A Conceptual Approach to Integrated Science—Now with MasteringPhysics®!

This best-selling introduction to the physical and life sciences emphasizes concepts over computation. It treats equations as guides to thinking so the reader can connect ideas across the sciences. This text covers physics, chemistry, earth science, astronomy, and biology at a level appropriate for non-science students. This book is now supported in MasteringPhysics—an unrivaled homework, tutorial, and assessment system.

Hallmark Strengths of the Book

Integrated Science 18A
PHYSICS AND CHEMISTRY

Moving Water Up a Tree

EXPLAIN THIS How can a plant get water by losing water?

LEARNING OBJECTIVE
Explain how water moves up from a plant's roots to its shoots.

The tallest trees are as tall as 30-story skyscrapers. Their highest branches and leaves need water, just like the rest of the plant. How do trees and other plants transport water, against gravity, all the way up to their highest points? Let's take a look.

In a plant, there are continuous columns of water molecules extending all the way through the xylem—from the leaves to the roots. These water molecules stick to one another and to the walls of the xylem. The attachment of water molecules to other water molecules is called *cohesion*; the attachment of water molecules to other molecules, such as those of the xylem wall, is called *adhesion*. Both cohesion and adhesion are the result of hydrogen bonds, which form when the positively charged end of one molecule sticks to the negatively charged end of another molecule. (Figure 12.30 in Chapter 12 shows cohesion in water molecules.)

UNIFYING CONCEPT
- *The Gravitational Force* Section 5.3

Cohesion and adhesion maintain the continuous columns of water molecules in the xylem. But how do these columns of water move up the plant? The process starts at the leaves, when a plant loses water through transpiration. *Transpiration* occurs when water evaporates from the moist cells inside a leaf and diffuses through the stomata to the outside air. As water is lost from the leaves, a tension is transferred all the way down the water column; water molecules in the leaf pull on water molecules in the nearby xylem, which pull on water molecules farther down the xylem, and so on all the way down to the roots. Transpiration pulls water up the xylem almost the way sucking on a straw pulls water up the straw. This mechanism of moving water up a plant is called the *transpiration-cohesion-tension mechanism* (Figure 18.20).

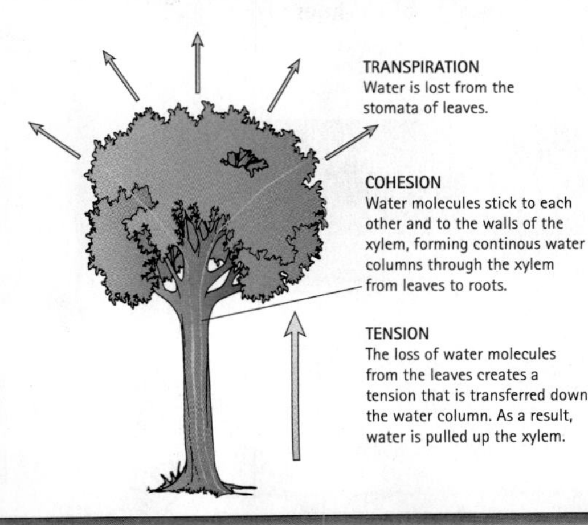

TRANSPIRATION
Water is lost from the stomata of leaves.
COHESION
Water molecules stick to each other and to the walls of the xylem, forming continous water columns through the xylem from leaves to roots.
TENSION
The loss of water molecules from the leaves creates a tension that is transferred down the water column. As a result, water is pulled up the xylem.

FIGURE 18.20 The transpiration-cohesion-tension mechanism describes how plants move water from their roots to their shoots.

Learning Objectives
A new Learning Objective has been added to each section to help students focus on the most important concepts in each chapter. Instructors using MasteringPhysics can assign content that is associated with these book-specific Learning Objectives.

Integrated Science (IS) Sections
These sections are found in every chapter and show how the foundational ideas developed in a chapter can find application across the different sciences. End-of-chapter questions are associated with most IS sections.

Unifying Concepts Icons
These marginal features highlight and cross reference a unifying concept developed in another part of the book.

NEW! Readiness Assurance Test
A Readiness Assurance Test (RAT) has been added to the end of each chapter. The RAT is a set of multiple-choice questions designed to help students assess their understanding of the material in that chapter.

MasteringPhysics®

www.masteringphysics.com

Now Available with MasteringPhysics for ***Conceptual Integrated Science,* Second Edition!**

The Mastering platform is the most widely used and effective online homework, tutorial, and assessment system for the sciences. It delivers self-paced tutorials that provide individualized coaching, focus on your course objectives, and respond to each student's progress.

Outstanding Content Accompanied by Unparalleled Tutoring

The Mastering system provides tutorials and coaching activities covering content relevant to the integrated science course and motivates students to learn outside of class and arrive prepared for lecture.

Video Activities ask students to answer multiple-choice questions based on the content of Paul Hewitt's popular classroom demonstrations.

Feedback and **hints** coach students back onto the right track, emulating how an instructor works with students during an office-hour visit.

Interactive Figure Activities help students master important topics by interacting with key figures, bringing principles to life. Hints and specific wrong answer feedback help guide students toward understanding the scientific principles.

Survey data show that the immediate feedback and tutorial assistance in MasteringPhysics motivate students to do more homework. The result is that students learn more and improve their test scores.

MasteringPhysics®
www.masteringphysics.com

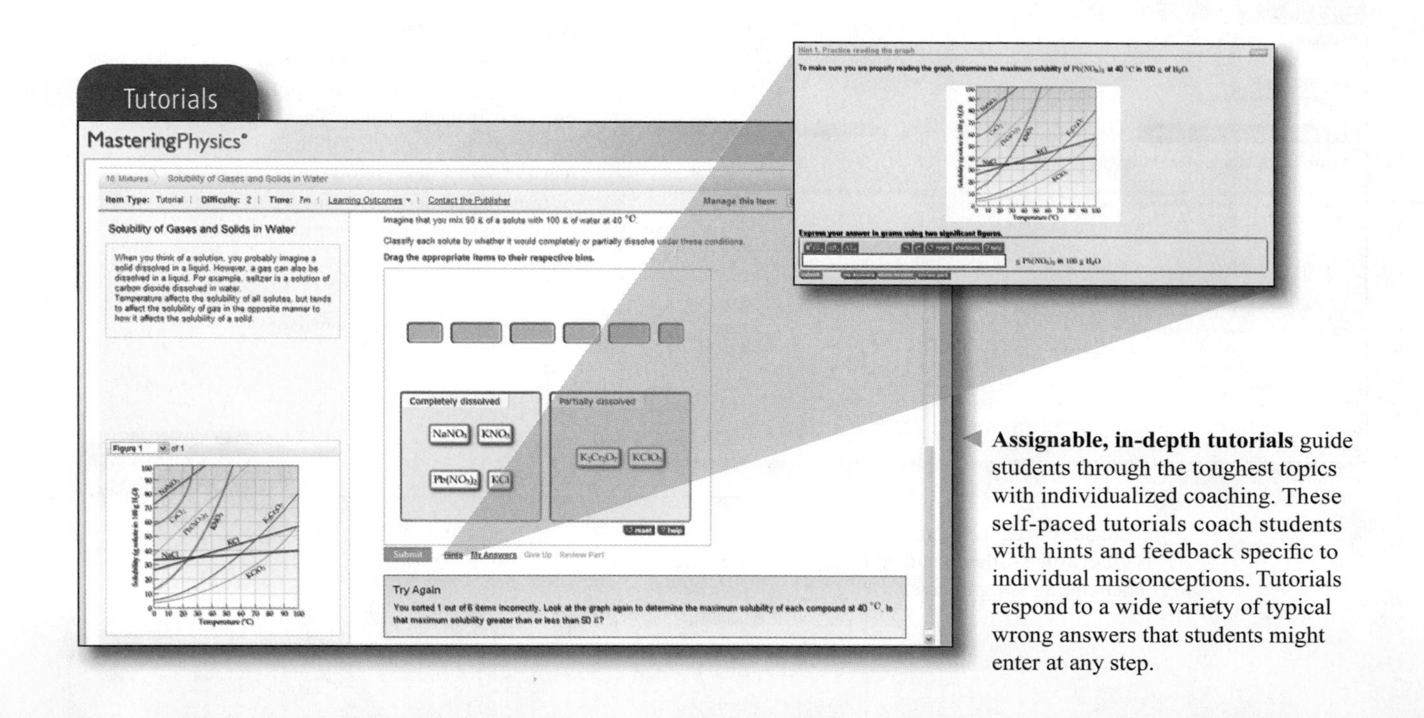

Assignable, in-depth tutorials guide students through the toughest topics with individualized coaching. These self-paced tutorials coach students with hints and feedback specific to individual misconceptions. Tutorials respond to a wide variety of typical wrong answers that students might enter at any step.

Coaching Activities have students interact with content, and available hints and/or feedback promote comprehension of the concepts.

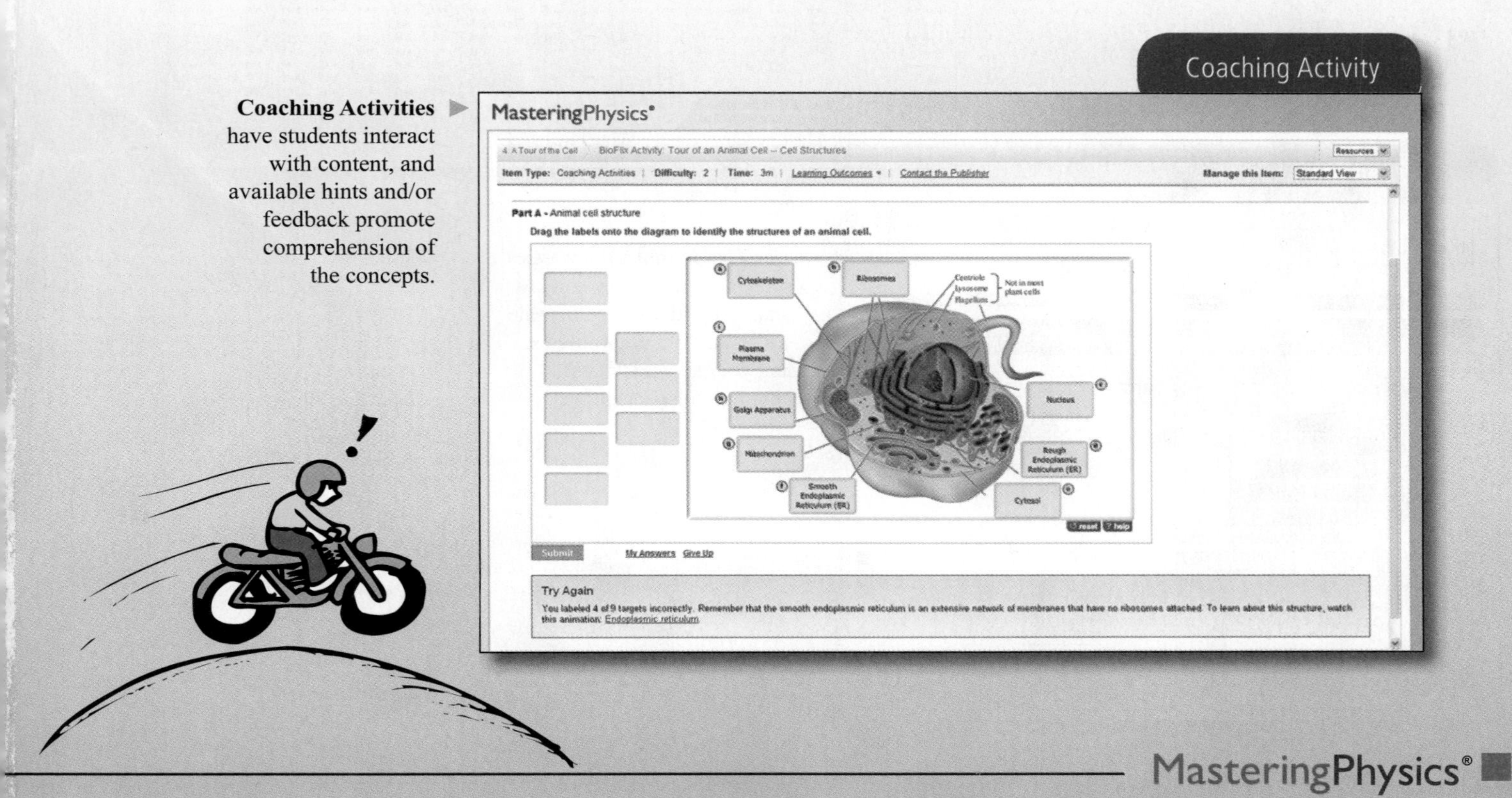

Student Results

The Mastering platform was developed by scientists for science students and instructors, and has a proven history with over 11 years of student use. Mastering currently has more than 1.5 million active registrations with active users in 50 states and in 41 countries.

Learning Outcomes

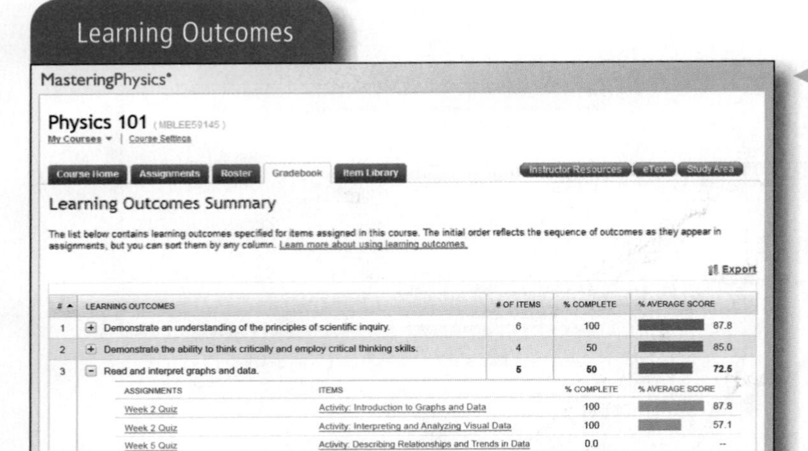

A learning objective has been added to each section of the text to help the students focus on the most important concepts in each chapter. These learning outcomes are associated with content in MasteringPhysics®, allowing the work of tracking student performance against course learning outcomes to be done automatically.

Gradebook

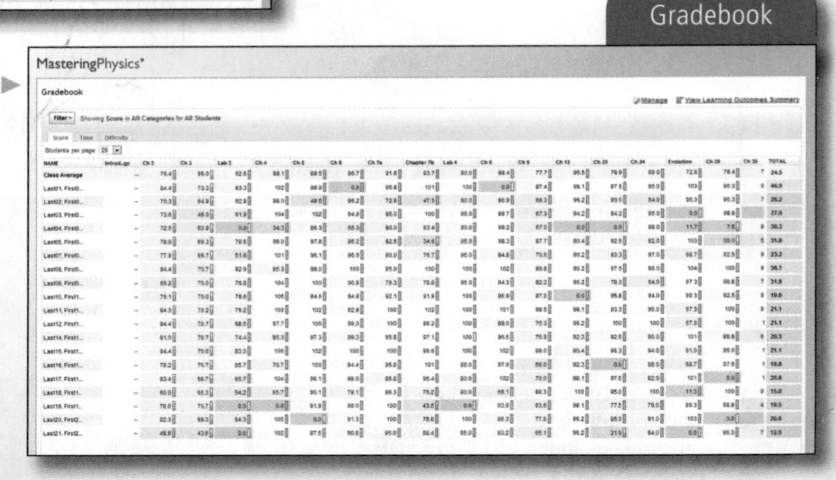

Shades of red **highlight vulnerable students** and challenging assignments.

Gradebook Diagnostics

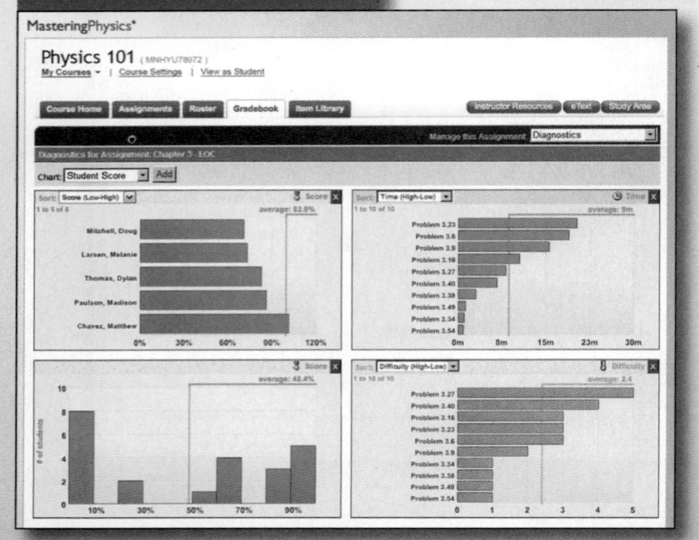

The Gradebook Diagnostics screen allows you to quickly and easily identify vulnerable students, difficult problems, and your students' most common misconceptions.

CONCEPTUAL Integrated Science

SECOND EDITION

Paul G. Hewitt
City College of San Francisco

Suzanne Lyons
California State University, Sacramento

John Suchocki
Saint Michael's College

Jennifer Yeh
University of California, San Francisco

PEARSON

Boston Columbus Indianapolis New York San Francisco Upper Saddle River
Amsterdam Cape Town Dubai London Madrid Milan Munich Paris Montréal Toronto
Delhi Mexico City São Paulo Sydney Hong Kong Seoul Singapore Taipei Tokyo

Publisher: Jim Smith
Project Manager: Chandrika Madhavan
Editorial Manager: Laura Kenney
Editorial Assistant: Kyle Doctor
Marketing Manager: Will Moore
Senior Program Manager: Corinne Benson
Media Producer: Kate Brayton
Production Service and Composition: Cenveo Publisher Services
Project Manager, Production Service: Cindy Johnson, Cenveo Publisher Services
Copyeditor: Carol Reitz
Design Manager: Derek Bacchus
Text Designer: Naomi Schiff, Seventeenth Street Studios
Cover Designer: Naomi Schiff, Seventeenth Street Studios
Illustrations: Rolin Graphics, Inc.
Photo Researcher: Sarah Bonner, Bill Smith Group
Image Lead: Maya Melenchuk
Manufacturing Buyer: Jeffrey Sargent
Printer and Binder: R.R. Donnelley
Cover Printer: Lehigh Phoenix
Cover Photo Credit: age fotostock / SuperStock

Credits and acknowledgments borrowed from other sources and reproduced, with permission, in this textbook appear on page C-1.

Library of Congress Cataloging-in-Publication Data
Hewitt, Paul G.
Conceptual integrated science / Paul G. Hewitt, City College of San Francisco, Suzanne Lyons, California State University, Sacramento, John Suchocki, Saint Michael's College, Jennifer Yeh, University of California, San Francisco. -- Second edition.
p. cm.
Includes index.
ISBN 978-0-321-81850-8 (student edition) -- ISBN 978-0-321-82745-6 (exam copy) -- ISBN 978-0-13-310526-1 (NASTA)
1. Interdisciplinary approach to knowledge. 2. Science--Philosophy. 3. Science--History. I. Title.
Q175.32.K45C66 2012
500--dc23

2012028366

ISBN 10: **0-321-81850-4;** ISBN 13: **978-0-321-81850-8** (Student Edition)
ISBN 10: **0-321-82745-7;** ISBN 13: **978-0-321-82745-6** (Exam Copy)
ISBN 10: **0-321-82287-0;** ISBN 13: **978-0-321-82287-1** (Books a la Carte Edition)

www.pearsonhighered.com

1 2 3 4 5 6 7 8 9 10—RRD—16 15 14 13 12

This book is dedicated to inquiring minds devoted to protecting this beautiful planet we call home.

Brief Contents

Detailed Contents

PART TWO

Chemistry 213

PART THREE

Biology 409

Special Features

MATH CONNECTION

SCIENCE AND SOCIETY

HISTORY OF SCIENCE

TECHNOLOGY

The *Conceptual Integrated Science* Photo Album

THIS BOOK IS VERY PERSONAL to the authors, family undertakings shown in the many photographs throughout. Author Paul is seen with wife Lillian on page 58, and Lil again on pages 176, and 210. Lil's dad, Wai Tsan Lee, is on page 176. Paul's grown children begin with son Paul with his mother Millie on page 138, and daughter Leslie in her student days on page 240. Paul's sister (and John's mom) Marjorie Hewitt Suchocki (pronounced Su-hock-ee, with a silent *c*), a leading process theologian, is shown reflectively on page 188. Paul's brother Steve shows Newton's third law with his daughter Gretchen on page 67. Paul's grandchild Emily Abrams opens Part One on page 17. Paul's friends include Tenny Lim pages 1 and 79, Will Maynez on page 77, Burl Grey on page 25, Dan Johnson on page 146 and Bay Johnson (Dan's grandson) on page 185, John Hubisz on page 128, Mike and Jane Jukes on pages 34 and 68, Cassy Cosme on page 73. Little kids that Paul loves include Carlos Vasquez on page 195, Miriam Dijamco and Michelle Anna Wong on page 178, Francesco Ming Giovannuzzi on page 125, and Andrea Wu on pages 94 and 191.

Chemistry author John, who in his "other life" is John Andrew, singer and song-writer walks barefoot on red-hot coals on page 137. John's wife, Tracy, is seen again with oldest son Ian on page 216 and with their second son Evan on page 283. Their third child, Maitreya Rose, is featured in the Chapter 14 opening photograph on page 376. Exploring the microscopic realm with the uncanny resolution of electron waves is cousin George Webster, who is seen on page 234. Friend Rinchen Trashi looks through the spectroscope on page 230. John's former students and co-stars of the Kai and Maile Show from *Conceptual Chemistry Alive!* appear on page 370. Coauthor Leslie Hewitt is seen as a 16 year old inspecting a molecular model on page 240. Nephews and niece Liam, Bo, and Neve Hopwood are seen together in the chemistry part opener of page 213.

Paul's granddaughter Grace Hewitt is featured in the astronomy part opener on page 819. The left two photographs on page 842 show our lab manual author Dean Baird surrounded by crescent shaped images of the Sun during the 2012 solar eclipse. Colleague Paul Doherty, on the far right on page 842, reveals the rarely seen sun circles that formed during same eclipse at the time and place of annular totality.

Biology author Jennifer's daughter Daphne and her prized sunflower open up Biology on page 409. A special thank you to Daphne's aunt Anita Sherman for bringing Daphne the sunflower all the way from her Baltimore garden! Jennifer's kids Io and Pico show off their sunscreen and sun hats on page 472. Io also appears on page 552, where she engages in her favorite pastime. Jennifer's husband Nils shoots a basket on page 591. And Jennifer's sister Pam, who is also a biologist, shows a specimen she caught (and later released) during an ecological study in Nepal on page 609.

Earth-Science author Suzanne Lyons with her children Simone and Tristan are shown on page 211. Tristan is on page 32 demonstrating friction and Simone is on page 192 pondering the color of a rose. Suzanne's husband Pete demonstrates thermodynamics in everyday life on page 126.

These photographs are of people very dear to the authors, which makes *Conceptual Integrated Science* all the more our labor of love.

Paul G. Hewitt

Former silver-medal boxing champion, sign painter, uranium prospector, and soldier, Paul began college at the age of 27, with the help of the GI Bill. He pioneered the conceptual approach to teaching physics at the City College of San Francisco. He has taught as a guest teacher at various middle schools and high schools, the University of California at both the Berkeley and Santa Cruz campuses, and the University of Hawaii at both the Manoa and Hilo campuses. He also taught for 20 years at the Exploratorium in San Francisco, which honored him with its Outstanding Educator Award in 2000. He is the author of *Conceptual Physics* and a co-author of *Conceptual Physical Science* and *Conceptual Physical Science Explorations* (with John Suchocki and Leslie Hewitt Abrams).

Suzanne Lyons

Suzanne received her B.A. in physics from the University of California, Berkeley. She earned her M.A. in education and her California teaching credential both from Stanford University. She earned another M.A. degree in Integrated Earth Sciences from California State University Sacramento. Suzanne was editor of *Conceptual Physics* and other books in the Conceptual series for 16 years and has authored 7 books on physics, hands-on science activities, and other topics in science and education. She has taught science and education courses to students of diverse ages and ability levels, from elementary school through college. She is always interested in developing new ways to teach and to that end, she founded the small business CooperativeGames.com.

John A. Suchocki

John is the author of *Conceptual Chemistry* and coauthor of *Conceptual Physical Science*. John obtained his Ph.D. in organic chemistry from Virginia Commonwealth University where he also completed a postdoctoral fellowship in pharmacology. As a tenured professor at Leeward Community College his interests turned to science education, the development of distance learning programs, and student-centered learning curricula. Currently an adjunct professor at Saint Michael's College in Vermont, John also produces science multimedia through his company Conceptual Productions. His popular tutorial video lessons, as well as those of his coauthors, are freely available at ConceptualAcademy.com.

Jennifer Yeh

Jennifer earned a Ph.D. in integrative biology from the University of Texas, Austin, for her work on frog skeleton evolution. She obtained her B.A. in physics and astronomy from Harvard University. Following her graduate work, Jennifer was a postdoctoral fellow at the University of California, San Francisco, where she studied the genetics of breast cancer. Jennifer has taught courses in physics, cell biology, human embryology, vertebrate anatomy, and ecology and evolution. She is the author of various scientific papers as well as the book *Endangered Species: Must They Disappear?* (Thomson/Gale, © 2002, 2004). Jennifer continues to work on a wide variety of introductory biology materials, including various online materials.

To the Student

WELCOME TO *Conceptual Integrated Science.* The science you'll learn here is INTEGRATED. That means we'll explore the individual science disciplines of physics, chemistry, biology, Earth science, and astronomy PLUS the areas where these disciplines overlap. Most of the scientific questions you're curious about, or need to know about, involve not just one discipline, but several of them in an overlapping way. How did the universe originate? That's astronomy + physics. How are our bodies altered by the foods we eat, the medicines we take, and the way we exercise? That's chemistry + biology. What's the greenhouse effect? Will it trigger irreversible global warming, threatening life on our planet? Physics, chemistry, biology, and Earth science are all needed to understand the answers.

We're convinced that the CONCEPTUAL orientation of this book is the way in which students best learn science. That means that we emphasize concepts *before* computation. Although much of science is mathematical, a firm qualitative grasp of concepts is also important. Too much emphasis on mathematical problem solving early in your science studies can actually distract you from the concepts and prevent you from fully comprehending them. If you continue in science, you may follow up with classes requiring advanced mathematical methods. Whether you do or don't, we think you'll be glad you learned the concepts first with just enough math to make them clearer.

This course provides plenty of resources beyond the text as well. For example, the interactive figures, interactive tutorials, and demonstration videos on www.masteringphysics.com will help you visualize science concepts, particularly processes that vary over time, such as the velocity of an object in free fall, the phases of the Moon, or the formation of chemical bonds. The activities in the *Laboratory Manual* will build your gut-level feeling for concepts and your analytical skills. Ponder the puzzlers in the *Conceptual Integrated Science Practice Book* and work through the simple review questions—all this will increase your confidence and mastery of science.

As with all things, what you get out of this class depends on what you put into it. So study hard, ask all the questions you need to, and most of all enjoy your scientific tour of the amazing natural world!

To the Instructor

THIS SECOND EDITION OF *Conceptual Integrated Science* with its important ancillaries provides your students an enjoyable and readable introductory coverage of the natural sciences. As with the previous edition, the 29 chapters are divided into five main parts—Physics, Chemistry, Biology, Earth Science, and Astronomy. We begin with physics, the basic science that provides a foundation for chemistry, which in turn underlies biology, which extends to Earth science and astronomy.

For the nonscience student, this book affords a means of viewing nature perceptively. One can see that a surprisingly few relationships make up its rules, most of which are the laws of physics presented in Part One. Physics laws are nature's secret codes. Here they are expressed both in words and in equation form. We view equations as *guides to thinking*. Even students who shy away from mathematics can learn to read equations to guide their thinking—to see how concepts connect. The symbols in equations are akin to musical notes that guide musicians.

For the science student, this same foundation affords a springboard to further study. For quantitatively oriented students, ample end-of-chapter material provides problem-solving activity through the *Think and Solve* problems.

Physics begins with static equilibrium so that students can start with forces before studying velocity and acceleration. After success with simple forces, the coverage touches lightly on kinematics, enough preparation for Newton's laws of motion. The pace picks up with the conventional order of mechanics topics followed by heat, thermodynamics, electricity and magnetism, sound, and light. Physics chapters lead to the realm of the atom—a bridge to chemistry.

The chemistry chapters begin with a look at the submicroscopic world of the atom, which is described in terms of subatomic particles and the periodic table. Students are then introduced to the atomic nucleus and its relevance to radioactivity, nuclear power, as well as astronomy. Subsequent chemistry chapters follow a traditional approach covering chemical changes, bonding, molecular interactions, and the formation of mixtures. With this foundation students are then set to learn the mechanics of chemical reactions and the behavior of organic compounds. As with previous editions, chemistry is related to the student's familiar world—the fluorine in their toothpaste, the Teflon on their frying pans, and the flavors produced by various organic molecules. The environmental aspects of chemistry are also highlighted—from how our drinking water is purified to how atmospheric carbon dioxide influences the pH of rainwater and our oceans.

The biology section begins by asking—what constitutes life? Each of the first three chapters focuses on a key feature of living things. We begin with a discussion of cells, move on to genes, and finally, tackle evolution and the origin of life. From here, we proceed to an overview of the different kinds of living things found on Earth. This overview is followed by two chapters on humans, our own species. In these chapters, we study the human body and how it works. Finally, we look at ecology, the study of how living organisms interact with their environments.

The Earth science chapters begin with plate tectonics, the theory that establishes the underlying framework of the geosciences. The next chapter is about rocks and minerals, the principal materials that make up the solid Earth. Then comes a tour of Earth's landforms, surface features, and geography followed by a chapter on surficial processes—those processes of weathering, erosion, and deposition that originate at Earth's surface and shape the planet's contours. Plate tectonics is about Earth's interior, and the chapters on rock, landforms, and surficial processes describe Earth's surface. The next chapter in the sequence rises higher still—into the atmosphere—with weather. The subject of weather is broken down into elements from atmospheric pressure to wind to precipitation that can be learned separately but then applied to complex phenomena such as weather systems. The Earth science unit concludes with a chapter on environmental geology, which is new to the second edition. It provides an updated review of earthquakes, tsunami, hurricanes, volcanic eruptions, and other geologic hazards. Most importantly, it features expanded coverage of our changing climate including extensive discussion of natural and anthropogenic climate change.

The applications of physics, chemistry, biology, and the Earth sciences applied to other massive bodies in the universe culminate in Part Five—astronomy. This unit introduces the basic structure of the universe from our local solar system and the stars we see at night to galaxies and superclusters of galaxies. Focus is given to modern theories describing how this structure evolved and is continuing to evolve. Many recent discoveries are featured in this edition, illustrating that science is more than a growing body of knowledge; it is an arena in which humans actively and systematically reach out to learn more about our place in the universe.

What's New to This Edition

Conceptual Integrated Science now comes with a powerful media package including **MasteringPhysics®**, the most widely used, educationally proven, and technologically advanced tutorial and homework system available.

MasteringPhysics®
www.masteringphysics.com

MasteringPhysics contains:

- A **library of assignable and automatically graded content**, including tutorials, visual activities, end-of-chapter problems, and test bank questions so instructors can create the most effective homework assignments with just a few clicks. A **color-coded gradebook** instantly identifies vulnerable students or topic areas that are challenging for students in the class.
- **A student study area** with practice quizzes, Interactive Figures, self-guided tutorials, flashcards, videos, access to the Pearson eText version of the book, and more.
- An **instructor resources section** with PowerPoint lectures, clicker questions, Instructor Manual files and more.

Another significant revision for this second edition lies with the development of the end-of-chapter review. New questions were added while older ones were either discarded or reworded for improved quality. All questions were then organized following Bloom's taxonomy of learning as follows:

Summary of Terms (Knowledge)

These key terms match the definitions given within the chapter and are now listed in alphabetical order so that they appear as a mini-glossary for the chapter.

Reading Check Questions (Comprehension)

These questions frame the important ideas of each section in the chapter. They are for review and a check of reading comprehension. They are simple questions and all answers can easily be discovered in the chapter.

Think Integrated Science

Questions pertaining to the Integrated Science sections of each chapter are contained in this section. Questions range from straightforward, reading-check type questions to critical-thinking exercises.

Think and Do (Hands-On Application)

The Think and Do items are easy-to-perform hands-on activities designed to help students experience physical science concepts for themselves.

Think and Solve (Mathematical Application)

The *Think and Solve* questions blend simple mathematics with concepts. They allow students to apply problem-solving techniques, many of which are featured in the Math Connection boxed features.

Think and Rank (Analysis)

The *Think and Rank* questions ask students to make comparisons of quantities. For example, when asked to rank quantities such as momentum or kinetic energy, appreciably more judgment is called for than in providing numerical answers. Some *Think and Rank* analyze trends, as in ranking atoms in order of increasing size based upon student understanding of the periodic table. This feature elicits critical thinking that goes beyond *Think and Solve.*

Think and Explain (Synthesis)

The *Think and Explain* questions, by a notch or two, are the more challenging questions at the end of each chapter. Many require critical thinking while others are designed to prompt the application of science to everyday situations. All students wanting to perform well on exams should be directed to the *Exercises* because these are the questions that directly assess student understanding. Accordingly, many of the *Exercises* have been adapted to a multiple-choice format and integrated into the *Conceptual Integrated Science, 2e* test bank. This will hopefully allow the instructor to reward those students who put time and effort into the *Exercises.*

Think and Discuss (Evaluation)

The *Think and Discuss* topics provide students the opportunity to apply science concepts to real-life situations, such as whether a cup of hot coffee served to you in a restaurant cools faster when cream is added promptly or a few minutes later. Other discussion questions allow students to present their educated opinions on a number of science-related hot topics, such as the appearance of pharmaceuticals in drinking water.

Readiness Assurance Test

Each chapter review concludes with a set of 10 multiple choice questions that students can take for self-assessment. They are advised to study further if they score less than 7 correct answers.

Also new to this edition are the solutions to the odd-numbered end-of-chapter questions in the back of this book. As before, solutions to all end-of-chapter questions are available to instructors through the Instructor Manual for *Conceptual Integrated Science,* which is found in the Instructor Resource Center and in the Instructor Resource area of MasteringPhysics.

This second edition features a new and, we think, refreshing page layout design. Integrated into this design are **learning objectives** that appear alongside each chapter section head. Each learning objective begins with an active verb that specifies what the student should be able to do after studying that section, such as "Calculate the energy released by a chemical reaction." These section-specific learning objectives are further integrated into the new MasteringPhysics online tutorial/assessment tool.

Also within the design, appearing beneath each section head is another new feature, which we call an **"Explain This"** question. An ET question would be fairly difficult for the student to answer without having read the chapter section. Some require that the student recall earlier material. Others reveal interesting applications of concepts. In all cases the ET question should serve well as a launching point for classroom discussions. The answers to these ET questions appear only within the Instructor Manual.

The text of all chapters has been edited for accuracy, better readability and also updated to reflect current events, such as the nuclear power plant disaster following the 2011 Japanese earthquake and tsunami, and the discovery of Fermi clouds arising from the center of our Milky Way galaxy.

The scope and sequence of chapters is revised for this second edition. The material on the atom has been folded into the chemistry unit so that the atomic theory is explained at the point of use. In Part Three—Biology, the genetics chapter has been reorganized, and much new material has been added on DNA technology, its uses, and its potential dangers. The ecology chapter has also been reorganized with new attention to human population growth and human ecological footprints. The Earth science material has been reorganized such that the geography material is now separated from the discussion of surficial processes, allowing for more discussion of the oceans. A chapter on Historical Geology was eliminated with the most important concepts (such as the geologic time scale, Cretaceous extinction, and the nature of the rock record) being integrated into other chapters. The elimination of Historical Geology allowed the new chapter on Environmental Geology to be added with in-depth coverage of climate change. In Part Five—Astronomy, aside from updates from recent discoveries, the first section of Chapter 28 has been heavily revised in its presentation of nebular theory and the second chapter of this unit is expanded greatly to include discussions of cosmology.

Ancillary Materials

MasteringPhysics®
www.masteringphysics.com

Most significantly, *Conceptual Integrated Science* is now available with MasteringPhysics—a homework, tutorial, and assessment system based on years of research into how students work problems and precisely where they need help. Studies show that students who use MasteringPhysics significantly increase their scores compared to hand-written homework. MasteringPhysics achieves this improvement by providing students with instantaneous feedback specific to their wrong answers and simpler sub-problems upon request when they get stuck. Instructors can also assign End-of-Chapter (EOC) problems from every chapter including multiple-choice questions, section-specific exercises, and general problems. Quantitative problems can be assigned with numerical answers and randomized values or solutions.

The Pearson eText of *Conceptual Integrated Science* is available through MasteringPhysics. Allowing students access to the text wherever they have access to the Internet, the Pearson eText comprises the full text, including figures that can be enlarged for better viewing, popup definitions and terms, a note-taking feature, and more.

Tutorial video lessons and screencasts featuring the authors are now freely available to students at ConceptualAcademy.com. This is a must-visit website for any student who needs a bit of extra help and it is also a great tool for the online component of any course. Many author-created resources for the instructor are also available on this website, which is a great place to communicate directly with the authors as well as other instructors using *Conceptual Integrated Science*.

The ***Instructor Manual for Conceptual Integrated Science*** (ISBN 0-321-82743-0), which you'll find to be different from most instructors' manuals, allows for a variety of course designs to fit your taste. It contains many lecture ideas and topics not treated in the textbook as well as teaching tips and suggested step-by-step lectures and demonstrations. It has full-page answers to all the end of chapter material in the text.

The ***Conceptual Integrated Science Practice Book*** (ISBN 0-321-82298-6), our most creative work, guides your students to a sometimes computational way of developing concepts. It spans a wide use of analogies and intriguing situations, all with a user-friendly tone.

The ***Test Bank for Conceptual Integrated Science*** (ISBN 0-321-82276-5) has more than 2400 multiple choice questions as well as short answer and essay questions. The questions are categorized according to level of difficulty. The Test Bank allows you to edit questions, add questions, and create multiple test versions.

The ***Laboratory Manual for Conceptual Integrated Science*** (ISBN 0-321-82297-8) is written by the authors and Dean Baird. In addition to interesting laboratory experiments, it includes a range of activities similar to the activities in the textbook. These guide students to experience phenomena before they quantify the same phenomena in a follow-up laboratory experiment. Answers to the lab manual questions are in the Instructor Manual.

Another valuable media resource available to you is the ***Instructor Resource DVD for Conceptual Integrated Science*** (ISBN 0-321-82744-9). This cross-platform DVD set provides instructors with the largest library available of purpose-built, in-class presentation materials, including all the images from the book in high-resolution JPEG format; interactive figures™ and videos; PowerPoint® lecture outlines and clicker questions in PRS-enabled format for each chapter, all of which are written by the authors; and Hewitt's acclaimed Next-Time Questions in PDF format. The *Instructor Resource DVD* provides you with everything you need to prepare for dynamic, engaging lectures in no time.

Go to it! Your conceptual integrated science course really can be the most interesting, informative, and worthwhile science course your students will ever take.

Acknowledgments

THE AUTHORS WISH TO express their sincere appreciation to the many talented and generous people who helped make *Conceptual Integrated Science*, now in its Second Edition, come to life. To the teachers and professors who reviewed the manuscript, giving generously of their time, we express heartfelt appreciation.

We thank all the contributors to the first and second editions of *Conceptual Integrated Science*, as well as the many people who contributed to the other books in the *Conceptual* series: *Conceptual Physics*, *Conceptual Chemistry*, *Conceptual Physical Science*, and *Conceptual Integrated Science—Explorations*. For helping to shape the physics content over the years, in the student editions as well as our many supplements, we thank: Dean Baird, Tsing Bardin, Howie Brand, Alexi Cogan, Paul Doherty, Marshall Ellenstein, Ken Ford, Lillian Lee Hewitt, Jim Hicks, David Housden, John Hubisz, Will Maynez, Fred Myers, Bruce Novak, Ron Perkins, Diane Reindeau, David Williamson, Larry Weinstein, Phil Wolf, and Dean Zollman.

For development of chemistry chapters, thanks go to Adedoyin Adeyiga, John Bonte, Emily Borda, Charles Carraher, Natashe Cleveland, Sara Devo, Andy Frazer, Kenneth French, Marcia Gillette, Chu-Ngi Ho, Frank Lambert, Jeremy Mason, Daniel Predecki, Britt Price, Jeremy Ramsey, Kathryn Rust, William Scott, Anne Marie Sokol, Jason Vohs, Bob Widing, and David Yates.

For advice and wide-ranging contributions to the biology section, we thank Pamela Yeh, Sarah Ying, Nina Shapley, Nils Gilman, Todd Schlenke, Howard Ying, Brian West, Robert Dudley, Vivianne Ding, Mike Fried, W. Bryan Jennings, Rachel Zierzow, Ernie Brown, and Lil Hewitt.

For Earth science inspiration, input and advice, we are grateful to Leslie Hewitt Abrams and Bob Abrams. They gave generously of their time to share insights gained from their work authoring and teaching Earth science. We are also grateful to Tsing Bardin, Judy Kusnick, Bruce Gervais, Lil Hewitt, Diane Carlson, and Lynne Cherry.

For space science we are grateful to Jeffrey Bennett, Megan Donahue, Nicholas Schneider, and Mark Volt for permission to use many of the graphics that appear in their textbook *The Cosmic Perspective*, 6th edition. Also, for reviews of the astronomy chapters we remain grateful to Richard Crowe, Bjorn Davidson, Stacy McGaugh, Michelle Mizuno-Wiedner, John O'Meara, Neil de Grass Tyson, Joe Wesney, Lynda Williams, and Erick Zackrisson.

Our colleagues at Pearson have been our partners in this project and given us much support and guidance. For the second edition, we thank Jim Smith, seasoned Publisher, for wise and sensible overall direction. We thank our project editor Chandrika Madhavan for being great to work with, patient, and wonderfully competent in her role at the crossroads of communication. We thank Cindy Johnson for graceful and intelligent handling of the production process.

The *Conceptual Integrated Science* authors are fortunate to have helpful and loving spouses who have supported us through the long hours and contributed to our efforts. Thanks go to Lillian Lee Hewitt, Pete Lang (Suzanne's husband), Tracy Suchocki, and Nils Gilman (Jennifer's husband). And to our kids, ranging now from preschool to high school, we send our love and gratitude: Tristan and Simone Lyons Lang; Ian, Evan, and Maitreya Suchocki; and Io, Pico, and Daphne Yeh Gilman.

Reviewers

Leila Amiri, University of South Florida
Leanne Avery, Indiana University of Pennsylvania
Bambi Bailey, Midwestern State University
Dirk Baron, California State University, Bakersfield
Daniel Berger, Bluffton University
Reginald Blake, City Tech University of New York
Derrick Boucher, King's College
Martin Brock, Eastern Kentucky University
Linda Brown, Gainsville College
Mary Brown, Lansing Community College
Steven Burns, St. Thomas Aquinas College
Erik Burtis, Northern Valley Community College, Woodbridge
Gerry Clarkson, Howard Payne University
Anne Coleman, Cabrini College
Gary Courts, University of Dayton
Red Chasteen, Sam Houston State University
Randy Criss, St. Leo University
Jason Dahl, Bemidji State University
Terry Derting, Murray State University
David DiMattio, St. Bonaventure University
Gary Neil Douglas, Berea College
S. Keith Dunn, Centre College
George Econ, Jackson Community College
Michael S. Epstein, Mount St. Mary's University
Charles Figura, Wartburg College
Lori K. Garrett, Danville Area Community College
David Goldsmith, Westminster College
Brian Goodman, Lakeland College
Nydia R. Hannah, Georgia State University
Carole Hillman, Elmhurst University
James Houpis, California State University, Chico
Thomas Hunt, Jackson Community College
David T. King, Jr., Auburn University
Jeremiah K. Jarrett, Central Connecticut State University
Peter Jeffers, State University of New York, Cortland
Charles Johnson, South Georgia College
Richard Jones, Texas Women's University
Carl Klook, California State University, Bakersfield
Kenneth Laser, Edison Community College
Jeffrey Laub, Rogers State University
Holly Lawson, State University of New York, Fredonia
David Lee, Biola University
Steven Losh, State University of New York, Cortland
Ntungwa Maasha, Coastal Georgia Community College
Kingshuk Majumdar, Berea College
Lynette McGregor, Wartburg College
Preston Miles, Centre College
Frank L. Misiti, Bloomburg University
Matthew Nehring, Adams State College
Marlene Morales, Miami Dade College
Douglas Nelson, Coastal Carolina University
Jan Oliver, Troy State University
Treva Pamer, New Jersey City University
Todd Pedlar, Luther College
Denice Robertson, Northern Kentucky University
Judy Rosovsky, Johnson State College
Steven Salaris, All Saints Christian School
Terry Shank, Marshall University
Sedonia Sipes,
Southern Illinois University, Carbondale
Ran Sivron, Baker University
Priscilla Skalac, Olivet Nazarene University
Stanley Sobolewski, Indiana University of Pennsylvania
John Snyder, Lansing Community College
Stuart Snyder, Montana State University
Anne Marie Sokol, Buffalo State College
Pamela Stephens, Midwestern State University
Laura Stumpe, Holy Cross College, Notre Dame
Karen Swanson, William Patterson University
Timothy Swindle, University of Arizona
Rachel Teasdale, California State University, Chico
Diana Treahy, Anderson University
Christos Valiotis, Antelope Valley College
Robin van Tine, St. Leo University
Daniel Vaughn, Southern Illinois University
Stephen Webb, Brescia College
Karen Wehn, Buffalo State University
Adam Wenz, Montana State University, Great Falls
William Wickun, Montana State University, Billings
Bonnie S. Wood, University of Maine, Presque Isle
Robert Zdor, Andrews University

CHAPTER 1
About Science

MODERN CIVILIZATION is built on science. Nearly all forms of technology—from medicine to space travel—are applications of science. One such application is *Curiosity*, the latest vehicle to explore the surface of Mars. Tenny Lim, lead designer of its descent stage, stands in front of a model of Curiosity in the photo above to show its size. Tenny's science and engineering career was ignited when she was in Paul Hewitt's conceptual physics class.

Science is a way of seeing the world and making sense of it. Science is also a human endeavor, as Tenny well knows when she teams with other investigators at Jet Propulsion Laboratory in California. Science is the culmination of centuries of human effort from all parts of the world, making it the legacy of countless thinkers and experimenters of the past.

LEARNING OBJECTIVE
Acknowledge contributions to science by various cultures.

1.1 A Brief History of Advances in Science

EXPLAIN THIS How did the advent of the printing press affect the growth of science?

When a light goes out in your room, you ask, "How did that happen?" You might check to see if the lamp is plugged in or if the bulb is burned out, or you might look at homes in your neighborhood to see if there has been a power outage. When you think and act like this, you are searching for *cause-and-effect* relationships—trying to find out what events cause what results. This type of thinking is *rational thinking*, applied to the physical world. It is basic to science.

Today, we use rational thinking so much that it's hard to imagine other ways of interpreting our experiences. But it wasn't always this way. In other times and places, people relied heavily on superstition and magic to interpret the world around them. They were unable to analyze the *physical* world in terms of *physical* causes and effects.

The ancient Greeks used logic and rational thought in a systematic way to investigate the world around them and make many scientific discoveries. They learned that Earth is round and determined its circumference. They discovered why things float and suggested that the apparent motion of the stars throughout the night is due to the rotation of Earth. The ancient Greeks founded the science of botany—the systematic study and classification of plants—and even proposed an early version of the principle of natural selection. Such scientific breakthroughs, when applied as technology, greatly enhanced the quality of life in ancient Greece. For example, engineers applied principles articulated by Archimedes and others to construct an elaborate public waterworks, which brought fresh water into the towns and carried sewage away in a sanitary manner.

When the Romans conquered ancient Greece, they adopted much of Greek culture, including the scientific mode of inquiry, and spread it throughout the Roman Empire. When the Roman Empire fell in the 5th century AD, advancements in science came to a halt in Europe. Nomadic tribes destroyed much in their paths as they conquered Europe and brought in the Dark Ages. While religion held sway in Europe, science continued to advance in other parts of the world.

FIGURE 1.1
A view of the Acropolis, or "high city," in ancient Greece. The buildings that make up the Acropolis were built as monuments to the achievements of the residents of the area.

The Chinese and Polynesians were charting the stars and the planets. Arab nations developed mathematics and learned to make glass, paper, metals, and certain chemicals. Finally, during the 10th through 12th centuries, Islamic people brought the spirit of scientific inquiry back into Europe when they entered Spain. Then universities sprang up. When the printing press was invented by Johannes Gutenberg in the 15th century, science made a great leap forward. People were able to communicate easily with one another across great distances. The printing press did much to advance scientific thought, just as computers and the Internet are doing today.

Up until the 16th century, most people thought Earth was the center of the universe. They thought that the Sun circled the stationary Earth. This thinking was challenged when the Polish astronomer Nicolaus Copernicus quietly published a book proposing that the Sun is stationary and Earth revolves around it. These ideas conflicted with the powerful institution of the Church and were banned for 200 years.

Modern science began in the 17th century, when the Italian physicist Galileo Galilei revived the Copernican view. Galileo used experiments, rather than speculation, to study nature's behavior (we'll say more about Galileo in chapters that follow). Galileo was arrested for popularizing the Copernican theory and for his other contributions to scientific thought. But, a century later, his ideas and those of Copernicus were accepted by most educated people.

Scientific discoveries are often opposed, especially if they conflict with what people want to believe. In the early 1800s, geologists were condemned because their findings differed from religious accounts of creation. Later in the same century, geology was accepted, but theories of evolution were condemned. Every age has had its intellectual rebels who have been persecuted, vilified, condemned, or suppressed but then later regarded as harmless and even essential to the advancement of civilization and the elevation of the human condition. "At every crossway on the road that leads to the future, each progressive spirit is opposed by a thousand men appointed to guard the past."*

1.2 Mathematics and Conceptual Integrated Science

LEARNING OBJECTIVE
Recount how mathematics is a key in formulating good science.

EXPLAIN THIS What is meant by "Equations are guides to thinking"?

Pure mathematics is different from science. Math studies relationships among numbers. When math is used as a tool of science, the results can be astounding. Measurements and calculations are essential parts of the powerful science we practice today. For example, it would not be possible to send missions to Mars if we were unable to measure the positions of spacecraft or to calculate their trajectories.

You will make some calculations in this course, especially when you make measurements in lab. In this book, we don't make a big deal about math. Our focus is on understanding concepts in everyday language. We use equations as guides to thinking rather than as recipes for "plug-and-chug" computational work. We believe that focusing on computations too early, especially on math-based problem solving, is a poor substitute for learning the concepts. That's why the emphasis in this book is on building concepts. Only when concepts are understood does computational problem solving make sense.

1.3 The Scientific Method—A Classic Tool

LEARNING OBJECTIVE
List the steps in the classic scientific method, and cite other processes that advance science.

EXPLAIN THIS What other processes besides the classic scientific method advance science?

The practice of **science** usually encompasses keen observations, rational analysis, and experimentation. In the 17th century, Galileo and the English philosopher Francis Bacon were the first to formalize a particular method

*From Count Maurice Maeterlinck's "Our Social Duty."

MATH CONNECTION

Equations as Guides to Thinking

In this book we recognize the value of equations as guides to thinking. What we mean by this is that simple equations tell you immediately how one quantity is related to another. In Chapter 5, when you study gravity, you will learn about the inverse-square relationship—a mathematical form that comes up over and over again in science. In Appendix D, you can study exponential relations in general. But, to start off, let's consider two basic mathematical relationships:

The direct proportion The more you study for this course, the better you'll do. That's a direct proportion. Similarly, the more coffee you drink, the more nervous you'll feel. The longer time you drive at a constant speed, the farther you travel. If you're paid by the hour, the longer you work, the more money you make. All these examples show relationships between two quantities, and in each case, the relationship is a direct proportion. A direct proportion has the mathematical form $x \sim y$. Direct proportions have graphs of the form shown here.

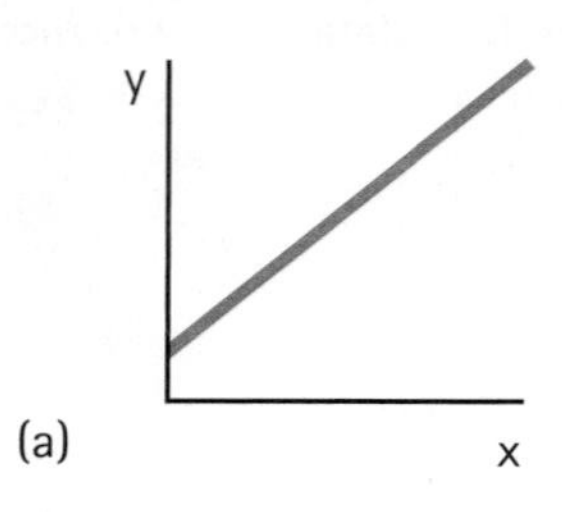

(a) The direct proportion.

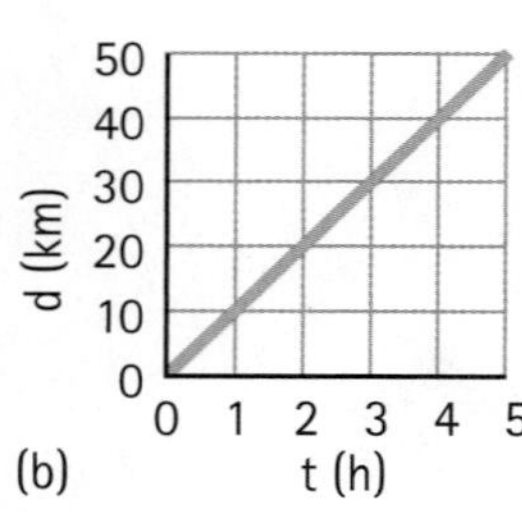

(b) A car travels at a constant speed. The more time it travels, the farther it goes. Distance is directly proportional to time.

The inverse proportion Some quantities are related to each other so that as one increases, the other decreases. The *more* you compress an air-filled balloon, the *smaller* it becomes. The *more* massive a grocery cart, the *less* it accelerates when you push it. These quantities are related through the inverse proportion, which has the mathematical form $y \sim \frac{1}{x}$. Inverse proportions have graphs of the form shown here.

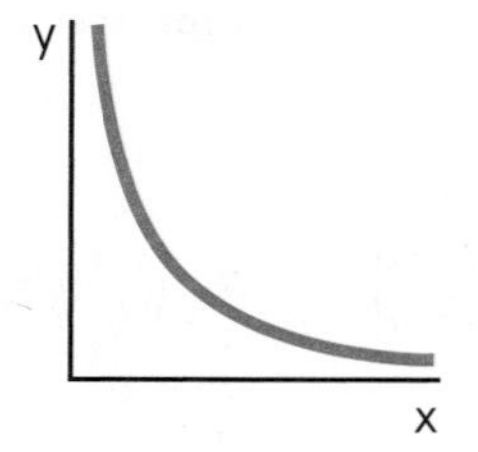

Graph of an inverse proportion.

Note that these mathematical relations have been stated as proportional relations, rather than as *equations*. For a proportional relation to be stated as an equation, the numbers and units on both sides must be the same. We can state a direct or indirect proportion as an exact equation by inserting a *proportionality constant, k,* into the relation. Proportionality constants make the numbers and units on both sides of an equation match up.

For example, consider Hooke's law. Hooke's law tells us about springs and other stretchy, elastic objects. Imagine a spring, such as a Slinky. According to Hooke's law, the more a Slinky is stretched, the harder it is to stretch it further. Written as a direct proportion, Hooke's law is

$$F \sim x$$

where F is your pulling force and x is the distance the spring is stretched beyond its resting length. But F, a force, has units of newtons (N) and x, a distance, has units such as centimeters (cm). We convert Hooke's law into an equation by inserting k into the relation. The value of k in this case depends on the shape and material of the spring. For a common Slinky-type metal coil, the proportionality constant k is about 2.5 N/cm. Now we can state Hooke's law as an exact equation:

$$F = kx$$

(a) An unstretched spring. (b) The spring is stretched past its resting length. (c) Stretching the spring further takes more force. This is Hooke's law, $F = kx$.

Problems

1. A freely falling object picks up a speed of 10 m/s during each second of fall. This is expressed as $v = gt$, where v is the speed picked up, g is the acceleration of free fall, and t is the time of fall. What type of mathematical relation is this?
2. A spring stretches and compresses according to Hooke's law, which states $F = kx$, where F is the force, k is the spring constant, and x is the stretched or compressed distance. For a certain spring with a spring constant of 3 N/cm, how much force is needed to stretch this spring 4 cm past its resting length?

Solutions

1. This relationship is of the form $x \sim y$, a direct proportion.
2. $F = kx = (3\,\frac{\text{N}}{\text{cm}}) \times (4\text{ cm}) = 12\text{N}.$

for doing science. What they outlined has come to be known as the classic **scientific method**. It essentially includes the following steps:

1. **Observe** Closely observe the physical world around you.
2. **Question** Recognize a question or a problem.
3. **Hypothesize** Make an educated guess—a *hypothesis*—to answer the question.
4. **Predict** Predict consequences that can be observed if the hypothesis is correct. The consequences should be *absent* if the hypothesis is not correct.
5. **Test predictions** Do experiments to see if the consequences you predicted are present.
6. **Draw a conclusion** Formulate the simplest general rule that organizes the hypothesis, predicted effects, and experimental findings.

UNIFYING CONCEPT

- ***The Scientific Method***

Science is a way to teach how something gets to be known, what is not known, to what extent things are known (for nothing is known absolutely), how to handle doubt and uncertainty, what the rules of evidence are, how to think about things so that judgments can be made, how to distinguish truth from fraud, and from show.—*Richard Feynman*

Although the scientific method is powerful, good science is often done differently, in a less systematic way. In the Integrated Science feature at the end of the chapter, "An Investigation of Sea Butterflies," you will see a recent application of the classic scientific method. However, many scientific advances involve trial and error, experimenting without guessing, or just plain accidental discovery. More important than a particular method, the success of science has to do with an attitude common to scientists. This attitude is one of inquiry, experimentation, and humility before the facts.

1.4 The Scientific Hypothesis

LEARNING OBJECTIVE
Describe the value of testing for furthering scientific knowledge.

EXPLAIN THIS Why do scientists work only with hypotheses that are testable?

A scientific **hypothesis** is an educated guess that tentatively answers a question or solves a problem in regard to the physical world. Typically, experiments are done to test hypotheses.

The cardinal rule in science is that all hypotheses must be testable—in other words, they must, at least in principle, be capable of being shown wrong. In science, it is more important that there be a means of proving an idea wrong than that there be a means of proving it right. This is a major feature that distinguishes science from nonscience. The idea that scientific hypotheses must be capable of being proven wrong is a pillar of the philosophy of science, and it is stated formally as the **principle of falsifiability**:

For a hypothesis to be considered scientific it must be testable—it must, in principle, be capable of being proven wrong.

At first this principle may seem strange, for when we wonder about most things, we concern ourselves with ways of finding out whether they are true. Scientific hypotheses are different. In fact, if you want to determine whether a hypothesis is scientific or not, look to see whether there is a test for proving it wrong. If there is no test for possible wrongness, then the hypothesis is not scientific. Albert Einstein put it well: "No number of experiments can prove me right; a single experiment can prove me wrong."

For example, Einstein hypothesized that light is bent by gravity. This idea might be proven wrong if starlight that grazed the Sun and could be seen during an eclipse were not deflected from a normal path. But starlight *is* found to bend as it passes close to the Sun, just as Einstein's hypothesis would have predicted. If and when a hypothesis or scientific claim is confirmed, it is regarded as useful and as a steppingstone to additional knowledge.

Consider another hypothesis: "The alignment of planets in the sky determines the best time for making decisions." Many people believe it, but this hypothesis is not scientific. It cannot be proven wrong, nor can it be proven right. It is speculation. Likewise, the hypothesis "Intelligent life exists on planets somewhere in the universe besides Earth" is not scientific.* Although it can be proven correct by the verification of a single instance of life existing elsewhere in the universe, there is no way to prove it wrong if no life is ever found. If we searched the far reaches of the universe for eons and found no life, we would not prove that it doesn't exist "around the next corner." A hypothesis that is capable of being proven right but not capable of being proven wrong is not a scientific hypothesis. Many such statements are quite reasonable and useful, but they lie outside the domain of science.

CHECK YOURSELF

Which statements are *scientific* hypotheses?

(a) Better stock market decisions are made when the planets Venus, Earth, and Mars are aligned.
(b) Atoms are the smallest particles of matter that exist.
(c) The Moon is made of Swiss cheese.
(d) Outer space contains a kind of matter whose existence can't be detected or tested.
(e) Albert Einstein was the greatest physicist of the 20th century.

CHECK YOUR ANSWER

All these statements are hypotheses, but only statements a, b, and c are scientific hypotheses because they are testable. Statement a can be tested (and proven wrong) by researching the performance of the stock market during times when these planets were aligned. Not only can statement b be tested; it has been tested. Although the statement has been found to be untrue (many particles smaller than atoms have been discovered), the statement is nevertheless a scientific one. Likewise for statement c, where visits to the Moon have proven that the statement is wrong. Statement d, on the other hand, is easily seen to be unscientific because it can't be tested. Last, statement e is an assertion that has no test. What possible test, beyond collective opinion, could prove that Einstein was the greatest physicist? How could we know? Greatness is a quality that cannot be measured in an objective way.

LEARNING OBJECTIVE
Discuss how experimentation helps prevent the acceptance of false ideas.

1.5 The Scientific Experiment

EXPLAIN THIS Why do experiments trump philosophical discussion in science?

A well-known scientific hypothesis that turned out to be incorrect was that of the greatly respected Greek philosopher Aristotle (384–322 BC), who claimed that heavy objects naturally fall faster than light objects. This hypothesis was considered to be true for nearly 2000 years—mainly because

*The search for intelligent life in the universe is, however, ongoing. This search is based on the *question*: Might there be intelligent life somewhere besides on Earth? This question is the starting point for scientific observations of the physical world, but strictly speaking it is not a scientific hypothesis. A hypothesis is a sharper scientific tool than a question—a better, more finely honed instrument for separating scientific fact from fiction.

nearly everyone who knew of Aristotle's conclusions had such great respect for him as a thinker that they simply assumed he couldn't be wrong. Also, in Aristotle's time, air resistance was not recognized as an influence on how quickly an object falls. We've all seen that stones fall faster than leaves fluttering in the air. Without investigating further, we can easily accept false ideas.

Galileo very carefully examined Aristotle's hypothesis. Then he did something that caught on and changed science forever. He *experimented.* Galileo showed the falseness of Aristotle's claim with a single experiment—dropping heavy and light objects from the Leaning Tower of Pisa. Legend tells us that the objects fell at equal speeds. In the scientific spirit, one experiment that can be reproduced outweighs any authority, regardless of reputation or the number of advocates.

Experiment, not philosophical discussion, decides what is correct in science.

Scientists must accept their experimental findings even when they would like them to be different. They must strive to distinguish between the results they see and those they wish to see. This is not easy. Scientists, like most people, are capable of fooling themselves. People have always tended to adopt general rules, beliefs, creeds, ideas, and hypotheses without thoroughly questioning their validity. And sometimes we retain these ideas long after they have been shown to be meaningless, false, or at least questionable. The most widespread assumptions are often the least questioned. Too often, when an idea is adopted, great attention is given to the instances that support it. Contrary evidence is often distorted, belittled, or ignored.

The fact that scientific statements will be thoroughly tested before they are believed helps to keep science honest. Sooner or later, mistakes (or deceptions) are found out. A scientist exposed for cheating doesn't get a second chance in the community of scientists. Honesty, so important to the progress of science, thus becomes a matter of self-interest to scientists. There is relatively little bluffing in a game where all bets are called.

1.6 Facts, Theories, and Laws

LEARNING OBJECTIVE
Distinguish among facts, theories, and laws.

EXPLAIN THIS How does a theory relate to a collection of facts?

When a scientific hypothesis has been tested over and over again and has not been contradicted, it may become known as a **law** or *principle.* A scientific **fact**, on the other hand, is generally something that competent observers can observe and agree to be true. For example, it is a fact that an amputated limb of a salamander can grow back. Anyone can watch it happen. It is not a fact—yet—that a severed limb of a human can grow back.

Scientists use the word *theory* in a way that differs from its use in everyday speech. In everyday speech, a theory is the same as a hypothesis—a statement that hasn't been tested. But scientifically speaking, a **theory** is a synthesis of facts and well-tested hypotheses. Physicists use quantum theory to explain the behavior of light. Chemists have theories about how atoms bond to form molecules. The theory of evolution is key to the life sciences. Earth scientists use the theory of plate tectonics to explain why the continents move, and astronomers speak of the theory of the Big Bang to account for the observation that galaxies are moving away from one another.

Facts are revisable data about the world.
Theories interpret facts.

Theories are a foundation of science, but they are not fixed. Rather, they evolve. They pass through stages of refinement. For example, since the theory of

SCIENCE AND SOCIETY

Pseudoscience

For a claim to qualify as "scientific" it must meet certain standards. For example, the claim must be reproducible by others who have no stake in whether the claim is true or false. The data and subsequent interpretations are open to scrutiny in a social environment where it's okay to have made an honest mistake but not okay to have been dishonest or deceiving. Claims that are presented as scientific but do not meet these standards are what we call **pseudoscience**, which literally means "fake science." In the realm of pseudoscience, skepticism and tests for possible wrongness are downplayed or flatly ignored.

Examples of pseudoscience abound. Astrology is an ancient belief system that supposes a person's future is determined by the positions and movements of planets and other celestial bodies. Astrology mimics science in that astrological predictions are based on careful astronomical observations. Yet astrology is not a science because there is no validity to the claim that the positions of celestial objects influence the events of a person's life. After all, the gravitational force exerted by celestial bodies on a person is smaller than the gravitational force exerted by objects making up the earthly environment: trees, chairs, other people, bars of soap, and so on. Further, the predictions of astrology do not hold true; there just is no evidence that astrology works.

For more examples of pseudoscience, turn on the television. You can find advertisements for a plethora of pseudoscientific products. Watch out for remedies for ailments from baldness to obesity to cancer, for air-purifying mechanisms, and for "germ-fighting" cleaning products in particular. While many such products do operate on solid science, others are pure pseudoscience. Buyer beware!

Humans are very good at denial, which may explain why pseudoscience is such a thriving enterprise. Many pseudoscientists themselves do not recognize their efforts as pseudoscience. A practitioner of "absent healing," for example, may truly believe in her ability to cure people she will never meet except through e-mail and credit card exchanges. She may even find anecdotal evidence to support her contentions. The placebo effect, as discussed in Chapter 20, can mask the ineffectiveness of various healing modalities. In terms of the human body, what people believe *will* happen often *can* happen because of the physical connection between the mind and the body.

That said, consider the enormous downside of pseudoscientific practices. Today there are more than 20,000 practicing astrologers in the United States. Do people listen to these astrologers just for the fun of it? Or do they base important decisions on astrology? You might lose money by listening to pseudoscientific entrepreneurs or, worse, you could become ill. Delusional thinking, in general, carries risk.

Meanwhile, the results of science literacy tests given to the general public show that most Americans lack an understanding of the basic concepts of science. Some 63% of American adults are unaware that the mass extinction of the dinosaurs occurred long before the first human evolved; 75% do not know that antibiotics kill bacteria but not viruses; 57% do not know that electrons are smaller than atoms. What we find is a rift—a growing divide—between those who have a realistic sense of the capabilities of science and those who do not understand the nature of science and its core concepts or, worse, think that scientific knowledge is too complex for them to understand. Science is a powerful method for understanding the physical world—and a whole lot more reliable than pseudoscience as a means for bettering the human condition.

Those who can make you believe absurdities can make you commit atrocities.—*Voltaire*

the atom was proposed 200 years ago, it has been refined many times in light of new evidence. Those who know only a little about science may argue that scientific theories can't be taken seriously because they are always changing. Those who understand science, however, see it differently: Theories grow stronger and more precise as they evolve to include new information.

LEARNING OBJECTIVE
Distinguish between the natural and the supernatural.

1.7 Science Has Limitations

EXPLAIN THIS What is the fundamental difference between the natural and the supernational?

Science deals with only hypotheses that are testable. Its domain is therefore restricted to the observable natural world. Although scientific methods can be used to debunk various paranormal claims, they have no way of accounting for testimonies involving the supernatural. The term *supernatural* literally means "above nature." Science works within nature, not above it. Likewise, science is unable to answer philosophical questions, such as What is the purpose of life?, or religious questions, such as What is the nature of the human spirit? Although these questions are valid and may have great importance to us, they rely on subjective personal experience and do not lead to testable hypotheses. They lie outside the realm of science.

We each need a *knowledge filter* to tell the difference between what is true and what only pretends to be true. The best knowledge filter ever invented for explaining the physical world is science.

1.8 Science, Art, and Religion

LEARNING OBJECTIVE
Discuss some similarities and differences among science, art, and religion.

EXPLAIN THIS When are science and religion compatible, and when are they incompatible?

The search for a deeper understanding of the world around us has taken different forms, including science, art, and religion. Science is a system by which we discover and record physical phenomena and think about possible explanations for such phenomena. The arts are concerned with personal interpretation and creative expression. Religion addresses the source, purpose, and meaning of it all. Simply put, science asks *how*, art asks *who*, and religion asks *why*.

Science and the arts have certain things in common. In the art of literature, we find out about what is possible in human experience. We can learn about emotions from rage to love, even if we haven't yet experienced them. The arts describe these experiences and suggest what may be possible for us. Similarly, knowledge of science tells us what is possible in nature. Scientific knowledge helps us to predict possibilities in nature even before they have been experienced. It provides us with a way of connecting things, of seeing relationships between and among them, and of making sense of the great variety of natural events around us. While art broadens our understanding of ourselves, science broadens our understanding of our environment.

Science and religion have similarities also. For example, both are motivated by curiosity about the natural world. Both have great impact on society. Science, for example, leads to useful technological innovations, while religion provides a foothold for many social services. Science and religion, however, are basically different. Science is concerned with understanding the physical universe, whereas many religions are concerned with spiritual matters, such as belief and faith. Scientific truth is a matter of public scrutiny; religion is a deeply personal matter. In these respects, science and religion are as different as apples and oranges and yet do not contradict each other.

Art is about cosmic beauty. Science is about cosmic order. Religion is about cosmic purpose.

Ultimately, in learning more about science, art, and religion, we find that they are not mutually exclusive. Rather, they run parallel to each other like strings on a guitar, each resonating at its own frequency. When played together, they can produce a chord that is profoundly rich. Science, art, and religion can work very well together, which is why we should never feel forced into choosing one over another.

That science and religion can work very well together deserves special emphasis. When we study the nature of light later in this book, we will treat light as both a wave and a particle. At first, waves and particles may appear to be contradictory. You might believe that light can be only one or the other, and that you must choose between them. What scientists have discovered, however, is that light waves and light particles *complement* each other and that, when these two

No wars have ever been fought over science.

ideas are taken together, they provide a deeper understanding of light. In a similar way, it is mainly people who are either uninformed or misinformed about the deeper natures of both science and religion who feel that they must choose between believing in religion and believing in science. Unless one has a shallow understanding of either or both, there is no contradiction in being religious in one's belief system and being scientific in one's understanding of the natural world.* What your religious beliefs are and whether you have any religion at all are, of course, private matters for you to decide. The tangling up of science and religion has led to many unfortunate arguments over the course of human history.

CHECK YOURSELF

Which of the following activities involves the utmost human expression of passion, talent, and intelligence?

(a) painting and sculpture
(b) literature
(c) music
(d) religion
(e) science

CHECK YOUR ANSWER

All of them. In this book, we focus on science, which is an enchanting human activity shared by a wide variety of people. With present-day tools and know-how, scientists are reaching further and finding out more about themselves and their environment than people in the past were ever able to do. The more you know about science, the more passionate you feel toward your surroundings. There is science in everything you see, hear, smell, taste, and touch!

LEARNING OBJECTIVE
Relate technology to the furthering of science, and science to the furthering of technology.

1.9 Technology—The Practical Use of Science

EXPLAIN THIS What does it mean to say that technology is a double-edged sword?

Science and technology are also different from each other. Science is concerned with gathering knowledge and organizing it. **Technology** enables humans to use that knowledge for practical purposes, and it provides the instruments scientists need to conduct their investigations.

Technology is a double-edged sword. It can be both helpful and harmful. We have the technology, for example, to extract fossil fuels from the ground and then burn the fossil fuels to produce useful energy. Energy production from fossil fuels has benefited society in countless ways. On the flip side, the burning of fossil fuels damages the environment. It is tempting to blame technology itself for such problems as pollution, resource depletion, and even overpopulation. These problems, however, are not the fault of technology any more than a stabbing is the fault of the knife. It is humans who use the technology, and humans who are responsible for how it is used.

*Of course, this does not apply to certain religious extremists, who steadfastly assert that one cannot embrace both their brand of religion and science, and aspects of some religions, including the world's largest ones, that are distinctly anti-science.

Remarkably, we already possess the technology to solve many environmental problems. This 21st century will likely see a switch from fossil fuels to more sustainable energy sources. We recycle waste products in new and better ways. In some parts of the world, progress is being made toward limiting human population growth, a serious threat that worsens almost every problem faced by humans today. Difficulty solving today's problems results more from social inertia than from failing technology. Technology is our tool. What we do with this tool is up to us. The promise of technology is a cleaner and healthier world. Wise applications of technology *can* improve conditions on planet Earth.

There are many paths scientists can follow in doing science. Scientists who explore the ocean floor or who chart new galaxies, for example, are focused on making and recording new observations.

1.10 The Natural Sciences: Physics, Chemistry, Biology, Earth Science, and Astronomy

LEARNING OBJECTIVE
Compare the fields of physics, chemistry, biology, Earth science, and astronomy.

EXPLAIN THIS Why is physics considered to be the basic science?

Science is the present-day equivalent of what used to be called *natural philosophy.* Natural philosophy was the study of unanswered questions about nature. As the answers were found, they became part of what is now called *science.* The study of science today branches into the study of living things and nonliving things: the life sciences and the physical sciences. The *life sciences* branch into such areas as molecular biology, microbiology, and ecology. The *physical sciences* branch into such areas as physics, chemistry, the Earth sciences, and astronomy. In this book, we address the life sciences and physical sciences and the ways in which they overlap—or *integrate.* This gives you a foundation for more specialized study in the future and a framework for understanding science in everyday life and in the news, from the greenhouse effect to tsunamis to genetic engineering.

A few words of explanation about each of the major divisions of science: Physics is the study of such concepts as motion, force, energy, matter, heat, sound, light, and the components of atoms. Chemistry builds on physics by telling us how matter is put together, how atoms combine to form molecules, and how the molecules combine to make the materials around us. Physics and chemistry, applied to Earth and its processes, make up Earth science—geology, meteorology, and oceanography. When we apply physics, chemistry, and geology to other planets and to the stars, we are speaking about astronomy. Biology is more complex than the physical sciences because it involves matter that is alive. Underlying biology is chemistry, and underlying chemistry is physics. So physics is basic to both the physical sciences and the life sciences. That is why we begin this book with physics, then follow with chemistry and biology, and finally investigate Earth science and conclude with astronomy. All are treated conceptually, with the twin goals of enjoyment and understanding.

1.11 Integrated Science

LEARNING OBJECTIVE
Relate learning integrated science to an increased appreciation of nature.

EXPLAIN THIS Who gets the most out of something—one who has an understanding of it or one without understanding?

Just as you can't enjoy a ball game, computer game, or party game until you know its rules, so it is with nature. Because science helps us learn the rules of nature, it also helps us appreciate nature. You may see beauty in a tree, but you'll see more beauty in that tree when you understand how trees and other plants trap solar energy and convert it into the chemical energy that sustains nearly all life on Earth. Similarly, when you look at the stars, your sense of their beauty is enhanced if you know how stars are born from mere clouds of gas and dust—with a little help from the laws of physics, of course. And how much richer it is, when you look at the myriad objects in your environment, to know that they are all composed of atoms—amazing, ancient, invisible systems of particles regulated by an eminently knowable set of laws.

Understanding the physical world—to appreciate it more deeply or to have the power to alter it—requires concepts from different branches of science. For example, the process by which a tree transforms solar energy to chemical energy—photosynthesis—involves the ideas of radiant energy (physics), bonds in molecules (chemistry), gases in the atmosphere (Earth science), the Sun (astronomy), and the nature of life (biology). Thus, for a complete understanding of photosynthesis and its importance, concepts beyond biology are required. And so it is for most of the real-world phenomena we are interested in. Put another way, the physical world integrates science, so to understand the world we need to look at science in an integrated way.

If the complexity of science intimidates you, bear this in mind: All the branches of science rest upon a relatively small number of basic ideas. Some of the most important unifying concepts are identified at the back of this book and in the page margins where they come up. Learn these underlying ideas, and you will have a tool kit to bring to any phenomenon you wish to understand.

Go to it—we live in a time of rapid and fascinating scientific discovery!

LEARNING OBJECTIVE
Apply the scientific method to a biological investigation.

Integrated Science 1A
CHEMISTRY AND BIOLOGY

An Investigation of Sea Butterflies

EXPLAIN THIS How does reproducibility relate to the validity of research?

Let's consider an example of a recent scientific research project that shows how the scientific method can be put to work. Along the way, we'll get a taste of how biology and chemistry are integrated in the physical world.

The Antarctic research team headed by James McClintock, Professor of Biology at the University of Alabama at Birmingham, and Bill Baker, Professor of Chemistry at the University of South Florida, was studying the toxic chemicals Antarctic marine organisms secrete to defend themselves against predators (Figure 1.2). McClintock and Baker observed an unusual relationship between two animal species, a sea butterfly and an amphipod—a relationship that led to a question, a scientific hypothesis, a prediction, tests concerning the chemicals involved in the relationship, and finally a conclusion. The research generally proceeded according to the steps of the classic scientific method.

UNIFYING CONCEPT
- *The Scientific Method*
Section 1.3

1. **Observe** The sea butterfly *Clione Antarctica* is a brightly colored, shell-less snail with winglike extensions used in swimming (Figure 1.3a), and the amphipod *Hyperiella dilatata* resembles a small shrimp. McClintock and Baker observed a large percentage of amphipods carrying sea butterflies on their backs, with the sea butterflies held tightly by the legs of the amphipods (Figure 1.3b). Any amphipod that lost its sea butterfly would quickly seek another—the amphipods were actively abducting the sea butterflies!

FIGURE 1.2
The Chemical Ecology of Antarctic Marine Organisms Research Project was initiated in 1988 by James McClintock, shown here (fifth from left) with his team of colleagues and research assistants. In 1992, he was joined by Bill Baker (second from right). Baker is shown in the inset dressing for a dive into the icy Antarctic water. Like many other science projects, this one was interdisciplinary, involving the efforts of scientists from a wide variety of backgrounds.

2. **Question** McClintock and Baker noted that amphipods carrying butterflies were slowed considerably, making the amphipods more vulnerable to predators and less adept at catching prey. Why then did the amphipods abduct the sea butterflies?
3. **Hypothesize** Given their experience with the chemical defense systems of various sea organisms, the research team hypothesized that amphipods carry sea butterflies to produce a chemical that deters a predator of the amphipod.
4. **Predict** Based on their hypothesis, they predicted (a) that they would be able to isolate this chemical and (b) that an amphipod predator would be deterred by it.
5. **Test predictions** To test their hypothesis and predictions, the researchers captured several predator fish species and conducted the test shown in Figure 1.4. The fish were presented with solitary sea butterflies, which they took into their mouths but promptly spat back out. The fish readily ate uncoupled amphipods but spit out any amphipod coupled with a sea butterfly. These are the results expected if the sea butterfly was secreting some sort of chemical deterrent. The same results would be obtained, however, if a predator fish simply didn't like the feel of a sea butterfly in its mouth. The results of this simple test were therefore ambiguous. A conclusion could not yet be drawn.

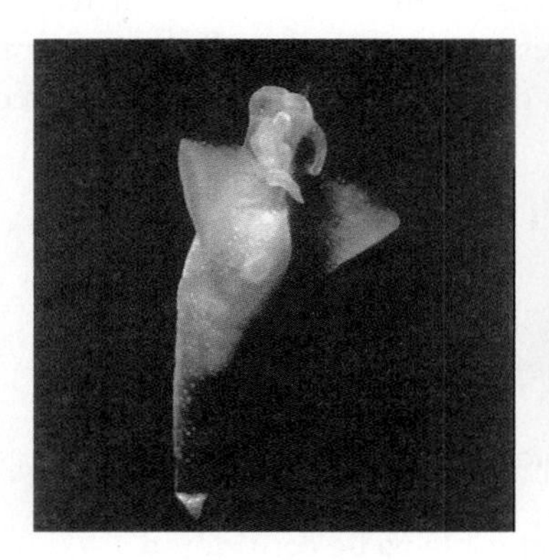

(a)

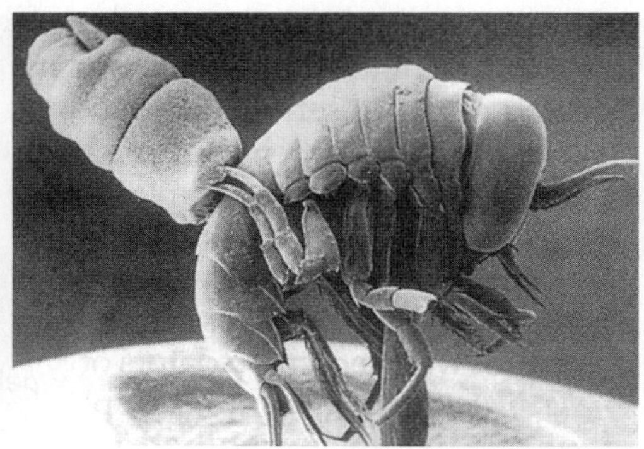

(b)

FIGURE 1.3
(a) The graceful Antarctic sea butterfly is a species of snail that does not have a shell. (b) The shrimplike amphipod attaches a sea butterfly to its back even though doing so limits the amphipod's mobility.

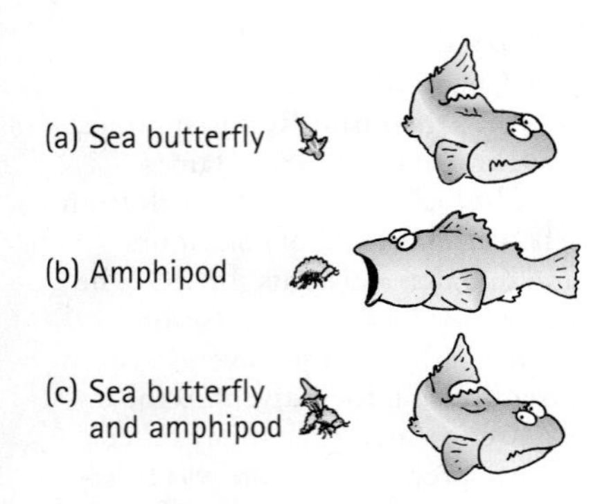

FIGURE 1.4
In McClintock and Baker's initial experiment, a predatory fish (a) rejected the sea butterfly, (b) ate the free-swimming amphipod, and (c) rejected the amphipod coupled with a sea butterfly.

All scientific tests need to minimize the number of possible conclusions. Often this is done by running an experimental test along with a **control**. Ideally, the experimental test and the control should differ by only one variable. Any differences in results can then be attributed to how the experimental test differed from the control.

To confirm that the deterrent was chemical and not physical, the researchers made one set of food pellets containing both fish meal and sea butterfly extract (the experimental pellets). For their control test, they made a physically identical set containing only fish meal (the control pellets). As shown in Figure 1.5, the predator fish readily ate the control pellets but not the experimental ones. These results strongly supported the chemical hypothesis.

Further processing of the sea butterfly extract yielded five major chemical compounds, only one of which deterred the predator fish from eating the pellets. Chemical analysis of this compound revealed it to be the previously unknown molecule shown in Figure 1.6, which they named pteroenone.

6. **Draw a conclusion** In addition to running control tests, scientists confirm experimental results by repeated testing. In this case, the Antarctic researchers made many food pellets, both experimental and control, so that each test could be repeated many times. Only after obtaining consistent results in repeated tests can a scientist draw a conclusion. McClintock and Baker were thus able to conclude that amphipods abduct sea butterflies in order to use the sea butterflies' secretion of pteroenone as a defense against predator fish.

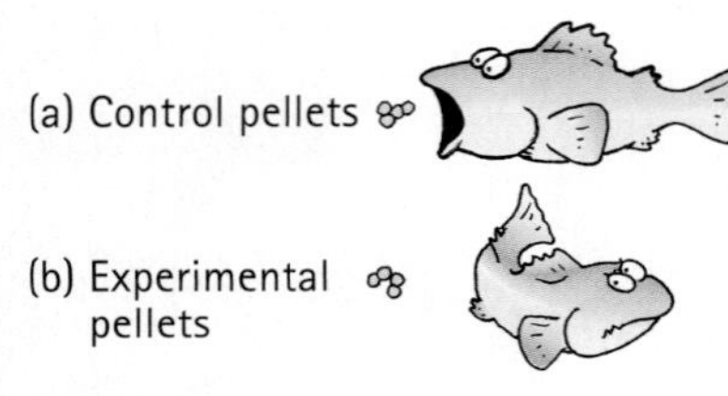

FIGURE 1.5
The predator fish (a) ate the control pellets but (b) rejected the experimental pellets, which contained sea butterfly extract.

Yet, this conclusion would still be regarded with skepticism in the scientific community. Why? There is a great potential for unseen error in any experiment. A laboratory may have faulty equipment that leads to consistently wrong results, for example. Because of the potential for unseen error from any particular research group, experimental results must be *reproducible* to be considered valid. This means that other scientists must be able to reproduce the same experimental findings in separate experiments. Thus you can see that it is a long road from bright idea to accepted scientific finding! The plodding, painstaking nature of this process is beneficial, though—it is the reason that scientific knowledge is highly trustworthy.

As frequently happens in science, McClintock and Baker's results led to new questions. What are the properties of pteroenone? Does this substance have applications—for example, can it be used as a pest repellent? Could it be useful for treating human disease? In fact, a majority of the chemicals we use were originally discovered in natural sources. This illustrates that there is an important reason for preserving marine habitats, tropical rainforests, and the other diverse natural environments on Earth—they are storehouses of countless yet-to-be-discovered substances.

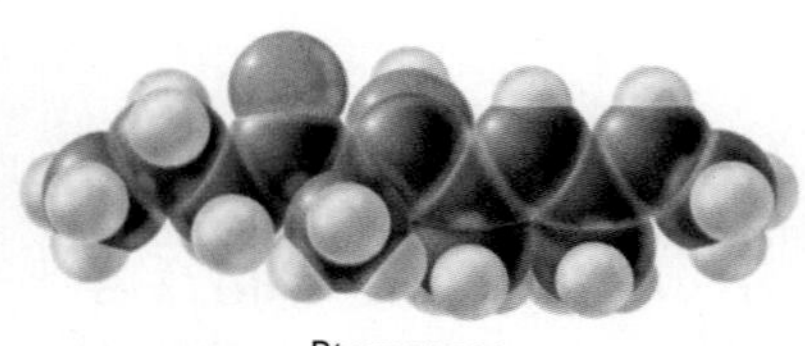

FIGURE 1.6
Pteroenone is a molecule produced by sea butterflies as a chemical deterrent against predators. Its name is derived from *ptero-*, which means "winged" (for the sea butterfly), and *-enone*, which describes information about the chemical structure. The black spheres represent carbon atoms, the white spheres hydrogen atoms, and the red spheres oxygen atoms.

CHECK YOURSELF

1. **What variable did the experimental fish pellets contain that was not found in the control pellets?**
2. **If the fish had eaten the experimental pellets, what conclusion could the scientists have drawn?**
3. **Why must experimental findings be reproducible to be considered valid?**

CHECK YOUR ANSWERS

1. Sea butterfly extract.
2. The scientists would have had to conclude that the predator fish were not deterred by the sea butterfly secretions and thus that the amphipods did not capture the sea butterflies for this reason.
3. Reproducibility of results is essential because every research project may contain unseen errors.

For instructor-assigned homework, go to www.masteringphysics.com 

SUMMARY OF TERMS (KNOWLEDGE)

Control A test that excludes the variable being investigated in a scientific experiment.

Fact A phenomenon about which competent observers can agree.

Hypothesis An educated guess or a reasonable explanation. When the hypothesis can be tested by experiment, it qualifies as a *scientific hypothesis.*

Law A general hypothesis or statement about the relationship of natural quantities that has been tested over and over again and has not been contradicted; also known as a *principle.*

Principle of falsifiability For a hypothesis to be considered scientific, it must be testable—it must, in principle, be capable of being proven wrong.

Pseudoscience A theory or practice that is considered to be without scientific foundation but purports to use the methods of science.

Science The collective findings of humans about nature, and the process of gathering and organizing knowledge about nature.

Scientific method An orderly method for gaining, organizing, and applying new knowledge.

Technology The means of solving practical problems by applying the findings of science.

Theory A synthesis of a large body of information that encompasses well-tested hypotheses about certain aspects of the natural world.

READING CHECK QUESTIONS (COMPREHENSION)

1.1 A Brief History of Advances in Science

1. What launched the era of modern science in the 17th century?

1.2 Mathematics and Conceptual Integrated Science

2. Why do we believe that focusing on math too early is a mistake in an introductory science course?

1.3 The Scientific Method—A Classic Tool

3. Specifically, what do we mean when we say that a scientific hypothesis must be testable?

1.4 The Scientific Hypothesis

4. Is any hypothesis that is not scientific necessarily unreasonable? Explain.

1.5 The Scientific Experiment

5. How did Galileo disprove Aristotle's idea that heavy objects fall faster than light objects?

1.6 Facts, Theories, and Laws

6. Distinguish among a scientific fact, a hypothesis, a law, and a theory.
7. How does the definition of the word *theory* differ in science versus in everyday life?

1.7 Science Has Limitations

8. Your friend says that scientific theories cannot be believed because they are always changing. What can you say to counter this argument?

1.8 Science, Art, and Religion

9. What is meant by the term *supernatural*, and why doesn't science deal with the supernatural?
10. Why do religious questions—such as What is the nature of the human spirit?—lie outside of the domain of science?

1.9 Technology—The Practical Use of Science

11. Clearly distinguish between science and technology.

1.10 The Natural Sciences: Physics, Chemistry, Biology, Earth Science, and Astronomy

12. In what sense does physics underlie chemistry?
13. In what sense is biology more complex than the physical sciences?

1.11 Integrated Science

14. Why should we study integrated science?

THINK INTEGRATED SCIENCE

1A—An Investigation of Sea Butterflies

15. What two scientific disciplines were needed to understand the curious behavior of the Antarctic amphipods?
16. When was a control used in the investigation of the amphipods and sea butterflies? Why was a control necessary?
17. What was McClintock and Baker's hypothesis? Was it a scientific hypothesis? Why?

THINK AND DO (HANDS-ON APPLICATION)

18. Use the scientific method: (1) Based on your observations of your environment, (2) develop a question, (3) hypothesize the answer, (4) predict the consequences if your hypothesis is correct, (5) test your predictions, and (6) draw a conclusion. On a sheet of paper, describe in detail how you performed each step of the method from (1) through (6).

THINK AND SOLVE (MATHEMATICAL APPLICATION)

19. The more candy bars you add to your diet per day, the more weight you gain (all other factors, such as the amount of exercise you get, being equal). Is this an example of a direct proportion or an inverse proportion?
20. State the relation in Exercise 19 in mathematical form. (Hint: Don't forget to include a proportionality constant with appropriate units.)
21. Give an example of two quantities that are related in an inverse proportion that you have observed in your daily life? Express this relation in mathematical form.

THINK AND EXPLAIN (SYNTHESIS)

22. Are the various branches of science separate, or do they overlap? Give several examples to support your answer.
23. In what way is the printing press like the Internet in the history of science?
24. Which of the following are scientific hypotheses? (a) Chlorophyll makes grass green. (b) Earth rotates about its axis because living things need an alternation of light and darkness. (c) Tides are caused by the Moon.

THINK AND DISCUSS (EVALUATION)

25. Discuss the value Galileo placed on experimentation over philosophical discussions.
26. What do science, art, and religion have in common? How are they different?
27. Can a person's religious beliefs be proven wrong? Can a person's understanding of a particular scientific concept be proven wrong?
28. In what sense is science grand and breathtaking? In what sense is it dull and painstaking?

1

PART ONE

Physics

What I enjoy most is discovering that I understand things. Like learning that our Earth is round for the same reason the Moon and Sun are round—gravity. Every bit of mass inside them pulls on every other bit of mass, all pulling inward and making them ball-shaped. If Earth had corners, gravity would pull them in too. Gravity produces ocean tides, the curved paths of baseballs, and the motion of satellites—which orbit because they continually fall around Earth. Physics tells me that electricity and magnetism connect to become light. Since physics is everywhere, it gives me a foundation for integrating chemistry, biology, Earth science, and astronomy. Best of all, learning science conceptually, starting with physics, is phun!

CHAPTER 2
Describing Motion

EVERYTHING MOVES—atoms and molecules; jumping dolphins, all living things and parts of living things; mountains and clouds; stars, planets, galaxies, and even the universe itself. Understanding motion, therefore, is important in all areas of science. Do objects *start* moving spontaneously, or is a force required to make them move? Do they *stop* moving on their own, or is a force required to make them stop? How do objects interact through forces? How does friction operate? By answering questions like these, we can begin to understand many types of motion—from the blowing wind, to your car on a freeway, to the path of a crawling bug.

2.1 Aristotle on Motion

LEARNING OBJECTIVE
Establish Aristotle's influence on classifying motion.

EXPLAIN THIS How did Aristotle classify motion?

Some two thousand years ago, Greek scientists understood some of the physics we understand today. They had a good grasp of the physics of floating objects and of some properties of light, but they were confused about motion. One of the first to study motion seriously was Aristotle, the most outstanding philosopher-scientist in ancient Greece. Aristotle attempted to clarify motion by classification. He classified all motion into two kinds: *natural motion* and *violent motion*. We shall briefly consider each, not as study material but as a background to modern ideas about motion.

In Aristotle's view, natural motion proceeds from the "nature" of an object. He believed that all objects were some combination of four elements—earth, water, air, and fire—and he asserted that motion depends on the particular combination of elements an object contains. He taught that every object in the universe has a proper place, which is determined by its "nature"; any object not in its proper place will "strive" to get there. For example, an unsupported lump of clay, being of the earth, properly falls to the ground; an unimpeded puff of smoke, being of the air, properly rises; a feather properly falls to the ground, but not as rapidly as a lump of clay, because it is a mixture of air and earth. Aristotle stated that heavier objects strive harder and fall faster than lighter ones.

Natural motion was understood to be either straight up or straight down, as in the case of all things on Earth. Natural motion beyond Earth, such as the motion of celestial objects, was circular. Both the Sun and Moon seemed to circle Earth in paths without beginning or end. Aristotle taught that different rules apply in the heavens and that celestial bodies are perfect spheres made of a perfect and unchanging substance, which he called *quintessence*.*

FIGURE 2.1
Does a force keep the cannonball moving after it leaves the cannon?

Violent motion, Aristotle's other class of motion, is produced by pushes and pulls. Violent motion is imposed motion. A person pushing a cart or lifting a heavy boulder imposes motion, as does someone hurling a stone or winning at tug-of-war. The wind imposes motion on ships. Floodwaters impose it on boulders and tree trunks. Violent motion is externally caused and is imparted to objects, which move not of themselves, not by their nature, but because of impressed **forces**—pushes or pulls.

2.2 Galileo's Concept of Inertia

LEARNING OBJECTIVE
Establish Galileo's influence in understanding motion.

EXPLAIN THIS Does a moving object need a force to keep it moving?

Aristotle's ideas were accepted as fact for nearly 2000 years. Then, in the early 1600s, the Italian scientist Galileo demolished Aristotle's belief that heavy objects fall faster than light ones. According to the legend mentioned in Chapter 1, Galileo dropped both heavy and light objects from the Leaning Tower of Pisa (Figure 2.2). He showed that, except for the effects of air friction, objects of different weights fell to the ground in the same amount of time.

FIGURE 2.2
Galileo's famous demonstration.

Galileo made another discovery: He showed that Aristotle was wrong about forces being necessary to keep objects in motion. Although a force is needed to

*Quintessence is the fifth essence, the other four being earth, water, air, and fire.

start an object moving, Galileo showed that, once the object is moving, no force is needed to *keep* it moving—except for the force needed to overcome friction. (We will learn more about friction in Section 2.8.) When friction is absent, a moving object needs no force to keep it moving. It will remain in motion all by itself.

Rather than philosophizing about ideas, Galileo did something that was quite remarkable at the time. As mentioned earlier, Galileo tested his revolutionary idea by *experiment.* This was the beginning of modern science. He rolled balls down inclined planes and observed and recorded the gain in speed as rolling continued (Figure 2.3). On downward-sloping planes, the force of gravity increases a ball's speed. On an upward slope, the force of gravity decreases a ball's speed. What about a ball rolling on a level surface? While rolling on a level surface, the ball rolls neither with nor against the vertical force of gravity—it neither speeds up nor slows down. The rolling ball maintains a constant speed. Galileo reasoned that a ball moving horizontally would move forever if friction were entirely absent (Figure 2.4). Such a ball would move all by itself—of its own *inertia.*

Slope downward–
Speed increases

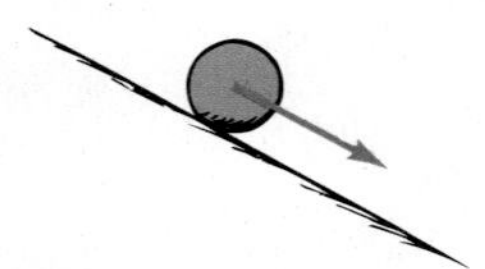

Slope upward–
Speed decreases

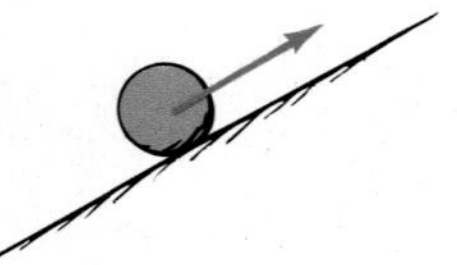

No slope–
Does speed change?

FIGURE 2.3
Motion of balls on different planes.

CHECK YOURSELF

A ball rolling along a level surface slowly comes to a stop. How would Aristotle explain this behavior? How would Galileo explain it? How would you explain it? (Throughout this book think about the Check Yourself questions *before* you read the answers. When you formulate your own answers first, you'll find yourself learning more—much more.)

HISTORY OF SCIENCE

Aristotle (384–322 BC)

Aristotle was the foremost philosopher, scientist, and educator of his time. Born in Greece, he was the son of a physician who personally served the king of Macedonia. At the age of 17, Aristotle entered the Academy of Plato, where he worked and studied for 20 years until Plato's death. He then became the tutor of young Alexander the Great. Eight years later, Aristotle formed his own school. His aim was to systematize existing knowledge, just as Euclid had systematized geometry. Aristotle made critical observations, collected specimens, and gathered, summarized, and classified almost all of the existing knowledge of the physical world. His systematic approach became the method from which Western science later arose. After Aristotle's death, his voluminous notebooks were preserved in caves near his home and were later sold to the library at Alexandria. Scholarly activity ceased in most of Europe throughout the Dark Ages, and the works of Aristotle were forgotten and lost in the scholarship that continued in the Byzantine and Islamic empires. Various texts were reintroduced to Europe during the 11th and 12th centuries and were translated into Latin. The Church, the dominant political and cultural force in Western Europe, at first prohibited the works of Aristotle but later accepted and incorporated them into Christian doctrine.

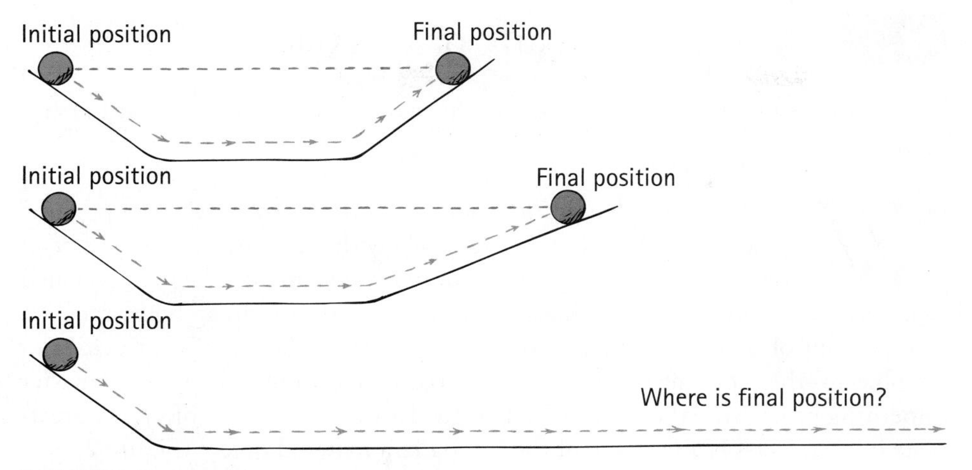

FIGURE 2.4
A ball rolling down an incline on the left tends to roll up to its initial height on the right. The ball must roll a greater distance as the angle of incline on the right is reduced.

CHECK YOUR ANSWERS
Aristotle would probably say that the ball stops because it seeks its natural state of rest. Galileo would probably say that friction overcomes the ball's natural tendency to keep rolling—that friction overcomes the ball's inertia and brings it to a stop. Only you can answer the last question.

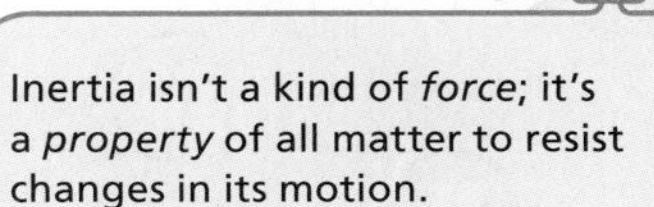

Inertia isn't a kind of *force;* it's a *property* of all matter to resist changes in its motion.

Galileo noted that moving objects tend to remain moving, without the need of an imposed force. Objects at rest tend to remain at rest. This property of objects to maintain their state of motion is called **inertia**.

HISTORY OF SCIENCE

Galileo Galilei (1564–1642)

Galileo was born in Pisa, Italy, in the same year in which Shakespeare was born and Michelangelo died. He studied medicine at the University of Pisa and then changed to mathematics. He developed an early interest in motion and was soon in opposition to others around him, who held to Aristotelian ideas about falling bodies. Galileo left Pisa to teach at the University of Padua, where he became an advocate of the new theory of the solar system advanced by the Polish astronomer Copernicus. Galileo was one of the first to build a telescope, and he was the first to direct it to the nighttime sky and discover mountains on our Moon and the moons of Jupiter. Because he published his findings in Italian instead of Latin, which was expected of so reputable a scholar, and because of the recent invention of the printing press, his ideas reached many people. He soon ran afoul of the Church, however, and was warned not to teach, and not to hold to, Copernican views. Galileo restrained himself publicly for nearly 15 years. Then he defiantly published his observations and conclusions, which opposed Church doctrine. The outcome was a trial in which he was found guilty, and he was forced to renounce his discoveries. By then an old man and broken in health and spirit, Galileo was sentenced to perpetual house arrest. Nevertheless, he completed his studies on motion, and his writings were smuggled out of Italy and published in Holland. His eyes had been damaged earlier by viewing the Sun through a telescope, and that led to his blindness at age 74. He died 4 years later.

LEARNING OBJECTIVE
Describe and distinguish between mass and weight.

2.3 Mass—A Measure of Inertia

EXPLAIN THIS Why is your mass, but not your weight, the same on Earth as on the Moon?

When an object changes its state of motion—by speeding up, slowing down, or changing course—we say that it undergoes *acceleration*. How much acceleration it will undergo depends on the forces applied to it and on the **inertia** of the object—how much it resists changing its motion. The amount of inertia an object possesses depends on the amount of matter in the object—the more matter, the more inertia. In speaking of how much matter something has, we use the term *mass*. The greater the mass of an object, the greater its inertia. Mass is a measure of the inertia in a material object (Figure 2.5).

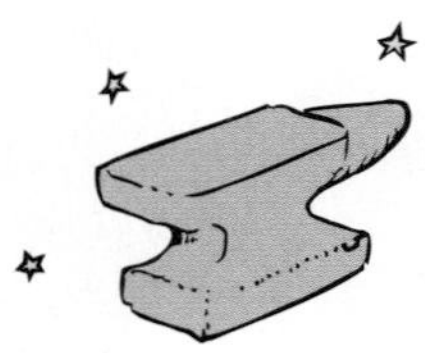

FIGURE 2.5
An anvil in outer space—beyond the Sun, for example—may be weightless, but it still has mass.

Mass corresponds to our intuitive notion of weight. We casually say that something contains a lot of matter if it weighs a lot. But there is a difference between mass and weight. We can define each as follows:

Mass: The quantity of matter in an object. It is also the measure of the inertia or sluggishness that an object exhibits in response to any effort to start it, stop it, or change its state of motion in any way.

Weight: The force on an object due to gravity.

Mass and weight are directly proportional to each other.* If the mass of an object is doubled, its weight is also doubled; if the mass is halved, the weight is halved. Because of this, the concepts of mass and weight are often interchanged. Also, mass and weight are sometimes confused because it is customary to measure the quantity of matter in things (their mass) by their gravitational attraction to Earth (their weight). But mass doesn't depend on gravity. Gravity on the Moon, for example, is much less than it is on Earth. Whereas your weight on the surface of the Moon would be much less than it is on Earth, your mass would be the same in both locations. Mass is a fundamental quantity that completely escapes the notice of most people.

FIGURE 2.6
The astronaut in space finds it is just as difficult to shake the "weightless" anvil as it would be on Earth. If the anvil is more massive than the astronaut, which shakes more—the anvil or the astronaut?

You can sense how much mass is in an object by sensing its inertia. When you shake an object back and forth, you can feel its inertia. If it has a lot of mass, it's difficult to change the object's direction. If it has a small mass, shaking the object is easier. To-and-fro shaking requires the same force even in regions where gravity is different—on the Moon, for example. An object's inertia, or mass, is a property of the object itself and not of its location (Figure 2.6).

Mass is measured in **kilograms** (kg). If an object has a large mass, it may or may not have a large volume. Volume is a measure of space, measured in such units as cubic centimeters, cubic meters, or liters. How many kilograms of matter an object contains and how much space the object occupies are two different things. Mass is different from volume.

A nice demonstration that distinguishes mass from weight is provided by the massive ball suspended on the string that is shown in Figure 2.7. The top string breaks when the lower string is pulled with a gradual increase in force,

*Directly proportional means directly related. If you change one quantity, the other quantity changes proportionally. The constant of proportionality is g, the acceleration due to gravity. As we shall soon see, weight $= mg$ (or mass $\times$ acceleration due to gravity), so 10 N $=$ (1 kg)(10 m/s^2). Later, in Chapter 5, we'll refine our definition of weight to be the gravitational force of a body pressing against a support (such as against a weighing scale).

but the bottom string breaks when the string is jerked. Which break illustrates the weight of the ball, and which illustrates the mass of the ball? Note that only the top string bears the weight of the ball. So, when the lower string is gradually pulled, the tension supplied by the pull is transmitted to the top string. The total tension in the top string is the result of the pull plus the weight of the ball. The top string breaks when the breaking point is reached. But, when the bottom string is jerked, the mass of the ball—its tendency to remain at rest—is responsible for the break of the bottom string.

FIGURE 2.7
Why will a slow, continuous increase in downward force break the string above the massive ball, whereas a sudden increase in downward force will break the lower string?

CHECK YOURSELF

1. **Does a 2-kg block of iron have twice as much *inertia* as a 1-kg block of iron? Twice as much *mass*? Twice as much *weight* when weighed in the same location? Twice as much *volume*?**
2. **Does a 2-kg iron block have twice as much *inertia* as a 1-kg bunch of bananas? Twice as much *mass*? Twice as much *weight* when weighed in the same location? Twice as much *volume*?**
3. **How does the mass of a gold bar vary with location?**

CHECK YOUR ANSWERS

1. The answer is *yes* to all questions. A 2-kg block of iron has twice as many iron atoms and therefore twice the inertia, mass, and weight. The blocks consist of the same material, so the 2-kg block also has twice the volume.
2. Two kg of anything has twice the inertia and twice the mass of 1 kg of anything else. Because mass and weight are proportional at the same location, 2 kg of anything will weigh twice as much as 1 kg of anything. Except for volume, the answer to all questions is *yes*. Volume and mass are proportional only when the materials are identical—when they have the same density. Iron is much more dense than bananas, so 2 kg of iron must occupy less volume than 1 kg of bananas.
3. The mass of a gold bar does not vary. It consists of the same number of atoms no matter what its location. Although its weight may vary with location, it has the same mass everywhere. This is why mass is preferred to weight in scientific studies.

Mass (quantity of matter) and weight (force due to gravity) are directly proportional to each other.

One Kilogram Weighs 10 Newtons

The standard unit of mass is the kilogram, abbreviated kg. The standard unit of force is the **newton**, abbreviated N. The abbreviation N is written with a capital letter because the unit is named after a person. A 1-kg bag of any material at Earth's surface has a weight of 10 N in standard units. Away from Earth's surface, where the force of gravity is less (on the Moon, for example), the bag would weigh less. (In the laboratory when precision is needed, 10 N or 10 m/s^2 will be replaced with the more precise value 9.8 N or 9.8 m/s^2.) In any event, 1 kg of something weighs about 10 N. If you know the mass in kilograms and want the weight in newtons, multiply the number of kilograms by 10. Or, if you know the weight in newtons, divide by 10 and you'll have the mass in kilograms. Weight and mass are proportional to each other.

The relationship between kilograms and pounds is that 1 kg weighs 2.2 lb at Earth's surface. (That means that 1 lb is equal to 4.45 N.)

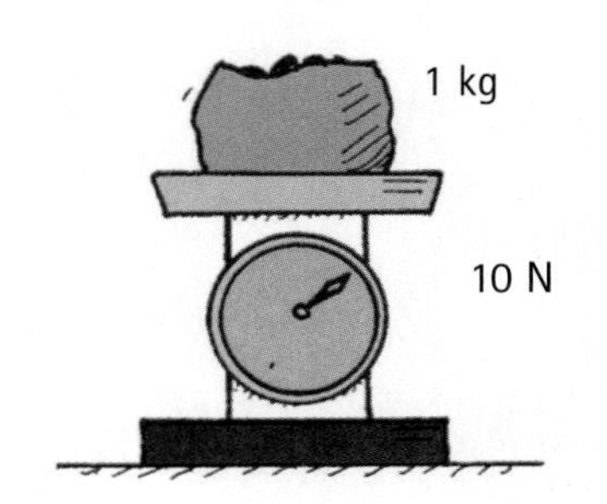

FIGURE 2.8
One kg of nails weighs 10 N, which is equal to 2.2 lb.

Density Is Mass Divided by Volume

UNIFYING CONCEPT
- *Density*

An important property of a material, whether solid, liquid, or gas, is the measure of its compactness: **density**. Density is a measure of how much mass is squeezed into a given space; it is the amount of matter per unit volume:

$$\text{Density} = \frac{\text{mass}}{\text{volume}}$$

A pillow is bigger than a car battery, but which has more matter? Which has more *inertia*? Which has more *mass*?

Like mass and weight, density has to do with the "lightness" or "heaviness" of materials. But the distinction is that density also involves the volume of an object, the space it occupies. For example, a kilogram of lead has the same mass as a kilogram of feathers, and, at the surface of Earth, both of them have the same weight—2.2 lb. But their densities are very different. A kilogram of lead is very dense and would fit into a tennis ball, while the same mass of feathers has a very low density and could adequately stuff the shell of a down-filled sleeping bag. Volume is often measured in cubic centimeters (cm^3) or cubic meters (m^3), so density is most typically expressed in units of g/cm^3 or kg/m^3. Water has a density of 1 g/cm^3. Mercury's density of 13.6 g/cm^3 means that it has 13.6 times as much mass as an equal volume of water. Although different masses of a given material have different volumes, the given materials have the same density.

CHECK YOURSELF

1. **Would 1 kg of gold have the same density on the Moon as on Earth?**
2. **Which has the greater density—an entire candy bar or half a candy bar?**

CHECK YOUR ANSWERS

1. Yes. Since mass and volume remain the same despite gravitational variations, the ratio of mass to volume remains constant as well.
2. Both half a candy bar and an entire candy bar have the same density.

LEARNING OBJECTIVE
Distinguish between force and net force, and give examples.

2.4 Net Force

EXPLAIN THIS When and how can two people push on something to produce a net force of zero?

In simplest terms, a force is a push or a pull. An object doesn't speed up, slow down, or change direction unless a force acts on it. When we say "force," we imply the total force, or **net force**, acting on the object. Often more than one force may be acting on an object. For example, when you throw a softball, the force of gravity and the pushing force you apply with your muscles both act on the ball. When the ball is sailing through the air, the force of gravity and air resistance both act on it. The net force on the ball is the combination of forces. It is the net force that changes an object's state of motion.

For example, suppose you pull on a shoebox with a force of 5 N (slightly more than 1 lb). If your friend also pulls with 5 N in the same direction, the net force on the box is 10 N. If your friend pulls on the box with the same force as you but in the opposite direction, the net force on the box is zero. Now if you increase your pull to 10 N and your friend pulls oppositely with a force of 5 N, the net force is 5 N in the direction of your pull. You can see these examples in Figure 2.9.

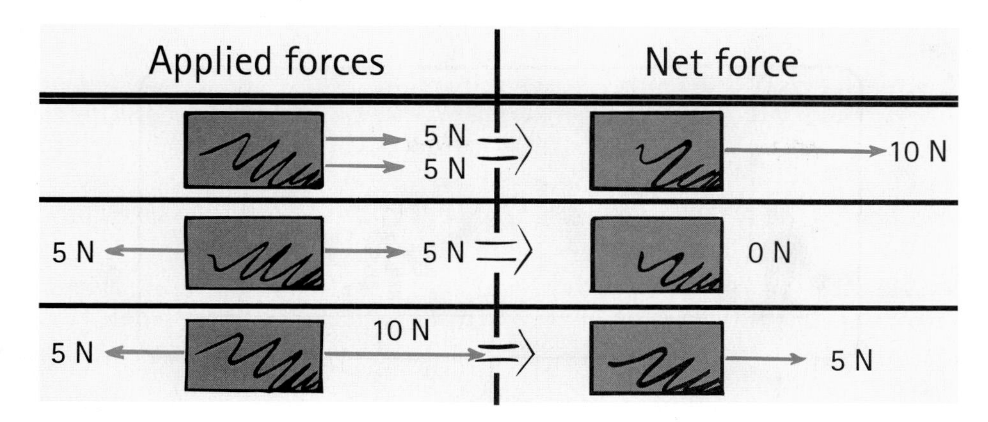

FIGURE 2.9
Net force.

The forces in Figure 2.9 are shown by arrows. Forces are vector quantities. A **vector quantity** has both magnitude (how much) and direction (which way). When an arrow represents a vector quantity, the arrow's length represents magnitude and its direction shows the direction of the quantity. Such an arrow is called a *vector*.

2.5 The Equilibrium Rule

LEARNING OBJECTIVE
Describe the rule $\Sigma F = 0$, and give examples.

EXPLAIN THIS How can the sum of real forces result in no force at all?

If you tie a string around a 2-lb bag of flour and suspend it on a weighing scale, a spring in the scale stretches until the scale reads 2 lb. The stretched spring is under a "stretching force" called *tension*. A scale in a science lab (Figure 2.10) is likely calibrated to read the same force as 9 N. Both pounds and newtons are units of weight, which, in turn, are units of force. The bag of flour is attracted to Earth with a gravitational force of 2 lb—or, equivalently, 9 N. Suspend twice as much flour from the scale and the reading will be 18 N.

Two forces are acting on a bag of flour—tension force acting upward and weight acting downward. The two forces on the bag are equal and opposite, and they cancel to zero. Hence, the bag remains at rest.

When the net force on something is zero, we say that something is in mechanical equilibrium.* Anything in mechanical equilibrium obeys an interesting rule: In mathematical notation, the **equilibrium rule** is

$$\Sigma F = 0$$

The symbol Σ is the capital Greek letter sigma, which stands for "the vector sum of"; F stands for "forces." For a suspended body at rest, like the bag of flour, the equilibrium rule states that the forces acting upward on the body must be balanced by other forces acting downward to make the vector sum equal zero. (Vector quantities take direction into account, so, if upward forces are positive, downward forces are negative, and when summed they equal zero. See Figure 2.11.)

FIGURE 2.10
Burl Grey, who taught the author about tension forces, suspends a 2-lb bag of flour from a spring scale, showing the weight and tension in the string of about 9 N.

*We'll see in Appendix B that another condition for equilibrium is that the net torque is zero.

Everything that isn't undergoing a change in motion is in mechanical equilibrium.
That's because $\Sigma F = 0$.

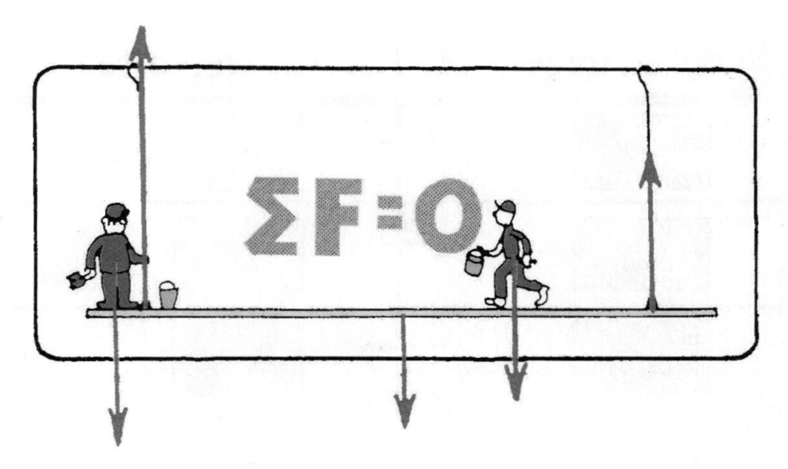

FIGURE 2.11
The sum of the upward vectors equals the sum of the downward vectors. $\Sigma F = 0$, and the scaffold is in equilibrium.

CHECK YOURSELF

Consider the gymnast hanging from the rings.

1. **If Nellie hangs with her weight evenly divided between the two rings, how do the scale readings in both supporting ropes compare with her weight?**
2. **Suppose Nellie hangs with slightly more of her weight supported by the left ring. What does the scale on the right read?**

CHECK YOUR ANSWERS

1. The reading on each scale will be *half her weight*. The sum of the readings on both scales then equals her weight.
2. When more of her weight is supported by the left ring, the reading on the right is *less than half her weight*. No matter how she hangs, the sum of the scale readings equals her weight. For example, if one scale reads two-thirds her weight, the other scale will read one-third her weight. Get it?

LEARNING OBJECTIVE
Define the support force, and describe its relationship to weight.

2.6 The Support Force

EXPLAIN THIS How does the support force relate to weight?

Consider a book lying at rest on a table. It is in equilibrium. What forces act on the book? One is the force due to gravity—the weight of the book. Since the book is in equilibrium, there must be another force acting on it to produce a net force of zero—an upward force opposite to the force of gravity. The table exerts this upward force, called the **support force**. This upward support force, often called the *normal force*, must equal the weight of the book.*

Can you see evidence of $\Sigma F = 0$ in bridges and other structures around you?

*This force acts at right angles to the surface. When we say "normal to," we are saying "at right angles to," which is why this force is called a normal force.

SCIENCE AND SOCIETY

Paul Hewitt and the Origin of Conceptual Integrated Science

Paul Hewitt, the founding author of this book, wrote the physics textbook *Conceptual Physics* when he was a young instructor at City College of San Francisco. *Conceptual Physics* has been the leading physics book for nonscience majors in America for more than 40 years, and it's had a major impact on how science is taught—concepts first, with computations and technical details brought in later. The following is Paul's personal story about how he discovered the fascination of science by observing physics principles at work in everyday life:

When I was in high school, my counselor advised me not to enroll in science and math classes but instead to focus on my interest in art. I took this advice. For a while, my major interests were drawing comic strips and boxing, but neither of these earned me much success. After a stint in the army, I tried my luck at sign painting, and the cold Boston winters drove me south to Miami, Florida. There, at age 26, I got a job painting billboards and met a new friend who became a great intellectual influence for me, Burl Grey. Like me, Burl had never studied physics in high school. But he was passionate about science in general, and he shared his passion with many questions as we painted together.

I remember Burl asking me about the tensions in the ropes that held up the scaffold we were standing on. The scaffold was simply a heavy horizontal plank suspended by a pair of ropes. Burt twanged the rope nearest his end of the scaffold and asked me to do the same with mine. He was comparing the tensions in both ropes—to determine which was greater. Like a more tightly stretched guitar string, the rope with greater tension twangs at a higher pitch. The finding that Burl's rope had a higher pitch seemed reasonable because he was heavier and his rope supported more of the load.

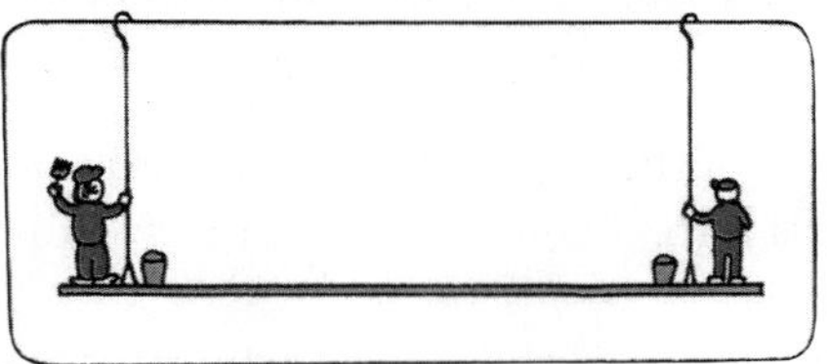

When I walked toward Burl to borrow one of his brushes, he asked if the tensions in the ropes had changed. Did the tension in his rope change as I moved closer? We agreed that it should have, because even more of the load was supported by Burl's rope. How about my rope? Did its tension decrease? We agreed that it would, for it would be supporting less of the total load. I was unaware at the time that I was discussing physics.

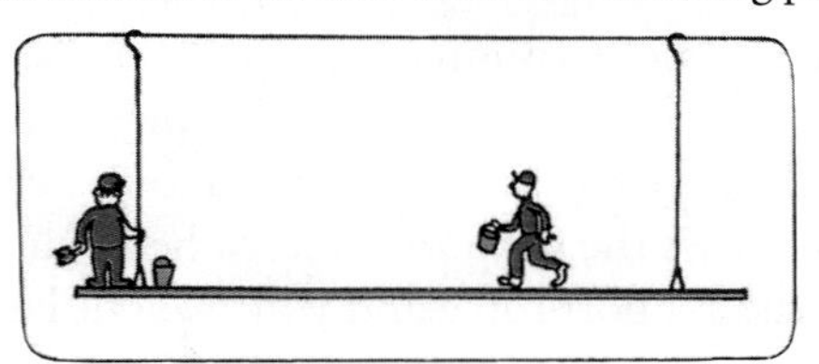

Burl and I used exaggeration to bolster our reasoning (just as physicists do). If we both stood at an extreme end of the scaffold and leaned outward, it was easy to imagine the opposite end of the scaffold rising like the end of a seesaw, with the opposite rope going limp. Then there would be no tension in that rope. We reasoned that the tension in my rope would gradually decrease as I walked toward Burl. It was fun posing such questions and seeing if we could answer them.

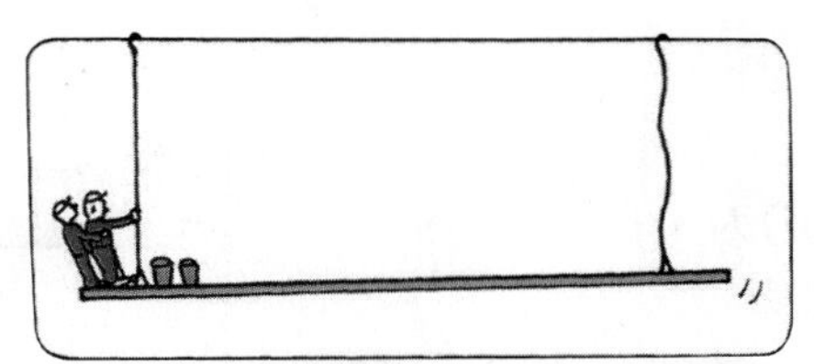

A question we couldn't answer was whether or not the decrease in tension in my rope when I walked away from it would be exactly compensated by a tension increase in Burl's rope. For example, if my rope underwent a decrease of 50 N, would Burl's rope gain 50 N? (We talked about pounds back then, but here we use the scientific unit of force, the *newton*—abbreviated N.) Would the gain be exactly 50 N? And, if so, would this be a grand coincidence? I didn't know the answer until more than a year later, when Burl's stimulation resulted in my leaving full-time painting and going to college to learn more about science.

There I learned that any object at rest, such as the sign-painting scaffold I had worked on with Burl, is said to be in equilibrium. That is, all the forces that act on it balance to zero $(\Sigma F = 0)$. So the upward forces supplied by the supporting ropes indeed do add up to our weights plus the weight of the scaffold. A 50-N loss in one would be offset by a 50-N gain in the other.

I tell this story to make the point that one's thinking is very different when there is a rule to guide it. Now, when I look at any motionless object, I know immediately that all the forces acting on it cancel out. We see nature differently when we know its rules. Nature becomes simpler and easier to understand. Without the rules of physics, we tend to be superstitious and to see magic where there is none. Quite wonderfully, everything is beautifully connected to everything else by a surprisingly small number of rules. Physics is a study of nature's rules.

MATH CONNECTION

Applying the Equilibrium Rule

Use the physics you've learned so far to solve these practice problems.

Problems

1. When Burl stands alone in the exact middle of his scaffold, the reading on the left scale is 500 N. Fill in the reading on the right scale. The total weight of Burl and the scaffold is ______N.

2. Burl moves farther from the left side. Fill in the reading on the right scale.

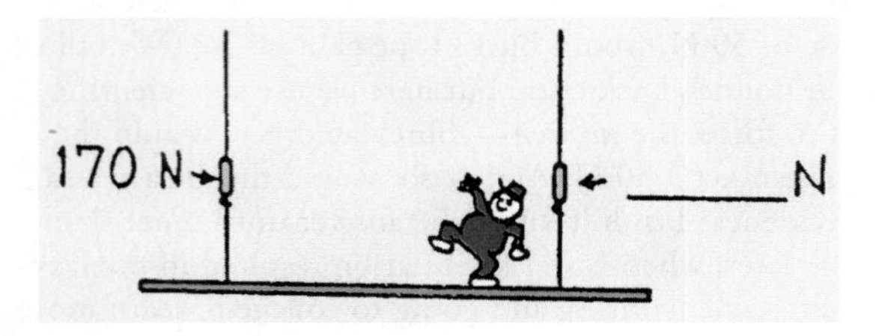

3. In a silly mood, Burl dangles from the right end. Fill in the reading on the right scale.

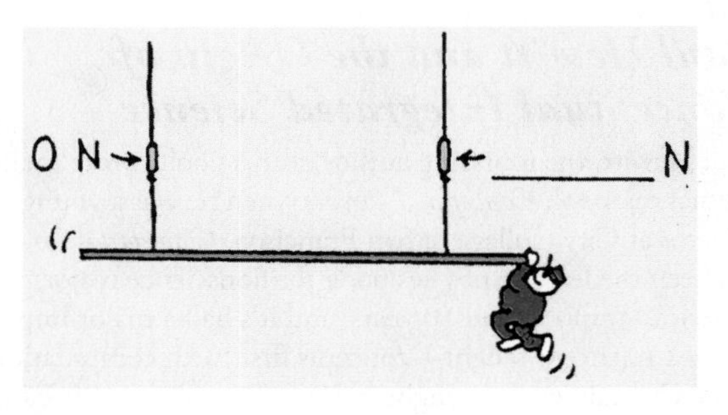

Solutions

1. The total weight is 1000 N. The right rope must be under 500 N of tension because Burl is in the middle and both ropes support his weight equally. Since the sum of the tensions is 1000 N, the total weight of Burl and the scaffold must be 1000 N. Let's call the upward tension forces +1000 N. Then the downward weights are −1000 N. What happens when you add +1000 and −1000? The answer is that they equal zero. So we see that $\Sigma F = 0$.
2. Did you get the correct answer of 830 N? Reasoning: We know from question 1 that the sum of the rope tensions equals 1000 N, and since the left rope has a tension of 170 N, the other rope must make up the difference—that is, 1000 N − 170 N = 830 N. Get it? If so, great. If not, discuss it with your friends until you do. Then read further.
3. The answer is 1000 N. Do you see that this illustrates $\Sigma F = 0$?

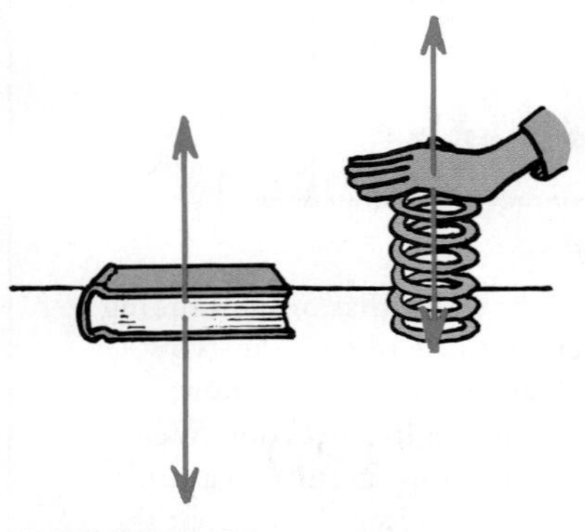

FIGURE 2.12
The table pushes up on the book with as much force as the downward force of gravity on the book. The spring pushes up on your hand with as much force as you exert to push down on the spring.

If we designate the upward force as positive, then the downward force (weight) is negative, and the sum of the two is zero. The net force on the book is zero. Stated another way, $\Sigma F = 0$.

To understand better that the table pushes up on the book, compare the case of compressing a spring (Figure 2.12). If you push the spring down, you can feel the spring pushing up on your hand. Similarly, the book lying on the table compresses the atoms in the table, which behave like microscopic springs. The weight of the book squeezes downward on the atoms, and they squeeze upward on the book. In this way, the compressed atoms produce the support force.

When you step on a bathroom scale, two forces act on the scale. One is the downward pull of gravity (your weight) and the other is the upward support force of the floor. These forces compress a spring that is calibrated to show your weight (Figure 2.13). In effect, the scale shows the support force. When you weigh yourself on a bathroom scale at rest, the support force and your weight have the same magnitude.

CHECK YOURSELF

1. **What is the net force on a bathroom scale when a 150-lb person stands on it?**
2. **Suppose you stand on two bathroom scales with your weight evenly distributed between the scales. What is the reading on each of the scales? What happens when you stand with more of your weight on one foot than on the other?**

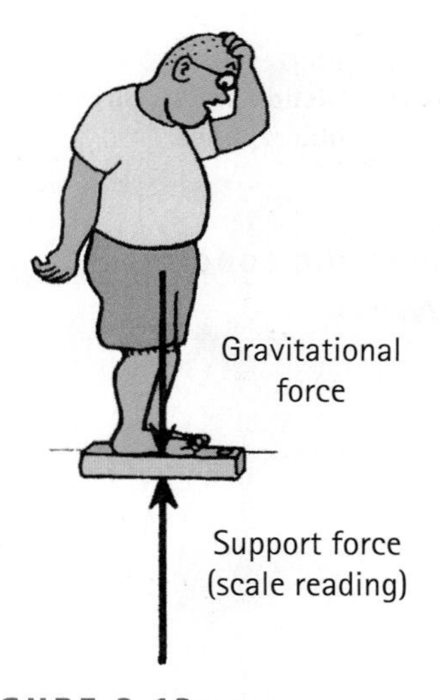

FIGURE 2.13
The upward support force is as much as the downward force of gravity, your weight.

CHECK YOUR ANSWERS

1. Zero, because the scale remains at rest. The scale reads the support force (which has the same magnitude as weight), not the net force.
2. The reading on each scale is half your weight, because the sum of the scale readings must balance your weight, so that the net force on you will be zero. If you lean more on one scale than on the other, more than half your weight will be read on that scale, but less on the other, so they will still add up to your weight. Like the example of the gymnast hanging by the rings, if one scale reads two-thirds of your weight, the other scale will read one-third of your weight.

2.7 Equilibrium of Moving Things

LEARNING OBJECTIVE
Distinguish between equilibrium at rest and when moving.

EXPLAIN THIS How can a crate pushed across a factory floor slide while no net force acts?

Equilibrium is a state of no change. Rest is only one form of equilibrium. An object moving at a constant speed in a straight-line path is also in equilibrium. A bowling ball rolling at a constant speed in a straight line is also in equilibrium—until it hits the pins. Whether at rest or steadily rolling in a straight-line path, $\Sigma F = 0$.

An object under the influence of only one force cannot be in equilibrium. The net force couldn't be zero. Only when two or more forces act on it can the object be in equilibrium. We can test whether or not something is in equilibrium by noting whether or not it undergoes changes in its state of motion.

Consider a crate being pushed across a factory floor. If it moves at a steady speed in a straight-line path, it is in equilibrium. This indicates that more than one force is acting on the crate. Another force exists—likely the force of friction between the crate and the floor. The fact that the net force on the crate equals zero tells us that the force of friction must be equal to, and opposite to, the pushing force (Figure 2.14).

FIGURE 2.14
When the push on the crate is as great as the force of friction between the crate and the floor, the net force on the crate is zero, and the crate slides at an unchanging speed.

LEARNING OBJECTIVE
Describe friction and its direction when an object slides.

UNIFYING CONCEPT
- *Friction*

2.8 The Force of Friction

EXPLAIN THIS How much friction acts on a solid block when it is pushed with constant velocity?

Friction occurs when one object rubs against something else.* Friction occurs for solids, liquids, and gases. An important rule of friction is that it always acts in a direction to oppose motion. If you pull a solid block along a floor to the left, the force of friction on the block will be to the right. A boat propelled to the east by its motor experiences water friction to the west. When an object falls downward through the air, the force of friction, **air resistance**, acts upward. Friction always acts in a direction to oppose motion.

CHECK YOURSELF
An airplane flies through the air at a constant velocity. In other words, it is in equilibrium. Two horizontal forces act on the plane. One is the thrust of the propeller that pushes it forward. The other is the force of air resistance that acts in the opposite direction. Which force is greater?

CHECK YOUR ANSWER
Both horizontal forces have the same magnitude. If you call the forward force exerted by the propeller positive, then the air resistance is negative. Since the plane is in equilibrium, can you see that the two forces combine to zero?

The amount of friction between two surfaces depends on the kinds of material and how much they are pressed together. Friction is due to surface bumps and also to the "stickiness" of the atoms on the surfaces of the two materials (Figure 2.15). The friction between a crate and a smooth wooden floor is less than the friction between the same crate and a rough floor. And, if the crate is full, the friction is more than it would be if the crate were empty because the crate presses down harder on the floor when it weighs more.

When you push horizontally on a crate and it slides across a factory floor, both your force and the opposite force of friction affect the crate's motion. When you push hard enough on the crate to match the friction, the net force on the crate is zero, and it slides at a constant velocity. Notice that we are talking about what we recently learned—that no change in motion occurs when $\Sigma F = 0$.

FIGURE 2.15
Friction results from the mutual contact of irregularities in the surfaces of sliding objects. Even surfaces that appear to be smooth have irregular surfaces when viewed at the microscopic level.

CHECK YOURSELF
1. **Suppose you exert a 100-N horizontal force on a heavy crate resting motionless on a factory floor. The fact that it remains at rest indicates that 100 N isn't great enough to make it slide. How does the force of friction between the crate and the floor compare with your push?**
2. **You push harder—say, 110 N—and the crate still doesn't slide. How much friction acts on the crate?**

*Even though it may not seem so yet, most of the concepts in physics are not really complicated. But friction is different; it is a very complicated phenomenon. The findings are empirical (gained from a wide range of experiments), and the predictions are approximate (also based on experiments).

3. **You push still harder, and the crate moves. Once the crate is in motion, you push with 115 N, which is just sufficient to keep it sliding at a constant velocity. How much friction acts on the crate?**
4. **What net force does a sliding crate experience when you exert a force of 125 N and the friction between the crate and floor is 115 N?**

CHECK YOUR ANSWERS

1. The force of friction is 100 N in the opposite direction. Friction opposes the motion that would occur otherwise. The fact that the crate is at rest is evidence that $\Sigma F = 0$.
2. The friction increases to 110 N; again $\Sigma F = 0$.
3. 115 N, because when the crate is moving at a constant velocity, $\Sigma F = 0$.
4. 10 N, because $\Sigma F = 125\text{ N} - 115\text{ N}$. In this case, the crate *accelerates*.

MasteringPhysics®
VIDEO: Friction

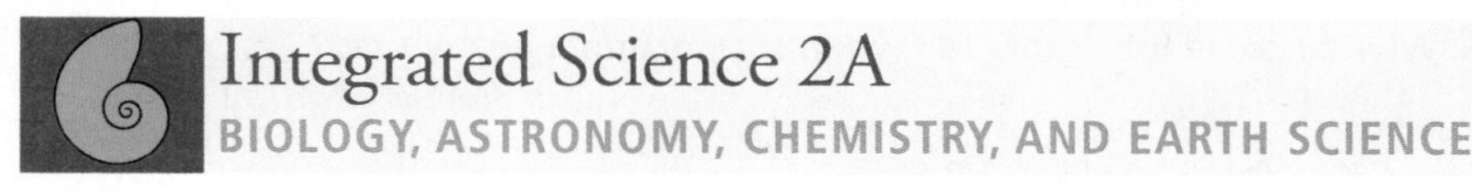

Integrated Science 2A
BIOLOGY, ASTRONOMY, CHEMISTRY, AND EARTH SCIENCE

LEARNING OBJECTIVE
Relate friction to different areas of scientific study.

Friction Is Universal

EXPLAIN THIS How does friction relate to shooting stars?

Friction is the opponent of all motion. If an object moves to the left, friction acts on it to the right. If an object moves upward, friction pushes it downward. Whenever two objects are in contact, friction acts in such a way as to prevent or slow their relative motion.

Your body is well adapted to a friction-filled environment. The fingerprint ridges in your palms and fingers increase surface roughness and so enhance friction between your hands and the things they touch. When your hands are wet and water partially fills in the troughs between the ridges, friction is reduced—and it's easy to drop a glass or a plate. Your toes and the soles of your feet are similarly patterned with grooves and ridges that help you grip the surface of the ground. If not for the friction between your feet and the ground, your feet would slip out from under you like smooth-soled shoes on ice when you tried to walk. It's friction between her hands and the rock that holds the climber to the nearly vertical mountain face (Figure 2.16). Can you see the reason why rock climbers often rub chalk on their hands to absorb hand perspiration before a climb?

FIGURE 2.16
The rock climber grips the sheer rock face with her hands. Why is it important that her hands are dry?

Astronomy has its share of interesting friction effects as well. Shooting stars or *meteors* are bits of material falling through Earth's atmosphere. They are heated to incandescence by friction with the gas particles that make up the atmosphere. For some tiny dust grains, *micrometeoroids*, air resistance is enough to slow them sufficiently so that they do not burn up. Instead, they fall gently to Earth and accumulate, adding hundreds or thousands of tons to Earth's mass every day!

Although friction is often useful, there are many situations in which it reduces efficiency, and minimizing it would save energy—for example, inside most machines. So industry employs chemists to develop lubricants that minimize friction. Lubricants reduce friction by separating two contacting surfaces with an intermediate layer of more slippery material. Then, instead of rubbing against each other, the surfaces rub against the lubricant. Most lubricants are oils or greases. Currently, chemists are trying to develop lubricants that won't evaporate at high temperatures or freeze at low temperatures, won't lock when called upon to carry heavy loads, and won't leak through gaskets and seals when spun at high speeds.

Earthquakes are an Earth-science phenomenon that depends in an obvious way on friction. Earthquakes happen when adjoining, massive blocks of rock are pushed or pulled in different directions. The blocks of rock, locked together by friction, resist motion until the stress becomes too great. At that point, friction is overcome, the blocks of rock let go, and they slip into new positions, releasing energy that vibrates the Earth.

CHECK YOURSELF

1. **Tristan is holding a brick, as shown, by pressing his hands inward on the brick to supply the contact force. In what direction does gravity act? In what direction does friction act? What is the net force on the brick? What does the texture of Tristan's hands have to do with the amount of friction acting on the brick?**
2. **Why do some lubricants fail when supporting a heavy load?**

CHECK YOUR ANSWERS

1. Gravity pulls the brick downward. Friction opposes gravity, so it acts upward. The net force on the brick is zero, as evidenced by no change in its state of motion. Tristan's hands are covered with grooves and ridges that, like treads on a tire, increase friction and improve his grip.
2. Lubricants fail when they too readily evaporate at high temperatures, or freeze at low temperatures, or leak through gaskets and seals when spun at high speeds.

LEARNING OBJECTIVE
Distinguish different kinds of speed and velocity.

2.9 Speed and Velocity

EXPLAIN THIS Under what circumstance can you drive at constant speed while your velocity changes?

Speed

Before the time of Galileo, people described moving things as "slow" or "fast." Such descriptions were vague. Galileo was the first to measure speed by comparing the distance covered with the time it takes to move that distance. He defined **speed** as the distance covered per amount of travel time:

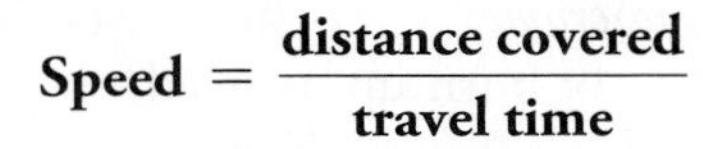

$$\textbf{Speed} = \frac{\textbf{distance covered}}{\textbf{travel time}}$$

For example, if a bicyclist covers 20 km in 1 h, her speed is 20 km/h. Or, if she runs 6 m in 1 s, her speed is 6 m/s.

Any combination of units for distance and time can be used for speed—kilometers per hour (km/h), centimeters per day (the speed of a sick snail), or whatever is useful and convenient. The slash (/) is read as "per" and means "divided by." In science, the preferred unit of speed is meters per second (m/s). Table 2.1 compares some speeds in different units.

FIGURE 2.17
A cheetah can maintain a very high speed, but only for a short time.

TABLE 2.1 APPROXIMATE SPEEDS IN DIFFERENT UNITS

12 mi/h = 20 km/h = 6 m/s (bowling ball)
25 mi/h = 40 km/h = 11 m/s (very good sprinter)
37 mi/h = 60 km/h = 17 m/s (sprinting rabbit)
50 mi/h = 80 km/h = 22 m/s (tsunami)
62 mi/h = 100 km/h = 28 m/s (sprinting cheetah)
75 mi/h = 120 km/h = 33 m/s (batted softball)
100 mi/h = 160 km/h = 44 m/s (batted baseball)

Instantaneous Speed

Moving things often have variations in speed. A car, for example, may travel along a street at 50 km/h, slow to 0 km/h at a red light, and then speed up to only 30 km/h because of traffic. At any instant, you can tell the speed of the car by looking at its speedometer. The speed at any given instant is the *instantaneous speed.*

FIGURE 2.18
A common automobile speedometer. Note that speed is shown both in km/h and in mi/h.

Average Speed

In planning a trip by car, the driver wants to know the travel time. The driver is concerned with the *average speed* for the trip. How is average speed defined?

$$\textbf{Average speed} = \frac{\textbf{total distance covered}}{\textbf{travel time}}$$

Average speed can be calculated rather easily. For example, if you drive a distance of 80 km in 1 h, your average speed is 80 km/h. Likewise, if you travel 320 km in 4 h,

$$\text{Average speed} = \frac{\text{total distance covered}}{\text{travel time}} = \frac{320 \text{ km}}{4 \text{ h}} = 80 \text{ km/h}$$

Note that when a distance in kilometers (km) is divided by a time in hours (h), the answer is in kilometers per hour (km/h).

Since average speed is the entire distance covered divided by the total time of travel, it doesn't indicate the various instantaneous speeds that may have occurred along the way. At any moment on most trips, the instantaneous speed is often quite different from the average speed.

If we know average speed and travel time, the distance traveled is easy to find. A simple rearrangement of the definition above gives

$$\textbf{Total distance covered} = \textbf{average speed} \times \textbf{travel time}$$

For example, if your average speed on a 4-h trip is 80 km/h, then you cover a total distance of 320 km.

If you get a traffic ticket for speeding, is it because of your *instantaneous speed* or your *average speed?*

MasteringPhysics®
VIDEO: Definition of Speed
VIDEO: Average Speed

CHECK YOURSELF

1. **What is the average speed of a cheetah that sprints 100 m in 4 s? How about if it sprints 50 m in 2 s?**
2. **If a car travels at an average speed of 60 km/h for 1 h, it will cover a distance of 60 km. (a) How far would it travel if it moved at this rate for 4 h? (b) For 10 h?**

3. **In addition to the speedometer on the dashboard of every car, there is an odometer, which records the distance traveled. If the initial reading is set at zero at the beginning of a trip and the reading is 40 km after 0.5 h, what was the average speed?**
4. **Would it be possible to attain the average speed in question 3 and never go faster than 80 km/h?**

CHECK YOUR ANSWERS

(Are you reading this before you have thought about the answers in your mind? As mentioned earlier, *think* before you read the answers. You'll not only learn more, but you'll enjoy learning more.)

1. In both cases, the answer is 25 m/s:

$$\text{Average speed} = \frac{\text{total distance covered}}{\text{travel time}} = \frac{100\text{ m}}{4\text{ s}} = \frac{50\text{ m}}{2\text{ s}} = 25\text{ m/s}$$

2. The distance traveled is average speed × time of travel, so:
   (a) Distance $= 60\text{ km/h} \times 4\text{ h} = 240\text{ km}$.
   (b) Distance $= 60\text{ km/h} \times 10\text{ h} = 600\text{ km}$.
3. $\text{Average speed} = \frac{\text{total distance covered}}{\text{travel time}} = \frac{40\text{ km}}{0.5\text{ h}} = 80\text{ km/h}$.
4. No, not if the trip starts from rest and ends at rest. During the trip, there are times when the instantaneous speeds are less than 80 km/h, so the driver must at some time drive faster than 80 km/h in order to average 80 km/h. In practice, average speeds are usually much less than high instantaneous speeds.

FIGURE 2.19
Although Mike Jukes in his 1935 Ford Sprint can maintain a constant speed while rounding the curved part of the track, he cannot maintain a constant velocity on the curve. Why?

Velocity

When we know both the speed and direction of an object, we know its **velocity**. For example, if a vehicle travels at 60 km/h, we know its speed. But, if we say how fast it moves at 60 km/h to the north, we specify its *velocity*. Speed is a description of how fast; velocity is a description of how fast and in what direction. Velocity is a vector quantity.

Constant speed means steady speed, neither speeding up nor slowing down. Constant velocity, on the other hand, means both constant speed and constant direction. Constant direction is a straight line—the object's path doesn't curve. So constant velocity means motion in a straight line at a constant speed—motion, as we shall soon see, with no acceleration.

MasteringPhysics®
VIDEO: Velocity
VIDEO: Changing Velocity

CHECK YOURSELF

"She moves at a constant speed in a constant direction." Restate this sentence in just a few words.

CHECK YOUR ANSWER

"She moves at a constant velocity."

Motion Is Relative

Everything is always moving. Even when you think you're standing still, you're actually speeding through space. You're moving relative to the Sun and stars, although you are at rest relative to Earth. At this moment, your speed relative to the Sun is about 100,000 km/h, and it is even faster relative to the center of our galaxy.

When we discuss the speed or velocity of something, we mean the speed or velocity relative to something else. For example, when we say that a space vehicle travels at 30,000 km/h, we mean relative to Earth. Unless stated otherwise, all speeds discussed in this book are relative to the surface of Earth. Motion is relative (Figure 2.20).

FIGURE 2.20
When you are sitting on a chair, your speed is zero relative to Earth but 30 km/s relative to the Sun.

2.10 Acceleration

EXPLAIN THIS Why is the word *change* important in describing acceleration?

LEARNING OBJECTIVE
Define acceleration, and distinguish it from velocity and speed.

Most moving things usually experience variations in their motion. We say they undergo **acceleration**. The first to formulate the concept of acceleration was Galileo, who developed the idea in his experiments with inclined planes. He found that balls rolling down inclined planes rolled faster and faster. Their velocities changed as they rolled. Further, the balls gained the same amount of speed in equal time intervals (Figure 2.21).

Galileo defined the rate of change of velocity, or acceleration, as:*

$$\textbf{Acceleration} = \frac{\textbf{change of velocity}}{\textbf{time interval}}$$

You experience acceleration when you're in a moving bus. When the driver steps on the gas pedal, the bus picks up speed. We say that the bus accelerates. We can see why the gas pedal is called the "accelerator." When the bus driver applies the brakes, the vehicle slows down. This is also acceleration because the velocity of the vehicle is changing. When something slows down, we often call this *deceleration*.

MasteringPhysics®
VIDEO: Definition of Acceleration
VIDEO: Force Causes Acceleration

Consider driving a car that steadily increases in speed. Suppose that, in 1 s, you steadily increase your velocity from 30 km/h to 35 km/h. In the next second, you steadily go from 35 km/h to 40 km/h, and so on. You change your velocity by 5 km/h each second. Thus we can see that

$$\text{Acceleration} = \frac{\text{change in velocity}}{\text{time interval}} = \frac{5\text{ km/h}}{1\text{ s}} = 5\text{ km/h·s}$$

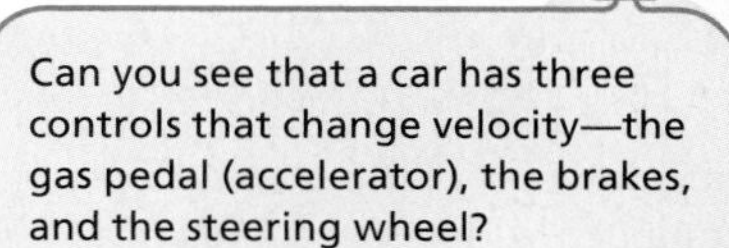

Can you see that a car has three controls that change velocity—the gas pedal (accelerator), the brakes, and the steering wheel?

In this example, the acceleration is described as "5 kilometers per hour-second" (abbreviated as 5 km/h·s). Note that a unit for time enters twice: once for the unit of velocity and again for the interval of time in which the velocity is changing. Also note that acceleration is not just the change in velocity; it is the change

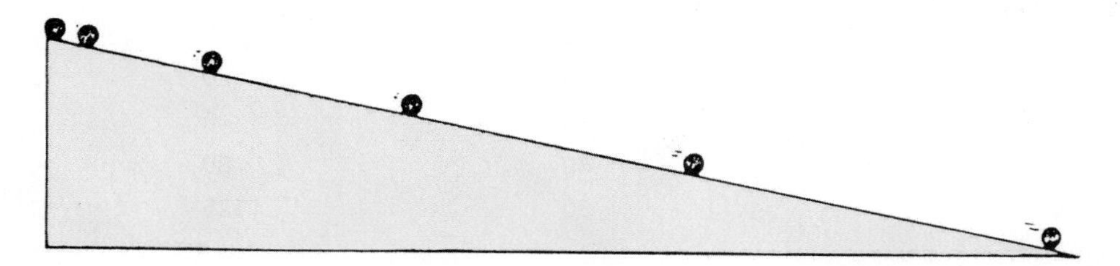

FIGURE 2.21
A ball gains the same amount of speed in equal intervals of time. It undergoes acceleration.

*The capital Greek letter Δ (delta) is often used as a symbol for "change in" or "difference in." In "delta" notation, $a = \Delta v/\Delta t$, where Δv is the change in velocity and Δt is the change in time (the time interval). From this we can see that $\Delta v = a\Delta t$. See the further development of linear motion in Appendix B.

FIGURE 2.22
We say that a body undergoes acceleration when there is a change in its state of motion.

in velocity per second. If either speed or direction changes, or if both change, then velocity changes.

When a car makes a turn, even if its speed does not change, it is accelerating. Can you see why? Acceleration often occurs because the car's direction is changing. Acceleration refers to a change in velocity. So acceleration involves a change in speed, a change in direction, or changes in both speed and direction. Figure 2.22 illustrates this.

CHECK YOURSELF

In 2.0 s, a car increases its speed from 60 km/h to 65 km/h while a bicycle goes from rest to 5 km/h. Which has the greater acceleration?

CHECK YOUR ANSWER

Both have the same acceleration, since both gain the same amount of speed in the same time. Both accelerate at 2.5 km/h·s.

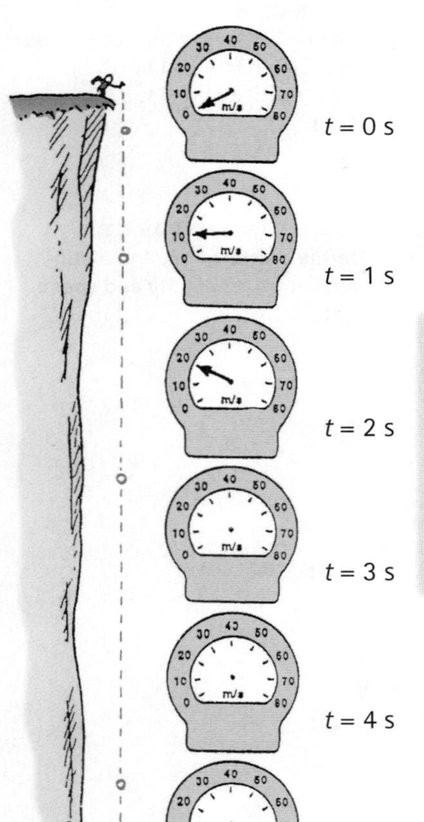

FIGURE 2.23
Imagine that a falling boulder is equipped with a speedometer. In each succeeding second of fall, you'd find the boulder's speed increasing by the same amount: 10 m/s. Sketch in the missing speedometer needle at $t = 3$ s, $t = 4$ s, and $t = 5$ s.

Hold a stone at a height above your head and drop it. It accelerates during its fall. When the only force that acts on a falling object is that due to gravity, when air resistance doesn't affect its motion, we say that the object is in **free fall**. All freely falling objects in the same vicinity undergo the same acceleration. At Earth's surface, a freely falling object gains speed at the rate of 10 m/s each second, as shown in Table 2.2.

$$\text{Acceleration} = \frac{\text{change in velocity}}{\text{time interval}} = \frac{10 \text{ m/s}}{1 \text{ s}} = 10 \text{ m/s}^2$$

TABLE 2.2 FREE FALL

Time of Fall (s)	Speed of Fall (m/s)	Distance of Fall (m)
0	0	0
1	10	5
2	20	20
3	30	45
4	40	80
5	50	125
.	.	.
.	.	.
.	.	.
t	$10t$	$\frac{1}{2}10t^2$

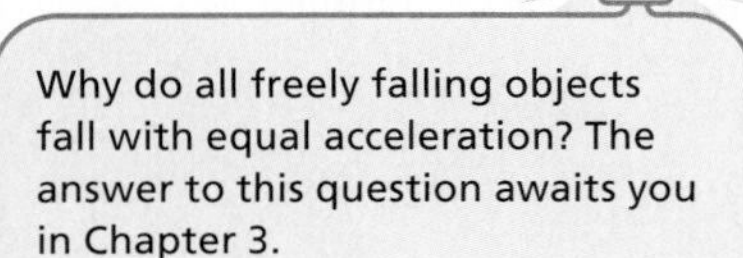

Why do all freely falling objects fall with equal acceleration? The answer to this question awaits you in Chapter 3.

MasteringPhysics®
VIDEO: Free Fall: How Far?

We read the acceleration of free fall as "10 meters per second squared" (more precisely, 9.8 m/s^2). This is the same as saying that acceleration is "10 meters per second per second." Note again that the unit of time, the second, appears twice. It appears once in the units of velocity and once for the time during which the velocity changes.

In Figure 2.23, we imagine a freely falling boulder with a speedometer attached. As the boulder falls, the speedometer shows that the boulder goes 10 m/s faster each second. This 10-m/s gain each second is the boulder's acceleration. (The acceleration of free fall is further developed in Appendix B.)

Up-and-down motion is shown in Figure 2.24. The ball leaves the thrower's hand at 30 m/s. Call this the initial velocity. The figure uses the convention of up as positive and down as negative, indicated by a minus sign (−). Notice that the 1-s interval positions correspond to changes in velocity of 10 m/s.

Aristotle used logic to establish his ideas of motion. Galileo used experiment. Galileo showed that experiments are superior to logic in testing knowledge. Galileo was concerned with how things move rather than why they move. The path was paved for Isaac Newton to make further connections of concepts in motion.

3 s Velocity = 0
2 s v = 10 m/s
4 s v = −10 m/s
1 s v = 20 m/s
5 s v = −20 m/s
0 s v = 30 m/s
6 s v = −30 m/s
7 s v = −40 m/s

FIGURE 2.24
INTERACTIVE FIGURE MP

The rate at which velocity changes each second is the same.

Integrated Science 2B
BIOLOGY

Hang Time

EXPLAIN THIS Describe the relationship between jumping height and time.

LEARNING OBJECTIVE
Apply the equations of motion to the act of jumping.

Some athletes and dancers have great jumping ability. Leaping straight up, they seem to defy gravity, hanging in the air for what feels like at least 2 or 3 s. In reality, however, the "hang time" of even the best jumpers is almost always less than 1 s. What determines hang time? Just what you'd expect—how high you jump.

Just as we tend to overestimate how long the best jumpers stay in the air, we can easily overestimate how high people jump. You can test your own jumping ability by performing what's called a standing vertical jump: Stand facing a wall with your feet flat on the floor and your arms extended upward. Make a mark on the wall at the top of your reach. Then jump, arms outstretched, and make another mark on the wall at your peak. The distance between the two marks measures your vertical jump. If it's more than 0.6 m (2 ft), you're exceptional.

A standing vertical jump of 1.25 m is a record breaker. The best basketball stars can't top this. At the top of a jump, right when you stop going up and are about to start coming down, your speed is zero and you are at rest. As we show in Appendix B, the relationship between the time it takes to reach the ground and the vertical height for a uniformly accelerating object starting from rest is

$$d = \frac{1}{2}gt^2$$

We can rearrange this equation to calculate time:

$$t = \sqrt{\frac{2d}{g}}$$

For a record-breaking jump, we use 1.25 m for d and 10 m/s^2 for g and see that:

$$t = \sqrt{\frac{2d}{g}} = \sqrt{\frac{2(1.25\ \text{m})}{10\ \text{m/s}^2}} = 0.50\ \text{s}$$

The hang time is actually double this, since t is the time for only one way of an up-and-down round trip—so the total hang time would be an impressive 1 s. It's safe to say you're not acquainted with anybody who can do a 1-s standing jump! (You can win bets on this!)

What determines jumping ability? Your jumping ability increases with the length of your legs and with the strength of your leg muscles. If you look at the bodies of animals that specialize in jumping—frogs, kangaroos, and rabbits, for example—you'll see that they all have elongated and very muscular hind legs.

CHECK YOURSELF

The red kangaroo has a vertical jump height of about 1.8 m (6 ft). What is the kangaroo's hang time?

CHECK YOUR ANSWER

The time it takes for the kangaroo to reach the ground from a height of 1.8 m is

$$t = \sqrt{\frac{2d}{g}} = \sqrt{\frac{2(1.8\ \text{m})}{10\ \text{m/s}^2}} = 0.60\ \text{s}$$

The hang time is double this, or 1.2 s.

For instructor-assigned homework, go to www.masteringphysics.com 

SUMMARY OF TERMS (KNOWLEDGE)

Acceleration The rate at which velocity changes with time; the change in velocity may be in magnitude, or in direction, or in both. It is usually measured in m/s^2.

Air resistance The force of friction acting on an object due to its motion through air.

Density A measure of mass per volume for a substance.

Equilibrium rule The vector sum of forces acting on a non-accelerating object equals zero: $\Sigma F = 0$.

Force Simply stated, a push or a pull.

Free fall Motion under the influence of gravitational pull only.

Friction The resistive force that opposes the motion or attempted motion of an object through a fluid or past another object with which it is in contact.

Inertia The property of things to resist changes in motion.

Kilogram The unit of mass. One kilogram (symbol kg) is the mass of 1 liter (symbol L) of water at 4°C.

Mass The quantity of matter in an object. More specifically, mass is a measure of the inertia or sluggishness that an object exhibits in response to any effort made to start it, stop it, deflect it, or change its state of motion in any way.

Net force The combination of all forces that act on an object.

Newton The scientific unit of force.

Speed The distance traveled per unit of time.

Support force The force that supports an object against gravity; often called the normal force.

Vector quantity A quantity that specifies direction as well as magnitude.

Velocity A vector quantity that specifies both the speed of an object and its direction of motion.

Weight Simply stated, the force of gravity on an object. More specifically, the gravitational force with which a body presses against a supporting surface.

READING CHECK QUESTIONS (COMPREHENSION)

2.1 Aristotle on Motion

1. What were Aristotle's two main classifications of motion?
2. Did Aristotle believe that forces are necessary to keep moving objects moving, or did he believe that, once moving, they would move by themselves?

2.2 Galileo's Concept of Inertia

3. What two main ideas of Aristotle did Galileo discredit?
4. Which dominated Galileo's way of extending knowledge—philosophical discussion or experiment?
5. What is the name of the property of objects to maintain their states of motion?

2.3 Mass—A Measure of Inertia

6. Which depends on gravity—weight or mass?
7. Where would your weight be greater—on Earth or on the Moon? Where would your mass be greater?
8. What are the units of measurement for weight and mass?
9. One kg weighs 10 N on Earth. Would it weigh more or less on the Moon?
10. Which has the greater density—1 kg of water or 10 kg of water?

2.4 Net Force

11. What is the net force on a box that is being pushed to the left with a force of 50 N while it is also being pushed to the right with a force of 60 N?
12. What two quantities are necessary to determine a vector quantity?

2.5 The Equilibrium Rule

13. What is the name given to a force that occurs in a rope when both ends are pulled in opposite directions?
14. How much tension is there in a rope that holds a 20-N bag of apples at rest?
15. What does $\Sigma F = 0$ mean?

2.6 The Support Force

16. Why is the support force on an object often called the normal force?
17. When you weigh yourself, how does the support force of the scale acting on you compare with the gravitational force between you and Earth?

2.7 Equilibrium of Moving Things

18. What test tells us whether or not a moving object is in equilibrium?
19. If we push a crate at constant velocity, how does friction acting on the crate compare with our pushing force?

2.8 The Force of Friction

20. How does the direction of a friction force compare with the direction of motion of a sliding object?
21. If you push on a heavy crate to the right and it slides, what is the direction of friction on the crate?
22. Suppose you push on a heavy crate, but not hard enough to make it slide. Does a friction force act on the crate?

2.9 Speed and Velocity

23. What equation shows the relationship among speed, distance, and time.
24. Why do we say that velocity is a vector and speed is not a vector?
25. Does the speedometer on a vehicle show the average speed or the instantaneous speed?
26. How can you be at rest and also moving at 100,000 km/h at the same time?

2.10 Acceleration

27. What equation shows the relationship among velocity, time, and acceleration?
28. What is the acceleration of an object in free fall at Earth's surface?
29. Why does the unit of time appear twice in the definition of acceleration?
30. When you toss a ball upward, by how much does its upward speed decrease each second?

THINK INTEGRATED SCIENCE

2A—Friction Is Universal

31. *Joints* are places where bones meet. Many of them, such as the ball-and-socket joints in your shoulders and hips, are bathed with *synovial fluid*, a viscous substance resembling the white of an egg. Speculate about what the purpose of the synovial fluid might be.
32. Describe one phenomenon from each of the major natural sciences—physics, chemistry, biology, Earth science, and astronomy—in which friction plays a major and interesting role.
33. Is it more correct to say that friction prevents earthquakes or that friction causes earthquakes? Justify your answer.

2B—Hang Time

34. When during a jump is your speed zero?
35. What are some anatomical features that affect an animal's jumping ability?

THINK AND DO (HANDS-ON APPLICATION)

36. Roll cans of different masses across the floor at equal initial speeds and notice the effect of inertia and friction on how far each can rolls.
37. By any method you choose, determine both your walking speed and your running speed.

PLUG AND CHUG (FORMULA FAMILIARIZATION)

These are "plug-in-the-number" tasks to familiarize you with the main formulas that link the physics concepts of this chapter. They are one-step substitutions, much less challenging than the Think and Solve problems that follow.

$$\textbf{Average speed} = \frac{\textbf{total distance covered}}{\textbf{travel time}}$$

$$\textbf{Acceleration} = \frac{\textbf{change of velocity}}{\textbf{time interval}}$$

38. Show that the average speed of a rabbit that runs a distance of 30 m in a time of 2 s is 15 m/s.
39. Calculate your average walking speed when you step 1.0 m in 0.5 s.
40. Show that the acceleration of a hamster is 5 m/s^2 when it increases its velocity from rest to 10 m/s in 2 s.
41. Show that the acceleration of a car that can go from rest to 100 km/h in 10 s is 10 km/h·s.
42. Show that the acceleration of a rock that reaches a speed of 40 m/s in 4 s is 10 m/s^2.

THINK AND COMPARE (ANALYSIS)

43. The weights of Burl, Paul, and the scaffold produce tensions in the supporting ropes. Rank the tensions in the left rope, from greatest to least, in the three situations A, B, and C.

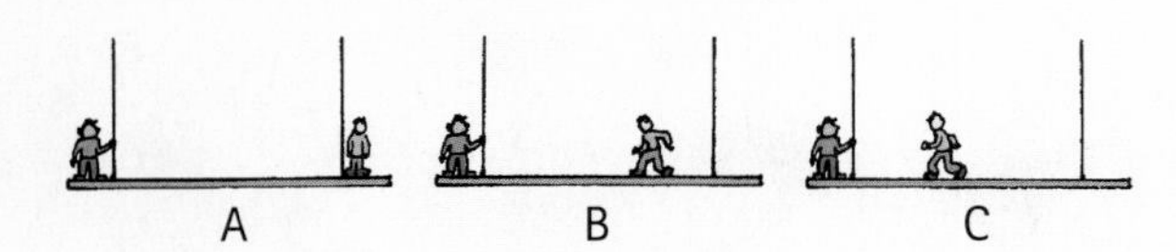

44. Rank the net forces on the block from greatest to least in the four situations A, B, C, and D.

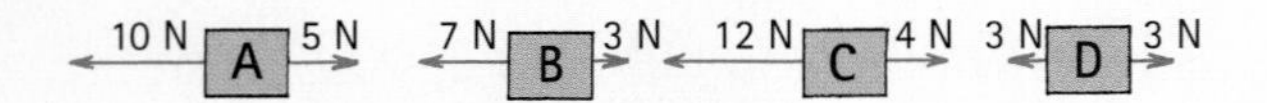

45. Different materials, A, B, C, and D, rest on a table.
    (a) From greatest to least, rank them by how much they resist being set in motion.
    (b) From greatest to least, rank them by the support (normal) force the table exerts on them.

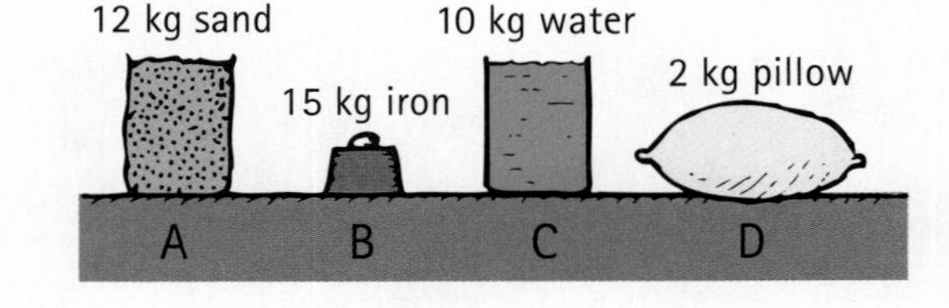

46. Three pucks, A, B, and C, are sliding across ice at the given speeds. The forces of air and ice friction are negligible.
    (a) From greatest to least, rank the pucks by the force needed to keep them moving.
    (b) From greatest to least, rank the pucks by the force needed to stop them in the same time interval.

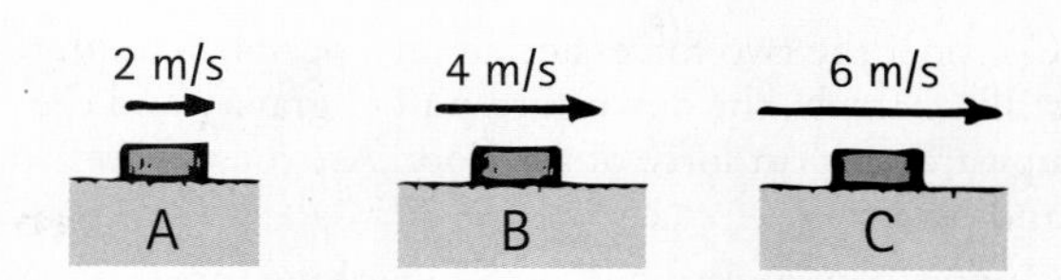

THINK AND SOLVE (MATHEMATICAL APPLICATION)

47. Suppose that a 30-N force and a 20-N force act on an object.
    (a) Show that when both forces act in the same direction the resultant force is 50 N.
    (b) Show that when the forces act in opposite directions the resultant force is 10 N. Is the resultant force in the direction of the 30-N force or the 20-N force?
48. A horizontal force of 100 N is required to push a box across a floor at a constant velocity.
    (a) Show that the net force acting on the box is zero.
    (b) Show that the friction force that acts on the box is 100 N.
49. A firefighter with a mass of 100 kg slides down a vertical pole at a constant speed. Show that the force of friction provided by the pole is about 1000 N.
50. The ocean's level is currently rising at about 1.5 mm per year. Show that at this rate the sea level will be 3 m higher in 2000 years.
51. A vehicle changes its velocity from 90 km/h to a dead stop in 10 s. Show that its acceleration in doing so is $-2.5\ m/s^2$.
52. A ball is thrown straight up with an initial speed of 40 m/s.
    (a) Show that its time in the air is about 8 s.
    (b) Show that the ball's maximum height, neglecting air resistance, is about 80 m.
53. Extend Table 2.2 (which gives values from 0 to 5 s) from 6 to 10 s, assuming no air resistance.
54. A ball is thrown straight up with enough speed so that it is in the air for several seconds.
    (a) What is the velocity of the ball when it reaches its highest point?
    (b) What is the ball's velocity 1 s before it reaches its highest point?
    (c) What is the change in its velocity during this 1-s interval?
    (d) What is the ball's velocity 1 s after it reaches its highest point?
    (e) What is the change in its velocity during this 1-s interval?
    (f) What is the change in the ball's velocity during the 2-s interval from 1 s before the highest point to 1 s after the highest point? (Caution: We are asking for velocity, not speed.)
    (g) What is the acceleration of the ball during any of these time intervals and at the moment the ball has zero velocity at the top of its path?

THINK AND EXPLAIN (SYNTHESIS)

55. What Aristotelian idea did Galileo discredit in his fabled experiment at the Leaning Tower of Pisa? In his experiments with inclined planes?
56. What physical quantity is a measure of how much inertia an object has?
57. Does a person on a diet lose mass or lose weight?
58. One gram of lead has a mass of $11.3\ g/cm^3$. What is its density? Two grams of aluminum have a mass of $5.4\ g/cm^3$. What is the density of aluminum?
59. Which has the greater density—5 kg of lead or 10 kg of aluminum?
60. Consider a pair of forces, one with a magnitude of 25 N and the other with a magnitude of 15 N. What maximum net force is possible for these two forces? What is the minimum net force possible?
61. The sketch shows a painter's scaffold in mechanical equilibrium. The person in the middle weighs 250 N, and the tension in each rope is 200 N. What is the weight of the scaffold?

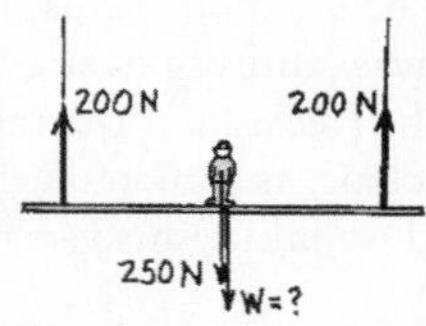

62. A different scaffold that weighs 300 N supports two painters, one weighing 250 N and one weighing 300 N. The reading on the left scale is 400 N. What is the reading on the right scale?

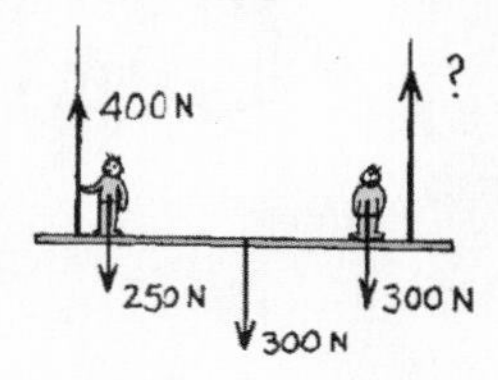

63. Consider the two forces acting on a person standing still—namely, the downward pull of gravity and the upward support force of the floor. Are these forces equal and opposite?
64. Can we accurately say that, if something moves at constant velocity, there are no forces acting on it? Explain.
65. At the moment an object that has been tossed upward into the air reaches its highest point, is it in equilibrium? Defend your answer.
66. If you push horizontally on a crate and it slides across the floor, slightly gaining speed, how does the friction acting on the crate compare with your push?
67. What is the impact speed when a car moving at 100 km/h bumps into the rear of another car traveling in the same direction at 98 km/h?
68. Harry Hotshot can paddle a canoe in still water at 8 km/h. How successful will he be in canoeing upstream in a river that flows at 8 km/h?
69. A destination 120 mi away is posted on a highway sign, and the speed limit is 60 mi/h. If you drive at the posted speed, will you reach the destination in 2 h or in more than 2 h?
70. Suppose that a freely falling object were somehow equipped with a speedometer. By how much would its speed reading increase with each second of fall?
71. Suppose that the freely falling object in Exercise 70 were also equipped with an odometer. Would the readings of distance fallen each second indicate equal or unequal distances of fall for successive seconds? Explain.
72. Someone standing on the edge of a cliff (as in Figure 2.24) throws one ball straight up at a certain speed and another ball straight down with the same initial speed. If air resistance is negligible, which ball has the greater speed when it strikes the ground below?
73. For a freely falling object dropped from rest, what is its acceleration at the end of the fifth second of fall? At the end of the tenth second? Defend your answer (and distinguish between velocity and acceleration).

THINK AND DISCUSS (EVALUATION)

74. A bowling ball rolling along a lane gradually slows. Discuss how Aristotle would interpret this observation. How would Galileo interpret it?
75. Can an object be in mechanical equilibrium when only a single force acts on it? Discuss.
76. Nellie Newton hangs at rest from the ends of the rope as shown. How does the reading on the scale compare with her weight?

77. Harry the painter swings year after year from his bosun's chair. His weight is 500 N, and the rope, unknown to him, has a breaking point of 300 N. Why doesn't the rope break when he is supported as shown on the left? One day, Harry was painting near a flagpole, and, for a change, he tied the free end of the rope to the flagpole instead of to his chair, as shown on the right. Discuss why Harry ended up taking his vacation early.

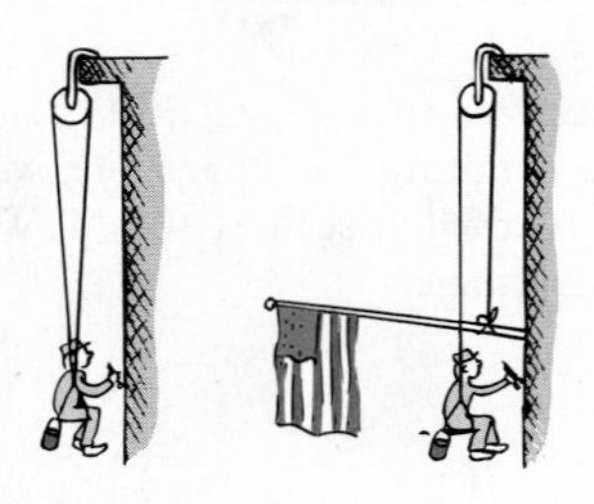

78. When a ballplayer throws a ball straight up, by how much does the speed of the ball decrease each second while it is ascending? In the absence of air resistance, by how much does the ball's speed increase each second while it is descending? How much time elapses during the ball's ascent compared with its descent?
79. Two balls, A and B, are released simultaneously from rest at the left end of the equal-length tracks, as shown. Which ball will reach the end of its track first?

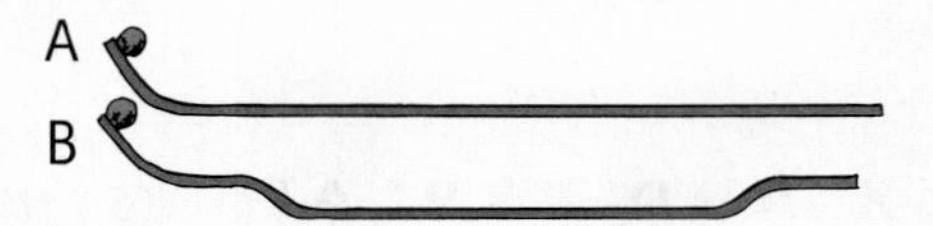

80. Refer to the tracks in Exercise 79.
    (a) Does ball B roll faster along the lower part of track B than ball A rolls along track A?
    (b) Is the speed gained by ball B going down the extra dip the same as the speed it loses going up near the right-hand end? Does this mean that the speeds of balls A and B will be the same at the ends of the tracks?
    (c) For ball B will the average speed dipping down and up be greater than the average speed of ball A during the same time?
    (d) So, overall, does ball A or ball B have the greater average speed? (Do you wish to change your answer to Exercise 79?)

READINESS ASSURANCE TEST (RAT)

If you have a good handle on this chapter, you should be able to score at least 7 out of 10 on this RAT. If you score less than 7, you need to study further before moving on.

Choose the BEST answer to each of the following:

1. According to Galileo, inertia is a
   (a) force like any other force.
   (b) special kind of force.
   (c) property of all matter.
   (d) concept opposite to force.
2. If gravity between the Sun and Earth suddenly vanished, Earth would continue moving in
   (a) a curved path.
   (b) an outward spiral path.
   (c) an inward spiral path.
   (d) a straight line.
3. To be in mechanical equilibrium, an object must be
   (a) at rest.
   (b) moving at constant velocity.
   (c) either at rest or moving at constant velocity.
   (d) neither at rest nor moving at constant velocity.
4. When you stand on two bathroom scales, one foot on each scale with your weight evenly distributed, each scale will read
   (a) your weight.
   (b) half your weight.
   (c) zero.
   (d) actually more than your weight.
5. If an object moves along a straight-line path at constant speed, then it must be
   (a) accelerating.
   (b) acted on by a force.
   (c) both of these
   (d) neither of these
6. What is the net force on a box that is being pushed to the left with a force of 40 N while it is also being pushed to the right with a force of 50 N?
   (a) 10 N to the left
   (b) 10 N to the right
   (c) 90 N to the left
   (d) 90 N to the right
7. Neglecting air resistance, when you toss a rock upward, by about how much does its upward speed decrease each second?
   (a) 10 m/s
   (b) 10 m/s^2
   (c) The answer depends on the initial speed.
   (d) none of these
8. During each second of free fall, the speed of an object
   (a) increases by the same amount.
   (b) changes by increasing amounts.
   (c) remains constant.
   (d) doubles.
9. In 2.0 s, a car increases its speed from 30 km/h to 35 km/h while a bicycle goes from rest to 6 km/h. Which has the greater acceleration?
   (a) the car
   (b) the bicycle
   (c) The accelerations are equal.
   (d) It is impossible to know from the information provided.
10. A freely falling object has a speed of 40 m/s at one instant. Exactly 1 s later its speed will be
    (a) the same.
    (b) 10 m/s.
    (c) 45 m/s.
    (d) greater than 45 m/s.

Answers to RAT

1. c, 2. d, 3. c, 4. b, 5. d, 6. b, 7. a, 8. a, 9. b, 10. d

CHAPTER 3
Newton's Laws of Motion

A HEAVY WINGSUIT flyer falls faster than a lighter one, and even after parachuting, has a rougher landing—but why? Have you tried the party trick where you pull a tablecloth out from under place settings and the dishes stay put? How does this "trick" work, and what law of motion does it demonstrate? Have you heard the expression "You can't touch without being touched"? Does this statement about the objective world of physics have a corollary in the world of human emotions? How did Newton's laws get us to the Moon? How do birds fly, rockets take off, and people walk? How do Newton's laws of motion interface with modern discoveries about motion gained from relativity and quantum mechanics? You'll learn all this and much more in this chapter.

3.1 Newton's First Law of Motion

LEARNING OBJECTIVE
Define Newton's first law of motion and relate it to inertia.

EXPLAIN THIS Why isn't inertia a kind of force?

Galileo's work on inertia set the stage for Isaac Newton, who was born shortly after Galileo's death in 1642. By the time Newton was 23, he had developed his famous three laws of motion, which completed the overthrow of Aristotelian ideas about motion. These three laws first appeared in one of the most famous books of all time, Newton's *Philosophiae Naturalis Principia Mathematica,** often known as simply the *Principia*. The first law is a restatement of Galileo's concept of inertia, the second law relates acceleration to its cause—force, and the third is the law of action and reaction. **Newton's first law** is:

MasteringPhysics®
VIDEO: Newton's Law of Inertia

UNIFYING CONCEPT
● *Newton's First Law*

> **Every object continues in its state of rest, or a uniform speed in a straight line, unless acted on by a nonzero force.**

The key word in this law is *continues*; an object continues to do whatever it happens to be doing unless a force is exerted on it. If the object is at rest, it continues in a state of rest. This is nicely demonstrated when a tablecloth is skillfully whipped out from beneath dishes sitting on a tabletop, leaving the dishes in their initial state of rest (Figure 3.1).** On the other hand, if an object is moving, it continues to move without changing its speed or direction, as evidenced by space probes that continuously move in outer space. As stated in Chapter 2, this property of objects to resist changes in motion is called *inertia* (Figures 3.1 and 3.2).

FIGURE 3.1
Inertia in action.

CHECK YOURSELF
When a space vehicle travels in a nearly circular orbit around Earth, is a force required to maintain its high speed? If the force of gravity were suddenly cut off, what type of path would the space vehicle follow?

CHECK YOUR ANSWER
There is no force in the direction of the vehicle's motion, which is why it coasts at a constant speed by its own inertia. The only force acting on it is the force of gravity, which acts at right angles to its motion (toward Earth's center). We'll see later that this right-angled force holds the vehicle in a circular path. If the force of gravity were cut off, the vehicle would fly off in a straight line at a constant velocity.

FIGURE 3.2
Rapid deceleration is sensed by the driver, who lurches forward—inertia in action!

3.2 Newton's Second Law of Motion

LEARNING OBJECTIVE
Relate the three concepts acceleration, force, and mass.

EXPLAIN THIS What happens to the acceleration of a sliding brick when you increase the force of your push on it?

Isaac Newton was the first to recognize the connection between force and mass in producing acceleration, which is one of the central rules of nature, as expressed in his second law of motion.

MasteringPhysics®
TUTORIAL: Parachutes and Newton's Second Law

*The Latin title means "mathematical principles of natural philosophy."
**Close inspection reveals that brief friction between the dishes and the fast-moving tablecloth starts the dishes moving, but then friction between the dishes and the tabletop stops the dishes before they slide very far. If you try this, use unbreakable dishes!

UNIFYING CONCEPT

● *Newton's Second Law*

Newton's second law states:

The acceleration produced by a net force on an object is directly proportional to the net force, is in the same direction as the net force, and is inversely proportional to the mass of the object.

Or, in shorter notation,

$$\textbf{Acceleration} \sim \frac{\textbf{net force}}{\textbf{mass}}$$

By using consistent units such as newtons (N) for force, kilograms (kg) for mass, and meters per second squared (m/s^2) for acceleration, we produce the exact equation:

$$\textbf{Acceleration} = \frac{\textbf{net force}}{\textbf{mass}}$$

In its briefest form, where a is acceleration, F is net force, and m is mass:

$$a = \frac{F}{m}$$

Acceleration equals the net force divided by the mass. If the net force acting on an object is doubled, the object's acceleration will be doubled. Suppose instead that the mass is doubled. Then the acceleration will be halved. If both the net force and the mass are doubled, then the acceleration will be unchanged (Figures 3.4 and 3.5).

An object accelerates in the direction of the net force acting on it. Speed changes when the net force acts in the direction of the object's motion. When the net force acts at right angles to the object's motion, the direction of the object changes. A net force acting in any other direction results in a combination of speed change and deflection (Figure 3.6).

If one thing is inversely proportional to another, then, as one gets bigger, the other gets smaller.

FIGURE 3.3
INTERACTIVE FIGURE MP
The greater the mass, the greater a force must be for a given acceleration.

Force of hand accelerates the brick

Twice as much force produces twice as much acceleration

Twice the force on twice the mass gives the same acceleration

FIGURE 3.4
Acceleration is directly proportional to force.

Force of hand accelerates the brick

The same force accelerates 2 bricks 1/2 as much

3 bricks, 1/3 as much acceleration

FIGURE 3.5
Acceleration is inversely proportional to mass.

FIGURE 3.6
When you accelerate in the direction of your velocity, you speed up; when you accelerate against your velocity, you slow down; when you accelerate at an angle to your velocity, your direction changes.

HISTORY OF SCIENCE

The Moving Earth

In 1543, the Polish astronomer Nicolaus Copernicus caused a great controversy when he published a book proposing that the Earth revolved around the Sun.* This idea conflicted with the popular view that the Earth was the center of the universe. Copernicus's concept of a Sun-centered solar system was the result of years of studying the planets. He had kept his theory from the public for two reasons. First, he feared persecution: A theory so completely different from common opinion would surely be taken as an attack on the established order. Second, he had reservations about it himself: He could not reconcile the idea of a moving Earth with the prevailing ideas of motion. The concept of inertia was unknown to him and to others of his time. In the final days of his life, at the urging of close friends, Copernicus sent his manuscript, *De Revolutionibus Orbium Coelestium*,† to the printer. The final copy of his famous exposition reached him on the day he died—May 24, 1543.

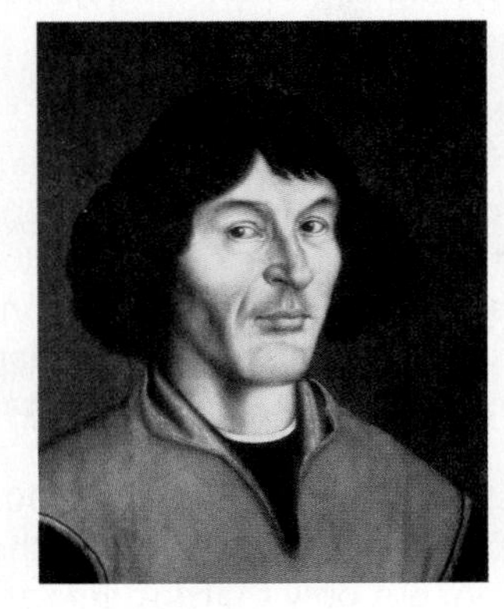

Nicolaus Copernicus, 1473–1543

The idea of a moving Earth was much debated. Europeans thought about the universe much as Aristotle had, and the existence of a force big enough to keep Earth moving was beyond their imagination. They had no concept of inertia. One of the arguments against a moving Earth was the following:

Consider a bird sitting at rest on a branch of a tall tree. On the ground below is a fat, juicy worm. The bird sees the worm and drops vertically below and catches it. It was argued that this would be impossible if Earth were moving. A moving Earth would have to travel at an enormous speed to circle the Sun in one year. While the bird would be in the air descending from its branch to the ground below, the worm would be swept far away along with the moving Earth. It seemed that catching a worm on a moving Earth would be an impossible task. The fact that birds do catch worms from tree branches seemed to be clear evidence that Earth must be at rest.

Can the bird drop down and catch the worm if Earth moves at 30 km/s?

Can you see the mistake in this argument? You can if you use the concept of inertia. You see, not only is Earth moving at a great speed, but so are the tree, the branch of the tree, the bird that sits on it, the worm below, and even the air in between. Things in motion remain in motion if no unbalanced forces are acting on them. So when the bird drops from the branch, its initial sideways motion remains unchanged. It catches the worm quite unaffected by the motion of its total environment.

We live on a moving Earth. If you stand next to a wall and jump up so that your feet are no longer in contact with the floor, does the moving wall slam into you? Why not? It doesn't because you are also traveling at the same horizontal speed—before, during, and after your jump. The speed of Earth relative to the Sun is not the speed of the wall relative to you.

Four hundred years ago, people had difficulty with ideas like these. One reason is that they didn't yet travel in high-speed vehicles. Rather, they experienced slow, bumpy rides in horse-drawn carts. People were less aware of the effects of inertia. Today, we can flip a coin in a high-speed car, bus, or plane and catch the vertically moving coin as easily as we could if the vehicle were at rest. We see evidence of the law of inertia when the horizontal motion of the coin before, during, and after the catch is the same. The coin always keeps up with us.

When you flip a coin in a high-speed airplane, it behaves as if the airplane were at rest. The coin keeps up with you—inertia in action!

* Copernicus was certainly not the first to think of a Sun-centered solar system. In the 5th century, for example, the Indian astronomer Aryabhatta taught that Earth circles the Sun, not the other way around (as the rest of the world believed).
† The Latin title means "On the Revolution of Heavenly Spheres."

CHECK YOURSELF

1. **In Chapter 2, we defined acceleration as the time rate of change of velocity—that is, $a =$ (change in v)/time. Are we now saying that acceleration is instead the ratio of force to mass—that is, $a = F/m$? Which is it?**

MasteringPhysics®
VIDEO: Newton's Second Law
VIDEO: Free Fall: Acceleration Explained
VIDEO: Air Resistance and Falling Objects

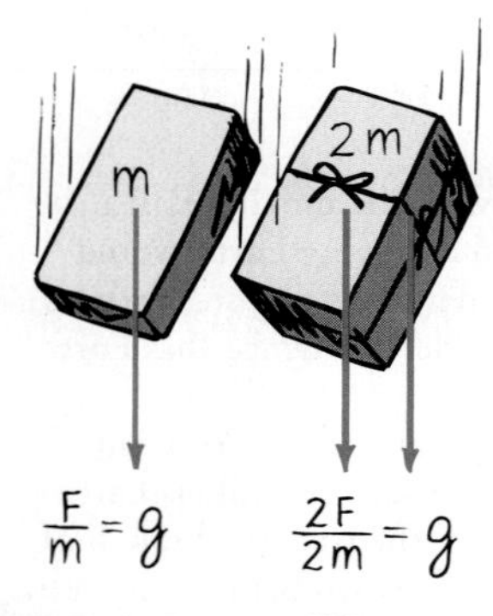

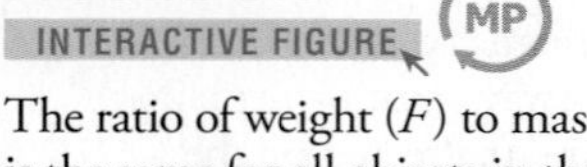

FIGURE 3.7

INTERACTIVE FIGURE 

The ratio of weight (F) to mass (m) is the same for all objects in the same locality; hence, their accelerations are the same in the absence of air resistance.

2. A jumbo jet cruises at a constant velocity of 1000 km/h when the thrusting force of its engines is a constant 100,000 N. What is the acceleration of the jet? What is the force of air resistance on the jet?

3. Suppose you apply the same amount of force to two separate carts, one cart of mass 1 kg and the other of mass 2 kg. Which cart will accelerate more, and with how much greater acceleration?

CHECK YOUR ANSWERS

1. Acceleration is *defined* as the time rate of change of velocity, and it is *produced* by a force. The magnitude of force/mass (often the cause) determines the rate change in velocity/time (often the effect).
2. The acceleration is zero, as evidenced by a constant velocity. Newton's second law tells us that zero acceleration means that the net force is zero, which tells us that the force of air resistance must just equal the thrusting force of 100,000 N and that it must act in the opposite direction. So the air resistance on the jet is 100,000 N. This is in accord with $\Sigma F = 0$. (Note that we don't need to know the velocity of the jet to answer this question, but only that the velocity is *constant*. In a nutshell, zero acceleration means that the net force is also zero.)
3. The 1-kg cart will have greater acceleration—twice as much, in fact—because it has half as much mass, which means it has half as much resistance to a change in motion.

Only a single force acts on something in free fall—the force of gravity.

When Acceleration Is g—Free Fall

Although Galileo articulated both concepts of inertia and acceleration and was the first to measure the acceleration of falling objects, he was unable to explain why objects of different masses fall with equal accelerations. Newton's second law provides the explanation.

We know that a falling object accelerates toward Earth because of the gravitational force of attraction between the object and Earth. As discussed in Chapter 2, when the force of gravity is the only force—that is, when air resistance is negligible—we say that the object is in a state of **free fall**. An object in free fall accelerates toward Earth at 10 m/s^2 (or, more precisely, at 9.8 m/s^2).

The greater the mass of an object, the stronger the gravitational pull between it and Earth. The double brick in Figure 3.7, for example, has twice the gravitational attraction to Earth as the single brick. Why, then, doesn't the double brick fall twice as fast, as Aristotle supposed it would? The answer is evident in Newton's second law: The acceleration of an object depends not only on the force (weight, in this case) but also on the object's resistance to motion—its inertia (mass). Whereas a force produces an acceleration, inertia is a *resistance* to acceleration. So twice the force exerted on twice the inertia produces the same acceleration as half the force exerted on half the inertia. Both accelerate equally. The acceleration due to gravity is symbolized by g. We use the symbol g, rather than a, to denote that acceleration is due to gravity alone.

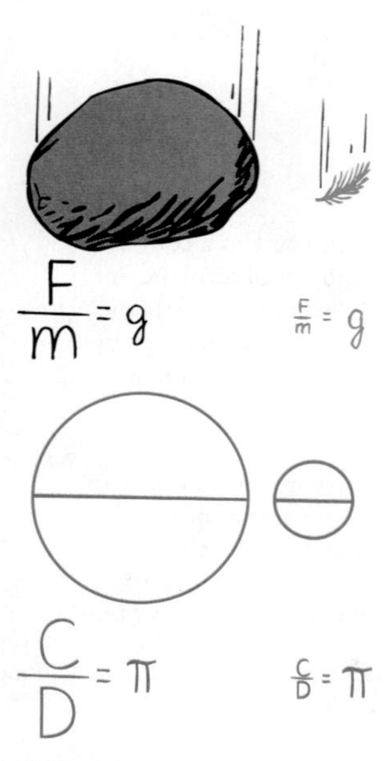

FIGURE 3.8

The ratio of weight (F) to mass (m) is the same for the large rock and the small feather; similarly, the ratio of the circumference (C) to the diameter (D) is the same for the large and small circles.

So the ratio of weight to mass for freely falling objects equals a constant, g. This is similar to the ratio of a circle's circumference to its diameter, which equals the constant π. The ratio of weight to mass is identical for both heavy objects and light objects, just as the ratio of circumference to diameter is the same for both large and small circles (Figure 3.8).

MATH CONNECTION

Equations as Guides to Thinking: $a = \frac{F}{m}$

Newton's second law is not only simple in form but also widely applicable, so it's a highly useful tool in many problem-solving situations—it is well worth your while to become adept at using it.

For example, if we know the mass of an object in kilograms (kg) and its acceleration in meters per second squared (m/s^2), then the force will be expressed in newtons (N). One newton is the force needed to give a mass of 1 kilogram an acceleration of 1 meter per second squared. We can rearrange Newton's law to read

$$\text{Force} = \text{mass} \times \text{acceleration}$$
$$1\ \text{N} = (1\ \text{kg}) \times (1\ \text{m/s}^2)$$

We can see that

$$1\ \text{N} = 1\ \text{kg}\cdot\text{m/s}^2$$

The centered dot between "1 kg" and "m/s^2" means that one expression is multiplied by the other.

If we know two of the quantities in Newton's second law, we can calculate the third. For example, how much force, or thrust, must a 20,000-kg jet plane develop to achieve an acceleration of 1.5 m/s^2? Using the equation, we can calculate

$$\begin{aligned} F &= ma \\ &= (20{,}000\ \text{kg}) \times (1.5\ \text{m/s}^2) \\ &= 30{,}000\ \text{kg}\cdot\text{m/s}^2 \\ &= 30{,}000\ \text{N} \end{aligned}$$

Suppose we know the force and the mass and we want to find the acceleration. For example, what acceleration is produced by a force of 2000 N applied to a 1000-kg car? Using Newton's second law, we find that

$$\begin{aligned} a &= \frac{F}{m} = \frac{2000\ \text{N}}{1000\ \text{kg}} \\ &= \frac{2000\ \text{kg}\cdot\text{m/s}^2}{1000\ \text{kg}} = 2\ \text{m/s}^2 \end{aligned}$$

If the force is 4000 N, the acceleration is

$$\begin{aligned} a &= \frac{F}{m} = \frac{4000\ \text{N}}{1000\ \text{kg}} \\ &= \frac{4000\ \text{kg}\cdot\text{m/s}^2}{1000\ \text{kg}} = 4\ \text{m/s}^2 \end{aligned}$$

Doubling the force on the same mass simply doubles the acceleration.

Physics problems are often more complicated than these. We don't focus on solving complicated problems in this book; instead, we emphasize equations, such as Newton's second law, in which the relationships among physical quantities are clear. Such equations serve as guides to thinking, rather than recipes for mathematical problem solving. Remember, mastering concepts first makes problem solving more meaningful.

We now understand that the acceleration of free fall is independent of an object's mass. A boulder 100 times more massive than a pebble falls at the same acceleration as the pebble because, although the force on the boulder (its weight) is 100 times that of the pebble, the greater force offsets the equally greater mass.

When Galileo tried to explain why all objects fall with equal accelerations, wouldn't he have loved to know the rule $a = \frac{F}{m}$?

CHECK YOURSELF

In a vacuum, a coin and a feather fall at an equal rate, side by side. Would it be correct to say that equal forces of gravity act on both the coin and feather in the vacuum?

CHECK YOUR ANSWER

No, no, no—a thousand times no! These objects accelerate equally not because of equal forces of gravity acting on them but because the ratios of their weights to their masses are equal. Although air resistance is not present in a vacuum, gravity is. (You'd know this if you placed your hand in a vacuum chamber and a cement truck rolled over it.) If you answered *yes* to this question, let this be a signal for you to be more careful when you think about physics.

When Acceleration Is Less Than g—Non–Free Fall

UNIFYING CONCEPT

- *Friction*
Section 2.8

Most often, air resistance is not negligible for falling objects. Then the acceleration of the object's fall is less. Air resistance, which is the force of friction acting between an object and the surrounding air, depends on two quantities: speed and surface area. When a skydiver steps from a high-flying plane, the air resistance on the skydiver's body builds up as the falling speed increases. The result is reduced acceleration. The acceleration can be reduced further by increasing the surface area. A skydiver does this by orienting his or her body so that more air is encountered by its surface—by spreading out like a flying squirrel. So air resistance depends both on speed and on the surface area encountered by the air.

For free fall, the downward net force is weight—only weight. But, when air is present, the downward net force = weight − air resistance. Can you see that the presence of air resistance reduces the net force? And that less force means less acceleration? So, as a skydiver falls faster and faster, the acceleration of the fall decreases.* What happens to the net force if air resistance builds up to equal the weight of the skydiver? The answer is that the net force becomes zero. Does this mean the skydiver comes to a stop? No! What it means is that the skydiver no longer gains speed. Acceleration terminates—it no longer occurs. We say that the skydiver has reached **terminal speed**. If we are concerned with direction—down, for falling objects—we say that the skydiver has reached *terminal velocity*.

FIGURE 3.9
In a vacuum, a feather and a coin fall with equal accelerations.

Terminal speed for a human skydiver varies from about 150 to 200 km/h, depending on the weight, size, and orientation of the body. A heavier person has to fall faster for air resistance to balance weight.** The greater weight is more effective in "plowing through" the air, resulting in a higher terminal speed for a heavier person. Increasing the frontal area reduces the terminal speed. That's where a parachute is useful. A parachute increases the frontal area, which greatly increases the air resistance, reducing the terminal speed to a safe 15 to 25 km/h.

When the force of gravity and air resistance act on a falling object, the object is not in free fall.

Consider the interesting demonstration of the falling coin and the feather in the glass tube. When air is inside the tube, the feather falls more slowly because of air resistance. The feather's weight is very small, so the feather reaches terminal speed very quickly because it doesn't have to fall very far or very fast before air resistance builds up to equal its small weight. The coin, on the other hand, doesn't have a chance to fall fast enough for air resistance to build up to equal its weight. If you were to drop a coin from a very high location, such as from the top of a tall building, its terminal speed would be reached when the speed of the coin is greater than 100 km/h. This is a much, much higher terminal speed than that of a falling feather.

When Galileo allegedly dropped objects of different weights from the Leaning Tower of Pisa, they didn't actually hit the ground at the same time. They almost did, but, because of air resistance, the heavier one hit slightly before the lighter one. But this still contradicted the much longer time difference expected by the followers of Aristotle. The behavior of falling objects was never really understood until Newton announced his second law of motion.

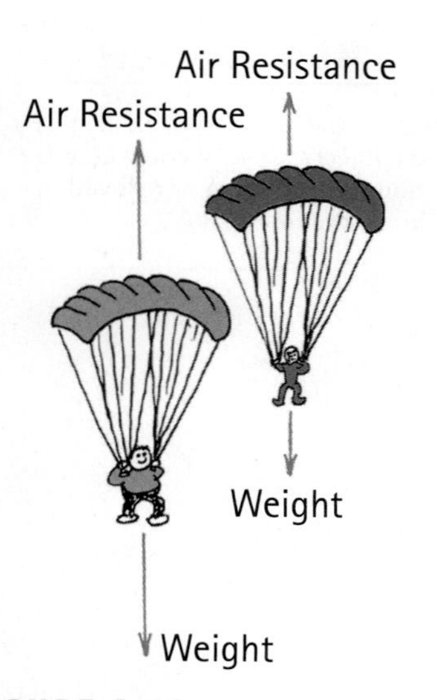

FIGURE 3.10
The heavier parachutist must fall faster than the lighter parachutist for air resistance to cancel her greater weight.

*In mathematical notation, $a = \frac{F_{net}}{m} = \frac{mg - R}{m}$, where mg is the weight and R is the air resistance. Note that, when $R = mg$, $a = 0$; then, with no acceleration, the object falls at a constant velocity. With elementary algebra, we can proceed another step and get $a = \frac{F_{net}}{m} = \frac{mg - R}{m} = g - \frac{R}{m}$. We see that the acceleration a will always be less than g if air resistance R impedes falling. Only when $R = 0$ does $a = g$.

** A skydiver's air resistance is proportional to speed squared.

CHECK YOURSELF

Consider two parachutists, a heavy person and a light person, who jump from the same altitude with parachutes of the same size.

1. **Which person reaches terminal speed first?**
2. **Which person has the higher terminal speed?**
3. **Which person reaches the ground first?**
4. **If there were no air resistance, like on the Moon, how would your answers to the questions differ?**

CHECK YOUR ANSWERS

To answer these questions correctly, think of a coin and a feather falling in air.

1. Just as a feather reaches terminal speed very quickly, the light person reaches terminal speed first.
2. Just as a coin falls faster than a feather through air, the heavy person falls faster and reaches a higher terminal speed.
3. Just as in the race between a falling coin and a falling feather, the heavy person falls faster and reaches the ground first.
4. If there were no air resistance, there would be no terminal speed at all. Both parachutists would be in free fall, and both would hit the ground at the same time.

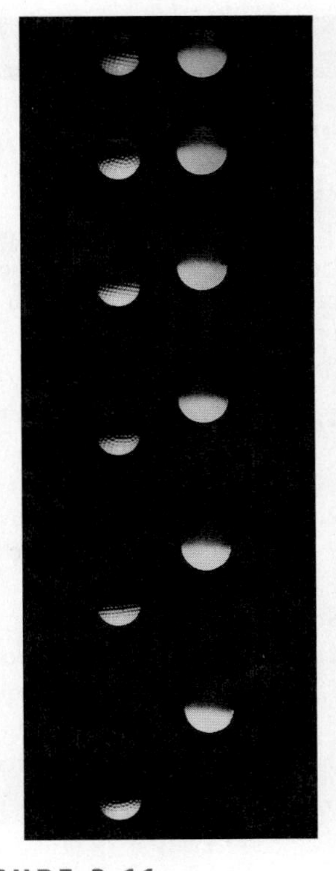

FIGURE 3.11
A stroboscopic study of a golf ball (left) and a Styrofoam ball (right) falling in air. The air resistance is negligible for the heavier golf ball, and its acceleration is nearly equal to *g*. Air resistance is not negligible for the lighter Styrofoam ball, however, and it reaches its terminal velocity sooner.

Integrated Science 3A
BIOLOGY

Gliding

EXPLAIN THIS How is a wingsuit flyer like a flying squirrel?

LEARNING OBJECTIVE
Relate how living creatures are able to glide.

Only three groups of living organisms—birds, bats, and insects—can truly fly. Gliding, however, has evolved many times in the biological world. Gliding is a mode of locomotion in which animals move through the air in a controlled fall. There are gliding squirrels, gliding lizards, gliding snakes, gliding frogs, even gliding ants. Although gliders cannot generate the forward thrust that enables fliers to power through the air, many gliders nevertheless have remarkable control—many, for example, are able to execute sharp turns in midair.

How does gliding work? When an animal jumps out of a tree, it falls toward the ground due to the downward force of gravity. Air resistance slows the animal's fall, just as it slows the motion of any object moving through air. The more air resistance an animal encounters, the slower and more controllable its fall. Since the amount of air resistance a falling object encounters depends on the object's surface area, all gliding animals have evolved special structures that increase their surface area. "Flying" squirrels have large flaps of skin between their front and hind legs. We now see humans emulating flying squirrels (see Figure 3.12). "Flying dragons" (gliding lizards of the genus *Draco*) have long extendable ribs that support large gliding membranes. "Flying" frogs have very long toes with extensive webbing between them. Gliding geckos have skin flaps along their sides and tails in addition to webbed toes. Gliding tree snakes spread out their ribs and suck in their stomachs when they leap off a branch, creating a concave parachute to slow their descent.

MATH CONNECTION

When Air Resistance Slows Acceleration

The effect of air resistance on acceleration can be made clearer with some problem-solving practice. Examine the problems and solutions. You'll have more opportunities for practice at the end of the chapter.

Problems

1. A skydiver jumps from a high-flying helicopter. As she falls faster and faster through the air, does her *acceleration* increase, decrease, or remain the same?
2. What will be her acceleration if her weight is mg and air resistance builds up to be equal to half her weight?

Solutions

1. Acceleration decreases because the net force on her decreases. Net force is equal to her weight minus her air resistance, and, because air resistance increases with increasing speed, net force and therefore acceleration

also decrease. According to Newton's second law,

$$a = \frac{F_{\text{net}}}{m} = \frac{mg - R}{m}$$

where mg is her weight and R is the air resistance she encounters. As R increases, a decreases. Note that, if she falls fast enough so that $R = mg$, $a = 0$; then, with no acceleration, she falls at terminal speed.

2. We find the acceleration from

$$a = \frac{F_{\text{net}}}{m} = \frac{mg - R}{m} = \frac{(mg - mg/2)}{m} = \frac{(mg/2)}{m} = \frac{g}{2}$$

Gliding locomotion is particularly common in certain types of habitat. For example, the presence of tall trees without much "clutter" between them has resulted in the evolution of many gliding species in the rainforests of Southeast Asia. Gliding offers a number of advantages. First, it allows rapid, energetically efficient descent, which is useful in many contexts, such as escaping from predators. Second, gliding allows animals to move from one tree to another without descending all the way to the ground and climbing back up. This is, again, energy efficient. And gliding allows gliders to avoid potentially dangerous forest understories.

(a)

(b)

FIGURE 3.12
(a) A flying squirrel increases its frontal area by spreading out. The result is greater air resistance and a slower fall. (b) Likewise for a wingsuit flyer.

3.3 Forces and Interactions

LEARNING OBJECTIVE
Identify forces in pairs.

EXPLAIN THIS When you push, what pushes back?

So far, we've treated force in its simplest sense—as a push or a pull. In a broader sense, a force is not a thing in itself but an **interaction** between one thing and another. If you push a wall with your fingers, more is happening than you pushing on the wall. You're interacting with the wall, and the wall is simultaneously pushing on you. The fact that your fingers and the wall push on each other is evident in your bent fingers (Figure 3.13). Your push on the wall and the wall's push on you are equal in magnitude (amount) and opposite in direction. The pair of forces constitutes a single interaction. In fact, you can't push on the wall unless the wall pushes back.*

When pushing my fingers together I see the same discoloration on each of them. Aha —evidence that each experiences the same amount of force!

In Figure 3.14, we see a boxer's fist hitting a massive punching bag. The fist hits the bag (and dents it) while the bag hits back on the fist (and stops its motion). This force pair is fairly large. But what if the boxer were hitting a piece of tissue paper? The boxer's fist can exert only as much force on the tissue paper as the tissue paper can exert on the boxer's fist. Furthermore, the fist can't exert any force at all unless what is being hit exerts the same amount of reaction force. An interaction requires a pair of forces acting on two different objects.

MasteringPhysics®
VIDEO: Forces and Interactions

When a hammer hits a stake and drives it into the ground, the stake exerts an equal amount of force on the hammer, and that force brings the hammer to an abrupt halt. And when you pull on a cart, the cart pulls back on you, as evidenced, perhaps, by the tightening of the rope wrapped around your hand. One thing interacts with another: The hammer interacts with the stake, and you interact with the cart.

FIGURE 3.13
When you lean against a wall, you exert a force on the wall. The wall simultaneously exerts an equal and opposite force on you. Hence, you don't topple over.

FIGURE 3.14
The boxer can hit the massive bag with considerable force. But, with the same punch, he can exert only a tiny force on the tissue paper in midair.

FIGURE 3.15
In the interaction between the hammer and the stake, each exerts the same amount of force on the other.

*We tend to think of only living things pushing and pulling, but inanimate things can also push and pull. So please don't be troubled by the idea of the inanimate wall pushing back on you. It does push back, just as another person pushing back on you would.

LEARNING OBJECTIVE
Define Newton's third law of motion, and give examples.

3.4 Newton's Third Law of Motion

EXPLAIN THIS How does Newton's third law account for rocket propulsion?

Newton's third law states:

> **Whenever one object exerts a force on a second object, the second object exerts an equal and opposite force on the first.**

We can call one force the action force, and we can call the other the reaction force. The important thing is that they are coequal parts of a single interaction and that neither force exists without the other. Action and reaction forces are equal in strength and opposite in direction. They occur in pairs, and they make up a single interaction between two things (Figure 3.16).

When walking, you interact with the floor. Your push against the floor is coupled to the floor's push against you. The pair of forces occurs simultaneously. Likewise, the tires of a car push against the road while the road pushes back on the tires—the tires and the road push against each other. In swimming, you interact with the water that you push backward, while the water pushes you forward—you and the water push against each other. The reaction forces are what account for our motion in these cases. These simultaneous forces depend on friction; a person or a car on ice, for example, may not be able to exert the action force necessary to produce the desired reaction force.

UNIFYING CONCEPT
- *Newton's Third Law*

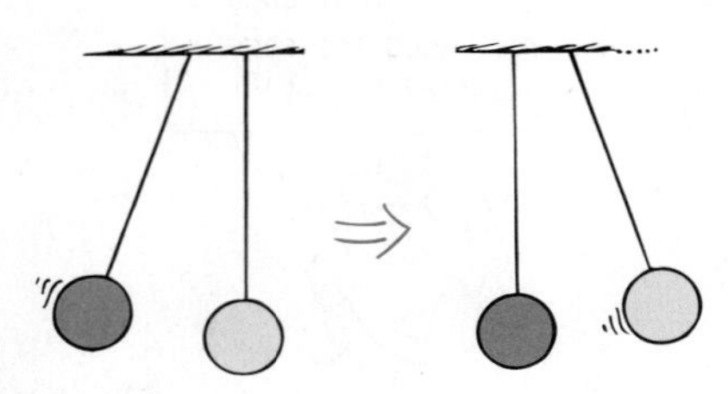

FIGURE 3.16
The impact forces between the blue and yellow balls move the yellow ball and stop the blue ball.

You can't pull on something unless that something simultaneously pulls back on you. That's the law!

MasteringPhysics®
TUTORIAL: Newton's Third Law

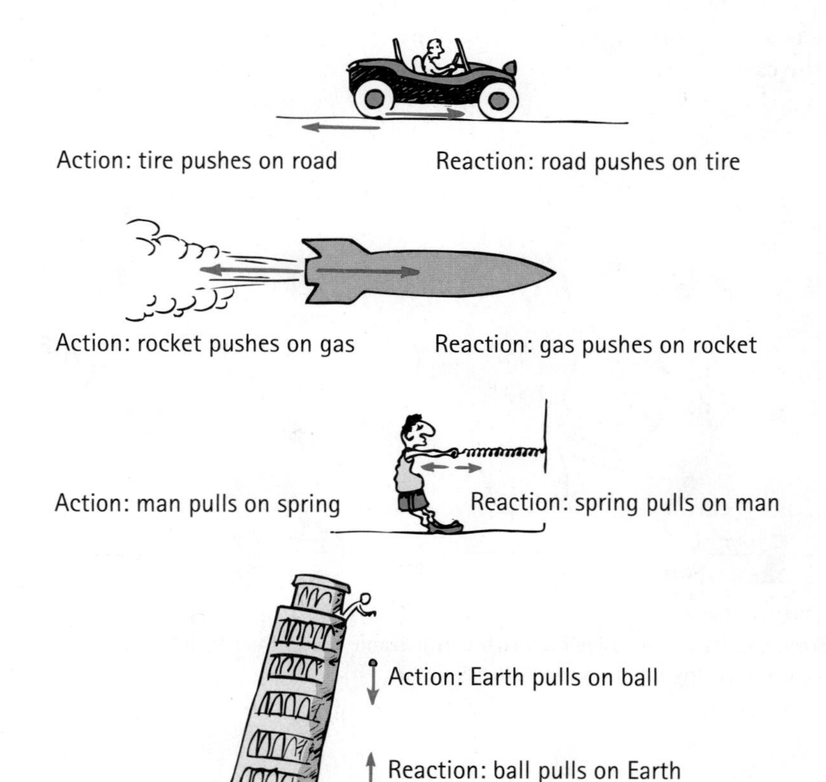
Action: tire pushes on road
Reaction: road pushes on tire
Action: rocket pushes on gas
Reaction: gas pushes on rocket
Action: man pulls on spring
Reaction: spring pulls on man
Action: Earth pulls on ball
Reaction: ball pulls on Earth

FIGURE 3.17
Action and reaction forces. Note that, when the action is "A exerts force on B," the reaction is simply "B exerts force on A."

Simple Rule to Identify Action and Reaction

There is a simple rule for identifying action and reaction forces. First, identify the interaction—one thing (object A) interacts with another (object B). Then, action and reaction forces can be stated in the following form:

Action: Object A exerts a force on object B.

Reaction: Object B exerts a force on object A.

The rule is easy to remember. If action is A acting on B, then reaction is B acting on A. We see that A and B are simply switched around. Consider the case of your hand pushing on the wall. The interaction is between your hand and the wall. We'll say the action is your hand (object A) exerting a force on the wall (object B). Then the reaction is the wall exerting a force on your hand.

MasteringPhysics®
VIDEO: Action and Reaction on Different Masses

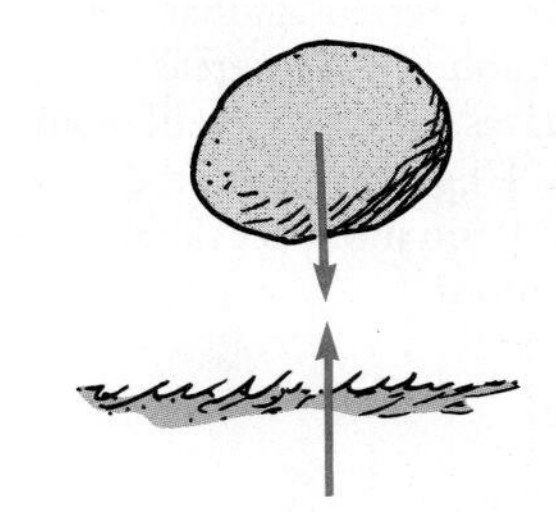

FIGURE 3.18
Earth is pulled up by the boulder with just as much force as the boulder is pulled down by Earth.

CHECK YOURSELF

1. **A car accelerates along a road. Identify the force that moves the car.**
2. **Identify the action and reaction forces for the case of an object in free fall (with no air resistance).**

CHECK YOUR ANSWERS

1. The force is the road that pushes the car along. Except for air resistance, only the road provides a horizontal force on the car. How does the road do this? The rotating tires of the car push back on the road (action). The road simultaneously pushes forward on the tires (reaction). How about that!
2. To identify a pair of action–reaction forces in any situation, first identify the pair of interacting objects. In this case, Earth interacts with the falling object via the force of gravity. So Earth pulls the falling object downward (call it action). Then the reaction is the falling object pulling Earth upward. It is only because of Earth's enormous mass that you don't notice its upward acceleration.

Action and Reaction on Different Masses

Quite interestingly, a falling object pulls upward on Earth with as much force as Earth pulls downward on the object. The resulting acceleration of the falling object is evident, while the upward acceleration of Earth is too small to detect (Figure 3.18).

Consider the exaggerated examples of the two planetary bodies shown in Figure 3.19. The forces between bodies A and B are equal in magnitude and oppositely directed in each case. If the acceleration of Planet A is unnoticeable in (a), then it is more noticeable in (b), where the difference between the masses is less extreme. In (c), where both bodies have equal mass, the acceleration of Planet A is as evident as it is for Planet B. Continuing, we see that the acceleration of Planet A becomes even more evident in (d) and even more so in (e). So, strictly speaking, when you step off the curb, the street rises ever so slightly to meet you.

When a cannon is fired, there is an interaction between the cannon and the cannonball (Figure 3.20). The sudden force that the cannon exerts on the cannonball is exactly equal and opposite to the force the cannonball exerts on the cannon. This is why the cannon recoils (kicks). But the effects of these equal

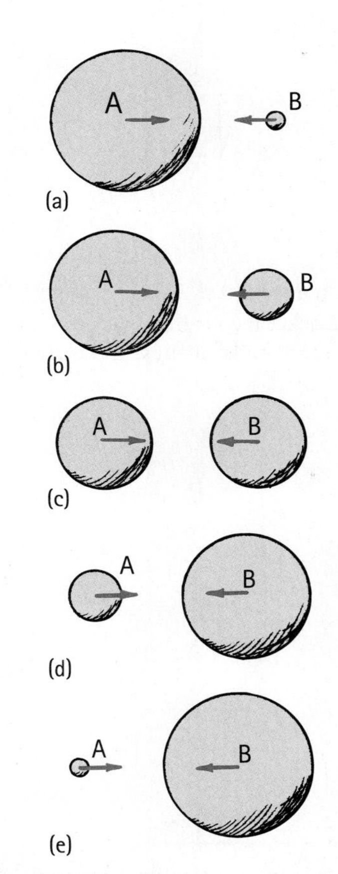

FIGURE 3.19
Which planet falls toward the other, A or B? Although the magnitudes of the forces are the same, how do the accelerations of each planet relate to their relative masses?

FIGURE 3.20
INTERACTIVE FIGURE 

The force exerted against the recoiling cannon is just as great as the force that drives the cannonball along the barrel. Why, then, does the cannonball undergo more acceleration than the cannon?

forces are very different because the forces act on different masses. The different accelerations are evident in Newton's second law,

$$a = \frac{F}{m}$$

Let F represent both the action and reaction forces, $\Large m$ the mass of the cannon, and m the mass of the cannonball. Different-sized symbols are used here to indicate the relative masses and the resulting accelerations. Then the acceleration of the cannonball and cannon can be represented in the following way:

$$\text{cannonball: } \frac{F}{m} = \Large a$$

$$\text{cannon: } \frac{F}{\Large m} = \scriptstyle a$$

Thus we see why the change in velocity of the cannonball is so large compared with the change in velocity of the cannon. A given force exerted on a small mass produces a large acceleration, while the same force exerted on a large mass produces a small acceleration.

We can extend the idea of a cannon recoiling from the ball it fires to understanding rocket propulsion. Consider an inflated balloon recoiling when air is expelled. If the air is expelled downward, the balloon accelerates upward. The same principle applies to a rocket, which continuously "recoils" from the ejected exhaust gas. Each molecule of exhaust gas is like a tiny cannonball shot from the rocket (Figure 3.21).

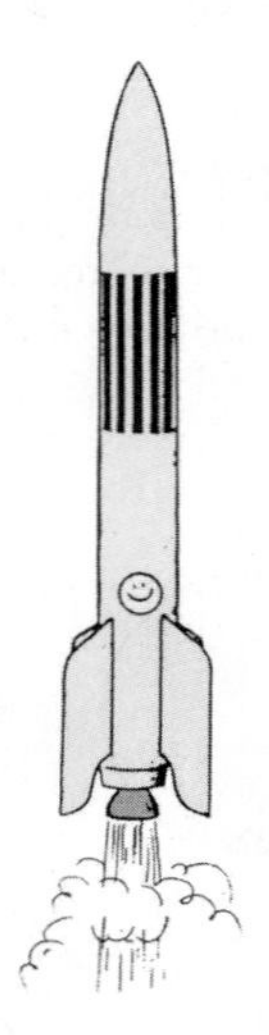

FIGURE 3.21
The rocket recoils from the "molecular cannonballs" it fires, and it rises.

A common misconception is that a rocket is propelled by the impact of exhaust gases against the atmosphere. In fact, before the advent of rockets, it was commonly thought that sending a rocket to the Moon was impossible. Why? Because there is no air above Earth's atmosphere for the rocket to push against. But this is like saying a cannon wouldn't recoil unless the cannonball had air to push against. Not true! Both the rocket and the recoiling cannon accelerate because of the reaction forces by the material they fire, not because of any pushes on the air. In fact, a rocket operates better above the atmosphere, where there is no air resistance.

CHECK YOURSELF

1. **We know that Earth pulls on the Moon. Does it follow that the Moon also pulls on Earth?**
2. **Which pulls harder—Earth on the Moon or the Moon on Earth?**
3. **A high-speed bus and an unfortunate bug have a head-on collision. The force of the bus on the bug splatters it all over the windshield. Is the corresponding force of the bug on the bus greater than, less than, or the same as the force of the bus on the bug? Is the resulting deceleration of the bus greater than, less than, or the same as that of the bug?**

CHECK YOUR ANSWERS

1. Yes, both pulls make up an action–reaction pair of forces associated with the gravitational interaction between Earth and the Moon. We can say that (a) Earth pulls on the Moon, and (b) the Moon likewise pulls on Earth; but it is more insightful to think of this as a single interaction—Earth and Moon simultaneously pulling on each other, each with the *same* amount of force.

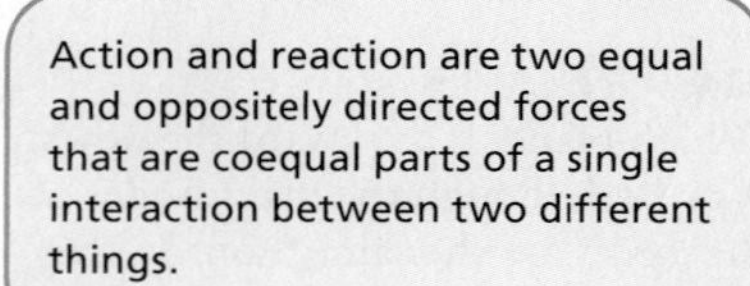

Action and reaction are two equal and oppositely directed forces that are coequal parts of a single interaction between two different things.

2. Both pull with the same strength. This is like asking which distance is greater—from San Francisco to New York, or from New York to San Francisco? Both distances, like both forces in the Moon–Earth pulls, are the same.
3. The magnitudes of the forces are the same because they constitute an action–reaction force pair that makes up the interaction between the bus and the bug. The accelerations, however, are very different because the masses are different. The bug undergoes an enormous and lethal deceleration, while the bus undergoes a very tiny deceleration—so tiny that the very slight slowing of the bus is unnoticed by its passengers. But, if the bug were more massive—as massive as another bus, for example—the slowing down would be very apparent.

Defining Your System

An interesting question often arises: Since action and reaction forces are equal and opposite, why don't they cancel to zero? To answer this question, we must consider the system. Consider, for example, a system consisting of the single orange shown in Figure 3.22. The dashed line surrounding the orange encloses and defines the system. The vector that pokes outside the dashed line represents an external force on the system. The system accelerates in accord with Newton's second law. In Figure 3.23, we see that this force is provided by the apple, which doesn't change our analysis. The apple is outside the system. The fact that the orange simultaneously exerts a force on the apple, which is external to the system, may affect the apple (another system) but not the orange. You can't cancel a force on the orange with a force on the apple. So, in this case, the action–reaction forces don't cancel.

Now let's consider a larger system, enclosing both the orange and the apple. We see the system bounded by the dashed line in Figure 3.24. Notice that the force pair is internal to the orange–apple system. These forces do cancel each other. They play no role in accelerating the system. A force external to the system is needed for acceleration. That's where friction with the floor comes in (Figure 3.25). When the apple pushes against the floor, the floor simultaneously pushes on the apple—an external force on the system. The system accelerates to the right.

Inside a baseball are trillions and trillions of interatomic forces at play. They hold the ball together, but they play no role in accelerating the ball. Although every one of the interatomic forces is part of an action–reaction pair inside the ball, they combine to zero, no matter how many of them there are. A force external to the ball, such as a swinging bat provides, is needed to accelerate the ball.

If this is confusing, it may be well to note that Newton had difficulties with the third law himself.

FIGURE 3.22

INTERACTIVE FIGURE

A force acts on the orange, and the orange accelerates to the right.

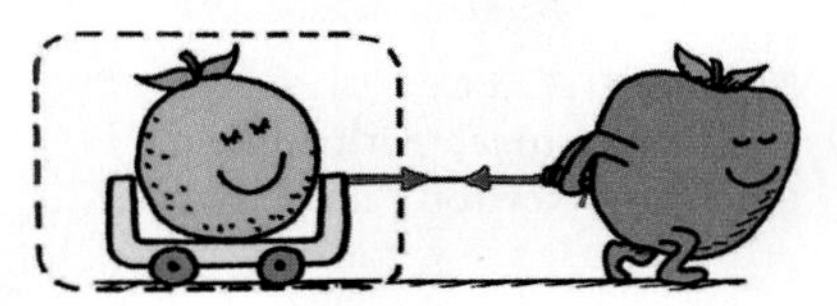

FIGURE 3.23

INTERACTIVE FIGURE

The force on the orange, provided by the apple, is not canceled by the reaction force on the apple. The orange still accelerates.

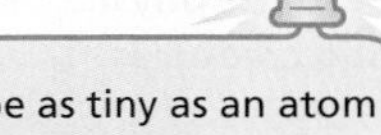

A system may be as tiny as an atom or as large as the universe.

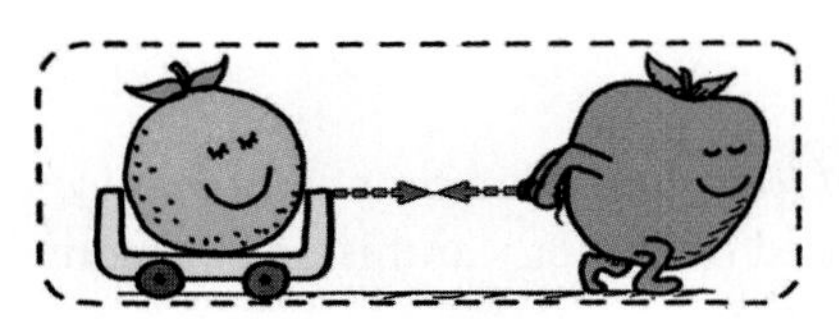

FIGURE 3.24

INTERACTIVE FIGURE

In the larger system of orange + apple, action and reaction forces are internal and cancel. If these are the only horizontal forces, with no external force, then no acceleration of the system occurs.

FIGURE 3.25

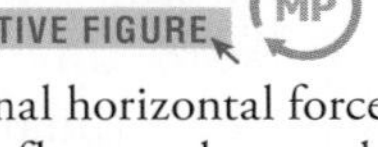

INTERACTIVE FIGURE

An external horizontal force occurs when the floor pushes on the apple (reaction to the apple's push on the floor). The orange + apple system accelerates.

CHECK YOURSELF

1. **On a cold, rainy day, your car battery is dead, and you must push the car to move it and get it started. Why can't you move the car by remaining comfortably inside and pushing against the dashboard?**
2. **Does a fast-moving softball possess force?**

CHECK YOUR ANSWERS

1. In this case, the system to be accelerated is the car. If you remain inside and push on the dashboard, the force pair you produce acts and reacts within the system. These forces cancel out, as far as any motion of the car is concerned. To accelerate the car, there must be an interaction between the car and something external—for example, you on the outside pushing against the road.
2. No, a force is not something an object has, like mass; a force is an interaction between one object and another. A speeding softball may possess the capability of exerting a force on another object when interaction occurs, but it does not possess force as a thing in itself. As we will see in the next chapter, moving things possess momentum and kinetic energy.

FIGURE 3.26
You cannot touch without being touched—Newton's third law.

We see Newton's third law in action everywhere. A fish propels water backward with its fins, and the water propels the fish forward. The wind caresses the branches of a tree, and the branches caress back on the wind to produce whistling sounds. Forces are interactions between different things. Every contact requires at least a twoness; there is no way that an object can exert a force on nothing. Forces, whether large shoves or slight nudges, always occur in pairs, each opposite to the other. Thus, we cannot touch without being touched (Figure 3.26).

LEARNING OBJECTIVE
Relate the physics of motion to living creatures.

Animal Locomotion

EXPLAIN THIS When you push backward on the ground while walking, what pushes you forward?

The study of how animals move, *animal locomotion*, is a branch of *biophysics*. Biophysics draws both on physics and on biology and is one of many crossover science disciplines thriving today.

Much of the study of animal locomotion is based on Newton's third law. To move forward, an animal pushes back on something else; the reaction force pushes the animal forward. For example, birds fly by pushing against air. When they flap their wings, birds push the air downward. The air, in turn, pushes the bird upward. When a bird is soaring, its wing is shaped so that moving air particles are deflected downward; the upward reaction force is lift. A fish swims by pushing against water—the fish propels water backward with its fins, and the water propels the fish forward. Likewise, land animals such as humans push against the ground, and the ground in turn pushes them forward.

Let's consider an example of animal locomotion on land in more detail. Specifically, how does Newton's third law come into play when you walk? First, note that when you are standing still, you are not accelerating. The forces that act on you, gravity and the normal force, balance as shown in Figure 3.27a. To walk, you must accelerate horizontally—the vertical forces of gravity and the normal force don't help. The forces involved in walking are horizontal *frictional* forces (Figure 3.27b). Because your feet are firmly pressed to the floor, there is friction when you push your foot horizontally against the floor. By Newton's third law, the floor pushes back on you in the opposite direction—forward. So the reaction force that allows you to walk is the friction force that the floor applies to your foot and thereby to your mass as a whole. (Don't be confused by all the internal forces within your body that are involved in walking, such as the rotation of your bones and the stretching of your muscles and tendons. An *external* force must act on your body to accelerate it; friction is that force.)

After friction nudges you forward from a standstill, your step is like a controlled fall. You step forward, and your body drops a short distance until your front foot becomes planted in front of you. Friction, as shown in Figure 3.27c, acts in the opposite direction now as it prevents your front foot from sliding forward.

Locomotion is important for many life functions (eating, finding mates, escaping predators, and so on). Biophysical research in this area, therefore, has beneficial applications for countless animals—human and otherwise—that have impaired locomotion.

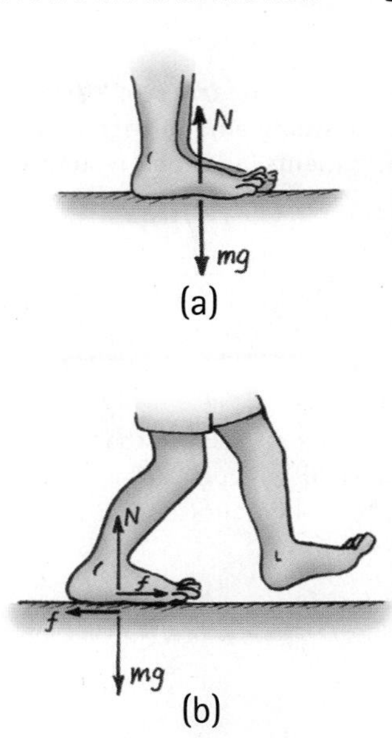

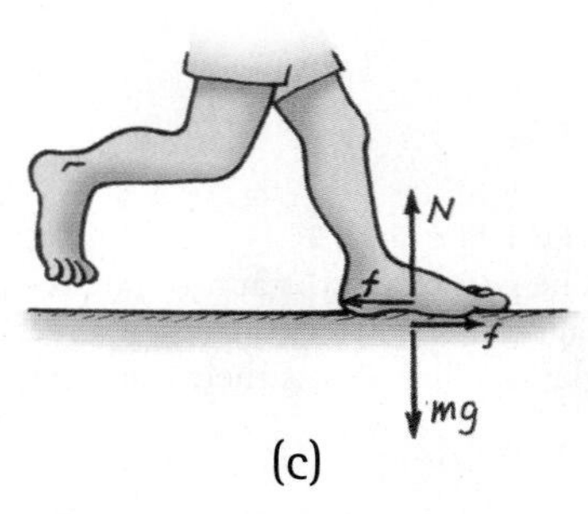

FIGURE 3.27
(a) Standing still, you push against the floor with a force equal to your weight; the normal force pushes you back equally—action and reaction. (b) Your lifted foot doesn't accelerate you horizontally. It's that frictional force you apply to the floor and its frictional push back on you that move you forward. (c) When your forward foot lands, friction acts again but in the opposite direction. Friction stops your foot from slipping forward as the rest of your body catches up.

CHECK YOURSELF

1. **In what way is the study of animal locomotion an integrated science?**
2. **Why is Newton's third law such a necessary piece of information for understanding animal locomotion?**
3. **Why don't the force interactions among your muscles, bones, and other internal organs—or, for that matter, the forces among the atoms and molecules in your body—move your body as a whole?**
4. **Why is walking in a puddle of grease so much more difficult than walking on carpet?**

CHECK YOUR ANSWERS

1. It combines biology and physics.
2. Generally speaking, animals move by pushing back on some medium, and the reaction force pushes them forward.
3. Forces internal to a system cannot accelerate a system.
4. Grease is so smooth that it offers little friction to your feet and therefore insufficient reaction force to get you walking.

LEARNING OBJECTIVE
Differentiate among vectors, their components, and their resultants.

3.5 Vectors

EXPLAIN THIS What is the difference between a vector quantity and a scalar quantity?

We have learned that quantities such as velocity, force, and acceleration require both magnitude and direction for a complete description. Such a quantity is a **vector quantity**. By contrast, a quantity that can be described by magnitude only, a quantity not involving direction, is called a **scalar quantity**. Mass, volume, and speed are scalar quantities.

FIGURE 3.28
This vector, scaled so that 1 cm equals 20 N, represents a force of 60 N to the right.

A vector quantity is nicely represented by an arrow. When the length of the arrow is scaled to represent the quantity's magnitude and the direction of the arrow indicates the direction of the quantity, we refer to the arrow as a **vector** (Figure 3.28).

Adding vectors is quite simple when they act along parallel directions: If they are in the same direction, they add; if they are in opposite directions, they subtract. The sum of two or more vectors is called their **resultant**. To find the resultant of nonparallel vectors, we use the parallelogram rule.* Construct a parallelogram in which the two vectors are adjacent sides—the diagonal of the parallelogram shows the resultant. In Figure 3.29, the parallelograms are rectangles.

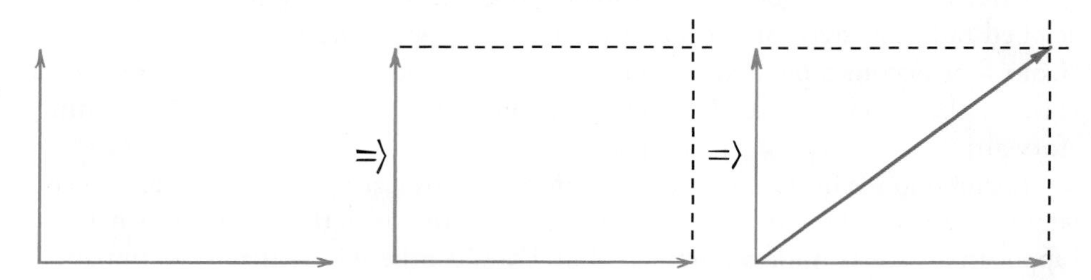

FIGURE 3.29
The pair of vectors at right angles to each other form two sides of a rectangle; the diagonal is their resultant.

In the special case of two perpendicular vectors that are equal in magnitude, the parallelogram is a square (Figure 3.30). Since, for any square, the length of a diagonal is $\sqrt{2}$, or 1.41 times the length of one of the sides, the resultant is $\sqrt{2}$ times one of the vectors. So, the resultant of two equal vectors of magnitude 100 acting at a right angle to each other is 141. Consider the **force vectors** shown in Figure 3.31. The figure shows the top view of a pair of horizontal forces acting on the box. One is 30 N and the other is 40 N. Simple measurement shows that the resultant is 50 N. (Vectors are developed in more detail in Appendix C.)

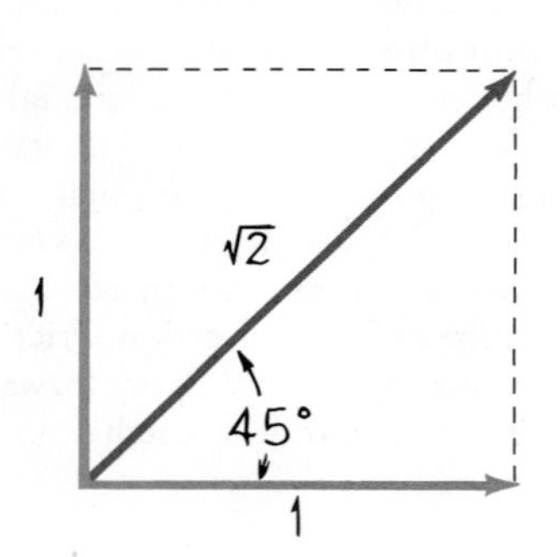

FIGURE 3.30
When a pair of equal-length vectors at right angles to each other are added, they form a square. The diagonal of the square is $\sqrt{2}$ times the length of the other side.

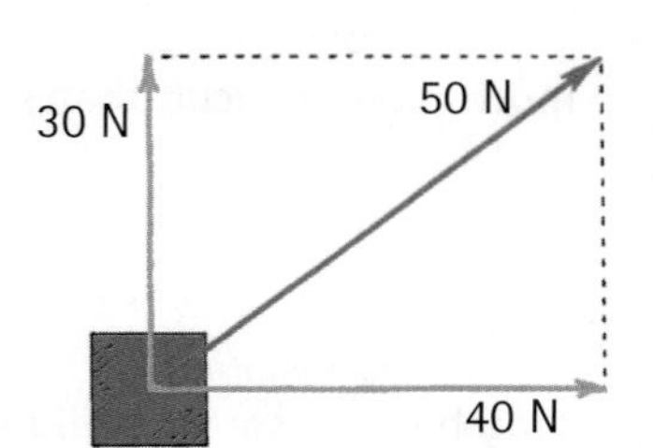

FIGURE 3.31
INTERACTIVE FIGURE
MP
The resultant of the 30-N and 40-N forces is 50 N.

*A parallelogram is a four-sided figure with opposite sides equal in length and parallel to each other. You can determine the length of the diagonal by measuring, but, in the special case in which the two vectors **V** and **H** are perpendicular, forming a square or rectangle, you can apply the Pythagorean theorem, $\mathbf{R}^2 = \mathbf{V}^2 + \mathbf{H}^2$, to give the resultant: $\mathbf{R} = \sqrt{(\mathbf{V}^2 + \mathbf{H}^2)}$.

MATH CONNECTION

Vector Components

Just as two vectors at right angles can be resolved into one resultant vector, any vector can be "resolved" into two component vectors perpendicular to each other. These two vectors are known as the **components** of the given vector they replace. The process of determining the components of a vector is called *resolution*. Any vector drawn on a piece of paper can be resolved into vertical and horizontal components.

Vector resolution is illustrated in the accompanying figure. A vector **V** is drawn in the proper direction to represent a vector quantity. Then vertical and horizontal lines (axes) are drawn at the tail of the vector. Next, a rectangle is drawn that has **V** as its diagonal. The sides of this rectangle are the desired components, vectors **X** and **Y**. In reverse, note that the vector sum of vectors **X** and **Y** is **V**. We'll return to vector components when we treat projectile motion in Chapter 5.

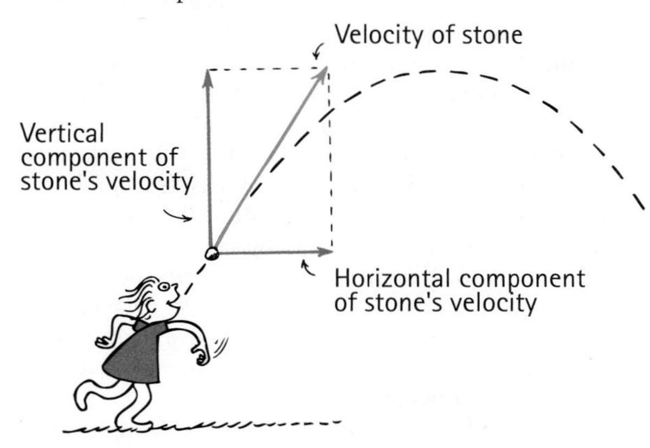

The horizontal and vertical components of a ball's velocity.

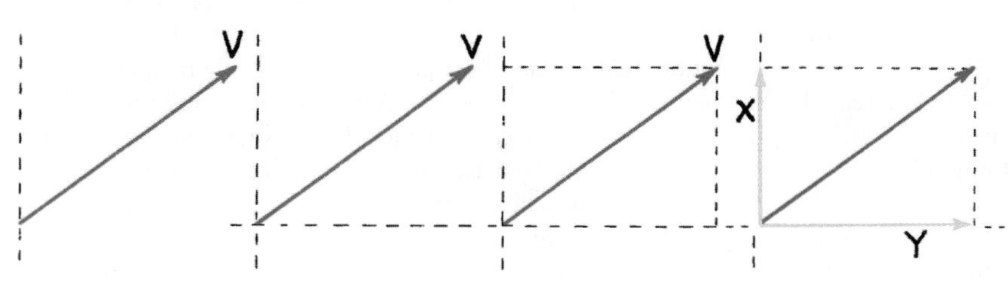

Construction of the vertical and horizontal components of a vector.

Problems

1. With a ruler, draw the horizontal and vertical components of the two vectors shown here. Measure the components.

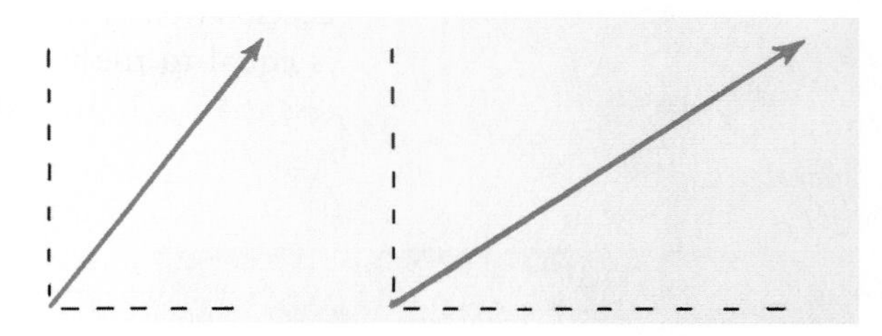

2. Consider an airplane that sets its course north but is blown off course by a crosswind blowing eastward at 60 km/h. The resultant velocity of the plane is 100 km/h relative to the ground in a direction between north and northeast. Find the plane's airspeed (its speed if there were no wind).

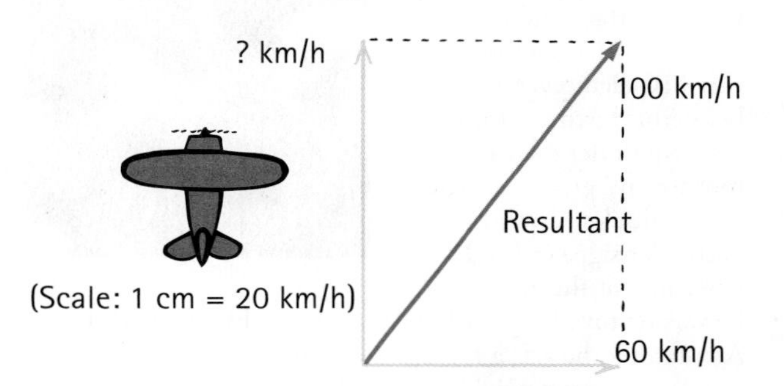

Solutions

1. Left vector: The horizontal component is 2 cm; the vertical component is 2.6 cm.
   Right vector: The horizontal component is 3.9 cm; the vertical component is 2.6 cm.
2. The plane's airspeed is 80 km/h.

3.6 Summary of Newton's Three Laws

LEARNING OBJECTIVE
Summarize and contrast Newton's three laws of motion.

EXPLAIN THIS If the *action* is the force acting on a dropped ball, identify the *reaction*.

Newton's first law, the law of inertia: An object at rest tends to remain at rest; an object in motion tends to remain in motion at constant speed along a straight-line path. This property of objects to resist change in motion is called inertia. Objects will undergo changes in motion only in the presence of a net force.

Newton's second law, the law of acceleration: When a net force acts on an object, the object will accelerate. The acceleration is directly proportional to the net force and inversely proportional to the mass. Symbolically, $a \sim F/m$. Acceleration is always in the direction of the net force. When an object falls in a vacuum, the net force is simply the weight and the acceleration is g (the symbol g denotes that acceleration is due to gravity alone). When an object falls in air, the net force is equal to the weight minus the force of air resistance, and the acceleration is less than g. If and when the force of air resistance equals the weight of a falling

HISTORY OF SCIENCE

Isaac Newton (1642–1727)

Isaac Newton was born prematurely (and barely survived) on Christmas Day, 1642, at his mother's farmhouse in Woolsthorpe, England, the same year that Galileo died. Newton's father had died several months before his birth, and Isaac grew up under the care of his mother and grandmother. As a child, he showed no particular signs of brightness, and, at the age of $14\frac{1}{2}$, he was removed from school to work on his mother's farm. As a farmer, he was a failure, preferring to read books he borrowed from a neighborhood druggist. An uncle sensed the potential in young Isaac and prompted him to study at Cambridge University, which he did for 5 years, graduating without particular distinction.

A plague swept through London, and Newton retreated to his mother's farm—this time to continue his studies. While on the farm, at age 23, he laid the foundations for the work that was to make him immortal. Seeing an apple fall to the ground led him to consider the force of gravity extending to the Moon and beyond. He formulated the law of universal gravitation (which he later proved). He invented the calculus, a very important mathematical tool in science. He extended Galileo's work and formulated the three fundamental laws of motion. He also formulated a theory of the nature of light and demonstrated with prisms that white light is composed of all colors of the rainbow. It was his experiments with prisms that initially brought him fame.

When the plague subsided, Newton returned to Cambridge and soon established a reputation for himself as a first-rate mathematician. His mathematics teacher resigned in his favor, and Newton was appointed the Lucasian Professor of Mathematics, a post he held for 28 years. In 1672, he was elected to the Royal Society, where he exhibited the world's first reflector telescope. It can still be seen, preserved at the library of the Royal Society in London, with this inscription: "The first reflecting telescope, invented by Sir Isaac Newton, and made with his own hands."

It wasn't until Newton was 42 that he began to write what is generally acknowledged as the greatest scientific book ever written, the *Philosophiae Mathematica Principia Naturalis*. He wrote the work in Latin and completed it in 18 months. It appeared in print in 1687, but an English translation wasn't printed until 1729, two years after his death. When asked how he was able to make so many discoveries, Newton replied that he solved his problems by continually thinking very long and hard about them—and not by sudden insight.

At the age of 46, Newton was elected a member of Parliament for two years and never gave a speech. One day he rose and the House fell silent to hear the great man speak. Newton's "speech" was very brief; he simply requested that a window be closed because of a draft.

A further turn from his work in science was his appointment as warden, and then as master, of the mint. Newton resigned his professorship and directed his efforts toward greatly improving the workings of the mint, to the dismay of counterfeiters who flourished at that time. He maintained his membership in the Royal Society and was elected president, then was reelected each year for the rest of his life. At the age of 62, he wrote *Opticks*, which summarized his work on light. Nine years later, he wrote a second edition of his *Principia*.

Although Newton's hair turned gray when he was 30, it remained full, long, and wavy all his life. Unlike others of his time, he did not wear a wig. He was a modest man, very sensitive to criticism, and never married. He remained healthy in body and mind into old age. At 80, he still had all his teeth, his eyesight and hearing were sharp, and his mind was alert. In his lifetime, he was regarded by his countrymen as the greatest scientist who ever lived. In 1705, he was knighted by Queen Anne. Newton died at the age of 85, and he was buried in Westminster Abbey, along with England's kings and heroes.

Newton showed that the universe ran according to natural laws—a knowledge that provided hope and inspiration to people of all walks of life and that ushered in the Age of Reason. The ideas and insights of Isaac Newton truly changed the world and elevated the human condition.

object, acceleration terminates and the object falls at a constant speed (called the *terminal speed*).

Newton's third law, the law of action–reaction: Whenever one object exerts a force on a second object, the second object simultaneously exerts an equal and opposite force on the first. Forces occur in pairs: One is an action and the other is a reaction, and together they constitute the interaction between one object and the other. Action and reaction always act on different objects. Neither force exists without the other.

Are Newton's laws *always* accurate? From falling apples to rising smoke to the orbits of planets, Newton's laws seem to explain all motion. However, there are limitations to these laws. Modern experiments show that Newton's laws are valid over the range of phenomena we normally observe; however, they are *not* valid for:

1. Objects moving near the speed of light: To understand the motion of very fast moving objects, we must use Einstein's principles of special relativity.
2. Objects that are very small—on the scale of an atom: Objects that consist of just a few atoms move according to the principles of *quantum mechanics* rather than Newtonian physics.
3. Objects under the influence of very strong gravitational forces: To understand the motions of such astronomical objects, we invoke Einstein's theory of *general relativity,* which appears valid for all gravitational forces.

Though they have limitations, Newton's laws remain the major tools of scientists working today. Most objects are slow enough, big enough, and not subject to extraordinary gravitational forces, so Newton's laws work just fine to describe their motion. Also, the mathematics of special relativity, quantum mechanics, and general relativity are so much more complicated than the mathematics of Newtonian physics that these modern theories are usually reserved for extreme situations where Newton's laws do not apply. After all, it was primarily Newton's laws that got us to the Moon!

For instructor-assigned homework, go to www.masteringphysics.com 

SUMMARY OF TERMS (KNOWLEDGE)

Free fall Motion under the influence of gravitational pull only.

Force vector An arrow drawn to scale so that its length represents the magnitude of a force and its direction represents the direction of the force.

Interaction Mutual action between objects during which one object exerts an equal and opposite force on the other object.

Newton's first law of motion Every object continues in a state of rest, or of uniform speed in a straight line, unless acted on by a nonzero force.

Newton's second law of motion The acceleration produced by a net force on an object is directly proportional to the net force, is in the same direction as the net force, and is inversely proportional to the mass of the object.

Newton's third law of motion Whenever one object exerts a force on a second object, the second object exerts an equal and opposite force on the first object.

Resultant The net result of a combination of two or more vectors.

Scalar quantity A quantity, such as mass, volume, speed, and time, that can be completely specified by its magnitude.

Terminal speed The speed at which the acceleration of a falling object terminates when air resistance balances weight.

Vector An arrow whose length represents the magnitude of a quantity and whose direction represents the direction of the quantity.

Vector components Parts into which a vector can be separated and that act in different directions from the vector.

Vector quantity A quantity whose description requires both magnitude and direction.

READING CHECK QUESTIONS (COMPREHENSION)

3.1 Newton's First Law of Motion

1. What is Newton's first law of motion?
2. What kind of path would the planets follow if suddenly their attraction to the Sun no longer existed?

3.2 Newton's Second Law of Motion

3. What is Newton's second law of motion?
4. (a) Is acceleration *directly* proportional to force, or is it *inversely* proportional to force? Give an example.
   (b) Is acceleration *directly* proportional to mass, or is it *inversely* proportional to mass? Give an example.
5. If the mass of a sliding block is tripled at the same time the net force on it is tripled, how does the resulting acceleration compare with the original acceleration?
6. What is the acceleration of a 10-N freely falling object with no air resistance?
7. Why doesn't a heavy object accelerate more than a light object when both are freely falling?
8. What is the acceleration of a falling object that has reached its terminal velocity?
9. What two quantities does air resistance depend on?
10. Who falls faster when wearing the same-size parachute—a heavy person or a light person—or do both fall at the same speed?

3.3 Forces and Interactions

11. How many forces are required for a single interaction?
12. When you push against a wall with your fingers, they bend because they experience a force. What is this force?
13. A boxer can hit a heavy bag with a great force. Why can't he hit a sheet of newspaper in midair with the same amount of force?

3.4 Newton's Third Law of Motion

14. What is Newton's third law of motion?
15. Consider hitting a baseball with a bat. If we call the force of the bat against the ball the action force, what is the reaction force?
16. Do action and reaction forces act in succession or simultaneously?
17. If the forces that act on a cannonball and the recoiling cannon from which it is fired are equal in magnitude, why do the cannonball and cannon have very different magnitudes?
18. What is needed to accelerate a system?

3.5 Vectors

19. Cite three examples of a vector quantity. Then cite three examples of a scalar quantity.
20. How great is the resultant of two equal-magnitude vectors at right angles to each other?
21. According to the parallelogram rule, what does the diagonal of a constructed parallelogram represent?
22. Can it be said that, when two vectors are at right angles to each other, the resultant is greater than either of the vectors separately?
23. When a vector at an angle is resolved into horizontal and vertical components, can it be said that each component has less magnitude than the original vector?

THINK INTEGRATED SCIENCE

3A—Gliding

24. What is gliding locomotion?
25. Why is having a large surface area important for effective gliding?
26. Describe some of the physical characteristics that gliding organisms have evolved to increase their surface area.

3B—Animal Locomotion

27. Explain how Newton's third law underlies many forms of animal locomotion—from fish, to birds, to humans.
28. A squid propels itself forward by pushing water backward. Why does this occur?
29. When you walk, what is the force that pushes you forward?
30. Why does a duck in an oil spill find it difficult to walk?

THINK AND DO (HANDS-ON APPLICATION)

31. The net force acting on an object and the resulting acceleration are always in the same direction. You can demonstrate this with a spool. If the spool is pulled horizontally to the right, in which direction will it roll? (Some of your classmates may be surprised.)
32. Hold your hand with the palm down like a flat wing outside the window of a moving automobile. Then slightly tilt the front edge of your hand upward and notice the lifting effect as air bounces from the bottom of your hand. Which of Newton's laws is illustrated here?

PLUG AND CHUG (FORMULA FAMILIARIZATION)

Do these simple one-step calculations and familiarize yourself with the equation that links the concepts of force, mass, and acceleration.

$$\textbf{Acceleration} = \frac{\textbf{net force}}{\textbf{mass}} = \frac{F_{net}}{m}$$

33. In Chapter 2 acceleration is defined as $a = \frac{\Delta v}{\Delta t}$. Use this formula to show that the acceleration of a cart on an inclined plane that gains 6.0 m/s each 1.2 s is 5.0 m/s^2.
34. In this chapter we learned that the cause of acceleration is given by Newton's second law: $a = F_{net}/m$. Show that the 5.0-m/s^2 acceleration in Exercise 33 can result when a 15-N net force is exerted on a 3.0-kg cart. (Note: The unit N/kg is equivalent to m/s^2.)
35. If you know that a 1-kg object weighs 10 N, confirm that the acceleration of a 1-kg stone in free fall is 10 m/s^2.
36. A simple rearrangement of Newton's second law gives $F_{net} = ma$. Show that a net force of 84 N is needed to give a 12-kg package an acceleration of 7.0 m/s^2. (Note: The units kg·m/s^2 and N are equivalent.)

THINK AND COMPARE (ANALYSIS)

37. Four boxes of different masses are on a friction-free level table. From greatest to least, rank the (a) net forces on the boxes and (b) accelerations of the boxes.

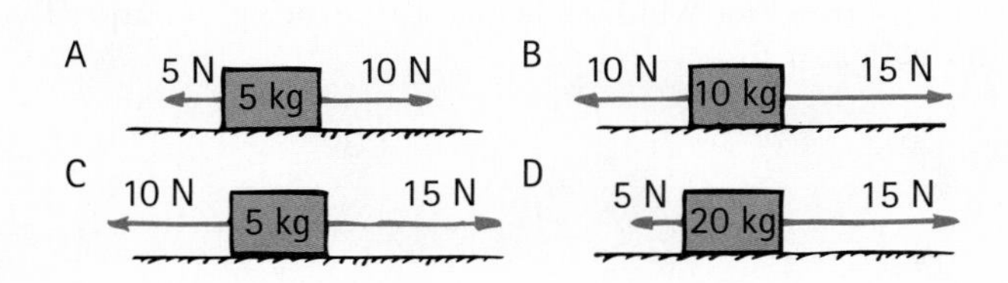

38. In cases A, B, and C, the crate is in equilibrium (no acceleration). From greatest to least, rank the amounts of friction between the crate and the floor.

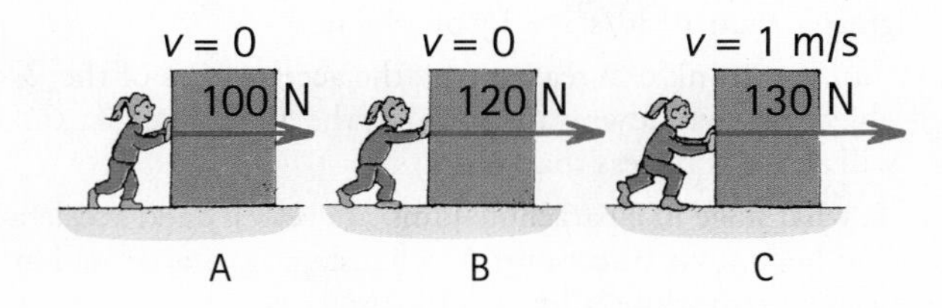

39. Consider a 100-kg box of tools in locations A, B, and C. Rank from greatest to least the (a) masses of the 100-kg box of tools and (b) weights of the 100-kg box of tools.

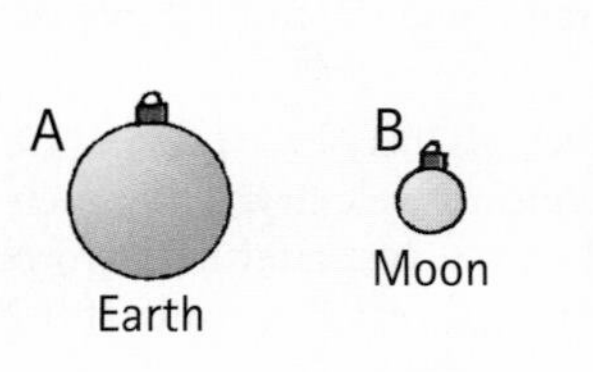

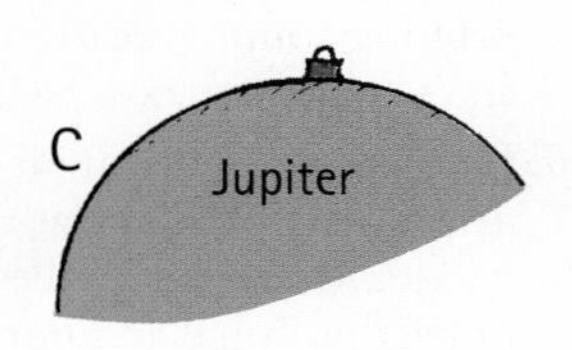

40. Three parachutists, A, B, and C, each have reached terminal velocity at the same altitude. (a) From fastest to slowest, rank their terminal velocities. (b) From longest to shortest times, rank their order in reaching the ground.

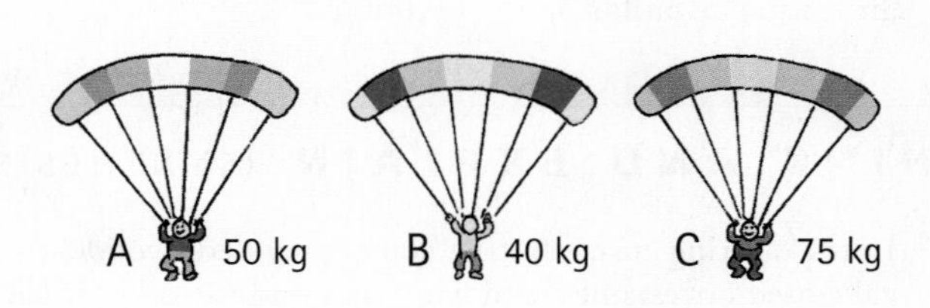

41. The strong man is pulled in the three situations shown. Rank the amounts of tension in the rope in his right hand (the one attached to the tree in B and C) from least to greatest.

THINK AND SOLVE (MATHEMATICAL APPLICATION)

42. A 400-kg bear grasping a vertical tree slides down at a constant velocity. Show that the friction force that acts on the bear is about 4000 N.
43. When two horizontal forces are exerted on a cart, 600 N forward and 400 N backward, the cart undergoes acceleration. Show that the additional force needed to produce nonaccelerated motion is 200 N.
44. You push with a 20-N horizontal force on a 2-kg mass resting on a horizontal surface. The horizontal friction force is 12 N. Show that the acceleration is 4 m/s^2.
45. You push with a 40-N horizontal force on a 4-kg mass resting on a horizontal surface. The horizontal friction force is 12 N. Show that the acceleration is 7 m/s^2.
46. A cart of mass 1 kg is accelerated at 1 m/s^2 by a force of 1 N. Show that a 2-kg cart pushed with a 2-N force would also accelerate at 1 m/s^2.
47. A rocket of mass 100,000 kg undergoes an acceleration of 2 m/s^2. Show that the force being developed by the rocket engines is 200,000 N.
48. A 747 jumbo jet of mass 30,000 kg experiences a 30,000-N thrust for each of four engines during takeoff. Show that its acceleration is 4 m/s^2.
49. Suppose the jumbo jet in Exercise 48 flies against an air resistance of 90,000 N while the thrust of all four engines is 100,000 N. Show that its acceleration will be about 0.3 m/s^2. What will the acceleration be when the air resistance builds up to 100,000 N?
50. A boxer punches a sheet of paper in midair, bringing it from rest to a speed of 25 m/s in 0.05 s. If the mass of the paper is 0.003 kg, show that the force the boxer exerts on it is only 1.5 N.
51. Suppose that you are standing on a skateboard near a wall and you push on the wall with a force of 30 N. (a) How hard does the wall push on you? (b) Show that if your mass is 60 kg your acceleration while pushing will be 0.5 m/s^2.
52. If raindrops fall vertically at a speed of 3 m/s and you are running horizontally at 4 m/s, show that the drops will hit your face at a speed of 5 m/s.
53. Horizontal forces of 3 N and 4 N act at right angles on a block of mass 5 kg. Show that the resulting acceleration is 1 m/s^2.
54. Suzie Skydiver with her parachute has a mass of 50 kg.
    (a) Before Suzie opens her chute, show that the force of air resistance she encounters when reaching terminal velocity is about 500 N.
    (b) After her chute is open and she reaches a lower terminal velocity, show that the force of air resistance she encounters is also about 500 N.
    (c) Discuss why your answers are the same.
55. An airplane with an airspeed of 120 km/h encounters a 90-km/h crosswind. Show that the plane's groundspeed is 150 km/h.

THINK AND EXPLAIN (SYNTHESIS)

56. In the orbiting space shuttle, you are handed two identical closed boxes, one filled with sand and the other filled with feathers. How can you tell which is which without opening the boxes?
57. Your empty hand is not hurt when it bangs lightly against a wall. Why is it hurt if it is carrying a heavy load when it bangs against the wall? Which of Newton's laws is most applicable here?
58. As you stand on a floor, does the floor exert an upward force against your feet? How much force does it exert? Why aren't you moved upward by this force?
59. A rocket becomes progressively easier to accelerate as it travels through space. Why? (Hint: About 90% of the mass of a newly launched rocket is fuel.)
60. As you are leaping upward from the ground, how does the force you exert on the ground compare with your weight?
61. On which of these hills does the ball roll down with increasing speed and decreasing acceleration along the path? (Use this example if you wish to explain to someone the difference between speed and acceleration.)

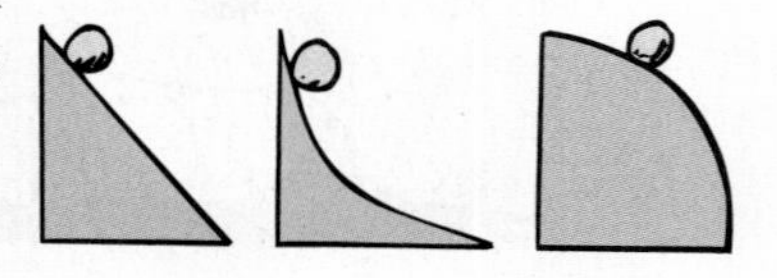

62. Neglecting air resistance, if you drop an object, its acceleration toward the ground is 10 m/s^2. If you throw it down instead, will its acceleration after throwing be greater than 10 m/s^2? Why or why not?
63. Can you think of a reason why the acceleration of the object thrown downward through the air in Exercise 62 will actually be less than 10 m/s^2?
64. At what stage in a parachute jump are velocity and acceleration in opposite directions? At what stage does acceleration become zero while falling continues?
65. You hold an apple over your head. (a) Identify all the forces acting on the apple and their reaction forces. (b) When you drop the apple, identify all the forces acting on it as it falls and the corresponding reaction forces.
66. If Earth exerts a gravitational force of 1000 N on an orbiting communications satellite, how much force does the satellite exert on Earth?
67. If you exert a horizontal force of 200 N to slide a crate across a factory floor at a constant velocity, how much friction is exerted by the floor on the crate? Is the force of friction equal and oppositely directed to your 200-N

push? Does the force of friction make up the reaction force to your push? Why?

68. If a Mack truck and a motorcycle have a head-on collision, on which vehicle is the impact force greater? Which vehicle undergoes the greater change in motion? Explain your answers.

69. Two people of equal mass attempt a tug-of-war with a 12-m rope while standing on frictionless ice. When they pull on the rope, each person slides toward the other. How do their accelerations compare, and how far does each person slide before they meet?

70. Suppose that one person in Exercise 69 has twice the mass of the other. How far does each person slide before they meet?

71. Which team wins in a tug-of-war—the team that pulls harder on the rope or the team that pushes harder against the ground? Explain.

72. The photo shows Steve Hewitt and his daughter Gretchen. Is Gretchen touching her dad or is he touching her? Explain.

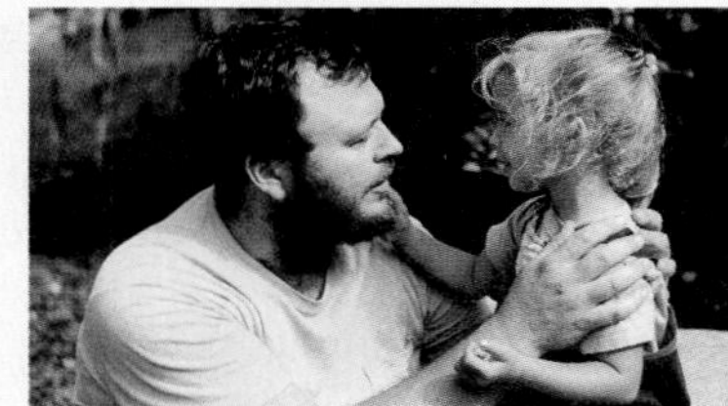

73. How does the weight of a falling body compare with the air resistance it encounters just before it reaches terminal velocity? Just after?

74. Free fall is motion in which gravity is the only force acting. (a) Is a skydiver who has reached terminal speed in free fall? (b) Is a satellite circling Earth above the atmosphere in free fall?

75. Why is it that a cat that falls from the top of a 50-story building will hit the safety net below no faster than if it fell from the 20th story?

76. You tell your friend that the acceleration of a skydiver before the chute opens decreases as falling progresses. Your friend then asks if this means the skydiver is slowing down. What is your response?

77. Which is more likely to break—the ropes supporting a hammock stretched tightly between a pair of trees or the ropes supporting a hammock that sags more when you sit on it? Defend your answer.

78. A stone is shown at rest on the ground. (a) The vector shows the weight of the stone. Complete the vector diagram by showing another vector that results in zero net force on the stone. (b) What is the conventional name of the vector you have drawn?

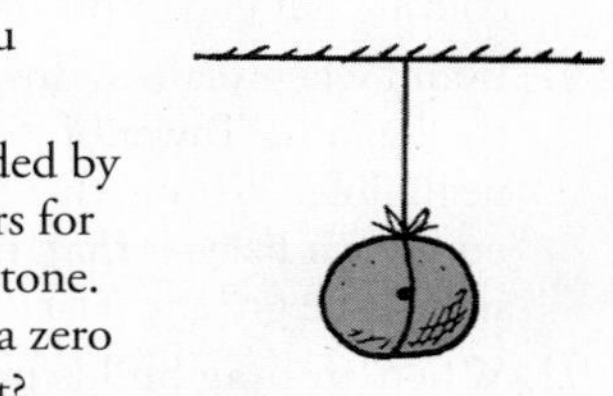

79. Here a stone at rest is suspended by a string. (a) Draw force vectors for all the forces that act on the stone. (b) Should your vectors have a zero resultant? (c) Why or why not?

80. Here the same stone is being accelerated vertically upward. (a) Draw force vectors to some suitable scale showing the relative forces acting on the stone. (b) Which is the longer vector, and why?

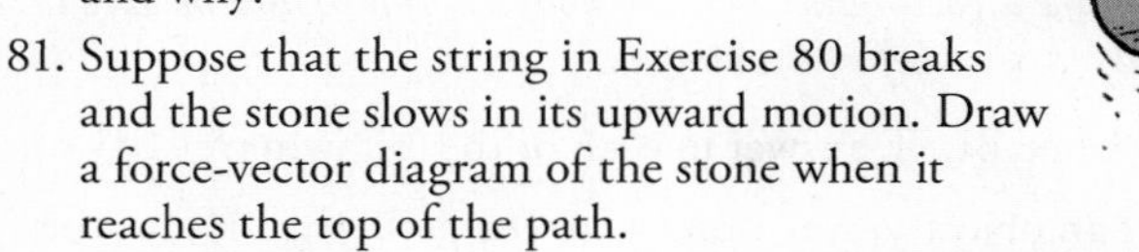

81. Suppose that the string in Exercise 80 breaks and the stone slows in its upward motion. Draw a force-vector diagram of the stone when it reaches the top of the path.

82. What is the net force on the stone in Exercise 81 when it is at the top of its path? What is its instantaneous velocity? What is its acceleration?

83. Here is the stone sliding down a friction-free incline. (a) Identify the forces that act on it, and draw appropriate force vectors. (b) By the parallelogram rule, construct the resultant force on the stone (carefully showing that it has a direction parallel to the incline—the same direction as the stone's acceleration).

84. Here is the stone at rest, interacting with both the surface of the incline and the block. (a) Identify all the forces that act on the stone, and draw appropriate force vectors. (b) Show that the net force on the stone is zero. (Hint 1: There are two normal forces *on* the stone. Hint 2: Be sure the vectors you draw are for forces that act on the stone, not for forces that act on the surfaces *by* the stone.)

THINK AND DISCUSS (EVALUATION)

85. Each of the vertebrae forming your spine is separated from its neighbors by disks of elastic tissue. What happens, then, when you jump heavily on your feet from an elevated position? Can you think of a reason why you are a little taller in the morning than you are at the end of the day? (Hint: Think about how Newton's first law of motion applies in this case.)

86. To pull a wagon across a lawn at a constant velocity, you must exert a steady force. Discuss and reconcile this fact with Newton's first law, which states that motion with a constant velocity indicates no force.

87. A common saying goes, "It's not the fall that hurts you; it's the sudden stop." Translate this statement into Newton's laws of motion.

88. Does a stick of dynamite contain force? Discuss and defend your answer.
89. Can a dog wag its tail without the tail in turn "wagging the dog"? (Consider a dog with a relatively massive tail.)
90. When your hand turns the handle of a faucet, water comes out. Do your push on the handle and the water coming out constitute an action–reaction pair? Discuss.
91. If and when Galileo dropped two balls from the top of the Leaning Tower of Pisa, air resistance was not really negligible. Assume that both balls are the same size yet one much heavier than the other. Discuss which ball struck the ground first—and why.
92. When air drag builds up to equal the combined weight of Dick and Jane in their tandem skydive, a terminal velocity of nearly 200 km/h is reached. How would this terminal velocity compare for each if they fell separately?

93. If you simultaneously drop a pair of tennis balls from the second story of a building, they will strike the ground at the same time. If one of the tennis balls is filled with lead pellets, will it fall faster and hit the ground first? Which of the two will encounter more air resistance? Defend your answers.

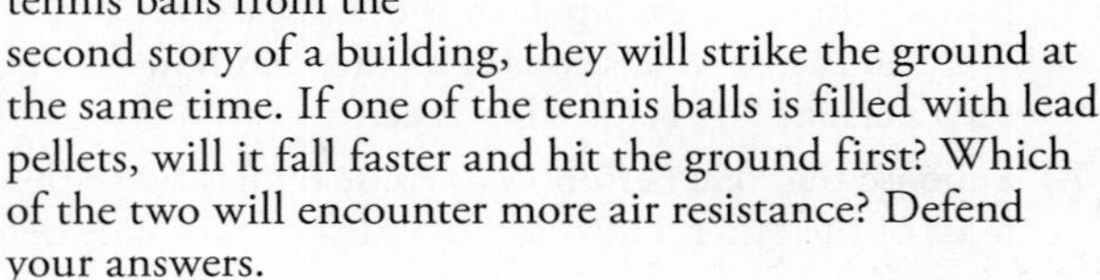

READINESS ASSURANCE TEST (RAT)

If you have a good handle on this chapter, you should be able to score at least 7 out of 10 on this RAT. If you score less than 7, you need to study further before moving on.

Choose the BEST answer to each of the following:

1. If an object moves along a curved path, then it must be
   (a) accelerating.
   (b) acted on by a force.
   (c) both of these
   (d) none of these
2. As mass is added to a pushed object, its acceleration
   (a) increases.
   (b) decreases.
   (c) remains constant.
   (d) quickly reaches zero.
3. A ball rolls down a curved ramp as shown. As its speed increases, its rate of gaining speed
   (a) increases.
   (b) decreases.
   (c) remains unchanged.
   (d) none of these
4. A heavy rock and a light rock in free fall (zero air resistance) have the same acceleration. The *reason* the heavy rock doesn't have a greater acceleration is that the
   (a) force due to gravity is the same on each.
   (b) air resistance is always zero in free fall.
   (c) inertia of both rocks is the same.
   (d) ratio of force to mass is the same.
   (e) none of these
5. You drop a basketball off the edge of the tallest building on your campus. While the ball falls, its speed
   (a) and acceleration both increase.
   (b) increases and its acceleration decreases.
   (c) and acceleration both decrease.
   (d) decreases and its acceleration increases.
6. A karate chop delivers a force of 3000 N to a board that breaks. The force that the board exerts on the hand during this event is
   (a) less than 3000 N.
   (b) 3000 N.
   (c) greater than 3000 N.
   (d) More information is needed.
7. Two parachutists, a heavy person and a light person, jumping from the same altitude have the same size parachute. Which reaches the ground first?
   (a) The heavy person hits the ground first.
   (b) The light person hits the ground first.
   (c) They reach the ground at the same time.
   (d) More information is needed.
8. The amount of air resistance that acts on a wingsuit flyer (and a flying squirrel) depends on the flyer's
   (a) area.
   (b) speed.
   (c) area and speed.
   (d) acceleration.
9. When you push a marble with a 0.5-N force, the marble
   (a) accelerates at 10 m/s^2.
   (b) resists being pushed with its own 0.5 N.
   (c) will likely not move.
   (d) pushes on you with a 0.5-N force.
10. The force that propels a rocket is provided by
   (a) gravity.
   (b) Newton's laws of motion.
   (c) its exhaust gases.
   (d) the atmosphere against which the rocket pushes.

Answers to RAT:

1. d, 2. b, 3. b, 4. d, 5. b, 6. b, 7. a, 8. c, 9. d, 10. c

CHAPTER 4
Momentum and Energy

MOVING OBJECTS have a quantity that objects at rest don't have. More than a hundred years ago, this quantity was called *impedo.* A boulder at rest had no impedo, while the same boulder rolling down a steep incline possessed impedo. The faster an object moved, the greater the impedo. The change in impedo depended on force and, more important, on how long the force acted. Apply a force to a cart and you give it impedo. Apply a long force and you give it more impedo.

But what do we mean by "long"? Does "long" refer to time or distance? When this distinction was made, the term *impedo* gave way to two more precise ideas—*momentum* and *kinetic energy.* And these two ideas are related to a cluster of other concepts, including work, power, efficiency, potential energy, and impulse—all of which, as we shall see in later chapters, nicely relate to the "machinery" of living organisms.

LEARNING OBJECTIVE
Describe the relationship between mass and velocity.

4.1 Momentum

EXPLAIN THIS When does a moving car have more momentum than a moving ship?

MasteringPhysics®
VIDEO: Definition of Momentum

We all know that a heavy truck is more difficult to stop than a lighter car moving at the same speed. We state this fact by saying that the truck has more momentum than the car. By **momentum**, we mean inertia in motion—or, more specifically, the product of the mass of an object and its velocity; that is,

$$\textbf{Momentum} = \textbf{mass} \times \textbf{velocity}$$

Or, in shorthand notation,

$$\textbf{Momentum} = mv$$

When direction is not an important factor, we can say momentum = mass × speed, which we still abbreviate mv.

We can see from the definition that a moving object can have a large momentum if either its mass or its velocity is large or both its mass and its velocity are large. A truck has more momentum than a car moving at the same velocity because it has a greater mass. A huge ship moving at a low velocity can have a large momentum, as can a small bullet moving at a high velocity. A massive truck moving down a steep hill with no brakes has a large momentum, whereas the same truck at rest has no momentum at all.

FIGURE 4.1
The boulder, unfortunately, has more momentum than the runner.

CHECK YOURSELF
When do a 1000-kg car and a 2000-kg truck have the same momentum?

CHECK YOUR ANSWER
They have the same momentum when the car travels twice as fast as the truck. Then (1000 kg × 2v) for the car equals (2000 kg × v) for the truck. Or, if they are both at rest, they have the same momentum—zero.

LEARNING OBJECTIVE
Describe the relationship between force and time.

4.2 Impulse

EXPLAIN THIS What is the role of force in the concept of impulse?

MasteringPhysics®
VIDEO: Changing Momentum

Changes in momentum may occur when there is a change in the mass of an object, or a change in its velocity, or both. If the momentum changes while the mass remains unchanged, as is most often the case, then the velocity changes. Acceleration occurs. And what produces acceleration? The answer is *force.* The greater the net force on an object, the greater will be the change in velocity and, hence, the change in momentum.

But something else is important also: time—how long the force acts. Apply a force briefly to a stalled automobile and you produce a small change in its momentum. Apply the same force over an extended period of time, and a greater momentum change results (Figure 4.2). A long sustained force produces more change in momentum than the same force applied briefly. So, for changing an object's momentum, both force and the time interval during which the force acts are important.

FIGURE 4.2
When you push with twice the force for twice the time, you impart twice the impulse and produce twice the momentum.

The quantity force × time interval is called **impulse**.

4.3 Impulse–Momentum Relationship

LEARNING OBJECTIVE
Relate the impulse on a body to the resulting change in its momentum.

EXPLAIN THIS If the impulse on a body is doubled, what change in its momentum occurs?

The greater the impulse exerted on something, the greater will be its change in momentum. This is known as the **impulse–momentum relationship**. Mathematically, the exact relationship is

$$\textbf{Impulse} = \textbf{change in momentum}$$

or

$$\boldsymbol{Ft} = \textbf{change in } \boldsymbol{mv}$$

which reads "force multiplied by the time during which the force acts equals change in momentum."*

We can express all terms in this relationship in shorthand notation and use the delta symbol Δ (a capital letter in the Greek alphabet signifying "change in" or "difference in"):

$$\boldsymbol{Ft} = \boldsymbol{\Delta(mv)}$$

Different forces exerted over different time intervals can produce the same impulse:

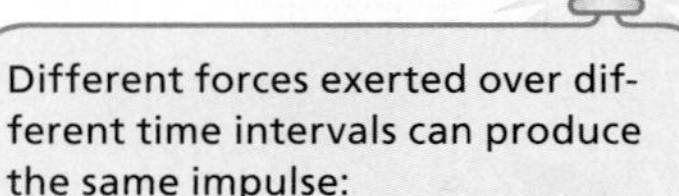

So, whenever you exert a force on something, you also exert an impulse. Recall that a net force produces acceleration. Now we are also saying that a net force multiplied by the time during which that force acts produces a change in an object's momentum.

CHECK YOURSELF

1. **Does a moving object possess impulse?**
2. **Does a moving object possess momentum?**

CHECK YOUR ANSWERS

1. No. Recall that an object cannot possess force. Similarly, an object cannot possess impulse. Just as a force is something an object can provide when it changes velocity, an impulse is something an object can provide, or something it can experience, only when it interacts with another object.
2. Yes, a moving object can possess momentum but, like velocity, only in a relative sense—that is, with respect to a frame of reference, such as Earth's surface. For example, a fly inside a fast-moving airplane cabin may have a large momentum relative to Earth below, but it has very little momentum relative to the cabin.

Impulse may be viewed as causing momentum change, or momentum change may be viewed as causing impulse. It doesn't matter which way you think about it. The important thing to know is that impulse and change of momentum are always linked. Here we will consider some ordinary examples in which impulse is related to (1) increasing momentum and (2) decreasing momentum.

*In Newton's second law ($F/m = a$), we can insert the definition of acceleration (a = change in v/t) and get F/m = (change in v)/t. Then multiplying both sides by mt gives Ft = change in (mv), or, in delta notation, $Ft = \Delta(mv)$.

FIGURE 4.3
The force of contact on a golf ball varies throughout duration of the contact.

Increasing Momentum

If you wish to produce the maximum increase in the momentum of something, you not only apply the greatest force but also extend the time of application as much as possible (hence the different results obtained by pushing briefly on a stalled automobile and by giving it a sustained push).

Long-range cannons have long barrels. The longer the barrel, the higher the velocity of the emerging cannonball or shell. Why? The force of exploding gunpowder in a long barrel acts on the cannonball for a longer time, increasing the impulse on it, which increases its momentum. Of course, the force that acts on the cannonball is not steady—it is strong at first and weaker as the gases expand. Most often the forces involved in impulses vary over time. The force that acts on the golf ball in Figure 4.3, for example, increases rapidly as the ball is distorted and then decreases as the ball gets up to speed and returns to its original shape. When we speak of any force that makes up impulse in this chapter, we mean the *average* force.

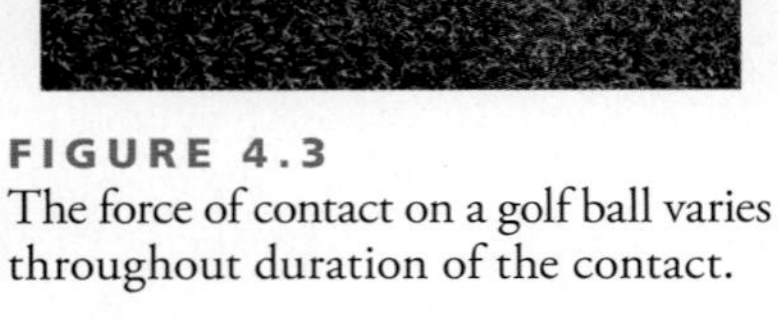

Timing is important—especially when you're changing momentum.

Decreasing Momentum

Imagine that you are in a car that is out of control, and you're faced with a choice of slamming into either a concrete wall or a haystack. You don't need much physics knowledge to make the better decision, but knowing some physics aids you in understanding why hitting something soft is entirely different from hitting something hard. Whether you hit the wall or the haystack, your momentum will be decreased by the same amount, and this means that the impulse required to stop you is the same. The same impulse means the same product of force and time, not the same force or the same time. You have a choice. By hitting the haystack instead of the wall, you extend the time of contact—you extend the time during which your momentum is brought to zero. The longer time is compensated for by a lesser force. If you extend the time of contact 100 times, you reduce the force of contact to a hundredth of what it might have been. So, whenever you wish the force of a contact to be small, extend the time of the contact (Figure 4.4).

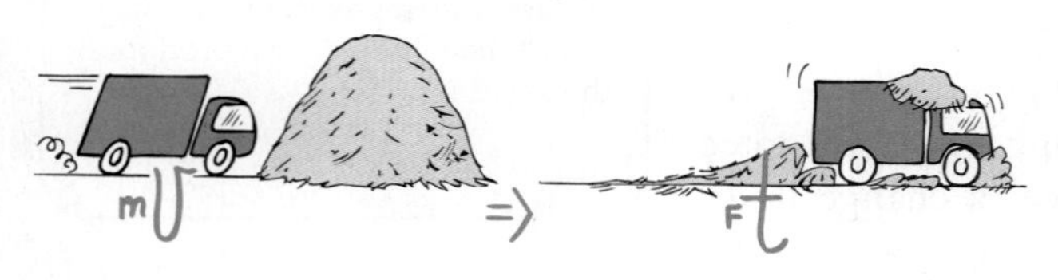

FIGURE 4.4
If the change in momentum occurs over a long time, the hitting force is small.

Conversely, if the time over which the force acts is short, the force itself will be comparatively large, for a given change in momentum. Going back to the example of the car, you can see why you are in much more trouble if you hit the concrete wall rather than the haystack. Your time of contact is short, so the force on you is large, as your momentum decreases to zero.

FIGURE 4.5
If the change in momentum occurs over a short time, the hitting force is large.

LEARNING OBJECTIVE
Relate the impulse–momentum relationship to living creatures.

Integrated Science 4A
BIOLOGY

The Impulse–Momentum Relationship in Sports

EXPLAIN THIS How do the amount of force and its time of application affect a change in momentum?

MasteringPhysics®
VIDEO: Decreasing Momentum Over a Short Time

In sports, we use the impulse–momentum relationship to our advantage. In many cases, we try to decrease momentum over an extended period of time. For example, a wrestler who is being thrown to the floor tries to prolong the length of his fall to the floor by relaxing his muscles and spreading the crash into a series of contacts, as foot, knee, hip, ribs, and shoulder fold onto the floor in turn. The increased time of contact reduces the force of contact.

When you jump from an elevated position to a floor below, you bend your knees when you land, which extends the time during which your momentum is reduced. Because this extended time can be 10 or 20 times longer than the time of an abrupt, stiff-legged landing, the forces on your bones can be reduced by 10 to 20 times. Of course, falling on a mat is preferable to falling on a solid floor, and this also increases the time of contact.

Ballet dancers much prefer a wooden floor with "give" to a hard floor with little or no "give." The wooden floor allows a slightly longer time of contact whenever the dancer lands, thus decreasing the force of contact and reducing the chance of injury. A safety net used by acrobats provides an obvious example of a small contact force spread out over a long time to provide the required impulse to reduce the momentum of a fall.

Bungee jumping puts the impulse–momentum relationship to a thrilling test. The momentum gained during the fall must be decreased to zero by an impulse equal to the gain in momentum. The long stretching time of the bungee cord ensures a small average force to bring the jumper to a safe halt. Bungee cords typically stretch to about twice their original length during the fall.

If you're about to catch a fast baseball with your bare hand, you extend your hand forward so you'll have plenty of space to allow your hand to move backward after making contact with the ball. You extend the time of contact and thereby reduce the force of the contact. Similarly, a boxer rides or rolls with the punch to reduce the force of the contact (Figure 4.6).

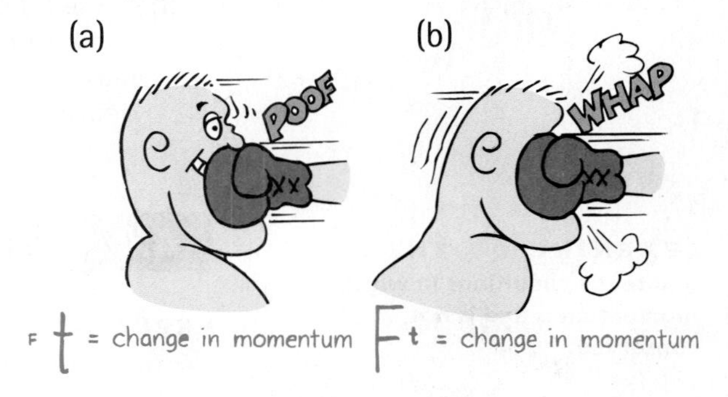

FIGURE 4.6
In both cases, the boxer's jaw provides an impulse that changes the momentum of the punch. (a) When the boxer moves away (that is, when he "rides with the punch"), he extends the time and diminishes the force. (b) If the boxer moves into the punch, the time is reduced and he must withstand a greater force.

Sometime it's advantageous to decrease momentum over a short time—this is how a karate expert is able to break a stack of bricks with the blow of her bare hand (Figure 4.7). She brings her hand swiftly against the bricks with considerable momentum. This momentum is quickly reduced when she delivers an impulse to the bricks. The impulse is the force of her hand against the bricks multiplied by the time her hand makes contact with the bricks. By swift execution, her time of contact is very brief, with a correspondingly huge force of contact. If her hand is made to bounce upon contact, the force is even greater. Why does the bounce increase the force?

Impulses are greater when an object bounces. The impulse required to bring an object to a stop and then to "throw it back again" is greater than the impulse required merely to bring the object to a stop. Suppose you catch a soccer ball with your hands, for example. You provide an impulse to reduce its momentum to zero. If you throw the ball back again, you have to provide additional impulse. It takes more impulse to catch it and throw it back than merely to catch it. If you're a serious soccer player, you may have mastered the technique of allowing the ball to bounce off your head. Your head supplies the increased amount of impulse needed to make the ball bounce. This trick is best reserved for the professionals—and only for those pros willing to risk a painful and even potentially damaging blow to the head.

FIGURE 4.7
Cassy imparts a large impulse to the bricks in a short time and produces considerable force.

CHECK YOURSELF

1. **If the boxer in Figure 4.6 makes the time of contact three times longer by riding with the punch, by how much is the force reduced?**
2. **If the boxer instead moves into the punch and shortens the contact time by half, by how much is the force increased?**
3. **A boxer being hit with a punch tries to extend the time of contact for best results, whereas a karate expert delivers a blow in a short time for best results. Isn't there a contradiction here?**

CHECK YOUR ANSWERS

1. The force will be one-third as great as it would have been if he hadn't pulled back.
2. The force will be two times greater than it would have been if he had held his head still. Forces of this kind account for many knockouts.
3. There is no contradiction because the best results for each are different. The best result for the boxer is reduced force, which is accomplished by maximizing time. The best result for the karate expert is increased force delivered in the shortest possible time.

MasteringPhysics®
VIDEO: Bowling Ball and Conservation of Energy

LEARNING OBJECTIVE
Relate the conditions in which momentum is and is not conserved.

4.4 Conservation of Momentum

EXPLAIN THIS What stays the same when a rolling pool ball stops upon hitting another ball at rest?

Newton's second law tells us that, if we want to accelerate an object, we must apply a force to it. The force must be an external force. If we want to accelerate a car whose engine can't be started, we must push the car from the outside. Inside forces don't qualify—sitting inside the automobile and pushing against the dashboard, with the dashboard pushing back, has no effect in accelerating the automobile. Likewise, if we want to change the momentum of an object, the force must be external. In the impulse–momentum relationship, $Ft = \Delta mv$, internal forces have no influence. If no external force is present, then no change in momentum is possible.

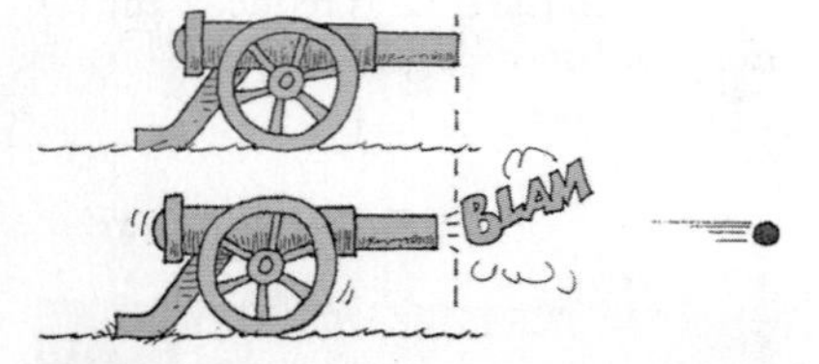

FIGURE 4.8
INTERACTIVE FIGURE (MP)
The momentum before firing is zero. After firing, the net momentum is still zero, because the momentum of the cannon is equal and opposite to the momentum of the cannonball.

Consider a cannon being fired. The force that drives the cannonball and the force that makes the cannon recoil are equal and opposite (Newton's third law). To the system consisting of the cannon and the cannonball, they are internal forces. No external net force acts on the cannon–cannonball system, so the momentum of the system undergoes no net change. Before the firing, the momentum is zero; after the firing, the net momentum is still zero (Figure 4.8). Like velocity, momentum is a vector quantity. The momentum gained by the cannonball is equal to and opposite to the momentum gained by the recoiling cannon.* They cancel. No momentum is gained, and no momentum is lost.

Whenever a physical quantity remains unchanged during a process, that quantity is said to be *conserved*. We say momentum is conserved.

The concept that momentum is conserved when no external force acts is elevated to a central law of mechanics, called the law of **conservation of momentum**:

In the absence of an external force, the momentum of a system remains unchanged.

UNIFYING CONCEPT
- *The Law of Conservation of Momentum*

* Here we neglect the momentum of ejected exhaust gases, which can be considerable.

Newton's first law tells us that a body in motion remains in motion when no external forces act. We say the same thing in a different context when we say the momentum of a body doesn't change when no external forces act. Whatever momentum a system may have, in the absence of external force, that momentum remains unchanged. In any system in which all forces are internal—whether cars colliding or stars exploding—the net momentum of the system before and after the event is the same. Whether forces are external or internal depends on the system being considered, as Figure 4.9 illustrates.

(a) 8-ball system

(b) cue-ball system

(c) cue-ball + 8-ball system

FIGURE 4.9
A cue ball hits an 8 ball head-on. Consider this event in three systems: (a) An external force acts on the 8-ball system, and its momentum increases. (b) An external force acts on the cue-ball system, and its momentum decreases. (c) No external force acts on the cue-ball-plus-8-ball system, and momentum is conserved (simply transferred from one part of the system to the other).

CHECK YOURSELF

If you toss a ball horizontally while standing on a skateboard, you will roll backward with the same amount of momentum that you have given to the ball. Will you roll backward if you go through the motions of tossing the ball but instead hold onto it?

CHECK YOUR ANSWER

No, you will not roll backward without immediately rolling forward to produce no net rolling. In third-law fashion, if no net force acts on the ball, no net force acts on you. In terms of momentum, if no net momentum is imparted to the ball, no net momentum will be imparted to you. Try it and see.

Collisions

The conservation of momentum is especially useful in collisions, where the forces involved are internal forces. In any collision, we can say that

Net momentum before collision = net momentum after collision.

When a moving billiard ball hits another billiard ball at rest head-on, the first ball comes to rest and the second ball moves with the initial velocity of the first ball. We call this an **elastic collision**; the colliding objects rebound without lasting deformation or the generation of heat. In this collision, momentum is transferred from the first ball to the second (Figure 4.10). Momentum is conserved. Billiard balls approximate perfectly elastic collisions, while collisions between molecules in a gas are perfectly elastic.

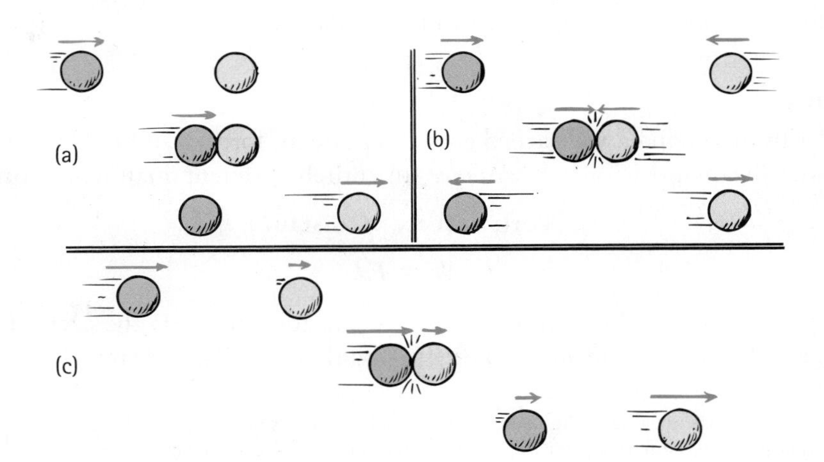

FIGURE 4.10
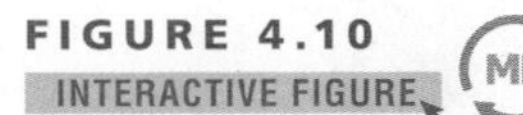

Elastic collisions of equally massive balls. (a) A green ball strikes a yellow ball at rest. (b) A head-on collision. (c) A collision of balls moving in the same direction. In each case, momentum is transferred from one ball to the other.

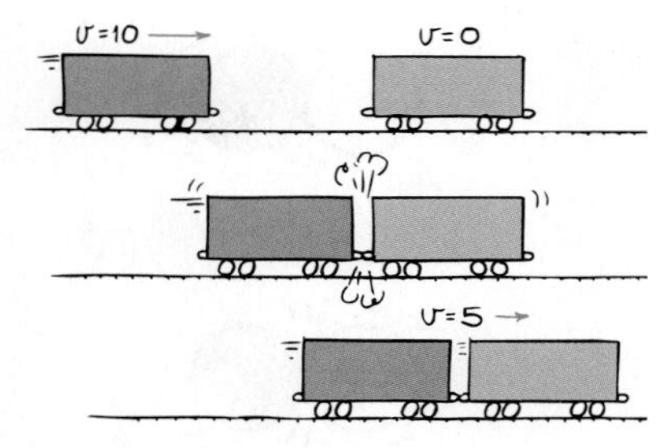

FIGURE 4.11
INTERACTIVE FIGURE MP

Inelastic collision. The momentum of the freight car on the left is shared with the freight car on the right after the collision.

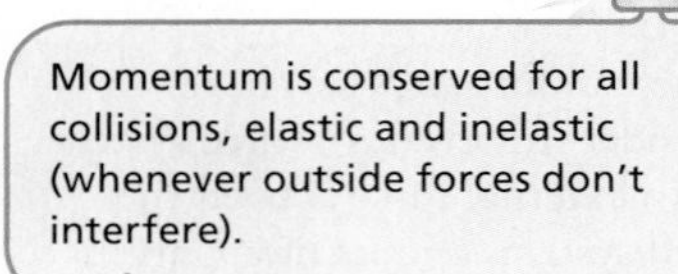

Momentum is conserved for all collisions, elastic and inelastic (whenever outside forces don't interfere).

Momentum is conserved even when the colliding objects don't rebound. This is an **inelastic collision**, characterized by deformation, generation of heat, or both. Sometimes an inelastic collision results in the coupling of colliding objects. Consider, for example, the case of a freight car moving along a track and colliding with another freight car at rest (Figure 4.11). If the freight cars have equal masses and are coupled by the collision, can we predict the velocity of the coupled cars after contact?

Suppose the moving car has a velocity of 10 meters per second, and we consider the mass of each car to be m. Then, from the conservation of momentum,

$$(\text{net } mv)_{\text{before}} = (\text{net } mv)_{\text{after}}$$

$$(m \times 10)_{\text{before}} = (2m \times v)_{\text{after}}$$

By simple algebra, $v = 5$ m/s. This makes sense because, since twice as much mass is moving after the collision, the velocity must be half as much as the velocity before the collision. Both sides of the equation are then equal.

So we see that changes in an object's motion depend both on force and on how long the force acts. When "how long" means time, we refer to the quantity force × time as *impulse*. But "how long" can mean distance also. When we consider the quantity force × *distance*, we are talking about something entirely different—the concept of *energy*.

LEARNING OBJECTIVE
Describe how work done on an object relates to its change in energy.

4.5 Energy

EXPLAIN THIS What is the relationship between work and energy?

Energy is perhaps the most central concept in science. The combination of energy and matter makes up the universe; matter is substance, and energy is the mover and changer of substance. The idea of matter is easy to grasp—it is stuff that we can see, smell, and feel. It has mass and it occupies space. Energy, on the other hand, is abstract. We cannot see, smell, or feel most forms of energy. Energy isn't even noticeable unless it is undergoing a change of some kind—being transferred or transformed. Surprisingly, the idea of energy was unknown to Isaac Newton, and its existence was still being debated in the 1850s. Energy comes from the Sun in the form of sunlight, it is in the food you eat, and it sustains all life. It's in heat, sound, electricity, and radiation. Even matter itself is condensed, bottled-up energy, as set forth in Einstein's famous formula $E = mc^2$, which we'll return to in Chapter 10.

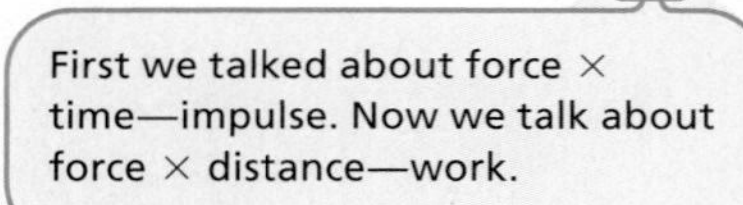

First we talked about force × time—impulse. Now we talk about force × distance—work.

Work

Earlier in this chapter, we learned about impulse, or force × time. Now we will consider the quantity force × *distance*, an entirely different quantity—**work**:

$$\textbf{Work = force} \times \textbf{distance}$$

$$W = Fd$$

In every case in which work is done, two factors enter: (1) the exertion of a force and (2) the movement of something by that force.* For example, if we lift

FIGURE 4.12
The man may expend energy when he pushes the wall, but, if it doesn't move, no work is performed on the wall.

* Force and distance must be in the same direction. When force is not along the direction of motion, work equals the component of force in the direction of motion × distance moved.

MATH CONNECTION

Quantifying Collisions

Billiard balls, cars, molecules, football players—collisions between objects are everywhere. You can use the conservation of momentum to analyze them. For example, consider this practice problem.

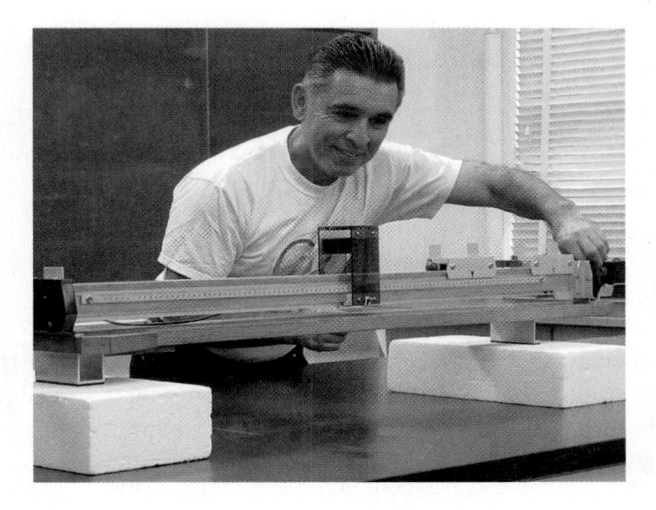

Problem Consider the air track in the photo. Suppose that a gliding cart with a mass of 0.5 kg bumps into, and sticks to, a stationary cart that has a mass of 1.5 kg. If the speed of the gliding cart before contact is 4 m/s, how fast will the coupled carts glide after the collision?

Solution According to momentum conservation, the momentum of the cart of mass m and velocity v before the collision will equal the momentum of both carts stuck together after the collision:

$$(\text{total } mv)_{\text{before}} = (\text{total } mv)_{\text{after}}$$

$$0.5\text{ kg}(4\text{ m/s}) = (0.5\text{ kg} + 1.5\text{ kg})v$$

$$v = \frac{(0.5\text{ kg})(4\text{ m/s})}{(0.5\text{ kg} + 1.5\text{ kg})}$$

$$= \frac{(2.0\text{ kg}\cdot\text{m/s})}{2.0\text{ kg}} = 1\text{ m/s}$$

This makes sense, because four times as much mass will be moving after the collision, so the coupled carts will glide more slowly. In keeping the momentum equal, four times the mass glides one-quarter as fast.

two loads one story high, we do twice as much work as we do when lifting one load the same distance, because the *force* needed to lift twice the weight is twice as much. Similarly, if we lift a load two stories high instead of one story high, we do twice as much work, because the *distance* is twice as much.

In raising a heavy barbell, a weightlifter does work on the barbell and gives energy to it. Interestingly, when a weightlifter simply holds a barbell overhead, no work is done on the barbell. Work involves not only force but motion as well. The weightlifter may get tired holding the barbell still, but, if the barbell is not moved by the force exerted, no work is done *on the barbell.* Work may be done on the weightlifter's muscles, as they stretch and contract, which is force × distance on a biological scale. But this work is not done *on the barbell. Lifting* the barbell is different from *holding* the barbell.

FIGURE 4.13
Work is done in lifting the barbell.

You do work when you push and move things horizontally. Imagine that you push a box across a very well-waxed and slippery floor. The amount of work you do on the box is your average push × the distance moved. The work you do sets the box in motion.

The unit of work combines the unit of force (N) with the unit of distance (m), the newton-meter (N·m). We call a newton-meter the **joule** (J), which rhymes with "cool." One joule of work is done when a force of 1 newton is exerted over a distance of 1 meter in the direction of the force, as in lifting an apple over your head. For larger values, we speak of kilojoules (kJ), which are thousands of joules, or megajoules (MJ), which are millions of joules. The weightlifter in Figure 4.13 does work that can be measured in kilojoules. The work done to vertically hoist a heavily loaded truck can be measured in kilojoules.

The word *work*, in common usage, means physical or mental exertion. Don't confuse the science definition of work with the everyday notion of work.

MATH CONNECTION

Work Practice Problems

Solve these practice problems relating to work to be sure you can apply this concept.

Problems

1. How much work is needed to lift an object that weighs 500 N to a height of 2 m?
2. How much work is needed to lift the object twice as high?
3. How much work is needed to lift a 1000-N object to a height of 4 m?

Solutions

1. $W = Fd = (500 \text{ N})(2 \text{ m}) = 1000 \text{ J}$.
2. Twice the height requires twice the work; that is, $W = Fd = (500 \text{ N})(4 \text{ m}) = 2000 \text{ J}$.
3. Lifting twice the load twice as high requires four times the work; that is, $Fd = (1000 \text{ N})(4 \text{ m}) = 4000 \text{ J}$.

LEARNING OBJECTIVE
Specify the relationship between work and power.

4.6 Power

EXPLAIN THIS Why do you run out of breath when you run up stairs but not when you walk up?

The definition of work says nothing about how long it takes to do the work. The same amount of work is done when we carry a load up a flight of stairs, whether we walk up or run up. So, why are we more tired after running upstairs in a few seconds than after walking upstairs in a few minutes? To understand this difference, we need to talk about a measure of how fast the work is done—power. **Power** is the rate at which work is done—the amount of work done per time it takes to do the work:

$$\textbf{Power} = \frac{\textbf{work done}}{\textbf{time interval}}$$

The work done in climbing stairs requires more power when the worker is running up rapidly than when the worker is climbing up slowly. A high-power automobile engine does work rapidly. An engine that delivers twice the power of another, however, does not necessarily move a car twice as fast or twice as far. Twice the power means that the engine can do twice the work in the same amount of time—or it can do the same amount of work in half the time. A powerful engine can produce greater acceleration.

Power is also the rate at which energy is changed from one form to another. The unit of power is the joule per second, called the watt. This unit was named in honor of James Watt, the eighteenth-century developer of the steam engine. One watt (W) of power is used when 1 joule of work is done in 1 second. One kilowatt (kW) equals 1000 watts. One megawatt (MW) equals 1 million watts.

Your heart uses slightly more than 1 watt of power in pumping blood through your body.

LEARNING OBJECTIVE
Specify the relationship between potential energy and change of location.

4.7 Potential Energy

EXPLAIN THIS How much more energy is required to lift an object twice as high?

An object can store energy because of its position, shape, or state. Such stored energy is called **potential energy** (PE) because, in the stored state, it has the potential to do work. For example, a stretched or compressed

MATH CONNECTION

Power Practice Problems

Power calculations are straightforward. Try these problems to reinforce your understanding of the concept.

Problems

1. How much power is expended when lifting a 1000-N load a vertical distance of 4 m in a time of 2 s?
2. How much power is needed to perform the same job in 1 s?

Solutions

1. $$\text{Power} = \frac{\text{work done}}{\text{time interval}} = \frac{(1000\ \text{N})(4\ \text{m})}{2\ \text{s}}$$
$$= \frac{4000\ \text{J}}{2\ \text{s}} = 2000\ \text{W (or 2 kW)}$$
2. Twice the power is needed to do the same job in half the time; that is,
$$\text{Power} = \frac{\text{work done}}{\text{time interval}} = \frac{4000\ \text{J}}{1\ \text{s}}$$
$$= 4000\ \text{W (or 4 kW)}$$

spring has the potential for doing work. Or, when an archer draws a bow with an arrow, energy is stored in the fibers of the bent bow (Figure 4.14). When the bowstring is released, much of the energy in the bow is transferred to the arrow (the rest warms the bow).

The chemical energy in fuels is potential energy; it is the energy of position from a microscopic point of view. Such energy characterizes fossil fuels, electric batteries, and the food we eat. This energy is available when atoms are rearranged—that is, when a chemical change occurs. Any substance that can do work through chemical action possesses potential energy.

The form of potential energy that is easiest to visualize is when work is done to elevate objects against Earth's gravity. The potential energy due to elevated positions is called gravitational potential energy. Water in an elevated reservoir and the elevated ram of a pile driver have gravitational potential energy. The amount of gravitational potential energy possessed by an elevated object is equal to the work done against gravity in lifting it. The work done equals the force required to move it upward times the vertical distance it has been moved ($W = Fd$). The upward force equals the weight mg of the object. So, the work done in lifting an object to a height h is given by the product mgh:

$$\text{Gravitational potential energy} = \text{weight} \times \text{height}$$
$$\text{PE} = mgh$$

Note that the height h is the distance above some reference level, such as the ground or the floor of a building. The potential energy mgh is relative to that level, and it depends on only mg and height h. You can see in Figure 4.15 that the potential energy of the ball at the top of the structure depends on the vertical displacement and not on the path taken.

FIGURE 4.14
The potential energy of Tenny's drawn bow equals the work (average force × distance) she did in drawing the arrow back into position. When the bowstring is released, most of the potential energy of the drawn bow will become the kinetic energy of the arrow.

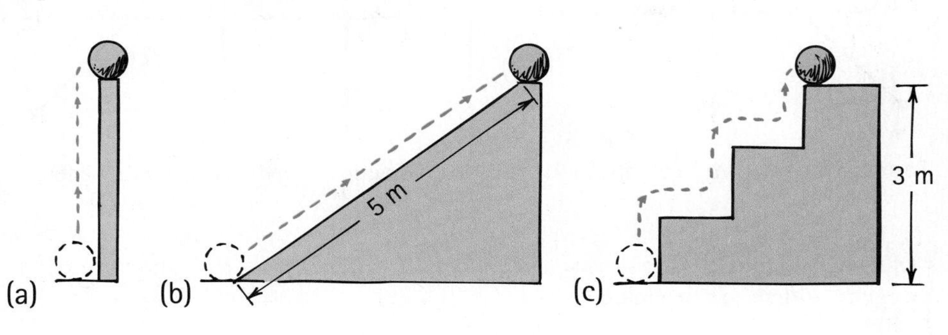

FIGURE 4.15
The potential energy of the 10-N ball is the same (30 J) in all three cases because the work done in elevating the ball 3 m is the same whether it is (a) lifted vertically with 10 N of force, (b) pushed with 6 N of force up the 5-m incline, or (c) lifted with 10 N up each 1-m stair. No work is done in moving the ball horizontally (neglecting friction).

MATH CONNECTION

Calculating Potential Energy

Why are ramps used? Solve these practice problems to appreciate why.

Problems

1. How much work is done in lifting the 100-N block of ice a vertical distance of 2 m, as shown in Figure 4.16?
2. How much work is done in pushing the same block of ice up the 4-m-long ramp? The force needed is only 50 N (which is the reason inclines are used).
3. What is the increase in the block's potential energy in both cases?

Solutions

1. $W = Fd = (100\text{ N})(2\text{ m}) = 200\text{ J}$.
2. $W = Fd = (50\text{ N})(4\text{ m}) = 200\text{ J}$.
3. In both cases the block's potential energy is increased by 200 J. The ramp simply makes this work easier to perform.

LEARNING OBJECTIVE
Describe how work done on an object relates to its change in energy.

FIGURE 4.16
Both do the same work in elevating the same-mass blocks. When both blocks are raised to the same vertical height, both possess the same potential energy.

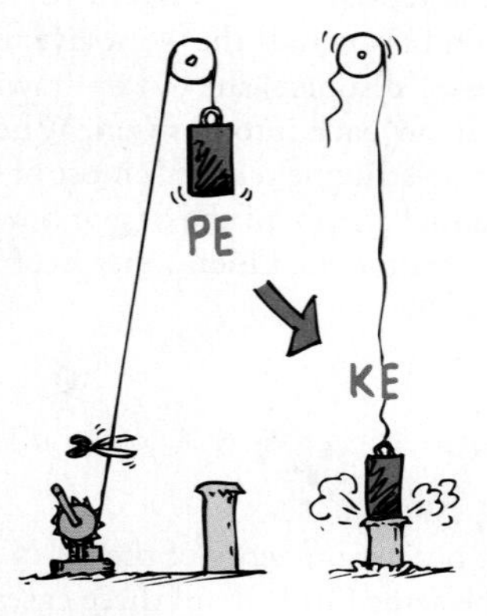

FIGURE 4.17
The potential energy of the elevated ram is converted into kinetic energy when released.

4.8 Kinetic Energy

EXPLAIN THIS How much more energy is required to push an object twice as fast?

When you push on an object, you can make it move. If it moves, the object can apply a force to something else and move it through a distance—the object you pushed can do work. If a moving object has the capacity to do work, it must have energy, but what kind of energy? The energy associated with a moving body, by virtue of its motion alone, is called **kinetic energy** (*KE*).

Energy can be *transferred* from one object to another, such as when a rolling bowling ball transfers some of its kinetic energy to the pins and sets them in motion. Energy also *transforms*, or changes form. For example, the gravitational potential energy of a raised ram transforms into kinetic energy when the ram is released from its elevated position, as shown in Figure 4.17. And, when you raise a pendulum bob against the force of gravity, you do work on it. That work is stored as potential energy until you let the pendulum bob go. Then its potential energy transforms into kinetic energy as it picks up speed and loses elevation.

The kinetic energy of an object depends on its mass and its speed. The kinetic energy is equal to the mass multiplied by the square of the speed, multiplied by the constant $\frac{1}{2}$.

$$\text{Kinetic energy} = \tfrac{1}{2}\,\text{mass} \times \text{speed}^2$$
$$\text{KE} = \tfrac{1}{2}mv^2$$

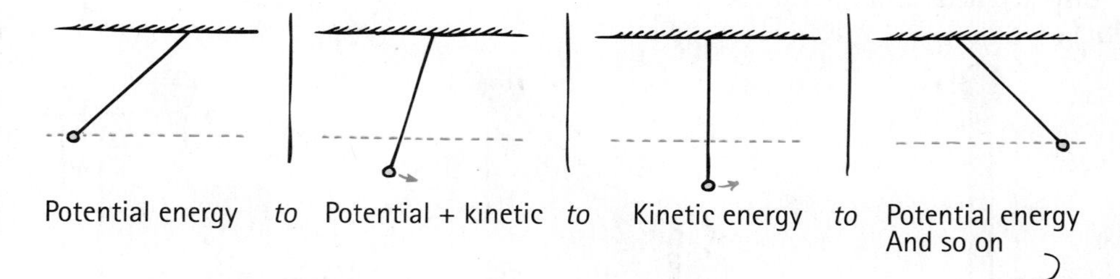

FIGURE 4.18
Energy transitions in a pendulum.

A car moving along a road has kinetic energy. A car that is twice as heavy and moving at the same speed has twice the kinetic energy. That's because a car that is twice as heavy has twice the mass. Kinetic energy depends on mass. But note that kinetic energy also depends on speed—not just plain speed but speed multiplied by itself—speed squared. If you double the speed of a car, you'll increase its kinetic energy by four ($2^2 = 4$). Or, if you drive three times as fast, you will have nine times the kinetic energy ($3^2 = 9$). The fact that kinetic energy depends on the square of the speed means that small changes in speed can produce large changes in kinetic energy. The squaring of speed means that kinetic energy can be only zero or positive—never negative. Now, let's relate these ideas to work.

Which of these does a speeding baseball *not* possess—force, momentum, or energy? (Hint: The correct answer begins with *F*.)

4.9 The Work–Energy Theorem

LEARNING OBJECTIVE
Describe how the work done on an object relates to the object's change in energy.

EXPLAIN THIS What is the relationship between work and kinetic energy?

To increase the kinetic energy of an object, work must be done on it. The change in kinetic energy is equal to the work done. This important relationship is called the **work–energy theorem**. We abbreviate "change in" with delta Δ and write

$$\text{Work} = \Delta KE$$

Work equals change in kinetic energy. The work in this equation is net work—that is, the work based on the net force.

Recall from Section 4.3 that a cannonball fired from a cannon with a longer barrel has a higher velocity because of the longer time of the impulse. The greater speed is also evident from the work–energy theorem because of the longer *distance* through which the force acts. The work done on the cannonball is the force exerted on it multiplied by the distance through which the force acts: $Fd = \Delta KE$.

The work–energy theorem emphasizes the role of change. If there were no change in an object's energy, then we know that no net work was done on it. This theorem applies to changes in potential energy as well. Recall our earlier example of the weightlifter raising the barbell. When work was being done on the barbell, its potential energy was being changed. But, when the barbell was held stationary, no further work was being done on it, as evidenced by no further change in its energy.

FIGURE 4.19
The work required in raising the roller-coaster car against gravity converts into kinetic energy as the car begins to fall.

Similarly, if you push against a box on a floor and it doesn't slide, then you are not doing work on the box. There is no change in kinetic energy. But, if you push harder and the box slides, then you're doing work on it. When the amount of work done to overcome friction is small, the amount of work done on the box is practically matched by its gain in kinetic energy.

The work–energy theorem applies to decreasing speed as well. Energy is required to reduce the speed of a moving object and to bring it to a halt. When we apply the brakes to slow a moving car, we do work on it. This work is the friction force supplied by the brakes multiplied by the distance over which the friction force acts. The more kinetic energy something has, the more work is required to stop it.

Interestingly, the friction supplied by the brakes is very nearly the same whether the car moves slowly or quickly. Friction doesn't depend on speed. The variable that makes a difference is the braking distance. A car moving at twice the speed of another takes four times ($2^2 = 4$) as much work to stop. Therefore, it takes four times as much distance in which to stop. Accident investigators are well aware that an automobile going 100 kilometers per hour has four times the kinetic energy as it would have if it were going 50 kilometers per hour. So a car going 100 kilometers per hour will skid four times as far when its brakes are applied as it would if it were going 50 kilometers per hour. Kinetic energy depends on speed *squared*.

Automobile brakes convert kinetic energy into heat. Professional drivers are familiar with another way to slow a vehicle—shift to low gear and allow the engine to do the braking. Hybrid cars do the same and divert braking energy to electric storage batteries where it is used to complement the energy produced by gasoline combustion.

CHECK YOURSELF

1. **If the brakes of a bicycle become locked and the bike skids to a stop, how much farther will the bike skid if it's moving three times as fast?**
2. **Can an object possess energy?**
3. **Can an object possess work?**

CHECK YOUR ANSWERS

1. Nine times farther. The bicycle has nine times as much energy when it travels three times as fast: $\frac{1}{2}m(3v)^2 = \frac{1}{2}m9v^2 = 9(\frac{1}{2}mv^2)$. The friction force will ordinarily be the same in either case. Therefore, to do nine times the work to stop requires nine times as much stopping distance.
2. Yes, but only in a relative sense. For example, an elevated object may possess PE relative to the ground, but no energy relative to a point at the same elevation. Similarly, the kinetic energy of an object is relative to a frame of reference, usually Earth's surface.
3. No; unlike energy, work is not something an object has. Work is something an object *does* to some other object. An object can *do* work only if it has energy. Or, stated another way, an object spends energy when it does work on something else.

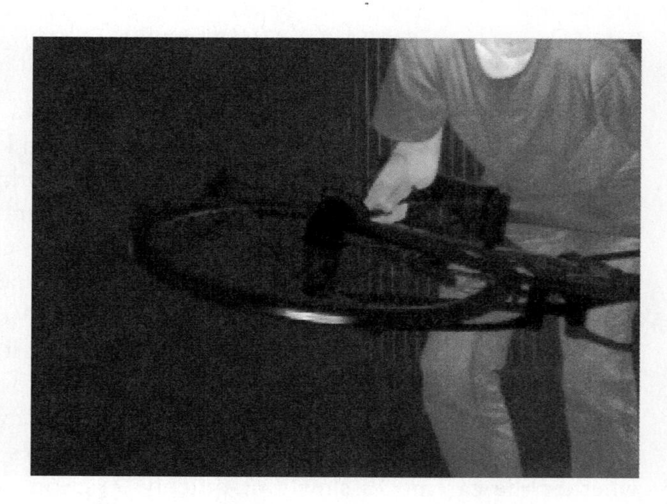

FIGURE 4.20
Due to the friction, energy is transferred both to the floor and to the tire when the bicycle skids to a stop. An infrared camera reveals (*left*) the heated tire track (*red streak on the floor*) and (*right*) the warmth of the tire. (Courtesy of Michael Vollmer.)

MATH CONNECTION

Applying the Work–Energy Theorem

Determine the work done on an object even though you don't know the forces or distances involved.

Problems

1. Calculate the change in kinetic energy when a 50-kg shopping cart moving at 2 m/s is pushed to a speed of 6 m/s.
2. How much work is required to make this change in kinetic energy?

Solutions

1. $\Delta KE = \frac{1}{2}m(v_f^2 - v_o^2)$

$$= \frac{1}{2}(50\text{ kg})[(6\text{ m/s})^2 - (2\text{ m/s})^2]$$

$$= 800\text{ J}$$

2. $W = \Delta KE = 800$ J because the change in kinetic energy equals the work done on the shopping cart.

Kinetic energy underlies other seemingly different forms of energy, such as heat, sound, and light. Random molecular motion is sensed as heat: When fast-moving molecules bump into the molecules in the surface of your skin, they transfer kinetic energy to your molecules, much as the balls in a game of pool or billiards transfer energy to each other. Sound consists of molecules vibrating in rhythmic patterns: Shake a group of molecules in one place and, in cascade fashion, they disturb neighboring molecules that, in turn, disturb others, preserving the rhythm of shaking throughout the medium. Electrons in motion produce electric currents. Even light energy originates from the motion of electrons within atoms. Kinetic energy is far-reaching.

Understanding the distinction between momentum and kinetic energy is high-level physics.

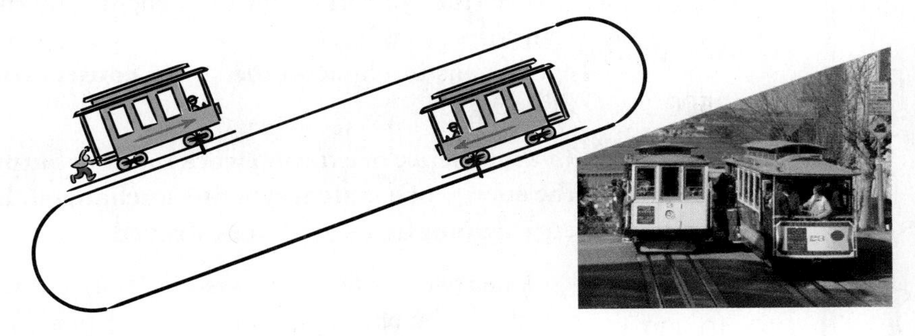

FYI Cable cars on the steep hills of San Francisco nicely transfer energy to one another via the cable beneath the street. The cable forms a complete loop that connects cars going both downhill and uphill. In this way a car moving downhill does work on a car moving uphill. So the increased gravitational PE of an uphill car is due to the decreased gravitational PE of a car moving downhill.

Comparison of Kinetic Energy and Momentum

Momentum and kinetic energy are properties of moving things, but they differ from each other. Like velocity, momentum is a vector quantity and is therefore directional and capable of being canceled entirely. But kinetic energy is a nonvector (scalar) quantity, like mass, and can never be canceled. The momenta (plural of *momentum*) of two firecrackers approaching each other may cancel, but, when they explode, there is no way their energies can cancel. Energies transform to other forms, but momenta do not.

Another difference is the velocity dependence of the two. Whereas momentum depends on velocity (mv), kinetic energy depends on the square of velocity ($\frac{1}{2}mv^2$). An object that moves with twice the velocity of another object of the same mass has twice the momentum but four times the kinetic energy. So, when a car traveling twice as fast crashes, it crashes with four times the energy.

If the distinction between momentum and kinetic energy isn't really clear to you, you're in good company. Failure to make this distinction, when impedo was in vogue, resulted in disagreements and arguments among the best British and French physicists for two centuries.

LEARNING OBJECTIVE
Relate conservation of energy to physics and science in general.

4.10 Conservation of Energy

EXPLAIN THIS When is energy most evident?

Whenever energy is transformed or transferred, none is lost and none is gained. In the absence of work input or output, the total energy of a system before some process or event is equal to the total energy after.

Consider the system of a bow, arrow, and target. In the process of drawing the arrow in the bow, we do work in bending the bow, and we give the arrow and the bow potential energy. When the bowstring is released, most of this potential energy is transferred to the arrow as kinetic energy (the rest slightly warms the bow). The arrow, in turn, transfers this energy to its target, perhaps a bale of hay. The distance the arrow penetrates into the hay multiplied by the average force of contact doesn't quite match the kinetic energy of the arrow. The energy score doesn't balance. But, if we investigate further, we discover that both the arrow and the hay are a bit warmer. By how much? By the energy difference. In these transformations of energy, taking the form of thermal energy into account, we find energy transforms without net loss or gain. Quite remarkable!

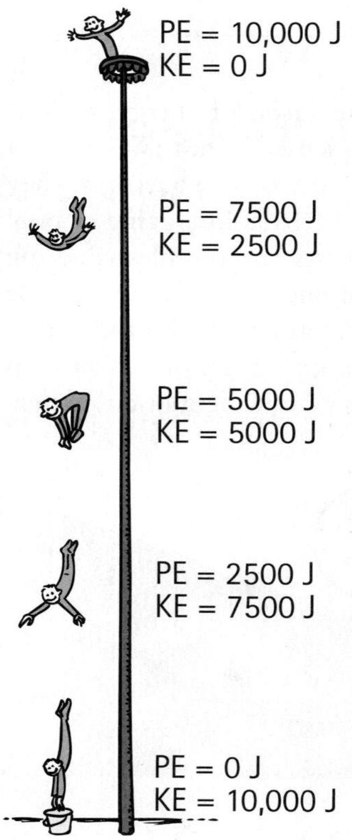

PE = 10,000 J
KE = 0 J

PE = 7500 J
KE = 2500 J

PE = 5000 J
KE = 5000 J

PE = 2500 J
KE = 7500 J

PE = 0 J
KE = 10,000 J

FIGURE 4.21 INTERACTIVE FIGURE (MP)

A circus diver at the top of a pole has a potential energy of 10,000 J. As he dives, his potential energy converts into kinetic energy. Notice that, at successive positions (one-fourth, one-half, three-fourths, and all the way down), the total energy is constant.

The study of various forms of energy and their transformations has led to one of the greatest generalizations in physics—the law of **conservation of energy**:

In the absence of external work input or output, the energy of a system remains unchanged. Energy cannot be created or destroyed.

UNIFYING CONCEPT

- ***The Law of Conservation of Energy***

When we consider any system in its entirety, whether it be as simple as a swinging pendulum or as complex as an exploding galaxy, there is one quantity that doesn't change: energy. It may change form or it may simply be transferred from one part of the system to another, but, as far as we can tell, the total energy score remains the same. This energy score takes into account the fact that the atoms that make up matter are themselves concentrated bundles of energy. When the nuclei (cores) of atoms rearrange themselves, enormous amounts of energy can be released. The Sun shines because some of this energy is transformed into radiant energy. In nuclear reactors, much of this energy is transformed into heat. Enormous gravitational forces in the deep, hot core of the Sun push hydrogen nuclei together to form helium. This welding together of atomic nuclei is called thermonuclear fusion (see Chapter 10). This process produces radiant energy, some of which reaches Earth. Part of this energy falls on plants and part, in turn, later becomes coal. Another part supports life in the food chain that begins with plants, and part of this energy later becomes petroleum. Part of the energy from the Sun powers the evaporation of water from the ocean, and part returns to Earth as rain that may be trapped behind a dam. By virtue of its position, the water behind the dam has energy that may be used to power a generating plant below, where it will be transformed into electric energy. The energy travels through wires to homes, where it is used for lighting, heating, cooking, and operating electric gadgets. How wonderful that energy is transformed from one form into another!

MasteringPhysics®
VIDEO: Conservation of Energy

CHECK YOURSELF

Rows of wind-powered generators are used in various windy locations to generate electric power. Does the power generated affect the speed of the wind? Would locations behind the windmills be windier if the windmills weren't there?

CHECK YOUR ANSWERS

Windmills generate power by taking kinetic energy from the wind, so the wind is slowed by interaction with the windmill blades. So, yes, it would be windier behind the windmills if they weren't present.

Integrated Science 4B
BIOLOGY AND CHEMISTRY

LEARNING OBJECTIVE
Relate energy conservation to living things.

Glucose: Energy for Life

EXPLAIN THIS What role does glucose play in the energy of a living system?

Your body is in many ways a machine—a fantastically complex machine. It is made up of smaller machines, the living cells (Figure 4.22). Like any machine, a living cell needs a source of energy. The principal energy source used by most living things is the sugar glucose, $C_6H_{12}O_6$. One glucose molecule contains six atoms of carbon (C), twelve atoms of hydrogen (H), and six atoms of oxygen (O). The glucose molecule is rich in stored energy (chemical potential energy). Organisms break glucose down in their cells and harvest the energy it contains to power the chemical and physical processes that sustain life, as discussed in more detail in Chapter 15.

You obtain glucose from the food you eat indirectly, by way of some rather complex chemical reactions. A few super-sweet foods contain glucose, but most consist of other, more complex carbohydrates, such as starch, or some combination of carbohydrates, fats, and proteins. Your body must break down these nutrients to produce glucose, a raw fuel that is then passed on to your cells. Glucose molecules are taken apart inside your cells, where energy is liberated from them. The actual energy harvesting typically takes place through *cellular respiration*, a process that occurs in specialized structures within the cell—mitochondria, the "power plants" of the cells. Cells use the released energy to do all the tasks they must do to stay alive and to perform their specialized functions.

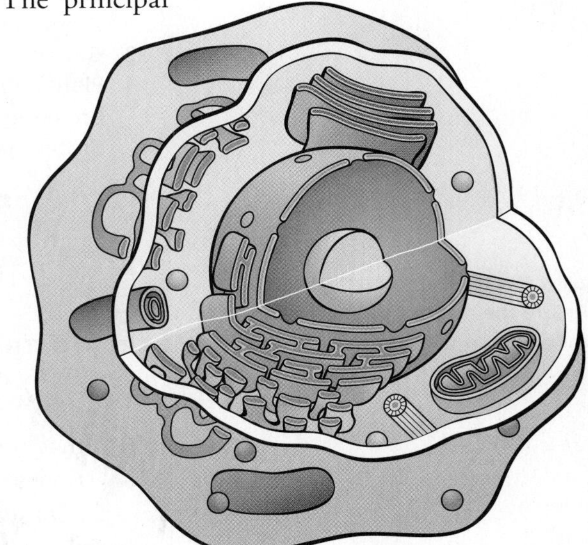

FIGURE 4.22
This cutaway view of a generalized animal cell shows various specialized structures, including the orange and yellow nucleus in the center. The pink structure at the lower right and the others scattered throughout the cell are *mitochondria*, which provide the cell with energy through cellular respiration.

Green plants, on the other hand, manufacture glucose directly during *photosynthesis*. Photosynthesis is the process by which plants, algae, and certain kinds of bacteria convert light energy from the Sun into chemical energy in sugar

FIGURE 4.23
Plants capture solar energy and transform it into chemical energy, which is stored in large molecules. When other organisms consume the plants, they obtain the energy they need for life.

molecules. Almost all life on Earth is either directly or indirectly dependent on photosynthesis. The overall chemical reaction for photosynthesis is

$$6CO_2 + 6H_2O + \text{sunlight} \rightarrow C_6H_{12}O_6 + 6O_2$$

Carbon dioxide, water, and sunlight go in; glucose and oxygen come out. Glucose is typically converted by plant tissues into complex carbohydrates, which are long molecules built of glucose units. Some plants, of course, don't have the opportunity to consume the glucose they make for themselves, instead donating it to the animals that consume them. A potato, for instance, is crammed with glucose stored in a thicket of starch molecules. The potato's starch molecules are broken down to glucose in your mouth and small intestine, and the glucose is transported to your cells, powering their lives—and yours.

With the exception of nuclear power, all Earth's energy comes from the Sun.

CHECK YOURSELF

1. **Why do cells need energy?**
2. **What is the ultimate source of the energy that powers most life on Earth? Explain.**
3. **What is the principal energy source used by most living things?**
4. **How do plants obtain glucose? How do animals obtain it?**

CHECK YOUR ANSWERS

1. Cells need energy to do all those things that require work. For example, to move, to change their shape, to reproduce, to maintain and repair cellular structures, or to create new cells requires considerable energy.
2. The Sun; it is the source of the energy plants need to perform photosynthesis. Plants form the base of most food chains, and so the energy they obtain from the Sun is transferred to other organisms that consume them.
3. Glucose.
4. Plants obtain glucose through photosynthesis. Animals obtain it by breaking down nutrients in the foods they eat, such as proteins, carbohydrates, and fats.

LEARNING OBJECTIVE
Relate the concept of energy conservation to machines.

4.11 Machines

EXPLAIN THIS Should you invest your savings in a machine that creates energy? Why or why not?

A machine is a device for multiplying forces or simply changing the direction of forces. Underlying every machine is the conservation of energy. Consider one of the simplest machines, the lever (Figure 4.24). At the same time we do work on one end of the lever, the other end does work on the load. We see that the direction of force is changed: If we push down, the load is lifted up. If the heat from friction forces is small enough to neglect, the work input will be equal to the work output:

$$\text{Work input} = \text{work output}$$

Since work equals force times distance, we have

$$(\text{Force} \times \text{distance})_{\text{input}} = (\text{force} \times \text{distance})_{\text{output}}$$

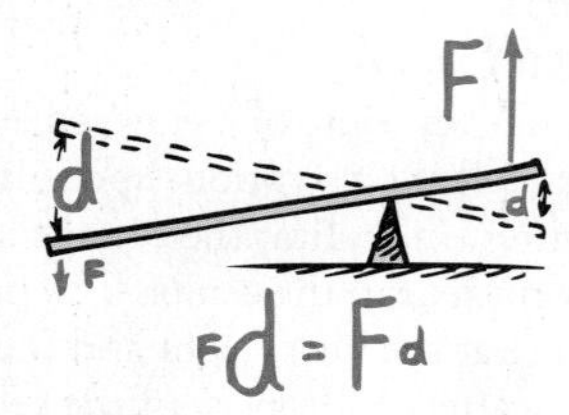

FIGURE 4.24
The lever.

FIGURE 4.25
Applied force × applied distance = output force × output distance.

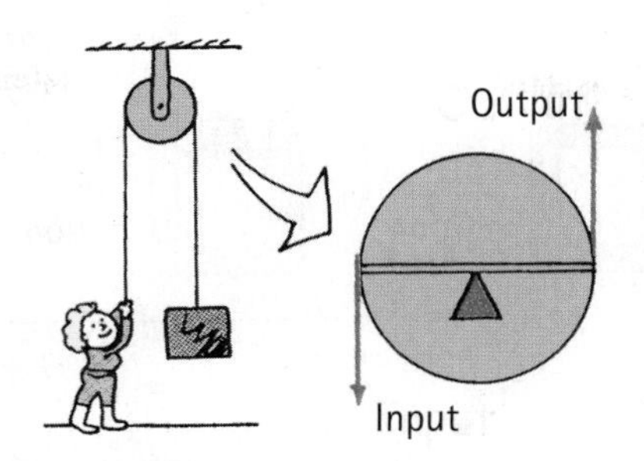

FIGURE 4.26
The pulley acts like a lever. It changes only the direction of the force.

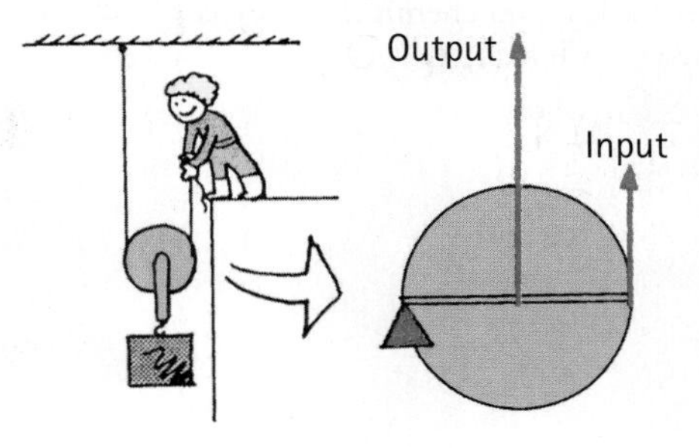

FIGURE 4.27
In this arrangement, a load can be lifted with half the input force.

If the pivot point, or fulcrum, of the lever is relatively close to the load, then a small input force will produce a large output force. This is because the input force is exerted through a large distance and the load is moved over a correspondingly short distance. In this way, a lever can multiply forces. But no machine can multiply work or multiply energy. That's a conservation of energy no-no!

You use the principle of the lever in jacking up the front of an automobile. By exerting a small force through a large distance, you are able to provide a large force acting through a small distance. Consider the ideal example illustrated in Figure 4.25. Every time the jack handle is pushed down 25 cm, the car rises 0.25 cm—only a hundredth as far, but with 100 times the force.

A block and tackle, or system of pulleys, is a simple machine that multiplies force at the expense of distance. One can exert a relatively small force through a relatively large distance and lift a heavy load through a relatively short distance. With the ideal pulley shown in Figure 4.28, the man pulls 10 m of rope with a force of 50 N and lifts 500 N through a vertical distance of 1 m. The work done by the man in pulling the rope is numerically equal to the increased potential energy of the 500-N block.

Any machine that multiplies force does so at the expense of distance. Likewise, any machine that multiplies distance, even that of your forearm and elbow, does so at the expense of force. No device or machine can put out more energy than is put into it. No machine can create energy; a machine can only transfer energy from one place to another or transform it from one form into another.

FIGURE 4.28
Applied force × applied distance = output force × output distance.

UNIFYING CONCEPT

- ***The Law of Conservation of Energy***
Section 4.10

MasteringPhysics®
VIDEO: Machines: Pulleys

CHECK YOURSELF
If a lever is arranged so that the input distance is twice the output distance, can we predict that the energy output will be doubled?

CHECK YOUR ANSWER
No, no, a thousand times no! We can predict that the output force will be doubled, but never the output energy. Work and energy remain the same, which means force × distance remains the same. Shorter distance means greater force, and vice versa. Be careful to distinguish between the concepts of *force* and *energy*.

A machine can multiply force, but never energy. No way!

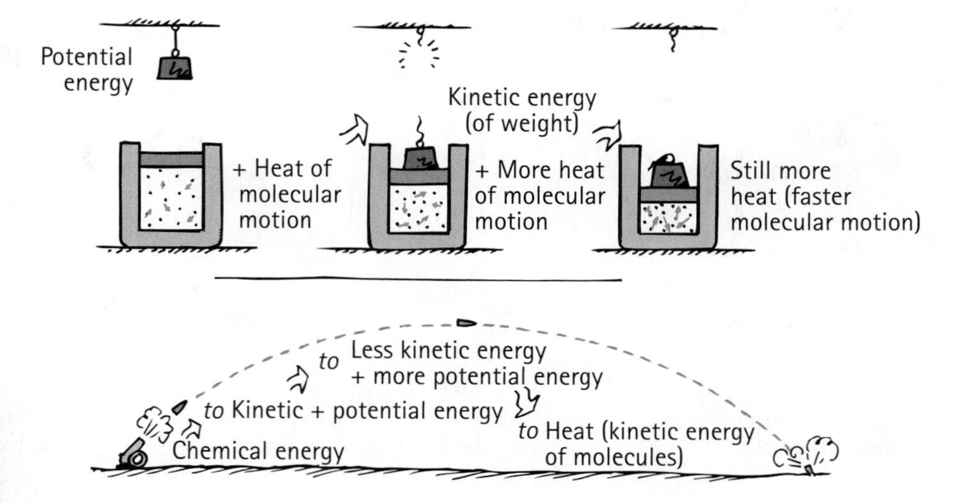

FIGURE 4.29
Energy transitions. The graveyard of mechanical and chemical energy is strewn with heat.

FIGURE 4.30
Power harvested by photovoltaic cells, such as these on railroad track ties, can be used to extract hydrogen for fuel-cell pollution-free transportation (see www.SuntrainUSA.com).

FYI The most efficient transportation is by bicycle, which is much more efficient than train and car travel and even that of fish and animals. Hooray for bicyclists!

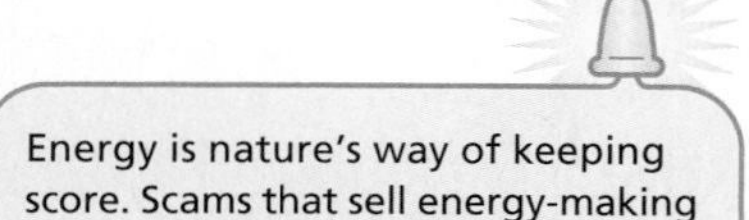
Energy is nature's way of keeping score. Scams that sell energy-making machines rely on funding from deep pockets and shallow brains!

Efficiency

Given the same energy input, some machines can do more work than others. Machines that can perform more work are said to be more efficient. **Efficiency** can be expressed by a ratio:

$$\text{Efficiency} = \frac{\text{work done}}{\text{energy used}}$$

Even a lever converts a small fraction of input energy into heat when it rotates about its fulcrum. We may do 100 J of work but get out only 98 J of productive work. The lever is then 98% efficient, and we waste 2 J of work input as heat. In a pulley system, a larger fraction of input energy goes into heat. If we do 100 J of work, the forces of friction acting through the distances in which the pulleys turn and rub about their axles may dissipate 60 J of energy as heat. So the work output is only 40 J, and the pulley system has an efficiency of 40%. The lower the efficiency of a machine, the greater the amount of energy wasted as heat.

An automobile engine is a machine that transforms chemical energy stored in gasoline into mechanical energy. But only a fraction of the energy stored in the fuel actually moves the car. Nearly half is wasted in the friction of the moving engine parts. Some goes out in the hot exhaust gases as waste. In addition to these inefficiencies, some of the fuel doesn't even burn completely and goes unused.

TECHNOLOGY

Junk Science

Scientists have to be open to new ideas. That's how science grows. But there is a body of established knowledge that can't be easily overthrown. That includes energy conservation, which is woven into every branch of science and supported by countless experiments on the atomic to the cosmic scale. Yet no concept has inspired more "junk science" than energy. Wouldn't it be wonderful if we could get energy for nothing, to build a machine that gives out more energy than is put into it? That's what many practitioners of junk science offer. Gullible investors put their money into some of these schemes. But none of them pass the test of being real science. Perhaps some day a flaw in the law of energy conservation will be discovered. If that ever happens, scientists will rejoice at the breakthrough. But, so far, energy conservation is as solid as any knowledge we have. Don't bet against it.

MATH CONNECTION

Efficiency Calculations

Energy efficiency is a consideration when choosing appliances, insulating your home, or driving a car, and in countless other practical matters of daily life. Hone your understanding of the concept with a couple practice problems.

Problems

1. Consider an imaginary dream car that has a 100%-efficient engine and burns fuel that has an energy content of 40 MJ/L. If the air resistance plus frictional forces on the car traveling at highway speed is 500 N, what is the maximum distance the car can go on 1 L of fuel?
2. One can only dream of a car with a 100%-efficient engine. More realistically, a conventional car engine is about 30% efficient. With the same air resistance and the same fuel as the car in problem 1, what is the maximum distance per liter for the realistic car?

Solutions

1. From the definition work = force × distance, simple rearrangement gives distance = work/force. If all 40 MJ of energy in 1 L is used to do the work of overcoming air resistance and frictional force, the distance covered is:

$$\text{Distance} = \frac{\text{work}}{\text{force}} = \frac{40{,}000{,}000\ \text{J}}{500\ \text{N}}$$
$$= 80{,}000\ \text{m} = 80\ \text{km}$$

The important point here is that, even with a perfect engine, there is an upper limit of fuel economy dictated by the conservation of energy.

2. The realistic distance per liter is 30% of 80 km = 0.3(80 km) = 24 km/L.

For instructor-assigned homework, go to www.masteringphysics.com 

SUMMARY OF TERMS (KNOWLEDGE)

Conservation of energy In the absence of external work input or output, the energy of a system remains unchanged. Energy cannot be created or destroyed.

Conservation of energy and machines The work output of any machine cannot exceed the work input. In an ideal machine, where no energy is transformed into heat,

$$\text{work}_{\text{input}} = \text{work}_{\text{output}}$$

and

$$(Fd)_{\text{input}} = (Fd)_{\text{output}}$$

Conservation of momentum In the absence of an external force, the momentum of a system remains unchanged. Hence, the momentum before an event involving only internal forces is equal to the momentum after the event:

$$mv_{(\text{before})} = mv_{(\text{after})}$$

Efficiency The percentage of the work put into a machine that is converted into useful work output. (More generally, efficiency is useful energy output divided by total energy input.)

Elastic collision A collision in which colliding objects rebound without lasting deformation or the generation of heat.

Energy The property of a system that enables it to do work.

Impulse The product of the force acting on an object and the time during which it acts.

Impulse–momentum relationship Impulse is equal to the change in the momentum of the object upon which the impulse acts. In symbolic notation,

$$Ft = \Delta mv$$

Inelastic collision A collision in which the colliding objects become distorted, generate heat, and possibly stick together.

Joule The SI unit of energy and work, equivalent to a newton-meter.

Kinetic energy The energy of motion, described by the relationship

$$\text{Kinetic energy} = \tfrac{1}{2}mv^2$$

Momentum The product of the mass of an object and its velocity.

Potential energy The stored energy that a body possesses because of its position.

Power The time rate of work:

$$\text{Power} = \frac{\text{work}}{\text{time}}$$

(More generally, power is the rate at which energy is expended.)

Work The product of the force and the distance through which the force moves:

$$W = Fd$$

Work–energy theorem The work done on an object equals the change in kinetic energy of the object:

$$\text{Work} = \Delta KE$$

READING CHECK QUESTIONS (COMPREHENSION)

4.1 Momentum

1. Which has a greater momentum—a heavy truck at rest or a moving skateboard?
2. How can a huge ship have an enormous momentum when it moves relatively slowly?

4.2 Impulse

3. How does impulse differ from force?
4. What are the two ways in which the impulse exerted on something can be increased?

4.3 Impulse–Momentum Relationship

5. For the same force, which cannon imparts the greater speed to a cannonball—a long cannon or a short one? Explain.
6. Consider a baseball that is caught and thrown at the same speed. Which case illustrates the greatest change in momentum—the baseball (a) being caught, (b) being thrown, or (c) being caught and then thrown back?
7. In the preceding question, which case requires the greatest impulse?

4.4 Conservation of Momentum

8. Can you produce a net impulse on an automobile by sitting inside and pushing on the dashboard? Defend your answer.
9. What does it mean to say that a quantity is conserved?
10. Distinguish between an elastic collision and an inelastic collision. For which type of collision is momentum conserved?
11. Railroad car A rolls at a certain speed and collides elastically with car B of the same mass. After the collision, car A is at rest. How does the speed of B after the collision compare with the initial speed of A?
12. If the equally massive cars of the preceding question stick together after colliding inelastically, how does their speed after the collision compare with the initial speed of car A?

4.5 Energy

13. When is energy most evident?
14. What do we call the quantity force × distance, and what quantity does it change?
15. In what units are work and energy measured?

4.6 Power

16. True or false: One watt is the unit of power equivalent to 1 joule per second.
17. How many watts of power are expended when a force of 6 N moves a book 2 m in a time interval of 3 s?

4.7 Potential Energy

18. A car is lifted a certain distance in a service station and therefore has potential energy with respect to the floor. If the car were lifted twice as high, how much potential energy would it have?
19. Two cars, one twice as heavy as the other, are lifted to the same height in a service station. How do their potential energies compare?

4.8 Kinetic Energy

20. When a car travels at 50 km/h, it has kinetic energy. How much more kinetic energy does it have at 100 km/h?

4.9 The Work–Energy Theorem

21. What is the evidence for saying whether or not work is done on an object?
22. The brakes do a certain amount of work to stop a car that is moving at a particular speed. How much work must the brakes do to stop a car that is moving four times as fast?

4.10 Conservation of Energy

23. Cite the law of energy conservation.
24. What is the source of energy that powers a hydroelectric power plant?

4.11 Machines

25. Can a machine multiply input force? Input distance? Input energy? (If your three answers are the same, seek help. The last question is especially important.)
26. A force of 50 N applied to the end of a lever moves that end a certain distance. If the other end of the lever is moved half as far, how much force does it exert?
27. Is it possible to design a machine that has an efficiency greater than 100%? Discuss.

THINK INTEGRATED SCIENCE

4A—The Impulse–Momentum Relationship in Sports

28. (a) Why is it a good idea to have your hand extended forward when you are getting ready to catch a fast-moving baseball with your bare hand? (b) In boxing, why is it advantageous to roll with the punch? (c) In karate, why is it advantageous to apply a force for a very brief time?
29. In Figure 4.7, how does the force that Cassy exerts on the bricks compare with the force exerted on her hand?
30. How will the impulse differ if Cassy's hand bounces back when striking the bricks?

4B—Glucose: Energy for Life

31. The word *burn* is often used to describe the process of cellular respiration, in which cells release energy from the chemical bonds in food molecules. How is the "burning" that goes on in cells different from literal burning—for example, burning a log on a campfire?
32. In what sense are you powered by solar energy?

THINK AND DO (HANDS-ON APPLICATION)

33. Pour some dry sand into a tin can that has a cover. Compare the temperature of the sand before and after vigorously shaking the can for more than a minute. Explain your observations.
34. Place a small rubber ball on top of a basketball, and then drop them together. How high does the smaller ball bounce? Can you reconcile this with energy conservation? (What if the basketball was not elastic?)

PLUG AND CHUG (FORMULA FAMILIARIZATION)

These one-step calculations are to familiarize you with the equations of the chapter.

Momentum = mv

35. Calculate the momentum of a 10-kg bowling ball rolling at 3 m/s.
36. Show that the momentum of a 50-kg carton that slides at 3 m/s across an icy surface is 150 kg·m/s.

Impulse = Ft

37. Calculate the impulse that occurs when an average force of 10 N is exerted on a cart for 5 s.
38. Show that an impulse of 100 N·s occurs when the same 10-N force acts on the cart for twice the time.

Work $W = Fd$ (where F and d are in same direction)

39. Calculate the work done when a force of 2 N moves a book 3 m.
40. Show that 45 J of work is done when a 15-N force pushes a cart 3 m.

Power = work done/time interval, $P = W/t$

41. Calculate the watts of power expended when a force of 1 N moves a book 2 m in a time interval of 1 s.
42. Show that 140 W of power is expended when a 20-N force pushes a cart 3.5 m in a time of 0.5 s.

Gravitational potential energy = weight × height, PE = mgh

43. How many joules of potential energy does a 1.5-kg book gain when it is elevated 2 m? When it is elevated 4 m? (Let $g = 10$ N/kg.)
44. Show that there is a 600-J increase in potential energy when a 20-kg block of ice is lifted a vertical distance of 3 m.

Kinetic energy: KE = $\frac{1}{2}mv^2$

45. Calculate the number of joules of kinetic energy a 1-kg parrot has when it flies at 6 m/s.
46. Show that the kinetic energy of a 3-kg dog running at a speed of 4 m/s is 24 J.

Work–energy theorem: Work = ΔKE, $Fd = \Delta\frac{1}{2}mv^2$

47. How much work is required to increase the kinetic energy of a motor scooter by 4000 J?
48. Show that a 50-J change in kinetic energy occurs when a model airplane on takeoff is moved a distance of 5 m by a sustained net force of 10 N.

THINK AND COMPARE (ANALYSIS)

Compare amounts for each of the following situations. If some are equal, put them in parentheses. For example, if A and B are tied, say (A&B tied).

49. The mass and speed of three vehicles, A, B, and C, are shown. Rank them from greatest to least for (a) momentum, (b) kinetic energy, (c) work done to bring them up to their respective speeds from rest.

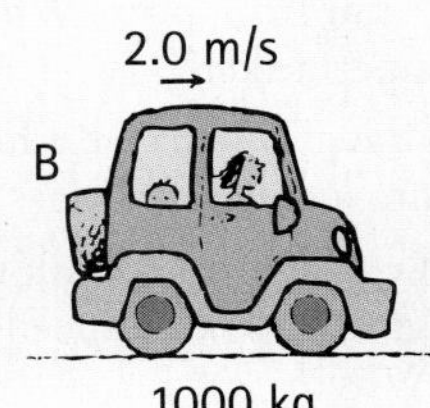

50. A ball is released at the left end of the metal track. Assume it has only enough friction to roll, but not to lessen its speed. Rank these quantities from greatest to least at each point: (a) momentum, (b) kinetic energy, (c) potential energy.

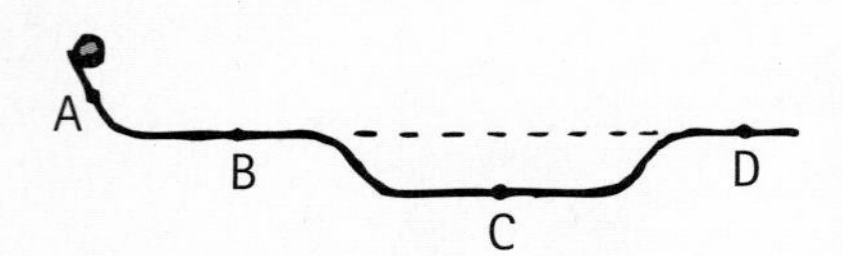

51. The roller coaster starts from rest at point A. Rank these quantities from greatest to least at each point: (a) speed, (b) kinetic energy, (c) potential energy.

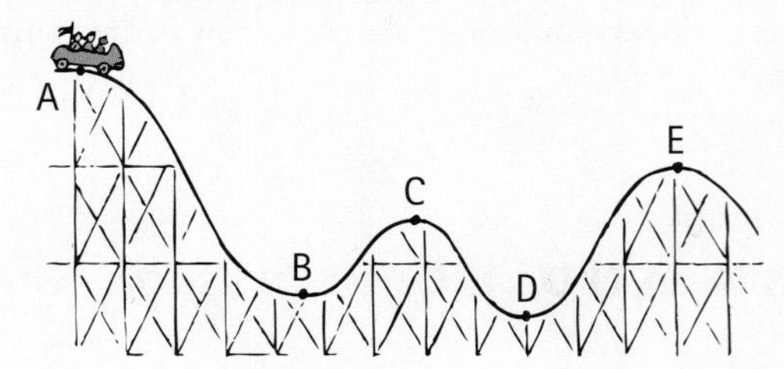

THINK AND SOLVE (MATHEMATICAL APPLICATION)

52. A car with a mass of 1000 kg moves at 20 m/s. Show that the braking force needed to bring the car to a halt in 10 s is 2000 N.
53. A railroad diesel engine weighs four times as much as a freight car. If the diesel engine coasts at 5 km/h into a freight car that is initially at rest, show that the two coast at 4 km/h after they couple together.
54. A 5-kg fish swimming at 1 m/s swallows an absent-minded 1-kg fish at rest. (a) Show that the speed of the larger fish after lunch is –5 m/s. (b) What would the speed of the larger fish be if the smaller fish were swimming toward it at 4 m/s?
55. Comic-strip hero Superman meets an asteroid in outer space and hurls it at 800 m/s, as fast as a bullet. The asteroid is a thousand times more massive than Superman. In the strip, Superman is seen at rest after the throw. Taking physics into account, what would be his recoil velocity?
56. Consider the inelastic collision between the two freight cars in Figure 4.11. The momentum before and after the collision is the same. The kinetic energy, however, is less after the collision than before the collision. How much less, and what has become of this energy?
57. This question is typical on some driver's license exams: A car moving at 50 km/h skids 15 m with locked brakes. How far will the car skid with locked brakes at 150 km/h?
58. In the hydraulic machine shown, it is observed that, when the small piston is pushed down 10 cm, the large piston is raised 1 cm. If the small piston is pushed down with a force of 100 N, show that the large piston is capable of exerting 1000 N of force.

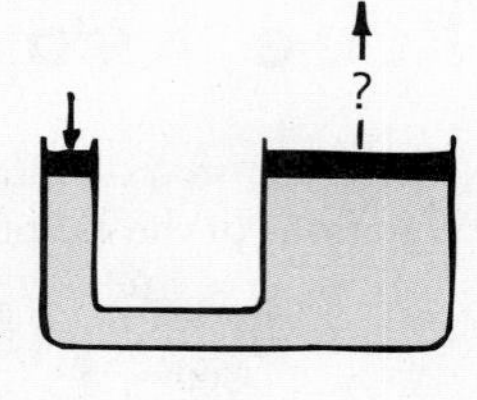

59. Consider a car with a 25%-efficient engine that encounters an average retarding force of 500 N. Assume that the energy content of the gasoline is 40 MJ/L. Show that the car will get 20 km per liter of fuel.
60. When a cyclist expends 1000 W of power to deliver mechanical energy to her bicycle at a rate of 100 W, show that the efficiency of her body is 10%.
61. The decrease in PE for a freely falling object equals its gain in KE, in accord with the conservation of energy. (a) Using simple algebra, find an equation for an object's speed v after falling a vertical distance h. Do this by equating KE to the object's change in PE. (b) Then figure out how much farther a freely falling object must fall to have twice the speed when it hits the ground.

THINK AND EXPLAIN (SYNTHESIS)

62. What is the purpose of a "crumple zone" (which has been manufactured to collapse in a crash) in the front section of an automobile?
63. To bring a supertanker to a stop, its engines are typically cut off about 25 km from port. Why is it so difficult to stop or turn a supertanker?
64. Why might a wine glass survive a fall onto a carpeted floor but not onto a concrete floor?
65. If you throw an egg against a wall, the egg will break. If you throw an egg at the same speed into a sagging sheet, it won't break. Why?
66. Railroad cars are loosely coupled so that there is a noticeable delay from the time the first car is moved and the time the last cars are moved from rest by the locomotive. Discuss the advisability of this loose coupling and slack between cars from an impulse–momentum point of view.

67. You're on a small raft next to a dock, and you jump from the raft only to fall into the water. What physics principle did you fail to take into account?

68. Freddy Frog drops vertically from a tree onto a horizontally moving skateboard. The skateboard slows. Give two reasons for the slowing, one in terms of a horizontal friction force between Freddy's feet and the skateboard, and one in terms of momentum conservation.

69. When a cannon with a long barrel is fired, the force of expanding gases acts on the cannonball for a long distance. What effect does this have on the velocity of the emerging cannonball?
70. You and a flight attendant toss a ball back and forth in an airplane in flight. Does the KE of the ball depend on the speed of the airplane? Carefully explain.
71. Can something have energy without having momentum? Explain. Can something have momentum without having energy? Defend your answer.
72. An inefficient machine is said to "waste energy." Does this mean that energy is actually lost? Explain.
73. A child can throw a baseball at 20 mph. Some professional ball players can throw a baseball at 100 mph, which is five times as fast. How much more energy does the pro ball player give to the faster ball?
74. If a golf ball and a Ping-Pong ball both move with the same kinetic energy, can you say which has the greater speed? Explain in terms of KE. Similarly, in a gaseous mixture of massive molecules and light molecules with the same average KE, can you say which have the greater speed?
75. Consider a pendulum swinging to and fro. At what point in its motion is the KE of the pendulum bob at a maximum? At what point is its PE at a maximum? When the pendulum's KE is half its maximum value, how much PE does it have?
76. A physics instructor demonstrates energy conservation by releasing a heavy pendulum bob, as shown in the sketch, allowing it to swing to and fro. What would happen if, in his exuberance, he gave the bob a slight shove as it left his nose? Defend your answer.

77. If an automobile had an engine that was 100% efficient, would it be warm to your touch? Would its exhaust heat the surrounding air? Would it make any noise? Would it vibrate? Would any of its fuel remain unused?

THINK AND DISCUSS (EVALUATION)

78. Why is a punch more forceful with a bare fist than with a boxing glove?
79. A boxer can punch a heavy bag for more than an hour without tiring but will tire more quickly when boxing with an opponent for a few minutes. Why? (Hint: When the boxer's punches are aimed at the bag, what supplies the impulse to stop them? When aimed at the opponent, what (or who) supplies the impulse to stop the punches that are missed?)
80. A fully dressed person is at rest in the middle of a pond on perfectly frictionless ice and must reach the shore. Discuss how this can be done.
81. A high-speed bus and an innocent bug have a head-on collision. Is the sudden change in momentum of the bus greater than, less than, or the same as the change in momentum of the unfortunate bug? Discuss the distinction between momentum and a *change* in momentum.
82. Why is it difficult for a firefighter to hold a hose that ejects large quantities of water at high speed?
83. Your friend says that conservation of momentum is violated when you step off a chair and gain momentum as you fall. A discussion follows. What do you say?
84. If a Mack truck and a Honda Civic have a head-on collision, which vehicle will experience the greater force of contact? The greater impulse? The greater change in momentum? The greater acceleration? Discuss how a failure to distinguish between these concepts leads to difficulties in understanding physics!
85. Would a head-on collision between two cars be more damaging to the occupants if the cars stuck together or if the cars rebounded upon contact?
86. In Chapter 3, rocket propulsion was explained in terms of Newton's third law; that is, the force that propels a rocket is from the exhaust gases pushing against the rocket, the reaction to the force the rocket exerts on the exhaust gases. Discuss rocket propulsion in terms of momentum conservation.
87. Suppose three astronauts are outside a spaceship and two of them decide to play catch using the third man. All the astronauts weigh the same on Earth and are equally strong. The first astronaut throws the second one toward the third one and the game begins. Discuss the motion of the astronauts as the game proceeds. In terms of the number of throws, how long will the game last?

88. Discuss how it is possible that a flock of birds in flight can have a momentum of zero but not have zero kinetic energy.
89. Discuss the design of the roller coaster shown in the sketch in terms of the conservation of energy.

90. In an effort to combat wasteful habits, we often urge others to "conserve energy" by turning off lights when they are not in use, for example, or by setting thermostats at a moderate level. In this chapter, we also speak of "energy conservation." Distinguish between these two expressions.

91. Consider the identical balls released from rest on tracks A and B as shown. When each ball has reached the right end of its track, which will have the greater speed? Why is this question easier to answer than the similar question in Exercise 79 in Chapter 2?

92. Strictly speaking, does a car burn more gasoline when the lights are turned on? Does the overall consumption of gasoline depend on whether or not the engine is running? Defend your answer.

93. Consider the swinging-balls apparatus shown. If two balls are lifted and released, momentum is conserved as two balls pop out the other side with the same speed as the released balls at impact. But momentum would also be conserved if one ball popped out at twice the speed. Think about "energy" and discuss why this never happens.

READINESS ASSURANCE TEST (RAT)

If you have a good handle on his chapter, you should be able to score at least 7 out of 10 on this RAT. If you score less than 7, you need to study further before moving on.

Choose the BEST answer to each of the following:

1. A freight train rolls along a track with considerable momentum. If it rolls at the same speed but has twice as much mass, its momentum is
   (a) zero.
   (b) doubled.
   (c) quadrupled.
   (d) unchanged.

2. When you are in the way of a fast-moving object and can't escape, you will suffer a smaller force if the collision time is
   (a) long.
   (b) short.
   (c) The force is the same either way.

3. A 1-kg glider and a 2-kg glider slide toward each other at 1 m/s on an air track. They collide and stick. The combined gliders move at
   (a) 0 m/s.
   (b) ½ m/s.
   (c) ⅓ m/s.
   (d) ⅔ m/s.

4. How much work is done on a 200-kg crate that is hoisted 2 m in a time of 4 s?
   (a) 400 J
   (b) 1000 J
   (c) 1600 J
   (d) 4000 J

5. When an increase in speed doubles the momentum of a moving body, its kinetic energy
   (a) increases, but less than doubles.
   (b) doubles.
   (c) more than doubles.
   (d) depends on factors not stated.

6. Exert 100 J in 50 s, and your power output is
   (a) ¼ W.
   (b) ½ W.
   (c) 2 W.
   (d) 4 W.

7. You lift a barbell a certain distance from the floor. If you lift it twice as high, its potential energy is
   (a) twice as great.
   (b) half as great.
   (c) the same.
   (d) none of the above

8. A car moving at 50 km/h skids 20 m with locked brakes. How far will it skid if its initial speed is 100 km/h?
   (a) 60 m
   (b) 100 m
   (c) 120 m
   (d) 180 m

9. Compared to a recoiling rifle, the bullet that is fired has a greater
   (a) momentum.
   (b) kinetic energy.
   (c) both
   (d) neither

10. True or false: A hydraulic press properly arranged, like a simple lever, is capable of multiplying energy input.
   (a) true
   (b) sometimes true
   (c) false

Answers to RAT

1. b, 2. a, 3. c, 4. d, 5. c, 6. c, 7. a, 8. b, 9. b, 10. c

CHAPTER 5
Gravity

IT WOULD be erroneous to say that Newton discovered gravity. The discovery of gravity goes back much further than Newton's era, to earlier times when Earth dwellers fell from trees and from ledges inside their caves, or when they discovered the consequences of tripping. What Newton discovered was that gravity is universal—that it is not a phenomenon unique to Earth, as his contemporaries had assumed. Further, Newton demonstrated that gravity is universal and can be described by a simple law. What is Newton's universal law of gravitation? How does it unify phenomena as seemingly diverse as falling apples and orbiting planets? What is the relationship between weight and gravity? How is the motion of a thrown basketball essentially similar to the motion of a satellite? We'll discover answers to these and other gravity-related questions in this chapter.

LEARNING OBJECTIVE
Describe the setting where Newton linked gravity to the Moon.

5.1 The Legend of the Falling Apple

EXPLAIN THIS What connection did Newton discover between a falling apple and the Moon?

According to popular legend, Newton was sitting under an apple tree when he made a connection that changed the way we see the world. He saw an apple fall. Perhaps he then looked up through the branches toward the origin of the falling apple and noticed the Moon. In any event, Newton had the insight to realize that the force pulling on the apple was the same force that pulls on the Moon. Newton realized that Earth's gravity reaches to the Moon (Figure 5.1).

FIGURE 5.1
Could the gravitational pull on the apple reach to the Moon?

LEARNING OBJECTIVE
Link the high tangential velocity of the Moon to its continuous fall around Earth.

5.2 The Fact of the Falling Moon

EXPLAIN THIS How do higher tangential speeds affect the curved path of a projectile?

If Earth's gravity is pulling the Moon toward it, why doesn't the Moon fall toward Earth, like an apple falls from a tree? If you hold an apple above your head and drop it, the apple will fall in a vertical straight-line path. But, rather than dropping the apple from rest, move your hand sideways as you drop it. In effect, you're tossing it, and you see that it follows a curved path. The greater the initial sideways motion, the wider the curve. Later in this chapter, we'll see that, if the apple or anything else moves fast enough so that its curved path matches Earth's curvature, it becomes a satellite.

As the Moon traces out its orbit around Earth, it maintains a **tangential velocity**—a velocity parallel to Earth's surface (Figure 5.2). Newton realized that the Moon's tangential velocity keeps it falling *around* Earth instead of directly into it. Newton further realized that the Moon's path around Earth is similar to the paths of the planets around the Sun.

From the time of Aristotle, the circular motions of heavenly bodies were regarded as natural. The ancients believed that stars, planets, and the Moon moved in divine circles, free from the forces that dictate motion here on Earth. They believed there were two sets of laws, one for earthly events and a different set for motions in the heavens. Newton's stroke of intuition—that the force between Earth and apples is the same force that pulls moons, planets, and everything else—was a revolutionary break with prevailing notions. Newton synthesized terrestrial and cosmic laws.

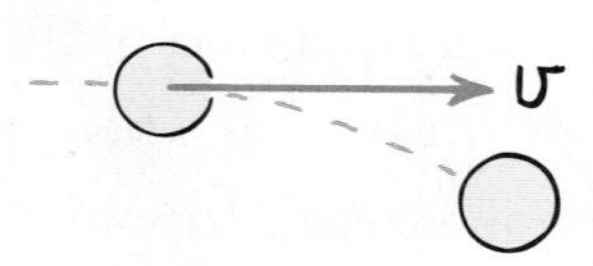

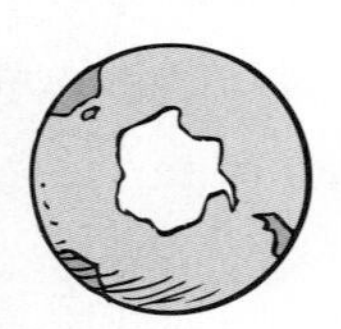

FIGURE 5.2
The tangential velocity of the Moon about Earth allows it to fall around Earth rather than directly into it. If this tangential velocity were reduced to zero, the Moon would then crash into Earth.

LEARNING OBJECTIVE
Define and describe Newton's law of universal gravitation.

5.3 Newton's Law of Universal Gravitation

EXPLAIN THIS What exactly did Newton discover about gravity?

Newton further realized that *everything* pulls on *everything* else. He discovered that a **gravitational force** acts on all things in a beautifully simple way—a way that involves only mass and distance. According to Newton, every mass pulls on every other mass with a force that is directly proportional to the product of the two interacting masses. The force is inversely proportional to the square of the distance separating them. This statement is known as the **law of universal gravitation**:

$$\text{Force} \sim \frac{(\text{mass}_1 \times \text{mass}_2)}{\text{distance}^2}$$

Expressed in symbolic shorthand,

$$F \sim \frac{m_1 m_2}{d^2}$$

where m_1 and m_2 are the masses, and d is the distance between their centers. Thus, the greater the masses m_1 and m_2, the greater the force of attraction between them. The greater the distance of separation d, the weaker is the force of attraction—weaker as the inverse square of the distance between their centers (Figure 5.3).

FIGURE 5.3
As the rocket gets farther from Earth, the strength of the gravitational force between the rocket and Earth decreases.

CHECK YOURSELF

1. **According to the equation for gravitational force, what happens to the force between two bodies if the mass of only one body is doubled?**
2. **What happens to the force if the masses of both bodies are doubled?**
3. **What happens to the force if the mass of one body is doubled and the mass of the other is tripled?**
4. **Gravitational force acts on all bodies in proportion to their masses. Why, then, doesn't a heavy body fall faster than a light body?**

CHECK YOUR ANSWERS

1. When one mass is doubled, the force between them doubles.
2. The force is four times as much.
3. Double × triple = six, so the force is six times as much.
4. The answer goes back to Chapter 3. Recall Figure 3.9, in which a feather and a coin fall with the same acceleration because both have the same ratio of weight to mass. Newton's second law ($a = \frac{F}{m}$) reminds us that greater force acting on greater mass does not result in greater acceleration.

UNIFYING CONCEPT

- *The Gravitational Force*

5.4 Gravity and Distance: The Inverse-Square Law

LEARNING OBJECTIVE
Describe the rule by which gravity decreases with distance.

EXPLAIN THIS How much smaller is the force of gravity on an object when it is twice as far away?

Gravity gets weaker with distance the same way a lit lamp gets dimmer as you move farther from it. Consider the candle flame in Figure 5.4. Light from the flame travels in all directions in straight lines. A patch is shown 1 meter from the flame. Notice that at a distance of 2 meters away, the light rays that fall on the patch spread to fill a patch twice as tall and twice

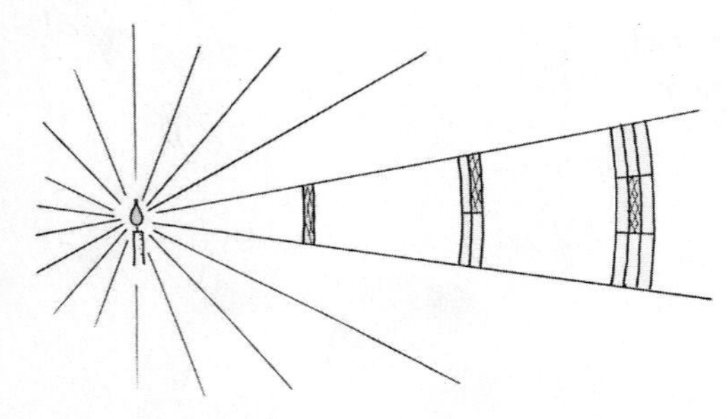

FIGURE 5.4
Light from the flame spreads in all directions. At twice the distance, the same light is spread over 4 times the area; at 3 times the distance, it is spread over 9 times the area.

as wide. In other words, the same light falls on a patch with 4 times the area. If the same light were 3 meters away, it would spread to fill a patch 3 times as tall and 3 times as wide, and it would fill a patch 9 times the area.

As the light spreads out, its brightness decreases. Can you see that when you're twice as far away, the light appears $\frac{1}{4}$ as bright? And can you see that when you're 3 times as far away, the light appears $\frac{1}{9}$ as bright? There is a rule here: The intensity of the light decreases as the inverse square of the distance. This is the **inverse-square law**.

The inverse-square law applies to gravity. The greater the distance from Earth's center, the less the gravitational force on an object. In Newton's equation for gravity, the distance term d is the distance between the centers of the masses of objects attracted to each other. Note that the girl at the top of the ladder in Figure 5.5 weighs only $\frac{1}{4}$ as much as she weighs on Earth's surface. That's because she is *twice* the distance from Earth's center. (Recall from Chapter 2 that *weight* is the force due to gravity on a body.)

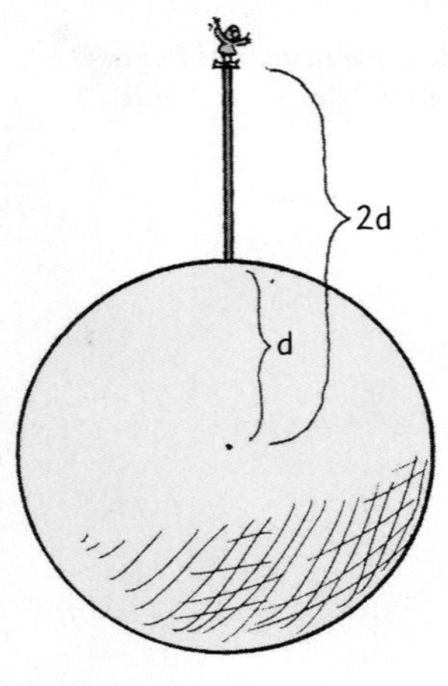

FIGURE 5.5
At the top of the ladder, the girl is twice as far from Earth's center, and she weighs only $\frac{1}{4}$ as much as she did at the bottom of the ladder.

CHECK YOURSELF

1. **How much does the force of gravity change between Earth and a receding rocket when the distance between them is doubled? When the distance is tripled? When it is increased tenfold?**
2. **Consider an apple at the top of a tree. The apple is pulled by Earth's gravity with a force of 1 N. If the tree were twice as tall, would the force of gravity be only $\frac{1}{4}$ as strong? Defend your answer.**

CHECK YOUR ANSWERS

1. When the distance is doubled, the force is only $\frac{1}{4}$ as much. When the distance is tripled, the force is only $\frac{1}{9}$ as much. When the distance is increased tenfold, the force is only $\frac{1}{100}$ as much.
2. No, because the twice-as-tall apple tree is not twice as far from Earth's center. The taller tree would have to be 6370 km tall (as tall as Earth's radius) for the apple's weight to reduce to $\frac{1}{4}$ N. For a decrease in weight of 1%, an object must be raised 32 km—nearly four times the height of Mt. Everest. So, as a practical matter, we disregard the effects of everyday changes in elevation for gravity. The apple has practically the same weight at the top of the tree as it has at the bottom.

MasteringPhysics®
VIDEO: Inverse-Square Law

So, the gravitational attraction between two objects gets appreciably weaker as the objects get farther apart. But, no matter how great the distance, gravity approaches, but never quite reaches, zero. There is still a gravitational attraction between any two masses, no matter how far apart they are. Even if you were removed to the far reaches of the universe, the gravitational influence of Earth would still remain with you. It may be overwhelmed by the influences of nearer or more massive bodies, but its presence is still there. The gravitational influence of every material object, however small or however far, is exerted through all of space.

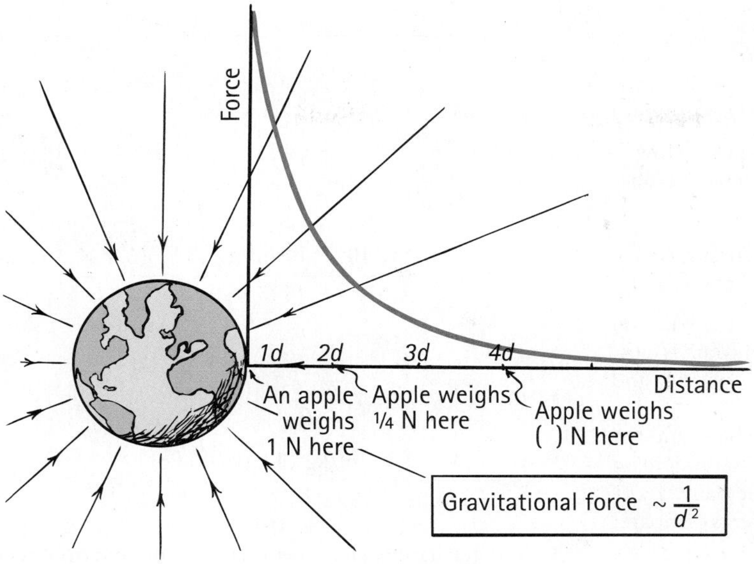

FIGURE 5.6

INTERACTIVE FIGURE

If an apple weighs 1 N at Earth's surface, it would weigh only $\frac{1}{4}$ N twice as far from the center of Earth. At three times the distance, it would weigh only $\frac{1}{9}$ N. What would the apple weigh at four times the distance? At five times the distance? As the distance approaches infinity, the gravitational force approaches zero. The graph of gravitational force versus distance is plotted in red.

5.5 The Universal Gravitational Constant, *G*

> **LEARNING OBJECTIVE**
> **Describe the role of *G* in the law of universal gravitation.**

EXPLAIN THIS How do the units of *G* affect the units of force in the equation for gravitational force?

The universal law of gravitation can be written as an exact equation when the **universal constant of gravitation**, *G*, is used:

$$F = G\frac{m_1 m_2}{d^2}$$

The units of *G* make the force come out in newtons. The magnitude of *G* is the same as the gravitational force between two 1-kilogram masses that are 1 meter apart: 0.0000000000667 newton,

$$G = 6.67 \times 10^{-11} \frac{\text{N}\cdot\text{m}^2}{\text{kg}^2}$$

This is an extremely small number. It shows that gravity is a very weak force compared with electrical forces. The large net gravitational force that we feel as weight occurs because of the immensity of the number of bits of mass in the planet Earth that are pulling on us.

Just as π relates the circumference of a circle to its diameter, *G* relates gravitational force with mass and distance. *G*, like π, is a proportionality constant.

Integrated Science 5A
BIOLOGY

> **LEARNING OBJECTIVE**
> **Relate gravity to balance.**

Your Biological Gravity Detector

EXPLAIN THIS In relation to balance, what do vestibular organs detect?

How many senses do you have? The answer is *not* five. Beyond sight, taste, smell, hearing, and touch, there are other senses—for example, hunger and thirst. You have a sense of how your body is oriented in

MATH CONNECTION

Comparing Gravitational Attractions

Every mass gravitationally attracts every other mass. How do different pulls between different pairs of interacting objects compare? Try a few examples.

Problems A 3-kg newborn baby at Earth's surface is gravitationally attracted to Earth with a force of about 30 N.

1. Calculate the force of gravity with which the baby on Earth is attracted to the planet Mars, when Mars is closest to Earth. (The mass of Mars is 6.4×10^{23} kg, and its closest distance is 5.6×10^{10} m.)
2. Calculate the force of gravity between the baby and the physician who delivers her. Assume that the physician has a mass of 100 kg and is 0.5 m from the baby.
3. How do these forces in the preceding problems compare?

Solutions

1. Mars:

$$F = G\frac{m_1 m_2}{d^2}$$

$$= \frac{[6.67 \times 10^{-11}\ \text{N}\cdot\text{m}^2/\text{kg}^2(3\ \text{kg})(6.4 \times 10^{23}\ \text{kg})]}{(5.6 \times 10^{10}\ \text{m})^2}$$

$$= 4.1 \times 10^{-8}\ \text{N}$$

2. Physician: $F = G\frac{m_1 m_2}{d^2}$

$$\frac{[6.67 \times 10^{-11}\ \text{N}\cdot\text{m}^2/\text{kg}^2(3\ \text{kg})(100\ \text{kg})]}{(0.5\ \text{m})^2} = 8.0 \times 10^{-8}\ \text{N}$$

3. The gravitational force between the baby and the physician is about twice the force between the baby and Mars.

space, too. This is called your *vestibular sense*, and it depends on your ability to detect both your acceleration and your orientation with respect to Earth's gravity. You do this by means of organs in your inner ear called *vestibular organs*.

There are two kinds of vestibular organs, as shown in Figure 5.7—the *semicircular canals* and the *otolith organs*. While it's the job of the semicircular canals to detect rotational motion of the head, the otolith organs detect linear acceleration of the head as well as whether or not the head is tilted with respect to Earth's gravity. Otolith organs contain small sensory areas about 2 mm in diameter known as *maculas*. Each macula contains thousands of receptor cells called *hair cells*. The hair cells have stalklike cilia that are embedded in a gelatinous matrix called the *otolithic membrane*. And, on top of the membrane, there are small piles of calcium carbonate ($CaCO_3$) crystals, which are called *otoliths*. The anatomy of a macula is shown in Figure 5.8.

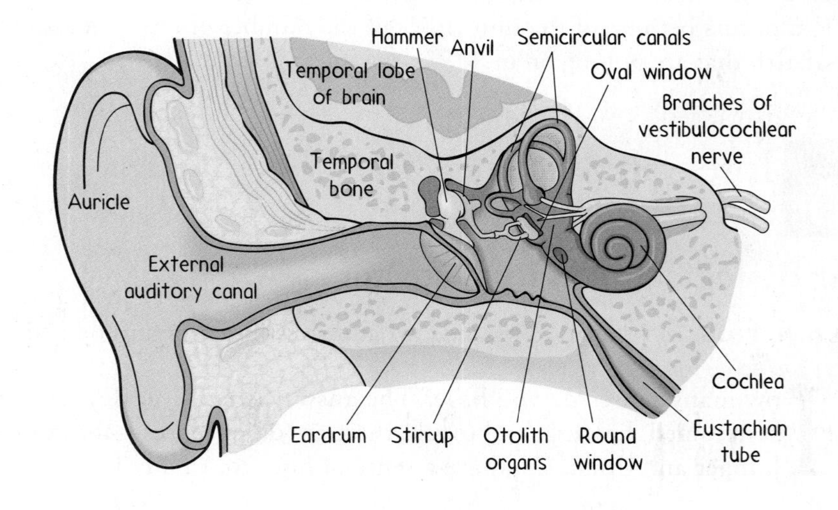

FIGURE 5.7
The vestibular system consists of the semicircular canals and the otolith organs.

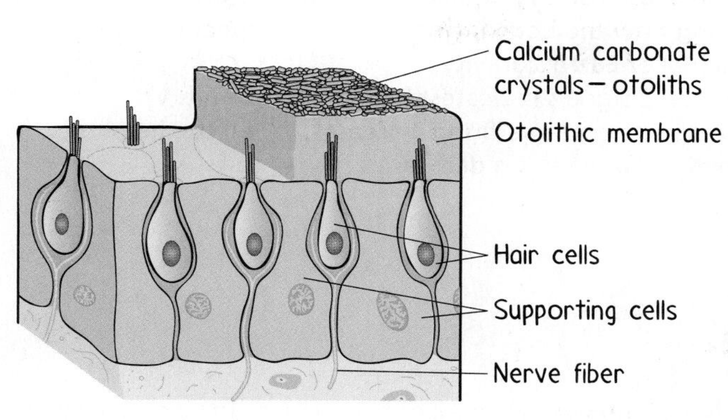

FIGURE 5.8
Anatomical features of a macula—the structure that contains receptor cells that detect acceleration of the head as well as its orientation with respect to gravity. Each macula contains several thousand hair cells.

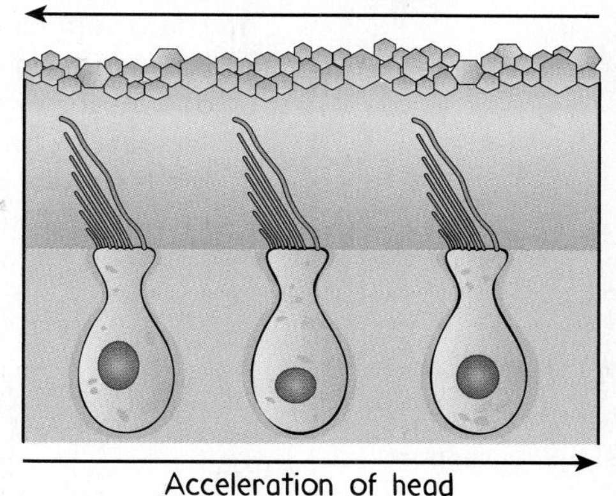

FIGURE 5.9
The greater density of the otoliths and otolithic membrane compared with the surrounding tissue pulls and bends the cilia of receptor cells when you accelerate.

Because the otoliths are more dense than the surrounding tissue, they have greater inertia. This enables them to indicate linear acceleration: When you accelerate in one direction, the mass of the otolithic membrane and its otoliths causes the hair-cell cilia to bend. This bending of the cilia stimulates the cells to send signals in a particular pattern that your brain interprets as acceleration. When you move with a constant velocity, the otoliths reach equilibrium and you no longer have the perception that you are moving (Figure 5.9).

The otoliths detect head tilt in a way similar to how they sense linear acceleration. When your head tilts, the planes of your maculae change with respect to the direction of gravity's pull, and this bends the cilia of your hair cells. When the cilia bend in one direction, they increase the rate at which they fire nerve signals. When the cilia bend in the opposite direction, they fire fewer nerve signals, as shown in Figure 5.10. The overall pattern of signal firing from tilted macula indicates the direction and degree of head tilt.

By the way, the literal meaning of the word *otoliths* is "ear stones." So you really do have rocks in your head!

CHECK YOURSELF

1. **What is the vestibular sense, and what does it have to do with gravity?**
2. **How do you sense the direction in which your head tilts?**

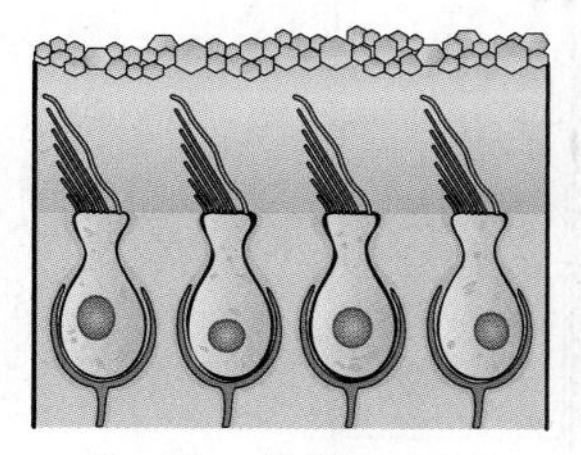

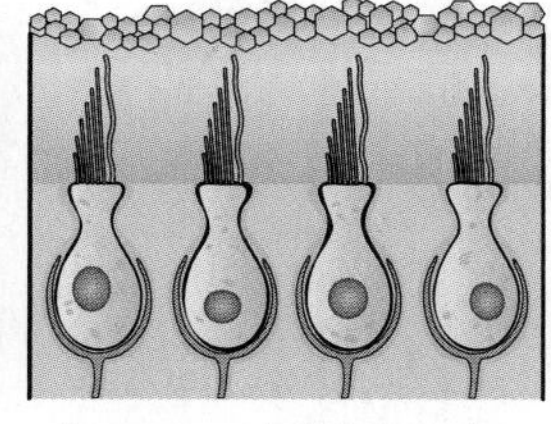

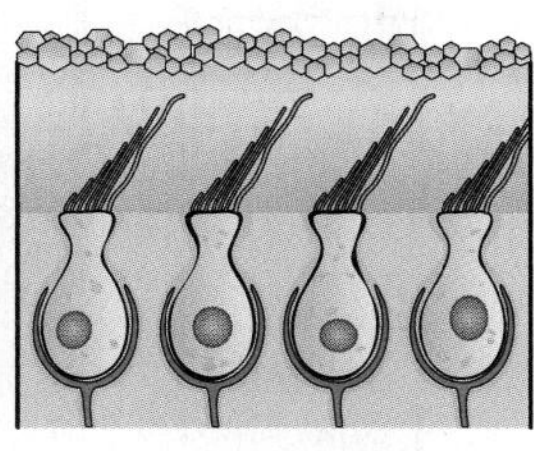

FIGURE 5.10
Maculae respond to tilting by varying the frequency of neural signals.

CHECK YOUR ANSWERS

1. The vestibular sense is the sense that indicates your body's orientation. It depends on knowing how the body is oriented with respect to the downward-directed pull of Earth's gravity.
2. The rate of firing of nerve signals indicates the direction of tilt. When the head tilts in one direction, nerve firing increases. When the head tilts in the opposite direction, nerve firing decreases.

LEARNING OBJECTIVE
Describe how weight is a support force.

5.6 Weight and Weightlessness

EXPLAIN THIS How does your weight change when you're inside an accelerating elevator?

MasteringPhysics®
VIDEO: Weight and Weightlessness
VIDEO: Apparent Weightlessness

When you step on a bathroom scale, in effect, you compress a spring inside it that is affixed to a pointer. When the pointer stops moving, the elastic force of the deformed spring balances the gravitational force between you and Earth—you and the scale are in equilibrium. The pointer is calibrated to show your weight (Figure 5.11). If you stand on a bathroom scale in a moving elevator, you'll find variations in your weight. If the elevator accelerates upward, the springs inside the bathroom scale are more compressed and your weight reading is higher. If the elevator accelerates downward, the springs inside the scale are less compressed and your weight reading is lower. If the elevator cable breaks and the elevator falls freely, the reading on the scale goes to zero. According to the scale's reading, you would be weightless. Would you really be weightless? We can answer this question only if we agree on what we mean by *weight*.

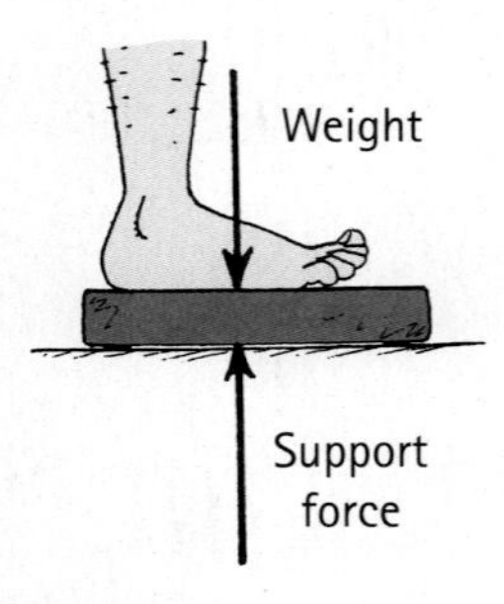

FIGURE 5.11
When you step on a weighing scale, two forces act on it: a downward force of gravity (your ordinary weight, mg, if there is no acceleration) and an upward support force. These equal and opposite forces squeeze a springlike device inside the scale that is calibrated to show your weight.

In Chapter 2, we defined weight as the force due to gravity on a body, mg. Your weight has the value of mg if you're not accelerating. To generalize, we can refine this definition by saying that the weight of something is the force it exerts against a supporting floor or weighing scale. According to this definition, you are as heavy as you feel. So, in an elevator that accelerates downward, the supporting force of the floor is less and, therefore, you weigh less. If the elevator is in free fall, your weight is zero (Figure 5.12). Even in this weightless condition, however, there is still a gravitational force acting on you, causing your downward acceleration. But gravity now is not felt as weight because there is no support force.

FIGURE 5.12
Your weight equals the force with which you press against the supporting floor. If the floor accelerates up or down, your weight varies (even though the gravitational force mg that acts on you remains the same).

Consider an astronaut in orbit. The astronaut is weightless because he is not supported by anything (Figure 5.13). There would be no compression in the springs of a bathroom scale placed beneath his feet because the bathroom scale is falling as fast as he is. Any objects that are released fall together with him and remain in his vicinity, unlike what occurs on the ground. All the local effects of gravity are eliminated. The body organs respond as if gravity forces were absent, and this gives the sensation of **weightlessness**. The astronaut experiences the same sensation in orbit that he would feel falling in an elevator—a state of free fall.

FIGURE 5.13
Both are weightless.

On the other hand, if the astronaut were in a spacecraft undergoing acceleration, even in deep space and far removed from any attracting objects, he *would* have weight. Like the girl in the upward accelerating elevator, the astronaut would be pressed against a scale or supporting surface.

The International Space Station in Figure 5.14 provides a weightless environment. The station and the astronauts inside all accelerate equally toward Earth, at somewhat less than 1 *g* because of their altitude. This acceleration is not sensed at all. With respect to the station, the astronauts experience zero *g*.

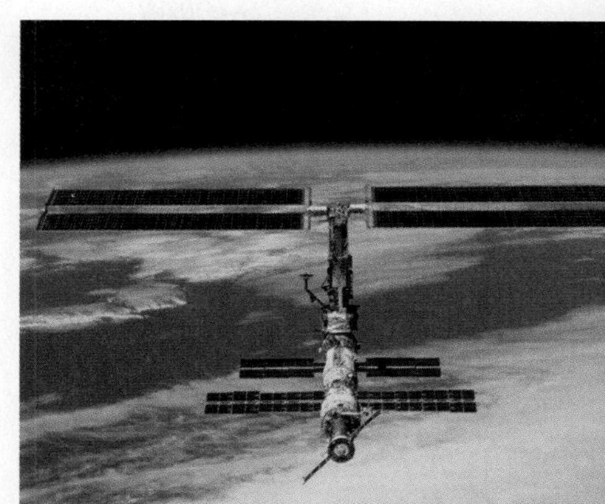

FIGURE 5.14
The inhabitants of this laboratory and docking facility continuously experience weightlessness. They are in free fall around Earth. Does a force of gravity act on them?

Integrated Science 5B
BIOLOGY

LEARNING OBJECTIVE
To relate center of gravity to balance.

Center of Gravity of People

EXPLAIN THIS What is the relationship between the support base of an object and balance and toppling?

The *center of gravity* (CG) of an object is the point located at the object's average position of weight. For a symmetrical object, this point is at the geometric center. But an irregularly shaped object, such as a baseball bat, has more weight at one end, so its CG is toward the heavier end. A piece of tile cut into the shape of a triangle has its CG one-third of the way up from its base (Figure 5.15).

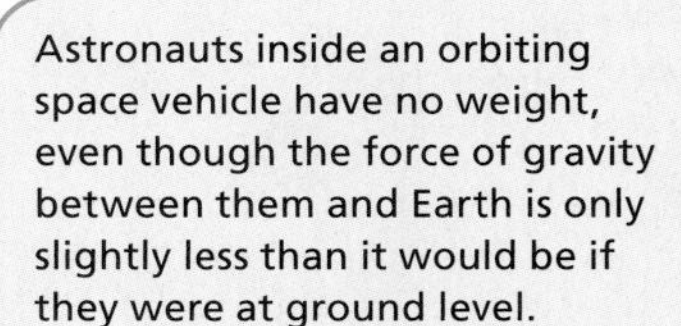

Astronauts inside an orbiting space vehicle have no weight, even though the force of gravity between them and Earth is only slightly less than it would be if they were at ground level.

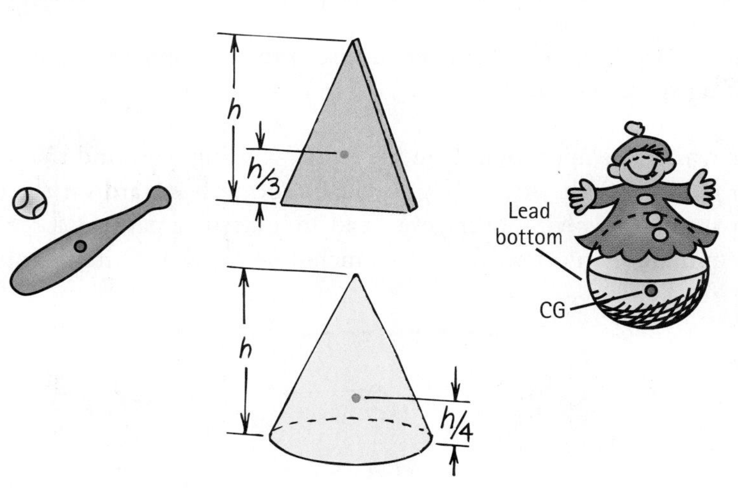

FIGURE 5.15
The center of gravity for each object is shown by a red dot.

FIGURE 5.16
The Leaning Tower of Pisa does not topple over because its CG lies above its base.

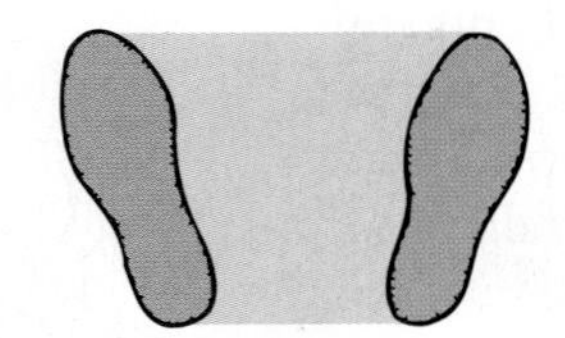

FIGURE 5.17
Your support base is the shaded area bounded by your feet.

FIGURE 5.18
You can lean over and touch your toes only if your CG is above your base.

The position of an object's CG relative to its base of support determines the object's stability. The rule for stability is this: If the CG of an object is above the area of support, the object will remain upright. If the CG extends outside the area of support, the object will topple.

This is why the Leaning Tower of Pisa doesn't topple. Its CG does not extend beyond its base. If the tower leaned far enough over so that its CG extended beyond its base, it would topple (Figure 5.16).

When you stand erect with your arms hanging at your sides, your CG is typically 2 to 3 cm below your navel and midway between your front and back. The CG is slightly lower in women than in men because women tend to be proportionally larger in the pelvis and smaller in the shoulders.

MasteringPhysics®
VIDEO: Locating the Center of Gravity
VIDEO: Toppling

When you stand, your CG is somewhere above your support base, the area bounded by your feet (Figure 5.17). In unstable situations, you place your feet farther apart to increase this area. Standing on one foot greatly decreases the area of your support base. A baby must learn to coordinate and position its CG above one foot. Many birds—pigeons, for example—do this by jerking their heads back and forth with each step.

LEARNING OBJECTIVE
Describe the way that centripetal force produces weight in a rotating space station.

5.7 Centripetal Force Can Simulate Gravity

EXPLAIN THIS How does centripetal force provide support for an astronaut in a rotating space station?

If you whirl an empty tin can on the end of a string, you find that you must keep pulling on the string (Figure 5.19). You pull inward on the string to keep the can revolving over your head in a circular path. A force of some kind is required for any circular motion, including the nearly circular motions of

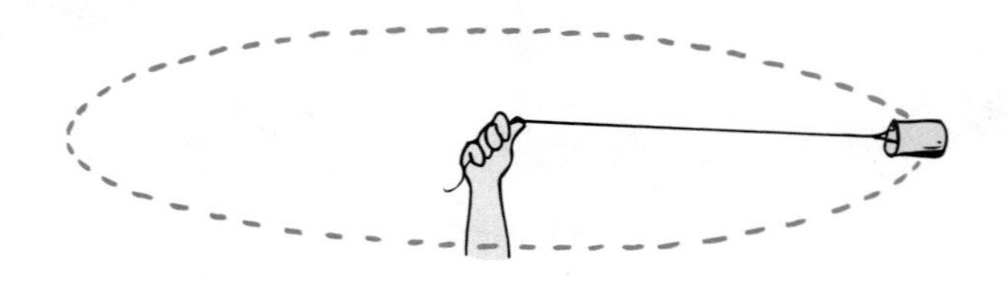

FIGURE 5.19
The only force that is exerted on the whirling can (neglecting the downward pull of gravity) is directed *toward* the center of circular motion. It is called a centripetal force. No *outward* force is exerted on the can.

the planets around the Sun. Any force that causes an object to follow a circular path is called a **centripetal force**. *Centripetal* means "center-seeking" or "toward the center." Centripetal force is not a new kind of force. It is simply a name given to any force that is directed at right angles to the path of a moving object and that tends to produce circular motion. The gravitational force acting across space is a centripetal force that keeps the Moon in Earth's orbit. Likewise, electrons that revolve about the nucleus of an atom are held by an electrical force that is directed toward the central nucleus.

Centripetal force could rescue future space travelers from the effects of weightlessness. Today's space habitats may someday be replaced by lazily rotating giant wheels (Figure 5.20). Occupants in the rotating habitat will travel in circular paths and feel a centripetal force, which is directed inward toward the axis of rotation. The centripetal force is the support force, and it is sensed as weight. If the space habitat rotates at just the right speed, the support force will simulate normal Earth weight. Any space explorers in rotating systems "out there" will experience the gravity of their home planets.

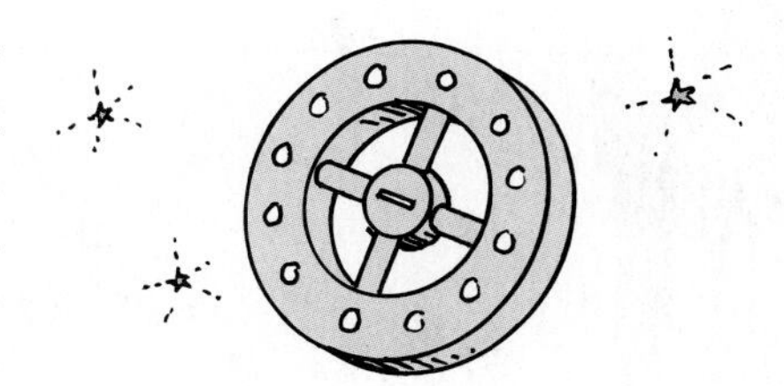

FIGURE 5.20
Occupants in the rotating space habitat will experience simulated weight and can "stand up" inside. In upright positions, their feet are pressed by the outer rim and their heads point toward the center.

MasteringPhysics®
VIDEO: Simulated Gravity
VIDEO: Centripetal Force

5.8 Projectile Motion

LEARNING OBJECTIVE
Apply the independence of horizontal and vertical motion to projectiles.

EXPLAIN THIS Why does a dropped ball and a ball thrown horizontally hit the ground in the same time?

UNIFYING CONCEPT
- ***Newton's First Law***
Section 3.1

Without gravity, you could toss a rock upward at an angle and it would follow a straight-line path. But, due to gravity, the path curves. A tossed rock, a cannonball, or any object that is projected by some means and continues in motion by its own inertia is called a **projectile**. To the cannoneers of earlier centuries, the curved paths of projectiles seemed very complex. Today, we see that these paths are surprisingly simple when the horizontal and vertical components of velocity are considered separately. We'll first consider the vertical part of a projectile's motion, the component that is affected by gravity.

A very simple projectile is a falling stone, as shown in Figure 5.21. (This is a version of Figure 2.23, which we studied in Chapter 2.) The stone gains speed as it falls straight down, as indicated by a speedometer. Remember that a freely falling object gains 10 meters per second during each second of fall. This is the acceleration due to gravity, 10 m/s^2. If the stone begins its fall from rest, 0 m/s, then at the end of the first second of fall its speed is 10 m/s. At the end of 2 seconds, its speed is 20 m/s—and so on. It keeps gaining 10 m/s each second it falls.

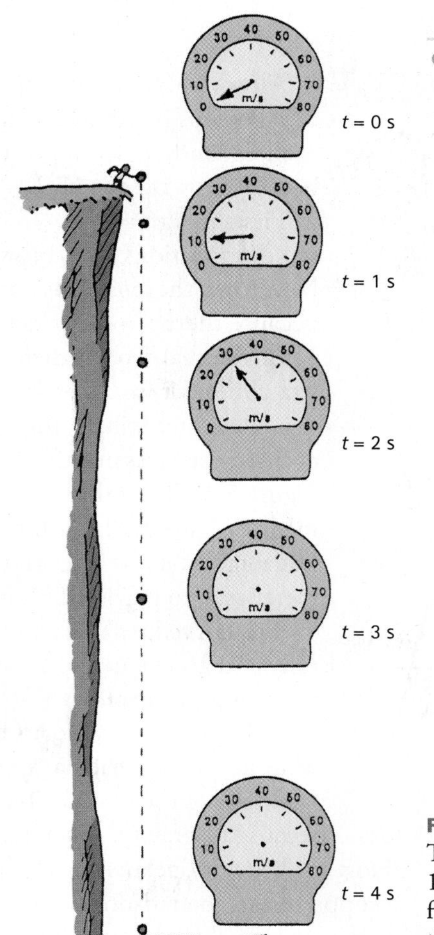

FIGURE 5.21
The falling stone gains a speed of 10 m/s each second. How would you fill in the speedometer readings for the times 3 seconds and 4 seconds?

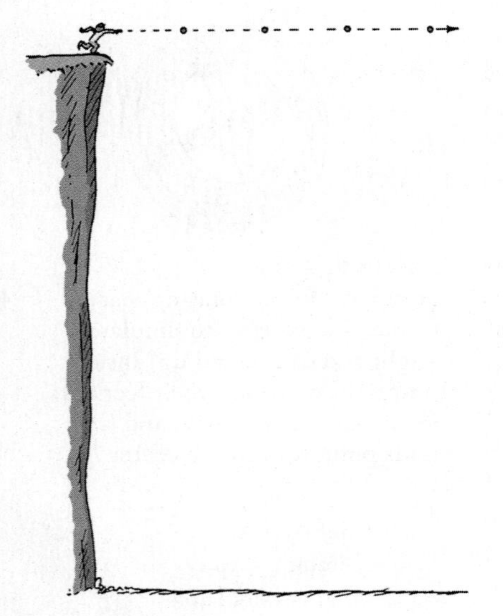

FIGURE 5.22
If there were no gravity, a stone thrown horizontally would move in a straight-line path and cover equal distances in equal time intervals.

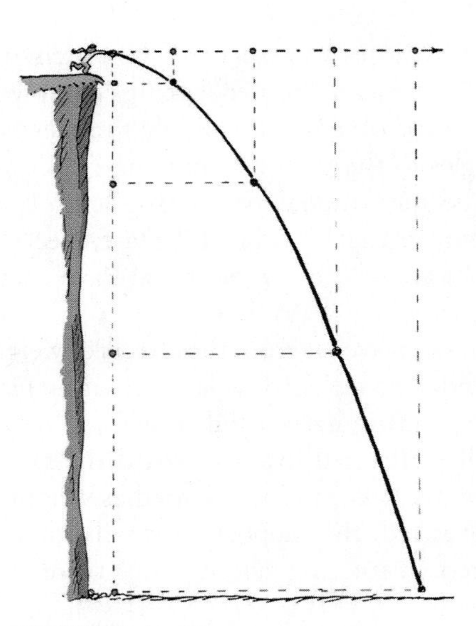

FIGURE 5.23
The vertical dashed line is the path of a stone dropped from rest. The horizontal dashed line would be its path if there were no gravity. The curved solid line shows the resulting trajectory that combines horizontal and vertical motions.

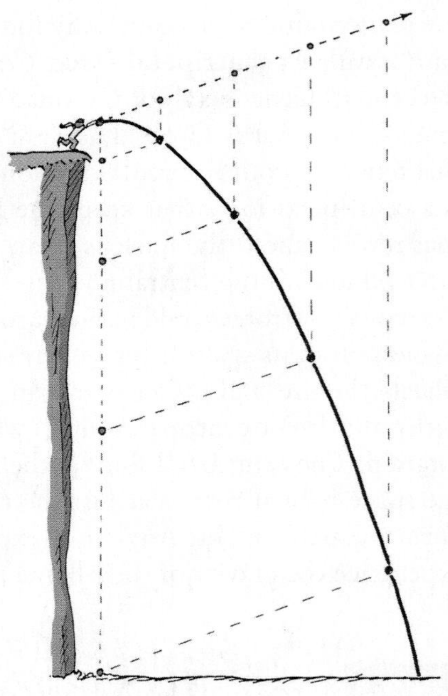

FIGURE 5.24
A stone thrown at an upward angle would follow the dashed line in the absence of gravity. Because of gravity, however, it falls beneath this line and describes the parabola shown by the solid curve.

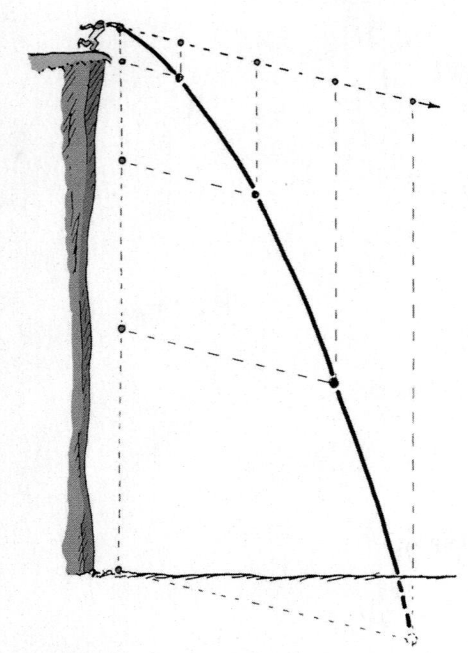

FIGURE 5.25
A stone thrown at a downward angle follows a somewhat different parabola.

Although the change in speed is the same each second, the distance of fall keeps increasing. That's because the average speed of fall increases each second. Let's apply this to a new situation—throwing the stone horizontally off the cliff.

First, imagine that gravity doesn't act on the stone. In Figure 5.22 we see the positions the thrown stone would have if there were no gravity. Note that the positions each second are the same distance apart. That's because there's no force acting on the stone.

In the real world, there is gravity. The thrown stone falls beneath the straight line it would follow with no gravity (Figure 5.23). The stone curves as it falls. Interestingly, this familiar curve is the result of two kinds of motion occurring at the same time. One kind is the straight-down vertical motion of Figure 5.21. The other is the horizontal motion of constant velocity, as imagined in Figure 5.22. Both motions occur simultaneously. As the stone moves horizontally, it also falls straight downward—beneath the place it would be if there were no gravity. This is indicated in Figure 5.23.

The curved path of a projectile is the result of constant motion horizontally and accelerated motion vertically under the influence of gravity. This curve is a **parabola**.

In Figure 5.24, we consider a stone thrown upward at an angle. If there were no gravity, the path would be along the dashed line with the arrow. Positions of the stone at 1-second intervals along the line are shown by red dots. Because of gravity, the actual positions (dark dots) are below these points. How far below? The answer is, the same distance an object would fall if it were dropped from the red-dot positions. When we connect the dark dots to plot the path, we get a different parabola.

In Figure 5.25, we consider a stone thrown at a downward angle. The physics is the same. If there were no gravity, it would follow the dashed line with the arrow. But, because of gravity, it falls beneath this line, just as in the preceding cases. The path is a somewhat different parabola.

The curved path of a projectile is a combination of horizontal and vertical motions. Consider the girl throwing the stone in Figure 5.26. The velocity she gives the stone is shown by the light blue vector. Notice that this vector has horizontal and vertical components. These components, interestingly, are completely independent of each other. Each of them acts as if the other didn't exist. Combined, they produce the resultant velocity vector.

A typical projectile path would have the velocity vectors and their components as shown in Figure 5.27. Notice that the horizontal component remains the same at all points. That's because no horizontal force exists to change this component of velocity (assuming negligible air resistance). The vertical component changes because of the vertical influence of gravity.

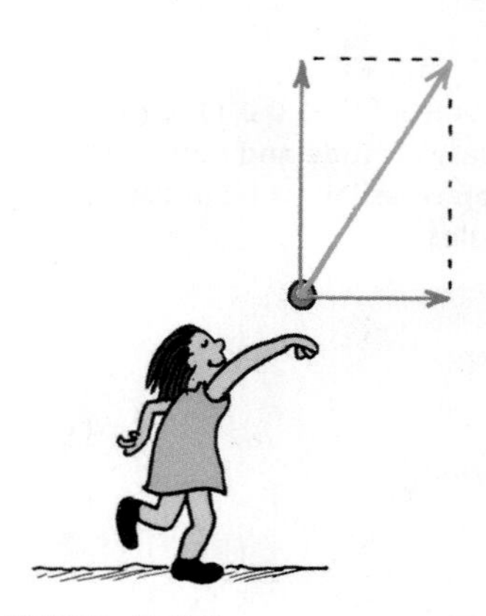

FIGURE 5.26
The velocity of the ball (light blue vector) has vertical and horizontal components. The vertical component relates to how high the ball will go. The horizontal component relates to the horizontal range of the ball.

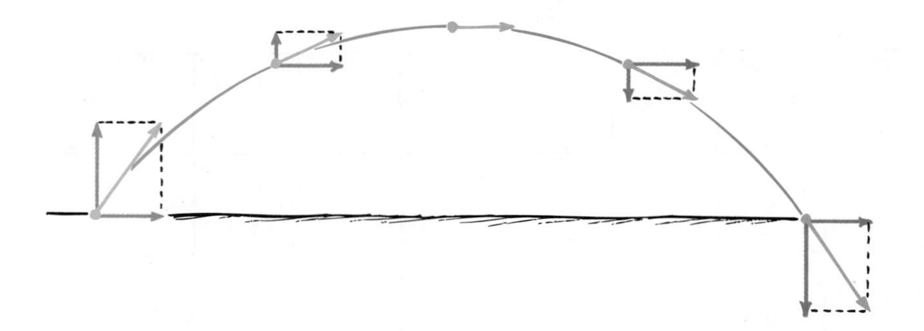

FIGURE 5.27

The velocity of a projectile at various points along its trajectory. Note that the vertical component changes while the horizontal component is the same everywhere.

CHECK YOURSELF

1. **At what part of its trajectory does a projectile have its minimum speed?**
2. **A tossed ball changes speed along its parabolic path. When the Sun is directly overhead, does the shadow of the ball across the field also change speed?**

CHECK YOUR ANSWERS

1. The speed of a projectile is at a minimum at the top of its path. If it is launched vertically, its speed at the top is zero. If it is projected at an angle, the vertical component of speed is zero at the top, leaving only the horizontal component. So, the speed at the top is equal to the horizontal component at any point.
2. No; the shadow moves at constant velocity across the field, showing exactly the motion due to the horizontal component of the ball's velocity.

MasteringPhysics®
TUTORIAL: Projectile Motion
VIDEO: Projectile Motion Demonstration

PRACTICING PHYSICS

Hands-On Dangling Beads

Make your own model of projectile paths. Divide a ruler or a stick into five equal spaces. At position 1, hang a bead from a string that is 1 cm long, as shown. At position 2, hang a bead from a string that is 4 cm long. At position 3, do the same with a 9-cm length of string. At position 4, use 16 cm of string, and for position 5, use 25 cm of string. If you hold the stick horizontally, you will have a version of Figure 5.23. Hold it at a slight upward angle to show a version of Figure 5.24. Hold it at a downward angle to show a version of Figure 5.25.

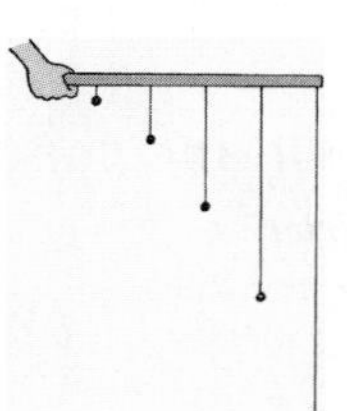

LEARNING OBJECTIVE
Relate altitude and range to projectiles launched at various angles.

5.9 Projectile Altitude and Range

EXPLAIN THIS How does the altitude and range compare for a pair of projectiles launched at 30° and 60° above the horizontal?

In Figure 5.28, we see the paths of several projectiles in the absence of air resistance. All of them have the same initial speed but different projection angles. Notice that these projectiles reach different *altitudes*, or heights above the ground. They also have different *ranges*, or distances traveled horizontally. The remarkable thing to note is that the same range is obtained from two different projection angles—a pair whose projection angles add up to 90°! An object thrown into the air at an angle of 60°, for example, will have the same range as if it were thrown at the same speed at an angle of 30°. For the smaller angle, of course, the object remains in the air for a shorter time.

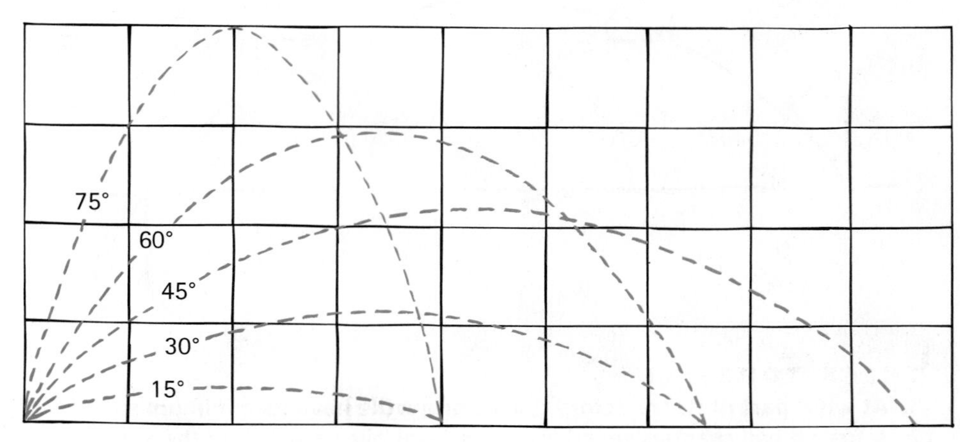

FIGURE 5.28
The paths of projectiles launched with equal speeds but different projection angles. Note that the same range occurs for pairs whose projection angles add up to 90°.

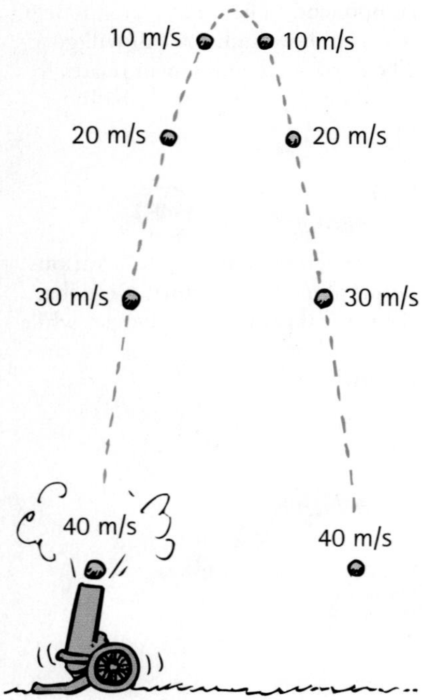

FIGURE 5.29
Without air drag, the speed lost while going up equals the speed gained while going down.

When air drag is low enough to be negligible, a projectile will rise to its maximum height in the same time it takes to fall from that height to the ground. This is because the speed it loses while going up is the same speed it had when it was projected from the ground (Figure 5.29).

LEARNING OBJECTIVE
Explain the effects of air drag on projectile motion.

5.10 The Effect of Air Drag on Projectiles

EXPLAIN THIS How does air drag affect altitude and range for projectiles?

UNIFYING CONCEPT
- ***Friction***
Section 2.8

We have considered projectile motion without air drag. You can neglect air drag for a ball you toss back and forth with your friends because the speed is low. But, at higher speeds, air resistance matters. Air drag is an important factor for high-speed projectiles. The result of air drag is that both range and altitude are decreased (Figure 5.30).

Consider baseball: Air drag greatly affects the range of balls batted and thrown in baseball games. Without air drag, a ball normally batted to the middle of center field would be a home run. If baseball were played on the Moon (not scheduled in the near future!), the range of balls would be considerably farther—about six times the ideal range on Earth. This is because there is no atmosphere on the Moon, so air drag on the Moon is completely absent. In addition, gravity is one-sixth as strong on the Moon, so the paths are higher and longer.

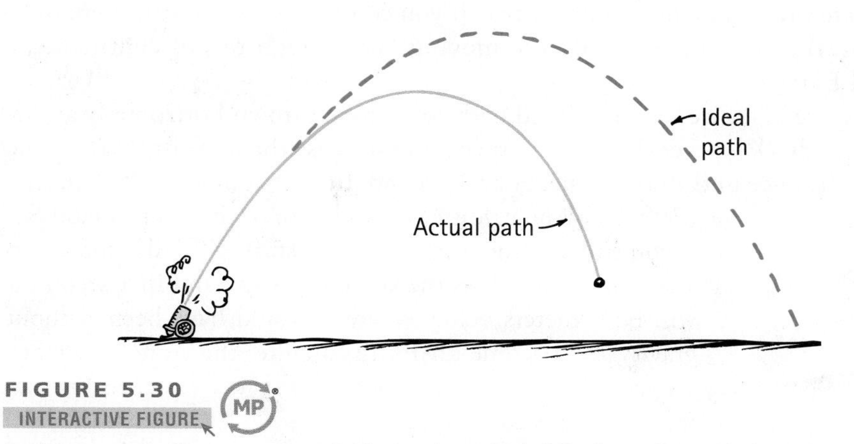

FIGURE 5.30
INTERACTIVE FIGURE MP

In the presence of air resistance, a high-speed projectile falls short of a parabolic path. The dashed line shows an ideal path with no air resistance; the solid line indicates an actual path.

CHECK YOURSELF

If the boy in Figure 5.31 simply drops a baseball a vertical distance of 5 m, it will hit the ground in 1 s. Suppose instead that he throws the ball horizontally as shown. The ball lands 20 m downrange. What is his pitching speed?

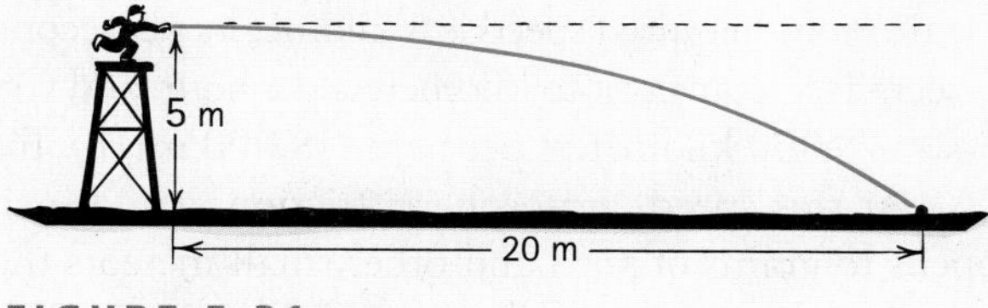

FIGURE 5.31
How fast is the ball thrown?

CHECK YOUR ANSWER

The ball is thrown horizontally, so the pitching speed is the horizontal distance divided by the time. A horizontal distance of 20 m is given. How about the time? Isn't the time along the parabola the same time it takes to fall vertically 5 m? Isn't this time 1 s? So the pitching speed $v = \frac{d}{t} = \frac{(20 \text{ m})}{(1 \text{ s})} = 20 \text{ m/s}$.

Back here on Earth, baseball games normally take place on level ground. Baseballs curve over a flat playing field. The speeds of baseballs are not great enough for Earth's curvature to affect the ball's path. For very long-range projectiles, however, the curvature of Earth's surface must be taken into account. As we will now see, when an object is projected fast enough, it can fall all the way around Earth and become a *satellite*.

MasteringPhysics®
VIDEO: More Projectile Motion

LEARNING OBJECTIVE
Relate a projectile trajectory that matches Earth's curvature to satellite motion.

5.11 Fast-Moving Projectiles—Satellites

EXPLAIN THIS What does Earth's curvature have to do with Earth satellites?

Suppose a cannon fires a cannonball so fast that its curved path matches the curvature of Earth. Then, without air drag, it would be an Earth satellite! The same would be true if you could throw a stone fast enough. Any **satellite** is simply a projectile moving fast enough to fall continuously around Earth.

In Figure 5.32, we see the curved paths of a stone thrown horizontally at different speeds. Whatever the pitching speed, in each case the stone drops the same vertical distance in the same time. For a 1-second drop, that distance is 5 meters. (Perhaps by now you have made use of this fact in lab.) So, if you simply drop a stone from rest, it will fall 5 meters in 1 second of fall. Toss the stone sideways, and in 1 second it will be 5 meters below where it would have been without gravity. To become an Earth satellite, the stone's horizontal velocity must be great enough for its falling distance to match Earth's curvature.

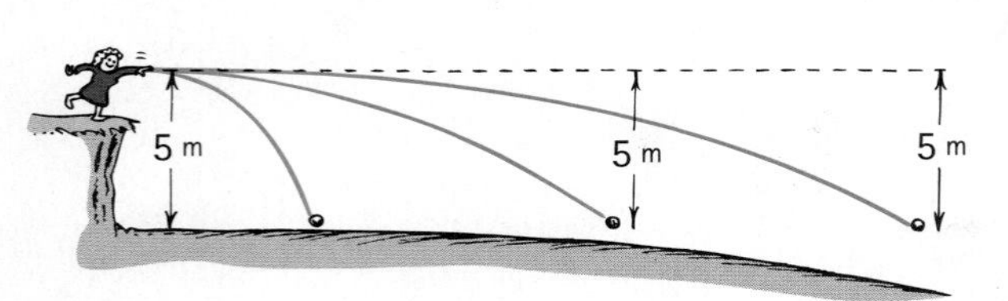

FIGURE 5.32
If you throw a stone at any speed, 1 second later it will have fallen 5 meters below where it would have been if there were no gravity.

It is a geometric fact that the surface of Earth drops a vertical distance of 5 meters for every 8000 meters tangent to the surface (Figure 5.33). A tangent to a circle or to Earth's surface is a straight line that touches the circle or surface at only one place. With this amount of Earth's curvature, if you were floating in a calm ocean, you would be able to see only the top of a 5-meter mast on a ship 8000 meters (8 kilometers) away. We live on a round Earth.

What do we call a projectile that moves fast enough to travel a horizontal distance of 8 kilometers during 1 second? We call it a satellite. With no air drag, it would follow the curvature of Earth. A little thought tells you that the minimum required speed is 8 kilometers per second. If this doesn't seem fast, convert it to kilometers per hour, and you get an impressive 29,000 kilometers per hour (18,000 mi/h). Fast, indeed!

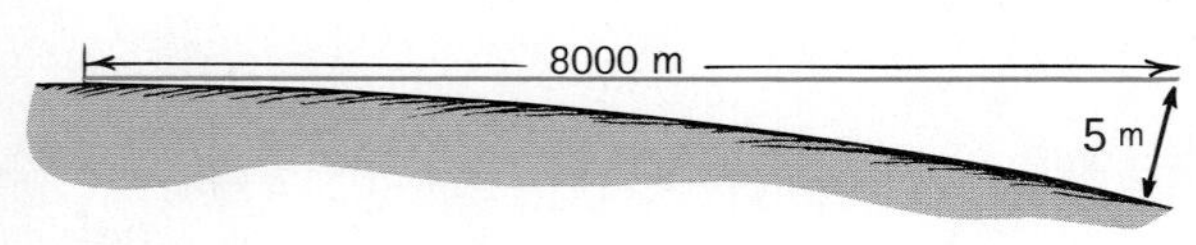

FIGURE 5.33
Earth's curvature drops a vertical distance of 5 m for each 8000-m tangent (not to scale).

At this speed, atmospheric friction would incinerate the projectile. This happens to grains of sand and other small meteors that graze Earth's atmosphere, burn up, and appear as "falling stars." That is why Earth satellites are launched to altitudes higher than 150 kilometers—to be above the atmosphere.

It is a common misconception that satellites orbiting at high altitudes are free of gravity. Nothing could be further from the truth. The force of gravity on a satellite 150 kilometers above Earth's surface is nearly as great as it would be at the surface. If there were no gravity, the motion of the satellite would be along a straight-line path instead of curving around Earth. High altitude puts the satellite beyond Earth's *atmosphere*, but not beyond its *gravity*. As you know, Earth's gravity goes on forever, getting weaker and weaker with distance, but never reaching zero.

FIGURE 5.34
If the speed of the object and the curvature of its trajectory are great enough, the stone may become a satellite.

Isaac Newton understood satellite motion. He reasoned that the Moon is simply a projectile circling Earth under gravitational attraction. This concept is illustrated in Figure 5.35, which is an actual drawing by Newton. He compared the Moon's motion to a cannonball fired from atop a high mountain. He imagined that the mountaintop was above Earth's atmosphere, so that air drag would not slow the motion of the cannonball. If a cannonball were fired with a low horizontal speed, it would follow a curved path and soon hit Earth below. If it were fired faster, its path would be wider and it would hit a place on Earth farther away. If the cannonball were fired fast enough, Newton reasoned, the curved path would circle Earth indefinitely. The cannonball would be in orbit.

Newton calculated the speed for a circular orbit around Earth. However, since such a cannon-muzzle velocity was clearly impossible, he did not foresee humans launching satellites (and it is quite likely that he didn't foresee multistage rockets).

Both the cannonball and the Moon continuously circle with a tangential velocity parallel to Earth's surface. This velocity is enough to ensure motion around Earth rather than into it. Without air drag to reduce speed, the Moon or any Earth satellite "falls" around and around Earth indefinitely. Similarly, the planets continuously fall around the Sun in closed paths.

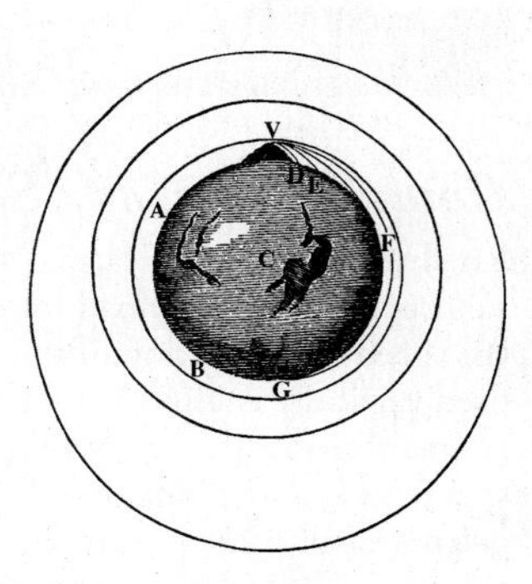

FIGURE 5.35
Newton's mountain: "The greater the velocity . . . with which [an object] is projected, the farther it goes before it falls to the Earth. We may therefore suppose the velocity to be so increased that it would describe an arc of 1, 2, 5, 10, 100, 1000 miles before it arrived at the Earth, till at last, exceeding the limits of the Earth, it should pass into space without touching." —Isaac Newton, *System of the World*

CHECK YOURSELF

Can we say that a satellite stays in orbit because it's above the pull of Earth's gravity?

CHECK YOUR ANSWER

No, no, no! No satellite is "above" Earth's gravity. If the satellite were not in the grip of Earth's gravity, it would not orbit but would follow a straight-line path instead.

Why don't the planets crash into the Sun? Because of their tangential velocities. What would happen if their tangential velocities were reduced to zero? The answer is simple enough: Their motion would be straight toward the Sun, and they would indeed crash into it. Any objects in the solar system without sufficient tangential velocities have long ago crashed into the Sun. What remains is the harmony we observe.

5.12 Elliptical Orbits

LEARNING OBJECTIVE
Describe the role of speeds in circular and elliptical orbits, and the role of speeds greater than 11.2 km/s.

EXPLAIN THIS What is the fate of a projectile launched above the atmosphere with a speed greater than 8 km/s (but less than 11.2 km/s)? Greater than 11.2 km/s?

If a projectile just above the resistance of the atmosphere is given a horizontal speed somewhat greater than 8 kilometers per second, it will overshoot a circular path and trace an oval path called an **ellipse**.

An ellipse is a specific curve—the closed path taken by a point that moves in such a way that the sum of its distances from two fixed points (called *foci*, the plural of *focus*) is constant. For a satellite orbiting a planet, one focus is at the center of the planet; the other focus could be inside or outside of the planet. An ellipse can be constructed easily by using a pair of tacks (one at each focus), a loop of string, and a pen; see Figure 5.36. The nearer the foci are to each other, the closer the ellipse is to a circle. When both foci are together, the ellipse *is* a circle. So we see that a circle is a special case of an ellipse.

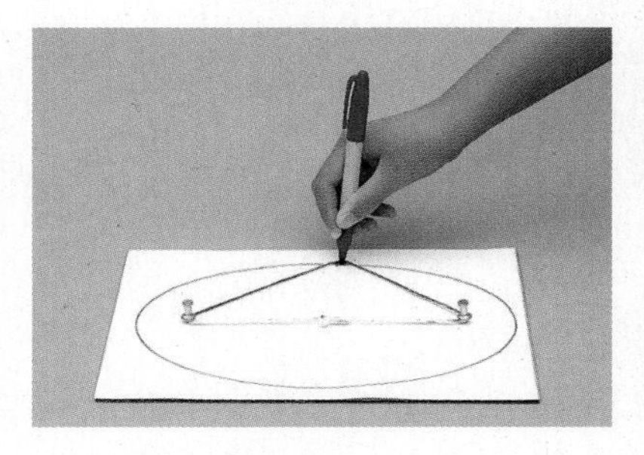

FIGURE 5.36
A simple method for constructing an ellipse.

Whereas the speed of a satellite is constant in a circular orbit, the speed of a satellite varies in an elliptical orbit. When the initial speed is greater than 8 km/s, the satellite overshoots a circular path and moves away from Earth, against the

TECHNOLOGY

Communications Satellites

Satellites are payloads carried above the atmosphere by rockets. Putting a payload into orbit requires control over the speed and direction of the rocket. A rocket initially fired vertically is intentionally tipped from the vertical course as it rises. Then, once above the drag of the atmosphere, it is aimed horizontally, and then the payload is given a final thrust to orbital speed.

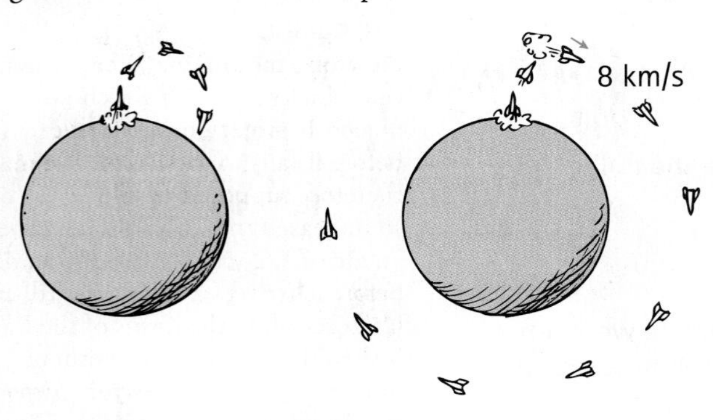

For a satellite orbiting close to Earth, the period (the time for a complete orbit about Earth) is about 90 minutes. For satellites orbiting at higher altitudes, gravitation is less and the orbital speed is lower, so the period is longer. For example, communications satellites orbiting at an altitude of 5.5 Earth radii have a period of 24 hours. This period matches the period of daily Earth rotation. A satellite in orbit around the equator stays above the same point on the ground; that is, it is in *geosynchronous* orbit.

Satellite television employs communications satellites. Satellite TV is much like traditional broadcast television, but it has a larger range. Both systems use electromagnetic signals (radio waves) to send programming to your home. Broadcast stations transmit the waves from powerful land-based antennas, and viewers pick up the signals with smaller antennas.

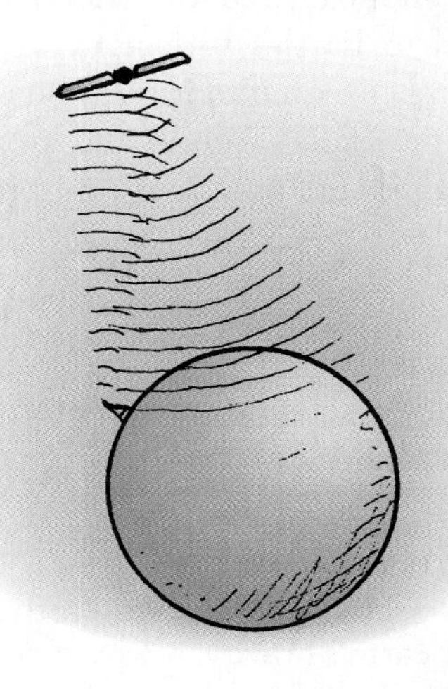

The problem with this technology, especially in pre-satellite times, is that radio signals travel away from a broadcast antenna in a straight line. To receive the signals, you need to be in the direct "line of sight" of the antenna. Earth's curvature interrupts the line of sight, so the broadcast signals can be sent over only a short distance.

Satellite TV solves the problem by transmitting the signals from satellites in orbit high above Earth. This way, Earth's curvature doesn't interrupt the line of sight. Since the satellite is in geosynchronous orbit, the relative positions of the satellite and receiving dish are fixed. You don't need to readjust your dish—just grab the remote.

GPS systems rely on multiple satellites in lower orbits.

MasteringPhysics®
VIDEO: Circular Orbits
TUTORIAL: Orbits and Kepler's Laws

force of gravity. It therefore loses speed. The speed it loses in receding is gained as it falls back toward Earth, and it finally rejoins its path with the same speed it had initially (Figure 5.37). The procedure repeats over and over, and the process repeats each cycle.

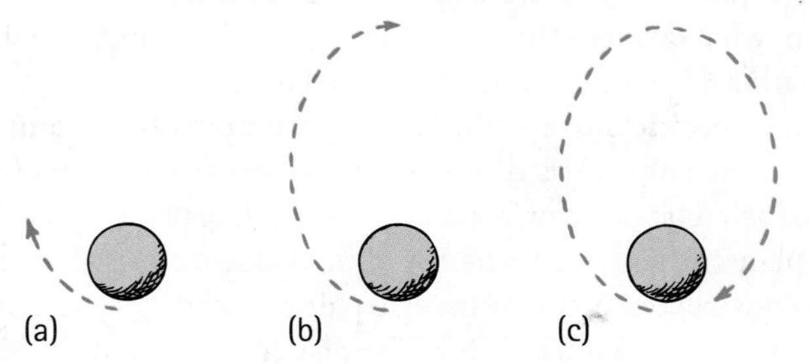

FIGURE 5.37
Elliptical orbit. (a) An Earth satellite that has a speed somewhat greater than 8 km/s overshoots a circular orbit and travels away from Earth. (b) Gravitation slows it to a point at which it no longer moves farther away from Earth. (c) The satellite falls toward Earth, gaining the speed that it lost in receding, and it follows the same path as before in a repetitious cycle.

CHECK YOURSELF

The orbital path of a satellite is shown in the sketch. In which of the marked positions, A through D, does the satellite have the highest speed? In which position does it have the lowest speed?

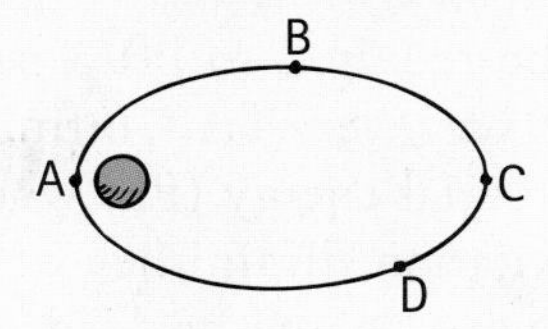

CHECK YOUR ANSWERS

The satellite has its highest speed as it whips around A, and it has its lowest speed at C. After passing C, the satellite gains speed as it falls back to A to repeat the cycle.

Integrated Science 5C

ASTRONOMY

Escape Speed

LEARNING OBJECTIVE
Relate orbital speed and escape speed.

EXPLAIN THIS What is escape speed from the surface of Earth?

We know that a cannonball fired horizontally at 8 kilometers per second from Newton's mountain will go into orbit. But what would happen if the cannonball were instead fired at the same speed vertically? It would rise to some maximum height, reverse direction, and then fall back to Earth. Then the old saying "What goes up must come down" would hold true, just as surely as a stone tossed skyward will be returned by gravity (unless, as we shall see, its speed is great enough.)

In this age of space travel, it is more accurate to say "What goes up *may* come down" because there is a critical starting speed that allows a projectile to outrun gravity and to escape the Earth. This critical speed is called the **escape speed** or,

TABLE 5.1 ESCAPE SPEEDS FROM THE SURFACES OF BODIES IN THE SOLAR SYSTEM

Astronomical Body	Mass (in Earth masses)	Radius (in Earth radii)	Escape Speed (km/s)
Sun	333,000	109	620
Sun (at a distance of Earth's orbit)		23,500	42.2
Jupiter	318.0	11.0	60.2
Saturn	95.2	9.2	36.0
Neptune	17.3	3.47	24.9
Uranus	14.5	3.7	22.3
Earth	1.00	1.00	11.2
Venus	0.82	0.95	10.4
Mars	0.11	0.53	5.0
Mercury	0.055	0.38	4.3
Moon	0.0123	0.27	2.4

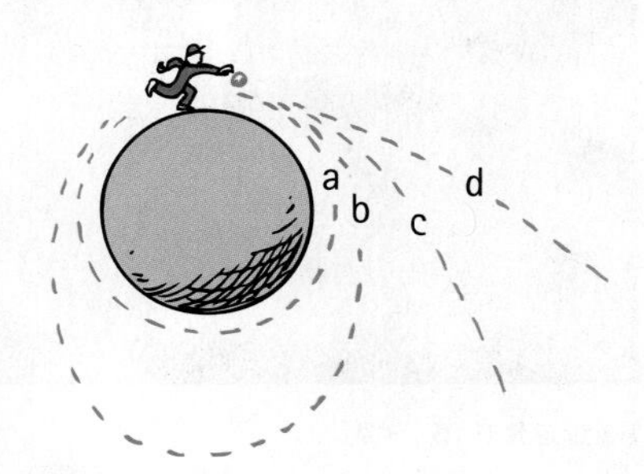

FIGURE 5.38 

INTERACTIVE FIGURE

If Superman were to toss a ball 8 km/s horizontally from the top of a mountain high enough to be just above air resistance (a), then, about 90 minutes later, he could turn around and catch it (neglecting Earth's rotation). Tossed slightly faster (b), the ball would take an elliptical orbit and return in a slightly longer time. Tossed faster than 11.2 km/s (c), it would escape Earth's gravitational pull. Tossed faster than 42.5 km/s (d), it would escape from the solar system.

If you were to drop a candy bar to Earth from a distance as far away from Earth as Earth is from Pluto, its speed of impact would be about 11.2 km/s.

if direction is involved, the *escape velocity*. From the surface of Earth, the escape speed is 11.2 kilometers per second. Launch a projectile at any speed greater than that and it will leave Earth, traveling slower and slower, but never stopping due to Earth's gravity.* We can understand the magnitude of this speed from an energy point of view.

How much work would be required to lift a payload against the force of Earth's gravity to a distance very, very far ("infinitely far") away? We might think that the change of potential energy (PE) would be infinite because the distance is infinite. But gravitation diminishes with distance by the inverse-square law. The force of gravity on the payload would be strong only near Earth. It turns out that the change of PE of a 1-kilogram body moved from the surface of Earth to an infinite distance is 62 million joules, or 62 megajoules (62 MJ). So, to put a payload infinitely far from Earth's surface requires at least 62 megajoules of energy per kilogram of load. We won't go through the calculation here, but 62 megajoules per kilogram corresponds to a speed of 11.2 kilometers per second, whatever the total mass involved. This is the escape speed from the surface of Earth.**

If we give the payload any more energy than 62 megajoules per kilogram at the surface of Earth—or, equivalently, any speed higher than 11.2 kilometers per second—then, neglecting air resistance, the payload will escape from Earth, never to return. As it continues outward, its PE increases and its kinetic energy (KE) decreases. Its speed becomes lower and lower, although it is never reduced to zero. The payload outruns Earth's gravity. It escapes.

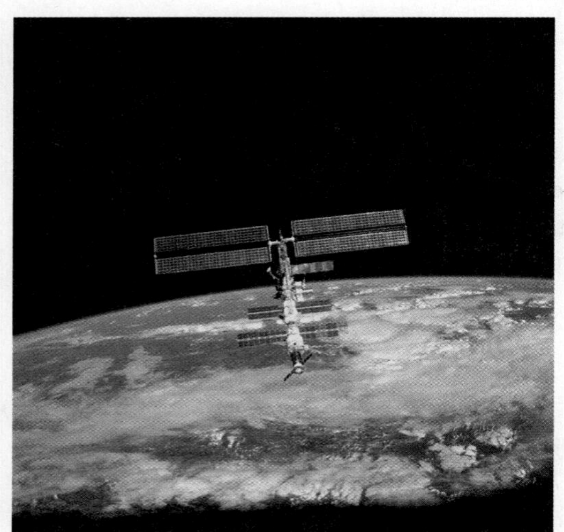

FIGURE 5.39
The International Space Station (ISS) is a projectile in a continuous state of free fall. It follows an elliptical orbit and falls around Earth rather than into it.

The escape speeds from various bodies in the solar system are listed in Table 5.1. Note that the escape speed from the surface of the Sun is 620 km/s. Even at a 150,000,000-km distance from the Sun (Earth's distance), the escape speed needed to break free of the Sun's influence is 42.5 km/s, which is considerably greater than the escape speed of Earth. An object projected from Earth at a speed greater than 11.2 km/s but less than 42.5 km/s will escape Earth, but it will not escape the Sun. Rather than receding forever, it will occupy an orbit around the Sun.

The first probe to escape the solar system, *Pioneer 10*, was launched from Earth in 1972 with a speed of only 15 km/s. The escape was accomplished by directing the probe into the path of oncoming Jupiter. The probe was whipped about by Jupiter's strong gravitation, gaining speed in the process—similar to the increase in speed of a baseball encountering an oncoming bat. Its speed of departure from Jupiter was increased enough to exceed the escape speed from the Sun at the distance of Jupiter. *Pioneer 10* passed the orbit of Pluto in 1984. Unless it collides with another body, it will wander indefinitely through interstellar space. Like a bottle cast into the sea with a note inside, *Pioneer 10* contains information about Earth that might be of interest to extraterrestrials, information put there in hopes that it will one day wash up and be found on some distant "seashore" (Figure 5.40). Since we've recently learned that there are more planets than stars in our galaxy, that seashore seems more plausible than when *Pioneer 10* was launched.

*The escape speed from any planet or any body is given by $v = \frac{2GM}{d}$, where G is the universal gravitational constant, M is the mass of the attracting body, and d is the distance from its center. (At the surface of the body, d would be simply the radius of the body.) Compare this formula with the one for orbital speed $v = \sqrt{\frac{GM}{d}}$.

**Interestingly enough, this might be called the maximum falling speed. Any object, however far from Earth, if it is released from rest and allowed to fall to Earth only under the influence of Earth's gravity, would not exceed 11.2 km/s. (With air resistance, it would be somewhat slower.)

It is important to point out that the escape speed of a body is the initial speed given by an initial thrust, after which there is no force to assist motion. One could escape Earth at any sustained speed greater than zero, given enough time. For example, suppose a rocket is launched to a destination such as the Moon. If the rocket engines burn out when the rocket is still close to Earth, the rocket needs a minimum speed of 11.2 kilometers per second. But, if the rocket engines can be sustained for long periods of time, the rocket could go to the Moon without ever attaining 11.2 kilometers per second.

It is interesting to note that the accuracy with which an unmanned rocket reaches its destination is not determined by its remaining on a preplanned path or by its getting back on that path if it strays off course. No attempt is made to return the rocket to its original path. Instead, the control center asks, in effect, Where is it now, and what is its velocity? What is the best way to reach its destination, given its present situation? With the aid of high-speed computers, the answers to these questions are used to find a new path. Corrective thrusters put the rocket on this new path. This process is repeated over and over again all the way to the goal.*

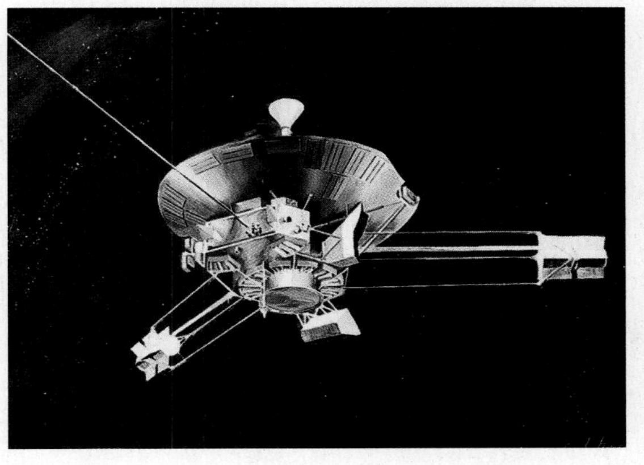

FIGURE 5.40
Pioneer 10, launched from Earth in 1972, left the solar system in 1984 and is now wandering in our galaxy.

CHECK YOURSELF

1. **What is the minimum speed a molecule in Earth's upper atmosphere must have if it is to escape Earth's gravity and wander into space?**
2. **The ISS orbits at an average distance of 370 km above Earth's surface. What evidence do you have that its orbital speed is less than 11.2 km/s?**

CHECK YOUR ANSWERS

1. The minimum speed is 11.2 km/s.
2. The fact that the ISS remains in Earth orbit is startling evidence that its speed is less than escape speed. Seeing is believing, for the ISS can be seen with the naked eye.

Just as planets fall around the Sun, stars fall around the centers of galaxies. Those with insufficient tangential speeds are pulled into, and are gobbled up by, the galactic nucleus, which is usually a black hole.

*Is there a personal lesson to be learned here? Suppose you find that you are off course. You may, like the rocket, find it more fruitful to take a course that will lead you to your goal, as best plotted from your present position and circumstances, rather than to try to get back on the course you plotted from the earlier position and, perhaps, under different circumstances.

For instructor-assigned homework, go to www.masteringphysics.com

SUMMARY OF TERMS (KNOWLEDGE)

Centripetal force Any force that is directed at right angles to the path of a moving object and that causes the object to follow a circular path.

Ellipse A closed curve of oval shape for which the sum of the distances from any point on the curve to two internal focal points is a constant.

Escape speed The speed that a projectile, space probe, or similar object must reach in order to escape the gravitational influence of Earth or of another celestial body to which it is attracted.

Gravitational force The attractive force between objects due to mass.

Inverse-square law A law relating the intensity of an effect to the inverse square of the distance from the cause:

$$\text{Intensity} \sim \frac{1}{\text{distance}^2}$$

Law of universal gravitation Every body in the universe attracts every other body with a mutually attracting force. For two bodies, this force is directly proportional to the product of their masses and inversely proportional to the square of the distance separating them: $F = G\dfrac{m_1 m_2}{d^2}$.

Parabola The curved path followed by a projectile near Earth under the influence of gravity only.

Projectile Any object that moves through the air or through space under the influence of gravity.

Satellite A projectile or small body that orbits a larger body.

Tangential velocity Velocity that is parallel (tangent) to a curved path.

Universal constant of gravitation, *G* The proportionality constant in Newton's law of gravitation.

Weightlessness A condition encountered in free fall in which a support force is lacking.

READING CHECK QUESTIONS (COMPREHENSION)

5.1 The Legend of the Falling Apple

1. What connection did Newton make between a falling apple and the Moon?

5.2 The Fact of the Falling Moon

2. What does it mean to say that something moving in a curved path has a tangential velocity?
3. In what sense does the Moon "fall"?

5.3 Newton's Law of Universal Gravitation

4. State Newton's law of gravitation in words. Then state the law in one equation.

5.4 Gravity and Distance: The Inverse-Square Law

5. How does the force of gravity between two bodies change when the distance between them is doubled?
6. How does the brightness of light on a surface change when a point source of light is brought twice as far away?
7. At what distance from Earth is the gravitational force on an object zero?

5.5 The Universal Gravitational Constant, *G*

8. What is the magnitude of gravitational force between two 1-kg bodies that are 1 m apart?
9. What is the magnitude of the gravitational force between Earth and a 1-kg body?

5.6 Weight and Weightlessness

10. Would the springs inside a bathroom scale be more compressed or less compressed if you weighed yourself in an elevator that was accelerating upward? That was accelerating downward?
11. Answer the preceding questions for the case of an elevator moving upward and then downward at *constant velocity*.
12. When is your weight equal to mg?
13. When is your weight greater than mg?
14. When is your weight zero?

5.7 Centripetal Force Can Simulate Gravity

15. When you whirl a can at the end of a string in a circular path, what is the direction of the force that acts on the can?
16. How can weight be simulated in a space habitat?

5.8 Projectile Motion

17. What exactly is a projectile?
18. How much speed does a freely falling object gain during each second of fall?
19. With no gravity, a horizontally moving projectile follows a straight-line path. With gravity, how far below the straight-line path does it fall compared with the distance of free fall?
20. A ball is batted upward at an angle. What happens to the vertical component of its velocity as it rises? With no air drag, what happens to the horizontal component for the ball?

5.9 Projectile Altitude and Range

21. A projectile is launched at an angle of 75° above the horizontal and strikes the ground downrange. For what other angle of launch at the same speed would this projectile land at the same distance?
22. A projectile is launched vertically at 30 m/s. If air drag can be neglected, at what speed will it return to its initial level?

5.10 The Effect of Air Drag on Projectiles

23. What is the effect of air drag on the height and range of batted baseballs?

5.11 Fast-Moving Projectiles—Satellites

24. Why will a projectile that moves horizontally at 8 km/s follow a curve that matches the curvature of Earth?
25. Why is it important that the projectile in the preceding question be above Earth's atmosphere?
26. Is it correct to say that the planets of the solar system are simply projectiles falling around the Sun?

5.12 Elliptical Orbits

27. Why does the force of gravity on a satellite moving in an elliptical orbit vary?
28. Why does the speed of a satellite moving in an elliptical orbit vary?

THINK INTEGRATED SCIENCE

5A—Your Biological Gravity Detector

29. In what location are the sense organs in humans that allow them to sense gravity?
30. Speculate on how the vestibular system might be involved in "space sickness"—the feeling of nausea and disorientation that astronauts experience.

5B—Center of Gravity of People

31. Why doesn't the Leaning Tower of Pisa topple?
32. Why does spreading feet apart help a surfer stay on the board?

5C—Escape Speed

33. What is the minimum speed for orbiting Earth in a close orbit? What is the maximum speed? What happens above this speed?
34. How was *Pioneer 10* able to escape the solar system with an initial speed less than escape speed?

THINK AND DO (HANDS-ON APPLICATION)

35. Hold your hand outstretched with one hand twice as far from your eyes as the other. Make a casual judgment about which hand looks bigger. Most people see the hands as about the same size, and many see the nearer hand as slightly bigger. Very few people see the hand as four times as big, but by the inverse-square law, the nearer hand should appear twice as tall and twice as wide. Twice times twice means four times as big. That's four times as much of your visual field as is occupied by the more distant hand. It is likely that your belief that your hands are the same size is so strong that your brain overrules this information. Try it again, only this time overlap your hands slightly and view them with one eye closed. Aha! Do you now see more clearly that the nearer hand is bigger? This raises an interesting question: What other illusions do you experience that can't be checked so easily?
36. Repeat the eye-balling experiment in Exercise 35, only this time use two dollar bills—one regular and the other folded in half lengthwise and then widthwise so it has $\frac{1}{4}$ the area. Now hold the two bills in front of your eyes. Where do you hold the folded bill so that it appears to be the same size as the unfolded one? Share this experiment with a friend.
37. Ask a friend to stand facing a wall with her toes against the wall. Then ask her to stand on the balls of her feet without toppling backward. Your friend won't be able to do it. Explain to her why it can't be done.
38. With stick and strings, make a "trajectory stick" as shown on page 107.
39. With your friends, whirl a bucket of water in a vertical circle fast enough so the water doesn't spill out. As it happens, the water in the bucket is falling, but with less speed than you give to the bucket. Tell your friends how your bucket swing is like satellite motion—that satellites in orbit continuously fall toward Earth, but not with enough vertical speed to get closer to the curved Earth below. Remind your friends that physics is about finding the connections in nature!

PLUG AND CHUG (FORMULA FAMILIARIZATION AND CALCULATOR EXERCISE)

$$F = G\frac{m_1 m_2}{d^2}$$

40. Using the formula for gravity, show that the force of gravity on a 1-kg mass at Earth's surface is nearly 10 N. You need to know that the mass of Earth is 6×10^{24} kg and its radius is 6.4×10^6 m.
41. Calculate the force of gravity on the same 1-kg mass if it were 6.4×10^6 m above Earth's surface (that is, if it were two Earth radii from Earth's center).
42. Calculate the force of gravity between Earth (mass = 6.0×10^{24} kg) and the Moon (mass = 7.4×10^{22} kg). The average Earth–Moon distance is 3.8×10^8 m.
43. Calculate the force of gravity between Earth and the Sun (Sun's mass = 2.0×10^{30} kg; average Earth–Sun distance = 1.5×10^{11} m).

THINK AND COMPARE (ANALYSIS)

44. The planet and its moon gravitationally attract each other. Rank the forces of attraction between the pairs, A to D, from greatest to least.

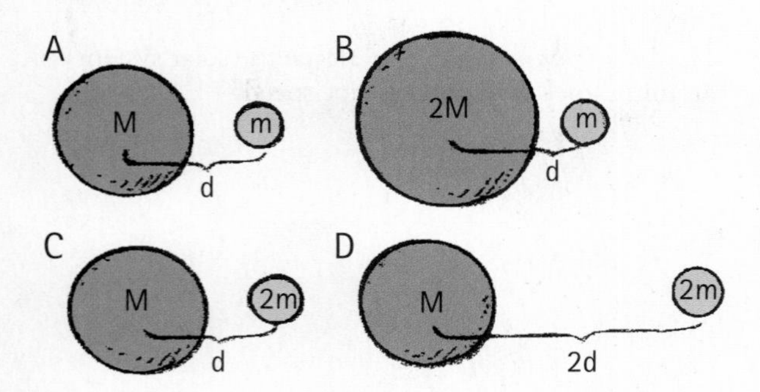

45. Rank the average gravitational forces from greatest to least between (a) the Sun and Mars, (b) the Sun and the Moon, (c) the Sun and Earth.
46. A ball is thrown upward at the velocities and angles shown in A to D. From greatest to least, rank them by their (a) vertical components of velocity, (b) horizontal components of velocity, (c) accelerations when they reach the top of their paths.

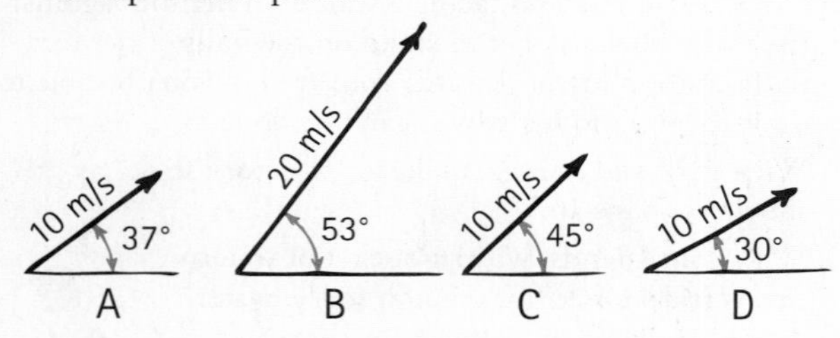

47. The dashed lines show three circular orbits, A, B, and C, about Earth. Rank these quantities from greatest to least: (a) their orbital speed, (b) their time to orbit Earth.

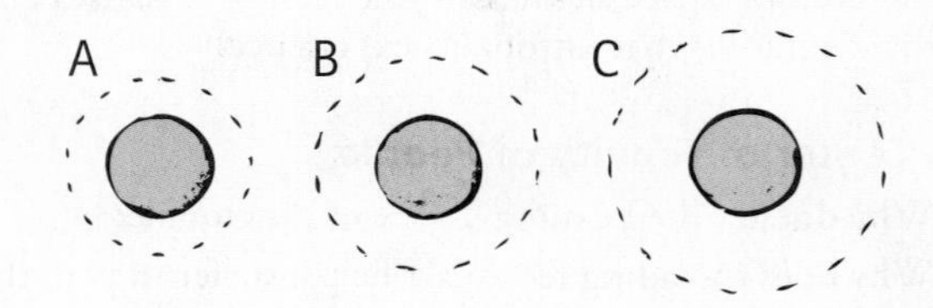

48. The positions of a satellite in elliptical orbit are indicated by A through D. Rank these quantities from greatest to least: (a) gravitational force, (b) speed, (c) momentum, (d) KE, (e) PE, (f) total energy (KE + PE), (g) acceleration.

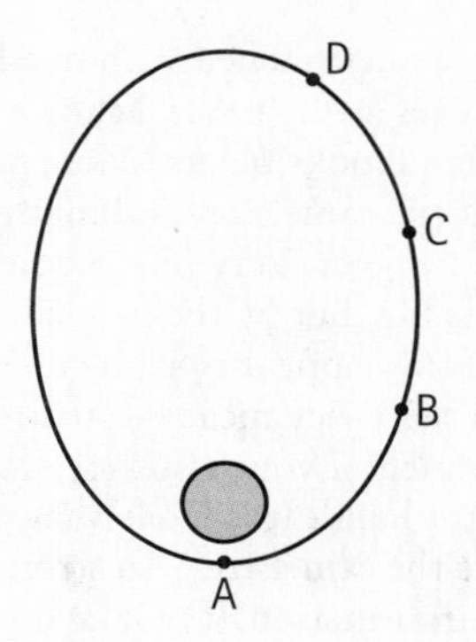

THINK AND SOLVE (MATHEMATICAL APPLICATION)

49. If you stood atop a super-tall ladder three times as far from Earth's center as at Earth's surface, how would your weight compare with it present value?
50. Find the change in the force of gravity between two planets when the masses of both planets are doubled but the distance between them stays the same.
51. Find the change in the force of gravity between two planets when their masses remain the same but the distance between them is increased tenfold.
52. Find the change in the force of gravity between two planets when the distance between them is *decreased* to a tenth of the original distance.
53. Find the change in the force of gravity between two planets when the masses of the planets don't change but the distance between them is decreased to a fifth of the original distance.
54. By what factor would your weight change if Earth's diameter were doubled and its mass were doubled?
55. Find the change in the force of gravity between two objects when both masses are doubled and the distance between them is also doubled.
56. Consider a bright point light source located 1 m from a square opening with an area of 1 m^2. Light passing through the opening illuminates an area of 4 m^2 on a wall 2 m from the opening. (a) Find the areas that would be illuminated if the wall were moved to distances of 3 m, 5 m, and 10 m from the opening. (b) How can the same amount of light illuminate more area as the wall is moved farther away?

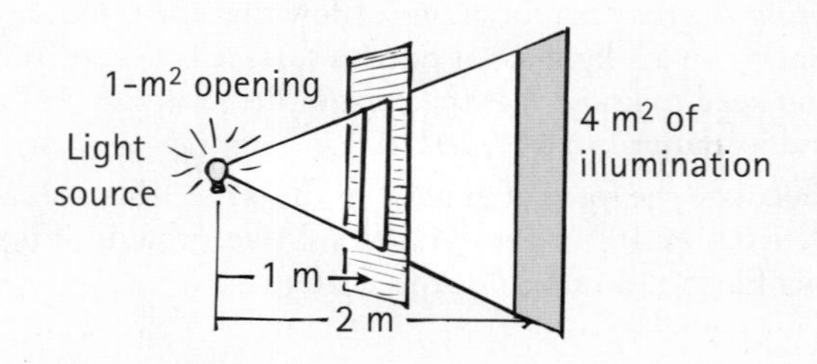

57. Calculate the force of gravity between Earth (6.4×10^{24} kg) and the Sun (2×10^{30} kg). The average distance between the two is 1.5×10^{11} m.
58. Students in a lab roll a steel ball off the edge of a table and measure its speed to be 4.0 m/s. They also know that if they simply dropped the ball from rest off the edge of the table it would take 0.5 s to hit the floor. How far from the bottom of the table should they place a small piece of paper so that the ball will hit it when it lands?

THINK AND EXPLAIN (SYNTHESIS)

59. What would be the path of the Moon if somehow all gravitational force on the Moon vanished to zero?
60. Is the force of gravity stronger on a piece of iron than it is on a piece of wood if both have the same mass? Defend your answer.
61. Is the force of gravity on a piece of paper stronger when it is crumpled? Defend your answer.
62. What are the magnitude and direction of the gravitational force that acts on a professor who weighs 1000 N at the surface of Earth?
63. Is gravitational force acting on a person who falls off a cliff? Is it acting on an astronaut inside an orbiting space vehicle?
64. If you were in a freely falling elevator and you dropped a pencil, it would hover in front of you rather than falling to the floor. Is there a force of gravity that is acting on the pencil? Defend your answer.
65. Are the planets of the solar system simply projectiles falling around the Sun?
66. What path would you follow if you fell off a rotating merry-go-round? What force prevents you from following that path while you're on the merry-go-round?
67. A heavy crate accidentally falls from a high-flying airplane just as it flies over a shiny red sports car parked in a parking lot. Relative to the location of the car, where will the crate crash?

68. How does the vertical component of motion for a ball kicked off a high cliff compare with the motion of vertical free fall?
69. In the absence of air drag, why doesn't the horizontal component of the ball's motion change while the vertical component does change?
70. At what point in its trajectory does a batted baseball have its minimum speed? If air drag can be neglected, how does this compare with the horizontal component of its velocity at other points?
71. Each of two golfers hits a ball at the same speed, one at 60° and the other at 30° above the horizontal. Which ball goes farther? Which hits the ground first? (Ignore air resistance.)
72. Does the speed of a falling object depend on its mass? Does the speed of a satellite in orbit depend on its mass? Defend your answers.
73. If you have ever watched the launching of an Earth satellite, you may have noticed that the rocket starts vertically upward, then veers from a vertical course, and continues its rise at an angle. Why does it start vertically? Why doesn't it continue vertically?
74. A satellite can orbit at 5 km above the Moon but not at 5 km above Earth. Why?

THINK AND DISCUSS (EVALUATION)

75. To begin your wingsuit flight, you step off the edge of a high cliff. Why are you then momentarily weightless? At that point, is gravity acting on you? Discuss why some people are confused by this.
76. Earth and the Moon are attracted to each other by gravitational force. Does the more massive Earth attract the less massive Moon with a force that is greater than, smaller than, or the same as the force with which the Moon attracts Earth?
77. What would you say to a friend who says that, if gravity follows the inverse-square law, the effect of gravity on you when you are on the 20th floor of a building should be one-fourth as much as it would be if you were on the 10th floor?
78. Why do passengers in high-altitude jet planes feel the sensation of weight but passengers in an orbiting space vehicle do not?
79. A park ranger shoots a monkey hanging from a branch of a tree with a tranquilizing dart. The ranger aims directly at the monkey, not realizing that the dart will follow a parabolic path and, therefore, will fall below the monkey's position. The monkey, however, seeing the dart leave the gun, lets go of the branch to avoid being hit. Will the monkey be hit anyway? Defend your answer.

80. Since the Moon is gravitationally attracted to Earth, why doesn't it simply crash into Earth?

81. Newton knew that if a cannonball were fired from a tall mountain, gravity would change its speed all along its trajectory. But if it were fired fast enough to attain circular orbit, gravity would not change its speed. Discuss.

82. How could an astronaut in orbit "drop" an object so that it would fall vertically to Earth's surface?

83. A new member of your discussion group says the primary reason that astronauts in orbit are weightless is because they are being pulled by other planets and stars. Why do you agree or disagree?

READINESS ASSURANCE TEST (RAT)

If you have a good handle on this chapter, then you should be able to score at least 7 out of 10 on this RAT. If you score less than 7, you need to study further before moving on.

Choose the BEST answer to each of the following:

1. The Moon falls toward Earth in the sense that it falls
   (a) with an acceleration of 10 m/s^2, as do apples on Earth.
   (b) beneath the straight-line path it would follow without gravity.
   (c) both of these
   (d) neither of these

2. The force of gravity between two planets depends on their
   (a) planetary compositions.
   (b) planetary atmospheres.
   (c) rotational motions.
   (d) none of these

3. Inhabitants of the International Space Station do not have a
   (a) force of gravity on their bodies.
   (b) sufficient mass.
   (c) support force.
   (d) condition of free fall.

4. A spacecraft on its way from Earth to the Moon is pulled equally by Earth and the Moon when it is
   (a) closer to Earth's surface.
   (b) closer to the Moon's surface.
   (c) halfway from Earth to Moon.
   (d) at no point, since Earth always pulls more strongly.

5. If you tossed a baseball horizontally and with no gravity, it would continue in a straight line. With gravity it falls about
   (a) 1 m below that line.
   (b) 5 m below that line.
   (c) 10 m below that line.
   (d) none of these

6. When no air resistance acts on a projectile, its horizontal acceleration is
   (a) *g*.
   (b) at right angles to *g*.
   (c) centripetal.
   (d) zero.

7. Without air resistance, a ball tossed at an angle of 40° with the horizontal goes as far downrange as a ball tossed at the same speed at an angle of
   (a) 45°.
   (b) 50°.
   (c) 60°.
   (d) none of these

8. When you toss a projectile sideways, it curves as it falls. It will become an Earth satellite if the curve it makes
   (a) matches the curve of Earth's surface.
   (b) results in a straight line.
   (c) spirals out indefinitely.
   (d) none of these

9. A satellite in elliptical orbit about Earth travels fastest when it moves
   (a) close to Earth.
   (b) far from Earth.
   (c) It travels at the same speed everywhere.
   (d) halfway between the near and far points from Earth.

10. A satellite in Earth's orbit is mainly above Earth's
    (a) atmosphere.
    (b) gravitational field.
    (c) both of these
    (d) neither of these

Answers to RAT

1. b, 2. d, 3. c, 4. b, 5. b, 6. d, 7. b, 8. a, 9. a, 10. a

6

CHAPTER 6

Heat

LAVA HAS an average temperature of about 1000°C—ten times the boiling point of water. White-hot sparks from a Fourth of July sparkler are about 2000°C, twice as hot as typical lava. Why, then, would lava severely burn your skin on contact, but the sparkler's sparks leave you unhurt? Can an object get colder and colder or hotter and hotter forever—or are there limits to coldness and hotness? Why does a tile floor feel colder to bare feet than a carpeted floor at the same temperature? Another question . . . Why does air in a balloon, the concrete of a sidewalk, and almost everything else expand as it warms? Why does ice water do the opposite, contracting instead of expanding as its temperature rises? And why does ice form on the top of a pond rather than at the bottom, thus enabling fish to swim comfortably all winter? This chapter will answer these and other questions relating to heat.

LEARNING OBJECTIVE
Distinguish between the forms of energy that make up thermal energy.

6.1 The Kinetic Theory of Matter

EXPLAIN THIS What is thermal energy?

According to the *kinetic theory of matter*, matter is made up of tiny particles—atoms and molecules. The particles are always moving, and they move in a number of ways. They rotate, vibrate, and move in straight lines between collisions (Figure 6.1).

The energy that atoms and molecules have relates to their motion and their position. They have translational kinetic energy due to their translational (straight-line) motion as well as rotational and vibrational kinetic energy. Also, the particles have potential energy from the attractions between them or from their mutual repulsion when they are at close range. The *total* of all these forms of energy in a particular substance is its **thermal energy**. (Physicists usually refer to thermal energy as *internal energy* because it is internal to a substance.)

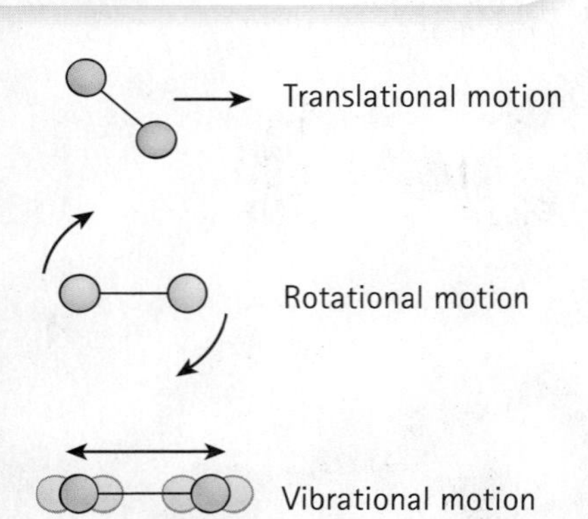

FIGURE 6.1
Types of motion of particles in matter.

LEARNING OBJECTIVE
Distinguish between thermal energy and temperature.

6.2 Temperature

EXPLAIN THIS What are two ways to state the freezing point of water?

When you strike a nail with a hammer, the nail becomes warm. Why? Because the hammer's blow makes the nail's atoms move faster. When you put a flame to a liquid, the liquid becomes warmer as its molecules move faster. When you rapidly compress air in a tire pump, the air becomes warmer. In these cases, the molecules are made to race back and forth faster. They gain kinetic energy. In general, the warmer an object, the more kinetic energy its atoms and molecules possess. **Temperature**, the degree of "hotness" or "coldness" of an object, is proportional to the average translational kinetic energy of the atoms or molecules that make up the object. So, the translational motion of particles contributes to the temperature of an object (see Figure 6.1, top).

FIGURE 6.2
Can we trust our sense of hot and cold? Will both fingers feel the same temperature when they are put in the warm water? Try this and see (feel) for yourself.

MasteringPhysics®
VIDEO: Low Temperatures with Liquid Nitrogen
VIDEO: How a Thermostat Works

It is important to note that temperature is not a measure of the total kinetic energy in a substance. For example, there is twice as much molecular kinetic energy in 2 liters of boiling water as in 1 liter of boiling water—but both volumes of water have the same temperature because the *average* kinetic energy per molecule is the same.

We express temperature quantitatively by a number that corresponds to the degree of hotness on some chosen scale. A common thermometer takes advantage of the fact that most substances expand as temperature rises. Such a thermometer measures temperature by comparing the expansion and contraction of a liquid (usually mercury or colored alcohol) to increments on a scale. Some modern thermometers avoid contact and measure temperature by the infrared radiation emitted by substances, as we'll discuss in the section on heat transfer and radiation later in this chapter.

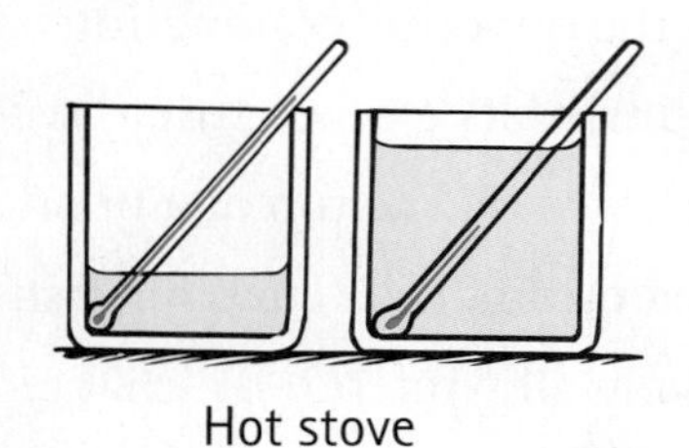

FIGURE 6.3
Although the same quantity of heat is added to both containers, the temperature increases more in the container with the smaller amount of water.

On a worldwide basis, the thermometer most often used is the Celsius thermometer. This thermometer is named in honor of the Swedish astronomer Anders Celsius (1701–1744), who first suggested the scale of 100 degrees between the freezing point and boiling point of water. Zero (0) is assigned to the temperature at which water freezes, and the number 100 is assigned to the temperature at which water boils (at standard atmospheric pressure). In between freezing and boiling temperatures are 100 equal parts called *degrees*.

In the United States, the number 32 is traditionally assigned to the temperature at which water freezes, and the number 212 is assigned to the temperature at which water boils. Such a scale makes up the Fahrenheit thermometer, named after its inventor, the German physicist G. D. Fahrenheit (1686–1736). Although the Fahrenheit scale is the one most commonly used in the United States, the Celsius scale is standard in scientific applications.

Using arithmetic formulas to convert from one temperature scale to the other is common in classroom exams. Because such arithmetic conversions aren't really physics, we won't be concerned with them here. Besides, the conversion between Celsius and Fahrenheit temperatures is closely approximated in the side-by-side scales shown in Figure 6.4.*

It is interesting that a thermometer actually registers its own temperature. When a thermometer is in thermal contact with something whose temperature we wish to know, thermal energy flows between the two until their temperatures are equal. At this point, thermal equilibrium is established. So, when we look at the temperature of the thermometer, we learn about the temperature of the substance with which it reached thermal equilibrium.

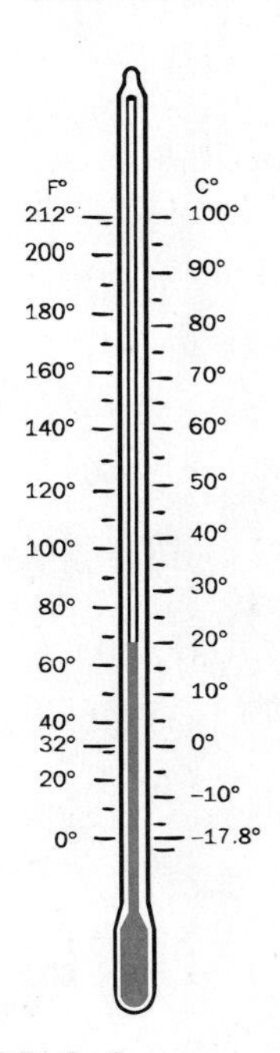

FIGURE 6.4
Fahrenheit and Celsius scales on a thermometer.

6.3 Absolute Zero

LEARNING OBJECTIVE
Describe the meaning of the lowest possible temperature in nature.

EXPLAIN THIS How cold is absolute zero?

In principle, there is no upper limit to temperature. As thermal motion increases, a solid object first melts and then vaporizes. As the temperature is further increased, molecules dissociate into atoms, and atoms lose some or all of their electrons, thereby forming a cloud of electrically charged particles—a plasma. Plasmas exist in stars, where the temperature is many millions of degrees Celsius. Temperature has no upper limit.

In contrast, there is a definite limit at the opposite end of the temperature scale. Gases expand when heated, and they contract when cooled. Nineteenth-century experimenters found that all gases, regardless of their initial pressures or volumes, change by $\frac{1}{273}$ of their volume at 0°C for each drop in temperature of 1 degree Celsius, provided that the pressure is held constant. So, if a gas at 0°C were cooled down by 273°C, it would contract $\frac{273}{273}$ volumes and be reduced to zero volume. Clearly, we cannot have a substance with zero volume (Figure 6.5).

The same is true of pressure. The pressure of a gas of fixed volume decreases by $\frac{1}{273}$ for each drop in temperature of 1 degree Celsius. If it were cooled to 273°C below zero, it would have no pressure at all. In practice, every gas converts to a liquid before becoming this cold. Nevertheless, these decreases by $\frac{1}{273}$ increments suggested the idea of lowest temperature: −273°C. That's the lower limit of temperature, **absolute zero**. At this temperature, molecules have lost all available

*Okay, if you really want to know, the formulas for temperature conversion are $C = \frac{5}{9}(F - 32)$ and $F = \frac{9}{5}C + 32$, where C is the Celsius temperature and F is the corresponding Fahrenheit temperature.

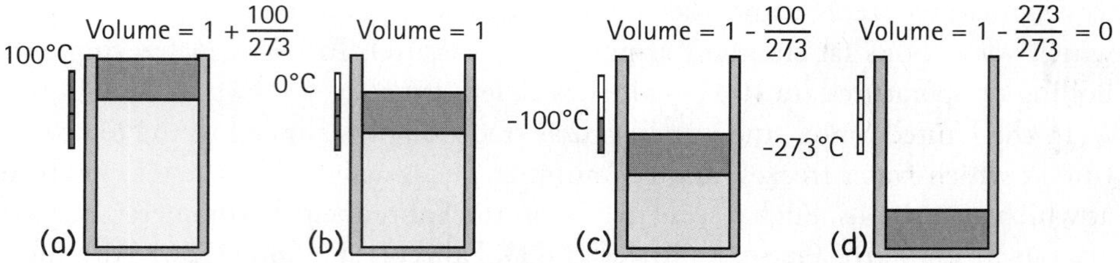

FIGURE 6.5
The gray piston in the vessel goes down as the volume of gas (blue) shrinks. The volume of gas changes by $\frac{1}{273}$ of its volume at 0°C with each 1°C change in temperature when the pressure is held constant. (a) At 100°C, the volume is $\frac{100}{273}$ greater than it is at (b), when its temperature is 0°C. (c) When the temperature is reduced to −100°C, the volume is reduced by $\frac{100}{273}$. (d) At −273°C, the volume of the gas would be reduced by $\frac{273}{273}$ and so would be zero.

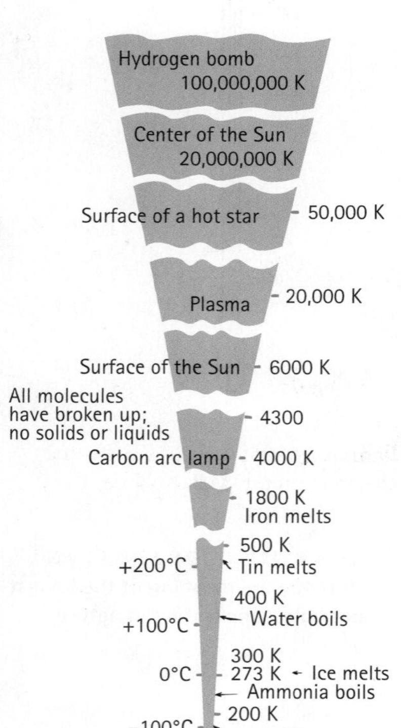

FIGURE 6.6
Some absolute temperatures.

kinetic energy.* No more energy can be removed from a substance at absolute zero. It can't get any colder (Figure 6.6).

The absolute temperature scale is called the Kelvin scale, named after the famous British mathematician and physicist William Thomson, First Baron Kelvin. Absolute zero is 0 K (short for "zero kelvin"; note that the word *degrees* is not used with Kelvin temperatures). There are no negative numbers on the Kelvin scale. Its temperature divisions are identical to the divisions on the Celsius scale. Thus, the melting point of ice is 273 K, and the boiling point of water is 373 K.

CHECK YOURSELF

1. **Which is larger—a Celsius degree or a kelvin?**
2. **A sample of hydrogen gas has a temperature of 0°C. If the sample were heated until it had twice the thermal energy, what would its temperature be?**

CHECK YOUR ANSWERS

1. Neither; they are equal.
2. The 0°C gas has an absolute temperature of 273 K. Twice the thermal energy means that it has twice the absolute temperature, or 2 × 273 K. This would be 546 K, or 273°C.

LEARNING OBJECTIVE
Distinguish between heat and temperature.

6.4 What Is Heat?

EXPLAIN THIS Why do we say that no substances contain heat?

If you touch a hot stove, thermal energy enters your hand because the stove is warmer than your hand. When you touch a piece of ice, however, thermal energy passes out of your hand and into the colder ice. The direction of energy flow is always from a warmer thing to a neighboring cooler thing. A scientist defines **heat** as the thermal energy transferred from one thing to another due to a temperature difference.

Which has a higher temperature—a red-hot tack or a lake? Which has more thermal energy?

*Even at absolute zero, molecules still possess a small amount of kinetic energy, called the zero-point energy. Helium, for example, has enough motion at absolute zero to prevent it from freezing. The explanation for this involves quantum theory.

According to this definition, matter does not contain heat. Matter contains thermal energy, not heat. Heat is the flow of thermal energy due to a temperature difference. Once thermal energy has been transferred to an object or substance, it ceases to be heat. Heat is thermal energy in transit.

For substances in thermal contact, thermal energy flows from the higher-temperature substance to the lower-temperature substance until thermal equilibrium is reached. This does not mean that thermal energy necessarily flows from a substance with more thermal energy into one with less thermal energy. For example, there is more thermal energy in a bowl of warm water than there is in a red-hot thumbtack. If the tack is placed into the water, thermal energy doesn't flow from the warm water to the tack. Instead, it flows from the hot tack to the cooler water. Thermal energy never flows unassisted from a lower-temperature substance to a higher-temperature substance.

If heat is thermal energy that transfers in a direction from hot to cold, what is cold? Does a cold substance contain something opposite to thermal energy? No. An object is not cold because it contains something but because it lacks something. It lacks thermal energy. When outdoors on a near-zero winter day, you feel cold not because something called cold gets to you. You feel cold because you lose heat. That's the purpose of your coat—to slow down the heat flow from your body. Cold is not a thing in itself; it is the result of reduced thermal energy.

FIGURE 6.7
The temperature of the sparks is very high, about 2000°C. That's a lot of thermal energy per molecule of spark. However, because there are few molecules per spark, the thermal energy is safely small. Temperature is one thing; transfer of energy is another.

Just as dark is the absence of light, cold is the absence of thermal energy.

CHECK YOURSELF

1. **Suppose you apply a flame to 1 liter of water for a certain time and the water's temperature rises by 2°C. If you apply the same flame for the same time to 2 liters of water, by how much will the water's temperature rise?**
2. **If a fast marble hits a random scatter of slow marbles, does the fast marble usually speed up or slow down? Which lose(s) kinetic energy and which gain(s) kinetic energy—the initially fast-moving marble or the initially slow ones? How do these questions relate to the direction of heat flow?**

CHECK YOUR ANSWERS

1. Its temperature will rise by only 1°C, because there are twice as many molecules in 2 liters of water, and each molecule receives only half as much energy on the average. So, the water's average kinetic energy, and thus its temperature, increases by half as much.
2. A fast-moving marble slows when it hits slower-moving marbles. It gives up some of its kinetic energy to the slower ones. Likewise with heat: Molecules with higher kinetic energy that make contact with molecules that have lower kinetic energy give up some of their excess kinetic energy to the slower ones. The direction of energy transfer is from hot to cold. For both the marbles and the molecules, however, the total energy before and after contact is the same.

Quantity of Heat

Heat is a form of energy, and it is measured in joules. It takes about 4.2 joules of heat to change 1 gram of water by 1 Celsius degree. A unit of heat still common in the United States is the **calorie.*** A calorie is defined as the amount of heat

Temperature is measured in degrees. Heat is measured in joules or calories.

*Another common unit of heat is the British thermal unit (Btu). The Btu is defined as the amount of heat required to change the temperature of 1 pound of water by 1 degree Fahrenheit.

needed to change the temperature of 1 gram of water by 1 Celsius degree. (The relationship between calories and joules is 1 calorie = 4.19 joules.)

The energy ratings of foods and fuels are measured by the energy released when they are burned. (Metabolism is really "burning" at a slow rate.) The heat unit for labeling food is the kilocalorie, which is 1000 calories (the heat needed to change the temperature of 1 kilogram of water by 1°C). To differentiate this unit and the smaller calorie, the food unit is usually called a Calorie, with a capital C. So, 1 Calorie is really 1000 calories.

FIGURE 6.8
To the weight watcher, the peanut contains 10 Calories; to the physicist, it releases 10,000 calories (or 41,900 joules) of energy when burned or digested.

CHECK YOURSELF

Which will raise the temperature of water more—adding 4.19 joules or 1 calorie?

CHECK YOUR ANSWER

They are the same. This is like asking which is longer—a 1-mile-long track or a 1.6-kilometer-long track. They're the same length in different units.

LEARNING OBJECTIVE
Describe the three laws of thermodynamics.

6.5 The Laws of Thermodynamics

EXPLAIN THIS How does thermodynamics relate to the conservation of energy?

What we've learned thus far about heat and thermal energy is summed up in the laws of **thermodynamics**. The word *thermodynamics* stems from Greek words meaning "movement of heat." It is the study of heat and its transformation to different forms of energy.

When thermal energy transfers as heat, it does so without net loss or gain. The energy lost in one place is gained in another. When the law of conservation of energy (which we discussed in Chapter 4) is applied to thermal systems, we have the **first law of thermodynamics**:

Whenever heat flows into or out of a system, the gain or loss of thermal energy equals the amount of heat transferred.

A *system* is any substance, device, or well-defined group of atoms or molecules. The system may be the steam in a steam engine, the entire Earth's atmosphere, or even the body of a living creature. When we add heat to any of these systems, we increase its thermal energy. The added energy enables the system to do work. The first law makes good sense.

The first law is nicely illustrated when you put an airtight can of air on a hot stove and warm it. The energy that is put in increases the thermal energy of the enclosed air, so its temperature rises. If the can is fitted with a movable piston, then the heated air can do *mechanical work* as it expands and pushes the piston outward. This ability to do mechanical work is energy that comes from the energy you put in to begin with. The first law says you don't get energy from nothing.

FIGURE 6.9
When Pete pushes down on the piston, he does work on the air inside. What happens to the air's temperature?

The **second law of thermodynamics** restates what we've learned about the direction of heat flow:

Heat never spontaneously flows from a lower-temperature substance to a higher-temperature substance.

UNIFYING CONCEPT

- *The Second Law of Thermodynamics*

When heat flow is spontaneous—that is, when no external work is done—the direction of flow is always from hot to cold. In winter, heat flows from inside a warm home to the cold air outside. In summer, heat flows from the hot air outside into the cooler interior. Heat can be made to flow the other way only when work is done on the system or when energy is added from another source. This occurs with heat pumps and air conditioners. In these devices, thermal energy is pumped from a cooler to a warmer region. But without external effort, the direction of heat flow is from hot to cold. The second law, like the first law, makes logical sense.*

The **third law of thermodynamics** restates what we've learned about the lowest limit of temperature:

No system can reach absolute zero.

As investigators attempt to reach this lowest temperature, it becomes more difficult to get closer to it. Physicists have been able to record temperatures that are less than a millionth of 1 kelvin, but never as low as 0 K.

Entropy

LEARNING OBJECTIVE
Relate the dispersal of energy to the concept of entropy.

EXPLAIN THIS When does entropy decrease in a system?

Energy tends to disperse. It flows from where it is localized to where it is spread out. For example, consider a hot pan once you have taken it off the stove. The pan's thermal energy doesn't stay localized in the pan. Instead, it disperses outward, away from the pan into its surroundings. As the pan is heated, energetic molecules transfer energy to the air by molecular collisions as well as by radiation. Consider a second example: The chemical energy in gasoline burns explosively in a car engine when the molecules combust. Some of this energy disperses through the transmission to get the car moving. The rest of the energy disperses as heat into the metal of the engine, into the coolant that flows through the engine and the radiator, and into the gases that flow through the engine and out the exhaust pipe. The energy, once localized in the small volume of the gasoline, is now spread out through a larger volume of space. Or, witness the dispersion of energy when you pick up an object—a marble, for example—and drop it. When you lift the marble, you give it potential energy. Drop it and that potential energy converts to kinetic energy, pushing air aside as it falls (therefore spreading out the marble's kinetic energy a bit), before hitting the ground. When the marble hits, it disperses energy as sound and as heat (when it heats the ground a bit). The potential energy you put into the marble by lifting it, which was once localized in the marble, is now spread out and dispersed in a little air movement plus the heating of the air and ground. The marble bounces before it finally comes to rest; in each bounce, energy spreads out from the marble to its surroundings.

*The laws of thermodynamics were all the rage back in the 1800s. At that time, horses and buggies were yielding to steam-driven locomotives. There is the story of the engineer who explained the operation of a steam engine to a peasant. The engineer cited in detail the operation of the steam cycle—how expanding steam drives a piston that in turn rotates the wheels. After some thought, the peasant said, "Yes, I understand all that. But where's the horse?" This story illustrates how hard it is to abandon our way of thinking about the world when a newer method comes along to replace established ways. Are we any different today?

The tendency of energy to spread out is one of the central driving forces of nature. Processes that disperse energy tend to occur spontaneously—they are favored. The opposite holds true as well. Processes that result in the concentration of energy tend not to occur—they are not favored. Heat from the room doesn't spontaneously flow into the frying pan to heat it up. Likewise, the lower-energy molecules of a car's exhaust won't on their own come back together to re-form the higher-energy molecules of gasoline. And, needless to say, dropped marbles don't jump back into your hand. The natural flow of energy is always a one-way trip from where it is concentrated to where it is spread out. The second law of thermodynamics states this principle for heat flow. But now we can see that the second law of thermodynamics can be generalized and stated this way:

FIGURE 6.10
Entropy: Order in nature tends to progress to disorder.

Natural systems tend to disperse from concentrated and organized-energy states toward diffuse and disorganized states.

The least concentrated form of energy is thermal energy. So, since organized forms of energy tend to become less organized, they ultimately degrade into the environment as thermal energy. Further, when energy is dispersed, it is less able to do useful work than when it was concentrated—in effect, it becomes diluted. So, thermal energy is the graveyard of useful energy.

The measure of energy dispersal is a quantity known as **entropy**.* More entropy means more degradation of energy. Since energy tends to degrade and disperse with time, the total amount of entropy in any system tends to increase with time. Things wear down (Figure 6.10). The same is true for the largest system, the universe. The net entropy in the universe is continuously increasing (the universe is continuously running "downhill").

Here's another way of stating the laws of thermodynamics: You can't win (because you can't get any more energy out of a system than you put into it), you can't break even (because you can't get as much useful energy out as you put in), and you can't get out of the game (entropy in the universe is always increasing).

We say *net* entropy because there are some regions where energy is actually being organized and concentrated. Work input from outside of an isolated system can decrease entropy in the system, with energy proceeding toward organization and concentration in that system. For example, diffuse thermal energy in the air can be concentrated in a heat pump. And living things seem to defy the second law of thermodynamics with their highly organized and concentrated energy. But, on closer examination, the orderliness we observe among life forms is a result of energy input. Ultimately, the energy that builds and maintains orderly biological systems comes mostly from the Sun when plants build energy-rich sugar molecules from disorderly gases and liquids during photosynthesis. The *spontaneous* processes that occur within organisms actually do increase entropy—consider, for example, the diffusion of nutrients across a cell membrane. Without some outside energy input, processes in which entropy decreases are not observed in nature (Figure 6.11).

Interestingly, the direction of time's passage is linked to increasing entropy; examples are a leaf falling from a tree, wood burning in a fire, and even the hands of a clock moving. As these occur, energy is dispersed, and we gain the sense that time moves forward. To put it another way, consider the likelihood of a burned log in a fire becoming whole, or a leaf on the ground spontaneously moving upward to join the branch from which it came. These cases involve the opposite of the dispersion of energy, which would be perceived as time moving backward. Hence, entropy is both a gauge for the dispersal of energy and time's arrow.

*Entropy can be expressed mathematically. The increase in entropy ΔS of a thermodynamic system is equal to the amount of heat added to the system ΔQ divided by the temperature T at which the heat is added: $\Delta S = \Delta Q/T$.

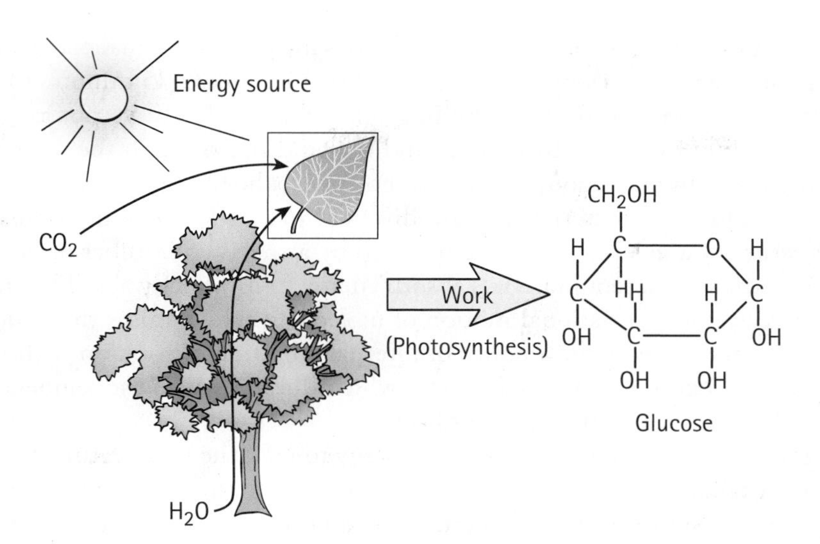

FIGURE 6.11
Work is needed to transform a state of disorderly, diffuse energy into a state of more concentrated energy. The Sun supplies the energy to do this work when plants transform liquids and vapors into sugar molecules—a plant's storehouse of usable, concentrated energy.

CHECK YOURSELF

1. **As energy disperses, where does it ultimately go?**
2. **Which has greater entropy—the molecules of perfume in a closed perfume bottle or the molecules of perfume when the perfume bottle is opened?**
3. **In the formation of molecular hydrogen H_2 from atomic hydrogen H, there is a net increase in entropy. Will this chemical reaction proceed on its own? Justify your answer.**
4. **A tree takes in carbon dioxide from the air, water from the soil, and a small amount of water vapor from the air. Structures in the tree's leaves, called chloroplasts, convert these disorderly building materials into sugar molecules, which are highly concentrated forms of energy. Does this violate the second law of thermodynamics? Explain your answer.**

CHECK YOUR ANSWERS

1. Ultimately the energy disperses into thermal energy.
2. The molecules diffuse when the bottle is opened, spreading their thermal energy over a larger volume of space. Entropy is increased.
3. Yes; by the second law of thermodynamics, processes that increase entropy are favored.
4. No; the radiant energy of the Sun provides the energy input needed to convert less-concentrated energy into a more concentrated, usable form.

6.6 Specific Heat Capacity

LEARNING OBJECTIVE
Relate the specific heat capacity of substances to thermal inertia.

EXPLAIN THIS Why does a hot frying pan cool faster than the same mass of equally hot water?

You've likely noticed that some foods remain hotter much longer than others. Whereas the filling of a hot apple pie can burn your tongue, the crust does not, even when the pie has just been removed from the oven (Figure 6.12). Or a piece of toast may be comfortably eaten a few seconds after coming from the hot toaster, whereas you must wait several minutes before eating soup that initially had the same temperature.

FIGURE 6.12
The filling of a hot apple pie may be too hot to eat, even though the crust is not.

Different substances have different capacities for storing thermal energy. If we heat a pot of water on a stove, we might find that it requires 15 minutes to raise it from room temperature to its boiling temperature. But if we put an equal mass of iron on the same stove, we'd find it would rise through the same temperature range in only about 2 minutes. For silver, the time would be less than a minute. Different materials require different quantities of heat to raise the temperature of a given mass of the material by a specified number of degrees. This is because different materials absorb energy in different ways. The energy may increase the translational motion of molecules, which raises the temperature; or it may increase the amount of internal vibration or rotation within the molecules and go into potential energy, which does not raise the temperature. Generally, there is a combination of both.

A gram of water requires 1 calorie of energy to raise the temperature 1 degree Celsius. It takes only about one-eighth as much energy to raise the temperature of a gram of iron by the same amount. Water absorbs more heat than iron for the same change in temperature. We say water has a higher **specific heat capacity** (sometimes called *specific heat*).

The specific heat capacity of any substance is the quantity of heat required to change the temperature of a unit mass of the substance by 1 degree Celsius.

If you add 1 calorie of heat to 1 gram of water, you'll raise its temperature by 1°C.

We can think of specific heat capacity as thermal inertia. Recall that *inertia* is a term used in mechanics to signify the resistance of an object to a change in its state of motion. Specific heat capacity is like thermal inertia because it signifies the resistance of a substance to a change in temperature.

CHECK YOURSELF

Which has a higher specific heat capacity—water or sand? In other words, which takes longer to warm in sunlight (or longer to cool at night)?

CHECK YOUR ANSWER

Water has the higher specific heat capacity. In the same sunlight, the temperature of water increases more slowly than the temperature of sand. And water will cool more slowly at night. The low specific heat capacity of sand and soil, as evidenced by how quickly they warm in the morning Sun and how quickly they cool at night, affects local climates.

Water has a much higher capacity for storing energy than almost all other substances. A lot of heat energy is needed to change the temperature of water. This explains why water is very useful in the cooling systems of automobiles and other engines. It absorbs a great quantity of heat for small increases in temperature. Water also takes longer to cool.

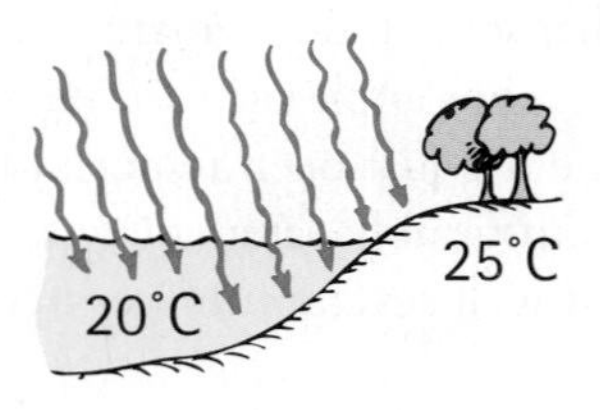

FIGURE 6.13
Because water has a high specific heat capacity and is transparent, it takes more energy to warm a body of water than to warm the land. Solar energy incident on the land is concentrated at the surface, but solar energy hitting the water extends beneath the surface and is diluted.

Integrated Science 6B
EARTH SCIENCE

LEARNING OBJECTIVE
Relate thermal inertia to specific heat capacity.

Specific Heat Capacity and Earth's Climate

EXPLAIN THIS Relate specific heat capacity to the oceans.

Water's high specific heat capacity affects the world's climate. Look at a globe or a map of the Northern Hemisphere and notice the high latitude of Europe. Water's high specific heat keeps Europe's climate appreciably milder than regions of the same latitude in northeastern Canada. Both Europe and Canada receive about the same amount of sunlight per square kilometer. What happens is that the Atlantic Ocean current known as the Gulf Stream carries warm water northeastward from the Caribbean Sea. The water retains much of its thermal energy long enough to reach the North Atlantic Ocean off the coast of Europe. Then it cools, releasing 4.19 joules of energy for each gram of water that cools by 1°C. The released energy is carried by westerly winds over the European continent.

MATH CONNECTION

The Heat-Transfer Equation

We can use specific heat capacity to write a formula for the quantity of heat Q involved when a mass m of a substance undergoes a change in temperature: $Q = cm\Delta T$. In words, heat transferred into or out of a substance = specific heat capacity of the substance × mass of the substance × the substance's temperature change. This equation is valid for a substance that gets warmer as well as for one that cools. When a substance is warming up, the heat transferred into it, Q, is positive. When a substance is cooling off, Q has a minus sign.

Let's apply this equation to a few examples.

Problems

1. A 2.0-kg aluminum pan is heated on the stove from 20°C to 110°C. How much heat had to be transferred to the aluminum? The specific heat capacity of aluminum is 900 J/kg·°C.
2. What will be the final temperature of a mixture of 50 g of 20°C water and 50 g of 40°C water?
3. What will be the final temperature when 100 g of 25°C water is mixed with 75 g of 40°C water?
4. Radioactive decay of granite and other rocks in Earth's interior provides enough energy to keep the interior hot, to produce magma, and to provide warmth to natural hot springs. This is due to the average release of about 0.03 J per kilogram each year. How many years are required for a chunk of thermally insulated granite to increase 500°C in temperature, assuming that the specific heat of granite is 800 J/kg·°C?

Solutions

1. $Q = cm\Delta T = (900\text{ J/kg}\cdot°\text{C})(2.0\text{ kg})(110°\text{C} - 20°\text{C}) = 1.62 \times 10^5\text{ J}$. The sign of Q is positive because the pan is absorbing heat.
2. The heat gained by the cooler water = the heat lost by the warmer water. Since the masses of the water are the same, the final temperature is midway between the two, 30°C. So, we'll end up with 100 g of 30°C water.
3. Here we have different masses of water that are mixed together. We equate the heat gained by the cool water to the heat lost by the warm water. We can express this equation formally, and then we can let the expressed terms lead to a solution:

   Heat gained by cool water = heat lost by warm water

   $$cm_1\Delta T_1 = cm_2\Delta T_2$$

   ΔT_1 doesn't equal ΔT_2 as in Problem 2 because of different masses of water. We can see that ΔT_1 will be the final temperature T minus 25°C, since T will be greater than 25°C; ΔT_2 is 40°C minus T, because T will be less than 40°C. Then, showing magnitude only, we see

   $$c(100)(T - 25) = c(75)(40 - T)$$

   $$100T - 2500 = 3000 - 75T$$

   $$T = 31.4°\text{C}$$

4. Here, we switch to rock, but the same concept applies. No particular mass is specified, so we'll work with quantity of heat/mass (our answer should be the same for a small chunk of rock or a huge chunk). From $Q = cm\Delta T$, $\frac{Q}{m} = c\Delta T = (800\text{ J/kg}\cdot°\text{C}) \times (500°\text{C}) = 400{,}000\text{ J/kg}$. The time required is $(400{,}000\text{ J/kg})/(0.03\text{ J/kg}\cdot\text{yr}) =$ 13.3 million years. Small wonder it remains hot down there!

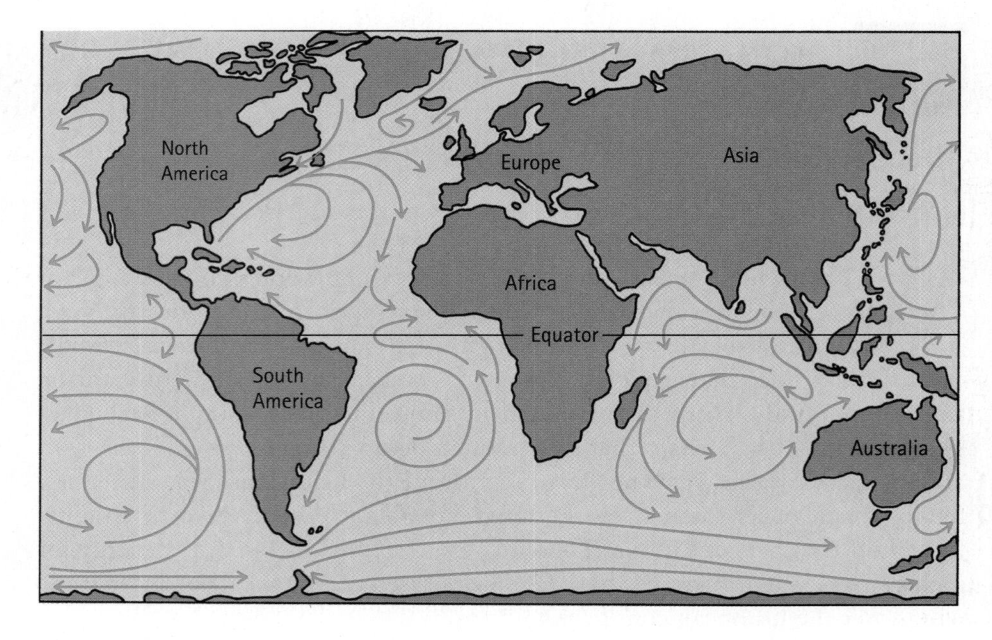

FIGURE 6.14
Many ocean currents, shown in blue, distribute heat from the warmer equatorial regions to the colder polar regions.

A similar effect occurs in the United States. The winds in North America are mostly westerly. On the West Coast, air moves from the Pacific Ocean to the land. In winter months, the ocean water is warmer than the air. Air blows over the warm water and then moves over the coastal regions. This produces a warmer climate. In summer, the opposite occurs. The water cools the air, and the coastal regions are cooled. The East Coast does not benefit from the moderating effects of water because the direction of the prevailing wind is eastward from the land to the Atlantic Ocean. Land, with a lower specific heat capacity, gets hot in the summer but cools rapidly in the winter.

Islands and peninsulas do not have the extremes of temperatures that are common in the interior regions of a continent. The high summer and low winter temperatures common in Manitoba and the Dakotas, for example, are largely due to the absence of large bodies of water. Europeans, islanders, and people living near ocean air currents should be glad that water has such a high specific heat capacity. San Franciscans certainly are!

See Chapter 26 on weather for more discussion of the effects of ocean currents on global climate.

CHECK YOURSELF

1. **Bermuda is close to North Carolina, but, unlike North Carolina, it has a tropical climate year-round. Why?**
2. **How is the thermal energy that is given up by the northern Atlantic Ocean off the coast of Europe carried to the European continent?**
3. **If the winds at the latitude of San Francisco and Washington, DC, were from the east rather than from the west, why might cherry trees grow in San Francisco and palm trees grow in Washington, DC?**

CHECK YOUR ANSWERS

1. Bermuda is an island. The surrounding water warms it when it might otherwise be too cold, and cools it when it might otherwise be too warm.

2. Thermal energy moves from the Atlantic Ocean into the air and is carried over Europe by westerly winds.
3. As the ocean off the coast of San Francisco cools in the winter, the heat it loses warms the atmosphere it comes in contact with. This warmed air blows over the California coastline to produce a relatively warm climate. If the winds were easterly instead of westerly, the climate of San Francisco would be chilled by winter winds from dry and cold Nevada. The climate would also be reversed in Washington, DC, because air warmed by the Atlantic Ocean would blow over Washington, DC, and produce a warmer climate there in the winter.

6.7 Thermal Expansion

LEARNING OBJECTIVE
Relate the open crystalline structure of ice to water's maximum density at 4°C.

EXPLAIN THIS Why does ice float?

Molecules in a hot substance jiggle faster and move farther apart than molecules in a colder substance. The result is thermal expansion. Most substances expand when heated and contract when cooled. Sometimes the changes aren't noticed, and sometimes they are. Telephone wires are longer and sag more on a hot summer day than they do in winter. Railroad tracks laid on cold winter days tend to expand and buckle during the hot summer (Figure 6.15). Metal lids on glass fruit jars can often be loosened by heating them under hot water. If one part of a piece of glass is heated or cooled more rapidly than adjacent parts, the resulting expansion or contraction may break the glass. This is especially true of thick glass. Pyrex glass is an exception because it is specially formulated to expand very little with increasing temperature.

Thermal expansion must be taken into account in structures and devices of all kinds. A dentist uses filling material that has the same rate of expansion as teeth. A civil engineer uses reinforcing steel that has the same expansion rate as concrete. A long steel bridge usually has one end anchored while the other rests on rockers (Figure 6.16). Notice also that many bridges have tongue-and-groove gaps called expansion joints (Figure 6.17). Similarly, concrete roadways and sidewalks are intersected by gaps, which are sometimes filled with tar, so that the concrete can expand freely in summer and contract in winter.

FIGURE 6.15
Thermal expansion gone wild. Extreme heat on a July day caused the buckling of these railroad tracks.

FIGURE 6.16
One end of the bridge rides on rockers to allow for thermal expansion. The other end (not shown) is anchored.

FIGURE 6.17
The gap in the roadway of a bridge is called an expansion joint. It allows the bridge to expand and contract. (Was this photo taken on a warm day or a cold day?)

Thermal expansion accounts for the creaky noises often heard in the attics of old houses on cold nights. The construction materials expand during the day and contract at night, creaking as they grow and shrink.

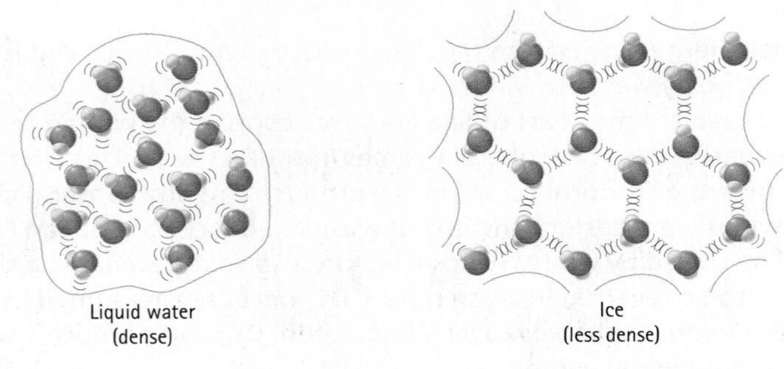

FIGURE 6.18
Liquid water is more dense than ice because water molecules in a liquid are closer together than water molecules frozen in ice, where they have an open crystalline structure.

With increases in temperature, liquids expand more than solids. We notice this when gasoline overflows from a car's tank on a hot day. If the tank and its contents expanded at the same rate, the gas would not overflow. This is why you shouldn't "top off" a gas tank when you fill it, especially on a hot day.

Expansion of Water

Water, like most substances, expands when it is heated, except in the temperature range between 0°C and 4°C. Something fascinating happens in this range. Ice has a crystalline structure, with open-structured crystals. Water molecules in this open structure occupy a greater volume than they do in the liquid phase (Figures 6.18 and 6.19). This means that ice is less dense than water.

FIGURE 6.19
The six-sided structure of a snowflake is a result of the six-sided ice crystals that make it up.

When ice melts, not all the six-sided crystals collapse. Some of them remain in the ice-water mixture, making up a microscopic slush that slightly "bloats" the water, increasing its volume (Figure 6.20). This results in ice water being less dense than slightly warmer water. As the temperature of water at 0°C is increased, more of the remaining ice crystals collapse. This further decreases the volume of the water. This contraction occurs only up to 4°C. That's because two things occur at the same time—contraction and expansion. Volume tends to decrease as ice crystals collapse, while volume tends to increase due to greater molecular motion. The collapsing effect dominates until the temperature reaches 4°C. After that, expansion overrides contraction because most of the ice crystals have melted (Figure 6.21).

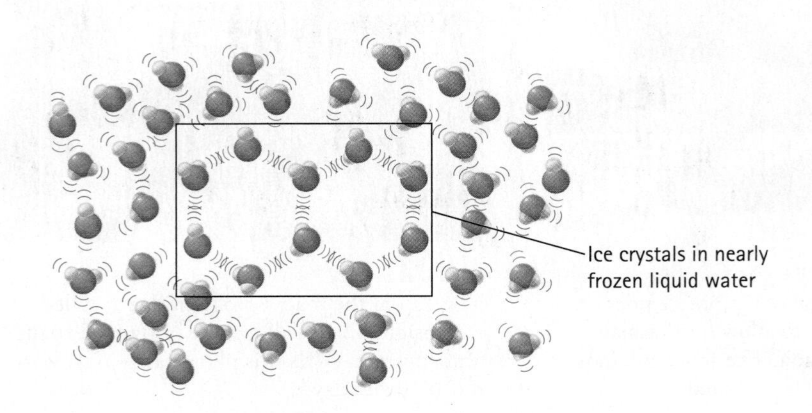

FIGURE 6.20
Close to 0°C, liquid water contains crystals of ice. The open structure of these crystals increases the volume of the water slightly.

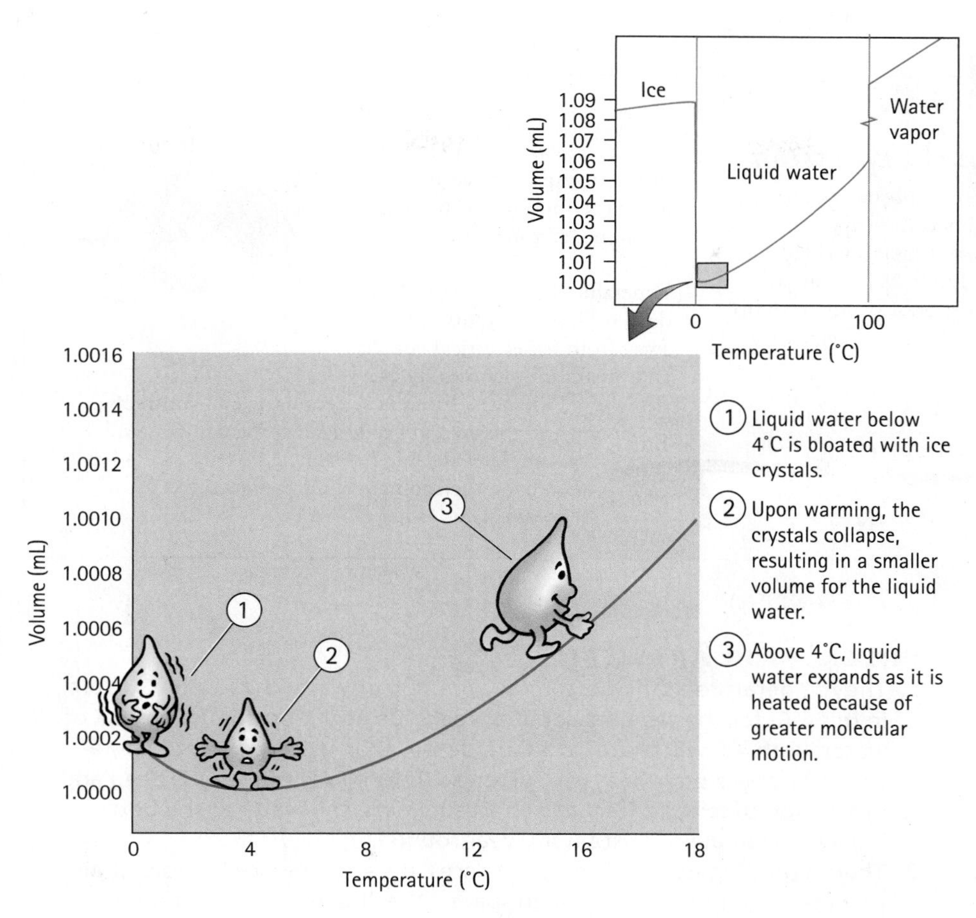

FIGURE 6.21
Between 0°C and 4°C, the volume of liquid water decreases as temperature increases. Above 4°C, water behaves the way other substances do. Its volume increases as its temperature increases. The volumes shown here are for a 1-gram sample.

When ice water freezes to become solid ice, its volume increases tremendously—and its density is therefore much lower. That's why ice floats on water. Like most other substances, solid ice contracts without further cooling. This behavior of water is very important in nature. If water were most dense at 0°C, it would settle to the bottom of a lake or pond. Because water at 0°C is less dense, it floats at the surface. That's why ice forms at the surface (Figure 6.22).

UNIFYING CONCEPT
- ***Density***
Section 2.3

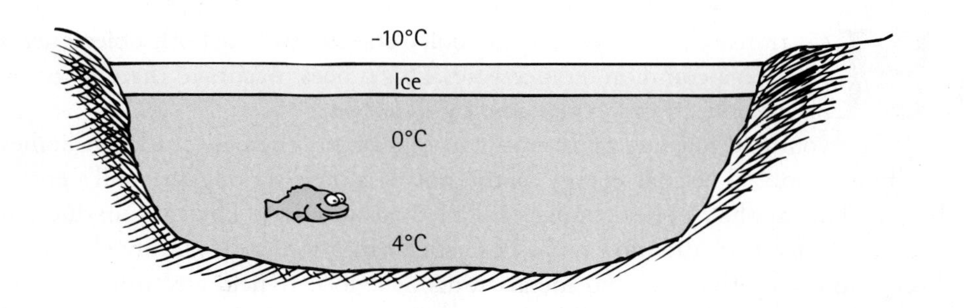

FIGURE 6.22
As water cools, it sinks until the entire lake is 4°C. Then, as the water at the surface is cooled further, it floats on top and can freeze. Once ice is formed, temperatures lower than 4°C can extend down into the lake.

CHECK YOURSELF

1. **What was the precise temperature at the bottom of Lake Michigan on New Year's Eve in 1901?**
2. **What's inside the open spaces of the ice crystals shown in Figure 6.20? Is it air, water vapor, or nothing?**

TECHNOLOGY

Engineering for Thermal Expansion

An illustration of the fact that different substances expand at different rates is provided by a bimetallic strip. This device is made of two strips of different metals welded together, one of brass and the other of iron. When the strip is heated, the greater expansion of the brass bends the strip. This bending may be used to turn a pointer, to regulate a valve, or to close a switch.

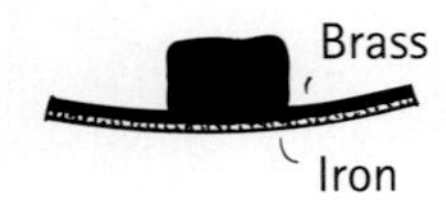

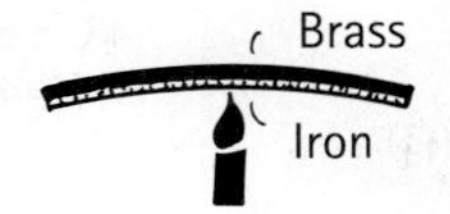

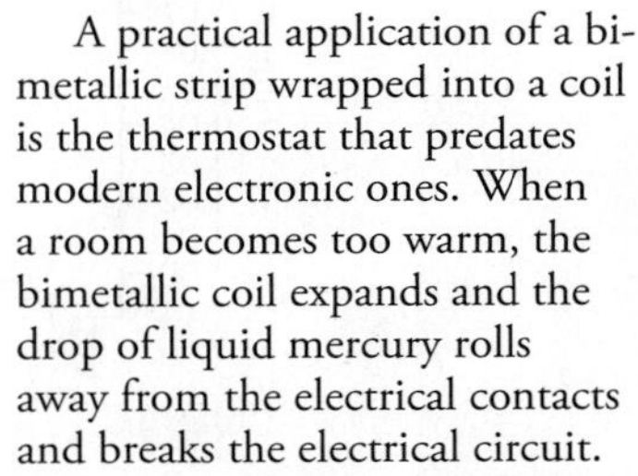

A practical application of a bimetallic strip wrapped into a coil is the thermostat that predates modern electronic ones. When a room becomes too warm, the bimetallic coil expands and the drop of liquid mercury rolls away from the electrical contacts and breaks the electrical circuit. When the room is too cool, the coil contracts and the mercury rolls against the contacts and completes the circuit. Bimetallic strips are used in oven thermometers, refrigerators, electric toasters, and other devices.

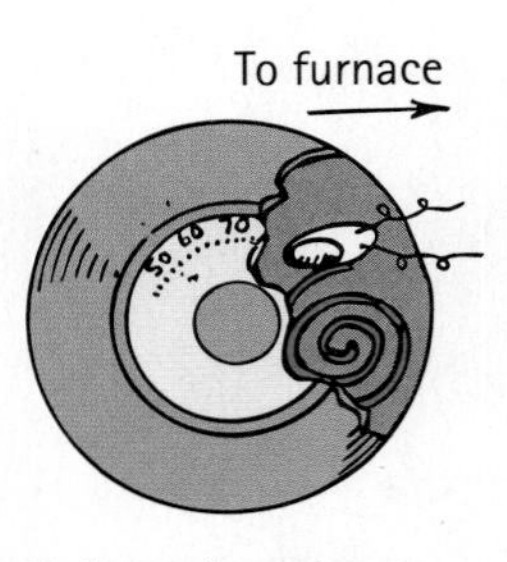

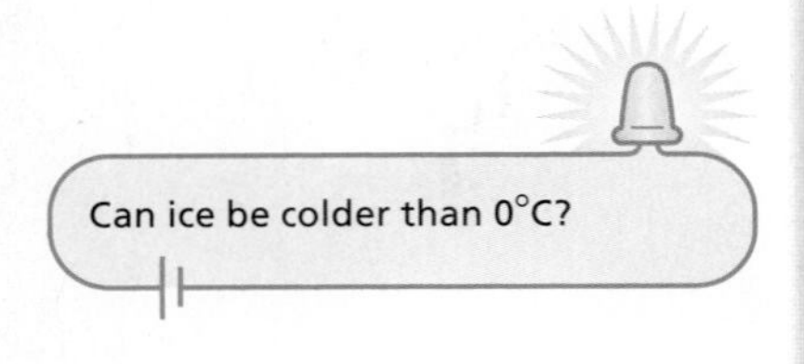

CHECK YOUR ANSWERS

1. The temperature at the bottom of any body of water that has 4°C water in it is 4°C, for the same reason that rocks are at the bottom of a body of water. Both 4°C water and rocks are denser than water at any temperature. Water is a poor heat conductor, so, if the body of water is deep and in a region of long winters and short summers, the water at the bottom is likely to remain a constant 4°C year round.
2. There's nothing at all in the open spaces. It's empty space—a void. If air or water vapor were in the open spaces, the illustration should show molecules there—oxygen and nitrogen for air and H_2O for water vapor.

LEARNING OBJECTIVE
Describe the nature of conduction in solids.

6.8 Heat Transfer: Conduction

EXPLAIN THIS Why does a tile floor feel colder to your feet than a rug of the same temperature?

Heat transfers from warmer to cooler objects, so that both objects tend to reach a common temperature. This process occurs in three ways: by *conduction*, by *convection*, and by *radiation*.

When you hold one end of an iron nail in a flame, the nail quickly becomes too hot to hold. Thermal energy at the hot end travels along the nail's entire length. This method of heat transfer is called **conduction**. Thermal conduction occurs by means of the movement of particles in a material, mainly electrons. Every atom has electrons, and metal atoms have loosely held electrons that are free to migrate in the metal. We shall see, in Chapter 7, that metals are good electrical conductors for the same reason. Thermal conduction occurs by atomic particles colliding inside the heated object.

Solids whose atoms or molecules have loosely held electrons are good conductors of heat. Metals have the loosest electrons, and they are excellent conductors of heat. Silver is the best, copper is next, and then, among the common metals, aluminum and iron. Poor conductors include wool, wood, paper, cork, and plastic foam. Molecules in these materials have electrons that are firmly attached to them. Poor conductors are called *insulators*.

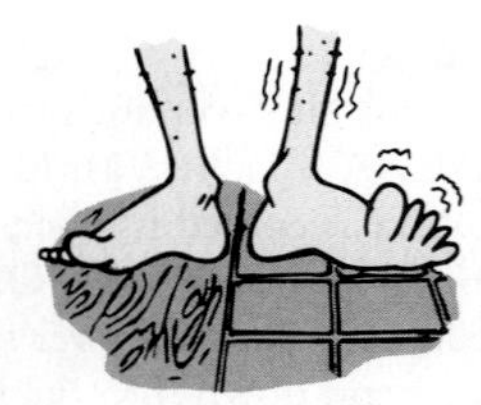

FIGURE 6.23
The tile floor feels colder than the wooden floor, even though both floor materials are at the same temperature. This is because tile is a better conductor of heat than wood, and so heat is more readily conducted out of the foot touching the tile.

MasteringPhysics®
VIDEO: The Secret to Walking on Hot Coals
VIDEO: Air is a Poor Conductor

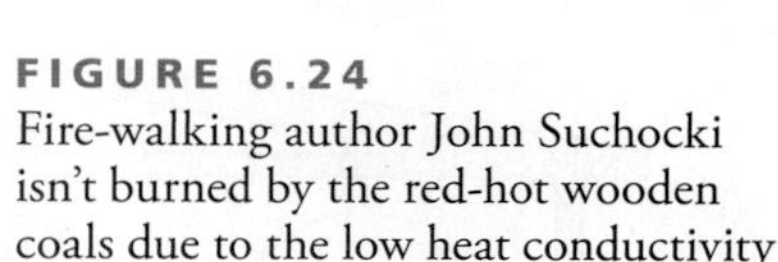

FIGURE 6.24
Fire-walking author John Suchocki isn't burned by the red-hot wooden coals due to the low heat conductivity of wood.

FIGURE 6.25
When you touch a nail stuck in ice, does cold flow from the ice to your hand, or does energy flow from your hand to the ice?

FIGURE 6.26
Snow patterns on the roof of a house show areas of conduction and insulation. Bare parts show where heat from inside has leaked through the roof and melted the snow.

Wood is a great insulator, and it is often used for cookware handles. Even when a pot is hot, you can briefly grasp the wooden handle with your bare hand without harm. An iron handle of the same temperature would surely burn your hand. Wood is a good insulator even when it's red hot. This explains how fire-walking coauthor John Suchocki can walk barefoot on red-hot wood coals without burning his feet (Figure 6.24). (Caution: Don't try this on your own; even experienced fire walkers sometimes receive bad burns when conditions aren't just right.) The main factor here is the poor conductivity of wood—even red-hot wood. Although its temperature is high, very little energy is conducted to the feet. A fire walker must be careful that no iron nails or other good conductors are among the hot coals. Ouch!

Air is a very poor conductor as well. You can briefly put your hand into a hot pizza oven without harm. The hot air doesn't conduct thermal energy well. But don't touch the metal in the hot oven. Ouch again! The good insulating properties of such things as wool, fur, and feathers are largely due to the air spaces they contain. Porous substances are also good insulators because of their many small air spaces. Be glad that air is a poor conductor; if it weren't, you'd feel quite chilly on a 20°C (68°F) day!

Snow is a poor conductor of thermal energy. Snowflakes are formed of crystals that trap air and provide insulation. That's why a blanket of snow keeps the ground warm in winter. Animals in the forest find shelter from the cold in snow banks and in holes in the snow. The snow doesn't provide them with thermal energy—it simply slows down the loss of body heat generated by the animals. Then there are igloos, Arctic dwellings built from compacted snow to shield those inside from the bitter cold of Arctic winters.

LEARNING OBJECTIVE
Describe the nature of convection in fluids.

6.9 Heat Transfer: Convection

EXPLAIN THIS Why does warm air rise?

Liquids and gases transfer thermal energy mainly by **convection**, which is heat transfer due to the actual motion of the fluid itself. Unlike conduction (in which heat is transferred by successive collisions of electrons and atoms), convection involves the motion of a fluid—currents. Convection occurs in all fluids, whether liquids or gases. Whether we heat water in a pan or heat

Convection ovens are simply ovens with a fan inside. Cooking is speeded up by the circulation of heated air.

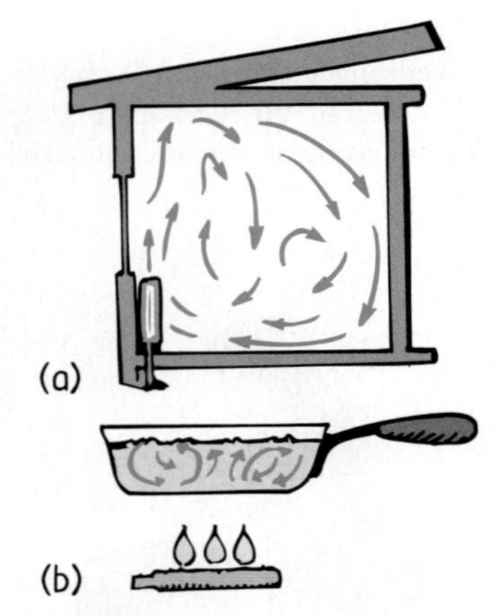

FIGURE 6.27
(a) Convection currents in air.
(b) Convection currents in liquid.

air in a room, the process is the same (Figure 6.27). As the fluid is heated from below, the molecules at the bottom move faster, spread apart more, become less dense, and are buoyed upward. Denser, cooler fluid moves in to take their place. In this way, convection currents keep the fluid stirred up as it heats—warmer fluid moving away from the heat source and cooler fluid moving toward the heat source.

We can see why warm air rises. When warmed, it expands, becomes less dense, and is buoyed upward in the cooler surrounding air like a balloon buoyed upward. When the rising air reaches an altitude at which the air density is the same, it no longer rises. We see this occurring when smoke from a fire rises and then settles off as it cools and its density matches that of the surrounding air. To see for yourself that expanding air cools, right now do the experiment shown in Figure 6.28. Expanding air really does cool.*

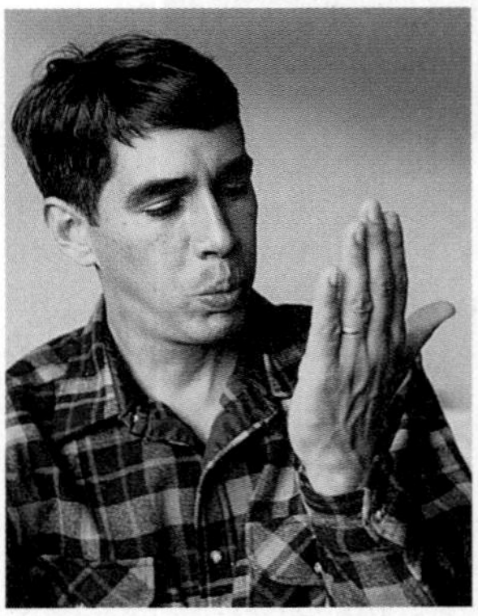

FIGURE 6.28
Blow warm air onto your hand from your wide-open mouth. Now reduce the opening between your lips so that the air expands as you blow. Try it now. Do you notice a difference in the air temperature?

A dramatic example of cooling by expansion occurs when steam expands through the nozzle of a pressure cooker (Figure 6.29). The combined cooling effects of expansion and rapid mixing with cooler air will allow you to hold your hand comfortably in the jet of condensed vapor. (Caution: If you try this, be sure to place your hand high above the nozzle at first and then lower it slowly to a comfortable distance above the nozzle. If you put your hand directly at the nozzle where no steam is visible, watch out! Steam is invisible and is clear of the nozzle before it expands and cools. The cloud of "steam" you see is actually condensed water vapor, which is much cooler than live steam.)

FIGURE 6.29
The hot steam expands from the pressure cooker and is cool to Millie's touch.

Cooling by expansion is the opposite of what occurs when air is compressed. If you've ever compressed air with a tire pump, you probably noticed that both air and pump became quite hot. Compression of hot air warms it.

Convection currents stir the atmosphere and produce winds. Some parts of Earth's surface absorb thermal energy from the Sun more readily than others. This results in uneven heating of the air near the ground. We see this effect at the seashore, as Figure 6.30 shows. In the daytime, the ground warms up more than the water. Then the warmed air close to the ground rises and is replaced by cooler air that moves in from above the water. The result is a sea breeze. At night, the process reverses because the shore cools off more quickly than the water, and then the warmer air is over the sea. If you build a fire on the beach, you'll see that the smoke sweeps inland during the day and then seaward at night.

As something expands, it spreads its energy over a greater area and therefore it cools.

*Where does the energy go in this case? It goes to work done on the surrounding air as the expanding air pushes outward.

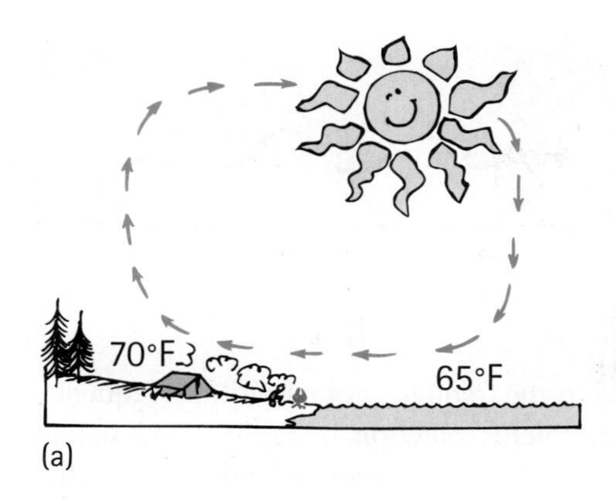

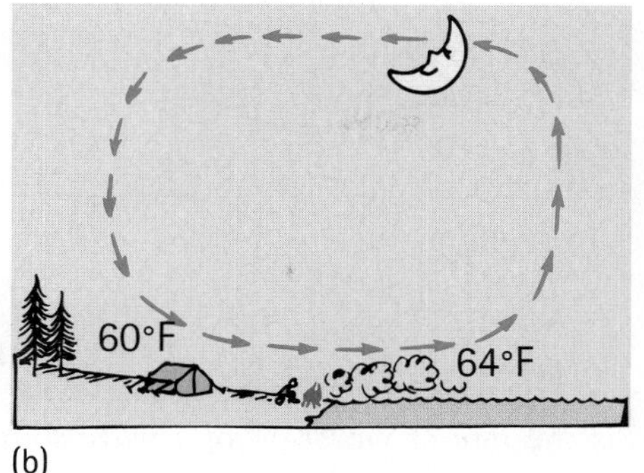

FIGURE 6.30
Convection currents produced by unequal heating of land and water. (a) During the day, warm air above the land rises, and cooler air over the water moves in to replace it. (b) At night, the direction of the airflow is reversed because now the water is warmer than the land.

CHECK YOURSELF

Explain why you can hold your fingers beside the candle flame without harm, but not above the flame.

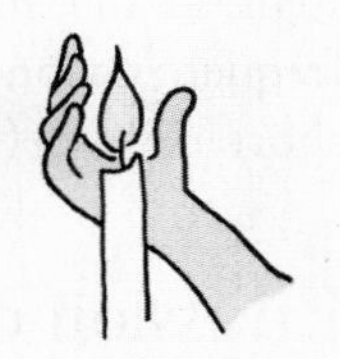

CHECK YOUR ANSWER

Thermal energy travels upward by convection. Since air is a poor conductor, very little energy travels sideways to your fingers.

UNIFYING CONCEPT

- *Convection*

6.10 Heat Transfer: Radiation

LEARNING OBJECTIVE
Describe the nature of radiant energy.

EXPLAIN THIS How do we know the temperatures of stars?

Thermal radiation from the Sun travels through space and then through Earth's atmosphere and warms Earth's surface. This energy cannot pass through the empty space between the Sun and Earth by conduction or convection because there is no medium for doing so. Energy must be transmitted some other way—by **radiation**.* The energy so radiated is called radiant energy.

Radiant energy exists in the form of electromagnetic waves. It includes a wide span of waves, ranging from longest to shortest: radio waves, microwaves, infrared waves (invisible waves below red in the visible spectrum), visible waves, then to waves that can't be seen by the eye, including ultraviolet waves, X-rays, and gamma rays. (We'll discuss electromagnetic waves further in Chapter 7 and other types of waves in Chapter 8.)

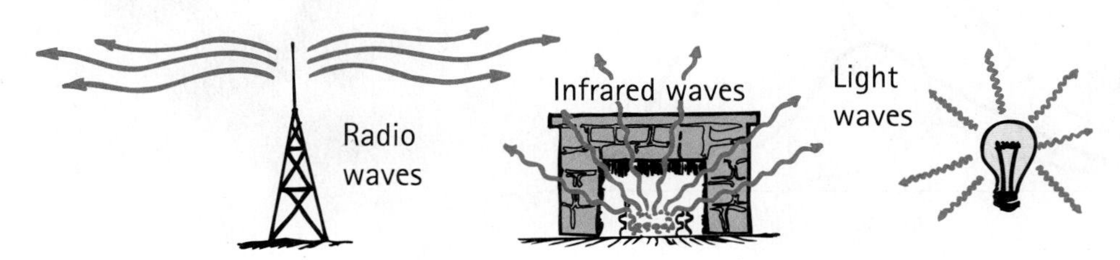

FIGURE 6.31
Types of radiant energy (electromagnetic waves).

*The radiation we are talking about here is electromagnetic radiation, including visible light. Don't confuse this with radioactivity, a process of the atomic nucleus that we'll discuss in Chapter 10.

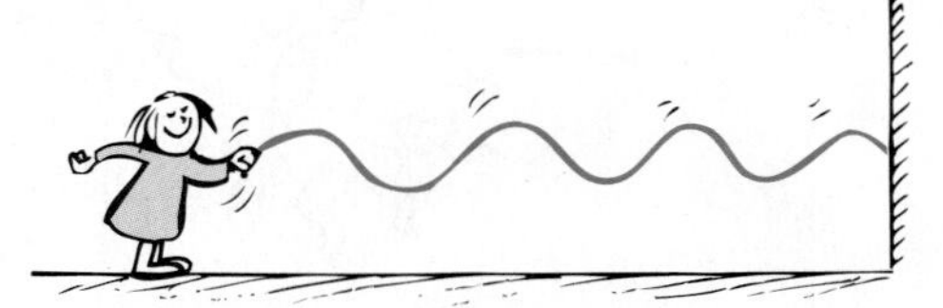

FIGURE 6.32
A wave of long wavelength is produced when the rope is shaken gently (at a low frequency). When it is shaken more vigorously (at a high frequency), a wave of shorter wavelength is produced.

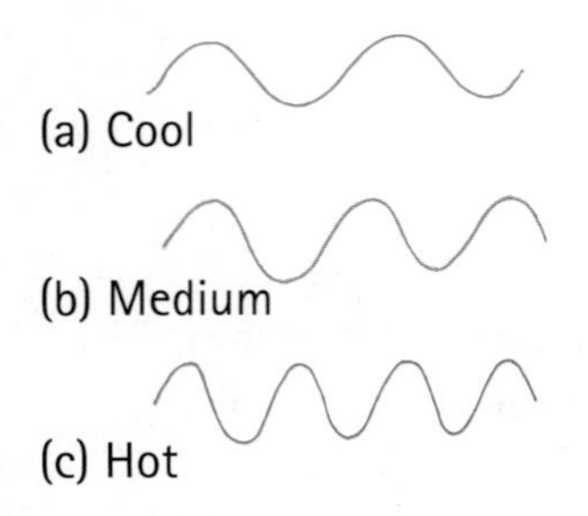

FIGURE 6.33
(a) A low-temperature (cool) source emits primarily low-frequency, long-wavelength waves. (b) A medium-temperature source emits primarily medium-frequency, medium-wavelength waves. (c) A high-temperature (hot) source emits primarily high-frequency, short-wavelength waves.

The wavelength of radiation is related to the frequency of vibration. Frequency is the rate of vibration of a wave source. Nellie Newton in Figure 6.32 shakes a rope at both a low frequency (left) and a higher frequency (right). Note that shaking at a low frequency produces a long, lazy wave, and the higher-frequency shake produces shorter waves. This is also true with electromagnetic waves. We shall see in Chapter 8 that vibrating electrons emit electromagnetic waves. Low-frequency vibrations produce long waves, and high-frequency vibrations produce shorter waves (Figure 6.33).

Emission of Radiant Energy

All substances at any temperature above absolute zero emit radiant energy. The average frequency f of the radiant energy is directly proportional to the absolute temperature T of the emitter: $f \sim T$.

Figure 6.34 shows radiation curves for an object at three sample temperatures. At these temperatures, the peak radiation frequencies are in the infrared part of the electromagnetic spectrum.

The fact that all objects in our environment continuously emit infrared radiation underlies the increasingly common infrared thermometers (Figure 6.35). Quite remarkably, you simply point the thermometer at something whose temperature you want, press a button, and a digital temperature reading appears. There is no need to touch the thermometer to whatever is being measured. The radiation it emits provides the reading. Typical classroom infrared thermometers operate in the range of about −30°C to 200°C.

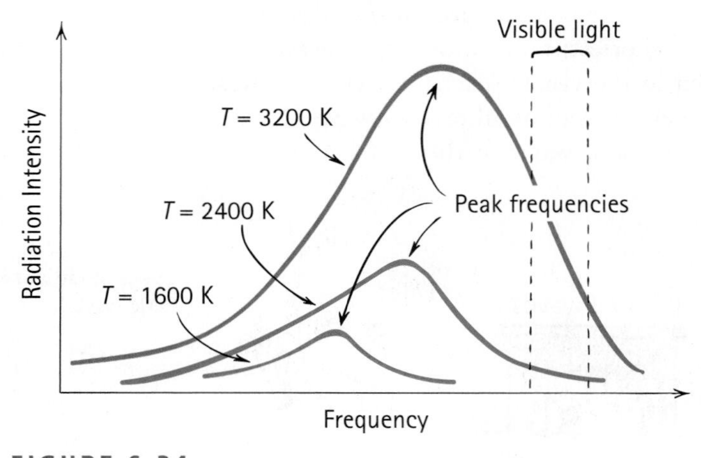

FIGURE 6.34
Radiation curves for different temperatures. The peak frequency of radiant energy is directly proportional to the absolute temperature of the emitter.

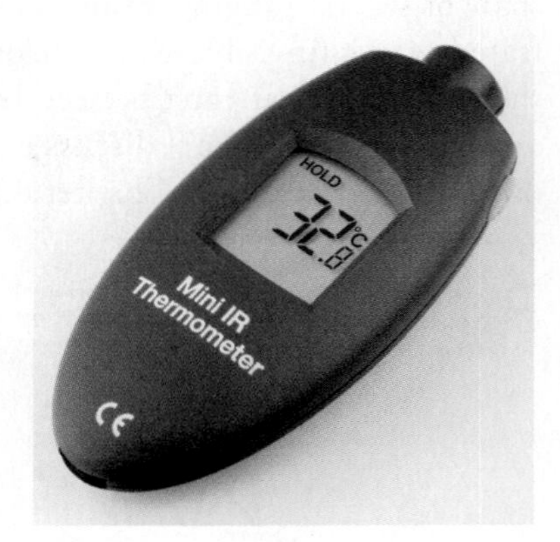

FIGURE 6.35
An infrared thermometer measures the infrared radiant energy emitted by a body and converts it into temperature.

If an object is hot enough, some of the radiant energy it emits is in the range of visible light. At a temperature of about 500°C, an object begins to emit the longest waves we can see, red light. Higher temperatures produce a yellowish light. At about 1200°C all the different waves to which the eye is sensitive are emitted and we see an object as "white hot." A blue-hot star is hotter than a white-hot star, and a red-hot star is less hot. Since a blue-hot star has twice the light frequency of a red-hot star, it therefore has twice the surface temperature of a red-hot star.

The surface of the Sun has a high temperature (by earthly standards) and therefore emits radiant energy at a high frequency—much of it in the visible portion of the electromagnetic spectrum. Earth's surface, by comparison, is relatively cool, and so the radiant energy it emits has a frequency lower than that of visible light. The radiation emitted by Earth is in the form of infrared waves—below our threshold of sight. As you will learn in Chapter 25, radiant energy emitted by Earth is called *terrestrial radiation*.

Radiant energy is emitted by both the Sun and Earth, and it differs only in the range of frequencies and the amount (Figure 6.36). When we study meteorology in Chapter 25, we'll learn how the atmosphere is transparent to the high-frequency solar radiation but opaque to much of the lower-frequency terrestrial radiation. This produces the greenhouse effect, which plays a role in global warming.

All objects—you, your instructor, and everything in your surroundings—continuously emit radiant energy in a mixture of frequencies (because temperature corresponds to a mixture of different molecular kinetic energies). Objects with everyday temperatures mostly emit low-frequency infrared waves. When your skin absorbs the higher-frequency infrared waves, you feel the sensation of heat. So it is common to refer to infrared radiation as heat radiation.

Common infrared sources that give the sensation of heat are the burning embers in a fireplace, a lamp filament, and the Sun. All of these sources emit both infrared radiation and visible light. When this radiant energy falls on other objects, it is partly reflected and partly absorbed. The part that is absorbed increases the thermal energy of the objects on which it falls.

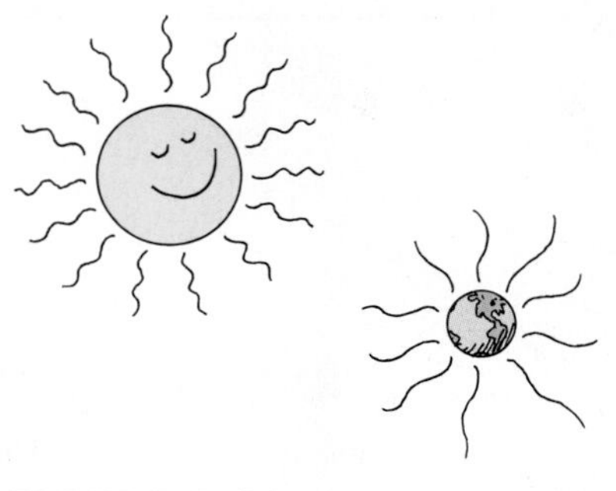

FIGURE 6.36
Both the Sun and Earth emit the same kind of radiant energy. The Sun's glow consists of longer waves and isn't visible to the eye.

CHECK YOURSELF

Which of these do not emit radiant energy: (a) the Sun, (b) lava from a volcano, (c) red-hot coals, (d) this textbook?

CHECK YOUR ANSWER

All emit radiant energy—even your textbook, which, like the other substances, has a temperature. According to the relation $f \sim T$, the book emits radiation whose average frequency f is quite low compared with the radiation frequencies emitted by the other substances. Everything with any temperature above absolute zero emits radiant energy. That's right—everything!

Absorption of Radiant Energy

If everything is radiating energy, why doesn't everything finally run out of energy? The answer is that everything is also absorbing energy. Good emitters of radiant energy are also good absorbers; poor emitters are poor absorbers. For

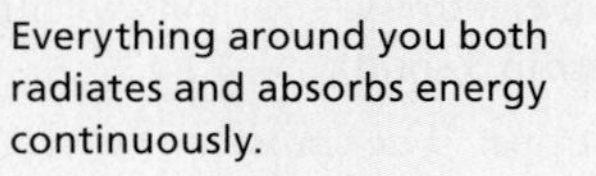

Everything around you both radiates and absorbs energy continuously.

FIGURE 6.37
When the black, rough-surfaced container and the shiny, polished one are filled with hot (or cold) water, the blackened one cools (or warms) faster.

example, a radio dish antenna constructed to be a good emitter of radio waves is also, by its very design, a good receiver (absorber) of them. A poorly designed transmitting antenna is also a poor receiver.

It's interesting to note that, if a good absorber were not also a good emitter, black objects would remain warmer than lighter-colored objects, and the two would never reach a common temperature. Objects in thermal contact, given sufficient time, will reach the same temperature. A blacktop pavement may remain hotter than the surroundings on a hot day, but, at nightfall, it cools faster. Sooner or later, all objects come to thermal equilibrium. So, a dark object that absorbs a lot of radiant energy must emit a lot as well (Figure 6.37).

The surface of any material, hot or cold, both absorbs and emits radiant energy. If the surface absorbs more energy than it emits, it is a net absorber and its temperature rises. If it emits more than it absorbs, it is a net emitter and its temperature drops. Whether a surface plays the role of net absorber or net emitter depends on whether its temperature is higher or lower than that of its surroundings. In short, if it's hotter than its surroundings, the surface will be a net emitter and will cool; if it's colder than its surroundings, it will be a net absorber and will warm.

A hot pizza put outside on a winter day is a net emitter. The same pizza placed in a hotter oven is a net absorber.

CHECK YOURSELF

1. **If a good absorber of radiant energy were a poor emitter, how would its temperature compare with the temperature of its surroundings?**
2. **A farmer turns on the propane burner in his barn on a cold morning and heats the air to 20°C (68°F). Why does he still feel cold?**

CHECK YOUR ANSWERS

1. If a good radiator were not also a good emitter, there would be a net absorption of radiant energy and the temperature of the absorber would remain higher than the temperature of the surroundings. Things around us approach a common temperature only because good absorbers are, by their very nature, also good emitters.
2. The walls of the barn are still cold. The farmer radiates more energy to the walls than the walls radiate back at him, and he feels chilly. (On a winter day, you are comfortable inside your home or classroom only if the walls are warm—not just the air.)

For instructor-assigned homework, go to www.masteringphysics.com 

SUMMARY OF TERMS (KNOWLEDGE)

Absolute zero The lowest possible temperature that a substance may have—the temperature at which molecules of the substance have their minimum kinetic energy.

Calorie The amount of heat needed to change the temperature of 1 gram of water by 1 Celsius degree.

Conduction The transfer of thermal energy by molecular and electronic collisions within a substance (especially within a solid).

Convection The transfer of thermal energy in a gas or liquid by means of currents in the heated fluid. The fluid flows, carrying energy with it.

Entropy The measure of the energy dispersal of a system. Whenever energy freely transforms from one form into another, the direction of transformation is toward a state of greater disorder and, therefore, toward greater entropy.

First law of thermodynamics A restatement of the law of energy conservation, usually as it applies to systems involving changes in temperature: Whenever heat flows into or out of a system, the gain or loss of thermal energy equals the amount of heat transferred.

Heat The thermal energy that flows from a substance of higher temperature to a substance of lower temperature, commonly measured in calories or joules.

Radiation The transfer of energy by means of electromagnetic waves.

Second law of thermodynamics Heat never spontaneously flows from a lower-temperature substance to a higher-temperature substance. Also, all systems tend to become more and more disordered as time goes by.

Specific heat capacity The quantity of heat required to raise the temperature of a unit mass of a substance by 1 degree Celsius.

Temperature A measure of the hotness or coldness of substances, related to the average translational kinetic energy per molecule in a substance; measured in degrees Celsius, in degrees Fahrenheit, or in kelvins.

Thermal energy The total energy (kinetic plus potential) of the submicroscopic particles that make up a substance (often called *internal energy*).

Thermodynamics The study of heat and its transformation into different forms of energy.

Third law of thermodynamics No system can reach absolute zero.

READING CHECK QUESTIONS (COMPREHENSION)

6.1 The Kinetic Theory of Matter

1. What kinds of particle motion account for thermal energy?
2. Why does a penny become warmer when it is struck by a hammer?

6.2 Temperature

3. What are the temperatures for freezing water on the Celsius and Fahrenheit scales? What are the temperatures for boiling water on those scales?
4. Is the temperature of an object a measure of the total translational kinetic energy of the molecules in the object or a measure of the average translational kinetic energy per molecule in the object?
5. What is meant by this statement: "A thermometer measures its own temperature"?

6.3 Absolute Zero

6. What pressure would you expect in a rigid container of 0°C gas if you cooled it by 273 Celsius degrees?
7. How much energy can be taken from a system at a temperature of 0 K?

6.4 What Is Heat?

8. When you touch a cold surface, does cold travel from the surface to your hand or does energy travel from your hand to the cold surface? Explain.
9. (a) Distinguish between temperature and heat.
   (b) Distinguish between heat and thermal energy.
10. What determines the direction of heat flow?
11. Distinguish between a calorie and a Calorie, and between a calorie and a joule.

6.5 The Laws of Thermodynamics

12. How does the law of conservation of energy relate to the first law of thermodynamics?
13. What happens to the thermal energy of a system when mechanical work is done on the system? What happens to the temperature of the system?
14. How does the second law of thermodynamics relate to the direction of heat flow?

6.6 Specific Heat Capacity

15. Which warms up faster when heat is applied—iron or silver?
16. Does a substance that heats up quickly have a high or low specific heat capacity?
17. How does the specific heat capacity of water compare with the specific heat capacities of other common materials?

6.7 Thermal Expansion

18. Which generally expands more for an equal increase in temperature—solids or liquids?
19. What is the reason ice is less dense than water?
20. Why does ice form at the surface of a pond instead of at the bottom?

6.8 Heat Transfer: Conduction

21. What is the role of "loose" electrons in heat conduction?
22. Distinguish between a heat conductor and a heat insulator.
23. Why is a barefoot fire walker able to walk safely on red-hot wooden coals?
24. Why are such materials as wood, fur, and feathers—and even snow—good insulators?

6.9 Heat Transfer: Convection

25. Describe how convection transfers heat.
26. What happens to the temperature of air when it expands?
27. Why does the direction of coastal winds change from day to night?

6.10 Heat Transfer: Radiation

28. (a) What exactly is radiant energy? (b) What is heat radiation?
29. How does the frequency of radiant energy relate to the absolute temperature of the radiating source?
30. Since all objects continuously radiate energy to their surroundings, why don't the temperatures of all objects continuously decrease?

THINK INTEGRATED SCIENCE

6A—Entropy

31. What does it mean to say that energy becomes less useful when it is transformed?
32. What is the physicist's term for the measure of energy dispersal?
33. Consider the decomposition of water (H_2O) to form hydrogen (H_2) and oxygen (O_2). At room temperature, the products of this reaction have less entropy than the reactants. Is this reaction thermodynamically favored? Justify your answer.
34. A deer is a more concentrated form of energy than the grass it feeds on. Does this imply that the second law of thermodynamics is violated as the deer converts its food into tissue? Explain.

6B—Specific Heat Capacity and Earth's Climate

35. Northeastern Canada and much of Europe receive about the same amount of sunlight per unit area. Why, then, is Europe generally warmer in the winter months?
36. Iceland, so named to discourage conquest by expanding empires, is not ice-covered like Greenland and parts of Siberia, even though it is nearly on the Arctic Circle. The average winter temperature of Iceland is considerably higher than the temperatures of regions at the same latitude in eastern Greenland and central Siberia. Why is this so?
37. Why does the presence of large bodies of water tend to moderate the climate of nearby land—to make it warmer in cold weather and cooler in warm weather?

THINK AND DO (HANDS-ON APPLICATION)

38. Hold the bottom end of a test tube full of cold water in your hand. Heat the top part in a flame until the water boils. The fact that you can still hold the bottom shows that water is a poor conductor of heat. This is even more dramatic when you wedge chunks of ice at the bottom; then the water above can be brought to a boil without melting the ice. Try it and see.

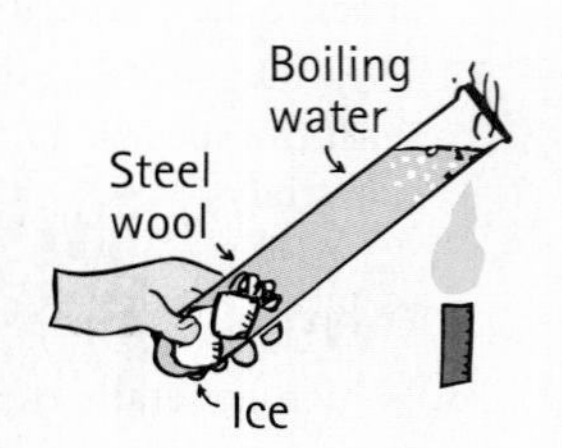

39. Wrap a piece of paper around a thick metal bar and place it in a flame. Note that the paper will not catch fire. Can you figure out why? (Paper will generally not ignite until its temperature reaches 233°C.)
40. Watch the spout of a teakettle of boiling water. Notice that you cannot see the steam that issues from the spout. The cloud that you see farther away from the spout is not steam but condensed water droplets. Now hold the flame of a candle in the cloud of condensed steam. Can you explain your observations?

PLUG AND CHUG (FORMULA FAMILIARIZATION)

$$Q = cm\Delta T$$

41. Show that 5000 cal is required to increase the temperature of 50 g of water from 0°C to 100°C. The specific heat capacity of water is 1 cal/g·°C.
42. Calculate the quantity of heat absorbed by 20 g of water that warms from 30°C to 90°C.

THINK AND COMPARE (ANALYSIS)

43. Rank the magnitudes of these units of thermal energy from greatest to least: (a) 1 calorie, (b) 1 Calorie, (c) 1 joule.
44. The precise volume of water in a beaker depends on the temperature of the water. Rank from greatest to least the volumes of water at these temperatures: (a) 0°C, (b) 4°C, (c) 10°C.
45. From best to worst, rank these materials as heat conductors: (a) copper wire, (b) snow, (c) a glass rod.
46. From greatest to least, rank the frequencies of radiation of these emitters of radiant energy: (a) red-hot star, (b) blue-hot star, (c) the Sun.

THINK AND SOLVE (MATHEMATICAL APPLICATION)

47. Pounding a nail into wood makes the nail warmer. Consider a 5-g steel nail 6 cm long and a hammer that exerts an average force of 500 N on the nail when it is being driven into a piece of wood. Show that the increase in the nail's temperature will be 13.3°C. (Assume that the specific heat capacity of steel is 450 J/kg·°C.)
48. If you wish to warm 100 kg of water by 20°C for your bath, show that 8370 kJ of heat is required.

49. The specific heat capacity of copper is 0.092 calorie per gram per degree Celsius. Show that the heat required to raise the temperature of a 10-g piece of copper from 0°C to 100°C is 92 calories.
50. When 100 g of 40°C iron nails is submerged in 100 g of 20°C water, show that the final temperature of the water will be 22.1°C. (The specific heat capacity of iron is 0.12 J/kg·°C. Here, you should equate the heat gained by the water to the heat lost by the nails.)
51. A 10-kg iron ball is dropped onto a pavement from a height of 100 m. If half the heat generated goes into warming the ball, show that the temperature increase of the ball will be 1.1°C. (In SI units, the specific heat capacity of iron is 450 J/kg·°C.) Why is the answer the same for an iron ball of any mass?

THINK AND EXPLAIN (SYNTHESIS)

52. Which is greater—an increase in temperature of 1°C or an increase of 1°F?
53. Which has the greater amount of thermal energy—an iceberg or a hot cup of coffee? Explain.
54. When air is rapidly compressed, why does its temperature increase?
55. What happens to the gas pressure within a sealed gallon can when it is heated? What happens to the pressure when the can is cooled? Why?
56. After a car has been driven for some distance, why does the air pressure in the tires increase?
57. Why doesn't adding the same amount of heat to two different objects necessarily produce the same increase in temperature?
58. Why will a watermelon stay cool for a longer time than sandwiches when both are removed from a cooler on a hot day?
59. Cite an exception to the claim that all substances expand when heated.
60. An old method for breaking boulders was to put them into a hot fire and then douse them with cold water. Why does this fracture the boulders?
61. A metal ball is just able to pass through a metal ring. When the ball is heated, however, it will not pass through the ring. What would happen if the ring, rather than the ball, were heated? Would the size of the hole increase, stay the same, or decrease?

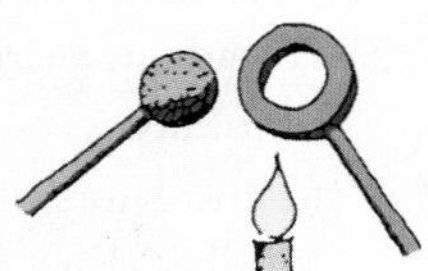

62. After a machinist very quickly slips a hot, snugly fitting iron ring over a very cold glass cylinder, there is no way that the two can be separated intact. Why is this so?
63. Suppose you cut a small gap in a metal ring. If you heat the ring, will the gap become wider or narrower?
64. Why is it important to protect water pipes so that they don't freeze?
65. If you wrap a fur coat around a thermometer, will its temperature rise?
66. If you hold one end of a nail against a piece of ice, the end in your hand soon becomes cold. Does cold flow from the ice to your hand? Explain.
67. From the rules that a good absorber of radiation is a good radiator and a good reflector is a poor absorber, state a rule relating the reflecting and radiating properties of a surface.
68. Suppose that, at a restaurant, you are served coffee before you are ready to drink it. So that it will be as warm as possible when you are ready to drink it, would you be wiser to add cream right away or to add it just before you are ready to drink it?

THINK AND DISCUSS (EVALUATION)

69. In your room, there are tables, chairs, other people, and other things. Discuss which of these things has a temperature (1) lower than, (2) higher than, (3) equal to the temperature of the air.
70. Discuss why you can't establish whether you are running a high temperature by touching your own forehead.
71. Use the laws of thermodynamics to defend the statement that 100% of the electrical energy that goes into lighting an incandescent lamp is converted into thermal energy.
72. If you drop a hot rock into a pail of water, the temperature of the rock and the water will change until both are equal. The rock will cool and the water will warm. Does this hold true if the hot rock is dropped into the Atlantic Ocean? Discuss.
73. On cold winter nights in the old days it was common to bring a hot object to bed. Which would do a better job of keeping you warm through the cold night—a 10-kg iron brick or a 10-kg jug of hot water at the same temperature? Discuss.
74. Would you or the gas company gain by having gas warmed before it passes through your gas meter? Discuss.
75. Suppose that water instead of mercury is used in a thermometer. If the temperature is at 4°C and then changes, why can't the thermometer indicate whether the temperature is rising or falling?

76. In terms of physics, speculate as to why some restaurants serve baked potatoes wrapped in aluminum foil.
77. Wood is a better insulator than glass, yet fiberglass is commonly used as an insulator in wooden buildings. Discuss.
78. Visit a snow-covered cemetery and note that the snow does not slope upward against the gravestones but, instead, forms depressions around them, as shown. Can you think of a reason for this?

79. Why is it that you can safely hold your bare hand in a hot pizza oven for a few seconds, but, if you were to touch the metal inside, you'd burn yourself?
80. In a still room, smoke from a candle will sometimes rise only so far, not reaching the ceiling. Discuss.
81. After boiling a bit of water in a gallon can and then sealing the can when water vapor has driven out most of the air, Dan Johnson and his class watch the can slowly crumple. Discuss the role of condensation of water vapor inside the can. What does the crumpling?

READINESS ASSURANCE TEST (RAT)

If you have a good handle on this chapter, then you should be able to score at least 7 out of 10 on this RAT. If you score less than 7, you need to study further before moving on.

Choose the BEST answer to each of the following:

1. When scientists discuss kinetic energy per molecule, the concept being discussed is
   (a) temperature.
   (b) heat.
   (c) thermal energy.
   (d) entropy.
2. In a mixture of hydrogen gas, oxygen gas, and nitrogen gas, the molecules with the greatest average speed are those of
   (a) hydrogen.
   (b) oxygen.
   (c) nitrogen.
   (d) All have the same speed.
3. Your garage gets messier every day. In this case entropy is
   (a) decreasing.
   (b) increasing.
   (c) holding steady.
   (d) none of these
4. The specific heat capacity of aluminum is more than twice that of copper. If equal quantities of heat are added to equal masses of aluminum and copper, the metal that more rapidly increases in temperature is
   (a) aluminum.
   (b) copper.
   (c) Both increase at the same rate.
   (d) none of these
5. A bimetallic strip used in thermostats relies on the fact that different metals have different
   (a) specific heat capacities.
   (b) thermal energies at different temperatures.
   (c) rates of thermal expansion.
   (d) all of these
6. Water at 4°C will expand when it is
   (a) slightly cooled.
   (b) slightly warmed.
   (c) both cooled and warmed.
   (d) neither cooled nor warmed.
7. The principal reason one can walk barefoot on red-hot wood coals without burning the feet has to do with
   (a) the low temperature of the coals.
   (b) the low thermal conductivity of the coals.
   (c) mind-over-matter techniques.
   (d) none of these
8. Thermal convection is linked mostly to
   (a) radiant energy.
   (b) fluids.
   (c) insulators.
   (d) all of these
9. Which of these electromagnetic waves has the lowest frequency?
   (a) infrared
   (b) visible
   (c) ultraviolet
   (d) gamma rays
10. Compared with terrestrial radiation, the radiation from the Sun has
   (a) a longer wavelength.
   (b) a lower frequency.
   (c) both a longer wavelength and a lower frequency.
   (d) neither a longer wavelength nor a lower frequency.

Answers to RAT

1. a, 2. a, 3. b, 4. b, 5. c, 6. c, 7. b, 8. b, 9. a, 10. d

CHAPTER 7
Electricity and Magnetism

ELECTRICITY UNDERLIES almost everything around us. It's in the lightning from the sky, it powers devices from computers to flashlights, and it's what holds atoms together to form molecules. But what *is* electricity? Why can you feel a spark when you grab a doorknob after scuffing your feet along the carpet? What's the difference between electric current and voltage—and which of these gives us electric shocks? How do electric circuits work? Is electricity related to magnetism—if so, how? How do the magnetic strips on credit cards, timing systems for traffic lights, and metal detectors at airports tap both magnetic and electrical forces? We begin our study with a fundamental idea—the concept of electric charge.

LEARNING OBJECTIVE
Distinguish between negative and positive electric charges.

7.1 Electrical Force and Charge

EXPLAIN THIS What is meant by saying that electric charge is conserved?

UNIFYING CONCEPT
- ***The Electric Force***

Suppose that the universe consisted of two kinds of particles—say, positive and negative. Suppose that positives repelled positives but attracted negatives, and that negatives repelled negatives but attracted positives. What would the universe be like? The answer is simple: It would be like the one we are living in. For there *are* such particles and there *is* such a force—the *electric force.*

The terms *positive* and *negative* refer to electric charge, the fundamental quantity that underlies all electric phenomena. The positively charged particles in ordinary matter are protons, and the negatively charged particles are electrons. Charge is not something added to protons and electrons. It is a basic attribute of them, just as gravitational attractiveness is an attribute of any mass.

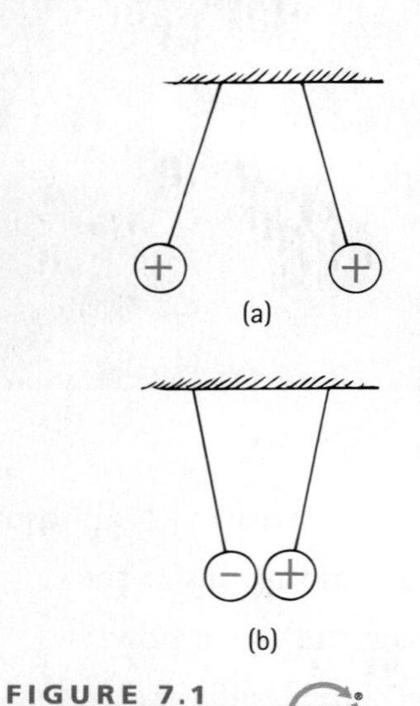

FIGURE 7.1
INTERACTIVE FIGURE

(a) Like charges repel. (b) Unlike charges attract.

The attraction between protons and electrons pulls them together into tiny units—atoms. (Atoms also contain neutral particles called neutrons. More interesting details about atoms are presented in Part Two of this book.) In order to understand the basic principles of electricity, however, we must preview some fundamental facts about atoms:

1. Every atom is composed of a positively charged nucleus surrounded by negatively charged electrons.
2. Each of the electrons in any atom has the same quantity of negative charge and the same mass. Electrons are identical to one another.
3. Protons and neutrons make up the nucleus. (The only exception is the most common form of hydrogen atom, which has no neutrons.) Protons are about 1800 times more massive than electrons, but each proton carries an amount of positive charge equal to the negative charge of the electrons. Neutrons have slightly more mass than protons, and they have no net charge.

MasteringPhysics®
TUTORIAL: Electrostatics

Normally, an atom has as many electrons as protons. When an atom loses one or more electrons, it has a positive net charge; when it gains one or more electrons, it has a negative net charge. A charged atom is called an *ion.* A *positive ion* has a net positive charge. A *negative ion,* with one or more extra electrons, has a net negative charge.

Material objects are made of atoms, which in turn are composed of electrons and protons (and neutrons as well). Although the innermost electrons in an atom are attracted very strongly to the oppositely charged atomic nucleus, the outermost electrons of many atoms are attracted more loosely and can be dislodged. The amount of work required to tear an electron away from an atom varies for different

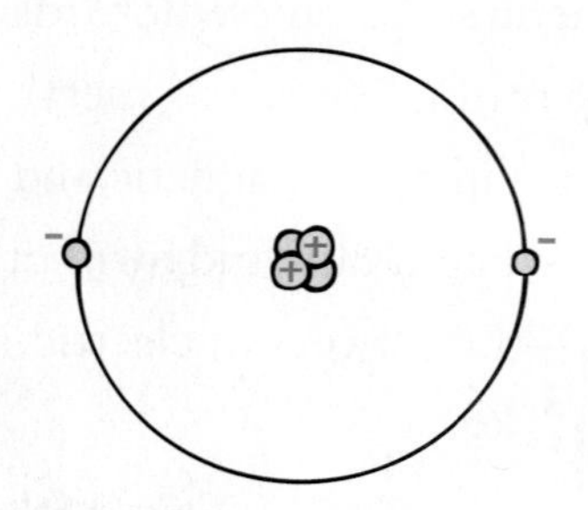

FIGURE 7.2
INTERACTIVE FIGURE

Model of a helium atom. The atomic nucleus is made up of two protons and two neutrons. The positively charged protons attract two negatively charged electrons. What is the net charge of this atom?

FIGURE 7.3
Fur has a greater affinity for electrons than plastic. So, when a plastic rod is rubbed with fur, electrons are transferred from the fur to the rod. The rod is then negatively charged. By how much compared with the rod? Positively or negatively?

substances. Electrons are held more firmly in rubber or plastic than in your hair, for example. Thus, when you rub a comb against your hair, electrons transfer from the hair to the comb. The comb then has an excess of electrons and is said to be *negatively charged*. Your hair, in turn, has a deficiency of electrons and is said to be *positively charged*.

So, protons attract electrons and we have atoms. Electrons repel electrons and we have matter—because atoms don't mesh into one another. This pair of rules is the guts of electricity.

Negative and *positive* are just names given to opposite charges. The names chosen could just as well have been *east* and *west*, or *up* and *down*, or *Mary* and *Larry*. Positive charge is not "better" than negative charge—the two kinds of charge are just opposites of each other.

Conservation of Charge

Another basic fact of electricity is that, whenever something is charged, no electrons are created or destroyed. Electrons are simply transferred from one material to another. Charge is *conserved*. In every event, whether large-scale or at the atomic and nuclear level, the principle of *conservation of charge* has always been found to apply. No case of the creation or destruction of charge has ever been found (Figure 7.4).

FIGURE 7.4
Why will you get a slight shock from the doorknob after scuffing across the carpet?

CHECK YOURSELF
If you walk across a rug and scuff electrons from your feet, are you negatively or positively charged?

CHECK YOUR ANSWER
You have fewer electrons after you scuff your feet, so you are positively charged (and the rug is negatively charged).

7.2 Coulomb's Law

LEARNING OBJECTIVE
Relate the inverse-square law to electrical forces.

EXPLAIN THIS What do Newton's law of gravity and Coulomb's law have in common?

The electric force, like the gravitational force, decreases inversely as the square of the distance between the charges. This relationship, which was discovered by Charles Coulomb in the 18th century, is called **Coulomb's law**. It states that, for two charged objects that are much smaller than the distance between them, the force between them varies directly as the product of their charges and inversely as the square of the distance between them. The force acts along a straight line from one charge to the other. Coulomb's law can be expressed as

$$F = k\frac{q_1 q_2}{d^2}$$

where d is the distance between the charged particles, q_1 is the charge on one particle, q_2 is the charge on the second particle, and k is the proportionality constant.

The unit of electric charge is called the **coulomb**, abbreviated C. It turns out that a charge of 1 C is equal in magnitude to the total charge of 6.25 billion billion electrons. This might seem like a great number of electrons, but it represents only the amount of charge that flows through a common 100-watt lightbulb in a little more than a second. Dividing the value of 1 C by the number of electrons with this much charge, we find the charge on a single electron: 1.60×10^{-19} C. An electron and a proton carry this same magnitude of charge.

Because all charged objects carry multiples of this charge, 1.60×10^{-19} C is considered the *fundamental charge*.

The proportionality constant k in Coulomb's law is similar to G in Newton's law of gravity. Instead of being a very small number, like G, however, k is a very large number—approximately

$$k = 9{,}000{,}000{,}000\ \text{N}\cdot\text{m}^2/\text{C}^2$$

In scientific notation, $k = 9.0 \times 10^9\ \text{N}\cdot\text{m}^2/\text{C}^2$. The unit $\text{N}\cdot\text{m}^2/\text{C}^2$ is not central to our interest here; it simply converts the right-hand side of the equation to the unit of force, the newton (N). What is important is the large magnitude of k. If, for example, a pair of charges of 1 coulomb each were 1 meter apart, the force of repulsion between the two would be 9 billion N. That would be about ten times the weight of a battleship. Obviously, such quantities of unbalanced charge do not usually exist in our everyday environment.

So, Newton's law of gravitation for masses is similar to Coulomb's law for electrically charged bodies. The most important difference between gravitational and electrical forces is that electrical forces may be either attractive or repulsive, whereas gravitational forces are only attractive.

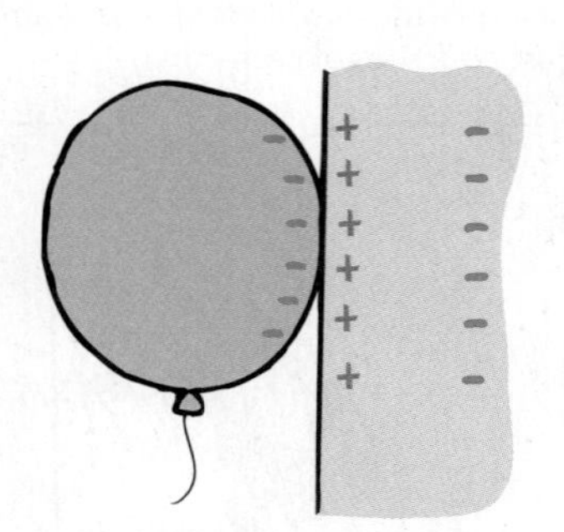

FIGURE 7.5
The negatively charged balloon polarizes molecules in the wooden wall and creates a positively charged surface, so the balloon sticks to the wall.

CHECK YOURSELF

1. **The proton is the nucleus of the hydrogen atom, and it attracts the electron that orbits it. Relative to this force, does the electron attract the proton with less force, more force, or the same amount of force?**
2. **If a proton at a particular distance from a charged particle is repelled with a given force, by how much will the force decrease when the proton is three times as distant from the particle? When it is five times as distant?**
3. **What is the sign of the charge on the particle in this case?**

CHECK YOUR ANSWERS

1. The same amount of force, in accord with Newton's third law—basic mechanics! Recall that a force is an interaction between two things—in this case, between the proton and the electron. They pull on each other equally.
2. In accord with the inverse-square law, it decreases to $\frac{1}{9}$ of its original value when it is three times farther away and to $\frac{1}{25}$ of its original value when it is five times farther away.
3. The sign is positive.

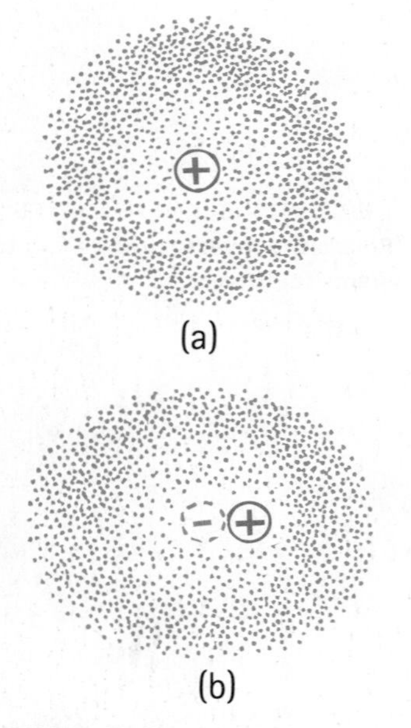

FIGURE 7.6
An electron buzzing around the atomic nucleus makes up an electron cloud. (a) The center of the negative "cloud" of electrons coincides with the center of the positive nucleus in an atom. (b) When an external negative charge is brought nearby, as on a charged balloon, the electron cloud is distorted so that the centers of negative and positive charge no longer coincide. The atom is electrically polarized.

Charge Polarization

If you charge an inflated balloon by rubbing it on your hair and then place the balloon against a wall, it sticks (Figure 7.5). This is because the charge on the balloon alters the charge distribution in the atoms or molecules in the wall, effectively inducing an opposite charge on the wall. The molecules cannot move from their relatively stationary positions, but their "centers of charge" are moved. The positive part of the atom or molecule is attracted toward the balloon while the negative part is repelled. This has the effect of distorting the atom or molecule (Figure 7.6). The atom or molecule is said to be **electrically polarized**. (We will treat electrical polarization further in Chapter 12.)

7.3 Electric Field

LEARNING OBJECTIVE
Relate the gravitational field to the electric field.

EXPLAIN THIS How does the magnitude of an electric field relate to the force that acts on a positive test charge placed in the field?

Electrical forces, like gravitational forces, can act between things that are not in contact with each other. How do bodies that are not touching exert forces on one another? To model "action-at-a-distance" forces such as gravity and the electric force, we use the concept of the *force field*. Every mass is surrounded by a gravitational field, while an electric field surrounds any charged object. Since you are more familiar with the gravitational force than the electric force at this point, first consider how the gravitational field works. The space surrounding any mass is altered such that another mass introduced into this region experiences a force. This "alteration in space" is called its *gravitational field*. We can think of any other mass as interacting with the field and not directly with the mass that produces it. For example, when an apple falls from a tree, we say that it is interacting with the mass of Earth, but we can also think of an apple as responding to the gravitational field of Earth at that point.

MasteringPhysics®
VIDEO: Van de Graaff Generator

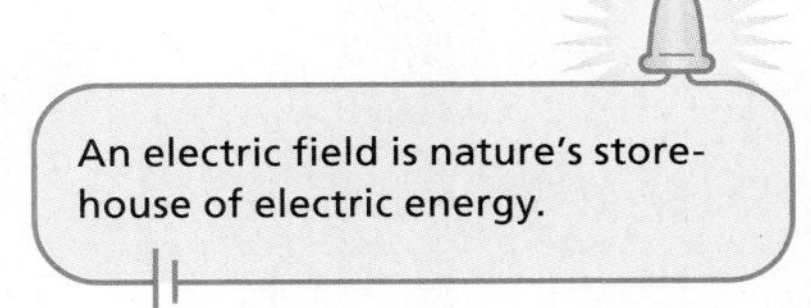

An electric field is nature's storehouse of electric energy.

Similarly, the space around every electric charge is filled with an **electric field**—an energetic aura that extends through space. An electric field is a vector quantity, having both magnitude and direction. The magnitude of the field at any point is simply the force per unit charge. If a charge q experiences a force F at some point in space, then the electric field at that point is $E = F/q$. The direction of the electric field is away from positive charge and toward negative charge.

If you place a charged particle in an electric field, it will experience a force. The direction of the force on a positive charge is the same direction as the field. The electric field can best be visualized with *field lines*. Field lines show the direction of the electric field—away from positive charge and toward negative charge. Field lines also show the intensity of the electric field—where the field lines are most tightly bunched together, the field is strongest. The electric field and field lines about an electron point toward the electron (Figure 7.7). The electric field and field lines about a proton point in the opposite direction—radially away from the proton. As with the electric force, the electric field about a particle obeys the inverse-square law. Some electric-field configurations are shown in Figure 7.8.

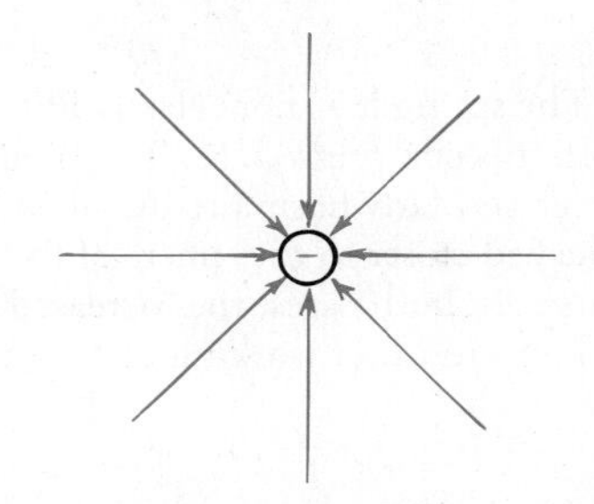

FIGURE 7.7
INTERACTIVE FIGURE (MP)

Electric-field representations about a negative charge.

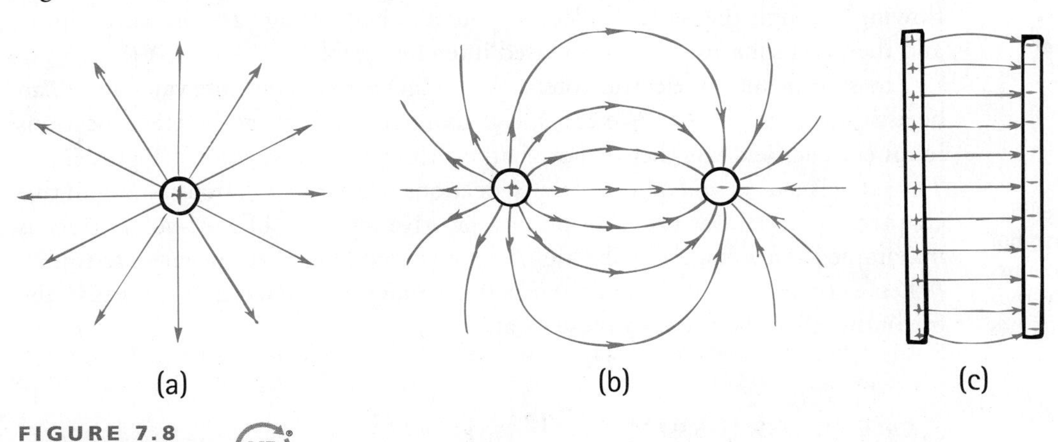

FIGURE 7.8
INTERACTIVE FIGURE (MP)

Some electric-field configurations. (a) Field lines about a single positive charge. (b) Field lines for a pair of equal but opposite charges. (c) Field lines between two oppositely charged parallel plates.

LEARNING OBJECTIVE
Distinguish between electric potential energy and electric potential.

7.4 Electric Potential

EXPLAIN THIS Why aren't you harmed when you touch a 5000-V party balloon?

In our study of energy in Chapter 4, we learned that an object has gravitational potential energy due to its location in a gravitational field. Similarly, a charged object has potential energy due to its location in an electric field. Just as work is required to lift a massive object against the gravitational field of Earth, work is required to push a charged particle against the electric field of a charged body. This work changes the electric potential energy (PE) of the charged particle. This is similar to the work done in compressing a spring, increasing the potential energy of the spring (Figure 7.9a). Likewise, the work done in pushing a positively charged particle closer to the positively charged sphere in Figure 7.9b increases the potential energy of the charged particle. We call this energy possessed by the charged particle that is due to its location **electric potential energy**. If the particle is released, it accelerates in a direction away from the sphere, and its electric potential energy changes to kinetic energy.

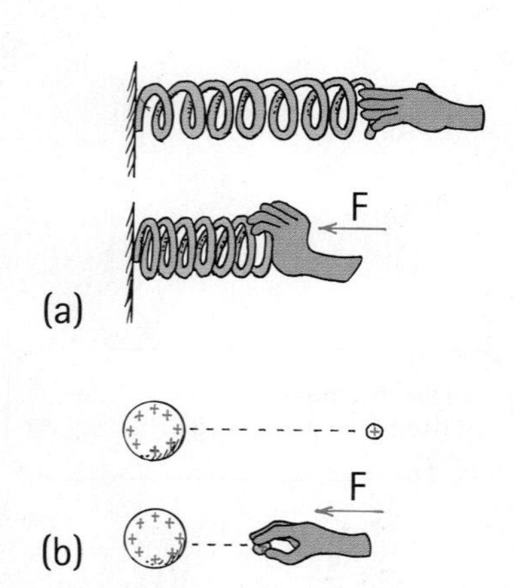

FIGURE 7.9
(a) The spring has more elastic PE when it is compressed. (b) The small charge similarly has more PE when it is pushed closer to the sphere of like charge. In both cases, the increased PE is the result of work input.

If we push a particle with twice the charge, we do twice as much work. Twice the charge in the same location has twice the potential energy; with three times the charge, there is three times as much potential energy; and so on. When working with electricity, rather than dealing with the total potential energy of a charged body, we find it is convenient to consider the electric potential energy *per charge*. We simply divide the amount of energy by the amount of charge. The electric potential energy per amount of charge is called **electric potential**; that is,

$$\text{Electric potential} = \frac{\text{electric potential energy}}{\text{charge}}$$

The unit of measurement for electric potential is the volt, so electric potential is often called *voltage*. A potential of 1 volt (V) equals 1 joule (J) of energy per coulomb (C) of charge:

$$1 \text{ volt} = \frac{1 \text{ joule}}{\text{coulomb}}$$

Thus, a 1.5-volt battery gives 1.5 joules of energy to every 1 coulomb of charge flowing through the battery. Electric potential and voltage are the same thing, and the two terms are commonly used interchangeably.

The significance of electric potential (voltage) is that a definite value for it can be assigned to a location. We can speak about the voltages at different locations in an electric field whether or not charges occupy those locations (Figure 7.10). The same is true of voltages at various locations in an electric circuit. Later in this chapter, we'll see that the location of a positive terminal of a 12-volt battery is maintained at a voltage 12 volts higher than the location of the negative terminal. When a conducting medium connects this voltage difference, any charges in the medium will move between these locations.

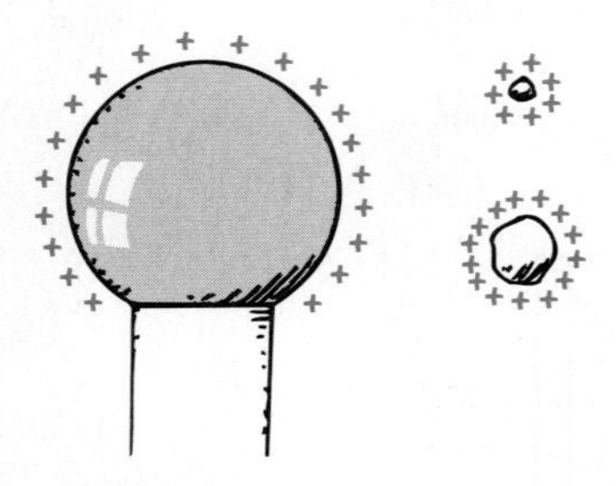

FIGURE 7.10
The larger test charge has more PE in the field of the charged dome, but the electric potential of any amount of charge at the same location is the same.

CHECK YOURSELF

1. **Suppose there were twice as many coulombs in one of the test charges near the charged sphere in Figure 7.10. Would the electric potential energy of this test charge relative to the charged sphere be the same,**

or would it be twice as great? Would the electric potential of the test charge be the same, or would it be twice as great?

2. What does it mean to say that your car has a 12-volt battery?

FIGURE 7.11
Although the voltage of the charged balloon is high, the electric potential energy is low because of the small amount of charge.

CHECK YOUR ANSWERS

1. The result of twice as many coulombs is twice as much electric potential energy because it takes twice as much work to put the charge there. But the electric potential would be the same. Twice the energy divided by twice the charge gives the same potential as one unit of energy divided by one unit of charge. Electric potential is not the same thing as electric potential energy. Be sure you understand this before you read any further.
2. It means that one of the battery terminals is 12 V higher in potential than the other one. Soon we'll see that it also means that, when a circuit is connected between these terminals, each coulomb of charge in the resulting current will be given 12 J of energy as it passes through the battery (and 12 J of energy is "spent" in the circuit).

High voltage at low energy is very similar to the harmless high-temperature sparks emitted by a Fourth of July sparkler. Temperature, the ratio of energy/molecule, means a lot of energy only if a lot of molecules are involved. High voltage means a lot of energy only if a lot of charge is involved.

Rub a balloon on your hair, and the balloon becomes negatively charged—perhaps to several thousand volts. That would be several thousand joules of energy, if the charge were 1 coulomb. However, 1 coulomb is a fairly respectable amount of charge. The charge on a balloon rubbed on hair is typically much less than a millionth of a coulomb. Therefore, the amount of energy associated with the charged balloon is very, very small (Figure 7.11).

7.5 Conductors and Insulators

LEARNING OBJECTIVE
Distinguish between electric conductors and insulators.

EXPLAIN THIS What is the role of electrons in conducting materials and in insulating materials?

Electric **conductors** are materials that allow charged particles (usually electrons) to pass through them easily. Copper, silver, and other metals are good electric conductors for the same reason they are good heat conductors: Atoms of metals have one or more outer electrons that are loosely bound to their nuclei. These are called free electrons. It is these free electrons that conduct through a metallic conductor when an electrical force is applied to them, making up a current. (A *current* is a flow of charged particles, usually electrons, which is discussed in Section 7.7.)

There are great differences in the electrical conductivity of conductors and insulators. For example, in a common appliance cord, electrons flow through several meters of metal wire rather than through a small fraction of a centimeter of vinyl or rubber insulation.

The electrons in other materials—rubber and glass, for example—are tightly bound and belong to particular atoms. Consequently, it isn't easy to make them flow. These materials are poor electric conductors for the same reason they are generally poor heat conductors. Such a material is called a good **insulator**.

All substances can be arranged in order of their electrical conductivity. Those at the top of the list are conductors and those at the bottom are insulators. The ends of the list are very far apart. The conductivity of a metal, for example, can be more than a million trillion times greater than the conductivity of an insulator such as glass.

Miniaturized semiconducting electronic components take up less space, and they are faster and require less energy to operate than old-fashioned circuit components.

Some materials are neither good conductors nor good insulators; they are **semiconductors**. These materials fall into the midrange of conductivity, since they possess few electrons that are free to move. However, the number of free electrons in a semiconductor can be adjusted by introducing small amounts of another element. This process is called *doping*. By doping a semiconductor, a scientist can create a conductor with a specific conductivity. Silicon and germanium are the most common semiconductors. These elements, once they have been doped, serve as the basic material for computer chips and miniaturized electronic components—transistors, for example. Transistors are essential components of computers and other electronic devices. A transistor can act as a conductor or as an insulator depending on the applied voltage, controlling the flow of charge to different parts of a circuit.

LEARNING OBJECTIVE
Recognize how a potential difference is necessary for electric current.

7.6 Voltage Sources

EXPLAIN THIS Why is an electric battery often called an *electric pump*?

When the ends of an electric conductor are at different electric potentials—when there is a **potential difference**—charges in the conductor flow from the higher potential to the lower potential. The flow of charges persists until both ends reach the same potential. Without a potential difference, no flow of charge will occur.

FIGURE 7.12
An unusual source of voltage. The electric potential between the head and tail of the electric eel (*Electrophorus electricus*) can be as high as 650 V.

To attain a sustained flow of charge in a conductor, some arrangement must be provided to maintain a difference in potential while charge flows from one end to another. The situation is analogous to the flow of water from a higher reservoir to a lower one (Figure 7.13a). Water will flow in a pipe that connects the reservoirs only as long as a difference in the water levels exists. The flow of water in the pipe, like the flow of charge in a wire, will cease when the pressures at each end are equal. A continuous flow is possible if the difference in the water levels—and hence the difference in the water pressures—is maintained with the use of a suitable pump (Figure 7.13b).

A sustained electric current requires a suitable pumping device to maintain a difference in electric potential—to maintain a voltage. Chemical batteries or generators are "electrical pumps" that can maintain a steady flow of charge. These devices do work to pull negative charges apart from positive ones. In chemical batteries, this work is done by the chemical disintegration of lead or zinc in acid, and the energy stored in the chemical bonds is converted into electric potential energy. (See Chapter 13 for a discussion of how batteries work.) Generators separate charge by electromagnetic induction, a process described later in this chapter.

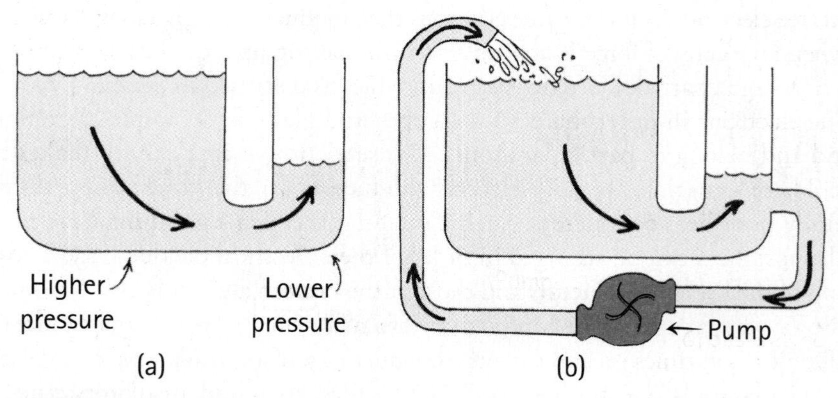

FIGURE 7.13
(a) Water flows from the reservoir of higher pressure to the reservoir of lower pressure. The flow will cease when the difference in pressure ceases. (b) Water continues to flow because the pump maintains a difference in pressure.

The work that is done (by whatever means) in separating the opposite charges is available at the terminals of the battery or generator. This energy per charge provides the difference in potential (voltage) that provides the "electrical pressure" to move electrons through a circuit joined to those terminals. A common automobile battery provides an electrical pressure of 12 volts to a circuit connected across its terminals. Then 12 joules of energy are supplied to each coulomb of charge that is made to flow in the circuit. A simple battery that can be made with a lemon provides about 1 V. (Instructions for making a lemon battery are given in Exercise 36 at the end of the chapter.)

In ac circuits, 120 volts is what is called the "root-mean-square" average of the voltage. The actual voltage in a 120-volt ac circuit varies between +170 volts and −170 volts, delivering the same power to an iron or a toaster as a 120-volt dc circuit.

7.7 Electric Current

LEARNING OBJECTIVE
Relate the speed of electrons in a circuit to dc and ac.

EXPLAIN THIS What kind of current is produced by a battery? By a generator?

Just as water current is a flow of H_2O molecules, **electric current** is a flow of charged particles. In circuits of metal wires, electrons make up the flow of charge. In metals, one or more electrons from each atom are free to move throughout the atomic lattice. These charge carriers are called *conduction electrons*. Protons, on the other hand, do not move because they are bound within the nuclei of atoms that are more or less locked in fixed positions. In fluids, however, positive ions as well as electrons may make up the flow of an electric charge. This occurs inside a common automobile battery.

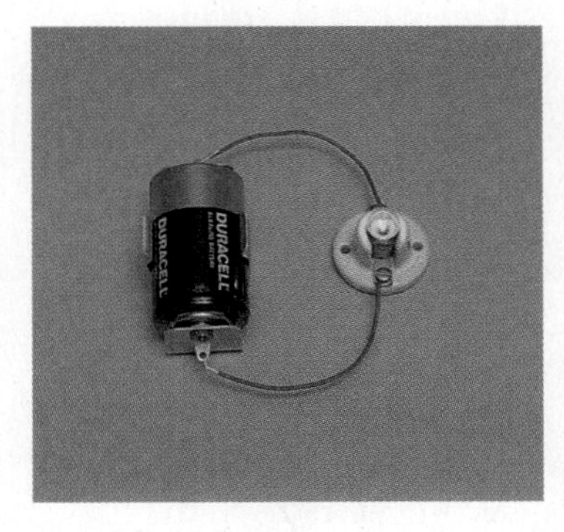

FIGURE 7.14
Each coulomb of charge that is made to flow in a circuit that connects the ends of this 1.5-V flashlight cell is energized with 1.5 J.

An important difference between water flow and electron flow has to do with their conductors. If you purchase a water pipe at a hardware store, the clerk doesn't sell you the water to flow through it. You provide that yourself. By contrast, when you buy an "electric pipe" (that is, an electric wire), you also get the electrons. Every bit of matter, wires included, contains enormous numbers of electrons that swarm about in random directions. When a source of voltage sets them moving, we have an electric circuit.

The rate of electrical flow is measured in amperes. An ampere is the rate of flow of 1 coulomb of charge per second. (That's a flow of 6.25 billion billion electrons per second.) For example, in a wire that carries 4 amperes to a car headlight bulb, 4 coulombs of charge flow past any cross section in the wire each second.

It is interesting to note that the speed of electrons as they drift through a wire is surprisingly slow. This is because electrons continually bump into atoms in the wire. The net speed, or *drift speed*, of electrons in a typical circuit is much less than 1 centimeter per second. The electrical signal, however, travels at nearly the speed of light. That's the speed at which the electric *field* in the wire is established.

We often think of current flowing through a circuit, but we don't say this around somebody who is picky about grammar, because the expression "current flows" is redundant. More properly, charge flows—which *is* current.

Direct Current and Alternating Current

Electric current may be dc or ac. By dc, we mean **direct current**, which refers to charges flowing in only one direction. A battery produces direct current in a circuit because the terminals of the battery always have the same sign. Electrons move from the repelling negative terminal toward the attracting positive terminal, and they always move through the same circuit in the same direction.

FYI André-Marie Ampère is often called the "Newton of electricity." In the 1820s he showed that parallel wires carrying current in the same direction attract each other, and he postulated that circulating charge is responsible for magnetism. In his honor, the unit for current is the ampere, often shortened to *amps*.

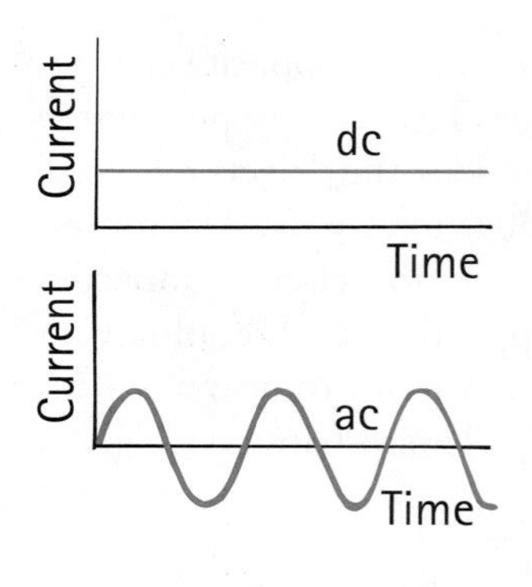

FIGURE 7.15
Time graphs of direct and alternating current.

MasteringPhysics®
VIDEO: Alternating Current
VIDEO: Caution on Handling Wires

Alternating current (ac) acts as the name implies. Electrons in the circuit are moved first in one direction and then in the opposite direction, alternating to and fro about relatively fixed positions (Figure 7.15). This is accomplished in a generator or alternator by periodically switching the sign at the terminals. Nearly all commercial ac circuits involve currents that alternate back and forth at a frequency of 60 cycles per second. This is 60-hertz current. (One cycle per second is called a hertz.) In some countries, 25-hertz, 30-hertz, or 50-hertz current is used.

LEARNING OBJECTIVE
Relate the length and width of wires to electrical resistance.

7.8 Electrical Resistance

EXPLAIN THIS What distinguishes a conductor from a *superconductor*?

How much current is in a circuit depends not only on voltage but also on the **electrical resistance** of the circuit. Just as narrow pipes resist water flow more than wide pipes, thin wires resist electric current more than thicker wires. And length contributes to resistance also. Just as long pipes have more resistance than short ones, long wires offer more electrical resistance. Most important is the material from which the wires were made. Copper has a low electrical resistance, while a strip of rubber has an enormous resistance. Temperature also affects electrical resistance. The greater the jostling of atoms within a conductor (the higher the temperature), the greater its resistance. The resistance of some materials reaches zero at very low temperatures. These materials are referred to as *superconductors*.

Electrical resistance is measured in units called *ohms*. The Greek letter *omega*, Ω, is commonly used as the symbol for the ohm. The unit was named after Georg Simon Ohm, a German physicist who, in 1826, discovered a simple and very important relationship among voltage, current, and resistance.

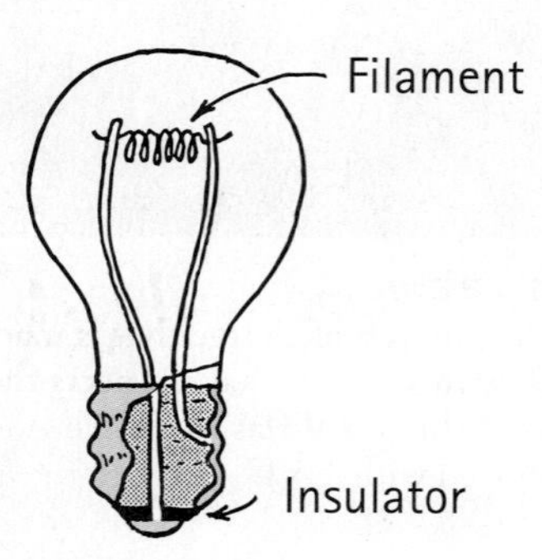

FIGURE 7.16
The conduction electrons that surge to and fro in the filament of the incandescent lamp do not come from the voltage source. They are in the filament to begin with. The voltage source simply provides them with surges of energy. When the voltage source is switched on, the very thin tungsten filament heats up to 3000°C and roughly doubles its resistance.

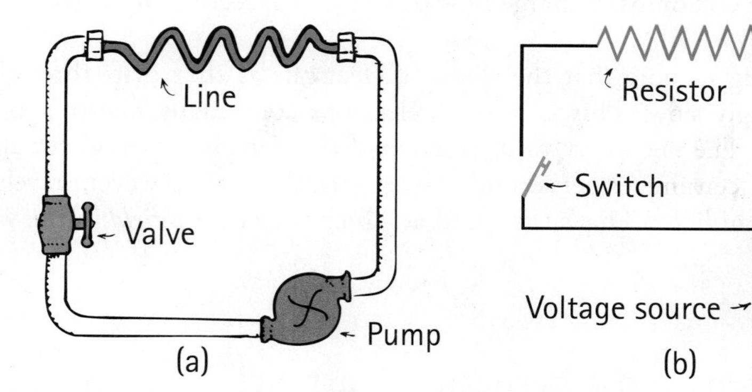

FIGURE 7.17
Analogy between a simple hydraulic circuit and an electric circuit. (a) In a hydraulic circuit, a narrow pipe (green) offers resistance to water flow. (b) In an electric circuit, a lamp or other device (shown by the zigzag symbol for resistance) offers resistance to electron flow.

HISTORY OF SCIENCE

110 Volts

In the early days of electric lighting, high voltages burned out electric filaments, so low voltages were more practical. The hundreds of power plants built in the United States prior to 1990 adopted 110 volts (or 115 volts, or 120 volts) as their standard. The tradition of 110 volts was decided upon because it made the bulbs of the day glow as brightly as a gas lamp. By the time electric lighting became popular in Europe, engineers had figured out how to make lightbulbs that would not burn out so fast at higher voltages. Power transmission is more efficient at higher voltages, so Europe adopted 220 volts as the standard. The United States remained with 110 volts (today, it is officially 120 volts) because of the initial huge expense necessary to install 110-volt equipment.

7.9 Ohm's Law

LEARNING OBJECTIVE
Relate current, voltage, and resistance in electric circuits.

EXPLAIN THIS What is the source of electrons in a body undergoing electric shock?

The relationship among voltage, current, and resistance is summarized by a statement known as **Ohm's law**. Ohm discovered that the amount of current in a circuit is directly proportional to the voltage established across the circuit and is inversely proportional to the resistance of the circuit:

$$\textbf{Current} = \frac{\textbf{voltage}}{\textbf{resistance}}$$

Or, in the form of units,

$$\text{Amperes} = \frac{\text{volts}}{\text{ohms}}$$

And, in symbolic form [since *V* stands for voltage (in volts), *I* for current (in amperes), and *R* for resistance (in ohms)], we express Ohm's law as

$$I = \frac{V}{R}$$

So, for a given circuit of constant electrical resistance, current and voltage are proportional to each other. This means that we'll get twice the current for twice the voltage. The greater the voltage, the greater the current. But, if the resistance is doubled for a circuit, the current will be half what it would have been otherwise. The greater the resistance, the smaller the current. Ohm's law makes good sense.

The resistance of a typical lamp cord is much less than 1 ohm, and a typical lightbulb has a resistance greater than 100 ohms. An iron or an electric toaster has a resistance of 15 to 20 ohms. The current inside these and all other electrical devices is regulated by circuit elements called *resistors* (Figure 7.18), whose resistance may be a few ohms or millions of ohms.

The unit of resistance is the ohm, Ω—like the old cowboy song, "Ω, Ω on the Range."

MasteringPhysics®
VIDEO: Ohm's Law

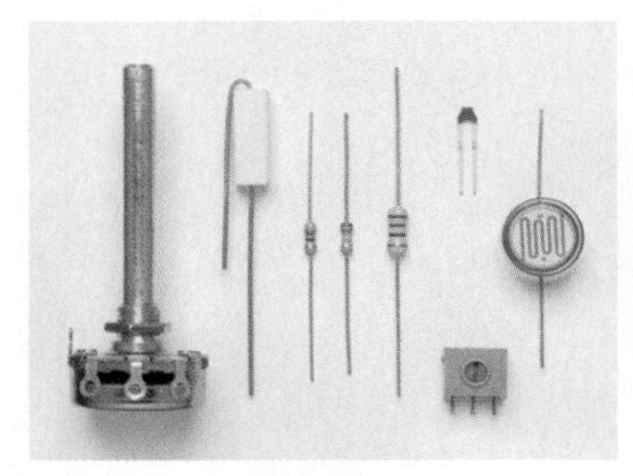

FIGURE 7.18
Resistors. The graphic symbol for a resistor in an electric circuit is —⋀⋀⋀—.

Current is a flow of charge, pressured into motion by voltage and hampered by resistance.

MATH CONNECTION

Ohm's Law

We can rearrange Ohm's law to show that $I = \frac{V}{R}$ can be expressed $R = \frac{V}{I}$. If any two variables are known, the third can be found.

Problems

1. How much current is in a lamp of resistance 60 when the voltage across the lamp is 12 V?
2. What is the resistance of a toaster that draws a current of 12 A when connected to a 120-V circuit?

Solutions

1. From Ohm's law: $\text{Current} = \frac{\text{voltage}}{\text{resistance}} = \frac{12\text{ V}}{60\ \Omega} = 0.2\text{ A}$
2. Rearranging Ohm's law, we get

$$\text{Resistance} = \frac{\text{voltage}}{\text{current}} = \frac{120\text{ V}}{12\text{ A}} = 10\ \Omega$$

LEARNING OBJECTIVE
Relate electrical phenomena to electric shock.

Integrated Science 7A
BIOLOGY

Electric Shock

EXPLAIN THIS What is the source of the electrons that shock a person?

The damaging effects of electric shock are the result of current passing through the human body. But what causes electric shock in the body—current or voltage? From Ohm's law, we can see that this current depends on the voltage that is applied and also on the electrical resistance of the human body. The resistance of one's body depends on its condition, and the resistance ranges from about 100 ohms, if the body is soaked with salt water, to about 500,000 ohms, if the skin is very dry. If we touch the two electrodes of a battery with dry fingers, completing the circuit from one hand to the other, we offer a resistance of about 100,000 ohms. We usually cannot feel 12 volts, and 24 volts just barely tingles. If our skin is moist, 24 volts can be quite uncomfortable. Table 7.1 lists the effects that different amounts of current have on the human body.

To receive a shock, there must be a difference in electric potential between one part of your body and another part. Most of the current will pass along the path of least electrical resistance connecting these two points. Suppose you fell from a bridge and managed to grab onto a high-voltage power line, halting your fall. So long as you touch nothing that has a different potential, you will receive no shock at all. Even if the wire is a few thousand volts above ground potential and you hang by it with two hands, no appreciable charge will flow from one hand to the other. This is because there is no appreciable difference in potential between your two hands. If, however, you reach over with one hand and grab onto a wire that has a different

TABLE 7.1 EFFECTS OF ELECTRIC CURRENTS ON THE BODY

Current (A)	Effect
0.001	Can be felt
0.005	Is painful
0.010	Causes involuntary muscle contractions (spasms)
0.15	Causes loss of muscle control
0.70	Goes through the heart; causes serious disruption; probably fatal if current lasts longer than 1 s

potential—zap! We have all seen birds perched on high-voltage wires. Every part of their bodies is at the same potential as the wire, so they feel no ill effects (Figure 7.19).

Many people are killed each year from common 120-volt electric circuits. If your hand touches a faulty 120-volt light fixture while your feet are on the ground, there's likely a 120-volt "electrical pressure" between your hand and the ground. Resistance to current is usually greatest between your feet and the ground, and so the current is usually not enough to do serious harm. But if your feet and the ground are wet, there is a low-resistance electrical path between you and the ground. Pure water is not a good conductor. But the ions that normally are found in water make it a fair conductor. Dissolved materials in water, especially small quantities of salt, lower the resistance even more. There is usually a layer of salt remaining on your skin from perspiration, which, when wet, lowers your skin resistance to a few hundred ohms or less. Handling electrical devices when wet is a definite no-no.

Injury by electric shock occurs in three forms: (1) burning of tissues by heating, (2) contraction of muscles, and (3) disruption of cardiac rhythm. These conditions are caused by the delivery of excessive power for too long a time in critical regions of the body.

Electric shock can upset the nerve center that controls breathing. In rescuing shock victims, the first thing to do is remove them from the source of the electricity. Use a dry wooden stick or some other nonconductor so that you don't get electrocuted yourself. Then apply artificial respiration. It is important to continue artificial respiration. There have been cases of victims of lightning who did not breathe without assistance for several hours but who were eventually revived and completely regained good health.

FIGURE 7.19
The bird can stand harmlessly on one wire of high potential, but it had better not reach over and touch a neighboring wire. Why not?

FYI While Alexandra Volta was experimenting with metals and acids in 1791 he touched a silver spoon and a piece of tin to his tongue (saliva is slightly acidic) and connected them with a piece of copper wire. The sour taste indicated electricity. He went on to assemble a pile of cells to form a battery. In Volta's honor, electric potential is measured in units of "volts." (Touch the two terminals of a 9-volt battery to your tongue to experience this for yourself.)

CHECK YOURSELF

1. **What causes electric shock—current or voltage?**
2. **At 100,000 Ω, how much current will flow through your body if you touch the terminals of a 12-V battery?**
3. **If your skin is very moist, so that your resistance is only 1000 Ω, and you touch the terminals of a 12-V battery, how much current do you receive?**

CHECK YOUR ANSWERS

1. Electric shock occurs when current is produced in the body, but the current is *caused* by impressed voltage.
2. $\text{Current} = \dfrac{\text{voltage}}{\text{resistance}} = \dfrac{12\text{ V}}{100{,}000\ \Omega} = 0.00012\text{ A}.$
3. $\text{Current} = \dfrac{\text{voltage}}{\text{resistance}} = \dfrac{12\text{ V}}{1000\ \Omega} = 0.012\text{ A}.$ Ouch!

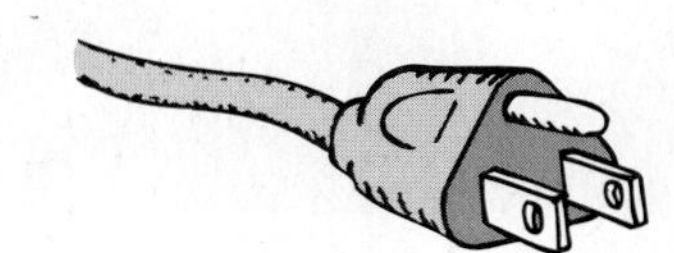

FIGURE 7.20
The third prong connects the body of the appliance directly to ground. Any charge that builds up on an appliance is therefore conducted to the ground.

Most electrical plugs and sockets are wired with three connections. The principal two flat prongs on an electrical plug are for the current-carrying double wire, one part "live" and the other neutral, while the third round prong is grounded—connected directly to Earth (Figure 7.20). Appliances such as stoves, washing machines, and dryers are connected with these three wires. If the live wire accidentally comes into contact with the metal surface of the appliance and you touch the surface of the appliance, you could receive a dangerous shock. This won't occur when the appliance casing is grounded via the ground wire, which assures that the appliance casing is at zero ground potential. A lamp has an insulating body and doesn't need the third (ground) wire.

A battery doesn't supply electrons to a circuit; it instead supplies energy to electrons that already exist in the circuit.

LEARNING OBJECTIVE
Distinguish between series and parallel circuits.

7.10 Electric Circuits

EXPLAIN THIS How can a circuit be connected so that the current in each device is the same?

Any path along which electrons can flow is a circuit. For a continuous flow of electrons, there must be a complete circuit with no gaps. A gap is usually provided by an electric switch that can be opened or closed either to cut off energy or to allow energy to flow. Most circuits have more than one device that receives electric energy. These devices are commonly connected to a circuit in one of two ways: in *series* or in *parallel.*

A simple **series circuit** is shown in Figure 7.21. Three lamps are connected in series with a battery. The same current exists almost immediately in all three lamps when the switch is closed. The current does not "pile up" or accumulate in any lamp but flows through each lamp. Electrons that make up this current leave the negative terminal of the battery, pass through each of the resistive filaments in the lamps in turn, and then return to the positive terminal of the battery. (The same amount of current passes through the battery.) This is the only path of the electrons through the circuit. A break anywhere in the path results in an open circuit, and then the flow of electrons ceases. Such a break occurs when the switch is opened, when the wire is accidentally cut, or when one of the lamp filaments burns out.

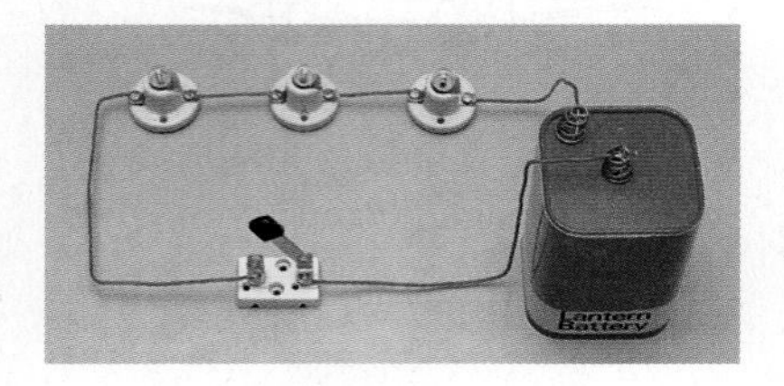

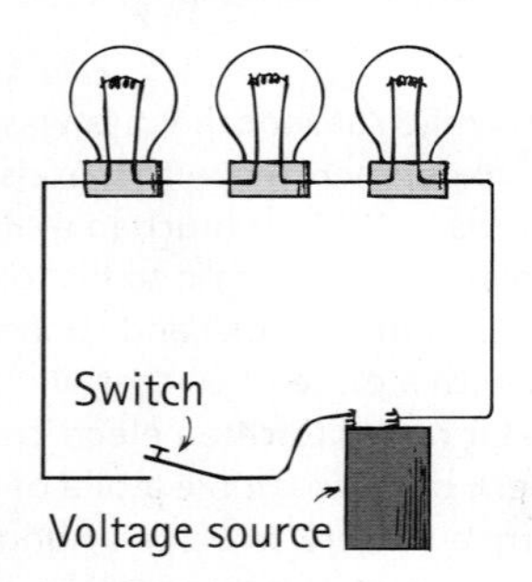

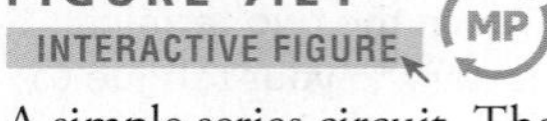

FIGURE 7.21
INTERACTIVE FIGURE (MP)
A simple series circuit. The 6-V battery provides 2 V across each lamp.

MasteringPhysics®
VIDEO: Electric Circuits

It is easy to see the main disadvantage of the series circuit; if one device fails, current in the entire circuit ceases. Some cheap Christmas-tree lights are connected in series. When one bulb burns out, it's fun and games (or frustration) trying to locate which one to replace.

Most circuits are wired so that it is possible to operate several electrical devices, each independently of the other. In your home, for example, a lamp can be turned on or off without affecting the operation of other lamps or electrical devices. This is because these devices are connected not in series but in parallel with one another. A simple **parallel circuit** is shown in Figure 7.22. Three lamps are connected to the same two points, A and B. Electrical devices connected to the same two points of an electric circuit are said to be *connected in parallel.* Electrons leaving the negative terminal of the battery need to travel through only one lamp filament before returning to the positive terminal of the battery. In this case, the current branches into three separate pathways from A to B. A break in any one path does not interrupt the flow of charge in the other paths. Each device operates independently of the other devices.

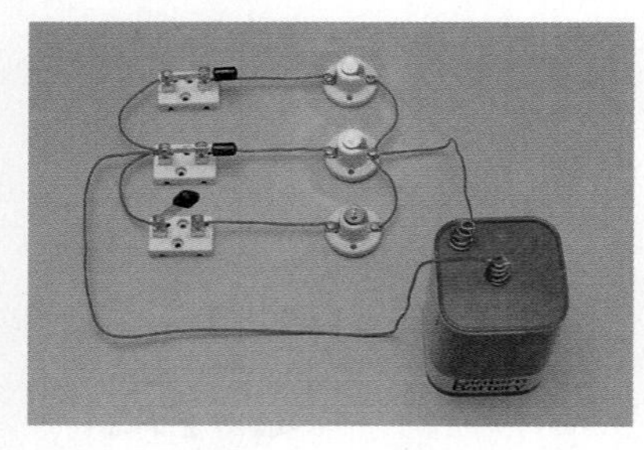

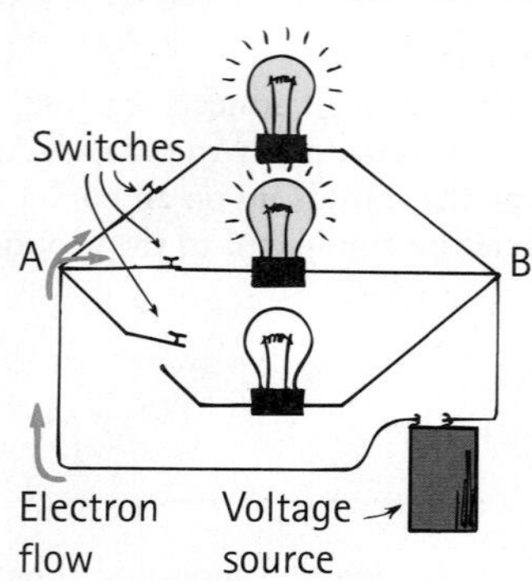

FIGURE 7.22
INTERACTIVE FIGURE

A simple parallel circuit. A 6-V battery provides 6 V across each lamp.

CHECK YOURSELF

1. **In a circuit consisting of two lamps connected in series, if the current in one lamp is 1 A, what is the current in the other lamp?**
2. **In a circuit consisting of two lamps connected in parallel, if there is 6 V across one lamp, what is the voltage across the other lamp?**

CHECK YOUR ANSWERS

1. 1 A; current does not "pile up" anywhere in the circuit.
2. 6 V; the voltages in all branches of a parallel circuit are the same.

7.11 Electric Power

LEARNING OBJECTIVE
Relate current and voltage to power with their units of measurement.

EXPLAIN THIS Why shouldn't you connect a 120-V hairdryer to 240 V?

The moving charges in an electric current do work. This work, for example, can heat a circuit or turn a motor. The rate at which this work is done—that is, the rate at which electric energy is converted into another form, such as mechanical energy, heat, or light—is called **electric power**. Electric power is equal to the product of current and voltage:*

$$\text{Power} = \text{current} \times \text{voltage} = IV$$

If a lamp rated at 120 volts operates on a 120-volt line, you can figure that it will draw a current of 1 ampere (120 watts = 1 ampere × 120 volts). A 60-watt lamp draws 1/2 ampere on a 120-volt line. This relationship becomes a practical matter when you wish to know the cost of electric energy, which is usually a small fraction of a dollar per kilowatt-hour, depending on the locality. A kilowatt is 1000 watts, and a kilowatt-hour represents the amount of energy consumed in 1 hour at the rate of 1 kilowatt.** Therefore, in a locality where electric energy costs 25 cents per kilowatt-hour, an incandescent 100-watt electric lightbulb can operate for 10 hours at a cost of 25 cents, or a half nickel for each hour (Figure 7.23). Compact fluorescent lamps (CFLs) are much less expensive to operate (Figure 7.24).

FIGURE 7.23
The power-and-voltage designation on the incandescent lightbulb reads "100 W 120 V." How many amperes will flow through the bulb?

It is interesting that a 26-W CFL (see Figure 7.24) provides about the same amount of light as a 100-W incsandescent bulb. That's only one-quarter of the power for the same light!† In addition to significantly greater efficiencies, CFLs also have longer bulb lifetimes.†† Incandescent bulbs are now being replaced by CFLs.

FIGURE 7.24
The power-and-voltage designation on the compact fluorescent lamp (CFL) reads "26 W 120 V." It glows as bright as a more power-consuming 100-W incandescent bulb.

Another up and coming light source is the light-emitting diode (LED), the most primitive being the little red lights that tell you whether your electronic devices are on or off. LEDs that emit a full range of colors were developed in the 1960s. Today, LEDs have advanced beyond indicator displays on electronic appliances and are common in television screens, traffic lights, automobile lights, airport runway lighting, and even billboards. LEDs are compact, efficient, require

FYI A bulb's brightness depends on how much power it uses. An incandescent bulb that uses 100 W is brighter than one that uses 60 W. Because of this, many people mistakenly think that a watt is a unit of brightness, but it isn't. A 26-W CFL is as bright as a 100-W incandescent bulb. Does that mean that an incandescent bulb wastes electricity? Yes. The extra electric power used just heats the bulb, which is why incandescent bulbs are much hotter to touch than equally bright CFLs.

*Recall from Chapter 4 that power $= \frac{\text{work}}{\text{time}}$; 1 watt = 1 J/s. Note that the units for mechanical power and electric power agree (work and energy are both measured in joules):

$$\text{Power} = \frac{\text{charge}}{\text{time}} \times \frac{\text{energy}}{\text{charge}} = \frac{\text{energy}}{\text{time}}$$

If the voltage is expressed in volts and the current is expressed in amperes, then the power is expressed in watts. So, in units form,

$$\text{Watts} = \text{amperes} \times \text{volts}$$

Using Ohm's law to substitute IR for V, we have an alternative statement for power: power $= I^2R$.

** Since power = energy/time, simple rearrangement gives energy = power × time; thus energy can be expressed in the unit *kilowatt-hours* (kWh).

† It turns out that the power formula $P = IV$ doesn't apply to CFLs, because in a CFL the alternating voltage and current are out of step with each other (out of phase), and the product of current and voltage is greater than the actual power consumption. How much greater? Check the printed data at the base of a CFL.

†† A disadvantage of CFLs is the trace amounts of mercury sealed in their glass tubing, some 4 mg. But the single largest source of mercury emissions in the environment is coal-fired power plants. According to the EPA, when coal power is used to illuminate a single incandescent lamp, more mercury is released in the air than exists in a comparably luminous CFL.

MATH CONNECTION

Solving Power Problems

You'll find that you can answer many practical, everyday questions related to electricity if you know the relationships among power, voltage, and current. Here are two examples.

Problems

1. If a 120-V line to a socket is limited to 15 A by a safety fuse, will it operate a 1200-W hair dryer?
2. At 30¢/kWh, show that it costs 36¢ to operate the 1200-W hair dryer for 1 h.

Solutions

1. Yes. From the expression watts = amperes × volts, we can see that current $= \frac{1200\text{ W}}{120\text{ V}} = 10$ A, so the hair dryer will operate when connected to the circuit. But two hair dryers on the same circuit will blow the fuse.
2. 1200 W = 1.2 kW; 1.2 kW × 1 h × 30¢/kWh = 36¢.

no filament, and are long-lasting (about 100 times longer-lasting than incandescent bulbs). They emit 15 times as much light per watt as an incandescent bulb. And, they do not contain the traces of harmful mercury found in CFLs. With competition from CFLs, then LEDs, watch for the commonly used incandescent bulbs to become history.

LEARNING OBJECTIVE
Establish the rule for the attraction and repulsion of magnetic poles.

7.12 The Magnetic Force

EXPLAIN THIS Why are refrigerator magnets short range?

Anyone who has played around with magnets knows that magnets exert forces on one another. **Magnetic forces** are similar to electrical forces, in that magnets can both attract and repel without touching (depending on which ends of the magnets are held near one another), and the strength of their interaction depends on the distance between them. Whereas electric charges produce electrical forces, regions called *magnetic poles* give rise to magnetic forces.

If you suspend a bar magnet at its center by a piece of string, you've got a compass. One end, called the north-seeking pole, points northward. The opposite end, called the south-seeking pole, points southward. More simply, these are called the *north* and *south* poles. All magnets have both a north and a south pole (some have more than one of each). Refrigerator magnets have narrow strips of alternating north and south poles. These magnets are strong enough to hold sheets of paper against a refrigerator door, but they have a very short range because the north and south poles cancel a short distance from the magnet. In a simple bar magnet, the magnetic poles are located at the two ends. A common horseshoe magnet is a bar magnet bend into a U shape. Its poles are also located at its two ends (Figure 7.25).

FIGURE 7.25
A horseshoe magnet.

When the north pole of one magnet is brought near the north pole of another magnet, they repel each other. The same is true of a south pole near a south pole. If opposite poles are brought together, however, attraction occurs. We find the following rule:

Like poles repel; opposite poles attract.

This rule is similar to the rule for the forces between electric charges, where like charges repel each other and unlike charges attract. But there is a very important difference between magnetic poles and electric charges. Whereas electric charges can be isolated, magnetic poles cannot. Electrons and protons are entities by themselves, but the north and south poles of a magnet are like the head and tail of the same coin.

If you break a bar magnet in half, each half still behaves as a complete magnet. Break the pieces in half again, and you have four complete magnets. You can continue breaking the pieces in half and never isolate a single pole. Even if your pieces were one atom thick, there would still be two poles on each piece, which suggests that the atoms themselves are magnets (Figure 7.26).

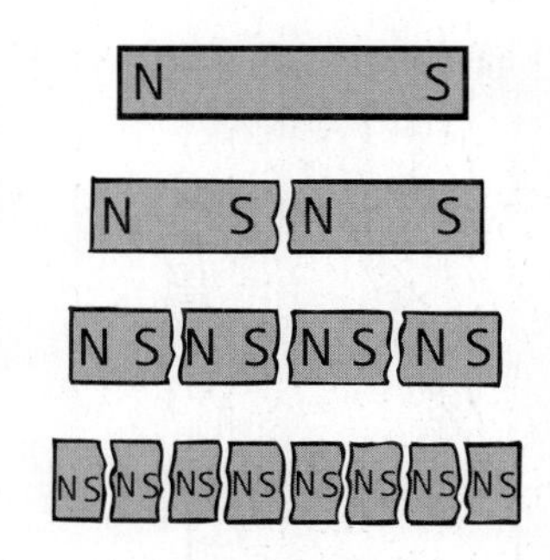

FIGURE 7.26
If you break a magnet in half, you will have two magnets. If you break these two magnets in half, you will have four magnets, each with a north and a south pole. If you continue breaking the pieces, you will find that you always get the same results. Magnetic poles exist in pairs.

7.13 Magnetic Fields

EXPLAIN THIS What is the origin of all magnetic fields?

LEARNING OBJECTIVE
Relate magnetic field strength to magnetic field patterns.

You have learned that the space around every electric charge is filled with an electric field. Similarly, the space around a magnet contains a **magnetic field**—an energetic aura that extends through space. If you sprinkle some iron filings on a sheet of paper placed on a magnet, you'll see that the filings move in response to the magnetic field and trace out an orderly pattern of lines that surround the magnet. The shape of the field is revealed by magnetic field lines that spread out from one pole and return to the other pole. It is interesting to compare the field patterns in Figures 7.27 and 7.28 with the electric field patterns in Figure 7.8b.

The direction of the field outside the magnet is from the north pole to the south pole. Where the lines are closer together, the field is stronger. We can see that the magnetic field strength is greater at the poles. If we place another magnet or a small compass anywhere in the field, its poles will tend to align with the magnetic field (Figure 7.29).

A magnetic field is produced by moving electric charges. Where, then, is this motion in a common bar magnet? The answer is, in the electrons of the atoms that make up the magnet. These electrons are in constant motion. Two kinds of electron motion produce magnetism: electron spin and electron revolution. In most common magnets, electron spin is the main contributor to magnetism.

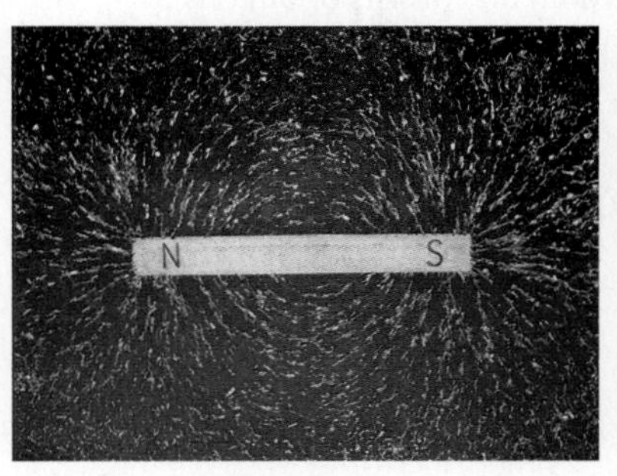

FIGURE 7.27
INTERACTIVE FIGURE MP
Top view of iron filings sprinkled on a sheet of paper on top of a magnet. The filings trace out a pattern of magnetic field lines in the surrounding space. Interestingly enough, the magnetic field lines continue inside the magnet (not revealed by the filings) and form closed loops.

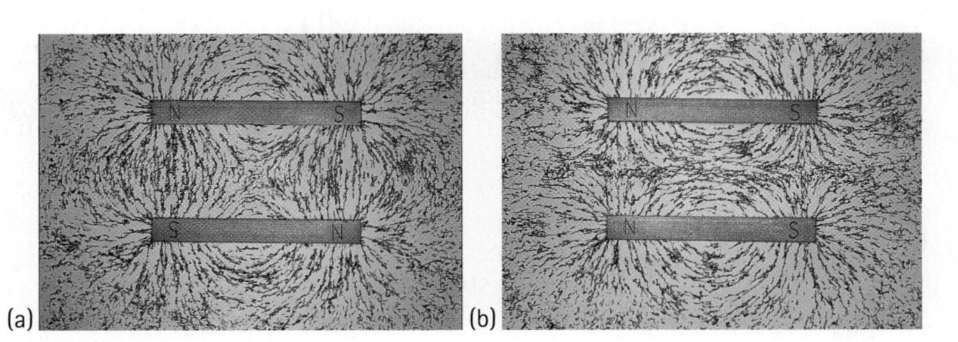

FIGURE 7.28
The magnetic field patterns for a pair of magnets. (a) Opposite poles are nearest each other, and (b) like poles are nearest each other.

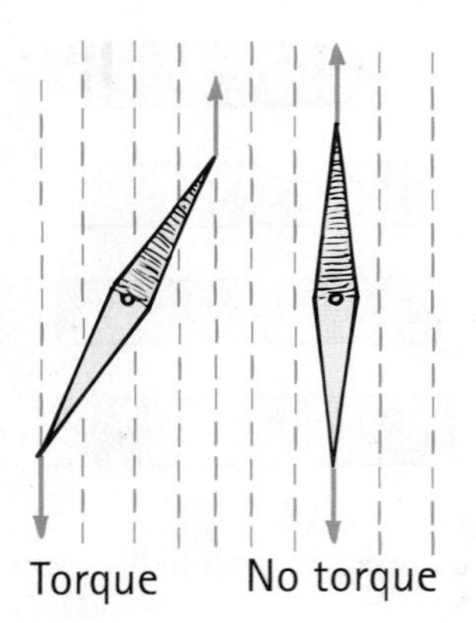

FIGURE 7.29
When the compass needle is not aligned with the magnetic field, the oppositely directed forces produce a pair of *torques* (a torque is a turning force) that twist the needle into alignment.

Every spinning electron is a tiny magnet. A pair of electrons spinning in the same direction create a stronger magnet. A pair of electrons spinning in opposite directions, however, work against each other, and the magnetic fields cancel. This is why most substances are not magnets. In most atoms, the various fields cancel one another because the electrons spin in opposite directions. In such materials as iron, nickel, and cobalt, however, the fields do not cancel entirely. Each iron atom has four electrons whose spin magnetism is uncanceled. Each iron atom, then, is a tiny magnet. The same is true, to a lesser extent, of nickel and cobalt atoms. Most common magnets are therefore made from alloys containing iron, nickel, cobalt, and aluminum in various proportions.

Magnetic Domains

The magnetic field of an individual iron atom is so strong that interactions among adjacent atoms cause large clusters of them to line up with one another. Regions with these clusters of aligned atoms are called **magnetic domains**. Each domain is perfectly magnetized and is made up of billions of aligned atoms. The domains are microscopic (Figure 7.30), and there are many of them in a crystal of iron.

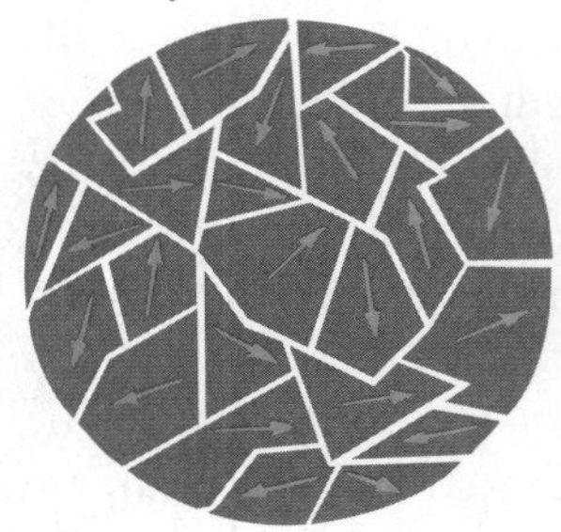

FIGURE 7.30
A microscopic view of magnetic domains in a crystal of iron. Each domain consists of billions of aligned atoms.

Not every piece of iron is a magnet because the domains in ordinary iron are not aligned. In a common nail, for example, the domains are randomly oriented. But, when you bring a magnet nearby, the domains can be induced into alignment. When you remove the nail from the magnet, ordinary thermal motion causes most or all of the domains in the nail to return to a random arrangement.

A magstripe on a credit card contains millions of tiny magnetic domains held together by a resin binder. Data is encoded in binary code, with zeroes and ones distinguished by the frequency of domain reversals. It is quite amazing how quickly your name pops up when an airline reservationist swipes your card.

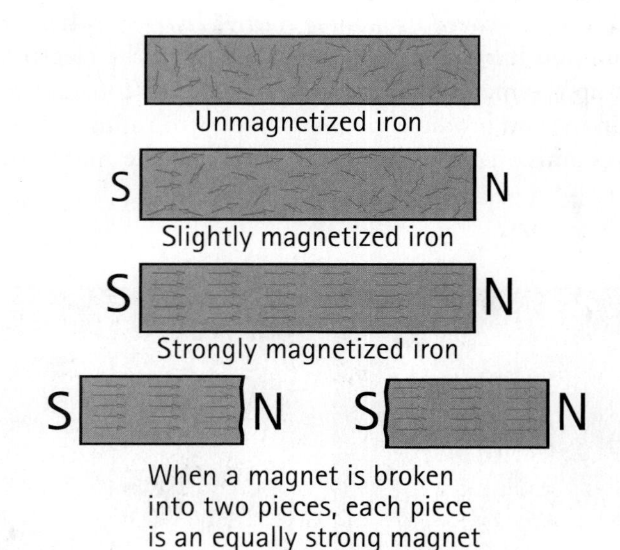

FIGURE 7.31
Pieces of iron in successive stages of magnetization. The arrows represent domains; the head is a north pole and the tail is a south pole. Poles of neighboring domains neutralize each other's effects, except at the ends.

Electric Currents and Magnetic Fields

A moving electric charge produces a magnetic field. A current of charges, then, also produces a magnetic field. The magnetic field that surrounds a current-carrying wire can be demonstrated by arranging an assortment of compasses around the wire (Figure 7.32). The magnetic field about the current-carrying wire makes up a pattern of concentric circles. When the current reverses direction, the compass needle turns around, showing that the direction of the magnetic field changes also.

If the wire is bent into a loop, the magnetic field lines become bunched up inside the loop (Figure 7.33). If the wire is bent into another loop that overlaps the first, the concentration of magnetic field lines inside the loops is doubled. It follows that the magnetic field intensity in this region is increased as the number of loops is increased. The magnetic field intensity is appreciable for a current-carrying coil that has many loops (Figure 7.34).

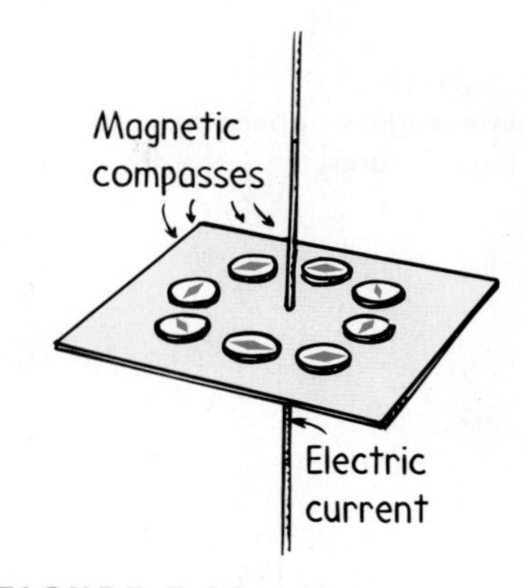

FIGURE 7.32
The compasses show the circular shape of the magnetic field surrounding the current-carrying wire.

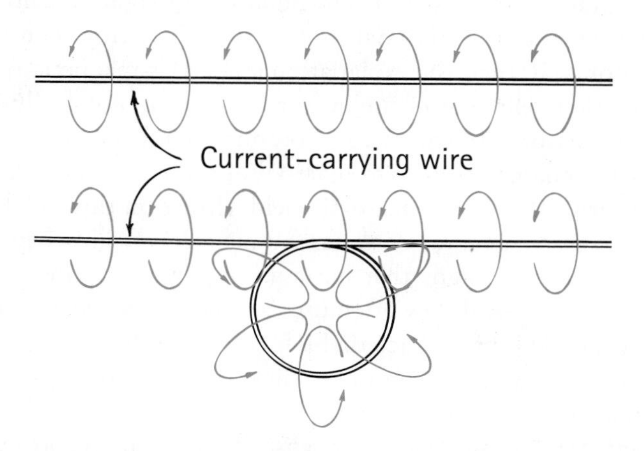

FIGURE 7.33
Magnetic field lines about a current-carrying wire become bunched up when the wire is bent into a loop.

If a piece of iron is placed in a current-carrying coil of wire, the alignment of magnetic domains in the iron produces a particularly strong magnet known as an **electromagnet**. The strength of an electromagnet can be increased simply by increasing the current through the coil or by increasing the number of loops in the coil. Electromagnets powerful enough to lift automobiles are a common sight in junkyards.

There's much bunk about magnetism—hence the need for a knowledge filter to tell the difference between what's true and what's not true. The best knowledge filter ever invented is science.

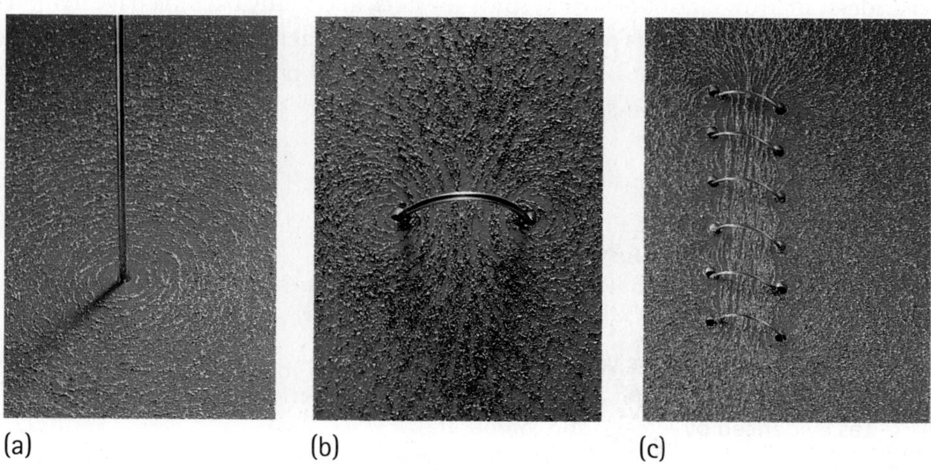

FIGURE 7.34
Iron fillings sprinkled on paper reveal the magnetic-field configurations about (a) a current-carrying wire, (b) a current-carrying loop, and (c) a current-carrying coil of loops.

LEARNING OBJECTIVE
Relate magnetic phenomena to living creatures.

Integrated Science 7B

BIOLOGY AND EARTH SCIENCE

Earth's Magnetic Field and the Ability of Organisms to Sense It

EXPLAIN THIS What serves as a compass in creatures that can sense Earth's magnetic field?

A suspended magnet or compass points northward because Earth itself is a huge magnet and the compass aligns with Earth's magnetic field (Figure 7.35).

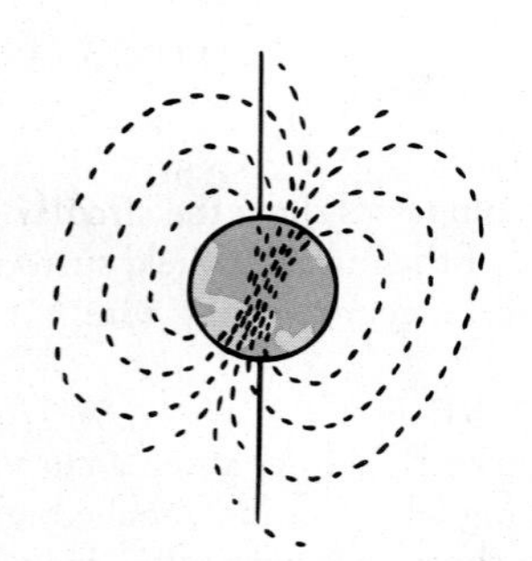

FIGURE 7.35
Earth is a magnet.

The configuration of Earth's magnetic field is similar to that of a strong bar magnet placed near the center of Earth. But Earth is not a magnetized chunk of iron like a bar magnet. Temperatures rise quickly underground, and they are too high for individual atoms to stabilize in magnetic domains with the proper orientation. Random thermal motion would destroy such an organized alignment of atoms. Rather, the explanation of Earth's magnetic field has to do with the convection cells that occur in Earth's interior, especially in the liquid outer core. Accelerating electric charges produce magnetic fields, as we know. As the electrically charged, iron-rich material of the hot outer core circulates in convection patterns, it creates a magnetic field. (More on this in Chapter 22.)

Many organisms are able to sense Earth's magnetic field. They use it, rather than the maps and street signs that humans employ, to figure out how to get where they need to go—and they often use their magnetic sense with remarkable accuracy. For example, certain bacteria biologically produce single-domain-sized grains of magnetite (an iron-containing magnetic material) that they string together to form magnetic compasses. They use these compasses to detect the dip in Earth's magnetic field. Equipped with a sense of direction, the organisms are able to locate food supplies. Pigeons have been found to have multiple-domain magnetite magnets within their skulls that are connected by a large number of nerves to their brains. Magnetic material has also been found in the abdomens of bees, whose behavior is affected by small magnetic fields and fluctuations in Earth's magnetic field. Monarch butterflies and sea turtles are known to employ their magnetic sense to migrate vast distances. Organisms as diverse as hamsters, salamanders, sparrows, rainbow trout, spiny lobsters, and many bacteria have demonstrated a magnetic sense in laboratory experiments. Magnetite crystals resembling those found in bacteria have been found in human brains. No one knows how these may be useful, but it seems likely that humans, too, have a magnetic sense.

FIGURE 7.36
The pigeon may well be able to sense direction because of a built-in magnetic "compass" in its skull.

CHECK YOURSELF

1. **Is Earth a magnet? Justify your answer.**
2. **How does the presence of magnetite in an organism's body help it navigate?**

CHECK YOUR ANSWERS

1. Yes; Earth is a magnet that produces a net magnetic field outside itself, as evidenced by a magnetic compass.
2. The domains in magnetite line up in an external magnetic field much like a compass, providing organisms that contain magnetite with an internal compass to indicate orientation with respect to Earth's magnetic field.

7.14 Magnetic Forces on Moving Charges

LEARNING OBJECTIVE
Show how the relative direction, field, and motion of charged particles affect the force on those particles.

EXPLAIN THIS How does Earth's magnetic field protect us from cosmic radiation?

A charged particle has to be moving to interact with a magnetic field. Charges at rest don't respond to magnets. But, when they are moving, charged particles experience a deflecting force.* The force is greatest when the particles move at right angles to the magnetic field lines. At other angles, the force is less, and it becomes zero when the particles move parallel to the field lines. The force is always perpendicular to the magnetic field lines and perpendicular to the velocity of the charged particle (Figure 7.37). This means that a moving charge is deflected when it crosses through a magnetic field, but *not* when it travels parallel to the field.

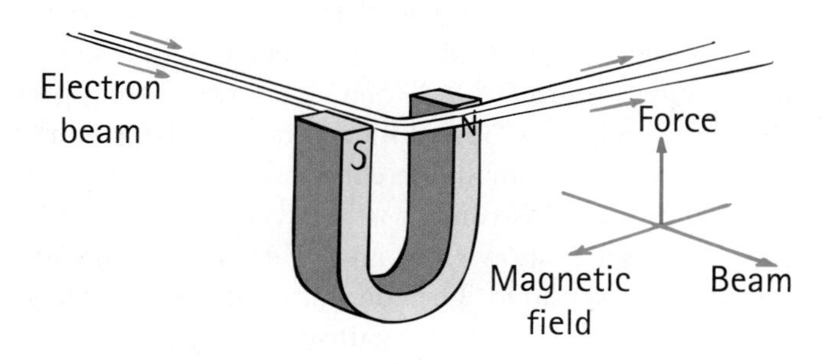

FIGURE 7.37
A beam of electrons is deflected by a magnetic field.

This deflecting force is very different from the forces that occur in other interactions. Gravitation acts in a direction parallel to the line between masses, and electrical forces act in a direction parallel to the line between charges. But magnetic force acts at right angles to the magnetic field and the velocity of the charged particle.

We are fortunate that charged particles are deflected by magnetic fields. This fact is used when Earth's magnetic field deflects charged particles from outer space. Without this magnetic deflection, the harmful cosmic rays that bombard Earth's surface would be much more intense (Figure 7.38).

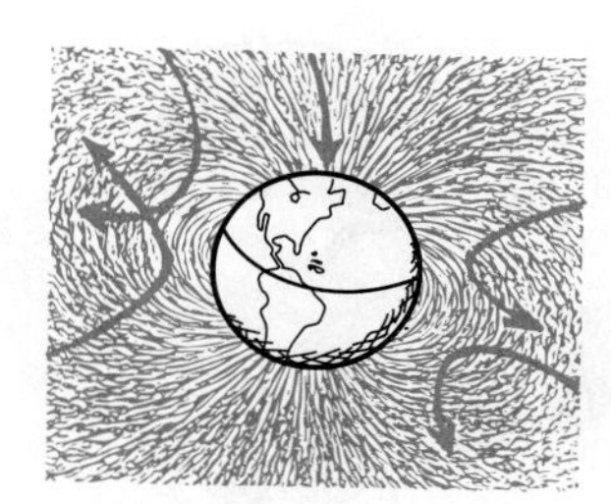

FIGURE 7.38
Earth's magnetic field picks up many charged particles that make up cosmic radiation.

Magnetic Force on Current-Carrying Wires

Simple logic tells you that, if a charged particle moving through a magnetic field experiences a deflecting force, then a current of charged particles moving through a magnetic field also experiences a deflecting force. If the particles are deflected while moving inside a wire, the wire is also deflected (Figure 7.39).

MasteringPhysics®
VIDEO: Magnetic Force on Current-Carrying Wire

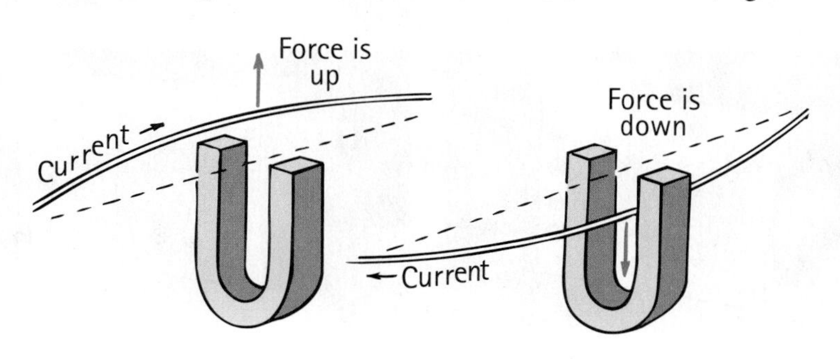

FIGURE 7.39
A current-carrying wire experiences a force in a magnetic field.

*When particles of electric charge q and velocity v move perpendicularly into a magnetic field of strength B, the force F on each particle is simply the product of the three variables: $F = qvB$. For nonperpendicular angles, v in this relationship must be the component of velocity perpendicular to B.

If we reverse the direction of the current, the deflecting force acts in the opposite direction. The force is strongest when the current is perpendicular to the magnetic field lines. The direction of force is neither along the magnetic field lines nor along the direction of current. The force is perpendicular both to the field lines and to the current. It is a sideways force.

We see that, just as a current-carrying wire will deflect a magnet, such as a compass needle, a magnet will deflect a current-carrying wire. When discovered, these complementary links between electricity and magnetism created much excitement. Almost immediately, people began harnessing the electromagnetic force for useful purposes—with great sensitivity in electric meters and with great force in electric motors.

Electric Meters

FIGURE 7.40
A very simple galvanometer.

The simplest meter to detect electric current is a magnetic compass. The next simplest meter is a compass in a coil of wires (Figure 7.40). When an electric current passes through the coil, each loop produces its own effect on the needle, so even a very small current can be detected. Such a current-indicating instrument is called a galvanometer.

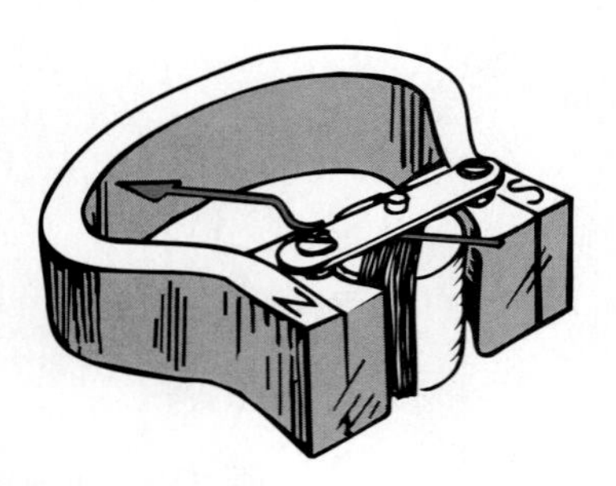

FIGURE 7.41
A common galvanometer design.

A more common design is shown in Figure 7.41. It employs more loops of wire and is therefore more sensitive. The coil is mounted for movement, and the magnet is held stationary. The coil turns against a spring, so the greater the current in its windings, the greater its deflection. A galvanometer may be calibrated to measure current (amperes), in which case it is called an *ammeter*, or it may be calibrated to measure electric potential (volts), in which case it is called a *voltmeter* (Figure 7.42).

Electric Motors

If we change the design of the galvanometer slightly, so that deflection makes a complete turn rather than a partial rotation, we have an electric motor. The principal difference is that the current in a motor is made to change direction each time the coil makes a half rotation. This happens in a cyclic fashion to produce continuous rotation, which has been used to run clocks, operate gadgets, and lift heavy loads.

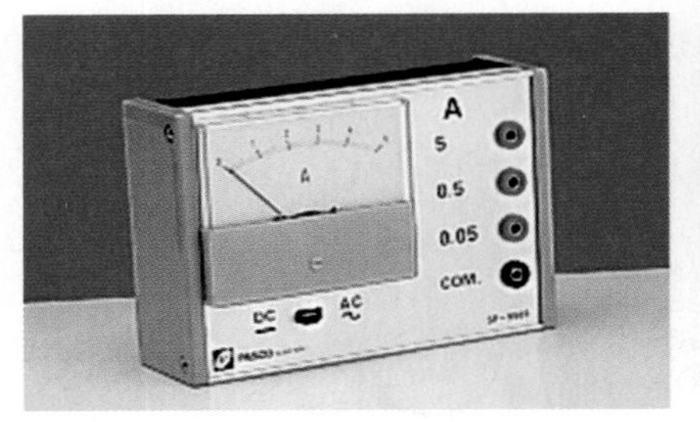

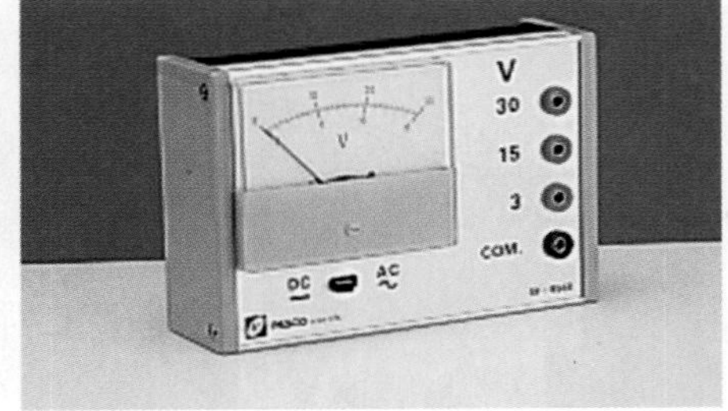

FIGURE 7.42
Both the ammeter and the voltmeter are basically galvanometers. (The electrical resistance of the instrument is designed to be very low for the ammeter and very high for the galvanometer.)

In Figure 7.43, we see the principle of the electric motor in bare outline. A permanent magnet produces a magnetic field in a region where a rectangular loop of wire is mounted to turn about the axis shown by the dashed line. When a current passes through the loop, it flows in opposite directions in the upper and lower sides of the loop. (It must do this because, if charge flows into one end of the loop, it must flow out the other end.) If the upper portion of the loop is forced to the left, then the lower portion is forced to the right, as if it were a galvanometer. But, unlike a galvanometer, the current is reversed during each half revolution by means of stationary contacts on the shaft. The parts of the wire that brush against these contacts are called brushes. In this way, the current in the loop alternates so that the forces in the upper and lower regions do not change directions as the loop rotates. The rotation is continuous as long as current is supplied.

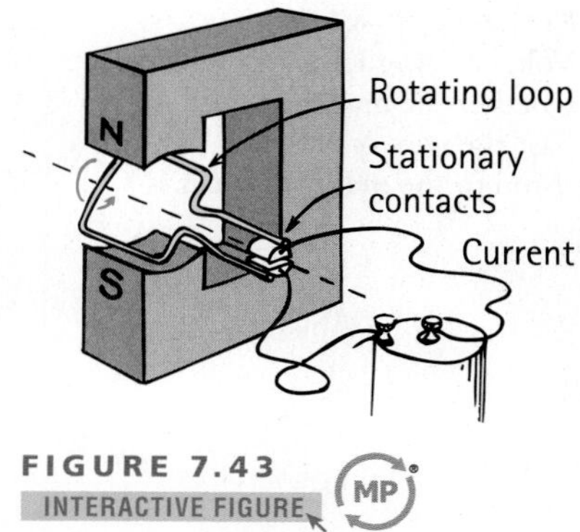

FIGURE 7.43
INTERACTIVE FIGURE
A simple motor.

We have described here only a very simple dc motor. In larger motors, dc or ac, the permanent magnet is usually replaced by an electromagnet that is energized by the power source. Of course, more than a single loop is used. Many loops of wire are wound about an iron cylinder, called an armature, which then rotates when the wire carries current.

A generator is a motor in reverse (Figure 7.44). In a motor, electric energy is the input and mechanical energy is the output. In a generator, mechanical energy is the input and electric energy is the output. Both devices transform energy from one form to another. To understand how a generator works, you need to understand electromagnetic induction, the subject of the next section.

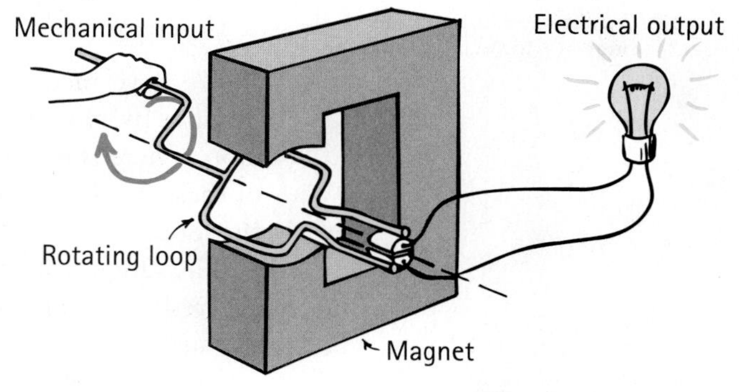

FIGURE 7.44

INTERACTIVE FIGURE
A simple generator. Voltage is induced in the loop when it is rotated in the magnetic field.

7.15 Electromagnetic Induction

LEARNING OBJECTIVE
Describe how Faraday's law is central to an industrial age.

EXPLAIN THIS What is the role of *motion* in electromagnetic induction? Motion of what?

In the early 1800s, when electricity and magnetism were topics of much scientific research, the question arose as to whether electricity could be produced from magnetism. The answer was provided in 1831 by two physicists, Michael Faraday in England and Joseph Henry in the United States—each working without knowledge of the other. Their discovery changed the world by making electricity commonplace.

Faraday and Henry both discovered that electric current could be produced in a wire simply by moving a magnet into or out of a coil of wire (Figure 7.45). No battery or other voltage source was needed—only the motion of a magnet in a coil or a wire loop. They discovered that voltage is caused, or *induced*, by the relative motion between a wire and a magnetic field. Whether the magnetic field moves near a stationary conductor or vice versa, voltage is induced (Figure 7.46).

FIGURE 7.45
When the magnet is plunged into the coil, charges in the coil are set in motion, and voltage is induced in the coil.

The greater the number of loops of wire moving in a magnetic field, the greater the induced voltage (Figure 7.47). Pushing a magnet into a coil with twice as many loops induces twice as much voltage; pushing into a coil with ten times

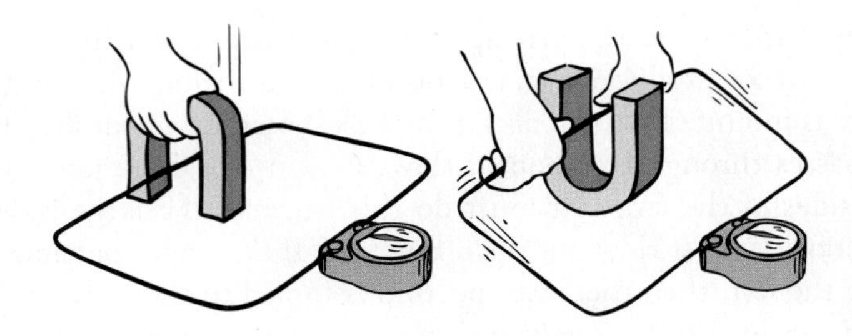

FIGURE 7.46
Voltage is induced in the wire loop whether the magnetic field moves past the wire or the wire moves through the magnetic field.

as many loops induces ten times as much voltage; and so on. **Electromagnetic induction** can be summarized by **Faraday's law:**

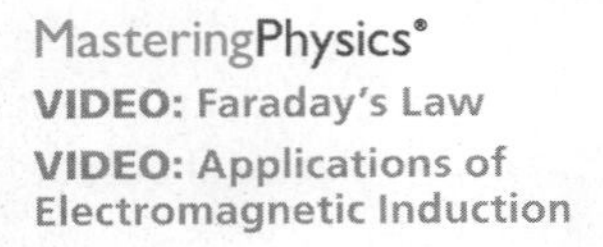

MasteringPhysics®
VIDEO: Faraday's Law
VIDEO: Applications of Electromagnetic Induction

> **The induced voltage in a coil is proportional to the number of loops multiplied by the rate at which the magnetic field changes within those loops.**

We have mentioned two ways in which voltage can be induced in a loop of wire: by moving the loop near a magnet and by moving a magnet near the loop. There is a third way: by changing the current in a nearby loop. All three of these ways have the same essential ingredient—a changing magnetic field in the loop.

The amount of current produced by electromagnetic induction depends on the resistance of the coil and the circuit it connects, as well as the induced voltage.

Electromagnetic induction explains the induction of voltage in a wire, as we have seen. However, the more basic concept of fields is at the root of the induced voltages and currents we observe in coils. The modern view of electromagnetic induction is that electric and magnetic *fields* are induced—and these in turn produce the voltages we have considered. So, induction occurs whether or not a conducting wire or any material medium is present. In this more general sense, Faraday's law states:

> **An electric field is induced in any region of space in which a magnetic field is changing with time.**

There is a second effect, an extension of Faraday's law. It is the same except that the roles of electric and magnetic fields are interchanged. It is one of nature's many symmetries. This effect, which was introduced by British physicist James Clerk Maxwell in about 1860 is known as **Maxwell's counterpart to Faraday's law:**

> **A magnetic field is induced in any region of space in which an electric field is changing with time.**

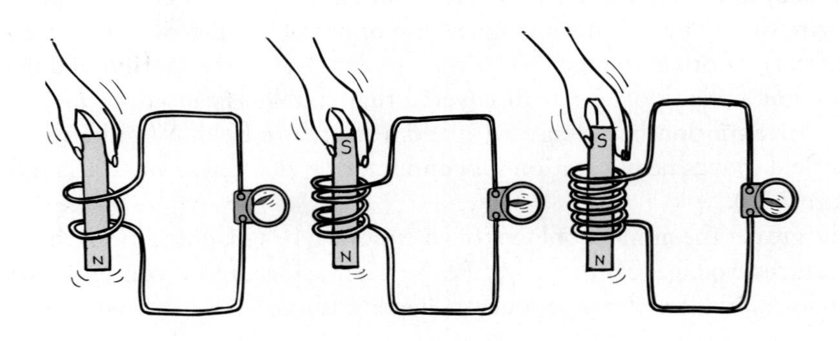

FIGURE 7.47
When a magnet is plunged into a coil with twice as many loops as another, twice as much voltage is induced. If the magnet is plunged into a coil with three times as many loops, three times as much voltage is induced.

HISTORY OF SCIENCE

Generators and Power Production

Fifty years after Faraday and Henry discovered electromagnetic induction, Nikola Tesla and George Westinghouse put those findings to practical use. They showed the world that electricity could be generated reliably and in sufficient quantities to light entire cities.

Tesla built generators much like those still in use. Tesla's generators were more complicated than the simple model we have discussed. His generators had armatures made up of bundles of copper wires. The armatures were forced to spin within strong magnetic fields by a turbine, which, in turn, was spun by the energy of either steam or falling water. The rotating loops of wire in the armature cut through the magnetic field of the surrounding electromagnets. In this way, they induced alternating voltage and currents.

It is important to emphasize that the energy put out by a turbine is the energy provided by the rushing water in a waterfall or by steam. In traditional steam power plants, the water is heated to steam by burning fossil fuels. In nuclear power plants, the energy to heat water to make steam is supplied by nuclear fission reactions.

It's important to know that generators don't create energy—they simply convert energy from some other form into electric energy. Energy from a source—whether fossil or nuclear fuel, or wind, or water—is converted into mechanical energy to drive the turbine, and then converted into electricity, which then carries the energy to where it can be used.

In each case, the strength of the induced field is proportional to the rates of change of the inducing field. The induced electric and magnetic fields are at right angles to each other.

Maxwell saw the link between electromagnetic waves and light.* If electric charges are set into vibration in the range of frequencies that match those of light, waves are produced that *are* light! Maxwell discovered that light is simply electromagnetic waves in the range of frequencies to which the eye is sensitive. You will learn much more about electromagnetic waves in the next chapter.

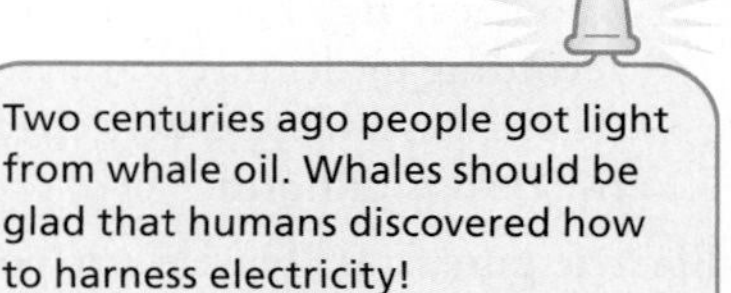

Two centuries ago people got light from whale oil. Whales should be glad that humans discovered how to harness electricity!

TECHNOLOGY

Electromagnetic Induction in Everyday Technologies

We see electromagnetic induction all around us. On the road, we see it operate when we drive over buried coils of wire to activate a nearby traffic light. When the iron parts of a car move over the buried coils, the effect of Earth's magnetic field is changed, which induces a voltage to trigger the changing of the traffic lights. Similarly, when you walk through the upright coils in the security system at an airport, any metal you carry slightly alters the magnetic field in the coils. This change induces voltage, which sounds an alarm. When the magnetic strip on the back of a credit card is scanned, induced voltage pulses identify the card. Electromagnetic induction is everywhere. As we shall see in the next chapter, it underlies the electromagnetic waves we call light.

*On the eve of his discovery, story has it that Maxwell had a date with a young woman he was later to marry. While walking in a garden, his date remarked about the beauty and wonder of the stars. Maxwell asked how she would feel to know that she was walking with the only person in the world who knew what starlight really was. And that was true. At that time, James Clerk Maxwell was the only person in the world who knew that light of any kind is energy carried in waves of electric and magnetic fields that continuously regenerate each other.

SUMMARY OF TERMS (KNOWLEDGE)

Alternating current (ac) An electric current that repeatedly reverses its direction; the electric charges vibrate about relatively fixed positions, 60 Hz in the United States.

Conductor Any material having free charged particles that easily flow through it when an electrical force acts on them.

Coulomb The SI unit of electric charge. One coulomb (symbol C) is equal in magnitude to the total charge of 6.25×10^{18} electrons.

Coulomb's law The relationship among force, charge, and distance:

$$F = k\frac{q_1 q_2}{d^2}$$

If the charges are alike in sign, the force is repelling; if the charges are unlike, the force is attractive.

Direct current (dc) An electric current that flows in one direction only.

Electric current A flow of charged particles that transports energy from one place to another, measured in amperes, where 1 A is the flow of 6.25×10^{18} electrons per second, or 1 coulomb per second.

Electric field Defined as force per unit charge, it can be considered an energetic aura surrounding charged objects. About a charged point, the field decreases with distance according to the inverse-square law, like a gravitational field. Between oppositely charged parallel plates, the electric field is uniform.

Electric potential The electric potential energy per amount of charge, measured in volts, and often called *voltage*:

$$\text{Electric potential} = \frac{\text{electric potential energy}}{\text{charge}}$$

Electric potential energy The energy a charge possesses by virtue of its location in an electric field.

Electric power The rate of energy transfer, or rate of doing work; the amount of energy per unit time, which can be measured by the product of current and voltage:

$$\text{Power} = \text{current} \times \text{voltage}$$

Electric power is measured in watts (or in kilowatts).

Electrical resistance The property of a material that resists the flow of electric charge through it; measured in ohms (Ω).

Electrically polarized Description of an atom or molecule in which the charges are aligned so that one side has a slight excess of positive charge and the other side a slight excess of negative charge.

Electromagnet A magnet whose field is produced by an electric current. It is usually in the form of a wire coil with a piece of iron inside the coil.

Electromagnetic induction The induction of voltage when a magnetic field changes with time.

Faraday's law An electric field is induced in any region of space in which a magnetic field is changing with time. The magnitude of the induced electric field is proportional to the rate at which the magnetic field changes. The direction of the induced field is at right angles to the changing magnetic field.

Insulator Any material with very little or no free charged particles and through which current does not easily flow.

Magnetic domains Clustered regions of aligned magnetic atoms. When these regions themselves are aligned with one another, the substance containing them is a magnet.

Magnetic field The region of magnetic influence around a magnetic pole or around a moving charged particle.

Magnetic force (1) Between magnets, it is the attraction of unlike magnetic poles for each other and the repulsion between like magnetic poles. (2) Between a magnetic field and a moving charge, it is a deflecting force due to the motion of the charge; the deflecting force is perpendicular to the velocity of the charge and perpendicular to the magnetic field lines.

Maxwell's counterpart to Faraday's law A magnetic field is induced in any region of space in which an electric field is changing with time.

Ohm's law The amount of current in a circuit varies in direct proportion to the potential difference or voltage and inversely with the resistance:

$$\text{Current} = \frac{\text{voltage}}{\text{resistance}}$$

Parallel circuit An electric circuit with two or more devices connected in such a way that the same voltage acts across each one, and any single one completes the circuit independently of all the others.

Potential difference The difference in electric potential between two points, measured in volts; often called *voltage difference*.

Semiconductor A material that can be made to behave sometimes as an insulator and sometimes as a conductor.

Series circuit An electric circuit with devices connected in such a way that the same electric current exists in all of them.

READING CHECK QUESTIONS (COMPREHENSION)

7.1 Electrical Force and Charge

1. Which part of an atom is positively charged, and which part is negatively charged?
2. What is meant by saying that charge is conserved?

7.2 Coulomb's Law

3. How is Coulomb's law similar to Newton's law of gravity? How is it different?
4. How does a coulomb of charge compare with the charge of a single electron?

7.3 Electric Field

5. Give two examples of common force fields.
6. How is the direction of an electric field defined?

7.4 Electric Potential

7. In terms of the units that measure them, distinguish between *electric potential energy* and *electric potential.*
8. A balloon may easily be charged to several thousand volts. Does that mean it has several thousand joules of energy? Explain.

7.5 Conductors and Insulators

9. What is the difference between a conductor and an insulator? Between a conductor and a semiconductor?
10. What kind of materials are the best conductors? Why are they so good at conducting electricity?

7.6 Voltage Sources

11. What condition is necessary for heat energy to flow from one end of a metal bar to another? For electric charge to flow?
12. What condition is necessary for a sustained flow of electric charge through a conducting medium?

7.7 Electric Current

13. Why do electrons, rather than protons, make up the flow of charge in a metal wire?
14. Distinguish between dc and ac.

7.8 Electrical Resistance

15. Which has more resistance—a thick wire or a thin wire of the same length? A short wire or a long wire?
16. What is the unit of electrical resistance?

7.9 Ohm's Law

17. What is the effect on the current through a circuit of steady resistance when the voltage is doubled? What is the effect on the current if both voltage and resistance are doubled?
18. How much current does a radio speaker with a resistance of 8 Ω draw when 12 V is impressed across it?

7.10 Electric Circuits

19. Which type of circuit is favored for operating several electrical devices, each independently of the other—series or parallel? Defend your answer.
20. How does the sum of the currents through the branches of a simple parallel circuit compare with the current that flows through the voltage source?

7.11 Electric Power

21. What is the relationship among electric power, current, and voltage?
22. Between a kilowatt and a kilowatt-hour, which is a unit of energy and which is a unit of power?

7.12 The Magnetic Force

23. How is the rule for the interaction between magnetic poles similar to the rule for the interaction between electric charges?
24. How are magnetic poles different from electric charges?

7.13 Magnetic Fields

25. What produces a magnetic field?
26. Why is iron magnetic and wood is not?

7.14 Magnetic Forces on Moving Charges

27. What is the direction of the magnetic force that acts on a moving charged particle?
28. What is a galvanometer? What is it called when it is calibrated to read current? To read voltage?

7.15 Electromagnetic Induction

29. What are the three ways in which voltage can be induced in a wire?
30. (a) What is induced by the rapid alternation of a magnetic field? (b) What is induced by the rapid alternation of an electric field?

THINK INTEGRATED SCIENCE

7A—Electric Shock

31. High voltage by itself does not produce electric shock. What does?
32. What is the source of the electrons that shock you when you touch a charged conductor?
33. If a current of one-tenth or two-tenths of an ampere were to flow into one of your hands and out the other, you would probably be electrocuted. But if the same current were to flow into your hand and out at the elbow above the same hand, you could survive, even though the current might be large enough to burn your flesh. Explain.

7B—Earth's Magnetic Field and the Ability of Organisms to Sense It

34. People have wondered about the "mystery" of animal migration for generations. Give one possible, very general explanation as to how animals find their way during migration.
35. What is the likely cause of Earth's magnetic field?

THINK AND DO (HANDS-ON APPLICATION)

36. An electric cell is made by placing two plates of different materials that have different affinities for electrons in a conducting solution. You can make a simple 1.5-V cell by placing a strip of copper and a strip of zinc in a tumbler of salt water. The voltage of a cell depends on the materials and the solution they are placed in, not the size of the plates. A battery is actually a series of cells.

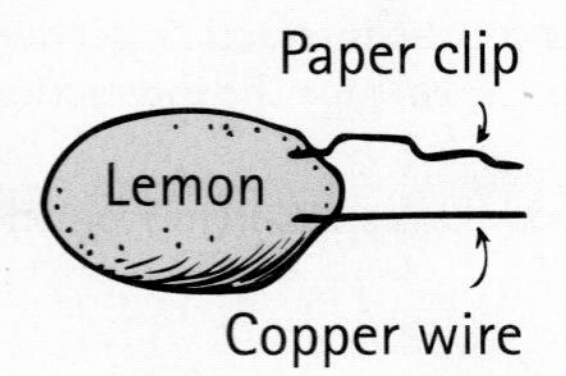

An easy cell to construct is the lemon cell. Stick a straightened paper clip and a piece of copper wire into a lemon. Hold the ends of the paper clip and the wire close together, but not touching, and place the ends on your tongue. The slight tingle you feel and the metallic taste you experience result from a small current of electricity pushed by the citrus cell through the paper clip and the wire when your moist tongue closes the circuit.

37. An iron bar can be magnetized easily by aligning it with the magnetic field lines of Earth and striking it lightly a few times with a hammer. This works best if the bar is tilted down to match the dip of Earth's magnetic field. The hammering jostles the magnetic domains in the bar so that they can fall into a better alignment with Earth's magnetic field. The bar can be demagnetized by striking it when it is oriented in an east-west direction.
38. Text or write a letter to Grandma and convince her that whatever electric shocks she may have received over the years have been due to the movement of electrons already in her body—not electrons coming from somewhere else.

PLUG AND CHUG (FORMULA FAMILIARIZATION)

Coulomb's law: $F = k\dfrac{q_1 q_2}{d^2}$

39. Two point charges, each with 0.1 C of charge, are 0.1 m apart. Given that k is $9 \times 10^9 \text{ N} \cdot \text{m}^2/\text{C}^2$ (the proportionality constant for Coulomb's law), show that the force between the charges is 9×10^9 N.

Ohm's law: $I = \dfrac{V}{R}$

40. A toaster has a heating element of 15 Ω and is connected to a 120-V outlet. Show that the current drawn by the toaster is 8 A.
41. When you touch your fingers (resistance 1000 Ω) to the terminals of a 6-V battery, show that the small current moving through your fingers is 0.006 A.
42. Calculate the current in the 240-Ω filament of a bulb connected to a 120-V line.

Power $= IV$

43. An electric toy draws 0.5 A from a 120-V outlet. Show that the toy consumes 60 W of power.
44. Show that the power consumed by a 120-V device that draws 5 A of current is 500 W.

THINK AND COMPARE (ANALYSIS)

45. The three pairs of metal same-size spheres have different charges on their surfaces as indicated. Each pair is brought together, allowed to touch, and then separated. Rank from greatest to least the total amounts of charge on the pairs of spheres after separation.

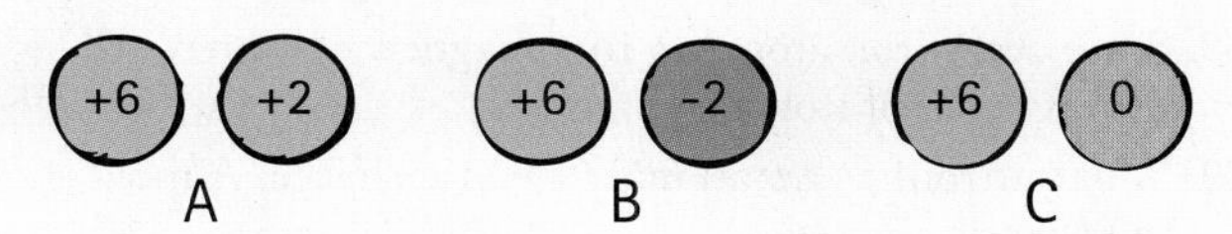

46. Rank circuits A, B, and C according to the brightness of the identical bulbs, from brightest to dimmest.

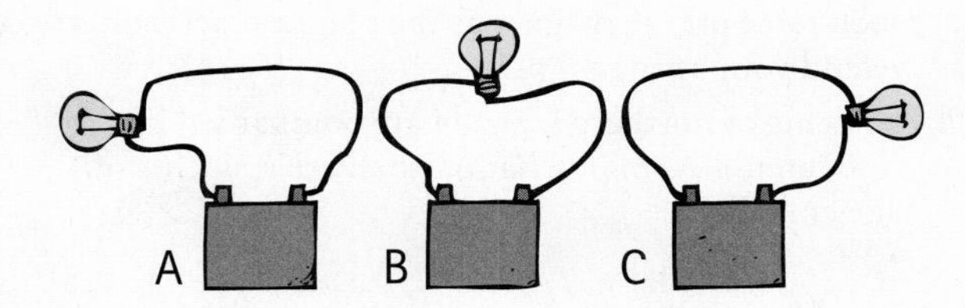

47. The bulbs in circuits A, B, and C are identical. An ammeter is placed in different locations, as shown. Rank the current readings in the ammeter from greatest to least.

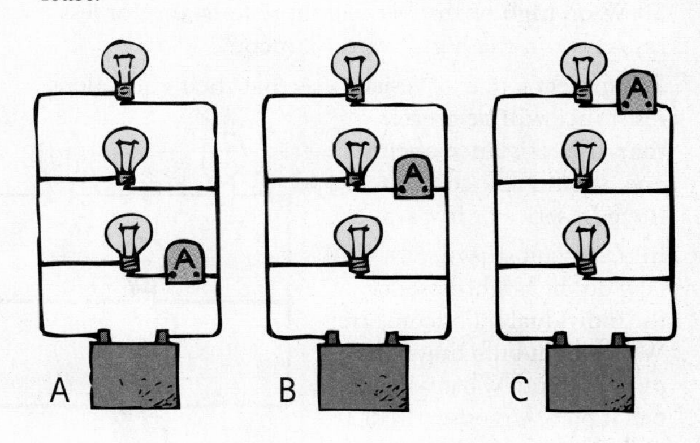

THINK AND SOLVE (MATHEMATICAL APPLICATION)

48. Two point charges are separated by 6 cm. The attractive force between them is 20 N. Show that the force between them when they are separated by 12 cm is 5 N.
49. A droplet of ink in an industrial ink-jet printer carries a charge of 1.6×10^{-10} C and is deflected onto paper by a force of 3.2×10^{-4} N. Show that the strength of the electric field that is required to produce this force is 2×10^{6} N/C.
50. Find the voltage change (a) when an electric field does 12 J of work on a charge of 0.0001 C and (b) when the same electric field does 24 J of work on a charge of 0.0002 C.
51. Rearrange this equation

$$\text{Current} = \frac{\text{voltage}}{\text{resistance}}$$

to express resistance in terms of current and voltage. Then consider the following: A certain device in a 120-V circuit has a current rating of 20 A. Show that the resistance of the device is 6 Ω.

52. Use the formula

$$\text{Power} = \text{current} \times \text{voltage}$$

to find that the current drawn by a 1200-W hair dryer connected to 120 V is 10 A. Then, using the method from Exercise 51, show that the resistance of the hair dryer is 12 Ω.

53. Show that it costs \$3.36 to operate a 100-W lamp continuously for a week if the power utility rate is 20¢/kWh.
54. An electric iron connected to 120 V draws 9 A of current. Show that the amount of heat generated in 1 minute is nearly 65,000 J.

THINK AND EXPLAIN (SYNTHESIS)

55. With respect to forces, how are electric charge and mass alike? How are they different?
56. When combing your hair, you scuff electrons from your hair onto the comb. Is your hair then positively or negatively charged? How about the comb?
57. The 5000 billion billion freely moving electrons in a penny repel one another. Why don't they fly out of the penny?
58. Two equal charges exert equal forces on each other. What if one charge has twice the magnitude of another? How do the forces they exert on each other compare?
59. Suppose that the strength of the electric field around an isolated point charge has a certain value at a distance of 1 m. How will the electric field strength compare at a distance of 2 m from the point charge? What law guides your answer?
60. Why is a good conductor of electricity also a good conductor of heat?
61. When a car is moved into a painting chamber, a mist of paint is sprayed around it. When the body of the car is given a sudden electric charge and the mist of paint is attracted to it, presto—the car is quickly and uniformly painted. What does the phenomenon of polarization have to do with this?

62. Will the current in a lightbulb connected to a 220-V source be greater or less than the current in the same bulb when it is connected to a 110-V source?
63. Are automobile headlights wired in parallel or in series? What is your evidence?
64. A car's headlights dissipate 40 W on low beam and 50 W on high beam. Is there more resistance or less resistance in the high-beam filament?
65. To connect a pair of resistors so that their equivalent resistance will be greater than the resistance of either one, should you connect them in series or in parallel?
66. In the circuit shown, how do the brightnesses of the individual bulbs compare? Which light bulb draws the most current? What will happen if bulb A is unscrewed? If bulb C is unscrewed?

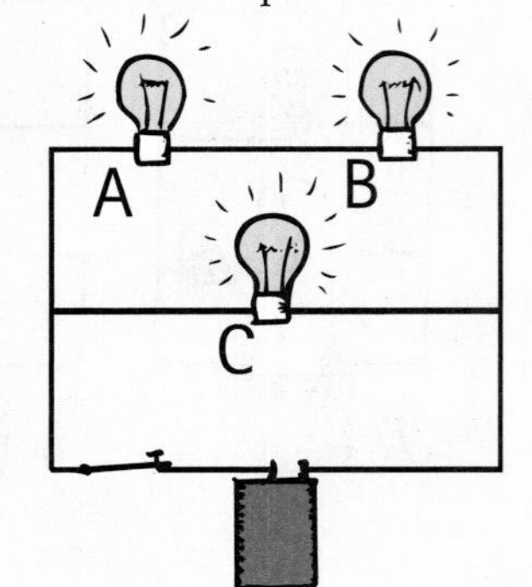

67. Why might the wingspans of birds be a consideration in determining the spacing between parallel wires on power poles?
68. As more and more bulbs are connected in series to a flashlight battery, what happens to the brightness of each bulb? Assuming that heating inside the battery is negligible, what happens to the brightness of each bulb when more and more bulbs are connected in parallel?
69. Since every iron atom is a tiny magnet, why aren't all things made of iron also magnets?
70. What surrounds a stationary electric charge? A moving electric charge?
71. A strong magnet attracts a paper clip to itself with a certain force. Does the paper clip exert a force on the strong magnet? If not, why not? If so, does it exert as much force on the magnet as the magnet exerts on it? Defend your answers.
72. Residents of northern Canada are bombarded by more intense cosmic radiation than are residents of Mexico. Why?

THINK AND DISCUSS (EVALUATION)

73. An electroscope is a simple device consisting of a metal ball that is attached by a conductor to two thin leaves of metal foil protected from air disturbances in a jar, as shown. When the ball is touched by a charged body, the leaves, which normally hang straight down, spread apart. Why?
74. If you place a free electron and a free proton in the same electric field, how will the forces acting on them compare? How will their accelerations compare? How will their directions of travel compare?
75. You are not harmed by contact with a charged metal ball, even though its voltage may be very high. Is the reason for this similar to the reason you are not harmed by the sparks from a Fourth-of-July sparkler, even though the temperature of each of those sparks is higher than 1000°C? Discuss and defend your answer in terms of the energies that are involved.
76. Lillian is charged to some 50,000 V but is unharmed. Her hair, however, stands out. Why is she not harmed, and why does her hair stand out?

77. Discuss the circuits shown. In which of these circuits does a current exist to light the bulb?

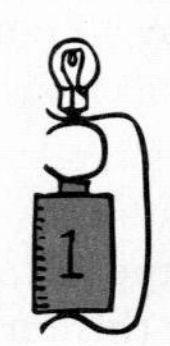

78. Sometimes you hear someone say that a particular appliance "uses up" electricity. What is it that the appliance actually uses up, and what becomes of it?
79. Wai Tsan Lee shows iron nails that have become induced magnets. Is the physics of this situation similar to that of the sticking balloon in Figure 7.5? Discuss and defend your answer.

80. Can an electron at rest in a magnetic field be set into motion by the magnetic field? What if it were at rest in an electric field?
81. A magician places an aluminum ring on a table, underneath which is hidden an electromagnet. When the magician says "abracadabra" (and pushes a switch that starts current flowing through the coil under the table), the ring jumps into the air. Discuss and explain his "trick."

82. Only a small percentage of the electric energy fed into an incandescent lamp is transformed into light. What happens to the remaining energy?

83. In terms of heat generated, why are compact fluorescent lamps (CFLs) more efficient than incandescent lamps?

84. A person in your discussion group says that changing electric and magnetic fields generate each other and this gives rise to visible light when the frequency of the change matches the frequency of light. Do you agree? Explain.

85. Another person in your discussion group says that a battery provides not a source of electrons, but a means of moving the electrons already in the circuit. Do you agree? Defend your answer.

86. Still another person in your discussion group says that adding bulbs in series to a circuit provides more opposition to the flow of charge and results in less current in more bulbs. But adding bulbs in parallel provides more pathways so that there is more current in the circuit. Do you agree? Defend your answer.

READINESS ASSURANCE TEST (RAT)

If you have a good handle on this chapter, then you should be able to score at least 7 out of 10 on this RAT. If you score less than 7, you need to study further before moving on.

Choose the BEST answer to each of the following:

1. The electric force of attraction between an electron and a proton is greater on
   (a) the proton.
   (b) the electron.
   (c) neither the proton nor the electron; both are the same.
   (d) none of these

2. When a pair of charged particles are brought twice as close to each other, the force between them becomes
   (a) twice as strong.
   (b) four times as strong.
   (c) half as strong.
   (d) one-quarter as strong.

3. An electric field surrounds
   (a) the electric charge.
   (b) all the electrons.
   (c) all the protons.
   (d) all of these

4. When you double the voltage in a simple electric circuit, you double
   (a) the current.
   (b) the resistance.
   (c) both the current and the resistance.
   (d) neither the current nor the resistance.

5. In a simple circuit consisting of a single lamp and a single battery, when the current in the lamp is 2 A, the current in the battery is
   (a) 1 A.
   (b) 2 A.
   (c) dependent on the internal battery resistance.
   (d) More information is needed.

6. In a circuit with two lamps in parallel, if the current in one lamp is 2 A, the current in the battery is
   (a) 1 A.
   (b) 2 A.
   (c) greater than 2 A.
   (d) none of these

7. If both the current and the voltage in a circuit are doubled, the power
   (a) remains unchanged if the resistance is constant.
   (b) is halved.
   (c) is doubled.
   (d) is quadrupled.

8. The essential physics concept in an electric generator is
   (a) Coulomb's law.
   (b) Ohm's law.
   (c) Faraday's law.
   (d) Newton's second law.

9. If you change the magnetic field in a closed loop of wire, you induce a(n)
   (a) current.
   (b) voltage.
   (c) electric field.
   (d) all of these

10. The mutual induction of electric and magnetic fields can produce
    (a) light.
    (b) sound.
    (c) both light and sound.
    (d) neither light nor sound.

Answers to RAT

1. c, 2. b, 3. d, 4. a, 5. b, 6. c, 7. d, 8. c, 9. d, 10. a

CHAPTER 8

Waves—Sound and Light

SOUND IS the only thing we can really hear, as young violinists Michelle and Miriam remind us. But what is sound? And light is the only thing we can really see. But what *is* light? You may know that light is an electromagnetic wave and that it is part of the electromagnetic spectrum—a continuum of waves including X-rays, radio waves, microwaves, and others. Where do electromagnetic waves come from? What are their properties? Do electromagnetic waves of various lengths really permeate our environment like vibrating strands of invisible spaghetti? Sound, like light, is a wave phenomenon. How do sound waves differ from light waves? Does the speed of sound differ in different materials? Can sound travel in a vacuum? Can one sound wave cancel another, so that two loud noises combine to make zero noise? Are there technological applications for this idea? Many things vibrate—from musical instruments to atoms to vocal chords—and when doing so, they produce waves. Waves, the subject of this chapter, are everywhere!

8.1 Vibrations and Waves

EXPLAIN THIS How do vibrating electrons produce radio waves?

LEARNING OBJECTIVE
Distinguish among amplitude, wavelength, frequency, and period.

Anything—from your vocal chords to a pendulum—that moves back and forth, to and fro, in and out, or up and down is vibrating. A *vibration* is a wiggle. A wiggle that travels is a *wave*. A wave extends from one location to another. Light and sound are both vibrations that propagate throughout space as waves, but waves of two very different kinds. Sound is the propagation of mechanical vibrations through a material medium—a solid, a liquid, or a gas. If there is no medium to vibrate, then no sound is possible. Sound cannot travel in a vacuum. But light can, because light is a vibration of nonmaterial electric and magnetic fields—a vibration of pure energy. Although light can pass through many materials, it does not require a material. This is evident when light propagates through the near vacuum between the Sun and Earth.

MasteringPhysics®
TUTORIAL: Vibrations and Waves

UNIFYING CONCEPT
- *Waves*

The relationship between a vibration and a wave is shown in Figure 8.1. A marking pen on a bob attached to a vertical spring vibrates up and down and traces a waveform on a sheet of paper that is moved horizontally at constant speed. The waveform is actually a *sine curve*, a graphical representation of a wave. Like a water wave, the high points of a sine wave are called *crests* and the low points are the *troughs*. The straight dashed line represents the "home" position, or midpoint, of the vibration. The term **amplitude** refers to the distance from the midpoint to the crest (or to the trough) of the wave. So the amplitude equals the maximum displacement from equilibrium. Waves carry energy from one place to another. The amount of energy a wave carries depends on its amplitude. The larger the amplitude of a wave, the more energy it has.

Waves carry energy, as anyone who has witnessed a pounding ocean wave upon the shore can see.

The **wavelength** of a wave is the distance from the top of one crest to the top of the next crest or, equivalently, the distance between successive identical parts of the wave. The wavelengths of waves at the beach are measured in meters, the wavelengths of ripples in a pond in centimeters, and the wavelengths of light in billionths of a meter (nanometers). All waves have a vibrating source.

How frequently a vibration occurs is described by its frequency. The **frequency** of a vibrating pendulum, or of an object on a spring, specifies the number of to-and-fro vibrations it makes in a given time (usually in 1 second). A complete to-and-fro oscillation is one vibration. If the oscillation occurs in one second, the frequency is one vibration per second. If two vibrations occur in one second, the frequency is two vibrations per second.

The unit of frequency is called the **hertz** (Hz), after Heinrich Hertz, who demonstrated the existence of radio waves in 1886. One vibration per second is 1 hertz, two vibrations per second is 2 hertz, and so on. Higher frequencies are measured in kilohertz (kHz) and still higher frequencies in megahertz (MHz).

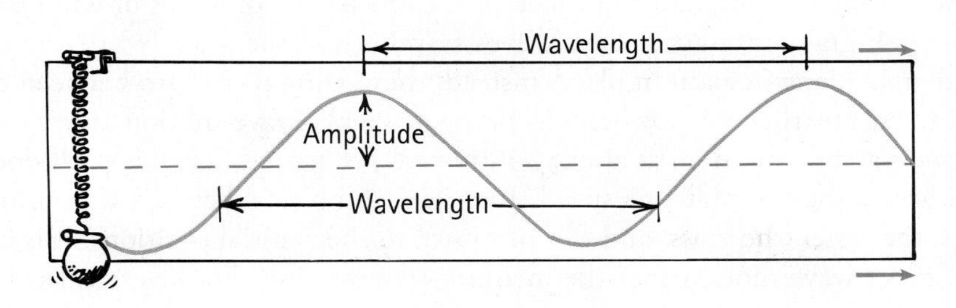

FIGURE 8.1
INTERACTIVE FIGURE
When the bob vibrates up and down, a marking pen traces out a sine curve on paper that is moved horizontally at constant speed.

TECHNOLOGY

Broadcasting Radio Waves

AM radio waves are usually measured in kilohertz, while FM radio waves are measured in megahertz. A station at 960 kHz on the AM radio dial, for example, broadcasts radio waves that have a frequency of 960,000 vibrations per second. A station at 101.7 MHz on the FM dial broadcasts radio waves with a frequency of 101,700,000 hertz. These radio-wave frequencies are the frequencies at which electrons are forced to vibrate in the antenna of a radio station's transmitting tower. The frequency of the vibrating electrons and the frequency of the wave produced are the same.

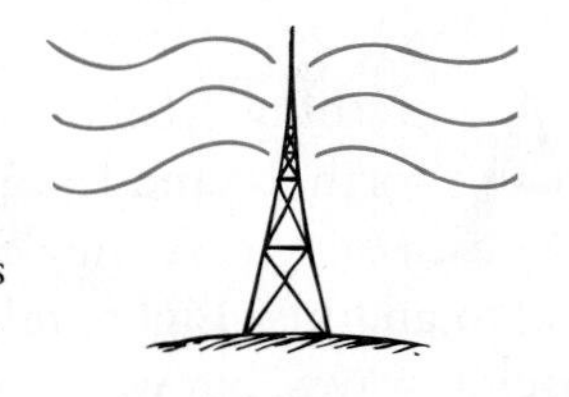

The **period** of a wave or vibration is the time it takes for a complete vibration—for a complete cycle. Period can be calculated from frequency, and vice versa. Suppose, for example, that a pendulum makes two vibrations in one second. Its frequency is 2 Hz. The time needed to complete one vibration—that is, the period of vibration—is $\frac{1}{2}$ second. Or, if the vibration frequency is 3 Hz, then the period is $\frac{1}{3}$ second. The frequency and period are the inverse of each other:

$$\text{Frequency} = \frac{1}{\text{period}} \quad \text{or, vice versa,} \quad \text{Period} = \frac{1}{\text{frequency}}$$

A wave transfers energy without transferring matter. If matter were to move along with the energy in a wave, the oceans would be emptied as ocean waves travel to the shore.

CHECK YOURSELF

1. **An electric razor completes 60 cycles every second. (a) What is its frequency? (b) What is its period?**
2. **Gusts of wind cause the Willis Tower in Chicago to sway back and forth, completing a cycle every 10 seconds. (a) What is its frequency? (b) What is its period?**

CHECK YOUR ANSWERS

1. (a) 60 cycles per second, or 60 Hz. (b) $\frac{1}{60}$ second.
2. (a) $\frac{1}{10}$ Hz. (b) 10 seconds.

LEARNING OBJECTIVE
Describe how energy is carried in waves.

8.2 Wave Motion

EXPLAIN THIS How does wave speed relate to frequency and wavelength?

If you drop a stone into a calm pond, waves will travel outward in expanding circles. Energy is carried by the wave, traveling from one place to another. The water itself goes nowhere. This can be seen by waves encountering a floating leaf. The leaf bobs up and down, but it doesn't travel with the waves. The waves move along, not the water. The same is true of waves of wind over a field of tall grass on a gusty day. Waves travel across the grass, while the individual grass plants remain in place; instead, they swing to and fro between definite limits, but they go nowhere. When you speak, wave motion through the air travels across the room at about 340 meters per second. The air itself doesn't travel across the room at this speed. In these examples, when the wave motion ceases, the water, the grass, and the air return to their initial positions. It is characteristic of wave motion that the medium transporting the wave returns to its

FIGURE 8.2
Water waves.

initial condition after the disturbance has passed. Putting all the information about waves together, we can now specifically define what a wave is: A **wave** is a disturbance that travels from one place to another transporting energy, but not necessarily matter, along with it.

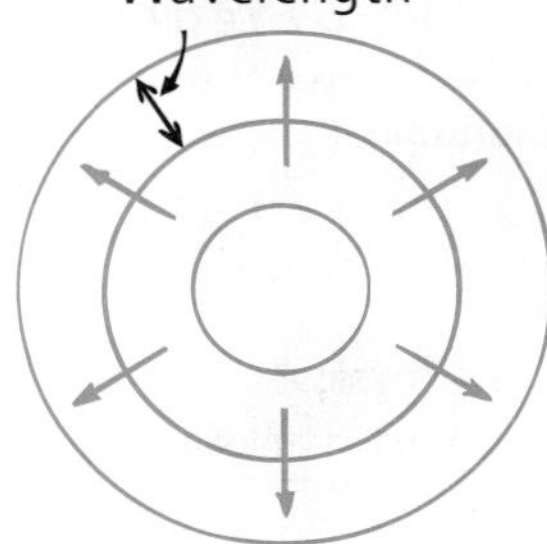

FIGURE 8.3
A top view of water waves.

Wave Speed

The speed of periodic wave motion is related to the frequency and wavelength of the waves. Consider the simple case of water waves (Figures 8.2 and 8.3). Imagine that we fix our eyes on a stationary point on the water's surface and observe the waves passing by that point. We can measure how much time passes between the arrival of one crest and the arrival of the next one (the period), and we can also observe the distance between crests (the wavelength). We know that speed is defined as distance divided by time. In this case, the distance is one wavelength and the time is one period, so wave speed = wavelength/period.

For example, if the wavelength is 10 meters and the time between crests at a point on the surface is 0.5 second, then the wave is traveling 10 meters in 0.5 second and its speed is 10 meters divided by 0.5 second, or 20 meters per second. Since period is equal to the inverse of frequency, the formula wave speed = wavelength/period = wavelength × frequency. We usually say:

$$\text{Wave speed} = \text{frequency} \times \text{wavelength}$$

This relationship applies to all kinds of waves, whether they are water waves, sound waves, or light waves.

The speed of light waves in a vacuum, approximately 3.0×10^8 m/s and denoted as c, is nature's speed limit. No material objects in the universe travel faster than this.

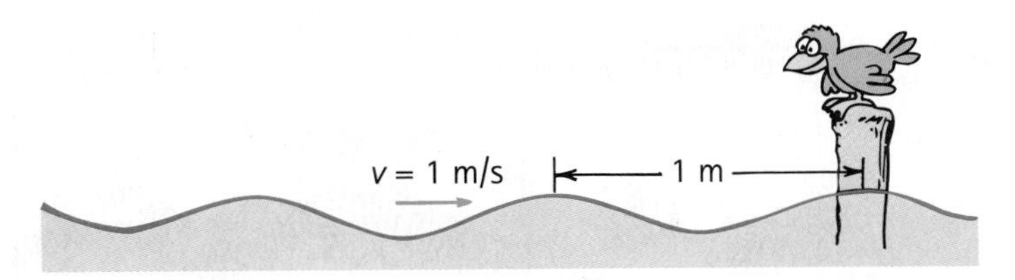

FIGURE 8.4

INTERACTIVE FIGURE

If the wavelength is 1 m, and one wavelength per second passes the pole, then the speed of the wave is 1 m/s.

MATH CONNECTION

Frequency and Wave Speed

The frequency of a wave is the same as the frequency of the vibrating source that produces it. The speed of a wave equals the product of frequency and wavelength. It is customary to express this speed by the equation $v = f\lambda$, where v is wave speed, f is wave frequency, and λ (the Greek letter lambda) is wavelength.

Problems

1. If a train of freight cars, each 10 m long, rolls by you at the rate of three cars each second, what is the speed of the train?
2. If a water wave oscillates up and down three times each second and the distance between wave crests is 2 m, what are the wave's (a) frequency, (b) wavelength, and (c) wave speed?
3. The sound from a 60-Hz electric razor spreads out at 340 meters per second. (a) What is the frequency of the sound waves? What are (b) their period, (c) their speed, and (d) their wavelength?

Solutions

1. The answer is 30 m/s. We can see this in two ways. (a) According to the definition of speed in Chapter 2, $v = \frac{d}{t} = \frac{3 \times 10\text{ m}}{1\text{ s}} = 30$ m/s, since 30 m of train passes you in 1 s. (b) If we compare our train to wave motion, where wavelength corresponds to 10 m and frequency is 3 Hz, then

$$\begin{aligned}\text{Speed} &= \text{frequency} \times \text{wavelength}\\ &= 3\text{ Hz} \times 10\text{ m} = 30\text{ m/s}\end{aligned}$$

2. (a) 3 Hz. (b) 2 m. (c) Wave speed = frequency × wavelength = $\frac{3}{s} \times 2$ m = 6 m/s. Do you see that parts (a) and (b) are given in the question itself?
3. (a) 60 Hz, as given in the question. (b) Period = $\frac{1}{\text{frequency}} = \frac{1}{60}$ s. (c) 340 m/s, which is given in the question. (d) From $v = f\lambda$, $\lambda = \frac{v}{f} = \frac{340\text{ m/s}}{60/\text{s}} = 5.7$ m.

LEARNING OBJECTIVE
Distinguish between transverse and longitudinal waves.

8.3 Transverse and Longitudinal Waves

EXPLAIN THIS Exactly what is transmitted in all kinds of waves?

Fasten one end of a rope to a wall and hold the free end in your hand. If you shake the free end up and down, you will produce vibrations that are at right angles to the direction of wave travel. The right-angled, or sideways, motion is called *transverse motion*. This type of wave is called a **transverse wave.** Waves in the stretched strings of musical instruments are transverse waves. We will see later that electromagnetic waves, such as radio waves and light waves, are also transverse waves.

MasteringPhysics®
VIDEO: Longitudinal vs. Transverse Waves

A **longitudinal wave** is one in which the direction of wave travel is along the direction in which the source vibrates. A Slinky with one end fastened to a wall nicely illustrates this. Shake the Slinky back and forth along its long axis and you produce a longitudinal wave (Figure 8.5). The vibrations are then parallel to the direction of energy transfer. Part of the Slinky is compressed, and a wave of compression travels along it. In between successive compressions is a stretched region, called a rarefaction. Both compressions and rarefactions travel in the same direction along the Slinky. Together they make up the longitudinal wave. You produce a transverse wave when you shake the Slinky up and down.

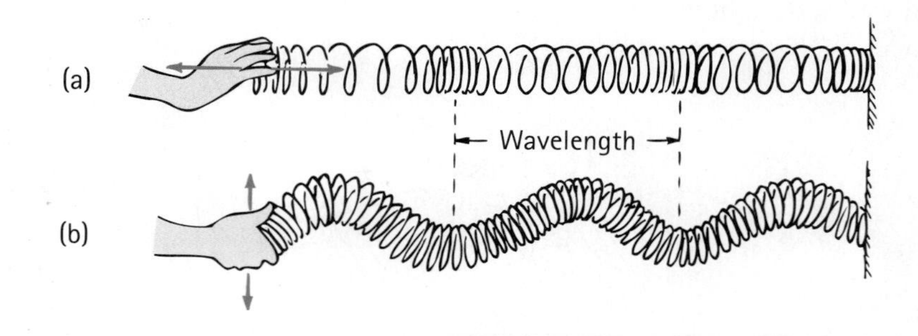

FIGURE 8.5
INTERACTIVE FIGURE

Both waves transfer energy from left to right. (a) When the end of the Slinky is pushed and pulled rapidly along its length, a longitudinal wave is produced. (b) When its end is shaken up and down (or from side to side), a transverse wave is produced.

LEARNING OBJECTIVE
Identify compressions and rarefactions in a sound wave.

8.4 The Nature of Sound

EXPLAIN THIS Why doesn't sound travel in a vacuum?

Think of the air molecules in a room as tiny, randomly moving Ping-Pong balls. If you vibrate a Ping-Pong paddle in the midst of the balls, you'll set them vibrating to and fro (Figure 8.6). The balls will vibrate in rhythm with your vibrating paddle. In some regions, they will be momentarily bunched up (compressions), and in other regions in between, they will be momentarily spread out (rarefactions). The vibrating prongs of a tuning fork do the same to air molecules. Vibrations made up of compressions and rarefactions spread from the tuning fork throughout the air, and a sound wave is produced (Figure 8.7).

The wavelength of a sound wave is the distance between successive compressions or, equivalently, the distance between successive rarefactions. Each molecule in the air vibrates to and fro about some equilibrium position as the waves move by.

FIGURE 8.6
If you vibrate a Ping-Pong paddle in the midst of a lot of Ping-Pong balls, the balls will vibrate also.

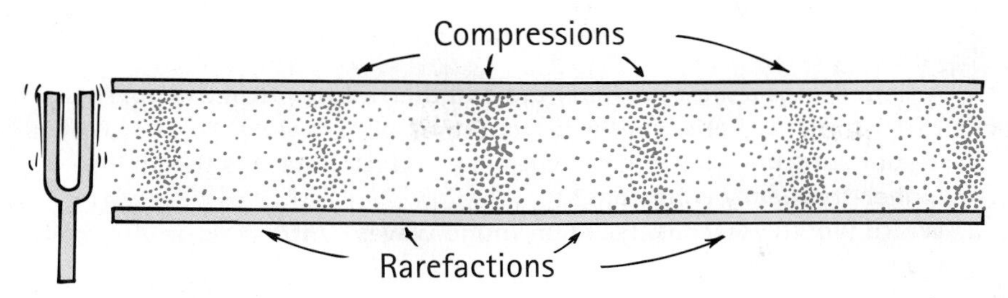

FIGURE 8.7
Compressions and rarefactions travel (both at the same speed and in the same direction) from the tuning fork through the air in the tube.

Integrated Science 8A
BIOLOGY

Sensing Pitch

LEARNING OBJECTIVE
Relate the frequency of sound to pitch.

EXPLAIN THIS How does pitch relate to the frequency of sound?

Our subjective impression about the frequency of sound is described as *pitch.* A high-pitched sound, like the sound from a tiny bell, has a high vibration frequency. The sound from a large bell has a low pitch because its vibrations are of a low frequency.

The human ear can normally hear pitches from sounds ranging from about 20 hertz to about 20,000 hertz. As we age, this range shrinks. Sound waves with frequencies lower than 20 hertz are called *infrasonic* waves, and those with frequencies higher than 20,000 hertz are called *ultrasonic* waves. We cannot hear infrasonic or ultrasonic sound waves, but dogs and some other animals can.

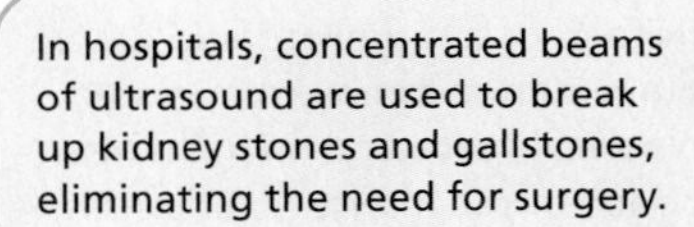

In hospitals, concentrated beams of ultrasound are used to break up kidney stones and gallstones, eliminating the need for surgery.

Hearing any sound occurs because air molecules next to a vibrating object are themselves set into vibration. These molecules, in turn, vibrate against neighboring molecules, which, in turn, do the same, and so on. As a result, rhythmic patterns of compressed and rarefied air emanate from the sound source. The resulting vibrating air sets your eardrum into vibration, which, in turn, sends cascades of rhythmic electrical impulses along nerves in the cochlea of your inner ear and into the brain. Thus, when you hear a high-pitched sound, a high-frequency wave from a quickly vibrating source sets your eardrum into fast vibration. Bass guitars, foghorns, and deep-throated bullfrogs vibrate slowly, making low-pitched waves that set your eardrums into slow vibration.

Most sound is transmitted through air, but any elastic substance—solid, liquid, or gas—can transmit sound.* Air is a poor conductor of sound compared with solids and liquids. You can hear the sound of a distant train clearly by placing your ear against the rail. When swimming, have a friend at a distance click two rocks together beneath the surface of the water while you are submerged. You will experience how well water conducts the sound.

*An elastic substance is "springy," has resilience, and can transmit energy with little loss. Steel, for example, is elastic, whereas lead and putty are not.

Sound requires a medium. It can't travel in a vacuum because then there's nothing to compress and stretch.

CHECK YOURSELF

1. **A singer sings a high-pitched note and then a low-pitched note. For which note are her vocal chords vibrating more rapidly? For which note is your eardrum vibrating more rapidly? Which note sets the air into higher-frequency vibrations?**
2. **Which would you consider to be sound waves—infrasonic or ultrasonic waves?**

CHECK YOUR ANSWERS

1. A high-pitched note is the answer to all these questions. A high-frequency sound wave arises from a rapidly vibrating source, setting the air into higher-frequency oscillations than a low-pitched sound. This then vibrates your eardrum at a higher frequency.
2. Both; although we can't hear them. Infrasonic and ultrasound waves are outside the range of human hearing—but not outside the range of hearing for certain other creatures.

Speed of Sound

If, from a distance, we watch a person chopping wood or hammering, we can easily see that the blow occurs a noticeable time before its sound reaches our ears. We often hear thunder seconds after we see a flash of lightning. These common experiences show that sound requires time to travel from one place to another. The speed of sound depends on wind conditions, temperature, and humidity. It does not depend on the loudness or the frequency of the sound; all sounds travel at the same speed in a given medium. The speed of sound in dry air at 0°C is about 330 meters per second, which is nearly 1200 kilometers per hour. Water vapor in the air increases this speed slightly. Sound also travels faster through warm air than cold air. This is to be expected because the faster-moving molecules in warm air bump into each other more frequently and, therefore, can transmit a pulse in less time.** For each degree rise in temperature above 0°C, the speed of sound in air increases by 0.6 meter per second. Thus, in air at a normal room temperature of about 20°C, sound travels at about 340 meters per second. In water, sound speed is about four times its speed in air; in steel, it's about 15 times its speed in air.

FYI Small fractal antennas have replaced the long antennas of mobile phones. The fractal antenna, like other fractals, has a shape that repeats itself. In any fractal, similar shapes appear at all levels of magnification. Common fractals in nature include snowflakes, shorelines, and even cauliflower and broccoli. Due to its folded, self-similar design, a fractal antenna can be compressed into a small space and be built into the body of the phone. Yum to this technology.

CHECK YOURSELF

1. **Do compressions and rarefactions in a sound wave travel in the same direction or in opposite directions from one another?**
2. **Approximately how far away is a thunderstorm when you note a 3-s delay between the flash of lightning and the sound of thunder?**

CHECK YOUR ANSWERS

1. They travel in the same direction.
2. Assuming that the speed of sound in air is about 340 m/s, in 3 s it will travel (340 m/s)(3 s) = 1020 m. There is no appreciable time delay for the flash of light, so the storm is slightly more than 1 km away.

**The speed of sound in a gas is about $\frac{3}{4}$ the average speed of its molecules.

8.5 Resonance

LEARNING OBJECTIVE
Distinguish between forced vibration and resonance.

EXPLAIN THIS Why can a metal surface be set into resonance but not a handkerchief?

If you strike an unmounted tuning fork, its sound is rather faint. Repeat with the handle of the tuning fork held against a table after striking it, and the sound is louder. This is because the table is forced to vibrate, and its larger surface sets more air in motion. The table is forced into vibration by a fork of any frequency. This is an example of **forced vibration**. The vibration of a factory floor caused by the running of heavy machinery is another example of forced vibration. A more pleasing example is the sounding boards of stringed instruments.

If you drop a wrench and a baseball bat onto a concrete floor, you can easily notice the difference in their sounds. This is because each vibrates differently when it strikes the floor. The objects are not forced to vibrate at a particular frequency, but, instead, each vibrates at its own characteristic frequency. Any object composed of an elastic material will, when disturbed, vibrate at its own special set of frequencies, which together form its special sound. We speak of an object's **natural frequency**, which depends on such factors as the elasticity and the shape of the object. Bells and tuning forks, of course, vibrate at their own characteristic frequencies. Interestingly, most things, from atoms to planets and almost everything else in between, have a springiness to them, and they vibrate at one or more natural frequencies.

FIGURE 8.8
Each time Bay Johnson strikes a piano key, a string in the piano is tapped. Piano strings for low notes are heavier, have more inertia, and vibrate at a lower frequency—a lower pitch than lighter strings of the same string tension. Loudness involves how hard the keys are struck, which affects the amplitudes of the vibrating strings. The touch sensitivity of the piano distinguishes it from earlier keyboard instruments such as the harpsichord.

When the frequency of forced vibrations on an object matches the object's natural frequency, a dramatic increase in amplitude occurs. This phenomenon is called **resonance**. Putty doesn't resonate because it isn't elastic, and a dropped handkerchief is too limp to resonate. In order for something to resonate, it needs both a force to pull it back to its starting position and enough energy to maintain its vibration.

Resonance is not restricted to wave motion. It occurs whenever successive impulses are applied to a vibrating object in rhythm with its natural frequency. Cavalry troops marching across a footbridge near Manchester, England, in 1831 inadvertently caused the bridge to collapse when they marched in rhythm with the bridge's natural frequency. Since then, it has been customary to order troops to "break step" when crossing bridges. A more recent bridge disaster was caused by wind-generated resonance, as shown in Figure 8.9.

FIGURE 8.9
In 1940, four months after it was completed, the Tacoma Narrows Bridge in the state of Washington was destroyed by wind-generated resonance. A mild gale produced a fluctuating force in resonance with the natural frequency of the bridge, steadily increasing the amplitude until the bridge collapsed.

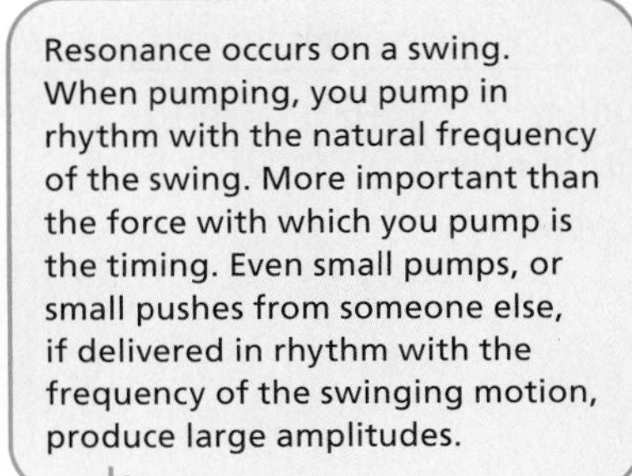

Resonance occurs on a swing. When pumping, you pump in rhythm with the natural frequency of the swing. More important than the force with which you pump is the timing. Even small pumps, or small pushes from someone else, if delivered in rhythm with the frequency of the swinging motion, produce large amplitudes.

The effects of resonance are all around us. Resonance underscores not only the sound of music but also the color of autumn leaves, the height of ocean tides, the operation of lasers, and a vast multitude of phenomena that add to the beauty of the world.

LEARNING OBJECTIVE
Describe the nature and range of electromagnetic waves.

8.6 The Nature of Light

EXPLAIN THIS How are electromagnetic waves produced?

During the day, the primary source of light is the Sun. Other common sources are flames, white-hot filaments in lightbulbs, and glowing gases in glass tubes. What these sources emit, and what we perceive as light, are electromagnetic waves with frequencies that fall within a certain range. Recall from Chapter 7 that an **electromagnetic wave** is a wave of energy produced when an electric charge oscillates (Figure 8.10). Light is only a tiny part of a larger whole—the wide range of electromagnetic waves called the **electromagnetic spectrum** (Figure 8.11).

FIGURE 8.10
If you shake an electrically charged object to and fro, you produce an electromagnetic wave.

Light originates from the accelerated motion of electrons. If you shake the end of a stick back and forth in still water, you'll create waves on the water's surface. Similarly, if you shake an electrically charged rod to and fro in empty space, you'll create electromagnetic waves in space. This is because the moving charge on the rod is an electric current. Recall from Chapter 7 that a magnetic field surrounds an electric current and the field changes as the current changes. Recall also that a changing magnetic field induces an electric field—electromagnetic induction. And what does the changing electric field do? It induces a changing magnetic field. The vibrating electric and magnetic fields regenerate each other to make up an electromagnetic wave (Figure 8.12).

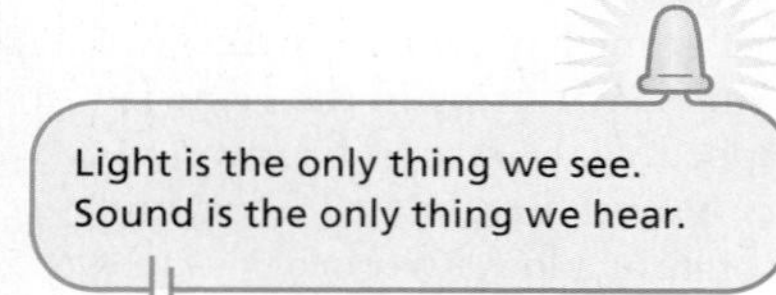

Light is the only thing we see.
Sound is the only thing we hear.

In a vacuum, all electromagnetic waves move at the same speed—the speed of light, 300,000 kilometers per second. We call this speed of light c. These waves

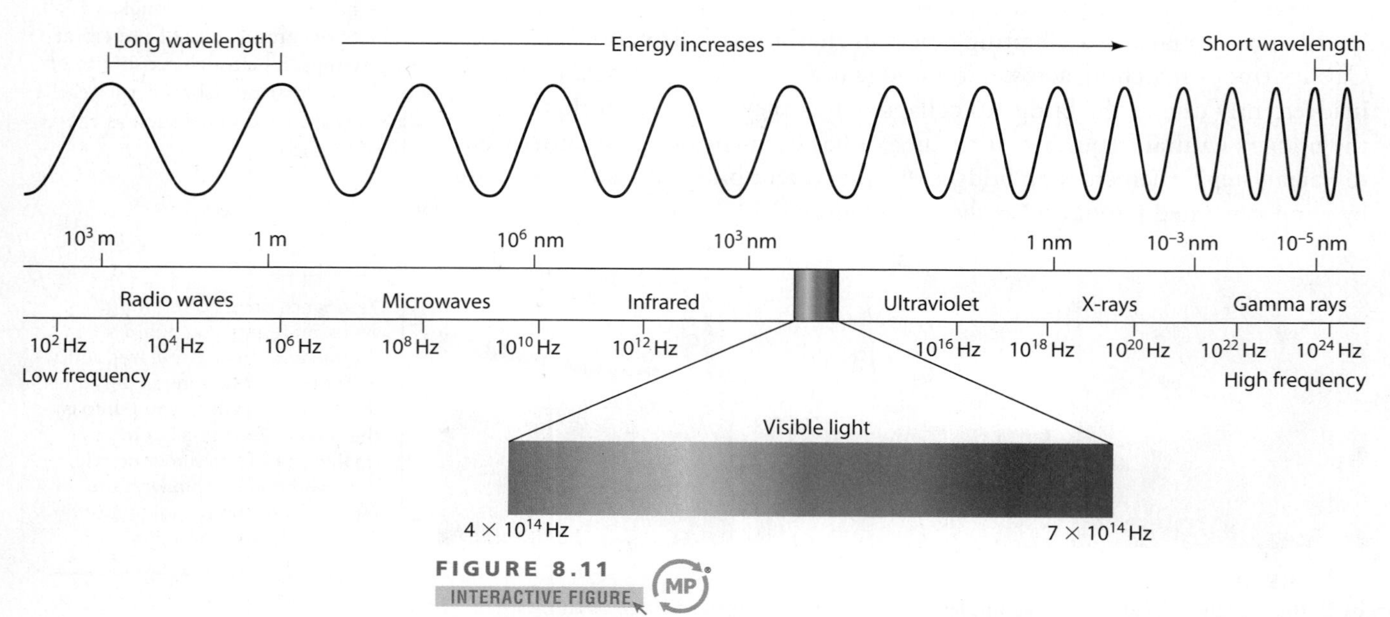

FIGURE 8.11
INTERACTIVE FIGURE MP

The electromagnetic spectrum is a continuous range of electromagnetic waves extending from radio waves to gamma rays. The descriptive names of the sections are merely a historical classification because all the waves are the same in nature, differing principally in frequency and wavelength; all travel at the same speed.

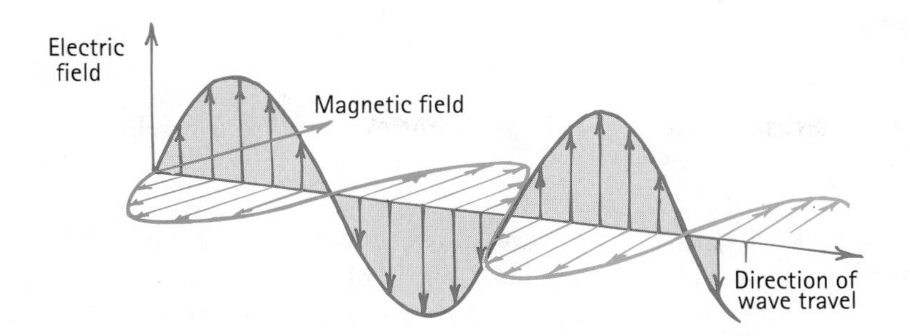

FIGURE 8.12
INTERACTIVE FIGURE MP

The electric and magnetic fields of an electromagnetic wave are perpendicular to each other and to the direction of the wave.

differ from one another in terms of their frequency. Electromagnetic waves have been detected with a frequency as low as 0.01 hertz (Hz). Others, with frequencies of several thousand hertz (kHz), are classified as low-frequency radio waves. One million hertz (1 MHz) lies in the middle of the AM radio band. The very high frequency (VHF) television band of waves begins at about 50 million hertz (MHz), and FM radio frequencies extend from 88 to 108 MHz. Then come ultrahigh frequencies (UHF), followed by microwaves, beyond which are infrared waves. Farther still is visible light, which makes up less than a millionth of 1% of the electromagnetic spectrum.

Light is energy carried in an electromagnetic wave emitted by vibrating electrons in atoms. In air, light travels a million times faster than sound.

The lowest frequency of light we can see with our eyes appears red. The highest visible frequencies, which are nearly twice the frequency of red light, appear violet. Still higher frequencies are ultraviolet, which are invisible to the human eye. These higher-frequency waves are more energetic and can cause sunburns. Beyond ultraviolet light are the X-ray and gamma-ray regions. There is no sharp boundary between regions of the spectrum; they actually grade continuously into one another. The spectrum is divided into these arbitrary regions for the sake of classification.

The frequency of the electromagnetic wave as it vibrates through space is identical with the frequency of the oscillating electric charge that generates it. Different frequencies result in different wavelengths—low frequencies produce long wavelengths and high frequencies produce short wavelengths. The higher the frequency of the vibrating charge, the shorter the wavelength of the radiation.*

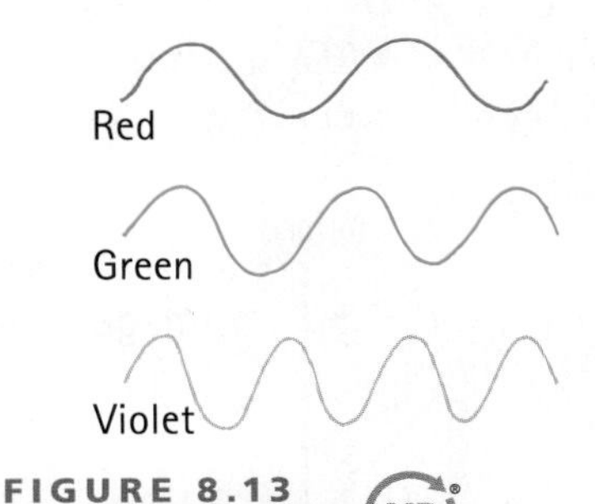

FIGURE 8.13
INTERACTIVE FIGURE MP

Relative wavelengths of red, green, and violet light. Violet light has nearly twice the frequency of red light and half the wavelength.

CHECK YOURSELF

Are we correct in saying that a radio wave is a low-frequency light wave and that a radio wave is a sound wave?

CHECK YOUR ANSWER

Yes and no; both radio waves and light waves are electromagnetic waves emitted by vibrating electrons. Radio waves have lower frequencies than light waves, so a radio wave might be considered a low-frequency light wave (and a light wave might be considered a high-frequency radio wave). But a sound wave is a mechanical vibration of matter and is not electromagnetic. A sound wave is fundamentally different from an electromagnetic wave. So, a radio wave is definitely not a sound wave. (Don't confuse a radio wave with the sound that a loudspeaker emits.)

*The relationship is $c = f\lambda$, where c is the speed of light (constant), f is the frequency, and λ is the wavelength. It is common to describe sound and radio by frequency and light by wavelength. In this book, however, we'll use the single concept of frequency in describing light.

LEARNING OBJECTIVE
State the law of reflection.

8.7 Reflection

EXPLAIN THIS Where is your image when you look at yourself in a plane mirror?

If you look carefully at the waves in a swimming pool, you will see that they bounce back when they hit the side of the pool. This is an example of **reflection**—the returning of a wave to the medium from which it came when it strikes a barrier.

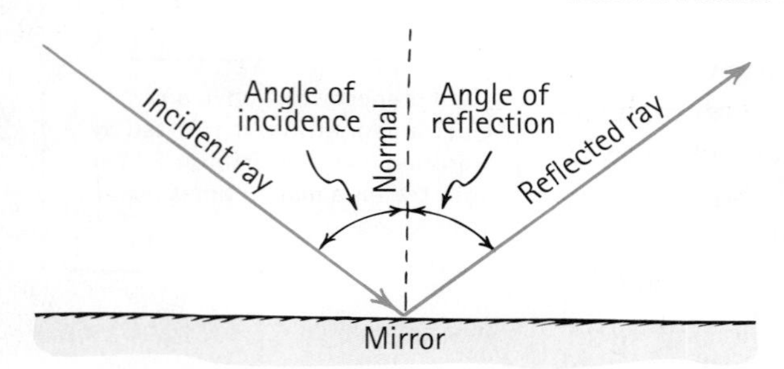

FIGURE 8.14
INTERACTIVE FIGURE
The law of reflection.

Water waves reflect off a surface in much the same way that a ball bounces back when it strikes a surface. When the ball hits a surface and bounces back, the angle of incidence (the angle at which the ball strikes the surface) equals the angle of rebound. Waves behave the same way. This is the **law of reflection**, and it holds for all angles:

The angle of reflection equals the angle of incidence.

The law of reflection is illustrated with light rays (with arrows representing the direction of light wave travel) in Figure 8.14. Instead of measuring the angles of incident and reflected rays from the reflecting surface, it is customary to measure them from a line perpendicular to the plane of the reflecting surface. This imaginary line is called the *normal*. The incident ray, the normal, and the reflected ray all lie in the same plane.

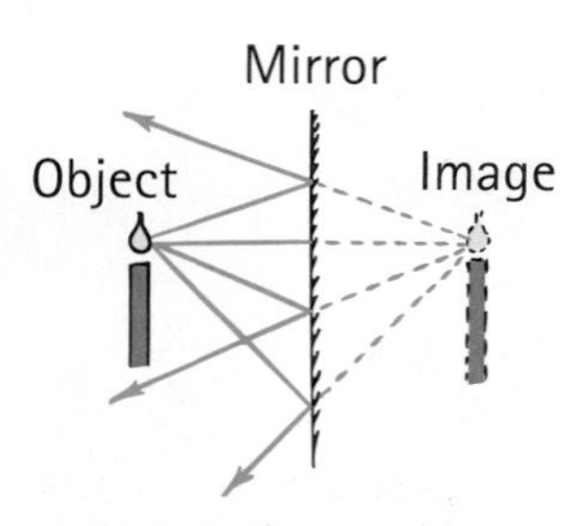

FIGURE 8.15
A virtual image is located behind the mirror and is at the position where the extended reflected rays (dashed lines) converge.

If you place a candle in front of a mirror, rays of light radiate from the flame in all directions. Figure 8.15 shows only four of the infinite number of rays leaving one of the infinite number of points on the candle. When these rays meet the mirror, they reflect at angles equal to their angles of incidence. The rays diverge from the flame. Note that they also diverge when reflecting from the mirror. These divergent rays appear to emanate from behind the mirror (dashed lines). You see an image of the candle at this point. The light rays do not actually come from this point, so the image is called a *virtual image*. The image is as far behind the mirror as the object is in front of the mirror, and image and object are the same size. When you view yourself in a mirror, for example, the size of your

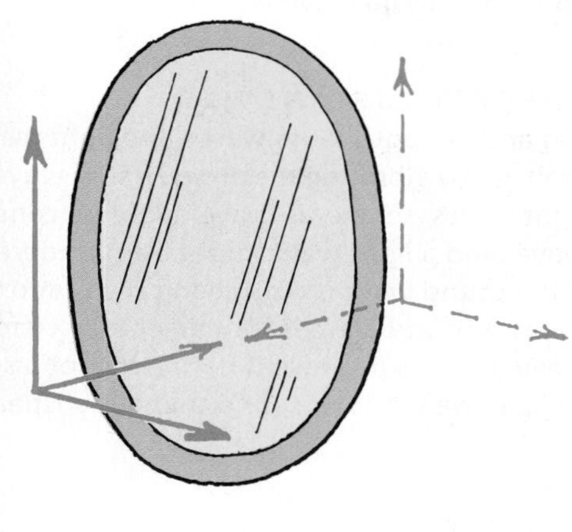

FIGURE 8.16
Marjorie's image is as far behind the mirror as she is in front of it. Note that she and her image have the same color of clothing—evidence that light doesn't change frequency upon reflection. Interestingly, her left–right axis is no more reversed than her up-and-down axis. The axis that is reversed, as shown to the right, is her front–back axis. That's why it appears that her left hand faces the right hand of her image.

image is the same as the size your twin would appear to be if located as far behind the mirror as you are in front, as long as the mirror is flat.

Only part of the light that strikes a surface is reflected. For example, on a surface of clear glass and for normal incidence (light perpendicular to the surface), only about 4% is reflected from each surface. On a clean and polished aluminum or silver surface, however, about 90% of the incident light is reflected. The light that is not reflected by a surface penetrates the surface and is either *absorbed* there (if the material is opaque) or *transmitted*—absorbed and reemitted (if the material is transparent).

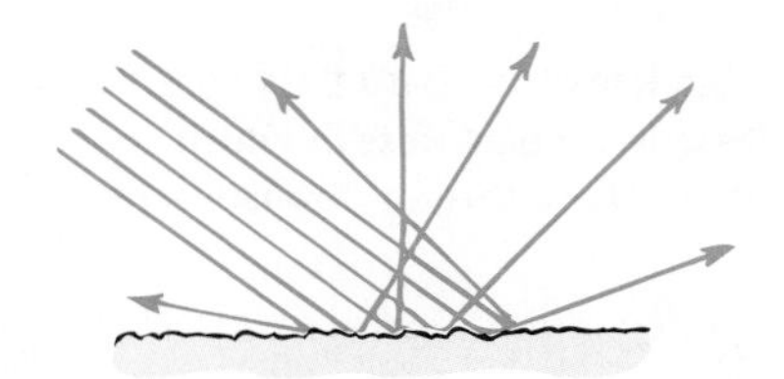

FIGURE 8.17
Diffuse reflection. Although reflection of each single ray obeys the law of reflection, the many different surface angles that the light rays encounter in striking a rough surface produce reflection in many directions.

Diffuse Reflection

When light is incident on a rough or granular surface, the light is reflected in many directions. This is called *diffuse reflection* (Figure 8.17). If the surface is so smooth that the distances between successive elevations on the surface are less than about one-eighth the wavelength of the light, there is very little diffuse reflection and the surface is said to be *polished*. A surface therefore may be polished for radiation of long wavelengths but rough for light of short wavelengths. The wire-mesh "dish" shown in Figure 8.18 is very rough for light waves and is hardly mirrorlike. But, for long-wavelength radio waves, it is polished and is an excellent reflector.

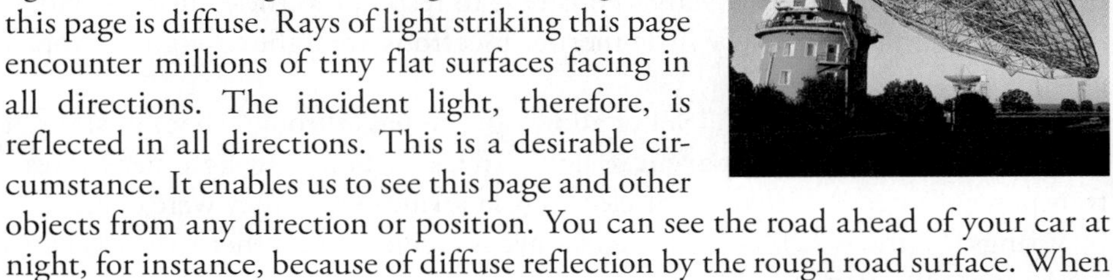

FIGURE 8.18
The open-mesh parabolic dish is a diffuse reflector for short-wavelength light but a polished reflector for long-wavelength radio waves.

This page is smooth to a radio wave, but to a light wave it is rough, so the light reflecting from this page is diffuse. Rays of light striking this page encounter millions of tiny flat surfaces facing in all directions. The incident light, therefore, is reflected in all directions. This is a desirable circumstance. It enables us to see this page and other objects from any direction or position. You can see the road ahead of your car at night, for instance, because of diffuse reflection by the rough road surface. When the road is wet, however, it is smoother with less diffuse reflection and therefore more difficult to see. Most of our environment is seen by diffuse reflection.

FIGURE 8.19
A magnified view of the surface of ordinary paper.

Reflection of Sound

We call the reflection of sound an *echo*. The fraction of sound energy reflected from a surface is large if the surface is rigid and smooth, but it is less if the surface is soft and irregular. The sound energy that is not reflected is transmitted or absorbed.

Sound reflects from a smooth surface in the same way that light reflects—the angle of incidence is equal to the angle of reflection. Sometimes, when sound reflects from the walls, ceiling, and floor of a room, the surfaces are too reflective and the sound becomes garbled. This is due to multiple reflections called reverberations. In contrast, if the reflective surfaces are too absorbent, the sound level is low and the room may sound dull and lifeless. Reflected sound in a room makes it sound lively and full, as you have probably experienced while singing in the shower. In the design of an auditorium or concert hall, a balance must be found between reverberation and absorption. The study of sound properties is called *acoustics*.

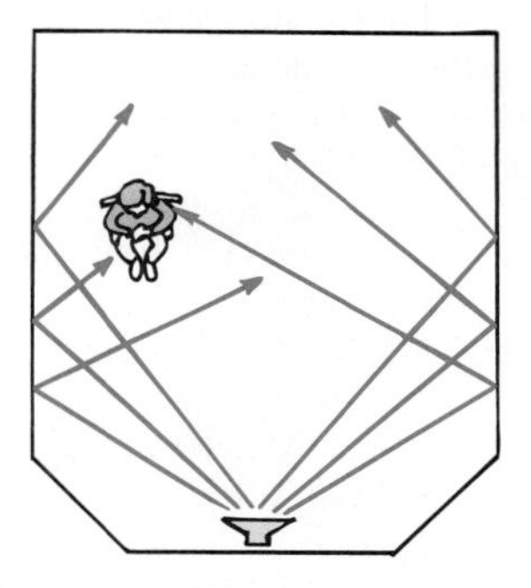

FIGURE 8.20
The angle of incident sound is equal to the angle of reflected sound.

LEARNING OBJECTIVE
Describe the process in which light travels through transparent materials.

8.8 Transparent and Opaque Materials

EXPLAIN THIS Why don't the photons that strike a pane of glass travel through it?

Light is energy carried in an electromagnetic wave emitted by vibrating electrons in atoms. When light is incident upon matter, some of the electrons in the matter are forced into vibration. In this way, vibrations in the emitter are transformed into vibrations in the receiver. This is similar to the way in which sound is transmitted (Figure 8.21).

Thus the way a receiving material responds when light is incident upon it depends on the frequency of the light and on the natural frequency of the electrons in the material. Visible light vibrates at a very high rate, some 100 trillion times per second (10^{14} hertz). If a charged object is to respond to these ultrafast vibrations, it must have very, very little inertia. Electrons are light enough to vibrate at this rate.

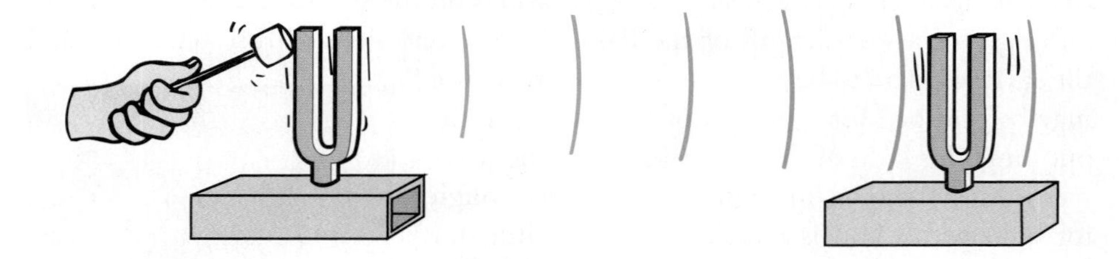

FIGURE 8.21
Just as a sound wave can force a sound receiver into vibration, a light wave can force the electrons in materials into vibration.

Such materials as glass and water allow light to pass through them in straight lines. We say they are **transparent** to light. To understand how light penetrates a transparent material, visualize the electrons in an atom as if they were connected by springs (Figure 8.22).* When a light wave is incident upon them, the electrons are set into vibration.

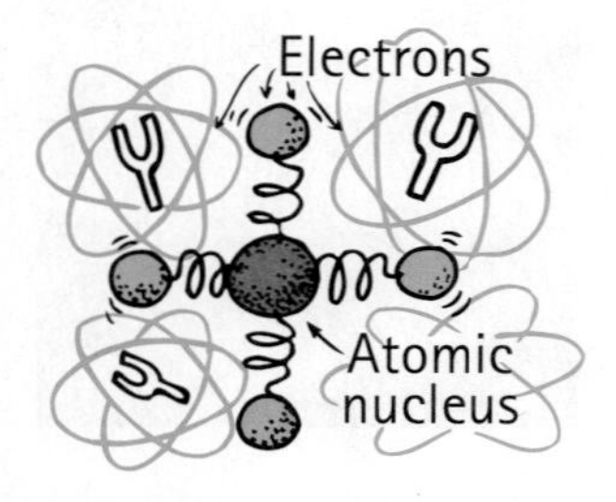

FIGURE 8.22
The electrons of atoms have certain natural frequencies of vibration, and these can be modeled as particles connected to the atomic nucleus by springs. As a result, atoms and molecules behave somewhat like optical tuning forks.

Materials that are springy (elastic) respond more to vibrations at some frequencies than to vibrations at other frequencies. Bells ring at a particular frequency, tuning forks vibrate at a particular frequency, and so do the electrons of atoms and molecules. The natural vibration frequencies of an electron depend on how strongly it is attached to its atom or molecule. Different atoms and molecules have different "spring strengths." Electrons in glass have a natural vibration frequency in the ultraviolet range. When ultraviolet rays shine on glass, resonance occurs as the wave builds and maintains a large amplitude of vibration of the electron, just as pushing someone at the resonant frequency on a swing builds a large amplitude. The energy that atoms in the glass receive may be passed on to neighboring atoms by collisions, or the energy may be reemitted. Resonating atoms in the glass can hold onto the energy of the ultraviolet light for quite a long time (about 100 millionths

*Electrons, of course, are not really connected by springs. We simply present a "spring model" of the atom to help us understand the interaction of light and matter. Scientists devise such conceptual models to understand nature, particularly at the submicroscopic level. The value of a model lies not in whether it is "true" but whether it is useful—in explaining observations and in predicting new ones. If predictions are contrary to new observations, then the model is usually either refined or abandoned.

FIGURE 8.23
A light wave incident upon a pane of glass sets up vibrations in the molecules that produce a chain of absorptions and reemissions that pass the light energy through the material and out the other side. Because of the time delay between absorptions and reemissions, the light travels through the glass more slowly than through empty space.

of a second). During this time, the atom makes about 1 million vibrations, collides with neighboring atoms, and transfers absorbed energy as heat. Thus, glass is not transparent to ultraviolet. Glass absorbs ultraviolet.

At lower wave frequencies, such as those of visible light, electrons in the glass are forced into vibrations of lower amplitudes. The atoms or molecules in the glass hold the energy for less time, with less chance of collision with neighboring atoms and molecules, and with less of the energy being transformed into heat. The energy of vibrating electrons is reemitted as light. Glass is transparent to all the frequencies of visible light. The frequency of the reemitted light that is passed from molecule to molecule is identical to the frequency of the light that produced the vibration originally. However, there is a slight time delay between absorption and reemission.

FIGURE 8.24
When a ball hits the array of balls, the ball that emerges at the opposite side is not the ball that initiates the transfer of energy. Likewise, light that emerges from a pane of glass is not the same light that was incident on the glass.

It is this time delay that results in a lower average speed of light through a transparent material (Figure 8.23). Light travels at different average speeds through different materials. We say average speeds because the speed of light in a vacuum, whether in interstellar space or in the space between molecules in a piece of glass, is c (a constant 300,000 km/s or 186,000 mi/s). The speed of light in the atmosphere is slightly less than it is in a vacuum, but it is usually rounded off as c. In water, light travels at 75% of its speed in a vacuum, or $0.75c$. In glass, light travels at about $0.67c$, depending on the type of glass. In a diamond, light travels at less than half its speed in a vacuum, only $0.41c$. When light emerges from these materials into the air, it travels at its original speed.

Infrared waves, which have frequencies lower than those of visible light, vibrate not only the electrons but also the entire molecules in the structure of the glass and in many other materials. This molecular vibration increases the thermal energy and temperature of the material, which is why infrared waves produce temperature increases in these materials. Glass is transparent to visible light, but not to ultraviolet and infrared light (Figure 8.25).

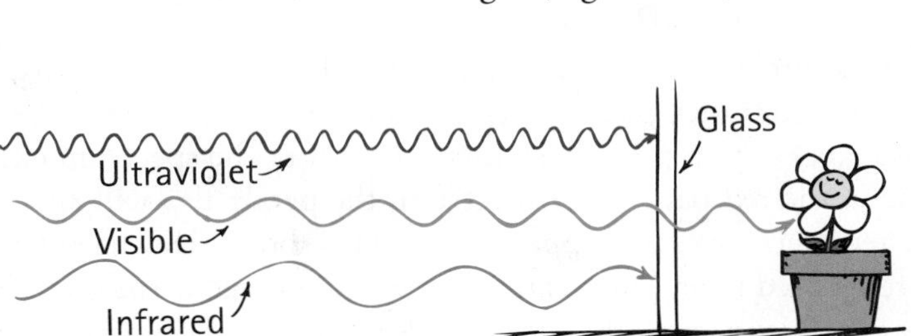

FIGURE 8.25
Clear glass blocks both infrared and ultraviolet light, but it is transparent to all frequencies of visible light.

Different substances have different molecular structures and therefore absorb or reflect light differently over various spectral ranges.

CHECK YOURSELF

1. **Pretend that, while you are at a social gathering, you make several momentary stops across the room to greet people who are "on your wavelength." How is this analogous to light traveling through glass?**
2. **In what way is your behavior at a party not analogous to light traveling through glass?**

CHECK YOUR ANSWERS

1. Your average speed across the room is slowed because of the time delays associated with your momentary stops. Likewise, the speed of light in glass is slowed because of the time delays caused by interactions with atoms along its path.
2. As you walk across the room, you are the one who both begins the walk and completes the walk. This is not analogous to the similar case of light because (according to our model for light passing through a transparent material) the light that is absorbed by an electron that has been made to vibrate is not the same light that is reemitted—even though the two, like identical twins, are indistinguishable.

Most objects around us are **opaque**—they absorb light without reemitting it. Books, desks, chairs, and people are opaque. Vibrations given by light to their atoms and molecules are turned into random kinetic energy—into thermal energy. They become slightly warmer.

LEARNING OBJECTIVE
Describe how color depends on the frequency of light.

8.9 Color

EXPLAIN THIS Why do red, green, and blue combined make white on your TV screen?

Roses are red and violets are blue; colors intrigue artists and scientists too. The colors we see depend on the frequency of the light we see. Different frequencies of light are perceived as different colors; the lowest frequency we detect appears, to most people, as the color red, and the highest appears as violet. Between them range the infinite number of hues that make up the color spectrum of the rainbow. By convention, these hues are grouped into seven colors: red, orange, yellow, green, blue, indigo, and violet. These colors combined appear white. The white light from the Sun is a composite of all the visible frequencies.

Selective Reflection

A rose, for example, doesn't emit light; it reflects light. If we pass sunlight through a prism and then place a deep-red rose in various parts of the spectrum, the rose will appear brown or black in all regions of the spectrum except in the red region. In the red part of the spectrum, the petals also will appear red, but the green stem and leaves will appear black. This shows that the red rose has the ability to reflect red light, but it cannot reflect other colors; the green leaves have the ability to reflect green light and, likewise, cannot reflect other colors. When the rose is held in white light, the petals appear red and the leaves appear green because the petals reflect the red part of the white light and the leaves reflect the green part of the white light.

FIGURE 8.26
The rose appears red because it reflects light in this frequency range.

Usually, a material will absorb light of some frequencies and reflect the rest. If a material absorbs most of the light and reflects red, for example, the material appears red. If it reflects light of all the visible frequencies, like the white part of this page, it will be the same color as the light that shines on it. If a material absorbs light and reflects none, then it is black (Figures 8.27 and 8.28).

FIGURE 8.27
The square on the left reflects all the colors illuminating it. In sunlight, it is white. When illuminated with blue light, it is blue. The square on the right absorbs all the colors illuminating it. In sunlight, it is warmer than the white square.

FIGURE 8.28
The bunny's dark fur absorbs all the incoming sunlight and therefore appears black. Fur on other parts of its body reflects light of all frequencies and therefore appears white.

Interestingly, the petals of most yellow flowers, like daffodils, reflect red and green as well as yellow. Yellow daffodils reflect a broad band of frequencies. The reflected colors of most objects are not pure single-frequency colors but rather a mixture of frequencies.

An object can reflect only those frequencies that are present in the illuminating light. The appearance of a colored object, therefore, depends on the kind of light that illuminates it. An incandescent lamp, for instance, emits light of lower average frequencies than sunlight, enhancing any reds viewed in this light. In a fabric that has only a little bit of red in it, the red is more apparent under an incandescent lamp than it is under a fluorescent lamp. Fluorescent lamps are richer in the higher frequencies, and so blues are enhanced in their light. For this reason, it is difficult to tell the true color of objects viewed in artificial light (Figure 8.29).

FIGURE 8.29
Color depends on the light source.

Selective Transmission

The color of a transparent object depends on the color of the light it transmits. A red piece of glass appears red because it absorbs all the colors of white light except red, so red light is transmitted. Similarly, a blue piece of glass appears blue because it transmits primarily blue and absorbs the other colors that illuminate it (Figure 8.30). Colored pieces of glass contain dyes or pigments—fine particles that selectively absorb light of particular frequencies and selectively transmit others. Ordinary window glass doesn't have a color because it transmits light of all visible frequencies equally well.

MasteringPhysics®
TUTORIAL: Color

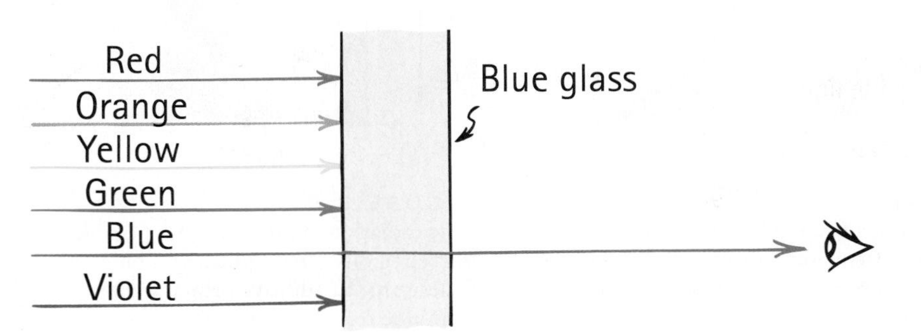

FIGURE 8.30
Only energy that has the frequency of blue light is transmitted; energy of the other frequencies is absorbed and warms the glass.

CHECK YOURSELF

1. **When illuminated with green light, why do the petals of a red rose appear black?**
2. **If you hold a match, a candle flame, or any small source of white light between you and a piece of red glass, you'll see two reflections from the glass: one from the front surface of the glass and one from the back surface. What color reflections will you see?**

CHECK YOUR ANSWERS

1. The petals absorb rather than reflect the green light. Since green is the only color illuminating the rose, and green contains no red to be reflected, the rose reflects no color at all and appears black.
2. You will see white reflected from the top surface. You'll see red reflected from the back surface because only red reaches the back surface and reflects from there.

LEARNING OBJECTIVE
Relate colors to light frequencies.

Integrated Science 8B
BIOLOGY

Mixing Colored Lights

MasteringPhysics®
VIDEO: Colored Shadows
VIDEO: Yellow-Green Peak of Sunlight

EXPLAIN THIS What three colors of light combine to produce white light?

You can see that white light from the Sun is composed of all the visible frequencies when you pass sunlight through a prism. The white light is dispersed into a rainbow-colored spectrum. The distribution of solar frequencies is uneven (Figure 8.31), and the light is most intense in the yellow-green part of the spectrum. How fascinating it is that our eyes have evolved to have maxim um sensitivity in this range. That's why many fire engines are painted yellow-green, particularly at airports, where visibility is vital. Our sensitivity to yellow-green light is also why we see better under the illumination of yellow sodium-vapor lamps at night than we do under incandescent lamps of the same brightness.

Every artist knows that if you mix red, green, and blue paint, the result will not be white but a muddy dark brown. The mixing of pigments in paints and dyes is a process called subtractive color mixing and is entirely different from mixing light of different colors.

All the colors combined produce white. Interestingly, we also see white from the combination of only red, green, and blue light. We can understand this by dividing the solar radiation curve into three regions, as in Figure 8.32. Three

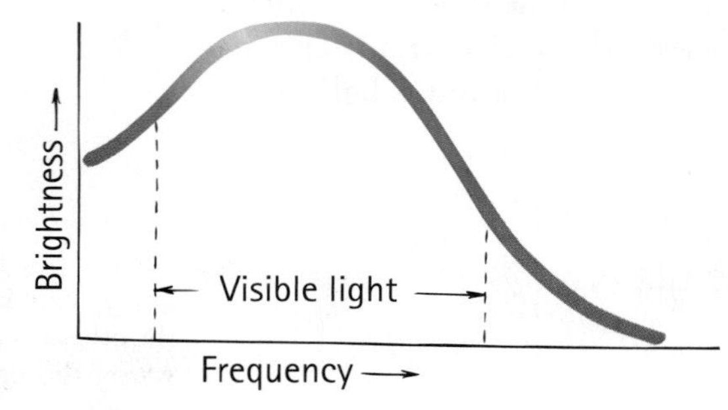

FIGURE 8.31
The radiation curve of sunlight is a graph of brightness against frequency. Sunlight is brightest in the yellow-green region, which is in the middle of the visible range.

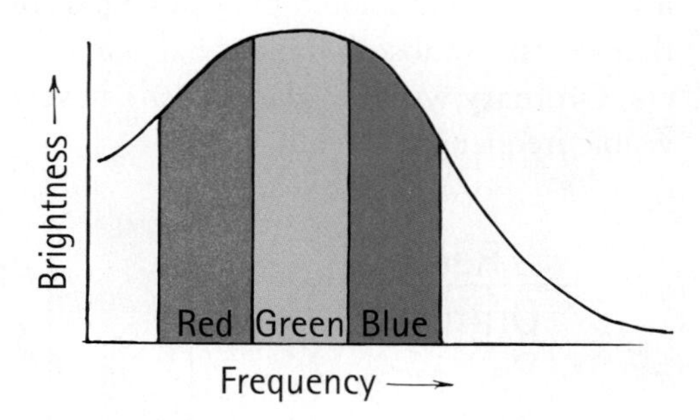

FIGURE 8.32
The radiation curve of sunlight divided into three regions—red, green, and blue (RGB). These are the additive primary colors, which combine to produce white.

types of cone-shaped receptor cells in our eyes ("cones") perceive color. (For more on this, see Chapter 19.) Each cone cell is stimulated only by certain frequencies of light. Light of lower visible frequencies stimulates the cones that are sensitive to low frequencies and appears red. Light of middle frequencies stimulates the cones that are sensitive to midfrequencies and appears green. Light of higher frequencies stimulates the cones that are sensitive to high frequencies and appears blue. When all three types of cones are stimulated equally, we see white.

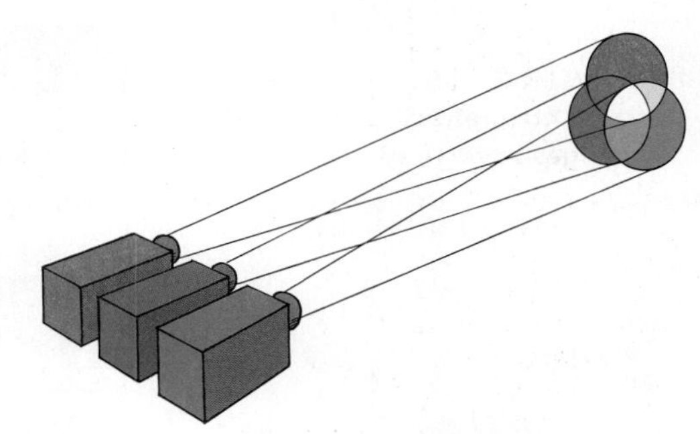

FIGURE 8.33
INTERACTIVE FIGURE (MP)
Color addition. When three projectors shine red, green, and blue light onto a white screen, the overlapping parts produce different colors. Red and blue light combine to make magenta; red and green light combine to make yellow; and blue and green light combine to make cyan. White is produced where all three overlap.

Project red, green, and blue lights onto a screen and, where they all overlap, white is produced. If two of the three colors overlap, or are added, then another color sensation is produced (Figure 8.33). By adding various amounts of red, green, and blue, the colors to which each of our three types of cones are sensitive, we can produce any color in the spectrum. For this reason, red, green, and blue are called the *additive primary colors* (known as RGB in color monitors). Here's what happens when any two of the primary colors are combined:

Red + Blue = Magenta
Red + Green = Yellow
Blue + Green = Cyan

We say that magenta is the opposite of green, cyan is the opposite of red, and yellow is the opposite of blue. The addition of any color to its opposite color results in white. Two colors that add together to produce white are known as complementary colors:

Magenta + Green = White (= Red + Blue + Green)
Cyan + Red = White (= Blue + Green + Red)
Yellow + Blue = White (= Red + Green + Blue)

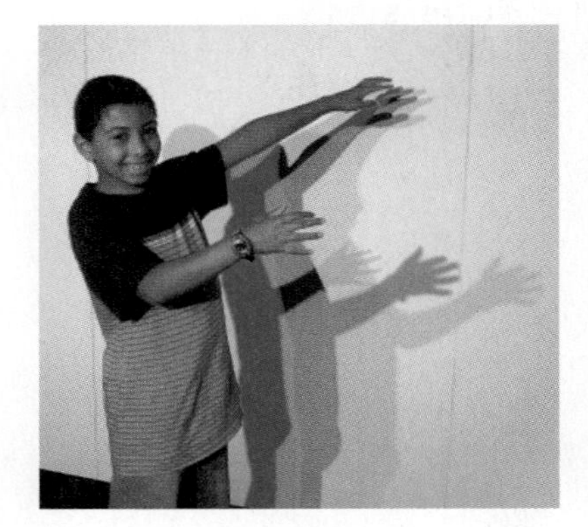

FIGURE 8.34
Carlos Vasquez displays a variety of colors when he is illuminated by only red, green, and blue lamps. Can you see where complementary colors appear in the shadows?

Note the complementary colors in the shadows produced by the boy when he is illuminated by red, green, and blue lamps (Figure 8.34). Cyan, yellow, and magenta are the *subtractive primary colors* because they result when light of particular frequencies subtract from white light. Inks of the subtractive primaries are used in printing. Ink-jet printers, for example, deposit various combinations of cyan, magenta, yellow, and black inks. This is CMYK printing (K indicates black).

CHECK YOURSELF

1. **From Figure 8.33, find the complements of cyan, of yellow, and of red.**
2. **Red + blue = ?**
3. **White − red = ?**
4. **White − blue = ?**

CHECK YOUR ANSWERS

1. Red, blue, cyan
2. Magenta
3. Cyan
4. Yellow

FIGURE 8.35
The color green on a printed page consists of cyan and yellow dots.

LEARNING OBJECTIVE
Describe how refraction is caused by changes in wave speed.

8.10 Refraction

EXPLAIN THIS Why does a fish in water appear closer to the surface than it actually is?

The speed of a wave changes when the medium changes. This change in speed can cause the wave to bend. The bending of waves due to a change in wave speed is called **refraction**.

A simple way to understand refraction is to compare the motion of a wave to the motion of a pair of toy cart wheels, as shown in Figure 8.36. The wheels roll from a smooth sidewalk onto a grass lawn. If the wheels meet the grass at an angle, as the figure shows, they are deflected from their straight-line course. Note that the left wheel slows first when it comes in contact with the grass on the lawn. The right wheel maintains its higher speed while on the sidewalk. It pivots about the slower-moving left wheel because it travels farther in the same time. So, the direction of the rolling wheels is bent toward the normal, the black dashed line perpendicular to the grass–sidewalk border in Figure 8.36.

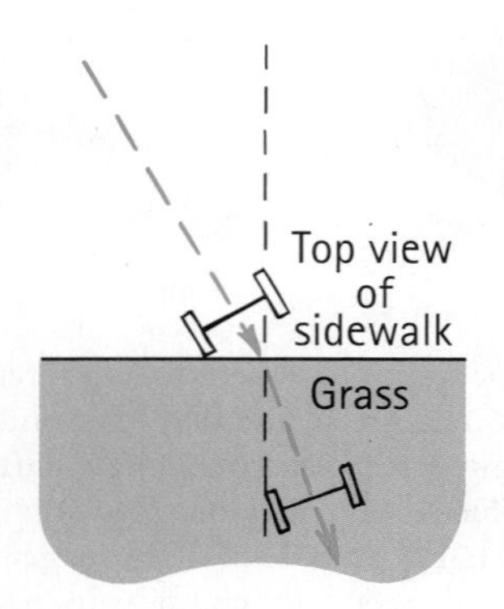

FIGURE 8.36
The direction of rolling wheels changes when one wheel slows before the other one does.

Figure 8.37 shows how a light wave bends in a similar way. Note the direction of light, indicated by the blue arrow (the light ray). Also note the wavefronts drawn at right angles to the ray. (If the light source were close, the wavefronts would appear circular, but if the distant Sun is the source, the wavefronts are practically straight lines.) The wavefronts are everywhere at right angles to the light rays. In the figure, the wave meets the water surface at an angle. This means that the left portion of the wave slows down in the water while the remainder in the air travels at the full speed of light, c. The light ray remains perpendicular to the wavefront and therefore bends at the surface. It bends like the wheels bend when they roll from the sidewalk onto the grass. In both cases, the bending is caused by a change of speed.*

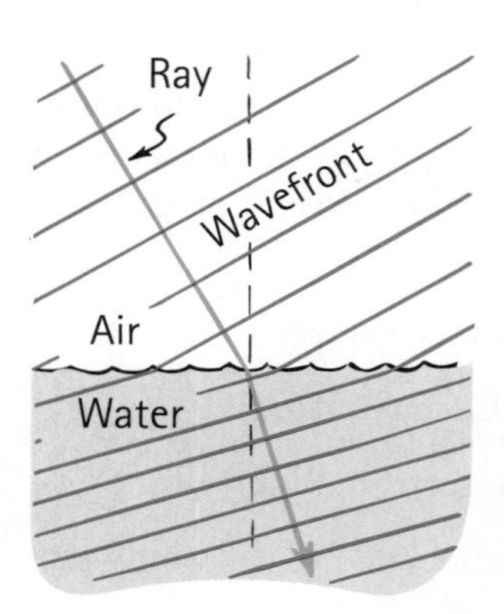

FIGURE 8.37
The direction of light waves changes when one part of the wave slows before the other part slows.

Figure 8.38 shows a beam of light entering water at the left and exiting at the right. The path would be the same if the light entered from the right and exited at the left. The light paths are reversible for both reflection and refraction. If you see someone's eyes by way of a reflective or refractive device, such as a mirror or a prism, then that person can see you by way of the device also.

Refraction causes many illusions. One of them is the apparent bending of a stick that is partially submerged in water. The submerged part appears closer to the surface than it actually is. The same is true when you look at a fish in water. The fish appears nearer to the surface and closer than it really is (Figure 8.39). If we look straight down into water, an object submerged 4 meters beneath the surface appears to be only 3 meters deep. Because of refraction, submerged objects appear to be magnified.

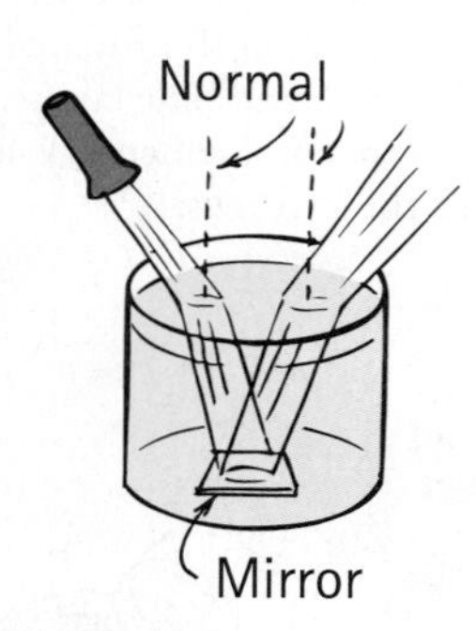

FIGURE 8.38
When light slows down in going from one medium to another, as it does in going from air to water, it bends toward the normal. When it speeds up in traveling from one medium to another, as it does in going from water to air, it bends away from the normal.

*The quantitative law of refraction, called *Snell's law*, is credited to Willebrord Snell, a 17th-century Dutch astronomer and mathematician: $n_1 \sin \theta_1 = n_2 \sin \theta_2$, where n_1 and n_2 are the indices of refraction of the media on either side of the surface, and θ_1 and θ_2 are the angles of incidence and refraction, respectively. If three of these values are known, the fourth can be calculated from this relationship.

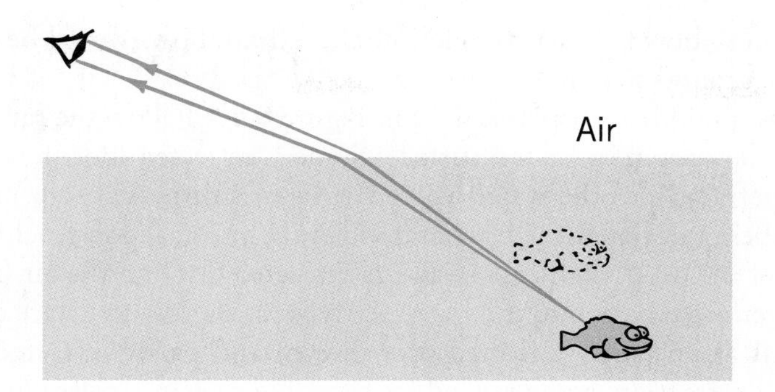

FIGURE 8.39
Because of refraction, a submerged object appears to be nearer to the surface than it actually is.

Sound waves refract when parts of the wave fronts travel at different speeds. This may happen when sound waves are affected by uneven winds, or when sound is traveling through air of different temperatures. On a warm day, the air near the ground may be appreciably warmer than the air above, so the speed of sound near the ground increases. Sound waves therefore tend to bend away from the ground, resulting in sound that does not seem to transmit well (Figure 8.40).

MasteringPhysics®
VIDEO: Refraction of Sound

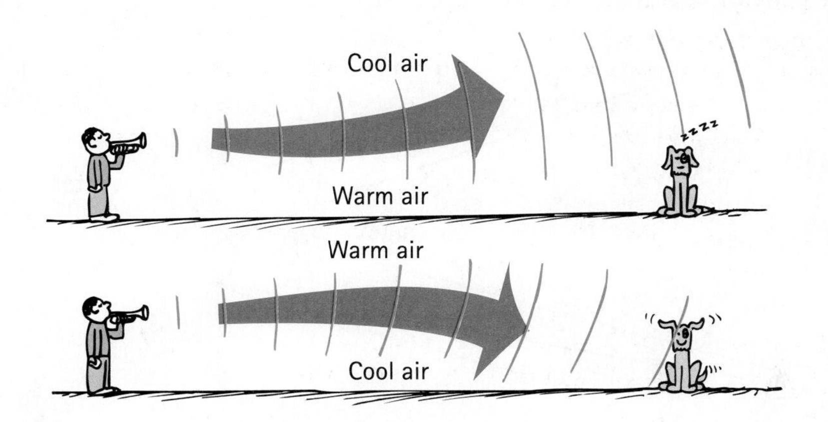

FIGURE 8.40
Sound waves are bent in air as speed changes with different temperatures.

Dispersion

Recall that light that resonates with the electrons of atoms and molecules in a material is absorbed. Such a material is opaque to light. Also recall that transparency occurs for light of frequencies near (but not at) the resonant frequencies of the material. Light is slowed due to the absorption/reemission sequence, and the closer to the resonant frequencies, the slower the light, as was shown in Figure 8.23. The grand result is that high-frequency light in a transparent medium travels slower than low-frequency light. Violet light travels about 1% slower in ordinary glass than red light. Light of colors between red and violet travel at their own respective speeds in glass.

Because light of various frequencies travels at different speeds in transparent materials, different colors of light refract by different amounts. When white light is refracted twice, as in a prism, the separation of light by colors is quite noticeable. This separation of light into colors arranged by frequency is called **dispersion** (Figure 8.41). Because of dispersion, there are rainbows!

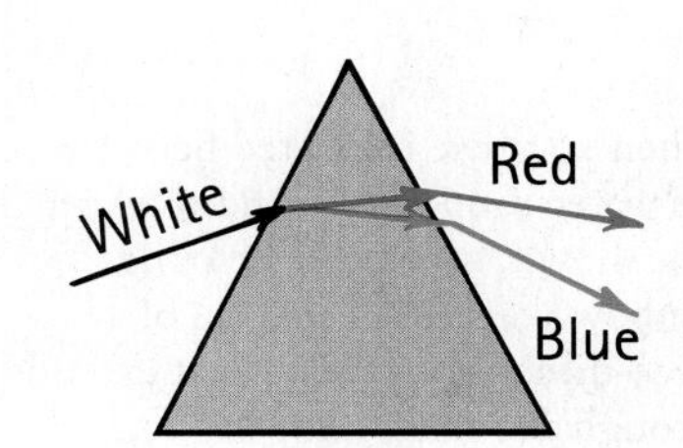

FIGURE 8.41
Dispersion by a prism makes the components of white light visible.

Rainbows

We see a rainbow when the Sun shines on drops of water in a cloud or in falling rain. The drops act as prisms that disperse light. When you face a rainbow, the Sun is behind you, in the opposite part of the sky. Seen from an airplane near

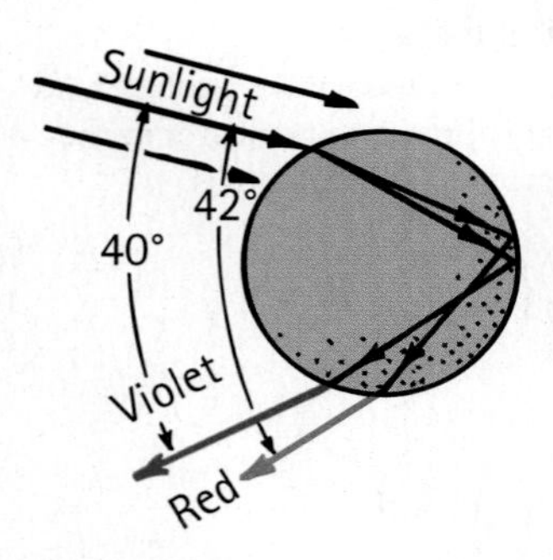

FIGURE 8.42
Dispersion of sunlight by a single raindrop.

midday, the rainbow forms a complete circle. All rainbows would be completely round if the ground were not in the way.

Note how a raindrop disperses light in Figure 8.42. Follow the ray of sunlight as it enters the drop near its top surface. Some is reflected (not shown) and the remainder refracts into the water where the light is dispersed into its spectrum colors, red being deviated the least and violet the most. Upon reaching the opposite side of the drop, each color is partly refracted out into the air (not shown) and partly reflected back into the water. Arriving at the lower surface of the drop, each color is again partly reflected (not shown) and partly refracted back into the air. This refraction at the second surface, like that in a prism, increases the dispersion already produced at the first surface.

Although each drop disperses a full spectrum of colors, an observer is in a position to see only a single color from any one drop (Figure 8.43). If violet light from a single drop reaches an observer's eye, red light from the same drop is incident elsewhere toward the observer's feet. To see red light, one must look to a drop higher in the sky. The color red will be seen where the angle between a beam of sunlight and the dispersed light is 42°. The color violet is seen where the angle between the sunbeams and the dispersed light is 40°.

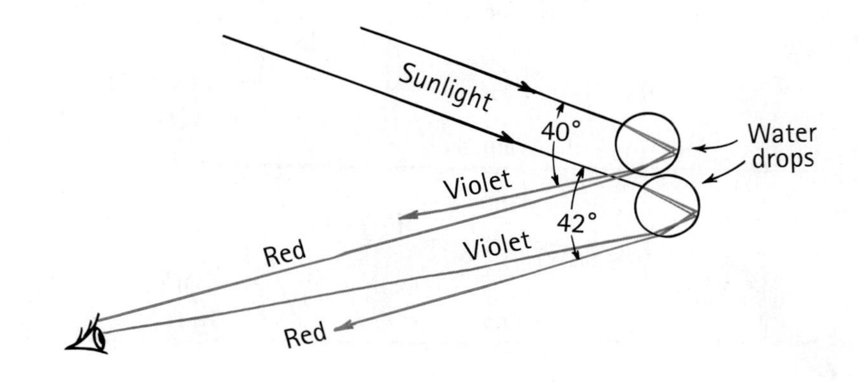

FIGURE 8.43
Sunlight incident on two raindrops, as shown, emerges from them as dispersed light. The observer sees the red light from the upper drop and the violet light from the lower drop. Millions of drops produce the entire spectrum of visible light.

A rainbow is not the flat two-dimensional arc it appears to be, but is actually a three-dimensional cone of dispersed light (Figure 8.44). The apex of this cone is at your eye. The thicker the region containing water drops, the thicker the conical edge you look through and the more vivid the rainbow.

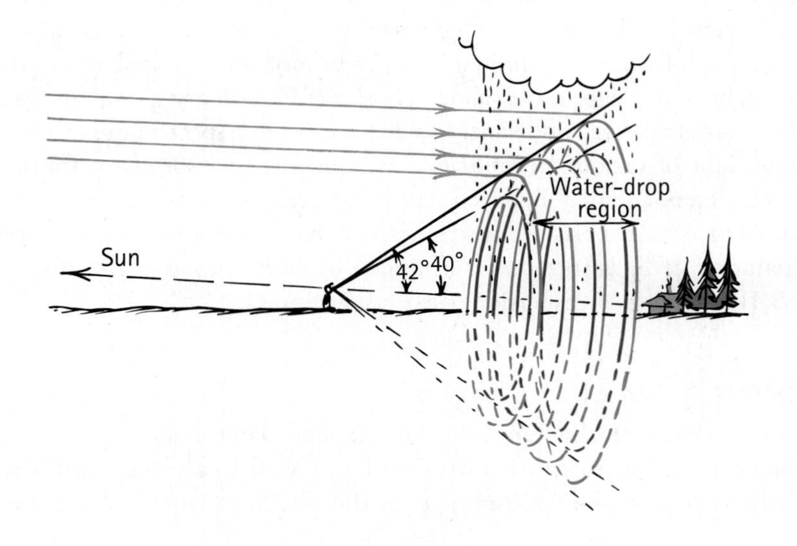

FIGURE 8.44
When your eye is located between the water-drop region and the Sun (not shown, off to the left), the rainbow you see is the edge of a three-dimensional cone that extends through the water-drop region. Violet is dispersed by drops that form a 40° conical surface; red is seen from drops along a 42° conical surface, with other colors in between. (Innumerable layers of drops form innumerable two-dimensional arcs, like the four sets suggested here.)

FIGURE 8.45
Two refractions and a reflection in water droplets produce light at all angles up to about 42°, with the intensity concentrated where we see the rainbow at 40° to 42°. Light doesn't exit the water droplet at angles greater than 42° unless it undergoes two or more reflections inside the drop. Thus the sky is brighter inside the rainbow than outside it. Notice the weak secondary rainbow.

It is interesting that a rainbow always faces you squarely. When you move, the rainbow appears to move with you. So you can never approach the side of a rainbow or see it end-on as in the exaggerated view of Figure 8.44. You *can't* reach its end. Thus the saying "looking for the pot of gold at the end of the rainbow" means pursuing something you can never reach.

Often a larger, secondary bow with its colors reversed can be seen arching at a larger angle around the primary bow. We won't treat this secondary bow except to say that it is formed by similar circumstances and is a result of double reflection within the raindrops (Figure 8.46). Because of this extra reflection (and extra refraction loss), the secondary bow is much dimmer and reversed.

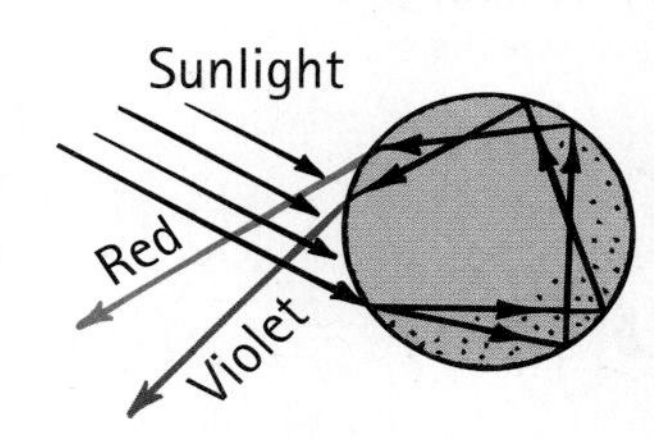

FIGURE 8.46
Double reflection in a drop produces a secondary bow.

Lenses

When you think of lenses, think of sets of glass prisms arranged as shown in Figure 8.47. They refract incoming parallel light rays so that the rays converge to (or diverge from) a point. The arrangement shown in Figure 8.47a converges the light, and we have a *converging lens*. Notice that it is thicker in the middle. In the arrangement shown in Figure 8.47b, the middle is thinner than the edges. Because this lens diverges the light, we have a *diverging lens*. Note that the prisms in (b) diverge the incident rays in a way that makes them appear to originate from a single point in front of the lens.

In both lenses, the greatest deviation of rays occurs at the outermost prisms because they have the largest angle between the two refracting surfaces. No deviation occurs exactly in the middle because in that region the two surfaces of

Your cone of vision intersects the cloud of drops and creates your rainbow. It is ever so slightly different from the rainbow seen by a person nearby. So, when a friend says, "Look at the pretty rainbow," you can reply, "Okay, move aside so I can see it, too." Everybody sees his or her own personal rainbow.

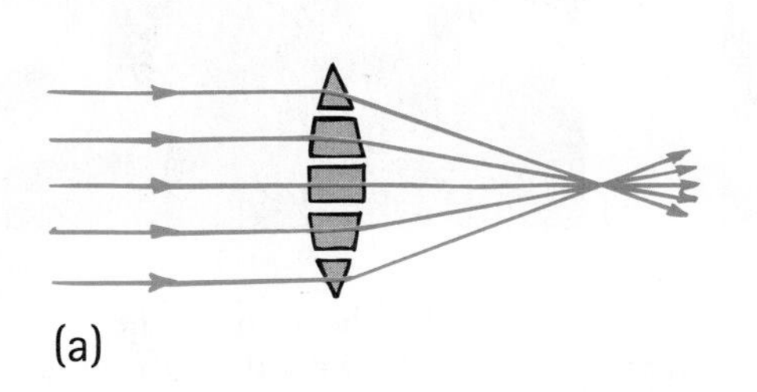

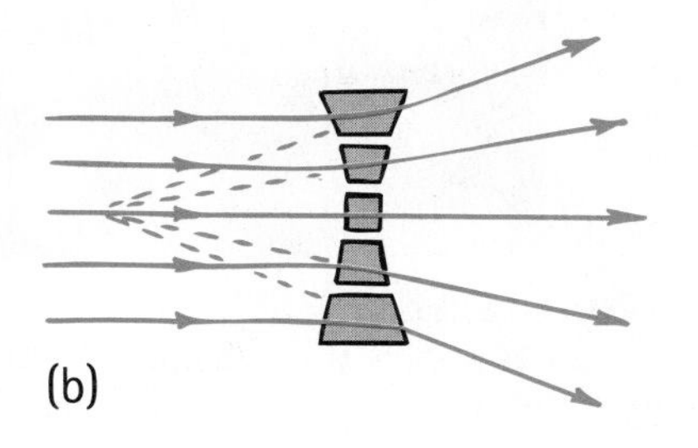

FIGURE 8.47
A lens may be thought of as a set of blocks and prisms. (a) A converging lens. (b) A diverging lens.

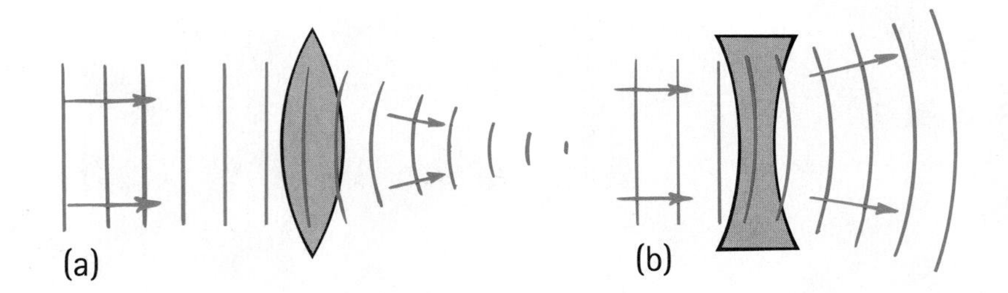

FIGURE 8.48
Wave fronts travel more slowly in glass than in air. (a) The waves slow more through the center of the lens, and convergence results. (b) The waves slow more at the edges, and divergence results.

the glass are parallel to each other (light doesn't deviate when it passes through glass with parallel surfaces, like window glass). A real lens is not made of prisms, of course. It is made of a solid piece of glass with surfaces ground usually to a circular curve. In Figure 8.48, we see how smooth lenses refract waves.

LEARNING OBJECTIVE
Describe the spreading of waves that pass through openings of various widths.

8.11 Diffraction

EXPLAIN THIS Do small openings or large openings produce more diffraction?

When you touch your finger to the surface of still water, circular ripples are produced. When you touch the surface with a straightedge, such as a horizontally held meterstick, you produce a *plane wave*. You can produce a series of plane waves by successively dipping a meterstick into the surface of the water (Figure 8.49).

FIGURE 8.49
The oscillating meterstick makes plane waves in the tank of water. Water oscillating in the opening acts as a source of waves that fan out on the other side of the opening. Water diffracts through the opening.

The photographs in Figure 8.50 are top views of water ripples in a shallow glass tank (called a ripple tank). A barrier with an adjustable opening is in the tank. When plane waves meet the barrier, they continue through with some distortion. In the left image, where the opening is wide, the waves continue through the opening almost without change. At the two ends of the opening, however, the waves bend. This bending is called **diffraction**. (Any bending of light by means other than reflection and refraction is diffraction.) Notice in Figure 8.50 that as the opening is narrowed the waves spread more. Smaller openings produce greater diffraction. Diffraction is a property of all kinds of waves, including sound and light waves.

Diffraction is not confined to narrow slits or to openings in general but can be seen around the edges of all shadows (Figure 8.51). On close examination, even the sharpest shadow is blurred slightly at its edges (Figure 8.52).

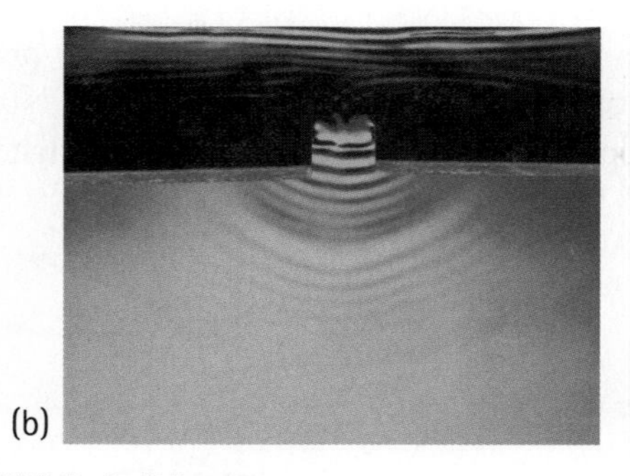

(a) (b) (c)

FIGURE 8.50
Plane waves passing through openings of various sizes. The smaller the opening, the greater the bending of the waves at the edges—in other words, the greater the diffraction.

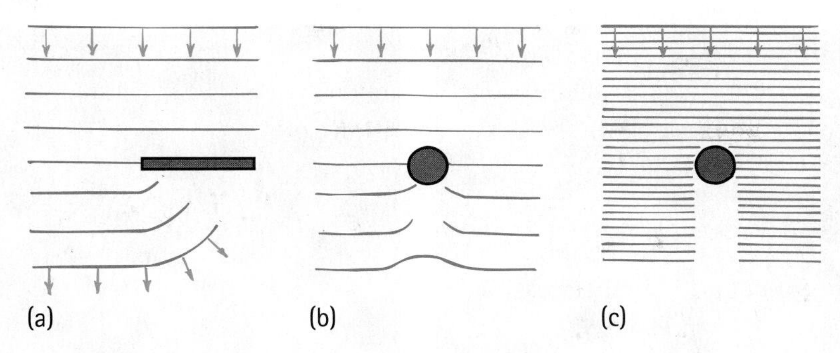

FIGURE 8.51
(a) Waves tend to spread into the shadow region. (b) When the wavelength is about the size of the object, the shadow is soon filled in. (c) When the wavelength is short compared with the object, a sharp shadow is cast.

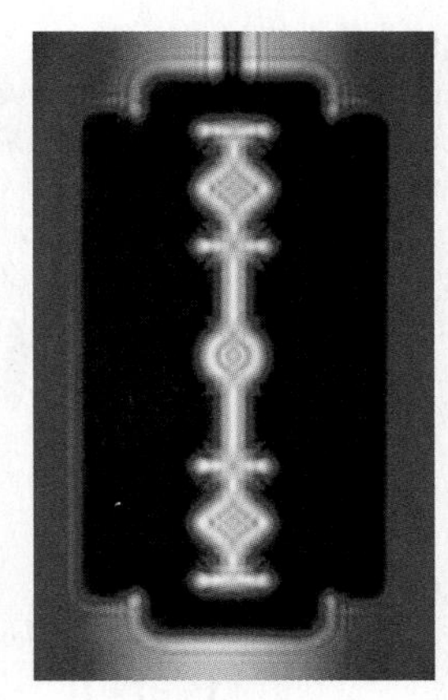

FIGURE 8.52
Diffraction fringes are evident in the shadows of monochromatic (single-frequency) light. These fringes would be filled by multitudes of other fringes if the source were white light.

Diffraction poses a problem when you are viewing very small objects with a light microscope. If the size of an object is about the same as the wavelength of light, diffraction blurs the image. If the object is smaller than the wavelength of light, no structure can be seen. The entire image is lost, due to diffraction. No amount of magnification or perfection of microscope design can defeat this fundamental diffraction limit. However, the electron microscope is useful for viewing ultratiny objects, as described in Chapter 15.

8.12 Interference

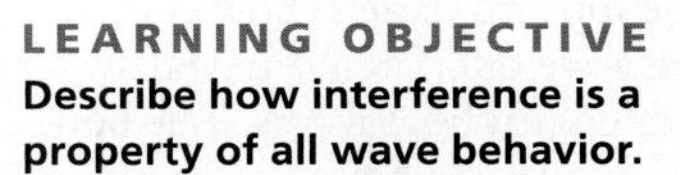

EXPLAIN THIS In what way can both sound and light be canceled?

An intriguing property of all waves is **interference**—the combined effect of two or more waves overlapping. Consider transverse waves: When the crest of one wave overlaps the crest of another, their individual effects add together. The result is a wave of increased amplitude. This is constructive interference (Figure 8.53, left). When the crest of one wave overlaps the trough of another, their individual effects are reduced. The high part of one wave simply fills in the low part of the other. This is destructive interference (Figure 8.53, right).

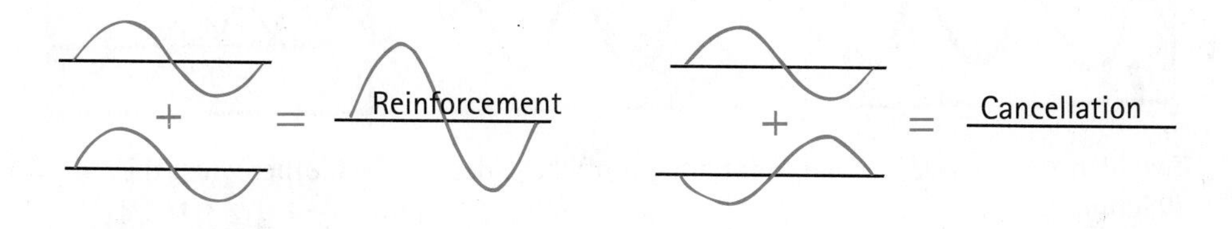

FIGURE 8.53
Constructive and destructive interference in a transverse wave.

Wave interference is easiest to observe in water. In Figure 8.54, we see the interference pattern produced when two vibrating objects touch the surface of water. We can see the regions in which the crest of one wave overlaps the trough of another to produce a region of zero amplitude. At points along such regions, the waves arrive out of step. We say they are out of phase with one another.

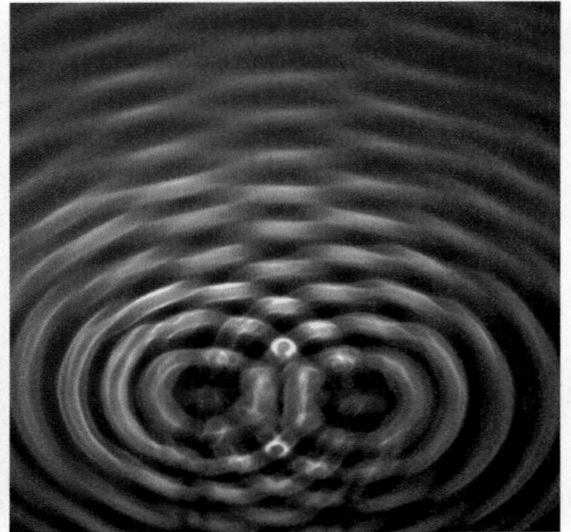

FIGURE 8.54
Two sets of overlapping water waves produce an interference pattern. The left image is an idealized drawing of the expanding waves from two sources. The right image is a photograph of an actual interference pattern.

Interference is a property of all wave motion, whether the waves are water waves, sound waves, or light waves. We see a comparison of interference for transverse and for longitudinal waves in Figure 8.55. In the case of sound, the crest of a wave corresponds to a compression and the trough of a wave corresponds to a rarefaction.

FIGURE 8.55
The top two panels show constructive wave interference in transverse and longitudinal waves. The bottom panels illustrate destructive interference.

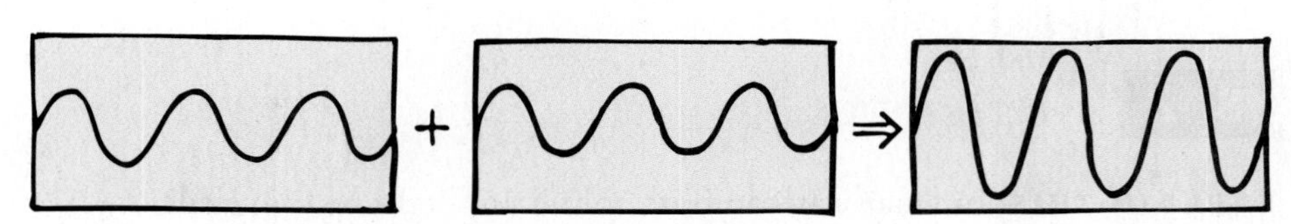

The superposition of two identical transverse waves in phase produces a wave of increased amplitude.

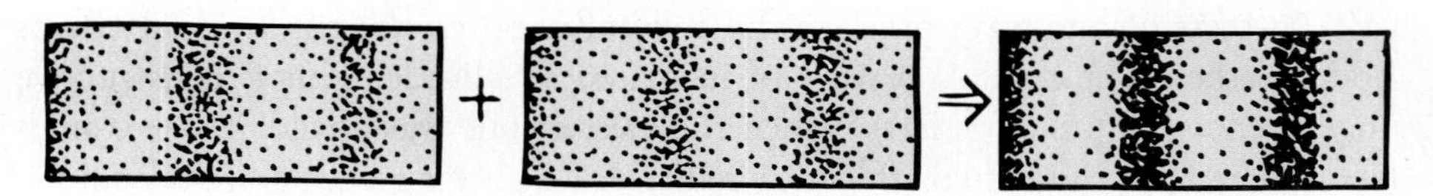

The superposition of two identical longitudinal waves in phase produces a wave of increased intensity.

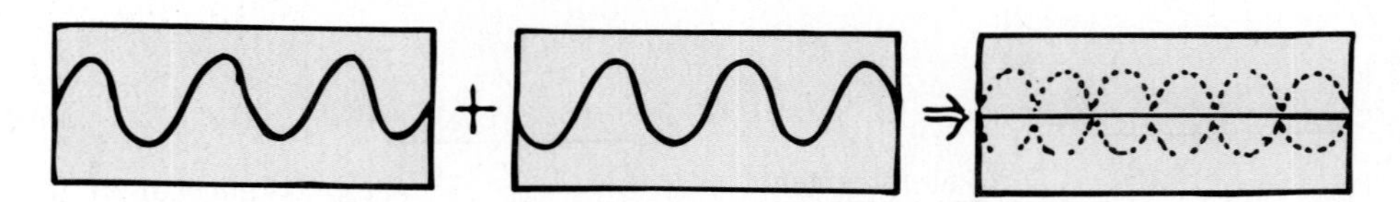

Two identical transverse waves that are out of phase destroy each other when they are superimposed.

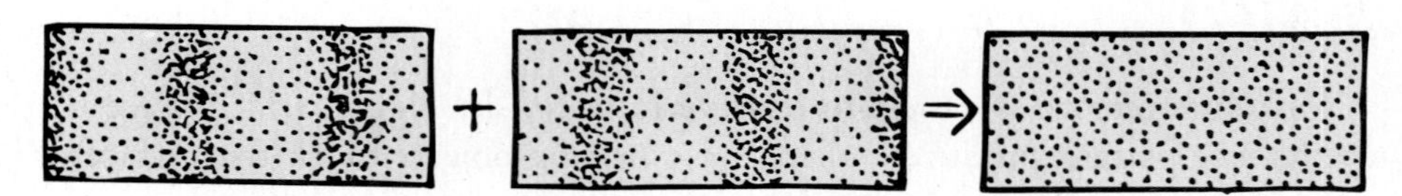

Two identical longitudinal waves that are out of phase destroy each other when they are superimposed.

8.13 The Doppler Effect

LEARNING OBJECTIVE
Relate the shortening and lengthening of waves due to motion to the Doppler effect.

EXPLAIN THIS When does the pitch of an ambulance siren change?

Consider a bug in the middle of a quiet pond. A pattern of water waves is produced when it jiggles its legs and bobs up and down (Figure 8.56). The bug is not traveling anywhere but merely treads water in a stationary position. The waves it creates are concentric circles because wave speed is the same in all directions. If the bug bobs in the water at a constant frequency, the distance between the wave crests (the wavelength) is the same in all directions. Waves encounter point A as frequently as they encounter point B. Therefore, the frequency of wave motion is the same at points A and B, or anywhere in the vicinity of the bug. This wave frequency remains the same as the bobbing frequency of the bug.

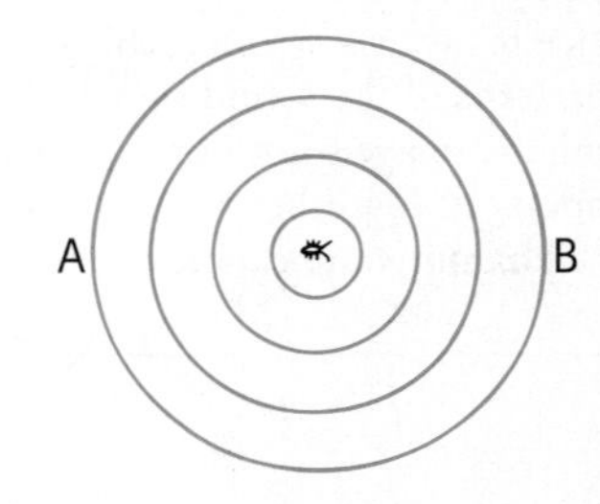

FIGURE 8.56
Top view of water waves made by a stationary bug jiggling on the surface of still water.

Suppose the jiggling bug moves across the water at a speed less than the wave speed. In effect, the bug chases part of the waves it has produced. The wave pattern is distorted and is no longer concentric circles (Figure 8.57). The center of the outer wave originated when the bug was at the center of that circle. The center of the next smaller wave originated when the bug was at the center of that circle, and so forth. The centers of the circular waves move in the direction of the swimming bug. Although the bug maintains the same bobbing frequency as before, an observer at B would see the waves coming more often. The observer would measure a higher frequency. This is because each successive wave has a shorter distance to travel and therefore arrives at B sooner than if the bug weren't moving toward B. An observer at A, on the other hand, measures a lower frequency because of the longer time between wave-crest arrivals. This occurs because each successive wave travels farther to A as a result of the bug's motion. This change in frequency of a wave as measured by an observer due to the motion of the source (or due to the motion of the observer) is called the **Doppler effect** (after the Austrian physicist and mathematician Christian Johann Doppler, 1803–1853).

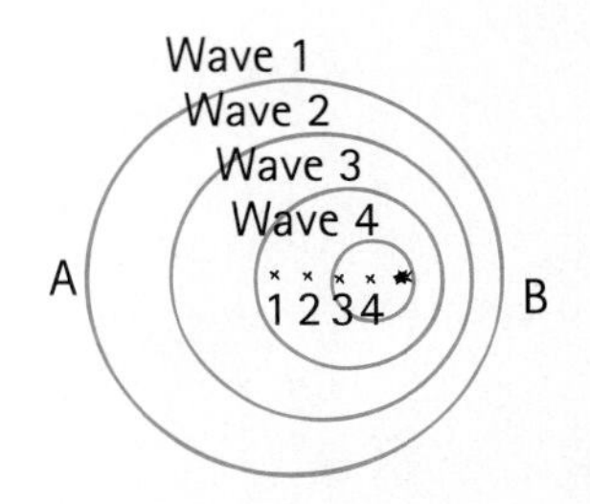

FIGURE 8.57
Water waves made by a bug swimming on the surface of still water toward point B.

Water waves spread over the flat surface of the water. In contrast, sound and light waves travel in three-dimensional space in all directions, like an expanding balloon. Just as circular waves are closer together in front of the swimming bug, spherical sound or light waves ahead of a moving source are closer together and reach an observer more frequently. The Doppler effect holds for all types of waves.

The Doppler effect is evident when you hear the changing pitch of an ambulance or fire-engine siren. When the siren is approaching you, the crests of the sound waves encounter your ear more frequently, and the pitch is higher than normal. And when the siren passes you and moves away, the crests of the waves encounter your ear less frequently, and you hear a drop in pitch (Figure 8.58).

MasteringPhysics®
TUTORIAL: The Doppler Effect
VIDEO: The Doppler Effect

FIGURE 8.58
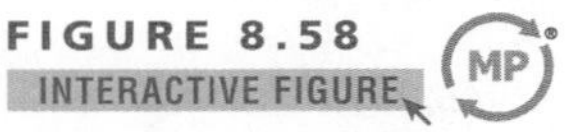

The pitch of sound increases when the source of the sound moves toward you, and it decreases when the source moves away.

TECHNOLOGY

Antinoise Technology

Destructive sound interference is at the heart of antinoise technology. Such noisy devices as jackhammers are being equipped with microphones that send the sound of the device to electronic microchips that create mirror-image wave patterns of the sound signals. For the jackhammer, this mirror-image sound signal is fed to earphones worn by the operator. Sound compressions (or rarefactions) from the jackhammer are canceled by mirror-image rarefactions (or compressions) in the earphones. The combination of signals cancels the jackhammer noise. Antinoise devices are becoming more common in aircraft, which today are much quieter inside than before this technology was introduced. Are automobiles next, perhaps eliminating the need for mufflers?

LEARNING OBJECTIVE
Relate the Doppler effect to astronomy.

Integrated Science 8C
ASTRONOMY

The Doppler Effect and the Expanding Universe

EXPLAIN THIS What is meant by a red shift and a blue shift for astronomical objects?

The Doppler effect also occurs for light waves. When a light source approaches, there is an increase in its measured frequency; when it recedes, there is a decrease in its frequency. An increase in light frequency is called a *blue shift* because the increase is toward a higher frequency, or toward the blue end of the color spectrum. A decrease in frequency is called a *red shift*, referring to a shift toward a lower frequency, or toward the red end of the color spectrum. A rapidly spinning star, for example, shows a red shift on the side turning away from us and a relative blue shift on the side turning toward us. This shifting of frequencies enables us to calculate the star's spin rate.

Galaxies, too, show a red shift in the light they emit. This observation was first made by American astronomer Edwin Hubble. When Hubble observed galaxies through his telescope, he noticed that the colors of the light emitted by their elements seemed to be red-shifted. This implied that the galaxies must be moving away from Earth. Further, Hubble's observations established that the farther a galaxy is from Earth, the faster it is moving away. This is the basis of our current belief that the universe is ever-expanding. (You will learn more about Hubble and his fundamentally important contributions to astronomy in Chapter 28.)

CHECK YOURSELF

1. **If Hubble had observed that the light from distant galaxies was blue-shifted, would this be evidence for an expanding or a shrinking universe? Explain.**
2. **Explain why, in terms of the bunching together of waves, light from a receding source is red-shifted.**

CHECK YOUR ANSWERS

1. Blue-shifted atomic spectra would indicate that the source was approaching the observer; if all surrounding galaxies showed blue-shifted atomic spectra, this would be evidence for a shrinking universe.
2. Light from a receding source should be red-shifted because, just like the bug on the pond, wave crests grow successively farther apart and thus have a lower frequency.

8.14 The Wave–Particle Duality

LEARNING OBJECTIVE
Relate how light can be seen as composed of particles and as composed of waves.

EXPLAIN THIS Which experiments favor a particle view of light and which favor a wave view?

We have described light as a wave. The earliest ideas about the nature of light, however, held that light was composed of tiny particles. In ancient times, Plato and other Greek philosophers believed in a particle view of light, as did Isaac Newton, who first became famous for his experiments with light in the early 1700s. A hundred years later, the wave nature of light was demonstrated by Thomas Young in his interference experiments. The wave view was reinforced in 1862 by Maxwell's finding that light is energy carried in oscillating electric and magnetic fields of electromagnetic waves. The wave view of light was confirmed experimentally by Heinrich Hertz 25 years later.

Then, in 1905, Albert Einstein published a Nobel Prize–winning paper that challenged the wave theory of light. Einstein stated that light in its interactions with matter was confined not in continuous waves, as Maxwell and others had envisioned, but in tiny particles of energy called *photons*. Einstein's particle model of light explained a perplexing phenomenon of that time—*the photoelectric effect*.

The Photoelectric Effect

When light shines on certain metal surfaces, electrons are ejected from those surfaces. This is the photoelectric effect, which is put to use in electric eyes, in light meters, and in motion-picture sound tracks. What perplexed investigators at the turn of the 20th century was that ultraviolet and violet light imparted sufficient energy to knock electrons from those metal surfaces, while lower-frequency light did not—even when the lower-frequency light was very bright. Ejection of electrons depended only on the frequency of light, and the higher the frequency of the light, the greater the kinetic energy of the ejected electrons. Very dim high-frequency light ejected fewer electrons, but it ejected each of them with the same kinetic energy of electrons ejected in brighter light of the same frequency (Figure 8.59).

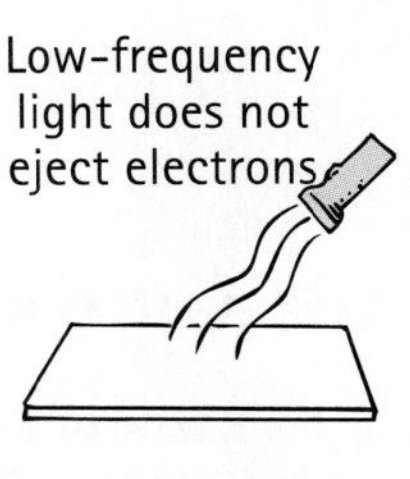
Low-frequency light does not eject electrons
High-frequency light *does* eject electrons

FIGURE 8.59
The photoelectric effect depends on frequency.

Einstein's explanation was that the electrons in the metal were being bombarded by "particles of light"—by photons. Einstein stated that the energy of each photon was proportional to its frequency—that is,

$$E \sim f$$

So Einstein viewed light as a hail of photons, each carrying energy proportional to its frequency. One photon is completely absorbed by each electron ejected from the metal.

All attempts to explain the photoelectric effect by waves failed. A light wave has a broad front, and its energy is spread out along this front. For the light wave to eject a single electron from a metal surface, all its energy would somehow have to be concentrated on that one electron. But this is as improbable as an ocean wave hitting a beach and knocking only one single seashell far inland with an energy equal to the energy of the whole wave. Therefore, the photoelectric effect suggests that, instead of thinking of light encountering a surface as a continuous train of waves, we should conceive of light encountering a surface, or any detector, as a succession of particle-like photons. The energy of each photon is proportional to the frequency of light, and that energy is given completely to a single electron in the metal's surface. The number of ejected electrons has to do with the number of photons—the brightness of the light.

Experimental verification of Einstein's explanation of the photoelectric effect was made 11 years later by the American physicist Robert Millikan. Every aspect of Einstein's interpretation was confirmed. The photoelectric effect proves conclusively that light has particle properties. A wave model of light is inconsistent with the photoelectric effect. On the other hand, interference demonstrates convincingly that light has wave properties, and a particle model of light is inconsistent with interference.

Evidently, light has both a wave nature and a particle nature—a **wave–particle duality**. It reveals itself as a wave or particle depending on how it is being observed. Simply stated, light behaves as a stream of photons when it interacts with a sheet of metal or other detector, and it behaves as a wave in traveling from a source to the place where it is detected. Light travels as a wave and hits as a stream of photons. The fact that light exhibits both wave and particle behavior is one of the most interesting surprises that physicists discovered in the 20th century.

SUMMARY OF TERMS (KNOWLEDGE)

Amplitude For a wave or a vibration, the maximum displacement on either side of the equilibrium (midpoint) position.

Diffraction Any bending of light by means other than reflection and refraction.

Dispersion The separation of light into colors arranged by frequency.

Doppler effect The change in frequency of a wave due to the motion of the source (or due to the motion of the receiver).

Electromagnetic spectrum The continuous range of electromagnetic waves that extends in frequency from radio waves to gamma rays.

Electromagnetic wave An energy-carrying wave produced by an oscillating electric charge.

Forced vibration The setting up of vibrations in an object by a vibrating source.

Frequency For a vibrating body, the number of vibrations per unit time. For a wave, the number of crests that pass a particular point per unit time.

Hertz The SI unit of frequency; one hertz (symbol Hz) equals one vibration per second.

Interference The combined effect of two or more waves overlapping.

Law of reflection The angle of reflection equals the angle of incidence.

Longitudinal wave A wave in which the medium vibrates in a direction parallel (longitudinal) to the direction in which the wave travels.

Natural frequency A frequency at which an elastic object naturally tends to vibrate.

Opaque Description of materials that absorb light without reemission.

Period The time required for a vibration or a wave to make one complete cycle.

Reflection The returning of a wave to the medium from which it came when it hits a barrier.

Refraction The bending of waves due to a change in wave speed.

Resonance A dramatic increase in the amplitude of a wave that results when the frequency of forced vibrations matches an object's natural frequency.

Transparent Description of materials that allow light to pass through them in straight lines.

Transverse wave A wave in which the medium vibrates in a direction perpendicular (transverse) to the direction in which the wave travels.

Wave A disturbance that travels from one place to another transporting energy, but not necessarily matter, along with it.

Wavelength The distance from the top of one crest to the top of the next crest or, equivalently, the distance between successive identical parts of the wave.

Wave-particle duality The exhibition of both a wave nature and a particle nature of light.

READING CHECK QUESTIONS (COMPREHENSION)

8.1 Vibrations and Waves

1. Distinguish among amplitude, wavelength, frequency, and period.
2. What is the source of all waves?

8.2 Wave Motion

3. In one word, what is it that moves from source to receiver in wave motion?
4. Does the medium in which a wave travels move with the wave?
5. What is the relationship among frequency, wavelength, and wave speed?

8.3 Transverse and Longitudinal Waves

6. In what direction are the vibrations relative to the direction of wave travel in a transverse wave? In a longitudinal wave?
7. Distinguish between a compression and a rarefaction.

8.4 The Nature of Sound

8. Define the wavelength of sound in terms of successive compressions of air.
9. Can sound travel through a vacuum? Why or why not?

8.5 Resonance

10. Why does a struck tuning fork sound louder when its handle is held against a table?
11. Distinguish between forced vibrations and resonance.

8.6 The Nature of Light

12. What is the principal difference between a radio wave and light? Between light and X-rays?
13. How does the frequency of an electromagnetic wave compare with the frequency of the vibrating electrons that produces it?

8.7 Reflection

14. What is the law of reflection?
15. Does the law of reflection hold for diffuse reflection? Explain.

8.8 Transparent and Opaque Materials

16. The sound coming from one tuning fork can force another to vibrate. What is the analogous effect for light?
17. (a) What is the fate of the energy in ultraviolet light incident on glass? (b) What is the fate of the energy in visible light incident on glass?
18. How does the average speed of light in glass compare with its speed in a vacuum?

8.9 Color

19. What is the relationship between the frequency of light and its color?
20. Distinguish between the white of this page and the black of this ink, in terms of what happens to the white light that falls on both.

8.10 Refraction

21. What causes the bending of light in refraction?
22. Does a single raindrop illuminated by sunlight deflect light of a single color, or does it disperse a spectrum of colors?
23. Does a viewer see a single color or a spectrum of colors coming from a single faraway drop?
24. Distinguish between a converging lens and a diverging lens.

8.11 Diffraction

25. For an opening of a given size, is diffraction more pronounced for a longer wavelength or a shorter wavelength?
26. Does diffraction help or hinder viewing with a light microscope?

8.12 Interference

27. What kinds of waves exhibit interference?
28. Distinguish between constructive interference and destructive interference.

8.13 The Doppler Effect

29. Why does an observer measure waves from an approaching source as having a higher frequency than if the source were stationary?

8.14 The Wave–Particle Duality

30. What evidence can you cite for the particle nature of light? The wave nature of light?
31. Which are more successful in dislodging electrons from a metal surface—photons of violet light or photons of red light? Why?
32. When does light behave as a particle? When does it behave as a wave?

THINK INTEGRATED SCIENCE

8A—Sensing Pitch

33. How does the pitch of sound relate to its frequency?
34. A cat can hear sound frequencies up to 70,000 Hz. Bats send and receive ultrahigh-frequency squeaks up to 120,000 Hz. Which hears sound of shorter wavelengths—cats or bats?

8B—Mixing Colored Lights

35. Explain how you are able to see a wide range of colors even though there are only three kinds of light-sensitive cells ("cones") in your eyes.
36. Stare at a piece of colored paper for 45 seconds or so. Then look at a plain white surface. What do you see now? Can you explain what you see? (Hint: To figure this out, you need to know that the cones in your retina that are receptive to the color of the paper become fatigued.)

8C—The Doppler Effect and the Expanding Universe

37. Swing a buzzer of any kind over your head in a circle. Ask some friends to stand off to the side, listen to the buzzer, and report their observations. Then switch places so you too can hear the buzzer as it moves. How does the sound of the buzzer change? Why?
38. How does the Doppler effect provide evidence that we live in an expanding universe?

THINK AND DO (HANDS-ON APPLICATION)

39. Suspend a wire grill from a refrigerator or an oven from a string, holding the ends of the string to your ears. Let a friend gently stroke the grill with pieces of broom straw and with other objects. The effect is best appreciated when you are in a relaxed condition with your eyes closed. Be sure to try this!

40. Set up two pocket mirrors at right angles, and place a coin between them. You'll see four coins. Change the angle of the mirrors, and see how many images of the coin you can see. With the mirrors at right angles, look at your face. Then wink. What do you see? You now see yourself as others see you. Hold a printed page up to the double mirrors and compare its appearance with the reflection of a single mirror.

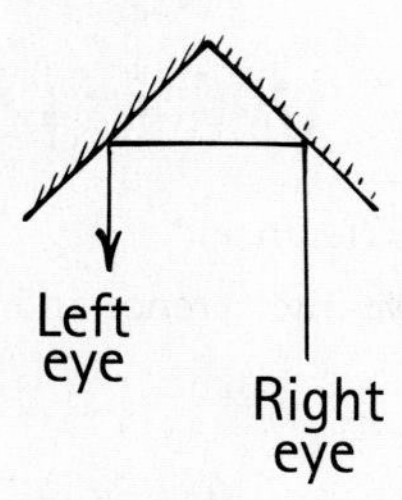

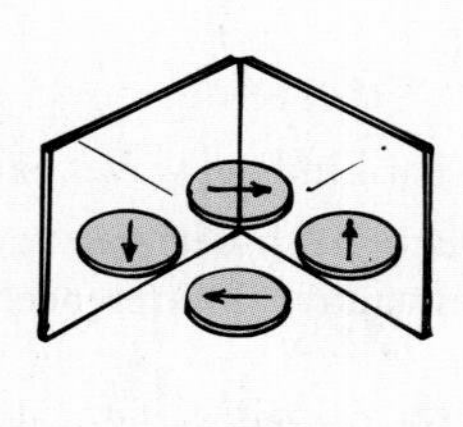

41. Make a pinhole camera, as illustrated here. Cut out one end of a small cardboard box, and cover the end with tissue or wax paper. Make a clean-cut pinhole at the other end. (If the cardboard is thick, make the hole through a piece of aluminum foil placed over an opening in the cardboard.) Aim the camera at a bright object in a darkened room, and you see an upside-down image on the tissue paper. When photographic film was in vogue, students replaced the tissue paper with unexposed photographic film, covering the back so it was light tight, and covering the pinhole with a removable flap, all ready to take a picture. Exposure times differed, depending principally on the kind of film and the amount of light. Lenses on today's commercial cameras are much bigger than pinholes and therefore admit more light in less time—hence the term *snapshot.* For now it will be enough to view images on the tissue or wax paper. Point your camera toward the Sun. And if you do so during a solar eclipse, you'll marvel at the clear crescents on your viewing screen.

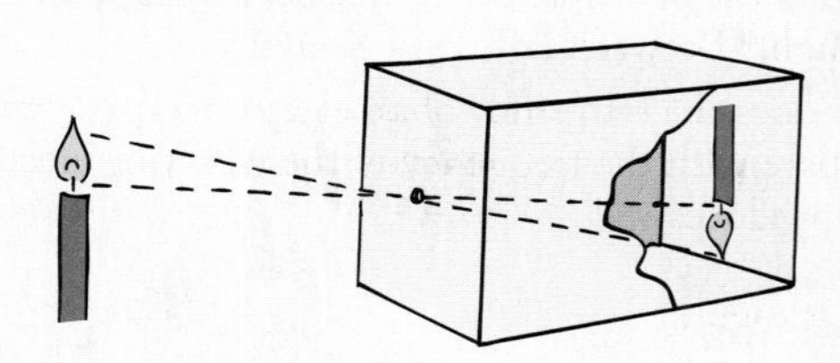

42. Write a letter to Grandpa explaining why we now say that light is not just a particle, and not just a wave, but in fact is both—a "wavicle"!

PLUG AND CHUG (FORMULA FAMILIARIZATION)

$$\textbf{Frequency} = \frac{1}{\textbf{period}}; \quad \textbf{Period} = \frac{1}{\textbf{frequency}}$$

43. A pendulum swings to and fro every 3 s. Show that its frequency of swing is $\frac{1}{3}$ Hz.
44. Another pendulum swings to and fro at a regular rate of 2 times per second. Show that its period is 0.5 s.

$$\textbf{Wave speed} = \textbf{frequency} \times \textbf{wavelength} = f\lambda$$

45. A 3-m-long wave oscillates 1.5 times each second. Show that the speed of the wave is 4.5 m/s.
46. Show that a certain 1.2-m-long wave with a frequency of 2.5 Hz has a wave speed of 3.0 m/s.
47. A tuning fork produces a sound with a frequency of 256 Hz and a wavelength in air of 1.33 m. Show that the speed of sound in the vicinity of the fork is 340 m/s.

THINK AND COMPARE (ANALYSIS)

48. The siren of a fire engine is heard when the fire engine is traveling in three different situations: A, toward the listener at 30 km/h; B, toward the listener at 50 km/h; and C, away from the listener at 20 km/h. Rank the pitches heard, from highest to lowest.
49. Three shock waves are produced by supersonic aircraft. Rank their speeds from highest to lowest.

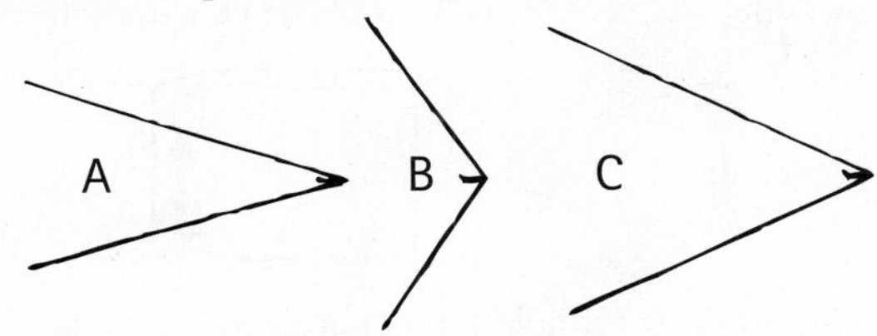

50. A woman looks at her face in the handheld mirror. Rank the amount of her face she sees in the three situations A, B, and C, from greatest to least (or does she see the same amount in all positions?).

51. Wheels from a toy cart are rolled from a concrete sidewalk onto three surfaces: A, a paved driveway; B, a grass lawn; and C, a close-cropped grass on a golf-course putting green. Each set of wheels bends at the boundary due to slowing and is deflected from its initial straight-line course. Rank the surfaces according to the amount each set of wheels bends at the boundary, from greatest amount of bending to least.

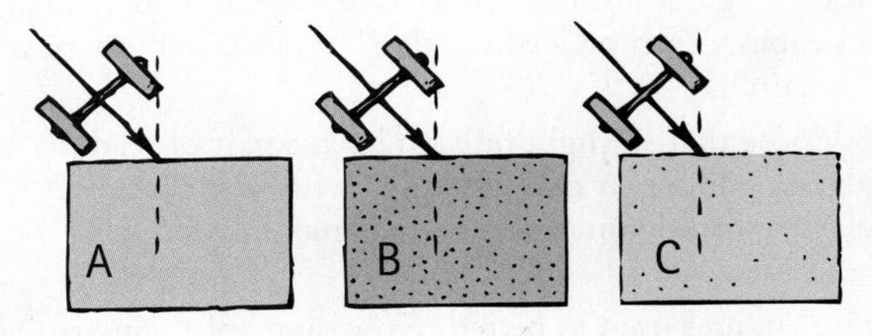

52. Identical rays of light enter three transparent blocks composed of different materials. Light slows down when it enters the blocks. Rank the blocks according to the speed light travels in each, from highest to lowest.

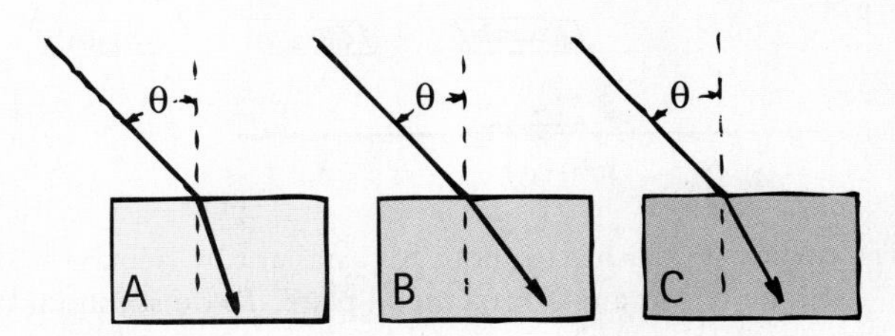

THINK AND SOLVE (MATHEMATICAL APPLICATION)

53. What is the frequency, in hertz, that corresponds to each of these periods: (a) 0.10 s, (b) 5 s, (c) $\frac{1}{60}$ s?
54. The nearest star beyond the Sun is Alpha Centauri, which is 4×10^{16} m away. If we were to receive a radio message from this star today, show that it would have been sent 1.4×10^8 s ago (4.4 yr ago).
55. Blue-green light has a frequency of about 6×10^{14} Hz. Use the relationship $c = f\lambda$ to show that the wavelength of this light in air is 5×10^{-7} m. How does this wavelength compare with the size of an atom, which is about 10^{-10} m?
56. A certain blue-green light has a wavelength of 600 nm in air. Show that its wavelength in water, where light travels at 75% of its speed in air, is 450 nm. Show that its speed in Plexiglas, where light travels at 67% of its speed in air, is 400 nm. (1 nm = 10^{-9} m)
57. A certain radar installation that is used to track airplanes transmits electromagnetic radiation with a wavelength of 3 cm. (a) Show that the frequency of this radiation, measured in billions of hertz, is 10 GHz. (b) Show that the time required for a pulse of radar waves to reach an airplane 5 km away and return is 3.3×10^{-5} s.
58. Suppose you walk toward a mirror at 2 m/s. Show that you and your image approach each other at a speed of 4 m/s (and not 2 m/s).

THINK AND EXPLAIN (SYNTHESIS)

59. What kind of motion should you impart to a stretched coiled spring (or to a Slinky) to produce a transverse wave? To produce a longitudinal wave?
60. What does it mean to say that a radio station is "at 101.1 on your FM dial"?
61. In the stands at a racetrack, you notice smoke from the starter's gun before you hear it fire. Explain.
62. What is the danger posed by people in the balcony of an auditorium stamping their feet in a steady rhythm?
63. What is the fundamental source of electromagnetic radiation?
64. Which has the shorter wavelengths—ultraviolet or infrared? Which has the higher frequencies?
65. Do radio waves travel at the speed of sound, at the speed of light, or at some speed in between?
66. What determines whether a material is transparent or opaque?
67. You can get a sunburn on a cloudy day, but you can't get a sunburn even on a sunny day if you are behind glass. Explain.
68. Suppose that sunlight falls on both a pair of reading glasses and a pair of dark sunglasses. Which pair of glasses would you expect to become warmer? Defend your answer.
69. Fire engines used to be red. Now many of them are yellow-green. Why was the color changed?
70. A woman's eye at point P looks into the mirror. Which of the numbered cards can she see reflected in the mirror?

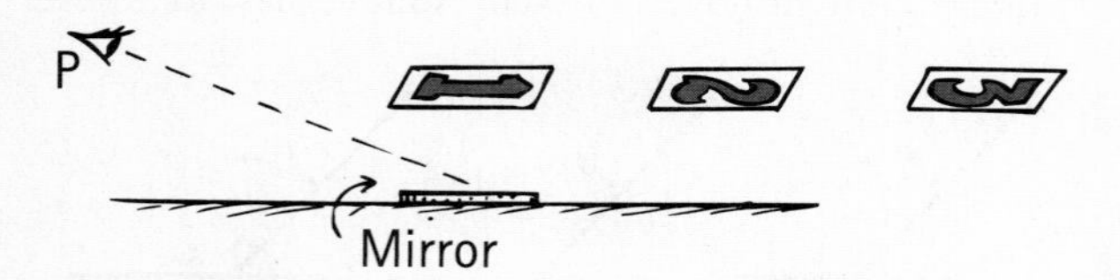

71. Cowboy Joe wishes to shoot his assailant by ricocheting a bullet off a mirrored metal plate. To do so, should he simply aim at the mirrored image of his assailant? Explain.
72. What happens to light of a certain frequency when it is incident on a material whose natural frequency is the same as the frequency of the light?
73. Two observers standing apart from each other do not see the "same" rainbow. Explain.
74. An ocean wave is cyan. What color(s) of light does it absorb? What colors does it reflect?
75. A rainbow viewed from an airplane may form a complete circle. Where will the shadow of the airplane appear? Explain.
76. A bat flying in a cave emits a sound and receives its echo 0.1 s later. How far away is the cave wall?
77. Why do radio waves diffract around buildings whereas light waves do not?
78. Suntanning produces cellular damage in the skin. Why is ultraviolet radiation capable of producing this damage whereas visible radiation is not?
79. Why doesn't the sharpness of the image in a pinhole camera depend on the position of the viewing screen?

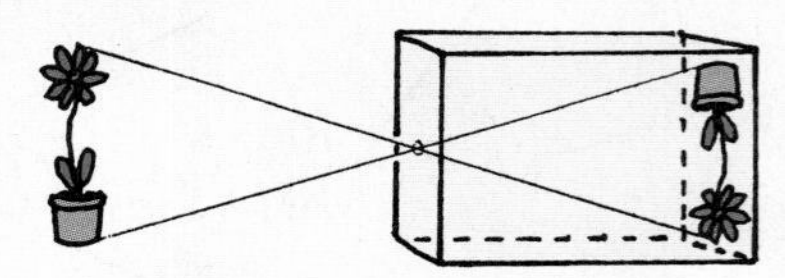

80. If you point the pinhole camera of Exercise 79 at the Sun, you will see a clear and bright solar image on the viewing screen. How does this relate to the circular spots that surround Lillian beneath the sunlit tree shown in the photo?

THINK AND DISCUSS (EVALUATION)

81. The wavelength of light changes as light goes from one medium to another, while the frequency remains the same. Is the wavelength longer or shorter in water than in air? Explain in terms of the equation: Speed $=$ frequency $\times$ wavelength.
82. The sitar, an Indian musical instrument, has a set of strings that vibrate and produce music, even though the player never plucks those strings. These "sympathetic strings" are identical to the plucked strings and are mounted below them. What is your explanation?
83. A railroad locomotive is at rest with its whistle shrieking, and then it starts moving toward you. (a) Does the frequency that you hear increase, decrease, or stay the same? (b) How about the wavelength reaching your ear? (c) How about the speed of sound in the air between you and the locomotive?
84. The radiation curve of the Sun (see Figure 8.31) shows that the brightest light from the Sun is yellow-green. Why then do we see the Sun as whitish instead of yellow-green?
85. If, while standing on the bank of a stream, you wished to spear a fish swimming in the water out in front of you, would you aim above, below, or directly at the observed fish to make a direct hit? If you decided instead to zap

the fish with a laser, would you aim above, below, or directly at the observed fish? Defend your answers.

86. A rule of thumb for estimating the distance in kilometers between an observer and a lightning strike is to divide the number of seconds in the interval between the flash and the sound by 3. Show that this rule is correct.
87. If a single disturbance some unknown distance away sends out both transverse and longitudinal waves that travel at distinctly different speeds in the medium, such as the ground during earthquakes, how could the origin of the disturbance be located?
88. The photo at right shows Earth science author Suzanne Lyons with her son Tristan wearing red and her daughter Simone wearing green. Below that is the negative of the photo, which shows these colors differently. What is your explanation?
89. Explain briefly how the photoelectric effect is used in the operation of at least two of the following: an electric eye, a photographer's light meter, the sound track of a motion picture.

90. Does the photoelectric effect prove that light is made up of particles? Do interference experiments prove that light is composed of waves? (Is there a distinction between what something is and how it behaves?)

READINESS ASSURANCE TEST (RAT)

If you have a good grasp of this chapter, if you really do, then you should be able to score at least 7 out of 10 on this RAT. If you score less than 7, consider studying further before moving on.

Choose the BEST answer to each of the following:

1. Which of these does NOT belong in the family of electromagnetic waves?
   (a) light
   (b) sound
   (c) radio waves
   (d) X-rays
2. The source of electromagnetic waves is vibrating
   (a) electrons.
   (b) atoms.
   (c) molecules.
   (d) energy fields.
3. The slowing of light in transparent materials has to do with
   (a) the time for absorption and reemission of the light.
   (b) the density of materials.
   (c) different frequency ranges in materials.
   (d) the fundamental difference between light and sound.
4. Whether a particular surface acts as a polished reflector or a diffuse reflector depends on the
   (a) color of reflected light.
   (b) brightness of reflected light.
   (c) wavelength of light.
   (d) angle of incoming light.
5. When a light ray passes at an angle from water into the air, the ray in the air bends
   (a) toward the normal.
   (b) away from the normal.
   (c) either away from or toward the normal.
   (d) parallel to the normal.
6. Refracted light that bends away from the normal is light that has
   (a) slowed down.
   (b) speeded up.
   (c) bounced.
   (d) diffracted.
7. The colors on the cover of your physics book are due to
   (a) color addition.
   (b) color subtraction.
   (c) color interference.
   (d) scattering.
8. The redness of a sunup or sunset is due mostly to light that hasn't been
   (a) absorbed.
   (b) transmitted.
   (c) scattered.
   (d) polarized.
9. A rainbow is the result of light in raindrops that undergoes
   (a) internal reflection.
   (b) dispersion.
   (c) refraction.
   (d) all of these
10. Light has both a wave nature and a particle nature. Light behaves primarily as a particle when it
    (a) travels from one location to another.
    (b) interacts with matter.
    (c) both
    (d) neither

Answers to RAT

1. b, 2. a, 3. a, 4. c, 5. b, 6. b, 7. b, 8. c, 9. d, 10. b

2 PART TWO

Chemistry

Hey, Liam, like everyone, I'm made of atoms, which are so small and numerous that I inhale billions of trillions with each breath. Many of these atoms stay and become a part of my body. When I exhale, I'm releasing into the air many of the atoms that were once a part of me.

Gee, Bo, the atoms you exhale are the very ones both our baby sister Neve and I inhale. So your atoms then become a part of us, just as the atoms we exhale eventually become a part of you!

In each breath we inhale, we recycle atoms that once were a part of every person who lived. Hey, in this sense, we're all one!

CHAPTER 9

Atoms and the Periodic Table

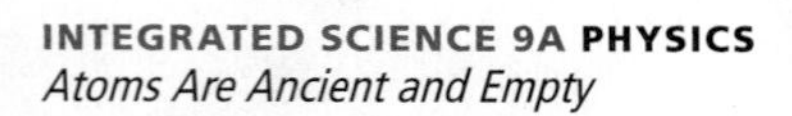

WE HUMANS have long tinkered with the materials around us and used them to our advantage. Once we learned how to control fire, we were able to create many new substances. Moldable wet clay, for example, was found to harden to ceramic when heated by fire. By 5000 BC, pottery fire pits gave way to furnaces hot enough to convert copper ores to metallic copper. By 1200 BC, even hotter furnaces were converting iron ores into iron. This technology allowed for the mass production of metal tools and weapons and made possible the many achievements of ancient Chinese, Egyptian, and Greek civilizations.

Fast-forward to the 21st century, and we've since learned that all the materials around us are made of remarkably small particles called atoms. We have learned how to manipulate these atoms to produce a vast array of new and useful modern materials. In this chapter, we will explore both the nature of atoms and the amazing chart that tells their story—the periodic table.

Integrated Science 9A

PHYSICS

Atoms Are Ancient and Empty

LEARNING OBJECTIVE
Describe the origin of atoms and the empty nature of their internal structure.

EXPLAIN THIS If atoms are empty, why can't we walk through walls?

Hydrogen, H, the lightest atom, makes up more than 90% of the atoms in the known universe. Most of these hydrogen atoms were formed during the beginning of our universe about 13.7 billion years ago. Heavier atoms are produced in stars, which are massive collections of hydrogen atoms pulled together by gravitational forces. The great pressures deep in a star's interior cause hydrogen atoms to fuse into heavier atoms. With the exception of hydrogen, therefore, all the atoms that occur naturally on Earth—including those in your body—are the products of stars. You are made of stardust, as is everything that surrounds you.

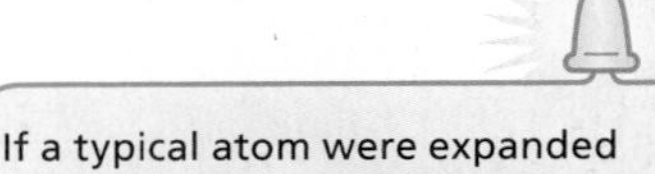

If a typical atom were expanded to a diameter of 3 km, about as big as a medium-sized airport, the nucleus would be about the size of a basketball. Atoms are mostly empty space.

So, atoms are ancient. They have existed through imponderable ages, recycling through the universe in innumerable forms, both nonliving and living. In this sense, you don't "own" the atoms that make up your body—you are simply their present caretaker. Many more caretakers will follow.

Atoms are so small that each breath you exhale contains more than 10 billion trillion of them. This is more than the number of breaths in Earth's atmosphere. Within a few years, the atoms of your breath are uniformly mixed throughout the atmosphere. What this means is that anyone anywhere on Earth inhaling a breath of air takes in numerous atoms that were once part of you. And, of course, the reverse is true: You inhale atoms that were once part of everyone who has ever lived. We are literally breathing one another.

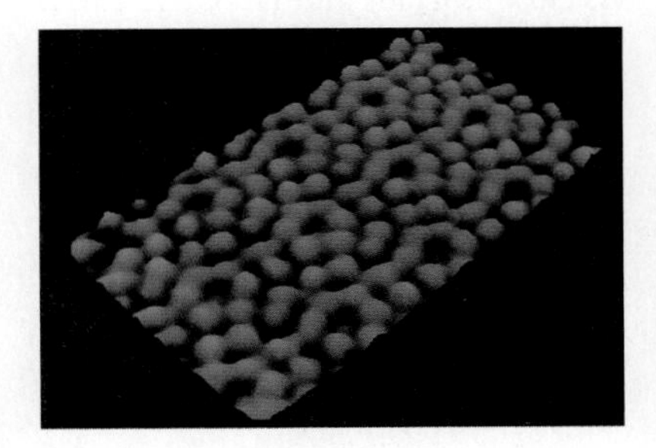

FIGURE 9.1
An image of carbon atoms obtained with a scanning probe microscope.

Atoms are so small that they can't be seen with visible light. That's because they are even smaller than the wavelengths of visible light. We could stack microscope on top of microscope and never "see" an atom. Photographs of atoms, such as in Figure 9.1, are obtained with a scanning probe microscope. Discussed further in Integrated Science 9B, this is an imaging device that bypasses light and optics altogether.

Today we know that the atom is made up of smaller, subatomic particles—*electrons*, *protons*, and *neutrons*. We also know that atoms differ from one another only in the number of subatomic particles they contain. Protons and neutrons are bound together at the atom's center to form a larger particle—the **atomic nucleus**. The nucleus is a relatively heavy particle that makes up most of an atom's mass. Surrounding the nucleus are the tiny **electrons** in an electron cloud, as shown in Figure 9.2.

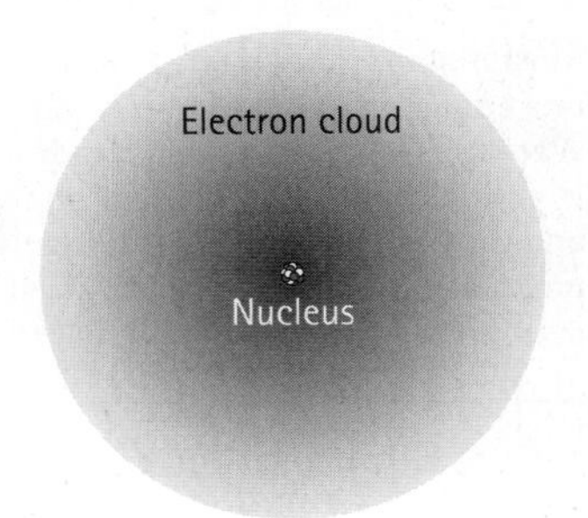

FIGURE 9.2
Electrons whiz around the atomic nucleus, forming what can be best described as a cloud that is more dense where the electrons tend to spend most of their time. Electrons, however, are invisible to us. Hence, such a cloud can only be imagined. Furthermore, if this illustration were drawn to scale, the atomic nucleus with its protons and neutrons would be too small to be seen. In short, atoms are not well suited to graphical depictions.

CHECK YOURSELF
A friend claims there are atoms in his brain that were once in the brain of Albert Einstein. Is your friend's claim likely correct or nonsense?

CHECK YOUR ANSWER
Your friend is correct! In addition, there are atoms in your friend's and everyone else's body that were once part of Einstein and everybody else too! The arrangements of these atoms, however, are now quite different. What's more, the atoms of which you and your friend are composed will be found in the bodies of all the people on Earth who are yet to be.

UNIFYING CONCEPT

- *Newton's Third Law* Section 3.4

We and all materials around us are mostly empty space. How can this be? Electrons move about the nucleus in an atom, defining the volume of space that the atom occupies. But electrons are very small. If an atom were the size of a baseball stadium, one of its electrons would be smaller than a grain of rice. Furthermore, all the electrons of an atom are widely spaced apart. Atoms are indeed mostly empty space.

MasteringPhysics®
VIDEO: Evidence for Atoms
VIDEO: Atoms are Recyclable

So, why don't atoms simply pass through one another? How is it that we are supported by the floor despite the empty nature of its atoms? Although subatomic particles are much smaller than the volume of the atom, the range of their electric field is several times larger than that volume. In the outer regions of any atom are electrons, which repel the electrons of neighboring atoms. Two atoms therefore can get only so close to each other before they start repelling (provided they don't join in a chemical bond, as will be discussed in Chapter 12).

FIGURE 9.3
As close as Tracy and Ian are in this photograph, none of their atoms meet.

When the atoms of your hand push against the atoms of a wall, electrical repulsions between electrons in your hand and electrons in the wall prevent your hand from passing through the wall. These same electrical repulsions prevent us from falling through the solid floor. They also allow us the sense of touch. Interestingly, when you touch someone, your atoms and those of the other person do not meet. Instead, atoms from the two of you get close enough so that you sense an electrical repulsion. A tiny, though imperceptible, gap still exists between the two of you (Figure 9.3).

9.1 The Elements

LEARNING OBJECTIVE
Recognize the elements of the periodic table as the fundamental building blocks of matter.

EXPLAIN THIS Why isn't water an element?

You know that atoms make up the matter around you, from stars to steel to chocolate ice cream. Given all these different types of material, you might think that there must be many different kinds of atoms. But the number of different kinds of atoms is surprisingly small. The great variety of substances results from the many ways a few kinds of atoms can be combined. Just as the three colors red, green, and blue can be combined to form any color on a television screen or the 26 letters of the alphabet make up all the words in a dictionary, only a few kinds of atoms combine in different ways to produce all substances. To date, we know of slightly more than 100 distinct atoms. Of these, about 90 are found in nature. The remaining atoms have been created in the laboratory.

Most materials are made from more than one kind of atom. Water, H_2O, for example, is made from the combination of hydrogen and oxygen atoms. These materials are called *compounds*, and we will discuss them further in Chapter 11.

Any material made of only one type of atom is classified as an **element**. Three examples are shown in Figure 9.4. Pure gold, for example, is an element— it contains only gold atoms. Nitrogen gas is an element because it contains only nitrogen atoms. Likewise, the graphite in your pencil is an element—carbon.

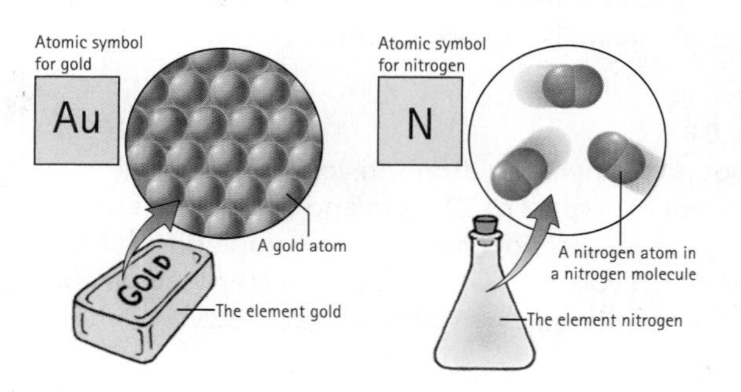

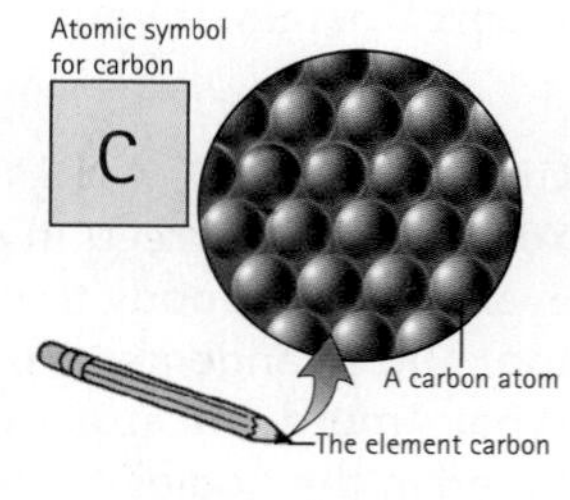

FIGURE 9.4
Any element consists of only one kind of atom. Gold consists of only gold atoms, a flask of gaseous nitrogen consists of only nitrogen atoms, and the carbon of a graphite pencil consists of only carbon atoms.

1 H																	2 He
3 Li	4 Be											5 B	6 C	7 N	8 O	9 F	10 Ne
11 Na	12 Mg											13 Al	14 Si	15 P	16 S	17 Cl	18 Ar
19 K	20 Ca	21 Sc	22 Ti	23 V	24 Cr	25 Mn	26 Fe	27 Co	28 Ni	29 Cu	30 Zn	31 Ga	32 Ge	33 As	34 Se	35 Br	36 Kr
37 Rb	38 Sr	39 Y	40 Zr	41 Nb	42 Mo	43 Tc	44 Ru	45 Rh	46 Pd	47 Ag	48 Cd	49 In	50 Sn	51 Sb	52 Te	53 I	54 Xe
55 Cs	56 Ba	57 La	72 Hf	73 Ta	74 W	75 Re	76 Os	77 Ir	78 Pt	79 Au	80 Hg	81 Tl	82 Pb	83 Bi	84 Po	85 At	86 Rn
87 Fr	88 Ra	89 Ac	104 Rf	105 Db	106 Sg	107 Bh	108 Hs	109 Mt	110 Ds	111 Rg	112 Cn	113 Uut	114 Fl	115 Uup	116 Lv	117 Uus	118 Uuo
			58 Ce	59 Pr	60 Nd	61 Pm	62 Sm	63 Eu	64 Gd	65 Tb	66 Dy	67 Ho	68 Er	69 Tm	70 Yb	71 Lu	
			90 Th	91 Pa	92 U	93 Np	94 Pu	95 Am	96 Cm	97 Bk	98 Cf	99 Es	100 Fm	101 Md	102 No	103 Lr	

FIGURE 9.5
The periodic table lists all the known elements.

Graphite is made up solely of carbon atoms. All of the elements are listed in a chart called the **periodic table**, shown in Figure 9.5.

As you can see from the periodic table, each element is designated by its **atomic symbol**, which comes from the letters of the element's name. For example, the atomic symbol for carbon is C, and the symbol for chlorine is Cl. In many cases, the atomic symbol is derived from the element's Latin name. Gold has the atomic symbol Au after its Latin name, *aurum*. Lead has the atomic symbol Pb after its Latin name, *plumbum* (Figure 9.6). Elements with symbols derived from Latin names are usually those that were discovered earliest.

Note that only the first letter of an atomic symbol is capitalized. The symbol for the element cobalt, for instance, is Co, but CO is a combination of two elements: carbon, C, and oxygen, O.

UNIFYING CONCEPT
- ***The Atomic Theory of Matter***

The terms *element* and *atom* are often used in a similar context. You might hear, for example, that gold is an element made of gold atoms. Generally, *element* is used in reference to an entire macroscopic or microscopic sample, and *atom* is used when speaking of the submicroscopic particles in the sample. The important distinction is that elements are made of atoms and not the other way around.

The number of atoms that arrange themselves in a unit of an element is shown by the *elemental formula*. For elements in which two or more atoms are bonded into molecules, the **elemental formula** is the chemical symbol followed by a

FIGURE 9.6
A plumb bob, a heavy weight attached to a string and used by carpenters and surveyors to establish a straight vertical line, gets it name from the lead (*plumbum*, Pb) that is still sometimes used as the weight. Plumbers got their name because they once worked with lead pipes. Because of lead's toxicity, copper or PVC (polyvinyl chloride) pipes are now used.

numeral subscript indicating the number of atoms in each molecule. For example, elemental nitrogen, as was shown in Figure 9.4, consists of molecules containing two nitrogen atoms per molecule. (As we discuss further in Chapter 11, a molecule is a group of atoms bonded together in some particular arrangement.) Thus, N_2 is the elemental formula for atmospheric nitrogen. Similarly, atmospheric oxygen has the elemental formula O_2, while the elemental formula for sulfur is S_8. For elements in which the basic units are individual atoms (no molecules), the elemental formula is simply the chemical symbol. This is the case for most elements. To name two examples, Au is the elemental formula for gold and Li is the elemental formula for lithium.

LEARNING OBJECTIVE
Describe the structure of the atomic nucleus and how the atomic mass of an element is calculated.

9.2 Protons and Neutrons

EXPLAIN THIS Why aren't we harmed by drinking heavy water, D_2O?

Let us take a closer look at the atom and investigate the particles found in the atomic nucleus. A **proton** carries a positive charge and is relatively heavy—nearly 2000 times as massive as an electron. The proton and electron have the same quantity of charge, but the opposite sign. The number of protons in the nucleus of any atom is equal to the number of electrons whirling about the nucleus. So, the opposite charges of protons and electrons balance each other, producing a zero net charge. For example, an oxygen atom has a total of eight protons and eight electrons and is thus electrically neutral.

Scientists have agreed to identify elements by **atomic number**, which is the number of protons in each atom of a given element. The modern periodic table lists the elements in order of increasing atomic number. Hydrogen, with one proton per atom, has atomic number 1; helium, with two protons per atom, has atomic number 2; and so on.

CHECK YOURSELF
How many protons are there in an iron atom, Fe (atomic number 26)?

CHECK YOUR ANSWER
The atomic number of an atom and its number of protons are the same. Thus, there are 26 protons in an iron atom. Another way to put this is that all atoms that contain 26 protons are, by definition, iron atoms.

If we compare the electric charges and masses of different atoms, we see that the atomic nucleus must be made up of more than just protons. Helium, for example, has twice the electric charge of hydrogen but four times the mass. The added mass is due to another subatomic particle found in the nucleus, the neutron. The **neutron** has about the same mass as the proton, but it has no electric charge. Any object that has no net electric charge is said to be electrically neutral, and that is where the neutron got its name. (We will discuss the important role that neutrons play in holding the atomic nucleus together in Chapter 10.)

Both protons and neutrons are called **nucleons**, a generic term that denotes their location in the atomic nucleus. Table 9.1 summarizes the basic facts about electrons, protons, and neutrons.

TABLE 9.1 SUBATOMIC PARTICLES

	Particle	Charge	Mass Compared to Electron	Actual Mass* (kg)
	Electron	−1	1	9.11×10^{-31}**
Nucleons	Proton	+1	1836	1.673×10^{-27}
	Neutron	0	1841	1.675×10^{-27}

*Not measured directly but calculated from experimental data.
**9.11×10^{-31} kg = 0.000000000000000000000000000000911 kg.

Isotopes and Atomic Mass

MasteringPhysics®
TUTORIAL: Atoms and Isotopes

For any element, the number of neutrons in the nucleus may vary. For example, most hydrogen atoms (atomic number 1) have no neutrons. A small percentage, however, have one neutron, and a smaller percentage have two neutrons. Similarly, most iron atoms (atomic number 26) have 30 neutrons, but a small percentage have 29 neutrons. Atoms of the same element that contain different numbers of neutrons are **isotopes** of one another.

We identify isotopes by their mass number, which is the total number of protons and neutrons (in other words, the number of nucleons) in the nucleus. As Figure 9.7 shows, a hydrogen isotope with only one proton is called hydrogen-1, where 1 is the mass number. A hydrogen isotope with one proton and one neutron is therefore hydrogen-2, and a hydrogen isotope with one proton and two neutrons is hydrogen-3. Similarly, an iron isotope with 26 protons and 30 neutrons is called iron-56, and one with only 29 neutrons is iron-55.

An alternative method of indicating isotopes is to write the mass number as a superscript and the atomic number as a subscript to the left of the atomic symbol. For example, an iron isotope with a mass number of 56 and an atomic number of 26 is written:

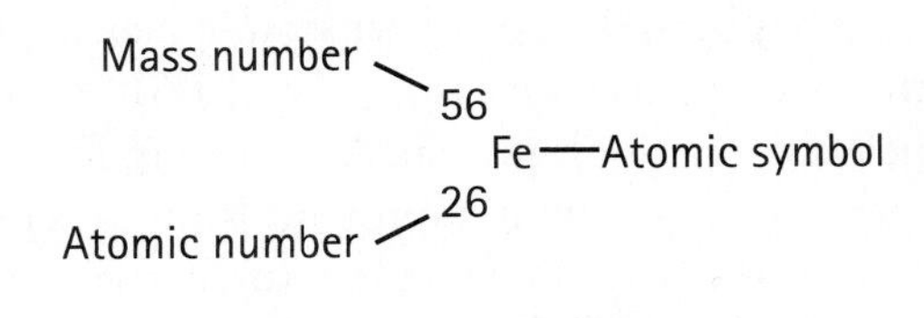

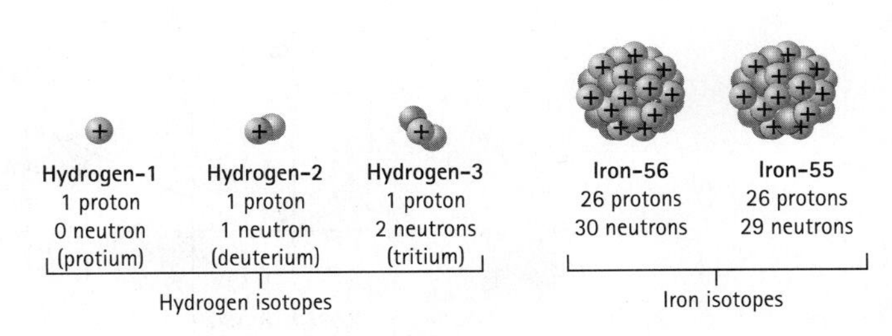

FIGURE 9.7
Isotopes of an element have the same number of protons but different numbers of neutrons and hence different mass numbers. The three hydrogen isotopes have special names: protium for hydrogen-1, deuterium for hydrogen-2, and tritium for hydrogen-3. Of these three isotopes, hydrogen-1 is most common. For most elements, such as iron, the isotopes have no special names and are indicated merely by mass number.

The total number of neutrons in an isotope can be calculated by subtracting its atomic number from its mass number:

$$\begin{array}{r} \text{mass number} \\ -\ \text{atomic number} \\ \hline \text{number of neutrons} \end{array}$$

For example, uranium-238 has 238 nucleons. The atomic number of uranium is 92, which tells us that 92 of these 238 nucleons are protons. The remaining 146 nucleons must be neutrons:

$$\begin{array}{l} \text{238 protons and neutrons} \\ -\ \text{92 protons} \\ \hline \text{146 neutrons} \end{array}$$

Atoms interact with one another electrically. Therefore the way any atom behaves in the presence of other atoms is determined largely by the charged particles it contains, especially its outer electrons. Isotopes of an element differ only by mass, not by electric charge. For this reason, isotopes of an element share many characteristics—in fact, as chemicals they cannot be distinguished from one another. For example, a sugar molecule containing carbon atoms with seven neutrons is digested no differently from a sugar molecule containing carbon atoms with six neutrons. Interestingly, about 1% of the carbon we consume in food is the carbon-13 isotope containing seven neutrons per nucleus. The remaining 99% of the carbon in our diet is the more common carbon-12 isotope containing six neutrons per nucleus.

UNIFYING CONCEPT

- ***The Atomic Theory of Matter***
  Section 9.1

Most water molecules, H_2O, consist of hydrogen atoms with no neutrons. The few that have neutrons, however, are heavier, and because of this difference they can be isolated. Such water is appropriately called "heavy water," which has the formula D_2O.

The total mass of an atom is called its **atomic mass**. This is the sum of the masses of all the atom's components (electrons and the nucleus). Because electrons are so much less massive than the nucleus, their contribution to atomic mass is negligible. A special unit has been developed for atomic masses. This is the *atomic mass unit* (amu), where 1 atomic mass unit is equal to 1.661×10^{-24} gram, which is slightly less than the mass of a single proton. As shown in Figure 9.8, the atomic masses listed in the periodic table are in atomic mass units. The atomic mass of an element as presented in the periodic table is actually the *average* atomic mass of its various isotopes. The Math Connection describes how these averages are calculated.

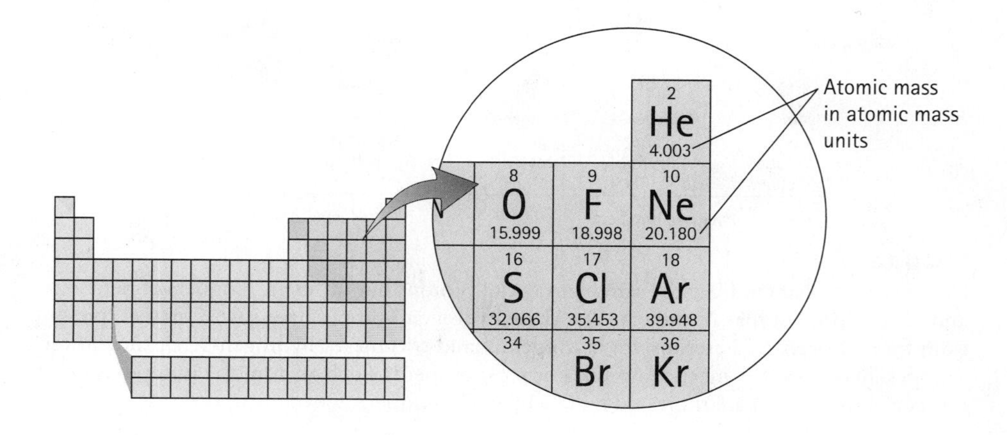

FIGURE 9.8
Helium, He, has an atomic mass of 4.003 amu, and neon, Ne, has an atomic mass of 20.180 amu.

MATH CONNECTION

Calculating Atomic Mass

About 99% of all carbon atoms are the isotope carbon-12, and most of the remaining 1% are the heavier isotope carbon-13. This small amount of carbon-13 raises the average mass of carbon from 12.000 amu to the slightly greater value of 12.011 amu.

To arrive at the atomic mass presented in the periodic table, first multiply the mass of each naturally occurring isotope of an element by the fraction of its abundance and then add up all the fractions.

Problem Carbon-12 has a mass of 12.0000 amu and makes up 98.89% of naturally occurring carbon. Carbon-13 has a mass of 13.0034 amu and makes up 1.11% of naturally occurring carbon. Use this information to show that the atomic mass of carbon shown in the periodic table, 12.011 amu, is correct.

Solution Recognize that 98.89% and 1.11% expressed as decimals are 0.9889 and 0.0111, respectively.

	Contributing Mass of ^{12}C	Contributing Mass of ^{13}C	
Fraction of Abundance	0.9889	0.0111	step 1
Mass (amu)	× 12.0000	× 13.0034	
	11.867	0.144	

$$\text{Atomic mass} = 11.867 + 0.144 = 12.011 \qquad \text{step 2}$$

Try the following exercise. Chlorine-35 has a mass of 34.97 amu, and chlorine-37 has a mass of 36.95 amu. Determine the atomic mass of chlorine, Cl (atomic number 17), if 75.53% of all chlorine atoms are the chlorine-35 isotope and 24.47% are the chlorine-37 isotope. (See Figure 9.8 for the answer.)

CHECK YOURSELF

Distinguish between mass number and atomic mass.

CHECK YOUR ANSWER

Both terms include the word *mass* and so are easily confused. Focus your attention on the second word of each term, however, and you'll get it right every time. Mass number is a count of the *number* of nucleons in an isotope. An atom's mass number requires no units because it is simply a count. Atomic mass is a measure of the total *mass* of an atom, which is given in atomic mass units.

9.3 The Periodic Table

LEARNING OBJECTIVE

Interpret how elements are organized in the periodic table.

EXPLAIN THIS How is the periodic table like a dictionary?

The periodic table is a listing of all the known elements with their atomic masses, atomic numbers, and atomic symbols. But the periodic table contains much more information. The way the table is organized, for example, tells us much about the elements' properties. Let's look at how the elements are grouped as metals, nonmetals, and metalloids.

As shown in Figure 9.9, most of the known elements are metals, which are defined as elements that are shiny, opaque, and good conductors of electricity and heat. Metals are *malleable*, which means they can be hammered into different shapes or bent without breaking. They are also *ductile*, which means they can be drawn into wires. All but a few metals are solid at room temperature. The exceptions are mercury, Hg; gallium, Ga; cesium, Cs; and francium, Fr, which are all liquids at a warm room temperature of 30°C (86°F). Another interesting exception is hydrogen, H, which takes on the properties of a liquid metal only at very high pressures (Figure 9.10). Under normal conditions, hydrogen behaves like a nonmetallic gas.

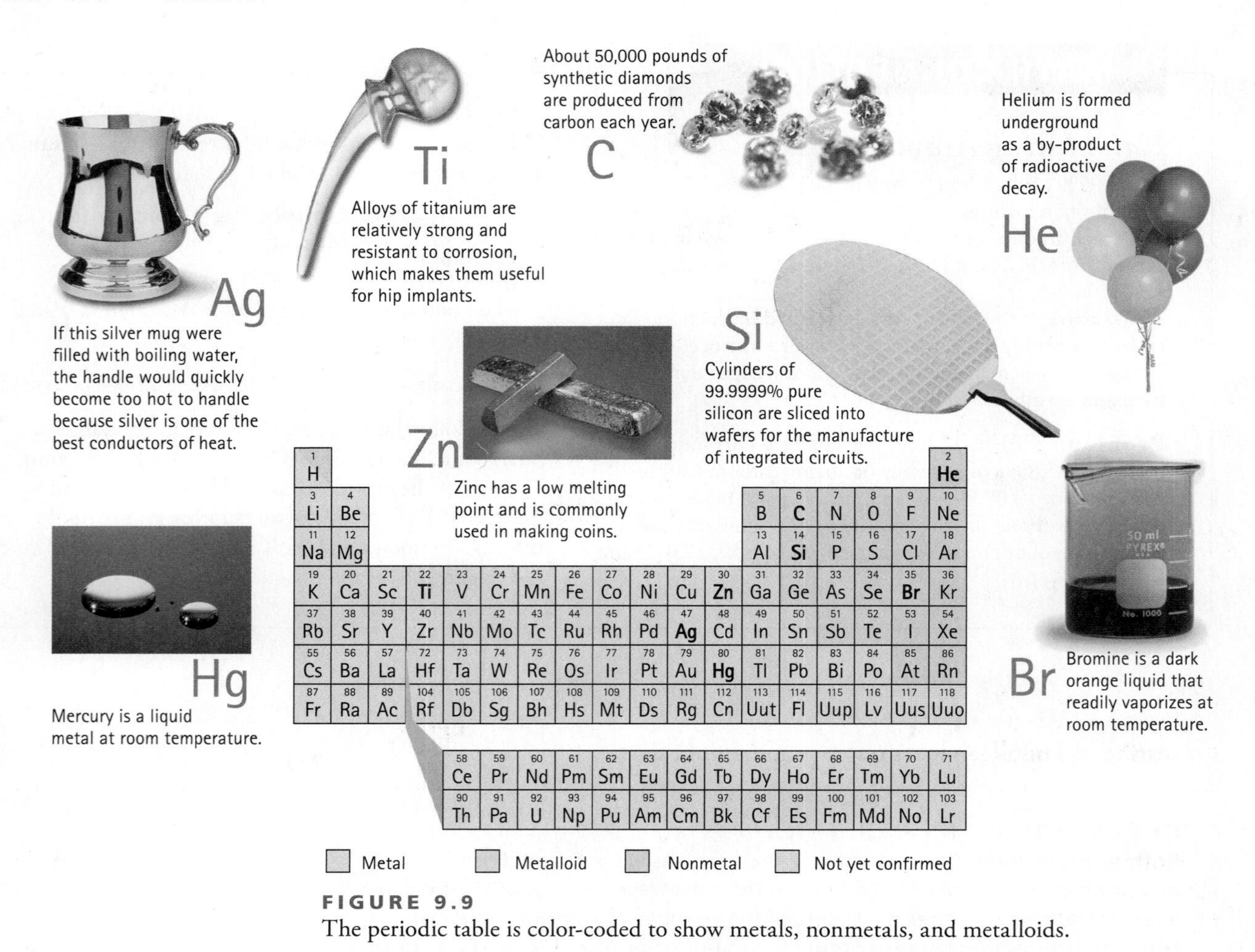
About 50,000 pounds of synthetic diamonds are produced from carbon each year.
C
Ti
Alloys of titanium are relatively strong and resistant to corrosion, which makes them useful for hip implants.
Helium is formed underground as a by-product of radioactive decay.
He
Ag
If this silver mug were filled with boiling water, the handle would quickly become too hot to handle because silver is one of the best conductors of heat.
Si
Cylinders of 99.9999% pure silicon are sliced into wafers for the manufacture of integrated circuits.
Zn
Zinc has a low melting point and is commonly used in making coins.
Hg
Mercury is a liquid metal at room temperature.
Br
Bromine is a dark orange liquid that readily vaporizes at room temperature.
1 H
2 He
3 Li 4 Be 5 B 6 C 7 N 8 O 9 F 10 Ne
11 Na 12 Mg 13 Al 14 Si 15 P 16 S 17 Cl 18 Ar
19 K 20 Ca 21 Sc 22 Ti 23 V 24 Cr 25 Mn 26 Fe 27 Co 28 Ni 29 Cu 30 Zn 31 Ga 32 Ge 33 As 34 Se 35 Br 36 Kr
37 Rb 38 Sr 39 Y 40 Zr 41 Nb 42 Mo 43 Tc 44 Ru 45 Rh 46 Pd 47 Ag 48 Cd 49 In 50 Sn 51 Sb 52 Te 53 I 54 Xe
55 Cs 56 Ba 57 La 72 Hf 73 Ta 74 W 75 Re 76 Os 77 Ir 78 Pt 79 Au 80 Hg 81 Tl 82 Pb 83 Bi 84 Po 85 At 86 Rn
87 Fr 88 Ra 89 Ac 104 Rf 105 Db 106 Sg 107 Bh 108 Hs 109 Mt 110 Ds 111 Rg 112 Cn 113 Uut 114 Fl 115 Uup 116 Lv 117 Uus 118 Uuo
58 Ce 59 Pr 60 Nd 61 Pm 62 Sm 63 Eu 64 Gd 65 Tb 66 Dy 67 Ho 68 Er 69 Tm 70 Yb 71 Lu
90 Th 91 Pa 92 U 93 Np 94 Pu 95 Am 96 Cm 97 Bk 98 Cf 99 Es 100 Fm 101 Md 102 No 103 Lr
Metal
Metalloid
Nonmetal
Not yet confirmed

FIGURE 9.9
The periodic table is color-coded to show metals, nonmetals, and metalloids.

FIGURE 9.10
Geoplanetary models suggest that hydrogen exists as a liquid metal deep beneath the surfaces of Jupiter (shown here) and Saturn. These planets are composed mostly of hydrogen. Inside them, the pressure exceeds 3 million times Earth's atmospheric pressure. At this tremendously high pressure, hydrogen is pressed to a liquid-metal phase.

The nonmetallic elements, with the exception of hydrogen, are on the right side of the periodic table. Nonmetals are very poor conductors of electricity and heat, and may also be transparent. Solid nonmetals are neither malleable nor ductile. Rather, they are brittle and shatter when hammered. At 30°C (86°F), some nonmetals are solid (carbon, C), others are liquid (bromine, Br), and still others are gaseous (helium, He).

Six elements are classified as metalloids: boron, B; silicon, Si; germanium, Ge; arsenic, As; antimony, Sb; and tellurium, Te. Situated between the metals and the nonmetals in the periodic table, the metalloids have both metallic and nonmetallic characteristics. For example, these elements are weak conductors of electricity, which makes them useful as semiconductors in the integrated circuits of computers. Note from the periodic table how germanium, Ge (atomic number 32), is closer to the metals than to the nonmetals. Because of this positioning, we can deduce that germanium has more metallic properties than silicon, Si (atomic number 14), and is a slightly better conductor of electricity. So, we find that integrated circuits fabricated with germanium operate faster than those fabricated with silicon. Because silicon is much more abundant and less expensive to obtain, however, silicon computer chips remain the industry standard.

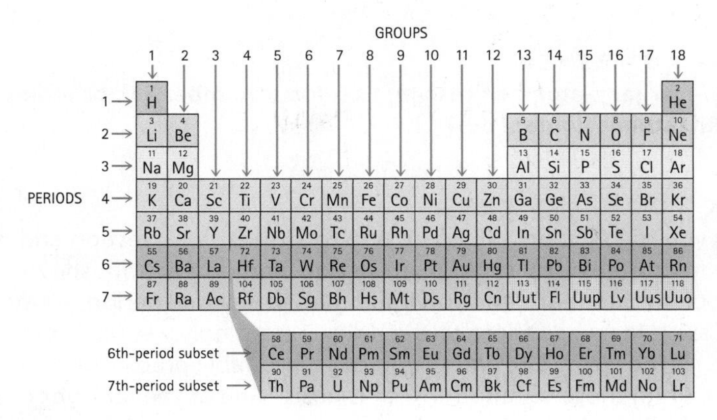

Please put to rest any fear you may have about needing to memorize the periodic table, or even parts of it—better to focus on the many great concepts behind its organization.

FIGURE 9.11
The 7 periods (horizontal rows) and 18 groups (vertical columns) of the periodic table. Note that not all periods contain the same number of elements. Also note that, for reasons explained later, the sixth and seventh periods each include a subset of elements, which are listed apart from the main body.

Periods and Groups

Two other important ways in which the elements are organized in the periodic table are by horizontal rows and vertical columns. Each horizontal row is called a **period**, and each vertical column is called a **group** (or sometimes a *family*). As shown in Figure 9.11, there are 7 periods and 18 groups.

Across any period, the properties of elements gradually change. This gradual change is called a periodic trend. As is shown in Figure 9.12, one periodic trend is that atomic size tends to decrease as you move from left to right across any period. Note that the trend repeats from one horizontal row to the next. This phenomenon of repeating trends is called *periodicity*, a term used to indicate that the trends recur in cycles. Each horizontal row is called a period because it corresponds to one full cycle of a trend.

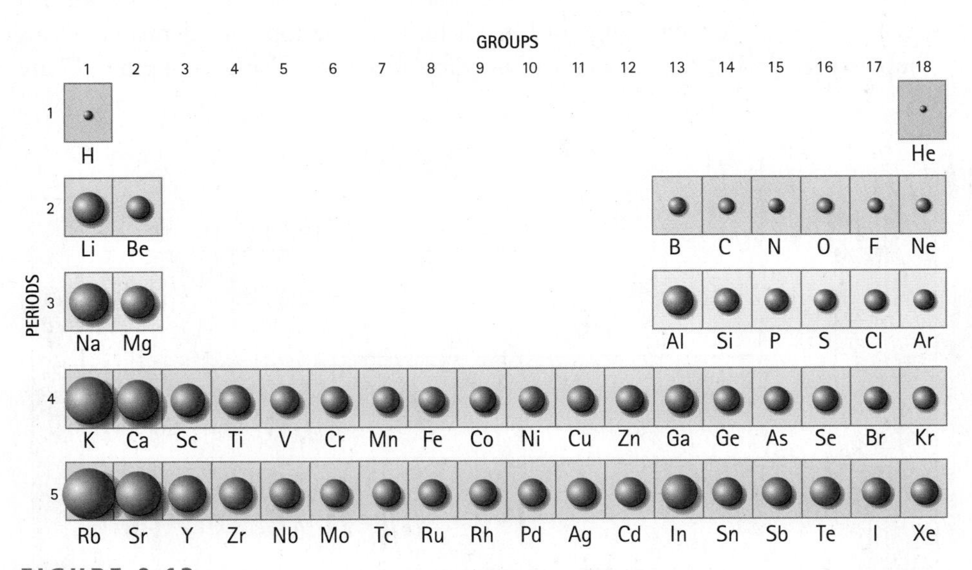

FIGURE 9.12
The size of atoms gradually decreases as we move from left to right across any period. Atomic size is a periodic (repeating) property.

FYI The *carat* is the common unit used to describe the mass of a gem. A 1-carat diamond, for example, has a mass of 0.20 g. The *karat* is the common unit used to describe the purity of a precious metal, such as gold. A 24-karat gold ring is as pure as can be. A gold ring that is 50% pure is 12 karat.

CHECK YOURSELF

Which are larger—atoms of cesium, Cs (atomic number 55), or atoms of radon, Rn (atomic number 86)?

CHECK YOUR ANSWER

Perhaps you tried looking at Figure 9.12 to answer this question and quickly became frustrated because the sixth-period elements are not shown. Well, relax. Look at the trends and you'll see that, in any one period, all atoms to the left are larger than those to the right. Accordingly, cesium is positioned at the far left of period 6, and so you can reasonably predict that its atoms are larger than those of radon, which is positioned at the far right of period 6. The periodic table is a road map to understanding the elements.

FYI As the tungsten filament inside a lightbulb is heated, minute particles of tungsten evaporate. Over time, these particles are deposited on the inner surface of the bulb, causing the bulb to blacken. As it loses its tungsten, the filament eventually breaks and the bulb "burns out." A remedy is to replace the air inside the bulb with a halogen gas, such as iodine or bromine. In such a *halogen* bulb, the evaporated tungsten combines with the halogen rather than depositing on the bulb, which remains clear. Furthermore, the tungsten becomes unstable and splits from the halogen when it touches the hot filament. The halogen returns as a gas while the tungsten is deposited onto the filament, thereby restoring the filament. This is why halogen lamps have such long lifetimes.

Down any group (vertical column), the properties of elements tend to be remarkably similar, which is why these elements are said to be "grouped" or "in a family." As Figure 9.13 shows, several groups have traditional names that describe the properties of their elements. Early in human history, people discovered that ashes mixed with water produce a slippery solution useful for removing grease. By the Middle Ages, such mixtures were described as being *alkaline*, a term derived from the Arabic word for ashes, *al-qali*. Alkaline mixtures found many uses, particularly in the preparation of soaps (Figure 9.14). We now know that alkaline ashes contain compounds of group 1 elements, most notably potassium carbonate, also known as potash. Because of this history, group 1 elements, which are metals, are called the *alkali metals*.

Elements of group 2 also form alkaline solutions when mixed with water. Furthermore, medieval alchemists noted that certain minerals (which we now know are made up of group 2 elements) do not melt or change when put into fire. These fire-resistant substances were known to the alchemists as "earth." As a holdover from these ancient times, group 2 elements are known as the *alkaline-earth metals*.

Toward the right side of the periodic table, elements of group 16 are known as the *chalcogens* ("ore-forming" in Greek) because the top two elements of this group, oxygen and sulfur, are so commonly found in ores. Elements of group 17 are

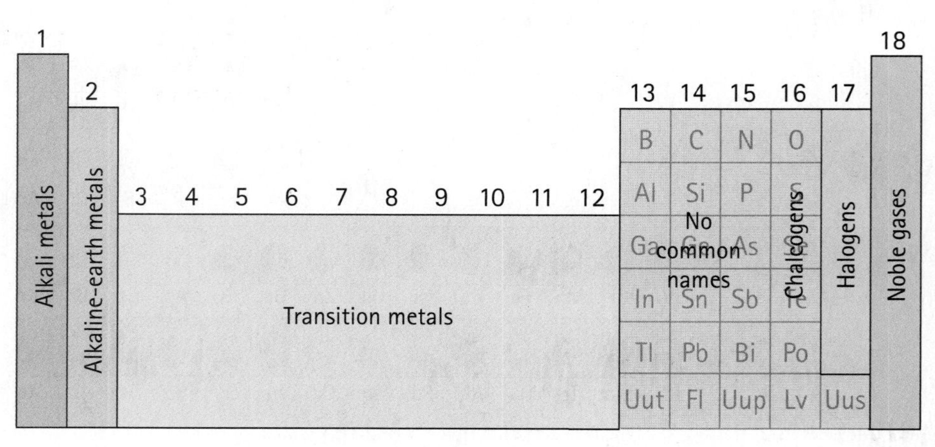

FIGURE 9.13
The common names for various groups of elements.

known as the *halogens* ("salt-forming" in Greek) because of their tendency to form various salts. Group 18 elements are all unreactive gases that tend not to combine with other elements. For this reason, they are called the *noble gases*, presumably because the nobility of earlier times were above interacting with common folk.

The elements of groups 3 through 12 are all metals that do not form alkaline solutions with water. These metals tend to be harder than the alkali metals and less reactive with water; hence they are used for structural purposes. Collectively they are known as the *transition metals*, a name that denotes their central position in the periodic table. The transition metals include some of the most familiar and important elements—iron, Fe; copper, Cu; nickel, Ni; chromium, Cr; silver, Ag; and gold, Au. They also include many lesser-known elements that are nonetheless important in modern technology. People with hip implants appreciate the transition metals titanium, Ti; molybdenum, Mo; and manganese, Mn, because these noncorrosive metals are used in implant devices.

FIGURE 9.14
Ashes and water make a slippery alkaline solution once used to clean hands.

CHECK YOURSELF

The elements copper, Cu; silver, Ag; and gold, Au, are three of the few metals that can be found naturally in their elemental state. These three metals have found great use as coins and jewelry for a number of reasons, including their resistance to corrosion and their remarkable colors. How is the fact that these metals have similar properties reflected in the periodic table?

CHECK YOUR ANSWER

Copper (atomic number 29), silver (atomic number 47), and gold (atomic number 79) are all in the same group in the periodic table (group 11), which suggests they should have similar—though not identical—properties.

A uranium atom is 40 times as heavy as a lithium atom, but only slightly larger because its more highly charged nucleus pulls harder on its electrons. But it has more electrons to pull, a balancing act that barely changes the atom's size.

Within the sixth period is a subset of 14 metallic elements (atomic numbers 58 to 71) that are quite unlike any of the other transition metals. A similar subset (atomic numbers 90 to 103) is found within the seventh period. These two subsets are the *inner transition metals*. Inserting the inner transition metals into the main body of the periodic table as in Figure 9.15 results in a long and cumbersome table. So that the table can fit nicely on a standard paper size, these elements are commonly placed below the main body of the table, as shown in Figure 9.16.

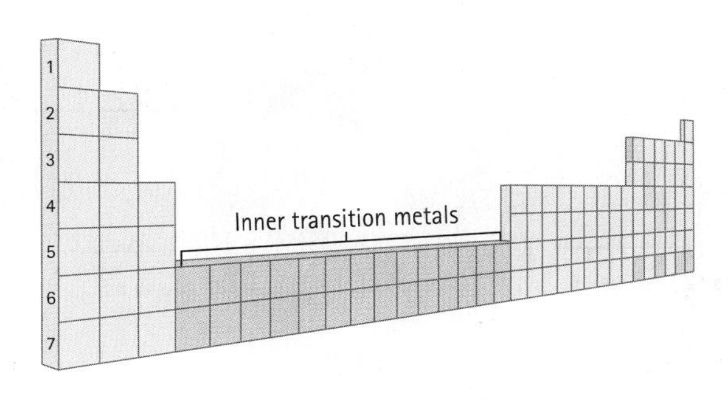

FIGURE 9.15
Inserting the inner transition metals between atomic groups 3 and 4 results in a periodic table that is not easy to fit on a standard sheet of paper.

FIGURE 9.16
The typical display of the inner transition metals. The count of elements in the sixth period goes from lanthanum (La, 57) to cerium (Ce, 58) on through to lutetium (Lu, 71) and then back to hafnium (Hf, 72). A similar jump is made in the seventh period.

PERIODS																		
1	1 H																	2 He
2	3 Li	4 Be											5 B	6 C	7 N	8 O	9 F	10 Ne
3	11 Na	12 Mg											13 Al	14 Si	15 P	16 S	17 Cl	18 Ar
4	19 K	20 Ca	21 Sc	22 Ti	23 V	24 Cr	25 Mn	26 Fe	27 Co	28 Ni	29 Cu	30 Zn	31 Ga	32 Ge	33 As	34 Se	35 Br	36 Kr
5	37 Rb	38 Sr	39 Y	40 Zr	41 Nb	42 Mo	43 Tc	44 Ru	45 Rh	46 Pd	47 Ag	48 Cd	49 In	50 Sn	51 Sb	52 Te	53 I	54 Xe
6	55 Cs	56 Ba	57 La	72 Hf	73 Ta	74 W	75 Re	76 Os	77 Ir	78 Pt	79 Au	80 Hg	81 Tl	82 Pb	83 Bi	84 Po	85 At	86 Rn
7	87 Fr	88 Ra	89 Ac	104 Rf	105 Db	106 Sg	107 Bh	108 Hs	109 Mt	110 Ds	111 Rg	112 Cn	113 Uut	114 Fl	115 Uup	116 Lv	117 Uus	118 Uuo

Inner transition metals

Lanthanides	58 Ce	59 Pr	60 Nd	61 Pm	62 Sm	63 Eu	64 Gd	65 Tb	66 Dy	67 Ho	68 Er	69 Tm	70 Yb	71 Lu
Actinides	90 Th	91 Pa	92 U	93 Np	94 Pu	95 Am	96 Cm	97 Bk	98 Cf	99 Es	100 Fm	101 Md	102 No	103 Lr

FYI From 1943 to 1986, the Hanford nuclear facility in central Washington state produced 72 tons of plutonium, nearly two-thirds the nation's supply. Creating this much plutonium generated an estimated 450 billion gallons of radioactive and hazardous liquids, which were discharged into the local environment. Today, some 53 million gallons of high-level radioactive and chemical wastes are stored in 177 underground tanks, many of them leaking into the groundwater.

The sixth-period inner transition metals are called the *lanthanides* because they fall after lanthanum, La. Because of their similar physical and chemical properties, they tend to occur mixed together in the same locations on Earth. Also because of their similarities, lanthanides are unusually difficult to purify. Recently, the commercial use of lanthanides has increased. Several lanthanide elements, for example, are used in the fabrication of the light-emitting diodes (LEDs) of computer monitors and flat-screen televisions.

The seventh-period inner transition metals are called the *actinides* because they fall after actinium, Ac. They, too, all have similar properties and hence are not easily purified. The nuclear power industry faces this obstacle because it requires purified samples of two of the most publicized actinides: uranium, U, and plutonium, Pu. Actinides heavier than uranium are not found in nature but are synthesized in the laboratory.

LEARNING OBJECTIVE
Distinguish between models that describe physical attributes and models that describe the behavior of a system.

Integrated Science 9B
BIOLOGY AND EARTH SCIENCE

Physical and Conceptual Models

EXPLAIN THIS How do we predict the behavior of atoms?

Atoms are so small that the number of them in a baseball is roughly equal to the number of Ping-Pong balls that could fit inside a hollow sphere as big as Earth, as Figure 9.17 illustrates. This number is incredibly large—beyond our intuitive grasp. Atoms are so incredibly small that we can never see

FIGURE 9.17
If Earth were filled with nothing but Ping-Pong balls, the number of balls would be roughly equal to the number of atoms in a baseball. Put differently, if a baseball were the size of Earth, one of its atoms would be the size of a Ping-Pong ball.

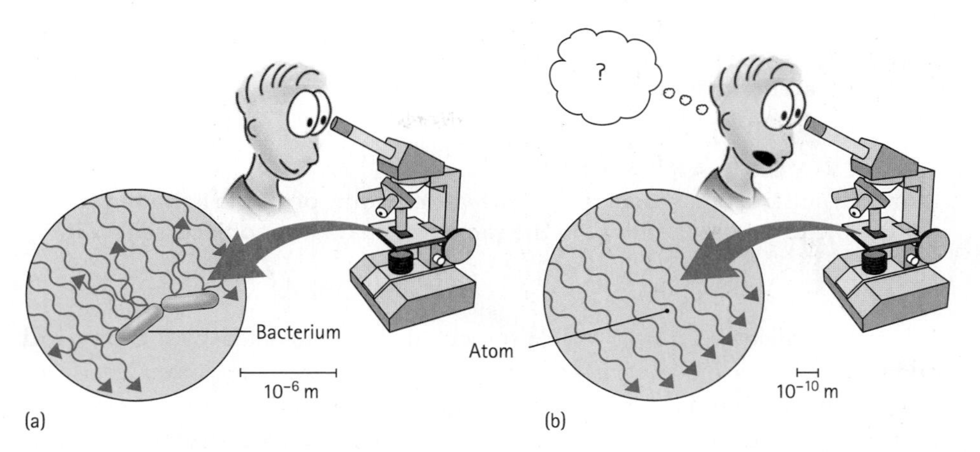

FIGURE 9.18
Microscopic objects can be seen through a microscope that works with visible light, but submicroscopic particles cannot. (a) A bacterium is visible because it is larger than the wavelengths of visible light. We can see the bacterium through the microscope because the bacterium reflects visible light. (b) An atom is invisible because it is smaller than the wavelengths of visible light and so does not reflect the light toward our eyes.

them in the usual sense. This is because light travels in waves, and atoms are smaller than the wavelengths of visible light, which is the light that allows the human eye to see things. As illustrated in Figure 9.18, the diameter of an object visible under the highest magnification of a microscope must be larger than the wavelengths of visible light.

Although we cannot see atoms *directly*, we can generate images of them *indirectly*. In the mid-1980s, researchers developed the *scanning probe microscope*, which produces images by dragging an ultrathin needle back and forth over the surface of a sample. Bumps the size of atoms on the surface cause the needle to move up and down. This vertical motion is detected and translated by a computer into a topographical image that corresponds to the positions of atoms on the surface (Figure 9.19). A scanning probe microscope can also be used to push individual atoms into desired positions. This ability opened the field of nanotechnology, which we will discuss further in Chapter 11.

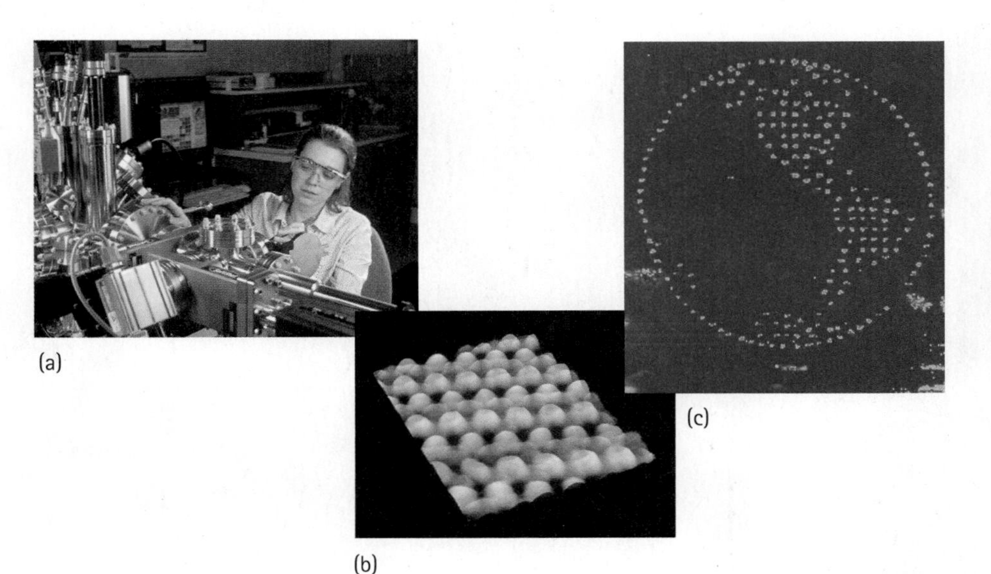

FIGURE 9.19
(a) Scanning probe microscopes are relatively simple devices used to create submicroscopic imagery. (b) An image of gallium and arsenic atoms obtained with a scanning probe microscope. (c) Each dot in the world's tiniest map consists of a few thousand gold atoms, each atom moved into its proper place by a scanning probe microscope.

CHECK YOURSELF
Why are atoms invisible?

CHECK YOUR ANSWER
An individual atom is smaller than the wavelengths of visible light and so is unable to reflect that light in the same way that larger objects reflect it.

A very small or very large visible object can be represented with a **physical model**, which is a model that replicates the object at a more convenient scale. Figure 9.20a, for instance, shows a large-scale physical model of a microorganism that a biology student uses to study the microorganism's internal structure. Because atoms are invisible, however, we cannot use a physical model to represent them. In other words, we cannot simply scale up the atom to a larger size, as we might with a microorganism. (A scanning probe microscope merely shows the *positions* of atoms and not actual images of atoms, which do not have the solid surfaces implied in the scanning probe image of Figure 9.19b.) So, rather than describing the atom with a physical model, chemists use what is known as a **conceptual model**, which describes a system. The more accurate a conceptual model, the more accurately it predicts the behavior of the system. The weather is best described using a conceptual model like the one shown in Figure 9.20b. Such a model shows how the various components of the system—humidity, atmospheric pressure, temperature, electric charge, the motion of large masses of air—interact with one another. Other systems that can be described by conceptual models are the economy, population growth, the spread of diseases, and team sports.

CHECK YOURSELF
A basketball coach describes a playing strategy to her team by way of sketches on a game board. Do the illustrations represent a physical model or a conceptual model?

FIGURE 9.20
(a) This large-scale model of a microorganism is a physical model. (b) Weather forecasters rely on conceptual models such as this one to predict the behavior of weather systems.

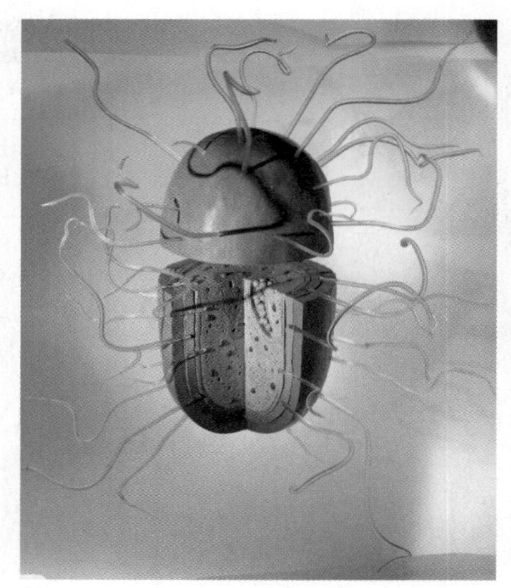

(a)

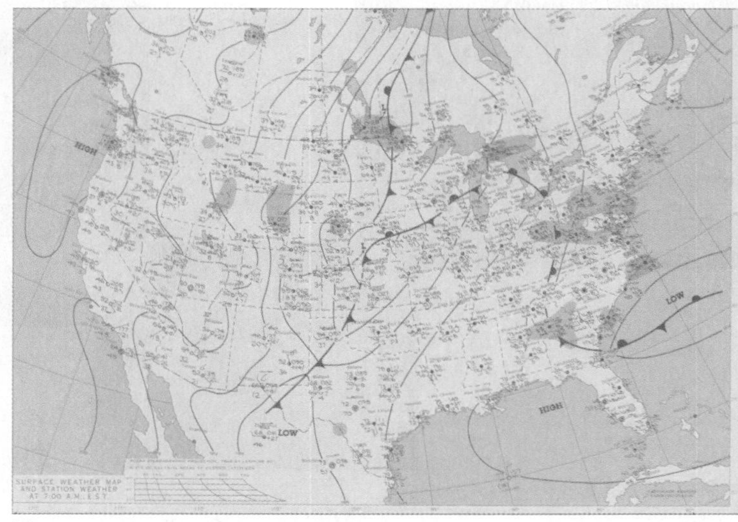

(b)

CHECK YOUR ANSWER
The sketches are a conceptual model the coach uses to describe a system (the players on the court), with the hope of predicting an outcome (winning the game).

Like the weather, the atom is a complex system of interacting components, and it is best described with a conceptual model. Thus, you should be careful not to interpret any visual representation of an atomic conceptual model as a re-creation of an actual atom. In Section 9.5, for example, you will be introduced to the planetary model of the atom, in which electrons are shown orbiting the atomic nucleus much as planets orbit the Sun. This planetary model is limited, however, in that it fails to explain many properties of atoms. Thus newer and more accurate (and more complicated) conceptual models of the atom have since been introduced. In these models, electrons appear as a cloud hovering around the atomic nucleus, but even these models have their limitations. Ultimately, the best models of the atom are purely mathematical. We can't "see" an atom because it is too small. We can't see the farthest star either. There's much that we can't see. But that doesn't prevent us from thinking about such things or even collecting indirect evidence.

UNIFYING CONCEPT
- ***The Scientific Method***
Section 1.3

In this book, our focus is on conceptual atomic models that are easily represented by visual images, including the planetary model and a model in which electrons are grouped in units called shells. Despite their limitations, such images are excellent guides to learning about the behavior of atoms, especially for the beginning student. As we discuss in the following sections, scientists developed these models to help explain how atoms emit light.

9.4 Identifying Atoms Using the Spectroscope

LEARNING OBJECTIVE
Describe how an atom reveals its identity by the light it emits.

EXPLAIN THIS How is it possible to tell what stars are made of when they are so very far away?

Recall from Chapter 8 that we see white light when all the frequencies of visible light reach our eye at the same time. By passing white light through a prism or through a diffraction grating, we can separate the color components of the light, as shown in Figure 9.21. (Remember—each color

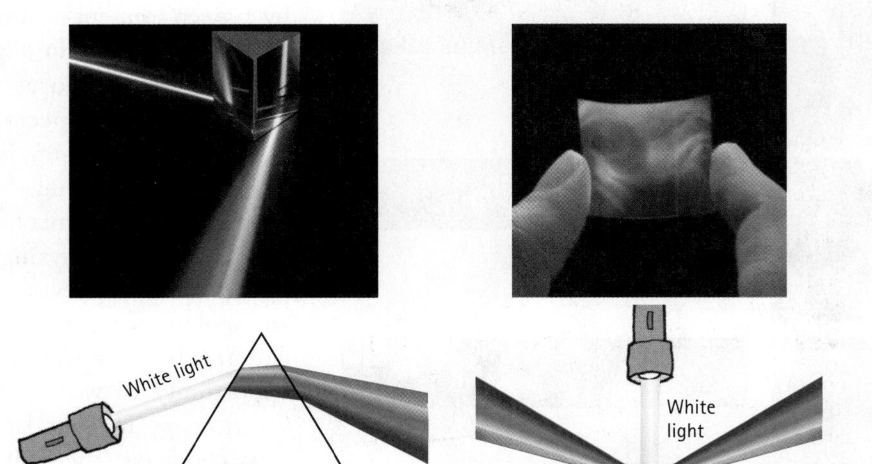

FIGURE 9.21
White light is separated into its color components by (a) a prism and (b) a diffraction grating.

FIGURE 9.22

INTERACTIVE FIGURE 

(a) In a spectroscope, light emitted by atoms passes through a narrow slit before being separated into particular frequencies by a prism or (as shown here) a diffraction grating. (b) This is what the eye sees when the slit of a diffraction-grating spectroscope is pointed toward a white-light source. Spectra of colors appear to the left and right of the slit.

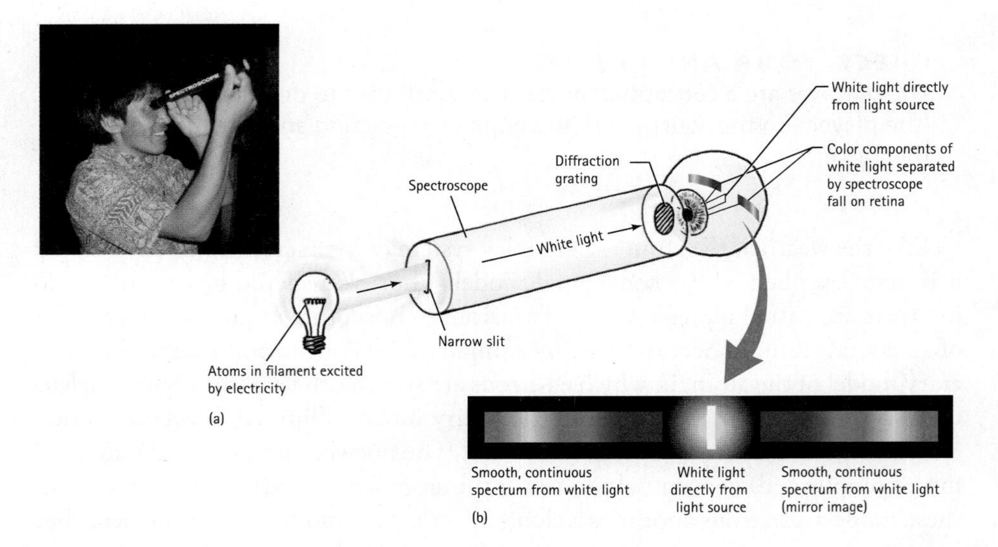

of visible light corresponds to a different frequency.) A **spectroscope**, shown in Figure 9.22, is an instrument used to observe the color components of any light source. The spectroscope allows us to analyze the light emitted by elements as they are made to glow.

Light is given off by atoms subjected to various forms of energy, such as heat or electricity. The atoms of a given element emit only certain frequencies of light, however. As a consequence, each element emits a distinctive glow when energized. Sodium atoms emit bright yellow light, which makes them useful as the light source in streetlamps because our eyes are very sensitive to yellow light. To name just one more example, neon atoms emit a brilliant red-orange light, which makes them useful as the light source in neon signs.

When we view the light from a glowing element through a spectroscope, we see that the light consists of a number of discrete (separate from one another) frequencies rather than a continuous spectrum like the one shown in Figure 9.22. The pattern of frequencies formed by a given element—some of which are shown in Figure 9.23—is referred to as that element's **atomic spectrum**. The atomic spectrum is an element's fingerprint. You can identify the elements in a light source by analyzing the light through a spectroscope and looking for characteristic patterns. If you don't have the opportunity to work with a spectroscope in your laboratory, check out the activities at the end of this chapter.

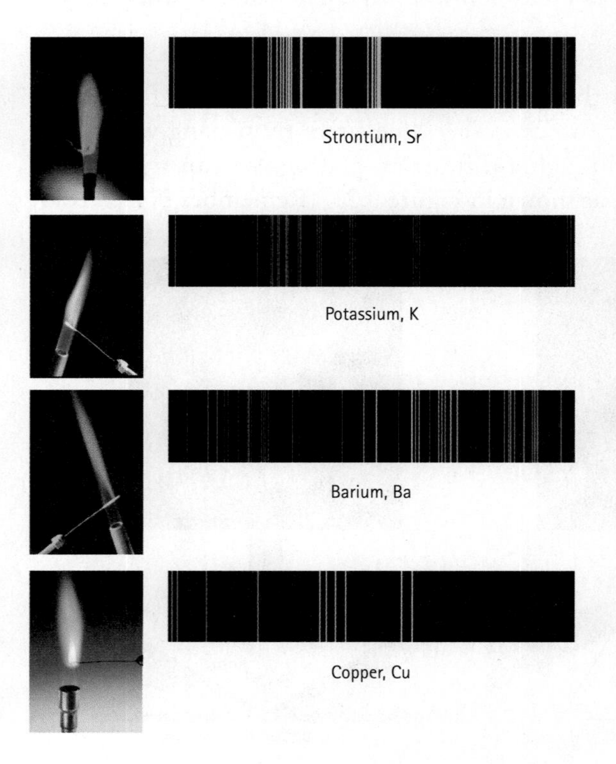

FIGURE 9.23

Elements heated by a flame glow their characteristic color. This is commonly called a flame test and is used to test for the presence of an element in a sample. When viewed through a spectroscope, the color of each element is revealed to consist of a pattern of distinct frequencies known as an atomic spectrum.

CHECK YOURSELF
How could you deduce the elemental composition of a star?

CHECK YOUR ANSWER
Aim a well-built spectroscope at the star and study its spectral patterns. In the late 1800s, this was done with our own star, the Sun. Spectral patterns of hydrogen and some other known elements were observed, in addition to one pattern that could not be identified. Scientists concluded that this unidentified pattern belonged to an element not yet discovered on Earth. They named this element helium after the Greek word for "sun," *helios*.

FYI A star's age is revealed by its elemental makeup. The first and oldest stars were composed of hydrogen and helium because those were the only elements available at that time. Heavier elements were produced after many of these early stars exploded in supernovae. Later stars incorporated these heavier elements in their formation. In general, the younger a star, the greater amounts of these heavier elements it contains.

9.5 The Quantum Hypothesis

LEARNING OBJECTIVE
Recount how the quantum nature of energy led to Bohr's planetary model of the atom.

EXPLAIN THIS Why do atomic spectra contain only a limited number of light frequencies?

An important step toward our present understanding of atoms and their spectra was taken by the German physicist Max Planck (1858–1947). In 1900, Planck hypothesized that light energy is *quantized* in much the same way matter is. The mass of a gold brick, for example, equals some whole-number multiple of the mass of a single gold atom. Similarly, an electric charge is always some whole-number multiple of the charge on a single electron. Mass and electric charge are therefore said to be *quantized* in that they consist of some number of fundamental units.

Planck identified each discrete parcel of light energy as a **quantum**, which is represented in Figure 9.24. A few years later, Einstein recognized that a quantum of light energy behaves much like a tiny particle of matter. To emphasize its particulate nature, each quantum of light was called a *photon*, a name coined because of its similarity to the word *electron*.

MasteringPhysics®
TUTORIAL: Bohr's Shell Model

Using Planck's quantum hypothesis, the Danish scientist Niels Bohr (1885–1962) explained the formation of atomic spectra as follows. First, an electron has more potential energy when it is farther from the nucleus. This is analogous to the greater potential energy an object has when it is held higher above the ground. Second, Bohr recognized that when an atom absorbs a photon of light, it is absorbing energy. This energy is acquired by one of the electrons. Because this electron has gained energy, it must move away from the nucleus.

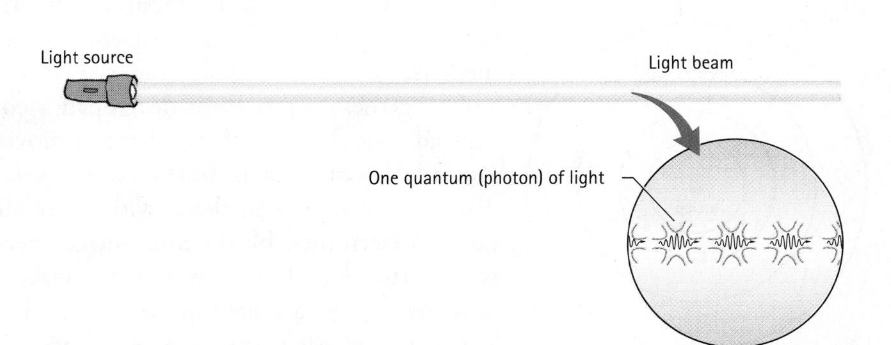

FIGURE 9.24
Light is quantized, which means it consists of a stream of energy packets. Each packet is called a quantum, also known as a photon.

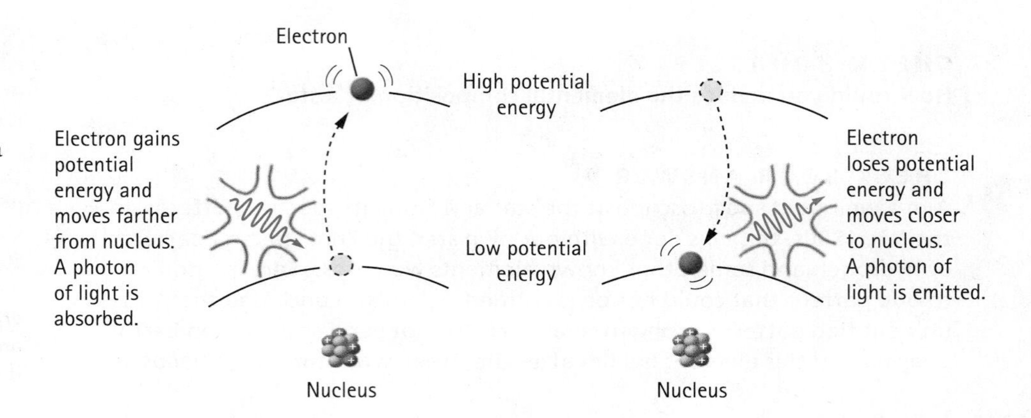

FIGURE 9.25
An electron is lifted away from the nucleus as the atom it is in absorbs a photon of light and drops closer to the nucleus as the atom releases a photon of light.

Bohr also realized that the opposite is true: When a high-potential-energy electron in an atom loses some of its energy, the electron moves closer to the nucleus and the energy lost from the electron is emitted from the atom as a photon of light. Both absorption and emission are illustrated in Figure 9.25.

CHECK YOURSELF
What is released as an electron makes a transition from a higher to a lower energy level?

CHECK YOUR ANSWER
A photon of light is released.

Bohr reasoned that because light energy is quantized, the energy of an electron in an atom must also be quantized. In other words, an electron cannot have just any amount of potential energy. Rather, within the atom there must be a number of distinct energy levels, analogous to steps on a staircase. Where you are on a staircase is restricted to where the steps are—you cannot stand at a height that is, say, halfway between any two adjacent steps. Similarly, an atom has only a limited number of permitted energy levels, and an electron can never have an amount of energy between these permitted energy levels. Bohr gave each energy level a **quantum number** n, where n is always some integer. The lowest energy level has a principal quantum number $n = 1$. An electron for which $n = 1$ is as close to the nucleus as possible, and an electron for which $n = 2$, $n = 3$, and so forth is farther away, in a stepwise fashion, from the nucleus.

Using these ideas, Bohr developed a conceptual model in which an electron moving around the nucleus is restricted to certain distances from the nucleus, with these distances determined by the amount of energy the electron has. Bohr saw this as similar to how the planets are held in orbit around the Sun at given distances from the Sun. The

FIGURE 9.26
Bohr's planetary model of the atom, in which electrons orbit the nucleus much as planets orbit the Sun, is a graphical representation that helps us understand how electrons can possess only certain quantities of energy.

allowed energy levels for any atom, therefore, could be graphically represented as orbits around the nucleus, as shown in Figure 9.26. Bohr's quantized model of the atom thus became known as the *planetary model.*

Bohr used his planetary model to explain why atomic spectra contain only a limited number of light frequencies, as shown in Figure 9.27. According to the model, photons are emitted by atoms as electrons move from higher-energy outer orbits to lower-energy inner orbits. The energy of an emitted photon is equal to the difference in energy between the two orbits. Because an electron is restricted to discrete orbits, only particular light frequencies are emitted, as atomic spectra show.

Interestingly, any transition between two orbits is always instantaneous. In other words, the electron doesn't "jump" from a higher to a lower orbit the way a squirrel jumps from a higher branch in a tree to a lower one. Rather, an electron takes no time to move between two orbits. Bohr was serious when he stated that electrons could never exist between permitted energy levels!

CHECK YOURSELF

Is the Bohr model of the atom a physical model or a conceptual model?

CHECK YOUR ANSWER

The Bohr model is a conceptual model. It is not a scaled-up version of an atom, but instead is a representation that accounts for the atom's behavior.

Bohr's planetary atomic model proved to be a tremendous success. By utilizing Planck's quantum hypothesis, Bohr's model solved the mystery of atomic spectra. Despite its successes, though, Bohr's model was limited because it did not explain why energy levels in an atom are quantized. Bohr was quick to point out that his model was to be interpreted only as a crude beginning, and the picture of electrons whirling about the nucleus like planets about the Sun was not to be taken literally (a warning to which popularizers of science paid no heed).

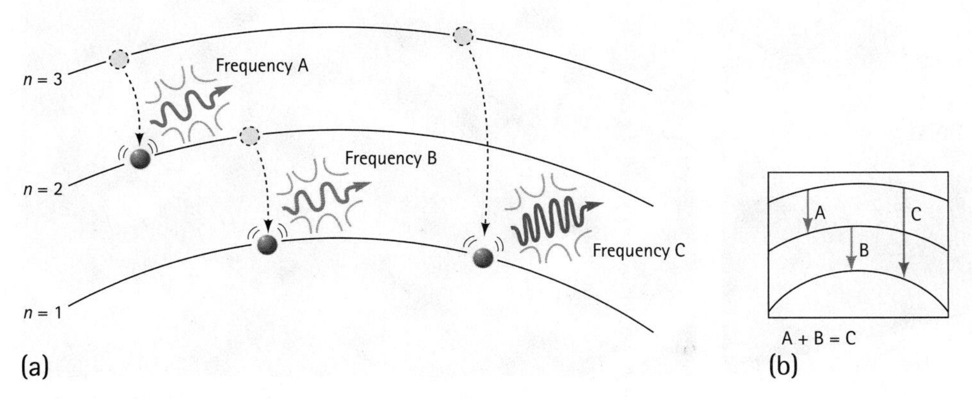

FIGURE 9.27

(a) The frequency of light emitted (or absorbed) by an atom is proportional to the energy difference between electron orbits. Because the energy differences between orbits are discrete, the frequencies of light emitted (or absorbed) are also discrete. The electron here can emit only three discrete frequencies of light—A, B, and C. The greater the transition, the higher the frequency of the photon emitted. (b) The sum of the energies (and frequencies) for transitions A and B equals the energy (and frequency) of transition C.

LEARNING OBJECTIVE
Summarize how electrons, when confined to an atom, behave like self-reinforcing wavelike entities.

9.6 Electron Waves

EXPLAIN THIS How is a plucked guitar string like an electron in an atom?

If light has both wave properties and particle properties, why can't a material particle, such as an electron, also have both? This question was posed by the French physicist Louis de Broglie (1892–1987) while he was still a graduate student in 1924. His revolutionary answer was that every moving particle of matter is endowed with the characteristics of a wave.

UNIFYING CONCEPT
- *Waves*
Section 8.1

We now speak of waves as an essential feature of any bit of matter. An electron, or any particle, can show itself as a wave or as a particle, depending on how we examine it. This is called the *wave–particle duality*. Just a few years after de Broglie's suggestion, researchers in Great Britain and the United States confirmed the wave nature of electrons by observing diffraction and interference effects when electrons bounced from crystals.

A practical application of the wave properties of electrons is the electron microscope, which focuses not visible-light waves but rather electron waves. Because electron waves are much shorter than visible-light waves, electron microscopes can show far greater detail than optical microscopes, as Figure 9.28 shows.

FYI Electron waves are three-dimensional, which makes them difficult to visualize, but scientists have come up with ways of visualizing them, including *probability clouds* and *atomic orbitals*. (You may learn more about these in a follow-up course on chemistry.)

An electron's wave nature can be used to explain why electrons in an atom are restricted to particular energy levels. Permitted energy levels are a natural consequence of electron waves having to form closed circular patterns around the atomic nucleus.

As an analogy, consider the wire loop shown in Figure 9.29. This loop is affixed to a mechanical vibrator that can be adjusted to create waves of different wavelengths in the wire. Waves that are some multiple of the length of the wire are able to meet up with themselves after traveling around the wire. This results

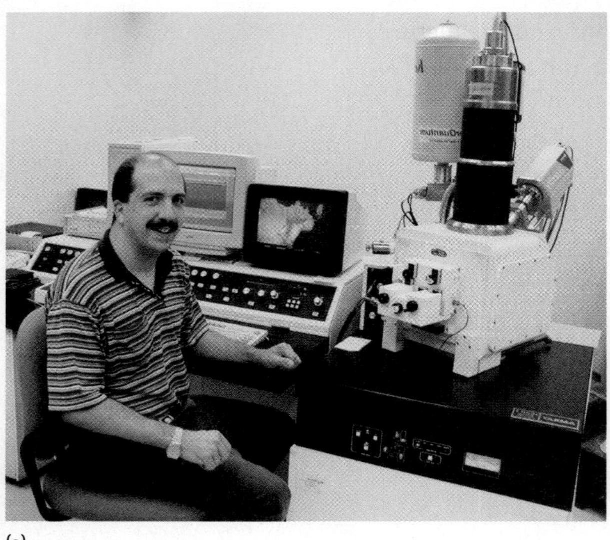

(a)

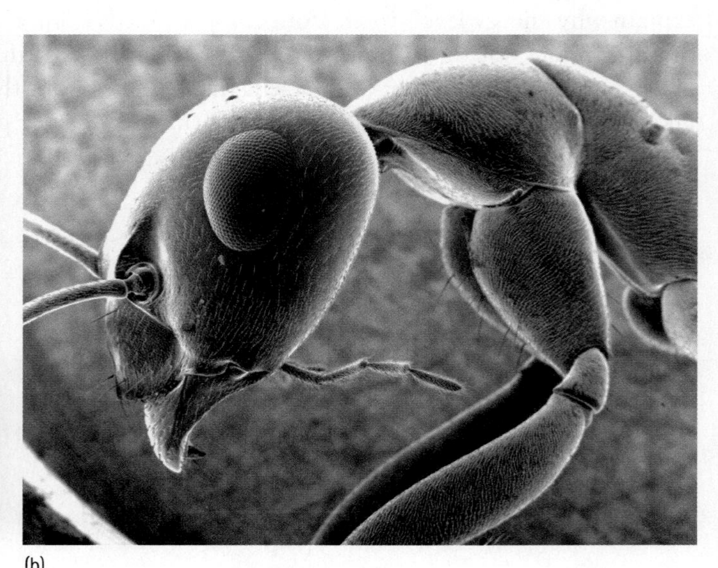

(b)

FIGURE 9.28
(a) An electron microscope makes practical use of the wave nature of electrons. The wavelengths of electron beams are typically shorter than the wavelengths of visible light by a factor of a thousand, and so the electron microscope can distinguish detail not visible with optical microscopes. (b) Detail of an ant as seen with an electron microscope at a "low" magnification of 200×. Note the remarkable resolution.

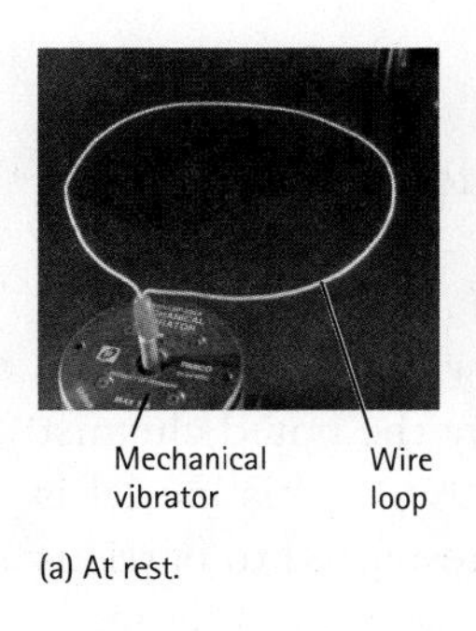

(a) At rest.

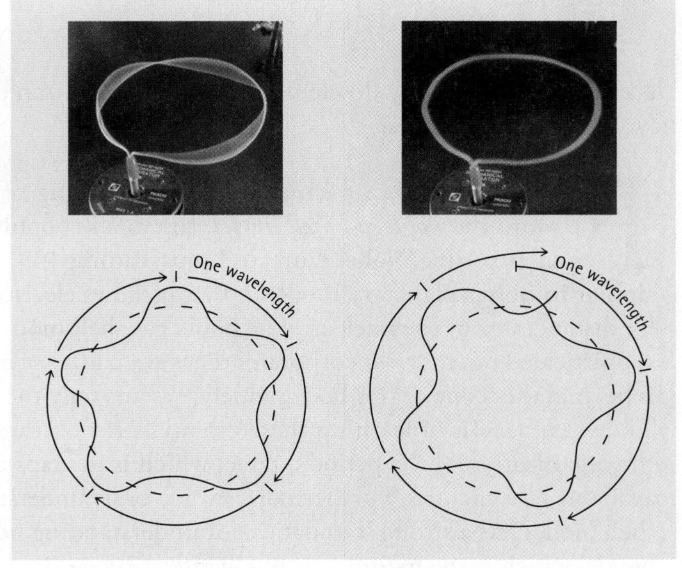

(b) Wavelength is self-reinforcing.

(c) Wavelength produces chaotic motion.

FIGURE 9.29
For the fixed circumference of a wire loop, only some wavelengths are self-reinforcing. (a) The loop affixed to the post of a mechanical vibrator at rest. Waves are sent through the wire when the post vibrates. (b) Waves created by vibration at particular rates are self-reinforcing. (c) Waves created by vibration at other rates are not self-reinforcing.

in a stationary wave pattern called a *standing wave,* as shown in Figure 9.29b. This pattern results because the peaks and valleys of successive waves are perfectly matched. With other wavelengths, as shown in Figure 9.29c, successive waves are not synchronized. As a result, the waves do not build to great amplitude.

The only waves that an electron exhibits while confined to an atom are those that are self-reinforcing. These resemble a standing wave centered on the atomic nucleus. Each standing wave corresponds to one of the permitted energy levels. Only the frequencies of light that match the difference between any two of these permitted energy levels can be absorbed or emitted by an atom.

The wave nature of electrons also explains why they do not spiral closer and closer to the positive nucleus that attracts them. By viewing each electron orbit as a self-reinforcing wave, we see that the circumference of the smallest orbit can be no smaller than a single wavelength.

CHECK YOURSELF
What must an electron be doing in order to have wave properties?

CHECK YOUR ANSWER
According to de Broglie, particles of matter behave like waves by virtue of their motion. An electron must therefore be moving in order to have wave properties.

LEARNING OBJECTIVE
Show how electrons behave as though they are arranged in a series of shells centered around the atomic nucleus.

9.7 The Noble Gas Shell Model

EXPLAIN THIS Why do elements in the same group of the periodic table have similar properties?

For the purposes of a simplified understanding of how atoms behave, we turn to the *noble gas shell model*, first made popular by the noted chemist and two-time Nobel laureate Linus Pauling (1901–1994). This model is similar to Bohr's planetary model in that it shows electrons restricted to particular distances from the nucleus. The noble gas shell model, however, is a bit more sophisticated because it incorporates the wave nature of electrons. How it does so is beyond the scope of this book, which presents only the essentials of this model. The great benefit of learning these essentials is that they help us to understand the organization of the periodic table, which is perhaps the most important tool available to a chemist. Furthermore, even a basic understanding of the noble gas shell model sets a strong foundation for understanding how atoms form chemical bonds, which is the main focus of Chapter 12.

The quality of a song depends on the arrangement of musical notes. In a similar fashion, the properties of an element depend on the arrangements of electrons in its atoms.

According to the noble gas shell model, electrons behave as though they are arranged in a series of concentric shells. A single **noble gas shell** is a region of space around the atomic nucleus upon which electrons may reside.* An important aspect of this model is that there are at least seven shells and each shell can hold only a limited number of electrons. As shown in Figure 9.30, the innermost shell can hold 2 electrons, the second and third shells 8 each, the fourth and fifth shells 18 each, and the sixth and seventh shells 32 each. When a shell is filled, it represents a noble-gas element, which is why we refer to this model as the "noble gas" shell model.

What do poets and scientists have in common? They both use metaphors to help us understand abstract concepts and relationships. The "shell," for example, is a metaphor that helps us visualize an invisible reality. Scientific models are essentially equivalent to the metaphorical language used in poetry.

A series of seven such concentric noble gas shells accounts for the seven periods of the periodic table. Furthermore, the number of elements in each period is equal to the shell's capacity for electrons. The first shell, for example, can hold only two electrons. That's why we find only two elements, hydrogen and helium, in the first period (Figure 9.31). Hydrogen is the element whose

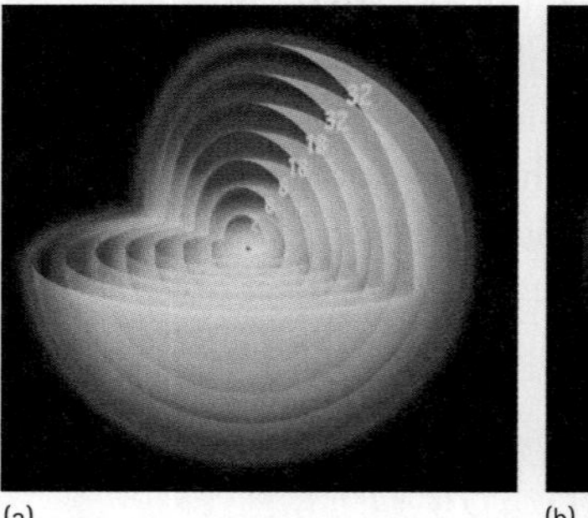
(a)

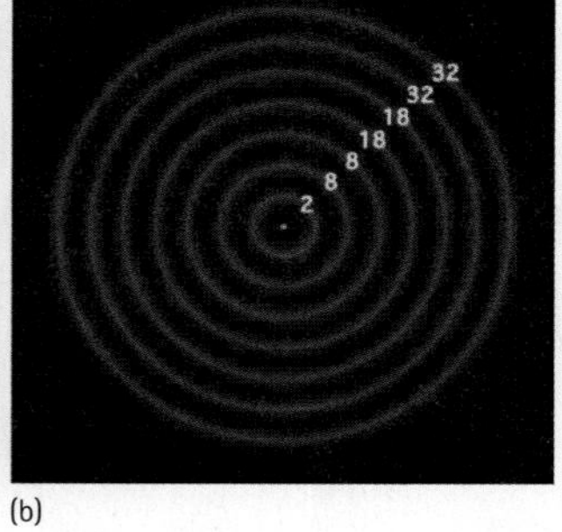

(b)

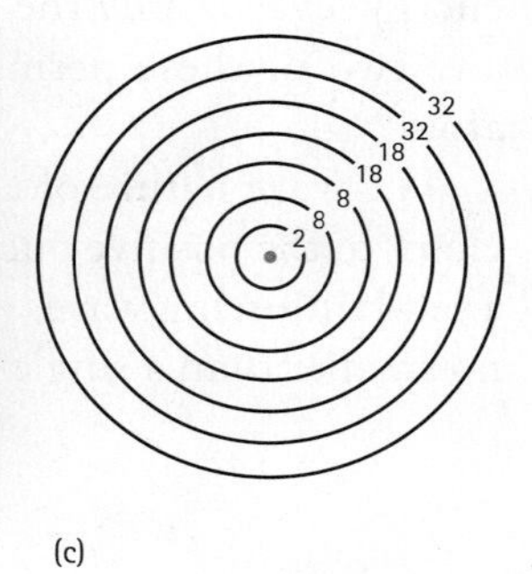

(c)

FIGURE 9.30
(a) A cutaway view of the seven shells, showing the number of electrons each shell can hold. (b) A two-dimensional, cross-sectional view of the shells. (c) An easy-to-draw cross-sectional view that resembles Bohr's planetary model.

*Technical note: As used here, a *noble gas shell* means a set of orbitals in a multielectron atom grouped by similar energy levels rather than by principal quantum number. Historically, these are the "argonian" shells developed by Linus Pauling to explain chemical bonding and the organization of the periodic table.

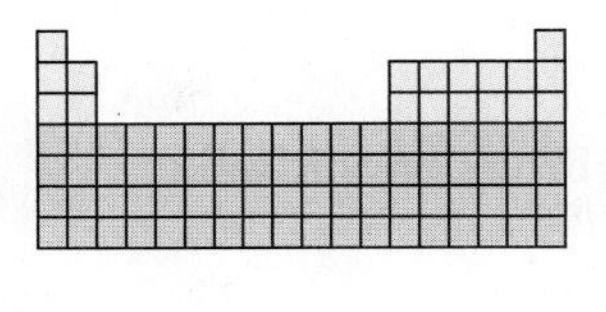

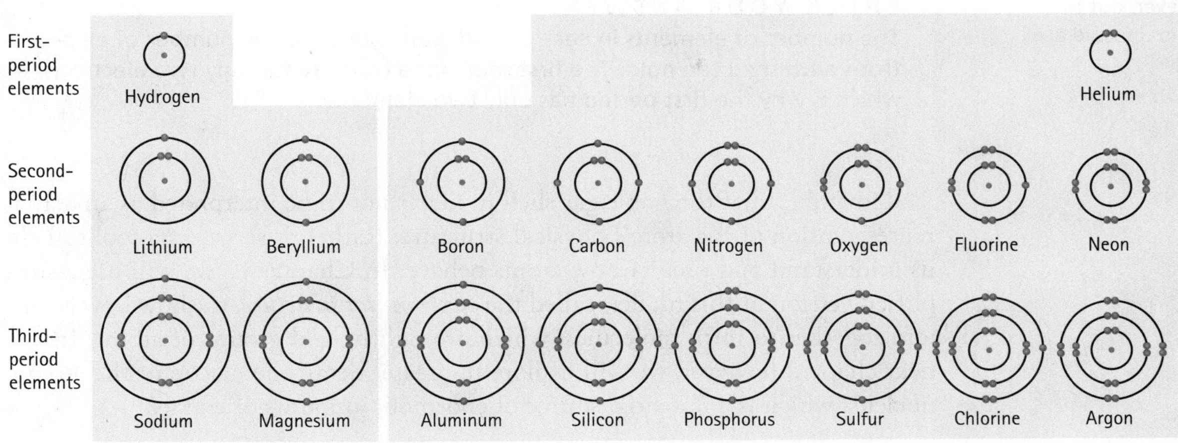

FIGURE 9.31
The first three periods of the periodic table according to the shell model. Elements in the same period (horizontal row) have electrons in the same shells. Elements in the same period differ from one another by the number of electrons in the outermost shell. Elements in the same group (vertical column) have the same number of electrons in the outermost shell. These outer-shell electrons determine the character of the atom, which is why elements in the same group have similar properties.

atoms have only one electron. This one electron resides within the first shell, which is the shell closest to the nucleus. Each helium atom has two electrons, both of which are also within the first shell, which is thus filled to its maximum capacity. Similarly, the second and third shells each have the capacity for eight electrons, so eight elements are found in both the second and the third periods.

The electrons of the outermost occupied shell in any atom are directly exposed to the external environment and are the first to interact with other atoms. Most notably, they are the ones that participate in chemical bonding, as we will discuss in Chapter 12. The electrons in the outermost shell, therefore, are quite important.

The quality of a song depends on the arrangement of musical notes. In a similar fashion, the properties of an element depend on the arrangements of electrons in its atoms, especially the outer-shell electrons. Look carefully at Figure 9.31. Can you see that the outer-shell electrons of atoms above and below one another on the periodic table—that is, within the same group—are similarly organized? For example, atoms of the first group, which include hydrogen, lithium, and sodium, each have a single outer-shell electron. The atoms of the second group, including beryllium and magnesium, each have two outer-shell electrons. Similarly, atoms of the last group, including helium, neon, and argon, each have their outermost shell filled to capacity with electrons—two for helium and eight for both neon and argon. In general, the outer-shell electrons of atoms in the same group of the periodic table are similarly organized. This explains why elements of the same group have similar properties—a concept first presented in Section 9.3.

FIGURE 9.32
Linus Pauling was an early proponent of teaching beginning chemistry students a shell model, from which the organization of the periodic table could be described. In 1954, Pauling won the Nobel Prize in Chemistry for his research into the nature of the chemical bond. In 1962, he was awarded the Nobel Peace Prize for his campaign against the testing of nuclear bombs, which introduced massive amounts of radioactivity into the environment.

FYI According to Einstein's theory of special relativity, at 60% of the speed of light, gold's innermost electrons experience only 52 seconds for each one of our minutes. A diamond may be forever, but the innermost electrons of gold are 8 s/min slow!

CHECK YOURSELF

Why are there only two elements in the first period of the periodic table?

CHECK YOUR ANSWER

The number of elements in each period corresponds to the number of electrons each shell can hold. The first shell has a capacity for only two electrons, which is why the first period has only two elements.

Remember that the noble gas shell model is not to be interpreted as an actual representation of the atom's physical structure. Rather, it serves as a tool to help us understand and predict how atoms behave. In Chapter 12 we will use a simplified version of this model, called the *electron-dot structure*, to show how atoms join together to form molecules, which are tightly held groups of atoms. In the next chapter, however, we will explore in greater detail the nature of the atomic nucleus, which is a potential source of enormous amounts of energy.

For instructor-assigned homework, go to www.masteringphysics.com 

SUMMARY OF TERMS (KNOWLEDGE)

Atomic mass The mass of an element's atoms; listed in the periodic table as an average value based on the relative abundance of the element's isotopes.

Atomic nucleus The dense, positively charged center of every atom.

Atomic number A count of the number of protons in the atomic nucleus.

Atomic spectrum The pattern of frequencies of electromagnetic radiation emitted by the atoms of an element; considered to be an element's fingerprint.

Atomic symbol An abbreviation for an element or atom.

Conceptual model A representation of a system that helps us predict how the system behaves.

Electron An extremely small, negatively charged subatomic particle found in a cloud outside the atomic nucleus.

Element Any material that is made up of only one type of atom.

Elemental formula The chemical symbol of an element followed by a numeral subscript indicating the number of atoms in each molecule of the element.

Group A vertical column in the periodic table; also known as a family of elements.

Isotope Any member of a set of atoms of the same element whose nuclei contain the same number of protons but different numbers of neutrons.

Neutron An electrically neutral subatomic particle in the atomic nucleus.

Noble gas shell A region of space around the atomic nucleus within which electrons may reside.

Nucleon Any subatomic particle found in the atomic nucleus; another name for either a proton or a neutron.

Period A horizontal row in the periodic table.

Periodic table A chart in which all known elements are listed in order of atomic number.

Physical model A representation of an object on some convenient scale.

Proton A positively charged subatomic particle in the atomic nucleus.

Quantum A small, discrete packet of light energy.

Quantum number An integer that specifies the quantized energy level of an atomic orbital.

Spectroscope A device that uses a prism or diffraction grating to separate light into its color components.

READING CHECK QUESTIONS (COMPREHENSION)

9.1 The Elements

1. How many types of atoms can you expect to find in a pure sample of any element?
2. Distinguish between an atom and an element.
3. What is the atomic symbol for the element cobalt?

9.2 Protons and Neutrons

4. What role does atomic number play in the periodic table?
5. Distinguish between atomic number and mass number.
6. Distinguish between mass number and atomic mass.

9.3 The Periodic Table

7. Are most elements metallic or nonmetallic?
8. How many periods are in the periodic table? How many groups?
9. What happens to the properties of elements as you go across any period of the periodic table?

9.4 Identifying Atoms Using the Spectroscope

10. What does a spectroscope do to the light coming from an atom?
11. What causes an atom to emit light?
12. Why do we say that atomic spectra are like the fingerprints of elements?

9.5 The Quantum Hypothesis

13. What was Planck's quantum hypothesis?
14. Which has more potential energy—an electron close to an atomic nucleus or one far from an atomic nucleus?
15. Did Bohr think of his planetary model as an accurate representation of what an atom looks like?

9.6 Electron Waves

16. Who first proposed that electrons exhibit the properties of a wave?
17. What is a practical application of the wave nature of an electron?
18. An electron confined to an atom has waves that are what?

9.7 The Noble Gas Shell Model

19. Does the periodic table explain the noble gas shell model, or does the noble gas shell model explain the periodic table?
20. Which electrons are most responsible for the properties of an atom?
21. What is the relationship between the maximum number of electrons each shell can hold and the number of elements in each period of the periodic table?

THINK INTEGRATED SCIENCE

9A—Atoms Are Ancient and Empty

22. Which is the oldest element?
23. Is it possible to see an atom using visible light?
24. What is at the center of every atom?

9B—Physical and Conceptual Models

25. If a baseball were the size of Earth, about how large would its atoms be?
26. When we use a scanning probe microscope, do we see atoms directly or do we see them only indirectly?
27. What is the difference between a physical model and a conceptual model?

THINK AND DO (HANDS-ON APPLICATION)

28. Fluorescent lights contain spectral lines from the light emission of mercury atoms. Special coatings on the inner surface of the bulb help to accentuate visible frequencies, which can be seen through the diffraction grating reflection of a compact disc (CD). Cut a narrow slit through some thick paper (or thin cardboard) and place it over a bright fluorescent bulb. View this slit at an oblique angle against a CD and look for spectral lines. Place the slit over an incandescent bulb and you'll see a smooth continuous spectrum (no lines) because the incandescent filament glows at all visible frequencies. Try looking at different brands of fluorescent bulbs. You'll also be able to see spectral lines in streetlights and fireworks. For those it is best to use "rainbow" glasses, available from a nature, toy, or hobby store.

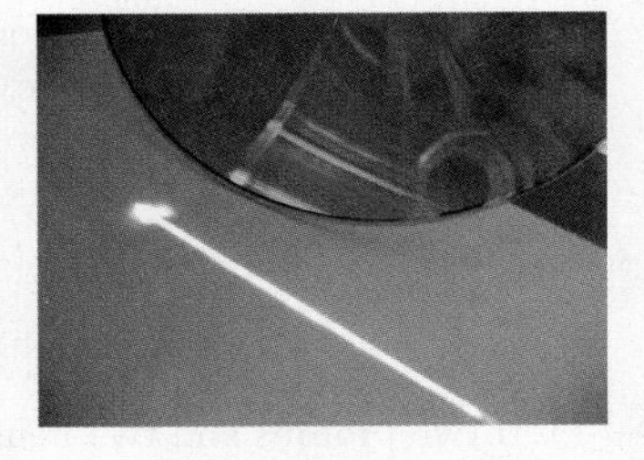

29. You can "quantize" your whistle by whistling down a long tube, such as the tube from a roll of wrapping paper. First, without the tube, whistle from a high pitch to a low pitch. Do it loudly and in a single breath. (If you can't whistle, find some one who can.) Next, try the same thing while holding the tube to your lips. Aha! Note that some frequencies simply cannot be whistled, no matter how hard you try. These frequencies are forbidden because their wavelengths are not a multiple of the length of the tube.

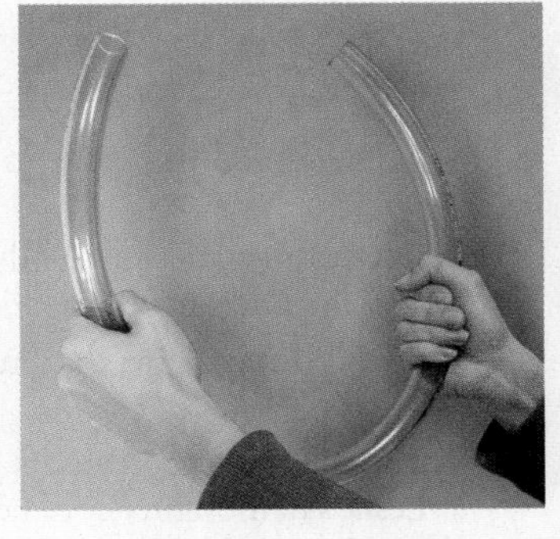

Try experimenting with tubes of different lengths. To hear yourself more clearly, use a flexible plastic tube and twist the outer end toward your ear.

When your whistle is confined to the tube, the consequence is a quantization of its frequencies. When an electron wave is confined to an atom, the consequence is a quantization of the electron's energy.

30. Stretch a rubber band between your two thumbs and then pluck it with your index finger. Better yet, stretch the rubber band in front of a windy fan to get it vibrating. Note that the area of greatest oscillation is always at the midpoint. This is a self-reinforcing wave that occurs as overlapping waves bounce back and forth from thumb to thumb.

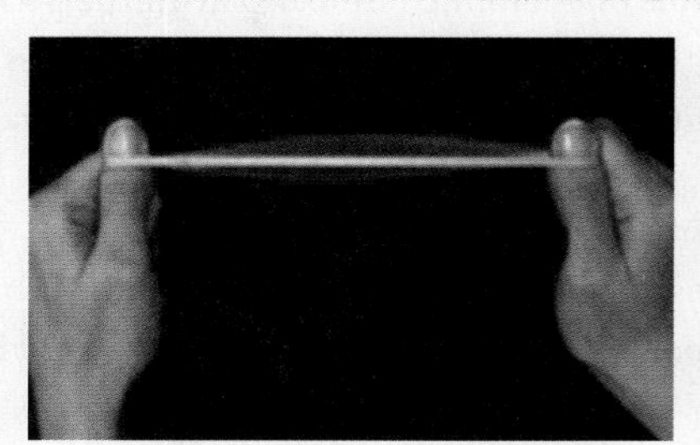

THINK AND COMPARE (ANALYSIS)

31. Rank these three subatomic particles in order of increasing mass: (a) neutron, (b) proton, (c) electron.
32. Consider these atoms: helium, He; chlorine, Cl; and argon, Ar. Rank them by their atomic number, from smallest to largest.
33. Consider three 1-gram samples of matter: (a) carbon-12, (b) carbon-13, (c) uranium-238. Rank the samples by the number of atoms in each, from most to least.
34. Consider these atoms: helium, He; aluminum, Al; argon, Ar. Rank them, from smallest to largest, in order of (a) size, (b) number of protons in the nucleus, (c) number of electrons.
35. Rank these atoms by the number of electrons they tend to lose, from fewest to most: (a) sodium, Na; (b) magnesium, Mg; (c) aluminum, Al.
36. Rank these noble-gas atoms by their number of filled shells, from least to most: (a) argon, (b) radon, (c) helium, (d) neon.

THINK AND SOLVE (MATHEMATICAL APPLICATION)

37. A class of 20 students takes an exam and every student scores 80%. What is the class average? Would the class average be slightly lower, the same, or slightly higher if one of the students instead scored 100%? How is this similar to how we calculated the atomic masses of elements?
38. The isotope lithium-7 has a mass of 7.0160 atomic mass units, and the isotope lithium-6 has a mass of 6.0151 atomic mass units. Given the information that 92.58% of all lithium atoms found in nature are lithium-7 and 7.42% are lithium-6, show that the atomic mass of lithium, Li (atomic number 3), is 6.941 amu.
39. The element bromine, Br (atomic number 35), has two major isotopes of similar abundance, both about 50%. The atomic mass of bromine is reported in the periodic table as 79.904 atomic mass units. Choose the most likely set of mass numbers for the bromine isotopes: (a) Br-79, Br-81; (b) Br-79, Br-80; (c) Br-80, Br-81.
40. Chlorine (atomic number 17) is composed of two principal isotopes, chlorine-35, which has a mass of 34.9689 atomic mass units, and chlorine-37, which has a mass of 36.9659 atomic mass units. Assume that 75.77% of all chlorine atoms are the chlorine-35 isotope and 24.23% are the chlorine-37 isotope. Show that the atomic mass of natural chlorine is 35.45 amu.

THINK AND EXPLAIN (SYNTHESIS)

41. If all the molecules of a body remained part of that body, would the body have any odor?
42. Where did the atoms that make up a newborn baby originate?
43. Where did the carbon atoms in Leslie's hair originate? (The photo shows Leslie at age 16.)

44. Is the head of a politician really made of 99.99999999% empty space?
45. If two protons and two neutrons are removed from the nucleus of an oxygen-16 atom, a nucleus of which element remains?
46. An atom has 43 electrons, 56 neutrons, and 43 protons. What is its approximate atomic mass? What is the name of this element?
47. The nucleus of an electrically neutral iron atom contains 26 protons. How many electrons does this iron atom have?
48. Evidence for the existence of neutrons did not come until many years after the discoveries of the electron and the proton. Give a possible explanation.

49. Which has more atoms—a 1-gram sample of carbon-12 or a 1-gram sample of carbon-13? Explain.
50. Why aren't the atomic masses listed in the periodic table whole numbers?
51. Which contributes more to an atom's mass—electrons or protons? Which contributes more to an atom's size?
52. What is the approximate mass of an oxygen atom in atomic mass units? What is the approximate mass of two oxygen atoms? How about an oxygen molecule?
53. What is the approximate mass of a carbon atom in atomic mass units? How about a carbon dioxide molecule?
54. Which is heavier—a water molecule, H_2O, or a carbon dioxide molecule, CO_2?
55. When we breathe we inhale oxygen, O_2, and exhale carbon dioxide, CO_2, plus water vapor, H_2O. Which likely has more mass—the air that we inhale or the same volume of air that we exhale? Does breathing cause you to lose or gain weight?
56. As a tree respires it takes in carbon dioxide, CO_2, and water vapor, H_2O, from the air while also releasing oxygen, O_2. Does the tree lose or gain weight as it respires? Explain.
57. Which of these diagrams best represents the size of the atomic nucleus relative to the size of the atom?

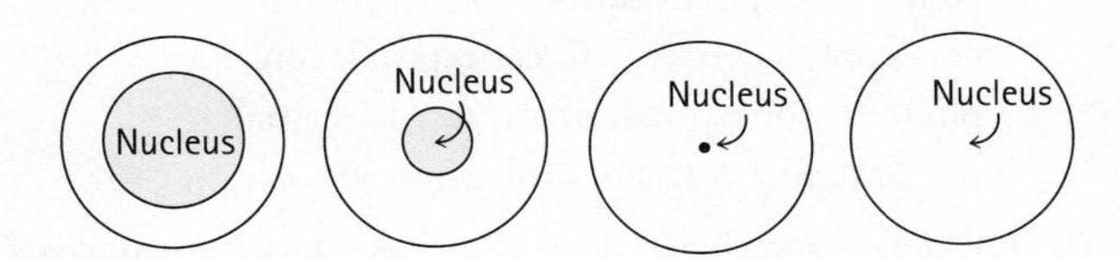

58. A beam of protons and a beam of neutrons of the same energy are both harmful to living tissue. The beam of neutrons, however, is less harmful. Suggest why.
59. Germanium, Ge (number 32), computer chips operate faster than silicon, Si (number 14), computer chips. So, how might a gallium, Ga (number 31), chip compare with a germanium chip?
60. Helium, He, is a nonmetallic gas and the second element in the periodic table. Rather than being placed adjacent to hydrogen, H, however, helium is placed on the far right of the table. Why?
61. Name ten elements that you have access to macroscopic samples of as a consumer here on Earth.
62. Strontium, Sr (number 38), is especially dangerous to humans because it tends to accumulate in calcium-dependent bone marrow tissues (calcium, Ca, number 20). How does this fact relate to what you know about the organization of the periodic table?
63. As depicted in Figure 9.19, are gallium atoms really red and arsenic atoms green?
64. With scanning probe microscopy technology, we see not actual atoms but rather images of them. Explain.
65. Why isn't it possible for a scanning probe microscope to make images of the inside of an atom?
66. What do the components of a conceptual model have in common?
67. Would you use a physical model or a conceptual model to describe the following: gold coin, dollar bill, car engine, air pollution, virus, spread of sexually transmitted disease?
68. What is the function of an atomic model?
69. What is the relationship between the light emitted by an atom and the energies of the electrons in the atom?
70. How might you distinguish a sodium-vapor streetlight from a mercury-vapor streetlight?
71. What particle within an atom vibrates to generate electromagnetic radiation? This particle is vibrating back and forth between what?
72. How can a hydrogen atom, which has only one electron, create so many spectral lines?
73. Which color of light comes from a greater energy transition—red or blue?
74. How does the wave model of electrons orbiting the nucleus account for the fact that the electrons can have only discrete energy values?
75. Some older cars vibrate loudly when driven at particular speeds. For example, at 65 mph the car is very quiet, but at 60 mph the car rattles uncomfortably. How is this analogous to the quantized energy levels of an electron in an atom?
76. Does a noble gas shell have to contain electrons in order to exist?
77. Place the correct number of electrons in each shell.

Sodium, Na
Rubidium, Rb
Krypton, Kr
Chlorine, Cl

78. Use the noble gas shell model to explain why a potassium atom, K, is larger than a sodium atom, Na.
79. Neon, Ne (atomic number 10) cannot attract any additional electrons. Why?
80. Use the shell model to explain why a lithium atom, Li, is larger than a beryllium atom, Be.

THINK AND DISCUSS (EVALUATION)

81. In what sense can you truthfully say that you are a part of every person around you?
82. Considering how small atoms are, discuss the chances that at least one of the atoms exhaled in your first breath will be in your last breath.
83. The atoms that constitute your body are mostly empty space, and structures such as the chair you're sitting on are composed of atoms that are also mostly empty space. Discuss why it is that you don't fall through the chair.
84. How might the spectrum of an atom appear if the atom's electrons were not restricted to particular energy levels?
85. If matter is made of atoms and atoms are made of subatomic particles, what comes together to create subatomic particles? Where might you find such information?

READINESS ASSURANCE TEST (RAT)

If you have a good handle on this chapter, if you really do, then you should be able to score 7 out of 10 on this RAT. If you score less than 7, you need to study further before moving on.

Choose the BEST answer to each of the following:

1. Which are older—the atoms in the body of an elderly person or the atoms in the body of a baby?
   (a) A baby's atoms are older because this is surely a trick question.
   (b) An elderly person's are older because they have been around much longer.
   (c) The atoms are the same age, which is appreciably older than the solar system.
   (d) The atoms' ages depend on the person's diet.
2. You could swallow a capsule of germanium, Ge (atomic number 32), without significant ill effects. If a proton were added to each germanium nucleus, however, you would not want to swallow the capsule because the germanium would
   (a) become arsenic.
   (b) become radioactive.
   (c) expand and likely lodge in your throat.
   (d) have a change in flavor.
3. Why aren't the atomic masses listed in the periodic table whole numbers?
   (a) Scientists have yet to make the precise measurements.
   (b) That would be too much of a coincidence.
   (c) The atomic masses are average atomic masses.
   (d) Today's instruments are able to measure the atomic masses to many decimal places.
4. If an atom has 43 electrons, 56 neutrons, and 43 protons, what is its approximate atomic mass? What is the name of this element?
   (a) 137 amu; barium
   (b) 99 amu; technetium
   (c) 99 amu; radon
   (d) 142 amu; einsteinium
5. An element found in another galaxy exists as two isotopes. If 80.0% of the atoms have an atomic mass of 80.00 atomic mass units and the other 20.0% have an atomic mass of 82.00 atomic mass units, what is the approximate *atomic mass* of the element?
   (a) 80.4 amu
   (b) 81.0 amu
   (c) 81.6 amu
   (d) 64.0 amu
   (e) 16.4 amu
6. List the following atoms in order of increasing atomic size: thallium, Tl; germanium, Ge; tin, Sn; phosphorus, P.
   (a) Ge $<$ P $<$ Sn $<$ Tl
   (b) Tl $<$ Sn $<$ P $<$ Ge
   (c) Tl $<$ Sn $<$ Ge $<$ P
   (d) P $<$ Ge $<$ Sn $<$ Tl
7. Which of these elements has chemical properties the most similar to those of chlorine (Cl, atomic number 17)?
   (a) O
   (b) Na
   (c) S
   (d) Ar
   (e) Br
8. Would you use a physical model or a conceptual model to describe each of the following: the brain, the mind, the solar system, the beginning of the universe?
   (a) conceptual, physical, conceptual, physical
   (b) conceptual, conceptual, conceptual, conceptual
   (c) physical, conceptual, physical, conceptual
   (d) physical, physical, physical, physical
9. How does the wave model of electrons orbiting the nucleus account for the fact that the electrons can have only discrete energy values?
   (a) Electrons are able to vibrate only at particular frequencies.
   (b) When an electron wave is confined, it is reinforced only at particular frequencies.
   (c) The energy values of an electron occur only where its wave properties have a maximum amplitude.
   (d) The wave model accounts for the shells an electron may occupy, not its energy levels.
10. How many electrons are in the third shell of sodium, Na (atomic number 11)?
    (a) none
    (b) 1
    (c) 2
    (d) 3

Answers to RAT

1. c, 2. a, 3. c, 4. b, 5. a, 6. d, 7. e, 8. c, 9. b, 10. b

10

CHAPTER 10

The Atomic Nucleus and Radioactivity

THE ATOMIC nucleus and nuclear processes are one of the most misunderstood and controversial areas of science. Distrust of anything *nuclear*, or anything *radioactive*, is much like the fears of electricity more than a century ago. Indeed, electricity can be dangerous, and even lethal, when improperly handled. But with safeguards and well-informed consumers, society has determined that the benefits of electricity outweigh its risks. Today we are making similar decisions about nuclear technology's risks and benefits. The risks became most evident with the 2011 earthquake and tsunami that destroyed Japan's Fukushima Daiichi nuclear power plant. The benefits, however, include the large-scale production of electric energy with no emission of carbon dioxide, which is a potent greenhouse gas. Should society continue to invest in nuclear energy? Now, more than ever, it is important that we "know nukes"!

LEARNING OBJECTIVE
Identify three forms of radioactivity.

10.1 Radioactivity

EXPLAIN THIS Why is it both impractical and impossible to prevent our exposure to radioactivity?

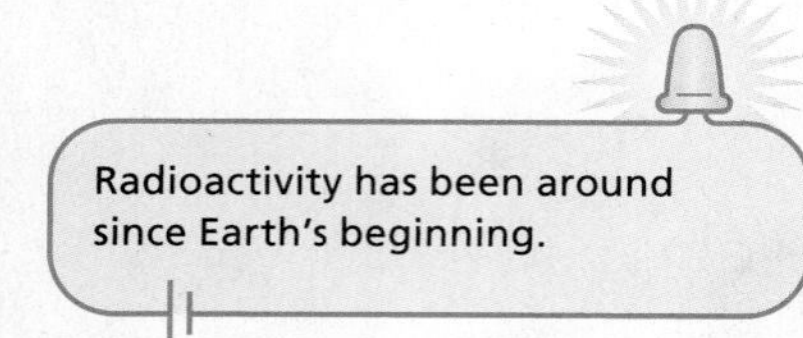

Elements with unstable nuclei are said to be *radioactive.* They eventually break down and eject energetic particles and emit high-frequency electromagnetic radiation. The emission of these particles and radiation is called **radioactivity**. Because the process involves the decay of the atomic nucleus, it is often called *radioactive decay.*

A common misconception is that radioactivity is new in the environment, but it has been around far longer than the human race. Interestingly, the deeper you go below Earth's surface, the hotter it gets. At a depth of only 30 km the temperature is hotter than 500°C. At greater depths it is so hot that rock melts into magma, which can rise to Earth's surface to escape as lava. Superheated subterranean water can escape violently to form geysers or more gently to form a soothing natural hot spring. The main reason it gets hotter down below is that Earth contains an abundance of radioactive isotopes and is heated as it absorbs radiation from these isotopes. So volcanoes, geysers, and hot springs are all powered by radioactivity. Even the drifting of continents (see Chapter 22) is related to Earth's internal radioactivity. Radioactivity is as natural as sunshine and rain.

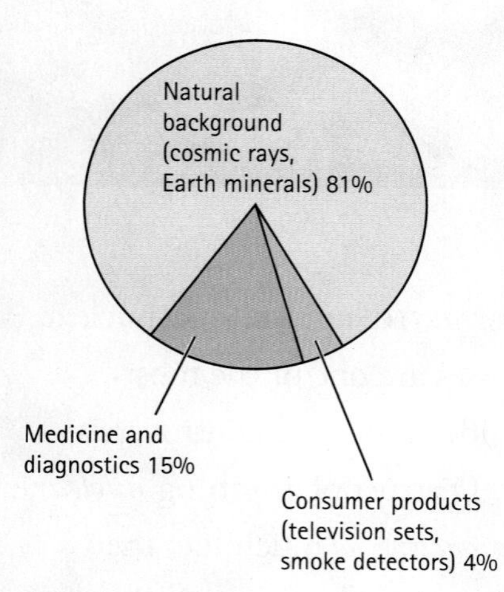

FIGURE 10.1
Sources of radiation exposure for an average individual in the United States.

MasteringPhysics®
TUTORIAL: Radiation and Biological Effects

Alpha, Beta, and Gamma Rays

All isotopes of elements with an atomic number higher than 83 (bismuth) are radioactive. These isotopes, and certain lighter radioactive isotopes, emit three distinct types of radiation, named by the first three letters of the Greek alphabet, α, β, and γ—*alpha*, *beta*, and *gamma*. Alpha rays carry a positive electric charge, beta rays carry a negative charge, and gamma rays carry no charge. The three rays can be separated by placing a magnetic field across their paths (Figure 10.2).

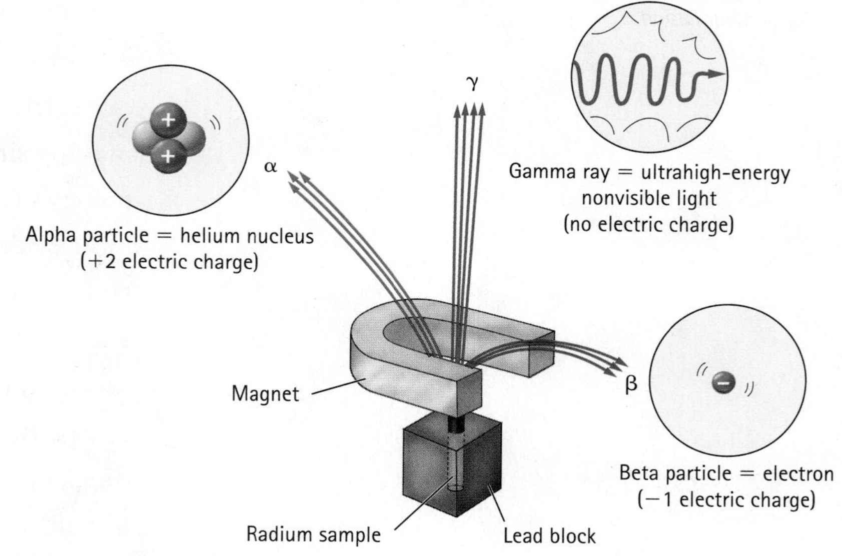

FIGURE 10.2
INTERACTIVE FIGURE MP

In a magnetic field, alpha rays bend one way, beta rays bend the other way, and gamma rays don't bend at all. Note that the alpha rays bend less than do the beta rays. This is because alpha particles have more inertia (mass) than beta particles.

An **alpha particle** is the combination of two protons and two neutrons (in other words, it is the nucleus of the helium atom, atomic number 2). Alpha particles are relatively easy to shield because of their relatively large size and their double positive charge (+2). For example, they do not normally penetrate through light materials such as paper or clothing. Because of their high kinetic energies, however, alpha particles can cause significant damage to the surface of a material, especially living tissue. When traveling through only a few centimeters of air, alpha particles pick up electrons and become nothing more than harmless helium. As a matter of fact, that's where the helium in a child's balloon comes from—practically all of Earth's helium atoms were at one time energetic alpha particles.

A **beta particle** is an electron ejected from a nucleus. Once ejected, it is indistinguishable from an electron in a cathode ray or electric circuit, or one orbiting the atomic nucleus. The difference is that a beta particle originates inside the nucleus—from a neutron. As we shall soon see, the neutron becomes a proton when it loses the electron that is a beta particle. A beta particle is normally faster than an alpha particle and carries only a single negative charge (−1). Beta particles are not as easy to stop as alpha particles are, and they can penetrate light materials such as paper or clothing. They can penetrate fairly deeply into skin, where they have the potential for harming or killing living cells. But they are not able to penetrate deeply into denser materials such as aluminum. Beta particles, once stopped, simply become part of the material they are in, like any other electron.

Gamma rays are high-frequency electromagnetic radiation emitted by radioactive elements. Like visible light, a gamma ray is pure energy. The amount of energy in a gamma ray, however, is much greater than in visible light, ultraviolet light, or even X-rays. Because they have no mass or electric charge and because of their high energies, gamma rays can penetrate most materials. However, they cannot penetrate unusually dense materials such as lead, which absorbs them. Delicate molecules inside cells throughout our bodies that are zapped by gamma rays suffer structural damage. Hence, gamma rays are generally more harmful to us than alpha or beta particles (unless the alphas or betas are ingested).

FYI Once alpha and beta particles slow down, they combine to form harmless helium. This happens primarily deep underground. As the newly formed helium seeps toward the surface, it becomes concentrated within natural gas deposits. Some natural gas deposits, such as those in Texas, contain as much as 7% helium. This helium is isolated and sold for various applications, such as blimps and helium balloons. Interestingly, natural gas fields within the United States contain about two-thirds of the world's supply of helium.

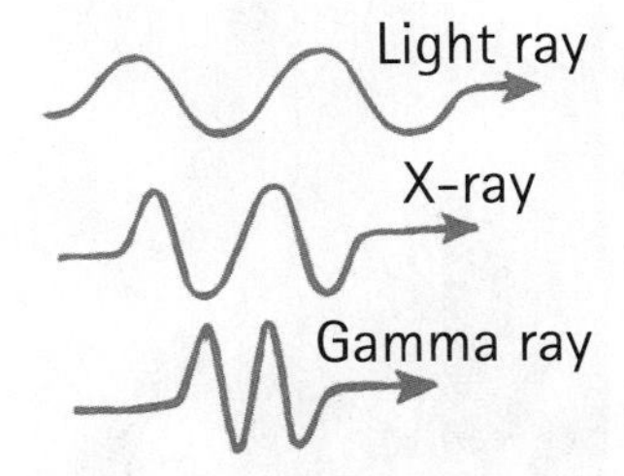

FIGURE 10.3
A gamma ray is simply electromagnetic radiation, much higher in frequency and energy than light and X-rays.

CHECK YOURSELF

Imagine that you are given three radioactive rocks—one an alpha emitter, one a beta emitter, and one a gamma emitter. You can throw away one, but of the remaining two, you must hold one in your hand and place the other in your pocket. What can you do to minimize your exposure to radiation?

CHECK YOUR ANSWER

Hold the alpha emitter in your hand because the skin on your hand shields you. Put the beta emitter in your pocket because beta particles are likely stopped by the combined thickness of your clothing and skin. Throw away the gamma emitter because gamma rays penetrate your body from either of these locations. Ideally, of course, you should keep as much distance as possible between you and all of the rocks.

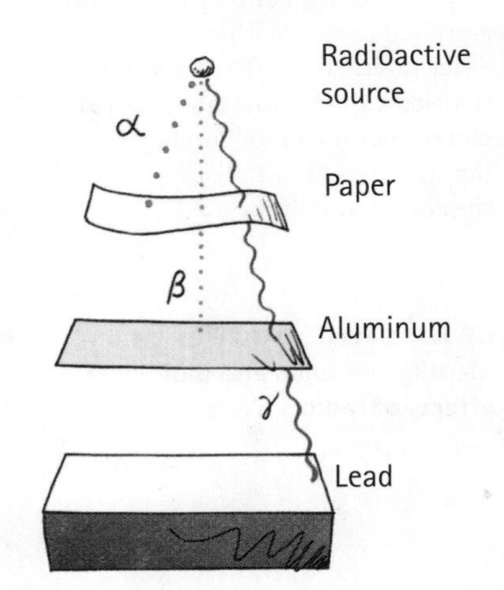

FIGURE 10.4
INTERACTIVE FIGURE MP

Alpha particles are the least penetrating and can be stopped by a few sheets of paper. Beta particles readily pass through paper, but not through a sheet of aluminum. Gamma rays penetrate several centimeters into solid lead.

FIGURE 10.5
The shelf life of fresh strawberries and other perishables is markedly lengthened when the food is subjected to gamma rays from a radioactive source. The strawberries on the right were treated with gamma radiation, which kills the microorganisms that normally lead to spoilage. The food is only a receiver of radiation and is not transformed into an emitter of radiation, as can be confirmed with a radiation detector.

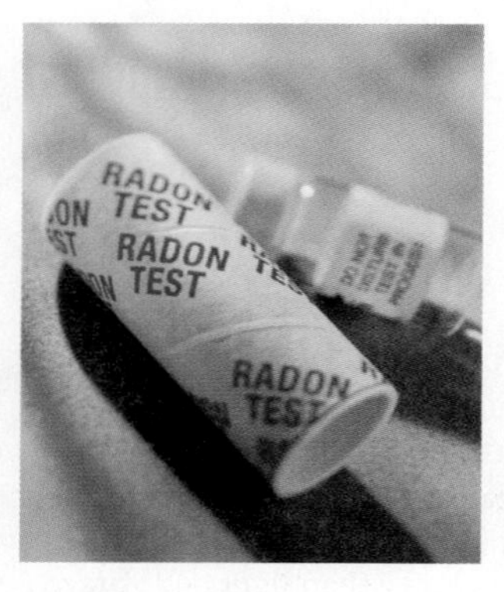

FIGURE 10.6
A commercially available radon test kit for the home. The canister is unsealed in the area to be sampled. Radon seeping into the canister is adsorbed by activated carbon within the canister. After several days, the canister is resealed and sent to a laboratory that determines the radon level by measuring the amount of radiation emitted by the adsorbed radon.

Common rocks and minerals in our environment contain significant quantities of radioactive isotopes because most of them contain trace amounts of uranium. People who live in brick, concrete, or stone buildings are exposed to greater amounts of radiation than people who live in wooden buildings.

The leading source of naturally occurring radiation is radon-222, an inert gas arising from uranium deposits. Radon is a heavy gas that tends to accumulate in basements after it seeps up through cracks in the floor. Levels of radon vary from region to region, depending on local geology. You can check the radon level in your home with a radon detector kit (Figure 10.6). If levels are abnormally high, corrective measures such as sealing the basement floor and walls and maintaining adequate ventilation are recommended.

About one-fifth of our annual exposure to radiation comes from nonnatural sources, primarily medical procedures. Television sets, fallout from nuclear testing, and the coal and nuclear power industries are also contributors. The coal industry far outranks the nuclear power industry as a source of radiation. The global combustion of coal annually releases about 13,000 tons of radioactive thorium and uranium into the atmosphere. Both these minerals are found naturally in coal deposits, so their release is a natural consequence of burning coal. Worldwide, the nuclear power industries generate about 10,000 tons of radioactive waste each year. Most of this waste, however, is contained and *not* released into the environment.

LEARNING OBJECTIVE
Identify the units and biological effects of radioactivity.

Integrated Science 10A
BIOLOGY

Radiation Dosage

EXPLAIN THIS Why are household smoke detectors radioactive?

Radiation received by living tissue is commonly measured in *rads* (*r*adiation *a*bsorbed *d*ose), a unit of absorbed energy. One **rad** is equal to 0.01 J of radiant energy absorbed per kilogram of tissue. The capacity for nuclear radiation to cause damage to living tissue, however, is not just a function of its level of energy. Some forms of radiation are more harmful than others. For example, suppose you have two arrows, one with a pointed tip and one with a suction cup at its tip. Shoot the two of them at an apple at the same speed and both have the same kinetic energy. The one with the pointed tip, however, invariably does more damage to the apple than the one with the suction cup. Similarly, some forms of radiation cause greater harm than other forms, even when we receive the same number of rads from both forms.

The unit of measure for radiation dosage based on potential damage is the **rem** (*r*oentgen *e*quivalent *m*an).* In calculating the dosage in rems, we multiply the number of rads by a factor that corresponds to different health effects of different types of radiation as determined by clinical studies. For example, 1 rad of alpha particles has the same biological effect as 10 rads of beta particles.** We call both of these dosages 10 rems:

Particle	Radiation Dosage		Factor		Health Effect
alpha	1 rad	×	10	=	10 rems
beta	10 rad	×	1	=	10 rems

CHECK YOURSELF

Would you rather be exposed to 1 rad of alpha particles or 1 rad of beta particles?

CHECK YOUR ANSWER

Multiply these quantities of radiation by the appropriate factor to get the dosages in rems. Alpha: 1 rad × 10 = 10 rems. Beta: 1 rad × 1 = 1 rem. The factors show us that, physiologically speaking, alpha particles are 10 times as damaging as beta particles.

Lethal doses of radiation begin at 500 rems. A person has about a 50% chance of surviving a dose of this magnitude received over a short period of time. During radiation therapy, a patient may receive localized doses in excess of 200 rems each day for a period of weeks (Figure 10.7).

All the radiation we receive from natural sources and from medical procedures is only a fraction of 1 rem. For convenience, the smaller unit *millirem* is used, where 1 millirem (mrem) is 1/1000 of a rem.

The average person in the United States is exposed to about 360 mrem a year, as Table 10.1 indicates. About 80% of this radiation comes from natural sources, such as cosmic rays and Earth itself. A typical chest X-ray exposes a person to 5–30 mrem (0.005–0.030 rem), less than 1/10,000 of the lethal dose. The human body itself is a significant source of natural radiation, primarily from the potassium we ingest. Our bodies contain about 200 g of potassium. Of this quantity, about 20 mg is the radioactive isotope potassium-40, which is a gamma ray emitter. Radiation is indeed everywhere.

When radiation encounters the intricately structured molecules in the watery, ion-rich brine that makes up our cells, the radiation can create chaos on the atomic scale. Some molecules are broken, and this change alters other molecules, which can be harmful to life processes.

Cells can repair most kinds of molecular damage caused by radiation if the radiation is not too severe. A cell can survive an otherwise lethal dose of radiation if the dose is spread over a long period of time to allow intervals for healing. When

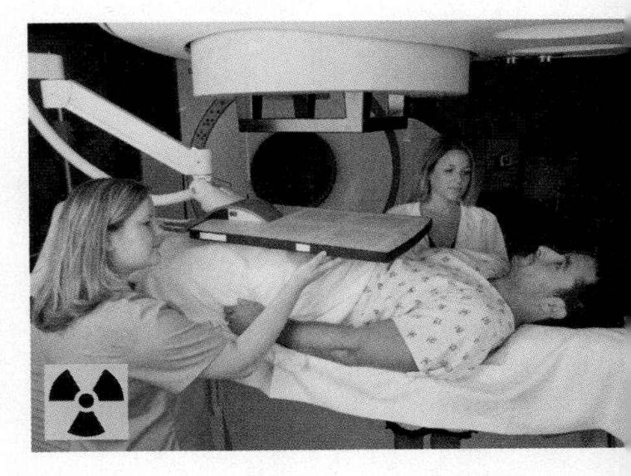

FIGURE 10.7
Nuclear radiation is focused on harmful tissue, such as a cancerous tumor, to selectively kill or shrink the tissue in a technique known as radiation therapy. This application of nuclear radiation has saved millions of lives—a clear-cut example of the benefits of nuclear technology. The inset shows the internationally used symbol to indicate an area where radioactive material is being handled or produced.

*This unit is named for the discoverer of X-rays, Wilhelm Roentgen.
**This is true even though beta particles have more penetrating power, as mentioned earlier.

TABLE 10.1 ANNUAL RADIATION EXPOSURE

Source	Typical Amount Received in 1 Year (Millirems)
Natural Origin	
Cosmic radiation	26
Ground	33
Air (radon-222)	198
Human tissues (K-40; Ra-226)	35
Human Origin	
Medical procedures	
Diagnostic X-rays	40
Nuclear medicine	15
TV tubes and other consumer products	11
Weapons-test fallout	1

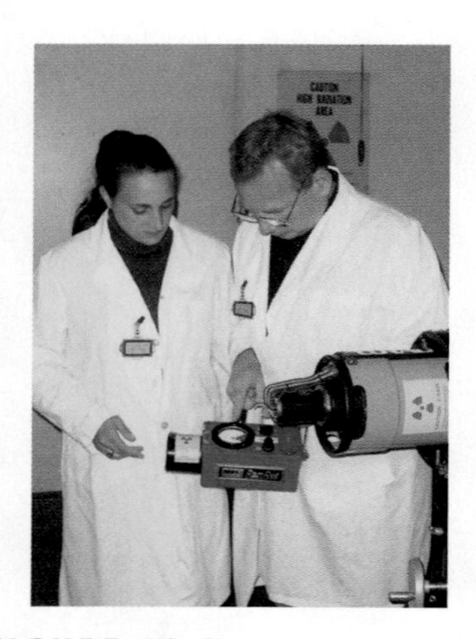

FIGURE 10.8
The film badges worn by Tammy and Larry contain audible alerts for both radiation surge and accumulated exposure. Information from the individualized badges is periodically downloaded to a database for analysis and storage.

radiation is sufficient to kill cells, the dead cells can be replaced by new ones. Sometimes a radiated cell survives with a damaged DNA molecule. New cells arising from the damaged cell retain the altered genetic information, producing a *mutation*. Usually the effects of a mutation are insignificant, but occasionally the mutation results in cells that do not function as well as unaffected ones, sometimes leading to a cancer. If the damaged DNA is in an individual's reproductive cells, the genetic code of the individual's offspring may retain the mutation.

FYI The only element beyond uranium to find a commercial application is americium, Am, which is a key component of almost all household smoke detectors. This element completes an electric circuit by ionizing air within a chamber. Smoke particles interfere with this ionization, thus breaking the circuit and triggering the alarm.

Radioactive Tracers

Radioactive samples of all the elements have been made in scientific laboratories. This is accomplished by bombardment with neutrons or other particles. Radioactive materials are extremely useful in scientific research and industry. To check the action of a fertilizer, for example, researchers combine a small amount of radioactive material with the fertilizer and then apply the combination to a few plants. The amount of radioactive fertilizer taken up by the plants can be easily measured with radiation detectors. From such measurements, scientists can inform farmers of the proper amount of fertilizer to use. Radioactive isotopes used to trace such pathways are called *tracers*.

In a technique known as medical imaging, tracers are used to diagnose internal disorders. This technique works because the path the tracer takes is influenced only by its physical and chemical properties, not by its radioactivity. The tracer may be introduced alone or along with some other chemical that helps target the tracer to a particular type of tissue in the body.

FIGURE 10.9
Tracking fertilizer uptake with a radioactive isotope.

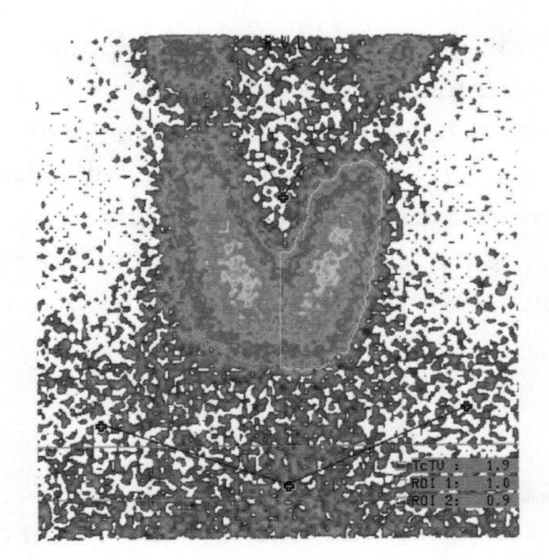

FIGURE 10.10
The thyroid gland, located in the neck, absorbs much of the iodine that enters the body through food and drink. Images of the thyroid gland, such as the one shown here, can be obtained by giving a patient the radioactive isotope iodine-131. These images are useful in diagnosing metabolic disorders.

10.2 The Strong Nuclear Force

LEARNING OBJECTIVE
Describe how the strong nuclear force acts to hold nucleons together in the atomic nucleus.

EXPLAIN THIS Why are larger nuclei less stable than smaller nuclei?

As described in Chapter 9, the atomic nucleus occupies only a tiny fraction of the volume of an individual atom, leaving most of the atom as empty space. The nucleus is composed of *nucleons*, which, as discussed in Chapter 9, is the collective name for protons and neutrons.

We know that electric charges of like sign repel one another. So, how do positively charged protons in the nucleus stay clumped together? This question led to the discovery of an attraction called the **strong nuclear force**, which acts between all nucleons. This force is very strong but over only extremely short distances (about 10^{-15} m, the diameter of a typical atomic nucleus). Repulsive electric interactions, on the other hand, are relatively long-ranged. Figure 10.11 suggests a comparison of the strengths of these two forces over distance. For protons that are close together, as in small nuclei, the attractive strong nuclear force easily overcomes the repulsive electric force.

Without the strong nuclear force there would be no atoms beyond hydrogen.

UNIFYING CONCEPT
• ***The Electric Force***
Section 7.1

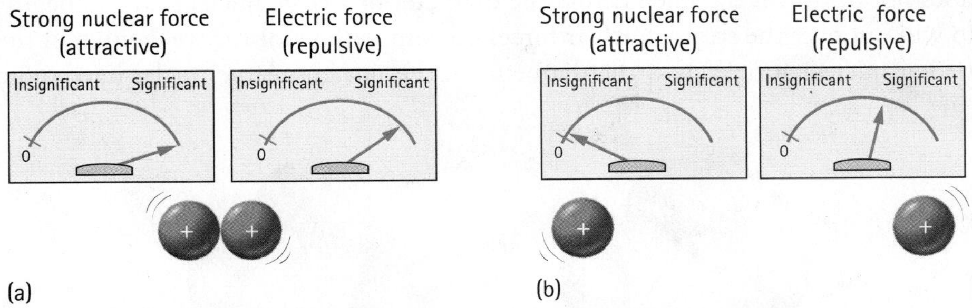

Strong nuclear force (attractive) | Electric force (repulsive) | Strong nuclear force (attractive) | Electric force (repulsive)
Insignificant Significant | Insignificant Significant | Insignificant Significant | Insignificant Significant
(a) (b)

FIGURE 10.11
INTERACTIVE FIGURE

(a) Two protons near each other experience both an attractive strong nuclear force and a repulsive electric force. At this tiny separation distance, the strong nuclear force overcomes the electric force, and the protons stay together. (b) When the two protons are relatively far from each other, the electric force is more significant and the protons repel each other. This proton–proton repulsion in large atomic nuclei reduces nuclear stability.

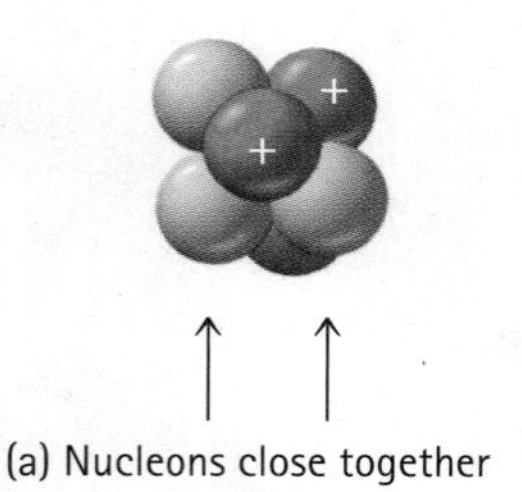

(a) Nucleons close together

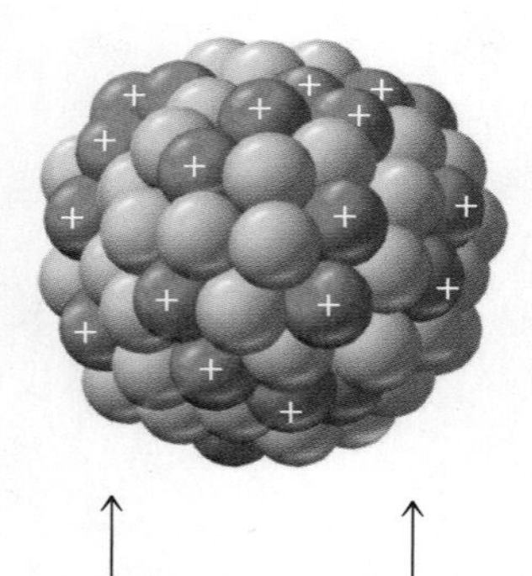

(b) Nucleons far apart

FIGURE 10.12
(a) All nucleons in a small atomic nucleus are close to one another; hence, they experience an attractive strong nuclear force. (b) Nucleons on opposite sides of a larger nucleus are not as close to one another, and so the attractive strong nuclear forces holding them together are much weaker. The result is that the large nucleus is less stable.

But for protons that are far apart, such as those on opposite edges of a large nucleus, the attractive strong nuclear force may be weaker than the repulsive electric force.

A large nucleus is not as stable as a small one. In a helium nucleus, which has two protons, each proton feels the repulsive effect of only one other proton. In a uranium nucleus, however, each of the 92 protons feels the repulsive effects of the other 91 protons! The nucleus is unstable. We see that there is a limit to the size of the atomic nucleus. For this reason, all nuclei with more than 83 protons are radioactive.

CHECK YOURSELF

Two protons in the atomic nucleus repel each other, but they are also attracted to each other. Why?

CHECK YOUR ANSWER

Although two protons repel each other by the electric force, they also attract each other by the strong nuclear force. Both of these forces act simultaneously. As long as the attractive strong nuclear force is stronger than the repulsive electric force, the protons remain together. When the electric force overcomes the strong nuclear force, however, the protons fly apart from each other.

Neutrons serve as "nuclear cement" holding the atomic nucleus together. Protons attract both protons and neutrons by the strong nuclear force. Protons also repel other protons by the electric force. Neutrons, on the other hand, have no electric charge and so attract other protons and neutrons only by the strong nuclear force. The presence of neutrons therefore adds to the attraction among nucleons and helps hold the nucleus together (Figure 10.13).

The more protons there are in a nucleus, the more neutrons are needed to help balance the repulsive electric forces. For light elements, it is sufficient to have about as many neutrons as protons. The most common isotope of carbon, C-12, for instance, has equal numbers of each—six protons and six neutrons. For large nuclei, more neutrons than protons are needed. Because the strong nuclear force diminishes rapidly over distance, nucleons must be practically touching in order for the strong nuclear force to be effective. Nucleons on opposite sides of a large atomic nucleus are not as attracted to one another. The electric force, however, does not diminish by much across the diameter of a large nucleus and so begins to win out over the strong nuclear force. To compensate for the weakening of the strong nuclear force across the diameter of the nucleus, large nuclei have more

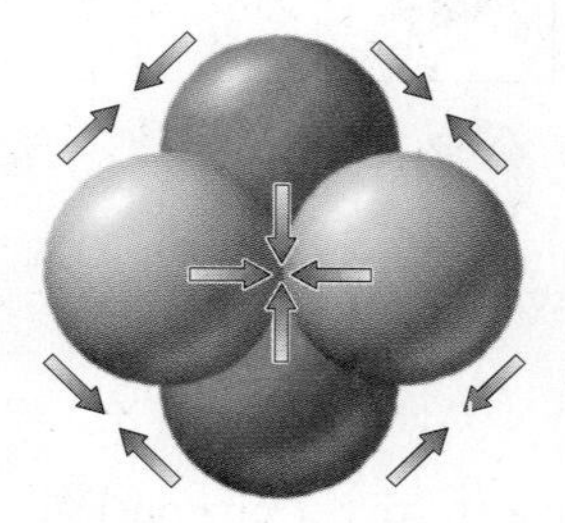

All nucleons, both protons and neutrons, attract one another by the strong nuclear force.

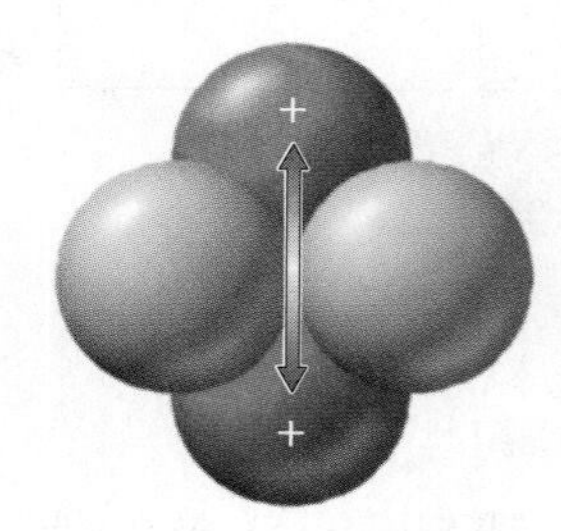

Only protons repel one another by the electric force.

FIGURE 10.13
The presence of neutrons helps hold the nucleus together by increasing the effect of the strong nuclear force, represented by the single-headed arrows.

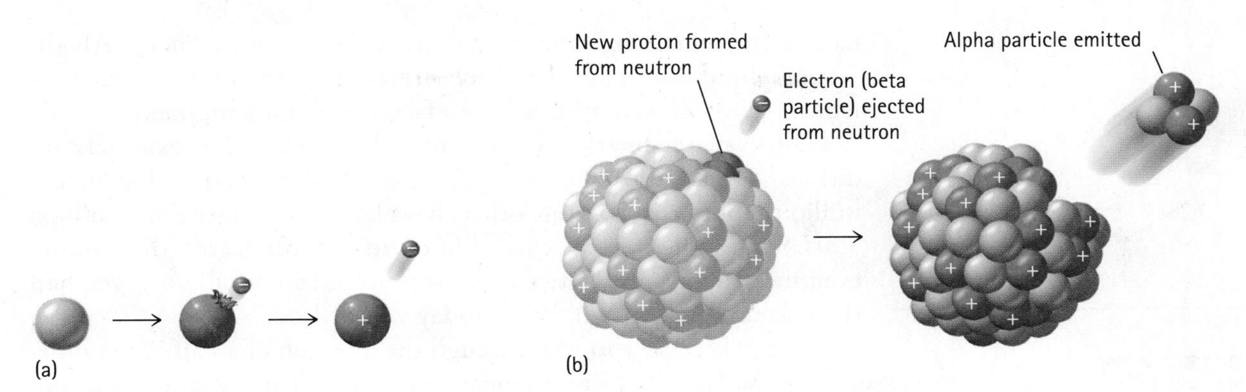

FIGURE 10.14
(a) A neutron near a proton is stable, but a neutron by itself is unstable and decays to a proton by emitting an electron. (b) Destabilized by an increase in the number of protons, the nucleus begins to shed fragments, such as alpha particles.

neutrons than protons. Lead, for example, has about one and a half times as many neutrons as protons.

So, we see that neutrons are stabilizing and large nuclei require an abundance of them. But neutrons are not always successful in keeping a nucleus intact. Interestingly, neutrons are not stable when they are by themselves. A lone neutron is radioactive and spontaneously transforms into a proton and an electron (Figure 10.14a). A neutron seems to need protons around to keep this from happening. After the size of a nucleus reaches a certain point, the neutrons so outnumber the protons that there are not enough protons in the mix to prevent the neutrons from turning into protons. As neutrons in a nucleus change into protons, the stability of the nucleus decreases because the repulsive electric force becomes increasingly significant. The result is that pieces of the nucleus fragment away in the form of radiation, as shown in Figure 10.14b.

CHECK YOURSELF
What role do neutrons serve in the atomic nucleus? What is the fate of a neutron when alone or distant from one or more protons?

CHECK YOUR ANSWERS
Neutrons serve as nuclear cement in nuclei and add to nuclear stability. But when alone or away from protons, a neutron becomes radioactive and spontaneously transforms to a proton and an electron.

10.3 Half-Life and Transmutation

LEARNING OBJECTIVE
Recognize how radioactive elements can be identified by the rate at which they decay and how this decay results in the formation of new elements.

EXPLAIN THIS How is the rate of transmutation related to half-life?

The rate of decay for a radioactive isotope is measured in terms of a characteristic time, the **half-life**. This is the time it takes for half of an original quantity of an element to decay. For example, radium-226 has a half-life of 1620 years, which means that half of a radium-226 sample will be converted into other elements by the end of 1620 years. In the next 1620 years,

MasteringPhysics®
VIDEO: Radioactive Decay
VIDEO: Half-Life

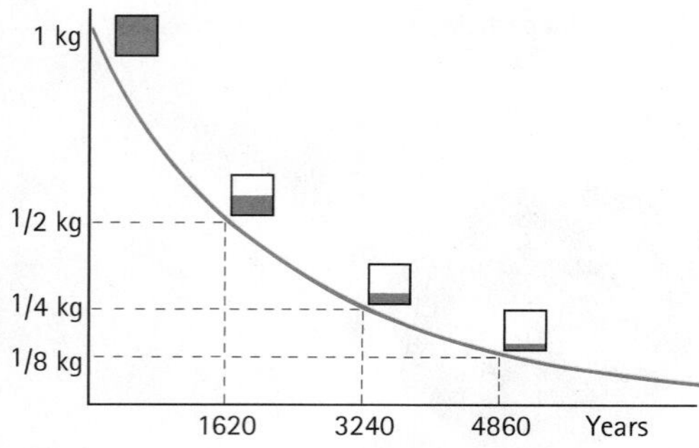

FIGURE 10.15
INTERACTIVE FIGURE

Every 1620 years the amount of radium decreases by half.

half of the remaining radium will decay, leaving only one-fourth the original amount of radium. (After 20 half-lives, the initial quantity of radium-226 will be diminished by a factor of about 1 million.)

Half-lives are remarkably constant and not affected by external conditions. Some radioactive isotopes have half-lives that are less than a millionth of a second, while others have half-lives longer than a billion years. Uranium-238 has a half-life of 4.5 billion years. All uranium eventually decays in a series of steps to lead. In 4.5 billion years, half the uranium presently in Earth today will be lead.

It is not necessary to wait through the duration of a half-life in order to measure it. The half-life of an element can be calculated at any given moment by measuring the rate of decay of a known quantity. This is easily done using a radiation detector (Figure 10.16). In general, the shorter the half-life of a substance, the faster it disintegrates and the more radioactivity per amount is detected.

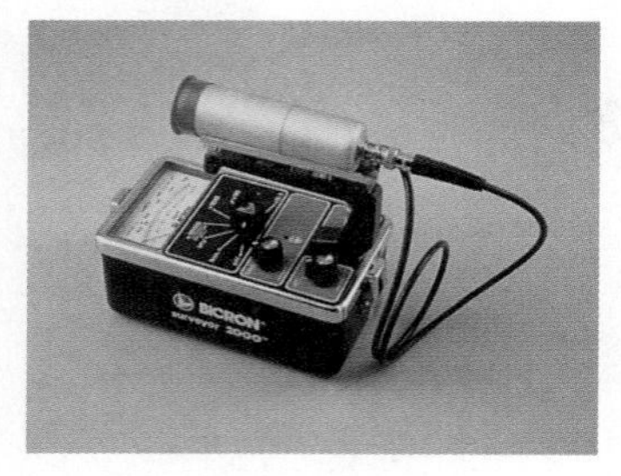

(a)

(b)

FIGURE 10.16
Radiation detectors. (a) A Geiger counter detects incoming radiation by its ionizing effect on enclosed gas in the tube. (b) A scintillation counter detects incoming radiation by flashes of light that are produced when charged particles or gamma rays pass through it.

CHECK YOURSELF

1. **If a radioactive isotope has a half-life of 1 day, how much of an original sample is left at the end of the second day? The third day?**
2. **Which gives a higher counting rate on a radiation detector—a radioactive material with a short half-life or a radioactive material with a long half-life?**

CHECK YOUR ANSWERS

1. One-fourth of the original sample is left at the end of the second day—the three-fourths that underwent decay is then a different element altogether. At the end of three days, one-eighth of the original sample remains.
2. The material with the shorter half-life is more active and shows a higher counting rate on a radiation detector.

When a radioactive nucleus emits an alpha or a beta particle, there is a change in atomic number, which means that a different element is formed. (Recall from Chapter 9 that an element is defined by its atomic number, which is the number of protons in the nucleus.) The changing of one chemical element into another is called **transmutation**. Transmutation occurs in natural events and is also initiated artificially in the laboratory.

The radioactive half-life of a material is also the time for its decay rate to reduce to half.

Natural Transmutation

Consider uranium-238, the nucleus of which contains 92 protons and 146 neutrons. When an alpha particle is ejected, the nucleus loses two protons and two neutrons. Because an element is defined by the number of protons in its nucleus, the 90 protons and 144 neutrons left behind are no longer identified as being uranium. Instead we have the nucleus of a different element—thorium. This transmutation can be written as a nuclear equation:

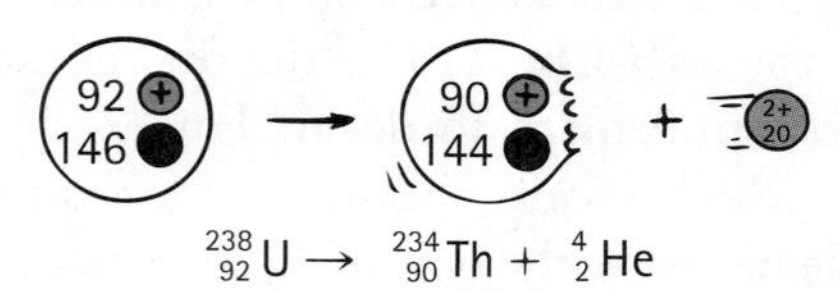

$$^{238}_{92}\text{U} \rightarrow\ ^{234}_{90}\text{Th} +\ ^{4}_{2}\text{He}$$

We see that $^{238}_{92}U$ transmutes to the two elements written to the right of the arrow. When this transmutation occurs, energy is released, partly in the form of kinetic energy of the alpha particle ($^{4}_{2}He$), partly in the kinetic energy of the thorium atom, and partly in the form of gamma radiation. In this and all such equations, the mass numbers at the top balance ($238 = 234 + 4$) and the atomic numbers at the bottom also balance ($92 = 90 + 2$).

Thorium-234, the product of this reaction, is also radioactive. When it decays, it emits a beta particle. Because a beta particle is an electron, the atomic number of the resulting nucleus is *increased* by 1. So, after beta emission by thorium with 90 protons, the resulting element has 91 protons. It is no longer thorium, but the element protactinium. Although the atomic number has increased by 1 in this process, the mass number (protons + neutrons) remains the same. The nuclear equation is

90 ⊕ 144 ● → 91 ⊕ 143 ● + ⊖

$$^{234}_{90}Th \rightarrow \, ^{234}_{91}Pa + \, ^{0}_{-1}e$$

We write an electron as $^{0}_{-1}e$. The superscript 0 indicates that the electron's mass is insignificant relative to that of protons and neutrons. The subscript -1 is the electric charge of the electron.

We see that when an element ejects an alpha particle from its nucleus, the mass number of the resulting atom is decreased by 4 and its atomic number is decreased by 2. The resulting atom is an element two places back in the periodic table of the elements. When an element ejects a beta particle from its nucleus, the mass of the atom is practically unaffected, meaning there is no change in mass number, but its atomic number increases by 1. The resulting atom belongs to an element one place forward in the periodic table. Gamma radiation results in no change in either the mass number or the atomic number. So, we see that radioactive elements can decay backward or forward in the periodic table.

The successions of radioactive decays of $^{238}_{92}U$ to $^{206}_{82}Pb$, an isotope of lead, are shown in Figure 10.17. Each gray arrow shows an alpha decay, and each red arrow shows a beta decay. Notice that some of the nuclei in the series can decay in both ways. This is one of several similar radioactive series that occur in nature.

MasteringPhysics®
TUTORIAL: Nuclear Chemistry
VIDEO: Plutonium

FYI Beta emission is also accompanied by the emission of a neutrino, which is a neutral particle with nearly zero mass that travels at about the speed of light. Neutrinos are hard to detect because they interact very weakly with matter—a piece of lead about 8 light-years thick would be needed to stop half the neutrinos produced in typical nuclear decays. Thousands of neutrinos are flying through you every second of every day because the universe is filled with them. Only occasionally, one or two times a year or so, does a neutrino or two interact with the matter of your body.

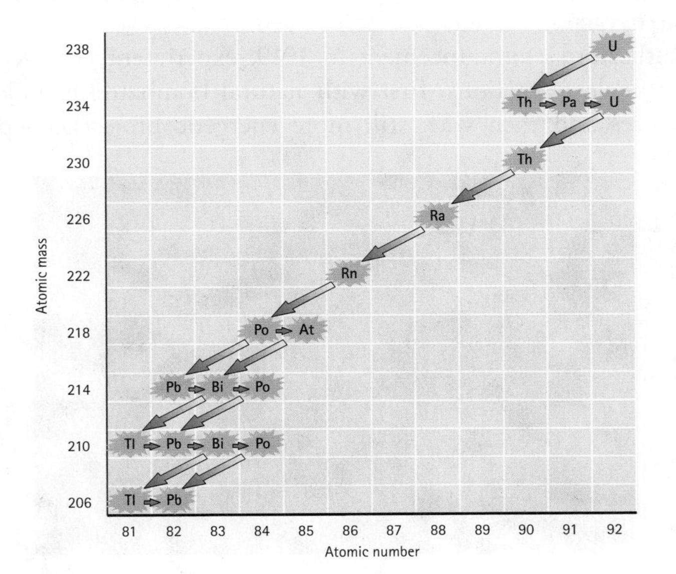

FIGURE 10.17
U-238 decays to Pb-206 through a series of alpha and beta decays.

The alchemists of old tried in vain to cause the transmutation of one element into another. Despite their fervent efforts and rituals, they never came close to succeeding. Ironically, natural transmutations were going on all around them.

CHECK YOURSELF

1. Complete the following nuclear reactions:

a. $^{226}_{88}\text{Ra} \rightarrow {}^{?}_{?}? + {}^{0}_{-1}e$

b. $^{209}_{84}\text{Po} \rightarrow {}^{205}_{82}\text{Pb} + {}^{?}_{?}?$

2. What finally becomes of all the uranium that undergoes radioactive decay?

CHECK YOUR ANSWERS

1. a. $^{226}_{88}\text{Ra} \rightarrow {}^{226}_{89}\text{Ac} + {}^{0}_{-1}e$

   b. $^{209}_{84}\text{Po} \rightarrow {}^{205}_{82}\text{Pb} + {}^{4}_{2}\text{He}$

2. All uranium ultimately becomes lead. On the way to becoming lead, it exists as a series of elements, as indicated in Figure 10.17.

Artificial Transmutation

Ernest Rutherford, in 1919, was the first of many investigators to succeed in transmuting a chemical element. He bombarded nitrogen gas with alpha particles from a piece of radioactive ore. The impact of an alpha particle on a nitrogen nucleus transmutes nitrogen into oxygen:

$$^{4}_{2}\text{He} + {}^{14}_{7}\text{N} \rightarrow {}^{17}_{8}\text{O} + {}^{1}_{1}\text{H}$$

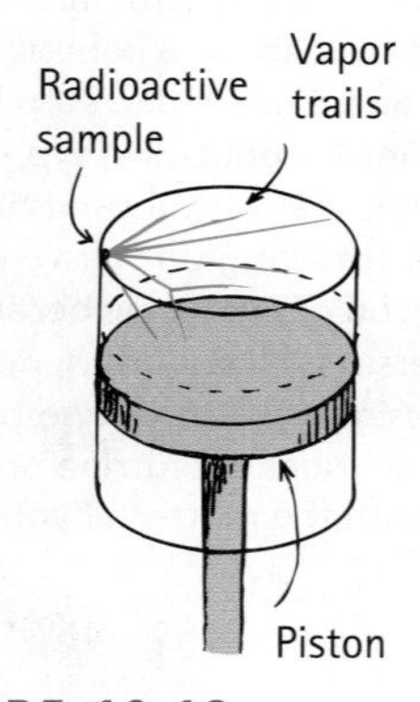

FIGURE 10.18
A cloud chamber. Charged particles moving through supersaturated vapor leave trails. When the chamber is in a strong electric or magnetic field, the bending of the tracks provides information about the charge, mass, and momentum of the particles.

Rutherford used a device called a *cloud chamber* to record this event (Figure 10.18). In a cloud chamber, moving charged particles show a trail of ions along their path in a way similar to the ice crystals that show the trails of jet planes high in the sky. From a quarter of a million cloud-chamber tracks photographed on movie film, Rutherford showed seven examples of atomic transmutation. Analysis of tracks bent by a strong external magnetic field showed that when an alpha particle collided with a nitrogen atom, a proton bounced out and the heavy atom recoiled a short distance. The alpha particle disappeared. The alpha particle was absorbed in the process, transforming nitrogen into oxygen.

Since Rutherford's announcement in 1919, experimenters have carried out many other nuclear reactions, first with natural bombarding projectiles from radioactive ores and then with still more energetic projectiles—protons and

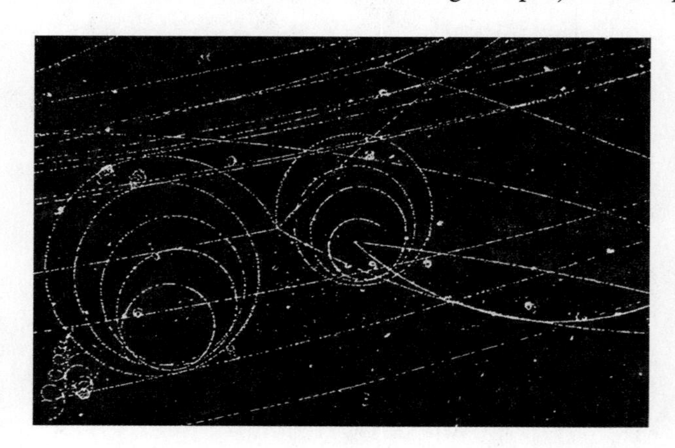

FIGURE 10.19
Tracks of elementary particles in a bubble chamber, a similar yet more complicated device than a cloud chamber. Two particles have been destroyed at the points where the spirals emanate, and four have been created in the collision.

electrons hurled by huge particle accelerators. Artificial transmutation produces the hitherto unknown synthetic elements at the upper end of the periodic table. All of these artificially made elements have short half-lives. If they ever existed naturally when Earth was formed, they have long since decayed.

Integrated Science 10B
BIOLOGY AND EARTH SCIENCE

Radiometric Dating

LEARNING OBJECTIVE
Review how the age of ancient artifacts can be determined by measuring the amounts of remaining radioactivity they contain.

EXPLAIN THIS How does radioactivity enable archaeologists to measure the age of ancient artifacts?

Earth's atmosphere is continuously bombarded by cosmic rays, and this bombardment causes many atoms in the upper atmosphere to transmute. These transmutations result in many protons and neutrons being "sprayed out" into the environment. Most of the protons are stopped as they collide with the atoms of the upper atmosphere, stripping electrons from these atoms to become hydrogen atoms. The neutrons, however, keep going for longer distances because they have no electric charge and therefore do not interact electrically with matter. Eventually, many of them collide with the nuclei in the denser lower atmosphere. A nitrogen nucleus that captures a neutron, for instance, becomes an isotope of carbon by emitting a proton:

$$^{1}_{0}n + {}^{14}_{7}N \rightarrow {}^{14}_{6}C + {}^{1}_{1}H$$

This carbon-14 isotope, which makes up less than one-millionth of 1% of the carbon in the atmosphere, is radioactive and has eight neutrons. (The most common isotope, carbon-12, has six neutrons and is not radioactive.) Because both carbon-12 and carbon-14 are forms of carbon, they have the same chemical properties. Both these isotopes can chemically react with oxygen to form carbon dioxide, which is taken in by plants. This means that all plants contain a tiny bit of radioactive carbon-14. All animals eat plants (or at least plant-eating animals) and therefore have a little carbon-14 in them. In short, all living things on Earth contain some carbon-14.

FYI A 1-g sample of carbon from recently living matter contains about 50 trillion billion (5×10^{22}) carbon atoms. Of these carbon atoms, about 65 billion (6.5×10^{10}) are the radioactive C-14 isotope. This gives the carbon a beta disintegration rate of about 13.5 decays per minute.

Carbon-14 is a beta emitter and decays back to nitrogen by the following reaction:

$$^{14}_{6}C \rightarrow {}^{14}_{7}N + {}^{0}_{-1}e$$

Because plants continue to take in carbon dioxide as long as they live, any carbon-14 lost by decay is immediately replenished with fresh carbon-14 from the atmosphere. In this way, a radioactive equilibrium is reached at which there is a constant ratio of about one carbon-14 atom to every 100 billion carbon-12 atoms. When a plant dies, replenishment of carbon-14 stops. Then the percentage of carbon-14 decreases

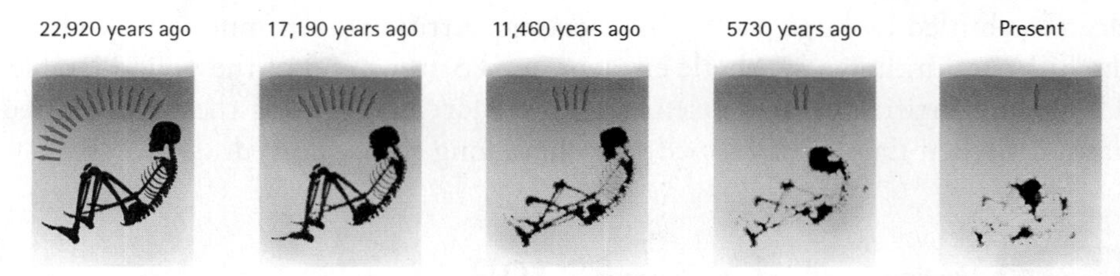

FIGURE 10.20
The amount of radioactive carbon-14 in the skeleton diminishes by half every 5730 years, with the result that today the skeleton contains only a fraction of the carbon-14 it originally had. The red arrows symbolize relative amounts of carbon-14.

at a constant rate given by its half-life. The longer a plant or other organism is dead, therefore, the less carbon-14 it contains relative to the constant amount of carbon-12.

The half-life of carbon-14 is about 5730 years. This means that half of the carbon-14 atoms that are now present in a plant or animal that dies today will decay over the next 5730 years. Half of the remaining carbon-14 atoms will then decay over the following 5730 years, and so forth.

With this knowledge, scientists can calculate the age of carbon-containing artifacts, such as wooden tools or skeletons, by measuring their current level of radioactivity. This process, known as **carbon-14 dating**, enables us to probe as much as 50,000 years into the past. Beyond this time span, too little carbon-14 remains to permit accurate analysis.

Carbon-14 dating would be an extremely simple and accurate dating method if the amount of radioactive carbon in the atmosphere had been constant over the ages. But it hasn't been. Fluctuations in the Sun's magnetic field as well as changes in the strength of Earth's magnetic field affect cosmic-ray intensities in Earth's atmosphere, which in turn produce fluctuations in the production of C-14. In addition, changes in Earth's climate affect the amount of carbon dioxide in the atmosphere. The oceans are great reservoirs of carbon dioxide. When the oceans are warm, they release more carbon dioxide into the atmosphere than when they are cold. (We'll return to the oceans and their important interplay with carbon dioxide in Chapter 25.)

One ton of ordinary granite contains about 9 g of uranium and 20 g of thorium. Basalt rocks contain 3.5 g and 7.7 g of the same elements, respectively.

CHECK YOURSELF

Suppose an archaeologist extracts a gram of carbon from an ancient axe handle and finds it one-fourth as radioactive as a gram of carbon extracted from a freshly cut tree branch. About how old is the axe handle?

CHECK YOUR ANSWER

Assuming the ratio of C-14 to C-12 was the same when the axe was made, the axe handle is as old as two half-lives of C-14, or about 11,460 years old.

The dating of older, but nonliving, materials is accomplished with radioactive minerals, such as uranium. The naturally occurring isotopes U-238 and U-235 decay very slowly and ultimately become isotopes of lead—but not the common lead isotope Pb-208. For example, U-238 decays through several stages to finally become Pb-206, whereas U-235 finally becomes the isotope Pb-207. Lead isotopes 206 and 207 that now exist were at one time uranium. The older the rock, the higher the percentage of these remnant isotopes.

From the half-lives of uranium isotopes and the percentage of lead isotopes in uranium-bearing rock, it is possible to calculate the date at which the rock was formed. (We'll return to isotopic dating when we investigate Earth's dynamic interior in Chapter 22.)

10.4 Nuclear Fission

LEARNING OBJECTIVE
Describe the process by which large atomic nuclei can split in half, leading to the production of energy.

EXPLAIN THIS Why isn't it possible for a nuclear power plant to explode like a nuclear bomb?

MasteringPhysics®
VIDEO: Nuclear Fission

In 1938, two German scientists, Otto Hahn and Fritz Strassmann, made an accidental discovery that was to change the world. While bombarding a sample of uranium with neutrons in the hope of creating new, heavier elements, they were astonished to find chemical evidence for the production of barium, an element with about half the mass of uranium. Hahn wrote of this news to his former colleague Lise Meitner, who had fled from Nazi Germany to Sweden because of her Jewish ancestry. From Hahn's evidence, Meitner concluded that the uranium nucleus, activated by neutron bombardment, had split in half. Soon thereafter, Meitner, working with her nephew Otto Frisch, also a physicist, published a paper in which the term *nuclear fission* was first coined.

In the nucleus of every atom is a delicate balance between attractive nuclear forces and repulsive electric forces between protons. In all known nuclei, the nuclear forces dominate. In certain isotopes of uranium, however, this domination is tenuous. If a uranium nucleus stretches into an elongated shape (Figure 10.21), the electric forces may push it into an even more elongated shape. If the elongation passes a certain point, the electric forces overwhelm the strong nuclear forces, and the nucleus splits. This is **nuclear fission**.

The energy released by the fission of one U-235 nucleus is enormous—about 7 million times the energy released by the explosion of one TNT molecule. This energy is mainly in the form of kinetic energy of the fission fragments that fly apart from one another, with some energy given to ejected neutrons and the rest to gamma radiation.

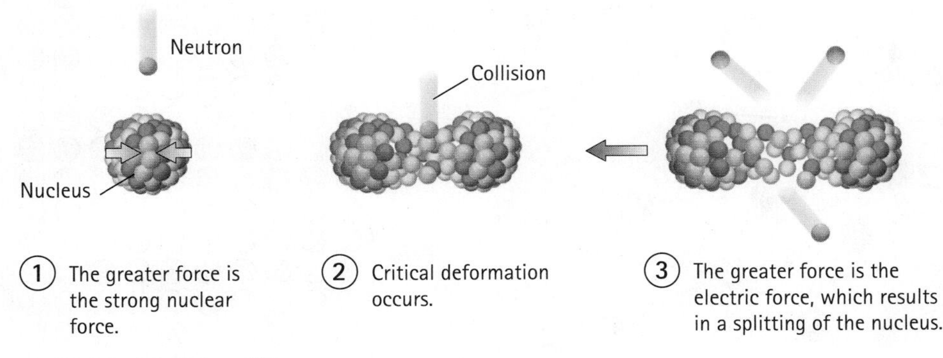
Neutron
Collision
Nucleus
1 The greater force is the strong nuclear force.
2 Critical deformation occurs.
3 The greater force is the electric force, which results in a splitting of the nucleus.

FIGURE 10.21
INTERACTIVE FIGURE (MP)

Nuclear deformation may result in repulsive electric forces overcoming attractive nuclear forces, in which case fission occurs.

A typical uranium fission reaction is

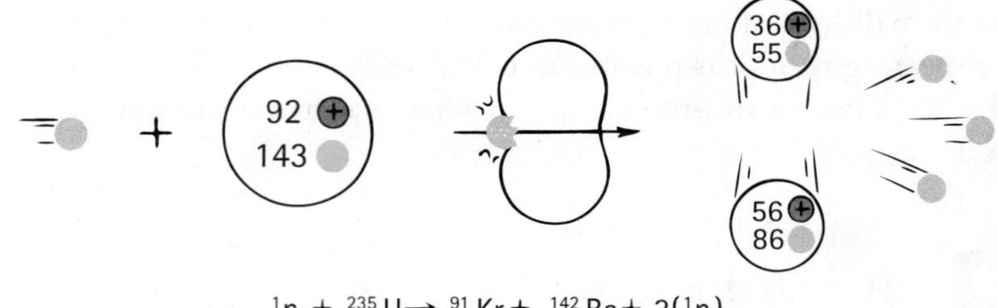

$$^{1}_{0}n + ^{235}_{92}U \rightarrow ^{91}_{36}Kr + ^{142}_{56}Ba + 3(^{1}_{0}n)$$

FYI Otto Hahn, rather than Lise Meitner, received the Nobel Prize for the work on nuclear fission. Notoriously, Hahn didn't even acknowledge Meitner's role. See more about this in the readable book $E = mc^2$ by David Bodanis.

Note in this reaction that 1 neutron starts the fission of a uranium nucleus and that the fission produces 3 neutrons. (A fission reaction may produce fewer or more than 3 neutrons.) These product neutrons can cause the fissioning of 3 other uranium atoms, releasing 9 more neutrons. If each of these 9 neutrons succeeds in splitting a uranium atom, the next step in the reaction produces 27 neutrons, and so on. Such a sequence, illustrated in Figure 10.22, is called a **chain reaction**—a self-sustaining reaction in which the products of one reaction event stimulate further reaction events.

Why don't chain reactions occur in naturally occurring uranium ore deposits? They would if all uranium atoms fissioned so easily. Fission occurs mainly for the rare isotope U-235, which makes up only 0.7% of the uranium in naturally occurring uranium metal. When the more abundant isotope U-238 absorbs neutrons created by fission of U-235, the U-238 typically does not undergo fission. So, any chain reaction is snuffed out by the neutron-absorbing U-238 as well as by the rock in which the ore is imbedded.

UNIFYING CONCEPT

- ***Exponential Growth and Decay***
Appendix D

If a chain reaction occurred in a baseball-size chunk of pure U-235, an enormous explosion would result. If the chain reaction were started in a smaller chunk of pure U-235, however, no explosion would occur. This is because of geometry: The ratio of surface area to mass is higher in a small piece than in a large one (just as there is more skin on six small potatoes with a combined mass of 1 kg than there is on a single 1-kg potato). So there is more surface area on a bunch of small pieces of uranium than on a large piece. In a small piece of U-235, neutrons leak through the surface before an explosion can occur. In a bigger piece, the chain

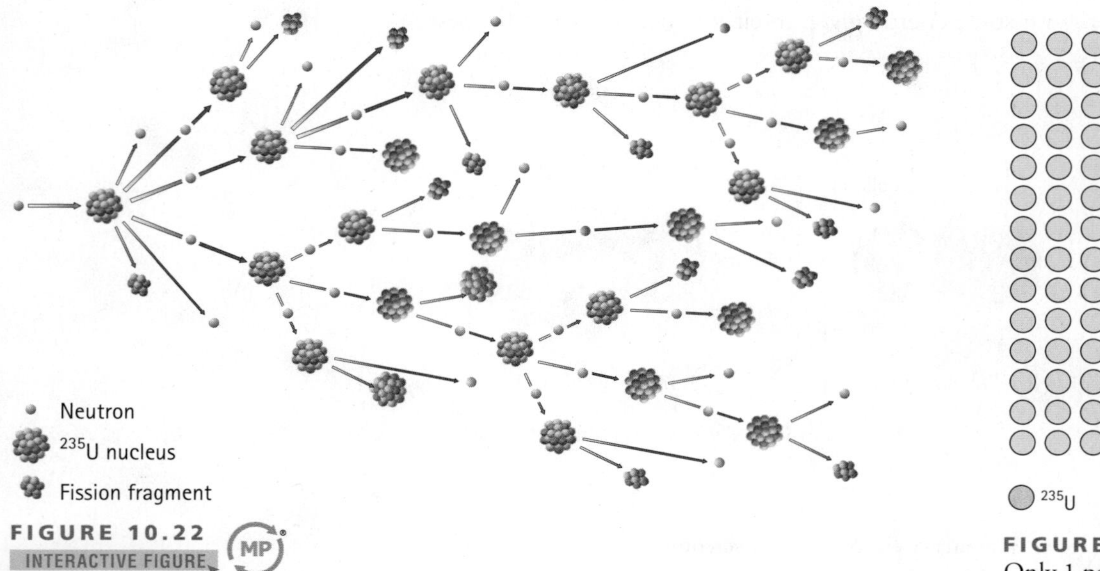

FIGURE 10.22
INTERACTIVE FIGURE
A chain reaction.

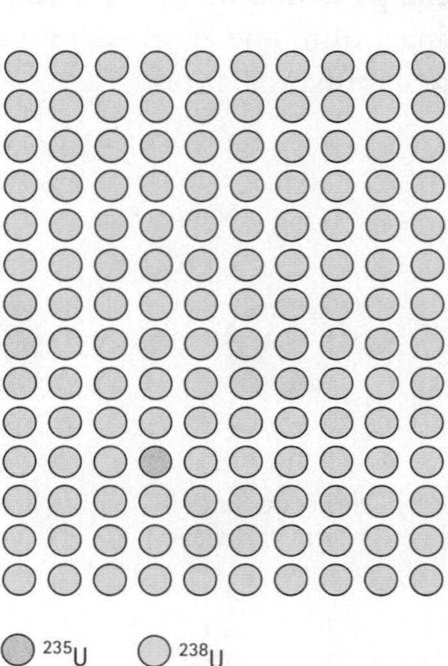

FIGURE 10.23
Only 1 part in 140 of naturally occurring uranium is U-235.

reaction builds up to enormous energies before the neutrons get to the surface and escape (Figure 10.24). For masses greater than a certain minimum amount, called the **critical mass**, an explosion of enormous magnitude may take place.

Consider a large quantity of U-235 divided into two pieces, each with a mass less than critical. The units are *subcritical*. Neutrons in either piece readily reach a surface and escape before a sizable chain reaction builds up. But, if the pieces are suddenly driven together, the total surface area decreases. If the timing is right and the combined mass is greater than critical, a violent explosion takes place. This is what happens in a nuclear fission bomb (Figure 10.25).

Constructing a fission bomb is a formidable task. The difficulty is separating enough U-235 from the more abundant U-238. Scientists took more than two years to extract enough U-235 from uranium ore to make the bomb that was detonated at Hiroshima in 1945. To this day uranium isotope separation remains a difficult process.

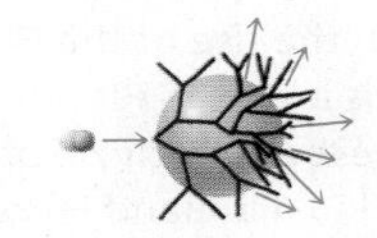

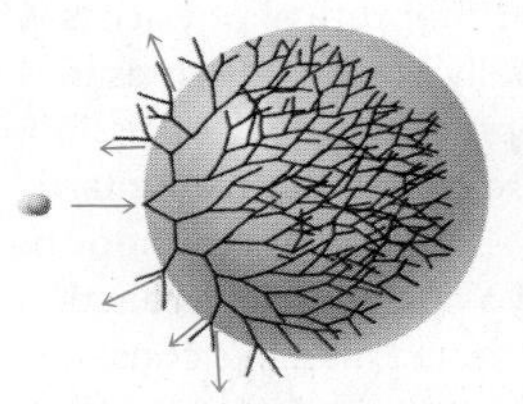

FIGURE 10.24
The exaggerated view shows that a chain reaction in a small piece of pure U-235 runs its course before it can cause a large explosion because neutrons leak from the surface too soon. The surface area of the small piece is large relative to the mass. In a larger piece, more uranium and a smaller surface area are presented to the neutrons.

CHECK YOURSELF
A 1-kg ball of U-235 is at critical mass, but the same ball broken up into small chunks is not. Explain.

CHECK YOUR ANSWER
The small chunks have a larger combined surface area than the ball from which they came (just as the combined surface area of gravel is larger than the surface area of a boulder of the same mass). Neutrons escape via the surface before a sustained chain reaction can build up.

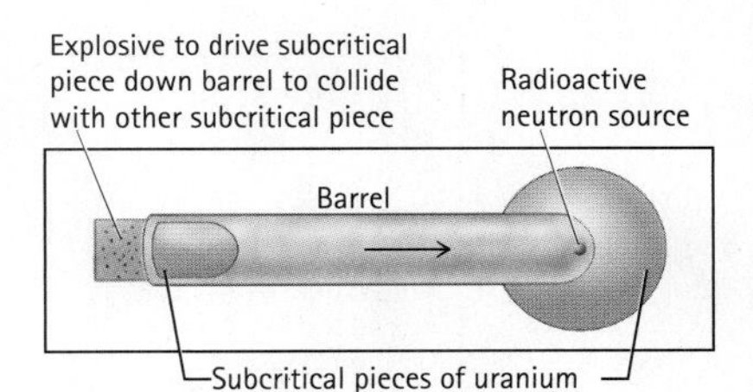

FIGURE 10.25
Simplified diagram of a uranium fission bomb.

Nuclear Fission Reactors

The awesome energy of nuclear fission was introduced to the world in the form of nuclear bombs, and this violent image still colors our thinking about nuclear power, making it difficult for many people to recognize its potential usefulness. Currently, about 20% of electric energy in the United States is generated by *nuclear fission reactors* (whereas in some other countries most of the electric power is nuclear—about 75% in France). These reactors are simply nuclear furnaces. They, like fossil fuel furnaces, do nothing more elegant than boil water to produce steam for a turbine (Figure 10.26). The greatest practical difference is the amount

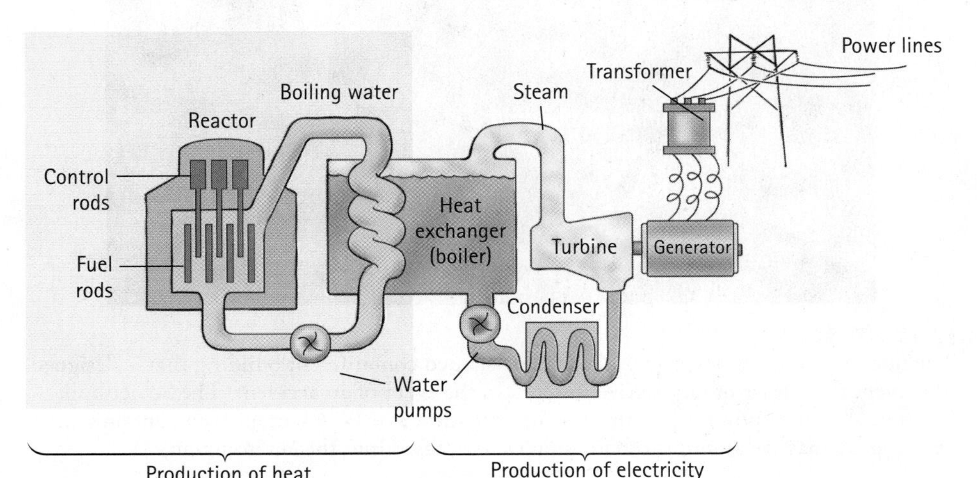

FIGURE 10.26
Diagram of a nuclear fission power plant. Note that the water in contact with the fuel rods is completely contained, and radioactive materials are not involved directly in the generation of electricity.

FYI With the rise of the German Nazis in the 1930s, many scientists, especially those of Jewish ancestry, fled mainland Europe to America. They included dozens of brilliant theoretical physicists who eventually played key roles in the development of nuclear fission. Of these physicists, Leo Szilard (1898–1964) first envisioned the idea of a chain nuclear reaction. With Albert Einstein's consent, Szilard drafted a letter that was signed by Einstein and delivered to President Roosevelt in 1939. This letter outlined the possibility of the chain reaction and its implications for a nuclear bomb. Within six years the first test nuclear bomb was exploded in the desert in New Mexico. In 1945, Szilard generated a petition in which 68 of the scientists involved in the nuclear program asked President Truman not to drop the atomic bomb on a populous Japanese city, such as Nagasaki.

of fuel involved: A mere kilogram of uranium fuel, smaller than a baseball, yields more energy than 30 freight-car loads of coal.

A fission reactor contains four components: nuclear fuel, control rods, moderator (to slow neutrons, which is required for fission), and liquid (usually water) to transfer heat from the reactor to the turbine and generator. The nuclear fuel is primarily U-238 plus about 3% U-235. Because the U-235 isotopes are so highly diluted with U-238, an explosion like that of a nuclear bomb is not possible. The reaction rate, which depends on the number of neutrons that initiate the fission of other U-235 nuclei, is controlled by rods inserted into the reactor. The control rods are made of a neutron-absorbing material, usually the metal cadmium or boron.

Heated water around the nuclear fuel is kept under high pressure to keep it at a high temperature without boiling. It transfers heat to a second lower-pressure water system, which operates the turbine and electric generator in a conventional fashion. In this design, two separate water systems are used so that no radioactivity reaches the turbine or the outside environment.

A significant disadvantage of fission power is the generation of radioactive waste products. Light atomic nuclei are most stable when composed of equal numbers of protons and neutrons, as discussed earlier, and heavy nuclei need more neutrons than protons for stability. For example, U-235 has 143 neutrons but only 92 protons. When uranium fissions into two medium-weight elements, the extra neutrons in their nuclei make them unstable. They are radioactive, most with very short half-lives, but some with half-lives of thousands of years. Safely disposing of these waste products as well as materials made radioactive in the production of nuclear fuels requires special storage casks and procedures. Although fission has been successfully producing electricity for a half century, disposing of radioactive wastes in the United States remains problematic.

The designs for nuclear power plants have progressed over the years. The earliest designs from the 1950s through the 1990s are called the Generation I, II, and III reactors. The safety systems of these reactors are "active" in that they rely on

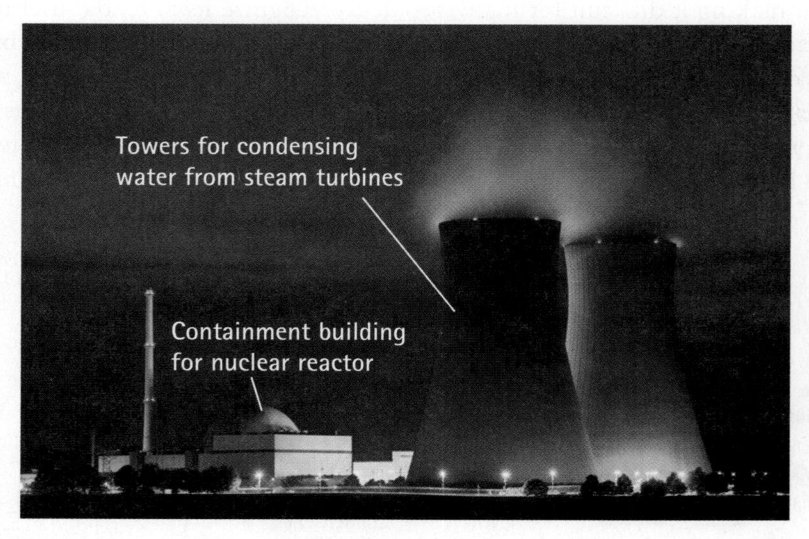

FIGURE 10.27
The nuclear reactor is housed within a dome-shaped containment building that is designed to prevent the release of radioactive isotopes in the event of an accident. The Soviet-built Chernobyl nuclear power plant that reached meltdown in 1986 had no such containment building, so massive amounts of radiation were released into the environment.

active measures, such as water pumps, to keep the reactor core cool in the event of an accident. Notably, these active measures failed when Japan's Generation II Fukushima Daiichi nuclear plant was hit by a powerful earthquake and tsunami in 2011. Though not yet operational, the latest Generation IV nuclear reactors will have fundamentally different designs. For example, they will incorporate passive safety measures that cause the reactor to shut down by itself in the event of an emergency. The fuel source may be the depleted uranium stockpiled from earlier reactors. Furthermore, these reactors can be built as small modular units that generate between 150 and 600 megawatts of power rather than the 1500 megawatts that is the usual output of today's reactors. Smaller reactors are easier to manage and can be used to build a generating capacity suited to the community being served.

A nuclear power plant "meltdown" occurs when the fissioning nuclear fuels are no longer submerged within a cooling fluid, such as water. The temperature rises to the point that the solid nuclear fuel, and the reaction vessel itself, melt into a liquid phase that has the potential of penetrating through the floor of the containment building.

The benefits of fission power include plentiful electricity and the conservation of many billions of tons of fossil fuels. Every year these fuels are turned to heat, smoke, and megatons of poisonous gases such as sulfur oxides. Notably, fossil fuels are far more precious as sources of organic molecules, which, as we will discuss in Chapter 19, can be used to create medicines, clothing, automobiles, and much more.

FYI Recent evidence discovered by neutrino research in 2011 indicates that a major source of Earth's internal energy, perhaps half, is due to nuclear fission within Earth's core. This heat-generating process is occurring at great depths beneath your feet right now! Indeed, power from the atomic nuclei is as old as Earth itself.

CHECK YOURSELF

Coal contains tiny quantities of radioactive materials, enough that more environmental radiation surrounds a typical coal-fired power plant than a fission power plant. What does this indicate about the shielding typically surrounding the two types of power plants?

CHECK YOUR ANSWER

Coal-fired power plants are as American as apple pie, with no required (and expensive) shielding to restrict the emissions of radioactive particles. Nukes, on the other hand, are required to have shielding to ensure strict low levels of radioactive emissions.

The Breeder Reactor

One of the fascinating features of fission power is the breeding of fission fuel from nonfissionable U-238. This breeding occurs when small amounts of fissionable isotopes are mixed with U-238 in a reactor. Fission liberates neutrons that convert the relatively abundant nonfissionable U-238 to U-239, which beta-decays to Np-239, which in turn beta-decays to fissionable plutonium—Pu-239. So, in addition to the abundant energy produced, fission fuel is bred from the relatively abundant U-238 in the process.

An average ton of coal contains 1.3 parts per million (ppm) of uranium and 3.2 ppm of thorium. That's why the average coal-burning power plant produces much more airborne radioactive material than a nuclear power plant.

Breeding occurs to some extent in all fission reactors, but a reactor specifically designed to breed more fissionable fuel than is put into it is called a *breeder reactor*. Using a breeder reactor is like filling your car's gas tank with water, adding some gasoline, then driving the car and having more gasoline after the trip than at the beginning! The basic principle of the breeder reactor is very attractive: After a few years of operation a breeder-reactor power plant can produce vast amounts of power while breeding twice as much fuel as its original fuel.

The downside is the enormous complexity of successful and safe operation. The United States gave up on breeders about two decades ago, and only Russia,

France, Japan, and India are still investing in them. Officials in these countries point out that the supplies of naturally occurring U-235 are limited. At present rates of consumption, all natural sources of U-235 may be depleted within a century. If countries then decide to turn to breeder reactors, they may well find themselves digging up the radioactive wastes they once buried.

LEARNING OBJECTIVE
Show how the mass of a nucleon depends on the identity of the nucleus within which it is contained.

10.5 Mass–Energy Equivalence

EXPLAIN THIS Why does it get easier to pull nucleons away from nuclei heavier than iron?

In the early 1900s, Albert Einstein discovered that mass is actually "congealed" energy. Mass and energy are two sides of the same coin, as stated in his celebrated equation $E = mc^2$. In this equation E stands for the energy that any mass has at rest, m stands for the mass, and c is the speed of light. This relationship between energy and mass is the key to understanding why and how energy is released in nuclear reactions.

UNIFYING CONCEPT
- ***Mass–Energy Equivalence***

Is the mass of a nucleon inside a nucleus the same as the mass of the same nucleon outside a nucleus? This question can be answered by considering the work that would be required to separate nucleons from a nucleus. From physics we know that work, which is expended energy, equals *force* × *distance*. Think of the amount of force required to pull a nucleon out of the nucleus through a sufficient distance to overcome the attractive strong nuclear force, comically indicated in Figure 10.28. Enormous work would be required. This work is energy added to the nucleon that is pulled out.

FIGURE 10.28
Work is required to pull a nucleon from an atomic nucleus. This work increases the energy and hence the mass of the nucleon outside the nucleus.

According to Einstein's equation, this newly acquired energy reveals itself as an increase in the nucleon's mass. The mass of a nucleon outside a nucleus is greater than the mass of the same nucleon locked inside a nucleus. For example, a carbon-12 atom—the nucleus of which is made up of six protons and six neutrons—has a mass of exactly 12.00000 atomic mass units (amu). Therefore on average, each nucleon contributes a mass of 1 amu. However, outside the nucleus, a proton has a mass of 1.00728 amu and a neutron has a mass of 1.00867 amu. Thus we see that the combined mass of six free protons and six free neutrons—$(6 \times 1.00728) + (6 \times 1.00867) = 12.09570$—is greater than the mass of one carbon-12 nucleus. The greater mass reflects the energy required to pull the nucleons apart from one another. Thus, what mass a nucleon has depends on where the nucleon is.

A graph of the nuclear masses for the elements from hydrogen through uranium is shown in Figure 10.29. The graph slopes upward with increasing atomic number as expected: Elements are more massive as atomic number increases. The slope curves because there are proportionally more neutrons in the more massive atoms.

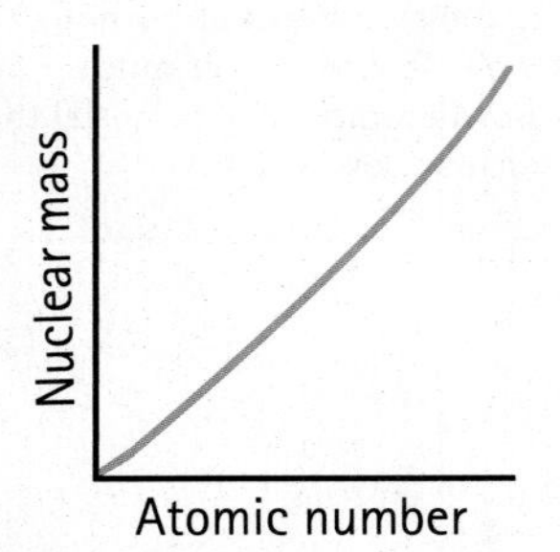

FIGURE 10.29
The plot shows how nuclear mass increases with increasing atomic number.

A more important graph results from the plot of nuclear mass *per nucleon* from hydrogen through uranium (Figure 10.30). This is perhaps the most important graph in this book because it is the key to understanding the energy associated with nuclear processes.

Note that the masses of the nucleons are different when combined in different nuclei. The greatest mass per nucleon occurs for the proton alone, hydrogen, because it has no binding energy to pull its mass down. Progressing beyond hydrogen, the mass per nucleon is less, and is least for a nucleon in the nucleus of the iron

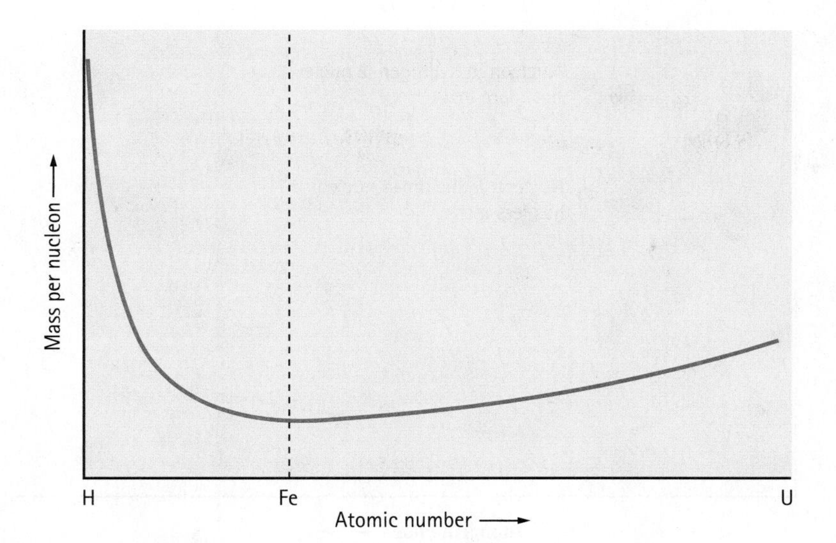

FIGURE 10.30
This graph shows that the average mass of a nucleon depends on which nucleus it is in. Individual nucleons have the greatest mass in the lightest nuclei, the least mass in iron, and intermediate mass in the heaviest nuclei.

atom. Beyond iron, the process reverses itself as nucleons have progressively greater and greater mass in atoms of increasing atomic number. This continues all the way to uranium and elements heavier than uranium.

From Figure 10.30 we can see how energy is released when a uranium nucleus splits into two nuclei of lower atomic number. Uranium, being toward the right-hand side of the graph, is shown to have a relatively large amount of mass per nucleon. When the uranium nucleus splits in half, however, smaller nuclei of lower atomic numbers are formed. As shown in Figure 10.31, these nuclei are lower on the graph than uranium, which means that they have a smaller amount of mass per nucleon. Thus, nucleons lose mass in their transition from being in a uranium nucleus to being in one of its fragments. When this decrease in mass is multiplied by the speed of light squared (c^2 in Einstein's equation), the product is equal to the energy yielded by each uranium nucleus as it undergoes fission.

MasteringPhysics®
VIDEO: Controlling Nuclear Fusion

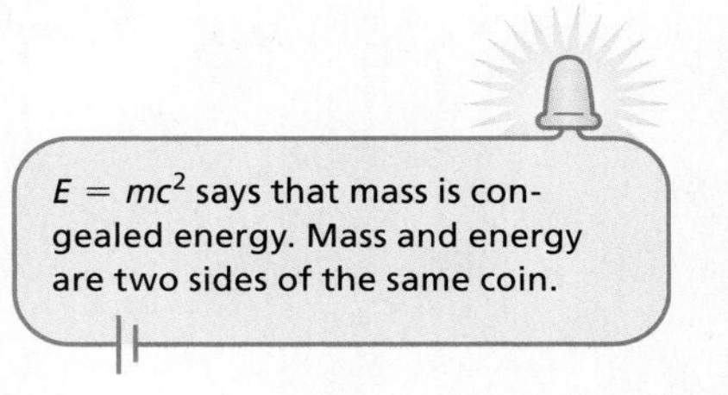

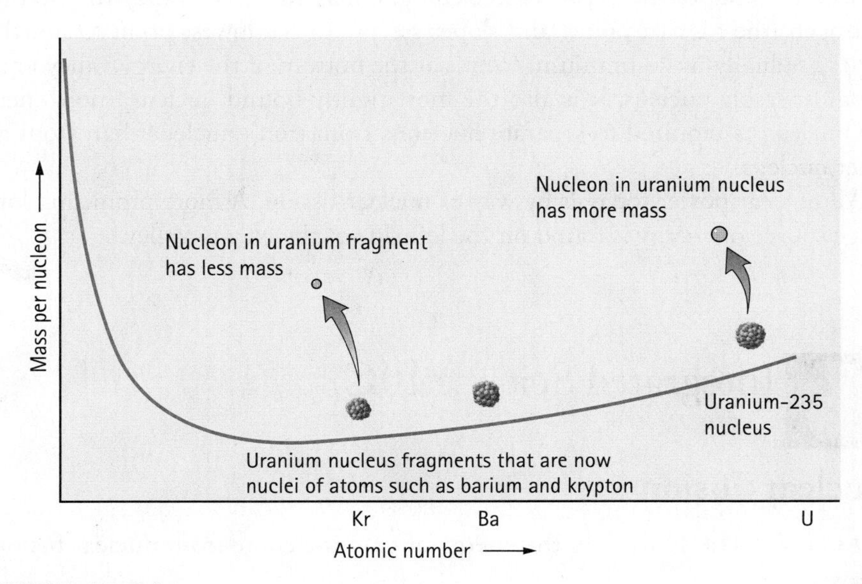

FIGURE 10.31
The mass of each nucleon in a uranium nucleus is greater than the mass of each nucleon in any one of its nuclear fission fragments. This lost mass is mass that has been transformed into energy, which is why nuclear fission is an energy-releasing process.

FIGURE 10.32
INTERACTIVE FIGURE

The mass of each nucleon in a hydrogen-2 nucleus is greater than the mass of each nucleon in a helium-4 nucleus, which results from the fusion of two hydrogen-2 nuclei. This lost mass has been converted into energy, which is why nuclear fusion is an energy-releasing process.

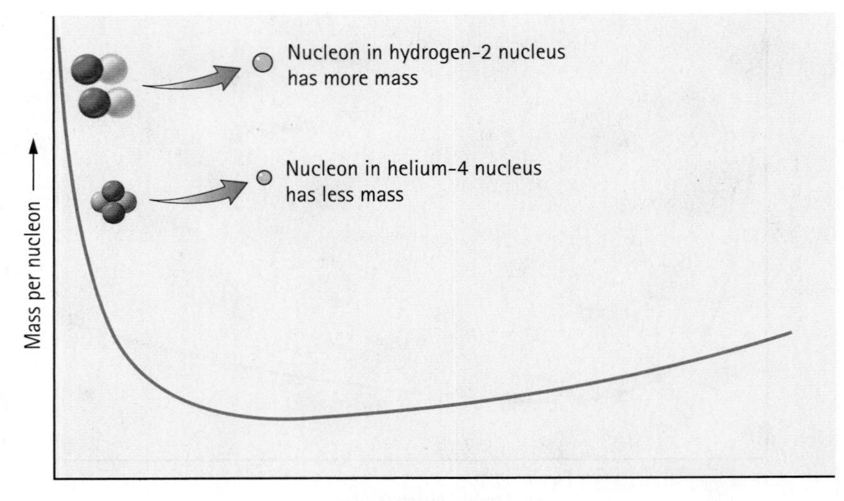

CHECK YOURSELF

Correct the following incorrect statement: When a heavy element such as uranium undergoes fission, there are fewer nucleons after the reaction than before.

CHECK YOUR ANSWER

When a heavy element such as uranium undergoes fission, there aren't fewer nucleons after the reaction. Instead, there's *less mass* in the same number of nucleons.

The graph of Figure 10.30 (and Figures 10.31 and 10.32) reveals the energy of the atomic nucleus, a primary source of energy in the universe—which is why it can be considered the most important graph in this book.

We can think of the mass-per-nucleon graph as an energy valley that starts at hydrogen (the highest point) and slopes steeply to the lowest point (iron), then slopes gradually up to uranium. Iron is at the bottom of the energy valley and is the most stable nucleus. It is also the most tightly bound nucleus; more energy per nucleon is required to separate nucleons from iron's nucleus than from any other nucleus.

All nuclear power today is by way of nuclear fission. A more promising long-range source of energy is found on the left side of the energy valley.

LEARNING OBJECTIVE
Describe the process by which small nuclei can join together, leading to the production of energy such as occurs in the Sun.

Nuclear Fusion

EXPLAIN THIS How does the energy of gasoline come from nuclear fusion?

Notice in the graph of Figure 10.30 that the steepest part of the energy valley goes from hydrogen to iron. Energy is released as light nuclei combine. This combining of nuclei is **nuclear fusion**—the opposite of nuclear fission. We see from Figure 10.32 that, as we move along the list of elements from hydrogen to iron, the average mass per nucleon decreases. Thus, when two small nuclei fuse—say, two hydrogen isotopes—the mass of the resulting helium-4 nucleus is less than the mass of the two small nuclei before fusion. Energy is released as smaller nuclei fuse.

For a fusion reaction to occur, the nuclei must collide at a very high speed in order to overcome their mutual electric repulsion. The required speeds correspond to the extremely high temperatures found in the core of the Sun and other stars. Fusion brought about by high temperatures is called **thermonuclear fusion**. In the high temperatures of the Sun, approximately 657 million tons of hydrogen are converted into 653 million tons of helium *each second*. The missing 4 million tons of mass are discharged as radiant energy.

Such reactions are, quite literally, nuclear burning. Thermonuclear fusion is analogous to ordinary chemical combustion. In both chemical and nuclear burning, a high temperature starts the reaction; the release of energy by the reaction maintains a high enough temperature to spread the fire. The net result of the chemical reaction is a combination of atoms into more tightly bound molecules. In nuclear fusion reactions, the net result is more tightly bound nuclei.

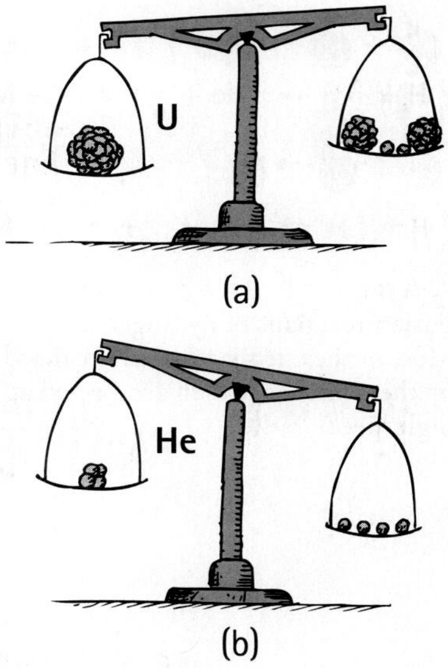

FIGURE 10.33

The mass of a nucleus is not equal to the sum of the masses of its parts. (a) The fission fragments of a heavy nucleus such as uranium are less massive than the uranium nucleus. (b) Two protons and two neutrons are more massive in their free states than when combined to form a helium nucleus.

CHECK YOURSELF

1. **Fission and fusion are opposite processes, yet each releases energy. Isn't this contradictory?**
2. **To get nuclear energy released from the element iron, should iron be fissioned or fused?**
3. **Predict whether the temperature of the core of a star increases or decreases when iron and elements of higher atomic number than iron in the core are fused.**

CHECK YOUR ANSWERS

1. No, no, no! This is contradictory only if the same element is said to release energy by the processes of both fission and fusion. Only the fusion of light elements and the fission of heavy elements result in a decrease in nucleon mass and a release of energy.
2. Neither, because iron is at the very bottom of the "energy valley." Fusing a pair of iron nuclei produces an element to the right of iron on the curve, where the mass per nucleon is greater. If you split an iron nucleus, the products lie to the left of iron on the curve—also a greater mass per nucleon. So no energy is released. For energy release, "decrease mass" is the name of the game—any game, chemical or nuclear.
3. In the fusion of iron and any nuclei beyond, energy is absorbed and the star core cools at this late stage of its evolution. This, however, leads to the star's collapse, which then greatly increases its temperature. Interestingly, elements beyond iron are not manufactured in normal fusion cycles in stellar sources but are manufactured when stars violently explode—supernovae.

FYI A common reaction is the fusion of H-2 and H-3 nuclei to become He-4 plus a neutron. Most of the energy released is in the kinetic energy of the ejected neutron, with the rest of the energy in the kinetic energy of the recoiling He-4 nucleus. Interestingly, without the neutron energy carrier, a fusion reaction won't occur. The intensity of fusion reactions is measured by the accompanying neutron flux.

Before the development of the atomic bomb, the temperatures required to initiate nuclear fusion on Earth were unattainable. When researchers found that the temperature inside an exploding atomic bomb is four to five times the temperature at the center of the Sun, the thermonuclear bomb was but a step

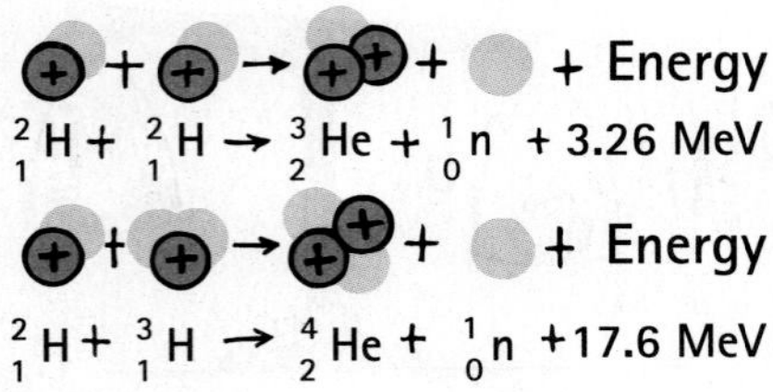

$$^{2}_{1}H + ^{2}_{1}H \rightarrow ^{3}_{2}He + ^{1}_{0}n + 3.26\ MeV$$

$$^{2}_{1}H + ^{3}_{1}H \rightarrow ^{4}_{2}He + ^{1}_{0}n + 17.6\ MeV$$

FIGURE 10.34
Fusion reactions of hydrogen isotopes. Most of the energy released is carried by the neutrons, which are ejected at high speeds.

away. This first thermonuclear bomb, a hydrogen bomb, was detonated in 1952. Whereas the critical mass of fissionable material limits the size of a fission bomb (atomic bomb), no such limit is imposed on a fusion bomb (thermonuclear or hydrogen bomb). Just as there is no limit to the size of an oil-storage depot, there is no theoretical limit to the size of a fusion bomb. Like the oil in the storage depot, any amount of fusion fuel can be stored safely until ignited. Although a mere match can ignite an oil depot, nothing less energetic than an atomic bomb can ignite a thermonuclear bomb. We can see that there is no such thing as a "baby" hydrogen bomb. A typical thermonuclear bomb stockpiled by the United States today, for example, is about 1000 times as destructive as the atomic bomb detonated over Hiroshima at the end of World War II.

The hydrogen bomb is another example of a discovery used for destructive rather than constructive purposes. The potential constructive possibility is the controlled release of vast amounts of clean energy.

Controlling Fusion

Carrying out fusion reactions under controlled conditions requires temperatures of millions of degrees. A variety of techniques exist for attaining high temperatures. No matter how the temperature is produced, a problem is that all materials melt and vaporize at the temperatures required for fusion. One solution is to confine the reaction in a nonmaterial container.

A nonmaterial container is a magnetic field that can exist at any temperature and can exert powerful forces on charged particles in motion. "Magnetic walls" of sufficient strength provide a kind of magnetic straitjacket for hot gases called plasmas. Magnetic compression further heats the plasma to fusion temperatures. At this writing, fusion by magnetic confinement has been only partially successful— a sustained and controlled reaction has so far been out of reach.

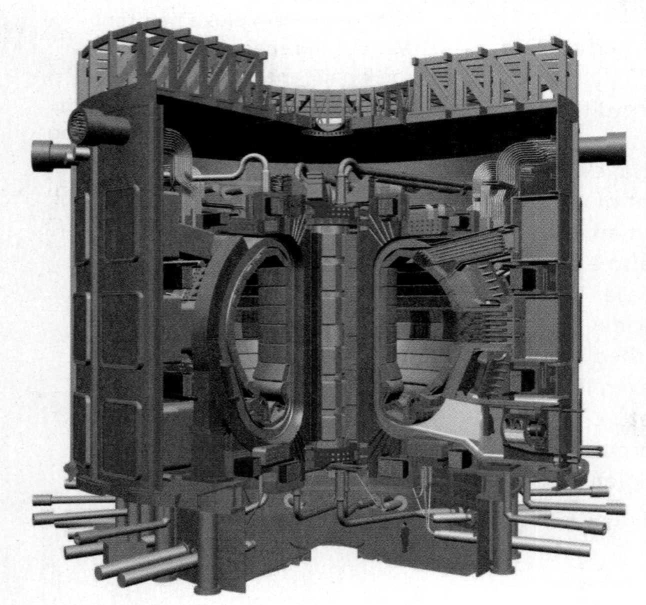

FIGURE 10.35
A cross-sectional view of the ITER (rhymes with "fitter") planned to be built and operating in Cadarache, France, before 2020.

Although no nuclear fusion power plants are currently operating, an international project now exists whose goal is to prove the feasibility of nuclear fusion power in the near future. This fusion power project is the International Thermonuclear Experimental Reactor (ITER). After construction at the chosen site in Cadarache, France, the first sustainable fusion reaction may begin as early as 2015 (Figure 10.35). The reactor will house electrically charged hydrogen gas (plasma) heated to more than 100 million °C, which is hotter than the center of the Sun. In addition to producing about 500 MW of power, the reactor could be the energy source for the creation of hydrogen, H_2, which could be used to power fuel cells such as those incorporated into automobiles.

If people are one day to dart about the universe in the same way we jet about Earth today, their supply of fuel is ensured. The fuel for fusion—hydrogen—is found in every part of the universe, not only in the stars but also in the space between them. About 91% of the atoms in the universe are estimated to be hydrogen. For people of the future, the supply of raw materials is also ensured because all the elements known to exist result from the fusing of more and more hydrogen nuclei. Future humans might synthesize their own elements and produce energy in the process, just as the stars have always done.

For instructor-assigned homework, go to www.masteringphysics.com 

SUMMARY OF TERMS (KNOWLEDGE)

Alpha particle A subatomic particle made up of the combination of two protons and two neutrons ejected by a radioactive nucleus. The composition of an alpha particle is the same as that of the nucleus of a helium atom.

Beta particle An electron emitted during the radioactive decay of a radioactive nucleus.

Carbon-14 dating The process of estimating the age of once-living material by measuring the amount of radioactive carbon-14 present in the material.

Chain reaction A self-sustaining reaction in which the products of one reaction event initiate further reaction events.

Critical mass The minimum mass of fissionable material needed for a sustainable chain reaction.

Gamma rays High-frequency electromagnetic radiation emitted by radioactive nuclei.

Half-life The time required for half the atoms in a sample of a radioactive isotope to decay.

Nuclear fission The splitting of the atomic nucleus into two smaller halves.

Nuclear fusion The combining of nuclei of light atoms to form heavier nuclei.

Rad A quantity of radiant energy equal to 0.01 J absorbed per kilogram of tissue.

Radioactivity The high-energy particles and electromagnetic radiation emitted by a radioactive substance.

Rem The unit for measuring radiation dosage in humans based on harm to living tissue.

Strong nuclear force The attractive force between all nucleons, effective at only very short distances.

Thermonuclear fusion Nuclear fusion brought about by high temperatures.

Transmutation The changing of an atomic nucleus of one element into an atomic nucleus of another element through a decrease or increase in the number of protons.

READING CHECK QUESTIONS (COMPREHENSION)

10.1 Radioactivity

1. Which type of radiation—alpha, beta, or gamma—results in the greatest change in mass number? The greatest change in atomic number?
2. Which of the three rays—alpha, beta, or gamma—has the greatest penetrating power?

10.2 The Strong Nuclear Force

3. Why doesn't the repulsive electric force of protons in the atomic nucleus cause the protons to fly apart?
4. Which have more neutrons than protons—large nuclei or small nuclei?
5. What role do neutrons play in the atomic nucleus?

10.3 Half-Life and Transmutation

6. In what form is most of the energy released by atomic transmutation?
7. What change in atomic number occurs when a nucleus emits an alpha particle? A beta particle?
8. What is the long-range fate of all the uranium that exists in the world today?
9. What is meant by the half-life of a radioactive sample?
10. What is the half-life of uranium-238?

10.4 Nuclear Fission

11. What happens to the uranium-235 nucleus when it is stretched out?
12. Is a chain reaction more likely to occur in two separate pieces of uranium-235 or in the same pieces stuck together?
13. How is a nuclear reactor similar to a conventional fossil-fuel power plant? How is it different?

10.5 Mass–Energy Equivalence

14. Who discovered that energy and mass are two different forms of the same thing?
15. In which atomic nucleus do nucleons have the least mass?
16. How does the mass per nucleon in uranium compare with the mass per nucleon in the fission fragments of uranium?

THINK INTEGRATED SCIENCE

10A—Radiation Dosage

17. What is the origin of most of the natural radiation we encounter?
18. Which produces more radioactivity in the atmosphere—coal-fired power plants or nuclear power plants?
19. Is radioactivity on Earth something relatively new? Defend your answer.

10B—Radiometric Dating

20. What happens to a nitrogen atom in the atmosphere that captures a neutron?
21. Why is there more carbon-14 in living bones than in once-living ancient bones of the same mass?
22. Why is lead found in all deposits of uranium ores?

10C—Nuclear Fusion

23. How does the mass of a pair of atoms that have fused compare to the sum of their masses before fusion?
24. What kind of containers are used to hold plasmas at temperatures of millions of degrees?
25. What kind of nuclear power is responsible for sunshine?

THINK AND DO (HANDS-ON APPLICATION)

26. Throw ten coins onto a flat surface. Move aside all the coins that landed tails-up. Collect the remaining coins. After tossing them once again, remove all the coins that landed tails-up. Repeat this process until all the coins have been removed. Can you see how this relates to radioactive half-life? In units of "tosses" what is the average half-life of 25 coins? 50 coins? 1 million coins?
27. Repeat Exercise 26 but use 10 dimes and 25 pennies. Let the dimes represent a radioactive isotope, such as carbon-14, while the pennies represent a nonradioactive isotope, such as carbon-12. Remove only the dimes when they land heads-up. Collect all the pennies and add them to the dimes that landed heads-up. Does the number of pennies affect the behavior of the dimes? Someone gives you two sets of coins. The first set contains 10 dimes and 25 pennies. The second set contains 2 dimes and 25 pennies. Which set of coins has gone through a greater number of tosses? Which set provides the most "radioactivity" after a toss? Which set is analogous to a sample of once-living ancient material?
28. Calculate your estimated annual dose of radiation using the EPA's radiation dose calculator available at http://www.epa.gov/radiation/understand/calculate.html.
29. Stand one domino upright so that when it topples it hits two other upright dominos, which also each hit two other upright dominos, and so forth. Arrange as many upright dominos as you can in this fashion so that they fan out as shown in the photograph. Your challenge is to arrange the dominos so that every one of them falls.

    Topple the first domino and observe your chain reaction. Focus your attention on the sound.

    This dominoes chain reaction occurs on a two-dimensional flat surface. What is the dimensional geometry of a nuclear chain? What would happen if a Ping-Pong ball were tossed into a room in which the floor was covered with thousands of set-to-kill spring-action mouse traps? Such an explosive event can be seen by using the keywords "mouse trap chain reaction" for an Internet video search.

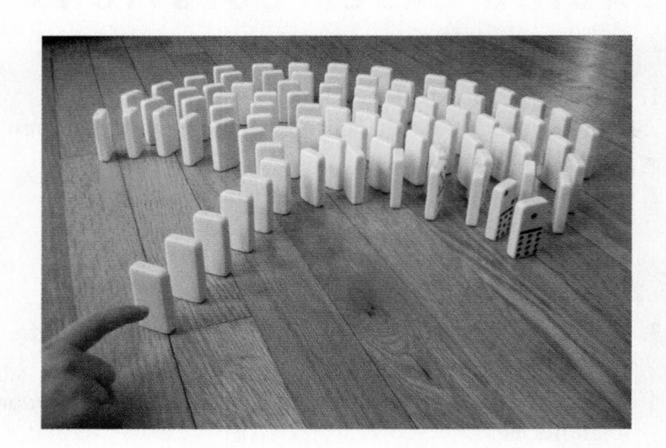

THINK AND COMPARE (ANALYSIS)

30. Rank these three types of radiation by their ability to penetrate a page of a book, from best to worst: (a) alpha particle, (b) beta particle, (c) gamma ray.
31. Consider the atoms C-12, C-14, and N-14. From greatest to least, rank them by the number of (a) protons in the nucleus, (b) neutrons in the nucleus, (c) nucleons in the nucleus.
32. Rank these isotopes in order of their radioactivity, from the most radioactive to the least radioactive: (a) nickel-59, half-life 75,000 years; (b) uranium-238, half-life 4.5 billion years; (c) actinium-225, half-life 10 days.
33. Rank the following in order from the most energy released to the least energy released: (a) uranium-235 splitting into two equal fragments, (b) uranium-235 splitting into three equal fragments, (c) uranium-235 splitting into 92 equal fragments.

THINK AND SOLVE (MATHEMATICAL APPLICATION)

34. Radiation from a point source follows an inverse-square law, where the amount of radiation received is proportional to $1/d^2$, where d is distance. If a Geiger counter that is 1 m away from a small source reads 100 counts per minute, what will be its reading 2 m from the source? 3 m from the source?
35. Consider a radioactive sample with a half-life of one week. How much of the original sample will be left at the end of the second week? The third week? The fourth week?
36. A radioisotope is placed near a radiation detector, which registers 80 counts per second. Eight hours later, the detector registers 5 counts per second. What is the half-life of the radioactive isotope?
37. Uranium-238 absorbs a neutron and then emits a beta particle. Show that the resulting nucleus is neptunium-239.

THINK AND EXPLAIN (SYNTHESIS)

38. Just after an alpha particle leaves the nucleus, would you expect it to speed up? Defend your answer.
39. A pair of protons in an atomic nucleus repel each other, but they are also attracted to each other. Explain.
40. Why do different isotopes of the same element have the same chemical properties?
41. In bombarding atomic nuclei with proton "bullets," why must the protons be given large amounts of kinetic energy in order to make contact with the target nuclei?
42. Why is lead found in all deposits of uranium ores?
43. What do the proportions of lead and uranium in rock tell us about the age of the rock?
44. What are the atomic number and atomic mass of the element formed when $^{218}_{84}Po$ emits a beta particle? What are they if the polonium emits an alpha particle?
45. Elements heavier than uranium in the periodic table do not exist in any appreciable amounts in nature because they have short half-lives. Yet there are several elements below uranium in the periodic table that have equally short half-lives but do exist in appreciable amounts in nature. How can you account for this?
46. People who work around radioactivity wear film badges to monitor the amount of radiation that reaches their bodies. Each badge consists of a small piece of photographic film enclosed in a lightproof wrapper. What kind of radiation do these devices monitor? How can they determine the amount of radiation the people receive?
47. When food is irradiated with gamma rays from a cobalt-60 source, does the food become radioactive? Defend your answer.
48. Radium-226 is a common isotope on Earth, but it has a half-life of about 1620 years. Given that Earth is some 5 billions years old, why is there any radium at all?
49. Is carbon dating advisable for measuring the age of materials a few years old? How about a few thousand years old? A few million years old?
50. Why is carbon-14 dating not accurate for estimating the age of materials more than 50,000 years old?
51. The age of the Dead Sea Scrolls was determined by carbon-14 dating. Could this technique have worked if they had been carved on stone tablets? Explain.
52. If you make an account of 1000 people born in the year 2000 and find that half of them are still living in 2060, does this mean that one-quarter of them will be alive in 2120 and one-eighth of them will be alive in 2180? What is different about the death rates of people and the "death rates" of radioactive atoms?
53. The uranium ores of the Athabasca Basin deposits of Saskatchewan, Canada, are unusually pure, containing up to 70% uranium oxides. Why doesn't this uranium ore undergo an explosive chain reaction?
54. "Strontium-90 is a pure beta source." How could a physicist test this statement?
55. Why will nuclear fission probably never be used directly for powering automobiles? How could it be used indirectly?
56. Why is carbon better than lead as a moderator in nuclear reactors?
57. How does the mass per nucleon in uranium compare with the mass per nucleon in the fission fragments of uranium?
58. Why doesn't iron yield energy if it undergoes fusion or fission?
59. Uranium-235 releases an average of 2.5 neutrons per fission, while plutonium-239 releases an average of 2.7 neutrons per fission. Which of these elements might you therefore expect to have the smaller critical mass?
60. Which process would release energy from gold—fission or fusion? From carbon? From iron?
61. If a uranium nucleus were to fission into three fragments of approximately equal size instead of two, would more energy or less energy be released? Defend your answer using Figure 10.31.

62. Is the mass of an atomic nucleus greater or less than the sum of the masses of the nucleons composing it? Why don't the nucleon masses add up to the total nuclear mass?
63. The original reactor built in 1942 was just "barely" critical because the natural uranium that was used contained less than 1% of the fissionable isotope U-235 (half-life 713 million years). If, in 1942, the Earth had been 9 billion years old instead of 4.5 billion years old, would this reactor have reached critical stage with natural uranium?
64. Heavy nuclei can be made to fuse—for instance, by firing one gold nucleus at another one. Does such a process yield energy or cost energy? Explain.
65. Which produces more energy—the fission of a single uranium nucleus or the fusing of a pair of deuterium nuclei? The fission of a gram of uranium or the fusing of a gram of deuterium? (Why do your answers differ?)
66. If a fusion reaction produces no appreciable radioactive isotopes, why does a hydrogen bomb produce significant radioactive fallout?
67. Explain how radioactive decay has always warmed the Earth from the inside and how nuclear fusion has always warmed the Earth from the outside.
68. What percentage of nuclear power plants in operation today are based on nuclear fusion?
69. Sustained nuclear fusion has yet to be achieved and remains a hope for abundant future energy. Yet the energy that has always sustained us has been the energy of nuclear fusion. Explain.
70. Oxygen and two hydrogen atoms combine to form a water molecule. At the nuclear level, if one oxygen and two hydrogen were fused, what element would be produced?
71. If a pair of carbon nuclei were fused and the product emitted a beta particle, what element would be produced?
72. Ordinary hydrogen is sometimes called a perfect fuel because of its almost unlimited supply on Earth, and when it burns, harmless water is the product of the combustion. So why don't we abandon fission energy and fusion energy, not to mention fossil-fuel energy, and just use hydrogen?

THINK AND DISCUSS (EVALUATION)

73. Why might some people consider it a blessing in disguise that fossil fuels are such a limited resource? Centuries from now, what attitudes about the combustion of fossil fuels are our descendants likely to have?
74. The 1986 accident at Chernobyl, in which dozens of people died and thousands more were exposed to cancer-causing radiation, created fear and outrage worldwide and led some people to call for the closing of all nuclear plants. Yet many people choose to smoke cigarettes in spite of the fact that 2 million people die every year from smoking-related diseases. The risks posed by nuclear power plants are involuntary, risks we must all share like it or not, whereas the risks associated with smoking are voluntary because a person chooses to smoke. Why are we so unaccepting of involuntary risk but accepting of voluntary risk?
75. Your friend Paul says that the helium used to inflate balloons is a product of radioactive decay. Your mutual friend Steve says no way. Then there's your friend Alison, fretful about living near a fission power plant. She wishes to get away from radiation by traveling to the high mountains and sleeping out at night on granite outcroppings. Still another friend, Michele, has journeyed to the mountain foothills to escape the effects of radioactivity altogether. While bathing in the warmth of a natural hot spring, she wonders aloud how the spring gets its heat. What do you tell these friends?
76. Speculate about some worldwide changes that are likely to follow the advent of successful fusion reactors.

READINESS ASSURANCE TEST (RAT)

If you have a good handle on this chapter, then you should be able to score at least 7 out of 10 on this RAT. If you score less than 7, you need to study further before moving on.

Choose the BEST answer to each of the following:

1. Which type of radiation from cosmic sources predominates on the inside of a high-flying commercial airplane?
   (a) alpha
   (b) beta
   (c) gamma
   (d) None of these predominates; all three are abundant.
2. Is it possible for a hydrogen nucleus to emit an alpha particle? Why?
   (a) yes, because alpha particles are the simplest form of radiation
   (b) no, because it would require the nuclear fission of hydrogen, which is impossible
   (c) yes, but it does not occur very frequently
   (d) no, because the nucleus does not contain enough nucleons

3. A sample of radioactive material is usually a little warmer than its surroundings because
   (a) it efficiently absorbs and releases energy from sunlight.
   (b) its atoms are continuously being struck by alpha and beta particles.
   (c) it is radioactive.
   (d) it emits alpha and beta particles.
4. What evidence supports the contention that the strong nuclear force is stronger than the electrical interaction at short internuclear distances?
   (a) Protons are able to exist side by side in an atomic nucleus.
   (b) Neutrons spontaneously decay into protons and electrons.
   (c) Uranium deposits are always slightly warmer than their immediate surroundings.
   (d) Radio interference arises adjacent to any radioactive source.
5. When the isotope bismuth-213 emits an alpha particle, what new element results?
   (a) lead
   (b) platinum
   (c) polonium
   (d) thallium
6. A certain radioactive element has a half-life of 1 hour. If you start with a 1-g sample of the element at noon, how much of this same element will be left at 3:00 PM?
   (a) 0.5 g
   (b) 0.25 g
   (c) 0.125 g
   (d) 0.0625 g
7. The isotope cesium-137, which has a half-life of 30 years, is a product of nuclear power plants. How long will it take for this isotope to decay to about one-half its original amount?
   (a) 0 yr
   (b) 15 yr
   (c) 30 yr
   (d) 60 yr
   (e) 90 yr
8. If uranium were to split into 90 pieces of equal size instead of two, would more energy or less energy be released? Why?
   (a) less energy, because of less mass per nucleon
   (b) less energy, because of greater mass per nucleon
   (c) more energy, because of less mass per nucleon
   (d) more energy, because of greater mass per nucleon
9. Which process would release energy from gold—fission or fusion? From carbon?
   (a) gold: fission; carbon: fusion
   (b) gold: fusion; carbon: fission
   (c) gold: fission; carbon: fission
   (d) gold: fusion; carbon: fusion
10. If an iron nucleus split in two, its fission fragments would have
    (a) greater mass per nucleon.
    (b) less mass per nucleon.
    (c) the same mass per nucleon.
    (d) either greater or less mass per nucleon.

Answers to RAT

1. c, 2. d, 3. b, 4. a, 5. d, 6. c, 7. c, 8. b, 9. a, 10. a

11

CHAPTER 11

Investigating Matter

As you progress through this science course, you will note an accumulating list of key terms. For example, we say that there are more than 100 kinds of *atoms* and that any material consisting of a single kind of atom is an *element*. Atoms can link together to form a *molecule*, and a molecule consisting of atoms from different elements is a *compound*. And on and on, one term building on another, as we attempt to describe the nature of matter beyond its casual appearance.

Instead of just memorizing the formal definitions of terms, you will serve yourself better by making sure that you understand the underlying concepts. Practice articulating and paraphrasing those concepts—aloud to yourself or to a friend without looking at the book. When you are able to express these concepts in your own words—not memorized words—you will have the insight to do well in this course and beyond.

11.1 Chemistry: The Central Science

LEARNING OBJECTIVE
Define chemistry as a central science that has had a great impact on society.

EXPLAIN THIS How has chemistry influenced our modern lifestyles?

When you wonder what the land, sky, or ocean is made of, you are thinking about chemistry. When you wonder how a rain puddle dries up, how a car acquires energy from gasoline, or how your body extracts energy from the food you eat, you are again thinking about chemistry. By definition, **chemistry** is the study of matter and the transformations it can undergo. Matter is anything that occupies space. It is the stuff that makes up all material things; anything you can touch, taste, smell, see, or hear is matter. The scope of chemistry, therefore, is very broad.

Chemistry is often described as a central science because it touches all the other sciences. It springs from the principles of physics, and it serves as the foundation for the most complex science of all—biology. Indeed, many of the great advances in the life sciences today, such as genetic engineering, are applications of some very exotic chemistry. Chemistry sets the foundation for the major Earth sciences—geology, oceanography, meteorology. It is also an important component of space science (Figure 11.1). Just as we learned about the origin of the Moon from the chemical analysis of moon rocks in the early 1970s, we are now learning about the history of Mars and other planets from the chemical information gathered by space probes.

FIGURE 11.1
Special materials of chemistry, such as rocket fuels, metals for spaceships, and fabrics for the space suits, were required to enable astronauts to reach and explore the surface of the Moon.

Progress in science is made as scientists conduct research. Research is any activity aimed at the systematic discovery and interpretation of new knowledge. Many scientists focus on **basic research**, which leads us to a greater understanding of how the natural world operates. The foundation of knowledge laid down by basic research frequently leads to useful applications. Research that focuses on developing these applications is known as **applied research**. Most chemists choose applied research as their major focus. Applied research in chemistry has provided us with medicine, food, water, shelter, and many of the material goods that characterize modern life. Just a few examples are shown in Figure 11.2.

MasteringPhysics®
TUTORIAL: What Is Chemistry?

Over the course of the past century, we excelled at manipulating atoms and molecules to create materials to suit our needs. At the same time, however, we made mistakes in caring for the environment. Waste products were dumped into

FIGURE 11.2
Most of the materials in any modern house are shaped by some human-devised chemical process.

rivers, buried in the ground, or vented into the air without regard for possible long-term consequences. Many people believed that Earth was so large that its resources were virtually unlimited and that it could absorb wastes without being significantly harmed.

Most nations now recognize this as a dangerous attitude. As a result, government agencies, industries, and concerned citizens are involved in extensive efforts to clean up toxic-waste sites. Such regulations as the international ban on ozone-destroying chlorofluorocarbons have been enacted to protect the environment. Members of the American Chemistry Council, who produce 80–85% of the chemicals manufactured in the United States, have adopted a program called Responsible Care®, in which they have pledged to continuously improve their environmental, health, safety, and security performance. The Responsible Care program emblem is shown in Figure 11.3. If we use chemistry wisely, most waste products can be minimized, recycled, engineered into salable commodities, or rendered environmentally benign.

FIGURE 11.3
The Responsible Care symbol of the American Chemistry Council; go to responsiblecare.org.

Chemistry has influenced our lives in profound ways, and it will continue to do so in the future. For this reason, it is in everyone's interest to become acquainted with the basic concepts of chemistry.

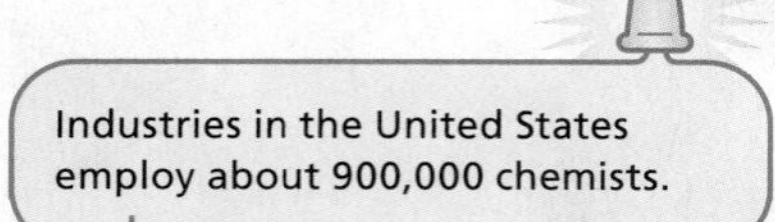

Industries in the United States employ about 900,000 chemists.

CHECK YOURSELF

Chemists have learned how to produce aspirin using petroleum as a starting material. Is this an example of basic or applied research?

CHECK YOUR ANSWER

This is an example of applied research because the primary goal was to develop a useful commodity. However, the ability to produce aspirin from petroleum depended on an understanding of atoms and molecules developed from many years of basic research.

LEARNING OBJECTIVE
Introduce the molecule as a fundamental unit of matter.

11.2 The Submicroscopic View of Matter

EXPLAIN THIS What is found between two adjacent molecules of a gas?

From afar, a sand dune appears to be a smooth, continuous material. Up close, however, the dune reveals itself to be made of tiny particles of sand. In a similar fashion, as discussed in Chapter 9, everything around us—no matter how smooth it may appear—is made of the basic units you know as *atoms*. Atoms are so small, however, that a single grain of sand contains on the order of 125 million trillion of them. There are roughly 250,000 times more atoms in a single grain of sand than there are grains of sand in the dunes shown in Figure 11.4.

UNIFYING CONCEPT
- ***The Atomic Theory of Matter*** **Section 9.1**

As small as atoms are, there is much we have learned about them. We know, for example, that there are more than 100 different types of atoms; they are listed in the widely recognized periodic table, which was introduced in Section 9.3. Some atoms link together to form larger but still incredibly small basic units of matter called **molecules**. As shown in Figure 11.4, for example, two hydrogen atoms and one oxygen atom link together to form a single molecule of water, also known as H_2O. Water molecules are so small that an 8-oz glass of water contains about a trillion trillion of them.

FIGURE 11.4
There are far more atoms in a glass of water than there are grains of sand in this towering sand dune.

Our world can be studied at different levels of magnification. At the *macroscopic* level, matter is large enough to be seen, measured, and handled. A handful of sand and a glass of water are macroscopic samples of matter. At the *microscopic* level, physical structure is so fine that it can be seen only with a microscope. A biological cell is microscopic, as is the detail on a dragonfly's wing. Beyond the microscopic level is the **submicroscopic**—the realm of atoms and molecules and an important focus of chemistry.

The Phase of a Material

One of the most evident ways we can describe matter is by its physical form, which may be one of three phases (also sometimes described as physical states): *solid*, *liquid*, or *gas*. A **solid** material, such as a rock, occupies a constant amount of space and does not readily deform when pressure is applied to it. In other words, a solid has both definite volume and definite shape. A **liquid** also occupies a constant amount of space (it has a definite volume), but its form changes readily (it has an indefinite shape). A liter of milk, for example, may take the shape of its carton or the shape of a puddle, but its volume is the same in both cases. A **gas** is diffuse, having neither definite volume nor definite shape. Any sample of gas assumes both the shape and the volume of the container it occupies. A given amount of air, for example, may assume the volume and shape of a toy balloon or the volume and shape of a bicycle tire. Released from its container, a gas diffuses into the atmosphere, which is a collection of various gases held to our planet only by the force of gravity.

FYI Coffee and tea are decaffeinated using carbon dioxide in a fourth phase of matter known as a supercritical fluid. This phase in which carbon dioxide behaves like a gaseous liquid is attained by adding lots of pressure and heat. Supercritical carbon dioxide is relatively easy to produce. To get water to form a supercritical fluid, however, requires pressures in excess of 217 atmospheres and a temperature of 374°C.

On the submicroscopic level, the solid, liquid, and gas phases are distinguished by the extent of interaction between the submicroscopic particles (the atoms or molecules). This is illustrated in Figure 11.5. In solid matter, the attractions between particles are strong enough to hold all the particles together in some fixed three-dimensional arrangement. The particles are able to vibrate about fixed positions, but they cannot move past one another. Adding heat causes these vibrations to increase until, at a certain temperature, the vibrations are rapid enough to disrupt the fixed

FIGURE 11.5
The familiar bulk properties of a solid, liquid, and gas. (a) The particles of the solid phase vibrate about fixed positions. (b) The particles of the liquid phase slip past one another. (c) The fast-moving particles of the gas phase are separated by large average distances.

arrangement. The particles can then slip past one another and tumble around much like a bunch of marbles in a bag. This kind of motion is representative of the liquid phase of matter, and it is the mobility of the particles that gives rise to the liquid's fluid character—its ability to flow and take on the shape of its container.

Further heating causes the particles in the liquid to move so fast that the attractions they have for one another are unable to hold them together. They then separate from one another, forming a gas. Moving at an average speed of 500 meters per second (1100 miles per hour), the particles of a gas are widely separated from one another. Matter in the gas phase therefore occupies much more volume than it does in the solid or liquid phase, as Figure 11.6 shows. Applying pressure to a gas squeezes the gas particles closer together, which reduces their volume. Enough air for an underwater diver to breathe for many minutes, for example, can be squeezed (compressed) into a tank small enough to be carried on the diver's back.

Although gas particles move at high speeds, the speed at which they can travel from one side of a room to the other is relatively slow. This is because the gas particles are continuously hitting one another, and the paths they end up taking are circuitous. At home, you get a sense of how long it takes for gas particles to migrate each time someone opens the oven door after baking, as Figure 11.7 shows. A shot of aromatic gas particles escapes from the oven, but there is a notable delay before the aroma reaches the nose of someone sitting in the next room.

CHECK YOURSELF
Why are gases so much easier to compress into smaller volumes than are solids and liquids?

CHECK YOUR ANSWER
Because there is a lot of space between gas particles. The particles of a solid or liquid, on the other hand, are already close to one another, meaning there is little room left for a further decrease in volume.

(a)

(b)

(c)

FIGURE 11.6
The gaseous phase of any material occupies significantly more volume than either its solid or liquid phase. (a) Solid carbon dioxide (dry ice) is broken up into powder form. (b) The powder is funneled into a balloon. (c) The balloon expands as the contained carbon dioxide becomes a gas as the powder warms up.

FIGURE 11.7
In traveling from point A to point B, the typical gas particle travels a circuitous path because of numerous collisions with other gas particles—about eight billion collisions every second! The changes in direction shown here represent only a few of these collisions. Although the particle travels at very high speeds, it takes a relatively long time for them to cross between two distant points because of these numerous collisions. This net movement of gas molecules from an area of high concentration to low concentration is called diffusion.

The net migration of aromatic gas molecules away from a baking pie to a person in an adjacent room is an example of diffusion. **Diffusion** is the tendency of molecules to move from an area of high concentration to one of low concentration. Over time, a gas will diffuse to completely fill its container. Because all gases do this, a mixture of gases eventually will become thoroughly and evenly mixed.* This is not surprising when you think about it. The gases that make up the air in your classroom, for example, have an even composition throughout the classroom. You wouldn't expect the oxygen molecules to concentrate on one side of the classroom while leaving the other half of the class gasping for breath!

UNIFYING CONCEPT

- *The Second Law of Thermodynamics*
Section 6.5

Diffusion is not limited to gases. Liquids diffuse readily as well, and even solids diffuse gradually. If you put a bar of gold and a bar of silver side by side, leave them alone for several months, and then chemically analyze their compositions, you will find that some of the gold particles have diffused into the silver and some of the silver particles have diffused into the gold. (In Chapter 15, you will learn how the diffusion of materials across cellular membranes is a principal mechanism for moving nutrients into and wastes out of a cell.)

11.3 Phase Changes

LEARNING OBJECTIVE
Describe phase changes from a molecular point of view.

EXPLAIN THIS What gas is found within a bubble of boiling water?

Figure 11.8 illustrates that you must either add heat to a substance or remove heat from it if you want to change its phase. The process of a solid transforming into a liquid is called **melting**. To visualize what happens when

*All gases do not diffuse at the same rate, however. At a constant temperature, molecules with less mass diffuse more rapidly than molecules of greater mass. Graham's law states the quantitative relationship: The rate of diffusion of a gas is inversely proportional to the square root of its formula mass.

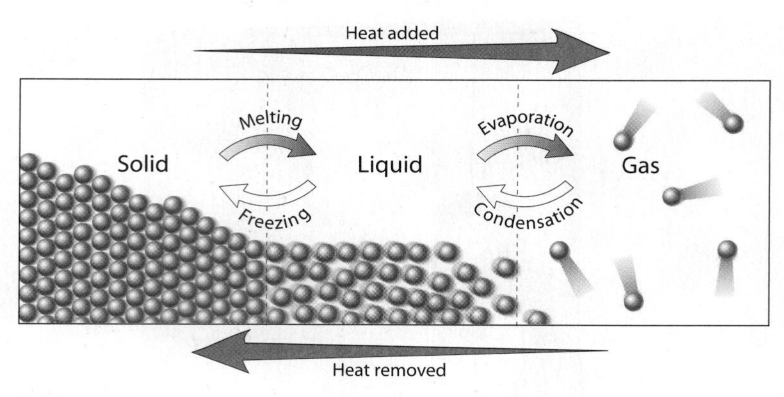

FIGURE 11.8
Melting and evaporation involve the addition of heat; condensation and freezing involve the removal of heat.

heat begins to melt a solid, imagine that you are holding hands with a group of people and all of you start jumping around randomly. The more violently you jump, the more difficult it is to hold onto one another. If everyone jumps violently enough, keeping hold is impossible. Something like this happens to the particles of a solid when it is heated. As heat is added to the solid, the particles vibrate more and more violently. If enough heat is added, the attractive forces between the particles are no longer able to hold them together. The solid melts.

A liquid can be changed to a solid by the removal of heat. This process is called **freezing**, and it is the reverse of melting. As heat is withdrawn from the liquid, particle motion decreases until the particles, on average, are moving slowly enough for attractive forces between them to take permanent hold. The only motion the particles are capable of then is vibration about fixed positions, which means the liquid has solidified, or frozen. Of course, freezing is just the reverse of melting. The temperature at which a substance freezes or melts is the same. For pure water, this freezing/melting point is 0°C. Interestingly, 0°C is the temperature at which water can remain in both a liquid and a solid phase indefinitely. If you raise the temperature, even slightly, then all the ice will eventually melt. Similarly, if you lower the temperature, then all the liquid water will freeze.

FIGURE 11.9
Evaporation is a cooling process. Our sweat glands take advantage of this by producing perspiration to cool us down when we begin to overheat. As the water molecules in perspiration evaporate from the surface of our skin, they carry away unwanted energy, which cools us down.

A liquid can be heated so that it becomes a gas—a process called **evaporation**. As heat is added, the particles of the liquid acquire more kinetic energy and move faster. Particles at the liquid surface eventually gain enough energy to jump out of the liquid and enter the air. In other words, they enter the gas phase. As more and more particles absorb the heat being added, they too acquire enough energy to escape from the liquid surface and become gas particles. Because a gas results from evaporation, this phase is also sometimes referred to as *vapor*. Water in the gaseous phase, for example, may be referred to as water vapor.

The rate at which a liquid evaporates increases with temperature. A puddle of water, for example, evaporates from a hot pavement more quickly than it does from your cool kitchen floor. When the temperature is hot enough, evaporation occurs beneath the surface of the liquid. As a result, bubbles form and are buoyed up to the surface. We say that the liquid is **boiling**. A substance is often characterized by its *boiling point*, which is the temperature at which it boils. At sea level, the boiling point of fresh water is 100°C.

The transformation from gas to liquid—the reverse of evaporation—is called **condensation**. This process can occur when the temperature of a gas decreases. The water vapor held in the warm daylight air, for example, may condense to form a wet dew in the cool of the night.

The submicroscopic particles of a solid are able to change directly into the gas phase—a process called **sublimation**. The molecules that make up mothballs, for example, sublime quite readily, which is why mothballs are so smelly and also why they disappear if not kept in a sealed container. Frozen water also sublimes and this accounts for the loss of much snow and ice, especially on high, sunny mountaintops. It's also why ice cubes left in the freezer for a long time tend to get smaller. In the opposite process of sublimation, a gas transforms directly into a solid; this process is called **deposition**. On chilly mornings water vapor in the air may *deposit* as ice crystals on cold surfaces, such as vegetation or windows, to form frost.

FIGURE 11.10
Patches of frost crystals betray the hidden entrance to a mouse burrow. Each cluster of crystals is the frozen water vapor coming from the mouse's breath!

Note that the underlying cause of phase changes is the transfer of energy. Energy must be added to melt ice into liquid water or vaporize liquid water into water vapor. Energy must be removed to condense water vapor back into liquid water or freeze liquid water into ice. We call the energy that is absorbed or released in a change of phase **latent heat**. The word *latent* means "hidden." The heat energy involved in changing the phase of a material is *hidden* in the sense that the material does not change its temperature as it absorbs or releases this heat. For example, water boiling in a pot on the stove will remain at 100°C during its phase change to water vapor. The heat pumped into the pot of water goes into disrupting the attractive forces between molecules in the liquid state, rather than changing the water's temperature. A thermometer will register 100°C until all the water has evaporated.

More specifically, the amount of energy needed to change any substance from solid to liquid (and vice versa) is called the substance's **heat of fusion**. For water, this is 334 joules per gram. And the amount of energy required to change any substance from liquid to gas (and vice versa) is called the substance's **heat of vaporization**. For water, this is a whopping 2256 joules per gram. Water's high heat of vaporization allows you to briefly touch your wetted finger to a hot clothes iron without being burned. Firefighters also take advantage of water's high heat of vaporization, as shown in Figure 11.11.

FIGURE 11.11
Liquid water extinguishes a flame by absorbing much of the heat that the fire needs to sustain itself, as well as by wetting, which blocks oxygen from reaching the burning material.

11.4 Physical and Chemical Properties

LEARNING OBJECTIVE
Describe how materials can be identified by their physical and chemical properties.

EXPLAIN THIS Why are physical changes typically easier to reverse than chemical changes?

Properties that describe the look or feel of a substance, such as color, hardness, density, texture, and phase, are called **physical properties**. Every substance has its own set of characteristic physical properties that we can use to identify that substance (Figure 11.12).

The physical properties of a substance can change when conditions change, but that does not mean that a different substance is created. Cooling liquid water to below 0°C causes the water to transform into solid ice, but the substance is still water, no matter what the phase. The only difference is the relative orientation of the H_2O molecules to one another. In the liquid phase, the

Gold	Diamond	Water
Opacity: opaque	**Opacity:** transparent	**Opacity:** transparent
Color: yellowish	**Color:** colorless	**Color:** colorless
Phase at 25°C: solid	**Phase at 25°C:** solid	**Phase at 25°C:** liquid
Density: 19.3 g/mL	**Density:** 3.5 g/mL	**Density:** 1.0 g/mL

FIGURE 11.12
Gold, diamond, and water can be identified by their physical properties. If a substance has all the physical properties listed under gold, for example, then it must be gold.

water molecules tumble around one another, whereas in the ice phase, they vibrate about fixed positions. The freezing of water is an example of what chemists call a physical change. During a **physical change**, a substance changes its phase or some other physical property, but not its chemical composition. As shown in Figure 11.13, water in either the liquid or solid phase is still made of water molecules. Likewise, the density of elemental mercury, Hg, decreases with increasing temperature because its atoms become spaced farther apart—but the mercury is still made of mercury atoms.

UNIFYING CONCEPT
- *Density*
Section 2.3

CHECK YOURSELF
The melting of gold is a physical change. Why?

CHECK YOUR ANSWER
During a physical change, a substance changes only one or more of its physical properties; its chemical identity does not change. Because melted gold is still gold but in a different form, its melting represents only a physical change.

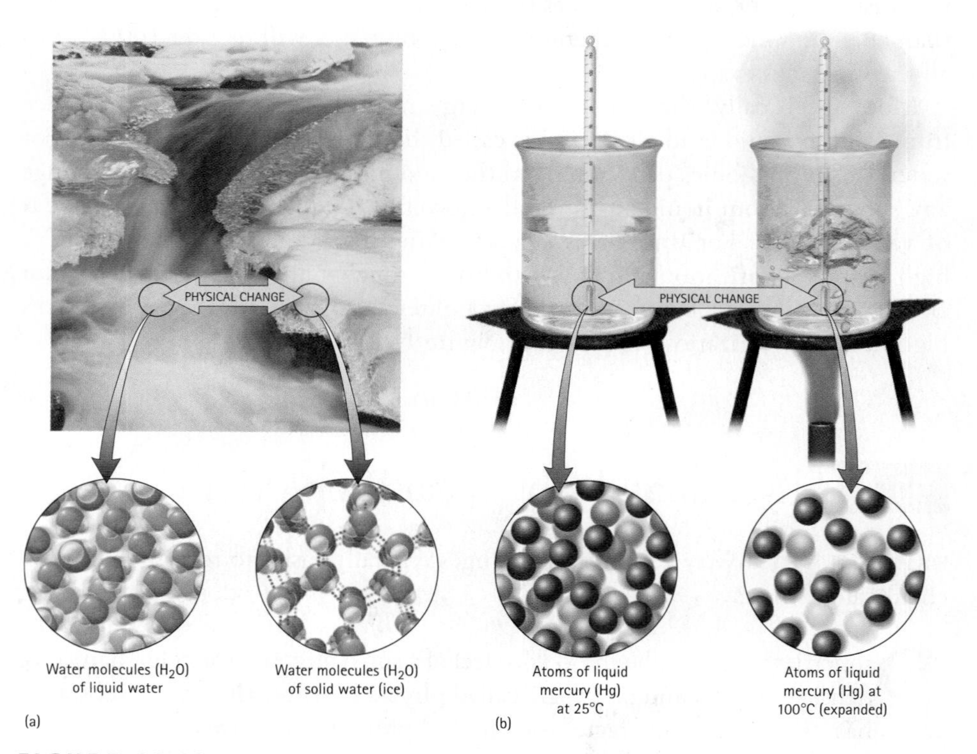

FIGURE 11.13
Two physical changes. (a) Liquid water and ice may appear to be different substances, but a submicroscopic view shows that both consist of water molecules. (b) At 25°C, the atoms in a sample of mercury are a certain distance apart, yielding a density of 13.53 g/mL. At 100°C, the atoms are farther apart, meaning that each milliliter now contains fewer atoms than at 25°C, and the density is now 13.35 g/mL. The physical property we call density has changed with temperature, but the identity of the substance remains unchanged: Mercury is mercury.

Methane
Reacts with oxygen to form carbon dioxide and water, giving off lots of heat during the reaction.

Baking soda
Reacts with vinegar to form carbon dioxide and water, absorbing heat during the reaction.

Copper
Reacts with carbon dioxide and water to form the greenish-blue substance called patina.

FIGURE 11.14
The chemical properties of substances determine the ways in which they can change into new substances. Natural gas and baking soda, for example, can both undergo chemical reactions in which they are transformed into carbon dioxide and water. Similarly, copper can be transformed into patina.

Chemical properties characterize the ability of a substance to react with other substances or to transform from one substance into another. Figure 11.14 shows three examples. The methane of natural gas has the chemical property of reacting with oxygen to produce carbon dioxide and water, along with appreciable heat energy. Similarly, baking soda has the chemical property of reacting with vinegar to produce carbon dioxide and water while absorbing a small amount of heat energy. Copper has the chemical property of reacting with carbon dioxide and water to form a greenish-blue solid known as *patina*. Copper statues exposed to the carbon dioxide and water in the air become coated with patina. The patina is not copper, it is not carbon dioxide, and it is not water. It is a new substance formed by the reaction of these chemicals with one another.

All three of these transformations involve a change in the way the atoms in the molecules are *chemically bonded* to one another. A **chemical bond** is the force of attraction between two atoms that holds them together within a compound. A methane molecule, for example, is made of a single carbon atom bonded to four hydrogen atoms, and an oxygen molecule is made of two oxygen atoms bonded to each other. Figure 11.15 shows the chemical change in which the atoms in a methane molecule and those in two oxygen molecules first pull apart and then form new bonds with different partners, resulting in the formation of molecules of carbon dioxide and water.

Any change in a substance that involves a rearrangement of the way atoms are bonded is called a **chemical change**. Thus the transformation of methane to carbon dioxide and water is a chemical change, as are the other two transformations shown in Figure 11.14.

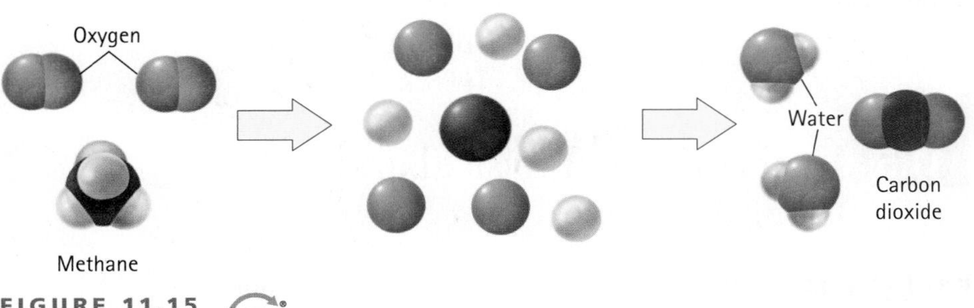

FIGURE 11.15
INTERACTIVE FIGURE MP

The chemical change in which molecules of methane and oxygen transform into molecules of carbon dioxide and water, as atoms break old bonds and form new ones. Although the actual mechanism of this transformation is more complicated than depicted here, the idea that new materials are formed by the rearrangement of atoms is accurate.

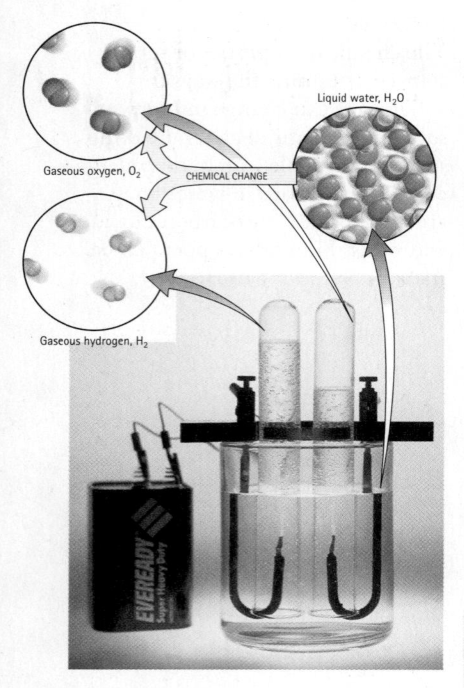

FIGURE 11.16
Water can be transformed into hydrogen gas and oxygen gas by applying the energy of an electric current. This is a chemical change because new materials (the two gases) are formed as the atoms originally found in the water molecules are rearranged.

The chemical change shown in Figure 11.16 occurs when an electric current is passed through water. The energy of the current is sufficient to pull bonded atoms away from each other. Loose atoms then form new bonds with different atoms, which results in the formation of new molecules. Thus, water molecules are changed into molecules of hydrogen and oxygen, two substances that are very different from water. The hydrogen and oxygen are both gases at room temperature, and they can be seen as bubbles rising to the surface.

In the language of chemistry, materials undergoing a chemical change are said to be *reacting*. Methane reacts with oxygen to form carbon dioxide and water. Water reacts when exposed to electricity to form hydrogen gas and oxygen gas. Thus, the term *chemical change* means the same thing as *chemical reaction*. During a **chemical reaction**, new materials are formed by a change in the way atoms are bonded together. Again for emphasis, *during a chemical reaction, new materials are formed by a change in the way atoms are bonded together.* (We shall explore chemical bonds and the reactions in which they are formed and broken in Chapters 12 and 13.)

A substance's tendency to change into another substance is one of its chemical properties. For example, it is a chemical property of iron to transform into rust.

CHECK YOURSELF

Each sphere in the following diagrams represents an atom. Joined spheres represent molecules. One set of diagrams shows a physical change, and the other shows a chemical change. Which is which?

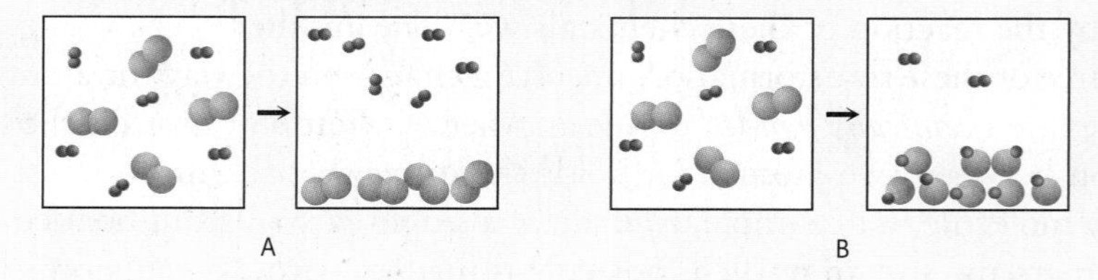

CHECK YOUR ANSWER

Remember that a chemical change (also known as a chemical reaction) involves molecules breaking apart so that the atoms are free to form new bonds with new partners. Be careful to distinguish this breaking apart from a mere change in the relative positions of a group of molecules. In set A, the molecules are the same both before and after the change. They differ only in their positions relative to one another. Set A, therefore, represents only a physical change. In set B, new molecules, consisting of bonded red and blue spheres, appear after the change. These molecules represent a new material, and so set B represents a chemical change.

LEARNING OBJECTIVE
Spell out the difficulty involved in distinguishing between physical and chemical properties.

11.5 Determining Physical and Chemical Changes

EXPLAIN THIS Why is the air over a campfire always moist?

How can you determine whether an observed change is physical or chemical? This can be tricky because in both cases there are changes in physical appearance. Water, for example, looks quite different after it freezes, just as a car looks quite different after it rusts (Figure 11.17). The

FIGURE 11.17
The transformation of water into ice and the transformation of iron into rust both involve changes in physical appearance. The formation of ice is a physical change, whereas the formation of rust is a chemical change.

freezing of water is a physical change because liquid water and frozen water are both forms of water—only the orientation of the water molecules to one another changes. The rusting of a car, by contrast, is the result of the transformation of iron into rust. This is a chemical change because iron and rust are two different materials, each consisting of a different arrangement of atoms. As we shall see in the next two sections, iron is an element and rust is a compound consisting of iron and oxygen atoms.

Two powerful guidelines can help you assess physical and chemical changes. First, in a physical change, a change in appearance is the result of a new set of conditions imposed on the same material. Restoring the original conditions restores the original appearance: frozen water melts upon warming. Second, in a chemical change, a change in appearance is the result of the formation of a new material that has its own unique set of physical properties. The more evidence you have suggesting that a different material has been formed, the greater the likelihood that the change is a chemical change. Iron is a moldable metal that can be used to build cars. Rust is a reddish material that tends to fall apart. This suggests that the rusting of iron is a chemical change.

CHECK YOURSELF

Evan, shown to the right, has grown an inch in height over the past year. Is this best described as a physical or a chemical change?

CHECK YOUR ANSWER

Are new materials being formed as Evan grows? Absolutely—created out of the food he eats. His body is very different from, say, the peanut butter sandwich he ate yesterday. Yet, through some very advanced chemistry, his body is able to absorb the atoms of that peanut butter sandwich and rearrange them into new materials. Biological growth, therefore, is best described as a chemical change.

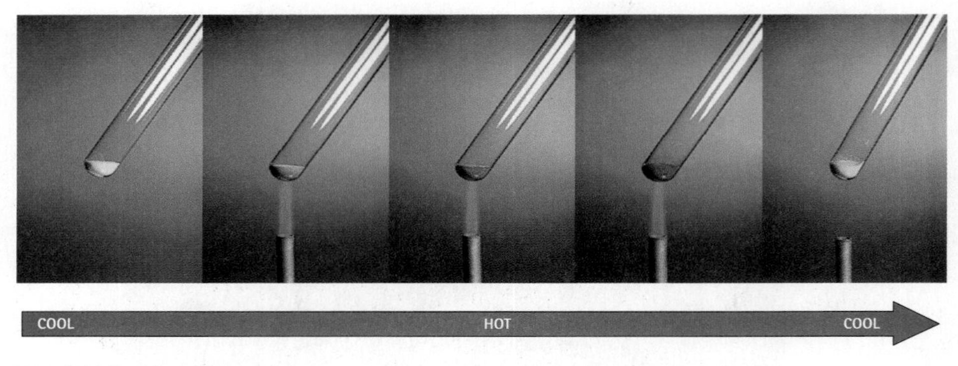

FIGURE 11.18
Potassium chromate changes color as its temperature changes. This change in color is a physical change. A return to the original temperature restores the original bright yellow color.

Figure 11.18 shows potassium chromate, a material whose color depends on its temperature. At room temperature, potassium chromate is a bright canary yellow. At higher temperatures, it is a deep reddish orange. Upon cooling, the canary color returns, suggesting that the change is physical. With a chemical change, reverting to the original conditions does not restore the original appearance. Ammonium dichromate, shown in Figure 11.19, is an orange material that, when heated, explodes into ammonia, water vapor, and green chromium (III) oxide. When the test tube is returned to the original temperature, there is no trace of orange ammonium dichromate. In its place are new substances that have completely different physical properties.

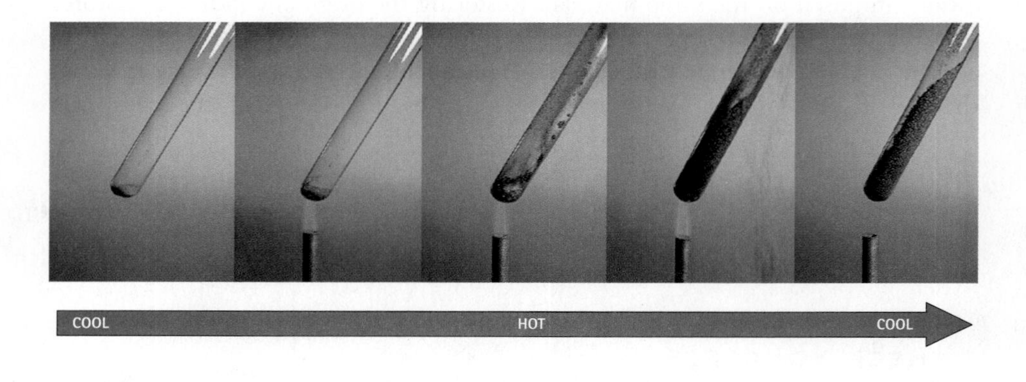

FIGURE 11.19
When heated, orange ammonium dichromate undergoes a chemical change to ammonia, water vapor, and chromium (III) oxide. A return to the original temperature does not restore the orange color because the ammonium dichromate is no longer there.

LEARNING OBJECTIVE
Contrast compounds with the elements from which they are created.

11.6 Elements to Compounds

EXPLAIN THIS How are compounds different from elements?

The terms *element* and *atom* are often used in a similar context. You might hear, for example, that gold is an element made of gold atoms. Generally, *element* is used in reference to an entire macroscopic or microscopic sample, and *atom* is used when speaking of the submicroscopic particles in the sample. The important distinction is that elements are made of atoms and not the other way around.

The fundamental unit of an element is indicated by its **elemental formula**. For elements in which the fundamental units are individual atoms, the elemental formula is simply the chemical symbol: Au is the elemental formula for gold, and

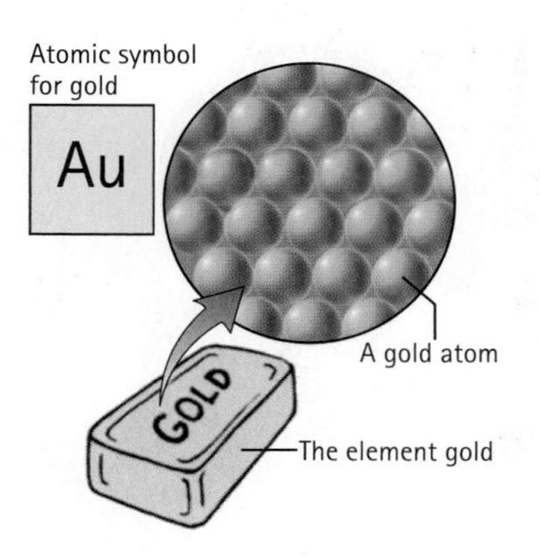

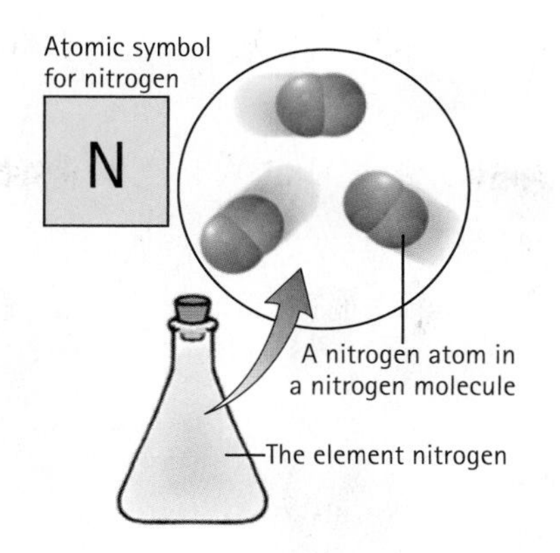

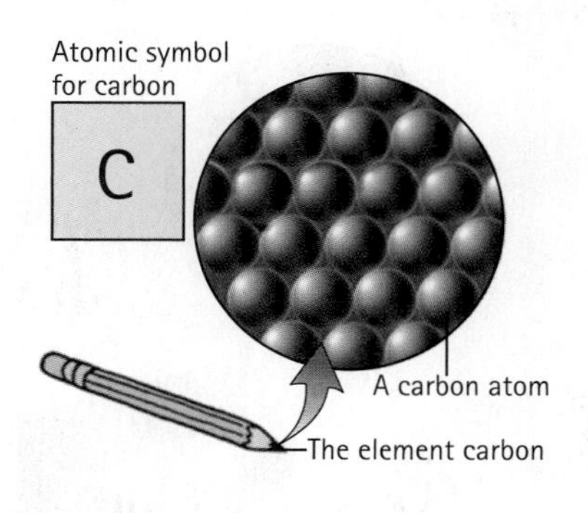

FIGURE 11.20
Any element consists of only one kind of atom. Gold consists of only gold atoms, a flask of gaseous nitrogen consists of only nitrogen atoms, and the carbon of a graphite pencil consists of only carbon atoms.

Li is the elemental formula for lithium, to name just two examples. For elements in which the fundamental units are two or more atoms bonded into molecules, the elemental formula is the chemical symbol followed by a subscript indicating the number of atoms in each molecule. For example, elemental nitrogen, shown in Figure 11.20, commonly consists of molecules containing two nitrogen atoms per molecule. Thus, N_2 is the usual elemental formula given for nitrogen. Similarly, O_2 is the elemental formula for the oxygen we breathe, and S_8 is the elemental formula for sulfur.

Physical change? Chemical change? It's not always easy to distinguish between the two. Because of many subtleties that are recognized only after years of study and laboratory experience, you'll not soon have a firm handle on how to categorize many observed changes. It's okay to learn a little now, and to entrust a lot that remains for some future time or perhaps to others who chose to specialize in this field.

CHECK YOURSELF

The oxygen we breathe, O_2, is converted into ozone, O_3, in the presence of an electric spark. Is this a physical or chemical change?

CHECK YOUR ANSWER

When atoms regroup, the result is an entirely new substance, and that is what happens here. The oxygen we breathe, O_2, is odorless and life-giving. Ozone, O_3, can be toxic; it has a pungent smell commonly associated with electric motors. The conversion of O_2 into O_3 is therefore a chemical change. However, both O_2 and O_3 are elemental forms of oxygen.

FYI Carbon is the only element that can form bonds with itself indefinitely. Sulfur's practical limit is S_8 and nitrogen's limit is about N_{12}. The elemental formula for a 1-carat diamond, however, is about $C_{10,000,000,000,000,000,000,000}$.

When atoms of different elements bond to one another, they make a **compound**. Sodium atoms and chlorine atoms, for example, bond to make the compound sodium chloride, commonly known as table salt. Nitrogen atoms and hydrogen atoms join to make the compound ammonia, which is a common household cleaner.

A compound is represented by its **chemical formula**, in which the symbols for the elements are written together. The chemical formula for sodium chloride is NaCl, and the formula for ammonia is NH_3. Numerical subscripts indicate the ratio in which the atoms combine. By convention, the subscript 1 is understood and omitted. So, the chemical formula NaCl tells us that the compound sodium chloride has one sodium atom for every chlorine atom; the chemical formula NH_3 tells us that the compound ammonia has one nitrogen atom for every three hydrogen atoms, as Figure 11.21 shows.

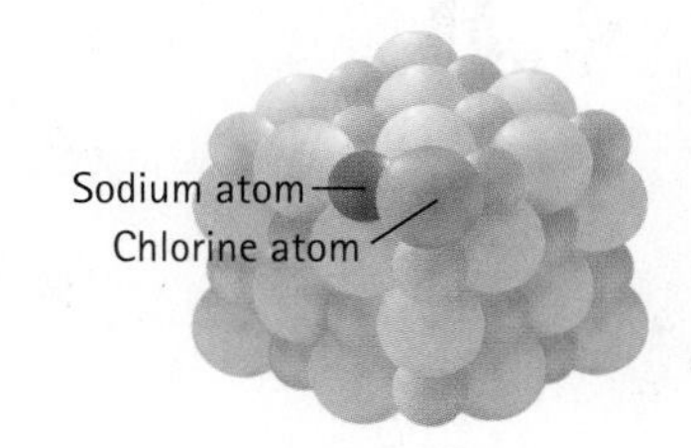

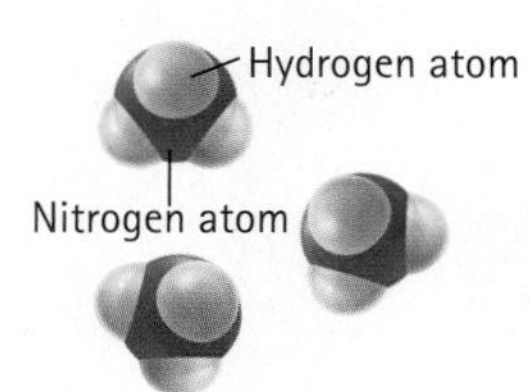

FIGURE 11.21
The compounds sodium chloride and ammonia are represented by their chemical formulas, NaCl and NH_3. A chemical formula shows the ratio of atoms that make up the compound.

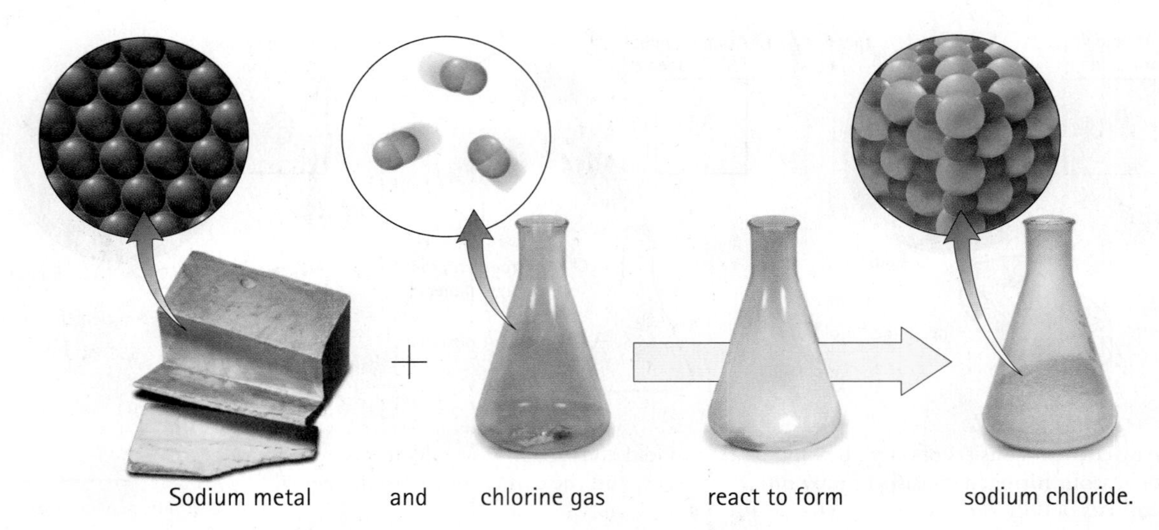
Sodium metal
and
chlorine gas
react to form
sodium chloride.

FIGURE 11.22
Sodium metal and chlorine gas react together to form sodium chloride. Although the compound sodium chloride is composed of sodium and chlorine, the physical and chemical properties of sodium chloride are very different from the physical and chemical properties of either sodium metal or chlorine gas.

Compounds have physical and chemical properties that are completely different from the properties of their elemental components. The sodium chloride, NaCl, shown in Figure 11.22 is very different from the elemental sodium and the elemental chlorine used in its formation. Elemental sodium, Na, consists of nothing but sodium atoms, which form a soft, silvery metal that can be cut easily with a knife. Its melting point is 97.5°C, and it reacts violently with water. Elemental chlorine, Cl_2, consists of chlorine molecules. This material, a yellow-green gas at room temperature, is very toxic; it was used as a chemical warfare agent during World War I. Its boiling point is −34°C. The compound sodium chloride, NaCl, is a translucent, brittle, colorless crystal with a melting point of 800°C. Sodium chloride does not react chemically with water the way sodium does; not only is it not toxic to humans, which chlorine is, but the very opposite is true—it is an essential component of all living organisms. Sodium chloride is not sodium, nor is it chlorine; it is uniquely sodium chloride, a tasty chemical when sprinkled lightly over popcorn. A compound is uniquely different from the elements from which it is made. For example, water is a liquid, while the elements that are in it, hydrogen and oxygen, are gases. The harmless compound known as table salt is composed of two very dangerous chemicals: metallic sodium and chlorine gas.

CHECK YOURSELF

Hydrogen sulfide, H_2S, is one of the smelliest compounds. Rotten eggs get their characteristic bad smell from the hydrogen sulfide they release. Can you infer from this information that elemental sulfur, S_8, is just as smelly?

CHECK YOUR ANSWER

No, you cannot. In fact, the odor of elemental sulfur is negligible compared with that of hydrogen sulfide. Compounds are truly different from the elements from which they are formed. Hydrogen sulfide, H_2S, is as different from elemental sulfur, S_8, as water, H_2O, is from elemental oxygen, O_2.

11.7 Naming Compounds

LEARNING OBJECTIVE
List three guidelines used to name compounds.

EXPLAIN THIS What information is found within the name of a compound?

A system for naming the countless possible compounds has been developed by the International Union of Pure and Applied Chemistry (IUPAC). This system is designed so that a compound's name reflects the elements it contains and how those elements are joined. Anyone familiar with the system, therefore, can deduce the chemical identity of a compound from its systematic name.

As you might imagine, this system is very intricate. There is no need for you to learn all its rules. Instead, learning some guidelines will prove most helpful. These guidelines alone will not enable you to name every compound. However, they will acquaint you with how the system works for many simple compounds.

As we will discuss in Chapter 12, atoms are held together in a chemical compound by their electric charges. Whether an atom in a compound takes on a positive or negative charge can be predicted by its place in the periodic table. The atom of an element located closer to the left side of the periodic table tends to take on a positive charge, while the atom of an element closer to the right side of the periodic table takes on a negative charge. This applies to the naming of compounds in that the more positively charged atom is, by convention, listed first. That's why we have "sodium chloride" rather than "chloride sodium" as spelled out in guideline 1—note that sodium, Na, is on the left side of the periodic table and chlorine, Cl, is on the right side.

GUIDELINE 1 The name of the element farther to the left in the periodic table is followed by the name of the element farther to the right, with the suffix *-ide* added to the name of the latter:

NaCl	sodium chloride	HCl	hydrogen chloride
Li_2O	lithium oxide	MgO	magnesium oxide
CaF_2	calcium fluoride	Sr_3P_2	strontium phosphide

GUIDELINE 2 When two or more compounds have different numbers of the same elements, prefixes are added to remove the ambiguity. This occurs primarily with compounds of nonmetals. The first four prefixes are *mono-* (one), *di-* (two), *tri-* (three), and *tetra-* (four). The prefix *mono-*, however, is commonly omitted from the beginning of the first word of the name:

Carbon and oxygen
CO carbon monoxide
CO_2 carbon dioxide

Nitrogen and oxygen
NO_2 nitrogen dioxide
N_2O_4 dinitrogen tetroxide

Sulfur and oxygen
SO_2 sulfur dioxide
SO_3 sulfur trioxide

GUIDELINE 3 Atoms can clump together to form a molecular unit that acts as single electrically charged group, called a *polyatomic ion*. For example, a carbon atom can join with three oxygen atoms to form what is known as a carbonate ion,

TABLE 11.1
COMMON POLYATOMIC IONS

Name	Formula
Acetate ion	$CH_3CO_2^-$
Ammonium ion	NH_4^+
Bicarbonate ion	HCO_3^-
Carbonate ion	CO_3^{2-}
Cyanide ion	CN^-
Hydroxide ion	OH^-
Nitrate ion	NO_3^-
Phosphate ion	PO_4^{3-}
Sulfate ion	SO_4^{2-}

CO_3^{2-}. Some commonly encountered polyatomic ions are listed in Table 11.1. Note that most of them are negatively charged. We'll explore the nature of these polyatomic ions in greater detail in Chapter 12. For now it suffices to recognize that positively charged polyatomic ions are placed first in the name (without the word *ion*). An example is ammonium chloride, NH_4Cl. Negatively charged polyatomic ions are placed at the end of the name. An example is lithium nitrate, $LiNO_3$.

A polyatomic ion may appear more than once within a compound. This is indicated by placing the polyatomic ion in parentheses. A subscript just outside the parentheses indicates the number of times the polyatomic ion appears. To keep it simple, the prefixes *mono-*, *di-*, *tri-*, and *tetra-* are commonly not included for polyatomic ions:

K_2CO_3	potassium carbonate
$AuPO_4$	gold phosphate
$Mg(CN)_2$	magnesium cyanide
$Al_2(SO4)_3$	aluminum sulfate

GUIDELINE 4 Many compounds are not usually referred to by their systematic names. Instead, they are assigned common names that are more convenient or have been used traditionally for many years. Some common names are water for H_2O, ammonia for NH_3, and methane for CH_4.

CHECK YOURSELF

What is the systematic name for $Ca(CH_3CO_2)_2$? How many oxygen atoms does it have?

CHECK YOUR ANSWER

The systematic name for this compound, which consists of calcium and the polyatomic acetate ion, is calcium acetate. Each acetate ion has two oxygen atoms. This compound has two acetate ions, which means it has a total of four oxygen atoms.

LEARNING OBJECTIVE
Show how nanotechnology is a novel and promising application of science.

Integrated Science 11A
PHYSICS AND BIOLOGY

The Advent of Nanotechnology

EXPLAIN THIS Is nanotechnology the result of basic or applied research?

The age of microtechnology was ushered in some 60 years ago with the invention of the solid-state transistor, a device that serves as a gateway for electronic signals. Physicists and engineers were quick to grasp the idea of integrating many transistors together to create logic boards that could perform calculations and run programs. The more transistors they could squeeze into a circuit, the more powerful the logic board. The race thus began to squeeze more and more transistors together into tinier and tinier circuits. The scales achieved were in the realm of the micron (10^{-6} m): thus the term *micro*technology. At the time of the transistor's invention, few people realized the impact microtechnology would have on society—from personal computers to cell phones to the Internet.

Today, we are at the beginning of a similar revolution. Technological advances have recently brought us past the realm of microns to the realm of the nanometer (10^{-9} m), which is the scale of individual atoms and molecules—a realm where we have reached the basic building blocks of matter. Technology that works on this scale where we engineer materials by manipulating individual atoms or molecules is known as **nanotechnology**. No one knows exactly what impact nanotechnology will have on society, but we are quickly coming to realize its vast potential, which is likely to be much greater than that of microtechnology.

FYI Before he died in 2005, Nobel laureate Rick Smalley advocated that carbon nanotubes, if developed into wires, could be an ideal material for efficiently transporting electricity over vast distances. If such an infrastructure were in place, the wind energy of the Great Plains of the United States would be sufficient to supply the electrical needs of the entire country.

Nanotechnology generally concerns the manipulations of objects from 1 to 100 nanometers in size. For perspective, a DNA molecule is about 2.0 nm wide, while a water molecule is only about 0.2 nm wide. Like microtechnology, nanotechnology is interdisciplinary, requiring the cooperative efforts of physicists, engineers, chemists, molecular biologists, and many others. Interestingly, there are already many products on the market that contain components developed through nanotechnology. These include sunscreens, mirrors that don't fog, dental bonding agents, automotive catalytic converters, stain-free clothing, water filtration systems, the heads of computer hard drives, and many more. Nanotechnology, however, is still in its infancy, and it will likely be decades before its potential is fully realized (Figure 11.23). Consider, for example, that personal computers didn't blossom until the 1990s, some 40 years after the first solid-state transistor.

There are two main approaches to building nanoscale materials and devices: top down and bottom up. The top-down approach is an extension of microtechnology techniques to smaller and smaller scales. A nanosize circuit board, for example, might be carved from a larger block of material. The bottom-up approach involves building nanosized objects atom by atom. A most important tool for either of these approaches is the **scanning probe microscope**, which was first discussed in Chapter 9. This device detects and characterizes the surface atoms of materials using an ultrathin probe tip, as shown in Figure 11.24. The tip is mechanically dragged over the surface. Interactions between the tip

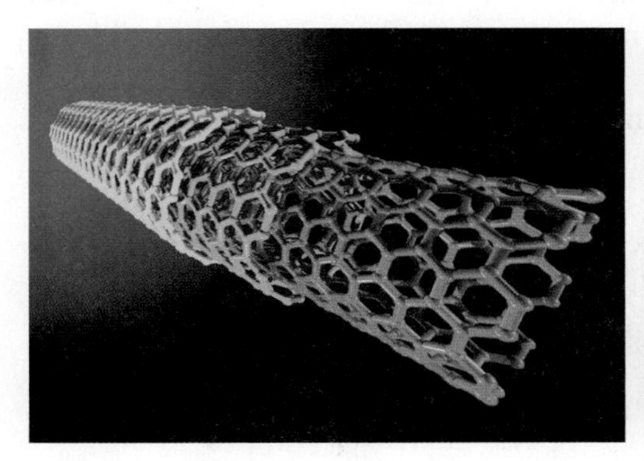

FIGURE 11.23
Carbon nanotubes can be nested within each other to provide the strongest fiber known—a thread 1 mm in diameter can support a weight of about 13,000 lb. A network of such strong fibers could be used to build the once science-fictional space elevator.

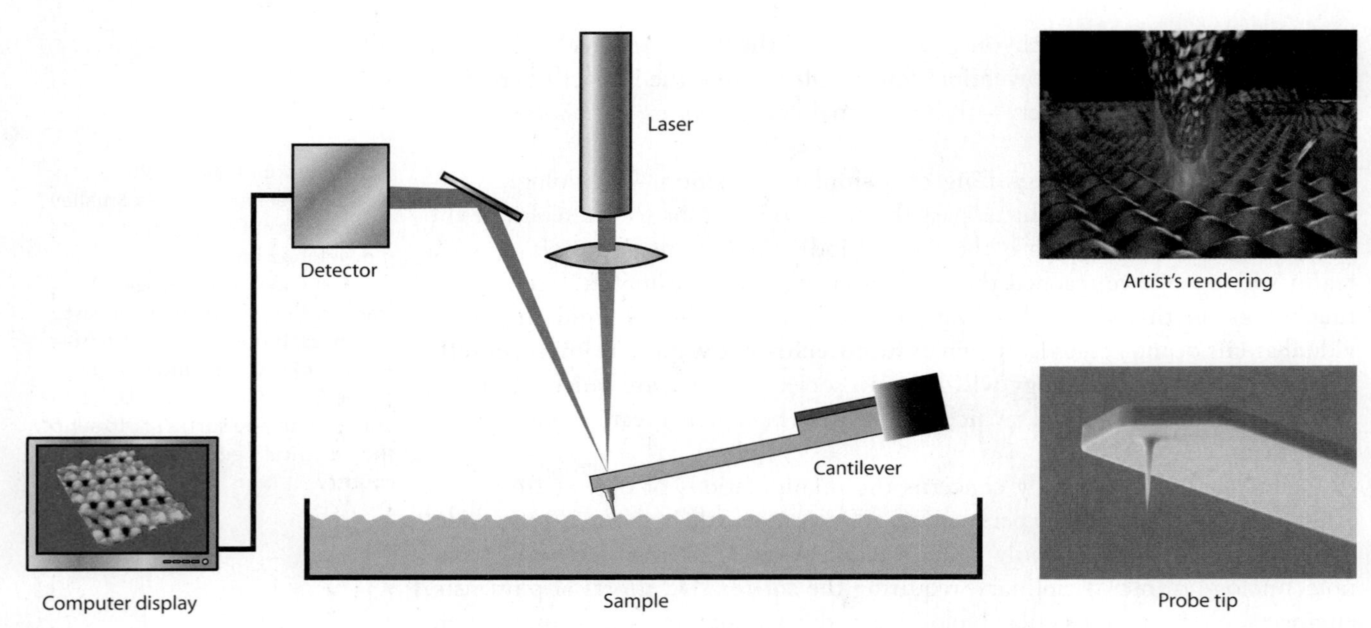

FIGURE 11.24
A schematic of a scanning probe microscope.

FYI An interesting discovery of nanoscience is that the properties of a material at the level of its atoms can be different from its properties in bulk quantities. A bar of gold, for example, is gold in color. A thin sheet of gold atoms, by contrast, is dark red. Much research is currently being directed toward the discovery of the unique nano properties of materials. Many novel applications of these nano properties are sure to follow.

and the surface atoms cause movements in a cantilever attached to the tip that are detected by a laser beam and translated by a computer into a topographical image. Scanning probe microscopes can also be used to move individual atoms into desired positions.

Nanotechnology allows the continued miniaturization of integrated circuits needed for ever smaller and more powerful computers. But a computer need not rely on an integrated circuit of nanowires for processing power. A wholly new approach involves designing logic boards in which molecules (not electric circuits) read, process, and write information. One molecule that has proved most promising for such *molecular computation* is DNA, the same molecule that holds our genetic code. An advantage that molecular computing has over conventional computing is that it can run a massive number of calculations in parallel (at the same time). Because of such fundamental differences, molecular computing may one day outshine even the fastest integrated circuits. Molecular computing, in turn, may then be eclipsed by other novel approaches, such as quantum or photon computing, also made possible by nanotechnology.

The ultimate expert on nanotechnology is nature. Living organisms, for example, are complex systems of interacting biomolecules all functioning on the scale of nanometers. In this sense, the living organism is nature's nanomachine. We need look no farther than our own bodies to find evidence of the feasibility and power of nanotechnology. With nature as our teacher, we have much to learn. Such knowledge will be particularly applicable to medicine. By becoming nanotechnology experts ourselves, we would be well equipped to understand the exact causes of nearly any disease or disorder (aging included) and empowered to develop innovative cures.

What are the limits of nanotechnology? As a society, how will we deal with the impending changes nanotechnology may bring? Consider the possibilities: wall paint that can change color or be used to display video; smart dust that the military could use to seek out and destroy an enemy; solar cells that capture sunlight so efficiently that they render fossil fuels obsolete; robots with so much processing power that we begin to wonder whether they experience consciousness; nanobots that roam our circulatory systems destroying cancerous tumors or arterial plaque; nanomachines that can "photocopy" three-dimensional objects, including living organisms; medicines that more than double the average human life span. Stay tuned for an exciting new revolution in human capabilities.

CHECK YOURSELF

How believable would our present technology be to someone living 200 years ago? How believable might the technology of 200 years in the future be to us right now?

CHECK YOUR ANSWER

Hindsight is 20-20. It's always easy to look back over time and see the progression of events that led to our present state. Much more difficult is to think forward and project possible scenarios. The technology of 200 years from now may be just as unbelievable to us as our present technology would be unbelievable to someone living 200 years ago.

For instructor-assigned homework, go to www.masteringphysics.com 

SUMMARY OF TERMS (KNOWLEDGE)

Applied research Research that focuses on developing applications of knowledge gained through basic research.

Basic research Research that leads us to a better understanding of how the natural world operates.

Boiling Evaporation in which bubbles form beneath the liquid surface.

Chemical bond The force of attraction between two atoms that holds them together within a compound.

Chemical change The formation of a new substance(s) by the rearrangement of the atoms of the original material(s).

Chemical formula A notation that indicates the composition of a compound, consisting of the atomic symbols for the elements in the compound and numerical subscripts indicating the ratio in which the atoms combine.

Chemical property A property that characterizes the ability of a substance to react with other substances or to change into a different substance under specific conditions.

Chemical reaction A term synonymous with chemical change.

Chemistry The study of matter and the transformations it can undergo.

Compound A structure in which atoms of different elements are bonded to one another.

Condensation The transformation of a gas to a liquid.

Deposition The transformation of a gas to a solid.

Diffusion The tendency of the particles of a material, such as molecules, to move from an area of high concentration to one of low concentration.

Elemental formula A notation that uses the atomic symbol and (sometimes) a numerical subscript to denote how many atoms are bonded in one unit of an element.

Evaporation The transformation of a liquid to a gas.

Freezing The transformation of a liquid to a solid.

Gas Matter that has neither a definite volume nor a definite shape, always filling any space available to it.

Heat of fusion The amount of energy needed to change any substance from a solid to liquid or liquid to solid phase.

Heat of vaporization The amount of energy needed to change any substance from a liquid to gas or gas to liquid phase.

Latent heat The energy absorbed or released in order to cause a change of phase.

Liquid Matter that has a definite volume but not definite shape, assuming the shape of its container.

Melting The transformation of a solid to a liquid.

Molecule An extremely small fundamental structure built of atoms.

Nanotechnology Technology on the scale of a nanometer, where materials are engineered by the manipulation of individual atoms or molecules.

Physical change A change in which a substance's physical properties are altered but with no change in its chemical identity.

Physical property Any physical attribute of a substance, such as color, density, or hardness.

Scanning probe microscope A tool of nanotechnology that detects and characterizes the surface atoms of materials using an ultrathin probe tip, which is detected by laser light as it is mechanically dragged over the surface.

Solid Matter that has a definite volume and a definite shape.

Sublimation The transformation of a solid to a gas.

Submicroscopic The realm of atoms and molecules, where objects are smaller than can be detected by optical microscopes.

READING CHECK (COMPREHENSION)

11.1 Chemistry: The Central Science

1. Why is chemistry often called the central science?
2. What is the difference between basic research and applied research?
3. What pledge have many members of the American Chemistry Council made through the Responsible Care program?

11.2 The Submicroscopic View of Matter

4. Is a biological cell macroscopic, microscopic, or submicroscopic?
5. Are atoms made of molecules or are molecules made of atoms?
6. How are the particles in a solid arranged differently from the particles in a liquid?
7. Which occupies the greatest volume—1 gram of ice, 1 gram of liquid water, or 1 gram of water vapor?

11.3 Phase Changes

8. What is it called when evaporation takes place beneath the surface of a liquid?
9. How is sublimation different from evaporation?
10. Why doesn't the temperature of melting ice rise as the ice is heated?
11. How much heat is needed to melt 1 gram of ice? Give your answer in joules.

11.4 Physical and Chemical Properties

12. What happens to the chemical identity of a substance during a physical change?
13. What is a physical property? A chemical property?
14. What is a chemical bond?
15. What changes during a chemical reaction?

11.5 Determining Physical and Chemical Changes

16. Why is the freezing of water considered to be a physical change?
17. Why is it sometimes difficult to determine whether an observed change is physical or chemical?
18. Why is the rusting of iron considered to be a chemical change?
19. What are some clues that help us to determine whether an observed change is physical or chemical?

11.6 Elements to Compounds

20. Distinguish between an atom and an element.
21. What is the difference between an element and a compound?
22. How many atoms are in one molecule of H_3PO_4? How many atoms of each element are in one molecule of H_3PO_4?
23. What does the chemical formula of a substance tell us about that substance?
24. Are the physical and chemical properties of a compound necessarily similar to those of the elements from which it is composed?

11.7 Naming Compounds

25. Which element within a compound is given first in the compound's name?
26. What is the IUPAC systematic name for the compound KF?
27. What is the chemical formula for the compound titanium dioxide?
28. Why are common names often used for chemical compounds instead of systematic names?

THINK INTEGRATED SCIENCE

11A—The Advent of Nanotechnology

29. How soon will nanotechnology give rise to commercial products?
30. What are the two main approaches to building nanoscale materials and devices?
31. Who is the ultimate expert at nanotechnology?

THINK AND DO (HANDS-ON APPLICATION)

32. A TV screen looked at from a distance appears as a smooth continuous flow of images. Up close, however, we see that this is an illusion. What really exists are a series of tiny dots (pixels) that change color in a coordinated way to produce images. Use a magnifying glass to examine closely the screen of a computer monitor or television set.
33. Add a pinch of red Kool-aid crystals to a still glass of hot water. Add the same amount of crystals to a second still glass of cold water. With no stirring, which would you expect to become uniform in color first—the hot water or the cold water? Why?
34. Air molecules stuck inside an inflated balloon are perpetually colliding with the inner surface of the balloon. Each collision provides a little push outward on the balloon. All the many collisions working together is what keeps the balloon inflated. To get a "feel" for what's happening here, add about a tablespoon of tiny beads to a large balloon (pellets, beans, BBs, grains of rice, etc., also work). Inflate the balloon to its full size and tie it shut. Hold the balloon in the palms of both hands and shake rapidly. Can you feel the collisions? As you shake the balloon wildly, the flying beads represent the gaseous phase. How should you move the balloon so that the beads represent the liquid phase? The solid phase?
35. This activity is for those who have access to a gas stove. Place a large pot of cool water on top of the stove, and set the burner on high. What product from the combustion of the natural gas do you see condensing on the outside of the pot? Where did it come from? Would more or less of this product form if the pot contained ice water? Where does this product go as the pot gets warmer? What physical and chemical changes can you identify?

36. When you pour a solution of hydrogen peroxide, H_2O_2, over a cut, an enzyme in your blood decomposes it to produce oxygen gas, O_2, as evidenced by the bubbling that takes place. It is this oxygen at high concentrations at the site of the injury that kills off microorganisms. A similar enzyme is found in baker's yeast.

    Wear safety glasses and remove all combustibles, such as paper towels, from a clear countertop area. Pour a small packet of baker's yeast into a tall glass. Add a couple capfuls of 3% hydrogen peroxide and watch oxygen bubbles form. Test for the presence of oxygen by holding a lighted match with tweezers and putting the flame near the bubbles. Look for the flame to glow more brightly as the escaping oxygen passes over it. Describe oxygen's physical and chemical properties.

THINK AND COMPARE (ANALYSIS)

37. Rank the following in order of increasing volume: (a) bacterium, (b) virus, (c) water molecule.
38. Rank these substances in order of increasing strength of attraction between submicroscopic particles: (a) sugar, (b) water, (c) air.
39. Rank the following physical and chemical changes in order of the amount of energy released, from least to greatest: (a) the condensation of rain in a thunderstorm, (b) the burning of a gallon of gasoline in a car engine, (c) the explosion of a firecracker.
40. Rank these compounds in order of increasing number of atoms: (a) $C_{12}H_{22}O_{12}$, (b) DNA, (c) $Pb(C_2H_3O_2)_2$.

THINK AND EXPLAIN (SYNTHESIS)

41. While visiting a foreign country, a foreign-speaking citizen tries to give you verbal directions to a local museum. After multiple attempts he is unsuccessful. An onlooker sees your frustration and concludes that you are not smart enough to understand simple directions. What is another reason you do not understand the directions?
42. What is the best way to really prove to yourself that you understand an idea?
43. Of physics, chemistry, and biology, which is the most fundamental science? Why?
44. Why might biology be considered a more complex science than chemistry?
45. Is chemistry the study of the submicroscopic, the microscopic, the macroscopic, or all three? Defend your answer.

46. You combine 50 mL of small BBs with 50 mL of large BBs and get a total of 90 mL of BBs of mixed size. Explain.
47. You combine 50 mL of water with 50 mL of purified alcohol and get a total of 98 mL of the mixture. Explain.
48. Red Kool-aid crystals are added to a still glass of water. The crystals sink to the bottom. Twenty-four hours later the entire solution is red even though no one stirred the water. Explain.
49. With no one looking, you add 5 mL of a cinnamon solution to a blue balloon, which you tie shut. You also add 5 mL of fresh water to a red balloon, which you also tie shut. You heat the two balloons in a microwave until they each inflate to about the size of a grapefruit. Your brother then comes along, examines the inflated balloons, and tells you that the blue balloon is the one that contains the cinnamon. How did he know?
50. Which has stronger attractions among its submicroscopic particles—a solid at 25°C or a gas at 25°C? Explain.
51. Humidity is a measure of the amount of water vapor in the atmosphere. Why is humidity always very low inside your kitchen freezer?
52. Why is perfume typically applied behind the ear rather than on the ear?
53. When liquid water freezes, is heat released to the surroundings or absorbed from the surroundings?
54. Why does it take so much more energy to boil 10 grams of liquid water than to melt 10 grams of ice?
55. The leftmost diagram below shows the moving particles of a gas in a rigid container. Which of the three boxes on the right—(a), (b), or (c)—best represents this material when heat is added?

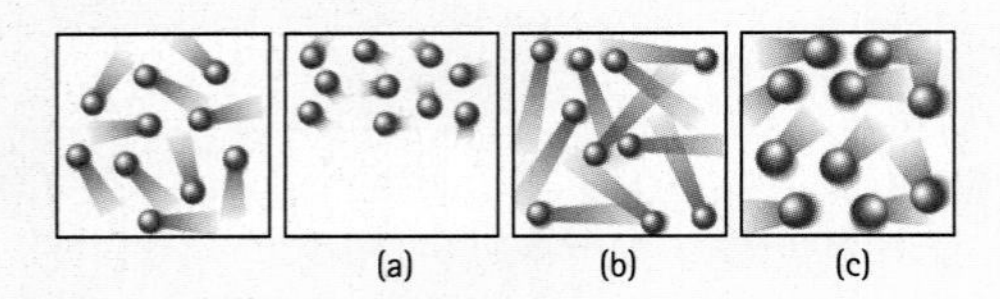

56. The leftmost diagram below shows two phases of a single substance. In the middle box, draw what these particles would look like if heat were taken away. In the box on the right, show what they would look like if heat were added. If each particle represents a water molecule, what is the temperature of the box on the left?

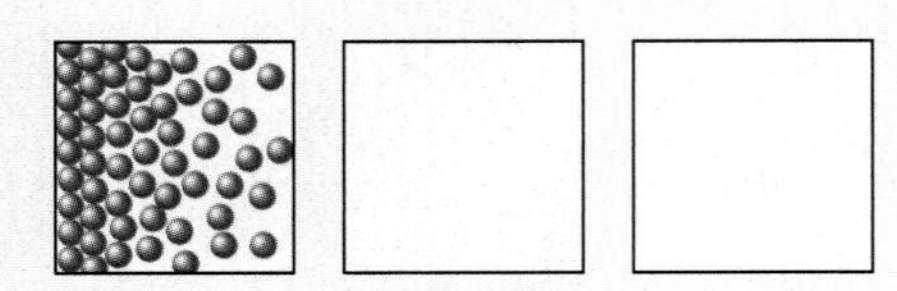

57. A cotton ball is dipped in alcohol and wiped across a tabletop. Explain what happens to the alcohol molecules that are deposited on the tabletop? Is this a physical or chemical change?
58. A skillet is lined with a thin layer of cooking oil followed by a layer of unpopped popcorn kernels. Upon heating the kernels all pop, thereby escaping the skillet. Identify any physical or chemical changes.
59. A cotton ball dipped in alcohol is wiped across a tabletop. Would the resulting smell of the alcohol be more or less noticeable if the tabletop were much warmer? Explain.
60. Use Exercise 58 as an analogy to describe what occurs in Exercise 59. Why does it make sense to think that the alcohol is made of very tiny particles (molecules) rather than being an infinitely continuous material?
61. In the winter Vermonters make a tasty treat called "sugar on snow" in which they pour boiled-down maple syrup onto a scoop of clean fresh snow. As the syrup hits the snow it forms a delicious taffy. Identify the physical changes involved in the making of sugar on snow. Identify any chemical changes.
62. Oxygen, O_2, has a boiling point of 90 K (−183°C), and nitrogen, N_2, has a boiling point of 77 K (−196°C). Which is a liquid and which is a gas at 80 K (−193°C)?
63. Is aging primarily an example of a physical or chemical change?
64. Each sphere in the diagrams below represents an atom. Joined spheres represent molecules. Which box contains a substance in the liquid phase? Why can't you assume that box B represents a lower temperature?

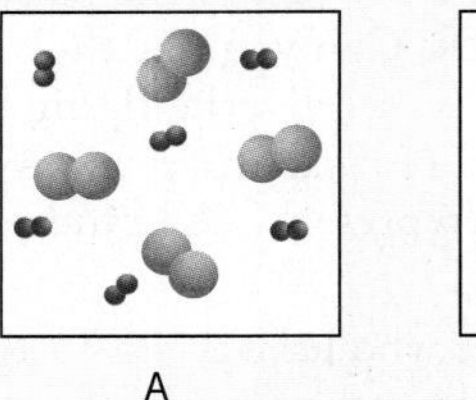

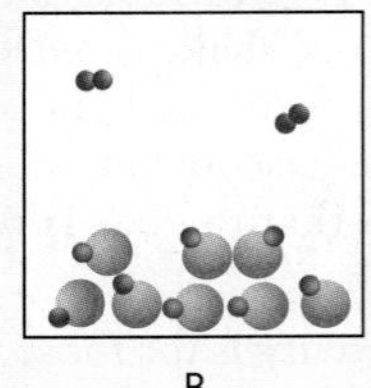

65. State whether each of the following is an example of a physical or chemical property of matter: (a) Graphite conducts electricity. (b) Bismuth, Bi, loses its iridescence upon melting. (c) A copper penny is smushed into an embossed souvenir.
66. State whether each of the following is an example of a physical or chemical property of matter: (a) Carbon dioxide escapes when a soda can is opened. (b) A bronze statue turns green. (c) A silver spoon tarnishes.
67. Classify the following changes as physical or chemical. Even if you are incorrect in your assessment, you should be able to defend your choice. (a) Grape juice turns to wine. (b) Wood burns to ashes. (c) Water begins to boil. (d) A broken leg mends itself.
68. Classify the following changes as physical or chemical. Even if you are incorrect in your assessment, you should be able to defend your choice. (a) Grass grows. (b) An infant gains 10 pounds. (c) A rock is crushed to powder. (d) A tire is inflated with air.
69. Oxygen atoms are used to make water molecules. Does this mean that oxygen, O_2, and water, H_2O, have similar properties? Why do we drown when we breathe in water despite all the oxygen atoms present in water?
70. Oxygen, O_2, is certainly good for you. Does it follow that if small amounts of oxygen are good for you, then large amounts of oxygen would be especially good for you?
71. Why isn't water classified as an element?

72. Which of the boxes contains only an element? Which contains only a compound? How many different types of molecules are shown altogether in all three boxes?

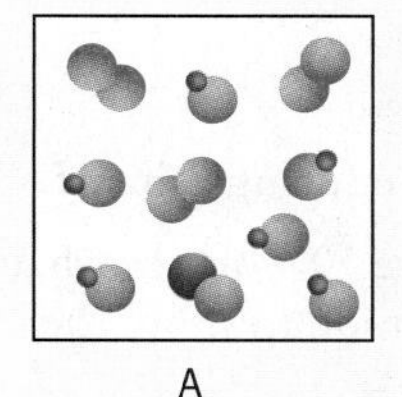
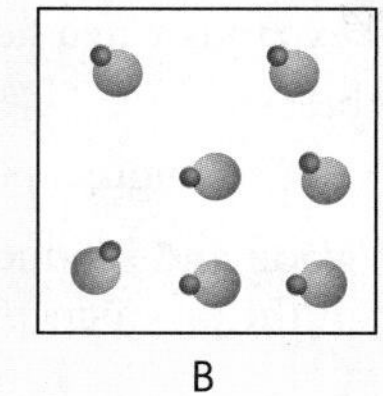
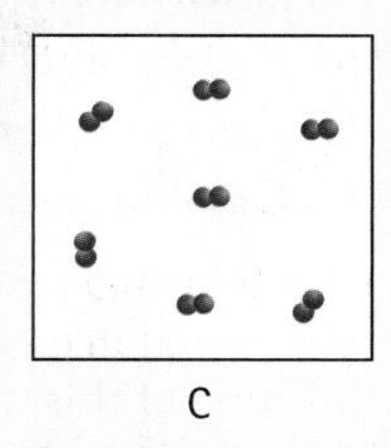

A B C

73. Elemental copper, Cu, is copper colored. Elemental sulfur, S_8, is yellow. What does this tell you about the color of the compound copper sulfide, CuS?
74. What is the chemical formula for the compound dihydrogen sulfide?
75. What is the chemical name for a compound with the formula Ba_3N_2?
76. What is the common name for trioxygen?
77. What is the name of the compound with the formula $Sr_3(PO_4)_2$?
78. Give the name and the formula for a compound that results from the combination of aluminum, sulfur, and oxygen?
79. How does a scanning probe microscope differ from an optical microscope?
80. People often behave differently when they are in a group than when they are by themselves. Explain how this is similar to the behavior of atoms. Is this good news or bad news for the development of nanotechnology?
81. How is chemistry similar to nanotechnology? How is it different?

THINK AND DISCUSS (EVALUATION)

82. Your friend smells cinnamon coming from an inflated rubber balloon that contains cinnamon extract. You tell him that the cinnamon molecules are passing through the micropores of the balloon. He accepts the idea that the balloon contains micropores but insists that he is simply smelling cinnamon-flavored air. You explain that scientists have discovered that gases are made of molecules, but that's not good enough for him. He needs to see the evidence for himself. How might you lead him to accept the concept of molecules?
83. British diplomat, physicist, and environmentalist John Ashton, in speaking to a group of scientists, stated (paraphrased): "There has to be much better communication between the world of science and the world of politics. Consider the different meaning of the word *uncertainty*. To scientists, it means uncertainty over the strength of a signal. To politicians it means 'go away and come back when you're certain.'" Pretend you are a scientist with strong but inconclusive evidence in support of impending climate change. How might you best persuade politicians to take action?
84. The famous 20th-century physicist Richard Feynman (1918–1988) noted: "The laws of science do not limit our ability to manipulate single atoms and molecules." What does?
85. A calculator is useful, but certainly not exciting. Why would someone from 100 years ago vehemently disagree with this statement? We often marvel at a new technology, but how long does this marveling last? How soon before a new technology becomes assumed? Think of other examples. Is technology the source of happiness?

READINESS ASSURANCE TEST (RAT)

If you have a good handle on this chapter, if you really do, then you should be able to score 7 out of 10 on this RAT. If you score less than 7, you need to study further before moving on.

Choose the BEST answer to each of the following:

1. Chemistry is the study of
   (a) matter.
   (b) transformations of matter.
   (c) only microscopic phenomena.
   (d) only macroscopic phenomena.
   (e) both matter and transformations of matter.
2. Imagine that you can see individual molecules. You watch a small collection of molecules that are moving around slowly while vibrating and bumping against one another. The slower-moving molecules then start to line up, but as they do so their vibrations increase. Soon all the molecules are aligned and vibrating about fixed positions. The sample is being
   (a) cooled and the material is freezing.
   (b) heated and the material is melting.
   (c) cooled and the material is condensing.
   (d) heated and the material is boiling.
3. The phase in which atoms and molecules no longer move is the
   (a) solid phase.
   (b) liquid phase.
   (c) gas phase.
   (d) none of these

4. What chemical change occurs when a wax candle burns?
   (a) The wax near the flame melts.
   (b) Carbon soot collects above the flame.
   (c) The molten wax is pulled upward through the wick.
   (d) The heated wax molecules combine with oxygen molecules.
   (e) Two of these are signs of chemical change.
5. Based on the information given in the diagrams, which substance has the lower boiling point—one made from molecule A, , or one made from molecule B, ?

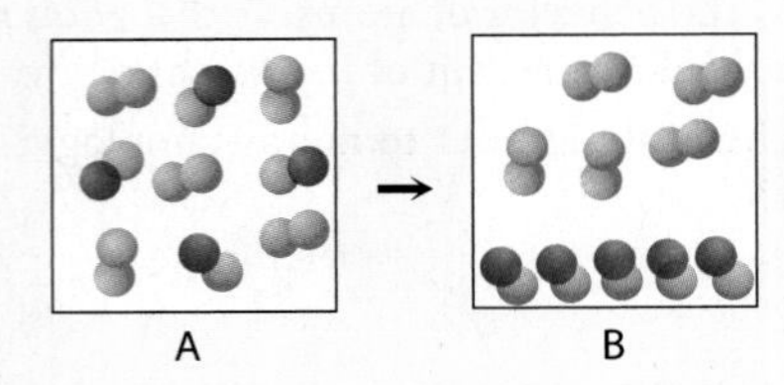

   (a) molecule A, which is the first to transform into a liquid
   (b) molecule B, which is the first to transform into a liquid
   (c) molecule A, which remains in the gas phase
   (d) molecule B, which remains in the gas phase
6. Does this transformation represent a physical change or a chemical change? Why?

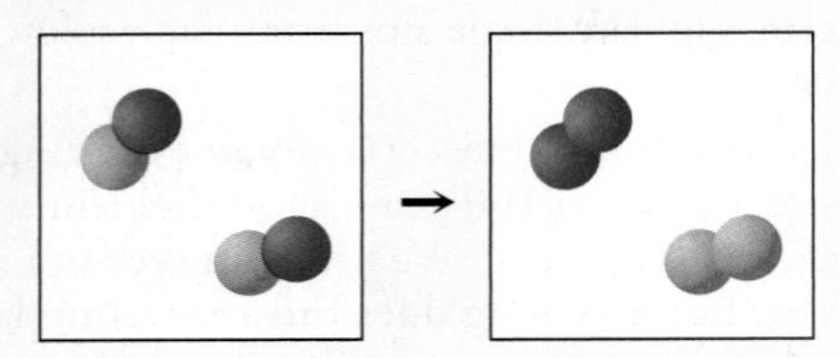

   (a) chemical, because of the formation of elements
   (b) physical, because a new material has been formed
   (c) chemical, because the atoms are connected differently
   (d) physical, because of a change in phase
7. Which is an example of a chemical change?
   (a) Water freezes into ice crystals.
   (b) Aftershave or perfume on your skin generates a smell.
   (c) A piece of metal expands when heated.
   (d) A glass window breaks.
   (e) Gasoline burns in a car engine, producing exhaust.
8. If you burn 50 g of wood and produce 10 g of ash, what is the total mass of all the products produced from the burning of this wood?
   (a) more than 50 g
   (b) 10 g
   (c) less than 10 g
   (d) 50 g
   (e) none of these
9. If you have one molecule of TiO_2, how many molecules of O_2 does it contain? Why?
   (a) one, because TiO_2 is a mixture of Ti and O_2
   (b) none, because O_2 is a different molecule than TiO_2
   (c) two, because TiO_2 is a mixture of Ti and 2 O
   (d) three, because TiO_2 contains three molecules
   (e) none of these
10. What is the name of the compound $CaCl_2$?
   (a) carbon chloride
   (b) dichlorocalcium
   (c) calc two
   (d) dicalcium chloride
   (e) calcium chloride

Answers to RAT

1. e, 2. a, 3. d, 4. e, 5. c, 6. c, 7. e, 8. a, 9. e, 10. e

CHAPTER 12
Chemical Bonds and Mixtures

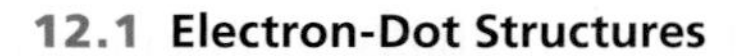

WHY DO salt crystals have a distinct cubic shape? As we will see in this chapter, the macroscopic properties of any substance can be traced to how its submicroscopic parts are held together. The sodium and chloride ions in a salt crystal, for example, hold together in a cubic orientation and, as a result, the macroscopic object we know as a salt crystal is also cubic.

The force of attraction that holds ions or atoms together is the electric force. Chemists refer to this atom-binding force as a *chemical bond*. In this chapter, we explore three types of chemical bonds: the *ionic bond*, which holds ions together in a crystal; the *metallic bond*, which holds atoms together in a piece of metal; and the *covalent bond*, which holds atoms together in a molecule. Then, later in this chapter, we explore how the behavior of ions and molecules gives rise to everyday phenomena, such as why oil and water don't mix and how fish get the oxygen they need from water.

LEARNING OBJECTIVE
Identify paired and unpaired electrons in an electron-dot structure.

12.1 Electron-Dot Structures

EXPLAIN THIS Why do the atoms of group 18 resist forming chemical bonds?

An atomic model is needed to help us understand how atoms bond. We begin this chapter with a brief overview of the noble gas shell model presented in Section 9.7. Recall how electrons are arranged around the atomic nucleus. Rather than moving in neat orbits like planets around the Sun, electrons are wavelike entities that vibrate within regions of space called *noble gas shells.*

As was shown in Figure 9.30, seven noble gas shells are available to the electrons in an atom, and the electrons fill these shells in order, from innermost to outermost. Furthermore, the maximum number of electrons allowed in the first shell is 2, and for the second and third shells it is 8. The fourth and fifth shells can each hold 18 electrons, and the sixth and seventh shells can each hold 32 electrons.* These numbers match the number of elements in each period (horizontal row) of the periodic table. Figure 12.1 shows how this model applies to the first three elements of group 18.

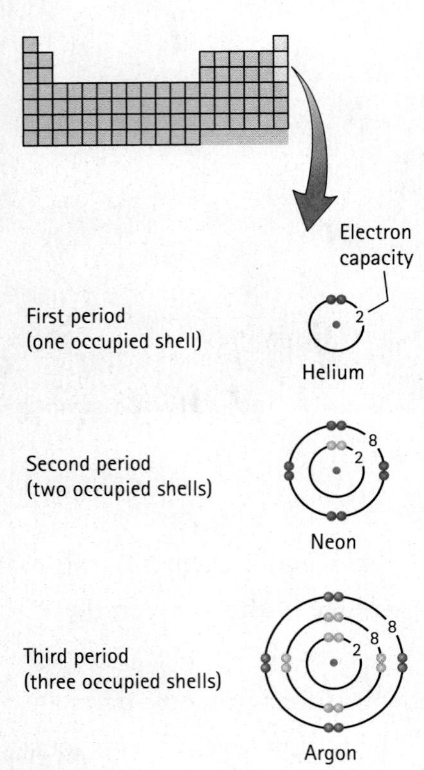

FIGURE 12.1
Occupied shells in the group 18 elements helium through argon. Each of these elements has a filled outermost occupied shell, and the number of electrons in each corresponds to the number of elements in the period to which a particular group 18 element belongs.

Electrons in the outermost occupied shell of any atom may play a significant role in that atom's chemical properties, including its ability to form chemical bonds. To indicate their importance, we call these electrons **valence electrons**, and we call the shell they occupy the **valence shell**. Valence electrons can be conveniently represented as a series of dots surrounding an atomic symbol. This notation is called an **electron-dot structure**, or sometimes a *Lewis dot symbol* (in honor of the American chemist G. N. Lewis, who first proposed the concepts of shells and valence electrons).

Figure 12.2 shows the electron-dot structures for the atoms that are important in our discussions of ionic and covalent bonds. Electron-dot structures are not so useful in describing metallic bonds, however. This is because within a metallic bond, the bonding electrons readily flow from one atom to the next, as is discussed further in Section 12.3. Therefore the metallic groups 3–12 are not included in Figure 12.2.

When you look at the electron-dot structure of an atom, you immediately know two important things about that element. You know how many valence electrons it has and how many of these electrons are *paired.* Chlorine, for example, has three sets of paired electrons and one unpaired electron, and carbon has four unpaired electrons:

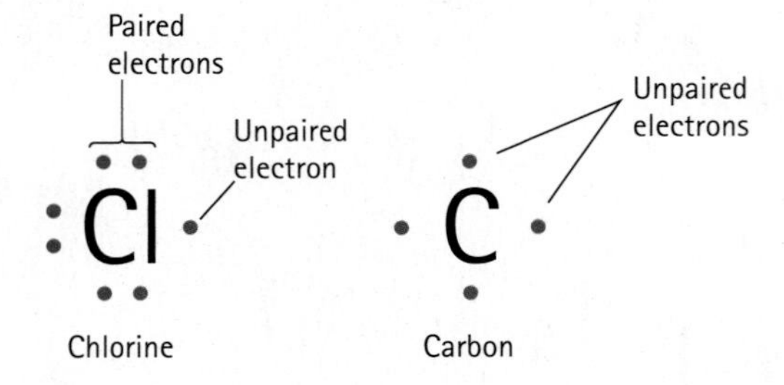

"The rapid progress true Science now makes occasions my regretting sometimes that I was born so soon. It is impossible to imagine the heights to which may be carried, in a thousand years, the power of man over matter. O that moral Science were in as fair a way of improvement, that men would cease to be wolves to one another, and that human beings would at length learn what they now improperly call humanity."—Benjamin Franklin, in a letter to chemist Joseph Priestley, 8 February 1780.

*As a point of reference for physicists reading this text, these are shells of orbitals grouped by similar energy levels rather than by principal quantum numbers. They are the "argonian" shells developed by Linus Pauling in the 1930s to explain the periodic table. This is an old atomic model, but it works well for a simple description of chemical bonding.

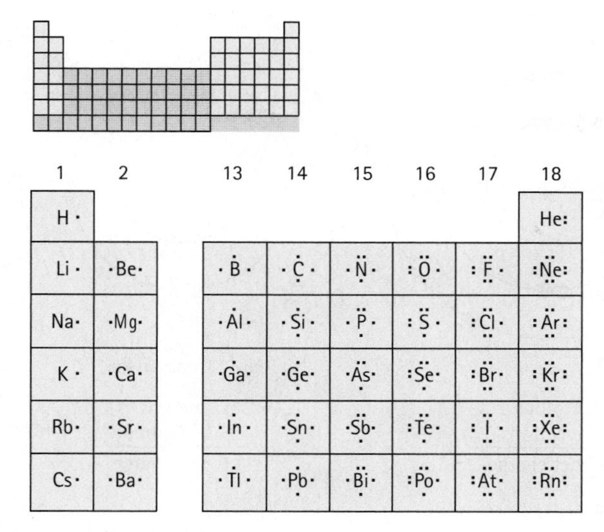

FIGURE 12.2
The valence electrons of an atom are shown in its electron-dot structure. Note that the first three periods here parallel Figure 9.31. Also note that for larger atoms, not all the electrons in the valence shell are valence electrons. Krypton, Kr, for example, has 18 electrons in its valence shell, but only 8 of these are classified as valence electrons. The reasons for this are best left to a follow-up course in chemistry.

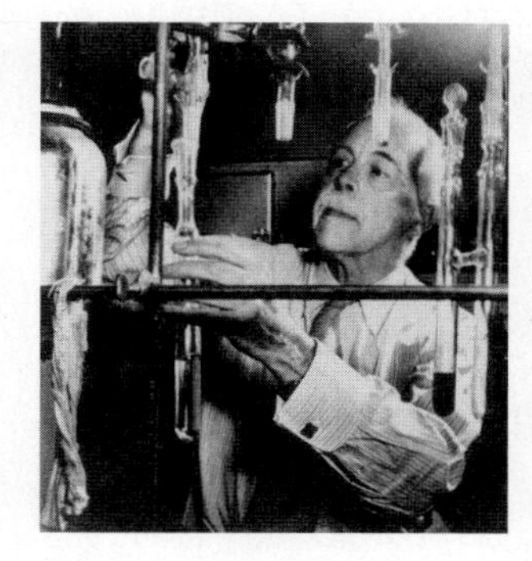

FIGURE 12.3
Gilbert Newton Lewis (1875–1946) revolutionized chemistry with his theory of chemical bonding, which he published in 1916. He worked most of his life in the chemistry department of the University of California, Berkeley, where he was not only a productive researcher but also an exceptional teacher. Among his teaching innovations was the idea of providing students with problem sets as a follow-up to lectures and readings.

Paired valence electrons are relatively stable. In other words, they do not readily form chemical bonds with other atoms. For this reason, electron pairs in an electron-dot structure are called **nonbonding pairs**.

Valence electrons that are *unpaired*, by contrast, have a strong tendency to participate in chemical bonding. By doing so, they become paired with an electron from another atom. The ionic and covalent bonds discussed in this chapter all result from either a transfer or a sharing of unpaired valence electrons.

CHECK YOURSELF
Where are valence electrons located, and why are they important?

CHECK YOUR ANSWER
Valence electrons are located in the outermost occupied shell of an atom. They are important because they play a leading role in determining the chemical properties of the atom.

Too much detail to learn? What would the scientists of 200 years ago have given for the information that today is so readily available to you?

12.2 The Ionic Bond

LEARNING OBJECTIVE
Describe how ions combine to form ionic compounds.

EXPLAIN THIS Why do ionic compounds have very high melting points?

When the number of protons in the nucleus of an atom equals the number of electrons in the atom, the charges balance and the atom is electrically neutral. If one or more electrons are lost or gained, as illustrated in Figures 12.4 and 12.5, the balance is upset and the atom takes on a net electric charge. Any atom that has a net electric charge is an **ion**. When

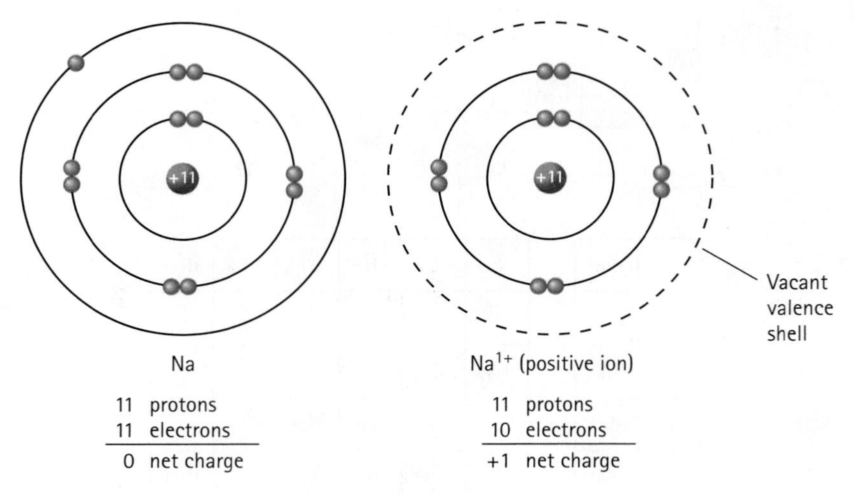

FIGURE 12.4
An electrically neutral sodium atom contains 11 negatively charged electrons surrounding the 11 positively charged protons of the nucleus. When this atom loses an electron, the result is a positive ion.

electrons are lost, protons outnumber electrons and the ion has a positive net charge. When electrons are gained, electrons outnumber protons and the ion has a negative net charge.

UNIFYING CONCEPT

- ***The Electric Force***
  Section 7.1

Chemists use a superscript to the right of the atomic symbol to indicate the magnitude and sign of an ion's charge. Thus, as shown in Figures 12.4 and 12.5, the positive ion formed from the sodium atom is written Na^{1+} and the negative ion formed from the fluorine atom is written F^{1-}. Usually the numeral 1 is omitted when indicating either a 1^{+} or 1^{-} charge. Hence, these two ions are most frequently written Na^{+} and F^{-}. To give two more examples, a calcium atom that loses two electrons is written Ca^{2+}, and an oxygen atom that gains two electrons is written O^{2-}. (Note that the convention is to write the numeral before the sign, not after it: 2+, not +2.)

You can use the periodic table as a quick reference when determining the type of ion an atom tends to form. As Figure 12.6 shows, each atom of any group 1 element, for example, has only one valence electron and so tends to form the 1+ ion. Each atom of any group 17 element has room for one additional electron in its valence shell and therefore tends to form the 1− ion. Atoms of the noble-gas elements tend not to form ions of any type because their valence shells are already filled to capacity.

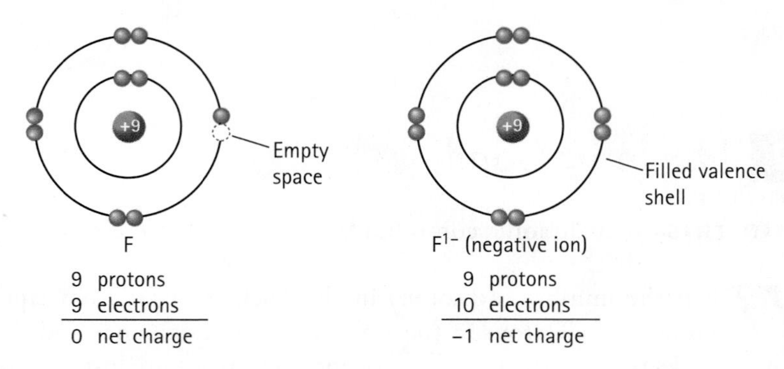

FIGURE 12.5
An electrically neutral fluorine atom contains 9 protons and 9 electrons. When this atom gains an electron, the result is a negative ion.

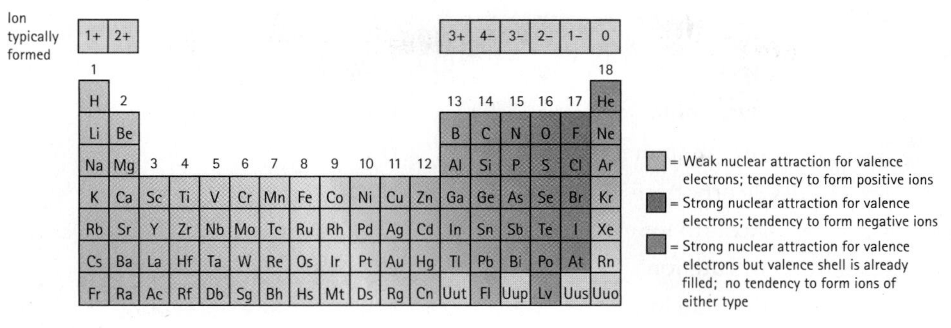

FIGURE 12.6
The periodic table is your guide to the types of ions that atoms tend to form within an ionic compound.

CHECK YOURSELF
What type of ion does the magnesium atom, Mg, tend to form?

CHECK YOUR ANSWER
The magnesium atom (atomic number 12) is found in group 2 and has two valence electrons to lose (see Figure 12.2). Therefore, it tends to form the 2+ ion.

Electrons are negatively charged. So, gaining an electron results in a negative ion, and losing an electron results in a positive ion.

The noble gas shell model we've presented is too simplified to work well for the transition metals of groups 3 through 12, or for the inner transition metals. In general, these metal atoms tend to form positive ions, but the number of electrons lost varies. For example, depending on conditions, an iron atom may lose two electrons to form the Fe^{2+} ion, or it may lose three electrons to form the Fe^{3+} ion.

Atoms form ions by losing or gaining electrons. Interestingly, molecules can also become ions. In most cases, this occurs when a molecule loses or gains a proton—equivalent to the hydrogen ion, H^+. For example, a water molecule, H_2O, can gain a hydrogen ion, H^+ (a proton), to form the hydronium ion, H_3O^+:

$$H_2O + H^+ \longrightarrow H_3O^+$$

Water | Hydrogen ion (proton) | Hydronium ion

Similarly, the carbonic acid molecule, H_2CO_3, can lose two protons to form the carbonate ion, CO_3^{2-}:

$$H_2CO_3 \longrightarrow CO_3^{2-} + 2\,H^+$$

Carbonic acid | Carbonate ion | Hydrogen ions (protons)

FYI What do the ions of the following elements have in common: calcium, Ca; chromium, Cr; cobalt, Co; copper, Cu; iodine, I; iron, Fe; magnesium, Mg; manganese, Mn; molybdenum, Mo; nickel, Ni; phosphorus, P; potassium, K; selenium, Se; sodium, Na; sulfur, S; zinc, Zn? They are all dietary minerals that are essential for good health but that can be harmful, even lethal, when consumed in excessive amounts.

TABLE 12.1 COMMON POLYATOMIC IONS

Name	Formula
Hydronium ion	H_3O^+
Ammonium ion	NH_4^+
Bicarbonate ion	HCO_3^-
Acetate ion	$CH_3CO_2^-$
Nitrate ion	NO_3^-
Cyanide ion	CN^-
Hydroxide ion	OH^-
Carbonate ion	CO_3^{2-}
Sulfate ion	SO_4^{2-}
Phosphate ion	PO_4^{3-}

How these reactions occur will be explored in later chapters. For now, you should understand that the hydronium and carbonate ions are examples of **polyatomic ions**, which are molecules that carry a net electric charge. Table 12.1 lists some commonly encountered polyatomic ions.

Ionic Bond Formation

When an atom that tends to lose electrons is placed in contact with an atom that tends to gain them, the result is an electron transfer and the formation of two oppositely charged ions. This occurs when sodium and chlorine are combined. As shown in Figure 12.7, the sodium atom loses one of its electrons to the chlorine atom, resulting in the formation of a positive sodium ion and a negative chloride ion. The two oppositely charged ions are attracted to each other by the electric force, which holds them close together. This electric force of attraction between two oppositely charged ions is called an **ionic bond**.

A sodium ion and a chloride ion together make the chemical compound sodium chloride, NaCl, commonly known as table salt. This and all other chemical compounds containing ions are referred to as **ionic compounds**. All ionic compounds are completely different from the elements from which they are made. As discussed in Section 11.6, sodium chloride is not sodium, nor is it chlorine.

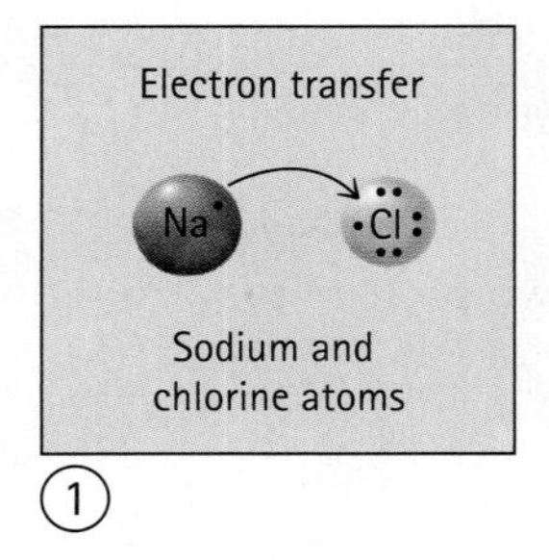

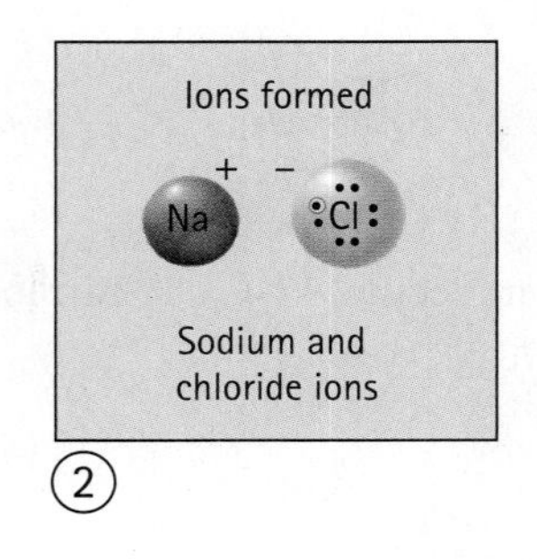

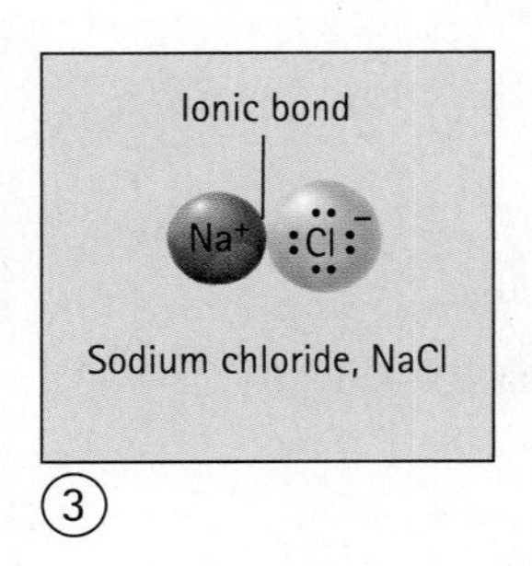

FIGURE 12.7
(1) An electrically neutral sodium atom loses its valence electron to an electrically neutral chlorine atom. (2) This electron transfer results in two oppositely charged ions. (3) The ions are then held together by an ionic bond. The spheres drawn around these and subsequent illustrations of electron-dot structures indicate the relative sizes of the atoms and ions.

Rather, it is a collection of sodium and chloride ions that form a unique material with its own physical and chemical properties.

CHECK YOURSELF

Is the transfer of an electron from a sodium atom to a chlorine atom a physical change or a chemical change?

CHECK YOUR ANSWER

Recall from Chapter 11 that only a chemical change involves the formation of new material. Thus, this or any other electron transfer, because it results in the formation of a new substance, is a chemical change.

Ionic compounds typically consist of elements that are found on opposite sides of the periodic table.

For all ionic compounds, the positive and negative charges must balance. In sodium chloride, for example, there is one sodium 1+ ion for every chloride 1−. Charges must also balance in compounds containing ions that carry multiple charges. The calcium ion, for example, carries a charge of 2+, but the fluoride ion carries a charge of only 1−. Because two fluoride ions are needed to balance each calcium ion, the formula for calcium fluoride is CaF_2. Calcium fluoride occurs naturally in the drinking water of some communities, where it is a good source of the tooth-strengthening fluoride ion, F^-.

An aluminum ion carries a 3+ charge, and an oxide ion carries a 2− charge. Together, these ions make the ionic compound aluminum oxide, Al_2O_3, the main component of such gemstones as rubies and sapphires. Figure 12.8 illustrates the formation of aluminum oxide. The three oxide ions in Al_2O_3 carry a total charge of 6−, which balances the total 6+ charge of the two aluminum ions. Rubies and sapphires differ in color because of the impurities they contain. Rubies are red because of minor amounts of chromium ions, and sapphires are blue because of minor amounts of iron and titanium ions.

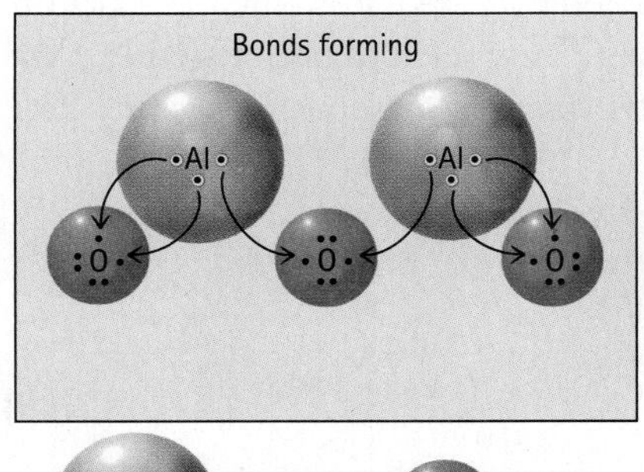

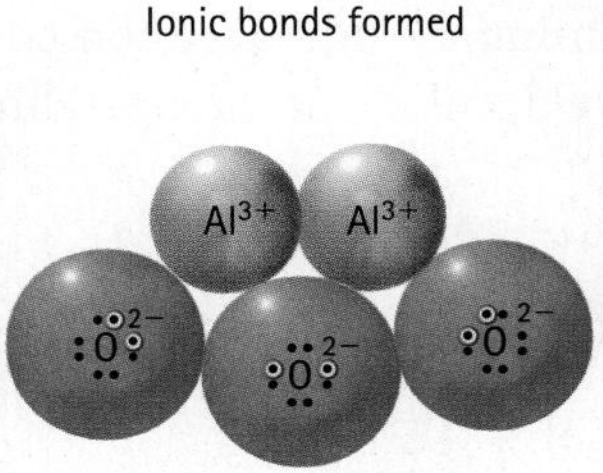

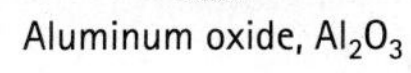

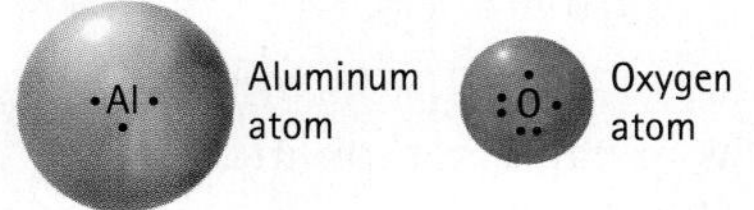

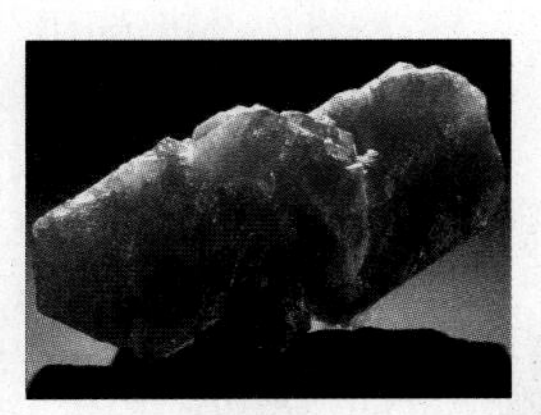

Ruby

Sapphire

FIGURE 12.8

Two aluminum atoms lose a total of six electrons to three oxygen atoms. In the process, the aluminum atoms become aluminum ions, Al^{3+}, and the oxygen atoms become oxide ions, O^{2-}. The oppositely charged ions join to form the ionic compound aluminum oxide, Al_2O_3. Certain gemstones are crystalline aluminum oxide with trace amounts of impurities, such as chromium, which makes ruby, and titanium, which makes sapphire.

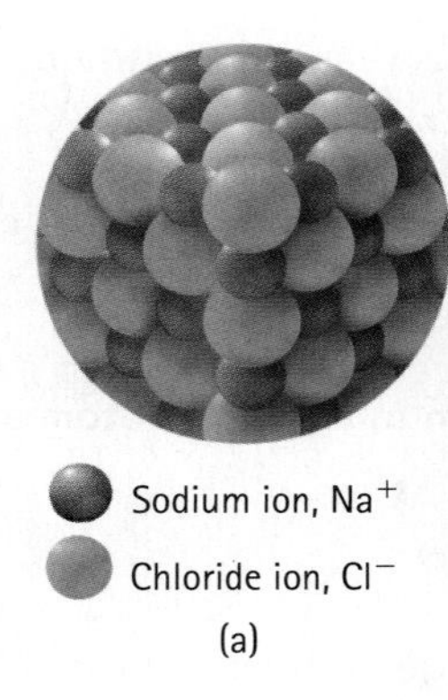

FIGURE 12.9
(a) Sodium chloride, as well as other ionic compounds, forms ionic crystals in which every internal ion is surrounded by ions of the opposite charge. (For simplicity, only a small portion of the ion array is shown here. A typical NaCl crystal involves millions of ions.) (b) A view of crystals of table salt through a microscope shows their cubic structure. The cubic shape is a consequence of the cubic arrangement of sodium and chloride ions.

CHECK YOURSELF
What is the chemical formula for the ionic compound magnesium oxide?

CHECK YOUR ANSWER
Because magnesium is a group 2 element, you know that a magnesium atom must lose two electrons to form a Mg^{2+} ion. Because oxygen is a group 16 element, an oxygen atom gains two electrons to form an O^{2-} ion. These charges balance in a one-to-one ratio, and so the formula for magnesium oxide is MgO.

An ionic compound typically contains a multitude of ions grouped together in a highly ordered three-dimensional array. In sodium chloride, for example, each sodium ion is surrounded by six chloride ions, and each chloride ion is surrounded by six sodium ions (Figure 12.9). Overall, there is one sodium ion for each chloride ion, but there are no identifiable sodium–chloride pairs. Such an orderly array of ions is known as an ionic crystal. As mentioned at the beginning of this chapter, on the atomic level, the crystalline structure of sodium chloride is cubic, which is why macroscopic crystals of table salt are also cubic. Smash a large cubic sodium chloride crystal with a hammer, and what do you get? Smaller cubic sodium chloride crystals!

Similarly, the crystalline structures of other ionic compounds, such as calcium fluoride and aluminum oxide, are a consequence of how the ions pack together. (We will go into more detail about the crystalline structures of minerals in Chapter 23.)

LEARNING OBJECTIVE
Relate the properties of a metal to how the atoms of that metal are chemically bonded.

Integrated Science 12A
EARTH SCIENCE

Metals from Earth

EXPLAIN THIS Why aren't alloys described as metallic compounds?

In Section 11.4 you learned about physical properties. Among the physical properties of metals are that they conduct electricity and heat, are opaque to light, and deform—rather than fracture—under pressure. Because of these properties, metals are used to build homes, appliances, cars, bridges, airplanes, and skyscrapers. Metal wires across the landscape transmit communication

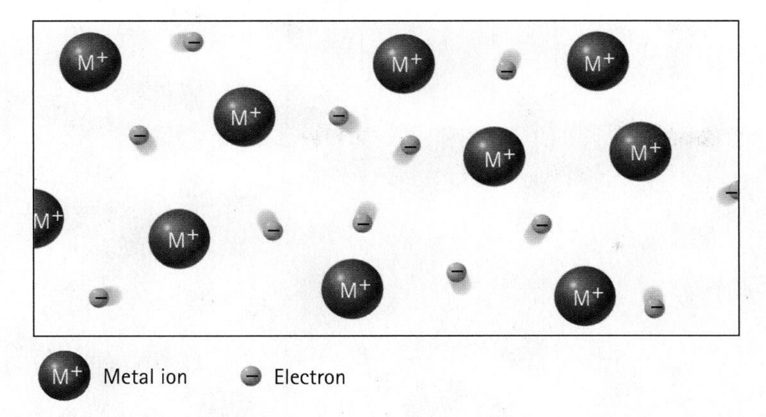

FIGURE 12.10
Metal ions are held together by freely flowing electrons. These loose electrons form a kind of "electronic fluid," which flows through the lattice of positively charged ions.

signals and electric power. We wear metal jewelry, exchange metal currency, and drink from metal cans. Yet what gives a metal its metallic properties? We can answer this question by looking at the behavior of its atoms.

The outer electrons of most metal atoms tend to be weakly held to the atomic nucleus. Consequently, these electrons are easily dislodged, leaving behind positively charged metal ions. The many electrons dislodged from a large group of metal atoms flow freely through the resulting metal ions, as is depicted in Figure 12.10. This "fluid" of electrons holds the positively charged metal ions together in the type of chemical bond known as a **metallic bond**.

The mobility of electrons in a metal accounts for the metal's significant ability to conduct electricity and heat. Also, metals are opaque and shiny because the free electrons easily vibrate to the oscillations of any light falling on them, reflecting most of it. Furthermore, the metal ions are not rigidly held to fixed positions, as ions are in an ionic crystal. Rather, because the metal ions are held together by a "fluid" of electrons, these ions can move into various orientations relative to one another, which is what happens when a metal is pounded, pulled, or molded into a different shape.

Two or more different metals can be bonded to each other by metallic bonds. This occurs, for example, when molten gold and molten palladium are blended to form a homogeneous solution known as white gold. The quality of the white gold can be modified simply by changing the proportions of gold and palladium. White gold is an example of an **alloy**, which is any mixture composed of two or more metallic elements. By playing around with proportions, metal workers can readily modify the properties of an alloy. For example, in designing the Sacagawea dollar coin, shown in Figure 12.11, the U.S. Mint needed a metal that had a gold color—so that it would be easy to recognize—and also the same electrical characteristics as the Susan B. Anthony dollar coin—so that the new coin could substitute for the Anthony coin in vending machines.

FIGURE 12.11
The gold color of the Sacagawea U.S. dollar coin is achieved by an outer surface made of an alloy of 77% copper, 12% zinc, 7% manganese, and 4% nickel. The interior of the coin is pure copper.

Only a few metals—gold and platinum are two examples—appear in nature in metallic form. Deposits of these natural metals, also known as *native metals*, are quite rare. For the most part, metals found in nature are chemical compounds. Iron, for example, is most frequently found as iron oxide, Fe_2O_3, and copper is found as chalcopyrite, $CuFeS_2$. Geologic deposits containing relatively high concentrations of metal-containing compounds are called *ores*. The metals industry mines these ores from the ground, as shown in Figure 12.12, and then

FIGURE 12.12
The world's biggest open-pit mine is the copper mine at Bingham Canyon, Utah.

processes them into metals. Although metal-containing compounds occur just about everywhere, only ores are concentrated enough to make the extraction of the metal economical.

Metal ores contain ionic compounds in which the metal atoms have lost electrons to become positive ions. To convert the ores into metals requires that electrons be given back to the metal ions. This is done by heating the ore with electron-releasing materials, such as carbon, in hot furnaces that reach about 1500°C, as shown in Figure 12.13. The metal emerges in a molten state that can be cast into a variety of useful shapes.

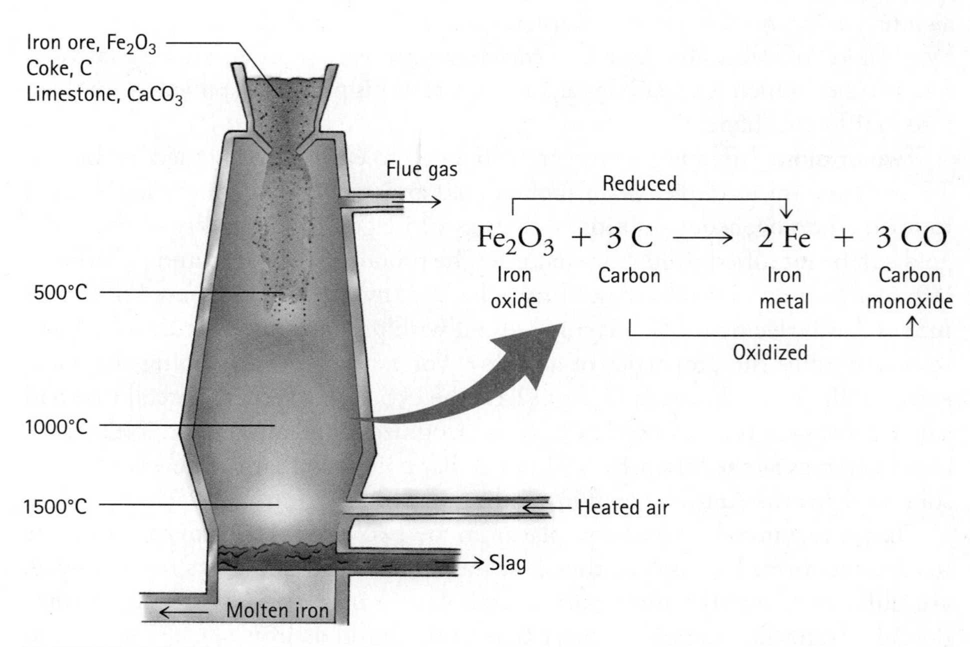

FIGURE 12.13
A mixture of iron oxide, carbon, and limestone is dropped into a blast furnace, where the iron ions gain electrons to form iron atoms, which collect to form molten iron. The by-products of this process, called slag, float above the molten iron and are easily removed.

12.3 The Covalent Bond

LEARNING OBJECTIVE
Describe how atoms combine to form covalent compounds.

EXPLAIN THIS A lone proton encounters the lone pair of electrons of an ammonia molecule and forms what?

UNIFYING CONCEPT
- ***The Electric Force*** Section 7.1

Imagine two children playing together and sharing their toys. Perhaps a force that keeps the children together is their mutual attraction to the toys they share. In a similar fashion, two atoms can be held together by their mutual attraction for the electrons they share. A fluorine atom, for example, has a strong attraction for one additional electron to fill its outermost occupied shell. As shown in Figure 12.14, a fluorine atom can obtain an additional electron by holding onto the unpaired valence electron of another fluorine atom. This results in a situation in which the two fluorine atoms are mutually attracted to the same two electrons. This type of electrical attraction in which atoms are held together by their mutual attraction for shared electrons is called a **covalent bond**, in which *co-* signifies sharing and *-valent* indicates that valence electrons are being shared.

MasteringPhysics®
TUTORIAL: Covalent Bonds

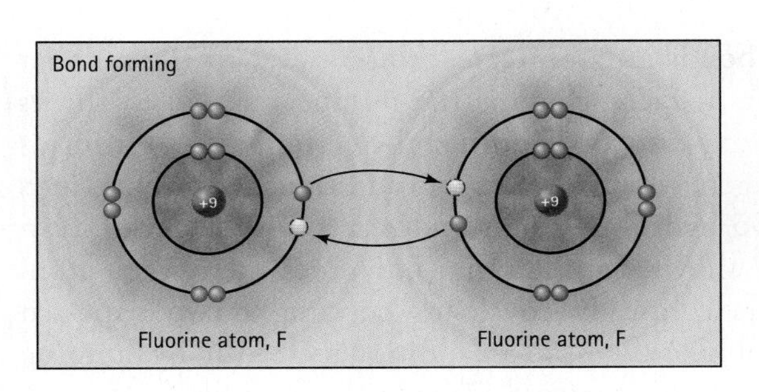

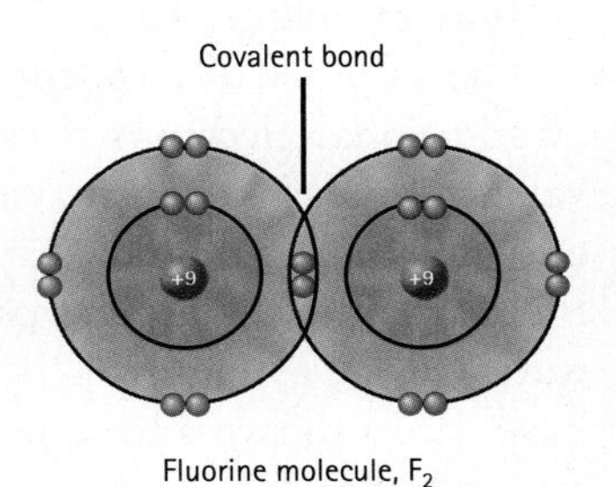

FIGURE 12.14
The effect of the positive nuclear charge (represented by red shading) of a fluorine atom extends beyond the atom's outermost occupied shell. This positive charge can cause the fluorine atom to become attracted to the unpaired valence electron of a neighboring fluorine atom. Then the two atoms are held together in a fluorine molecule by the attraction they both have for the two shared electrons. Each fluorine atom achieves a filled valence shell.

A substance composed of atoms held together by covalent bonds is a **covalent compound**. The fundamental unit of most covalent compounds is a **molecule**, which we can now formally define as any group of atoms held together by covalent bonds. Figure 12.15 uses the element fluorine to illustrate this principle.

When writing electron-dot structures for covalent compounds, chemists often use a straight line to represent the two electrons involved in a covalent bond. In some representations, the nonbonding electron pairs are ignored. This occurs in instances where these electrons play no significant role in the process being illustrated. Here are two frequently used ways of showing a fluorine molecule without using spheres to represent the atoms:

$$:\ddot{\underset{..}{F}}-\ddot{\underset{..}{F}}: \qquad F-F$$

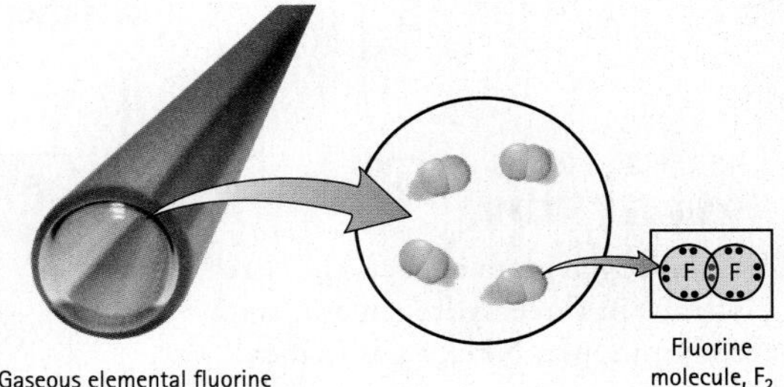

FIGURE 12.15
Molecules are the fundamental units of the gaseous covalent compound fluorine, F_2. Notice that in this model of a fluorine molecule, the spheres overlap, whereas the spheres shown earlier for ionic compounds do not. Now you know that this difference in representation is because of the difference in bond types.

Remember—the straight line in both versions represents two electrons, one from each atom. Thus, we now have two types of electron pairs to keep track of. The term *nonbonding pair* refers to any electron pair in the electron-dot structure of an individual atom, and the term *bonding pair* refers to any pair that results from the formation of a covalent bond.

Recall from Section 12.2 that an ionic bond is formed when an atom that tends to lose electrons makes contact with an atom that tends to gain them. A covalent bond, by contrast, is formed when two atoms that tend to gain electrons are brought into contact with each other. Atoms that tend to form covalent

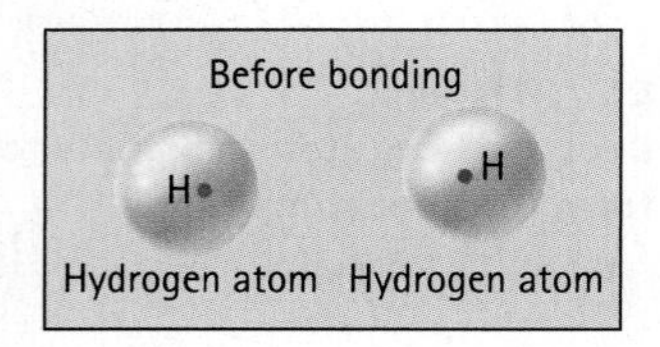

Covalent bond formed

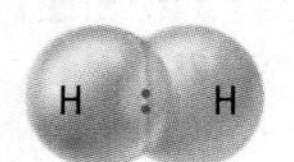

Hydrogen molecule, H_2

FIGURE 12.16

Two hydrogen atoms form a covalent bond as they share their unpaired electrons.

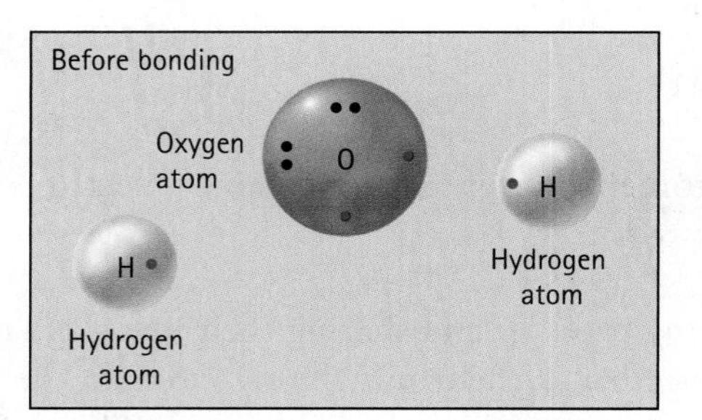

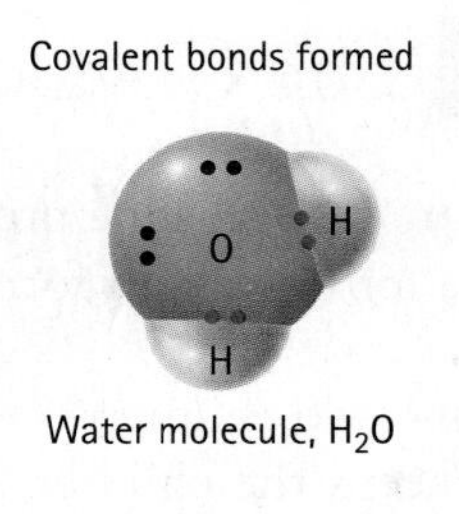

FIGURE 12.17

INTERACTIVE FIGURE

The two unpaired valence electrons of oxygen pair with the unpaired valence electrons of two hydrogen atoms to form the covalent compound water.

FYI Spectroscopic studies of interstellar dust within our galaxy have revealed the presence of more than 120 kinds of molecules, such as water, H_2O; acetylene, H_2C_2; formic acid, HCO_2H; methanol, CH_3OH; methyl amine, NH_2CH_3; acetic acid, CH_3CO_2H; and even the biomolecule glycine, $NH_2CH_2CO_2H$. As discussed in Chapter 10, the atoms originated from the nuclear fusion of ancient stars. How interesting that these atoms then join together to form molecules across the universe. Chemistry is truly everywhere.

bonds are therefore primarily atoms of the nonmetallic elements (including hydrogen) because these elements all tend to gain electrons (with the exception of the noble-gas elements, which are very stable and tend not to form bonds). Two hydrogen atoms, for example, covalently bond to form a hydrogen molecule, H_2, as shown in Figure 12.16.

The number of covalent bonds an atom can form is equal to the number of additional electrons it can attract, which is the number needed to fill its valence shell. Hydrogen attracts only one additional electron, and so it forms only one covalent bond. Oxygen, which attracts two additional electrons, finds them when it encounters two hydrogen atoms and reacts with them to form water, H_2O, as Figure 12.17 shows. In water, not only does the oxygen atom have access to two additional electrons by covalently bonding to two hydrogen atoms, but each hydrogen atom has access to an additional electron by bonding to the oxygen atom. Each atom thus achieves a filled valence shell.

Nitrogen attracts three additional electrons and thus can form three covalent bonds, as occurs in ammonia, NH_3, shown in Figure 12.18. Likewise, a carbon atom can attract four additional electrons and is thus able to form four covalent bonds, as occurs in methane, CH_4. Note that the number of covalent bonds formed by these and other nonmetallic elements parallels the type of negative ions they tend to form (see Figure 12.6). This makes sense because covalent-bond

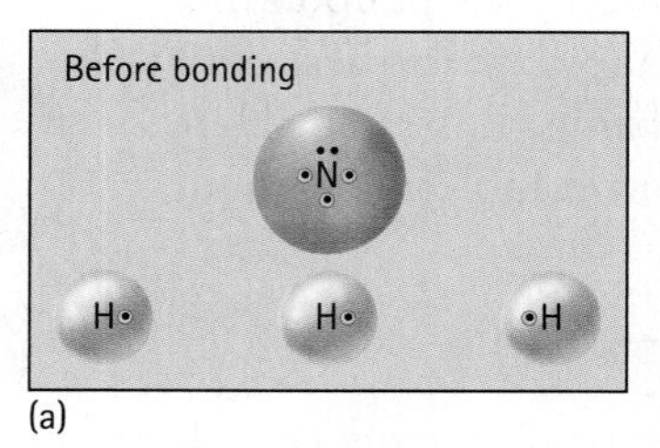

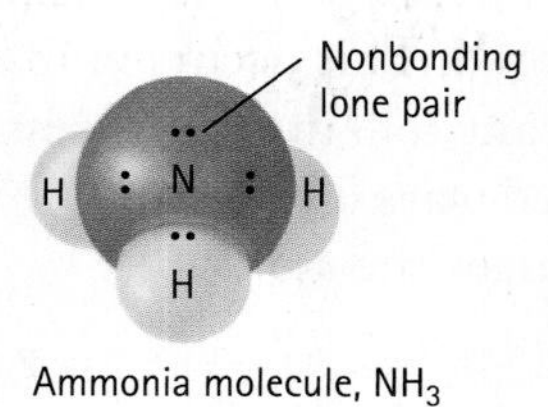

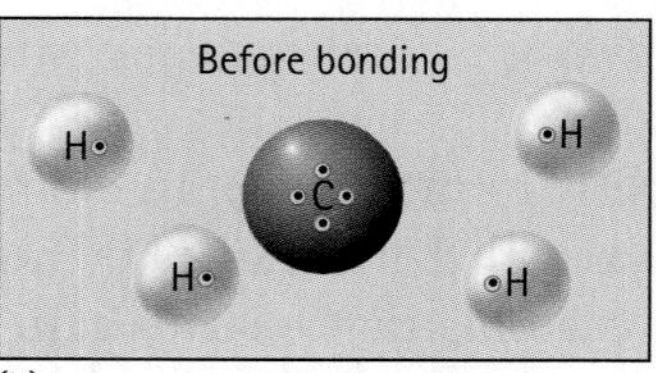

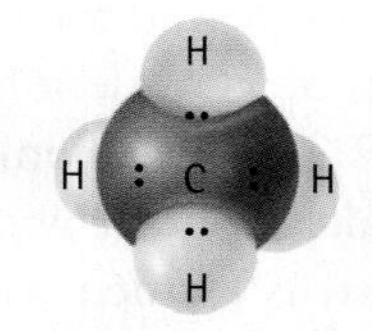

FIGURE 12.18

INTERACTIVE FIGURE

(a) A nitrogen atom attracts the three electrons in three hydrogen atoms to form ammonia, NH_3, a gas that can dissolve in water to make an effective cleanser. (b) A carbon atom attracts the four electrons in four hydrogen atoms to form methane, CH_4, the primary component of natural gas. In these and most other cases of covalent-bond formation, the result is a filled valence shell for all the atoms involved.

FIGURE 12.19
The crystalline structure of diamond is nicely illustrated with sticks to represent the covalent bonds. The molecular nature of a diamond is responsible for its extreme hardness.

formation and negative-ion formation are both applications of the same concept: Nonmetallic atoms tend to gain electrons until their valence shells are filled.

Diamond is a very unusual covalent compound consisting of carbon atoms covalently bonded to one another in four directions. The result is a covalent crystal, which, as shown in Figure 12.19, is a highly ordered, three-dimensional network of covalently bonded atoms. This network of carbon atoms forms a very strong and rigid structure, which is why diamonds are so hard.

FYI Astronomers have recently discovered an expired star that has a solid core made of diamond. This star-sized diamond is about 4000 km wide, which amounts to about 10 billion trillion trillion carats. The star has been named "Lucy" after the Beatles song "Lucy in the Sky with Diamonds." In about 7 billion years, our own star, the Sun, is also likely to crystallize into a huge diamond ball.

CHECK YOURSELF
How many electrons make up a covalent bond?

CHECK YOUR ANSWER
Two—one from each participating atom.

It is possible to have more than two electrons shared between two atoms, and Figure 12.20 shows three examples. Molecular oxygen, O_2, consists of two oxygen atoms connected by four shared electrons. This arrangement is called a *double covalent bond* or, for short, a *double bond*. As another example, the covalent compound carbon dioxide, CO_2, consists of two double bonds connecting two oxygen atoms to a central carbon atom.

Some atoms can form triple covalent bonds, in which six electrons—three from each atom—are shared. One example is molecular nitrogen, N_2. Any double or triple bond is often referred to as a multiple covalent bond. Multiple bonds higher than these, such as the quadruple covalent bond, are not commonly observed.

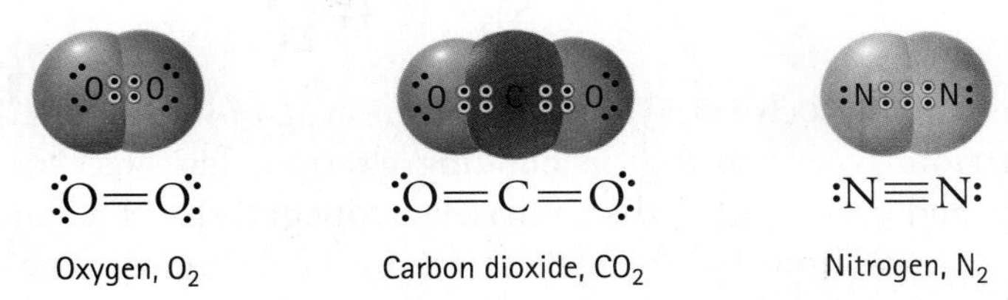

FIGURE 12.20
Double covalent bonds in molecules of oxygen, O_2, and carbon dioxide, CO_2, and a triple covalent bond in a molecule of nitrogen, N_2.

LEARNING OBJECTIVE
Differentiate ionic, polar covalent, and nonpolar covalent chemical bonds.

MasteringPhysics®
TUTORIAL: Bonds and Bond Polarity

12.4 Polar Covalent Bonds

EXPLAIN THIS How is the chemical bond between sodium and chlorine mostly ionic but partially covalent?

If the two atoms in a covalent bond are identical, then their nuclei have the same positive charge and therefore the electrons are shared evenly. We can represent these electrons as being centrally located by using an electron-dot structure with the electrons situated exactly halfway between the two atomic symbols. Alternatively, we can draw a cloud in which the positions of the two bonding electrons over time are shown as a series of dots. Where the dots are most concentrated is where the electrons have the greatest probability of being located:

H : H H H

There are always two electrons per covalent bond. A double bond, therefore, consists of four electrons, while a triple bond consists of six electrons.

In a covalent bond between nonidentical atoms, the nuclear charges are different, and consequently the bonding electrons may be shared unevenly. This occurs in a hydrogen–fluorine bond, where electrons are more attracted to fluorine's greater nuclear charge:

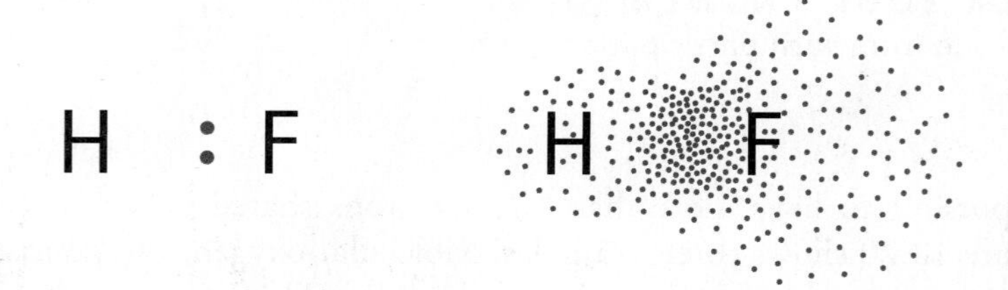

The bonding electrons spend more time around the fluorine atom. For this reason, the fluorine side of the bond is slightly negative and, because the bonding electrons have been drawn away from the hydrogen atom, the hydrogen side of the bond is slightly positive. This separation of charge is called a **dipole** (pronounced *die*-pole) and is represented either by the characters $\delta-$ and $\delta+$ (read "slightly negative" and "slightly positive," respectively) or by a crossed arrow pointing to the negative side of the bond:

$$\overset{\delta+}{\mathrm{H}}-\overset{\delta-}{\mathrm{F}} \qquad \overset{+\!\!\longrightarrow}{\mathrm{H}-\mathrm{F}}$$

So, atoms forming a chemical bond engage in a tug-of-war for electrons. How strongly an atom is able to tug on bonding electrons has been measured experimentally and quantified as the atom's **electronegativity**. The range of electronegativities runs from 0.7 to 3.98, as Figure 12.21 shows. The greater an atom's electronegativity, the greater its ability to pull electrons toward itself when bonded. Thus, in hydrogen fluoride, fluorine has a greater electronegativity, or pulling power, than hydrogen.

H 2.2																	He –
Li 0.98	Be 1.57											B 2.04	C 2.55	N 3.04	O 3.44	F 3.98	Ne –
Na 0.93	Mg 1.31											Al 1.61	Si 1.9	P 2.19	S 2.58	Cl 3.16	Ar –
K 0.82	Ca 1.0	Sc 1.36	Ti 1.54	V 1.63	Cr 1.66	Mn 1.55	Fe 1.83	Co 1.88	Ni 1.91	Cu 1.90	Zn 1.65	Ga 1.81	Ge 2.01	As 2.18	Se 2.55	Br 2.96	Kr –
Rb 0.82	Sr 0.95	Y 1.22	Zr 1.33	Nb 1.6	Mo 2.16	Tc 1.9	Ru 2.2	Rh 2.28	Pd 2.20	Ag 1.93	Cd 1.69	In 1.78	Sn 1.96	Sb 2.05	Te 2.1	I 2.66	Xe –
Cs 0.79	Ba 0.89	La 1.10	Hf 1.3	Ta 1.5	W 2.36	Re 1.9	Os 2.2	Ir 2.20	Pt 2.8	Au 2.54	Hg 2.00	Tl 2.04	Pb 2.33	Bi 2.02	Po 2.0	At 2.2	Rn –
Fr 0.7	Ra 0.9	Ac 1.1	Rf –	Db –	Sg –	Bh –	Hs –	Mt –	Ds –	Rg –	Cn –	Uut –	Fl –	Uup –	Lv –	Uus –	Uuo –

FIGURE 12.21
The experimentally measured electronegativities of elements.

Electronegativity is greatest for elements at the upper right of the periodic table and lowest for elements at the lower left. Noble gases are not considered in electronegativity discussions because, as previously mentioned, they rarely participate in chemical bonding.

When the two atoms in a covalent bond have the same electronegativity, no dipole is formed (as is the case with H_2) and the bond is classified as a **nonpolar** bond. When the electronegativities of the atoms differ, a dipole may form (as with HF) and the bond is classified as a **polar** bond. Just how polar a bond is depends on the difference between the electronegativity values of the two atoms—the greater the difference, the more polar the bond.

As can be seen in Figure 12.21, the greater the distance between two atoms in the periodic table, the greater the difference in their electronegativities, and hence the greater the polarity of the bond between them. So, a chemist can predict which bonds are more polar than others without reading the electronegativities. Bond polarity can be inferred by looking at the relative positions of the atoms in the periodic table—the farther apart they are, especially when one is at the lower left and one is at the upper right, the greater the polarity of the bond between them.

CHECK YOURSELF

Which bond should be more polar: P—F or Ge—F? (F, fluorine, atomic number 9; P, phosphorus, atomic number 15; Ge, germanium, atomic number 32).

CHECK YOUR ANSWER

The greater the difference in electronegativities between two bonded atoms, the greater the polarity of the bond. Thus, Ge—F is more polar than P—F. Note that this answer can be obtained by looking only at the relative positions of these elements in the periodic table rather than by calculating the differences in their electronegativities.

The magnitude of bond polarity is sometimes indicated by the size of the crossed arrow or the $\delta-$ and $\delta+$ symbols used to depict a dipole.

What is important to understand here is that there is no black-and-white distinction between ionic and covalent bonds. Rather, there is a gradual change from one to the other as the atoms that bond are farther apart in the periodic table. This continuum is illustrated in Figure 12.22. Atoms on opposite sides of the periodic table have great differences in electronegativity, and hence the bonds

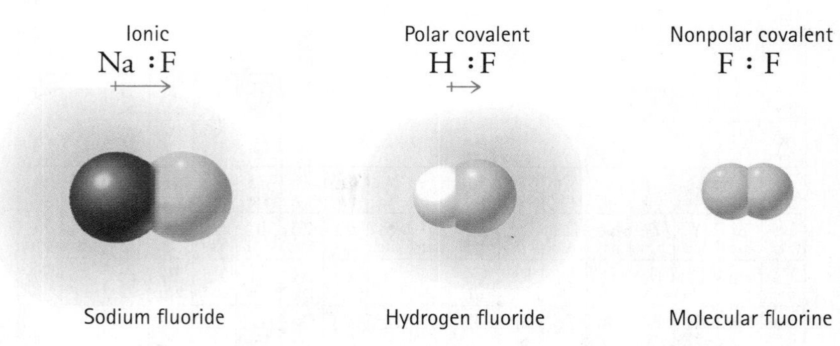

FIGURE 12.22
The ionic bond and the nonpolar covalent bond represent the two extremes of chemical bonding. The ionic bond involves a transfer of one or more electrons, and the nonpolar covalent bond involves the equitable sharing of electrons. The character of a polar covalent bond falls between these two extremes.

between them are highly polar—in other words, ionic. Nonmetallic atoms of the same type have the same electronegativities, and so their bonds are nonpolar covalent. The polar covalent bond with its uneven sharing of electrons and slightly charged atoms is between these two extremes.

LEARNING OBJECTIVE
Show how the shape of a molecule affects the molecule's polarity.

12.5 Molecular Polarity

EXPLAIN THIS Which is heavier—carbon dioxide or water?

When all the bonds in a molecule are nonpolar, the molecule as a whole is also nonpolar—as is the case with H_2, O_2, and N_2. When a molecule consists of only two atoms and the bond between them is polar, the polarity of the molecule is the same as the polarity of the bond—as with HF, HCl, and ClF.

Complexities arise when assessing the polarity of a molecule that contains more than two atoms. Consider carbon dioxide, CO_2, shown in Figure 12.23. The cause of the dipole in either one of the carbon–oxygen bonds is oxygen's greater pull on the bonding electrons (because oxygen is more electronegative than carbon). At the same time, however, the oxygen atom on the opposite side of the carbon pulls those electrons back to the carbon. The net result is an even distribution of bonding electrons around the entire molecule. So dipoles that are of equal strength but pull in opposite directions in a molecule effectively cancel each other, with the result that the molecule as a whole is nonpolar.

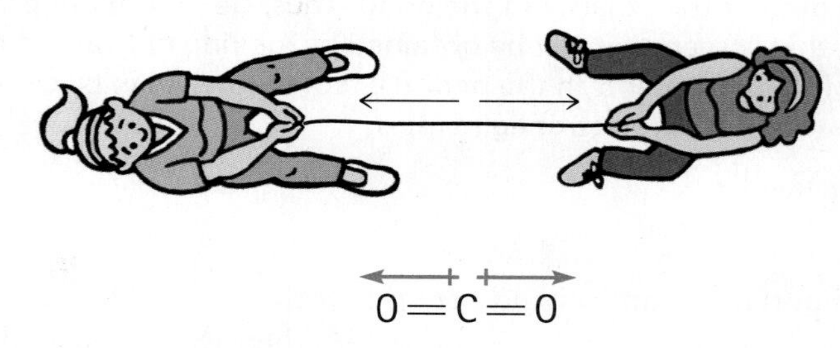

FIGURE 12.23
There is no net dipole in a carbon dioxide molecule, and so the molecule is nonpolar. This is analogous to two people in a tug-of-war. As long as they pull with equal forces but in opposite directions, the rope remains stationary.

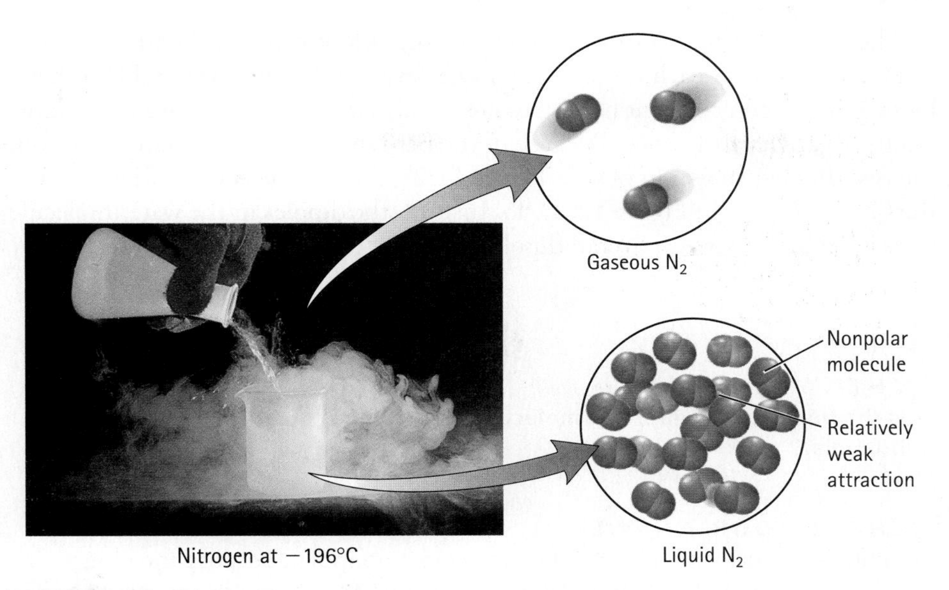

FIGURE 12.24
Nitrogen is a liquid at temperatures below its chilly boiling point of −196°C. Nitrogen molecules are not very attracted to one another because they are nonpolar. As a result, the small amount of heat energy available at −196°C is enough to separate them and allow them to enter the gaseous phase.

Nonpolar molecules have only relatively weak attractions to other nonpolar molecules. The covalent bonds in a carbon dioxide molecule, for example, are many times stronger than any forces of attraction that might occur between two adjacent carbon dioxide molecules. This lack of attraction between nonpolar molecules explains the low boiling points of many nonpolar substances. Recall from Section 11.3 that boiling is the process in which the molecules of a liquid separate from one another as they go into the gaseous phase. When there are only weak attractions between the molecules of a liquid, less heat energy is required to liberate the molecules from one another and allow them to enter the gaseous phase. This translates into a relatively low boiling point for the liquid, as, for example, in molecular nitrogen, N_2, shown in Figure 12.24. The boiling points of hydrogen (H_2), oxygen (O_2), and carbon dioxide (CO_2) are also quite low for the same reason (see Table 12.2).

TABLE 12.2 BOILING POINTS OF SOME POLAR AND NONPOLAR SUBSTANCES

Substance	Boiling point (°C)
Polar	
Hydrogen fluoride, HF	20
Water, H_2O	100
Ammonia, NH_3	−33
Nonpolar	
Hydrogen, H_2	−253
Oxygen, O_2	−183
Nitrogen, N_3	−196
Boron trifluoride, BF_3	−100
Carbon dioxide, CO_2	−79

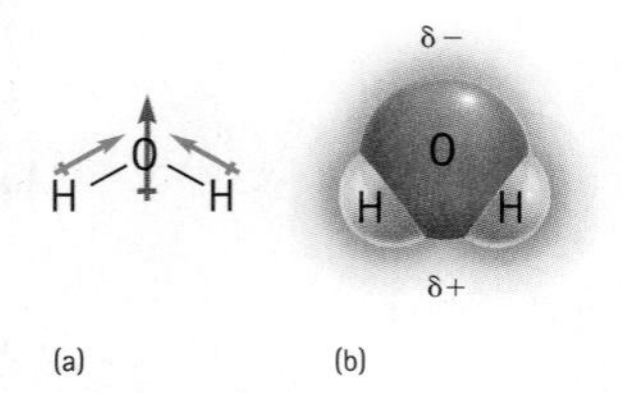

FIGURE 12.25
(a) The individual dipoles in a water molecule add together to give a large overall dipole for the whole molecule, shown in purple. (b) The region around the oxygen atom is therefore slightly negative, and the region around the two hydrogen atoms is slightly positive.

There are many instances in which the dipoles of different bonds in a molecule do not cancel each other. The most relevant example is water, H_2O. Each hydrogen–oxygen covalent bond has a relatively large dipole because of the large electronegativity difference. Because of the bent shape of the molecule, however, the two dipoles, shown in blue in Figure 12.25, do not cancel each other the way the C—O dipoles in Figure 12.23 do. Instead, the dipoles in the water molecule work together to give an overall dipole, shown in purple, for the molecule.

CHECK YOURSELF
Is the boron trifluoride, BF_3, molecule (shown here) polar or nonpolar?

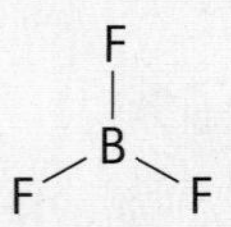

CHECK YOUR ANSWER
Symmetry usually means that the molecule is nonpolar. Because the molecule shown here is symmetrical, the dipoles on the three sides cancel one another. This molecule is therefore nonpolar. Note its low boiling point in Table 12.2.

A water molecule is a natural dipole—a bit positive on one end and negative on the other. What's the net charge of a dipole?

Figure 12.26 illustrates how polar molecules electrically attract one another and, as a result, are relatively difficult to separate. In other words, polar molecules can be thought of as being "sticky," which is why it takes more energy to separate them—to change phase. For this reason, substances composed of polar molecules typically have higher boiling points than substances composed of nonpolar molecules, as Table 12.2 shows.

Water boils at 100°C, whereas carbon dioxide boils at −79°C. This 179°C difference is quite dramatic when you consider that a carbon dioxide molecule is more than twice as massive as a water molecule.

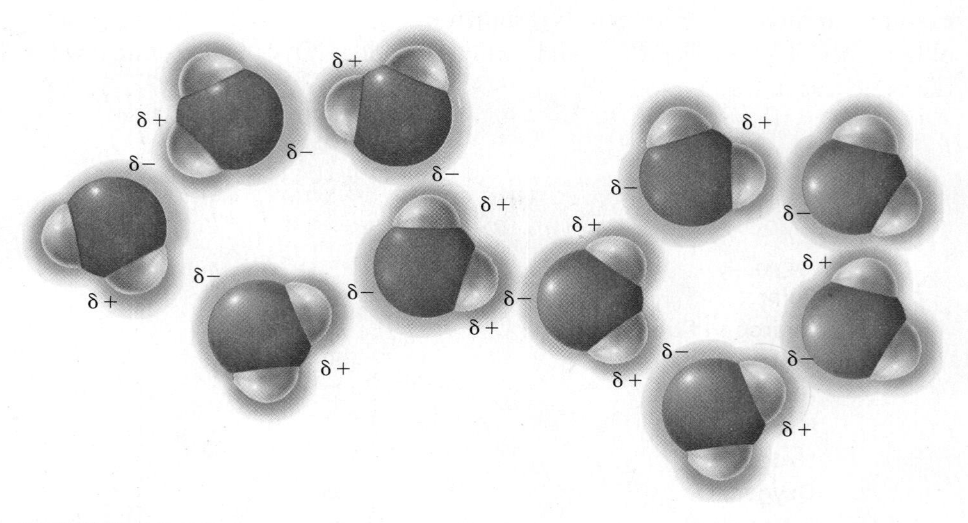

FIGURE 12.26
Water molecules attract one another because each contains a slightly positive side and a slightly negative side. The molecules position themselves such that the positive side of one faces the negative side of a neighbor.

FIGURE 12.27
Oil and water are difficult to mix, as is evident from this oil spill. It's not the case, however, that oil and water repel each other. Rather, water molecules are so attracted to themselves because of their polarity that they pull themselves together. The nonpolar oil molecules are thus excluded and left to themselves. Being less dense than water, oil floats on the surface, where it poses great danger to birds and other wildlife.

Because molecular "stickiness" can play a leading role in determining a substance's macroscopic properties, molecular polarity is a central concept of chemistry. Figure 12.27 describes an interesting example.

12.6 Molecular Attractions

LEARNING OBJECTIVE
Recognize the important role that molecular interactions play in determining the physical properties of a material.

EXPLAIN THIS Is it possible for a fish to die from drowning?

So far you have learned that the atoms of a molecule are held together by covalent bonds. Furthermore, the molecule, behaving as a fundamental unit, may have electrical attractions with neighboring molecules. As discussed in the preceding section, the greater the polarity of the molecule, the greater its attraction to neighboring molecules. This explains why water has such a high boiling point—the water molecules, being quite polar, are so attracted to one another that a lot of energy is required to separate them from one another into the gaseous phase. In this section, we explore further how the physical properties of a material, such as boiling point, can be deduced from the polarity of its molecules. In addition to discussing the attractions among the molecules within a single substance, we explore the attractions between the fundamental units of different substances, such as water and salt.

In the first part of this chapter, we talked about how molecules form. Now we see how molecules mix together.

As listed in Table 12.3, there are four types of electrical attractions involving molecules. The strength of even the strongest of these attractions is much weaker than any chemical bond. The attraction between two adjacent water molecules, for example, is only about $\frac{1}{20}$ as strong as the chemical bonds holding the hydrogen

TABLE 12.3 TYPES OF MOLECULAR ATTRACTIONS

Attraction	Relative Strength
Ion–dipole	Strongest
Dipole–dipole	
Dipole–induced dipole	
Induced dipole–induced dipole	Weakest

MasteringPhysics®
TUTORIAL: Polar Attraction
TUTORIAL: Intermolecular Forces

and oxygen atoms together in the water molecule. Although molecule-to-molecule attractions are relatively weak, their effects on the physical properties of substances are most significant.

Ions and Dipoles

Recall from Section 12.5 that a *polar* molecule is one in which the bonding electrons are unevenly distributed. One side of the molecule carries a slight negative charge, and the opposite side carries a slight positive charge. This separation of charge makes up a *dipole.*

So, what happens to polar molecules, such as water molecules, when they are near an ionic compound, such as sodium chloride? The opposite charges electrically attract one another. A positive sodium ion attracts the negative side of a water molecule, and a negative chloride ion attracts the positive side of a water molecule. This phenomenon is illustrated in Figure 12.28. Such an attraction between an ion and the dipole of a polar molecule is called an *ion–dipole* attraction.

Ion–dipole attractions are much weaker than ionic bonds. However, a large number of ion–dipole attractions can act collectively to disrupt ionic bonds. This is what happens to sodium chloride in water. Attractions exerted by the water molecules break the ionic bonds and pull the ions away from one another. The result, represented in Figure 12.29, is a solution of sodium chloride in water. (A solution in water is called an *aqueous solution.*)

An attraction between two polar molecules is called a *dipole–dipole* attraction. An unusually strong dipole–dipole attraction is the **hydrogen bond**. This attraction occurs between molecules that have a hydrogen atom covalently bonded to a highly electronegative atom, usually nitrogen, oxygen, or fluorine. Recall from Section 12.4 that the electronegativity of an atom describes how well that atom is able to pull bonding electrons toward itself. The greater the atom's electronegativity, the better it is able to gain electrons and thus the more negative is its charge.

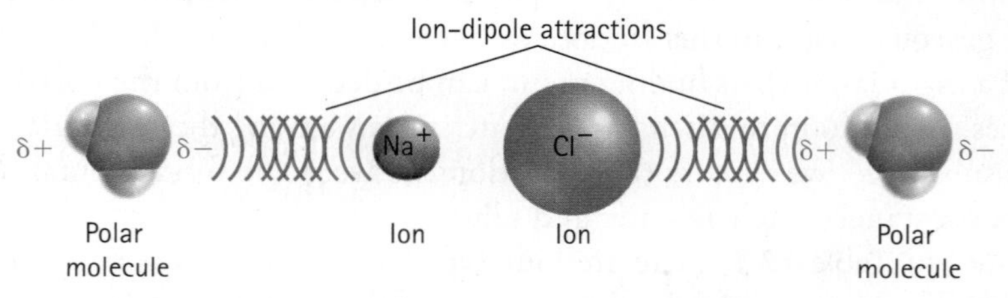

FIGURE 12.28
Electrical attractions are shown as a series of overlapping arcs. The blue arcs indicate negative charge, and the red arcs indicate positive charge.

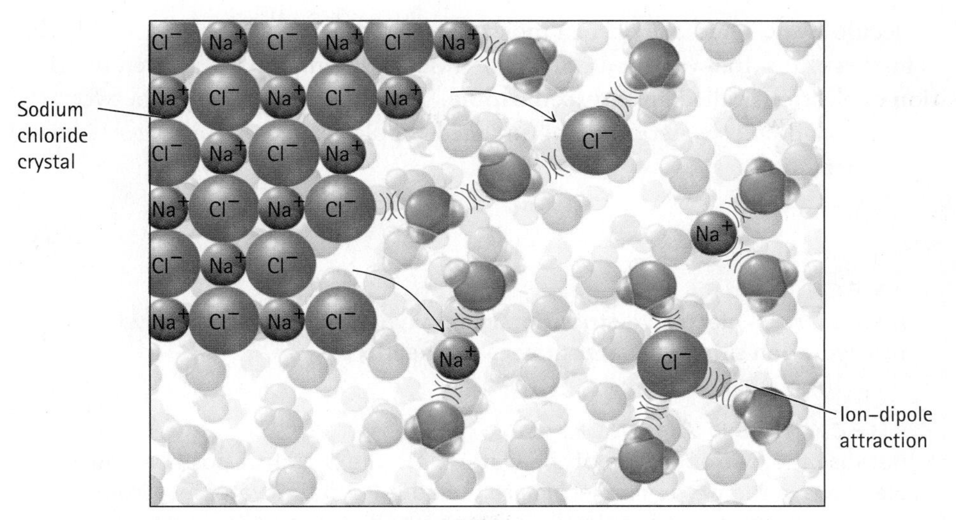

FIGURE 12.29
Sodium and chloride ions tightly bound in a crystal lattice are separated from one another by the collective attraction exerted by many water molecules to form an aqueous solution of sodium chloride.

Look at Figure 12.30 to see how hydrogen bonding works. The hydrogen side of a polar molecule (water, in this example) has a positive charge because the more electronegative oxygen atom pulls more strongly on the electrons of the covalent bond. The hydrogen is therefore electrically attracted to a pair of non-bonding electrons on the negatively charged atom of another molecule (in this case, another water molecule). This mutual attraction between hydrogen and the negatively charged atom of another molecule is a *hydrogen bond.*

Even though the hydrogen bond is much weaker than any covalent or ionic bond, the effects of hydrogen bonding can be very pronounced. For example, water owes many of its properties to hydrogen bonds. The hydrogen bond is also of great importance in determining the properties of biological molecules, such as carbohydrates, proteins, and even DNA, which is the molecule that holds an individual's genetic code.

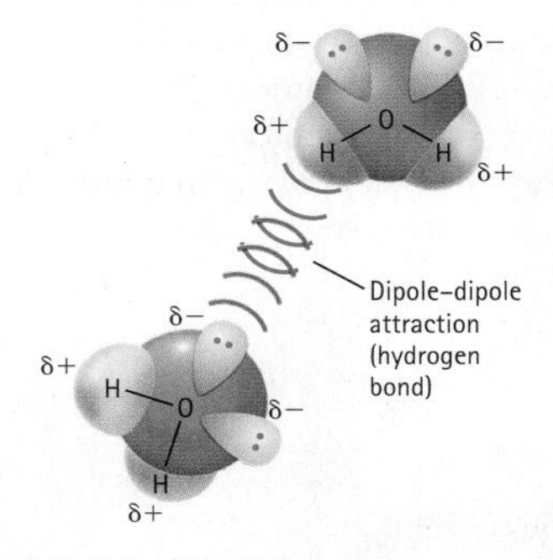

FIGURE 12.30
The dipole–dipole attraction between two water molecules is a hydrogen bond because it involves hydrogen atoms bonded to highly electronegative oxygen atoms.

Induced Dipoles

In many molecules, the electrons are distributed evenly, and so there is no dipole. The oxygen molecule, O_2, is an example. Such a nonpolar molecule can be induced to become a temporary dipole, however, when it is brought close to a water molecule (or to any other polar molecule), as Figure 12.31 illustrates. The slightly negative side of the water molecule pushes the electrons in the oxygen

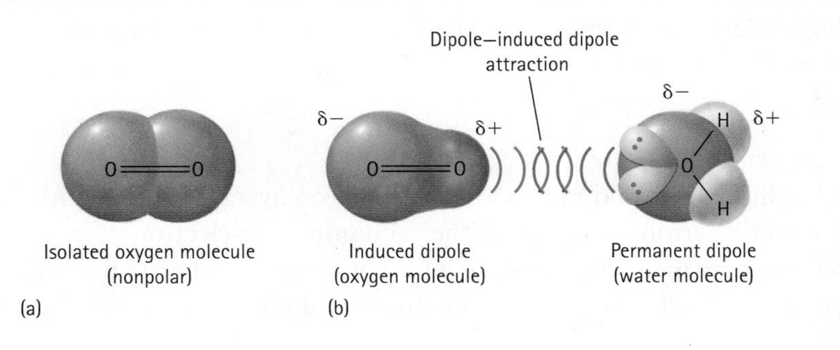

FIGURE 12.31
INTERACTIVE FIGURE MP

(a) An isolated oxygen molecule has no dipole; its electrons are distributed evenly. (b) An adjacent water molecule induces a redistribution of electrons in the oxygen molecule. (The slightly negative side of the oxygen molecule is shown larger than the slightly positive side because the slightly negative side contains more electrons.)

molecule away. Thus, the oxygen molecule's electrons are pushed to the side that is farthest from the water molecule. The result is a temporarily uneven distribution of electrons called an **induced dipole**. The resulting attraction between the permanent dipole (water) and the induced dipole (oxygen) is a *dipole–induced dipole* attraction.

FIGURE 12.32
The electrical attraction between water and oxygen molecules is relatively weak, which explains why not much oxygen is able to dissolve in water. For example, water fully aerated at room temperature contains only about 1 oxygen molecule for every 200,000 water molecules. The gills of a fish, therefore, must be highly efficient at extracting molecular oxygen from water.

CHECK YOURSELF

How does the electron distribution in an oxygen molecule change when the hydrogen side of a water molecule is nearby?

CHECK YOUR ANSWER

Because the hydrogen side of the water molecule is slightly positive, the electrons in the oxygen molecule are pulled toward the water molecule, inducing in the oxygen molecule a temporary dipole in which the larger side is nearest the water molecule (rather than as far away as possible, as it was in Figure 12.31).

Remember—induced dipoles are only temporary. If the water molecule in Figure 12.31b were removed, the oxygen molecule would return to its normal, nonpolar state. In general, dipole–induced dipole attractions are much weaker than dipole–dipole attractions. But dipole–induced dipole attractions are strong enough to hold relatively small quantities of oxygen dissolved in water, as depicted in Figure 12.32. This attraction between water and molecular oxygen is vital for fish and other forms of aquatic life that rely on molecular oxygen dissolved in water.

Dipole–induced dipole attractions are also responsible for holding plastic wrap to glass, as shown in Figure 12.33. These wraps are made of very long nonpolar molecules that are induced to have dipoles when placed in contact with glass, which is highly polar. As we discuss next, the molecules of a nonpolar material, such as plastic wrap, can also induce dipoles among themselves. This explains why plastic wrap sticks not only to polar materials such as glass but also to itself.

FIGURE 12.33
Temporary dipoles induced in the normally nonpolar molecules in plastic wrap make it stick to glass.

CHECK YOURSELF

Distinguish between a dipole–dipole attraction and a dipole–induced dipole attraction.

CHECK YOUR ANSWER

The dipole–dipole attraction is stronger and involves two permanent dipoles. The dipole–induced dipole attraction is weaker and involves a permanent dipole and a temporary one.

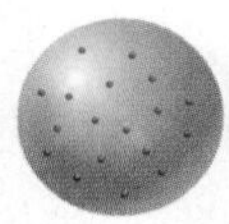

FIGURE 12.34
The electron distribution in an atom is normally even. At any given moment, however, the electron distribution may be somewhat uneven, resulting in a temporary dipole.

Individual atoms and nonpolar molecules, on average, have a fairly even distribution of electrons. Because of the randomness of electron motion, however, at any given moment the electrons in an atom or a nonpolar molecule may be bunched to one side. The result is a temporary dipole, as shown in Figure 12.34.

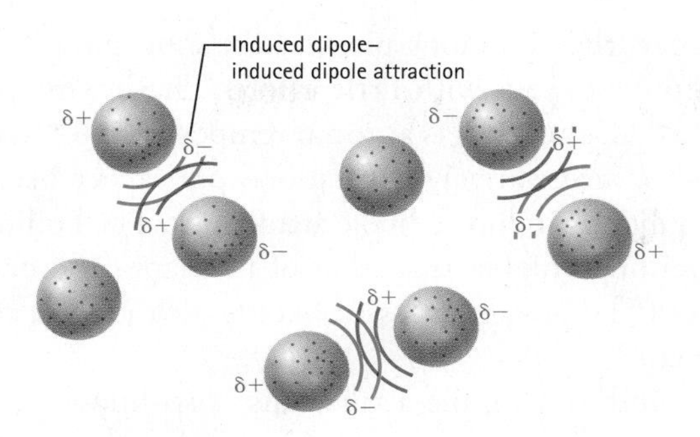

FIGURE 12.35
Because the normally even distribution of electrons in atoms can momentarily become uneven, atoms can be attracted to one another through induced dipole–induced dipole attractions.

Just as the permanent dipole of a polar molecule can induce a dipole in a nonpolar molecule, a temporary dipole can do the same thing. This gives rise to the relatively weak *induced dipole–induced dipole* attraction, illustrated in Figure 12.35.

CHECK YOURSELF

What is the distinction between a dipole–induced dipole attraction and an induced dipole–induced dipole attraction?

CHECK YOUR ANSWER

The dipole–induced dipole attraction is stronger and involves a permanent dipole and a temporary one. The induced dipole–induced dipole attraction is weaker and involves two temporary dipoles.

Induced dipole–induced dipole attractions help explain why natural gas is a gas at room temperature but gasoline is a liquid. The major component of natural gas is methane, CH_4, and one of the major components of gasoline is octane, C_8H_{18}. We can see in Figure 12.36 that the number of induced dipole–induced dipole attractions between two methane molecules is appreciably less than the number between two octane molecules. You know that two small pieces of Velcro

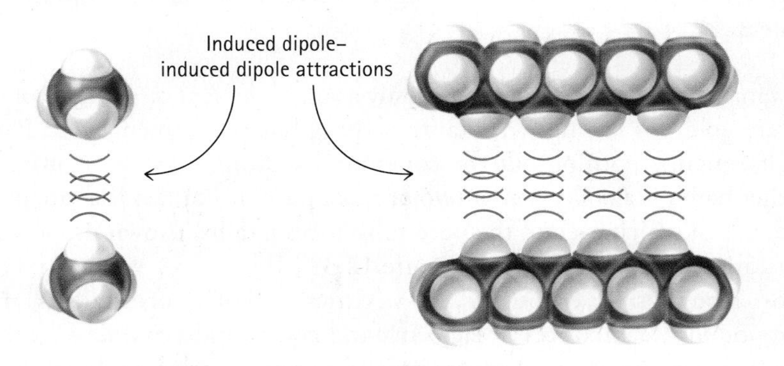

FIGURE 12.36
(a) Two nonpolar methane molecules are attracted to each other by induced dipole–induced dipole attractions, but there is only one attraction per molecule. (b) Two nonpolar octane molecules are similar to methane, but they are longer. The number of induced dipole–induced dipole attractions between these two molecules is therefore greater.

FIGURE 12.37
If the gecko's foot is so sticky, how does the gecko keep its feet clean? Answer: the gecko's foot is extremely nonpolar. Dirt may stick to it briefly, but after a few steps, the dirt sticks better to the surface on which the gecko walks. Of course, there is at least one surface a gecko finds very difficult to climb—Teflon.

are easier to pull apart than two long pieces. Like short pieces of Velcro, methane molecules can be pulled apart with little effort. That's why methane has a low boiling point, −161°C, and is a gas at room temperature. Octane molecules, like long strips of Velcro, are relatively difficult to pull apart because of the larger number of induced dipole–induced dipole attractions. The boiling point of octane, 125°C, is therefore much higher than that of methane, and octane is a liquid at room temperature. (The greater mass of octane also plays a role in making its boiling point higher.)

Induced dipole–induced dipole attractions, also known as *dispersion forces*, also explain how the gecko can race up a glass wall and support its entire body weight with only a single toe. A gecko's feet are covered with billions of microscopic hairs called *spatulae*, each of which is about $\frac{1}{300}$ as thick as a human hair. The force of attraction between these hairs and the wall is the weak induced dipole–induced dipole attraction. But because there are so many hairs, the surface area of contact is relatively large, and hence the total force of attraction is enough to prevent the gecko from falling (Figure 12.37). Research is currently under way to develop a synthetic dry glue based on gecko adhesion. Velcro, watch out!

CHECK YOURSELF

Methanol, CH_3OH, which can be used as a fuel, is not much larger than methane, CH_4, but it is a liquid at room temperature. Suggest why.

CHECK YOUR ANSWER

The polar oxygen–hydrogen covalent bond in each methanol molecule leads to hydrogen bonding between molecules. These relatively strong interparticle attractions hold methanol molecules together as a liquid at room temperature.

LEARNING OBJECTIVE
Classify the states of matter into the categories of pure and mixture.

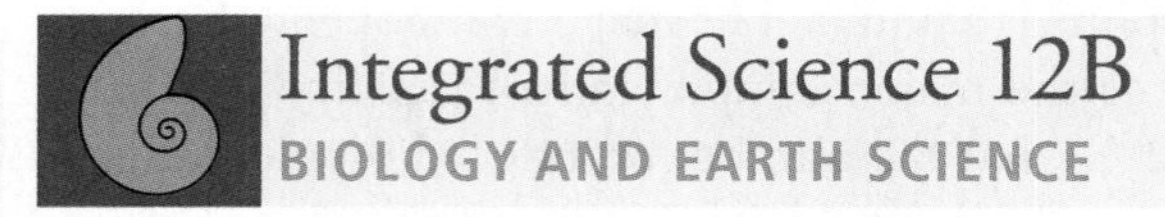

Most Materials Are Mixtures

EXPLAIN THIS Does 500 mL of sugar-sweetened water also contain 500 mL of water?

If a material is **pure**, it consists of only a single element or a single compound. Pure gold, for example, contains nothing but the element gold. Pure table salt contains nothing but the compound sodium chloride. A **mixture**, on the other hand, is a collection of two or more pure substances that are physically mixed and in which each of the pure substances retains its own properties. This classification scheme is shown in Figure 12.38.

Most materials we encounter are mixtures: mixtures of elements, mixtures of compounds, or mixtures of elements and compounds. Stainless steel, for example, is a mixture of the elements iron, chromium, nickel, and carbon. Seltzer water is a mixture of a liquid compound, water, and a gaseous compound, carbon

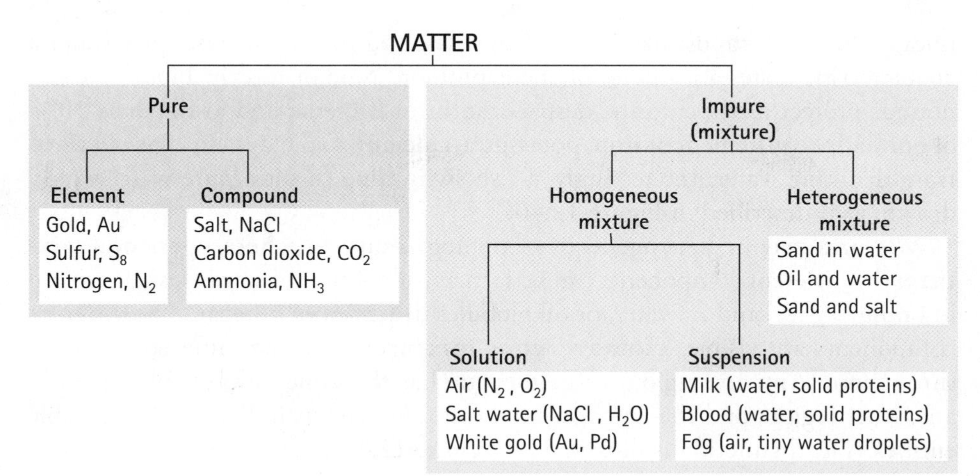

FIGURE 12.38
A chemical classification of matter.

dioxide. Our atmosphere, as Figure 12.39 illustrates, is a mixture of the elements nitrogen, oxygen, and argon, plus small amounts of such compounds as carbon dioxide and water vapor.

Because atoms and molecules are so small, it is impractical to prepare a sample that is truly pure—that is, truly 100% of a single material. For example, if just one atom or molecule out of a trillion trillion were different, then the 100% pure status would be lost. Samples, however, can be "purified" by various methods such as distillation. *Pure* is understood to be a relative term. When we compare the purity of two samples, the purer one contains fewer impurities. A sample of water that is 99.9% pure has a greater proportion of impurities than does a purer sample of water that is 99.9999% pure.

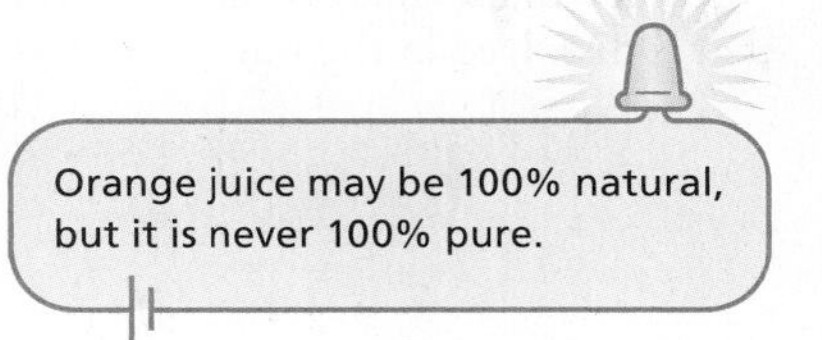

Orange juice may be 100% natural, but it is never 100% pure.

Tap water is a mixture containing mostly water but also many other compounds. Depending on your location, your water may contain compounds of calcium, magnesium, chlorine, fluorine, iron, and potassium; trace amounts of compounds of lead, mercury, and cadmium; organic compounds; and dissolved oxygen, nitrogen, and carbon dioxide. Although minimizing any toxic components in your drinking water is important, completely removing every impurity is unnecessary,

FIGURE 12.39
Earth's atmosphere is a mixture of gaseous elements and compounds. Some of them are shown here. (You will learn much more about the composition of the atmosphere and its effect on the weather in Chapter 26.)

FIGURE 12.40
Most of the oxygen in the air bubbles produced by an aquarium aerator escapes into the atmosphere. Some of the oxygen, however, mixes with the water. Fish depend on this oxygen to survive. Without this dissolved oxygen, which fish extract from the water with their gills, the fish would promptly drown. So, fish don't "breathe" water. They breathe the O_2 that is dissolved in the water.

undesirable, and impossible. Some of the dissolved solids and gases give water a characteristic taste, and many of them promote human health: Fluoride compounds protect teeth, chlorine destroys harmful bacteria, and as much as 10% of our daily requirement of iron, potassium, calcium, and magnesium is obtained from drinking water. Interestingly, a fish swimming in ultra pure water would drown, as is described in Figure 12.40.

Mixtures may be heterogeneous or homogeneous. In a **heterogeneous mixture**, the different components can be seen as individual substances, such as pulp in orange juice, sand in water, or oil globules dispersed in vinegar. The different components are visible. **Homogeneous mixtures** have the same composition throughout. Any one region of the mixture has the same ratio of substances as any other region, and the components cannot be seen as individual identifiable entities. The distinction is illustrated in Figure 12.41.

A homogeneous mixture may be either a solution or a suspension. In a **solution**, all components are in the same phase. The atmosphere is a gaseous solution consisting of the gaseous elements nitrogen and oxygen as well as minor amounts of other gaseous materials. Salt water is a liquid solution because both the water and the dissolved sodium chloride are found in a single liquid phase. An example of a solid solution is the alloy white gold, which is a homogeneous mixture of the elements gold and palladium.

A **suspension** is a homogeneous mixture in which the different components are in different phases, such as solids in liquids or liquids in gases. In a

Chemists have many ways of separating the components of a mixture, such as distillation and filtration. Most of these techniques employ the simple principle of separating the components by differences in their physical properties.

Granite

"Snow" in snow globe

Pizza

(a) Heterogeneous mixtures

Air

Clear seawater

White gold

(b) Homogeneous mixtures

FIGURE 12.41
(a) In heterogeneous mixtures, the different components can be seen with the naked eye. (b) In homogeneous mixtures, the different components are mixed at a much finer level and so are not readily distinguished.

suspension, the mixing is so thorough that the different phases cannot be readily distinguished. Milk is a suspension because it is a homogeneous mixture of proteins and fats finely dispersed in water. Blood is a suspension composed of finely dispersed blood cells in water. Another example of a suspension is clouds, which are homogeneous mixtures of tiny water droplets suspended in air. Shining a light through a suspension, as in Figure 12.42, results in a visible cone as the light is reflected by the suspended components.

FIGURE 12.42
The path of light becomes visible when the light passes through a suspension.

CHECK YOURSELF

Impure water can be purified by which of these?

(a) removing the impure water molecules
(b) removing everything that is not water
(c) breaking down the water into its simplest components
(d) adding some disinfectant such as chlorine

CHECK YOUR ANSWER

The answer is (b): When something other than water molecules is found in the water, we say that the water is impure.

12.7 Describing Solutions

LEARNING OBJECTIVE
Describe the components of a solution, and calculate a solution's concentration.

EXPLAIN THIS What is the solvent in brown sugar?

What happens to table sugar, known chemically as *sucrose*, as it is stirred into water? The answer is that the sucrose simply loses its crystalline form. Each crystal consists of billions of sucrose molecules packed neatly together. When the crystals are exposed to water, as illustrated in Figure 12.43, an even greater number of water molecules pull on the sucrose molecules via hydrogen bonds. With a little stirring, the sucrose molecules soon mix throughout the water to form a homogeneous mixture. Because this mixture is *homogeneous*, the sweetness of the first sip is the same as the sweetness of the last sip.

To most people, solutions mean the answers. To chemists, however, solutions are things that are still all mixed up.

FIGURE 12.43
Water molecules pull the sucrose molecules in a sucrose crystal away from one another. This pulling away does not, however, affect the covalent bonds within each sucrose molecule, which is why each dissolved sucrose molecule remains intact as a single molecule.

In describing solutions, the component present in the largest amount is the **solvent**, and any other components are **solutes**. For example, when a teaspoon of table sugar is mixed with 1 L of water, we identify the sugar as the solute and the water as the solvent.

The process of mixing a solute with a solvent is called **dissolving**. To make a solution, a solute must dissolve in a solvent; that is, the solute and solvent must form a homogeneous mixture. Whether one material dissolves in another is a function of their electrical attractions for each other.

CHECK YOURSELF

What is the solvent in the gaseous solution we call air?

CHECK YOUR ANSWER

Nitrogen is the solvent because it is the component that is present in the greatest quantity.

There is a limit to how much of a given solute can be dissolved in a given solvent, as Figure 12.44 illustrates. We know that when you add table sugar to a glass of water, for example, the sugar readily dissolves. As you continue to add sugar, however, there comes a point when it no longer dissolves. Instead, it collects at the bottom of the glass, even after stirring. At this point, the water is *saturated* with sugar, meaning that the water cannot accept any more sugar. When this happens, we have a **saturated solution**, defined as one in which no more solute can be dissolved. A solution that has not reached the limit of solute that will dissolve is called an **unsaturated solution**.

The quantity of solute dissolved in a solution is described in mathematical terms by the solution's **concentration**, which is the amount of solute dissolved per amount of solution:

$$\text{Concentration} = \frac{\text{amount of solute}}{\text{amount of solution}}$$

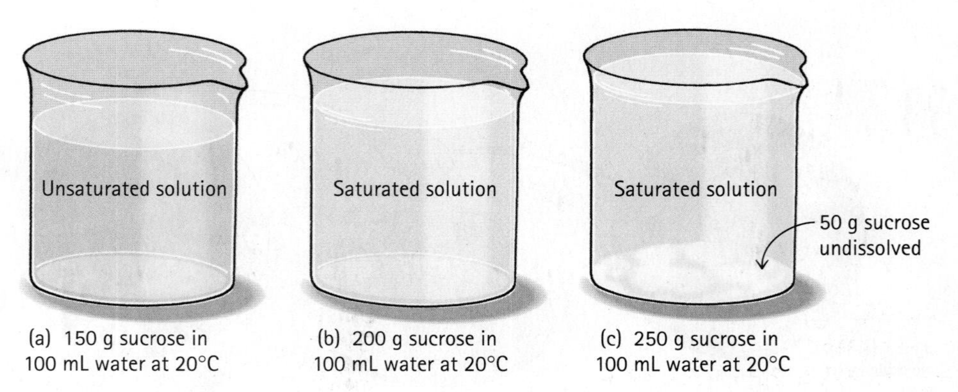

FIGURE 12.44

A maximum of 200 g of sucrose dissolves in 100 mL of water at 20°C. (a) Mixing 150 g of sucrose produces an unsaturated solution. (b) Mixing 200 g produces a saturated solution. (c) If 250 g of sucrose is mixed, then 50 g of sucrose remains undissolved. (As we discuss later, the saturation point varies with temperature.)

For example, a sucrose–water solution may have a concentration of 1 g of sucrose for every liter of solution. This can be compared with concentrations of other solutions. A sucrose–water solution containing 2 g of sucrose per liter of solution, for example, is more *concentrated*, and one containing only 0.5 g of sucrose per liter of solution is less concentrated, or more *dilute*.

Chemists are often more interested in the number of solute particles in a solution than in the number of grams of solute. Submicroscopic particles, however, are so small that the number of them in any observable sample is incredibly large. To avoid awkwardly large numbers, scientists use a unit called the *mole*. One **mole** of any type of particle is, by definition, 6.02×10^{23} particles:

$$\begin{aligned} 1 \text{ mole} &= 6.02 \times 10^{23} \text{ particles} \\ &= 602{,}000{,}000{,}000{,}000{,}000{,}000{,}000 \text{ particles} \end{aligned}$$

One mole of gold atoms, for example, is 6.02×10^{23} gold atoms, and 1 mole of sucrose molecules is 6.02×10^{23} sucrose molecules. This number 6.02×10^{23} is known as *Avogadro's number* in honor of the 19th-century Italian scientist who played a key role in the discovery of the particulate nature of matter.

Even if you've never heard the term *mole* in your life before now, you are already familiar with the basic idea. Saying "one mole" is just a shorthand way of saying "six point oh two times ten to the twenty-third particles." Just as "a couple of" means 2 of something and "a dozen of" means 12 of something, "a mole of" means 6.02×10^{23} of some elementary unit, such as atoms, molecules, or ions.

The term *mole* is derived from the Latin word *moles*, meaning heap, mass, or pile. A stack containing "1 mole" of pennies would reach a height of about 860 quadrillion km, which is roughly equal to the diameter of our galaxy, the Milky Way.

Sucrose molecules are so small that there are 6.02×10^{23} of them in only 342 g of sucrose, which is about a cupful. Thus, because 342 g of sucrose contains 6.02×10^{23} molecules of sucrose, we can use our shorthand wording and say that 342 g of sucrose contains 1 mole of sucrose. As Figure 12.45 shows, therefore, an aqueous solution that has a concentration of 342 g of sucrose per liter of solution also has a concentration of 6.02×10^{23} sucrose molecules per liter of solution or, by definition, a concentration of 1 mole of sucrose per liter of solution. The number of grams tells you the mass of solute in a given solution, and the number of moles indicates the actual number of molecules.

A common unit of concentration used by chemists is **molarity**, which is the solution's concentration expressed in moles of solute per liter of solution:

$$\text{Molarity} = \frac{\text{number of moles of solute}}{\text{liter of solution}}$$

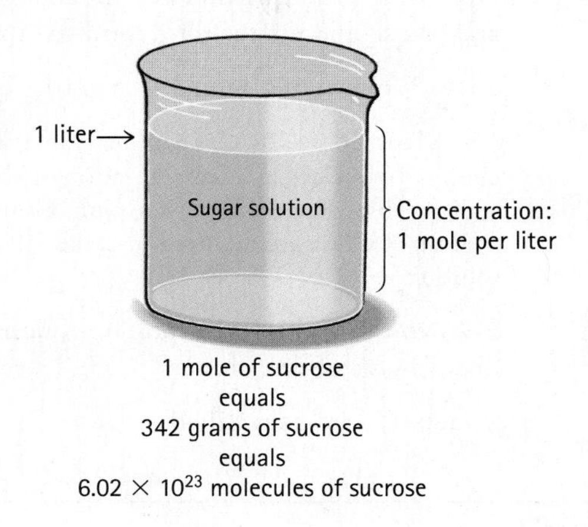

FIGURE 12.45
An aqueous solution of sucrose that has a concentration of 1 mole of sucrose per liter of solution contains 6.02×10^{23} sucrose molecules (342 g) in every liter of solution.

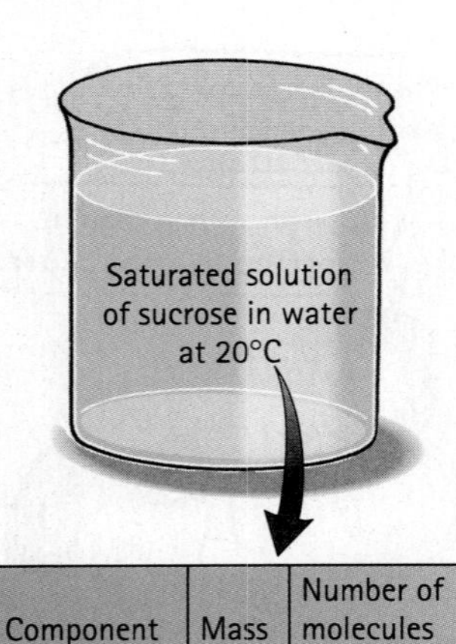

Component	Mass	Number of molecules
Sucrose	200 g	3.5×10^{23}
Water	100 g	3.3×10^{24}

FIGURE 12.46
Although 200 g of sucrose is twice as massive as 100 g of water, there are about 10 times as many water molecules in 100 g of water as there are sucrose molecules in 200 g of sucrose. How can this be? Each water molecule is about 1/20 as massive as (and much smaller than) each sucrose molecule, which means that about 10 times as many water molecules can fit in half the mass.

A solution containing 1 mole of solute per liter of solution is a 1-molar solution, which is often abbreviated 1 M. A 2-molar (2 M) solution contains 2 moles of solute per liter of solution. The difference between referring to the number of molecules of solute and referring to the number of grams of solute can be illustrated by the following question. A saturated aqueous solution of sucrose contains 200 g of sucrose and 100 g of water. Which is the solvent—sucrose or water?

As shown in Figure 12.46, there are 3.5×10^{23} molecules of sucrose in 200 g of sucrose, but there are almost 10 times as many molecules of water in 100 g of water—3.3×10^{24} molecules. As defined earlier, the solvent is the component present in the largest amount, but what do we mean by *amount*? If *amount* means number of molecules, then water is the solvent. If *amount* means mass, then sucrose is the solvent. So, the answer depends on how you look at it. From a chemist's point of view, *amount* typically means the number of molecules, and so water is the solvent in this case.

CHECK YOURSELF

Does 1 L of a 1 M solution of sucrose in water contain 1 L of water, less than 1 L of water, or more than 1 L of water?

CHECK YOUR ANSWERS

The definition of molarity refers to the number of liters of solution, not to the number of liters of solvent. When sucrose is added to a given volume of water, the volume of the solution increases. So, if 1 mole of sucrose is added to 1 L of water, the result is more than 1 L of solution. Therefore, 1 L of a 1 M solution contains less than 1 L of water.

MATH CONNECTION

Calculating for Solutions

From the formula for the concentration of a solution, we can derive equations for the amount of solute and the amount of solution:

Concentration of solution = amount of solute/volume of solution

Amount of solute = concentration of solution × volume of solution

Volume of solution = amount of solute/concentration of solution

In solving for any of these values, the units must always match. If concentration is given in grams per liter of solution, for example, then the amount of solute must be in grams and the amount of solution must be in liters.

Problem 1 How many grams of sucrose are in 3 L of an aqueous solution that has a concentration of 2 g of sucrose per liter of solution?

Solution This question asks for the amount of solute, and so you should use the second of the three formulas given above:

$$\text{Amount of solute} = 2\text{ g}/1\text{ L} \times 3\text{ L} = 6\text{ g}$$

Problem 2 A solution you are using in an experiment has a concentration of 10 g of solute per liter of solution. If you pour enough of this solution into an empty laboratory flask to make the flask contain 5 g of the solute, how many liters of the solution have you poured into the flask?

Solution This question asks for amount of solution, and you should use the third formula given above:

$$\text{Volume of solution} = 5\text{ g}/10\text{ g/L} = 0.5\text{ L}$$

Problem 3 At 20°C, a saturated solution of sodium chloride in water has a concentration of about 380 g of sodium chloride per liter of solution. How many grams of sodium chloride are required to make 3 L of a saturated solution?

Solution Multiply the solution concentration by the final volume of the solution:

$$\text{Amount of solute required} = 380\text{ g/L} \times 3\text{ L} = 1140\text{ g}$$

12.8 Solubility

LEARNING OBJECTIVE
Discuss how solutes dissolve in solvents and how solubility changes with temperature.

EXPLAIN THIS How can oxygen be removed from water?

The **solubility** of a solute is its *ability* to dissolve in a solvent. As can be expected, this ability mainly depends on the attractions between solute particles and solvent particles. If a solute has any appreciable solubility in a solvent, then that solute is said to be **soluble** in that solvent.

Solubility depends on the attractions of solute particles for one another and the attractions of solvent particles for one another. Sucrose molecules, for example, can form multiple hydrogen bonds with one another, which is why sucrose is a solid at room temperature. In order for sucrose to dissolve in water, the water molecules must first pull the tightly held sucrose molecules away from one another. This puts a limit on the amount of sucrose that can dissolve in water—eventually a point is reached where there are not enough water molecules to separate the sucrose molecules from one another. This is the point of saturation, and any additional sucrose added to the solution does not dissolve.

Grease is soluble in paint thinner, which is why paint thinner can be used to clean grease off one's hands. But body oils are also soluble in paint thinner, which is why hands cleaned with paint thinner feel dry and chapped.

You probably know from experience that water-soluble solids usually dissolve better in hot water than in cold water. Heating water to boiling, for example, allows you to make a highly concentrated sugar solution. This is how syrups and hard candy are made. Solubility increases with increasing temperature because water molecules have greater kinetic energy. Therefore, in their random thermal motions, higher-temperature molecules are able to collide with the solid solute more vigorously. The vigorous collisions facilitate the disruption of electrical particle-to-particle attractions in the solid.

MasteringPhysics®
TUTORIAL: Solubility

Although the solubilities of many solid solutes—sucrose, to name just one example—are greatly increased by temperature increases, the solubilities of other solid solutes, such as sodium chloride, are only mildly affected, as Figure 12.47 shows. This difference involves a number of factors, including the strength of the chemical bonds in the solute molecules and the way those molecules are packed together. Some chemicals, such as calcium carbonate, $CaCO_3$, actually become *less* soluble as the water temperature increases. This explains why the inner surface of tea kettles are often coated with calcium carbonate residues. This also explains why coral reefs, which are made of calcium carbonate, form in only warm tropical regions of our oceans—in colder climates the calcium carbonate is more soluble and thus dissolves.

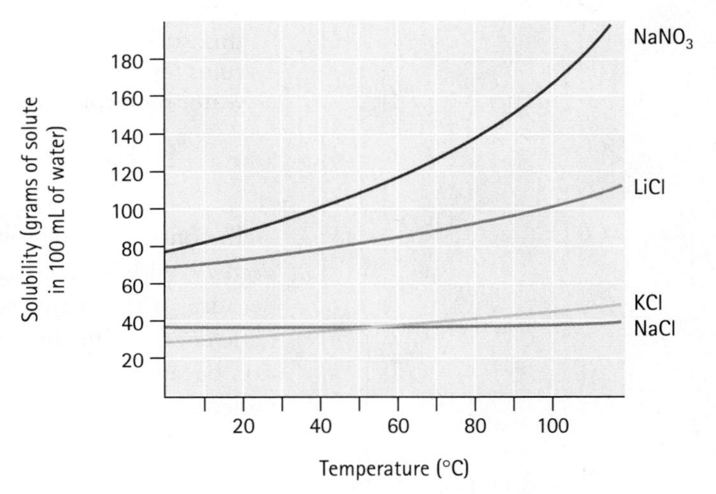

FIGURE 12.47
The solubility of many water-soluble solids increases with temperature, while the solubility of others is only very slightly affected by temperature.

When a sugar solution saturated at a high temperature is allowed to cool, some of the sugar usually comes out of solution and forms a **precipitate**. When this occurs, the solute—sugar in this case—is said to have *precipitated* from the solution.

Let's put on our quantitative thinking caps and consider another example. At 100°C, the solubility of sodium nitrate, $NaNO_3$, in water is 165 g per 100 mL of water. As we cool this solution, the solubility of $NaNO_3$ decreases, as shown in Figure 12.48, and this change in solubility causes some of the dissolved $NaNO_3$ to precipitate (come out of solution). At 20°C, the solubility of $NaNO_3$

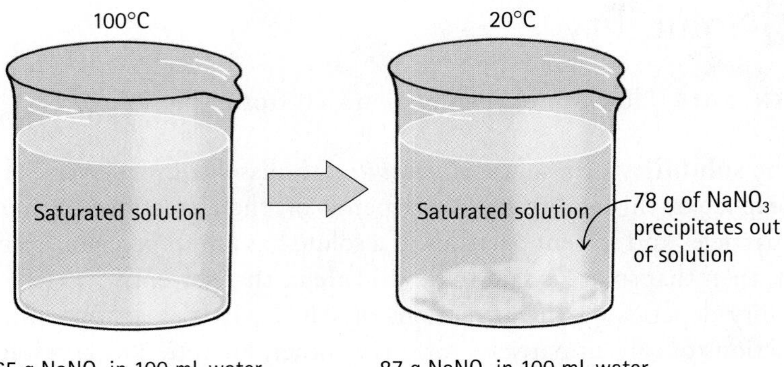

FIGURE 12.48
The solubility of sodium nitrate is 165 g per 100 mL of water at 100°C but only 87 g per 100 mL at 20°C. Cooling a 100°C saturated solution of $NaNO_3$ to 20°C causes 78 g of the solute to precipitate.

is only 87 g per 100 mL of water. So, if we cool the 100°C solution to 20°C, 78 g (165 g − 87 g) precipitates, as shown in Figure 12.48.

CHECK YOURSELF
Why isn't sucrose infinitely soluble in water?

CHECK YOUR ANSWER
The attraction between two sucrose molecules is much stronger than the attraction between a sucrose molecule and a water molecule. Because of this, sucrose dissolves in water only as long as the number of water molecules far exceeds the number of sucrose molecules. When there are too few water molecules to dissolve any additional sucrose, the solution is saturated.

In contrast to the solubilities of most solids, the solubilities of gases in liquids *decrease* with increasing temperature, as Table 12.4 shows. This effect occurs because, with an increase in temperature, the solvent molecules have more kinetic energy. This makes it more difficult for a gaseous solute to remain in solution because the solute molecules are literally ejected by the high-energy solvent molecules.

TABLE 12.4 TEMPERATURE-DEPENDENT SOLUBILITY OF OXYGEN GAS IN WATER AT A PRESSURE OF 1 ATMOSPHERE

Temperature (°C)	O_2 Solubility (g O_2/L H_2O)
0	0.0141
10	0.0109
20	0.0092
25	0.0083
30	0.0077
35	0.0070
40	0.0065

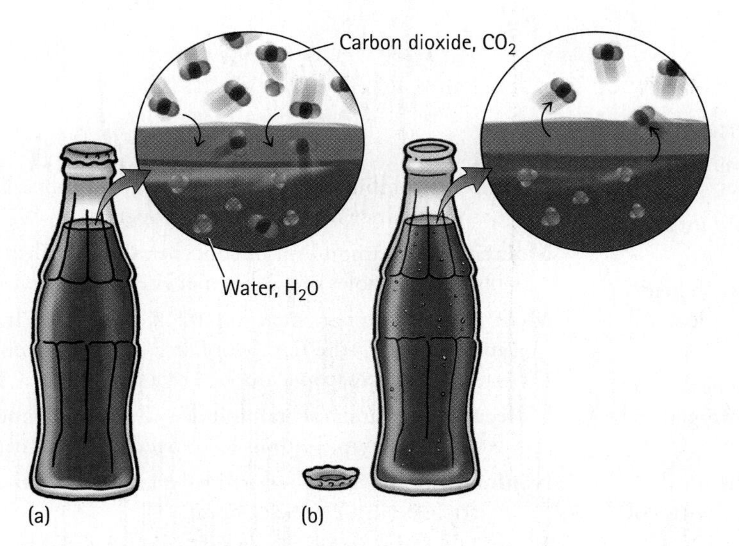

FIGURE 12.49
(a) The carbon dioxide gas above the liquid in an unopened soft drink bottle consists of many tightly packed carbon dioxide molecules that are forced by pressure into solution. (b) When the bottle is opened, the pressure is released, and carbon dioxide molecules originally dissolved in the liquid can escape into the air.

Perhaps you have noticed that warm carbonated beverages go flat faster than cold ones. The higher temperature causes the molecules of carbon dioxide gas to leave the liquid solvent at a higher rate.

The solubility of a gas in a liquid also depends on the pressure of the gas immediately above the liquid. In general, a higher gas pressure above the liquid means more of the gas dissolves. A gas at a high pressure has many, many gas particles crammed into a given volume. The "empty" space in an unopened soft drink bottle, for example, is crammed with carbon dioxide molecules in the gaseous phase. With nowhere else to go, many of these molecules dissolve in the liquid, as shown in Figure 12.49. Alternatively, we might say that the great pressure forces the carbon dioxide molecules into solution. When the bottle is opened, the "head" of highly pressurized carbon dioxide gas escapes. Now the gas pressure above the liquid is lower than before. As a result, the solubility of the carbon dioxide drops, and the carbon dioxide molecules that were once squeezed into the solution begin to escape into the air above the liquid.

FYI It is not just dipole–induced dipole attractions that keep carbon dioxide dissolved in water. As we'll discuss in Chapter 13, carbon dioxide reacts with water to form carbonic acid, which is much more soluble in water. When a can of carbonated soda is opened, much of this carbonic acid quickly transforms back into water and carbon dioxide, which quickly bubbles out of solution because of its low solubility.

CHECK YOURSELF

You open two cans of soda, one from a warm kitchen shelf and the other from the coldest depths of your refrigerator. Which fizzes more in your mouth?

CHECK YOUR ANSWER

The solubility of carbon dioxide in water decreases with increasing temperature. The warm drink, therefore, fizzes more in your mouth than the cold one does.

SUMMARY OF TERMS (KNOWLEDGE)

Alloy A mixture of two or more metallic elements.

Concentration A quantitative measure of the amount of solute in a solution.

Covalent bond A chemical bond in which atoms are held together by their mutual attraction for two or more shared electrons.

Covalent compound A substance, such as an element or a chemical compound, in which atoms are held together by covalent bonds.

Dipole A separation of charge that occurs in a chemical bond because of differences in the electronegativities of the bonded atoms.

Dissolving The process of mixing a solute in a solvent to produce a homogeneous mixture.

Electron-dot structure A shorthand notation of the noble gas shell model of the atom, in which valence electrons are shown as dots surrounding an atomic symbol. The electron-dot structure for an atom or ion is sometimes called a *Lewis dot symbol*, and the electron-dot structure for a molecule or polyatomic ion is sometimes called a *Lewis structure*.

Electronegativity The ability of an atom to attract a bonding pair of electrons to itself when bonded to another atom.

Heterogeneous mixture A mixture in which the different components can be seen as individual substances.

Homogeneous mixture A mixture in which the components are so finely mixed that any one region of the mixture has the same ratio of substances as any other region; the components cannot be seen as identifiable individual substances.

Hydrogen bond An unusually strong dipole–dipole attraction between molecules that have a hydrogen atom covalently bonded to a highly electronegative atom, usually nitrogen, oxygen, or fluorine.

Induced dipole A temporarily uneven distribution of electrons in an otherwise nonpolar atom or molecule.

Ion An atom that has a net electric charge because of either a loss or gain of electrons.

Ionic bond A chemical bond created by an electrical attraction between two oppositely charged ions.

Ionic compound A chemical compound containing ions.

Metallic bond A chemical bond in which positively charged metal ions are held together within a "fluid" of loosely held electrons.

Mixture A combination of two or more substances in which each substance retains its own properties.

Molarity A common unit of concentration equal to the number of moles of a solute per liter of solution.

Mole A large number equal to 6.02×10^{23}; usually used in reference to the number of atoms, ions, or molecules within a macroscopic amount of a material.

Molecule The fundamental unit of a chemical compound, which is a group of atoms held together by covalent bonds.

Nonbonding pairs Two paired valence electrons that do not participate in a chemical bond.

Nonpolar Description of a chemical bond or molecule that has no dipole; the electrons are distributed evenly.

Polar Description of a chemical bond or molecule that has a dipole; the electrons are congregated on one side, which makes that side slightly negative while the opposite side (lacking electrons) becomes slightly positive.

Polyatomic ion A molecule that carries a net electric charge.

Precipitate A solute that has come out of solution.

Pure The state of a material that consists of only a single element or compound.

Saturated solution A solution containing the maximum amount of solute that will dissolve in its solvent.

Solubility The ability of a solute to dissolve in a given solvent.

Soluble Capable of dissolving to an appreciable extent in a given solvent.

Solute Any component in a solution that is not the solvent.

Solution A homogeneous mixture in which all components are in the same phase.

Solvent The component in a solution that is present in the largest amount.

Suspension A homogeneous mixture in which the various components are thoroughly mixed but remain in different phases.

Unsaturated solution A solution that is capable of dissolving additional solute.

Valence electron The electrons in the outermost occupied shell of an atom.

Valence shell The outermost occupied shell of an atom.

READING CHECK (COMPREHENSION)

12.1 Electron-Dot Structures

1. How many electrons can occupy the first shell? How many can occupy the second shell?
2. Which electrons are represented by an electron-dot structure?
3. How do the electron-dot structures of elements in the same group in the periodic table compare with one another?

12.2 The Ionic Bond

4. How does an ion differ from an atom?
5. To become a negative ion, does an atom lose or gain electrons?
6. Which elements tend to form ionic bonds?

12.3 The Covalent Bond

7. Which elements tend to form covalent bonds?
8. How many electrons are shared in a double covalent bond?
9. Within a polyatomic ion, how many covalent bonds does a negatively charged oxygen atom form?

12.4 Polar Covalent Bonds

10. What is a dipole?
11. Which element in the periodic table has the greatest electronegativity? Which has the least electronegativity?
12. Which is more polar—a carbon–oxygen bond or a carbon–nitrogen bond?

12.5 Molecular Polarity

13. Why do nonpolar substances boil at relatively low temperatures?
14. Which has a greater degree of symmetry—a polar molecule or a nonpolar molecule?
15. Why don't oil and water mix?

12.6 Molecular Attractions

16. What is the primary difference between a chemical bond and an attraction between two molecules?
17. What is a hydrogen bond?
18. Are induced dipoles permanent?

12.7 Describing Solutions

19. What happens to the volume of a sugar solution as more sugar is dissolved in it?
20. Distinguish between a solute and a solvent.
21. Is concentration typically given with the volume of solvent or the volume of solution?

12.8 Solubility

22. Why does the solubility of a gas solute in a liquid solvent decrease with increasing temperature?
23. Why do sugar crystals dissolve faster when crushed?
24. Is sugar a polar or nonpolar substance?

THINK INTEGRATED SCIENCE

12A—Metals from Earth

25. Do metals more readily gain or lose electrons?
26. What is an alloy?
27. What is a native metal?

12B—Most Materials Are Mixtures

28. What defines a material as being a mixture?
29. Why isn't it practical to have a macroscopic sample that is 100% pure?
30. How is a solution different from a suspension?

THINK AND DO (HANDS-ON APPLICATION)

31. View crystals of table salt through a magnifying glass or, better yet, a microscope if one is available. If you have a microscope, crush the crystals with a spoon and examine the resulting powder. Purchase some sodium-free salt, which is potassium chloride, KCl, and examine these ionic crystals, both intact and crushed. Sodium chloride and potassium chloride both form cubic crystals, but there are significant differences. What are they?
32. To see the action of the ion–dipole attraction, create a static charge on a rubber balloon by rubbing it across your hair. Hold this charged balloon up close to, but not touching, a thin stream of water running from a faucet. Watch the charged balloon divert the path of the falling water. Your balloon is negatively charged because it picks up electrons from your hair. Why would a balloon that was positively charged also attract the stream of water?
33. To see the gases dissolved in your water, fill a clean cooking pot with water and let it stand at room temperature for several hours. Note the tiny bubbles that adhere to the inner sides of the pot. Where did these tiny bubbles come from? What do you suppose they contain? For further experimentation, repeat this activity in two pots side by side. In one pot, use warm water from the kitchen faucet. In the second pot, use boiled water that has cooled down to the same temperature.
34. Put on your safety glasses and add several cups of tap water to a cooking pot. Boil the water to dryness. (Turn off the burner before the water is all gone. The heat from the pot will finish the evaporation. Watch out for splattering!) Examine the resulting residue by scraping it with the knife. These are the solids you ingest with every glass of water you drink.

35. Here's a quick recipe for rock candy. In a cooking pot make a hot saturated solution of sugar in water. Start by mixing sugar and water in a 2:1 ratio by volume. Add more sugar or water as necessary to obtain a clear, runny syrup. Let the mixture cool for 10 minutes. Roll a wet wooden skewer stick in some granulated sugar. Pour the warm sugar syrup into a jar. Submerge the skewer into the sugar syrup. Cover the top and store in a cool place. The longer you wait, the larger the crystals will be.

THINK AND COMPARE (ANALYSIS)

36. Rank the bonds in order of increasing polarity: (a) C—H, (b) O—H, (c) N—H.
37. Rank the compounds in order of increasing boiling point: (a) fluorine, F_2; (b) hydrogen fluoride, HF; (c) hydrogen chloride, HCl.
38. Rank the following in order of increasing symmetry: (a) CH_4, (b) NH_3, (c) H_2O.
39. Rank the following in order of decreasing boiling point: (a) CH_4, (b) NH_3, (c) H_2O.
40. Rank these solutions in order of increasing concentration: (a) 0.5 mole of sucrose in 2.0 L of solution; (b) 1.0 mole of sucrose in 3.0 L of solution; (c) 1.5 moles of sucrose in 4.0 L of solution.
41. Rank the compounds in order of increasing solubility in water:

CH_3CH_2—OH	$CH_3CH_2CH_2CH_2$—OH	$CH_3CH_2CH_2CH_2CH_2CH_2$—OH
Ethanol	Butanol	Hexanol

THINK AND SOLVE (MATHEMATICAL APPLICATION)

42. Ores of manganese, Mn, sometimes contain the mineral rhodochrosite, $MnCO_3$, which is an ionic compound of manganese ions and carbonate ions. How many electrons has each manganese atom lost to make this compound?
43. Magnesium ions carry a 2+ charge, and chloride ions carry a 1− charge. What is the chemical formula for the ionic compound magnesium chloride?
44. Barium ions carry a 2+ charge, and nitrogen ions carry a 3− charge. What would be the chemical formula for the ionic compound barium nitride?
45. How much sodium chloride, in grams, is needed to make 15 L of a solution that has a concentration of 3.0 g of sodium chloride per liter of solution?
46. If water is added to 1 mole of sodium chloride in a flask until the volume of the solution is 1 L, what is the molarity of the solution? What is the molarity when water is added to 2 moles of sodium chloride to make 0.5 L of solution?

THINK AND EXPLAIN (SYNTHESIS)

47. How many more electrons can fit in the valence shell of a fluorine atom?
48. How many more electrons can fit in the valence shell of a hydrogen atom?
49. The valence electron of a sodium atom does not sense the full 11+ charge of the sodium nucleus. Why not?
50. How is the number of unpaired valence electrons in an atom related to the number of bonds that the atom can form?
51. Why is it so easy for a magnesium atom to lose two electrons?
52. Why doesn't the neon atom tend to lose or gain any electrons?
53. Sulfuric acid, H_2SO_4, loses two protons to form what polyatomic ion? What molecule loses a proton to form the hydroxide ion, OH^-?
54. Which should be larger—the potassium ion, K+, or the potassium atom, K? Which should be larger—the potassium ion, K^+, or the argon atom, Ar? Explain.
55. Which should be more difficult to pull apart—a sodium ion from a chloride ion or a potassium ion from a chloride ion? Explain.

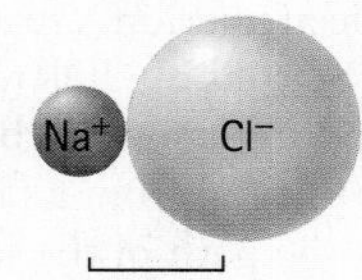

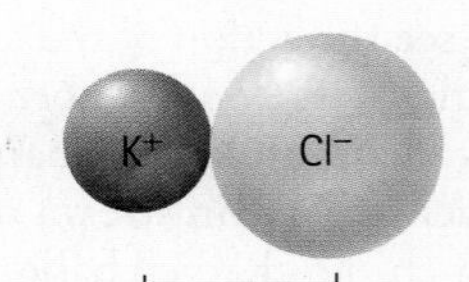

56. Given that the total number of atoms on our planet remains fairly constant, how is it ever possible to deplete a natural resource such as a metal?

57. An artist wants to create a metal sculpture using a mold so that his artwork can be readily mass produced. He wants his sculpture to be exactly 6 inches tall. Should the mold also be 6 inches tall? Why or why not?
58. Which are closer together—the two nuclei within potassium fluoride, KF, or the two nuclei within molecular fluorine, F_2? Explain.
59. Two fluorine atoms join together to form a covalent bond. Why don't two potassium atoms do the same thing?
60. What drives an atom to form a covalent bond—its nuclear charge or the need to have a filled outer shell? Explain.
61. Atoms of nonmetallic elements form covalent bonds, but they can also form ionic bonds. How is this possible?
62. Examine the three-dimensional geometries of PF_5 and SF_4 shown here. Which compound do you think is more polar?

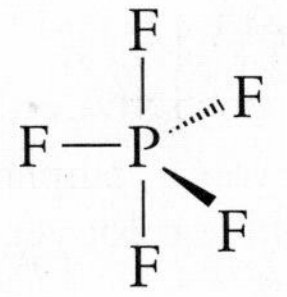

```
      F                 F
      |   F             |   F
  F—P                 : S
      |  F              |  F
      F                 F

     PF5               SF4
```

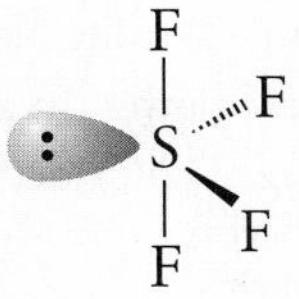

63. In each of these molecules, which atom carries the greater positive charge: H—Cl, Br—F, C≡O, Br—Br?
64. Which is more polar—a sulfur–bromine bond (S—Br) or a selenium–chlorine bond (Se—Cl)?
65. True or false: The greater the nuclear charge of an atom, the greater its electronegativity. Explain.
66. Water, H_2O, and methane, CH_4, have about the same mass and differ by only one type of atom. Why is the boiling point of water so much higher than that of methane?
67. Which molecule from each pair should have a higher boiling point (atomic numbers: Cl = 17, O = 8, C = 6, H = 1)?

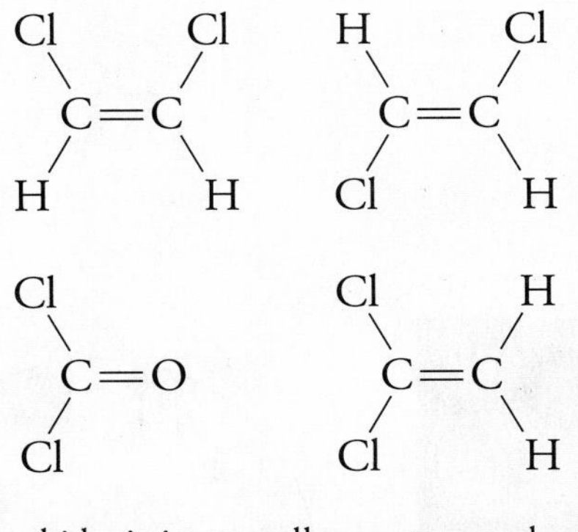

```
Cl      Cl      H       Cl
  \    /          \    /
   C=C             C=C
  /    \          /    \
H       H       Cl      H

Cl              Cl      H
  \               \    /
   C=O             C=C
  /               /    \
Cl              Cl      H
```

68. Three kids sitting equally apart around a table are sharing jelly beans. One of the kids, however, tends only to take jelly beans and only rarely gives one away. If each jelly bean represents an electron, who ends up being slightly negative? Who ends up being slightly positive? Is the negative kid just as negative as one of the positive kids is positive? Would you describe this as a polar or nonpolar situation? How about if all three kids were equally greedy?
69. Which is stronger—the covalent bond that holds atoms together within a molecule or the electrical attraction between two neighboring molecules? Explain.
70. The charges in sodium chloride are all balanced—for every positive sodium ion there is a corresponding negative chloride ion. Since its charges are balanced, how can sodium chloride be attracted to water, and vice versa?
71. Why are ion–dipole attractions stronger than dipole–dipole attractions?
72. Why is calcium fluoride, CaF_2, a high-melting-point crystalline solid while stannic chloride, $SnCl_4$, is a volatile liquid?
73. Of the two structures shown, one is a typical gasoline molecule and the other is a typical motor oil molecule. Which is which? Base your reasoning not on memorization but rather on what you know about electrical attractions between molecules and the various physical properties of gasoline and motor oil.

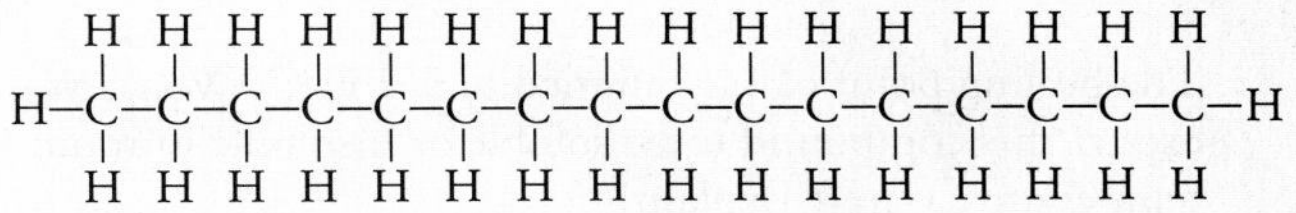

```
  H H H H H H H H H H H H H H H H
  | | | | | | | | | | | | | | | |
H-C-C-C-C-C-C-C-C-C-C-C-C-C-C-C-C-H
  | | | | | | | | | | | | | | | |
  H H H H H H H H H H H H H H H H
```

Structure A

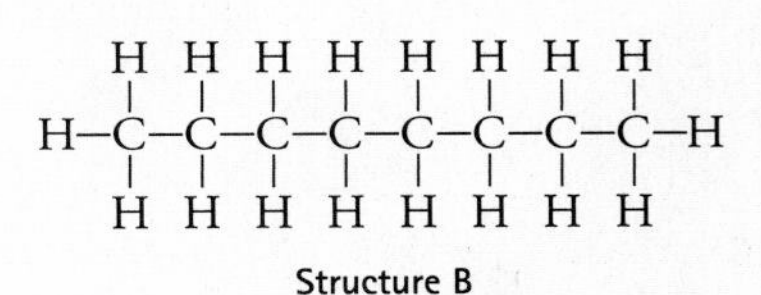

Structure B

74. Why can't the elements of a compound be separated from one another by physical means?
75. Classify the following as an element, compound, or mixture, and justify your classifications: salt, stainless steel, tap water, sugar, vanilla extract, butter, maple syrup, aluminum, ice, milk, cherry-flavored cough drops.
76. Which of the boxes below best represents a suspension?

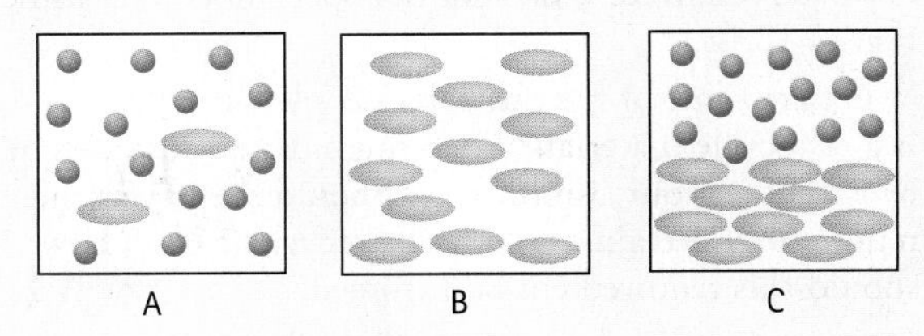

77. Which of the boxes best represents a solution?
78. Which of the boxes best represents a compound?
79. Which should weigh more—100 mL of fresh water or 100 mL of fresh sparkling seltzer water? Why? Which should weigh more—100 mL of flat seltzer water at 20°C or the same 100 mL of flat seltzer water brought to 80°C? Why?
80. The volume of many liquid solvents expands with increasing temperature. What happens to the concentration of a solution made with such a solvent as the temperature of the solution is increased?
81. Some bottled water is now advertised as containing extra quantities of "Vitamin O," which is a marketing

gimmick for selling oxygen, O_2. Might this bottled water actually contain extra quantities of oxygen, O_2? How much more might one find in regular bottled water? How might the amount of oxygen we absorb through our lungs compare to the amount we might absorb through our stomach—after burping?

82. Explain why, for these three substances, the solubility in 20°C water goes down as the molecules get larger but the boiling point goes up.

Substance	Boiling point/ Solubility
$CH_3{-}O{-}H$	65°C infinite
$CH_3CH_2CH_2CH_2{-}O{-}H$	117°C 8 g/100 mL
$CH_3CH_2CH_2CH_2CH_2{-}O{-}H$	138°C 2.3 g/100 mL

83. The boiling point of 1,4-butanediol is 230°C. Would you expect this compound to be soluble or insoluble in room-temperature water? Explain.

$H{-}O{-}CH_2CH_2CH_2CH_2{-}O{-}H$

1,4-Butanediol

84. Account for the observation that ethanol, C_2H_5OH, dissolves readily in water but dimethyl ether, CH_3OCH_3, which has the same number and kinds of atoms, does not.

$H{-}CH_2{-}CH_2{-}O{-}H$ (Ethanol) $\qquad$ $H_3C{-}O{-}CH_3$ (Dimethyl ether)

Ethanol

Dimethyl ether

85. At 10°C, which is more concentrated—a saturated solution of sodium nitrate, $NaNO_3$, or a saturated solution of sodium chloride, NaCl? (See Figure 12.47.)

86. Why are both rain and snow called precipitation?

87. Hydrogen chloride, HCl, is a gas at room temperature. Would you expect it to be very soluble or not very soluble in water?

THINK AND DISCUSS (EVALUATION)

88. What should be done with mining pits after all their ore has been removed? Consider the open-pit copper mine of Figure 12.12.

89. What are some of the obstacles people face when trying to recycle materials? How might these obstacles be overcome in your community? Should the government require that certain materials be recycled? If so, how should this requirement be enforced?

90. Oxygen, O_2, dissolves quite well within a class of compounds known as liquid perfluorocarbons—so well that oxygenated perfluorocarbons can be inhaled in a liquid phase, as is demonstrated by the rodent shown below the water-bound goldfish. Do you suppose perfluorocarbon molecules are polar or nonpolar? Why would the rodent drown if it were brought up to the water layer, and why would the goldfish die if they swam down into the perfluorocarbon layer? How might perfluorocarbons be used to clean our lungs or serve as an artificial blood? When is it acceptable to sacrifice the lives of animals for scientific research?

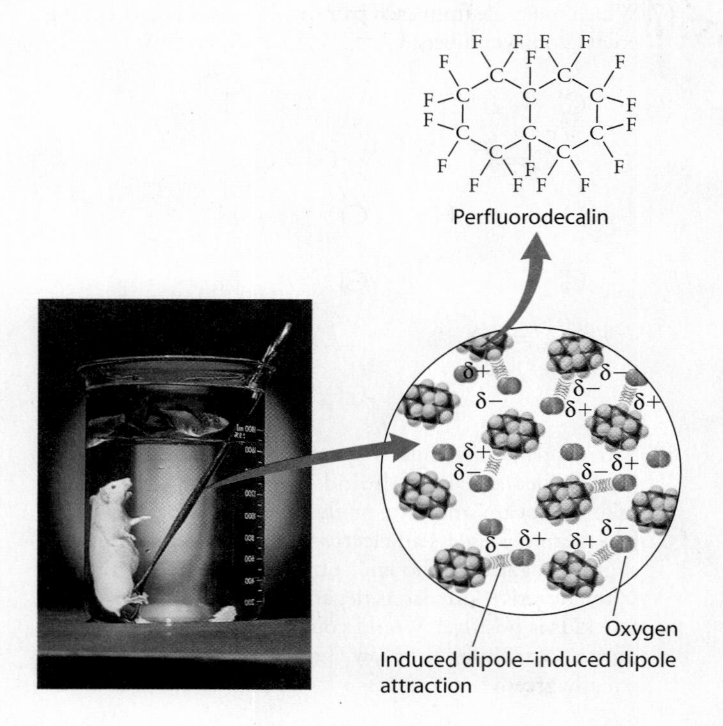

READINESS ASSURANCE TEST (RAT)

If you have a good handle on this chapter, if you really do, then you should be able to score 7 out of 10 on this RAT. If you score less than 7, you need to study further before moving on.

Choose the BEST answer to each of the following:

1. An atom loses an electron to another atom. Is this an example of a physical or chemical change?
   (a) chemical change involving the formation of ions
   (b) physical change involving the formation of ions
   (c) chemical change involving the formation of covalent bonds
   (d) physical change involving the formation of covalent bonds

2. Aluminum ions carry a 3+ charge, and chloride ions carry a 1− charge. What is the chemical formula for the ionic compound aluminum chloride?
   (a) Al_3Cl
   (b) $AlCl_3$
   (c) Al_3Cl_3
   (d) $AlCl$

3. Why are ores so valuable?
   (a) They are sources of naturally occurring gold.
   (b) Metals can be efficiently extracted from them.
   (c) They tend to occur in scenic mountainous regions.
   (d) They hold many clues to Earth's natural history.

4. In terms of the periodic table, is there an abrupt or gradual change between ionic and covalent bonds?
   (a) An abrupt change occurs across the metalloids.
   (b) Actually, any element of the periodic table can form a covalent bond.
   (c) There is a gradual change; the farther apart the elements, the more ionic.
   (d) Whether an element forms an ionic or a covalent bond depends on nuclear charge, not on the relative positions in the periodic table.

5. When nitrogen and fluorine combine to form a molecule, the most likely chemical formula is:
   (a) N_3F.
   (b) N_2F.
   (c) NF_4.
   (d) NF.
   (e) NF_3.

6. A substance consisting of which of the two molecules shown below should have a higher boiling point? Why?

   S=C=O   O=C=O

   (a) the molecule on the left, SCO, because it comes later in the periodic table
   (b) the molecule on the left, SCO, because it has less symmetry
   (c) the molecule on the right, OCO, because it has greater symmetry
   (d) the molecule on the right, OCO, because it has more mass

7. Someone argues that you shouldn't drink tap water because it contains hundreds of molecules of some impurity in each glass. How would you respond in defense of the water's purity, if it indeed does contain hundreds of molecules of some impurity per glass?
   (a) Impurities aren't necessarily bad; in fact, they may be good for you.
   (b) The water contains water molecules, and each water molecule is pure.
   (c) There's no defense. If the water contains impurities, it should not be drunk.
   (d) Relatively speaking, a hundred molecules is practically nothing.

8. What is the difference between a compound and a mixture?
   (a) Both consist of atoms of different elements.
   (b) The difference is the way in which their atoms are bonded together.
   (c) The components of a mixture are not chemically bonded together.
   (d) One is a solid and the other is a liquid.

9. Fish don't live very long in water that has just been boiled and brought back to room temperature. Why?
   (a) There is now a higher concentration of dissolved CO_2 in the water.
   (b) The nutrients in the water have been destroyed.
   (c) Since some of the water was evaporated while boiling, the salts in the water are now more concentrated, which has a negative effect on the fish.
   (d) The boiling process removes the air that was dissolved in the water. Upon cooling the water is void of its usual air content and hence the fish drown.

10. How many moles of sugar (sucrose) are there in 5 L of sugar water that has a concentration of 0.5 M?
   (a) 5.5 moles
   (b) 5.0 moles
   (c) 2.5 moles
   (d) 1.5 moles

Answers to RAT

1. a, 2. b, 3. b, 4. c, 5. e, 6. b, 7. d, 8. c, 9. d, 10. c

CHAPTER 13
Chemical Reactions

THE HEAT of a lightning bolt causes atmospheric nitrogen and oxygen to react, leading to the formation of nitric acid, HNO_3, and nitrous acid, HNO_2. As part of the *nitrogen cycle*, these acids are carried by rain into the ground, where they transform into nitrate ions that plants use for growing. We, in turn, eat the plants, or plant-eating animals, to support life-sustaining chemical reactions within ourselves.

Countless chemical reactions take place continuously in the outside environment, such as the reactions of the nitrogen cycle. Many more occur within our bodies, such as the chemical reactions that help us digest our food. How do chemicals store and release energy? What is pH, and how do we measure it? What do our bodies have in common with the burning of a campfire or the rusting of old farm equipment? We will address such questions in this chapter as we investigate the dynamic submicroscopic world of reacting chemicals.

13.1 Chemical Equations

LEARNING OBJECTIVE
Identify whether a chemical equation is balanced or not balanced.

EXPLAIN THIS How can 50 g of wood burn to produce more than 50 g of products?

As we discussed in Chapter 11, during a chemical reaction, atoms rearrange to create one or more new compounds. This activity is neatly summed up in written form as a **chemical equation**. A chemical equation shows the reacting substances, called **reactants**, to the left of an arrow that points toward the newly formed substances, called **products**:

$$\text{reactants} \longrightarrow \text{products}$$

Typically, reactants and products are represented by their elemental or chemical formulas. Sometimes molecular models or simply names may be used instead. Phases are also often shown: (*s*) for solid, (*l*) for liquid, and (*g*) for gas. Compounds dissolved in water are designated (*aq*) for aqueous solution. Lastly, numbers are placed in front of the reactants or products to show the ratio in which they either combine or form. These numbers are called *coefficients*, and they represent numbers of individual atoms and molecules. For instance, two hydrogen gas molecules, H_2, react with one oxygen gas molecule, O_2, to produce two molecules of water, H_2O, in the gaseous phase:

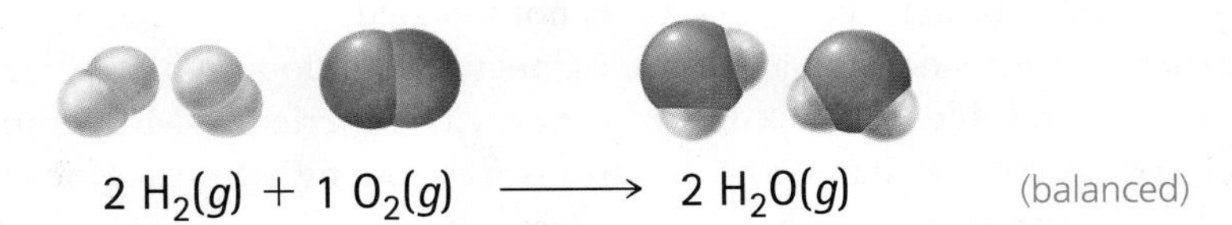

$$2\ H_2(g) + 1\ O_2(g) \longrightarrow 2\ H_2O(g) \qquad \text{(balanced)}$$

One of the most important principles of chemistry is the **law of mass conservation**. The law of mass conservation states that matter is neither created nor destroyed during a chemical reaction.* The atoms present at the beginning of a reaction merely rearrange to form new molecules. This means that no atoms are lost or gained during any reaction. The chemical equation must therefore be *balanced*. In a balanced equation, each atom must appear the same number of times on both sides of the arrow. The equation for the formation of water shown earlier is balanced because each side shows four hydrogen atoms and two oxygen atoms. You can count the number of atoms in the models to see this for yourself.

UNIFYING CONCEPT
- ***The Atomic Theory of Matter***
Section 9.1

A coefficient in front of a chemical formula tells us the number of times that element or compound must be counted. For example, 2 H_2O indicates two water molecules, which contain a total of four hydrogen atoms and two oxygen atoms. By convention, the coefficient 1 is omitted, so the chemical equation for water is typically written:

$$2\ H_2(g) + O_2(g) \longrightarrow 2\ H_2O(g) \qquad \text{(balanced)}$$

*For all practical purposes this law holds true. Technically, however, any energy released or absorbed by a chemical reaction arises from the transformation of matter into energy, or vice versa. The amount of matter lost or gained in a chemical reaction is so small that, for all practical purposes, we can ignore this detail. This is not the case for the nuclear reactions discussed in Chapter 10. For nuclear reactions, matter energy conversions are much greater.

MasteringPhysics®
TUTORIAL: Chemical Reactions and Equations

CHECK YOURSELF

How many oxygen atoms are shown in this balanced equation?

$$3\ O_2(g) \longrightarrow 2\ O_3(g)$$

CHECK YOUR ANSWER

Before the reaction, there are six oxygen atoms in three O_2 molecules. After the reaction, these same six atoms are found in two O_3 molecules.

FYI Chemical explosions typically involve the transformation of an unstable solid or liquid chemical into more stable gases that occupy much more volume. Upon detonation, 1 mole of nitroglycerin, $C_3H_5N_3O_9$, produces 7.25 moles of gases including carbon dioxide, CO_2; nitrogen, N_2; oxygen, O_2; and water vapor, H_2O. The volume change is dramatic—from less than 0.3 L to about 170 L, which is an increase of about 600%. For nitroglycerin and similar high explosives, these gases expand at supersonic speeds, creating a powerful and destructive shock wave.

An unbalanced chemical equation shows the reactants and products without the correct coefficients. For example, the equation

$$NO(g) \longrightarrow N_2O(g) + NO_2(g) \qquad \text{(not balanced)}$$

is not balanced because there are one nitrogen atom and one oxygen atom to the left of the arrow, but three nitrogen atoms and three oxygen atoms on the right.

You can balance unbalanced equations by adding or changing the coefficients to produce correct ratios. (It's important not to change the subscripts, however, because to do so changes the compound's identity—H_2O is water, but H_2O_2 is hydrogen peroxide!) For example, to balance the equation above, add a 3 before the NO:

$$3\ NO(g) \longrightarrow N_2O(g) + NO_2(g) \qquad \text{(balanced)}$$

Now there are three nitrogen atoms and three oxygen atoms on each side of the arrow, and the law of mass conservation is not violated.

Practicing chemists develop a skill for balancing equations. This skill involves creative energy and, like other skills, improves with experience. More important than being an expert at balancing equations is knowing why they need to be balanced. And the reason is the law of mass conservation, which tells us that atoms are neither created nor destroyed in a chemical reaction—they are simply rearranged. So, every atom present before the reaction must be present after the reaction, even though the groupings of atoms are different.

CHECK YOURSELF

Write a balanced equation for the reaction showing hydrogen gas, H_2, and nitrogen gas, N_2, forming ammonia gas, NH_3:

$$\_\ H_2(g) + \_\ N_2(g) \longrightarrow \_\ NH_3(g)$$

CHECK YOUR ANSWER

Initially, we see two hydrogen atoms to the left of the reaction arrow and three on the right. This can be remedied by placing a coefficient of 3 by the hydrogen, H_2, and a coefficient of 2 by the ammonia, NH_3. This makes for six hydrogen atoms both before and after the reaction arrow. Meanwhile, the coefficient of 2 by the ammonia makes for two nitrogen atoms to the right of the arrow, which balances out the two nitrogen atoms that appear to the left. The full balanced equation is therefore:

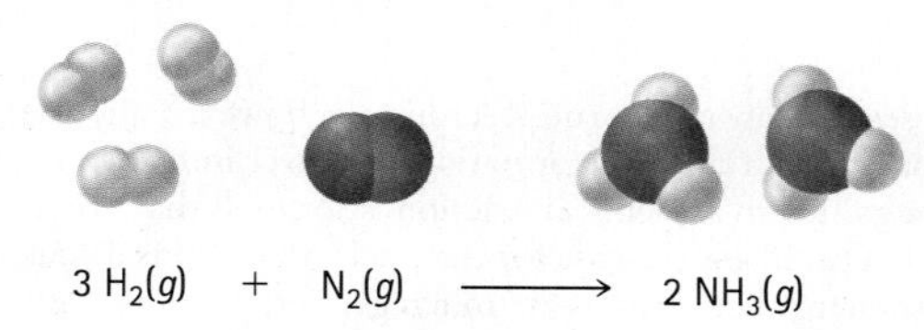

$$3\ H_2(g) + N_2(g) \longrightarrow 2\ NH_3(g)$$

Chemists use many methods to balance equations. Your instructor may share with you his or her favorite methods. For more practice balancing equations, see the questions at the end of this chapter.

13.2 Energy and Chemical Reactions

LEARNING OBJECTIVE
Calculate the amount of energy released or absorbed by a chemical reaction using the bond energies of reactants and products.

EXPLAIN THIS What changes during a chemical reaction?

Once a reaction is complete, there may be either a net release or a net absorption of energy. Reactions in which there is a net release of energy are called **exothermic**. Rocket ships lift off into space and campfires glow red hot as a result of exothermic reactions. Reactions in which there is a net absorption of energy are called **endothermic**. Photosynthesis, for example, involves a series of endothermic reactions that are driven by the energy of sunlight. Both exothermic and endothermic reactions, illustrated in Figure 13.1, can be understood through the concept of bond energy.

During a chemical reaction, chemical bonds are broken and atoms rearrange to form new chemical bonds. Such breaking and forming of chemical bonds involve changes in energy. As an analogy, consider a pair of magnets. To separate them requires an input of "muscle energy." Conversely, when the two magnets come together, they accelerate toward each other. Upon colliding, they become slightly warmer as their kinetic energy is transformed into heat. So, the magnets must absorb energy if they are to break apart, while they release energy as they come together. The same principle applies to atoms. To pull bonded atoms apart requires an energy input. When atoms combine, there is an energy output, usually in the form of faster-moving atoms and molecules, electromagnetic radiation, or both.

I must supply energy to these magnets in order to pull them apart.
Energy is released when they come together!

The amount of energy required to pull two bonded atoms apart is the same as the amount released when they are brought together. This energy, whether it is the energy that is absorbed as a bond breaks or the energy that is released as a bond forms, is called **bond energy**. Each chemical bond has its own characteristic

FIGURE 13.1
Chemical reactions that occur when wood is burning have a net release of energy. Chemical reactions that occur in a photosynthetic plant have a net absorption of energy.

TABLE 13.1 SELECTED BOND ENERGIES

Bond	Bond Energy (kJ/mole)	Bond	Bond Energy (kJ/mole)
H—H	436	N—N	159
H—C	414	O—O	138
H—N	389	Cl—Cl	243
H—O	464	C=O	803
H—F	569	N=O	631
C—O	351	O=O	498
H—Cl	431	C≡C	837
C—C	347	N≡N	946

bond energy. The hydrogen–hydrogen bond energy, for example, is 436 kJ/mole. This means that 436 kJ of energy is absorbed as 1 mole of hydrogen–hydrogen bonds break apart, and 436 kJ of energy is released upon the formation of 1 mole of hydrogen–hydrogen bonds. Different bonds involving different elements have different bond energies, as Table 13.1 shows.

By convention, a positive bond energy represents the amount of energy absorbed as a bond breaks, and a negative bond energy represents the amount of energy released as a bond forms. Thus, when you are calculating the net energy released or absorbed during a reaction, you need to be careful about plus and minus signs. It is standard practice when doing such calculations to assign a plus sign to energy absorbed and a minus sign to energy released. For instance, when dealing with a reaction in which 1 mole of H—H bonds are broken, we write +436 kJ to indicate energy absorbed—this is the energy *gained* by the atoms. When dealing with the formation of 1 mole of H—H bonds, we write −436 kJ to indicate energy released—this is the energy *lost* by the atoms. We'll do some sample calculations in a moment.

Remember, in a chemical reaction, the bonds being formed are different from the bonds that were broken. The bond energies of the bonds being formed, therefore, are also different from the energies of the bonds that were broken.

CHECK YOURSELF

Do all covalent single bonds have the same bond energy?

CHECK YOUR ANSWER

No. Bond energy depends on the types of atoms bonding. The H—H single bond, for example, has a bond energy of 436 kJ/mole, but the H—O single bond has a bond energy of 464 kJ/mole.

UNIFYING CONCEPT

- ***The Law of Conservation of Energy***
  Section 4.10

Exothermic Reaction: Net Release of Energy

For any chemical reaction, the total amount of energy absorbed in breaking bonds in the reactants is always different from the total amount of energy released as bonds form in the products. Consider the reaction in which hydrogen and oxygen react to form water:

$$\text{H—H} + \text{H—H} + \text{O=O} \longrightarrow \text{H—O—H} + \text{H—O—H}$$

In the reactants, hydrogen atoms are bonded to hydrogen atoms, and oxygen atoms are double-bonded to oxygen atoms. The total amount of energy absorbed as these bonds break is +1370 kJ.

Type of Bond	Number of Moles	Bond Energy	Total Energy
H—H	2	+436 kJ/mole	+872 kJ
O=O	1	+498 kJ/mole	+498 kJ
		Total energy absorbed:	+1370 kJ

In the products there are four hydrogen–oxygen bonds. The total amount of energy released as these bonds form is −1856 kJ.

Type of Bond	Number of Moles	Bond Energy	Total Energy
H—O	4	−464 kJ/mole	−1856 kJ
		Total energy released:	−1856 kJ

The amount of energy released in this reaction exceeds the amount of energy absorbed. The net energy of the reaction is found by adding the two quantities:

$$\begin{aligned}\text{Net energy of reaction} &= \text{energy absorbed} + \text{energy released}\\ &= +1370\text{ kJ} + (-1856\text{ kJ})\\ &= -486\text{ kJ}\end{aligned}$$

The negative sign on the net energy indicates that there is a net release of energy, and so the reaction is exothermic. For any exothermic reaction, energy can be considered a product and is thus sometimes included after the arrow of the chemical equation:

$$2\,H_2 + O_2 \longrightarrow 2\,H_2O + \text{energy}$$

In an exothermic reaction, the potential energy of atoms in the product molecules is lower than their potential energy in the reactant molecules. This is illustrated in the reaction profile shown in Figure 13.2. The lowered potential energy of the atoms in the product molecules is due to their being held together more tightly. This is analogous to two attracting magnets whose potential energy decreases as they come closer together. The loss of potential energy is balanced by a gain in kinetic energy. Like two free-floating magnets coming together and

FYI NASA scientists routinely test various materials for their durability against atomic oxygen, O, which is abundant in the low orbit of the space shuttle. They discovered that atomic oxygen effectively transforms surface organic materials into gaseous carbon dioxide. The scientists realized atomic oxygen's usefulness for restoring paintings damaged by smoke or other organic contaminants. Together with art conservationists they used atomic oxygen to restore certain damaged paintings, and it worked spectacularly.

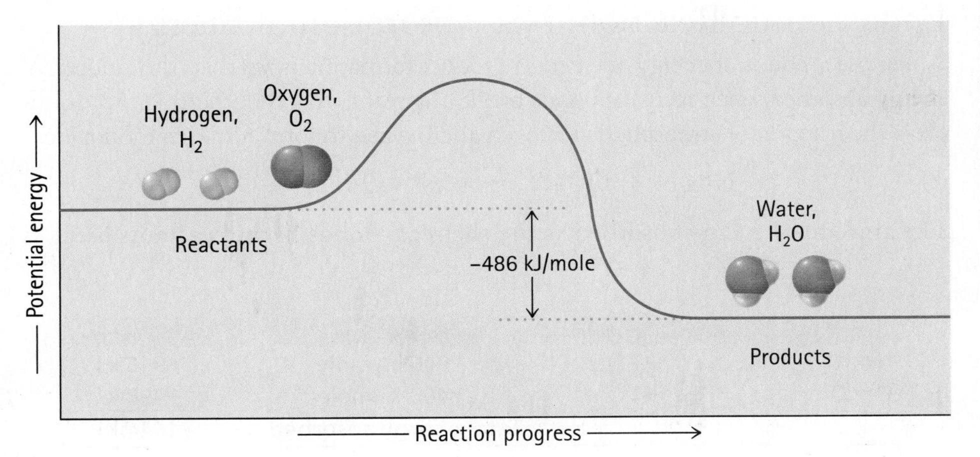

FIGURE 13.2
In an exothermic reaction, the product molecules are at a lower potential energy than the reactant molecules. The net amount of energy released by the reaction is equal to the difference in potential energies of the reactants and products.

accelerating to higher speeds, the potential energy of the reactants is converted into faster-moving atoms and molecules, electromagnetic radiation, or both. This kinetic energy released by the reaction is equal to the difference between the potential energy of the reactants and the potential energy of the products, as is indicated in Figure 13.2.

It is important to understand that the energy released by an exothermic reaction is not created by the reaction. This agrees with the *law of conservation of energy*, which tells us that energy is neither created nor destroyed in a chemical reaction (or any process). Instead, energy is merely converted from one form into another. During an exothermic reaction, energy that was once in the form of the potential energy of chemical bonds is released as the kinetic energy of fast-moving molecules and/or as electromagnetic radiation.

FIGURE 13.3
A space shuttle uses exothermic chemical reactions to lift off from Earth's surface.

The amount of energy released in an exothermic reaction depends on the amounts of the reactants. The reaction of large amounts of hydrogen and oxygen, for example, provides the energy to lift the space shuttle shown in Figure 13.3 into orbit. There are two compartments in the large central tank to which the orbiter is attached—one filled with liquid hydrogen and the other filled with liquid oxygen. Upon ignition, these two liquids mix and react chemically to form water vapor, which produces the needed thrust as it is expelled out the rocket cones. Additional thrust is obtained by a pair of solid-fuel rocket boosters that contain a mixture of ammonium perchlorate, NH_4ClO_4, and powdered aluminum. On ignition, these chemicals react to form products that are expelled at the rear of the rocket. The balanced equation representing this reaction is

$$3\,NH_4ClO_4 + 3\,Al \longrightarrow Al_2O_3 + AlCl_3 + 3\,NO + 6\,H_2O + \text{energy}$$

CHECK YOURSELF
Where does the net energy released in an exothermic reaction go?

CHECK YOUR ANSWER
This energy goes into increasing the speeds of reactant atoms and molecules and often into electromagnetic radiation.

Recall from Chapter 3 that for every action there is an opposite and equal reaction. A rocket is thrust upward, for example, only as its exhaust chemicals are thrust downward.

Endothermic Reaction: Net Absorption of Energy

When the amount of energy released in product formation is *less* than the amount of energy absorbed when reactant bonds break, the reaction is endothermic. An example is the reaction of atmospheric nitrogen and oxygen to form nitrogen monoxide:

$$N{\equiv}N + O{=}O \longrightarrow N{=}O + N{=}O$$

The amount of energy absorbed as the chemical bonds in the reactants break is

Type of Bond	Number of Moles	Bond Energy	Total Energy
N≡N	+1	+946 kJ/mole	+946 kJ
O=O	+1	+498 kJ/mole	+498 kJ
		Total energy absorbed:	+1444 kJ

The amount of energy released upon the formation of bonds in the products is

Type of Bond	Number of Moles	Bond Energy	Total Energy
N=O	2	−631 kJ/mole	−1262 kJ
		Total energy released:	−1262 kJ

As before, the net energy of the reaction is found by adding the two quantities:

$$\begin{aligned}\text{Net energy of reaction} &= \text{energy absorbed} + \text{energy released}\\ &= +1444\text{ kJ} + (-1262\text{ kJ})\\ &= +182\text{ kJ}\end{aligned}$$

The positive sign indicates a net *absorption* of energy, meaning the reaction is endothermic. For any endothermic reaction, energy can be considered a reactant and is thus sometimes included to the left of the arrow in the chemical equation:

$$\text{Energy} + N_2 + O_2 \longrightarrow 2\text{ NO}$$

In an endothermic reaction, the potential energy of atoms in the product molecules is higher than their potential energy in the reactant molecules. This is illustrated in the reaction profile shown in Figure 13.4. Raising the potential energy of the atoms in the product molecules requires a net input of energy, which must come from some external source, such as electromagnetic radiation, electricity, or heat. Thus, nitrogen and oxygen react to form nitrogen monoxide only with the application of much heat, as occurs adjacent to a lightning bolt or in an internal-combustion engine.

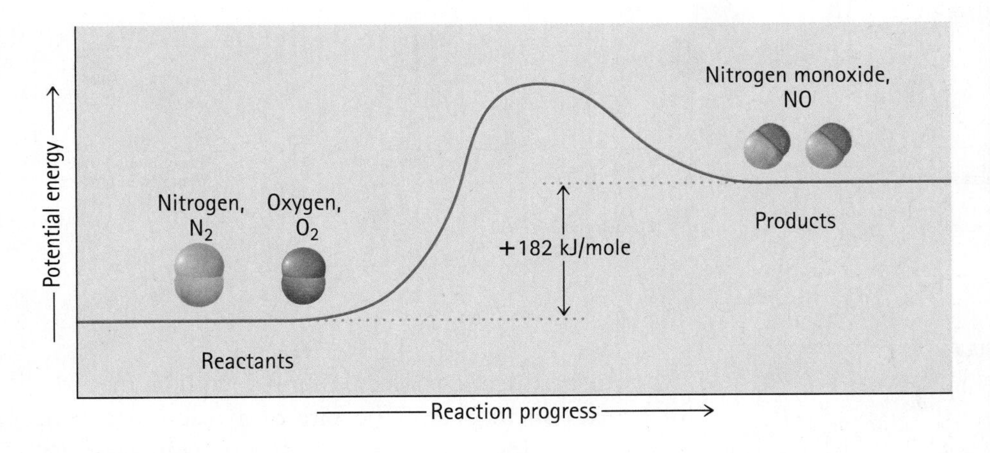

FIGURE 13.4
In an endothermic reaction, the product molecules are at a higher potential energy than the reactant molecules. The net amount of energy absorbed by the reaction is equal to the difference in potential energies of the reactants and products.

13.3 Reaction Rates

LEARNING OBJECTIVE
Describe the requirements that must be met in order for a chemical reaction to occur.

EXPLAIN THIS Why does blowing into a campfire make the fire burn brighter?

Some chemical reactions, such as the rusting of iron, are slow, while others, such as the burning of gasoline, are fast. The speed of any reaction is indicated by its *reaction rate,* which is an indicator of how quickly the reactants transform to products. As shown in Figure 13.5, initially a flask may contain only reactant molecules. Over time, these reactants form product

FIGURE 13.5
Over time, the reactants in this reaction flask may transform to products. If this happens quickly, the reaction rate is high. If this happens slowly, the reaction rate is low.

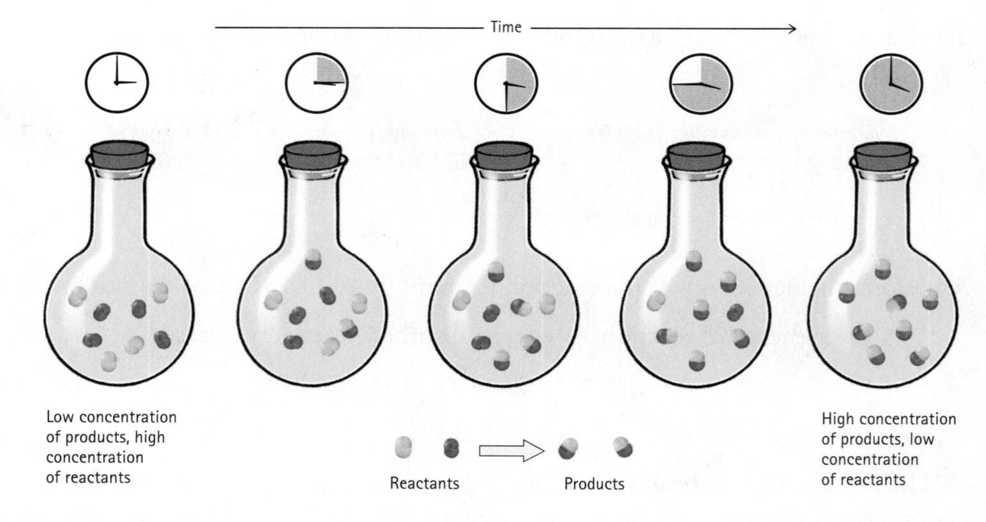

molecules and, as a result, the concentration of product molecules increases. The **reaction rate**, therefore, can be defined either as how quickly the concentration of products increases or as how quickly the concentration of reactants decreases.

What determines the rate of a chemical reaction? The answer is complex, but one important factor is that reactant molecules must physically come together. Because molecules move rapidly, this physical contact is appropriately described as a collision. We can illustrate the relationship between molecular collisions and reaction rate by considering the reaction of gaseous nitrogen and gaseous oxygen to form gaseous nitrogen monoxide, as shown in Figure 13.6.

UNIFYING CONCEPT

- ***The Law of Conservation of Momentum***
  Section 4.4

FIGURE 13.6
INTERACTIVE FIGURE 
During a reaction, reactant molecules collide with one another.

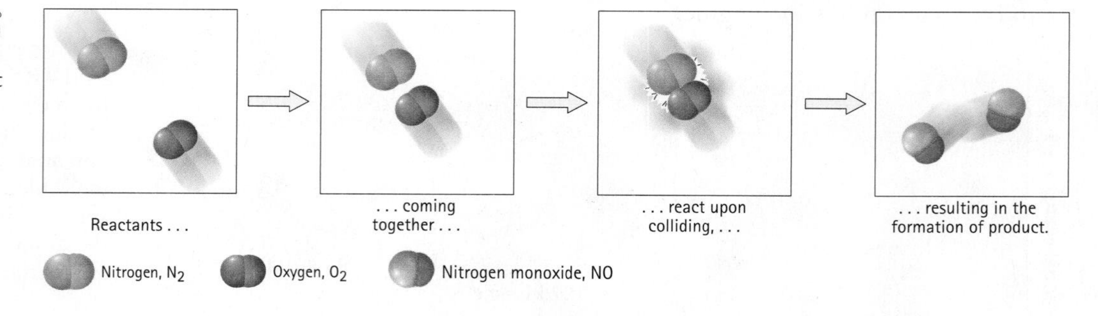

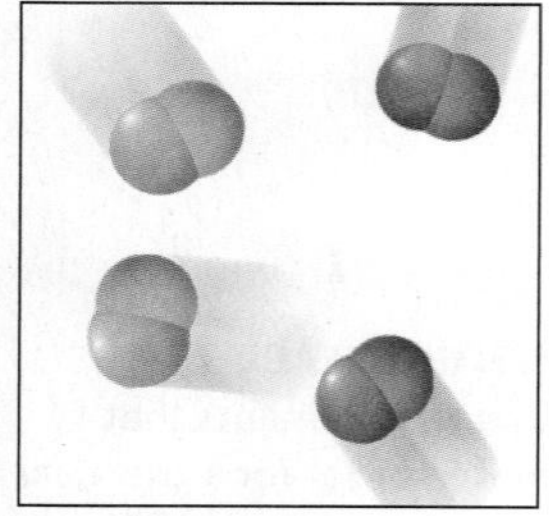

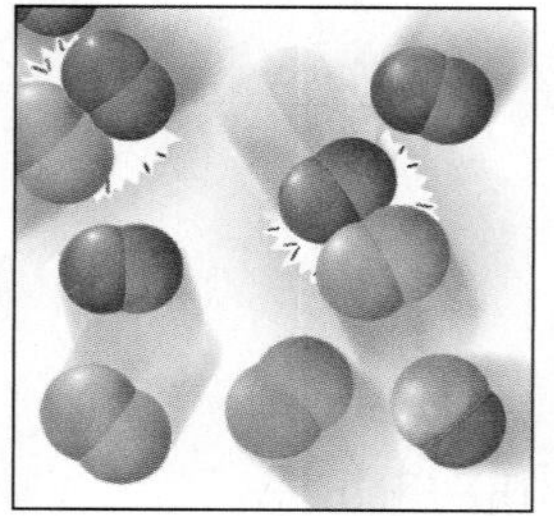

FIGURE 13.7
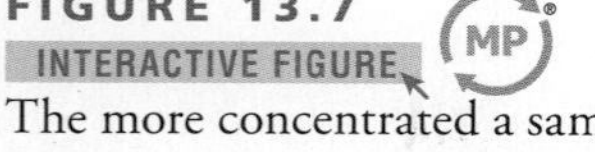

The more concentrated a sample of nitrogen and oxygen, the greater the probability that N_2 and O_2 molecules will collide and form nitrogen monoxide.

Because reactant molecules must collide in order for a reaction to occur, the rate of a reaction can be increased by increasing the number of collisions. An effective way to increase the number of collisions is to increase the concentration of the reactants. Figure 13.7 shows that with higher concentrations, more molecules are in a given volume, which makes collisions between molecules more probable. As an analogy, consider a group of people on a dance floor—as the number of people increases, so does the rate at which they bump into one another. An increase in the concentration of nitrogen and oxygen molecules, therefore, leads to a greater number of collisions between these molecules; hence, a greater number of nitrogen monoxide molecules form in a given period of time.

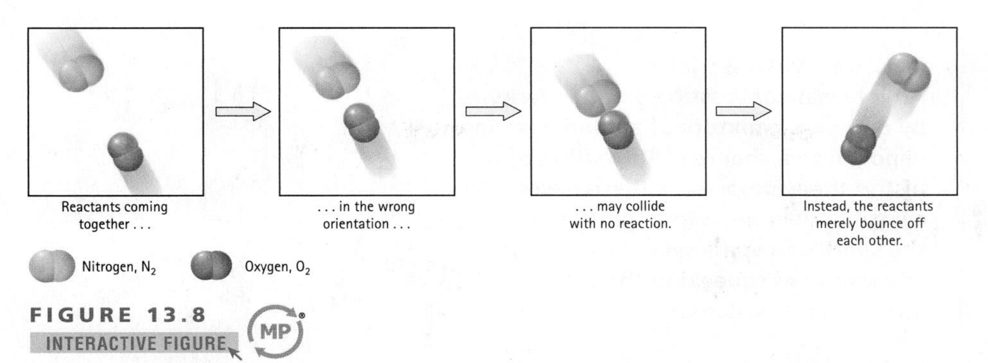
Reactants coming together . . .
. . . in the wrong orientation . . .
. . . may collide with no reaction.
Instead, the reactants merely bounce off each other.
Nitrogen, N_2
Oxygen, O_2

FIGURE 13.8
INTERACTIVE FIGURE

The orientation of reactant molecules in a collision can determine whether a reaction occurs. A perpendicular collision between N_2 and O_2 tends not to result in the formation of a product molecule.

Not all collisions between reactant molecules lead to products, however, because the molecules must collide in a certain orientation in order to react. Nitrogen and oxygen, for example, are much more likely to form nitrogen monoxide when the molecules collide in the parallel orientation shown in Figure 13.6. When they collide in the perpendicular orientation shown in Figure 13.8, nitrogen monoxide does not form. For larger molecules, which can have numerous orientations, this orientation requirement is even more restrictive.

A second reason that not all collisions lead to the formation of products is that the reactant molecules must collide with enough kinetic energy to break their bonds. Only then is it possible for the atoms in the reactant molecules to change bonding partners and form product molecules. The bonds in N_2 and O_2 molecules, for example, are quite strong. In order for these bonds to be broken, collisions between the molecules must contain enough energy to break the bonds. As a result, collisions between slow-moving N_2 and O_2 molecules, even those that collide in the proper orientation, may not form NO, as is shown in Figure 13.9.

The higher the temperature of a material, the faster its molecules move and the more forceful the collisions between them. Higher temperatures, therefore, increase reaction rates. The nitrogen and oxygen molecules that make up our atmosphere, for example, are continuously colliding with one another. At the ambient temperatures of our atmosphere, however, these molecules do not generally have sufficient kinetic energy for the formation of nitrogen monoxide. The heat of a lightning bolt, however, dramatically increases the kinetic energy of these molecules to the point that a large portion of the collisions in the vicinity of the bolt result in the formation of nitrogen monoxide. The nitrogen monoxide formed in this manner undergoes further reactions to form nitrate ions that plants depend on for survival, as was discussed in the opening of this chapter. This is an example of *nitrogen fixation*, which we explore further in Chapter 21.

The life sciences involve fantastic applications of chemistry, with nitrogen fixation being just one example. Others are photosynthesis, cellular respiration, and molecular genetics. So, there are distinct advantages to learning about chemistry *before* advancing to the study of the life sciences.

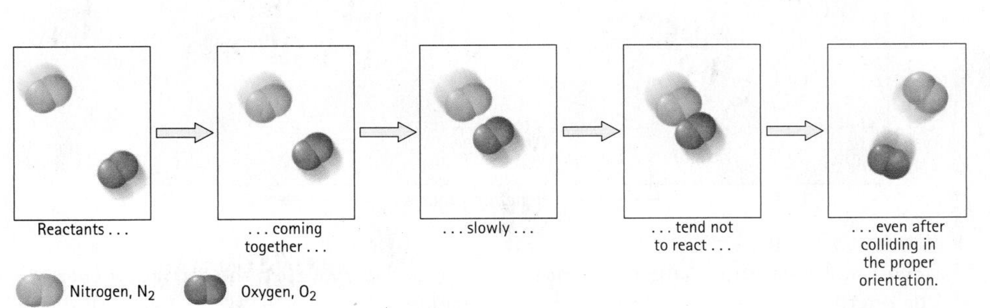
Reactants . . .
. . . coming together . . .
. . . slowly . . .
. . . tend not to react . . .
. . . even after colliding in the proper orientation.
Nitrogen, N_2
Oxygen, O_2

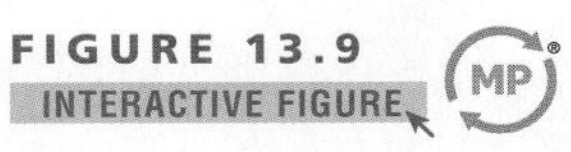
FIGURE 13.9
INTERACTIVE FIGURE

Slow-moving molecules may collide with insufficient force to break their bonds. As a result, they cannot react to form product molecules.

CHECK YOURSELF

An internal-combustion engine works by drawing a mixture of air and gasoline vapors into a chamber. The action of a piston then compresses these gases into a smaller volume before ignition by the spark of a spark plug. What is the advantage of squeezing the vapors into a smaller volume?

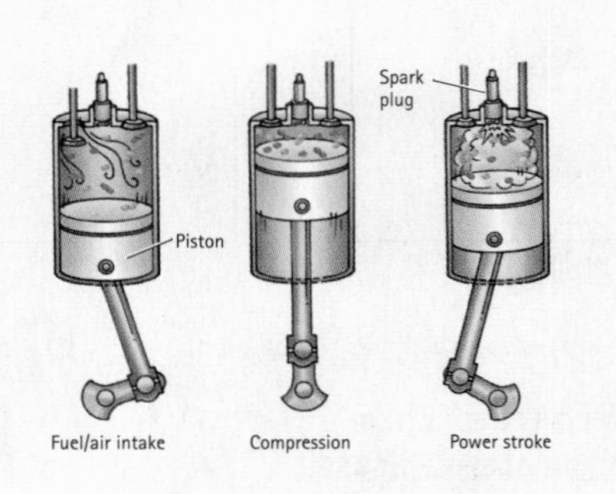

CHECK YOUR ANSWER

Squeezing the vapors into a smaller volume effectively increases their concentration and, hence, the number of collisions between molecules. This, in turn, promotes the chemical reaction. As discussed in Chapter 6, compression also increases the temperature, which further favors the chemical reaction.

The energy required to break bonds can also come from the absorption of electromagnetic radiation. As the radiation is absorbed by reactant molecules, the atoms in the molecules may start to vibrate so rapidly that the bonds between them are easily broken. In many instances, direct absorption of electromagnetic radiation is sufficient to break chemical bonds and to initiate a chemical reaction. The common atmospheric pollutant nitrogen dioxide, NO_2, for example, may transform into nitrogen monoxide and atomic oxygen merely on exposure to sunlight:

$$NO_2 + \text{sunlight} \longrightarrow NO + O$$

Whether they result from collisions, absorption of electromagnetic radiation, or both, broken bonds are a necessary first step in most chemical reactions. The energy required for this initial breaking of bonds can be viewed as an *energy barrier*. The minimum energy required to overcome this energy barrier is known as the **activation energy** (E_a).

In the reaction between nitrogen and oxygen to form nitrogen monoxide, the activation energy is so high (because the bonds in N_2 and O_2 are strong) that only the fastest-moving nitrogen and oxygen molecules possess sufficient energy to react. Figure 13.10 shows the activation energy in this chemical reaction as a vertical hump.

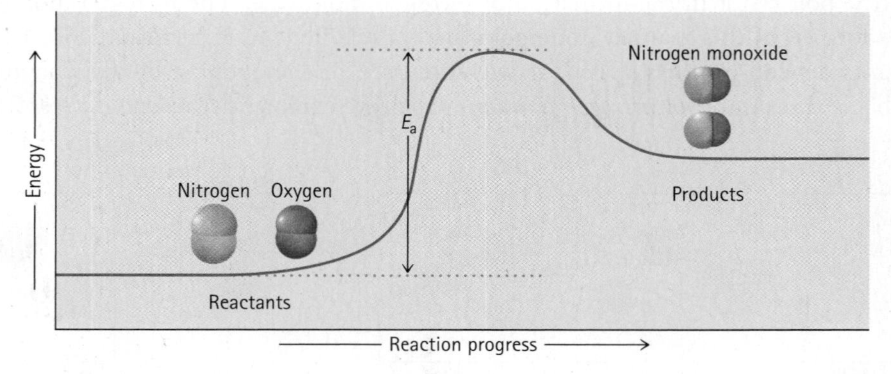

FIGURE 13.10

Reactant molecules must gain a minimum amount of energy, called the activation energy, E_a, before they can transform into product molecules.

The activation energy of a chemical reaction is analogous to the energy a car needs to drive over the top of a hill. Without sufficient energy to climb to the top of the hill, the car cannot get to the other side. Likewise, reactant molecules can transform into product molecules only if the reactant molecules possess an amount of energy equal to or greater than the activation energy.

Kinetic energies not sufficient to overcome energy barrier

Kinetic energies sufficient to overcome energy barrier

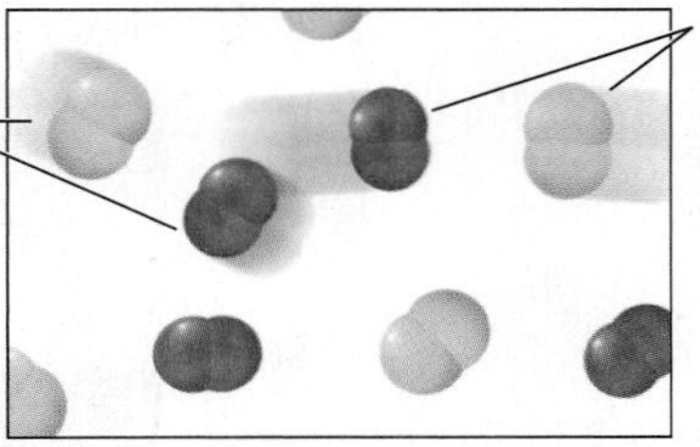

FIGURE 13.11
Because fast-moving reactant molecules possess sufficient energy to pass over the energy barrier, they are the first ones to transform into product molecules.

At any given temperature, there is a wide distribution of kinetic energies in reactant molecules. Some are moving slowly, and others are moving quickly. As we discussed in Chapter 6, the temperature of a material is related to the average of all these kinetic energies. The few fast-moving reactant molecules in Figure 13.11 have enough energy to pass over the energy barrier and are the first to transform into product molecules.

When the temperature of the reactants is increased, the number of reactant molecules possessing sufficient energy to pass over the barrier also increases, which is why reactions are generally faster at higher temperatures. Conversely, at lower temperatures, fewer molecules have sufficient energy to pass over the barrier. Hence, reactions are generally slower at lower temperatures.

CHECK YOURSELF

What kitchen device is used to lower the rate at which microorganisms grow on food?

CHECK YOUR ANSWER

The refrigerator! Microorganisms, such as bread mold, are everywhere and difficult to avoid. By lowering the temperature of microorganism-contaminated food, the refrigerator decreases the rate of the chemical reactions that these microorganisms depend on for growth, thereby increasing the food's shelf life.

In order for two chemicals to be able to react, they must first collide in the proper orientation. Second, they must have sufficient kinetic energy to initiate the breaking of chemical bonds so that new bonds can form.

Most chemical reactions are influenced by temperature in this manner, including reactions that occur in living bodies. The body temperature of animals that regulate their internal temperature, such as humans, is fairly constant. However, the body temperature of some animals, such as the alligator shown in Figure 13.12,

FIGURE 13.12
This alligator became immobilized on the pavement after being caught in the cold night air. By midmorning, shown here, the temperature had warmed sufficiently to allow the alligator to get up and walk away.

rises and falls with the temperature of the environment. On a warm day, the chemical reactions occurring in an alligator are "up to speed," and the animal is more active. On a chilly day, however, the chemical reactions proceed at a lower rate and, as a consequence, the alligator's movements are unavoidably sluggish.

LEARNING OBJECTIVE
Discuss how a catalyst can speed up a chemical reaction, using the destruction of stratospheric ozone as an example.

Catalysts

EXPLAIN THIS Chew a salt-free soda cracker for a few minutes and the cracker begins to taste sweet. Why?

As we have discussed, increasing the concentration or the temperature of the reactants can cause a chemical reaction to proceed faster. A third way to increase the rate of a reaction is to add a **catalyst**, which is any substance that increases the rate of a chemical reaction by lowering its activation energy. The catalyst may participate as a reactant, but it is then regenerated as a product and is thus available to catalyze subsequent reactions.

The conversion of ozone, O_3, to oxygen, O_2, is normally sluggish because the reaction has a relatively high activation energy, as shown in Figure 13.13a. However, when chlorine atoms act as a catalyst, the energy barrier is lowered, as shown in Figure 13.13b, and the reaction can proceed faster.

Chlorine atoms lower the energy barrier of this reaction by providing an alternative pathway involving intermediate reactions, each having a lower activation energy than the uncatalyzed reaction. This alternative pathway involves two steps. Initially, the chlorine reacts with the ozone to form chlorine monoxide and oxygen:

$$\underset{\text{Chlorine}}{Cl} + \underset{\text{Ozone}}{O_3} \longrightarrow \underset{\text{Chlorine monoxide}}{ClO} + \underset{\text{Oxygen}}{O_2}$$

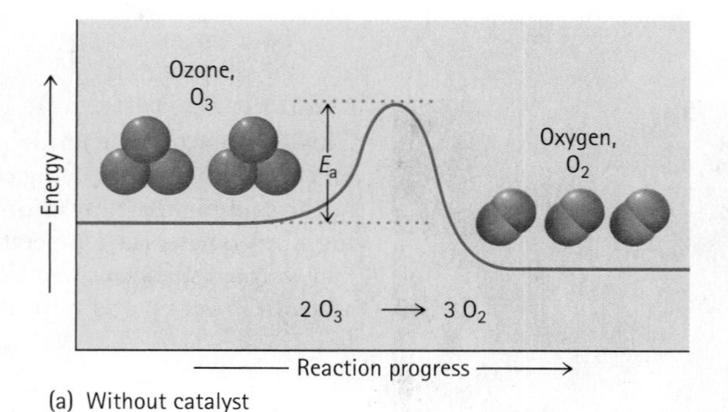

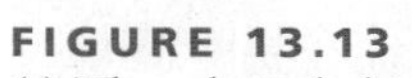

(a) Without catalyst

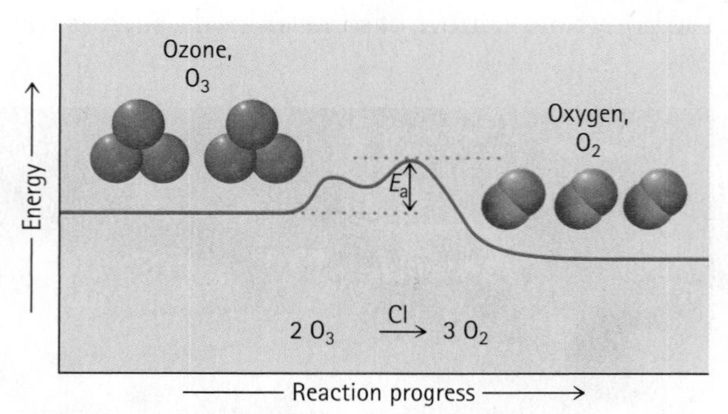

(b) With chlorine catalyst

FIGURE 13.13
(a) The relatively high activation energy (energy barrier) means that only the most energetic ozone molecules can react to form oxygen molecules. (b) The presence of chlorine atoms lowers the activation energy, which means that more reactant molecules have sufficient energy to form product. The chlorine allows the reaction to proceed in two steps, and the two smaller activation energies correspond to these steps. (Note that the convention is to write the name of the catalyst above the reaction arrow.)

The chlorine monoxide then reacts with another ozone molecule to re-form the chlorine atom as well as produce two additional oxygen molecules:

$$\underset{\text{Chlorine monoxide}}{ClO} + \underset{\text{Ozone}}{O_3} \longrightarrow \underset{\text{Chlorine}}{Cl} + \underset{\text{Oxygen}}{2\,O_2}$$

Although chlorine is depleted in the first reaction, it is regenerated in the second reaction. As a result, there is no net consumption of chlorine. At the same time, however, because of the lower energy of activation for these reactions, ozone molecules are rapidly converted into oxygen molecules.

Chlorine atoms in the stratosphere catalyze the destruction of Earth's ozone layer. Evidence indicates that chlorine atoms are generated in the stratosphere as a by-product of human-made chlorofluorocarbons (CFCs), once widely produced as the cooling fluid of refrigerators and air conditioners. Destruction of the ozone layer is a serious concern because of its role in protecting us from the Sun's harmful ultraviolet rays. One chlorine atom in the ozone layer is estimated to catalyze the transformation of 100,000 ozone molecules into oxygen molecules in the one or two years before the chlorine atom is removed by natural processes.

Chemists have been able to harness the power of catalysts for numerous beneficial purposes. The exhaust that comes from an automobile engine, for example, contains a wide assortment of pollutants, such as nitrogen monoxide, carbon monoxide, and uncombusted fuel vapors (hydrocarbons). To reduce the amount of these pollutants entering the atmosphere, most automobiles are equipped with *catalytic converters*, as shown in Figure 13.14. Metal catalysts in a converter speed up reactions that convert exhaust pollutants into less toxic substances. Nitrogen monoxide is transformed into nitrogen and oxygen, carbon monoxide is transformed into carbon dioxide, and unburned fuel is converted into carbon dioxide and water vapor. Because catalysts are not consumed by the reactions they facilitate, a single catalytic converter may continue to operate effectively for the lifetime of the car.

Catalytic converters, along with microchip-controlled fuel–air ratios, have led to a significant drop in the per-vehicle emission of pollutants. This improvement,

FYI Before the fall of the Soviet Union, numerous oil-drilling sites in Siberia were allowed to vent natural gas freely into the atmosphere, presumably because the natural gas had no commercial value. After the fall of the Soviet Union, the wells were capped to prevent this venting. Within weeks, instruments at the Mauna Loa weather observatory on the other side of the planet noted a significant drop in atmospheric levels of methane and its by-product, carbon dioxide. The effect that we humans have on global atmospheric conditions is measurable.

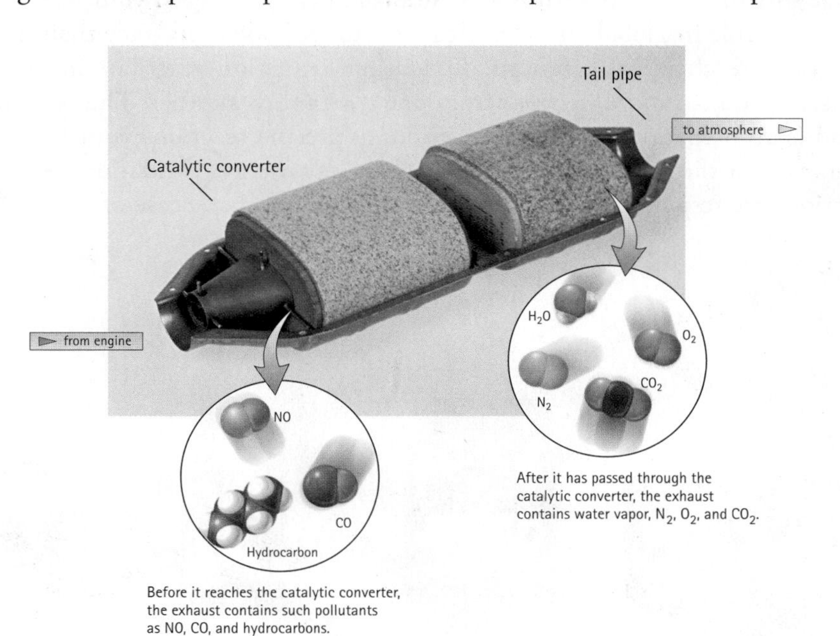
Tail pipe
to atmosphere
Catalytic converter
from engine
H_2O
O_2
CO_2
N_2
NO
CO
Hydrocarbon
After it has passed through the catalytic converter, the exhaust contains water vapor, N_2, O_2, and CO_2.
Before it reaches the catalytic converter, the exhaust contains such pollutants as NO, CO, and hydrocarbons.

FIGURE 13.14
A catalytic converter reduces the pollution caused by automobile exhaust by converting such harmful combustion products as NO, CO, and hydrocarbons to harmless N_2, O_2, and CO_2. The catalyst is typically platinum, Pt; palladium, Pd; or rhodium, Rd.

FIGURE 13.15
The exhaust from automobiles today is much cleaner than before the advent of the catalytic converter, but many more cars are on the road. In 1960, there were about 74 million registered motor vehicles in the United States; by 2008, there were about 250 million.

however, has been offset by an increase in the number of cars being driven, as exemplified by the traffic jam shown in Figure 13.15. It is also offset by the large number of SUVs, which bypass pollution requirements.

The chemical industry depends on catalysts because they lower manufacturing costs by lowering required temperatures and by providing greater product yields without being consumed. Indeed, more than 90% of all manufactured goods are produced with the assistance of catalysts. Without catalysts, the price of gasoline would be much higher, as would be the prices of such consumer goods as rubber, plastics, pharmaceuticals, automobile parts, clothing, and food grown with chemical fertilizers. Living organisms rely on special types of catalysts known as *enzymes*, which allow complex biochemical reactions to occur with ease. This includes the digestion of food and the genetic activity of DNA. (You will learn more about the nature and behavior of enzymes in Chapter 15.)

CHECK YOURSELF
How does a catalyst lower the activation energy of a chemical reaction?

CHECK YOUR ANSWER
The catalyst provides an alternative and easier-to-achieve pathway along which the chemical reaction can proceed.

LEARNING OBJECTIVE
Identify when a chemical behaves like an acid or a base.

13.4 Acids Donate Protons; Bases Accept Them

EXPLAIN THIS Why are many pharmaceuticals treated with hydrogen chloride?

MasteringPhysics®
TUTORIAL: The Nature of Acids and Bases

For the rest of this chapter, we explore two main classes of chemical reactions: acid–base reactions and oxidation–reduction reactions. Acid–base reactions involve the transfer of *protons* from one reactant to another. These sorts of reactions within your stomach help you digest your food. They play a key role in global climate. Most consumer goods can trace their origins to acid–base chemical reactions. Oxidation–reduction reactions involve the transfer of one or more *electrons* from one reactant to another. The burning of wood is an oxidation–reduction reaction, as are the reactions your body uses to transform the food you eat into biochemical energy. Oxidation–reduction reactions are responsible for the rusting of a car and the release of energy from a fuel cell.

(a)

(b)

(c)

(d)

FIGURE 13.16
Examples of acids. (a) Citrus fruits contain many types of acids, including ascorbic acid, $C_6H_8O_6$, which is vitamin C. (b) Vinegar contains acetic acid, $C_2H_4O_2$, and can be used to preserve foods. (c) Many toilet bowl cleaners are formulated with hydrochloric acid, HCl. (d) All carbonated beverages contain carbonic acid, H_2CO_3; many also contain phosphoric acid, H_3PO_4.

The term *acid* comes from the Latin *acidus*, which means "sour." The sour taste of vinegar and citrus fruits is due to the presence of acids. Acids are essential in the chemical industry. For example, more than 85 billion pounds of sulfuric acid are produced annually in the United States, making this the number-one manufactured chemical. Sulfuric acid is used to make fertilizers, detergents, paint dyes, plastics, pharmaceuticals, and storage batteries, as well as to produce iron and steel. Sulfuric acid is so important in the manufacturing of goods that its production is considered a standard measure of a nation's industrial strength. Figure 13.16 shows only a few of the acids we commonly encounter.

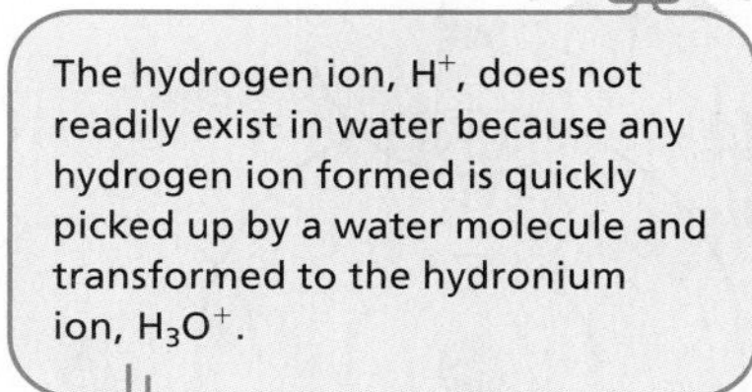

The hydrogen ion, H^+, does not readily exist in water because any hydrogen ion formed is quickly picked up by a water molecule and transformed to the hydronium ion, H_3O^+.

Bases are characterized by their bitter taste and slippery feel. It is interesting that bases themselves are not slippery. Rather, they cause skin oils to transform into slippery solutions of soap. Most commercial preparations for unclogging drains contain sodium hydroxide, NaOH (also known as lye), which is extremely basic and hazardous when concentrated. Bases are also heavily used in industry. Each year in the United States, about 25 billion pounds of sodium hydroxide are manufactured for use in the production of various chemicals and in the pulp and paper industry. Solutions containing bases are often called *alkaline*, a term derived from the Arabic *al-qali* ("the ashes"). Ashes are slippery when wet because of the presence of the base potassium carbonate, K_2CO_3. Figure 13.17 shows some familiar bases.

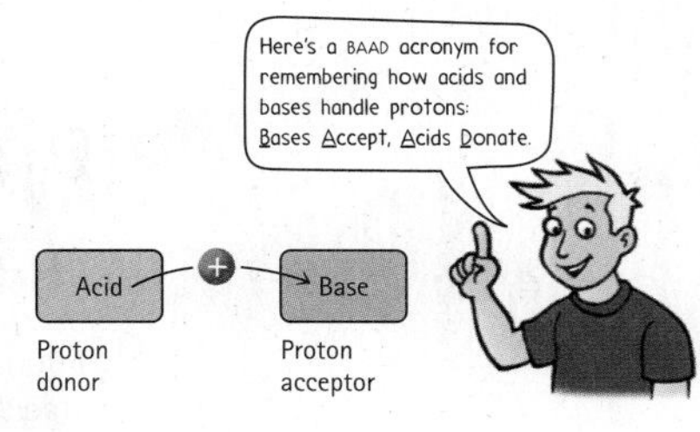

Here's a BAAD acronym for remembering how acids and bases handle protons: Bases Accept, Acids Donate.
Acid
Base
Proton donor
Proton acceptor

We can define an **acid** as any chemical that donates a hydrogen ion, H^+, and a **base** as any chemical that accepts a hydrogen ion. Recall from Chapter 12 that, because a hydrogen atom consists of one electron surrounding a one-proton nucleus, a hydrogen ion formed from the loss of an electron is nothing more than a lone proton. Thus, it is also sometimes said that an acid is a chemical that donates a proton and a base is a chemical that accepts a proton. Consider what happens when hydrogen chloride is mixed into water:

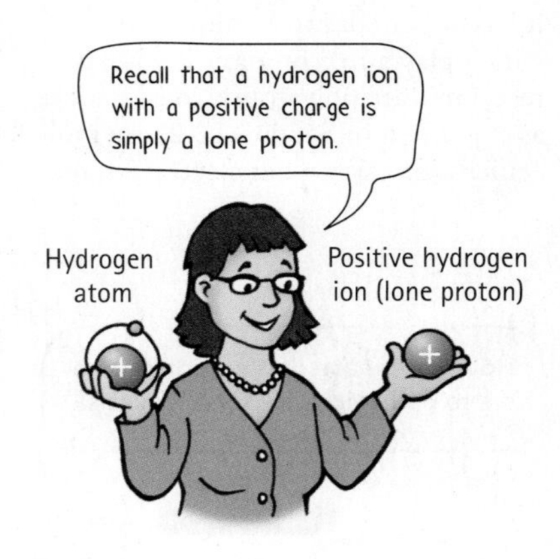

Recall that a hydrogen ion with a positive charge is simply a lone proton.
Hydrogen atom
Positive hydrogen ion (lone proton)

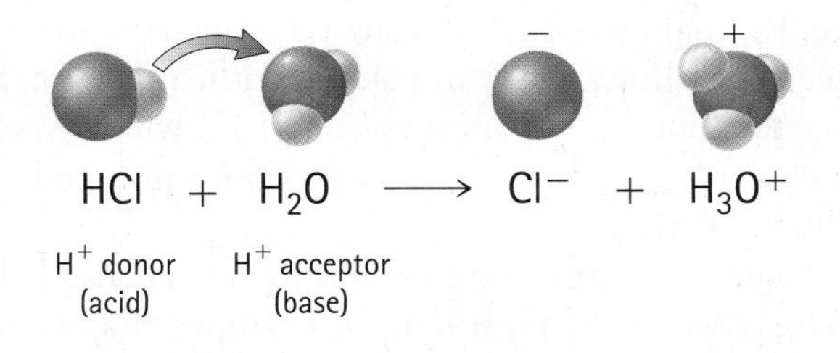

$$HCl + H_2O \longrightarrow Cl^- + H_3O^+$$

(a) (b) (c) (d)

FIGURE 13.17
Examples of bases. (a) Reactions involving sodium bicarbonate, $NaHCO_3$, cause baked goods to rise. (b) Ashes contain potassium carbonate, K_2CO_3. (c) Soap is made by reacting bases with animal or vegetable oils. The soap itself, then, is slightly alkaline. (d) Powerful bases, such as sodium hydroxide, NaOH, are used in drain cleaners.

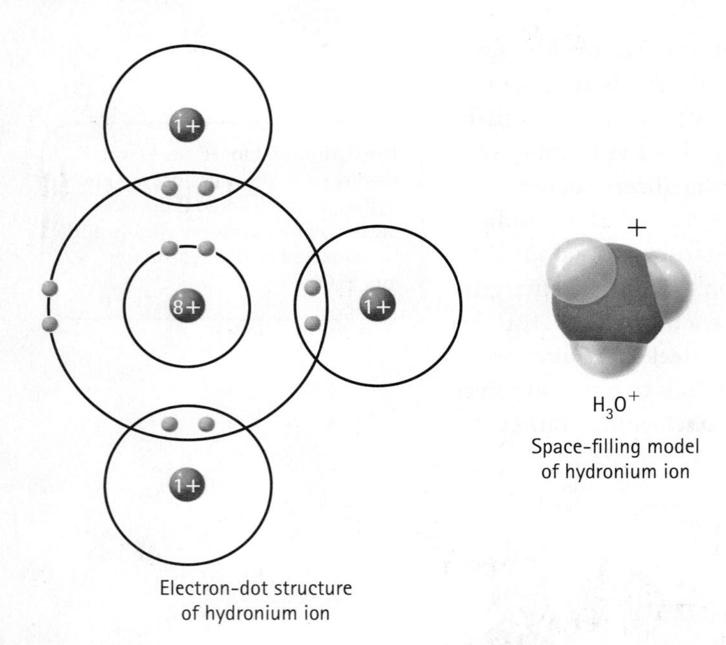

Total protons	11+
Total electrons	10−
Net charge	1+

FIGURE 13.18
The hydronium ion's positive charge is a consequence of the extra proton it has acquired. Hydronium ions, which play a role in many acid–base reactions, are polyatomic ions, which, as explained in Section 12.2, are molecules that carry a net electric charge.

Hydrogen chloride donates a hydrogen ion to one of the nonbonded electron pairs on a water molecule, resulting in a third hydrogen bonded to the oxygen. In this case, hydrogen chloride behaves as an acid (proton donor) and water behaves as a base (proton acceptor). The products of this reaction are a chloride ion and a **hydronium ion**, H_3O^+, which, as Figure 13.18 shows, is a water molecule with an extra proton.

When added to water, ammonia behaves as a base as its nonbonded electrons accept a hydrogen ion from water, which, in this case, behaves as an acid:

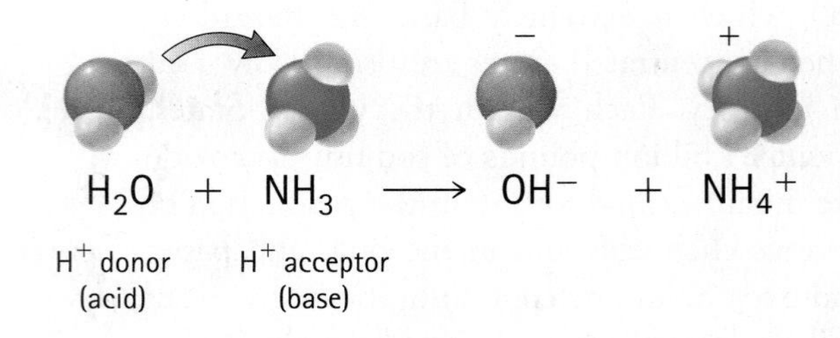

This reaction results in the formation of an ammonium ion and a **hydroxide ion**, which, as shown in Figure 13.19, is a water molecule without the nucleus of one of the hydrogen atoms.

Note that a substance is defined as an acid or a base according to its *behavior*. We say, for example, that hydrogen chloride *behaves* as an acid when mixed with water, which *behaves* as a base. Similarly, ammonia *behaves* as a base when mixed with water, which under this circumstance *behaves* as an acid. Because acid–base is seen as a behavior, there is really no contradiction when a chemical like water behaves as a base in one instance but as an acid in another instance. By analogy, consider yourself. You are who you are, but your behavior changes depending on whom you are with. Likewise, it is a chemical property of water to behave as a base (to accept H^+) when mixed with hydrogen chloride and to behave as an acid (to donate H^+) when mixed with ammonia.

How we behave depends on whom we're with. Likewise for chemicals.

The products of an acid–base reaction can also behave as acids or as bases. An ammonium ion, for example, may donate a hydrogen ion back to a hydroxide ion to re-form ammonia and water:

8+
1+
Electron-dot structure
of hydroxide ion

OH^-
Space-filling model
of hydroxide ion

Total protons	9+
Total electrons	10−
Net charge	1−

FIGURE 13.19
Hydroxide ions have a net negative charge, which is a consequence of having lost a proton. Like hydronium ions, they play a part in many acid–base reactions.

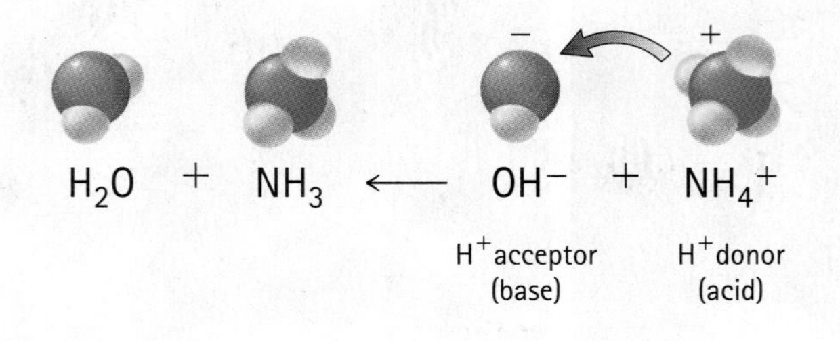

Forward and reverse acid–base reactions proceed simultaneously and can therefore be represented as occurring at the same time by using two oppositely facing arrows:

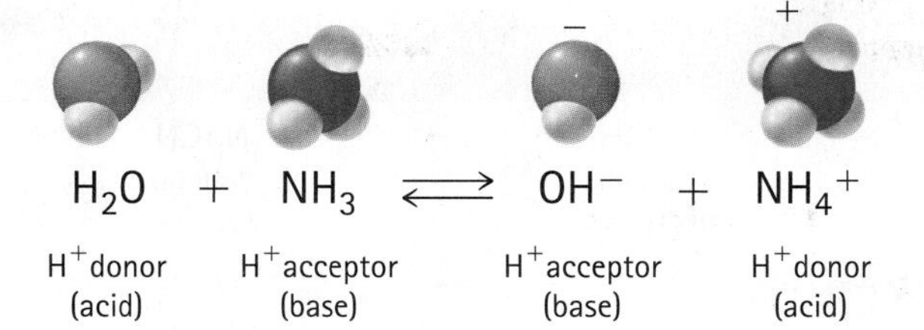

When the equation is viewed from left to right, the ammonia behaves as a base because it accepts a hydrogen ion from the water, which therefore acts as an acid. Viewed in the reverse direction, the equation shows that the ammonium ion behaves as an acid because it donates a hydrogen ion to the hydroxide ion, which therefore behaves as a base.

CHECK YOURSELF

Identify the acid or base behavior of each participant in the reaction:

$$H_2PO_4^- + H_3O^+ \rightleftharpoons H_3PO_4 + H_2O$$

CHECK YOUR ANSWER

In the forward reaction (left to right), $H_2PO_4^-$ gains a hydrogen ion to become H_3PO_4. In accepting the hydrogen ion, $H_2PO_4^-$ is behaving as a base. It gets the hydrogen ion from the H_3O^+, which is behaving as an acid. In the reverse direction, H_3PO_4 loses a hydrogen ion to become $H_2PO_4^-$ and is thus behaving as an acid. The recipient of the hydrogen ion is H_2O, which is behaving as a base as it transforms into H_3O^+.

A Salt Is the Ionic Product of an Acid–Base Reaction

In everyday language, the word *salt* implies sodium chloride, NaCl, table salt. In the language of chemistry, however, **salt** is a general term meaning any ionic compound formed from the reaction between an acid and a base. Hydrogen chloride and sodium hydroxide, for example, react to produce the salt sodium chloride and water:

HCl	+	NaOH	⟶	NaCl	+	H_2O
Hydrogen chloride (acid)		Sodium hydroxide (base)		Sodium chloride (salt)		Water

Similarly, the reaction between hydrogen chloride and potassium hydroxide yields the salt potassium chloride and water:

HCl	+	KOH	⟶	KCl	+	H_2O
Hydrogen chloride (acid)		Potassium hydroxide (base)		Potassium chloride (salt)		Water

Potassium chloride is the main ingredient in "salt-free" table salt, as noted in Figure 13.20.

Salts are generally far less corrosive than the acids and bases from which they are formed. A corrosive chemical has the power to disintegrate a material or wear

FIGURE 13.20
"Salt-free" table salt substitutes contain potassium chloride in place of sodium chloride. Caution is advised in using these products, however, because excessive quantities of potassium salts can lead to serious illness. Furthermore, sodium ions are a vital component of our diet and should never be totally excluded. For a good balance of these two important ions, you might inquire about commercially available half-and-half mixtures of sodium chloride and potassium chloride, such as the one shown here.

TABLE 13.2 ACID–BASE REACTIONS AND THE SALTS FORMED

Acid		Base		Salt		Water
HCN Hydrogen cyanide	+	NaOH Sodium hydroxide	$\longrightarrow$	NaCN Sodium cyanide	+	H_2O
HNO_3 Nitric acid	+	KOH Potassium hydroxide	$\longrightarrow$	KNO_3 Potassium nitrate	+	H_2O
2 HCl Hydrogen chloride	+	$Ca(OH)_2$ Calcium hydroxide	$\longrightarrow$	$CaCl_2$ Calcium chloride	+	2 H_2O
HF Hydrogen fluoride	+	NaOH Sodium hydroxide	$\longrightarrow$	NaF Sodium fluoride	+	H_2O

away its surface. Hydrogen chloride is a remarkably corrosive acid, which makes it useful for cleaning toilet bowls and etching metal surfaces. Sodium hydroxide is a very corrosive base used for unclogging drains. Mixing hydrogen chloride and sodium hydroxide together in equal portions, however, produces an aqueous solution of sodium chloride—salt water, which is not nearly as destructive as either starting material.

There are as many salts as there are acids and bases. Sodium cyanide, NaCN, is a deadly poison. "Saltpeter," which is potassium nitrate, KNO_3, is useful as a fertilizer and in the formulation of gunpowder. Calcium chloride, $CaCl_2$, is commonly used to de-ice walkways, and sodium fluoride, NaF, helps prevent tooth decay. The acid–base reactions that form these salts are shown in Table 13.2.

The reaction between an acid and a base is called a **neutralization** reaction. As can be seen in the color-coding of the neutralization reactions in Table 13.2, the positive ion of a salt comes from the base and the negative ion comes from the acid. The remaining hydrogen and hydroxide ions join to form water.

FYI What makes one acid strong and another weak? Briefly, it involves the stability of the negative ion that remains after the proton has been donated. Hydrogen chloride is a strong acid because the chloride ion can accommodate the negative charge rather well. Acetic acid, however, is a weaker acid because the resulting oxygen ion is less able to accommodate the negative charge.

Not all neutralization reactions result in the formation of water. In the presence of hydrogen chloride, for example, the drug pseudoephedrine behaves as a base by accepting H^+ from a hydrogen chloride. The negative Cl^- then joins the pseudoephedrine$-H^+$ ion to form the salt pseudoephedrine hydrochloride, which is a nasal decongestant, shown in Figure 13.21. This salt is soluble in water and can be absorbed through the digestive system.

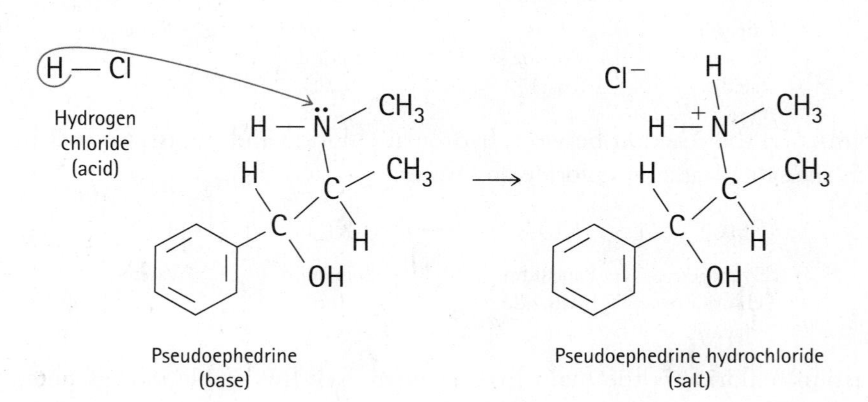

FIGURE 13.21
Hydrogen chloride and pseudoephedrine react to form the salt *pseudoephedrine hydrochloride*, which, because of its solubility in water, is readily absorbed into the body.

CHECK YOURSELF

Is a neutralization reaction best described as a physical change or a chemical change?

CHECK YOUR ANSWER

New chemicals are formed during a neutralization reaction, which means that the reaction is a chemical change.

13.5 Acidic, Basic, and Neutral Solutions

LEARNING OBJECTIVE
Calculate the pH of a solution given its hydronium ion concentration.

EXPLAIN THIS Why can't water be absolutely pure?

A substance whose ability to behave as an acid is about the same as its ability to behave as a base is said to be **amphoteric**. Water is a good example. Because it is amphoteric, water has the ability to react with itself. In behaving as an acid, a water molecule donates a hydrogen ion to a neighboring water molecule, which in accepting the hydrogen ion is behaving as a base. This reaction produces a hydroxide ion and a hydronium ion, which react together to re-form the water:

$$H_2O + H_2O \rightleftharpoons OH^- + H_3O^+$$

Water | Water | Hydroxide ion | Hydronium ion

From this reaction we can see that, in order for a water molecule to gain a hydrogen ion, a second water molecule must lose a hydrogen ion. This means that for every one hydronium ion formed, there is also one hydroxide ion formed. In pure water, therefore, the total number of hydronium ions must be the same as the total number of hydroxide ions. Experiments reveal that the concentration of hydronium and hydroxide ions in pure water is extremely low—about 0.0000001 M for each, where M stands for molarity or moles per liter (see Chapter 12). Water by itself, therefore, behaves as a very weak acid as well as a very weak base.

FYI The outer surface of hair is made of microscopic scale-like structures called cuticles that, like window shutters, can open and close. A beautician can control how long hair retains artificial coloring by modifying the pH of the hair-coloring solution. With an acidic solution, the cuticles close shut so that the dye binds to only the outside of each shaft of hair. This results in a temporary hair coloring, which may come off with the next hair washing. With an alkaline solution, the cuticles open up so that the dye can penetrate into the hair for a more permanent effect.

CHECK YOURSELF

Do water molecules react with one another?

CHECK YOUR ANSWER

Yes, but not to any large extent. When water molecules do react, they form hydronium and hydroxide ions. (*Note*: Make sure you understand this point because it serves as a basis for understanding the pH scale.)

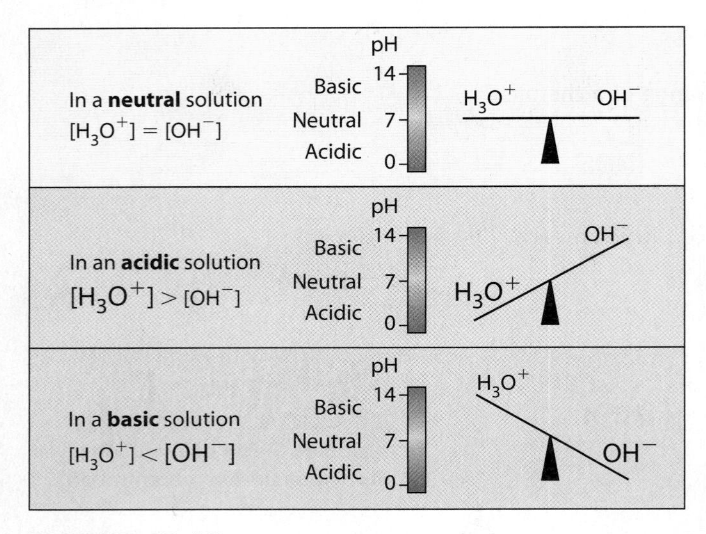

FIGURE 13.22
The relative concentrations of hydronium and hydroxide ions determine whether a solution is acidic, basic, or neutral.

An aqueous solution can be described as neutral, acidic, or basic. A **neutral** solution is one in which the hydronium ion concentration equals the hydroxide ion concentration. Pure water is an example of a neutral solution—not because it contains so few hydronium and hydroxide ions, but because it contains *equal* numbers of them. A neutral solution is also obtained when equal quantities of acid and base are combined, which is why acids and bases are said to *neutralize* each other.

An **acidic** solution is one in which the hydronium ion concentration is higher than the hydroxide ion concentration. An acidic solution is made by adding an acid to water. The effect of this addition is to increase the concentration of hydronium ions. Interestingly, the excess amounts of hydronium ions have the effect of neutralizing the hydroxide ions. Thus, as the hydronium ion concentration increases, the hydroxide ion concentration necessarily decreases, as is depicted in Figure 13.22.

A **basic** solution is one in which the hydroxide ion concentration is higher than the hydronium ion concentration. A basic solution is made by adding a base to water. This addition increases the concentration of hydroxide ions. Notably, the excess amounts of hydroxide ions have the effect of neutralizing the hydronium ions. Thus, as the hydroxide ion concentration increases, the hydronium ion concentration necessarily decreases, as is depicted in Figure 13.22.

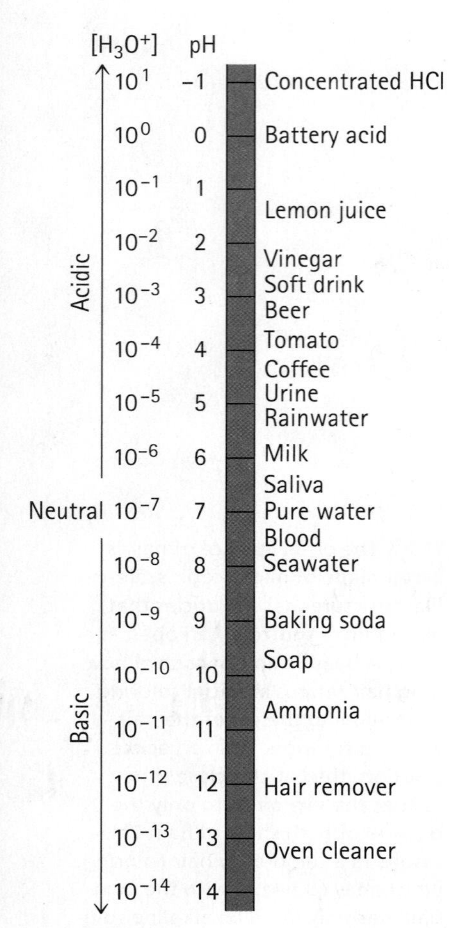

FIGURE 13.23
The pH values of some common solutions.

CHECK YOURSELF

How does adding ammonia, NH_3, to water make a basic solution when there are no hydroxide ions in the formula for ammonia?

CHECK YOUR ANSWER

Ammonia indirectly increases the hydroxide ion concentration by reacting with water:

$$NH_3 + H_2O \longrightarrow NH_4^+ + OH^-$$

This reaction raises the hydroxide ion concentration, which has the effect of lowering the hydronium ion concentration. With the hydroxide ion concentration now higher than the hydronium ion concentration, the solution is basic.

The pH Scale Is Used to Describe Acidity

The *pH scale* is a numeric scale used to express the acidity of a solution. Mathematically, **pH** is equal to the negative logarithm of the hydronium ion concentration:

$$pH = -\log[H_3O^+]$$

Note that brackets are used to represent molar concentrations, so $[H_3O^+]$ is read "the molar concentration of hydronium ions." For help with understanding the logarithm function, see the Math Connection.

Consider a neutral solution that has a hydronium ion concentration of 1.0×10^{-7} M. To find the pH of this solution, we first take the logarithm of this value, which is −7. The pH is by definition the negative of this value, which means $-(-7) = 7$. Hence, in a neutral solution, in which the hydronium ion concentration equals 1.0×10^{-7} M, the pH is 7.

Acidic solutions have pH values lower than 7. For an acidic solution in which the hydronium ion concentration is 1.0×10^{-4} M, for example, $pH = -\log(1.0 \times 10^{-4}) = 4$. The more acidic a solution is, the higher its hydronium ion concentration and the lower its pH.

Basic solutions have pH values higher than 7. For a basic solution in which the hydronium ion concentration is 1.0×10^{-8} M, for example, $pH = -\log(1.0 \times 10^{-8}) = 8$. The more basic a solution is, the lower its hydronium ion concentration and the higher its pH.

Figure 13.23 shows typical pH values of some familiar solutions, and Figure 13.24 shows two common ways of determining pH values.

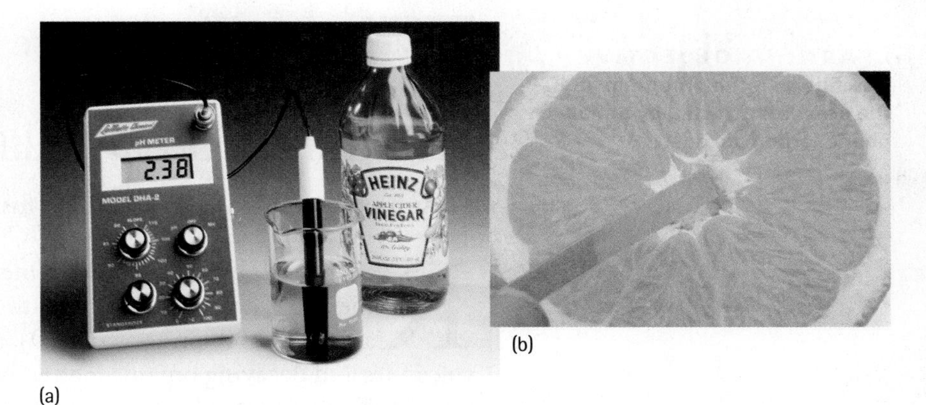

FIGURE 13.24
(a) The pH of a solution can be measured electronically using a pH meter. (b) A rough estimate of the pH of a solution can be obtained with pH paper, which is coated with a dye that changes color with pH.

MasteringPhysics®
TUTORIAL: The pH Scale

MATH CONNECTION

Logarithms and pH

The logarithm of a number can be found on any scientific calculator by keying in the number and pressing the [log] button. The calculator finds the power to which 10 is raised to give the number. The logarithm of 10^2, for example, is 2 because that is the power to which 10 is raised to give the number 10^2. If you know that 10^2 is equal to 100, then you'll understand that the logarithm of 100 also is 2. Check this out on your calculator. Similarly, the logarithm of 1000 is 3 because 10 raised to the third power, 10^3, equals 1000.

Any positive number, including a very small one, has a logarithm. The logarithm of 0.0001, which equals 10^{-4}, for example, is −4 (the power to which 10 is raised to equal this number).

Problem 1 What is the logarithm of 0.01?

Solution The number 0.01 is 10^{-2}, the logarithm of which is −2 (the power to which 10 is raised).

The concentration of hydronium ions in most solutions is typically much lower than 1 M. Recall, for example, that in neutral water the hydronium ion concentration is 0.0000001 M (10^{-7} M). The logarithm of any number less than 1 (but greater than zero) is a negative number. The definition of pH includes the minus sign so as to transform the logarithm of the hydronium ion concentration to a positive number.

When a solution has a hydronium ion concentration of 1 M, the pH is 0 because 1 M = 10^0 M. A 10 M solution has a pH of −1 because 10 M = 10^1 M.

Problem 2 What is the pH of a solution that has a hydronium ion concentration of 0.001 M?

Solution The number 0.001 is 10^{-3}, so

$$\begin{aligned} pH &= -\log[H_3O^+] \\ &= -\log 10^{-3} \\ &= -(-3) = 3 \end{aligned}$$

Problem 3 What is the logarithm of 10^5?

Solution What is the logarithm of 10^5? can be restated as To what power is 10 raised to give the number 10^5? The answer is 5.

Problem 4 What is the logarithm of 100,000?

Solution You should know that 100,000 is the same as 10^5. Thus the logarithm of 100,000 is 5.

Problem 5 What is the pH of a solution that has a hydronium ion concentration of 10^{-9} M? Is this solution acidic, basic, or neutral?

Solution The pH is 9, which means this is a basic solution:

$$\begin{aligned} pH &= -\log[H_3O^+] \\ &= -\log 10^{-9} \\ &= -(-9) \\ &= 9 \end{aligned}$$

LEARNING OBJECTIVE
Describe how the pH of rain and the oceans is affected by atmospheric carbon dioxide.

Integrated Science 13B
EARTH SCIENCE

Acid Rain and Ocean Acidification

EXPLAIN THIS How does burning fossil fuels lower the pH of the ocean?

Rainwater is naturally acidic. One source of this acidity is carbon dioxide, the same gas that gives fizz to soft drinks. The atmosphere contains about 829 billion tons of CO_2, most of it from such natural sources as volcanoes and decaying organic matter, but a growing amount (about 230 billion tons) is generated from human activities.

Water in the atmosphere reacts with carbon dioxide to form *carbonic acid*:

$$\underset{\text{Carbon dioxide}}{CO_2(g)} + \underset{\text{Water}}{H_2O(l)} \longrightarrow \underset{\text{Carbonic acid}}{H_2CO_3(aq)}$$

Carbonic acid, as its name implies, behaves as an acid and lowers the pH of water. The CO_2 in the atmosphere brings the pH of rainwater to about 5.6—noticeably lower than the neutral pH value of 7. Because of local fluctuations, the normal pH of rainwater varies between 5 and 7. This natural acidity of rainwater may accelerate the erosion of land, and under certain circumstances it can lead to the formation of underground caves.

By convention, *acid rain* is a term used for rain with a pH lower than 5. Acid rain is created when airborne pollutants, such as sulfur dioxide, are absorbed by atmospheric moisture. Sulfur dioxide is readily converted into sulfur trioxide, which reacts with water to form *sulfuric acid*:

$$\underset{\text{Sulfur dioxide}}{2\,SO_2(g)} + \underset{\text{Oxygen}}{O_2(g)} \longrightarrow \underset{\text{Sulfur trioxide}}{2\,SO_3(g)}$$

$$\underset{\text{Sulfur trioxide}}{SO_3(g)} + \underset{\text{Water}}{H_2O(l)} \longrightarrow \underset{\text{Sulfuric acid}}{H_2SO_4(aq)}$$

Each year about 20 million tons of SO_2 are released into the atmosphere by the combustion of sulfur-containing coal and oil. Sulfuric acid is much stronger than carbonic acid, and, as a result, rain laced with sulfuric acid eventually corrodes metal, paint, and other exposed substances. Each year, the damage costs billions of dollars. The cost to the environment is also high (Figure 13.25). Many rivers and lakes receiving acid rain become less capable of sustaining life. Much vegetation that receives acid rain doesn't survive. This is particularly evident in heavily industrialized regions.

Acid rain from industrial air pollution remains a serious problem in many regions of the world. Significant progress, however, has been made toward fixing the problem. In the United States, for example, sulfur dioxide and nitrogen oxide emissions have been reduced by nearly half since 1980. Also, in 2005 the U.S. Environmental Protection Agency (EPA) implemented the Clean Air Interstate Rule (CAIR), which is designed to reduce levels of these pollutants even further, especially for areas downwind of heavily industrialized regions. An ultimate long-term solution, however, would be a shift from using fossil fuels to developing more energy sources such as nuclear and solar.

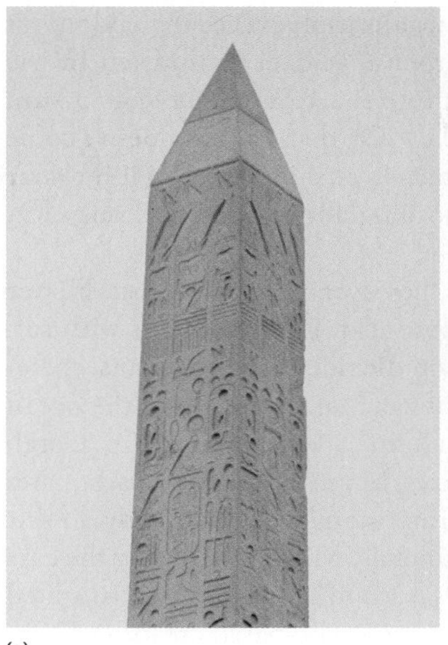

(a)

(b)

FIGURE 13.25
(a) These two photographs show the same obelisk in New York City's Central Park before and after the effects of acid rain. (b) Many forests downwind from heavily industrialized areas, such as in the northeastern United States and in Europe, have been noticeably hard hit by acid rain.

CHECK YOURSELF

When sulfuric acid, H_2SO_4, is added to water, what makes the resulting aqueous solution corrosive?

CHECK YOUR ANSWER

Because H_2SO_4 is a strong acid, it readily forms hydronium ions when dissolved in water. Hydronium ions are responsible for the corrosive action.

When speaking of atmospheric pollutants such as sulfur dioxide, we talk in terms of millions of tons. The amount of carbon dioxide we pump into the atmosphere, however, is measured in *billions* of tons. This carbon dioxide will never lower the pH of rain as much as sulfur dioxide does over a small area. But because there is so much of it, the impact around the entire world can be significant, especially for the oceans.

Consider that the amount of carbon dioxide put into the atmosphere by human activities is growing, as shown in Figure 13.26. Surprisingly, however, the atmospheric concentration of CO_2 is not increasing proportionately. One explanation has to do with the oceans, as described in Figure 13.27. When atmospheric CO_2 dissolves in any body of water—a raindrop, a lake, or the ocean—it forms carbonic acid, H_2CO_3. In fresh water, this carbonic acid transforms back to water and carbon dioxide, which is released back into the atmosphere. In the ocean, however, the carbonic acid is quickly neutralized by dissolved alkaline minerals. (The ocean is

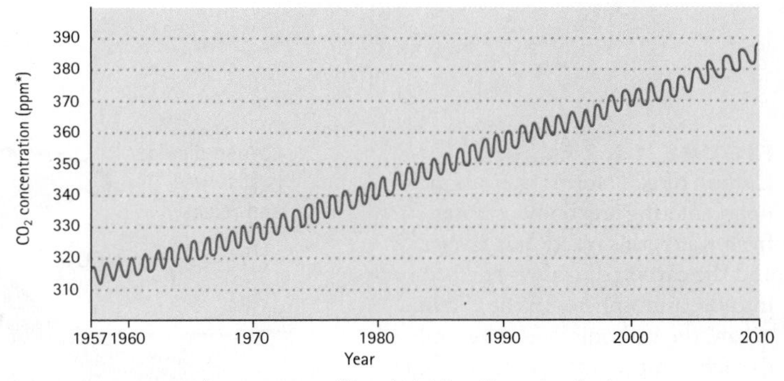

*ppm = parts per million, which tells us the number of carbon dioxide molecules for every million molecules of air

FIGURE 13.26
Researchers at the Mauna Loa Weather Observatory in Hawaii have recorded increasing concentrations of atmospheric carbon dioxide since they began collecting data in the 1950s.

alkaline, pH $\approx$ 8.1.) The products of this neutralization eventually end up on the ocean floor as insoluble solids. Thus, carbonic acid neutralization in the ocean prevents CO_2 from being released back into the atmosphere. The ocean, therefore, is a carbon dioxide *sink*—most of the CO_2 that goes in doesn't come out. Pushing more CO_2 into our atmosphere means pushing more of it into our vast oceans. So far, the oceans have been absorbing about a third of our CO_2 emissions.

The movement of our CO_2 into the oceans, however, comes at a cost. Notice from Figure 13.27 that the CO_2 derived carbonic acid, H_2CO_3, reacts with calcium carbonate, $CaCO_3$. The addition of carbon dioxide into our oceans, therefore, has the effect of decreasing the amount of calcium carbonate in the ocean water, as well as other carbonates, such as magnesium carbonate, Mg_2CO_3. Coral, shelled organisms, and many other marine species, however, use these carbonates to build and maintain their bodily structures. Importantly, they can only do this when the ocean water is saturated with these minerals, which is currently the case in most surface waters. But the addition of CO_2 leads to ocean water unsaturated in carbonates. In regions where this occurs, the carbonate-based creatures begin to dissolve, which means they perish. This, in turn, can have a significant impact throughout the marine ecosystem, which would affect us as well. Consider this one example: Pink salmon off the southern coast of Alaska live on a diet of sea snails, also known as pteropods, which are dependent upon carbonate minerals. The destruction of this pteropod population by ocean acidification would also mean an end to the Alaskan pink salmon fishing industry.

UNIFYING CONCEPT

- ***The Ecosystem***
  Section 21.1

CHECK YOURSELF

How does adding CO_2 to the oceans cause a decrease in the concentration of carbonate ions, CO_3^{2-}?

CHECK YOUR ANSWER

The CO_2 reacts with water, H_2O, to form carbonic acid, H_2CO_3, which effectively removes the carbonate ion, CO_3^{2-}, by reacting with it to form the bicarbonate ion, HCO_3^-.

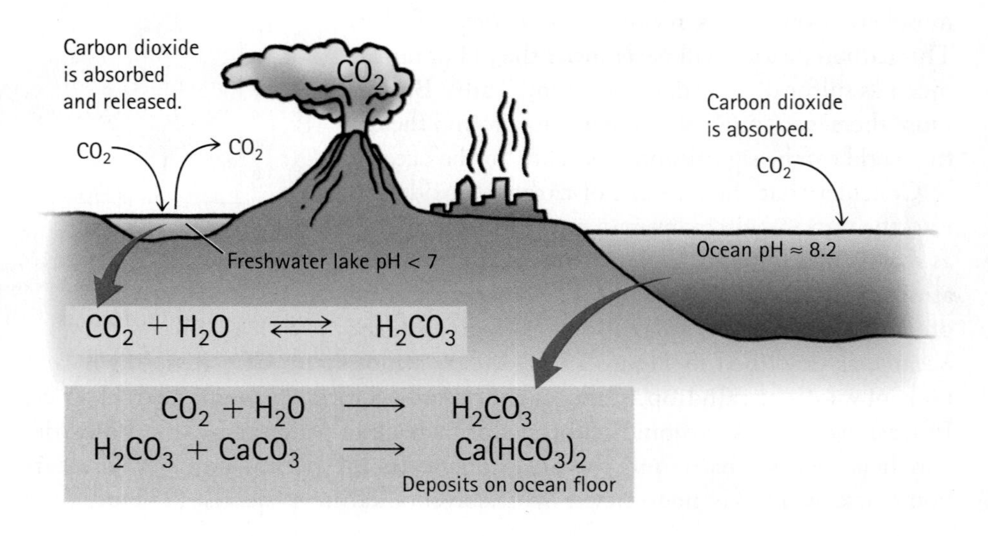

FIGURE 13.27
Carbon dioxide forms carbonic acid upon entering any body of water. In fresh water, this reaction is reversible, and the carbon dioxide is released back into the atmosphere. In the alkaline ocean, the carbonic acid is neutralized to such compounds as calcium bicarbonate, $Ca(HCO_3)_2$, which precipitate to the ocean floor. As a result, most of the atmospheric carbon dioxide that enters our oceans remains there.

Is the effect of human-produced CO_2 on ocean pH measurable? Absolutely and it is significant. Over the past 100 years, increases in atmospheric carbon dioxide—primarily due to the burning of fossil fuels—has lowered the average pH of the ocean by about 0.1 pH units. On a geologic time scale, this is lightning fast. The last comparable decrease in ocean pH occurred about 56 million years ago. At that time, major increases in atmospheric carbon dioxide caused the pH of the ocean to decrease by about 0.45 units. These changes, however, took place over 5000 years for an average decrease in ocean pH of about 0.01 units per century. The result was a huge die-off of marine organisms, which can be seen as a distinct layer of brown sediment within core samplings of the ocean floor. We are on track for surpassing this event at a rate that is about 10 times as fast.

There is much attention in the media given to the role atmospheric carbon dioxide plays on global climate, which we discuss in Chapter 27. You should understand, however, that our production of copious amounts of carbon dioxide is a two-fold problem. One is the potential for a not-so-predictable change in global climate. The other is a very predictable change in ocean chemistry. Both need to be considered carefully.

13.6 Losing and Gaining Electrons

LEARNING OBJECTIVE
Identify when a chemical undergoes oxidation or reduction.

EXPLAIN THIS Why is the chlorine atom such a strong oxidizing agent?

Oxidation is the process whereby a reactant loses one or more electrons. **Reduction** is the opposite process, whereby a reactant gains one or more electrons. Oxidation and reduction are complementary processes that occur at the same time. They always occur together; you cannot have one without the other. The electrons lost by one chemical in an oxidation reaction don't simply disappear; they are gained by another chemical in a reduction reaction.

An oxidation–reduction reaction occurs when sodium and chlorine react to form sodium chloride, as shown in Figure 13.28. The equation for this reaction is

$$2\,Na(s) + Cl_2(g) \longrightarrow 2\,NaCl(s)$$

FIGURE 13.28
In the exothermic formation of sodium chloride, sodium metal is oxidized by chlorine gas, and chlorine gas is reduced by sodium metal.

To see how electrons are transferred in this reaction, we can look at each reactant individually. Each electrically neutral sodium atom changes to a positively charged ion. At the same time, we can say that each atom loses an electron and is therefore oxidized:

$$2\,Na(s) \longrightarrow 2\,Na^+ + 2e^- \qquad \text{Oxidation}$$

Each electrically neutral chlorine molecule changes into two negatively charged ions. Each of these atoms gains an electron and is therefore reduced:

$$Cl_2 + 2e^- \longrightarrow 2\,Cl^- \qquad \text{Reduction}$$

The net result is that the two electrons lost by the sodium atoms are transferred to the chlorine atoms. Therefore, each of the two equations shown above actually represents one-half of an entire process, which is why each is called a **half reaction**. In other words, an electron won't be lost from a sodium atom without the presence of a chlorine atom available to pick up that electron. Both half reactions are required to represent the *whole* oxidation–reduction process. Half reactions are useful for showing which reactant loses electrons and which reactant gains them, which is why we use half reactions in our discussion.

FYI Scientists have experimented with ways of enhancing the ocean's ability to absorb atmospheric carbon dioxide. Adding powdered iron to a small plot of the ocean, they found, has the effect of fostering the growth of microorganisms that speed up the rate at which carbon dioxide is absorbed. Might this be a solution to the problem of global warming? Might adding too much iron initiate another ice age or alter the ocean's ecology?

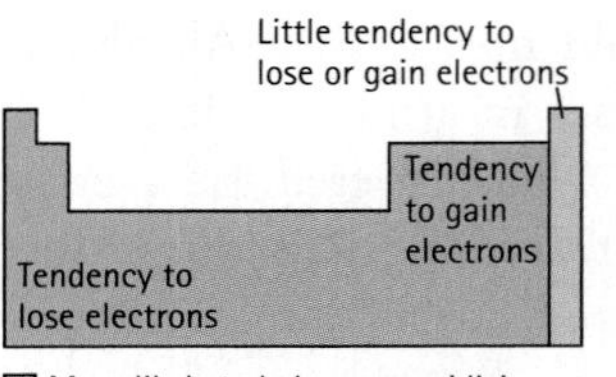

FIGURE 13.29
The ability of an atom to gain or lose electrons is indicated by its position in the periodic table. Those at the upper right tend to gain electrons, and those at the lower left tend to lose them.

Because the sodium causes reduction of the chlorine, the sodium is acting as a *reducing agent*. A reducing agent is any reactant that causes another reactant to be reduced. Note that sodium is oxidized when it behaves as a reducing agent—it loses electrons. Conversely, chlorine causes oxidation of the sodium and so is acting as an *oxidizing agent*. Because it gains electrons in the process, an oxidizing agent is reduced. Just remember that **l**oss of **e**lectrons is **o**xidation, and **g**ain of **e**lectrons is **r**eduction. Here is a helpful mnemonic adapted from a once-popular children's story: **Leo** the lion went "**ger**."

Different elements have different oxidation and reduction tendencies—some lose electrons more readily, while others gain electrons more readily, as Figure 13.29 illustrates.

CHECK YOURSELF

True or false?

1. **Reducing agents are oxidized in oxidation–reduction reactions.**
2. **Oxidizing agents are reduced in oxidation–reduction reactions.**

CHECK YOUR ANSWERS

Both statements are true.

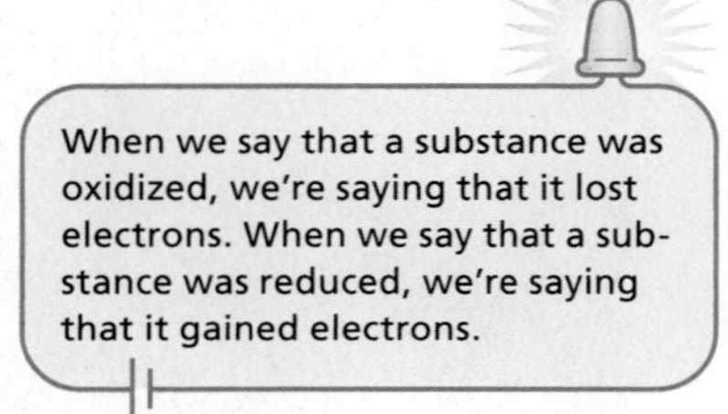

When we say that a substance was oxidized, we're saying that it lost electrons. When we say that a substance was reduced, we're saying that it gained electrons.

Whether a reaction is classified as an oxidation–reduction reaction is not always immediately apparent. The chemical equation, however, can provide some important clues. First, look for changes in the ionic states of elements. Sodium metal, for example, consists of neutral sodium atoms. In the formation of sodium chloride, these atoms transform into positively charged sodium ions, which occurs as sodium atoms lose electrons (oxidation). A second way to identify a reaction as an oxidation–reduction reaction is to look to see whether an element is joining with (gaining) or departing from (losing) *oxygen* atoms. As the element gains the oxygen, it is losing electrons to that oxygen because of the oxygen's high electronegativity. The gain of oxygen, therefore, is oxidation (loss of electrons), while the loss of oxygen is reduction (gain of electrons). For example, hydrogen, H_2, reacts with oxygen, O_2, to form water, H_2O, as follows:

$$\mathrm{H{-}H + H{-}H + O{=}O \longrightarrow H{-}O{-}H + H{-}O{-}H}$$

Note that the element hydrogen becomes attached to an oxygen atom through this reaction. The hydrogen, therefore, is oxidized.

A third way to identify a reaction as an oxidation–reduction reaction is to see whether an element is gaining or losing hydrogen atoms. The gain of a hydrogen atom is reduction, while the loss of a hydrogen atom is oxidation. In the formation of water shown above, we see that the element oxygen is gaining hydrogen atoms, which means that the oxygen is being reduced—that is, the oxygen is gaining electrons from the hydrogen, which is why the oxygen atom in water is slightly negative, as discussed in Section 12.5. The three ways of identifying a reaction as an oxidation–reduction type of reaction are summarized in Figure 13.30.

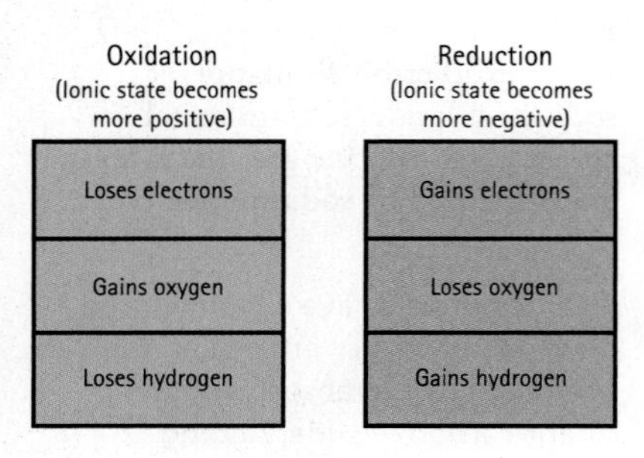

Oxidation
(Ionic state becomes more positive)
Reduction
(Ionic state becomes more negative)
Loses electrons
Gains oxygen
Loses hydrogen
Gains electrons
Loses oxygen
Gains hydrogen

FIGURE 13.30
Oxidation results in a greater positive charge, which can be achieved by losing electrons, gaining oxygen atoms, or losing hydrogen atoms. Reduction results in a greater negative charge, which can be achieved by gaining electrons, losing oxygen atoms, or gaining hydrogen atoms.

CHECK YOURSELF

In this equation, is carbon oxidized or reduced?

$$CH_4 + 2\,O_2 \rightarrow CO_2 + 2\,H_2O$$

CHECK YOUR ANSWER

As the carbon of methane, CH_4, forms carbon dioxide, CO_2, it is losing hydrogen and gaining oxygen, which tells us that the carbon is being oxidized.

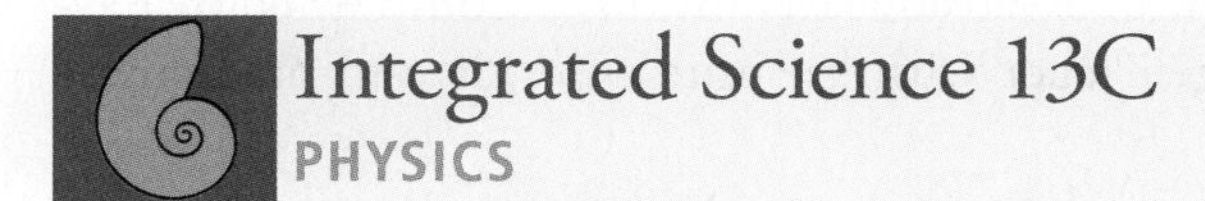

Integrated Science 13C
PHYSICS

LEARNING OBJECTIVE
Describe the key differences between a battery and a fuel cell.

Batteries and Fuel Cells

EXPLAIN THIS Why is lithium a preferred metal for the making of batteries?

An oxidation–reduction reaction can produce electric energy. This typically occurs when a material that tends to lose electrons (oxidation) is put into contact with a material that tends to gain electrons (reduction). What happens initially is that electrons flow between these two materials. Soon, however, the material that is gaining electrons acquires a negative charge. The electrons then stop flowing to that material because they are repelled by its negative charge. It is possible, however, to engineer a system so that this buildup of negative charge is alleviated by the flow of positively charged ions found within a salty medium. This is how a battery works.

Let's look at the common *dry-cell battery*, which was invented in the 1860s and is still used today as a disposable energy source for flashlights, toys, and the like. The basic design consists of a zinc cup filled with a thick paste of ammonium chloride, NH_4Cl; zinc chloride, $ZnCl_2$; and manganese dioxide, MnO_2. Immersed in this paste is a porous stick of graphite that projects out the top of the battery, as shown in Figure 13.31.

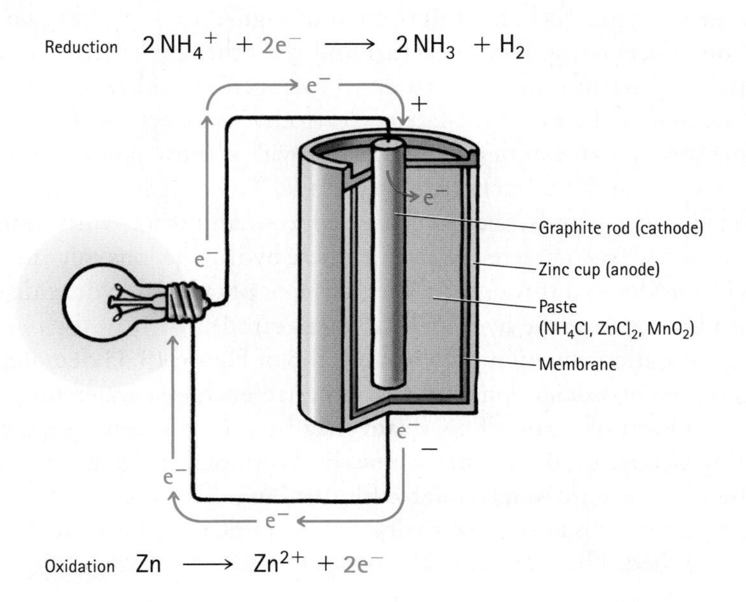

FIGURE 13.31
A common dry-cell battery with a graphite rod immersed in a paste of ammonium chloride, manganese dioxide, and zinc chloride.

The zinc is the source of electrons according to the reaction

$$Zn(s) \longrightarrow Zn^{2+}(aq) + 2e^- \quad \text{Oxidation}$$

These electrons flow through an external circuit and are then received by the graphite, which is a good conductor of electricity. From the graphite, the electrons flow into the paste, where they react with ammonium ions as follows:

$$2\,NH_4^+(aq) + 2e^- \longrightarrow 2\,NH_3(g) + H_2(g) \quad \text{Reduction}$$

The reduction of ammonium ions in a dry-cell battery produces two gases—ammonia, NH_3, and hydrogen, H_2—that need to be removed to avoid a pressure buildup and a potential explosion. Removal is accomplished by having the ammonia and hydrogen react with the zinc chloride and manganese dioxide.

The life of a dry-cell battery is relatively short. Oxidation causes the zinc cup to deteriorate, and eventually the contents leak out. Even while the battery is not operating, the zinc corrodes as it reacts with ammonium ions. As discussed in Section 13.3, chemical reactions slow down with decreasing temperature. Chilling a battery, therefore, slows down the rate at which the zinc corrodes, which increases the life of the battery.

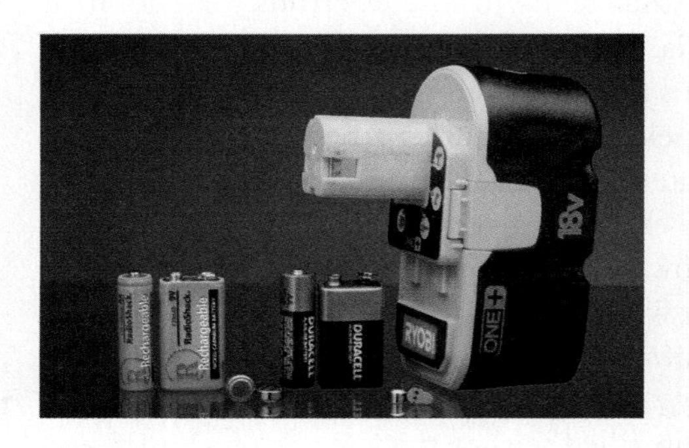

FIGURE 13.32
Alkaline batteries last much longer than dry-cell batteries and give a steadier voltage, but they are more expensive.

Another type of disposable battery, the more expensive *alkaline battery*, avoids many of the problems of dry-cell batteries by operating in a strongly alkaline paste (Figure 13.32). The small lithium disposable batteries are variations of the alkaline battery. In the lithium battery, lithium metal, rather than zinc, is used as the source of electrons. Not only can lithium maintain a higher voltage than zinc, but it also is about $\frac{1}{13}$ as dense, which allows for a lighter battery.

A *fuel cell* is a device that converts the chemical energy of a fuel into electric energy. Fuel cells are by far the most efficient means of generating electricity. A hydrogen–oxygen fuel cell is illustrated in Figure 13.33. It has two compartments, one for entering hydrogen fuel and the other for entering oxygen fuel, separated by a set of porous electrodes. Hydrogen is oxidized on contact with hydroxide ions at the hydrogen-facing electrode. The electrons from this oxidation flow through an external circuit and provide electric power before meeting up with oxygen at the oxygen-facing electrode. The oxygen readily picks up the electrons (in other words, the oxygen is reduced) and reacts with water to form hydroxide ions. To complete the circuit, these hydroxide ions migrate across the porous electrodes and through an ionic paste of potassium hydroxide, KOH, to join with hydrogen at the hydrogen-facing electrode.

As the oxidation equation shown at the top of Figure 13.33 demonstrates, the hydrogen and hydroxide ions react to produce energetic water molecules that arise in the form of steam. This steam may be used for heating systems or for generating electricity in a steam turbine. Furthermore, the water that condenses from the steam is pure water, suitable for drinking!

Although fuel cells are similar to dry-cell batteries, they don't run down as long as fuel is supplied. The International Space Station uses hydrogen–oxygen fuel cells

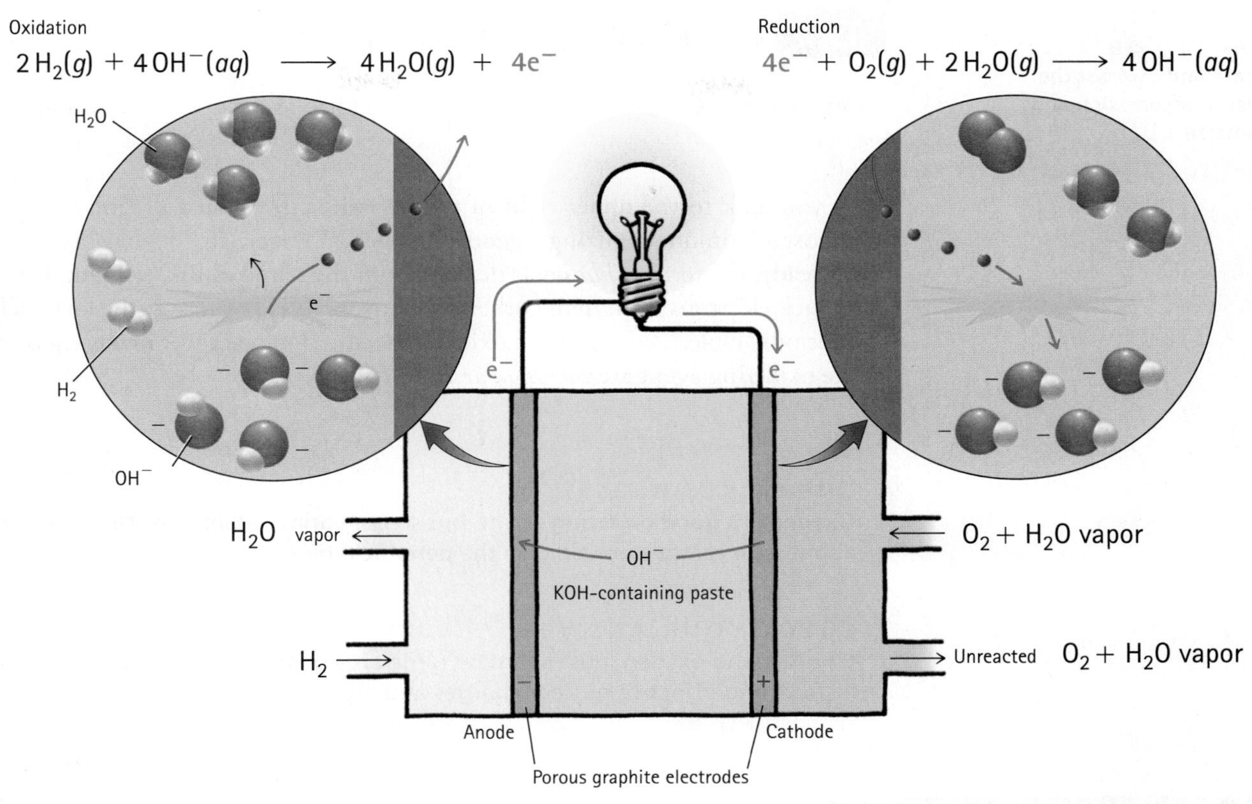

FIGURE 13.33
The hydrogen–oxygen fuel cell.

to meet its electrical needs. Back on Earth, researchers are developing fuel cells for buses and automobiles. As shown in Figure 13.34, experimental fuel-cell buses are already operating in several cities, such as Vancouver, British Columbia, and Chicago, Illinois. These vehicles produce very few pollutants and can run much more efficiently than vehicles that burn fossil fuels.

Commercial buildings as well as individual homes are now being outfitted with fuel cells as an alternative to receiving electricity (and heat) from regional power stations. Researchers are also working on miniature fuel cells that could replace the batteries used for portable electronic devices, such as cell phones and laptop computers. Such devices could operate for extended periods of time on a single "ampoule" of fuel available at your local supermarket.

FIGURE 13.34
Because this bus is powered by a fuel cell, its tailpipe emits mostly water vapor.

CHECK YOURSELF

As long as fuel is available to it, a given fuel cell can supply electric energy indefinitely. Why can't batteries do the same?

CHECK YOUR ANSWER

Batteries generate electricity as the chemical reactants they contain are reduced and oxidized. Once these reactants are consumed, the battery can no longer generate electricity. A rechargeable battery can be made to operate again, but only after the energy flow is interrupted so that the reactants can be replenished.

LEARNING OBJECTIVE
Compare and contrast the processes of corrosion and combustion.

13.7 Corrosion and Combustion

EXPLAIN THIS Do our bodies gradually oxidize or reduce the food molecules we eat?

If you look to the upper right of the periodic table, you will find one of the most common oxidizing agents—oxygen. In fact, if you haven't guessed already, the term *oxidation* is derived from the name of this element. Oxygen can pluck electrons from many other elements, especially those at the lower left of the periodic table. Two common oxidation–reduction reactions involving oxygen as the oxidizing agent are *corrosion* and *combustion*.

CHECK YOURSELF
Oxygen is a good oxidizing agent, but so is chlorine. What does this indicate about their relative positions in the periodic table?

CHECK YOUR ANSWER
Chlorine and oxygen must lie in the same area of the periodic table. Both have strong effective nuclear charges and are strong oxidizing agents.

FIGURE 13.35
Rust itself does not harm the iron structures on which it forms, but the loss of metallic iron ruins the structural integrity of these objects.

Corrosion is the process by which a metal deteriorates. Corrosion caused by atmospheric oxygen is a widespread and costly problem. About one-quarter of the steel produced in the United States, for example, goes into replacing corroded iron, at a cost of billions of dollars annually. Iron corrodes when it reacts with atmospheric oxygen and water to form iron oxide trihydrate, which is the naturally occurring reddish-brown substance you know as rust, shown in Figure 13.35:

$$\underset{\text{Iron}}{4\,Fe} + \underset{\text{Oxygen}}{3\,O_2} + \underset{\text{Water}}{3\,H_2O} \longrightarrow \underset{\text{Rust}}{2\,Fe_2O_3{\cdot}3\,H_2O}$$

Another common metal oxidized by oxygen is aluminum. The product of aluminum oxidation is aluminum oxide, Al_2O_3, which is not water soluble. Because of its insolubility, aluminum oxide forms a protective coat that shields the metal from further oxidation. This coat is so thin that it's transparent, which is why aluminum maintains its metallic shine.

A protective, water-insoluble oxidized coat is the principle underlying a process called *galvanization*. Zinc has a slightly greater tendency to oxidize than does iron. For this reason, many iron objects, such as the nail pictured in Figure 13.36, are *galvanized* by coating them with a thin layer of zinc. The zinc oxidizes to zinc oxide, an inert, insoluble substance that protects the iron underneath it from rusting.

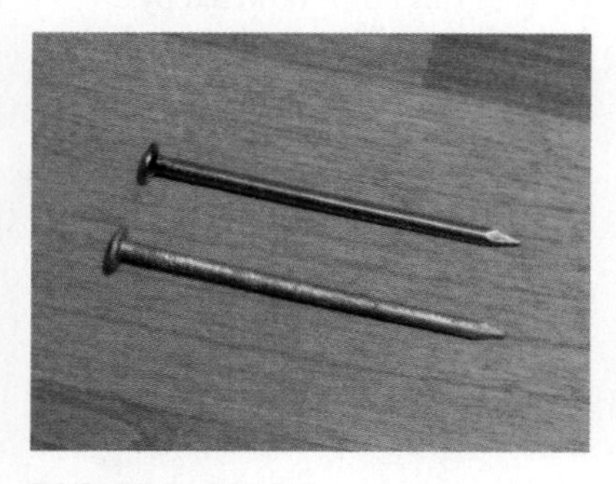

FIGURE 13.36
The galvanized nail (*bottom*) is protected from rusting by the sacrificial oxidation of zinc.

In a technique called *cathodic protection*, iron structures can be protected from oxidation by placing them in contact with certain metals, such as zinc or magnesium, that have a greater tendency to oxidize. This forces the iron to accept electrons. Ocean tankers, for example, are protected from corrosion by strips of zinc affixed to their hulls, as shown in Figure 13.37. Similarly, outdoor steel pipes are protected by being connected to magnesium rods inserted into the ground.

FIGURE 13.37
Zinc strips help protect the iron hull of an oil tanker from oxidizing. The zinc strips shown here are attached to the hull's exterior surface.

Combustion is a rapid oxidation–reduction reaction between a material and molecular oxygen. Combustion reactions are characteristically exothermic (energy-releasing). A violent combustion reaction is the formation of water from hydrogen and oxygen. As discussed in Section 13.2, the energy from this reaction is used to power rockets into space. More common examples of combustion include the burning of wood and fossil fuels. The combustion of these and other carbon-based chemicals forms carbon dioxide and water. Consider, for example, the combustion of methane, the major component of natural gas:

$$CH_4 \quad + \quad 2\,O_2 \quad \longrightarrow \quad CO_2 \quad + \quad 2\,H_2O \quad + \quad \text{energy}$$

Methane	Oxygen dioxide	Carbon	Water

In combustion, electrons are transferred when polar covalent bonds are formed in place of nonpolar covalent bonds, or vice versa. (This is in contrast with the other examples of oxidation–reduction reactions presented in this chapter, which involve the formation of ions from atoms or, conversely, atoms from ions.) This concept is illustrated in Figure 13.38, which compares the electronic structures of the combustion starting material, molecular oxygen, and the combustion product, water. Molecular oxygen is a nonpolar covalent compound. Although each oxygen atom in the molecule has a fairly strong electronegativity, the four bonding electrons are pulled equally by both atoms and thus cannot congregate on one side or the other. After combustion, however, the electrons are shared between the oxygen and hydrogen atoms in a water molecule and are pulled toward the oxygen. This gives the oxygen a slight negative charge, which is another way of saying it has gained electrons and has thus been reduced. At the same time, the hydrogen atoms in the water molecule develop a slight positive charge, which is another way of saying they have lost electrons and have thus been oxidized. This gain of electrons by oxygen and loss of electrons by hydrogen are an energy-releasing process. Typically, the energy is released either as molecular kinetic energy (heat) or as light (the flame).

Interestingly, combustion oxidation–reduction reactions occur throughout your body. You can visualize a simplified model of your metabolism by reviewing Figure 13.38 and substituting a food molecule for the methane. Food

FYI There are two kinds of matches: the "strike anywhere" type, which usually has a "bull's-eye" tip, and the "safety match," which requires you to strike the match on a strip on the packaging. Both involve the burning of sulfur in the tip of the match. Getting the sulfur to burn using only the oxygen in the air, however, is difficult, which is why the sulfur is blended with an oxidizing agent, such as potassium chlorate, $KClO_3$. For the strike anywhere match, a third ingredient, red phosphorus, P_4, is included. The heat of friction causes the red phosphorus to convert into white phosphorus—an alternative form of phosphorus that burns rapidly in air. This initiates the reduction–oxidation reaction between the sulfur and the potassium chlorate, which in turn ignites the burning of the matchstick. Safety matches work the same way, except that the red phosphorus is embedded within the striking strip, which is the only place where the match can be lit.

FIGURE 13.38
(a) Neither atom in an oxygen molecule can preferentially attract the bonding electrons. (b) The oxygen atom of a water molecule pulls the bonding electrons away from the hydrogen atoms on the water molecule, making the oxygen atom slightly negative and the two hydrogen atoms slightly positive.

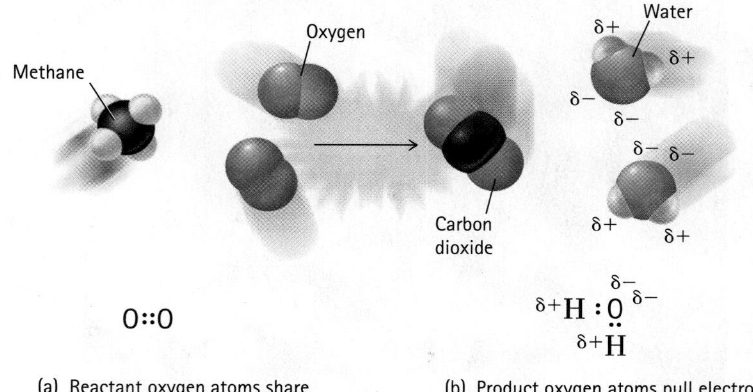
Methane
Oxygen
Water
Carbon dioxide
δ+
δ−
O::O
δ+H : O δ−
δ+H
(a) Reactant oxygen atoms share electrons equally in O_2 molecules.
(b) Product oxygen atoms pull electrons away from H atoms in H_2O molecules and are reduced.

molecules relinquish their electrons to the oxygen molecules you inhale. The products are carbon dioxide, water vapor, and energy. You exhale the carbon dioxide and water vapor, but much of the energy from the reaction is used to keep your body warm and to drive the many other biochemical reactions necessary for life.

For instructor-assigned homework, go to www.masteringphysics.com 

SUMMARY OF TERMS (KNOWLEDGE)

Acid A substance that donates hydrogen ions.

Acidic Description of a solution in which the hydronium ion concentration is higher than the hydroxide ion concentration.

Activation energy The minimum energy required in order for a chemical reaction to proceed.

Amphoteric Description of a substance that can behave as either an acid or a base.

Base A substance that accepts hydrogen ions.

Basic Description of a solution in which the hydroxide ion concentration is higher than the hydronium ion concentration; also sometimes called *alkaline.*

Bond energy The amount of energy required to pull two bonded atoms apart, which is the same as the amount of energy released when the two atoms are brought together into a bond.

Catalyst Any substance that increases the rate of a chemical reaction without itself being consumed by the reaction.

Chemical equation A representation of a chemical reaction in which reactants are written to the left of an arrow that points toward the products on the right.

Combustion A rapid exothermic oxidation–reduction reaction between a material and molecular oxygen.

Corrosion The deterioration of a metal, typically caused by atmospheric oxygen.

Endothermic Description of a chemical reaction in which there is a net absorption of energy.

Exothermic Description of a chemical reaction in which there is a net release of energy.

Half reaction One half of an oxidation–reduction reaction, represented by an equation showing electrons as either reactants or products.

Hydronium ion A polyatomic ion made by adding a proton (hydrogen ion) to a water molecule.

Hydroxide ion A polyatomic ion made by removing a proton (hydrogen ion) from a water molecule.

Law of mass conservation Matter is neither created nor destroyed during a chemical reaction; atoms are merely rearranged, without any apparent loss or gain of mass, to form new molecules.

Neutral Description of a solution in which the hydronium ion concentration is equal to the hydroxide ion concentration.

Neutralization A reaction between an acid and a base.

Oxidation The process by which a reactant loses one or more electrons.

pH A measure of the acidity of a solution; equal to the negative logarithm of the hydronium ion concentration.

Products The new materials formed in a chemical reaction.

Reactants The reacting substances in a chemical reaction.

Reaction rate A measure of how quickly the concentration of products in a chemical reaction increases or the concentration of reactants decreases.

Reduction The process by which a reactant gains one or more electrons.

Salt An ionic compound commonly formed from the reaction between an acid and a base.

READING CHECK (COMPREHENSION)

13.1 Chemical Equations

1. What is the purpose of coefficients in a chemical equation?
2. How many oxygen atoms are indicated on the right side of this balanced chemical equation:

$$4\ Cr(s) + 3\ O_2(g) \longrightarrow 2\ Cr_2O_3(g)$$

3. Why is it important that a chemical equation be balanced?

13.2 Energy and Chemical Reactions

4. If it takes 436 kilojoules to break a bond, how many kilojoules are released when the same bond is formed?
5. What is released by an exothermic reaction?
6. What is absorbed by an endothermic reaction?

13.3 Reaction Rates

7. Why don't all collisions between reactant molecules lead to product formation?
8. What generally happens to the rate of a chemical reaction with increasing temperature?
9. Which reactant molecules are the first to pass over the energy barrier?

13.4 Acids Donate Protons; Bases Accept Them

10. How is an acid different from a base?
11. When an acid is dissolved in water, what ion does the water form?
12. How many salts are there?

13.5 Acidic, Basic, and Neutral Solutions

13. Are there many hydronium ions in neutral water?
14. What is true about the relative concentrations of hydronium and hydroxide ions in an acidic solution? How about in a neutral solution? A basic solution?
15. What does the pH of a solution indicate?

13.6 Losing and Gaining Electrons

16. Which elements have the greatest tendency to behave as oxidizing agents?
17. What elements have the greatest tendency to behave as reducing agents?
18. What happens to a reducing agent as it reduces?

13.7 Corrosion and Combustion

19. What metal coats a galvanized nail?
20. What is iron forced to accept during cathodic protection?
21. What happens to the polarity of oxygen atoms as they transform from molecular oxygen, O_2, into water molecules, H_2O?

THINK INTEGRATED SCIENCE

13A—Catalysts

22. What catalyst is effective in the destruction of atmospheric ozone, O_3?
23. What does a catalyst do to the energy of activation for a reaction?
24. What net effect does a chemical reaction have on a catalyst?

13B—Acid Rain and Ocean Acidification

25. What is the product of the reaction between carbon dioxide and water?
26. What does sulfur dioxide have to do with acid rain?
27. As carbon dioxide is absorbed by the ocean, it reacts with water and then carbonate ions to form what?

13C—Batteries and Fuel Cells

28. A material that tends to lose electrons is put into contact with a material that tends to gain electrons. Why won't electrons flow continuously between these two materials?
29. What is the primary difference between a battery and a fuel cell?
30. What else do fuel cells produce besides electricity?

THINK AND DO (HANDS-ON APPLICATION)

31. Hold some room-temperature water in the cupped palm of your hand over a sink. Pour an equal amount of room-temperature rubbing alcohol into the water. Is this mixing an exothermic or endothermic process? What's going on at the molecular level?

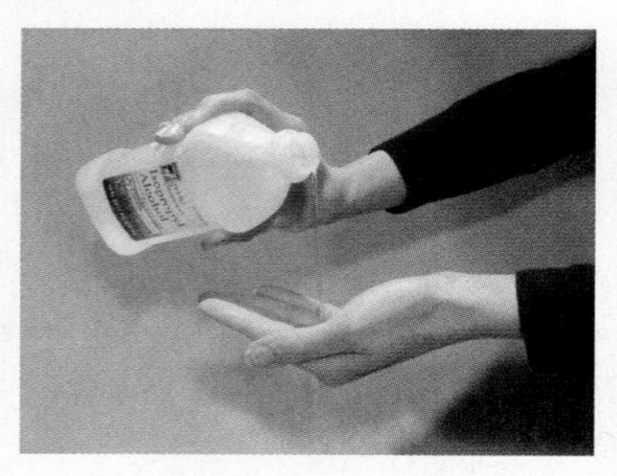

32. Add lukewarm water to two plastic cups. (Do *not* use insulating Styrofoam cups.) Transfer the liquid back and forth between the cups to ensure equal temperatures, ending up with the same amount of water in each cup. Add several tablespoons of table salt to one cup and stir. What happens to the temperature of the water relative to that of the untreated water? (Hold the cups up to your cheeks to tell.) Is this an exothermic or endothermic process? What's going on at the molecular level?
33. An Alka-Seltzer antacid tablet reacts vigorously with water. But how does this tablet react in a solution of half water and half corn syrup? Propose an explanation involving the relationship between reaction speed and the frequency of molecular collisions.
34. Baker's yeast contains a biological catalyst known as *catalase*, which catalyzes the transformation of hydrogen peroxide, H_2O_2, into oxygen, O_2, and water, H_2O. Write a balanced equation for this reaction. Add a couple milliliters of 3% hydrogen peroxide to a glass containing a small amount of baker's yeast. What happens? Why?

35. The pH of a solution can be approximated with a *pH indicator*, which is any chemical whose color changes with pH. Many pH indicators are found in plants; the pigment of red cabbage is a good example. This pigment is red at low pH values (acidic), light purple at slightly acidic pH values, blue at neutral pH values, light green at moderately alkaline pH values, and dark green at very alkaline pH values.

    Boil shredded red cabbage in water for about five minutes. Strain the broth from the cabbage and allow it to cool. Add the cooled blue broth to at least three clear cups so that each cup is less than half-filled or to three

white porcelain bowls. Add a teaspoon of white vinegar to one cup and a teaspoon of baking soda to a second cup. Watch for color changes. Add nothing to the third cup so that it remains blue.

    What color is the red cabbage before being boiled? Are the juices in red cabbage more or less acidic than vinegar? What would happen to the baking soda solution if you were to slowly add vinegar to it? What color would you get if a teaspoon of concentrated broth were added to a glass of water?
36. Add about an inch of water to a large test tube and then add a couple drops of phenolphthalein pH indicator, which you will likely need to obtain from your classroom. Add a small pinch of washing soda, which contains sodium carbonate, Na_2CO_3. Upon mixing, the washing soda turns the solution basic, as evidenced by the pink color that forms. Neutralize this base by adding an acid, but not just any acid—use the acid of your breath. Bubble your breath into the solution through a straw until the pink color disappears. What acid are you adding? How does this activity relate to the acidity of rain? Why do you want to add only a small pinch of washing soda and not a tablespoon?
37. Copper metal slowly reacts with the oxygen in air to form reddish copper (I) oxide, Cu_2O, which is a compound that coats the surface of older pennies and makes them look tarnished. When such a penny is placed in a solution of salt in vinegar, the copper (I) oxide acts as a base and reacts with the vinegar to form copper salts. This effectively cleans the penny. The copper salts can then be transformed back into copper metal when exposed to an iron nail.

    Stir about half a teaspoon of salt into about half a cup of white distilled vinegar. Use a nonmetal container, such as a ceramic or plastic bowl. Dip a tarnished penny halfway into the solution and notice the rapid cleaning effect. Add at least a dozen tarnished pennies to the solution. As they get cleaned this increases the concentration of copper ions in solution. Sandpaper an iron nail to give it a clean surface and then rest the nail in the vinegar solution for about 10 minutes. Watch for the formation of copper metal on the nail.

Are copper ions positively or negatively charged? What is the charge on the copper atoms in Cu_2O? What is the charge on the oxygen? What must copper ions gain in order to transform back into a metallic form? What was the iron of the nail able to do for the copper ions?

38. Silver tarnishes because it reacts with the small amounts of smelly hydrogen sulfide, H_2S, we put into the air as we digest our food. In this reaction, the silver loses electrons to the sulfur. You can reverse this reaction by allowing the silver to get its electrons back from aluminum.

    Flatten some aluminum foil on the bottom of a cooking pot. Fill the pot halfway with water and bring the water to a boil. When the water is boiling, remove the pot from its heat source. Add a couple tablespoons of baking soda to the hot water. Slowly immerse a tarnished piece of silver into the water and allow the silver to touch the aluminum foil. You should see an immediate effect when the silver and aluminum make contact. (Add more baking soda if you don't.) If your silver piece is very tarnished, you may notice the unpleasant odor of hydrogen sulfide as it is released back into the air.

    As silver tarnishes, is it oxidized or reduced? Is baking soda, $NaHCO_3$, an ionic compound? Why is it needed in this activity? Does aluminum behave as an oxidizing agent or a reducing agent as it restores the silver to its untarnished state?

THINK AND COMPARE (ANALYSIS)

39. Rank these reaction profiles in order of increasing reaction speed:

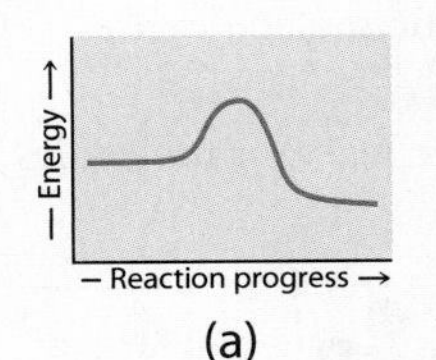

(a)

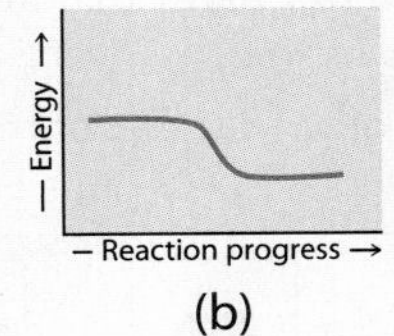

(b)

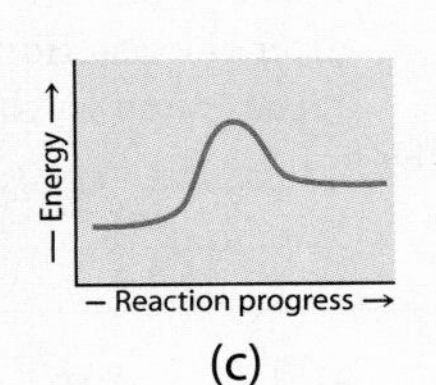

(c)

40. Rank the covalent bonds in order of increasing bond strength: (a) C≡C, (b) C=C, (c) C—C.

41. Review the concept of electronegativity in Section 12.4 and rank the elements from the weakest to strongest oxidizing agent: (a) chlorine, Cl; (b) sulfur, S; (c) sodium, Na.

42. Review the concept of electronegativity in Section 12.4 and rank the elements from the weakest to strongest reducing agent: (a) chlorine, Cl; (b) sulfur, S; (c) sodium, Na.

43. Rank these molecules from least oxidized to most oxidized:

Ethane ($H_3C{-}CH_3$)

Ethanol ($H_3C{-}CH_2{-}O{-}H$)

Acetaldehyde ($H_3C{-}C(=O){-}H$)

Acetic aid ($H_3C{-}C(=O){-}O{-}H$)

THINK AND SOLVE (MATHEMATICAL APPLICATION)

44. Use the bond energies in Table 13.1 and the accounting format shown in Section 13.2 to determine whether these reactions are exothermic or endothermic:

    (a) $H_2 + Cl_2 \rightarrow 2\ HCl$

    (b) $2\ HC{\equiv}CH + 5\ O_2 \rightarrow 4\ CO_2 + 2\ H_2O$

45. Use the bond energies in Table 13.1 and the accounting format shown in Section 13.2 to determine whether these reactions are exothermic or endothermic:

(a)

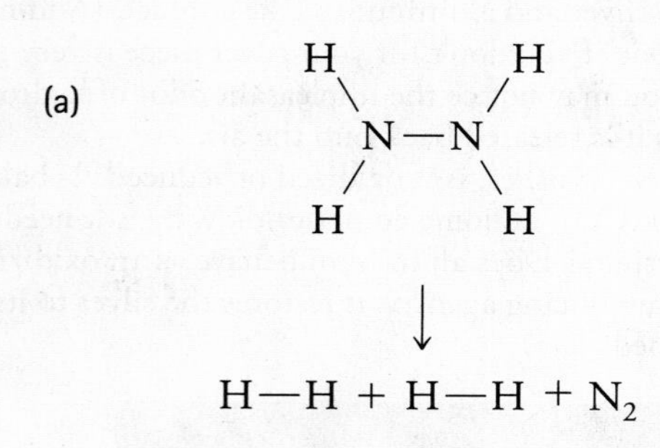

(b) H—O—O—H + H—O—O—H

↓

O=O + H—O—H + H—O—H

46. When the hydronium ion concentration of a solution is 1×10^{-10} M, what is the pH of the solution? Is the solution acidic or basic? When the hydronium ion concentration of a solution is 1×10^{-4} M, what is the pH of the solution? Is the solution acidic or basic?

47. When the pH of a solution is 1, the concentration of hydronium ions is 10^{-1} M $= 0.1$ M. Assume that the volume of this solution is 500 mL. What is the pH after 500 mL of water is added? You will need a calculator with a logarithm function to answer this question.

48. Show that the pH of a solution is -0.301 when its hydronium ion concentration equals 2 moles per liter. Is the solution acidic or basic?

THINK AND EXPLAIN (SYNTHESIS)

49. Balance these equations:
   (a) __ $Fe(s)$ + __ $O_2(g)$ → __ $Fe_2O_3(s)$
   (b) __ $H_2(g)$ + __ $N_2(g)$ → __ $NH_3(g)$
   (c) __ $Cl_2(g)$ + __ $KBr(aq)$ → __ $Br_2(l)$ + __ $KCl(aq)$
   (d) __ $CH_4(g)$ + __ $O_2(g)$ → __ $CO_2(g)$ + __ $H_2O(l)$

50. Balance these equations:
   (a) __ $Fe(s)$ + __ $S(s)$ → __ $Fe_2S_3(s)$
   (b) __ $P_4(s)$ + __ $H_2(g)$ → __ $PH_3(g)$
   (c) __ $NO(g)$ + __ $Cl_2(g)$ → __ $NOCl(g)$
   (d) __ $SiCl_4(l)$ + __ $Mg(s)$ → __ $Si(s)$ + __ $MgCl_2(s)$

51. Which of these two chemical equations is balanced?

$$2\,C_4H_{10}(g) + 13\,O_2(g) \rightarrow 8\,CO_2(g) + 10\,H_2O(l)$$

$$4\,C_6H_7N_5O_{16}(s) + 19\,O_2(g) \rightarrow 24\,CO_2(g) + 20\,NO_2(g) + 14\,H_2O(g)$$

Use the following illustration for Exercises 52–54.

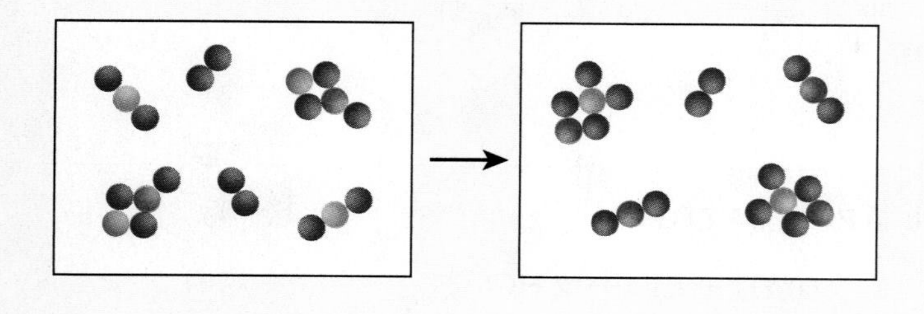

52. Assume the illustrations at left are two frames of a movie—one from before the reaction and the other from after the reaction. How many diatomic molecules are represented in this movie?

53. There is an excess of at least one of the reactant molecules. Which one?

A

B

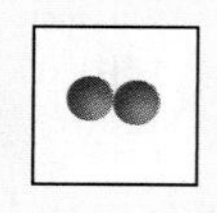
C

D

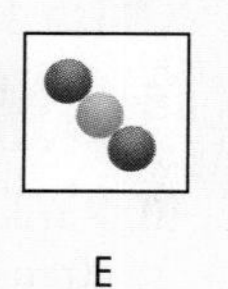
E

54. Which equation best describes this reaction?
   (a) $2\,AB_2 + 2\,DCB_3 + B_2 \rightarrow 2\,DBA_4 + 2\,CA_2$
   (b) $2\,AB_2 + 2\,CDA_3 + B_2 \rightarrow 2\,C_2A_4 + 2\,DBA$
   (c) $2\,AB_2 + 2\,CDA_3 + A_2 \rightarrow 2\,DBA_4 + 2\,CA_2$
   (d) $2\,BA_2 + 2\,DCA_3 + A_2 \rightarrow 2\,DBA_4 + 2\,CA_2$

55. The reactants shown schematically on the left represent methane, CH_4, and water, H_2O. Write out the full balanced chemical equation that is depicted.

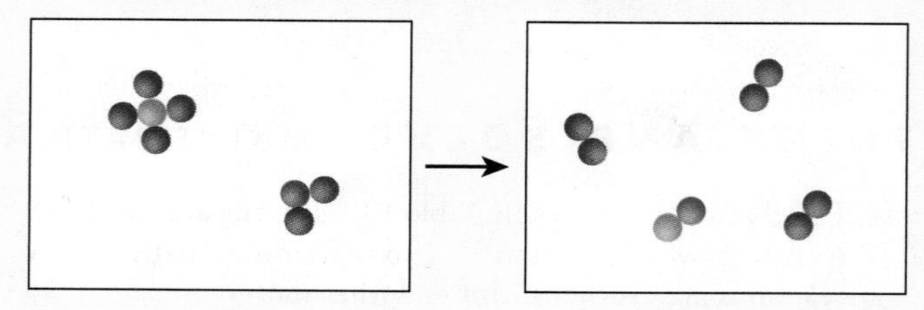

56. The reactants shown schematically on the left represent iron oxide, Fe_2O_3, and carbon monoxide, CO. Write out the full balanced chemical equation that is depicted.

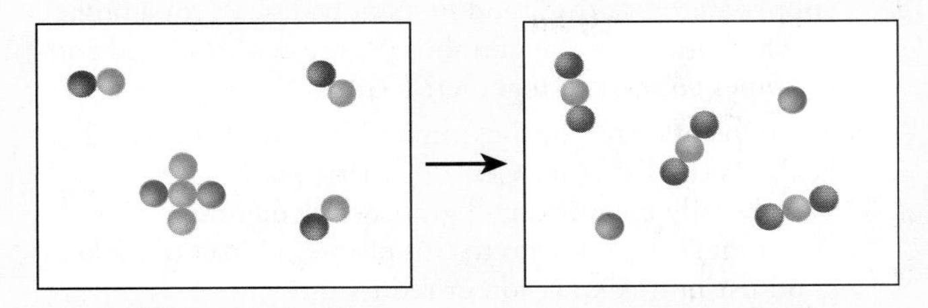

57. Are the chemical reactions that take place in a disposable battery exothermic or endothermic? What evidence supports your answer? Is the reaction going on in a rechargeable battery while it is recharging exothermic or endothermic?
58. Is photosynthesis (see Chapter 15) an exothermic or endothermic process? Please explain.
59. How is it possible for a jet airplane carrying 110 tons of jet fuel to emit 340 tons of carbon dioxide?
60. Does a refrigerator prevent or delay the spoilage of food? Explain.
61. Why does a glowing splint of wood burn only slowly in air but burst into flames when placed in pure oxygen?
62. Give two reasons why heat is often added to chemical reactions performed in the laboratory.
63. Explain the connection between photosynthetic life on Earth and the ozone layer.
64. Does the ozone pollution from automobiles help to diminish the ozone hole over the South Pole? Defend your answer.
65. Chlorine is put into the atmosphere by volcanoes in the form of hydrogen chloride, HCl, but this form of chlorine does not remain in the atmosphere for very long. Why?
66. In the following reaction sequence for the catalytic formation of ozone from molecular oxygen, which chemical compound is the catalyst—nitrogen monoxide or nitrogen dioxide?

$$O_2 + 2\,NO \rightarrow 2\,NO_2$$

$$2\,NO_2 \rightarrow 2\,NO + 2\,O$$

$$2\,O + 2\,O_2 \rightarrow 2\,O_3$$

67. An acid and a base react to form a salt, which consists of positive and negative ions. Which forms the positive ions—the acid or the base? Which forms the negative ions?
68. Identify the acid or base behavior of each substance in these reactions:

(a) $H_3O^+ + Cl^- \rightleftharpoons H_2O + HCl$

____ ____ ____ ____

(b) $H_2PO_4^- + H_2O \rightleftharpoons H_3O^+ + HPO_4^{2-}$

____ ____ ____ ____

(c) $HSO_4^- + H_2O \rightleftharpoons OH^- + H_2SO_4$

____ ____ ____ ____

(d) $O^{2-} + H_2O \rightleftharpoons OH^+ + OH^-$

____ ____ ____ ____

69. Many of the smelly molecules of cooked fish are alkaline compounds. How might these smelly molecules be conveniently transformed into less smelly salts just prior to eating the fish?
70. The main component of bleach is sodium hypochlorite, NaOCl, which consists of sodium ions, Na^+, and hypochlorite ions, ^-OCl. What products are formed when this compound is reacted with the hydrochloric acid, HCl, of toilet bowl cleaner?
71. The amphoteric reaction between two water molecules is endothermic, which means the reaction requires the input of heat energy in order to proceed:

$$\text{Energy} + H_2O + H_2O \Leftrightarrow H_3O^+ + OH^-$$

The warmer the water, the more heat energy is available for this reaction, and the more hydronium and hydroxide ions are formed.

(a) Which has a lower pH—pure water that is hot or pure water that is cold?

(b) Is it possible for water to be neutral but have a pH lower or higher than 7.0?

72. Water at 374°C and 218 atmospheres of pressure transforms into what is called a supercritical fluid, which behaves both as a liquid and as a gas. In a neutral solution of supercritical water the pH is about 2. What is the concentration of hydronium ions in this neutral solution? What is the concentration of hydroxide ions? Why is supercritical water so corrosive?
73. What is the concentration of hydronium ions in a solution that has a pH of −3? Why is such a solution impossible to prepare?
74. Can an acidic solution be made less acidic by adding an acidic solution?
75. Bubbling carbon dioxide into water causes the pH of the water to go down (become more acidic) because of the formation of carbonic acid. Will the pH also drop when carbon dioxide is bubbled into a solution of 1 M hydrochloric acid, HCl?
76. What happens to the pH of soda water as it loses its carbonation?
77. How does burning fossil fuels lower the pH of the ocean?
78. What element oxidized in this equation and what element is reduced?

$$I_2 + 2\,Br^- \rightarrow 2\,I^- + Br_2$$

79. What element behaves as the oxidizing agent in this equation and what element behaves as the reducing agent?

$$Sn^{2+} + 2\,Ag \rightarrow Sn + 2\,Ag^+$$

80. Hydrogen sulfide, H_2S, burns in the presence of oxygen, O_2, to produce water, H_2O, and sulfur dioxide, SO_2. Through this reaction, is sulfur oxidized or reduced?

$$2\,H_2S + 3\,O_2 \rightarrow 2\,H_2O + 2\,SO_2$$

81. Unsaturated fatty acids, such as $C_{12}H_{22}O_2$, react with hydrogen gas, H_2, to form saturated fatty acids, such as $C_{12}H_{24}O_2$. Are the unsaturated fatty acids being oxidized or reduced through this process?

82. The general chemical equation for photosynthesis is shown here. Through this reaction are the oxygen atoms of the water molecules, H_2O, oxidized or reduced?

$$6\ CO_2 + 6\ H_2O \rightarrow C_6H_{12}O_6 + 6\ O_2$$

83. A chemical equation for the combustion of propane, C_3H_8, is shown. Through this reaction is the carbon oxidized or reduced?

$$C_3H_8 + 5\ O_2 \rightarrow 3\ CO_2 + 4\ H_2O$$

84. Pennies manufactured after 1982 are made of zinc metal, Zn, with a coat of copper metal, Cu. Zinc is more easily oxidized than copper. Why, then, don't these pennies quickly corrode?

85. Water is 88.88% oxygen by mass. Oxygen is exactly what a fire needs to grow brighter and stronger. So, why doesn't a fire grow brighter and stronger when water is added to it?

86. Iron atoms have a greater tendency to oxidize than do copper atoms. Is this good news or bad news for a home in which much of the plumbing consists of iron and copper pipes connected together? Explain.

87. The type of iron that the human body needs for good health is the Fe^{2+} ion. Cereals fortified with iron, however, usually contain small grains of elemental iron, Fe. What must the body do to this elemental iron to make good use of it? Oxidation or reduction?

88. Why is it easier for the body to excrete a polar molecule than to excrete a nonpolar molecule? What chemistry does the body use to get rid of molecules it no longer needs?

THINK AND DISCUSS (EVALUATION)

89. Many people hear about atmospheric ozone depletion and wonder why we don't simply replace the ozone that has been destroyed. Knowing about chlorofluorocarbons and knowing how catalysts work, explain how this would not be a lasting solution.

90. Can industries be trusted to self-regulate the amount of pollution they produce? Is government really necessary to enforce these regulations? Shouldn't the mind of the consumer and the economic advantages of sustainable practices be sufficient to motivate industries to protect the environment?

READINESS ASSURANCE TEST (RAT)

If you have a good handle on this chapter, if you really do, then you should be able to score 7 out of 10 on this RAT. If you score less than 7, you need to study further before moving on.

Choose the BEST answer to each of the following:

1. What coefficients balance this equation?

   $\_\ P_4(s) + \_\ H_2(g) \rightarrow \_\ PH_3(g)$

   (a) 4, 2, 3
   (b) 1, 6, 4
   (c) 1, 4, 4
   (d) 2, 10, 8

2. Is the synthesis of ozone, O_3, from oxygen, O_2, an example of an exothermic or endothermic reaction? Why?

   (a) exothermic, because ultraviolet light is emitted during its formation
   (b) endothermic, because ultraviolet light is emitted during its formation
   (c) exothermic, because ultraviolet light is absorbed during its formation
   (d) endothermic, because ultraviolet light is absorbed during its formation

3. How much energy, in kilojoules, is released or absorbed from the reaction of 1 mole of nitrogen, N_2, with 3 moles of molecular hydrogen, H_2, to form 2 moles of ammonia, NH_3? Consult Table 13.1 for bond energies.

   (a) +899 kJ/mol
   (b) −993 kJ/mol
   (c) +80 kJ/mol
   (d) −80 kJ/mol

4. The yeast in bread dough feeds on sugar to produce carbon dioxide. Why does the dough rise faster in a warmer area?

   (a) There is a greater number of effective collisions among the reacting molecules.
   (b) Atmospheric pressure decreases with increasing temperature.
   (c) The yeast tends to "wake up" with warmer temperatures, which is why baker's yeast is best stored in the refrigerator.
   (d) The rate of evaporation increases with increasing temperature.

5. What role do CFCs play in the catalytic destruction of ozone?

   (a) Ozone is destroyed upon binding to a CFC molecule that has been energized by ultraviolet light.
   (b) There is no strong scientific evidence that CFCs play a significant role in the catalytic destruction of ozone.
   (c) CFC molecules activate chlorine atoms into their catalytic action.
   (d) CFC molecules migrate to the upper stratosphere, where they generate chlorine atoms upon being destroyed by ultraviolet light.

6. What is the relationship between the hydroxide ion and a water molecule?
   (a) A hydroxide ion is a water molecule plus a proton.
   (b) A hydroxide ion and a water molecule are the same thing.
   (c) A hydroxide ion is a water molecule minus a hydrogen nucleus.
   (d) A hydroxide ion is a water molecule plus two extra electrons.

7. When the hydronium ion concentration equals 1 mole per liter, what is the pH of the solution? Is the solution acidic or basic?
   (a) pH = 0; acidic
   (b) pH = 1; acidic
   (c) pH = 10; basic
   (d) pH = 7; neutral

8. When lightning strikes, nitrogen molecules, N_2, and oxygen molecules, O_2, in the air react to form nitrates, NO_3^{1-}, which come down in the rain to help fertilize the soil. Is this an example of oxidation or reduction?
   (a) The formation of nitrates is an example of reduction.
   (b) The formation of nitrates is an example of oxidation.
   (c) Both. The nitrogen is oxidized as it reacts with the oxygen while the oxygen is reduced.
   (d) Neither. Although the bonds of both the N_2 and O_2 molecules are broken to form the NO_3^{1-}, neither oxidation nor reduction occurs.

9. Why does a battery that has thick zinc walls last longer than a battery that has thin zinc walls?
   (a) Thick zinc walls prevent the battery from overheating.
   (b) Thick zinc walls prevent electrons from being lost into the surrounding environment.
   (c) Thick zinc walls are chemically resistant to battery acid.
   (d) The zinc walls are transformed into zinc ions as the battery provides electricity.

10. What element is oxidized in this equation and what element is reduced?

$$Sn^{2+} + 2\ Ag \rightarrow Sn + 2\ Ag^{+}$$

   (a) The tin ion, Sn^{2+}, is oxidized; the silver, Ag, is reduced.
   (b) The tin ion, Sn^{2+}, is reduced; the silver, Ag, is oxidized.
   (c) Both the tin ion, Sn^{2+}, and the silver, Ag, are reduced.
   (d) Both the tin ion, Sn^{2+}, and the silver, Ag, are oxidized.

Answers to RAT

1. b, 2. d, 3. d, 4. a, 5. d, 6. c, 7. a, 8. c, 9. d, 10. b

14

CHAPTER 14

Organic Compounds

CARBON ATOMS are perhaps the most versatile of all atoms. Add to this the fact that carbon atoms can also bond with atoms of other elements, and you see the possibility of an endless number of different carbon-based molecules. Each molecule has its own unique set of physical, chemical, and biological properties. The flavor of vanilla, for example, is perceived when the compound *vanillin* is absorbed by the sensory organs in the nose. The flavor of chocolate is generated when a selection of compounds, such as *tetramethylpyrazine*, are absorbed in the nose.

The study of carbon-containing compounds has come to be known as **organic chemistry**. Because organic compounds are so closely tied to living organisms and because they have many applications—flavorings, fuels, polymers, medicines, agriculture, and more—it is important to have a basic understanding of them. We begin with the simplest organic compounds—those consisting of only carbon and hydrogen.

14.1 Hydrocarbons

LEARNING OBJECTIVE
Identify the structures of hydrocarbons.

EXPLAIN THIS How is a road like an oil spill?

Organic compounds that contain only carbon and hydrogen atoms are called **hydrocarbons**, which differ from one another by the number of carbon and hydrogen atoms they contain. The simplest hydrocarbon is methane, CH_4, with only one carbon atom per molecule. Methane is the main component of natural gas. The hydrocarbon octane, C_8H_{18}, has eight carbon atoms per molecule and is a component of gasoline. The hydrocarbon polyethylene contains hundreds of carbon and hydrogen atoms per molecule. Polyethylene is a plastic used to make many familiar items, such as milk containers and plastic bags.

Methane, CH_4

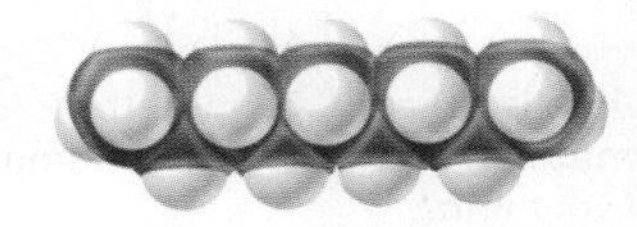

Octane, C_8H_{18}

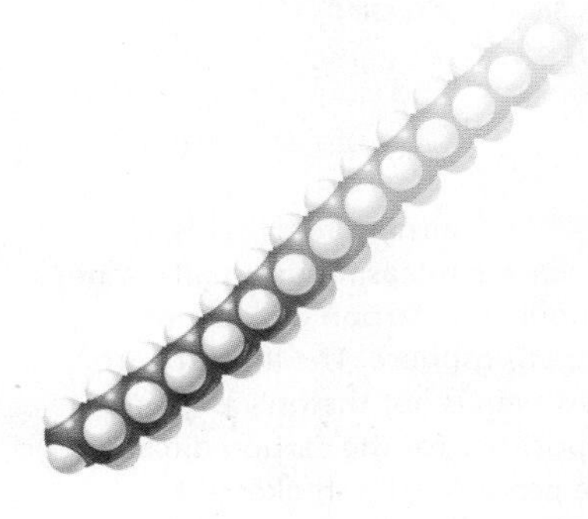

Polyethylene

Hydrocarbons also differ in the way the carbon atoms connect to one another. Figure 14.1 shows the three hydrocarbons pentane, isopentane, and neopentane. These hydrocarbons have the same molecular formula, C_5H_{12}, but they are structurally different from one another.

We can see the different structural features of pentane, isopentane, and neopentane more clearly by drawing the molecules in two dimensions, as shown in the middle row of Figure 14.1. Alternatively, we can represent them by the *stick structures* shown in the bottom row. A stick structure is a commonly used shorthand notation for representing an organic molecule. Each line (stick) represents a covalent bond, and carbon atoms are understood to exist at the end of any line or wherever two or more straight lines meet (unless another type of atom is drawn at the end of the line). Any hydrogen atoms bonded to the carbons are also typically not shown. Instead, their presence is only implied, so that the focus can remain on the skeletal structure that is formed by the carbon atoms.

Molecules such as pentane, isopentane, and neopentane have the same molecular formula, which means they have the same numbers of the same kinds of atoms. The way these atoms are put together, however, is different. We say that each has its own **configuration**, which refers to how the atoms are connected.

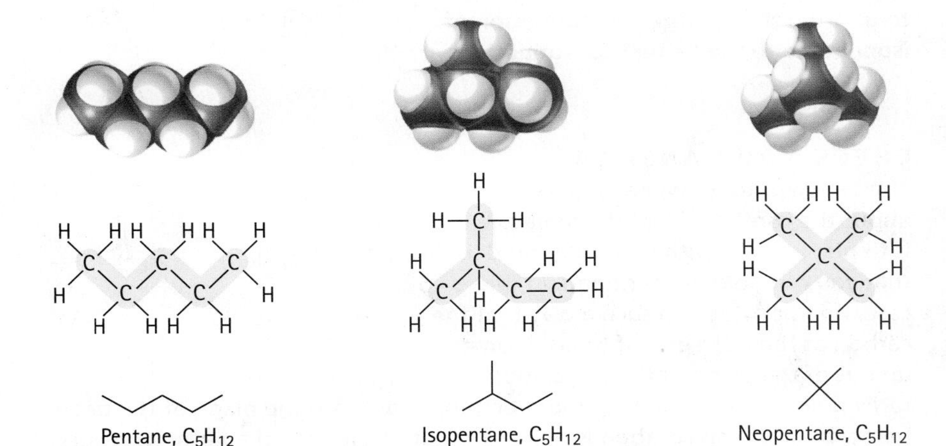

FIGURE 14.1
These three hydrocarbons all have the same molecular formula. We can see their different structural features by highlighting the carbon framework in two dimensions. Easy-to-draw stick structures that use lines for all carbon–carbon covalent bonds are also useful.

When you look at the stick structures, remember that each corner or end represents a carbon atom and that each carbon atom must be bonded four times. Because hydrogen atoms are assumed, they're not usually depicted.

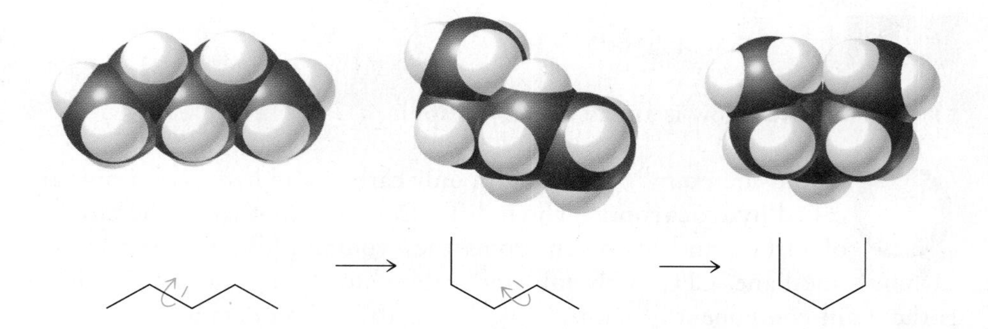

FIGURE 14.2
Three conformations for a molecule of pentane. The molecule looks different in each conformation, but the five-carbon framework is the same in all three conformations. In a sample of liquid pentane, the molecules are found in all conformations—not unlike a bucket of worms.

Different configurations result in different chemical structures. Molecules with the same molecular formula but different configurations (and hence different structures) are known as **structural isomers**. Structural isomers differ from each other and have different physical and chemical properties. For example, pentane has a boiling point of 36°C, isopentane's boiling point is 30°C, and neopentane's boiling point is 10°C.

The number of possible structural isomers for a chemical formula increases rapidly as the number of carbon atoms increases. There are three structural isomers for compounds that have the formula C_5H_{12}, 18 for C_8H_{18}, 75 for $C_{10}H_{22}$, and a whopping 366,319 for $C_{20}H_{42}$!

A carbon-based molecule can have different spatial orientations called **conformations**. Flex your wrist, elbow, and shoulder joints, and you'll find your arm passing through a range of conformations. Likewise, organic molecules can twist and turn about their carbon–carbon single bonds and thus have a range of conformations. The structures in Figure 14.2, for example, are different conformations of pentane. In the language of organic chemistry, we say that the *configuration* of a molecule, such as pentane, has a broad range of *conformations*. Change the *configuration* of pentane, however, and you no longer have pentane. Rather, you have a different structural isomer, such as isopentane, which has its own range of different conformations.

Which is easier to change—the conformation or the configuration of your arm?

MasteringPhysics®
TUTORIAL: Organic Molecules and Isomers

FYI By burning fossil fuels, we are releasing unusually large amounts of carbon dioxide into the atmosphere. The atmosphere, however, is not the only possible repository for the carbon dioxide we produce. The smokestacks of power plants, for example, can be modified to capture CO_2, which is then liquefied and pumped kilometers deep into the ground. This idea is not so far-fetched. Carbon dioxide is already being stored underground at the Salah natural gas refinery in Algeria. Such a system would have its costs. The price of electricity from a CO_2-capturing coal-fired power plant would rise by about 20%. The long-term costs of not implementing such systems, however, may be even higher.

CHECK YOURSELF

Which carbon–carbon bond was rotated to go from the "before" conformation of isopentane to the "after" conformation?

a b c d
Before

a b c d
After

CHECK YOUR ANSWER

The best way to answer any question about the conformation of a molecule is to play around with molecular models that you can hold in your hand. In this case, bond c rotates in such a way that the carbon at the right end of bond d comes up out of the plane of the page, momentarily points straight at you, and then plops back into the plane of the page below bond c. This rotation is similar to that of the arm of an arm wrestler who, with the arm just above the table while on the brink of losing, suddenly gets a surge of strength and swings the opponent's arm (and his or her own) through a half-circle arc and wins.

a b c d
Before

a b c d
After

Hydrocarbons are obtained primarily from coal and petroleum. Most of the coal and petroleum that exists today was formed between 280 million and 395 million years ago when plant and animal matter decayed in the absence of oxygen. At that time, Earth was covered with extensive swamps that, because they were close to sea level, periodically became submerged. The organic matter of the swamps was buried beneath layers of marine sediments and was eventually transformed into either coal or petroleum.

Coal is a solid material that contains many large, complex hydrocarbon molecules. Most of the coal mined today is used to produce steel and to generate electricity at coal-burning power plants.

Petroleum, also called crude oil, is a liquid readily separated into its hydrocarbon components through a process known as *fractional distillation*, shown in Figure 14.3. The crude oil is heated in a pipe still to a temperature high enough to vaporize most of the components. The hot vapor flows into the bottom of a fractionating tower, which is warmer at the bottom than at the top. As the vapor rises in the tower and cools, the various components begin to condense. Hydrocarbons that have high boiling points, such as tar and lubricating stocks, condense first at warmer temperatures. Hydrocarbons that have low boiling points, such as gasoline, travel to the cooler regions at the top of the tower before condensing. Pipes drain the various liquid hydrocarbon fractions from the tower. Natural gas, which is primarily methane, does not condense. It remains a gas and is collected at the top of the tower.

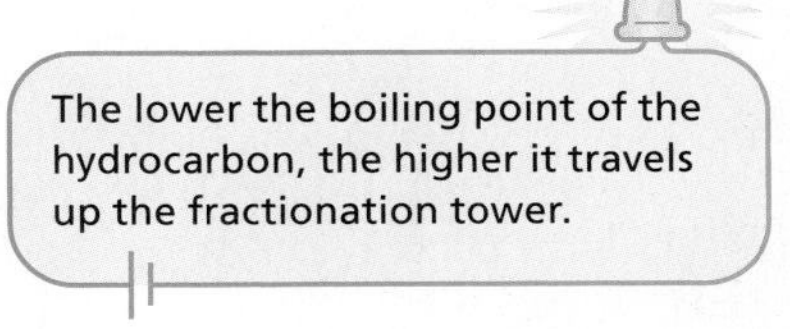

The lower the boiling point of the hydrocarbon, the higher it travels up the fractionation tower.

Differences in the strength of molecular attractions explain why different hydrocarbons condense at different temperatures. As discussed in Section 12.6, in our comparison of induced dipole–induced dipole attractions in methane and octane, larger hydrocarbons experience many more of these attractions than smaller hydrocarbons do. For this reason, the larger hydrocarbons condense readily at high temperatures and so are found at the bottom of the tower. Smaller molecules, because they experience fewer attractions to their neighbors, condense only at the cooler temperatures found at the top of the tower.

UNIFYING CONCEPT

- ***The Electric Force***
Section 7.1

The gasoline obtained from the fractional distillation of petroleum consists of a wide variety of hydrocarbons that have similar boiling points. Some of these components burn more efficiently than others in a car engine. The straight-chain hydrocarbons, such as heptane, burn too quickly, causing what is called

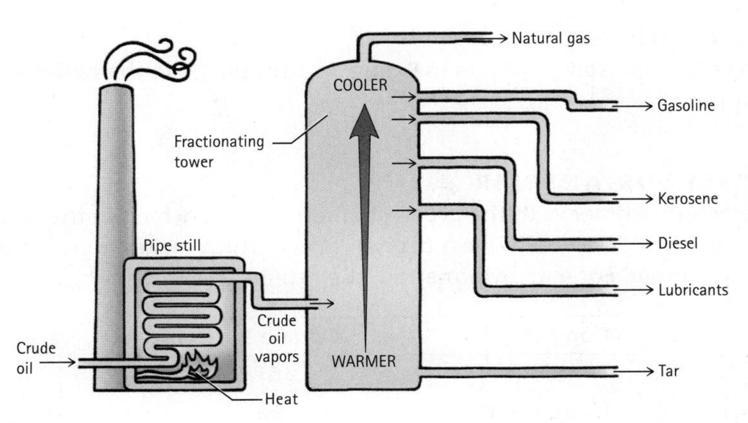

FIGURE 14.3
A schematic for the fractional distillation of petroleum into its useful hydrocarbon components.

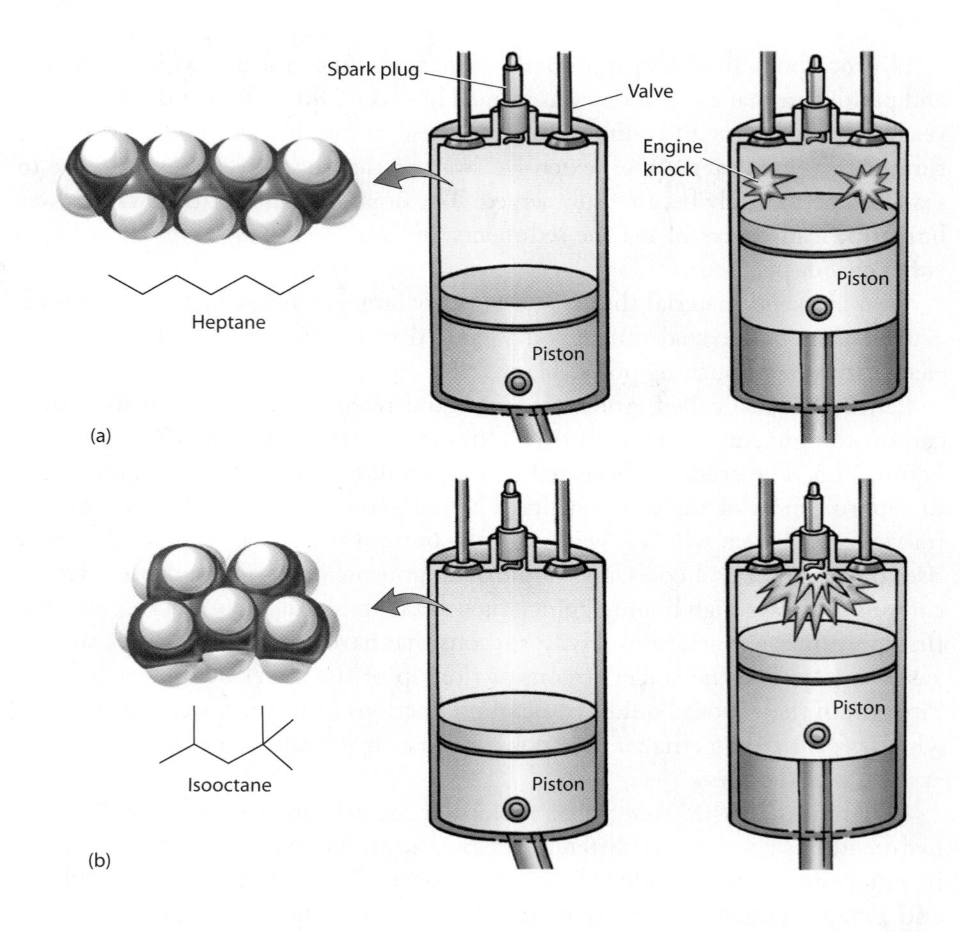

FIGURE 14.4
(a) A hydrocarbon, such as heptane, can be ignited from the heat generated as gasoline is compressed by the piston—before the spark plug fires. This upsets the timing of the engine cycle, giving rise to a knocking sound. (b) Branched hydrocarbons, such as isooctane, burn less readily and are ignited not by compression alone but only when the spark plug fires.

engine knock, as illustrated in Figure 14.4. Gasoline hydrocarbons that have more branching, such as isooctane, burn slowly and result in the engine running more smoothly. These two compounds, heptane and isooctane, are used as standards in assigning *octane ratings* to gasoline. An octane rating of 100 is arbitrarily assigned to isooctane, and heptane is assigned an octane rating of 0. The antiknock performance of a particular gasoline is compared with that of various mixtures of isooctane and heptane, and an octane rating is assigned. Figure 14.5 shows the octane information that appears on a typical gasoline pump.

FIGURE 14.5
Octane ratings are posted on gasoline pumps. The engines of modern cars are designed to run best on 87 octane grade fuel. With higher-octane fuels, not only do you lose performance, but you lose your money as well.

CHECK YOURSELF

Which structural isomer shown in Figure 14.1 should have the highest octane rating?

CHECK YOUR ANSWER

The structural isomer with the greatest amount of branching in the carbon framework likely has the highest octane rating, which makes neopentane the clear winner. For your information, the ratings are as follows:

Compound	Octane Rating
Pentane	61.7
Isopentane	92.3
Neopentane	116

14.2 Unsaturated Hydrocarbons

LEARNING OBJECTIVE
Identify the structures of unsaturated hydrocarbons.

EXPLAIN THIS With four unpaired valence electrons, how can carbon bond to only three adjacent atoms?

Recall from Section 12.1 that a carbon atom has four unpaired valence electrons. As shown in Figure 14.6, each of these electrons is available for pairing with an electron from another atom, such as hydrogen, to form a covalent bond.

In all the hydrocarbons discussed so far, including the methane shown in Figure 14.6, each carbon atom is bonded to four neighboring atoms by four single covalent bonds. Such hydrocarbons are known as **saturated hydrocarbons**. The term *saturated* means that each carbon has as many atoms bonded to it as possible. We now explore cases in which one or more carbon atoms in a hydrocarbon are bonded to fewer than four neighboring atoms. This occurs when at least one of the bonds between a carbon and a neighboring atom is a multiple bond. (See Section 12.3 for a review of multiple bonds.)

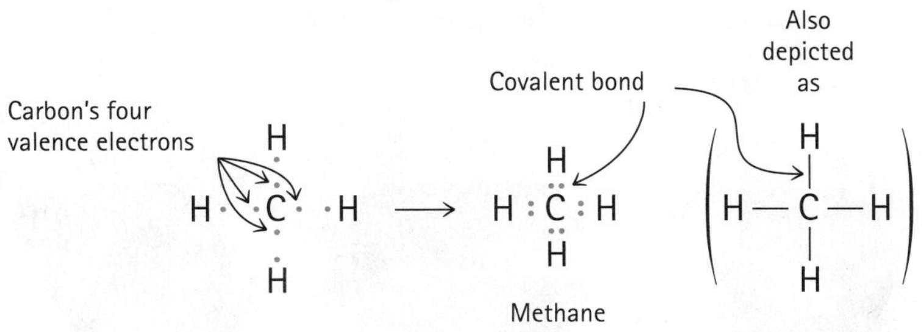

Carbon's four valence electrons
Covalent bond
Also depicted as
Methane

FIGURE 14.6
Carbon has four valence electrons. Each electron pairs with an electron from a hydrogen atom in the four covalent bonds of methane.

A hydrocarbon that contains a multiple bond—either double or triple—is known as an **unsaturated hydrocarbon**. Because of the multiple bond, two of the carbons are bonded to fewer than four other atoms. These carbons are thus said to be *unsaturated.*

Figure 14.7 compares the saturated hydrocarbon butane with the unsaturated hydrocarbon 2-butene. The number of atoms that are bonded to each of the two middle carbons of butane is four, whereas each of the two middle carbons of 2-butene is bonded to only three other atoms—a hydrogen and two carbons.

FYI Saturated fats in our diet have long carbon chains that contain single bonds. These chains tend to be straight, so they pack well together (like wooden matches in a box). Hence, saturated fats, such as lard, are solid at room temperature. The carbon chains of unsaturated fats have double bonds and take on a bent shape, so they don't pack well together. Unsaturated fats, such as vegetable oils, are liquid at room temperature.

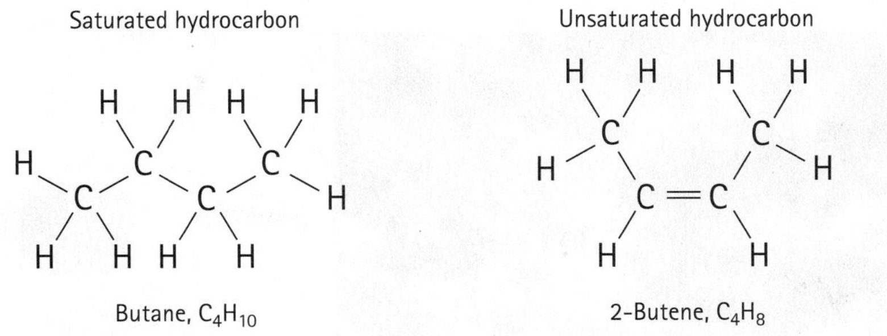

Saturated hydrocarbon
Butane, C_4H_{10}

Unsaturated hydrocarbon
2-Butene, C_4H_8

FIGURE 14.7
The carbons of the hydrocarbon butane are saturated, each being bonded to four other atoms. Because of the double bond, two of the carbons of the unsaturated hydrocarbon 2-butene are bonded to only three other atoms, which makes the molecule an unsaturated hydrocarbon.

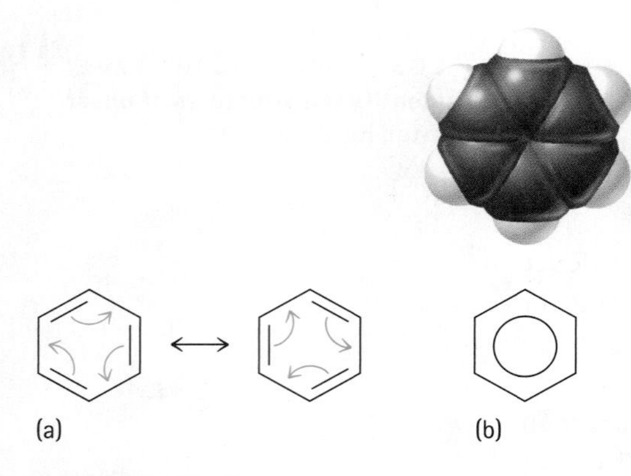

FIGURE 14.8
(a) The double bonds of benzene, C_6C_6, can migrate around the ring. (b) For this reason, they are often represented by a circle inside the ring.

An important unsaturated hydrocarbon is benzene, C_6H_6, which may be drawn as three double bonds contained within a flat hexagonal ring, as shown in Figure 14.8a. Unlike the double-bond electrons in most other unsaturated hydrocarbons, the electrons of the double bonds in benzene are not fixed between any two carbon atoms. Instead, these electrons can move freely around the ring. The double bonds are commonly represented by a circle drawn inside the ring, as shown in Figure 14.8b, rather than by individual double bonds.

Many organic compounds contain one or more benzene rings in their structure. Because many of these compounds are fragrant, any organic molecule that contains a benzene ring is classified as an **aromatic compound** (even if it is not particularly fragrant). Figure 14.9 shows three examples. Toluene, a common solvent used as a paint thinner, is toxic and gives airplane glue its distinctive odor. Some aromatic compounds, such as naphthalene, contain two or more benzene rings fused together. At one time, mothballs were made of naphthalene. Most mothballs sold today, however, are made of the less toxic 1,4-dichlorobenzene.

FIGURE 14.9
The structures for three odoriferous organic compounds that contain one or more benzene rings: toluene, naphthalene, and 1,4-dichlorobenzene.

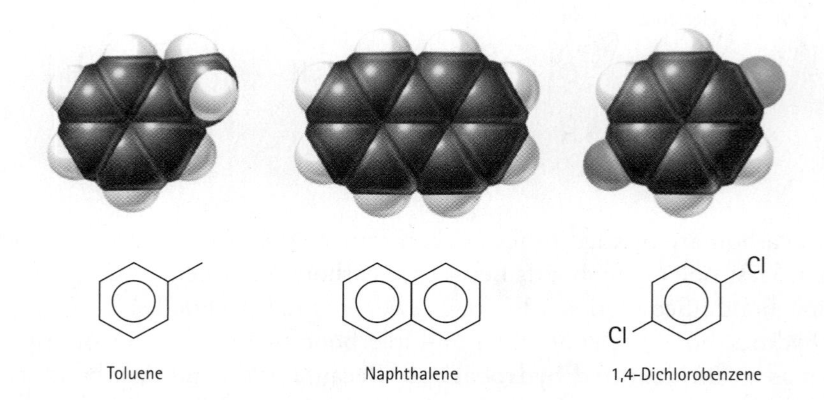

An example of an unsaturated hydrocarbon that contains a triple bond is acetylene, C_2H_2. A confined flame of acetylene burning in oxygen is hot enough to melt iron, which makes acetylene a choice fuel for welding (Figure 14.10).

FIGURE 14.10
The unsaturated hydrocarbon acetylene, C_2H_2, when burned in this torch, produces a flame that is hot enough to melt iron.

CHECK YOURSELF

Prolonged exposure to benzene increases a person's risk of developing certain cancers. The structure of aspirin contains a benzene ring. Does this mean that prolonged exposure to aspirin increases a person's risk of developing cancer?

CHECK YOUR ANSWER

No. Although benzene and aspirin both contain a benzene ring, these two molecules have different overall structures and quite different chemical properties. Each carbon-containing organic compound has its own set of unique physical, chemical, and biological properties. Although benzene may cause cancer, aspirin works as a safe remedy for headaches.

14.3 Functional Groups

LEARNING OBJECTIVE
Discuss the significance of heteroatoms in organic compounds.

EXPLAIN THIS Why are there so many different organic compounds?

Carbon atoms can bond to one another and to hydrogen atoms in many ways, which results in an incredibly large number of hydrocarbons. But carbon atoms can bond to atoms of other elements as well, further increasing the number of possible organic molecules. In organic chemistry, any atom other than carbon or hydrogen in an organic molecule is called a **heteroatom**, where *hetero-* indicates that the atom is different from either carbon or hydrogen.

MasteringPhysics®
TUTORIAL: Introduction to Organic Molecules

A hydrocarbon structure can serve as a framework for the attachment of various heteroatoms. This is analogous to a Christmas tree serving as the scaffolding on which ornaments are hung. Just as the ornaments give character to the tree, so do heteroatoms give character to an organic molecule. Heteroatoms have profound effects on the properties of an organic molecule.

Consider ethane, C_2H_6, and ethanol, C_2H_6O, which differ by only a single oxygen atom. Ethane has a boiling point of −88°C, making it a gas at room temperature, and it does not dissolve in water very well. Ethanol, by contrast, has a boiling point of +78°C, making it a liquid at room temperature. It is infinitely soluble in water, and it is the active ingredient of alcoholic beverages. Consider further ethylamine, C_2H_7N, which has a nitrogen atom on the same basic two-carbon framework. This compound is a corrosive, pungent, highly toxic gas—very different from either ethane or ethanol.

The chemistry of hydrocarbons is surely interesting, but start adding heteroatoms to these organic molecules and the chemistry becomes extraordinarily interesting. The organic chemicals of living organisms, for example, all contain heteroatoms.

Organic molecules are classified according to the functional groups they contain. A **functional group** is defined as a combination of atoms that behave as a unit. Most functional groups are distinguished by the heteroatoms they contain; some common groups are listed in Table 14.1.

The remainder of this chapter introduces the classes of organic molecules shown in Table 14.1. The role heteroatoms play in determining the properties

TABLE 14.1 FUNCTIONAL GROUPS IN ORGANIC MOLECULES

General Structure	Class	General Structure	Class
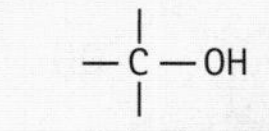 —C—OH Hydroxyl group	Alcohols	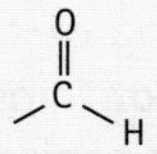 O=C—H Aldehyde group	Aldehydes
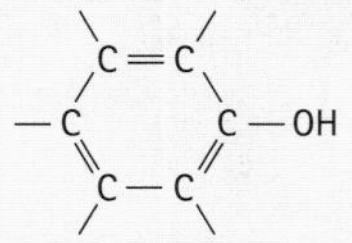 C=C, C—C ring, C—OH Phenolic group	Phenols	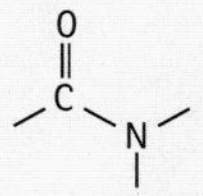 O=C—N Amide group	Amides
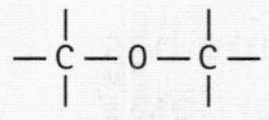 —C—O—C— Ether group	Ethers	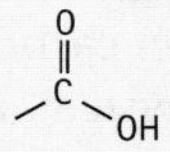 O=C—OH Carboxyl group	Carboxylic acids
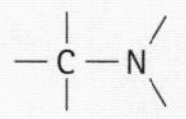 —C—N Amine group	Amines	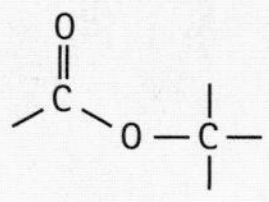 O=C—O—C— Ester group	Esters
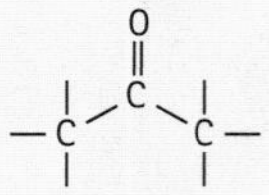 —C—C(=O)—C— Ketone group	Ketones		

of each class is the underlying theme. As you study this material, focus on understanding the chemical and physical properties of the various classes of compounds; doing so will give you a greater appreciation of the remarkable diversity of organic molecules and their many applications.

CHECK YOURSELF

What is the significance of heteroatoms in an organic molecule?

CHECK YOUR ANSWER

Heteroatoms largely determine an organic molecule's "personality."

LEARNING OBJECTIVE

Review the general properties of alcohols, phenols, and ethers.

14.4 Alcohols, Phenols, and Ethers

EXPLAIN THIS What do alcohols, phenols, and ethers have in common?

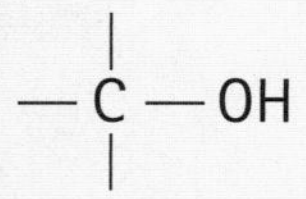

—C—OH

Hydroxyl group

Alcohols are organic molecules in which a *hydroxyl group* is bonded to a saturated carbon. The hydroxyl group consists of an oxygen bonded to a hydrogen. Because of the polarity of the oxygen–hydrogen bond, low-mass alcohols are often soluble in water, which is itself very polar. Three common alcohols and their melting and boiling points are listed in Table 14.2.

More than 11 billion pounds of methanol, CH_3OH, are produced annually in the United States. Most of it is used for making formaldehyde and acetic acid, important starting materials in the production of plastics. In addition, methanol

TABLE 14.2 SIMPLE ALCOHOLS

Structure	Scientific Name	Common Name	Melting Point (°C)	Boiling Point (°C)
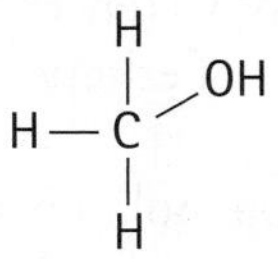 CH_3OH (structural formula)	Methanol	Methyl alcohol	−97	65
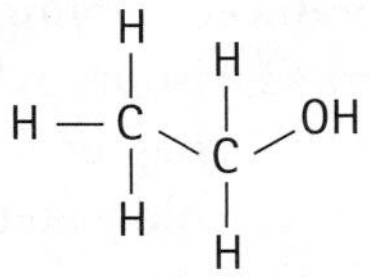 CH_3CH_2OH (structural formula)	Ethanol	Ethyl alcohol	−115	78
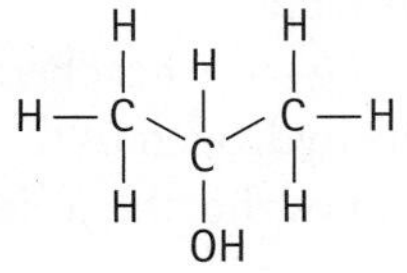 $CH_3CH(OH)CH_3$ (structural formula)	2-Propanol	Isopropyl alcohol	−126	97

is used as a solvent, an octane booster, and an anti-icing agent in gasoline. Sometimes called *wood alcohol* because it can be obtained from wood, methanol should never be ingested because, in the body, it is metabolized to formaldehyde and formic acid. Formaldehyde is harmful to the eyes, can lead to blindness, and was once used to preserve dead biological specimens. Formic acid, the active ingredient in an ant bite, can lower the pH of the blood to dangerous levels. Methanol has its own inherent toxicities. Ingesting only about 15 mL (about 3 Tbsp) of methanol may lead to blindness, and about 30 mL can cause death.

Ethanol, C_2H_5OH, on the other hand, is the "alcohol" of alcoholic beverages, and it is one of the oldest chemicals manufactured by humans. Ethanol is prepared by feeding the sugars of various plants to certain yeasts, which produce ethanol through a biological process known as *fermentation*. Ethanol is also widely used as an industrial solvent. For many years, ethanol intended for this purpose was made by fermentation, but today industrial-grade ethanol is more cheaply manufactured from petroleum by-products, such as ethene, as Figure 14.11 illustrates.

The liquid produced by fermentation has an ethanol concentration no higher than about 12% because at this concentration the yeast cells begin to die. This is why most wines have an alcohol content of about 12%—they are produced solely by fermentation. To attain the higher ethanol concentrations found in "hard" alcoholic beverages such as gin and vodka, the fermented liquid must be distilled. In the United States, the ethanol content of distilled alcoholic beverages

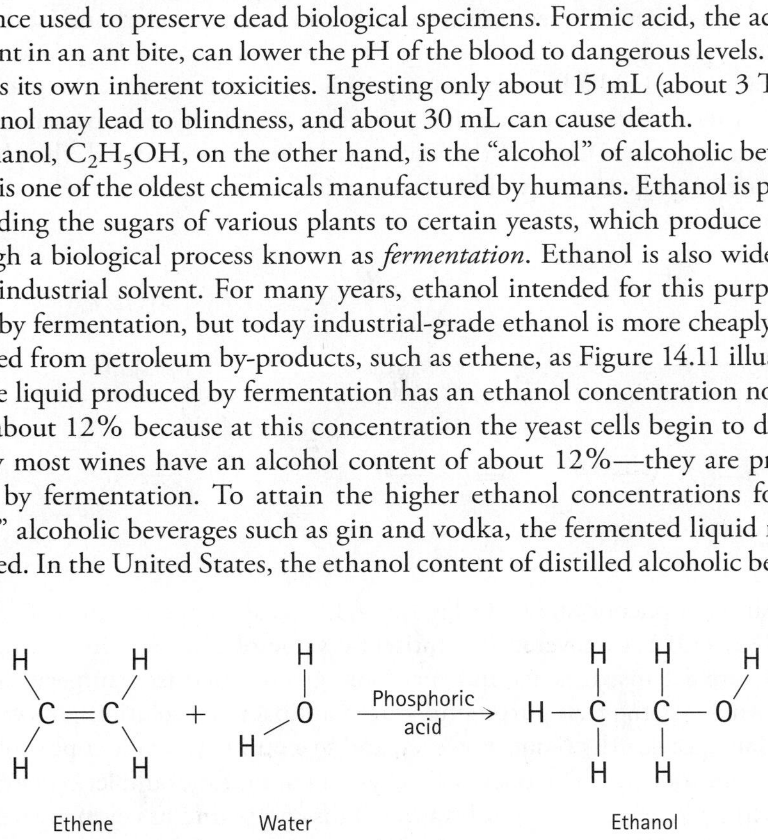

FYI Just as the body metabolizes methanol into formaldehyde, HCOH, it metabolizes ethanol into acetaldehyde, CH_3COH. Acetaldehyde won't cause you to go blind, but it does provide for some painful side effects, which people who drink too much experience as part of their "hangover."

$$H_2C{=}CH_2 + H_2O \xrightarrow{\text{Phosphoric acid}} CH_3CH_2OH$$

Ethene Water Ethanol

FIGURE 14.11
Ethanol can be synthesized from the unsaturated hydrocarbon ethene, with phosphoric acid as a catalyst.

is measured as *proof,* which is twice the percentage of ethanol. An 86-proof whiskey, for example, is 43% ethanol by volume. The term *proof* evolved from a crude method once used to test alcohol content. Gunpowder was wetted with a beverage of suspect alcohol content. If the beverage was primarily water, the powder did not ignite. If the beverage contained a significant amount of ethanol, the powder burned, thus providing "proof" of the beverage's worth.

A third well-known alcohol is isopropyl alcohol, also called 2-propanol. This is the rubbing alcohol you buy at the drugstore. Although 2-propanol has a relatively high boiling point, it evaporates readily, leading to a pronounced cooling effect when it is applied to skin—an effect once used to reduce fevers. (Isopropyl alcohol is very toxic if ingested. See the activities at the end of this chapter to understand why. In place of isopropyl alcohol, washcloths wetted with cold water are nearly as effective in reducing fever, and they are far safer.) You are probably most familiar with the use of isopropyl alcohol as a topical disinfectant.

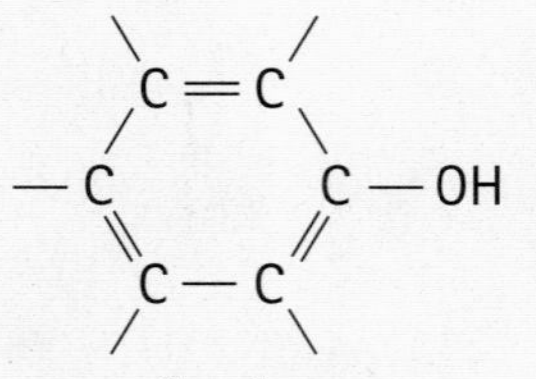

Phenolic group

Phenols contain a phenolic group, which consists of a hydroxyl group attached to a benzene ring. Because of the presence of the benzene ring, the hydrogen of the hydroxyl group is readily lost in an acid–base reaction, which makes the phenolic group mildly acidic.

The reason for this acidity is illustrated in Figure 14.12. How readily an acid donates a hydrogen ion is a function of how well the acid can accommodate the resulting negative charge it gains after donating the hydrogen ion. After phenol donates the hydrogen ion, it becomes a negatively charged phenoxide ion. The negative charge of the phenoxide ion, however, is not restricted to the oxygen atom. Recall that the electrons of the benzene ring can migrate around the ring. In a similar manner, the electrons responsible for the negative charge of the phenoxide ion can migrate around the ring, as shown in Figure 14.12. Just as several people can easily hold a hot potato by quickly passing it around, the phenoxide ion can easily hold the negative charge because the charge gets passed around. Because the negative charge of the ion is so nicely accommodated, the phenolic group is more acidic than it would be otherwise.

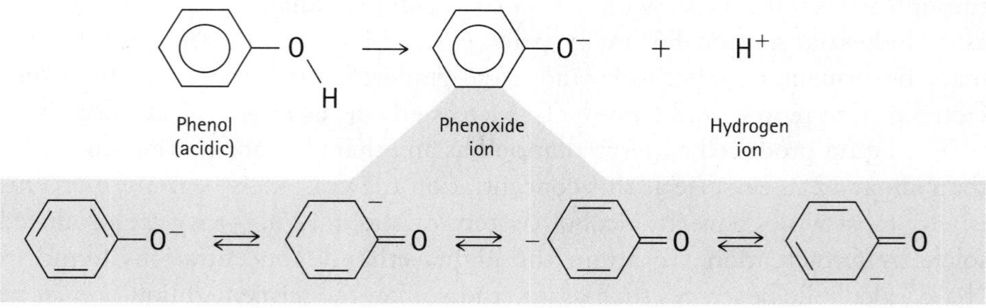

Phenol (acidic) → Phenoxide ion + Hydrogen ion

FIGURE 14.12
The negative charge of the phenoxide ion can migrate to select positions on the benzene ring. This mobility helps accommodate the negative charge, which is why the phenolic group readily donates a hydrogen ion.

We're classifying organic molecules based on the functional groups they contain. As you will see shortly, however, organic molecules may contain more than one type of functional group. A single organic molecule, therefore, might be classified as both a phenol and an ether.

The simplest phenol, shown in Figure 14.13, is called phenol. In 1867, Joseph Lister (1827–1912) discovered the antiseptic value of phenol, which, when applied to surgical instruments and incisions, greatly increased surgery survival rates. Phenol was the first purposefully used antibacterial solution, or *antiseptic.* Phenol damages healthy tissue, however, and so a number of milder phenols have since been introduced. The phenol 4-hexylresorcinol, for example, is commonly used in throat lozenges and mouthwashes. This compound has even greater antiseptic properties than phenol, and yet it does not damage tissue. Listerine brand mouthwash (named after Joseph Lister) contains the antiseptic phenols thymol and methyl salicylate.

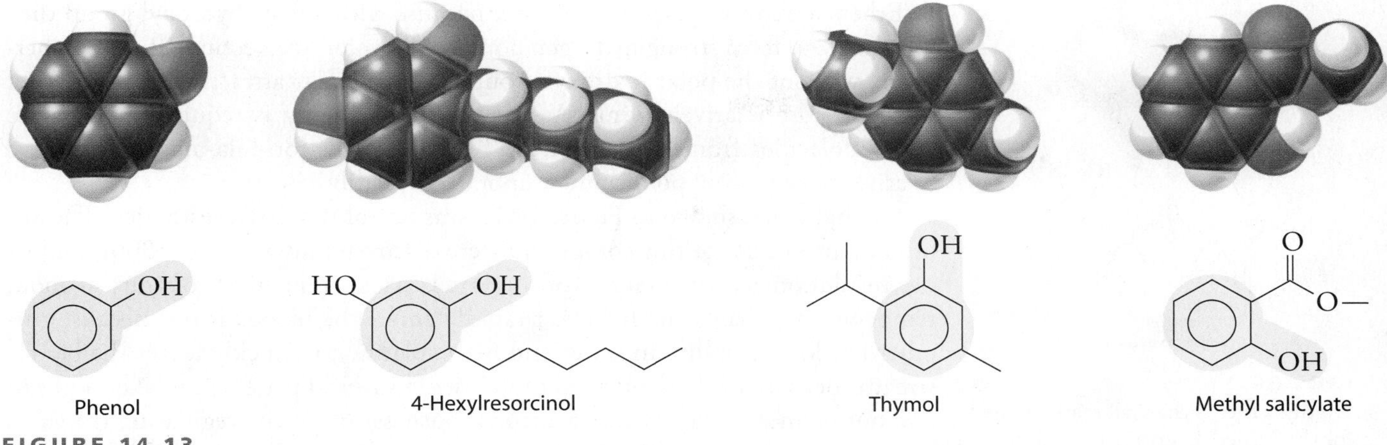

FIGURE 14.13
Every phenol contains a phenolic group (highlighted in blue).

CHECK YOURSELF
Why are alcohols less acidic than phenols?

CHECK YOUR ANSWER
An alcohol does not contain a benzene ring adjacent to the hydroxyl group. If the alcohol were to donate the hydroxyl hydrogen, the result would be a negative charge on the oxygen. Without an adjacent benzene ring, this negative charge has nowhere to go. As a result, an alcohol behaves as only a very weak acid, much the way water does.

Ethers are organic compounds structurally related to alcohols. The oxygen atom in an ether group, however, is bonded not to a carbon and a hydrogen but rather to two carbons. As we see in Figure 14.14, ethanol and dimethyl ether have the same chemical formula, C_2H_6O, but their physical properties are vastly different. Whereas ethanol is a liquid at room temperature (boiling point 78°C) and mixes quite well with water, dimethyl ether is a gas at room temperature (boiling point −25°C) and is much less soluble in water.

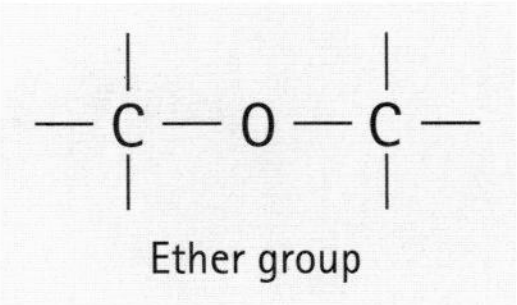

—C—O—C—

Ether group

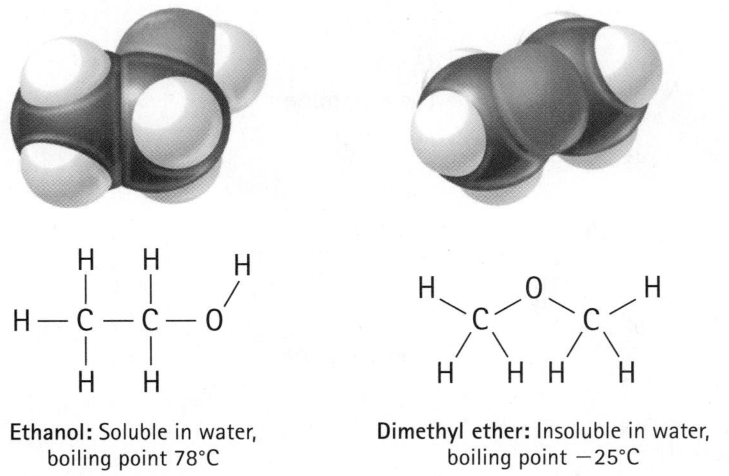

Ethanol: Soluble in water, boiling point 78°C

Dimethyl ether: Insoluble in water, boiling point −25°C

FIGURE 14.14
The oxygen in an alcohol, such as ethanol, is bonded to one carbon atom and one hydrogen atom. The oxygen in an ether, such as dimethyl ether, is bonded to two carbon atoms. Because of this difference, alcohols and ethers of similar molecular mass have vastly different physical properties.

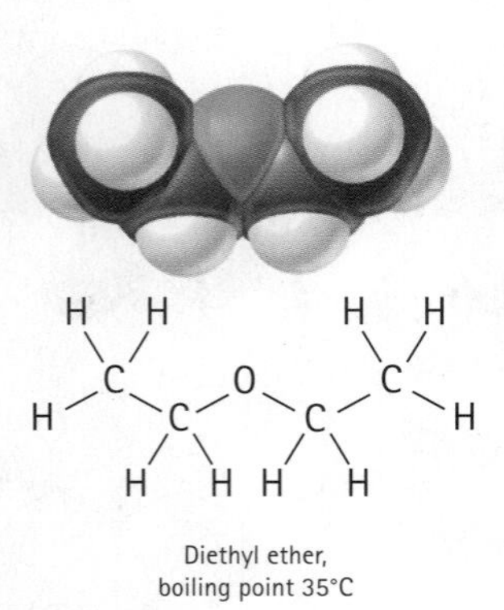

FIGURE 14.15
Diethyl ether is the systematic name for the "ether" historically used as an anesthetic.

Ethers are not very soluble in water because without the hydroxyl group they are unable to form strong hydrogen bonds with water (see Section 12.6). Furthermore, without the polar hydroxyl group, the molecular attractions among ether molecules are relatively weak. As a result, little energy is required to separate ether molecules from one another. This is why low-formula-mass ethers have relatively low boiling points and evaporate so readily.

Diethyl ether, shown in Figure 14.15, was one of the first anesthetics. The anesthetic properties of this compound were discovered in the early 1800s, and its use revolutionized the practice of surgery. Because of its high volatility at room temperature, inhaled diethyl ether rapidly enters the bloodstream. Because this ether has low solubility in water and high volatility, it quickly leaves the bloodstream. Because of these physical properties, a surgical patient can be brought in and out of anesthesia (a state of unconsciousness) simply by regulating the gases breathed. Modern gaseous anesthetics have fewer side effects than diethyl ether, but they operate on the same principle.

LEARNING OBJECTIVE
Review the general properties of amines and alkaloids.

14.5 Amines and Alkaloids

EXPLAIN THIS Why are rainforests of great interest to pharmaceutical companies?

Amine group

Amines are organic compounds that contain the amine group—a nitrogen atom bonded to one, two, or three saturated carbons. Amines are typically less soluble in water than are alcohols because the nitrogen–hydrogen bond is not quite as polar as the oxygen–hydrogen bond. The lower polarity of amines also means their boiling points are typically somewhat lower than those of alcohols of similar formula mass. Table 14.3 lists three simple amines.

TABLE 14.3 SIMPLE AMINES

Structure	Name	Melting Point (°C)	Boiling Point (°C)
	Ethylamine	−18	17
	Diethylamine	−50	55
	Triethylamine	−7	89

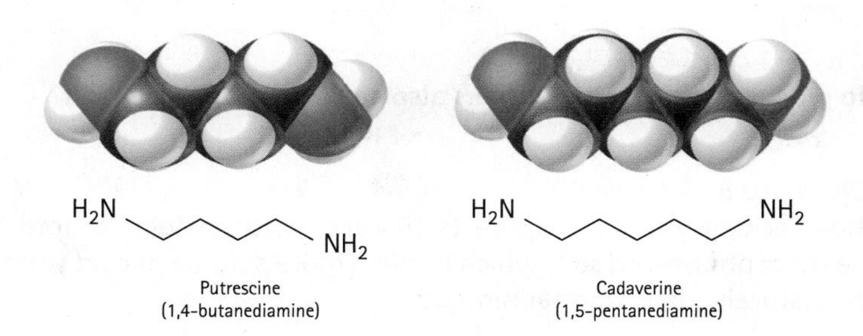

FIGURE 14.16
Low-formula-mass amines such as these tend to have offensive odors.

One of the most notable physical properties of many low-formula-mass amines is their offensive odor. Figure 14.16 shows two appropriately named amines, putrescine and cadaverine, which are partly responsible for the odor of decaying flesh.

Amines are typically alkaline because the nitrogen atom readily accepts a hydrogen ion from water, as Figure 14.17 illustrates.

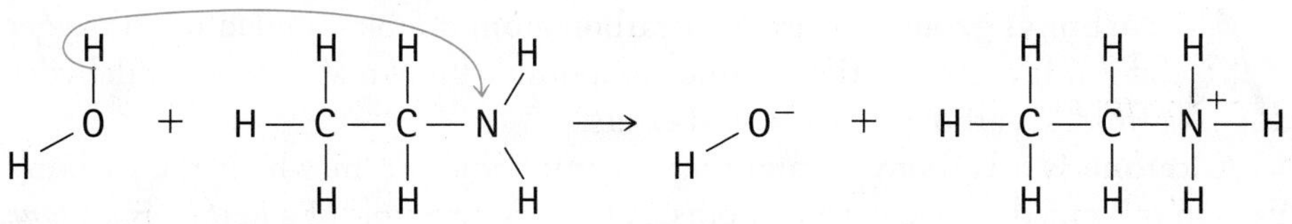

Water (acid) + Ethylamine (base) ⟶ Hydroxide ion + Ethylammonium ion

FIGURE 14.17
Ethylamine acts as a base and accepts a hydrogen ion from water to become the ethylammonium ion. This reaction generates a hydroxide ion, which increases the pH of the solution.

A group of naturally occurring complex molecules that are alkaline because they contain nitrogen atoms are often called *alkaloids*. Because many alkaloids have medicinal value, there is great interest in isolating these compounds from plants or marine organisms that contain them. As shown in Figure 14.18, an alkaloid reacts with an acid to form a salt that is usually quite soluble in water. This is in contrast to the non-ionized form of the alkaloid, known as a *free base*, which is typically insoluble in water.

Most alkaloids exist in nature not in their free-base form but rather as the salts of naturally occurring acids known as *tannins*, a group of phenol-based organic acids that have complex structures. The alkaloid salts of these acids are usually much more soluble in hot water than in cold water. The caffeine in coffee and tea exists in the form of the tannin salt, which is why coffee and tea are more effectively brewed in hot water. As Figure 14.19 relates, tannins are also responsible for the stains caused by these beverages.

FIGURE 14.19
Tannins are responsible for the brown stains in coffee mugs or on a coffee drinker's teeth. Because tannins are acidic, they can be readily removed with an alkaline cleanser. Use a little laundry bleach on the mug, and brush your teeth with baking soda.

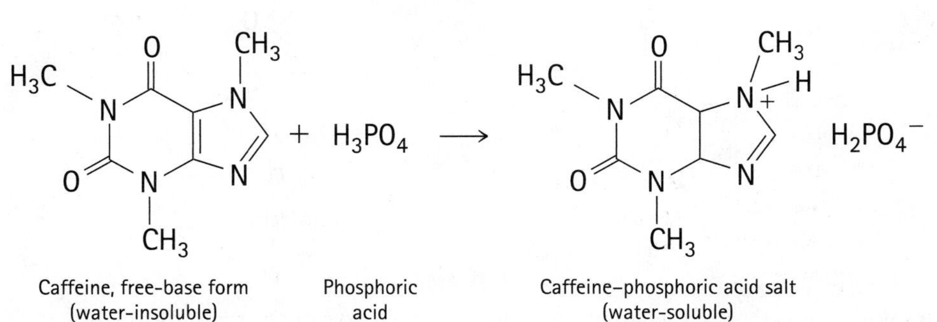

FIGURE 14.18
All alkaloids are bases that react with acids to form salts. An example is the alkaloid caffeine, shown here reacting with phosphoric acid.

Most pharmaceuticals that can be administered orally contain nitrogen heteroatoms in the water-soluble salt form.

CHECK YOURSELF
Why do most caffeinated soft drinks also contain phosphoric acid?

CHECK YOUR ANSWER
Phosphoric acid, as shown in Figure 14.18, reacts with caffeine to form the caffeine–phosphoric acid salt, which is much more soluble in cold water than the naturally occurring tannin salt.

LEARNING OBJECTIVE
Review the general properties of carbonyl compounds.

14.6 Carbonyl Compounds

EXPLAIN THIS Why does the carbon of the carbonyl usually have a slightly positive charge?

Ketone group

Aldehyde group

A **carbonyl group** consists of a carbon atom double-bonded to an oxygen atom. It occurs in the organic compounds known as ketones, aldehydes, amides, carboxylic acids, and esters.

A **ketone** is a carbonyl-containing organic molecule in which the carbonyl carbon is bonded to two carbon atoms. A familiar example of a ketone is *acetone*, which is often used in fingernail-polish remover and is shown in Figure 14.20a. In an **aldehyde**, the carbonyl carbon is bonded either to one carbon atom and one hydrogen atom, as in Figure 14.20b, or, in the special case of formaldehyde, to two hydrogen atoms.

Many aldehydes are particularly fragrant. A number of flowers, for example, owe their pleasant odor to the presence of simple aldehydes. The smells of lemons, cinnamon, and almonds are due to the aldehydes citral, cinnamaldehyde, and benzaldehyde, respectively. The structures of these three aldehydes are shown in Figure 14.21. Another aldehyde, vanillin, which was

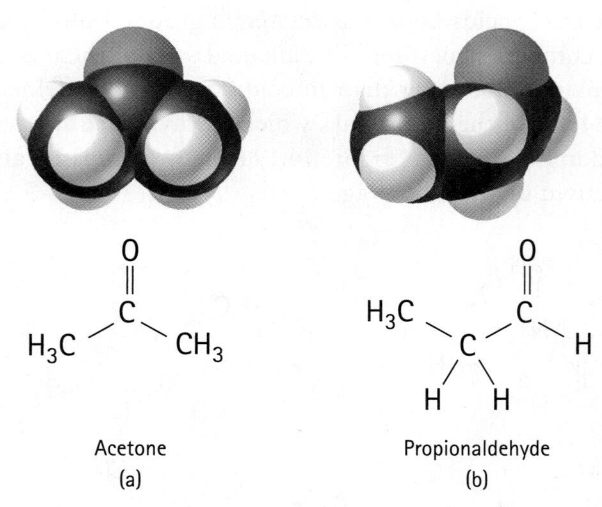

FIGURE 14.20
(a) When the carbon of a carbonyl group is bonded to two carbon atoms, the result is a ketone. An example is acetone. (b) When the carbon of a carbonyl group is bonded to at least one hydrogen atom, the result is an aldehyde. An example is propionaldehyde.

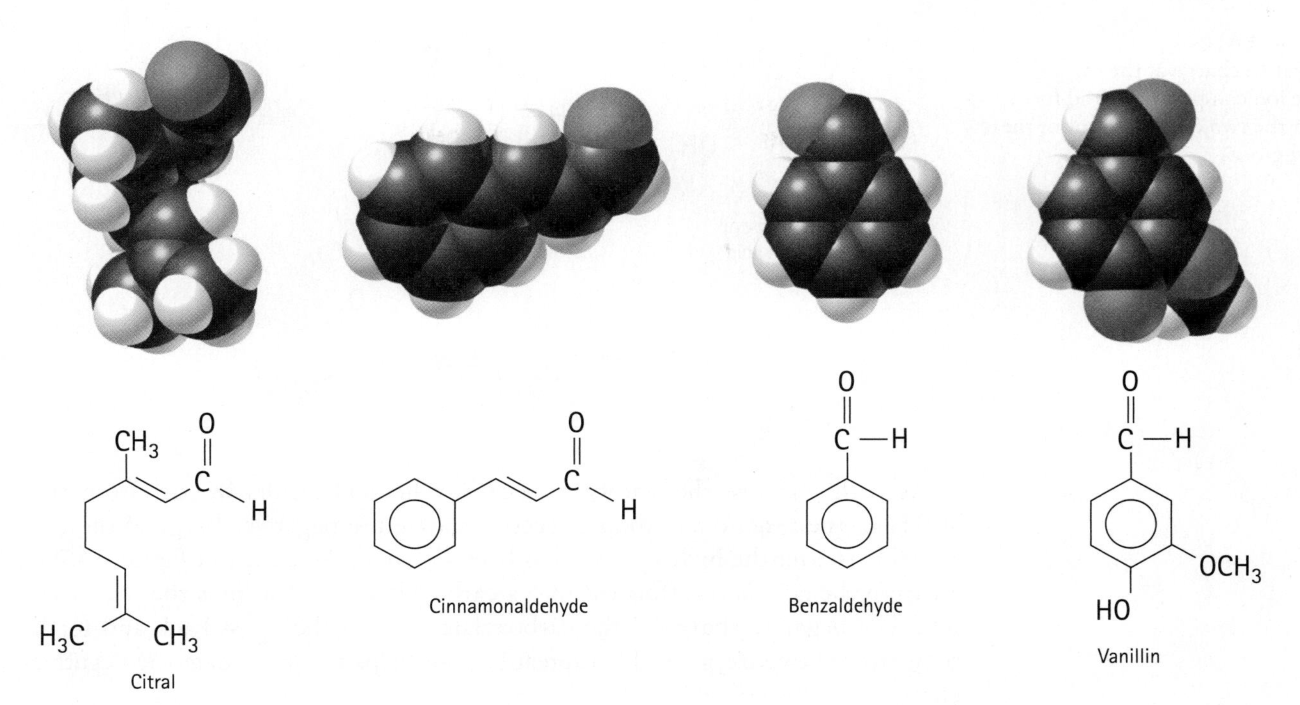

FIGURE 14.21
Aldehydes are responsible for many familiar fragrances.

introduced at the beginning of this chapter, is the key flavoring molecule derived from seed pods of the vanilla orchid. You may have noticed that vanilla seed pods and vanilla extract are fairly expensive. Imitation vanilla flavoring is less expensive because it is merely a solution of the compound vanillin, which is economically synthesized from waste chemicals from the wood-pulp industry. Imitation vanilla does not taste the same as natural vanilla extract, however, because, in addition to vanillin, many other flavorful molecules contribute to the complex taste of natural vanilla. Many books manufactured in the days before "acid-free" paper smell of vanilla because of the vanillin formed and released as the paper ages, a process that is accelerated by the acids the paper contains.

An **amide** is a carbonyl-containing organic molecule in which the carbonyl carbon is bonded to a nitrogen atom. The active ingredient of most mosquito repellents is an amide whose chemical name is *N,N*-diethyl-*m*-toluamide but is commercially known as DEET, shown in Figure 14.22. This compound is actually not an insecticide. Rather, it causes certain insects, especially mosquitoes, to lose their sense of direction, which effectively protects DEET wearers from being bitten.

A **carboxylic acid** is a carbonyl-containing organic molecule in which the carbonyl carbon is bonded to a hydroxyl group. As its name implies, this functional group can donate hydrogen ions. Organic molecules that contain it are therefore acidic. An example is acetic acid, $C_2H_4O_2$, which, after water, is the main ingredient of vinegar.

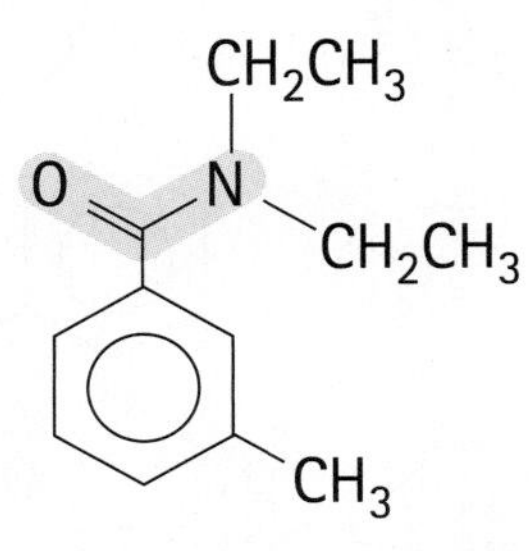

N,N-Diethyl-*m*-toluamide
(DEET)

FIGURE 14.22
N,N-diethyl-*m*-toluamide is an example of an amide. Amides contain the amide group, shown highlighted in blue.

FIGURE 14.23
The negative charge of the carboxylate ion can pass back and forth between the two oxygen atoms of the carboxyl group.

$$H_3C{-}C({=}O){-}OH \longrightarrow H_3C{-}C({=}O){-}O^- + H^+$$

Carboxyl group in acetic acid

Carboxylate ion in acetate ion

Hydrogen ion

$$-C({=}O){-}O^- \rightleftharpoons -C(O^-){=}O$$

As with phenols, the acidity of a carboxylic acid results in part from the ability of the functional group to accommodate the negative charge of the ion that forms after the hydrogen ion has been donated. As shown in Figure 14.23, a carboxylic acid is transformed into a carboxylate ion as it loses the hydrogen ion. The negative charge of the carboxylate ion can then pass back and forth between the two oxygens. This spreading out helps accommodate the negative charge.

An interesting example of an organic compound that contains both a carboxylic acid and a phenol is salicylic acid, found in the bark of willow trees and illustrated in Figure 14.24a. At one time brewed for its antipyretic (fever-reducing) effect, salicylic acid is an important analgesic (painkiller), but it causes nausea and stomach upset because of its relatively high acidity, a result of the presence of two acidic functional groups. In 1899, Friedrich Bayer and Company, in Germany, introduced a chemically modified version of this compound in which the acidic

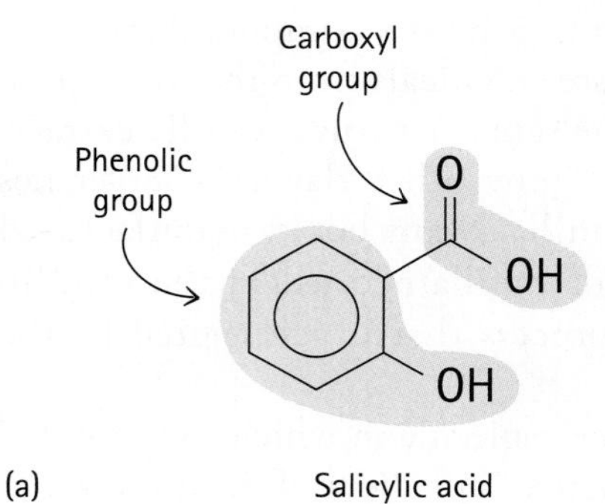

(a)

Carboxyl group

O

OH

O

O

Ester

Aspirin (acetylsalicylic acid)

(b)

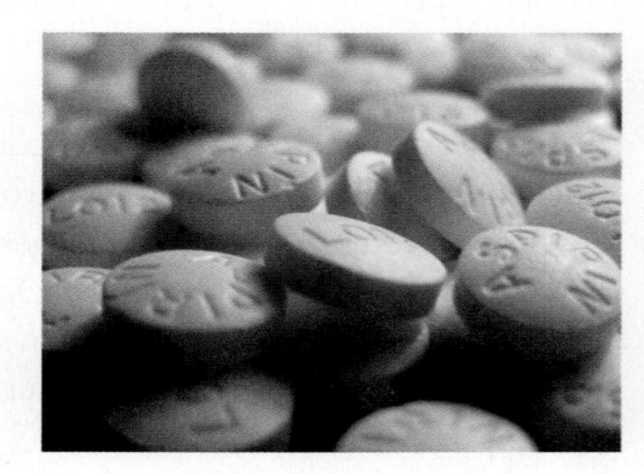

FIGURE 14.24
(a) Salicylic acid, which is found in the bark of willow trees, is an example of a molecule that contains both a carboxyl group and a phenolic group. (b) Aspirin, acetylsalicylic acid, is less acidic than salicylic acid because it no longer contains the acidic phenolic group, which has been converted into an ester.

phenolic group was transformed into an ester functional group. The result was the less acidic and more tolerable acetylsalicylic acid, the chemical name for aspirin, shown in Figure 14.24b.

An **ester** is an organic molecule similar to a carboxylic acid except that in the ester, the hydroxyl hydrogen is replaced by a carbon. Unlike carboxylic acids, esters are not acidic because they lack the hydrogen of the hydroxyl group. Like aldehydes, many simple esters have notable fragrances and are often used as flavorings. Four familiar ones are listed in Table 14.4.

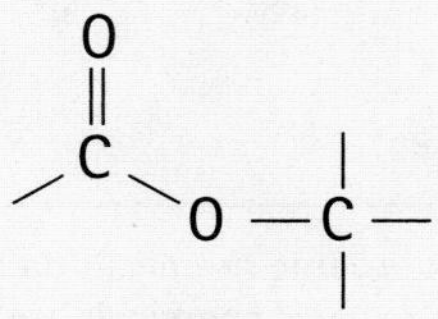

$$-C(=O)-O-C-$$

Ester group

Esters are fairly easy to synthesize by dissolving a carboxylic acid in an alcohol and then bringing the mixture to a boil in the presence of a strong acid, such as sulfuric acid, H_2SO_4. The synthesis of methyl salicylate from salicylic acid and methanol is one example. Methyl salicylate is responsible for the smell of wintergreen and is a common ingredient of hard candies.

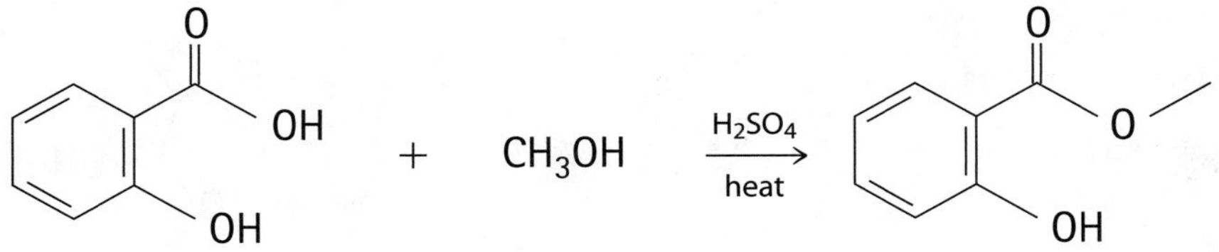

$$\text{Salicylic acid} + CH_3OH \xrightarrow[\text{heat}]{H_2SO_4} \text{Methyl salicylate}$$

Salicylic acid | Methanol | Methyl salicylate (wintergreen)

FYI In the 1800s most salicylic acid used by people was produced not from willow bark but from coal tar. Tar residues within the salicylic acid had a nasty taste. This, combined with salicylic acid's stomach irritation, led many to view the salicylic acid cure as worse than the disease. Felix Hoffman, a chemist working at Bayer, added the acetyl group to the phenol group of salicylic acid in 1897. According to Bayer, Hoffman was inspired by his father, who had been complaining about salicylic acid's side effects. To market the new drug, Bayer invented the name *aspirin*, in which "a" is for acetyl; "spir" is for the spirea flower, another natural source of salicylic acid; and "in" is used as a common suffix for medications. After World War I, Bayer, a German company, lost the rights to use the name *aspirin*. Bayer didn't regain these rights until 1994 for a steep price of $1 billion.

There is much more to organic chemistry than just learning functional groups and their general properties. Many, if not most, practicing organic chemists dedicate much of their time to the synthesis of organic molecules that have practical applications, such as for agriculture or pharmaceuticals. Often these target molecules are organic compounds that have been isolated from nature, where they

TABLE 14.4 SOME ESTERS AND THEIR FLAVORS AND ODORS

Structure	Name	Flavor/Odor
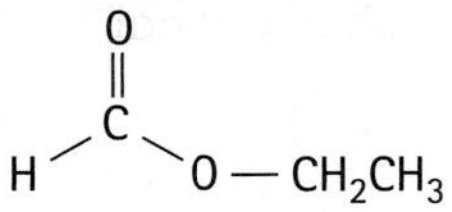 $H-C(=O)-O-CH_2CH_3$	Ethyl formate	Rum
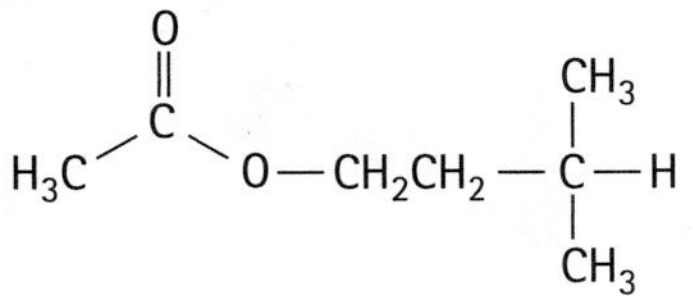 $H_3C-C(=O)-O-CH_2CH_2-CH(CH_3)-CH_3$	Isopentyl acetate	Banana
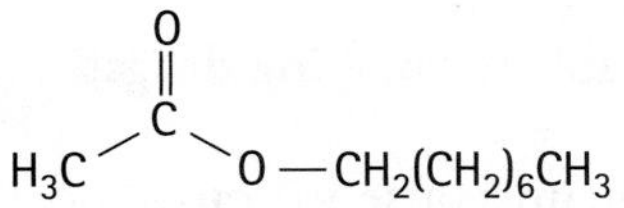 $H_3C-C(=O)-O-CH_2(CH_2)_6CH_3$	Octyl acetate	Orange
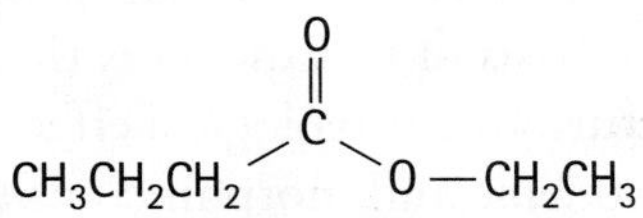 $CH_3CH_2CH_2-C(=O)-O-CH_2CH_3$	Ethyl butyrate	Pineapple

can be found in only small quantities. To create large amounts of these chemicals, the organic chemist devises a pathway through which the compound can be synthesized in the laboratory from readily available smaller compounds. Once synthesized, the compound produced in the laboratory is chemically identical to that compound found in nature. In other words, it has the same physical and chemical properties and also has the same biological effects, if any.

We eat organic chemicals daily. In fact, organic chemicals are the only things we eat, except for some important minerals, such as the ions of sodium and calcium.

CHECK YOURSELF

Identify all the functional groups in these four molecules (ignore the sulfur group in penicillin G).

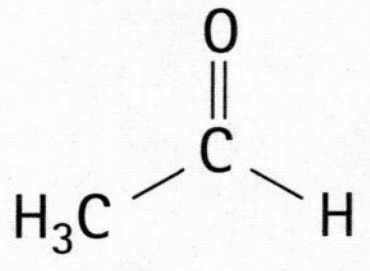

Acetaldehyde

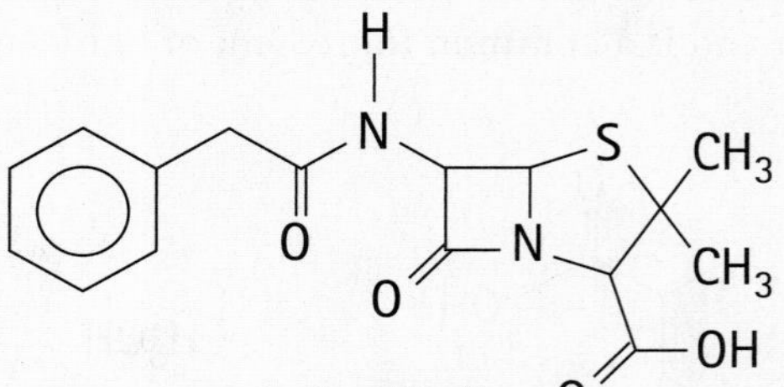

Penicillin G

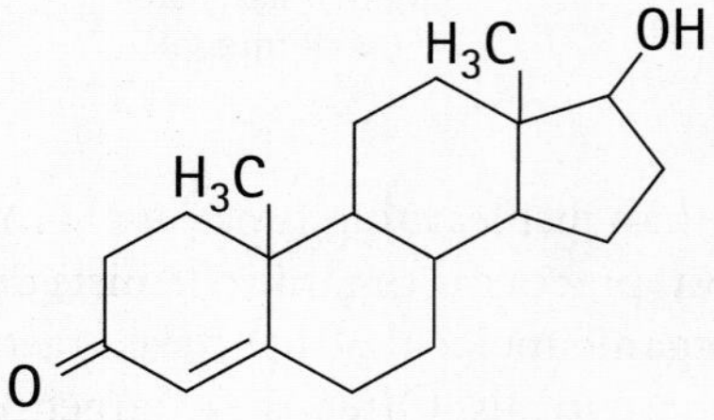

Testosterone

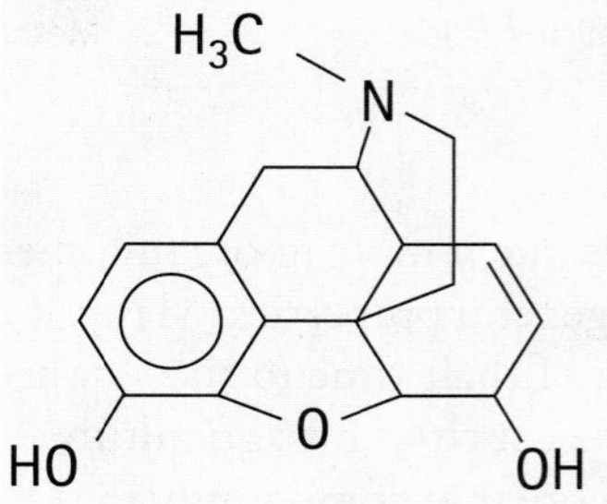

Morphine

CHECK YOUR ANSWERS

Acetaldehyde: aldehyde. Penicillin G: amide (two amide groups) and carboxylic acid. Testosterone: alcohol and ketone. Morphine: alcohol, phenol, ether, and amine.

LEARNING OBJECTIVE
Use the lock-and-key model of drug action to demonstrate how the chemical structure of a drug affects its biological activity.

Integrated Science 14A
BIOLOGY

Drug Action

EXPLAIN THIS Why are organic chemicals so suitable for making drugs?

To find new and more effective medicines, chemists use various models that describe how drugs work. By far, one of the most useful models of drug action is the **lock-and-key model**. The basis of this model is that there is a connection between a drug's chemical structure and its biological effect. For example, related pain-relieving opioids, such as codeine and morphine, have the T-shaped structure shown in Figure 14.25.

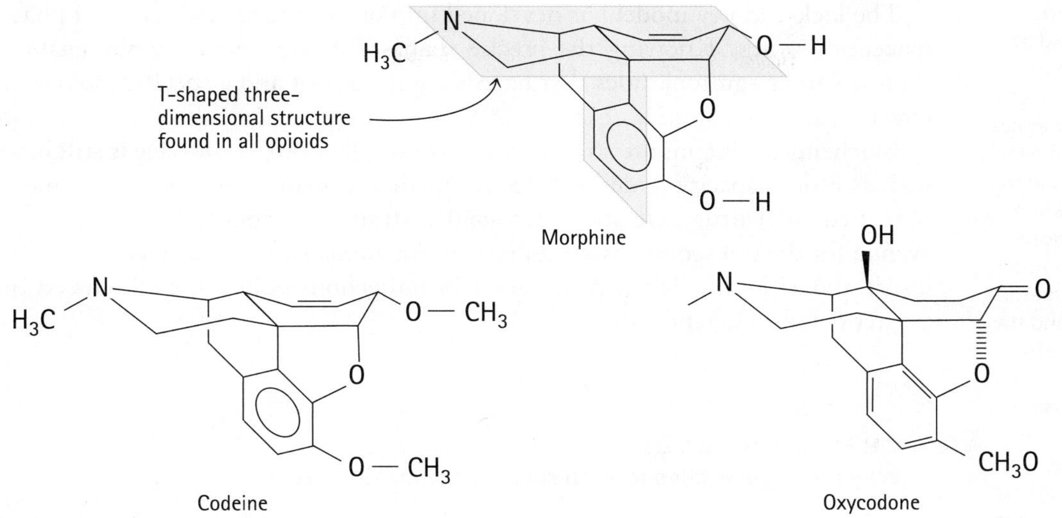

T-shaped three-dimensional structure found in all opioids

Morphine

Codeine

Oxycodone

FIGURE 14.25
All drugs that act like morphine have the same basic three-dimensional shape as morphine.

According to the lock-and-key model, illustrated in Figure 14.26, biologically active molecules function by fitting into *receptor sites* on proteins in the body, where they are held by molecular attractions, such as hydrogen bonding (see Chapter 12). When a drug molecule fits into a receptor site the way a key fits into a lock, a particular biological event is triggered, such as a nerve impulse, a change in the shape of a protein, or even a chemical reaction. In order to fit into a particular receptor site, however, a molecule must have the proper shape, just as a key must have properly shaped notches in order to fit a lock.

Another facet of this model is that the molecular attractions holding a drug to a receptor site are easily broken. (Recall from Chapter 12 that most molecular attractions are many times weaker than chemical bonds.) A drug is therefore held to a receptor site only temporarily. Once the drug is removed from the receptor site, body metabolism destroys the drug's chemical structure and the effects of the drug are said to have worn off. Using this model, we can also understand why some drugs are more potent than others. The painkiller methadone, for example, works by binding to morphine's receptor site. It is less potent than morphine, however, because its structure lacks all the pieces needed to make a perfect fit, as shown in Figure 14.27.

Drug molecule (morphine)

Receptor site

FIGURE 14.26
Many drugs act by fitting into receptor sites on molecules found in the body, much as a key fits in a lock.

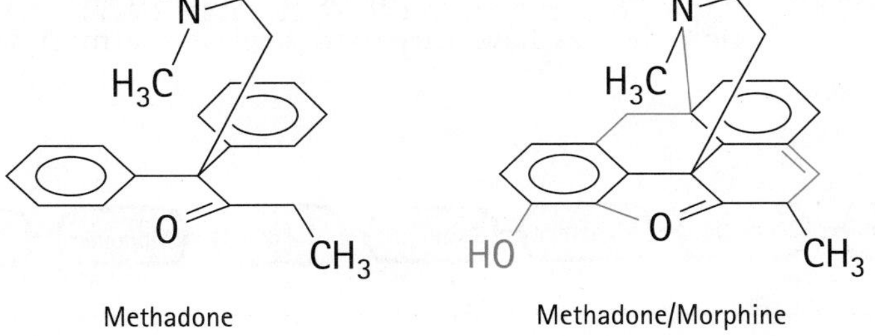

Methadone

Methadone/Morphine

FIGURE 14.27
The structure of methadone (black) superimposed on that of morphine (blue and black).

FYI The most widely used approach to treating opioid addiction is methadone maintenance. *Methadone*, shown in Figure 14.27, is a synthetic opioid derivative that has most of the effects of other opioids, including euphoria, but differs in that it retains much of its activity when taken orally. This means that doses are very easy to control and monitor. The withdrawal symptoms of methadone are also far less severe, and the addict may be slowly weaned off the opioid without excessive stress. An addict may be freed of physical dependence in a matter of months. The psychological dependence, however, usually persists throughout the individual's life, which is why the relapse rate is so high.

The lock-and-key model has developed into one of the central tenets of pharmaceutical study. Knowing the precise shape of a target receptor site enables chemists to design molecules that have an optimum fit and a specific biological effect.

Biochemical systems are so complex, however, that our knowledge is still limited, as is our capacity to design effective medicinal drugs. For this reason, most new medicinal drugs are still discovered instead of designed. One important avenue for drug discovery is ethnobotany. An *ethnobotanist* is a researcher who learns about the medicinal plants used in indigenous cultures, as discussed in greater detail in Chapter 18.

CHECK YOURSELF
Why are organic chemicals so suitable for making drugs?

CHECK YOUR ANSWER
The vast diversity of organic chemicals permits the manufacture of the many different types of medicines needed to combat the many different types of illnesses humans are subject to.

LEARNING OBJECTIVE
Describe how polymers are synthesized from monomers.

14.7 Polymers

EXPLAIN THIS Why are plastics generally so inexpensive?

MasteringPhysics®
TUTORIAL: Polymers

Polymers are exceedingly long molecules made up of repeating molecular units called **monomers**, as Figure 14.28 illustrates. Monomers have relatively simple structures consisting of anywhere from 4 to 100 atoms per molecule. When monomers are chained together, they can form polymers with hundreds of thousands of atoms per molecule. These large molecules are still too small to be seen with the unaided eye. They are, however, giants in the submicroscopic world—if a typical polymer molecule were as thick as a kite string, it would be 1 km long.

Many of the molecules that constitute living organisms are polymers, including DNA, proteins, the cellulose of plants, and the complex carbohydrates of starchy foods. For now, we focus on the human-made polymers, also known as synthetic polymers, that make up the class of materials that are commonly known as plastics.

We will begin by exploring the two major types of synthetic polymers used today: *addition polymers* and *condensation polymers*. As shown in Table 14.5, addition and condensation polymers have a wide variety of uses. Solely the product of human design, these polymers pervade modern living. In the United States, for example, synthetic polymers have surpassed steel as the most widely used material.

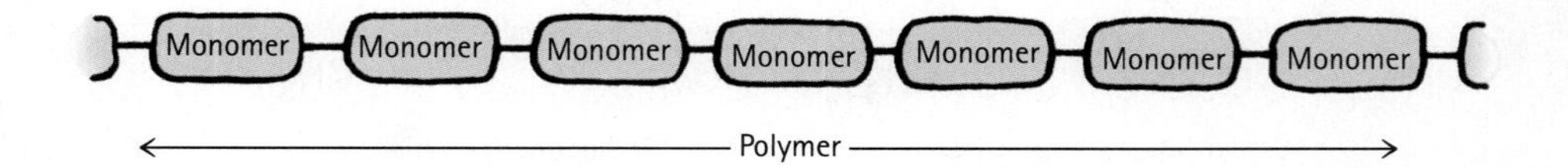

FIGURE 14.28
A polymer is a long molecule consisting of many smaller monomer molecules linked together.

TABLE 14.5 ADDITION AND CONDENSATION POLYMERS

Polymers	Repeating Unit	Common Uses	Recycling Code
Addition			
Polyethylene (PE)	$\cdots CH_2{-}CH_2 \cdots$	Plastic bags, bottles	2 HDPE, 4 LDPE
Polypropylene (PP)	$\cdots CH_2{-}CH(CH_3) \cdots$	Indoor–outdoor carpets	5 PP
Polystyrene (PS)	$\cdots CH_2{-}CH(C_6H_5) \cdots$	Plastic utensils, insulation	6 PS
Polyvinyl chloride (PVC)	$\cdots CH_2{-}CHCl \cdots$	Shower curtains, tubing	3 V
Condensation			
Polyethylene terephthalate	$\cdots C(=O){-}C_6H_4{-}C(=O){-}O{-}CH_2CH_2{-}O \cdots$	Clothing, plastic bottles	1 PET
Melamine–formaldehyde resin (Melmac, Formica)	triazine ring (C_3N_3) with three NH groups, each bonded to $\cdots C{=}O \cdots$	Dishes, countertops	Not recycled

Addition Polymers

Addition polymers form simply by the joining of monomer units. For this to happen, each monomer must contain at least one double bond. As shown in Figure 14.29, polymerization occurs when two of the electrons from each double bond split away from each other to form new covalent bonds with neighboring monomer molecules. During this process, no atoms are lost, so the total mass of the polymer is equal to the sum of the masses of all the monomers.

Nearly 12 million tons of polyethylene is produced annually in the United States; that's about 90 lb per U.S. citizen. The monomer from which it is synthesized, ethylene, is an unsaturated hydrocarbon produced in large quantities from petroleum.

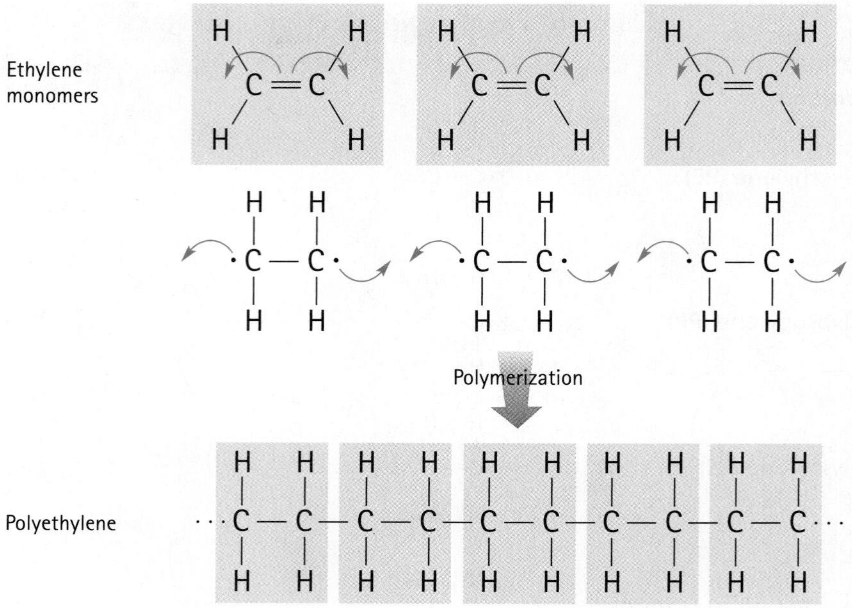

Ethylene monomers

Polymerization

Polyethylene

FIGURE 14.29
The addition polymer polyethylene is formed as electrons from the double bonds of ethylene monomer molecules split away and become unpaired valence electrons. Each unpaired electron then joins with an unpaired electron of a neighboring carbon atom to form a new covalent bond that links two monomer units.

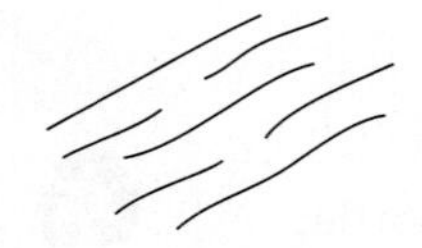

(a) Molecular strands of HDPE

(b) Molecular strands of LDPE

FIGURE 14.30
(a) The polyethylene strands of HDPE can pack closely together, much like strands of uncooked spaghetti. (b) The polyethylene strands of LDPE are branched, which prevents them from packing well.

Two principal forms of polyethylene are produced by using different catalysts and reaction conditions. High-density polyethylene (HDPE), shown schematically in Figure 14.30a, consists of long strands of straight-chain molecules packed closely together. The tight alignment of neighboring strands makes HDPE a relatively rigid, tough plastic useful for such things as bottles and milk jugs. Low-density polyethylene (LDPE), shown in Figure 14.30b, is made of strands of highly branched chains, an architecture that prevents the strands from packing closely together. This makes LDPE more bendable than HDPE and gives it a lower melting point. HDPE holds its shape in boiling water; LDPE deforms. LDPE is most useful for such items as plastic bags, photographic film, and electrical-wire insulation.

Other addition polymers are created by using different monomers. The only requirement is that the monomer must contain a double bond. The monomer propylene, for example, yields polypropylene, as shown in Figure 14.31. Polypropylene is a tough plastic material useful for pipes, hard-shell suitcases, and appliance parts. Fibers of polypropylene are used for upholstery, indoor–outdoor carpets, and even thermal underwear.

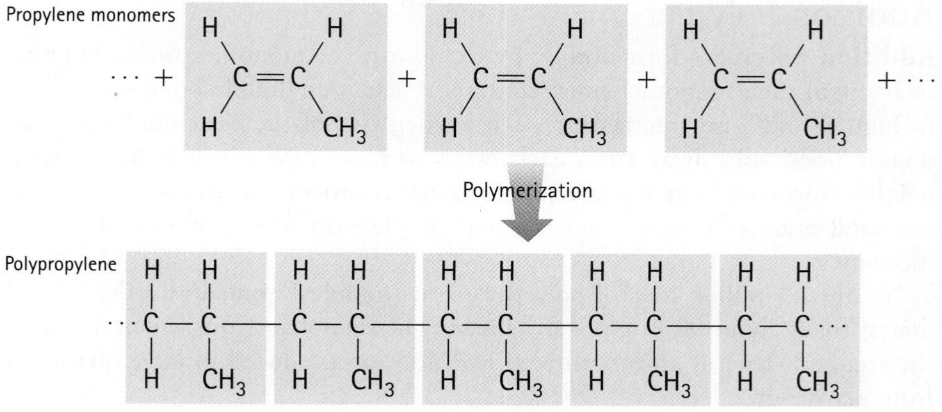

Propylene monomers

Polymerization

Polypropylene

FIGURE 14.31
Propylene monomers polymerize to form polypropylene.

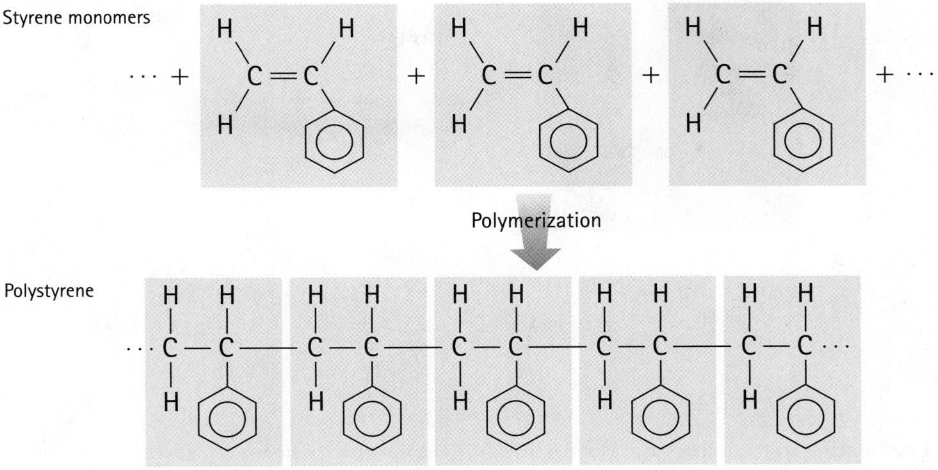

Styrene monomers

Polymerization

Polystyrene

FIGURE 14.32
Styrene monomers polymerize to form polystyrene. Another important addition polymer is polyvinyl chloride (PVC), which is tough and easily molded. Floor tiles, shower curtains, and pipes are most often made of PVC, shown in Figure 14.33.

Figure 14.32 shows that using styrene as the monomer yields polystyrene. Transparent plastic cups are made of polystyrene, as are thousands of other household items. Blowing gas into liquid polystyrene generates Styrofoam, which is widely used for coffee cups, packing material, and insulation.

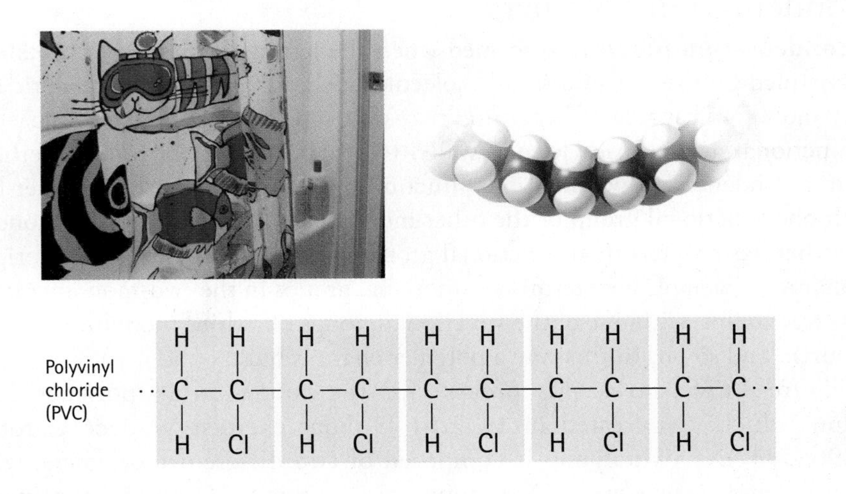

FIGURE 14.33
Another important addition polymer is polyvinyl chloride (PVC), which is used to fabricate many household items.

The addition polymer polytetrafluoroethylene, shown in Figure 14.34, is what you know as Teflon. In contrast to the chlorine-containing Saran, fluorine-containing Teflon has a nonstick surface because the fluorine atoms tend not to experience any molecular attractions. In addition, because carbon–fluorine bonds are unusually strong, Teflon can be heated to high temperatures before decomposing. These properties make Teflon an ideal coating for cooking surfaces. It is also relatively inert, which is why many corrosive chemicals are shipped or stored in Teflon containers.

FYI So, if nothing sticks to Teflon, how is Teflon made to adhere to a pan as a coating? That's a trade secret, but rumor has it that there are microscopic pits in the metal pan that help the Teflon adhere physically. Of course, we all know that the Teflon is fairly easy to scrape out of the pan, which is why manufacturers recommend that you stir-fry with a wooden utensil.

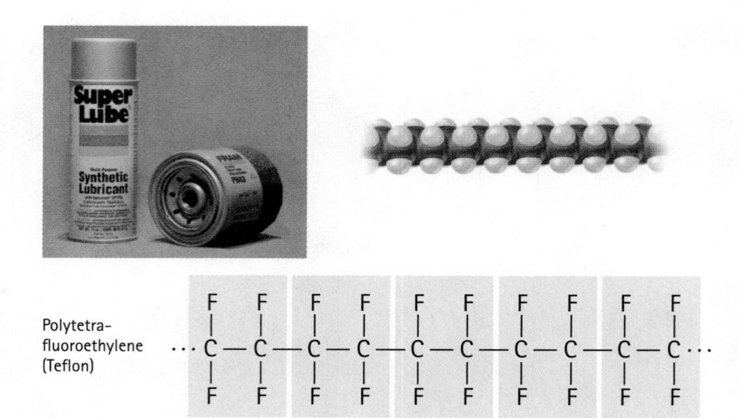

FIGURE 14.34
The fluorine atoms in polytetrafluoroethylene tend not to experience molecular attractions, which is why this addition polymer is used as a nonstick coating and lubricant.

CHECK YOURSELF

What do all monomers that are used to make addition polymers have in common?

CHECK YOUR ANSWER

They have a double covalent bond between two carbon atoms.

Condensation Polymers

A **condensation polymer** is formed when the joining of monomer units is accompanied by the loss of a small molecule, such as water or hydrochloric acid. Any monomer capable of becoming part of a condensation polymer must have a functional group on each end. When two such monomers come together to form a condensation polymer, one functional group of the first monomer links with one functional group of the other monomer. The result is a two-monomer unit that has two terminal functional groups, one from each of the two original monomers. Each of these terminal functional groups in the two-monomer unit is now free to link with one of the functional groups of a third monomer, and then a fourth, and so on. In this way a polymer chain is built.

Figure 14.35 shows this process for the condensation polymer called nylon, which was created in 1937 by DuPont chemist Wallace Carothers (1896–1937). This polymer is composed of two different monomers, which classifies it as a *copolymer*. One monomer is adipic acid, which contains two reactive end groups, both carboxyl groups. The second monomer is hexamethylenediamine, in which two amine groups are the reactive end groups. One end of an adipic acid molecule and one end of a hexamethylenediamine molecule can be made to react with each other, splitting off a water molecule in the process. After two monomers have joined, reactive ends still remain for further reactions, which leads to a growing polymer chain. Aside from its use in hosiery, nylon also has important uses in the manufacture of ropes, parachutes, clothing, and carpets.

Another widely used condensation polymer is polyethylene terephthalate (PET), which is formed from the copolymerization of ethylene glycol and terephthalic

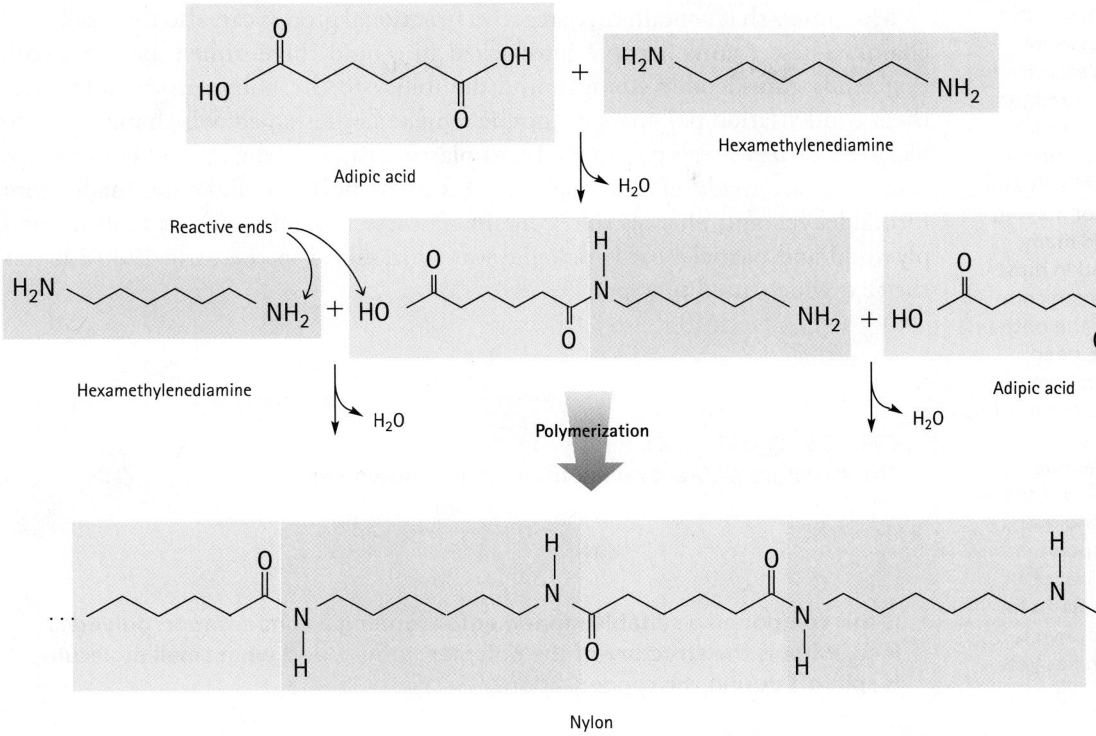

Adipic acid
Hexamethylenediamine
Reactive ends
Hexamethylenediamine
Adipic acid
Polymerization
Nylon

FIGURE 14.35
Adipic acid and hexamethylenediamine polymerize to form the condensation copolymer nylon.

acid, as shown in Figure 14.36. Plastic soda bottles are made from this polymer. Also, PET fibers are sold as Dacron polyester, a product used in clothing and stuffing for pillows and sleeping bags. Thin films of PET, which are called Mylar, can be coated with metal particles to make magnetic recording tape or those metallic-looking balloons that are sold at most grocery store checkout counters.

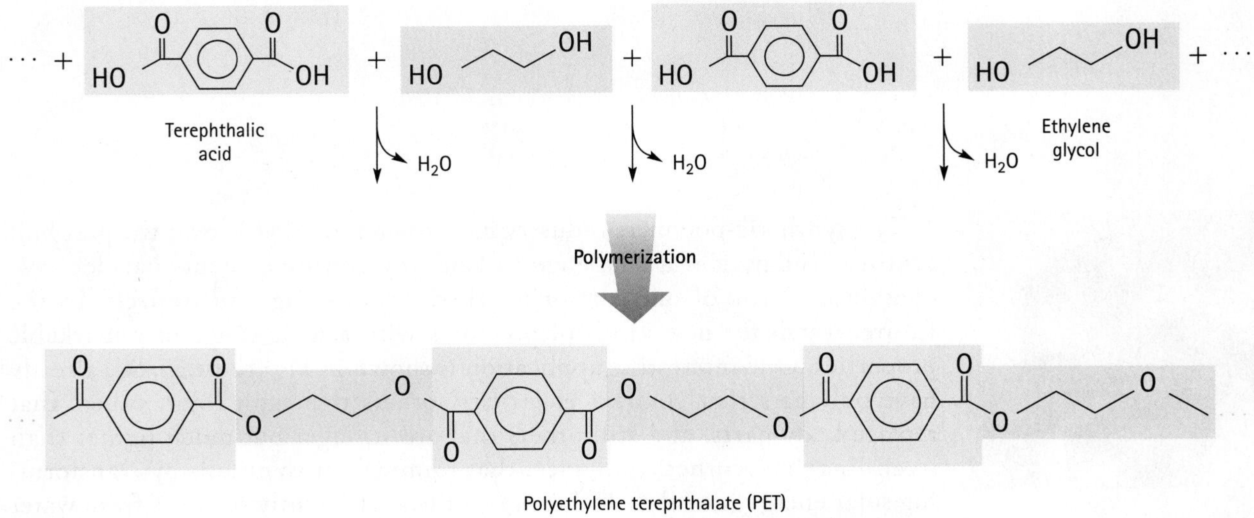

Terephthalic acid
Ethylene glycol
Polymerization
Polyethylene terephthalate (PET)

FIGURE 14.36
Terephthalic acid and ethylene glycol polymerize to form the condensation copolymer polyethylene terephthalate.

FYI Rigid polymers such as PVC can be made soft by incorporating small molecules called plasticizers. Pure PVC, for example, is a tough material great for making pipes. Mixed with a plasticizer, however, the PVC becomes soft and flexible and thus useful for making shower curtains, toys, and many other products now found in most households. One of the more commonly used plasticizers is the phthalates, some of which have been shown to disrupt the development of reproductive organs, especially in the fetus and in growing children. Governments and manufacturers are now working to phase out these plasticizers. But some phthalates, such as DINP, have been shown to be safe. For social and political simplicity, should all phthalates be banned or just the ones shown to be harmful? This question has yet to be resolved.

Monomers that contain three reactive functional groups can also form polymer chains. These chains become interlocked in a rigid three-dimensional network that lends considerable strength and durability to the polymer. Once formed, these condensation polymers cannot be remelted or reshaped, which makes them hard-set, or *thermoset*, polymers. Hard plastic dishes (Melmac) and countertops (Formica) are made of this material. A similar polymer, Bakelite, made from formaldehyde and phenols that contain multiple oxygen atoms, is used to bind plywood and particleboard. Bakelite was synthesized in the early 1900s; it was the first widely used polymer.

CHECK YOURSELF

The structure of 6-aminohexanoic acid is shown here:

H_2N — (chain) — C(=O)OH

Is this compound a suitable monomer for forming a condensation polymer? If so, what is the structure of the polymer formed, and what small molecule is split off during the condensation?

CHECK YOUR ANSWERS

The compound is a suitable monomer because the molecule has two reactive ends. You know that both ends are reactive because they are the ends shown in Figure 14.35. The only difference here is that both types of reactive ends are on the same molecule. Monomers of 6-aminohexanoic acid combine by splitting off water molecules to form the polymer known as nylon-6:

··· N(H) — (chain) — C(=O) — N(H) — (chain) — C(=O) — N(H) — (chain) — C(=O) ···

FIGURE 14.37
Flexible and flat video displays, also known as OLEDs, can now be fabricated from polymers.

The synthetic-polymers industry has grown remarkably over the past half century. Today, it is a challenge to find any consumer item that does *not* contain a plastic of one sort or another. Try finding one yourself. In the future, watch for new kinds of polymers with a wide range of remarkable properties. One interesting application is shown in Figure 14.37. We already have polymers that conduct electricity, others that emit light, others that replace body parts, and still others that are stronger but much lighter than steel. Imagine synthetic polymers that mimic photosynthesis by transforming solar energy into chemical energy, or that efficiently separate fresh water from the oceans. These are not dreams. They are realities that chemists have already been demonstrating in the laboratory. Polymers hold a clear promise for the future.

For instructor-assigned homework, go to www.masteringphysics.com 

SUMMARY OF TERMS (KNOWLEDGE)

Addition polymer A polymer formed by the joining together of monomer units with no atoms being lost as the polymer forms.

Alcohol An organic molecule that contains a hydroxyl group bonded to a saturated carbon.

Aldehyde An organic molecule that contains a carbonyl group, the carbon of which is bonded either to one carbon atom and one hydrogen atom or to two hydrogen atoms.

Amide An organic molecule that contains a carbonyl group, the carbon of which is bonded to a nitrogen atom.

Amine An organic molecule that contains a nitrogen atom bonded to one or more saturated carbon atoms.

Aromatic compound Any organic molecule that contains a benzene ring.

Carbonyl group A carbon atom double-bonded to an oxygen atom; found in ketones, aldehydes, amides, carboxylic acids, and esters.

Carboxylic acid An organic molecule that contains a carbonyl group, the carbon of which is bonded to a hydroxyl group.

Condensation polymer A polymer formed by the joining together of monomer units accompanied by the loss of small molecules, such as water.

Configuration The way the atoms within a molecule are connected. For example, two structural isomers have the same numbers and the same kinds of atoms, but in different configurations.

Conformation One of a wide range of possible spatial orientations of a particular configuration.

Ester An organic molecule that contains a carbonyl group, the carbon of which is bonded to one carbon atom and one oxygen atom bonded to another carbon atom.

Ether An organic molecule that contains an oxygen atom bonded to two carbon atoms.

Functional group A specific combination of atoms that behaves as a unit in an organic molecule.

Heteroatom Any atom other than carbon or hydrogen in an organic molecule.

Hydrocarbon A chemical compound that contains only carbon and hydrogen atoms.

Ketone An organic molecule that contains a carbonyl group, the carbon of which is bonded to two carbon atoms.

Lock-and-key model A conceptual model explaining how drugs function by interacting with receptor sites on proteins in the body.

Monomers The small molecular units from which a polymer is formed.

Organic chemistry The study of carbon-containing compounds.

Phenol An organic molecule in which a hydroxyl group is bonded to a benzene ring.

Polymer A long organic molecule made up of many repeating units.

Saturated hydrocarbon A hydrocarbon that contains no multiple covalent bonds, with each carbon atom bonded to four other atoms.

Structural isomers Molecules that have the same molecular formula but different chemical structures.

Unsaturated hydrocarbon A hydrocarbon that contains at least one multiple covalent bond.

READING CHECK (COMPREHENSION)

14.1 Hydrocarbons

1. How do two structural isomers differ from each other?
2. How are two structural isomers similar to each other?
3. What physical property of hydrocarbons is used in fractional distillation?
4. What types of hydrocarbons are more abundant in higher-octane gasoline?

14.2 Unsaturated Hydrocarbons

5. To how many atoms is a saturated carbon atom bonded?
6. What is the difference between a saturated hydrocarbon and an unsaturated hydrocarbon?
7. How many multiple bonds must a hydrocarbon have in order to be classified as unsaturated?
8. Aromatic compounds contain what kind of ring?

14.3 Functional Groups

9. What is a heteroatom?
10. Why do heteroatoms make such a difference in the physical and chemical properties of an organic molecule?

14.4 Alcohols, Phenols, and Ethers

11. Why are low-formula-mass alcohols soluble in water?
12. What distinguishes an alcohol from a phenol?
13. What distinguishes an alcohol from an ether?

14.5 Amines and Alkaloids

14. Which heteroatom is characteristic of an amine?
15. Do amines tend to be acidic, neutral, or basic?
16. Are alkaloids found in nature?
17. What are some examples of alkaloids?

14.6 Carbonyl Compounds

18. Which elements make up the carbonyl group?
19. How are ketones and aldehydes related to each other? How are they different from each other?
20. How are amides and carboxylic acids related to each other? How are they different from each other?
21. From what naturally occurring compound is aspirin prepared?

14.7 Polymers

22. What happens to the double bond of a monomer that participates in the formation of an addition polymer?
23. What is released in the formation of a condensation polymer?
24. Why is plastic wrap made of polyvinylidene chloride stickier than plastic wrap made of polyethylene?
25. What is a copolymer?

THINK INTEGRATED SCIENCE

14A—Drug Action

26. In the lock-and-key model, is a drug viewed as the lock or the key?
27. What holds a drug to its receptor site?
28. Which fits better into the opioid receptor site—morphine or methadone?

THINK AND DO (HANDS-ON APPLICATION)

29. Two carbon atoms connected by a single bond can rotate relative to each other. This ability to rotate can give rise to numerous conformations (spatial orientations) of an organic molecule. Is it also possible for two carbon atoms connected by a double bond to rotate relative to each other?

    Hold two toothpicks side by side and attach one jellybean to each end such that each jellybean has both toothpicks poked into it. Hold one jellybean while rotating the other. What kind of rotations are possible? Relate what you observe to the carbon–carbon double bond. Which structure of Figure 14.7 do you suppose has more possible conformations—butane or 2-butene? What do you suppose is true about the ability of atoms connected by a carbon–carbon triple bond to twist relative to each other?

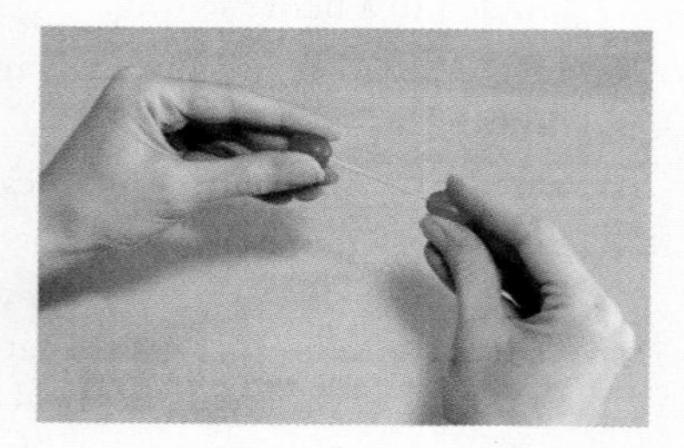

30. A property of polymers is their glass transition temperature, T_g, which is the approximate temperature below which the polymer is hard and rigid, but above which the polymer is soft and flexible. The T_g of polyethylene is a chilly −125°C, which is why polyethylene food wrap is flexible at ambient temperatures. Consider the two polymers polyethylene terephthalate (PETE) and polystyrene (PS). Which do you suppose has the higher T_g? Dip some plastics of these two polymers in boiling water to find out. A common polymer used to make chewing gum is polyvinyl acetate, which has a T_g of about 28°C, below body temperature but above room temperature. That's why most chewing gums are hard until they soften up in your warm mouth. Drink ice water while chewing and note how the gum quickly hardens.
31. Isopropyl alcohol, also known as rubbing alcohol, is very toxic if ingested. This is because it acts to destroy the digestive proteins and other important biomolecules in your stomach. Do this activity to see firsthand the destructive action of isopropyl alcohol on proteins. Crack open an egg, and put the egg white and the yolk into two separate bowls. Pour a capful of isopropyl alcohol into the egg white and observe what happens. In the second bowl, stir the yolk with a fork. Add another capful of isopropyl alcohol to the stirred yolk and observe what happens. The same sort of destruction would occur to your own stomach proteins, as well as various tissues, if you ingested the isopropyl alcohol. Not good! Our skin, however, is more impervious to the destructive powers of isopropyl alcohol, which therefore serves as a good topical antiseptic.

THINK AND COMPARE (ANALYSIS)

32. Rank these molecules in order of the phase they form at room temperature: solid, liquid, gas.

(a) $H_3C-C_6H_4-CH(CH_3)-C(=O)OH$

(b) $CH_3CH_2CH_2CH_3$

(c) $CH_3CH_2CH_2CH_2-OH$

33. Rank these hydrocarbons in order of increasing number of hydrogen atoms:

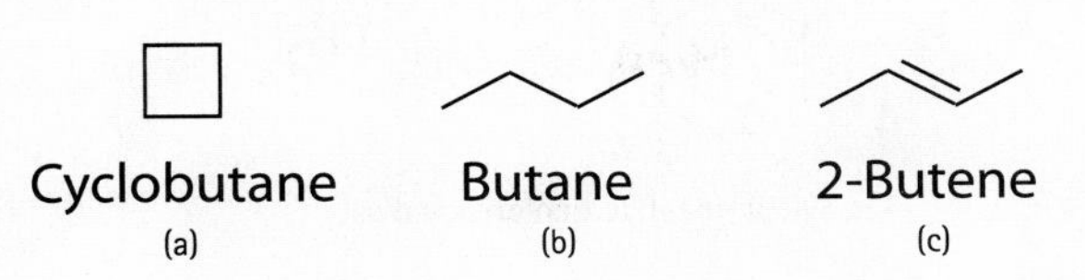

34. Rank these hydrocarbons in order of increasing number of hydrogen atoms:

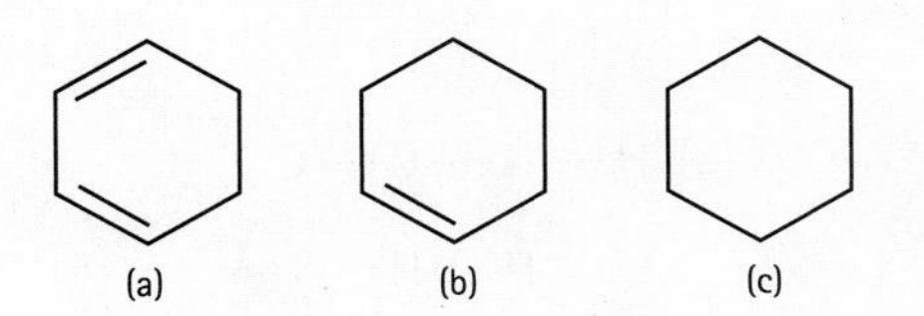

35. Rank the organic molecules in order of increasing solubility in water:

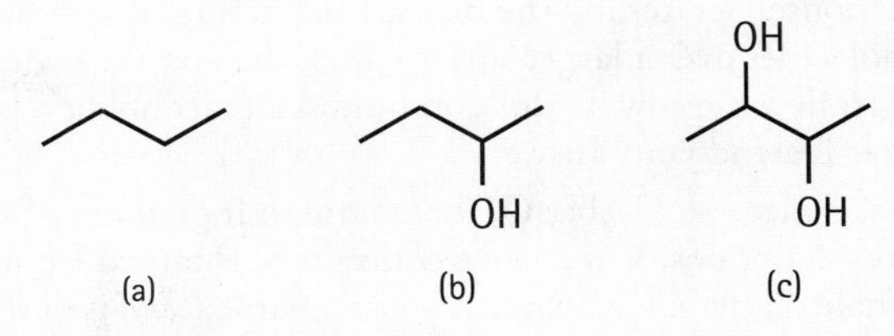

36. Rank the organic molecules in order of increasing solubility in water:

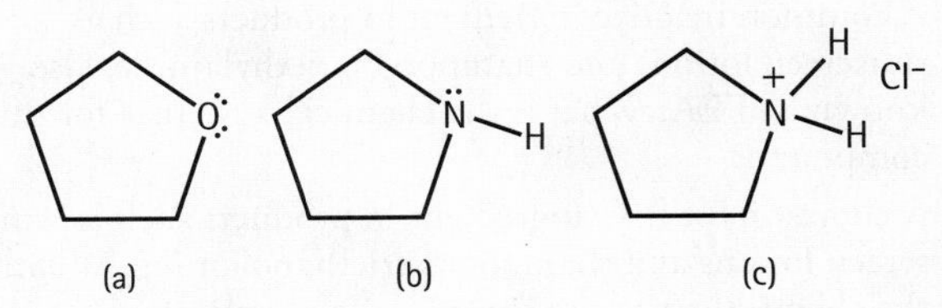

THINK AND EXPLAIN (SYNTHESIS)

37. What property of carbon allows for the formation of so many different organic molecules?

38. Why does the melting point of hydrocarbons get higher as the number of carbon atoms per molecule increases?

39. Draw all the structural isomers for hydrocarbons that have the molecular formula C_6H_{14}.

40. How many structural isomers are shown here?

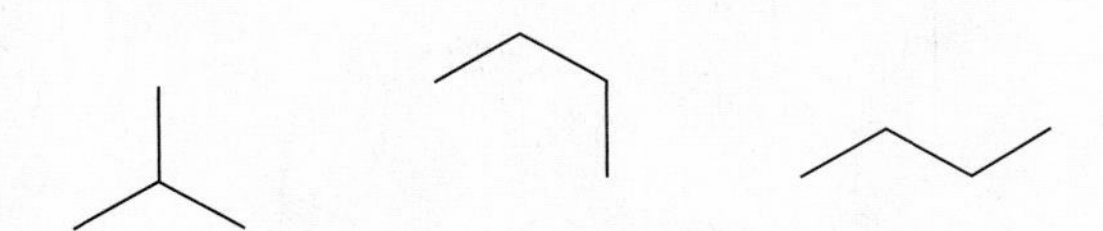

41. What do the compounds cyclopropane and propene have in common?

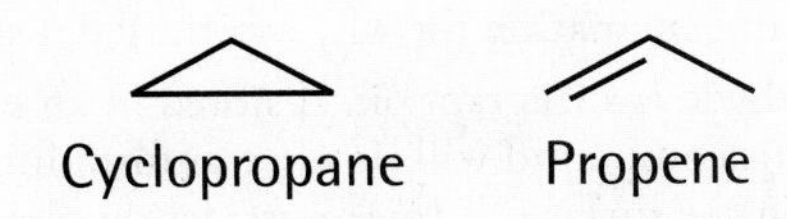

42. According to Figure 14.3, which has a higher boiling point—gasoline or kerosene?

43. The temperatures in a fractionating tower at an oil refinery are important, but so are the pressures. Where might the pressure in a fractionating tower be greatest—at the bottom or at the top? Defend your answer.

44. There are five atoms in the methane molecule, CH_4. One of these five is a carbon atom, which is $1/5 \times 100 = 20\%$ carbon. What is the percent carbon in ethane, C_2H_6? Propane, C_3H_8? Butane, C_4H_{10}?

45. Do heavier hydrocarbons tend to produce more or less carbon dioxide upon combustion compared with lighter hydrocarbons? Why?

46. What are the chemical formulas for these structures?

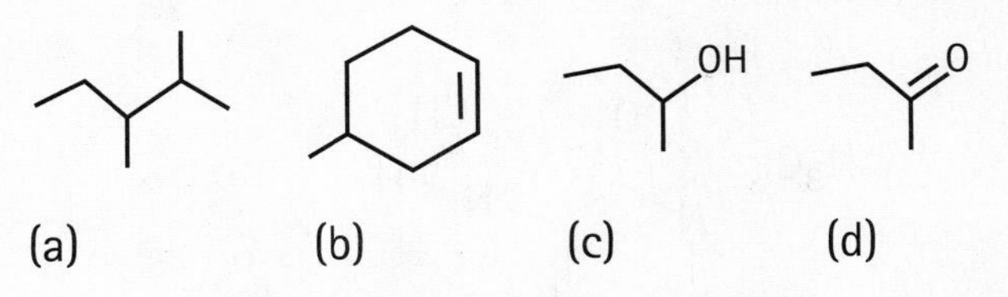

47. What do these two structures have in common?

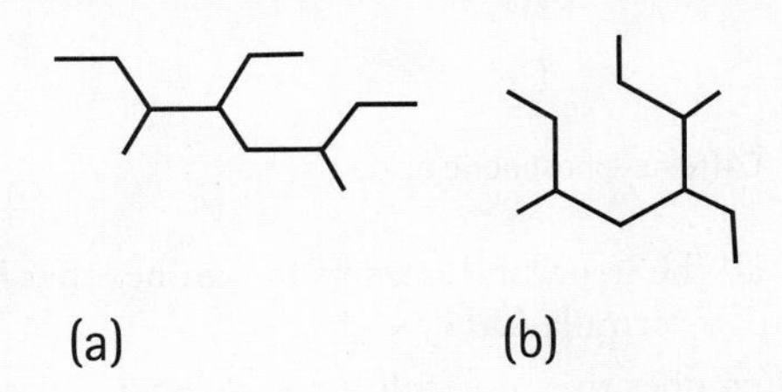

48. Remember that carbon–carbon single bonds can rotate but carbon–carbon double bonds cannot rotate. How many different structures are shown here?

49. Why do ethers typically have lower boiling points than alcohols?

50. What is the percent volume of water in 80-proof vodka?

51. One of the skin-irritating components of poison oak is tetrahydrourushiol:

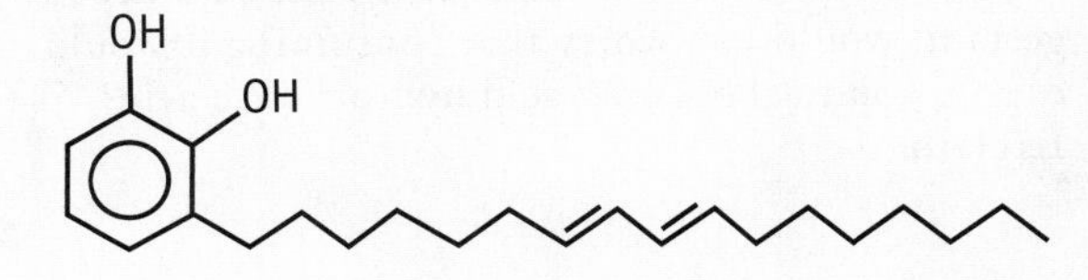

The long, nonpolar hydrocarbon tail embeds itself in a person's oily skin, where the molecule initiates an allergic response. Scratching the itch spreads tetrahydrourushiol molecules over a larger surface area, causing the zone of irritation to grow. Is this compound an alcohol or a phenol? Defend your answer.

52. Cetyl alcohol, $C_{16}H_{34}O$, is a common ingredient of soaps and shampoos. It was once commonly obtained from whale oil, which is where it gets its name (*cetyl* is derived from "cetacean"). There is no branching in the chemical structure of this molecule. Draw a likely structure.

53. A common inactive ingredient in products such as sunscreen lotions and shampoo is triethylamine, also known as TEA. What is the chemical structure for this compound?

54. A common inactive ingredient in products such as sunscreen lotions and shampoo is triethanolamine. What is the chemical structure for this tri-alcohol?

55. The phosphoric acid salt of caffeine has the structure shown below. This molecule behaves as an acid in that it can donate a hydrogen ion, created from the hydrogen atom bonded to the positively charged nitrogen atom. What are all the products formed when 1 mole of this salt reacts with 1 mole of sodium hydroxide, NaOH, a strong base?

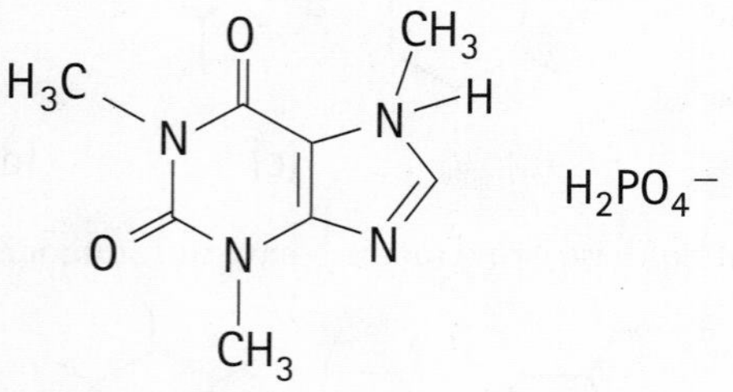

$H_2PO_4^-$

Caffeine–phosphoric acid salt

56. Draw all the structural isomers for amines that have the molecular formula C_3H_9N.

57. In water, does this molecule act as an acid, a base, neither, or both?

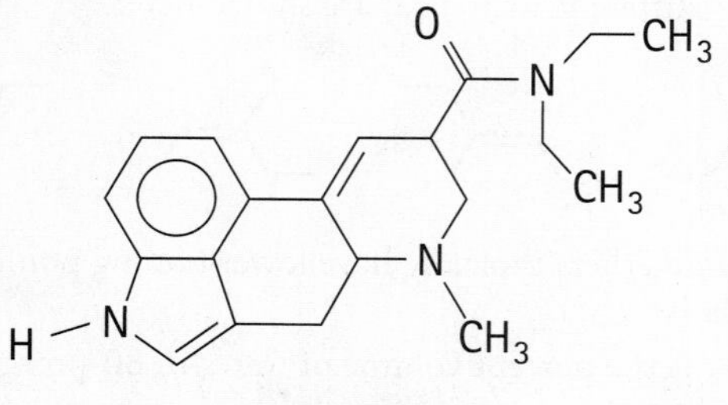

Lysergic acid diethylamide

58. If you saw the label phenylephrine HCl on a decongestant, would you worry that consuming it would expose you to the strong acid hydrochloric acid? Explain.

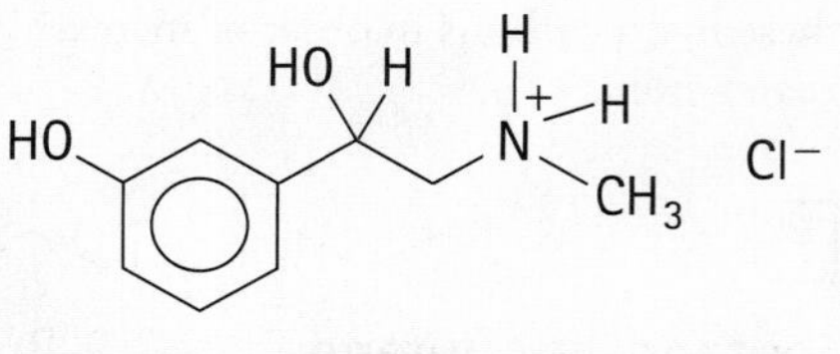

Cl^-

Phenylephrine–hydrochloric acid salt

59. An amino acid is an organic molecule that contains both an amine group and a carboxyl group. At an acidic pH, which structure is more likely? Explain your answer.

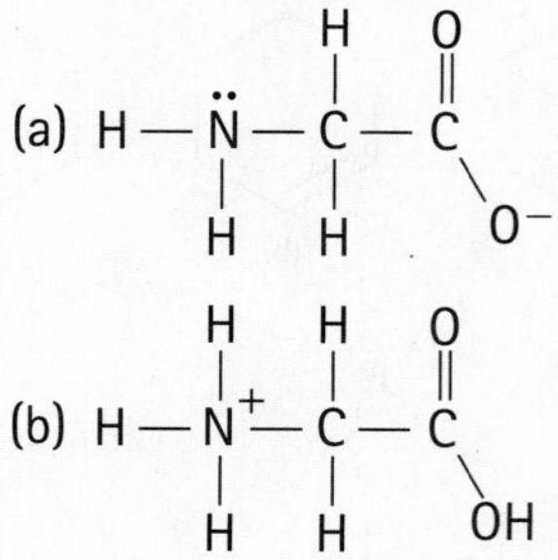

(a)

(b)

60. Identify the following functional groups in this organic molecule: amide, ester, ketone, ether, alcohol, aldehyde, amine.

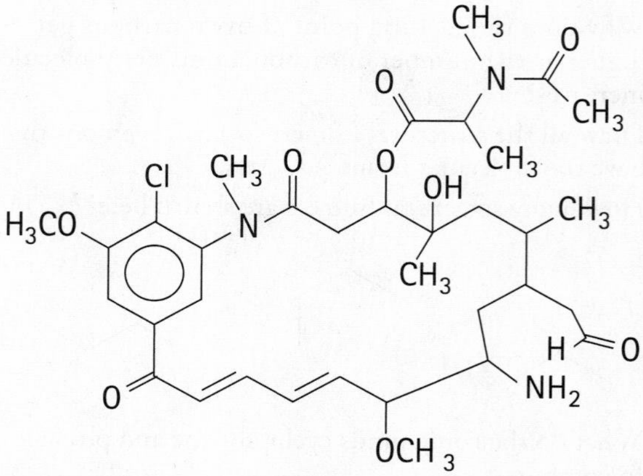

61. Suggest an explanation for why aspirin has a sour taste.

62. Benzaldehyde is a fragrant oil. If stored in an uncapped bottle, this compound will slowly tranform into benzoic acid along the surface. Is this an oxidation or a reduction?

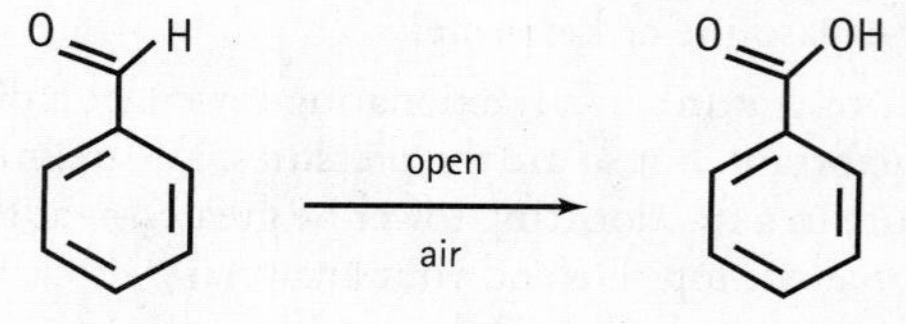

Benzaldehyde —open air→ Benzoic acid

63. What products are formed upon the reaction of benzoic acid with sodium hydroxide, NaOH? One of these products is a common food preservative. Can you name it?

64. The amino acid lysine is shown here. What functional group must be removed in order to produce cadaverine as shown in Figure 14.16?

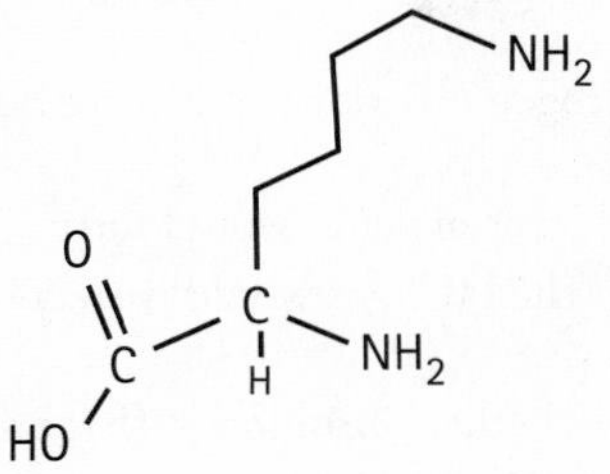

Lysine

65. Aspirin can cure a headache, but when you pop an aspirin tablet, how does the aspirin know to go to your head rather than to your big toe?

66. Which is better for you—a drug that is a natural product or one that is synthetic?

67. Would you expect polypropylene to be denser or less dense than low-density polyethylene? Why?

68. Hydrocarbons release a lot of energy when ignited. Where does this energy come from?

69. The polymer styrene-butadiene rubber (SBR), shown here, is used for making tires as well as bubble gum. Is it an addition polymer or a condensation polymer?

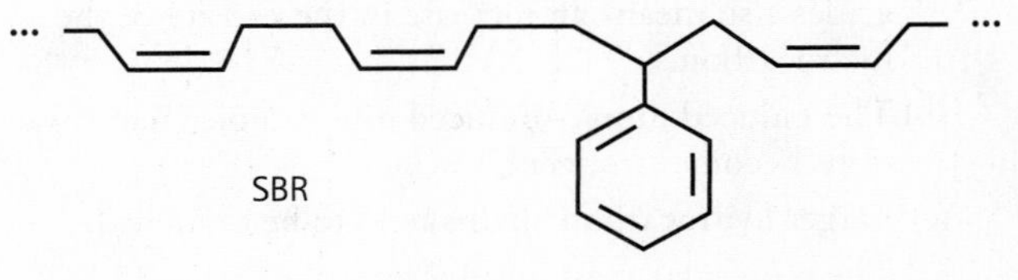

SBR

70. Citral and camphor are both 10-carbon odoriferous natural products made from the joining of two isoprene units plus the addition of a carbonyl functional group. Their chemical structures are shown here. Find and circle the two isoprene units in each of these molecules.

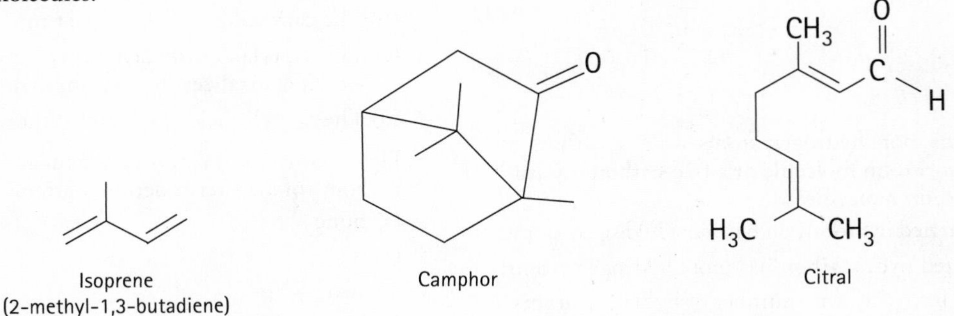

Isoprene (2-methyl-1,3-butadiene) — Camphor — Citral

71. Many of the natural product molecules synthesized by photosynthetic plants are formed by the joining together of isoprene monomers via an addition polymerization. A good example is the nutrient beta-carotene. How many isoprene units are needed to make one beta-carotene molecule? Find and circle these units in the beta-carotene structure shown here.

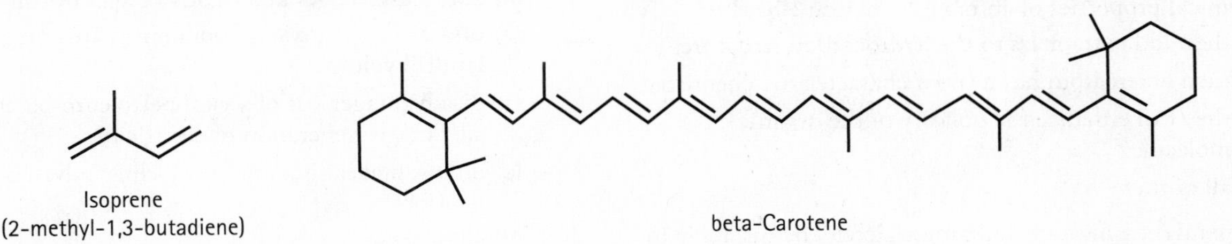

Isoprene (2-methyl-1,3-butadiene) — beta-Carotene

THINK AND DISCUSS (EVALUATION)

72. The solvent diethyl ether can be mixed with water but only by shaking the two liquids together. After the shaking is stopped, the liquids separate into two layers, much like oil and vinegar. The free-base form of the alkaloid caffeine is readily soluble in diethyl ether but not in water. Suggest what might happen to the caffeine of a caffeinated beverage if the beverage was first made alkaline with sodium hydroxide and then shaken with some diethyl ether.

73. Alkaloid salts are not very soluble in the organic solvent diethyl ether. What might happen to the free-base form of caffeine dissolved in diethyl ether if gaseous hydrogen chloride, HCl, were bubbled into the solution?

74. Go online and look up the total synthesis of the anticancer drug Taxol. With this major accomplishment in mind, discuss the relative merits of specializing in a single area versus becoming an expert in many different areas. When in life do we have the opportunity to simultaneously narrow our focus while also expanding our horizons?

75. Medicines, such as pain relievers and antidepressants, are being found in the drinking water supplies of many municipalities. How did these medicines get there? Does it matter that they are there? Should something be done about it. If so, what?

READINESS ASSURANCE TEST (RAT)

If you have a good handle on this chapter, if you really do, then you should be able to score 7 out of 10 on this RAT. If you score less than 7, you need to study further before moving on.

Choose the BEST answer to each of the following:

1. Why does the melting point of hydrocarbons get higher as the number of carbon atoms per molecule increases?
   (a) An increase in the number of carbon atoms per molecules also means an increase in the density of the hydrocarbon.
   (b) The induced dipole–induced dipole molecular attractions become stronger.
   (c) Larger hydrocarbon chains tend to be branched.
   (d) The molecular mass also increases.

2. How many structural isomers are there for hydrocarbons that have the molecular formula C_4H_{10}?
   (a) none
   (b) one
   (c) two
   (d) three

3. Which contains more hydrogen atoms—a five-carbon saturated hydrocarbon molecule or a five-carbon unsaturated hydrocarbon molecule?
   (a) The unsaturated hydrocarbon has more hydrogen atoms.
   (b) The saturated hydrocarbon has more hydrogen atoms.
   (c) They both have the same number of hydrogen atoms.
   (d) It depends on whether the unsaturation is due to a double or triple bond.

4. Heteroatoms make a difference in the physical and chemical properties of an organic molecule because
   (a) they add extra mass to the hydrocarbon structure.
   (b) each heteroatom has its own characteristic chemistry.
   (c) they can enhance the polarity of the organic molecule.
   (d) all of these

5. Why might a high-formula-mass alcohol be insoluble in water?
   (a) A high-formula-mass alcohol is too attracted to itself to be soluble in water.
   (b) The bulk of a high-formula-mass alcohol likely consists of nonpolar hydrocarbons.
   (c) Such an alcohol would likely be in a solid phase.
   (d) In order for two substances to be soluble in each other, their molecules need to be of comparable mass.

6. Alkaloid salts are not very soluble in the organic solvent diethyl ether. What might happen to the free-base form of caffeine (an alkaloid) dissolved in diethyl ether if gaseous hydrogen chloride, HCl, were bubbled into the solution?
   (a) A second layer of water would form.
   (b) Nothing; the HCl gas would merely bubble out of solution.
   (c) The diethyl ether–insoluble caffeine salt would form as a white precipitate.
   (d) The acid–base reaction would release heat, which would cause the diethyl ether to start evaporating.

7. Explain why caprylic acid, $CH_3(CH_2)_6COOH$, dissolves in a 5% aqueous solution of sodium hydroxide but caprylaldehyde, $CH_3(CH_2)_6CHO$, does not dissolve.
   (a) With two oxygens, the caprylic acid is about twice as polar as the caprylaldehyde.
   (b) The caprylaldehyde is a gas at room temperature.
   (c) The caprylaldehyde behaves as a reducing agent, which neutralizes the sodium hydroxide.
   (d) The caprylic acid reacts to form the water-soluble salt.

8. How many oxygen atoms are bonded to the carbon of the carbonyl of an ester functional group?
   (a) none
   (b) one
   (c) two
   (d) three

9. One solution to the problem of our overflowing landfills is to burn plastic objects instead of burying them. What are some advantages and disadvantages of this practice?
   (a) disadvantage: toxic air pollutants; advantage: reduced landfill volume
   (b) disadvantage: loss of a vital petroleum-based resource; advantage: generation of electricity
   (c) disadvantage: discourages recycling; advantage: provides new jobs
   (d) all of these

10. Which would you expect to be more viscous—a polymer made of long molecular strands or a polymer made of short molecular stands? Why?
    (a) long strands, because they tend to tangle among themselves
    (b) short strands, because of a higher density
    (c) long strands, because of a greater molecular mass
    (d) short strands, because their ends are typically polar

Answers to RAT

1. b, 2. c, 3. b, 4. d, 5. b, 6. c, 7. d, 8. c, 9. d, 10. a

3

PART THREE

Biology

Look at the hundreds and hundreds of seeds on this sunflower! Each of these seeds has the potential to grow into a new sunflower plant. Of course, if every seed did that, the world would quickly become filled with sunflowers. In fact, only a tiny fraction of sunflower seeds will end up growing into plants. That's part of the reason why this sunflower makes so many seeds. The beautiful spiral arrangement of seeds on the head of the flower allows the plant to pack in as many seeds as possible, giving it a better chance at successfully reproducing. Nature loves spirals, as the cover of this book attests. Even scientists working on solar power recently discovered that the Sun's energy is captured more efficiently using spiral shapes. But how do living things end up with such wonderful solutions to life's challenges? It's time to delve into biology—the study of life.

CHAPTER 15

The Basic Unit of Life—The Cell

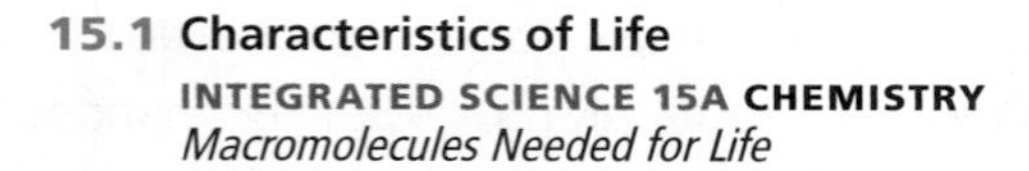

IT'S A rock. Or wait—could it be a well-disguised stonefish waiting for a juicy crab to come near? We know a living thing when we see one, even if our eyes fool us sometimes. But what, exactly, is a living thing? Do living things all share characteristics that differentiate them from nonliving things? Do they all reproduce? Use energy? Evolve? Are all living things made up of one or more cells? What are cells? How do cells "talk" with each other? How do cells make new cells? How do cells obtain energy, and how do we take advantage of this process when we bake bread and brew beer? How do plant cells use sunlight, air, and water to build living tissue, and how is this process the basis of life as we know it? In this chapter, we will explore the nature of life and the world of the cell.

15.1 Characteristics of Life

LEARNING OBJECTIVE
List and describe the characteristics of living things.

EXPLAIN THIS How can you tell a plant is alive even though it doesn't talk or run around?

Biology is the study of life and living organisms. But what is a living organism? What distinguishes living things from nonliving things?

Living things share certain characteristics. For one thing, they use energy. Living things, such as the sunflowers and lions in Figure 15.1, take energy from the environment and convert it into other forms of energy for their own use. Plants take electromagnetic energy from sunlight and convert it into chemical energy, which they can use to build their stems and leaves or fuel their activities. Animals eat, converting the energy they get from food into chemical energy, which they store in their bodies. This chemical energy is eventually converted again into kinetic and potential energy and heat as animals crawl, or fly, or grow. Of course, all the ways in which living things convert energy are consistent with the laws of physics. This means, first, that energy is always conserved and, second, that in any energy conversion, some energy is lost to the environment as heat.

UNIFYING CONCEPT
- *The Law of Conservation of Energy* Section 4.10
- *The Second Law of Thermodynamics* Section 6.5

Another characteristic of living things is that they develop and grow. When chicks hatch, they are small and covered with downy yellow feathers. Over time, they grow bigger, and their downy feathers are replaced by stiff adult feathers (Figure 15.2).

Living things maintain themselves. They generate structures, such as stems and leaves or skin and bones, and they repair damage done to those structures. When you scrape your knee, your blood clots to stop the bleeding, and the wounded skin scabs over and heals. Living things also maintain their internal environment, keeping it stable in the face of changing external conditions. Whether it is freezing cold or blisteringly hot, your body temperature stays right around 37C (98.6°F).

Living things have the capacity to reproduce. They make offspring that are exact or inexact copies of themselves. Figure 15.3 shows the two ways living things reproduce, asexually and sexually. In *asexual reproduction* a living organism reproduces all by itself, such as by dividing into two. Bacteria and sea anemones are organisms that are able to reproduce asexually. In *sexual reproduction*

(a)

(b)

FIGURE 15.1
Living things take energy from the environment and convert it into other forms of energy. (a) Plants such as these sunflowers convert energy from sunlight into chemical energy, which can be used to build tissues or fuel activity. (b) Animals such as these lions convert the chemical energy stored in food into motion or other activity, or use it for growth and reproduction.

FIGURE 15.2
Living things grow and develop over time.

(a)

(b)

FIGURE 15.3
Living things have the capacity to reproduce. (a) A sea anemone reproduces asexually by dividing. (b) Penguins reproduce sexually.

organisms form special sex cells, such as sperm and eggs, that join to develop into new individuals. Humans, penguins, beetles, and oak trees reproduce sexually.

Finally, living things are parts of populations that evolve. Populations do not remain constant from one generation to the next but change over time. Often, populations change in response to their environments. During the Industrial Revolution, when cities became polluted and blackened with soot, peppered moth populations evolved so that better-camouflaged dark-winged moths became more prevalent than light-winged moths. After antipollution laws were passed and cities were cleaned up, light-winged moths again became more common.

CHECK YOURSELF

1. **We know that cars are not living things. Check cars against the list of characteristics of living organisms. Which characteristics do cars have and which are they lacking?**
2. **Do all living things have all the characteristics of life we have listed?**

CHECK YOUR ANSWERS

1. Cars use energy, converting the energy in gasoline into motion. We might be able to argue that cars "develop" over time, acquiring nicks and dents and wearing down the treads on their tires. However, cars definitely do not maintain themselves, do not reproduce, and do not evolve.
2. There are some exceptions; for example, mules are sterile and unable to reproduce, but they are certainly alive.

LEARNING OBJECTIVE
Describe the structure and functions of the four types of molecules that make up living things: proteins, carbohydrates, lipids, and nucleic acids.

Integrated Science 15A
CHEMISTRY

Macromolecules Needed for Life

EXPLAIN THIS Do the same ingredients go into both you and a fried egg sandwich?

Living things are made up of four main types of macromolecules, or "big molecules." Some of these may already be familiar to you if you've heard that protein helps you grow or that a healthy diet should not include too much fat—or if you've ever "carbo-loaded" before a long run. The four main types of macromolecules in living organisms are proteins, carbohydrates, lipids, and nucleic acids. You will see that they are called macromolecules for a reason—each of these "big molecules" consists of multiple smaller molecules joined together.

Proteins perform a wide range of functions in living organisms. The protein keratin provides structure—it is a major component of skin, hair, and feathers. Insulin is a protein that acts as a hormone, enabling one type of cell in the body to communicate with other types. Actin and myosin are proteins that allow muscles to contract. Hemoglobin, a protein found in red blood cells, transports oxygen to body tissues. Proteins called antibodies protect the body from disease. And proteins known as digestive enzymes break down food during digestion.

What are proteins? Proteins are folded chains of organic molecules called *amino acids*. All amino acids include a central carbon (C) atom bonded to an amino group (NH_2), a carboxyl group (COOH), a hydrogen atom (H), and a side chain (called R) that varies from one amino acid to another. The amino acid leucine is shown in Figure 15.4a. Although only 20 different amino acids are found in living organisms, they can be strung together and folded in practically countless ways to create proteins with unique three-dimensional structures. This is why proteins are able to perform such a wide variety of functions.

Carbohydrates store energy in living organisms. Simple sugars, such as glucose (Figure 15.4b) and fructose, are carbohydrates. More complex carbohydrates

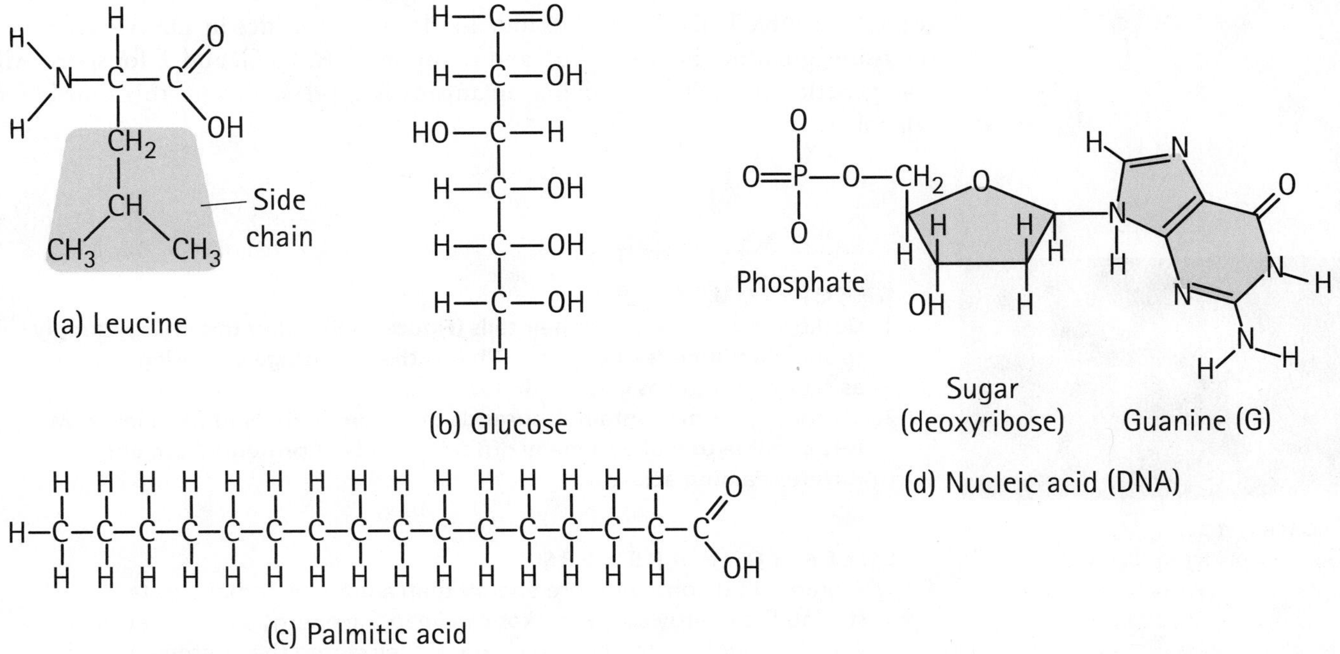

FIGURE 15.4
(a) Proteins are made up of amino acids. Leucine is one of the 20 amino acids found in living organisms. (b) Carbohydrates are made up of one or more simple sugars. This structure is glucose, a primary energy source for many living things. (c) Lipids have a variety of chemical structures, but many include fatty acids. Palmitic acid is a fatty acid found in lard and butter. (d) Nucleic acids are made up of nucleotides. This guanine nucleotide is one of the four nucleotides that make up DNA.

are made up of chains of simple sugars. Starch and glycogen—the primary energy-storage substances in plants and animals, respectively—consist of linked glucose molecules. Carbohydrates can also have structural functions. Cellulose, found in plant cell walls, is a structural carbohydrate built from glucose subunits—and it is the most abundant organic compound in the world. Carbohydrates are made up of carbon, hydrogen, and oxygen atoms, generally in the form of $(CH_2O)_n$.

Lipids serve diverse functions in living organisms. As fats or oils, lipids are used by many living organisms to store energy. Lipids store energy much more efficiently than carbohydrates—that is, 1 gram of fat or oil contains a lot more energy than 1 gram of carbohydrate. For this reason, lipids are used for long-term energy storage by many organisms, including humans. Lipids can have structural functions as well; for example, phospholipids are an essential component of cell membranes. One of the most familiar lipids is cholesterol, which the body uses to make such hormones as estrogen and testosterone. Lipids have a variety of chemical structures, but many include fatty acids—strings of hydrocarbons (carbon and hydrogen atoms) with a carboxyl (COOH) group at one end—as a major component (Figure 15.4c). Lipids are hydrophobic; that is, they are not soluble in water.

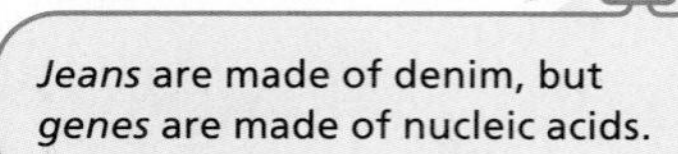

Jeans are made of denim, but *genes* are made of nucleic acids.

The role of the fourth type of macromolecule, **nucleic acids**, is to store genetic information in living organisms—all our genes are made up of nucleic acids. (In Chapter 16, we will see that genes carry the information that cells need to build proteins.) Nucleic acids are made up of strands of smaller units called nucleotides. A *nucleotide* includes a sugar molecule, a phosphate group, and a nitrogenous base. The two kinds of nucleic acids found in living organisms are *deoxyribonucleic acid* (DNA) and *ribonucleic acid* (RNA). DNA consists of two nucleic acid strands twisted into a spiral, which is why it is sometimes called a double helix. There are four kinds of nucleotides in DNA—adenine, cytosine, guanine (Figure 15.4d), and thymine, or A, C, G, and T for short. All the genetic information in living organisms is expressed using this four-letter alphabet.

FIGURE 15.5
Geckos store fat in their tails.

CHECK YOURSELF

1. **Gecko lizards store fat in their tails (Figure 15.5), using this energy supply to survive during lean periods. What is the advantage of storing energy as fat rather than as carbohydrate?**
2. **Living organisms contain thousands of different kinds of proteins. How is it possible to make so many different kinds of proteins from only 20 different amino acids?**

CHECK YOUR ANSWERS

1. A gram of fat contains more energy than a gram of carbohydrate. By storing the energy as fat, geckos can keep the energy supply they need without weighing themselves down. If they stored the energy as carbohydrate, they would need a much heavier tail.
2. The 20 amino acids that make up proteins can be strung together and then folded in a huge number of different ways, thus creating many different proteins.

15.2 Cell Types: Prokaryotic and Eukaryotic

LEARNING OBJECTIVE
Describe the key differences between the two basic types of cells—prokaryotic and eukaryotic.

EXPLAIN THIS Why are you more like yeast than like bacteria?

Now that you know what molecules living organisms are made of, let's look at living things themselves. **Cells** are the basic units of life, the same way that atoms are the basic units of matter. All living organisms are made up of one or more cells.* Organisms made up of just one cell are *unicellular*—bacteria are unicellular. Organisms made up of many cells are *multicellular*.

MasteringPhysics®
TUTORIAL: Comparing Prokaryotic and Eukaryotic Cells

You yourself have more than 10 trillion (10^{13}) cells in your body, and the diversity of things they do is amazing. Right now, muscle cells are moving your eyeballs as you follow the text on this page, sensory cells in your eyes are taking in the shapes of the letters, cells in your ears are absorbing nearby sounds, red blood cells are carrying oxygen to all the other cells in your body, and digestive cells are making the enzymes that will break down your last snack. And some extremely impressive cells—the neurons in your brain—are producing your thoughts about how amazing cells are.

Two distinct types of cells are found in different living organisms today: prokaryotic cells and eukaryotic cells. They are distinguished primarily by the presence or absence of a **nucleus**, a distinct structure within the cell that contains the cell's DNA. *Prokaryotic cells* do not have a nucleus (*pro* means "before" and *karyote* refers to "nut" or "nucleus"). *Eukaryotic cells* ("true nucleus") have a nucleus as well as other structures not present in prokaryotic cells. Organisms with prokaryotic cells are called **prokaryotes**, and organisms with eukaryotic cells are called **eukaryotes**. Figure 15.6 compares typical prokaryotic and eukaryotic cells.

Prokaryotes have existed on Earth far longer than eukaryotes. Prokaryotes first evolved 3.5 billion to 4 billion years ago and were the only living things on Earth until about 2 billion years ago, when the first eukaryotes appeared.

*Viruses, which possess some of the characteristics of life and straddle the line between living and nonliving, are not composed of cells. We will discuss viruses further in Chapter 18.

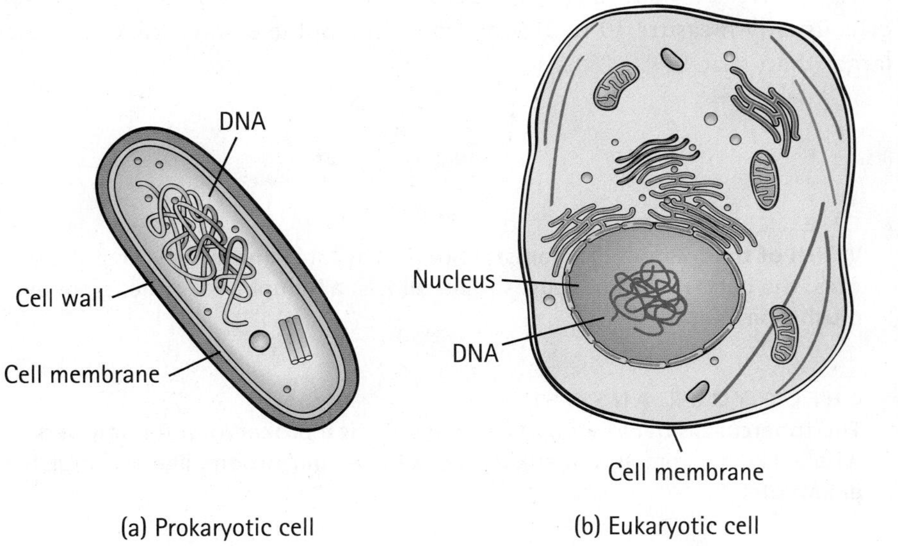

DNA
Cell wall
Cell membrane

(a) Prokaryotic cell

Nucleus
DNA
Cell membrane

(b) Eukaryotic cell

FIGURE 15.6
(a) Prokaryotic cells have no nucleus. (b) Eukaryotic cells have a nucleus as well as structures called organelles that are not found in prokaryotes.

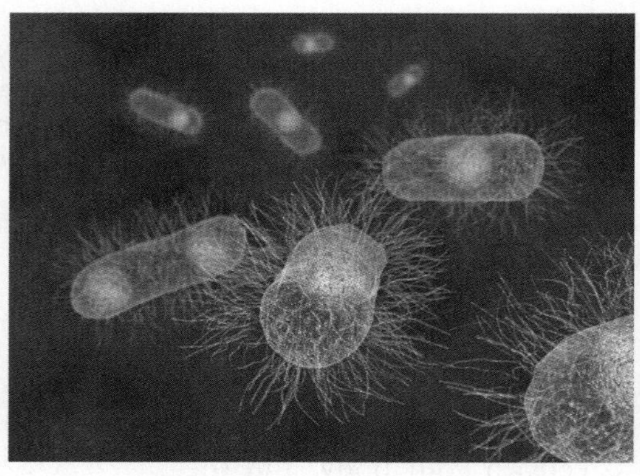

FIGURE 15.7
Escherichia coli (commonly referred to as *E. coli*) is a prokaryote that lives in the human digestive tract.

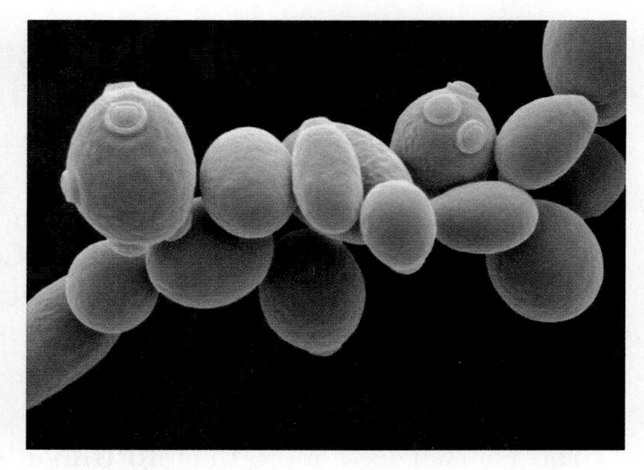

FIGURE 15.8
Saccharomyces cerevisiae, commonly known as baker's yeast or brewer's yeast, is a single-celled eukaryote.

You just learned that you are made up of more than 10 trillion (10^{13}) cells. But did you know that you are outnumbered even in your own body? There are more than 100 trillion (10^{14}) prokaryotic bacteria living in your body right now! How nice that they are—for the most part—friendly!

Prokaryotes now include two major lineages, the bacteria and the archaea (see Chapter 18). Prokaryotes are single-celled organisms and are very small, ranging from about 0.1 to 10 micrometers (10^{-6}meter) in diameter. Their structure is simpler than that of eukaryotes. The DNA of prokaryotes is found in a single circular structure and is not contained within a nucleus. Most prokaryotes have an outer *cell wall* that helps protect the cell. The prokaryote *Escherichia coli*, an occupant of the human digestive tract and one of the best-studied organisms in the world, is shown in Figure 15.7.

Eukaryotes can be single-celled, like prokaryotes, or they can be composed of many cells. The fungus known as baker's yeast, commonly used in baking and brewing, is a single-celled eukaryote (Figure 15.8). Humans are multicellular eukaryotes. Eukaryotes include all animals, plants, fungi, and protists. Eukaryotic cells have their DNA inside a distinct nucleus, a feature that distinguishes them from prokaryotes. In addition, the DNA of eukaryotic cells is found in linear, rather than circular, **chromosomes**. Eukaryotic cells also have numerous **organelles**, structures that perform specific functions for the cell. Finally, eukaryotic cells are larger than prokaryotic cells—whereas prokaryotic cells measure 0.1 to 10 micrometers, eukaryotic cells usually measure 10 to 100 micrometers. Some eukaryotic cells are even larger than that.

An egg yolk is a single cell. That makes the ostrich egg yolk, which measures 10 to 15 centimeters in diameter, the largest cell in the world.

Ostrich eggs may be the largest cells in the world, but nerve cells are the longest. The longest cell in the human body is a neuron that runs all the way from the spinal cord to the toes. You can try this cell out right now by wiggling your toes.

CHECK YOURSELF

Which of the following organisms are prokaryotes and which are eukaryotes: the bacterium that causes tuberculosis, a humpback whale, a honey mushroom?

CHECK YOUR ANSWER

The tuberculosis bacterium, like all bacteria, is a prokaryote. A humpback whale, like all animals, is a eukaryote. A honey mushroom, like all fungi, is a eukaryote.

HISTORY OF SCIENCE

Cell Theory

Cell theory, the idea that the cell is the basic unit of life, was several centuries in the making. In 1665, Robert Hooke, an English scientist, coined the term *cell* and published the first description of cells in his book *Micrographia*. Hooke examined a piece of cork under a microscope and saw a series of small boxlike chambers. He called these chambers "cells" because they reminded him of monks' cells. We now know that the chambers Hooke saw were not actually living cells but the cell walls that remain in dried plant matter.

It was not until the 1800s that the central importance of the cell was established. In 1838, careful studies of plants led German scientist Matthias Schleiden to conclude that all plants are made of cells. The following year, another German scientist, Theodor Schwann, came to the same conclusion about animals. The cell theory was finally completed in 1855, when German scientist Rudolph Virchow observed that all living cells come from other living cells.

In summary, the cell theory says:

1. All living things are made up of one or more cells.
2. All cells come from other cells.

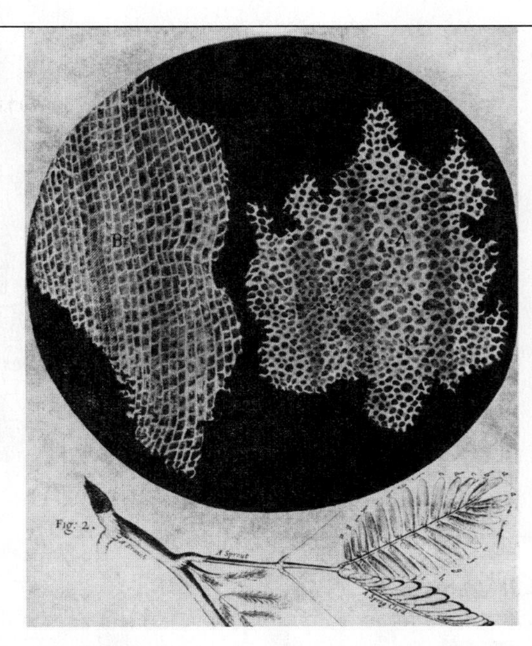

Robert Hooke examined cork under a microscope and called the small chambers he saw "cells." This is Hooke's original drawing of what he saw.

Integrated Science 15B
PHYSICS

The Microscope

LEARNING OBJECTIVE
Explain how light and electron microscopes work, describing the uses of each for biology.

EXPLAIN THIS Can you use a microscope to look at an atom?

Microscopes are high-tech magnifying glasses that allow us to see very small objects with a fine level of detail. One type of microscope, the light microscope, has been around for centuries—in fact, Robert Hooke discovered the existence of cells while using one. Light microscopes work by passing visible light through a specimen and then through a series of lenses. The lenses refract, or bend, the light in order to produce a magnified image of the specimen (Figure 15.9a).

Light microscopes are able to resolve objects on the order of a micrometer (10^{-6} meter) in size. In other words, two lines closer together than 10^{-6} meter appear as a single line. (The resolving power of the human eye is about 1/10 millimeter, or 10^{-4} meter.) This is because diffraction blurs the image if the size of an object is about the same as the wavelength of light (see Chapter 8). Furthermore, if an object is *smaller* than the wavelength of light, no structure can be seen at all. The entire image is lost due to diffraction. No amount of magnification or perfection of microscope design can overcome this fundamental diffraction limit.

UNIFYING CONCEPT
- ***Waves***
Section 8.1

With a resolving power of 10^{-6} meter, light microscopes allow us to view cells and to make out the larger features within them, such as the nucleus.

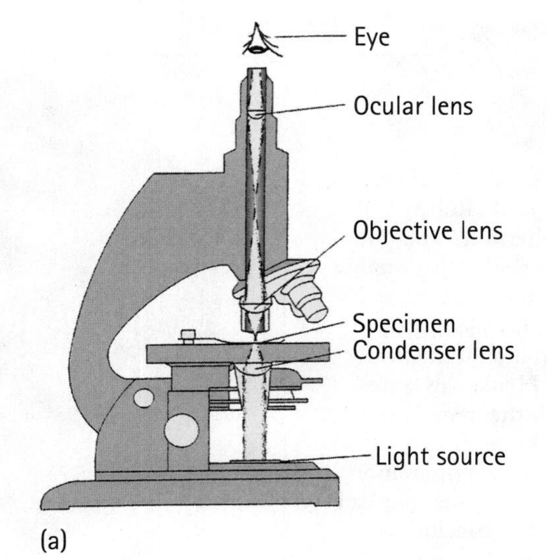

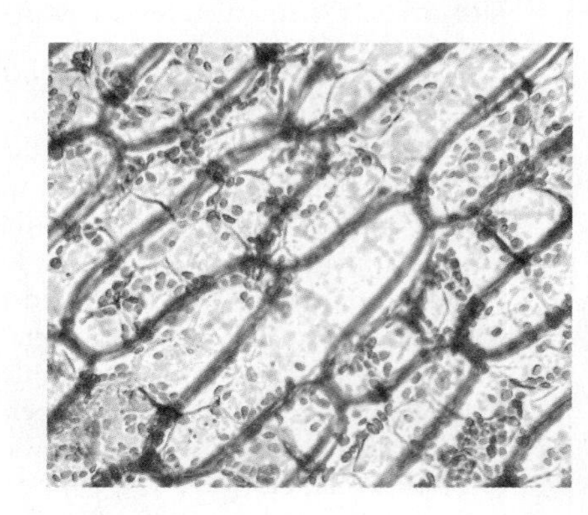

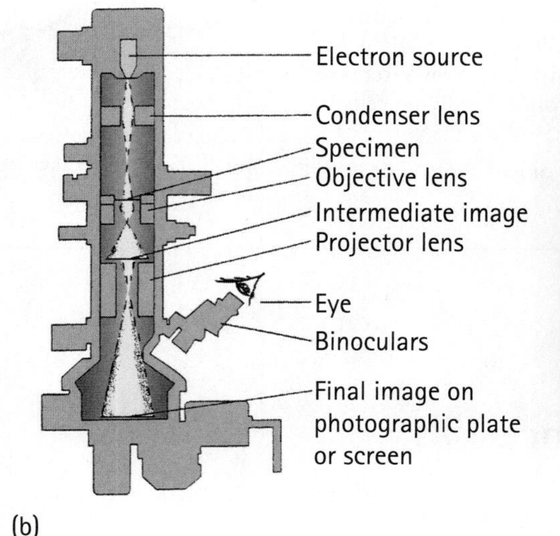

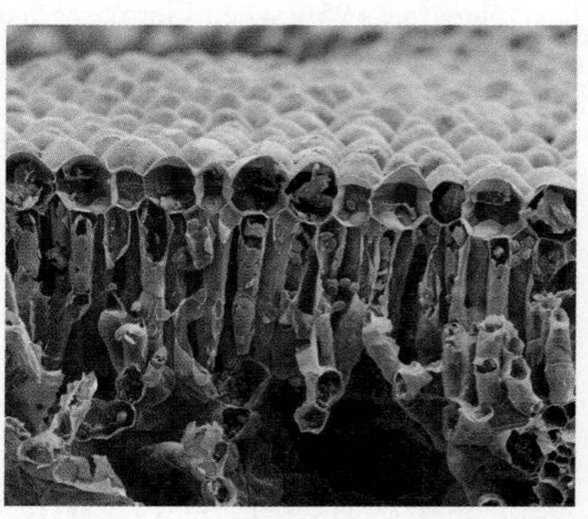

FIGURE 15.9
Microscopes enable us to examine objects that are too small for the human eye to see. (a) Light microscopes use glass lenses to bend light and magnify specimens. The photo shows plant cells seen through a light microscope. The small circular forms inside the cells are plant structures called chloroplasts. (b) Electron microscopes use electric and magnetic fields to magnify specimens. This is an electron microscope photo of a leaf in cross-section. Chloroplasts can be seen inside the cells.

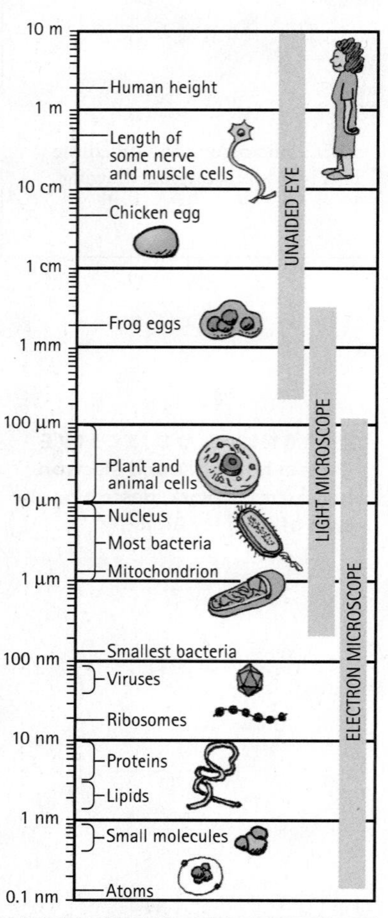

FIGURE 15.10
Microscopes allow us to look into the world of the cell. Depending on the size of a feature, the naked eye, a light microscope, or an electron microscope may be used.

However, they do not really allow us to see organelles and other cellular structures in detail.

To get around this problem, scientists illuminate very tiny objects with electron beams rather than with light. All matter has wave properties, and, compared with light waves, electron beams have extremely short wavelengths, allowing them to resolve much smaller objects. In an electron microscope, electric and magnetic fields, rather than optical lenses, are used to focus electron beams (Figure 15.9b). Electron microscopes are able to resolve objects about a nanometer (10^{-9} meter) in size, which covers just about everything of biological interest (Figure 15.10).

There are two types of electron microscopes. Scanning electron microscopes create a three-dimensional image of the surface of a specimen (see Figure 15.9b). Transmission electron microscopes image thin sections.

CHECK YOURSELF

Tiny bacteria called mycoplasmas have diameters as small as 0.1 micrometer. Can they be studied with light microscopes? With electron microscopes?

CHECK YOUR ANSWER

Mycoplasmas are too small to be examined with light microscopes, but they can be studied with electron microscopes.

15.3 Tour of a Eukaryotic Cell

LEARNING OBJECTIVE
Describe the main features of eukaryotic cells.

EXPLAIN THIS How are a cell's cytoskeleton and organelles like a person's skeleton and organs?

You, your tulips, and your dog are all eukaryotes, organisms composed of eukaryotic cells. In fact, most of the living things we encounter on a daily basis—all the plants, animals, and fungi—are eukaryotes.

MasteringPhysics®
TUTORIAL: Build an Animal and a Plant Cell

All eukaryotic cells are surrounded by a cell membrane (Figure 15.11). The **cell membrane** separates the inside of the cell from the outside and is responsible for controlling what goes into and out of the cell. Plant cells also have a rigid cell wall outside the cell membrane made of cellulose and other materials. The cell wall helps to protect and support the cell.

Eukaryotic cells have a nucleus, a structure within the cell that contains the cell's DNA. The nucleus is surrounded by a double membrane. The portion of

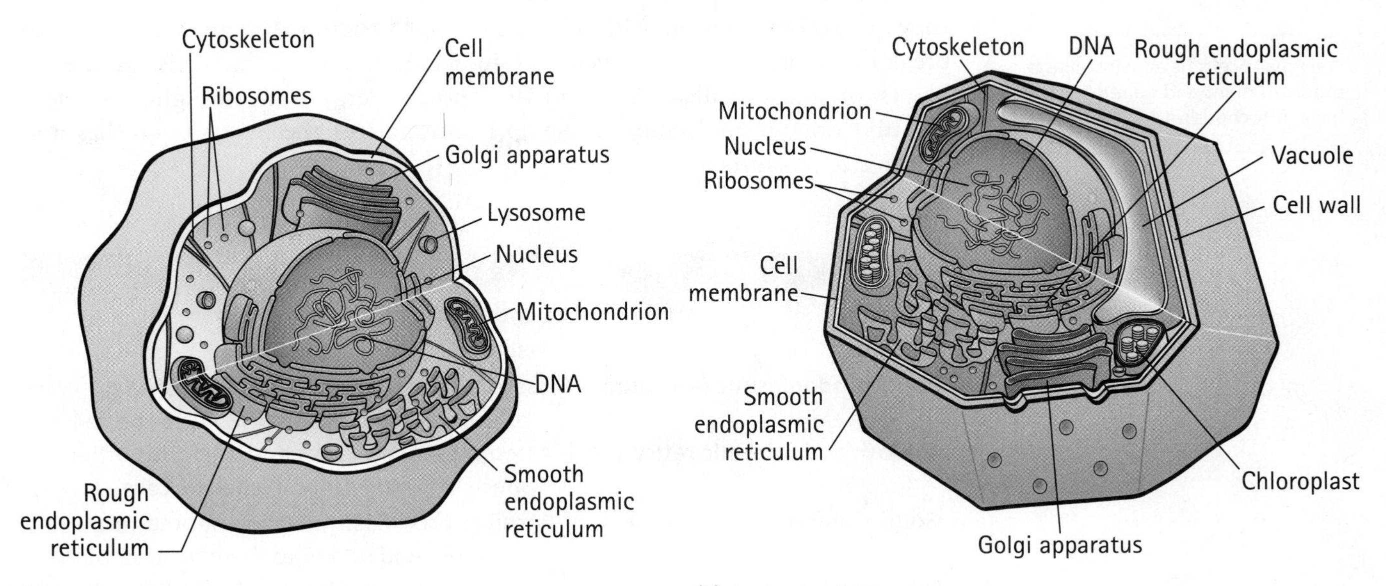

FIGURE 15.11
Typical plant and animal cells have a cell membrane, a nucleus, and many different organelles. (a) This is a typical animal cell. The lysosomes in animal cells are not found in plant cells. (b) This is a typical plant cell. Plants cells have a cell wall outside the cell membrane, a large central vacuole, and chloroplasts, features that are absent from animal cells.

the cell that is inside the cell membrane but outside the nucleus is called the cell's **cytoplasm**. The cytoplasm is crisscrossed by fibers of the *cytoskeleton*, which helps the cell hold its shape.

The cytoplasm of eukaryotic cells contains many organelles that are attached to the cytoskeleton. These are called organelles because, like the organs of the body, each performs a specific function in the cell. Each organelle, except for the ribosomes, is also surrounded by a membrane. Let's consider the organelles one at a time. *Ribosomes* are organelles that assemble proteins. Some ribosomes are suspended in the fluid of the cytoplasm. These make proteins that will remain inside the cell. Other ribosomes are attached to an organelle called the *rough endoplasmic reticulum*. These ribosomes assemble proteins that will go to the cell membrane or be exported from the cell. The rough endoplasmic reticulum appears "rough" because of the ribosomes embedded within it. The *smooth endoplasmic reticulum* assembles membranes and, depending on the cell, may have additional functions. For example, the smooth endoplasmic reticulum of liver cells detoxifies drugs and other poisons. In cells that secrete such steroid hormones as estrogen or testosterone, it is the smooth endoplasmic reticulum that synthesizes the hormones. The *Golgi apparatus* is sometimes described as the "post office" of a cell. It receives products from the endoplasmic reticulum, modifies them, and packages them in membrane-bound vesicles for transport within or out of the cell. *Lysosomes* are the garbage disposals of a cell. These organelles break down organic materials, such as damaged or worn-out organelles. Lysosomes are also used for other purposes: Certain white blood cells of the immune system use lysosomes to destroy the bacteria they have engulfed. *Vacuoles* are sacs surrounded by membrane. Plant cells usually have a single large vacuole that can be used to store nutrients or other materials. In flowers, vacuoles store the pigments that provide color. Animal cells typically have smaller vacuoles, sometimes called *vesicles*, that are used to hold or transport a wide array of products. For example, neurons have many vesicles that contain the chemicals they use to communicate with other neurons. **Mitochondria** are organelles that break down organic molecules to obtain energy in a form that cells can use. In plants, organelles called **chloroplasts** capture energy from sunlight and use it to build organic molecules. Table 15.1 summarizes the major organelles and features of eukaryotic cells.

Why does our tolerance for many drugs and medications increase over time, so that we require higher doses to obtain the same effect? Regular exposure to a drug causes the amount of smooth endoplasmic reticulum in liver cells to increase. This enables the liver to detoxify, or break down, the drug molecules more efficiently. As a result, a higher dose is necessary to make the same amount of medication available to the body.

Only plant cells have chloroplasts, but both plant and animal cells have mitochondria.

TABLE 15.1 MAJOR FEATURES OF EUKARYOTIC CELLS

Nucleus	Contains the cell's DNA
Ribosome	Assembles proteins for the cell
Rough endoplasmic reticulum	Assembles proteins destined either to go to the cell membrane or to leave the cell
Smooth endoplasmic reticulum	Assembles membranes and performs other specialized functions in specific cells
Golgi apparatus	Receives products from the endoplasmic reticulum and packages them for transport
Lysosome	Breaks down organic material
Mitochondrion	Obtains energy for the cell to use
Chloroplast	In plant cells, captures energy from sunlight to build organic molecules
Cytoskeleton	Helps cell hold its shape

15.4 The Cell Membrane

LEARNING OBJECTIVE
Describe the structure and function of the cell membrane.

EXPLAIN THIS Why is the cell membrane called a "fluid mosaic"?

The cell membrane defines a cell's boundary, separating the inside of the cell from the outside. One of its main functions is to serve as a gatekeeper, controlling what goes into the cell and what comes out of it. To see how the cell membrane performs this function, let's look at its structure. The three primary components of the cell membrane are phospholipids, proteins, and short carbohydrates.

Phospholipids are part hydrophilic and part hydrophobic. Have you ever noticed the way oil and water separate after they have been combined? The oil floats on top of the water in a distinct layer, rather than mixing with it (Figure 15.12). This is because oil is *hydrophobic*, or insoluble in water. (*Hydrophobic* literally means "afraid of water.") The opposite of hydrophobic is *hydrophilic*, or soluble in water. (*Hydrophilic* literally means "water-loving.") Phospholipids have hydrophilic "heads" and hydrophobic "tails" (Figure 15.13). The hydrophilic heads are naturally drawn to the watery environment inside and outside the cell, whereas the hydrophobic tails naturally try to avoid it. The result is that the phospholipids form a double layer, or bilayer, with the hydrophobic tails pointing inward and the hydrophilic heads pointing outward (Figure 15.14). You can think of the phospholipids as making a sandwich, with the heads as two slices of bread and the tails as the peanut butter inside.

The cell membrane also includes a large number of *membrane proteins* embedded like toothpicks in the phospholipid sandwich. Membrane proteins serve a variety of functions: They help cells communicate with other cells, control transport into and out of cells, and join cells to one another. Because different cells have different functions, membrane proteins also differ from one cell type to another.

FIGURE 15.12
Water and oil do not mix when they are combined.

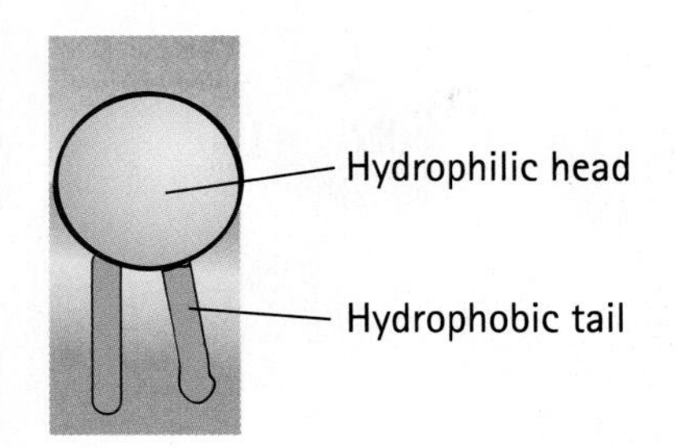

FIGURE 15.13
A phospholipid has a hydrophilic head and hydrophobic tails.

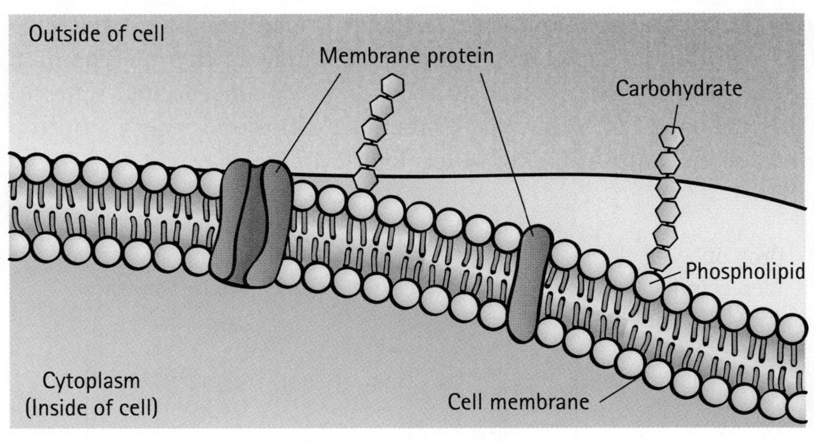

(a)

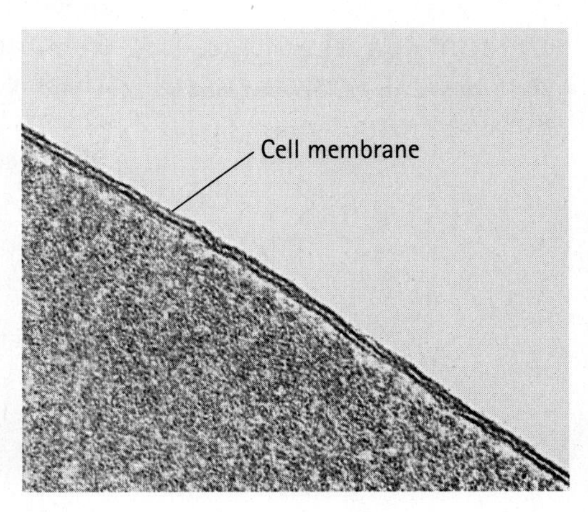

(b)

FIGURE 15.14
(a) The cell membrane includes a phospholipid bilayer, with hydrophobic tails pointed inward and hydrophilic heads pointing toward the inside and outside of the cell. Proteins are embedded in the membrane, and short carbohydrates are attached to the outside of the membrane. (b) This photograph shows the cell membrane of a red blood cell. Note the double layer of phospholipids.

Short carbohydrates are attached to the membrane proteins and phospholipids on the outside surface of the cell. These carbohydrates play an important role in cell recognition, the ability to distinguish one type of cell from another. For example, certain immune-system cells use these short carbohydrates to identify foreign cells, such as disease-causing bacteria (see Chapter 20).

MasteringPhysics®
TUTORIAL: Membrane Structure

Because the cell membrane includes a mosaic of phospholipids and proteins, and because the phospholipids and many membrane proteins slide freely around the cell surface, the cell membrane is often described as *fluid mosaic.* The fluidity of the cell membrane is essential to its ability to control the flow of materials into and out of the cell.

15.5 Transport Into and Out of Cells

LEARNING OBJECTIVE
Describe the different ways molecules move into and out of cells.

EXPLAIN THIS Why does oxygen go right into your cells, whereas potassium ions have to be pumped in?

Cells need to take in a variety of resources, including water, oxygen, and organic molecules. Cells also generate wastes that they must dispose of. Now that we know the structure of the cell membrane, we can discuss how it performs the essential task of controlling how things move into and out of cells. Transport across the cell membrane occurs in a number of ways—through diffusion, facilitated diffusion, active transport, endocytosis, and exocytosis.

Diffusion

MasteringPhysics®
TUTORIAL: Diffusion

Some molecules are able to cross the phospholipid bilayer of the cell membrane directly. Hydrophobic molecules, such as the gases oxygen and carbon dioxide, can pass directly through the double layer of hydrophobic tails. Certain small hydrophilic molecules—such as water—can also cross the cell membrane this way. What governs the way these substances move into and out of cells? A process known as **diffusion**, the movement of molecules from an area of high concentration to an area of low concentration—that is, down a concentration gradient (see Chapter 11). Diffusion is a direct result of the second law of thermodynamics, which states that natural systems tend to degrade from concentrated states to diffuse states (see Chapter 6). A familiar example of diffusion is the way a drop of food coloring spreads in a beaker of water (Figure 15.15).

UNIFYING CONCEPT
- ***The Second Law of Thermodynamics***
Section 6.5

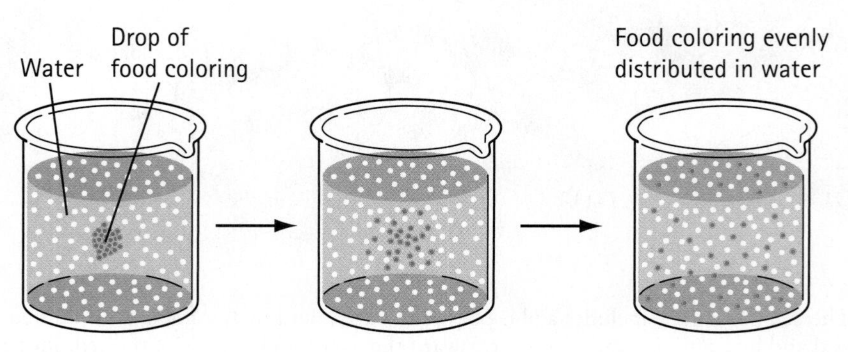

FIGURE 15.15
A drop of food coloring diffuses in a beaker of water. Over time, molecules diffuse from where they are more concentrated to where they are less concentrated. Eventually, the water becomes uniformly colored.

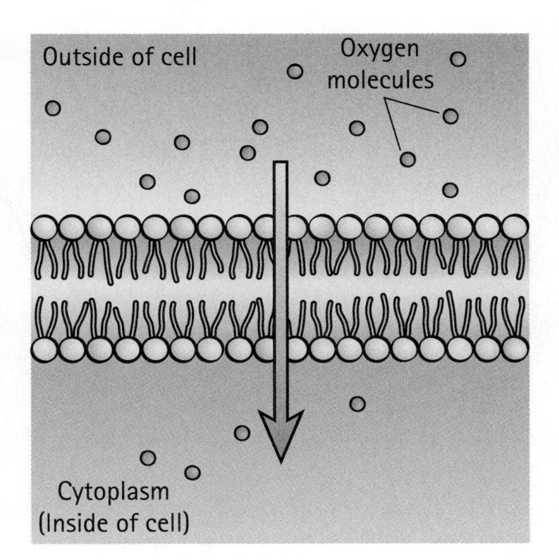

FIGURE 15.16
Molecules that diffuse freely across the cell membrane move from an area of higher concentration to an area of lower concentration.

Let's look at how molecules diffuse across the cell membrane. For example, consider how oxygen diffuses from the fluid surrounding our cells into the cells themselves (Figure 15.16). Oxygen molecules are found both inside and outside the cell, and they move around randomly. Sometimes an oxygen molecule drifts from outside the cell to inside, and sometimes an oxygen molecule drifts from inside the cell to outside. However, because there is a higher concentration of oxygen molecules outside the cell than inside, the net effect of diffusion is to move oxygen molecules into the cell. (Note that there are fewer oxygen molecules inside cells because cells use up oxygen in cellular respiration, a process we will discuss later in this chapter.)

FIGURE 15.17
The Chinese giant salamander is the largest amphibian in the world. Some individuals reach lengths of up to six feet. These salamanders live in fast-flowing mountain streams and obtain oxygen from diffusion across their wrinkly skin. The species is in danger of extinction.

Diffusion works best over small distances. Because of this, bodily processes that depend on diffusion require very thin structures. The walls of our capillaries are very thin so that oxygen can diffuse efficiently from our bloodstream to our tissues. Processes that depend on diffusion also require large surface areas. The intricate branching of our lungs creates a lot of surface area so that a lot of oxygen can diffuse into the bloodstream. If you know that amphibians get much of their oxygen from diffusion across the skin, can you develop a hypothesis for why the giant salamander in Figure 15.17 has such wrinkly skin?

MasteringPhysics®
TUTORIAL: Osmosis and Water Balance in Cells

The diffusion of water has a special name—*osmosis.* Like other substances, water diffuses from an area with a high concentration of water molecules to an area with a low concentration of water molecules. Because a higher concentration of water molecules means a lower concentration of solutes, and vice versa (see Chapter 12), another way to say this is that diffusion moves water from an area of lower solute concentration to an area of higher solute concentration (Figure 15.18). This is important because controlling water flow is important to all cells—with too much water, they could burst; with too little, they shrivel.

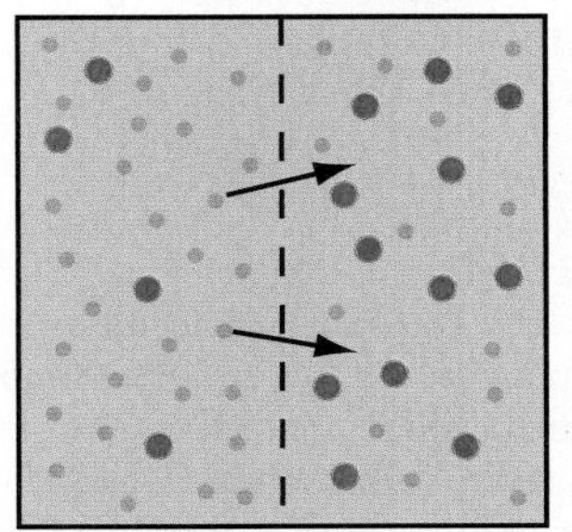

FIGURE 15.18
In osmosis, water molecules move from an area of lower solute concentration to an area of higher solute concentration. Note that the solute molecules cannot cross the barrier, but the water molecules can.

Facilitated Diffusion

Many of the molecules that cells need, including ions and large hydrophilic molecules such as proteins and carbohydrates, cannot pass freely across the phospholipid bilayer of the cell membrane. How do these molecules get into

FIGURE 15.19
Transport proteins and the molecules they transport fit together like a lock and key.

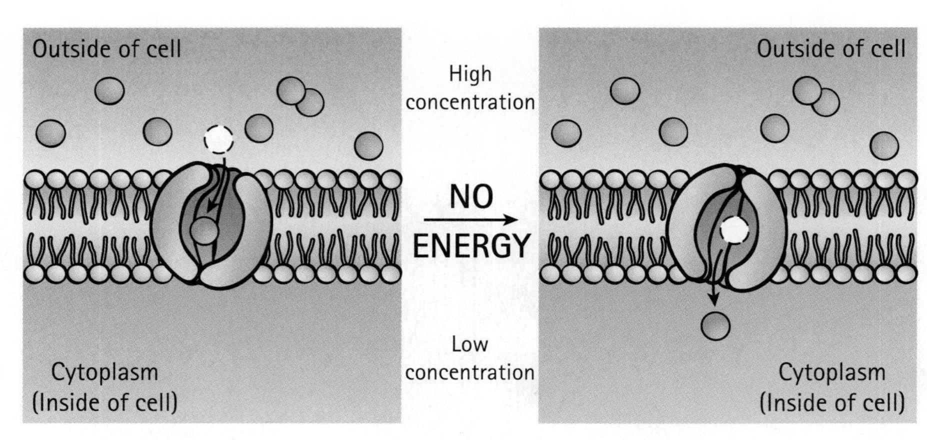

FIGURE 15.20
In facilitated diffusion, molecules move from an area of high concentration to an area of low concentration using a "gate" provided by a transport protein.

and out of cells? They use special "gates" in the cell membrane. These gates are membrane proteins called *transport proteins*. Transport proteins are very specific about the molecules they let through the cell membrane. A molecule fits into its transport protein the way a key fits into a lock—only the right key will work in a given lock (Figure 15.19).

MasteringPhysics®
TUTORIAL: Facilitated Diffusion

In *facilitated diffusion*, a transport protein moves molecules down a concentration gradient, from an area of high concentration to an area of low concentration (Figure 15.20). One example of facilitated diffusion is the movement of the sugar glucose (the basic fuel that cells burn for energy) into red blood cells. Water, in addition to diffusing directly across the phospholipid bilayer, can also use facilitated diffusion to cross the cell membrane. The transport proteins used by water are called aquaporins. Aquaporins allow water to move more quickly across the cell membrane than it can through diffusion alone. Like diffusion, facilitated diffusion requires no energy from the cell. For this reason, diffusion and facilitated diffusion are both examples of **passive transport**.

Active Transport

MasteringPhysics®
TUTORIAL: Active Transport

In **active transport**, a transport protein moves molecules *against* a concentration gradient, from an area of low concentration to an area of high concentration (Figure 15.21). In this case, the second law of thermodynamics tells us that energy

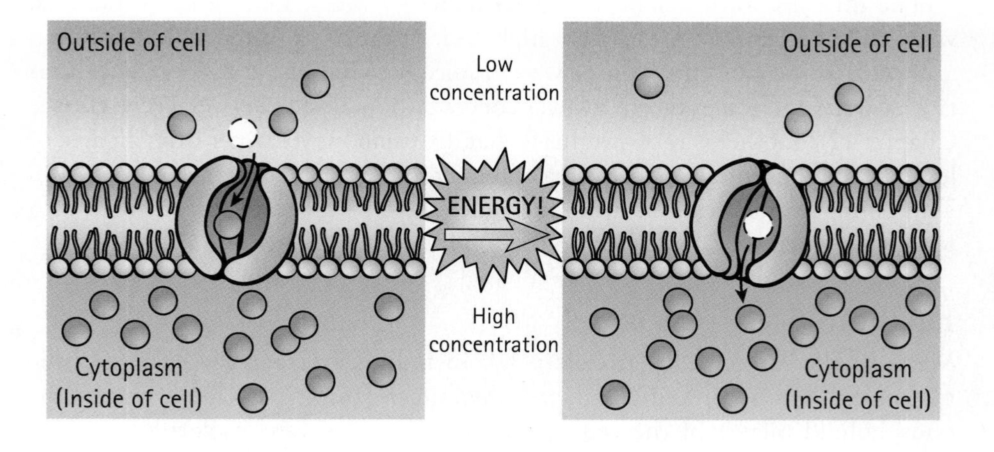

FIGURE 15.21
In active transport, molecules move from an area of low concentration to an area of high concentration—that is, against a concentration gradient. This requires both a transport protein and energy.

is required, since natural systems do not move spontaneously from more diffuse states to more concentrated states. Active transport is used to move many organic molecules, including most proteins, into cells. Active transport is also used to control the concentration of many ions inside and outside cells. An example of active transport, the movement of sodium and potassium ions by the sodium-potassium pump, is described later in this chapter.

UNIFYING CONCEPT

- *The Second Law of Thermodynamics*
Section 6.5

Endocytosis and Exocytosis

Larger amounts of material can be moved into and out of cells through endocytosis and exocytosis (Figure 15.22). In endocytosis, a portion of the cell membrane folds inward and pinches off, enclosing material within a vesicle inside the cell. Endocytosis is used by certain white blood cells of the human immune system to engulf invading bacteria. In exocytosis, the opposite process occurs—a vesicle fuses its membrane with the cell membrane and dumps its contents outside the cell. Many endocrine cells use exocytosis to release hormones into the bloodstream. Neurotransmitters—the chemicals that neurons use to signal one another—are also released through exocytosis (see Chapter 19).

CHECK YOURSELF

Insects don't have lungs. Instead, they get oxygen from a series of small branching tubules in their bodies that are connected to the outside air. Oxygen diffuses through the tubules to reach their tissues. Can you use this information to explain why a mosquito could never grow to be 12 feet tall?

CHECK YOUR ANSWER

Because diffusion works well only at small distances, the insect respiratory system constrains insects to relatively small body sizes. A 12-foot mosquito would not be able to get enough oxygen to the tissues inside its body.

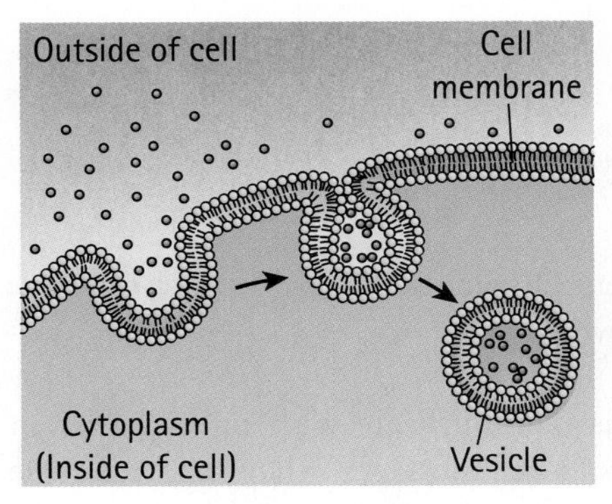

(a) Endocytosis

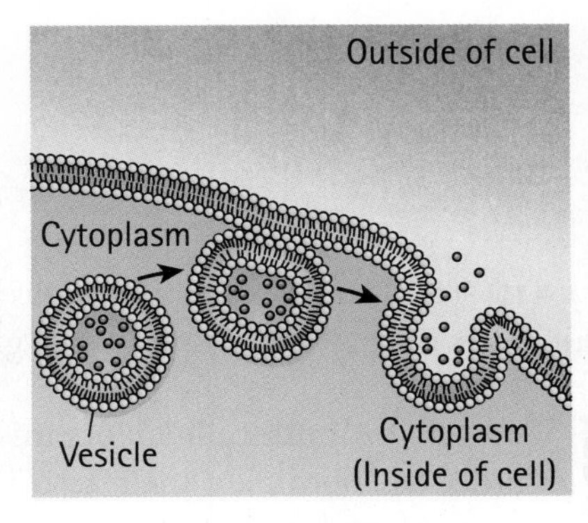

(b) Exocytosis

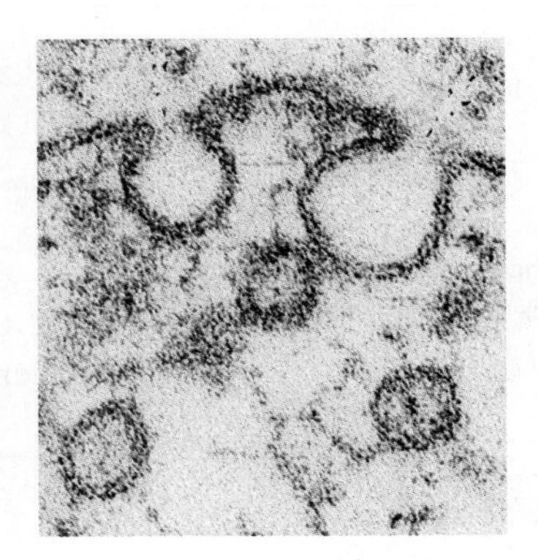

(c)

FIGURE 15.22
(a) In endocytosis, a portion of the cell membrane pinches off to form a vesicle that brings materials into the cell. (b) In exocytosis, a vesicle inside the cell fuses with the cell membrane, dumping its contents outside the cell. (c) This photograph shows vesicles dumping their contents through exocytosis.

MATH CONNECTION

Why Does Diffusion Limit the Size of Cells?

Cells rely on diffusion to obtain many crucial resources. The amount of any given resource that a cell needs usually depends on the cell's volume. The larger the volume, the more resources the cell needs. But, the rate at which molecules diffuse into the cell depends on the cell's surface area—the total area of the cell membrane. This means that for diffusion to work well, a cell needs to have a lot of surface area relative to its volume. What implication does this have for cell size?

Let's look at how well diffusion works in cells of different sizes. We'll assume that cells are spherical. The surface area of a sphere is $4\pi r^2$, where π is the constant equal to approximately 3.14 and r is the radius of the sphere. The volume of a sphere is $\frac{4}{3}\pi r^3$. Let's look at how surface area compares to volume in three cells with radii 1, 2, and 3 micrometers, respectively. For a cell with a radius of 1 micrometer,

$$\frac{\text{Surface area}}{\text{Volume}} = \frac{4\pi r^2}{4/3\pi r^3} = \frac{4\pi(1)^2}{4/3\pi(1)^3} = \frac{4\pi}{4/3\pi} = 3$$

That is, the amount of surface area the cell has is three times greater than the volume of the cell. For a cell with a radius of 2 micrometers,

$$\frac{\text{Surface area}}{\text{Volume}} = \frac{4\pi r^2}{4/3\pi r^3} = \frac{4\pi(2)^2}{4/3\pi(2)^3} = \frac{4\pi(4)}{4/3\pi(8)} = 1.5$$

For a cell with a radius of 3 micrometers,

$$\frac{\text{Surface area}}{\text{Volume}} = \frac{4\pi r^2}{4/3\pi r^3} = \frac{4\pi(3)^2}{4/3\pi(3)^3} = \frac{4\pi(9)}{4/3\pi(27)} = 1$$

What's happening here? Even though a cell's surface area increases as it gets bigger, its volume increases even more quickly. As a result, bigger cells have a *smaller* surface-area-to-volume ratio, making it harder for them to meet their needs through diffusion.

For this reason, it was long thought that bacteria, which rely on diffusion to obtain nutrients, could not grow very large. The discovery of two giant species of bacteria—large enough to be visible to the naked eye—was startling. A close examination of these organisms reveals how they do it. One giant, *Thiomargarita namibiensis*, has all its cytoplasm in a thin layer just under its cell membrane. Most of the interior of the cell is occupied by a giant vacuole. The other giant, *Epulopiscium fishelsoni*, has an extremely wrinkled cell membrane. The wrinkles increase the surface area available for diffusion. In order to use diffusion effectively for transport, these species have evolved ways of compensating for their large size.

Problem The smallest bacteria, called mycoplasmas, have a radius of about 0.1 micrometer, and the largest bacteria (the giant *Thiomargarita namibiensis* mentioned above) have a radius close to 500 micrometers. What is the surface-area-to-volume ratio for each of these?

Solution For the mycoplasma:

$$\frac{\text{Surface area}}{\text{Volume}} = \frac{4\pi r^2}{4/3\pi r^3} = \frac{4\pi(0.1)^2}{4/3\pi(0.1)^3}$$

$$= \frac{4\pi(0.01)}{4/3\pi(0.001)} = 30$$

For the giant bacteria:

$$\frac{\text{Surface area}}{\text{Volume}} = \frac{4\pi r^2}{4/3\pi r^3} = \frac{4\pi(500)^2}{4/3\pi(500)^3}$$

$$= \frac{4\pi(250{,}000)}{4/3\pi(125{,}000{,}000)} = 0.006$$

LEARNING OBJECTIVE
Explain how cells can send chemical messages to other cells.

15.6 Cell Communication

EXPLAIN THIS If growth hormone is sent all over the body, why do only certain parts of the body respond and grow?

The cells of multicellular organisms communicate with one another in order to coordinate their activities. The "messages" they send take the form of molecules. For example, nerve cells send special molecules to muscle cells, telling the muscle cells to contract. The pituitary gland sends a different molecule, called growth hormone, to many different cells in the body, telling them to grow.

MasteringPhysics®
TUTORIAL: Signal Transduction

In animals and plants, special structures allow very local messages to pass directly from one cell to an adjacent cell (Figure 15.23). In animal cells, the

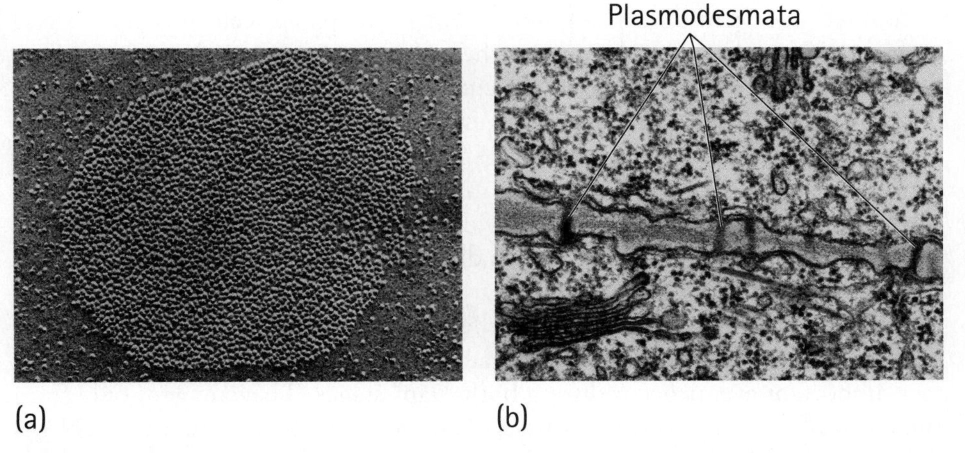

FIGURE 15.23
(a) Gap junctions are tiny channels between adjacent animal cells that allow small molecules to pass. This is a cross-sectional view of a gap junction between two rat cells. (b) This figure shows plasmodesmata, narrow passages that link the cytoplasm of adjacent plant cells. (The green elements are part of the cytoskeleton, and the large red structure is the Golgi apparatus.)

structures are gap junctions, tiny channels between cells. Gap junctions are found in almost every cell in the body. In the heart, for example, communication via gap junctions allows muscle cells to contract simultaneously to produce the heartbeat. Plasmodesmata in plant cells serve a function similar to gap junctions in animal cells. Plasmodesmata are slender threads of cytoplasm that link adjacent plant cells.

Message molecules may also travel through the bloodstream to faraway target cells. When a message molecule reaches a target cell, it binds to a protein called a *receptor*. Some receptors are membrane proteins, whereas others are inside the cell. Receptors are extremely specific about the molecules they bind to. This is because a message molecule and its receptor fit together like a key in a lock—only the right combination will work. This lock-and-key fit also means that only cells with the appropriate receptors will be able to "receive" a specific message molecule. The binding of a message molecule to its receptor sets off a series of chemical reactions that results ultimately in the target cell's response to the message. As just one example, a cell may receive a message that tells it to grow and divide (Figure 15.24).

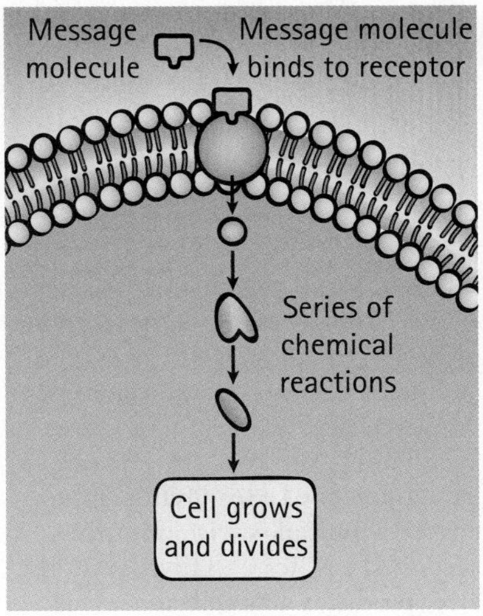

Problems with cell communication can be dangerous. In some cancer cells, a problem in the communication process causes cells to receive the "grow and divide" message continuously and divide out of control. We will present several more examples of message molecules and receptors when we discuss the nervous, sensory, and endocrine systems in Chapter 19.

FIGURE 15.24
A cell receives a message when a message molecule binds to a receptor in its cell membrane. This begins a chain reaction that ends with the cell's response to the message.

15.7 How Cells Reproduce

LEARNING OBJECTIVE
Describe the process of cell division.

EXPLAIN THIS When a cell reproduces, does it produce two identical cells?

Cells reproduce by dividing. Cell division enables single-celled organisms to reproduce themselves and multicellular organisms to develop, grow, and maintain their tissues (Figure 15.25).

MasteringPhysics®
TUTORIAL: Cell Cycle
TUTORIAL: Mitosis and Cytokinesis

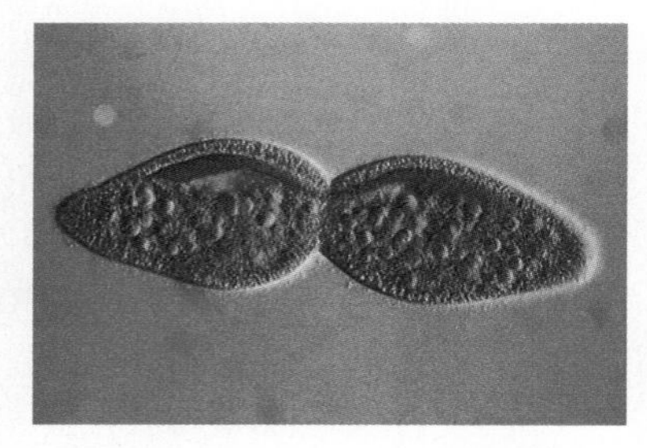

(a)

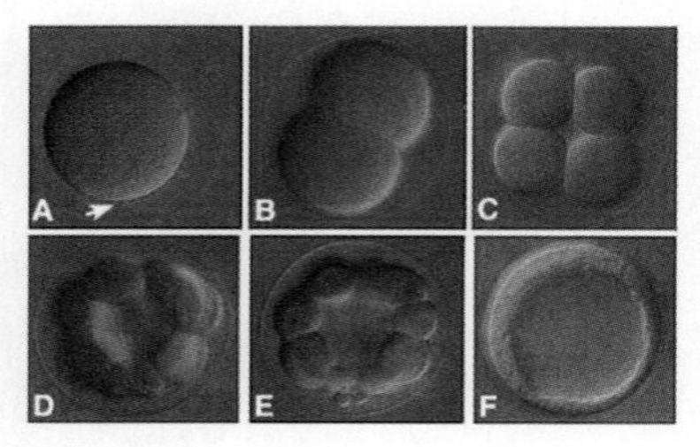

(b)

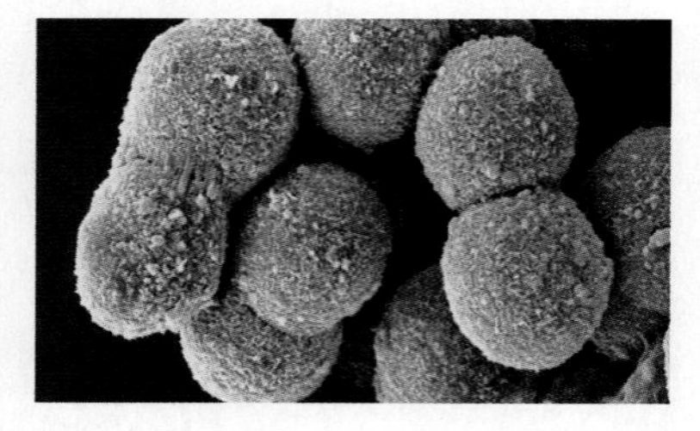

(c)

FIGURE 15.25
Cell division is essential for reproduction, growth and development, and maintenance. (a) A paramecium, a single-celled organism, reproduces by dividing into two. (b) The early development of a sea urchin embryo involves multiple divisions of the fertilized egg. (c) Cell division in the liver produces new cells to replace old, worn-out cells.

Mnemonics are little sayings that help us remember things. Here's a mnemonic for the cell cycle: "***G****o*, ***S****ally*, ***G****o*! ***M****ake* ***C****ake*!" (Gap 1, Synthesis, Gap 2, Mitosis and Cytokinesis).

Mitosis is a form of cell division in which one parent cell divides into two daughter cells, each of which contains the same genetic information as the parent cell. Cells that are preparing to divide enter the *cell cycle*, shown in Figure 15.26. The cell cycle is divided into four stages—gap 1, synthesis, gap 2, and mitosis and cytokinesis. Gap 1, synthesis, and gap 2 are collectively known as interphase. During interphase, the cell makes the necessary preparations for division. During mitosis and cytokinesis, the cell divides.

During gap 1 (G_1), a cell prepares to divide by growing to approximately double its original size. All the important components of the cytoplasm, including the mitochondria and other organelles, also double in number. (Calling this stage a "gap" is a little misleading in that it suggests that nothing is going on. In fact, important events occur during both "gap" stages. They are gaps only from the point of view of someone focused exclusively on whether the cell's DNA is doing anything interesting.)

During synthesis (S), the cell creates an exact copy of its genetic material—its DNA. (The process of duplicating DNA is described in Chapter 16.)

During gap 2 (G_2), the cell builds the machinery necessary for division. This includes the structures that will separate the two copies of the genetic material and divide the cell into two daughter cells.

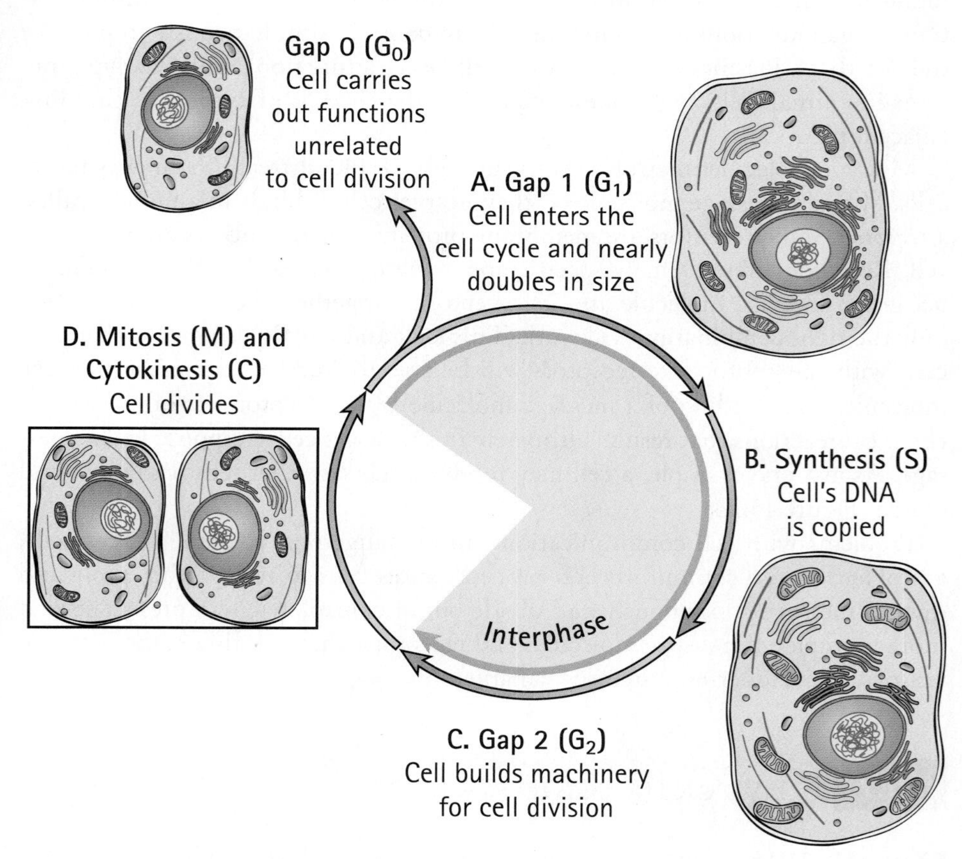

FIGURE 15.26
The cell cycle has four stages. (A) During G_1, the cell grows to about double its original size. (B) During S, an exact copy of the cell's DNA is made. (C) During G_2, the cell builds the machinery required for mitosis. (D) During mitosis and cytokinesis, the cell divides.

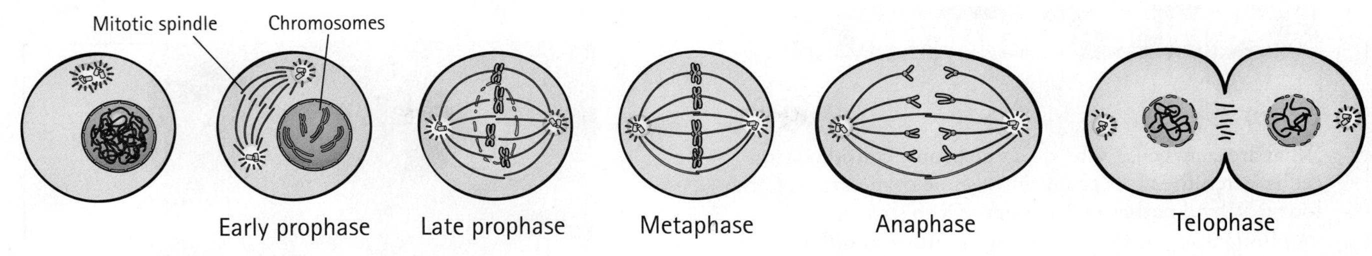

FIGURE 15.27
During mitosis, the nucleus divides in four phases: prophase, metaphase, anaphase, and telophase.

During mitosis and cytokinesis (M), the cell divides. Mitosis describes the division of the nucleus, which takes place in four steps, or "phases" (Figure 15.27). During *prophase*, the normally loosely packed chromosomes condense and the membranes surrounding the nucleus break down. When the chromosomes condense, it becomes clear that each consists of two identical sister chromatids attached at a point called the centromere. The mitotic spindle also forms during prophase. The mitotic spindle, which consists of a series of fibers that attach to the duplicated chromosomes, is responsible for splitting the genetic material between the two daughter cells. During *metaphase*, the chromosomes line up at the equatorial plane, the plane that passes through the imaginary "equator" of the cell. During *anaphase*, the two sister chromatids are pulled apart by the shortening of the mitotic-spindle fibers and move to opposite poles of the cell. During *telophase*, new nuclear membranes form around each set of chromosomes, and the chromosomes return to their loosely packed state. The division of the nucleus is followed by *cytokinesis*, the division of the cytoplasm to yield two separate daughter cells.

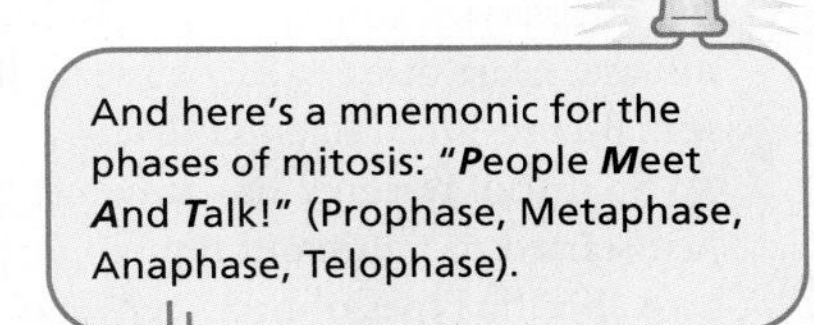

And here's a mnemonic for the phases of mitosis: "***P***eople ***M***eet ***A***nd ***T***alk!" (Prophase, Metaphase, Anaphase, Telophase).

Isn't biology neat?
That's just what I was going to say!

Cells are not always in the cell cycle. Many cells are neither dividing nor preparing to divide, but simply carrying out their regular functions. These cells are said to be in gap 0 (G_0). Some cells are in G_0 temporarily and then eventually reenter the cell cycle. Other cells, such as many neurons, are in permanent G_0 and will never divide again.

CHECK YOURSELF

During which stages of the cell cycle does the cell have twice the genetic material it normally has?

CHECK YOUR ANSWER

The cell has twice the normal amount of genetic material during G_2 and M stages—that is, after S stage and continuing up until cytokinesis is complete. The cell has double the amount of genetic material after S because it copies its genetic material during S. During M, all that genetic material is still in one big cell. Only after cytokinesis, when two new cells are formed, does each new cell have the normal amount of genetic material.

SCIENCE AND SOCIETY

Stem Cells

What are stem cells? Why do some people consider stem-cell research the most promising avenue of medical research today? Why do other people oppose it so strongly?

Humans are made up of a huge number of different kinds of cells—skin cells, muscle cells, liver cells, brain cells, and so on. Embryonic stem cells come from human embryos that have yet to differentiate into distinct types of cells. Because they are still undifferentiated, embryonic stem cells have the capacity to develop into all the different kinds of cells in the body. The hope of stem-cell research is to use embryonic stem cells to grow healthy cells that can then be used to replace defective or diseased cells in the body. Embryonic stem cells provide great promise for treating a variety of conditions, including diabetes, Parkinson's disease, heart disease, Alzheimer's disease, arthritis, strokes, spinal-cord injuries, and burns.

Stem-cell research has only just begun. Scientists have just started to figure out how to direct stem cells to develop into specific types of body cells. For example, one group of researchers recently made embryonic stem cells develop into insulin-producing cells. These cells could one day be used to treat type 1 diabetes, a disease in which the insulin-producing cells of the pancreas are destroyed by the body's immune system. Other researchers have made embryonic stem cells develop into special immune-system cells that attack tumors. These cells may eventually be effective in treating cancer. California-based Geron Corporation is testing the use of cells derived from embryonic stem cells to treat spinal-cord injuries. The cells are injected into patients' bodies, where they differentiate into the supportive cells that sheathe nerve cells and allow them to function. This study, which began in 2009, is the first clinical trial of a stem cell therapy to be approved by the U.S. Food and Drug Administration (FDA).

Beyond the challenge of directing stem-cell differentiation, scientists have to figure out how to prevent a patient's immune system from identifying transplanted cells as foreign and attacking them. In addition, the great capacity for cell division that characterizes stem cells has to be carefully controlled. Otherwise, the cells may form tumors after they have been transplanted.

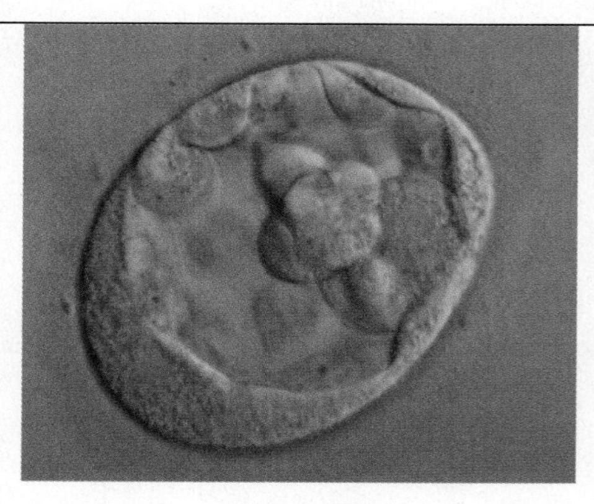

Embryonic stem cells are isolated from three- to five-day-old human embryos, such as the one shown here.

Why is stem-cell research controversial? Embryonic stem cells are grown from cells that are removed from three- to five-day-old human embryos. These embryos were created at fertility clinics to aid infertile couples and were then donated to stem-cell research, with donor consent, when they were no longer needed. Because removing the stem cells kills the embryo, however, some people are opposed to stem-cell research. (The extra embryos generated at fertility clinics are discarded if they are not donated to another couple or used for research.) Although "stem cells" also exist in adults, adult stem cells are very different from embryonic stem cells. Adult stem cells can give rise to only a limited range of cell types. They are also much harder to grow in the lab.

In the United States, federal funding of research that involves the destruction of human embryos has been banned since 1996. However, it is sometimes permissible to use federal funds to work on stem-cell lines that were initially established without the use of federal funds. Just what is permissible is currently being fought out in court battles. In the meantime, much of the progress in stem-cell research is occurring in the private sector as well as abroad.

LEARNING OBJECTIVE
Describe the role of enzymes in cells.

15.8 How Cells Use Energy

EXPLAIN THIS Why is lead toxic?

At any moment, countless chemical reactions are occurring in cells. These reactions sustain life by enabling cells to carry out such essential functions as building macromolecules, transporting molecules across membranes, and dividing. But what determines exactly which chemical reactions occur?

We learned in Chapter 13 that in order for a chemical reaction to occur, two conditions must be met. First, the reaction must obey the law of conservation of energy. Second, the reacting molecules must collide with enough energy to get the reaction going—they must have the *activation energy* necessary for the initial breaking of bonds.

Let's look at conservation of energy first. Remember from Chapter 13 that *exothermic* reactions release energy, and *endothermic* reactions absorb energy. Because of conservation of energy, exothermic reactions tend to happen all by themselves. On the other hand, endothermic reactions can occur only if energy from an external source is provided. In cells, usable energy is found in molecules of adenosine triphosphate, or **ATP**.

UNIFYING CONCEPT

- ***The Law of Conservation of Energy*** **Section 4.10**

ATP and Chemical Reactions in Cells

LEARNING OBJECTIVE
Explain how ATP provides cells with the energy needed to run chemical reactions.

EXPLAIN THIS If ATP is like a rechargeable battery, how is its energy used up, and how is it recharged?

ATP provides energy for cellular processes. One way to think of ATP is as the cell's equivalent of a dollar bill. Dollar bills can be used to buy all sorts of things, just as ATP can be used to power all sorts of chemical reactions. ATP also provides an appropriate amount of energy for the types of chemical reactions that occur in living organisms. Other energy storehouses—glucose, for example, or starches, oils, or fats—are like gold nuggets or Van Gogh paintings: They are worth a lot, but you can't use them to pay for milk at the grocery store.

ATP consists of an adenosine molecule and three phosphate groups. Cells get energy from ATP when one of its phosphate groups is removed, leaving adenosine diphosphate, or ADP (Figure 15.28). Cells eventually turn ADP back into ATP by adding a phosphate group during cellular respiration (Figure 15.29). ATP is like a rechargeable battery that can be repeatedly used up and recharged.

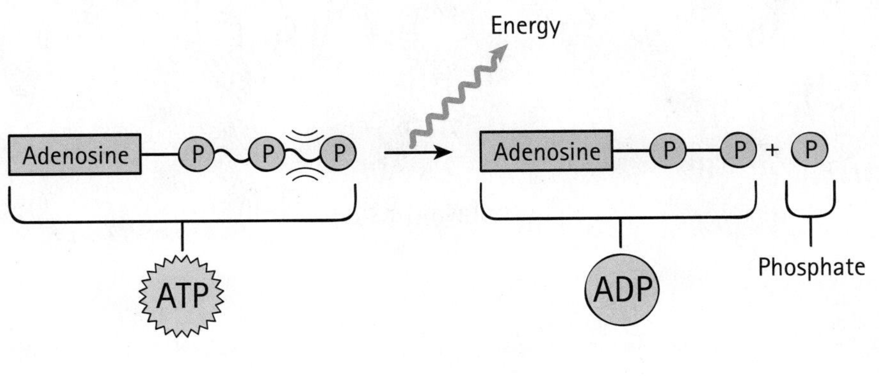

FIGURE 15.28
Energy is obtained from ATP when one of its phosphate groups is removed, leaving ADP.

FIGURE 15.29
Cells obtain energy from ATP when one of its phosphate groups is removed, leaving ADP. ADP is eventually turned back into ATP through the addition of a phosphate group during cellular respiration.

UNIFYING CONCEPT

- *The Second Law of Thermodynamics* Section 6.5

Of course, the second law of thermodynamics (see Chapter 6) tells us that it will take more energy to make ATP than cells eventually get out of it. This means that not all the energy we take in through food gets turned into growth, reproduction, or activity; a large proportion of it is lost to the environment as heat. The implications of this energy loss extend all the way to the functioning of ecosystems, as we will see in Chapter 21.

An example of a biological process that requires exactly one molecule of ATP to run is the sodium-potassium pump. The sodium-potassium pump uses active transport to maintain a low concentration of sodium ions (Na^+) and a high concentration of potassium ions (K^+) inside animal cells. It does this by using ATP to pump sodium ions out and to pump potassium ions in. Maintaining appropriate ion concentrations helps to regulate water flow into and out of the cell and is also critical for such processes as the firing of neurons and the contraction of muscles (see Chapter 19).

How does the pump work? The sodium-potassium pump is a transport protein in the cell membrane (Figure 15.30). It behaves like a swinging door, shuffling ions into and out of the cell. In its default state, the protein is open to the inside of the cell. There, it binds three sodium ions. An ATP reaction, in which a molecule of ATP transfers a phosphate group to the protein, causes the protein to shift and open to the outside of the cell. The sodium ions are released, and two potassium ions are bound, causing the phosphate group to be released from the transport protein. The loss of the phosphate group causes the protein to shift back to its original position and to release the potassium ions inside the cell. Amazingly, more than a third of all the ATP consumed by animals is used to run this pump.

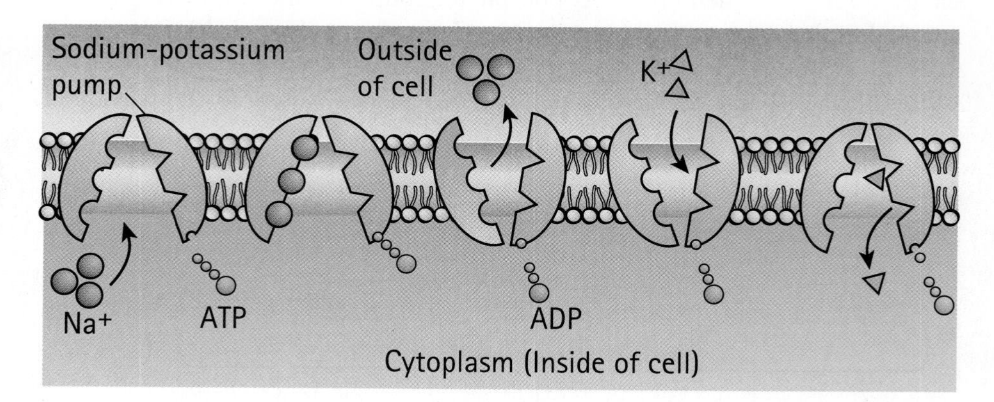

FIGURE 15.30
The sodium-potassium pump moves sodium ions out of cells and potassium ions into cells. One molecule of ATP allows the movement of three sodium ions out and two potassium ions in.

CHECK YOURSELF

1. **Each "pump" of the sodium-potassium pump moves three sodium ions out and two potassium ions in. If you have five molecules of ATP, how many sodium and potassium ions can be transferred from one side of the cell membrane to the other?**
2. **If ATP were temporarily unavailable, what would happen to a sodium-potassium pump that was just about to release sodium ions outside the cell?**

CHECK YOUR ANSWERS

1. Fifteen sodium ions can be moved out of the cell, and ten potassium ions can be moved into the cell.
2. The sodium-potassium pump would release the sodium ions, bind potassium ions, shift, release the potassium ions inside the cell, and bind more sodium ions. (None of these activities requires ATP.) At this point, the pump would be stuck, since ATP is required for it to proceed to the next step: shifting to open to the cell's exterior.

We now see how cells use energy from ATP to power chemical reactions. But a second condition must be met before a chemical reaction can occur: The reacting molecules must collide with enough energy to get the reaction going—this activation energy is needed for the initial breaking of bonds. Unfortunately for living organisms, the activation energy for many essential chemical reactions is very high. Some important reactions would take 100 years to happen on their own. Because of this, cells rely on catalysts to lower the activation energy of reactions and allow them to happen more quickly. The catalysts in cells are large, complex proteins called **enzymes**. Enzymes help specific chemical reactions proceed more quickly by lowering the activation energy. With the help of enzymes, important chemical reactions can happen in milliseconds instead of taking 100 years.

How does an enzyme work? An enzyme binds the reactants of a reaction—the enzyme's *substrate*—at its active site and then releases the products (Figure 15.31). Enzymes, like other catalysts, are not altered or destroyed in the reactions they catalyze. This means they are available to catalyze the same reaction over and over again. Enzymes are involved in nearly all of the chemical reactions in cells; in fact, several thousand unique enzymes have been identified in living organisms. Each of these enzymes is highly specific for a certain reaction. That is, each enzyme is very picky about the substrate to which it binds and therefore about the reaction it catalyzes.

Cells regulate enzymes carefully in order to control chemical reactions. Regulation takes place in different ways. First, cells control the synthesis and degradation of enzymes. (Enzymes are not used up when they catalyze a reaction, so they must be actively degraded when they are no longer needed.) Second, how

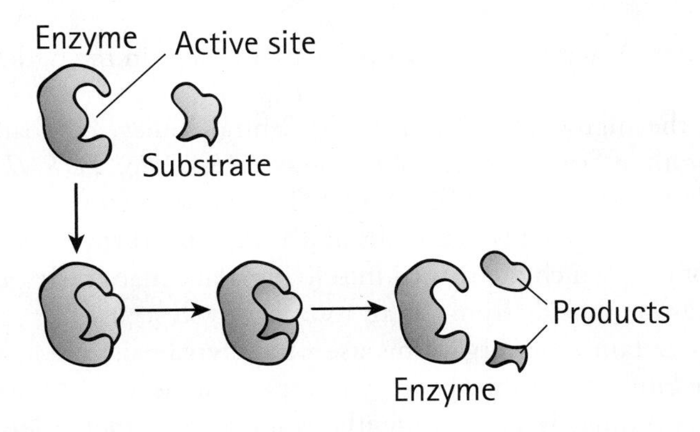

FIGURE 15.31
Enzymes are large proteins that function as catalysts. They are very specific, binding to only specific molecules and catalyzing only specific reactions. Note that the enzyme is not used up in the reaction it catalyzes, and so it is available to catalyze the same reaction again.

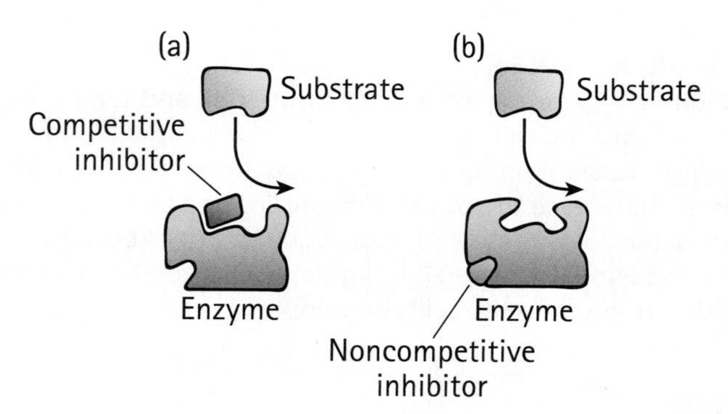

FIGURE 15.32
(a) In competitive inhibition, the inhibitor binds to the enzyme's active site and blocks the substrate from binding. (b) In noncompetitive inhibition, the inhibitor binds to the enzyme somewhere other than at the active site. This changes the active site in such a way that the enzyme is no longer able to bind to the substrate.

well an enzyme works depends on the chemical features of its environment—the temperature, pH, and other factors. For example, if the temperature is too high or too low, an enzyme's shape may change and the reactants may not fit. Finally, enzymes can be blocked by inhibitors. In competitive inhibition (Figure 15.32a), an inhibitor binds to the active site of an enzyme, preventing it from binding to its substrate. (If a substrate fits into an enzyme like a key in a lock, the inhibitor is like bubble gum stuck in the keyhole.) This type of inhibition is called competitive inhibition because the inhibitor and the substrate compete for the active site. Cyanides are toxic because they competitively inhibit enzymes that are critical for cellular respiration. In noncompetitive inhibition (Figure 15.32b), an inhibitor binds to an enzyme somewhere other than at the active site, changing the enzyme so that it can no longer bind to its substrate. Lead is toxic because it binds to a number of enzymes in the body and disrupts their structures.

Many medications work by inhibiting enzymes. The antibiotic penicillin kills bacteria by inhibiting an enzyme that bacteria need to build cell walls. Aspirin works as a painkiller because it inhibits an enzyme needed to make prostaglandins, molecules that make us more sensitive to pain (see Chapter 19).

LEARNING OBJECTIVE
Explain how plant cells use light energy from the Sun, carbon dioxide, and water to make sugars.

15.9 Photosynthesis

EXPLAIN THIS What happens to a plant that is kept in the dark?

Along the highways of California's Central Valley, billboards celebrate the work of farmers. Some of them say *FARMERS: TURNING WATER AND SUNLIGHT INTO FOOD*. This is what crops do and, in fact, what all plants do. They take sunlight and water and turn them into organic molecules, some of which are turned into living plant matter. (It turns out that plants need carbon dioxide from the air too, as we will see.)

Plants are constantly removing carbon dioxide from the air as they photosynthesize. You may also know that the increasing amount of carbon dioxide in the atmosphere is causing global warming. This is why conserving forests is an important part of the battle against global warming.

Plants and certain other organisms use **photosynthesis** to convert light energy from the Sun into chemical energy in organic molecules. Almost all life on Earth depends ultimately on photosynthesis for organic molecules and energy. This is because plants and other photosynthesizers (such as algae and certain bacteria) are food for herbivores, and herbivores are food for carnivores. So, photosynthesizers are the ultimate source of all food.

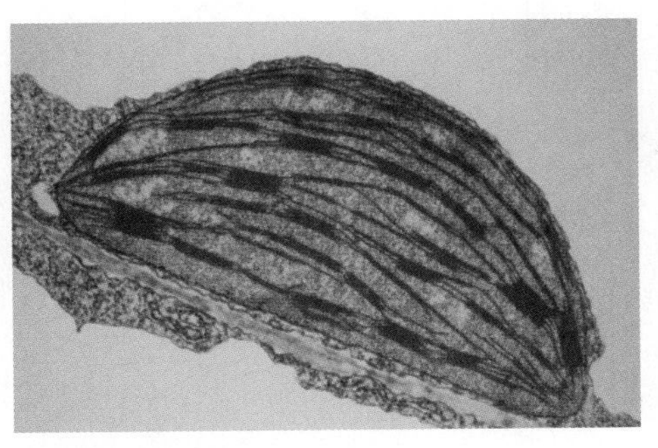

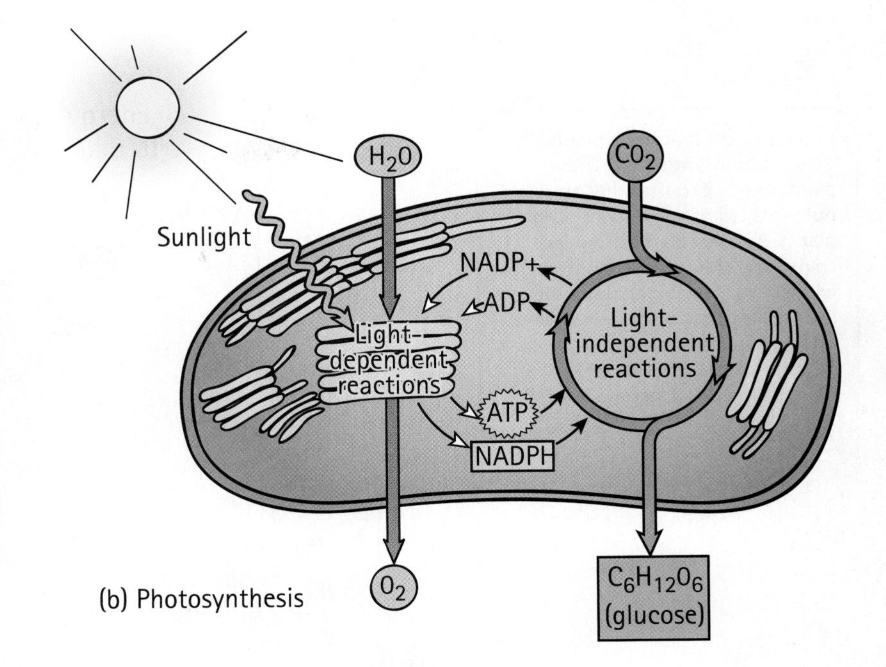

FIGURE 15.33
(a) Photosynthesis takes place in plant organelles called chloroplasts, found primarily in stems and leaves. This photo shows a chloroplast in a lilac leaf. (b) Photosynthesis occurs in two steps: the light-dependent and light-independent reactions.

The chemical reaction for photosynthesis is:

$$6\,CO_2 + 6\,H_2O + \text{sunlight} \longrightarrow C_6H_{12}O_6 + 6\,O_2$$

Carbon dioxide, water, and sunlight go in; glucose and oxygen come out. (Keep in mind, though, that this is a summary of the overall process of photosynthesis. In fact, a large number of chemical reactions are needed before the reactants in photosynthesis are converted into the products.) The oxygen released during photosynthesis is the source of the oxygen we breathe—yet another way photosynthesis is fundamental to life as we know it.

Photosynthesis takes place in the chloroplasts of plant cells (Figure 15.33a). It occurs in two steps: *light-dependent reactions* and *light-independent reactions* (Figure 15.33b). During the light-dependent reactions, energy is captured from sunlight. During the light-independent reactions, carbon is fixed—that is, carbon atoms are moved from atmospheric carbon dioxide to the organic molecule glucose.

The light-dependent reactions begin when sunlight strikes a chlorophyll molecule inside a chloroplast. *Chlorophyll* is a pigment, meaning it is a molecule that absorbs light. (Remember from Chapter 8 that visible light is just one part of the electromagnetic spectrum, which also includes radio waves, ultraviolet light, X-rays, and gamma rays.) Chlorophyll absorbs blue-violet light and red-orange light best, as you can see in Figure 15.34. The light that chlorophyll does not absorb—green light—is reflected by plants. This is why plants are green.

When sunlight strikes a chlorophyll molecule, the energy in sunlight excites an electron in the chlorophyll molecule and knocks it out (Figure 15.35). The chlorophyll molecule, which is now missing an electron, extracts an electron

UNIFYING CONCEPT

- ***Waves***
Section 8.1

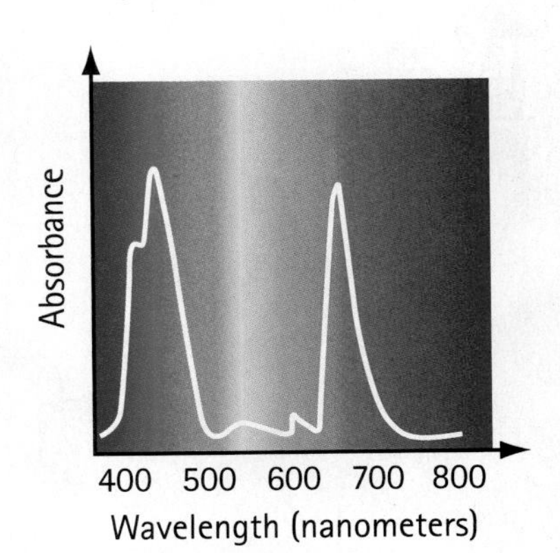

FIGURE 15.34
Light is a type of electromagnetic radiation. Like other types of electromagnetic radiation, it takes the form of waves. Different colors of visible light have different wavelengths (the wavelength is the distance from the top of one crest to the top of the next), ranging from about 400 nanometers for blue-violet light to 800 nanometers for red light. This graph shows that chlorophyll absorbs blue-violet light and red-orange light best. It does not absorb green light well.

Did you know that some plants do not photosynthesize? These plants live as parasites that absorb nutrients from their hosts—other plants. Parasitic plants often lack chlorophyll. Many do not bother with stems, leaves, or roots either. Instead, they focus their energies on reproduction, and it shows—one parasitic plant is known for having a huge flower a full meter across!

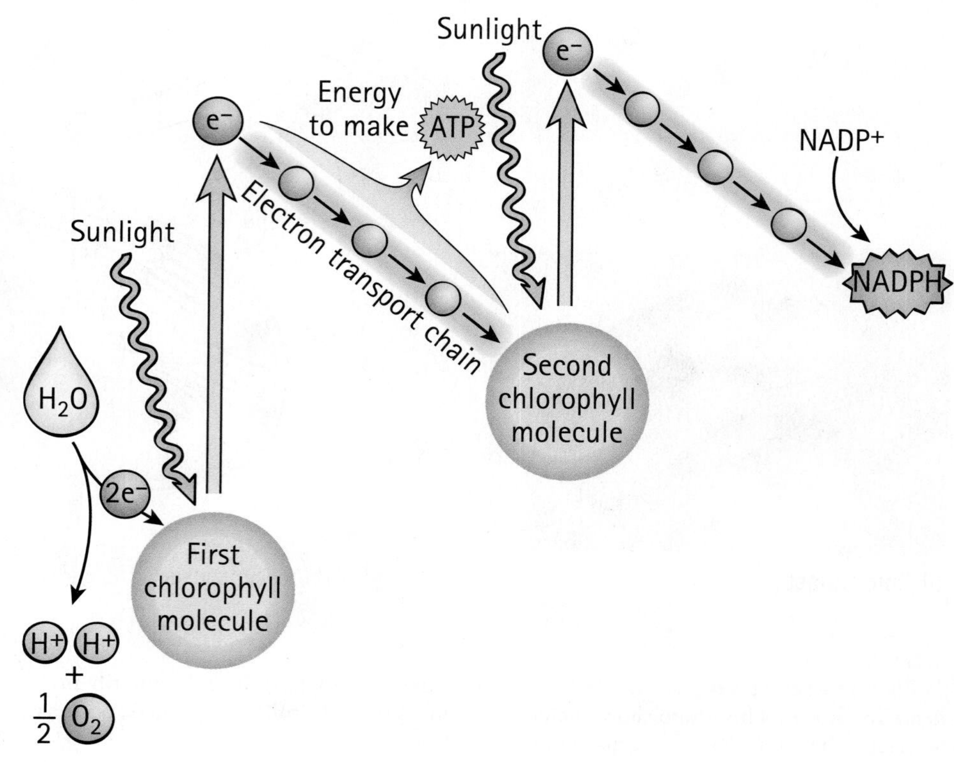

FIGURE 15.35
In the light-dependent reactions of photosynthesis, two electron transport chains capture energy from sunlight and store it in ATP and NADPH. Oxygen is released.

FIGURE 15.36
In an electron transport chain, the electron is passed from one carrier to the next the way a bucket of water is passed from one person to the next in a cartoon fire brigade. The electron loses energy as it is passed down the chain.

from water (H_2O), producing oxygen (O_2). Meanwhile, the high-energy electron that was knocked out of the chlorophyll molecule passes down an electron transport chain. The *electron transport chain* consists of a series of carrier molecules, each of which receives the electron and then passes it on to the next carrier. You can imagine the electron transport chain as a cartoon fire brigade, in which a bucket of water is passed from one person to the next until the last person tosses it over the flames (Figure 15.36). The electron is the bucket of water, passed from one carrier to the next. As the electron passes down the transport chain, it loses energy. This energy is used to convert ADP into ATP. At the end of the transport chain, a second, different chlorophyll molecule receives the electron. This chlorophyll molecule is struck by sunlight, and the electron is knocked out again and passed down a second electron transport chain. This time, the energy lost by the electron as it passes down the transport chain is used to convert a molecule called $NADP^+$ into NADPH.

To summarize, two important things happen during the light-dependent reactions. First, energy from sunlight is converted into chemical energy in the form of ATP and NADPH. Second, oxygen is released.

In the light-independent reactions, also known as the *Calvin cycle*, the cell uses the energy stored in ATP and NADPH to fix carbon (see Figure 15.33b). During the Calvin cycle, six molecules of carbon dioxide (CO_2) are taken in and used to make a molecule of glucose ($C_6H_{12}O_6$). Cells use glucose as a starting point for making other carbohydrates, lipids, and, with the addition of nitrogen, amino acids and nucleic acids—in short, all the macromolecules of life.

MasteringPhysics®
TUTORIAL: Overview of Photosynthesis

CHECK YOURSELF

1. **What three things does a plant need to photosynthesize? What two products are made during photosynthesis?**
2. **Is oxygen released during the light-dependent or the light-independent reactions of photosynthesis?**

CHECK YOUR ANSWERS

1. Plants need sunlight, water, and carbon dioxide for photosynthesis. The two products are glucose and oxygen.
2. Oxygen is released during the light-dependent reactions.

15.10 Cellular Respiration and Fermentation

LEARNING OBJECTIVE
Explain how cells break down glucose to obtain energy in the form of ATP.

EXPLAIN THIS Why can't you live without oxygen?

What do the following things have in common: heavy breathing during exercise, bread dough rising, and alcohol in wine and beer? They are all related to how cells obtain energy.

All cells need ATP, which provides the energy required for many essential cellular processes. In order to obtain ATP, cells break down glucose and other organic molecules. The *aerobic*, or oxygen-using, breakdown of glucose is known as **cellular respiration**. The equation for cellular respiration is

MasteringPhysics®
TUTORIAL: Overview of Cellular Respiration

$$C_6H_{12}O_6 + 6\ O_2 + \text{about 38 molecules of ADP}$$
$$\rightarrow 6\ CO_2 + 6\ H_2O + \text{about 38 molecules of ATP}$$

Glucose, oxygen, and ADP go in; carbon dioxide, water, and ATP come out. This equation summarizes the process of cellular respiration. The many chemical reactions that occur from the beginning to the end of cellular respiration can be divided into three steps: glycolysis, the Krebs cycle, and electron transport (Figure 15.37).

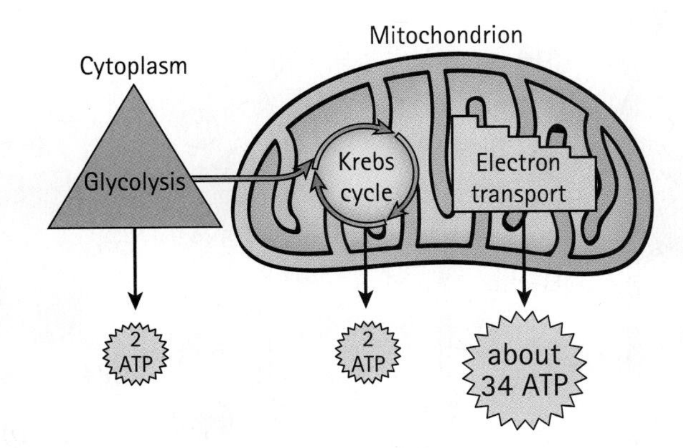

FIGURE 15.37
Cellular respiration takes place in three steps: glycolysis, the Krebs cycle, and electron transport. Glycolysis occurs in the cytoplasm. The Krebs cycle and electron transport occur in the mitochondria.

Glycolysis

The first step in breaking down glucose is *glycolysis* (which literally means "sugar splitting"). Glycolysis takes place in the cell cytoplasm. During glycolysis, the six-carbon glucose molecule is split into two molecules of pyruvic acid, each of which contains three carbon atoms. Two molecules of ATP are produced in the process.

The Krebs Cycle and Electron Transport

The Krebs cycle and electron transport occur in the mitochondria. Before entering the Krebs cycle, the pyruvic acid that is produced during glycolysis is converted into acetic acid and bound to a molecule of coenzyme A. This entire complex is called acetyl-CoA. During the Krebs cycle, acetyl-CoA is broken down into carbon dioxide. Two molecules of ATP are produced, and additional energy is stored in two other molecules, NADH and $FADH_2$.

During electron transport, electrons carried by NADH and $FADH_2$ are sent down electron transport chains (Figure 15.38). As electrons are passed from one carrier in the transport chain to the next, they lose energy. This energy is used to pump hydrogen ions (H^+) across a membrane inside the mitochondrion. At the end of the electron transport chain, the electrons combine with an oxygen molecule to generate water. (This is what all the oxygen we breathe is for—it is needed to catch those electrons at the end of electron transport chains.) The concentration gradient of hydrogen ions across the inner mitochondrial membrane is then used to make ATP. As hydrogen ions move back across the inner mitochondrial membrane (down their concentration gradient), they pass through a protein complex called ATP synthase and turn ADP into ATP. This process generates the bulk of the ATP harvested during cellular respiration.

Fermentation

In certain cells, glycolysis is sometimes followed by an *anaerobic* (non–oxygen-using) process known as **fermentation** instead of the aerobic process just described. Fermentation yields no ATP, but it does regenerate the molecules

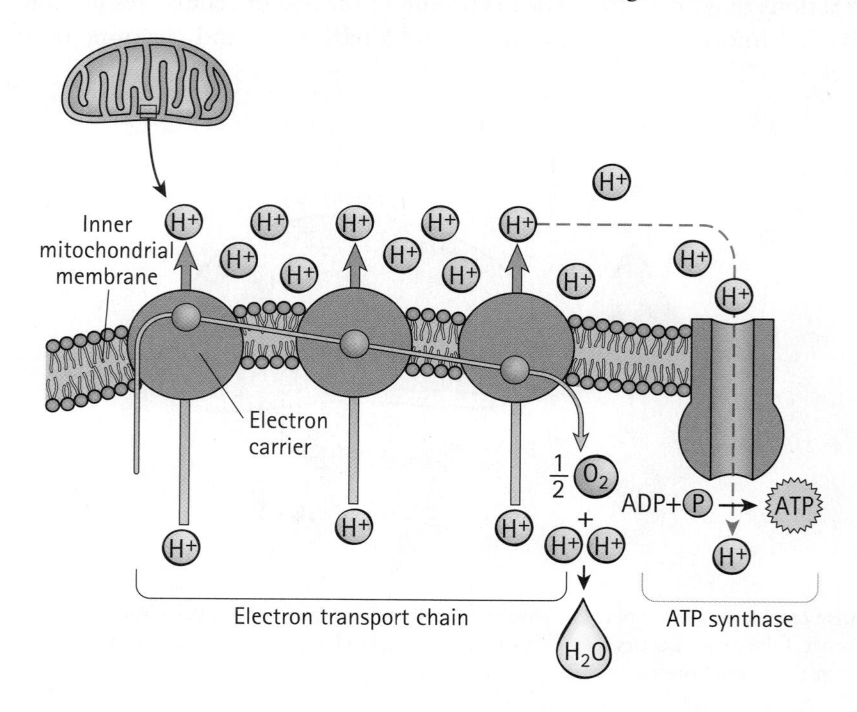

FIGURE 15.38
As electrons move down electron transport chains, hydrogen ions are pumped across the membrane inside mitochondria. When the hydrogen ions move back across the membrane, they generate ATP.

necessary to keep glycolysis going. In this way, cells can continue to obtain ATP through glycolysis.

Alcoholic fermentation is an anaerobic process used by yeast. It takes place in the cytoplasm of yeast cells. In alcoholic fermentation, the pyruvic acid from glycolysis is broken down into ethanol and carbon dioxide. Alcoholic fermentation is essential both to making wine and to baking bread. Yeast cells ferment the sugar in grape juice to turn it into wine. The same process makes bread rise—yeast cells ferment the sugars in bread dough, releasing tiny bubbles of carbon dioxide gas. (And, yes, fermentation also produces ethanol in bread dough, but it evaporates during baking. This is why you don't have to be 21 to purchase dinner rolls.)

As yeasts ferment the sugars in bread dough, they release tiny bubbles of carbon dioxide gas, which make the dough rise.

Lactic acid fermentation is an anaerobic process that occurs in some animal cells as well as in certain species of bacteria and fungi. It takes place in the cytoplasm of these cells. In lactic acid fermentation, the pyruvic acid from glycolysis is broken down to lactic acid. In animal muscle cells, lactic acid fermentation occurs during strenuous exercise, when the oxygen supply—despite heavy breathing—can't quite meet the demand. By regenerating the molecules required for glycolysis, lactic acid fermentation allows muscle cells to continue to make ATP without oxygen. The lactic acid produced during strenuous exercise causes a burning sensation in the muscles. Red blood cells, which lack mitochondria, also rely on lactic acid fermentation to obtain ATP. Finally, lactic acid fermentation by certain species of bacteria and fungi is used to make cheese and yogurt.

Why do you continue to breathe hard even after strenuous activity ends? The liver needs oxygen to convert lactic acid back into pyruvic acid, which can then be converted back into glucose.

Table 15.2 compares cellular respiration and fermentation.

CHECK YOURSELF

1. **About 38 molecules of ATP are obtained from one glucose molecule. Of these 38, how many come from glycolysis, the Krebs cycle, and electron transport?**
2. **How are the chemical reactions for photosynthesis and cellular respiration similar? How are they different?**

CHECK YOUR ANSWERS

1. Of the approximately 38 ATP molecules, 2 come from glycolysis, 2 from the Krebs cycle, and about 34 from electron transport.
2. The reactions are similar in that the products of one are the reactants of the other, and vice versa. They are different in that photosynthesis takes energy and small molecules to build a larger molecule (glucose), whereas cellular respiration takes a large molecule and breaks it down into smaller molecules, releasing energy in the process.

Your body relies on cellular respiration for endurance activities such as long-distance running or cycling, but on lactic acid fermentation for activities that require a burst of energy, such as sprinting or hitting a baseball.

TABLE 15.2 COMPARING CELLULAR RESPIRATION AND FERMENTATION

Cellular Respiration	Fermentation
Aerobic (uses oxygen)	Anaerobic (does not use oxygen)
Steps: glycolysis, Krebs cycle, electron transport	Steps: glycolysis, fermentation
Takes place in cytoplasm (glycolysis) and mitochondria (Krebs cycle and electron transport)	Takes place in cytoplasm only
Makes 38 molecules of ATP	Makes 2 molecules of ATP (during glycolysis)

For instructor-assigned homework, go to www.masteringphysics.com MP

SUMMARY OF TERMS (KNOWLEDGE)

Active transport Transport across the cell membrane that requires energy from the cell.

ATP Adenosine triphosphate, the molecule that provides the energy for most cellular processes.

Biology The study of life and living organisms.

Carbohydrates Organic molecules that consist of a single simple sugar or chains of simple sugar molecules.

Cell membrane The structure that separates the inside of the cell from the outside of the cell.

Cells The basic units of life that make up all living organisms.

Cellular respiration The aerobic breakdown of glucose to produce ATP.

Chloroplasts The organelles within plant cells in which photosynthesis occurs.

Chromosomes The DNA-containing structures within cells.

Cytoplasm The part of a cell that is inside the cell membrane but outside the nucleus.

Diffusion The passive movement of molecules from an area of high concentration to an area of low concentration.

Enzymes Complex proteins that catalyze chemical reactions in living organisms.

Eukaryotes Organisms whose cells have a true nucleus, such as protists, animals, plants, and fungi.

Fermentation The anaerobic breakdown of glucose that results in the production of ethanol and carbon dioxide (alcoholic fermentation) or lactic acid (lactic acid fermentation).

Lipids Hydrophobic organic molecules, many of which include fatty acids as a primary component.

Mitochondria Eukaryotic organelles that break down organic molecules to produce ATP.

Mitosis A form of cell division in which one cell divides into two daughter cells, each of which contains the same genetic information as the original cell.

Nucleic acids Organic molecules that are composed of chains of nucleotides and store genetic information.

Nucleus A structure within eukaryotic cells that is surrounded by a double membrane and that contains the cell's DNA.

Organelles Structures within the cytoplasm of eukaryotic cells that perform specific functions in the cell.

Passive transport Transport across the cell membrane that does not require energy from the cell.

Photosynthesis The process in plants and some other organisms in which light energy from the Sun is converted into chemical energy in organic molecules.

Prokaryotes Single-celled organisms, such as bacteria and archaea, whose cells lack a nucleus.

Proteins Organic molecules composed of folded chains of amino acids.

READING CHECK (COMPREHENSION)

15.1 Characteristics of Life

1. What are some of the characteristics that living organisms share?
2. Describe what it means to say that living things are parts of populations that evolve.

15.2 Cell Types: Prokaryotic and Eukaryotic

3. What are some examples of prokaryotes? What are some examples of eukaryotes?
4. Describe three or more differences between prokaryotic cells and eukaryotic cells.
5. How is the DNA of prokaryotes packaged differently from the DNA of eukaryotes?

15.3 Tour of a Eukaryotic Cell

6. What is the nucleus of a cell? What does the nucleus contain?
7. Describe the functions of the following organelles: mitochondria, ribosomes, lysosomes, chloroplasts.

15.4 The Cell Membrane

8. What are three components of the cell membrane?
9. How are phospholipids arranged in the cell membrane?
10. What are some of the functions of membrane proteins?

15.5 Transport Into and Out of Cells

11. Name three molecules that are able to move directly across the phospholipid bilayer of the cell membrane.
12. What is diffusion?
13. What distinguishes passive transport from active transport?
14. How do endocytosis and exocytosis move materials into and out of cells?

15.6 Cell Communication

15. What are plasmodesmata? What function do they serve?
16. Describe what happens when a message molecule binds to a receptor on the cell membrane.
17. Does one receptor bind to many different kinds of message molecules?

15.7 How Cells Reproduce

18. What are the stages of the cell cycle? What happens during synthesis (S)?
19. Describe the phases of mitosis.
20. What are the end products of mitosis?

15.8 How Cells Use Energy

21. Why do cells need catalysts? What are the catalysts in cells?
22. How do cells regulate enzymes?
23. How does penicillin kill bacteria?

15.9 Photosynthesis

24. Why is almost all life on Earth dependent either directly or indirectly on photosynthesis?
25. What happens during the light-dependent reactions of photosynthesis? What happens during the light-independent reactions?
26. In the chemical reaction for photosynthesis, what are the reactants and what are the products?

15.10 Cellular Respiration and Fermentation

27. Describe the process of glycolysis. Is ATP produced during glycolysis?
28. About how many ATP molecules does a cell obtain from one molecule of glucose through cellular respiration? What other products result from cellular respiration?
29. What are the products of alcoholic fermentation?
30. Give two examples of cells in the human body that use lactic acid fermentation. Why does each of these use lactic acid fermentation?

THINK INTEGRATED SCIENCE

15A—Macromolecules Needed for Life

31. What are some of the different functions of proteins?
32. Give an example of (a) a carbohydrate that functions in energy storage and (b) a carbohydrate that has a structural function.
33. Describe the structure of DNA.
34. Explain this statement: Proteins, carbohydrates, lipids, and nucleic acids are called macromolecules because they are composed of smaller units joined together.

15B—The Microscope

35. Why are light microscopes of limited use to cell biologists?
36. Why are electron microscopes particularly useful to cell biologists?

15C—ATP and Chemical Reactions in Cells

37. How does ATP provide energy for cells?
38. Is the amount of energy it takes to make ATP the same as the amount of energy that cells eventually get out of ATP?
39. Describe how the sodium-potassium pump works. Which step in the active transport of sodium and potassium ions requires ATP?

THINK AND DO (HANDS-ON APPLICATION)

40. Does a pound of lipid (such as fat or oil) really store more energy than a pound of carbohydrate? Examine the nutrition labels on a bag of sugar and a stick of butter or a bottle of cooking oil. How many calories are there in a pound of sugar? What about in a pound of butter or cooking oil?
41. Carefully place a teabag in a mug of very still hot water and observe what happens. Why is the water darker closer to the teabag than farther away from it? If you remove the teabag and wait, will your tea eventually become uniformly colored? If so, explain why. Which cellular transport mechanism does this exploration examine?
42. Find a recipe and mix up some bread dough. Is yeast one of the ingredients? (You may also be able to buy some fresh dough at a supermarket; some supermarkets sell dough for making pizza crusts.) Place a ball of dough in a bowl or other container and note how big the ball is. Let the dough sit for an hour. Has the dough risen? Pull it apart. Do you see any evidence of yeast at work?

THINK AND COMPARE (ANALYSIS)

43. Rank these three living things from largest to smallest: (a) a tuberculosis bacterium, a prokaryote; (b) baker's yeast, a single-celled eukaryote; (c) a human, a multicellular eukaryote.
44. Rank the molecules in order of the amount of energy they contain, from most to least: (a) ATP, (b) starch, (c) glucose.
45. Rank the processes in order of how much ATP they produce, from most to least: (a) glycolysis, (b) electron transport, (c) fermentation.

THINK AND SOLVE (MATHEMATICAL APPLICATION)

46. As energy-storage substances, carbohydrates produce about 4 kilocalories of energy per gram, whereas fats produce about 9 kilocalories of energy per gram. The American black bear hibernates for as long as seven months in the winter, during which it does not eat. Before hibernating, black bears put on a lot of weight, often spending 20 hours a day eating and storing as much as 50 kilograms of fat. Show that the bear would have to gain 112.5 kilograms if it stored energy as carbohydrate instead of as fat.
47. A typical cell in the body makes about 10 million molecules of ATP per second. Show that the cell breaks down about 263,158 molecules of glucose per second.
48. Two different bacteria have radii of 1 micrometer and 5 micrometers. What is the surface area of each cell? How does the surface area compare with the volume for each cell—that is, what is the surface-area-to-volume ratio? Why is the larger cell able to obtain more molecules through diffusion? Why is it nonetheless more challenging for the larger cell to meet its needs through diffusion? Recall that the surface area of a sphere is $4\pi r^2$ and the volume of a sphere is $\frac{4}{3}\pi r^3$.
49. Proteins are folded chains of amino acids. All the proteins in living organisms are made up of only 20 different amino acids. Show that there are 400 different ways to make a string of two amino acids, and 8000 different ways to make a string of three amino acids. How many different ways are there to make a string of 10 amino acids? Do you see why the number of proteins that living organisms can make is practically countless?

THINK AND EXPLAIN (SYNTHESIS)

50. What are some features of living organisms? Describe how human beings show each of these features.
51. Bacteria reproduce by dividing in two. Is this an example of asexual reproduction or sexual reproduction? Defend your answer.
52. When birds migrate south for the winter, they may fly hundreds or even thousands of kilometers in a relatively short period of time. Why do birds put on a layer of fat before their annual migration? Why don't they store this energy in the form of carbohydrate?
53. DNA uses only four different kinds of nucleotides. How can only four nucleotides code for all the different kinds of genes that are found in different living things?
54. You look at a cell under a microscope and discover that the cell has mitochondria and chloroplasts. Is the cell a eukaryotic cell or a prokaryotic cell? Is the cell from a plant or from an animal? Defend your answers.
55. What organelle is found only in plants? What does it do? Does this explain why animals have to eat but plants don't?
56. Some cells in the body have more mitochondria than others. For example, nerve cells, muscle cells, and liver cells have lots of mitochondria. Bone cells and fat cells have very few mitochondria. Why do muscle cells have more mitochondria than fat cells?
57. How is the function of a cell wall different from the function of a cell membrane? Do you have cell walls, cell membranes, both, or neither?
58. Certain cells in the body make protein hormones that they release into the bloodstream via exocytosis. Trace the path of one such hormone from its assembly in the cell to its release.
59. Transport proteins and the molecules they transport are described as fitting together like a lock and key. Why is it important that they have such a specific fit?
60. You learned that all cells face the challenge of controlling water flow. You also learned that water is able to cross the cell membrane directly or via aquaporins. Are organisms that occupy freshwater habitats likely to have the problem of too little water entering their cells or too much water entering their cells? Why?
61. Glucose moves into many cells through facilitated diffusion. This works because there is usually a higher concentration of glucose molecules outside the cell than inside the cell. Why is this the case? (Hint: Do cells use up the glucose inside them?)
62. Why does oxygen diffuse into cells rather than out of them? Why does carbon dioxide diffuse out of cells rather than into them?
63. In plants, roots absorb water (among other functions). Why are the roots of many plants highly branched?
64. What is the difference between endocytosis and using a transport protein to cross the cell membrane?
65. How are gap junctions and plasmodesmata similar? How do they differ?
66. Message molecules and their receptors are described as fitting together like a lock and key. Why is it important that they have such a specific fit?
67. Imagine that a cell goes through all the stages in the cell cycle, except that cytokinesis doesn't happen. How will that cell be different from normal cells?

68. The figure below shows a cell in the process of cell division. In which stage of the cell cycle is it?

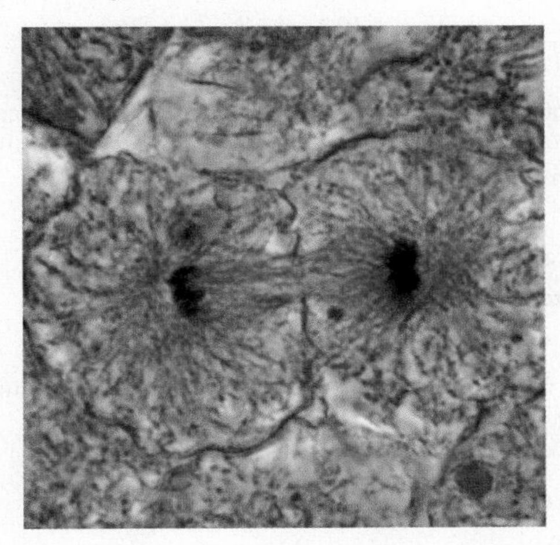

69. The deadly nerve gas sarin binds to an enzyme called acetylcholinesterase, which breaks down acetylcholine in the body. If acetylcholine is not broken down, muscles are unable to relax after contracting. Without prompt treatment, respiratory collapse and death follow. Sarin works by binding to acetylcholinesterase at the site where acetylcholine normally binds. What form of enzyme regulation does this represent?
70. Enzymes and their substrates are described as having a lock-and-key fit. Why is it important that they have such a specific fit?
71. Global warming has occurred because of the large amounts of carbon dioxide released by the burning of fossil fuels. Carbon dioxide traps heat. Why might the loss of forests also contribute to global warming?
72. What are some differences between fermentation and cellular respiration? Which process produces more ATP? Why do some cells in the human body use fermentation?
73. Where do the bubbles in champagne come from? (Hint: In nonbubbly wines, fermentation happens with the grape juice exposed to air. In champagne, there is an extra round of fermentation during which the bottles are capped tight.)
74. Some animals that live in desert environments, such as this kangaroo rat, never drink water. Kangaroo rats live entirely on the starches and lipids in the dry seeds they eat. Yet we know that all living organisms need water, and, in fact, the bodies of kangaroo rats have about the same water content as those of other animals. How do kangaroo rats get their water?

75. Write a letter to Grandma telling her about stem cells. Be sure to tell her how stem cells are different from all the other cells in our bodies and why scientists believe they may be helpful for treating certain diseases.

THINK AND DISCUSS (EVALUATION)

76. A friend in your study group claims that a change in just one amino acid in an important protein can cause a serious disease. Another friend says this doesn't make sense because a protein with a slight change should still be able to function. Both friends look at you. What do you say?
77. Membrane proteins often have hydrophilic regions as well as hydrophobic regions. Based on your understanding of the structure of the cell membrane, which part of a membrane protein would you predict is hydrophilic? Which part would you predict is hydrophobic?
78. A friend in your class is reading about cells that store energy in the form of tiny lipid droplets. He says, "It says here that the membranes around the droplets are just a single phospholipid layer, instead of a double layer like everywhere else. That doesn't make sense, does it?" What is your reply?
79. You and a friend are eating lunch in the cafeteria. Suddenly, you both smell very strong perfume. "Wow," your friend says, "I wish they would turn the fan off. It's blowing that perfume right at us." You say, "Even if they turn the fan off, there would still be diffusion. You would still smell the perfume." Which of you is correct?
80. You and a friend go to a garden shop to buy fertilizer for your new plants. Your friend says, "Didn't we just learn that plants make their own organic molecules through photosynthesis? Plants shouldn't need fertilizer." Another customer overhears your discussion and looks at the bag of fertilizer you are holding, "Well," she says, "it looks like the main ingredients in fertilizer are nitrogen and phosphorus. I don't know if plants really need that stuff or not." What do you say? Do you buy the fertilizer?
81. You are talking with a winemaker about yeast. "I know that yeast can survive under both aerobic and anaerobic conditions," he says, "but what I don't understand is why they seem to need so much more sugar under anaerobic conditions. I think they gobble a hundred times as much sugar." Can you explain to him why yeast might need more sugar under anaerobic conditions than under aerobic conditions? Is his estimation of 100 times as much sugar reasonable?

READINESS ASSURANCE TEST (RAT)

If you have a good handle on this chapter, if you really do, then you should be able to score 7 out of 10 on this RAT. If you score less than 7, you need to study further before moving on.

Choose the BEST answer to each of the following:

1. Which of the following is *NOT* a characteristic of living things?
   (a) use energy
   (b) maintain themselves
   (c) have the capacity to reproduce
   (d) are part of populations that remain constant from one generation to the next
2. The macromolecules made from folded chains of amino acids are
   (a) proteins.
   (b) carbohydrates.
   (c) lipids.
   (d) nucleic acids.
3. One difference between prokaryotic and eukaryotic cells is that
   (a) prokaryotic cells have a nucleus, whereas eukaryotic cells do not.
   (b) eukaryotic cells have existed on Earth far longer than prokaryotic cells.
   (c) the DNA of eukaryotic cells is found in linear chromosomes, whereas the DNA of prokaryotic cells is found in a single circular chromosome.
   (d) eukaryotic cells are usually smaller than prokaryotic cells.
4. In plant cells, which organelles break down organic molecules to obtain energy in a form that cells can use?
   (a) mitochondria
   (b) ribosomes
   (c) chloroplasts
   (d) lysosomes
5. Which of the following is not a component of the cell membrane?
   (a) phospholipids
   (b) proteins
   (c) nucleic acids
   (d) carbohydrates
6. A transport protein moves a molecule across the cell membrane against its concentration gradient. This is an example of
   (a) diffusion.
   (b) facilitated diffusion.
   (c) active transport.
   (d) endocytosis.
7. During which stage of the cell cycle does a cell duplicate its genetic material?
   (a) gap 1
   (b) gap 2
   (c) synthesis
   (d) mitosis and cytokinesis
8. Which of these statements about enzymes is true?
   (a) Enzymes provide energy for specific chemical reactions in cells.
   (b) Enzymes are catalysts that enable specific chemical reactions in cells to happen more quickly than they would otherwise.
   (c) Both statements are true.
   (d) Neither statement is true.
9. The products of photosynthesis are
   (a) carbon dioxide and water.
   (b) carbon dioxide, water, and sunlight.
   (c) glucose.
   (d) glucose and oxygen.
10. Which of the following processes requires oxygen?
    (a) glycolysis
    (b) Krebs cycle and electron transport
    (c) fermentation
    (d) none of these

Answers to RAT

1. d, 2. a, 3. c, 4. a, 5. c, 6. c, 7. c, 8. b, 9. d, 10. b

CHAPTER 16

Genetics

HE HAS his mother's eyes and his father's nose . . . but why those, and not his *father's* eyes and his *mother's* nose? Why does a chicken egg always hatch a chick? Why not a flamingo, or a bullfrog, or—just once!—a *Tyrannosaurus rex*? The offspring of living organisms are like their parents because they inherit their parents' genes. But what are genes? What are genes made of? How do genes determine what organisms are like? How are changes in genes responsible for the remarkable diversity of life on Earth—as well as such devastating diseases as cancer? How can genetics research revolutionize everything from what we eat to how long we live? We'll learn the answers to these questions in this chapter.

LEARNING OBJECTIVE
Define a gene, and explain why an organism's genes determine many of its characteristics.

16.1 What Is a Gene?

EXPLAIN THIS How do your genes determine what kinds of chemical reactions occur in your cells?

Genes determine many traits in living organisms—the colors of an orchid's flowers, the length of a cat's tail, the substances that make up a crab's shell or a bacterium's cell wall. In humans, genes affect our eye color, whether we are tall or short, and whether our hair is straight or curly. Genes even influence our personalities. But what is a gene, and how does a gene determine a trait?

A **gene** is a section of DNA that contains the instructions for building a protein. An organism's genes, found in its DNA, make up its **genotype**. The traits of an organism make up its **phenotype**. For example, you might have two genes for eye color, one brown gene and one blue gene. If so, you will have brown eyes. Your genotype is one brown gene and one blue gene, but your phenotype is brown eye color. How genotype becomes phenotype—how genes become traits—is one subject of this chapter.

But, if genes contain instructions for building proteins, you might wonder: Why are proteins so important? Why do so many of our traits depend on proteins? The answer is that proteins do many important things in living organisms. Proteins provide structure, act as hormones, transport molecules, function in cell communication, and protect organisms from disease. In addition, enzymes are proteins, and enzymes are needed for practically every chemical reaction that occurs in living things.

LEARNING OBJECTIVE
Describe how DNA is packed into chromosomes.

16.2 Chromosomes: Packages of Genetic Information

EXPLAIN THIS Why do some of your cells have only 23 chromosomes instead of 46?

Where are our genes? In eukaryotes, DNA is found in the cell nucleus, where it is packaged in *chromosomes* (Figure 16.1). Each chromosome consists of a long DNA molecule wrapped around small proteins called histones. The histones keep DNA from getting tangled when it is not being used. DNA is unwound from the histones when the cell needs to use it. Chromosomes are loosely packed most of the time, but they become condensed during cell division.

A single human cell contains about 7 feet of DNA.

Most cells have two of each kind of chromosome, like a pair of matched shoes. These cells are **diploid**, and their matched chromosomes are called *homologous chromosomes*. Some cells—such as sperm and eggs—have only one of each kind of chromosome. These cells are **haploid**. Different organisms have different numbers of chromosomes. Chickens have 78 (39 pairs), mosquitoes have 6 (3 pairs), and yeast have 32 (16 pairs). In humans, there are 46 chromosomes, or 23 pairs (Figure 16.2). The last pair are called *sex chromosomes* because they determine sex. Females have two X chromosomes, whereas males have one X and one Y chromosome. The rest of the chromosomes are known as *autosomes*.

MasteringPhysics®
TUTORIAL: DNA and RNA Structure
TUTORIAL: DNA Double Helix

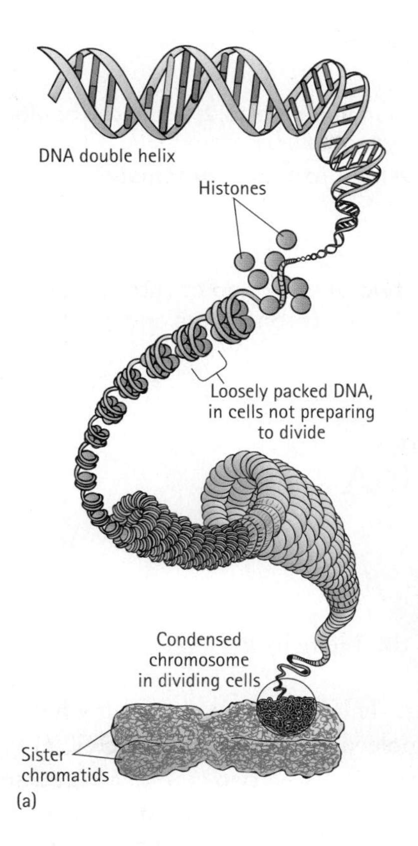

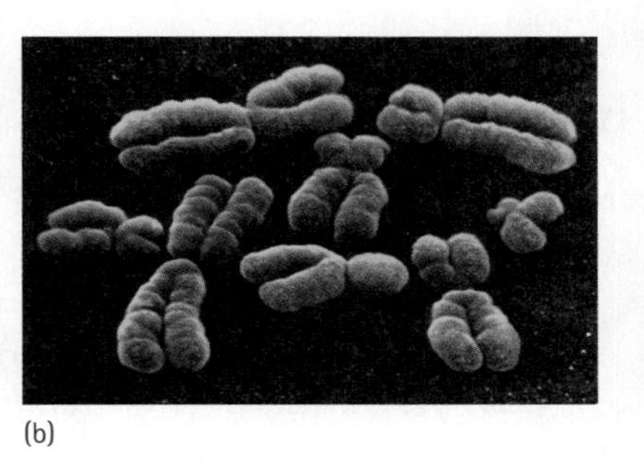

FIGURE 16.1
(a) Chromosomes consist of DNA wrapped around histone proteins. Chromosomes are loosely packed most of the time but become condensed during cell division. (b) These chromosomes are condensed in preparation for cell division.

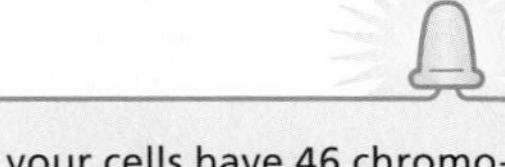

Most of your cells have 46 chromosomes, but eggs and sperm have 23, and red blood cells have 0. Red blood cells don't even have a nucleus. This makes them smaller, so they can move through blood vessels more easily. But it also means that red blood cells can't make the proteins they need to repair and maintain themselves. Because of this, red blood cells do not live very long and must be replaced frequently.

Most of our cells (except our haploid sperm and eggs) have the same 46 chromosomes, with the same DNA and the same genes. This might make you wonder: If all our cells have the same genes, what makes different kinds of cells so different? For example, what makes a brain cell so different from a muscle cell? The answer is that different genes are *expressed* in the two cells. The proteins a brain cell makes from its DNA are very different from the proteins a muscle cell makes. To see how proteins are made from DNA, let's start at the beginning. What is the structure of DNA?

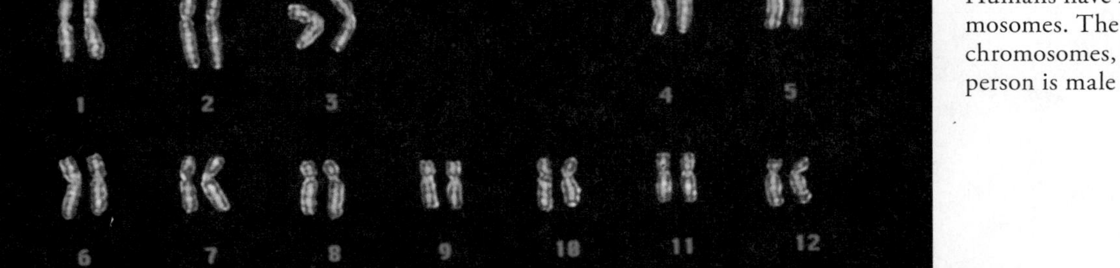

FIGURE 16.2
Humans have 23 pairs of chromosomes. The last pair, the sex chromosomes, determine whether a person is male or female.

CHECK YOURSELF

1. **If a cell contains the chromosomes shown in Figure 16.2, is the cell haploid or diploid?**
2. **Do the chromosomes in Figure 16.2 belong to a male or female?**

CHECK YOUR ANSWERS

1. The cell is diploid because it contains two of each kind of chromosome.
2. They belong to a male because there is an X chromosome and a Y chromosome.

LEARNING OBJECTIVE
Describe the structure of DNA.

The Structure of DNA

EXPLAIN THIS Why is DNA called a "double helix"?

Now that we know where genes are, let's look more closely at what genes are made of, at DNA itself. A molecule of **deoxyribonucleic acid**, or **DNA**, consists of two strands. Together, the two strands make a spiral ladder with two "sides" and a series of "rungs" (Figure 16.3). Because DNA is made up of two strands twisted into a spiral or helix, it is called a *double helix.*

Let's look at a single strand of DNA, and then consider how DNA's two strands fit together. You already know that a DNA strand is a chain of nucleotides, and that a nucleotide includes a nitrogenous base, a sugar molecule, and a phosphate group (see Chapter 15). The sugar in DNA nucleotides is deoxyribose. So, a DNA strand has a backbone (or "side" of the ladder) made up of alternating molecules of deoxyribose sugar and phosphate. Sticking out from this backbone is a series of nitrogenous bases. Each nitrogenous base is half of a "rung" of the DNA ladder. The four nitrogenous bases found in DNA are adenine (A), guanine (G), cytosine (C), and thymine (T).

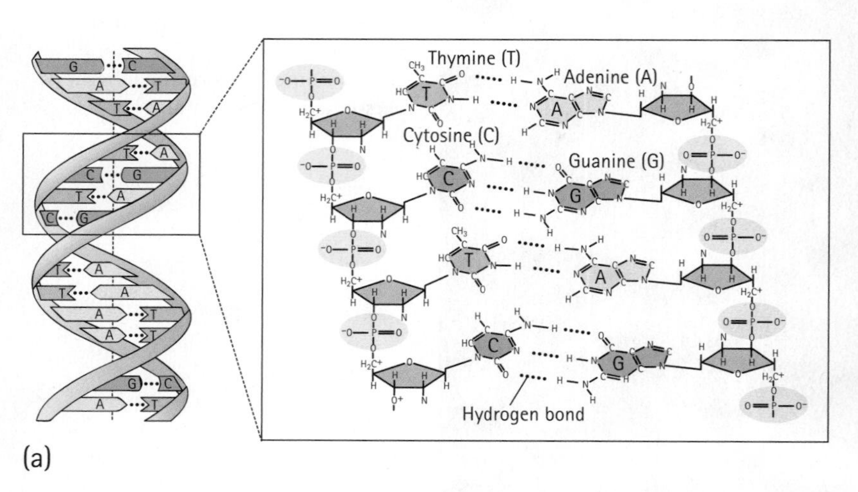

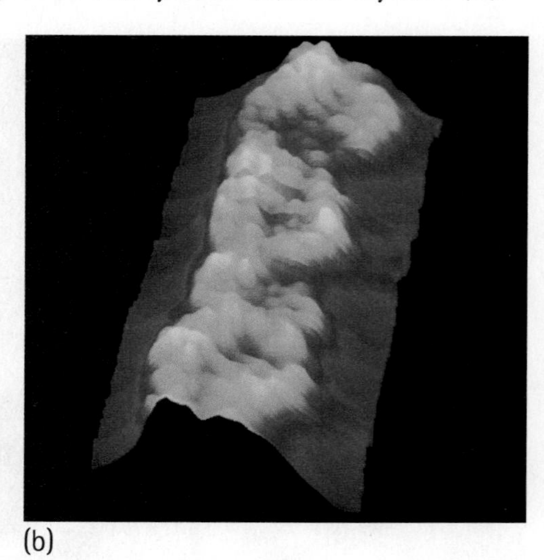

FIGURE 16.3
(a) DNA is shaped like a spiral ladder with two sugar–phosphate strands as the "sides" of the ladder and paired bases as the "rungs." (b) This photograph of DNA shows that it is a double helix.

Now let's put two strands of DNA together. Each nitrogenous base binds with a base on the other strand using chemical bonds called hydrogen bonds (see Chapter 12). Because each nitrogenous base can best form hydrogen bonds with another specific nitrogenous base, the binding of bases always happens in a specific way. Adenine always pairs with thymine (A–T), and guanine always pairs with cytosine (G–C).

CHECK YOURSELF

If one strand of DNA contains the nucleotides ACCTGA, what are the nucleotides on the opposite strand?

CHECK YOUR ANSWER

Because of the way the nitrogenous bases pair, the opposite strand must have TGGACT.

UNIFYING CONCEPT

● ***The Scientific Method***
Section 1.3

HISTORY OF SCIENCE

Discovery of the Double Helix

By the early 1950s, scientists knew that the genetic material of eukaryotic organisms was contained in their chromosomes. Scientists also knew that, of the two types of molecules found in chromosomes—proteins and DNA—the genetic material was DNA. The structure of DNA was still a mystery, however, and it represented the biggest unsolved problem in biology. In 1951, Francis Crick and James D. Watson began to tackle the problem. Their hope was to build a model of DNA that would be consistent with available experimental evidence. Meanwhile, both Rosalind Franklin and Maurice Wilkins were taking X-ray photos of DNA.

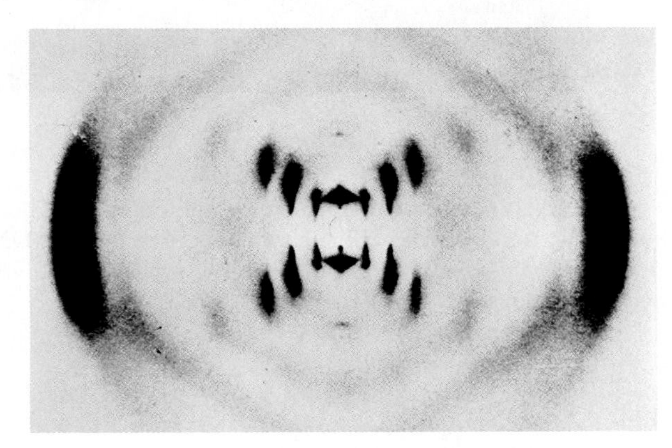

Rosalind Franklin's X-ray photo may look like a blurry "X" to you, but to Watson and Crick it showed that DNA is a double helix.

The breakthrough came in January 1953, when Wilkins shared Franklin's photograph of DNA with Watson. A report on Franklin's experimental findings also made its way to Watson and Crick. Watson and Crick began to build DNA models and to test them against Franklin's photographs and data. Within a few weeks, they had it—DNA was a double helix. Furthermore, adenine and thymine, and guanine and cytosine, paired up between the strands. The Watson–Crick model of DNA was published in April 1953, along with two papers offering supporting evidence, one by Wilkins and his collaborators and one by Franklin and her assistant. Acceptance of the Watson–Crick model was immediate and widespread.

(a) (b)

(a) James D. Watson and Francis Crick figured out the structure of DNA in 1953. (b) Rosalind Franklin took the famous X-ray photo that led Watson and Crick to the structure of DNA.

For their discovery of the double helix, Watson, Crick, and Wilkins were awarded the 1962 Nobel Prize in Physiology or Medicine. Franklin had died by then—of ovarian cancer that probably resulted from radiation exposure related to her work. She never knew how important her results had been in enabling Watson and Crick to develop their model. In fact, Franklin's importance to the discovery of DNA structure was not fully appreciated until Watson wrote his memoir *The Double Helix*.

LEARNING OBJECTIVE
Explain how DNA is copied.

MasteringPhysics®
TUTORIAL: DNA Replication

During DNA replication, 50 nucleotides are added every second. This sounds fast—and it is—but we have 3 billion pairs of nucleotides in each cell! It's a good thing copying occurs simultaneously at many different points in each chromosome. As a result, a human cell can copy all of its DNA in only a few hours.

16.3 DNA Replication

EXPLAIN THIS When DNA is copied, why don't we get an "old" molecule and a "new" molecule?

DNA must be copied in order for cells to divide and reproduce. This process is called **DNA replication**. During DNA replication, DNA's two strands are separated as if the spiral ladder were unzipped down the middle. Each strand then serves as a template for building a new partner. This is possible because of the way the nitrogenous bases pair—because A always pairs with T, and G always pairs with C. As free nucleotides pair up with nucleotides on the template strand, they are attached to the new DNA strand by enzymes called *DNA polymerases*. Each new DNA molecule includes one old strand and one new strand, and the new DNA molecules are identical to the original (Figure 16.4).

DNA replication always begins at fixed spots within chromosomes. In eukaryotes, replication begins simultaneously at many different points in a chromosome, allowing the job to be completed efficiently.

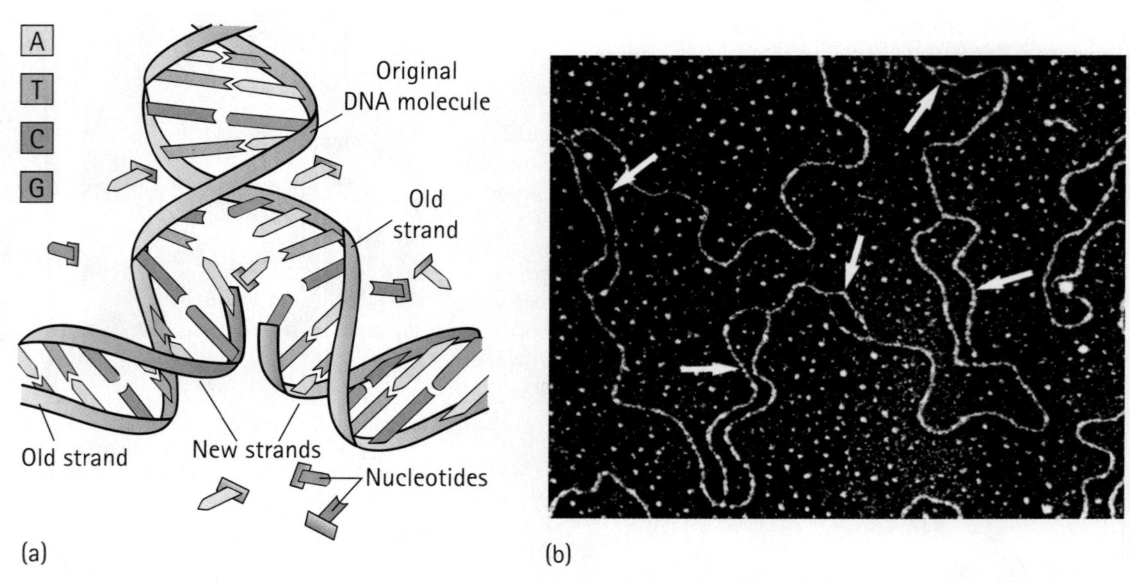

FIGURE 16.4
(a) When DNA is copied, the two strands are separated. Each old strand (orange backbone) serves as a template for building a new strand (green backbone). (b) This photo shows DNA replication in the cell of a fruitfly (*Drosophila melanogaster*). Arrows show where the DNA strands have been separated and replicated.

LEARNING OBJECTIVE
Explain how information contained in DNA is used to build a protein.

16.4 How Proteins Are Built

EXPLAIN THIS What is the genetic code?

We now know the structure of DNA and how DNA is copied. We are almost ready to look at how DNA provides cells with the instructions for building proteins. Before we do that, though, we have to introduce one more molecule—RNA.

RNA

Like DNA, **ribonucleic acid**, or **RNA**, consists of a sugar–phosphate backbone attached to nitrogenous bases (Figure 16.5). But RNA differs from DNA in three ways: (1) RNA has only one strand instead of two; (2) RNA uses the sugar ribose instead of deoxyribose; and (3) RNA uses the nitrogenous base uracil (U) instead of thymine (T).

During **transcription**, DNA is used as a template for building a molecule of RNA. During **translation**, this RNA molecule is used to assemble a protein (Figure 16.6).

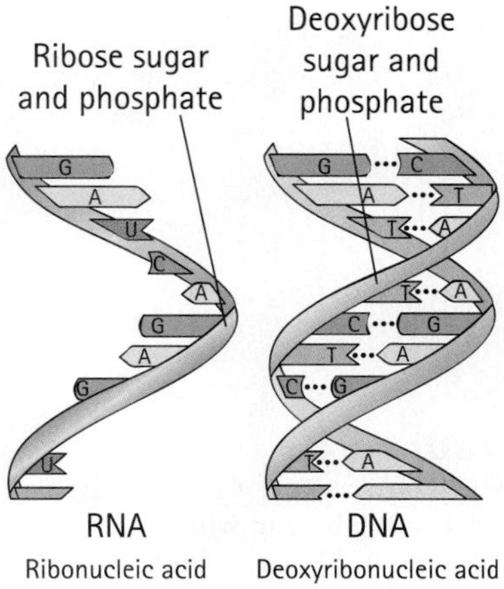

Ribose sugar and phosphate
Deoxyribose sugar and phosphate
RNA
Ribonucleic acid
DNA
Deoxyribonucleic acid

FIGURE 16.5
RNA is a nucleic acid, like DNA. However, RNA is usually single-stranded, uses the sugar ribose instead of deoxyribose, and uses the nitrogenous base uracil instead of thymine.

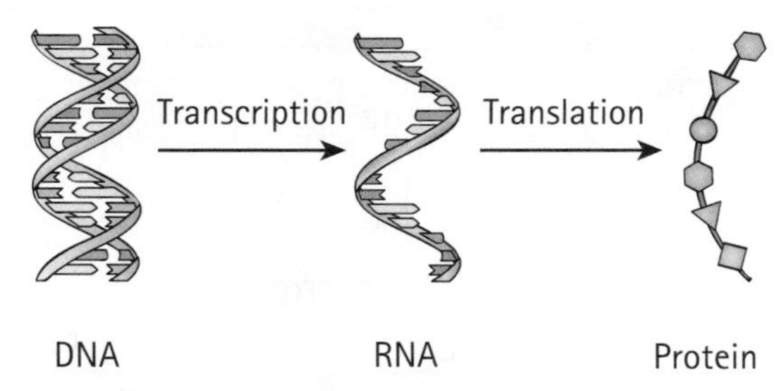

FIGURE 16.6
DNA is used to build a protein through the processes of transcription and translation.

Transcription

In eukaryotes, transcription occurs in the cell nucleus. The two strands of DNA are separated, and one strand serves as a template for building an RNA transcript. Special nucleotide sequences in the DNA indicate where transcription should begin and end, marking the beginning and end of a gene. The construction of the RNA transcript follows the same base-pairing rules we saw for DNA, except that RNA uses uracil (U) instead of thymine (T). So, where DNA has the nucleotides A, C, G, and T, the RNA transcript will have the nucleotides U, G, C, and A, respectively (Figure 16.7). As free RNA nucleotides pair up with complementary nucleotides on the DNA strand, an enzyme called *RNA polymerase* adds them to the growing RNA molecule. Once transcription is complete, the DNA zips back up and the RNA transcript begins a processing phase.

MasteringPhysics®
TUTORIAL: Nucleic Acid Structure
TUTORIAL: Transcription

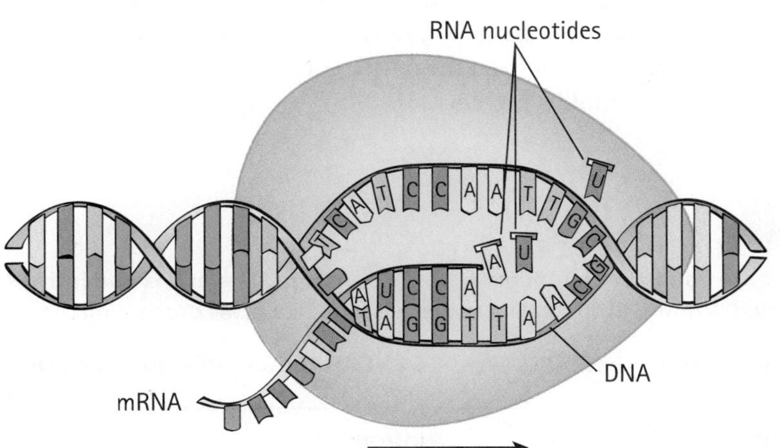

FIGURE 16.7
During transcription, information in DNA is stored in a molecule of RNA.

CHECK YOURSELF

If a stretch of DNA has the nucleotide sequence ACCTGAT, what sequence will the RNA transcript have?

CHECK YOUR ANSWER

The RNA transcript will have the nucleotide sequence UGGACUA.

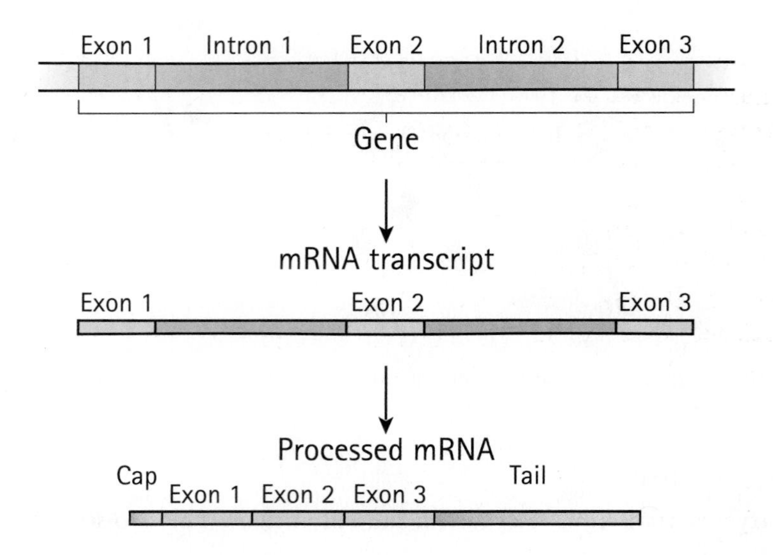

FIGURE 16.8
The RNA transcript goes through a processing phase in which introns are removed and a cap and tail are added. The result is an mRNA molecule that is ready for translation.

If you have trouble remembering which are exons and which are introns, just think: "EXons EXpressed, INtrons IN the trash."

During the processing phase, the RNA transcript becomes a mature **messenger RNA** molecule, or **mRNA** molecule, that is ready for translation. First, stretches of nucleotides that are not needed to build the protein, called *introns*, are removed (Figure 16.8). The stretches of nucleotides that are used to build the protein, called *exons*, remain. It is as if a starting transcript of Shakespeare's *Hamlet* contained the line "*aggfr uidosa to be dfjklsdf or rewerwe not to be*," and the irrelevant parts had to be removed in order to leave the coherent "*to be or not to be*." Scientists are not sure why introns exist. However, a single RNA transcript can sometimes be processed in multiple ways, resulting in different mRNA molecules. Second, a "cap" and "tail" are added to the beginning and end of the RNA molecule. The cap and tail allow the cell to recognize the molecule as mRNA. Once processing is complete, the mRNA molecule moves from the nucleus of the cell to the cytoplasm, where it will be translated to build a protein.

CHECK YOURSELF

Suppose you have the following RNA transcript. The exons and introns are labeled. What will the mRNA molecule look like after processing?

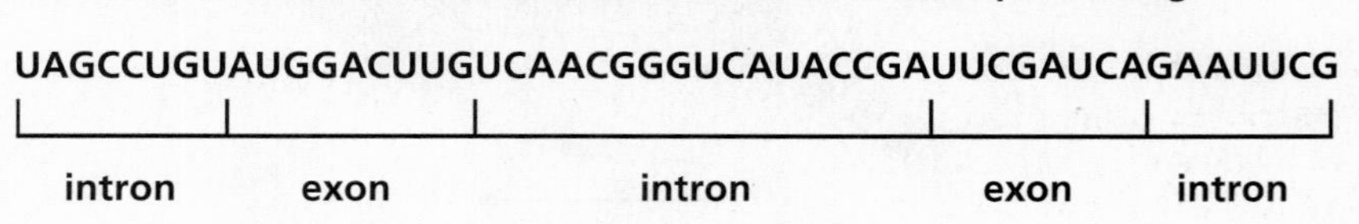

CHECK YOUR ANSWER

The introns will be removed, and a cap and tail will be added, resulting in:
cap–AUGGACUUGUUCGAUCA–tail

Translation

During translation, mRNA is used to build a protein. Translation takes place in the cell cytoplasm at organelles called ribosomes (see Chapter 15). Ribosomes are composed of proteins and a type of RNA called *ribosomal RNA*, or *rRNA*.

How does translation work? Remember that a protein is a folded chain of amino acids. In translation, three nucleotides along the mRNA strand form a three-letter "word" called a **codon**. During translation, codons are "read" from the mRNA molecule one at a time. Most of these codons tell the ribosome to add a specific amino acid to the protein being built. A few codons tell the ribosome that there are no more amino acids in the protein and that translation is complete. The *genetic code* tells us what each codon stands for (Table 16.1). For example, the codon AGU translates to the amino acid serine, and the codon GUG translates to the amino acid valine. Because there are more codons than amino acids, multiple codons can translate to the same amino acid. CGU, CGG, and CGC all translate to arginine. Certain codons, such as UAA, are called *stop codons* because they tell the ribosome that there are no more amino acids in the protein. The first codon to be translated from mRNA is always AUG—methionine. So, you read the codons in mRNA, string together amino acids in the right order, and presto! You have a protein.

Now let's look at how translation actually happens in the ribosome. Translation requires another type of RNA, **transfer RNA**, or **tRNA**. A tRNA molecule includes a sequence of three nucleotides called an *anticodon* and carries a single, specific amino acid.

MasteringPhysics®
TUTORIAL: Translation

With very few exceptions, the genetic code is the same for all living organisms. Animals, fungi, plants, bacteria—even viruses use the same genetic code! This suggests that the genetic code originated very early in the evolution of life and was then passed on to all living species.

TABLE 16.1 THE GENETIC CODE

First base	Second base: U	Second base: C	Second base: A	Second base: G	Third base
U	UUU, UUC: Phenylalanine (Phe) UUA, UUG: Leucine (Leu)	UCU, UCC, UCA, UCG: Serine (Ser)	UAU, UAC: Tyrosine (Tyr) UAA: Stop UAG: Stop	UGU, UGC: Cysteine (Cys) UGA: Stop UGG: Tryptophan (Trp)	U C A G
C	CUU, CUC, CUA, CUG: Leucine (Leu)	CCU, CCC, CCA, CCG: Proline (Pro)	CAU, CAC: Histidine (His) CAA, CAG: Glutamine (Gln)	CGU, CGC, CGA, CGG: Arginine (Arg)	U C A G
A	AUU, AUC, AUA: Isoleucine (Ile) AUG: Methionine or start	ACU, ACC, ACA, ACG: Threonine (Thr)	AAU, AAC: Asparagine (Asn) AAA, AAG: Lysine (Lys)	AGU, AGC: Serine (Ser) AGA, AGG: Arginine (Arg)	U C A G
G	GUU, GUC, GUA, GUG: Valine (Val)	GCU, GCC, GCA, GCG: Alanine (Ala)	GAU, GAC: Aspartic acid(Asp) GAA, GAG: Glutamic acid (Glu)	GGU, GGC, GGA, GGG: Glycine (Gly)	U C A G

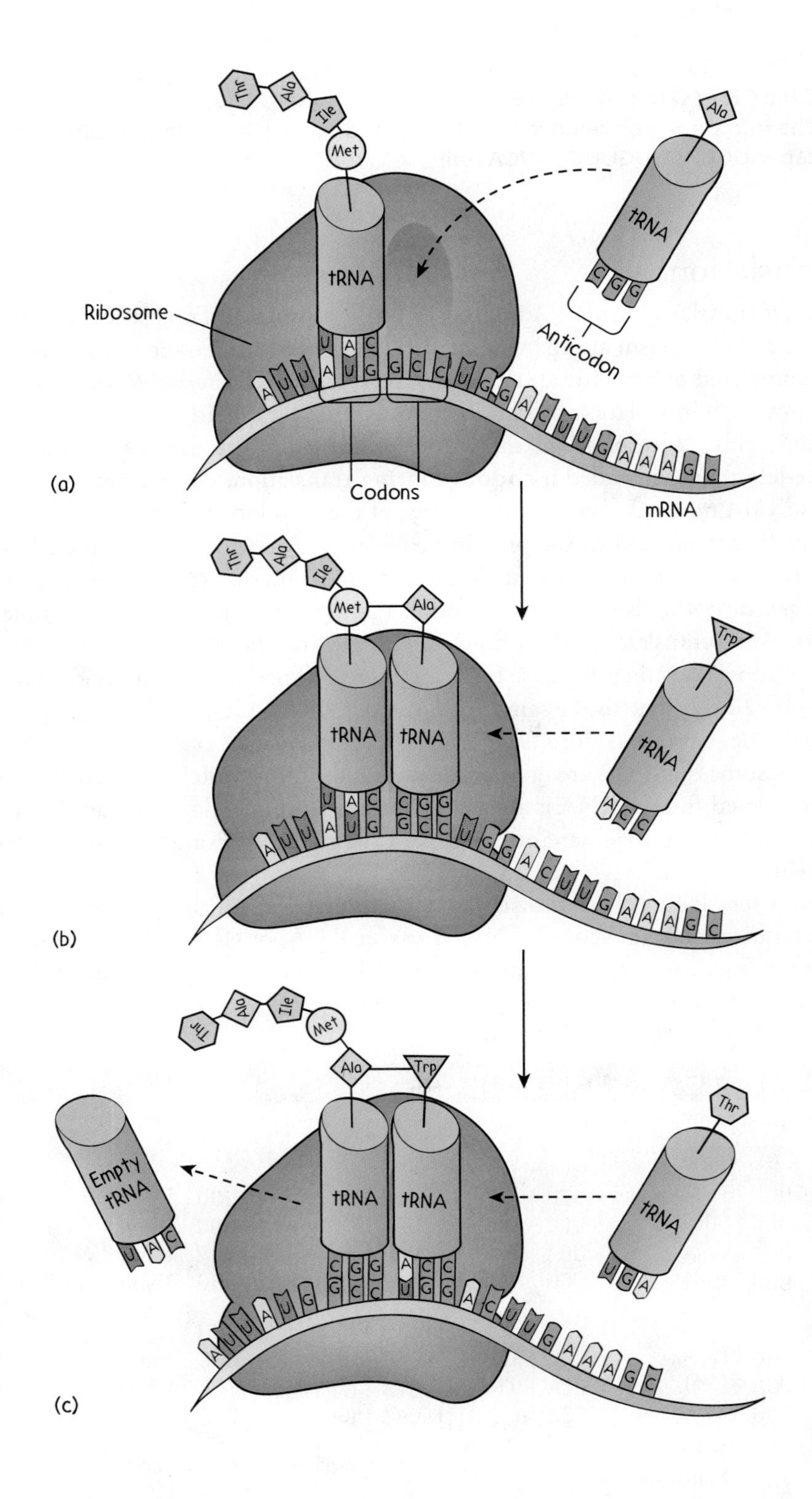

FIGURE 16.9
During translation, an mRNA molecule is translated to build a protein. (a) The anticodon of a tRNA molecule binds to a codon in the mRNA molecule. (b) The amino acid carried by the anticodon is added to the growing protein. (c) The mRNA molecule is shifted so that a new codon can be translated.

During translation, the mRNA molecule binds to a ribosome. The codon being translated is positioned at a specific site. A tRNA molecule with the appropriate anticodon binds to the codon (Figure 16.9a). The binding of anticodon to codon follows the usual rules—A binds with U, and G binds with C. The amino acid carried by the tRNA is then added to the growing protein (Figure 16.9b). After this, the ribosome shifts the mRNA so that a new codon can be translated (Figure 16.9c). This process is repeated until a stop codon is reached. As translation proceeds, the growing protein folds up into its appropriate three-dimensional shape.

CHECK YOURSELF

1. **A molecule of mRNA has the codon GUC. What is the anticodon on the tRNA that binds to it during translation? What amino acid does the tRNA carry?**
2. **Consider the mRNA sequence AUGAGCCUGUAC. What string of amino acids does this sequence code for?**

CHECK YOUR ANSWERS

1. The tRNA has the anticodon CAG. Looking at the genetic code table, we see that this tRNA molecule carries the amino acid valine. (When you use the genetic code table, remember that you need to look up the codon: GUC translates to valine.)
2. Dividing the sequence into codons and using the genetic code table, we have AUG = methionine, AGC = serine, CUG = leucine, and UAC = tyrosine. So, the amino acid string is methionine–serine–leucine–tyrosine.

A ribosome builds a typical protein in less than one minute.

16.5 Genetic Mutations

LEARNING OBJECTIVE
Define a genetic mutation, and describe the potential consequences of genetic mutations.

EXPLAIN THIS Why is a frameshift mutation more likely to disrupt a protein's function than a point mutation?

A **genetic mutation** occurs when the sequence of nucleotides in an organism's DNA is changed. Genetic mutations may result from errors during DNA replication or from exposure to things that damage DNA, such as ultraviolet light, X-rays, and chemicals.

A genetic mutation can have no effect at all, or it can have a huge effect. A mutation in a gene is likely to have a bigger effect than one that strikes DNA that is not part of a gene. A mutation in an exon is likely to have a bigger effect than one in an intron. A mutation in an egg or sperm may have a bigger effect than one that strikes a skin cell on the big toe—this is because a mutation in an egg or sperm can be passed on to offspring and then appear in every cell in the offspring's body.

The majority of genetic mutations have no effect on proteins or organisms. Other mutations interfere with how proteins function, and so are disadvantageous for organisms. Every once in a while, however, a mutation will produce something new and advantageous. Mutations are the original source of all genetic diversity, and they provide the raw materials for evolution (see Chapter 17). For this reason, mutations are ultimately responsible for the diversity of life on Earth.

Mutations can interfere with how our proteins function. It is definitely wise to avoid exposure to chemicals and other things that cause mutations.

Point Mutations

A *point mutation* occurs when one nucleotide is substituted for another, such as when a C becomes a G. A point mutation may change the sequence of amino acids in a protein. For example, a point mutation that changes AAC to AAG changes the amino acid asparagine to lysine. (Refer to the genetic code in Table 16.1.) Not

(a) Original sequence:	AGC	CUG	UAC	UGG	ACA	UUG	CCA
	serine	leucine	tyrosine	tryptophan	threonine	leucine	proline
(b) Point mutation 1:	AGC	CUG	UAC	UGG	ACU	UUG	CCA
	serine	leucine	tyrosine	tryptophan	threonine	leucine	proline
(c) Point mutation 2:	AGG	CUG	UAC	UGG	ACA	UUG	CCA
	arginine	leucine	tyrosine	tryptophan	threonine	leucine	proline
(d) Nonsense mutation:	AGC	CUG	UAG	UGG	ACA	UUG	CCA
	serine	leucine	STOP				
(e) Frameshift mutation:	AGC	UCU	GUA	CUG	GAC	AUU	GCC A
	serine	serine	valine	leucine	aspartic acid	isoleucine	alanine

FIGURE 16.10
Mutations can affect proteins in many different ways. (a) This is a gene's original nucleotide sequence. (b) This point mutation does not change the amino acid sequence of the protein. (c) This point mutation changes one amino acid into another. (d) This nonsense mutation produces a stop codon, which results in the production of a shorter protein. (e) The insertion of a nucleotide causes a frameshift mutation that completely changes the sequence of amino acids in a protein.

all point mutations change amino acids, however. A point mutation that changes GCA to GCC has no effect because both codons stand for the same amino acid—alanine. You might wonder: Does changing one amino acid in a protein really matter? The answer is sometimes. Some amino acid changes do not affect protein function. Others have a significant effect—a single amino acid change is responsible for several serious human diseases, including sickle cell anemia, which we will discuss later in this chapter.

A *nonsense mutation* is a point mutation that creates a stop codon in the middle of a gene. Nonsense mutations cause translation to stop before all the amino acids in a protein have been added. Nonsense mutations result in short, often nonfunctional, proteins.

Frameshift Mutations

Nucleotides can also be inserted into or removed from DNA. If this happens in a gene, the codons that are "read" during translation can be shifted, producing a *frameshift mutation*. Frameshift mutations completely change a protein's amino acid sequence and usually result in nonfunctional proteins.

Figure 16.10 shows examples of different mutations. Figure 16.11 compares mutations to changes in English sentences—this may help you understand how different mutations affect proteins.

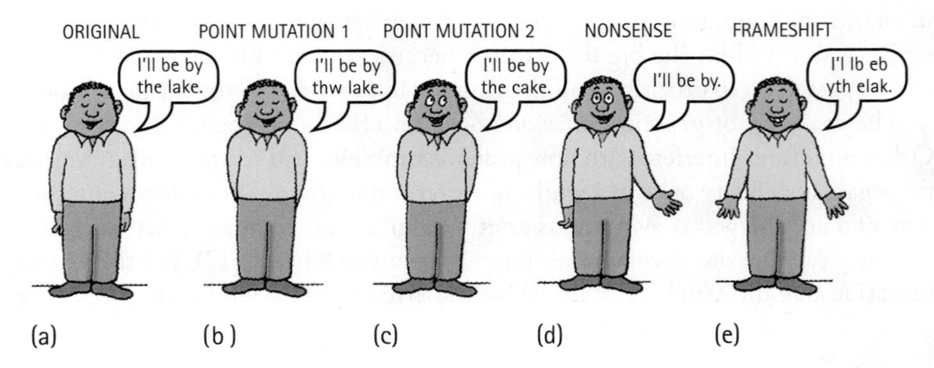

FIGURE 16.11
If a DNA nucleotide sequence were a sentence, different mutations would have different effects on the meaning of the sentence. (a) This is the original sentence (DNA nucleotide sequence). (b) This point mutation has only a minimal effect on the meaning (protein). (c) This point mutation affects the meaning (protein) significantly. (d) This is a nonsense mutation that causes the meaning to be lost (the protein is nonfunctional). (e) This is a frameshift mutation that causes the meaning to be lost (the protein is nonfunctional).

CHECK YOURSELF

Two point mutations are shown below. What effect does each mutation have on the protein that is produced? (Use the genetic code in Table 16.1.)
Original sequence: AGC CUG UAC UGG ACA UUG CCA

1. **AGC CUG UAC UGG ACC UUG CCA**
2. **AGC CUG UAG UGG ACA UUG CCA**

CHECK YOUR ANSWERS

1. ACC and ACA code for the same amino acid (threonine), so this mutation has no effect on the protein.
2. This is a nonsense mutation—it produces a stop codon, shortening the protein to just serine–leucine. This is a serious mutation because it has a severe effect on the protein. A mutation like this in an important gene could cause illness or death.

Integrated Science 16B
PHYSICS

LEARNING OBJECTIVE
Explain the effect of ionizing radiation on DNA.

How Radioactivity Causes Genetic Mutations

EXPLAIN THIS What is dangerous about a free radical?

The atomic bombs dropped on Hiroshima and Nagasaki in 1945 caused immediate death and destruction. But they also affected survivors in ways that became apparent only years later. For example, survivors of the bombs developed many forms of cancer at very high rates. This was attributed to genetic mutations caused by radioactivity. How does radioactivity cause mutations?

Radioactive materials (see Chapter 10) release ionizing radiation—gamma rays, beta particles, and alpha particles. When these forms of radiation strike electrons in the body with enough energy, they free the electrons from the atoms they were orbiting. Sometimes, the free electrons damage DNA directly. More frequently, the damage occurs indirectly when the free electrons strike water molecules in a cell and produce free radicals. A *free radical* is a group of atoms that has an unpaired electron and is consequently highly reactive. Free radicals will react with many molecules in the body, including DNA. Their interactions with DNA damage it, causing genetic mutations (Figure 16.12).

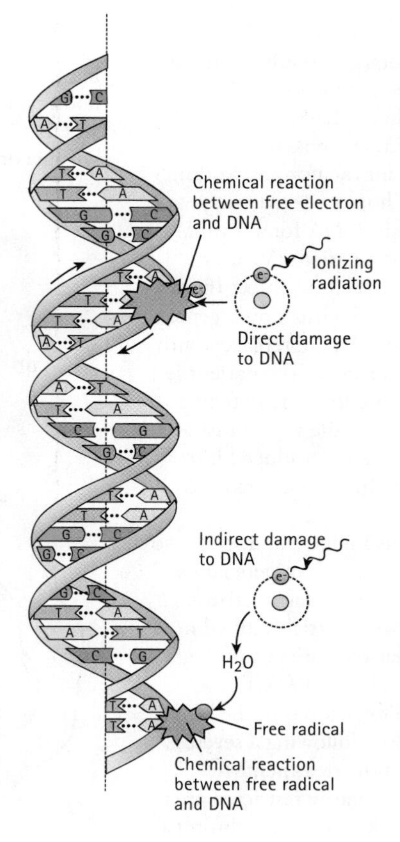

FIGURE 16.12
Radioactivity causes genetic mutations by releasing ionizing radiation. Ionizing radiation strikes and frees electrons in the body. These free electrons can strike and damage DNA directly or produce free radicals that interact with and damage DNA.

Cells can repair DNA damage, but they vary in their ability to do so. Cells in the body that divide frequently often replicate their DNA and pass on mutations before they have time to repair DNA damage. As a result, frequently dividing cells are especially vulnerable to radiation damage. Frequently dividing cells include cells in the bone marrow (where blood cells are made), in the lining of the digestive tract, in the testes, and in a developing fetus. Cancer cells also divide frequently, which is why radiation is used to treat some tumors.

CHECK YOURSELF

1. **When radiation is used to treat cancer, why does it kill many tumor cells while many normal cells survive?**
2. **Why does radiation therapy for cancer cause such side effects as skin damage and hair loss?**

CHECK YOUR ANSWERS

1. Tumor cells divide frequently, so they are less able to repair DNA damage caused by radiation. Excessive DNA damage is often fatal to cells.
2. The side effects are the result of inadvertent damage to healthy cells. Skin and hair cells divide frequently compared to other cells in the body, so they are more vulnerable to radiation damage.

TECHNOLOGY

Gene Therapy

You now know that some genetic mutations result in proteins that are unable to carry out their normal functions. Many genetic diseases occur when people do not have a working gene for making an important protein. Is it possible to cure a genetic disease by introducing DNA for the normal, working gene into a person's cells? That is the hope of *gene therapy*.

In gene therapy, researchers obtain DNA for a normal gene and introduce this DNA into the patient. Viruses, which naturally insert their DNA into cells to infect them, are often used to perform the transfer. A virus's own genes (the ones that give us a cold, for example) are replaced with the DNA that codes for the normal protein. The patient is infected with the virus, and the DNA enters the patient's cells. (Sometimes, cells are infected with the virus outside the patient's body, and the cells are then introduced into the patient.) The DNA begins to produce functional proteins, and the patient is cured.

However, gene therapy is still in its infancy, and so far, no gene therapy treatments have been approved for general use. Unfortunately, the field experienced several tragic setbacks early in its history. In 1999, a gene therapy trial resulted in the death of Jesse Gelsinger, an 18-year-old patient suffering from ornithine transcarboxylase deficiency (OTCD), a disease in which ammonia climbs to toxic levels in the blood. Gelsinger died of multiple organ failure following a severe immune response to the virus used in the treatment. In 2003, all gene therapy trials were temporarily restricted after several children developed leukemia-like symptoms during a trial for severe combined immunodeficiency disease (SCID, also known as "bubble baby syndrome").

But there have been promising results as well. In 2008, doctors from Moorfields Eye Hospital and University College in London used gene therapy to treat a type of inherited childhood blindness. The trials suggest that the treatment is safe and that it improves sight. And in 2011, University of Pennsylvania doctors reported several cases in which gene therapy was successfully used to treat certain kinds of leukemia.

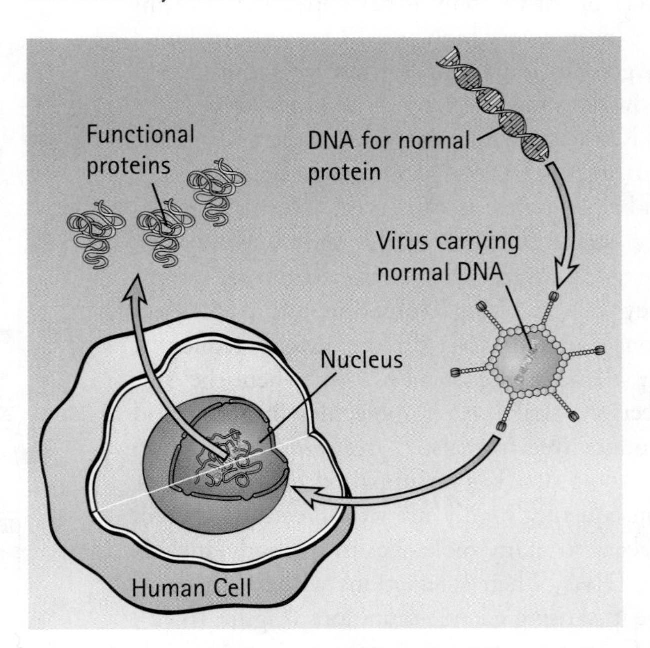

In gene therapy, DNA from a normal gene can help a patient make a working protein.

16.6 Meiosis and Genetic Diversity

LEARNING OBJECTIVE
Describe meiosis, and explain how it contributes to genetic diversity.

EXPLAIN THIS How can the same two parents produce children that are all different?

In Chapter 15, we learned that some cells reproduce through a type of cell division called mitosis. In mitosis, one cell divides into two daughter cells, each of which contains the same genetic information as the original cell. **Meiosis** is another type of cell division. Meiosis is used to make haploid cells, such as our eggs and sperm. Only the cells that produce our eggs and sperm go through meiosis. In meiosis, one diploid cell, with two of each kind of chromosome, divides into four haploid cells, each with only one of each kind of chromosome. The normal diploid chromosome number is restored during sexual reproduction when sperm and egg fuse at fertilization (Figure 16.13).

MasteringPhysics®
TUTORIAL: Meiosis Animation

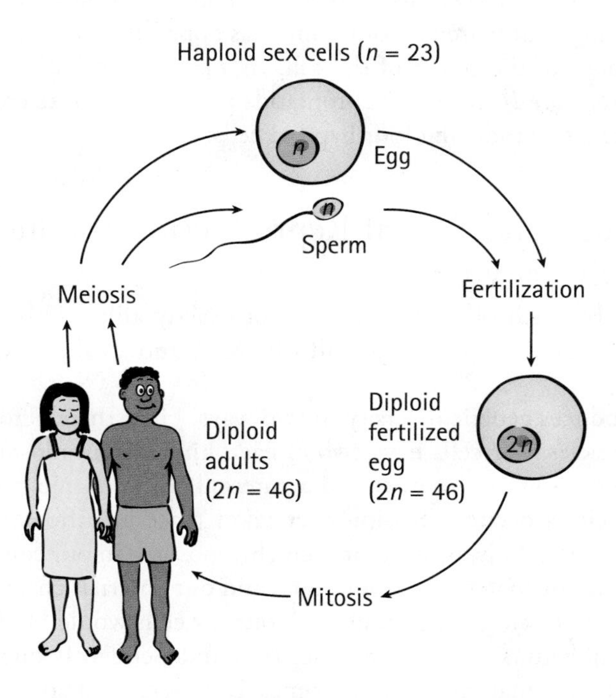

FIGURE 16.13
Sperm and eggs are haploid cells produced during meiosis. At fertilization, sperm and egg fuse to make a diploid cell. This diploid cell develops into a diploid individual.

When meiosis begins, the diploid cell has already copied its DNA. Meiosis takes place in two steps, *meiosis I* and *meiosis II* (Figure 16.14). During meiosis I, the original cell divides into two cells. During meiosis II, these two cells divide again to produce four haploid cells.

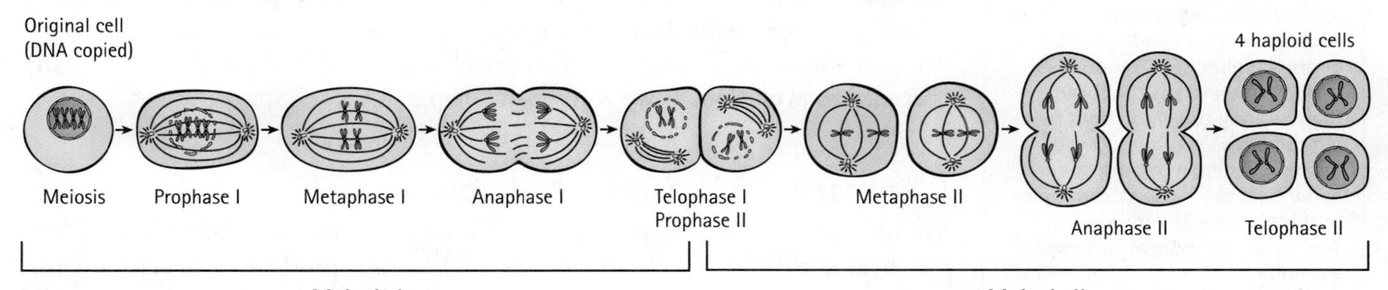

FIGURE 16.14
Meiosis produces four haploid cells.

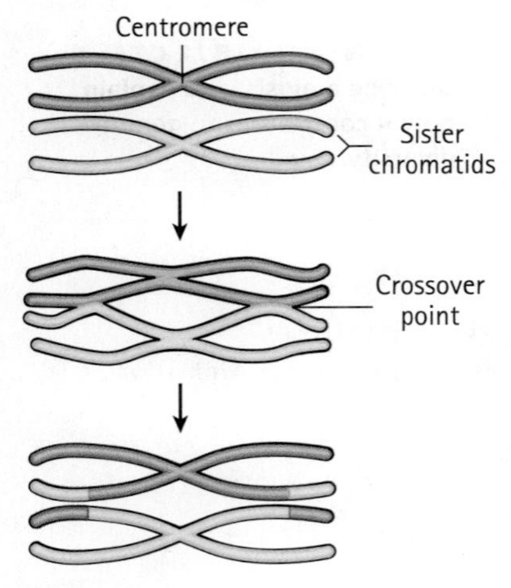

FIGURE 16.15
During meiosis, homologous chromosomes exchange genetic material in a process called crossing over. The result of crossing over is that chromosomes end up with parts of each of the two original homologous chromosomes—you can see that after crossing over, the chromosomes are part purple and part yellow.

Meiosis I begins with prophase I. During *prophase I*, the chromosomes condense and the membranes of the nucleus break down. As in mitosis, each chromosome includes two identical sister chromatids. Homologous chromosomes then line up with each other, and *crossing over* occurs. In crossing over, a chromosome exchanges parts with its homologous chromosome (Figure 16.15). As a result, the chromosomes in the dividing cell are no longer identical to the ones in the original cell. Instead, many chromosomes now include parts of *each* of the two original homologous chromosomes. On a genetic level, crossing over results in **recombination**, the production of new combinations of genes different from those found in the original chromosomes. We'll see why this is important when we consider how meiosis produces genetic diversity. In *metaphase I*, homologous chromosomes line up in the middle of the cell. In *anaphase I*, the homologous chromosome pairs separate. In *telophase I*, the chromosomes move to opposite ends of the cell. Cytokinesis occurs, producing two cells.

Meiosis II is similar to meiosis I, except for a few key differences. During *metaphase II*, the single, unpaired chromosomes (as opposed to homologous chromosome pairs) move to the center of the cell. In *anaphase II*, the sister chromatids separate. In *telophase II*, the sister chromatids move to opposite ends of the cell. Cytokinesis occurs, producing four haploid cells.

How Meiosis and Sexual Reproduction Produce Genetic Diversity

Siblings resemble each other, but they're not exactly alike. This is because the eggs and sperm of their parents are all different, and so they receive different genes.

Meiosis produces genetic diversity in two ways. First, there's crossing over. At the start of meiosis, our cells have two of each kind of chromosome—a maternal chromosome from our mother and a paternal chromosome from our father. Every time meiosis occurs, crossing over takes place at different points along these chromosomes. This means that each chromosome in our eggs and sperm is a unique mix of our maternal chromosome and our paternal chromosome.

Even without crossing over, though, all our sex cells would be different. This is because homologous chromosomes separate independently during meiosis I. One egg might get maternal chromosomes 1, 3, 4, 5, and so on and paternal chromosomes 2, 6, 7, 8, and so on. A second egg is almost certain to get different chromosomes, perhaps maternal chromosomes 2, 3, 5, 6, and so on and paternal chromosomes 1, 4, 7, 8, and so on (Figure 16.16). There are a huge number of possibilities!

Finally, during fertilization, the joining of an egg and sperm, genetic material is brought together in different ways. Each of the many different possible eggs can join with each of many different possible sperm. All of this genetic diversity is crucial to evolution, as we will see in Chapter 17.

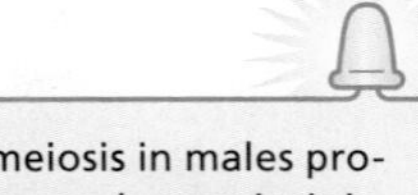

In humans, meiosis in males produces four sperm, but meiosis in females produces only one egg. This is because one of the four haploid cells—the future egg—receives almost all of the cytoplasm. This ensures that the future egg will have the resources it needs to develop. The other three cells receive almost no cytoplasm and quickly degenerate.

A mistake during meiosis can cause a sperm or egg to end up with the wrong number of chromosomes. In humans, embryos produced by these sex cells usually do not survive, but there are exceptions. One of the most common chromosomal abnormalities is having three copies of chromosome 21—trisomy 21. Trisomy 21 causes Down syndrome, a condition characterized by mental retardation and defects of the heart and respiratory system.

Comparing Mitosis and Meiosis

We have now looked at two types of cell division: mitosis and meiosis. There are three key differences between them: (1) Mitosis produces two cells; meiosis produces four cells. (2) Mitosis produces diploid cells; meiosis produces haploid cells. (3) The cells produced by mitosis are identical to one another and to the original cell; the cells produced by meiosis are all different.

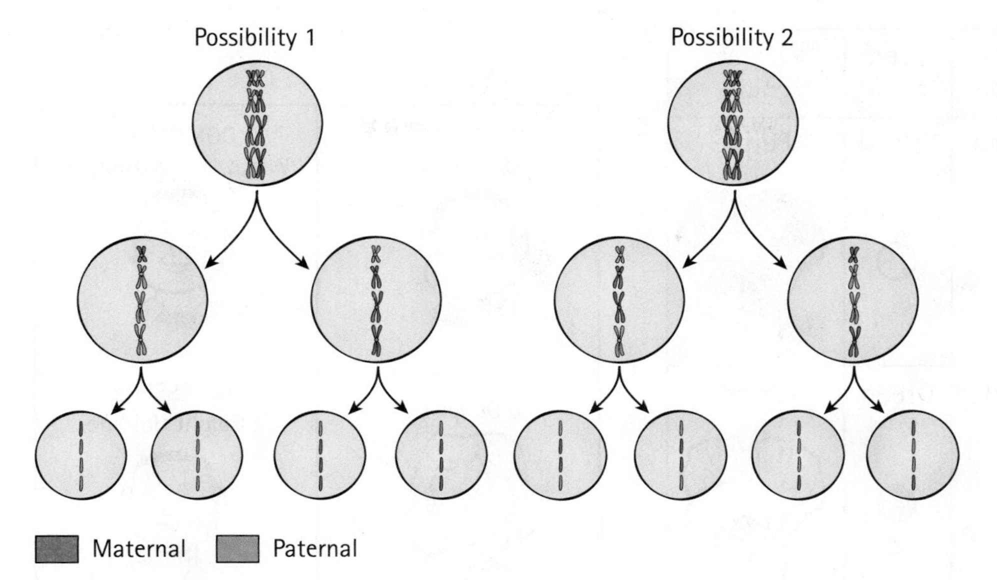

FIGURE 16.16
Each time meiosis occurs, the resulting cells receive different combinations of maternal and paternal chromosomes. This figure shows just two possibilities when meiosis occurs in an organism with 4 pairs of chromosomes. With the 23 pairs found in humans, there are even more possibilities.

16.7 Mendelian Genetics

LEARNING OBJECTIVE
Describe Mendel's laws, and explain how they account for his breeding results.

EXPLAIN THIS Why do parents with dimples sometimes have children without dimples?

We look like our parents because we inherit our parents' genes. But why do we resemble our parents in certain ways but not others? For example, why do we have our mom's dimples but not her curly hair? Why do we have our dad's brown eyes but not his freckles? Why do brown-eyed parents sometimes have a blue-eyed child? And, strangest of all, why do traits sometimes skip generations? For example, a crook in Grandpa's nose might not show up in any of his children and then suddenly appear in his grandson! How can we explain that?

Gregor Mendel (Figure 16.17) discovered the answers to these questions. Mendel did not make his discoveries by studying people, though. He studied peas.

Mendel's First Law

Mendel bred pea plants, which vary in a number of traits. They have round or wrinkled peas, yellow or green peas, purple or white flowers, and other variations. Mendel started by finding plants that "bred true," meaning these plants always produced offspring that looked like them. He then bred two plants that differed in a single trait—for example, a round-pea plant and a wrinkled-pea plant.

What did Mendel find? Mendel found that when he bred two plants that differed in a single trait, all of the offspring resembled *one* of the two parents. For example, when a round-pea plant was bred with a wrinkled-pea plant, all the offspring had round peas. Traits that were expressed in the offspring (such as round peas) Mendel

FIGURE 16.17
Gregor Mendel was the founder of modern genetics. Here, he is examining a plant.

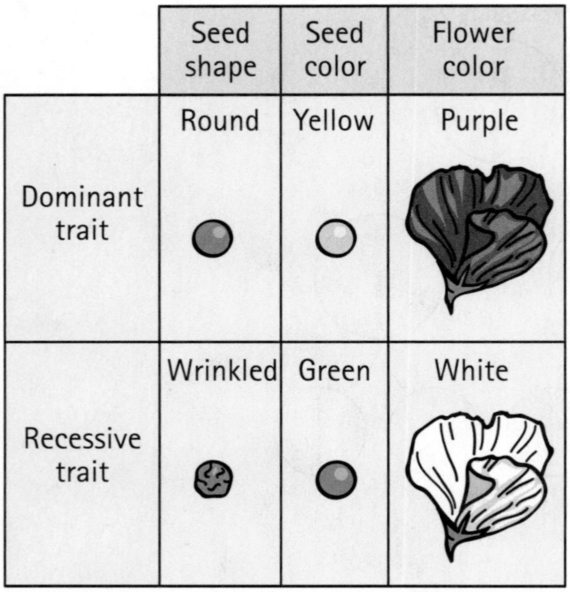

	Seed shape	Seed color	Flower color
Dominant trait	Round	Yellow	Purple
Recessive trait	Wrinkled	Green	White

FIGURE 16.18
Mendel bred pea plants that varied in a number of traits. For every pair of traits he looked at, one trait was dominant and the other was recessive.

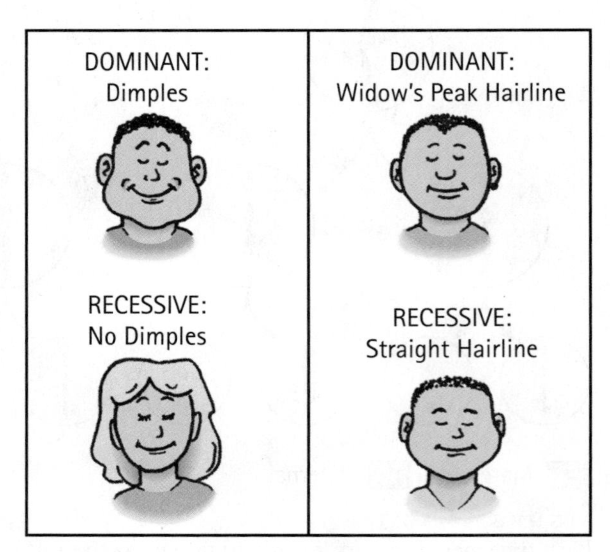

FIGURE 16.19
Dimples are a dominant human trait. So is a hairline that comes to a peak in the middle—a widow's peak hairline.

called *dominant.* Traits that were not expressed in the offspring (such as wrinkled peas) Mendel called *recessive.* In every pair of traits Mendel looked at, one was dominant and the other was recessive (Figure 16.18). Humans have dominant and recessive traits too (Figure 16.19).

CHECK YOURSELF
When Mendel bred green-pea plants and yellow-pea plants, all of the offspring had yellow peas. Which trait is dominant? Which trait is recessive?

CHECK YOUR ANSWER
Yellow peas are dominant. Green peas are recessive.

What just happened with Mendel's peas is a lot like the story about the crook in Grandpa's nose at the beginning of this section. The crook disappears in Grandpa's children and reappears in Grandpa's grandson, just the way the wrinkled peas disappeared in the first generation and reappeared in the next generation.

After breeding two plants that differed in a single trait, Mendel bred the offspring with themselves. (Many plants can breed with themselves, something humans can't do.) For example, after Mendel bred round-pea plants and wrinkled-pea plants together, he took their offspring (all of which were round-pea) and bred them with themselves. Mendel was surprised to find that the recessive trait, which had disappeared in the first generation, reappeared in the second generation. Moreover, in the second generation, the ratio of plants expressing the dominant trait to plants expressing the recessive trait was 3:1. That is, there were three times as many plants expressing the dominant trait as there were plants expressing the recessive trait.

How did Mendel explain his results? Mendel postulated that the genes that determine traits are made up of two separate **alleles**, or versions of the gene. One allele is inherited from each parent. Then Mendel came up with his first law, the principle of segregation. The *principle of segregation* says that when an individual makes sex cells (sperm or eggs), half the sex cells carry one allele and the other half carry the other allele.

UNIFYING CONCEPT

- *The Scientific Method* **Section 1.3**

Let's see how the principle of segregation explains Mendel's breeding results. Mendel looked at plants that differed in a single trait—say, round peas or wrinkled peas. The round-pea plant had two round alleles, a genotype we write as *RR*. The wrinkled-pea plant had two wrinkled alleles, or genotype *rr*. (Notice that a capital letter is used for a dominant trait and a lowercase letter for a recessive trait.) The *RR* and *rr* plants are homozygotes for this trait—a **homozygote** has two identical alleles for a given trait.

When Mendel bred the *RR* (round-pea) and *rr* (wrinkled-pea) plants together, the first-generation offspring inherited an *R* allele from the round parent and an *r* allele from the wrinkled parent. These first-generation offspring had the genotype *Rr*. An *Rr* plant is a **heterozygote** for this trait because it has two different alleles. Which trait does an *Rr* plant have? An *Rr* plant has round peas because the *R* allele is dominant. The **dominant** allele is expressed in the heterozygote. The **recessive** allele is not expressed in the heterozygote; it is hidden.

CHECK YOURSELF

1. **In humans, "dimples" is dominant and "no dimples" is recessive. You can use *D* and *d* to represent the two alleles. Which allele represents "dimples"? Which allele represents "no dimples"?**
2. **What is the genotype of a person who is a heterozygote for dimples? What is this person's phenotype—that is, does this person have dimples?**
3. **If a person has dimples, what are the two possibilities for his or her genotype? If a person has no dimples, what is his or her genotype?**

CHECK YOUR ANSWERS

1. We use capital letters for dominant traits and lowercase letters for recessive traits. So *D* represents "dimples" and *d* represents "no dimples."
2. A heterozygote has two different alleles, so a heterozygote has the genotype *Dd*. This person has dimples because *D* is dominant.
3. A person with dimples must have at least one dominant allele, so his or her genotype is either *DD* or *Dd*. A person with no dimples must have two recessive alleles, so his or her genotype is *dd*.

Then Mendel bred the *Rr* plants with themselves. According to the principle of segregation, *Rr* plants make equal numbers of *R* and *r* sperm and eggs. What alleles will the offspring inherit? It is equally likely that an offspring will get (1) *R* egg and *R* sperm, (2) *R* egg and *r* sperm, (3) *r* egg and *R* sperm, or (4) *r* egg and *r* sperm. This means that a quarter of the offspring are *RR*, half are *Rr*, and a quarter are *rr* (Figure 16.20). *RR* and *Rr* plants have round peas; *rr* plants have wrinkled peas. So, three-quarters of the offspring have round peas, and one-quarter have wrinkled peas—the ratio of round-pea plants to wrinkled-pea plants is 3:1.

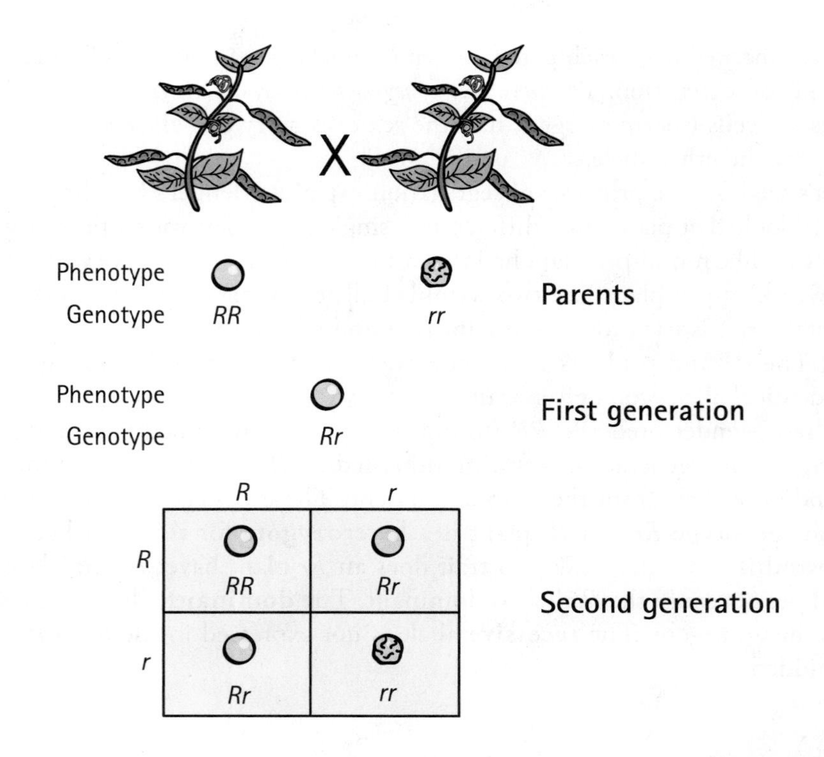

FIGURE 16.20
Mendel bred a round-pea plant with a wrinkled-pea plant. In the first generation, all the offspring had round peas. Then Mendel bred the offspring with themselves. The result was round-pea plants and wrinkled-pea plants in a ratio of 3:1.

CHECK YOURSELF

1. **You breed a tall pea plant (*TT*) with a short pea plant (*tt*). What allele is found in the sex cells of each plant? What alleles do the offspring inherit? Are the offspring tall or short?**
2. **When you breed *Tt* plants with themselves, what alleles do the offspring have? Are they tall or short? In what ratio?**

CHECK YOUR ANSWERS

1. The tall plant produces sex cells that carry the *T* allele. The short plant produces sex cells that carry the *t* allele. The offspring are *Tt*. They are all tall.
2. A quarter of the offspring are *TT*, half are *Tt*, and a quarter are *tt*. Three-quarters are tall, and one-quarter are short.

Mendel's Second Law

Mendel's next experiment was to breed plants that differed in two traits. He bred plants with round yellow peas (*RRYY*) to plants with wrinkled green peas (*rryy*). In the first generation, all the offspring inherited *R* and *Y* alleles from the first parent and *r* and *y* alleles from the second parent. The offspring were *RrYy*. They had round yellow peas.

Next, Mendel bred the *RrYy* plants with themselves. He wanted to test two possibilities. The first possibility is that alleles for the two genes stay together when the offspring make sex cells. If this is the case, *RrYy* plants should make only two kinds of sex cells: *RY* and *ry* (because *R* and *Y* were inherited together from one parent, and *r* and *y* were inherited together from the other parent). The second possibility is that alleles for the two genes behave independently when sex cells are made. In this case, *RrYy* plants should make sex cells with all combinations of alleles—*RY*, *Ry*, *rY*, and *ry*.

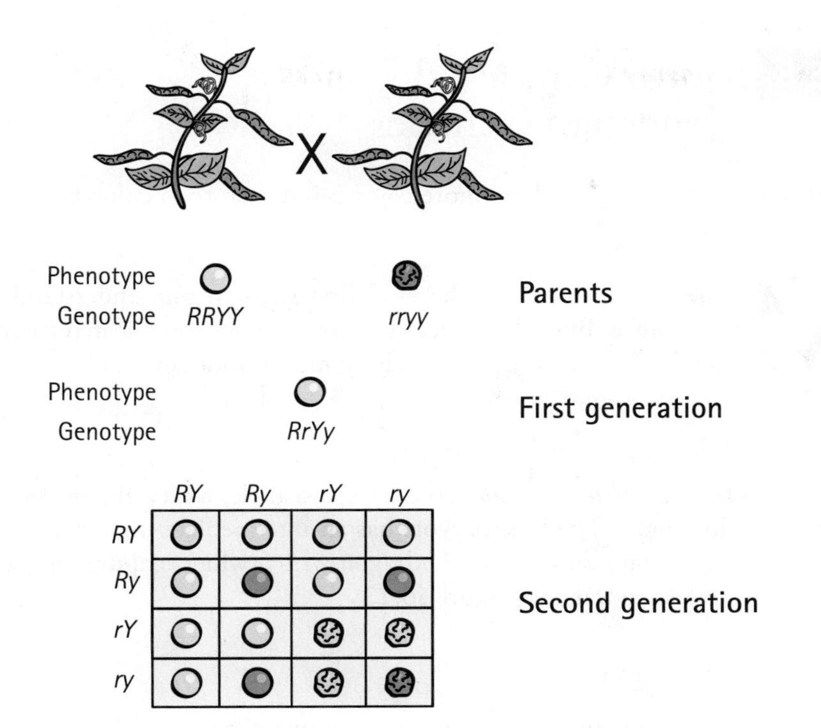

FIGURE 16.21
When Mendel bred plants that differed in two traits, the traits were inherited independently. Breeding plants with round yellow peas and plants with wrinkled green peas produced plants with round yellow peas in the first generation. In the second generation, Mendel saw all the different combinations of traits in a ratio of 9:3:3:1.

When the *RrYy* plants were bred with themselves, the offspring had every type of pea: round yellow, round green, wrinkled yellow, and wrinkled green. This couldn't happen if all the sex cells had been either *RY* or *ry*, because then all the offspring would have either round yellow peas or wrinkled green peas. You could never get wrinkled yellow peas or round green peas. Also, the ratio of round yellow : round green : wrinkled yellow : wrinkled green peas was 9:3:3:1, exactly what you expect if the sex cells carry all combinations of alleles (Figure 16.21). The inheritance of one trait is independent of the inheritance of a second trait. This is Mendel's second law, the *principle of independent assortment.*

UNIFYING CONCEPT
- ***The Scientific Method***
Section 1.3

Mendel knew nothing about meiosis when he developed his ideas on inheritance. However, we now know that both of Mendel's laws are true because of the way sex cells are produced during meiosis.

CHECK YOURSELF

Your father gave you an allele for no dimples (*d*) and an allele for a widow's peak (*W*). Your mother gave you a dimples allele (*D*) and a straight hairline allele (*w*).

1. **What is your genotype? Are you a homozygote or a heterozygote for each of these traits?**
2. **What is your phenotype? That is, do you have dimples? Do you have a widow's peak?**
3. **What will the genotypes of your sex cells be?**

CHECK YOUR ANSWERS
1. Your genotype is *DdWw.* You are a heterozygote for both traits.
2. You have dimples, and you have a widow's peak. Both those traits are dominant.
3. You make sex cells with all possible combinations of alleles: *DW, Dw, dW,* and *dw.*

LEARNING OBJECTIVE
Explain how traits can be inherited in ways that are different from Mendelian genetics.

16.8 More Wrinkles: Beyond Mendelian Genetics

EXPLAIN THIS Why are there more color-blind men than color-blind women?

Mendel's work provided the vital first steps in our understanding of inheritance. But inheritance can also be more complicated than what Mendel described. Let's consider some additional wrinkles.

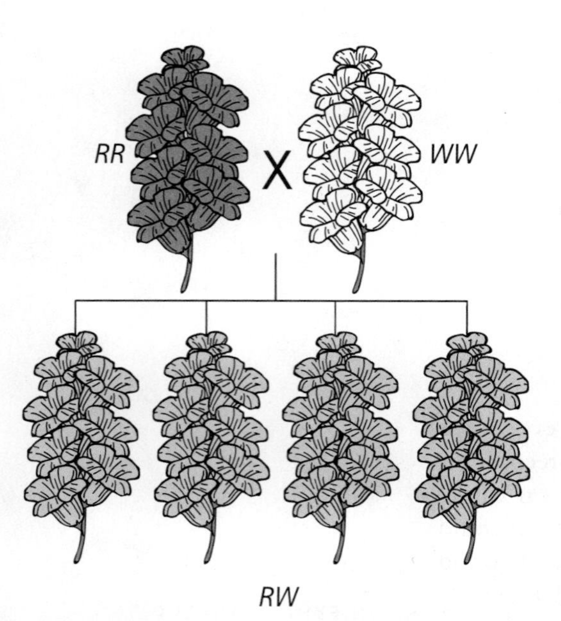

FIGURE 16.22
Flower color in snapdragons shows incomplete dominance. When a snapdragon with red flowers (*RR*) is bred with a snapdragon with white flowers (*WW*), the offspring (*RW*) have pink flowers.

Incomplete Dominance

In *incomplete dominance*, there are two alleles for a trait, and neither is dominant. The heterozygote has an intermediate trait. For example, when you breed a red snapdragon with a white snapdragon, you get pink snapdragons (Figure 16.22).

Codominance

In *codominance*, a heterozygote with two different alleles expresses the traits of *both* alleles. An example can be found in human blood type. Your blood type describes molecules on the surface of your red blood cells. You can have the A molecule (blood type A), the B molecule (blood type B), neither (blood type O), or both (blood type AB). There are three blood type alleles: *A*, *B*, and *O*. A person with genotype *AA* or *AO* has A molecules (blood type A). A person with genotype *BB* or *BO* has B molecules (blood type B). A person with genotype *OO* has neither molecule (blood type O). A person with genotype *AB* has both A and B molecules (blood type AB)—both the *A* trait and the *B* trait are expressed. The *A* and *B* alleles are codominant (Figure 16.23).

Polygenic Traits

Polygenic traits are determined by more than one gene. (*Poly* means "many," so *polygenic* refers to "many genes.") Human height and skin color are both polygenic. Polygenic traits show more of a continuum than traits determined by a single gene. To see why, consider human height.

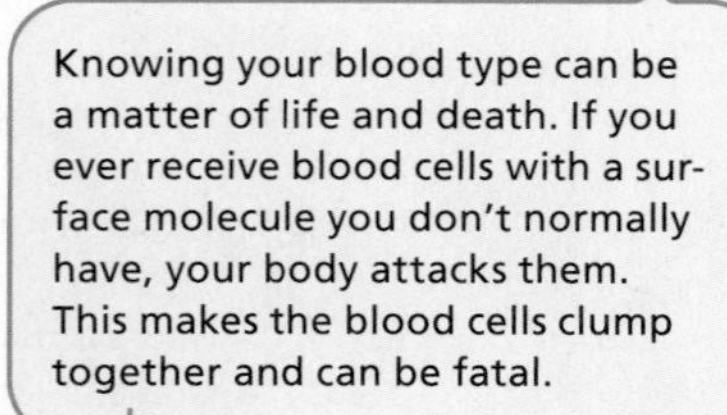

Knowing your blood type can be a matter of life and death. If you ever receive blood cells with a surface molecule you don't normally have, your body attacks them. This makes the blood cells clump together and can be fatal.

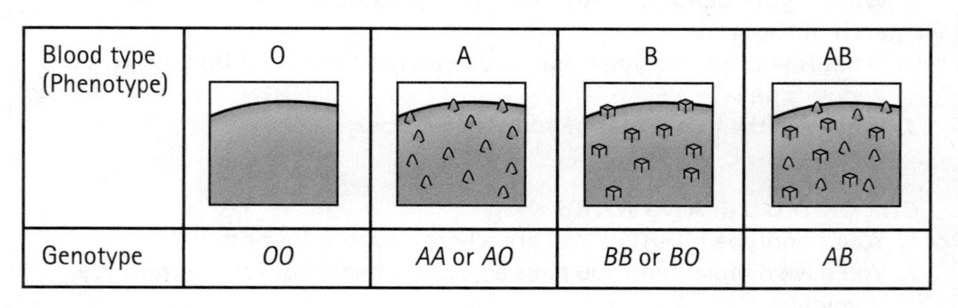

Blood type (Phenotype)
O
A
B
AB
Genotype
OO
AA or AO
BB or BO
AB

FIGURE 16.23
Human blood type offers an example of codominance. Blood type is determined by combinations of the *A*, *B*, and *O* alleles. *A* and *B* are both dominant to *O*, but *A* and *B* are codominant.

FIGURE 16.24
Human height is a polygenic trait. Taller people have more tall alleles. (Height is also affected by nongenetic factors such as diet and nutrition.)

Suppose three genes A, B, and C determine height. (More than three genes are actually involved.) Each gene has a tall allele and a short allele, which we will indicate with *T* and *S*. The shortest people have all short alleles—genotype $A_SA_SB_SB_SC_SC_S$. People with genotype $A_SA_SB_TB_SC_SC_S$ (five short alleles and one tall allele) are somewhat taller, those with four short alleles and two tall alleles are even taller, and so on. In this way, multiple genes produce a gradation of heights (Figure 16.24).

Pleiotropy

In *pleiotropy*, a single gene affects more than one trait. The disease sickle cell anemia provides an example of pleiotropy in humans. The sickle-cell allele for hemoglobin makes red blood cells turn sickle-shaped under certain conditions, causing tissue damage and pain (Figure 16.25). Homozygotes (people who have two sickle-cell alleles) are severely affected. Heterozygotes (people who have one sickle-cell allele and one normal allele) have only mild symptoms. Given the sickle-cell allele's harmful effects, scientists were puzzled by how common it is, especially among people of African descent. Why hadn't evolution through natural selection (which we will study in Chapter 17) caused the sickle-cell allele to disappear from human populations?

Pleiotropy provides the answer. It turns out that the sickle-cell allele also protects people from malaria. This is an example of pleiotropy because the sickle-cell allele affects more than one trait—it affects the shape of red blood cells *and* resistance to malaria. The fact that the sickle-cell allele protects against malaria—plus its mild symptoms in heterozygotes—explains why the allele is common in populations where malaria has been a danger in the past.

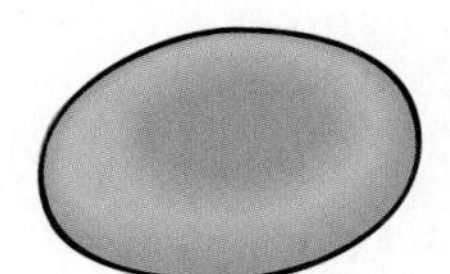

FIGURE 16.25
The sickle-cell allele offers an example of pleiotropy in humans. The allele can cause normally round red blood cells to assume a sickle shape, resulting in pain and tissue damage. However, the allele also offers protection from malaria.

Linked Genes

Linked genes are frequently inherited together, in apparent contradiction to Mendel's principle of independent assortment. How can genes be linked? The key is that independent assortment occurs when alleles found on *different* chromosomes separate independently at meiosis. But, if the alleles for two genes are on the *same* chromosome, they are often inherited together—though not quite all the time, because sometimes crossing over shuffles them up. In fact, the closer two alleles are to each other on a chromosome, the more likely they are to be inherited together. (The genes for Mendel's pea traits happened to be on different chromosomes—that's why he concluded that genes sort independently.) Body color and wing size in fruit flies are linked genes.

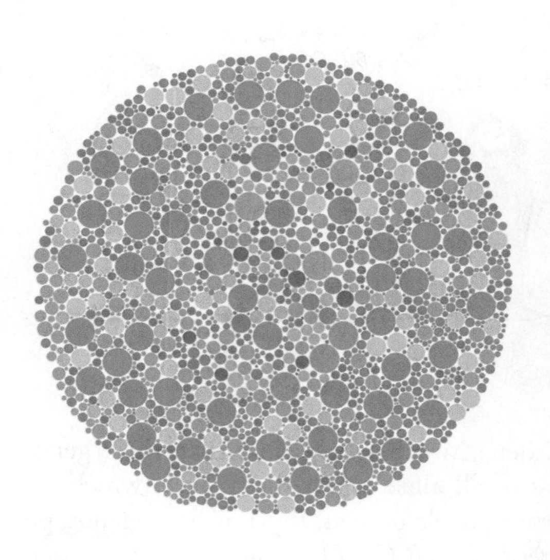

(a)

(b)

FIGURE 16.26
Red-green color-blindness is a sex-linked trait that affects many more males than females. (a) This image is used to test for red-green color-blindness. Do you see the number? (b) A plate of fruits and vegetables as seen by a person with red-green color-blindness.

Sex-Linked Traits

Sex-linked traits are determined by genes found on the X chromosome. Because males have only one X chromosome, they have only one allele for these traits. This means that males need only one recessive allele to express a recessive sex-linked trait, whereas females need two recessive alleles. Because of this, recessive sex-linked traits are seen more often in males than in females. In humans, two recessive sex-linked traits are red-green color-blindness (Figure 16.26) and the blood disease hemophilia. Both conditions affect far more males than females.

CHECK YOURSELF

1. **You breed a plant with two blue-flower alleles to a plant with two white-flower alleles. If the blue allele is incompletely dominant to the white allele, what color flowers will the offspring have?**
2. **Your blood type is O, your mother's blood type is A, and your father's blood type is B. What is your genotype? What about your mother's and your father's? Explain your reasoning.**

CHECK YOUR ANSWERS

1. They will have pale blue flowers.
2. Because your blood type is O, your genotype must be *OO*. Your mother's blood type is A, so she must have the genotype *AO*. (She must have an *O* allele because she passed one to you.) Similarly, your father must be *BO*.

SCIENCE AND SOCIETY

Genetic Counseling

What is your risk of someday having a child with a genetic disease? Are there tests that show whether a developing fetus has a genetic disease? What is it like to raise a child who has a genetic disease? These are some of the questions genetic counselors help people answer.

Genetic counseling is especially relevant for couples with a family history of genetic disease. Genetic counselors often begin by creating a family *pedigree*—a family tree that shows which relatives are affected by genetic disease. They then use information about the inheritance pattern of the disease in question—for example, whether the disease allele is dominant or recessive—to assess a couple's risk of having an affected child.

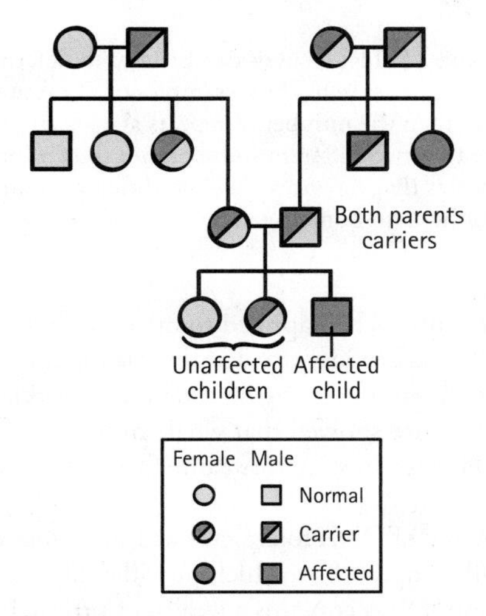

This three-generation pedigree shows the occurrence of Tay–Sachs disease within a family. The allele for Tay–Sachs disease is recessive.

For some diseases, medical tests can provide additional information. Those with a family history of Tay–Sachs disease (a fatal disease of the central nervous system) or cystic fibrosis (a serious disease characterized by mucus buildup in the lungs, digestive system, and other organs) can undergo simple tests to determine whether they are *carriers* of the disease—that is, whether they possess a disease allele in addition to a normal allele. Because these diseases affect only people who have two recessive alleles, a couple is at risk of having an affected child only if both parents are carriers. If both parents are in fact carriers, procedures such as *chorionic villus sampling* or *amniocentesis*, in which fetal cells are collected and examined during pregnancy, can be used to determine whether a fetus is affected. Couples may also have the option of undergoing *in vitro* fertilization, testing the embryos, and implanting only unaffected ones.

Genetic testing is also useful for families without a history of genetic disease. For Down syndrome, a condition associated with mental retardation and other health issues, risk depends primarily on the age of the mother. Older women have a greater risk of having children with Down syndrome. Down syndrome is caused by trisomy 21, the presence of three copies of chromosome 21 instead of two. Tests during pregnancy can look at the chromosomes in a fetus's cells to determine whether they show trisomy 21.

Down syndrome occurs in about one out of every 1000 births. It is caused by trisomy 21.

16.9 The Human Genome

LEARNING OBJECTIVE
Describe the key features of the human genome.

EXPLAIN THIS Does all your DNA code for proteins?

As soon as Crick and Watson discovered the structure of DNA, scientists began to develop tools for studying—as well as manipulating—genetic material. Advances in DNA technology now allow scientists to study the total genetic material of a living organism—the organism's **genome**. One of the first genomes scientists chose to study was the human genome.

SCIENCE AND SOCIETY

DNA Forensics

When detectives investigate a crime scene, they no longer stop at dusting for fingerprints. Now, they carefully collect all the hair, blood, saliva, and other bodily products they can find in order to look for the genetic version of fingerprints—DNA. The DNA collected can be compared with that of suspects and can provide compelling evidence as to whether a suspect was present at the scene of the crime.

Forensic scientists usually use *short tandem repeat (STR) analysis* to test whether DNA samples match. STR analysis looks at certain non–protein-coding sites in the human genome where a short sequence of DNA (such as AGAT) is repeated multiple times. The exact number of repeats varies enormously from one person to the next. If two DNA samples are from the same person, they must have the same number of repeats at every site. The FBI uses a standard set of 13 sites for which the odds of a false match are less than one in a billion. (However, identical twins share the same DNA and so have matching STR patterns.) DNA data are stored in an FBI database called CODIS (Combined DNA Index System).

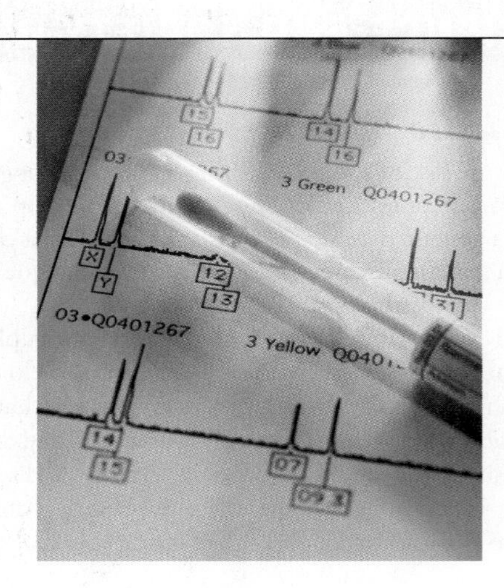

Short tandem repeat (STR) analysis can be used to determine whether DNA samples match. If two samples of DNA are from the same person, then the number of repeats at each site must be identical. (Note that two different numbers are shown for some sites—this indicates that the individual has different numbers of repeats on his or her two homologous chromosomes.)

DNA evidence has been used in criminal cases since 1986. Although DNA appears in only a small fraction of cases, it frequently plays a role when violent crimes such as murder and rape are involved. Although DNA can be used to help convict suspects, it can also exclude suspects who have been falsely accused and exonerate people imprisoned for crimes they did not commit. The Innocence Project reports that 272 prisoners were exonerated by DNA evidence between 1989 and 2011, including 17 who were on death row.

DNA forensics has uses beyond criminal justice. For example, it can be used to identify the victims of accidents or crimes. In the September 11, 2001, terrorist attacks at the World Trade Center, more than half of the victims were ultimately identified by DNA evidence alone. DNA tests are also commonly used to establish paternity and trace family relationships. When applied to other species, DNA forensics can be used to identify disease-causing microorganisms or the illegal harvesting of endangered species. In 2010, DNA evidence showed that whale meat served in Los Angeles–area Japanese restaurants came from endangered species.

Despite the uses of DNA forensics, there are serious ethical concerns. Unlike fingerprints, which provide only identification information, DNA contains a wealth of private information about a person's family relationships, susceptibility to diseases, and perhaps even behavioral tendencies. In some states, DNA can be collected not only from convicted felons but also from anyone who is arrested. This would certainly include many who are innocent of any crimes.

MasteringPhysics®
TUTORIAL: DNA Fingerprinting

The Human Genome Project, which was completed in 2003, determined the DNA sequence of the entire human genome. We now know that there are about 3.2 billion nucleotide pairs in the human genome. Amazingly, more than 99.9% of these are identical in all humans.

We also know that the human genome is not made up exclusively of genes. In fact, only 2% of the genome carries instructions for building proteins. Most of the genome—at least 50%—consists of repeat sequences that appear over and over. Although repeat sequences do not have a direct function, they may occasionally rearrange the human genome, creating new genes or reshuffling existing ones.

Humans have a total of about 22,000 genes. Genes are often found in clusters at random spots along our chromosomes, with large sections of non–protein-coding DNA between them. Although 22,000 genes is a lot fewer than the 100,000 that scientists originally postulated, many genes give rise to RNA transcripts that are processed in different ways. The result is that one gene can provide the instructions for building multiple proteins. The function of more than half of our genes is still completely unknown.

Scientists have also identified more than 3 million locations in the genome where the nucleotide sequence differs among human beings. These differences are called *single-nucleotide polymorphisms*, or *SNPs*. SNPs make every person unique, of course, but they may also help scientists identify genes related to human diseases.

CHECK YOURSELF

Scientists believe that most mutations have little effect on genes or organisms. Can you use what you learned about the human genome to support this statement?

CHECK YOUR ANSWER

Genes make up only about 2% of the human genome. This suggests that, just by chance, the majority of mutations that strike human DNA will not affect any genes.

16.10 Cancer: Genes Gone Awry

LEARNING OBJECTIVE
Explain how genetic mutations that occur during our lifetimes give rise to cancer.

EXPLAIN THIS How does a tumor become deadly?

Cell division is normally a carefully orchestrated process controlled by a large number of genes. Cancer occurs when mutations in these genes cause cells to divide out of control. A mutation in a single gene is not enough to produce cancer—mutations in many important genes are required.

Cancer tends to strike older people more often than younger people because mutations have had more time to accumulate in their cells. People who have been exposed to mutation-causing agents accumulate mutations more rapidly, making cancer more likely to develop. Finally, some people are prone to developing certain types of cancers because they have inherited mutations in cancer-related genes. For example, mutations in the *BRCA* genes (named for *BR*east *CA*ncer) are linked to aggressive breast and ovarian cancers.

Several types of genes have been implicated in cancer. Proto-oncogenes are genes that, when mutated, turn into *oncogenes* that stimulate abnormal cell division. *Tumor-suppressor genes* normally prevent cancer by inhibiting cell division. In cancer cells, it is typical to find that oncogenes have been activated and tumor-suppressor genes have become nonfunctional.

In a deadly cancer, tumor cells do more than divide out of control. Cancers are said to have *metastasized* if tumor cells have developed the ability to spread around the body and give rise to secondary tumors. Metastasis represents a crucial point in disease progression because cancers typically become much harder to treat once they reach that point.

LEARNING OBJECTIVE
List and describe some important environmental causes of cancer.

Integrated Science 16C
EARTH SCIENCE

Environmental Causes of Cancer

EXPLAIN THIS Why should you use sunscreen?

When people migrate from one place to another, their risk of developing cancer is determined largely by the place they move to, rather than the place they have moved from. This suggests that cancer is primarily an environmental, rather than an inherited, disease. In fact, a person's environment is probably responsible for about 80%–90% of the mutations that result in cancer, with inherited genetic factors accounting for the other 10%–20%.

FIGURE 16.27
Reducing exposure to UV light will help you avoid skin cancer. Here, Io and Pico demonstrate the use of hats and sunglasses. Also present, though not visible, is lots of sunscreen.

Environmental factors that increase the risk for cancer include smoking, diet, radiation, ultraviolet light, chemicals, and infection by certain viruses and bacteria. The roles of factors such as smoking and radiation are well understood: Scientists know how and why these things increase genetic mutations. The roles of other factors are still being studied. For example, scientists know that diet is important in the development of many cancers. However, people's diets are such complicated mixes of nutrients that it can be hard to connect cancer with any particular foods. Yet, most of the research shows that a diet rich in fruits and vegetables can help protect against cancer.

A well-known environmental risk factor for skin cancer is ultraviolet (UV) light. Staying out of sunlight and using sunscreen help reduce exposure to UV light (Figure 16.27). Unfortunately, the reduction of ozone in Earth's atmosphere allows more UV radiation to reach Earth's surface (see Chapter 26).

CHECK YOURSELF
If cancer is usually caused by environmental factors, why do we call it a genetic disease?

CHECK YOUR ANSWER
Cancer is a genetic disease because it is the result of genetic mutations. But, unlike other genetic diseases, cancer is not usually an *inherited* genetic disease. The mutations that cause cancer are not usually inherited from one's parents. Instead, they are caused mainly by environmental factors.

LEARNING OBJECTIVE
Use examples to describe some practical uses of transgenic organisms and cloning.

16.11 Transgenic Organisms and Cloning

EXPLAIN THIS How could transgenic plants help humans adjust to the effects of global warming?

Advances in DNA technology have led to many practical applications as well as some serious concerns. In this section, we will consider two topics on the edge of DNA technology today: transgenic organisms and cloning.

Transgenic Organisms

A *transgenic organism* is an organism that contains a gene from another species. Transgenic organisms are usually engineered for some practical purpose. Scientists produced one of the first useful transgenic organisms when they placed the gene for human insulin into the bacterium *Escherichia coli*, a common inhabitant of the human digestive tract. At the time, insulin, which is used to treat some forms of diabetes, had to be collected from the carcasses of pigs and cows. When transgenic *E. coli* began to crank out human insulin, the advantages were clear—human insulin could now be produced easily and in large amounts. In 1982, insulin became the world's first genetically engineered pharmaceutical product. Transgenic bacteria now produce many other products, and scientists continue to work to expand this list (Figure 16.28). Figure 16.29 shows a typical process for developing transgenic bacteria.

FIGURE 16.28
Transgenic bacteria (and other microorganisms) make a wide variety of important pharmaceutical products. This is an insulin production facility in Russia.

Transgenic plants have been developed for a variety of purposes. Some produce useful products; for example, transgenic tobacco plants are used to make medicines for certain autoimmune diseases. Other transgenic plants have genes that offer resistance to pests, disease, or herbicides. Transgenic varieties of corn, potato, and cotton contain a gene from the soil bacterium *Bacillus thuringiensis*; the gene codes for a toxic protein that kills certain insects and makes the crops resistant to these pests. A transgenic soybean carries foreign DNA that makes it resistant to the herbicide Roundup. Farmers who plant these "Roundup Ready" soybeans can use Roundup to kill weeds without damaging their crops. Still other transgenic plants

Another product made by transgenic bacteria is human growth hormone (HGH). Supplemental HGH can help children who have a deficiency of the hormone achieve normal size. But the easy availability of HGH has led to its abuse by some adults, including some athletes. When healthy adults take HGH, they often experience dangerous side effects. It is a banned substance in most sports.

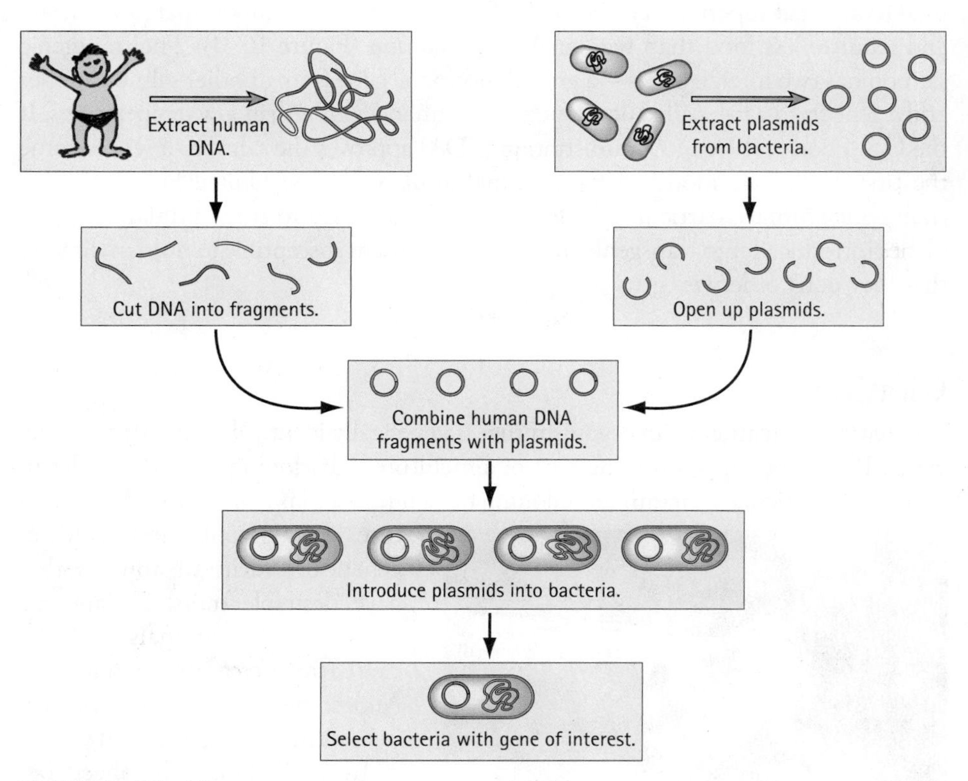

FIGURE 16.29
This figure shows how transgenic bacteria are developed. Transgenic bacteria often carry their foreign genes in plasmids, small circles of DNA separate from the main bacterial chromosome. Special enzymes are used to cut and reseal DNA when new genes are introduced.

FIGURE 16.30
Naturally drought-resistant plants such as the resurrection fern provide genes that can be transferred to crop plants, enabling them to survive the high temperatures and severe droughts that accompany global warming. These ferns appear dead, but they will revive as soon as it rains.

have been engineered to grow in harsh environments. Transgenic varieties of tomato, canola, and rice grow in salty soils with the help of a special ion transport protein from a naturally salt-tolerant plant. (Salty soils are a common side effect of intensive agriculture.) Considerable effort is now being made to develop plants that are resistant to drought and high temperatures, both of which have worsened with global warming. Drought-tolerant species such as wild barley and the resurrection fern (Figure 16.30) have special genes that can be transferred to crops such as wheat. Transgenic wheat carrying drought-related wild barley genes needs only one-eighth as much water as regular wheat. Other research is focusing on the development of transgenic plants for biofuels and for use in cleaning up pollutants.

FIGURE 16.31
This transgenic salmon may become the first transgenic animal approved for human consumption in the United States. The transgenic salmon (*rear*) grows faster and requires less food than the regular salmon (*forward*). These two animals are the same age.

Transgenic animals make proteins for human use, provide agricultural products, and aid in the study of human diseases. Certain transgenic sheep and goats make foreign proteins in their milk, and researchers are working on a transgenic chicken that can produce foreign proteins in its eggs. In agriculture, researchers increased wool production in sheep by inserting special DNA sequences from mice. Other scientists are trying to use a roundworm gene to engineer healthier pork—transgenic pigs that carry the gene produce higher amounts of healthy omega-3 fatty acids. The first transgenic animal to arrive at supermarkets, however, may be an Atlantic salmon that grows faster and requires less food than regular Atlantic salmon (Figure 16.31). The transgenic salmon has two foreign genes—a growth hormone gene from another salmon species and a gene from a fish called ocean pout that enhances growth at low temperatures. If the U.S. Food and Drug Administration (FDA) approves the salmon, it will become the first genetically modified food animal available in the United States. Finally, transgenic animals, particularly transgenic mice, are used to study human diseases. Genes introduced into transgenic mice can make them susceptible to human diseases that they don't otherwise get.

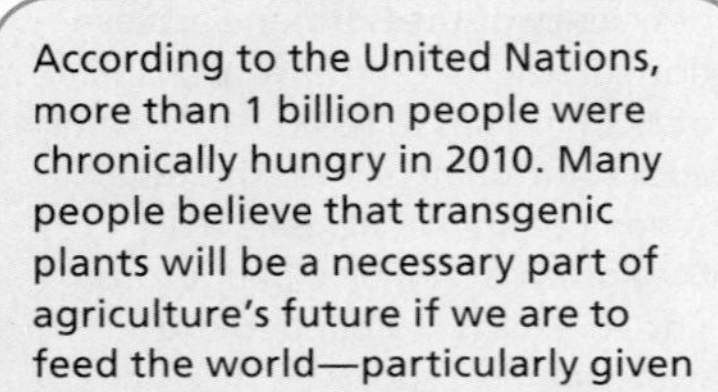

According to the United Nations, more than 1 billion people were chronically hungry in 2010. Many people believe that transgenic plants will be a necessary part of agriculture's future if we are to feed the world—particularly given the changes that global warming is bringing.

Cloning

Cloning is the creation of an organism that is genetically identical to one that already exists. Plant cloning is a routine part of agriculture, but cloning is much harder in animals—particularly mammals. Cloning has a number of potential uses. In agriculture, cloning could create herds of genetically identical animals that have desirable traits. Endangered or even extinct animals could be cloned to increase their numbers. And some people have expressed interest in cloning deceased pets.

FIGURE 16.32
Dolly the sheep was the first cloned mammal.

In 1997, Dolly the sheep became the first cloned mammal (Figure 16.32). Cloning was done through a process called *nuclear transplantation*, in which the nucleus of a cell from the animal being cloned is placed into an egg cell that has had its own nucleus removed. The resulting embryo is then implanted into a surrogate mother. Since Dolly, other mammals have been cloned, including dogs, cats, mice, and horses as well as rare species such as the European mouflon (a small wild sheep) and the banteng (a type of wild cattle). Unfortunately, cloned animals often suffer from developmental abnormalities or other health problems. Human cloning has never been performed and would certainly run into serious ethical obstacles. Laws currently prevent federal funds from being used to work on human cloning, and it is also illegal in some states.

A group of Japanese, Russian, and American scientists are trying to use tissue from a frozen mammoth to clone a live woolly mammoth. The plan is to insert the nucleus of a mammoth cell into an elephant egg cell that has had its own nucleus removed. If an embryo results, it could then be implanted into a female elephant.

CHECK YOURSELF

Is an animal produced through nuclear transplantation genetically identical to the animal that provides the nucleus, the animal that provides the egg, or the surrogate mother? Explain.

CHECK YOUR ANSWER

The animal is genetically identical to the animal that provides the nucleus. DNA is contained within the nucleus.

16.12 DNA Technology—What Could Possibly Go Wrong?

LEARNING OBJECTIVE
Explain some of the safety, social, and ethical concerns associated with DNA technology.

EXPLAIN THIS What is a "superweed"?

Although DNA technology has been the source of great excitement, it has also given rise to serious concerns. Could scientists accidentally release a deadly bacteria or virus? Are transgenic plants and animals safe to eat? Are transgenic crops damaging the environment and giving rise to dangerous "superweeds"? Also, which DNA technologies could one day be applied to humans, and what impact will this have on our society and culture? Finally, who will have access to DNA technologies?

Some genetically engineered bacteria and viruses are dangerous to human health or natural habitats. To help keep these organisms from leaving the lab, scientists give them genetic mutations that prevent them from performing important natural functions. This makes it harder for them to survive in the wild. (In the lab, scientists keep them alive by providing conditions that make up for the genetic deficiencies.) Despite these precautions, accidental releases remain a legitimate concern.

Hey mom, I just cloned a *T. rex* with two heads!
Okay, but don't let it bother your brother's woolly mammoth.

Genetically modified (GM) crops, including transgenic crops, are probably the most controversial product of DNA technology today. In the United States, GM crops are very common—for example, they include most of our soybeans, corn, and cotton. In many European countries, however, GM crops are banned or strictly controlled. One basic question is: Are GM plants safe to eat? In 2010, the European Commission reviewed 25 years of research and concluded that GM plants are not riskier for human health than plants produced through conventional breeding methods. However, some scientists

FIGURE 16.33
Should genetically modified foods be labeled? This photo shows a labeled GM food product sold in Europe. Labeling is not currently required in the United States.

and consumer groups worry that safety is not always adequately tested. In 2009, a group of two dozen scientists from public research institutions in 17 states warned the U.S. Environmental Protection Agency that the companies that develop GM plants routinely act to "inhibit public scientists from pursuing their mandated role on behalf of the public good" and that these companies had made independent analyses of GM crops impossible. A related issue is the labeling of GM foods (Figure 16.33). Labeling is not currently required in the United States, although a large majority of consumers favor it. Labeling is opposed by industry, which believes that consumers may avoid products that are labeled.

Another concern about GM crops is their effect on the environment. For example, crops that contain *Bacillus thuringiensis* genes, which are toxic to pest species, also kill nontarget species such as the Monarch butterfly. Heavy use of the herbicide Roundup on "Roundup Ready" crops (described in Section 16.11) has caused weeds to evolve resistance, turning them into "superweeds." Farmers respond by using more and more Roundup until it is no longer effective, at which point they switch to more toxic herbicides or pull weeds by hand. The result is more toxic chemicals in the environment, rising costs for farmers, and higher food prices for consumers.

A further worry is contamination by transgenic plants or their genes. In a 2007 test of "Roundup Ready" creeping bentgrass, developed for use on golf courses, some plants escaped from a field in Idaho and ended up in irrigation canals miles away in Oregon. Hundreds of escaped plants were found, including many that had already produced their own seeds. The plants must now be eradicated using special herbicides, a process expected to take years. Another worry is that GM crops will breed with wild plants and transfer their herbicide-resistance genes to the wild varieties, creating instant "superweeds." One such plant, a resistant mustard weed, was discovered in 2005 in a field used for trials of GM oilseed rape (a plant used to produce canola oil). Before the discovery, such transfers were described as nearly impossible. Genes from GM crops can also contaminate traditional crops. In 2004, the Union of Concerned Scientists studied corn, canola, and soybean and found that the traditional seed supplies for all three were "pervasively contaminated with low levels of DNA sequences derived from transgenic varieties." Contamination is a particular worry for organic farmers because organic produce cannot include GM varieties. In 2011, an organic farmer in Australia sued his neighbor after GM canola blew onto his fields, resulting in the loss of his organic license.

FIGURE 16.34
In the 1993 movie *Jurassic Park*, things don't go according to plan at a theme park filled with cloned dinosaurs.

A final concern about GM crops is cost. GM seeds are too expensive for farmers in many parts of the world. In India, many farmers were encouraged to take out large loans to buy "insect-proof" GM seeds. When the crops failed due to drought (some GM crops require more water than traditional crops), many of the farmers went bankrupt.

GM animal foods have sparked new concerns. For example, consumer groups charge that dangerously high levels of growth hormone are found in transgenic Atlantic salmon. Ecologists and environmentalists also worry about effects on native populations—although the transgenic fish are supposed to be sterile, a small percentage could remain fertile; although they are supposed to be isolated from natural habitats, some could escape. In 2011, a group of senators asked the FDA to stop considering the transgenic salmon for approval, even threatening to strip the agency of funding to study the fish. But was it science or politics that influenced them? Most of the senators represent states with salmon fisheries, where constituents worry about competition from transgenic varieties.

Finally, what about DNA technology and human society? What will happen when we apply DNA technologies to ourselves? Curing a genetic disease, as gene therapy attempts to do, is one thing, but what about enhancing our abilities? Also, could DNA technology blur the distinctions between species, including the line between humans and nonhuman animals? How would we protect the rights of any such organisms that came into existence? More generally, what constraints should be placed on DNA technology? In addition, as genetic technologies are developed, who will control access to them? Should important technologies be available only to the wealthy and privileged? On the other hand, could the ability to manipulate DNA become too readily available—in easy, do-it-yourself kits? Will we have enough time to think through the social and ethical implications of a technology or to legislate effective controls? And what if something unexpected occurs (see Figure 16.34)? We will be facing questions like these for years to come.

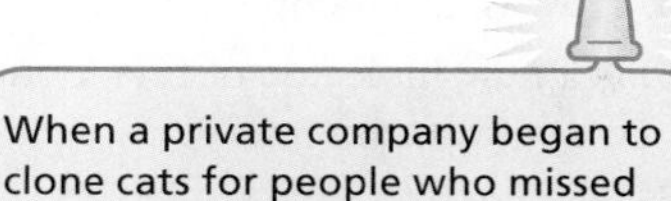

When a private company began to clone cats for people who missed their deceased pets, the price tag was $50,000.

For instructor-assigned homework, go to www.masteringphysics.com

SUMMARY OF TERMS (KNOWLEDGE)

Alleles Versions of a gene.

Codon A sequence of three nucleotides in an mRNA molecule that either stands for an amino acid or ends translation.

Deoxyribonucleic acid (DNA) The cell's genetic material: a double-stranded molecule, consisting of sugar–phosphate backbones attached by pairs of matched nitrogenous bases, in the form of a double helix.

Diploid Description of a cell that has two of each kind of chromosome.

DNA replication The process through which a DNA molecule is copied in order for cells to divide and reproduce.

Dominant Description of an allele that is expressed in a heterozygote.

Gene A section of DNA that contains the instructions for building a protein.

Genetic mutation A change in the nucleotide sequence of an organism's DNA.

Genome The total genetic content of an organism.

Genotype The genetic makeup of an organism.

Haploid Description of a cell that has one of each kind of chromosome.

Heterozygote An organism that has two different alleles for a given gene.

Homozygote An organism that has two identical alleles for a given gene.

Meiosis A form of cell division in which one diploid cell divides to produce four haploid cells.

Messenger RNA (mRNA) An RNA molecule made during transcription that carries genetic information from DNA to the ribosomes.

Phenotype The traits of an organism.

Recessive Description of an allele that is not expressed in a heterozygote.

Recombination The production of new combinations of genes that differ from combinations found in the parental chromosomes resulting from crossing over during meiosis.

Ribonucleic acid (RNA) A single-stranded molecule consisting of a sugar–phosphate backbone attached to a series of nitrogenous bases.

Transcription The first step in building a protein, in which a molecule of RNA is assembled from information contained in DNA.

Transfer RNA (tRNA) An RNA molecule that transfers an amino acid to a growing protein during translation.

Translation The assembly of a protein based on information contained in an mRNA molecule.

READING CHECK (COMPREHENSION)

16.1 What Is a Gene?

1. What is a gene?
2. Why do proteins, built from instructions contained in an organism's genes, determine many of an organism's traits?

16.2 Chromosomes: Packages of Genetic Information

3. How is DNA packaged into chromosomes?
4. What is the difference between a diploid cell and a haploid cell? What types of cells are haploid?

16.3 DNA Replication

5. How is DNA copied?
6. Is a new molecule of DNA put together using two newly made strands?

16.4 How Proteins Are Built

7. How does RNA differ from DNA?
8. What base-pairing rules are followed in making an RNA transcript from a DNA template?
9. Name two things that happen during RNA processing.
10. What is a codon?
11. Describe the role of tRNA in translation.

16.5 Genetic Mutations

12. What is a point mutation?
13. What is a frameshift mutation, and what is its effect on a protein?

16.6 Meiosis and Genetic Diversity

14. What is crossing over? Why is crossing over important to genetic diversity?
15. What are the products of meiosis?

16.7 Mendelian Genetics

16. When Mendel bred pea plants that differed in a single trait, what did he see in the offspring? When Mendel allowed these offspring to self-fertilize, what did he see?
17. What does Mendel's principle of independent assortment tell us?

16.8 More Wrinkles: Beyond Mendelian Genetics

18. What is codominance? Provide an example of codominance.
19. Why does Mendel's law of independent assortment hold for his pea traits but not for all genes?
20. Why are recessive sex-linked traits seen more often in males than in females?

16.9 The Human Genome

21. How many genes does a human have?
22. What is an SNP?

16.10 Cancer: Genes Gone Awry

23. Is cancer usually the result of a single genetic mutation?
24. What is the difference between an oncogene and a tumor-suppressor gene?

16.11 Transgenic Organisms and Cloning

25. What is a transgenic organism?
26. Use examples to describe how transgenic organisms are useful to humans.
27. Explain how a mammal is cloned.

16.12 DNA Technology—What Could Possibly Go Wrong?

28. Describe some of the safety, social, and ethical concerns that relate to genetically modified (GM) crops.

THINK INTEGRATED SCIENCE

16A—The Structure of DNA

29. Why is DNA often described as a double helix?
30. How is DNA like a ladder? What are the "sides" of the ladder, and what are the "rungs"?
31. What are the four nucleotides found in DNA? How do they pair?

16B—How Radioactivity Causes Genetic Mutations

32. Explain how DNA is damaged by exposure to radioactive materials.
33. Which cells are most vulnerable to ionizing radiation?
34. Why is radiation used to treat some forms of cancer?

16C—Environmental Causes of Cancer

35. What are some of the most important cancer-related environmental risk factors?
36. How is the loss of ozone from Earth's atmosphere related to cancer?

THINK AND DO (HANDS-ON APPLICATION)

37. Trace the inheritance of blood type in your immediate family. What is your blood type? What blood-type alleles do you have? (Note: You may be able to say for sure what alleles you have, or you may have to list multiple possibilities.) What are your parents' blood types? What alleles might they have? (Again, you may or may not be able to say for sure. What are the possibilities?) If you are a heterozygote, can you tell which allele you received from your mother and which allele you received from your father?
38. Answer the questions in Exercise 37 for the presence versus absence of dimples and for straight hairline versus widow's peak hairline.

39. How different is mouse growth hormone from human growth hormone? Nucleotide sequences for the two genes are available at the National Center for Biotechnology Information (NCBI) Web site at http://www.ncbi.nlm.nih.gov/. Perform a "Nucleotide" search for accession numbers NM_008117 and NM_000515. Scroll down to look at the mRNA sequences. You may wish to focus on a limited portion of the gene—say, the first 30 nucleotides. Are there differences between the mouse mRNA and the human mRNA? How many differences are there? Now use the genetic code (Table 16.1) to translate the mRNA sequences into amino acids. Do the differences in mRNA sequences result in different amino acid sequences?

THINK AND COMPARE (ANALYSIS)

40. Rank these types of human cells in order of the number of chromosomes they have, from most to least: (a) a sperm cell, (b) a brain cell, (c) a red blood cell
41. Let's assume that human height is a polygenic trait determined by three genes: A, B, and C. For each gene, there is a tall allele and a short allele. We'll use T and S to indicate these. Seven people have the following genotypes: $A_SA_SB_SB_SC_SC_S$, $A_SA_TB_T B_SC_SC_T$, $A_TA_TB_TB_TC_TC_S$, $A_SA_TB_SB_SC_SC_S$, $A_TA_T B_TB_TC_TC_T$, $A_SA_TB_SB_TC_SC_S$, $A_SA_TB_TB_TC_TC_S$. Rank them from tallest to shortest.
42. The human genome includes all the DNA found in all of our chromosomes. Rank these parts of the human genome from highest percentage of our total DNA to lowest percentage of our total DNA: (a) DNA that is identical in all humans, (b) DNA that is not identical in all humans, (c) DNA that contains instructions for making proteins, (d) repeat sequences.

THINK AND SOLVE (MATHEMATICAL APPLICATION)

43. If an organism's diploid cells have 64 chromosomes, how many chromosomes will its haploid cells have?
44. One strand of DNA has the nucleotide sequence CTGAGGTCAGGA. What are the nucleotides on the opposite strand?
45. A section of DNA with the nucleotide sequence CTGAGGTCAGGA is transcribed. What will the nucleotide sequence of the RNA transcript be?
46. Suppose an mRNA molecule with the nucleotide sequence AGUCGUUGGCAGGAAGUA is translated. What sequence of amino acids will be produced?
47. Suppose an mRNA molecule has the nucleotide sequence AGUCGUUGGCAGGAAGUA. What point mutation in this sequence will produce a nonsense mutation?
48. Suppose an mRNA molecule has the nucleotide sequence AGUCGUUGGCAGGAAGUA. Give two examples of point mutations in this sequence that will not affect the protein that the sequence codes for.
49. You have a pea plant with round seeds. Can you say for sure what pea-shape alleles the plant carries? What are the two possibilities? You want to distinguish between these possibilities, so you decide to let the plant self-fertilize. What kind of offspring do you expect in each case?
50. A woman carries an allele for red-green color-blindness on one of her X chromosomes. Her husband is not red-green color-blind. Show that her daughters are not at risk for red-green color-blindness but her sons are at risk.

THINK AND EXPLAIN (SYNTHESIS)

51. Look at your finger. Is it made of diploid cells or haploid cells?
52. What kind of sex chromosomes do you have? Where in your body are sex chromosomes found?
53. Explain why every new DNA molecule has one old strand and one new strand.
54. Can RNA replicate the way DNA replicates? Why or why not?
55. How is transcription similar to DNA replication? How is it different?
56. We compared mRNA processing to editing "*aggfr uidosa to be dfjklsdf or rewerwe not to be*" to obtain "*to be or not to be.*" In this comparison, is "*aggfr*" an exon or an intron? Is "*not*" an exon or an intron?
57. Do all codons code for amino acids? If not, what else can they code for?
58. Examine the genetic code in Table 16.1. Are point mutations in the first, second, and third positions of a codon equally likely to cause a change in the amino acid sequence of a protein? What type of point mutation is least likely to change the amino acid sequence?
59. If there were no such thing as crossing over or recombination, would all the offspring of two parents be identical? Why or why not?
60. Describe three differences between mitosis and meiosis.

61. The figure below shows a set of human chromosomes. What is unusual about this person's genetic makeup? What health issues might this person suffer from? How does meiosis relate to this condition?

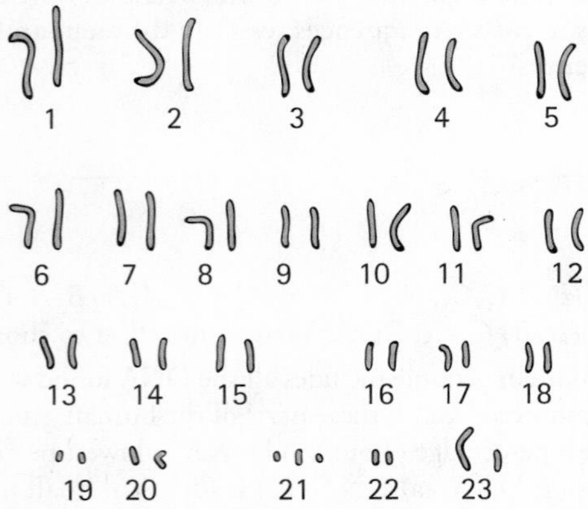

62. Explain how a trait can skip generations.
63. Can you tell what alleles a pea plant with round seeds has? Can you tell what alleles a red snapdragon has? Why are your answers different?
64. If you have dimples, will all your children have dimples? (Remember that "dimples" is a dominant trait.)
65. If you do not have dimples, will all your children lack dimples too? (Remember that "no dimples" is a recessive trait.)
66. Is it possible for two parents with widow's peaks to have a child that has a straight hairline? Is it possible for two parents with straight hairlines to have a child with a widow's peak? (Remember that a widow's peak hairline is dominant, and a straight hairline is recessive.)
67. You are in an accident and you need a blood transfusion, but you have forgotten your blood type. Which type of blood should you be given? Does this explain why people with type O blood are called universal donors?
68. Universal receivers are people who can safely receive any blood type during a blood transfusion. Which blood type do universal receivers have? Explain your answer.
69. Suppose you are studying traits in mice, and you notice that two traits seem to be found together more often than you would expect under Mendel's law of independent assortment. For example, you keep noticing that mice with blue eyes are often deaf too. Can you think of two possible explanations for this?
70. Duchenne's muscular dystrophy is a condition that affects many more males than females. Knowing only this, how do you think the condition is inherited?
71. At their genetic counseling session, a couple learns that one partner is a carrier of cystic fibrosis but the other is not. Are their children at risk for inheriting the condition? Could their children be carriers of the disease?
72. Suppose that you are studying two different mutations in a gene that codes for a protein. In the first, a nonsense mutation occurs near the beginning of the gene. In the second, a nonsense mutation occurs near the end of the gene. Which mutation is more likely to disrupt protein function?
73. Suppose that you are studying two different mutations in a gene that codes for a protein. In one mutation, a single nucleotide is inserted near the beginning of the gene. In the other mutation, three nucleotides are inserted near the beginning of the gene. Which mutation is more likely to disrupt protein function? Why?
74. Although offspring resemble their parents in many ways, it is also possible for a child to differ from both of her parents in some trait. Think of at least two possible explanations for this.
75. If two genes are found on the same chromosome, are they always inherited together? Why or why not?
76. Give an example of a trait that is determined partly by an organism's genes and partly by the organism's environment. Give an example of a trait that is determined primarily by genes—that is, a trait on which the environment has little effect.
77. Cancer is caused by "genes gone awry"—yet cancer is not usually an inherited genetic condition. Explain why.
78. Write a letter to Grandpa telling him about transgenic organisms. Tell him why the development of this technology could result in many practical applications. Also tell him about the potential dangers of this technology. Are you more excited or more worried about the possible consequences of DNA technology?

THINK AND DISCUSS (EVALUATION)

79. What are the three types of RNA, and what is the function of each type?
80. A friend in your study group says that an RNA molecule can sometimes stick to itself, forming a stem and loop that sort of look like a hairpin. "But I thought RNA was single-stranded!" another friend says. You think about it for a moment, mutter "base pairing," and then grab a piece of paper. "I think I know how it works," you say, and you begin to draw. What do you draw?
81. Explain how a mistake during meiosis can result in trisomy 21.
82. Does the process of meiosis explain Mendel's two laws: the principle of segregation and the principle of independent assortment?
83. In the case of linked genes, explain why two genes are more likely to be inherited together the closer together they are on a chromosome.
84. "Listen to this," your friend says, reading from a newspaper article. "'Scientists believe that most genetic mutations have no effect on organisms, a small number have a disadvantageous effect, and an even smaller number have an advantageous effect.'" She looks at you. "Do you agree?"
85. You learned that when red snapdragons are bred with white snapdragons, the offspring are pink snapdragons—an example of incomplete dominance. What happens when you breed a pink snapdragon with another pink snapdragon?

86. If humans had only 2 chromosomes instead of 46 chromosomes (1 homologous pair instead of 23 homologous pairs), would all the children of two parents be identical?
87. Artemisinin is a powerful antimalaria drug. Unfortunately, it is expensive and in short supply because it must be extracted from the leaves of the sweet wormwood tree, a rare plant with a very limited distribution. As a result, many of the populations that suffer most from malaria cannot afford artemisinin. Could DNA technology be used to address this problem?
88. If you were a lawmaker, what constraints would you place on DNA technology? Should there be different rules regarding applying DNA technology to humans and applying it to other species?

READINESS ASSURANCE TEST (RAT)

If you have a good handle on this chapter, if you really do, then you should be able to score 7 out of 10 on this RAT. If you score less than 7, you need to study further before moving on.

Choose the BEST answer to each of the following:

1. If an organism's haploid cells have 10 chromosomes, then its diploid cells have
   (a) 5 chromosomes.
   (b) 10 chromosomes.
   (c) 20 chromosomes.
   (d) none of these
2. One strand of DNA has the sequence AGCCTG. The opposite strand has the sequence
   (a) AGCCTG.
   (b) GATTCA.
   (c) CTAAGT.
   (d) TCGGAC.
3. When DNA is replicated in eukaryotes,
   (a) the new molecule of DNA consists of two newly synthesized strands.
   (b) a molecule of messenger RNA (mRNA) is constructed using DNA as a template.
   (c) each strand of DNA serves as a template for building a new partner strand.
   (d) replication begins at one end of the chromosome and continues to the other end.
4. Which of these statements regarding transcription and translation is false?
   (a) mRNA includes the bases A, C, G, and U.
   (b) The RNA transcript may contain sections that are removed before translation.
   (c) A codon consists of a sequence of three nucleotides.
   (d) The tRNA molecule includes a codon sequence and carries a single amino acid.
5. A mutation that creates a stop codon in the middle of a protein-coding sequence is
   (a) a nonsense mutation.
   (b) a frameshift mutation.
   (c) both a nonsense and a frameshift mutation.
   (d) neither a nonsense nor a frameshift mutation.
6. Crossing over occurs during which phase of meiosis?
   (a) metaphase I
   (b) prophase I
   (c) anaphase II
   (d) prophase II
7. If having dimples is dominant and not having dimples is recessive, then
   (a) all the children of two nondimpled parents will have no dimples.
   (b) some of the children of two dimpled parents will have no dimples.
   (c) all the children of two dimpled parents will have dimples.
   (d) some of the children of two nondimpled parents will have dimples.
8. The sickle-cell allele for hemoglobin, which affects both red blood cell shape and resistance to malaria, provides an example of
   (a) codominance.
   (b) pleiotropy.
   (c) incomplete dominance.
   (d) a polygenic trait.
9. Which of these statements about the human genome is true?
   (a) About 50% of nucleotides are identical in all humans.
   (b) Humans have about 100,000 genes.
   (c) About 90% of the human genome carries instructions for making proteins.
   (d) SNPs identify locations in the human genome where the nucleotide sequence differs among individuals.
10. Cancer
   (a) is usually caused by a single mutation in a key gene.
   (b) is more likely to develop in younger people or people who have been exposed to mutagens (mutation-causing agents).
   (c) requires a mutation in the *BRCA* genes.
   (d) is primarily an environmental, and not an inherited, disease.

Answers to RAT

1. c, 2. d, 3. c, 4. d, 5. a, 6. b, 7. a, 8. b, 9. d, 10. d

CHAPTER 17
The Evolution of Life

MARINE IGUANAS swim through seawater with their long, flattened tails. Flies taste food with the hairs on their feet. Bats catch insects in midair. Cactuses grow sharp spines that protect them from animals. These adaptations, and the countless other ways in which organisms are structured to survive and reproduce, make up the incredible story of evolution. How do living things change over time in response to their environments? After all, a giraffe can't grow a long neck just because it *wants* to. So, how do adaptations (such as a giraffe's long neck) actually come about? Does the same process explain how new types of living things—new species—originate? Also, if all organisms today evolved from earlier organisms, then how did life get started in the first place? Read on to discover these secrets of life.

17.1 The Origin of Life

LEARNING OBJECTIVE
Explain how scientists think life originated, and describe the evidence that supports these ideas.

EXPLAIN THIS Why do scientists think the first living things had genes made of RNA rather than DNA?

How did life originate? For thousands of years, the answer to this question was thought to be *spontaneous generation*, the sudden emergence of living organisms from nonliving materials. For "simple" life forms, spontaneous generation was believed to occur regularly, and evidence for it was everywhere—frogs leaped out of the mud when rain fell, mice appeared in grain stores, maggots squirmed suddenly in rotting meat. More complex organisms such as walruses or human beings were thought to have sprung into existence through spontaneous generation as well, though not as often.

The idea that larger organisms appeared through spontaneous generation lost favor in the 1600s, when experiments with maggots—lowly creatures universally considered to originate that way—showed that they did not appear when rotting meat was kept isolated from flies. However, many people saw spontaneous generation at work in the way microscopic organisms appeared in huge numbers, as if from nowhere, in places like meat broths. It wasn't until 1862, with the experiments of Louis Pasteur, that spontaneous generation was dealt a fatal blow. Pasteur designed a flask that kept out dust and other airborne particles, filled the flask with sterile meat broth, let it sit, and waited for life to emerge (Figure 17.1). It never did, and Pasteur concluded that life did not arise from nonlife.

FIGURE 17.1
Louis Pasteur demonstrated that life did not arise from nonlife.

UNIFYING CONCEPT
- ***The Scientific Method***
Section 1.3

So, how did life originate? We know from fossils that life has existed for at least 3.5 billion years. This means that the Earth on which life evolved was very different from the Earth of today. It contained vast, lifeless oceans, violent volcanoes, and a turbulent atmosphere filled with lightning storms (see Figure 23.40). The atmosphere of the early Earth included no oxygen, which, we will see, was produced by the activity of living things. Although this environment seems a hostile place for life today, it may have been appropriate for producing the first life. Why do we think that the early Earth environment could produce life from nonliving materials? A famous experiment suggests exactly this.

In 1953, Stanley Miller and Harold Urey built a model of the early Earth in a chemistry lab (Figure 17.2). A flask containing a mixture of simple compounds—including water vapor, ammonia, methane, and hydrogen gas—simulated Earth's early atmosphere. Liquid water was added to represent Earth's oceans. Electric sparks sent through the gases simulated lightning. When this model of early Earth was assembled, an amazing thing happened. Many complex organic molecules were formed, including amino acids, the building blocks of proteins. Not only had these molecules formed quickly, they formed in huge numbers. Further experiments showed that all the important organic molecules that make up life—not just amino acids but also sugars, lipids, and the nitrogenous bases found in RNA and DNA—can be generated in a similar way.

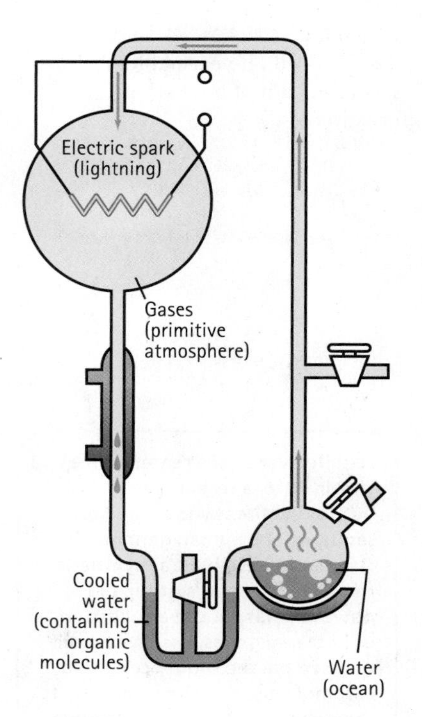

FIGURE 17.2
Stanley Miller and Harold Urey built a model of the early Earth and showed that complex organic molecules could be formed during lightning storms.

However, some scientists today question the importance of the Miller–Urey experiment. They think that Earth's early atmosphere was actually quite different from the model atmosphere Miller and Urey used, and that organic molecules may not have been so easy to generate. They have proposed two other hypotheses for how Earth got its first organic building blocks. One

UNIFYING CONCEPT

- *The Scientific Method*
Section 1.3

hypothesis is that organic molecules were brought to Earth by meteorites. Earth was steadily bombarded by meteorites during its early history, and some of the meteorites recovered here do in fact contain a wide variety of complex organic molecules, presumably formed in outer space. For example, a meteorite found in Australia in 1969 contained nearly one hundred different amino acids. A second hypothesis is that large numbers of organic molecules were formed in deep-sea environments on Earth, similar to the hydrothermal vent habitats of today (see Chapter 22).

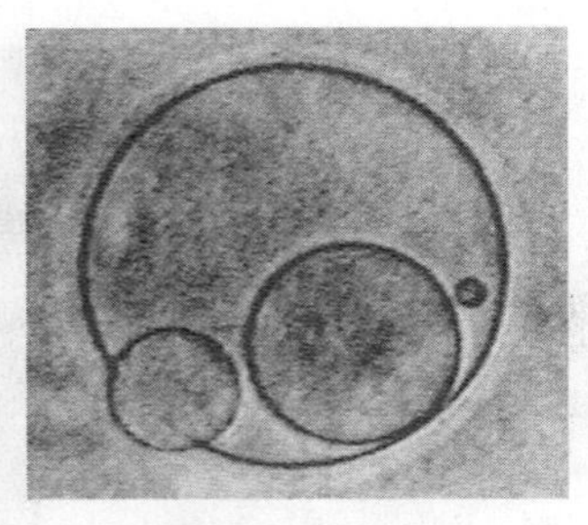

FIGURE 17.3
Liposomes show some cell-like behaviors. Here, a large liposome produces smaller "offspring."

The next question is how these many separate organic molecules advance to become living cells. Scientists do not know the entire story, but they have discovered some clues. For example, when certain lipids are added to water, they spontaneously form tiny hollow spheres called liposomes. Liposomes have double membranes similar to cell membranes. Although they are not alive, liposomes sometimes behave like living cells—they grow, shrink, and divide (Figure 17.3). Liposomes also run chemical reactions inside their membranes and control what molecules move into and out of them, two key features of living cells.

A lipid sphere with captured nucleic acids—the ultimate ancestor of all life today?

Some liposomes may have eventually captured nucleic acids—that is, primitive genes. These early genes were probably made of RNA, not DNA. This is because, even in the absence of cells and enzymes, short strands of RNA can spontaneously assemble from individual nucleotides and even reproduce themselves. With a few more changes, RNA-containing liposomes may have become the very first cells—the first organisms on Earth. However it occurred, the transition to living cells was complete by 3.5 billion years ago, the time of the earliest known fossil organisms (Figure 17.4).

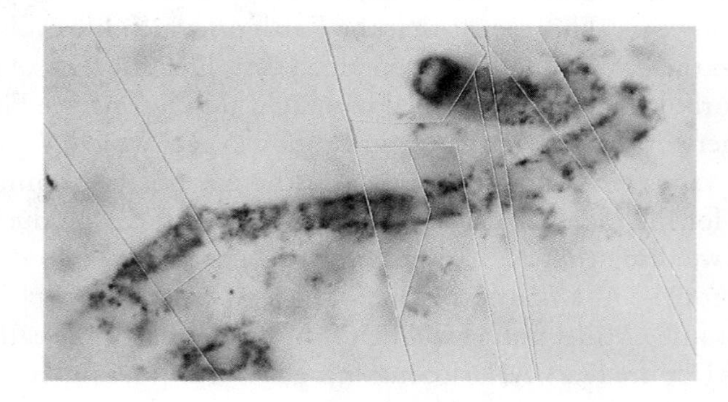

FIGURE 17.4
This 3.5-billion-year-old fossil shows a chain of microscopic prokaryotes.

Has life ever been created in the lab? In 2010, a research team led by Craig Venter, who helped to sequence the human genome, claimed to have built a "synthetic cell." What Venter's group did, however, was choose a collection of bacterial genes, use a yeast cell to assemble the sequences into a chromosome, and then transplant the chromosome into a bacterium. It was not a case of creating a new life form or of creating a living organism from scratch.

CHECK YOURSELF

Miller and Urey found that organic molecules are easily formed in large quantities from non-organic materials. But Pasteur had already shown that life does not come from nonlife. Why aren't these results contradictory?

CHECK YOUR ANSWER

Pasteur's experiment took place under conditions on Earth today, whereas life originated on a very different, younger Earth. Miller and Urey modeled conditions that may have been present on this early Earth. Also, it is a long way from organic molecules of the sort that Miller and Urey obtained to the microscopic organisms Pasteur was looking for.

Integrated Science 17A
ASTRONOMY

Did Life on Earth Originate on Mars?

LEARNING OBJECTIVE
Describe the two sides of the debate over whether a meteorite from Mars contains fossil bacteria.

EXPLAIN THIS If Mars once had life, how could that life have moved from Mars to Earth?

Is that bacteria?! In 1996, National Aeronautics and Space Administration (NASA) scientists found what looked like fossils of tiny bacteria in a Martian meteorite (Figure 17.5). The potential fossils were found very close to complex organic molecules and carbonate minerals that, on Earth, are associated with living organisms. The fossil-like structures are tube-shaped and measure 20 to 100 nanometers across, or less than 1/100th the width of a human hair. The meteorite is about 3.6 billion years old, so it dates from a time when Mars was a much warmer and wetter planet.

This discovery fueled tremendous speculation about whether life on Earth could possibly have originated on Mars. Some scientists suggested that life may have found its way to Earth in Martian dust that was set adrift in space when a comet collided with Mars. Skeptics were quick to point out, however, that the proposed fossils are much smaller than the tiniest bacteria on Earth and that they are likely to be too small to contain all the DNA, proteins, and other molecules a bacterium needs to function. Since the original report, no further evidence has been uncovered. However, ongoing NASA missions to Mars continue to collect data from the Red Planet, so eventually the mystery may be solved.

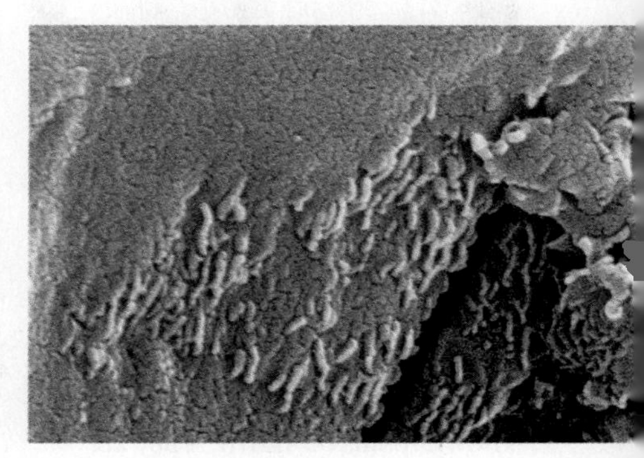

FIGURE 17.5
A startling possibility, or just science fiction? Could the tubelike structures on this Martian meteorite be fossilized bacteria?

CHECK YOURSELF
1. **What evidence is there that life on Earth could have originated on Mars?**
2. **What is the primary objection to this hypothesis?**

CHECK YOUR ANSWERS
1. A meteorite from Mars contains structures that could be the fossils of very small bacteria-like creatures. That these potential fossils were found very close to complex organic molecules and carbonate minerals associated with living organisms on Earth is particularly intriguing.
2. The primary objection is the supposed fossils' small size—they are smaller than any bacteria found on Earth and perhaps too small to contain all the molecules necessary for a living organism to function.

17.2 Early Life on Earth

LEARNING OBJECTIVE
Describe some of the key events in the early history of life on Earth.

EXPLAIN THIS What role did prokaryotes play in the origin of the eukaryotes?

We now know how life on Earth may have originated. In this section, we will look at some key events in the early history of life on Earth.

The Origin of Autotrophs

The earliest living organisms were marine prokaryotes living in a world with no free oxygen. They were **heterotrophs**, organisms that obtain energy and organic molecules from outside sources, as humans and other animals do today. Earth's early heterotrophs found a ready supply of food in the organic molecules that had accumulated in the oceans. As the number of heterotrophs increased through reproduction, however, the supply of organic molecules dwindled.

At some point, **autotrophs**, organisms able to convert inorganic molecules into food and organic molecules, evolved. The origin of autotrophic organisms was a crucial event in Earth history—without autotrophs, heterotrophs would have eaten through their food supply and died out. Plants are present-day examples of autotrophs: Plants use light energy from the Sun, water, and carbon dioxide to build organic molecules during photosynthesis (see Chapter 15). Some of Earth's early autotrophs also used energy from the Sun to build organic molecules. Others were *chemoautotrophs* that used energy from certain inorganic chemicals. Today, the large majority of autotrophs photosynthesize. However, some living prokaryotes are chemoautotrophs (see Chapter 18).

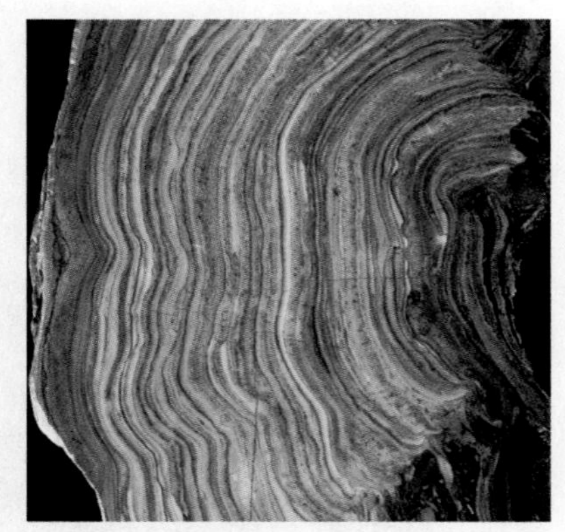

FIGURE 17.6
Stromatolites are among the oldest known fossils on Earth. They are formed by mats of photosynthetic cyanobacteria, the prokaryotes that changed the history of life on Earth by creating an atmosphere rich in oxygen.

The Oxygenation of the Atmosphere

Oxygen is essential to life as we know it. Specifically, oxygen is essential for cellular respiration (see Chapter 15), a process used by most living things today to obtain energy. As we saw, though, Earth's early atmosphere contained no oxygen. This changed with the rise of the cyanobacteria, a group of photosynthetic bacteria, about 2.7 billion years ago. Cyanobacteria release oxygen as a by-product of photosynthesis, and it was their incredible success that first introduced oxygen into Earth's atmosphere. Evidence of an oxygenated atmosphere comes from the presence of banded iron formations in old sedimentary rocks. These formations are produced when atmospheric oxygen combines with iron dissolved in Earth's oceans. Fossilized cyanobacteria, called stromatolites, are shown in Figure 17.6.

Oxygen is essential to life as we know it in more ways than one. Atmospheric ozone, another form of oxygen, shields the Earth from dangerous mutation-causing ultraviolet radiation. Without this protective ozone layer, life might never have been able to move onto land.

The First Eukaryotes

The living organisms we are most familiar with—animals, plants, and fungi—are all eukaryotes. Unlike the prokaryotes, which are at least 3.5 billion years old, eukaryotes first appeared on Earth only about 2 billion years ago.

Eukaryotes differ from prokaryotes in many ways. Most important, eukaryotic cells contain a nucleus and many organelles (see Chapter 15). The nucleus and most eukaryotic organelles probably originated from inward foldings of the cell membrane. Mitochondria and chloroplasts, however, appear to have a different origin. (Remember that mitochondria, which are found in most eukaryotic cells, function in cellular respiration. Chloroplasts are found in plant cells and are responsible for photosynthesis.) Scientists believe that mitochondria and chloroplasts evolved from prokaryotes that started to live inside the earliest eukaryotic cells (Figure 17.7). This *endosymbiotic theory* (*endo* means "in" and *symbiotic* means "to live with") is supported by several observations. First, mitochondria and chloroplasts have their own membranes and their own DNA. Furthermore, this DNA is in the form of a circular chromosome, just like prokaryotic DNA. Finally, both mitochondria and chloroplasts make their own proteins, using ribosomes that resemble those of prokaryotes. So, which prokaryotes did mitochondria and chloroplasts evolve from? By studying their

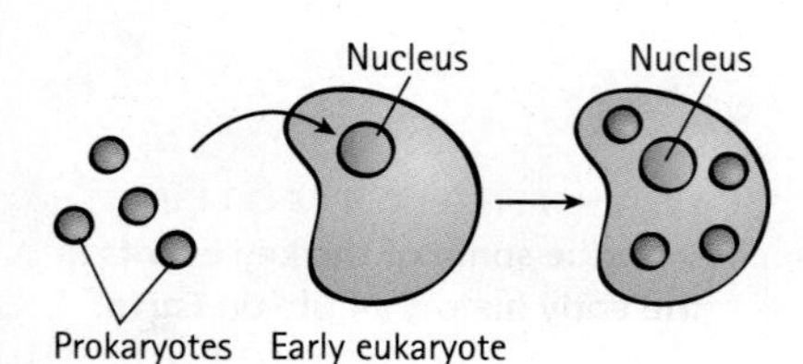

FIGURE 17.7
The mitochondria and chloroplasts in eukaryotic cells evolved from prokaryotes that started living inside early eukaryotes.

structures, scientists have concluded that mitochondria are most likely descended from a group of oxygen-breathing bacteria and that chloroplasts most likely originated from photosynthesizing cyanobacteria. This makes sense given the functions of mitochondria and chloroplasts today.

17.3 Charles Darwin and *The Origin of Species*

LEARNING OBJECTIVE
Describe some of the influences and events that brought Darwin to his theory of evolution through natural selection.

EXPLAIN THIS How did the Galápagos finches contribute to Darwin's ideas about evolution?

How has life on Earth changed over time? For example, how did we get from tiny, primitive cells to humans, hippos, redwoods, and all the amazing diversity of life on Earth today?

For thousands of years, people believed that life on Earth did not change. They believed that Earth had always had the same species, and always would. Then fossils were discovered in Earth's rocks, and people began to wonder. Fossils suggested that the kinds of species living on Earth changed over time—old species disappeared, and new species appeared. Also interesting was that fossil organisms sometimes showed a distinct resemblance to modern species (Figure 17.8). Could some fossils actually be the ancestors of modern species?

French naturalist Jean-Baptiste Lamarck (1744–1829) was one of the first to argue that this was the case. Lamarck believed that modern species were descended from ancestors that had evolved—changed over time—to become better adapted to the environments they lived in. According to Lamarck, organisms acquired new characteristics during their lifetimes and then passed these characteristics to their offspring. For example, ancestral giraffes stretched their necks to grab the high leaves on a tree, and their necks became longer. They then passed these longer necks to their offspring. The offspring reached for even higher leaves, stretching their necks even further, and so on (Figure 17.9a). Lamarck's theory for how change occurs, called the *inheritance of acquired characteristics*, proved to be incorrect: Organisms cannot pass characteristics acquired during their lifetimes to their offspring because these acquired characteristics are not genetic. However, Lamarck's fierce support for the idea that organisms evolve set the stage for Charles Darwin.

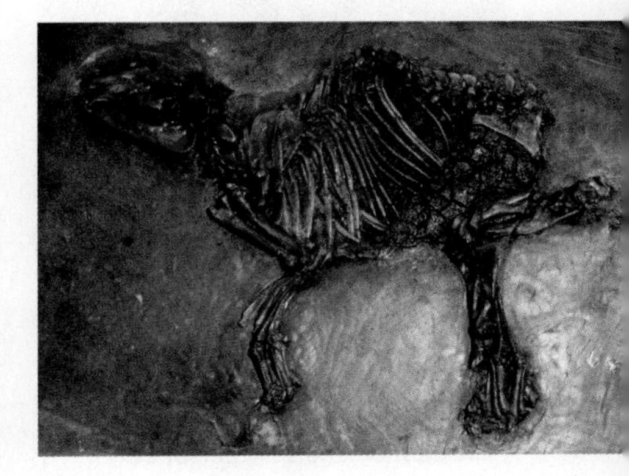

FIGURE 17.8
Could fossils be the ancestors of modern species? This fossil, found in Germany, is about 50 million years old. It has a clear resemblance to a horse, yet is only the size of a fox.

UNIFYING CONCEPT
- ***The Scientific Method***
Section 1.3

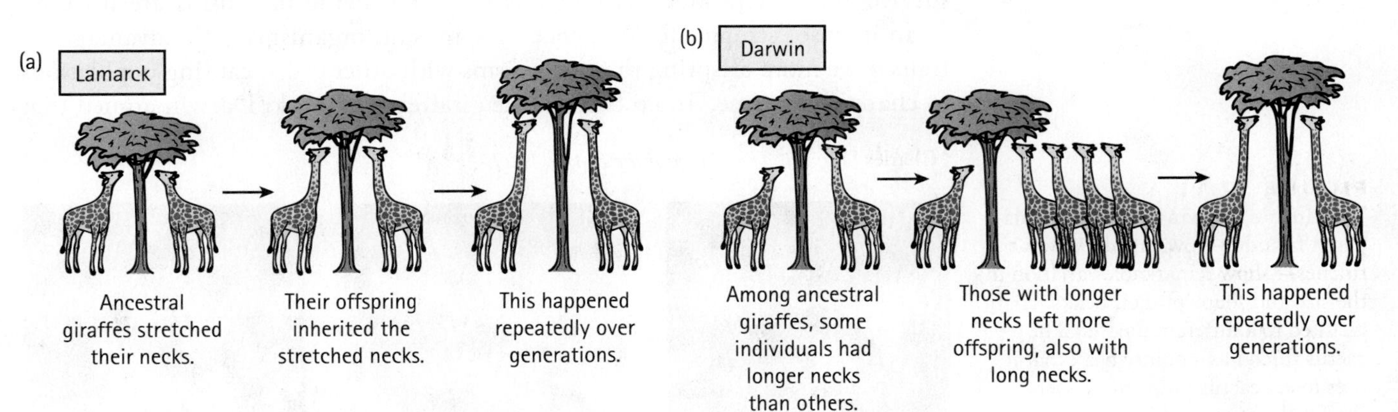
(a)
Lamarck
Ancestral giraffes stretched their necks.
Their offspring inherited the stretched necks.
This happened repeatedly over generations.
(b)
Darwin
Among ancestral giraffes, some individuals had longer necks than others.
Those with longer necks left more offspring, also with long necks.
This happened repeatedly over generations.

FIGURE 17.9
(a) Lamarck believed that organisms acquired new characteristics during their lifetimes and passed these characteristics to their offspring. (b) In his theory of evolution by natural selection, Darwin argued that organisms with advantageous traits left more offspring than organisms with other traits. As a result, advantageous traits became more common in a population.

FIGURE 17.10
Charles Darwin developed the theory of evolution by natural selection.

English naturalist Charles Darwin (1809–1882), shown in Figure 17.10, set forth the theory of evolution in his book *The Origin of Species by Means of Natural Selection*, published in 1859. Darwin proposed that **evolution**—inherited changes in populations of organisms over time—had produced all the living forms on Earth.

Darwin's theory of evolution grew out of the observations he made as the official naturalist aboard the H.M.S. *Beagle*, which sailed around South America from 1831 to 1836. During these years, Darwin studied South American species, collecting large numbers of plants, animals, and fossils. Darwin became increasingly intrigued by the question of how species got to be the way they were. He was particularly struck by the living things he encountered on the Galápagos Islands, 950 kilometers from the South American continent. Darwin took particular note of the 13 species of Galápagos finches—now known as Darwin's finches. Darwin's finches showed remarkable variation in the size and shape of their beaks, with each beak being suited to, and used for, a different diet (Figure 17.11). How had the beaks of these finches come to differ in this way? Darwin wrote, "Seeing this gradation and diversity of structure in one small, intimately related group of birds, one might really fancy that from an original paucity of birds in this archipelago, one species had been taken and modified for different ends."*

There's an expression: Genius is 1% inspiration and 99% perspiration. Darwin's genius reflects a lot of perspiration. While on the *Beagle*, Darwin collected 1529 alcohol-preserved specimens and 3907 skins, bones, and dried specimens. He also took 2000 pages of notes on plants, animals, and geology. It's no wonder that when he wrote down his theory, he was able to support it with a wide variety of well-considered examples.

Darwin was also inspired by the work of two of his contemporaries, Charles Lyell and Thomas Malthus. Lyell, a geologist, argued that Earth's geological features were created not by major catastrophic events—the favored theory of the time—but by gradual processes that produced their effects over long time periods. For example, the formation of a deep canyon did not require a cataclysmic flood, but could result from a river's slow erosion of rock over millennia. Darwin realized this could be true for organisms as well: The accumulation of gradual changes over long periods could produce all the diversity of living organisms as well as all their remarkable features.

The economist Thomas Malthus was a second important influence for Darwin, and the one who led Darwin to his great idea on the cause of evolutionary change. Malthus observed that human populations grow much faster than available food supplies, and he concluded, with despair, that famine was an inevitable feature of human existence. Darwin applied Malthus's idea to the natural world and argued that, because there are not enough resources for all organisms to survive and to reproduce as much as they can, living organisms are involved in an intense "struggle for existence." As a result, organisms with advantageous traits leave more offspring than organisms with other traits, causing populations to change over time. To go back to the giraffe's long neck: Darwin argued that

*Charles Darwin, *The Voyage of the Beagle*, 1909.

(a)

(b)

(c)

FIGURE 17.11
The finches Darwin saw on the Galápagos Islands—now called Darwin's finches—show remarkable variation in the size and shape of their beaks. Each is suited to a different diet. (a) The cactus finch has a pointy beak that it uses to eat cactus pulp and flowers. (b) The large ground finch has a blunt, powerful beak that it uses to crack seeds. (c) The woodpecker finch has a woodpecker-like beak that it uses to drill holes in wood. It then uses a cactus spine to pry out insects.

ancestral giraffes with longer necks were better at reaching the high leaves on trees. Because longer-necked giraffes got more food, they were able to survive and leave more offspring than ancestral giraffes with shorter necks. This happened repeatedly over generations. Over time, there were more longer-necked giraffes in the giraffe population (Figure 17.9b). This process, which Darwin called **natural selection**, is the major driving force behind evolution.

MasteringPhysics®
TUTORIAL: Darwin and the Galapagos Islands
VIDEO: Galapagos Islands Overview
VIDEO: Galapagos Marine Iguana

CHECK YOURSELF

1. **If Lamarck had been correct and evolutionary change occurred through the inheritance of acquired characteristics, what trait might a bodybuilder pass to his offspring?**
2. **Many animals that live in the Arctic, such as Arctic hares, have white fur. How could natural selection explain the evolution of their white fur color?**

CHECK YOUR ANSWERS

1. If Lamarck were correct, the bodybuilder's children would inherit the increased muscle mass that the bodybuilder had acquired over a lifetime of weightlifting. Because Lamarck's theory turned out to be incorrect, however, the children will have to do their own bodybuilding.
2. Animals that were harder to see in their snowy environments had an advantageous trait—predators were less likely to spot them. Arctic hares with whiter fur were more likely to survive to adulthood, reproduce, and leave offspring. These offspring would also have inherited whiter fur. As a result, whiter fur became more common in the Arctic hare population. Over many generations, natural selection produced a white coat that matches the Arctic snow.

17.4 How Natural Selection Works

LEARNING OBJECTIVE
Explain how natural selection results in populations becoming adapted to their environments.

EXPLAIN THIS What does it mean to say that one rabbit has greater fitness than another?

Rabbits were introduced into Australia in 1859, when a man named Thomas Austin released 24 individuals onto his property in the southeastern part of the continent. The rabbits quickly became pests, devastating farmlands and natural habitats (Figure 17.12). Breeding "like rabbits," they spread across the continent in such large numbers that they were described as a "gray blanket" that covered the land. Many attempts were made to control the rabbit population, including the construction of an 1822-kilometer-long "rabbit-proof" fence—still the longest fence in the world. Unfortunately, by the time the fence was completed in 1907, the rabbits had already passed through. (The fence wouldn't have worked anyway—even after it was completed, rabbits would pile up so thickly behind it that some were eventually able to walk right over their companions' backs to the other side.)

MasteringPhysics®
TUTORIAL: Causes of Microevolution

In the early 1950s, the government decided to try to control the rabbit population by releasing myxoma virus, a virus deadly to rabbits. Initially, the virus was a wonder, killing more than 99.9% of infected rabbits. Within a few years, however, fewer rabbits were dying. What had happened? Within the original rabbit population, a small number of individuals happened to be resistant to the

(a)

(b)

FIGURE 17.12
(a) Rabbits introduced into Australia caused widespread destruction, including here on Phillip Island. (b) This photo shows the same area after rabbits were eradicated. The vegetation has grown back.

myxoma virus. These resistant individuals survived the disease and reproduced, producing more disease-resistant offspring (Figure 17.13). Over time, the number of disease-resistant rabbits increased, and the virus became less and less effective. The rabbit population had evolved resistance to the myxoma virus through natural selection.

Natural selection occurs when organisms with advantageous traits leave more offspring than organisms with other traits, causing populations to change over time. Let's look more carefully at the process of natural selection.

1. *Variation.* In any population of organisms, individuals have many traits that show **variation**—that is, they vary from individual to individual. In humans, some variable traits are height, hair color, hairstyle, foot size, and blood type.
2. *Heritability.* Many traits are determined at least partly by genes and so are **heritable**—that is, they are passed from parents to offspring. Which of the human traits listed above are heritable? All of them are heritable except hairstyle. Hairstyle is not heritable because it is not genetically determined.
3. *Natural selection.* Some variable heritable traits are advantageous. The organisms that possess these advantageous traits are able to leave more offspring than organisms without the advantageous traits. The **fitness** of an organism describes the number of offspring it leaves over its lifetime compared to other individuals in the population. An organism that leaves more offspring than other individuals in the population is said to have greater fitness.
4. *Adaptation.* Because organisms with advantageous traits leave more offspring, advantageous traits are "selected for" and become more common in a population. What is the result? The population evolves to become better adapted to its environment.

Figure 17.14 summarizes the process of natural selection. Note that, although natural selection acts on individuals within a population, allowing some individuals to leave more offspring than others, it is the population as a whole that evolves and becomes adapted to its environment.

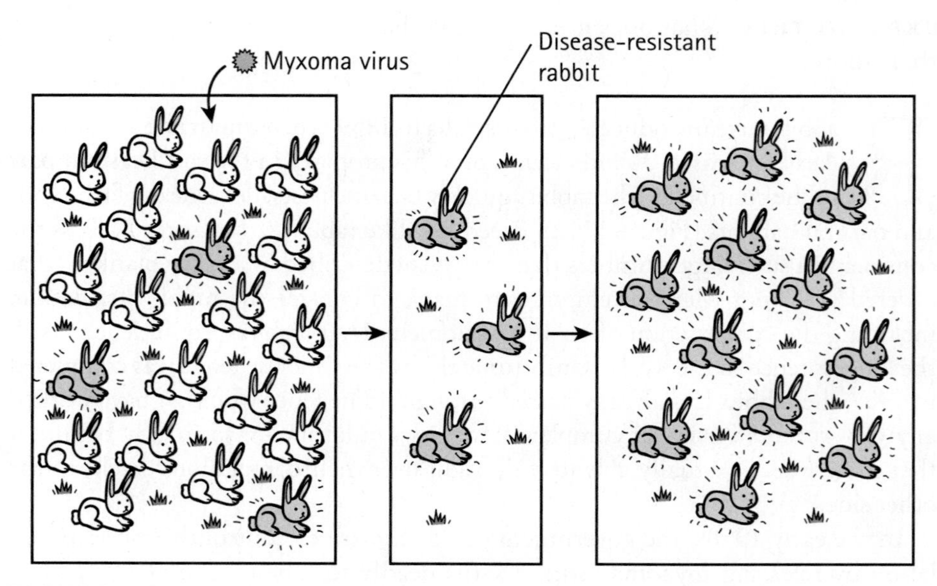

FIGURE 17.13
At first, the myxoma virus killed 99.9% of infected rabbits. However, a small number of naturally disease-resistant rabbits (blue) survived and reproduced, passing their myxoma-resistant genes to their offspring. The population became more resistant, and the virus became less effective.

(1) VARIATION

Organisms have lots of traits, many of which show variation.

(2) HERITABILITY

Some traits are heritable. They are determined by genes and so are passed from parents to offspring.

(3) NATURAL SELECTION

Variation in heritable traits can result in some organisms leaving more offspring than others. This is called natural selection.

(4) ADAPTATION

Natural selection causes advantageous traits to become more common in a population. In this way, entire populations become adapted to their environments.

FIGURE 17.14
How natural selection works.

CHECK YOURSELF

1. **(a) Which of these traits are variable in cats: fur color, tail length, number of eyes? (b) Which of the traits are heritable?**
2. **The cheetah is the fastest land animal on Earth. It can run 112 kilometers/hour (70 miles/hour)! Cheetahs prey on Thomson's gazelles that can run almost as fast, 80 kilometers/hour (50 miles/hour). How might natural selection have produced the cheetah's fast running speed?**

CHECK YOUR ANSWERS

1. (a) Fur color varies among cats—there are tabby cats, black cats, gray cats, and so on. Tail length also varies—not all cats' tails are exactly the same length. But there is no variation in the number of eyes—all cats have two eyes. (b) All three traits are heritable because all are determined genetically.
2. Faster cheetahs were better at catching Thomson's gazelles. Being better at catching food made faster cheetahs better at surviving and reproducing. As a result, faster cheetahs left more offspring, which were also fast. This resulted in a cheetah population with faster individuals. Over many generations, natural selection produced the remarkably fast cheetah we know today.

HISTORY OF SCIENCE

The Peppered Moth

During the Industrial Revolution, coal was the primary fuel in England. Burning coal slathered dark soot on trees, rocks, and ground. And then a startling thing happened to the moths.

Peppered moths in England had always been light in color, with the scattering of dark peppery flecks that gave them their name. Their coloration made them hard to see in a habitat of lichen-covered trees and rocks. (Lichens are fungi that grow with photosynthetic algae or bacteria; they form crustlike growths on rocks, trees, and other surfaces.) It was believed that this camouflage protected the moths from birds, their main predators.

As the Industrial Revolution progressed, pollution killed the lichens, leaving the trees first bare and then darkened with soot. In 1848, the first dark peppered moth was found in the industrial center of Manchester, England. Dark moths had probably always existed in the population, but they had been extremely rare. Over the next decades, as more coal burned and the environment became increasingly sooty, more and more dark moths were seen. By 1895, 98% of peppered moths in industrialized areas were dark. Then, in the second half of the 20th century, antipollution laws were passed and soot disappeared. Light moths increased in number, and today the dark moths have all but disappeared.

Did natural selection cause the coloration shifts in the peppered moth? Biologists hypothesized that in lichen-covered habitats, natural selection favored light moths because they were better camouflaged. In sooty habitats, natural selection favored dark moths. A series of experiments by Bernard Kettlewell tested this hypothesis. Kettlewell released equal numbers of marked dark and light moths in polluted and unpolluted areas. After a while, he tried to recapture the moths. In polluted areas, Kettlewell recaptured more dark moths than light moths, which suggested that dark moths had survived better. The opposite was true in unpolluted habitats, where he recaptured more light moths. Kettlewell also placed moths on tree trunks and filmed birds eating the moths. He found that birds ate what they could see: Birds ate more light moths in polluted habitats and more dark moths in unpolluted habitats.

Can you find the moths? Light peppered moths are well camouflaged on lichen-covered trees.

Kettlewell's work became a classic example of natural selection. Eventually, however, certain aspects of his experiments were challenged. For example, moth experts pointed out that peppered moths don't usually sit on tree trunks, where Kettlewell had placed them. Instead, they usually rest on the undersides of branches. In addition, Kettlewell released the normally nocturnal moths during the daytime. This may have affected the moths' ability to find resting spots. Finally, Kettlewell used a mix of lab-raised and wild-caught moths, which could differ in their behavior. These doubts led Michael Majerus of Cambridge University to conduct a new set of experiments between 2001 and 2007. Majerus's work confirmed that bird predation was the key factor affecting the relative numbers of light and dark peppered moths. It is also interesting that a shift from light to dark forms in polluted areas (and back again, as pollution is cleaned up) has been reported in more than 70 other moth species in England and the United States alone.

LEARNING OBJECTIVE
Use examples to describe different kinds of adaptations found in living organisms.

17.5 Adaptation

EXPLAIN THIS Why do some birds have bright feathers despite the fact that the vivid colors make them more visible to predators?

Natural selection leads to the evolution of **adaptations**—traits that make organisms well suited to living and reproducing in their environments. The Check Yourself question in the preceding section gave an example of an adaptation—the cheetah's speed. The cheetah's speed helps it catch the food it needs to survive and reproduce.

Adaptations can relate to various aspects of an organism's life. Some adaptations help organisms survive. Survival is, after all, usually an important first step in successful reproduction. Survival requires that organisms be able to acquire food and other necessary resources. It also requires that organisms avoid becoming food for someone else (Figure 17.15). Anti-predator adaptations include camouflage, toxicity, or just the ability to hide or run away.

(a)

(b)

FIGURE 17.15
Almost every organism has adaptations that help prevent it from becoming food for someone else. (a) The spines of this cactus prevent most animals from eating it. (b) When threatened, this octopus releases a cloud of dark ink that may confuse a predator long enough for the octopus to escape.

The peacock may be the organism with the most famous adaptation for attracting mates. The male peacock's great fan of colorful tail feathers not only is admired by people but, more important, impresses peahens.

And speaking of bright colors—the bold colors of organisms such as wasps, coral snakes, and poison dart frogs evolved to warn potential predators that they are dangerous.

Other adaptations have evolved to help organisms acquire mates. These include the beautiful feathers of male peacocks and birds of paradise (Figure 17.16a), the sexy "rib-bits" of male frogs, and the enchanting songs of many male birds. Males have evolved these "sexy" traits because females of the species find them attractive. In other species, females don't choose their mates based on attractive traits. Instead, males fight with other males to obtain mates. The adaptations of these males may include large size, great strength, or fighting structures such as antlers (Figure 17.16b). Natural selection that favors individuals best able to acquire mates is also called *sexual selection*.

(a) (b)

FIGURE 17.16
Some adaptations for acquiring mates. (a) The beautiful feathers of this male bird of paradise (shown here displaying his wings) help attract female mates. (b) These male deer are fighting for control of territory as well as mates.

Finally, some adaptations relate to bearing and raising young. Figure 17.17 shows one such adaptation—parental care. Parental care evolved because natural selection favored organisms that were able to help their offspring survive and thrive. Parental care is found in many animals, including humans.

Natural selection has produced remarkable adaptations over time. Nature does not plan ahead—it does not plan to make a falcon or a polar bear. Instead, adaptations are built step by step, through the never-ending selection of the most successful forms.

FIGURE 17.17
Parental care occurs in many species. This male poison dart frog is carrying his tadpoles on his back.

When a male praying mantis (the smaller insect on top) mates with a female, he is in danger of having his head bitten off.

CHECK YOURSELF

Mating is very dangerous for a male praying mantis. Quite often, the female will eat him as he mates with her.

1. **What advantage does the female get from eating the male?**
2. **Would it be more advantageous ("adaptive") for the male not to mate at all?**

CHECK YOUR ANSWERS

1. The female gets nutrients when she eats the male.
2. A male praying mantis that never mates is more likely to survive to old age. But, if he doesn't mate, he won't leave any offspring. Remember

SCIENCE AND SOCIETY

Antibiotic-Resistant Bacteria

A patient is ill with pneumonia and gets a prescription for penicillin. After three days, he feels better and stops taking his pills. A few days later, his symptoms return. He quickly finds his pills and starts taking them again, but this time they have no effect. What happened? This frightening phenomenon is called *antibiotic resistance.* Antibiotic resistance is caused by natural selection: Penicillin killed most of the pneumonia bacteria, but a few penicillin-resistant bacteria survived. These bacteria multiplied, and the patient's infection came back—only this time, the bacteria are resistant to penicillin.

Antibiotics are wonder drugs. When penicillin, the first antibiotic, appeared, it dramatically cut the number of illnesses and deaths resulting from bacterial infections. After only a decade of use, however, the first penicillin-resistant bacterial strains appeared. Since then, antibiotic resistance has spread, with more and more bacterial populations becoming resistant to more and more different antibiotics. Diseases once easy to treat—tuberculosis, pneumonia, even common childhood ailments such as ear infections—are now often resistant to multiple antibiotics. In 2011 the World Health Organization reported that about 440,000 new cases of multi-drug-resistant tuberculosis appear each year, resulting in at least 150,000 deaths.

Some of the most dangerous antibiotic-resistant bacteria are found in hospitals, where the use of many different types of antibiotics allows widely resistant strains to evolve. The Centers for Disease Control reported that in 2005, methicillin-resistant *Staphylococcus aureus* (MRSA), a bacterial strain that is resistant to most of the antibiotics currently available, was responsible for more than 94,000 life-threatening infections and 18,650 deaths in the United States alone. And, some MRSA strains are beginning to show resistance to the antibiotic vancomycin, often considered "the drug of last resort." Another worrisome development is the emergence of MRSA in the wider community. Community-based MRSA infections usually start as skin infections and spread through skin-to-skin contact. Some of these cases turn into "flesh-eating" disease, and others are halted only by drastic measures such as amputation. Environments with a higher risk for community-based MRSA infections include athletic facilities, dorms, prisons, and day-care centers. Compared to people whose infections respond to antibiotics, people who have antibiotic-resistant infections require longer hospital stays and are more likely to die from their infections.

All antibiotic use has the potential of contributing to resistance. However, resistance has been greatly accelerated by the overuse of antibiotics. Under pressure from patients, physicians may prescribe antibiotics for illnesses that are not caused by bacteria. (Many common illnesses, such as colds, flus, and most sore throats, are caused by viruses.) These antibiotics select for resistance in the normal (non–disease-causing) bacterial populations in our bodies, making it possible for resistant genes to be transferred to disease-causing bacteria that later invade the body. The fact that patients sometimes stop taking their medications too soon contributes to the problem; this selects for antibiotic-resistant strains without providing the sustained dose that would actually kill all the bacteria. Antibiotics are also used heavily in the livestock industry, where animals are given antibiotics regularly—even when they are healthy—to promote growth. Unfortunately, this practice greatly promotes the evolution of antibiotic resistance. In recent years, reports of food-borne illnesses caused by antibiotic-resistant bacteria have become regular items in the news. For example, in August 2011, an outbreak of antibiotic-resistant salmonella in ground turkey caused at least 79 illnesses and one death.

What can be done about antibiotic resistance? First, humans must learn to use antibiotics wisely, taking them only when they are needed—that is, for bacterial infections—and then taking the entire course of treatment. Second, physicians and veterinarians can promote a socially responsible approach to antibiotics by educating patients and agriculturalists on the proper application of these drugs. Third, antibiotics should not be used to promote growth in livestock. In 2012, steps were finally taken to ban the agricultural use of certain antibiotics. Finally, since many antibiotics are less effective now because of resistance, scientists must search for new antibiotics to take the place of those that no longer do the job.

that adaptations are traits that make organisms good at living *and reproducing* in their environments. It's not enough to survive—you also have to reproduce! This male praying mantis may not have long to live, but at least he has a good chance of leaving offspring.

Integrated Science 17B
PHYSICS

Staying Warm and Keeping Cool

LEARNING OBJECTIVE
Explain how the need to thermoregulate affects the size and shape of animals that live in very cold and very hot habitats.

EXPLAIN THIS Why do Arctic mammals have relatively big bodies, short legs, and small ears?

In this section, we will see how being the right size and shape can be an adaptation. Most mammals maintain a fairly constant body temperature. We humans, for example, have a body temperature that always stays around 37°C (98.6°F). Animals that live in extremely hot or extremely cold habitats need to be able to maintain appropriate body temperatures in those environments—to *thermoregulate*. In deserts, animals have to be able to lose heat to avoid overheating. In cold habitats, animals have to be able to retain heat.

A key factor in thermoregulation is an animal's surface-area-to-volume ratio. (This ratio was also discussed in Chapter 15 in the context of cells.) The heat an animal produces depends on its volume. The heat it loses to its environment depends on its surface area because heat is lost through the body's surface. As a result, animals are better able to *lose* heat if they have a high surface-area-to-volume ratio, and they are better able to *retain* heat if they have a low surface-area-to-volume ratio. This has consequences for both the size and shape of animals that live in extreme habitats.

A large animal tends to have a *lower* surface-area-to-volume ratio because volume increases more quickly than surface area as organisms get bigger (see Chapter 15). For this reason, animals that live in cold habitats are often larger than related species that live in warm habitats. For example, the smallest bear in the world is the sun bear, found in the tropical forests of Southeast Asia (Figure 17.19). Adult sun bears weigh between 27 and 65 kilograms (60 to 140 pounds). The largest bear in the

FIGURE 17.18
The amount of *caramel* on this caramel-covered apple is determined by its surface area. The amount of *apple* in this caramel-covered apple is determined by its volume. The surface-area-to-volume ratio is the amount of caramel divided by the amount of apple. A small caramel-covered apple has a high surface-area-to-volume ratio. A big caramel-covered apple has a low surface-area-to-volume ratio.

Good cooks know that there are a lot more peelings from several small potatoes than from one large potato of the same volume. This is because little things have more skin per volume than big things!

(a)

(b)

FIGURE 17.19
(a) The small sun bear is found in tropical forests in Southeast Asia. A small animal is better able to lose heat because of its high surface-area-to-volume ratio. (b) The polar bear, the largest terrestrial carnivore in the world, is found throughout the Arctic. A large animal is better able to retain heat because of its low surface-area-to-volume ratio.

(a) (b)

FIGURE 17.20
(a) This black-tailed jackrabbit lives in a California desert. Extensive blood vessels in its ears help it dissipate heat. It also has long legs. (b) This arctic hare is a relative of the black-tailed jackrabbit. Its ears are much smaller.

world is the polar bear, which ranges throughout the Arctic. Adult polar bears weigh between 200 and 800 kilograms (440 to 1760 pounds).

Animals adapted to hot versus cold climates also vary in shape. Desert species often have long legs and large ears that increase the surface area available for heat dissipation. Their legs and ears are also covered with extensive blood vessels that carry heat from the core of the body to the skin, where convection, the transfer of heat by moving air, cools the animal. Arctic species often have short appendages and small ears that help conserve heat. An example is the rabbits shown in Figure 17.20.

UNIFYING CONCEPT
- *Convection*
Section 6.9

CHECK YOURSELF
On cold days, people often bundle up babies and small children carefully. Are babies more likely to need the extra bundling than adults? Why or why not?

CHECK YOUR ANSWER
Babies have higher surface-area-to-volume ratios than adult humans because they are smaller. So, yes, they are likely to appreciate the extra bundling.

LEARNING OBJECTIVE
Explain how an understanding of genetics produced insights about the mechanisms of evolution and the origin of genetic diversity.

17.6 Evolution and Genetics

EXPLAIN THIS Can chance cause a population to evolve?

So far, we've seen how natural selection acts on organisms' traits—giraffe neck length, cheetah speed, peppered moth color, and so on. Traits are only part of the story, though, because what gets passed from parents to offspring are not traits, but genes. The incorporation of modern genetics (see Chapter 16) into Darwin's theory of evolution took place in the middle of the twentieth century and produced many new insights about how populations evolve.

The focus on genes led to a description of evolution as *changes in the allele frequencies of genes over time. Allele frequencies* describe how common different alleles are in a population. For example, the peppered moths we discussed earlier have a light allele (*a*) and a dark allele (*A*) for color. A population with many light moths and few dark moths might have allele frequencies of 92% *a* and 8% *A*. As the habitat becomes more polluted, dark moths become more common, and the dark allele increases in frequency. In a polluted area, the allele frequencies might change to 5% *a* and 95% *A*.

We can describe natural selection in terms of allele frequencies as well: (1) There is variation in a gene when multiple alleles for that gene exist within a population. For example, in peppered moths there are two alleles for color, *A* and *a*. (2) A specific allele may give an organism an advantage that allows it to reproduce more than other organisms in the population. In a polluted habitat, for example, the *A* allele is advantageous. (3) As a result, more copies of the advantageous allele are passed to the next generation, and the frequency of the advantageous allele increases in the population. In a polluted habitat, the frequency of the *A* allele increases.

Notice that, although natural selection *affects* genes and allele frequencies, natural selection does not act *directly* on genes. Another way to say this is: Natural selection acts on an organism's phenotype (traits), not on its genotype (genes). To see why, let's go back to the peppered moth. In peppered moths, the dark allele (*A*) is dominant and the light allele (*a*) is recessive. This means that both *AA* moths and *Aa* moths have dark wings (Figure 17.21). Whether a bird is likely to eat the moth depends on the moth's phenotype (whether it is dark or light), not its genotype. A bird is equally likely to eat a dark moth whether it has genotype *AA* or *Aa*.

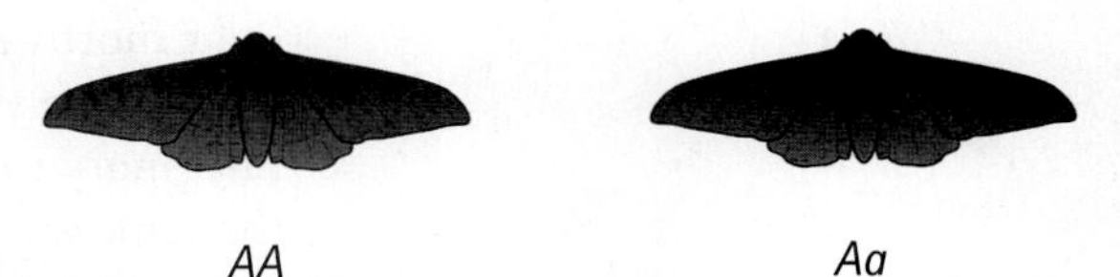

FIGURE 17.21
Natural selection acts on phenotype, not genotype. In the case of these two dark moths, it's the phenotype (dark color) that matters, not the genotype (*AA* versus *Aa*).

Mechanisms of Evolution

Natural selection is the driving force behind evolution, and it causes populations to become adapted to their environments. However, natural selection is not the only mechanism that causes populations to evolve. Populations also change over time because of mutation pressure, genetic drift, and gene flow. In order to understand these processes, let's consider peppered moths again.

Mutation pressure exists if the alleles responsible for peppered moth color are more likely to mutate in one direction than the other. For example, a genetic mutation may be more likely to turn a dark allele into a light allele than vice versa. If so, then over time, the frequency of the light allele will increase.

Genetic drift is the evolution of populations due to chance. Imagine a half-polluted town where light moths and dark moths are equally successful—neither allele is advantageous. Now suppose a sudden storm wiped out part of the town's peppered moth population. It *just might happen*—just by chance—that more dark moths survive. If so, then the frequency of the dark allele will increase. Notice that genetic drift produces evolution (heritable changes in a population over time) but that this evolution is not the result of natural selection—the dark allele was not advantageous. Genetic drift can also occur when, just by chance, more alleles of one type are transmitted to the next generation than alleles of the other type. For example, even if light and dark moths have equal fitness,

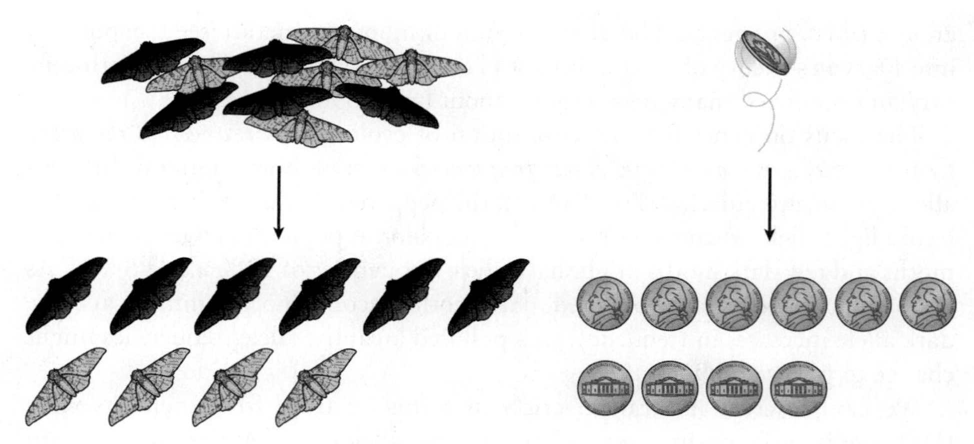

FIGURE 17.22
If light and dark moths have equal fitness, we would expect them to leave the same number of offspring. However, just by chance, one group could leave more offspring than the other, resulting in genetic drift. This is true for the same reason that, even though you expect the same number of heads and tails when you flip a coin 10 times, you might get more heads or more tails.

light moths might just happen to leave more offspring (and therefore more light alleles) one year. Genetic drift works a lot like a coin flip (Figure 17.22). In our imaginary town, light and dark moths are equally likely to survive and reproduce the same way you are equally likely to get heads or tails when you flip a coin. But, if you flip a coin 100 times, you won't always get *exactly* 50 heads and 50 tails. Similarly, of 100 moths born in the next generation, there won't necessarily be *exactly* 50 light ones and 50 dark ones.

Genetic drift is particularly important in small populations because chance is more likely to change allele frequencies significantly in a small population. To see why, consider flipping a coin 10 times (a small population) versus 1000 times (a large population). With 10 flips, it is not at all unlikely that you'll get heads 60% of the time—that is, 6 heads and 4 tails. On the other hand, a similar result with 1000 flips—600 heads and 400 tails—is practically impossible.

Gene flow describes changes in allele frequencies that result from a net movement of alleles into or out of a population. For example, our half-polluted town may be next to a clean woodland that is home to a population of light moths. If a few of these light moths migrate from the woodland into town, the frequency of the light allele in the town will increase.

Where Variation Comes From

Natural selection cannot happen without variation. Furthermore, populations with more variation have a better chance of adapting to a changing environment. This is because with more variation, it is more likely that somewhere in the population there are alleles that will allow some individuals to survive under the new conditions. For instance, what would have happened to peppered moths during the Industrial Revolution if all the moths had been light and none were dark? In polluted areas, populations with only light moths might have died out. (In Chapter 21, we'll see that having many kinds of *species* in a habitat also increases the chance that at least some organisms will survive major changes in the environment.)

But where does variation come from? An understanding of genetics enabled biologists to answer this question. Genetic mutations (see Chapter 16) constantly create new variations within populations. For example, when a genetic mutation changes the amino acids in a protein, it may produce a new allele for a given gene. Sexual reproduction also contributes to variation by bringing together alleles for different traits in new combinations.

CHECK YOURSELF

1. **In a peppered moth population, genetic drift causes the frequency of the dark allele to increase one year. Will genetic drift have the same effect the following year?**
2. **Twenty dark moths migrate into a peppered moth population, and 30 light moths migrate out. What effect does gene flow have on this population?**

CHECK YOUR ANSWERS

1. Genetic drift is evolution due to chance, and there is no guarantee that chance will have the same effect the following year. The situation is the same as flipping coins—when you flip a coin 100 times and then do it again, you may get more heads the first time and more tails the second time.
2. Gene flow causes the frequency of the dark allele to increase in this population.

17.7 How Species Form

LEARNING OBJECTIVE
Explain how new species arise.

EXPLAIN THIS Why is speciation often associated with the introduction of a geographic barrier?

We have seen how evolution through natural selection and other mechanisms causes populations to change over time. Can evolution also explain how different kinds of living things came to live on Earth? How does evolution produce new species?

A **species** is a group of organisms whose members can breed with one another but not with members of other species. (Notice that this definition works only for organisms that reproduce sexually. For asexually reproducing organisms, species are usually recognized by their similar characteristics and ways of life.) This means that the key to **speciation**—the formation of new species—is the evolution of *reproductive barriers* that prevent two groups of organisms from interbreeding.

Humans may vary in significant ways from one part of the world to another, but we all belong to the same species—all humans are able to interbreed!

There are two kinds of reproductive barriers: prezygotic and postzygotic. (A *zygote* is a fertilized egg, so *prezygotic* means "before fertilization" and *postzygotic* means "after fertilization.") *Prezygotic reproductive barriers* prevent individuals of different species from mating in the first place or prevent fertilization from occurring if they do mate. There are many types of prezygotic barriers—organisms may differ in when they breed, where they breed, or in the details of their courtship rituals. Their sex organs may not fit together properly, preventing successful sperm transfer, or other factors may prevent fertilization if sperm is transferred. Figure 17.23 shows an example of a prezygotic reproductive barrier. *Postzygotic reproductive barriers* act after fertilization has taken place. Postzygotic barriers

FIGURE 17.23
During courtship in red-crowned cranes, the birds dance around each other, bob their heads, stretch their necks, extend their wings, and leap straight into the air, singing in unison. Unless you can perform all these behaviors just right, you have little hope of convincing a red-crowned crane to mate with you.

MasteringPhysics®
VIDEO: Albatross Courtship
VIDEO: Blue-Footed Booby Courtship

occur when mating produces hybrids that either don't survive or are sterile—unable to breed themselves. The mule, the offspring of a horse and a donkey, is sterile and cannot reproduce. Likewise, a liger (Figure 17.24), the product of the mating of a lion and a tiger, is sterile.

Now let's consider how reproductive barriers—and therefore new species—evolve. In **allopatric speciation**, new species are formed after a geographic barrier divides a single population into two isolated populations (Figure 17.25). A geographic barrier could be a mountain range, a river, an ocean, a canyon, or—for aquatic organisms—a piece of land. Once two populations are geographically isolated from each other, they evolve independently. Over time, natural selection and genetic drift may contribute to the evolution of key differences that prevent interbreeding. If a reproductive barrier evolves, the different populations become separate species.

Numerous instances of allopatric speciation have been recorded. The rise of the Isthmus of Panama, 3 million years ago, divided the Caribbean Sea from the Pacific Ocean, splitting hundreds of types of marine organisms into separate Caribbean and Pacific populations. Most of these populations subsequently

FIGURE 17.24
A postzygotic reproductive barrier—the liger, a lion–tiger hybrid, is sterile.

FIGURE 17.25
Geographic barriers isolate populations and allow them to evolve independently. Sometimes, a reproductive barrier will evolve, resulting in allopatric speciation. In this example, the courtship song of birds divided by a mountain range diverges, resulting in a prezygotic reproductive barrier.

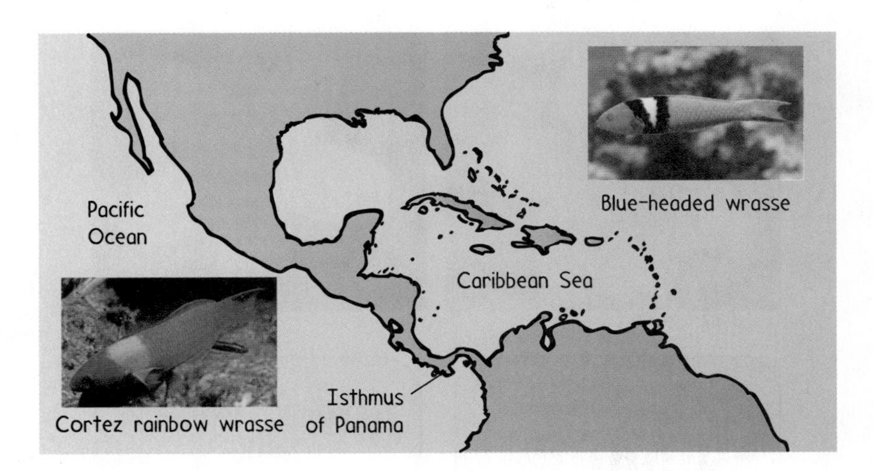

FIGURE 17.26
The formation of the Isthmus of Panama 3 million years ago isolated Pacific and Caribbean marine populations, producing numerous instances of allopatric speciation. The blue-headed wrasse (Caribbean) and the Cortez rainbow wrasse (Pacific) are descended from a single ancestral species that formerly spanned Pacific and Caribbean waters.

speciated by evolving reproductive barriers, including the wrasses in Figure 17.26. *Adaptive radiations* are spectacular examples of allopatric speciation where many new species, each adapted to a distinct way of life, evolve from a single ancestor. Many adaptive radiations have occurred on island archipelagos, which have abundant opportunities for geographic isolation. Adaptive radiations often occur after a new habitat is colonized. Examples of adaptive radiations include Darwin's finches, representing 13 species on the Galápagos Islands, and the Hawaiian honeycreepers, which include more than 30 species that differ in plumage, beak shape and size, and diet (Figure 17.27).

Sympatric speciation occurs without geographic isolation. Sympatric speciation is less common than allopatric speciation and often results from a sudden chromosomal change. One such chromosomal change is *polyploidy*, which occurs when organisms inherit more than the usual two sets of chromosomes, usually as a result of improper meiosis (see Chapter 16). Figure 17.28a shows a species of anemone that arose through polyploidy—it has four copies of each chromosome instead of two copies. Sympatric speciation can also result from hybridization. *Hybridization* occurs when two species interbreed and produce fertile offspring (Figure 17.28b). In both polyploidy and hybridization, chromosomal differences between the new species and the parent species prevent interbreeding. These types of speciation are more common in plants than in animals.

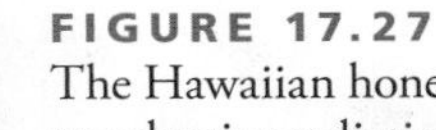

FIGURE 17.27
The Hawaiian honeycreepers represent an adaptive radiation consisting of more than 30 bird species. The honeycreepers differ in plumage, beak shape and size, and diet. Unfortunately, many species are extinct or endangered because of habitat destruction and the introduction of nonnative species such as rats, pigs, mongooses, cats, and mosquitoes. (Mosquitoes are harmful because some carry avian malaria.)

(a)

(b)

FIGURE 17.28
Sympatric speciation often occurs through an abrupt chromosomal change. (a) A new species of anemone (right) was produced through polyploidy. The parental species is shown on the left. Note the doubling of chromosomes. (b) The sunflower *Helianthus anomalus* (right) originated through the hybridization of two other sunflower species (left and center).

CHECK YOURSELF

1. **A small river forms, dividing a group of moles into two isolated populations. After many years, a biologist puts moles from opposite sides of the river together and finds that they will not mate. Has speciation occurred? If so, what type of speciation was it?**
2. **Do you think the same river would cause a population of birds to become two separate species?**
3. **Two species of frogs do not interbreed because one species breeds in the spring and the other breeds in the fall. Is this a prezygotic or postzygotic reproductive barrier?**

CHECK YOUR ANSWERS

1. The moles on the two sides of the river now represent two different species because they don't interbreed. This was allopatric speciation because it occurred after a geographic barrier (the river) separated the populations.
2. Probably not, since a small river is not much of a geographic barrier for flying animals.
3. Prezygotic because it prevents mating.

17.8 Evidence of Evolution

LEARNING OBJECTIVE
List and provide examples of the main kinds of evidence that support the theory of evolution.

EXPLAIN THIS How do corn on the cob, a dog's dewclaw, and the human hand provide evidence for evolution?

All scientific theories make predictions about what we should observe in nature (see Chapter 1). If these predictions are confirmed, the theory is supported. The theory of evolution has been tested repeatedly against observations of the natural world, and the evidence for evolution is overwhelming. Eight main kinds of evidence support the idea that evolution produced the diversity of life on Earth: (1) observations of natural selection in action, (2) artificial selection, (3) similarities in body structures, (4) vestigial organs, (5) DNA and molecular evidence, (6) patterns of development, (7) hierarchical organization of living things, (8) biogeography, and (9) fossils. We will look at the first eight topics here, and then consider fossils in Integrated Science 17C.

UNIFYING CONCEPT
- *The Scientific Method*
Section 1.3

1. *Observations of natural selection in action.* In many cases, scientists have seen natural selection produce evolutionary changes in populations; they have observed and measured the actual changes in populations. Examples include some of the cases we have looked at: Australian rabbits evolved resistance to the myxoma virus, so that over time a smaller and smaller fraction of individuals died from the disease. Peppered moths evolved to become better camouflaged in their environments—dark moths became more and more common as habitats became polluted, and then became less and less common as pollution was cleaned up. Bacteria evolved resistance to certain antibiotics, so that these antibiotics no longer controlled infections. Scientists have also studied how the beaks of Darwin's finches evolve after a drought, how insects evolve resistance to pesticides, and natural selection in a wide variety of other populations.
2. *Artificial selection.* **Artificial selection** is the selective breeding of organisms with desirable traits in order to obtain organisms with similar traits. Humans artificially select for desirable traits in domesticated animals and crops all the time: We breed fast racehorses to try to get faster racehorses; different types of dogs to produce superior hunters, herders, or sled-pullers (Figure 17.29); and varieties of strawberries to grow the largest and sweetest fruit. In artificial selection, humans control the reproductive success of different organisms and bring about distinct evolutionary changes in populations over time. These changes can be dramatic—think how much a Chihuahua differs from the animal it is descended from, the wolf. Or look at Figure 17.30 to see the difference between the corn we eat today and teosinte, the plant from which corn was bred. Artificial selection has produced countless forms of domestic animals and crops, all with traits valued by humans.
3. *Similarities in body structures.* We see evidence of the evolutionary histories of species in the structures of their bodies. Consider, for example, the limbs of different mammals. Different mammals use their front limbs for different purposes: Humans use theirs as arms and hands for manipulating tools, cats use theirs to walk on, whales use theirs as flippers, and bats use theirs as wings. If each of these animals had originated independently, we would expect their limbs to look completely different. Each limb would have been designed from scratch to best perform its function. But, despite the different functions of human hands, cat legs, whale flippers, and bat wings, all these limbs show the same arrangement of bones (Figure 17.31). This suggests that the limbs were inherited from a common ancestor and then modified through natural selection for different functions.

FIGURE 17.29
Artificial selection has produced great diversity in dogs.

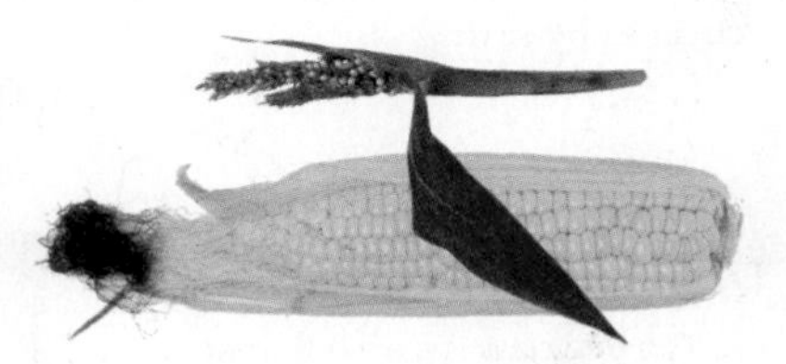

FIGURE 17.30
Corn (*below*), one of the most important agricultural crops in the world, was laboriously bred through artificial selection from teosinte (*above*). Teosinte has tiny cobs, only a few rows of kernels, and inedible hard coverings on its seeds.

A mouse and a whale are about as different as two mammals can be. Yet just about every bone in a mouse corresponds to a specific bone in a whale. These similarities suggest that mice and whales had a common ancestor and that their skeletons were modified over time by natural selection to fit different environments and ways of life.

MasteringPhysics®
TUTORIAL: Reconstructing Forelimbs

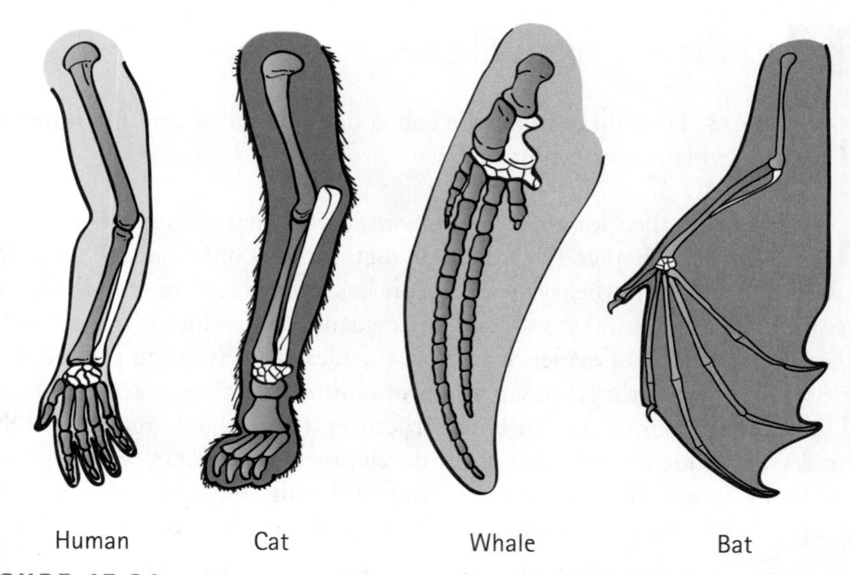

FIGURE 17.31
Although these mammalian limbs are used for different activities, they are composed of the same set of bones, evidence that they were inherited from a common ancestor.

4. *Vestigial organs.* An organism's evolutionary history often leaves traces in its body. Some organisms have vestigial organs. *Vestigial organs* are not functional—they are just the remains of an organ found in the organism's ancestor. For example, we think of snakes as legless. But did you know that certain snakes actually have tiny, partial hind legs? The tiny stubs have no purpose—they are just the remains of what once were bigger limbs. A snake's vestigial hind legs provide evidence that snakes evolved from animals with legs. In the same way, many blind cave species lack functional eyes in their lightless habitats but retain vestigial eyes (Figure 17.32). These vestigial organs suggest that cave species evolved from animals with eyes.
5. *DNA and molecular evidence.* The DNA of related species have similar nucleotide (ACGT) sequences. In fact, the more closely related two species are, the more similar their DNA sequences tend to be. This is true not only for DNA sequences that code for proteins, but even for sequences that have no known function. If each species on Earth had originated independently, would we expect to see similar noncoding DNA in related species? DNA similarity suggests that DNA did not originate independently in each species but was inherited from a common ancestor and then modified during evolution.
6. *Patterns of development.* Related species develop in similar ways. If each species on Earth had originated independently, we wouldn't expect these similarities in development. For example, even though humans have no tails, we go through a tailed stage, just like other vertebrates (Figure 17.33).
7. *Hierarchical organization of living things.* Darwin's theory of evolution explains Earth's diversity of species as originating through numerous speciation events. If this is the case, then we expect living things to be organized into hierarchical sets of "nested groups"—that is, "groups within groups." Each living species should have fewer traits in common with more distant relatives, and more traits in common with species that it split off from more recently. This is in fact how living things on Earth are organized. Humans, for example, share a backbone with other vertebrates such as fishes, amphibians, reptiles, and mammals; they share four limbs with terrestrial vertebrates such as amphibians, reptiles, and mammals but not with fish, which are more distantly related; they share a waterproof skin with reptiles and mammals

FIGURE 17.32
The Texas blind salamander lives in lightless caves. It has tiny vestigial eyes (dark dots in the photo) that are covered by skin.

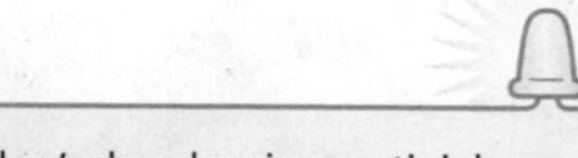

A dog's dewclaw is a vestigial organ. The dewclaw is a digit that appears on the inside of the front paws. It does not reach the ground and has no function. It is just what remains of a formerly functional toe.

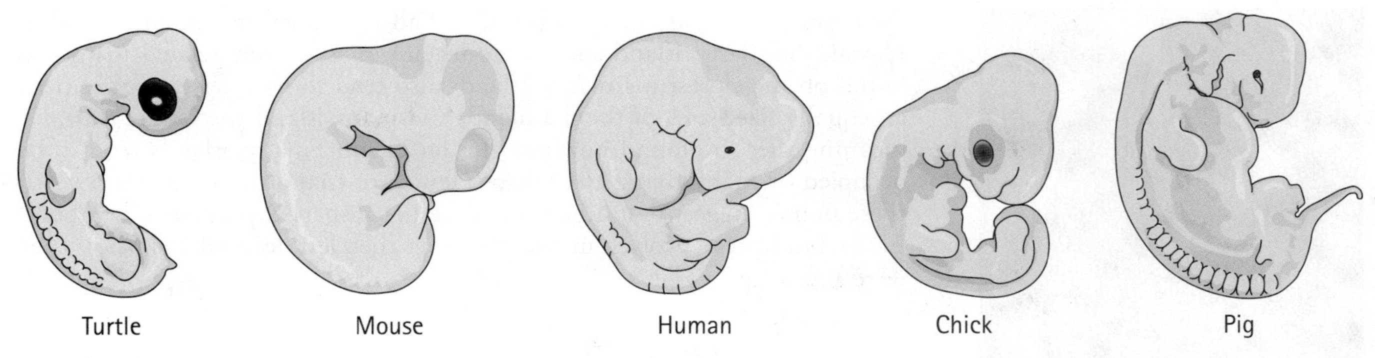

FIGURE 17.33
Related species go through similar stages in their development. The human embryo goes through a tailed stage just like the other vertebrates, even though humans don't have tails.

but not with amphibians, which are more distantly related; and they share the trait of nursing their young with milk with other mammals but not with the more distantly related reptiles. Living things fit into a hierarchical organization, as predicted by evolution. We do not see traits scattered across living things. For example, we do not see a backbone in vertebrates plus some worms and some insects and some snails. The characteristics that organisms have make sense based on their evolutionary history and relationships.

8. *Biogeography. Biogeography* is the study of how species are distributed on Earth. Biogeography is consistent with evolution: It supports the idea that organisms evolved in a certain place and then left descendants in the places where they were able to spread. Biogeography does not support the idea that organisms were specially designed to fit into a specific type of habitat and then distributed where these habitats occur on Earth. For example, even though the Arctic and Antarctic have similar environments, they are occupied by entirely different species (Figure 17.34). The same is true for New World tropical forests and Old World tropical forests.

What biogeography does show is that the ranges of many species are bounded by geographic barriers such as oceans or mountain ranges. For example, many organisms are restricted to a single continent. In addition, closely related species tend to be found close together, suggesting that they evolved in one place and then spread. For example, all of Darwin's finches

FIGURE 17.34
The Arctic and Antarctic, which have similar habitats, are occupied by very different species. Polar bears are found in the Arctic but not the Antarctic. Penguins are found in the Antarctic but not the Arctic.

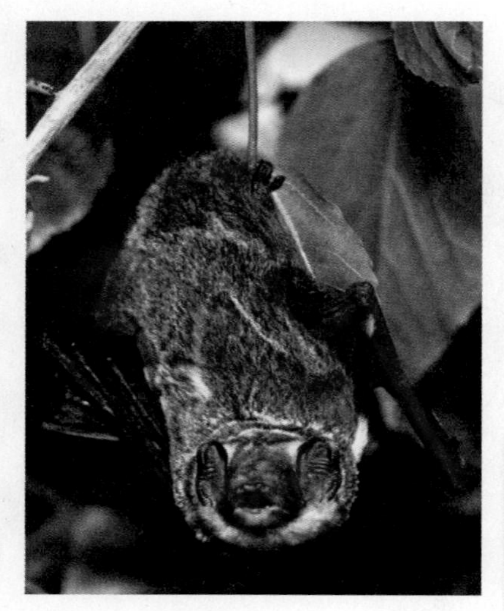

FIGURE 17.35
Why are terrestrial vertebrates rare or absent from islands, whereas flying species are common? This is the Hawaiian hoary bat, the only mammal found on Hawaii prior to human colonization of the islands.

are found in or near the Galápagos, and all the honeycreepers are found in Hawaii. Similarly, island species are usually most closely related to species found on the closest mainland. Islands also tend to have fewer species than an equally sized area of the mainland, and many island species are *endemic*, meaning they are found nowhere else on Earth. Finally, islands tend to be occupied by many flying animals but few terrestrial ones (Figure 17.35). All these points suggest that organisms were not dispersed purposefully around Earth, but instead evolved in one place and then left descendants where they were able to spread.

CHECK YOURSELF
Why is the fact that many species found on islands resemble species found on the nearest mainland evidence for evolution?

CHECK YOUR ANSWER
This pattern suggests that island species evolved when some mainland individuals colonized the island and then evolved in isolation, rather than that species were distributed purposefully around the Earth.

LEARNING OBJECTIVE
Explain how fossils provide evidence of evolution.

Integrated Science 17C
EARTH SCIENCE

Fossils: Earth's Tangible Evidence of Evolution

EXPLAIN THIS Why do fossil whales have legs?

Evolution has left a record in Earth's rocks—fossils. Because we can date fossils from the age of the rock formations they belong to, we can follow the evolution of certain groups of organisms over time. For example, fossil whales show that whales are descended from hoofed mammals. Fossil whales also tell us how many key whale traits evolved. In Figure 17.36a, we can see how, over time, whale nostrils moved from the front of the skull to the top of the skull, forming a blowhole. Fossil whales also show how whales lost their hind legs as they became more and more adapted to an aquatic existence. The oldest whale fossils, such as the 50-million-year-old *Ambulocetus*, have large hind legs that were used both on land and for swimming (Figure 17.36b). *Ambulocetus* also has small hooves on its front legs, providing clear evidence that whales are descended from hoofed mammals. *Rhodocetus*, a 46-million-year-old fossil whale, shows reduced hind legs—these are not attached to the backbone and so could not have supported much weight. *Rhodocetus* also shows prominent tail muscles that would have been effective for swimming. In the 40-million-year-old *Dorudon*, hind limbs are present, but they are tiny: *Dorudon* was clearly a fully aquatic species. In modern whales, there is no evidence of hind limbs on the outside of the body, although tiny remnants of the pelvis and sometimes femurs remain inside the body.

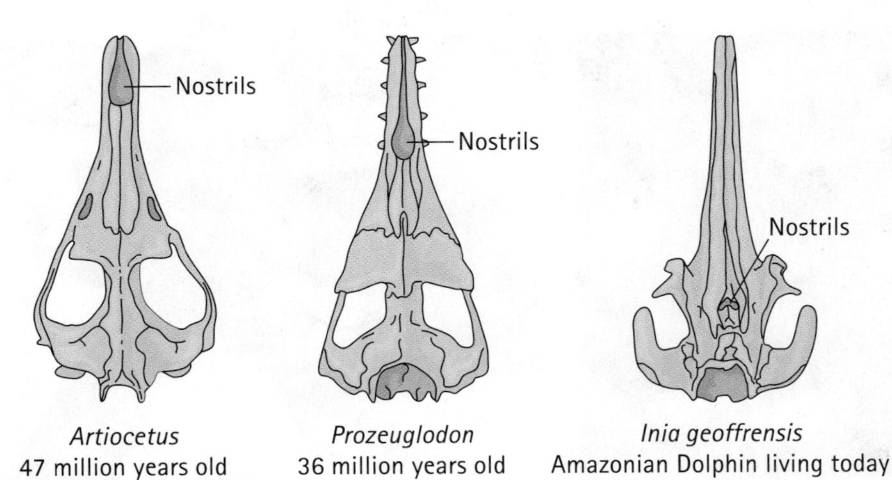

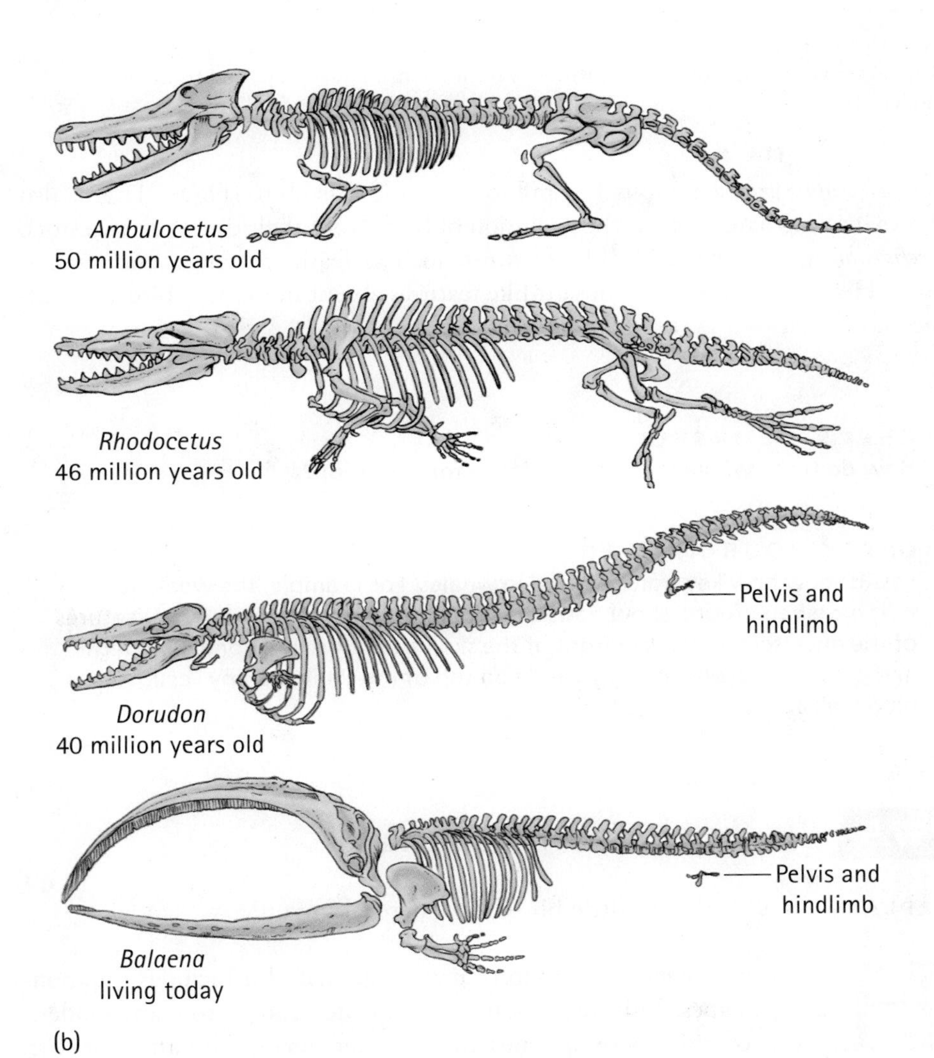

FIGURE 17.36
Fossil whales show how key features of these marine creatures evolved over time. (a) These fossil skulls show that the location of the nostrils shifted over time, from a position in front of the skull to a position on top of the skull—the "blowhole" seen in modern species. (b) Fossil whales also show the reduction and loss of hind legs over time.

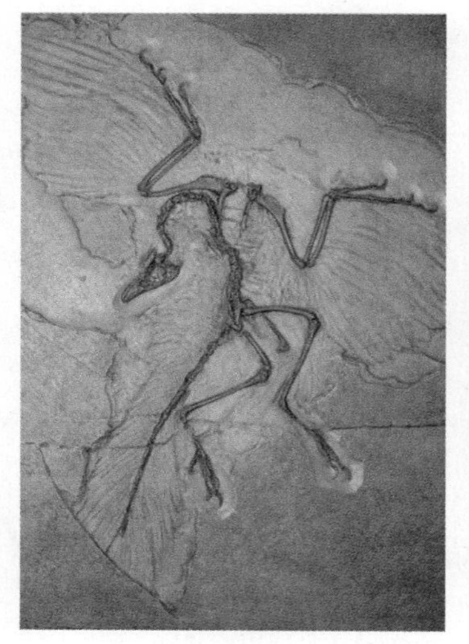

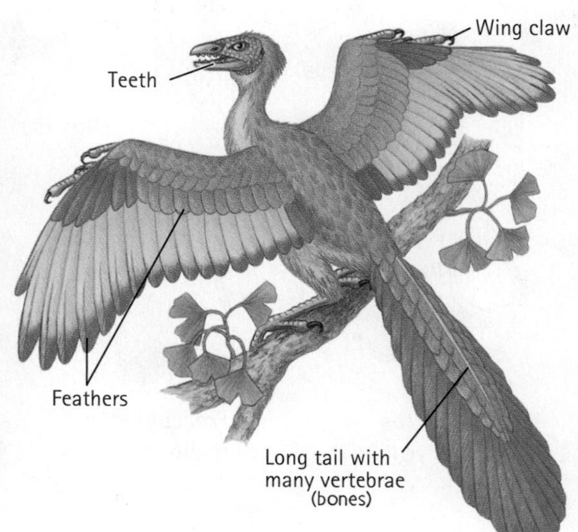

FIGURE 17.37
Archaeopteryx, an early bird, has features of both the dinosaurs it evolved from and modern birds.

You just learned that *Archaeopteryx*, the ancient bird, had clawed wings. Most birds today do not have claws on their wings, but there are a few exceptions—the most famous may be the hoatzin, which lives in tropical forests in the Amazon. Hoatzin chicks use their claws to move along branches. In addition, when threatened, they may drop from one tree, swim or move to another tree trunk, and then climb back up using their claws.

Archaeopteryx, the famous 150-million-year-old fossil bird (Figure 17.37), also shows intermediate traits in the evolution of birds from their dinosaur ancestors. *Archaeopteryx* has many birdlike features, such as feathers, wings, and a wishbone. However, it also has dinosaur-like features absent in modern birds, including claws on its wings, bones in its tail, and teeth.

CHECK YOURSELF
How do fossil whales provide evidence for evolution?

CHECK YOUR ANSWER
Fossils show how key traits evolved in whales. For example, the whale fossils that have been found show traits that are intermediate between the features of the ancestors (nostrils in front of the skull and large functional hind legs) and present-day whales (a blowhole on top of the skull and tiny vestigial hind limbs).

LEARNING OBJECTIVE
Describe some fossil hominids and what they reveal about the evolution of humans.

17.9 The Evolution of Humans

EXPLAIN THIS Is there a little bit of Neanderthal in you?

Humans are *primates*, a group of mammals that also includes the monkeys and apes. This does not mean we are descended from any modern species of monkey or ape, just that we share a common ancestor with these species more recently than we do with a dog, or a lizard, or a plant. Humans are also *hominids*, the group within the primates that includes modern *Homo sapiens* (our species) as well as some of our extinct relatives. Although humans are the only hominids in existence today, fossil hominids provide clues as to how humans evolved. A timeline of human evolution is shown in Figure 17.38.

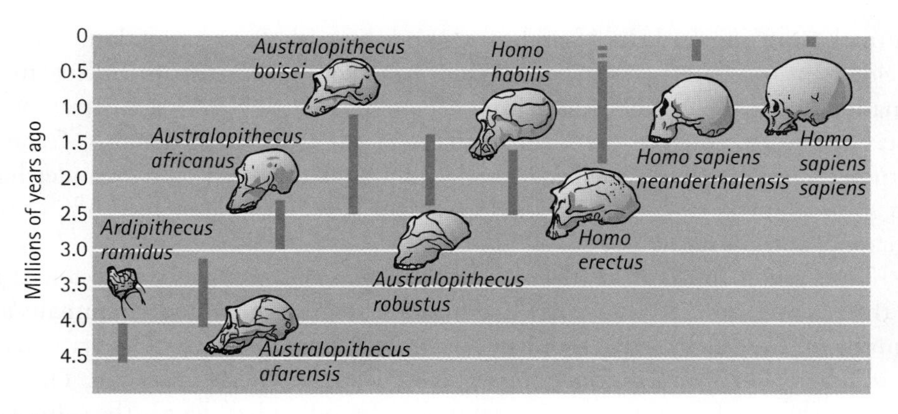

FIGURE 17.38
This timeline shows when certain hominid species existed on Earth. The skulls are all drawn to the same scale to show relative brain sizes.

Humans are not descended from any species of monkey or ape that is living today. However, we are more closely related to monkeys and apes than we are to other animals. *Descended from* and *related to* are entirely different.

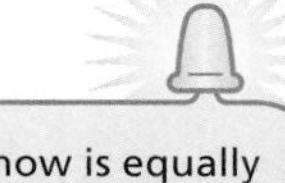

Every creature alive now is equally evolved. Every creature alive today is the product of at least 3.5 billion years of evolution. Humans are not "more evolved" than any other species.

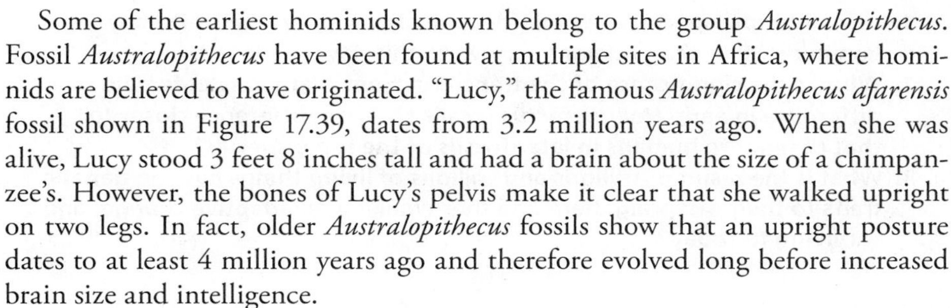

Some of the earliest hominids known belong to the group *Australopithecus*. Fossil *Australopithecus* have been found at multiple sites in Africa, where hominids are believed to have originated. "Lucy," the famous *Australopithecus afarensis* fossil shown in Figure 17.39, dates from 3.2 million years ago. When she was alive, Lucy stood 3 feet 8 inches tall and had a brain about the size of a chimpanzee's. However, the bones of Lucy's pelvis make it clear that she walked upright on two legs. In fact, older *Australopithecus* fossils show that an upright posture dates to at least 4 million years ago and therefore evolved long before increased brain size and intelligence.

Homo habilis is the earliest known species that belongs to the group *Homo*, which includes the species most closely related to modern humans. Some *Homo habilis* fossils are 2.2 million years old. *Homo habilis* had a larger brain than *Australopithecus*. *Homo habilis* also made stone tools—in fact, its scientific name means "handy man." Male *Homo habilis* were much larger than females. This is interesting because in other primates, such as gorillas and baboons, a big size difference between males and females is a sign that males fight each other for female mates.

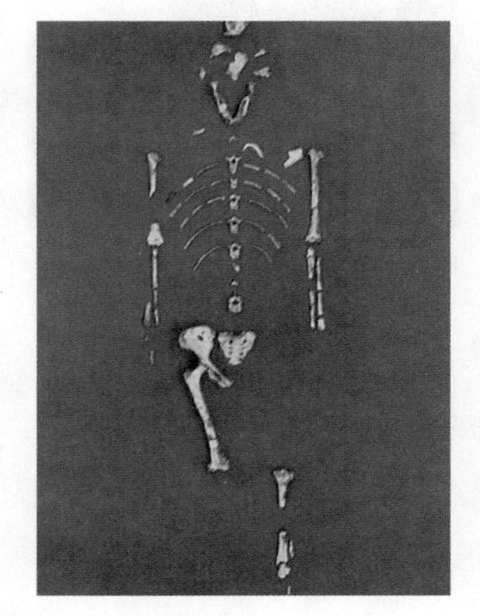

FIGURE 17.39
"Lucy," a fossil *Australopithecus afarensis*, stood upright and walked on two feet.

Homo erectus lived from about 2 million years ago to about 400,000 years ago. *Homo erectus* had an even larger brain than *Homo habilis*. In fact, the brain of *Homo erectus* was not much smaller than that of modern humans. *Homo erectus* was a skilled toolmaker as well as the first hominid species to migrate out of Africa and spread into much of what is now Europe and Asia. Like *Homo habilis*, older *Homo erectus* fossils show that males were much larger than females. However, later fossils of the same species show a male–female size difference closer to that present in modern humans, suggesting the development of a more humanlike social system.

The Neanderthals—*Homo sapiens neanderthalensis*—are closely related to modern humans (Figure 17.40). They lived from about 200,000 years ago to about 30,000 years ago. Neanderthals had very thick arms and legs, and their brains were as large as those of modern humans. Archaeological finds show that Neanderthals were effective hunters, had complex burial rituals, and made use of medicinal plants. One question that remains unanswered is whether the Neanderthals had language. For thousands of years, modern humans coexisted with Neanderthals. However, Neanderthal populations disappeared as modern humans spread. Scientists are not sure why, although it seems likely that modern humans outcompeted the Neanderthals and drove them to extinction. The development of modern genetic techniques has allowed scientists to collect information about the Neanderthals from a new source—DNA. Scientists are now attempting to sequence the Neanderthal genome from fossil remains. Genetic studies have already revealed that modern humans interbred with Neanderthals at some point in time; Neanderthal DNA accounts for at least 1%–4% of the genome of most humans.

The earliest fossils of modern humans, *Homo sapiens sapiens*, were found in Ethiopia and are 195,000 years old. Although anatomically modern humans are quite old, the cultural traits we associate with humans—things like art, music, and religion—are more recent, appearing only about 50,000 years ago. The reason for this gap between modern anatomy and modern behavior is the subject of continued debate.

In one of the most spectacular archaeological finds in centuries, skeletons of a tiny human relative were discovered on a remote Indonesian island in 2004. Nicknamed "hobbits," *Homo floresiensis* adults had skulls the size of grapefruits and were no bigger than 3-year-old modern children. *Homo floresiensis* lived alongside pygmy elephants, giant rodents, and Komodo dragons. Most amazing is the fact that they still occupied the island only 13,000 years ago, which means that they coexisted with our own species.

FIGURE 17.40
Neanderthals coexisted with modern humans and interbred with them. These reconstructions appear at the Neanderthal Museum in Mettmann, Germany.

CHECK YOURSELF

1. **Have multiple species of hominids ever coexisted on Earth? Do any hominids other than humans survive to this day?**
2. **What is the significance of the transition from a large male–female size difference in early *Homo erectus* fossils to a size difference closer to that of modern humans in later fossils of the same species?**
3. **What is the result of trillions and trillions of living things passing genetic traits to their offspring, here and there making an adaptive change, and surviving to today?**

CHECK YOUR ANSWERS

1. The timeline of hominid evolution shows that multiple species of hominids coexisted during much of hominid history. Today, however, humans are the only species of hominids in existence. The others have all died out.
2. A large size difference between males and females is a sign that males fought each other for female mates. This may have been true in early *Homo erectus.* More equal body sizes in later *Homo erectus* suggests that males and females had longer-term bonds, perhaps as they raised offspring together.
3. We and Earth's other living things are the result of this long and astounding journey!

For instructor-assigned homework, go to www.masteringphysics.com 

SUMMARY OF TERMS (KNOWLEDGE)

Adaptations Evolved traits that make organisms well suited to living and reproducing in their environments.

Allopatric speciation Speciation that occurs after a geographic barrier divides a group of organisms into two isolated populations.

Artificial selection The selective breeding of organisms with desirable traits in order to produce offspring with the same traits.

Autotrophs Living organisms that convert inorganic molecules into food and organic molecules.

Evolution Inherited changes in populations of organisms over time.

Fitness The number of offspring an organism produces in its lifetime compared to other organisms in the population.

Gene flow The evolution of a population due to the movement of alleles into or out of the population.

Genetic drift The evolution of a population due to chance.

Heritable Description of traits that are passed from parents to offspring because they are at least partially determined by genes.

Heterotrophs Living organisms that obtain energy and organic molecules from other living organisms or other organic materials.

Natural selection The process in which organisms with heritable, advantageous traits leave more offspring than organisms with other traits, causing these advantageous traits to become more common in a population over time.

Speciation The formation of new species.

Species A group of organisms whose members can breed with one another but not with members of other species.

Sympatric speciation Speciation that occurs without geographic isolation.

Variation Differences in a trait from one individual to another.

READING CHECK (COMPREHENSION)

17.1 The Origin of Life

1. How did Pasteur disprove the idea of spontaneous generation?
2. What experiment did Miller and Urey perform? What were their results?
3. How do liposomes resemble real cells?
4. Why is RNA, rather than DNA, believed to be the first genetic material?

17.2 Early Life on Earth

5. Why was the evolution of autotrophs an important event in the history of life on Earth?
6. What fundamental change in Earth's environment is attributed to the cyanobacteria?
7. How did the mitochondria and chloroplasts of eukaryotic cells originate?

17.3 Charles Darwin and *The Origin of Species*

8. What was Lamarck's theory about how evolutionary change occurred?
9. What impressed Darwin about the finches on the Galápagos Islands?
10. How did the work of Thomas Malthus influence Darwin?
11. How did Charles Lyell's work influence Darwin?

17.4 How Natural Selection Works

12. What is variation?
13. What is a heritable trait?
14. Describe how natural selection occurs.

17.5 Adaptation

15. From the point of view of natural selection, why is it important for an organism to survive?
16. Define sexual selection, and provide some examples of adaptations that evolved as a result of sexual selection.
17. Why is parental care adaptive in certain species?

17.6 Evolution and Genetics

18. Does natural selection act on genotype or phenotype?
19. Define genetic drift, and provide an example of how genetic drift can cause a population to evolve.
20. What is gene flow? Use an example to explain how gene flow can cause a population to evolve.
21. Why are genetic mutations and sexual reproduction important to creating and maintaining variation in populations?

17.7 How Species Form

22. What is a species?
23. What is the difference between a prezygotic reproductive barrier and a postzygotic reproductive barrier? Give an example of each.
24. Explain the difference between allopatric speciation and sympatric speciation. Which is more common?
25. What is an adaptive radiation, and when does it most commonly occur?

17.8 Evidence of Evolution

26. What is artificial selection? Why does artificial selection provide evidence for evolution?
27. Why does the similarity of the mammalian limb in all different species of mammals provide evidence for evolution?
28. How does biogeography provide evidence for evolution?

17.9 The Evolution of Humans

29. What important feature of modern humans can already be seen in 4-million-year-old *Australopithecus* fossils?
30. What was the first species of hominid to leave Africa and spread throughout Europe and Asia?
31. How old is our species, the modern humans known as *Homo sapiens sapiens*?

THINK INTEGRATED SCIENCE

17A—Did Life on Earth Originate on Mars?

32. Why do some NASA scientists think that life on Earth could originally have come from Mars?
33. If life on Earth did originate on Mars, how did it get here?
34. Why are some people skeptical that the supposed Martian fossils are of bacteria?

17B—Staying Warm and Keeping Cool

35. Why is the surface-area-to-volume ratio important in thermoregulation?
36. Recall that the amount of caramel on a caramel-covered apple is determined by the apple's surface area, and the amount of apple is determined by the apple's volume. If you like to eat a lot more caramel than apple, should you choose one large caramel apple or two smaller caramel apples with the same total volume? Defend your answer.
37. You are studying a species of tropical goat and comparing it with a related Arctic species. Based on your knowledge of thermoregulation in mammals and its effect on the size and shape of organisms, predict some of the differences you might see between the tropical and Arctic species.

17C—Fossils: Earth's Tangible Evidence of Evolution

38. Explain how the fossil whales that have been discovered support Darwin's theory of evolution.
39. What does the fossil *Archaeopteryx* tell us about bird evolution?

THINK AND DO (HANDS-ON APPLICATION)

40. Look at these photos, taken from Kettlewell's original publication on peppered moths. If *you* were the primary predator of peppered moths, would there be natural selection for color? Why or why not? (How many moths do you see in each photo? Which one did you see first?)

41. Take a hike or a walk in your neighborhood, and examine some of the plants, insects, birds, and other organisms that you come across. For each organism, note one or two traits that make it adapted to its environment. Did you notice any adaptations that keep organisms from being eaten by potential predators? What types of adaptations were these—ones that allowed them to flee or ones that allow them to remain camouflaged? Did you notice any adaptations related to finding mates? What about adaptations related to raising offspring?
42. Here's one more thing to think about during your walk. The species you see (whether they are pigeons, squirrels, mice, robins, sow bugs, or others) all share the ability to coexist with humans in a human-created environment. What adaptations allow these species to thrive in human communities?

THINK AND COMPARE (ANALYSIS)

43. Three bears are shown. Rank the bears according to how well suited they are to living (and thermoregulating) in a cold climate, from best suited to least suited.

44. Peppered moths collected from three different areas are shown. Rank the habitats the moths live in, from most sooty to least sooty.

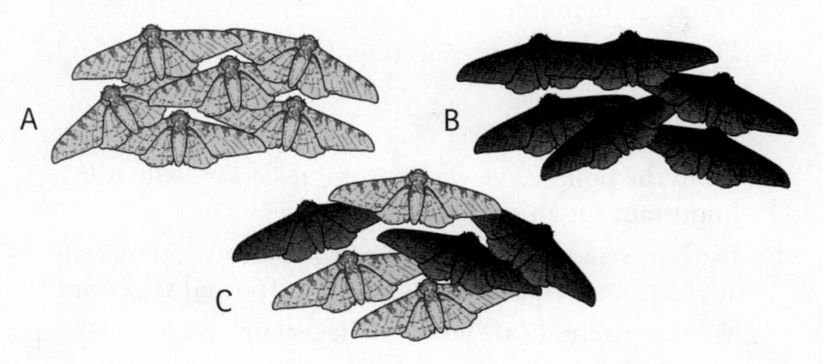

THINK AND SOLVE (MATHEMATICAL APPLICATION)

45. Let's look at how natural selection causes advantageous traits to become more common in populations. Suppose there is a population of bugs in which some individuals are green and some individuals are brown. Suppose that, because brown bugs are better camouflaged against predators, each brown bug leaves two brown offspring per generation (on average) and each green bug leaves one green offspring per generation (on average). (Is this natural selection? Why?) You start with two brown bugs and two green bugs in generation 1. How many brown and green bugs are there in generation 2? Calculate the number of brown and green bugs there are in generations 1 to 10. Show that 50% of the bugs in generation 1 are brown, 94% in generation 5 are brown, and more than 99% in generation 10 are brown. What is happening?
46. Let's consider a very small population of snapdragons, one with only two individuals. One snapdragon has two red alleles for flower color—it is *RR*. The other snapdragon has a red allele and a white allele for flower color—it is *RW*. (You may wish to review the inheritance of snapdragon flower color in Chapter 16.) Show that the frequency of the red allele *R* in the population is 0.75 and the frequency of the white allele *W* is 0.25.
47. Now let's assume that the two snapdragons in our tiny population mate and produce a single offspring with genotype *RR*. We now have a snapdragon population with only one individual. Calculate the allele frequencies of the red and white alleles in the population. Is this an example of genetic drift?

THINK AND EXPLAIN (SYNTHESIS)

48. What types of experiments showed that living organisms are not spontaneously generated in nonliving matter? Why do you think the idea of spontaneous generation survived so long—that is, why was spontaneous generation so difficult to disprove?
49. Why do scientists consider Miller and Urey's experiment important?
50. How are liposomes similar to cells? How are they different from real cells?
51. What are some human traits that do not show variation? What are some that do show variation? What are some heritable human traits? Some nonheritable human traits?
52. How would you determine whether a trait you were interested in studying is heritable?
53. How is the story of the myxoma virus and Australian rabbits similar to the story of antibiotic resistance in bacteria?
54. Nancy Burley of the University of California, Irvine, ran the following experiment: She placed red color bands on the feet of some male birds and green color bands on the feet of other male birds. Females preferred to mate with males that had red color bands. Is this an example of natural selection? Why or why not?
55. In recent decades, average human height has increased in many parts of the world. Do you think this is an example of evolution?
56. On islands, many large animals—such as elephants—evolve to become miniaturized. On the other hand, many small animals—including some rodents—evolve to be exceptionally large. Why might natural selection produce these results? Do you think this phenomenon sheds light on *Homo floresiensis*, the miniature relative of humans?
57. Male birds of many species have brighter feathers than females. Bright colors on males are often adaptations for winning mates, as in the case of birds of paradise and peacocks, discussed in this chapter. Is being less colorful adaptive for female birds? Defend your answer.
58. Two species of foxes are shown here. One is a kit fox in Arizona. The other is an Arctic fox. Which is which? How can you tell? Describe at least two traits that make each animal well adapted to its environment.

59. You are studying a population of beetles that includes some red individuals and some yellow individuals. You know that color is a heritable trait in the population. By counting the number of red and yellow beetles over a period of 5 years, you notice that the proportion of red individuals is increasing over time while the proportion of yellow individuals is decreasing over time. How could you determine whether this is a result of natural selection? Are there other potential explanations?
60. In a population of mice that you are studying, tail length appears to be increasing over time. However, you find no evidence that natural selection is acting on tail length. What are two alternative explanations for your observation?
61. Individuals of two different fish species sometimes mate, but their offspring die soon after hatching. Is this an example of a prezygotic or postzygotic reproductive barrier?

62. Finches on two closely situated islands look different; on one island, they have brown tail feathers, and on the other island, they have black tail feathers. Can you conclude that these are two different species? How could you determine whether they are in fact distinct species?
63. At your field site, there are butterflies with yellow wings and butterflies with orange wings. After observing them carefully, you notice that the yellow butterflies always mate in shady areas under trees, whereas the orange butterflies always mate in sunny meadows. Can you conclude that they are different species?
64. Many of the living organisms in Hawaii are found nowhere else on Earth. Hawaii has numerous unique species of plants, birds, insects, mammals, mushrooms, and other living things. Why?
65. What are some examples of artificial selection? How are artificial selection and natural selection similar? How are they different?
66. Islands tend to have fewer species than the mainlands they resemble. Furthermore, island species often include many flying organisms and few terrestrial ones. Do these biogeographic patterns support evolution or the purposeful distribution of organisms? Why?
67. Laura says she doesn't believe that humans were at one time chimpanzees or gorillas. Jeff says he doesn't believe it either. Explain why biologists also don't believe that humans are descended from chimps or gorillas.
68. Write a letter to Grandma telling her about drug resistance in living organisms. Explain to her why drug resistance is such a common phenomenon—including why insects become more resistant to pesticides over time, and why diseases such as tuberculosis and malaria have become harder to treat in recent years.

THINK AND DISCUSS (EVALUATION)

69. During a drought, the supply of seeds available to a finch population decreases. The smaller, softer seeds, which are easier to crack, are quickly eaten up. Finches with larger, stronger beaks are better able to crack the larger seeds that remain. What evolutionary changes do you expect to see in this finch population?
70. Caterpillars of the Monarch butterfly eat plants that are toxic to other animals so that their tissues become toxic. Birds that try to eat Monarchs vomit and then avoid the striking orange-and-black pattern in the future. The viceroy is another species of butterfly. Viceroys resemble monarchs, but they are not toxic. Is the appearance of the viceroy adaptive? How could you test this hypothesis?

Can you tell which is the Monarch and which is the Viceroy? Viceroys have a black stripe in the hind wing that goes across the other stripes. Monarchs do not.

71. You are eating a salad when you almost bite down on a green insect hidden among the lettuce leaves. The friend who is eating with you says, "That would have been gross, but I don't think it would have poisoned you." Do you agree?
72. Bird eggs vary tremendously in color. Do you think the color of a bird's eggs is adaptive? What factors may have shaped the evolution of egg color in different species?

73. A population of beetles that includes both sandy and green individuals is introduced into a grassy environment. How do you expect the population to evolve due to natural selection? Now suppose that another beetle population lives in a nearby sandbank. Most of the individuals in this population are sandy colored. If beetles regularly migrate from one population to the other, what will be the effect on each population? Does gene flow make it easier or harder for these beetle populations to adapt to their environments?
74. Islands tend to have fewer species than an equally sized area of the mainland. Is this consistent with the idea that species were spread around Earth purposefully? Is it consistent with evolution?
75. Scientific theories must be falsifiable. Is evolution falsifiable? For example, can you imagine some biogeographic evidence that would not be consistent with evolution? Has any evidence of this sort been found?

76. Scientists who are searching for new fossils of early hominids usually look in Africa. Does this make sense, or should they expand their search?
77. Broad-spectrum antibiotics are effective against a wide variety of bacteria. Narrow-spectrum antibiotics are effective against only certain types of bacteria. Public health officials have suggested that one way to combat antibiotic resistance is to use narrow-spectrum antibiotics whenever possible. Do you agree? Explain how the use of narrow-spectrum antibiotics instead of broad-spectrum antibiotics could slow the evolution of resistance in bacteria.

READINESS ASSURANCE TEST (RAT)

If you have a good handle on this chapter, if you really do, then you should be able to score 7 out of 10 on this RAT. If you score less than 7, you need to study further before moving on.

Choose the BEST answer to each of the following:

1. Which of these statements regarding the origin of life is false?
   (a) Life originated on an Earth whose atmosphere contained high levels of oxygen.
   (b) Miller and Urey obtained amino acids and other organic molecules when they sent electric sparks through a model of Earth's early atmosphere.
   (c) The first genes were probably made of RNA.
   (d) When certain lipids are added to water, they spontaneously form structures that resemble cell membranes.
2. The primary problem with the hypothesis that life on Earth originated on Mars is that
   (a) Mars has never had water.
   (b) the proposed Martian fossils are much smaller than the tiniest bacteria on Earth.
   (c) life on Mars would have had no way to get to Earth.
   (d) life has never been found on Mars.
3. Photosynthesizing plants are
   (a) heterotrophs.
   (b) autotrophs.
   (c) chemoautotrophs.
   (d) archaeans.
4. Organisms with heritable, advantageous traits leave more offspring than organisms with other traits, which causes these advantageous traits to become more common in a population over time. This describes
   (a) the inheritance of acquired characteristics.
   (b) speciation.
   (c) natural selection.
   (d) genetic drift.
5. Which of these adaptations is the result of sexual selection?
   (a) the spines on cactus plants
   (b) parental care in male poison dart frogs
   (c) the songs of male birds
   (d) dark wings in peppered moths
6. If we compare related rabbit species in desert and Arctic environments, we would expect
   (a) the desert species to have shorter legs.
   (b) the Arctic species to be smaller.
   (c) the desert species to have larger ears.
   (d) the Arctic species to be the same size as the desert species.
7. Which of the following mechanisms of evolution consistently causes populations to become more adapted to their environments?
   (a) natural selection
   (b) mutation pressure
   (c) genetic drift
   (d) gene flow
8. When a lion mates with a tiger, the offspring are sterile. This is an example of
   (a) allopatric speciation.
   (b) sympatric speciation.
   (c) speciation by hybridization.
   (d) a postzygotic reproductive barrier.
9. Which of the following provides evidence for evolution?
   (a) changes in the coloration of peppered moth populations over time
   (b) the presence of vestigial eyes in cave salamanders
   (c) the fact that island species tend to most closely resemble species found on the nearest mainland
   (d) all of these
10. Which statement about human evolution is true?
   (a) The earliest fossils of modern humans are almost 200,000 years old.
   (b) Humans are descended from chimpanzees.
   (c) Modern humans are the only hominids that used tools.
   (d) During human evolution, large brains evolved before upright posture.

Answers to RAT

1. a, 2. b, 3. b, 4. c, 5. c, 6. c, 7. a, 8. d, 9. d, 10. a

CHAPTER 18
Diversity of Life on Earth

MORE THAN 1.5 million known species live on Earth today. Scientists believe that many more—somewhere between 10 and 100 million—have yet to be discovered. Earth's living species show remarkable diversity in their adaptations and ways of life. What are the major groups of living things? Is it true that humans are more closely related to bread mold than to daisies? Is it true that life on Earth would die out without that tiniest of life forms—bacteria? Why do some flowers smell sweet, while others smell like, well . . . a dead horse? Could evolution one day produce a 10-foot mosquito? And are there really thousands of species of dinosaurs flying around on Earth today? Read on to learn all about the diversity of life.

18.1 Classifying Life

LEARNING OBJECTIVE
Compare Linnaean classification with evolutionary classification.

EXPLAIN THIS Why is a bird also a reptile?

The desire to classify living things is an ancient one. Thousands of years ago, the Greek philosopher Aristotle arranged a "Chain of Being" that proceeded from minerals to plants, animals, man, and God. Other thinkers built on Aristotle's ideas, arranging organisms from "simple" things like plants and worms to humans, the most "complex."

Linnaean Classification

In the 18th century, Swedish naturalist Carolus Linnaeus developed a new system of classification that emphasized the shared similarities of organisms. Under the Linnaean system, different *species* are grouped together into a *genus*. Different genera (plural of *genus*) are grouped together into a *family*. Different families are grouped into an *order*. Different orders are grouped into a *class*. Different classes are grouped together into a *phylum*. Different phyla (plural of *phylum*) are grouped into a *kingdom*. And, more recently, different kingdoms have been grouped into a *domain*. At every level in the Linnaean system, species are grouped together based on shared similarities. For example, species in the class Mammalia (all mammals) have hair and nurse their young with milk.

Mnemonics are little sayings that help people remember things. You can use this mnemonic to remember the levels of classification in the Linnaean system: In his *domain, King Phillip* called the *class* to *order*—the *family genius* will now *speak*.

Let's look at the classification of humans in Figure 18.1. Humans are *Homo sapiens*. Our genus, *Homo*, includes humans as well as some of our extinct relatives (see Chapter 17). The family we belong to is Hominidae, which includes all the hominids. Humans are the only hominid species in existence today, but we know from fossils that there were other hominid species that are now extinct. Humans belong to the order Primates, which also includes monkeys and apes. We are in the class Mammalia, which includes all mammals. We belong to the phylum Chordata, which includes mammals, other vertebrates (birds, frogs, fish, etc.), and a few other organisms. We belong to the kingdom Animalia, which includes all animals. Finally, we are in the domain Eukarya, which includes all the eukaryotes—animals, plants, fungi, and other living things.

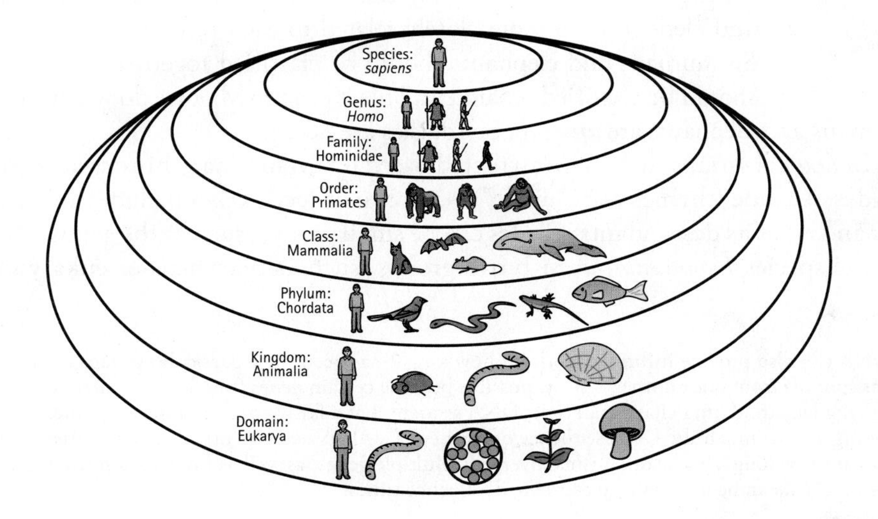

FIGURE 18.1
In the Linnaean system, humans belong to the domain Eukarya (eukaryotes), kingdom Animalia (animals), phylum Chordata (chordates), class Mammalia (mammals), order Primates (monkeys and apes), family Hominidae (humans and some of our extinct relatives), genus *Homo*, and species *sapiens*.

Linnaeus also came up with rules about how species are named. Species have a two-part scientific name made up of their genus name and species name. For example, humans are *Homo sapiens* ("wise human") and dogs are *Canis familiaris* ("intimate dog"). Genus and species names are Latin; by convention, the names are in italics with the genus name capitalized. Sometimes the genus name is abbreviated as a single letter, as in *E. coli* for the human gut bacterium *Escherichia coli*.

Evolutionary Classification

Since Linnaeus's time, science has expanded our understanding of the history of life on Earth. Darwin's theory of evolution showed that the wealth of species on Earth is the result of numerous instances of speciation followed by the independent evolution of new lineages (see Chapter 17). Biologists now classify living things based on this evolutionary history. Specifically, they group species together based on how closely related they are to one another. How does this work?

The first step in an evolutionary classification is to reconstruct the history of speciation events among a group of organisms. This allows biologists to determine which species split off from each other more recently and are therefore more closely related. Biologists use fossils as well as information on the anatomy, behavior, and genetics of existing species to try to reconstruct this history. DNA sequences have proved to be a particularly valuable source of information.*

Once biologists have a hypothesis about how speciation events caused different species to diverge, they create an **evolutionary tree**. A simple evolutionary tree is shown in Figure 18.2. This tree includes three species—daisies, elephants, and humans. Notice the dot labeled Speciation Event 1. Up until the time specified by this dot, daisies, elephants, and humans had a shared history. Then a speciation event occurred, producing two separate lineages. (A *lineage* is a line of descendants from an ancestor.) One of the lineages eventually gave rise to daisies (and many other species, which are not shown). The other lineage eventually gave rise to humans and elephants (as well as other species not shown). At a later point in time, Speciation Event 2 caused the lineage that led to elephants to split from the lineage that led to humans. This evolutionary tree tells us that, because humans and elephants split from each other more recently than they split from daisies, humans and elephants are more closely related to each other than to daisies. So, humans and elephants should be classified together, and daisies should be classified separately. This is exactly what we do when we say humans and elephants are *animals*, but daisies are *plants*.

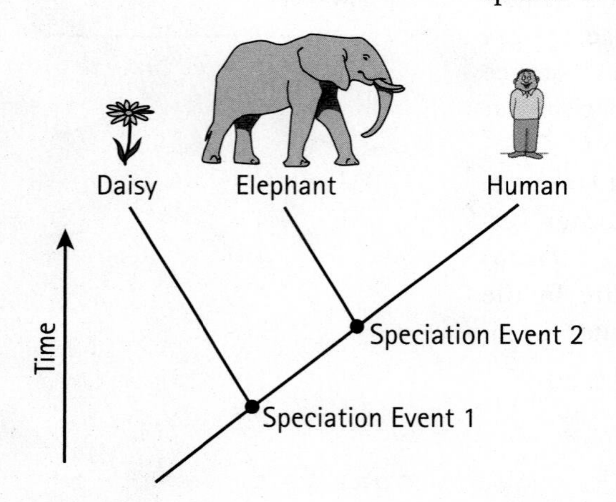

FIGURE 18.2
An evolutionary tree shows how species or other biological groups are related. This evolutionary tree for humans, elephants, and daisies shows that humans and elephants are more closely related to each other than either is related to daisies.

Biological groups that are constructed based on evolutionary history are called clades. A **clade** (rhymes with "made") is a group of species that includes an ancestor and all of its descendants. Clades can be small groups, such as the genus *Homo* or the species *Homo sapiens*, or broad groups, such as mammals or eukaryotes.

* DNA can also provide information about how long it has been since particular groups of organisms split off from one another. This is possible because certain genes function as *molecular clocks*, evolving (accumulating changes in their DNA sequence) at a fairly constant rate over time. By looking at how much the DNA sequence of a gene differs between two organisms, scientists can estimate how long ago the organisms diverged. Multiple genes, as well as information from the fossil record, can be used to help estimate divergence times.

How does evolutionary classification compare to Linnaean classification? In many cases, Linnaean groups *are* clades and continue to be used by biologists. For example, the Linnaean kingdom Animalia—all animals—is a clade. The Linnaean kingdoms Plantae and Fungi are also clades. Mammals, amphibians, primates, birds, frogs—these familiar Linnaean groups are all clades.

Some Linnaean groups are not clades, however, and have had to be reconsidered. For example, the Linnaean class Reptilia—the reptiles—grouped together turtles, lizards, snakes, and crocodiles. Birds were placed in a separate class, Aves. But, reconstructing the evolutionary relationships among these organisms showed that birds are descended from the last common ancestor of the reptiles, and so they are reptiles too (Figure 18.3). Classifying turtles, lizards, snakes, and crocodiles together while excluding birds is like classifying elephants and daisies together while excluding humans. What is the reason for the discrepancy between Linnaean and evolutionary classification? Recall that, under the Linnaean system, species are grouped together based on shared similarities. Turtles, lizards, snakes, and crocodiles were grouped together based on shared features such as "cold-bloodedness" and the possession of scales. Birds were placed in a different group because they are "warm-blooded" and have feathers. An evolutionary classification groups species together based on their evolutionary relationships and places birds squarely among the reptiles.

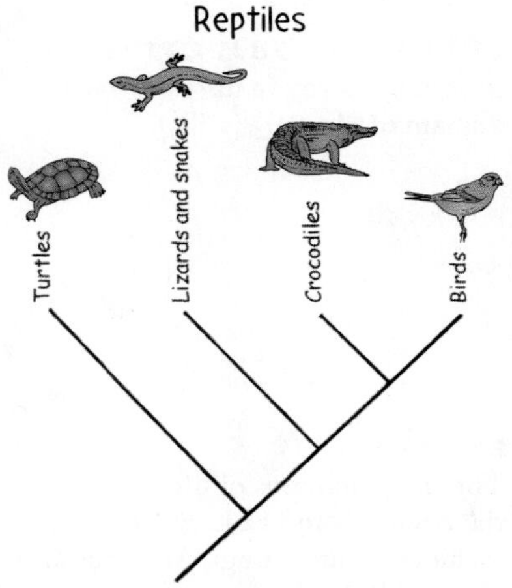

FIGURE 18.3
This evolutionary tree shows that birds are descended from the last common ancestor of reptiles, and so they are reptiles also.

One advantage of an evolutionary classification is that it is much less arbitrary than trying to figure out which species are most "similar" to one another. What does "similar" mean exactly? For example, are humans more similar to bread mold or to cherries? You could argue about that for a long time. On the other hand, biologists are pretty sure they know which one is more closely related to humans. (Read on to find out!)

Biological classification is always a work in progress. As biologists learn more about the evolution of species, they sometimes need to draw new evolutionary trees. Classification then changes to reflect this new understanding of species relationships.

UNIFYING CONCEPT
- ***The Scientific Method***
Section 1.3

CHECK YOURSELF

1. **The scientific name of the endangered orangutan of Sumatra is *Pongo abelii*. What is its genus name? Its species name?**
2. **Why is evolutionary classification more useful for biologists than Linnaean classification?**

CHECK YOUR ANSWERS

1. The genus name is *Pongo*; the species name is *abelii*.
2. Evolutionary classification reveals, rather than obscures, the evolutionary history of species. For example, knowing that birds are reptiles allows biologists to ask appropriate questions about their traits, such as: If birds are "warm-blooded" rather than "cold-blooded" like other reptiles, how did they evolve warm-bloodedness? If birds have feathers rather than the scales seen in other reptiles, could scales have been modified to form feathers? Under Linnaean classification, reptiles and birds are separate classes, and there is no link between them. Only with an evolutionary classification do we understand that birds evolved from a "cold-blooded," scaly ancestor.

Birds aren't just reptiles—they're dinosaurs! This is because birds are descended from the last common ancestor of all the dinosaurs, which makes them dinosaurs too. So, dinosaurs didn't *all* go extinct: Birds survived, and they are certainly alive and well today, with nearly 10,000 known species.

LEARNING OBJECTIVE
Describe the key features of each domain of life.

18.2 The Three Domains of Life

EXPLAIN THIS Which domain of life do humans belong to?

Life is classified into three domains: Bacteria, Archaea, and Eukarya. Early in the history of life—probably 2.5 to 3.5 billion years ago—living organisms split into two separate lineages: one that produced the Bacteria and one that produced the Archaea and Eukarya (Figure 18.4).

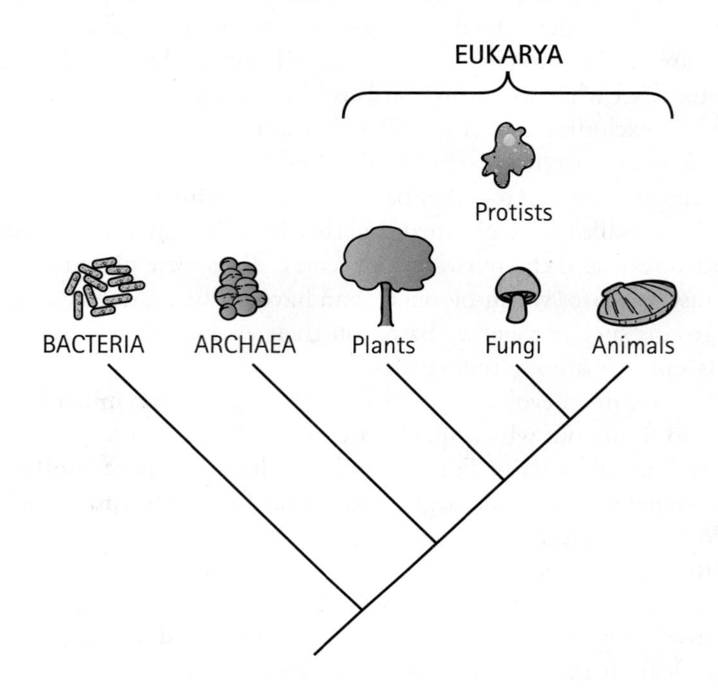

FIGURE 18.4
The three domains of life are Bacteria, Archaea, and Eukarya. Eukarya includes Plants, Fungi, Animals, and Protists. Protists are not shown in the tree because they are not a clade and do not have a clear place in the tree. (They would have to be drawn as multiple "twigs" in various spots along the Eukarya branch.)

Bacteria and Archaea consist of prokaryotic organisms, organisms whose cells do not have a nucleus (see Chapter 15). **Eukarya** includes all eukaryotic organisms, living things whose cells have a nucleus. The domain Eukarya is further divided into four kingdoms: Protists, Plants, Fungi, and Animals. Some key characteristics of each group are shown in Table 18.1.

TABLE 18.1 KEY CHARACTERISTICS OF THE MAJOR GROUPS OF LIVING THINGS

Group	Type of Cells	Single-Celled or Multicellular	Way of Obtaining Nutrition
Bacteria	Prokaryotic	Mostly single-celled	Heterotrophs or autotrophs
Archaea	Prokaryotic	Single-celled	Heterotrophs or autotrophs
Eukarya			
Protists	Eukaryotic	Single-celled or multicellular	Heterotrophs or autotrophs
Plants	Eukaryotic	Multicellular	Autotrophs
Fungi	Eukaryotic	Single-celled or multicellular	Heterotrophs
Animals	Eukaryotic	Multicellular	Heterotrophs

The Protist kingdom is problematic because it includes all eukaryotes that aren't plants, animals, or fungi—that is, it's a hodgepodge of species that don't represent a clade. Amoebas, kelp, and diatoms are all protists, athough they have little in common besides the fact that they are eukaryotes. Until a more accurate classification emerges, however, we are stuck with the term *Protists*.

CHECK YOURSELF

Are humans more closely related to bread mold or to cherries?

CHECK YOUR ANSWER

Humans are animals, bread mold is a fungus, and cherries are plants. In the evolutionary tree in Figure 18.4, we see that animals and fungi are more closely related to each other than either is to plants. So, we are more closely related to bread mold!

18.3 Bacteria

LEARNING OBJECTIVE
Describe the key features and ecological significance of bacteria.

EXPLAIN THIS Why would life on Earth be impossible without bacteria?

They live on your body by the trillions, occupy habitats where no other organisms can survive (Figure 18.5), and devastate human populations with diseases such as plague and tuberculosis. Yet, life on Earth would quickly end without them. *They* are bacteria, one of the most ancient lineages on Earth. Earth's oldest fossils, 3.5 billion years old, are of bacteria (see Chapter 17).

Bacteria are prokaryotes so diverse that it is hard to make generalizations about them. Some bacteria are autotrophs that, like plants, make their own food through photosynthesis. Others are chemoautotrophs that make food using chemical energy rather than energy from sunlight. Still others are heterotrophs that obtain food from organic matter. Heterotrophic bacteria are so diverse in what they eat that just about any type of organic molecule is food for some species of bacteria. Most bacteria are single-celled, but others gather in multicellular clusters. Bacteria come in varied shapes, including spheres, rods, and spirals. Many can move by using whiplike structures called *flagella*.

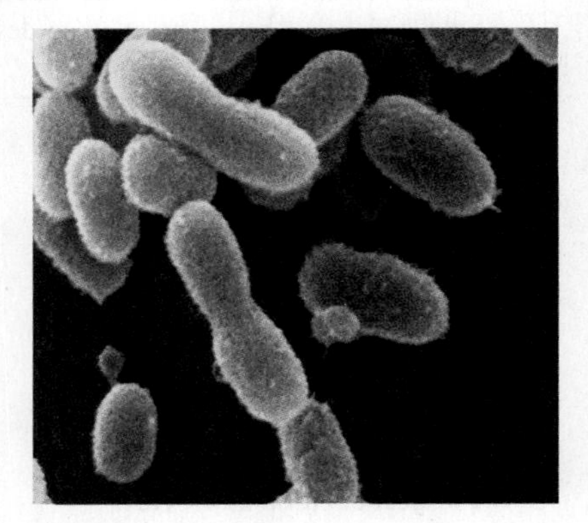

FIGURE 18.5
Bacteria can live in habitats where no other organisms can survive. These tiny bacteria were found in a 120,000-year-old ice core 3000 meters beneath a glacier in Greenland. They live in a habitat with sub-zero temperatures, high pressure, and little oxygen or nutrients.

Bacteria typically reproduce asexually by dividing. However, most species exchange genetic material at least occasionally—when they take up small pieces of naked DNA from the environment, when bacterial viruses inadvertently transfer DNA between organisms, or when two bacteria join together and one passes DNA to the other. Under favorable conditions, bacteria can divide very quickly, as often as every 20 minutes. This allows bacterial populations to grow rapidly when food is plentiful and conditions are good. In poor conditions, many bacteria form *spores*—tough, thick-walled structures that stay dormant until conditions improve.

Life on Earth would be impossible without bacteria because they play an important role in *decomposition*, the breaking down of organic matter. Without bacterial decomposition, carbon would stay trapped in dead organic matter, all the carbon dioxide in the atmosphere would eventually be used up, and photosynthesis would stop. Bacteria also help cycle other nutrients; some bacteria fix

TECHNOLOGY

Bacterial Clean-Up Crews

In 2010, the Deepwater Horizon offshore oil-drilling rig exploded in the Gulf of Mexico. The result was the worst marine spill in history—more than 5 million barrels of oil escaped into the Gulf. Response workers and volunteers tried to contain the oil, skim it from the water, and burn it away. They also tried to assist the living organisms that would ultimately deal with the mess—bacteria. All oceans have natural communities of bacteria that consume oil. Dozens of species feed off the hydrocarbons in oil, consuming oxygen and releasing carbon dioxide in the process. After the Deepwater Horizon spill, scientists measured a 30% decrease in oxygen levels in Gulf waters, a sign that bacteria were hard at work.

Chemical dispersants are spread over an oil slick after the Deepwater Horizon explosion. The dispersants break up the oil slick into smaller droplets, making it easier for bacteria to consume the oil.

The use of living organisms to clean up polluted soil, air, or water is called *bioremediation*. Bioremediation has been around for a long time—for example, bacteria have been used for many years to decompose human wastes in sewage treatment plants. But scientists are realizing that, with a little management, bacteria can be remarkably effective at clearing away a variety of messes, everything from oil spills to nuclear waste.

In the case of oil spills, the application of chemical dispersants may aid bacteria in their work. Dispersants break up large oil slicks into smaller droplets, so that it is easier for bacteria to reach the oil. However, the use of dispersants is controversial because they are toxic to humans and other species, and may also harm the bacteria they are trying to help. Scientists have also tried to develop genetically engineered bacteria that consume oil more effectively. So far, though, no genetically engineered strain can match the capabilities of naturally occurring species.

What about nuclear waste? Can bacteria help us with that? The remarkable answer is yes. Some unusual bacteria in the genus *Geobacter* attach to the radioactive uranium in nuclear waste. They use long extensions called pili to obtain energy from the waste, which simultaneously prevents the waste from escaping into the groundwater. Some researchers are hoping to genetically engineer *Geobacter* strains with more pili, which could make them even better at cleaning up.

nitrogen, transforming it from its inorganic atmospheric form to varieties that can be used by living organisms (see Chapter 21).

Countless bacteria live in and on our bodies, particularly on the skin and in the mouth, respiratory tract, and intestines. A few of these are potentially harmful, but others benefit us by producing vitamins and by keeping more dangerous bacteria from invading our bodies. Bacteria are used to make foods such as cheese and yogurt, and some genetically engineered strains produce human insulin and other medically important molecules (see Chapter 16). Of course, other bacteria cause diseases, including tuberculosis, syphilis, and Lyme disease. The development of *antibiotics*, substances that kill bacteria, was a huge step forward in medicine.

Most bacteria are very small. However, a few are actually big enough to see with the naked eye! The largest bacteria known are *Thiomargarita namibiensis*, which means "sulfur pearls of Namibia." (Discovered in Namibia, they grow in long strands like strings of pearls.) These giant bacteria are about the size of the period at the end of this sentence.

CHECK YOURSELF

Some people get yeast infections after taking antibiotics. Why?

CHECK YOUR ANSWER

Antibiotics kill "bad" bacteria as well as the normal "good" bacteria that live in our bodies. The "good" bacteria help keep yeast in check. With the good bacteria out of the way, yeast have a chance to grow.

18.4 Archaea

LEARNING OBJECTIVE
Describe the key features and ecological significance of archaea.

EXPLAIN THIS Why are some archaea described as "lovers of the extreme"?

Once considered a group of funny-looking bacteria, **archaea** ("OUR-kee-uh") are now recognized as a distinct domain of prokaryotic organisms more closely related to eukaryotes than to bacteria. Some features of archaean genetics in particular link archaea to eukaryotes—their ribosomes are like those of eukaryotes, their genes contain introns like those of eukaryotes, and their DNA is associated with histone proteins, like that of eukaryotes.

Many archaea are adapted to extreme environments, such as very salty ponds or the scalding waters of hot springs and hydrothermal vents (Figure 18.6). These archaea are called "extremophiles"—lovers of the extreme. Biologists are interested in extremophiles because they live in conditions similar to those found on the young Earth. Because of this, extremophiles may provide clues about what the earliest living organisms were like. For example, certain archaea thrive in the hydrothermal vent habitats (see Chapter 22) where life may have first evolved. These archaea obtain energy from a chemical that is abundant there, hydrogen sulfide, and form the basis of remarkable vent communities that are entirely independent of sunlight.

FIGURE 18.6
Large colonies of extremophile archaea—the orange and yellow layers—live in the scalding waters of this Nevada geyser.

Because methane is a powerful greenhouse gas, keeping large numbers of cows, sheep, and other livestock contributes significantly to global warming.

Not all archaea are extremophiles, though—many live in more familiar places. Some are found in the open ocean, and others live in the digestive tracts of termites, cows, and other herbivores. These archaea help their hosts digest plant material, and they release methane as a waste product.

18.5 Protists

LEARNING OBJECTIVE
Describe the key features and ecological significance of protists.

EXPLAIN THIS What kind of protist is sometimes found in toothpaste?

Eukaryotes that are not plants, animals, or fungi are lumped together in a group called **Protists**. This group includes autotrophs, heterotrophs, and even species that use both strategies to obtain nutrition. Protists may be single-celled or multicellular. Certain protists, the slime molds, are actually somewhere between single-celled and multicellular—they go from one condition to the other during the course of their lives. Many protists reproduce asexually, but others use sexual reproduction. Because protists do not form a clade, some biologists are in the process of splitting protists into separate groups.

Many protists are autotrophs that get their food from photosynthesis. *Diatoms* are single-celled protists that float in the open ocean. They perform the bulk of oceanic photosynthesis and are a critical part of many marine food chains. Diatoms have elaborate shells made of silica (Figure 18.7). These shells are sometimes used in human-made products—for example, they provide the gritty texture of some toothpastes.

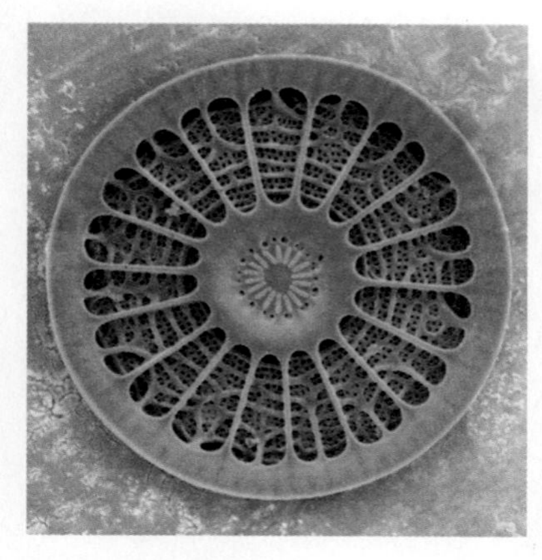

FIGURE 18.7
This microscopic view of a diatom shows its silica shell.

MasteringPhysics®
VIDEO: Diatoms Moving
VIDEO: Dinoflagellates

Dinoflagellates are another group of single-celled marine protists. Some dinoflagellates are autotrophs, and others are heterotrophs. When sunlight and nutrients are plentiful, dinoflagellate populations can explode, producing "red tides." Believe it or not, the discoloration that gives red tides their name is caused by the huge number of dinoflagellates in the water! Some red tides are toxic; shellfish that eat the dinoflagellates become contaminated and poisonous to humans.

FIGURE 18.8
Kelp forms large oceanic "forests" that are home to diverse species. This one is off the west coast of the United States.

Red tides aren't always red. The water may be pink, purple, green, orange, brown, or blue, depending on the dinoflagellate that is responsible.

Some photosynthetic protists are multicellular and can grow quite large. For example, all the different kinds of seaweeds are protists. Kelp forms huge oceanic forests that are home to many unique species (Figure 18.8). Red algae are the source of some of the seaweed we eat, including Japanese nori. Green algae are a group of multicellular protists that likely gave rise to terrestrial plants.

Heterotrophic protists are typically active, single-celled hunters with special cell vacuoles for digesting prey. *Amoebas* move by extending part of their body forward and then pulling the rest of the body behind (Figure 18.9). The extensions are called pseudopodia ("false feet"). Amoebas surround and engulf their prey. *Ciliates* move by beating numerous hairlike projections called cilia. *Flagellates* move by whipping a single long flagellum. Both ciliates and flagellates have openings that function as "mouths." One group of flagellates, called the choanoflagellates, probably gave rise to animals.

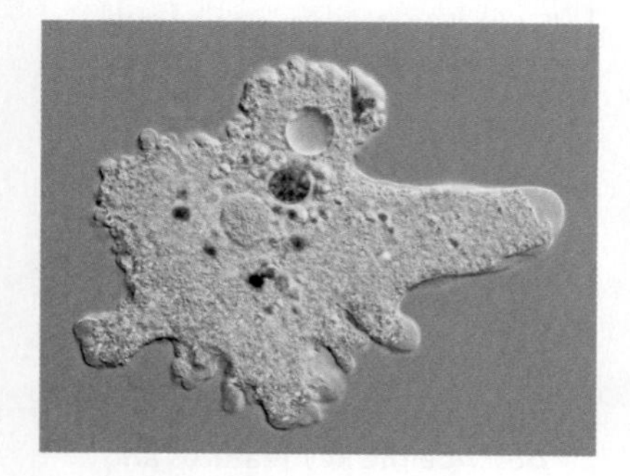

FIGURE 18.9
This freshwater amoeba feeds on bacteria and smaller protists. It uses extensions of its body to move as well as engulf food.

Protists cause a number of serious human diseases, including malaria, African sleeping sickness, and amoebic dysentery. Malaria is caused by *Plasmodium* protists that divide their life cycle between mosquitoes and humans. Humans contract the disease when infected mosquitoes bite them. The protists then move into our red blood cells (Figure 18.10) and reproduce in huge numbers. The synchronized emergence of protists from host red blood cells causes chills, fever, and vomiting.

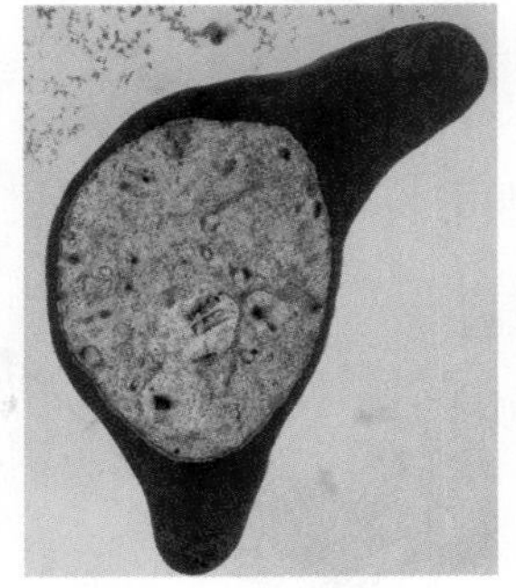

FIGURE 18.10
This misshapen red blood cell has been infected by the malaria-causing protist *Plasmodium* (shown in green).

MasteringPhysics®
VIDEO: *Paramecium* Cilia
VIDEO: *Paramecium* Vacuole
VIDEO: Amoeba
VIDEO: Amoeba Pseudopodia

18.6 Plants

EXPLAIN THIS Why do many flowers smell sweet?

LEARNING OBJECTIVE
Describe the key features and ecological significance of the major groups of plants.

Photos of Earth from space show large green patches stretching across wide areas of the continents (see photo on page 644). Much of Earth's land surface is green because it is covered with plants. **Plants** are terrestrial, multicellular, autotrophic eukaryotes that obtain energy through photosynthesis. Plants are green because they contain chlorophyll, a pigment used in photosynthesis (see Chapter 15).

UNIFYING CONCEPT
- ***The Second Law of Thermodynamics***
Section 6.5

Plants have a variety of adaptations for living in terrestrial environments. Roots anchor them to the ground and absorb water and nutrients from the soil. Shoots, the stems and leaves of a plant, conduct photosynthesis. The leaves of plants have a large surface area for catching sunlight, which powers photosynthesis. Carbon dioxide, which is also needed for photosynthesis, diffuses from the air into leaves through small pores called *stomata* (Figure 18.11).

FIGURE 18.11
Stomata are tiny pores in plant leaves that allow carbon dioxide to enter.

Most plants also have a *vascular system*, a sort of plant "circulatory system" that distributes water and other resources. The plant vascular system consists of two types of tissue: xylem and phloem. The *xylem* is made up of dead, tube-shaped cells through which water and nutrients move up from the roots. The *phloem* consists of living cells that pass the sugars produced during photosynthesis down from the leaves. The liquid that flows in a plant's vascular system is known as *sap*. One sap you may be familiar with is maple syrup, which is made by boiling down liquid collected from the vascular systems of maple trees.

Plant reproduction occurs through an **alternation of generations**, in which the life cycle alternates between a haploid stage called a *gametophyte* and a diploid stage called a *sporophyte*. (Recall from Chapter 16 that haploid cells contain a single set of chromosomes, whereas diploid cells contain two sets of chromosomes.) The details of this life cycle vary among the three major groups of plants—mosses, ferns, and seed plants.

FIGURE 18.12
Mosses growing in the forest understory in Great Smoky Mountains National Park, Tennessee.

Mosses

Mosses are small plants with no vascular systems. Instead, every part of a moss obtains water directly from the environment through diffusion. Because of this, mosses can live only in moist habitats such as bogs or the shady parts of forests (Figure 18.12).

The moss life cycle is shown in Figure 18.13. Mosses are unique among plants in that the gametophyte is much larger than the sporophyte—when you see a moss in the forest, you are looking at a haploid gametophyte. In most

UNIFYING CONCEPT
- ***The Second Law of Thermodynamics***
Section 6.5

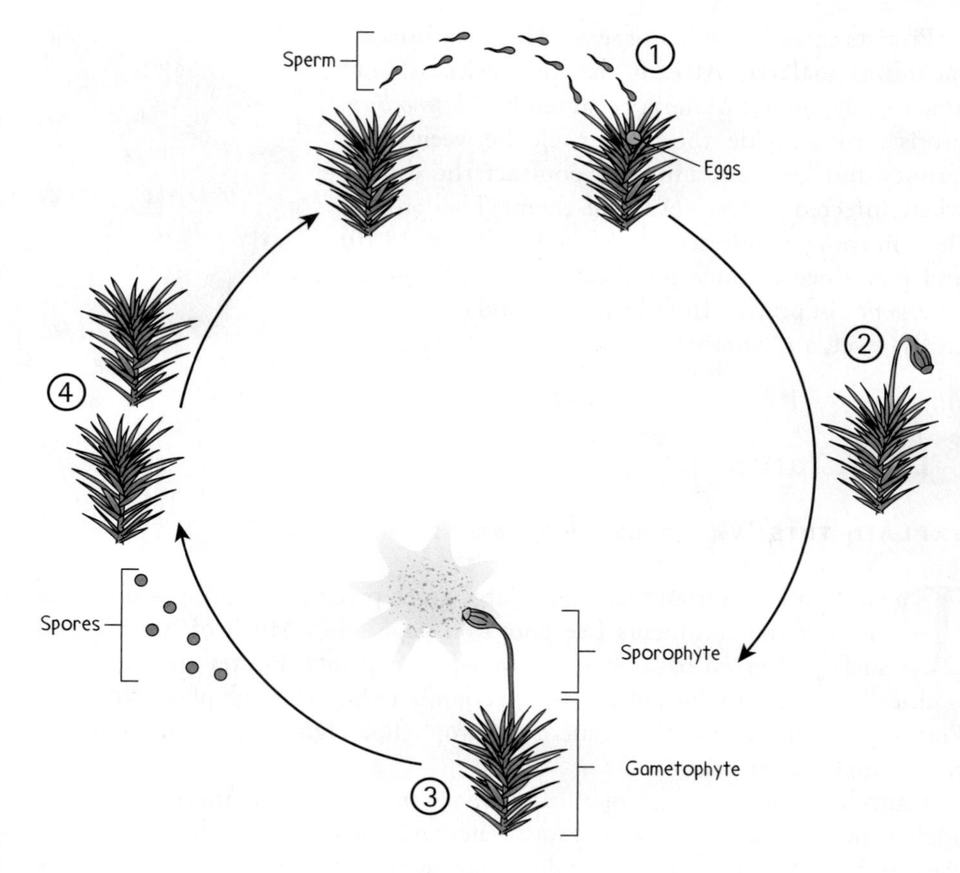

FIGURE 18.13
Mosses, like other plants, are characterized by a life cycle that alternates between a haploid gametophyte stage and a diploid sporophyte stage. (1) The mature gametophytes (haploid) produce haploid sperm and eggs. Sperm swim through a film of water to fertilize eggs. (2) Sporophytes (diploid) grow from fertilized eggs. (3) The mature sporophyte produces haploid spores through meiosis. (4) Gametophytes (haploid) grow from the spores.

mosses, male and female gametophytes are separate. The male gametophyte produces sperm and releases them directly into the environment. The sperm use flagella to swim through a film of water to eggs in the female gametophyte. This aspect of moss reproduction is another reason mosses can live only in moist habitats. Sperm and egg fuse and grow into a tiny diploid sporophyte that is completely dependent on the female gametophyte for nutrients. Eventually, cells in the sporophyte undergo meiosis to produce haploid spores that scatter and grow into new gametophytes—new moss plants. Moss spores have a tough outer coating that allows them to survive under difficult conditions for some time.

Ferns

Ferns are seedless plants with distinctive feathery leaves. They are often found in the forest understory, where they thrive in the shade of large trees (Figure 18.14). Unlike mosses, ferns have a vascular system for transporting water and nutrients. However, ferns are similar to mosses in that their sperm swim through the environment to fertilize eggs. Because of this, ferns can live only in moist habitats.

In ferns, the diploid sporophyte is much larger than the haploid gametophyte (Figure 18.15)—when you see a fern in the forest, you are looking at a diploid sporophyte. Mature fern sporophytes form haploid spores on the underside of special leaves. These spores grow into tiny, but independent, haploid gametophytes. The gametophytes produce eggs and sperm that fuse and grow into new sporophytes.

FIGURE 18.14
Ferns (*right*) grow in a forest on Vancouver Island in British Columbia, Canada.

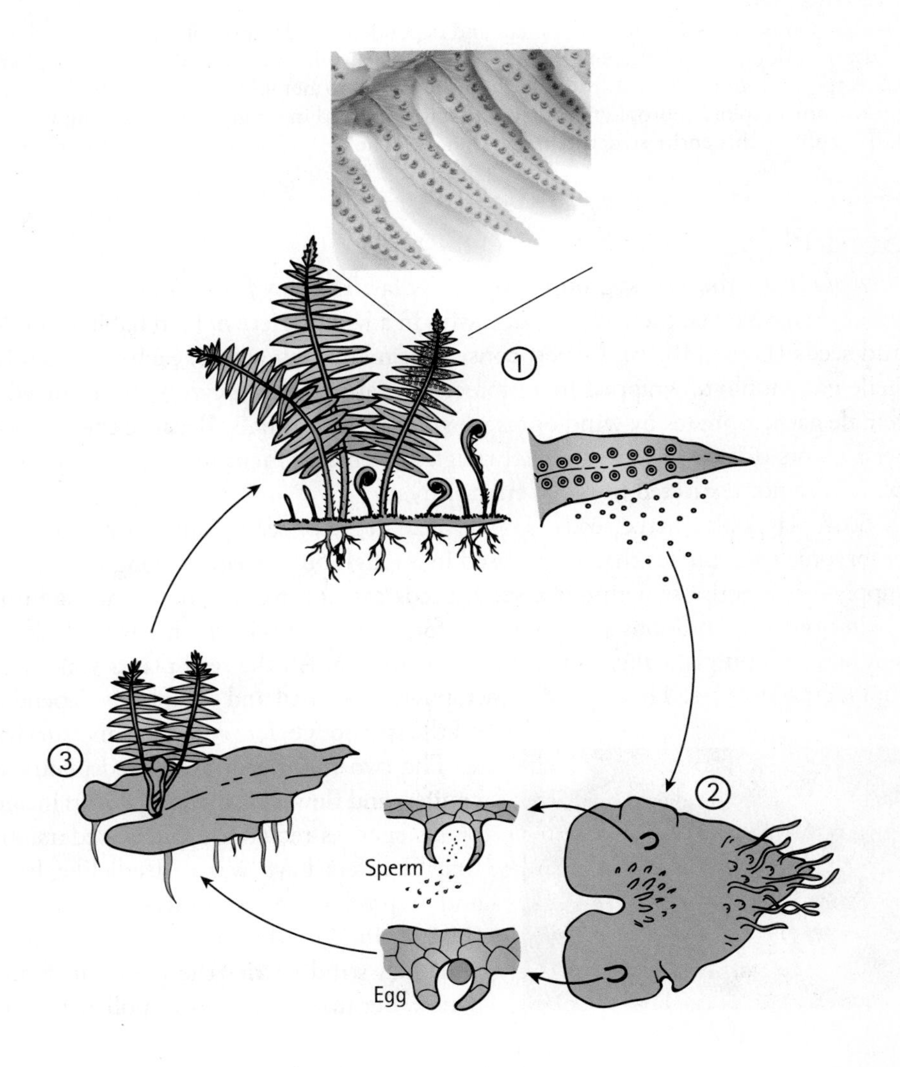

FIGURE 18.15
In the fern life cycle, the sporophyte stage is much larger than the gametophyte stage. The gametophyte is very small, but independent. (1) A mature sporophyte releases haploid spores from structures on the underside of special leaves (see photo). (2) The spores grow into tiny gametophytes, which release sperm that swim to and fertilize eggs. (3) New diploid sporophytes grow from the fertilized eggs.

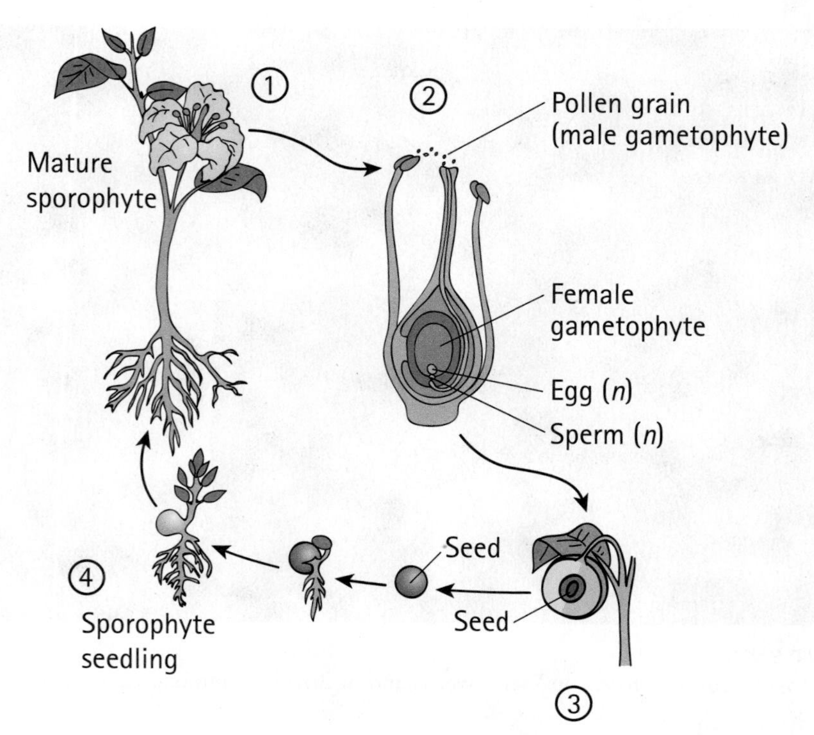

FIGURE 18.16
In seed plants, the gametophyte is small and dependent on the sporophyte for protection and survival. (1) A mature sporophyte produces pollen, which contains male gametophytes. (2) A sperm from pollen fertilizes an egg in the female gametophyte. (3) The fertilized egg grows into a diploid sporophyte embryo, which is encased in a tough outer coating with a food supply—this entire structure is the seed. (4) The seed grows into a new sporophyte.

The tallest tree in the world is a conifer, a coast redwood named "Hyperion." Hyperion stands just over 115 meters tall. According to one team of researchers, no tree can grow past a theoretical limit of 130 meters, the height of a 35-story skyscraper. Because of gravity and friction between water and the vessels of a tree's vascular system, it would be just too hard to transport water any higher.

Seed Plants

Seed plants are the largest group of plants by far. Two key features of the seed plant life cycle have made these plants successful in a wide variety of land habitats: pollen and seeds (Figure 18.16). **Pollen** consists of many tiny grains, each of which is a male gametophyte wrapped in a protective coating. Pollen can be transported to female gametophytes by wind or (as we will see) by animals. Because the sperm of seed plants do not have to swim through the environment to fertilize eggs, seed plants are not restricted to moist environments.

Seed plants also make seeds. The fertilized eggs of seed plants grow into small embryonic sporophytes that are encased in a tough outer coating along with a food supply—this entire structure is a **seed**. Seeds can survive in a dormant state until environmental conditions are appropriate for growth. This is why many seeds do not sprout until you plant them in soil and water them. All the seed plants you see are diploid sporophytes. The haploid gametophytes are small and completely dependent on the sporophyte for protection and survival.

FIGURE 18.17
Conifers are seed plants with reproductive structures called cones. These are cones from a bristlecone pine.

The two main groups of seed plants are conifers and flowering plants. *Conifers* include plants such as redwoods, pines, cedars, and firs. Conifers have waxy, needlelike leaves and reproductive structures called *cones* (Figure 18.17). Male cones release pollen, and then wind carries the pollen to female cones. Because wind blows pollen all over the place, conifers make large amounts of pollen—this makes it more likely that some of the pollen will reach female cones. Fertilization occurs in the female cones, which eventually drop the mature seeds.

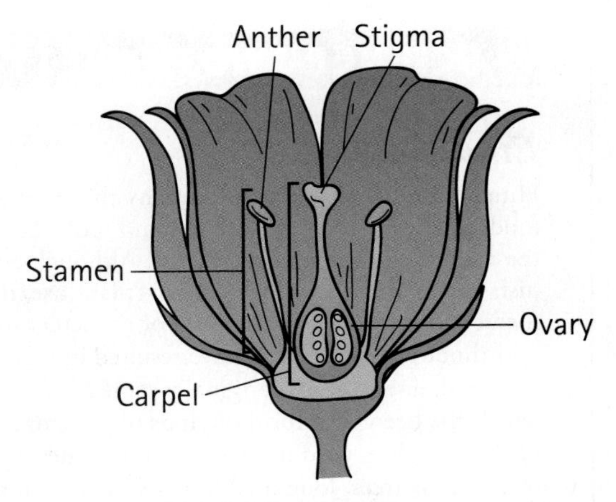

FIGURE 18.18
Flowers contain the reproductive structures of flowering plants.

Flowering plants are the largest and most successful group of seed plants. Flowering plants have two important features absent in conifers: flowers and fruit. A **flower** functions in reproduction—it contains the male structures that produce pollen and the female structures that produce eggs (Figure 18.18). The *stamen* is the male reproductive structure. It consists of a stalk capped with an *anther* where pollen develops. The *carpel* is the female reproductive structure. It includes an *ovary* where eggs develop and a stalk capped by the *stigma*, a sticky structure that traps pollen. In many flowering plants, insects or other animals transport pollen from one flower to another. The petals, scent, and nectar of many flowers have evolved to attract specific animal pollinators. In fact, flowers and their pollinators can be so perfectly suited to each other that you can often look at one and predict features of the other. For example, after Charles Darwin studied a night-blooming orchid in Madagascar, he predicted that there should exist a nocturnal moth with a tongue 30 centimeters long. Forty years later, that moth was finally discovered! Figure 18.19 shows more examples of flowers and their pollinators.

Not all flowers smell sweet! Flowers of the "dead horse arum lily" smell like rotting meat. What are their pollinators? Flies that like to lay their eggs in rotting meat!

Flowering plants surround their seeds with a structure called a **fruit**. A fruit is an adaptation for spreading seeds. When an animal eats a fruit, the seeds pass through its digestive system and eventually come out far from the parent plant. Tasty fruits evolved in certain plants because animals were more likely to eat them. But not all fruits have evolved to be eaten. The burrs that catch on your socks during a hike are also fruits. These fruits hitch a ride until you pull them off and drop them on the ground—again, far from the parent plant.

Seed plants are very important to human societies. We use many different trees for wood, and much of our food comes from the roots, stems, leaves, and fruits of flowering plants.

(a)

(b)

(c)

FIGURE 18.19
Many flowering plants are pollinated by insects or other animals. (a) Flowers pollinated by bees are often blue or yellow, the colors bees see best. Bee-pollinated flowers may also give off a pleasant scent and provide suitable bee landing spots. (b) Flowers pollinated by hummingbirds are often red. Hummingbirds have a poor sense of smell, and the flowers they pollinate are usually odorless. Hummingbirds hover while feeding, so hummingbird-pollinated flowers may be trumpet-shaped. (c) Flowers pollinated by bats, which are active at night, tend to be white, a color that is easy to see at night. Bat-pollinated flowers also have strong odors and grow at the tops of plants for easy access.

SCIENCE AND SOCIETY

Ethnobotany

Human societies use plants for many things, including food, medicine, shelter, clothing, and tools. *Ethnobotany* is the study of how people use plants. Although ethnobotanists are concerned with all types of plant use, the study of medicinal plants has always been of particular interest.

Ethnobotanical studies have resulted in many important medicines. Aspirin originally came from willow bark, which has been used for thousands of years to relieve pain. Quinine, a drug used to treat malaria, comes from the bark of cinchona trees, long used in native Peruvian medicine to treat fever, digestive ailments, and malaria. Artemisinin, another powerful antimalarial drug, is extracted from the leaves of the sweet wormwood tree, a plant that has been used to treat malarial fevers for more than a thousand years in China. More recently, Madagascar periwinkle, a plant used by native peoples for diabetes and other conditions, provided two new cancer drugs.

Scientists looking for new medicines rely on the knowledge of local healers to find promising plants. But developing modern drugs from medicinal plants leads to a difficult ethical issue. What are the rights of indigenous peoples, and how can these rights be protected? The cancer drugs developed from Madagascar periwinkle produced more than a billion dollars in profit for the pharmaceutical giant Eli Lilly, but nothing for traditional societies in Madagascar. Some drug developers now attempt to ensure that local peoples share in the profits. When the AIDS Research Alliance found an anti-HIV compound in a Samoan medicinal tree, the group made direct contributions to the village where the tree was known and also promised 20% of the profits from any drugs that are developed.

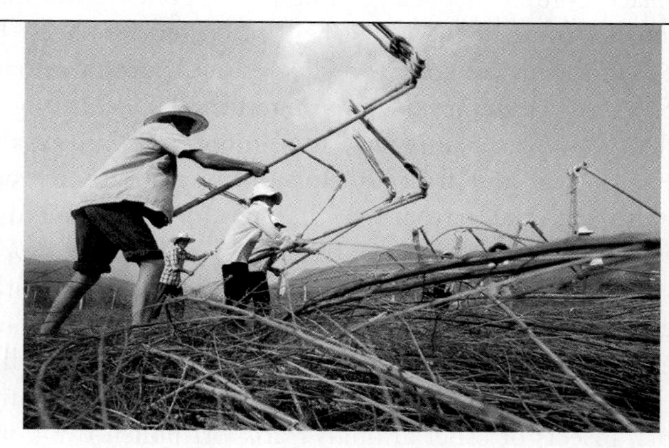

Farmers in Youyang, China, beat the stalks of sweet wormwood to remove the leaves, which will later be used to make the antimalarial drug artemisinin.

Unfortunately, both native plants and invaluable knowledge about them are disappearing rapidly. Native cultures are being steadily lost through "modernization," and countless plant species are going extinct because of habitat destruction and deforestation. One tragic consequence is that many medically useful plants will never be known.

The most massive living organism on Earth is a quaking aspen that has more than 47,000 separate tree trunks covering over 100 acres in Utah. The aspen is named "Pando"—Latin for "I spread." Although Pando looks like a forest of separate trees, its trunks are all connected to a single giant root system—and they all have the same DNA. Pando may be 80,000 years old or more!

CHECK YOURSELF

1. **How does the alternation of generations differ among mosses, ferns, and seed plants?**
2. **Fruits are adaptations that help flowering plants spread their seeds. Why is it adaptive for a plant to spread its seeds?**

CHECK YOUR ANSWERS

1. In mosses, the gametophyte is larger than the sporophyte, and the sporophyte is dependent on the gametophyte for water and nutrients. In ferns, the sporophyte is larger than the gametophyte; the gametophyte is small but independent. In seed plants, the sporophyte is larger than the gametophyte, and the gametophyte is dependent on the sporophyte. In other words, seed plants are sort of the "opposites" of mosses.
2. If plants are able to spread their seeds to a variety of environments, their offspring have a better chance of encountering environments that are well suited to their survival and reproduction.

Integrated Science 18A
PHYSICS AND CHEMISTRY

LEARNING OBJECTIVE
Explain how water moves up from a plant's roots to its shoots.

Moving Water up a Tree

EXPLAIN THIS How can a plant get water by losing water?

The tallest trees are as tall as 30-story skyscrapers. Their highest branches and leaves need water, just like the rest of the plant. How do trees and other plants transport water, against gravity, all the way up to their highest points? Let's take a look.

UNIFYING CONCEPT
- *The Gravitational Force*
Section 5.3

In a plant, there are continuous columns of water molecules extending all the way through the xylem—from the leaves to the roots. These water molecules stick to one another and to the walls of the xylem. The attachment of water molecules to other water molecules is called *cohesion*; the attachment of water molecules to other molecules, such as those of the xylem wall, is called *adhesion*. Both cohesion and adhesion are the result of hydrogen bonds, which form when the positively charged end of one molecule sticks to the negatively charged end of another molecule. (Figure 12.30 in Chapter 12 shows cohesion in water molecules.)

Cohesion and adhesion maintain the continuous columns of water molecules in the xylem. But how do these columns of water move up the plant? The process starts at the leaves, when a plant loses water through transpiration. *Transpiration* occurs when water evaporates from the moist cells inside a leaf and diffuses through the stomata to the outside air. As water is lost from the leaves, a tension is transferred all the way down the water column; water molecules in the leaf pull on water molecules in the nearby xylem, which pull on water molecules farther down the xylem, and so on all the way down to the roots. Transpiration pulls water up the xylem the way sucking on a straw pulls water up the straw. This mechanism of moving water up a plant is called the *transpiration-cohesion-tension mechanism* (Figure 18.20).

UNIFYING CONCEPT
- *The Second Law of Thermodynamics*
Section 6.5

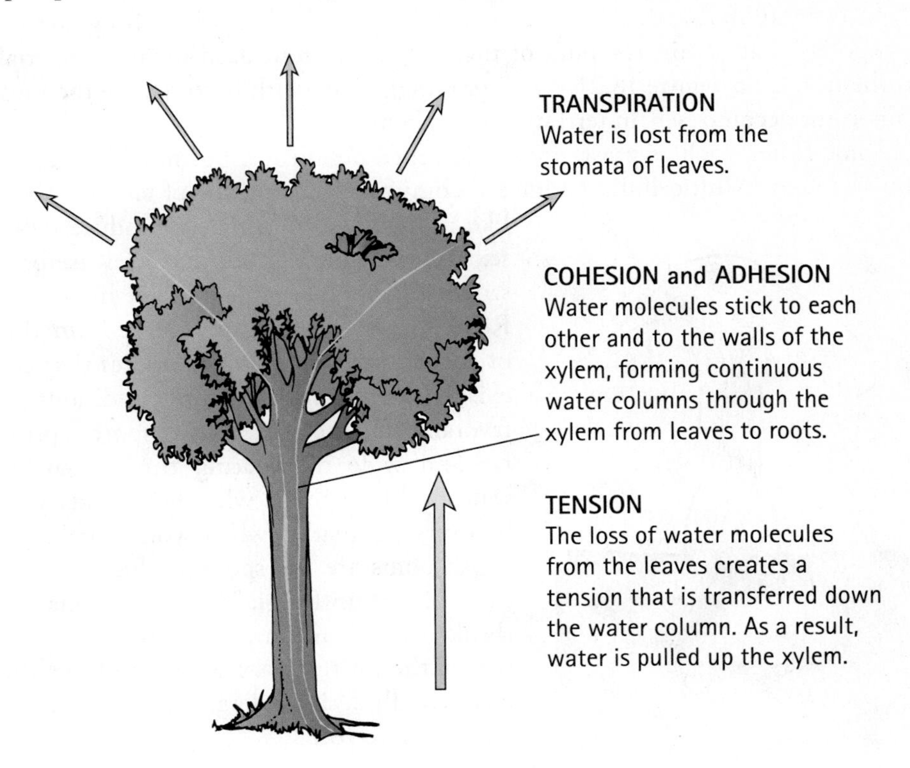

TRANSPIRATION
Water is lost from the stomata of leaves.
COHESION and ADHESION
Water molecules stick to each other and to the walls of the xylem, forming continuous water columns through the xylem from leaves to roots.
TENSION
The loss of water molecules from the leaves creates a tension that is transferred down the water column. As a result, water is pulled up the xylem.

FIGURE 18.20
The transpiration-cohesion-tension mechanism describes how plants move water from their roots to their shoots.

Not only does this mechanism move water up tall trees, it does so without any energy input from the plant!

CHECK YOURSELF

Explain why plants do not need to expend energy to move water from their roots to their shoots.

CHECK YOUR ANSWER

Transpiration does not require energy from the plant because it is the result of diffusion, a form of passive transport (see Chapter 15). The hydrogen bonds required for the cohesion and adhesion of water, which allow transpiration to pull water up the xylem, form on their own and also do not require energy from the plant.

MasteringPhysics®
VIDEO: ***Phlyctochytrium*** **Zoospore Release**

LEARNING OBJECTIVE
Explain the key features and ecological significance of fungi.

18.7 Fungi

EXPLAIN THIS What are you eating when you eat a mushroom?

When you keep a loaf of bread too long, you often end up with some fuzzy stuff called mold. Mold is a fungus, a living organism that belongs to the group Fungi. And it is doing what you once meant to do—it is eating your food!

Fungi were once grouped with plants because of their stationary way of life, but they are actually more closely related to animals. Like animals, fungi are heterotrophs that obtain food from other organisms. Fungi release digestive enzymes over organic matter and then absorb the nutrients. This distinguishes them from animals, which digest food inside their bodies. Many fungi are decomposers that obtain the bulk of their nutrients from dead organic material, as the fungus in Figure 18.21 is doing. Fungi, along with bacteria, are the most important decomposers in terrestrial ecosystems.

FIGURE 18.21
Many fungi are decomposers and obtain their nutrients from dead organic material. This fungus is growing on dead wood. The mushrooms are reproductive structures.

Some fungi, such as yeast, are single-celled organisms, but most species are multicellular. Multicellular fungi are composed of bunches of small thread-like filaments. Fungi may reproduce either sexually or asexually; most species use both strategies at some point in their life cycles. Reproduction occurs through the formation of spores, tiny reproductive bodies that can exist in a dormant state until conditions are favorable for growth. Fungal spores spread far and wide by floating through air or water—this explains why mold finds your leftovers no matter where you hide them. Mushrooms are the spore-producing structures of certain fungi. Notice that what we think of as a "mushroom" is only a small part of the entire organism. Most of the fungus actually lives underground, as you can see in Figure 18.22.

FIGURE 18.22
Mushrooms are spore-producing structures in certain fungi. Most of the organism actually lives underground.

Fungi are essential to the survival and growth of many, perhaps most, plants. This is because in most plant species, the roots form close associations with fungi. These associations, called *mycorrhizae* ("my-kuh-RYE-zuh"), benefit both fungus and plant. The fungus receives nutrients from the plant while helping the roots absorb water and minerals from the soil.

Fungi important to humans include yeast, which is used in baking and brewing, and edible mushrooms. Fungi are also used to make blue cheeses such as Roquefort and gorgonzola (the blue stuff is actually zillions of tiny fungal spores—enjoy!). Penicillin, the first antibiotic, was originally found in a fungus. Finally, human fungal diseases include yeast infections, ringworm, and athlete's foot.

Pando the tree may be the most *massive* living organism in the world, but the one with the largest overall *size* is a fungus—an underground honey mushroom in Oregon that measures 5.6 kilometers (3.5 miles) across. The fungus covers more than 2200 acres! It was discovered when scientists were trying to figure out why large groves of trees were dying. The fungus was eating them.

18.8 Animals

LEARNING OBJECTIVE
Explain the key features and ecological significance of the major groups of animals.

EXPLAIN THIS Can a mammal lay an egg?

Animals include creatures as varied as starfish, beetles, coral, and antelope. **Animals** are multicellular, heterotrophic eukaryotes that obtain nutrients by eating other organisms. Animals *ingest* food, taking it into their bodies for digestion. Most animals reproduce sexually and are diploid during most of their life cycle. The gametes—sperm and eggs—are the only haploid stage. Many animals go through a juvenile period as a *larva* that is markedly different from the adult in form and ecology; examples are butterfly caterpillars and frog tadpoles. Most animals also have muscles for moving, sense organs for collecting information from their environments, and nervous systems for controlling their actions. The evolutionary tree in Figure 18.23 shows a tentative hypothesis of the relationships among the major groups of animals.

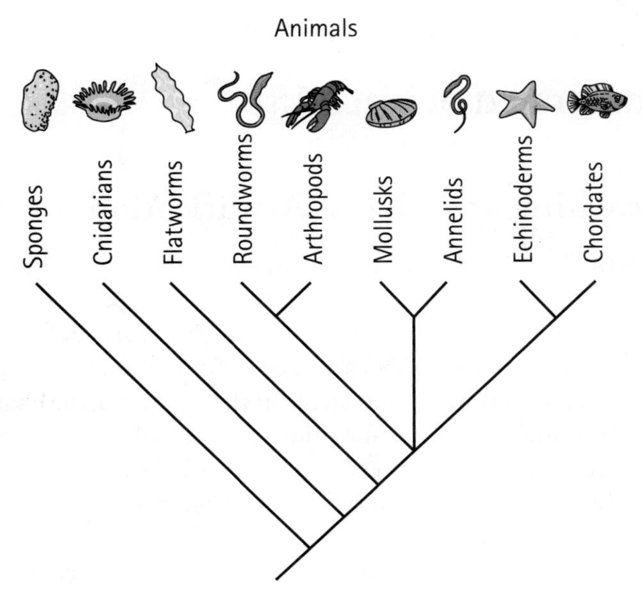

FIGURE 18.23
This evolutionary tree shows how the major groups of animals are related to one another.

FIGURE 18.24
This is a purple tube sponge.

Sponges

Sponges are sedentary marine animals (Figure 18.24). Most sponges have a tube-like shape with a large central cavity. Special cells in the sponge beat their flagella to produce a constant flow of water through the animal. Water enters through numerous pores, flows into the sponge's central cavity, and goes out the top. This constant current allows the sponge to catch food. Sponge cells trap bacteria from the water, digest them, and then distribute the nutrients to other cells.

Sponges are the only animals that lack tissues, groups of similar cells that perform a certain function. This allows sponges to do unusual things—for example, if you separate a sponge's cells by passing it through a sieve, the cells will reassemble on the other side, forming a new sponge. No other animals can do that.

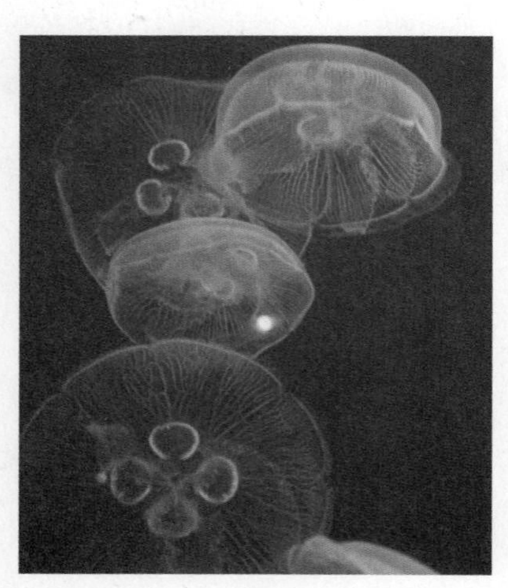

FIGURE 18.25
These are moon jellyfish, a species found in coastal waters all over the world. The moon jellyfish is a cnidarian in its medusa form.

Cnidarians

Cnidarians ("nye-DARE-ee-uhns") include animals such as jellyfish, sea anemones, and corals. Unlike sponges, cnidarians have two distinct tissue layers: an outer layer that protects the body and an inner layer that digests food. These layers are separated by a jellylike middle layer. Cnidarians use tentacles armed with barbed stinging cells to catch prey. In many species, the stinging cells release powerful toxins. (This is why jellyfish can be a danger to ocean swimmers.) Prey are digested in a *gastrovascular cavity* that has a single opening that serves as both mouth and anus. Many cnidarians alternate between a sedentary polyp stage and a mobile, bell-shaped, medusa stage. Cnidarians such as sea anemones and corals spend most of their lives as polyps. Cnidarians such as jellyfish spend most of their lives as medusas (Figure 18.25).

LEARNING OBJECTIVE
Explain how coral bleaching and ocean acidification threaten marine life.

Integrated Science 18B
EARTH SCIENCE AND CHEMISTRY

Coral Bleaching and Ocean Acidification

EXPLAIN THIS Why is ocean acidification dangerous for animals with shells?

UNIFYING CONCEPT
- ***The Ecosystem***
  Section 21.1

Coral reefs are found in tropical oceans, in clear, shallow waters with temperatures between 20°C and 28°C (68°F and 82°F). Coral reefs are among the most diverse ecosystems in the world (see the chapter-opening photo). Numerous marine species, including commercially important fish, spend all or part of their lives in coral reefs. Reefs also help protect shorelines from ocean waves. But, as humans burn more and more fossil fuels, the amount of carbon dioxide in Earth's atmosphere has increased, resulting in global warming and more acidic oceans (see Chapter 27). How will this affect corals and other marine species?

Corals are cnidarians. They live in colonies of tiny polyps, each wrapped in a calcium carbonate skeleton. Unlike many other cnidarians, though, corals are not hunters. They get most of their nutrients from photosynthesizing dinoflagellates that live within their cells. Coral bleaching occurs when corals kick out their dinoflagellates. The corals literally turn white because it is their dinoflagellates that give them their colors (Figure 18.26). Coral bleaching is most often triggered by an increase in seawater temperature. High temperatures interfere with photosynthesis in dinoflagellates, causing toxic molecules to build up in them. This in turn causes the corals to eject them. Corals can survive for a short time without their dinoflagellates. If water temperatures decrease again, dinoflagellates move back into bleached corals, and the corals survive. But, if warm temperatures continue for too long, the corals starve to death.

FIGURE 18.26
These bleaching corals are kicking out their dinoflagellates.

Mass coral bleaching has become common in recent years due to the high water temperatures associated with global warming. For example, in 2005, the United States lost half of its Caribbean corals to a massive bleaching event. Data showed that thermal stress in 2005 was worse than in all the preceding 20 years combined. As temperatures continue to rise, bleaching is likely to become more and more widespread. Will coral reefs disappear? Corals do vary in their preferred temperatures and in their susceptibility to bleaching. For example, certain corals have special fluorescent pigments that they use, like sunscreen, to shield their dinoflagellates. These fluorescent corals have survived mass bleaching episodes better than nonfluorescent corals.

High carbon dioxide levels also lead to ocean acidification. As we discussed in Chapter 13, atmospheric carbon dioxide (CO_2) is absorbed by the ocean where it reacts to form carbonic acid (H_2CO_3), which lowers the pH of the ocean while also transforming carbonate minerals, such as calcium carbonate ($CaCO_3$), into bicarbonate compounds:

$$CO_2 + H_2O \rightarrow H_2CO_3$$

$$H_2CO_3 + CaCO_3 \rightarrow Ca(HCO_3)_2$$

Many marine animals—including corals, echinoderms, crustaceans, and mollusks—need calcium carbonate to build their shells. In addition, acidified seawater increases the rate of dissolution of their shells. For these reasons, ocean acidification causes shelled animals to grow more slowly and have weaker shells. Scientists have observed that coral growth slows in acidified waters, and that the larvae of bivalves such as oysters and mussels are smaller. Some plankton, including certain photosynthetic species, also have calcium carbonate shells, and plankton mass decreases as seawater becomes more acidified. Because photosynthetic plankton form a crucial part of marine food chains, the consequences could extend across entire marine communities.

CHECK YOURSELF

1. **Global warming also causes sea levels to rise, mainly because seawater expands as its temperature increases. How do rising sea levels affect corals?**
2. **As global warming continues, how might the species composition of coral reefs change?**

CHECK YOUR ANSWERS

1. Corals are found only in clear, shallow waters, where there is enough sunlight for their dinoflagellates to photosynthesize. Rising sea levels will require corals to shift to shallower ground or to grow upward quickly. Whether corals can find appropriate habitat and keep pace with sea-level changes remains to be seen.
2. Corals that are adapted to warmer temperatures may spread as global warming continues. In addition, coral reefs may come to be dominated by corals that are more resistant to bleaching. For example, species that have fluorescent pigments may survive, while species that lack these sunscreens may die out.

Flatworms

UNIFYING CONCEPT

- ***The Second Law of Thermodynamics*** **Section 6.5**

Flatworms have distinct "head" and "tail" ends as well as "back" and "belly" sides. A single body opening serves as both mouth and anus, and an elaborately branched digestive tract transports nutrients to the entire body. The flat shape of flatworms allows oxygen to be absorbed efficiently across the skin via diffusion. Flatworms include many parasites as well as some nonparasitic species. The most familiar flatworms are tapeworms (Figure 18.27)—long, ribbonlike worms that live as parasites in the intestines of humans and other animals.

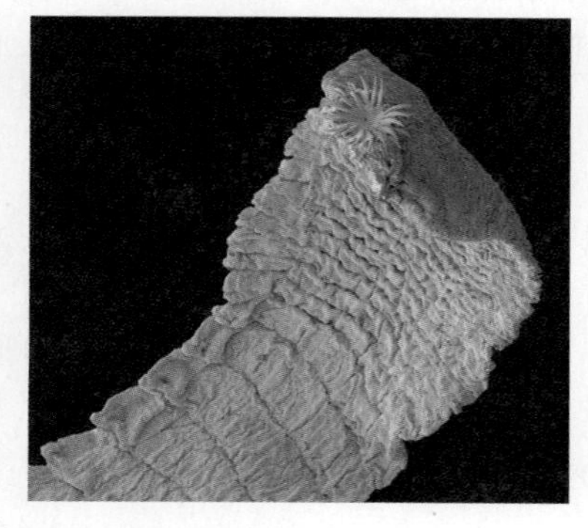

FIGURE 18.27
A tapeworm has hooks at the front end that help it attach to its host's intestines.

Roundworms

Roundworms (not to be confused with the more familiar earthworms) have small, slender bodies with tapered ends and a round cross section (Figure 18.28). They have both a mouth and an anus. Like their relatives the arthropods, roundworms have a tough outer cuticle that is shed periodically during growth. Roundworms eat a variety of things—bacteria, plants, fungi, other animals—but many specialize on decaying organic material. In many habitats, roundworms are important decomposers. Roundworms have muscles that run from head to tail. As a result, these worms move like flailing whips as muscles on alternate sides of the body contract. Roundworms are responsible for several human diseases, including hookworm, pinworm, elephantiasis, and trichinosis.

MasteringPhysics®
VIDEO: C. Elegans Crawling

FIGURE 18.28
A roundworm moves through rotting vegetables in a compost heap.

Arthropods

Arthropods include lobsters, barnacles, spiders, scorpions, ticks, centipedes, insects, and many other species. They are found in just about every known habitat on Earth. All arthropods have an external skeleton called an **exoskeleton** that protects and supports the body. The exoskeleton is incapable of growth and must be shed periodically as animals grow. The bodies of arthropods are divided into different segments, and their legs have bendable joints. Some arthropod legs have been modified during evolution to function as mouthparts, antennae, and reproductive organs. Arthropods have a brain and well-developed sense organs. Many species pass through a distinct larval stage during their growth and development.

The major groups of arthropods are the crustaceans, the chelicerates, and the uniramians (Figure 18.29). *Crustaceans* are mostly aquatic and include species such as lobsters, crabs, shrimp, krill, and barnacles. *Chelicerates* are eight-legged animals such as horseshoe crabs, spiders, scorpions, ticks, and mites. *Uniramians*

(a)

(b)

(c)

FIGURE 18.29
Arthropods have segmented bodies and jointed legs. (a) This barnacle is a crustacean. Its shell may make you think of mollusks, but its feathery jointed legs, which it uses to filter food from the water, are an arthropod feature. (b) This jumping spider is a chelicerate. (c) This damselfly is an insect and a uniramian. Here, the adult is shedding its exoskeleton.

MasteringPhysics®
VIDEO: Lobster Mouthparts

include centipedes, millipedes, and insects. Insects are the most diverse group of living organisms on Earth, with more than a million known species and perhaps as many as ten times that number waiting to be discovered. All insects have three body parts—a head, thorax, and abdomen—and three pairs of legs. Most also have two pairs of wings. Many insects are important to humans as plant pollinators. Others affect us because they transmit disease (mosquitoes carry malaria and West Nile virus) or are agricultural pests.

Despite their diversity and success, the one thing insects have not achieved is large size: Why are insects so small? The answer has to do with how they obtain oxygen. Insects obtain oxygen via branched tubules connected to the outside air. Oxygen must diffuse through the tubules to reach the tissues, a strategy that works in only extremely small bodies. However, there were much bigger insects during the Carboniferous period, 300 million years ago, when atmospheric oxygen levels were much higher than they are today. One Carboniferous dragonfly had a $2\frac{1}{2}$-foot wingspan!

The earliest flying insects weren't able to fold their wings back on their bodies. That evolutionary innovation came later in insect history. Dragonflies are an example of a group that still retains the original trait. Have you noticed how their wings stick straight out?

UNIFYING CONCEPT
- ***The Second Law of Thermodynamics***
  Section 6.5

Mollusks

Mollusks are soft-bodied animals such as clams, oysters, squids, octopuses, snails, and slugs. Most mollusks have a protective shell, although the shell is tiny in some species (squids) and entirely absent in others (octopuses and slugs). All mollusks have a muscular "foot" responsible for locomotion, a visceral mass that holds the digestive and reproductive organs, and a mantle that secretes the shell. There are three main groups of mollusks. *Bivalves* have two hinged shells and include species such as clams, oysters, mussels, and scallops. Most bivalves are sedentary and feed by filtering small particles from the water. *Cephalopods* such as squids and octopuses (Figure 18.30) are active predators that use arms (eight in octopuses and ten in squids) to capture prey. Cephalopods also have well-developed brains

FIGURE 18.30
This common octopus is a cephalopod, a type of mollusk.

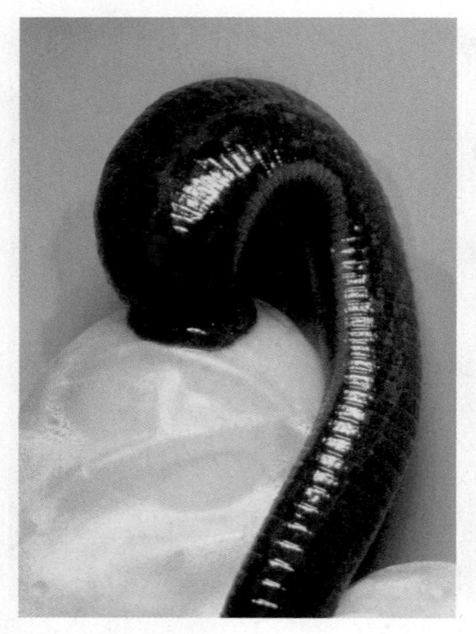

FIGURE 18.31
The leech is an annelid that feeds on blood. The leech shown here is sometimes used for medical purposes.

MasteringPhysics®
VIDEO: Earthworm Locomotion
VIDEO: Echinoderm Tube Feet

At one time, "bleeding" a sick person with leeches was common. Then this practice was viewed as medical quackery. Now, the leech is back! Leeches make powerful anticoagulants that are better than anything humans have invented. During certain delicate surgeries, leeches are used to keep blood flowing, so that blood vessels don't get clogged.

FIGURE 18.32
Two echinoderms in the Adriatic Sea: A brittle star and a sea cucumber.

and eyes. *Gastropods* have a single, spiral shell and include species such as snails, abalone, and limpets. Most gastropods are herbivores.

Annelids

Annelids are a group of segmented worms that include earthworms, leeches (Figure 18.31), and other species. Annelids have some muscles that go around the body and others that go head-to-tail. Their muscles allow for great flexibility of motion—for example, annelids can contract one part of their body while keeping the rest still. Earthworms are important decomposers that feed by ingesting large amounts of soil and absorbing the available nutrients. Their burrowing activity also helps to aerate soil, supplying it with oxygen. Leeches are parasites that feed on blood. They have bladelike teeth for cutting through their host's skin and secrete anticoagulants that prevent blood from clotting while they eat.

Echinoderms

Echinoderms include starfish, sea urchins, and sea cucumbers (Figure 18.32). All echinoderms have an internal skeleton, or **endoskeleton**, made of small, interlocking plates. They use small, suckerlike appendages called *tube feet* to move and stick to rocks. Starfish also use their tube feet to pry open the shells of bivalves. Most echinoderms live on the ocean floor and move very . . . very . . . slowly.

Chordates

The *chordates* include tunicates, lancelets, and vertebrates, the group to which humans belong. Chordates share four key features: a brain and a spinal cord that runs along the back of the body; a *notochord*—a stiff but bendable rod that supports the back; gill slits; and a tail that extends beyond the anus. These features are not apparent in all adult chordates (for example, humans don't have tails) but are generally present at some stage of development (human embryos do go through a tailed stage).

Tunicates are sedentary marine species also known as sea squirts. They feed by filtering small particles from the water. *Lancelets* are small, blade-shaped, swimming marine species that often bury themselves in sand. Like tunicates, lancelets filter food from the water.

Vertebrates are animals with backbones. The vertebrates include several groups of fishes as well as amphibians, reptiles, and mammals. Did you know that the earliest vertebrates had mouths but not hinged jaws? Only a few jawless vertebrates are still in existence. Lampreys, parasites that suck blood from other fish, are shown in Figure 18.33.

Cartilaginous fishes include sharks, skates, and rays. These fishes have skeletons made of cartilage rather than bone. Sharks are famous predators. Their unusual *electroreceptive organs* help them detect the tiny electric currents in the muscles and nerves of nearby prey. (We'll see why nerves and muscles have electric currents in Chapter 19.)

Ray-finned fishes include most of the animals we think of as "fish"—tuna, salmon, bass, and so on. Ray-finned fishes have a sac in their bodies called a *swim bladder*. The swim bladder is filled with just enough gas to make the fish's density the same as the density of water. This means that a ray-finned fish doesn't sink or float, which gives it great mobility in the water. Cartilaginous fishes, which do not have swim bladders, will actually sink if they stop swimming.

UNIFYING CONCEPT

- ***Density***
  Section 2.3

At first glance, *lungfishes* (Figure 18.34) and *coelacanths* may look like ray-finned fishes. However, they are actually more closely related to terrestrial vertebrates such as amphibians, reptiles, and mammals. The bones in their fins are arranged serially, like those in the limbs of terrestrial vertebrates, rather than as rays emanating from the base of the fin. Although only a few species remain, these species are of great interest to biologists because they provide clues about how terrestrial vertebrates evolved.

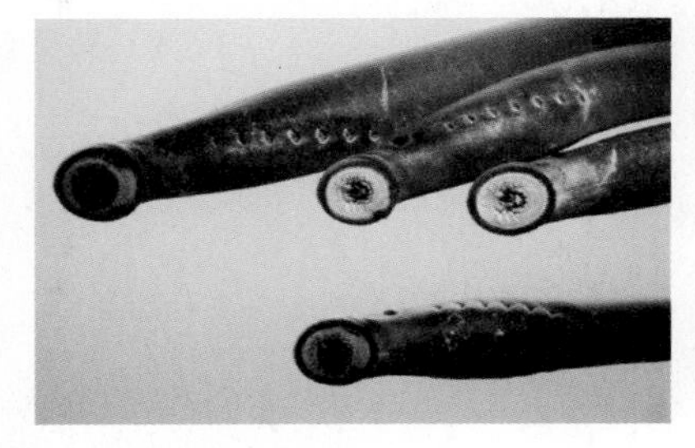

FIGURE 18.33
The earliest vertebrates did not have hinged jaws. Modern lampreys retain this condition. These individuals are clinging to glass with their circular mouths.

Amphibians include salamanders, frogs, and caecilians, a group of limbless species. The name *amphibian* refers to the fact that many species use both aquatic and terrestrial habitats. In many amphibians, aquatic larvae metamorphose into terrestrial adults. However, other amphibians are entirely aquatic, and many are entirely terrestrial. Amphibians are restricted to moist habitats because their skins are made of living cells that dry out. Amphibian eggs do not have shells and also require moisture (Figure 18.35). In the last few decades, dozens of amphibian species have suddenly gone extinct, and hundreds have experienced huge population declines. What is especially worrisome is that many species, such as the remarkable golden toad (Figure 18.36), have disappeared from protected nature reserves.

Reptiles (including the feathered ones—birds) and mammals are *amniotes*. Amniotes have two traits that enable them to live in diverse terrestrial habitats. First, their skin is made of dead cells, which helps to prevent water loss. Second, their eggs have shells, which keep them from drying out.

FIGURE 18.34
Lungfishes may look like other ray-finned fishes, but they are more closely related to terrestrial vertebrates. Note the unusual fins.

FIGURE 18.35
Amphibian eggs have no shells, and so require moisture to keep from drying out. These are frog eggs.

FIGURE 18.36
The golden toad is one of many amphibian species that has disappeared in the last few decades.

(a) (b)

FIGURE 18.37
Snakes and birds are reptiles, a group that also includes turtles, lizards, and crocodilians. (a) Many snakes have adaptations for swallowing large prey whole. This rat snake is eating a bird. (b) Birds are reptiles that have evolved the ability to fly. This is a great blue heron.

Flight has evolved three separate times in vertebrates—in birds, bats, and the extinct pterodactyls.

UNIFYING CONCEPT

- ***The Second Law of Thermodynamics*** **Section 6.5**

FIGURE 18.38
A baby koala sleeps in its mother's pouch.

Reptiles include turtles, lizards and snakes, crocodilians, and birds (Figure 18.37). Turtles have protective shells that are actually modified ribs. (Imagine squeezing your entire body inside your rib cage!) Lizards are a large and diverse group found in a wide array of terrestrial environments. Snakes are a group of lizards that have lost their legs and evolved adaptations for subduing large prey and swallowing them whole. Birds are exceptional in that they have evolved the ability to fly. In birds, adaptations for flight include wings, feathers, and hollow bones (for lightness).

All reptiles except birds are ectotherms. An **ectotherm** uses behavior to regulate its body temperature. For example, lizards bask in the sunlight to warm up and retreat to the shade to cool down. The body temperatures of ectotherms vary, to some degree, depending on environmental conditions. Birds and mammals are endotherms. An **endotherm** maintains a constant, high body temperature by breaking down food; this process releases heat, which warms the body. Because an endotherm warms itself by breaking down food, it has to eat more than an ectotherm of the same size. This is why you can feed a pet snake once a week, but a pet dog has to eat several times a day.

Mammals have hair and feed their young milk. The majority of mammals live on land, but bats fly, and seals and whales are partly or fully aquatic. There are three major groups of mammals. *Monotremes*, such as the platypus and spiny echidna, lay eggs. *Marsupials*, such as possums, koalas, and kangaroos, give birth to very immature live young. Newborn marsupials crawl up into their mother's pouch and attach to a nipple where they eat and continue to develop (Figure 18.38). *Placentals*, which include most living mammal species, also give birth to live young; however, newborn placentals are more mature than newborn marsupials.

HISTORY OF SCIENCE

Classifying the Platypus

The first dried platypus specimen reached England from Australia in 1799 and created an immediate sensation. The animal had a ducklike bill, webbed front legs, clawed hind legs, and a covering of thick fur. It did not have nipples, but, like birds, it had a cloaca—a single opening for the reproductive, excretory, and digestive systems. The platypus seemed to be an impossible cross between a bird and a mammal. Very quickly, attention focused on one key question: How did the platypus reproduce? Did it lay eggs or give birth to live young?

The platypus is an egg-laying mammal.

In 1821, naval surgeon Patrick Hill reported that certain Australian Aborigines were familiar with the platypus and described it as laying eggs in nests on the water surface. Then, in 1824, the anatomist Johann Meckel found mammary glands on a platypus—the glands opened directly to the skin without nipples. But this led to another puzzle: How could a platypus with a ducklike bill suckle? In 1831, platypus nests containing broken eggshells were found, and milk was seen to ooze from the skin of a female platypus's abdomen. In 1834, the anatomist Richard Owen examined the mouth of a baby platypus and concluded that it could suckle in the usual way. Plus, he found milk in the baby's stomach. Finally, in 1884, a female platypus was shot while in the process of laying eggs. To those who had yet to be convinced, this was decisive. The platypus was indeed a mammal that lays eggs.

Today, the platypus is classified in the mammalian clade Monotremata, which means "one hole" and refers to the animal's single exit for the reproductive, excretory, and digestive systems. Only one other creature belongs to this group—the spiny echidna, the platypus's closest relative left on Earth.

CHECK YOURSELF

Dispersal is important for most living organisms. How do animals disperse? How do animal dispersal strategies compare with those of plants?

CHECK YOUR ANSWER

Most animals are able to move. The ones that are sedentary (sponges, some bivalves, barnacles, and so forth) release their eggs into the water, where water currents disperse them. Some sedentary animals also have a mobile larval stage. Plants, which are sedentary, are able to disperse by forming spores or seeds that are carried by wind, water, or animals. Flowering plants have fruits that help them disperse their seeds.

Ectotherms and endotherms were once called "cold-blooded" and "warm-blooded," but these terms just aren't accurate. For example, some desert lizards have body temperatures much higher than 100°F, hotter than the body temperatures of most birds and mammals.

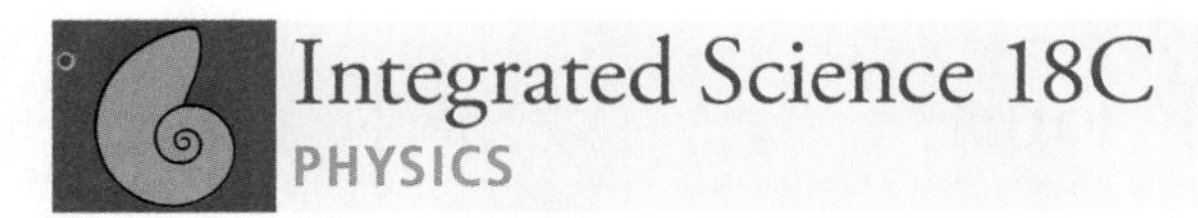

Integrated Science 18C
PHYSICS

LEARNING OBJECTIVE
Explain how birds fly.

How Birds Fly

EXPLAIN THIS Why are tiny hummingbirds such good fliers?

Bird flight has always fascinated humans. Without birds as a model, would we ever have dreamed up the airplane? But how can something heavier than air stay aloft?

Shape is the key. The wings of birds are *airfoils*—their curved shape causes air to flow faster over the top of the wing than under the wing (Figure 18.39). This is because, as a bird cuts through the air, air molecules have a greater distance to travel over the wing and so move faster. Bernoulli's principle (Appendix E) tells us that

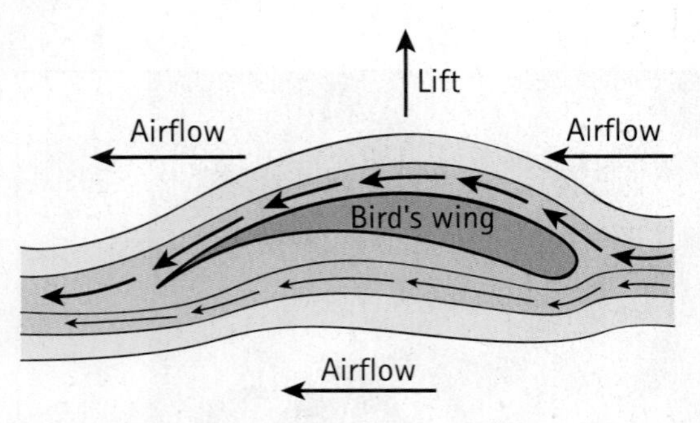

FIGURE 18.39
The shape of birds' wings causes air to flow faster over the top of the wing than under the wing. This produces lift, an upward force. Lift balances gravity, enabling birds to stay in the air.

UNIFYING CONCEPT

- ***Newton's Third Law***
  Section 3.4

when the speed of air over a surface increases, its pressure decreases. This means that air pressure above the wing is lower than air pressure below the wing. The result is an upward force called *lift*. Lift balances gravity and keeps birds in the air.

In order to generate lift, birds must move forward, which they do by flapping their wings. When a bird pulls its wings downward, it pushes air down and back. According to Newton's third law, if a bird pushes air backward, the air pushes it forward. As birds pull their wings back up, however, they produce *drag*, a force that drives them backward. Birds twist their wings during the upstroke to reduce drag. (You do something similar when you swim the breaststroke. On the downstroke, you push your arms back and your body moves forward. On the upstroke, you pull your arms in to reduce drag.)

Some birds are able to stay in the air for a long time without flapping their wings. This is called *soaring*. You often see eagles and vultures soar. Soaring is possible because the birds are floating on top of a *thermal*, a mass of rising hot air. It's sort of like sitting on top of a geyser but (luckily for the birds) much more controlled.

CHECK YOURSELF

1. **Which produces more lift as it flies: an eagle or a sparrow?**
2. **Given the way lift is produced, do you think it is harder for large birds or small birds to stay in the air?**

CHECK YOUR ANSWERS

1. For a bird to stay in the air, lift has to balance gravity. Because the eagle weighs more, it has to produce more lift to stay in the air.
2. Large birds have a harder time staying in the air. Because lift comes from air pressure under the wings, the amount of lift a bird generates depends on the surface area of its wings. Lift must counter gravity, which is proportional to the volume of the bird. Since the surface-area-to-volume ratio is smaller in larger animals (see Chapter 17), larger birds have more difficulty generating the necessary lift. This explains why large birds have relatively larger wings than small birds. It also explains why small birds are better fliers—think of the maneuverability of a hummingbird! Many of the largest birds actually spend most of their time soaring on thermals.

LEARNING OBJECTIVE
Describe how viruses and prions cause disease.

18.9 Viruses and Prions

EXPLAIN THIS Why does the flu come back in a different form every year?

Are any viruses beneficial to their hosts? Not that we know of. However, certain viruses are useful to humans. For example, viruses are used in many DNA technologies, including gene therapy and genetic engineering.

A **virus** is a small piece of genetic material wrapped in a protein coat. Viruses are on the border between living and nonliving. They are not made of cells, and they can reproduce only within a host cell. However, they have genes and they evolve. Where did viruses come from? Scientists believe viruses originate when little pieces of host DNA or RNA somehow evolve the ability to move from one cell to another. This means that viruses have probably originated many times.

Many viruses have normal double-stranded DNA, but others have genes made of single-stranded DNA, single-stranded RNA, or double-stranded RNA. (You can review DNA and RNA in Chapter 16.) Viruses reproduce by infecting a host cell and then using the cell's resources to copy their genetic material and build viral proteins. These are then assembled to form new viruses. All forms of life—from bacteria to plants and animals—are infected by viruses. Human immunodeficiency virus (HIV), the virus that causes acquired immunodeficiency syndrome (AIDS), is shown infecting a human immune cell in Figure 18.40.

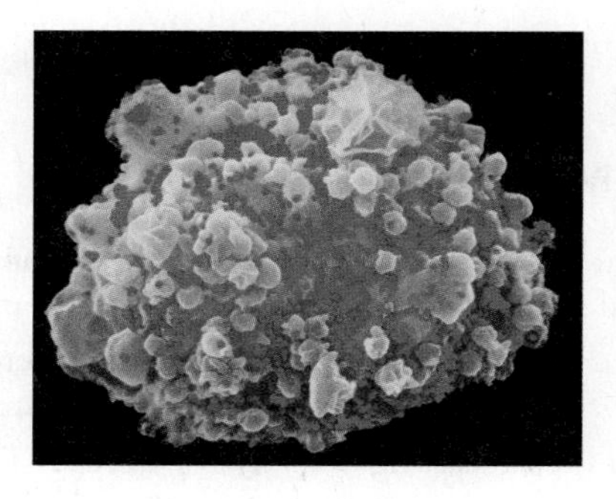

FIGURE 18.40
This human immune cell is infected with HIV, the virus that causes AIDS. The cell is covered with viral particles that will soon spread and infect other cells.

Viruses are responsible for many human diseases, including the common cold, flu, AIDS, and smallpox. One feature of viruses is that their genes mutate very quickly. This explains why there are always plenty of colds to catch and why the flu comes back, in a different form, year after year.

Prions are proteins that are incorrectly folded. (Remember that proteins are folded strings of amino acids. In prions, the folding has somehow gone wrong.) Prions cause mad cow disease and the related Creutzfeldt–Jakob disease in humans. Both these diseases cause severe damage to the brain. Prions infect cells and then "reproduce" by converting normal proteins into the incorrectly folded form. Prions are spread when we eat infected meat. Cooking, which destroys the nucleic acids found in bacteria or viruses, has no effect on prions.

Viruses can be used to make food safer. The Food and Drug Administration has approved a mixture of six viruses that can be sprayed on ready-to-eat meats such as cold cuts and sausages. These viruses attack dangerous *Listeria* bacteria that cause food poisoning.

SCIENCE AND SOCIETY

Swine Flu

The "swine flu" or H1N1 epidemic started in March 2009, when some unusual flu cases appeared in Mexico. Health officials became alarmed because this new flu appeared to cause many severe illnesses and deaths, including among young, otherwise healthy people. As H1N1 spread throughout the world, the World Health Organization (WHO) declared a global pandemic, the first in 41 years. People were encouraged to stay home from school or work when sick, to telecommute if possible, and to avoid large gatherings. Hundreds of schools were closed as a result of confirmed swine flu cases or as a precautionary measure.

Students in Nashville, Tennessee wear masks as a precaution against swine flu.

H1N1 was called "swine flu" because the virus originated in pigs. The H1N1 virus combined DNA from a Eurasian swine flu virus with DNA from North American swine, human, and bird flu viruses. (The deadly result of combining genes from different animal viruses later captured the imagination of Hollywood: The 2011 movie *Contagion* imagines a very deadly virus resulting from a cross between a bat virus and a pig virus.) Although H1N1 generally spread from human to human, several countries banned the import of pork or ordered the mass slaughter of pig herds. Because of this, the name "swine flu" was later dropped in favor of "H1N1," a more technical name describing the variants of two proteins (H and N) on the virus's surface.

Vaccinations were initially in short supply during the swine flu epidemic.

A vaccine for H1N1 was quickly developed, but manufacturing sufficient quantities took longer than expected. The resulting shortages highlighted the problems with old vaccine technologies as well as the risks of relying on foreign vaccine makers. Nonetheless, as the flu season wore on, vaccination efforts helped, and the H1N1 epidemic did not turn out to be as deadly as initially feared. As the number of cases waned, some critics even accused the WHO of overplaying the dangers and starting an unnecessary panic. Regardless, the H1N1 epidemic provided a good trial run, so that when the next global pandemic arrives, we will be better prepared.

For instructor-assigned homework, go to www.masteringphysics.com 

SUMMARY OF TERMS (KNOWLEDGE)

Alternation of generations The plant life cycle, which alternates between haploid and diploid stages.

Animals A clade of multicellular, heterotrophic eukaryotes that take food into their bodies for digestion.

Archaea One of the three domains of life, consisting of many different kinds of prokaryotic organisms that are more closely related to eukaryotes than bacteria; some archaea are adapted to extreme environments.

Bacteria One of the three domains of life, consisting of an extremely diverse array of prokaryotic organisms.

Clade A group of species that includes an ancestor and all of its descendants.

Ectotherm An organism that uses behavior to regulate its body temperature, which may vary to some degree.

Endoskeleton An internal skeleton, such as that found in echinoderms and chordates.

Endotherm An organism that maintains a constant, high body temperature by breaking down food.

Eukarya One of the three domains of life, consisting of eukaryotic organisms whose cells have a nucleus: animals, fungi, plants, and protists.

Evolutionary tree A diagram that shows the relationships among a set of organisms.

Exoskeleton An external skeleton, such as that found in arthropods.

Flower The reproductive structure of flowering plants.

Fruit In flowering plants, a structure surrounding the seeds that typically helps to spread the seeds.

Fungi A clade of heterotrophic eukaryotes that obtain food by secreting digestive enzymes over organic matter and then absorbing the nutrients.

Plants A clade of autotrophic, multicellular, terrestrial eukaryotes that make food through photosynthesis.

Pollen In seed plants, immature male gametophytes (reproductive cells) wrapped in protective coatings.

Protists A miscellaneous group of eukaryotic organisms that includes all the eukaryotes that are not plants, animals, or fungi.

Seed In seed plants, a structure consisting of a plant embryo, a food supply, and a tough outer coating.

Virus A small piece of genetic material wrapped in a protein coat that infects and reproduces within a host cell.

READING CHECK (COMPREHENSION)

18.1 Classifying Life

1. What criteria are used to classify species in the Linnaean system?
2. What criteria are used to classify species in an evolutionary classification system?
3. What is a clade?

18.2 The Three Domains of Life

4. What are the three domains of life?
5. Which domain of life do eukaryotes belong to?

18.3 Bacteria

6. Explain how bacteria that are autotrophs and chemoautotrophs obtain food.
7. How do bacteria reproduce? Do bacteria ever exchange genetic material?
8. Why is bacterial decomposition important?

18.4 Archaea

9. Which features of archaeans suggest they are more closely related to eukaryotes than to bacteria?
10. Why are some archaea described as “extremophiles”?

18.5 Protists

11. What are protists?
12. What is a diatom, and why are diatoms important to many other marine organisms?
13. Name three kinds of multicellular photosynthetic protists.
14. Heterotrophic protists move to capture prey. Describe some of the different ways they do this.
15. Describe how *Plasmodium*, the protist that causes malaria, infects humans. Also describe the symptoms of the disease.

18.6 Plants

16. What are the two components of the plant vascular system? What is the function of each?
17. How does the alternation of generations differ between mosses and all other plants?
18. How is pollen transferred from one plant to another in conifers? In flowering plants?

18.7 Fungi

19. How do fungi obtain food?
20. What are fungal spores?
21. Why are fungi essential to the growth and survival of most plants?

18.8 Animals

22. How do animals obtain nutrients?
23. What are some features of arthropods?
24. Describe the three major groups of mollusks.
25. Why must amphibians live in moist habitats?
26. What is the difference between an ectotherm and an endotherm? Which vertebrates are ectotherms, and which are endotherms?
27. How do monotremes differ from other mammals?

18.9 Viruses and Prions

28. What is a virus? How do viruses reproduce?
29. What are prions? How do they "reproduce"?

THINK INTEGRATED SCIENCE

18A—Moving Water up a Tree

30. What type of chemical bond is responsible for the cohesion and adhesion of water molecules in plants?
31. Why are cohesion and adhesion important to water transport in plants?
32. What is transpiration? How does transpiration drive the movement of water up the xylem?

18B—Coral Bleaching and Ocean Acidification

33. What single factor is most frequently responsible for coral bleaching?
34. Why do corals turn white during bleaching episodes?
35. Explain why ocean acidification threatens shelled marine species.

18C—How Birds Fly

36. How are birds able to stay in the air when they fly?
37. How is Newton's third law related to how birds fly?
38. How are some birds able to stay in the air without flapping their wings?

THINK AND DO (HANDS-ON APPLICATIONS)

39. Go for a walk and take notes on the flowers you see. Can you guess which pollinators they use? Look for yellow and blue flowers that may be pollinated by bees. Look for red, trumpet-shaped flowers that may be pollinated by hummingbirds. Do you see any white flowers that are closed during the day? These may be pollinated by nocturnal species such as moths or bats. If you wait long enough, you may see pollinators visit a flower! Is it the same species each time?
40. Let's explore water transport in plants. Place a few celery stalks in a beaker with some water. Add a few drops of blue food coloring to the water, and stir gently. Check the celery every few hours, and let the stalks sit overnight. What do you observe? Cut off the bottom of a celery stalk. Can you see the xylem?
41. Let's grow mold! Obtain some bread slices and plastic sandwich bags. Keep the bread in the bags so that you won't breathe in too many spores during the course of the experiment. How long does it take for mold to grow? Can you see the filaments? Can you see spores?
42. We know that birds stay in the air due to lift, an upward force produced by the difference in air pressure above and below the wings. This air-pressure difference is, in turn, the result of air flowing more quickly over the top of the wing than below the wing. Let's experiment with airflow and air pressure using a tissue. Hold the tissue vertically and blow on one side of it—the side you blow on is analogous to the top of the wing, where air flows faster. What happens to the tissue? Does it do what you expect? Can you produce enough lift to keep the tissue horizontal?

THINK AND COMPARE (ANALYSIS)

43. The groups in the Linnaean system are phylum, order, species, class, genus, kingdom, domain, and family. Rank these groups in order, from the one that includes the largest number of species to the one that includes the fewest species.
44. Examples of the three main groups of plants are shown here. Rank them in order, from most dependent on living in a moist habitat to least dependent on living in a moist habitat.

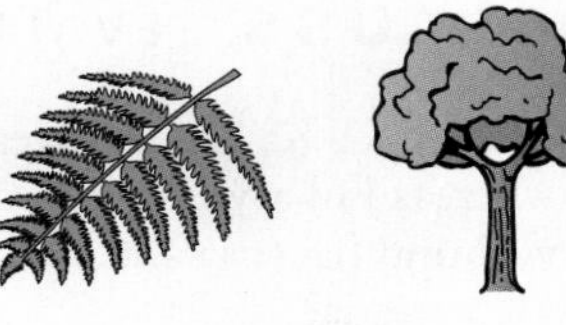

A B C

THINK AND SOLVE (MATHEMATICAL APPLICATION)

45. Suppose a species of bacteria divides once every 20 minutes. You start with a single bacterium on your unrefrigerated egg-and-baloney sandwich at 8:00 AM. Show that when you sit down to lunch at noon, there will be 4096 bacteria on your sandwich.
46. The lightest and heaviest flying birds are the bee hummingbird of Cuba, which weighs about 1.6 grams, and the great bustard of Europe and Asia, which can weigh as much as 21 kilograms. Show that the bee hummingbird produces about 0.016 newton of lift when it flies, whereas the great bustard produces about 205.8 newtons of lift. Which species would you expect to have proportionally larger wings? Why?

The bee hummingbird and the great bustard are the lightest and heaviest flying birds, respectively. The bee hummingbird weighs less than a penny!

THINK AND EXPLAIN (SYNTHESIS)

47. If two species belong to the same order, do they have to belong to the same class? Do they have to belong to the same genus?
48. What is the difference between a heterotroph and an autotroph? Name a clade of living things that consists exclusively of heterotrophs and one that consists exclusively of autotrophs. Name a clade that includes both heterotrophs and autotrophs.
49. What is a chemoautotroph? What does a chemoautotroph have in common with a plant? How does a chemoautotroph differ from a plant in how it obtains food?
50. What is the advantage of being able to produce spores, as many bacteria do?
51. Why is decomposition important to life on Earth?
52. We saw that life on Earth would be impossible without bacteria. Would life on Earth be impossible without eukaryotes?
53. Of the three major plant groups we discussed, which is most dependent on living in a moist habitat? Why? Which is least dependent?
54. You may have heard that moss grows on the north sides of trees—this is good to remember if you are lost in the woods! Why do mosses do best on the north side?
55. Which plants produce pollen? What strategies do plants use for pollination? What strategy do most flowering plants use?
56. Some plants, including many grasses, have small green flowers with no petals (see figure). How do you think these flowers are pollinated?
57. Some people are allergic to pollen. Do you think bee-pollinated plants or wind-pollinated plants are more likely to cause allergies? Why?
58. What do fungi and animals have in common? How do they differ?
59. Describe the function of fungal spores.
60. Name two different strategies used by cnidarians to obtain food.
61. How do the muscles of roundworms and earthworms differ? What does this mean about the way each animal moves?
62. Why is a salamander more dependent on living in a moist habitat than a lizard?
63. Many snakes can survive eating just once every few weeks. Why can't birds do this?
64. Birds and mammals are both endotherms, and they both have a four-chambered heart. Why are birds classified as reptiles rather than as mammals?
65. Birds have hollow bones, an adaptation for flight. How do hollow bones help birds fly?
66. Viruses straddle the line between living and nonliving. How do viruses resemble living things? How do they resemble nonliving things?
67. Write a letter to Grandpa telling him how global warming threatens marine life.

THINK AND DISCUSS (EVALUATION)

68. Of the three domains of life, Bacteria and Archaea both consist of prokaryotes, whereas Eukarya consists of eukaryotes. Why can't we lump Bacteria and Archaea together and call them all Bacteria?
69. Chemical dispersants are sometimes used to break up a large oil slick into tiny droplets. Explain the advantages of this strategy, using the concepts of surface area and volume in your answer. What are the drawbacks to using chemical dispersants?

70. Bacteria generally reproduce asexually. What are some advantages of asexual reproduction?
71. Most living organisms reproduce sexually sometimes or have some other mechanism for exchanging genetic material. What is the advantage of sexual reproduction or genetic exchange?
72. In humans, cells undergo meiosis to produce sperm and eggs. What kinds of cells are produced by meiosis in mosses?
73. Do plants have to use energy to obtain the carbon dioxide they need for photosynthesis?
74. How do mosses spread from one place to another? How do flowering plants spread from one place to another?
75. When you buy fresh flowers, you should cut the stems underwater to prevent an air bubble from forming in the xylem. What would happen if an air bubble entered the xylem?
76. Imagine that you are being interviewed by an ethnobotanist about how you use plants. What do you tell her? Remember that ethnobotanists are interested in everything related to plant use, so don't stop at the plants you eat. How else do you use plants and plant products? Clothing? Tools? Medicine? Shelter? Decoration?
77. We have mentioned the surface-area-to-volume ratio in the contexts of cell size (see Chapter 15) and thermoregulation in mammals (see Chapter 17). Explain how the surface-area-to-volume ratio also affects the size and shape of flying birds.

READINESS ASSURANCE TEST (RAT)

If you have a good handle on this chapter, if you really do, then you should be able to score 7 out of 10 on this RAT. If you score less than 7, you need to study further before moving on.

Choose the BEST answer to each of the following:

1. Under an evolutionary classification system, species are grouped together based on
   (a) their shared similarities.
   (b) how closely related they are to one another.
   (c) their position in a hierarchy from "simple" to "complex."
   (d) none of these
2. The scientific name of a species
   (a) consists of two parts, the family name and the species name.
   (b) is always underlined.
   (c) is always in Latin.
   (d) includes only the species name.
3. Why are birds considered reptiles in an evolutionary classification system?
   (a) Feathers evolved from scales.
   (b) Birds are more similar to reptiles than was previously thought.
   (c) Birds are descended from the last common ancestor of all reptiles.
   (d) Birds share certain similarities with crocodiles.
4. Life would be impossible without bacteria because
   (a) they photosynthesize.
   (b) they function in decomposition.
   (c) they reproduce quickly.
   (d) they occupy habitats where no other organisms can survive.
5. All protists are
   (a) eukaryotes.
   (b) autotrophs.
   (c) heterotrophs.
   (d) single-celled.
6. Which of the following is a characteristic of all plants?
   (a) seeds
   (b) pollen
   (c) swimming sperm
   (d) alternation of generations
7. Fruits help plants
   (a) attract animal pollinators.
   (b) obtain nutrients.
   (c) disperse seeds.
   (d) photosynthesize.
8. All fungi
   (a) are heterotrophs.
   (b) are multicellular.
   (c) reproduce asexually.
   (d) reproduce sexually.
9. Animals
   (a) are multicellular heterotrophs.
   (b) are more closely related to plants than fungi.
   (c) often have haploid larvae.
   (d) include some stationary, nonmobile organisms such as sponges, corals, and mushrooms.
10. Which of these statements is true?
   (a) All viruses have genes made of RNA.
   (b) A few viruses have evolved the ability to reproduce outside a host cell.
   (c) Viruses infect all other forms of life, from bacteria to plants and animals.
   (d) Prions are misfolded proteins that cause plant diseases.

Answers to RAT

1. b, 2. c, 3. c, 4. b, 5. a, 6. d, 7. c, 8. a, 9. a, 10. c

19

CHAPTER 19

Human Biology I—Control and Development

NOT EVERYONE can juggle with fire while walking a tightrope, but we all do things that are nearly as impressive on a daily basis. How can you ride a bike or fold laundry without even thinking about it? What part of the brain is working when you write a poem or solve a math problem? How can you tell that a strawberry is tasty but a rotten egg isn't? Why is it easy for you to touch your nose even with your eyes closed? In the next two chapters, we'll see how the body maintains exquisite control over its systems and allows us to function in a complex world.

19.1 Organization of the Human Body

LEARNING OBJECTIVE
Describe tissues, organs, and organ systems, and give examples of each.

EXPLAIN THIS Is the brain a tissue, an organ, or an organ system?

You wave to a friend, look at a traffic light, and step off the curb. As you cross the street, you consider whether you want tacos or a sandwich for lunch. You breathe in, bringing oxygen into your body, and then breathe out, removing carbon dioxide. Blood, pumped by your heart, moves through your blood vessels. At the same time, immune cells crowd around a small cut on your finger, eating up invading bacteria. At any moment, your body is involved in a huge number of different activities. To do their jobs, the cells in your body are organized into tissues, organs, and organ systems (Figure 19.1).

A **tissue** is a group of similar cells that performs a certain function. There are four main types of tissue in the body. *Epithelial tissue* consists of sheets of tightly packed cells that cover the internal and external surfaces of the body. Skin is an example of epithelial tissue. *Connective tissue* consists of cells scattered within an external matrix. Bone, cartilage, and blood are all connective tissues. *Muscle tissue* is made up of cells that are able to contract, or shorten. Three types of muscle tissue are found in the body. Skeletal muscle is responsible for all voluntary movements. Smooth muscle functions in the internal organs of the digestive system as well as in certain blood vessels. Cardiac muscle produces the heartbeat. *Nervous tissue* transmits information from one place in the body to another. Nervous tissue is found in the brain, spinal cord, and nerves.

Multiple tissues combine to make an **organ**, a structure in the body that has a specific function. The heart, stomach, and brain are examples of organs. Each of these organs is made up of multiple tissues. For example, the heart (1) is surrounded by epithelial tissue on the outside and lined with it on the inside; (2) contains blood vessels carrying blood, a connective tissue; (3) has walls made up largely of cardiac muscle; and (4) contains nerves, made of nervous tissue, that help to control its activity.

Multiple organs make up an **organ system**. An organ system is responsible for a particular bodily function. For example, the circulatory system moves nutrients, gases, and wastes throughout the body. The human body has ten major organ systems: nervous, sensory, endocrine, reproductive, muscular and skeletal, circulatory, respiratory, digestive, excretory, and immune. In the next two chapters, we'll look at what these organ systems do.

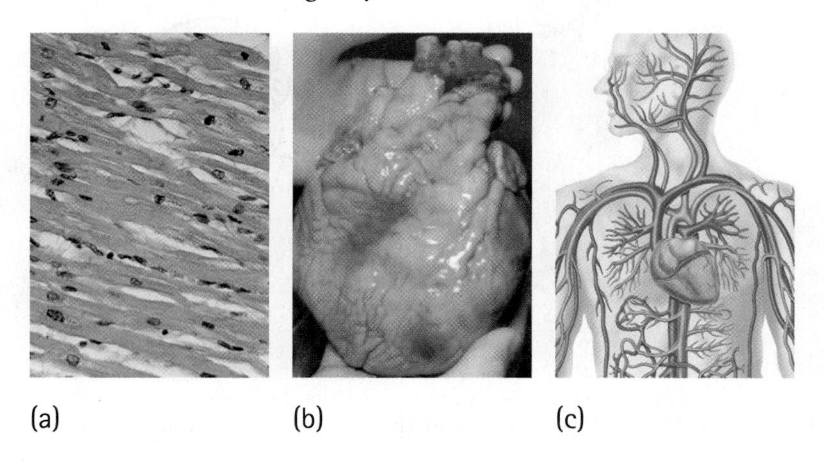

FIGURE 19.1
In the human body, cells are organized into tissues, organs, and organ systems. (a) Many muscle cells make up muscle tissue. This photo shows muscle tissue from a human heart. (b) The heart is an organ. (This heart has just been removed from a donor and will soon be transplanted into a recipient.) (c) The circulatory system is made up of the heart, blood, and blood vessels. The circulatory system transports nutrients, gases, and wastes to different parts of the body.

LEARNING OBJECTIVE
Use examples to describe homeostasis.

19.2 Homeostasis

EXPLAIN THIS Why do you shiver when you're cold?

Whether you are swimming in icy waters or hiking through scorching heat, your body temperature stays close to 37° Celsius (98.6° Fahrenheit). This consistency in body temperature is an example of **homeostasis**, the maintenance of a relatively stable internal environment. Homeostasis is a characteristic of all living organisms (see Chapter 15), and a huge amount of the body's activity goes toward maintaining it.

As just one example, your cells need a certain amount of oxygen to function. Your lungs and heart maintain a normal level of activity to supply this oxygen. If your activity level increases—say, because you run to catch a bus—your cells use up more oxygen. What happens? Your body responds by breathing harder to take in more oxygen and by increasing your heart rate to move that oxygen to your cells. Once your activity level returns to normal and your oxygen use decreases, your breathing and heart rate slow again.

How does increased activity trigger an increase in oxygen intake? The body doesn't respond to low oxygen levels directly. Instead, increased oxygen use during cellular respiration produces an increase in blood carbon dioxide levels. This carbon dioxide reacts with water to form carbonic acid, causing blood to become more acidic. And this pH change is detected by sensors, which cause the body to increase respiration and heart rate. (In Chapter 18, you learned that high levels of carbon dioxide cause ocean water to turn more acidic in just the same way. The same thing that happens in the ocean happens in your blood!)

To go back to body temperature, when it's cold outside, you *feel* cold and pile on more clothes, or wrap your arms around your body to reduce heat loss. In addition, less blood is sent to your limbs and extremities, which lose heat faster than the core of your body. (This explains why your fingers and toes often feel most cold when you're cold.) You may also shiver to generate heat (Figure 19.2). On the other hand, when it's hot outside, you take off your clothes, look for shade, and sweat to cool off. Also, more blood goes to the extremities and to the face, which are good at shedding heat. (This explains why your face turns red when you're hot.)

Controlling body temperature is an example of feedback regulation. In *feedback regulation*, changes in one variable affect a second variable, and changes in the second variable in turn affect the first variable. In this case, changes in body temperature trigger specific responses that in turn affect body temperature.

UNIFYING CONCEPT
- ***Feedback Regulation***

Oxygen supply and body temperature are only two of the many variables the body carefully regulates. The amount of water in the body, the concentration of nutrients and waste products in the blood, the concentrations of important ions inside and outside cells, and blood pH—these and many other variables are carefully regulated as part of maintaining homeostasis.

FIGURE 19.2
Your body responds to changing conditions in ways that help maintain homeostasis. (a) When you're cold, you shiver to warm up. (b) When you're hot, you sweat to cool down.

19.3 The Brain

LEARNING OBJECTIVE
Describe the structure and function of the brain.

EXPLAIN THIS Why is the surface of the brain wrinkled?

Your brain makes you who you are. All of your thoughts, feelings, and desires come from your brain. Your brain also controls all of your activities—both conscious ones such as choosing a shirt or kicking a soccer ball and unconscious ones such as breathing, blood circulation, and digestion. Your brain "heads up" your nervous system and is made up of five main parts (Figure 19.3):

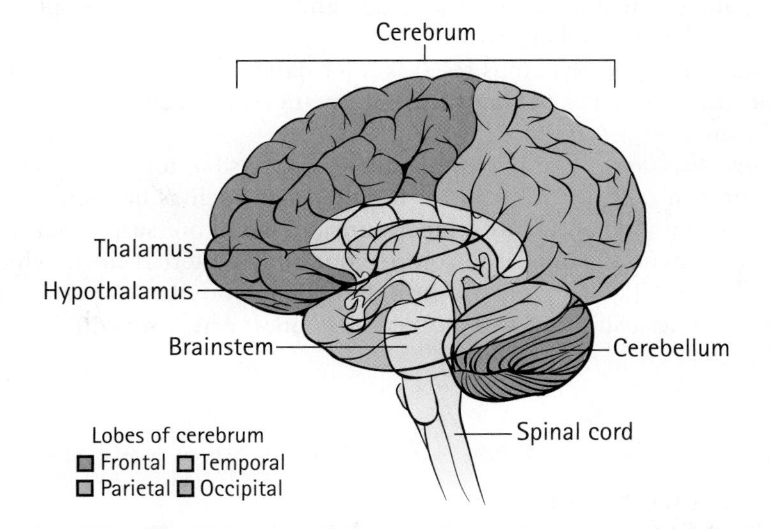

FIGURE 19.3
The major parts of the human brain are the brainstem, cerebellum, cerebrum, thalamus, and hypothalamus.

1. *The brainstem.* Can you imagine how much less you would get done if you had to remember to breathe every few seconds? It's a good thing the brainstem does this for you. The brainstem connects the spinal cord to the rest of the brain. It controls many of the body's basic, involuntary activities, such as heartbeat, respiration, and digestion. The brainstem also wakes you up every morning, bringing your body from sleep to wakefulness (Figure 19.4).
2. *The cerebellum.* The cerebellum controls balance, posture, coordination, and fine movements. It also controls all the motions you perform "without thinking." Consider bicycle riding. When you first learn how to ride a bike, you think very carefully about what your arms and legs are doing, and your motion is very awkward. With enough practice, though, you can ride a bike easily—"without thinking." When you do something "without thinking," it means the cerebellum has taken over.
3. *The cerebrum.* The cerebrum is the largest part of the brain. It is responsible for high-level functions such as reasoning, problem solving, language, and creativity. Your personality also resides in your cerebrum. The cerebrum collects information from the senses and controls all the conscious, voluntary activities of the body. The cerebrum has a right hemisphere (right side) and a left hemisphere (left side). The right hemisphere controls the left side of the body, and vice versa. This means that when you move your right hand, it's actually the left hemisphere of your cerebrum that controls that action.

   Most of the information processing that occurs in the cerebrum takes place in the cerebral cortex, the thin layer that covers the surface of the cerebrum. "Wrinkles" in the cerebral cortex give the brain its familiar convoluted appearance and increase the surface area available for information

FIGURE 19.4
The brainstem controls the transition between sleep and wakefulness.

processing. Each cerebral hemisphere consists of four lobes that are responsible for different activities. The *frontal lobes* deal with reasoning, voluntary movements, and speech. The *parietal lobes* take in sensory information about temperature, touch, taste, and pain. The *occipital lobes* process what you see—that is, visual information. The *temporal lobes* deal with sound and help you comprehend language.

The control of certain cognitive functions is dominated by either the right or left cerebral hemisphere. The left hemisphere is more adept at math, reasoning, language, and detail-oriented activities. The right hemisphere is more adept at spatial relationships, emotional processing, and music. This distinction has led to the popular conception of "left-brained" people who are organized, analytical, and attentive to detail, and "right-brained" people who are intuitive, flexible, and creative.

The cerebrum contains more than 10 billion neurons!

4. *The thalamus.* The thalamus receives information from many different parts of the brain. It sorts and filters this information and then passes it on to the cerebrum.
5. *The hypothalamus.* The hypothalamus is responsible for emotions such as pleasure and rage. It also controls bodily drives such as hunger, thirst, and sex drive and regulates body temperature and blood pressure. Another function of the hypothalamus is to control your body's internal clock, which tells you when it is day and when it is night. The hypothalamus performs some of its activities using molecules called hormones, which we will discuss later in this chapter.

CHECK YOURSELF

1. Io is reading a book. Which lobes of the cerebrum is she using?

2. In sports, it's called *choking*. It's a tense, decisive moment in a game, and the pressure's on. The ball comes toward you—it's an easy play, something you've done a million times before. But suddenly you become superaware of all your movements, and then it happens—the ball rolls right between your legs. What part of the brain normally controls practiced, automated movements? What part of your brain takes over when you choke?

CHECK YOUR ANSWERS

1. Io is using her occipital lobes to process visual information (the letters on the pages of her book) and her temporal lobes to comprehend language.
2. Automated movements that you can perform "without thinking" are normally controlled by the cerebellum. When you consciously control your movements and choke, you are using the frontal lobes of the cerebrum.

TECHNOLOGY

Mapping the Brain in Action: Functional MRIs

Traditional lie-detector tests are not always dependable. They rely on traits like pulse, blood pressure, breathing rate, and skin conductance (sweating) to give away the guilty. Someone who is nervous but telling the truth may fail, and a liar who stays calm may pass. But liars may find a new technique harder to fool. *Functional magnetic resonance imaging* (fMRI) reveals which parts of the brain are activated during different types of activity—and it shows that the brains of liars are doing different things from those of truth-tellers.

How does fMRI work? fMRI builds on the earlier technology of magnetic resonance imaging (MRI). MRI makes use of the fact that every hydrogen atom in the body—and there are two in every water molecule—is a tiny magnet. (This is because, as we learned in Chapter 7, every accelerating charged particle, including the spinning proton in the nucleus of a hydrogen atom, produces a magnetic field.) When living tissue is placed in the field of a strong magnet, all the hydrogen atoms line up the same way a compass lines up with Earth's magnetic field. A radio wave is then used to "bump" the atoms, knocking them slightly out of line. As the atoms bounce back to their natural alignment within the magnetic field, they release a small amount of energy that can be detected and recorded. Because body tissues vary in water concentration, different tissues release different amounts of energy, allowing a very detailed image to be constructed. Like MRI, fMRI constructs images based on different concentrations of water molecules in different parts of the body. With fMRI, however, the focus is on blood. Like other cells, neurons in the brain use more energy when they're active, and so require more oxygen. In order to accommodate this need, blood flow to active areas of the brain is increased. This increased flow can be detected and converted into an image of active brain areas.

So, what happens in the brain when people lie? In one study, volunteers were given a playing card—the five of clubs—and a $20 bill. They were told they could keep the money if they managed to fool the computer into thinking they had a different card. fMRI maps were made of the brain while volunteers lied by denying they had the five of clubs and while they told the truth by denying they had other cards. The maps were then compared. Lying caused increased activity in several areas of the brain (see the figure), including those responsible for attention, inhibiting actions, and monitoring errors. This suggests that lying requires the inhibition of a natural tendency to tell the truth as well as increased effort and attention. It is interesting that telling the truth did not increase activity in any part of the brain. Lying appears, overall, to be much harder work than telling the truth.

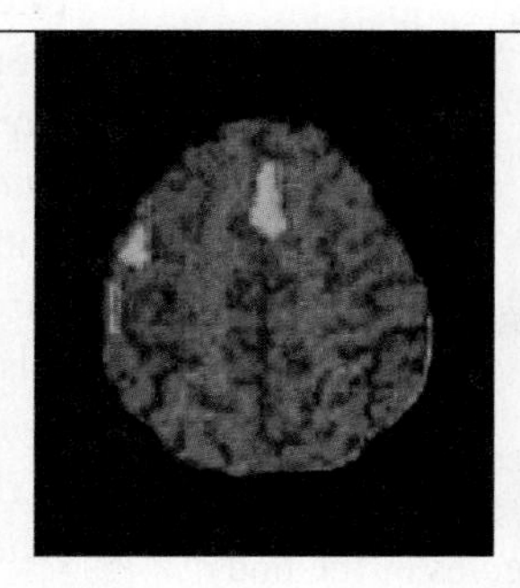

Functional magnetic resonance imaging (fMRI) allows scientists to compare activity levels in different parts of the brain during different activities. This image shows the areas of the brain that are activated during lying.

fMRI is too expensive to be used regularly for lie detection. However, it has been a valuable tool for figuring out which areas of the brain are responsible for different sensations, emotions, and activities. fMRI has already contributed to our understanding of how people remember information, feel love, gamble, recognize faces, and respond to pain. One group of researchers has even used fMRI to reconstruct the images that went through people's brains while they watched video clips. Researchers hope that one day a similar technique can be used to record what people see when they dream.

19.4 The Nervous System

LEARNING OBJECTIVE
Describe the structure and function of the nervous system.

EXPLAIN THIS What happens during the "fight or flight" response?

You're eating cornflakes and reading the sports page. You glance at the clock and see that you're late for class. Immediately, your body goes into overdrive. Your heart starts to pound and you begin to sweat. You dash out the door, run down the steps—and trip. Pain shoots through your knee, but you pull yourself up and continue your race to class. There, you drop into your seat. Your heart rate gradually slows, and your body relaxes. This entire sequence of actions and responses is controlled by your nervous system.

MasteringPhysics®
TUTORIAL: Neuron Structure

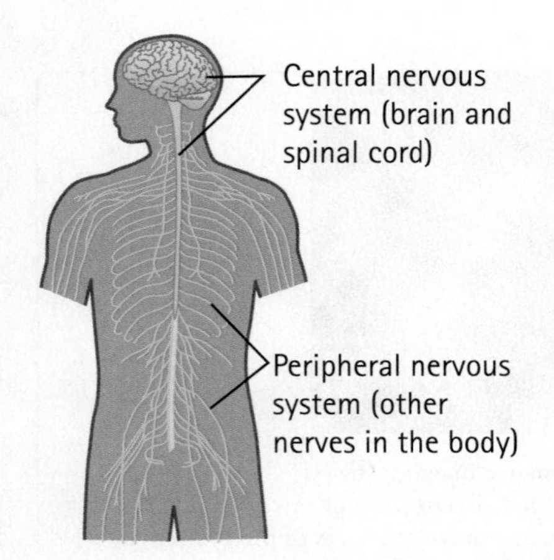

FIGURE 19.5
The central nervous system consists of the brain and spinal cord. The peripheral nervous system includes all the other nerves in the body.

The *nervous system* collects information about the body's internal and external environments and controls the body's activities. It can be divided into two parts. The *central nervous system* includes the brain and spinal cord, and the *peripheral nervous system* includes all the other nerves in the body (Figure 19.5).

The nervous system includes two different kinds of cells: neurons and glial cells (Figure 19.6). A **neuron** is a specialized cell that receives and transmits electrical signals from one part of the body to another. A typical neuron has three parts: (1) *dendrites* receive information from other neurons or cells, (2) the *cell body* contains the neuron's nucleus and organelles, and (3) the *axon* transmits information to other neurons or cells. *Glial cells* support, insulate, and protect neurons. The human brain actually has 10 to 50 times more glial cells than neurons.

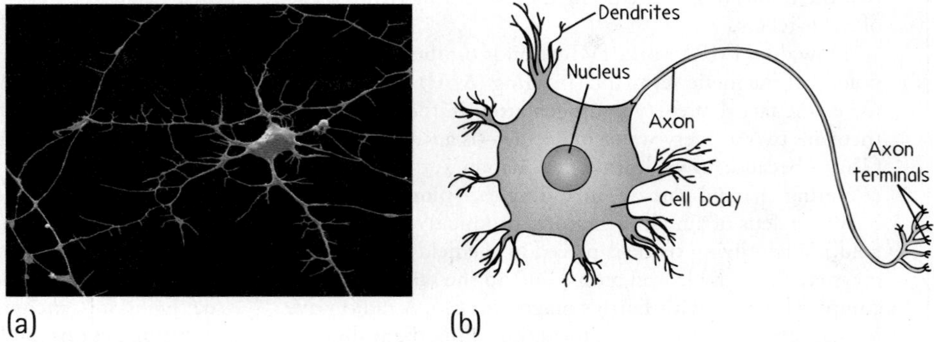

FIGURE 19.6
(a) Neurons transmit electrical signals from one part of the body to another. The red cells shown here are neurons. (b) Neurons have a cell body, dendrites that receive information from other cells, and an axon that transmits information to other cells.

Kinds of Neurons

There are three types of neurons: sensory neurons, interneurons, and motor neurons (Figure 19.7). *Sensory neurons* carry information from the senses to the central nervous system. The neurons that transmit visual information from your

FIGURE 19.7
Neurons are divided into different types based on their function.

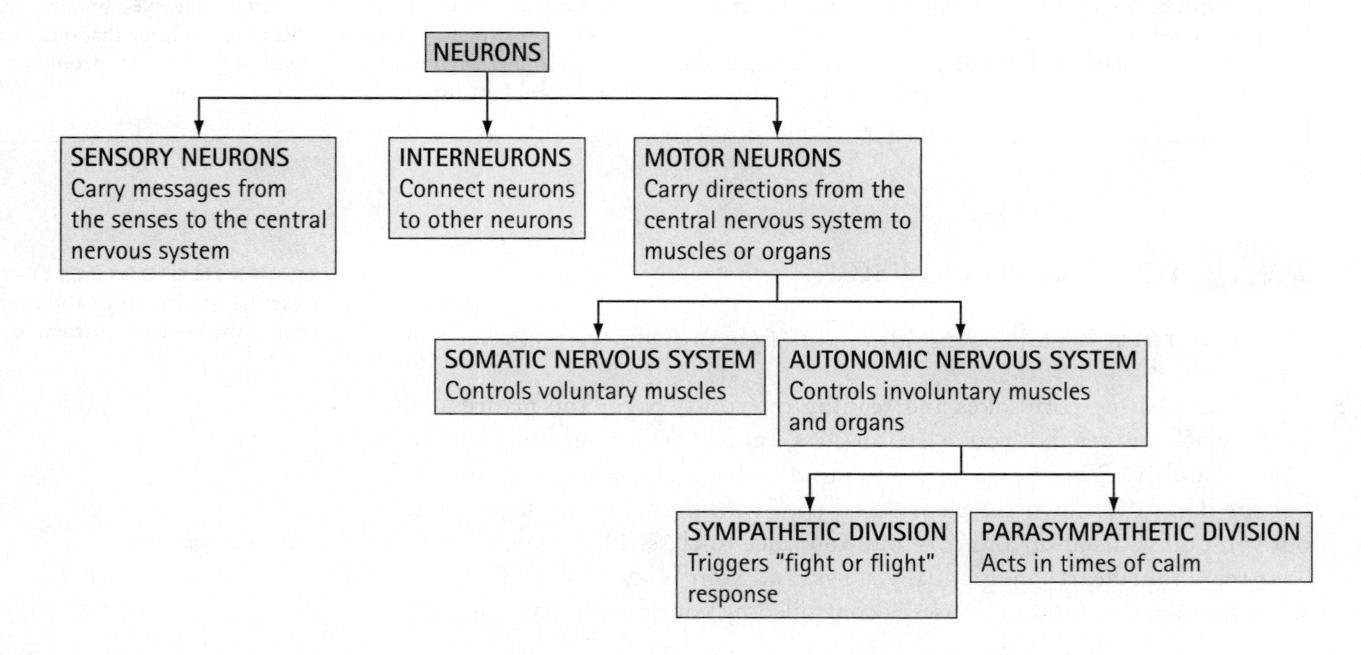

eyes to your brain are sensory neurons. So are the neurons that go from your fingertips to your central nervous system, telling you whether you are touching something cool or hot. In the story about your run to class, sensory neurons told your brain about the time on the clock and the pain of hitting your knee on concrete. *Interneurons* connect neurons to other neurons. Interneurons are found exclusively within the central nervous system. *Motor neurons* carry directions from the central nervous system to muscles or organs. During your run to class, motor neurons directed the motion of your legs. A motor neuron also made your heart beat faster.

You can decide to bend your elbow but you can't decide to make your heart beat faster. That's the difference between the somatic and autonomic nervous systems.

Motor neurons belong to either the somatic nervous system or the autonomic nervous system. Motor neurons in the *somatic nervous system* control voluntary actions, such as running to class. These neurons provide instructions to your voluntary muscles, such as the muscles of your arms, legs, and fingers. Motor neurons in the *autonomic nervous system* control involuntary actions. These neurons control organs and involuntary muscles (such as heart or stomach muscle). The neuron that made your heart beat faster is part of your autonomic nervous system.

The autonomic nervous system includes a sympathetic and a parasympathetic division (Figure 19.8). The *sympathetic nervous system* prepares your body for danger. It triggers a "fight or flight" response. "Fight or flight" refers to the fact that when you're threatened, you need to either run away ("flight") or stay and "fight." The sympathetic nervous system sped up your heartbeat

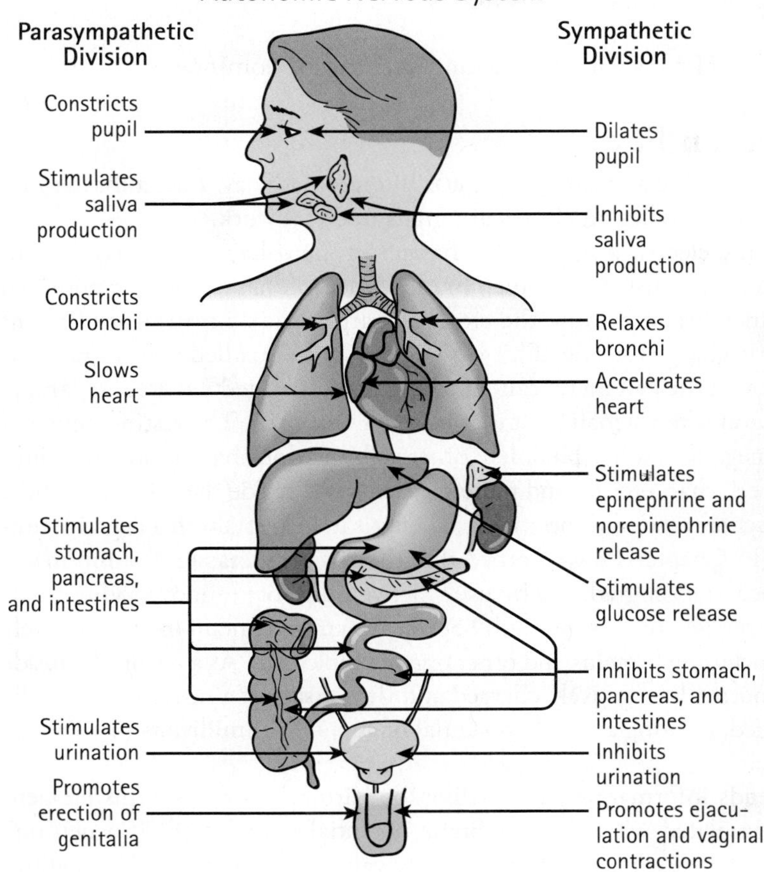

FIGURE 19.8
The autonomic nervous system has a sympathetic division and a parasympathetic division. (a) The sympathetic division prepares the body for danger—it initiates the "fight or flight" response. The parasympathetic division acts in a directly opposed way and operates in times of calm. (b) Is this man's sympathetic nervous system working hard right now?

when you realized you were late for class, speeding the transport of oxygen to your muscles. The sympathetic nervous system also slows down digestion and other activities that are unimportant during emergencies. The *parasympathetic nervous system* works during calmer times. Its effects are the opposite of those of the sympathetic nervous system; it calms you down. Your parasympathetic nervous system took over after you got to class and were able to relax.

CHECK YOURSELF

1. **You stand at the edge of a pool and wonder whether the water is too cold for a swim. What kinds of neurons do you use to find out?**
2. **What kind of motor neuron did you use to test the water—somatic or autonomic?**

CHECK YOUR ANSWERS

1. You use a motor neuron to dip your toe into the pool. Then a sensory neuron in the skin of your toe (one that senses temperature) tells you how cold the water is.
2. Somatic, because dipping your toe in the pool is a voluntary action.

LEARNING OBJECTIVE
Explain how a neuron transmits and receives information.

19.5 How Neurons Work

EXPLAIN THIS How is a neuron's axon like a row of dominoes?

How a Neuron Fires

Is the human body like a toaster? Well, it is in one small way. Like a toaster or a computer, your body relies on electrical signals to do its work.

In neurons, the electrical signals are changes in the voltage, or electric potential (see Chapter 7), across the cell membrane. A neuron has an electric potential across its cell membrane because the electric charge inside a neuron is different from the electric charge outside. This electric potential is called a *membrane potential* because it is the cell membrane that keeps the charged particles separate.

UNIFYING CONCEPT
- ***The Electric Force***
Section 7.1

When a neuron is not signaling, it is at its *resting potential*. The resting potential of a neuron is negative. Why? First, like other cells, neurons have more potassium ions inside the cell than outside and more sodium ions outside the cell than inside. The overall effect of these concentration gradients is to help establish a negative resting potential. (In Chapter 15, we learned that the sodium-potassium pump maintains these concentration gradients by using active transport to move sodium ions out of cells and potassium ions into cells.) Second, neurons contain many negatively charged ions, including proteins and other organic molecules. As a result, the inside of a neuron is normally negatively charged and the outside of a neuron is normally positively charged, creating a resting potential of about −70 millivolts (mV) across the cell membrane.

MasteringPhysics®
TUTORIAL: Nerve Signals
TUTORIAL: Neuron Communication

A neuron sends information by signaling, or "firing." How does this happen? A neuron is stimulated when its membrane potential is increased. If a neuron's membrane potential reaches a certain *threshold* value—about −55 mV—sodium channels in the neuron's cell membrane suddenly open, allowing positively charged sodium ions to move into the neuron. (Sodium ions move into the neuron because there are more sodium ions outside the cell than inside.) This influx of positively charged ions causes the membrane potential to spike and become positive. This spike is called an **action potential**. The action potential is the neuron's way of signaling, or "firing." Once the spike occurs, the sodium channels quickly close and the potassium channels open. Positively charged potassium ions flow out of the neuron (they flow out because there are more potassium ions inside the cell than outside), causing the membrane potential to return to resting potential. The sequence of events in an action potential is shown in Figure 19.9.

An action potential is an all-or-nothing event—a neuron either fires or it doesn't. A neuron cannot fire "harder" when a stimulus is more intense. However, it does fire more often. For example, touch sensors in your skin fire slowly when they feel a little pressure (tick . . . tick . . . tick . . .) and quickly when they feel lots of pressure (tick-tick-tick-tick-tick).

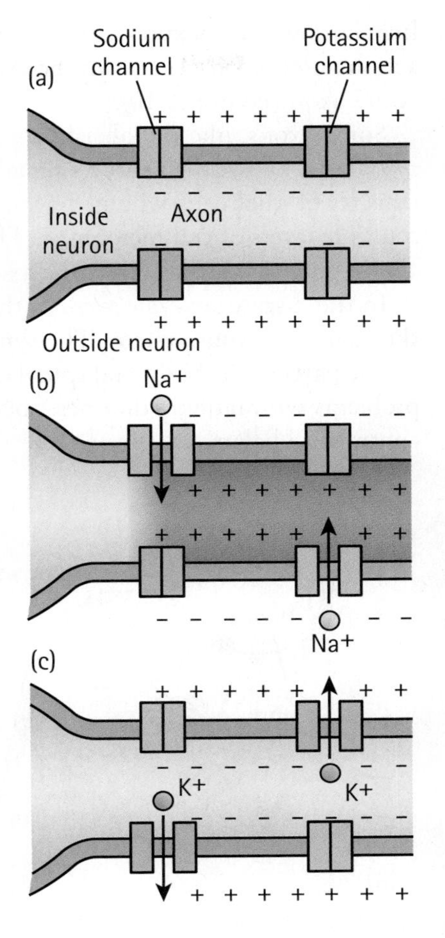

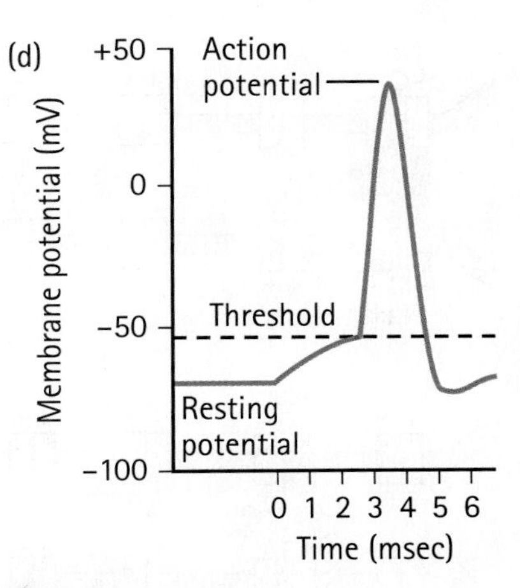

FIGURE 19.9
The action potential is the neuron's way of signaling. (a) Neurons have a negative resting potential. The inside of the neuron is negatively charged, and the outside is positively charged. (b) During an action potential, sodium channels in the neuron's cell membrane open. Sodium ions move into the neuron, causing the membrane potential to become positive. (c) Sodium channels close, and potassium channels open. Potassium ions move out of the neuron, causing the membrane potential to return to resting potential. (d) This graph shows how the membrane potential changes during an action potential. Resting potential is negative. When the neuron is stimulated, the membrane potential increases slowly until it reaches threshold. Then the action potential happens—sodium channels open, causing the membrane potential to spike. The membrane potential decreases when potassium channels open. It returns to resting potential.

CHECK YOURSELF

1. **What causes sodium channels to open, initiating an action potential?**
2. **What causes an action potential to end?**

CHECK YOUR ANSWERS

1. A stimulus that increases the membrane potential to its threshold value.
2. The opening of potassium channels, which returns the membrane potential to its resting potential.

How an Action Potential Is Propagated

An action potential doesn't spike everywhere on a neuron's cell membrane at once. It begins at one end of the axon, near the neuron's cell body, and then travels down the axon. How does this happen?

When an action potential begins, sodium ions enter the end of the axon that is closest to the cell body. These ions diffuse into adjacent areas along the axon. Because the sodium ions are positively charged, they cause the local membrane potential to increase. When the local membrane potential reaches threshold, a new action potential, farther along the axon, begins. The process is similar to the way a row of dominoes falls: The first domino knocks down the next one, which

UNIFYING CONCEPT

- ***The Second Law of Thermodynamics***
Section 6.5

knocks down the next one, and so forth. Similarly, an action potential sets off an action potential farther down the axon, and the action potential travels down the entire axon (Figure 19.10).

Some axons, like the one shown in Figure 19.11, are surrounded by a *myelin sheath* that allows the neuron's signal to be transmitted more quickly. The sheath consists of glial cells wrapped around and around the axon, insulating it with multiple layers of cell membrane. There are periodic gaps in the sheath, and the action potential jumps down the axon from one gap to the next.

In the disease *multiple sclerosis*, the body's immune system destroys myelin in the central nervous system. The symptoms of multiple sclerosis vary depending on the parts of the brain and spinal cord that are affected. Symptoms can include problems with fatigue, dizziness, bladder and bowel control, vision, and muscle control and balance.

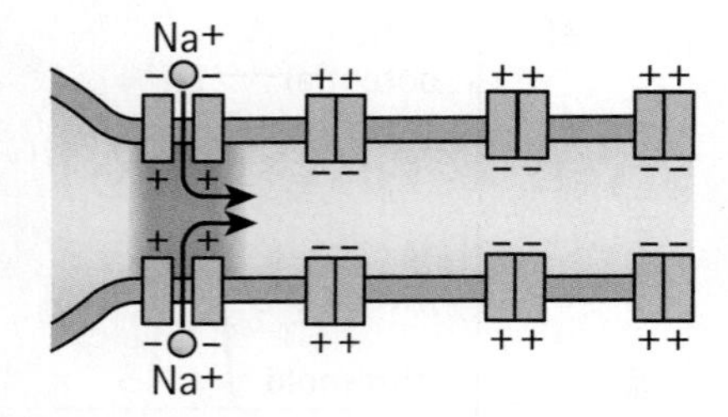

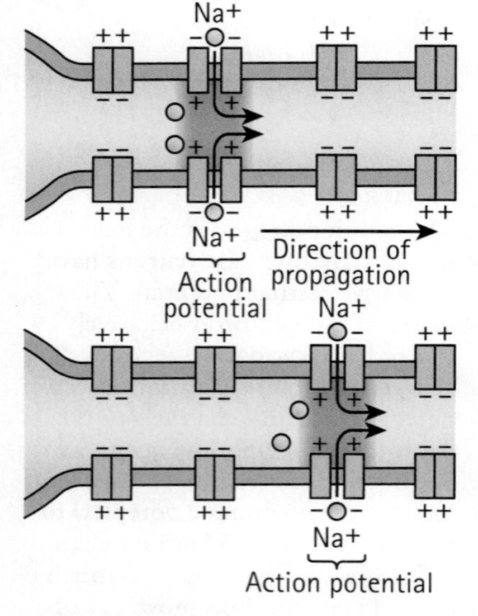

FIGURE 19.10
Action potentials are propagated down an axon. The action potential starts near the cell body. Sodium ions that enter the neuron diffuse down the axon, initiating an action potential farther along the axon. The action potential continues to move down the axon until it reaches the end of the axon.

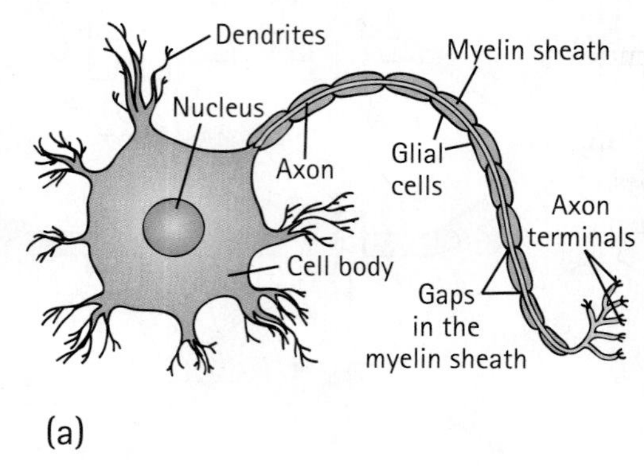

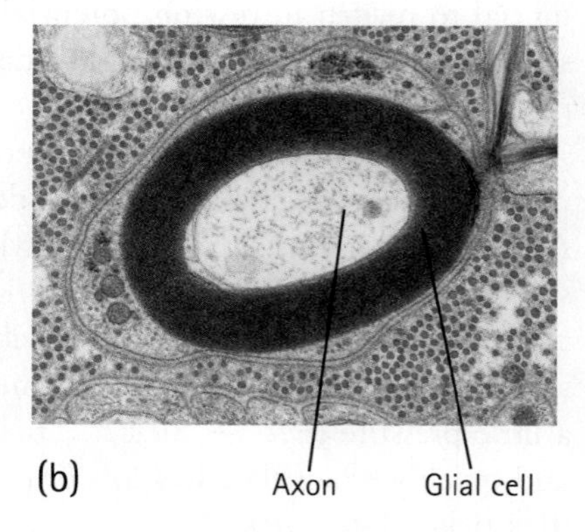

FIGURE 19.11
(a) A myelin sheath surrounds some axons, enabling the signals they transmit to travel faster. (b) The myelin sheath consists of glial cells wrapped around and around an axon, as shown in this cross-section. The brown ring is the myelin. (The green is the cytoplasm of the glial cell.)

LEARNING OBJECTIVE
Explain how axon size and myelination affect how quickly an action potential travels.

Integrated Science 19A
PHYSICS

How Fast Can Action Potentials Travel?

EXPLAIN THIS Squids have a giant axon, but humans do not. Why?

For a squid escaping from a hungry shark, speed is of the essence. How quickly the signal to *move!* gets from the brain to the muscles is the difference between life and death. How can you make an action potential go faster?

Action potentials travel faster if the sodium ions that produce them move more quickly down an axon. How quickly sodium ions move down an axon depends on Ohm's law, which says that current = voltage/resistance (see Chapter 7). The lower the resistance, the higher the current and the faster the action potential. Like any other material, an axon has lower resistance if it is thicker. This is because a thick axon resists current less than a thin axon, the same way a wide pipe resists water flow less than a thin pipe. So, one way to get a fast action potential is to build a thick axon. Thick axons have in fact evolved many times in different animals, including cockroaches, earthworms, and squid. The squid giant axon is nearly a millimeter in diameter and conducts action potentials very quickly—at a speed close to 100 meters/second. Compare this to a typical 1-micrometer axon, which conducts action potentials at a speed of about 2 meters/second.

Unfortunately, you can't pack very many giant axons into one animal; they simply take up too much space. Vertebrates and some other animals have evolved a different way to conduct action potentials quickly—myelination. In a myelinated axon, the action potential is not regenerated at every point along the axon. Instead, the action potential "jumps" from one gap in the myelin sheath to the next, saving time. A myelinated axon only 20 micrometers in diameter conducts action potentials as fast as the squid giant axon.

CHECK YOURSELF

Would an action potential move more quickly down a thick myelinated axon or down a thin myelinated axon?

CHECK YOUR ANSWER

A thick myelinated axon. A thick myelinated axon has the best of both worlds—it's big *and* myelinated.

How a Neuron Communicates with Another Cell

So far, we have seen what happens during an action potential and how an action potential travels down an axon. What happens when an action potential arrives at the end of an axon? At the end of an axon, a neuron connects with one or more target cells and passes information to them. A target cell can be another neuron or a cell that does something, such as a muscle cell or organ cell. A connection between a neuron and a target cell is called a **synapse**. There are two types of synapses: electrical synapses and chemical synapses.

In an *electrical synapse*, tiny channels in the cell membrane called gap junctions (see Chapter 15) connect a neuron to a target cell. When the neuron fires, ions move through these channels into the target cell, starting an action potential in the target cell (Figure 19.12a). Because the ions move very quickly, electrical

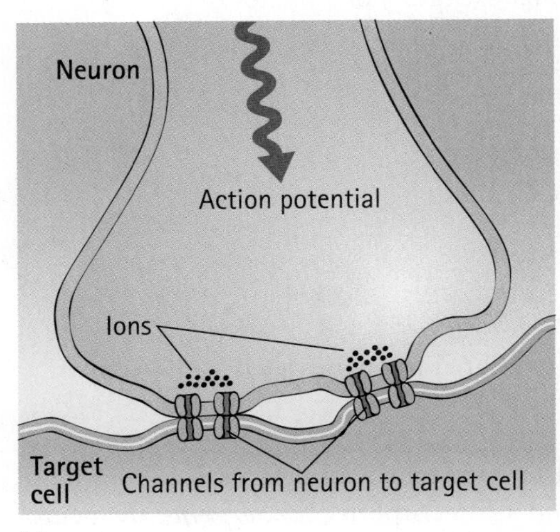

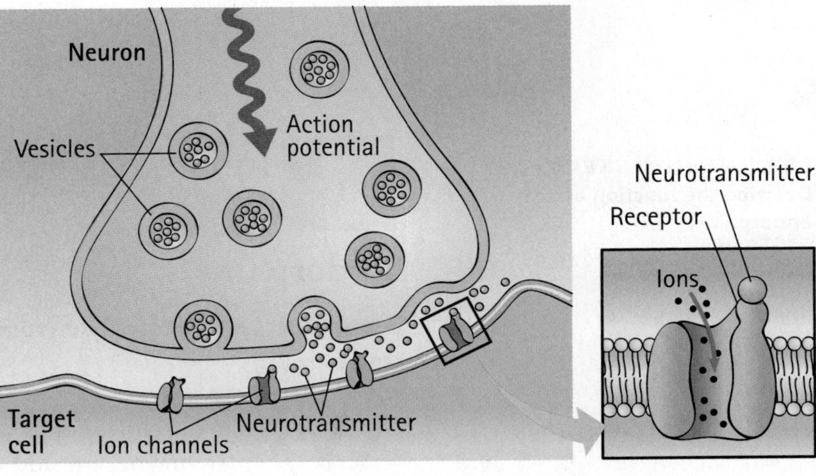

FIGURE 19.12
(a) In an electrical synapse, ions move through the channels that connect a neuron to a target cell and start an action potential in the target cell.
(b) In a chemical synapse, a neuron uses neurotransmitter to communicate with a target cell. When an action potential arrives at the end of the axon, neurotransmitter is released. The neurotransmitter diffuses to the target cell and binds to receptors on the target cell's membrane, causing ion channels to open. Ions enter the target cell and change its membrane potential.

UNIFYING CONCEPT

- *The Second Law of Thermodynamics* **Section 6.5**

synapses are very fast. Electrical synapses are found in places where speed is important. For example, in the heart, electrical synapses allow many muscle cells to contract nearly simultaneously to produce the heartbeat. Despite their speed, though, electrical synapses are fairly rare in the body.

Most synapses are *chemical synapses*. Chemical synapses are not as fast as electrical synapses, but they allow for a finer degree of control. In a chemical synapse, a neuron communicates with a target cell using molecules of *neurotransmitter*. When an action potential arrives at the end of an axon, vesicles containing neurotransmitter fuse to the neuron's cell membrane (Figure 19.12b). Neurotransmitter molecules are released into the narrow space between the axon and the target cell, and they diffuse to the target cell. There, the neurotransmitter molecules bind to receptors on the target cell's membrane. The binding causes ion channels to open, and ions enter the target cell.

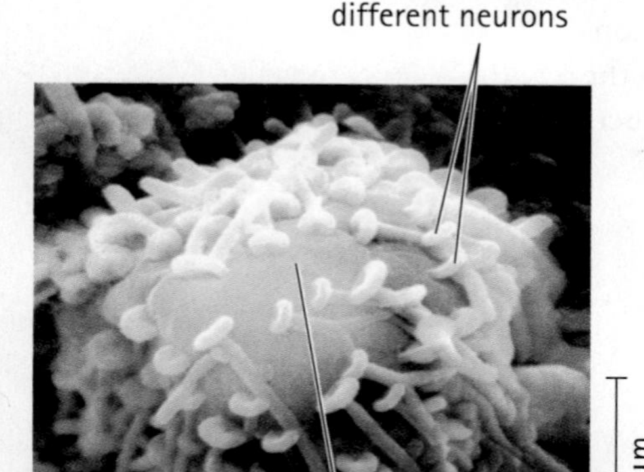

FIGURE 19.13
A target cell can receive signals from many neurons. Axons from many neurons have synapses with this target cell.

The membrane potential of the target cell may increase or decrease, depending on the type of ion that enters. An *excitatory signal* increases the target cell's membrane potential and makes the target cell more likely to fire. An *inhibitory signal* decreases the target cell's membrane potential and makes it less likely to fire. A target cell may receive signals from many other neurons (Figure 19.13). Depending on the sum total of their effects, the neuron will either reach its threshold and fire, or not.

The effect of a neurotransmitter on a target cell ends when the neurotransmitter is removed. Neurotransmitter molecules can be degraded by enzymes, taken up by the original neuron and repackaged into vesicles, or collected and broken down by glial cells.

CHECK YOURSELF

1. **What is the advantage of an electrical synapse over a chemical synapse?**
2. **What is the advantage of a chemical synapse over an electrical synapse?**

CHECK YOUR ANSWERS

1. An electrical synapse is faster than a chemical synapse.
2. A chemical synapse allows for finer control. A target cell can receive information from many neurons. Furthermore, each of these signals may be either strong or weak and either excitatory or inhibitory.

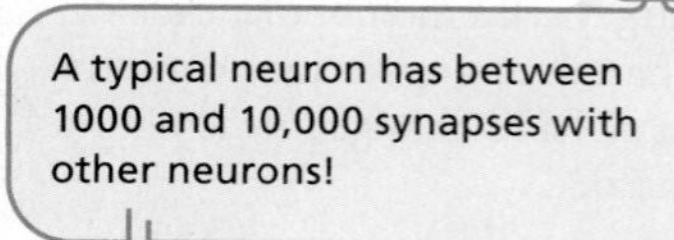

A typical neuron has between 1000 and 10,000 synapses with other neurons!

LEARNING OBJECTIVE
Describe the function of endorphins.

Integrated Science 19B
CHEMISTRY

Endorphins

EXPLAIN THIS How are endorphins related to opiates such as morphine and heroin?

What do chocoholics and long-distance runners have in common? Just maybe, an addiction to endorphins. *Endorphins* are neurotransmitters released by the brain during times of stress or pain.

Endorphins bind to opiate receptors on neurons, the same receptors used by drugs such as morphine, codeine, opium, and heroin. These drugs resemble endorphins in part of their chemical structure and so are able to bind to the same receptors (Figure 19.14). In fact, endorphins were discovered after scientists found the morphine receptor and realized that the body must make a molecule of its own that used the receptor.

Like opiates, endorphins decrease pain and bring on a feeling of euphoria. It is this euphoria that runners call "runner's high." Endorphin release is also associated with laughter, orgasm, acupuncture, massage, and deep meditation. Finally, certain foods—notably chocolate and chili peppers—increase endorphin release. That's comfort food indeed.

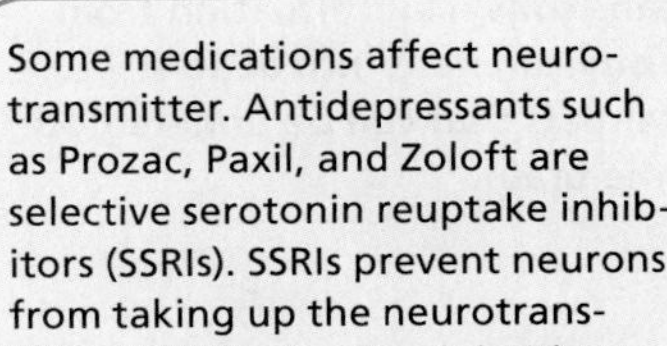

Some medications affect neurotransmitter. Antidepressants such as Prozac, Paxil, and Zoloft are selective serotonin reuptake inhibitors (SSRIs). SSRIs prevent neurons from taking up the neurotransmitter serotonin after it has been released at synapses. As a result, more serotonin remains to bind to and stimulate target neurons.

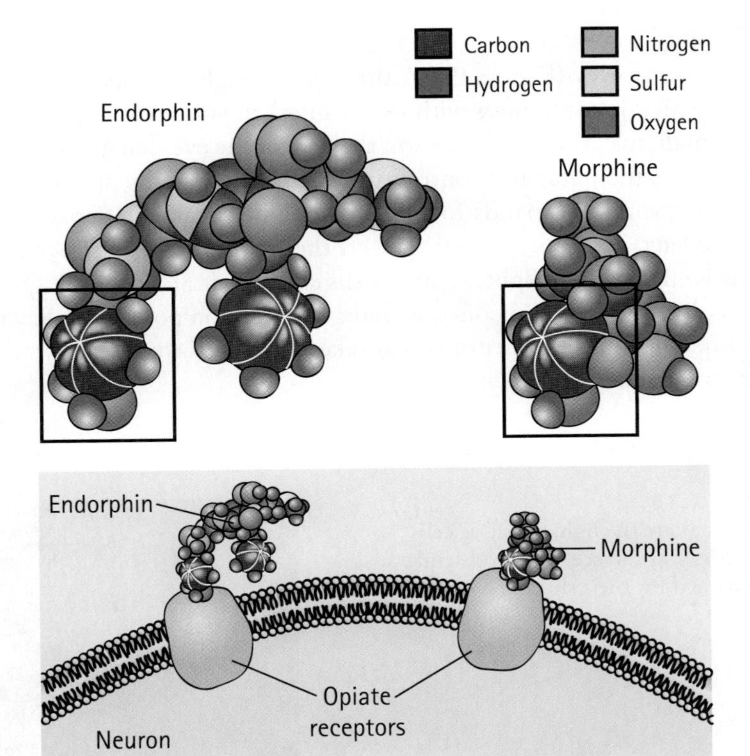

FIGURE 19.14
The boxed part of the endorphin molecule binds to opiate receptors in the nervous system. Morphine resembles endorphins in this part of its chemical structure and so is able to bind to the same receptors.

CHECK YOURSELF

In what way does the chemical structure of opiates such as morphine, codeine, and heroin resemble the structure of endorphins?

CHECK YOUR ANSWER

The part of an opiate that binds to the opiate receptor resembles the part of an endorphin molecule that binds to the same receptor. However, opiates do not resemble endorphins in their overall chemical structure, as you can see in Figure 19.14.

LEARNING OBJECTIVE
Describe how human sense organs convert information from the environment into action potentials that can be understood by the brain.

19.6 The Senses

EXPLAIN THIS How do you taste an apple?

Your senses are your connection to the world. What would happen if you couldn't see, hear, smell, touch, or taste? What if you couldn't tell, without looking at them, where your hands were? What if pain didn't warn you of danger or injury? Each of your senses takes information from the environment—light, sound, touch, or molecules—and converts it into action potentials that are sent to the brain. As you will see, each sense accomplishes this in its own way.

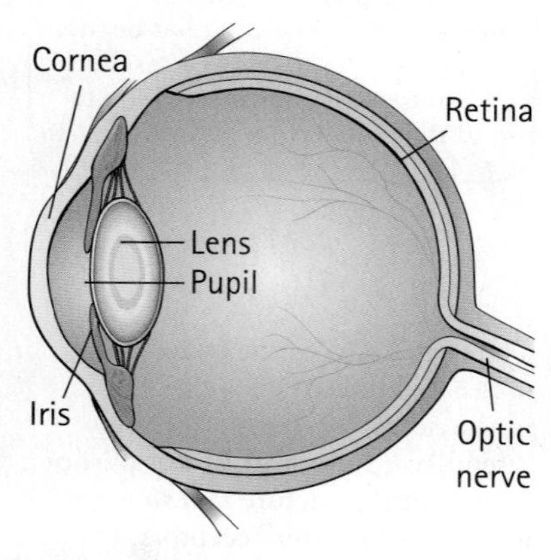

FIGURE 19.15
The eyes convert light into action potentials.

Vision

Light enters your eyes (Figure 19.15) through a tough, transparent layer called the *cornea*, which is continuous with the "whites" of your eyes. Light then passes through a small hole, the *pupil*. The *iris*, the part of the eye that gives you your eye color, surrounds the pupil and controls its size. In bright light, the pupil is small. In dim light, the pupil expands to let in more light. From the pupil, light passes through the lens. The *lens* focuses light on the *retina* at the back of the eyeball. The retina is covered with light-sensitive cells called rods and cones (Figure 19.16). When light hits the rods and cones, it changes the action potentials they transmit to the brain. The bundle of neurons that takes visual information from the retina to the brain is called the *optic nerve*.

Have you noticed that humans have more "whites" in their eyes than other mammals? Scientists believe that humans evolved this trait because it allows one person to see where another person is looking. This is adaptive in a social species like ours, where people live closely and cooperatively with other people.

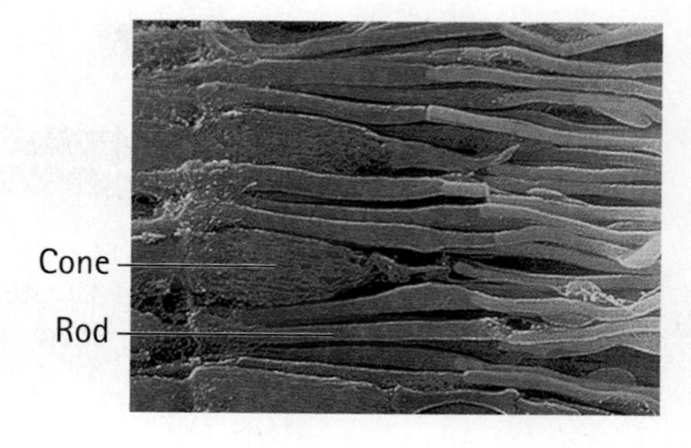

FIGURE 19.16
Rods and cones are the light-sensitive cells in the eyes. As you can see, both rods and cones are named for their shape.

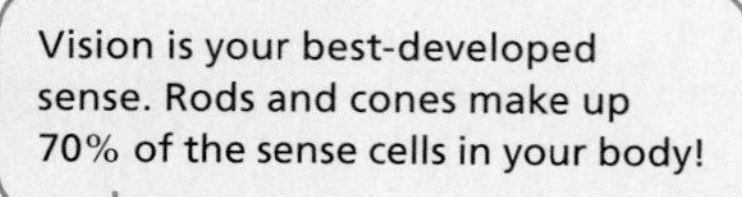

Vision is your best-developed sense. Rods and cones make up 70% of the sense cells in your body!

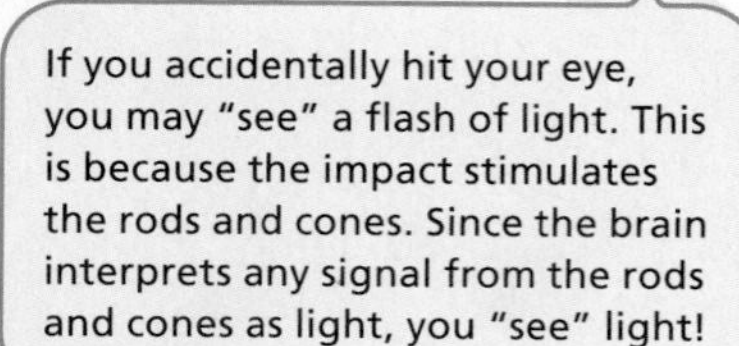

If you accidentally hit your eye, you may "see" a flash of light. This is because the impact stimulates the rods and cones. Since the brain interprets any signal from the rods and cones as light, you "see" light!

The two types of light-sensitive cells in your eyes, rods and cones, have different functions. *Rods* are very sensitive to light and are especially important for seeing in dim light. Rods cannot discriminate colors and allow you to see only black, white, and shades of gray. This is why, in a dark room, you cannot see the difference between a navy-blue shirt and a maroon shirt. (Most people are so used to this, however, that they don't even realize they've lost their color vision!) Rods also are not very good at making out fine details, which is why your night vision is grainy and not very sharp. *Cones* detect color. Your eyes have three types of cones that respond most strongly to red, green, and blue light. All the shades you see are made up of different combinations of these three colors. Color-blindness results from having a nonfunctioning version of one or more cone type.

Hearing

The ear consists of three parts: the outer ear, middle ear, and inner ear (Figure 19.17a). Sound waves move through the air to the *pinna*, the cartilaginous flap on the side of your head. The pinna funnels the waves in, and they move toward a thin membrane of skin—the *eardrum*. Sound waves make the eardrum vibrate, just the way blowing on a piece of paper makes it shake. The eardrum's vibrations move three middle ear bones—the hammer, the anvil, and the stirrup—in sequence. These bones amplify the vibrations, making them more powerful. The stirrup then transfers the vibrations to the fluid-filled inner ear.

In the inner ear, sound vibrations enter the cochlea, a coiled tube containing the organ of Corti. The *organ of Corti* contains the sensory cells responsible for hearing. Fluid vibrations in the inner ear move the organ of Corti's basilar membrane, causing sensory "hairs" embedded in it to brush against an overlying membrane and bend (Figure 19.17b). This bending causes ion channels to open, initiating action potentials that are transmitted to the brain. Note that the "hairs" of hearing cells are not like the hairs on your head; they are long extensions of sensory cells that happen to be shaped like hairs (Figure 19.17c). You can distinguish different noises—the high pitch of a siren versus the low pitch of a jackhammer—because different parts of the organ of Corti respond to different pitches.

UNIFYING CONCEPT

- ***Waves***
  Section 8.1

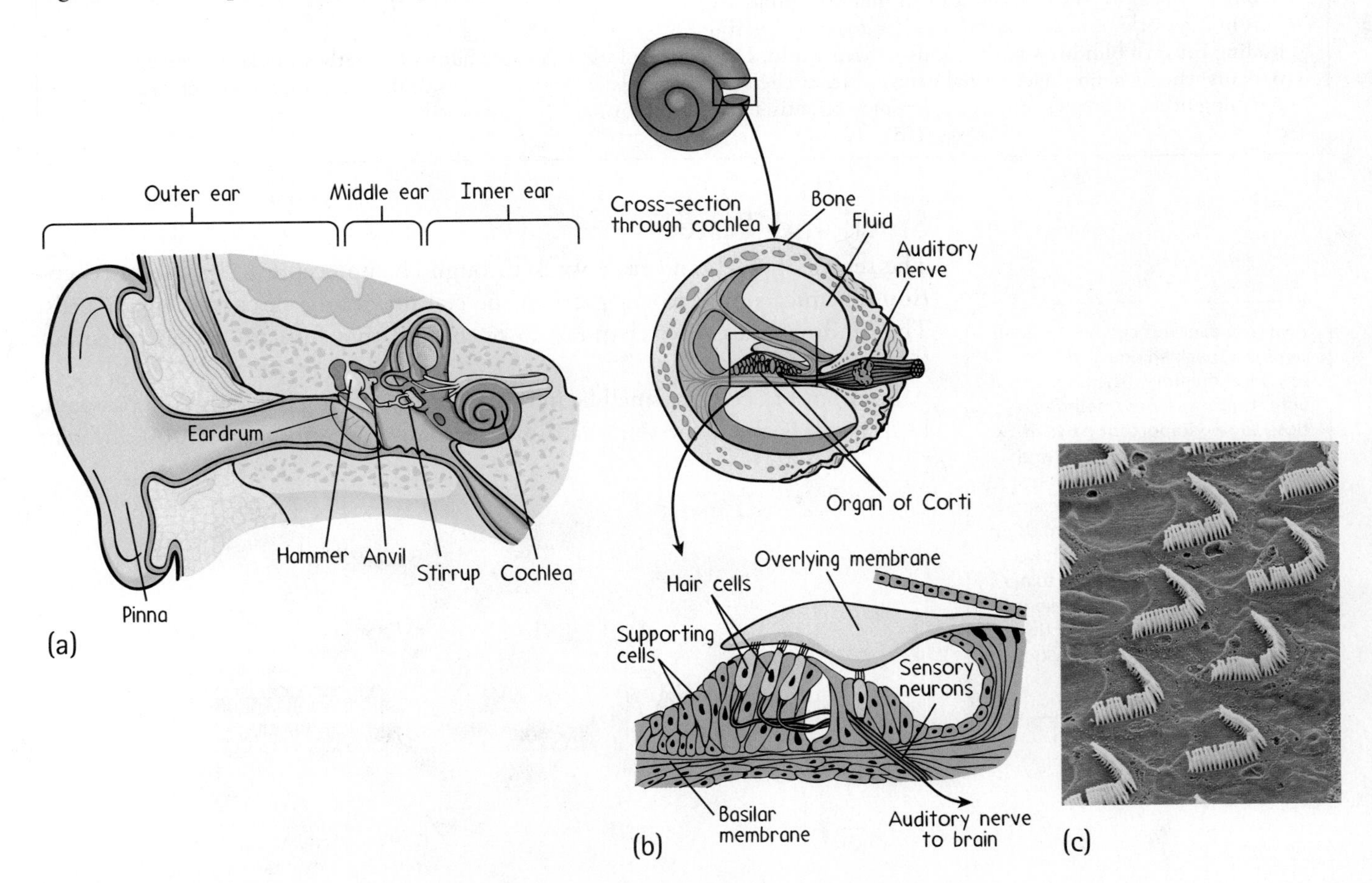

FIGURE 19.17
(a) The ear includes the structures of the outer ear, middle ear, and inner ear. (b) Sound-sensitive cells are contained in the organ of Corti. Fluid vibrations in the inner ear vibrate the basilar membrane of the organ of Corti, causing sensory "hairs" to brush against the overlying membrane and bend. The bending opens ion channels, starting action potentials. (c) This photo shows the "hairs" (yellow) in the organ of Corti.

TECHNOLOGY

Visual Prostheses for the Blind

Will blindness one day be as easy to treat as a toothache? Scientists are beginning to develop visual prostheses that bring vision to the blind. The most promising strategy at the moment involves implanting an array of *electrodes*—devices that conduct electric currents—behind the retina. These electrodes do the work of rods and cones: They receive wireless signals from a chip connected to a camera and stimulate intact neurons at the back of the retina. The neurons then pass visual information to the brain. Although retinal prostheses are still being tested, volunteers with the implants have been able to detect light, locate and count objects, and distinguish between simple objects such as a fork and a knife. As the technology advances, retinal prostheses may eventually allow patients to navigate unfamiliar locations, recognize faces, and read text.

Because retinal prostheses require functional retinal neurons, they are useful only in cases where blindness results from the loss of rods and cones. This is the case for *retinitis pigmentosa*, the leading cause of inherited blindness in the world, and *age-related macular degeneration*, the leading cause of blindness in the industrialized world. For patients who have no intact retinal neurons, researchers are pursuing other strategies. For example, some scientists are exploring the possibility of directly stimulating the optic nerve or even the parts of the brain that interpret visual information. Stay tuned.

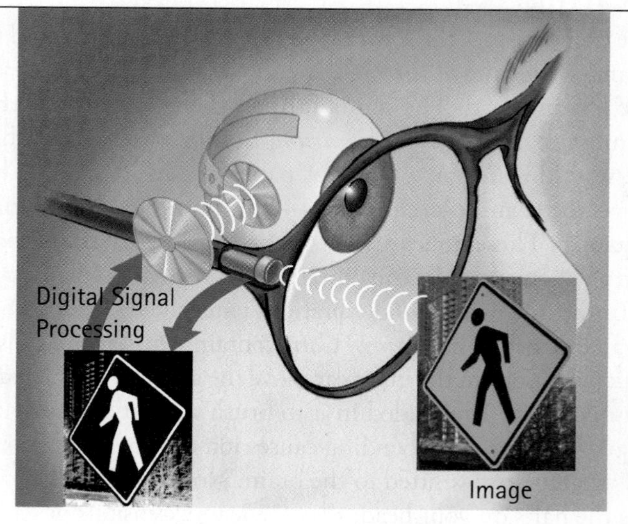

In this retinal prosthesis, a chip processes images from a camera connected to a pair of eyeglasses. Information is then wirelessly beamed to electrodes behind the retina.

Smell and Taste

The senses of smell and taste work through chemoreception. In **chemoreception**, chemicals bind to receptors in the cell membrane of chemosensory cells. The binding causes ion channels to open, and that initiates action potentials (Figure 19.18).

Your sensory cells for smell lie in two patches at the top of your nasal passages. Each patch is about the size of a dime. Humans have more than 1000 different

Have you ever noticed how a smell can bring back an emotionally powerful memory? The part of the brain that processes smells is very close to two important parts of the brain: the hippocampus and the amygdala. The hippocampus is responsible for memory. The amygdala is involved with emotion. Using fMRI experiments, scientists have shown that smells are linked to a person's memories and experiences in special ways. These connections have not been seen for other types of sensation.

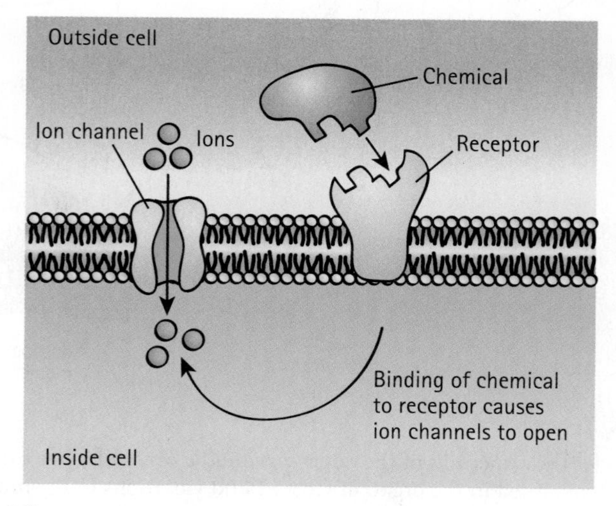

FIGURE 19.18
Smell and taste are examples of chemoreception. Chemicals bind to receptors on the cell membrane of chemosensory cells, causing ion channels to open. The ions initiate action potentials.

kinds of chemosensory cells for smell, each containing only one or a few different kinds of receptors. Because different chemicals trigger different combinations of sensory cells, however, you can distinguish well over 10,000 distinct odors.

Taste cells cluster in small bumps called taste buds. Taste buds are located on your tongue, on the insides of your cheeks, and on the roof of your mouth. Humans distinguish five basic tastes: sweet, salty, sour, bitter, and umami. *Umami* is the Japanese word for "delicious," and it describes the flavor found in foods such as meat, mushrooms, cheese, and asparagus. Monosodium glutamate, or MSG, has a strong umami taste. It is interesting that your experience of food—what you think of as "taste"—comes largely from your sense of smell. This is why food doesn't have nearly as much "taste" when you have a stuffy nose.

When you are hungry, your taste buds become more sensitive to salt and sugar, which makes food taste better. Hunger does not affect your ability to taste bitterness, which helps you identify toxic foods.

Touch

Your sense of touch is actually several different senses. These senses tell you about stimuli such as pressure, temperature, and pain. Your skin's sensory cells for touch are shown in Figure 19.19. Pressure causes the "hairs" on sensory cells to bend, opening ion channels and initiating action potentials. You have separate sensory cells for detecting light touch and heavy pressure. Temperature-sensing cells have ion channels that are affected by heat or cold. Some temperature sensors respond to heat, others to cold. The chemical menthol (found in peppermint) also stimulates cold receptors—it is this coincidence, not an actual cold temperature, that explains the cool feeling you get from eating a mint.

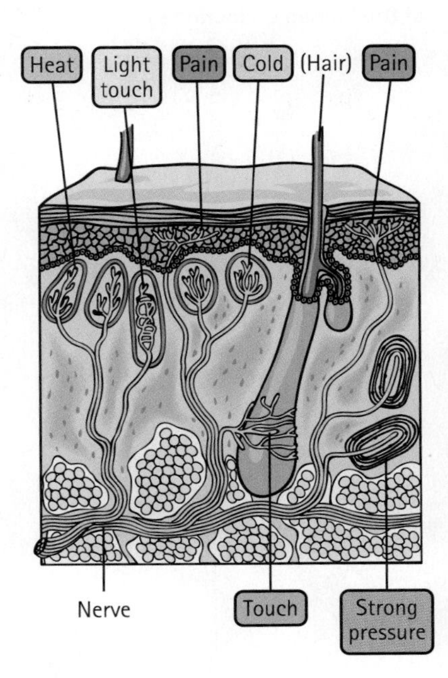

FIGURE 19.19
Sensory cells in the skin are responsible for the senses of touch.

Pain sensors respond to stimuli that damage the body. These sensory cells require strong stimulation before they respond. However, damaged tissues release chemicals called *prostaglandins* that increase the sensitivity of pain receptors. You might remember that aspirin provides pain relief by interfering with prostaglandin production (see Chapter 15). Pain sensors also become more sensitive with continued stimulation. This distinguishes pain receptors from most other sensory cells, which become less sensitive with repeated stimulation. (This is why you stop noticing the smells in your house or feeling the weight of your backpack as you walk.) Scientists believe that some types of chronic pain result from pain receptors that have become abnormally sensitive.

Other Senses

You have other senses in addition to the big five. *Proprioceptors* in your muscles, tendons, and joints tell you where different parts of your body are. You can try out this sense by closing your eyes and touching your nose with your finger. (You may be too used to this ability to be impressed, but consider the fact that you can't easily touch *another* person's nose with your eyes closed!) Finally, the *vestibular senses* tell you about body rotation and movement as well as which way is up (see Chapter 5).

CHECK YOURSELF

1. **When you are outside at night, looking at a starry sky, are you using your rods or your cones?**
2. **Deafness can result from a number of problems, including a ruptured eardrum or damaged sensory cells in the organ of Corti. Why would each of these problems make you unable to hear?**

CHECK YOUR ANSWERS

1. You are using your rods, which allow you to see in dim light.
2. If the eardrum is ruptured, sound waves cannot be conducted to the middle ear and no vibrations reach the organ of Corti. If sensory cells in the organ of Corti are damaged, they are unable to send signals to the brain.

LEARNING OBJECTIVE
Describe the structure and function of the human endocrine system.

19.7 Hormones

EXPLAIN THIS Why does your body release insulin after you eat a large meal?

MasteringPhysics®
TUTORIAL: Overview of Cell Signaling
TUTORIAL: Nonsteroid Hormone Action
TUTORIAL: Steroid Hormone Action

While the nervous system handles rapid actions and reactions, the hormones of the endocrine system regulate activities that take place over longer time periods. For example, hormones regulate your growth and development, prepare you for reproduction, determine how quickly you metabolize food, and tell you whether it is night or day. Hormones also play an important role in maintaining homeostasis in the body. So, what are hormones and how do they work?

A **hormone** is a molecule that gives instructions to the body. Hormones are produced in one place in the body, released into the blood, and then received by target cells elsewhere in the body. Hormones come in two types. *Protein hormones* are, as their name suggests, proteins or modified amino acids. *Steroid hormones* are made from cholesterol.

Protein hormones and steroid hormones work in different ways. Protein hormones bind to receptors on the cell membranes of their target cells. This binding starts a series of chemical reactions that result in the target cells' response to the hormone (Figure 19.20a). Steroid hormones cross the cell membrane and bind to receptors inside target cells—their receptors may be in a target cell's cytoplasm or nucleus. The hormone-receptor complex then binds to DNA in the nucleus and directly affects gene transcription (Figure 19.20b).

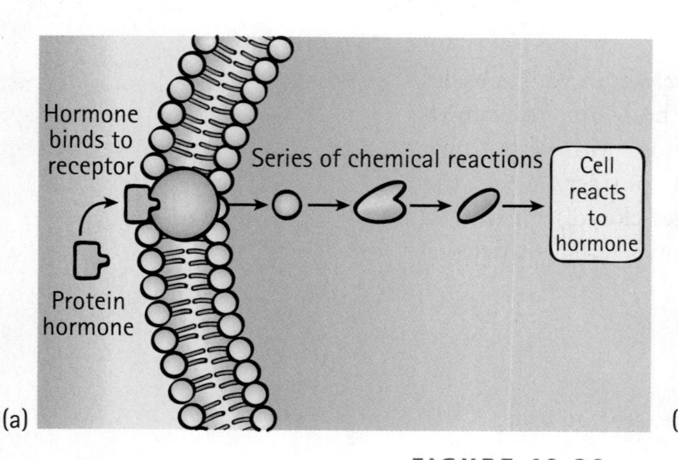

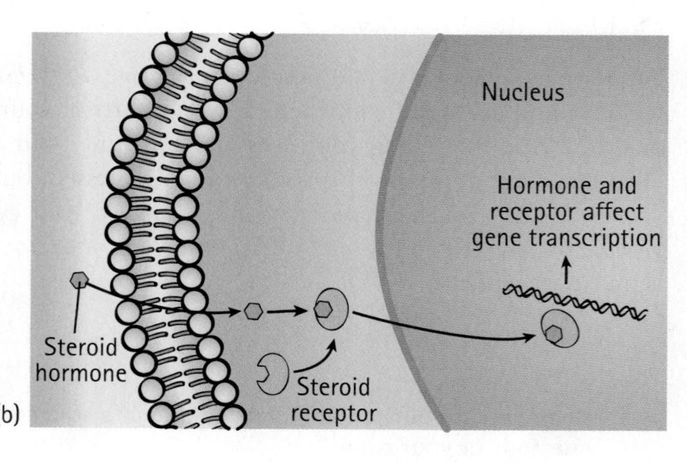

FIGURE 19.20
(a) Protein hormones bind to receptors in the cell membranes of target cells. This starts a sequence of events that result in the target cells' response to the hormone. (b) Steroid hormones enter target cells and bind to receptors in the cytoplasm or nucleus. Together, the hormone and receptor directly affect gene transcription.

Endocrine Organs and Their Hormones

Let's consider the human endocrine organs (Figure 19.21) and the hormones they make.

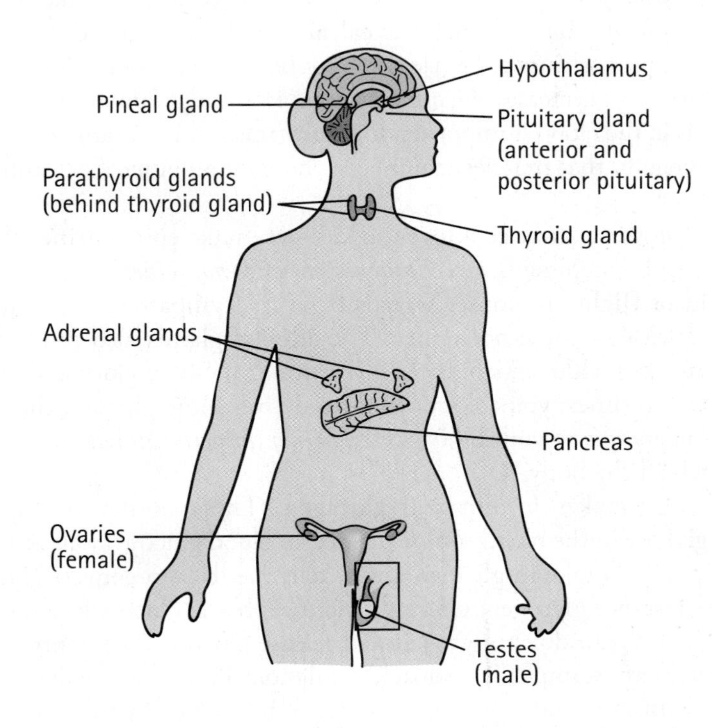

FIGURE 19.21
Human endocrine organs include the hypothalamus, pituitary gland (anterior and posterior pituitary), thyroid gland, parathyroid glands, adrenal glands, pancreas, ovaries and testes, and pineal gland.

The *hypothalamus* is a part of the brain; it is the endocrine system's control center. Many of the hypothalamus's hormones regulate the activity of another endocrine organ, the anterior pituitary. These hormones are released from the hypothalamus directly to the anterior pituitary, which is located just underneath the hypothalamus. The hypothalamus also makes hormones that are stored and released by the posterior pituitary (discussed below).

The *anterior pituitary* is sometimes called the "master gland." This is because many of its hormones regulate the activity of other endocrine organs. For example, anterior pituitary hormones tell the thyroid gland, sex organs, and adrenal glands to release their hormones. The anterior pituitary also makes growth hormone and prolactin. *Growth hormone* does what its name suggests—it promotes growth. Too little growth hormone results in dwarfism, and too much results in gigantism (Figure 19.22). *Prolactin* stimulates milk production in nursing mothers.

FIGURE 19.22
This composite picture shows dwarfism (*upper left*) and gigantism (*center*) relative to a person of normal size (*right*). Dwarfism and gigantism result from too little or too much growth hormone.

The *posterior pituitary* stores and controls the release of hormones made by the hypothalamus. *Antidiuretic hormone* helps regulate the amount of water in the body. Specifically, it helps the body conserve water by instructing the kidneys to produce urine that is more concentrated. (We will see how in Chapter 20.) Alcohol inhibits the release of antidiuretic hormone. This is why people produce more urine—and sometimes become dehydrated—when they consume alcohol. *Oxytocin* stimulates contraction of the uterus during childbirth. Women whose labor does not progress, or who need to have labor induced, may be given pitocin, a synthetic form of oxytocin, to stimulate contractions.

The *thyroid gland* makes thyroid hormones. *Thyroid hormones* are involved in metabolism, growth, and development. For example, thyroid hormones are necessary for proper brain development during childhood. Thyroid hormones contain iodine, and getting too little iodine can result in developmental problems. This is why iodine is added to many brands of table salt. The thyroid gland also makes *calcitonin*, which decreases calcium levels in the blood.

The *parathyroid glands*, which lie next to the thyroid gland, make parathyroid hormone. *Parathyroid hormone* increases calcium levels in the blood. It does this in three ways: It causes calcium to be released from bones, increases calcium absorption in the intestine, and decreases calcium excretion in urine. Maintaining appropriate calcium levels in the blood is important for many reasons. For example, you will see later in this chapter that muscles require just the right amount of calcium to function properly.

Have you noticed that some milk is labeled "rBGH Free"? rBGH stands for recombinant bovine growth hormone. rBGH is a genetically engineered growth hormone that is sometimes given to cows to increase milk production. Milk from cows that were given rBGH does not contain higher levels of growth hormone, but it does contain higher levels of other growth-promoting hormones. Whether these are dangerous for humans has been controversial, but many consumers are choosing to drink rBGH-free milk.

The *adrenal glands*, located above the kidneys, make epinephrine (also called adrenaline) and norepinephrine. *Epinephrine* and *norepinephrine* are involved in the "fight or flight" response. Signals from the sympathetic nervous system trigger the release of these hormones. The adrenal glands also produce glucocorticoids and mineralocorticoids. *Glucocorticoids* increase glucose levels in the blood; at certain times, your blood needs to deliver more glucose, the molecule that gives you energy, to your body's cells. *Mineralocorticoids* help regulate water and salt levels in the body.

The *pancreas* makes insulin and glucagon. These hormones regulate the amount of glucose in the blood. *Insulin* decreases blood glucose levels in two ways: It tells body cells to take in glucose, and it tells the liver to convert glucose into the storage substance glycogen. *Glucagon* increases blood glucose levels. Glucagon tells the liver to break down glycogen and release glucose. The control of blood glucose levels is an example of feedback regulation: Blood glucose levels regulate the release of insulin and glucagon, and these hormones in turn help regulate blood glucose levels (Figure 19.23).

UNIFYING CONCEPT

- ***Feedback Regulation***
Section 19.2

The disease *diabetes* results when the pancreas doesn't make enough insulin or when the body's cells do not respond to insulin. In either case, blood glucose levels become abnormally high. People who have diabetes have to control their diets and monitor their blood glucose levels. Some require regular injections of insulin.

The sex organs—*ovaries* in women and *testes* in men—make sex hormones. Both women and men make all three types of sex hormones: estrogens, progestins, and androgens. Women make more *estrogens* and *progestins*. These hormones

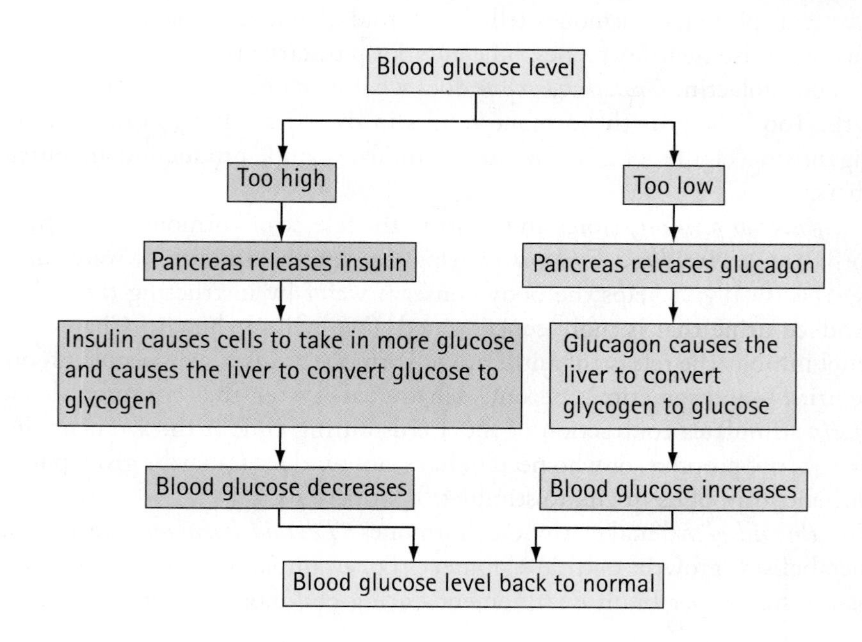

FIGURE 19.23
Feedback regulation enables the body to maintain appropriate blood glucose levels. If blood glucose levels are too high, insulin is released. Insulin causes blood glucose levels to decrease. If blood glucose levels are too low, glucagon is released. Glucagon causes blood glucose levels to increase.

SCIENCE AND SOCIETY

Fathers and Testosterone

After a man becomes a father, his testosterone levels fall. This was the conclusion of a study that measured testosterone in more than 600 men when they were 21 years old and single, and again 5 years later. Although testosterone levels dropped in all men, those who had children showed a greater drop. Moreover, fathers who spent the most time taking care of their children showed the largest drops. Researchers hypothesize that reducing testosterone is adaptive for fathers: Testosterone promotes aggressive behavior, and lower levels are likely to help maintain the family and promote effective child care. This evolved hormonal response also suggests that human fathers have long been involved in caring for their children. Mothers have not done it alone.

An evolved hormonal response—lower testosterone levels help fathers successfully raise families.

Data from the same study showed that men who had *higher* testosterone levels were more likely to become fathers in the first place. Apparently, high testosterone levels help a man find a partner, but low testosterone levels help him successfully raise a family.

regulate ovulation and the menstrual cycle and are involved in pregnancy. Estrogen also promotes breast development and fat storage in the hips and thighs. Men produce more *androgens*. Androgens such as testosterone are required for sperm production. Androgens also promote the development of male secondary sexual characteristics such as facial hair and increased muscle mass. It is this last effect of androgens that tempts some athletes to use anabolic steroids—synthetic versions of testosterone—to improve athletic performance. Unfortunately, steroids also have many negative side effects. They can cause aggressive behavior, mood swings, and irritability. In men, steroids also cause shrinking of the testicles, decreased sperm count, baldness, and breast development. In women, steroids disrupt the menstrual cycle, deepen the voice, and promote the growth of facial hair. In adolescents, steroids stunt growth and accelerate puberty.

The *pineal gland* produces the hormone melatonin. *Melatonin* regulates the body's internal clock, telling you when it is day and when it is night. Using light cues from the eyes, the pineal gland releases melatonin during the night hours. This is why some people use melatonin as a sleeping pill.

The major endocrine organs and the hormones they produce are summarized in Table 19.1.

A special group of light-sensitive cells in the retina tells the pineal gland whether it is day or night. This allows the pineal gland to set the body's internal clock. These cells are intact in some blind people but not in others. In people who are totally blind, the internal clock may shift out of sync with day and night, so that they are sleepy during the day and wide awake at night.

CHECK YOURSELF

Some hormones come in pairs with opposing effects. Each pair helps maintain homeostasis in the body by regulating the amount of an important substance or molecule—often through feedback regulation. What are two pairs of hormones with opposing effects?

CHECK YOUR ANSWER

Calcitonin and parathyroid hormone together regulate blood calcium levels. Calcitonin decreases blood calcium levels, and parathyroid hormone increases blood calcium levels. Insulin and glucagon together regulate blood glucose levels. Insulin decreases blood glucose levels, and glucagon increases blood glucose levels.

TABLE 19.1 MAJOR ENDOCRINE ORGANS AND THE HORMONES THEY PRODUCE

Endocrine Organ	Hormone	Hormone Type	Effect
Hypothalamus	Makes hormones that regulate the anterior pituitary and hormones released by the posterior pituitary (see below)		
Pituitary gland			
Anterior pituitary	Growth hormone	Protein	Stimulates growth
	Prolactin	Protein	Stimulates milk production
	Various hormones that stimulate other endocrine organs	Protein	Stimulate endocrine organs such as ovaries and testes, thyroid gland, and adrenal glands
Posterior pituitary (releases hormones made by hypothalamus)	Antidiuretic hormone	Protein	Promotes retention of water by kidneys
	Oxytocin	Protein	Stimulates contraction of uterus
Thyroid gland	Thyroid hormones	Protein	Stimulate growth and development; regulate metabolism
	Calcitonin	Protein	Decreases blood calcium levels
Parathyroid glands	Parathyroid hormone	Protein	Increases blood calcium levels
Adrenal glands	Epinephrine and norepinephrine	Protein	Promote "fight or flight" response
	Glucocorticoids	Steroid	Increase blood glucose levels
	Mineralocorticoids	Steroid	Regulate water and salt levels
Pancreas	Insulin	Protein	Decreases blood glucose levels
	Glucagon	Protein	Increases blood glucose levels
Sex organs			
Testes	Androgens	Steroid	Support sperm formation; promote male secondary sexual characteristics
Ovaries	Estrogens	Steroid	Maintain female reproductive system; promote female secondary sexual characteristics
	Progestins	Steroid	Promote uterine lining growth
Pineal gland	Melatonin	Protein	Regulates internal clock

LEARNING OBJECTIVE
Describe the key events in human reproduction and development.

19.8 Reproduction and Development

EXPLAIN THIS Does a mother's blood mix with her baby's during pregnancy?

Where do babies come from? Without reproduction, humans—or any other species—would quickly go extinct.

Human Reproduction

Human reproduction begins with the production of *gametes,* or sex cells—eggs in women and sperm in men. Eggs and sperm are produced through meiosis (see Chapter 16). They are haploid cells with only half the usual number of chromosomes. At **fertilization**, egg and sperm join to form a diploid cell that develops into a new human being. (You can review the human life cycle in Figure 16.13.) Let's look at the female and male reproductive systems.

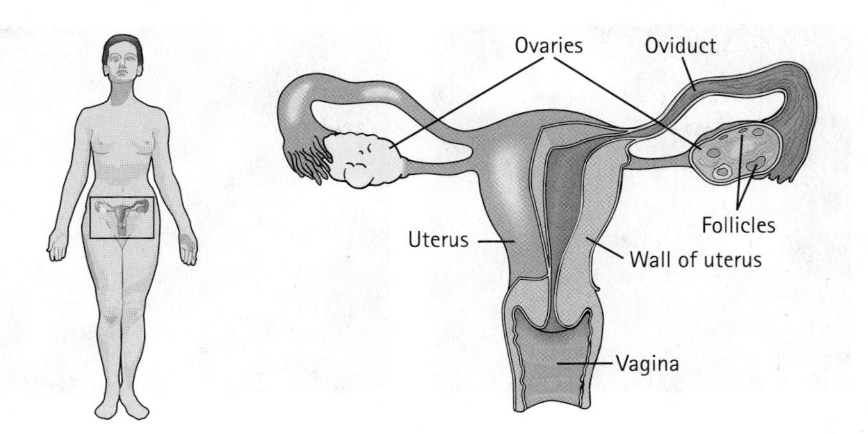

FIGURE 19.24
The female reproductive system.

In females, eggs are made in the *ovaries* (Figure 19.24). Each ovary is made up of many *follicles*, developing eggs surrounded by support cells. During each menstrual cycle, a single follicle matures and releases an egg in a process called **ovulation**. The egg is a large cell, with lots of nutrients stored in its cytoplasm. Eggs are large because they are the result of unequal meiosis: During cell division, the future egg receives almost all the cytoplasm, while the other cells (which quickly degenerate) receive very little cytoplasm (see Chapter 16). After ovulation, the egg enters the *oviduct*, where cilia sweep it toward the uterus. Fertilization typically takes place while the egg is still in the oviduct. The fertilized egg then continues to the *uterus*, where it implants and continues development.

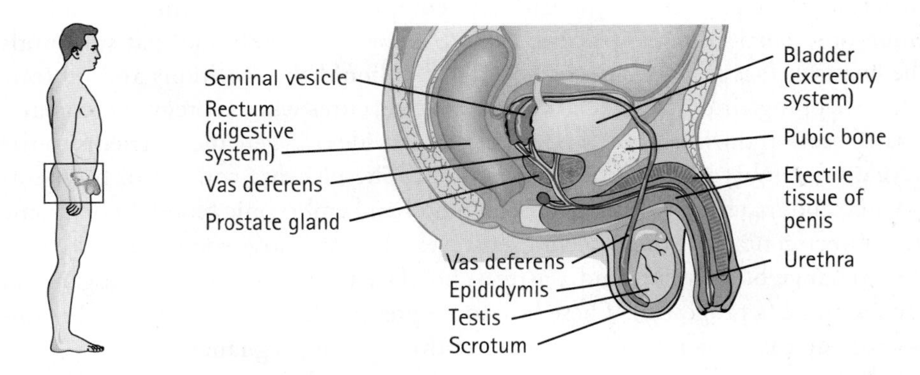

FIGURE 19.25
The male reproductive system.

If a man is trying to father a child, he should think about wearing boxers rather than briefs. It's not just superstition: Briefs can hold the scrotum too close to the body and make the testes too warm for sperm production.

In males, sperm are made in the *testes*, which are located in the scrotum (Figure 19.25). The scrotum hangs away from the body to keep the testes at a temperature lower than body temperature. This is essential for sperm production. From the testes, sperm enter the *epididymis*, where they complete development and become mobile. Each mature sperm cell has a head that contains DNA, mitochondria, and enzymes for penetrating the egg. Sperm also have a tail for swimming. During sexual intercourse, sperm travel along the *vas deferens* to the *urethra*, a tube inside the penis. Sperm are ejaculated from the urethra in *semen*. In addition to sperm, semen contains fluids from the *seminal vesicles* and *prostate gland* that nourish sperm and protect them from the acidic environment of the vagina. There are about half a billion sperm in each ejaculate.

After sexual intercourse, sperm swim up the oviduct to the egg (Figure 19.26). The egg is covered by a jellylike layer called the *zona pellucida*. Enzymes released from the heads of many sperm eat away at this cover. A single sperm finally reaches the egg's cell membrane, and the cell membranes of egg and sperm fuse. At this point, the zona pellucida undergoes changes that make it impenetrable to additional sperm, assuring that the fertilized egg doesn't end up with too many chromosomes.

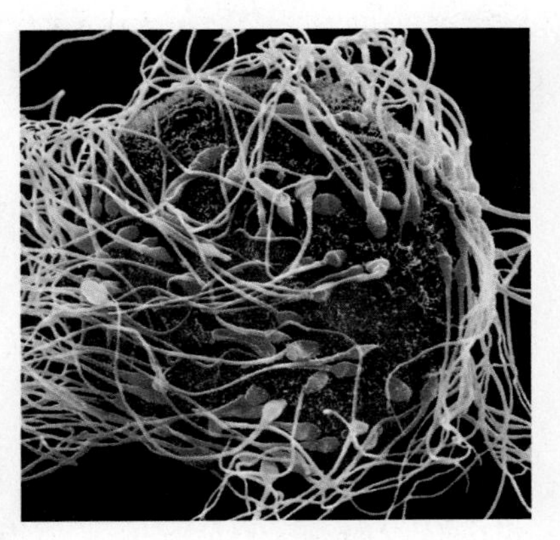

FIGURE 19.26
Sperm surround a human egg.

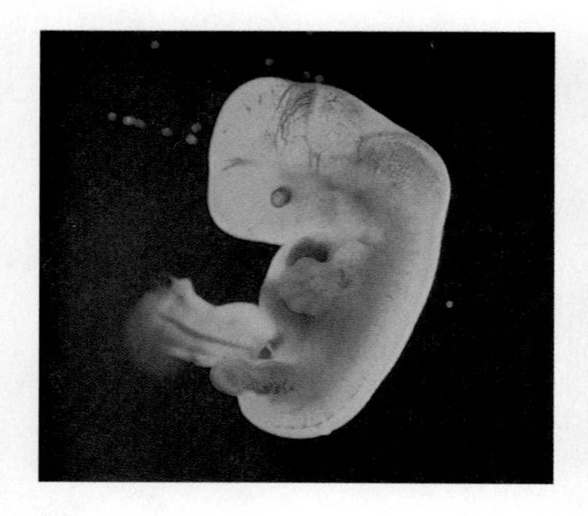

(a)

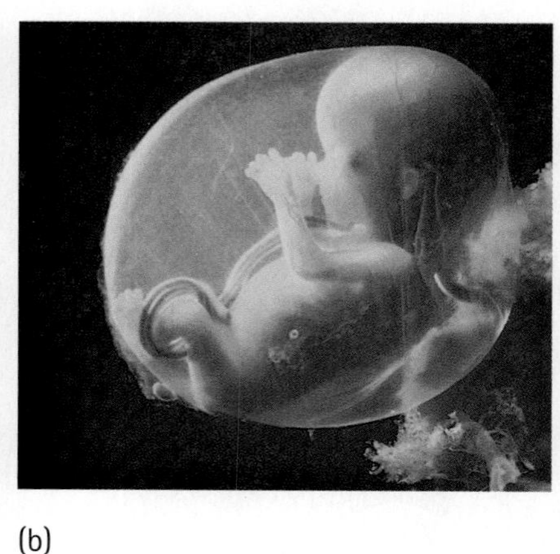

(b)

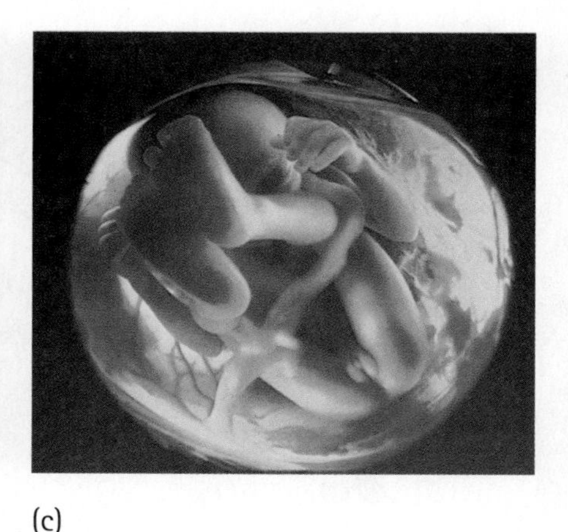

(c)

FIGURE 19.27
(a) This is a human embryo at 5 weeks. (b) This is a human embryo at 14 weeks. By the end of the first trimester, all the major organs and body parts have developed. (c) This is a human embryo at 20 weeks.

Human Development

After fertilization, the egg begins to divide through mitosis (see Chapter 15). By the time implantation occurs in the uterus, about 6 days after fertilization, the developing egg has become a hollow ball of cells called a *blastocyst*. Part of the blastocyst forms the embryo, the future baby. The rest of the blastocyst forms structures that protect and nourish the embryo, such as the amnion and the embryonic portion of the placenta. The *amnion* is a membrane that surrounds the embryo. It is filled with amniotic fluid, a liquid that cushions and protects the developing embryo. The amnion is what ruptures when a pregnant woman's "water breaks" during labor. The **placenta** provides oxygen and nutrients to the developing embryo and carries away wastes. The placenta consists of both embryonic and maternal tissues. Maternal blood and embryonic blood do not come into direct contact in the placenta; however, they are close enough to allow for the exchange of nutrients and wastes. The placenta also produces estrogen and progesterone (a progestin). These hormones prevent further ovulation and maintain the uterus in its nurturing condition throughout pregnancy.

The 9 months of pregnancy are divided into three 3-month trimesters. During the first trimester, all of the embryo's major organs and body parts develop. Further development, as well as most of the fetus's growth, occurs in the second and third trimesters. Three stages in human development are shown in Figure 19.27.

MasteringPhysics®
VIDEO: Ultrasound of a Fetus I
VIDEO: Ultrasound of a Fetus II

CHECK YOURSELF

If you place an unfertilized egg with its zona pellucida removed into a petri dish containing many sperm, what is the likely result?

CHECK YOUR ANSWER

Multiple sperm will probably fertilize the egg. Normally, the zona pellucida undergoes changes at fertilization to prevent additional sperm from entering.

19.9 The Skeleton and Muscles

LEARNING OBJECTIVE
Describe the structure and function of the skeleton and muscles.

EXPLAIN THIS If muscles can only pull, not push, how is it possible for you to both bend your arm and straighten it?

The Skeleton

The skeleton, shown in Figure 19.28, is made up of the bones and cartilages that protect and support the body. Human adults have 206 bones in all. Babies are born with more, but many of these fuse during growth. Besides bones, the skeleton includes several cartilages, including your external ears (pinnas) and the tip of your nose.

One function of the skeleton is to protect the body. For example, the skull protects the brain, the vertebrae protect the spinal cord, and the ribs protect the heart and lungs. The skeleton also supports the body and, working with the muscles, moves it. *Joints* are movable connections between bones. Some joints, like the elbow and knee, act like hinges, bending in only one direction. Other joints, like the one between the hip and thigh, resemble a ball and socket and allow for a greater range of motion. At a joint, the ends of connecting bones are covered with smooth cartilage and enclosed in a fluid-filled capsule. The fluid lubricates the joint so that the bones can move smoothly, without rubbing against each other. *Arthritis* is a condition where the tissues of the joint become inflamed and produce too much fluid. Both the inflammation and excess fluid cause painful bone damage.

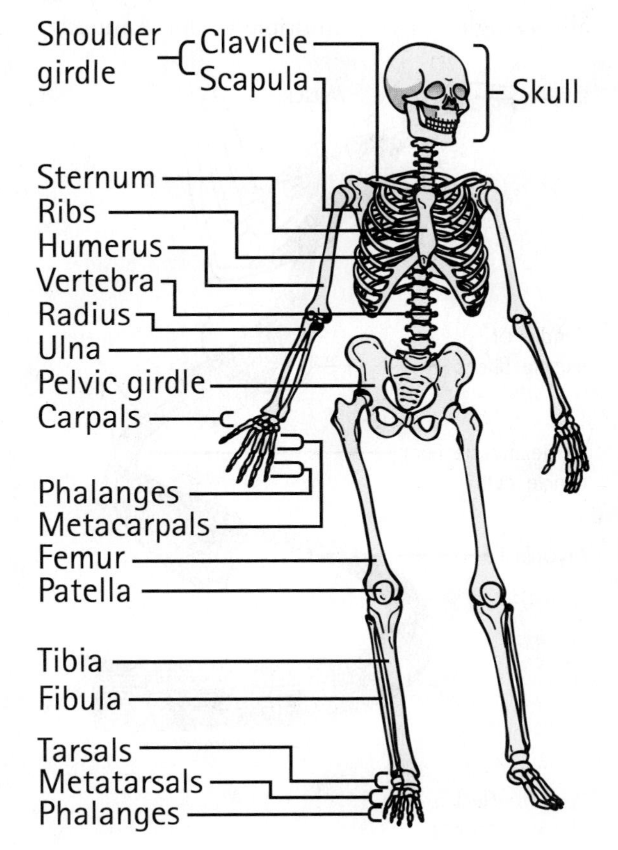

FIGURE 19.28
The human skeleton includes 206 bones (in adults) as well as a number of cartilages.

The largest bone in the body is the femur, or thigh bone. The smallest is the stirrup, a bone in the middle ear that is about a quarter of a centimeter long.

Bones are made up of three layers. First, there is a strong, hard outer layer of *compact bone.* Compact bone surrounds a lighter layer of *spongy bone.* Inside the spongy bone is a jellylike substance called bone marrow (Figure 19.29). *Red bone marrow* makes red and white blood cells for the circulatory system. *Yellow bone marrow* stores fat. Like other parts of the body, bones are made of living cells. These cells secrete the hard calcium-containing matrix that gives bones both strength and flexibility.

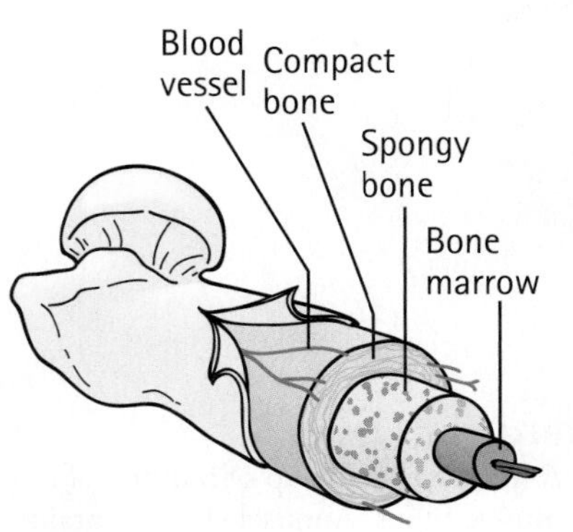

FIGURE 19.29
In bones, compact bone surrounds a layer of spongy bone, which surrounds the bone marrow.

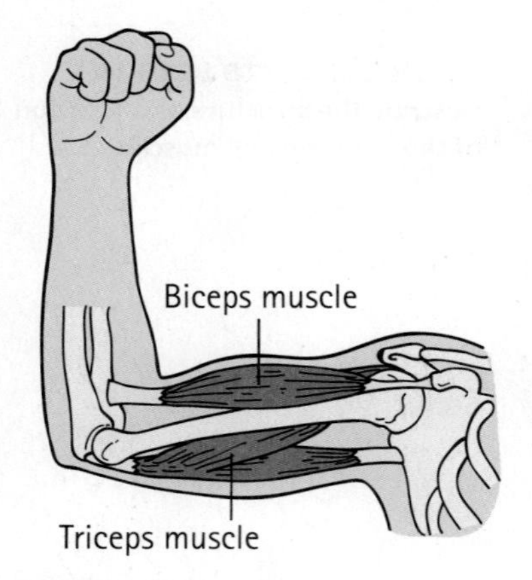

FIGURE 19.30
The biceps and triceps muscles move the forearm in opposite directions. The biceps bends the forearm, and the triceps straightens it.

MasteringPhysics®
TUTORIAL: Muscle Contraction

Muscles

You have more than 600 muscles in your body, including 30 in your face alone. You use these muscles for everything from smiling to making sandwiches. In fact, almost any time you say that you are "doing" something, you are doing it with your muscles.

Muscles work by contracting, or shortening. Many of your muscles are connected to bones via *tendons*. When these muscles contract, they pull at your bones, moving you. Because muscles can only pull, not push, you have many pairs of muscles with opposing effects. For example, your biceps muscle pulls on the inner part of your forearm, bending your forearm up toward your shoulder (Figure 19.30). Your triceps muscle pulls on the back end of your forearm (it attaches to the "funny bone" at the end of your elbow) and straightens the arm.

How does a muscle contract? Let's start by looking at its structure. A muscle (Figure 19.31) consists of bundles of long muscle fibers. Each muscle fiber is actually a single cell with multiple nuclei. A muscle fiber contains bundles of smaller elements called *myofibrils*. Each myofibril is made up of a series of contractile units called **sarcomeres**. Sarcomeres are made up of carefully arranged fibers of two proteins: thin filaments of *actin* and thick filaments of *myosin*. When a muscle contracts, the actin and myosin filaments slide past each other, shortening the length of each sarcomere. Let's look at how this happens.

A muscle contracts when it receives a signal from a motor neuron. An action potential in the motor neuron arrives at a chemical synapse connecting the neuron to a muscle cell. The action potential triggers the release of the neurotransmitter *acetylcholine* at the synapse (Figure 19.32). Acetylcholine then binds to receptors on the muscle cell's membrane, initiating an action potential in the muscle cell.

An action potential in a muscle cell causes calcium ions to be released from the muscle cell's endoplasmic reticulum (an organelle discussed in Chapter 15).

When you lift weights, you don't increase the number of muscle fibers in your muscles. That number stays the same. Instead, your muscles get bigger because the muscle fibers become thicker.

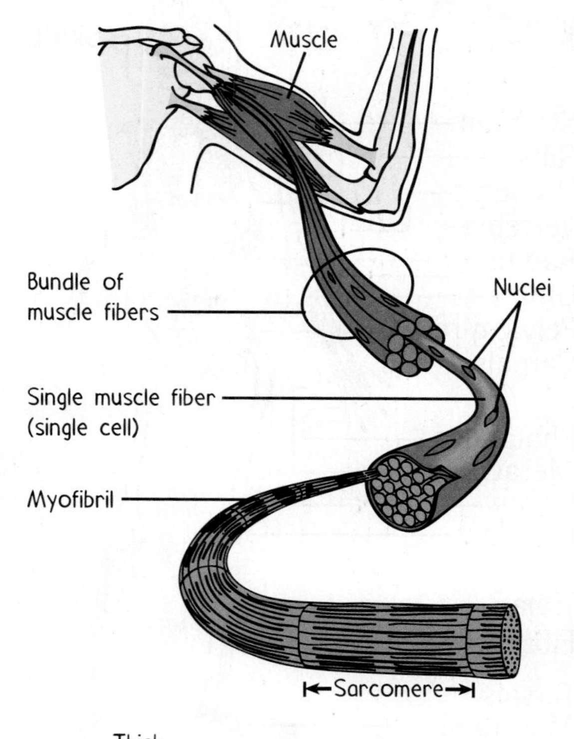

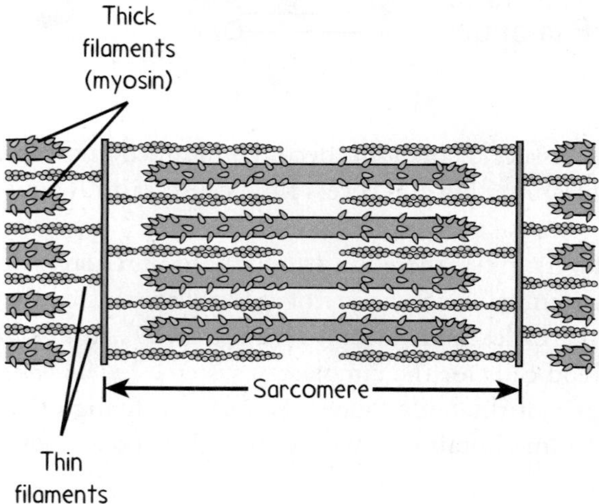

FIGURE 19.31
A muscle is made up of bundles of muscle fibers. A muscle fiber contains smaller fibers called myofibrils. A myofibril consists of sarcomeres arranged end to end. Two proteins in the sarcomere, actin and myosin, allow sarcomeres—and consequently muscles—to contract.

Calcium ions enable a series of "heads" on the myosin fibers to attach to actin (Figure 19.33). The myosin heads attach and pivot, pulling on the actin filaments. Each pull shortens the length of the sarcomere a tiny bit—about 10 nanometers—and, consequently, the length of the muscle as a whole. After pulling, the myosin heads release, extend, attach, and pull again. This cycle repeats until the signal to contract ends or until the muscle has fully contracted. Muscle contraction requires energy: ATP is needed for the myosin heads to release actin, an essential step in the contraction cycle.

Once the motor neuron stops signaling, acetylcholine stops binding to the muscle cell, and the muscle cell's endoplasmic reticulum stops releasing calcium. Without calcium, the myosin heads are unable to bind to actin, and the muscle relaxes.

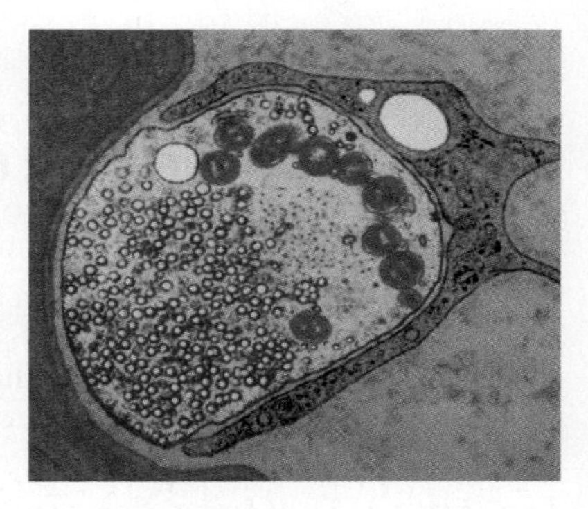

FIGURE 19.32
This photo shows a synapse between a motor neuron (blue) and a muscle cell (red). Small vesicles containing the neurotransmitter acetylcholine are clustered near the muscle cell. (The larger brown structures in the neuron are mitochondria. The green cell surrounding the neuron is a glial cell.)

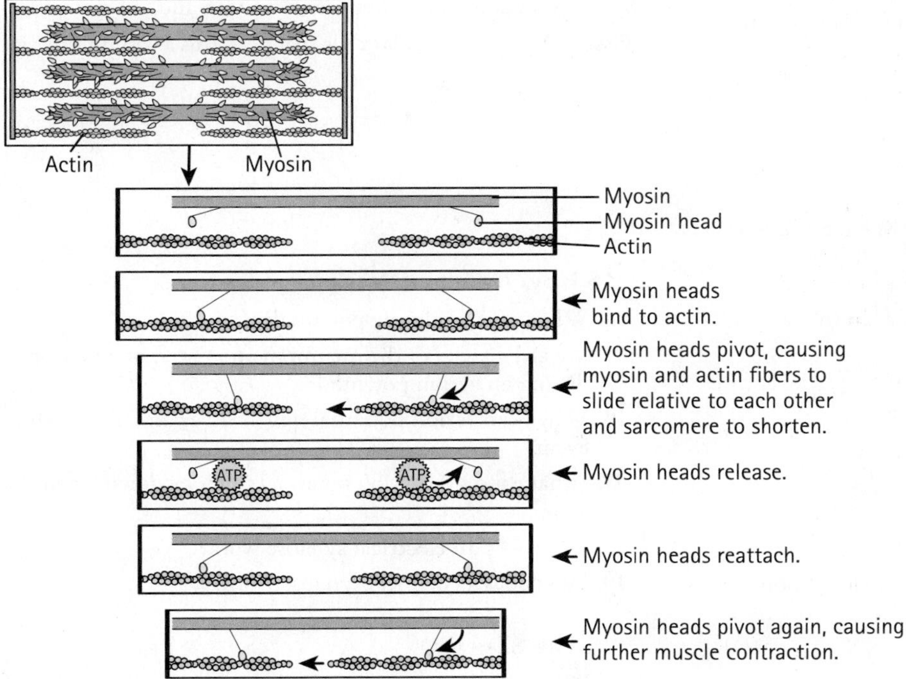
Relaxed sarcomere
Actin
Myosin
Myosin
Myosin head
Actin
Myosin heads bind to actin.
Myosin heads pivot, causing myosin and actin fibers to slide relative to each other and sarcomere to shorten.
ATP
ATP
Myosin heads release.
Myosin heads reattach.
Myosin heads pivot again, causing further muscle contraction.

FIGURE 19.33
During contraction, the myosin heads bind to actin, pivot (pulling the actin), release, reattach, and pull again. (For simplicity, only one myosin head is shown on each side. In reality, many myosin heads pull on actin.)

Some well-known toxins work by interfering with the neuron-to-muscle connection. Curare, an arrow poison used in the South American tropics for hunting, binds to acetylcholine receptors on muscle cells, preventing acetylcholine itself from binding. Curare causes paralysis and then death as the respiratory muscles become paralyzed. The powerful nerve gas sarin prevents acetylcholine from being broken down after muscles contract. Muscles are stimulated continuously and soon become exhausted. Again, death occurs through asphyxiation as the respiratory muscles stop working. Sarin is perhaps best known as the agent used in a terrorist attack on the Tokyo subway in 1995.

CHECK YOURSELF

1. **Why is calcium necessary for muscle contraction?**
2. **Why is ATP necessary for muscle contraction?**

CHECK YOUR ANSWERS

1. Calcium enables the myosin heads to attach to actin.
2. ATP is required for the myosin heads to release actin, an essential step in the contraction cycle.

What explains rigor mortis, the stiffness of the body that sets in after death? The mechanics of muscle contraction. After death, calcium ions leak from the endoplasmic reticulum of muscle cells. Myosin binds to actin, and muscles contract. Once the available ATP is used up, the myosin heads are unable to disengage, and the muscles remain contracted.

For instructor-assigned homework, go to www.masteringphysics.com 

SUMMARY OF TERMS (KNOWLEDGE)

Action potential A signal from a neuron or other cell that occurs when the cell's membrane potential becomes positive.

Chemoreception A way of sensing that occurs when chemicals bind to receptors on sensory cells, initiating action potentials.

Fertilization The joining of an egg and a sperm to form a diploid cell that can develop into a new organism.

Homeostasis The maintenance of a relatively stable internal environment.

Hormone A chemical molecule that gives instructions to the body; it is produced in one place in the body, released into the blood, and received by target cells elsewhere in the body.

Neuron A cell that receives and transmits electrical signals from one part of the body to another.

Organ A structure in the body that has a specific function.

Organ system A set of organs that work together to perform a particular bodily function.

Ovulation The release of a mature egg cell that occurs once during each menstrual cycle.

Placenta The organ that allows nutrients and wastes to be exchanged between a pregnant woman and a developing embryo.

Sarcomeres The contractile units of muscle cells.

Synapse A connection between a neuron and a target cell.

Tissue A group of similar cells that performs a certain function.

READING CHECK (COMPREHENSION)

19.1 Organization of the Human Body

1. What are the four main types of tissues in the human body? Give an example of each.
2. Multiple tissues combine to make a(n)_____, a structure in the body that has a specific function.
3. What is an organ system?

19.2 Homeostasis

4. What is homeostasis?
5. Give some examples of variables for which the body maintains homeostasis.

19.3 The Brain

6. What are some of the functions of the brainstem?
7. Which part of the brain is responsible for balance and posture?
8. Why can damage to one side of the cerebrum affect the functioning of the opposite side of the body?
9. Describe the functions of each of the four lobes of the cerebrum.

19.4 The Nervous System

10. Which structures make up the central nervous system?
11. A typical neuron has dendrites, a cell body, and an axon. What is the function of each of these parts?
12. What are the functions of sensory neurons, interneurons, and motor neurons?
13. What kinds of functions are controlled by the somatic nervous system and the autonomic nervous system?

19.5 How Neurons Work

14. What is a membrane potential?
15. What happens to the membrane potential of a neuron during an action potential?
16. Why is an action potential described as an all-or-nothing event?
17. What causes an action potential to travel down a neuron's axon?
18. How does an electrical synapse work?
19. How does a chemical synapse work?

19.6 The Senses

20. What are the two types of light-sensitive cells in the eyes? How do they differ from each other?
21. Describe how sound waves enter the ear and ultimately cause you to hear.
22. How are action potentials generated in chemosensory cells?
23. What are proprioceptors?

19.7 Hormones

24. What are the two types of hormones? How does each type of hormone produce an effect in a target cell?
25. Why is the anterior pituitary sometimes called the "master gland"?
26. What hormone made by the hypothalamus helps regulate the amount of water in the body? What is its effect?
27. Which endocrine organ produces the hormones associated with the "fight or flight" response?

19.8 Reproduction and Development

28. What is unusual about the process of meiosis that produces the egg?
29. How do sperm get past the zona pellucida that surrounds the egg?
30. What is the function of the placenta?
31. When during pregnancy do the major organs of the body develop?

19.9 The Skeleton and Muscles

32. What are the functions of the skeleton?
33. How does a signal from a motor neuron result in the contraction of a muscle?
34. At what point in the process of muscle contraction is ATP required?

THINK INTEGRATED SCIENCE

19A—How Fast Can Action Potentials Travel?

35. Why do action potentials travel more quickly down thicker axons than thinner ones?
36. What is the problem with achieving rapidly traveling action potentials through large numbers of giant axons?
37. How does myelination speed the propagation of action potentials?

19B—Endorphins

38. What are endorphins, and what causes cells in the brain to release them?
39. Why do endorphins have effects similar to those of drugs like morphine and heroin?
40. What is the effect of endorphins?

THINK AND DO (HANDS-ON APPLICATION)

41. You have a *blind spot* where your optic nerve exits your retina. Because of the presence of the optic nerve, there are no rods or cones in this spot. Why is there no obvious blind spot in your field of vision? The brain cleverly fills in this area using visual information from surrounding areas. Let's locate your blind spot. Draw a small dot on the left side of a piece of paper and a small x on the right side. The x should be about 6 inches from the dot. Now hold the paper in front of you at arm's length. Close your right eye and look at the x with your left eye. Slowly bring the sheet of paper closer to your face. At some point, the dot will disappear. Why? What is happening that prevents you from seeing the dot?
42. Is what you think of as "taste" largely smell? Cut up an apple and a pear, hold your nose, and chew. Can you tell which is which? Now release your nose and allow yourself to smell as you eat. Now can you tell which is which? This exploration can also be performed with different flavors of candy, jarred baby food, etc.

THINK AND COMPARE (ANALYSIS)

43. Rank the following from the smallest to the largest level of organization: organ, organ system, tissue, cell.
44. The membrane potential is the electric potential across the cell membrane of a neuron. Rank the following from smallest to largest electric potential: resting potential, membrane potential during action potential, threshold.
45. Rank the two types of light-sensitive cells, rods and cones, in terms of (a) ability to see in dim light, (b) ability to make out fine details, and (c) ability to distinguish different colors.

THINK AND SOLVE (MATHEMATICAL APPLICATION)

46. Two different types of neurons transmit pain signals to the central nervous system. The faster type transmits signals at 25 m/s. The slower type transmits signals at 0.5 m/s. The distance from your hand to your central nervous system is about 1 m. Suppose that you touch a hot stove. Show that you become aware of the first type of pain in 0.04 s, and that you become aware of the second type of pain after only 2 s. (You may have noticed that when you do something like touch a hot stove, you feel a flash of sharp pain first, followed by a slow throbbing pain.)
47. The human retina has an area of about 1000 mm^2. If 125 million rods and 6.5 million cones are found there, show that you have about 131,500 sensory cells per square millimeter in your retina.
48. You have about 1000 different kinds of smell receptors. Each of these receptors is a distinct protein coded for by a specific gene. In Chapter 16, you learned that the Human Genome Project revealed that humans have a total of about 22,000 genes. Show that about 4.5% of your genes are dedicated to helping you smell.

49. The egg is a large cell and contributes almost all the nutrients present in a fertilized egg. The sperm contributes little more than its set of chromosomes. Just how much bigger is a human egg than a human sperm? The human egg is about 100 micrometers in diameter. The head of a human sperm is about 4 micrometers in diameter (and, if you are curious, a human sperm is about 50 micrometers long). Show that the volume of a human egg is 15,625 times larger than the volume of a human sperm. Recall that the formula for the volume of a sphere is $\frac{4}{3}\pi r^3$.

THINK AND EXPLAIN (SYNTHESIS)

50. The brain is an organ and, like all organs, it is composed of multiple tissues. What are some of the tissues that make up the brain?
51. When you move your body, is your cerebrum in complete control? What other parts of your brain are involved? Explain.
52. The figure shows a map of the motor control area of the brain, found in the frontal lobe of the cerebrum. Why is such a large part of the brain responsible for controlling the actions of the hands and lips? Why is only a small part of the brain responsible for controlling the trunk?

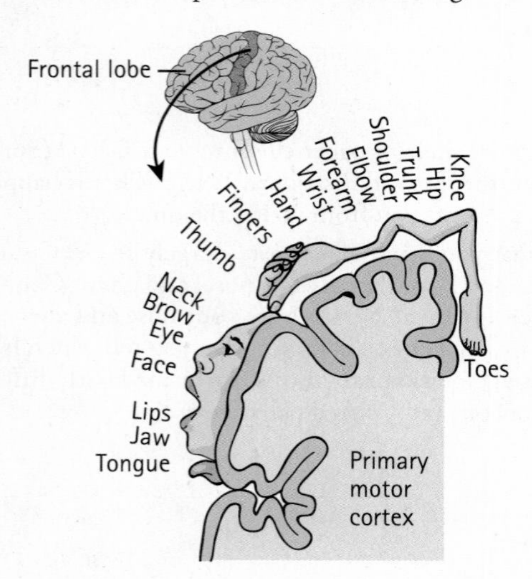

53. Of the three types of neurons (sensory neurons, motor neurons, and interneurons), which type goes to your biceps muscle and tells you to bend your elbow? Which type tells you whether your feet feel cold?
54. Is a neuron that slows your heartbeat part of the somatic or autonomic nervous system? Is this neuron part of the sympathetic or parasympathetic division?
55. What would be the effect of removing the myelin sheath from the axon of a neuron?
56. Do neurotransmitters enter the target cell? If not, how do they have an effect on the target cell?
57. What determines whether a target cell responds to a specific type of neurotransmitter?
58. Why do a lot of nocturnal animals have only rods in their retinas?
59. Are your rods or cones more important for reading a book?
60. In some people, the bones of the middle ear stiffen with age. This can result in deafness. Why?
61. Suppose you know that the receptor for a hormone you are studying is found in the target cell's cytoplasm. Are you studying a protein hormone or a steroid hormone?
62. On a brilliant, sunny day, you take a long hike through open country. You sweat a lot, losing a lot of water. What hormone does your body release? Why?
63. What are the roles of mitosis and meiosis in human reproduction and development?
64. Vasectomy is a form of male sterilization in which a section of each vas deferens is removed. How does this cause sterility, preventing a man from fathering children?
65. Does a fertilized human egg make anything other than the embryo?
66. What are the functions of bone marrow?

THINK AND DISCUSS (EVALUATION)

67. Does maintaining homeostasis of body temperature involve feedback regulation? Explain.
68. Animals vary in how "wrinkled" their brains are. Some animals have very wrinkled brains, while others have smooth brains. What would you predict about an animal that has a wrinkled brain rather than a smooth one?
69. You talk with a friend on the telephone. Which parts of your brain do you use?
70. If a signaling neuron has an excitatory effect on a target cell, does it increase or decrease the membrane potential of the target cell? Defend your answer.
71. Stars come in different colors depending on their surface temperatures (see Chapter 29). But when you look up at a starry night sky, all the stars look like they are the same color. Why?
72. Osteoporosis is a disease that primarily affects postmenopausal women, causing decreased bone density and brittle bones that are vulnerable to fracture. The hormone calcitonin is sometimes used to treat osteoporosis. Why?
73. Jet lag describes the fatigue and disorientation that result from flying across many time zones. What causes jet lag, and how does the body eventually adjust to a new time zone? Why is there no such thing as "train lag" or "bicycle lag"?

74. How has natural selection acted on testosterone levels in men? Is there an ideal testosterone level for all phases of a man's life?
75. Only one sperm fertilizes an egg. If this is the case, why can low sperm count be a factor in infertility?
76. Each time myosin heads pull on actin, the sarcomere contracts only about 10 nm (10^{-9} meter). Given that, how are you able to produce large motions with your muscles?
77. Explain what happens when you wiggle your toe. Start with the decision to wiggle your toe, which occurs in your brain, and end with a description of the activity of your muscles and bones.

READINESS ASSURANCE TEST (RAT)

If you have a good handle on this chapter, if you really do, then you should be able to score 7 out of 10 on this RAT. If you score less than 7, you need to study further before moving on.

Choose the BEST answer to each of the following:

1. Which part of the brain controls posture, balance, and fine movements?
   (a) brainstem
   (b) cerebellum
   (c) cerebrum
   (d) thalamus
2. Which part of the brain is responsible for reasoning, language, and the control of voluntary movement?
   (a) brainstem
   (b) cerebellum
   (c) cerebrum
   (d) thalamus
3. Which part of a neuron receives information from another cell or neuron?
   (a) dendrite
   (b) cell body
   (c) axon
   (d) myelin sheath
4. During an action potential,
   (a) the membrane potential becomes more negative until it hits threshold.
   (b) the membrane potential increases as potassium ions flow out.
   (c) the membrane potential increases as sodium ions flow in and then decreases as potassium ions flow out.
   (d) the membrane potential may or may not reach threshold.
5. Chemoreception characterizes
   (a) vision.
   (b) hearing.
   (c) touch.
   (d) taste.
6. The hormone insulin
   (a) helps regulate blood glucose levels.
   (b) is involved in the "fight or flight" response.
   (c) helps regulate the amount of water in the body.
   (d) works with another hormone, glucagon, to regulate calcium levels in the blood.
7. Which of these statements about human reproduction is false?
   (a) Cilia sweep the egg down the oviduct to the uterus.
   (b) Fertilization usually takes place in the uterus.
   (c) Semen contains fluids that help nourish sperm.
   (d) After fertilization, the zona pellucida becomes impenetrable to sperm.
8. The structure that provides oxygen and nutrients to the developing embryo is the
   (a) oviduct.
   (b) amnion.
   (c) blastocyst.
   (d) placenta.
9. The connection between a motor neuron and a muscle cell
   (a) involves an electrical synapse.
   (b) determines whether muscles contract or extend.
   (c) uses the neurotransmitter acetylcholine.
   (d) involves an opiate receptor.
10. During muscle contraction,
   (a) sodium ions are released by the muscle cell's endoplasmic reticulum.
   (b) calcium ions allow myosin heads to bind to actin.
   (c) ATP causes the myosin heads to pivot, pulling on actin.
   (d) actin heads repeatedly bind to myosin, pivot, release, extend, attach, and pivot again.

Answers to RAT

1. b, 2. c, 3. a, 4. c, 5. d, 6. a, 7. b, 8. d, 9. c, 10. b

CHAPTER 20

Human Biology II—Care and Maintenance

WHETHER YOU play world-class tennis—or just sit and read e-mail—your body is hard at work. In this chapter we focus on the care and maintenance of the human body. As we explore the body, we'll learn many things. What makes the "lub-dubb" sound of the heartbeat? Is it possible to forget to breathe? Which part of your body gives you the hiccups? Is swallowing a voluntary or involuntary action? Also, how does your body defend you against bacteria, viruses, and other organisms that cause disease? How do vaccines keep you from getting sick?

20.1 Integration of Body Systems

LEARNING OBJECTIVE
List examples of organ systems working together to perform important body functions.

EXPLAIN THIS How do the respiratory and circulatory systems work together to provide the body's tissues with oxygen?

The body's organ systems rarely act alone. Most of the body's major functions require the efforts of two or more organ systems. For example, the job of keeping the body supplied with oxygen is split between two organ systems: The respiratory system brings oxygen into the body, and the circulatory system transports it to the tissues. Similarly, getting rid of cellular wastes requires the coordinated efforts of the circulatory, respiratory, and excretory systems. The circulatory system collects wastes from the tissues, and the respiratory and excretory systems remove the wastes from the body via exhalation and urine production, respectively. In this chapter, we will look at how organ systems work together to maintain the body. Among their many tasks are two of the most essential: obtaining energy for the body's activities and protecting the body from disease.

20.2 The Circulatory System

LEARNING OBJECTIVE
Describe the structure and function of the circulatory system.

EXPLAIN THIS Why does a person's heart go "lub-dubb, lub-dubb, lub-dubb"?

Wouldn't it be nice if everything you needed came right to your door? Running low on milk? Here come a couple of gallons now. Hungry for oranges? There should be some along any minute. Broccoli? This is your lucky day. And no need to take the garbage out either. Just drop it, and it will make its own way to the dump.

Is this a couch potato's dream come true? Maybe, but it's also just a day in the life of your cells. Your circulatory system brings your cells everything they need, and it removes all the garbage they have to dispose of. The circulatory system delivers oxygen and nutrients, takes away wastes, and transports special items such as hormones and immune cells. It is like a food, garbage, and mail service rolled into one. The circulatory system consists of three parts: the heart, blood vessels, and blood.

The Heart

The heart (Figure 20.1) is a muscular pump that drives blood around the body. It is about the size of a clenched fist and has four separate chambers: right atrium, right ventricle, left atrium, and left ventricle. The right side of the heart pumps blood to the lungs, where oxygen is picked up. The left side pumps blood to the body.

Heart muscle does not need a neuron to tell it to contract, the way voluntary muscles do. Instead, the heart contracts on its own. Each heartbeat begins on its own in a part of the right atrium called the *sinoatrial node*, or *pacemaker*. The pacemaker initiates an action potential that sweeps quickly through the right and left atria, which contract simultaneously,

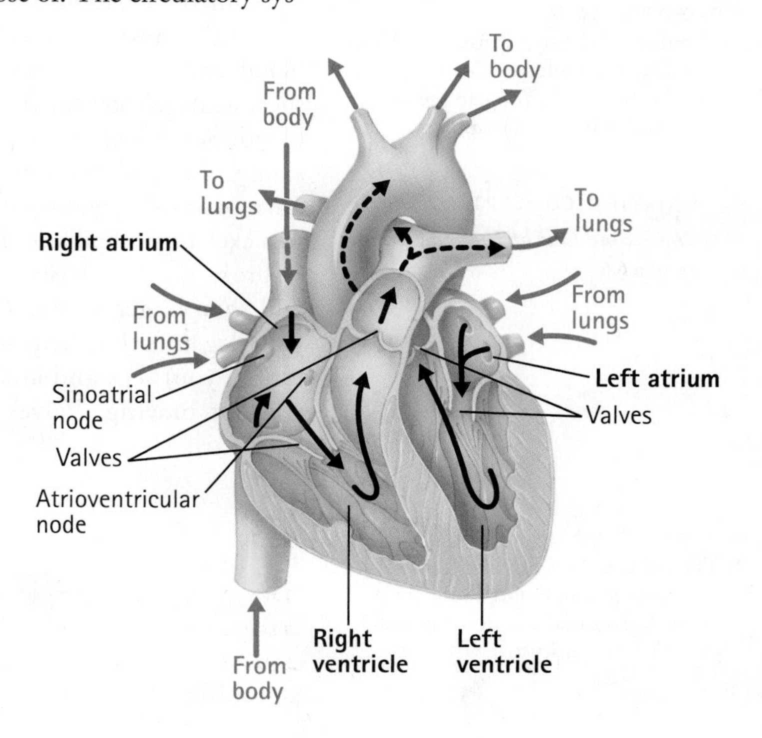

FIGURE 20.1
The heart has four chambers, labeled in bold. The arrows show how blood flows throughout the heart. Heart valves keep blood from flowing backward.

MasteringPhysics®
TUTORIAL: Path of Blood Flow in Animals

driving blood into the ventricles. The pacemaker's signal also travels to the *atrioventricular node.* From the atrioventricular node, the signal passes to the two ventricles, which also contract simultaneously, driving blood out through blood vessels.

Because the atrioventricular node conducts action potentials slowly, there is a delay between the contraction of the atria and the contraction of the ventricles. This is why each heartbeat consists of two separate sounds, the miraculous "lub-dubb . . . lub-dubb . . . lub-dubb . . ." that we associate with life. The "lub" occurs when the two atria contract; the "dubb" occurs when the two ventricles contract. Note, however, that the "lub-dubb" of the heartbeat doesn't come from muscle contractions. (After all, your arm and leg muscles don't thud when they contract!) The noises come from *valves* that snap shut after each contraction. Heart valves are flaps of tissue between the atria and the ventricles, and between the ventricles and blood vessels. These valves prevent blood from flowing backward.

The heart beats about 70 times a minute. This adds up quickly—to more than 100,000 heartbeats a day.

Blood Vessels

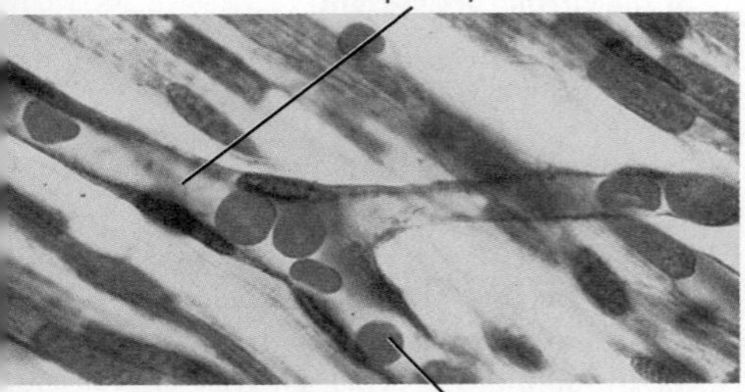

Blood travels throughout the body in tubes called *blood vessels.* **Arteries** are blood vessels that carry blood away from the heart. Arteries stretch out when blood is pumped, then bounce back. You can feel this stretch and recoil by feeling the pulse at your wrist or temple. From the arteries, blood moves into smaller vessels called *arterioles.* Each arteriole has a layer of smooth muscle around it that controls its diameter. Adjusting the diameters of different arterioles allows the body to control how much blood different tissues receive. Tissues need more blood when they are working harder. For example, a lot of blood goes to the tissues of your digestive system when you've just eaten a meal and are ready to digest it.

Capillary

Red blood cell

FIGURE 20.2
Capillaries are tiny, thin-walled blood vessels where molecules are exchanged with body tissues. You can see some red blood cells in this capillary.

From the arterioles, blood flows into capillaries (Figure 20.2). **Capillaries** are tiny, thin-walled blood vessels from which molecules are exchanged between blood and body tissues. Oxygen and nutrients move from blood to body tissues, while carbon dioxide and other waste products move from body tissues to blood. Some molecules, including oxygen and carbon dioxide, move via diffusion. Other molecules move by facilitated diffusion or active transport, or they are moved using endocytosis or exocytosis (see Chapter 15). In order to allow for this exchange, capillaries have very thin walls—often only a single cell thick!

UNIFYING CONCEPT
- ***The Second Law of Thermodynamics*** **Section 6.5**

From the capillaries, blood flows back toward the heart in small *venules* and then larger **veins**. The contractions of your voluntary muscles, such as your leg muscles, help squeeze blood along your veins (Figure 20.3). This is why your legs and ankles sometimes swell if you sit or stand for too long without moving. Valves in the veins help make sure blood does not flow backward.

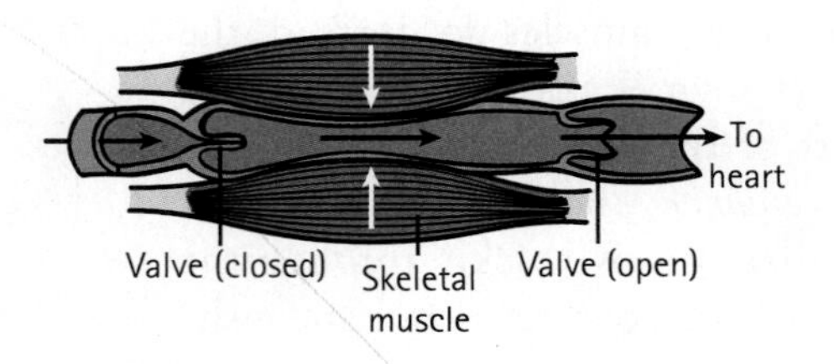

FIGURE 20.3
The contractions of your voluntary muscles squeeze blood along your veins. Valves make sure the blood doesn't flow backward.

How Blood Flows Around the Body

Like all good food, garbage, and mail services, the circulatory system does not move blood haphazardly around the body but follows a set route (Figure 20.4). The path of blood flow allows the circulatory system to efficiently carry out one of its primary tasks, delivering oxygen to tissues. How does blood flow? Let's start with blood returning from the body to the heart. Blood returning from the body contains low levels of oxygen—it is *deoxygenated.* This deoxygenated blood flows from veins into the right atrium of the heart. The right atrium pumps it to the right ventricle. The right ventricle pumps it out arteries that go to the lungs. There, blood picks up oxygen and drops off carbon dioxide. Blood is now *oxygenated.*

Oxygenated blood flows back to the heart through veins that go to the left atrium. The left atrium pumps blood to the left ventricle. The left ventricle then pumps it out arteries that go to tissues all over the body. The oxygen in the blood diffuses into the cells of all your tissues; every one of your cells needs oxygen. After carrying oxygen to the tissues, blood becomes deoxygenated again and returns to the heart. The entire circuit takes about one minute.

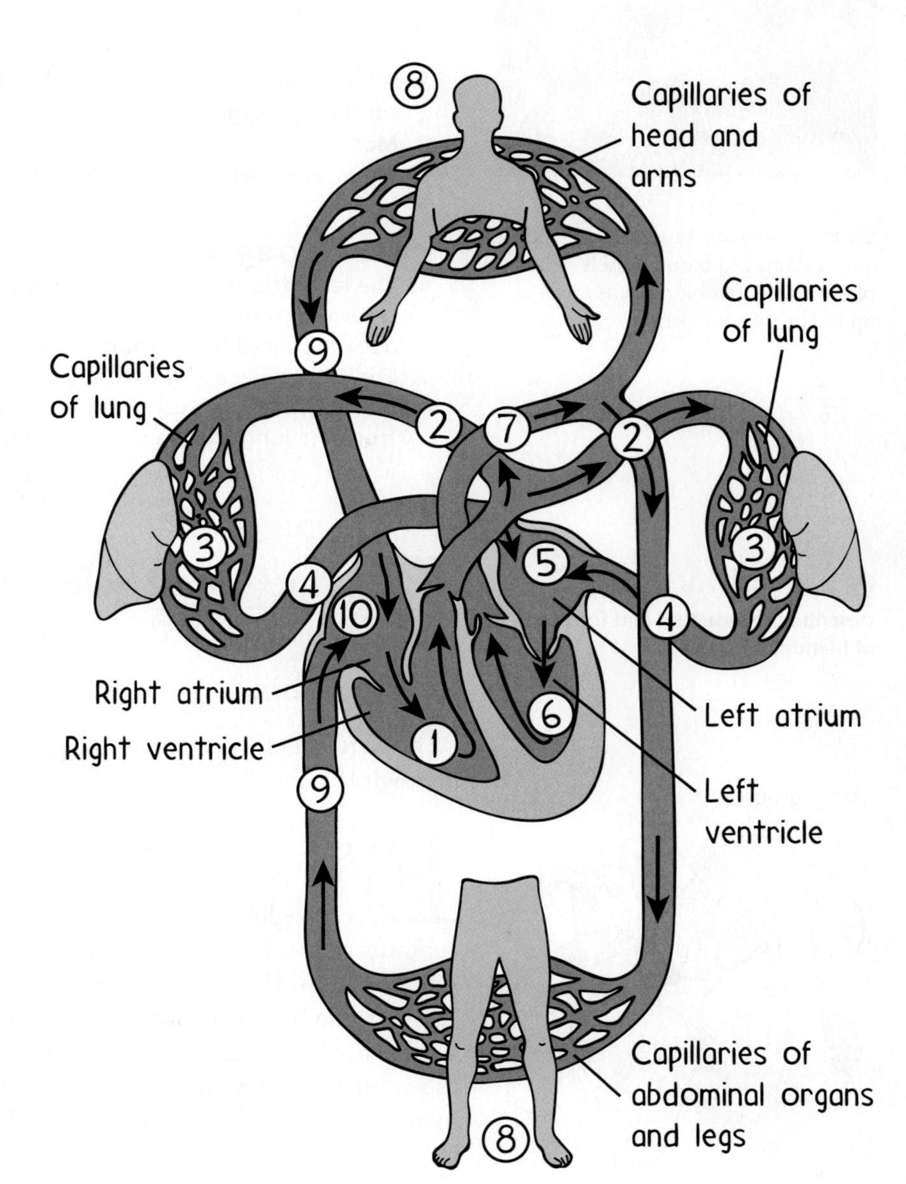

FIGURE 20.4
The path of blood flow around the body. Blue indicates deoxygenated blood, and red indicates oxygenated blood.

Blood

Now that we know how blood moves around the body, let's look more closely at blood itself. You have about 11 pints of blood in your body, which makes up about 8% of your body weight. A little more than half of this is *plasma.* Plasma is mostly water, but it also contains important molecules such as proteins, hormones, glucose, other nutrients, and wastes.

The rest of your blood is made up of cells—red blood cells, white blood cells, and platelets. **Red blood cells** transport oxygen to the body's tissues. Red blood cells contain numerous molecules of the protein **hemoglobin**, which binds to oxygen. There are as many as 300 million molecules of hemoglobin in a single red blood cell. Each hemoglobin molecule can carry up to four oxygen molecules. **White blood cells** are part of the immune system and help defend your body against disease. We'll learn more about white blood cells later. **Platelets** are involved in blood clotting. When body tissues are damaged, platelets attach to damaged blood vessels and release special clotting factors. These clotting factors convert proteins in blood plasma into a sticky substance

FIGURE 20.5
Clotting prevents blood from leaking out of damaged blood vessels. This photo shows red blood cells tangled up in threads of fibrin.

called *fibrin* (Figure 20.5). Fibrin prevents more blood from leaking out of the damaged blood vessels. The disease *hemophilia*, which is associated with a deficiency in blood clotting, is the result of genetic mutations that affect clotting factors.

CHECK YOURSELF

1. **Which chambers of the heart contain oxygenated blood? Which chambers contain deoxygenated blood?**
2. **Most arteries carry oxygenated blood, and most veins carry deoxygenated blood. Are there any exceptions?**

CHECK YOUR ANSWERS

1. The left atrium and left ventricle contain oxygenated blood that has just returned from the lungs. The right atrium and right ventricle contain deoxygenated blood that has just returned from the body.
2. There are only a few exceptions. The arteries that carry blood from the heart to the lungs carry deoxygenated blood. The veins that carry blood from the lungs to the heart carry oxygenated blood.

LEARNING OBJECTIVE
Describe the structure and function of hemoglobin.

Hemoglobin

EXPLAIN THIS How many oxygen molecules can a single molecule of hemoglobin carry?

Every human red blood cell contains about 300 million molecules of hemoglobin, the oxygen-carrying protein. What is the structure of hemoglobin? A hemoglobin molecule is made up of four smaller subunits. Each subunit contains a *heme group* that includes an iron atom at its center (Figure 20.6). It is this iron atom that binds to oxygen. Consequently, each hemoglobin molecule can carry up to four molecules of oxygen.

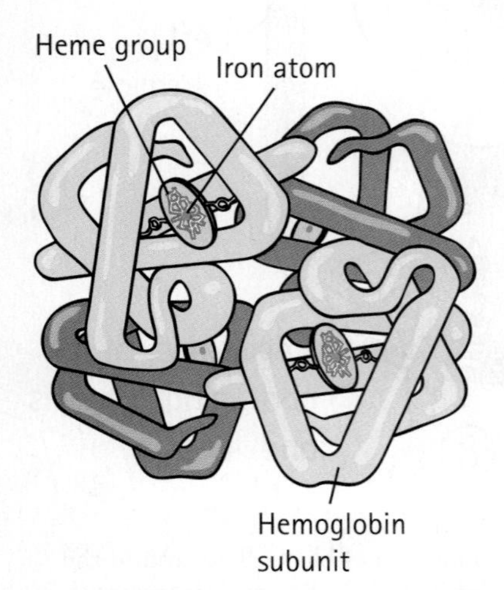

FIGURE 20.6
Hemoglobin is made up of four subunits. Each subunit includes a heme group with an iron atom at its center.

When oxygen binds to one of the four subunits in a hemoglobin molecule, the other three subunits are altered in such a way that their affinity for oxygen increases; that is, they become more likely to bind oxygen. As a result, most hemoglobin molecules will carry the maximum number of oxygen molecules away from the lungs. Similarly, when one subunit unloads an oxygen molecule at a body tissue, the other three subunits become more likely to give up their oxygen molecules as well, assuring that oxygen is passed efficiently to body tissues.

The oxygen affinity of a hemoglobin molecule also changes depending on its local environment. For example, lower blood pH (a more acidic environment) decreases hemoglobin's oxygen affinity. Why is this adaptive? An active, working tissue makes and uses more ATP and so releases more carbon dioxide during cellular respiration. Because carbon dioxide reacts with water in the blood to form carbonic acid, the presence of high levels of carbon dioxide decreases blood pH. Lower pH decreases the oxygen affinity of local hemoglobin molecules, making it easier for them to unload oxygen to the working tissue.

UNIFYING CONCEPT
● ***Feedback Regulation***
Section 19.2

TECHNOLOGY

Transplanting Bone Marrow in the Fight Against Cancer

Bone marrow, a spongy tissue found inside bones, contains the *blood-forming stem cells* that produce red blood cells, white blood cells, and platelets. In the disease leukemia, which describes cancers of the blood and bone marrow, blood cells grow out of control. One treatment for leukemia is a bone marrow transplant. In a bone marrow transplant, the patient's cancerous bone marrow is replaced with healthy bone marrow from a donor. Bone marrow transplants may also be useful in other types of cancer, when aggressive chemotherapy or radiation therapy destroys the patient's bone marrow as a side effect of treatment.

After a bone marrow transplant, it takes several years for a patient's immune system function to return to normal. This child is wearing a "biological isolation garment" to prevent infection during play.

During a bone marrow transplant, a patient receives either his or her own stem cells, harvested before cancer treatment, or cells from a donor. If stem cells are received from a donor, the donor's bone marrow must be a close genetic match to that of the patient. Without this close genetic match, one of two things could happen: The patient's immune system could destroy the transplanted cells. Or, white blood cells made by the new bone marrow could view the patient's body as foreign and attack it. For many patients, a sibling provides a good genetic match. In other cases, donors are found through an international donor registry.

The recipient of a bone marrow transplant receives stem cells directly into the bloodstream. These stem cells migrate to the bone marrow and begin to function a few weeks after the transplant. It can be some time before the new stem cell population is fully functional, however, and the patient's immune function usually does not return to normal for several years.

CHECK YOURSELF

Why is it important to consume enough iron? What happens if you don't consume enough iron?

CHECK YOUR ANSWER

Iron is an essential component of hemoglobin. People who do not consume enough iron may not have enough hemoglobin, resulting in a condition called *anemia.* In anemia, the blood does not carry enough oxygen. Because the tissues do not receive enough oxygen to support their activities, the result is weakness and fatigue.

Carbon monoxide gas is toxic to humans because it binds to hemoglobin even more strongly than oxygen binds. Extended exposure to carbon monoxide "fills up" hemoglobin, leaving no place for oxygen to bind. Carbon monoxide is found in car exhaust and cigarette smoke, and it is also released by some gas appliances. Because carbon monoxide is odorless, some people keep a carbon monoxide detector in their home.

20.3 Respiration

LEARNING OBJECTIVE

Describe the structure and function of the human respiratory system.

EXPLAIN THIS Why do you need your diaphragm to breathe?

Maybe it's only when you're huffing and puffing that you really think about breathing. Still, breathing in and out is something you do as many as 17,000 times a day! Breathing is your body's way of taking in oxygen, which is needed for cellular respiration, the process cells use to obtain ATP (see Chapter 15). Through breathing, your respiratory system moves oxygen from the air into your lungs. From the lungs, oxygen enters the circulatory system, which then delivers it to all your tissues. Breathing also allows your respiratory system to get rid of carbon dioxide, a side product of making ATP.

MasteringPhysics®

TUTORIAL: Human Respiratory System

TUTORIAL: Transport of Respiratory Gases

The Path of Air

The respiratory system is shown in Figure 20.7. When you inhale, air enters through your nose. (You can also breathe in through your mouth—this is a useful backup because otherwise a stuffy nose could be fatal.) Hairs in your nostrils trap dust and other particles. Air continues up your nasal passages, where it is moistened by mucus and warmed by a dense network of capillaries. In the nasal passages, your smelling cells also capture odor molecules present in the air (see Chapter 19). From the nasal passages, air moves through the *pharynx*, the part of the throat above the esophagus and windpipe. It then proceeds through the *larynx*, or voice box, and down the *trachea*, or windpipe. The trachea is a short tube stiffened by rings of cartilage. The rings keep the trachea open. The trachea branches into two tubes called *bronchi* that lead to the right and left lungs. Each bronchus branches into smaller and smaller tubules that finally end at tiny sacs called **alveoli**, where gas exchange occurs.

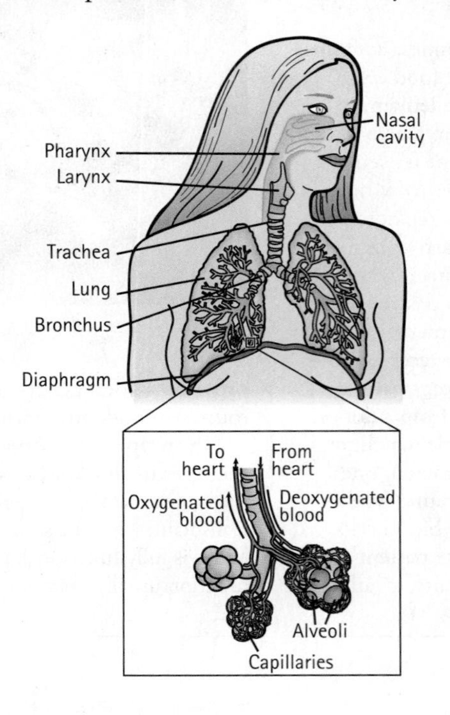

FIGURE 20.7
The respiratory system brings oxygen into the body and gets rid of carbon dioxide. Gas exchange occurs in the lungs, in tiny sacs called alveoli.

Your lungs contain about 300 million alveoli. Each alveolus is surrounded by a net of capillaries. Both the alveolus and the surrounding capillaries have extremely thin walls, consisting of only a single flattened cell. This allows gas exchange to happen through diffusion (see Chapter 15). Gas molecules diffuse down their concentration gradient—oxygen diffuses from the alveolus into the blood, and carbon dioxide diffuses from the blood into the alveolus.

UNIFYING CONCEPT

- ***The Second Law of Thermodynamics***
Section 6.5

When you exhale, air reverses its path. Air moves from the alveoli up through tubules to the bronchi, trachea, and nasal passages, and out through the nose. As air passes the larynx, it may vibrate your vocal cords (Figure 20.8). This makes the sound waves that allow you to talk, yell, and sing. In order to make different sounds, you control muscles that stretch your vocal cords, as well as other muscles in your lips, tongue, and cheeks.

UNIFYING CONCEPT

- ***Waves***
Section 8.1

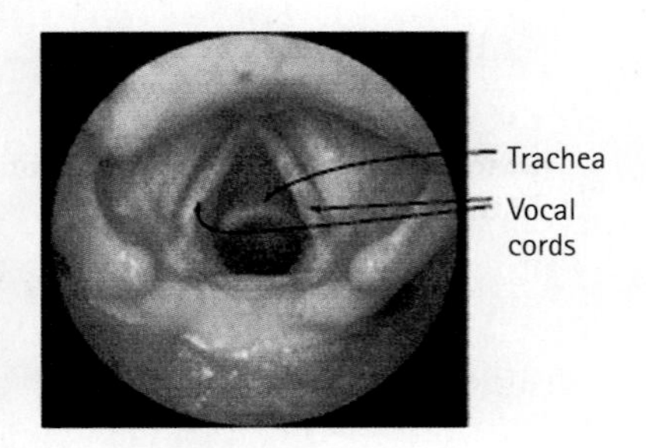

FIGURE 20.8
Vibrations in your vocal cords enable you to talk. You control the sounds you make by loosening or tightening your vocal cords and by moving the muscles of your lips, tongue, and cheeks.

Moving Air into and out of the Lungs

How do you inhale and exhale? Your lungs sit inside your rib cage in an air-filled pocket called the *thoracic cavity*. The bottom of the thoracic cavity is covered by a sheet of muscle called the **diaphragm**. When the diaphragm is relaxed, it is dome shaped (Figure 20.9). When you inhale, you contract your diaphragm. The diaphragm flattens, increasing the volume of the thoracic cavity. The muscles between your ribs also contract, pulling the rib cage up and out from the body. This further increases the volume of the thoracic cavity. So the volume of the thoracic cavity increases, while the amount of air inside it stays the same. What happens? The air pressure in the thoracic cavity drops. Air is sucked into the lungs, filling the alveoli. This process is similar to the way a bicycle pump sucks in air when you pull back its plunger.

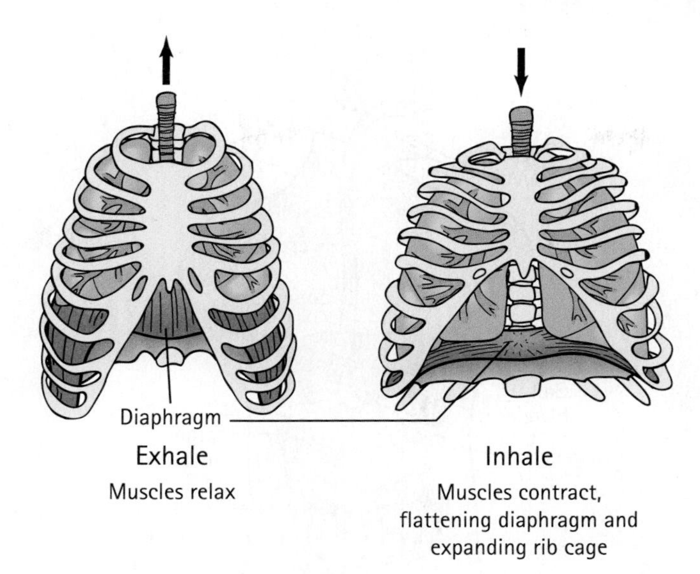

FIGURE 20.9
When you inhale, your diaphragm contracts and flattens and your rib cage expands. This increases the volume of your thoracic cavity, causing air to flow into your lungs.

When you exhale, the diaphragm and rib muscles relax, decreasing the volume of the thoracic cavity. This increases the air pressure in the thoracic cavity and pushes air out of the lungs.

You breathe about 12 times each minute. During a typical breath, only about 10% of the air in the lungs is exchanged with outside air—this is enough to keep your body tissues supplied with oxygen. Of course, you have some control over how often and how deeply you breathe. You don't have to worry about "forgetting" to breathe, though; respiration is controlled by the brainstem, along with many other involuntary activities (see Chapter 19).

Hiccups are sudden spasms of the diaphragm. Each spasm sucks air in, snapping the vocal cords shut and creating a "hic" noise. Hiccups may be caused by any irritation to the diaphragm, including eating too much or too quickly. Hiccups usually go away by themselves after a few minutes. However, Charles Osborne, who holds the world record for the longest bout of hiccups ever, hiccupped nonstop from 1922 to 1990, a total of 430 million spasms!

CHECK YOURSELF
Does it take energy to inhale, to exhale, or both?

CHECK YOUR ANSWER
It takes energy to inhale. Energy is needed to contract the muscles of the diaphragm and the rib cage. It does not take energy to exhale. Exhalation requires only that these muscles relax, a process that requires no energy.

20.4 Digestion

LEARNING OBJECTIVE
Describe the structure and function of the digestive system.

EXPLAIN THIS How does the epiglottis save you from choking?

If you go a few hours without eating, you're likely to start thinking about food. Food provides your body with organic molecules that can be broken down to make ATP, the molecule your cells use for energy. Food also provides many essential molecules that your body cannot produce on its own. Finally, the breakdown of food releases heat. This heat allows you to maintain a high, stable body temperature, the defining characteristic of endotherms (see Chapter 18).

MasteringPhysics®
TUTORIAL: Digestive System Function

Although you need food for many reasons, a chunk of food is way too big to travel in your bloodstream or move into your cells. During the process of **digestion**,

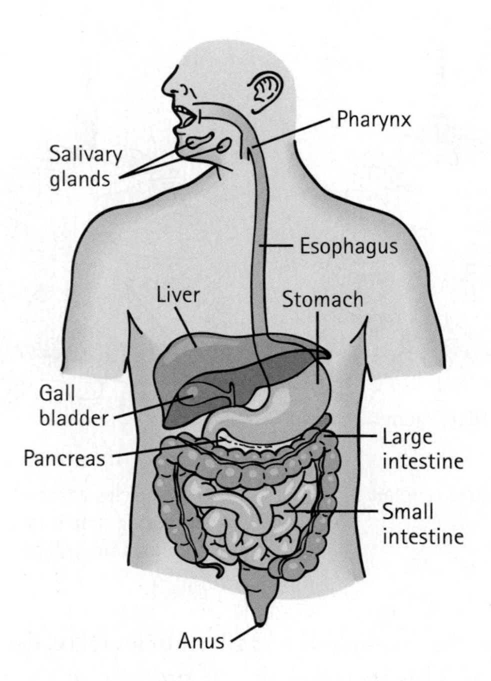

FIGURE 20.10
The digestive system breaks food down into small molecules that can be absorbed and used by the body.

FIGURE 20.11
Just thinking about food can cause the release of saliva.

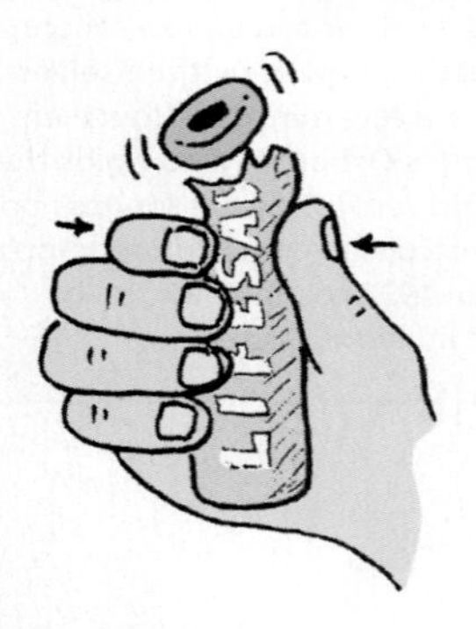

FIGURE 20.12
Peristalsis describes the moving wave of muscle contractions that moves food down the esophagus. It works a lot like the way you squeeze a LifeSavers® candy out of its wrapper.

food is broken down into small organic molecules that can be absorbed and used by the body. You may remember from Chapter 15 that living things are made up of different kinds of macromolecules—large molecules made up of smaller molecules linked together. During digestion, the macromolecules in your food are broken down into the small molecules that make them up. For example, the proteins in food are broken down into separate amino acids, and the carbohydrates in food are broken down into simple sugars. These small molecules are then absorbed and used by your body.

Digestion takes place in the digestive system, which is shown in Figure 20.10. Did you know that you begin to digest your food as soon as it enters your mouth? You chew food with your teeth, breaking it into little pieces. You also mix your food with *saliva.* Saliva contains a digestive enzyme that breaks down starches. Saliva also moistens food, so that the food can be moved easily around the mouth. Finally, because your taste buds can taste only molecules dissolved in liquid, saliva allows you to taste your food. Did you know that eating, or even just thinking about food, causes you to release saliva (Figure 20.11)?

After food is chewed and mixed with saliva, you swallow it. Food is pushed into the pharynx and then enters the esophagus. There are actually two openings in the pharynx, one to the esophagus and one to the trachea. You will choke if food goes down the trachea by mistake. A small flap of cartilage at the back of your tongue—the *epiglottis*—covers the trachea when you swallow so that food will not get into it. (This is why you cannot breathe while you are swallowing. Have you ever noticed that?)

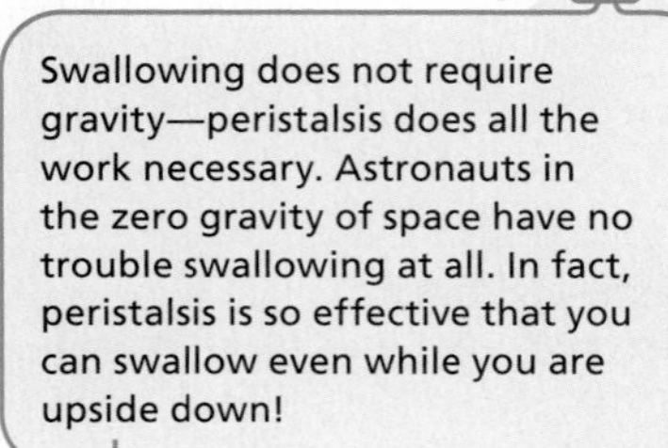

Swallowing does not require gravity—peristalsis does all the work necessary. Astronauts in the zero gravity of space have no trouble swallowing at all. In fact, peristalsis is so effective that you can swallow even while you are upside down!

Swallowing begins as a voluntary action. The muscles at the top of the esophagus are voluntary muscles. However, at a certain point, swallowing becomes involuntary. The lower part of the esophagus is made of involuntary muscle like that found in the rest of the digestive tract. In this part of the esophagus, food is pushed down by a moving wave of involuntary muscle contractions known as **peristalsis**. Peristalsis squeezes food down the esophagus by constricting behind it and pushing it along. You do something similar when you squeeze a LifeSavers® candy out of its paper-tube wrapper (Figure 20.12).

At the bottom of the esophagus, food moves through a *sphincter*, or ring-shaped muscle, into the stomach. Glands in the stomach wall release *gastric juice*, a highly acidic mixture of hydrochloric acid and digestive enzymes (Figure 20.13). These digestive enzymes, along with the churning of the stomach's muscular walls, convert food into a thick liquid called *chyme*. Meanwhile, the acidity of gastric juice kills any bacteria you swallow with your food. Gastric juice also contains a protective mucus that helps prevent the stomach from digesting its own tissues. Chyme leaves the stomach through a second sphincter and enters the small intestine. Usually, it takes the stomach about 4 hours to process a meal.

FIGURE 20.13
The stomach has numerous pits on its inner surface. These pits house the glands that release gastric juice.

The small intestine is about 20 feet long. In the first foot of the small intestine, digestion continues with the breakdown of proteins, fats, carbohydrates, and nucleic acids. Some of the digestive enzymes at work are made by the small intestine itself. Others are made by the pancreas and sent to the small intestine. Pancreatic enzymes also play an important role in neutralizing food, which is highly acidic when it arrives from the stomach. The small intestine also receives bile, a substance that is made by the liver and stored in the gall bladder. Bile is an emulsifier—it breaks fats into tiny droplets that are more easily attacked by enzymes.

When you vomit, the acidic nature of your stomach contents becomes immediately apparent—both from the taste and from the burning sensation left in your throat.

The rest of the small intestine is responsible for absorbing nutrients into the body. In order to do this job effectively, the small intestine has a huge surface area. The small intestine's inner surface is covered with fingerlike projections called *villi*. The villi are covered with even tinier projections called *microvilli* (Figure 20.14). Both the villi and microvilli increase the surface area of the small intestine. If your small intestine were flattened, it would fill the area of a tennis court! This impressive surface area enables your small intestine to absorb many digested nutrients at the same time. Most nutrients are absorbed into the body through facilitated diffusion or active transport. Absorbed nutrients enter capillaries located inside each villus and then are carried to all your tissues by the circulatory system.

After the nutrients are absorbed, what is left of your food moves into the large intestine. In the large intestine, some more water and minerals are absorbed into the body. Huge numbers of bacteria also live in the large intestine, where they feed on some of the undigested materials. Some of the bacteria make important vitamins, including vitamin K and some B vitamins. From the large intestine, feces are eliminated from the body through the anus. Feces are composed of living and dead bacteria as well as indigestible materials such as plant cellulose.

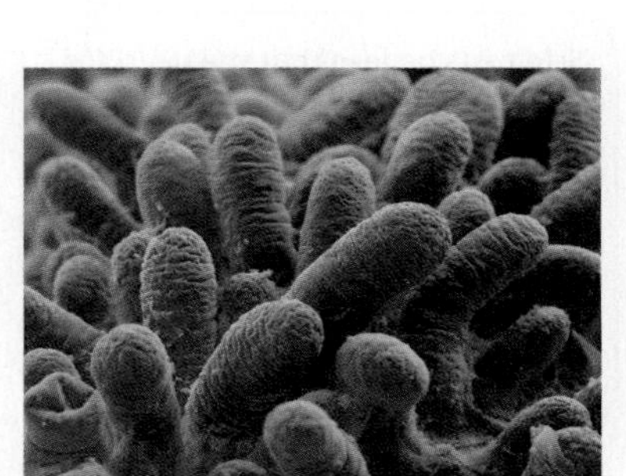

(a)

(b)

FIGURE 20.14
(a) Villi are fingerlike projections that increase the surface area of the small intestine. (b) Each villus in the small intestine is covered with even tinier projections called microvilli.

CHECK YOURSELF
How does bile help you digest your food?

CHECK YOUR ANSWER
Bile breaks fats into tiny droplets. Fats, which are hydrophobic, normally stay clumped together. By separating fats into tiny droplets, bile increases the surface area of the fats. This allows digestive enzymes to reach the fat molecules and break them down more quickly.

LEARNING OBJECTIVE
Explain how nutrition and exercise contribute to a healthy lifestyle.

20.5 Nutrition, Exercise, and Health

EXPLAIN THIS What is metabolic syndrome?

We now know how the body obtains the oxygen and nutrients it needs to fuel its activities. These demanding functions are performed day in and day out by the circulatory, respiratory, and digestive systems. How does lifestyle contribute to the functioning of these and other body systems? What can you do to live a healthy life?

Let's start with a healthy diet. You need a certain number of food Calories to support your activities. Recall that a food Calorie is actually a kilocalorie, or 1000 calories (the heat needed to change the temperature of 1 kg of water by 1°C). These food Calories provide the fuel that your cells use to make ATP. Most people need between 1500 and 3000 Calories a day—the exact number depends on your body size and activity level.

One of the latest trendy, healthy foods is quinoa. Quinoa is usually called a grain, but it is actually the seed of a plant closely related to spinach and chard. Quinoa is unusual in that it is a plant-based food that is a complete protein. (Another example is soybeans.)

You also need certain nutrients that your body can't make on its own. For example, humans cannot make 9 of the 20 amino acids needed to build proteins. This is why you need to eat a "complete protein"—one that contains all the amino acids—regularly. Meat is a complete protein and so is beans and rice, an important part of many vegetarian diets.

Humans also require a number of vitamins and minerals in small amounts. *Vitamins* are essential nonprotein components of certain enzymes. The body needs many different vitamins, but let's look at just a few. Vitamin C, found in citrus fruits, dark green leafy vegetables, and certain other foods, helps the body resist infections and repair wounds. Vitamin A, which you get from carrots, eggs, fortified milk, and other foods, is important for proper eye function. Vitamin A enables the eyes to adjust to dim light; insufficient amounts result in night blindness and other problems. Vitamin K, which is found in green leafy vegetables and synthesized by bacteria in the intestines, is essential for blood clotting. Insufficient amounts of vitamin K can result in hemorrhaging. Vitamin D is important for bone growth, immune system function, and other activities. Your body produces some vitamin D when ultraviolet light from the Sun strikes your skin. Vitamin D is not found naturally in many foods, however, and recent studies suggest that many people do not get enough vitamin D.

Scientists believe that the need for vitamin D drove the evolution of light skin in human populations that live far from the equator. In these areas, sunlight is weak during the long winter months. Light skin allows the body to absorb more ultraviolet light so that it is able to make more vitamin D.

Minerals are required as components of various body tissues. Important minerals include calcium for bones and teeth, phosphorus for ATP, and iron for hemoglobin. Mineral deficiencies can lead to a variety of health issues.

Beyond Calories and nutrients, does it matter what you eat? Yes! And, unfortunately, some of the things people like to eat, such as sweets and fats, aren't very good for them. Why would humans evolve a preference for foods that aren't good for them? Most likely, the explanation is that people's taste for these foods evolved under a different set of circumstances, when these types of food were much less readily available.

In fact, more than two-thirds of Americans today are either overweight or obese. This has resulted in an increasing number of people affected by metabolic syndrome, a set of characteristics that greatly increases the risk of heart disease, stroke, and type 2 diabetes. *Metabolic syndrome* is associated with two main features: carrying extra weight around the middle and upper parts of the body and insulin resistance, the inability of body tissues to adequately respond to the hormone insulin. Insulin resistance results in higher blood sugar levels as well as increased fat storage. In addition to diet, factors that may also contribute to metabolic syndrome include aging, genetics, and lack of exercise.

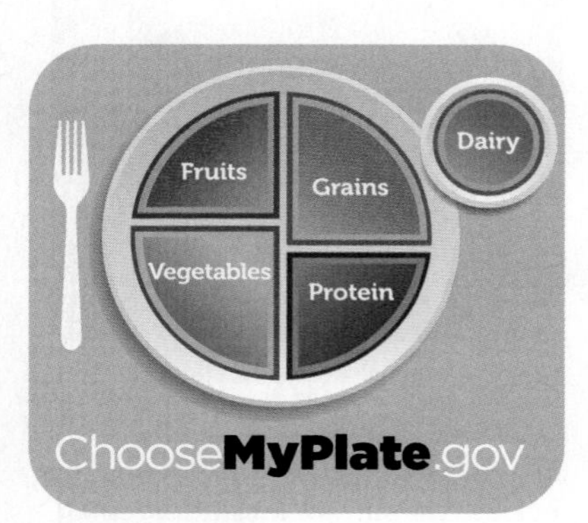

FIGURE 20.15
The USDA's "My Plate" emphasizes fruits and vegetables, whole grains, and healthy sources of protein.

So, what makes up a healthy diet? The U.S. Department of Agriculture's "MyPlate" offers some guidelines (Figure 20.15). The plate itself emphasizes the importance of sitting down to a well-planned meal. Half the plate is covered with fruits and vegetables, and another large segment is taken up by whole grains. Other recommendations include obtaining some protein from healthy sources, such as beans and seafood; cutting back on salt, sugar, and solid fats; eating an appropriate number of Calories; and exercising.

Exercise is a crucial part of any healthy lifestyle (Figure 20.16). For most of human history, daily life provided people with plenty of physical activity. This was true when most people belonged to hunting and gathering societies and when most people worked on farms. It is still true for a large fraction of the world's population, but not all of it. Desk jobs, cars, television, and the Internet have brought many people too close to a sedentary, couch-potato lifestyle. Many people now have to go out of their way to get the physical activity they need to stay healthy.

What are the benefits of exercise? Regular exercise reduces the risk of heart disease, high blood pressure, colon cancer, breast cancer, osteoporosis, diabetes, obesity, and many other health issues. Exercise improves the performance of the heart and lungs, maintains joints and tendons, and increases muscle mass and bone density. All of this contributes to strength and balance and helps people stay flexible as they age. Exercise is good for mental health too: It helps reduce depression, stress, and anxiety, and even helps you sleep better. There's no doubt that exercise is a good thing.

FIGURE 20.16
Even people with busy lives should make time for exercise. Here, Nils Gilman, busy father of three, shoots baskets in San Francisco's Golden Gate Park.

Integrated Science 20B
PHYSICS AND CHEMISTRY

Low-Carb Versus Low-Cal Diets

LEARNING OBJECTIVE
Explain how low-Calorie and low-carbohydrate diets help people lose weight.

EXPLAIN THIS What are some of the dangers of a low-carb diet?

A whopping 68% of Americans—more than two out of three—are either overweight or obese, according to the Centers for Disease Control and Prevention (CDC). At the same time, it has become more and more clear that maintaining a healthy body weight is a critical component of good health. And, in fact, many Americans are trying to lose weight. But losing weight is very difficult. How do you do it?

If you use more Calories than you take in, your body uses stored energy—such as fat—to support its activities. This is why a low-Calorie diet makes good sense. The importance of exercise is also clear. Exercise uses up a lot of Calories, and unused Calories are stored by the body as fat. By building muscle, exercise also makes your body use more Calories even when you're not exercising.

UNIFYING CONCEPT
- *The Law of Conservation of Energy* Section 4.10

Now what about low-carb diets, such as the Atkins diet or the South Beach diet? Low-carb diets have been extremely popular, and it is easy to see why. Studies have confirmed that, for many people, low-carb diets lead to weight loss more quickly and more consistently than low-Calorie diets. This appears to be because low-carb diets are easier to stick to, probably because they allow people to continue to eat fats. People on low-carb diets actually lose weight for the same reason that people on low-Calorie diets lose weight—they consume fewer total Calories. In addition, low-carb diets cause people to retain less water; this water is used during excretion to flush out the extra proteins consumed.

Are there any problems with low-carb diets? Yes—they tend to be high in saturated fats and cholesterol, which are associated with heart disease, and they

SCIENCE AND SOCIETY

What Are the Odds? Current Major Health Risks

According to the latest data, life expectancy at birth in the United States is 80.9 years for females and 76.0 years for males. The overall life expectancy of 78.5 years is the highest ever in this country. What are the current major health risks?

Heart disease is the leading cause of death in the United States, followed by cancer and strokes. The 15 leading causes of death in the United States are listed in the table, along with worldwide data for comparison.

Leading Causes of Death in the United States
(2010 Preliminary Data from the CDC)

	Total Deaths (thousands)	Percent of Total
1. Heart disease	595	24.1
2. Cancer	574	23.3
3. Chronic lung diseases (emphysema, chronic bronchitis, etc.)	138	5.6
4. Strokes	129	5.2
5. Accidents (unintentional injuries)	118	4.8
6. Alzheimer's disease	83	3.4
7. Diabetes	69	2.8
8. Kidney disease	50	2.0
9. Influenza and pneumonia	50	2.0
10. Intentional self-harm (suicide)	38	1.5
11. Septicemia (blood poisoning)	35	1.4
12. Chronic liver disease and cirrhosis	32	1.3
13. High blood pressure	27	1.1
14. Parkinson's disease	22	0.9
15. Pneumotitis (inflammation of lungs)	17	0.7
All other causes	489	19.9

Leading Causes of Death Worldwide
(2008 Data from the World Health Organization)

	Total Deaths (millions)	Percent of Total
1. Heart disease	7.25	12.8
2. Strokes	6.15	10.8
3. Lower respiratory infections (pneumonia)	3.46	6.1
4. Chronic lung diseases	3.28	5.8
5. Diarrheal diseases	2.46	4.3
6. HIV/AIDS	1.78	3.1
7. Lung cancers	1.39	2.4
8. Tuberculosis	1.34	2.4
9. Diabetes	1.26	2.2
10. Road traffic accidents	1.21	2.1
All other causes	27.1	48.0

tend to be short on whole grains and fruit, which are known to protect against many health problems. Finally, the long-term effects of low-carb diets are still unclear. For example, water loss and the processing of large amounts of proteins may be hard on organs such as the liver and kidneys.

CHECK YOURSELF

1. **How does a low-Calorie diet help you lose weight? How does a low-carb diet help you lose weight?**
2. **How does exercise help you lose weight?**

CHECK YOUR ANSWERS

1. In both kinds of diets you take in fewer Calories. Your body then uses stored energy, such as fats, to fuel its activities.
2. Exercise consumes Calories directly. Exercise also builds muscles, which use up significant numbers of Calories even when you're not exercising.

20.6 Excretion and Water Balance

LEARNING OBJECTIVE
Describe the structure and function of the excretory system.

EXPLAIN THIS Which structures in the excretory system help prevent the body from losing too much water in urine?

As cells go about their activities, they generate wastes. We already know what cells do with these wastes—let them diffuse into the bloodstream. But where do the wastes go from there? How does the body get rid of them? This is the job of the excretory system.

UNIFYING CONCEPT
- *The Second Law of Thermodynamics* **Section 6.5**

The excretory system, shown in Figure 20.17, filters your blood, removing wastes while leaving useful molecules in the blood. The excretory system also controls how much water, sodium, potassium, calcium, and other substances the body keeps. Did you know that your entire blood supply moves through your kidneys 16 times a day? The end result is something we're all familiar with—about 6 cups of urine.

One of the most important wastes found in urine is *urea*. When amino acids are broken down to make ATP, ammonia, a nitrogen-containing waste, is produced. Because ammonia is highly toxic, it is immediately converted into urea, a less toxic waste, by the liver. The liver releases urea back into the bloodstream for excretion.

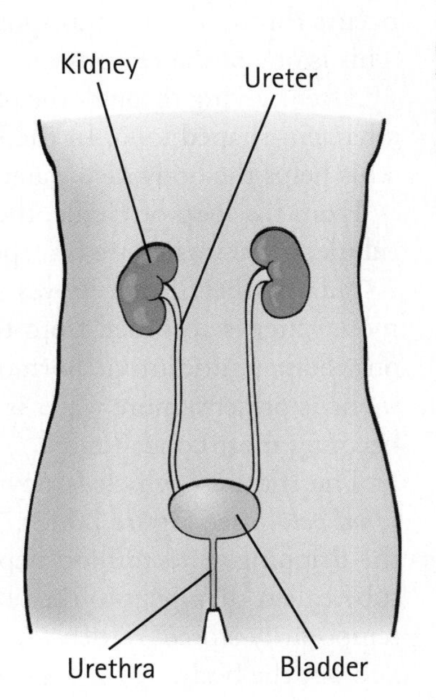

FIGURE 20.17
The excretory system consists of the kidneys, ureters, bladder, and urethra.

How does excretion happen? The process begins in the kidneys. Each kidney contains about a million structures called nephrons. The **nephron** is the functional unit of the kidney. What goes into the nephron is more or less blood plasma, and what comes out is urine. Let's look at how this happens.

Each nephron is associated with a cluster of capillaries (Figure 20.18). The first part of the nephron, a cup-shaped structure called *Bowman's capsule*, surrounds the capillaries. Blood pressure pushes fluid out of the capillaries and

FIGURE 20.18
The kidney is made up of about a million functional units called nephrons.

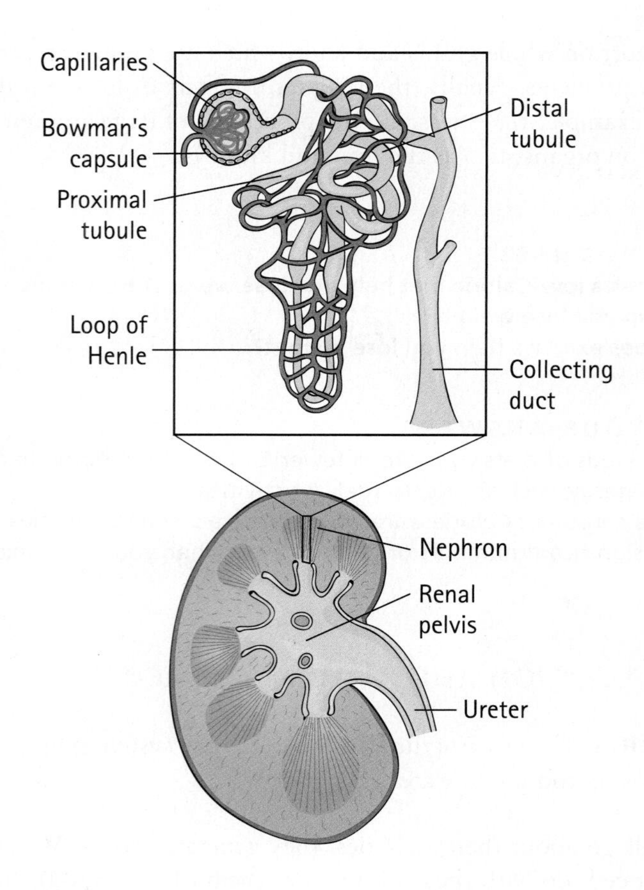

into Bowman's capsule. The fluid that enters Bowman's capsule is called the *filtrate*. At this point, the filtrate is very similar to blood plasma.

Did you know that diabetes used to be diagnosed by ants? Urine with a lot of sugar in it attracts ants. But urine becomes sugary only if glucose has built up to very high levels in the blood—that is, if a person either can't make or can't respond to insulin. Glucose is a "good" molecule that is normally absorbed by the kidneys, but when blood glucose levels are very high, the kidneys can't absorb all of it, and some ends up in urine.

From Bowman's capsule, the filtrate enters the *proximal tubule*. The proximal tubule is like a sorting machine. "Good" molecules in the filtrate—such as ions, glucose, vitamins, and amino acids—are transported back into the blood so that the body can keep them. "Bad" waste molecules are transported from the blood to the filtrate. The movement of molecules into and out of the proximal tubule occurs through active transport, a process that requires energy (see Chapter 15). This is one of the reasons excretion requires a significant amount of energy.

After moving through the proximal tubule, the filtrate enters the *loop of Henle*, a hairpin-shaped loop. In the loop of Henle, water is absorbed from the filtrate. This helps the body save water.

From the loop of Henle, the filtrate moves into the *distal tubule*. In the distal tubule, more wastes are transported into the filtrate.

Finally, the filtrate moves down the *collecting duct*. In the collecting duct, more water is absorbed from the filtrate. How much water is absorbed depends on whether antidiuretic hormone (see Chapter 19) is present. If antidiuretic hormone is present, more water is absorbed and kept by the body. As a result, urine becomes more concentrated.

The filtrate—which is now urine—flows from the collecting duct into the *renal pelvis* (see Figure 20.18). The renal pelvis is like a giant funnel that catches the drippings of a million nephrons. From the renal pelvis, urine goes down a tube called the *ureter* to the bladder. The *bladder* is a stretchy sac where urine is temporarily stored. When the bladder is emptied, urine flows down the *urethra* and out the body.

MasteringPhysics®
TUTORIAL: Structure of the Human Excretory System
TUTORIAL: Nephron Function

CHECK YOURSELF

We have now looked at all the organ systems that help the body make ATP. Which organ systems provide the raw materials for making ATP (such as glucose and oxygen)? Which organ systems get rid of the wastes (such as carbon dioxide and urea)?

CHECK YOUR ANSWER

Glucose—as well as other organic molecules that can be broken down to make ATP—comes from the digestive system, which breaks down the food you eat. Oxygen is brought into the body by the respiratory system. Carbon dioxide is removed from the body through the work of the respiratory system as well. Urea and other wastes are removed by the excretory system. The circulatory system is involved every step of the way—the circulatory system transports the raw materials and wastes to their destinations in the body.

20.7 Keeping the Body Safe: Defense Systems

LEARNING OBJECTIVE
Describe the structure and function of the human immune system.

EXPLAIN THIS How does a vaccine protect you from disease?

Without your body's defenses, you would quickly fall prey to the bacteria, viruses, and other **pathogens**—disease-causing organisms—around you. What protects you is your immune system (Figure 20.19). Your immune system includes barriers that keep pathogens out, such as your skin and mucous membranes. It also includes many different kinds of immune cells, also known as white blood cells. White blood cells are made in the bone marrow. They then mature in the bone marrow or thymus. White blood cells move throughout your body constantly, looking for pathogens. Most of your white blood cells are found in your circulatory system, lymphatic system, and spleen.

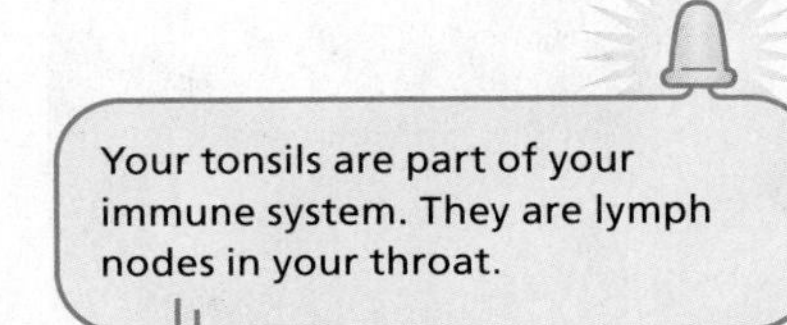

Your tonsils are part of your immune system. They are lymph nodes in your throat.

The *lymphatic system* is an important component of the immune system. The lymphatic system includes lymph vessels that, like blood vessels, travel all over the body. Lymph vessels carry a clear fluid called **lymph**. One function of the lymphatic system is to collect the fluid that leaks out of blood vessels and return it to the circulatory system. Another function of the lymphatic system is to carry white blood cells, which are found in large numbers in lymph. *Lymph nodes* are structures in the lymph vessels where many white blood cells are concentrated. Lymph nodes are found all over the body, including the throat, armpits, and groin. Swollen lymph nodes are a sure sign that the body is fighting an infection.

Let's look at the two parts of the immune system: the innate immune system and the acquired immune system.

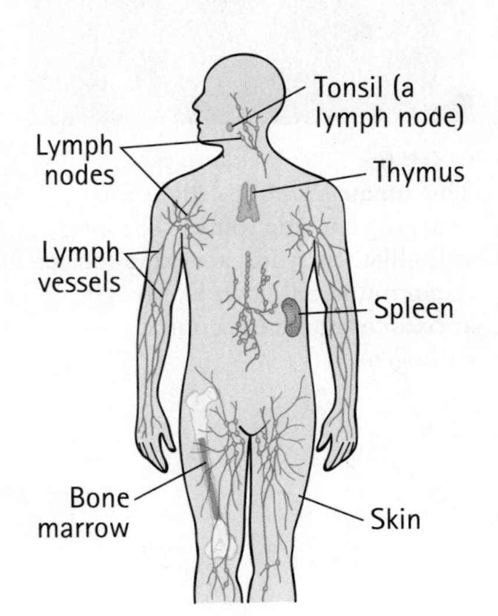

FIGURE 20.19
The immune system includes the skin, bone marrow, thymus, spleen, and lymph nodes and vessels. It also includes numerous immune cells found in the blood and tissues.

Innate Immunity

The **innate immune system** includes nonspecific body defenses that work against many different pathogens. These defenses are described as "innate" because they do not have to be activated by exposure to a pathogen—they are ready to go. Three key components of the innate immune system are the skin, the mucous membranes, and innate immune cells.

FIGURE 20.20
The lining of the respiratory tract includes mucus-secreting cells (yellow) as well as ciliated cells (pink). The cilia sweep mucus, along with trapped particles or pathogens, up to the pharynx, where it can be swallowed.

Skin The best way to deal with pathogens is to keep them out of the body in the first place. The skin is a crucial barrier, forming a tough outer layer that is hard to penetrate when intact. In addition, skin cells are shed frequently, making it hard for pathogens to establish a foothold. Hair follicles in skin also secrete special enzymes and acids that kill bacteria and fungi. In fact, many of your body secretions contain bacteria-killing enzymes—for example, saliva, tears, sweat, and milk.

Mucous Membranes The inside of your body is lined with *mucous membranes.* Mucous membranes are found inside your nose, mouth, and eyelids as well as in your respiratory, digestive, urinary, and reproductive tracts. Mucous membranes are not as tough as skin; however, all mucous membranes are covered by a layer of *mucus* that helps trap pathogens. The mucous membranes of the respiratory tract also include cilia that sweep mucus up to the pharynx (Figure 20.20). This mucus is swallowed, and stomach acid dispatches any pathogens. Many mucous membranes are also flushed by fluids such as tears, saliva, and urine. These fluids often contain pathogen-killing enzymes. In addition, the constant flushing makes it hard for pathogens to gain a foothold.

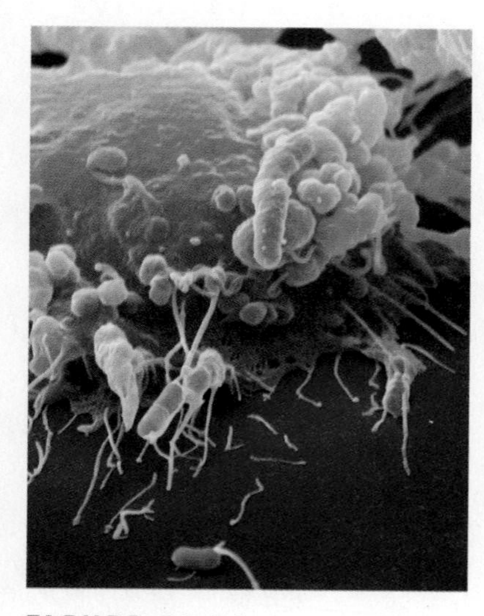

FIGURE 20.21
This innate immune cell (tan) is attacking bacteria (pink). The long stringlike structures are extensions of the immune cell's cytoplasm. They are searching the environment for pathogens.

Innate Immune Cells What happens when pathogens do get inside the body? Innate immune cells launch an immediate attack (Figure 20.21). A single innate immune cell is able to respond to many different pathogens. This is because the receptors of innate immune cells recognize molecules (usually carbohydrates, proteins, or nucleic acids) that are found in many different kinds of pathogens. All in all, the cells of the innate immune system include several hundred different receptors. If the innate immune system encounters the same pathogen more than once, its response is the same: The innate immune system retains no memory of pathogens it has encountered in the past.

One of the most important functions of the innate immune system is to initiate the *inflammatory response,* shown in Figure 20.22. You have no doubt experienced the inflammatory response many times—it is what makes your tissues swell and turn red when you cut your finger or scrape your knee. What happens during the inflammatory response? When a tissue is damaged, it releases chemicals called *histamines.* Histamines increase blood flow to the site of the injury and also cause local capillaries to leak fluid. The extra fluid causes swelling, which helps to isolate the wound from other body tissues. Histamines also attract innate immune cells. These cells squeeze out of the capillaries and attack any pathogens they find. Sometimes, the battle between innate immune cells and pathogens produces a whitish substance called pus. Pus consists of dead bacteria, dead tissue, and dead and living innate immune cells.

Acquired Immunity

The **acquired immune system** is a set of highly specific body defenses. These defenses are described as "acquired" because they become active only after specific pathogens are encountered. The acquired immune system differs from the innate immune system in several ways. First, a cell that belongs to the acquired immune system does not work against many different pathogens. Instead, each acquired immune cell has receptors that respond to a single **antigen**—a molecule or part of a molecule that belongs to a pathogen. Most often, antigens are parts of foreign proteins.

Second, the acquired immune system has more different receptors than the innate immune system. Whereas the innate immune system has several hundred different receptors, the acquired immune system has about 10 million receptors in all! With so

MasteringPhysics®
TUTORIAL: Immune Responses

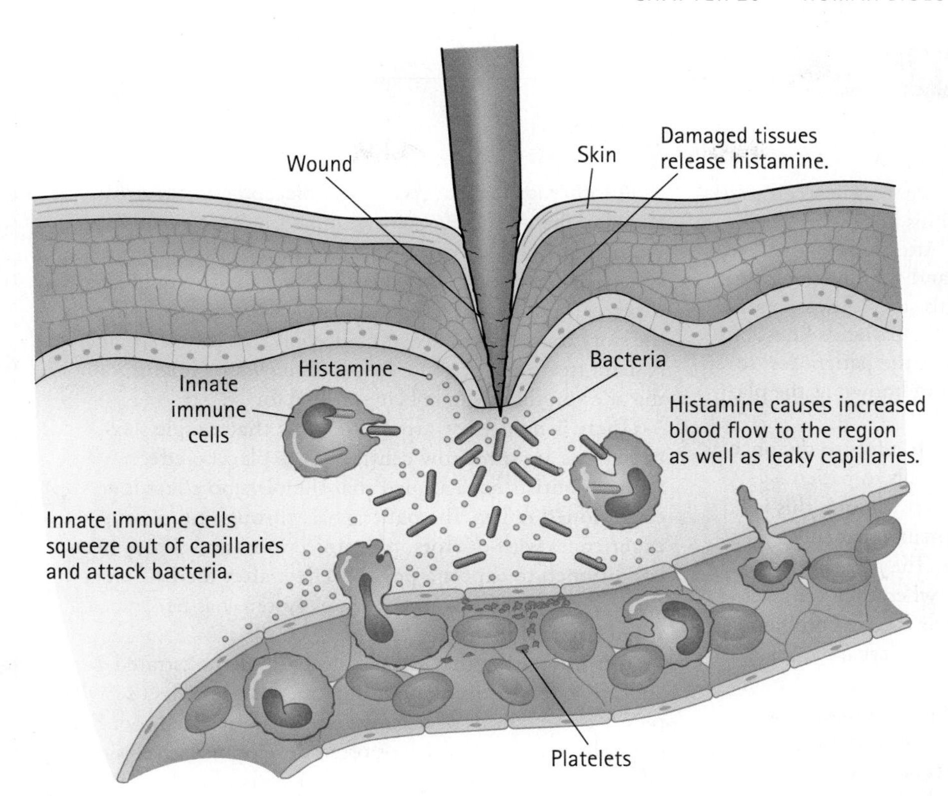

Wound
Skin
Damaged tissues release histamine.
Histamine
Bacteria
Innate immune cells
Histamine causes increased blood flow to the region as well as leaky capillaries.
Innate immune cells squeeze out of capillaries and attack bacteria.
Platelets

FIGURE 20.22
The inflammatory response begins when damaged tissues release histamines, which cause increased blood flow to a wound. Local capillaries become leaky and produce swelling. Innate immune cells squeeze out of the capillaries and attack bacteria or other microorganisms that enter the body.

many different receptors, it is just about guaranteed that any pathogen that enters the body will trigger a response by one or more acquired immune cells.

Third, the acquired immune system responds more slowly to pathogens than the innate immune system. Whereas the innate immune system responds immediately, the acquired immune response takes several days to reach full force. This time is needed to make multiple clones of the acquired immune cells that are needed. We will learn more about this process in a moment.

Finally, the acquired immune system retains a "memory" of pathogens it has encountered in the past. If it meets the same pathogen again, it reacts more quickly and more aggressively. The innate immune system does not have this kind of memory. Table 20.1 summarizes the differences between innate and acquired immunity.

TABLE 20.1 A COMPARISON OF INNATE IMMUNITY AND ACQUIRED IMMUNITY

Innate Immune System	Acquired Immune System
Each component can respond to **many different pathogens**	Each component responds to **a single antigen**
Includes **hundreds** of different receptors	Includes about **10 million** different receptors
Has an **immediate** response	Has a **delayed** response that reaches full force several days after the initial exposure to a pathogen
Has **no memory** of past encounters with a pathogen	Has a **memory** of past encounters with a pathogen; has a faster and more aggressive response to repeat exposure

SCIENCE AND SOCIETY

The Placebo Effect

A patient with knee pain goes in for surgery. He receives anesthesia, and cuts are made around his knee where the surgical instruments will be inserted. Afterward, the surgery appears to be successful—both pain and swelling are greatly diminished. What's unusual about this story? The operation was a sham. Cuts were made around the patient's knee, but nothing happened after that. Why does the patient feel so much better? Because of a phenomenon known as the placebo effect.*

Placebo is Latin for "I shall please." It refers to the once common practice of prescribing sugar pills to patients whom doctors otherwise couldn't help. Although this is now considered unethical, the fact remains that sugar pills frequently did have a beneficial effect. The *placebo effect* is the improvement patients experience when they are given a treatment that has no relevance to their medical problem.

Placebos appear to work for a wide variety of conditions and are usually far more effective than no treatment at all. In a study of patients with Parkinson's disease, a placebo worked just as well as medication in inducing the release of dopamine by the brain. Placebos have also been found to work as well as modern antidepressants in the treatment of depression. The placebo effect is certainly real, although placebos work better for some conditions than others. Placebos appear to be particularly effective for conditions related to the nervous system, including pain, depression, anxiety, headaches, fatigue, and gastrointestinal symptoms. For most of these conditions, placebos have about 30%–50% effectiveness—nearly as effective as many "real" treatments. Placebos are also believed to account for the "success" of certain alternative treatments that have no medical basis.

The placebo effect is one of the oddest phenomena in medicine. What causes it? Several possible mechanisms have been suggested. One is that the placebo effect operates through the release of endorphins (see Chapter 19). Release of the body's natural opiates would explain why placebos are so good at treating pain. In fact, placebos do become less effective as painkillers when patients are given a drug that blocks the opiate receptors. However, because the placebo effect works for many symptoms besides pain, this can't be the entire explanation.

Another idea is that receiving a placebo reduces stress, allowing the immune system to function more effectively. Numerous studies have shown that stress reduces the immune system's capabilities; consequently, stress relief would be expected to improve its function. Still, there must be more to the story than this because the placebo effect is very specific: A sham treatment does not help with *all* your ailments, only the one you think you're being treated for.

There is no getting around the fact that people's expectations are somehow central to the placebo effect. Some scientists have argued that the placebo effect is a conditioned reflex: The patient has, through numerous experiences with doctors, pills, injections, and so on, been conditioned to expect a positive effect after medical treatment. And, somehow, the nervous system has become wired to provide this.

It is interesting that studies have also demonstrated a "nocebo effect," sometimes called the placebo effect's "evil twin." Expectations of negative effects are realized too. For example, people on placebos often develop negative "side effects" from their treatments. Side effects of real medications may sometimes be caused by the nocebo effect as well. Studies have shown that patients who are warned of specific side effects tend to experience them more often than patients who are not warned. The nocebo effect can be even more dangerous. One study showed that women who believed they were vulnerable to heart disease were four times as likely to die of it as women who did not believe they were vulnerable but had similar risk factors. Never underestimate the power of the mind.

*Patients who participate in these types of studies—which are usually conducted to examine the effectiveness of a surgery, drug, or other treatment—are aware that they are enrolled in a clinical study and that they might be assigned to a "sham" treatment group. If the surgery or other treatment performs no better than the placebo (as occurred in this case), the conclusion is that the treatment is not effective. Testing treatments against placebos is important precisely because the placebo effect can be very powerful. If there is already a proven treatment for a particular condition, however, new treatments can be tested against the established one rather than against a placebo.

Let's look at how acquired immunity works. The acquired immune system has two kinds of cells: B cells and T cells. Both are made in the bone marrow, but each matures in a different organ—the *b*one marrow for B cells, and the *t*hymus for T cells.

B Cells *B cells* react to pathogens in bodily fluids such as blood or lymph. When a B cell encounters a pathogen and binds to an antigen on the pathogen, the cell

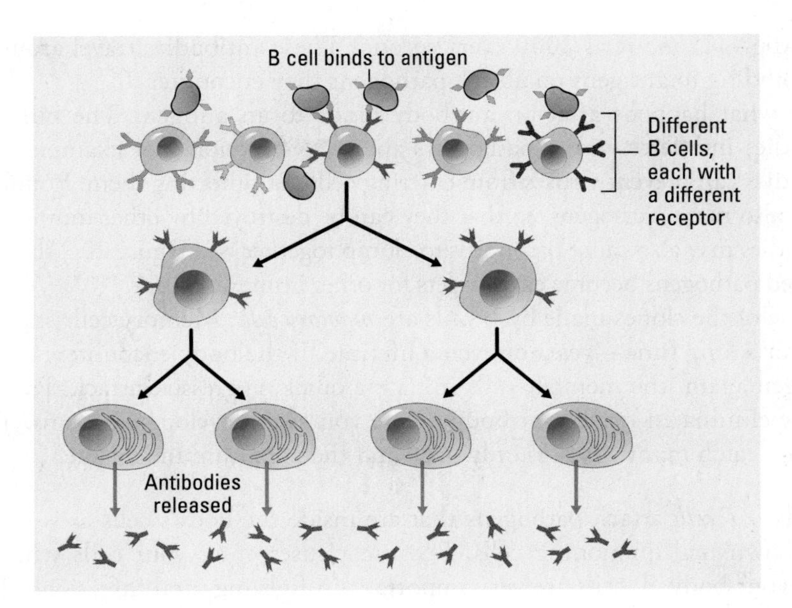

FIGURE 20.23
When a B cell binds to an antigen, the B cell begins to divide and make clones of itself. Mature clones then make and release large numbers of antibodies. A few of the clones are memory cells that remain in the body for a long time, ready to deal with subsequent infections by the same pathogen.

begins to divide. It makes many copies, or *clones*, of itself (Figure 20.23). The time it takes to make these clones is what delays the acquired immune response.

Once the clones are ready, they begin to make and release proteins called antibodies. An **antibody** is a large, Y-shaped protein that includes two antigen-binding sites (Figure 20.24a). An antibody binds to its antigen in a very specific way, with a fit like a lock and key. A single B cell can make a huge number of

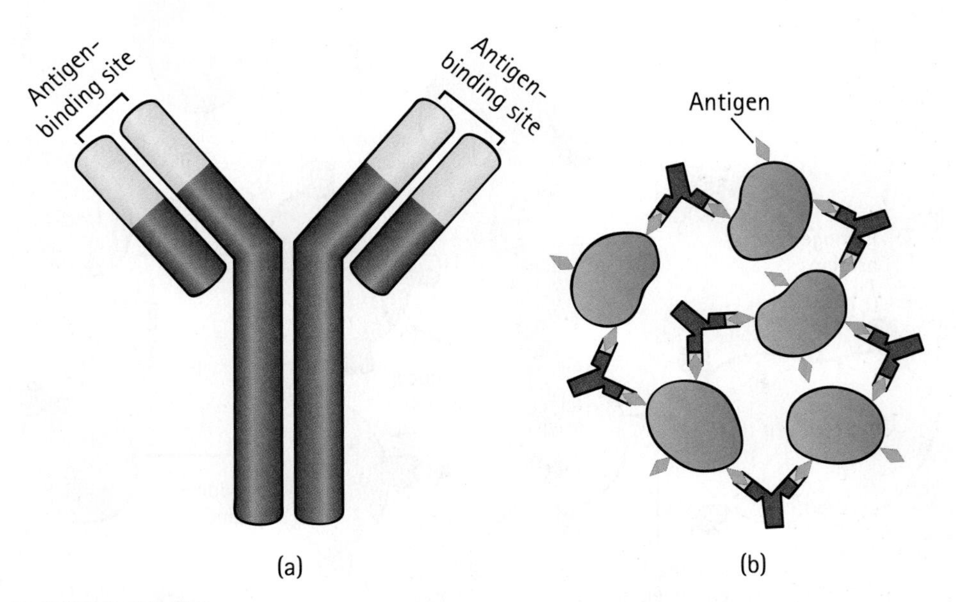

FIGURE 20.24
(a) Antibodies are large Y-shaped proteins that include two antigen-binding sites. Antibodies are very specific—an antibody binds to its antigen with a lock-and-key fit. (b) Because antibodies each have two binding sites, they can cause pathogens (shown here as green ovals) to clump together.

antibodies—as many as 2000 every second! These antibodies travel around the body binding to antigens on all the pathogens they encounter.

But what happens after an antibody binds to an antigen? The binding of antibodies interferes with a pathogen's ability to function. For example, bound antibodies can prevent viruses from entering cells and infecting them. Bound antibodies also mark pathogens, so that they can be destroyed by other immune cells. Antibodies may also cause pathogens to clump together (see Figure 20.24b)—these clumped pathogens become easy targets for other immune cells.

Some of the clones made by B cells are *memory cells*. Memory cells stay in the body for a long time—years or even a lifetime. If the body encounters the same pathogen again, the memory cells initiate a quick, aggressive attack. Pathogens may be eliminated from your body before you even develop symptoms. This is why you catch many diseases only once and then are immune for life.

T Cells *T cells* attack pathogens that are inside the body's cells as well as the body's own malfunctioning cells. Because viruses enter your cells when they infect your body, T cells are very important in fighting viral infections. T cells are also important in destroying cancer cells or other abnormal body cells.

When a body cell is infected by a pathogen, the cell displays some of the foreign proteins on its surface the way a ship might wave a distress flag (Figure 20.25). The display of foreign proteins serves as a signal to the body's T cells. A specific type of T cell called a *helper T cell* binds to an antigen on the pathogen protein and begins to divide and produce clones. Helper T cell clones then initiate many different immune activities: They stimulate B cells to produce clones that make antibodies against the pathogen. They also stimulate *killer T cells*, another group of T cells, to divide and produce clones. The killer T cells bind to displayed proteins and then, as their name suggests, kill infected cells. This prevents pathogens from coming

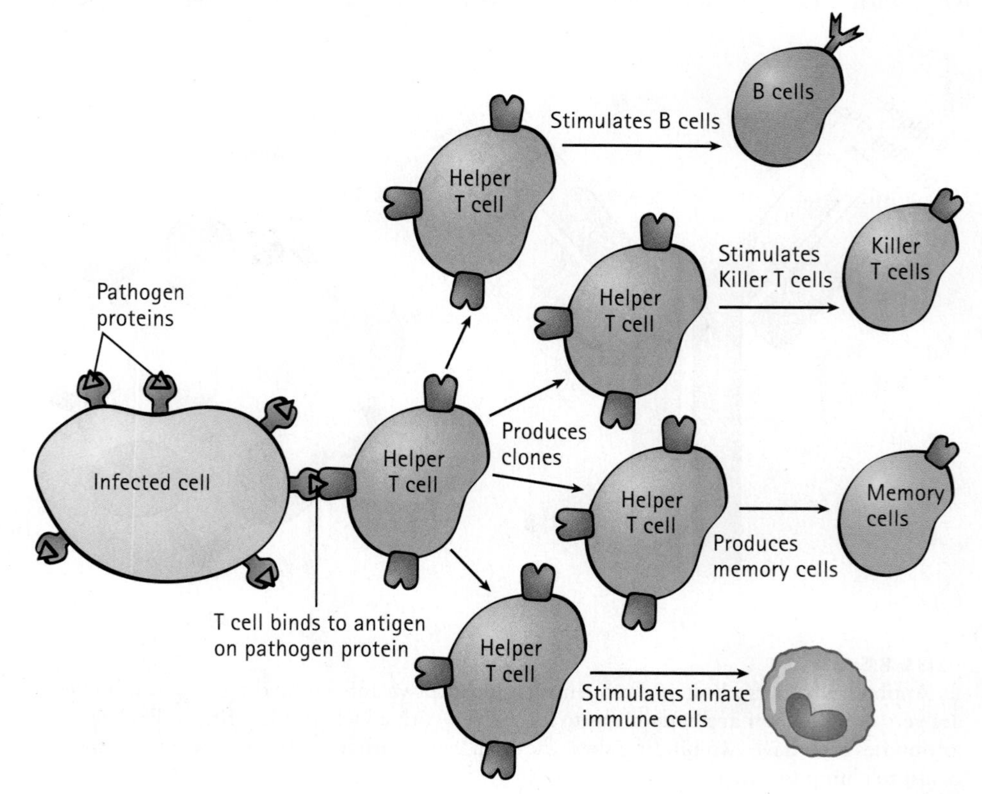

FIGURE 20.25
T cells target pathogens that are inside the body's cells. Infected cells display pathogen proteins the way a ship might display a distress flag. A helper T cell binds to an antigen on the pathogen protein and then produces clones that initiate a variety of immune activities.

out to infect new cells. Helper T cells also stimulate cells of the innate immune system. Finally, helper T cells produce memory cells that remain in the body for a long time and guard against a return of the pathogen.

Vaccines Vaccines protect you from disease by making use of your acquired immune system's "memory" for pathogens it has encountered in the past. A vaccine exposes your body to a pathogen's antigens, but—this is the key—it does not infect you with the pathogen itself!

Most vaccines contain either dead pathogens or weak strains of a pathogen. Or, a vaccine may use only part of a pathogen—perhaps a virus's protein coat or a bacterium's flagellum. Your acquired immune system reacts to antigens in the vaccine just as it would react to the real pathogen. It makes antibodies and—most critically—memory cells. If the real pathogen ever shows up, the acquired immune system is ready for it (Figure 20.26).

Vaccines are still the best way we have to fight most viruses. Although many drugs have been developed to attack pathogenic bacteria and fungi, it has been harder to develop drugs that work well against viruses. So, don't forget your vaccinations!

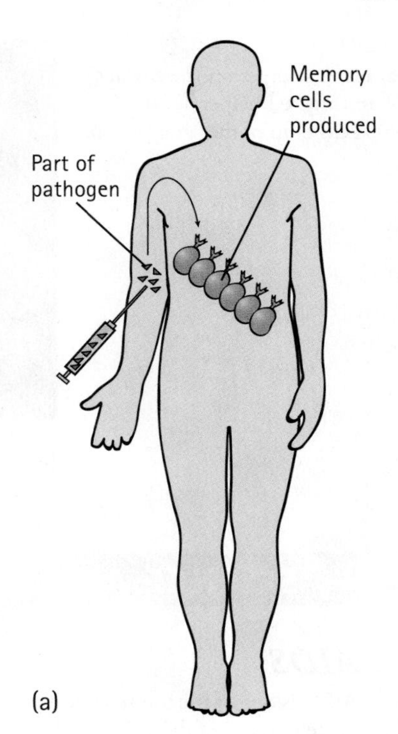

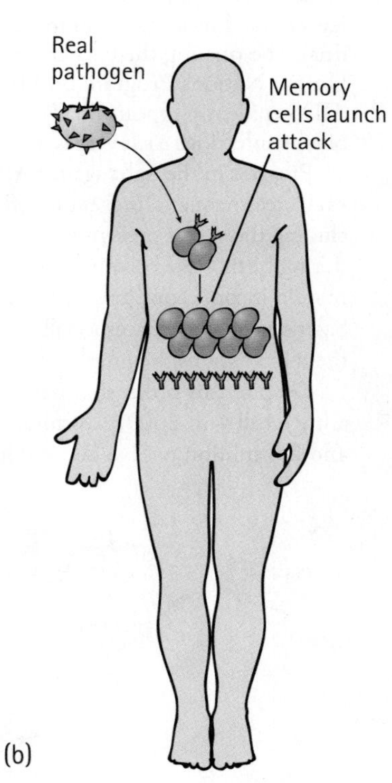

FIGURE 20.26
(a) Vaccines work by introducing dead pathogens, weak pathogens, or parts of a pathogen into the body. The acquired immune system reacts and produces memory cells. (b) If the actual pathogen is encountered later, the memory cells initiate a rapid, aggressive attack.

CHECK YOURSELF

1. **How does the inflammatory response help defend your body against disease?**
2. **Which immune system cells produce antibodies? How do antibodies help defend your body against pathogens?**

CHECK YOUR ANSWERS

1. During the inflammatory response, the tissues surrounding a wound swell, helping to isolate it from the rest of the body. The inflammatory response also brings innate immune cells to the site of injury so that they can attack bacteria or other pathogens.
2. Antibodies are produced by B cells in the acquired immune system. Antibodies work by binding to antigens on pathogens. This prevents pathogens from functioning or marks them for destruction by other immune cells. Antibodies may also cause pathogens to clump together, so that they become easy targets for immune cells.

Diseases of the Immune System

The immune system normally has no problem recognizing the body's own cells. In certain diseases, called *autoimmune diseases*, the immune system identifies certain body cells as foreign and attacks them. *Type 1 diabetes* occurs when the immune system attacks and destroys the insulin-producing cells of the pancreas. *Multiple sclerosis* occurs when immune cells destroy the myelin sheaths surrounding neurons. *Lupus* is a debilitating disease caused by the immune system attacking a wide range of healthy tissues.

Acquired immunodeficiency syndrome, or AIDS, is a disease caused by the *human immunodeficiency virus*, HIV. (The "acquired" in the name of the disease refers to the fact that immunodeficiency is acquired during a person's lifetime rather than being inborn.) HIV attacks immune cells, particularly helper T cells (Figure 20.27). With these crucial immune cells compromised, infections and cancers that are normally easy to fight off have a chance to overwhelm the body.

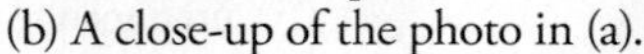

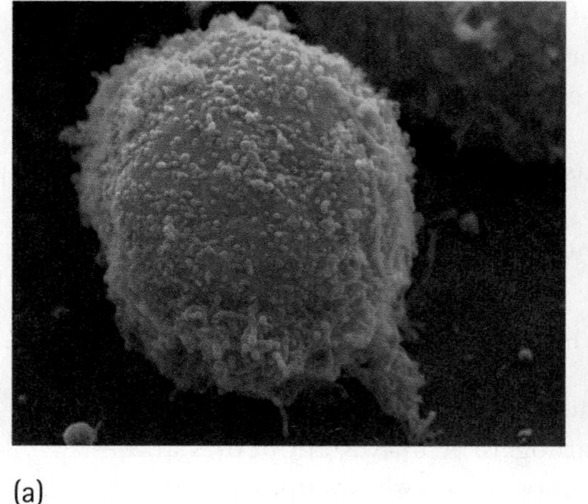

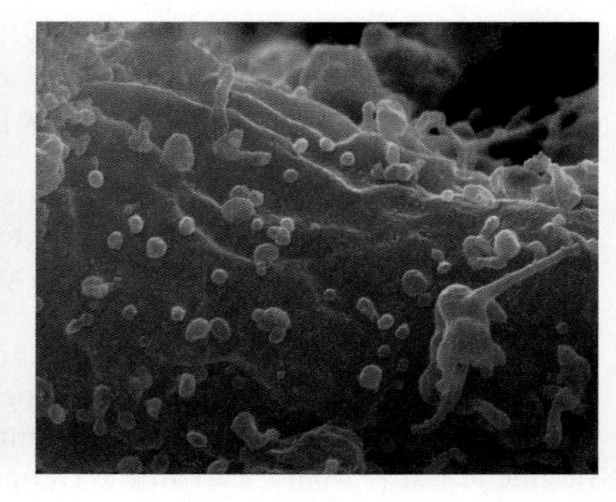

FIGURE 20.27
(a) HIV emerges on the surface of an infected helper T cell.
(b) A close-up of the photo in (a).

SCIENCE AND SOCIETY

AIDS

AIDS is one of the top ten leading causes of death worldwide. Since the start of the AIDS epidemic, in the 1980s, the disease has claimed more than 30 million lives. However, we may finally be turning the corner. In 2010, UNAIDS, the Joint United Nations Program on HIV/AIDS, reported that the AIDS epidemic appears finally to have stabilized. The number of new infections each year is now declining.

Progress in the fight against AIDS is credited largely to HIV treatment and prevention efforts. Anti-AIDS drugs, including those that prevent transmission from mother to child during birth, have become more widely available in low- and middle-income countries. AIDS education has also helped: Safer sexual behavior, especially among young people, is one of the main reasons the number of new infections has dropped.

Despite this promising news, the disease is still taking a huge toll—in 2009, 1.8 million people died of AIDS, and 2.6 million people became infected. There were also more than 33.3 million people living with HIV worldwide, including 2.5 million children. The prevalence of AIDS in different parts of the world is shown on the map.

Even if we continue to make progress in the battle against AIDS, the effects of the epidemic will be felt for a long time. In the countries that have been hardest hit by AIDS, many people were lost in the prime of their lives. This leaves fewer adults to care for children and the elderly. About 16 million children have been orphaned by AIDS, most of them in sub-Saharan Africa. The deaths of working-age adults have also significantly reduced the workforce, with important economic consequences.

In wealthy countries, where HIV treatment is widely available, AIDS rates have been stable for many years. Unfortunately, complacency about AIDS appears to have resulted in the return of some high-risk behaviors. In the United States, there are now worrisome hints of an AIDS resurgence.

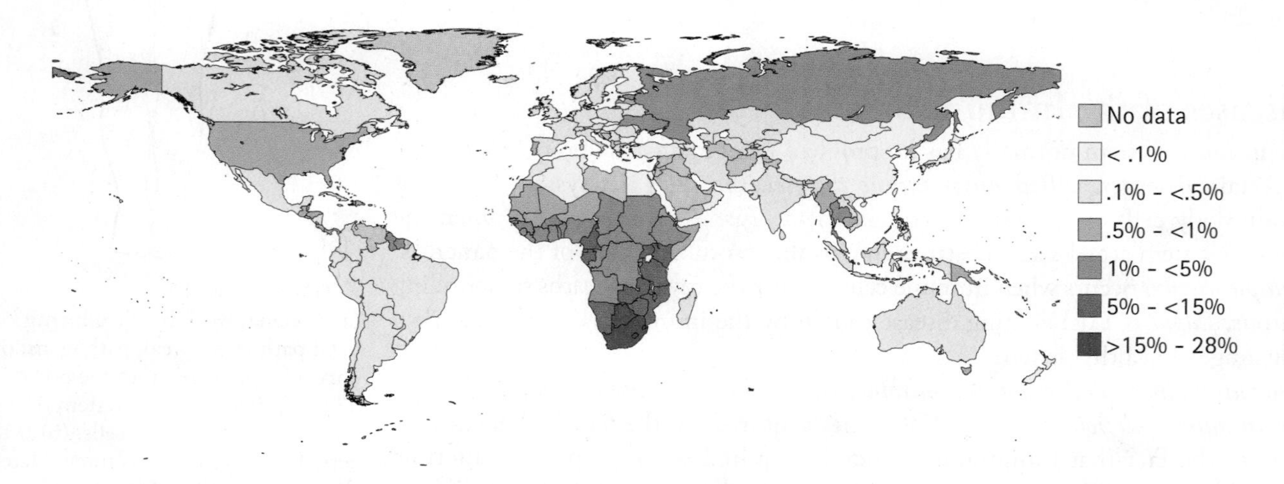

This map shows the percentages of the population infected with HIV throughout the world in 2009.

SUMMARY OF TERMS (KNOWLEDGE)

Acquired immune system A set of highly specific body defenses that respond to specific features of pathogens.

Alveoli The tiny sacs in the lungs where gas exchange occurs.

Antibody A large, Y-shaped protein that binds to a pathogen's antigen, disabling the pathogen or marking it for destruction.

Antigen A molecule or part of a molecule belonging to a pathogen that is recognized by cells of the acquired immune system.

Arteries Blood vessels that carry blood away from the heart.

Capillaries Tiny, thin-walled blood vessels from which molecules are exchanged between blood and body tissues.

Diaphragm The sheet of muscle that covers the bottom of the thoracic cavity; its contraction and relaxation cause inhalation and exhalation.

Digestion The breakdown of food into small organic molecules that can be absorbed and used by the body.

Hemoglobin A protein found in red blood cells that binds to and transports oxygen.

Innate immune system A set of nonspecific body defenses that work against many different pathogens.

Lymph A clear fluid inside lymph vessels that contains large numbers of immune cells.

Nephron The functional unit of a kidney; it filters waste molecules from blood to produce urine.

Pathogens Disease-causing agents such as bacteria, viruses, or other organisms.

Peristalsis A moving wave of muscular contractions that moves food down the digestive tract.

Platelets Blood cells that function in blood clotting.

Red blood cells Blood cells that transport oxygen to body tissues.

Veins Blood vessels that carry blood toward the heart.

White blood cells Blood cells that are part of the immune system.

READING CHECK (COMPREHENSION)

20.1 Integration of Body Systems

1. Which two organ systems work together to supply the body with oxygen?
2. How does getting rid of cellular wastes require the integrated efforts of multiple body systems?

20.2 The Circulatory System

3. What stimulates the heart to beat?
4. What makes the "lub-dubb" sound of the heartbeat?
5. Which blood vessels are responsible for nutrient and waste exchange with tissues?
6. Trace the path of blood through the body, beginning with blood returning from the tissues to the heart. Be sure to name each of the chambers of the heart.
7. What are the three types of blood cells, and what is the function of each?

20.3 Respiration

8. Trace the path of air as it moves to the alveoli.
9. How do the structures of the alveoli and their surrounding capillaries facilitate gas exchange?
10. How do you inhale?

20.4 Digestion

11. What does digestion accomplish?
12. What are the functions of saliva?
13. What prevents food from going into the trachea after it is swallowed?
14. What happens to food while it is in the stomach?
15. What structures increase the surface area available for nutrient absorption in the small intestine?

20.5 Nutrition, Exercise, and Health

16. Why is it important for you to eat a complete protein regularly?
17. What are some of the important minerals you obtain from your food?
18. What are some of the benefits of exercise?

20.6 Excretion and Water Balance

19. What is urea?
20. How does fluid move from the circulatory system into a nephron?
21. What happens to the filtrate in the proximal tubule?
22. What is the function of the loop of Henle?
23. How does antidiuretic hormone affect how concentrated urine becomes?

20.7 Keeping the Body Safe: Defense Systems

24. What are two functions of the lymphatic system?
25. What features of skin make it good at keeping pathogens out of the body?
26. Why is a single innate immune cell able to respond to many different pathogens?
27. What is an antigen?
28. Explain what happens when a B cell first encounters a pathogen and binds to an antigen on the pathogen.
29. What is the function of a memory cell?

THINK INTEGRATED SCIENCE

20A—Hemoglobin

30. What is the structure of hemoglobin? Which part of the hemoglobin molecule binds to oxygen?
31. How many oxygen molecules can one molecule of hemoglobin carry?
32. Explain how hemoglobin's oxygen affinity is affected by blood pH, and how this change in oxygen affinity is adaptive.

20B—Low-Carb Versus Low-Cal Diets

33. How does a low-Calorie diet help you lose weight?
34. How does exercise help you lose weight?
35. Do low-carb diets work? How do they help people lose weight?
36. What are some potential problems with low-carb diets?

THINK AND DO (HANDS-ON APPLICATION)

37. Your heart rate is the number of times your heart beats in a minute. It changes over the course of the day as your activity level increases or decreases. Begin by measuring your resting heart rate. Ideally, you should do this when you first wake up in the morning. Now measure your heart rate during different types of activities. What is it when you're reading quietly? When you're walking or doing chores around the house? Finally, measure your heart rate while you are exercising. Are you hitting your target zone for effective cardiovascular exercise? You can determine your target zone by following these directions:

    First calculate your maximal heart rate:

    $$\text{Maximal heart rate} = 220 - \text{your age}$$

    Then calculate your heart rate reserve:

    $$\text{Heart rate reserve} = \text{maximal heart rate} - \text{resting heart rate}$$

    You are in your target zone if you are using between 60% and 80% of your heart rate reserve:

    $$\text{Low end of target zone: } 60\%\ (\text{heart rate reserve}) + \text{resting heart rate}$$

    $$\text{High end of target zone: } 80\%\ (\text{heart rate reserve}) + \text{resting heart rate}$$

38. How healthy is your diet? Write down everything you eat and drink for a day. Note the number of Calories when you can. About how many Calories do you consume per day? How does your diet compare to that recommended by MyPlate? What are you eating too much of? Not enough of?

39. The body mass index (BMI) adjusts your weight for your height. You can calculate your BMI as follows:

    $$\text{BMI} = \frac{\text{your mass in kilograms}}{(\text{your height in meters})^2} \text{ or}$$

    $$= \frac{\text{your weight in pounds}}{(\text{your height in inches})^2} \times 703$$

    What does BMI mean? BMI tells you if you are underweight or overweight.

BMI	
Below 18.5	Underweight
18.5–24.9	Normal
25.0–29.9	Overweight
30.0 and above	Obese

Note, however, that BMI does not measure body fat, as the figure below should convince you.

6'3"
Height
6'3"
220 lb
Weight
220 lb
27.5
BMI
27.5

THINK AND COMPARE (ANALYSIS)

40. Rank the following from highest oxygen content to lowest: the air inside the alveoli, a working arm muscle, arteries to the arm, capillaries that supply the arm muscle.

THINK AND SOLVE (MATHEMATICAL APPLICATION)

41. Show that your blood supply can carry as many as 3×10^{22} molecules of oxygen. Here's some information you may find useful: You have 25 trillion red blood cells. Each red blood cell contains 300 million molecules of hemoglobin. Each molecule of hemoglobin can carry four molecules of oxygen.
42. A red blood cell has no nucleus and is therefore unable to make the proteins necessary to maintain itself. Because of this, red blood cells have a relatively short life span of about 120 days. Given that we have about 25 trillion red blood cells, show that more than 208 billion red blood cells die and are replaced each day. Also show that, in the 20 seconds it took you to read this problem, about 48 million red blood cells died and were replaced.
43. A typical person has a heart rate of 70 beats per minute and takes 12 breaths in a minute. Show that her heart beats about 4200 times an hour, 100,800 times a day, and 36.8 million times a year. Show also that she takes about 720 breaths per hour, 17,280 breaths per day, and 6.3 million breaths per year.

THINK AND EXPLAIN (SYNTHESIS)

44. How does reproduction require the integrated work of multiple organ systems?
45. In a developing embryo, the heart begins to beat long before developing nerves reach it. How is the embryonic heart able to beat without nervous system control?
46. Why do you think the atria of the heart are less muscular than the ventricles? Why is the left ventricle more muscular than the right ventricle?
47. The pumping of the heart does most of the work that is required to move blood around the body. How do your voluntary movements contribute?
48. Where in the body is blood most oxygenated?
49. How does the body control the amount of blood that different tissues receive?
50. Which functions, other than acquiring oxygen for the body, require the work of the respiratory system?
51. As food moves down your esophagus, your esophagus bulges. Why doesn't this cause the trachea, which lies adjacent to it, to become closed off?
52. Is breathing a voluntary or involuntary action?
53. Having the openings to both the trachea and the esophagus in the pharynx is problematic because it can lead to choking. Are there any advantages to this arrangement?
54. What are the functions of gastric juice?
55. What waste materials are produced during the process of making ATP? What body systems are responsible for removing the wastes from the body?
56. How does the endocrine system interact with the excretory system? Give examples.
57. What is the difference between elimination (feces) and excretion (urine)? What is the body getting rid of in each case?
58. How does the lymphatic system support the work of the circulatory system? How does it support the work of the immune system?
59. How do tears help defend your body against pathogens?
60. Why is the innate immune system described as "nonspecific"? Why is the acquired immune system described as "specific"?
61. What are some differences between the innate immune system and the acquired immune system?
62. Allergies occur when the immune system is abnormally sensitive to particular substances. Why do people sometimes take antihistamines for their allergies? Can you guess from the word *antihistamine* what an antihistamine does?
63. How are the antigens that B cells respond to different from the ones that T cells respond to?
64. Text or write a letter to Grandpa telling him about low-carb diets. Tell him why many people have been able to lose weight on low-carb diets, and explain some potential dangers of the diets as well.

THINK AND DISCUSS (EVALUATION)

65. Several of your senses provide examples of how multiple body systems work together to accomplish important tasks. What body systems are involved in hearing? In smelling? In tasting?
66. How do the arterioles react when you are running? When you are doing biceps curls? When you are sitting at your desk thinking and doing homework?
67. You use energy to contract your diaphragm and rib muscles when you breathe. Is energy also required for gas exchange—that is, for the transport of oxygen and carbon dioxide molecules between the air inside the alveoli and the blood in the surrounding capillaries?
68. Why shouldn't you talk with your mouth full (not just because it's impolite)?
69. If you hold a piece of cracker in your mouth without chewing it, the cracker will dissolve. However, this doesn't happen with a piece of meat. Why?
70. Why do you think people like to eat sweets, fats, and other foods that are not very good for them? Do you think that, over time, our taste for these and other foods might evolve?
71. Why might eating a high-protein diet be particularly hard on the liver and kidneys?
72. The liver is an organ that plays important roles in multiple organ systems. What role does the liver play in digestion? What role does it play in excretion?
73. A cell in your body breaks down an amino acid to make ATP and generates a molecule of ammonia, a nitrogen-containing waste. Describe what happens to this ammonia molecule.
74. What do you think explains the placebo effect?
75. The leading causes of death in low-income countries are listed in the table. Compare them with the leading causes of death in the United States given earlier in the chapter.

Leading Causes of Death in Low-Income Countries
(2008 data from the World Health Organization)

	Total Deaths (millions)	Percent of Total
1. Lower respiratory infections (pneumonia)	1.05	11.3
2. Diarrheal diseases	0.76	8.2
3. HIV/AIDS	0.72	7.8
4. Heart disease	0.57	6.1
5. Malaria	0.48	5.2
6. Strokes	0.45	4.9
7. Tuberculosis	0.40	4.3
8. Prematurity and low birth weight	0.30	3.2
9. Birth asphyxia and birth trauma	0.27	2.9
10. Neonatal infections	0.24	2.6
All other causes	9.2	43.5

READINESS ASSURANCE TEST (RAT)

If you have a good handle on this chapter, then you should be able to score 7 out of 10 on this RAT. If you score less than 7, you need to study further before moving on.

Choose the BEST answer to each of the following:

1. The right ventricle of the heart pumps blood to
   (a) the right atrium.
   (b) arteries going to the lungs.
   (c) arteries going to body tissues.
   (d) the left atrium.
2. The cells that carry oxygen to body tissues are
   (a) red blood cells.
   (b) white blood cells.
   (c) platelets.
   (d) hemoglobin.
3. Oxygen moves from the alveoli in the lungs into the bloodstream through the process of
   (a) active transport.
   (b) endocytosis.
   (c) diffusion.
   (d) exocytosis.
4. Air moves into the lungs when
   (a) the diaphragm contracts and flattens and the rib cage moves down and into the chest.
   (b) the diaphragm contracts and flattens and the rib cage moves up and out from the chest.
   (c) the diaphragm relaxes and the rib cage moves down and into the chest.
   (d) the diaphragm relaxes and the rib cage moves up and out from the chest.

5. Which of the following does not occur in the small intestine?
   (a) Bile from the gall bladder breaks fats into small droplets.
   (b) Nutrients are absorbed into the body.
   (c) Proteins are broken down.
   (d) A highly acidic environment kills bacteria present in food.
6. Aside from Calories, what else do you need to obtain through your diet?
   (a) a complete protein
   (b) a complete carbohydrate
   (c) vitamins and minerals
   (d) both a complete protein and vitamins and minerals
7. In which part of a nephron is water reabsorbed from the filtrate?
   (a) glomerulus
   (b) proximal tubule
   (c) loop of Henle
   (d) distal tubule
8. The stretchy sac where urine is temporarily stored is the
   (a) renal pelvis.
   (b) bladder.
   (c) ureter.
   (d) urethra.
9. Which of the following is not associated with the innate immune system?
   (a) skin
   (b) tears
   (c) the inflammatory response
   (d) antibodies
10. A difference between the innate immune system and the acquired immune system is
    (a) the innate immune system includes many more different types of receptors than the acquired immune system.
    (b) the acquired immune system response is immediate, whereas the innate immune response is delayed, peaking about 3 to 5 days after exposure.
    (c) the acquired immune system retains a memory for past exposure, with subsequent responses being faster and more aggressive; the innate immune system does not.
    (d) each acquired immune system cell responds to more different pathogens than each innate immune system cell.

Answers to RAT

1. b, 2. a, 3. c, 4. b, 5. d, 6. d, 7. c, 8. b, 9. d, 10. c

CHAPTER 21
Ecology

NO LIVING creature exists in a vacuum—all organisms interact with the living and nonliving things around them. The study of these interactions is called *ecology*. As we delve into ecology, we'll learn many things. Why do some animals, such as the elephant, produce only a single offspring every few years, while others, such as the razor clam, release more than 100 million eggs at a time? How can certain shrimp crawl inside an eel's mouth without being eaten? Why are there more zebras than lions living in the African savanna? Is it true that a disturbance can sometimes increase a habitat's biodiversity? What about the ecology of our own species? How many people live on Earth today? Will the global human population continue to grow, or is it likely to crash one day? Finally, could planting a tree be the key to world peace?

21.1 Organisms and Their Environments

LEARNING OBJECTIVE
Describe the different levels of ecological study.

EXPLAIN THIS What is the difference between a community and an ecosystem?

Ecology is the study of how organisms interact with their environments. An organism's environment includes nonliving, or *abiotic*, features, such as temperature, sunlight, rain, rocks, and ponds. An organism's environment also includes *biotic* features—other living things.

Ecology can be studied at many different levels, including those of the individual, population, community, and ecosystem. Studies that focus on individuals may ask how an organism's anatomy, physiology, and behavior help it function in its environment. For example, why do certain plants invest more energy in building root systems, whereas others invest more energy in their stems and leaves (anatomy)? Or, does the nutrient content of different seeds explain why finches prefer one type of seed to another (physiology)? Or, how do gulls divide their time between two different feeding strategies: begging for handouts from humans and foraging at sea (behavior)? Because individual-level studies in ecology usually focus on adaptation, there is often considerable overlap with evolutionary biology.

A **population** is a group of individuals of a single species that lives in a specific area. The raccoons in Berkeley, California, are a population. So are the snow leopards of Central Asia and the orange-flanked bush robins of Langtang National Park, Nepal (Figure 21.1). At the population level, ecologists are often interested in the size of a population and in how population size changes over time. In addition, population ecologists may study how a specific population uses resources, including the kinds of foods that are eaten and the types of habitats that are used.

FIGURE 21.1
Biologist Pamela Yeh holds an orange-flanked bush robin in Langtang National Park, Nepal. The bird will be measured, weighed, and given a unique identifying leg band before being released.

A **community** consists of all the organisms that live in a specific area. For example, the ecological community in an Idaho brush habitat might include sagebrush, prickly pear cactus, insects, squirrels, pronghorns, badgers, lizards, mountain lions, coyotes, and many kinds of microorganisms. At the community level, biologists often look at how species interact with one another. They may study who eats whom in a community or investigate how different species compete for resources.

An **ecosystem** consists of all the organisms that live in a specific area *and* all the abiotic features of their environment. At the ecosystem level, ecologists look at links between the biotic and abiotic worlds. For example, some ecosystem ecologists study the cycling of important resources such as water and carbon between living organisms and Earth. Others look at how energy moves through ecosystems. Additional examples of ecological studies at different levels of organization are shown in Figure 21.2.

UNIFYING CONCEPT
- *The Ecosystem*

CHECK YOURSELF

1. **An ecologist studies how hyenas and lions compete for food in the African savanna. Is this a study at the population, community, or ecosystem level?**
2. **An ecologist studies how the number of mountain lions in a certain park has changed over the past decade. Is this a study at the population, community, or ecosystem level?**

(a)

(b)

(c)

(d)

FIGURE 21.2
Ecologists study different kinds of questions. (a) An individual-level study: How do fireflies use light signals to attract mates? (b) A population-level study: How many giant pandas live in the bamboo forests of central China? Is this number increasing or decreasing? (c) A community-level study: Who eats whom in a forest in Wiltshire, England? (d) An ecosystem-level study: How does an important resource such as water cycle between the biotic and abiotic parts of an ecosystem?

CHECK YOUR ANSWERS

1. This study examines how two different species interact in a habitat, so it is a community-level study.
2. This study examines a group of individuals of a single species that lives in a specific area, so it is a population-level study.

LEARNING OBJECTIVE
Compare exponential and logistic population growth, and describe how these are connected to a population's life history.

21.2 Population Ecology

EXPLAIN THIS Why is a baby elephant considered an "expensive" offspring?

Population ecologists are often interested in how big a population is as well as in how population size changes over time. Sometimes, population ecologists count the total number of individuals in a population. Other times, they prefer to measure the *population density*, the number of organisms per unit area.

Population Growth

Four factors determine how a population's size changes over time: the birth rate, the death rate, the rate of immigration into the population, and the rate of emigration out of the population. Births and immigration increase population size. Deaths and emigration decrease population size. Whether a population is increasing or decreasing in size overall depends on the relative contributions of the four factors. Two models for how populations in the natural world grow are exponential growth and logistic growth.

UNIFYING CONCEPT

- ***Exponential Growth and Decay***
Appendix D

Exponential Growth **Exponential growth** occurs when a population grows at a fixed rate per amount of time (Figure 21.3). For example, a population that increases by 10% each year, or one that doubles each decade, grows exponentially (see Appendix D). As you can see from the graph in Figure 21.3a, a population that is growing exponentially increases in size more and more quickly as time goes on.

What kind of population grows exponentially? Exponential growth occurs when a population has unlimited resources. In this case, nothing limits population growth, and a bigger and bigger population continues to produce more and more offspring. Although resources are never truly unlimited in the real world, they are sometimes extremely plentiful. For example, organisms may find very plentiful resources when they colonize a new habitat, and their population may begin to grow exponentially. When reindeer were introduced on Saint Paul Island near Alaska, the population grew exponentially for many years (see Figure 21.3b). However, exponential growth cannot continue forever. On Saint Paul Island, the reindeer eventually ran out of food and the population crashed.

Exponential growth is often seen in populations that live in unstable environments. These populations may go through cycles of exponential growth and crash. Good conditions start exponential growth. When resources run out, the population crashes. When good conditions return, the population explodes again, and so on.

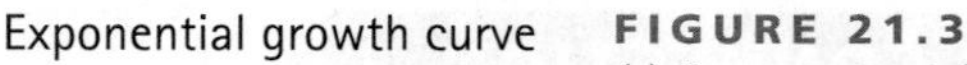
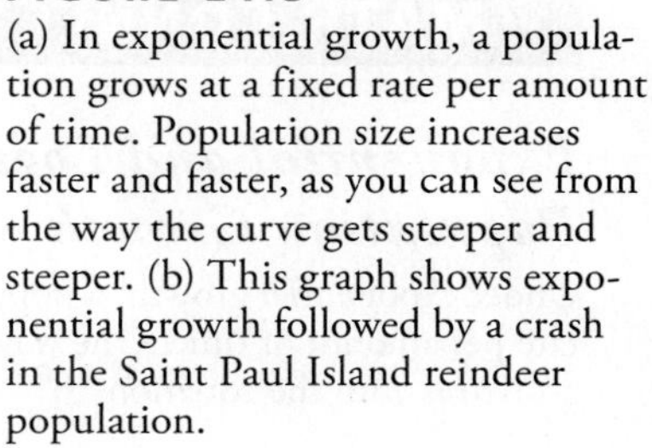

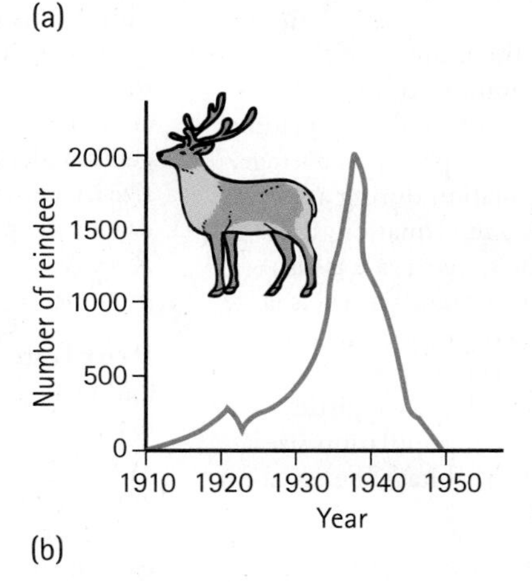

FIGURE 21.3
(a) In exponential growth, a population grows at a fixed rate per amount of time. Population size increases faster and faster, as you can see from the way the curve gets steeper and steeper. (b) This graph shows exponential growth followed by a crash in the Saint Paul Island reindeer population.

Logistic Growth **Logistic growth** occurs when population growth slows as the population approaches the habitat's carrying capacity (Figure 21.4). The **carrying capacity** is the maximum number of individuals or the maximum population density that a habitat can support. A certain lake may be able to support 250 trout, for example. Or, a certain grassland may be able to support 35 weasels per square kilometer.

Why would population growth slow as it approaches carrying capacity? There are many possible reasons. When a population is large, there may be more competition for resources such as food and space. This could make it more difficult to get enough resources to reproduce, causing birth rates to fall. Or, crowded conditions at large population sizes could increase the death rate.

In the real world, logistic growth is often seen in stable habitats. Logistic growth in the Tasmania Island sheep population is shown in Figure 21.4b.

CHECK YOURSELF

A population of shorebirds occupies a habitat where there are a limited number of safe nesting sites. What kind of population growth do you expect to see in this population?

CHECK YOUR ANSWER

Because only a limited number of nesting sites are available, the population's growth will slow as the population size increases. The population will grow logistically.

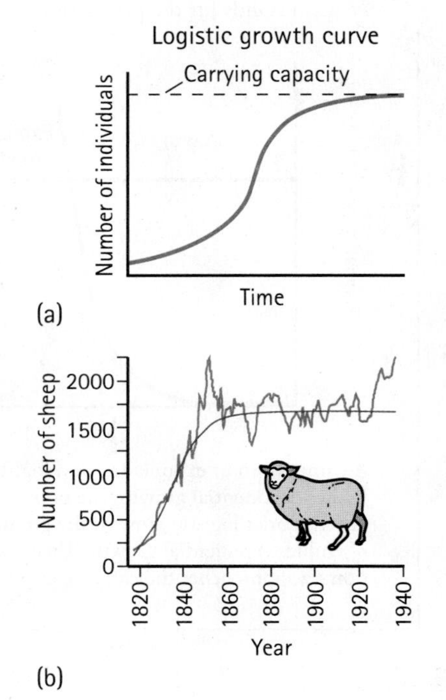

FIGURE 21.4
(a) In logistic growth, population growth slows as the population approaches the carrying capacity of its habitat. (d) This graph shows logistic growth in the Tasmanian sheep population.

MATH CONNECTION

Exponential and Logistic Population Growth

Under exponential growth, a population grows at a fixed rate per amount of time. One way to describe exponential growth is with the equation

$$\text{Growth rate} = rN$$

In this equation, the growth rate is the number of new individuals added during a specific amount of time, N stands for the population size, and r stands for the population's rate of increase. r measures how many offspring, on average, an individual contributes to the population during a specific amount of time. As you can see from the equation, the growth rate is higher if r is high. The growth rate is also higher if N is greater—that is, if the population size is large. This is why a population that is growing exponentially increases more and more quickly over time.

Now let's consider logistic growth. Under logistic growth, population growth slows as the population size approaches the carrying capacity of the habitat. The equation for logistic growth is

$$\text{Growth rate} = rN\frac{(K - N)}{K}$$

In this equation, r again stands for the rate of increase and N again stands for the population size. The new variable K stands for the carrying capacity, the maximum population size the habitat can support. Comparing the equations for exponential growth and logistic growth, we see that the only difference is that logistic growth has an extra $(K - N)/K$ term. What is the effect of this term? When the population size is very small compared to the carrying capacity (that is, when N is much smaller than K), $(K - N)/K$ is close to 1. This means that when the population size is small, logistic growth is similar to exponential growth. You can see this in the figure. However, as N gets larger (as N gets closer and closer to the carrying capacity K), the term $(K - N)/K$ gets smaller and smaller. This means that as the population size increases, the growth rate decreases. When $N = K$ (when the population size is equal to the carrying capacity), $K - N = 0$ and the growth rate is 0. The population stops growing.

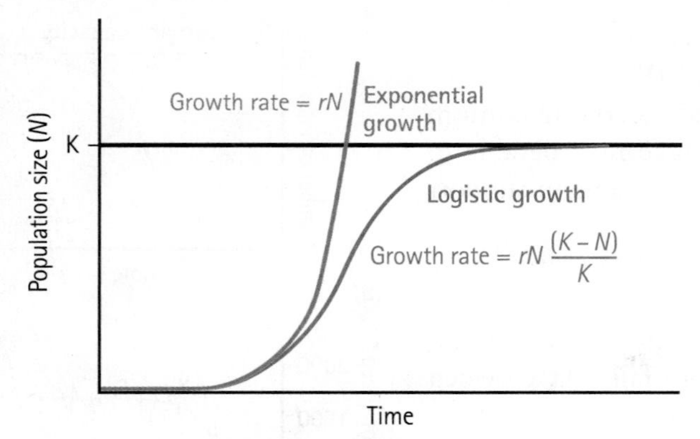

A comparison of exponential and logistic population growth. Under exponential growth, the population size grows faster and faster. Under logistic growth, the population growth initially resembles exponential growth. Then growth slows as the population size approaches the carrying capacity of the habitat.

Problem Let's compare growth rates under exponential and logistic growth. (a) Suppose a population is growing exponentially with rate of increase $r = 2$. What is the growth rate when population size $N = 1, 10, 20, 30, 40,$ and 50? (b) Now suppose a population is growing logistically with rate of increase $r = 2$ in a habitat where carrying capacity $K = 50$. What is the growth rate when population size $N = 1, 10, 20, 30, 40,$ and 50? (c) For both exponential and logistic growth, describe how the growth rate changes as the population size increases.

Solution For exponential growth, growth rate $= rN$. For logistic growth, growth rate $= rN(K - N)/K$.

Population size	(a) Exponential growth	(b) Logistic growth
1	2	1.96
10	20	16
20	40	24
30	60	24
40	80	16
50	100	0

(c) Under exponential growth, the growth rate increases as the population size increases. Under logistic growth, the growth rate starts out low, increases as the population size increases, and then decreases as the population size approaches the carrying capacity. When the population size equals the carrying capacity, the growth rate equals zero.

Life History

How a population grows depends in part on its *life history*—that is, on its schedule of survival and reproduction. A *survivorship curve* shows the proportion of individuals in a population that survive to a given age. Three different survivorship curves are shown in Figure 21.5.

MasteringPhysics®
TUTORIAL: Investigating Survivorship Curves

Type I organisms have low death rates early in life, with most individuals surviving until fairly late in life. Elephants have Type I survivorship—most elephant calves survive to adulthood. Type I organisms typically have large bodies and reach sexual maturity late. They produce a small number of offspring and devote significant resources to each offspring. For example, a female elephant takes about 10 years to reach sexual maturity, gives birth to a calf only every 5 years or so, and provides significant parental care.

Type III organisms have high death rates early in life, with few individuals surviving until late in life. Razor clams have Type III survivorship—most razor clams die early in life, and few survive to adulthood. Type III organisms typically reach sexual maturity early, produce large numbers of offspring, and devote few resources to each. For example, a female razor clam may reach sexual maturity in the first year of life and then produce 100 million eggs, releasing them into the water with no further attention or care. So, whereas elephants produce a small number of "expensive" offspring (expensive in the sense that the parents devote a lot of resources to each offspring), razor clams make many "inexpensive" offspring.

Type II organisms fall between Type I and Type III. Type II organisms experience a steady death rate throughout life—individuals are as likely to die early in life as late in life. Many songbirds show Type II survivorship.

How is life history related to population growth? The production of many "inexpensive" offspring is associated with exponential growth. In the unstable environments of exponential growth, life and death are often chance events. Producing many offspring is adaptive because it becomes more likely that at least some of the offspring will survive if conditions are poor. Also, when conditions are good, there will be many offspring to take advantage of plentiful resources. On the other hand, producing a few "expensive" offspring is associated with logistic population growth. In the stable environments of logistic growth, many populations are at or near carrying capacity. Consequently, an offspring that receives a lot of parental investment is more likely to be able to compete with other members of the population and successfully survive and reproduce. Table 21.1 summarizes some typical characteristics of Type I and Type III populations.

FIGURE 21.5
Three different survivorship curves. Type I organisms have low death rates early in life—most individuals live to an advanced age. Type II organisms have steady mortality throughout life. Type III organisms have high death rates early in life—few survive to adulthood.

TABLE 21.1 TYPE I VERSUS TYPE III POPULATIONS

Type I Populations	Type III Populations
Large bodies	Small bodies
Reach sexual maturity late	Reach sexual maturity early
Few "expensive" offspring	Many "inexpensive" offspring
Long life expectancy	Short life expectancy
Live in stable environments	Live in unstable environments
Logistic population growth	Exponential population growth

CHECK YOURSELF

What kind of life history characterizes humans?

CHECK YOUR ANSWER

Humans have a Type I survivorship curve. Most humans survive until late in life. Humans also have most of the other characteristics of Type I populations. For example, we have large bodies, reach sexual maturity late, and produce a small number of "expensive" offspring.

LEARNING OBJECTIVE

Describe how the global human population has grown over time, and how human population size is expected to change in the future.

21.3 Human Population Growth

EXPLAIN THIS What does it mean that the world human population is in "ecological overshoot"?

According to some scientists, humans came within a hair's breadth of extinction about 74,000 years ago, when the explosion of the Sumatran volcano Toba darkened Earth's skies for 6 years. Following the explosion, the world human population may have dropped to as few as 2000 individuals!

The World Population

More than 7 billion people live on Earth today—the end result of thousands of years of growth (Figure 21.6a). Where is the human population headed? Will it crash as resources run out? Or, will population growth slow down and slide toward carrying capacity?

According to United Nations projections, the world human population is expected to increase throughout the 21st century and exceed 10 billion by 2100. However, growth is expected to slow considerably during the second half of the century (Figure 21.6b). These projections assume that fertility will continue to decline in high-fertility countries and that death rates will continue to decline everywhere.

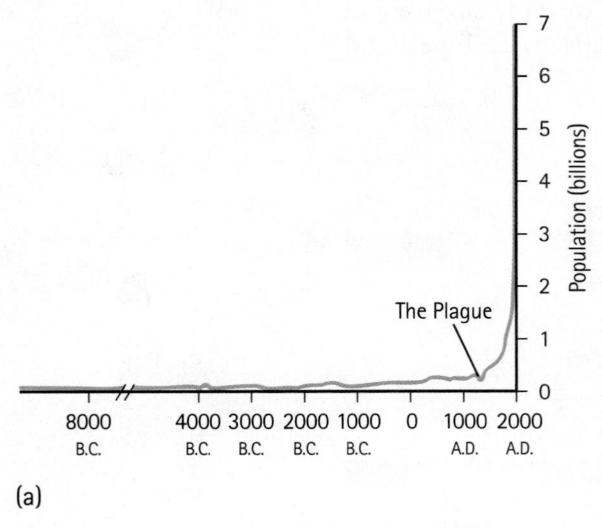

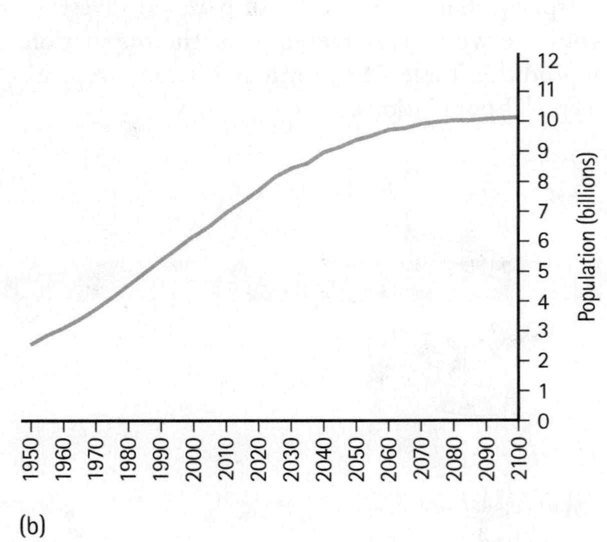

FIGURE 21.6

(a) The human population grew exponentially for thousands of years (with the exception of the occasional dip, such as during the bubonic plague in the 14th century). (b) Current projections suggest that population growth may slow considerably in the second half of the 21st century. However, the human population is expected to exceed 10 billion by the year 2100.

Source: United Nations, Department of Economic and Social Affairs, Population Division, *World Population Prospects: The 2010 Revision.*

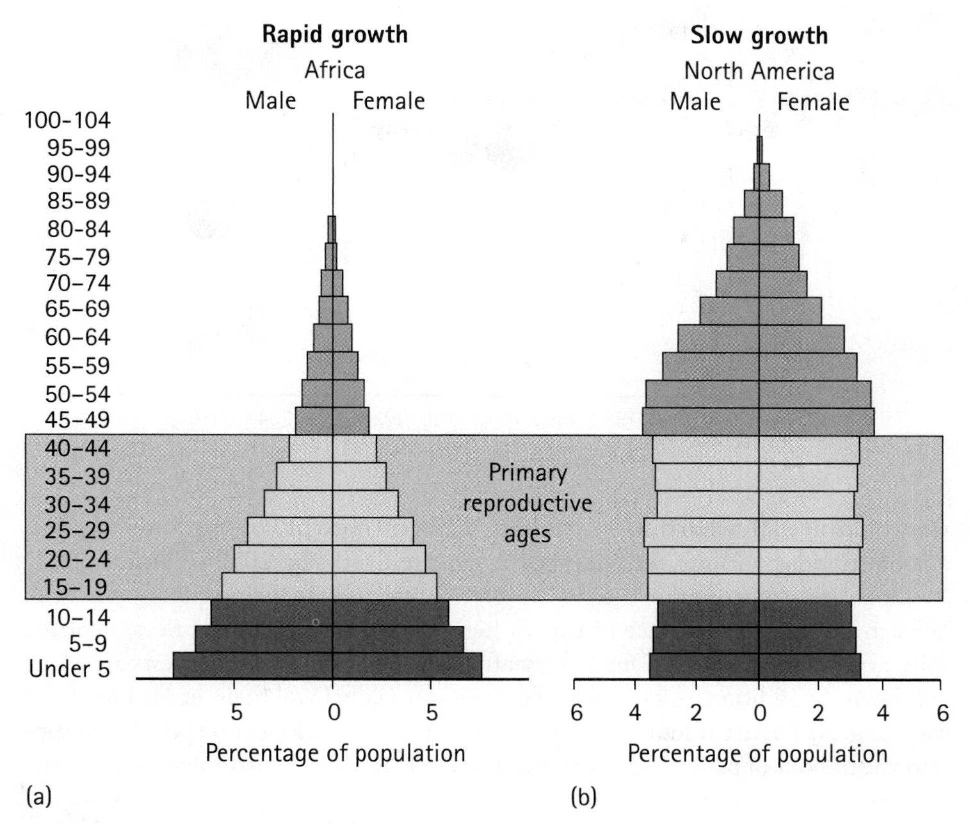

FIGURE 21.7
Age structures show the distribution of ages within a population. (a) Africa has a pyramid-shaped age structure. Most of the population is young, and the population is growing rapidly. (b) North America has a more even distribution of ages, and its population is growing slowly.
Source: United Nations, Department of Economic and Social Affairs, Population Division, *World Population Prospects: The 2010 Revision.*

We can learn how a population is growing from its *age structure*, the distribution of people's ages within the population. The age structures for Africa and North America are shown in Figure 21.7. A pyramid-shaped age structure, such as Africa's, is a sign of a rapidly growing population. Most of the population is young. North America has an age structure that is much more uniform—its population is growing slowly.

MasteringPhysics®
TUTORIAL: Analyzing Age Structure Diagrams

Many countries have experienced a *demographic transition*—a shift from high birth and death rates to low birth and death rates. The death rate typically decreases first, as a result of medical advances such as better care, improved nutrition, and immunization against childhood diseases. After some time, the birth rate also declines. During the period between the fall in the death rate and the fall in the birth rate, the combination of low death rate and high birth rate results in rapid population growth.

According to the U.S. Census Bureau, the population of the United States on May 31, 2012, was 313,647,717. One birth occurs in the country every 8 seconds. One death occurs every 13 seconds. Net migration into the country is one person every 46 seconds. This adds up to a net gain of one person every 14 seconds. (Check http://www.census.gov/population/www/popclockus.html for updates.)

Ecological Footprint

As the world human population grows, so does our demand for Earth's resources. An *ecological footprint* is a measure of how much land and water area a human population needs in order to produce the resources it consumes. The size of this footprint takes into account the population's use of water, soil, energy, food, and other resources. Measuring ecological footprints allows us to understand how resources are being consumed and how this compares with what Earth can provide. This is an important first step toward the goal of sustainable development.

For more than 30 years, the global human population has been in *ecological overshoot*, consuming more resources each year than Earth can provide. In fact,

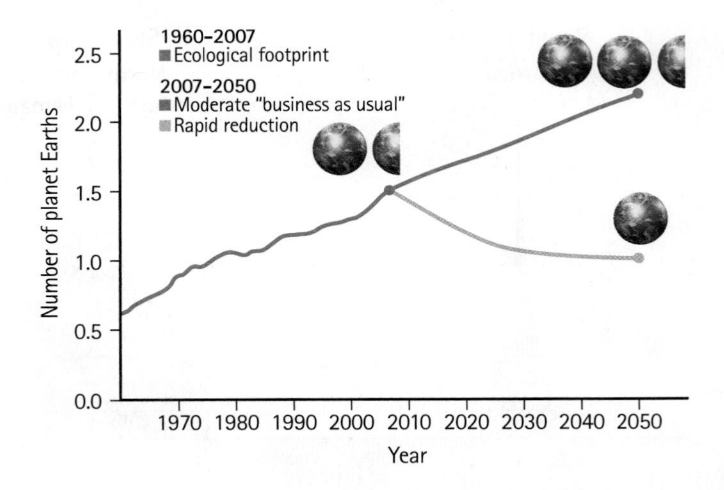

FIGURE 21.8
This graph shows how our global ecological footprint has increased in the last 50 years. We are now using the resources that it would take 1.5 Earths to provide. If present trends continue, we will soon be using the equivalent resources of 2 Earths.

we would need 1.5 Earths to sustain our current rate of consumption. And, if current trends continue, we will need 2.5 entire Earths by 2050 (Figure 21.8).

Global overconsumption means that Earth's resources are being quickly depleted. We can already see evidence of this in high food prices, high oil prices, depleted soils, and water shortages. One fishery after another is collapsing. Forests are being cut down. In addition, our overuse of energy, especially fossil fuels, has led to global warming. As Earth's resources begin to run out, it will no longer be possible to support the billions of people that live here. The likely result is catastrophic, deadly wars over resources as well as widespread starvation.

Of all the continents in the world, North America has the largest ecological footprint. In fact, if everyone in the world consumed as many resources as the average American, we would need the resources of *5.3 Earths*. It is vital for us to reduce consumption, increase efficiency, reduce energy use, and explore green energy alternatives.

LEARNING OBJECTIVE
Describe the ways species interact within communities.

21.4 Species Interactions

EXPLAIN THIS Can interspecific competition result in evolution through natural selection?

Lions eat zebras, eagles and storks fight over fish, and shrimp clean the mouths of eels. Within communities, species interact in different ways.

Food Chains and Food Webs

MasteringPhysics®
TUTORIAL: Interspecific Interactions
TUTORIAL: Food Webs

One important set of species interactions is who eats whom. These interactions can be shown in a *food chain* that contains multiple feeding levels (Figure 21.9).

At the bottom of a food chain are producers. A **producer** is an organism that makes organic molecules using inorganic molecules and energy. In almost all communities, the producers are photosynthesizing organisms such as plants.

A **consumer** obtains food by eating other organisms. A *primary consumer* eats producers. Herbivores, such as zebras or rabbits, are primary consumers. A *secondary consumer* eats primary consumers. Carnivores, such as lions and wolves, are secondary consumers. In some communities, there are consumers that eat at even higher levels in the food chain. The species at the top of a food chain is called the *top predator.*

Humans are not preyed on by other species. We are top predators.

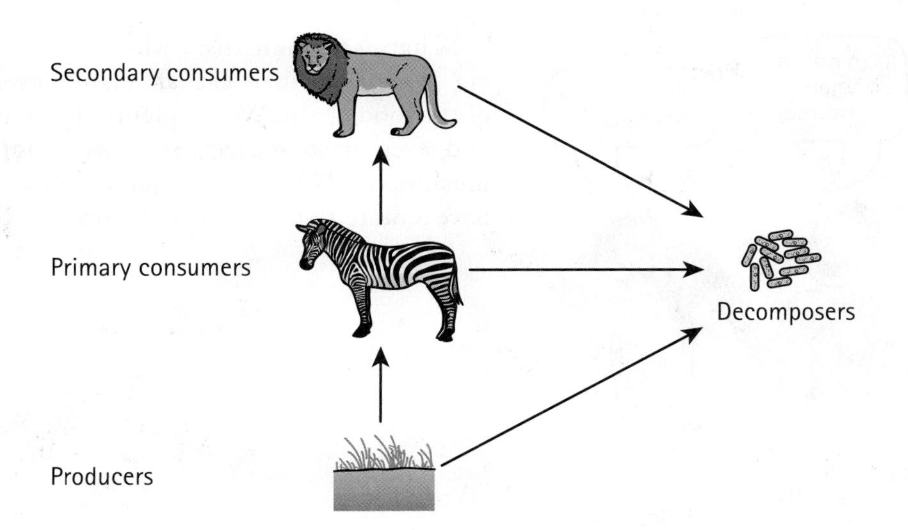

FIGURE 21.9
This food chain includes producers (grass), primary consumers (zebras), and secondary consumers (lions). Decomposers consume dead organisms at all levels of the food chain.

A **decomposer** obtains food by eating dead organic matter. Fungi and bacteria are important decomposers in many communities.

Most ecological communities are complicated enough that food "chains" are more accurately described as *food webs.* Figure 21.10 shows the food web in the Antarctic aquatic community.

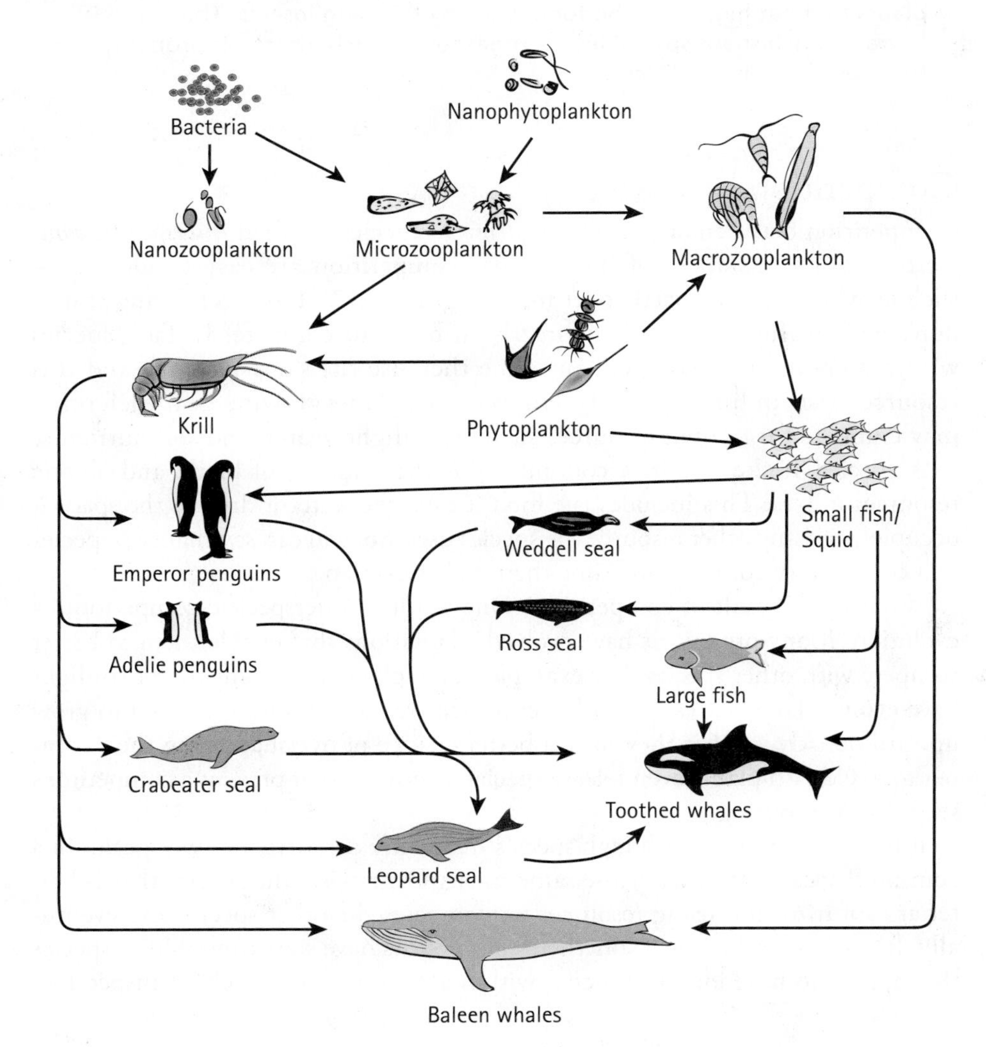

FIGURE 21.10
The Antarctic aquatic food web. Can you find the top predator in this community?

FIGURE 21.11
Most humans eat at many levels of the food chain, but some humans have a more restricted diet.

What are humans? Most humans are *omnivores* that eat at many levels of the food chain. We eat plants, animals, and even decomposers, such as certain mushrooms. Of course, some humans have a more restricted diet (Figure 21.11).

CHECK YOURSELF
Can an organism be a producer *and* a consumer?

CHECK YOUR ANSWER
Actually, yes. There are some parasitic plants that both photosynthesize and obtain nutrients from their hosts—other plants. This makes them both producers and primary consumers. There are even some predatory plants that eat higher up the food chain by trapping insects. The Venus flytrap, which snaps special leaves together to catch insects, is probably the most famous.

Competition

Competition between organisms of different species is called *interspecific competition*. Some instances of interspecific competition are easy to identify—such as when animals battle over food (Figure 21.12). However, living things don't have to fight each other directly in order to compete. In fact, species within a community compete any time they use the same resource and this resource exists in limited supply. For example, plants growing near each other may compete for limited resources such as sunlight, water, and soil nutrients.

A species' **niche** within a community is the total set of biotic and abiotic resources it uses. This includes the food it eats, the water it drinks, the space it occupies, and any other resource the species uses. So, you can say that two species in a community compete any time their niches overlap.

What is the result of competition? One result of interspecific competition is evolution. Many organisms have evolved adaptations that enable them to better compete with other species. For example, some plants that compete for sunlight have evolved large leaves with a large surface area. Others have evolved to grow upward quickly, so that they have a better chance of overshadowing their competitors. Certain plants even release special chemicals that prevent a competitor's seeds from sprouting.

It is interesting that, although species frequently compete, no two species in a community can have *exactly* the same niche. Otherwise, the species that is better at capturing and using resources outcompetes the other species and eventually drives it to extinction. Nonetheless, ecologists have sometimes found species that appear to have identical niches within a community. On closer inspection,

SCIENCE AND SOCIETY

Biodiversity, Nature's Insurance Policy

Biodiversity describes the number and variety of species found in a community. Are communities with greater biodiversity more stable? There are good reasons to expect this. A community that has only one producer is in big trouble if that producer has a bad year. A community that has three or four producers is much less likely to have them all do poorly at the same time.

Ecologist David Tilman tested the link between biodiversity and community stability. Tilman and his students marked off several hundred equal-sized plots of land in the Minnesota grasslands. For 11 years, they counted the number of plant species living on each plot. Tilman found that plots with more species were less affected by drought, showed less year-to-year variation in total plant cover, and were less vulnerable to invasion by new species. All of these findings suggest that greater biodiversity results in greater stability. Plots with more species also showed greater productivity—that is, they had a greater total aboveground mass of all plants combined. Both stability and productivity are considered good indicators of how well an ecosystem is functioning.

Why would a diverse community be more stable and productive? In a community with greater biodiversity, multiple species share a given role. Because different species respond differently to changing environmental conditions, one of these species may do well when another species is doing poorly. A diverse community might be more productive because different species make use of different resources. This means that available resources are likely to be used more completely in a diverse community. The more complete use of resources could also explain why new species have a harder time invading a diverse community.

however, it always turns out that the niches are not exactly identical. One of the most famous instances of this occurred in a group of birds now known as MacArthur's warblers. Ecologist Robert MacArthur studied five warbler species living in the coniferous forests of the Northeast United States. All five species lived in the same trees, and all five species ate insects. Did they share the same niche? By watching the warblers closely, MacArthur discovered that there were

FIGURE 21.12
An eagle and a stork fight over food.

FIGURE 21.13
The five species of MacArthur's warblers initially appeared to occupy the same niche—they all live in the same coniferous forests, and they all eat insects. However, closer inspection revealed that their use of resources differs in significant ways. For example, each species occupies a different part of the tree, as shown here (shaded in green). Each species also has a different way of hunting for insects.

crucial differences in how the species used resources: Each species of warbler occupied a different part of the tree, and each had a different way of hunting for insects (Figure 21.13). How did MacArthur's warblers come to have such similar, yet non-identical, niches? Ecologists believe that the niche differences were produced by evolution through natural selection. Individuals of each warbler species were more successful at surviving and reproducing if they competed less with individuals of other species. Over time, this caused the species' niches to diverge.

UNIFYING CONCEPT
- ***The Scientific Method***
Section 1.3

Symbiosis

In *symbiosis*, individuals of two species live in close association with one another. There are three kinds of symbiosis: parasitism, commensalism, and mutualism.

Parasitism describes a relationship that benefits one member of the interaction and harms the other. Fleas, ticks, and tapeworms are parasites that obtain nutrients from their hosts (Figure 21.14a). This relationship benefits the parasite,

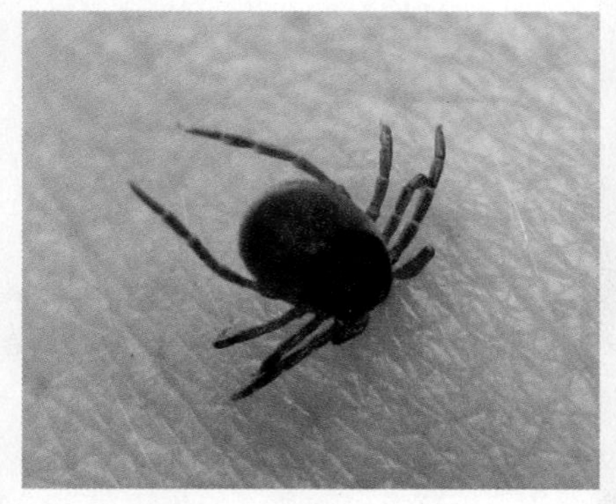

(a)

(b)

(c)

FIGURE 21.14
The three types of symbiosis are parasitism, commensalism, and mutualism. (a) Parasitism: A tick feeds on human blood. (b) Commensalism: A remora hitches a ride on a shark. (c) Mutualism: A cleaner shrimp removes parasites from the mouth of an eel.

SCIENCE AND SOCIETY

Invasive Species

When the brown tree snake arrived in Guam from its native New Guinea, it found a snake's paradise—plentiful food in the form of forest birds and their eggs and not a single natural enemy in sight. The snake quickly spread throughout the small island and now is found in startling numbers; in favorable habitats, there can be as many as 5000 individuals per square kilometer. As the brown tree snake thrives, however, it devastates Guam's unique bird fauna. A dozen species are now extinct, and many others are endangered.

An *invasive species* is a species that has moved from its native habitat to a new area, where it proceeds to do a lot of ecological damage. Although many species were purposely introduced in the past (such as the Australian rabbits discussed in Chapter 17), the introduction of nonnative species today is usually accidental, occurring when organisms hitch a ride on a ship or airplane. Many introduced species are never able to survive and become established in their new home. Others do become established, but have only a limited impact in their new environment. Sometimes, however, an introduced species will find that its new home is extremely well suited to its survival and reproduction. As the population explodes, the introduced species harms native species by preying on them or competing with them for food. This is when an introduced species becomes an invasive species.

A shopping cart left in zebra mussel–infested waters for a few months is completely covered with zebra mussels.

Invasive species are responsible for the decline of countless native species worldwide. More than a third of the species listed as endangered under the Endangered Species Act are threatened wholly or partly because of an invasive species. Some invasive species also have significant economic impact. The zebra mussel is an invasive species that has spread across the eastern United States since its accidental introduction in 1988. Zebra mussels clog water pipes at power plants and water treatment facilities, causing billions of dollars of damage annually. They also threaten native freshwater bivalves by competing with them for food. Most of these native species are now listed as endangered.

MasteringPhysics®
TUTORIAL: Fire Ants as an Invasive Species

which gets nutrients and a home, but harms the host. Pathogens such as bacteria or viruses are parasites as well.

Commensalism is a form of symbiosis that benefits one species while having no effect on the other. The remora is a small fish that attaches itself to a shark using a suction cup (Figure 12.14b). The remora feeds on small scraps of food left over when the shark eats. It also gets protection from its host. (Would you bother a fish hanging out under a shark?) The remora neither benefits nor harms the shark.

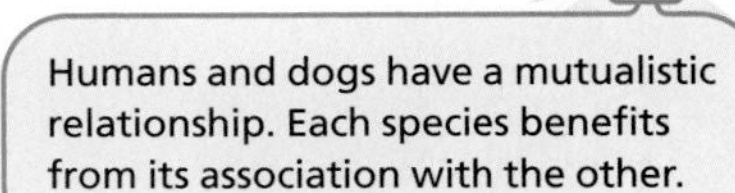

Humans and dogs have a mutualistic relationship. Each species benefits from its association with the other.

Mutualism is a form of symbiosis that benefits both species involved. Cleaner shrimp and eels have a mutualistic relationship. The eel opens its mouth, and a shrimp enters and eats the parasites there (Figure 21.14c). Both species benefit—the shrimp obtains food, and the eel is freed of its parasites. Sometimes, eels even "wait in line" for a shrimp's services! Another example of mutualism is mycorrhizae, the close association between fungi and plant roots. The fungi help the plant absorb water and minerals from the soil, and the plant provides nutrients and shelter for the fungi.

A species that lives in close association with another species evolves in response to that partner. Many parasites, for example, have evolved special adaptations for attaching to their hosts. Of course, hosts also evolve in response to parasites. As we saw, eels have evolved an amazing relationship with cleaner shrimp in order to get rid of their parasites.

CHECK YOURSELF

1. **This reed warbler is feeding a large cuckoo chick. Cuckoos lay their eggs in the nests of other species, and the parents then raise the cuckoo chick. What type of symbiotic relationship do these two species have?**
2. **In some places, mongooses and hornbills (a type of bird) eat together. Both look out for predators while they eat, and each can warn the other of danger. Hornbills are better at detecting predatory birds overhead. Mongooses, with their keen sense of smell, are better at detecting predators on land. What type of symbiosis do the mongoose and hornbill have?**

CHECK YOUR ANSWERS

1. This is an example of parasitism. The cuckoo is a parasite that benefits from its relationship with the reed warbler. The reed warbler is harmed.
2. The mongoose and hornbill have a mutualistic relationship—both species benefit from the interaction.

LEARNING OBJECTIVE
Describe the major types of terrestrial and aquatic ecosystems.

UNIFYING CONCEPT
- ***The Ecosystem***
  Section 21.1

MasteringPhysics®
TUTORIAL: Terrestrial Biomes

21.5 Kinds of Ecosystems

EXPLAIN THIS Which biome exhibits the greatest biodiversity?

Earth has many different kinds of ecosystems, and each is occupied by characteristic suites of organisms. We can divide Earth's ecosystems into two broad categories: terrestrial and aquatic.

Terrestrial Biomes

The land area of Earth is divided into eight major types of ecosystems, known as **biomes** (Figure 21.15). Each biome is occupied by specific types of biological communities and, particularly, by specific types of plant life. The same biome is often found in completely different parts of the world. For example, there are tropical forests—with their tall trees, dense vegetation, and astounding biodiversity—in the Amazon region of South America, in Southeast Asia, and in eastern and western Africa.

The type of biome found in a certain habitat depends mainly on climate, including factors such as temperature, rainfall, and the severity of seasonal variation. As a result, *latitude*—distance from the equator—and *altitude*—height above sea level—are major influences on the distribution of biomes on Earth. For example, sunlight strikes Earth most directly at the equator, warming the air there and causing it to rise. When the rising air cools, it releases a great deal of moisture as rain. This is why tropical forests, also known as rainforests, are found near the equator. The much dryer air then descends at around 30 degrees north and south latitude, creating many of Earth's largest deserts. Let's now survey Earth's biomes.

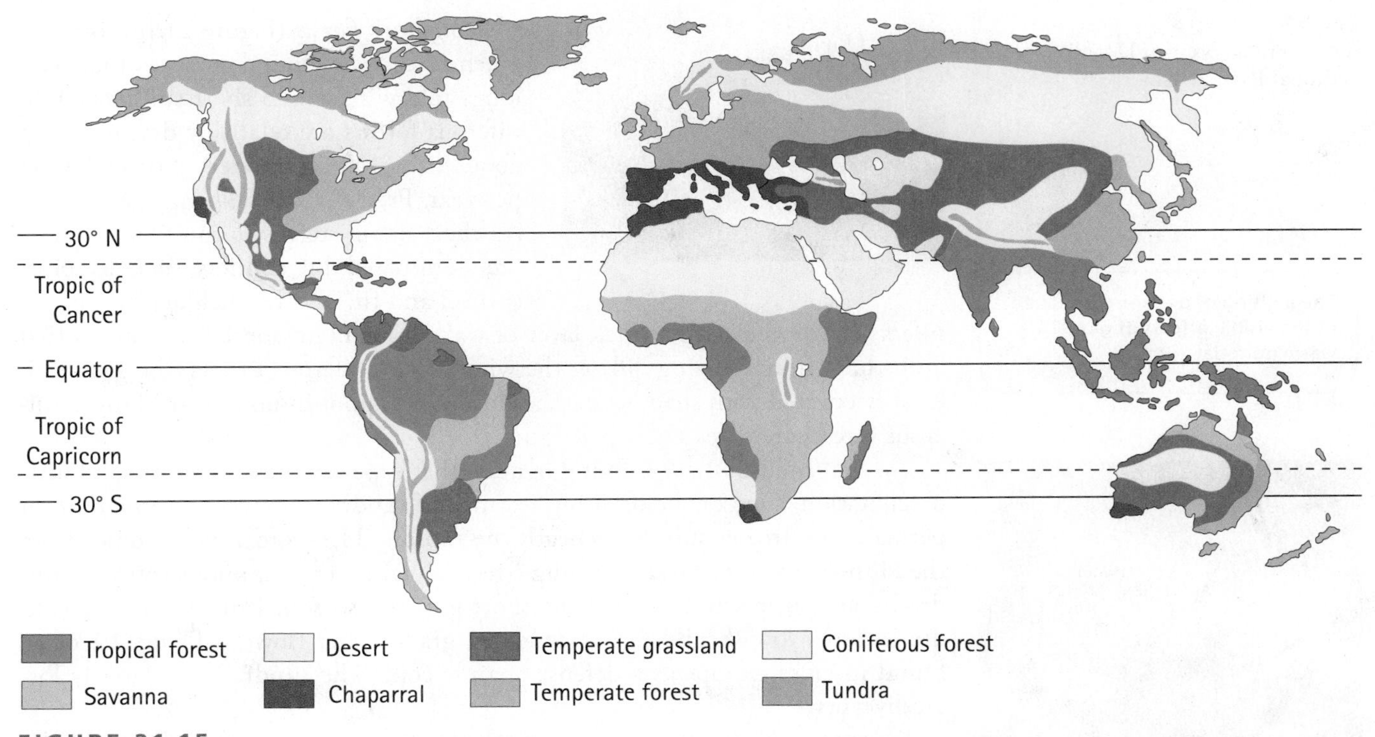

FIGURE 21.15
Earth's terrestrial biomes. Each biome is associated with a specific climate and with specific types of plant life.

Tropical forests (Figure 21.16), or rainforests, are found close to the equator. Temperatures are warm and fairly constant throughout the year. Tropical forests receive between 200 and 400 centimeters of rain a year, distributed between a wet season and a dry season. Tropical forests are famous for their biodiversity—more species are found in this biome than in all other biomes combined. A single square kilometer of tropical forest may contain as many as 100 different species of trees. The tallest trees form a dense canopy, shading the ground. Tropical forests have little leaf litter because all organic material is quickly decomposed. Most of the nutrients present in tropical forests are found in living organisms, and the soil tends to be poor. Tropical forests are rapidly being destroyed worldwide for timber and agriculture. Unfortunately, the poor quality of the soil means that forest that has been cleared for farmland can support crops for only a few years.

FIGURE 21.16
This tropical forest in Thailand shows the density and diversity of plant life found in all tropical forests.

Temperate forests (Figure 21.17) are found in areas that have four distinct seasons, including a warm growing season and a cold winter. Temperate forests receive between 75 and 150 centimeters of rainfall per year. There are usually three or four different species of trees per square kilometer. Some well-known temperate forest trees are elm, oak, beech, and maple. The leaves of temperate forest trees are unable to survive the freezing temperatures of winter. Consequently, these trees are *deciduous*—they drop their leaves in the autumn. Trees recover nutrients from their leaves before they are shed. The soil of temperate forests is fertile, and these forests make good farmland. In fact, many temperate forests have been cut down to make way for agriculture.

FIGURE 21.17
In this temperate forest, the season is autumn, and the trees are about to shed their leaves.

FIGURE 21.18
A coniferous forest in Denali National Park, Alaska.

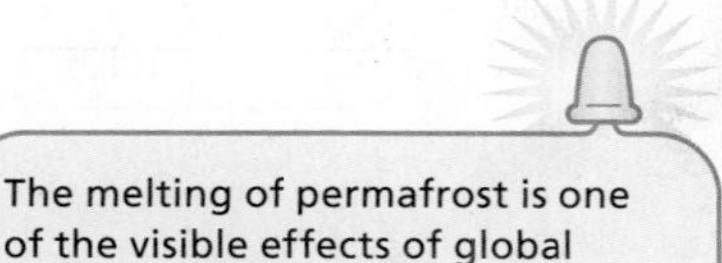

The melting of permafrost is one of the visible effects of global warming today.

Coniferous forests (Figure 21.18), or evergreen forests, are found in areas that have long, cold winters and short summers. Coniferous forests are relatively dry, receiving about 40 to 100 centimeters of precipitation per year. Precipitation falls mainly as snow. As their name suggests, coniferous forests are dominated by conifers such as pine, spruce, and fir. The needlelike leaves of conifers are covered with a thick layer of wax and contain special substances that keep them from freezing during the winter. In coniferous forests, the ground is usually covered with shed needles, and the soil is poor in nutrients. Many coniferous forests are threatened by logging.

FIGURE 21.19
A caribou in the Alaskan tundra.

Tundra (Figure 21.19) is found in areas that experience extreme cold and little precipitation. One of the defining features of tundra is a layer of *permafrost*, or permanently frozen subsoil, beneath the topsoil. The word *tundra* comes from the Finnish word *tunturia*, meaning "treeless plain." Trees cannot survive in tundra because of permafrost and the short growing season. Plant life in the tundra includes low shrubs, lichens, mosses, grasses, and flowers. Plants are often found in close clumps as a defense against cold. The tundra has relatively low biodiversity.

Savannas (Figure 21.20) are tropical grasslands with a warm climate and a long dry season. Savannas are covered with grass and occasional scattered trees. Plants that live in the savanna have long roots for coping with drought. Trees also have a thick bark that helps them survive the fires that occur during the dry season. Fires help maintain savannas by preventing tree growth. Living organisms also help maintain savanna habitats—for example, elephants eat and kill trees, and humans burn forests for agriculture. Without this activity, some savannas would grow into tropical forests. Savannas receive a moderate amount of rain, usually between 75 and 100 centimeters each year.

Temperate grasslands (Figure 21.21) are found in areas that have four distinct seasons, including a hot summer and a cold winter. Compared to savannas, temperate grasslands show more temperature variation and receive less rainfall—about 50 to 90 centimeters per year. Several factors prevent trees from taking over the habitat and turning it into temperate forest, including relatively low levels of precipitation, seasonal drought, fire, and grazing. The soil of temperate grasslands is extremely fertile, making for ideal farmland.

Chaparral (Figure 21.22) is found in places that have mild, rainy winters and hot, dry summers characterized by drought and fire. Chaparral habitats are

FIGURE 21.20
A savanna in Serengeti National Park, Tanzania.

FIGURE 21.21
A temperate grassland in South Dakota.

FIGURE 21.22
Chaparral in Southern California.

dominated by small trees and shrubs, many of which have small waxy leaves that help retain moisture. Chaparral plants also have extensive root systems that help them survive the summer drought.

Deserts (Figure 21.23) are habitats that receive very little precipitation, usually less than 50 centimeters per year. Although deserts are usually associated with hot climates, cold deserts also exist. In cold deserts, precipitation may fall as snow. Antarctica, the coldest and driest of the continents, is a desert. Many of the world's largest hot deserts, including the Sahara and the Australian desert, are found in a band around 30 degrees north latitude and 30 degrees south latitude. Desert plants usually have special adaptations for living in dry conditions, such as extensive root systems and the ability to store water. Desert soils are usually abundant in nutrients. This rich soil, combined with a long growing season and plentiful sunshine, makes deserts productive farmland if water can be brought in through irrigation.

FIGURE 21.23
A desert in Baja California, Mexico.

Aquatic Biomes

Life originated in the oceans, and many species continue to make use of aquatic habitats in both fresh water and salt water.

MasteringPhysics®
TUTORIAL: Aquatic Biomes

Freshwater Habitats Freshwater habitats include the still waters of lakes and ponds and the flowing waters of rivers and streams. Lakes and ponds vary tremendously in size and biodiversity. Nonetheless, all can be divided into three major zones, as illustrated in Figure 21.24. Habitats close to the water surface and to shore are part of the *littoral zone*. The littoral zone is relatively warm because of its exposure to sunlight. Organisms that live in the littoral zone include photosynthetic plants and algae, insects, mollusks, crustaceans, fishes, amphibians, ducks, and turtles. The *limnetic zone* includes habitats that are close to the water surface but far from shore. The limnetic zone is occupied mainly by *plankton*, organisms that float in the water rather than swimming actively through it. Phytoplankton are photosynthesizing plankton—they are the main producers in lakes and ponds. Zooplankton are heterotrophic plankton. In some lakes and ponds, fish also live in the limnetic zone. The *profundal zone* includes

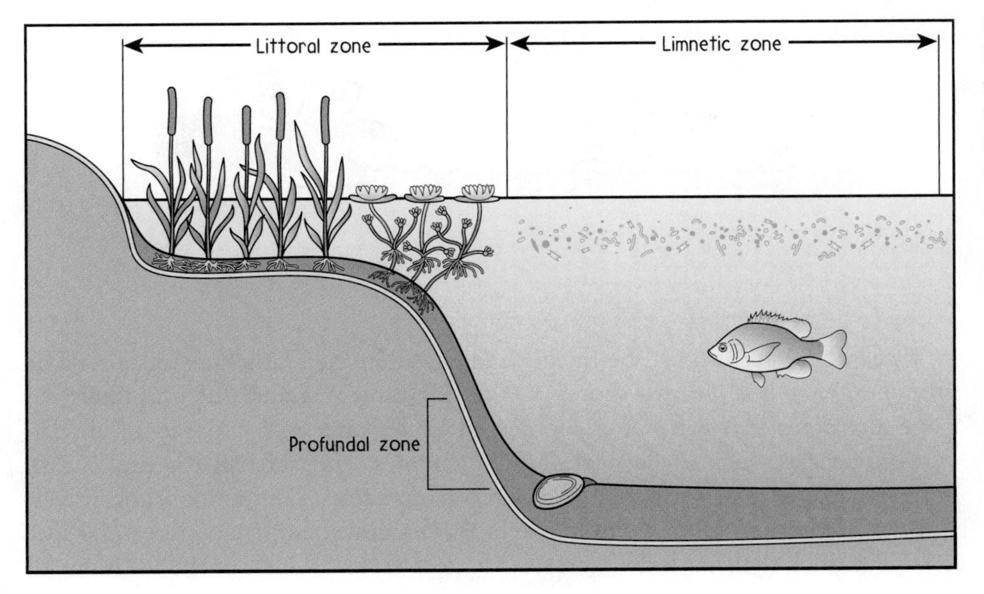

FIGURE 21.24
Lakes and ponds have three major zones: the littoral zone, limnetic zone, and profundal zone. Each is home to distinct kinds of living organisms.

FIGURE 21.25
An estuary in southern Florida.

the deep-water habitats in ponds and lakes. Most organisms in the profundal zone consume organic debris that drifts down from above.

Aquatic species that live in the flowing waters of rivers and streams usually have adaptations that keep them from being washed away. Some species have hooks or suckers for attaching to rocks. Others are strong swimmers. As in lakes and ponds, photosynthesizing plankton are the main producers in rivers and streams.

Estuaries are habitats where freshwater rivers flow into oceans. The plants found in estuaries, such as marsh grasses and mangroves (Figure 21.25), have adaptations for dealing with changing salinity conditions. Estuaries are also home to many fishes, invertebrates, and birds and are essential nursery habitats for many fishes and invertebrates.

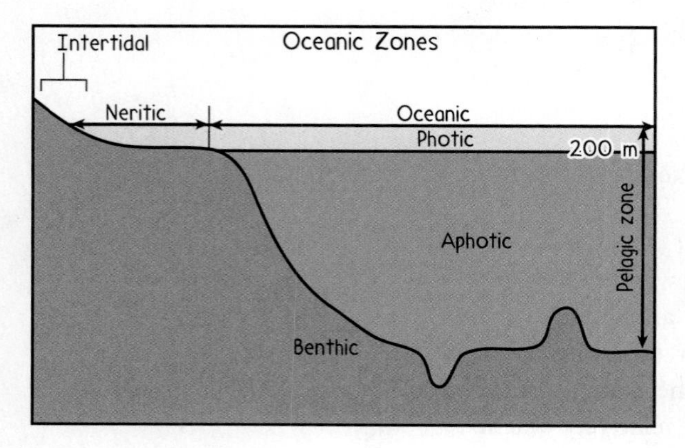

FIGURE 21.26
The ocean offers a variety of habitats to living organisms. Ocean habitats are divided into different zones based on depth and distance from shore.

Saltwater Habitats The oceans offer a wide range of habitats to living organisms, as shown in Figure 21.26. Many marine species are found in the *pelagic zone*—they live in the water column itself. The pelagic zone is further divided into two parts. The *photic zone* is close to the water surface and receives enough sunlight to power photosynthesis. The majority of pelagic species are found there. The deeper *aphotic zone* receives little sunlight and is much more limited in food availability and biodiversity.

Pelagic species use one of two modes of locomotion. Plankton float wherever water currents take them. Plankton include living things such as diatoms, dinoflagellates, and the larvae of animals such as clams, lobsters, and sea urchins. The phytoplankton, or photosynthetic plankton, are the main producers in oceanic food chains. *Nekton* swim actively through the water. Fish and sea turtles are examples of nekton. Humans aren't aquatic organisms, of course, but when we do get in the water, we can behave like plankton or nekton (Figure 21.27).

Most marine species live in the *benthic zone* on the ocean bottom. Benthic species may live on the surface of the ocean bottom, like lobsters, or burrow into the sediment, like clams and worms.

FIGURE 21.27
In the water, humans can behave like nekton or plankton.

FIGURE 21.28
Diverse organisms, including starfish, sea anemones, and sea urchins, occupy this tide pool.

Ocean habitats also can be categorized based on their closeness to shore. The *intertidal zone* is closest to shore. As the tide moves in and out, the intertidal zone alternates between being submerged underwater and exposed to air. Intertidal species, such as barnacles, sea anemones, and starfish (Figure 21.28), must be able to deal with exposure to air, temperature fluctuations, and waves. Special adaptations allow them to live in this environment. Many intertidal species have thick shells or hide in crevices to keep from drying out. In addition, all species can attach firmly to rocks or other surfaces so that they do not get washed up onto the beach. The *neritic zone* describes underwater marine habitats near the coasts. Coral reefs (see Chapter 18), the most diverse marine ecosystems on Earth, are found in the neritic zone. Habitats in the neritic zone contain high levels of nutrients, which wash into the water from land. Because of this, the neritic zone is richer in life than the deeper *oceanic zone* farther out.

CHECK YOURSELF

1. **Scallops live on shallow seafloors. They move by quickly opening and shutting their shells, shooting jets of water that propel them in spurts. Are scallops pelagic or benthic? Are they nekton or plankton?**
2. **The flounder has a very unusual feature—one of its eyes migrates from one side of the body to the other side during development. It ends up with two eyes on the same side! What type of aquatic habitat do you think the flounder occupies?**

CHECK YOUR ANSWERS

1. Scallops are benthic nekton.
2. Flounders are benthic fish that lie flat on the seafloor. One eye migrates from the "underside" of the body to the "top" side, where it is of more use.

Materials Cycling in Ecosystems

LEARNING OBJECTIVE
Describe how important substances such as water, carbon, and nitrogen cycle back and forth between living organisms and Earth.

EXPLAIN THIS How do living things return carbon to the atmosphere?

A water molecule in a cell on your cheek once bobbed on the Indian Ocean. A carbon atom on the skin over your knee once floated in the cold air above Antarctica. A nitrogen atom flowing through your veins once sat in the moist soil of a mountain in Peru. Parts of you have been all over the world, in places you have never seen with your own eyes! How is this possible?

Living organisms are made up of many substances, including water, carbon, nitrogen, phosphorus, sodium, calcium, sulfur, chlorine, and many others. All these substances move around Earth in a series of **biogeochemical cycles**, going back and forth between the tissues of living organisms and the abiotic world. The word *biogeochemical* emphasizes that substances cycle between living organisms (*bio*) and Earth (*geo*)—specifically, Earth's atmosphere, crust, and waters.

UNIFYING CONCEPT
- ***The Ecosystem***
  Section 21.1

As a molecule moves through its biogeochemical cycle, it passes through a series of *reservoirs*—that is, places where it stays. For example, a water molecule may sit in one reservoir, the ocean, until evaporation moves it to another reservoir, the atmosphere. It may then fall as precipitation into a third reservoir, a lake. The average amount of time a molecule spends in each reservoir—for example, the average amount of time a water molecule spends in the ocean before it moves to another reservoir—is known as its *residence time* in that reservoir. Although every substance used by living organisms has a biogeochemical cycle, we will focus on three of the most important substances: water, carbon, and nitrogen.

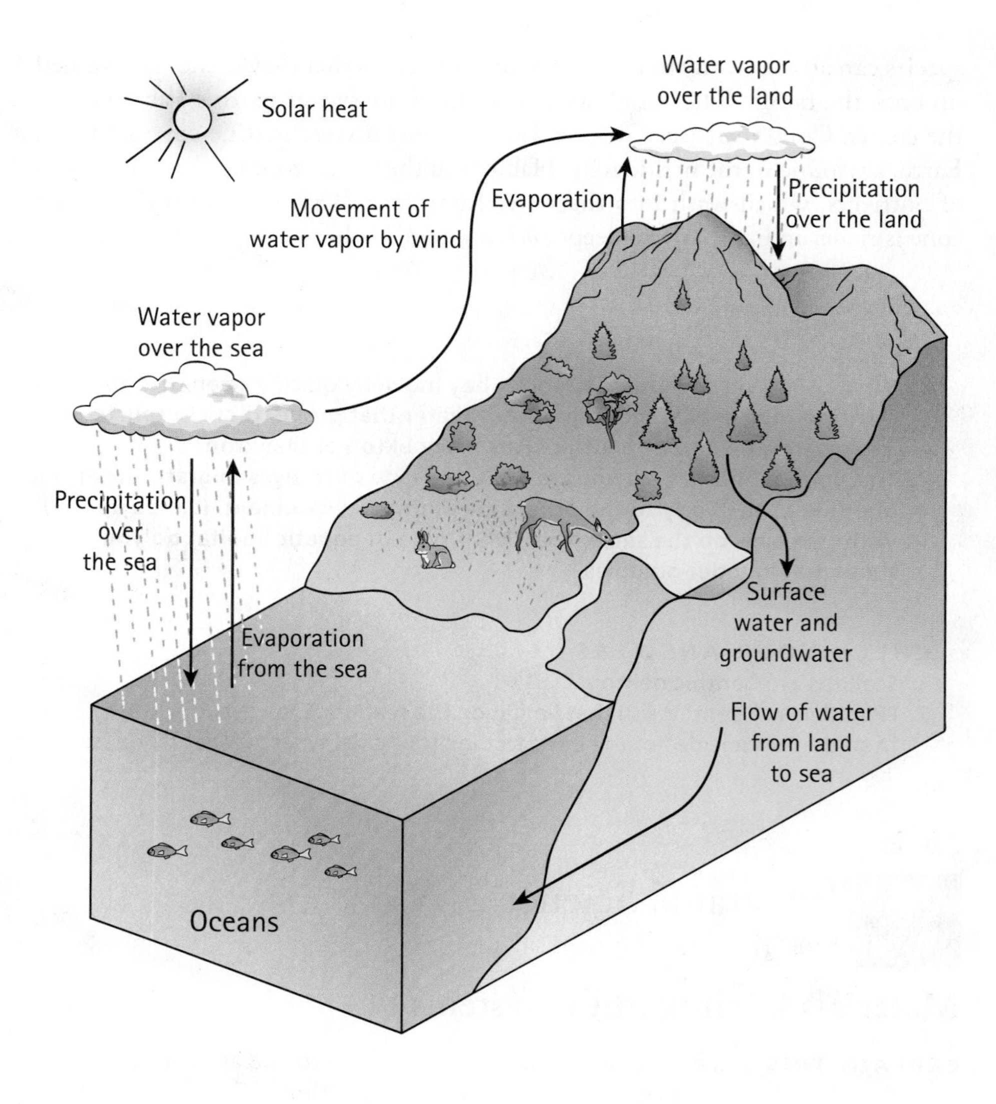

FIGURE 21.29
Water cycles through the oceans, atmosphere, and land. Some water enters the bodies of living organisms.

About 98% of the water on Earth is found in oceans, rivers, and lakes. The rest—only 2%—is found in glaciers, in the polar ice caps, in soil—and in living organisms!

Water

All living things need water. The global water cycle, shown in Figure 21.29, shows how water moves around Earth. Water evaporates from the oceans into the atmosphere, is moved around the atmosphere by winds, and then falls back to Earth as rain or snow. Water that falls on land flows back to the oceans through rivers, streams, and groundwater.

Water enters the biotic world when it is absorbed or swallowed by organisms. Some of this water may pass up the food chain. The rest is returned to the abiotic environment through processes such as respiration, perspiration, excretion, and elimination. Plants also lose water when it evaporates inside leaves and then escapes through the stomata to the atmosphere. Did you know that every time you exhale, you return water to the atmosphere? If you breathe onto a mirror or other piece of glass, you will create a foggy film of water vapor.

Carbon

MasteringPhysics®
TUTORIAL: Carbon Cycle

Carbon is an essential component of all organic molecules. The carbon cycle is shown in Figure 21.30. Most of the inorganic carbon on Earth exists as carbon dioxide and is found either in the atmosphere or dissolved in ocean waters.

Carbon enters the biotic world during photosynthesis, when plants and other producers use carbon dioxide to make the organic molecule glucose. This carbon becomes available to other organisms as it moves up the food chain. Carbon is returned to the abiotic world during cellular respiration, when organisms break down organic molecules (such as glucose) and release carbon dioxide.

An important part of Earth's carbon supply is found in fossil fuels such as coal and oil. Both coal and oil are formed over long time periods from the remains of dead organisms. (This is why they are called "fossil" fuels.) Human burning of fossil fuels has released so much carbon into the atmosphere that atmospheric carbon dioxide levels are now higher than they have been for 15 million years. Because atmospheric carbon dioxide traps heat on the planet, Earth has warmed noticeably as a result. This warming of Earth is called *global warming*.

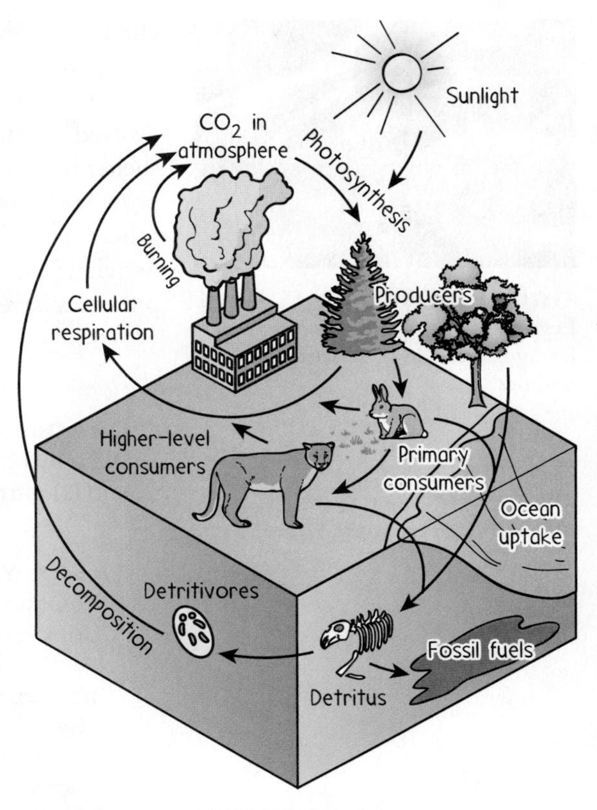

FIGURE 21.30
Carbon is found as carbon dioxide in the atmosphere and oceans. During photosynthesis, producers move carbon into the biotic world. Carbon then passes up the food chain. Carbon is returned to the abiotic world as carbon dioxide, a product of cellular respiration. Another important component of the carbon cycle is the burning of fossil fuels, which has released a huge amount of carbon previously trapped in the fuels.

Nitrogen

Nitrogen is an essential component of amino acids, DNA, and many other organic molecules. The nitrogen cycle is shown in Figure 21.31. Most of Earth's nitrogen exists as nitrogen gas (N_2) in the atmosphere. However, nitrogen gas is not a form of nitrogen that most living things can use. Bacteria convert nitrogen into a usable form. *Nitrogen-fixing bacteria* in soil convert nitrogen gas into ammonium (NH_4^+), and then *nitrifying bacteria* convert ammonium into nitrates (NO_3^-). Plants absorb nitrogen primarily in the form of nitrates, although they make some use of ammonium as well. Nitrogen then moves from plants up the food

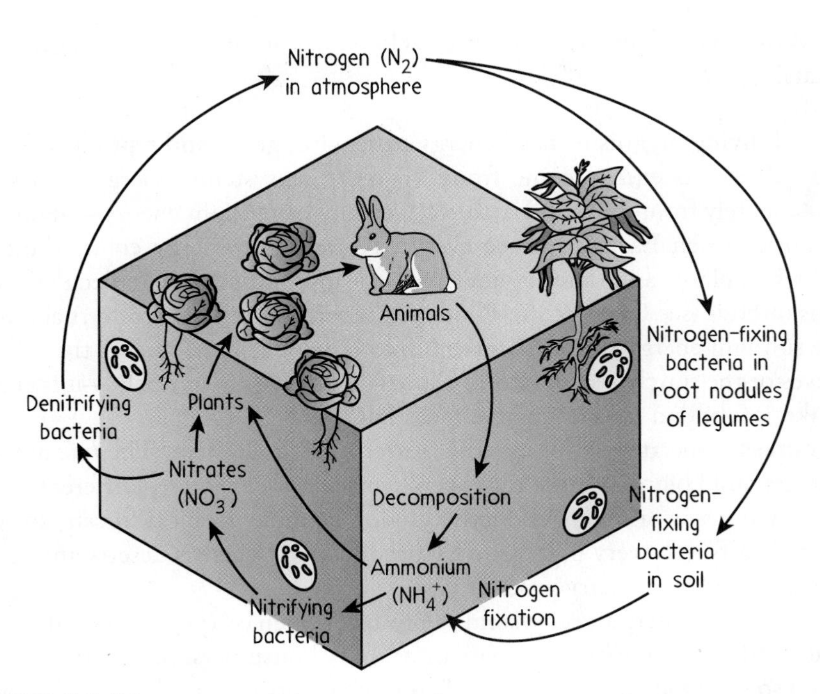

FIGURE 21.31
Nitrogen is found as nitrogen gas in the atmosphere. Bacteria convert atmospheric nitrogen into forms that can be used by living organisms.

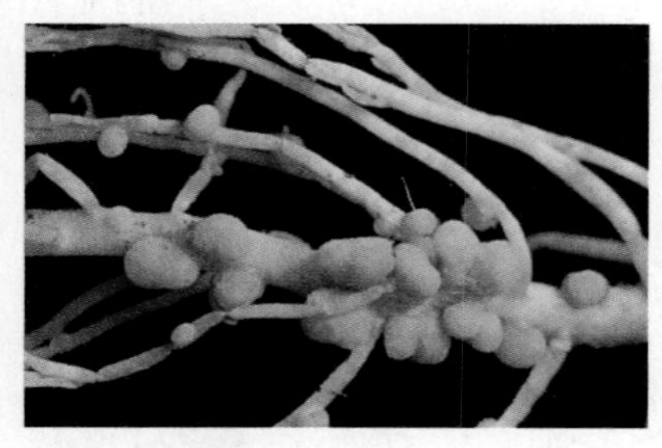

FIGURE 21.32
Legumes house symbiotic nitrogen-fixing bacteria in these special root nodules.

chain. Nitrogen returns to the abiotic environment when it is converted back into nitrogen gas by denitrifying bacteria.

Certain plants, including legumes such as peas, beans, clover, and alfalfa, have evolved a mutualistic relationship with nitrogen-fixing bacteria. These plants shelter the bacteria in special nodules in their roots (Figure 21.32), and the bacteria provide the plants with nitrogen. This is why farmers often grow legumes with other plants: The excess nitrogen enriches the soil and allows other crops to grow better. These days, however, not all plants rely on bacteria for nitrogen—industrial processes have been invented for making nitrogen-rich fertilizers.

MasteringPhysics®
TUTORIAL: Nitrogen Cycle
TUTORIAL: Greenhouse Effect

CHECK YOURSELF

What living things do you, as a human, rely on to obtain (1) water, (2) carbon, and (3) nitrogen?

CHECK YOUR ANSWERS

1. You don't have to rely on other living things to obtain water—you can drink it directly!
2. All the carbon in your body comes ultimately from plants and other photosynthesizers. This carbon comes to you through the food you eat.
3. The nitrogen in your body also comes through the food you eat. Nitrogen-fixing bacteria and nitrifying bacteria change nitrogen into a form that plants can use, and then you eat the plants (or the animals that eat the plants).

LEARNING OBJECTIVE
Describe how energy flows through an ecosystem.

21.6 Energy Flow in Ecosystems

EXPLAIN THIS Why are there more zebras than lions in African savanna habitats?

All living organisms need energy to survive, grow, and reproduce. Where does this energy come from? In most ecosystems, energy comes ultimately from the Sun. Earth receives a lot of sunlight energy—about 10^{19} kilocalories reach Earth's surface every day. Sunlight energy enters the biotic world when plants and other organisms use it to build organic molecules during photosynthesis (see Chapter 15). Photosynthesizing organisms convert about 1% of the sunlight energy that strikes them into organic matter, such as the cellulose in leaves or the starches in potatoes. Globally, this adds up quickly—it is enough to make 170 billion tons of organic material a year!

Biomass is the amount of organic matter in an ecosystem. The rate at which producers build biomass is the ecosystem's *primary productivity*. Different ecosystems vary in their primary productivity—for example, tropical forests, swamps, and coral reefs have very high primary productivity, whereas deserts and tundra have relatively low primary productivity.

UNIFYING CONCEPT
- ***The Ecosystem***
Section 21.1

Once energy enters an ecosystem as producer biomass (plant material), it goes up the food chain, creating biomass in primary consumers, secondary consumers, and so on. Does all the energy taken in by plants ultimately go into growing zebras, or lions? No. In all ecosystems, a huge amount of energy is lost as you move up the food chain. Only about 10% of the energy at one level of the food

MasteringPhysics®
TUTORIAL: Energy Flow and Chemical Cycling

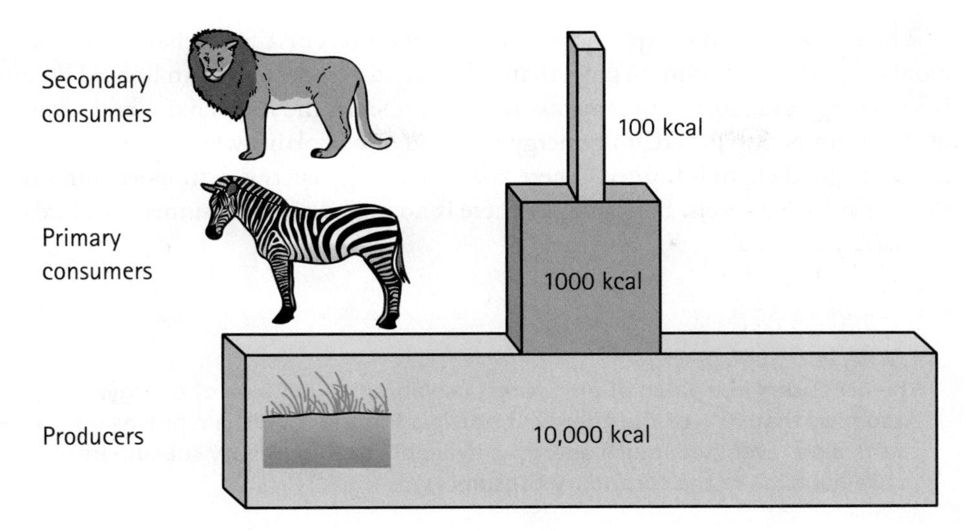

FIGURE 21.33
This energy pyramid shows that the amount of energy decreases as you move up the food chain.

chain moves up to the next level. For example, if the producers in one patch of African savanna use 10,000 kcal of energy, you would expect about 1000 kcal to be available to primary consumers (such as zebras), and 100 kcal to be available to secondary consumers (such as lions). This energy flow can be shown in an *energy pyramid*, such as the one in Figure 21.33. The shape of the pyramid shows the loss of energy as you go up the food chain.

MasteringPhysics®
TUTORIAL: Energy Pyramids

Why does only 10% of the energy at one level of the food chain move to the next level? What happens to the other 90%? First, some organisms at one level of the food chain are not eaten by organisms at the next level of the food chain. For example, some savanna plants are not eaten by zebras, and some zebras are not eaten by lions. If an organism is not eaten, then the energy in it cannot move up the food chain. Second, not all the energy an organism uses goes into building biomass—and only the energy that is converted into biomass can move up the food chain. For example, savanna plants use some of their energy to carry out cellular processes and other activities essential to life. Similarly, when a zebra eats, not all the energy in its food goes to building biomass. Some energy is lost in feces and some energy goes into maintenance. Feces contains organic matter that the zebra can't digest. Maintenance is the energy the zebra needs to live—the energy it takes to find food, run, breathe, heat its body, and so on. By the time feces and maintenance have taken their share of energy, only about 10% is left for growth and reproduction—for building new biomass (Figure 21.34).

UNIFYING CONCEPT

- ***The Law of Conservation of Energy***
  Section 4.10

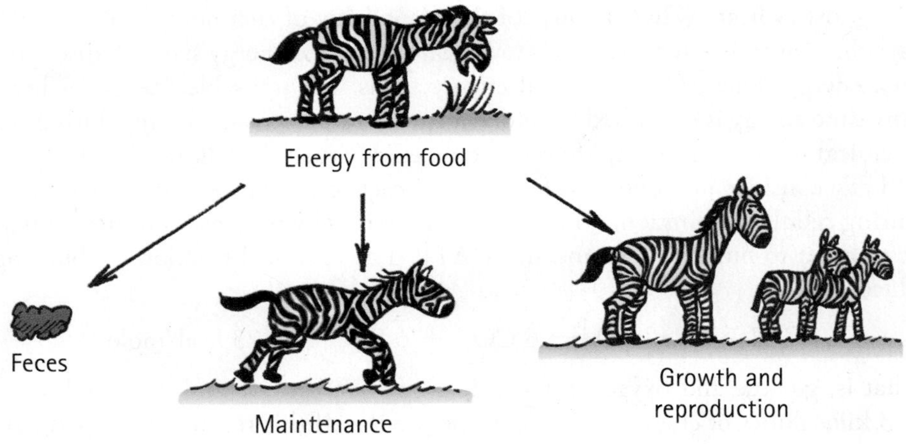

Energy from food
Feces
Maintenance
Growth and reproduction

FIGURE 21.34
The energy an organism takes in from food goes to three things: feces, maintenance, and growth and reproduction. Only growth and reproduction contribute to building biomass, and only energy that is converted into biomass can move up the food chain.

The energy pyramid explains why the higher levels of a food chain have less biomass—why there is more grass than zebras, and more zebras than lions: There is less energy available to organisms that eat higher in the food chain, and fewer of them can be supported. The energy pyramid also explains why the number of levels in a food chain is limited: There just isn't enough energy to support animals that eat at higher levels. For example, there is no higher-level consumer that feeds exclusively on lions.

Just as for other consumers, the energy in your food does not all go to building biomass. Despite all the Calories you eat, your weight doesn't change much from one week to the next. Where do these Calories go? The same place they go for other consumers—into feces and maintenance.

CHECK YOURSELF

The producers in a patch of pine forest use about 30,000 kcal of energy. Assuming that 10% of the energy at one level of the food chain moves up to the next level, how much energy is available to the primary consumers in the habitat? To the secondary consumers?

CHECK YOUR ANSWER

Energy available to primary consumers = 10%(30,000 kcal) = 3000 kcal.
Energy available to secondary consumers = 10%(3000 kcal) = 300 kcal.

LEARNING OBJECTIVE
Explain why the second law of thermodynamics is important to energy flow in ecosystems.

Integrated Science 21B
PHYSICS AND CHEMISTRY

Energy Leaks When Organisms Eat

EXPLAIN THIS How is the energy lost during cellular respiration related to the way mammals and birds maintain a warm, stable body temperature?

Why is an energy pyramid always a pyramid? That is, why does the amount of energy available always decrease as you move up a food chain? We have already seen two good reasons: First, some organisms at one level of a food chain are not eaten by organisms at the next level of the food chain. And, second, some energy from food is lost in feces or goes into maintenance.

A lot of the energy we call "maintenance" is energy lost to the environment as heat. This is because, as organisms carry out their activities, energy is constantly being lost as heat. Why? Because of the second law of thermodynamics, which says that natural systems tend to move from organized energy states to disorganized energy states. That is, useful energy turns into unusable energy, or heat. Any time energy is converted from one form into another—including during the chemical reactions in living things—some energy turns into heat.

UNIFYING CONCEPT
- *The Second Law of Thermodynamics* Section 6.5

Let's consider just one set of chemical reactions—the reactions that occur during cellular respiration. You may remember that living organisms use cellular respiration to burn glucose and make ATP. The chemical reaction for burning glucose is

$$C_6H_{12}O_6 + 6\ O_2 \rightarrow 6\ CO_2 + 6\ H_2O + 673\ \text{kcal/mole}$$

That is, glucose and oxygen react to form carbon dioxide and water, releasing 673 kilocalories of energy per mole in the process. If this reaction were perfectly efficient in organisms, the entire 673 kilocalories per mole released from burning glucose would be captured as ATP. Is it?

We know that about 38 molecules of ADP are converted into ATP as the result of burning a single glucose molecule. ATP then provides 7 kilocalories per mole when it is broken down into ADP and phosphate during cellular processes. But, $38 \times 7 = 266$, much less than 673. Clearly, a lot of energy is missing! What happened to it? It was lost to the environment as heat. In mammals and birds, heat lost this way contributes to the maintenance of stable, warm body temperatures. However, this heat is shed eventually to the environment as well. Once we consider the fact that *every* chemical reaction involves some energy loss to the environment, it becomes clear why so much energy leaks from one level in the food chain to the next (Figure 21.35). In fact, *all* the energy that Earth receives from the Sun is eventually lost as heat.

FIGURE 21.35
The second law of thermodynamics tells us that energy is lost to the environment as heat in every chemical reaction.

CHECK YOURSELF

1. **What ultimately happens to the energy that goes into the "maintenance" of a living organism?**
2. **Does energy move around Earth in cycles the way water, carbon, and nitrogen do?**

CHECK YOUR ANSWERS

1. It is ultimately lost to the environment as heat.
2. No, energy flows through ecosystems. Energy comes to Earth from the Sun. A fraction of this energy is captured by producers such as plants. Some of the energy captured by plants is used to build biomass, and some of this energy moves up the food chain. The rest is lost as heat. All living things lose energy to the environment as heat as a result of their activities. All of the energy Earth receives from the Sun is eventually lost to the environment as heat.

21.7 Change in an Ecosystem

LEARNING OBJECTIVE
Describe how a disturbance can affect an ecosystem.

EXPLAIN THIS Why are the earlier colonizers of a habitat displaced by later colonizers?

In 1883, a huge volcanic explosion tore through the tropical island of Krakatoa. Most of the island was destroyed. The fragment that remained was covered with ash, barren and lifeless. But within three years, a few living organisms, primarily photosynthetic bacteria and ferns, had colonized the island. By 1897, Krakatoa had become a savanna covered with tall grasses. Only a few species of animals were present—almost all were flying insects and birds. By 1919, scattered trees could be seen, and by 1931, a forest had developed. As more trees grew, animal diversity increased. Today, Krakatoa is once again tropical forest.

Krakatoa's history provides an example of ecological succession. **Ecological succession** describes how the community of species living in an ecosystem changes over time. There are two kinds of ecological succession: primary succession and secondary succession.

Primary succession occurs when bare land, devoid of soil, is colonized by successive waves of living organisms. For example, primary succession may begin

MasteringPhysics®
TUTORIAL: Primary Succession

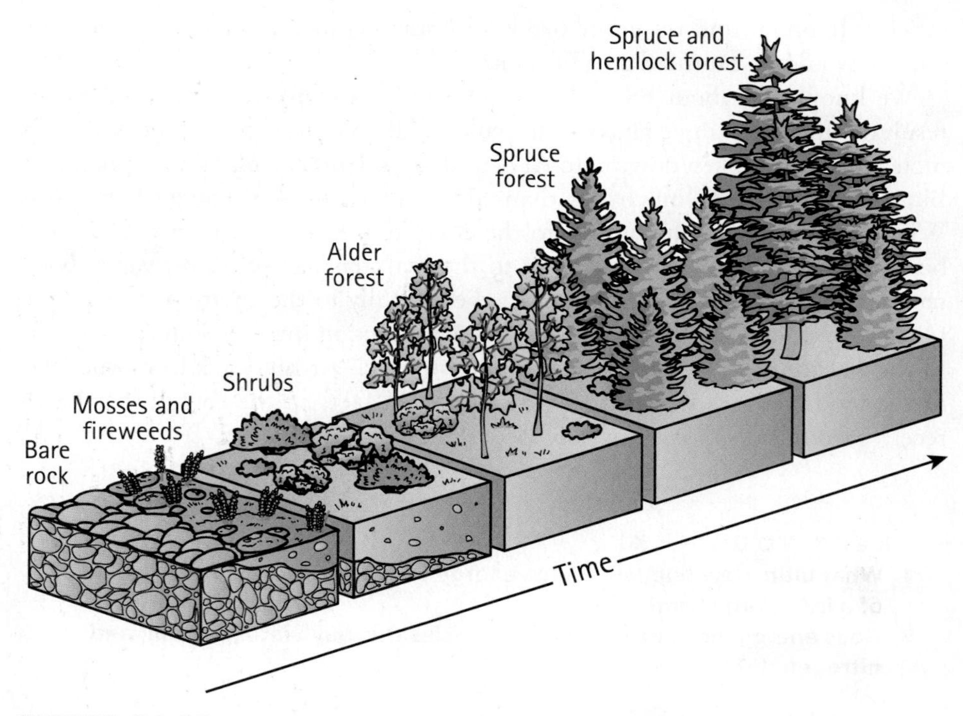

FIGURE 21.36
Primary succession after a glacier's retreat in Alaska: Bare rock was covered by mosses and fireweeds, then shrubs, an alder forest, a spruce forest, and finally a spruce and hemlock forest. The entire process took place over about 200 years.

when new land is formed by volcanic activity, or when a glacier's retreat reveals bare rock (Figure 21.36). The first species to colonize bare land are known as *pioneer species*. Pioneer species often include photosynthetic microorganisms as well as larger organisms such as lichens and mosses. Pioneer species must be able to survive with few nutrients and little existing organic matter. They also must be able to deal with direct sunlight and the variable temperatures that result from lack of cover. Pioneer species may be succeeded by grasses, shrubs, and finally trees. At each stage of succession, the activities of earlier waves of colonizers build up nutrients and organic matter, allowing later colonizers to thrive. Later colonizers eventually outcompete and displace earlier colonizers. Ecological succession ends with a *climax community* that is relatively stable. During the process of succession, the total biomass of the ecosystem typically increases, as does the number of species present.

Secondary succession occurs when a disturbance destroys existing life in a habitat but leaves the soil intact. Secondary succession may begin after a fire or when old farmland is abandoned. Because soil is already present, secondary succession proceeds more quickly than primary succession. As with primary succession, biomass and biodiversity typically increase as secondary succession progresses. For example, during secondary succession on abandoned farmland in the southeastern United States, the number of bird species increased as succession progressed (Figure 21.37).

All ecosystems experience change. Most ecosystems never suffer a cataclysm like Krakatoa's, but smaller disturbances, often affecting only a small area of habitat, are common. Fires, floods, and other natural events all damage habitats. Even the fall of a large tree can be a disturbance with significant consequences.

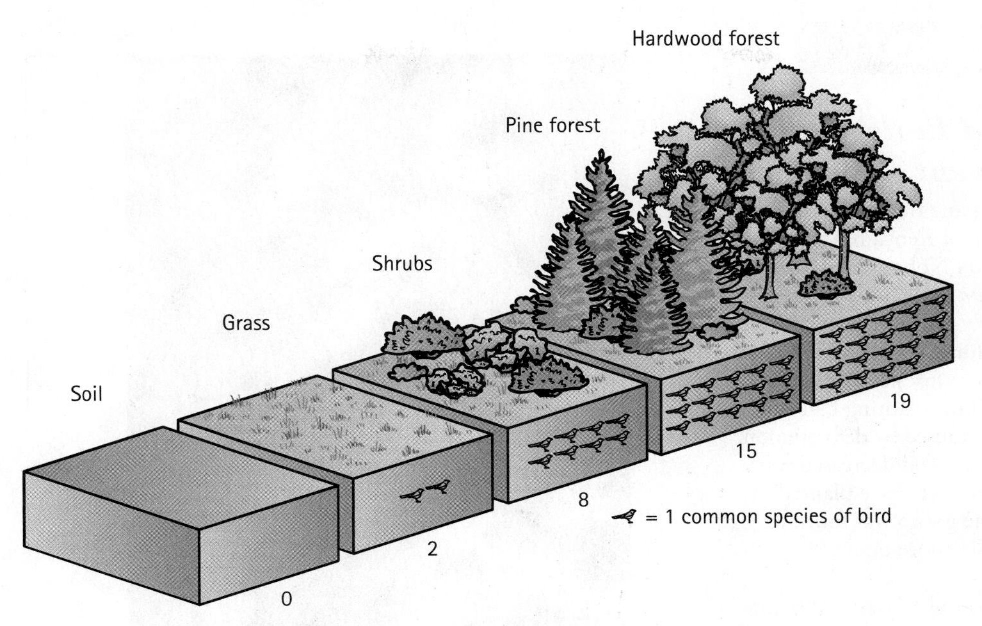

FIGURE 21.37
In secondary succession on abandoned farmland in the southeastern United States, empty fields progressed from grassland to shrubland, pine forest, and hardwood forest over the course of more than 100 years. Greater biodiversity—for example, more species of birds—was seen as succession progressed.

In tropical forests, for example, tree falls create holes in the canopy that make sunlight—a rare commodity—suddenly available to organisms in the understory.

According to the *intermediate disturbance hypothesis*, regular disturbances can actually contribute to biodiversity, as long as they are not too extreme. This is because an intermediate level of disturbance guarantees that there will always be habitat at varying stages of recovery, and different species make use of these different habitats. Of course, some disturbances have more lasting effects. Invasive species can devastate ecosystems, as we saw with the brown tree snake in Guam and the zebra mussel in the eastern United States. Human activity can convert one type of ecosystem into another, as when land is cleared for agriculture or for development. Finally, changes in climate, including global warming (see Chapter 27), can cause unpredictable changes to many ecosystems at the same time.

CHECK YOURSELF

How might global warming cause ecosystems to change?

CHECK YOUR ANSWER

This is a difficult question that many scientists are trying to answer. As Earth gets warmer, entire ecosystems are likely to shift to areas that have more appropriate climates. Usually, this means higher latitudes (locations farther away from the equator) and higher altitudes. However, many organisms have no appropriate habitat to shift to. For example, polar bears already live in the Arctic and cannot search for sea ice farther north. In addition, large areas of Earth have already been developed for human use. Even if there is appropriate new habitat somewhere on a warmer Earth, there may be no way for most organisms to move through human cities or farms to reach it.

SCIENCE AND SOCIETY

Wangari Maathai and Ecologically Sustainable Development

How much difference can planting one tree make? A lot, as Nobel laureate Wangari Maathai of Kenya has shown. Maathai, who was awarded the 2004 Nobel Peace Prize for her "contribution to sustainable development, democracy, and peace," founded the Green Belt Movement in 1977. This grassroots group organizes volunteers—primarily rural women—to plant indigenous trees in forests, wildlife reserves, farms, and public places. Tree planting helps to reverse some of the serious problems caused by deforestation, including erosion, loss of biodiversity, and decreased water supply. Green Belt Movement volunteers have planted more than 30 million trees since the group was founded, and the group continues to maintain more than 6000 tree nurseries throughout Kenya.

Maathai's Green Belt Movement also uses tree planting as an entry point for other activities. By enabling local communities to change their environmental history, the organization empowers people to address issues related to environmental conservation, community consciousness, equity, livelihood security, and accountability. Maathai's model has inspired similar groups in other countries, including Tanzania, Uganda, Ethiopia, and Zimbabwe.

Wangari Maathai plants a tree in Nyeri, Kenya. Maathai was the first African woman to win the Nobel Peace Prize, and it was also the first time the award was given for environmental work.

How important is environmentalism to the world? In Maathai's words, "Some people have asked what the relationship is between peace and environment, and to them I say that many wars are fought over resources, which are becoming increasingly scarce across the earth. If we did a better job of managing our resources sustainably, conflicts over them would be reduced. So, protecting the global environment is directly related to securing peace." Wangari Maathai died in 2011.

* From the Green Belt movement website, http://www.greenbeltmovement.org.

For instructor-assigned homework, go to www.masteringphysics.com 

SUMMARY OF TERMS (KNOWLEDGE)

Biogeochemical cycles The movement of substances such as water, carbon, and nitrogen between the bodies of living organisms and the abiotic world.

Biomass The amount of organic matter in an ecosystem or part of an ecosystem.

Biomes Earth's major types of terrestrial ecosystems.

Carrying capacity The maximum number of individuals or the maximum population density that a habitat can support.

Community All the organisms that live in a specific area.

Consumer An organism that obtains food by eating other organisms.

Decomposer An organism that obtains food by eating dead organic matter.

Ecological succession Changes over time in the community of species living in an ecosystem.

Ecology The study of how organisms interact with their environments.

Ecosystem All the organisms that live in a specific area and all the abiotic features of their environment.

Exponential growth A model of population growth in which a population grows at a fixed rate per amount of time.

Logistic growth A model of population growth in which growth slows as the population approaches the habitat's carrying capacity.

Niche The total set of biotic and abiotic resources a species uses within a community.

Population A group of individuals of a single species that lives in a specific area.

Producer An organism that makes organic molecules using inorganic molecules and energy.

READING CHECK (COMPREHENSION)

21.1 Organisms and Their Environments

1. What is ecology?
2. Explain the difference between the abiotic features and the biotic features of an organism's environment.
3. What is the difference between a community and an ecosystem?

21.2 Population Ecology

4. Describe exponential growth. Under what conditions do organisms grow exponentially?
5. Why do populations that live in unstable environments often grow exponentially and then crash?
6. Describe logistic growth. Under what conditions do populations experience logistic growth?
7. What are the differences between Type I, Type II, and Type III survivorship?

21.3 Human Population Growth

8. Explain how global human population size is expected to change during the 21st century.
9. What is the age structure of a population? What can you learn from a population's age structure?
10. What is an ecological footprint? What does the ecological footprint tell us about how the global human population is consuming resources today?

21.4 Species Interactions

11. What is the name for a diagram of who eats whom in a community?
12. Explain the difference between a producer and a consumer.
13. What is a decomposer? What organisms function as decomposers in most communities?
14. Can two species have the exact same niche in a community? Why or why not?
15. Define parasitism, and provide some examples.

21.5 Kinds of Ecosystems

16. What types of living things do scientists use to classify Earth's terrestrial habitats into biomes?
17. What climatic factors help determine the locations of different biomes?
18. Which biome includes more living things than all other biomes combined?
19. Why are fires important in savannas?
20. Describe the photic zone of ocean habitats.
21. What are some of the challenges that organisms face in the intertidal zone? What adaptations do intertidal organisms have for dealing with these challenges?

21.6 Energy Flow in Ecosystems

22. All organisms need energy in order to grow, reproduce, and perform the activities necessary for survival. Where does this energy ultimately come from?
23. How does sunlight energy enter the biotic world?
24. On average, how much of the energy at one level of the food chain becomes available to the next level? What happens to the rest of the energy?

21.7 Change in an Ecosystem

25. How does primary succession differ from secondary succession?
26. Why are the later colonizers of a habitat dependent on earlier waves of colonizers?
27. What usually happens to the total biomass in an ecosystem during succession? Does the number of species present in an ecosystem usually change as succession continues?
28. Why can regular disturbance sometimes contribute to the biodiversity of a habitat?

THINK INTEGRATED SCIENCE

21A—Materials Cycling in Ecosystems

29. What is a biogeochemical cycle?
30. How does carbon enter the biotic world? How is it returned to the abiotic world by organisms?
31. What role do nitrogen-fixing bacteria and nitrifying bacteria play in the nitrogen cycle?

21B—Energy Leaks When Organisms Eat

32. Explain how the second law of thermodynamics is related to energy loss between one level of a food chain and the next.
33. Is energy lost when glucose is converted into ATP during cellular respiration?

THINK AND DO (HANDS-ON APPLICATIONS)

34. Think about the ecology of *you*! What are some biotic and abiotic features of your environment? What species form part of your community? How do the abiotic features of your environment affect your life and activities?
35. Model exponential growth using pennies. Think of a penny as a single-celled organism that reproduces by dividing into two once each day. On day 1, you have one penny. On day 2, you have two pennies. Continue the exercise out to at least 10 days. How many pennies do you have at the end of that time? Now make a graph showing the number of pennies on the y-axis and the day on the x-axis. Does your graph look like the graph of exponential growth?
36. A number of online websites, such as http://www.footprintnetwork.org, allow you to calculate your personal ecological footprint. Use one of these sites to calculate your ecological footprint. What are five things you could do to reduce your ecological footprint?
37. Draw a food web that includes humans. What are some of the things we eat? Are we primary consumers? Secondary consumers? A top predator?

THINK AND COMPARE (ANALYSIS)

38. Rank the following from lowest in the food chain to highest in the food chain: secondary consumer, primary consumer, top predator, producer.
39. Rank the following biomes from least precipitation to most precipitation: coniferous forest, desert, tropical forest, temperate forest.

THINK AND SOLVE (MATHEMATICAL APPLICATION)

40. Suppose that you have a logistically growing population with rate of increase $r = 2$ in a habitat where carrying capacity $K = 50$. Suppose also that the habitat is overpopulated, with $N = 60$. (Note that N is larger than the carrying capacity, which means that the population size is larger than the habitat can support.) Use the equation for logistic growth, growth rate $= rN(K - N)/K$, to show that the growth rate of the population is negative—that is, the population size will decrease.
41. In a population of songbirds, 100 young are born in the year 2010. Each year, 10 individuals die. Make a table showing how many individuals are alive in each year from 2010 to 2020. Now draw a survivorship curve for the population. Does this songbird population have Type I, Type II, or Type III survivorship?
42. In a population of insects, 1 million young are born in the year 2010. Each year, 95% of the living individuals die. Make a table showing how many individuals are alive in each year from 2010 to 2015 (you can round off your answers). Now draw a survivorship curve for the population. Does this insect population have Type I, Type II, or Type III survivorship?
43. A patch of grassland is home to five carnivores that are secondary consumers in the ecosystem. Each of the carnivores uses 2000 kcal of energy over the course of a day. Assume that 10% of the energy available at each level of the food chain moves up to the next level of the food chain. Show that the primary consumers in the grassland use 100,000 kcal during a single day, and that grass and other producers use 1,000,000 kcal per day.

THINK AND EXPLAIN (SYNTHESIS)

44. A scientist studies how the number of coyotes in San Diego County has changed over the last decade. Is this a population-level study, a community-level study, or an ecosystem-level study? Defend your answer.
45. Another scientist examines how the presence of a nonnative bird species, the starling, affects other species of birds. Is this a population-level study, a community-level study, or an ecosystem-level study? Defend your answer.
46. Compare exponential growth and logistic growth. Under what circumstances can populations grow exponentially?
47. What factors could cause population growth to slow as population size increases?
48. The graph shows the growth of a population of single-celled *Paramecium* in the lab. Did the population grow exponentially or logistically? Can you estimate the carrying capacity of the habitat?

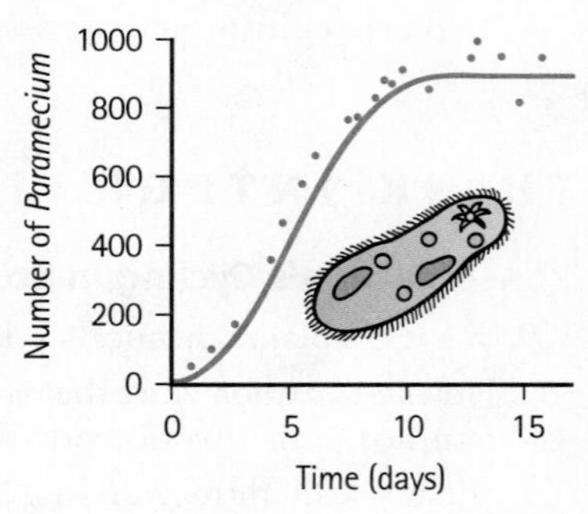

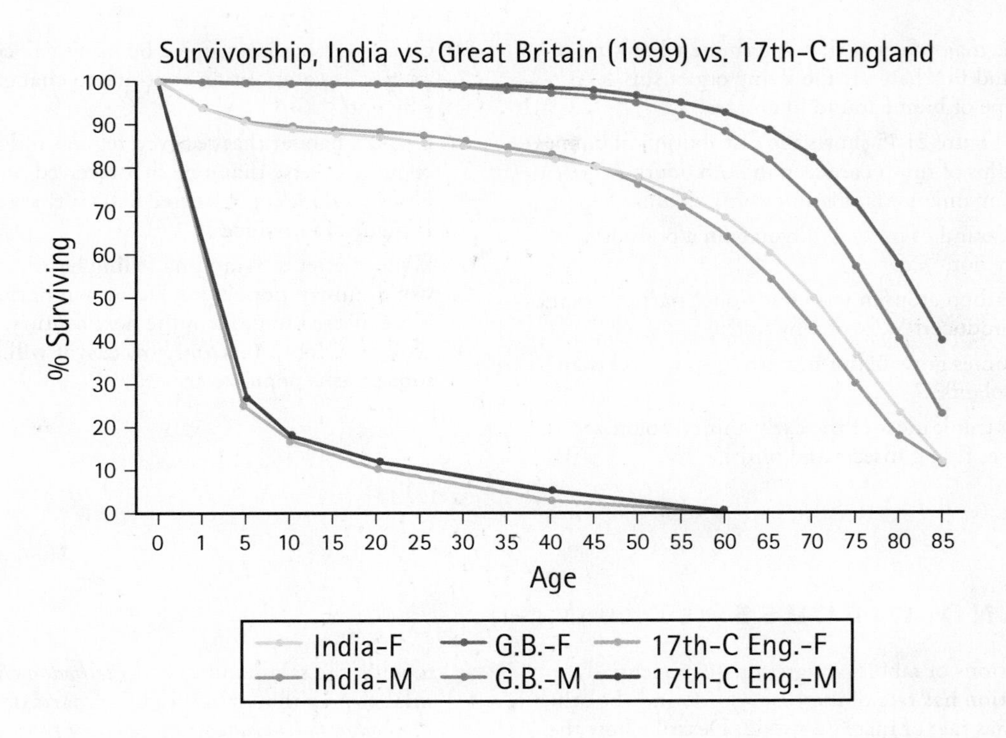

49. The graph above shows survivorship curves for males and females in Great Britain and India in 1999 as well as in 17th-century England. What differences do you see between Great Britain and India in 1999? What might explain these differences? How did survivorship in England change from the 17th century to 1999?

50. The figure below shows the age structures for the world human population in 1950 and 2010 as well as a projection for the year 2100. How did the age structure of the world population change between 1950 and 2010? How is it expected to change between 2010 and 2100?

51. All else being equal, will a person who eats a vegetarian diet or a person who eats meat regularly have a larger ecological footprint? Defend your answer.

52. List three examples of producers, primary consumers, and secondary consumers.

53. Some flowering plants rely on insects to carry pollen from male flowers to female flowers. The insects are dusted with pollen as they drink nectar produced by the flowers. Is this an example of parasitism, commensalism, or mutualism?

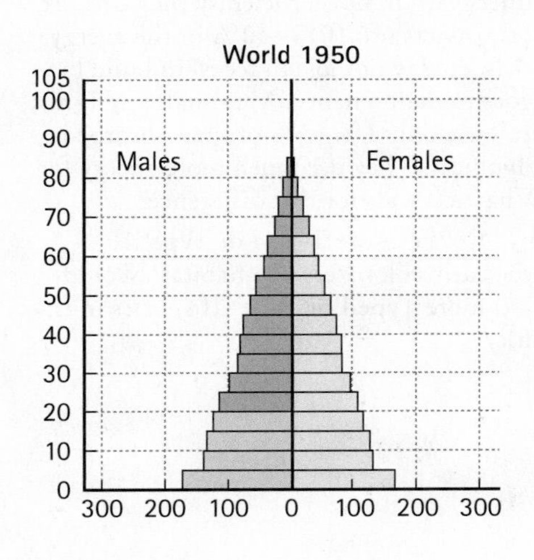

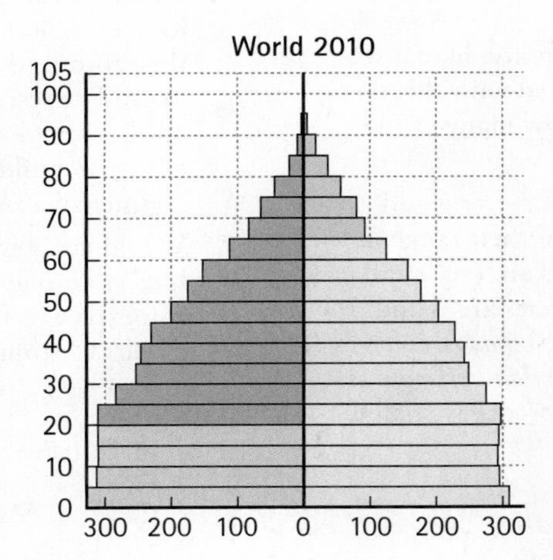

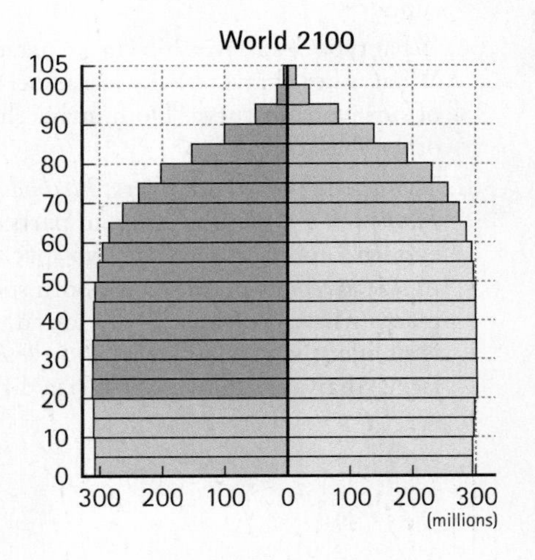

54. What are the major factors that determine what kind of biome is found in a habitat? Do living organisms ever affect the type of biome found in an area?
55. The map in Figure 21.15 shows the distribution of biomes on Earth. Strips of tundra are seen in both South America and the U.S. mainland. Is this an error? Explain.
56. If you eat a pound of pasta, will you gain a pound of weight? Why not?
57. Was every carbon atom in your body once part of a plant or another producer? Why or why not?
58. Why do legumes grow better in nutrient-poor soil than many other plants?
59. Why do you think most of the early animal colonizers of Krakatoa were flying insects and birds?
60. Once a habitat is occupied by its climax community, does its species composition continue to change? Why or why not?
61. Could a habitat that received regular mild disturbances be more diverse than one that received no disturbances? How would it compare to a habitat that received regular, extreme disturbances?
62. Write a letter to Grandma telling her about how the world human population has grown in the past and how it is expected to grow in the next century. Tell her about ecological footprints and how easy it will be for Earth to support this population.

THINK AND DISCUSS (EVALUATION)

63. Two populations of rabbits are growing exponentially. One population has rate of increase $r = 1$, and the other population has rate of increase $r = 2$. Describe how their population growth curves differ. What factors could cause one population to have a higher rate of increase than the other?
64. Two populations of monkeys are growing logistically. One population has carrying capacity $K = 1000$, and the other population has carrying capacity $K = 5000$. Describe how their population growth curves differ. What factors could cause one population to have a higher carrying capacity than the other?
65. A habitat's carrying capacity for a population can change over time. For example, a forest may be able to support a certain number of bears, but when the trees are cut down and the land is paved over, the carrying capacity drops (probably to zero). What other factors could cause the carrying capacity of a habitat to change? How has Earth's carrying capacity for humans changed over time? Does technology influence how many people Earth can support?
66. What type of survivorship curve characterizes humans? What other characteristics are associated with this type of survivorship curve? Do humans show many of these other characteristics?
67. Two species of salamanders, *Plethodon cinereus* and *Plethodon hoffmani*, overlap in parts of their range in the eastern United States. The two species are very similar in size *except* in places where both species are found. In places where both species are found, *Plethodon cinereus* is smaller than it usually is, and *Plethodon hoffmani* is larger than it usually is. This size difference is related to what the salamanders eat: *Plethodon cinereus* eats smaller prey than it eats in other parts of its range, and *Plethodon hoffmani* eats larger prey than it eats in other parts of its range. How can natural selection due to interspecific competition explain the evolution of size and diet in these salamanders?
68. Some acacia trees have evolved a special relationship with certain species of ants. The trees provide food and nesting sites (in the form of hollow thorns) for the ants. The ants attack insects and other species that try to eat the acacia, and they may also kill plants that grow in the immediate vicinity. Discuss the species interactions (food chain, competition, symbiosis, etc.) among the species in this community.
69. Why can't an ecosystem's energy pyramid be inverted (that is, upside down)?
70. A single tree can sometimes support many insects and birds. Does this contradict the idea that all ecosystems are characterized by energy *pyramids*?
71. Different consumers vary in how efficiently they convert food into biomass. Insects use 10%–40% of the energy they absorb (that is, energy not lost in feces) to build biomass. The rest goes to maintenance. Mammals and birds, on the other hand, use only 1%–3% of their absorbed energy to build biomass. They use much more energy for maintenance. What accounts for this difference?
72. Would you expect to find more Type I or Type III species among the early colonizers of a habitat? Would you expect to find more Type I or Type III species in the climax community?

READINESS ASSURANCE TEST (RAT)

If you have a good handle on this chapter, if you really do, then you should be able to score 7 out of 10 on this RAT. If you score less than 7, you need to study further before moving on.

Choose the BEST answer to each of the following:

1. All the organisms that live in a specific area make up a(n)
   (a) population.
   (b) community.
   (c) ecosystem.
   (d) food chain.
2. A Type III population is associated with
   (a) logistic population growth.
   (b) long life expectancy.
   (c) many "inexpensive" offspring.
   (d) large body size.
3. The world human population
   (a) is expected to grow exponentially between now and 2100.
   (b) is about 6 billion.
   (c) is expected to start to decline before 2100.
   (d) is expected to grow more slowly in the second half of this century.
4. Photosynthesizing plants are
   (a) producers.
   (b) primary consumers.
   (c) secondary consumers.
   (d) decomposers.
5. No two species in a community can occupy the same
   (a) ecosystem.
   (b) biome.
   (c) niche.
   (d) level of the food chain.
6. The relationship between a flea and its host is an example of
   (a) parasitism.
   (b) commensalism.
   (c) symbiosis.
   (d) parasitism and symbiosis.
7. Which biome is characterized by deciduous trees?
   (a) tropical forest
   (b) temperate forest
   (c) coniferous forest
   (d) chaparral
8. Which biogeochemical cycle(s) would be impossible without bacteria?
   (a) the water cycle
   (b) the carbon cycle
   (c) the nitrogen cycle
   (d) the carbon and nitrogen cycles
9. Which of these statements about energy flow in ecosystems is false?
   (a) Energy from sunlight enters the biotic world when plants and other photosynthesizers use it to build organic molecules.
   (b) Tropical forests and coral reefs have high primary productivity.
   (c) About 90% of the energy at one level of the food chain moves up to the next level of the food chain.
   (d) When a consumer eats, the energy it receives from food goes into feces, maintenance, and building biomass.
10. During ecological succession
    (a) the activities of earlier waves of colonizers allow later colonizers to thrive.
    (b) biodiversity typically decreases.
    (c) early colonizers typically remain in the habitat.
    (d) biomass typically decreases.

Answers to RAT

1. b, 2. c, 3. d, 4. a, 5. c, 6. d, 7. b, 8. c, 9. c, 10. a

4

PART FOUR

Earth Science

Some people think it's just "dirt," but soil is a precious Earth material. Plants need soil to grow. And people depend on plants for food and oxygen so without soil, there would be no people! A few centimeters of soil typically takes thousands of years to form. It forms from rock that's been *weathered*—broken down by physical, chemical, and biological processes. *Pedologists*—soil scientists—say there are over 100,000 kinds of soil on this planet. Each has its own developmental story, properties, and role to play in nature. To take care of soil—and other vital natural resources—we need to use some Earth science!

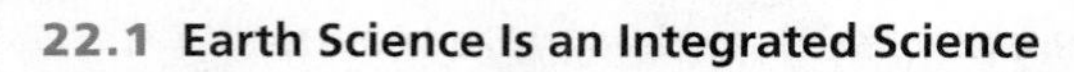

CHAPTER 22
Plate Tectonics

MOST OF us are curious about the big ball we live on: planet Earth. We wonder, for example, how do mountains form? If you could dig a hole all the way to the center of Earth, what would you find there? If Earth's inner core is almost as hot as the Sun, why doesn't the planet melt? Is Earth's surface really a jigsaw puzzle of moving pieces? How do we know?

Earth science, the study of Earth's history, structure, and natural processes, answers such questions. It also addresses questions of great concern such as these: Can earthquakes, tsunamis, and volcanic eruptions be predicted? Are Earth's resources being consumed too quickly? Must we sacrifice modern conveniences to protect the environment? How is the climate changing? The Earth science chapters in this book will help you understand these and many other issues that affect the planet and your life.

We begin our study with plate tectonics, the theory that explains how Earth's surface reacts to the huge forces acting below.

22.1 Earth Science Is an Integrated Science

LEARNING OBJECTIVE
Explain why Earth science is an integrated science.

EXPLAIN THIS Jim wants to major in geology but he doesn't want to take physics. Why is this a problem?

Earth is a rocky sphere. Almost three-quarters of it is covered by water. A thin *atmosphere* (or layer of gases) surrounds it. Earth orbits the Sun yearly, spins on its axis daily, and teems with life. Earth science is an integrated science: To understand our Earth, it is necessary to understand processes that are described in physics, chemistry, biology, and astronomy. That's because our home planet is a complex system of living and nonliving parts and processes.

Geology is the broadest Earth science discipline. It is concerned with the composition and structure of Earth. *Meteorology* is the study of weather. *Petrology* is the study of rocks, while *mineralogy* is the science of the minerals within those rocks. *Paleontology* is the study of Earth's early history, including the dinosaurs and other creatures whose fossilized remains fascinate us today. *Hydrology* investigates the flow of Earth's waters, while *oceanography* focuses on the oceans. *Volcanology* is the study of volcanoes. There are many other branches of Earth science too. Cross-disciplinary geosciences include geophysics and environmental science.

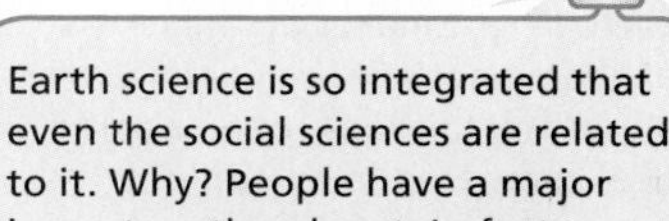

Earth science is so integrated that even the social sciences are related to it. Why? People have a major impact on the planet. In fact, we have become a geologic force. Our ideas, choices, and actions affect Earth in a big way!

FIGURE 22.1
Earth scientists at work.

LEARNING OBJECTIVE
Describe the basic chemical composition of each of Earth's compositional layers: crust, mantle, and core.

22.2 Earth's Compositional Layers

EXPLAIN THIS Why is Earth classified as one of the solar system's three "rocky" planets?

If you sliced Earth open and analyzed its chemical composition, you would find that it is layered like a hard-boiled egg. We recognize three main layers in an egg: the shell, egg white, and yolk. Similarly, Earth has three main layers: the crust, mantle, and core (Figure 22.2). Each of these layers consists of different materials.

Earth's crust makes up only 0.3% of the planet's entire volume, about the thickness of an apple peel compared to an apple.

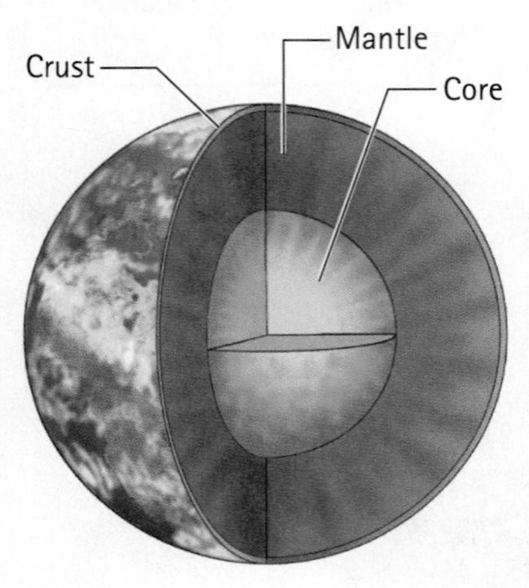

FIGURE 22.2
Earth has three layers that differ in their composition.

The Crust

The **crust** is Earth's surface layer. Like an eggshell, it is thin and brittle and can crack. The crust is composed mainly of rocks that contain minerals rich in silicon and oxygen. You might be surprised to learn that the most common element in the crust is actually oxygen. So, you see, oxygen not only fills the air but it is a major ingredient of solid Earth, too!

Earth's crust has two parts: *continental crust*, which makes up the landmasses, and *oceanic crust*, which underlies the oceans. The oceanic crust is made of mostly dense, dark gray or black, fine-grained rock called *basalt.* Continental crust is mostly lighter, less dense, *granitic* rock. Both the oceanic crust and the continental crust are composed mainly of silicon and oxygen with smaller amounts of other elements. The basaltic rock of the ocean floor contains a higher proportion of the dark minerals iron and magnesium than granitic rock. This accounts for basalt's striking dark color (Figure 22.3).

The oceanic crust is only 7 km thick on average, a distance you could walk in about an hour. Continental crust is thicker—on average, it's about 40 km from the top to the bottom.

As Figure 22.4 shows, mountains are like tips of icebergs. They are mostly buried beneath the surface. The mountaintops you see above ground are supported by deep "roots"—huge masses of continental crust that reach far into the mantle. When the tops of mountains are worn away by erosion, the roots move upward, poke above Earth's surface, and keep rising as the mountain loses mass

(a)

(b)

(c)

FIGURE 22.3
(a) Basalt is a fine-grained, dark rock rich in silicon, oxygen, iron, and magnesium. The oceanic crust is composed almost entirely of basaltic rock. (b) The continental crust is composed of chiefly light-colored, low-density granitic rock. Half Dome in Yosemite, shown here, is a classic example of the granitic rock of the continental crust. (c) Oceanic rock is found on land as well as under the sea. This basalt formed on the seafloor but was uplifted to a high altitude by tectonic forces.

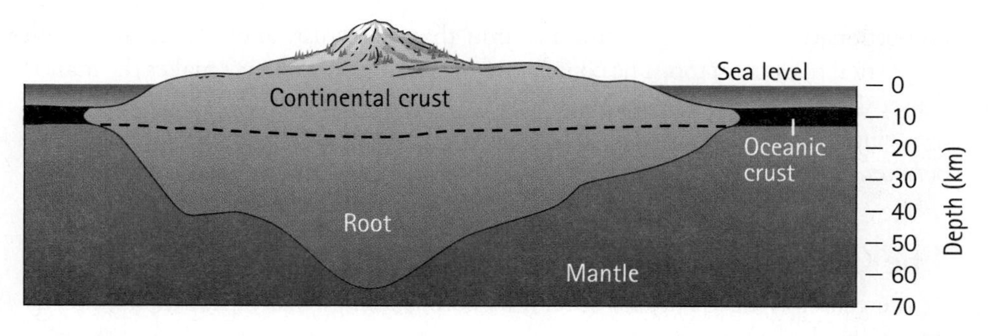

FIGURE 22.4
Mountains have "roots," blocks of continental crust that support them underground. The mountains we see are like the tips of icebergs.

at the top. So, as mountains wear down, they are continually replenished from below. The reason this happens is that the crust "floats" or buoys up and down in the mantle like a boat in the ocean. (Like all floating objects, the crust floats at the level where the buoyant force pushing it up equals the force of gravity pushing it down. Read more about this in Appendix E, the Physics of Fluids.) Just as a cargo ship rises higher in the water when it is unloaded, a mountain rises higher in the mantle when it "unloads" mass by erosion (Figure 22.5).

The vertical positioning of Earth's crust due to its flotation in the mantle is called *isostasy*. Isostasy explains why mountain roots rise up (and therefore why mountains take so long to wear away—they have reserve mass stored underground). Isostasy is also the reason that the oceanic crust sits lower in the mantle than the continental crust: The oceanic crust is made of denser rock, so, like the loaded cargo ship versus the empty one, the oceanic crust sinks deeper.

Now you know that Earth's crust moves up and down due to isostasy. But the crust also moves in far more complicated ways. Over geologic time, the entire crust churns, wrinkles, stretches, mixes up, and circulates. It's true that oceanic crust is mainly basaltic rock and continental crust is mainly granitic. Still, rocks of each type are intermixed throughout the crust. You can find ancient sea creatures embedded in black basaltic rock high in the mountains because, although these rocks formed on the seafloor, they were uplifted by great tectonic forces you will soon learn more about. Likewise, continental rock can be found on the seafloor among the basaltic rock because eroded continental sediments from the land eventually wash to the sea. The crust is a complex assemblage of rocks mixed over time like a very big pot of stew.

FIGURE 22.5
Isostasy: the vertical position of the crust is stable when gravity equals the buoyant force. Denser oceanic crust sits lower in the mantle than continental rock, just as the loaded ship sits lower in the water than the unloaded ship. Note that this principle of flotation governs the position of the crust in the mantle even though the mantle is solid. The reason is that the mantle actually flows very slowly over geologic time, as if it were a fluid.

The Mantle

Hidden beneath Earth's crust is the **mantle**, a thick layer of hot rock. Geologists know much less about the mantle than they know about the crust. We cannot see into the mantle because light does not travel through rock. Nor can we drill into the mantle because rock is so hard that drilling equipment breaks before it reaches the bottom of the crust. Scientists would like to drill all the way to the mantle someday because most of Earth—82% of its mass and 65% of its volume—*is* mantle.

The mantle's thickness is 2900 km from the bottom of the crust to the top of the core. Geologists have sampled molten (melted) rock that travels through volcanic vents in the ocean floor. Chemical analysis of this rock shows that the mantle consists mostly of rock rich in silicon and oxygen, like the crust. But mantle rock has

UNIFYING CONCEPT

- ***Density***
Section 2.3
- ***The Gravitational Force***
Section 5.3

MasteringPhysics®
TUTORIAL: The Earth's Mantle

UNIFYING CONCEPT

- *Density* Section 2.3

proportionately more magnesium and iron than the crust, and it contains a high proportion of calcium too. The contribution of the denser elements makes the mantle denser than the crust. The mantle's density is further increased by the weight of the overlying crust. The crust's weight bears down on the mantle, compressing the spaces between the rocks and also squeezing the rocks' molecular structures.

The Core

At the center of Earth lies its **core**. The core is a huge ball of hot metal, mostly iron with a lesser amount of nickel. Because iron is a very dense element, the core is much more dense than the mantle and the crust. The core has a radius of about 3500 km—approximately the distance from San Francisco to New York. Because we cannot observe the core, our knowledge of it is limited. What we do know about the core comes mainly from *seismology*, the study of earthquake waves.

CHECK YOURSELF
How is the Earth's crust like a ship?

CHECK YOUR ANSWERS
The crust floats on the mantle like a ship floats on the water. The crust's vertical position, as well as the ship's, is determined by its density.

LEARNING OBJECTIVE
Describe the process that divided Earth into layers of different density.

Earth Developed Layers When It Was Young, Hot, and Molten

EXPLAIN THIS Why does drilling equipment break before it reaches the mantle?

The density of Earth's crust, mantle, and core increases in concentric layers moving from the surface to the center. What's the reason for these density layers?

Earth, like the Moon and inner planets, is mostly rock. And, rock is composed of what all matter in the universe is made of—chemical elements. There are 112 naturally occurring elements, as you know from chemistry. Most chemical elements are rare on Earth. In fact, only eight elements make up about 98% of Earth's entire mass (Figure 22.6). All of the other elements combined make up the remaining 2% of Earth's mass.

Earth's elements are not distributed evenly. The heavier ones

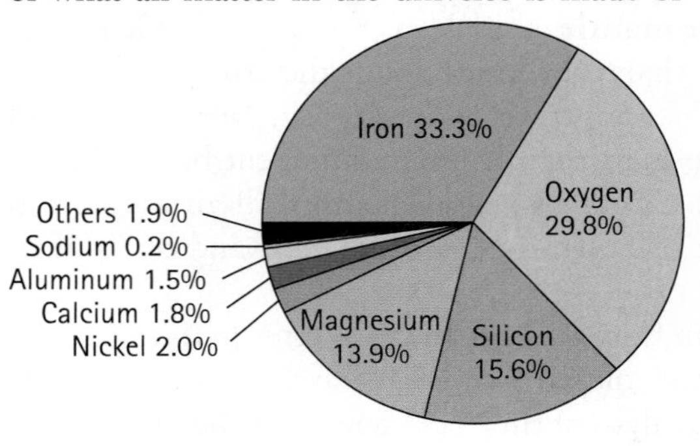

FIGURE 22.6
Only eight of the chemical elements are found in abundance on Earth.

are concentrated near the center of the planet, while the lighter elements are more abundant near the surface. The explanation for this has to do with Earth's early history. Earth formed 4.6 billion years ago from debris that was orbiting the newly forming Sun. Some of the debris collided and stuck together or "accreted" into a growing rocky mass, forming a rudimentary Earth. Earth grew as its gravity attracted more and more debris to it. Each time material from space collided with and stuck to Earth, the kinetic energy of the impacting bodies was transformed into heat, which warmed the developing planet. Earth was so hot, in fact, that it was *molten*—melted and able to flow. Under the influence of gravity, dense, iron-rich material sank to Earth's center to become the core. Less dense, silicon- and oxygen-rich material rose toward the surface by *differentiation*. **Differentiation** is the process by which gravity separates fluids according to density. (A mixture of oil and water, for example, undergoes differentiation when the denser water sinks to the bottom and the oil rises to the top.)

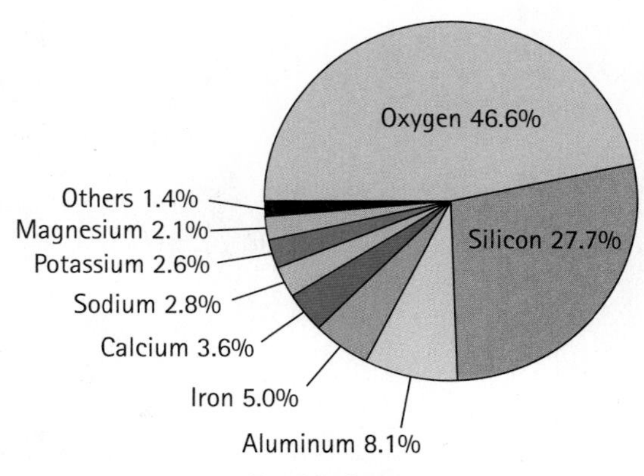

FIGURE 22.7
Percentages of the elements in Earth's crust by mass. Oxygen and silicon make up about 75% of Earth's crust.

Eventually, the frequency of impacts slowed and Earth cooled. Lighter materials near Earth's surface solidified to form minerals that clustered together to make rocks. Figure 22.7 shows the current composition of Earth's crust. Note that almost half (46.6%) of the crust is oxygen (O), and a little more than a fourth (27.7%) of it is silicon (Si), so together these two lightweight elements make up about three-quarters of Earth's solid surface.

UNIFYING CONCEPT

- ***Density***
  Section 2.3
- ***The Gravitational Force***
  Section 5.3

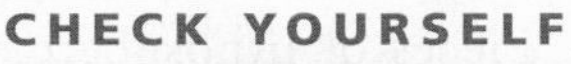

CHECK YOURSELF

Why was the early bombardment of Earth by space debris important to differentiation?

CHECK YOUR ANSWER

The kinetic energy of the space debris colliding with Earth was transformed into heat energy. The heat melted Earth's rock so it was molten and could flow. If the rock had been solid, it couldn't have settled into distinct layers.

22.3 Earth's Structural Layers

LEARNING OBJECTIVE
Name and describe Earth's five functional layers.

EXPLAIN THIS Why do geologists often speak of Earth's structural layers rather than the compositional layers?

To analyze Earth's interior, we can go beyond the three-layer, hard-boiled egg model. Geologists also divide Earth into five functional layers based on their physical properties. Physical properties include, for example, temperature, pressure, strength, and ability to flow. In terms of physical properties, Earth's five layers are the *lithosphere, asthenosphere, lower mantle, outer core,* and *inner core* (Figure 22.8).

The Lithosphere

The "rock sphere" we call the **lithosphere** ("LITH-uh-sphere") is a shell of cool, rigid rock. The lithosphere includes the crust and some of the upper portion of

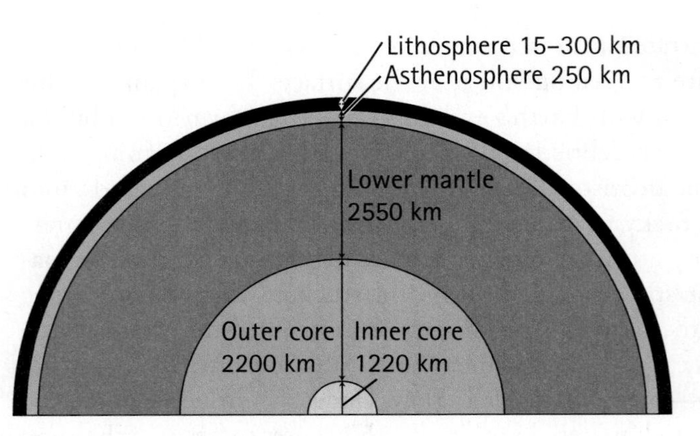

FIGURE 22.8
When we look at Earth in terms of physical properties, we see that it has a five-layered structure: the lithosphere, asthenosphere, lower mantle, outer core, and inner core.

Temperature and pressure both increase with depth in Earth's interior. This, along with Earth's varying composition, produces five layers with different physical properties.

Earth's mantle. On average the lithosphere is about 100 km thick. It is thickest below the continents and thinnest below the oceans. The lithosphere is not continuous like Earth's other layers. Instead, it is broken up into a patchwork of interlocking pieces. Pieces of lithosphere are called *tectonic plates.*

The Asthenosphere

The **asthenosphere** ("as-THEN-uh-sphere") lies under the lithosphere. It is made of mantle rock. The rock in the asthenosphere is soft and flows slowly like taffy or Silly Putty. We say it is "plastic"—that is, it is a solid that can flow slowly over immense spans of geologic time. Pieces of lithosphere are

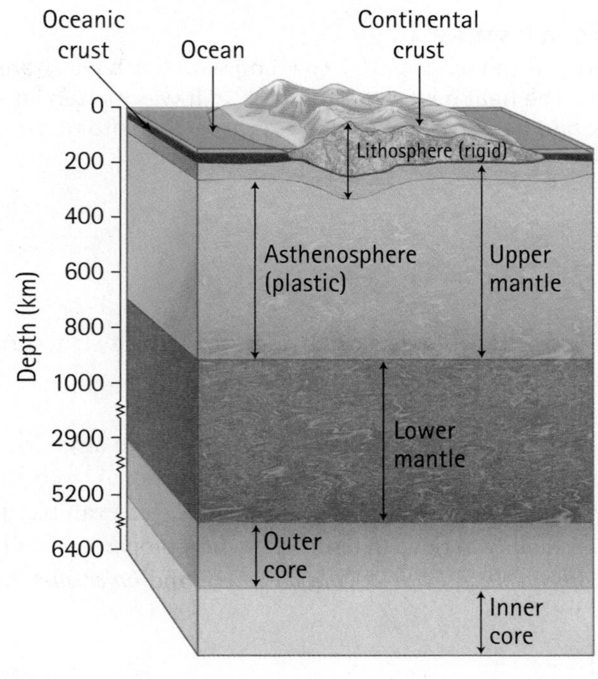

FIGURE 22.9
This diagram shows the relationship of the crust and mantle to the lithosphere and asthenosphere.

carried along on the flowing asthenosphere like rafts on a slowly flowing ocean.

When we say that the asthenosphere moves slowly, we mean *very* slowly. The hour hand on a clock moves about 10,000 times faster than the "flowing" asthenosphere. This layer flows so slowly that it would look like solid rock if you could see it.

FIGURE 22.10
This computer simulation shows the magnetic field generated by the motion of Earth's fluid outer core. Magnetic field lines are blue where the field is directed inward and yellow where it is directed outward. You can see that the field beyond the mantle–core boundary has a smooth overall structure, while the field at the core is highly complicated.

The Lower Mantle

The **lower mantle** sits between the asthenosphere and the outer core. It consists of strong, rigid mantle rock. Despite its extreme heat, the lower mantle is not nearly as plastic as the asthenosphere.

The Outer Core

The **outer core** is a shell of hot, liquid metal—mostly iron with some nickel. The outer core lies beneath the mantle and above the inner core. Since the outer core is liquid, it spins as Earth rotates. The outer core also swishes and flows because its enormous heat stirs up convection currents within it.

The outer core creates Earth's magnetic field. Do you remember from physics that moving charged particles make up an electric current and that an electric current always produces a magnetic field? The iron and nickel in the outer core supply charged particles, and the Earth's rotation and convection supply motion. The result is an immense magnetic field stretching thousands of kilometers into space. This "geomagnetic" field is essential to life because it shields Earth from the *solar wind*—the stream of harmful high-energy particles emanating from the Sun.

The Inner Core

The **inner core** is a solid sphere of hot metal. It is mostly iron, like the outer core. Its temperature is estimated to reach a maximum of over 7000°C. This is about as hot as the surface of the Sun!

UNIFYING CONCEPT
- ***Convection***
Section 6.9

With such a high temperature, how does the inner core remain solid? The answer is that intense pressure from the weight of the overlying rock keeps the inner core from melting. Pressure in the inner core packs the atoms so tightly that they cannot flow as they do in a liquid state.

LEARNING OBJECTIVE
Describe P-waves and S-waves and explain what each reveals about Earth's internal structure.

Integrated Science 22B
PHYSICS

Using Seismology to Explore Earth's Interior

EXPLAIN THIS Why do S-waves stop at the mantle–core boundary?

No one can observe Earth's internal structure. So, how do geologists know what's underground? Earthquakes are the key. An **earthquake** is the shaking or trembling of the ground that happens when big sections of rock underground shift or break. Earthquakes release vast amounts of energy.

Science knew very little about Earth's interior until very recently. In fact, Ford was manufacturing automobiles and there was a baseball World Series before scientists realized Earth has a core!

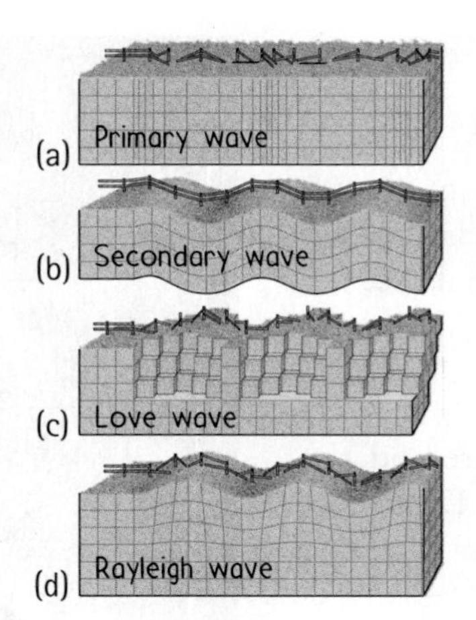

FIGURE 22.11
Block diagrams show the effects of seismic waves. The yellow portion of the left side of each diagram represents the undisturbed area. (a) Primary body waves alternately compress and expand Earth's crust, as shown by the different spacing between the vertical lines, similar to the action of a spring. (b) Secondary body waves cause the crust to oscillate up and down and side to side. (c) Love surface waves whip back and forth in a horizontal motion. (d) Rayleigh surface waves operate much like secondary body waves, but they affect only the surface of the Earth.

The energy radiates outward from the disturbance in the form of *seismic waves*. Seismic waves are mechanical waves, like sound waves, that travel through Earth. The study of seismic waves is called **seismology**, and seismology has provided most of what we know about Earth's interior.

Like any kind of wave, seismic waves may reflect or refract when they move from one material to another. Exactly how different kinds of seismic waves reflect and refract, as well as the changes in their speed and wavelength, reveals much about the medium the waves are traveling through.

Seismic waves come in two main varieties: *body waves*, which travel through Earth's interior, and *surface waves*, which travel on the surface (Figure 22.11). *Surface waves*, in turn, are divided into two types: *Rayleigh waves* and *Love waves* (each named after its discoverer). Rayleigh waves roll over and over in a tumbling motion similar to ocean waves. Love waves move in a snakelike, side-to-side transverse motion. Since Love waves shake things from side to side, they are particularly damaging to tall buildings. (Love waves are *not* loved by people who work in skyscrapers!) But it's the body waves, rather than the surface waves, that reveal Earth's inner structure because they travel deep inside Earth.

Body waves are classified as *primary waves* (P-waves) or *secondary waves* (S-waves). Primary waves are longitudinal waves; they compress and expand the material through which they move. P-waves are the fastest seismic waves, moving at speeds between 1.5 and 8 km per second through any type of material—solid rock, magma, water, and air. S-waves, on the other hand, are transverse waves; they vibrate the particles of their medium up and down and side to side, and they are slower than P-waves. Significantly, S-waves can travel through only solid materials, not liquids.

Near the turn of the 20th century, Irish geologist Richard Oldham was examining records of a massive earthquake in India when he discovered that its transverse S-waves traveled some distance through Earth and then stopped. He also observed that the longitudinal P-waves traveled to the same depth that the S-waves did but the P-waves then refracted and continued on at a lower speed. Since S-waves can't travel through liquids but P-waves can, though at a reduced speed, Oldham inferred that the earthquake waves had hit an internal boundary—a place where the solid Earth becomes liquid. In other words, he discovered that Earth has a distinct liquid center. The year was 1906.

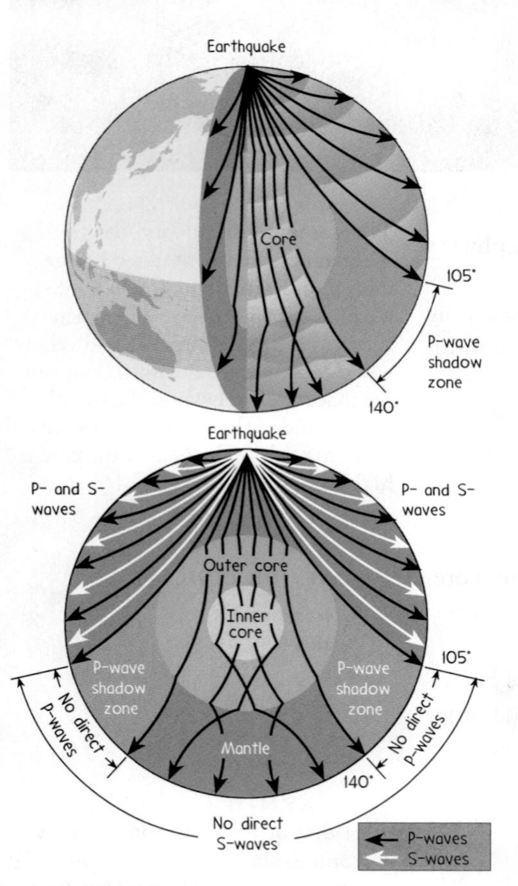

FIGURE 22.12
Cutaway and cross-sectional diagrams showing the change in wave paths at the major internal boundaries and the P-wave and S-wave shadows. The P-wave shadow between 105° and 140° from an earthquake is caused by the refraction of waves at the core–mantle boundary. The S-wave shadow is even more extensive. Any location more than 105° from an earthquake does not receive S-waves because the liquid outer core does not transmit them.

Three years later, seismologist Andrija Mohorovičić ("mo-huh-RO-vih-chich") analyzed seismic readings from an earthquake near his town of Zagreb, Croatia. He detected a sharp increase in the speed of seismic waves at another boundary, one that lay shallower within the Earth. Mohorovičić realized that the wave speed increased because the wave was passing from a low-density solid to a high-density solid. Thus, Mohorovičić discovered that Earth is composed of a light, thin, outer crust that sits upon a layer of denser material, the mantle. The dividing line between Earth's crust and mantle has been called the *Mohorovičić discontinuity*, or "Moho," ever since.

In 1913, Beno Gutenberg reinforced Oldham's earlier findings by showing that the mantle–core boundary is very distinct and is located at a depth of 2900 km. He observed that, when P-waves reach this depth, they are refracted so completely that the boundary actually casts a P-wave shadow over part of Earth (Figure 22.12). The shadow is a region where no waves are detected. Furthermore, the sharp boundary between the mantle and core casts an S-wave shadow that is even more extensive than the P-wave shadow. And, since S-waves are transverse and unable to pass through liquids, Gutenberg's work reinforced earlier findings that the core, or part of it, must be liquid. Taken together, the discoveries of Oldham, Mohorovičić, and Gutenberg established the three-layer, hard-boiled egg model of Earth.

This simple picture of Earth's structure was refined in 1936 by Ingre Lehmen, a Danish seismologist. Her research showed that P-waves refract not only at the core–mantle boundary but again at a certain depth within the core, where they gain speed. This suggested that the core actually has two parts: a liquid outer core and a solid inner core. Thus, Lehman's work helped to show that Earth is indeed layered—but layers exist according to physical properties as well as by composition.

Here's a tip for remembering body waves: P-waves are "push-pull waves" and S-waves are "side-to-side waves."

UNIFYING CONCEPT

- ***Waves***
  Section 8.1
- ***The Scientific Method***
  Section 1.3

CHECK YOURSELF

1. **What don't S-waves pass through Earth?**
2. **What is the Moho? How was it discovered?**

CHECK YOUR ANSWERS

1. S-waves can't travel through liquids, so they are interrupted by the liquid outer core.
2. The Moho is the boundary between Earth's mantle and crust. The Moho was detected by Mohorovičić, who observed an increase in the speed of P-waves across it.

22.4 Continental Drift—An Idea Before Its Time

LEARNING OBJECTIVE
Cite Wegener's evidence of continental drift.

EXPLAIN THIS Describe how scientists of the early 20th century viewed Wegener's ideas on continental drift. Why did they react this way?

If you've ever wondered what you would encounter if you dug a hole to the center of the Earth, now you know—thanks to seismology. But we live at Earth's surface, not underground. What's most crucial for us to understand is what's happening at Earth's surface and why it happens. Today we know that surface geology is largely determined by processes deep inside the Earth. For example, the continents are shaped and located as they are because the lithosphere floats in the asthenosphere. However, it took science many decades to figure out the connection between surface geology and Earth's internal structure.

Have you ever noticed on a world map that Africa and South America fit together like pieces of a giant jigsaw puzzle (Figure 22.13)? The first person to provide a detailed explanation of this was German scientist Alfred Wegener. Wegener's hypothesis, called **continental drift**, states that the world's continents

North America
Europe
Africa
South America
Continental shelf
Continent

FIGURE 22.13
The jigsaw puzzle fit of the continents suggests that they were once joined. The fit along the continental shelves is even better than the fit of the continents themselves.

move slowly over Earth's surface. Wegener hypothesized that the individual continents of today's world were once joined as a single, giant landmass. Wegener named this supercontinent *Pangaea* ("pan-GEE-uh"), which is Greek for "universal land."

Wegener supported his hypothesis of continental drift with impressive biological evidence. For example, fossils of the same species of *Glossopteris* plant, shown in Figure 22.15, were found in places that are separated today by large oceans—India, Australia, South America, Africa, and Antarctica. Seeds from these plants were too heavy to have been blown from continent to continent. Also, fossils of *Mesosaurus*, a reptile, appeared in both Africa and South America. That creature couldn't swim and could not have evolved in exactly the same way in these widely separated regions. So, why do we find fossil evidence of this plant and this reptile in these separate places?

To explain the matching of *Glossopteris* and other plant and animal fossils, geologists of Wegener's time proposed that prehistoric land bridges once connected the continents. Yet there was no physical evidence that such land bridges ever existed. Continental drift, on the other hand, was supported by evidence, *and* it provided a neat explanation of matching fossils. The continents had once been a single landmass, with a single community of animals and plants. When the continents later split along the present-day Mid-Atlantic Ocean, representatives of these species were divided among the new continents.

Wegener also used rock matching as evidence to support his hypothesis of continental drift. For example, diamond-bearing rock formations occur where Africa and South America would meet if the Atlantic Ocean did not separate them. And, in some places, mountain chains separated by oceans have patterns of folds that would be continuous if the mountains were brought together (Figure 22.16).

FIGURE 22.14
German naturalist Alfred Wegener (1880–1930). As a young man, Wegener yearned to explore Greenland. While training for this adventure, he established a record for the longest balloon flight (52 hours). Wegener's next passion was for astronomy, and he earned a doctoral degree in the subject. Later, he became a university professor of meteorology and geophysics. He proposed the hypothesis of continental drift, which led to the discipline of plate tectonics—a productive life indeed. In 1930, Wegener died while crossing an ice sheet on an expedition to Greenland.

(a) *Glossopteris*

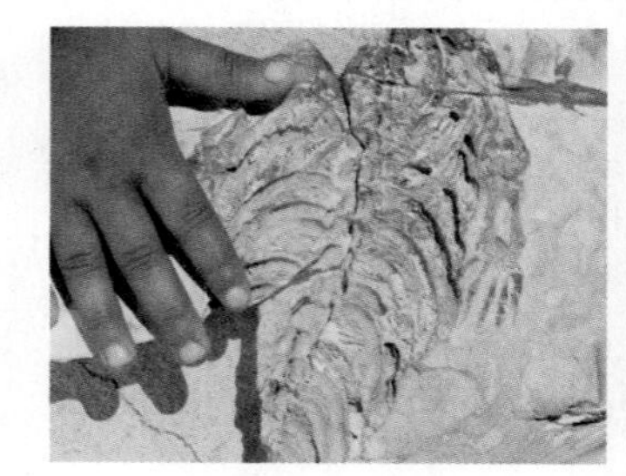

(b) *Mesosaurus*

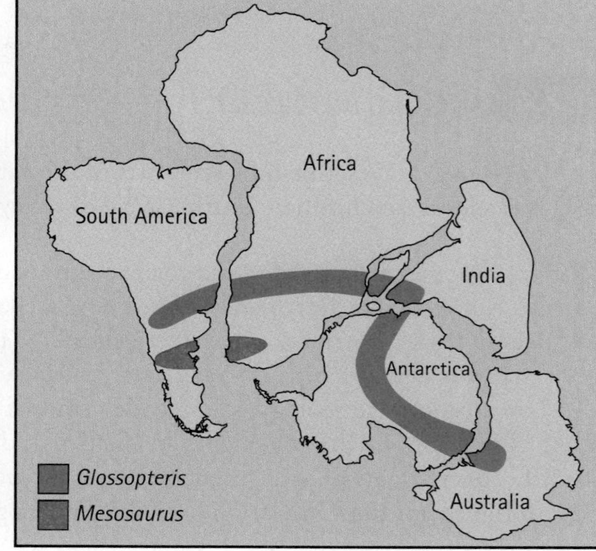

(c) Fossil distribution

FIGURE 22.15
(a) This rock contains a fossil of a *Glossopteris* fern, an ancient seed plant. Can you see its long leaves? (b) This is a fossilized reptile called *Mesosaurus*. (c) These species couldn't cross the oceans. Why, then, were *Mesosaurus* and *Glossopteris* fossils distributed as shown here?

Besides the matching of fossils and rock, ancient climate evidence supported Wegener's hypothesis. For example, continental drift explained the puzzling evidence of ancient ice sheets on regions now located near the equator. How did these ice sheets form on land that is in such a warm location today? Continental drift provides an explanation: When the ice sheets formed on these landmasses, the landmasses were located near the South Pole.

Despite the evidence, Wegener's hypothesis was dismissed in scientific circles. No one, including Wegener, could provide a mechanism for continental drift. How could massive continents of solid rock "drift" across Earth's surface? What force could be strong enough? These questions put acceptance of continental drift on hold for decades.

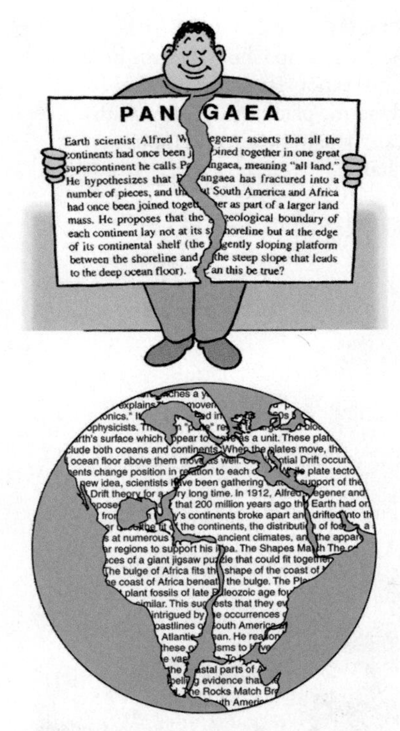

FIGURE 22.16
Wegener likened the fossil and rock matches to finding two pieces of torn newspaper with matching contours and lines of type. If the edges and the lines of type fit together, he said, the two pieces of newspaper must have originally been one.

CHECK YOURSELF

1. **What evidence might lead someone with no scientific knowledge to suspect that the continents were once connected?**
2. **Scientists of Wegener's day rejected his hypothesis of continental drift because they couldn't imagine how the continents could be made to move across the oceans. What didn't they know about the mantle?**

CHECK YOUR ANSWERS

1. The most obvious evidence is the matching of the edges of the African and South American continents, which can be seen on any world map or globe.
2. They did not know that the mantle has a plastic layer, the asthenosphere, over which "floating" continents can slide.

22.5 Seafloor Spreading—A Mechanism for Continental Drift

LEARNING OBJECTIVE
Explain how new lithosphere is created and old lithosphere is destroyed through the process of seafloor spreading.

EXPLAIN THIS Why is Earth's continental crust much older than its oceanic crust?

UNIFYING CONCEPT

- ***The Scientific Method*** **Section 1.3**

In the 1950s, many decades after Alfred Wegener proposed his hypothesis of continental drift, scientists used sonar technology to map the seafloor for the first time. They discovered that the ocean bottom is not flat, as had been presumed, but has deep canyons and trenches. It has many curious, flat-topped, underwater volcanoes.

An amazing discovery was the Mid-Atlantic Ridge, the longest and tallest mountain range in the world (Figure 22.17). The Mid-Atlantic Ridge is 19,312 km long. It winds through the center of the Atlantic Ocean basin parallel to the North American, South American, European, and African coastlines. Its highest peaks poke above sea level to form islands, including Iceland. In the center of the ridge and all along its length, there is a valley or *rift*. Continued investigation uncovered other midocean mountain ranges. Rocks of these midocean peaks were found to be the youngest rocks of the seafloor. We now know that an entire global system of underwater mountains called the **midocean ridge** winds all around Earth like the seam on a baseball.

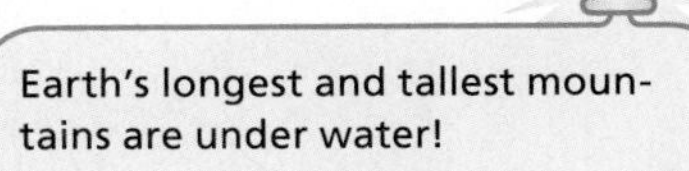

Earth's longest and tallest mountains are under water!

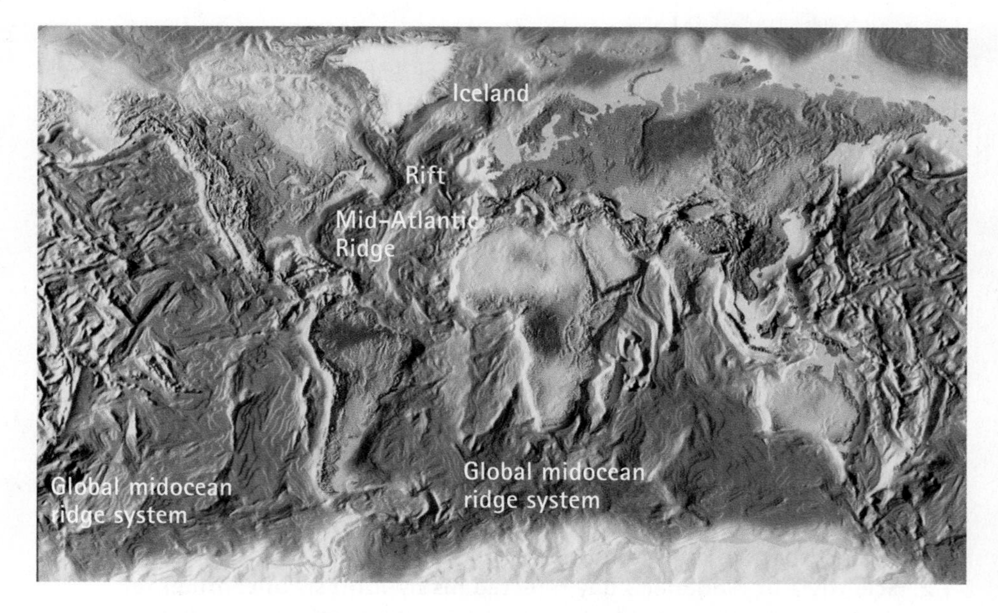

FIGURE 22.17
Seafloor maps show that the ocean floor is not flat and calm. It is a dynamic place, crisscrossed with canyons, trenches, and mountain chains.

FIGURE 22.18
The Mid-Atlantic Ridge runs down the center of the Atlantic Ocean floor. Its highest peaks poke above sea level, creating islands such as Iceland, shown here. You can see Iceland's rift valley in this photo.

Scientists also discovered ocean trenches, the deepest places on Earth. **Ocean trenches** are long, deep, steep troughs in the seafloor. Ocean trenches are found near continents, particularly around the edges of the Pacific Ocean. Also found at the ocean bottom were volcanic eruptions and zones of high-temperature water around underwater mountains. What kind of geologic activity could explain the seafloor's striking features?

American geologist Harry Hess answered this question with his theory of seafloor spreading. **Seafloor spreading** is the process by which new lithosphere is created at midocean ridges. Seafloor spreading works like a conveyor belt (Figure 22.19). Along the rift in a midocean ridge, slabs of lithosphere are slowly moving apart—too slowly for us to notice the movement. **Magma**, warm liquid rock from Earth's interior, wells up from the mantle and seeps out of the cracking rift, forming new lithosphere. Meanwhile, older lithosphere is pushed away from the rift zone. Gravity eventually pulls old oceanic lithosphere into

UNIFYING CONCEPT
- ***The Gravitational Force***
Section 5.3

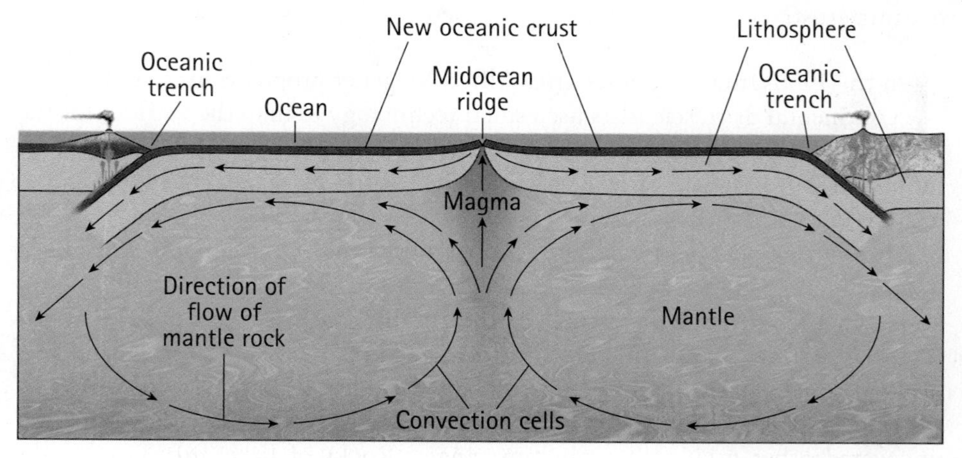

FIGURE 22.19
In conveyor-belt fashion, new lithosphere is formed at the rift zone of a midocean ridge while old lithosphere is recycled back into the mantle. As a result, the ocean floor is renewed about every 200 million years.

the mantle at a deep ocean trench. The sinking of lithosphere into the mantle is called **subduction**. The subducted lithosphere is heated by the hot mantle and eventually partially melts into it. So, in general, while seafloor spreading creates new lithosphere at the spreading center of a rift zone, old lithosphere is destroyed at an ocean trench in a subduction zone.

Earth has remained about the same size since it formed 4.6 billion years ago. This means that the lithosphere is being destroyed as fast as it is created. Can you see that the rate of seafloor spreading must be about equal to the rate of subduction?

CHECK YOURSELF

1. **Is the seafloor recycled? How?**
2. **If the seafloor is spreading and the continents are moving, why don't people notice it?**

CHECK YOUR ANSWERS

1. New lithosphere is created where magma wells up at a spreading center, and it is destroyed where it sinks back into the mantle at a subduction zone. Once lithosphere melts into the mantle, it can eventually well up as magma to become seafloor again. In this way, the seafloor is recycled.
2. The motion occurs so slowly in comparison with human lifetimes that it is not perceived.

Integrated Science 22C
PHYSICS, CHEMISTRY, AND ASTRONOMY

Magnetic Stripes Are Evidence of Seafloor Spreading

LEARNING OBJECTIVE
Describe what magnetic stripes are, and explain how they indicate that seafloor spreading has occurred.

EXPLAIN THIS The seafloor is covered with tiny compasses. Explain.

Seafloor spreading provides a simple mechanism for plate motion. What's the evidence for this appealing picture?

Some of the earliest evidence for seafloor spreading came from *magnetic stripes*. Earth is a huge magnet with north and south magnetic poles. Once in a great while, Earth's magnetic poles flip—the north and south poles exchange positions. This is called a *magnetic reversal*. In a magnetic reversal, the polarity of the magnetic field (positions of the north and south poles) reverses. Earth's magnetic field has had the same polarity for the past 700,000 years, but evidence shows that it's due for a reversal. There have been more than 300 magnetic reversals during the past 200 million years.

The new rock that forms at midocean ridges contains small crystals of the mineral magnetite. Magnetite crystals are magnetic. Like tiny compass needles, magnetite crystals align with the field set up by Earth's magnetic poles. When the new rock cools and solidifies, the magnetite crystals freeze in place. Thus, the alignment of the crystals becomes "locked in" when magma solidifies as oceanic crust. Rocks of the seafloor therefore hold a record of the polarity of Earth's magnetic field at the time that they solidify.

Paleo means old or ancient. As a prefix, paleo is used to describe things that occurred in the past. For example, *paleoclimate* refers to ancient climates. *Paleomagnetism* is the study of ancient magnetic data, such as the history of Earth's magnetic reversals.

TECHNOLOGY

Direct Measurement of Continental Drift

Today, continental drift is not just deduced from evidence—it can be directly measured. The *Very Long Baseline Interferometry System* (VLBI), for example, was the first system to directly measure the relative motion of Earth's tectonic plates and continents. The VLBI used radio telescopes to detect and record radio signals emitted from quasars. Quasars are so far from Earth (billions of light-years away) that they are virtually pointlike. Their radio emissions, therefore, can be used like a surveyor's beam from a stationary source. The same signal from a quasar arrives at slightly different times at different measuring sites. So, when the VLBI tracked changes in the arrival times of radio signals over a period of years, it showed the rate of movement of the sites relative to each other.

GPS satellites are used to measure continental drift directly.

The Global Positioning System (GPS) is currently used to measure the relative motion of different points on Earth. Because GPS results agree with the VLBI results, they provide a cross-check. The GPS system consists of 20 or so satellites that orbit Earth at an altitude of 20,000 km. These satellites transmit signals back to Earth continuously. Scientists at ground stations around the world use the signals to pinpoint their positions in terms of latitude, longitude, and altitude. Scientists repeatedly measure locations of ground stations, monitor changes in their relative positions, and thus track continental movement.

The different measurements of continental drift agree with one another and with theoretical predictions. For example, results show that Hawaii is moving in a northwesterly direction toward Japan at a rate of 8.3 cm per year. Maryland is moving away from England at a rate of 1.7 cm per year.

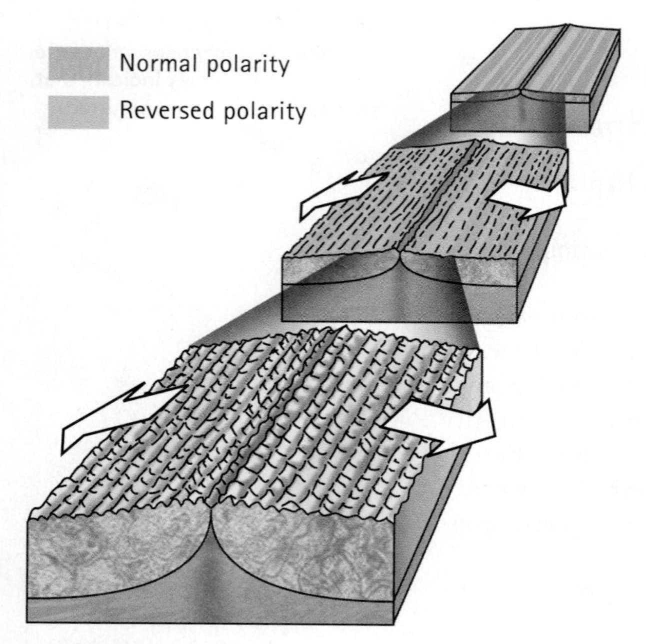

FIGURE 22.20
The symmetrical pattern of magnetic stripes on both sides of the spreading center shows that seafloor forms on both sides of a midocean ridge at about the same time.

Scientists surveying the seafloor in the 1950s noticed odd patterns in its magnetic character. They found magnetic "stripes," or bands of oppositely directed magnetic rock, on both sides of midocean ridges. One stripe had "normal" (present-day) polarity, while the band next to it had opposite or "reversed" polarity (Figure 22.20). Initially, scientists thought the alternating bands of normally and reversely magnetized rocks were so incredible that their measurements must be wrong. But other studies consistently found the same pattern. Then, in the 1960s, when the seafloor was revealed to be geologically active, scientists could see that magnetic stripes can be explained as a result of seafloor spreading. Matching magnetic stripes, symmetrically arranged on either side of an ocean ridge, strongly suggest that the seafloor spreads outward from a central rift.

The ages of the seafloor basalts reinforce the picture. Radiometric dating reveals that the rocks of the seafloor are youngest close to the ocean ridges and they become progressively older moving away from the ridges. Like the magnetic stripes, the age pattern in oceanic crust is symmetrical across a ridge. It thus appears that, as seafloor spreading progresses, previously formed rocks continually spread apart and move away from the ridge while fresh molten rock rises from the asthenosphere to form new lithosphere at the ridge.

CHECK YOURSELF

1. **Why doesn't the polarity of magnetite crystals reverse when Earth's magnetic field reverses?**
2. **Why is the pattern of magnetic stripes the same on both sides of a midocean ridge?**
3. **If there were a magnetic pole reversal tomorrow, in which direction would a compass needle point?**

CHECK YOUR ANSWERS

1. Magnetite crystals can align with Earth's magnetic field only when they are free to turn. They are free to turn when they are in liquid rock, but frozen in place when the rock cools to become solid.
2. The stripes are symmetrical because they consist of bands of rock laid down at the same time.
3. A compass needle would point south.

22.6 The Theory of Plate Tectonics

LEARNING OBJECTIVE
Summarize the theory of plate tectonics in one or two sentences.

EXPLAIN THIS Why was the history of planet Earth largely a mystery before the 1960s?

Plate tectonics is the far-reaching theory that explains the large-scale forces that act inside the Earth to shape its surface. It is founded on Wegener's hypothesis of continental drift but further refined and expanded by evidence gathered from the ocean floor in the 1950s and 1960s. When it was completed, the theory of plate tectonics was able to do for geology what the theory of evolution did for biology. As geology writer John McPhee once said, "The whole world suddenly makes sense."

UNIFYING CONCEPT
- *The Scientific Method* Section 1.3

Plate tectonics states that Earth's rigid outer shell, the lithosphere, is divided into a dozen or so large pieces plus a number of smaller ones (Figure 22.21). Each piece of lithosphere is called a **tectonic plate** or, simply, a *plate*. Plates are up to 100 km thick. Being slabs of lithosphere, tectonic plates consist of the uppermost mantle plus the crust. The plates ride atop the plastic asthenosphere below. Some

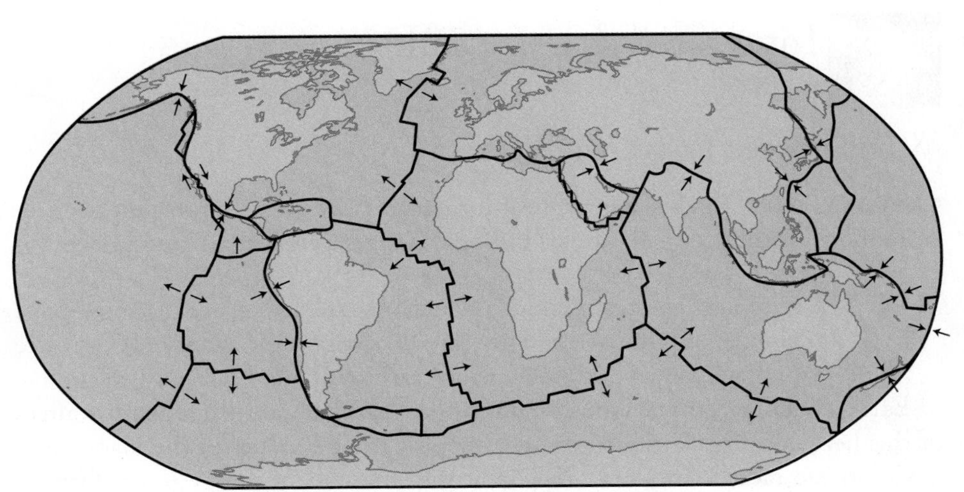

FIGURE 22.21
The lithosphere is broken into pieces called *tectonic plates*. There are a dozen or so major plates plus a number of smaller ones.

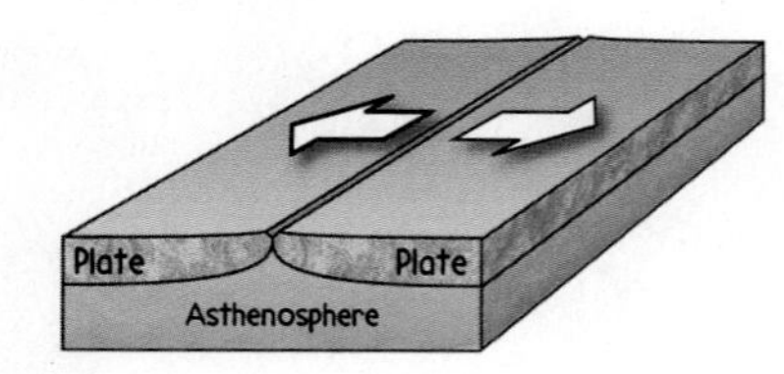

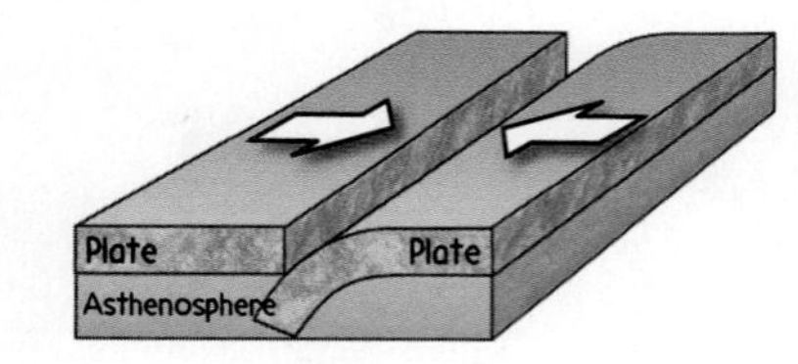

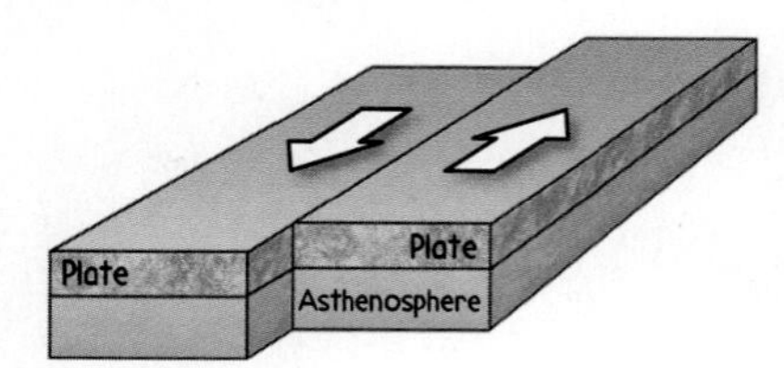

FIGURE 22.22
Plates interact at their boundaries, which results in intense geologic activity in those areas.

plates are large; others are small. Some are relatively immobile, while others are more active. Note that it is the plates that shuffle about Earth's surface. The continents move only because they are embedded in the plates. So, we can see that Wegener's hypothesis was a bit incomplete because whole plates are drifting—not merely the continents.

The continents don't have the same boundaries as the plates. The North American Plate, for example, is much larger than the North American continent. The western side of the North American Plate approximately traces the continent's West Coast, but the plate extends eastward halfway across the Atlantic Ocean to the Mid-Atlantic Ridge.

Earth's plates move in different directions at different speeds, ranging from 2 cm per year to about 15 cm per year. Plates that carry continents, such as the North American Plate, are slow movers, while oceanic plates, such as the Pacific Plate, tend to move much faster. This is because the continents act like heavy barges, with subsurface material—"roots"—extending farther down into the mantle than oceanic plates. Just as a barge dragging its bottom along a rocky river bed encounters resistance, so do continental plates when they drag their extended bottoms through the asthenosphere.

Plates pull apart, crash, merge, and separate from one another over geologic time. Because of all these interactions between plates, the edges of plates, *plate boundaries*, are regions of intense geologic activity (Figure 22.22). While interiors of plates are relatively quiet, most earthquakes, volcanic eruptions, and mountain building occur where plates meet.

CHECK YOURSELF

How does the theory of plate tectonics differ from the hypothesis of continental drift?

CHECK YOUR ANSWER

Plate tectonics states that plates, not just continents, move. Also, because the theory of plate tectonics includes seafloor spreading, it provides a way for continents to move—a *mechanism.*

LEARNING OBJECTIVE
Describe the role of mantle convection and the role of gravity in tectonic plate motion.

What Forces Drive the Plates?

EXPLAIN THIS Rock is a good heat insulator. Why is this important for plate motion still occurring today, billions of years after Earth's formation?

What forces are strong enough to push the seafloor apart and move huge chunks of lithosphere? There are two important factors: (1) the force of gravity and (2) the escape of Earth's internal heat by convection.

Earth contains a tremendous amount of heat. There are three main sources of this heat: The first is radioactivity. When unstable isotopes of the mantle and core undergo radioactive decay, they emit vast amounts of energy in the form of electromagnetic radiation and the kinetic energy of accelerated particles. When this energy is absorbed by rock, some of it is converted into heat, warming the rock. Second, heat remains stored in the Earth from the "Great Bombardment," the time in Earth's early history when chunks of space debris frequently crashed into Earth. The kinetic energy of the colliding particles was transformed into heat energy by friction when the particles crashed into the rock. The third source of Earth's heat also relates to Earth's early history. As denser material sank toward Earth's center through the process of differentiation, the gravitational potential energy of dense, sinking material was converted to kinetic energy then ultimately to heat.

As we know from the second law of thermodynamics, heat always flows from warmer to cooler places. This is why heat moves from Earth's interior to its surface. Heat flows from Earth's interior to its surface mostly through the process of convection. The convection cells in Earth are like the convection cells in a pot of boiling water (Figure 22.23). Hot material rises in the mantle and then moves horizontally while it slowly cools. The asthenosphere carries lithosphere with it, piggyback style. Cooled material sinks at subduction zones back into deeper parts of the mantle. The convection cells in the mantle that carry heat away from Earth's interior are the major driving force of plate tectonics.

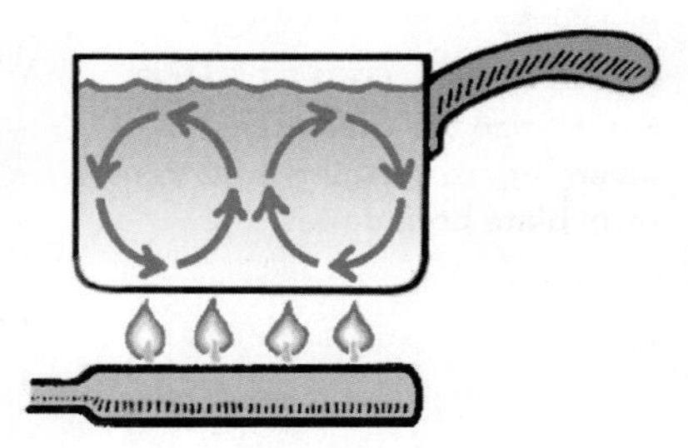

FIGURE 22.23
Convection cells in a pot of water.

UNIFYING CONCEPT

- ***The Second Law of Thermodynamics***
  Section 6.5
- ***Convection***
  Section 6.9

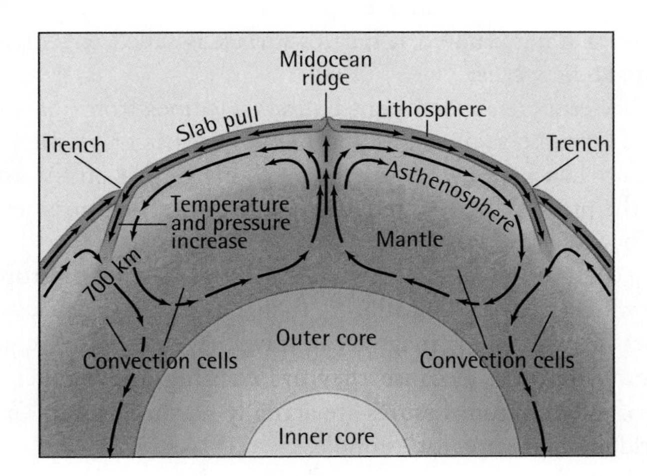

FIGURE 22.24
Simplified view of convection cells within the mantle.

UNIFYING CONCEPT

- ***The Gravitational Force***
  Section 5.3
- ***Density***
  Section 2.3

Gravity plays a role too. As the lithosphere moves away from spreading centers, it cools and becomes more dense. When the cool edge of the plate is dense enough to sink into the asthenosphere, it pulls the rest of the plate along with it. This process is called *slab pull*, and it contributes to the motion of tectonic plates.

CHECK YOURSELF

1. **Why does the asthenosphere flow?**
2. **What role does gravity play in plate motion?**

CHECK YOUR ANSWERS

1. Convection is the reason the asthenosphere flows. The asthenosphere is at the top of a convection cell that carries heat away from Earth's interior to its surface.
2. Gravity pulls the dense, cold edge of older lithosphere into the asthenosphere. This pulls the rest of the plate as well.

Recent evidence is refining our understanding of mantle convection. Many scientists think convecting rock may be more like the irregular blobs in a lava lamp than the more regular circulating currents in a pot of boiling water.

LEARNING OBJECTIVE
Summarize the characteristics of diverging, converging, and transform plate boundaries.

22.7 Plate Boundaries

EXPLAIN THIS Why should you live on a plate boundary if you want to experience earthquakes and other kinds of intense geologic activity?

Tectonic plates meet at *plate boundaries.* Because the plates interact with one another at their boundaries, a great deal of geologic activity occurs there. Mountain building, volcanoes, and earthquakes are much more common at plate boundaries than inside the plates. There are three types of tectonic plate boundaries:

- *Divergent boundaries*, where plates move away from each other
- *Convergent boundaries*, where plates move toward each other
- *Transform boundaries*, where plates slide past each other

Divergent Boundaries

Neighboring plates move away from one another along border zones called **divergent boundaries**. The plates are literally pulled apart at divergent boundaries as tensional forces tug them in opposite directions. As the plates separate, magma wells up from below and erupts into the widening gap between the plates. (Magma that has erupted at Earth's surface is called *lava*.) Lava cools and solidifies to make new crust.

The lava that erupts at a divergent boundary comes from magma in the asthenosphere. The asthenosphere, as you know, consists of plastic but still solid rock. However, when plates separate, the overlying weight on the asthenosphere decreases, so the pressure there is reduced. Rock in the asthenosphere then melts (partially) to become magma.

Lava can seep through fissures in Earth's surface or it can erupt through a central vent, which makes a volcano. (A fissure is simply a crack extending far into the planet through which magma can travel.) We are better acquainted with eruptions from volcanoes because they are exciting and dangerous, but the outpourings of magma from fissures are actually far more common—the rift at a midocean ridge is a fissure, for example.

Divergent boundaries in the ocean floor produce seafloor spreading. The Mid-Atlantic Ridge is a divergent boundary, dividing the North American Plate and the Eurasian Plate in the North Atlantic and the South American Plate and the African Plate in the South Atlantic. The rate of spreading at the Mid-Atlantic Ridge ranges between 1 cm and 6 cm per year. However slow this spreading may seem, over geologic time, the effect is tremendous. During the past 140 million years, seafloor spreading has transformed a tiny waterway through Africa, Europe, and the Americas into the vast Atlantic Ocean of today.

Divergent boundaries aren't only on the seafloor. They can also be in the middle of a continent, and then the continent tears apart. The East African Rift Zone is an example. If spreading continues there, the rift valley will lengthen and deepen and eventually extend out to the edge of the present-day African continent. The African continent will separate, and the rift will become a narrow sea as the Indian Ocean floods the area. The easternmost corner of Africa will then be a large island.

Convergent Boundaries

Earth has remained more or less its present size since it formed about 4.6 billion years ago. This means that lithosphere must be getting destroyed about as fast

MATH CONNECTION

Calculate the Age of the Atlantic Ocean

It's easy to calculate the age of the Atlantic Ocean if you can reasonably estimate the rate of seafloor spreading and the present width of the ocean. The Atlantic Ocean is currently about 7000 km or 7×10^8 cm wide. We assume that the average rate at which the plates diverge in the Atlantic Ocean is 5 cm per year, and we make the assumption that this rate has been constant over geologic time. We then apply the familiar equation that relates speed, time, and distance:

$$\text{Time} = \frac{\text{distance}}{\text{speed}}$$

$$= \frac{7 \times 10^8 \text{ cm}}{5 \text{ cm/year}}$$

$$= 1.4 \times 10^8 \text{ years}$$

$$= 140 \text{ million years}$$

Problem The Red Sea is presently a narrow body of water located over a divergent plate boundary. Based on current studies of the rate of seafloor spreading, the Red Sea will be as wide as the Atlantic Ocean in 200 million years. In order for this to be true, how fast is the seafloor spreading away at this divergent boundary?

Solution If we take the current width of the Red Sea as negligible, the seafloor will widen by a distance of 7×10^8 cm in 200 million years. The speed at which the seafloor spreads is then:

$$\text{Speed} = \frac{\text{distance}}{\text{time}}$$

$$= \frac{7 \times 10^8 \text{ cm}}{2 \times 10^8 \text{ cm}}$$

$$= 3.5 \text{ cm/yr}$$

as it is created. The destruction of Earth's lithosphere takes place at **convergent boundaries**. Here, plates come together in slow-motion collisions. Usually, one plate *subducts*, or descends, below another. The area around a subducting plate is called a **subduction zone**.

Plates converge in three different ways, depending on the kind of lithosphere that is involved. The three kinds of convergent plate boundaries are:

- Oceanic–oceanic convergence
- Oceanic–continental convergence
- Continental–continental convergence

In *oceanic–oceanic convergence*, both of the colliding plates are capped with oceanic crust. When these plates meet, the older (and therefore cooler and denser) plate bends and slides beneath the younger, less-dense plate. This creates a deep *ocean trench*, a long depression in the seafloor found in a subduction zone. Ocean trenches run parallel to the edges of convergent boundaries. They can be thousands of kilometers long, from 8 to 12 kilometers deep, and about 100 kilometers wide. Trenches are the deepest places on Earth, inhabited by strange organisms called *extremophiles* that can survive at extreme high pressure, low temperature, and no light.

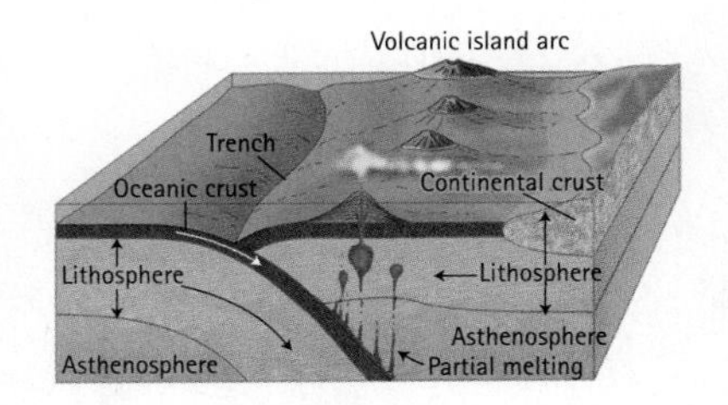

FIGURE 22.25
Oceanic–oceanic convergence occurs when plates that do not carry continental crust at their leading boundaries meet. A trench forms in the subduction zone, and an island arc of volcanoes grows from magma that is generated in the collision.

Figure 22.25 shows that mantle rock in the subduction zone partially melts to form magma. It buoyantly rises and erupts at the surface as lava. Over millions of years, the erupted lava and volcanic debris accumulate on the ocean floor until they grow tall enough to poke above sea level and form a volcanic island.

Volcanoes of this kind are grouped together in *island arcs* that parallel the trenches, such as the Aleutian Islands off the Alaskan Peninsula. Strong to moderate earthquakes are common along such boundaries as the descending plate sticks and slips by the overriding plate.

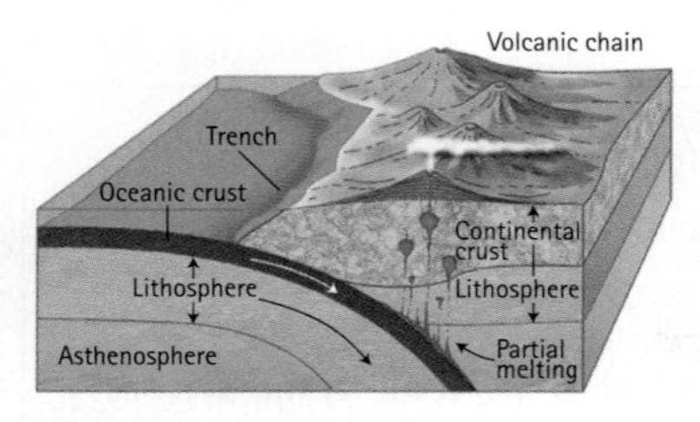

FIGURE 22.26
Oceanic–continental convergence occurs when a plate with oceanic crust at its leading edge is subducted beneath a plate with continental crust along its leading edge. A deep ocean trench and a coastal mountain range form as a result.

A second kind of plate collision is *oceanic–continental convergence* (Figure 22.26). In this case, a plate capped with continental crust on its leading edge slowly collides with a plate capped by oceanic crust. Being denser, the basaltic oceanic plate subducts beneath the less dense, granitic continental plate. A deep ocean trench forms offshore where the converging plates meet. As mantle rock partially melts, magma forms in the subduction zone and rises up and erupts at the surface as lava. Lava erupts, cools, and accumulates many times. Eventually, a volcanic mountain chain may develop.

But, why does rock partially melt in the subduction zone to form magma? It's not because the subducting plate is hot. Subducting lithosphere, remember, is cool and dense—it's relatively cold rock that sinks down into the asthenosphere. When a descending plate reaches a depth of 100–150 km, the heat and pressure of the surrounding environment drive water trapped in the subducting plate into the overlying mantle. The water acts like salt on ice or like flux at a foundry—it lowers the freezing point of the mantle rock. So, with the injection of water, the overlying mantle partially melts without changing temperature. Magma is generated and buoyantly rises. Often, it pools beneath the continental crust, where it may partially melt some of the neighboring crustal rocks. Eventually, this molten rock migrates to the surface, where it erupts—either calmly or explosively. The residents of the Pacific Northwest should appreciate this process because it has created the beautiful mountains of the Cascade Range.

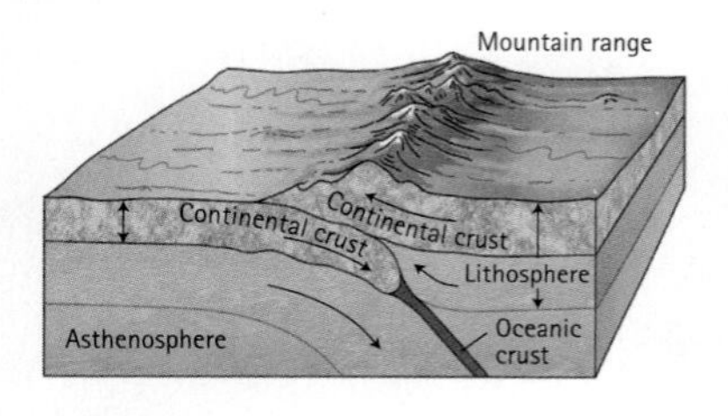

FIGURE 22.27
Continental–continental convergence occurs when continental crust caps the leading edge of each colliding plate. Mountains form where crust wrinkles and pushes upward.

Finally, when plates that are converging head-on have continental crust along their leading edges, they display *continental–continental convergence* (Figure 22.27). In this case, the blocks of colliding crust consist of the same type of buoyant granitic rock. Because the blocks of rock have the same density, neither sinks below the other when they collide—there is no subduction. Instead, the continents push one another upward, like crumpled cloth, and towering, jagged mountain chains result. For example, the Himalayas are the highest mountains in the world, rising to a majestic 8854 m (5.5 mi) above sea level. The Himalayas formed when the subcontinent of India rammed into Asia about 50 million years ago. By welding together along the plate boundary, India and Asia merged as one. The Himalayas are still growing at a rate of about 1 cm per year as the Eurasian and Indian plates continue to converge (Figure 22.28).

FIGURE 22.28
The continent-to-continent collision of India with Asia produced—and is still producing—the Himalayas.

Transform Boundaries

A **transform boundary** is a region where two tectonic plates meet, but rather than converging or separating, they slide past each other. Lithosphere is neither created nor consumed at these boundaries, since plates simply rub along each other as they move in opposite directions.

Transform boundaries are very large *faults*. A fault is any crack that divides two blocks of rock that have moved relative to each other. Faults can be much smaller than transform boundaries; networks of them often form near plate boundaries and branch into the interiors of plates.

Usually, transform boundaries join two segments of a midocean ridge. For example, the Mid-Atlantic Ridge is broken up into segments, which are connected by transform faults (Figure 22.29). These faults "transform" (or transfer) the motion from one ridge segment to another. As the figure shows, lithosphere at one ridge moves in a direction opposite that of the lithosphere of another ridge. In this way, slippage along the transform boundaries allows the tectonic plates to move.

As the rocky plates at a transform boundary grind past each other, their motion is usually steady and slow. But, in some locations friction is so great that entire sections of rock become stuck against each other. Plate motion continues, and the blocks of "stuck" rock become compressed or stretched. (Recall from earlier in this chapter that, even though rock *seems* brittle, it is not. It is capable of compressing as well as stretching and storing energy like a spring.) When the compressional or tensional forces pushing or pulling the rock exceed friction, the stressed rock suddenly breaks loose and slips, releasing its stored energy in a sudden jerk. This sticking and slipping of blocks of rock causes earthquakes.

Although most transform faults are located within the ocean basins, a few are found on continental plates. One continental transform fault that gets a lot of attention is the San Andreas Fault, which is shown in the drawing and photo in Figure 22.30. The San Andreas Fault is the thickest thread in a tangle of faults that collectively accommodate the motion between the North American Plate and the Pacific Plate in California.

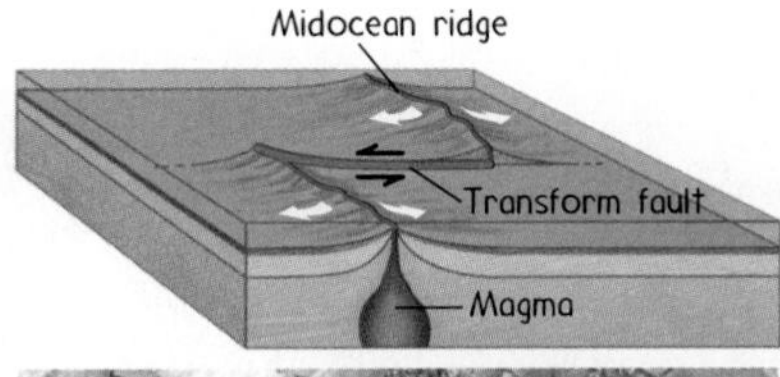

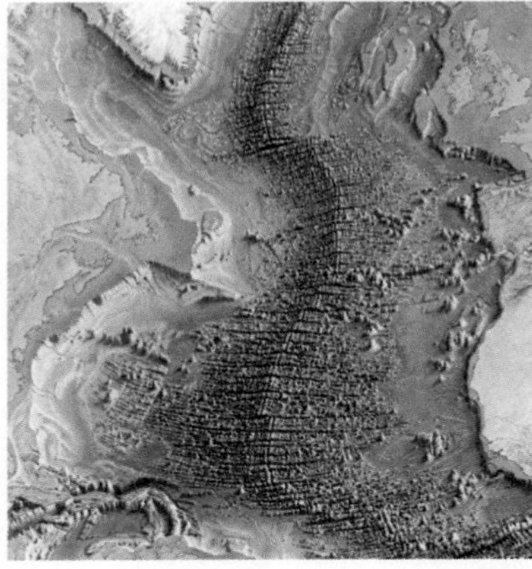

FIGURE 22.29
Most transform boundaries offset segments of a midocean ridge as shown, allowing plates to move laterally along Earth's surface. The plates stick and slip as they move along the boundary, causing earthquakes.

UNIFYING CONCEPT
- ***Friction***
  Section 2.8

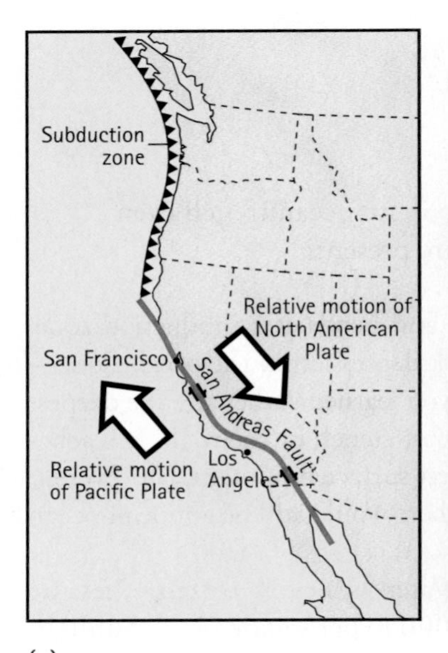

(a)

(b)

FIGURE 22.30
(a) The San Andreas Fault is a transform-fault plate boundary infamous for its earthquakes. The slice of California moving northwesterly lies on the Pacific Plate, while the rest of California sits on the North American Plate. (b) In this photo of the San Andreas Fault, notice the long valley created by many years of rock grinding along the fault.

The slice of California to the west of the fault is slowly moving northwest, while the rest of California is moving southeast. Contrary to popular opinion, you can see from the directions of the plates that Los Angeles is not going to split off and fall into the ocean. Instead, it will steadily advance northwesterly toward San Francisco while San Francisco moves in the opposite direction. In about 16 million years, the two cities will be side by side. In the meantime, California residents can expect plenty of earthquakes.

CHECK YOURSELF

1. **The African Plate is slowly moving northward toward the Eurasian Plate. Both plates have continental crust along the edges where they would meet. What is likely to happen if these plates collide with each other?**
2. **What types of plate boundary involve subduction? Which do not involve subduction?**
3. **Do volcanic island arcs and coastal mountain ranges form at spreading centers?**
4. **In one sentence, distinguish among the three major kinds of plate boundaries.**

CHECK YOUR ANSWERS

1. A mountain range may form along their continental–continental convergent boundary.
2. Plate boundaries that involve subduction are oceanic–continental and oceanic–oceanic convergent boundaries. The other kinds of boundaries do not involve subduction; these are transform boundaries, divergent boundaries, and continental–continental convergent boundaries.
3. No. Volcanic island arcs and coastal mountain ranges are associated with convergent plate boundaries.
4. Tectonic plates move apart at divergent boundaries, come together at convergent boundaries, and slide past one another at transform boundaries.

LEARNING OBJECTIVE
Describe the conditions in ocean trenches that make life difficult.

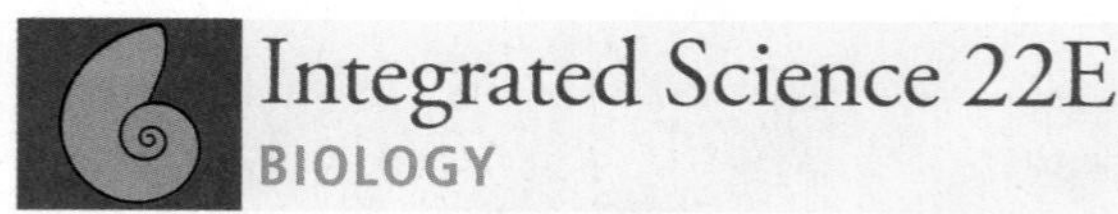

Integrated Science 22E
BIOLOGY

Life in the Trenches

EXPLAIN THIS How can food chains develop in an ocean trench even though the lack of light means that no plants are present?

When lithosphere is destroyed, it sinks and bends at a subduction zone. The bending plate creates a deep, dark depression in the ocean floor—an ocean trench. Ocean trenches, as you learned earlier, are the deepest places on Earth. The Mariana Trench is the deepest trench of them all. It reaches 11,000 m (7 mi) below sea level. If the world's tallest surface mountain, Mt. Everest, were sunk to the bottom of the Mariana Trench, there would still be a mile of ocean above it!

Conditions in the trenches are extreme. Water is nearly frozen. Pressure reaches more than 1000 atmospheres (about 16,000 lb per square inch). Sunlight is totally absent in the pitch-black depths.

Yet life thrives in the trenches. The trenches are inhabited by strange organisms called *extremophiles.* Trench extremophiles can survive at extreme high pressure, low temperature, and no light. Trench extremophiles include species of worms, shrimp, fish, crabs, and microbes. The world's deepest-dwelling fish, *Abyssobrotula galatheae,* was found in the Puerto Rican Trench at a depth of 8372 m (more than 5 mi). The adaptations of trench extremophiles are not well understood.

Some ocean trenches contain *hydrothermal vents.* Hydrothermal vents are cracks in the oceanic crust. Most often they are found in the newly formed oceanic crust in active rift zones. However, some hydrothermal vents occur where tectonic plates grind past one another in an ocean trench. Heat and chemicals such as hydrogen sulfide are released from Earth's interior at these vents. Pressure-adapted bacteria consume these chemicals and convert the chemical energy stored in them into the energy they need for life. These chemical-consuming bacteria also produce oxygen as a waste product. The process is called *chemosynthesis.* Other organisms eat the bacteria, and still other organisms eat the bacteria-eaters. And so a food chain develops in a hydrothermal vent despite the lack of light for photosynthesis (Figure 22.32).

FIGURE 22.31
The black anglerfish has a huge jaw and teeth that make it an effective predator. This fish needs to be an effective predator because food is so scarce in ocean trenches, where it lives. This fish also has a glowing or bioluminescent "lure" attached to its head to attract prey. Microbes that glow in the dark produce the light for the lure.

FIGURE 22.32
A hydrothermal vent community.

UNIFYING CONCEPT

- ***The Ecosystem***
  Section 21.1

CHECK YOURSELF

1. **What are the physical conditions in the trenches that make life difficult there?**
2. **Where does the heat that escapes from hydrothermal vents come from?**

CHECK YOUR ANSWERS

1. Lack of light, extreme cold, and extremely high pressure.
2. The heat comes from Earth's interior—more specifically, from the asthenosphere.

For instructor-assigned homework, go to www.masteringphysics.com MP

SUMMARY OF TERMS (KNOWLEDGE)

Asthenosphere One of Earth's structural layers—a layer of weak, warm rock that flows slowly over geologic time.

Continental drift The hypothesis that the world's continents move slowly over Earth's surface.

Convergent boundaries Places where tectonic plates come together.

Core Earth's innermost layer, which is mostly iron and includes the inner core and outer core.

Crust Earth's surface layer, consisting of oceanic and continental crust.

Differentiation The process by which Earth formed layers according to density.

Divergent boundaries Places where tectonic plates pull apart.

Earth science The study of the history, structure, and natural processes of planet Earth.

Earthquake The shaking of the ground that results when rock under Earth's surface moves or breaks.

Geology The Earth science that is concerned with the composition and structure of Earth.

Inner core One of Earth's structural layers—a solid sphere of hot metal, mostly iron, at the center of Earth.

Lithosphere Earth's outermost structural layer, consisting of cool, rigid rock.

Lower mantle One of Earth's structural layers—the lowest portion of the mantle, a zone of strong, rigid rock.

Magma Molten rock in Earth's interior.

Mantle The thick layer of dense, hot rock between Earth's crust and core.

Midocean ridge A global system of underwater mountains created by seafloor spreading.

Ocean trenches Long, deep, steep troughs in the seafloor where an oceanic plate sinks beneath an overlying plate.

Outer core One of Earth's structural layers—a shell of hot, liquid metal beneath the mantle and above the inner core.

Plate tectonics The theory that Earth's lithosphere is divided into large plates that move slowly around the globe.

Seafloor spreading The process by which new lithosphere is created at midocean ridges as older lithosphere moves away.

Seismology The study of seismic waves, waves that travel through Earth as a result of an earthquake or other disturbance.

Subduction The sinking of oceanic lithosphere into the mantle.

Subduction zone The region where an oceanic plate sinks into the asthenosphere at a convergent plate boundary.

Tectonic plates Separate pieces of lithosphere that move on top of the asthenosphere.

Transform boundaries Places where tectonic plates slide along beside one another as they move.

READING CHECK (COMPREHENSION)

22.1 Earth Science Is an Integrated Science

1. Name three or more branches of Earth science, and describe the focus of each.
2. Why is Earth science an integrated science?

22.2 Earth's Compositional Layers

3. In what way is Earth like a hard-boiled egg?
4. What kind of rock is most common in the oceanic crust? In the continental crust?

22.3 Earth's Structural Layers

5. Name and describe Earth's five structural layers.
6. Does the asthenosphere have the same composition throughout? What *is* uniform throughout the asthenosphere?
7. What are the large, interlocking pieces of lithosphere called?

22.4 Continental Drift—An Idea Before Its Time

8. Describe the fossil evidence that supported Wegener's hypothesis of continental drift.
9. Why was the hypothesis of continental drift dismissed by scientists for decades?
10. Describe evidence from the rock record that supports continental drift.

22.5 Seafloor Spreading—A Mechanism for Continental Drift

11. Where is lithosphere created? Where is it destroyed?
12. How does seafloor spreading relate to continental drift?
13. Earth has remained about the same size since it formed 4.6 billion years ago. What does this suggest about the rate of seafloor spreading compared to the rate of subduction?

22.6 The Theory of Plate Tectonics

14. In what way is the theory of plate tectonics like the theory of evolution, Newton's laws, and the periodic table?
15. Describe how tectonic plates move in terms of speed and direction.
16. How does plate tectonics differ from continental drift?

22.7 Plate Boundaries

17. Which is a more geologically stable place to live—along a plate boundary or in the interior of a plate? Explain.
18. Describe the three major kinds of plate boundaries.
19. How do plates interact when they collide? There are three possible ways; describe them all.
20. What type of plate boundary is the San Andreas Fault?

THINK INTEGRATED SCIENCE

22A—Earth Developed Layers When It Was Young, Hot, and Molten

21. What elements make up 98% of the Earth by weight?
22. In what way is Earth like a jar of water and oil left standing?
23. What two elements constitute about three-fourths of Earth's solid surface?

22B—Using Seismology to Explore Earth's Interior

24. Cite the seismic evidence that Earth has a liquid outer core.
25. What did studies of P-wave and S-wave shadows reveal about Earth's interior?
26. What is the difference between body waves and surface waves? Which of these seismic waves reveals key information about Earth's internal structure?

22C—Magnetic Stripes Are Evidence of Seafloor Spreading

27. What is a magnetic reversal, and how are magnetic reversals recorded in rock?
28. Where is the oldest rock—near a spreading center or far away from it? Explain your thinking.
29. How do magnetic stripes provide evidence of seafloor spreading?

22D—What Forces Drive the Plates?

30. In what way does the second law of thermodynamics relate to plate tectonics?
31. What are the two main factors in the movement of tectonic plates?
32. In what way is the asthenosphere like a pot of boiling water?

22E—Life in the Trenches

33. Photosynthesis cannot occur in a hydrothermal vent because there is no light there. How, then, can the microbes that serve as the lowest level of a food chain in a hydrothermal vent obtain the energy they need for life?
34. Where do most hydrothermal vents occur? Can they occur in other places as well? Explain.
35. Do all trench-dwelling organisms live in hydrothermal vent communities? Explain.

THINK AND DO (HANDS-ON APPLICATION)

36. Simulate motion along a transform fault with your hands. Press your palms together and move your hands in opposite directions, as shown in the figure. Notice that friction keeps them "locked" together until opposing forces no longer balance, friction no longer holds, and your palms slip suddenly past each other. This nicely models the sticking and slipping that occur along a transform fault.

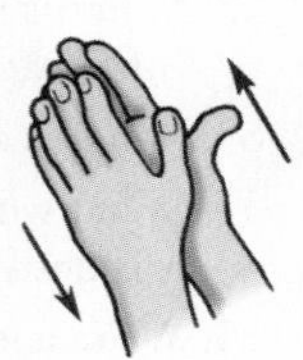

THINK AND SOLVE (MATHEMATICAL APPLICATION)

37. A sample of basalt has a mass of 5.6 g and a volume of 2 cm^3. Show that the density of basalt is 2.8 g/cm^3.
38. The San Andreas Fault separates the northwest-moving Pacific Plate, on which Los Angeles sits, from the North American Plate, on which San Francisco sits. If the plates slide past each other at the rate of 3.5 cm per year, show that the two cities will be neighbors in about 17 million years. (The distance from San Francisco to Los Angeles is 600 km.)
39. Suppose that a fence is built across the San Andreas Fault in 2013. Show that the fence will separate and the two sections of the broken fence will be 70 cm apart in 2033. Assume that the average rate of movement along the plate boundary is 3.5 cm (about 1½ in) per year.
40. The Nazca and Pacific Plates are spreading relatively fast—at the rate of 14.2 cm (5.6 in) per year. The East Pacific Rise is a midocean ridge system that divides these plates. If the central segment of the rise is 200 km wide, show that it is about 700,000 years old.

THINK AND EXPLAIN (SYNTHESIS)

41. Is Earth's inner core solid and the outer core liquid because the inner core is cooler than the outer core? Explain your answer.
42. How is magma generated at divergent plate boundaries?
43. What would happen if new lithosphere were created faster than it is destroyed?
44. Why do mountains rise in the mantle as they erode? What principle of Earth science explains this process?
45. What kinds of plate boundaries feature subduction zones?
46. Why are earthquakes common in subduction zones?
47. Briefly describe how an island arc forms.
48. Why are there expressions such as "solid earth" and "old as the hills," which suggest that Earth is unchanging?
49. How did seismic waves contribute to the discovery of Earth's deep internal layers?
50. Copy the diagram (which is not to scale) of Earth's interior. Add these labels: inner core, outer core, mantle, oceanic crust, continental crust, ocean.

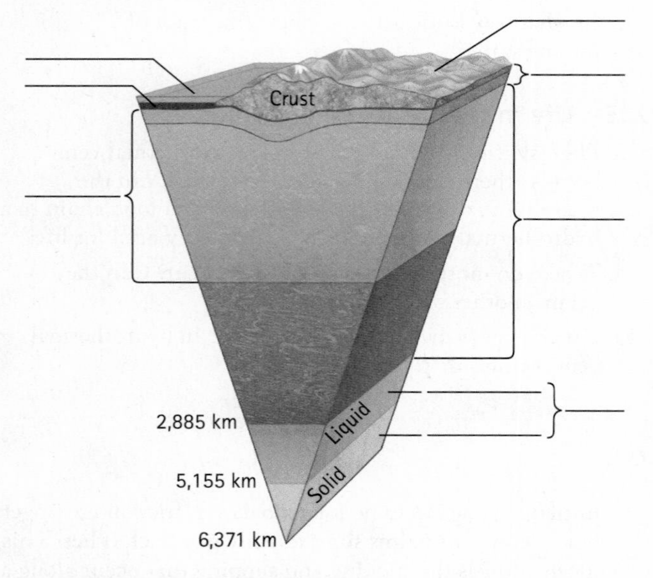

51. Why does Earth's crust "float" on the mantle?
52. Why are volcanic mountain ranges often found along oceanic–continental convergent boundaries?
53. What is the evidence that tectonic plates move?
54. Why are most earthquakes generated near plate boundaries?
55. The magnetic stripes that were laid down on the Pacific Ocean seafloor are wider than the magnetic stripes laid down over the same time period on the Atlantic Ocean seafloor. What does this tell you about the rate of seafloor spreading of the Pacific Ocean compared with that of the Atlantic Ocean?
56. How is Earth's lithosphere like a jigsaw puzzle? How is it NOT like a jigsaw puzzle?
57. Earth's surface is largely covered by volcanic rock or rock that is derived from volcanic rock. Use what you know about plate tectonics to explain why this is true.
58. Could Los Angeles fall into the ocean, as is popularly thought? Why or why not?
59. Is plate tectonics a theory based on integrated science? Why or why not?
60. The graph shows how the seafloor of the Atlantic Ocean varies with depth. Why was this picture surprising to scientists before Harry Hess? How does this graph support the hypothesis of continental drift?

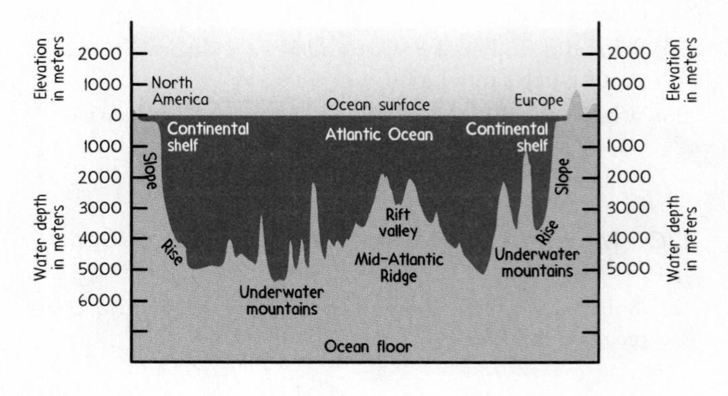

61. Evidence of ancient ice sheets has been found in areas near the equator. Give two possible explanations for this.
62. What is meant by magnetic pole reversals? What useful information do they provide about Earth's history?
63. What is a likely cause of Earth's magnetic field?
64. What would happen if new crust were created faster than it is destroyed?
65. Write a letter to your grandmother or grandfather with a one-paragraph summary of plate tectonics. (Plate tectonics was not an accepted theory until the 1960s, so people born much earlier than that missed out on this important theory in their education!)

THINK AND DISCUSS (EVALUATION)

66. Earth's Moon has a lithosphere that is continuous, not divided into continental and oceanic lithosphere as on Earth. Would you expect that there would be plate tectonic activity on the Moon similar to Earth? Why or why not?
67. The 1993 adventure film *The Core* is based on the premise that Earth's core has stopped spinning. Earth's magnetic field is therefore disrupted and humanity is in peril. Six scientists journey to the center of Earth to set off a nuclear device to set the outer core spinning again. Discuss the scientific flaws with this basic story line.
68. Play a game of "fortunately/unfortunately." First, finish the sentence: "Fortunately, because of plate tectonics . . . ." Then

have a partner finish a different sentence: "Unfortunately, because of plate tectonics . . . ." Go back and forth with your partner until you have a list of ways in which plate tectonics helps people and ways in which it makes life challenging.

69. What is the principal difference between the theory of plate tectonics and the continental drift hypothesis?
70. What law of physics relates heat flow to plate tectonics? Discuss how this law underlies plate tectonics and other everyday processes.
71. Are continents a permanent feature of our planet? Discuss why or why not.
72. Discuss the role of heat and the role of gravity in the motion of tectonic plates.
73. Where does the heat in Earth's interior come from?
74. How would GPS technology have been helpful to Alfred Wegener?
75. When the problem of disposing of radioactive waste first arose, some scientists suggested dumping it into subduction zones. What are two possible challenges with this method of disposal?

READINESS ASSURANCE TEST (RAT)

Choose the BEST answer to each of the following questions.

1. The lithosphere includes the crust and the top part of the
   (a) asthenosphere.
   (b) mantle.
   (c) core.
   (d) lower mantle.
2. The layer of Earth made up of soft rock that flows slowly over time is the
   (a) lithosphere.
   (b) lower mantle.
   (c) hydrosphere.
   (d) asthenosphere.
3. What type of plate boundary lies between two plates that are moving away from each other?
   (a) convergent boundary
   (b) divergent boundary
   (c) transform boundary
   (d) continental–continental convergent boundary
4. A long, deep, steep depression in the seafloor where subduction occurs is a(n)
   (a) midocean ridge.
   (b) rift zone.
   (c) ocean trench.
   (d) hydrothermal vent.
5. Is Earth's inner core a solid, liquid, or gas? Why?
   (a) a solid, because high pressure keeps it from melting
   (b) a liquid, because the temperature of the core is higher than the melting point of iron
   (c) a gas, because the inner core is nearly as hot as the surface of the Sun
6. Why is the outer core the source of Earth's magnetic field?
   (a) Moving liquid iron particles there form an electric current, and an electric current always produces a magnetic field.
   (b) It has north and south magnetic poles that reverse over time.
   (c) It contains iron, a magnetic material.
7. Why doesn't the heat in Earth's core melt Earth?
   (a) Heat tends not to move from place to place.
   (b) The core is a solid.
   (c) Heat escapes slowly to the surface because rock is a poor conductor.
   (d) Actually, Earth's core is not very hot.
8. Why is the inner core Earth's most dense region?
   (a) The inner core is made of mostly iron, which is very dense.
   (b) Dense material sank toward Earth's center when Earth was molten.
   (c) Pressure packs the atoms in the inner core tightly together.
   (d) all of these
9. Fossils such as *Glossopteris* and *Mesosaurus* have been found in widely separated continents because
   (a) *Mesosaurus* was a good swimmer and *Glossopteris* had lightweight seeds.
   (b) land bridges once connected all the continents.
   (c) all the continents were once joined in a supercontinent called Pangaea.
   (d) Alfred Wegener played a hoax on the scientific community.
10. Plate tectonics explains
    (a) how seafloor spreading and subduction account for the movement of tectonic plates.
    (b) why the continents move.
    (c) why plate boundaries are such active geologic regions.
    (d) all of these

Answers to RAT

1. b, 2. d, 3. b, 4. c, 5. a, 6. a, 7. c, 8. d, 9. c, 10. d

23

CHAPTER 23

Rocks and Minerals

ALMOST ALL manufactured products contain minerals or substances derived from minerals. From the aluminum in cans to the graphite in pencils to the halite on your French fries, minerals are practically everywhere. What exactly is a mineral? Where do all the minerals we consume come from? Minerals join together to create rocks, which are the dominant material of Earth. Rocks, like minerals, are economically important. To name just a few examples, coal is a rock burned for fuel, and marble and granite are popular ornamental stones. Rocks provide key information to science, too. They tell the story of Earth's past, if you know how to read them. The properties of rock affect geologic processes such as erosion, flooding, and earthquakes. In what sense does rock recycle? As "rock hounds" know, rocks are fascinating once you learn their story.

23.1 What Is a Mineral?

LEARNING OBJECTIVE
State the characteristics of a mineral.

EXPLAIN THIS Why are diamond and graphite classified as two different minerals even though both are made of pure carbon?

In everyday language, a *mineral* is a part of your diet ("vitamins and minerals"). But in Earth science, a *mineral* is something else. A **mineral** is a material that has these five characteristics:

1. *A mineral is naturally occurring.* In other words, a mineral is formed by nature rather than by people in a laboratory. Silver is a mineral, but nylon is not.
2. *Minerals are inorganic.* Minerals aren't produced just by living things. Wood isn't a mineral because living things—trees—are the only producers of wood.
3. *Minerals have a crystal structure.* Atoms, ions, or molecules are arranged in regular, repeating patterns in minerals. Halite (table salt) has the crystal structure of a cube repeated over and over again (Figure 23.1). Most solids—minerals and nonminerals alike—have a crystal structure. Two important exceptions are glass and plastic, which are *amorphous solids,* solids in which the particles are random and disorderly.
4. *Minerals are solid.* The ice in a glacier is a mineral, but liquid water is not.
5. *Minerals have a specific chemical composition.* For example, the mineral corundum always has the same chemical formula: Al_2O_3. This is corundum's formula whether it is in pure white form or has impurities that give it a red color (ruby) or a blue color (sapphire).

To be considered a mineral, a substance must have all five characteristics. For instance, sugar is solid and has a crystal structure. However, sugar is not a mineral because it is the product of a plant—sugar cane or sugar beets.

Sometimes two minerals contain the same elements in the same ratio but their atoms are arranged differently. As a result, they have different crystal structures

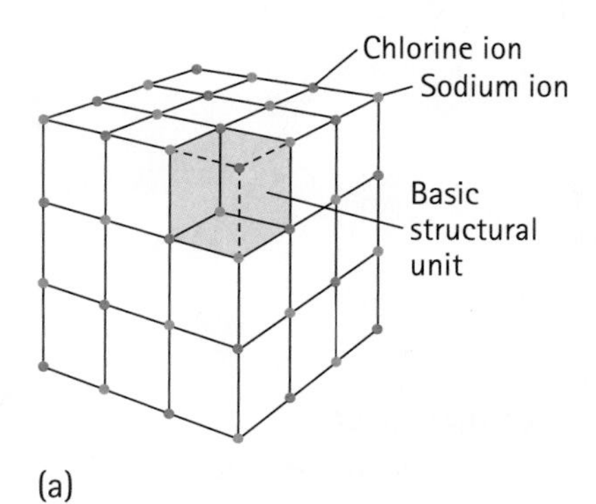

(a)

(b)

FIGURE 23.1
(a) Crystal structure of halite. (b) Grains of the mineral halite (table salt).

(a)

(b)

FIGURE 23.3
(a) Stone Age stone tools. (b) Marble is taken from this modern quarry and used for buildings and sculpture.

FIGURE 23.2
Fine "crystal" is not really crystal at all. Fine crystal is glass, an amorphous solid. In an amorphous solid, atoms are jumbled together in a random fashion. Gems, such as those in the lady's earrings, are true crystals—they have an orderly atomic structure.

FIGURE 23.4
Both graphite and diamond are pure carbon. (a) In diamond, each carbon atom is bonded to each of its neighbors. (b) In graphite, carbon atoms are bonded in two dimensions, forming sheets.

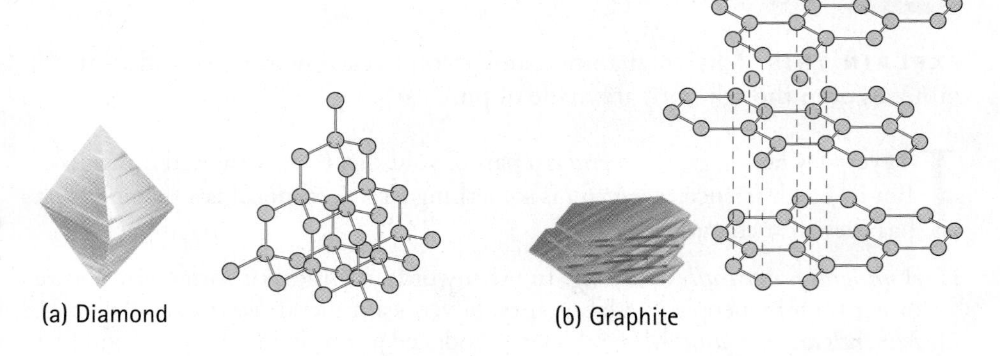

UNIFYING CONCEPT

- *The Atomic Theory of Matter* Section 9.1

and different properties. Such minerals are called *polymorphs*. Diamond and graphite (the "lead" in pencils) are polymorphs. They are both pure carbon, but diamond and graphite are different minerals because their atoms are arranged differently. In diamond, the carbon atoms form tight, three-dimensional networks (Figure 23.4). This makes diamond very hard. But, in graphite, the carbon atoms form sheets that slip past one another easily. When you write with a pencil, sheets of graphite atoms pull apart, and this leaves a mark on paper.

CHECK YOURSELF

1. **The graphite in pencil "lead" is a mineral. Why isn't the paper you write on a mineral?**
2. **Gold atoms bond together to make a tiny cube. This tiny cube is repeated over and over to make crystals of gold big enough to see. Halite (table salt) also has a cubic crystal structure. If gold and halite have the same crystal structure, why aren't they classified as the same mineral?**

CHECK YOUR ANSWERS

1. Paper is not a mineral for three reasons: One, it has no specific chemical composition. Two, it has an organic source—trees. Three, paper is made by people; it does not exist naturally in the environment. Pencils are also made by people, but the graphite in them has a specific chemical composition (carbon) and is a naturally occurring inorganic substance.
2. Gold and halite have different chemical compositions.

LEARNING OBJECTIVE
Name seven properties of minerals that are useful in identifying them.

23.2 Mineral Properties

EXPLAIN THIS Why do some minerals break into cubes when you hit them with a hammer?

About 4000 minerals are known today. Minerals can be classified into different groups by their physical properties. Seven important physical mineral properties are crystal form, hardness, cleavage, luster, color, specific gravity, and streak. Some minerals also have the property of being radioactive. Others are fluorescent or magnetic or toxic. Most of Earth's minerals look like everyday rocks, but some have vibrant colors and amazing forms (Figure 23.5).

FIGURE 23.5
A few of the many minerals with a striking appearance. (a) Hematite has a grape-cluster shape. (b) Amethyst is the purple variety of quartz, which grows in hexagonal (six-sided) crystals with pointed ends. (c) Pyrite, or "fool's gold," typically forms cubic crystals marked with parallel lines called "striations." (d) Rosasite has fibrous, blue-green crystals. (e) Rhodochrosite (whose name means "rose-colored") has rhombohedral crystals. (f) Asbestos forms strong, heat-resistant fibers.

Crystal Form

Each of the mineral samples shown in Figure 23.5 has a characteristic *crystal form*, or external shape. A mineral's crystal form is determined by its microscopic crystal structure. Sometimes a mineral's outward crystal form is *the same* as its internal structure of atoms. Figure 23.2 shows the cubic crystal structure of halite, which is mirrored in halite's cubic crystal form. Look at some salt crystals with a hand lens and you can see this cubic crystal form.

You are lucky if you find well-formed crystals in nature because they are rare. Usually, lots of crystals grow together in cramped locations. As they grow, each tiny crystal competes with neighboring crystals for space. The crystals that develop are so small and intermixed with one another that the mineral's crystal form can't be observed without a microscope.

Hardness

A mineral's *hardness* is its resistance to scratching. A harder mineral can scratch a softer one, but not the other way around. Why are some minerals harder than others? Hardness depends on the strength of a mineral's chemical bonds—the stronger its bonds, the harder the mineral. The carbon atoms in diamond are tightly bound, so diamond is especially hard. Mohs scale of hardness (Table 23.1) rates the hardness of different minerals on a scale of 1 to 10. Diamond has a rating of 10.

A piece of quartz can scratch feldspar, but feldspar cannot scratch quartz.

TABLE 23.1 MOHS SCALE OF HARDNESS

Mineral	Hardness	Object of Similar Hardness
Talc	1	
Gypsum	2.5	Fingernail
Calcite	3	Copper wire
Fluorite	4	
Apatite	5.5	
Feldspar	6	Window glass
Quartz	7	Steel file
Topaz	8	
Corundum	9	
Diamond	10	

CHECK YOURSELF

1. **Can quartz scratch topaz? Can topaz scratch quartz? Defend your answers.**
2. **Your friend finds a ring that looks like a diamond. But it could be a cubic zirconium—a "fake diamond." Zirconium has hardness 9. Your friend tests the ring and finds that it will scratch glass. Is the ring a diamond?**

CHECK YOUR ANSWERS

1. Quartz, with hardness 7, cannot scratch topaz, with hardness 8, because topaz is harder than quartz. Topaz can scratch quartz (and all other minerals with hardness less than 8).
2. You cannot tell because any mineral with hardness greater than 6 will scratch glass.

Cleavage and Fracture

Another property determined by a mineral's crystal structure and chemical bond strength is cleavage. *Cleavage* is the tendency of certain minerals to break along distinct planes of weakness, areas where the bonds holding the crystal together are weakest. Minerals that exhibit cleavage break along flat, even surfaces.

(a)

(b)

FIGURE 23.6
(a) Muscovite is a mineral of the mica group. It has the property of cleavage and easily breaks into flat sheets.
(b) Calcite has perfect cleavage in three directions. It breaks up into little cubes when you hit it with a hammer.

Minerals may display cleavage in one or more directions. For instance, mica displays cleavage in one direction. Mica's crystal structure is atoms arranged in stacks. If you hit a piece of mica with a hammer, the mica breaks into sheets (Figure 23.6). Other minerals have perfect cleavage in multiple directions, which allows them to be cut into neat blocks. Jewelers use the natural cleavage of diamonds, emeralds, and certain other minerals to cut them into beautiful gemstones.

A mineral that does not have cleavage breaks unevenly instead of along flat planes. We call this *fracture*. For instance, if you hit a piece of quartz with a hammer, it breaks along smooth, curved surfaces. This kind of fracture is called a *conchoidal* fracture (Figure 23.7).

FIGURE 23.7
This quartz sample does not exhibit cleavage. When it breaks, it develops a conchoidal fracture—a smooth, curved surface resembling a broken soda bottle.

Luster

The way a mineral reflects light is its *luster*. Geologists use a variety of terms to describe luster: Very shiny minerals have a "metallic" luster. Minerals that have a greasy or oily appearance have a "waxy" luster. Rough, dull minerals have an "earthy" luster. Other minerals have a "pearly" or "silky" luster.

Color

Sometimes a mineral's *color* can help you identify it—but only sometimes. This is because the same mineral can exist in several colors. Pure quartz is colorless, but quartz can also be purple, milky white, rose pink, yellow, or smoky brown if it contains impurities. Similarly, pure corundum is white. But if one out of every 5000 aluminum atoms is replaced with a chromium atom, we have the red variety of corundum prized as ruby (Figure 23.8).

Also, you can't use color to identify minerals because sometimes different minerals have the same color. Pyrite, or fool's gold, is gold in color, but so is the element gold. Is that mineral you found in the stream bed precious gold or merely pyrite? You must check other properties to be sure.

Ruby

Sapphire

FIGURE 23.8
Rubies and sapphires are varieties of the mineral corundum that occur when chemical impurities are present. As these examples show, color alone is not a reliable indicator of a mineral's identity.

Specific Gravity

Specific gravity is the ratio of the weight of a mineral to the weight of water. The specific gravity of copper is 9, so copper is nine times heavier than water for a given volume.

Although color isn't a good property for identifying minerals, specific gravity is. The specific gravity of gold is 19, while fool's gold (pyrite) has a specific gravity of 5. You can tell which is which by how heavy each mineral feels in relation to its size. Note that specific gravity is a lot like density: Both tell you how heavy a material is for its volume. The difference between the two is that specific gravity has no units.

UNIFYING CONCEPT

- ***Density***
  Section 2.3

Streak

Streak is the color of a mineral in its powdered form. To find a mineral's streak, you rub it on a piece of unglazed porcelain called a *streak plate.* The mineral's streak is not always the same as the color of the mineral itself, but it is characteristic of the mineral. For example, hematite comes in a variety of colors, but it always has a reddish-brown streak.

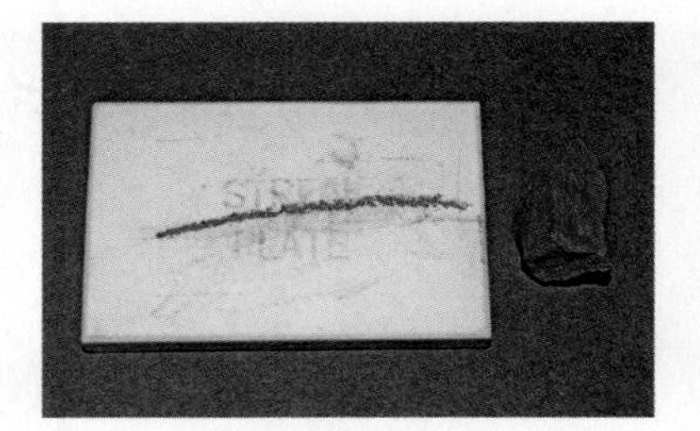

FIGURE 23.9
A mineral's streak is helpful in identifying it.

CHECK YOURSELF

1. **Why is crystal form alone not sufficient for identifying a mineral?**
2. **You work in the stockroom of a museum. A mineral sample has lost its label. What properties are the best ones to use to identify this mineral?**

CHECK YOUR ANSWERS

1. Different minerals can have the same crystal form. For example, both gold and halite have a cubic structure. Crystal form *and* chemical composition provide a better test.
2. Specific gravity, streak, and hardness are the best. Although each property cannot be used alone, when used together, they are reasonably accurate means of identification.

23.3 Types of Minerals

LEARNING OBJECTIVE
Identify the major differences between silicate and nonsilicate minerals.

EXPLAIN THIS Why are silicates the most common minerals in Earth's crust?

You learned in chemistry that there are 112 naturally occurring chemical elements. Most of these elements are rare. In fact, only eight elements make up 98% of Earth's entire mass! Just two chemical elements, silicon and oxygen, make up more than 75% of Earth's crust. It is no surprise, then, that most of Earth's minerals contain silicon and oxygen. Minerals that contain silicon and oxygen bonded together are called **silicates**. Minerals that do not contain silicon and oxygen bonded together are *nonsilicates.* Earth's 4000 or so minerals are divided into these two groups.

FIGURE 23.10
Although there are 112 naturally occurring elements, only a handful or so make up almost all of Earth's mass. More specifically, eight elements make up 98% of Earth's mass. Just two elements—silicon and oxygen—make up more than 75% of Earth's crust. This is why most minerals are silicates.

Silicates

Silicates, the large class of minerals that contains silicon and oxygen, is divided into two groups: *ferromagnesian* and *nonferromagnesian* silicates. Ferromagnesian silicates always contain iron or magnesium or both of these elements. (*Ferrum* is Latin for "iron," and *magnesian* refers to magnesium.) Ferromagnesian silicates tend to be dark, dense, and glassy and are often found in the oceanic crust. Olivine and hornblende are two examples of ferromagnesian silicates. Nonferromagnesian minerals are generally lighter in color and less dense, and they are extremely common in the continental crust. Feldspar is a nonferromagnesian silicate; feldspar minerals alone make up about half of Earth's crust!

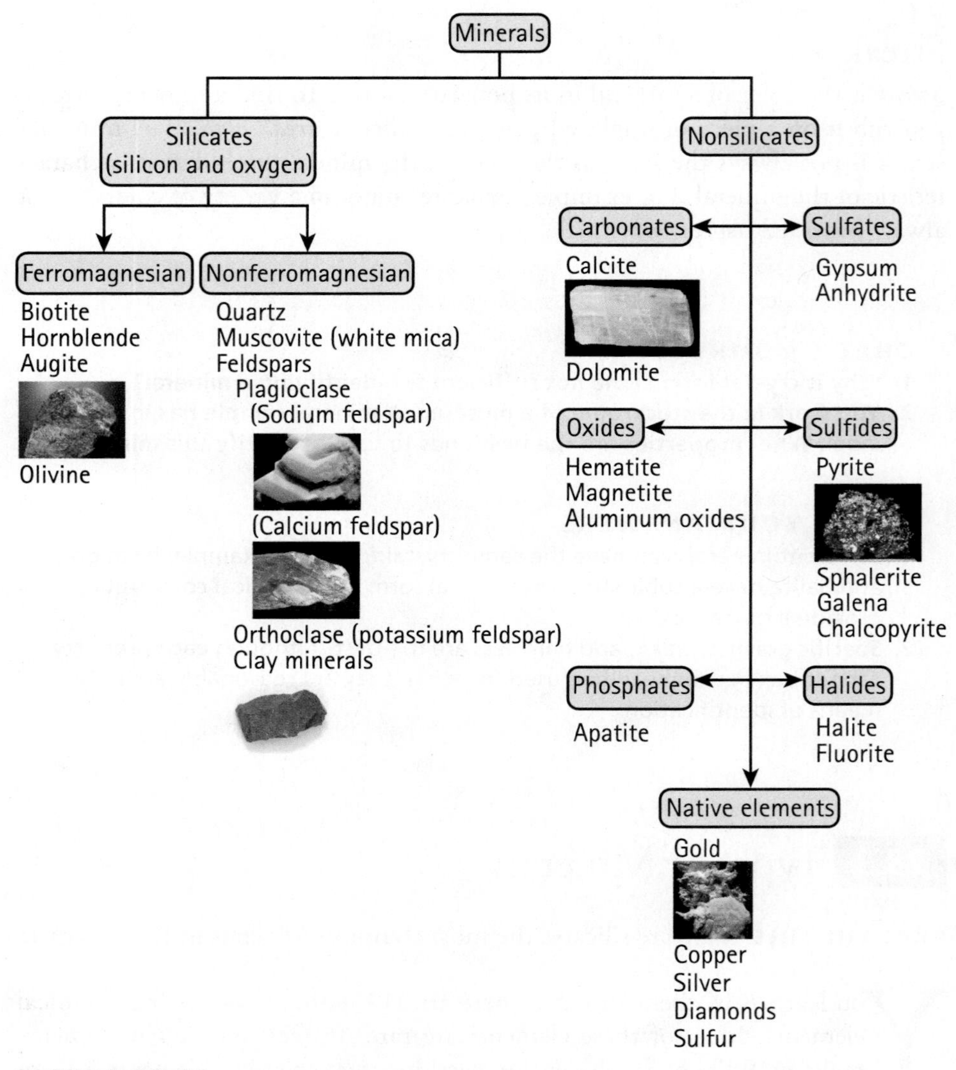

FIGURE 23.11
Classification of some common minerals.

Gold is truly a precious mineral. All the gold that has ever been discovered would fit into a cube with sides equal to those of a baseball diamond!

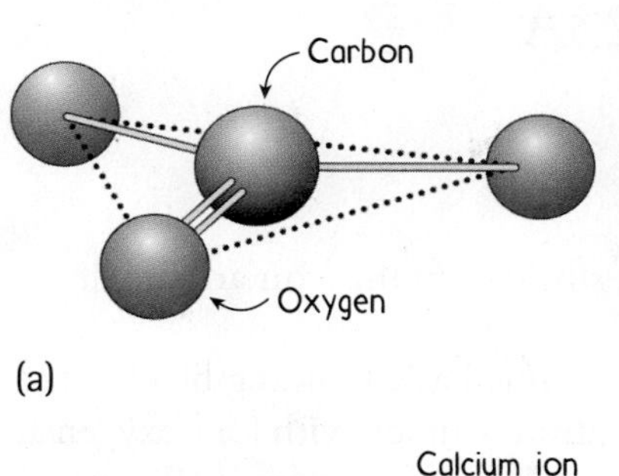

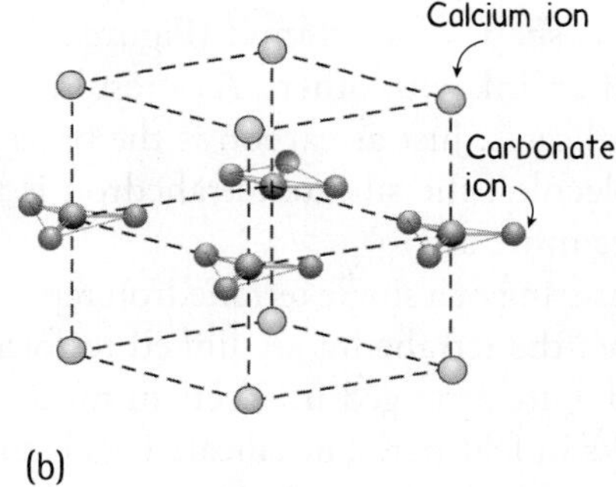

FIGURE 23.12
(a) The carbonate ion, CO_3^{2-}, has a triangular structure that features a central carbon atom bonded to three oxygen atoms. (b) Because carbonate minerals, such as calcite and dolomite, have a layered, sheetlike structure, they display the property of cleavage. (c) Dolomite rock (also called *dolostone*) is composed mainly of the mineral dolomite. (d) These cliffs along a stream in the Ozark Mountains of Missouri are made up of dolomite rock.

Nonsilicates

Nonsilicate minerals, including carbonates, sulfates, oxides, and native elements, do not contain silicon bonded to oxygen. Although they account for only about 8% of Earth's crust, they are quite important to nature's ecosystems as well as to civilization. The carbonates, for example, are used in cement. Oxide minerals are used in countless manufactured products such as medicines and airplane parts. The nonsilicate minerals also include the native elements, minerals made up of a single element. Some native elements are gold, silver, copper, carbon, and sulfur. It's easy to think of ways people use these elements—from jewelry to pencil leads!

UNIFYING CONCEPT
The Ecosystem
Section 21.1

CHECK YOURSELF

1. **Which of these photos could show ferromagnesian rock?**
2. **Which photo shows nonferromagnesian rock?**
3. **Which photo shows silicate rock?**

CHECK YOUR ANSWERS

1. Photo (a) could show ferromagnesian rock because it is black.
2. Photo (b) shows nonferromagnesian rock.
3. Both photos show silicate rocks, as Figure 23.11 indicates.

LEARNING OBJECTIVE
Explain why the silicate tetrahedron gives rise to so many different minerals.

The Silicate Tetrahedron

EXPLAIN THIS How are the atoms in a silicate tetrahedron arranged?

All of the silicates are made up of the same basic building block—the *silicate tetrahedron.* The silicate tetrahedron is an ion with four oxygen atoms joined to one silicon atom. It has the shape of a pyramid (Figure 23.13). This basic building block can occur by itself or link with others. As a result, there are many different kinds of silicates. Just as carbon is the basis for the vast array of organic molecules, the silicate tetrahedron is the basis for a vast array of silicate minerals.

Olivine has a structure consisting of a single tetrahedron repeated millions of times. In pyroxene, the tetrahedra are linked to form a single chain. Silicate tetrahedra are arranged in sheets in mica and in three-dimensional networks in feldspar. The silicate tetrahedron is a sturdy and versatile building block (Figure 23.14).

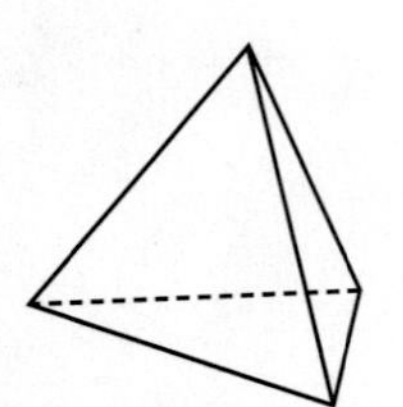

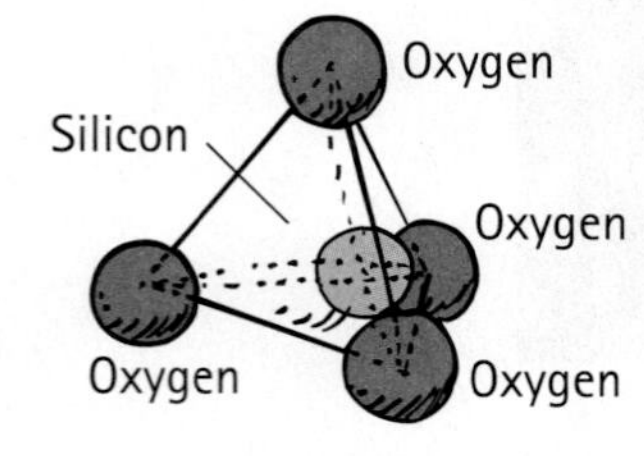

FIGURE 23.13
(a) The tetrahedron is a geometric shape with four triangular sides. (b) The silicate tetrahedron is an arrangement of four oxygen atoms with a silicon atom in the center.

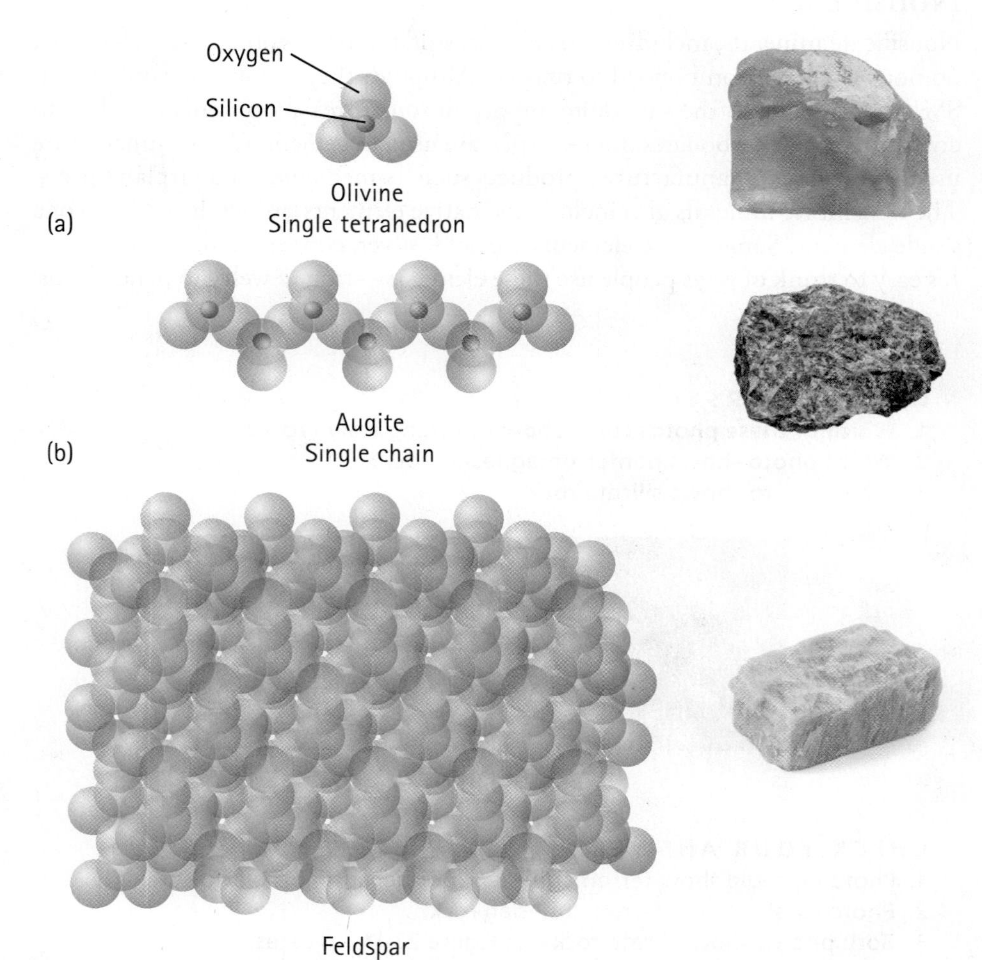

FIGURE 23.14
(a) Olivine, an olive-green mineral common in Earth's mantle, is made up of a single tetrahedron. (b) Augite, a member of the pyroxene group of minerals that are found in volcanic lava, is built of single chains of silicate tetrahedra. (c) In feldspar, the crust's dominant mineral group, silicate tetrahedra join together in three-dimensional networks.

CHECK YOURSELF

1. **The chemical formula for sapphire is Al_2O_3 (aluminum oxide). Do sapphire crystals contain silicate tetrahedra?**
2. **Why are there are more silicates in Earth's crust than other types of minerals?**

CHECK YOUR ANSWERS

1. No. Silicate tetrahedra require silicon atoms. These are missing in sapphire, as you can tell by its chemical formula. Sapphire is a nonsilicate oxide.
2. Silicates are common because they are made of the very abundant elements silicon and oxygen and because these elements form a versatile building block—the silicon tetrahedron.

23.4 How Do Minerals Form?

LEARNING OBJECTIVE
Describe the four pathways by which most minerals form.

EXPLAIN THIS Why are diamonds unstable at Earth's surface?

You now know what minerals are and how they are classified. You also know something about mineral properties and their structures. But you still may wonder: Where do minerals come from, and how do they form? All minerals arise through *crystallization*, the growth of a solid whose atoms come together in a specific structure. There are four major processes by which minerals crystallize: cooling magma, hydrothermal solutions, precipitation, and pressure and temperature changes.

Cooling Magma

At Earth's surface, rocks and minerals are solid. But, if the temperature is high enough, they can melt to become liquids. As miners know, Earth's interior heats up quickly with depth. On average, the temperature inside Earth increases about 30°C for every kilometer below the ground. The **geothermal gradient**, shown in Figure 23.15, illustrates this change. Rocks and minerals thus exist as liquids in large quantity deep underground. This naturally occurring, molten rocky material is called *magma*.

Magma has a texture like thick oatmeal. Because silicates are the most abundant minerals, magmas are usually rich in silica (SiO_2). Magma contains water and gases and usually contains some solids as well as atoms and molecules in the liquid state. Because magma is hot and less dense than solid rock, it rises buoyantly toward Earth's cooler surface. As magma rises and cools, solid mineral crystals begin to form. Most minerals are formed in this way by cooling magma. Some of the common minerals that form from cooling magma are feldspar, quartz, mica, and magnetite.

Different minerals crystallize at different temperatures. So, as magma cools, it keeps generating different minerals in a series according to the minerals' melting points. Feldspar crystallizes out of a cooling magma before quartz because feldspar has a higher melting point than quartz (see Figure 23.15d). That is, feldspar changes from the liquid to the solid phase at a higher temperature than

(a) Minerals form in areas once covered by salt water
(d) Minerals form where magma is cooling
(c) Minerals form where rock is undergoing metamorphosis
(b) Minerals form where crystals precipitate out of hot water solutions from groundwater

FIGURE 23.15
Minerals form through four basic pathways. (a) Minerals form where seawater evaporates and solids are left behind. Example: gypsum. (b) Minerals form when crystals precipitate out of the hot water solutions formed by groundwater. Example: copper. (c) Minerals form where rock is altered by changes in pressure or temperature (metamorphosis). Example: feldspar (d) Minerals form where magma is cooling. Example: feldspar.

quartz does. Because quartz solidifies at a lower temperature, it forms later than feldspar after the magma has cooled even more. Because minerals crystallize out of magma in series, one particular type of magma can be the "parent" of many different minerals.

Hydrothermal Solutions

Sometimes minerals crystallize out of hydrothermal solutions rather than from magma. A hydrothermal solution is a mixture of very hot water (over 100°C) and dissolved substances. Such a solution forms when water stored underground (*groundwater*) seeps into deep cracks in rock. If magma resides nearby, it heats

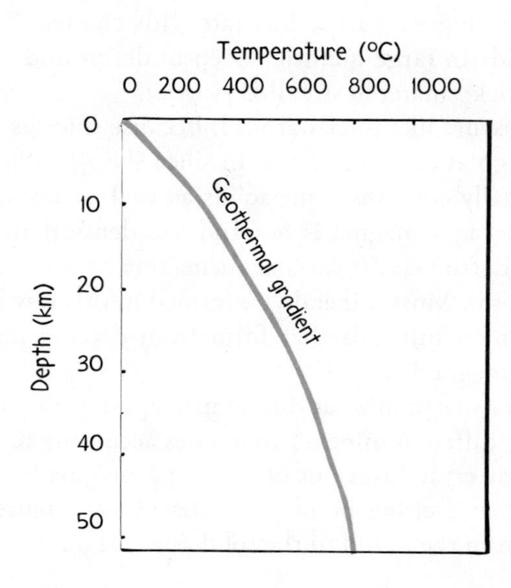
Temperature (°C)
0 200 400 600 800 1000
0
10
20
30
40
50
Depth (km)
Geothermal gradient

FIGURE 23.16
The temperature inside Earth increases about 30°C for each kilometer of depth from the surface. This increase of temperature with depth is known as the *geothermal gradient*.

the groundwater, which enhances the water's ability to dissolve or react with the minerals in the surrounding rock. Dissolved metals and other substances crystallize out of the solution when it becomes saturated. Minerals often formed this way include gold, copper, sulfur, and pyrite.

With all natural resources, including minerals, we should practice the basic three Rs: Reduce how much you consume, reuse mineral products rather than throw them away, and recycle minerals when it's time to discard them.

Evaporation

Surface water contains many dissolved substances. When this water evaporates, minerals can form. Such *evaporite* minerals, including halite, are left behind when a sea, lagoon, or other body of surface water dries up. Minerals also can form when water changes temperature because this can cause elements to precipitate out of solution.

Pressure and Temperature Changes

New minerals sometimes form inside rocks when the rock's environment changes—for example, when the temperature or pressure changes. You will learn more about this process, called *metamorphosis*, later in this chapter. Some minerals that form when rock undergoes metamorphosis are garnet, talc, and graphite.

Not every collection of atoms can make a mineral. A mineral will be *stable* only if its atoms hold together strongly enough. Temperature and pressure have a big effect on mineral stability. Diamonds illustrate this point. Diamonds form about 150 km underground, where high pressure pushes carbon atoms into tight networks. At Earth's surface, the pressure is much lower and a diamond is not really stable. The crystal structure of diamond eventually rearranges to become the mineral graphite. Fortunately, the conversion of diamond to graphite takes about a billion years, so diamonds are safe over the span of human lifetimes. It is only in this limited sense that we can say "diamonds are forever."

FIGURE 23.17
Diamonds aren't forever after all.

TECHNOLOGY

Synthetic Diamonds

In the laboratory, a sample of graphite can be sufficiently pressurized to turn it into a diamond. Until recently, achieving the required pressure was so costly that synthetic gem-quality diamonds were more expensive than mined diamonds. However, in 2004, two companies introduced synthetic gem-quality diamonds to the market. The synthetic diamonds cost roughly 30% less than mined diamonds. These new synthetic diamonds are visually and chemically identical to diamonds mined from Earth.

FIGURE 23.18
Which diamond is real? Which is synthetic? (Answer: The synthetic diamond is the one on the right.)

SCIENCE AND SOCIETY

Mining

People use minerals in millions of ways. Just look around. Iron, copper, aluminum, and other metals are used to make cars, refrigerators, and CD players. Building materials, such as concrete, sheet rock, cement, and brick, are made from minerals. So are paint, plastics, films, drugs, explosives, money, lubricants, nuclear reactor cores, steel, food additives, and more.

In the United States, several billion tons of mineral resources are needed each year for people to go about their normal everyday activities. Each man, woman, and child in the United States uses thousands of kilograms of new mineral resources per year, and the average American home contains slightly more than a quarter million pounds (more than 113,000 kg) of minerals and metals (Figure 23.19).

Where do all these minerals come from? They are mined from Earth's crust. Mineral deposits that are large enough and pure enough to be mined for profit are called **ore**. Have you ever seen a mining site (Figure 23.20)? Surface mining removes layers of Earth to obtain minerals that are at or near Earth's surface. Strip mines, open-pit mines, and quarries are surface mines. Obtaining ores such as coal and diamonds can require digging deep holes into the ground. Underground mines typically consist of networks of tunnels that branch out for many miles. Once ore is removed from a mine, the ore is processed to extract the needed minerals.

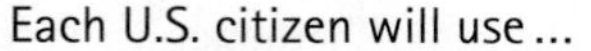

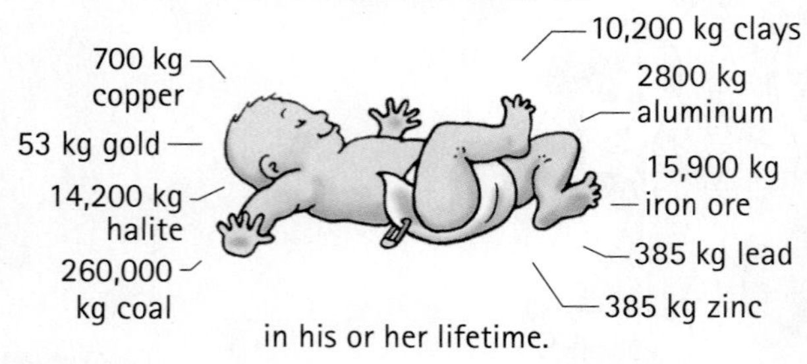

FIGURE 23.19
Each U.S. citizen, on average, will use large amounts of minerals for food, clothing, and shelter as well as transportation, manufacturing, and recreational activities.

FIGURE 23.20
The Kennecott Copper Mine in Utah is big enough to see from space. It is 1.2 km deep and 4 km wide. Billions of tons of rock have been removed from this mine in order to obtain copper and other minerals.

The extraction of great amounts of ore can do environmental damage. Because the land is disrupted, ecosystems may be destroyed. Removal of vegetation can cause erosion and landslides. Also, fossil fuels are consumed on a grand scale to process the minerals. Pollutants, such as carbon dioxide and sulfur dioxide, and toxic waste, such as lead and chromium, are generated. Once the ore is obtained, it must be refined to release its load of minerals. We need minerals, but how can we mine for them responsibly?

Many mining operations reduce environmental damage through reclamation. Reclamation is the practice of returning land to the condition it was in before the mining was done.

LEARNING OBJECTIVE
Distinguish between rocks and minerals.

23.5 What Is Rock?

EXPLAIN THIS How is a rock like a granola bar?

A **rock** is a solid, cohesive mixture of mineral crystals. The mineral crystals (grains) in rock are not necessarily all the same mineral; often many minerals are present, and sometimes glass as well. The minerals (and possibly glass) are held firmly together. Sand grains may be cemented or compacted together to make a rock, but if they are not consolidated, they are not a rock. The difference between minerals and rocks is that minerals are *chemical compounds*, whereas rocks are *physical mixtures*.

FIGURE 23.21
Rocks are solid mineral mixtures. This granite rock is a mixture of quartz, hornblende, and feldspar.

FIGURE 23.22
The three main classifications of rock. (a) Basalt and granite are igneous rocks. (b) Sandstone and limestone are igneous rocks. (c) Marble and slate are metamorphic rocks.

You can sometimes see the mineral crystals in a rock. Granite, the most common type of rock in the continental crust, is shown in Figure 23.21. It contains visible crystals of feldspar, quartz, and hornblende. But sometimes rock crystals are not so easily seen. Fine-grained rocks such as slate look smooth and even because they contain crystals that are too small to be seen with the naked eye.

Most surface rocks are covered with *soil* or what people call "dirt." Soil is a mixture of small bits of weathered rock, air, water, and *organic matter*—the decayed remains of plants, animals, and other organisms. But if you dig under the soil, you'll find rock. Wherever you dig, you're sure to find rock—some near Earth's surface and some deep down.

Rocks, like minerals, are as useful as they are abundant. People have been fashioning rock into useful tools since prehistoric times. Early humans didn't have many materials to choose from. Plastic, iron, steel, and such were not invented or discovered until modern times. The only materials available were trees, rocks, plants, minerals, and animal remains. Arrowheads made from obsidian could be sharpened enough to cut the skin of prey. Heavy granite rocks were used as blunt weapons. Large, flat sandstone rocks were used as grinding stones.

Another major use of rock is in building. The great cathedrals of Europe, the pyramids of Egypt, prehistoric stone temples, Mayan villages, ancient mosques, and even some entire ancient cities are built of rock. Modern buildings are often made of concrete, a material made from rock.

Rocks are classified into three major groups based on how they form (Figure 23.22). The three kinds of rocks are *igneous rocks*, *sedimentary rocks*, and *metamorphic rocks.*

FIGURE 23.23
People have been using rocks to make useful tools since the Stone Age.

LEARNING OBJECTIVE
Differentiate between intrusive and extrusive igneous rocks.

23.6 Igneous Rock

EXPLAIN THIS What determines the size of the crystals in an igneous rock?

The word *igneous* means "fire" in Latin. **Igneous rocks** ("IG-nee-us") form by the crystallization and solidification of magma that is cooling. Igneous rocks consist of tightly interlocking crystals, which makes them strong—able to withstand strong forces. The igneous rock granite is a desirable material for kitchen countertops because of its strength as well as its beauty (Figure 23.24).

FIGURE 23.24
Igneous rock, such as this granite, is strong because it consists of tightly interlocking crystals.

Igneous rocks vary in their mineral makeup and their texture. Mineral makeup is chemical composition, and texture refers to grain (crystal) size.

The mineral composition of an igneous rock reflects the chemical composition of the parent magma that generates the rock. Magmas around the world differ in composition according to the tectonic setting in which they form. The magmas generated at divergent plate boundaries and regions of oceanic crust are rich in the dark, iron- and magnesium-containing minerals as well as silica. When these magmas cool and crystallize, dark rocks including basalt and gabbro result (look ahead to Figure 23.28a). Magma generated in continental rock settings, on the other hand, has a higher silica content and less iron and magnesium. Light-colored rocks including granite and rhyolite solidify from high-silica magma.

Igneous rocks that form from magma cooling slowly underground are called **intrusive igneous rocks**. Intrusive igneous rocks are so named because the magma they derive from "intrudes" or pushes into the surrounding rock on its journey toward Earth's surface. And intrusive rocks are also sometimes called *plutonic* rocks, after Pluto, the Greek god of the underworld. Hence, underground igneous rock bodies are called *plutons*.

Plutons can be sheetlike or bulky depending on the rock they intrude into. *Dikes*, for example, are sheetlike igneous rock bodies that form when magma is injected into fractures cutting across rock layers (Figure 23.25a). A *batholith* is a large volume of plutonic rock that has been uplifted and exposed at Earth's surface after the overlying, softer rock has eroded away. Batholiths form the cores of many mountain systems. The Sierra Nevada mountain range is a gigantic batholith formed by the combined masses of many plutons (Figure 23.25b). It has a surface area of about 25,000 square miles!

(a)

(b)

FIGURE 23.25
(a) This sheetlike igneous intrusion is a dike in the Swiss Alps. (b) The core of the Sierra Nevada mountain range is a giant batholith, or intrusive igneous body, revealed at Earth's surface due to erosion.

When magma rises all the way to Earth's surface without solidifying, it's called **lava**. Lava may erupt violently from a volcano, or it may ooze out of cracks in the seafloor. Either way, it cools quickly to form solid rock. The igneous rocks that form at Earth's surface are called **extrusive igneous rocks** (Figure 32.26), or *volcanic* rocks. Whereas intrusive igneous rocks generally have big crystals, extrusive igneous rocks have small crystals or no crystals at all.

FIGURE 23.26
When lava cools enough to solidify, it forms extrusive igneous rock.

Classifying Igneous Rocks

It's convenient to classify igneous rocks by their *texture*—the size of their crystals. There are four igneous textures:

1. *Coarse-grained*—crystals that are visible without a microscope
2. *Fine-grained*—microscopic crystals
3. *Glassy*—no crystals
4. *Porphyritic*—large crystals embedded in a matrix of smaller crystals

The size of a crystal in an igneous rock depends on how rapidly the parent magma cools. Slow cooling produces large crystals, and rapid cooling produces small crystals (Figure 23.27). Extremely rapid cooling produces glassy-textured rocks, such as obsidian, that have no crystals at all. Why is this so?

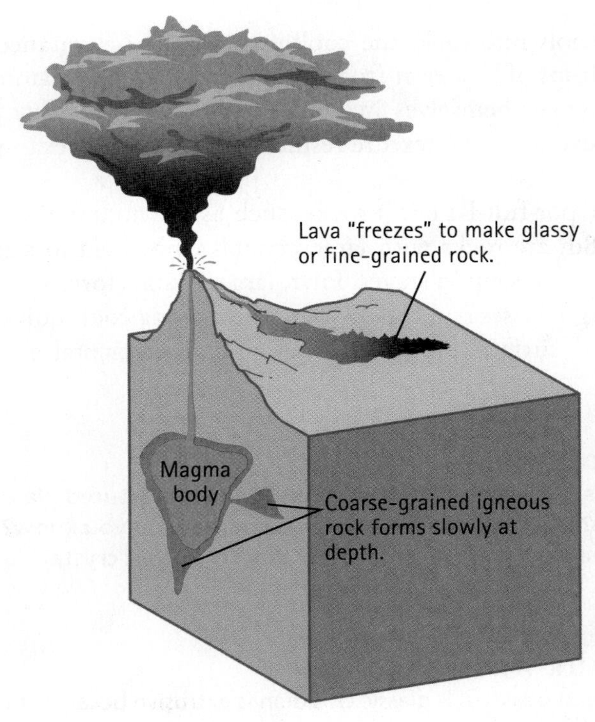
Lava "freezes" to make glassy or fine-grained rock.
Magma body
Coarse-grained igneous rock forms slowly at depth.

FIGURE 23.27
The texture of an igneous rock depends on how long it takes to cool. Coarse-grained igneous rocks have large crystals formed by slow cooling deep in Earth's interior. Fine-grained igneous rocks have small crystals and form closer to the surface.

(a)

(b)

(c)

(d)

FIGURE 23.28
Igneous rocks displaying different textures: (a) gabbro, a coarse-grained rock; (b) rhyolite, showing a fine-grained texture; (c) an obsidian sample with a glassy texture; and (d) andesite porphyry—a porphyritic rock.

A mineral crystal grows by adding atoms to its outside surface. When crystals growing side by side become large enough for their edges to touch, crystal growth stops because there is no more room. The entire magma body is solidified as a patchwork of interlocking crystals.

When magma cools slowly—over thousands or millions of years—crystals have time to get quite large before they have to compete with their neighbors for available atoms or for space. Thus, when you see a coarse-grained rock, such as the granite or the gabbro ("GAB-row") shown in Figure 23.28a, with its big, chunky crystals, you know that it cooled slowly from magma over a long period of time.

Sometimes cooling can occur rapidly, over days, weeks, or months. When cooling is this fast, lots of tiny crystals form at the same time. Each crystal doesn't have time to grow very large before its neighbors crowd in on it. As a result, rapid cooling produces a fine-grained rock, such as the rhyolite ("RI-o-lite") shown in Figure 23.28b.

When lava cools into rock, the cooling is almost instantaneous. This can happen when drops of lava eject from a violent volcano. The atoms lose energy before they arrange themselves in *any* crystal structure—they just freeze in place. An abrasive or glassy texture results. Obsidian is an example, shown in Figure 23.28c.

Porphyritic ("por-fuh-RIT-ic") rocks, such as the andesite porphyry shown in Figure 23.28d, are rocks with large crystals embedded in a matrix of tiny ones. These rocks develop in stages. First, large crystals form slowly at depth as a column of magma rises and cools. Then, the magma cools quickly as it erupts at or near Earth's surface, producing the fine-grained crystal matrix.

CHECK YOURSELF

1. **Obsidian is a volcanic glass that looks like manufactured glass. What is its texture? Is it intrusive or extrusive, and how do you know?**
2. **What is the texture of granite? How quickly does it crystallize from magma?**

CHECK YOUR ANSWERS

1. The texture of obsidian is glassy. Obsidian is extrusive because it is produced by the nearly instant cooling of lava.
2. Granite has a coarse-grained structure, which indicates that it crystallizes slowly from magma over a period of thousands to millions of years.

23.7 Sedimentary Rock

LEARNING OBJECTIVE
Describe how sedimentary rocks form.

EXPLAIN THIS How can sedimentary rock be the most common type of rock at Earth's surface yet the least common type of rock in Earth's crust?

Sedimentary rock is rock formed by the accumulation of weathered material. When a rock lies at or near Earth's surface, it gets broken down or *weathered* from the action of water, wind, organisms, reactive chemicals, or ice. The weathered rock fragments may be large or small, or they may even be tiny dissolved particles. Such fragments of weathered rock are called **sediments**.

Sediments don't stay in one spot forever. They are picked up and moved by streams, wind, glaciers, or other means of transportation in the process called *erosion*. Erosion ultimately moves sediments downhill and toward the sea because of gravity. Erosion comes to an end when the sediments are dumped at a new location, or *deposited*.

FIGURE 23.29
The strata of this sedimentary rock formation are visible in this photo. This formation is in Dorset, England. These strata were deposited more than 100 million years ago in the Jurassic period—a time when dinosaurs roamed the land.

Sediments mixed with organic matter make soil. But sediments that undergo the process of *lithification* (from the Greek *lithos*, meaning "stone") become sedimentary rock, which is more compact, dense, and cohesive than the original sediments. Lithification occurs when sediments are compacted, or compressed, by the weight of overlying layers. Or, the sediments may be held together by a natural cement that forms when mineral-rich water flows into empty spaces between sediments and evaporates. Sometimes cementation and compaction both occur.

Only about 5% of Earth's crust is composed of sedimentary rock. However, most of the rocks you come into contact with are sedimentary. They are spread out in a thin patchwork that covers about 75% of the continental crust. Sedimentary rock forms layers called **strata** that have been built up over thousands or millions of years. You may have noticed strata along cliffs or road cuts (Figure 23.29).

Sedimentary rocks sometimes contain **fossils**, the remains or traces of ancient life forms (Figure 23.30). Fossils are a record of the drama of life on Earth. Even without fossils, sedimentary rocks provide clues about Earth's geologic past—the positions of its continents, the rise and fall of sea level, and climate changes over time.

FIGURE 23.30
Hundreds of dinosaur fossils and skeletons have been found in the sandstone at Dinosaur Provincial Park in Canada. Sandstone is a common sedimentary rock.

Sedimentary rocks are classified into two main groups. **Clastic rocks** (from the Greek word *klastos*, which means "broken up") are formed when bits, pieces, or particles of weathered rock hold together. **Chemical rock** results when minerals precipitate out of a solution and then stick together.

Clastic sedimentary rocks are usually classified on the basis of the average size of their particles (Figure 23.31). *Sandstone* is composed of sand-sized sediment particles. *Shale* is made of finer sediments that cannot be seen with the naked eye. *Conglomerate* is a coarse-grained rock made of rounded particles larger than 2 mm in diameter. *Breccia* ("BRE-chee-a"), another coarse-grained clastic sedimentary rock, is made of large, angular pieces of rock held together by natural mineral cement.

Chemical sedimentary rocks are made when crystals precipitate out of a solution. Minerals can precipitate when water evaporates or when certain chemical reactions occur. For example, a chemical sedimentary rock can originate when rainwater dissolves rock. Mineral-rich water flows toward the ocean. Minerals remain dissolved in ocean water until so much water evaporates that its concentration of minerals exceeds its solubility limit. Then mineral crystals precipitate out of the water and sink to the ocean floor. Eventually, enough mineral crystals collect, compact, and join together to form a rock.

If you have to guess what kind of rock is in your backyard, guess sedimentary rock. Sedimentary rocks cover most of the continental crust.

FIGURE 23.31
Sedimentary rocks. (a) Shale, the most abundant sedimentary rock, is composed of fine particles. (b) Sandstone is made of medium-sized particles, about the size of sand grains. (c) Conglomerate consists of a variety of large, rounded particles. (d) Breccia consists of large, angular pieces of rock held together by cement.

Sometimes biological processes are involved in making chemical sedimentary rocks. Consider limestone: Marine animals, including clams and corals, chemically extract calcium carbonate ($CaCO_3$) from ocean water to build their shells and skeletons. When these organisms die, their shells and skeletons break up and pile up on the ocean bottom. Over time they can become buried, compacted, and cemented to form limestone. *Coquina* is a biochemical limestone composed of shells and shell fragments. *Chalk* is a fine-grained limestone that is also formed from the shells of marine organisms. *Travertine*, on the other hand, is a limestone formed inorganically—for example, by the precipitation of calcium carbonate (also called calcite) from mineral springs (Figure 23.32).

FIGURE 23.32
Coquina (a) and chalk (b) are two kinds of limestone. They are both chemical sedimentary rocks, formed from small marine animals, such as clams, that chemically extract calcium carbonate from seawater. Travertine is also limestone and a chemical sedimentary rock, although it is not derived from a biological source. The travertine specimen shown here (c) was formed when calcium carbonate precipitated out of a mineral spring in Turkey.

LEARNING OBJECTIVE
Explain how coal is formed.

Coal

EXPLAIN THIS Why does coal contain energy-rich organic molecules?

During the 19th century, coal replaced wood as the fuel of choice. Although coal was dangerous to mine, bulky to handle, and dirty to burn, it became widely available and produced more heat per volume than wood. When liquid and gaseous fossil fuels became available later in the 19th century, they began to replace coal, especially for home use. Coal is still an important energy source, however. Currently, about 20% of all U.S. energy is supplied by coal. And coal-fired power plants account for more than half of the electricity generated in the United States. As oil and gas supplies decline, there is increasing interest in expanding the role of coal as a fuel. On the other hand, there is concern about the environmental effects of coal, including greenhouse gas (carbon dioxide) emissions, air pollution from ash and toxic metals, and sulfur emissions that contribute to water pollution and acid rain.

(a)

(b)

FIGURE 23.33
(a) Anthracite coal. (b) A coal seam in New Zealand.

Coal is a black or brown combustible sedimentary rock (Figure 23.33). It occurs in rock strata called *coal seams* that are fairly common, particularly in the United States. It's estimated that more than 10 trillion tons of coal are stored in Earth's crust. The United States owns about 25% of the world's current reserves of this energy-rich rock.

Coal is derived from ancient land plants. Most coal was formed about 300 million years ago, when much of Earth was covered by steamy swamps. As the green plants died, their remains sank to the bottoms of the swamps. Plant material did not decay easily in the oxygen-poor swamp water. So the plant debris accumulated, layer upon layer, and the energy-rich molecules stored in the debris remained intact because they didn't react with oxygen. Eventually, the plant remains formed a dense, soggy, only partially decomposed organic material called *peat*.

Over geologic time, Earth's surface kept changing. Great rivers brought sand, clay, and other sediments to bury the peat. Sandstone and other sedimentary rock formed overlying strata, which heated and pressurized the buried peat even more. The spongy peat transformed into soft, brown coal (known as *lignite*) and then, with more time, heat, and pressure, into the harder coal *bituminous* (Figure 23.34). Bituminous that was squeezed and folded by tectonic forces during mountain building ultimately formed the highest-energy form of coal, *anthracite*. Lignite and bituminous are sedimentary rocks, but anthracite is a metamorphic rock.

Coal is a lot like petroleum. Both are fossil fuels. They contain hydrocarbons rich in chemical potential energy. Both coal and petroleum are nonrenewable resources that formed over millions of years when organisms died and were buried and then decayed in the absence of oxygen. Like all fossil fuels, coal and petroleum send carbon dioxide (a "greenhouse gas") into the atmosphere when burned, contributing to climate change. While petroleum is derived from ancient marine organisms, coal is the product of ancient land plants.

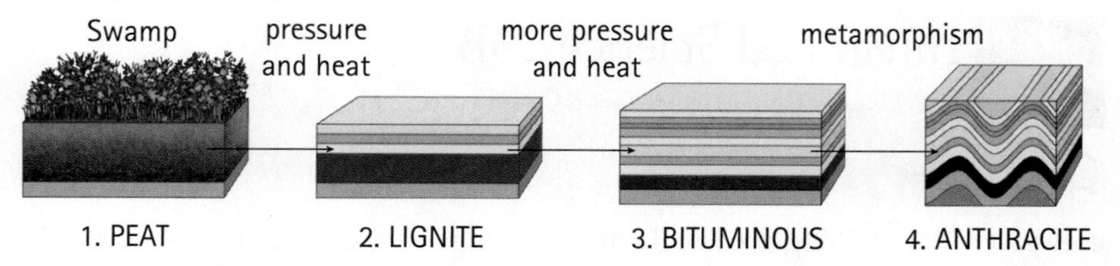

FIGURE 23.34
The formation of coal. (a) Peat forms from plant remains in a stagnant swamp. (b) The peat is buried by sediments and subjected to great heat and pressure over time. Lignite, a soft, brown, low-energy coal forms. (c) More time, burial, heat, and pressure lead to the formation of bituminous, a soft, black coal used for energy production. (d) Tectonic forces squeeze bituminous to anthracite, the hardest, highest-energy form of coal.

UNIFYING CONCEPT
- *The Law of Conservation of Energy* **Section 4.10**

Coal is sometimes called "buried sunshine." The ancient land plants that coal comes from, like all plants, captured solar energy through photosynthesis to manufacture the organic molecules of their tissues. When coal is combusted—chemically reacted with oxygen—its large reserves of chemical potential energy are unlocked and can be transformed into other kinds of useful energy.

CHECK YOURSELF
Why is coal sometimes called "buried sunshine"?

CHECK YOUR ANSWER
The ultimate source of the chemical energy in coal is the Sun.

LEARNING OBJECTIVE
Identify how metamorphic rocks differ from sedimentary and igneous rocks.

23.8 Metamorphic Rock

EXPLAIN THIS How are garnets, reddish-black gems, derived from plain sandstone?

How is a rock altered when it is subducted and sinks to a deeper location with higher temperature or pressure than the environment in which it formed? Or when it is subjected to the high stresses and temperatures associated with mountain building? How is a rock altered when it is exposed to the hot, reactive chemical fluids associated with magma? Such changes in the physical and chemical conditions to which a rock is exposed can transform rock. New rock is made from old. The new rock is stable under the new conditions, while the preexisting rock was not.

The word *metamorphism* literally means "changed form." Rocks that change form deep in Earth's interior to become more stable under new conditions are called **metamorphic rocks**. Igneous, sedimentary, and metamorphic rocks can all undergo metamorphism. During metamorphism, a rock's texture, size, shape, density, or mineral makeup may be altered.

The process occurs at depths ranging from a few kilometers below the surface all the way to the crust–mantle boundary, where elevated heat, pressure, and chemical fluids exist. *Regional metamorphism* affects large areas of rock. Regional metamorphism occurs deep in Earth's crust, where temperature and pressure are high enough to form huge regions of metamorphic rock. Regional metamorphism is usually produced from the heating and stresses that occur during mountain building and plate tectonic movement. On the other hand, when hot magma rises to shallow regions of the crust and heats the surrounding cooler rocks there, those rocks may metamorphose. This type of metamorphosis affects smaller regions of rock and is called *contact metamorphism*.

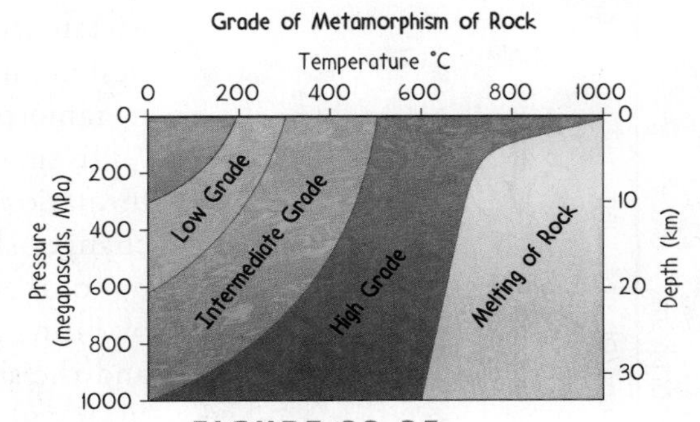

FIGURE 23.35
Varying degrees of rock metamorphism.

Note that minerals do not melt when a rock metamorphoses. Minerals may change their identity through recrystallization, but this does not require melting. *Recrystallization* occurs when a mineral's crystal structure collapses under the influence of elevated temperature or heat, and its atoms migrate and recombine to form new minerals. Recrystallization also happens when intruding chemical fluids supply new atoms that react with the minerals in preexisting rock, so that new mineral crystals form. Recrystallization can occur many times, but, once minerals actually melt, metamorphism is over and igneous activity has begun.

Metamorphism varies in degree. When a rock's environment changes only slightly and the rock is altered in a minor way, the change is referred to as *low-grade metamorphism* (Figure 23.35). The metamorphosis of the sedimentary rock shale into the metamorphic rock slate is an example of this—the rock changes mainly in that it becomes more dense as its mineral grains are aligned more compactly. Low-grade metamorphism takes place at just slightly higher temperatures and pressures than those that lithify sediments. High-grade metamorphism takes place under extreme conditions and can obliterate all features of the parent rock. An interesting example is "shock metamorphism," which yields the mineral stishovite, a high-density, tetragonal form of silica. The process occurs at ultrahigh pressures but not necessarily at high temperatures. Where might you find such conditions? At the site of a meteor impact. The first crystals of stishovite were discovered at Meteor Crater in Arizona (Figure 23.36). Today, the presence of stishovite is taken as diagnostic evidence of a meteor impact at craters of unknown origin.

FIGURE 23.36
Meteor Crater in Arizona, the site of "shock metamorphism."

Many metamorphic rocks are foliated. **Foliated rocks** contain minerals that are aligned in a particular direction, forming parallel layers that look like pages in a book. Usually the minerals are oriented perpendicular to the direction of applied pressure. Foliated metamorphic rocks, such as gneiss ("nice"), shown in Figure 23.37, often display distinctive banding from light and dark minerals.

(a) Macroscopic view of gneiss

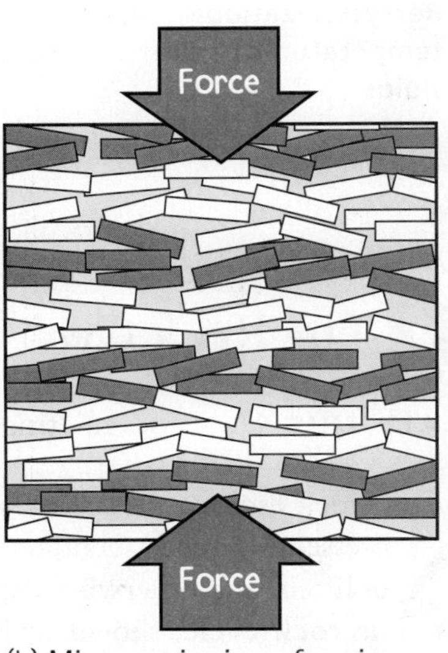

(b) Microscopic view of gneiss

FIGURE 23.37
(a) Foliated metamorphic rocks, such as the gneiss shown here, contain minerals aligned in a particular direction. (b) Usually they are aligned in a direction perpendicular to the applied force.

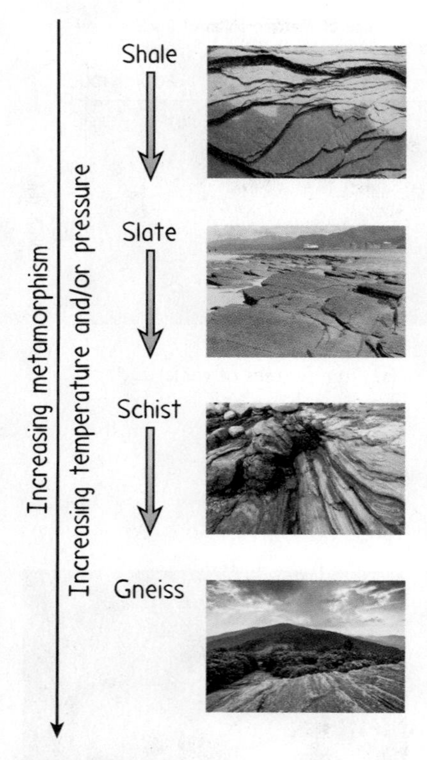

FIGURE 23.38
Shale, a sedimentary rock, recrystallizes to form gneiss in a series of metamorphic changes.

Metamorphic rocks that consist of just one non-platy mineral (except for impurities) are nonfoliated—that is, they don't exhibit layers. For example, limestone metamorphoses to marble, a nonfoliated metamorphic rock.

To summarize the metamorphosis of rock, consider an example—the transformation of shale to gneiss, which occurs through the series of metamorphic changes shown in Figure 23.38. If you start with clay and squeeze the water out of it at high pressure, you get the sedimentary rock shale. Compress shale by piling more rock and sediments on top or by the action of tectonic forces, and the shale metamorphoses into slate. Even higher temperatures and pressures can metamorphose slate into *schist*. Schist contains minerals, including beautiful reddish-black garnets, which form only under extremely high pressure and temperature. The garnet crystals were not in the original clay that created the shale, but as the rock was squeezed at higher and higher pressures, the garnet crystals began to grow at the expense of other minerals. Schist is not the end of the line for shale, however. With more pressure and temperature, schist metamorphoses into gneiss. Gneiss has a mineral structure that is completely different from that of the shale from which it began. The original atoms, those of the shale, are still there, but recrystallization has reformulated the original minerals and transformed the shale's visible texture. Yet even with gneiss the process of metamorphism is not over. If the temperature rises even higher, the gneiss begins to melt and becomes a partially melted rock called a *migmatite*.

You may wonder whether one rock can metamorphose into another and another indefinitely. It cannot. Once a rock melts completely, it becomes magma. And when the magma cools to form a rock, the new rock will, by definition, be an igneous rock.

CHECK YOURSELF

1. **When can a rock no longer undergo metamorphosis?**
2. **Why does recrystallization occur in a rock?**

CHECK YOUR ANSWERS

1. A rock cannot continue to undergo metamorphosis once it melts.
2. Recrystallization occurs because rock has been exposed to high temperatures or pressures or because it has been exposed to chemical fluids.

LEARNING OBJECTIVE
Draw the rock cycle.

23.9 The Rock Cycle

EXPLAIN THIS How does melting reset the rock cycle?

No rock escapes the forces of change. Rocks are broken, heated, compressed, stretched, bent, chemically altered, worn away, moved, melted, uplifted, and otherwise disturbed by biological, chemical, and physical agents. The **rock cycle**, shown in Figure 23.39, is a model that sums up the

MasteringPhysics®
TUTORIAL: The Rock Cycle

FIGURE 23.39
The rock cycle shows the processes that can change rock from one kind into another over and over again. Rocks can take more than one pathway through the cycle. The arrows represent the processes of change. The boxes represent the materials that undergo change.

formation, breakdown, and re-formation of rock as a result of igneous, sedimentary, and metamorphic processes.

Rocks can follow many different pathways around the rock cycle. Sometimes igneous rock, for example, is subjected to heat and pressure far below Earth's surface and becomes metamorphic rock. Or, metamorphic or sedimentary rocks at Earth's surface may decompose to become sediment that undergoes compaction and becomes new sedimentary rock. There are many possible variations of the cycle. Cycles even occur within cycles. Whatever the route, Earth's crust is formed when molten rock rises from the depths of Earth, cools, and solidifies to form a crust that, over time, is reworked by shifting and erosion, and it can eventually be returned to the interior, where it may be completely melted and once again become magma.

Many of Earth's materials move in cycles. Just as rock undergoes cycles of change, water changes phase and moves from the ocean, to the atmosphere, to the land, and back to the ocean in the *hydrologic* cycle. Both carbon and nitrogen move through Earth, to the atmosphere, and through living things in *biogeochemical* cycles.

LEARNING OBJECTIVE
State the age of Earth, the age of the oldest fossil, the age of the earliest fossilized human footprints, and the time of the Cretaceous extinction, along with the rock evidence for each of these ages.

Integrated Science 23C
BIOLOGY AND ASTRONOMY

Earth's History Is Written in Its Rocks

EXPLAIN THIS How is "shocked quartz" in the rock record suggestive of an asteroid impact?

How old is Earth? Was your home state once under the sea? Was your city a dinosaur's breeding ground? Amazing facts about Earth are revealed if you know how to interpret the *rock record*—the geologic clues recorded in rocks.

Radiometric dating is a laboratory method that can be used to determine the age of rock in years. Such testing indicates that the oldest known *rock* on Earth is a 4-billion-year-old gneiss formation in Canada (Figure 23.40). The oldest known natural *object* of terrestrial origin is a zircon mineral crystal enclosed in a metamorphosed sandstone rock in Australia. Radiometric dating shows that this crystal is 4.3 billion years old. These ages, coupled with astronomical evidence, put the age of Earth at about 4.6 billion years. A difficult stretch of time to comprehend! If the 4.6-billion-year age of Earth were condensed into a single year, all of recorded human history would take place in the last minute of New Year's Eve.

FIGURE 23.40
Radiometric techniques have shown that the Acasta Gneiss complex in Canada is 4 billion years old and is Earth's oldest rock outcropping.

Rock and mineral ages tell much about Earth's past—so do fossils. A fossil is any record of past life, such as bones, teeth, shells, impressions, and footprints. The earliest fossils known are stromatolites, large rocky structures built by *cyanobacteria* (blue-green algae) 3.5 billion years ago (Figure 23.41). The discovery of these fossils puts a lower limit on the beginning of life. Life must have existed on Earth *at least* 3.5 *billion* years ago. Another fossil find—this one only one thousandth the age of stromatolites at 3.6 *million* years old—displays the earliest known human footprints. From this fossil dig, scientists have inferred much about early hominids, even information about their social behavior (Figure 23.42).

FIGURE 23.41
This artwork of Precambrian Earth shows meteorites crashing into Earth's volcanic surface. Cyanobacteria thrive in the water. Stromatolites—the rocky structures built by cyanobacteria—are evidence of cyanobacteria's presence.

(a)

(b)

FIGURE 23.42
(a) This 3.6-million-year-old volcanic rock features two sets of adult footprints and a third of a child. These are the oldest known human-type footprints. They belong to early humanlike creatures called *Australopithecus afarensis*. (b) Paleontologists have determined that *Australopithecus afarensis* had the upright gait of modern humans.

FIGURE 23.43
These layers of the Grand Canyon, called the *Redwall Limestone*, are built of the rocky remains of marine animals, which indicates that the Redwall Limestone was once covered by a shallow sea.

Knowledge of the environments in which different kinds of sedimentary rock form helps scientists understand different geologic settings. For example, much of the world's limestone is formed from the calcium carbonate–rich skeletons and shells of marine organisms. So the presence of limestone suggests that a location was once under the sea. Figure 23.43 shows a section of the Grand Canyon called the Redwall Limestone. The limestone there indicates that this part of the Grand Canyon was once covered by seawater.

A dramatic example of the stories written in rock is the *Cretaceous extinction*. The Cretaceous period occurred between 150 and 65 million years ago. It ended with a mass extinction that killed more than 60% of Earth's species over a period of about 2 million years. Famously, the dinosaurs were devastated.

What caused the Cretaceous extinction? Scientists agree that climate change was the key factor. Mild Mesozoic climates suddenly turned cool, and sunlight grew scarce. But what was the underlying cause of the sudden climate change? This is where the geologic detective work gets interesting. Geologists support the *impact hypothesis*—the idea that a large asteroid struck Earth at the end of the Cretaceous period and triggered the climate changes that wiped out species (Figure 23.44).

There are several lines of evidence recorded in the rock record that support the impact hypothesis. Consider this: There is a thin layer of clay in the rock record that contains high levels of iridium. Iridium is a rare element on Earth, but it can occur in high concentrations in asteroids. The position of the iridium layer in the rock record matches the time of the Cretaceous extinction.

FIGURE 23.44
According to the impact hypothesis, a large asteroid triggered climate changes that caused a mass extinction at the end of the Cretaceous period.

FIGURE 23.45
The Chicxulub crater on the Yucatan Peninsula of Mexico, soon after it formed. Did the impact that produced this crater end the reign of the dinosaurs?

Could the iridium have been placed in the rock record by an asteroid crashing into Earth and exploding over a large area? The timing is right; the iridium was deposited about 65 million years ago—the time of the great Cretaceous extinction. A large object almost certainly did hit Earth then. An impact crater dating to this time, called *Chicxulub*, has been found in Mexico (Figure 23.45). The crater is more than 180 kilometers across; to produce a crater that size, an asteroid must have been larger than 10 kilometers in diameter. Such an asteroid could have triggered a mass explosion that ignited wildfires that blocked the Sun. The rock record shows a layer of soot in North American rocks that dates to the time of the proposed impact. Besides the iridium and soot, we also find "shocked" quartz, spheres of melted rock near Chicxulub. The crater, the soot, the shocked quartz, and the iridium—all of these point to a massive asteroid impact in the late Cretaceous. As you can see, rock is many things—including a record of Earth's amazing history.

For instructor-assigned homework, go to www.masteringphysics.com 

SUMMARY OF TERMS (KNOWLEDGE)

Chemical rocks Sedimentary rocks formed from minerals precipitated out of solution.

Clastic rocks Sedimentary rocks that are composed of pieces of preexisting rocks.

Extrusive igneous rock An igneous rock that forms at Earth's surface.

Foliated rocks Rocks that contain crystals arranged in parallel bands.

Fossils The remains or traces of ancient life forms.

Geothermal gradient The increase of temperature with depth in Earth's interior.

Igneous rock Rock that forms from magma that cools.

Intrusive igneous rock Igneous rock that forms from magma that cools underground.

Lava Magma that erupts at Earth's surface.

Metamorphic rock Rock that forms by the alteration of preexisting rock deep in Earth's interior due to the effects of temperature, pressure, or reactive chemicals.

Mineral A naturally occurring, solid, inorganic, crystal material with a specific chemical composition.

Ore Mineral deposits that are large enough and pure enough to be mined for profit.

Rock A solid, cohesive mixture of one or more kinds of mineral crystals.

Rock cycle A series of processes by which rock is formed, breaks down, and re-forms into different kinds of rock.

Sedimentary rock Rock that forms over time as bits and pieces of preexisting rock are consolidated through compaction and/or cementation.

Sediments Fragments of weathered rock or living organisms.

Silicates Minerals that contain silicon and oxygen and make up more than 75% of Earth's crust.

Strata Parallel layers of sedimentary rock.

READING CHECK (COMPREHENSION)

23.1 What Is a Mineral?

1. Diamond and graphite are minerals made of 100% carbon. Are diamond and graphite the same mineral or different minerals? Why?
2. Are plastic and nylon minerals?
3. Describe the microscopic structure of a mineral.

23.2 Mineral Properties

4. What is the chemical formula for rubies? Why are rubies red?
5. Identify six or more properties of minerals.
6. How does the hardness of a mineral relate to its chemical bonds?

23.3 Types of Minerals

7. How many minerals are known to exist? What are the two main groups into which minerals are classified?
8. Which is the largest group of minerals? What is the common feature of all minerals in this group?
9. Which are more common in Earth's crust and mantle—silicates or nonsilicates?

23.4 How Do Minerals Form?

10. List four ways that minerals can form.
11. A rock deep in Earth has a temperature higher than the melting point it has at Earth's surface. Why can this buried rock remain in the solid state even though its temperature is very high?
12. When can the rock described in Exercise 11 melt to become magma?

23.5 What Is Rock?

13. Earth's crust is made of rock, so why don't we see solid rock everywhere on the surface?
14. When can't we see the crystals in rocks?

23.6 Igneous Rock

15. What is the difference between extrusive and intrusive igneous rocks?
16. Why are most igneous rocks very hard?

23.7 Sedimentary Rock

17. Describe the process by which sedimentary rocks form.
18. Why are fossils found in sedimentary rocks rather than in metamorphic rocks?

23.8 Metamorphic Rock

19. Heat and pressure can make one rock change, or metamorphose, into another. Where do the heat and pressure that metamorphoses rock typically come from?
20. What kinds of rocks can undergo metamorphism?

23.9 The Rock Cycle

21. Are rocks permanent features of Earth? Explain.
22. How are all rocks related to one another?

THINK INTEGRATED SCIENCE

23A—The Silicate Tetrahedron

23. What do all of the most common minerals in Earth's crust share?
24. In what way is the silicate tetrahedron like a building block or a Lego?

23B—Coal

25. How is coal unlike other sedimentary rocks?
26. Describe the formation of coal.
27. What is coal made of?

23C—Earth's History Is Written in Its Rocks

28. How old is Earth? How old is the oldest rock formation? The oldest object of terrestrial origin?
29. What does the layer of iridium in the rock record tell about the demise of dinosaurs?
30. Fossils of ancient sea creatures have been found on modern mountains. Explain how this is possible.

THINK AND DO (HANDS-ON APPLICATION)

31. Look at some halite crystals under a magnifying glass or microscope and observe their shapes. They are not round or triangular but cubes. The salt factory has no machines designed to give salt these cubic shapes. The cubic shape occurs naturally and mirrors how the atoms of salt are organized—cubically.
32. Smash a few salt cubes and then look at them carefully again. What you'll see are smaller salt cubes. Explain why salt breaks into cubes in terms of the property of cleavage.

THINK AND SOLVE (MATHEMATICAL APPLICATION)

33. Concrete is made from sand, gravel, and cement (which is made from shale, limestone, and the minerals quartz, gypsum, iron, alumina, manganese, and clay). Every year, 4700 lb of concrete is produced for every person in the United States, so that our roads, homes, schools, businesses, factories, and so on can be built or maintained. Calculate how much concrete is produced for your entire class in one calendar year. (Source: Mineral Information Institute, www.mii.org.)
34. (a) What is the average temperature of Earth's interior 20 km below the surface? (b) How much warmer is it 30 km below the surface compared to 10 km below?

THINK AND EXPLAIN (SYNTHESIS)

35. A geologist finds an igneous rock that has large crystals embedded in a matrix of smaller crystals. What is the texture of this rock? How did it form?
36. Why are the ferromagnesian silicates often dark, dense, and magnetic?
37. Why do rocks made from slowly cooling magma have large crystals? Give an example of this kind of rock.
38. How does the atomic structure of glass differ from the atomic structure of the mineral calcite?
39. The chemical formula for quartz is SiO_2. Coesite has the same chemical formula. Do coesite and quartz have the same physical properties? Why or why not?
40. There are many types of igneous rocks. Are there as many types of magma? Explain why or why not.
41. Identify a natural cycle other than the rock cycle.
42. What is the difference between a clastic and a chemical sedimentary rock?
43. An impression is a type of fossil that is made by an organism that is buried quickly, before it can decompose. Is this impression of a fish contained in an igneous, sedimentary, or metamorphic rock? Defend your answer.

44. What is more plentiful on Earth—the group of minerals known as feldspars or the group of minerals known as silicates?
45. A tectonic plate is subducted. Why will its rocks and minerals change?
46. Why do some high-quality drills have diamond tips?
47. Toothbrushes and toothpastes usually consist of mica (it provides sparkles), limestone (it rubs away plaque), and plastic (it forms the handle and bristles). Which of these ingredients is a mineral? Which is a rock? Which is neither rock nor mineral?
48. "Minerals in Earth's crust generally do not contain oxygen because oxygen is a gas at surface temperatures." Is this statement right or wrong? Defend your answer.
49. Why are intrusive igneous rocks coarse grained? Why are extrusive igneous rocks fine grained?
50. Which of the three classes of rocks is formed at low temperatures?
51. Would you expect to find a fossil in marble? Why or why not?
52. Why should mines be air-conditioned?
53. The chemical formula for quartz is SiO_2. What is the chemical formula of coesite, a polymorph of quartz?
54. Why are metamorphic rocks formed underground?
55. Sedimentary rocks usually form under water. Why is this so? (*Hint:* Remember the role of gravity.)
56. Cycles in nature, such as the rock cycle, consist of both materials and processes. What are the processes of change that take place in the rock cycle? What are the materials that are affected by these processes?
57. How is the magma that crystallizes to make rhyolite different from the parent magma of basalt?
58. How can one magma body produce many different kinds of igneous rocks? (*Hint:* Remember that magma contains many different minerals.)
59. Why are quartz and diamond so much harder than gold?

60. Name two properties of minerals that are based on the mineral's crystal structure.
61. One of your friends thinks that all mining should be stopped because it can damage the environment. Another friend thinks that mining shouldn't be subject to any laws that would protect the environment. Which friend do you agree with—or don't you agree with either of them? Explain your thinking.
62. Earth's mineral resources are plentiful, but once extracted and used, they do not renew in a human lifetime. What are some possible environmental problems associated with the extraction of minerals?
63. Is the following rock a sedimentary rock, igneous rock, or metamorphic rock? Why do you think so?

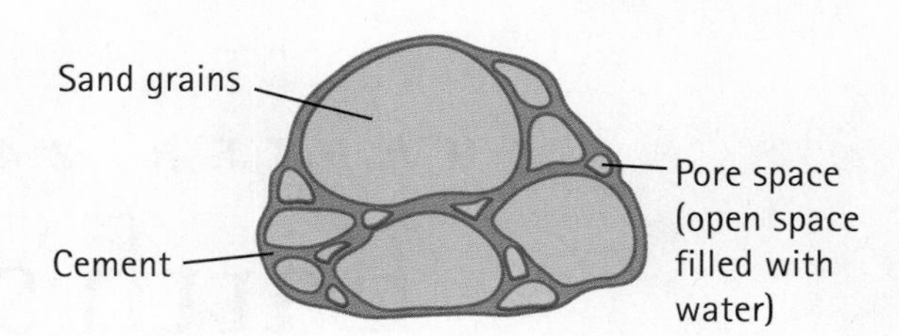

READINESS ASSURANCE TEST (RAT)

If you have a good handle on this chapter, then you should be able to score 7 out of 10 on this RAT. If you score less than 7, you need to study further before moving on.

Choose the BEST answer to each of the following:

1. The silicates are the largest mineral group because silicon and oxygen are
   (a) the hardest elements on Earth's surface.
   (b) the two most abundant elements on Earth's surface.
   (c) found in the common mineral quartz.
2. What physical change in metamorphic rock signals the end of metamorphism?
   (a) freezing
   (b) melting
   (c) crystallization
   (d) evaporation
3. Wow! You find a rock that contains a fossil. The rock could be
   (a) diamond.
   (b) sandstone.
   (c) marble.
   (d) basalt.
4. Which of these does not belong in your mineral collection?
   (a) table salt
   (b) diamond
   (c) quartz
   (d) coal
5. Igneous rocks are the most common of the three types of rocks. Yet they are not often observed because they are typically buried under
   (a) metamorphic rocks.
   (b) sedimentary rocks.
   (c) minerals.
6. Large crystals are usually associated with
   (a) intrusive igneous rocks.
   (b) metamorphic rocks.
   (c) extrusive igneous rocks.
   (d) minerals.
7. During the early stages of Earth's development, when the entire planet was molten, there were no rocks. What kind of rocks must have been the first to form?
   (a) igneous rocks
   (b) sedimentary rocks
   (c) metamorphic rocks
8. Magma transfers heat to the surrounding rock as it rises toward Earth's surface. The rocks that absorb this heat may undergo
   (a) compaction.
   (b) regional metamorphism.
   (c) contact metamorphism.
   (d) cleavage.
9. Conglomerate is a sedimentary rock that consists of large, rounded pieces of other rocks embedded in mineral "cement." What type of sedimentary rock is conglomerate?
   (a) foliated
   (b) chemical
   (c) clastic
   (d) extrusive
10. An igneous rock can be transformed into a metamorphic rock by
    (a) weathering, erosion, deposition, then compaction to become a sedimentary rock, then heat and pressure.
    (b) being squeezed by great tectonic forces such as occur at a convergent plate boundary.
    (c) being heated by magma flowing nearby.
    (d) all of these

Answers to RAT

1. b, 2. b, 3. b, 4. d, 5. b, 6. a, 7. a, 8. b, 9. c, 10. d

CHAPTER 24

Earth's Surface—Land and Water

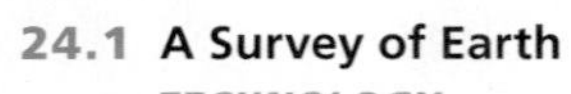

IMAGINE VIEWING Earth from a satellite and surveying its entire surface. The first thing you'd notice is that most of Earth—71%—is covered by ocean. The remaining 29% is taken up by seven continental landmasses. You'd see thousands of volcanoes on Earth's surface and notice that most of them are dotted along the rim of the Pacific Ocean. Earth's crust, in some places, is wrinkled by tight folds; in other places, it's etched by cracks and faults. Jagged peaks form mountain chains, many of which are located near coastlines. From your satellite view, you'd notice many other kinds of surface features, such as smoldering craters, snaking rivers, stark white glaciers, and fan-shaped deltas where rivers meet the sea. Looking at Earth as a whole, you'd see that its surface is quite varied—as if it had been broken and bent, twisted and pulled, built up and worn down. And, of course, over geologic time, it has! It's not so easy to board a satellite to survey Earth, but in this chapter you will nonetheless take a world tour of Earth's vast and varied surface.

24.1 A Survey of Earth

LEARNING OBJECTIVE
Describe Earth's major landforms.

EXPLAIN THIS If a sailor put a message in a bottle in the Atlantic Ocean, someone sailing on the Pacific Ocean might one day receive it. Explain how it's possible for an object to move from one ocean to another.

Almost three-quarters of Earth is covered by ocean. The land surface consists of scattered islands and seven large continents: Africa, Antarctica, Asia, Australia, Europe, North America, and South America. Earth's *topography*, or the shape of its surface, ranges from flat to bumpy to wrinkled. The most prominent topographic features of the continents are the linear mountain belts—for example, the Alpine-Himalayan Belt. The continents also feature expansive flat areas called *plains*, which are typically found at low elevations (Figure 24.1), and *plateaus*, which are uplifted flat areas composed of horizontal rock layers. On average, the continents lie 840 meters above sea level. Earth's continental elevation varies between the extremes of its highest point, Mt. Everest (8848 m, or 29,028 ft, above sea level), and its lowest point, on the shores of the Dead Sea (400 m, or 1312 ft, below sea level).

FIGURE 24.1
Plains in Kenya's Masai Mara National Reserve.

The three major oceans on Earth are the Pacific Ocean (Earth's largest, deepest, and oldest); the Atlantic Ocean (Earth's coldest and saltiest); and the Indian Ocean (the smallest). However, these oceans are connected, so, in truth, there is only one vast global ocean. Beneath the oceans, paralleling some coastlines, there are long, narrow ocean trenches. These are Earth's deepest places. The lowest point on Earth is in the Mariana Trench, with a depth of 11,033 m (36,198 ft)—that's deeper than any surface mountain is tall. Also under the ocean lies the global midocean ridge system, a continuous mountain belt 65,000 km (40,000 mi) long. Between the ocean trenches and ridges are deep undersea plains.

FIGURE 24.2
The Matterhorn in Switzerland. This mountain is named for its characteristic "horn" feature, which was produced by the action of massive glaciers.

About 10% of the world's land surface is covered by glaciers or other massive bodies of ice. One vast glacier nearly covers the entire continent of Antarctica.

Earth has a great diversity of *landforms* or surface features in addition to the plains, plateaus, trenches, and mountains (Figure 24.2). Some of these features—faults and folds, for example—were created largely through tectonic processes. Other landforms, such as valleys, canyons, deltas, and dunes, result mainly from the action of water, wind, gravity, and ice as they wear down and redistribute rock.

Also, much of Earth's present surface has been shaped not by nature but by people. In highly populated cities, more than 90% of the ground may be covered by paved streets, buildings, parking lots, and other human-made structures and materials. So, we see that humans are a major geologic force that is shaping Earth's surface in dramatic ways.

FIGURE 24.3
In modern times, people have changed much of Earth's surface. Industrialization, technology, agriculture, and the large human population all have major impacts.

TECHNOLOGY

Remote Sensing

Today, *remote sensing* is the primary tool used for mapping landforms and observing rapid or slow changes on Earth's surface. Remote sensing involves imaging Earth's surface with satellite-based cameras, radio receivers, scanners, thermal sensors, and other instruments. Such tools are used to create digital images, maps, and graphs of Earth's surface features. Most often, remote-sensing devices gather data in the form of electromagnetic waves—visible light, infrared radiation, microwaves, and so on.

There are many remote-sensing applications. Satellite images are used for mapmaking and surveying, to detect temperature variations on Earth's surface, and to track weather changes. Satellite images provide clues that point to subsurface deposits of mineral ores, oil, gas, and groundwater, and they help scientists and others to manage the land, the ocean, and the atmosphere.

Satellite pictures also allow us to compare landscapes before and after such natural events as floods, earthquakes, and fires, and to assess the damage done. In 2005, satellite images helped rescuers, homeowners, and the public identify the damage to New Orleans and the Gulf Coast after Hurricane Katrina. Satellite images, posted to the National Oceanographic and Atmospheric Administration (NOAA) Web site, were the first images of the devastated Gulf Coast that were made available to the public.

(a)

(b)

(c)

(a) This satellite image shows New Orleans, Louisiana, on March 9, 2004. At the time, New Orleans was a city with 1.3 million people and extensively developed buildings and roads. Because New Orleans lies below sea level, a system of levees was built to prevent flooding by surrounding waters. (b) New Orleans after Hurricane Katrina struck on August 31, 2005. The storm surge and rains led to the breaching of levees. Flooding, as shown here, occurred in more than 80% of the city. (c) An enlarged view of a portion of part (b), showing submerged roads and buildings.

24.2 Crustal Deformation—Folds and Faults

LEARNING OBJECTIVE
Identify the three types of stress that act on Earth's crust, and describe their effects.

EXPLAIN THIS Why does rock tend to break at Earth's surface but to bend below the surface?

Earth's surface is divided into a dozen or so massive patches of rock called tectonic plates. The movement of tectonic plates produces great **stresses**, or forces on Earth's rock. There are three types of stress:

1. *Compression* is the pushing together of masses of rock. Converging plates experience compressive stress.
2. *Tension* is the pulling apart of rock. Diverging plates undergo tensional stress.
3. *Shear stress* is produced when plates slide past one another.

A rock can respond in one of three ways to these applied stresses. It can *fracture* or break, as we commonly observe at Earth's surface. (What happens when you hit a rock with a hammer, for instance?) But rock responds to stresses in less familiar ways as well. It can *deform elastically*, in which case it bounces back to its initial size and shape when the applied stress is removed. Or, if the stress exceeds a level of stress called the "elastic limit" of the rock, the rock *deforms plastically*. The stressed rock permanently loses its original form—like Silly Putty™. Which of these three responses actually occurs depends on the nature of the rock, its temperature, the rate at which the stress is applied, and the confining pressure on the rock.

Colder surface rock is more brittle than warmer subsurface rock, which can slowly bend and flow. So, surface rocks typically fracture when their elastic limit is exceeded, but rock buried deep inside Earth often deforms plastically to produce folds.

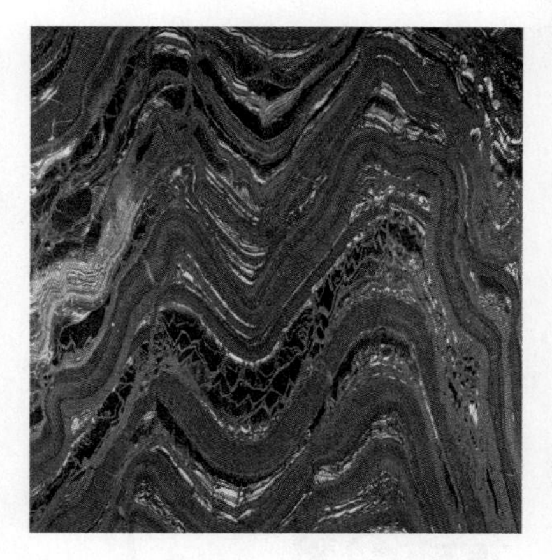

FIGURE 24.4
This banded iron formation (BIF) is a rock formed from layers of the iron oxide mineral hematite (red), tiger's eye, and jasper. At more than 3 billion years old, BIFs are among the oldest rocks on Earth. Deep burial of the rock and subsequent tectonic compression folded it. This sample is only about 2 cm across, so the folds in the rock are quite small.

Folds

Folds are bends in layered rock. Folds vary in size from tight, or even microscopic, folds in metamorphic rock (Figure 24.4) to broad undulations, such as those in folded sedimentary rock (Figure 24.5).

Folds in rock result from compressive forces. They are like the wrinkles you might find in a rug when you push on one end. Imagine the enormous forces required to wrinkle rock! Usually the force comes from the collision of tectonic plates, but sometimes other factors produce this compression, such as the movement of magma.

FIGURE 24.5
Folded strata are revealed in a roadside cliff along the San Andreas Fault in California. The North American Plate and the Pacific Plate are on opposite sides of the fault. The plates rub against each other as they move in opposite directions. This motion compresses and warps the rock at the plate boundary into folds.

Folds typically form series of arches and troughs. Arch-shaped folds, as shown in Figure 24.6, are called *anticlines*. *Synclines* are trough-shaped folds. The rocks at the center of a syncline are the youngest; that is, as you move horizontally away from the syncline's center, the rocks get older and older. In an anticline, the opposite is true: The rocks nearest the center of the fold are the oldest.

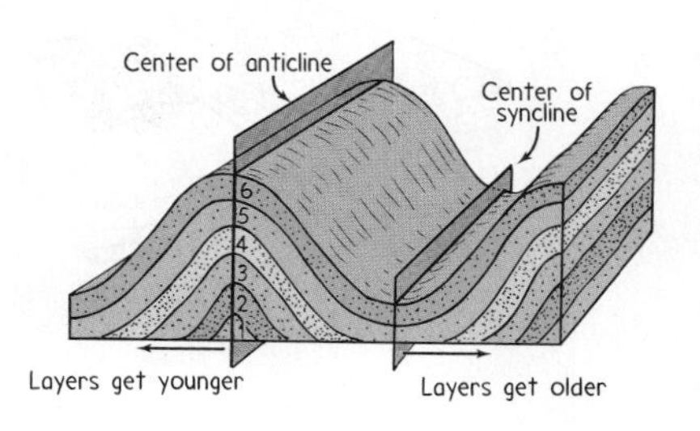

FIGURE 24.6
Anticline and syncline folds. Layer 1 is the oldest rock and layer 6 is the youngest. In a syncline, the youngest rock layers are near the center of the fold. In an anticline, the oldest rock is nearest the center.

CHECK YOURSELF

1. **Have you ever tossed a rock and seen it bend rather than break? What is your explanation for this?**
2. **When we can see a cross-section of Earth's surface—for example, at a road cut—layers of sedimentary rock sometimes look wavy. Why?**

CHECK YOUR ANSWERS

1. Not likely. Rock is brittle at Earth's surface because the temperature and pressure are low there, compared with the temperature and pressure in Earth's interior, where rock *can* bend.
2. The cross-section in such rock reveals synclines and anticlines—folds formed by compressive stress, most often related to the collision of tectonic plates.

Faults

A fracture is a crack in rock; a **fault** is a fracture where the rocks on either side have moved relative to each other. Movement along faults can occur rapidly in the form of a big earthquake, but such movement usually happens so gradually that we don't notice it.

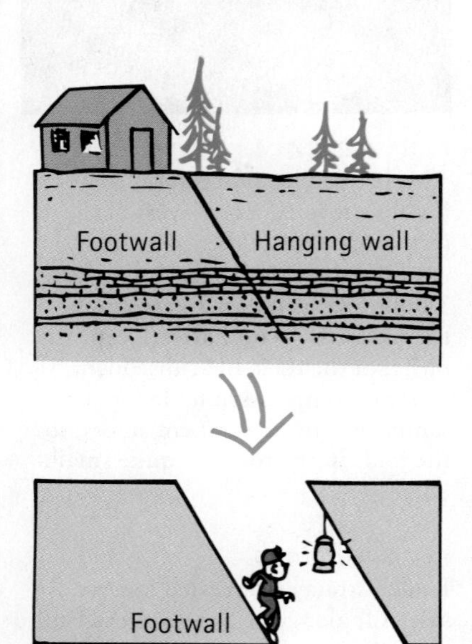

FIGURE 24.7
The terms *footwall* and *hanging wall* were commonly used by miners because a miner could hang a lamp on a hanging wall and stand on a footwall.

Faults mostly exist at plate boundaries where extreme tectonic stresses crack the crust, but faults can occur in the middle of a plate too. Faults vary in length from a few centimeters to hundreds of kilometers. The amount of displacement along faults also varies greatly, again in a range from centimeters to many kilometers. Faults are typically found in networks, rather than as single cracks, and at shallow depths, usually in the top 10–15 km of Earth's crust.

Faults are classified on the basis of the direction of displacement. There are three basic kinds: dip-slip faults, strike-slip faults, and oblique faults. To discuss these faults, it's useful to employ terminology coined by miners. Note the fault in Figure 24.7 (the slanted line in the top drawing). Imagine that you could pull the block diagram apart at the fault, as shown in the lower drawing. The half containing the fault surface where someone could stand is called the *footwall block*. The fault surface on the other block is also inclined, but in such a way as to make standing impossible; this is the *hanging-wall block*. (Miners could hang a lamp on a hanging wall and stand on a footwall.)

In a *dip-slip fault*, the hanging wall and footwall move vertically along the fault plane, as shown in Figures 24.8 and 24.9. Dip-slip faults are further classified as *normal faults* and *reverse faults*. In a reverse fault, the hanging wall moves up relative to the footwall. This fault motion, caused by compressional forces, is shown in Figure 24.8. In a normal fault, the hanging wall moves down relative to the footwall. (The name arises because it seems "normal" for the overhanging wall to slide down a sloping fault plane, since we are so accustomed to the effects of gravity.) The motion along a normal fault is due to tensional forces and is shown in Figure 24.9.

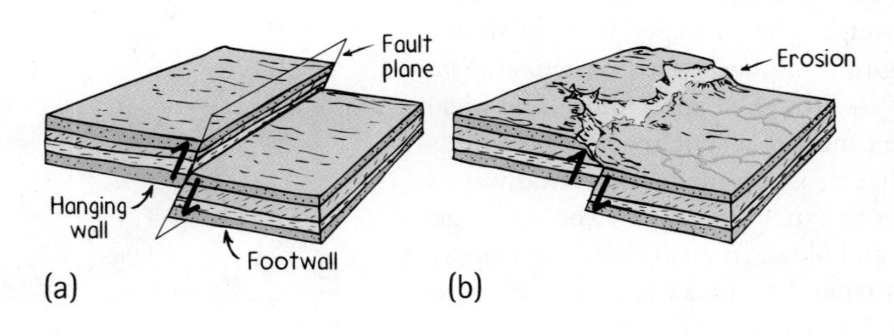

FIGURE 24.8
Reverse faulting occurs when compression pushes the hanging wall and footwall together and the rocks in the hanging wall are pushed upward relative to the rocks in the footwall. (a) A reverse fault before erosion. (b) A reverse fault after erosion.

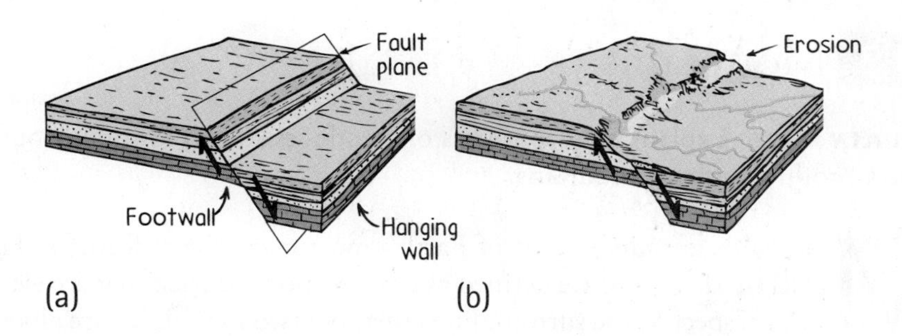

FIGURE 24.9
Normal faulting occurs when tension pulls the hanging wall and footwall apart and the rocks in the hanging wall drop down relative to the rocks in the footwall. (a) A normal fault before erosion. (b) A normal fault after erosion.

Strike-slip faults are characterized by horizontal motion (Figure 24.10). The world's most famous strike-slip fault, the San Andreas Fault, occurs at a transform-fault plate boundary that runs through California. Sticking and slipping of rock blocks along the San Andreas Fault (really a zone of networked faults) is responsible for many of the earthquakes California is noted for, including the great earthquake of 1906, which nearly destroyed San Francisco.

In an *oblique fault*, the blocks of rock have a combined motion. They move horizontally, as in a strike-slip fault, and vertically, as in a dip-slip fault. Oblique faulting occurs where there is a combination of tensional and compressional or shear forces.*

Networks of faults can be quite complex, and they can be used to read the history of tectonic events recorded in rock. For example, reverse faults correspond to compression. So if you find reverse faulting, your location could be a past or present convergent plate boundary. Normal faulting represents tension in rock. So if normal faults are present, you may be at a diverging plate boundary, a place where tectonic plates pull apart. Faults also have great economic importance because they can control the movement of groundwater and the distribution of minerals and fossil fuels. So faults are important for many reasons, not just because they are the locations of most earthquakes.

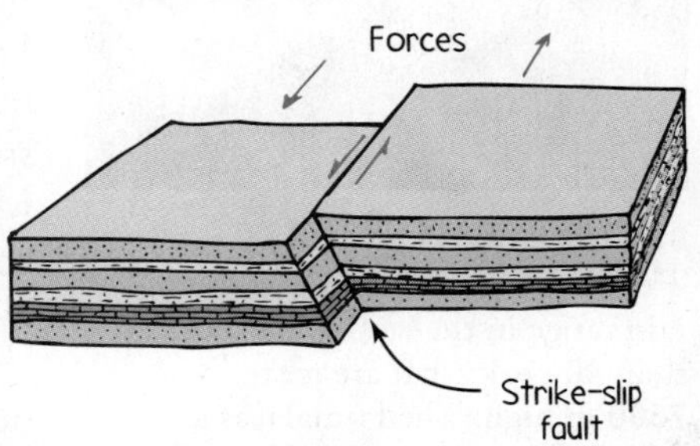

FIGURE 24.10
Blocks of rock rub and move past one another at a strike-slip fault.

CHECK YOURSELF

1. **You are walking across a desert landscape and note that the ground is cracked and dry. Should you conclude that the cracks you see are faults? Why or why not?**
2. **Why should a home builder be interested in the geologic study of faults?**
3. **What action occurs at a normal fault? Why is a normal fault so named?**

CHECK YOUR ANSWERS

1. No, faults are not simply cracks in Earth's surface. Faults are deep cracks along which sections of rock shift relative to one another. Actually, most faults cannot be seen at Earth's surface.
2. Since earthquake safety is a consideration in building design, home builders should be knowledgeable about earthquakes and therefore faults.
3. The hanging wall moves down relative to the footwall. This is a "normal" motion because it is the direction objects fall under the influence of gravity.

* Shear forces occur when one block of rock pushes on another block with a force that is at an angle to, rather than perpendicular to, the fault plane.

LEARNING OBJECTIVE
State the four major types of mountains, and explain how they form.

24.3 Mountains

EXPLAIN THIS Explain how compression produces some mountains but tension produces other mountains.

Mountains are surely some of Earth's most magnificent features (Figure 24.11). *Mountains* are defined as thick sections of crust that are elevated with respect to the surrounding crust. No two mountains are alike, but we can identify four basic types, which are classified according to their structural features: (1) folded mountains, (2) upwarped mountains, (3) fault-block mountains, and (4) volcanic mountains.

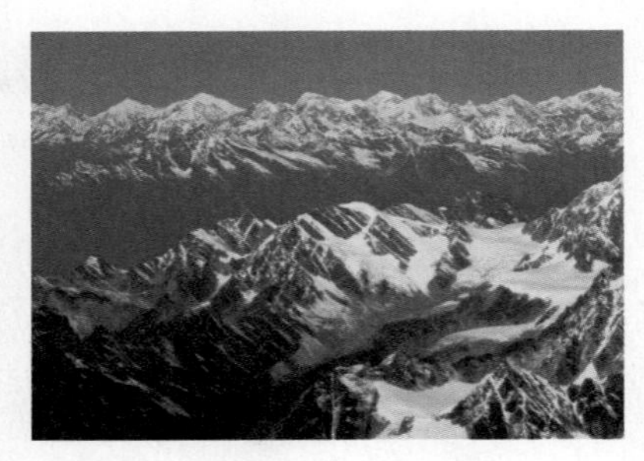

FIGURE 24.11
The Himalayas, the highest mountain range in the world, has more than 30 peaks that are more than 7600 m high. The Himalayas are folded mountains formed by colliding continental tectonic plates. They are located on the borders of Pakistan, India, Nepal, and Tibet.

Folded Mountains

Folded mountains are the most common type of mountain. They form over millions of years as tectonic plates collide. Tectonic collisions put compressive stress on rock so that it crumples to make folds. This is just like what happens when you push on a piece of paper or a rug from the edges and it wrinkles, except that the folds in rock are enormous. The world's tallest mountain ranges are folded mountains.

When you see large regions of folded mountains in the interiors of continents, you know that tectonic plates have collided there in the past. The Canadian Rockies are folded mountains resulting from recent plate convergence. The Himalayas are folded mountains too. They are still converging and growing as the Indian Plate keeps ramming into the Eurasian Plate. The Appalachian Mountains in the eastern United States were formed by a very ancient tectonic convergence 300 million years ago. They were once several kilometers taller than they are today, but erosion has worn them down.

The normal thickness of continental crust is about 35 km (22 miles). When a new folded mountain range forms, that thickness can be doubled. The Himalayas are about 80 km (50 miles) thick from the bottom of the crust to the summit of the highest mountains.

Upwarped Mountains

The Black Hills of South Dakota and the Adirondack Mountains of New York are **upwarped mountains**. Upwarped mountains have domelike shapes. Like folded mountains, they are produced by the compression of rock. However, upwarped mountains are usually single anticlines that form when Earth's crust heaves upward as a great amount of magma pushes its way up underneath. As Figure 24.12 shows, upwarped mountains are made up of older igneous and metamorphic bedrock that was once flat and overlain with sediment, but then was upwarped (pushed upward). As these regions were lifted, erosion removed

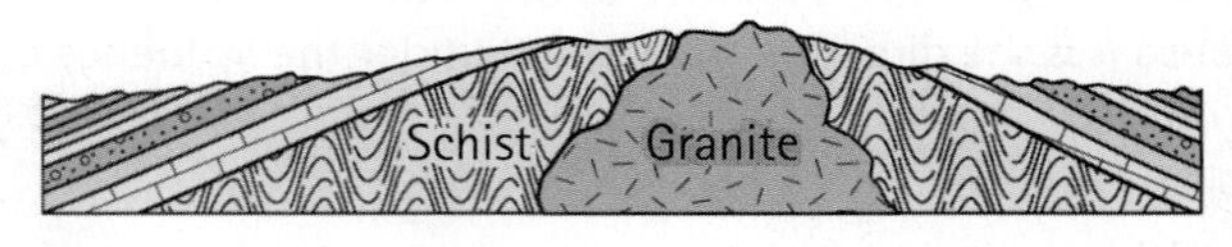

FIGURE 24.12
The Black Hills of South Dakota feature this large dome, an upwarped mountain. Sedimentary rock has eroded from the top, leaving an exposed dome of igneous and metamorphic rock.

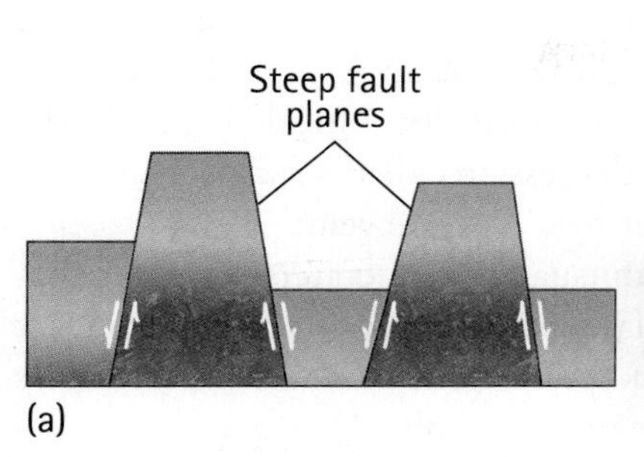

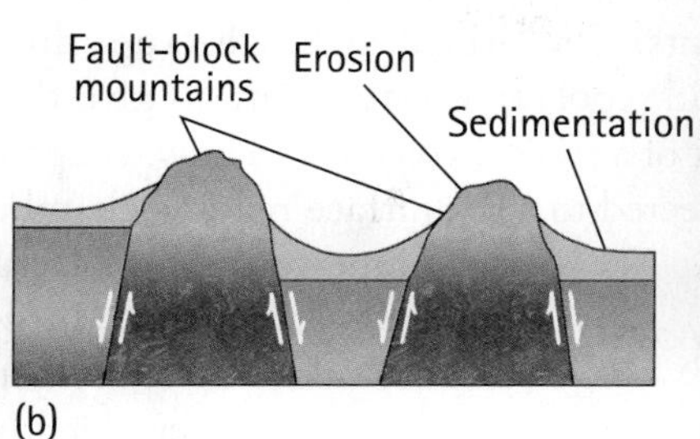

FIGURE 24.13
(a) Fault-block mountains are formed by blocks of rock that are uplifted along steep fault planes. (b) Erosion and deposition round out the fault planes that bound the mountains. (c) The Teton Range of Wyoming rises sharply above the surrounding land. This is a characteristic shared by fault-block mountains in general.

the easily weathered sedimentary surface rock. Domes of durable igneous and metamorphic rock were left behind.

Fault-Block Mountains

Mountain-size folds and domes result from compression, but tensional stress can produce mountains too. Mountains that form from tension and that have at least one side bounded by a normal fault are called **fault-block mountains**. Fault-block mountains occur when there has been broad uplifting over a large area. The uplifting stretches and elongates the crust so that it breaks along fault lines. Huge blocks of crust are sometimes thrust upward along steep fault planes while other sections of rock drop down, as Figure 24.13 shows. Blocks of rock end up stacked against one another like a series of toy block towers. Fault-block mountains rise steeply above the surrounding landscape. The western United States features many fault-block mountains, including the Teton Range in Wyoming and the Sierra Nevada in California.

CHECK YOURSELF

1. **Suppose huge blocks of crust drop down along normal faults. Between them, a block is left standing and over time it is uplifted to great heights. What kind of mountain is this?**
2. **The Sierra Nevada Range of California features fault-block mountains. Were these mountains produced by compressive or tensional forces?**
3. **A section of crust is compressed as it is pushed upward by forces acting on it from below. If the rock bends and arches upward to form a dome in response to this stress, what type of mountain will result? What kind of mountain will result if the rock breaks and moves upward along fault lines?**

CHECK YOUR ANSWERS

1. A fault-block mountain
2. Tensional forces
3. An upwarped mountain; a fault-block mountain

Volcanoes

A **volcano** is a hill or mountain formed by the extrusion of lava, ash, and rock fragments. The ejected materials erupt through the volcano's central vent, flow downhill, cool, and accumulate to give this landform its conical shape. At the summit of a typical volcano is a bowl-shaped depression called a *crater*. The crater is connected to a subsurface magma chamber by a pipelike vent.

Some volcanoes have unusually large craters. A crater that exceeds 1 kilometer in diameter is called a *caldera*. In most cases, calderas form when the summit of a volcano collapses into a partially emptied magma chamber; however, some calderas have been formed by explosive eruptions in which the top of the volcano was blown out, as shown in Figure 24.14. Yellowstone National Park is actually one huge caldera, about 40 miles in diameter!

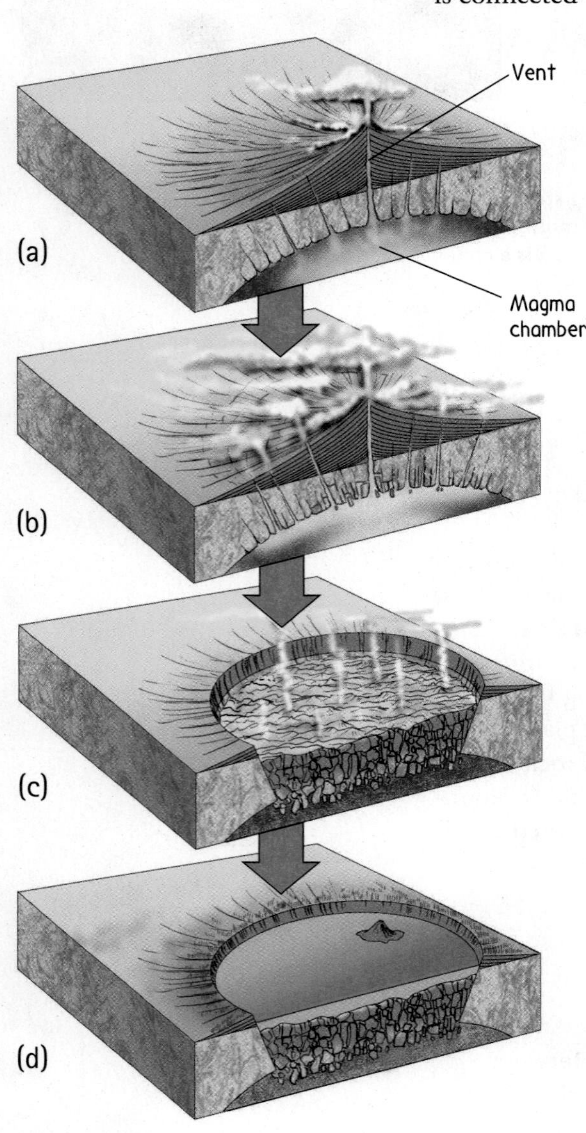

FIGURE 24.14
(a) Magma rising from a magma chamber into the vent of a volcano. (b) The formation of a caldera can involve a massive volcanic eruption, as shown here. (c) Material exploding from the volcano creates a much larger crater than the original volcanic vent. The magma chamber beneath the caldera eventually solidifies and, without a supply of magma, the volcano becomes dormant or extinct. (d) Rain and groundwater may eventually fill the caldera to form a lake.

There are three kinds of volcanoes, based on the types of material that erupt from them (Figure 24.15). Volcanoes that are built up from a steady supply of fluid basaltic lava are called *shield volcanoes*. Basaltic lava has a low viscosity—in other words, it flows easily. Shield volcanoes are shaped like a warrior's shield because the easily flowing lava that erupts from them pours out in all directions to form thin, gently sloping sheets.

Cinder cones are volcanoes that are very steep but rarely rise higher than 300 meters above ground level. Cinder cones are built from various ejected materials, including ash, cinders, glass and lava fragments, and rocks that erupt explosively from a single vent and pile up at a steep angle.

A *composite cone*, also known as a *stratovolcano*, is a volcano with a high, steep-sided summit and gently sloping flanks built of alternating layers of lava, ash, and mud. Mount St. Helens in Washington is an example of an active composite cone. In 1980, it erupted violently, disrupting the lives of thousands of people and transforming more than 200 square miles of lush forest into a burned, gray landscape.

Most volcanoes form near divergent or convergent plate boundaries. To see this, look at Figure 24.16, which shows a string of volcanoes in the *Ring of Fire*. The Ring of Fire encircles much of the Pacific Ocean. This is where 75% of the world's volcanoes are located. About 600 of these volcanoes are active. Tectonic plates meet at convergent boundaries all along the Ring of Fire.

Some volcanoes are not produced at the margins of tectonic plates but are instead created by hot spots. A **hot spot** is a stationary, exceptionally hot region deep in Earth's interior, usually originating at the mantle–core boundary. Mantle rock overlying a hot spot softens and is moved slowly upward by convection. Rising columns of hot, softened rock, called *mantle plumes*, make their way to the surface. This warmed rock melts under reduced pressure as it moves closer to the surface by convection, and then it erupts.

Most hot spots are located under the seafloor. Lava that erupts from a hot spot gradually builds a **seamount**, an undersea volcanic peak. Over geologic time, lava from countless eruptions enlarges a seamount. Ultimately, if the seamount gets

(a)

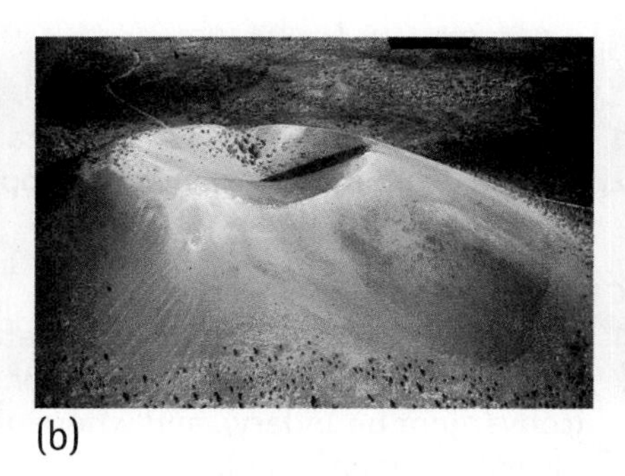
(b)

(c)

FIGURE 24.15
The three types of volcanoes. (a) Shield volcanoes, such as Mauna Loa, have broad, gentle slopes. (b) Cinder cones, such as Sunset Crater in Arizona, generally have smooth, steep sides and bowl-shaped summit craters. (c) Composite cones (also called *stratovolcanos*) such as Mount St. Helens (*foreground*) and Mount Rainier (*background*) are steep like cinder cones, but they are usually bigger because the mixture of lava and ash they erupt helps to hold them together. Note the volcanic ash on the southern flank of Mount St. Helens.

big enough, it will poke above the ocean as a volcanic island. Plates move, so that eventually a volcanic island loses its position over a hot spot. Because of this, the volcano will eventually lose its source of magma and become extinct. Over time, a new volcanic island can grow over the hot spot, and, if it does, a series of mountainous volcanic islands develops. This is how the Hawaiian Islands were created.

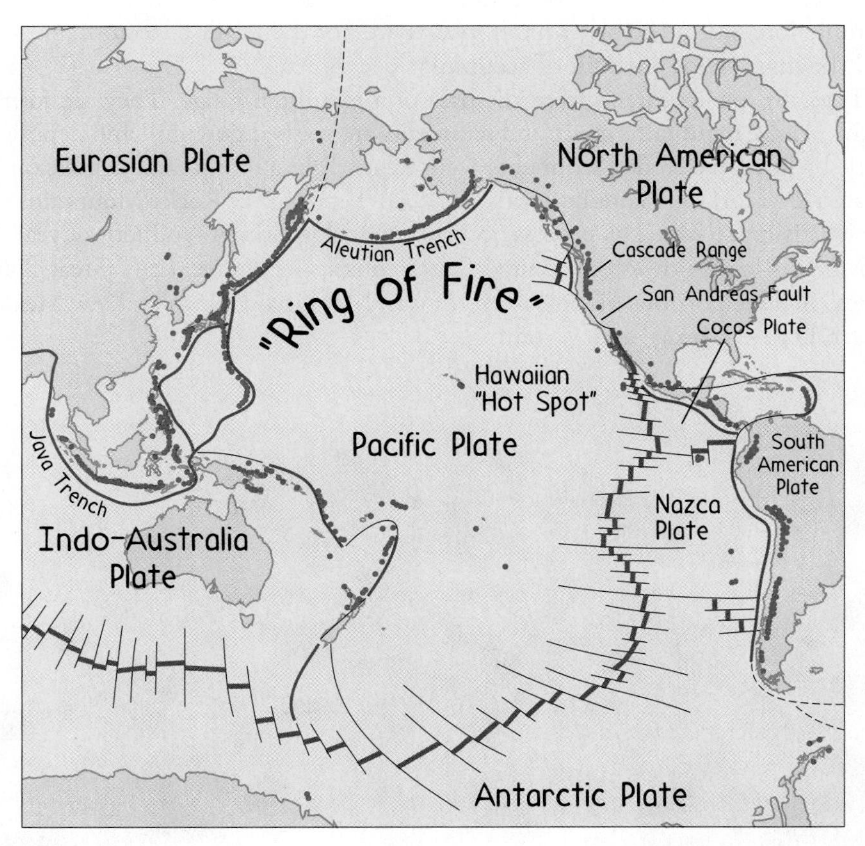

FIGURE 24.16
The Ring of Fire is the location of more than 75% of the world's volcanoes. Volcanoes are represented by red dots. Ocean trenches are shown as blue lines. Transform plate boundaries are designated by crosshatching. The plate boundaries along the Ring of Fire are convergent boundaries (not shown) along which subduction occurs.

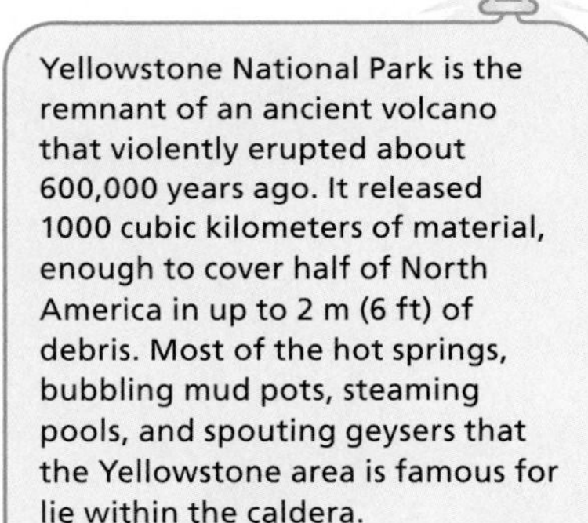
Yellowstone National Park is the remnant of an ancient volcano that violently erupted about 600,000 years ago. It released 1000 cubic kilometers of material, enough to cover half of North America in up to 2 m (6 ft) of debris. Most of the hot springs, bubbling mud pots, steaming pools, and spouting geysers that the Yellowstone area is famous for lie within the caldera.

CHECK YOURSELF

1. **Where are most of the world's volcanoes formed?**
2. **Why does a series of islands develop over a hot spot?**

CHECK YOUR ANSWERS

1. Most volcanoes are formed near plate boundaries. Volcanoes form where one plate subducts beneath another, triggering magma generation (convergent boundary), and where plates move away from one another (divergent boundaries).
2. The motion of the plates carries crust over a hot spot so that, over time, a chain of volcanoes may develop there.

LEARNING OBJECTIVE
Differentiate between plains and plateaus.

24.4 Plains and Plateaus

EXPLAIN THIS Why are plains located at the bases of mountain ranges?

Unlike mountains, plains and plateaus are flat. *Plains* are broad, flat areas that don't rise far above sea level. *Plateaus*, on the other hand, are flat areas uplifted more than 600 m above sea level (Figure 24.17).

Plateaus are usually massive sections of rock that have been uplifted by great tectonic forces. On the other hand, plains are not the result of tectonic heaving and shoving. Plains are built of accumulated sediment.

Typically, plains extend from the base of a mountain range. They are formed as the nearby mountains erode and sediments are washed downhill and deposited. Little by little, the eroded sediments accumulate, filling in dips and depressions to create a level surface. Plains built this way occur between the Rocky Mountains and the Mississippi River. The process occurs over geologic time—millions of years.

Much of the midwestern United States consists of plains. The "Great Plains states" include portions of Colorado, Kansas, Montana, Nebraska, New Mexico, North Dakota, Texas, and Wyoming.

(a)

(b)

FIGURE 24.17
Flat landforms: plains and plateaus. (a) Buffalo Gap National Grassland in South Dakota is a low, flat landform—a plain. (b) The Colorado Plateau covers a huge uplifted area in four states: Utah, New Mexico, Arizona, and Colorado. The Colorado River cuts the Colorado Plateau, forming the Grand Canyon in Arizona.

24.5 Earth's Waters

LEARNING OBJECTIVE
Describe how Earth's waters move in the water cycle.

EXPLAIN THIS Why is most of Earth's fresh water unavailable for drinking?

Water is vital to life, as we know. Water seems plentiful—most of Earth's surface is covered by it. However, consider the data in Figure 24.18. Almost all of Earth's water is salt water in the ocean. Less than 3% of Earth's water is fresh. Most fresh water is solid ice and snow in the mountains and in polar regions. The remainder of Earth's fresh water is groundwater; surface water in rivers, streams, and lakes; and water in the atmosphere and biosphere.

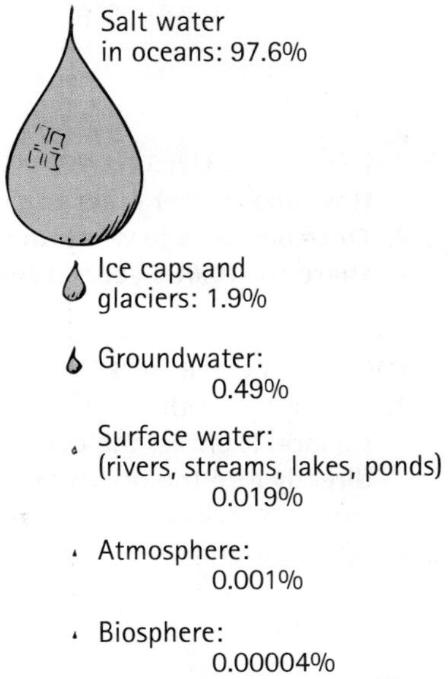

Salt water in oceans: 97.6%

Ice caps and glaciers: 1.9%

Groundwater: 0.49%

Surface water: (rivers, streams, lakes, ponds) 0.019%

Atmosphere: 0.001%

Biosphere: 0.00004%

FIGURE 24.18
Distribution of Earth's water. How much of Earth's water is available for drinking?

Water on Earth is constantly circulating, moving among the ocean, atmosphere, surface, and underground in a cycle, as shown in Figure 24.19. This cycle is called the **hydrologic cycle** or simply the *water cycle*. As water flows it also changes phase, evaporating and condensing in different parts of the cycle. The Sun gives water the energy it needs to evaporate. Gravity supplies the force that pulls water back to Earth's surface as *precipitation* (rain, snow, sleet, or hail). Gravity's constant pull moves water across Earth's surface downhill toward the oceans.

UNIFYING CONCEPT
- ***The Gravitational Force***
Section 5.3

Each of the places where water is stored as it moves through the water cycle is called a **reservoir**. Glaciers, the oceans, the atmosphere, cracks and pores in underground rock—and even the human body and other parts of the biosphere—are all reservoirs for Earth's total water supply. Water stays in different reservoirs for different lengths of time. An average water molecule will spend nine days in the atmosphere before falling as precipitation, but 3200 years in the ocean before leaving as evaporation. The average amount of time that a water molecule spends in each reservoir is called its **residence time**. Table 24.1 shows the residence times for a number of different reservoirs.

MasteringPhysics®
ACTIVITY: The Hydrologic Cycle

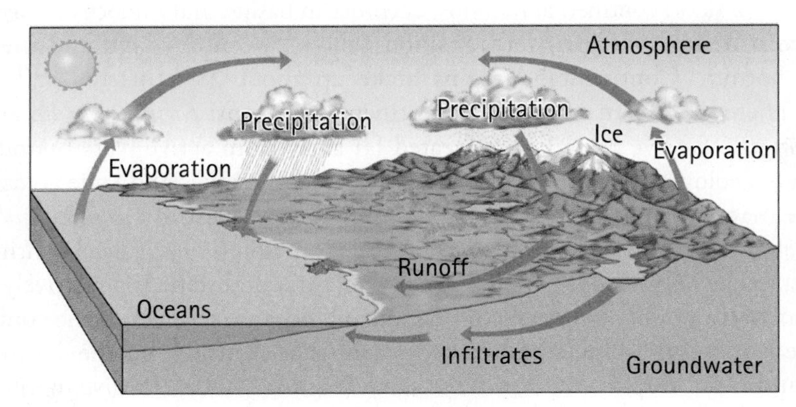

FIGURE 24.19
The major processes and reservoirs of the hydrologic cycle.

TABLE 24.1	RESIDENCE TIME OF WATER IN VARIOUS RESERVOIRS
Oceans	3200 years
Glaciers	20 to 100 years
Shallow groundwater	100 to 200 years
Deep groundwater	10,000 years
Lakes	50 to 100 years
Rivers	2 to 6 months
Atmosphere	9 days

CHECK YOURSELF

1. **How does water enter the atmosphere? How does water enter the ocean?**
2. **Once an average water molecule enters a certain reservoir, it will remain there for 10,000 years. Which reservoir is this?**

CHECK YOUR ANSWERS

1. Water enters the atmosphere by evaporating from the ocean or land surface. It enters the ocean by flowing across the surface, or by falling directly into the ocean, or by soaking into the ground and then flowing through rocks and cracks to the ocean.
2. Deep groundwater

There are many cycles in nature, including the rock cycle, the water cycle, and biogeochemical cycles. In any natural cycle, materials flow from one reservoir to another over and over again. Nature recycles!

LEARNING OBJECTIVE
Identify the regions of the ocean floor.

24.6 The Ocean

EXPLAIN THIS Why are the abyssal plains the flattest places on Earth?

The United States Geological Survey (USGS) estimates that there are 315 billion cubic miles of water in the ocean. What does Earth's surface look like under all this water? Oceanographers describe three main topographical units of the ocean floor: continental margins, deep-ocean basins, and midocean ridges.

A **continental margin** is a transition zone between dry land and the deep ocean bottom. Continental margins make up about 28% of the total ocean floor. There are two main kinds of continental margins: *passive margins* and *active margins.* Passive margins are located far away from active plate boundaries. They are geologically peaceful places with few earthquakes and little volcanism. Passive margins are wide and gently sloping. They consist of a *continental shelf, continental slope*, and *continental rise.* The shelf portion is nearly level continental crust that has been flooded by the sea. The continental slope is a relatively steep and narrow boundary between continental and oceanic crust. And the continental rise is a mound of accumulated sediment that eroded from the continental slope and is piled up at the foot of the slope (Figure 24.20a). Passive margins are found along the Atlantic Coast. Have you ever visited beaches on the East Coast of the United States? In many places, the continental shelf is wide and nearly flat, allowing plenty of space for swimming and boating near the shore.

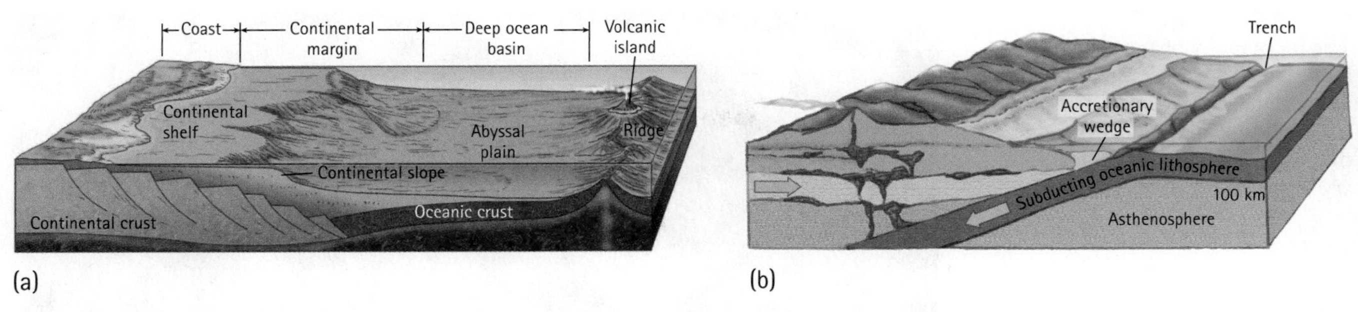

FIGURE 24.20
The continental margin is the transition zone between the coast and the ocean basin. The vertical drops are less steep in reality than shown here. (a) A passive margin is a place of geologic stability. (b) An active continental margin is associated with tectonic activity.

On the other hand, the West Coast lies along an active margin. An active margin is a place where the edge of a continent coincides with a tectonic plate boundary. Sections of the West Coast's active margin are involved in subduction or transform faulting. Active margins are common all along the Pacific Rim. The continental slope at an active margin can be the same feature as the landward wall of an ocean trench, as Figure 24.20b shows. Note that the active margin features an accumulation of sediment called an *accretionary wedge*. This is a mass of continental crust that is donated to the overlying plate by the descending plate. During subduction, sediments scrape off the descending plate. They crunch together and attach to the overriding plate as the accretionary wedge. Active margins are steeper, narrower, and more active than passive margins (Figure 24.20b).

The **ocean basin** is the deep depression in the lithosphere situated between continental margins and the midocean ridge. The ocean basin covers about 30% of Earth's surface. This is about the same area as all of Earth's exposed land. The ocean basin ranges between 3 and 5 km deep. It is composed of oceanic crust—dense, dark rock which is rich in basalt. The ocean basin contains abyssal plains, seamounts, and ocean trenches.

The **abyssal plains** are the flattest places on Earth because thick accumulations of sediment bury the uneven and rocky oceanic crust there. There is less than a 1-ft vertical change for every 1000 square feet of area in the abyssal plains—they are literally flatter than a pancake.

What would you find if you could deep dive below the ocean surface to visit the abyssal plains? The ocean changes with depth. The deeper you go, the less light there is. Sunlight can penetrate to only about 200 m, so the deep ocean is black. Temperature decreases with depth also. While the top 100–500 m of the ocean mixes and stays at about 25°C on average, the bottom of the ocean is dense and near freezing. Pressure increases with depth in the ocean. If you were to dive to the abyssal plains, which average about 4000 m below sea level, the force pressing in on you would be hundreds of times what atmospheric pressure is and you would immediately implode. Amazingly, abyssal organisms are adapted to this enormous pressure, total darkness, extreme cold, and scarcity of food. Most abyssal organisms eat dead organic matter that "snows" down on them from higher levels of the ocean.

Scattered about the ocean floor—particularly in the Pacific Basin—are the seamounts. As mentioned in Section 24.3, seamounts are isolated undersea volcanoes. They are steep-sided and tall; some tower thousands of meters above the abyssal plains.

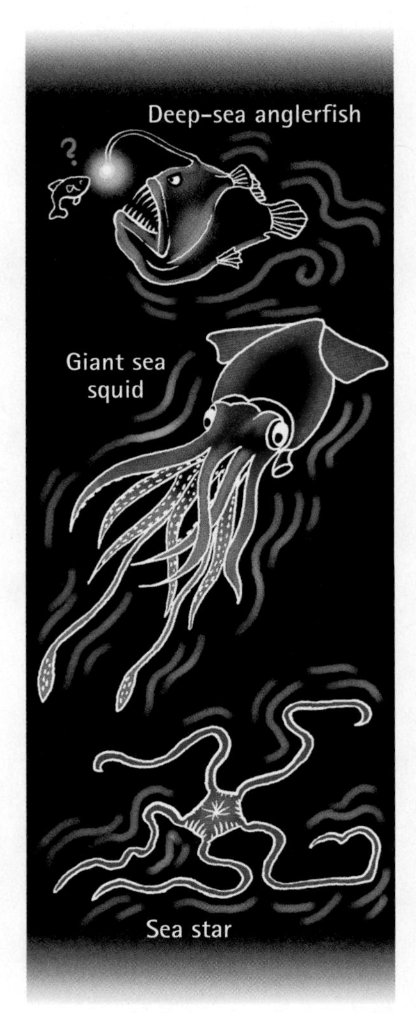

FIGURE 24.21
There are almost no light, little food, and extremely high pressure (5880–8820 pounds per square inch) in the abyssal plains. Nevertheless, life exists there. Residents of the abyssal plains include the giant sea squid, the largest of all invertebrates. It's more than 150 feet long and weighs more than a ton. Vicious anglerfish use light lures to attract their prey. Sea stars have luminescence on their arms that can be left behind to trick predators.

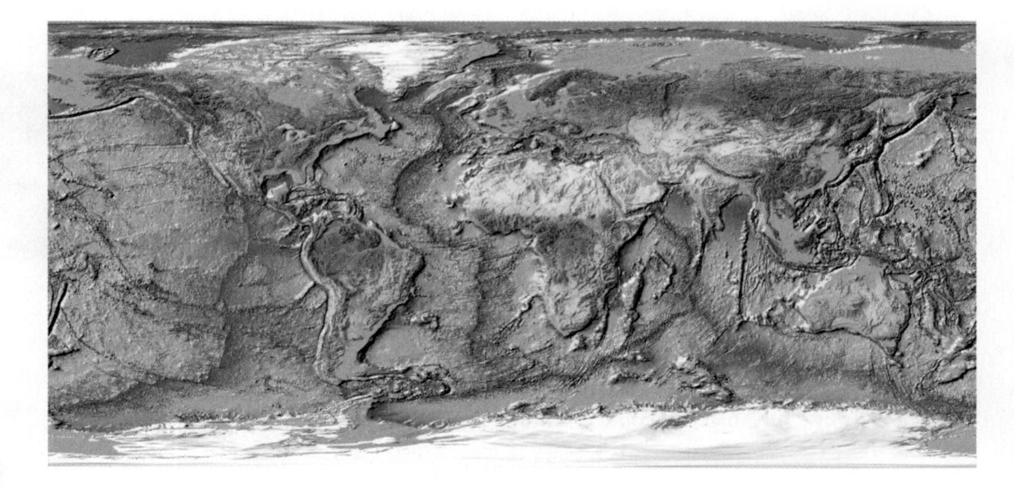

FIGURE 24.22
The topography of the sea floor. Note the midocean ridge, shown in darker blue. The Mid-Atlantic ridge is a section of the midocean ridge that marks a diverging plate boundary in the Atlantic basin.

UNIFYING CONCEPT

- ***Friction***
Section 2.8

Ocean trenches are deep furrows in the seafloor adjacent to active continental margins as discussed above. These are some of the sea's most hidden but fascinating places. Ocean trenches are the deepest places in the oceans, sometimes exceeding 10,000 m (more than 6 miles) in depth. That's much deeper than the height of most of Earth's surface mountains. Trenches are where the edge of one tectonic plate disappears as it slips under another plate and dives down into the mantle. The down-going plate bends as it slips under the trench, but it gets stuck along the way because of friction. When the force on the down-going plate exceeds the friction holding it in place, it slips, causing an earthquake. The friction caused by the motion of the down-going plate also creates heat, which contributes to the production of magma and the eventual formation of volcanoes.

The *midocean ridge* rises up from the ocean basin, typically demarcating seafloor spreading at a divergent plate boundary. It is a continuous chain of mountains that runs all around the globe like the seam on a baseball. It is about 65,000 m long, covering 21% of Earth's surface. Unlike surface mountains built of continental rock, the midocean ridge system is composed of dark, young basaltic rocks that erupted as magma, cooled, and were uplifted. This is Earth's longest topographic feature and very tall, rising 1000–3000 m above the surrounding plains. The term "ridge" is somewhat misleading, however, because the midocean ridge system is actually hundreds of meters wide. Broad rift valleys exist along the ridge crest in many locations.

The Mid-Atlantic Ridge is a portion of the midocean ridge that has been extensively studied because it sits in the middle of the Atlantic Ocean. The Mid-Atlantic Ridge stands between 2500 and 3000 m above the ocean floor but is still 2000–3000 m below sea level.

LEARNING OBJECTIVE
Describe where the salts in the ocean come from.

Integrated Science 24A
CHEMISTRY AND BIOLOGY

The Composition of Ocean Water

EXPLAIN THIS Why is the salinity of ocean water so constant?

People lost at sea can die of thirst even though they are surrounded by miles of water. Why? Ocean water is too salty to drink. Most of this salt is sodium chloride, the salt you safely sprinkle on your popcorn. Nevertheless,

TABLE 24.2 ABUNDANT SALTS OF THE SEA

Salt of Seawater	Weight per 1000 g
Sodium chloride ($NaCl$)	23.48 g
Magnesium chloride ($MgCl_2$)	4.98 g
Sodium sulfate (Na_2SO_4)	3.92 g
Calcium chloride ($CaCl_2$)	1.10 g
Sodium fluoride (NaF)	0.66 g
Total	34.8 g

if you were to drink ocean water, you could die of dehydration because your cells would pump water out of themselves, trying to dilute the excess salt in your bloodstream and becoming dehydrated as a result.

The salt in the ocean comes from minerals in Earth's crust that have been weathered and eroded for billions of years. Rivers and streams have carried dissolved elements and salts from the crust to the ocean, and they have been deposited there. Water, "the universal solvent," has a strong ability to dissolve salts, so salts exist as ions in the sea.

Salinity is defined as the proportion of dissolved salts to pure water. Seawater contains about 35 g of dissolved salts for every 1000 g of solution or 35 parts per thousand. Thus the salinity of seawater is about 35 g per 1 kg of seawater, which is represented as 35‰. How salty is seawater in everyday terms? One cubic foot of seawater contains 2.2 lb of salts.

It is a remarkable fact that the composition of seawater has remained virtually constant over millions of years—perhaps even a billion years. After all, there is a great deal of activity in the oceans. Salts and other compounds are continuously added to seawater by eroded continental rock, volcanic vents, dead marine organisms, and other sources.

To keep the composition of seawater constant, salts and other compounds must be removed from the sea as fast as they are deposited. Much salt is removed from seawater by chemical precipitation. Also, precipitates and other materials are taken into the mantle by subduction. Marine life has a strong influence on the composition of seawater by removing salts, dissolved gases, and other solutes. Multitudes of tiny *foraminifera* (marine protozoans) and various *crustaceans* (such as crabs and shrimp) remove calcium salts to build their bodies. Diatoms, which are microscopic marine algae, draw heavily on the ocean's dissolved silica to form their shells. Some animals concentrate elements that are present in seawater in minute, almost undetectable amounts. Lobsters extract copper and cobalt while certain seaweeds concentrate iodine. Vanadium is another rare element in seawater that serves a biological purpose. Sea cucumbers are blob-like animals that thrive in herds on the sea floor. They take vanadium from seawater and use it to make a blood pigment. Just as iron makes human blood red, vanadium makes sea cucumber blood yellow!

Water drains from Earth's surface and carries much of its material to the ocean. Besides salts, ocean waters hold decomposed biological material, dissolved gases, and waste from human activities. Ocean water has been called a weak solution of almost everything!

The Dead Sea—which is actually a lake—is about 35% salt, or 10 times as salty as the ocean. Why so salty? The Dead Sea is the lowest location on land, so water that runs into it has no outlet; the water stays there until it evaporates, leaving behind a salty mineral residue. We call it the Dead Sea because nothing can live in this extremely *hypersaline* lake.

FIGURE 24.23
A sea cucumber, one of the many organisms that removes substances from ocean water to maintain the ocean's salinity over time.

CHECK YOURSELF

1. **How could calcium salts taken into the mantle in a subduction zone ever become part of a crab's shell?**
2. **How many grams of salt are in 2 kg of seawater?**

CHECK YOUR ANSWERS

1. Crust that is pulled into the mantle through subduction may melt to form magma, and then flow back toward Earth's surface and one day erupt, forming a surface or underwater volcano. If the lava's minerals dissolve in the seawater, the crab could incorporate them in its shell.
2. Every 1000 g of water contains 35 g of salt. So 2000 g of water, or 2 kg, contains 70 g of salt.

LEARNING OBJECTIVE
Draw a diagram that shows Earth's freshwater reservoirs.

24.7 Fresh Water

EXPLAIN THIS Trace the route a water drop could take after it infiltrates the soil and then eventually emerges in a spring.

Fresh water is water that has a low concentration of minerals and salts. It's the water we and other animals drink and the water that most plants need, but less than 3% of Earth's water is fresh water. Most fresh water is frozen in ice sheets and glaciers, so it's not available for human use (Table 24.3). The largest supply of fresh water for human use is groundwater, which is water that has soaked underground. Surface water—liquid water in lakes, ponds, rivers, streams, springs, and puddles—is far less plentiful than groundwater, but it has a huge impact on life and surface geology. Rivers and streams, for example, contain only one one-thousandth of one percent (0.001%) of Earth's total water at any given time. Yet running water moves through them quickly, so rivers and streams play a major role in moving rock and sculpting Earth's surface. The ultimate source of Earth's fresh water is precipitation.

FIGURE 24.24
Ponds like this one form when water collects in depressions in Earth's surface. Less than 1% of Earth's total water supply is fresh liquid water available to animals for drinking.

Surface Water

When rain falls on land, most of it goes back up into the atmosphere through evaporation. Most of what doesn't evaporate soaks into the ground. This absorption of water by the ground is called **infiltration**. Whatever water the ground can't absorb becomes **runoff**, water that moves over Earth's surface. The proportion of rainfall that becomes runoff depends on the type of soil and on how wet or dry it is, as well as the steepness of the slope, the presence of plant life, and

TABLE 24.3 DISTRIBUTION OF EARTH'S FRESH WATER

Parts of the Hydrosphere	Volume of Fresh Water (km^3)	Percentage of Total Volume of Fresh Water
Ice sheets and glaciers	24,000,000	84.945
Groundwater	4,000,000	14.158
Lakes and reservoirs	155,000	0.549
Soil moisture	83,000	0.294
Water vapor in the atmosphere	14,000	0.049
River water	1,200	0.004
Total	28,253,200	100.0

(a)

(b)

(c)

FIGURE 24.25
A variety of streams. Streams merge to form bigger streams, and bigger streams merge to form rivers. Streams ultimately discharge their water to the sea.

the rate at which the rain falls—whether it's a sudden downpour or a sustained, gentle rain.

During and after a rainstorm, runoff collects in sheets that move downhill and merge to form *streams*. A stream is any body of flowing surface water, from the tiniest woodland creek to the mightiest river (a river is just a large stream). As streams move downhill, they merge with other streams. In this process, some streams can become quite large: Witness the Mississippi River in the United States and the Amazon River in South America. Whatever the size, most streams eventually discharge into the sea.

The area of land that drains into a stream is called the stream's *drainage basin* or **watershed**. Watersheds can be large or small, and every stream, tributary, and river has one. Large rivers that gather many streams also claim the streams' watersheds. In other words, the watershed of a large river is a patchwork of the many smaller watersheds that service the smaller streams.

Watersheds are separated from one another by **divides**, lines that trace the highest ground between streams. Under most circumstances, the separation is complete—rain that falls on one side of a divide cannot flow into an adjacent basin. A divide can be hundreds of kilometers long if it separates two large watersheds, or it can be only a short mountain ridge separating two small gullies. The Continental Divide, a continuous line running north to south down the length of North America, separates the Pacific basin on the west from the Atlantic basin on the east. Water to the west of the Continental Divide eventually flows to the Pacific Ocean, and water to the east of it flows to the Atlantic Ocean (Figure 24.26).

Surface runoff pauses on its way to the sea when it flows into a lake. A lake is formed by surface and subsurface waters flowing into a depression in Earth's surface. For a lake with a stable water level, evaporation plus drainage out of the lake balances the water flowing into it.

Why do you see a lot of runoff on city streets? Pavement reduces the ground's ability to soak up water.

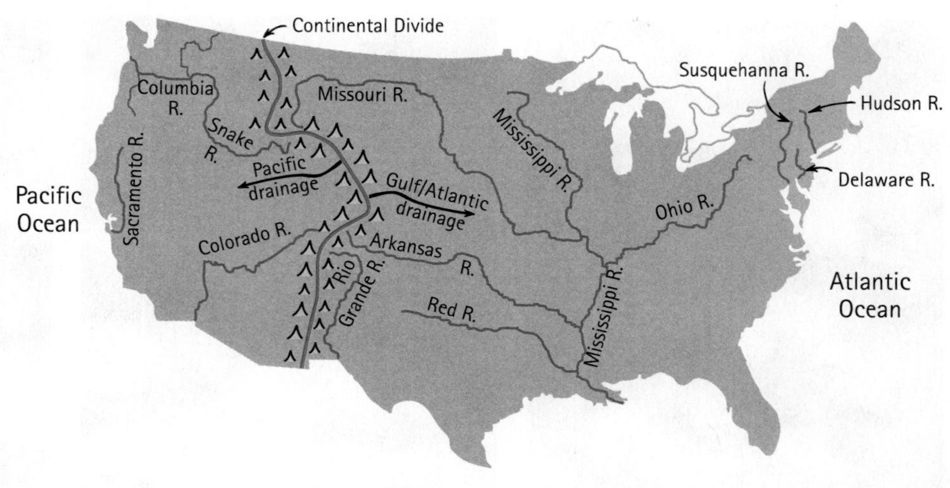

FIGURE 24.26
The Continental Divide in North America separates the Pacific basin on the west from the Atlantic basin on the east.

Groundwater

Rain and snowmelt absorbed by the ground may be retained by the soil. Some of this soil moisture is taken in by plants, and some of that water is transpired into the atmosphere. Water that isn't captured by soil percolates downward, moving between rocks and sediments and into the narrow joints, faults, and fractures in rock. Percolating water continues to move lower until it reaches the *saturation zone*. In the saturation zone, all the open spaces between sediments and rocks—and even the spaces between mineral grains *inside* rocks—are filled with water. Water that resides in the saturation zone is called **groundwater**. Groundwater continues to flow both downward and laterally in the saturation zone, eventually finding a stream or other outlet, often at the land surface (Figure 24.27).

Groundwater supplies streams, lakes, swamps, springs, and other surface waters. Most surface water is not obtained directly from runoff; instead, it flows into surface reservoirs from underground. During dry periods, when rain does not fall, it is groundwater that feeds surface waters. So, groundwater can provide needed water to the surface in time of drought.

Wells pump groundwater to the surface, where it is used for drinking, agriculture, and industry. In some regions, however, overuse of this resource has led to environmental problems, including land subsidence (lowering of the land surface), groundwater contamination, and streamflow depletion.

The upper boundary of the saturated zone is the **water table**. The level of the water table beneath Earth's surface varies with precipitation and climate. The water table is not flat like a kitchen table; rather, it tends to rise and fall with surface topography, as shown in Figure 24.27. Where the water table intersects the surface of the land, we find marshes, swamps, and springs. At lakes and perennial streams (streams that run all year), the water table is above the land surface.

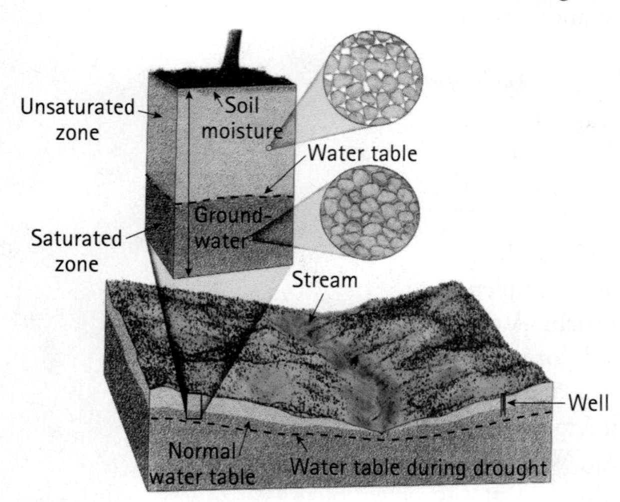

FIGURE 24.27
The unsaturated zone lies above the saturated zone. Groundwater resides in the saturated zone. The water table is the boundary between the two zones.

SCIENCE AND SOCIETY

Whose Water Is It?

If you have ever stood on the shore of one of the Great Lakes, or experienced the power of a waterfall, or been caught in a major downpour, it may seem to you that Earth's supply of fresh water is inexhaustible. From the perspective of one resident of the United States, water may be quite abundant. The world population, however, has grown to more than 7 billion people. If we were to spread ourselves evenly over all the habitable land, there would be more than 50 of us in every square kilometer. Thus it should come as no surprise that, collectively, we have a big impact on Earth's limited resources, including fresh water.

We are reminded that fresh water is a limited resource in the United States when farmers fight for the privilege to irrigate, or when our water utility bills rise, or when the water supply of downstream municipalities is endangered as upstream municipalities release sewage into the water. Globally, there are many nations in which the primary supply of fresh water is rivers that originate in some other nation. As the upstream nation diverts fresh water for its own expanding population, political tensions rise. Over the next decade, for example, it is projected that agricultural development in Ethiopia and Sudan will reduce the flow of the Nile River into Egypt by 15%. Similarly, Turkey is currently expanding its damming and irrigation projects along the headwaters of the Euphrates River. Once fully implemented, these projects could result in a 40% reduction of the river's flow into Syria and an 80% reduction of the river's flow into Iraq. It is not surprising that the issue of water is a major source of political tension among these nations.

You may have noticed that, during a rainstorm, sandy ground soaks up rain like a sponge. Sandy soils soak up water very easily; other soils, such as clay soils, do not. Rocky surfaces with little or no soil are the poorest absorbers of water. The amount of groundwater that a material can store depends on its porosity. **Porosity** is the percentage of the total volume of rock or sediment that consists of *pore spaces* (open spaces). Pore spaces are usually found between sediments, but fractures in rock, spaces between mineral grains within a rock, and pockets formed in soluble rock also contribute to porosity.

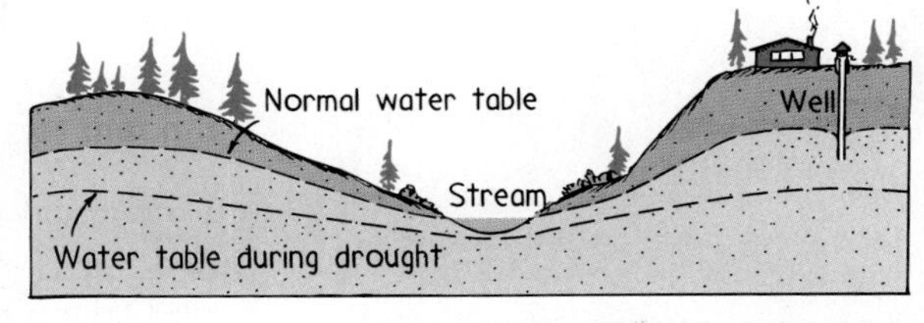

FIGURE 24.28
The water table roughly parallels the surface of the ground. If a well is drilled, the water level is at the water table. In times of drought, the water table falls, reducing streamflow and drying up wells. The water table also falls if the rate at which water is pumped out of a well exceeds the rate at which groundwater is replaced.

Rock or sediment may be very porous, but if the pore spaces are small and not interconnected, groundwater cannot freely move through it. **Permeability** is the ease with which fluids flow through the pore spaces in a rock or between rocks or sediments. Sand and gravel are highly permeable because they are composed of rounded particles that do not fit together tightly. Their pore spaces are large and connected, so water can flow easily from one pore space to the next. On the other hand, if the pore spaces are too small or poorly connected, water cannot flow through them at all. Clay sediments, for example, are composed of flat particles that fit tightly together. This is why clay, which can be quite porous, is practically impermeable (Figure 24.29).

An **aquifer** is a zone of water-bearing rock through which groundwater can flow. Aquifers generally have high porosity and high permeability. These reservoirs, which underlie the land surface in many places, contain an enormous amount of groundwater. They are important because wells can be drilled

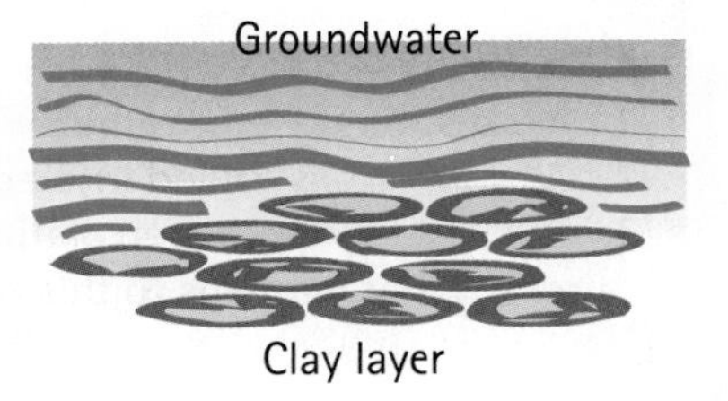

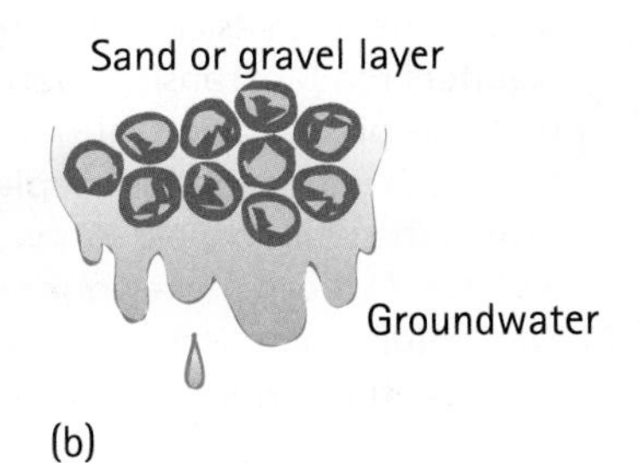

FIGURE 24.29
(a) The sediment particles in clay are flat and tightly packed, so the pore spaces are poorly connected. Clay is therefore not permeable. (b) The sediment particles in sand and gravel are relatively uniform in size and shape and they have large, well-connected pore spaces, so water can flow freely through sand and gravel. These are permeable materials.

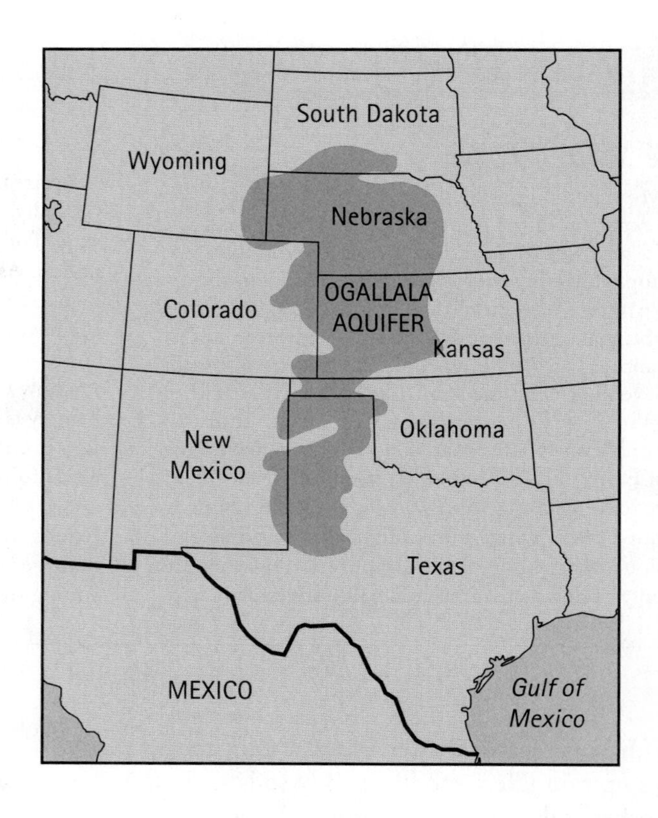

FIGURE 24.30
The Ogallala Aquifer is the biggest aquifer in the United States.

into them and water can be removed. More than half of the land surface in the United States is underlain by aquifers. The vast Ogallala Aquifer stretches from South Dakota to Texas and from Colorado to Kansas (Figure 24.30).

Groundwater, the largest supply of fresh water available for human use, has great economic importance. In the United States, more than 40% of the water used for everything except hydroelectric power production and power-plant cooling is groundwater.

Groundwater accumulates as precipitation percolates down from the surface. However, the process of restoring lost groundwater, called *recharging*, is slow. An aquifer constantly gains water from its recharge zone (the area of land from which the groundwater originates), but only a small amount of water reaches it each year. Completely recharging a depleted aquifer may take thousands or even millions of years. Thus, groundwater is considered a nonrenewable resource.

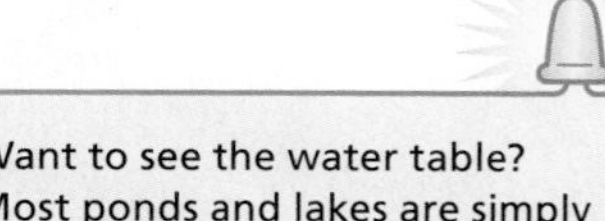

Want to see the water table? Most ponds and lakes are simply places where the land surface is below the water table. Swamps and marshes are places where the water table is level with the ground surface.

CHECK YOURSELF

1. **What's the difference between a saturation zone and an aquifer?**
2. **Why are wells drilled into aquifers? Can a well run dry?**
3. **At a certain depth, percolating subsurface water encounters impermeable rock. What happens to this water?**

CHECK YOUR ANSWERS

1. A saturation zone may consist of water-bearing but impermeable rock or sediments, such as clay. Aquifers have high porosity *and* high permeability.
2. Aquifers readily transmit water to wells. Withdrawn water is replaced eventually by precipitation, although the process can be very slow. Pumping out too much water too fast depletes the water in the aquifer. Eventually, this causes the well to yield less water and even to run dry. Pumping your own well too fast can make your neighbor's well run dry, too.
3. The water can no longer move downward, so it moves laterally—sideways—and eventually finds a surface outlet.

24.8 Glaciers

LEARNING OBJECTIVE
Distinguish between alpine and continental glaciers.

EXPLAIN THIS What glacier contains more than half of Earth's fresh water?

Near the poles and high in the mountains, there are places so cold that snow rarely melts. Instead, it piles up year after year. The weight of overlying snow crushes buried snow into ice crystals. A giant mass of ice the size of football fields or even a continent is produced. This is a **glacier**. There are two basic types of glaciers, as Figure 24.31 shows. *Alpine glaciers* form in the valleys of mountainous areas. *Continental glaciers*, which are also called *ice sheets*, are huge sheets of thick ice that cover a vast area such as a continent or an island. A single continental glacier called the Antarctic Ice Sheet almost covers the continent of Antarctica. It is so thick that it covers nearly everything except mountaintops. The Antarctic Ice Sheet contains about 90% of the glacial ice on the planet, or about 61% of Earth's fresh water.

(a)

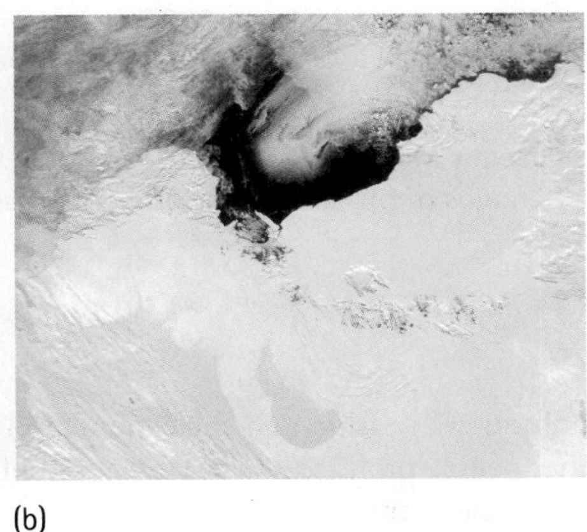
(b)

FIGURE 24.31
(a) An alpine glacier in the French Alps known as Glacier des Bossons. (b) As you can see in this satellite image, Antarctica is almost completely covered by a continental glacier called the Antarctic Ice Sheet.

Glaciers are sometimes called "rivers of ice" because they slowly glide across the land. High pressure at the base of the glacier can melt the ice there so that the glacier moves along a layer of liquid water. Gravity acts on a glacier too, constantly pulling it downhill. Alpine glaciers move much faster than continental glaciers because they are located on sloping land. As glaciers move, they carve the landscape like a giant bulldozer.

Integrated Science 24B
PHYSICS, CHEMISTRY, AND BIOLOGY

LEARNING OBJECTIVE
Distinguish between point pollution and nonpoint pollution.

Water Pollution

EXPLAIN THIS Describe the effects of large-scale industrialization and human population growth on water quality.

Water pollution is chemical, physical, or biological material introduced into water that harms organisms that depend on the water. The two biggest causes of water pollution are industrialization—the growth of large-scale factories and farms—and the human population explosion. Because

TABLE 24.4 COMMON KINDS OF WATER POLLUTANTS

Organic Chemicals
Fertilizers; pesticides; detergents; plastics; petroleum products, including gasoline and oil

Inorganic Chemicals
Metals; acids; salts

Toxic Chemicals
Chemicals that are poisonous to living things, including heavy metals such as arsenic and mercury; many industrial chemicals; some household chemicals, such as paint thinner and motor oil

Physical Pollutants
Heat; suspended particles, such as soil; litter, including fishing nets and plastic objects such as six-pack rings

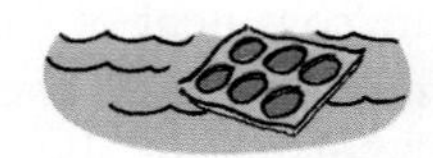

Pathogens
Organisms that cause disease, such as bacteria and viruses; pathogens in untreated sewage and animal feces that are washed into the water

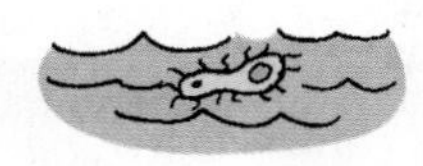

Organic Matter
Remains of organisms, including carcasses, feces, and plant material

of industrialization and the exponential growth of the human population, waste that people produce can't be disposed of as quickly as it is made. The waste ends up in the ocean, groundwater, rivers, lakes, and streams.

UNIFYING CONCEPT
• *The Ecosystem*
Section 21.1

Table 24.4 lists various kinds of water pollutants. Notice that some pollutants are disease-causing organisms, called *pathogens*, found in animal (including human) feces. Chemical water pollutants include toxic chemicals—poisons—and other chemicals that alter the chemistry of an ecosystem. One physical pollutant is heat. Are you surprised to see heat on the list? Some factories heat water and then discharge it into streams when they are finished with it. When the temperature of a stream rises, the amount of dissolved oxygen in it decreases. (Do you remember from chemistry that the solubility of oxygen decreases with temperature?) Fish and other aquatic organisms die when there is not enough dissolved oxygen for them to breathe.

The source of water pollution may be a particular factory that discharges excess heat or toxic chemicals. Or it may be an oil tanker that spills oil into the ocean, or a gas station that leaks chemicals into the groundwater. These are examples of *point pollution*—pollution due to a specific source. Point pollution is relatively easy to identify and therefore relatively easy to control.

Nonpoint pollution does not come from a single source. Nonpoint pollution comes from practically everywhere—homes, farms, freeways, city streets, and even acid rain. Nonpoint pollution is difficult to regulate and control because it comes from so many sources and because people may not see the connection between their activities and the polluted water. Therefore public awareness of nonpoint pollution is an important part of controlling water pollution. Table 24.5 shows some sources of point and nonpoint pollution.

TABLE 24.5 SOURCES OF WATER POLLUTION

Point Pollution
- Wastewater treatment plants
- Landfills
- Underground storage tanks, including gasoline tanks
- Septic tank systems

Nonpoint Pollution
- Salt applied to roadways
- Runoff from suburban and urban streets (contains litter, dog waste, oil, gasoline, etc.)
- Fertilizer
- Pesticides

CHECK YOURSELF

1. **How is studying integrated science helpful for understanding water pollution?**
2. **Why is public education important for controlling the problem of water pollution?**
3. **How can you help reduce point pollution? Nonpoint pollution?**

CHECK YOUR ANSWERS

1. Pollutants include biological, chemical, and physical agents.
2. People need to understand how certain activities cause pollution in order to develop ways of reducing pollution.
3. Your answer will depend on your particular circumstances. You can reduce the pollution that landfills create by being sure that you don't put any toxic chemicals, such as paint thinner, into the trash. Recycle electronics, glass, paper, and metals such as aluminum. To reduce nonpoint pollution, never litter and don't wash household chemicals down sewer drains. Tell a friend about the problem.

For instructor-assigned homework, go to www.masteringphysics.com 

SUMMARY OF TERMS (KNOWLEDGE)

Abyssal plains The flat regions of the ocean basin.

Aquifer A zone of water-bearing rock through which groundwater can flow.

Continental margin The transition zone between dry land and the deep ocean bottom.

Divide An imaginary line that is the highest ground separating two drainage basins.

Fault A fracture through a rock along which sections of the rock have moved.

Fault-block mountain A mountain that forms from tension and that has at least one side bounded by a normal fault.

Fold A layer or layers of bent (plastically deformed) rock.

Folded mountain A mountain that features extensive folding of rock layers.

Glacier A mass of dense ice that forms when snow on land is subjected to pressure from overlying snow, so that it is compacted and recrystallized.

Groundwater The water that resides in a saturation zone.

Hot spot A stationary, hot region deep in Earth's interior, usually originating near the mantle–core boundary.

Hydrologic cycle The cycle of evaporation and precipitation that controls the distribution of Earth's water.

Infiltration Absorption of water by the ground.

Ocean basin The deep portion of the sea floor between continental margins and the midocean ridge.

Permeability The ease with which fluids flow through the pore spaces in sediments and rocks.

Porosity The percentage of the volume of a rock or sediment that consists of open space.

Reservoir Each of the places where water is stored as it moves through the water cycle.

Residence time The length of time water exists in a reservoir.

Runoff Precipitation not absorbed by the ground that moves over Earth's surface.

Seamount An undersea volcanic peak.

Stress A force applied to rock.

Upwarped mountain A dome-shaped mountain produced by a broad arching of Earth's crust.

Volcano A hill or mountain formed by the extrusion of lava, ash, and rock fragments.

Watershed The area of land that drains into a stream.

Water table The upper boundary of the saturated zone.

READING CHECK (COMPREHENSION)

24.1 A Survey of Earth

1. What percentage of the Earth is covered with ocean?
2. What is Earth's highest point? How high is it? What is Earth's lowest point? How deep is it?

24.2 Crustal Deformation—Folds and Faults

3. (a) Identify and describe the three types of stress a rock may experience. (b) Describe the three ways in which rock can respond to applied stress.
4. Why does folding occur deep below Earth's surface?
5. Name three kinds of faults and distinguish among them.
6. Why are faults worth knowing about?

24.3 Mountains

7. (a) Name four types of mountains, classified by common structural features. (b) Give an example of each of the four types of mountains.
8. How do hot spots produce mountains?
9. Where are most of the volcanoes on Earth located?
10. What are the three different types of volcanoes?

24.4 Plains and Plateaus

11. What is the difference between a plain and a plateau?
12. Why do plains usually extend out from the base of a mountain?

24.5 Earth's Waters

13. (a) Where is most of Earth's water? What percentage of it resides there? (b) Where is most of Earth's fresh water?
14. (a) Describe the hydrologic cycle. (b) What part of the hydrologic cycle has particular relevance to the shaping of Earth's landforms?

24.6 The Ocean

15. (a) Describe the three parts of a passive continental margin. (b) Describe the parts of an active continental margin.
16. Describe the overall topography of the ocean floor.

24.7 Fresh Water

17. Approximately what percentage of Earth's fresh water is frozen in ice caps and glaciers? About what percentage is in groundwater? About what percentage is in streams and lakes?
18. What happens to rainwater when it falls to Earth?
19. Do you live in a watershed? Defend your answer.
20. Where are all the pore spaces in rocks and sediments filled with water?
21. In what way is the water table different from a table?
22. What is the water that resides in the saturation zone called?
23. Besides runoff from precipitation, what is the source of Earth's fresh water? Explain.

24.8 Glaciers

24. Why are glaciers called "rivers of ice"?
25. What percentage of the world's glacial ice is included in the Antarctic Ice Sheet?

THINK INTEGRATED SCIENCE

24A—The Composition of Ocean Water

26. Why do we infer that salts must be removed from seawater about as fast as they are deposited?
27. The salinity of seawater is almost constant over time. Provide an explanation.
28. If you drink too much ocean water you will die of dehydration. Explain how this is so.
29. Where does the salt in ocean water come from?

24B—Water Pollution

30. What is the difference between point and nonpoint pollution? Which is easier to control? Why?
31. Name one biological water pollutant, one physical water pollutant, and one chemical water pollutant.
32. There is a saying: "The solution to pollution is dilution." Do you think this is a good general rule? What might be some limitations of this approach to controlling water pollution?

THINK AND DO (HANDS-ON APPLICATION)

33. Compare the permeability and porosity of several soil samples. To do so, collect several samples of soil taken from various locations and put them in different containers. Look at each one, and predict through which sample the water will move the fastest. Slowly add 100 mL of water to each sample until it can hold no more. Note the time it takes for the water to reach the bottom of the container. Which sample had the greatest porosity? Which had the greatest permeability?

THINK AND SOLVE (MATHEMATICAL APPLICATION)

34. The volume of solids in a sediment sample is 975 cm^3, and the volume of open space is 325 cm^3. What is the porosity of the sediment? Describe what the result of your calculation means in physical terms.
35. Show that liquid fresh water makes up about 0.50% of the water on Earth based on the data in Figure 24.18.

THINK AND EXPLAIN (SYNTHESIS)

36. A factory emits steam into the air. How could those same water molecules eventually reside in the ocean? In a river? In groundwater? In a bear's body?
37. Some people "fold" under stress. Others "crack up." How is this like what happens to rock?
38. Where would you expect to see more runoff—along a city street or in a prairie meadow? Why?
39. Most glacial ice is stored in polar regions. Why, then, would a severe melting of glacial ice cause the sea level to rise off the coasts of countries located near the equator?
40. What immediately happens to rainwater when it falls to Earth? What eventually happens to it?
41. List five major landforms in the United States, and describe them in as much geologic detail as you can.
42. Why must aquifers consist of material that has both high permeability and high porosity?
43. What are the two different kinds of continental margins? How do they differ?
44. The soil under Samantha's home is rich in clay. The ground therefore has high porosity but low permeability. Would Samantha be able to build a good well in her backyard? Why or why not?
45. Your friend says that groundwater is a nonrenewable resource. Do you agree?
46. Rain falls on land. Then what happens to it?
47. Why does the area of the continental shelf change over geologic time? (Hint: Is sea level constant?)
48. If surface reservoirs such as lakes and streams suddenly dried up, would people have another source of drinking water? Explain.
49. What is the relationship between the level of the water table and the depth to which a well must be drilled?
50. Which surface features record tectonic compression acting on rock? Which features show tectonic tension?
51. The removal of groundwater can cause subsidence. If the removal of groundwater is stopped, will the land likely rise again to its original level? Defend your answer.
52. Is groundwater stored in underground rivers? What's your reasoning?
53. If you look at a map of any part of the world, you'll see that older cities are located either next to rivers or where rivers existed at the time the cities were built. Explain.
54. The oceans consist of salt water. Yet evaporation of ocean water makes clouds, and clouds make rain, which is fresh water. Why is there no salt in rain?
55. The Amazon River Basin is a giant watershed with an area of about 6 million square kilometers. What does runoff in the Amazon River Basin flow toward?
56. Relate the saying "What goes up, must come down" to the water cycle.
57. Why does kelp, a type of photosynthesizing seaweed, thrive in coastal waters but not in the vicinity of abyssal plains?
58. How is rock underground like a sponge?

59. Is the infiltration of water greatest on steep, rocky slopes or on gentle, sandy slopes? Defend your answer.
60. If the water table at location X is lower than the water table at location Y, does groundwater flow from X to Y or from Y to X?
61. (a) Refer to the data in Table 24.1. Compare how long a given water molecule remains in each of the following reservoirs, and rank the reservoirs in order from longest to shortest residence time: Atlantic Ocean, Lake Tahoe, the air over Miami, groundwater at the bottom of an aquifer. (b) Describe the process by which a water drop exits each one of these reservoirs.
62. Look at Table 24.4. What is easier to control—toxic chemicals leaching out of landfills or toxic chemicals in surface runoff? Defend your answer.

READINESS ASSURANCE TEST (RAT)

If you have a good handle on this chapter, then you should be able to score 7 out of 10 on this RAT. If you score less than 7, you need to study further before moving on.

Choose the BEST answer to each of the following:

1. Snow becomes glacial ice when it is subjected to
   (a) decreasing temperature.
   (b) pressure.
   (c) rain.
   (d) plastic deformation.
2. When a rock deforms plastically, it
   (a) fractures.
   (b) changes its size or shape temporarily.
   (c) changes its size or shape permanently.
   (d) changes its mineral composition.
3. Why are faults important?
   (a) They are the usual locations of earthquakes.
   (b) They control the movement of groundwater.
   (c) They control the subsurface accumulations of fossil fuels.
   (d) all of these
4. Approximately what percentage of Earth's liquid water is fresh water?
   (a) 3%
   (b) 25%
   (c) 0.50%
   (d) 75%
5. Which process in the water cycle requires enormous energy input from the Sun?
   (a) precipitation
   (b) evaporation
   (c) condensation
   (d) percolation
6. The ocean basin
   (a) is covered by undersea mountains.
   (b) is mostly flat but features mountains and trenches.
   (c) has not been mapped sufficiently to be understood.
   (d) is an area of low pressure.
7. The Ogallala Aquifer
   (a) underlies several states.
   (b) is being pumped at a high rate.
   (c) is a source of fresh water.
   (d) all of these
8. The salinity of seawater has remained about the same for millions of years—perhaps even a billion years. Why?
   (a) Salts stopped forming in the ocean long ago.
   (b) Salts are included in the molecular formula of seawater.
   (c) Salts are removed from seawater about as fast as they are deposited.
   (d) Actually, there is more salt in the ocean today because salts are continuously washed into the sea.
9. Underground water in the saturation zone is called
   (a) groundwater.
   (b) soil moisture.
   (c) the water table.
   (d) an artesian system.
10. Fault-block mountains
    (a) arise from the stretching of Earth's crust.
    (b) are formed by the convergence of tectonic plates.
    (c) originate over hot spots.
    (d) are produced by upwelling magma.

Answers to RAT

1. b, 2. c, 3. d, 4. c, 5. b, 6. b, 7. d, 8. c, 9. a, 10. a

CHAPTER 25
Surface Processes

How is a cave like a sponge that's been wrung out? Why do mountains eventually wash to the sea, bit by bit? What is soil made of—and why are there over 10,000 different kinds of it? How are glaciers like bulldozers, and what is the evidence that a gigantic glacier traveled through New York thousands of years ago? What caused the Dust Bowl? Could a similar disaster disrupt agriculture today? Why do many young rivers feature rapids while older rivers have meandering curves with levees along the banks? In this chapter, we entertain questions like these. We investigate *surficial processes*—those processes that remodel Earth by acting at its surface.

LEARNING OBJECTIVE
Identify three agents of change and three surface processes that reshape Earth's crust.

25.1 Processes That Sculpt Earth: Weathering, Erosion, and Deposition

EXPLAIN THIS Why are all the grains of sand on the beach usually about the same size and density?

The forces associated with plate tectonics act on a vast scale of time and space by building mountains, swallowing continental crust, and shoving entire continents against one another and pulling them apart. These mammoth tectonic forces that shape our environment originate underground. They are ultimately caused by heat convecting away from Earth's interior and toward its surface.

UNIFYING CONCEPT
- ***Newton's First Law*** Section 3.1
- ***The Gravitational Force*** Section 5.3
- ***Convection*** Section 6.9

There is also another entirely different set of forces that shape Earth's land surface, both its continental and oceanic crust. These geologic forces affect the land over small time scales (seconds, minutes, days, and years) as well as over thousands and millions of years. They act over small distances as well as over large expanses. Unlike tectonic forces, this set of forces doesn't originate in Earth's interior. **Surface processes** are the geologic events that originate at or near Earth's surface. Surface processes mostly involve the actions of water, ice (especially glaciers), wind, and gravity. Local climate significantly affects the relative influences of these four *agents of change*. The relationship between the Earth-sculpting surface processes and climate is yet another reason for the great interest in understanding possible future climate changes (Figure 25.1).

Agents of change—liquid water, ice, wind, and gravity—break down Earth's surface, move material, and build up new landforms through three surface processes: *weathering, erosion*, and *deposition*. The energy for these processes ultimately comes from the Sun (which drives the wind and runs the hydrologic cycle) and from the force of gravity (which constantly tries to pull all of Earth's surface features downhill).

Earth's landforms are built of rock. And rocks are aggregates of *minerals*—natural chemical compounds that are crystalline solids at the atomic level. The rocks and minerals that reside at Earth's surface are constantly subjected to weathering. **Weathering** is the breakdown of rock that happens at or near Earth's surface through chemical or physical means.

Once rock is decomposed or disintegrated by weathering into smaller bits (called *sediments*), it may be eroded. **Erosion** is the physical removal of weathered

(a)

(b)

(c)

(d)

FIGURE 25.1
The surface processes of weathering, erosion, and deposition create diverse landforms. (a) The spiky peaks of this rock formation in Switzerland are the result of glacial erosion. As the ice flowed downhill, it took rock with it, scooping out the mountain. (b) These channel islands were formed by erosion and deposition acting together along the Amazon River in Brazil. (c) Look at the extensive pockmarking of these limestone cliffs as well as the sea stack along this coastline in Portugal. These surface features are the result of erosive action by ocean waves. (d) These grassy plains at the Masai Mara National Reserve in Kenya are so flat because, over geologic time, sediments were eroded from distant slopes, transported downhill, and eventually deposited in depressions on the land surface.

particles from one place and their transport by water, ice, gravity, or wind to another place. Because gravity is a constant force, acting either alone or with a mobile agent such as running water, sediment always erodes downhill.

As sediments are transported, they continue to weather. Recently weathered sediments are normally quite angular and jagged. During transportation, especially by water, the chunks, pieces, and particles collide and break. This process rounds off the sharp edges. Rounding is one of the clues that tells you how extensively traveled and tumbled the sediment grains are.

Deposition occurs when eroded particles come to rest. Particles are deposited when the agent that is transporting them no longer has enough energy to carry them. For example, consider a rushing river high in the mountains. This stream has a lot of energy, mostly in the forms of kinetic and potential energy. Because the stream has so much energy, it can carry sediments in a range of sizes, even some as large as boulders. Some of the sediments are carried aloft, while others are pushed by the energetic rushing water along the stream bed. In contrast, the channels in a delta—the sprawling expanse of land where a river meets the sea—are almost flat. In these channels, water moves with very little energy of motion. Only the finest mud sediment can remain suspended in this water. Larger sediments are deposited progressively as the stream velocity slows, so they are deposited long before they reach the delta. In this way, sediments are sorted by size as they are transported. Sediment samples that are "well-sorted" contain particles of about the same size and density, whereas sediments that are poorly sorted contain particles of many different sizes (Figure 25.2). Sand dunes and beaches feature well-sorted sediments. The deposits along a seasonal stream are poorly sorted. Can you see why?

(a) Well-sorted sediments

(b) Poorly sorted sediments

FIGURE 25.2
A deposit that contains particles of similar sizes is called a "well-sorted" deposit. A poorly sorted deposit contains sediments of many sizes. In general, poorly sorted sediments traveled a short distance before they were deposited, and well-sorted sediments had a long journey before deposition.

Weathering, erosion, and deposition work together. New landforms are built up where weathered and eroded material is deposited. For example, deposition creates levees along *floodplains*. Floodplains are the locations along riverbanks where rivers overflow to during a flood. As the river flows outward from the stream channel onto the land, it slows down. And as the flooding water slows, sediments settle out of it progressively according to their mass. Over time and many cycles of flooding, sediments pile up. Levees are big accumulations of sediments built up over time on a floodplain. Natural levees help control floods. Their high walls prevent a river from overtopping its banks during a storm (Figure 25.3).

FIGURE 25.3
Natural levees are depositional landforms where a river overtops its banks and deposits sediment onto the floodplain.

CHECK YOURSELF

1. **A farmer is concerned that wind is eroding away his rich topsoil, so he plants hedges around his lettuce field. Why does this tactic work?**
2. **Granite is a strong and durable rock made of tightly interlocking crystals. Sandstone is a much softer rock formed by sand-sized grains stuck together but separated by larger pore spaces. Which rock is more subject to erosion? Why?**

CHECK YOUR ANSWERS

1. The hedges act as windbreaks, slowing the wind. As the wind loses energy, it loses its ability to carry sediment particles.
2. Sandstone is more subject to erosion because it is more easily weathered than granite. Water can flow into sandstone's pore spaces, and this causes the rock to weaken and come apart.

MATH CONNECTION

How Long Can a Mountain Exist?

Due to their impressive bulk, mountains are often considered to be symbols of permanence. But how long can a mountain really exist, given the steady erosion of its rock and soil? We can figure this out with a little math and a few approximations. The overall problem-solving strategy is to estimate the size of a typical mountain, estimate how much material erodes from the mountain per unit of time, and then divide the size of the mountain by its rate of erosion.

Mountains seem permanent, but weathering and erosion render them temporary—even short-lived by the standards of geologic time.

To estimate the volume of a mountain, we can approximate its shape as a rectangular box. Mountains come in an array of sizes and shapes, but a box 3 km long, 2 km wide, and 5 km high gives us the approximate volume of an average mountain:

$$\text{Volume} = \text{length} \times \text{width} \times \text{height}$$
$$= 3\text{ km} \times 2\text{ km} \times 5\text{ km} = 30\text{ km}^3$$

Expressing this in standard units of meters, we have

$$\text{Volume} = (3000\text{ m}) \times (2000\text{ m}) \times (5000\text{ m})$$
$$= 3.0 \times 10^{10}\text{ m}^3$$

Now we estimate how much rock, sand, and gravel are eroded in a typical day. The amount varies according to the kinds of rock that are present, the amount of precipitation, and other factors. But an average mountain could have four principal streams, and each one could easily carry a tenth of a cubic meter of material off the mountain per day. This is a conservative estimate—a tenth of a cubic meter is a box about a foot and a half on each side, about the size of a kitchen sink. So, the estimated rate of erosion per day is

$$\text{Rate of erosion per day} = (0.1\text{ m}^3/\text{stream}\cdot\text{day}) \times (4\text{ streams})$$
$$= 0.4\text{ m}^3/\text{day}$$

From this, we can calculate the rate of erosion per year:

$$\text{Rate of erosion per year} = (0.4\text{ m}^3/\text{day}) \times (365\text{ days/year})$$
$$= 146\text{ m}^3/\text{year}$$

Now we divide the volume of the mountain by the rate at which it is worn away in a year to find out how long the mountain can last:

$$\text{Duration of a mountain} = (3.0 \times 10^{10}\text{ m}^3)/(146\text{ m}^3/\text{year})$$
$$= 2.05 \times 10^8\text{ years}$$
$$= 205\text{ million years}$$

Thus, a typical mountain can exist for only about 200 million years, even with a conservative estimate for the rate of erosion. Compared with the age of Earth, which is 4.6 billion years, mountains are young features with short lifetimes. Erosion erases all landforms—even the largest ones.

LEARNING OBJECTIVE
Distinguish between chemical and mechanical weathering.

Integrated Science 25A
PHYSICS, CHEMISTRY, AND BIOLOGY

Weathering

EXPLAIN THIS Potholes form in the road during winter when water flows into asphalt and then freezes and expands. (Ice has a volume equal to 109% of liquid water.) So, ice residing in the asphalt pushes the asphalt apart. Explain why this is an example of mechanical weathering.

Weathering—the disintegration or decomposition of rock—is sometimes a chemical process, sometimes a physical process, and sometimes a combination of both chemical and physical forces.

Mechanical weathering, usually caused by water, physically breaks down rock into smaller pieces. *Ice wedging*, shown in Figures 25.4 and 25.5, is an example. In this process, liquid water seeps into rock and then freezes. When water freezes, it expands. So, the frozen water in the rock physically pushes the rock

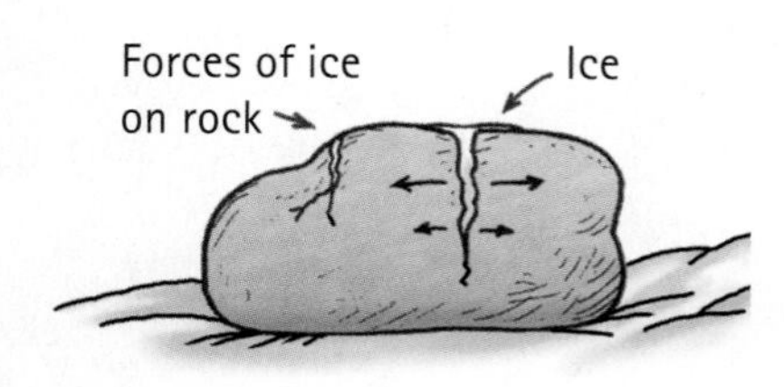

FIGURE 25.4
Water expands when it freezes. In ice wedging, water flows into a crack, freezes, and expands. This widens the crack so that when the ice melts, the rock is split into smaller pieces.

FIGURE 25.5
The rocks on this mountain peak in Wales have been split apart by ice wedging, a kind of mechanical weathering. In this cold environment, water in the cracks of the rocks has frozen and expanded and thereby caused the rocks to crack and break apart.

Time out for some vocabulary: In Newtonian physics, a *force* is a push or a pull. In geology, the word *force* is used more loosely. A *force*, also called a *driving force*, is anything that applies energy to the land. For example, Earth's internal heat and wind are driving forces. Now, back to Earth science: Landforms change when the driving forces on them exceed the resistance (strength) of their materials. For example, when wind on a sand dune overwhelms the friction between the sand particles, the sand dune can move. The driving force is greater than the resistance.

grains apart. When the ice melts, empty cracks are left behind. Biological agents, such as trees, also cause mechanical weathering. You have probably seen roots wedged into cracks in a rock. As the roots grow, they push the rock outward and enlarge the original cracks. Wind is another agent of mechanical weathering. It's not as forceful as ice or running water, but wind can be a significant force over time, especially on soft rock such as sandstone. Wind acts like sandpaper, blowing sand grains over surfaces and causing *abrasion*, or scraping. Abrasion smoothes rocky surfaces (Figure 25.6).

UNIFYING CONCEPT

- ***Newton's Second Law***
  Section 3.2
- ***Friction***
  Section 2.8

In contrast, **chemical weathering** is change in the chemical structure of the minerals in rock. During chemical weathering, the orignal rock decomposes into new substances that are stable in the surface environment.

Chemical weathering actually produces more sediment around the world than mechanical weathering does, although chemical weathering and mechanical weathering often work together. Mechanical weathering breaks up rock, which increases the surface area exposed to chemical attack. Water is the primary agent of chemical weathering. Pure water is not reactive but water is rarely pure in nature. Instead, it is a weak solution that contains many dissolved chemicals. The dissolving action of rainwater, in particular, weathers much of the world's rock by removing mineral elements from rock and putting them in water solution. Rainwater (H_20) is especially powerful when it reacts chemically with carbon dioxide (CO_2) in the air. This reaction forms carbonic acid (H_2CO_3), which makes rainwater slightly acidic and therefore a powerful agent of change.

FIGURE 25.6
Wind weathered this sandstone boulder in Utah, which led to the smoothing and rounding of its surface.

For example, consider how carbonic acid weathers granite, the most common continental rock. Granite is a speckled igneous rock that consists mostly of Earth's two most abundant minerals: quartz and potassium feldspar (Figure 25.7). Potassium feldspar reacts with carbonic acid to produce clay minerals. Because clay minerals are the end product

FIGURE 25.7
Granite, the most abundant continental rock, is composed mostly of quartz and potassium feldspar. Quartz is durable, but potassium feldspar chemically weathers to form clay.

FIGURE 25.8
Chemically weathered rock. Note that the exposed surface is reddish because iron-rich minerals have formed iron oxide, or rust.

of weathering, they are very stable under surface conditions and so are very common. If you have ever walked in nature after the rain and found yourself slipping and sliding in mud that feels as slimy as potter's clay, you are probably traversing a field of weathered feldspar from granite—otherwise known as clay. The quartz in granite is more resistant to weathering than feldspar. Quartz remains intact when attacked by carbonic acid. As a result, when granite weathers, its feldspar cystals become dull and slowly turn to clay but its quartz grains are released from the rock matrix still looking glassy and new. Indeed, quartz is so resistant to chemical weathering that it's commonly transported thousands of miles over rugged terrain all the way to the coast, where it is deposited in mass quantities to become the main constituent of sandy beaches.

Another very common example of chemical weathering is the formation of iron oxide—rust—on the surfaces of rocks. Oxygen dissolved in water will oxidize many materials—for example, rocks rich in iron. Have you ever noticed that the surface of a broken piece of rock is usually more red or brown than the inside of the rock? This is because iron oxide has formed on the rock's exposed surface (Figure 25.8).

CHECK YOURSELF

1. **In the process of hydration, rock minerals take water into their internal stucture. The added water expands the mineral structures, so the newly hydrated minerals become crowded. This overcrowding breaks up the rock. Is hydration an example of chemical or physical weathering?**
2. **Ocean waves pound a rocky shore over and over like a sledgehammer. This breaks the rocks there into smaller pieces. Is this an example of chemical or mechanical weathering?**

CHECK YOUR ANSWERS

1. Chemical weathering
2. Mechanical weathering

FIGURE 25.9
The rate of chemical weathering depends on the type of rock involved. The granite headstones (*left* and lying flat) are about the same age as the marble headstone (*right*). Marble is made of metamorphosed limestone, calcium carbonate, so, unlike granite, it is easily dissolved in even weak acids.

Many factors determine how fast weathering occurs at a given place. Climate is a crucial one. The ideal climate for chemical weathering is warm and wet. In cold locations, such as high altitudes and the polar regions, chemical weathering is slow and ineffective because water is locked up as ice. But, in a warmer climate, especially one that has lots of rain, chemical reactions are accelerated, so the weathering proceeds quickly. And, once the weathering is under way, erosion and deposition aren't far behind. Again, the current warming of Earth's climate portends big changes—even to the structure of the landforms around us.

Besides climate, the mineral composition of rock determines the rate and type of change that it can experience (Figure 25.9). Likewise, the number of cracks in rock, the rock's geometry and surface area, and many other factors all affect the rate and type of chemical weathering. The rocks in any location therefore typically show **differential weathering**, or weathering that is varied in its rate and extent. Differential weathering and subsequent differences in the rate of erosion can create weird and wonderful landforms and rock formations such as the natural bridges shown in Figure 25.10.

FIGURE 25.10
These natural bridges at Arches National Park were produced by differential weathering—differences in the rate of weathering at the same location.

CHECK YOURSELF

When sandstone rock at Arches National Park was uplifted several million years ago, tensional forces stretched the rock, producing many cracks and joints within it. Given this, develop a hypothesis to explain how curious structures such as the bridge in Figure 25.10 formed.

CHECK YOUR ANSWER

Groundwater is an agent of chemical weathering. Also, groundwater flows preferentially through cracks in rock. Therefore the portions of the ancient sandstone at Arches National Park that were compromised by joints and cracks experienced faster weathering and erosion than surrounding rock. The odd formations visible today are what remains after the differential weathering and erosion occurred.

LEARNING OBJECTIVE
Understand how soil develops.

UNIFYING CONCEPT
- ***The Ecosystem***
Section 21.1

Soil

EXPLAIN THIS Why do plants grow in topsoil rather than subsoil?

Soil is a mixture of organic and nonliving materials and is the end product of the weathering of rock. It is an essential part of almost every terrestrial ecosystem because plants require soil to grow.

Soil scientists have identified thousands of different kinds of soils throughout the world. All soils contain the same four ingredients—mineral matter, air, water, and **humus** (the decayed remains of animal and plant life)—although the proportions vary from soil to soil. As Figure 25.11 shows, fertile soil usually contains only about 5% humus. Still, humus is important. It provides

Soil is a mixture of rock, air, water, and organic matter. It reflects Earth as a system—atmosphere, biosphere, lithosphere, and hydrosphere—all working together.

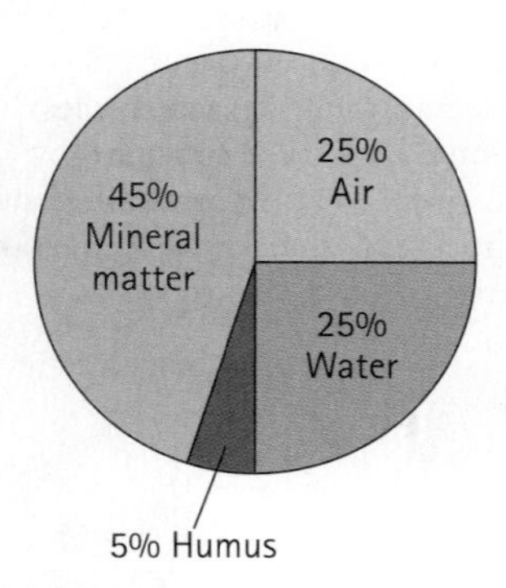

FIGURE 25.11
By volume, a fertile soil is about 45% mineral matter, 25% each air and water, and 5% humus. All of these components are needed for plant growth.

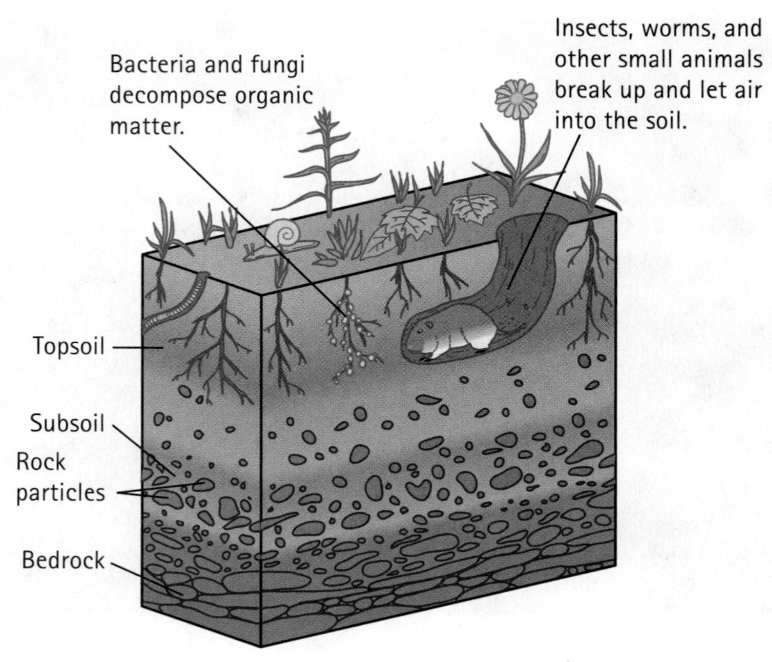

FIGURE 25.12
Soil consists of topsoil and subsoil. Diverse organisms inhabit the topsoil and help make it fertile. Dead and decaying organisms form humus, which also contributes to fertility.

Astounding numbers of organisms live in topsoil. In one acre of average farm soil there are about 270 million insects and 720,000 earthworms!

essential plant nutrients, which are minerals as well as organic compounds. Humus also contains a large proportion of pore space that holds the air and water plants need.

Soil has layers, as shown in Figure 25.12. The upper layer of soil, called **topsoil**, is where plants grow. It contains the humus. Beneath the topsoil is the *subsoil*, which most plant roots and animals cannot penetrate. Underneath the subsoil is a layer of rock particles from partially broken down parent material. And below that, typically, is bedrock. Soil forms from the surface downward because weathering occurs at Earth's surface. Minerals and other materials leach down from the top of a soil profile and accumulate below. That's why the top layer is the fluffiest and the bottom layers of soil are the most rocky.

In addition to the role that humus plays, living organisms help make soil fertile too. Fungi, bacteria, and other microorganisms do much of the job of decomposing dead animals and plants. Worms, insects, and other small animals add air by tunneling or even eating their way through soil. Topsoil is produced very slowly. It takes thousands of years for weathering and biological decay to produce a few centimeters of topsoil. So, topsoil, like groundwater and fossil fuels, is a nonrenewable resource.

Soils are classified according to their **texture**, which is the size of the particles they contain. Texture is an important soil property because it affects how well soil retains and transmits air and water to plants. The largest particles in soil are classified as *sand* (2–0.05 mm), medium-sized particles are *silt* (0.05–0.002 mm), and the smallest are *clay* (smaller than 0.002 mm). Most plants grow best in soil that has a mixture of particles of all sizes. Such soil, gardeners know, is called *loam* (Figure 25.13). Sandy soils transmit water quickly, so they dry out too fast for most plants. At the opposite extreme, clay-rich soils retain too much water and can become soggy.

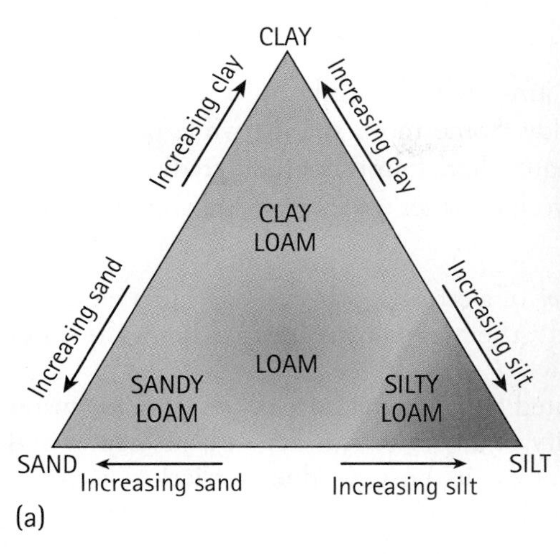

FIGURE 25.13
(a) Soil texture diagram. The texture of any soil can be represented by a point on this diagram. (b) Most plants, such as the vegetables in this garden, grow best in loam. (c) Some plants are adapted to extreme soils. These cacti have shallow roots to maximize water absorption from quickly draining sandy soil in the desert.

25.2 The Impact of Running Water

LEARNING OBJECTIVE
Describe the landforms created by erosion and deposition from running water.

EXPLAIN THIS Why are boulders and big rocks found at the base of a waterfall while smaller rocks and pebbles line the banks of tiny creeks?

A *stream* is any body of flowing water confined within a channel—from the tiniest creek to the mightiest river. As running water, streams transport tremendous loads of sediment across Earth's surface. The Mississippi River alone moves more than 300 million tons of sediment to the ocean each year. This sediment would fill a fleet of dump trucks that wraps around Earth nearly three times! Running water is the most widespread agent of erosion, sculpting more of Earth's surface area than all other forms of erosion.

Streams form where groundwater meets Earth's surface or from surface runoff. Runoff is precipitation that neither soaks underground nor evaporates. It collects at topographical low points and then flows downhill under the influence of gravity. Smaller streams called **tributaries** contribute to larger streams so streams tend to carry more water as they flow downstream. The land area that feeds a particular stream, including its tributaries, is the stream's **drainage basin** or *watershed* (Figure 25.14). The size of a stream at any point is related to the size (area) of its drainage basin upstream from that point because that is the area that determines how much rain or snow melt will flow into the stream. A stream's size is also influenced by other factors, including climate (especially precipitation and evaporation), vegetation, and the type of rocks that make up the stream channel.

In nature, a stream gradually carves its channel into the surrounding rock. The channel's size is proportional to the volume of water that the stream carries. Long-term changes in rainfall, snowmelt, land use, and other factors that change the volume of the water flowing in the stream change the channel's geometry over time. That is, stream channels, like the rest of Earth's surface, evolve in response to changing conditions.

FIGURE 25.14
This illustration shows the lower portion of a drainage basin. A drainage basin (or watershed) of a stream is the land area that drains into that stream. A drainage basin contains smaller streams (tributaries), as well as lakes possibly and other channels that drain into the main stream.

UNIFYING CONCEPT

- *The Gravitational Force* Section 5.3
- *Friction* Section 2.8

Properties of Streams

Streams occur all around the world: in jungles, at the seashore, in mountains, even in deserts. Some are clear, others muddy. Some move swiftly, while others move slowly. Despite their differences, all streams share four properties: gradient, velocity, discharge, and load. These properties work together to determine how powerfully a stream erodes the land.

FIGURE 25.15
This stream flows across nearly level plains—it has a very low gradient.

FIGURE 25.16
As the name suggests, "rapids" are *rapid*, or high-velocity, streams.

1. **Gradient** The **gradient** is the *slope* of a stream—the change in its elevation over a particular distance. Generally, the higher a stream's gradient, the faster it flows.
2. **Velocity** Stream velocity, represented by v, is a measure of how quickly water moves at a given point. The velocity of any stream is fastest at its center and slower at the insides of curves and near its bottom, due to friction with the stream bed. The faster a stream moves, the more energy it has and the better it is at picking up and transporting rock and soil.
3. **Discharge** **Discharge**, represented by Q, is the volume of water that a stream transports in a given amount of time. It depends on a stream's velocity (v) as well as its cross-sectional area, denoted by A. Discharge is measured in gallons per minute or cubic meters per second (m^3/s). It can be calculated with the simple equation

   $$Q = v \times A$$

   Stream discharge can be as low as 1 cubic foot per second for a small creek or as high as millions of cubic feet per second for a major river. Discharge is the principal index of a stream's erosive power.
4. **Load** The amount and type of sediment that a stream carries is its **load**. Stream load is directly related to velocity: Fast streams transport more sediment and bigger sediment grains than slow streams do. Also, load is related to discharge: More water flowing through a stream can carry a bigger sediment load.

There are three types of loads, which differ from one another in the sizes of their sediments. Dissolved sediments such as the minerals sodium and calcium are too small to be seen. They're called the *dissolved load*. Sediments that are carried in suspension are the stream's *suspended load*. These sediments move along with the water and make it look cloudy. Larger sediments, known collectively as the *bed load*, are moved downstream by bouncing and being dragged along the stream bed.

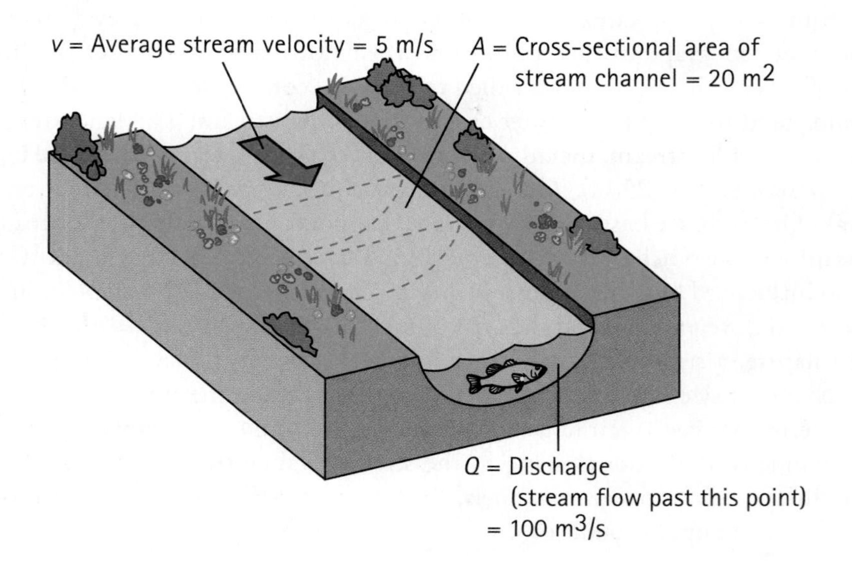

FIGURE 25.17
Discharge (Q) = stream cross-section (A) × stream velocity (v). Discharge indicates the ability of a stream to cause erosion.

CHECK YOURSELF

1. **A miner wants to set up camp by a stream, but he wonders whether the stream will provide enough fresh water for his horses every day. What property of the stream should he measure?**
2. **Why are speed and gradient factors in a stream's sediment load?**
3. **Trash such as old tires and plastic bags is illegally dumped in a stream channel that is upstream from a water-ski lake. Will more trash reach the lake in the spring or in the summer? Why?**

CHECK YOUR ANSWERS

1. Discharge
2. Discharge depends on the cross-sectional area of a stream and the stream's average speed. Speed depends on gradient. Thus, the amount of sediment in a stream depends on its discharge as well as its speed and its slope. Big streams on steep slopes carry the most sediment—they are powerful land movers!
3. Sediment load depends on discharge, the amount of water flowing per unit of time. In the spring, snowmelt and rain swell rivers, so discharge is at its peak. This is when the stream will carry the most trash and the biggest individual pieces of trash into the lake. If it is a really big, fast-moving stream with large sediments floating in it, even the old tires will end up in the lake.

Landforms Built by Running Water

Consider now the major landforms built by running water (Figure 25.18). We will see how stream properties influence their development.

The *headwaters* of a stream are its source region. Headwaters are often located high in the mountains where snow melts and runs off the ground. A mountain stream has a high gradient, so it flows quickly. By the time a stream reaches its end or *mouth*, its gradient is usually low. A stream's mouth is generally located near *base level*, which is the lowest elevation that a stream can erode to. Often base level is sea level. The closer a stream is to its base level, the lower is its gradient and the more slowly it tends to flow. Decreased velocity may be compensated for, however, by widened channels, so discharge may remain significant even in the flat channels near a stream's mouth. With sufficient discharge, the stream continues to cut deeper into its rocky channel, eroding more material. This erosion is counteracted all the while by sediments being deposited into the stream from the drainage basin. All in all, streams tend toward a balance or *equilibrium* in which erosion and deposition undo the effects of one another.

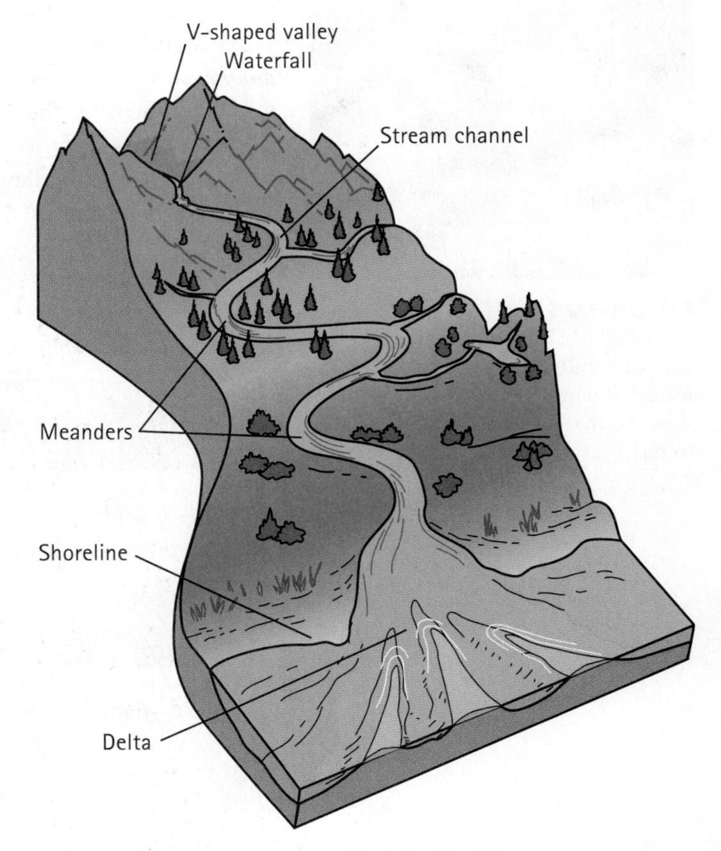

FIGURE 25.18
Some of the landforms created by weathering, erosion, and deposition facilitated by running water.

The stream profiles shown in Figure 25.19 are sketches of stream elevation from headwaters to mouth. A young stream undergoing active erosion features

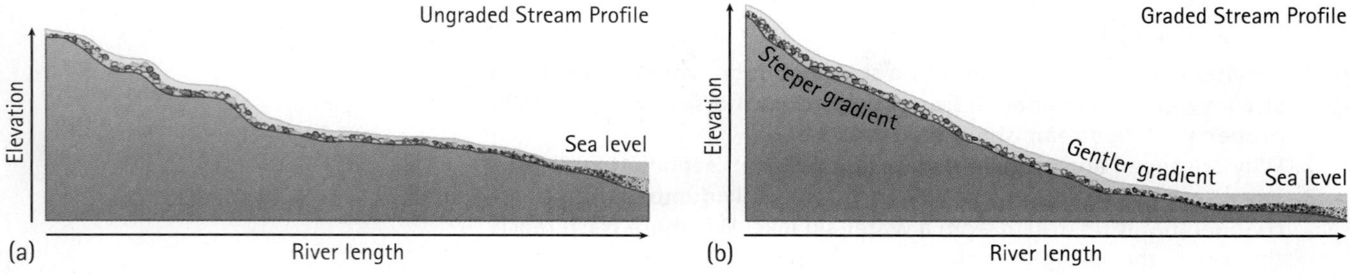

FIGURE 25.19
(a) Ungraded stream profile. Waterfalls and rapids are present. (b) Graded stream profile. The gradient decreases smoothly from the stream's source to its mouth. The sizes of sediments generally decrease with weathering and sorting during transportation.

rapids and waterfalls and is represented by the ungraded stream profile. Waterfalls develop where a stream crosses rock layers of varying hardness. The stream leaves behind the rock that is hard enough to resist erosion, but it erodes softer rock. The harder rock remains as the cliff that the waterfall tumbles over. Softer rock is weathered and washed downstream in pieces. Eventually the stream profile becomes graded, or smooth. Mature streams typically display the concave-upward shape of the graded stream profile; all the rough edges have been eroded away. The time it takes for a stream to change from an ungraded to a graded profile depends on the sediment load.

As a river erodes its channel ever more deeply over time—a process called *downcutting*—erosion begins to work on the exposed vertical sides of the channel. This widens the channel, and over geologic time it may create a valley (Figure 25.21). Valleys produced by river downcutting are called *V-shaped valleys*. V-shaped valleys have steep slopes. They are very common—and they are impressive evidence of the erosive power of streams (Figure 25.22).

FIGURE 25.20
Waterfall at Finger Lakes, New York. Boulders collect at the base of a waterfall where the stream suddenly slows so that larger sediments come to rest.

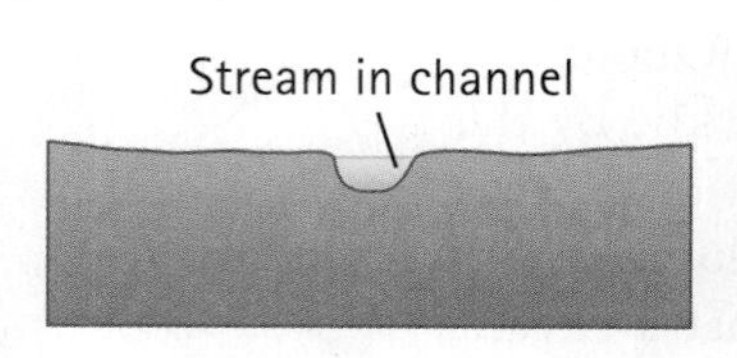

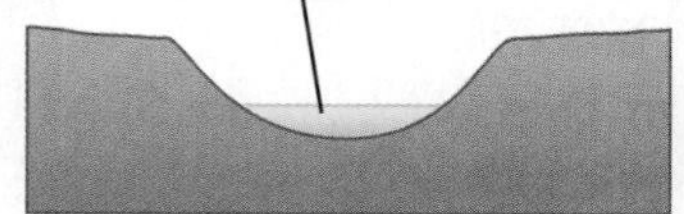

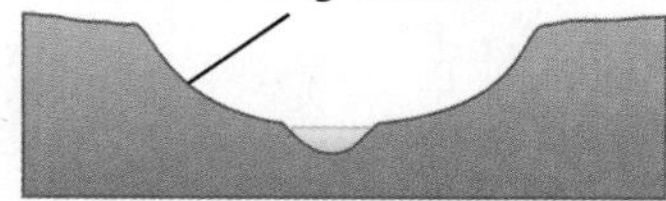

FIGURE 25.21
Downcutting stream profile.

FIGURE 25.22
A V-shaped valley in the Swiss Alps.

Many streams, especially older ones, don't follow straight paths. Instead, they bend and loop around like a snake (Figure 25.23). How does a curve in a stream, called a *meander*, develop? Imagine a stream that flows straight downhill under the influence of gravity. The channel will inevitably meet an irregularity or obstruction—a rock, a log, a plant, and so on. When water flows around the obstruction, it flows a little more quickly on the outside of the channel and a little slower closer to the blockage due to friction. Where the water moves faster, there's erosion; where it slows, there's deposition. This is how a meander gets started. Once it forms, a meander shifts to the side, gets bigger, and moves downstream. Meanders shift and grow because the water on the outside of the curve moves faster than the water on the inside of the curve; then, over time, the outside of the curve erodes while the inside of the curve is filled in by deposition. The original bend is sharpened. The site of erosion is called the *cut bank*, and the location of dumped sediment is called a *point bar*.

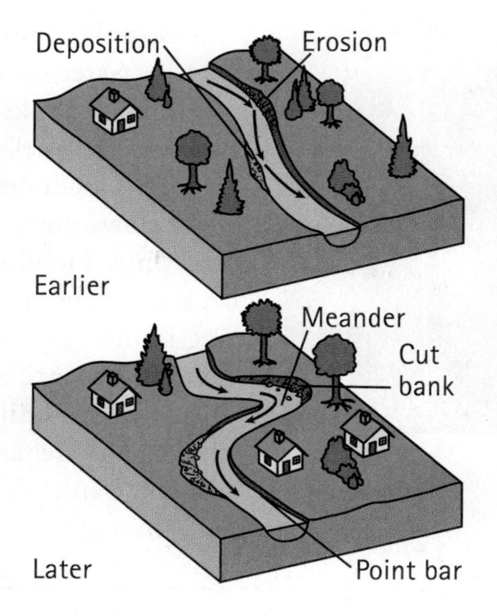

FIGURE 25.23
A meander forms because water on the outside of a stream has a higher velocity and therefore higher discharge and erosive ability compared to the inside of the curve.

UNIFYING CONCEPT

- ***Friction***
Section 2.8

When the sediment load is large, obstructions in the stream channel can also produce channel islands. *Channel islands* are mounds of sediment surrounded by flowing water (Figure 25.24). Deltas are also created by deposited sediment. A **delta** is a mass of sediment that is deposited where a river slows down as it enters a large body of water such as a lake or the ocean. It's easy to see how deltas form. When a stream reaches the ocean or a large lake, its gradient is low so its velocity is slow. Fine sediments, including gravel and sand, are deposited and build the delta. The Mississippi River Delta, shown in Figure 25.25, comprises more than 3 million acres—that's a lot of sediment! And it's a lot of habitat. Fish, mammals, birds, and all manner of other organisms congregate in the shallow coastal waters, swamps, and forests of the Mississippi Delta Region. Nearly 300 species of fish, reptiles, and amphibians and more than 300 species of migratory songbirds and waterfowl reside there (Figure 25.26).

FIGURE 25.24
Channel islands.

FIGURE 25.25
The Mississippi Delta—a depositional land feature covering over 3 million acres.

FIGURE 25.26
The Mississippi Delta is rich in biological diversity.

FIGURE 25.27
Floodplains are areas near a river that are subject to flooding. Floodplains are usually rich farmland because sediments are deposited there with each flood.

Floodplains are the flat areas along a stream that are covered by water only during a flood. When a stream floods, it overtops its banks, flowing ever more slowly as it flows away from its channel. Again, the deposition pattern comes into play: Finer sediments are deposited progressively as the moving water slows down. The deposited sediments build a landform over time. Floodplains have especially fertile soil for farming because of the minerals that are deposited there during recurring floods (Figure 25.27). This is why farmers have settled in floodplains since ancient times despite the risk of flooding.

Landforms built by deposition, such as floodplains, channel islands, and deltas, are called, reasonably enough, *depositional landforms*. Waterfalls, cut banks, V-shaped valleys, and other landforms created by erosion are called *erosional landforms*.

CHECK YOURSELF

1. **A stream erodes the land as it progresses to the sea. What landforms are likely to develop where the land is steep? What landforms are likely to develop where the land is more level?**
2. **Fine sediment grains are deposited in massive amounts along a shoreline, building up a beach. Why don't we see beaches high in the mountains?**
3. **A waterfall rushes over a granite cliff. Is granite a hard or soft rock? Why do you think so?**
4. **When kayaking downstream, you stop at a channel island to camp for the night. How did the island likely form?**

CHECK YOUR ANSWERS

1. The erosional features of steep landscapes include V-shaped valleys and waterfalls. More level landscapes may feature rivers with meanders, channel islands, and deltas.
2. Fine sediments are carried in fast-moving waters that run in mountain streams. The sediments aren't deposited until the water slows down along the shoreline, where the gradient is low.
3. The fact that the waterfall is rushing over it indicates that granite is a hard rock—it can withstand the force of fast-moving water and erodes slowly.
4. It formed from sediments deposited around an obstacle in the stream.

LEARNING OBJECTIVE
Describe the landforms created by glacial erosion and deposition.

25.3 Glaciers—Earth's Bulldozers

EXPLAIN THIS How are glaciers like bulldozers?

Most of Earth's fresh water is frozen in *glaciers*, enormous masses of moving ice. Glaciers are classified into two types: *alpine glaciers* and *continental glaciers*. The classification is based on size and where the glaciers occur. Alpine glaciers occupy mountainous terrain. Most of the 100,000 to 200,000 glaciers that are estimated to exist today are alpine glaciers. The rarer but bigger continental glaciers, also called *ice sheets*, are huge sheets of thick ice that can cover large portions of continents. Most of Greenland and Antarctica are covered by glaciers. All glaciers move downhill, pulled by gravity to the ocean.

Some glaciers, including some in Alaska, zip along at 150 ft (50 m) per day. These fast-moving glaciers threaten villages and oil pipelines in their path. Glaciers are moving faster now that the climate is warming.

Glaciers move for two reasons. First, they move because their solid ice crystals slip past one another. Also, the huge weight of a glacier produces pressure that can melt its base. The slippery bottom allows the glacier to slide downhill like a half-melted ice cube on a tilted tabletop.

Glacial speed varies. Most glaciers advance a sluggish few centimeters per day. Some move much faster, but even the fastest glaciers move slowly compared to running water. Yet, because of their great bulk, glaciers are the second most important cause of erosion and they are powerful agents of deposition as well.

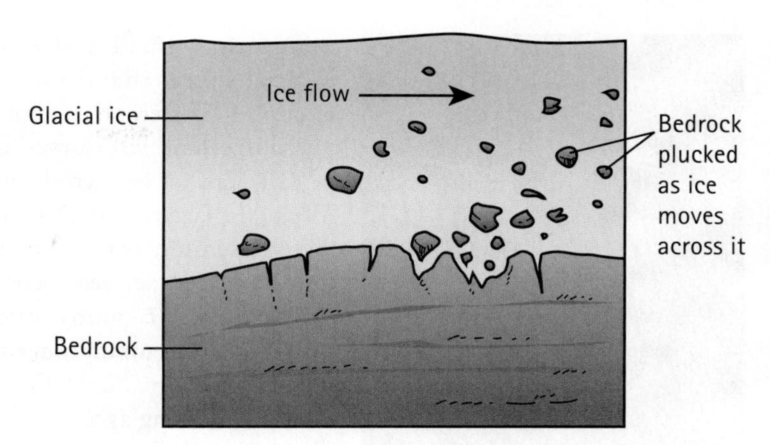

FIGURE 25.28
Glaciers lift and incorporate rock in the process of plucking.

Glaciers erode rock in two ways: plucking and abrasion. *Plucking* is illustrated in Figure 25.28. Plucking is based on ice wedging—ice forms in cracks in rocks and widens the cracks. As the glacier moves over the cracked bedrock, it picks up and carries large chunks of rocks with it. The debris plucked by the glacier is incorporated into it. So, a glacier is not just a huge mass of ice; it's a huge mass of ice and rock plowing over the land—a gigantic bulldozer.

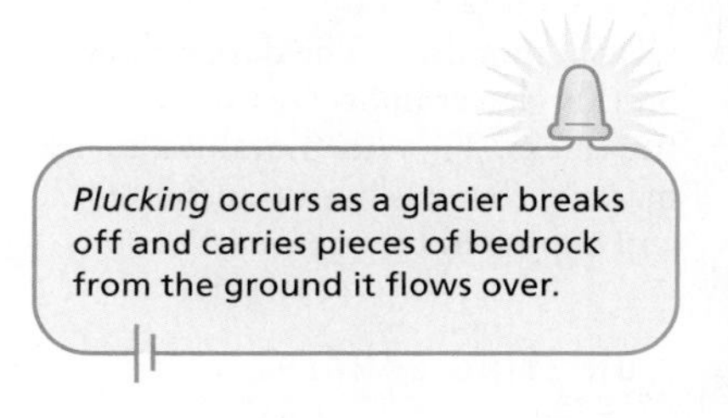

Plucking occurs as a glacier breaks off and carries pieces of bedrock from the ground it flows over.

Glaciers have roughened, abrasive surfaces because of the rocky debris they pick up as they move. These rough surfaces erode rock by *abrasion*. Sometimes glaciers leave parallel scratch marks on rock called *striations*. If you see striations on a rock formation, you know that a glacier at one time scoured its surface. You can also tell what direction the glacier was traveling by the direction of these scratch marks. For example, if you visit Central Park in New York City, you will see people playing among rounded rock formations and immense boulders—and some of these are marked by striations. Few of the park visitors realize it, but a sparkling shield of ice polished, gouged, scratched, and dumped car-sized masses of rock on this landscape 20,000 years ago as the Laurentide Ice Sheet traveled from northwest to southeast.

FIGURE 25.29
Glacial striations in Central Park, New York City, caused by a glacier moving through 20,000 years ago.

Glaciers leave many spectacular valleys, peaks, and ridges in their wake (Figure 25.30). When an alpine glacier flows through a sharp-sided valley cut by a stream, the glacier widens and deepens the valley by plucking. This can

(a)

(b)

(c)

FIGURE 25.30
Landforms created by the scooping action of glaciers. (a) This U-shaped valley is in Glacier National Park, Montana. (b) Lake Louise in the Rocky Mountains of Canada is a glacial lake. (c) The Matterhorn in Switzerland is a famous example of a glacial horn.

FIGURE 25.31
Glacial moraines. The dark stripes on this glacier in Switzerland are not tire tracks. They are moraines, rocky material that has been transported and deposited by a glacier.

UNIFYING CONCEPT

- ***The Gravitational Force*** **Section 5.3**

transform a small V-shaped valley into a grand one with a broad U-shape. Glacial valleys that fill with water become glacial lakes. When several glaciers scoop out different sides of a single peak, they create a *horn*. The Matterhorn in Switzerland was carved this way.

Glaciers not only erode powerfully but also deposit sediment on a grand scale. Because glaciers collect sediment as they move and because they usually don't deposit sediments until the ice melts, when a glacier finally makes a deposit, it's often a huge one. Sediment deposited directly from the melting of glacial ice is called *till*. Till is poorly sorted rock—it consists of particles with many sizes and shapes. An accumulation of till that is deposited in layers or ridges is called a *moraine* (Figure 25.31). Geologists use moraines to map the locations of glaciers that melted long ago.

CHECK YOURSELF

1. **You've been told that glaciers move and that they carry loads of sediment with them, but where are they going?**
2. **Why do glaciers carve broad U-shaped valleys rather than the narrow V-shaped valleys carved by running water?**

CHECK YOUR ANSWER

1. Glaciers, like all other agents of erosion, carry their load of sediment toward the ocean.
2. Glaciers erode rocks from the bottoms and sides of the valley. Running water, on the other hand, erodes the stream bed.

LEARNING OBJECTIVE
Distinguish among the types of mass movement, including falls, slides, flows, and creep.

25.4 Effects of Gravity Alone—Mass Movement

EXPLAIN THIS Why are mudflows a more hazardous type of mass movement than creep?

Gravity pulls you toward Earth's center, and it pulls on all of Earth's landforms too. Gravity is constantly trying to level everything. The pull of gravity causes a form of erosion known as *mass movement* or *mass wasting*. **Mass movement** is the downward movement of rock or soil due to gravity alone. Although most mass movement involves the transport of material down slopes, it can also occur as freely falling rock or as a broad lowering of the land surface (this is *subsidence*, which is associated with the extraction of oil or groundwater). Rock falls, landslides, avalanches, and mudflows are dramatic and common examples of mass movement.

Mass movements are classified on the basis of the kind of material that moves, how it moves, and the rate of movement. A *fall* is free-fall motion in which the moving material is not always in contact with the ground. For example, *rock falls* are common along rocky coastlines where slopes are undercut by wave erosion and along roadcuts as well (Figure 25.32a). A *slide* is the downhill motion of a cohesive mass of material along a clearly defined surface or plane. In the *landslide* shown in Figure 25.32b, a sheet of material has slipped downslope and left

UNIFYING CONCEPT

- ***Newton's First Law*** **Section 3.1**

(a) (b) (c)

FIGURE 25.32
Mass movement. (a) A rock fall and rockslide hazard along a coastal highway in Oregon. (b) A landslide in La Conchita, California. (c) This mudflow caught residents of Armero, Colombia, by surprise in 1985. A nearby volcano, Nevado del Ruiz, erupted in the middle of the night and melted the surrounding glacier. As melted snow and volcanic debris pushed downhill, rocks and soil were incorporated into the chaotic flowing mass. This *lahar* (a volcanically induced mudflow) reached a speed of 40 miles per hour and crossed six major rivers, picking up mass all the while, before it reached Armero. Tragically, more than 20,000 people died.

a broad scar of denuded rock (called a *scarp*) behind. Huge landslides can bury villages and housing developments. A *flow* is the chaotic movement of unconsolidated material that mixes together as it moves, like a fluid. An *avalanche* is a flow of snow and ice. *Earthflows* are flows involving fairly dry soil. *Mudflows* are flows of material saturated with water. They occur when heavy rains, volcanic ash or lava, or snowmelts mix with soil and rock to create a cementlike mud, which is carried by gravity downhill. Mudflows move fast and are powerful and destructive events, as Figure 25.32c shows.

Mass movement is sometimes so slow that you don't notice it. You know that it has happened only from the evidence left behind. *Creep* is the slow movement of soil down a slope. Creep is often the result of alternating freezing and thawing. Each time the ground freezes, ice pushes soil particles apart slightly. When the ice thaws, the loosened small particles of soil move a short distance downhill. Over time, creep makes objects that are stuck vertically into the ground lean over. Have you ever noticed utility poles tilting, or fences or tombstones leaning? These are likely signs of creep. Creep can also make a tree bend along its base, as shown in Figure 25.33.

FIGURE 25.33
An example of creep. As soil slips downhill, the tree is forced to tip. But the tree trunk rights itself by growing in the "pistol butt" shape.

Although mass movement is sometimes triggered by an event such as an earthquake or volcanic eruption, it's usually caused by the absorption of rainwater. Water makes soil heavy, which means that it experiences a stong gravitational pull. When a slope is sufficiently saturated, gravity pulls it downhill.

Factors that make a slope vulnerable include the steepness of the slope, weathering, and vegetation. When slopes are steep, gravity can easily overwhelm the friction holding rocks or soil in place, so movement occurs. Weathering is a factor because broken-up rock is easier to lift and transport. Vegetation affects mass movement too because plants that have extensive, shallow roots reduce erosion. When plants are removed from a hillside, soil is no longer anchored to the slope.

UNIFYING CONCEPT

- ***Friction***
  Section 2.8
- ***The Gravitational Force***
  Section 5.3

CHECK YOURSELF

What factors increase the chance for mass movement to occur on a slope?

CHECK YOUR ANSWER

Water saturation, gradient, weathering, and loss of vegetation increase the chance. Earthquakes and volcanic eruptions can also trigger mass movement.

LEARNING OBJECTIVE

Describe the process by which most caves form.

25.5 Groundwater Erodes Rock to Make Caves and Caverns

EXPLAIN THIS Why are sinkholes potentially dangerous?

Imagine hiking where cool, jagged walls press in on you. The ground is slick, so you move ahead with great caution. Holes and ledges are everywhere. It would be easy to slip into a void and never be seen again. Your headlamp is essential—with its light you exchange pitch blackness for a pastel, spike-studded alien world. But you are not in an alien world; you are on Earth. More specifically, you are *in* Earth. You are hiking in a cave.

A *cave* is an empty underground space large enough for a person to enter. There are many kinds of caves, including voids in glaciers and tunnels in solidified lava. However, most caves are created by the dissolving action of groundwater on rock, especially on limestone. Thus, most caves and caverns (caverns are simply large caves) are products of chemical weathering and erosion by groundwater.

Caves are a vivid example of water's ability to weather and erode rock. Groundwater contains small amounts of carbonic acid, formed when rainwater

FIGURE 25.34
Lechuguilla Cave in New Mexico. This spectacular cave was formed by weathering and erosion produced by acidified groundwater acting on limestone.

(a)

(b)

FIGURE 25.35
Stalactites are icicle-shaped mineral deposits that hang inside a cave, and stalagmites build up on the ground. (a) These stalactites and stalagmites are in the Blanchard Springs Caverns in Arkansas. (b) This spectacular stalagmite, called the Witch's Finger, is at Carlsbad Caverns in New Mexico.

Sinkholes are common in Florida, where limestone bedrock is common.

dissolves carbon dioxide from the air. Limestone (calcium carbonate) readily dissolves under the action of acidified groundwater. Usually caves form in the saturated zone beneath the water table. Over time, huge chunks of limestone go into solution. When the water table drops, holes in subsurface rock remain—these are caves.

Limestone caves are often decorated with *stalactites* and *stalagmites*. Stalactites hang from the roof of caves, and stalagmites form on the cave floor as water drips down from above. These cone-shaped features are made drop by drop as calcium carbonate or other minerals precipitate out of groundwater and are deposited on a cave's surfaces (Figure 25.35).

FIGURE 25.36
Sinkholes range from small, soil-lined depressions to bedrock-sided pits hundreds of meters wide.

Sometimes weathering and erosion by groundwater leave the roof of a cave very thin and fragile, and this is not visible above ground. At a certain point, the cavern's ceiling—with the soil and bedrock above it—may collapse. This creates a *sinkhole.* Sinkholes from collapsed caves can appear suddenly or gradually. They can swallow entire homes.

Landscapes where the effects of groundwater erosion can be seen above ground are said to have *karst topography*. Karst areas are often very scenic, featuring sinkholes, large rivers, and oddly carved, pitted, and pockmarked rock formations. Karst regions can contain large aquifers—in fact, more than 25% of the world's population either lives on or gets water from karst aquifers.

FIGURE 25.37
The Guilin region in China is famous for more than 5000 square kilometers of fantastic karst landscape. The towers of limestone look like giant green teeth. They are what's left behind after the surrounding limestone eroded away.

CHECK YOURSELF

1. **Stalactites and stalagmites are made of calcium carbonate that precipitates out of groundwater. How does the calcium carbonate get into the groundwater in the first place?**
2. **What must happen for a cave to be dry, rather than filled with groundwater?**

CHECK YOUR ANSWERS

1. Calcium carbonate is the mineral form of limestone. Calcium carbonate gets into the groundwater when acidified water dissolves limestone caves.
2. The water table must drop.

LEARNING OBJECTIVE
Describe the erosional effects of ocean waves.

UNIFYING CONCEPT
- ***Friction***
Section 2.8

25.6 Wave Effects

EXPLAIN THIS How is a wave like a sledgehammer?

If you've ever been in the ocean, and especially if you've surfed, you have felt the enormous force of ocean waves. Because waves carry so much energy, they are the most powerful agents of erosion along coastlines.

Waves are induced by wind flowing across the water surface. The wind sets the water in motion up and down, and through this motion the waves transport energy to shore. Note that it's energy, not individual water molecules, that move to shore with a wave. Individual water molecules move locally underneath a wave in a circular or "orbital" path that gets smaller with depth. As waves move closer to shore, orbits are disrupted as the water gets shallower and encounters friction with the sea bottom. Wavelength decreases and wave height steepens so the wave topples over with a crash. The crashing water moves into the surf zone—the area of frothy, foaming seawater that lies between the line of breakers and the shore at the beach (Figure 25.38).

Waves act like sledgehammers, pounding rock over and over again with each break of the surf. Ocean waves also weather rock by chemically dissolving it. Waves then pound into the cracks and break rock into smaller pieces. Erosion continues as waves carry rock fragments away. Smaller sediments carried in waves contribute to erosion when they act like sandpaper, abrading the rocks they crash against.

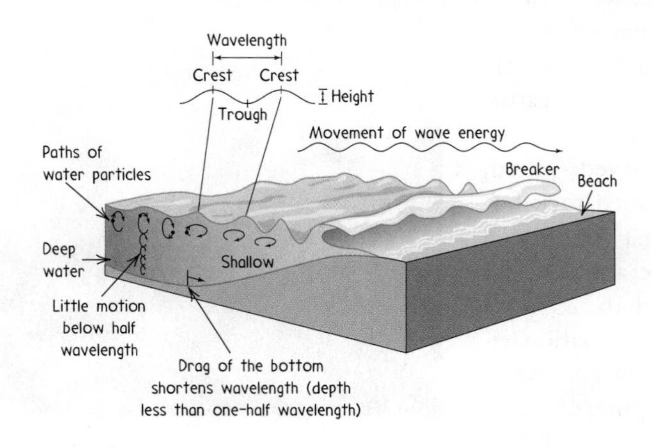

FIGURE 25.38
Waves change form as they travel from deep water through shallow water to the shore. In deep water, wave motion is circular. In shallow water, orbital motion becomes elliptical as the wave contacts the ocean floor. When waves reach a critical height, they break and crash onto the beach.

(a) (b) (c) (d)

FIGURE 25.39
Landforms created by ocean waves: (a) sea cave, (b) sea arch, (c) headlands, (d) sea stacks.

Wave erosion carves sea caves out of soft or fractured rock. Sea caves can further erode when waves cut all the way through them, leaving sea arches. When the arches eventually collapse, sea stacks may be left behind. Headlands, areas of land surrounded by water on three sides, form from cliffs that contain both hard and soft rock. The soft rock erodes to make beach sand, while the hard rock remains standing as headlands (Figure 25.39).

Waves don't actually transport water—they transport energy while the water molecules stay in place. Why, then, does a sailor put a message in a bottle hoping it will travel to a distant shore? In fact, ocean water does move, but it moves via *currents* rather than waves. Currents are like rivers of water in the larger sea. They are driven by tides, winds, or differences in water's density. Working along with waves, ocean currents are important agents of erosion and deposition along the coast. Waves and currents together carry sand along a beach in a zigzag pattern, redistributing it as sandbars, spits, and scalloped shorelines.

CHECK YOURSELF

Speaking very generally, are headlands and sea stacks formed from soft rock such as mudstone or from hard rock such as granite?

CHECK YOUR ANSWER

Headlands and sea stacks, like waterfalls, form from a rock formation that originally featured both hard and soft rock, but the softer rock eroded away first. Therefore, headlands and sea stacks are generally composed of hard rock, such as granite.

LEARNING OBJECTIVE
Describe wind erosion by abrasion.

25.7 Wind—Agent of Change

EXPLAIN THIS How do sand dunes form?

Wind, like water, shapes the land. However, because air is so much less dense than water, wind is a much less powerful agent of erosion. Still, wind can have pronounced effects in certain locations. Wind *abrades* rocks and other surfaces by grinding airborne sediments against them and carrying away small fragments. If you drive through a windstorm, the paint may be stripped from your car.

FIGURE 25.40
The ripple marks in this desert were generated by winds. They are small, elongated sand dunes.

Like water, wind erodes loose sediments more effectively than it erodes solid rock. So, wind-driven processes are most effective where sediments are loose, such as deserts, beaches, and poorly managed farmland.

Wind is like water in another way. The slower the wind's speed, the less sediment it can carry. A *sand dune* is a mound of sediment that forms when sediment-carrying winds slow down and drop their load of sediment. The coarser and heavier sediments drop out first. These deposited sediments form a windbreak that causes more deposition. Although a typical sand dune is between 3 and 100 m high, sand dunes 200 m high can develop.

SCIENCE AND SOCIETY

The Dust Bowl

A dramatic case of wind erosion took place in the midwestern United States in the 1930s. Farmers and ranchers on the plains of Colorado, New Mexico, Oklahoma, Kansas, and Texas had removed prairie grasses and wildflowers as they cleared land for crops and allowed cattle to eat the native vegetation. For years, normal amounts of rain fell and farms thrived.

But then several years of drought occurred. Crops died and left the soil bare. When strong winds blew, the soil was easily lifted up and carried away. It was wind erosion on a mass scale. Millions of tons of soil were blown through the air and dumped elsewhere. During this event, known as the Dust Bowl, dust storms called "black blizzards" hid the Sun. Dust blew as far away as Vermont. Tons of valuable topsoil were lost.

What caused the Dust Bowl? Although the native plants were adapted to the conditions of the Great Plains, the crops the farmers planted could not withstand the drought. When the crops died, and when livestock cleared out the native grasses, nothing was left to hold the soil in place. Farmers and ranchers lost their livelihoods. Thousands of people were forced to migrate as far away as California to find work. Eventually, rainfall increased to normal levels. Since the time of the Dust Bowl, soil conservation practices have prevented another recurrence of erosion on this scale. Still, topsoil erosion by wind and water remains a global concern.

FIGURE 25.41
Dust storms such as this were common during the Dust Bowl era.

SUMMARY OF TERMS (KNOWLEDGE)

Chemical weathering Change in the chemical structure of the minerals in rock.

Delta A mass of sediment that is deposited where a river slows down as it enters a large body of water such as a lake or the ocean.

Deposition The process of laying down eroded material in new locations.

Differential weathering Weathering that is varied in rate and degree due to climate, rock geometry, and other factors.

Discharge The volume of water that a stream transports in a given amount of time.

Drainage basin The area that supplies a stream with the water that flows through it.

Erosion The removal and transport of pieces of preexisting rock.

Floodplain A flat area along a stream that is covered by water only during a flood.

Gradient (of a stream) The slope; equivalently, the change in a stream's elevation over a particular distance.

Humus The decayed remains of animal and plant life in soil.

Load The amount and type of sediment that a stream carries.

Mass movement The downward movement of rock or soil due to gravity.

Mechanical weathering The breakdown of rock into smaller pieces through physical means.

Soil A mixture of weathered rock plus air, water, and organic matter that supports the growth of plants.

Surface processes Geologic processes that have causes and effects at or near Earth's surface.

Texture (of soil) The size of the particles in soil.

Topsoil The upper layer of soil, rich in humus, that supports plant life.

Tributary A smaller stream that flows into a larger one.

Weathering The disintegration or decomposition of rock at or near Earth's surface.

READING CHECK (COMPREHENSION)

25.1 Processes That Sculpt Earth: Weathering, Erosion, and Deposition

1. How are surface processes different from tectonic processes?
2. What are the four major agents of change that create and destroy landforms?
3. What are the two types of weathering?
4. What is the name of the geologic process that involves dumping sediment at a new location?
5. What is the name of the geologic process that involves lifting and transporting sediment?

25.2 The Impact of Running Water

6. How much sediment does the Mississippi River move each year?
7. How does the size of a stream's drainage basin relate to the size of the stream?
8. What is another name for a drainage basin?
9. There are three types of sediment loads. Describe each of them.
10. Describe a stream that causes a great deal of erosion. Give the four main properties of streams in your answer.
11. Identify one erosional landform produced by running water.
12. Identify one depositional landform produced by running water.
13. As a stream slows down, which sediments are deposited first—large ones or small ones?

25.3 Glaciers—Earth's Bulldozers

14. What type of glacier can cover vast portions of flat land? What is another name for this kind of glacier?
15. Why do glaciers move?
16. Why are glaciers powerful agents of erosion? Identify one landform produced by glacial erosion.
17. Why are glaciers powerful agents of deposition? Identify one landform produced by glacial deposition.

25.4 Effects of Gravity Alone—Mass Movement

18. What kind of mass movement occurs so slowly that you don't see it happening?
19. Why do trees sometimes bend at their base when they grow on sloping land?
20. What are the characteristics of a slide?
21. What are the characteristics of a flow?
22. What are the characteristics of a fall?
23. What's the most common cause of mass wasting events?
24. What factors predispose a slope to mass wasting?

25.5 Groundwater Erodes Rock to Make Caves and Caverns

25. Identify one landform produced by groundwater erosion.
26. How do caves form in subsurface limestone?
27. What is karst topography? Name one example somewhere in the world.
28. Why are sinkholes most common in locations where limestone makes up the bedrock?
29. How does a stalagmite form? A stalactite?
30. What do sinkholes and caverns have in common?

25.6 Wave Effects

31. Identify one landform produced by erosion from ocean waves.
32. Identify one landform produced by deposition from ocean waves.
33. Besides waves, what is an agent of change along the seashore?
34. How does a sea arch form?
35. How do headlands form?
36. A wave does not transport water molecules toward the shore. What does a wave transport shoreward?

25.7 Wind—Agent of Change

37. Why is wind a less powerful agent of erosion than water?
38. As wind speed is reduced, what happens to the sediment carried by wind?
39. Name one depositional landform produced by wind.
40. Name an environment where wind shapes the landscape to a significant degree.

THINK INTEGRATED SCIENCE

25A—Weathering

41. Distinguish between mechanical and chemical weathering.
42. Can ice wedging weather human-made materials? Give an example.
43. Water is unusual because it expands when it freezes. How does this behavior of water break down rocks? Explain.
44. Is wind usually a cause of mechanical or chemical weathering? Explain why you think so.
45. What are two end products that result from the weathering of granite?
46. What sort of climate accelerates chemical weathering?

25B—Soil

47. How does the acidification of rainwater contribute to the formation of soil?
48. What are the four ingredients in all soils?
49. Why is humus necessary for plant growth?
50. Describe the type of soil preferred by most plants.
51. How does silty soil differ from sandy soil?
52. Why is topsoil a nonrenewable resource?
53. Summarize the role of living things in the formation of fertile soil.

THINK AND DO (HANDS-ON APPLICATION)

54. Investigate the formation of stalactites and stalagmites. Dissolve the largest quantity of Epsom salt you can in a jar full of water. Fill two small cups with the solution, set them on paper towels, and suspend a thick string between them, with one end in each container. Let the string remain for several days. How do the deposits form from the solution? Where do they form? How is this similar to the formation of stalactites and stalagmites in caves?
55. The next time it rains, watch runoff move down a slope. Can you see streams forming and merging to make bigger ones? Can you see erosion occurring? Sketch the streams you see, and make notes about the stream properties you observe.
56. Go to a construction site and look for evidence of erosion. Record what you see.

THINK AND COMPARE (ANALYSIS)

57. Rank these soils in order of increasing particle size: silty loam, clay loam, sandy loam, sand.
58. Rank these agents of erosion from most powerful to least powerful, in terms of total sediment moved: wind, glaciers, running water, gravity alone.

THINK AND SOLVE (MATHEMATICAL APPLICATION)

59. A stream channel has a cross-sectional area of 30 m^2. The stream's velocity at that point is 5 m/s. Show that the stream's discharge is 150 m^3/s.
60. An alpine glacier is moving downhill at a speed of 10 m per day. Show that the glacier will reach a ranger station 250 m from the glacier's front edge in 25 days.
61. Show that a mountain that can be approximated by a box 4 km × 3 km × 4 km will exist for about 329 million years.

THINK AND EXPLAIN (SYNTHESIS)

62. Are karst mountains depositional or erosional landforms? Explain.
63. Which of these two valleys was carved by a stream? Which was carved by a glacier? What is your evidence?

(a)

(b)

64. Why are areas with fine sediments, such as beaches, so vulnerable to erosion?
65. Describe how the smallest sediments (those on the scale of atoms), medium-sized sediments, and large sediments move downstream.
66. In what way is a glacier like a bulldozer? Like sandpaper?
67. Sand dunes are not stationary—over time they move. What moves them? In what direction do they move?
68. You are hiking and you see a hazard sign that says, "Danger! Mass Movement." What is this sign warning you about? Draw the sign with the warning stated in everyday language.

69. Why is it smarter to build a house on a hillside with a gentle slope than on one with a steep slope? Include the concept of creep in your answer.
70. Why have farmers settled in floodplains since ancient times despite the risk of flooding there?
71. Which of these three agents of transportation—wind, water, or ice—transports the largest boulders? Why?
72. Name at least two factors that affect the rate of chemical weathering of rock.
73. How does rainwater cause weathering? Can it also cause erosion? If so, how?
74. Would you expect chemical weathering to occur faster in the Amazon rainforest or in the Antarctic Circle? Defend your answer.
75. How are the amount and size of the particles carried by a stream affected by the velocity of flow?
76. Why does water in a stream typically move slower as it nears the ocean?
77. You see a river with many large meanders. Do you suppose the river is an old or young geologic feature? What is the reason for your answer?
78. What is the major factor that determines a stream's ability to erode land?
79. This muskrat burrows and thereby adds silt and soil to a river. Does the muskrat increase the stream's bed load, dissolved load, or suspended load? Why do you think so?

80. In what way are glaciers like dirty snowballs?
81. In what way does wind erosion pose a threat to native people living in the desert?

THINK AND DISCUSS (EVALUATION)

82. A dam is built on a major river that flows into the ocean. The dam greatly reduces the flow of river water beyond it. Ten years after the dam is built, nearby beaches have noticeably less sand. Develop a hypothesis to explain what is happening at the beach.
83. Should you build a cabin at the foot of a slope? What should you consider to determine whether the slope presents an erosion hazard?

READINESS ASSURANCE TEST (RAT)

If you have a good handle on this chapter, then you should be able to score 7 out of 10 on this RAT. If you score less than 7, you need to study further before moving on.

Choose the BEST answer to each of the following questions.

1. Why is topsoil essential to conserve and protect?
   (a) Topsoil holds subsoil in place, preventing landslides.
   (b) Plants require topsoil to grow—and plants are vital because they provide food and oxygen and they absorb carbon dioxide, thereby protecting climate.
   (c) Topsoil is derived from fossil fuels, which means it is expensive and rare.
   (d) Plants grow poorly in the subsoil that remains when topsoil is eroded.
2. What landform is created by groundwater dissolving rock?
   (a) cave
   (b) beach
   (c) sea stack
   (d) sand dune
3. The most powerful agent of erosion in shaping Earth's surface is
   (a) running water.
   (b) wind.
   (c) glaciers.
   (d) waves.
4. Well-rounded sediment particles indicate
   (a) a quiet, low-energy environment.
   (b) mechanical weathering.
   (c) glacial deposits.
   (d) long transport distances.
5. A stream's sediment load can contain
   (a) dissolved sediments.
   (b) suspended sediments.
   (c) large sediments dragging along the stream bed.
   (d) all of these
6. A sinkhole is evidence of
   (a) wind erosion.
   (b) groundwater erosion.
   (c) ocean wave erosion.
   (d) stream erosion.
7. Identify the depositional feature for which wind is the agent of change.
   (a) morraine
   (b) levee
   (c) delta
   (d) sand dune
8. Glaciers are important agents of erosion because
   (a) they pull rocks out of the ground and carry them away.
   (b) they move slowly.
   (c) they are melting.
   (d) they contain enormous amounts of water.
9. Where is the world's largest glacier?
   (a) Greenland
   (b) Asia
   (c) Antarctica
   (d) Minnesota
10. Soft rock erodes
    (a) faster than hard rock.
    (b) slower than hard rock.
    (c) at the same speed as hard rock.

Answers to RAT

1. b, 2. a, 3. a, 4. d, 5. d, 6. b, 7. d, 8. a, 9. c, 10. a

CHAPTER 26
Weather

EARTH IS the "Goldilocks planet"—of all the planets in the solar system, Earth is neither too hot nor too cold for life. It's just right. Why? How does the movement of Earth's atmosphere—that big sea of air in which we live (visible in the photo above)—make the weather change? How do other factors, such as ocean currents, the angle of the Sun's rays, the greenhouse gases, and the distribution of land and water, contribute to weather?

Weather has a direct effect on health and comfort. It influences daily choices, from what to wear to what to eat. Large-scale and long-term weather patterns—known as *climate*—affect everything around us too, from the shape of Earth's rivers and mountains to the life cycles of penguins and mosquitos. Read on to find out about meteorology, the science of weather, and climatology, the study of climate.

LEARNING OBJECTIVE
Identify the six elements of weather.

26.1 The Atmosphere

EXPLAIN THIS State the reasons Earth's atmosphere is essential to life.

In science, a system is a set of interacting parts enclosed in a defined boundary. Within the Earth system, there are four interdependent parts or "spheres." These spheres include the atmosphere, lithosphere, hydrosphere, and biosphere. Earth's **atmosphere** is a thin envelope of gases, tiny solid particles, and liquid droplets that surrounds the solid planet. The hydrosphere is the sum of all the water at or near Earth's surface, the lithosphere is the outer shell consisting of the crust and mantle, and the biosphere is the portion of the planet that contains life. Earth's subsystems, or spheres, are not really separate. They interact in almost infinite ways. For example, the atmosphere weathers the lithosphere. And the biosphere affects the atmosphere when photosynthesizing organisms take carbon dioxide out of the air and add oxygen to it.

UNIFYING CONCEPT
- ***The Scientific Method***
Section 1.3

Even though Earth is a complex system of interacting parts that constantly changes in ways too complicated to completely track, science is good at detecting patterns and making predictions by breaking the system down into separate parts and cause-and-effect relations. To understand the extremely complex topics of weather and climate, we focus on the atmosphere. We look at how it is structured and what makes it change. **Weather** is defined as the state of the atmosphere at a particular time and place. Weather is made up of six elements of the atmosphere: atmospheric pressure, temperature, wind, precipitation, cloudiness, and

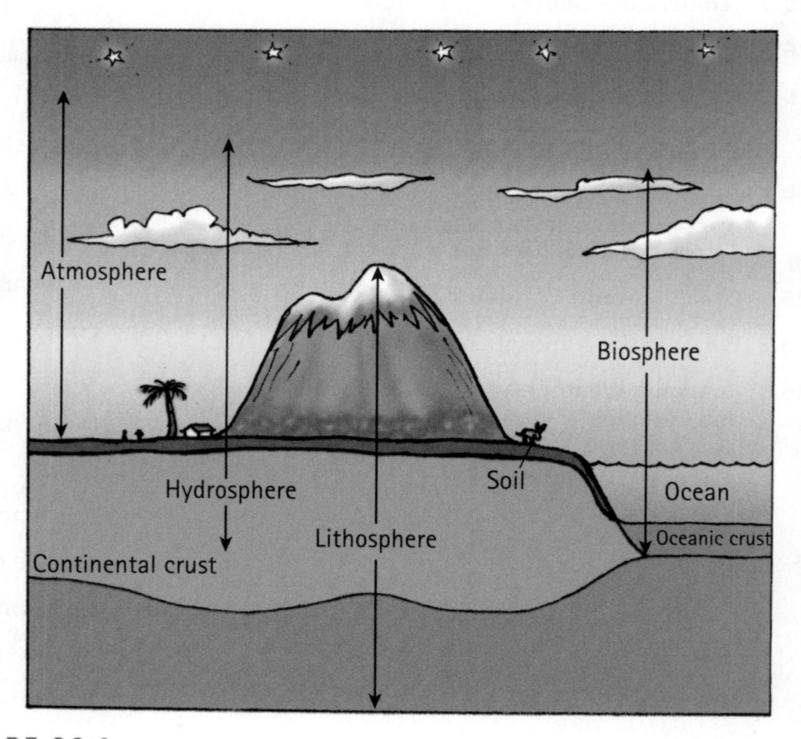

FIGURE 26.1
Earth is a system composed of subsystems or "spheres." Earth's environmental spheres are: the atmosphere, hydrosphere, lithosphere, and biosphere. When we study weather, we focus mainly on the atmosphere.

FIGURE 26.2
Changes in the atmosphere produce changes in the weather. On the left, the atmosphere is wet, cold, windy, and cloudy. On the right, the atmosphere is warm, clear, still, and dry.

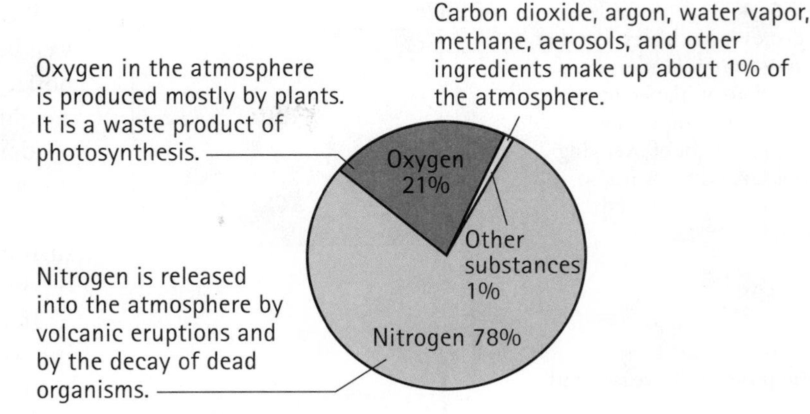

Carbon dioxide, argon, water vapor, methane, aerosols, and other ingredients make up about 1% of the atmosphere.
Oxygen in the atmosphere is produced mostly by plants. It is a waste product of photosynthesis.
Oxygen 21%
Other substances 1%
Nitrogen is released into the atmosphere by volcanic eruptions and by the decay of dead organisms.
Nitrogen 78%

FIGURE 26.3
The atmosphere consists of mainly nitrogen and oxygen gases with smaller amounts of various gases and aerosols.

humidity. Together, these elements create an overall atmospheric condition that we call "the weather."

Earth's present-day atmosphere is mostly a mixture of nitrogen and oxygen gas, with small percentages of water vapor, argon, methane, carbon dioxide, and other trace gases. Also, tiny suspended particles called *aerosols* are mixed into the atmosphere. The aerosols in the air include dust, ash, pollen, water droplets, and even air pollution such as soot. People typically call the atmosphere *air*, so we use the terms interchangeably here too.

The atmosphere is essential for life. It provides the oxygen and carbon dioxide that are required by living things. It provides the nitrogen that plants need. It protects organisms from the Sun's harmful high-energy ultraviolet waves, which damage not only the skin but also DNA. The atmosphere is an insulating blanket that traps heat and keeps temperatures sustainable for life—neither too hot nor too cold.

Climate is the general pattern of weather that occurs over a period of years. For instance, the weather in Anchorage, Alaska, may be clear, warm, and windy on a particular day, but the city has a cold, polar climate overall.

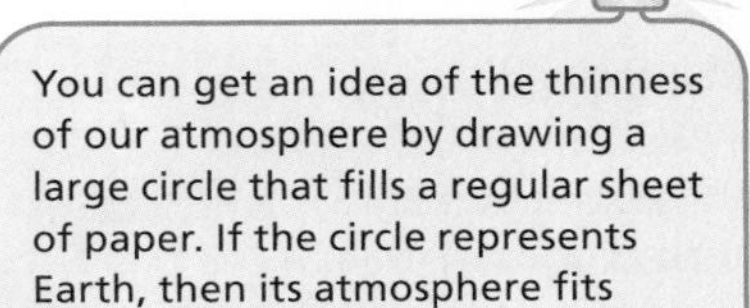

You can get an idea of the thinness of our atmosphere by drawing a large circle that fills a regular sheet of paper. If the circle represents Earth, then its atmosphere fits inside your pencil mark!

Integrated Science 26A
PHYSICS

LEARNING OBJECTIVE
Describe how air pressure relates to the weight of air molecules.

Atmospheric Pressure

EXPLAIN THIS Explain why air pressure decreases with altitude.

The atmosphere is more than 100 miles thick, and it contains many trillions of molecules. Since these molecules have weight, the atmosphere has weight, and it therefore pushes against Earth's surface with a certain force per unit area—a pressure. The pressure that the atmosphere exerts on a surface due to the weight of the molecules above that surface is known as **atmospheric pressure**, or simply *air pressure*. (Again, we use the terms *air* and *atmosphere* interchangeably.)

UNIFYING CONCEPT
- *The Gravitational Force*
Section 5.3

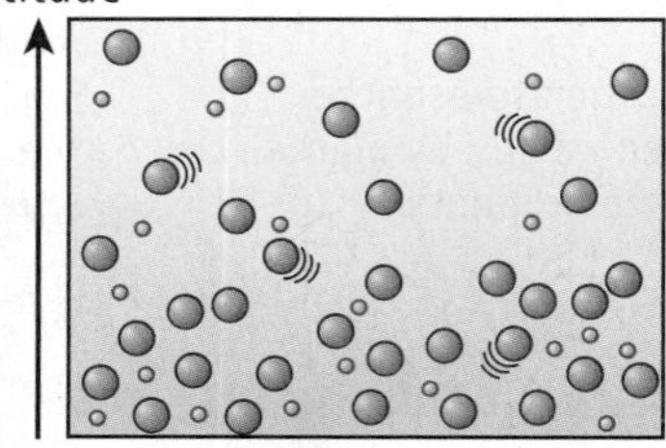

FIGURE 26.4
Air density decreases with altitude; that is, there are more air molecules closer to Earth's surface than at higher altitudes. This is because air is compressible—"squishable." The weight of overlying air presses molecules downward, so they pack relatively tightly together near the ground.

We have adapted so completely to the weight of the air around us that we don't notice it. Perhaps a fish "forgets" about the weight of water in the same way. The reason we don't feel the atmosphere's weight crushing against our bodies is that the pressure inside our bodies equals the pressure of the surrounding air. There is no net force acting on our skin.

If you have ever gone mountain climbing, you've noticed that the atmosphere gets thinner with altitude. There are fewer air molecules in a given volume of air up high, so it's harder to breathe. So, unlike the uniform density of water in a lake, the *density* of air in the atmosphere (its mass per volume) decreases with altitude (Figure 26.4).

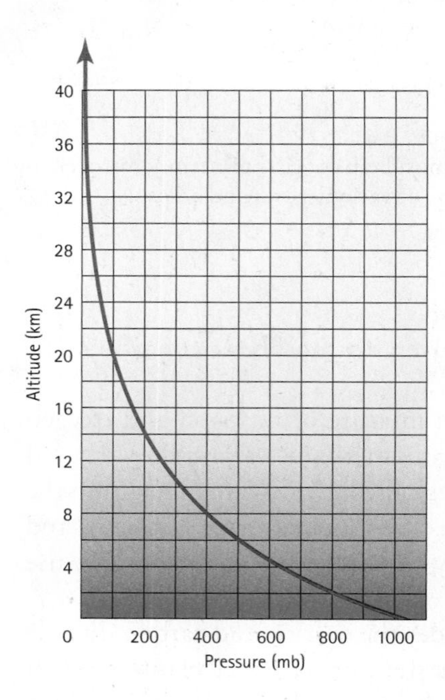

FIGURE 26.5
Atmospheric pressure decreases with altitude.

UNIFYING CONCEPT
- ***Density***
Section 2.3

Air pressure also decreases with altitude. This isn't surprising because air pressure is the weight of the overlying air. So, higher in the atmosphere, where the number of overlying molecules tapers off, air pressure also decreases (Figure 26.5). At sea level, air pressure is at its highest. What is the numerical value of air pressure at sea level? A column of air with a cross-sectional area of 1 square meter extending up through the atmosphere has a mass of about 10,000 kilograms. The weight of this air is about 100,000 newtons (10^5 N). This weight produces a pressure of 100,000 newtons per square meter, so we estimate the atmospheric pressure at sea level to be 10^5 N/m^2 (Figure 26.6).

MasteringPhysics®
VIDEO: Air Has Matter
VIDEO: Air Has Pressure
VIDEO: Air Has Weight

The SI unit of pressure is the pascal (Pa), and 1 Pa equals 1 N/m^2. Our estimate of atmospheric pressure thus may be stated as 100 kilopascals. Actually, the average atmospheric pressure at sea level is precisely 101.3 kilopascals (101.3 kPa), which is often expressed as 1 atmosphere (1 atm) of pressure. There are other popular units for atmospheric pressure. News programs report atmospheric pressure in *inches of mercury* or *millimeters of mercury*. These units refer to the mercury barometer, the instrument traditionally used to measure air pressure. Meteorologists often express air pressure in *bars* or *millibars* (mb). Conversion factors for air pressure are listed in Table 26.1.

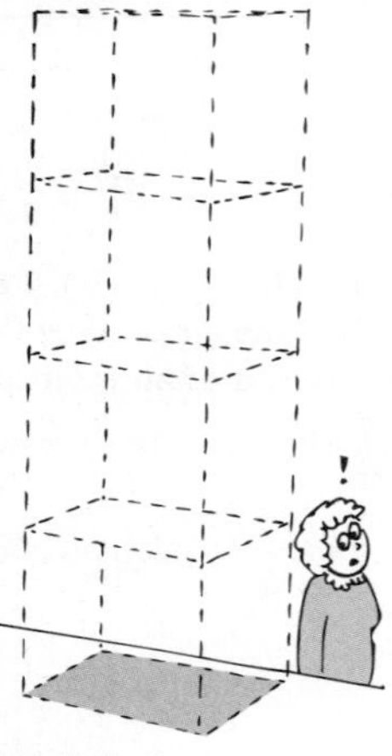

FIGURE 26.6
The weight of air that bears down on a 1-square-meter surface at sea level is about 100,000 N. So, the atmospheric pressure at sea level is about 10^5 N/m^2.

TABLE 26.1 EQUIVALENT MEASUREMENTS FOR ATMOSPHERIC PRESSURE
1 standard atmosphere (1 atm)
101,325 pascals
1.013 bars
14.7 pounds per square inch (14.7 psi or 14.7 lb/in^2)
760 torr
760 millimeters of mercury (760 mm Hg)

CHECK YOURSELF

1. **Why doesn't the pressure of the atmosphere break windows?**
2. **The density of atmospheric gas molecules diminishes with altitude. How does this affect atmospheric pressure?**

CHECK YOUR ANSWERS

1. Atmospheric pressure is exerted on both sides of a window, so no net force is exerted on the window. If, for some reason, the pressure is reduced on one side only, as in a strong wind, watch out!
2. Atmospheric pressure decreases due to the decreasing density of the atmospheric gas molecules.

UNIFYING CONCEPT

- ***Newton's Second Law*** Section 3.2
- ***Density*** Section 2.3

26.2 The Structure of the Atmosphere

LEARNING OBJECTIVE
State the layers of the atmosphere, and describe their principal characteristics.

EXPLAIN THIS Explain why weather occurs only in the troposphere.

The atmosphere has four layers based on temperature: the troposphere, stratosphere, mesosphere, and thermosphere.

The Troposphere

As you have learned, most of Earth's atmospheric gas molecules are held close to Earth. They are concentrated in Earth's lowest atmospheric layer—the *troposphere*. The troposphere is quite dense but thin. It contains almost 90% of the atmosphere's total mass and has an average thickness of only 12 km (8 mi).

Weather occurs in the troposphere. That's why commercial jets sometimes fly just above the troposphere. At this altitude, jets avoid turbulence caused by weather disturbances.

The troposphere contains almost all of the atmosphere's water vapor and suspended particles, and these create clouds and precipitation. For this reason, the troposphere is the layer where weather occurs.

The temperature in the troposphere decreases steadily (at 6°C per kilometer). At the top of the troposphere, the temperature averages a frigid −50°C. Certain gases of the troposphere absorb and retain warming energy from the Sun (we will give more details about this process in the next section). Therefore, as the density of air decreases, so does the troposphere's ability to retain heat (Figure 26.8). This is why high-altitude locations typically have cold climates.

FIGURE 26.7
Earth's weather occurs in the troposphere—the lowest, thinnest, and densest layer of the atmosphere. This satellite image shows the thick, spiraling clouds over the Atlantic Ocean that accompanied Tropical Storm Xina in 1985.

The Stratosphere

The *stratosphere* sits on top of the troposphere, from about 12 km to about 50 km above sea level. It is a thicker layer than the troposphere but much less dense. The bottom of the stratosphere is cold, only about

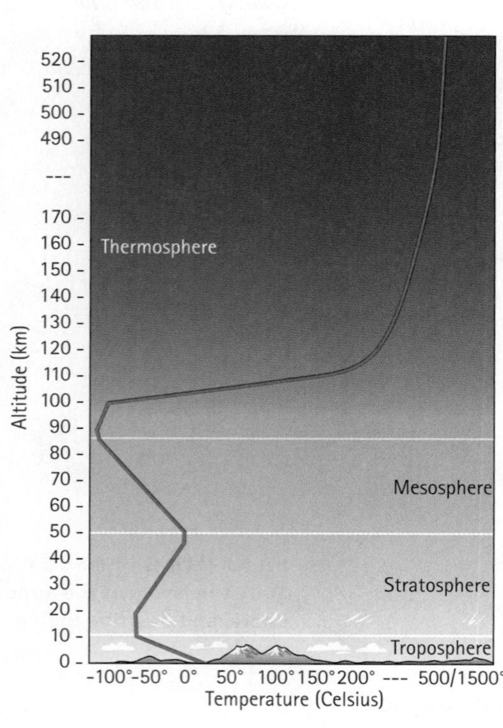

FIGURE 26.8
The average temperature of Earth's atmosphere varies in a zigzag pattern with altitude.

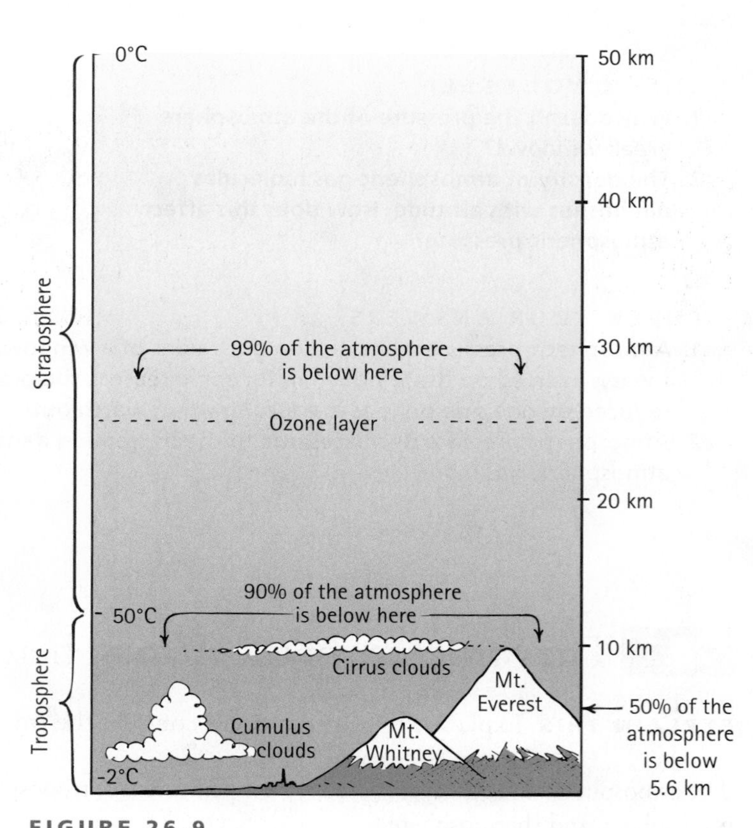

FIGURE 26.9
The two lowest layers of the atmosphere, the troposphere and the stratosphere, contain more than 99% of Earth's atmosphere.

MasteringPhysics®
ACTIVITY: Vertical Structure of the Atmosphere

−60°C. However, the temperature rises steadily to about 0°C at the top of the stratosphere.

The temperature rises in the stratosphere because of *ozone*. Ozone (O_3), a molecule made of three oxygen atoms, accumulates in the stratosphere and absorbs solar energy. Ozone absorbs warming infrared radiation as well as the Sun's higher-energy ultraviolet (UV) radiation. Atmospheric ozone is critical to life on Earth because it absorbs UV radiation. UV radiation has many harmful effects, including sunburn, skin cancer, and eye damage. UV radiation also disrupts DNA, which may cause genetic mutations.

If Earth's entire atmosphere were compressed to be as dense as concrete, it would only be about 15 feet thick.

The Mesosphere

The *mesosphere* extends upward from the top of the stratosphere to an altitude of about 80 km. The gases that make up the mesosphere absorb little solar radiation. As a result, the temperature decreases from 0°C at the bottom of the mesosphere to about −90°C at the top. Very chilly indeed!

The Thermosphere

There is little air in the *thermosphere*, the next layer up in the atmosphere. However, the air that is there readily absorbs solar radiation. For this reason, temperatures are surprisingly high in the thermosphere, ranging from 500°C to 1500°C depending on solar activity.

The Ionosphere

The *ionosphere* isn't an atmospheric layer in the same sense that the other layers are. Instead, it is a region of the upper mesosphere and lower thermosphere that contains ions—gas molecules that have become electrically charged by absorbing solar energy. Near Earth's magnetic poles, fiery light displays called *auroras* occur as the solar wind (high-speed charged particles ejected by the Sun) strikes and excites particles in the ionosphere.

The atmosphere rises up from Earth's surface, but where does it end and outer space begin? There is no distinct boundary. The atmosphere thins rapidly with distance from Earth, then fades gradually. It finally terminates where there are too few molecules to detect hundreds of kilometers above Earth's surface.

FIGURE 26.10
The aurora borealis, or northern lights, is a beautiful display of light that occurs in the ionosphere when charged particles are excited by incoming solar radiation.

CHECK YOURSELF

1. **In recent years, scientists have found "holes" in the ozone layer—areas where atmospheric ozone has become very thin. Why are scientists concerned about this?**
2. **Why do mountainous areas generally have cold climates?**

CHECK YOUR ANSWERS

1. Ozone protects life from damaging UV rays. Fortunately, the scientific and political communities have worked together and made great progress on the issue of the thinning ozone layer. By international agreement, the use of certain chemicals that damage the ozone layer has been restricted. Stratospheric ozone appears to be returning to healthy levels.
2. High-altitude places such as mountains have cold climates mainly because of the low density of the troposphere there.

Integrated Science 26B
PHYSICS

LEARNING OBJECTIVE
Describe how the greenhouse effect and solar radiation warm the atmosphere.

Solar Radiation and the Greenhouse Effect Drive Global Temperature

EXPLAIN THIS Explain why the visible radiation emitted by the Sun is not a cause of global warming.

Why does Earth have just the right temperature for life—neither too hot nor too cold? Two major factors work together to determine average global temperature: (1) the solar radiation that Earth absorbs, and (2) the concentration of greenhouse gases in its atmosphere.

The Sun emits **solar radiation**, which is energy in the form of electromagnetic waves. It moves outward from the Sun in all directions, like a light bulb. Earth intercepts a tiny part of it. We receive only about two-billionths of the total energy released by the Sun. This tiny fraction is the ultimate source of almost all of the energy used by everything on Earth. And it is also just the right amount of solar radiation required for sustaining temperatures suitable for life.

A bit less than half of the radiant energy from the Sun is in the form of visible light. As you know from physics, visible light is a portion of the

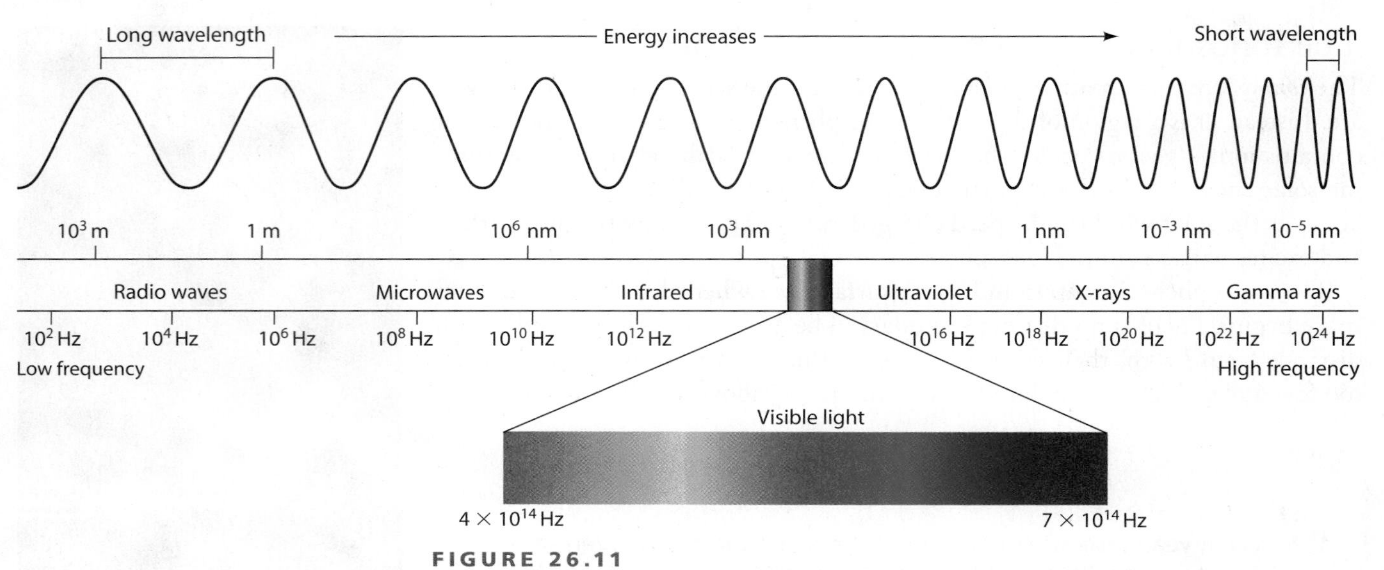

FIGURE 26.11
The electromagnetic spectrum consists of electromagnetic waves of different wavelengths.

UNIFYING CONCEPT
- ***Waves***
Section 8.1

electromagnetic spectrum, a continuum of radiant energy waves. Electromagnetic waves have wavelengths ranging from thousands of kilometers to fractions of an atom (Figure 26.11). The longest waves are radio waves, then microwaves, infrared radiation, visible light, ultraviolet light, X-rays, and gamma rays. Energy and wavelength are related *inversely*—the longer the wavelength of a wave, the less energy it carries. Visible light has higher energy and shorter wavelength compared to *infrared radiation*, which is the radiant energy that warms our bodies and other forms of matter too, including many molecules of the atmosphere.

The atmosphere is transparent to visible light. Most of the light transmitted through the atmosphere is absorbed by the ground, although some of it is reflected back to space before it ever reaches Earth's surface (Figure 26.12).

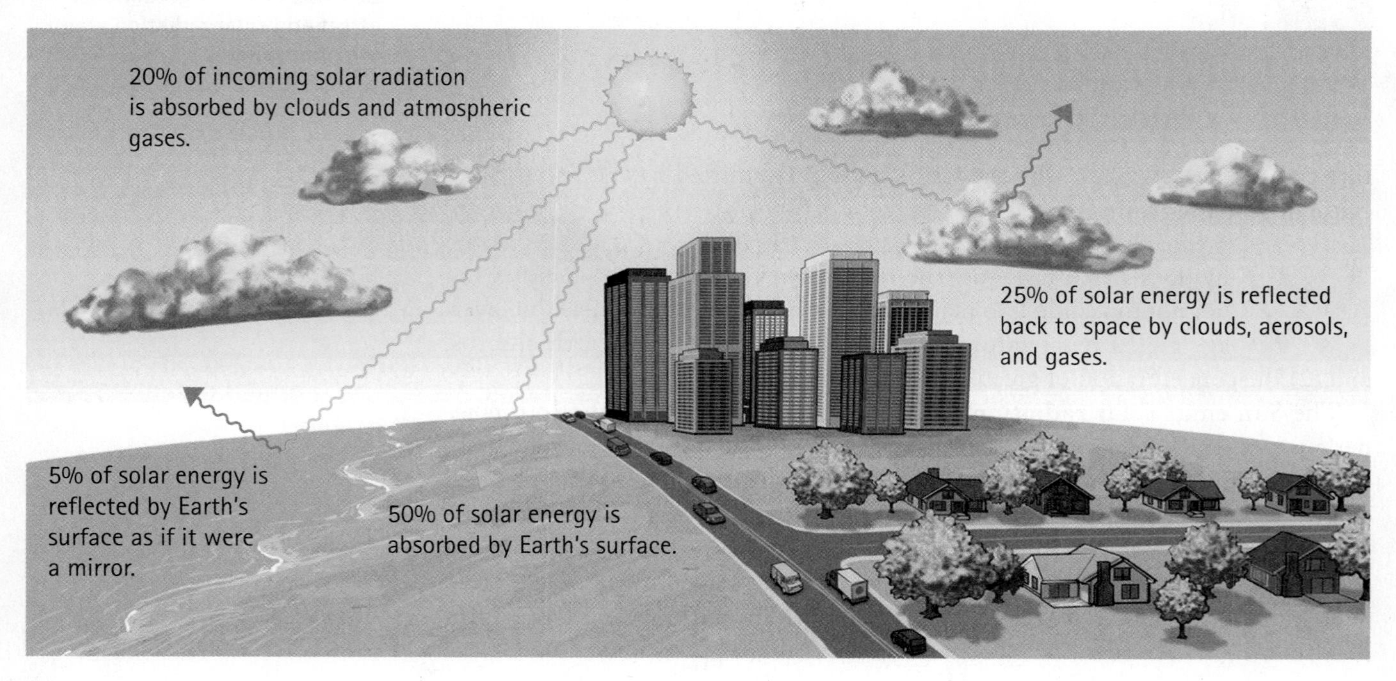

FIGURE 26.12
Some short-wavelength solar radiation is absorbed by the land, water, and atmosphere and eventually becomes heat. The rest is reflected back into space and doesn't affect temperature.

When the light energy that reaches the ground is absorbed, the ground in turn warms up and emits energy. But the energy that Earth's surface emits, which is called **terrestrial radiation**, has less energy and longer wavelengths than the incoming energy. Instead of being in the visible part of the spectrum, terrestrial radiation is infrared (Figure 26.13).

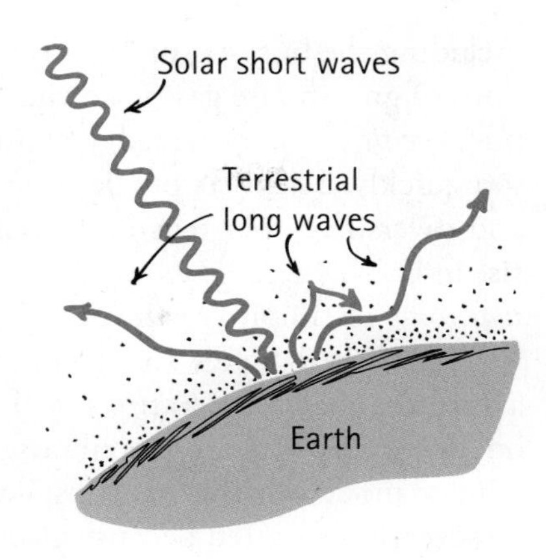

FIGURE 26.13
About 43% of the radiant energy from the Sun is in the visible part of the spectrum. The atmosphere is quite transparent to this radiation. Earth absorbs and Earth re-emits these waves as lower-energy, longer-wavelength infrared waves called terrestrial radiation.

Although the atmosphere is transparent to short-wave visible light, it does not allow the infrared waves to freely pass through. As Figure 26.14 shows, carbon dioxide, water vapor, methane, and other "greenhouse gases" in the atmosphere absorb the warming terrestrial radiation, re-emit it, and re-absorb it many times before it can escape into space. Thus, energy becomes trapped in the atmosphere by greenhouse gases, and this causes the temperature to rise. The process of trapping terrestrial radiation leading to increased atmospheric temperature is called the **greenhouse effect**. The greenhouse effect is named for a florist's greenhouse, as shown in Figure 26.15.

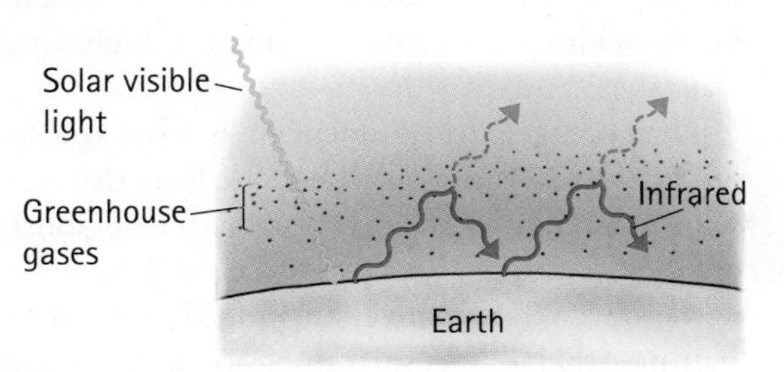

FIGURE 26.14
The greenhouse effect in Earth's atmosphere. Visible light is absorbed by the ground, which then emits infrared radiation. Carbon dioxide, water vapor, methane, and other greenhouse gases absorb and re-emit infrared radiation that would otherwise be radiated directly from Earth into space.

The greenhouse effect is natural and beneficial to life on Earth. If greenhouse gases weren't present and able to trap heat, the planet's average temperature would be a frigid −18°C. The present concern, however, is that human activities,

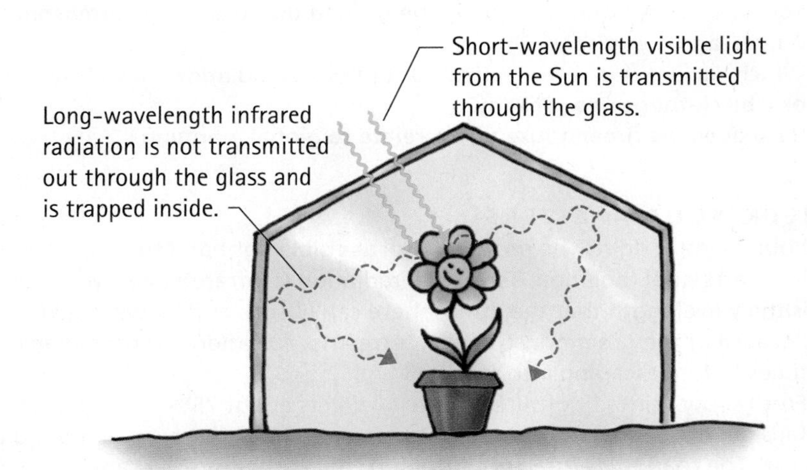

FIGURE 26.15
A florist's greenhouse is made of glass. Glass is transparent to visible light and opaque to infrared light just like the atmosphere is. Sunlight enters the greenhouse and warms the plants there. The plants then reradiate longer-wavelength infrared waves. The glass doesn't let the infrared waves pass through, so the infrared radiation builds up inside, increasing the temperature.

including the burning of fossil fuels and deforestation, produce high concentrations of greenhouse gases and thereby trap too much heat. As a result of this *anthropogenic* (human-caused) greenhouse effect, Earth is heating up too much and too quickly. Scientists predict that the complex ecosystems that have developed and diversified over tens of millions of years cannot be sustained. The sudden rise in Earth's average temperature, which threatens the biosphere, is called *global warming*, or simply *climate change*.

UNIFYING CONCEPT

- *The Ecosystem*
  Section 21.1

The United States Environmental Protection Agency (EPA) projects that average global temperatures will rise by 2°F to 11.5°F by 2100, depending on the level of future greenhouse gas emissions and the outcomes of various climate models. If this happens, ecosystems and human cultures worldwide will be severely disrupted because glaciers will melt and sea level will rise, there will be more frequent severe weather events, agriculture will de-stabilize, and there will be widespread economic impacts including the loss of habitable land along coastlines.

That being said, no one knows what the specific effects of global warming will be. Exactly how much will sea level rise and when? How severe and frequent will hurricanes be in different coastal regions? Where will drought, wildfires, and floods occur and how severe will they be compared to today? What will be the effect of climate change on disease-causing microbes and parasites when their natural habitats shift in basic ways? How will agricultural crops be affected when temperatures rise, soil dries out, and water for irrigation becomes scarce? There are many possible effects of climate change, and the problem is currently the focus of intense research and public interest. (More information on global climate change is provided in Chapter 27.)

CHECK YOURSELF

1. **Your friend says that it is mainly the ground that warms the atmosphere. Do you agree or disagree?**
2. **Which surface is a better emitter of terrestrial radiation—a snowy field or a black-topped road? Why?**
3. **How does the greenhouse effect relate to global warming?**

CHECK YOUR ANSWERS

1. Your friend is right. The ground absorbs visible light and then reradiates it as terrestrial radiation. Terrestrial radiation is infrared energy, which is the wavelength that the atmosphere can absorb and be warmed by. Because of the greenhouse effect, terrestrial radiation is reused many times before escaping into space.
2. Fresh snow shines brightly because it reflects about 75% of the light that falls on it. A road, by contrast, reflects about 10% of the incoming radiation. The road therefore is a much better absorber of solar radiation and emitter of terrestrial radiation than the snowy field.
3. Global warming is principally caused by the greenhouse effect. But it's not the natural greenhouse effect that is the problem. It's the intensified greenhouse effect that is caused by human activities that put excessive amounts of greenhouse gases into the atmosphere.

26.3 Temperature Depends on Latitude

LEARNING OBJECTIVE
Relate air temperature to solar intensity.

EXPLAIN THIS In terms of solar radiation, explain why you have a higher risk of getting a sunburn in Australia than in Canada.

Your house has a street address that identifies its unique location. Similarly, all points on Earth can be identified by their "address," a set of numbers called *longitude* and *latitude*.

Lines of longitude are imaginary lines that run north/south and pass through the poles (Figure 26.16). One of these lines, the one that passes through Greenwich, England, is called the *prime meridian*. The prime meridian, by international agreement, has a longitude of 0°. The longitude of any other location is its distance east or west from the prime meridian. Longitude is measured in degrees.

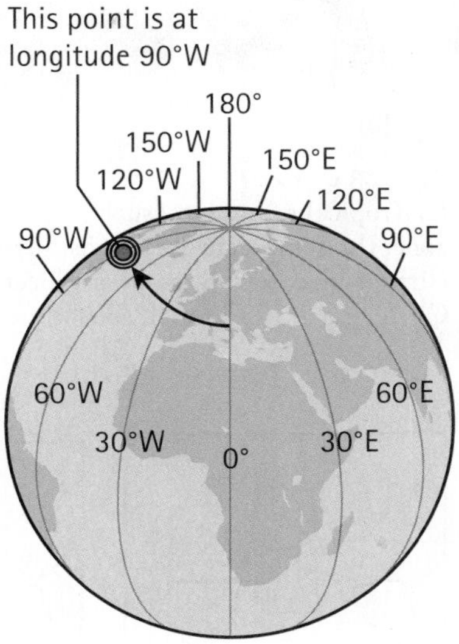

FIGURE 26.16
Longitude is measured up to 180° east or west from the line of 0°.

The equator is an imaginary circle that is halfway between the poles. It divides Earth into two halves: the Northern and Southern Hemispheres. Imaginary lines drawn around Earth parallel to the equator are called *lines of latitude*, or *parallels*. The equator has a latitude of 0°, and the latitudes of other locations are measured in degrees north or south from the equator. Figure 26.17 makes this clear.

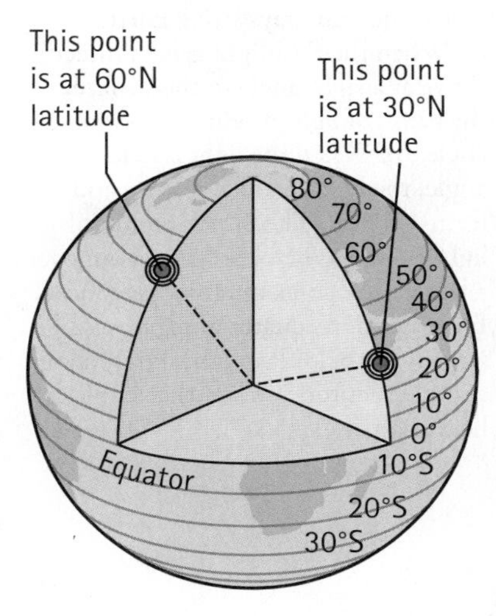

FIGURE 26.17
Latitude is a degree measurement that tells the angle between the center of Earth, the equator, and any particular point on Earth's surface.

Figure 26.18 shows how latitude relates to temperature. The world is divided into climate zones based on yearly temperature averages. Temperatures are highest in the latitudes nearest the equator, in the *tropical climate zone* (at sea level). In the tropical climate zone, temperature varies little throughout the year and averages higher than 18°C (or 64°F) even in the coldest months. At the middle latitudes, in the *temperate climate zone*, temperatures are neither very

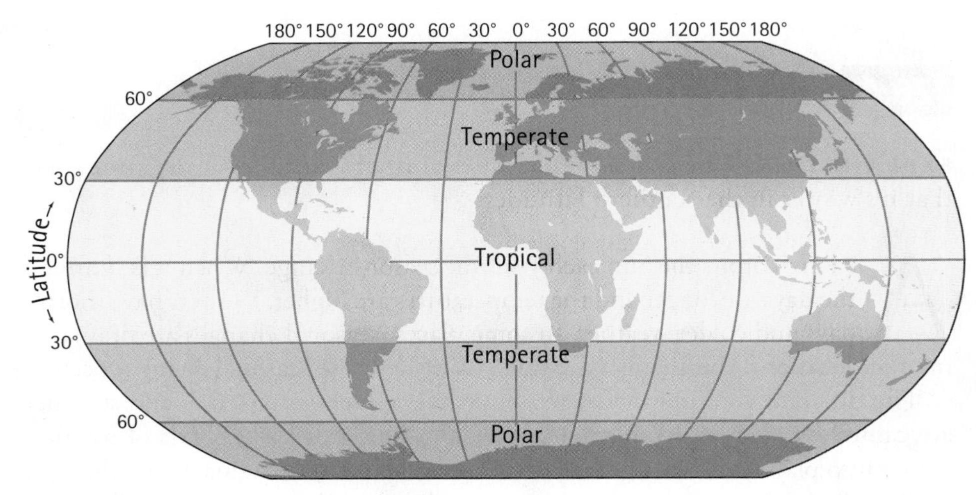

FIGURE 26.18
Yearly average temperatures vary in Earth's principal climate zones.

The Sun is directly overhead at noon only near the equator. Standing in the sunshine, you cast no shadow there. At locations farther from the equator, the Sun is never directly overhead at noon. The farther you stand from the equator, the longer is the shadow your body casts.

(a)

(b)

(c)

FIGURE 26.19
(a) Tropical rainforests, such as this one in Trinidad, are located in the tropical climate zone. (b) Temperate grasslands, such as the Oklahoma prairie, are found in the temperate climate zone. (c) The tundra is located in the polar climate zone. Here, within the Arctic Circle, grasses and tough shrubs grow in the frozen soil.

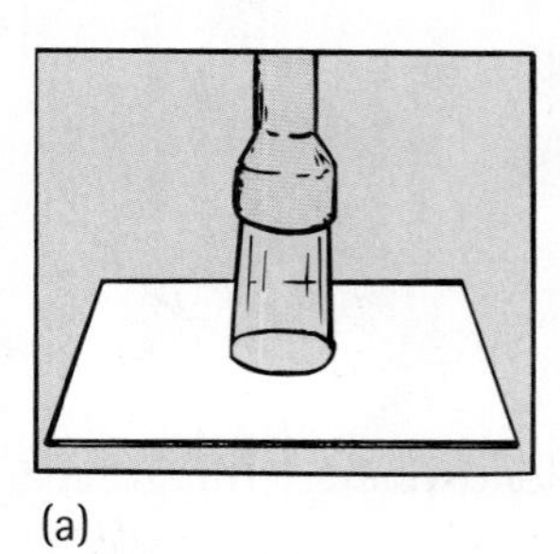
(a)

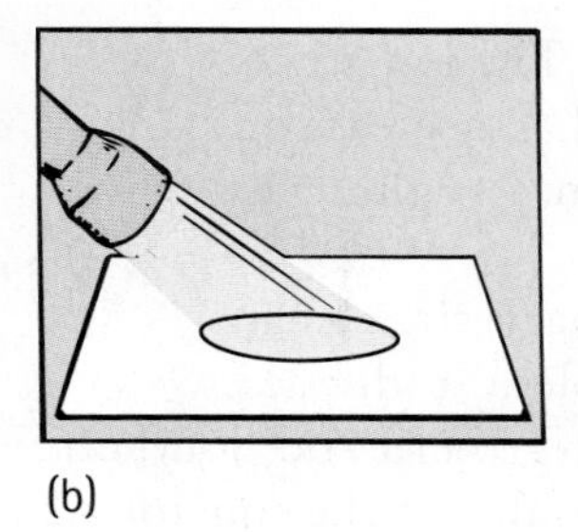
(b)

FIGURE 26.20
This simple demonstration shows why temperature depends on the angle at which the Sun's rays strike Earth. (a) When the flashlight is held directly above at a right angle to the surface, the beam of light produces a bright circle. (b) When the light is held at an angle, the light beam elongates and forms an oval. The same amount of light energy as before is spread out over a larger area, so its intensity is reduced. High noon in equatorial regions is like a vertically held flashlight. High noon at higher latitudes is like the flashlight held at an angle.

hot nor very cold. Average monthly temperatures vary between 10°C and 18°C (between 50°F and 64°F) throughout the year, with much seasonal variation. The *polar climate zone* is located at high latitudes. Monthly temperatures remain lower than 10°C (50°F) there even in the warmest months.

Why does the temperature depend on latitude? Latitude affects temperature because the Sun's rays strike different latitudes at different angles. While the Sun's rays strike regions near the equator dead-on perpendicularly, the rays hit polar latitudes at a steep angle. Figure 26.20 demonstrates this. As you can see, when rays of light strike a surface at right angles, the rays strike the smallest possible area—they are most concentrated. That area has maximum **solar intensity**, or solar radiation per area. Equatorial regions have high temperatures because they experience maximum solar intensity. Solar intensity is at a minimum at the polar latitudes, so they are cold places.

LEARNING OBJECTIVE
In words and diagrams, show how Earth's tilted axis of rotation produces the seasons.

26.4 Earth's Tilted Axis—The Seasons

EXPLAIN THIS Why is it summer in the northern latitudes at the same time that it's winter in the southern latitudes?

As Earth orbits the Sun each year, the seasons change. When it is summer, the days are longer and the temperatures are higher. Winters have shorter days and colder weather. In some places, seasonal changes are dramatic; in other locations, the changes are mild. Why do these seasonal changes occur?

The different conditions that we see in the winter, spring, fall, and summer are caused by changes in solar intensity. The tilting of Earth's axis of rotation comes into play here. Recall that Earth rotates about an imaginary line through the North and South Poles. This imaginary line is called Earth's *axis of rotation* (Figure 26.21). The axis is not straight up and down. Instead, it is tilted at an

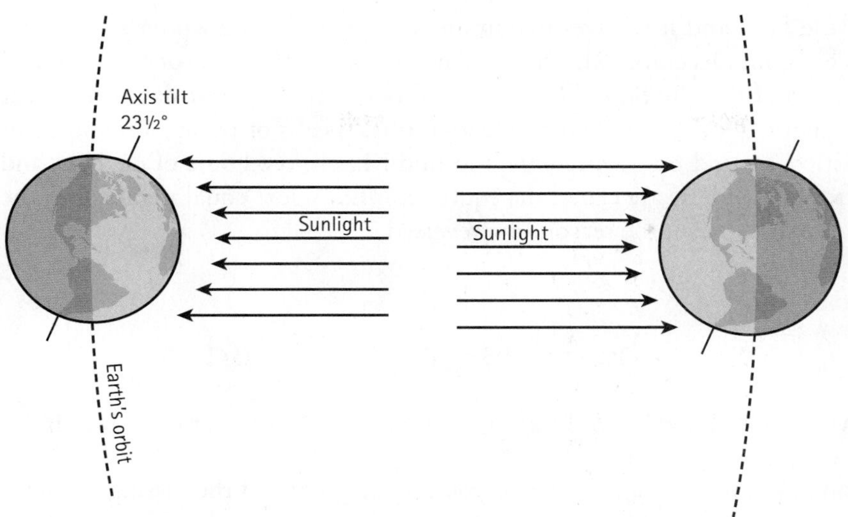

FIGURE 26.21
Earth spins about its axis of rotation while it orbits the Sun. The axis of rotation is not straight up and down—instead, it is tilted at an angle of $23\frac{1}{2}°$. When it is summer in the Northern Hemisphere, the Northern Hemisphere tilts toward the Sun.

FIGURE 26.22
When it is winter in the Northern Hemisphere, the Northern Hemisphere tilts away from the Sun. In the winter, it gets dark earlier and the weather is cooler.

angle of $23\frac{1}{2}°$. Because Earth's axis is tilted, different points on Earth face the Sun differently as Earth travels around the Sun.

During the summer in the Northern Hemisphere, the Northern Hemisphere is tilted toward the Sun. You can see this in Figure 26.21. In the winter, the Northern Hemisphere is tilted away from the Sun (Figure 26.22). The hemisphere that is tilted toward the Sun receives the Sun's energy more directly and has longer days. This combination of more direct rays and longer days creates the warmth of summer.

As Figure 26.23 shows, the summer solstice occurs in the Northern Hemisphere on or about June 21. This is the longest day of the year as well as the first day of summer. On this day, the Northern Hemisphere is pointed directly

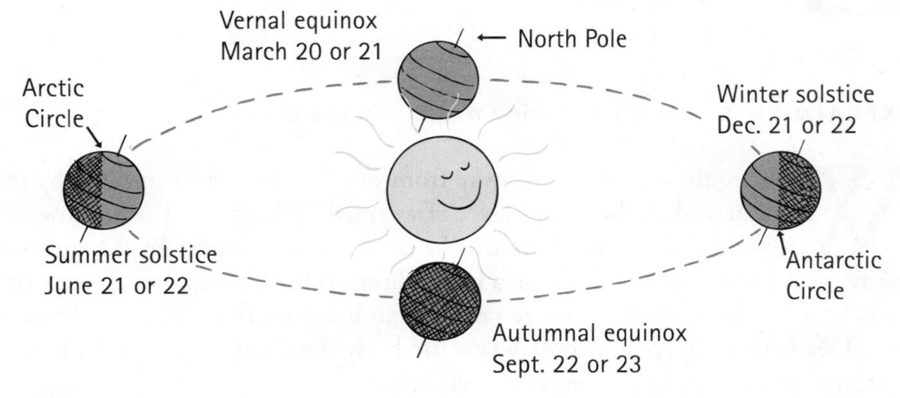

FIGURE 26.23
The tilt of Earth and the corresponding differences in the intensity of solar energy produce the yearly cycle of the seasons. Earth follows an elliptical (oval) path around the Sun and is farthest from the Sun when the Northern Hemisphere experiences summer. So, the angle of the Sun's rays, not the distance from the Sun, is most responsible for Earth's surface temperatures.

MasteringPhysics®
TUTORIAL: Seasons

toward the Sun, and it receives maximum solar energy. The winter solstice occurs on or about December 21. On this day, the Northern Hemisphere is pointed directly away from the Sun. The winter solstice is the shortest day of the year and the first day of winter. Halfway between the peaks of the winter and summer solstice, around mid-September and mid-March, the hours of daylight and night are equal. These are called the equinoxes (Latin for "equal nights"). In the Southern Hemisphere, the seasons are reversed.

LEARNING OBJECTIVE
Understand what wind is and how it relates to air pressure.

26.5 Flow of the Atmosphere—Wind

EXPLAIN THIS How is wind like air rushing from a tire that has a leak in it?

If you have ever seen a tire go flat, you probably noticed the hissing sound of air escaping. If your hand was in front of the tire, you also felt air rushing out of the tire. Air moves naturally from a region of high pressure to a region of low pressure. A punctured tire is just one example of this. A popped balloon is another example. Wind is a third example of the same phenomenon. **Wind** is air flowing horizontally from a region of high pressure to a region of lower pressure.

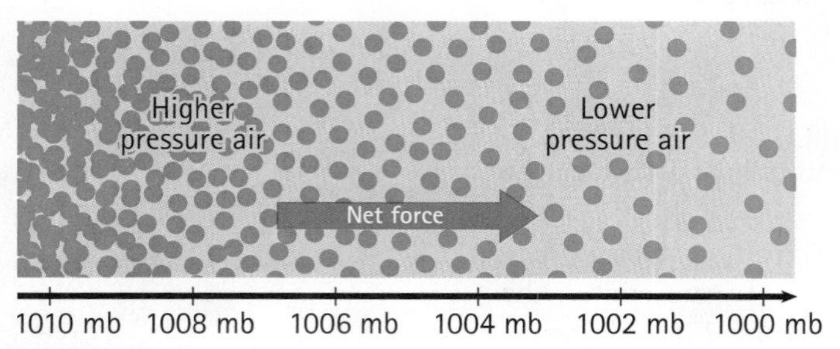

FIGURE 26.24
Air moves from a region of higher pressure to a region of lower pressure.

Winds are caused by differences in air pressure, and bigger differences in air pressure produce stronger, faster winds. A gentle breeze, one that just rustles the grass, is air moving at 10–15 km/h. A wind strong enough to rattle power lines moves at 40–50 km/h. A hurricane blows more than 120 km/h. Winds are named according to the direction from which they blow, so a "30-km/h northwesterly wind" blows from the northwest at a speed of 30 km/h.

LEARNING OBJECTIVE
Obtain and interpret the data provided on a wind chill chart.

Wind Chill

EXPLAIN THIS Explain why wind makes you feel cooler.

Warm air that transfers away from our bodies is held in place by our hair and clothes. This layer of warmed air keeps us warm. However, wind constantly blows this layer of warm air away. In the summer, the wind keeps us cool and this is a good thing. But, if it's cold outside, losing body heat can be dangerous. Extreme cold can bring on frostbite and hypothermia. Death from freezing occurs when the body does not have enough heat to perform the chemical reactions that sustain life.

The National Weather Service publishes a Wind Chill Index, a table that shows how air temperature and wind combine to produce a temperature that your body actually feels. A version of the Wind Chill Index is shown in Figure 26.25.

Wind (mph) \ Temperature (°F) Calm	40	35	30	25	20	15	10	5	0	-5	-10	-15	-20	-25	-30	-35	-40	-45
5	36	31	25	19	13	7	1	-5	-11	-16	-22	-28	-34	-40	-46	-52	-57	-63
10	34	27	21	15	9	3	-4	-10	-16	-22	-28	-35	-41	-47	-53	-59	-66	-72
15	32	25	19	13	6	0	-7	-13	-19	-26	-32	-39	-45	-51	-58	-64	-71	-77
20	30	24	17	11	4	-2	-9	-15	-22	-29	-35	-42	-48	-55	-61	-68	-74	-81
25	29	23	16	9	3	-4	-11	-17	-24	-31	-37	-44	-51	-58	-64	-71	-78	-84
30	28	22	15	8	1	-5	-12	-19	-26	-33	-39	-46	-53	-60	-67	-73	-80	-87
35	28	21	14	7	0	-7	-14	-21	-27	-34	-41	-48	-55	-62	-69	-76	-82	-89
40	27	20	13	6	-1	-8	-15	-22	-29	-36	-43	-50	-57	-64	-71	-78	-84	-91
45	26	19	12	5	-2	-9	-16	-23	-30	-37	-44	-51	-58	-65	-72	-79	-86	-93
50	26	19	12	4	-3	-10	-17	-24	-31	-38	-45	-52	-60	-67	-74	-81	-88	-95
55	25	18	11	4	-3	-11	-18	-25	-32	-39	-46	-54	-61	-68	-75	-82	-89	-97
60	25	17	10	3	-4	-11	-19	-26	-33	-40	-48	-55	-62	-69	-76	-84	-91	-98

Frostbite Times: 30 minutes; 10 minutes; 5 minutes

FIGURE 26.25
A wind chill chart. Wind speed and air temperature make the air feel colder to our bodies.

CHECK YOURSELF

1. **(a) If it is 40°F and there is a gentle, 5-mi/h breeze, what is the effective temperature your body experiences? (b) If the wind is blowing at 35 mi/h and it is 40°F, what is the effective temperature?**
2. **How long will it take for you to develop frostbite if it is 5°F and the wind is blowing at 30 mi/h?**

CHECK YOUR ANSWERS

1. (a) 36°F; (b) 28°F
2. 30 min

26.6 Local and Global Wind Patterns

LEARNING OBJECTIVE
Describe how and why temperature differences between polar and equatorial latitudes move the atmosphere.

EXPLAIN THIS Why do land breezes occur at night and sea breezes occur during the day?

There are two types of winds: *local winds* and *global winds.* Pressure differences over a small region create local winds. These winds blow short distances, and they can blow in any direction. Global winds, on the other hand, are part of a huge pattern of air circulation around the globe. Global winds blow consistently from the same direction. The wind where you live is a combination of local and global wind patterns.

Local Winds

Local temperature differences create wind where the geography varies—for example, where land meets water or where mountains meet valleys. Consider land and sea breezes (Figure 26.26). During the day, the land warms faster than the ocean because land has a lower specific heat capacity (specific heat capacity is "thermal inertia"—the resistance to a change in temperature). The hot air over the warmed land rises, creating an area of lower air pressure. The cooler,

During the day, a gentle, warm breeze moves from a valley floor upslope to mountain peaks. At night, a chilly breeze blows down the mountains as cold air sinks into the valley. The shapes of land features can produce local winds.

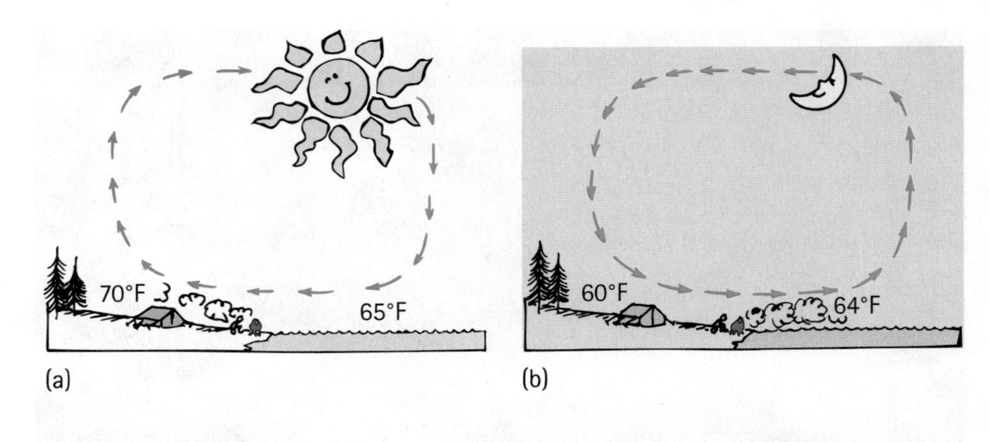

FIGURE 26.26
(a) Sea breeze. During the day, warm air above land rises, and cooler air moves in to replace it. (b) Land breeze. At night, the direction of airflow is reversed because then the water is warmer than the land.

higher-pressure air from over the ocean then blows from the sea to the land. This is a *sea breeze*. But at night, the land cools off faster than the sea, again because of its lower specific heat capacity relative to water. Cooler air descends over the land and creates an area of higher pressure. Wind blows from the land to the sea. This is a *land breeze*. You may have experienced a refreshing sea breeze if you have spent time along the coast in the summer, or even near the shores of large lakes.

Global Winds

UNIFYING CONCEPT

- ***Convection***
Section 6.9

All winds are created by differences in air pressure. Global winds are produced by planet-scale pressure differences that occur because of unequal heating of Earth at the equator and poles. At equatorial latitudes, Earth absorbs the Sun's rays directly. Air is warmed efficiently there, and it rises. The air flows toward the cooler poles but sinks back to Earth when it becomes too cold and dense to stay aloft. This creates an area of low pressure near the equator. At the poles, sinking cold air creates areas of high pressure. The unequal pressure causes the air to rise and sink, circulating in global patterns called *wind belts*. These wind belts are convection cells, and they move air in predictable patterns. As Figure 26.27 shows, there are six major convection cells: between the equator and latitudes of 30° north and south, between 30° and 60° north and south, and between 60° north and south and the poles. Global winds are the streams of horizontally moving air within these convection cells. Like land and sea breezes, global winds affect temperature patterns by transporting warming or cooling air.

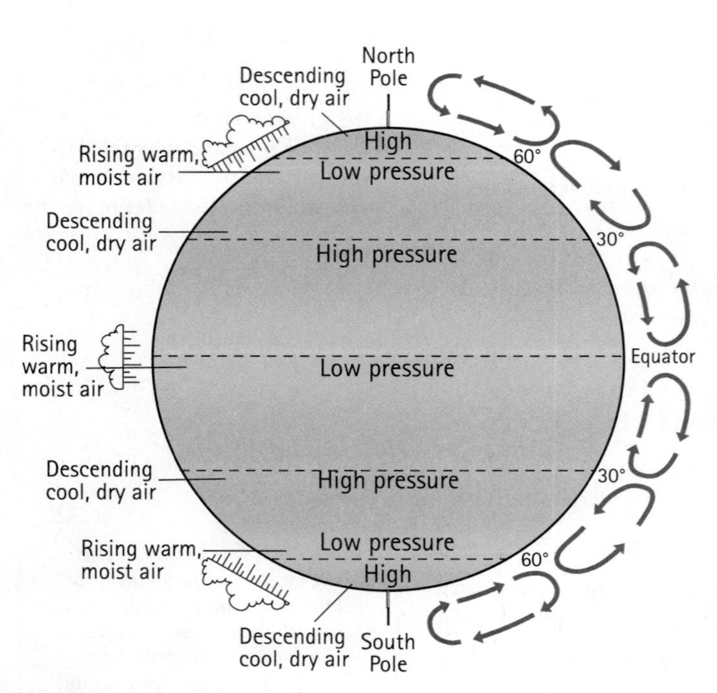

FIGURE 26.27
Earth is surrounded by a set of six major convection cells.

The global winds that flow between 0° and 30° latitude are called the *trade winds*. The trade winds are named for their role in propelling trading ships centuries ago. The winds between the equator and 30°N latitude are called the *northeast trade winds* because European sailing ships used them as they traveled toward the New World and the coast of the British Colonies. Then the sailors caught the global winds called the "westerlies" back to Europe. The *westerlies*, as Figure 26.28 shows, blow from west to east from 30° to 60° latitudes. Note that winds are named according to the direction from which they blow. The *easterlies* blow from east to west at the polar latitudes between 60° and 90° north and south.

Near the equator, where the trade winds die, there is a zone of still air. Sailors of long ago cursed the equatorial seas as their ships stalled for lack of wind, and they named the area the *doldrums*. Sailors were also frequently stalled in the calm air that occurs where convection cells meet at 30° north and south latitudes. According to legend, as food and water supplies dwindled, horses on board were either eaten or cast overboard to conserve fresh water. As a result, these regions became known as the *horse latitudes*.

There is another type of global wind—*jet streams*. Jet streams are narrow belts of high-speed wind that blow in the upper troposphere and lower stratosphere. Jet streams don't follow consistent paths around Earth as other global winds do. Instead, their altitude and latitude vary. Jet streams move very fast. In fact, jet streams were discovered by World War II pilots of high-altitude military aircraft. The pilots observed that they weren't flying forward because they were flying against winds blowing just as fast in the opposite direction!

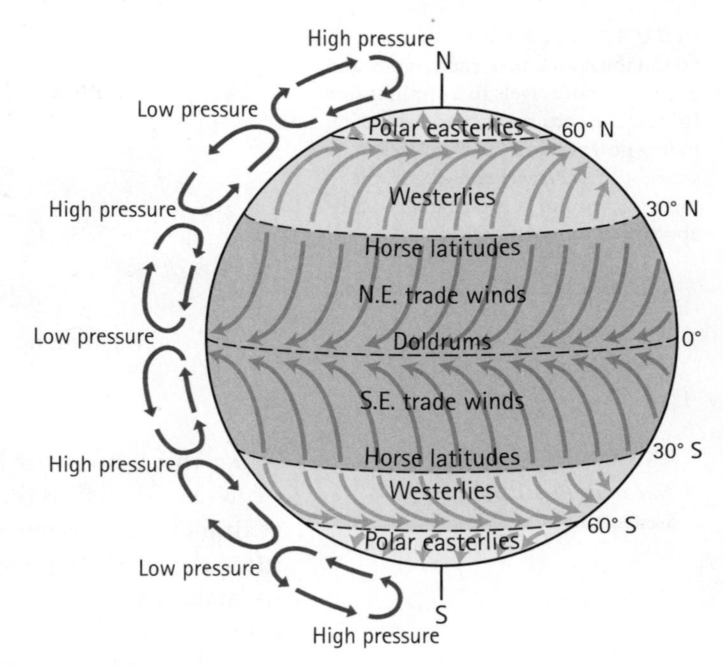

FIGURE 26.28
Note the locations of the major global winds: the trade winds, the westerlies, and the easterlies.

CHECK YOURSELF

1. **How are land and sea breezes alike? How are they different?**
2. **A migrating bird flies west to east across the United States and then back again. Which trip is easier for the bird? Why?**

CHECK YOUR ANSWERS

1. Both land breezes and sea breezes are local winds that occur where land is next to a large body of water. They are different in that a sea breeze blows from the sea to the land during the day, and a land breeze blows from the land to the sea at night.
2. The westerlies blow from west to east across the United States. This helps the bird when it flies from west to east but hinders the bird when it flies from east to west.

Why does Earth's atmosphere break up into six convection cells, instead of four, or nine, or 101? The number depends on how fast a planet rotates. Venus rotates slowly and has only two cells in each of its hemispheres. If Earth were to spin faster, like Jupiter, we would experience more air-circulation cells and faster mixing of our atmosphere.

Integrated Science 26D
PHYSICS

The Coriolis Effect

LEARNING OBJECTIVE
Describe the pattern of global air circulation that arises due to the Coriolis effect.

EXPLAIN THIS Why do global winds appear to blow along curved paths when in fact the air molecules that make them up are traveling in straight-line paths?

If Earth didn't rotate, the global winds would blow straight in a north–south direction. High-altitude winds would blow from the equator to the poles, while low-altitude winds would blow from the poles back to the equator.

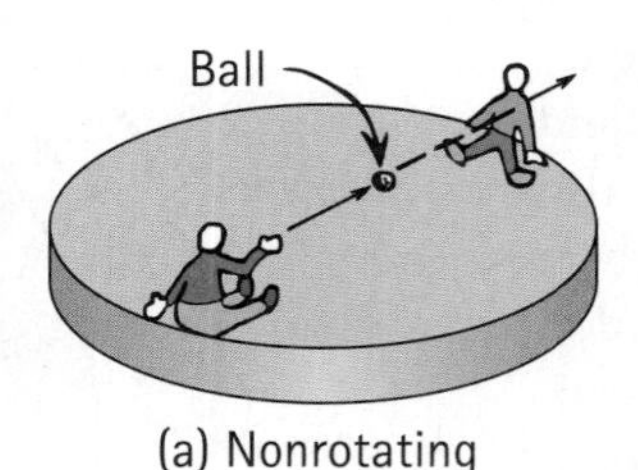

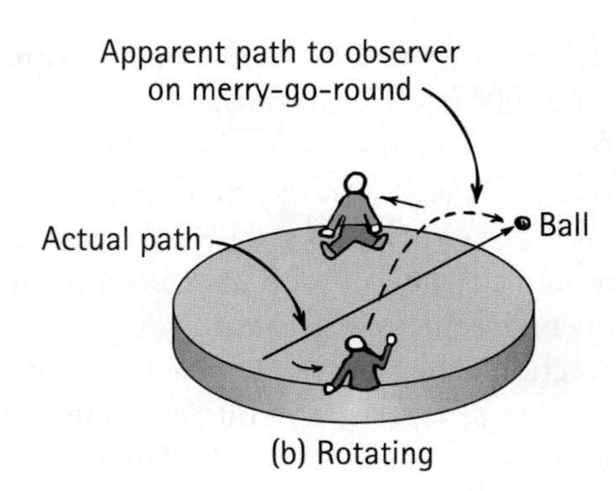

FIGURE 26.29
(a) On the nonrotating merry-go-round, a thrown ball travels in a straight line. (b) On the counterclockwise-rotating merry-go-round, the ball moves in a straight line. However, because the merry-go-round is rotating, the ball appears to move to the right.

UNIFYING CONCEPT

- *Newton's First Law*
  Section 3.1

But this is not what happens because Earth *does* rotate. The **Coriolis effect** ("core-ee-OH-lis") is the tendency of moving bodies not attached to Earth (such as air molecules) to move to their right in the Northern Hemisphere and to their left in the Southern Hemisphere. Because of the Coriolis effect, air in the planet's six major convection cells turns in the directions indicated by the arrows in Figure 26.28. The Coriolis effect varies according to the speed of the wind: The faster the wind, the more it appears to turn.

To see how the Coriolis effect works, consider an analogy. Think of Earth as a large merry-go-round rotating in a counterclockwise direction (in the same direction Earth spins, as viewed from the North Pole). You and a friend are playing catch on this merry-go-round. When you throw the ball to your friend, the circular movement of the merry-go-round affects the direction the ball appears to travel. Although the ball really travels in a straight-line path, it appears to curve to the right, as shown in Figure 26.29.

CHECK YOURSELF

1. **Why don't global winds flow in a straight direction between the equator and poles?**
2. **Which way does wind appear to turn in the Northern Hemisphere? In the Southern Hemisphere?**

CHECK YOUR ANSWERS

1. Winds turn because moving air molecules are subject to the Coriolis effect. Molecules of air are not attached to Earth. Because Earth rotates beneath them, air molecules turn relative to Earth.
2. In the Northern Hemisphere, wind appears to turn to the right. In the Southern Hemisphere, it appears to turn to the left.

LEARNING OBJECTIVE
Describe how global winds create surface currents.

26.7 Ocean Currents Distribute Heat

EXPLAIN THIS How does warm ocean water from the equator warm up the air temperature in Norway and Great Britain?

In addition to latitude, elevation, atmospheric gases, and wind effects, another important factor in atmospheric temperature is ocean currents.

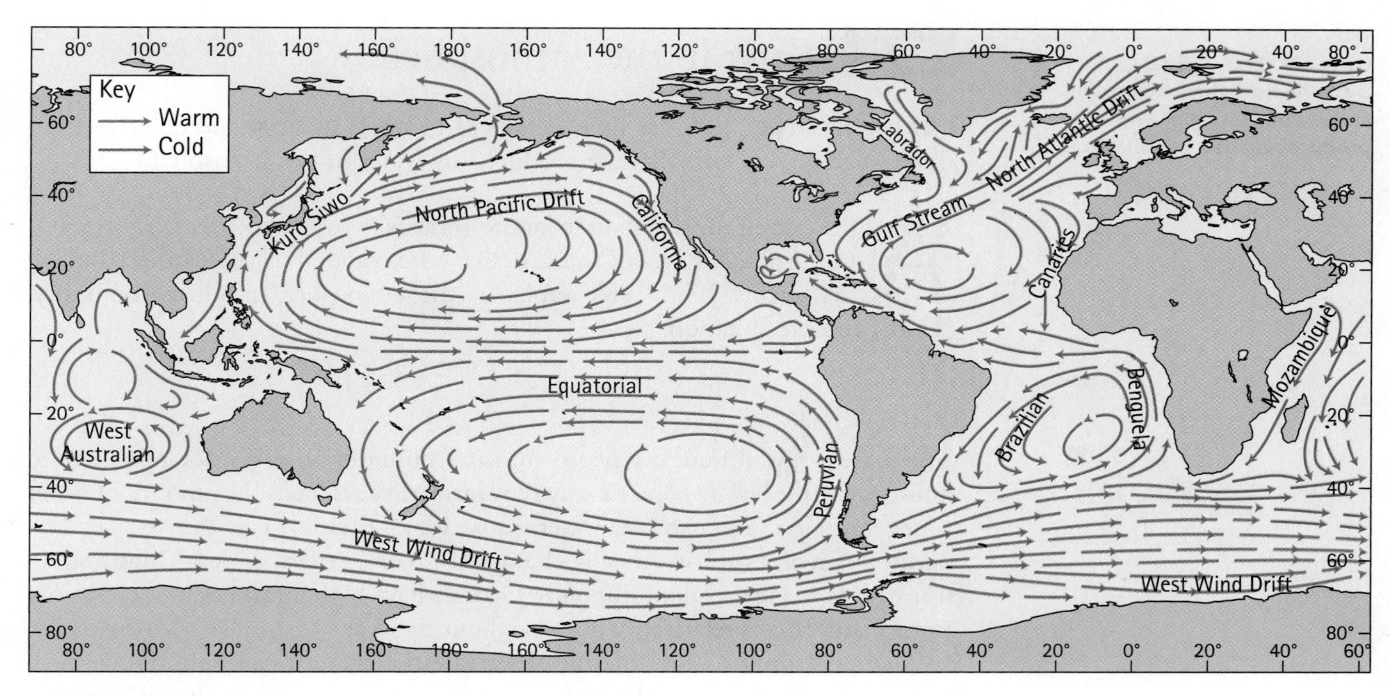

FIGURE 26.30
The ocean's major surface currents, shown by the arrows, closely match the pattern of global winds shown in Figure 26.31, except in regions where landmasses disrupt the water's flow.

Ocean water, like the fluid atmosphere, mixes and circulates. The oceans contain *currents*, streams of water that move relative to the larger ocean. **Surface currents** are usually created by global winds pushing water in the directions they blow. Compare Figure 26.30 closely with Figure 26.31. Notice that the surface currents match up quite well with the global winds if you take into account the interference of landmasses.

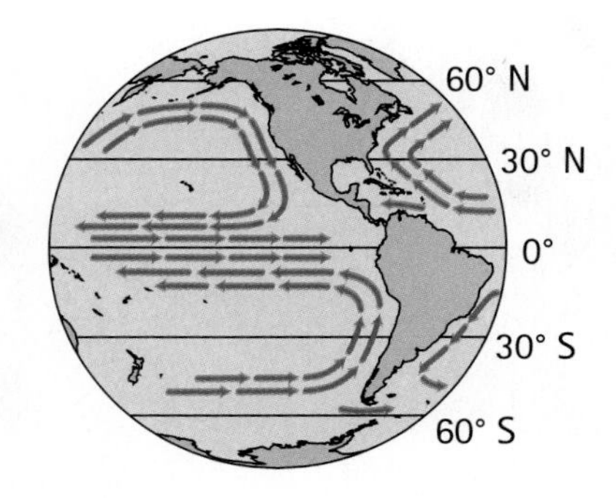

FIGURE 26.31
The world's major surface currents.

Surface currents, like the global winds, are turned and twisted along their path by Earth's rotation. Because of the Coriolis effect and other factors, surface water currents tend to form giant circular flow patterns called *gyres* (Figure 26.31).

Surface currents play a vital role in redistributing heat. The uneven heating of Earth's surface warms equatorial waters but leaves polar waters chilly. Surface currents redistribute heat by transporting waters far from their source. This circulation of the ocean affects not only water temperature but also atmospheric temperature. Consider the example of the Gulf Stream and how it warms Northern Europe.

Heat is transported in the North Atlantic Ocean as warm equatorial water flows westward into and around the Gulf of Mexico, then northward along the eastern coast of the United States. This warm-water current is called the *Gulf Stream*. As the Gulf Stream flows northward along the eastern coast of North America, global winds steer it toward Europe. When this warm water reaches the coasts of Great Britain and Norway, it cools and releases heat energy to the air. The heated air is carried by westerly winds over the lands of Great Britain and Norway.*

*Jet streams high in the atmosphere are also a major contributor to the warming of Europe.

LEARNING OBJECTIVE
Describe how the evaporation and condensation of water produce weather conditions.

26.8 Water in the Atmosphere

EXPLAIN THIS There are six elements of weather. Identify and describe the three weather elements that are produced by water in the atmosphere.

All air, even the driest air, contains some water vapor. *Water vapor* is water in the gaseous phase. You can't see water vapor, but you can feel and see its effects. The water vapor in the atmosphere produces humidity, clouds, and precipitation.

Humidity

There are a few different ways to measure humidity and it's not necessary to know them all. For now, just know that **humidity** refers to the amount of water vapor in the air. You can't see water vapor; it's an invisible gas. But you can certainly feel it. Humid weather is characteristic of Hawaii all year. Many states from Texas to New York to Illinois to Georgia have high humidity in the summer time. California is relatively dry throughout the year.

There is a limit to the amount of water vapor that air can hold. How much water vapor resides in air depends on the air's temperature. As the air temperature rises, the volume of the water vapor that it can hold increases too. That's why Earth's most humid locations are in the tropics and the dryest air is in the cold polar regions. Table 26.2 shows exactly how much air vapor can reside in air at various temperatures.

Weather reports describe humidity in terms of relative humidity. **Relative humidity** is the ratio of the amount of water vapor currently in the air to the largest amount of water vapor that it is possible for the air to hold at that temperature. You could say that relative humidity is a measurement of how full of water vapor the air is at a given temperature. Stated as an equation,

$$\text{Relative humidity} = \left(\frac{\text{water-vapor content}}{\text{water-vapor capacity}}\right) \times 100\%$$

Thus, if air contains 7 grams of water on a 25°C day, the relative humidity is expressed as 7/14, or 50%, because the water content of the air is half the amount that it can hold at a temperature of 25°C. Make sense?

The relative humidity of the air changes, of course, if the amount of water vapor in it changes or if the air's temperature changes. For instance, if the air temperature rises from 25°C to 40°C but 7 g of water vapor remains in the

TABLE 26.2 MAXIMUM AMOUNT OF WATER VAPOR THAT A 1-KG MASS OF AIR CAN HOLD AT DIFFERENT TEMPERATURES

Temperature (°C)	Grams of Water Vapor per kg of Air
−30	0.3
−20	0.75
−10	2
7	3.5
20	14
30	26.5
40	47

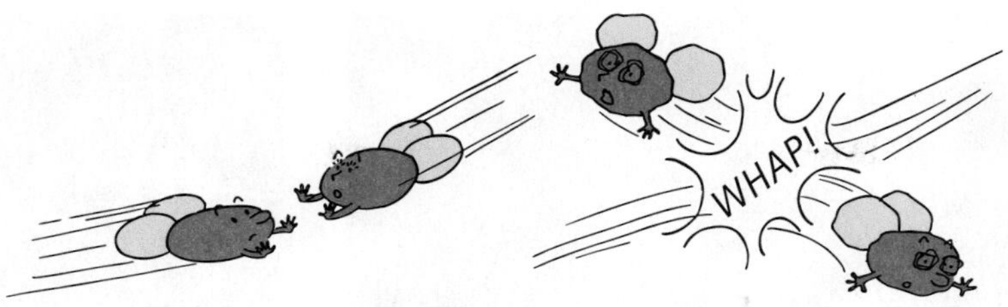

Fast-moving H_2O molecules rebound upon collision

Slow-moving H_2O molecules condense upon collision

FIGURE 26.32
Condensation of water molecules. Condensation occurs when water vapor reaches its dew point. At this temperature, the water molecules are moving slowly enough that they condense, rather than rebound, upon impact.

air, then the relative humidity decreases to (7 g/26.7 g) $\times$ 100% = 26% even though the actual quantity of water vapor in the air is unchanged.

When air contains as much water as it can possibly hold, the air is *saturated*. This is equivalent to saying that saturated air has 100% relative humidity. As an air mass cools, it can hold less and less water vapor before becoming saturated. If it cools down enough, the air mass reaches a point at which the water vapor present is the amount required to saturate the air at the lower temperature. This temperature, the temperature at which saturation occurs, is the **dew point**. Condensation occurs when the dew point is reached. Suppose, for example, that a certain mass of unsaturated air at 30°C is cooled to 15°C and that the air is saturated at that temperature. The dew point of this air is then 15°C; if the air is cooled further, its capacity for holding water vapor would be exceeded and the excess vapor would condense (Figure 26.32).

When perspiration evaporates from your skin, your perspiration takes the energy it needs to change state from liquid to gas from your skin. This energy, water's *latent heat of vaporization*, thus leaves your body, and you become cooler as a result. In humid weather, perspiration doesn't cool you as well as it does in dry air because evaporation is slowed. This is why humid air feels so much hotter than dry air of the same temperature.

Water vapor condenses high in the atmosphere to form clouds. It condenses close to the ground as well. When condensation in the air occurs near Earth's surface, we call it *dew*, *frost*, or *fog*. On cool, clear nights, objects near the ground cool down more rapidly than the surrounding air. As the air cools to its dew point, it cannot hold as much water vapor as it could when it was warmer. Water from the now-saturated air condenses on any available surface—a twig, a blade of grass, or the windshield of a car. We often call this type of condensation early-morning dew. When the dew point is at or below freezing, we have frost. When a large mass of air cools to its dew point, its relative humidity approaches 100%. And this produces a cloud. This usually happens high in the atmosphere where the air is cold and so can't hold much water vapor. But it can also happen closer to the ground—and then we have *fog* (Figure 26.33).

Clouds

As air rises, it expands and cools. As the air cools, water molecules move more slowly and condensation occurs. If there are larger and slower-moving particles or

(a) (b) (c) (d)

FIGURE 26.33
Water vapor condenses from a mass of air when the air's dew point is exceeded. This results in the formation of (a) clouds, (b) fog, (c) dew, or (d) frost.

ions present in the air, water vapor condenses on these particles, and this creates a **cloud**—a visible collection of minute water droplets or tiny ice crystals.

In 1803, British weather observer Luke Howard was the first to classify clouds according to their shapes. He recognized three cloud forms: *cirrus* (Latin for "curl"); *cumulus* (L., "piled up"); and *stratus* (L., "spread out") (Figure 26.34). If you have searched for these basic shapes in the sky, however, you know that clouds do not usually come in these simple forms; instead, they usually occur as composites of these forms. For this reason, Howard's simple classification has been modified so that clouds are generally classified by their form as well as their altitude. This results in ten basic cloud types, each of which belongs to one of the four major cloud groups (Table 26.3).

High clouds form at altitudes above 6000 meters and are denoted by the prefix *cirro-*. The air at this elevation is quite cold and dry, so clouds this high are made up almost entirely of ice crystals. The most common high clouds are *cirrus* clouds, which are blown by strong, high-altitude winds into their classic wispy shapes, such as the "mare's tail" and "artist's brush." *Cirrocumulus* clouds are arrays of rounded white puffs that rarely cover more than a small patch of the sky (Figure 26.35a). Small ripples and a wavy appearance make the cirrocumulus

(a)

(b)

(c)

FIGURE 26.34
Clouds classified by shape: (a) "wispy" cirrus clouds, (b) "piled-up" cumulus clouds, and (c) "spread-out" stratus clouds.

TABLE 26.3 THE FOUR MAJOR CLOUD GROUPS

1. High Clouds (above 6000 m)	2. Middle Clouds (2000–6000 m)	3. Low Clouds (below 2000 m)	4. Clouds of Vertical Development
Cirrus	Altostratus	Stratus	Cumulus
Cirrostratus	Altocumulus	Stratocumulus	Cumulonimbus
Cirrocumulus	Nimbostratus		

clouds look like the scaled body of a mackerel. Hence, cirrocumulus clouds make up what is often referred to as a "mackerel sky." Cirrus clouds usually indicate fair weather.

Middle clouds are denoted by the prefix *alto-*. They are made up of water droplets and, when temperature allows, ice crystals. *Altostratus* clouds are gray to blue-gray, and they often cover the sky for hundreds of square kilometers (Figure 26.35b). Altostratus clouds are often so thick that they diffuse incoming sunlight to the extent that objects on the ground don't produce shadows. Altostratus clouds often form before a storm. So, if you can't see your shadow when you're going on a picnic, cancel!

Low clouds are most often composed of water droplets, but they can contain ice crystals in colder climates. *Stratus* clouds tend to be the lowest of the low clouds. They are uniformly gray, and they cover the whole sky, often resembling a high fog. Stratus clouds are not associated with falling precipitation, but they sometimes generate a light drizzle or mist. *Nimbostratus* clouds are dark and foreboding (Figure 26.35c). They are a wet-looking cloud layer associated with rain and snow.

Clouds of vertical development do not fit into any of the three height categories. These clouds typically have their bases at low altitudes, but they reach up into the middle or high altitudes. Although cumulus clouds are fair-weather clouds, they are called "clouds of vertical development" because they can grow dramatically. Vertically moving air currents in them can produce a towering cloud with an anvil head—a *cumulonimbus* cloud (Figure 26.35d). Cumulonimbus clouds, popularly called "thunderheads," may produce heavy rain showers, thunder and lightning, and hail.

(a)

(b)

(c)

(d)

FIGURE 26.35
Clouds from each of the major cloud groups: (a) cirrocumulus, (b) altostratus, (c) nimbostratus, and (d) cumulonimbus.

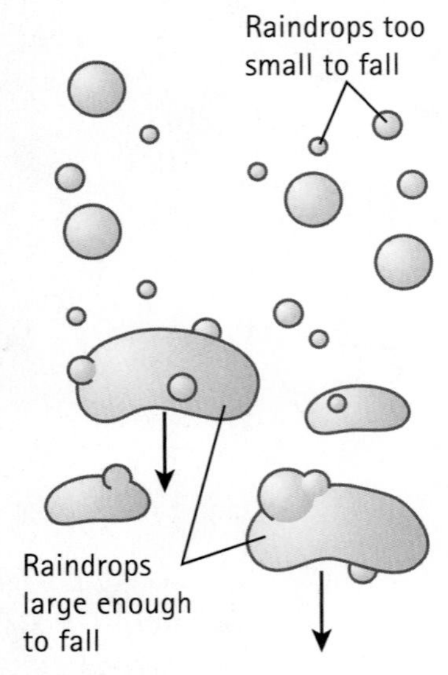

FIGURE 26.36
Raindrops form when tiny water droplets in clouds collide and join together. Before a raindrop falls, it must grow to at least 100 times the size of average water droplets in clouds.

FIGURE 26.37
Snowflakes are built of many microscopic ice crystals.

Precipitation

Precipitation is water in the liquid or solid phase that returns to Earth's surface from the atmosphere. It is an essential part of the water cycle, which means that precipitation controls everything from the weathering of rock to the sustenance of Earth's organisms. Precipitation occurs as snow and rain, but there are other less common forms too: mist, drizzle, hail, freezing rain, and sleet (in the form of solid ice crystals). Mist and drizzle generally fall from stratus clouds. Rain falls from nimbostratus and cumulonimbus clouds. Air temperature controls what kind of precipitation occurs.

Rain, the most common kind of precipitation, is liquid water that falls from clouds through the air at above-freezing temperature. Rain falls when the tiny water droplets in clouds collide and join together, becoming so big and heavy that they can no longer stay suspended in the air (Figure 26.36).

Here's a challenge question: If clouds are made of water droplets and ice crystals, which are denser than air, why don't we see them sinking to the ground? The gravitational force pulling a droplet down *is* enough to make it fall. So, why don't all droplets in clouds fall to the ground? The answer has to do with updrafts—rising air currents. A typical cumulus cloud has an updraft speed of at least 1 meter per second, which is faster than the droplet can fall. Small droplets don't fall to Earth because they are supported by upward-rising air. Raindrops, on the other hand, are huge compared with typical cloud droplets. A drop of rain big enough to reach the ground contains about a million times more water than a cloud droplet. Raindrops fall faster than most updrafts can push upward.

Snow is composed of tiny, six-sided ice crystals. It forms when water vapor goes directly to the solid state at temperatures below freezing. Snowflakes grow into large and fanciful shapes by bumping into one another as they fall. When the air is very cold, snow is usually light and fluffy. When the air is near freezing, snow is sticky and wet. All snowflakes are six-sided, like the ice crystals from which they form.

Hail is sometimes very damaging. Hail forms only in cumulonimbus, or thunder, clouds. Particles of hail are *hailstones*—layered balls of ice. Hailstones form when drops of water freeze in a cloud but strong upward air currents keep them aloft for a prolonged time. As more drops of water collide with the hailstone and freeze, new layers are added on. Hailstones as big as golf balls can easily dent cars and hurt people they hit!

(a)

(b)

FIGURE 26.38
(a) Hail damage. (b) Hailstones are made up of layers of ice.

CHECK YOURSELF

1. **Why is air in the tropics more humid than arctic air?**
2. **Why don't clouds fall to the ground?**
3. **In what way are snowflakes similar in shape to the ice crystals they contain?**

CHECK YOUR ANSWERS

1. Tropical air is often humid because it contains water that has evaporated from warm equatorial waters. Arctic air is drier because there is less evaporation in cold Arctic air.
2. Rising air keeps clouds from falling. (When clouds are at ground level, we call them fog.)
3. Both the flakes and the crystals have six sides.

26.9 Changing Weather—Air Masses, Fronts, and Cyclones

LEARNING OBJECTIVE
Compare and contrast: cyclones versus anticyclones, warm fronts versus cold fronts, and maritime versus continental air masses.

EXPLAIN THIS Describe what happens when a cold air mass meets a warm air mass.

When a meteorologist says a high-pressure system is moving to your area, what kind of weather can you expect? Is it time to grab an umbrella? To understand changing weather and weather forecasts, you need to know about the movement of *air masses*, including *fronts* and *cyclones*.

Air Masses

An **air mass** is a large pool of air that has similar temperature and moisture characteristics throughout. Air masses form when a huge body of air stays in one place for a long enough time to take on the properties of that region. For example, an air mass forming over the coast of Florida will be warm and humid because this region is warm and wet. Air masses aren't completely stationary, however, because global winds drive them. As they move, they carry their weather conditions with them. As this occurs, the properties of an air mass change as it interacts with new environments.

Air masses are classified according to the latitude over which they form and whether they form over land or water. Figure 26.39 shows the air masses that have the biggest influence on weather in North America. Each air mass is represented by a two-letter code. The first letter represents moisture content, and the second letter represents temperature. Thus, the air mass developing over Florida is mT for *maritime tropical.* Maritime tropical air masses bring moist warm air. Air masses labeled (cP) are *continental polar* air masses. These carry cold, dry air from Alaska and Canada.

Fronts

UNIFYING CONCEPT
- ***Convection***
Section 6.9

Recall the directions of global winds from Section 26.6. Remember that convection cells move warm air away from the equator toward the poles; cold air moves

FIGURE 26.39
The air masses that affect weather in North America, the regions they form over, and their typical directions of travel. Air masses keep their temperature and moisture characteristics as they move.

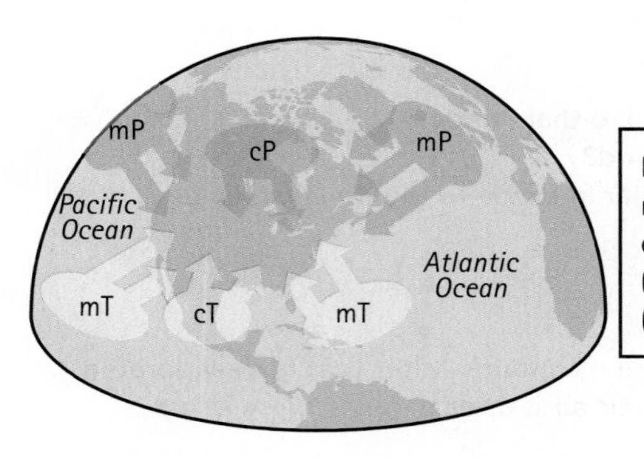

Legend:
maritime (m) – forms over water and is wet
continental (c) – forms over land and is dry
(T) – forms over the tropics and is warm
(P) – forms over the polar regions and is cold

UNIFYING CONCEPT
- *Density*
Section 2.3

from the poles to the equator. These movements set up global wind patterns that move air masses.

Just as global winds collide in the middle latitudes, so do air masses. And when air masses collide, they don't readily mix. Instead, air masses meet at a boundary that is called a *weather front*, or simply a **front**.

At a front, the less dense, warmer air mass flows upward over the denser, cooler air. The warmer air always rises vertically above the cooler air, but the horizontal movements of the air masses vary. Sometimes, the colder, denser air mass advances into and displaces a warm air mass. In this case, the contact zone between the air masses is called a *cold front* (Figure 26.40). But warm air can move into territory occupied by a cold air mass instead. In this case, the zone of contact is called a *warm front* (Figure 26.41). If neither of the air masses is moving, the contact zone is called a *stationary front*. An *occluded front* occurs when fast-moving cold air forces warm air up and traps it between two cold air masses. The warm air starts to cool and water vapor condenses. At the front—a region perhaps a few kilometers wide—there are often clouds, rain, winds, and storms.

You can generally predict that a cold front is moving in if you observe high cirrus clouds followed by alto and then stratus clouds, a shift in wind direction, and a drop in air pressure. As cold air moves into and displaces a warm air mass at the cold front, the warm air is forced upward, rises, and cools. If it is moist enough, the rising air can condense to form cumulonimbus clouds that produce thunderstorms with heavy showers and gusty winds. After the cold front passes, the lifted warm air cools and sinks, so pressure rises, and the rain stops. Except for a few fair-weather cumulus clouds, the skies clear and there is calm after the storm.

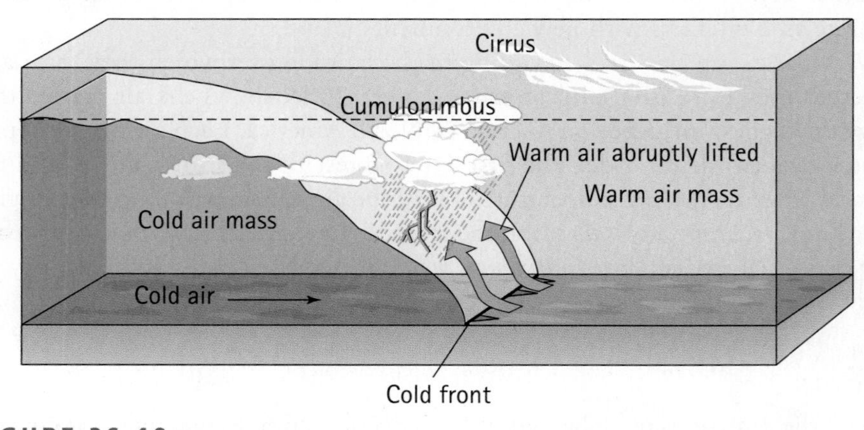

FIGURE 26.40
A cold front forms when a cold air mass moves into a warm air mass. The cold air forces the warm air upward, where it condenses to form clouds.

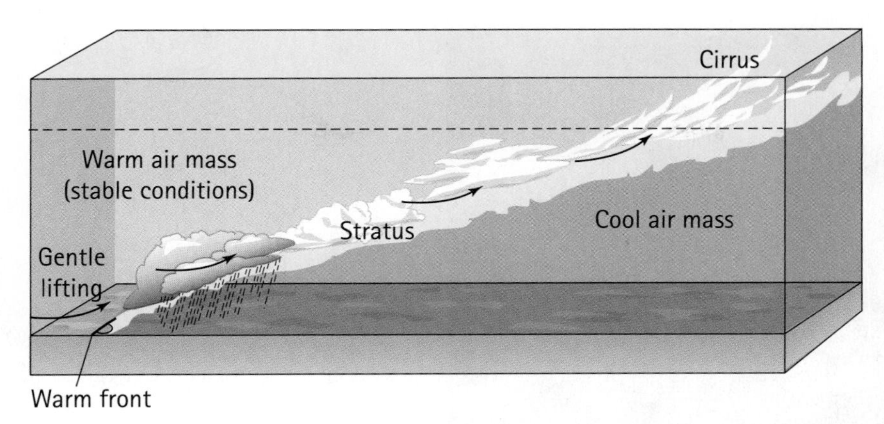

FIGURE 26.41
A warm front forms when a warm air mass moves into and displaces a cold air mass. The less-dense, warmer air rides up and over the colder, denser air.

A warm front may be coming if you see cirrus clouds in the sky followed by stratus clouds. A warm front separates an advancing mass of warm air from colder air ahead. As with a cold front, the different densities of cold and warm air discourage mixing of air masses so the warm air flows upward above the cold air mass. As the warm air rises and cools, water vapor in it may reach its dew point. Condensation and clouds ensue with light drizzle or fog along the front in summer and sleet at the front in the winter. Behind the front, the air is warm and the clouds scatter.

MasteringPhysics®
ACTIVITY: Cold Front
ACTIVITY: Warm Front

UNIFYING CONCEPT

• ***Density***
Section 2.3

Cyclones

Air masses belong to gigantic *weather systems*, organized systems of moving air spanning a thousand or more square kilometers. Weather systems are organized around a center of either high pressure or low pressure.

A system of low pressure is called a **cyclone** (a *low-pressure system* or simply a *low*). Cyclones are associated with rough weather (Figure 26.42). A cyclone's center is the region of lowest pressure, so air flows into it in a spiral, but then the air is forced upward.

The rising air can produce clouds and precipitation. In the summer, cyclones often bring rain and thunderstorms. In the winter, thunderstorms and snow are

(a)

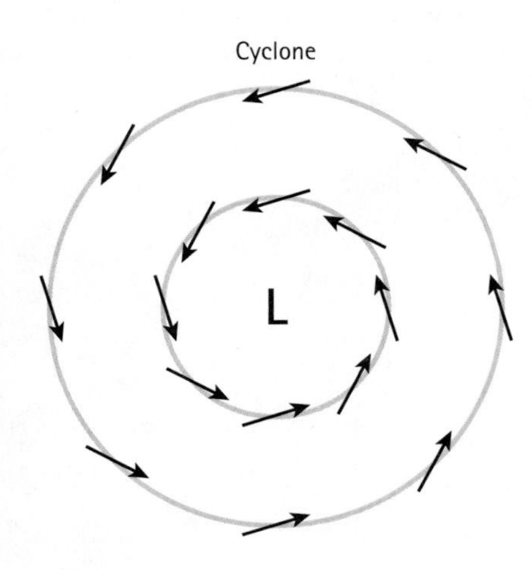

(b)

FIGURE 26.42
(a) A satellite image of a cyclone.
(b) In the Northern Hemisphere, winds blow counterclockwise into the cyclone. Air forced in and upward brings rough weather.

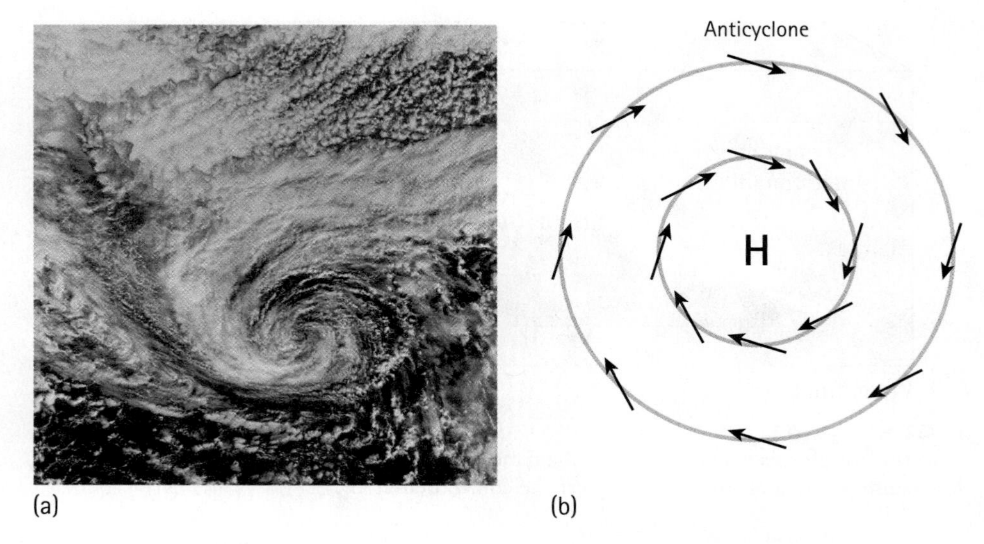

FIGURE 26.43
(a) A satellite image of an anticyclone. (b) In the Northern Hemisphere, winds blow clockwise away from the anticyclone. Anticyclones, or high-pressure systems, are associated with fair weather.

likely. Because of the Coriolis effect, the winds in a cyclone move counterclockwise in the Northern Hemisphere and clockwise in the Southern Hemisphere.

An **anticyclone** (a *high-pressure system* or simply a *high*) is an area of high pressure (Figure 26.43). Air moves from high pressure to low pressure, so the air moves outward from an anticyclone as well as downward. This sinking motion leads to generally fair skies and no precipitation near the high. (Does this make sense to you? Remember that clouds form where warm, moist air *rises* then condenses; *sinking* air is not the recipe for clouds and storms.) The Coriolis effect turns the moving air around a high-pressure center so that anticyclonic winds blow clockwise around a high in the Northern Hemisphere and counterclockwise around a high in the Southern Hemisphere.

Meteorologists plot the positions of fronts and pressure systems on weather maps to predict the weather (Figure 26.44). On a weather map, a high-pressure system is denoted with an *H*, and a low-pressure system is denoted with an *L*. The surface position of a warm front is shown as a line with semicircles on the side of cooler air. The surface position of a cold front is denoted by a line with triangles extending into the region of warmer air.

The weather map here, like the ones you see on television, is highly simplified. Weather forecasters build much more complex maps with computers in order to plot weather systems and calculate how fast the systems are moving. After all, it helps everyone—from farmers to astronauts—to know the weather!

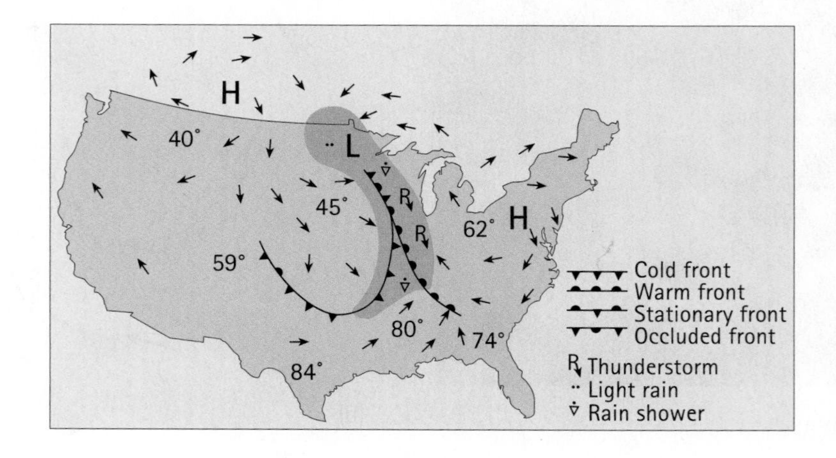

FIGURE 26.44
Weather maps show pressure systems and fronts. In this weather map, a cold front moves southward and a warm front moves eastward.

SUMMARY OF TERMS (KNOWLEDGE)

Air mass A large pool of air that has similar temperature and moisture characteristics throughout.

Anticyclone (*high-pressure system* or *high*) A weather system organized around an area of high pressure.

Atmosphere The thin envelope of gases, tiny solid particles, and liquid droplets surrounding the solid planet.

Atmospheric pressure The weight of all the air molecules in the atmosphere pressing down on Earth's surface.

Climate The general pattern of weather that occurs over a period of years.

Cloud A place in the atmosphere that contains enough water droplets or ice crystals to be visible.

Coriolis effect The tendency of moving bodies not attached to Earth (such as air molecules) to move to their right in the Northern Hemisphere and to their left in the Southern Hemisphere.

Cyclone (*low-pressure system* or *low*) A weather system organized around an area of low pressure.

Dew point The temperature at which the air becomes saturated.

Front (*weather front*) A boundary along which air masses meet.

Greenhouse effect The process by which certain gases warm the atmosphere by trapping infrared radiation that is emitted by Earth's surface.

Humidity The amount of water vapor in the air.

Precipitation Water in the liquid or solid state that returns to Earth's surface from the atmosphere.

Relative humidity The ratio of the water vapor actually in the air to the maximum water vapor the air can hold at that temperature.

Solar intensity Solar radiation per area.

Solar radiation Electromagnetic energy given off by the Sun.

Surface current A wind-driven, shallow ocean current.

Terrestrial radiation Infrared radiation emitted by Earth's surface.

Weather The state of the atmosphere at a particular time and place.

Wind Air flowing horizontally from an area of high pressure to an area of lower pressure.

READING CHECK (COMPREHENSION)

26.1 The Atmosphere

1. What is the difference between weather and climate?
2. What are the six elements of weather?
3. What two types of molecules make up more than 99% of the atmosphere?

26.2 The Structure of the Atmosphere

4. Name the four layers of the atmosphere.
5. Why does the stratosphere have a high temperature? Why does the mesosphere have a low temperature?

26.3 Temperature Depends on Latitude

6. Is San Francisco in the Northern or Southern Hemisphere? What is its approximate latitude?
7. How does latitude relate to temperature? Use the term *climate zone* in your answer.

26.4 Earth's Tilted Axis—The Seasons

8. Why are summer days warmer than winter days (on average)?
9. What is the winter solstice? The summer solstice? What are the equinoxes?

26.5 Flow of the Atmosphere—Wind

10. What is wind? What causes wind?
11. In what direction does wind blow?

26.6 Local and Global Wind Patterns

12. Give an example of a local wind pattern. Give an example of a global wind pattern.
13. How did the trade winds help traders in colonial America?

26.7 Ocean Currents Distribute Heat

14. What drives surface currents?
15. Do surface currents affect water temperature, air temperature, or both?

26.8 Water in the Atmosphere

16. What happens to the water vapor in the air when the air becomes saturated and its dew point is reached?
17. How do clouds form?
18. How do hailstones form?

26.9 Changing Weather—Air Masses, Fronts, and Cyclones

19. You hear of "low-pressure systems" on TV weather reports. What are two other names for a low-pressure system? What kind of weather are low-pressure systems associated with?
20. How do weather fronts develop?

THINK INTEGRATED SCIENCE

26A—Atmospheric Pressure

21. Why don't we feel atmospheric pressure?
22. Why is air pressure highest at sea level?

26B—Solar Radiation and the Greenhouse Effect Drive Global Temperature

23. About how much of solar radiation is intercepted by Earth? What portion of the solar energy intercepted by Earth is absorbed by the ground?
24. In what way is the greenhouse effect like a florist's greenhouse?
25. What is a greenhouse gas? Give three examples.
26. Distinguish between the natural greenhouse effect and global warming.

26C—Wind Chill

27. Why does wind generally make you feel cooler?
28. What is wind chill?

26D—The Coriolis Effect

29. What does the Coriolis effect do to the direction of global winds and ocean currents?
30. How is a ball tossed on a merry-go-round like the molecules in the atmosphere?

THINK AND DO (HANDS-ON APPLICATION)

31. Make a cloud—quickly. Simply open a can of soda and notice the cloud that forms for a moment over the opening. Why does the cloud form? The explanation is that bubbly drinks are held under pressure in cans. When you open the can, moist air escapes and quickly expands. Expanding air cools—to its dew point, as you can see.
32. Make a model of the greenhouse effect. You will need two nonmercury thermometers, one large plastic bag such as a produce bag, one small plastic bag such as a sandwich bag, and two twist ties. Place one thermometer in the small bag. Blow into the bag to inflate it, and tie it with a twist tie. Put the inflated bag inside the larger plastic bag. Inflate the large bag, and seal it with a twist tie. Now lay the bags containing the thermometer on a sunny windowsill or outside. Lay the unbagged thermometer next to it. Observe the temperature reading of both thermometers after 30 minutes. Explain your results.
33. The next time it rains, hold a sheet of black construction paper flat in the rain for a moment to catch a few raindrops. When you go inside, look at the marks the drops made. Are all raindrops the same size?

THINK AND COMPARE (ANALYSIS)

34. Suppose it's July 1. Rank the following locations in terms of the solar intensity they receive, from most to least: northern Africa, northern Canada, New York.
35. Rank the layers of the atmosphere in terms of density, from most dense to least dense.
36. Rank clouds classified as *alto*, *cirro*, and *stratus* according to their altitude, from highest to lowest.

THINK AND SOLVE (MATHEMATICAL APPLICATION)

37. Consider a house at sea level that has 2000 square feet of floor area. Show that the total force that the air inside the house exerts upward on the ceiling is 4.2×10^6 lb.
38. Suppose the air holds 75% of the water that it can hold at a given temperature before it becomes saturated. What is the relative humidity?
39. A tornado passes in front of a building, causing the pressure to drop there by 15% in 1 second. If a door on the side of the building is 6.5 feet tall and 3 feet wide, show that the net force on the closed door is 6200 lb.
40. At 50°C, the maximum amount of water vapor in the air is 9 g/m^3. If the relative humidity is 40%, show that the mass of water vapor in 1 m^3 of air is 3.6 g/m^3. [Hint: Relative humidity = (water-vapor content)/(water-vapor capacity) $\times$ 100%.]

THINK AND EXPLAIN (SYNTHESIS)

41. The summer solstice is the longest day of the year. Does the summer solstice occur because Earth is closest to the Sun on this day? Explain.
42. Sometimes the atmosphere's temperature doesn't decrease with altitude in a normal way. Instead, warmer air sits on top of colder air in a temperature inversion. What effect can this have on local air pollution?
43. Clouds are made of tiny water droplets or ice crystals. These are heavier than air, so why don't clouds fall to the ground?
44. According to the graph, what is average atmospheric pressure at sea level? Why does atmospheric pressure change with altitude as shown in this graph?

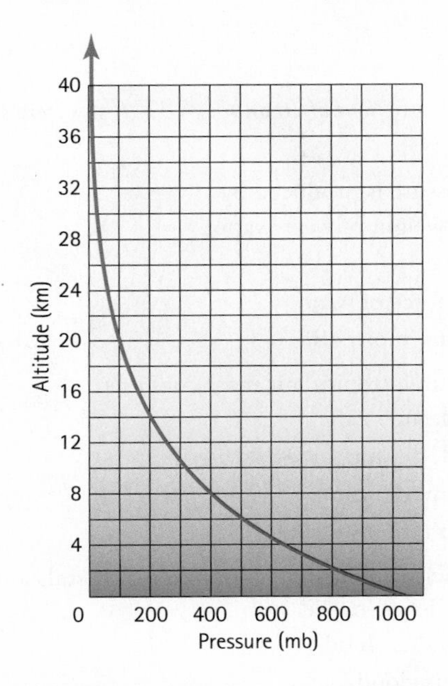

45. Identify the clouds shown in the photo. How are they formed? What kind of weather are they associated with?

46. After a day of skiing in the mountains, you decide to go indoors and get a cup of hot cocoa. As you enter the ski lodge, your eyeglasses fog up. Why?
47. Air is warmed and rises at the equator and then cools and sinks at the poles, as shown in the figure. Is this an accurate picture of the global circulation of air? Explain why or why not.
48. Why does warm, moist air blowing over cold water result in fog?
49. Why does atmospheric pressure typically drop before a storm comes?
50. What does convection in Earth's atmosphere produce? What does convection in Earth's mantle produce?
51. Why does the East Coast of the United States experience wider seasonal variation than the West Coast, even though both areas have oceans along their margins?
52. What role does the Sun play in ocean currents?
53. Explain why your ears pop when you climb to higher altitudes.
54. What is ozone? How might Earth be affected if there were no ozone layer in the stratosphere?
55. What are Earth's major climate zones? Describe each one.
56. Design an experiment to test the air pressure at different altitudes. What do you expect to observe?
57. When you go to school in the morning, the weather is sunny and warm. By lunchtime, it is cool, windy, and rainy. The weather front shown here has moved in. What kind of weather front is this—a cold front or a warm front?

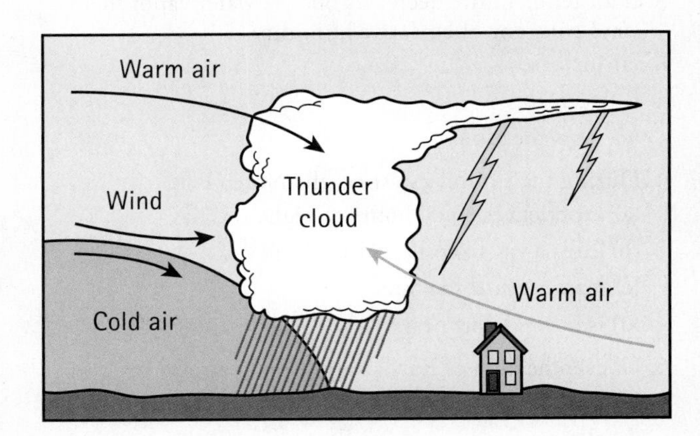

58. San Francisco, California, has mild winters. Springfield, Missouri, has cold winters. Yet San Francisco is farther north than Springfield. Is it surprising to learn then that Springfield is colder in the winter? Why? Why is San Francisco warm in the winter months?
59. At sea level, the air is about 23% oxygen. At the top of Mt. Everest, the air is still about 23% oxygen. So, why do almost all mountain climbers need to bring extra bottled oxygen to survive?

60. Identify which of these factors directly affect air temperature: altitude, latitude, proximity of water, ocean currents, the Coriolis effect.
61. As the air temperature decreases, does the relative humidity increase, decrease, or stay the same?
62. Why is it important that mountain climbers wear sunglasses and sunblock even when temperatures are below freezing?

THINK AND DISCUSS (EVALUATION)

63. When is the greenhouse effect a good thing? A bad thing?
64. The highest dew point ever recorded was 95°F, recorded in Saudi Arabia. Was the air humid or dry at that time? Explain your reasoning.
65. Do we see radiation emitted by the Earth? Do we feel it? Explain.

READINESS ASSURANCE TEST (RAT)

If you have a good handle on this chapter, if you really do, then you should be able to score 7 out of 10 on this RAT. If you score less than 7, you need to study further before moving on.

Choose the BEST answer to each of the following questions:

1. Earth's lower atmosphere is kept warm by
   (a) solar radiation.
   (b) terrestrial radiation.
   (c) shortwave radiation.
2. The wind blows because of
   (a) air pressure differences between different locations.
   (b) Earth's rotation.
   (c) the greenhouse effect.
   (d) differences in latitude.
3. If air temperature decreases but the water vapor in air stays constant, the relative humidity
   (a) increases.
   (b) decreases.
   (c) stays the same.
4. During the summer solstice, the North Pole
   (a) experiences equal hours of night and day.
   (b) leans away from the Sun.
   (c) leans toward the Sun.
   (d) is in total darkness.
5. Uneven heating of Earth's atmosphere produces
   (a) rain.
   (b) clouds.
   (c) air pressure.
   (d) wind.
6. The Gulf Stream redistributes heat from the Gulf of Mexico to
   (a) the North American coast, Great Britain, and Norway.
   (b) South America.
   (c) Japan.
   (d) Antarctica.
7. Air pressure is produced by
   (a) the weight of water vapor.
   (b) the weight of air.
   (c) the force of wind.
   (d) warm, moist air.
8. A maritime tropical air mass contains
   (a) cold, moist air.
   (b) cold, dry air.
   (c) warm, dry air.
   (d) warm, moist air.
9. High, wispy clouds that contain ice crystals and often bring rain are called
   (a) cumulus clouds.
   (b) cirrus clouds.
   (c) stratus clouds.
   (d) cumulonimbus clouds.
10. The relative humidity of air at its dew-point temperature is
    (a) 0%.
    (b) 10%.
    (c) 50%.
    (d) 100%.

Answers to RAT

1. b, 2. a, 3. a, 4. b, 5. d, 6. a, 7. b, 8. d, 9. b, 10. d

27

CHAPTER 27

Environmental Geology

WHEN DOES a natural event become a natural disaster? Floods, hurricanes, heat waves, volcanic eruptions, landslides, tsunami, etc. are natural events that become disasters when people are in the way. Sometimes it's possible to prevent natural disasters by staying out of the way. But sometimes, violent natural events occur where people live. In these cases, what can be done? Which violent natural events are preventable and how is this accomplished? When dangerous events can't be prevented, how can their impacts be mitigated? Disaster prediction, prevention, community preparedness, and emergency response all require state-of-the-art knowledge of environmental geology. In this chapter, we learn about the science that explains how environmental hazards work. We investigate climate change in particular depth because it's an environmental hazard of immediate and serious concern. This chapter is about serious science—the kind that saves lives!

LEARNING OBJECTIVE
Describe in detail what happens before, during, and after an earthquake.

27.1 Earthquakes

EXPLAIN THIS What happens when the friction holding huge blocks of rock in place is overcome by tectonic forces?

It was just after 5:00 PM on a warm October afternoon. Suzanne Lyons, one of the authors of this book, was leaving work in Oakland, California. Just as she walked through her office door, the ground started shaking. Furniture toppled. As she stood in the swaying doorway, she saw waves roll through the pavement like swells on the open ocean. Was this "the big one"—the massive earthquake all Californians are warned about? No, but it was serious. It was the Loma Prieta earthquake, which shook the San Francisco Bay Area in 1989. Freeways collapsed, buildings were destroyed, 63 people died, and $6 billion of property damage was done.

UNIFYING CONCEPT
- *The Law of Conservation of Energy* **Section 4.10**
- *Friction* **Section 2.8**
- *Waves* **Section 8.1**
- *Newton's Second Law* **Section 3.2**

An **earthquake**, as you learned in Chapter 22 on plate tectonics, happens when blocks of rock suddenly shift or break. Typically, earthquakes occur at plate boundaries. As the rocky plates grind beside one another, collide, or pull apart, their motion is usually steady and slow. In this case, the motion is called *creep*. But sometimes friction locks huge sections of rock together. Then, as force is applied to the rock, it cannot move freely and becomes compressed or stretched. When the force that is pushing or pulling the rock exceeds the friction that is holding it in place, or exceeds the strength of the rock, the stressed rock suddenly breaks loose and slips. Deformed rock snaps back elastically to its original position—a process called *elastic rebound* (Figure 27.2). The rock then releases its stored elastic energy in the form of **seismic waves**—mechanical waves that propagate through Earth. Geologists who study seismic waves and earthquakes are called *seismologists*.

When you bend a twig to the point that it breaks, you know that you have to put some energy into it. Imagine how much energy it takes to bend a rock. Now

FIGURE 27.1
This section of freeway collapsed during the 1989 Loma Prieta earthquake in California.

imagine the energy required to bend a slab of rock several kilometers long! This is the energy stored in rock before an earthquake. Where does this energy come from? The theory of *plate tectonics* provides the answer. Tectonic plates are huge sections of Earth's crust and upper mantle. Although they move slowly, their immense mass means that they carry enormous kinetic energy. (Recall that kinetic energy is proportional to mass times velocity squared: KE = $\frac{1}{2}mv^2$.) When huge zones of rock stick together because the plates are pushing against one another and locking, the rock's kinetic energy is converted into potential energy—energy that's stored in the rock.

The place where an earthquake starts is called the *focus* of the earthquake (Figure 27.3). Seismic waves radiate in all directions from the focus like sound from a ringing bell. The point at Earth's surface directly above the focus is the **epicenter**. Earthquakes generally occur at *faults*. A fault is a fracture where rocks on either side have moved relative to one another. Usually motion along a fault is smooth and slow; it's rapid only during an earthquake. Faults mostly exist at plate boundaries where extreme tectonic stresses crack the crust. But faults can occur in the middle of a plate too. Faults vary in length from a few centimeters to hundreds of kilometers. The amount of displacement along faults also varies greatly, again in a range of a few centimeters to many kilometers.

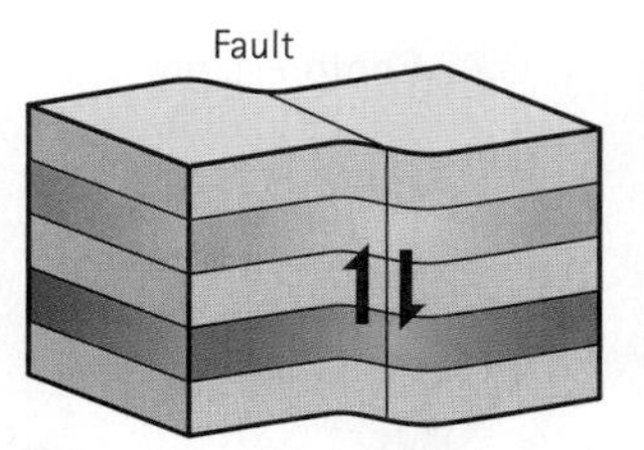

Sections of rock are forced in different directions though they are locked together by friction. The rock deforms.

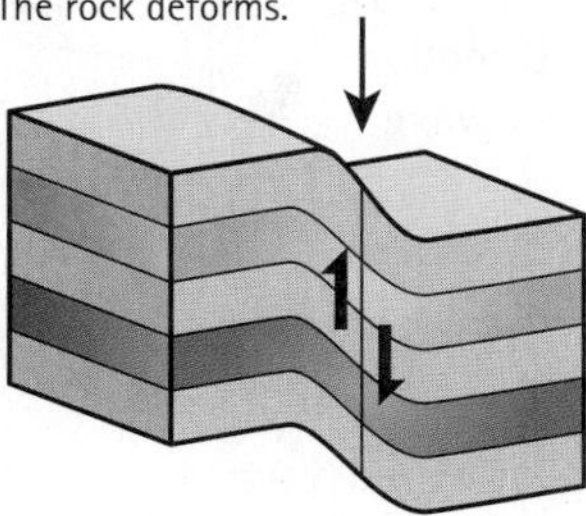

Stress builds in the rock as massive tectonic plates keep moving.

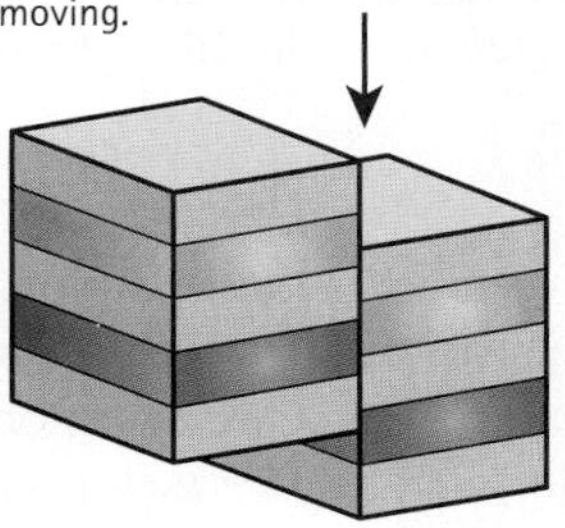

Rock slips along the fault and stress is released. Rocks rebound elastically–they return to their unstressed dimensions.

FIGURE 27.2
Huge blocks of rock moving under the influence of tectonic forces can become locked together through friction acting between them. When the force that is pushing or pulling the rocks exceeds the friction on them, the enormous zones of rock suddenly slip. The energy stored in them is released as seismic waves that travel through Earth and at its surface, shaking the ground.

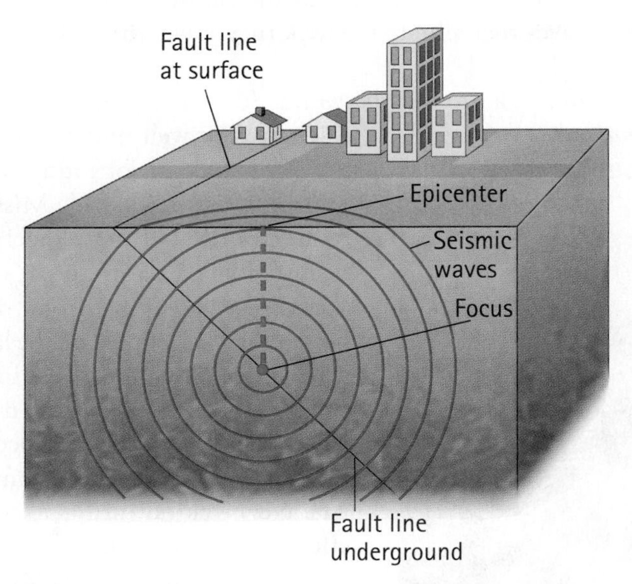

FIGURE 27.3
When rock suddenly moves in an earthquake, it is usually along a fault. The movement of rock generates seismic waves that radiate outward from the focus. The point at ground level directly above the focus is the epicenter.

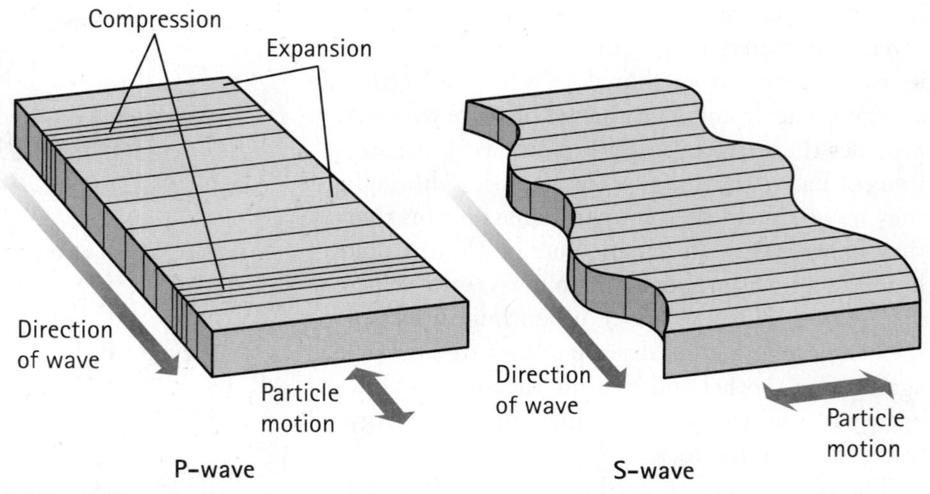

FIGURE 27.4
P-waves and S-waves are seismic waves known as *body waves*—they travel through Earth's interior. Notice that P-waves are longitudinal, whereas S-waves are transverse. They vibrate the rock they move through in different ways, as shown here. P-waves vibrate rock in the direction that they are traveling; S-waves vibrate the rock in a direction transverse (at right angles) to the direction of motion.

Faults are typically found in networks rather than as single cracks, and at shallow depths, usually within the top 10–15 km of Earth's crust.

Did you know that there are thousands of earthquakes every day? Most of them are minor—just the creeping of blocks of rock along a fault line. It's the sticking, bending, and slipping of giant blocks of rock that produce the big earthquakes that we can feel.

There are two types of seismic waves. *Body waves* travel through Earth's interior, and *surface waves* travel on Earth's surface. Because surface waves travel at ground level, these are the waves that damage surface structures. They can shake things up and down or side to side. Body waves are classified as either primary waves (P-waves) or secondary waves (S-waves). P-waves are longitudinal waves that move quickly through rock, water, or air, compressing and expanding it as they travel, like the squeezing and expanding of an accordion. S-waves are transverse body waves that vibrate the rock they travel through from side to side (Figure 27.4).

Earthquakes occur all around the world. Although most happen along plate boundaries, *intraplate* earthquakes (which are not well understood) occur away from plate boundaries. An infamous series of three major intraplate quakes occurred in New Madrid, Missouri, in the winter of 1811–1812. The first of them occurred just after midnight on December 16. People emerging from their shaking homes that night observed the land rolling in waves up to 1 m high. These intense surface waves opened deep cracks in the ground. The subsequent two quakes occurred within several weeks and were of similar intensity. The New Madrid earthquakes had a major effect on topography—new lakes were formed, the course of the Mississippi River was changed, more than 150,000 acres of forest were destroyed, and the contours of the land were reshaped over a wide area.

FIGURE 27.5
New Madrid, Missouri, was the site of more than 1000 small and large earthquakes in 1811–1812. The earthquakes ranged up to a Richter magnitude of 8.0, which is the highest recorded earthquake magnitude in the continental United States. The New Madrid earthquakes were intraplate quakes, which are rare in comparison to the earthquakes that occur at plate boundaries.

TABLE 27.1 EARTHQUAKE SEVERITY AND DEPTH RELATED TO TECTONIC SETTING (GENERAL TRENDS)

Plate Setting	Focus Depth and Earthquake Intensity
Diverging boundary	Shallow; mild
Subduction zone	Shallow to deep; high intensity
Continental–continental convergent boundary	Shallow to medium depth; moderate to high intensity
Transform boundary	Shallow; moderate to high intensity

Fortunately, because the region was sparsely settled at the time, there was little loss to human life and property.

Usually it's not the ground shaking itself but associated hazards that cause the most damage. Hazards associated with earthquakes include fire, landslides, tsunami, and *liquefaction*. What's liqufaction? When wet soil is shaken in an earthquake, soil particles are sometimes jarred apart and no longer locked together through friction. Then the soil is an unconsolidated slurry with insufficient strength to anchor buildings. Liquefaction is a major reason buildings topple over during earthquakes.

Intraplate quakes aside, most of the world's earthquakes—about 90%—are associated with movement along the boundaries of tectonic plates. In general, mild, shallow temblors occur at divergent plate boundaries. Earthquakes along transform plate boundaries are typically moderate. The strongest jolts mostly happen at convergent boundaries—especially at subduction zones where an oceanic plate slips under a plate that is less dense, as indicated in Table 27.1.

Most subduction occurs in the **Ring of Fire**, a horseshoe-shaped region of subduction zones that encircles much of the Pacific Ocean. About 80% of the world's big earthquakes occur in the Ring of Fire. Subduction produces volcanic activity as well as earthquakes. About 75% of the world's volcanoes are located in the Ring of Fire.

UNIFYING CONCEPT

● ***Friction***
Section 2.8

Earthquake Prediction

Can earthquakes be predicted? Yes, to some extent. Geologists can't predict exactly when or where an earthquake will occur, but they can measure the strain in rocks, and greater strain generally correlates with greater risk. Also, by mapping *seismic gaps*, scientists can identify which areas are most at risk within a timeframe of 30 to 100 years. Seismic gaps are stretches along major faults where little or no seismic activity has occurred for a long period of time even though the other regions of the fault are active. The dormant sections of the active fault—that is, the seismic gaps—are apparently dormant because they are locked by friction. In these areas, strain is building up rather than being released in small tremors. When the stress exceeds the friction holding the rock in place—snap! A large earthquake releases the strain. Figure 27.6 shows the locations of the major seismic gaps in the Western Hemisphere. Although the exact timing, location, and magnitude are impossible to predict, an understanding of seismic gaps allows scientists to prepare for future seismic activity. For example, scientists and policy officials used their knowledge of seismic gaps to prepare for the Loma Prieta earthquake by retrofitting buildings, enforcing new earthquake-safe building codes, and educating the local population about what to do when the quake hit.

FIGURE 27.6
This map shows the major seismic gaps of the Western Hemisphere (note the shaded bands). Seismic gaps are dormant sections of active faults where stress is presumed to be accumulating.

UNIFYING CONCEPT

- *The Law of Conservation of Energy* Section 4.10

CHECK YOURSELF

1. **It takes a lot of energy to deform huge slabs of rock and cause an earthquake. Where does this energy come from?**
2. **Where do most severe earthquakes occur? Why are they so common there?**
3. **Suppose you live along a major fault zone but no seismic activity has occurred in your location for hundreds of years, even though small earthquakes are common elsewhere along the fault. Should you be confident that you live in a very safe, earthquake-proof zone?**

CHECK YOUR ANSWERS

1. The energy that deforms rock comes from the motion of tectonic plates.
2. Most severe earthquakes occur along convergent plate boundaries where there is subduction. The Ring of Fire is where most of the subduction in the world is occurring.
3. No, you shouldn't feel especially safe. Check to see whether your location is in a seismic gap. If it is, you are in an especially hazardous portion of the fault. Practice earthquake preparedness!

Earthquake Measurement

The biggest earthquake to strike the United States occurred in Prince William Sound, Alaska, in 1964. It measured 9.2 on the Richter scale. By comparison, the Loma Prieta earthquake measured 7.0.

Geologists use a number of methods to determine the power of earthquakes. The *moment magnitude scale* measures *magnitude*, how much energy is released during an earthquake. This scale relies on actual measurements of a fault and the displacement of the plates along that fault. On this scale, each unit increase represents a 32-fold increase in the amount of energy released by the earthquake. In other words, a 4.0-magnitude earthquake has 32 times the energy of a 3.0-magnitude earthquake.

Although geologists use the moment magnitude scale, the media generally report the intensity of earthquakes in terms of the *Richter scale*. The Richter scale describes how much the ground shakes during an earthquake. On the Richter scale, each whole-number step represents a 10-fold increase in the amount of shaking that occurs. On the Richter scale, a 4.0 earthquake causes 10 times the shaking of a 3.0 earthquake.

Table 27.2 describes the intensity of earthquakes in terms of the Richter scale. Keep in mind, though, that two earthquakes assigned the same number may cause different amounts of damage. The amount of damage depends on whether the epicenter is near a populated area, how deep beneath the surface the focus is, the types of buildings located in the affected area, and other factors.

FIGURE 27.7
The standard seismometer records the shaking of the ground. Large-amplitude waves correspond to big shakes.

The moment magnitude scale is a more precise tool than the Richter scale, although both measure the size of an earthquake with a similar scale from 0 to 10.

The instrument used to measure the movement of Earth during an earthquake is called a *seismometer* (Figure 27.7). The stronger the earthquake, the harder the shaking and the greater the amplitude of the wave the seismometer records.

TABLE 27.2 THE RICHTER SCALE

Intensity	Description	Effects
Less than 2.0	Microquake	Can be recorded by a seismometer
2.0–2.9	Minor	Potentially detectable by average person
3.0–3.9	Minor	May be felt near the epicenter; rarely causes damage
4.0–4.9	Light	Strong shaking and rattling
5.0–5.9	Moderate	Possible structural damage
6.0–6.9	Strong	Widespread damage
7.0–7.9	Major	Can cause serious damage over a large area; approximately 20 per year worldwide
8.0–8.9	Great	Can cause severe damage hundreds of miles from the epicenter; approximately 1 per year worldwide
9.0–9.9	Catastrophic	Devastating across areas thousands of miles in diameter; recorded once every 20 years on average

27.2 Tsunami

LEARNING OBJECTIVE
Explain what causes tsunami and why tsunami are so destructive.

EXPLAIN THIS What is the best way to prevent loss of life due to tsunami?

Occasionally, a strong earthquake causes a *tsunami*, which is Japanese for "harbor wave." A massive subduction earthquake in the Ring of Fire on December 26, 2004, triggered a catastrophic tsunami that killed more than 200,000 people in 14 countries. The earthquake focus was under the Indian Ocean, off the coast of Sumatra, where two plates meet at a subduction zone. The earthquake's magnitude was 9.1 on the moment magnitude scale. Although the quake itself was damaging, the **tsunami**—a huge seismic sea wave—was even more destructive than the quake.

UNIFYING CONCEPT
- ***Waves*** **Section 8.1**

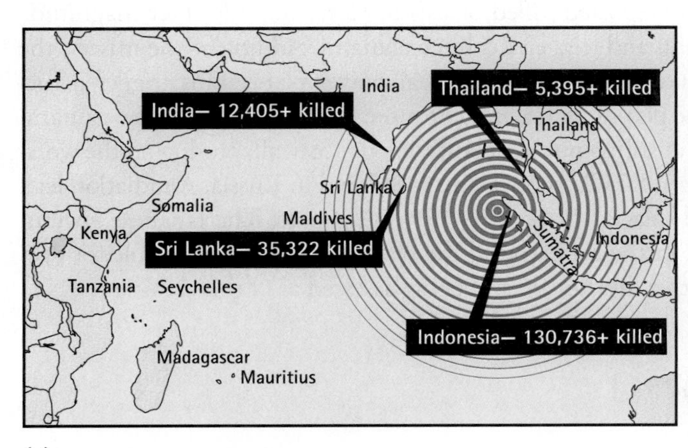

(a)

(b)

FIGURE 27.8
(a) The December 26, 2004, tsunami began with a strong subduction earthquake off the coast of Sumatra. (b) The tsunami strikes a coastline in Sri Lanka.

TABLE 27.3	CHARACTERISTICS OF A TSUNAMI VERSUS A TYPICAL OCEAN WAVE	
Wave Feature	**Ocean Wave**	**Tsunami**
Wave speed	8–10 km/h (5–60 mph)	800–1000 km/h (500–600 m/h)
Wave period	5–20 s	10–120 min
Wavelength	100–200 m (300–600 ft)	100–500 km (60–300 mi)
Initial disturbance	Wind	Earthquake, explosion, volcanic eruption, or meteor impact

Table 27.3 compares a tsunami with a typical ocean wave. As you can see, tsunami are long, fast-moving ocean waves triggered by a large disturbance—usually an earthquake. The 2004 tsunami was launched by a large but typical subduction earthquake: The subducting plate was sticking and bending along the overlying plate until friction no longer held it in place. When the bottom plate slipped, the overlying rock slab shot up like a piston with tremendous force. This force pushed a column of water upward above the surrounding sea level.

When a tsunami's initial bulge falls back down to sea level, it and subsequent ripples spread outward from the disturbance, traveling mostly underwater. The wave height of a tsunami is usually less than 1 m higher than the surface until it gets close to shore. However, close to shore, the shallower water and the shape of the shore compress the wave, which decreases its wavelength and greatly increases its amplitude. The Indonesian tsunami in 2004 reached a maximum height of nearly 30 feet above sea level.

Water is heavy, so a wall of water several stories high and moving at 500 miles per hour carries tremendous energy and packs giant force. The total energy released by the 2004 tsunami is estimated at five megatons of TNT. This is more than double the energy of all the bombs exploded in World War II, including the two atomic blasts. (Amazingly, the earthquake that triggered the tsunami possessed several orders of magnitude *more* energy than the tsunami!) Coastal buildings were smashed and seaside villages swept away when the tsunami reached coastal locations around the Pacific. And, since the tsunami inundated the land more than a mile past the shoreline, the killer wave had a significant impact inland as well as along the beach.

Another recent and catastrophic tsunami occurred in Japan on March 11, 2011. Over 20,000 people were reported killed or missing after a 9.0 Richter magnitude earthquake struck Honshu and triggered a huge tsunami. Adding to the misery, the tsunami flooded the Fukushima Daiichi nuclear power plant. This triggered fuel meltdowns at three of the plant's six reactors. At more than 40 feet high, the tsunami easily overtopped the Fukushima nuclear plant's 19-foot seawall, producing the worst nuclear disaster since the 1986 Chernobyl core meltdown in Russia. A radiation leak at Fukushima Daiichi displaced as many as 100,000 people. The Japanese government has said that some of the area around the plant will be uninhabitable for decades. The full effects of this disaster are still being assessed as of this writing.

LEARNING OBJECTIVE
Compare and contrast the three main kinds of volcanoes.

27.3 Volcanoes

EXPLAIN THIS Why are most volcanoes located along the Ring of Fire?

Temperature rises quickly with depth in Earth's interior. Between 30 and 150 miles below the surface, near the upper mantle, rock is so hot that it can exist as magma. Magma, a mixture of melted rock and trapped

(a)

(b)

(c)

FIGURE 27.9
(a) Kilauea is a shield volcano in Hawaii. Note the lava flowing gently down its broad sides. (b) Mount St. Helens, a composite volcano in Washington State erupted violently in 1980. (c) The Paricutin cinder cone in Mexico erupted in 1943.

gases, buoyantly rises toward Earth's surface, cooling all the while. Most of it crystallizes underground as igneous rock before it reaches the cool surface, but some magma makes it all the way to the surface without crystallizing. It then erupts through fissures and vents. Erupted magma, called *lava*, solidifies at cool surface temperatures to form volcanic rocks. Such rock, plus perhaps ash, cinders, and other erupted materials, can pile up in great masses to form a **volcano**.

There are three major types of volcanoes, each of which erupts different kinds of lava and other materials. The three types of volcanoes erupt in different ways, and they pose different hazards. The main categories of volcanoes are *shield volcanoes*, *composite volcanoes* (also known as *stratovolcanoes*), and *cinder cones* (Figure 27.9).

Shield volcanoes have broad, gentle slopes resembling a warrior's shield. Shield volcanoes erupt *basaltic* lava—a runny lava low in silica that hardens to make the iron- and magnesium-rich dark volcanic rock found on the ocean floor. Because shield volcanoes erupt smoothly flowing lava, they erupt quietly and steadily. On Hawaii, a shield volcano named *Kilauea* has been erupting almost continuously since 1983. Lava bubbles gently out of the ground at Kilauea and flows slowly enough downhill that you could outrun it most of the time.

In contrast, composite volcanoes erupt explosively. In 1980, *Mount St. Helens* in the state of Washington made history. Mount St. Helens is one of the majestic composite volcanoes that make up the Cascade Mountain Range. Most of the time, the snow-capped Cascade Range is a quiet place to hike or fish. However, on May 18, 1980, the top of Mount St. Helens blew off like a cork from a bottle in a giant eruption. A column of ash rose 60,000 feet into the air. Vegetation was charred 10 miles away. Burning, boulder-size **pyroclastics** ("pie-row-CLASS-ticks"), or debris fragments, were tossed into the air like popcorn. Composite volcanoes such as Mount St. Helens erupt thick, high-silica lava called *rhyolite* as well as a lava of intermediate consistency called *andesite*. Rhyolite and andesite are thick, chunky lavas that can clog up a volcano's vent. Trapped gases add to the pressure. When a composite volcano finally explodes, lava, toxic gases, ash, cinders, and pyroclastics hurl outward. Sometimes, exploding composite volcanoes trigger *lahars* ("LA-harz"), volcanic mudflows that can travel far and fast.

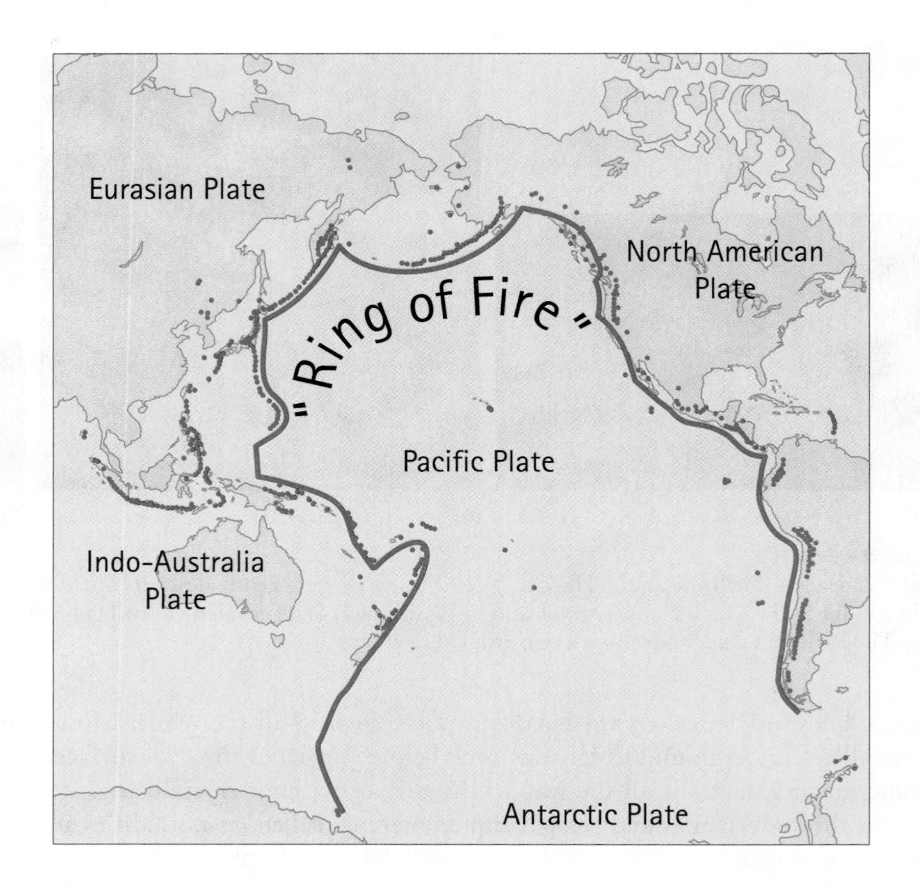

FIGURE 27.10
The Ring of Fire is a region of intense geologic activity. It is home to more than 75% of the world's volcanoes, and 90% of the world's earthquakes happen here. The reason? Convergent plate boundaries and subduction occur all along the Ring of Fire. Volcanoes are shown as red dots on this map.

Lahars have the consistency of wet cement as they flow and the consistency of dry cement when they come to rest.

A cinder cone is a small volcano with steep sides. Cinder cones are built up like heaps of assorted rubbish from basaltic, andesitic, or (less frequently) rhyolitic lava mixed together with various pyroclastics.

Most volcanoes are composite volcanoes. They form due to subduction, so they are mainly located in the Ring of Fire. As stated earlier, more than 75% of the world's volcanoes are composite volcanoes in the Ring of Fire (Figure 27.10). By contrast, shield volcanoes are usually formed over **hot spots**, which are randomly placed plumes of magma. Cinder cones are found anywhere volcanic activity occurs.

Sometimes cinder cones develop inside larger volcanoes. Wizard Island, for example, is a cinder cone sitting inside the composite volcano Mount Mazama, which erupted violently about 8000 years ago. Sometime after Mount Mazama blew, the Wizard Island cinder cone formed inside Mount Mazama's summit crater (or *caldera*). It is built of andesite lava, cinders, and other pyroclastics. Wizard Island, not surprisingly, sits within the Ring of Fire in Crater Lake National Park, Oregon (Figure 27.10).

A common misconception is that lava is the biggest hazard from volcanoes. In fact, lava is more threatening to structures than to people. Pyroclastics are more dangerous than lava. They are an impact hazard if they hit you, they can bury dwellings, and they create agricultural losses. Volcanic ash is also a burial hazard and an agricultural threat. Furthermore, volcanic ash causes lung damage. And, as you will read later in this chapter, ash can trigger climate change. The toxic gases that are released in volcanic eruptions are short-term hazards of volcanoes as well as causes of climate change.

FIGURE 27.11
Wizard Island, a cinder cone in Crater Lake National Park, Oregon. Wizard Island sits inside a caldera. Notice the dark, rugged lava rock encircling the island. That's Mount Mazama, the composite volcano that Wizard Island sits within.

SCIENCE AND SOCIETY

Disaster Warnings—When Do You Tell the People?

Figure 27.12 shows Mt. Vesuvius, a composite volcano that sits above Naples, Italy. This volcano erupted in AD 79, destroying the city of Pompeii. Mt. Vesuvius has erupted many times since. Today, geologists classify Mt. Vesuvius as one of the most dangerous volcanoes in the world. Like other composite volcanoes, it erupts explosively. A population of 3 million people has grown up around its base.

Predicting volcanic eruptions is far from certain. The most reliable approach is to constantly monitor them for signs of an impending eruption, such as bulging flanks. When there are clear signs of danger, disaster officials warn the population at risk. If there is sufficient evidence of an imminent eruption, officials order an evacuation.

Suppose you are a member of the International Volcanic Hazard Task Force. Your team has been monitoring Mt. Vesuvius carefully for many years. Also suppose that Mt. Vesuvius starts showing symptoms of an imminent eruption.

- When should your team warn people that they are at risk? What should you tell them? How will they react?

FIGURE 27.12
Mt. Vesuvius sits above Naples, a city of 3 million people.

- What will happen if you warn the people of an eruption but your indicators are wrong and the volcano goes quiet? What are some of the consequences of this "false alarm" to the local economy? To the people's trust in scientists and disaster officials?
- What will happen if you fail to issue a warning and the volcano blows sooner and more explosively than you predicted?

CHECK YOURSELF

1. **Where does lava come from?**
2. **Why are composite volcanoes more hazardous than shield volcanoes?**

CHECK YOUR ANSWERS

1. Lava is magma that erupts. Magma is melted rock that forms within the mantle.
2. Composite volcanoes erupt explosively. Both types of volcanoes erupt lava—which is a burn hazard. But in addition to lava, composite volcanoes emit toxic gases, pyroclastics, ash, and cinders. Also, composite volcanoes can trigger lahars—dangerous mudflows.

LEARNING OBJECTIVE
Explain the process by which hurricanes form.

27.4 Hurricanes

EXPLAIN THIS Why do hurricanes die when they move over dry land?

A **hurricane** is a large, rotating weather system with winds blowing at least 74 miles per hour (119 km/h). Hurricanes are the most powerful of all Earth's storms. They form over tropical seas and can be highly destructive if they reach land. As Figure 27.13 shows, hurricanes usually form between 5° and 20° north and south latitudes, where temperatures are warm. They vary from about 150 km to 1500 km in diameter, and they can travel for thousands of miles.

A hurricane starts as a storm over warm ocean water. The water needs to be at least 80°F down to a depth of 200 feet. As the Sun heats the ocean's surface, masses of warm, moist air move upward through the atmosphere, forming huge thunderclouds. More air is sucked into the warm, low-pressure air column beneath the storm clouds. The moving air is twisted by the Coriolis effect, so it turns in a spiral pattern around the low-pressure center. Figure 27.14 shows the pattern of

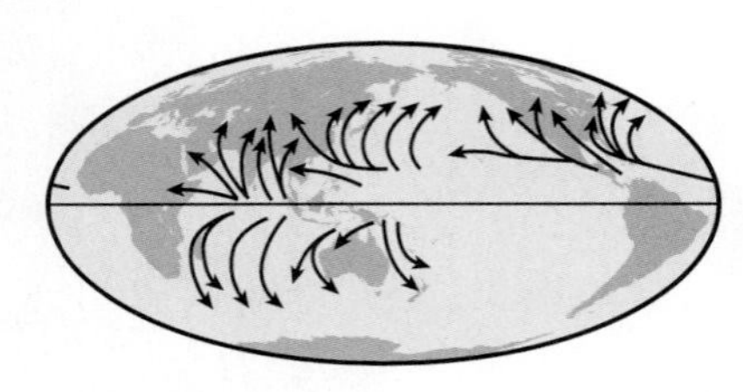

FIGURE 27.13
The map shows the places at risk for hurricanes. Hurricanes typically form at the tail of the arrow and track in the direction of the arrow to the arrowhead. The three regions that are most at risk for hurricane damage are the western Pacific Ocean, the Bay of Bengal (near India), and the Caribbean.

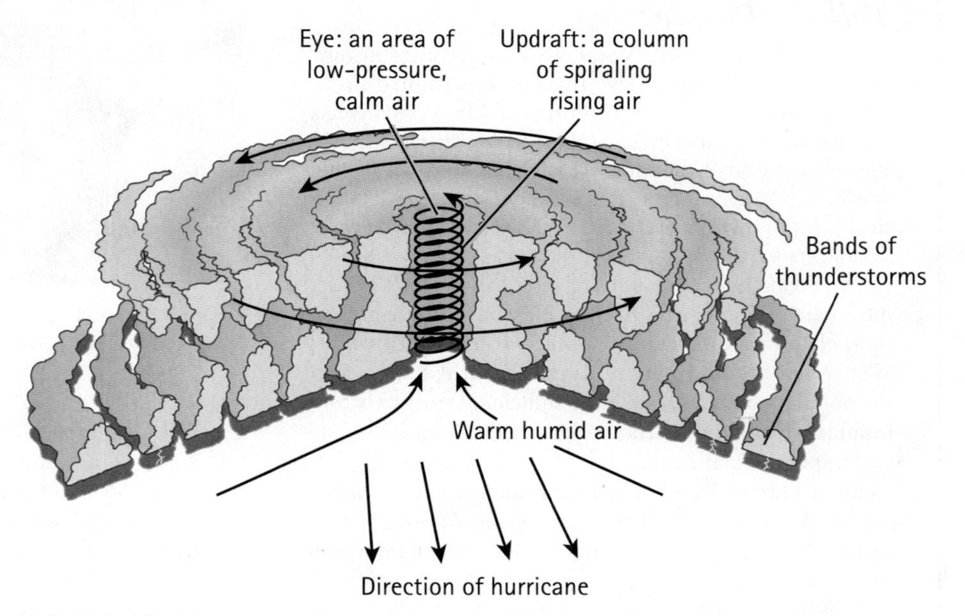

FIGURE 27.14
The low-pressure center of a hurricane is the eye. Moist air spirals around the eye and moves upward. Bands of rain and wind make it extend outward.

clouds and airflow that makes up a hurricane. Besides requiring warm water, warm air, and the Coriolis effect, hurricanes cannot exist in high-friction environments. So hurricanes are restricted to tropical seas.

UNIFYING CONCEPT

- ***Friction***
Section 2.8

The warm air that is sucked into the bottom of a hurricane is rich in water vapor. As the air rises, it cools and the water vapor in it condenses to form liquid water. Do you remember that as a gas cools to a liquid it gives off energy to its surroundings? The energy stored by the gas and released upon condensation is called its *latent heat*. The condensation of the water vapor in a hurricane releases latent heat—and this is the energy that fuels a hurricane. Being a dynamic system, a hurricane eventually moves away from the tropics. The cooler temperatures it then encounters reduce the supply of warm, moist air so the hurricane diminishes or dies. If the storm hits land, it is no longer fed by condensing water vapor from the ocean, so it will certainly lose its fuel and die.

Figure 27.15 shows Hurricane Katrina on August 28, 2005, as it twisted through the Gulf of Mexico. When a hurricane moves toward shore, it creates a wall of water called a *storm surge*. The storm surge is a hammer of water and often the most destructive part of a hurricane. The 20-foot storm surge created by Katrina overwhelmed New Orleans' levee and sea wall system, bringing mass flooding to the city and nearby communities. Nearly 2000 people lost their lives in Hurricane Katrina and in the floods that followed. The storm is estimated to have caused more than $81 billion in damage, making it the costliest natural disaster in U.S. history.

Hurricanes are known by different names in different parts of the world. People in the western Pacific call these storms *typhoons*. In the Indian Ocean and western South Pacific, they are called *tropical cyclones*. By whatever the name,

(a)

(b)

FIGURE 27.15
(a) Hurricane Katrina in the Gulf of Mexico on August 28, 2005. (b) Hurricane Katrina hammered the city of New Orleans with a 20-ft storm surge. New Orleans sits below sea level and has been protected from flooding by an engineered system of levees and sea walls. But, when the hurricane battered these barrier walls, the structures gave way and seawater rushed into the city. The floodwaters, filthy with floating debris, remained trapped like soup in a bowl within the city walls until the water could be drained.

TABLE 27.4 SAFFIR–SIMPSON HURRICANE SCALE

Category	Wind Speed (miles per hour)	Effects
No. 1	74–95	*Very dangerous winds produce some damage.* Examples: roofs damaged or destroyed; large tree branches snap; shallow-rooted trees topple; power lines and poles fall or break
No. 2	96–110	*Extremely dangerous winds cause extensive damage.* Examples: older mobile homes destroyed; some roofs removed; shallow-rooted trees uprooted blocking roads; long-term power loss
No. 3	111–130	*Devastating damage occurs.* Examples: electricity and water unavailable; extensive roof and siding damage; windows broken; high risk of injury or death due to flying and falling debris
No. 4	131–155	*Catastrophic damage occurs.* Examples: extensive structural damage to top floors; collapse of some masonry buildings; fallen trees and power poles isolate residential areas; long-term water shortages
No. 5	Greater than 155	*Catastrophic damage occurs.* Examples: people, livestock, and pets at very high risk of injury or death from falling or flying debris, even inside homes; many homes destroyed; nearly all trees uprooted; most of the area uninhabitable for weeks or months

hurricanes are by far most frequent in the Pacific Ocean. Hurricanes vary seasonally, as you might expect, because temperature varies seasonally. In the western Pacific, hurricane season peaks in late August/early September.

The severity of hurricanes can be compared using the *Saffir–Simpson Hurricane Scale* (Table 27.4). Hurricane Katrina was Category 5, an unusually severe event in terms of wind speed, height of the storm surge, and effects on land.

CHECK YOURSELF

1. **Why do winds flow inward toward the eye of a hurricane?**
2. **Do you live in a place that is at high risk for hurricanes?**

CHECK YOUR ANSWERS

1. The eye of a hurricane is a low-pressure zone. Air always moves from high to low pressure.
2. Look at the map in Figure 27.13. Is your home in the high-risk zone? If so, do you know what to do if a hurricane threatens your community?

Integrated Science 27A
PHYSICS, CHEMISTRY, BIOLOGY, AND ASTRONOMY

Climate Change

LEARNING OBJECTIVE
Describe how the use of fossil fuels over the past century has triggered a rise in global average temperature.

EXPLAIN THIS Explain why it's important that Edith's checkerspot butterfly has changed its range since the 1880s.

Signs of climate change—from melting sea ice, to increasingly frequent and severe hurricanes, to the relocation of native settlements away from flooded shorelines—are easy to find in the scientific literature and in the popular media. Climate change is here. But what exactly is climate change in the context of Earth's overall history? Is today's climate change natural or *anthropogenic* (human-caused)? What is the evidence that human activity is responsible for a recent spike in global temperature? What is planetary feedback? How is climate change affecting Earth systems, and what can be done in response? These questions frame our exploration of this important topic.

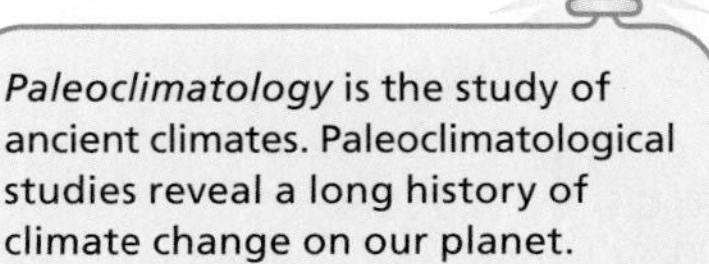

Paleoclimatology is the study of ancient climates. Paleoclimatological studies reveal a long history of climate change on our planet.

What Is Climate Change?

Weather is the state of the atmosphere at a particular time and place—its temperature, humidity, pressure, and precipitation, wind, and clouds. *Climate* is the weather pattern over several decades or longer. **Climate change**, then, is a lasting alteration to long-term weather patterns. One hot summer is not "global warming," nor is a single severe hurricane. Climate change is bigger than weather fluctuations—it's a lasting alteration to average weather conditions so new patterns of temperature, precipitation, storm intensity, and other weather features become established. As it's used in the popular media, the term *climate change* or *global warming* refers to the recent spike and projected future increase in average global temperature caused by an enhanced greenhouse effect. To understand climate change in a broad context, we begin by looking at how Earth's climate has varied throughout its history.

Eon	Era	Period	Sub-period	Epoch	MA
Phanerozoic	Cenozoic	Quaternary		Holocene	0.01
				Pleistocene	1.8
		Tertiary		Pliocene	5.3
				Miocene	23.8
				Oligocene	33.7
				Eocene	54.8
				Paleocene	65
	Mesozoic	Cretaceous			145.5
		Jurassic			199.6
		Triassic			248
	Paleozoic	Permian			290
		Carboniferous	Pennsylvanian		318
			Mississippian		354
		Devonian			417
		Silurian			443
		Ordovician			490
		Cambrian			542
Precambrian Time	Proterozoic				2500
	Archean				3800
	Hadean				4600

FIGURE 27.16
The geologic time scale divides Earth's history into time units of different size. The units on this scale are MA, or "millions of years ago." The Paleozoic era began about 542 Ma, or 542 million years ago, for example.

Earth's Climate Over Time Earth's climate has changed many times throughout its 4.6-billion-year history. The planet has withstood extremes. Figure 27.16 shows the geologic time scale, which divides Earth's long history into time units of different size. Today we are living in the Phanerozoic eon, the Cenozoic era, the Quaternary period, and the Holocene epoch. Most divisions of the geologic time scale mark major turning points in the history of geology or the history of life, such as mass extinctions. Many of these big turning points have involved, at least in part, a change in climate.

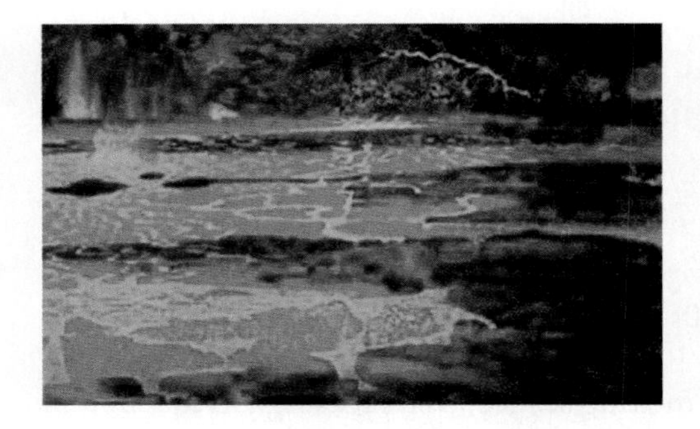

FIGURE 27.17
In the Hadean era, Earth's surface was mostly molten with a few fragments of crust.

In the Hadean eon, when Earth was newly formed, it was a scorching, lifeless sphere covered with churning, molten rock (Figure 27.17). About 600 million years ago, at the end of Precambrian time, the planet entered a frozen state called *Snowball Earth.* Ice completely covered it and life was sparse. In the Silurian period, about 400 million years ago, the global climate was stable and warm. Glaciers melted and warm shallow seas covered much of the globe, providing habitat for abundant species of marine life. A spectacular diversity of fishes evolved. Amphibians pushed the range of the animal kingdom onto the edge of land. Warm, moist forests developed as *gymnosperms,* plants bearing seeds, first appeared. The first animals to follow plants out of the water were wingless insects and scorpions. Later, in the Carboniferous period illustrated in Figure 27.18, reptiles appeared on land; they were able to survive in dry environments because they had developed the amniotic egg.

FIGURE 27.18
The Carboniferous period was marked by warm, moist, swampy forests. Reptiles first appeared in the Carboniferous period.

About 250 million years ago, at the end of the Permian period, a mass extinction destroyed almost 90% of the species living at the time. The Permian extinction is the most devastating extinction event in history. It coincided with the largest volcanic eruption ever—an eruption that lasted more than a million years and triggered two bouts of rapid climate change. Volcanic ash rapidly cooled the planet by blocking solar radiation (Figure 27.19). Then, as the

FIGURE 27.19
A million-year volcanic eruption at the end of the Permian period spewed huge amounts of ash into the air, thereby changing the world's climate.

eruption waned, Earth quickly warmed in a geologic whiplash. The two periods of rapid climate change—cooling and then warming—combined with disruption of the atmosphere's ozone layer, acid rain, the release of large amounts of carbon dioxide, and other factors are thought to have caused the huge die-off. Photosynthesizing species may have been most drastically affected first, with the effects of their death moving up the food chain. Although most species were destroyed in the Permian extinction, the reptiles that survived evolved and diversified to become the dominant animal species of the world—dinosaurs. Life flourished in many differentiated forms, driven by natural selection, until about 65 million years ago. Then, at the end of the Cretaceous period, about 60% of Earth's species were eliminated in the Cretaceous extinction. Famously, the dinosaurs were devastated. What caused this catastrophe?

The Mesozoic era is informally known as "The Age of the Dinosaurs" because of its famous reptilian species. Earth was warm and wet when the dinosaurs reigned.

Scientists agree that climate change was the key factor. Mild, warm Mesozoic climates suddenly turned cool. Sunlight dimmed. Ecosystems perished. According to most researchers, this round of climate change was triggered by an asteroid impact. In the rock record, there are widespread deposits of iridium, an element rare on Earth but characteristic of asteroids. The iridium layer in Earth's crust dates back to 65 million years ago—the time of the Cretaceous extinction. Apparently, a huge asteroid crashed into Earth and exploded over a large area, leaving its iridium "fingerprints" in rock strata. More important to the dinosaurs, though, the asteroid impact triggered wildfires that blanketed the atmosphere with soot and other particulates, and this blocked the Sun and altered the climate.

FIGURE 27.20
According to the impact hypothesis, a large asteroid triggered climate changes that caused a mass extinction at the end of the Cretaceous period.

CHECK YOURSELF

Among the survivors of the Cretaceous extinction are bottom-dwelling organisms such as clams. Why did they survive?

CHECK YOUR ANSWER

Bottom-dwelling communities were insulated from climate fluctuations that affected organisms closer to the atmosphere.

After the mass extinctions at the end of the Mesozoic era, many environmental niches were left vacant. They were filled in the Cenozoic era by new species. At the beginning of the Cenozoic era, tectonic events had cooled the climate. The supercontinent Pangaea had recently split apart. The breakup of Pangaea changed ocean circulation patterns by opening up seaways between landmasses. Cold-water currents could then cool the land. Subsequently, huge glaciers called *ice sheets* formed near the poles.

Ice Ages Geologic periods that are cold enough to support extensive glaciation are called **ice ages**. In an ice age, ice sheets and alpine glaciers are widespread.

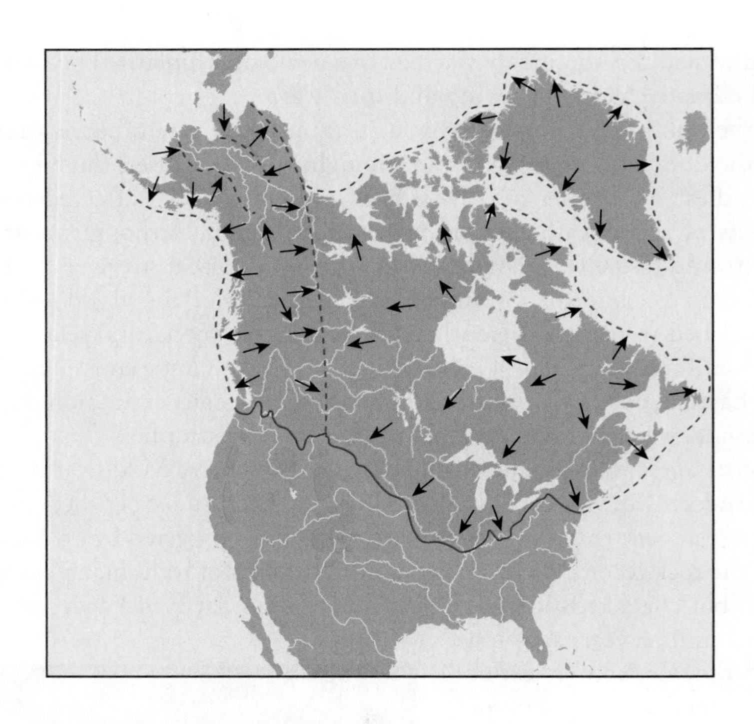

FIGURE 27.21
This map shows how far glaciers (in blue) reached into North America during the Pleistocene epoch. Arrows show the directions of ice movement. The Pleistocene spanned the time from about 2 million years ago to about 10,000 years ago.

Antarctica and Greenland are largely covered by ice sheets today, the Arctic Sea is frozen most of the year, and alpine glaciers remain at high altitudes. Accordingly, Earth is now in an ice age. Within an ice age, pulses of especially cold climate are called *glacials*. Glacials are the coldest of cold climates. In a glacial, glaciers advance toward the equator. Sea level drops because glaciers take their water from the ocean. Figure 27.21 shows that much of North America was covered by a glacier during the most recent glacial period. Today, we live in an **interglacial period**, a pulse of warmer temperatures that lasts thousands of years and occurs within an ice age. In an interglacial period, glaciers melt back to high latitudes. This causes sea level to rise worldwide. The oscillations between glacials and interglacials within an ice age are called *glacial/interglacial cycles*.

Ice ages are rare compared to the warm periods in geologic history. Earth was warmer than it is today for most of its past. All of human existence, however, has occurred within an ice age. Humans evolved during the Pleistocene epoch at about the same time that mammoths, rhinoceroses, bison, reindeer, and musk oxen appeared. These creatures all evolved with warm, woolly coats for protection against the frigid cold. Lacking such a coat, *Homo sapiens* relied on intelligence to cope with the hardships of cold weather.

You should have headed for Antarctica. You'd love the climate!

Following the Pleistocene epoch is the Holocene epoch. To see the Holocene environment, just look around. It is the most recent 10,000 years or so of Earth's history, including the present time. The transition from the Pleistocene to the Holocene, about 10,000 years ago, was marked by warming climate. It is this climate change, plus overhunting by humans, that apparently triggered the extinction of the large, cold-adapted mammals. From its beginning, the Holocene has been a warm and relatively stable interglacial period, allowing human culture to advance and population to grow exponentially.

To be sure, some climate variation has occurred even through the Holocene. Over the past 2000 years there have been three episodes of significant change: the Medieval Climate Anomaly (AD 900–1300), the Little Ice Age (AD 1500–1850), and the Industrial Era (AD 1900 to the present time). The Medieval Climate

Anomaly was a warmer-than-average phase that affected large geographic expanses; the American West was very dry. The Little Ice Age was—as its name suggests—a cold period: Alpine glaciers advanced in Europe, which shortened growing seasons there and decreased agricultural productivity. Over the past 100 years, in the Industrial Era, the temperature has changed again. It has spiked: The average global temperature has increased by about 1.3°F or about 0.7°C in the last century. This may sound like a small variation, but the change has been very rapid in geologic terms. The temperature is rising exponentially, in fact: The eight warmest years on record (since 1880) have all occurred since 2001. The ecosystems we are part of and depend upon are adapted to a narrow temperature range of only a few degrees Celsius. The 1-degree global temperature increase of the Industrial Era has had many environmental effects—including the shift in the range of Edith's checkerspot butterfly (Figure 27.22).

(a)

(b)

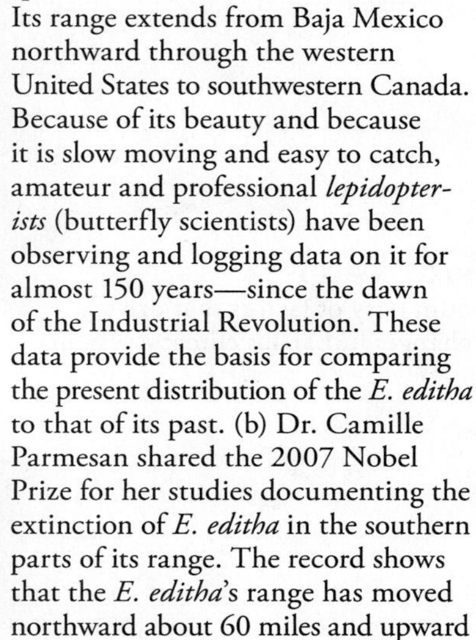

FIGURE 27.22
(a) Edith's checkerspot butterfly (*Euphydryas editha*) is a resident species of western North America. Its range extends from Baja Mexico northward through the western United States to southwestern Canada. Because of its beauty and because it is slow moving and easy to catch, amateur and professional *lepidopterists* (butterfly scientists) have been observing and logging data on it for almost 150 years—since the dawn of the Industrial Revolution. These data provide the basis for comparing the present distribution of the *E. editha* to that of its past. (b) Dr. Camille Parmesan shared the 2007 Nobel Prize for her studies documenting the extinction of *E. editha* in the southern parts of its range. The record shows that the *E. editha*'s range has moved northward about 60 miles and upward in elevation about 300 feet. In more recent research, Parmesan has found that 800 of the 1500 species she studied worldwide were responding in some way to a warming global climate.

UNIFYING CONCEPT

- ***Exponential Growth and Decay*** **Appendix D**

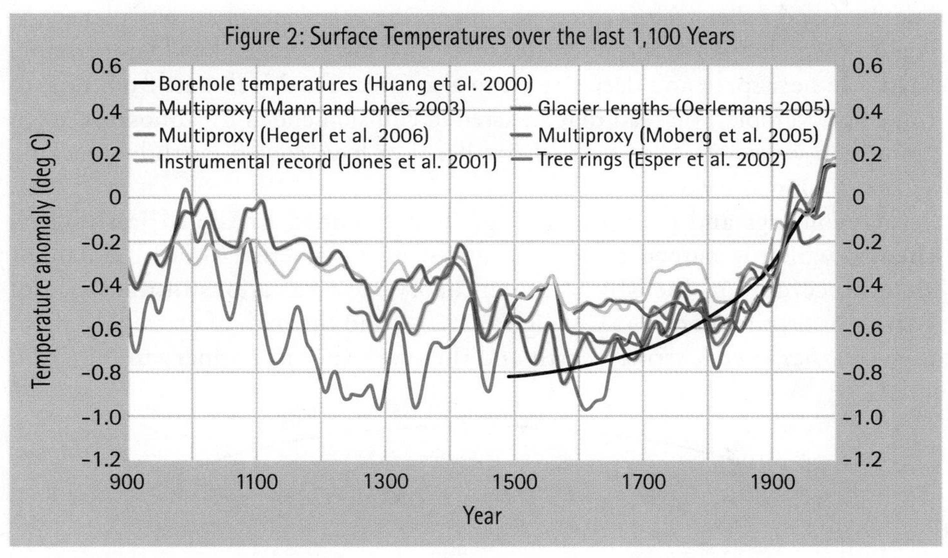

FIGURE 27.23
This graph tracks the average variations in surface temperatures in the Northern Hemisphere obtained from different research teams (in different colors). Can you find the Medieval Climate Anomaly, the Little Ice Age, and the Industrial Age on the graph?

Causes of Climate Change

What factors have driven Earth's climate swings? Paleoclimate researchers have identified several basic components. Some of the drivers are extraterrestrial in origin, whereas other factors originate here on Earth, as Figure 27.24 shows. Until recently, climate change occurred naturally. However, the temperature spike during the past hundred years, as well as projected near-term future climate warming, are mainly human-caused (*anthropogenic*). Anthropogenic climate change arises mostly from alterations to atmospheric chemistry that enhance the greenhouse effect.

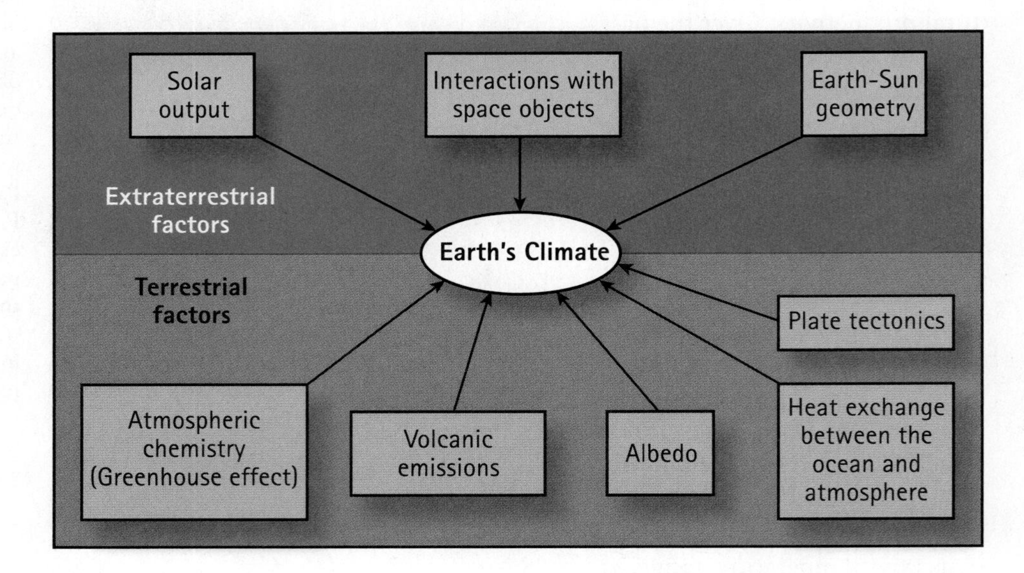

FIGURE 27.24
Summary of factors that drive changes in Earth's climatic system.

Natural Climate Change Before discussing anthropogenic climate change, let's examine the factors that would make the climate change even if people weren't around. This is called *natural climate change*. Note that the diagram of climate change drivers shows that Earth's temperature is controlled by the Sun's output. The amount of energy transmitted by the Sun is not truly constant. Sunspots and other factors affect the incoming solar energy to our planet. As you'd expect, low solar output corresponds with low average global temperature. Matter in near space and deep space can also affect Earth's climate from time to time. For example, you read that an asteroid collision ended the dinosaurs' reign 65 million years ago. Clouds of interstellar dust have cooled Earth by blocking solar radiation as well.

The comings and goings of ice ages are explained by the **Milankovitch theory**, which is named after its Yugoslavian discoverer, Milutin Milankovitch. According to the Milankovitch theory, slow variations in the shape of Earth's orbit, the tilt of its axis (Figure 27.25), and the path of its wobble affect the way solar energy strikes Earth and therefore alter its temperature.

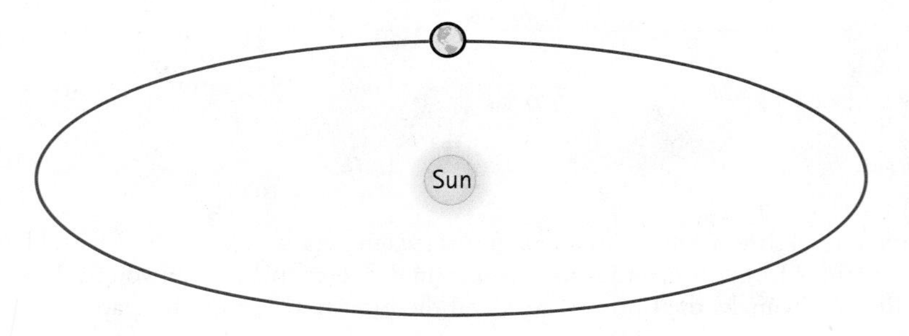

Earth's orbit around the Sun is an ellipse, an elongated loop. Over a 100,000-year period, the orbit changes from a more circular path to a more elliptical one and back again. Can you see that, the more elliptical Earth's orbit is, the farther away from the Sun it can be? This affects its climate.

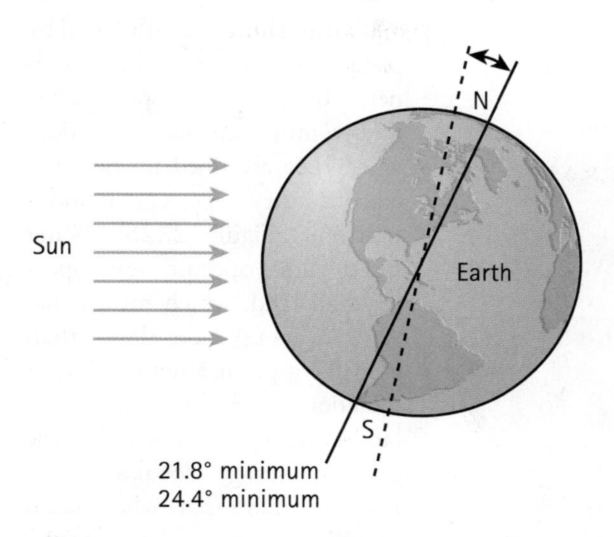

(a) The tilt of Earth's axis changes. Today, Earth is inclined at an angle of 23.5° but it varies between 22° and 25°every 41,000 years. When the tilt is 22°, Earth is most upright. Then, sunlight hits all parts of the Earth most evenly so seasonal intensity is reduced. This makes for cooler summers, which allows glaciers to build up year upon year.

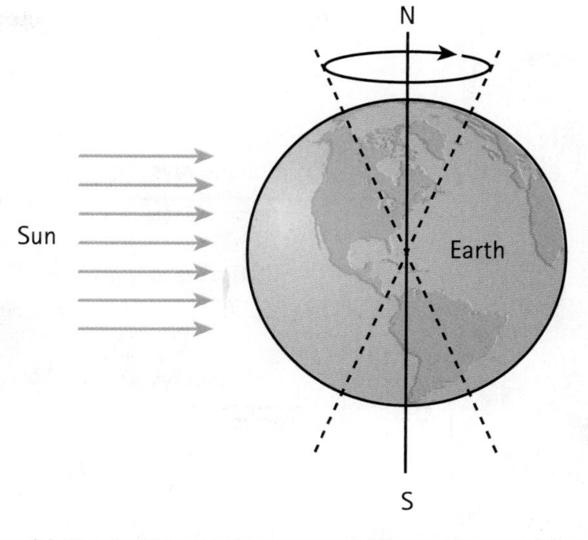

(b) The Earth's axis traces a complete circle, which affects its position in space and the solar energy it receives. Earth's axis completes a full circle every 26,000 years. The variation affects the amount of solar radiation it receives and therefore affects its climate.

FIGURE 27.25
Changes in Earth's orbital shape from elliptical to increasingly circular, (a) the tilt of its axis, and (b) the path of its wobble affect its climate and account for ice age changes, according to the Milankovitch theory.

The changes in Earth's orbit and axial spin occur in cycles of differing length. Sometimes changes that act toward a warming trend are canceled by other changes that act toward a cooling trend. But sometimes the effects all push climate in the same direction. Then the factors add together to cause extreme warming or cooling. Milankovitch proposed that ice ages occur when cooling effects occur at the same time and add up. Evidence shows that he was probably right.

CHECK YOURSELF

Do the changes in Earth's tilt, wobble, and orbital shape work together or separately to produce ice ages?

CHECK YOUR ANSWER

The variations work together—sometimes canceling and sometimes reinforcing one another. When several changes that produce cooling add up, ice ages occur.

Most of the causes of climate change originate here on Earth rather than in space. An important terrestrial component is volcanic emissions. Volcanic eruptions affect climate by filling the atmosphere with ash, dust, and gases that block the solar radiation received by Earth. Volcanic emissions can also work in the opposite direction—warming Earth rather than cooling it—because volcanoes emit carbon dioxide, a greenhouse gas.

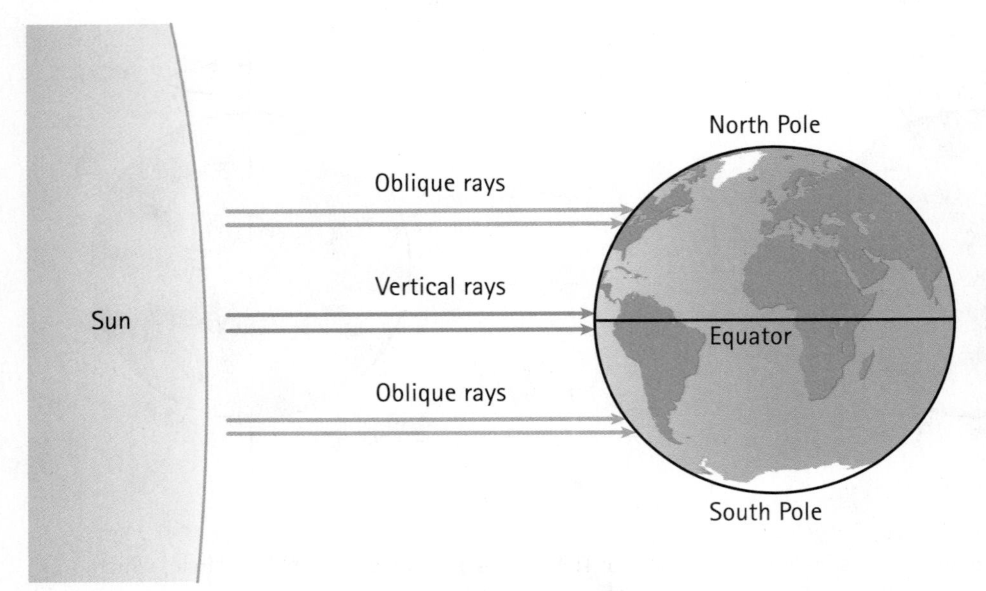

FIGURE 27.26
This diagram shows why latitude influences climate. Solar radiation is more intense at low latitudes because the Sun's rays strike Earth most directly there.

Earth's climate is influenced by plate tectonics, too. When continents drift near the poles, they receive minimum solar radiation; when they are lined up near the equator, they receive maximum radiation (Figure 27.26). Water has a higher specific heat capacity than land, which means that water heats up more slowly than land for a given amount of solar radiation. So when the continental landmasses are clustered near the equator, average global temperature rises more than when ocean occupies this space. Plate tectonics can also affect climate by shifting the path of ocean currents as they move around landmasses.

There is a continual exchange of heat between the oceans and atmosphere so the way in which ocean currents flow significantly affects the temperature. Surface currents move streams of warm and cold water around the world in response to global winds. There is another set of ocean currents, too. Cold, salty water sinks in the Arctic and is replaced by warm surface water circulating toward the poles from the tropics. This process is called *thermohaline circulation*. It is the basis of the "ocean conveyor belt," the circulation of seawater due to salinity and temperature differences that distributes heat around the globe. Perturbations in thermohaline circulation both result from and produce more changes to the global climate.

Reflectivity is another climate control. The fraction of solar radiation arriving at a surface and reflected is its **albedo**. Albedo depends on two factors: color and chemical composition, so it varies from surface to surface. The dark, blue ocean has a much lower albedo than the white ice caps of the poles, so the ocean absorbs a higher percentage of incoming energy and reflects less compared to

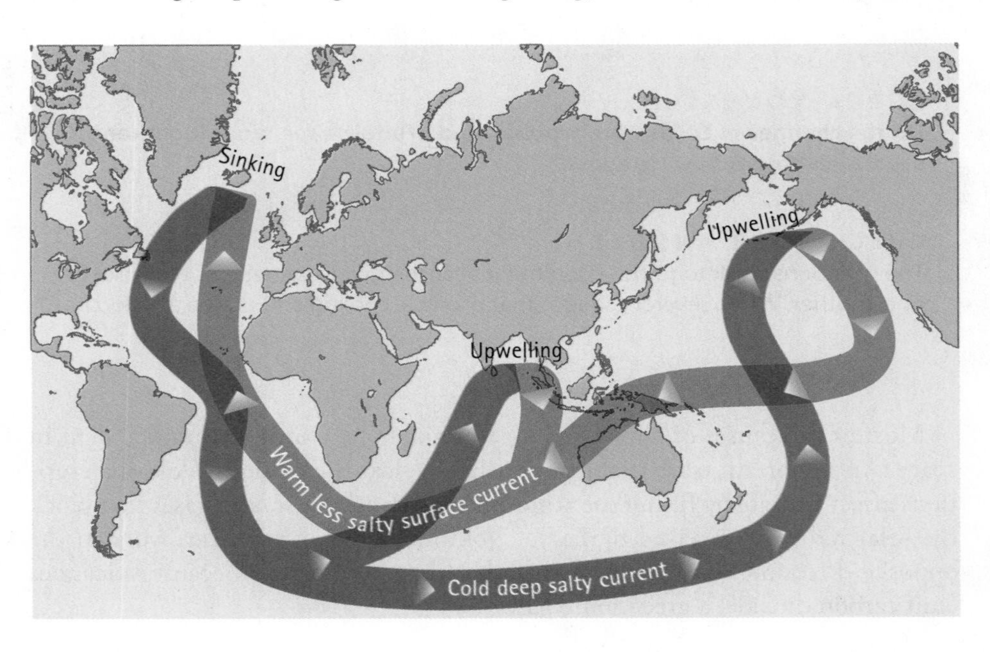

FIGURE 27.27
Thermohaline circulation: global ocean circulation pattern driven by density currents.

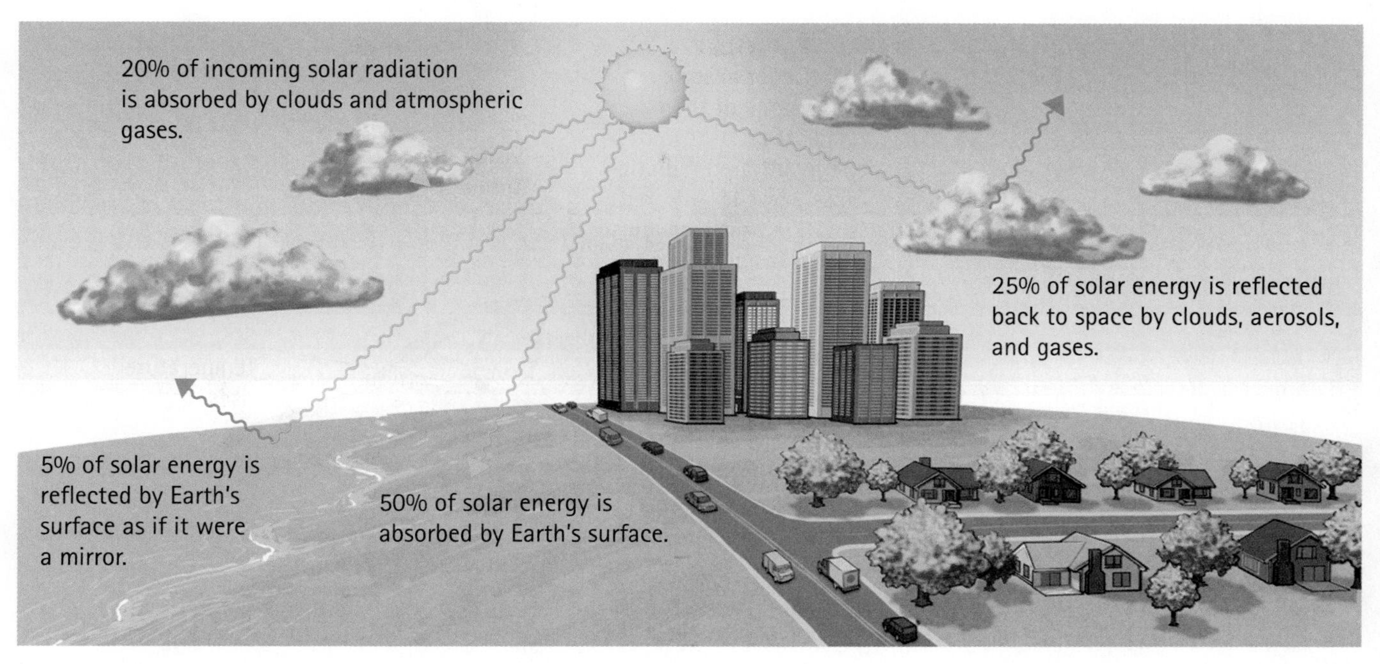

20% of incoming solar radiation is absorbed by clouds and atmospheric gases.
25% of solar energy is reflected back to space by clouds, aerosols, and gases.
5% of solar energy is reflected by Earth's surface as if it were a mirror.
50% of solar energy is absorbed by Earth's surface.

FIGURE 27.28
Earth's overall surface reflectivity or albedo is about 0.3, or 30%. This means that about 70% of incident radiation is absorbed by the land, water, and atmosphere and eventually becomes heat. The rest (30%) is reflected back to space and doesn't affect temperature.

the ice caps. Earth's total albedo varies mainly due to changes in cloud cover, snow and ice cover, human developments, and vegetation cover. Paving surfaces decrease their reflectivity. Asphalt absorbs nearly all of the energy incident on it. During ice ages, when glaciation is at its maximum, Earth's total albedo is very high. When albedo decreases due to, say, melting of the ice caps, Earth's global average temperature rises.

CHECK YOURSELF

1. **How does astronomy relate to climate change?**
2. **How do catastrophic events relate to climate change?**
3. **How does plate tectonics relate to climate change?**
4. **Which has a higher albedo—a mirror or Earth as a whole?**

CHECK YOUR ANSWERS

1. The intensity of solar radiation received, which affects Earth's temperature, depends on Earth's orbit and the tilt of its axis.
2. Asteroid impacts and volcanic eruptions throw gases and particles into the air that block sunlight and cool the climate.
3. There are many ways. A few of them are mentioned in this book. For example, the effects of continental drift include disruption of ocean currents and redistribution of continents to cold polar latitudes as well as increased reflection of incoming solar radiation by landmasses gathered along the equator.
4. A mirror's albedo is 100% because it's a virtually perfect reflector. Earth's albedo is only about 30%. Thus, about 70% of all incident solar radiation is absorbed by Earth and 30% is reflected.

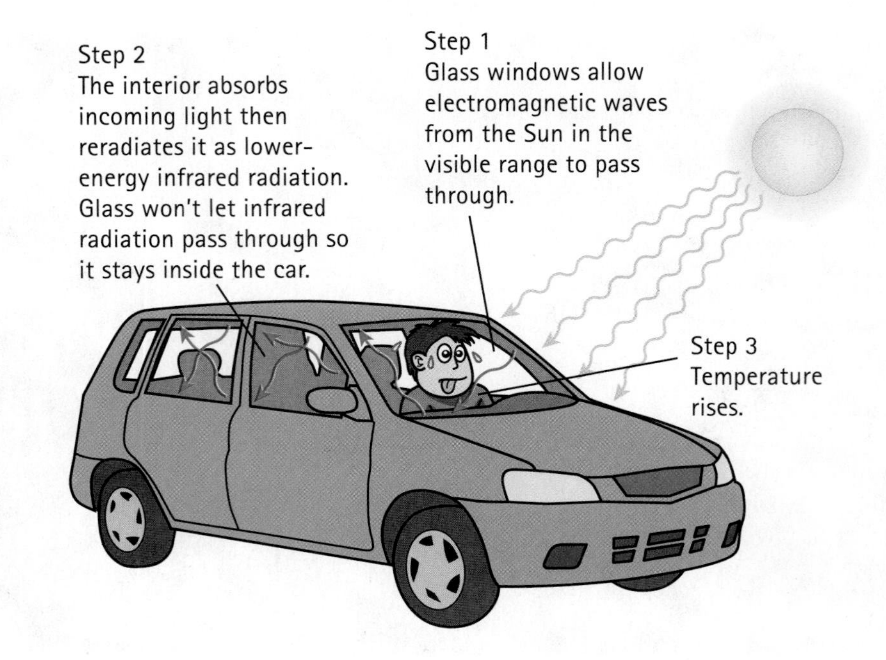

Step 2
The interior absorbs incoming light then reradiates it as lower-energy infrared radiation. Glass won't let infrared radiation pass through so it stays inside the car.
Step 1
Glass windows allow electromagnetic waves from the Sun in the visible range to pass through.
Step 3
Temperature rises.

FIGURE 27.29
The greenhouse effect in a car.

Last but certainly not least among natural climate controls is atmospheric chemistry. The atmosphere is an envelope of nitrogen gas (78%) and oxygen gas (21%) plus hundreds of species of trace gases; suspended particles (such as soot, ash, and dust); and water droplets and other liquids. Water vapor, carbon dioxide, and methane are important trace gases because they are potent *greenhouse gases.*

The greenhouse effect is a major mechanism by which Earth maintains its "Goldilocks climate"—not too hot, not too cold, but just right for life. Without the greenhouse effect, the planet's average temperature would be a frigid −18°C. The atmosphere's greenhouse effect works much like the glass of a florist's greenhouse or the glass windows of your car on a sunny day (Figure 27.29). Greenhouse gases in the atmosphere allow short-wave visible light from the Sun to pass through (Figure 27.30). The ground absorbs the light, then reradiates it. But the reradiated energy is no longer in the visible range. It's longer-wavelength *infrared radiation*—heat. When the infrared rays move away from the ground toward space, greenhouse gas molecules absorb and re-emit them many times before the waves finally escape. Every time the infrared radiation is absorbed, it heats the atmospheric gases a little more. The atmosphere is warmed up—just like the interior of your car on a sunny day. Eventually, the infrared radiation escapes to space, leaving a warmed Earth behind.

Solar visible light
Greenhouse gases
Infrared
Earth

FIGURE 27.30
The greenhouse effect in Earth's atmosphere. Visible light is absorbed by the ground, which emits infrared radiation. Carbon dioxide, water vapor, methane, and other greenhouse gases absorb and re-emit the infrared radiation that would otherwise be radiated directly from Earth into space.

Anthropogenic Climate Change There is more agreement about anthropogenic climate change among scientists than there is about almost any other scientific issue: temperature has risen sharply since the beginning of the Industrial Era and most of the increase has occurred during the past three decades due to the buildup of carbon dioxide (and, to a lesser extent, other greenhouse gases) in Earth's atmosphere. Carbon dioxide levels are higher today than at any time in the past 800,000 years. Carbon dioxide is released naturally by decaying organisms and volcanic emissions. But analysis of the isotopic composition of carbon dioxide in today's atmosphere shows that it comes mainly from burning fossil fuels, not from natural sources.

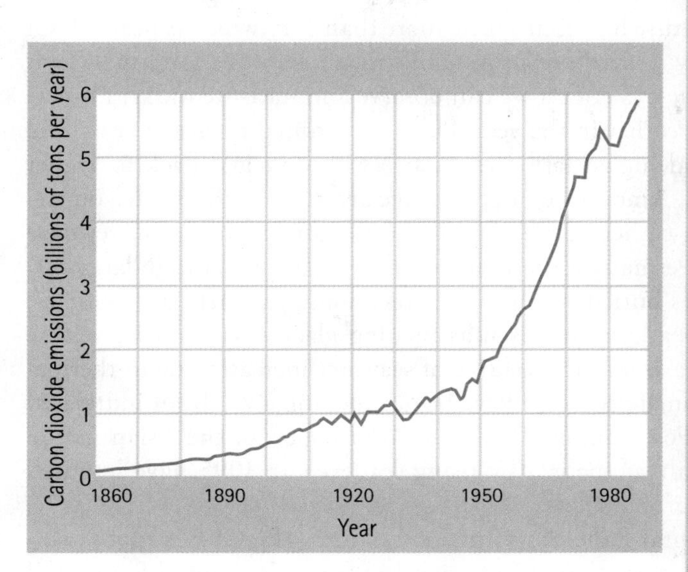

FIGURE 27.31
Carbon dioxide emissions from the burning of fossil fuels have increased exponentially since the time of the Industrial Revolution.

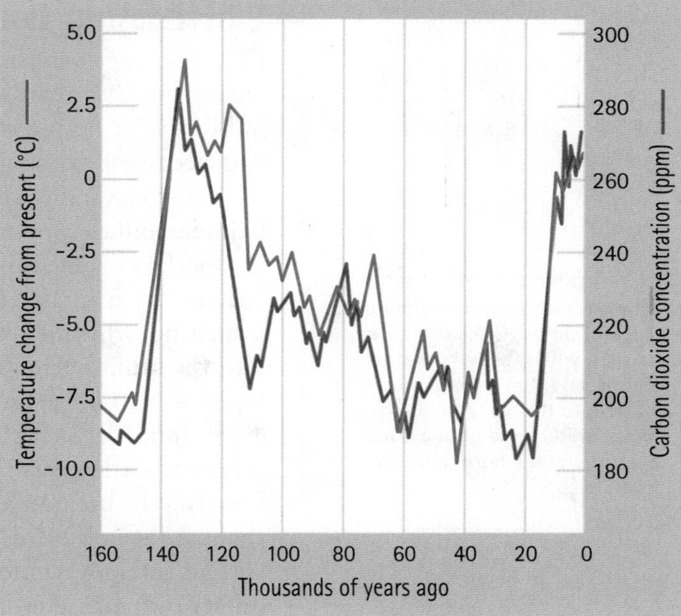

FIGURE 27.32
Carbon dioxide level and temperature have shown a close correlation for many thousands of years.

UNIFYING CONCEPT

- ***Exponential Growth and Decay***
Appendix D

How does burning fossil fuels put carbon dioxide into the air? Fossil fuels—by which we mean petroleum (crude oil and natural gas) and coal—are formed from animals and plants that died millions of years ago but did not decay completely. Instead, they were buried by sand and silt. Biomolecules, rich in chemical energy as well as carbon, remained buried. Over time, heat and pressure converted the biomass into fossil fuels. When these fossil fuels are extracted and then burned for energy, they release their stored carbon as carbon dioxide. Millions of tons of carbon dioxide are exhausted into the atmosphere every day from power plants that burn coal or oil, from trucks and cars that burn gas, and from other human activities.

Figure 27.31 shows the dramatic rise in carbon dioxide emissions since the mid-1800s, when humans began burning the fossil fuels that seemed abundant at the time. Burning coal, oil, and natural gas produces carbon dioxide as a waste product. Figure 27.32 shows that atmospheric carbon dioxide levels and temperatures have been closely correlated for many thousands of years. As CO_2 levels increase, so do temperatures.

Besides burning fossil fuels, people increase atmospheric carbon dioxide levels through *deforestation*. Deforestation is the clearing of forests, as shown in Figure 27.33. In many countries around the world, forests are being cleared to obtain firewood, build roads, and open up land for agriculture to feed a large and growing human population. Plants take carbon dioxide out of the air to perform photosynthesis, making them *carbon sinks*. So, a net loss of trees means a net gain of carbon dioxide in the atmosphere. Plus, if the trees are burned when they are cut down, they give off carbon dioxide right away through combustion.

Burning fossil fuels keeps many factories humming, but it also puts a lot of carbon dioxide into the air.

FIGURE 27.33
This deforested landscape was once rainforest. The forest was cut down to harvest palm oil in Indonesia.

Current and Future Effects of Climate Change

Earth's average temperature has risen a little more than 1°F (which is nearly 1°C) during the 20th century. That doesn't sound like much, but it's an unusual event in history. The rapid rise has effects we can observe and measure today: Flowers and trees are blooming earlier in the year. Plant and animal ranges have shifted to higher elevations and higher latitudes. Glaciers have melted back to higher latitudes and elevations. Amounts of arctic sea ice are at a record low. Ice on rivers and lakes is melting earlier in the year. Tropical storms and heat waves have become more intense. Permafrost (permanently frozen ground at high latitudes) is melting, which allows buried stores of the greenhouse gas methane to escape into the atmosphere. Sea level has risen as melting glaciers and sea ice release liquid water into the sea and as the volume of seawater increases due to thermal expansion. The first climate refugees have begun moving away from inundated shoreline settlements. For example, Bhola Island (a region of the Asian nation Bangladesh) became half-submerged by rising sea levels in 1995. Five hundred thousand people were left homeless.

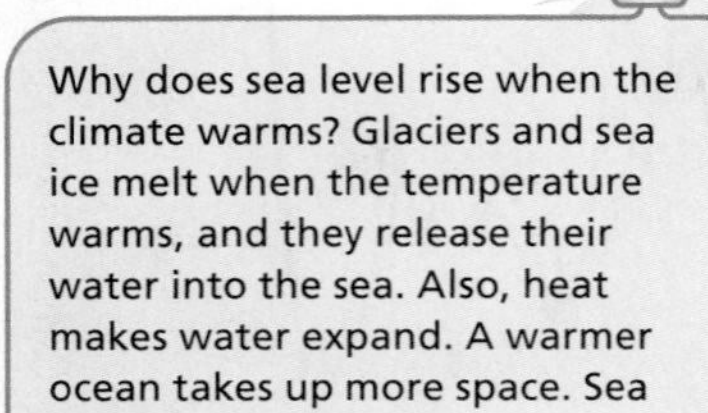

Why does sea level rise when the climate warms? Glaciers and sea ice melt when the temperature warms, and they release their water into the sea. Also, heat makes water expand. A warmer ocean takes up more space. Sea level rises.

The Intergovernmental Panel on Climate Change (IPCC) is a global volunteer organization of more than 1300 scientists from the United States and other countries. The IPCC reviews the credible scientific research on climate change and summarizes its findings in reports for policy makers and the public. The IPCC forecasts a temperature rise of 2.5°F to 10°F over the next century. The principal scientific agencies of the United States, including the Environmental Protection Agency (EPA) and the National Oceanic and Atmospheric Administration (NOAA), have published estimates that agree with the IPCC. There is high confidence and agreement among IPCC and other scientists that global temperatures will continue to rise precipitously, largely due to greenhouse gases produced by human activities. What will be the effects of this future warming?

It is impossible to predict what the particular future effects of global warming will be. There are too many factors involved to make specific forecasts. The IPCC has issued some predictions, however, and assigned probabilities to them; see Table 27.5.

One of the reasons it's so hard to predict future global warming and its effects is that climate is a *system*—a functioning unit made up of parts and processes

(a)

(b)

FIGURE 27.34

(a) A glacier-topped mountain. (b) A glacier that has almost completely melted. Where did the glacier's ice go? How does this relate to the rise in sea level that accompanies global warming?

TABLE 27.5 GLOBAL CLIMATE CHANGE: FUTURE TRENDS

Phenomena	Likelihood of Trend
Contraction of snow cover areas, increased thaw in permafrost regions, decrease in sea ice extent	Virtually certain
Increased frequency of hot extremes, heat waves, and heavy precipitation	Very likely to occur
Increase in hurricane intensity	Likely to occur
Precipitation increases in high latitudes	Very likely to occur
Precipitation decreases in subtropical land regions	Very likely to occur
Decreased water resources in many semi-arid areas, including western U.S. and Mediterranean basin	High confidence

Definitions of likelihood ranges used to express the assessed probability of occurrence: *virtually certain* > 99%, *very likely* > 90%, *likely* > 66%.

Source: Global Climate Change, Vital Signs of the Planet, NASA, 2011 and Summary for Policymakers, IPCC Synthesis Report, November 2007, http://www.ipcc.ch/

that interact with one another. There are many, many interacting components in Earth's overall climate system. Interacting variables aren't restricted to the atmosphere—the atmosphere, hydrosphere, biosphere, and lithosphere are all involved. The interconnections among the parts of a system are called *feedbacks* or **feedback loops**. In climatology, we call the interactions among climate variables *planetary feedbacks*.

The simplest feedback loop consists of just two variables, as shown in Figure 27.35. In a feedback loop, changes in one variable affect the other variable, *and* changes in the second variable in turn affect the first variable. Some feedbacks tend to inhibit change in the system, while other feedback relationships tend to support further changes to the system. Feedbacks that inhibit future change are called *negative feedbacks*. Negative feedback relationships among the variables in a system tend to stabilize the system because, when the second variable responds to changes in the first variable, it tends to suppress further change in the first variable. On the other hand, feedbacks that amplify the change in the initial variable are called *positive feedbacks*. Positive feedbacks are destabilizing—the changes in the interacting variables don't balance one another out; they amplify each other and change is exponential. In the real world, feedback loops are much more complicated than these simplified two-variable examples suggest. Planetary feedback is too complex even for today's sophisticated climate models to fully track.

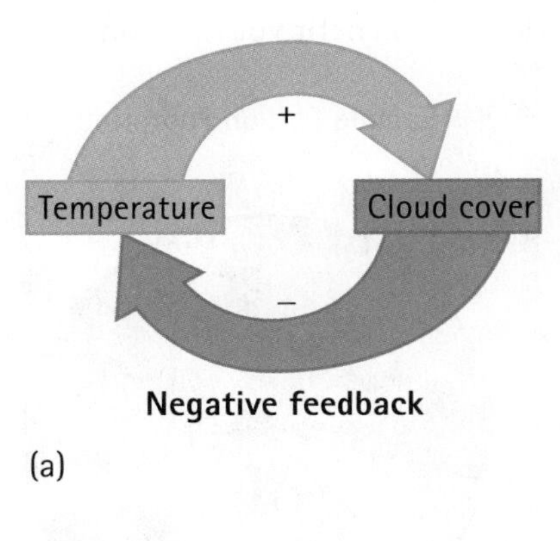

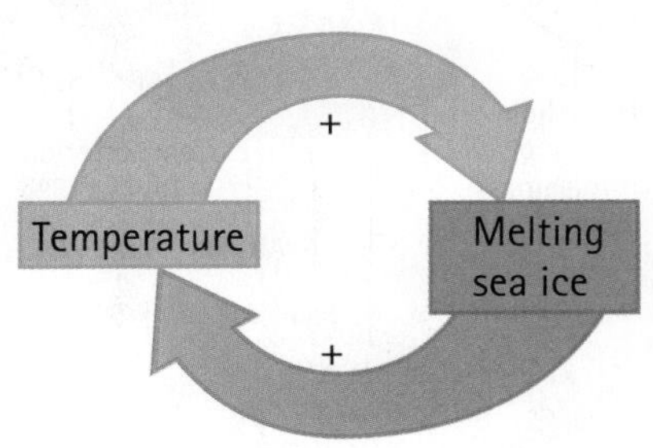

FIGURE 27.35
Diagram of a possible negative feedback loop between cloud cover and Earth's surface temperature. As the surface temperature increases, evaporation increases, which promotes the development of clouds. But the clouds can work against further temperature increases because they reflect sunlight, which cools the air. (b) Diagram of a positive feedback relationship between sea ice and temperature. As temperatures increase, sea ice melts. But, since sea ice is very reflective (it has a high albedo), melting sea ice drives Earth's temperature higher.

UNIFYING CONCEPT

- ***Feedback Regulation***
Section 19.2

To me and other animals who need cold temperatures, global warming just isn't cool.

Although numerous difficulties surely accompany climate change, there may be some positive effects for certain locations. Perhaps global warming will improve the potential for growing crops in areas that are now too cold. The effects simply cannot be predicted with confidence. But, on the whole, the forecast is for challenges ahead. Global warming is one risky experiment that we would rather not conduct.

Responding to Climate Change

There is no single solution to climate change. We need knowledge and we need wisdom too. We need Earth scientists to keep working hard and reporting their findings accurately. We need policy officials who understand the science and formulate laws and regulations that foster sustainability. We need citizens who care for and act on behalf of the planet and one another. Everyone has a part to play in protecting the big blue planet that is our home, our Mother Earth. What will be your role?

As you move ahead, here are three guidelines to help you exert a positive effect on Earth, particularly the changing climate. You'll find that these simple practices can benefit your personal life too.

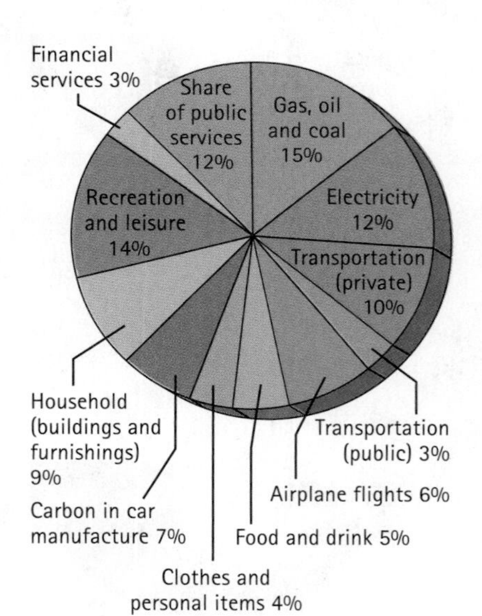

FIGURE 27.36
A "carbon footprint" is the total amount of greenhouse gas emissions that support human activities. It's often expressed as tons of carbon dioxide emitted during one year. This pie chart shows constituents of a typical American citizen's carbon footprint.

1. **Learn more.** You've studied this chapter—that's great. Now find out more! Tap whatever resources you have access to. Subscribe to NASA, NOAA, NSF, EPA, and other scientific and educational Web sites. Investigate citizen science programs to get involved in research projects right now. Find out about opportunities that appeal to you in your local community. Keep an open mind and ask questions until you get answers.
2. **Reduce your carbon footprint.** The average American emits 22 tons of carbon dioxide each year. Your transportation, dietary, and lifestyle choices do have an impact. Calculate your carbon footprint—a good measure of your impact on the climate—by visiting www.epa.gov/climatechange/ghgemissions/ind_calculator.html. Consider Figure 27.36, which represents a typical American citizen's carbon footprint. Develop a plan to reduce yours. Chances are, you'll be healthier and happier in the long run if you make Earth-wise choices.
3. **Stay connected to Earth.** Know enough science to understand and contribute to the public discussion on climate change. Learn more science still to empower yourself even more. Speak, vote, and live your values. Say what you think. Listen to others. Work with them too. Maintain your connection to nature through sports, the food you eat, and music and the arts. Use your talents and opportunities for the benefit of the biosphere. Enjoy a healthy and happy relationship with your home in space—beautiful planet Earth!

FIGURE 27.37
To live sustainably, remember these guidelines: Learn more. Understand your impact. Care for your home, planet Earth.

CHECK YOURSELF

1. **Your friend tells you not to worry about global warming because the greenhouse effect is a natural process that dates back to early Earth. What does he *not understand* about how global warming works?**
2. **Another friend tells you to just give up and plan on moving to Antarctica, where temperatures will rise and crops will thrive. What does she *not understand* about global warming?**
3. **A third friend tells you that people should find ways to reduce greenhouse gas emissions to minimize global warming. What does he likely *understand* about global warming?**

CHECK YOUR ANSWERS

1. In moderation, the greenhouse effect is beneficial to life. However, large concentrations of greenhouse gases in the atmosphere trap so much heat so quickly that ecosystems can't adapt to the temperature rise.
2. The regional effects of global warming are unknown.
3. When it comes to greenhouse gases, less is more. The less greenhouse gas people deposit in the air, the more likely it is that Earth's climate will continue to support a diverse biosphere.

For instructor-assigned homework, go to www.masteringphysics.com 

SUMMARY OF TERMS (KNOWLEDGE)

Albedo The fraction of solar radiation arriving at a surface and reflected by it.

Earthquake The sudden ground motion and energy release that occur when rocks slip along a fault.

Epicenter The point at Earth's surface directly above the focus of an earthquake.

Climate change (global warming) A long-term rise in average global temperature that results from an increase in the greenhouse effect.

Feedback loop A path between the parts of a system such that one part affects a second part and the second part in turn affects the initial part. Regarding climate change, positive feedback accelerates climate change, while negative feedback decelerates it.

Hot spots A place on Earth's surface that sits directly above a rising column of magma that extends to the lower mantle.

Hurricane A large rotating tropical storm with wind speeds of at least 119 km/h.

Ice age A time period that is cold enough so that glaciers collect at high latitudes and move toward lower latitudes.

Interglacial period A period within an ice age that is warm enough so that glaciers melt back to polar latitudes.

Milankovitch theory The idea that slow variations in Earth's orbit, the tilt of its axis, and the path of its wobble affect the way sunlight strikes Earth and therefore alter its temperature.

Pyroclastics Lava and rock fragments ejected into the atmosphere during a violent volcanic eruption.

Ring of Fire A horseshoe-shaped zone of frequent earthquakes and volcanic eruptions that encircles the basin of the Pacific Ocean.

Seismic waves Mechanical waves produced by an earthquake that propagate through Earth.

Tsunami A huge seismic sea wave.

Volcano A mountain that forms when lava or pyroclastic materials pile up around a volcano vent.

READING CHECK (COMPREHENSION)

27.1 Earthquakes

1. What does the Richter scale measure?
2. Why do earthquakes produce seismic waves?
3. Where do earthquakes occur?
4. What method uses records of past earthquakes to predict future earthquakes? What are the limitations of this method?

27.2 Tsunami

5. How is a tsunami similar to a ripple in a pond? How do they differ?
6. What causes a tsunami?
7. Describe two devastating tsunami that have occurred since the year 2000. Tell when they occurred and why they were costly in terms of lives and property.

27.3 Volcanoes

8. What are the three major kinds of volcanoes?
9. Which types of volcanoes erupt quietly? Which erupt most explosively?
10. What are the hazards associated with volcanoes?

27.4 Hurricanes

11. What is the source of a hurricane's energy?
12. Where do hurricanes form? Describe how they travel.
13. What was the costliest natural disaster in U.S. history?

THINK INTEGRATED SCIENCE

27A—Climate Change

14. What astronomical changes produce climate change on Earth?
15. Can continental drift cause climate change? Provide an example.
16. Cite two kinds of catastrophes that occur naturally and can alter climate.
17. Relate the greenhouse effect to global warming.
18. What are greenhouse gases?
19. Besides burning fossil fuels, what causes carbon dioxide buildup?
20. How do planetary feedbacks affect climate change?
21. What is it about the Industrial Era that has apparently caused the average global temperature to increase?
22. Explain why melting ice caps cause Earth's temperature to rise. (Hint: Use the term *albedo*.)
23. Give two examples of mass extinctions that apparently were related to climate change.
24. What does albedo depend upon?
25. Why is the greenhouse effect a good thing for life on Earth? Why is it a bad thing?
26. Why is melting permafrost a climate concern?
27. Why is sea level expected to rise as global temperature climbs?
28. By how much did the average global temperature rise during the past century?
29. Burning a plant releases carbon dioxide in two ways. What are they?
30. What is the IPCC? By how much does the IPCC predict climate will change over the next century?

THINK AND DO (HANDS-ON APPLICATION)

31. Use your hands to simulate motion along a fault. Press your palms together and move your hands in opposite directions, as shown in the figure. Notice that friction keeps them "locked" together until the opposing forces no longer balance, friction no longer holds, and your palms slip suddenly past each other. This demonstration nicely models the sticking and slipping that occur along a transform fault.

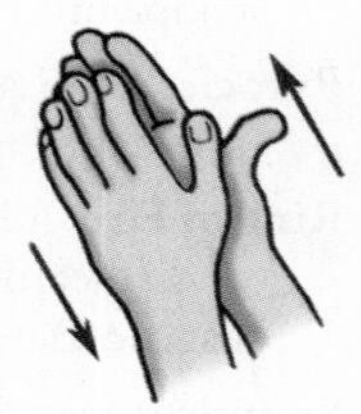

THINK AND SOLVE (MATHEMATICAL APPLICATION)

32. The Richter scale measures how much the ground shakes in an earthquake. Note that this is not equal to the energy released by an earthquake. However, there is a simple relationship between Richter magnitude and earthquake energy.

    A 1-point increase on the Richter scale corresponds to an energy increase of about 30 times. Thus a 5.0-magnitude earthquake releases about 30 times as much energy as a 4.0-magnitude earthquake. The 5.0-magnitude earthquake releases about 900 (30×30) times the energy of a 3.0-magnitude quake.

    (a) How much more energy does an 8.0-magnitude earthquake release than a 7.0-magnitude earthquake?
    (b) How much more energy does a 7.0-magnitude earthquake release than a 5.0-magnitude earthquake?
    (c) How much more energy does a 7.0-magnitude earthquake release than a 4.0-magnitude earthquake?
    (d) How much *less* energy does a 6.0-magnitude earthquake release than a 7.0-earthquake?

33. Investigate your carbon footprint. Go to the Environmental Protection Agency Web site (http://www.epa.gov/climatechange/kids/calc/index.html) and use the Global Climate Change Calculator. Calculate the impact you have through simple actions, such as turning off electric lights, on atmospheric carbon dioxide concentration.

THINK AND EXPLAIN (SYNTHESIS)

34. Why do shield volcanoes have broader bases than cinder cone volcanoes?
35. Why do hurricanes die when they reach land?
36. Have volcanoes been more of a threat in modern times or in ancient times? State your reasons.
37. How is rock before an earthquake like a compressed or stretched Slinky?
38. Briefly describe how a tsunami develops.
39. Heat waves kill more people in the United States than all of the other natural disasters combined. Why do heat waves mainly kill people who live in cities?
40. Name three natural hazards discussed in this chapter that affect coastal regions more than inland areas.
41. The use of fossil fuels contributes to global climate change. Are there other reasons to reduce our consumption of fossil fuels? If so, describe them.
42. Why have levels of carbon dioxide in the atmosphere increased sharply since the latter part of the 19th century?
43. What are some likely effects of global warming?
44. As technological society advances, humans use increasing amounts of energy. Does this violate the law of conservation of energy? Why or why not?
45. In what way do plants reduce the threat of global warming?
46. Can the problems associated with fossil fuels be cured by planting enough trees? Explain your reasoning.
47. How is a volcano like a shaken bottle of soda?
48. If you had to live near a composite volcano or a shield volcano, which type would you choose? Why?
49. Where is the Ring of Fire, and how did it earn its name?
50. What would eventually happen to a volcano that formed over a hot spot? (*Hint:* Think about plate movement.)
51. Some engineers have suggested burying radioactive wastes in ocean trenches at subduction zones. Why do you think they have proposed this? What hazards could this create?
52. Drowning causes the most deaths in a hurricane. Explain why, using the term *storm surge* in your answer.
53. Suppose geologists report that the stress in Earth's crust has increased where you live. Should you move? If so, how soon? Explain the reasons for your answer.
54. Draw a feedback loop with two variables: melting sea ice and temperature. Is this a positive or negative feedback loop? Does it enhance climate change or balance Earth's climatic system?
55. Which of these graphs represents feedback that destabilizes climate?

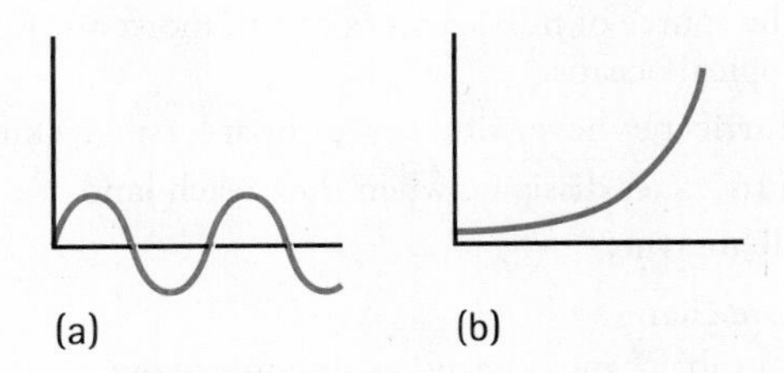

56. Do greenhouse gases truly trap the infrared radiation emitted by Earth or do they simply delay its transmission to space?
57. Why does paving the ground contribute to planetary warming?
58. Earth emits radiation. Is this the reason we can see Earth? Explain your answer.
59. Name one preventable natural disaster. Name one nonpreventable disaster.
60. What are three hazards associated with living in the Ring of Fire?
61. Why can't Arctic species use the same strategy to adapt to a warming climate that Edith's checkerspot butterfly used?
62. Earth has usually been much warmer than it is today. During the high point of the Mesozoic era, in fact, the sea surface temperature may have been 31°F warmer than at present. Yet life was abundant on planet Earth in the Mesozoic era. Why then do we worry about high surface temperatures today?
63. Why are hurricanes expected to become more frequent as the climate warms?
64. How do andesitic and basaltic lava differ from one another?
65. How does the type of magma in a volcano affect how the volcano erupts?
66. How did an earthquake in Japan in 2011 lead to the worst nuclear reactor accident since 1986?

THINK AND DISCUSS (EVALUATION)

67. Is climate change natural or anthropogenic? Give a thorough response.
68. How is the large human population a contributing cause to climate change?
69. Climate change is controversial even though the overwhelming majority of scientists agree that it is real and that it is human-caused. Why is the issue so controversial?
70. Seismic gap analysis shows that a densely populated region is at high risk for an earthquake. Should disaster officials warn the public? What are your reasons?

READINESS ASSURANCE TEST (RAT)

If you have a good handle on this chapter, if you really do, then you should be able to score 7 out of 10 on this RAT. If you score less than 7, you need to study further before moving on.

Choose the BEST answer to the following questions.

1. Which of these statements about hurricanes are true?
   (a) The source of hurricanes is warm, moist air over tropical oceans.
   (b) Hurricanes have wind speeds of at least 119 km/h.
   (c) Hurricanes dissipate when they reach land.
   (d) All are true.
2. Climate change
   (a) is a subject most scientists disagree about.
   (b) refers to the effect of greenhouse gases on atmospheric temperature.
   (c) refers to temperature increases in all of Earth's climate zones.
   (d) is not related to human activities.
3. Climates farther from the equator usually have
   (a) lower average temperatures.
   (b) higher solar intensity.
   (c) more precipitation.
   (d) more water.
4. Ice ages occur because of
   (a) chemicals people put into the atmosphere.
   (b) variations in Earth's tilt and orbit.
   (c) asteroids crashing into Earth.
   (d) changes in the temperature of Earth's core.
5. In an ice age, glaciers
   (a) advance.
   (b) retreat.
   (c) cause sea level to rise.
   (d) reflect incoming solar radiation.
6. Earthquakes occur because
   (a) tectonic plates shift as Earth cools.
   (b) tectonic plates shift as Earth warms.
   (c) seismic waves escape Earth's interior.
   (d) Pangaea continues to break up into smaller pieces.
7. Tsunami
   (a) are usually caused by hurricanes.
   (b) are usually caused by global warming.
   (c) are usually caused by the Milankovitch cycles.
   (d) are usually caused by earthquakes.
8. What can scientists do to predict earthquakes?
   (a) nothing
   (b) measure the strain in rocks and map inactive sections of active fault zones
   (c) measure the elasticity of rock
   (d) measure the distance from a certain location to the nearest subduction zone
9. Hurricanes are related to
   (a) heat, pressure, and water.
   (b) earthquakes.
   (c) tsunami.
   (d) volcanoes.
10. Which of the following is a sign that a volcano may soon erupt?
   (a) bulging of its flanks
   (b) climate change
   (c) warming of its base and flanks
   (d) odd animal behavior

Answers to RAT:

1. d, 2. b, 3. a, 4. b, 5. a, 6. a, 7. d, 8. b, 9. a, 10. a

5

PART FIVE

Astronomy

If I were standing on the Moon, I wonder what Earth would look like when the Moon passes into Earth's shadow—that is, during a lunar eclipse. I'm told Earth would look like a large dark orb surrounded by a ring of brilliant red as the light from the hidden Sun refracts through Earth's atmosphere—like a zillion sunsets all at once! Hmm. All those reddish sunsets would light up the Moon's surface. That must be why here on Earth the Moon appears red during a lunar eclipse!

28

CHAPTER 28

The Solar System

HOW DOES the Sun produce so much energy? How are the planets similar, and how are they different? How did our Moon form, and how does it go through phases? Why do we see only one side of the Moon? What are solar and lunar eclipses, and why are they rare? What are meteors, asteroids, and comets? How frequently do they collide with our planet, and why does a comet's tail always point away from the Sun?

For thousands of years, people have gazed into the night sky and pondered such questions. Thanks to scientific advances over the past century, however, the answers to these and other basic questions about the universe are now available to us. This chapter focuses on what we know about our solar system, which on a cosmic scale is Earth's own backyard. In the next chapter, we expand our view to galaxies and beyond, where deeper mysteries still remain.

28.1 The Solar System and Its Formation

LEARNING OBJECTIVE
Describe how nebular theory accounts for the main attributes of our solar system.

EXPLAIN THIS How is gravity responsible for solar energy?

Our solar system is the collection of objects gravitationally bound to the Sun. Along with the Sun itself, the solar system contains at least eight **planets**, which are large orbiting bodies massive enough for their gravity to make them spherical but small enough to avoid having nuclear fusion in their cores. The planets also have successfully cleared all debris from their orbital paths, which all lie roughly in the same plane. This plane, called the **ecliptic**, is defined as the plane of Earth's orbit. Within the solar system are also numerous moons (objects orbiting planets), asteroids (small, rocky bodies), comets (small, icy bodies), and a collection of miniature planets known as dwarf planets that orbit on the outer edges of the solar system. The most well-known dwarf planet is Pluto, which was downgraded from planet status in 2006. The planets and all these other objects are quite small compared to the Sun. Figure 28.1 shows the sizes of the planets relative to the Sun, which contains a whopping 99.86% of the solar system's mass.

MasteringPhysics®
TUTORIAL: Formation of the Solar System
VIDEO: History of the Solar System
VIDEO: Orbits in the Solar System

The vast distances between the Sun and the objects orbiting it can be grasped by imagining the Sun reduced to the size of a large beach ball 1 m in diameter. The closest planet, Mercury, would be an apple seed located about 40 m (130 ft) away. The next closest planet, Venus, would be the size of a pea about 80 m (255 ft) distant. Earth, also about the size of a pea, would be about 110 m away, which is greater than the length of a football field. The next planet, Mars, which is only a bit larger than Mercury, would be an apple seed almost two football field lengths away from our solar beach ball.

The ancients could tell the difference between planets and stars because of the difference in their movements in the sky. The stars remain relatively fixed in their patterns in the sky, but the planets wander. The term *planet* is derived from the Greek for "wandering star."

These first four planets—Mercury, Venus, Earth, and Mars—are called the **inner planets** because of their relatively close proximity to the Sun. All the inner planets are solid rocky planets. The **outer planets** are larger gaseous planets located much farther away. The first outer planet is Jupiter, which on the scale mentioned would be the size of a softball more than half a kilometer away. The second outer planet, Saturn, famous for its extensive ring system, would be the size of a baseball more than a kilometer away. Planets Uranus and Neptune would both be about the size of Ping-Pong balls located 2 and 3 km away, respectively. We see that solar system objects are mere specks in the vastness of the space about the Sun.

FIGURE 28.1
This illustration shows the order and relative sizes of planets. Moving away from the Sun, we have in order: Mercury, Venus, Earth, Mars, Jupiter, Saturn, Uranus, and Neptune. The planets range greatly in size, but the Sun dwarfs them all—containing more than 99% of the mass in the solar system. *Note:* Distances are not to scale in this illustration. If they were, Earth would be placed over 8 meters away. Neptune would be over 240 meters away!

TABLE 28.1 PLANETARY DATA

	Mean Distance from Sun (Earth distances, AU)	Orbital Period (years)	Diameter		Mass		Density (g/cm³)	Inclination to Ecliptic
			(km)	(Earth = 1)	(kg)	(Earth = 1)		
Sun			1,392,000	109.1	1.99×10^{30}	3.3×10^{5}	1.41	
Terrestrial								
Mercury	0.39	0.24	4,880	0.38	3.3×10^{23}	0.06	5.4	7.0°
Venus	0.72	0.62	12,100	0.95	4.9×10^{24}	0.81	5.2	3.4°
Earth	1.00	1.00	12,760	1.00	6.0×10^{24}	1.00	5.5	0.0°
Mars	1.52	1.88	6,800	0.53	6.4×10^{23}	0.11	3.9	1.9°
Jovian								
Jupiter	5.20	11.86	142,800	11.19	1.90×10^{27}	317.73	1.3	1.3°
Saturn	9.54	29.46	120,700	9.44	5.7×10^{26}	95.15	0.7	2.5°
Uranus	19.18	84.0	50,800	3.98	8.7×10^{25}	14.65	1.3	0.8°
Neptune	30.06	164.79	49,600	3.88	1.0×10^{26}	17.23	1.7	1.8°
Dwarf Planets								
Pluto	39.44	247.70	2,300	0.18	1.3×10^{22}	0.002	1.9	17°
Eris	67.67	557	2,400	0.19	1.6×10^{22}	0.002	1.9	44°

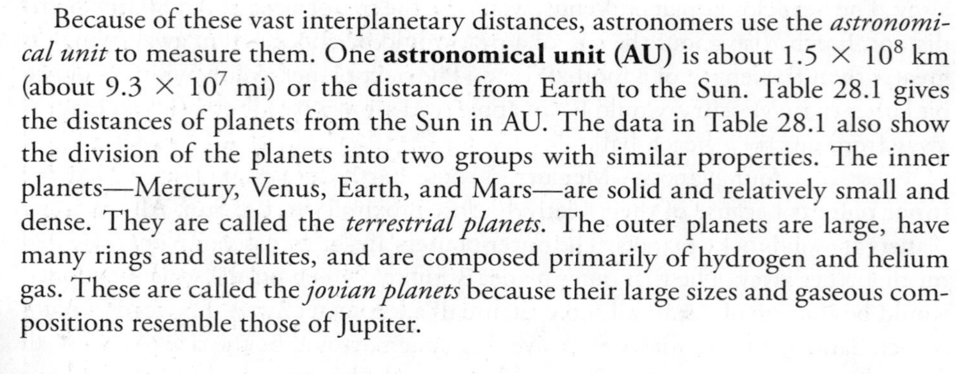

Because of these vast interplanetary distances, astronomers use the *astronomical unit* to measure them. One **astronomical unit (AU)** is about 1.5×10^8 km (about 9.3×10^7 mi) or the distance from Earth to the Sun. Table 28.1 gives the distances of planets from the Sun in AU. The data in Table 28.1 also show the division of the planets into two groups with similar properties. The inner planets—Mercury, Venus, Earth, and Mars—are solid and relatively small and dense. They are called the *terrestrial planets*. The outer planets are large, have many rings and satellites, and are composed primarily of hydrogen and helium gas. These are called the *jovian planets* because their large sizes and gaseous compositions resemble those of Jupiter.

FIGURE 28.2
This photograph, provided by the Hubble Telescope, shows the Orion Nebula. The Orion Nebula, like the nebula from which our solar system formed, is an interstellar cloud of gas and dust and the birthplace of stars.

Nebular Theory

Any theory of solar system formation must be able to explain two major regularities: (1) *the motions among large bodies of the solar system* and (2) *the division of planets into two main types*—terrestrial *and* jovian. Further, a viable theory of solar system formation must explain other known features of the solar system, including the existence of asteroids, comets, and moons.

The modern scientific theory that satisfies these requirements is called the **nebular theory**, which holds that the Sun and planets formed together from a cloud of gas and dust, a *nebula* (Latin for "cloud"). According to the theory, the solar system began to condense from the cloud of gas and dust about 5 billion years ago. The cloud would have been very diffuse and large, with a diameter thousands of times larger than Pluto's orbit. A nebula currently visible within the constellation of Orion is shown in Figure 28.2.

Within the nebula from which our solar system formed, the gravitational pull on particles exceeded the tendency of gas to disperse and fill all available space. Once the gravitational collapse of a cloud begins, gravity ensures that it continues. The universal law of gravity (see Chapter 5) is an inverse-square law: The strength of the gravitational force increases dramatically as the particles become

MATH CONNECTION

The Scale of the Solar System

Astronomical distances are mind-boggling. Try the following problems to better appreciate the sizes of bodies in the solar system and the distances between them. (Use the distance formula: distance = speed × time and data from Table 28.1.)

Problem The distance between Earth and the Moon is 384,401 km. How many Earth diameters would fit between Earth and the Moon?

Solution:

$$\frac{384{,}401 \text{ km}}{12{,}760 \text{ km}} \approx 30$$

About 30 Earth-sized planets would fit in the distance between Earth and the Moon.

Problem How long would it take to drive from Earth to the Moon if you drive at 55 mi/h? State your answer in years.

Solution: $d = vt \rightarrow t = \frac{d}{v}$

Convert all units to metric:
$(55 \text{ mi/h})(1 \text{ km}/0.62 \text{ mi}) \approx 89 \text{ km/h}$.
Then $t = 384{,}401 \text{ km}/89 \text{ km/h} =$
$4319 \text{ h } (1 \text{ day}/24 \text{ h})(1 \text{ yr}/365 \text{ days}) \approx 0.5 \text{ yr}$.
It would take about one-half year to drive to the Moon traveling at freeway speeds.

Problem If you could fly to the Sun on a jet that moves at 1000 km/h, how long would it take? State your answer in years.

Solution:

$$t = \frac{d}{v} = \frac{1.5 \times 10^8 \text{ km}}{1000 \text{ km/h}}$$

$$= 1.5 \times 10^5 \text{ h}$$

To convert to years, multiply by the following conversion factors:

$$(1.5 \times 10^5 \text{ h})\left(\frac{1 \text{ day}}{24 \text{ h}}\right)\left(\frac{1 \text{ yr}}{365 \text{ days}}\right) = 17 \text{ yr}$$

So if you could fly to the Sun in a jet without becoming vaporized, it would take about 17 years.

Problem The diameter of the Sun is 1,390,000 km. What is its diameter in AU? How much greater is the mean distance between Earth and the Sun compared to the diameter of the Sun?

Solution:

$$(1.39 \times 10^6 \text{ km}) \times (1 \text{ AU}/1.5 \times 10^8 \text{ km}) \approx 0.01 \text{ AU}$$

$$\frac{1 \text{ AU}}{0.01 \text{ AU}} = 100$$

The diameter of the Sun is approximately equal to 0.01 AU, so the distance between Earth and the Sun is about 100 times the diameter of the Sun. Big as the Sun is, the solar system is mostly empty space.

closer. The cloud maintained a constant mass as it shrank, but gravitational forces grew ever stronger, and the cloud came together, taking on a spherical shape.

UNIFYING CONCEPT
● ***The Gravitational Force***
Section 5.3

This nebula must also have had a slight net rotation, possibly due to the rotation of the galaxy itself. As the nebula shrank, it spun faster and faster (because of the *conservation of angular momentum* discussed in Appendix B). As any spinning object contracts, the speed of its spin increases such that angular momentum is conserved. A familiar example is an ice skater whose spin rate increases when her extended arms are pulled inward. A nebula does the same, as shown in Figure 28.3.

What happens to the shape of a sphere as it spins faster and faster? The answer is that it flattens. A familiar example is the chef who turns a ball of pizza dough into a disk by spinning it on his hands. Even planet Earth is a slightly "flattened" sphere because of its daily spin. Saturn, with a faster spin, noticeably departs from a purely spherical shape. So the initially spherical nebula progressed to a spinning disk, the center of which became the *protosun*.

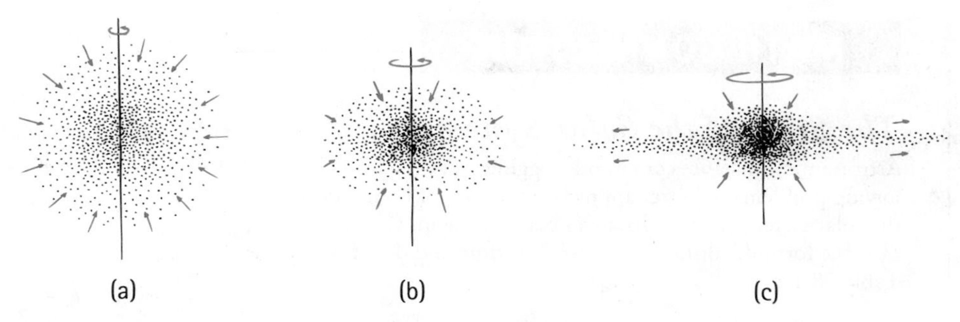

FIGURE 28.3
(a) The nebula from which the solar system formed was originally a large, diffuse cloud that rotated quite slowly. The cloud began to collapse under the influence of gravity. (b) As the cloud collapsed, it heated up as gravitational potential energy converted to heat. It spun faster by the conservation of angular momentum. (c) The cloud flattened into a disk as a result of its fast rotation. A spinning, flattened disk was produced whose mass was concentrated at its hot center.

UNIFYING CONCEPT
- *The Scientific Method* **Section 1.3**

The formation of the spinning disk explains the motions of our solar system today. All planets orbit the Sun in nearly the same plane because they formed from that same flat nebular disk. The direction in which the disk was spinning became the direction of the Sun's rotation and the orbits of planets.

CHECK YOURSELF
As a nebula contracts, its rate of spin increases. What rule of nature is at play here?

CHECK YOUR ANSWER
The rule that applies to all spinning bodies is the *conservation of angular momentum*. That rule and the conservation of energy are dominant players in the universe.

The surrounding disk was the source of material that became the planets. In the spinning disk, matter collected in some regions more densely than in others. Perhaps small particles of gas and dust stuck together via gravity or electrostatic attraction. Because of their extra mass, these clumps exerted a stronger gravitational force on one another than on neighboring regions of the disk, and so they pulled in even more material to them. This led to the accretion of the nebular disk into small objects called *planetesimals*, which ranged in size from boulders to objects several kilometers in diameter. Planetesimals grew larger through countless collisions until they gravitationally dominated surrounding matter and finally became full-grown planets.

All the while, compression due to gravity caused a rise in temperature, especially within the central portions of the solar nebula, which developed into a hot ball of gas called the *protosun*. The protosun transformed into the Sun once the temperature became sufficient to ignite thermonuclear fusion. Once thermonuclear fusion occurred and the Sun began to radiate energy, the nebular disk warmed, with the inner portions reaching higher temperatures than the outer portions. As a result, the inner and outer planets developed differently. The inner planets formed from materials that remained solid at high temperatures; hence, the inner planets are rocky. The outer planets, by contrast, consist mainly of hydrogen and helium gas that coalesced in the cold regions of the solar system far from the Sun. Thus, we can see that the nebular theory accounts for the formation of the planets and the neat division of them into two groups.

28.2 The Sun

LEARNING OBJECTIVE
Recognize the features of the Sun, including its interior, photosphere, sunspots, solar cycle, chromosphere, and corona.

EXPLAIN THIS Why is the Sun's surface much cooler than its inner core?

The Sun has a mass of about a trillion quadrillion tons (2×10^{30} kg). It produces energy from the thermonuclear fusion of hydrogen into helium. Each second, approximately 657 million tons of hydrogen is fused to 653 million tons of helium. The 4 million tons of mass lost is discharged as radiant energy. This conversion of hydrogen into helium in the Sun has been going on since it formed nearly 5 billion years ago, and it is expected to continue for another 5 billion years. A tiny fraction of the Sun's energy reaches Earth and is converted by photosynthesizing organisms into chemical energy stored in large molecules. These energy-rich molecules are the primary energy source for almost all the organisms of this planet. The Sun, Earth's nearest star, is the solar system's power supply.

MasteringPhysics®
TUTORIAL: The Sun

Solar energy is generated deep within the core of the Sun. The solar core constitutes about 10% of the Sun's total volume. It is very hot—more than 15,000,000 K. The core is also very dense, with more than 12 times the density of solid lead. Pressure in the core is 340 billion times Earth's atmospheric pressure! Because of these intense conditions, the hydrogen, helium, and minute quantities of other elements exist in the plasma state. *Plasma*, recall, is the phase of matter beyond gas, consisting of ions and electrons rather than atoms—electrons have been stripped from atoms by high energies. The nuclei of this plasma move fast enough to undergo nuclear fusion, as discussed in Chapter 10 (Integrated Science 10C). The energy released from this nuclear fusion rises to the surface, where it causes gases to emit a broad spectrum of electromagnetic radiation, centered in the visible region (see Figure 8.31).

The Sun's surface is a layer of glowing 5800 K plasma, which is much cooler than the Sun's core but hot enough to generate lots of light. This layer, called the *photosphere* (sphere of light), is about 500 km deep. Within the photosphere are relatively cool regions that appear as **sunspots** when viewed from Earth. Sunspots are cooler and darker than the rest of the photosphere and are caused by magnetic fields that impede hot gases from rising to the surface. As shown in Figure 28.4, sunspots can be seen by focusing the image of the Sun from a telescope or pair of binoculars onto a flat white surface. Sunspots are typically twice the size of Earth, move around because of the Sun's rotation, and

FIGURE 28.4
Never directly look at the Sun! Instead, you can get a nice view of the Sun by focusing the image of the Sun from a pair of binoculars onto a white surface. If the Sun is eclipsed by the Moon, which is a rare event, the Sun is seen as a crescent. More commonly, the Sun's image may reveal sunspots.

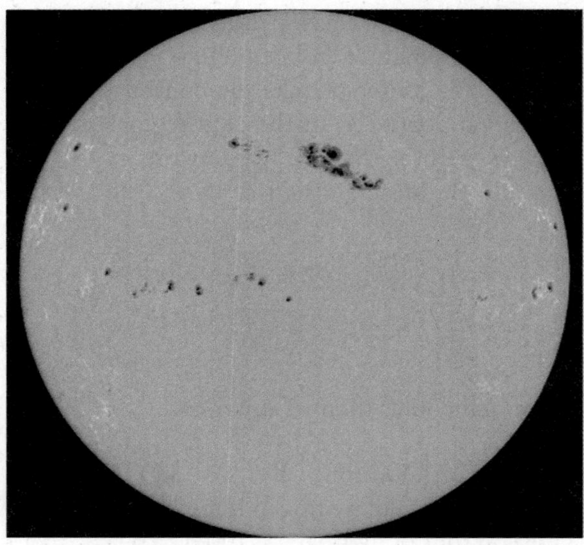

FIGURE 28.5
Sunspots on the solar surface are relatively cool regions. We say relatively cool because they are hotter than 4000 K. They look dark only in contrast with their 5800 K surroundings.

FIGURE 28.6
The pink chromosphere becomes visible when the Moon blocks most of the light from the photosphere during a solar eclipse.

FIGURE 28.7
The pearly white solar corona is visible only during a solar eclipse. Notice how this exceptional photo of the corona also captures some of the pink of the chromosphere as well as the face of the new Moon, which is faintly illuminated by light reflected from the full Earth. In other words, if you were standing on the Moon when this photo was taken, you would see your faint shadow cast by the fully lit Earth shining above you.

last about a week or so. Often, they cluster in groups as shown in Figure 28.5.

The Sun spins slowly on its axis. Because the Sun is a fluid rather than a solid, different latitudes of the Sun spin at different rates. Equatorial regions spin once in 25 days, but higher latitudes take up to 36 days to make a complete rotation. This differential spin means the surface near the equator pulls ahead of the surface farther north or south. The Sun's differential spin wraps and distorts the solar magnetic field, which bursts out to form the sunspots mentioned earlier. A reversal of magnetic poles occurs every 11 years, and the number of sunspots also reaches a maximum every 11 years (currently). The complete cycle of solar activity is 22 years.

Above the Sun's photosphere is a transparent 10,000-km-thick shell called the *chromosphere* (sphere of color), seen during an eclipse as a pinkish glow surrounding the eclipsed Sun. The chromosphere is hotter than the photosphere, reaching temperatures of about 10,000 K. Its beautiful pink color, as shown in Figure 28.6, arises from the emission of light from hydrogen atoms.

Beyond the chromosphere are streamers and filaments of outward-moving, high-temperature plasmas curved by the Sun's magnetic field. This outermost region of the Sun's atmosphere is the *corona*, which extends out several million kilometers (Figure 28.7). The temperature of the corona is amazingly high—on the order of 1 million K—and it is where most of the Sun's powerful X-rays are generated. Because the corona is not very dense, its brightness is not as intense as the Sun's surface, which makes the corona safe to observe during (and only during) a total solar eclipse. High-speed protons and electrons are cast outward from the corona to generate the *solar wind*, which powers the aurora on Earth and produces the tails of comets.

CHECK YOURSELF

1. **Was the Sun more massive 1000 years ago than it is today? Defend your answer.**
2. **Of the photosphere, chromosphere, and corona, which is thinnest? Which is hottest? Which is between the other two?**

CHECK YOUR ANSWER

1. Yes, although slightly compared with its great mass. The Sun loses mass as hydrogen nuclei combine to make helium nuclei.
2. The photosphere is the thinnest. The corona is the hottest. The chromosphere is the pinkish layer above the photosphere and below the vast corona.

FYI We have seven days in a week because ancient Europeans decided to name days after the seven wandering celestial objects they could observe. The English day names were derived from the language of the Teutonic tribes who lived in the region that is now Germany. In Teutonic, the Sun is *Sun* (Sunday), the Moon is *Moon* (Monday), Mars is *Tiw* (Tuesday), Mercury is *Woden* (Wednesday), Jupiter is *Thor* (Thursday), Venus is *Fria* (Friday), and Saturn is *Saturn* (Saturday).

28.3 The Inner Planets

LEARNING OBJECTIVE
Identify the major properties of the four inner planets: Mercury, Venus, Earth, and Mars.

EXPLAIN THIS Why does Venus, not Mercury, have the hottest surface of any planet in the solar system?

Compared with the outer planets, the four planets nearest the Sun are close together. These are Mercury, Venus, Earth, and Mars. Each of these rocky planets has a mineral-containing solid crust.

Mercury

Mercury (Figure 28.8) is about 1.4 times larger than Earth's Moon and similar in appearance. It is the closest planet to the Sun. Because of this closeness it is the fastest planet, circling the Sun in only 88 Earth days—which thus equals one Mercury "year." Mercury spins about its axis only three times for each two revolutions about the Sun. This makes its daytime very long and very hot, with temperatures as high as 430°C.

Because of Mercury's small size and weak gravitational field, it holds very little atmosphere. Mercury's atmosphere is only about a trillionth as dense as Earth's atmosphere—it's a better vacuum than laboratories on Earth can produce. So without a blanket of atmosphere, and because there are no winds to transfer heat from one region to another, nighttime on Mercury is very cold, about −170°C. Mercury is a fairly bright object in the nighttime sky and is best seen as an evening "star" during March and April or as a morning star during September and October. It is seen near the Sun at sunup or sunset.

FIGURE 28.8
Mercury is heavily cratered from the impacts of many meteorites. Mercury is a small planet, with only about 6% of the volume and mass of Earth. This photo was taken by the *Messenger* spacecraft, which was launched from Earth in 2004 and reached orbit around planet Mercury in 2011.

Venus

Venus is the second planet from the Sun and, like Mercury, has no moon. Venus is frequently the first starlike object to appear after the Sun sets, so it is often called the evening "star," as illustrated in Figure 28.9. Compared with other planets, Venus most closely resembles Earth in size, density, and distance from the Sun

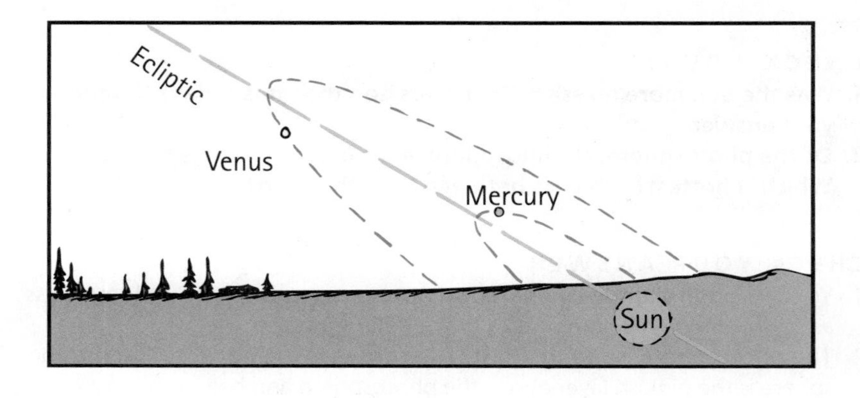

FIGURE 28.9
Because the orbits of Mercury and Venus lie inside the orbit of Earth, they are always near the Sun in our sky. Near sunset (or sunrise) they are visible as "evening stars" or "morning stars."

FIGURE 28.10
Venus is an Earth-sized planet barren of any oceans. The surface of Venus was first mapped in the early 1990s by the spacecraft *Magellan*, which used microwave radar ranging to "see through" the planet's thick atmosphere of carbon dioxide and sulfuric acid clouds.

FYI Ancient American cultures ran their lives using three calendars. Their secular calendar, which told them when to plant seeds and so forth, followed the 365-day orbit of Earth. Their religious calendar centered on the roughly 260-day orbit of Venus. Both of these calendars were cyclical and couldn't account for succeeding years. For that purpose they developed the "long count" calendar, which, interestingly enough, employed the concept of zero. They did this centuries before the concept of zero was recognized by accountants in India.

(Figure 28.10). However, Venus has a very dense atmosphere and opaque cloud cover that generate high surface temperatures (470°C)—too hot for oceans. The atmosphere of Venus is about 96% CO_2. Remember from Chapter 26 that carbon dioxide is a "greenhouse gas." By this we mean that CO_2 blocks the escape of infrared radiation from Earth's surface and contributes to the warming of our planet. The thick blanket of CO_2 surrounding Venus effectively traps heat near the Venusian surface. This and Venus's proximity to the Sun make Venus the hottest planet in the solar system.

Another difference between Venus and Earth is the way the two planets spin about their axes. Venus takes 243 Earth days to make one full spin and only 225 Earth days to make one revolution around the Sun. Furthermore, Venus spins in a direction opposite to the direction of Earth's spin. So, on Venus, the Sun rises in the west and sets in the east. But, because the cloud cover is so dense, a Venusian sunrise or sunset is never visible from its surface.

The slow spin of Venus means that the atmosphere is not disturbed by the Coriolis effect described in Chapter 26. As a result, there is very little wind and weather on the surface of Venus. Instead, the stifling hot dense air sits still throughout its long days and nights.

In recent years, 17 probes have landed on the surface of Venus. There have been 18 flyby spacecraft (notably *Pioneer Venus* in 1978 and *Magellan* in 1993). From spacecraft data, scientists have been able to account for why the atmospheres of Venus and Earth are so different. According to the generally accepted model, when the two planets first formed they had similar amounts of water. Venus, however, is a bit closer to the Sun than Earth is. It also rotates much more slowly. These two factors combined to make the Sun-facing side of Venus significantly warmer than Earth. With greater warmth, more of Venus's water evaporated into its atmosphere. Like carbon dioxide, water vapor is a powerful greenhouse gas. So more water vapor in the atmosphere caused even more warming, which caused more of the oceans to evaporate, causing further warming—runaway global warming! Venus's early oceans contained massive amounts of dissolved carbonates, just as on Earth. As the oceans evaporated, these carbonates transformed into carbon dioxide and moved into the atmosphere, increasing the intensity of the greenhouse effect. Water vapor in the early Venusian atmosphere was subject to the Sun's ultraviolet rays, which broke the water down into hydrogen and oxygen. The hydrogen escaped into space while the oxygen chemically reacted with minerals on the surface. In the end, the planet's supply of water was forever lost. All that remains we see today as a thick atmosphere of heat-trapping carbon dioxide, as well as other compounds, most notably, sulfuric acid, H_2SO_4.

CHECK YOURSELF

As the Sun gets older it also gets hotter. How will this affect the amount of water vapor in our atmosphere?

CHECK YOUR ANSWER

Initially, the additional heat from the Sun will increase the rate at which Earth's oceans evaporate. More water vapor in the atmosphere will then enhance the greenhouse effect, causing even warmer temperatures, which will provide for even more evaporation. As the oceans disappear, carbon dioxide levels will also rise dramatically, ensuring that the greenhouse effect keeps the water vaporized. Ultraviolet light from the Sun, however, will zap this water from the atmosphere. So, in the end, as the Sun gets hotter, the amount of water vapor in the atmosphere will drop to near nothing. We will share the same fate as our sister planet Venus. You can breathe a bit easy because the timeframe for this is about a billion years.

FYI Not all of the Venusian water was lost to space. Some instead reacted with volcano-generated sulfur dioxide to form sulfuric acid, which now laces the upper levels of the Venusian atmosphere. The water that stayed behind to react with sulfur dioxide was predominately water containing the heavier isotope of hydrogen, known as deuterium. This has been confirmed with direct measurements from space probes, which show an unusually high proportion of deuterium. This, in turn, supports the theory that Venus lost its water because of a runaway greenhouse effect. Without the greenhouse effect, the proportion of deuterium in the atmosphere would remain similar to that currently found on Earth.

Earth

Our home planet Earth resides within the Sun's *habitable zone*, which is a region not too close and not too far from the Sun so that water can exist predominately in the liquid phase, as shown in Figure 28.11. Earth has an abundant supply of liquid water covering about 70% of Earth's surface, which makes our planet the blue planet (Figure 28.12).

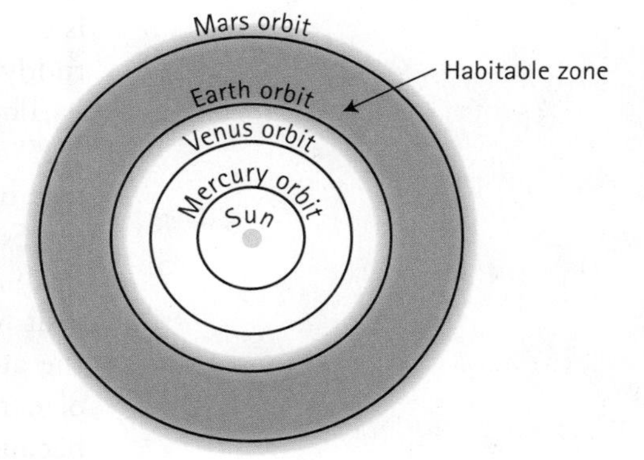

FIGURE 28.11
Earth resides on the inner side of the Sun's habitable zone, which is where conditions are favorable for life as we know it.

Earth's oceans support the *carbon dioxide cycle*, which acts as a thermostat to keep global temperatures from reaching harsh extremes. For example, if Earth froze over completely, carbon dioxide released by volcanoes would no longer be absorbed by the oceans. Instead the carbon dioxide would build up in the atmosphere, which would warm the atmosphere and hence melt the frozen Earth. Conversely, if Earth became hotter and hotter, more water would evaporate. This would lead to more precipitation, which would remove carbon dioxide from the atmosphere. With less carbon dioxide in the atmosphere, the greenhouse effect would be minimized and Earth would cool. So not only are we a nice distance from the Sun, but our atmosphere contains just enough water vapor and carbon dioxide to keep temperatures favorable for life. Furthermore, our relatively high daily spin rate allows only a brief and small lowering of temperature on the nighttime side of Earth. So temperature extremes of day and night are also kept moderate.

Moreover, recall from Chapters 7 and 22 that movements within Earth's molten core generate a strong magnetic field around our planet. This magnetic field, called the *magnetosphere*, extends thousands of kilometers into space and shields the Earth from *solar wind*, which is a flow of charged particles emanating from the Sun (see Figure 7.38). This form of cosmic radiation is harmful to living organisms. Furthermore, solar wind is capable of stripping a planet of its atmosphere. Planet Earth is indeed a sanctuary for life in an otherwise inhospitable universe.

FIGURE 28.12
Earth, the blue planet. This famous photo was taken by *Apollo 17* astronauts as they returned from the last manned mission to the Moon in 1972. It is the only existing photo of the full Earth from an appreciable distance. Can you see that it was taken in the summer months of the Southern Hemisphere?

CHECK YOURSELF

If Venus was also once protected by the carbon dioxide cycle, what went wrong?

CHECK YOUR ANSWER

The carbon dioxide cycle requires water. The Venusian carbon dioxide cycle, therefore, broke down as the Venusian water was split apart by the Sun's ultraviolet rays.

Mars

Mars captures our fancy as another world, perhaps even as a world with life because it resides on the outer fringes of the habitable zone. Mars is a little more than half Earth's size; its mass is about one-ninth that of Earth; and it has a core, a mantle, a crust, and a thin, nearly cloudless atmosphere. It has polar ice caps and seasons that are nearly twice as long as Earth's because Mars takes nearly two Earth years to orbit the Sun. Mars and Earth spin on their axes at about the same rate, which means the lengths of their days are about the same. When Mars is closest to Earth, a situation that occurs once every 15 to 17 years, its bright, ruddy color outshines the brightest stars.

The Martian atmosphere is about 95% carbon dioxide, with only about 0.15% oxygen. Yet because the Martian atmosphere is relatively thin, it doesn't trap heat via the greenhouse effect as much as Earth's and Venus's atmospheres do. So the temperatures on Mars are generally colder than on Earth, ranging from about 30°C in the day at the equator to a frigid −130°C at night. If you visit Mars, never mind your raincoat, for there is far too little water vapor in the atmosphere for rain. Even the ice at the planet's poles consists primarily of carbon dioxide. And don't give a second thought to waterproof footwear because the very low atmospheric pressure doesn't allow any puddles or lakes to form.

(a)

(b)

FIGURE 28.13

(a) A model of NASA's Mars Exploration Rover, *Spirit*, with cameras mounted on the white mast. (b) *Spirit* took photographs in June 2004 for this composite, true-color image of the region named Columbia Hills on Mars. The vehicle later traveled to the hills to analyze their composition.

Surface features on Mars, such as channels, indicate that liquid water was once abundant on this planet. This implies a distant past that was much warmer—a situation likely made possible by the greenhouse effect of a thicker atmosphere.

What happened to this thicker atmosphere? A currently accepted model is that because Mars is such a small planet, its molten core cooled down and solidified relatively fast—within a billion years of formation. This would have had at least two major consequences. The first is a decrease in the activity of volcanoes, which are a prime source of atmospheric gases. The second is the loss of the planet's magnetosphere. Without a magnetosphere, the bulk of the early Martian atmosphere—no longer replaced by volcanoes—was carried away into space by solar winds.

Today, landings on Mars show it to be a very dry and windy place. Because the Martian atmosphere has a very low density, its winds are about 10 times as fast as winds on Earth.

In 2004, spacecraft orbiting Mars detected signs of the organic compound methane, CH_4, in the atmosphere. This is unusual because methane decomposes fairly rapidly, which tells us that this compound is currently being produced. The likely source is ongoing residual volcanic activity, which could potentially melt underground ice into liquid water. Indeed, scientists have found evidence of the leakage of underground liquid water onto the surface occurring since we started surveying Mars from space. Once on the surface, this water would evaporate or freeze and sublime away. Underground pools of volcanically warmed liquid water, however, may harbor microscopic life forms.

Mars has two small moons—Phobos, the inner one, and Deimos, the outer. Both are potato-shaped and have cratered surfaces. Also, both orbit in the same easterly direction in which Mars spins (like our Moon). The eastward orbit of Phobos, however, is so fast that from the Martian surface, this moon is seen to rise in the west and set in the east. The slower moving and much more distant Deimos is only about half the size of Phobos and from the Martian surface appears as a barely visible point of light.

Why are planets round? All parts of a forming planet pull close together by mutual gravitation. No "corners" form because they're simply pulled in. So gravity is the cause of the spherical shapes of planets and other celestial bodies.

FYI Jupiter is near the point at which the addition of more matter would cause the size of this planet to contract. This is analogous to a stack of pillows. Start stacking pillows and the stack gets taller. Eventually, however, a point is reached at which the weight of upper pillows pushes down on lower pillows such that the column of pillows gets shorter. Interestingly, Jupiter is *larger* than the smallest stars, which, though smaller than Jupiter, are about 80 times as massive.

28.4 The Outer Planets

LEARNING OBJECTIVE
Identify the major properties of the four outer planets: Jupiter, Saturn, Uranus, and Neptune.

EXPLAIN THIS The exteriors of the outer planets are gaseous, but their interiors are mostly liquid. Why?

The outer planets—Jupiter, Saturn, Uranus, and Neptune—are gigantic, gaseous, low-density worlds. Each of them formed from rocky and metallic cores that were much more massive than the terrestrial planets. The gravitational forces of these cores were strong enough to sweep up gases of the early planetary nebula, primarily hydrogen and helium. The cores continued to collect gases until the Sun ignited and the solar wind blew away the remaining interplanetary gases. The core of Jupiter was the first to develop, and hence it had the longest time to collect gas before solar ignition. This is why Jupiter is the largest of the outer planets. Another commonality is that they all have ring systems, Saturn's being the most prominent. We will explore the outer planets in the order of their distance from the Sun.

Jupiter

FIGURE 28.14
Jupiter, with its moons Io (orange dot over planet) and Europa (white dot to right of planet), as seen from the *Voyager 1* spacecraft. The Great Red Spot (lower left), larger than Earth, is a cyclonic weather pattern.

Jupiter is the largest of all the planets. Its yellow light in our night sky is brighter than that of any star. Jupiter spins rapidly about its axis in about 10 hours, a speed that flattens it so that its equatorial diameter is about 6% greater than its polar diameter. As with the Sun, all parts do not rotate in unison. Equatorial regions complete a full revolution several minutes before nearby regions in higher and lower latitudes. This results in a pattern of stripes running parallel to the equator. The atmospheric pressure deep within Jupiter is more than a million times the atmospheric pressure of Earth. Jupiter's atmosphere is about 82% hydrogen, 17% helium, and 1% methane, ammonia, and other molecules.

Jupiter's average diameter is about 11 times Earth's, which means Jupiter's volume is more than 1000 times Earth's. Jupiter's mass is greater than the combined masses of all the other planets. Because of its low density, however—about one-fourth of Earth's—Jupiter's mass is more than 300 times Earth's. Investigations of Jupiter tell us that its core is a solid sphere about 15 times as massive as the entire Earth, and it is composed of iron, nickel, and other minerals.

More than half of Jupiter's volume is an ocean of liquid hydrogen. Beneath the hydrogen ocean lies an inner layer of hydrogen compressed into a liquid metallic state. In it are abundant conduction electrons that flow to produce Jupiter's enormous magnetic field.

FIGURE 28.15
This artist's rendering shows aurorae (pink) in the polar atmosphere of Jupiter. Aurorae, like the northern lights on Earth, are caused by charged particles from the solar wind exciting gas molecules in the upper atmosphere.

Jupiter has more than 60 moons in addition to a faint ring. The four largest moons were discovered by Galileo in 1610; Io and Europa are about the size of our Moon, and Ganymede and Callisto are about as large as Mercury (Figure 28.16). Jupiter's moon Io has more volcanic activity than any other body in the solar system. Perhaps most intriguing of all, however, are Ganymede and Europa, whose surfaces are made of frozen water. As shown in Figure 28.17, deep beneath this ice is likely an ocean of water kept warm by the strong tidal forces from nearby Jupiter.

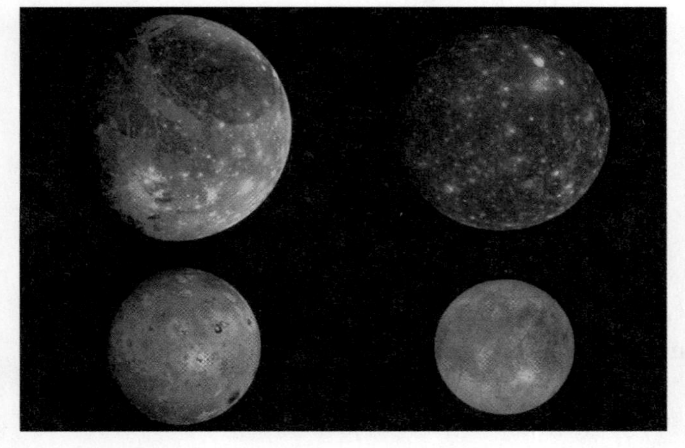

FIGURE 28.16
Galileo was the first to study the heavens with a telescope. He noted the changing positions of Jupiter's four largest moons and concluded they were orbiting Jupiter, which was a violation of the then widely held belief that all heavenly objects orbited Earth. In his honor, these four moons are known as the Galilean moons—from left to right starting with top row: Ganymede, Callisto, Io, and Europa.

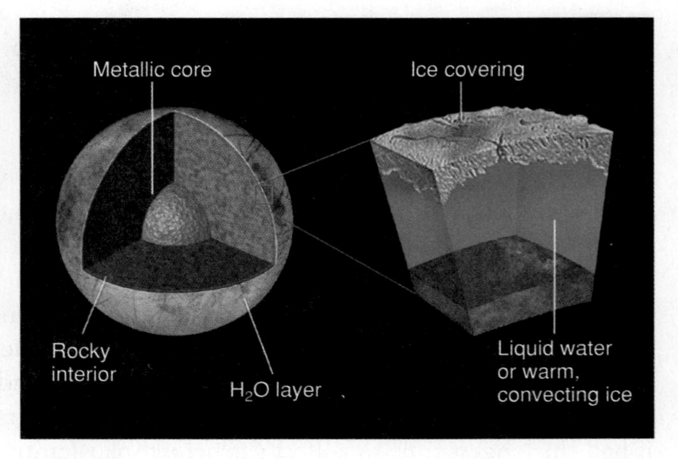

FIGURE 28.17
A model of the interior of Europa with a zoomed-in view of its ice-capped ocean, which, according to magnetic measurements, likely covers the entire sphere.

Saturn

Saturn is one of the most remarkable objects in the sky, with its rings clearly visible through a small telescope. It is quite bright—brighter than all but two stars—and it is second only to Jupiter in mass and size. Saturn is twice as far from Earth as Jupiter is. Its diameter, not counting its ring system, is nearly 10 times that of Earth, and its mass is nearly 100 times Earth's. It is composed primarily of hydrogen and helium, and it has the lowest density of any planet, only 0.7 times the density of water. These characteristics mean that Saturn would easily float in a bathtub, if the bathtub were large enough. Its low density and its 10.2-hour rapid spin produce more polar flattening than can be seen in the other planets. Notice its oblong shape in Figure 28.18.

FIGURE 28.18
Saturn surrounded by its famous rings, which are composed of rocks and ice.

Saturn's rings, only a few kilometers thick, lie in a plane coincident with Saturn's equator. Four concentric rings have been known for many years, and spacecraft missions have detected many others. The rings are composed of chunks of frozen water and rocks, believed to be the material of a moon that never formed or the remnants of a moon torn apart by tidal forces. All the rocks and bits of matter that make up the rings pursue independent orbits about Saturn. The inner parts of the ring travel faster than the outer parts, just as any satellite near a planet travels faster than a more distant satellite.

Saturn has about 50 moons beyond its rings. The largest is Titan, 1.6 times as large as our Moon and even larger than Mercury. It spins once every 16 days and has a methane atmosphere with atmospheric pressure that is likely greater than Earth's. Its surface temperature is cold, roughly −170°C. A space probe built by NASA and the European Space Agency landed on Titan in 2005. Remarkably, photos revealed a landscape similar to Earth's despite the fact that the materials are completely different (Figure 28.19). Lakes and streams are filled with not water but liquid methane. Rocks are made of ice. Instead of lava, Titan has a flowing slush of ice and liquid ammonia. No life is expected to be found on this moon because of the intensely cold temperatures. Titan, however, holds an intriguing soup of organic molecules whose chemistry may provide a clue to what Earth was like during the time before life arose here.

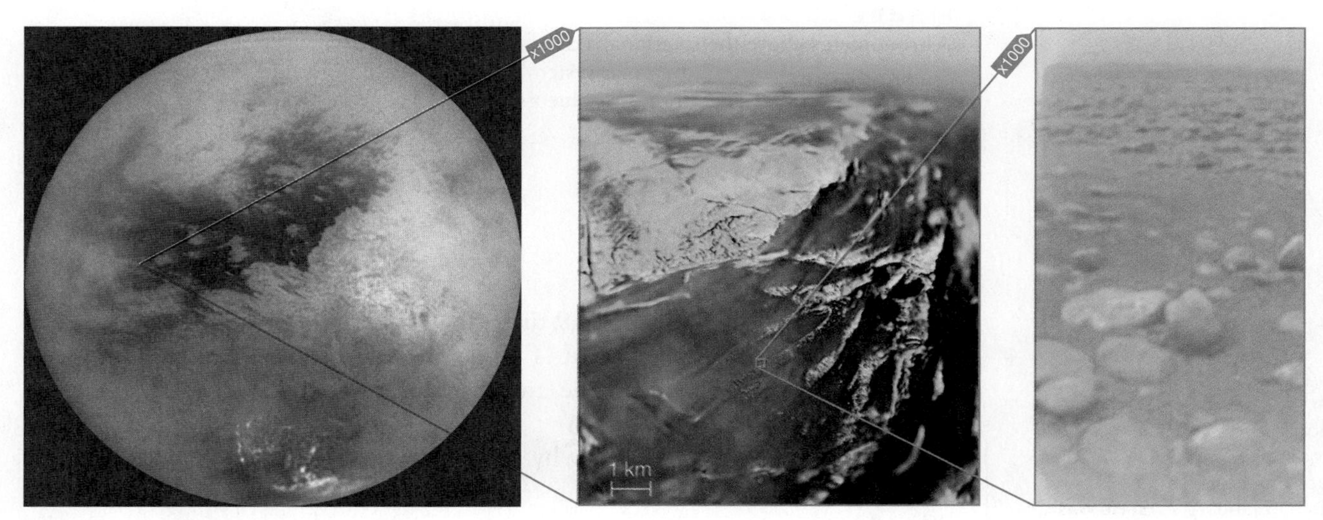

FIGURE 28.19
INTERACTIVE FIGURE MP

Images from Saturn's largest moon, Titan, taken by the *Cassini* spacecraft and its space probe, the *Huygens*, which successfully descended to the surface.

Uranus

Uranus ("YUR-uh-nus," accent on the first syllable) is twice as far from Earth as Saturn is, and it can barely be seen with the naked eye. Uranus was unknown to ancient astronomers and not discovered as a planet until 1781. The *Voyager 2* spacecraft first visited this planet in 1986. Uranus has a diameter four times that of Earth and a density slightly greater than that of water. So, if you could place Uranus in a giant bathtub, it would sink. The most unusual feature of Uranus is its tilt. Its axis is tilted 98° to the perpendicular of its orbital plane, so it lies on its side (Figure 28.20). Unlike Jupiter and Saturn, it appears to have no appreciable internal source of heat. Uranus is a cold place.

Uranus has at least 27 moons, in addition to a complicated faint ring system. Perturbations in the planet Uranus led to the discovery in 1846 of a farther planet, Neptune.

FIGURE 28.20
Astronomers believe the rotational axis of Uranus is tilted as a result of a collision it had with a large body early in the solar system's history. Methane in the upper atmosphere absorbs red light, giving Uranus its blue-green color.

Neptune

Neptune has a diameter about 3.9 times that of Earth, a mass 17 times as great, and a mean density about a third of Earth's. Its atmosphere is mainly hydrogen and helium, with some methane and ammonia, which makes Neptune bluer than Uranus (Figure 28.21).

Recent studies of Galileo's notebooks show that Galileo saw Neptune in December 1612 and again in January 1613. He was interested in Jupiter at the time, and so he merely plotted Neptune as a background star.

The *Voyager 2* spacecraft flew by Neptune in 1989. It showed that Neptune has at least 13 moons in addition to a ring system. The largest moon, Triton, orbits Neptune in 5.9 days in a direction opposite to the planet's eastward spin. This suggests that Triton is a captured object. Triton's diameter is three-quarters of our Moon's diameter, and yet Triton is twice as massive as Earth's Moon. It has bright polar caps and geysers of liquid nitrogen.

FIGURE 28.21
Cyclonic disturbances on Neptune in 1989 produced a great dark spot, which was even larger than Earth and similar to Jupiter's Great Red Spot. The spot has now disappeared. The gray horizon in the foreground of this computer-generated montage is a close-up of Neptune's moon Triton, which has a composition and size similar to those of Pluto.

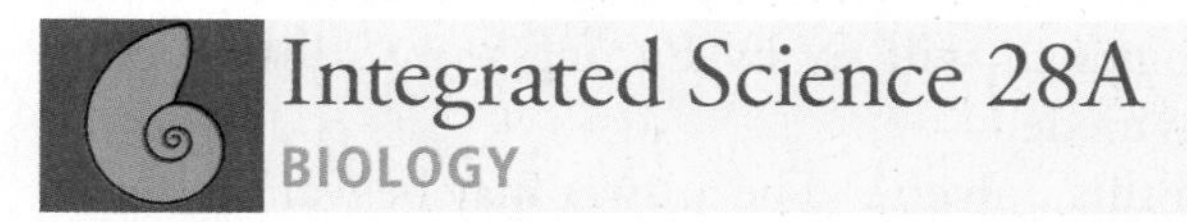

Integrated Science 28A
BIOLOGY

What Makes a Planet Suitable for Life?

LEARNING OBJECTIVE
Describe the conditions required for extraterrestrial life.

EXPLAIN THIS Jupiter's major moons keep getting stretched in different directions by tidal forces. What force causes these moons to become warm?

What does a planet need in order to support life? Astrobiologists, scientists who study the origin and distribution of life in the universe, have a number of ideas about what planetary characteristics are conducive to the evolution of life. First, most familiar life on Earth is carbon-based, and there are good reasons to expect life elsewhere to be carbon-based as well. Carbon has the unusual ability to bind to as many as four other atoms at a time, which enables it to serve as the basis for a wide variety of complex molecules. This versatility is probably necessary for life. So, the first attribute of a habitable planet is that it contain abundant amounts of carbon along with certain other elements such as hydrogen, nitrogen, and oxygen that combine with carbon to make biomolecules. Second, living things require energy. This energy can come from sunlight or from chemical reactions that take place on or inside planets or even tidal forces. Third, the evolution of life probably requires the presence of a liquid medium such as water. Molecules can move around and react with one another in liquids, and this is probably essential for any life form. The fourth condition needed is stability so that there is sufficient time for the life forms to develop and evolve.

Given these requirements, where in the solar system should we look for life? It is interesting that the first two conditions—the need for the right elements and the need for an energy source—are satisfied by most planetary bodies in the solar system. In contrast, the presence of water is relatively rare. There is abundant geologic evidence, however, that the planet Mars once had liquid water. Parts of the Martian surface appear to have been produced by flowing water, resembling floodplains or dry riverbeds, and the Martian poles are still covered with water ice. The one-time presence of liquid water also implies that Mars used to be significantly warmer. Mars almost certainly was habitable in the past. But was it inhabited? The discovery of bacteria-like "fossils" in a Martian meteorite (see Chapter 17) fueled speculation that Mars not only was inhabited but also was the source of life on Earth. However, these supposed bacteria are much tinier than any organisms found on Earth, and perhaps too tiny to contain all the cellular structures a bacterium needs to function. The question of whether life existed—or even exists—on Mars is still open, and further exploration is needed to provide definitive answers.

Other possible locations for extraterrestrial life in our solar system include the jovian moons Europa and Ganymede, which, as discussed in Section 28.4, may contain an ocean of liquid water beneath their icy surfaces. Any life on either of these moons may have originated adjacent to volcanic thermal vents on the ocean floor. Such extraterrestrial life forms may be similar to the bizarre forms of life recently discovered adjacent to deep thermal vents on Earth's ocean floor. Alternatively, they may be single-celled organisms, such as bacteria. Then again, there may be nothing. The European Space Agency is planning to develop a satellite scheduled to launch by 2022 and reach Jupiter's moons by 2030. Its mission is to probe beneath the surfaces of these moons with radar and other technologies. It will fly by Europa and Callisto before settling into orbit around Ganymede.

So, is life rare or common in this universe? The answer may be waiting for us in our own galactic backyard. Stay tuned.

CHECK YOURSELF

Many scientists believe that life is most likely to evolve on planets that contain liquid in their environments. One of the perceived advantages of water over other potential liquids (such as ammonia, methane, or ethane) is that frozen water—ice—floats. Why might this be important for life?

CHECK YOUR ANSWER

On Earth, a layer of floating ice insulates the underlying water and allows it to remain liquid. If ice sank, it would expose more liquid water to the environment, which would then also freeze and sink. In a cold period, lakes and oceans would eventually freeze solid—this is not good for the organisms living in them.

28.5 Earth's Moon

LEARNING OBJECTIVE
Summarize the Moon's origin, its phases, and the role it plays in both lunar and solar eclipses.

EXPLAIN THIS When the Moon rises at sunset, its phase is always full. Why?

Earth's Moon is puzzling. It is close to the size of Mercury, which is a planet and *not* a moon. The composition of Earth's Moon is nearly the same as Earth's mantle. Furthermore, the Moon possesses a rather small iron core. To explain these and a multitude of other facts about the Moon, scientists have pieced together the following possible scenario for its origin.

During the early history of the solar system, the young Earth had a Mercury-sized companion form within an orbit close to that of Earth. Normally, if the companion's orbit were a bit closer to the Sun, then it would orbit faster than Earth and move ahead of Earth. Upon passing through a special point, known as the *Lagrangian point*, however, the object would find that the gravitational pull of Earth was strong enough to hold it back so that it orbited with Earth in unison. So early Earth may have had a twin that paraded with Earth around the Sun much like two horses running side by side on a circular track.

We see the aftermath of meteoroid bombardment on the Moon because it wears no makeup. Similar bombardment on Earth has been long erased by erosion.

Eventually, a random event, such as the passing of an asteroid or comet, caused our companion to sway from the Lagrangian point and fall toward and collide

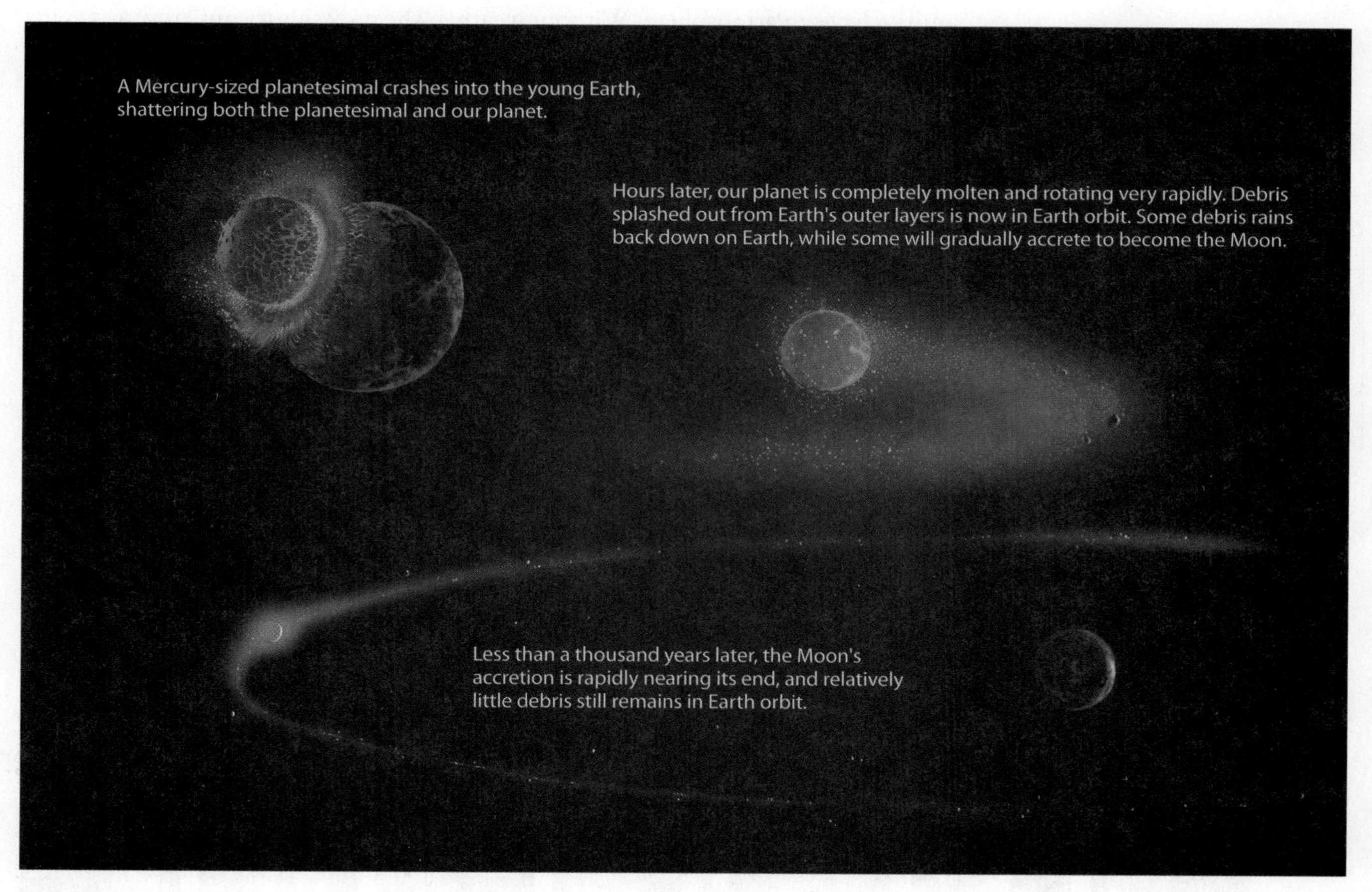

A Mercury-sized planetesimal crashes into the young Earth, shattering both the planetesimal and our planet.
Hours later, our planet is completely molten and rotating very rapidly. Debris splashed out from Earth's outer layers is now in Earth orbit. Some debris rains back down on Earth, while some will gradually accrete to become the Moon.
Less than a thousand years later, the Moon's accretion is rapidly nearing its end, and relatively little debris still remains in Earth orbit.

FIGURE 28.22
Three steps in the formation of Earth's Moon. A Mercury-sized object collides with Earth, which turns molten. Debris collects in a ring that accretes into the Moon, which is quite close to the rapidly rotating Earth. Over the next billion years tidal forces slow the rate of Earth's rotation while also causing the Moon to move farther away.

UNIFYING CONCEPT

- *The Scientific Method* Section 1.3

with Earth. The collision would have been massively spectacular, spewing debris everywhere while turning Earth fully molten. Hitting askew, the impact sent Earth into a wild spin rotating once every five hours. The debris soon collected as a ring around Earth, and then, within about 1000 years, the ring coalesced into the Moon. This scenario is known as the *giant impact theory* of the origin of Earth's Moon. It explains why the Moon is so large (we started out as twin planets), why its composition is similar to Earth's (it formed from our mantle and our mantle formed from it), why it has such a small iron core (Earth's iron core had already differentiated and was not sent up with the debris), and much more. This impact theory is still the subject of much research and is thus being continually refined. Although it was developed within only the past couple decades, scientists are excited by its explanatory powers.

FIGURE 28.23
Earth and the Moon as photographed in 1977 from the *Voyager 1* spacecraft on its way to Jupiter and Saturn.

From a distance, Earth and the Moon still resemble a twin planet system, as you can see in Figure 28.23. Compared to Earth, however, the Moon is relatively small, with a diameter of about the distance from San Francisco to New York City. It once had a molten surface, but it cooled too rapidly for the establishment of moving crustal plates, like those of Earth. In its early history, it was intensely bombarded by meteoroids (as was Earth). A little more than 3 billion years ago, meteoroid bombardment and volcanic activity filled basins with lava to produce its present surface. It has undergone very little change since then. Its igneous crust is thicker than Earth's. The Moon is too small with too little gravitational pull to have an atmosphere. Without weather, the only significant eroding agents have been meteoroid impacts.

The Phases of the Moon

Sunshine always illuminates half of the Moon's surface. The Moon shows us different amounts of its sunlit half as it circles Earth each month. These changes are the **Moon phases** (Figure 28.24). The Moon cycle begins with the **new Moon.** In this phase, its dark side faces us and we see darkness. This occurs when the Moon is between Earth and the Sun (position 1 in Figure 28.25).

Most planets wobble significantly as they spin about their axes. The Moon, however, helps keep Earth's wobble to a minimum. As a result, our weather patterns are fairly consistent through the ages, which makes our planet even more favorable for the development of life. Thank you, Moon!

During the next seven days, we see more and more of the Moon's sunlit side (position 2 in Figure 28.25). The Moon is going though its waxing crescent phase (*waxing* means "increasing"). At the first quarter, the angle between the Sun, the Moon, and Earth is 90°. At this time, we see half the sunlit part of the Moon (position 3 in Figure 28.25).

During the next week, we see more and more of the sunlit part. The Moon is going through its waxing gibbous phase (position 4 in Figure 28.25). (*Gibbous* means "more than half.") We see a **full Moon** when the sunlit side of the Moon faces us squarely (position 5 in Figure 28.25). At this time, the Sun, Earth, and the Moon are lined up, with Earth in between. To view this full Moon you need

FIGURE 28.24
The Moon in its various phases.

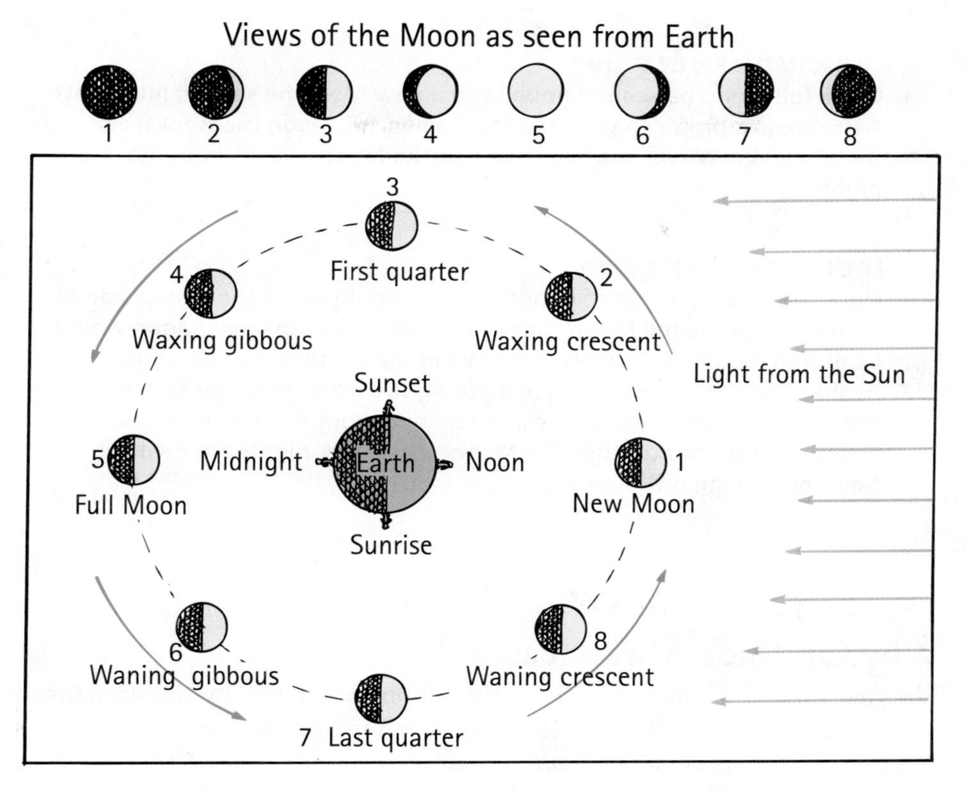

FIGURE 28.25
Sunlight always illuminates half of the Moon. As the Moon orbits Earth, we see varying amounts of its sunlit side. One lunar phase cycle takes 29.5 days.

to be on the nighttime side of Earth, at sunset when the full Moon rises from the east, or at sunrise when it sets in the west.

The cycle reverses during the following two weeks, as we see less and less of the sunlit side while the Moon continues in its orbit. This movement produces the waning gibbous, last quarter, and waning crescent phases. (*Waning* means "shrinking.") The time for one complete cycle is about 29.5 days.*

FYI During the time of the dinosaurs, a day was only about 19 hours. Today our days are about 24 hours. Billions of years from now, as Earth continues to slow down, a day will last about 47 hours. At that time, Earth and the Moon will be gravity locked such that the Moon will always appear in one location in the sky. To see the Moon, you will need to be on the Moon side of the planet, where perhaps real estate prices will be higher because of the view. One huge problem, however, is that by then, our Sun will have already gone through its dying phases, through which Earth will be subject to the fate of Venus, where it is forever cloudy. Nothing is permanent.

*The Moon actually orbits Earth once every 27.3 days relative to the stars. The 29.5-day cycle is relative to the Sun and is due to the motion of the Earth–Moon system as it revolves about the Sun.

FIGURE 28.26
Edwin E. Aldrin, Jr., one of the three *Apollo 11* astronauts, stands on the dusty lunar surface. To date, 12 people have stood on the Moon.

CHECK YOURSELF

1. **Can a full Moon be seen at noon? Can a new Moon be seen at midnight?**
2. **Astronomers prefer to view the stars when the Moon is absent from the night sky. When, and how often, is the Moon absent from the night sky?**

CHECK YOUR ANSWERS

1. Figure 28.25 shows that at noontime, you would be on the wrong side of Earth to view the full Moon. Likewise, at midnight, the new Moon would be absent. The new Moon is in the sky in the daytime, not at night.
2. At the time of the new Moon and during the week on either side of the new Moon, the night sky can be seen without the Moon. Unless an astronomer wishes to study the Moon, these dark nights are the best time for viewing other objects.

If someone shone a flashlight on a ball in a dark room, you could tell where the flashlight was by looking at the illumination on the ball. The Moon is similarly lit by the Sun.

Why One Side Always Faces Us

The first images of the back side of the Moon were taken by the unmanned Russian spacecraft *Lunik 3* in 1959. The first human witnesses of the Moon's back were *Apollo 8* astronauts, who orbited the Moon in 1968. From Earth, we see only a single lunar side. The familiar facial features of the "man in the Moon" are always turned toward us on Earth. Does this mean that the Moon doesn't spin about its axis as Earth does daily? No, but, relative to the stars, the Moon in fact does spin, although quite slowly—about once every 27 days. This monthly rate of spin matches the rate at which the Moon revolves about Earth. This explains why the same side of the Moon always faces Earth (Figure 28.27). The matching of monthly spin rate and orbital revolution rate is not a coincidence. After you answer the following question, we'll explore why.

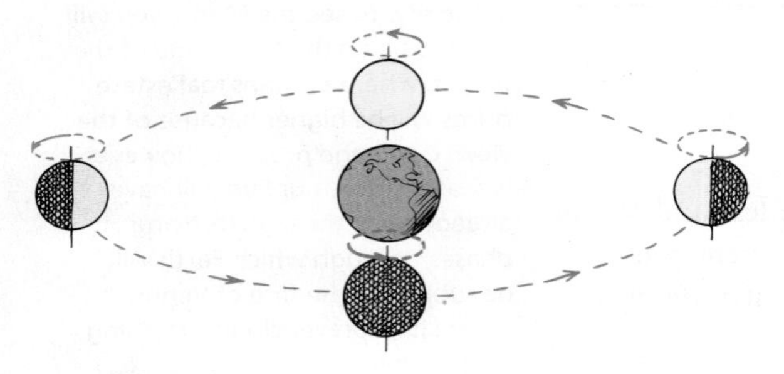

FIGURE 28.27
The Moon spins about its own polar axis just as often as it circles Earth. So as the Moon circles Earth, it spins so that the same side (shown in yellow) always faces Earth. In each of the four successive positions shown here, the Moon has spun one-quarter of a turn.

CHECK YOURSELF

A friend says that the Moon does not spin about its axis, and evidence for a nonspinning Moon is the fact that its same side always faces Earth. What do you say?

CHECK YOUR ANSWER

Place a quarter and a penny on a table. Pretend the quarter is Earth and the penny is the Moon. Keeping the quarter fixed, revolve the penny around it in such a way that Lincoln's head is always pointed to the center of the quarter. Ask your friend to count how many rotations the penny makes in one revolution (orbit) around the quarter. He or she will see that it rotates once with each revolution. The key concept is that the Moon takes the same amount of time to complete one rotation as it does to revolve around Earth.

Think of a compass needle that lines up with a magnetic field. This lineup is caused by a *torque*—a "turning force with leverage" (like that produced by the weight of a child at the end of a seesaw). The compass needle on the left in Figure 28.28 rotates because of a pair of torques. The needle rotates counterclockwise until it aligns with the magnetic field. In a similar manner, the Moon aligns with Earth's gravitational field.

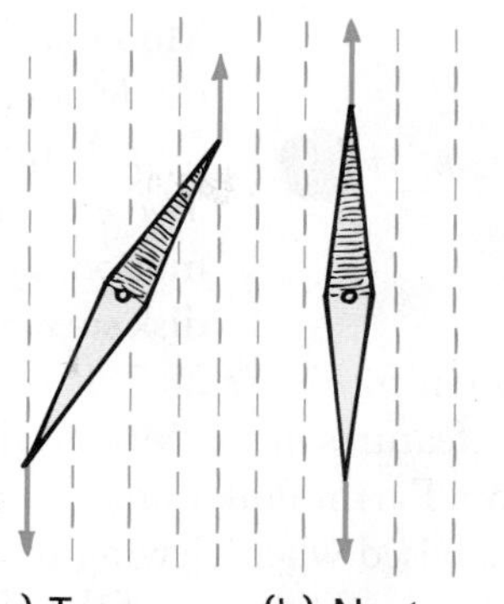

FIGURE 28.28
(a) When the compass needle is not aligned with the magnetic field (dashed lines), the forces represented by the blue arrows at either end produce a pair of torques that rotate the needle. (b) When the needle is aligned with the magnetic field, the forces no longer produce torques.

We know from the law of universal gravitation that gravity weakens with the inverse square of distance, so the side of the Moon nearer to Earth is gravitationally pulled more than the farther side. This stretches the Moon out slightly toward a football shape. (The Moon does the same to Earth and gives us tides.) If its long axis doesn't line up with Earth's gravitational field, a torque acts on it as shown in Figure 28.29. Like a compass in a magnetic field, it turns into alignment. So the Moon lines up with Earth in its monthly orbit. One hemisphere always faces us. It's interesting to note that for many moons orbiting other planets, a single hemisphere faces the planet. We say these moons are "tidally locked."

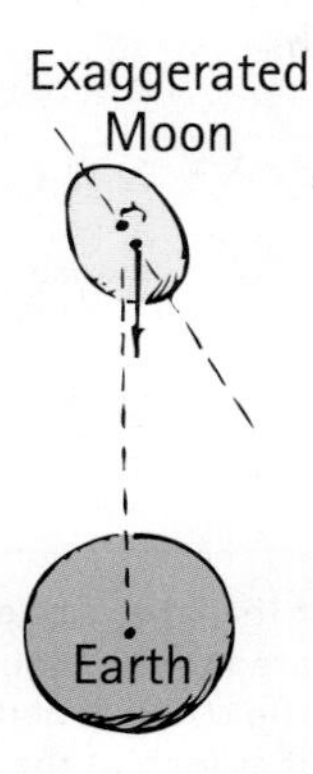

FIGURE 28.29
When the long axis of the Moon is not aligned with Earth's gravitational field, Earth exerts a torque that rotates the Moon into alignment.

Eclipses

Although the Sun is 400 times as large in diameter as the Moon, it is also 400 times as far away. So, from Earth, both the Sun and Moon measure the same angle (0.5°) and appear to be the same size in the sky. This coincidence allows us to see solar eclipses.

Both Earth and the Moon cast shadows from the sunlight shining on them. When the path of either of these bodies crosses into the shadow cast by the other, an eclipse occurs. A **solar eclipse** occurs when the Moon's shadow falls on Earth. Because of the large size of the Sun, the rays taper to provide an umbra and a surrounding penumbra, as shown in Figures 28.30 and 28.31.

An observer in the umbra part of the shadow experiences darkness during the day—a total eclipse, *totality*. Totality begins when the Sun disappears behind the

FIGURE 28.30
A solar eclipse occurs when the Moon passes in front of the Sun as seen from Earth. The Moon's shadow has two portions: a dark, central umbra that the Moon blocks completely from sunlight, surrounded by a lighter penumbra from which sunlight is only partly blocked. A total eclipse is seen from within the umbra and may last several minutes.

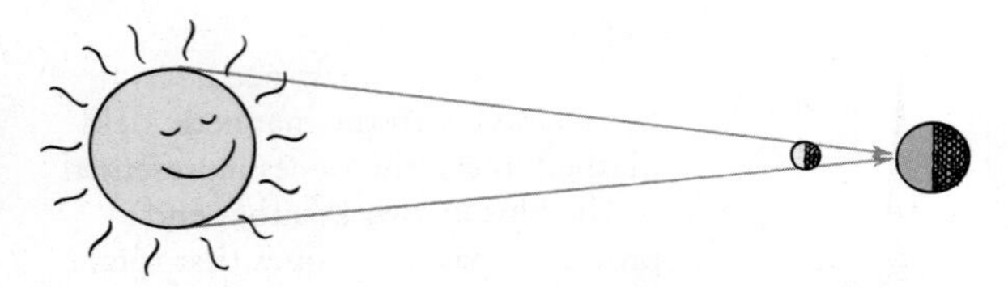

FIGURE 28.31
Geometry of a solar eclipse. During a solar eclipse, the Moon is directly between the Sun and Earth and the Moon's shadow is cast on Earth. Because of the small size of the Moon and tapering of the solar rays, a total solar eclipse occurs only on a small area of Earth.

Moon, and ends when the Sun reappears on the other edge of the Moon. The average time of totality at any location is about 2 or 3 minutes, with a maximum no longer than 7.5 minutes. The eclipse time in any location is brief because of the Moon's motion. During totality, what appears in the sky is an eerie black disk surrounded by the pearly white streams of the corona, as was shown in Figure 28.7. It is an experience one can never forget. With binoculars, the features of the Moon can be seen because they are lit by the sunlight reflected from Earth. Pink flares from the chromosphere may also appear. But great caution is advised when viewing the totality, which must be a totality of 100%. The moment the first edges of the photosphere appear, which is the moment you now have 99.99% totality, is the very moment when you can seriously damage your eyes if you continue to look.* At that moment you have entered the penumbra, where the eclipse is partial. An ideal way to view the partial solar eclipse is to focus the light of the eclipse onto a white surface, as was shown in Figure 28.4. Alternatively, you can view the crescent Sun under the shade of a tree, which casts pinhole images of the Sun onto the ground, as shown in Figure 28.32. Check the map shown in Figure 28.33 to see if a solar eclipse is coming to your area soon. Many solar eclipse enthusiasts travel the world to view this inspiring natural phenomenon.

The alignment of Earth, the Moon, and the Sun also produces a **lunar eclipse** when the Moon passes into the shadow of Earth, as shown in Figure 28.34. Usually a lunar eclipse precedes or follows a solar eclipse by two weeks. Just as all

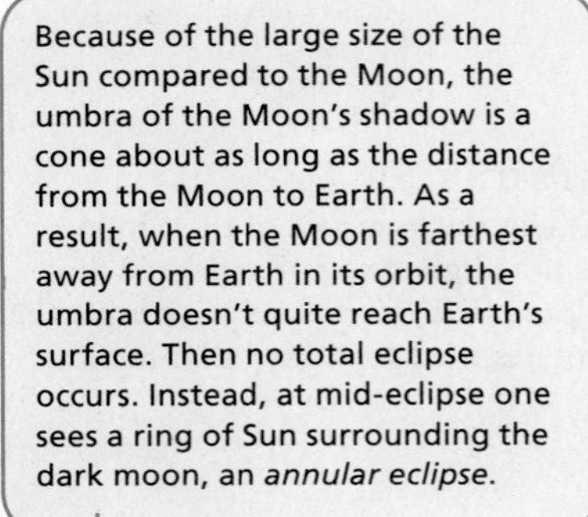

Because of the large size of the Sun compared to the Moon, the umbra of the Moon's shadow is a cone about as long as the distance from the Moon to Earth. As a result, when the Moon is farthest away from Earth in its orbit, the umbra doesn't quite reach Earth's surface. Then no total eclipse occurs. Instead, at mid-eclipse one sees a ring of Sun surrounding the dark moon, an *annular eclipse*.

* People are cautioned not to look at the Sun at the time of a solar eclipse because the brightness and the ultraviolet light of direct sunlight damage the eyes. This good advice is often misunderstood by those who then think that sunlight is more damaging at this special time. However, staring at the Sun when it is high in the sky is harmful whether or not an eclipse occurs. In fact, staring at the bare Sun is more harmful than when part of the Moon blocks it. The reason for special caution at the time of an eclipse is simply that more people are interested in looking at the Sun during this time.

(a) (b) (c)

FIGURE 28.32
(a) The spots of light on the wall and the ground, as well as on lab manual author Dean Baird, are pinhole images of the Sun cast through small openings between leaves on nearby trees. The spots are round because the Sun is round. (b) During a partial solar eclipse, the solar images are crescents because the Sun is then crescent shaped, as the Moon covers part of the Sun in the sky. (c) During an annular eclipse the Moon aligns with but doesn't fully cover the Sun, which results in the rarely seen Sun circles as demonstrated by Paul Doherty during the 2012 annular eclipse that passed over the western United States.

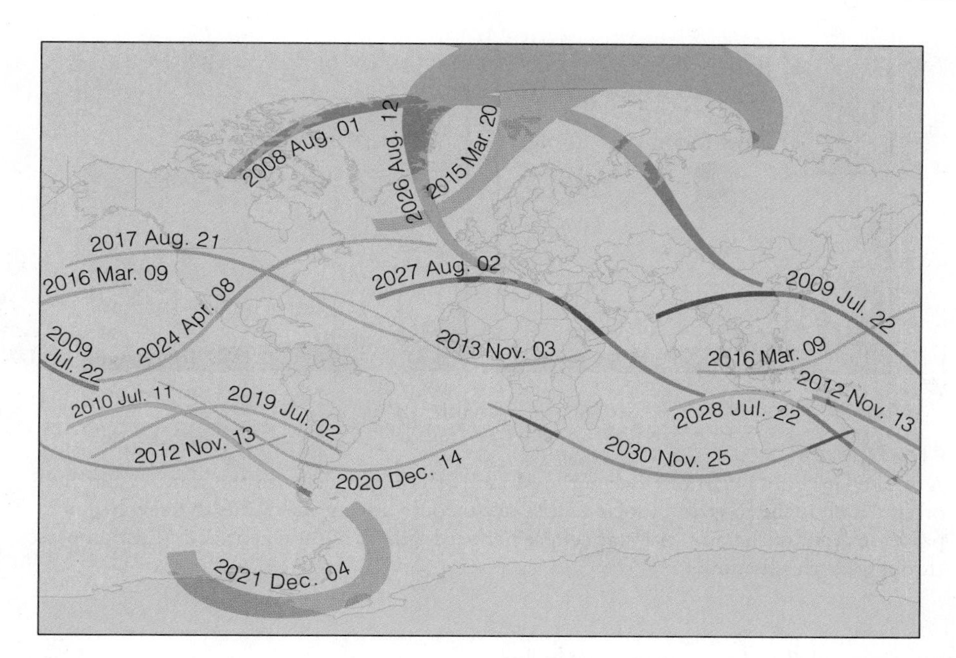

FIGURE 28.33
This map shows the paths of total solar eclipses from 2006 through 2030. More details about these and other future solar eclipses can be found on NASA's eclipse Web site, http://sunearth.gsfc.nasa.gov/eclipse/eclipse.html. If you live in the United States, August 21, 2017, is your special day.

solar eclipses involve a new Moon, all lunar eclipses involve a full Moon. They may be partial or total. All observers on the dark side of Earth see a lunar eclipse at the same time. Interestingly enough, when the Moon is fully eclipsed, it is still visible as is shown and discussed in Figure 28.35.

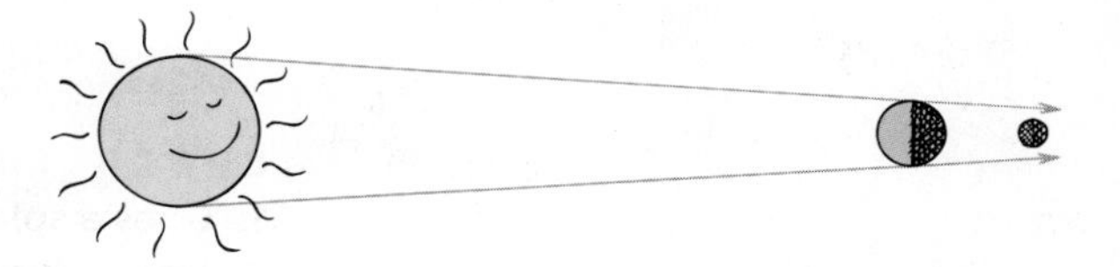

FIGURE 28.34
A lunar eclipse occurs when Earth is directly between the Moon and the Sun and Earth's shadow is cast on the Moon.

Why are eclipses relatively rare events? This has to do with the different orbital planes of Earth and the Moon. Earth revolves around the Sun in a flat planar orbit. The Moon similarly revolves about Earth in a flat planar orbit. But the planes are slightly tipped with respect to each other—a 5.2° tilt, as shown in Figure 28.36. If the planes weren't tipped, eclipses would occur monthly. Because of the tip, eclipses occur only when the Moon intersects the Earth–Sun plane at the time of a three-body alignment (Figure 28.37). This occurs about two times per year, which is why there are at least two solar eclipses

FIGURE 28.35
A fully eclipsed Moon is not completely dark in the shadow of Earth but is quite visible. This is because Earth's atmosphere acts as a lens and refracts light into the shadow region—sufficient light to faintly illuminate the Moon.

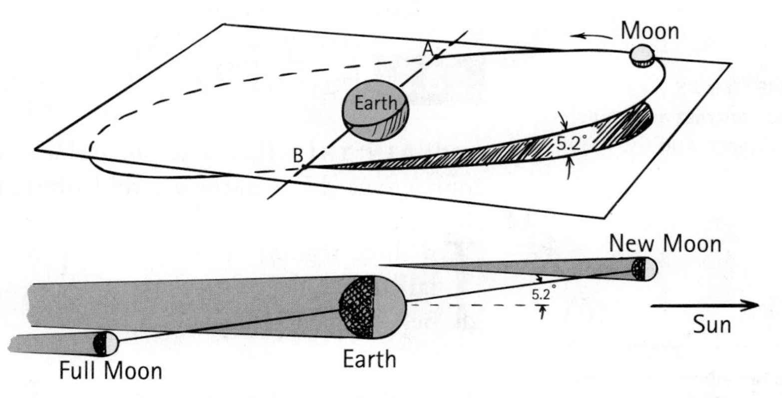

FIGURE 28.36
The Moon orbits Earth in a plane tipped 5.2° relative to the plane of Earth's orbit around the Sun. A solar or lunar eclipse occurs only when the Moon intersects the Earth–Sun plane (points A and B) at the precise time of a three-body alignment. Otherwise, the shadows are not aligned, as shown in the lower drawing.

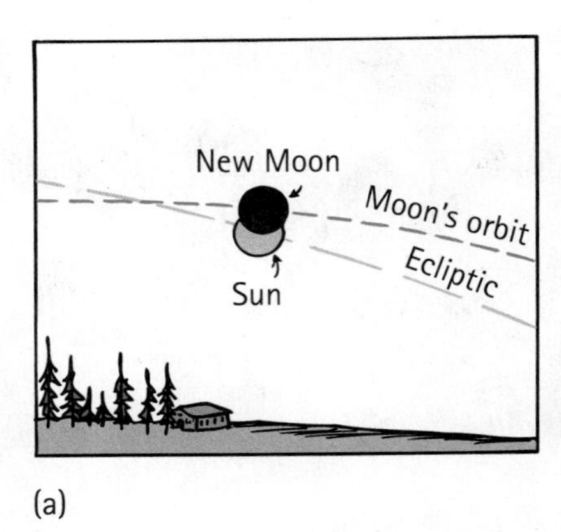

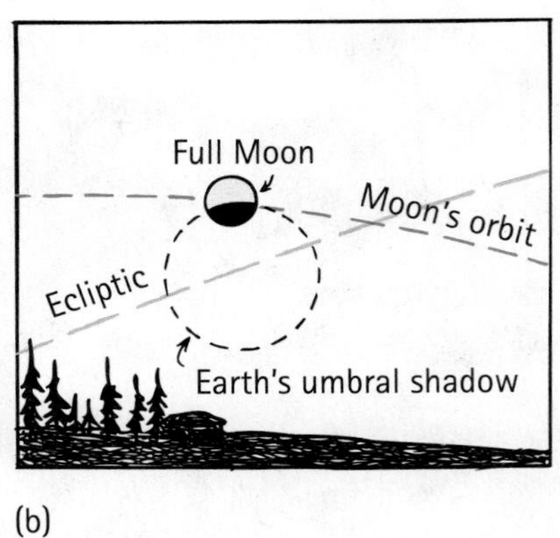

(a) (b)

FIGURE 28.37
A total eclipse can occur only when the Moon's orbit intersects with the plane of Earth's orbit, which is the ecliptic. A solar eclipse occurs only during the day as the new Moon passes in front of the Sun. A lunar eclipse happens only at night when the full Moon passes through Earth's shadow.

per year (visible from only certain locations on Earth). Sometimes there are as many as seven solar and lunar eclipses in a year.

CHECK YOURSELF

1. **Does a solar eclipse occur at the time of a full Moon or a new Moon?**
2. **Does a lunar eclipse occur at the time of a full Moon or a new Moon?**

CHECK YOUR ANSWERS

1. A solar eclipse occurs at the time of a new Moon, when the Moon is directly in front of the Sun. Then the shadow of the Moon falls on part of Earth.
2. A lunar eclipse occurs at the time of a full Moon, when the Moon and Sun are on opposite sides of Earth. Then the shadow of Earth falls on the full Moon.

LEARNING OBJECTIVE
Compare and contrast asteroids, Kuiper belt objects, and the Oort cloud.

28.6 Failed Planet Formation

EXPLAIN THIS If an asteroid and a comet of equal mass were on a collision course toward Earth, the asteroid would be easier to deflect. Why?

In three regions of our solar system, we find the remains of material that failed to collect into planets. These regions are the asteroid belt, the Kuiper belt, and the Oort cloud.

Meteorites fall all over our planet, but the easiest place to find them is on the icy white surfaces found in polar regions. Do you want to collect your own meteorites? Head south to Antarctica!

The Asteroid Belt and Meteors

The **asteroid belt** is a collection of rocks located between the orbits of Mars and Jupiter. More than 150,000 asteroids have been cataloged so far, but many more no doubt have yet to be discovered. They come in all shapes and sizes, but the largest asteroid, Ceres, is just less than a thousand kilometers in diameter.

Although Ceres is large enough to be fairly round, most asteroids are shaped more like a potato, as shown in Figure 28.38.

Evidence suggests that when the solar system was forming, the asteroid belt held much more mass than it does today. Massive Jupiter likely disrupted the orbits of this material, sending it off in many directions, including toward the inner planets and out of the solar system. The two moons of Mars, for example, may be former asteroids. What remains of the asteroid belt is small. If all the presently remaining asteroids were scrunched together, they would make a sphere less than half the size of our Moon.

Jupiter also causes the collisions of asteroids with asteroids, which then break apart into smaller fragments. So rather than building into a planet, this material is slowly ground down and pushed off course. Asteroid fragments known as **meteoroids** frequently find their way to Earth, where they are heated white-hot by friction with the atmosphere. As they descend with a fiery glow, they are called **meteors** (Figure 28.39).

If the meteoroid is large enough, it may survive to reach the surface, where it is called a **meteorite**. Most meteoroids, meteors, and meteorites are from asteroids, but many are also from comets, as we discuss later. Fortunately, smaller meteorites hit us more frequently than larger ones do. About 200 tons of small meteorites strike Earth every day. Every 10,000 years or so we are hit with a meteorite big enough to create a large crater, such as the one shown in Figure 28.40. Every 100 million years or so we are hit with one big enough—about 10 km in diameter—to cause a mass extinction, as occurred 65 million years ago at the end of the Cretaceous period discussed in Chapter 27. Accordingly, one of NASA's goals is to detect up to 90% of all large, near-Earth objects. If we can detect a dangerous space fragment early enough, we can take actions to alter its orbital path sufficiently to avoid impending disaster.

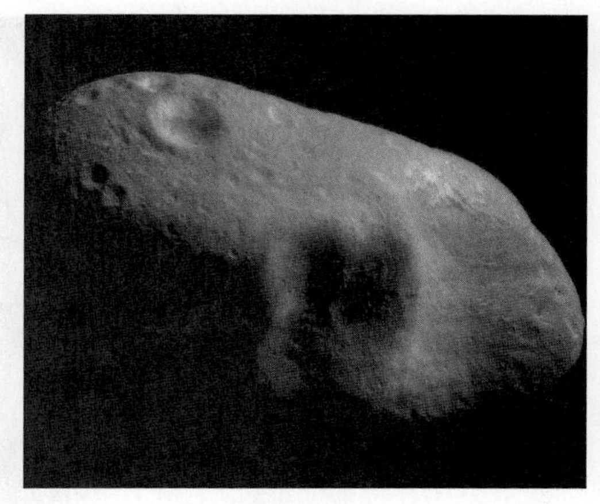

FIGURE 28.38
The asteroid Eros is about 40 km long, and like other small objects in the solar system, it is not spherical.

FIGURE 28.39
A meteor is produced when a meteoroid, usually about 80 km high, enters Earth's atmosphere. Most are sand-sized grains, which are seen as "falling" or "shooting" stars.

The Kuiper Belt and Dwarf Planets

Beyond Neptune at a distance from about 30 to 50 AU is a region known as the **Kuiper belt** ("KI-pur," rhymes with *hyper*). The Kuiper belt is occupied by many rocky, ice-covered objects. The most well-known Kuiper belt object is Pluto, which until recently was classified as a planet. Since its discovery in 1930, however, astronomers knew that Pluto was quite different from all the other known planets. For example, Pluto orbits at an angle to the plane of Earth's orbit—the ecliptic. Also, Pluto is quite small, being only one-seventh as massive as our Moon. Then, starting in the 1990s, astronomers began discovering many more Kuiper belt objects, some as large as or larger than Pluto. So in 2006 these Pluto-sized Kuiper belt objects were officially classified as **dwarf planets**. The main reason they do not meet full planet status is that they have yet to accrete all the material in their orbital paths. In the outer edges of our solar system, however, matter is simply too sparse for that to happen. Interestingly, if the Kuiper belt were more dense with material, then these dwarf planets could have served as cores for additional jovian planets. But that never happened, and so the Kuiper belt is another zone of failed planet formation.

Space probes have yet to visit any of the dwarf planets of the Kuiper belt. Pluto and its moon, Charon, however, are due to be visited by the *New Horizons*

FIGURE 28.40
The Barringer Crater in Arizona, made 25,000 years ago by an iron meteorite with a diameter of about 50 m. The crater extends 1.2 km across and reaches 200 m deep.

FIGURE 28.41
This infrared image of Pluto being orbited by its moon, Charon, is fuzzy because of the small size of these bodies and their great distance from Earth.

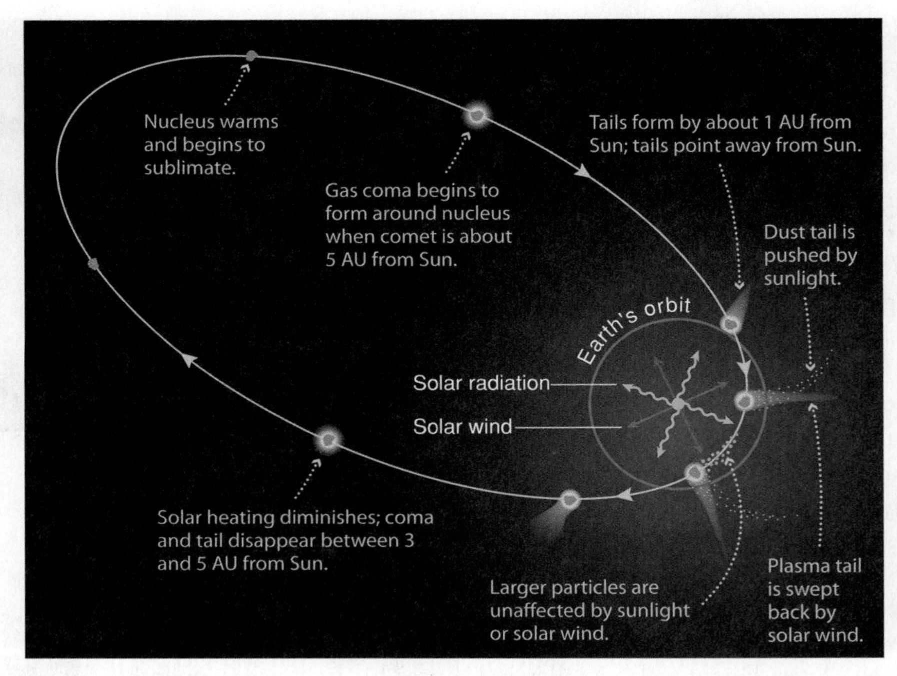

Nucleus warms and begins to sublimate.
Gas coma begins to form around nucleus when comet is about 5 AU from Sun.
Tails form by about 1 AU from Sun; tails point away from Sun.
Dust tail is pushed by sunlight.
Earth's orbit
Solar radiation
Solar wind
Solar heating diminishes; coma and tail disappear between 3 and 5 AU from Sun.
Larger particles are unaffected by sunlight or solar wind.
Plasma tail is swept back by solar wind.

FIGURE 28.42
A comet warms as it gets closer to the Sun and initially develops a coma, which is a halo of gases surrounding the comet nucleus. From this coma arises the tail, which is blown outward by the solar wind. Note how the tail always extends away from the Sun. Most Kuiper belt objects never make this journey and instead remain perpetually frozen within the outer reaches of our solar system.

spacecraft in 2015. We may have already had a preview, however, when the *Voyager 2* spacecraft took pictures of Neptune's moon Triton. Astronomers now suspect that Triton is a Kuiper belt dwarf planet pulled off course and captured into orbit around Neptune.

The larger Kuiper belt objects, such as Pluto, have a fair amount of inertia and so are not so easily thrown off course. Lighter Kuiper belt objects, however, are thrown off course quite frequently. Sometimes they are thrown toward the Sun, where the added heat and solar wind cause the ice and other volatile materials to be ejected, always in a direction away from the Sun. We see these objects as **comets**, which are characterized by their long and sometimes quite brilliant tails. Comets that come from the Kuiper belt tend to have orbital periods of less than 200 years. An example is Comet Halley, which returns to the inner solar system every 76 years—once in an average lifetime (Figure 28.43). Its next scheduled return is in 2061.

Comets apparently reside in at least two regions. The first is the Kuiper belt, which lies roughly within the same plane of the solar system. The second region lies much farther out and surrounds our entire solar system—like a cloud.

The Oort Cloud and Comets

As the jovian planets grew, their gravitational pulls became stronger, which made them even more effective at pulling in additional interplanetary debris. Not all debris, however, was pulled fully into the jovian planets. In many cases, a chunk of rock or ice just missing a planet was instead whipped around the planet and

FIGURE 28.43
Observations of Comet Halley have been recorded for thousands of years. Although it usually provides a brilliant display, its last visit in 1986 was not so spectacular when viewed from Earth. We were ready, though, with space probes that flew close enough to Halley to capture dramatic images of its nucleus.

then flung violently outward in some direction. Over billions of years this created a sphere of far-out objects just barely held to our solar system. We refer to this collection of far-out objects as the **Oort cloud** (*Oort* rhymes with *court*). Evidence suggests that the Oort cloud consists of trillions of objects extending as far out as 50,000 AU, which brings the cloud about a quarter of the way to the nearest star. A few of these objects occasionally fall toward and then around the Sun, where they appear as comets. The orbital periods of comets originating from the Oort cloud are on the order of thousands or even millions of years. They come from nearly any angle.

Most comets usually last only a couple of orbits before they break up. But if the solar system is billions of years old, shouldn't they be depleted by now? This very question led to the idea of the Oort cloud, which provides a continual supply of new comets, replacing those that are destroyed.

Whether the comet comes from the Kuiper belt or the Oort cloud, it still has the potential for colliding with a planet. In 1994, Comet Shoemaker-Levy

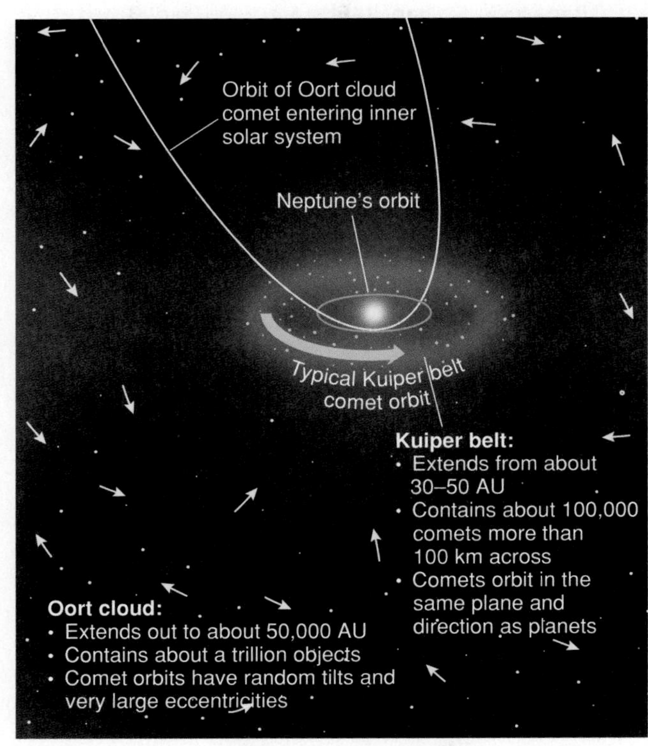

FIGURE 28.44
There are two major sources for comets: the Kuiper belt and the Oort cloud.

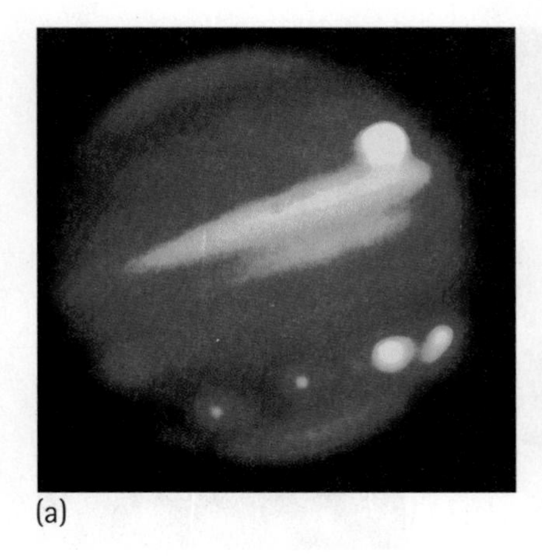

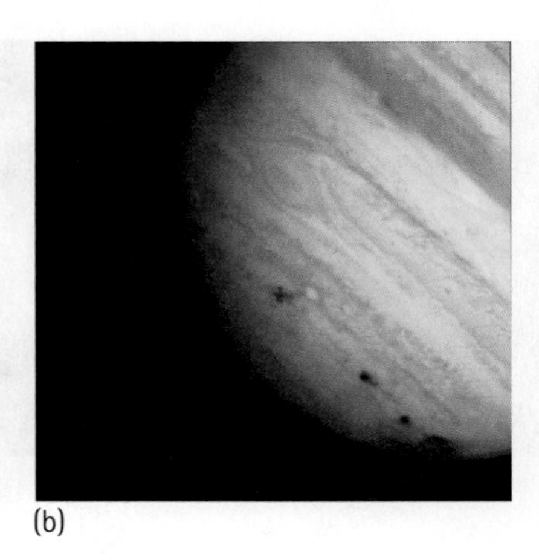

FIGURE 28.45
Comet Shoemaker-Levy was already broken up into a string of objects just before it collided with Jupiter in 1994. The image in part (a) is an infrared view of the collision, which produced much heat as well as visible scars (the black dots), as shown in the photograph in part (b).

collided spectacularly with Jupiter, as shown in Figure 28.45. Also, the large meteorite that collided with Earth 65 million years ago causing the mass extinction of the dinosaurs may have been a comet.

The tail of the comet leaves behind a wide trail of particles. Each year Earth passes through the remnants of comet tails that create annual meteor showers, as indicated in Table 28.2. Just go outside, look up at the sky, and with good eyesight every minute or so you will see a shooting star. Each streak is a tiny chip of a comet, once so very far away, that has fallen into Earth's neighborhood (Figure 28.46).

FIGURE 28.46
When the orbiting Earth intercepts the debris from a comet, we see a meteor shower.

TABLE 28.2 METEOR SHOWER DATA

Shower Name	Radiant*	Dates	Peak Dates	Meteors per Hour
Quadrantids	Pegasus	Jan. 1–6	Jan. 3	60
Eta Aquarids	Aquarius	May 1–10	May 6	35
Perseids	Perseus	Jul. 23–Aug. 20	Aug. 12	75
Orionids	Orion	Oct. 16–27	Oct. 22	25
Geminids	Gemini	Dec. 7–15	Dec. 13	75

*Meteors appear to radiate from a certain region of the sky, appropriately called a *radiant*. Radiants refer to constellations. See Chapter 29 for more on where the various constellations are located in the night sky.

CHECK YOURSELF

Of the asteroid belt, the Kuiper belt, and the Oort cloud:

1. **Which is closest to the Sun?**
2. **Which generates comets?**
3. **Which gives us the most meteorites?**
4. **Which gives us the brightest meteor showers?**
5. **Which consists of fragments that never coalesced into planets?**

CHECK YOUR ANSWERS

1. The asteroid belt
2. The Kuiper belt and the Oort cloud
3. The asteroid belt
4. The Kuiper belt and the Oort cloud
5. All of them

For instructor-assigned homework, go to www.masteringphysics.com (MP)

SUMMARY OF TERMS (KNOWLEDGE)

Asteroid belt A region between the orbits of Mars and Jupiter that contains small, rocky, planet-like fragments that orbit the Sun. These fragments are called *asteroids* ("small star" in Latin).

Astronomical unit (AU) The average distance between Earth and the Sun; about 1.5×10^8 km (about 9.3×10^7 mi).

Comet A body composed of ice and dust that orbits the Sun, usually in a very eccentric orbit, and that casts a luminous tail produced by solar radiation pressure when close to the Sun.

Dwarf planet A relatively large icy body, such as Pluto, that originated within the Kuiper belt.

Ecliptic The plane of Earth's orbit around the Sun. All major objects of the solar system orbit roughly within this same plane.

Full Moon The phase of the Moon when its sunlit side faces Earth.

Inner planets The four planets orbiting within 2 AU of the Sun: Mercury, Venus, Earth, and Mars—all are rocky and known as the *terrestrial* planets.

Kuiper belt The disk-shaped region of the sky beyond Neptune populated by many icy bodies and a source of short-period comets.

Lunar eclipse The phenomenon in which the shadow of Earth falls on the Moon, producing the relative darkness of the full Moon.

Meteor The streak of light produced by a meteoroid burning in Earth's atmosphere; a "shooting star."

Meteorite A meteoroid, or a part of a meteoroid, that has survived passage through Earth's atmosphere to reach the ground.

Meteoroid A small rock in interplanetary space, which can include a fragment of an asteroid or comet.

Moon phases The cycles of change of the "face" of the Moon, changing from *new*, to *waxing*, to *full*, to *waning*, and back to *new*.

Nebular theory The idea that the Sun and planets formed together from a cloud of gas and dust, a *nebula*.

New Moon The phase of the Moon when darkness covers the side facing Earth.

Oort cloud The region beyond the Kuiper belt populated by trillions of icy bodies and a source of long-period comets.

Outer planets The four planets orbiting beyond 2 AU of the Sun, Jupiter, Saturn, Uranus, and Neptune—all are gaseous and known as the *jovian* planets.

Planets The major bodies orbiting the Sun that are massive enough for their gravity to make them spherical and small enough to avoid having nuclear fusion in their cores. They also have successfully cleared all debris from their orbital paths.

Solar eclipse The phenomenon in which the shadow of the Moon falls on Earth, producing a region of darkness in the daytime.

Sunspots Temporary, relatively cool and dark regions on the Sun's surface.

READING CHECK (COMPREHENSION)

28.1 The Solar System and Its Formation

1. How many known planets are in our solar system?
2. What dwarf planet was downgraded from planetary status in 2006?
3. How are the outer planets different from the inner planets aside from their location?
4. Why does a nebula spin faster as it contracts?
5. According to the nebular theory, did the planets start forming before or after the Sun ignited?

28.2 The Sun

6. What happens to the amount of the Sun's mass as it "burns"?
7. What are sunspots?
8. What is the solar wind?
9. How does the rotation of the Sun differ from the rotation of a solid body?
10. What is the age of the Sun?

28.3 The Inner Planets

11. Why are the days on Mercury very hot and the nights very cold?
12. What two planets are evening or morning "stars"?
13. Why is Earth called "the blue planet"?
14. What gas makes up most of the Martian atmosphere?
15. What evidence tells us that Mars was at one time wetter than it presently is?

28.4 The Outer Planets

16. What surface feature do Jupiter and the Sun have in common?
17. Which move faster—Saturn's inner rings or the outer rings?
18. How tilted is Uranus's axis?
19. Why is Neptune bluer than Uranus?

28.5 Earth's Moon

20. Why doesn't the Moon have an atmosphere?
21. Where is the Sun located when you view a full Moon?
22. Where are the Sun and the Moon located at the time of a new Moon?
23. Why don't eclipses occur monthly, or nearly monthly?
24. How does the Moon's rate of rotation about its own axis compare with its rate of revolution around Earth?

28.6 Failed Planet Formation

25. Between the orbits of what two planets is the asteroid belt located?
26. What is the difference between a meteor and a meteorite?
27. What is the Kuiper belt?
28. What is the Oort cloud, and what is it noted for?
29. What is a falling star?
30. What causes comet tails to point away from the Sun?

THINK INTEGRATED SCIENCE

28A—What Makes a Planet Suitable for Life?

31. Why is carbon such a special atom?
32. Why does the evolution of life probably require the presence of a liquid on a planet?

THINK AND DO (HANDS-ON APPLICATION)

33. Find a Ping-Pong ball on the next clear day when the Moon is out. Hold the Ping-Pong ball with your arm stretched out toward where the Moon is so that the ball overlaps the Moon. Look carefully at how the ball is lit by the Sun. Notice that this is the same way the Moon is lit by the Sun! For an example, see the photograph accompanying Exercise 69. To see the different phases that the Moon would have if it were elsewhere in the sky, move your Ping-Pong ball around. Note that as you bring the ball closer to the Sun, the crescent on the ball gets thinner. The same thing happens with the Moon. This activity is a great way of really experiencing the roundness of the Moon.
34. Simulate the lunar phases. Insert a pencil into a Styrofoam ball. This will be the Moon. Position a lamp (representing the Sun) in another room near the doorway. Hold the ball in front and slightly above yourself. Slowly turn yourself around, keeping the ball in front of you as you move. Observe the patterns of light and shadow on the ball. Relate this to the phases of the Moon.
35. When viewed from the North Pole, Earth spins counterclockwise, which is toward the east. This means that the stars appear to move in the opposite direction, which is toward the west. This is just like when you're sitting in a train that begins moving eastward. The only way you know that you're moving eastward is because things outside your window give the appearance of moving westward. Just as Earth spins counterclockwise, the Moon revolves around us counterclockwise, though not as fast as we spin. Look where the Moon is located one night at, say, 11:00. Look for the Moon the next night at the same time, and you'll see that it has moved eastward (a counterclockwise direction) from where it was on the previous night.
36. The crescent Moon always points toward the Sun. You can use this fact to estimate your latitude. Down by the equator (0° latitude), the setting crescent Moon lies flat on the horizon, while up close to the North Pole (90° latitude), the crescent Moon stands on its end. Deviations to this can arise because of Earth's 23° tilt and because the Moon's orbit lies 5° outside the ecliptic. Nonetheless, those who live in Alaska see a crescent Moon that is much more upright than those who live in Hawaii. With this in mind, the next time you see the crescent Moon close to the horizon, look carefully at its angle and try correlating that angle to your local latitude.

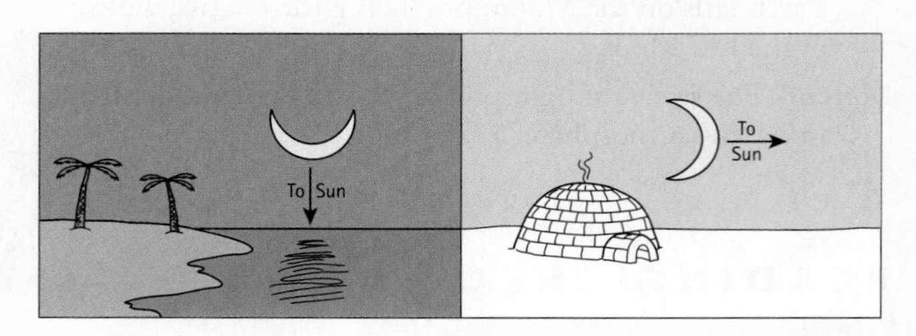

37. The planets of our solar system orbit in roughly the same plane, which is the plane of our solar system. We can identify the plane of our solar system in the night sky by noting the positions of the planets, which, from our point of view on Earth, appear in a roughly straight line relative to one another. This straight line, which is a cross-section of our solar system, always intersects with the Sun and often with the position of our Moon. On a clear evening after sunset, several planets can often be seen in the western sky forming a line that points directly toward the Sun. Upon finding this alignment, you are looking directly at the plane of our solar system. Check it out! How might this line look different if viewed from the North Pole versus the equator?
38. The dates of meteor showers are rather predictable as listed in Table 28.2. How intense the shower might be, however, is still somewhat of a guess. So keep your eye to the nighttime sky for these meteor showers. While doing

so, review in your mind the distinctions among meteoroids, meteors, meteorites, and comets. Also notable in our nighttime skies are the lunar eclipses, which can be seen by anyone who happens to be on the nighttime side of Earth during the eclipse (assuming skies are clear). Here are the dates for upcoming lunar eclipses viewable in North America.

Date of eclipse	Type
Apr. 15, 2014	Total
Oct. 8, 2014	Total
Sep. 28, 2015	Total
Jan. 21, 2019	Total

THINK AND COMPARE (ANALYSIS)

39. Rank these planets in order from longest to shortest year: (a) Mercury, (b) Venus, and (c) Earth.
40. Rank these planets in order of increasing number of moons: (a) Mars, (b) Venus, and (c) Earth.
41. Rank in order of increasing average density: (a) Jupiter, (b) Saturn, and (c) Earth.
42. Rank in order of increasing pressure at the center of each planet: (a) Jupiter, (b) Saturn, and (c) Earth.
43. Rank in order of decreasing number of people who have seen a: (a) solar eclipse, (b) lunar eclipse, and (c) new Moon.
44. Rank in order of increasing average distance from the Sun: (a) Kuiper belt objects, (b) asteroids, and (c) Oort cloud objects.

THINK AND SOLVE (MATHEMATICAL APPLICATION)

45. Knowing that the speed of light is 300,000 km/s, show that it takes about 8 min for sunlight to reach Earth.
46. How many days does sunlight take to travel the 50,000 AU from the Sun to the outer reaches of the Oort cloud?
47. The light-year is a standard unit of distance used by astronomers. It is the distance light travels in one Earth year. In units of light-years, what is the approximate diameter of our solar system, including the outer reaches of the Oort cloud? (Assume that 1 light-year equals 63,000 AU.)
48. The nearest star to our Sun is Alpha Centauri, which is about 4.4 light-years away. Assume that it too has an Oort cloud about 1.6 light-years in diameter. Show that there is enough space between us and it to fit about 1.75 solar systems.
49. If the Sun were the size of a beach ball, Earth would be the size of a green pea 110 m away. Show that the nearest star, Alpha Centauri (4.4 light-years away), would be about 30,000 km distant. (*Hint:* Find the distance to Alpha Centauri in units of AU.)

THINK AND EXPLAIN (SYNTHESIS)

50. According to nebular theory, what happens to a nebula as it contracts under the force of gravity?
51. What happens to the shape of a nebula as it contracts and spins faster?
52. A TV screen is normally light gray when not illuminated. How is the darkness of sunspots similar to the black parts of an image on a TV screen?
53. When a contracting ball of hot gas spins into a disk shape, it cools. Why?
54. If Earth didn't spin on its axis but still revolved around the Sun, how long would an Earth day be?
55. If Earth didn't spin on its axis but still revolved around the Sun, would the Sun set on the eastern or western horizon or not at all?
56. The greenhouse effect is very pronounced on Venus but doesn't exist on Mercury. Why?
57. What is the cause of winds on Mars (and also on almost every other planet)?
58. Why is there so little wind on the surface of Venus?
59. If Venus were somehow transported into the habitable zone, would conditions once again become favorable for life?
60. Mercury and Venus are never seen at night straight up toward the top of the sky. Why not?
61. What is the major difference between the terrestrial and jovian planets?
62. What does Jupiter have in common with the Sun that the terrestrial planets don't? What differentiates Jupiter from a star?
63. When it comes to celestial bodies, such as planets and stars, why doesn't a larger size necessarily mean a larger mass?
64. Why are the seasons on Uranus different from the seasons on any other planet?

65. Earth rotates much faster than Venus. How does the giant impact theory of the Moon account for this fact?
66. Why are many craters evident on the surface of the Moon but not on the surface of Earth?
67. Why is there no atmosphere on the Moon? Defend your answer.
68. Is the fact that we see only one side of the Moon evidence that the Moon spins or that it doesn't spin? Defend your answer.
69. Photograph (a) shows the Moon partially lit by the Sun. Photograph (b) shows a Ping-Pong ball in sunlight. Compare the positions of the Sun in the sky when each photograph was taken. Do the photos support or refute the claim that they were taken on the same day? Defend your answer.

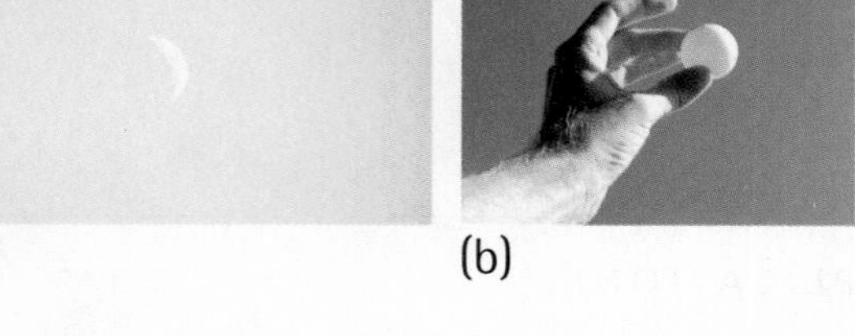

(a) (b)

70. We always see the same face of the Moon because the rotation of the Moon on its axis matches the rate at which it revolves around Earth. Does it follow that an observer on the Moon always sees the same face of Earth?
71. If we never see the back side of the Moon, would an observer on the back side of the Moon ever see Earth?
72. In what alignment of the Sun, the Moon, and Earth does a solar eclipse occur?
73. In what alignment of the Sun, the Moon, and Earth does a lunar eclipse occur?
74. What does the Moon have in common with a compass needle?
75. If you were on the Moon and you looked up and saw a full Earth, would it be nighttime or daytime on the Moon?
76. If you were on the Moon and you looked up and saw a new Earth, would it be nighttime or daytime on the Moon?
77. Earth takes 365.25 days to revolve around the Sun. If Earth took this same amount of time to spin on its axis, what might we note about the Sun's position in the sky?
78. Do astronomers make stellar observations during the full Moon part of the month or during the new Moon part of the month? Does it make a difference?
79. Nearly everybody has witnessed a lunar eclipse, but relatively few people have seen a solar eclipse. Why?
80. Because of Earth's shadow, the partially eclipsed Moon looks like a cookie with a bite taken out of it. Explain with a sketch how the curvature of the bite indicates the size of the Earth relative to the size of the Moon. How does the tapering of the Sun's rays affect the curvature of the bite?

Use the following illustration for Exercises 81–84.

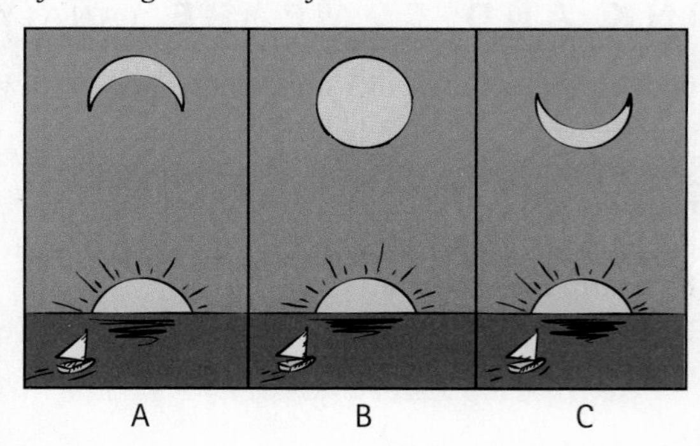

81. Which of the three orientations of the Moon at sunset is most correct?
82. Assuming the illustrations depict a sunset, within 24 hours of when this scene is depicted will the Moon appear to be farther from or closer to the Sun?
83. Is the sailboat sailing at a location closer to the North Pole or closer to the equator? How can you tell?
84. Where and how would the Moon be positioned if the scenes were close to the North Pole?
85. In what sense is Pluto a potential comet?
86. Smaller chunks of asteroids are sent hurling toward Earth much more frequently than larger chunks of asteroids. Why?
87. Why are meteorites so much more easily found in Antarctica than in other continents?
88. A meteor is visible only once, but a comet may be visible at regular intervals throughout its lifetime. Why?
89. What would be the consequence of a comet's tail sweeping across Earth?
90. Chances are about 50–50 that in any night sky there is at least one visible comet that has not been discovered. This keeps amateur astronomers busy looking night after night because the discoverer of a comet gets the honor of having it named for him or her. With this high probability of comets in the sky, why aren't more of them found?

THINK AND DISCUSS (EVALUATION)

91. Project what human civilization would be like if Earth had no Moon.
92. What are the chances that microbial life forms might one day be found elsewhere in our solar system? How much effort should we spend on searching for such life forms, and what precautions should we take upon such a discovery?

READINESS ASSURANCE TEST (RAT)

If you have a good handle on this chapter, if you really do, then you should be able to score 7 out of 10 on this RAT. If you score less than 7, you need to study further before moving on.

Choose the BEST answer to each of the following:

1. The Sun contains what percentage of the solar system's mass?
   (a) about 35%
   (b) 85%
   (c) the percentage varies over time
   (d) over 99%

2. The solar system is like an atom in that both
   (a) are governed principally through the electric force.
   (b) consist of a central body surrounded by objects moving in elliptical paths.
   (c) are composed of plasma.
   (d) are mainly empty space.

3. The nebular theory is based on the observation that the solar system
   (a) follows patterns indicating that it formed progressively from physical processes.
   (b) has a structure much like an atom.
   (c) is highly complex and appears to have been built by chaotic processes.
   (d) appears to be very old.

4. When a contracting hot ball of gas spins into a disk shape, it cools faster due to
   (a) increased radiation transfer.
   (b) increased surface area.
   (c) decreased insulation.
   (d) increased convection currents.
   (e) eddy currents.

5. Each second, the burning Sun's mass
   (a) increases.
   (b) remains unchanged.
   (c) decreases.

6. Compared to your weight on Earth, your weight on Jupiter would be about
   (a) 3000 times as much.
   (b) half as much.
   (c) 3 times as much.
   (d) 300 times as much.
   (e) 100 times as much.

7. When the Moon assumes its characteristic thin crescent shape, the position of the Sun is
   (a) almost directly in back of the Moon.
   (b) almost directly behind Earth, so that Earth is between the Sun and the Moon.
   (c) at right angles to the line between the Moon and Earth.

8. When the Sun passes between the Moon and Earth, we have
   (a) a lunar eclipse.
   (b) a solar eclipse.
   (c) met our end.

9. Asteroids orbit
   (a) the Moon.
   (b) Earth.
   (c) the Sun.
   (d) all of these
   (e) none of these

10. With each pass of a comet about the Sun, the comet's mass
    (a) remains virtually unchanged.
    (b) actually increases.
    (c) is appreciably reduced.

Answers to RAT

1. d, 2. d, 3. a, 4. b, 5. c, 6. c, 7. a, 8. c, 9. c, 10. c

29 CHAPTER 29 The Universe

On a moonless night and away from city lights the unaided eye sees no more than 3000 stars, horizon to horizon. Many more stars become visible with a telescope, especially when the telescope is pointed toward a cloudlike band of light that stretches north to south. The ancient Greeks called this diffuse band of light the Milky Way.

Today we know that the Milky Way is a vast collection of more than 100 billion stars. When viewed from afar, all of these stars—along with our own star, the Sun—appear as a great swirl of stars known as a *galaxy*.

In this chapter we will explore the nature of stars—how they form, how they die, and how they are organized within galaxies. We will explore how there are many different types of stars, just as there are many different types of galaxies. We will conclude with a discussion of cosmology, in which we attempt to answer such questions as How did the universe come into being? and What might be its ultimate fate?

29.1 Observing the Night Sky

EXPLAIN THIS When can winter constellations be seen in the summer?

Early astronomers divided the night sky into groups of stars called *constellations*. The names of the constellations today carry over mainly from the names assigned to them by early Greek, Babylonian, and Egyptian astronomers. The Greeks, for example, included the stars of the Big Dipper in a larger group of stars that outlined a bear. The large constellation Ursa Major (the Great Bear) is illustrated in Figure 29.1. The groupings of stars and the significance given to them have varied from culture to culture. To some cultures, the constellations stimulated storytelling and the making of great myths; to other cultures, the constellations honored great heroes, such as Hercules and Orion; to yet others, they served as navigational aids for travelers and sailors. To many cultures, including the African Bushmen and Masai, the constellations provided a guide for planting and harvesting crops because they were seen to move in the sky in concert with the seasons. Charts of this periodic movement became some of the first calendars. We can see in Figure 29.2 why the background of stars varies throughout the year.

The stars are at different distances from Earth. However, because all the stars are so far away, they appear equally remote. This illusion led the ancient Greeks and others to conceive of the stars as being attached to a gigantic sphere surrounding Earth, called the **celestial sphere**. Though we know it is imaginary, the celestial sphere is still a useful construction for visualizing the motions of the stars (Figure 29.3).

The stars appear to turn around an imaginary north–south axis. This is the *diurnal motion* of the stars. Diurnal motion is easy to visualize as a rotation of the celestial sphere from east to west. This motion is a consequence of the daily counterclockwise rotation of Earth on its axis. When we speak of the diurnal motion of the stars, we are referring to the motions of celestial objects as a

LEARNING OBJECTIVE
Distinguish between the diurnal and intrinsic motions of celestial objects, and express the vast distances in space in units of light-years.

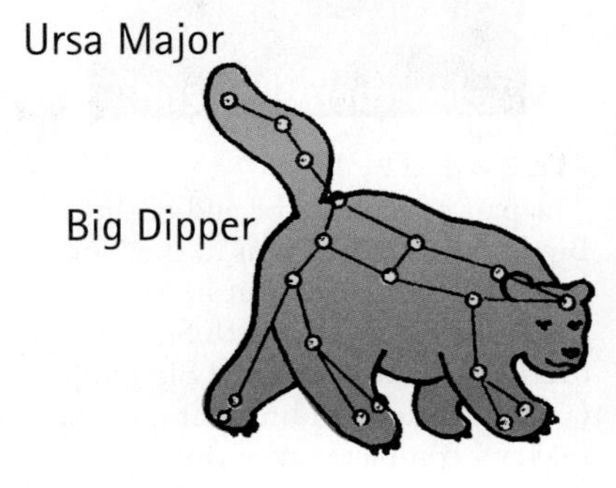

FIGURE 29.1
The constellation Ursa Major, the Great Bear. The seven stars in the tail and back of Ursa Major form the Big Dipper.

The grouping of stars into constellations tells us about the thinking of astronomers of earlier times, but it tells us nothing about the stars themselves.

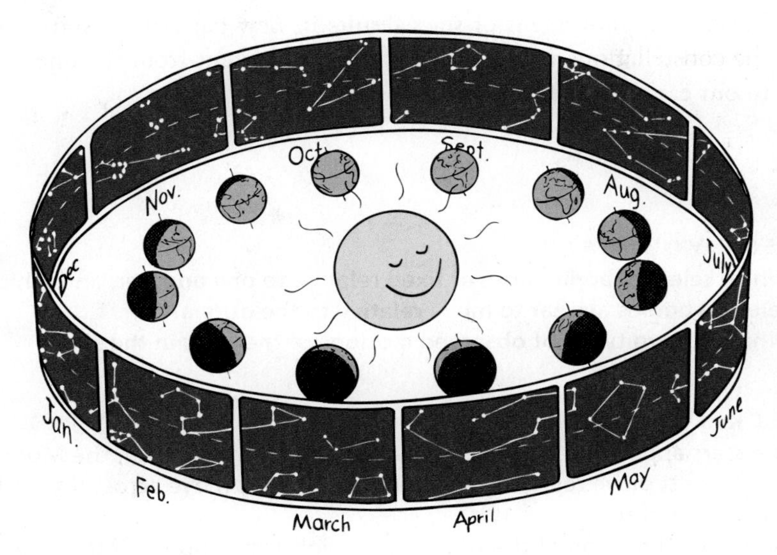

FIGURE 29.2
The night side of Earth always faces away from the Sun. As Earth circles the Sun, different parts of the universe are seen in the nighttime sky. Here the circle, representing 1 year, is divided into 12 parts—the monthly constellations. The stars in the nighttime sky change in a yearly cycle.

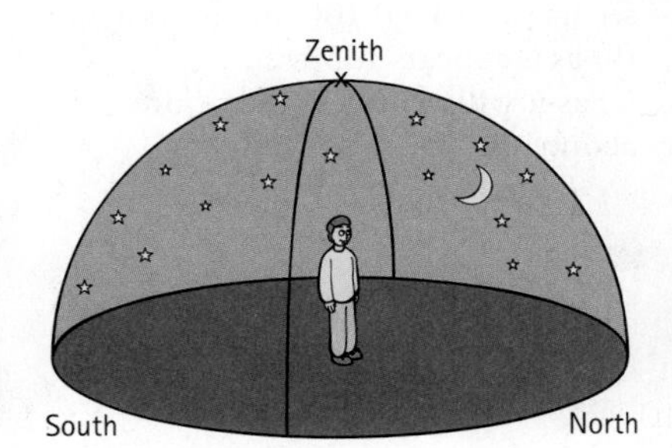

FIGURE 29.3
The celestial sphere is an imaginary sphere to which the stars are attached. We see no more than half of the celestial sphere at any given time. The point directly over your head at any time is called the *zenith*.

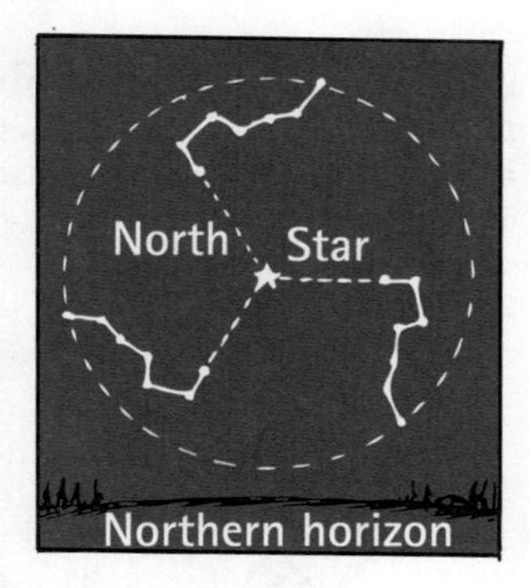

FIGURE 29.4
The pair of stars in the end of the Big Dipper's bowl point to the North Star. Earth rotates about its axis and therefore about the North Star, so over a 24-hour period the Big Dipper (and other surrounding star groups) makes a complete revolution.

FIGURE 29.5
A time exposure of the northern night sky.

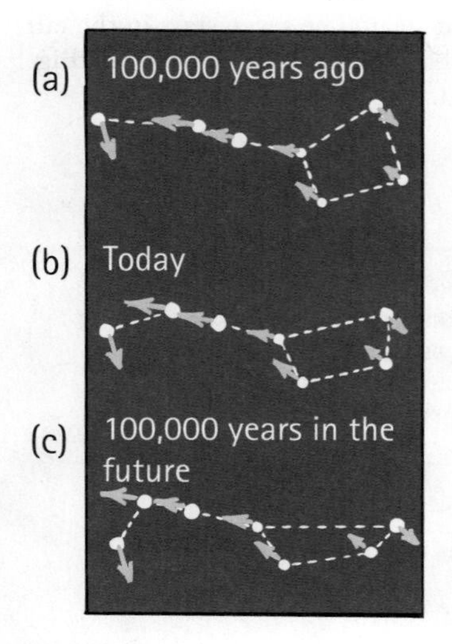

FIGURE 29.6
The present pattern of the Big Dipper is temporary. Here we can see its pattern (a) 100,000 years ago; (b) as it appears at present; and (c) as it will appear in the future, about 100,000 years from now.

whole; this motion does not change the relative positions of objects. Figure 29.4 shows the diurnal motion of the stars making up the Big Dipper. Time-exposure photographs show that the Big Dipper appears to move in circles around the North Star (Figure 29.5). The North Star appears nearly stationary as the celestial sphere rotates because it lies very close to the projection of Earth's rotational axis.

In addition to the diurnal motion of the sky, there is *intrinsic motion* of certain bodies that change their positions with respect to the stars. The Sun, the Moon, and planets, called "wanderers" by ancient astronomers, appear to migrate across the fixed backdrop of the celestial sphere. Interestingly, the stars themselves have intrinsic motion. They are so far away, however, that this motion is not apparent on the time scale of a human life. As shown in Figure 29.6, over thousands of years, the intrinsic movement of stars results in new patterns of stars. In other words, the constellations we see today are quite different from the ones that appeared to our earliest ancestors.

CHECK YOURSELF

1. **Which celestial bodies appear fixed relative to one another, and which celestial bodies appear to move relative to the others?**
2. **What are two types of observed motions of the stars in the sky?**

CHECK YOUR ANSWERS

1. The stars appear fixed as they move across the sky. The Sun, the Moon, and planets move relative to one another as they move across the backdrop of the stars.
2. One type of motion of the stars is their nightly rotation as if they were painted on a rotating celestial sphere; this is due to Earth's rotation on its own axis. Stars also appear to undergo a yearly cycle around the Sun because of Earth's revolution about the Sun.

Some stars on the celestial sphere are actually much farther away than others from Earth. Astronomers measure the vast distances between Earth and the stars using *light-years*. One **light-year** is the distance that light travels in 1 year, nearly 10 trillion km. For perspective, the diameter of Neptune's orbit is about 0.001 light-year. The distance from the Sun to the outer edges of the Oort cloud (the full radius of our solar system) is about 0.8 light-year. The star closest to our Sun, Proxima Centauri, is about 4.2 light-years away. The diameter of the Milky Way galaxy is about 100,000 light-years. The next closest major galaxy, the Andromeda galaxy, is about 2.3 million light-years distant. Figure 29.7 shows the distances to the seven stars making up the Big Dipper in light-years.

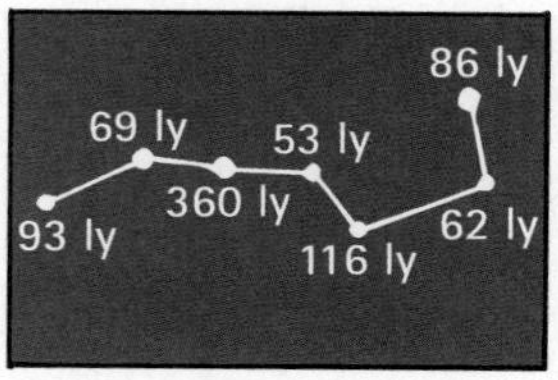

FIGURE 29.7
The seven stars of the Big Dipper are at distances from Earth. Note their distances in light-years (ly).

The speed of light (as we know from Chapter 8) is 3×10^8 m/s. Although this is very fast, it nevertheless takes light appreciable time to travel large distances. And so, when you see the light emitted by a very distant object, you are actually seeing the light it emitted long ago—you are looking back in time. Consider the example of Supernova 1987a (a supernova is the explosion of a star, as you will learn more about in Section 29.4). This supernova occurred in a galaxy 190,000 light-years from Earth. Although we witnessed the supernova in 1987, the light from this explosion took 190,000 years to reach our planet, so the explosion actually occurred 190,000 years earlier. "News" of the supernova took 190,000 years to reach Earth!

29.2 The Brightness and Color of Stars

LEARNING OBJECTIVE
Distinguish between a star's apparent brightness and its luminosity, and identify its temperature by its color.

EXPLAIN THIS How do astronomers gauge the temperature of a star?

Stars are born from clouds of interstellar dust with roughly the same chemical composition as the Sun (see Chapter 28). About three-fourths of the interstellar material from which a star forms is hydrogen; one-fourth is helium; and no more than 2% of the material from which a star forms consists of heavier chemical elements. Stars shine brilliantly for millions or billions of years because of the nuclear fusion reactions that occur in their cores. And all stars, the Sun included, ultimately exhaust their nuclear fuel and die. Yet not all stars are the same. If you look into the night sky, you will see that stars differ in two very visible ways: brightness and color.

Brightness relates to how much energy a star produces. However, although a star's brightness is related to its energy output, its brightness also depends on how far away it is from Earth. Recall from earlier chapters inverse-square laws; in this case, the intensity of light diminishes as the reciprocal of the square of the distance from the source. For example, the stars Betelgeuse and Procyon appear equally bright even though Betelgeuse emits about 5000 times as much light as Procyon. The reason? Procyon is much closer to Earth than Betelgeuse is.

To avoid confusing brightness with energy output, astronomers clearly distinguish between apparent brightness and the more important property, *luminosity*. Apparent brightness is the brightness of a star as it appears to our eyes. Luminosity, on the other hand, is the total amount of light energy that a star emits into space. Luminosity is usually expressed relative to the Sun's luminosity, which is noted L_{Sun}. For example, the luminosity of Betelgeuse is 38,000 L_{Sun}. This indicates that Betelgeuse is a very luminous star emitting about 38,000 times as much energy each second into space as the Sun. On the other hand, Proxima Centauri is quite dim, with a luminosity of 0.00006 L_{Sun}. Astronomers have

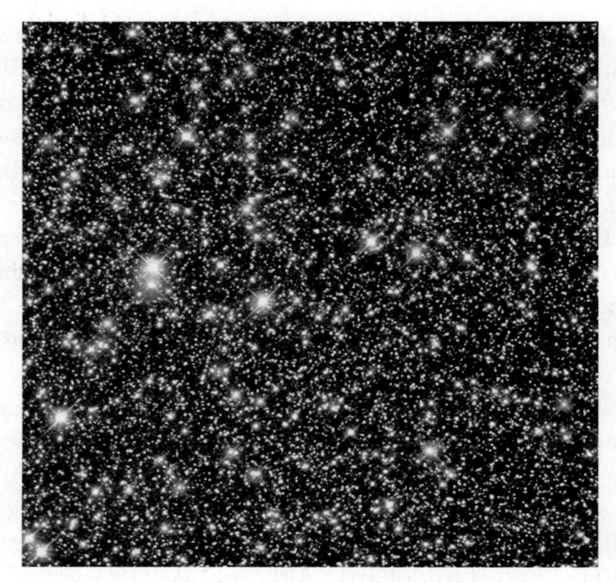

FIGURE 29.8
Most of the stars in this photograph are approximately the same distance—2000 light-years—from the center of the Milky Way galaxy. A star's color indicates its surface temperature—a blue star is hotter than a yellow star, and a yellow star is hotter than a red star. This photo was taken by the Hubble Telescope.

measured the luminosity of many stars and found that stars vary greatly in this respect. The Sun is somewhere in the middle of the luminosity range. The most luminous stars are about a million times as luminous as the Sun, while the dimmest stars produce about 1/10,000 as much energy per second as the Sun.

Besides apparent brightness, color is another property that varies widely among stars. Figure 29.8, a photograph of stars taken with the Hubble Telescope, shows that stars come in every color of the rainbow. A star's color directly tells you about its surface temperature—for example, a blue star is hotter than a yellow star, and a yellow star is hotter than a red star. In fact, astronomers use color to measure the temperatures of stars. Why is it that a star's color corresponds to its temperature?

Radiation Curves of Stars

As you learned in Chapters 6 and 8, all objects that have a temperature emit energy in the form of electromagnetic radiation. The peak frequency $\bar{f}$ of the radiation is directly proportional to the absolute temperature T of the emitter:

$$\bar{f} \sim T$$

Stars have different colors because they emit different frequencies of electromagnetic waves in the visible range. Our eyes sense different frequencies of visible radiation as different colors. Figure 29.9 shows the radiation curves, which are graphs of the intensity of emitted radiation versus wavelength, for two stars of the same size with different temperatures. The radiation curves show that the hotter a star is, the shorter the wavelength of its peak frequency and the bluer it looks. So the blue stars in the night sky have higher temperatures than the red ones. The Sun, for example, with its approximately 5800 K surface temperature, emits most strongly in the middle of the visible spectrum and so appears yellow.

FYI Interestingly, the Earth's atmosphere is transparent to a narrow band of light centered upon the Sun's peak frequency. Creatures here on Earth's surface evolved to be sensitive to these most abundant frequencies, which we now perceive as visible light. Within the spectrum of visible light we are most sensitive to a greenish-yellow, which is why many emergency vehicles are commonly painted greenish-yellow.

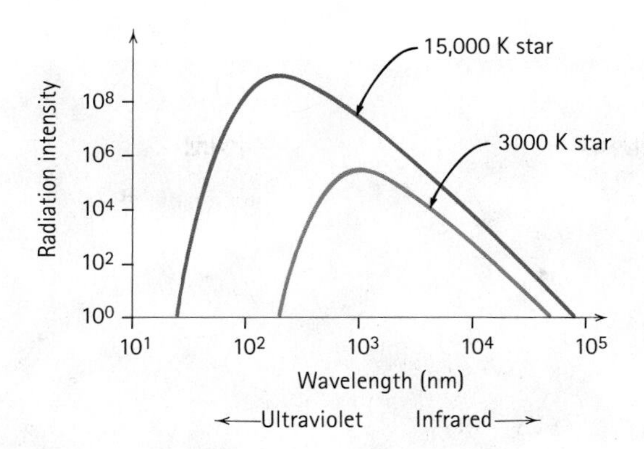

FIGURE 29.9
These radiation curves for stars of the same size and different surface temperatures show two important facts: (1) hotter stars emit radiation with higher average frequency than cooler stars, and (2) hotter stars emit more radiation per unit surface area at every frequency than cooler stars.

Betelgeuse, on the other hand, appears red because of its cooler surface temperature (about 3400 K). Betelgeuse emits more red light than blue light.

Notice also from Figure 29.9 that the hotter a star is, the more radiant energy it emits. Thus we see that hot blue stars are more luminous than cooler red stars of the same size.

CHECK YOURSELF
The temperature of Sirius is about 9400 K. What color is this star—and why?

CHECK YOUR ANSWER
Sirius has a slightly blue color. It emits more blue light than red light because of its high surface temperature.

29.3 The Hertzsprung–Russell Diagram

LEARNING OBJECTIVE
Describe the relationship between stellar luminosity and surface temperature as portrayed in the Hertzsprung–Russell diagram.

EXPLAIN THIS When is a cool star larger than a hot star?

When you compare the luminosity of stars to their temperature, interesting patterns emerge. Early in the 20th century, Danish astronomer Ejnar Hertzsprung and American astronomer Henry Norris Russell did just this. They produced a diagram known as the **Hertzsprung–Russell diagram**, or **H–R diagram**, which is of key importance in astronomy (Figure 29.10). The H–R diagram is a plot of the luminosity versus surface temperature of stars. Luminous stars are near the top of the diagram, and dim stars are toward the bottom. Hot bluish stars are toward the left side of the diagram, and cool reddish stars are toward the right side.

The H–R diagram shows several distinct regions of stars. Most stars are plotted on the band that stretches diagonally across the diagram. This band is called the **main sequence**. Stars on the main sequence, including our Sun, generate energy by fusing hydrogen to helium. As we would expect, the hottest main-sequence stars are the brightest and bluest stars and the coolest main-sequence

Because the giants and supergiants are so luminous, they are easy to see in the night sky even if they are not close to Earth. You can often identify them by their reddish color.

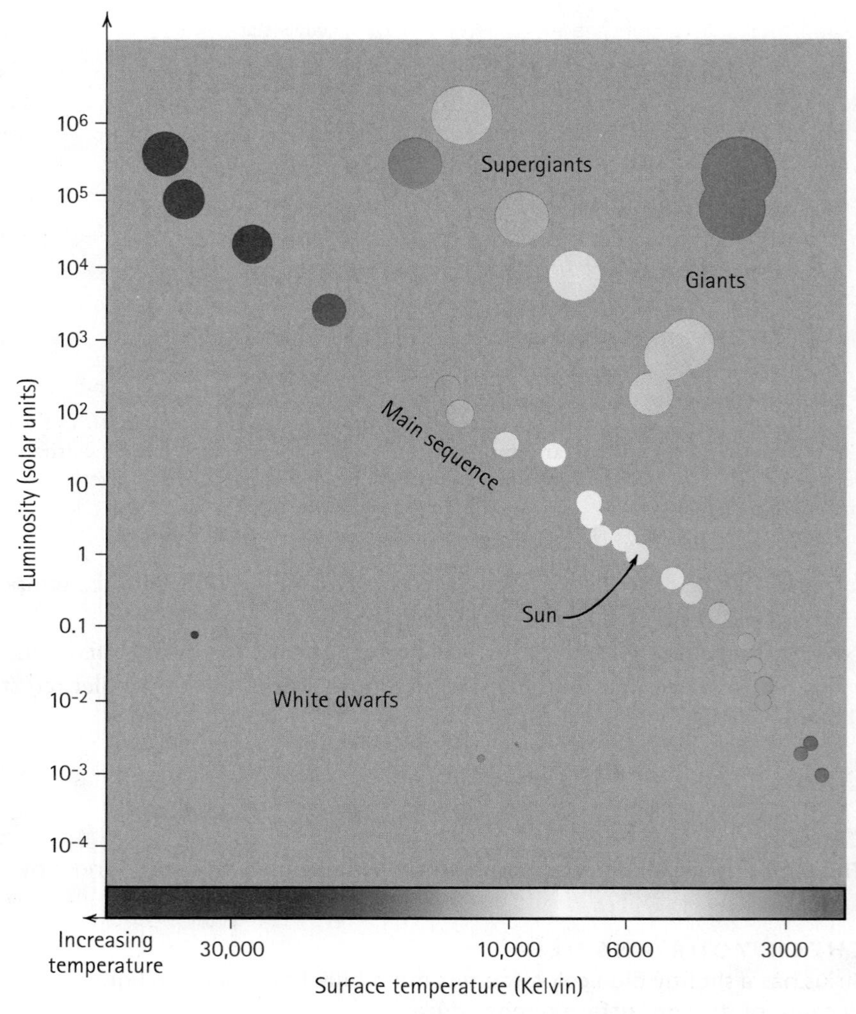

FIGURE 29.10
The H–R diagram shows a star's surface temperature on the horizontal axis and its luminosity on the vertical axis. The giant and supergiant stars shown here as circles are not drawn to scale. The red supergiant, Antares, for example, is so large that, if drawn to scale, it would reach the ceiling of your classroom. Interestingly, although the radius of Antares is 700 times that of our Sun, its mass is only about 15 times greater. So, Antares is much larger, but it is also much less dense.

FYI The H–R diagram is to astrophysicists what the periodic table is to chemists—an extremely important tool. A star's position on the H–R diagram can reveal its age. The age of our galaxy can be estimated by looking at the positions of our oldest stars and their white-dwarf remnants.

stars are the most dim and red stars. Take a moment to locate the Sun on the H–R diagram. Can you see that the Sun is a roughly average main-sequence star in terms of its luminosity and temperature?

Toward the upper right of the diagram is a distinct group of stars—the **giant stars**. These stars clearly do not follow the pattern of the hydrogen-burning main-sequence stars. Because these stars are red, we know they must have low surface temperatures. If they were main-sequence stars, the giants would be dim. Yet notice how high the giants are on the luminosity scale—they are very bright. The fact that the giants are both much cooler and much brighter than the Sun tells us that these stars must also be much larger than the Sun. (Hence the name *giant.*) Above the giants on the H–R diagram are a few rare stars, the *supergiants*. The supergiants are even larger and brighter than the giants. As you will see in the next section, the giants and supergiants are stars nearing the end of their lives because the fuel in their cores is running out.

Toward the lower left are some stars that are so dim they cannot be seen with the unaided eye. The surfaces of these stars can be hotter than the Sun, which makes them blue or white. Yet their luminosities are quite low—on the order of $0.1L_{Sun}$ to $0.0001L_{Sun}$. To be so hot and radiate so little light, these stars must be very small—they are called the **white dwarfs**. White dwarfs are typically the size of Earth or even smaller, yet they have mass comparable to the Sun. The density (or mass per volume) of a white dwarf is thus extremely high—about a million g/cm^3. For comparison, gold has a density of about 19 g/cm^3, while the average density of Earth is about 5.4 g/cm^3. As you will learn in the next section, white dwarfs are dead stars, the remnants of stars that have exhausted their nuclear fuel.

CHECK YOURSELF

1. **What characteristic do all main-sequence stars share?**
2. **Giant stars have cool surface temperatures yet are highly luminous. Does this mean that the frequency of light emitted by a giant star does not depend on its surface temperature as described by Figure 29.9?**

CHECK YOUR ANSWERS

1. All main-sequence stars generate energy by the nuclear fusion of hydrogen to helium.
2. No, radiation curves hold for a giant star as for any other radiating body. Giants do have a relatively low energy output per unit surface area; they are highly luminous only because they are very large.

SCIENCE AND SOCIETY

Astrology

There is more than one way to view the cosmos and its processes—astronomy is one and astrology is another. Astrology is a belief system that began more than 2000 years ago in Babylonia. Astrology has survived nearly unchanged since the second century AD, when some revisions were made by Egyptians and Greeks who believed that their gods moved heavenly bodies to influence the lives of people on Earth. Astrology today holds that the position of Earth in its orbit around the Sun at the time of birth, combined with the relative positions of the planets, has some influence over one's personal life. The stars and planets are said to affect such personal things as one's character, marriage, friendships, wealth, and death.

Could the force of gravity exerted by these celestial bodies be a legitimate factor in human affairs? After all, the ocean tides are the result of the Moon's and Sun's positions, and the gravitational pulls between the planets perturb one another's orbits. Because slight variations in gravity produce these effects, might not slight variations in the planetary positions at the time of birth affect a newborn? If the influence of stars and planets is gravitational, then credit must also be given to the effect of the gravitational pull between the newborn and Earth itself. This pull is enormously greater than the combined pull of all the planets, even when lined up in a row (as occasionally happens). The gravitational influence of the hospital building on the newborn far exceeds that of the distant planets. So planetary gravitation cannot be an underlying agent for astrology.

Astrology is not a science because it doesn't change with new information as science does, nor are its predictions borne out by experiment. Rather, its predictions depend on coincidence and also on the tendency of many people to seek external explanations for their fates or personal behaviors. Astrological beliefs are built on anecdotal evidence that is neither reproducible nor testable. Astrology means different things to different people, but in any case, it is outside the realm of science. It is a pseudoscience lying within the realm of superstition.

LEARNING OBJECTIVE
Use the Hertzsprung–Russell diagram to summarize the stages of stellar development from initial formation to an ultimate fate, such as supernovae.

29.4 The Life Cycles of Stars

EXPLAIN THIS Why doesn't a neutron star emit beta particles?

In Chapter 28, we discussed the nebular theory, which explains how the Sun formed from an expansive, low-density cloud of gas and dust called a *nebula* (Figure 29.11). Other stars are also thought to form in the same way. That is, over time, a nebula flattens, heats, and spins more rapidly as it gravitationally contracts. Furthermore, the center of the nebula becomes dense enough to trap infrared radiation so that this energy is no longer radiated away. The hot central bulge of a nebula is called a *protostar*.

MasteringPhysics®
TUTORIAL: Stellar Evolution
VIDEO: Lives of Stars

Mutual gravitation between the gaseous particles in a protostar results in an overall contraction of this huge ball of gas, and its density increases still further as matter is crunched together, with an accompanying rise in pressure and temperature. When the central temperature reaches about 10 million K, hydrogen nuclei begin fusing to form helium nuclei. This thermonuclear reaction, converting hydrogen into helium, releases an enormous amount of radiant and thermal energy, as discussed in Chapter 28. The ignition of nuclear fuel marks the change from protostar to star. Outward-moving radiant energy and the gas accompanying it exert an outward pressure called *thermal pressure* on the contracting matter. When nuclear fusion occurs fast enough, thermal pressure becomes strong enough to halt the gravitational contraction. At this point, outward thermal pressure balances inward gravitational pressure, and the star's size stabilizes.

UNIFYING CONCEPT
- ***The Gravitational Force***
Section 5.3

CHECK YOURSELF
What do the processes of thermonuclear fusion and gravitational contraction have to do with the physical size of a star?

CHECK YOUR ANSWER
The size of a star is the result of these two continually competing processes. Energy from thermonuclear fusion tends to blow the star outward like a hydrogen bomb explosion, and gravitation tends to contract its matter in an implosion. The outward thermonuclear expansion and inward gravitational contraction produce an equilibrium that accounts for the star's size.

FIGURE 29.11
This image of the Trifid Nebula was obtained by the Spitzer Space Telescope. This nebula is located 5400 light-years from Earth in the constellation Sagittarius. Within each of the red dust clouds are developing stars.

Though all stars are born in the same way from contracting nebulae, they do not all progress through their lives in the same way. A star's mass determines the stages a star will go through from birth to death. There are limits on the mass that a star can attain. A protostar with a mass less than 0.08 times the mass of the Sun (0.08 M_{Sun}) never reaches the 10 million K threshold needed for sustained fusion of hydrogen. On the other hand, stars with masses above 100 M_{Sun} would undergo fusion at such a furious rate that gravity could not resist thermal pressure and the star would explode. So stars exist within the limits of about a tenth of the mass of the Sun and 100 times the solar mass.*

*One solar mass, $1M_{\text{Sun}}$, is a unit of mass equivalent to that of the Sun: 2×10^{30} kg.

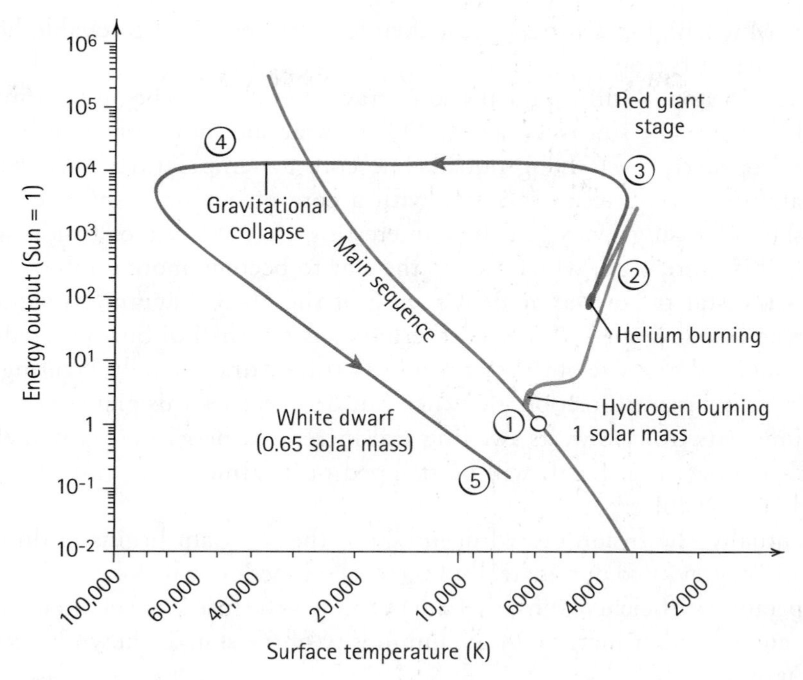

FIGURE 29.12
The stages of the Sun's life cycle are plotted on this H–R diagram. The short segment labeled Hydrogen burning lasts about 10 billion years. The later segments are much shorter.

Most stars have masses not very different from that of the Sun. Such stars inhabit a central place on the main sequence of the H–R diagram. If you plot the life-cycle stages of average stars on an H–R diagram, they trace a curve similar to the one for our Sun, which is shown in Figure 29.12. The Sun was born about 4.5 billion years ago at position 1, when the fusion of hydrogen ignited. The Sun will spend most of its lifetime—some 10 billion years—on the main sequence, with thermal pressure keeping gravity at bay. Speaking more generally, a star's hydrogen-burning lifetime lasts for a period of a few million to 50 billion years, depending on its mass.

More-massive stars have shorter lives than less-massive stars. This may sound counterintuitive, because if stars have more mass, they have more fuel to burn longer, right? High-mass stars, however, are more luminous than low-mass stars, meaning that they burn their hydrogen fusion fuel at a faster rate. Massive stars *must* be more luminous than small-mass stars so that the outward pressure of their nuclear fusion can offset the greater gravitational force of their contraction. Massive stars start out with more hydrogen fuel than small-mass stars, but they consume their fuel so much faster that they die billions of years younger than smaller stars.

A star's life cycle depends on its mass. The lowest-mass stars are brown dwarfs, dim but long-lived stars. Medium-mass stars progress from main-sequence stars to red giants or supergiants, then to white dwarfs. Very massive stars have short lives and die in massive explosions called *supernovae*.

No star lasts forever. In the old age of an average-mass star like our Sun, the nuclear fusion within the core eventually comes to a halt. This happens because the core has run out of hydrogen, which has been transformed into helium. But might this helium within the core start fusing to form heavier elements? The answer is no, because the temperatures within the core are not yet hot enough. Recall that each helium nucleus has two positively charged protons. Helium nuclei repel each other much more strongly than hydrogen nuclei. Much higher temperatures, therefore, are required to enable helium atoms to start fusing.

As fusion stops within the star's core, gravitational forces begin to dominate, and this causes the star to contract. This contraction, however, soon heats up the hydrogen that was lying outside the core to temperatures sufficient for nuclear fusion. The result is a star with a core of helium "ash" surrounded by a shell of fusing hydrogen. It is interesting that the rate of fusion within this shell is quite high, which causes the star to become more luminous. This pushes the star off of the main sequence of the H–R diagram, as shown by the green line in Figure 29.12. Furthermore, as the shell of burning hydrogen grows outward, the size of the star begins to inflate dramatically, creating what is called a red giant (position 2). When our Sun reaches this giant stage about 5 billion years from now, its swelling and increased energy output will elevate Earth's temperature. Earth will be stripped of its atmosphere and the oceans boiled dry. Ouch!

Eventually, the fusion of hydrogen within the red giant begins to diminish. This allows gravity to dominate, leading to a contraction, which heats the core to a temperature sufficient to ignite helium burning—the fusion of helium to carbon. This causes another increase in the luminosity of the star, as shown by position 3 in Figure 29.12.

For a star like our Sun, as helium fusion continues, carbon accumulates in the core, but temperatures will never become hot enough to allow the carbon to undergo fusion. Instead, carbon "ash" accumulates inside the star and fusion gradually tapers off. Then gravity predominates and the star contracts, which boosts its temperature. With higher temperatures, the color of the shrinking Sun will shift from red to blue and its position will shift to the left in the H–R diagram (position 4).

When our Sun turns into a red giant billions of years from now, its diameter will encompass the orbit of Venus.

CHECK YOURSELF

Why does a star shrink when its core runs out of nuclear fuel?

CHECK YOUR ANSWER

Outward thermal expansion and inward gravitational contraction produce an equilibrium that accounts for the star's size. As the heat from the inner thermonuclear reactions begins to die down, gravity predominates and the star shrinks. Upon shrinking, matter becomes compressed, which is an additional source of heat to ignite further nuclear fusion. For a star the size of our Sun, compression raises the temperature enough to fuse elements to carbon. The fusion of heavier elements in our Sun, however, is not possible.

Astronomers have found evidence suggesting that the carbon within the center of many white dwarfs crystallizes into diamond. They expect that when our Sun transforms into a white dwarf 5 billion years from now, its ember core will crystallize as well, leaving a planet-sized diamond at the center of our solar system.

Our fuel-exhausted Sun will continue to shrink until the electrons within the Sun are so squeezed that they resist any further compression. Having spent all of its nuclear fuel, our dead Sun, now quite small, will no longer be producing energy (position 5).

As our Sun goes through this final collapse, the layers of plasma and gas surrounding the core will be ejected in a brilliant display, forming what is called a **planetary nebula** (Figure 29.13). Despite its name, a planetary nebula has nothing to do with planets. The name is derived from the fact that the planetary nebula looks like a nebula from which planets could form. The planetary nebula, however, will disperse within a million years, leaving the Sun's cooling carbon core behind as a white dwarf. White dwarfs have the mass of a star but the volume of a planet, and are thus far more dense than anything on Earth. Because the nuclear fires of a white dwarf have burned out, it is not actually a star anymore, but is more accurately called a *stellar remnant*. In any case, a white dwarf cools for eons in space until it becomes too cold to radiate visible light (Figure 29.14).

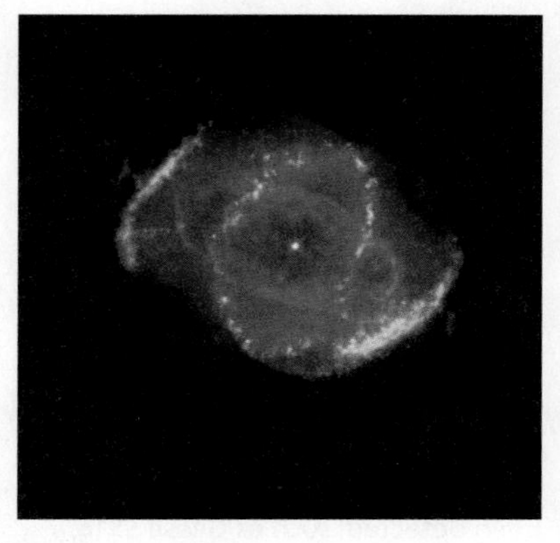

FIGURE 29.13
The Cat's Eye planetary nebula, seen here with the Hubble Space Telescope, measures about 1.2 light-years across, which is about a thousand times the diameter of Neptune's orbit. This planetary nebula is about 3000 light-years away, which places it within our galaxy. Clearly visible are the hot gases exploding away from the central Sun-sized star, which is in the process of transforming into a white dwarf.

Novae and Supernovae

There is another possible fate for a white dwarf, if it is part of a *binary star*. A binary star is a double star—a system of two stars that revolve about a common center, just as Earth and the Moon revolve about each other. If a white dwarf is a binary and close enough to its partner, the white dwarf may gravitationally pull hydrogen from its companion star. It then deposits this material on its own surface as a very dense hydrogen layer. Continued compacting increases the temperature of this layer, which ignites to embroil the white dwarf's surface in a thermonuclear blast that we see as a **nova**, which appears in the nighttime sky as a new star (*nova* is Latin for "new"). A nova is an event, not a stellar object. After a while, a nova subsides until enough matter accumulates to repeat the event. A given nova flares up at irregular intervals that may range from decades to hundreds of thousands of years.

Although low- and medium-mass stars become white dwarfs, the fate of stars more massive than about 10 M_{Sun} is quite different. When such a massive star contracts after its giant or supergiant phase, more heat is generated than in the contraction of a small star. Such a star does not shrink to become a white dwarf. Instead, carbon nuclei in its core fuse and liberate energy while synthesizing heavier elements, such as neon and magnesium. Thermal pressure halts further gravitational contraction until all the carbon is fused. Then the core of the star contracts again to produce even higher temperatures, and a new fusion series produces even heavier elements. The fusion cycles repeat until the element iron is formed.

Fusion of elements with atomic numbers higher than those of iron consumes energy rather than liberating energy. (The reason for this, as you may recall from Chapter 10, is that the average mass per nucleon is lower for iron than for any other element.) Once nuclei transform into iron, the fusion process stops. Thermal expansion that pushes against gravity, therefore, also stops. Gravity thus predominates and the entire star begins its final contraction.

FIGURE 29.14
A white dwarf, shown here in an artist's sketch, is the final stage in the evolution of low- and medium-mass stars. After a star has used all its nuclear fuel, its outer layers escape into space, leaving the dense core behind as a white dwarf. The strong gravitational field of a white dwarf causes it to attract matter from surrounding space to form an accretion disk. The disk is heated by friction where it meets the star, causing it to glow brightly.

Recall that with a dying medium-sized star, such as our Sun, contraction continues until gravity is counteracted by the resistance of electrons. With a supermassive supergiant, however, the gravitational forces are strong enough to overcome this resistance. The electrons, however, do not merge into one another. Instead, they combine with protons to form neutrons. What happens next is an astounding event called a **supernova**. Within minutes, the supergiant's iron core, about the size of Earth, collapses into a ball of neutrons only several kilometers in diameter. Massive amounts of energy are released—enough to outshine an entire galaxy. During this brief time of abundant energy, the heavy elements beyond iron are synthesized, as protons and neutrons outside the core mash into other nuclei to produce such elements as silver, gold, and uranium. These heavy elements are less abundant than the lighter elements because of the brief time available for synthesizing them.

FYI Do planets also orbit other stars? The answer is a resounding yes. These "exoplanets" reveal themselves by causing slight but detectable wobbles in the star they orbit. In some cases, the exoplanet transits in front of the star, which causes the star to become slightly dimmer, again at detectable levels. More than a thousand exoplanets have so far been discovered. Most are Jupiter-sized planets, but some near-Earth-sized planets have also been detected, such as Gliese 581g, which is about 20 light-years distant.

Most of the energy during the collapse of the iron core is released in the form of *neutrinos*—nearly massless subatomic particles that rarely interact with matter. Concentrations of neutrinos released from the collapse of the iron core are great enough to blow the outer shells of the star outward at speeds in excess of 10,000 km/s, which is fast enough to travel 1 AU in about four hours. Over time, this supernova wind of heavy elements spreads to far reaches of the galaxy where the elements are taken in by nebulae destined to become new stars. The gold and platinum we wear for jewelry here on Earth, as well as the bulk of Earth itself, are dust from supernovae that exploded many years before our solar system came to be.

CHECK YOURSELF

A star can undergo a nova more than once. Can a star also go through multiple supernovae? Why or why not?

CHECK YOUR ANSWER

A nova is a thermonuclear explosion that occurs when a white dwarf collects sufficient mass from a very close neighboring star. As long as the neighboring star provides mass, this explosion can be repeated multiple times. A supernova is such an energetic release of energy that it is an end-all event occurring never more than once for a particular supergiant star.

A supernova flares up to millions of times its former brightness. In AD 1054, Chinese astronomers recorded their observation of a star so bright that it could be seen by day as well as by night. This was a supernova (a "super new star"), its glowing plasma remnants now making up the spectacular Crab Nebula, shown in Figure 29.15. A less spectacular but more recent supernova was witnessed in 1987. The progress of this supernova, shown in Figure 29.16, has been very carefully monitored by modern scientific equipment.

FIGURE 29.15
The Crab Nebula is the remnant of a supernova explosion first observed on Earth in AD 1054. The explosion took place within our galaxy at a distance of about 7000 light-years from Earth. Had the blast occurred within 50 light-years, most life on Earth would have likely gone extinct. Is there any nemesis star right now within this limit ready to supernova? We know of none now, but check the Internet for information about Betelgeuse, which lies some 500 light-years away.

The superdense neutron core that remains after the supernova is called a **neutron star**. In accord with the law of conservation of angular momentum, these tiny bodies, with densities hundreds of millions times greater than those of white dwarfs, can spin at fantastic speeds. Neutron stars provide an explanation for the existence of *pulsars*. **Pulsars**, thought to be neutron stars, are rapidly varying sources of low-frequency radio emissions. As a pulsar spins, the beam of radiation it emits sweeps across the sky. If the beam sweeps over Earth, we detect its pulses. Of the approximately 300 known pulsars, only a few have been found emitting X-rays or visible light. One is in the center of the Crab Nebula (Figure 29.17). It has one of the highest rotational speeds of any pulsar studied, rotating more than 30 times per second. This thousand-year-old pulsar is relatively young. It is theorized that X-rays and visible light are emitted only during a pulsar's early history.

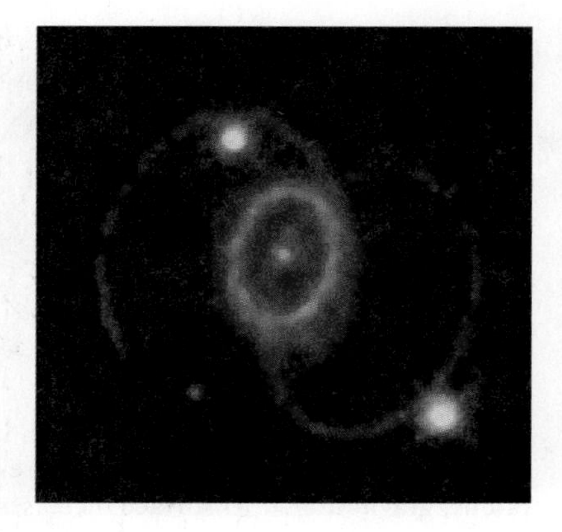

FIGURE 29.16
This image of the 1987A supernova was captured by the Hubble Telescope about 20 years after the initial explosion was sighted. Note the development of the ring systems, which continue to expand outward. This supernova occurred safely outside the Milky Way galaxy some 160,000 light-years away within a nearby smaller galaxy called the Large Magellanic Cloud. Some of its neutrinos were detected on Earth.

UNIFYING CONCEPT

- ***The Law of Conservation of Momentum***
  Section 4.4

We saw earlier that a medium-sized star, such as our Sun, can collapse no further than a white dwarf because the force of gravity is not strong enough to overcome the resistance of electrons, which refuse to trespass into the quantum states of neighboring electrons. Similarly, a neutron star stops collapsing because neutrons, like electrons, resist trespassing into their neighboring neutrons. For a dying star, however, the bigger they are, the harder they fall. When the collapsing star is the biggest of the big, gravitational forces can be strong enough to overcome even the resistance of neutrons. The collapse continues beyond the stage of a neutron star, and the star disappears altogether from the observable universe. What is left is a *black hole*.

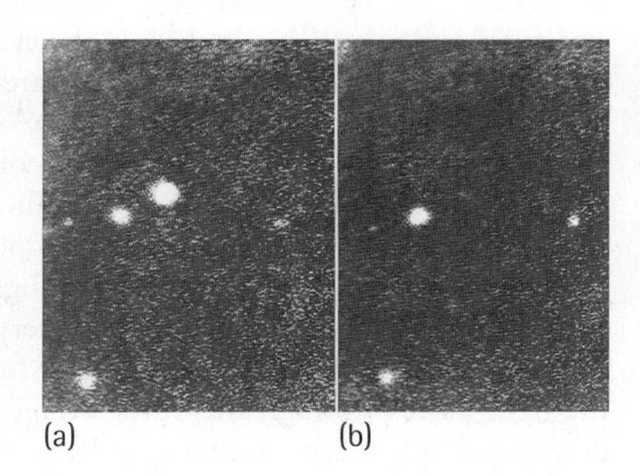

FIGURE 29.17
The pulsar in the Crab Nebula rotates like a searchlight, beaming visible light and X-rays toward Earth about 30 times per second, blinking on and off: (a) pulsar on, (b) pulsar off.

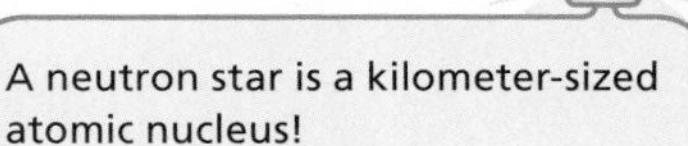
A neutron star is a kilometer-sized atomic nucleus!

29.5 Black Holes

LEARNING OBJECTIVE
Identify the major attributes of black holes, such as the photon sphere, event horizon, and singularity.

EXPLAIN THIS What happens to a light beam bouncing between two upright and perfectly parallel mirrors here on Earth?

A **black hole** is the remains of a supergiant star that has collapsed into itself. Upon this collapse, the force of gravity at the surface increases dramatically. Consider this from the perspective of Newton's law of gravity. According to this law, as discussed in Section 5.3, the force of gravity depends on the inverse square of the distance. If a star collapses to a tenth of its original size, the distance between the surface and the center of the star is one-tenth as much. The inverse square of one-tenth $(1/0.1^2)$ equals 100. The

MasteringPhysics®
TUTORIAL: Black Holes

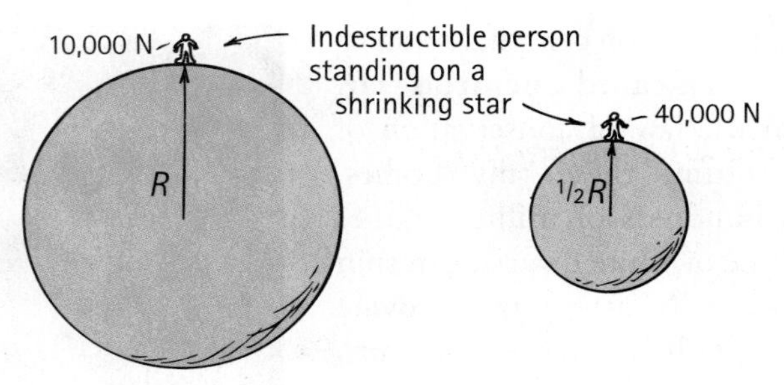

FIGURE 29.18
If a star collapses to half its radius with no mass change, gravitation at its surface increases fourfold (in accordance with the inverse-square law). If the star collapses to one-tenth its radius, gravitation at its surface increases a hundredfold.

FYI Contrary to stories about black holes, they're nonaggressive and don't reach out and swallow objects at a distance. Their gravitational fields are no stronger than the original fields about the stars before their collapse—except at distances less than the radius of the original star. If they don't get too close, black holes shouldn't worry future astronauts.

weight at the surface, therefore, is 100 times as much, as suggested in Figure 29.18. So the gravitational force at the surface of a collapsing star increases because the star is getting smaller.

As the force of gravity increases, so does the *escape speed*. Recall from Chapter 5 (Integrated Science 5C) that escape speed is the speed a moving object needs to fly away without ever falling back. For planet Earth, the escape speed is 11.2 km/s. This means that an object shot outward at 11.2 km/s (about 25,000 mi/h) will never fall back to Earth. The escape speed from the surface of our Sun is 618 km/s. For a supergiant star that has collapsed past the neutron star stage, the escape speed increases to the speed of light, which is 300,000 km/s.

In the early 20th century, Einstein proposed that light, like matter, is affected by gravity. We don't normally witness light being affected by gravity because light moves so fast, but with careful observations the influence is quite measurable. Starlight grazing the eclipsed Sun, for example, is seen to bend inward as the light passes through the Sun's strong gravitational field. So light is pulled downward by gravity. Sunlight can leave our Sun because the speed of light is much greater than the escape velocity. If a star such as our Sun, however, were to collapse to a radius of 3 km, the escape velocity from its surface would exceed the speed of light, and nothing—not even light—could escape. The Sun would be invisible. It would be a black hole.

The Sun, in fact, has too little mass to experience such a collapse, but when some stars with core masses over 40 times greater than the mass of the Sun reach the end of their nuclear resources, they undergo collapse; their collapse continues until the stars reach infinite densities. Gravitation near the surfaces of these shrunken stars is so enormous that light cannot escape from them. They have crushed themselves out of visible existence.

Except for material blown away during its formation, a black hole has the same amount of mass after its collapse as before its collapse. So, the gravitational field in regions at and beyond the star's original radius is no greater than before the star's collapse. An orbiting planet would keep on orbiting as though nothing happened. But closer distances near the vicinity of a black hole, beneath the star's original radius, are nothing less than the collapse of space itself, with a surrounding warp into which anything that passes too close—light, dust, or a spaceship—is drawn (Figure 29.19). Astronauts in a powerful spaceship could enter the fringes of this warp and still escape. At closer than a certain distance, however, they could not, and they would disappear from the observable universe.

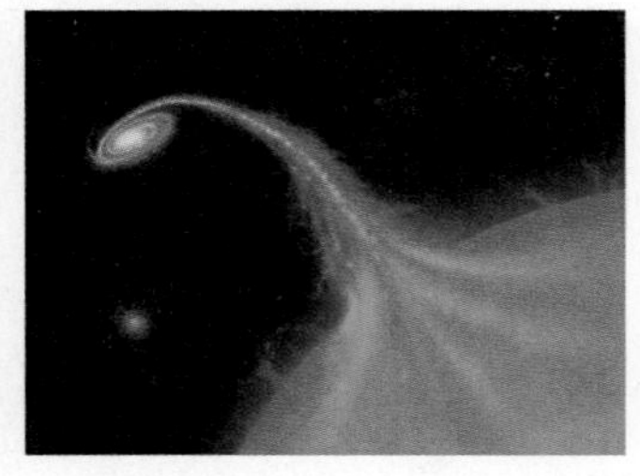

FIGURE 29.19
A rendering of a black hole stealing matter from a companion star.

CHECK YOURSELF

If the Sun somehow suddenly collapsed to a black hole, what change would occur in the orbital speed of Earth?

CHECK YOUR ANSWER

None. This is best understood classically; nothing in Newton's law of gravitation, $F = G\frac{mM}{d^2}$, changes. The fact that the Sun is compressed doesn't change its mass, M, or its distance, d, from Earth. Because Earth's mass, m, and G don't change either, the force, F, holding Earth in its orbit does not change.

Black Hole Geometry

As shown in Figure 29.20, a black hole can either deflect light or capture it. Also, there is one particular distance from the hole at which light can orbit in a circle. This distance is called the *photon sphere*. Light orbiting at that distance is highly unstable. The slightest disturbance will send it off into space or spiraling into the black hole.*

An indestructible astronaut with a powerful enough spaceship could venture into the photon sphere of a black hole and come out again. While inside the photon sphere, she could still send beams of light back into the outside universe as shown in Figure 29.21. If she directed her flashlight sideways and toward the black hole, the light would quickly spiral into the black hole, but light directed vertically and at angles close to the vertical would still escape. As she drew closer and closer to the black hole, however, she would need to shine the light beams closer and closer to the vertical for escape. Moving closer still, our astronaut would find a particular distance where *no* light can escape. No matter in what direction the flashlight pointed, all the beams would be deflected into the black hole. Our unfortunate astronaut would have passed within the **event horizon**, the boundary where no light within can escape. Once inside the event horizon, she could no longer communicate with the outside universe; neither light waves, radio waves, nor any matter could escape from inside the event horizon. Our astronaut would have performed her last experiment in the universe as we conceive it.

The event horizon surrounding a black hole is often called the *surface* of the black hole, the diameter of which depends on the mass of the hole. For example,

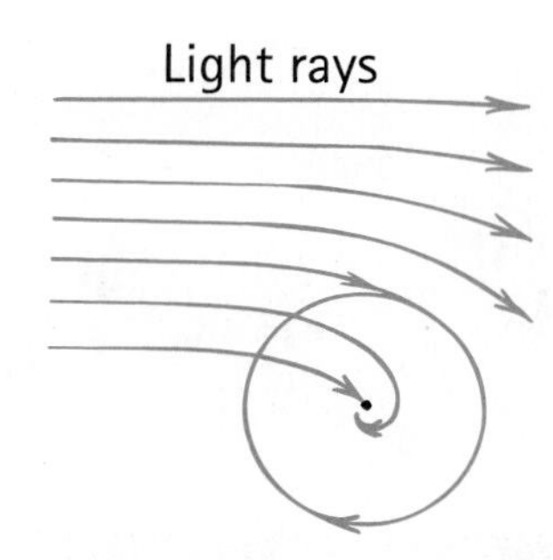

FIGURE 29.20
Light rays deflected by the gravitational field around a black hole. Light aimed far from the black hole is slightly deflected. Light aimed close to the black hole is drawn into the hole. In between, there is a particular radius at which photons can orbit the black hole.

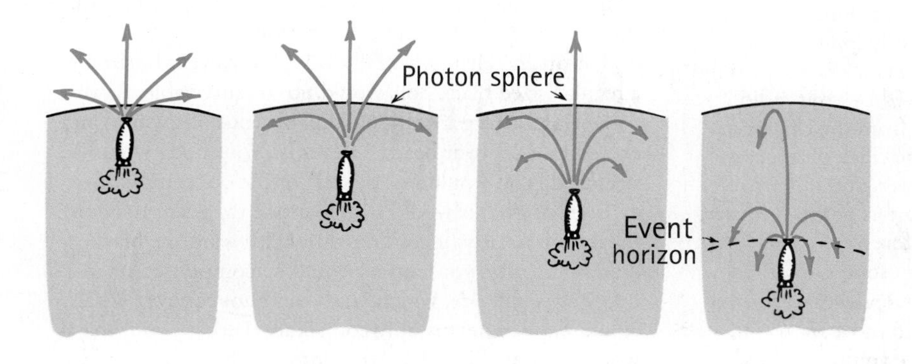

FIGURE 29.21
Just beneath the photon sphere, an astronaut can still shine light to the outside. But as she gets closer to the black hole, only light directed nearer to the vertical gets out, until finally even vertically directed light is trapped. This occurs at a distance called the event horizon.

*This discussion applies to a nonrotating black hole. The situation for a rotating one is more complicated.

TABLE 29.1 CALCULATED RADII OF EVENT HORIZONS FOR NONROTATING BLACK HOLES OF VARIOUS MASSES

Mass of Black Hole	Radius of Event Horizon
1 Earth mass	0.8 cm
1 Jupiter mass	2.8 m
1 solar mass	3 km
2 solar masses	6 km
3 solar masses	9 km
5 solar masses	15 km
10 solar masses	30 km
50 solar masses	148 km
100 solar masses	296 km
1000 solar masses	2961 km

a black hole resulting from the collapse of a star 10 times as massive as the Sun has an event-horizon diameter of about 30 km. Calculated radii of event horizons for black holes of various masses are shown in Table 29.1. The event horizon is not a physical surface. Falling objects pass right through it. The event horizon is simply the boundary of no return.

SCIENCE AND SOCIETY

Falling into a Black Hole

Imagine yourself exploring a black hole on some futuristic scientific mission. Your spacecraft is cruising within a safe orbit around the black hole. Your first experiment is to launch a clock-bearing probe toward the black hole. The clock consists of a large LED display of blue lights. Through a telescope you watch as the probe descends. Remarkably, the closer the probe gets to the black hole, the slower the clock appears to run. Furthermore, the light coming from the clock shifts from blue to a lower-frequency red. As the probe gets closer still, the clock runs even slower. Soon you can't make out the clock at all because it has shifted to the infrared. So you switch to your infrared telescope and see that as the probe gets closer to the black hole its clock slows to a creep. Furthermore, the probe seems to be taking an unusually long time to descend. Eventually, the light from the clock is visible only with your microwave telescope, followed by your radio telescope as the frequency of the light from the clock gets lower and lower. Ultimately, just as the clock disappears completely, emitting no light whatsoever, you note that the clock has frozen in time. To get to this point, however, would take, from your point of view, forever.

But you don't have forever, and you soon grow tired of watching the ultraslow clock as it creeps ever so slowly toward the black hole. So you decide to move on to the second experiment, for which you have volunteered to place yourself within a second probe equipped with a blue clock and an array of telescopes. As you descend toward the black hole you note that your clock runs perfectly normally without changing color. The clock on the mother ship, however, is running rather fast. Furthermore, its color is shifting toward ultraviolet and beyond. Your old shipmates are moving rather fast as well. As you descend even deeper, their peculiar speed grows even faster. Soon they grow impatient waiting for you and they leave you behind for dead. Before you know it, the remaining visible stars quickly pass through their life spans, their light coming to you in a flash of ultrahigh frequencies but through a narrowing field of view. And then there is nothing. At that moment you pass through the event horizon, which is a mathematical boundary, not a physical one. Singularity is still kilometers beneath you, but you are caught within its unrelenting grip. The universe you left behind has run through infinite time and exists no longer.

Unfortunately, such a fall through the event horizon of a regular-sized black hole would not be survivable. As you come closer, the gravitational pull on your feet would far exceed that on your head. As a result, your body would be stretched. You would be "ripped" of the opportunity to experience what it would be like inside the event horizon. Furthermore, as you continued to fall toward the black-hole singularity, your atoms would be compressed to an infinitely small size, which you would not survive. What would happen next is only conjecture. Perhaps your mass would explode like a Big Bang into another universe. Perhaps what happens to the mass that falls into the singularity is even stranger than we're capable of imagining. Maybe one day we crafty humans will come to understand such processes. If our species lasts that long!

When a collapsing star contracts within its own event horizon, the star still has substantial size. No known forces, however, can stop the continued contraction, and the star quickly shrinks until finally it is crushed, presumably to the size of a pinhead, then to the size of a microbe, and finally to a realm of size smaller than ever measured by humans. At this point, according to theory, what remains has infinite density. This point is the **black-hole singularity**.

Locating black holes is very difficult. One way to find them is to look for a binary system in which a single luminous star appears to orbit about an invisible companion, as was illustrated in Figure 29.19. If they are closely situated, matter ejected by the normal companion and accelerating into the neighboring black hole should emit X-rays. The first convincing candidate for a black hole, the X-ray star Cygnus X-1, was discovered in 1971. Many additional black hole candidates have since been found, which suggests that black holes are common. Studies of the center of our galaxy strongly suggest the presence of a black hole some 6 billion km in diameter, which is as large as our solar system! The origin of this mega black hole is likely related to the formation of the galaxy itself. It is currently thought that most, if not all, large galaxies contain central mega-sized black holes.

CHECK YOURSELF

What determines whether a star becomes a white dwarf, a neutron star, or a black hole?

CHECK YOUR ANSWER

The mass of a star is the principal factor that determines its fate. Stars that are about as massive as the Sun, and those that are less massive, evolve to become white dwarfs; stars with masses of $10M_{Sun}$ or greater evolve to become neutron stars; the most massive stars of about $40M_{Sun}$ or greater ultimately become black holes.

29.6 Galaxies

LEARNING OBJECTIVE

Describe the discovery of galaxies; their classification as elliptical, spiral, or irregular; and how they are organized into superclusters.

EXPLAIN THIS All the celestial objects discussed so far are located in what galaxy?

Look up into the clear nighttime sky away from the city lights and you will see plenty of stars. In between the stars you'll also see plenty of black. Before the early 20th century, the abundance of black in the night sky led many people to conclude that the universe consisted of an island of millions of stars nestled within a vast sea of emptiness. In addition to stars, however, are the cloudlike nebulae, some of them with a distinct spiral-shaped structure. As early as the 1750s, the German philosopher Immanuel Kant proposed that these spiral clouds were other islands of stars called *galaxies*. But without powerful telescopes, there was no way to tell whether that was true.

The debate about whether the universe consisted of one or many islands of stars was settled by the American astronomer Edwin Hubble. In 1927, working with the newly built largest telescope in the world at Mt. Wilson in California, Hubble made out individual stars within the Andromeda spiral nebula (Figure 29.22). Some of these stars he noticed to be *Cepheids,* which are stars that change their

FIGURE 29.22
Hubble showed that the great spiral nebula within the Andromeda constellation was not just a swirling cloud of gas, but a neighboring galaxy of stars, which is now called the Andromeda galaxy and cataloged as M31. You can see the Andromeda galaxy for yourself by looking between the constellations of Cassiopeia and Pegasus in the late fall nighttime sky. The galaxy appears huge, covering an area six times that of the full Moon. It is, of course, much dimmer than the Moon. Best viewing comes with a good pair of binoculars far away from city lights.

FYI Galaxies are cataloged by two systems. The first catalog is based on the work of Charles Messier, who in 1781 published a list of heavenly structures, such as galaxies, that are relatively easy to observe with small telescopes. The Andromeda galaxy, for example, is the 31st entry of this catalog and is thus listed as M31. A "New General Catalog" was begun in 1888 that was subsequently used to identify all structures, including the many more that became visible with the advent of more powerful telescopes. Under this system, the Andromeda galaxy is cataloged as NGC 224. You can use these catalog numbers in your Internet search engine to learn more about these objects, including their locations in the nighttime sky.

luminosity over short periods of time. Using photos of Cepheid variables in the Magellanic Clouds, Henrietta Leavitt had earlier discovered a relationship between Cepheids' periods and their luminosities. So, by measuring the rate at which they changed luminosity, Hubble estimated their distance, which he found to be much farther away than any star within our own galaxy. Spiral nebulae were not simply clouds—they were neighboring islands of stars within a vast emptiness that potentially extended forever.

But Hubble took his research a step further and discovered something even more amazing. He knew that the color of light emitted by a star or galaxy receding away from us shifts to the red because of the Doppler effect (see Chapter 8). The degree of redshift could be measured quantitatively by focusing on the line spectrum of hydrogen (see Chapter 9). The greater the shift in the lines of hydrogen's spectrum, the faster the receding speed. Using Cepheid variables and stellar spectra, Hubble's research team determined both the distances and redshifts of numerous galaxies and discovered that the farther the galaxy, the greater the redshift. This meant that the galaxies were not static islands. Rather, they were receding from us in every direction, which meant that the universe itself was expanding.

If distant galaxies were all moving away from one another, that could only mean that they were once much closer together. Running the cosmic movie backward would inevitably lead to a moment when all the galaxies were gathered together, perhaps within a single point. The universe as we know it, therefore, had a beginning. This moment has come to be known as the **Big Bang**, which we will discuss in more detail in Section 29.7. Now, however, we will describe the different kinds of galaxies and how they are organized within the observable universe.

A **galaxy** consists of a large assemblage of stars, interstellar gas, and dust. Galaxies are the breeding grounds of stars. Our own star, the Sun, is an ordinary star among more than 100 billion others in an ordinary galaxy known as the Milky Way galaxy (Figure 29.23). With unaided eyes, we see the Milky Way as a faint band of light that

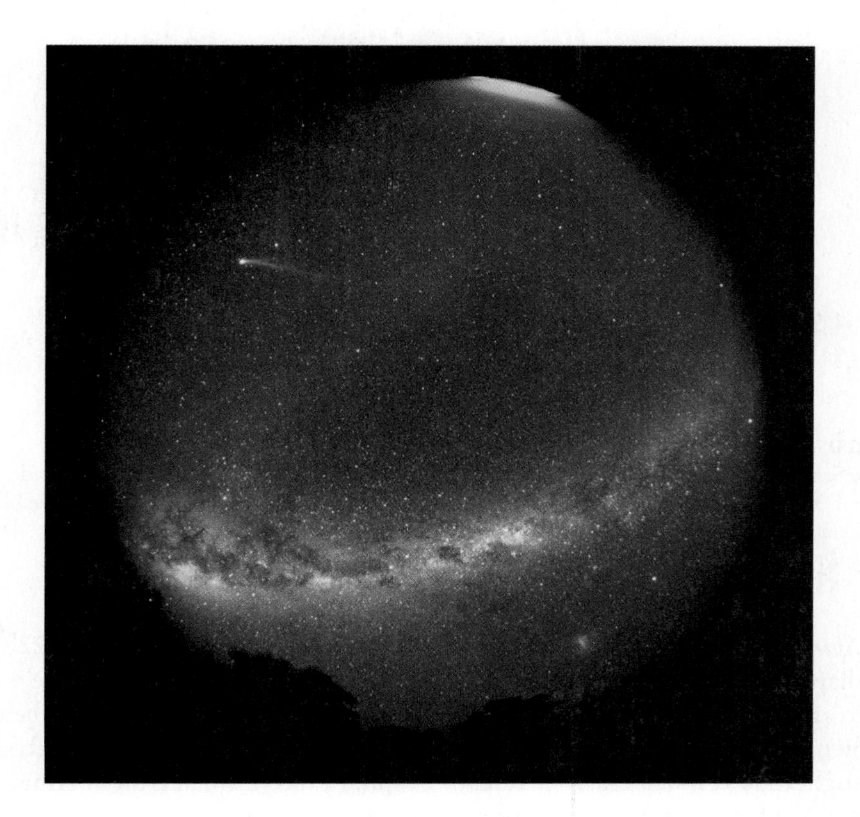

FIGURE 29.23
A wide-angle photograph of the Milky Way, which appears as a north–south cloudlike band of light. The dark lanes and blotches are interstellar gas and dust obscuring the light from the galactic center. If it weren't for this dust, the Milky Way would be a much more spectacular nighttime display. This photograph also shows Comet Hyakutake, which appeared in 1996.

stretches across the sky. The early Greeks called it the "milky circle" and the Romans called it the "milky road" or "milky way." The latter name has stuck.

The masses of galaxies range from about a millionth the mass of our galaxy to some 50 times more. Galaxies are calculated to have much more mass than can be seen with the telescope. A small proportion of the invisible mass is simply matter that has grown so cold that it doesn't emit enough light for us to see. The bulk of the invisible mass, however, is likely an unknown form of matter, called *dark matter*, that does not absorb or emit light. It does, however, possess mass and so its gravitational effects are quite measurable. In Section 29.8 we will describe how dark matter likely played a key role in the formation and distribution of galaxies.

FIGURE 29.24
This small elliptical galaxy, Leo I, found within the constellation Leo, is only about 2500 light-years in diameter. For comparison, the diameter of our Milky Way galaxy is about 100,000 light-years.

Elliptical, Spiral, and Irregular Galaxies

The millions of galaxies visible in photographs can be separated into three main classes—elliptical, spiral, and irregular. *Elliptical galaxies* are the most common galaxies in the universe. They are spherical, with the stars more crowded toward the center. Most contain little gas and dust, which makes them easy to see through. They also tend to be yellow, which tells us that they consist primarily of older stars—older stars are yellow, while hot young stars tend to be blue. Most ellipticals are small, consisting of fewer than a billion stars (Figure 29.24). An exception is the giant elliptical galaxy M87 (Figure 29.25). The largest ellipticals are about 5 times as large as our galaxy, and the smallest are 1/100 as large.

FIGURE 29.25
The giant elliptical galaxy M87, one of the most luminous galaxies in the sky, is located near the center of the Virgo cluster, some 50 million light-years from Earth. It is about 120,000 light-years across and about 40 times as massive as our own galaxy, the Milky Way.

FIGURE 29.26
The Sombrero galaxy, cataloged as M104, is about 80,000 light-years in diameter and about 32 million light-years from Earth. At its center is one of the most supermassive black holes measured in any nearby galaxy.

FYI The Andromeda galaxy is our closest spiral neighbor, being only some 2.3 million light-years away. It contains many more stars than the Milky Way, which makes it more luminescent. Also, its diameter is about 220,000 light-years, compared to the Milky Way's 100,000 light-years. Thus, our view of the Andromeda is likely more spectacular than the Andromeda's view of us.

Spiral galaxies, such as the Andromeda galaxy, shown in Figure 29.22, are perhaps the most beautiful arrangements of stars. Some spirals, such as the Sombrero galaxy of Figure 29.26, have a spheroid central hub. Others, like the one shown in Figure 29.27, have a hub shaped like a bar. The Milky Way galaxy is thought to look much like the NGC 6744 spiral galaxy, which is an intermediate between a barred and unbarred spiral (Figure 29.28).

Elliptical and spiral galaxies sometimes cross paths or even collide. In such cases, gravity causes the shape of the galaxy to become distorted. These distorted looking galaxies are called *irregular galaxies*. Most irregular galaxies are small and faint and are difficult to detect. They tend to contain large clouds of gas and dust mixed with both young (blue) and old (yellow) stars. The irregular galaxy first described by the navigator on Magellan's voyage around the world in 1521 is our nearest neighboring galaxy—the Magellanic Clouds. This galaxy consists of two "clouds," called the Large Magellanic Cloud (LMC) and the Small

FIGURE 29.27
The beautiful barred spiral galaxy NGC 1300 is about 100,000 light-years across and some 70 million light-years away.

FIGURE 29.28
The NGC 6744 galaxy is an intermediate between a barred and unbarred spiral galaxy. Studies of the Milky Way suggest that it too is an intermediate spiral. In other words, this is what we may look like from afar.

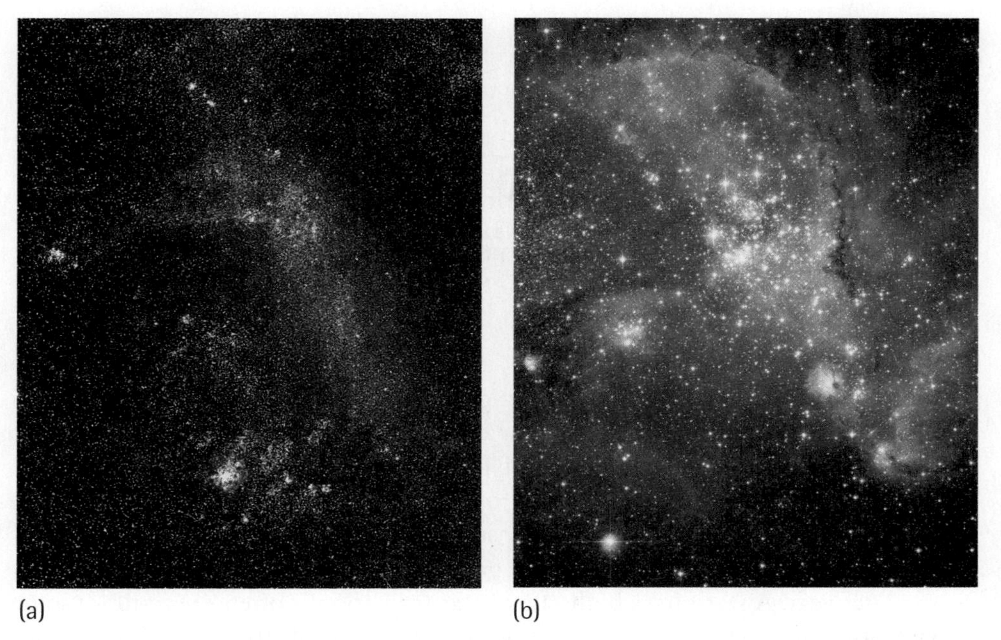

FIGURE 29.29
(a) The Large Magellanic Cloud and (b) the neighboring Small Magellanic Cloud are a pair of irregular galaxies. The Magellanic Clouds are our closest galactic neighbors, about 150,000 light-years distant. They likely orbit the Milky Way.

Magellanic Cloud (SMC), both of which are slowly being pulled into the Milky Way. The LMC is dotted with hot young stars with a combined mass of some 20 billion solar masses, and the SMC contains stars with a combined mass of about 2 billion solar masses (Figure 29.29). Some irregular galaxies, such as NGC 4038 shown in Figure 29.30, are the aftermaths of galactic collisions.

FIGURE 29.30
Shown in black and white is the ground-telescope view of an irregular galaxy resulting from the collision of two galaxies. Note the remnant arms that suggest two former spiral galaxies. The inset shows a close-up color view taken by the Hubble Telescope. Evident is the rapid formation of new stars (blue) occurring as the two galaxies combined.

CHECK YOURSELF
Is it possible for one type of galaxy to turn into another?

CHECK YOUR ANSWER
Yes, and this occurs as two symmetrically shaped galaxies collide to form an asymmetrically shaped irregular galaxy.

Active Galaxies

Galaxies differ greatly in the activity going on inside them. This activity may include star formation, supernova, or energy-releasing processes at the galactic core. Those galaxies with notable activity are sometimes called *active galaxies*.

When it comes to star formation, our Milky Way is a relatively calm place, producing on average about one new star per year. By comparison, one type of active galaxy, known as a **starburst galaxy**, can produce more than 100 new stars per year. A starburst's high rate of star formation is often the result of some violent disturbance, such as a collision between two galaxies. The irregular galaxy shown in Figure 29.30 is an example of a starburst galaxy. Another example is the Cigar galaxy, M82, which is being deformed by the tidal forces from its much larger neighbor, M81 (Figure 29.31). A starburst tends to die down once the disturbance is removed or after the starburst galaxy consumes all its interstellar fuel. Many elliptical galaxies are thought to be former starburst galaxies because of their low abundance of interstellar dust and gases.

Other active galaxies are active by virtue of their galactic core, which hosts a black hole more massive than millions or even billions of Suns. The event horizons of these black holes are about as large as our solar system! Most large galaxies, including the Milky Way, have such black holes in their centers, and these massive black holes can be a source of much activity.

For example, in 2010, using NASA's Fermi gamma-ray space telescope, astronomers discovered two massive gamma-ray–emitting bubbles extending north and south from the center of our galactic disk, as shown in Figure 29.32. Together, these bubbles, known as *Fermi bubbles*, extend about 50,000 light-years, which is about

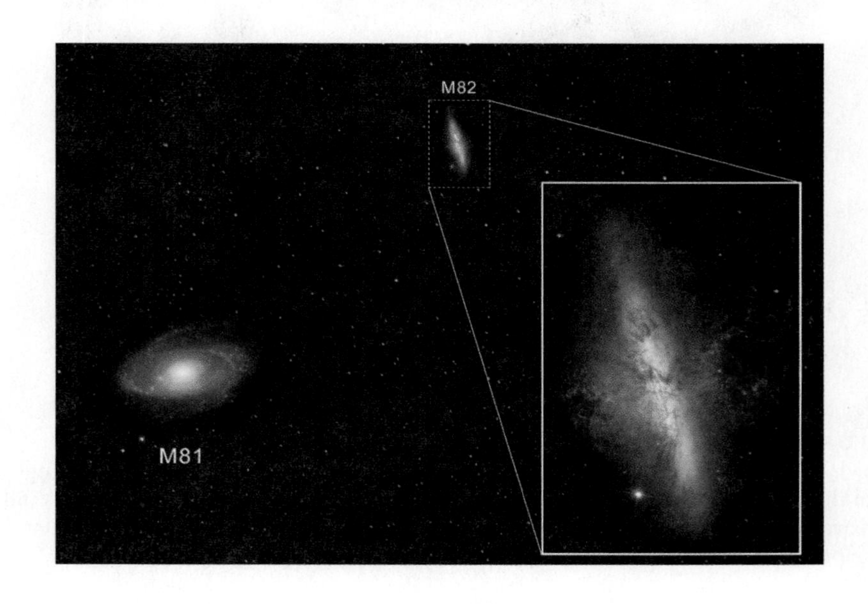

FIGURE 29.31
The Cigar galaxy, M82, is a spiral galaxy tilted away from us so that we see it from an edge-on view. Tidal forces from the nearby M81 galaxy disturb the distribution of matter within M82, which clumps, allowing for the formation of many new stars, as evidenced by M82's remarkable blue color. The red gases above and below the galactic plane are primarily hydrogen being pushed out by abundant stellar wind.

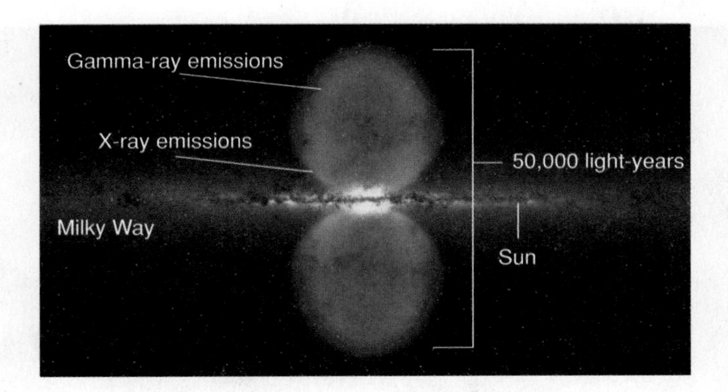

FIGURE 29.32
Vast gamma-ray–emitting bubbles, known as Fermi bubbles, extend north and south from the center of our galaxy. Though invisible to the naked eye, these bubbles span over half the sky when viewed from Earth through gamma-ray detectors.

half the diameter of our galaxy. Their edges are well defined (not diffuse), which suggests they are a fairly recent occurrence, perhaps arising only several million years ago. Their nature and origin are not yet fully understood. One model suggests they arose from large amounts of material falling into our galaxy's mega-size central black hole. Alternatively, they may have resulted from the outgassing of a brief but intense period of starburst activity occurring at the galactic core.

The Milky Way's Fermi bubbles, however, are minor compared to the activity we see arising from the centers of other large galaxies. In such cases, supermassive amounts of matter are likely falling into central supermassive black holes. Before falling into the black hole, the doomed mass forms a rapidly spinning disk, called an *accretion disk*, around the equator of the black hole. Charged particles in this hyperspinning disk create a narrow yet ultrastrong magnetic field that rises from the black hole's poles. Rather than falling into the black hole, some of the charged particles, such as electrons, are accelerated outward through these magnetic fields to nearly the speed of light. This results in two extremely long streams of particles, called *jets*, extending more than 100,000 light-years away from the galactic center, which is called an **active galactic nucleus** (AGN).

A relatively close AGN is found within the large elliptical galaxy M87, which was shown in Figure 29.25. High-resolution images of this galaxy, as shown in Figure 29.33, reveal a jet of material streaming away from the center of this galaxy. Interestingly, the jet is angled toward us. This plus the great speed of the jet (99.5% of the speed of light) helps make the jet appear more luminous. The opposite "counterjet" receding away from us at such great speeds is barely visible.

FIGURE 29.33
Material falling into the supermassive black hole in the center of M87 generates powerful jets that shoot out at near light speed.

Nearby active galactic nuclei, such as that of M87, give us a clue as to the possible nature of the most energetic galaxies of all—the *quasars*. Starting in the 1960s, astronomers began discovering extremely energetic bodies hundreds of times more luminous than our own galaxy, yet farther away than any observed object. Because they looked like radio-emitting stars, they were dubbed "quasi-stellar radio sources," which was shortened to **quasars**. Because all quasars are so very far away, they occurred a very long time ago—up to 13 billion years ago, which was close to the beginning of the universe. As we look to quasars, therefore, we are peering into the early lives of galaxies (Figure 29.34). During this galactic youth, it is thought that much material was still falling into the supermassive black holes found within their galactic cores. The dynamics of

FIGURE 29.34
Each disk in this deep-space image taken by the Hubble Telescope is a galaxy. The quasar shown in the center is billions of light-years behind this cluster of galaxies. Interestingly, the cluster's gravity has bent the light from the quasar like a lens so that multiple images of the quasar are also seen.

this process allow for an efficient conversion of mass into energy, which would provide for colossal jets of highly energetic particles and light. When one of these ancient jets faces our direction, the result is an unusually brilliant display of energy we call a *blazar*.

CHECK YOURSELF
Are there any quasars within the Milky Way galaxy?

CHECK YOUR ANSWER
No. A quasar is the active galactic nucleus of a galaxy as it appeared toward the beginning of the universe. All quasars are billions of light-years away from our galaxy.

Clusters and Superclusters

Galaxies are not the largest structures in the universe; they tend to cluster into distinguishable groups. Our Milky Way galaxy, for example, is part of a cluster of local galaxies that includes two other major spiral galaxies—namely, the Andromeda galaxy and the Triangulum galaxy. Also included are more than a dozen smaller elliptical galaxies, including the Leo I galaxy shown in Figure 29.24, and a few irregular galaxies, such as the Large Magellanic Cloud. Altogether this cluster of galaxies is called the **Local Group**. Their approximate distributions are shown in Figure 29.35. If drawn correctly to scale, Andromeda is only about 20 Milky Way diameters away from the Milky Way.

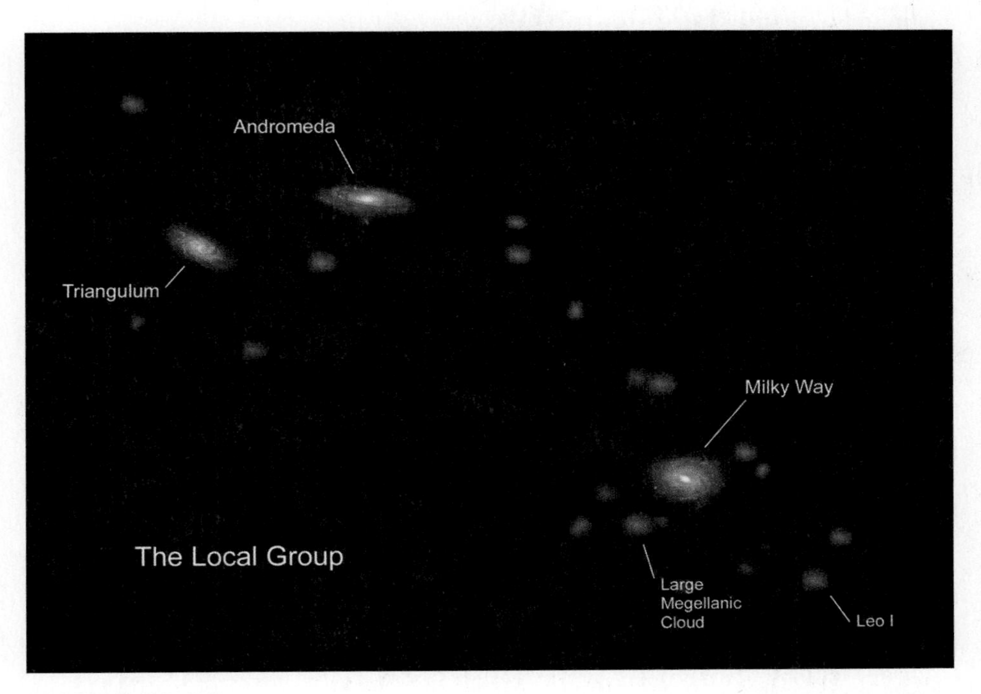

FIGURE 29.35
This two-dimensional composition shows the approximate relative distances between the members of our Local Group of galaxies. These galaxies are all moving toward each other and will one day collide into a larger supergalaxy.

The Triangulum galaxy, so named because it completes a triangle between the spirals, is even closer to Andromeda, but farther away from us.

Our Local Group of galaxies is also under the gravitational influence of neighboring galactic clusters. Our cluster plus all these other clusters make up what is called a *supercluster*, which is a cluster of galactic clusters. Our Local Group is actually a rather minor component of our **Local Supercluster**, as is illustrated in Figure 29.36.

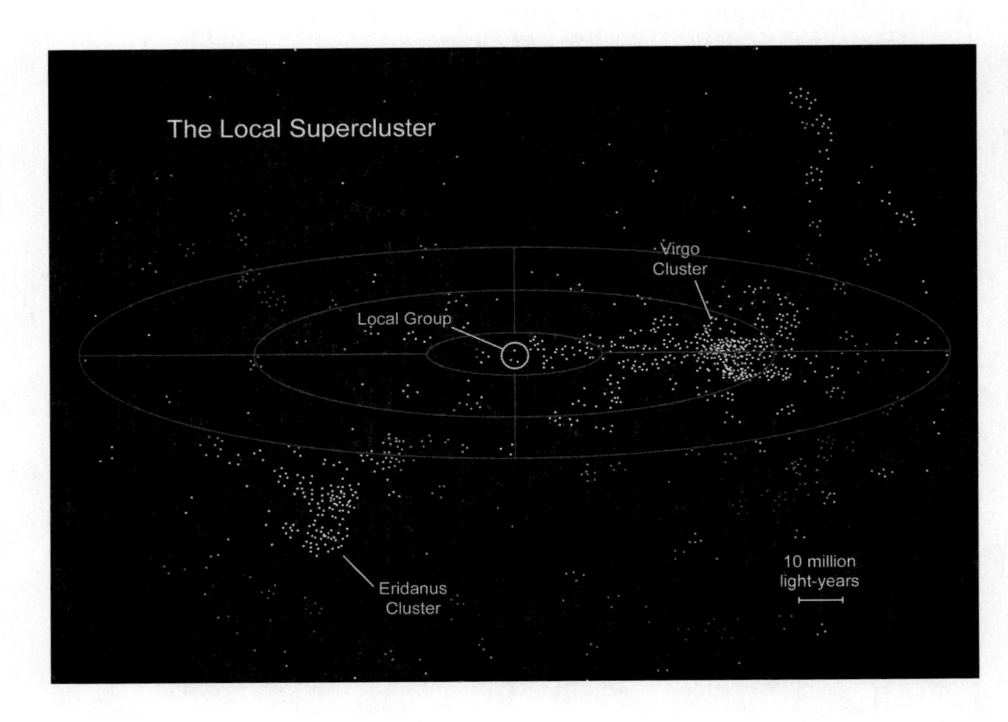

FIGURE 29.36
A supercluster is a cluster of galactic clusters. Each dot represents a galaxy. Note that our Local Group is found midway between two much larger clusters, the Virgo and Eridanus clusters.

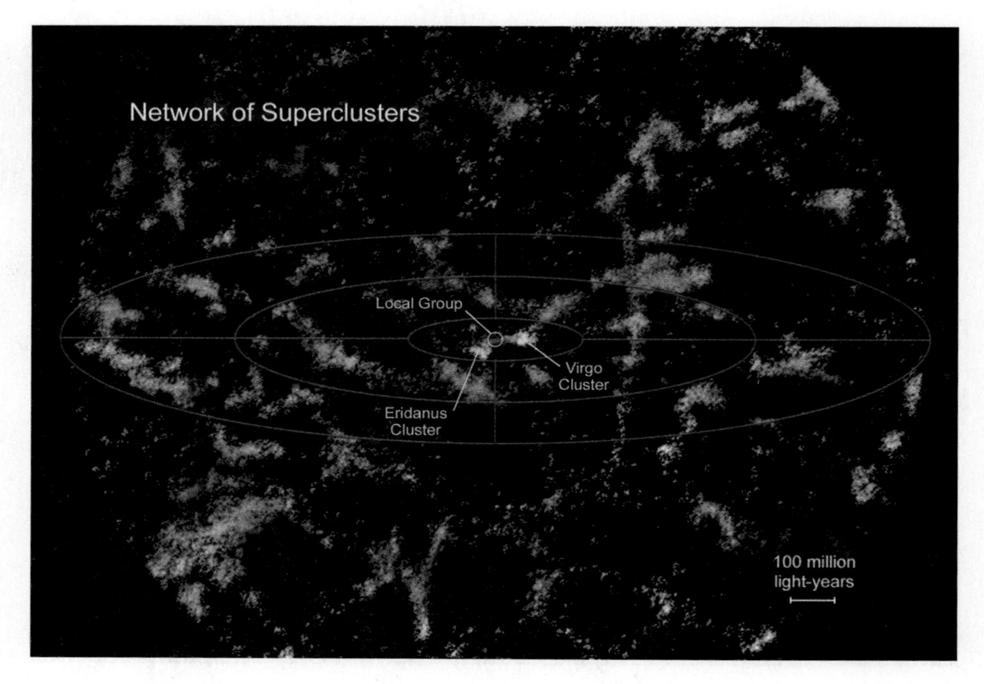

FIGURE 29.37
Each cloud represents a supercluster. Note that the superclusters are strung together as though on the surface of a foam.

Our Local Supercluster is tied in with an elaborate network of many other superclusters, as shown in Figure 29.37. Together, these superclusters appear as though they reside on the surface of a foam inside which are large voids of empty space. Zooming out farther we find that the network of superclusters extends to the edges of the *observable universe*, as illustrated in Figure 29.38. By "observable universe" we mean all that we are able to see given the fact that the universe is only about 14 billion years old.

Polls show that about half of American adults do not know that it takes one year for Earth to go around the Sun. Many of us, therefore, are still struggling with scientific ideas of 400 years ago. Your knowledge is much greater, and being aware of the amazing possibilities that science continues to reveal puts you in a very privileged minority. Celebrate that!

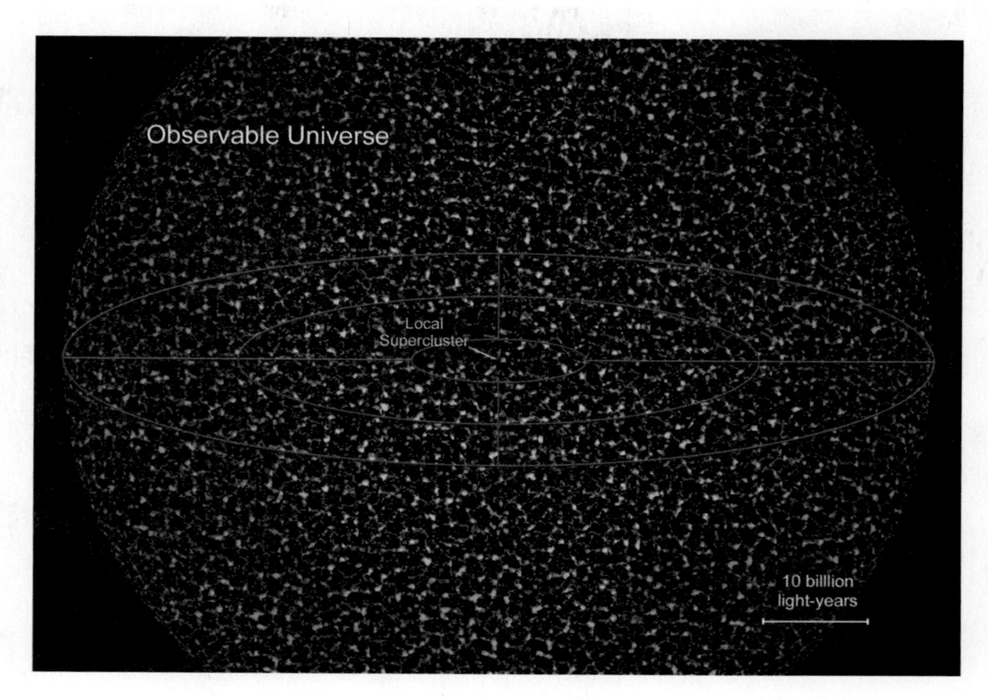

FIGURE 29.38
The network of superclusters extends to the edges of the observable universe, which is no farther than 14 billion light-years away. This illustration, however, shows a hypothetical bird's-eye view of this observable universe fully matured to the present moment, which, because of cosmic expansion, would place those most distant objects now some 42 billion light-years away.

MATH CONNECTION

The Scale of the Galaxy

Problem Earth is about 0.000016 light-year from the Sun and about 4.2 light-years from the next nearest star, which is Proxima Centauri. How far away is Proxima Centauri in Earth–Sun distances?

Solution Divide our distance from Proxima Centauri by our distance to the Sun:

$$\frac{4.2 \text{ light-years}}{0.000016 \text{ light-year}} = 262{,}500$$

So the nearest star is about 260,000 times as far from us as the Sun is.

Problem Our distance from the center of the Milky Way galaxy is about 26,000 light-years. How many Earth–Proxima Centauri distances is that?

Solution Divide the distance to the center of the galaxy by the distance to Proxima Centauri:

$$\frac{26{,}000 \text{ light-years}}{4.2 \text{ light-years}} = 6190$$

So the center of our galaxy is about 6200 times as far away as the nearest star.

Problem The Milky Way and Andromeda galaxies are about 2,300,000 light-years apart. The diameter of the Milky Way galaxy is about 100,000 light-years. How many Milky Way diameters distant is the Andromeda galaxy?

Solution Divide the distance to the Andromeda galaxy by the diameter of the Milky Way galaxy:

$$\frac{2{,}300{,}00 \text{ light-years}}{100{,}000 \text{ light-years}} = 23$$

So the Andromeda galaxy is about 23 Milky Way galaxy diameters away.

Problem The Andromeda galaxy is moving toward the Milky Way galaxy at a rate of about 300,000 mi/h. In how many years will these two galaxies collide?

Solution Convert 300,000 mi/h to light-years per year. First convert the 300,000 mi to light-years:

$$(300{,}000 \text{ mi})(1.61 \text{ km}/1 \text{ mi}) \times$$
$$(1 \text{ light-year}/9{,}460{,}000{,}000{,}000 \text{ km})$$
$$= 4.79 \times 10^{-8} \text{ light-year}$$

Second, convert hours to years:

$$(1 \text{ h})(1 \text{ day}/24 \text{ h})(1 \text{ yr}/365.25 \text{ days})$$
$$= 1.14 \times 10^{-4} \text{ year}$$

Put the two converted values together:

$$300{,}000 \text{ mi/h}$$
$$= 4.79 \times 10^{-8} \text{ light-year}/1.14 \times 10^{-4} \text{ yr}$$
$$= 4.20 \times 10^{-4} \text{ light-year/yr}$$

Use this equation for speed:

$$\text{Speed} = \text{distance/time}$$
$$4.20 \times 10^{-4} \text{ light-year/yr}$$
$$= 2{,}300{,}000 \text{ light-years}/x \text{ yr}$$

Solve for x:

$$x = 5{,}470{,}000{,}000 \text{ yr}$$

So in roughly 5.47 billion years, the Andromeda and Milky Way galaxies will collide. By this time our Sun will have exhausted almost all of its nuclear fuel, so this is not something we Earthlings will be around to witness. Collisions between galaxies, however, are fairly common, and astronomers have photographed many such occurrences now in progress.

CHECK YOURSELF

Which is greater—the number of stars in our galaxy or the number of galaxies in the observable universe?

CHECK YOUR ANSWER

There are far more galaxies in the entire universe than there are stars in our galaxy. Recall from the beginning of this chapter that astronomers estimate that there are about 100 billion stars in our galaxy and about 100 billion galaxies in our observable universe. If true, that means there are about 10^{22} stars in our observable universe, which is about the number of water molecules in a drop of water. As large as the observable universe is large is as small as the fundamental building blocks of our body are small. As humans we are nicely situated between these two extremes.

LEARNING OBJECTIVE
Describe how the Drake equation estimates the probability of extraterrestrial civilizations.

The Drake Equation

EXPLAIN THIS: How would finding microbial extraterrestrial life on Jupiter's moon Europa affect the value of N in the Drake equation?

The **s**earch for **e**xtra**t**errestrial **i**ntelligence (SETI) dates back to 1959, when physicists from Cornell University suggested that radio waves could be used for interstellar communication. The idea is that since human civilization leaks radio messages into space, the inhabitants of other planets might be doing the same thing. Large radio telescopes on Earth could detect such radio leaks from civilizations in nearby star systems, as well as stronger signals dispatched from planets thousands of light-years away. The young astronomer Frank Drake (who later became president of the SETI Institute) was the first to conduct a radio search throughout the galaxy. Although SETI telescopes have been analyzing signals for decades, the antenna have not yet picked up any unique frequencies that could signal distant civilizations, but there is plenty of sky yet to be searched.

Is it logical to search for extraterrestrial civilizations? What are the odds that such civilizations exist, and if they do, what are the chances that we could contact them? Frank Drake's answer to this question is the Drake equation. Since its introduction in 1961, this tool has framed the debate. The Drake equation estimates the odds of our ever making contact with another intelligent civilization by multiplying seven variables related to the possibility of life. The Drake equation is

$$N = R \times F \times Ne \times Fl \times Fi \times Fc \times L$$

where

R = the rate of formation of suitable stars—stars like the Sun—in our galaxy per year
F = the fraction of these stars that have planets
Ne = the number of Earth-like planets—meaning planets that have liquid water—within each planetary system
Fl = the fraction of Earth-like planets where life develops
Fi = the fraction of life sites where intelligent life develops
Fc = the fraction of intelligent life sites where communication develops
L = the "lifetime" (in years) of a communicative civilization
N = the number of civilizations with which we could possibly communicate

A problem with Drake's equation, however, is that we don't know the value of any of the terms! We can make rough estimates of some of them: Astronomers believe that the product $R \times F \times Ne$, which would give us the number of habitable planets in the Milky Way, may be in the neighborhood of 100 billion. The technology for detecting extrasolar planets—planets that orbit stars beyond our solar system—is rapidly advancing. Today, more than 1000 extrasolar planets have been found and new ones are rapidly being discovered.

But presently there is no way to know how many of these 100 billion hypothetical planets have produced life. Life arose quickly on Earth (see Chapter 17), which suggests that life develops fairly easily under the right conditions, but this is not definitive—N could be anywhere between 1 (if Earth is the only planet in the Milky Way with life) and 100 billion (if they all have life). The fact that life flourished on Earth some 4 billion years before humans developed suggests that the development of intelligent life is much more tenuous than the rise of microbial life. But how much more difficult? As to the fraction of intelligent life sites, there could be intelligent beings who haven't yet invented radio telescopes. (We humans belonged to this category until the 20th century.) Or, there could be other civilizations that have the means to communicate but don't—perhaps because they are not interested or they fear that they might endanger themselves by advertising where they live. And L, the lifetime of an intelligent, communicative civilization? Well, humans have been communicative for fewer than 100 years. Do you think we, or the average intelligent species, will remain willing and able to communicate with other star systems for a century? A thousand years? A million? If organizations of intelligent beings—civilizations—last long enough, the Milky Way may now be brimming with communicative, advanced beings. One day, we just may make contact with them.

29.7 Looking Back in Time

LEARNING OBJECTIVE
Describe the Big Bang and three lines of evidence that support it.

EXPLAIN THIS Where did the Big Bang occur?

Imagine some fantastic optical instrument through which we can see the actual history of human civilization. Tune into some 40,000 years ago and you are able to witness the migration of humans into the Australian subcontinent. Re-focus to some 4500 years ago and see the building of the Egyptian pyramids. With such a device we would have an amazing window into our past. The accuracy of history books would be assured.

MasteringPhysics®
TUTORIAL: Hubble's Law
VIDEO: From the Big Bang to Galaxies

When it comes to the history of the universe, we have exactly such a device. It is called the *telescope*. The speed of light is fast, but the universe is exceedingly large. The light from our nearest star, Proxima Centauri, takes 4.2 years to reach us. Thus, as we view this star, we are literally looking at the star as it appeared 4.2 years ago. Similarly, the Andromeda galaxy we see today is the Andromeda galaxy as it appeared 2.3 million years ago.

Want to see the history of our universe? All we need to do is look through our telescopes. The farther away the object, the older it is. Start cataloging the past as we see it today, and soon we come to an appreciable understanding of how the universe itself formed. The study of the overall structure and evolution of the universe is called **cosmology**. With the advent of modern technology, especially with the development of space telescopes, the field of cosmology is currently in a golden era of discovery.

UNIFYING CONCEPT
- ***The Scientific Method***
Section 1.3

In this section we will present some of the more well-established findings of cosmology. To provide a sense of the current excitement, we also dip into some speculations, such as the possible final fate of the universe. But because our discussions here are brief, you should consider exploring the latest news as it happens through online resources such as those listed in Table 29.2.

TABLE 29.2 WEB RESOURCES FOR COSMOLOGY

- The Harvard-Smithsonian Center for Astrophysics
  http://www.cfa.harvard.edu
- The Hubble Space Telescope
  http://Hubblesite.org
- The Wilkinson Microwave Anisotropy Probe
  http://map.gsfc.nasa.gov

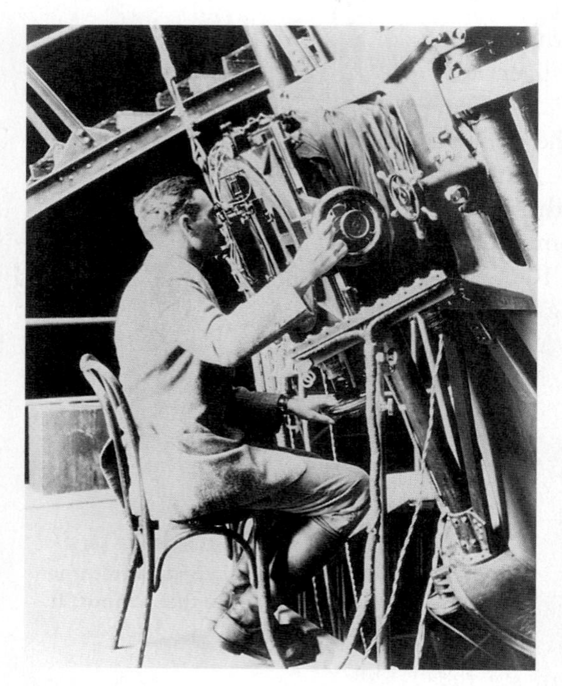

FIGURE 29.39
Edwin Hubble (1889–1953), astronomer and cosmologist, is shown here in 1923 using the 100-inch telescope at the Mt. Wilson Observatory, where he worked for most of his life. In 1929, he announced Hubble's law, which states that the farther apart galaxies are, the faster they move.

The Big Bang

Not so long ago it was commonly thought that our Milky Way galaxy made up the whole universe. Then, in the early 1920s, the astronomer Edwin Hubble, using the new Mt. Wilson telescope, discovered that the Andromeda "nebula" was, in fact, a separate galaxy farther away than the outermost stars of the Milky Way (Figure 29.39). This was a monumental discovery, but Hubble didn't stop there. As discussed earlier in this chapter, Hubble identified and measured the distances to numerous other galaxies. What he discovered next was astonishing—the galaxies are almost all receding from one another. Furthermore, the farther away the galaxy, the greater the velocity with which it is receding.

Hubble's observations had two major implications. The first was that the universe arose from an infinitesimal point that expanded rapidly in an event called the Big Bang. The second major implication was that the universe is not contained within a region of space. Rather, space is *in* the universe and expanding. This is peculiar because you may first think that the Big Bang occurred within an already existing infinite space and that matter and energy flew outward from this Big Bang to occupy this space. If this were the case, however, the distribution of galaxies we observe today and their relative motions would be quite different. To reemphasize this often confused concept: When we talk about the expansion of the universe, we are referring to an expansion of the very structure of space itself. A useful analogy is a group of ants on a balloon that is expanding, as shown in Figure 29.40. As the balloon is inflated, every ant sees every other ant moving farther away. Likewise, in an expanding universe, any observer sees all other galaxies moving away. So the Big Bang marked not only the beginning of time but also the beginning of space.

FIGURE 29.40 MP
INTERACTIVE FIGURE
Every ant on the expanding balloon sees all other ants moving farther away. Each ant may therefore think that it is at the center of the expansion. Not so! There is no center to the surface of the balloon, just as there are no edges.

Hubble suggested a simple relationship between the distance of an object from Earth and its velocity of recession. This relationship, which has been confirmed by numerous measurements over many decades, is known as **Hubble's law**:

$$\text{Hubble's law: } v = H \times d$$

where v is the velocity of a galaxy, H is a constant known as Hubble's constant, and d is the distance of the galaxy from Earth. This law tells us, for example, that if one galaxy is twice as far away as another, the farther galaxy recedes twice as fast from us. Furthermore, if a galaxy were to move from where we are now to its present location with a velocity v, then the time of this trip would be the distance it travels divided by its velocity:

$$t = \frac{d}{v}$$

We use Hubble's law to substitute for v:

$$t = \frac{d}{H \times d} = \frac{1}{H}$$

When we enter the value of H into this equation, we have an estimate of the interval of time for the expansion. Said another way, we have determined the age of the universe. Plugging the currently accepted value of H into the equation indicates that the universe is nearly 14 billion years old. Wow!

CHECK YOURSELF

Draw three dots equally spaced along the length of a rubber band. Assume the distance between these dots is 5 mm. In 1 s stretch the rubber band to 10 times its original length.

1. **Show that the second dot recedes from the first dot with a speed of 45 mm/s. (*Hint*: Speed = distance/time.)**
2. **Show that the third dot recedes from the first dot with a speed of 90 mm/s.**
3. **Why do farther galaxies recede from us at faster rates?**

CHECK YOUR ANSWERS

1. The original distance was 5 mm, which stretched to 50 mm. The change in distance, therefore, was 45 mm over a time of 1 s. So the speed equals 45 mm/s.
2. The difference between 100 mm and 10 mm is 90 mm over a time of 1 s, which equals 90 mm/s, or twice the speed of 45 mm/s.
3. For an expanding rubber band, we see that a dot twice as far away travels twice as fast. The same goes with the universe: A galaxy twice as far away from us is moving twice as fast. The reason for these receding velocities is the expansion of space itself. (A few nearby galaxies, including Andromeda, buck the trend and move toward us.)

Light from a distant galaxy is the light from glowing elements, which emit spectra of particular frequencies as discussed in Chapter 9. An examination of the full spectrum of a galaxy's light shows a pattern of peaks that are the sum of the spectra of all of the many glowing elements, primarily hydrogen and helium. If these peaks are shifted toward the red, then we know the galaxy must be receding away from us. The amount of redshift tells us how fast that galaxy is receding.

Where exactly did the Big Bang occur? Was it at some now far distant point from which we have long since traveled? The answer is an astounding NO! Rather, every point in the universe was present at the Big Bang. It's just that all of these points are now quite far away from one another. So if you want to point to the location of the Big Bang, just point your finger to the tip of your nose, or anywhere for that matter. You can't miss.

Cosmic Background Radiation

In addition to the expansion of the universe, a second line of evidence supporting the Big Bang theory is *cosmic background radiation*. In 1964, scientists Arno Penzias and Robert W. Wilson, working at Bell Labs in New Jersey, used a simple radio receiver to survey the heavens for radio signals (Figure 29.41). No matter which way they directed their receiver, they detected microwaves with a wavelength of 7.35 cm coming toward Earth. Penzias and Wilson were puzzled. With no specific source of the radiation, where were the microwaves coming from and why?

Remember that any object above absolute zero emits energy in the form of electromagnetic radiation. The frequency of this radiation is proportional to

FIGURE 29.41
Arno Penzias and Robert Wilson in front of the microwave receiver they used to detect the afterglow of the Big Bang.

FIGURE 29.42
This all-sky map of the cosmic background radiation taken by the Wilkinson Microwave Anisotropy Probe (WMAP) satellite reveals an average temperature of about 2.73 K everywhere. This is the cooled-off remnants of the Big Bang. The color shows minor temperature variation on the order of ± 0.0001 K.

the absolute temperature of the emitter. Theorists at Princeton, working around the same time as Penzias and Wilson, showed that if the universe began in a primordial explosion as described by the Big Bang, it would still be cooling off. Further, they showed that the temperature of the early universe would have cooled by today to an average of about 3 K. A universe of this temperature would be expected to emit microwave radiation of just the frequency observed by Penzias and Wilson. Thus the influx of microwave radiation that initially puzzled Penzias and Wilson was found to be emitted by the cooling universe itself. This faint microwave radiation is now referred to as **cosmic background radiation** and is taken as strong evidence of the Big Bang (Figure 29.42).

The Abundance of Hydrogen and Helium

The Big Bang answers another cosmic mystery involving the element helium. Measurements show that matter in the universe is about 75% hydrogen and 25% helium. (Heavier elements such as those found on Earth are very minor contributors to the total amount of matter in the universe.) Hydrogen is the simplest of all elements, consisting of a single proton nucleus. It makes sense that hydrogen was the original element. Helium, however, is a more complex element containing a nucleus of two protons and two neutrons. We know that helium is produced from the fusion of hydrogen in stars. But the number of stars is insufficient to account for all the observed helium—not more than 10% of the observed helium could have originated in stars. Most of the helium observed in the universe must have been created elsewhere. As described in the Science and Society box, the Big Bang model predicts that the early universe

SCIENCE AND SOCIETY

Big Bang Helium

As the universe expands, it cools. The cosmic background microwave radiation tells us that it has cooled to an average of about 3 K. Working backward from the present, scientists estimate a temperature in excess of 100 billion K within a few seconds after the Big Bang. At this extreme temperature, protons would be turning into neutrons and neutrons would be turning into protons. The rates of these transformations would be about the same, meaning the ratio of protons to neutrons in this super early universe would be about 1:1.

Over the next three minutes, the temperature dropped to less than 100 billion K, which favored the formation of protons.* Soon protons outnumbered neutrons. By the time the ratio of protons to neutrons was 7:1, the universe would have *cooled enough* to allow for nuclear fusion. We say "cooled enough" because the universe was still quite hot, but not as hot as before. At this point, the protons and neutrons would have begun fusing into deuterium nuclei (consisting of one proton and one neutron). Deuterium nuclei would have then fused to form helium. They would continue doing so until the deuterium grew sparse. (Incidentally, this explains why deuterium is such a rare isotope today.) Continued cooling would prevent the further fusion of helium into heavier elements, such as carbon. By the time the three minutes of helium synthesis were over, the universe would be left with about 75% hydrogen and 25% helium, which is what we observe in the universe today. The Milky Way is actually about 28% helium. The 3% excess helium is likely from the fusion of hydrogen in the stars. No galaxy detected so far, however, has a helium abundance of less than 25%, just as predicted by the Big Bang theory.

*Neutrons are slightly more massive than protons. The transformation of a proton into a neutron, therefore, requires a slightly greater input of energy in accordance with $E = mc^2$. At temperatures below 100 billion K, there is insufficient energy to allow this transformation. The conversion of neutrons into protons, however, releases energy at lower temperatures in accordance with the second law of thermodynamics.

would have been favorable to the formation of helium, but not the formation of other elements. A more detailed analysis shows that the amount of helium created just after the Big Bang should be about the same as the amount we observe in the universe today.

In summary, three major lines of evidence strongly support the Big Bang theory. The first is the current expansion of space, which causes galaxies to recede from one another. The second is the discovery of cosmic background radiation, which is thought to be the Big Bang's afterglow. The third is the Big Bang's ability to explain the observed proportions of elements. Because of these and other similar lines of evidence, the Big Bang has come to be widely accepted by the scientific community as the most viable explanation for the beginning of our universe.

An antenna-fed television (not cable or satellite) tuned into a channel with no local station shows a screen of static "snow." Interestingly, about 1% of this snow is due to photons from the cosmic background radiation.

29.8 Dark Matter and Dark Energy

LEARNING OBJECTIVE
Describe the evidence for dark matter and dark energy and their role in the evolution of the universe.

EXPLAIN THIS How is it possible to detect something in space that we can't see through our telescopes?

Evidence now suggests that the Big Bang generated matter in at least two different forms—one that we can see and another that we can't. The visible form of matter is the "ordinary matter" made of subatomic particles such as protons, neutrons, and electrons. As you learned in earlier chapters, these particles combine to make the atoms of the periodic table. Atoms then combine to make molecules, such as those that make our bodies. This matter we can touch. We interact with it directly. You, the planets, and the stars are made of this form of matter, which we will from here on refer to as **ordinary matter.**

The second form of matter generated from the Big Bang is quite unlike ordinary matter. This form of matter does not recognize the strong nuclear force, which means it cannot clump to form atomic nuclei. Neither does this second form of matter recognize the electromagnetic force, which makes it invisible to light as well as to our sense of touch. As was explained in Chapter 9, the electromagnetic force is responsible for the repulsion between electrons. The reason you can't walk through a wall is because of the repulsions between the electrons in your body and the electrons in the wall. If the wall were made of this invisible matter you would be able to walk right through it. Of course, you wouldn't be able to see the wall either. This invisible form of matter that we cannot see or touch is known as **dark matter.**

So, if dark matter is invisible to us, how do we know it's there? The answer is that this ghostly form of matter gives itself away by its gravitational effects. One of the first clues came to us as we were mapping the speeds at which stars orbit our galactic center. According to the laws of gravity, orbital speed is a function of the force of gravity between the orbiting object and the object being orbited—the greater the force of gravity, the greater the orbital speed. The inner planets of our solar system, for example, orbit the Sun much faster than the outer planets because they are closer to the Sun and experience stronger gravitational forces. Relative to our galaxy, we might expect the same trend—stars closest to the galactic center should have faster orbital speeds than stars farther out. Interestingly, that's not what we observe! Instead, stars closer to the galactic center and those farther out orbit with about the same speed. How can this be?

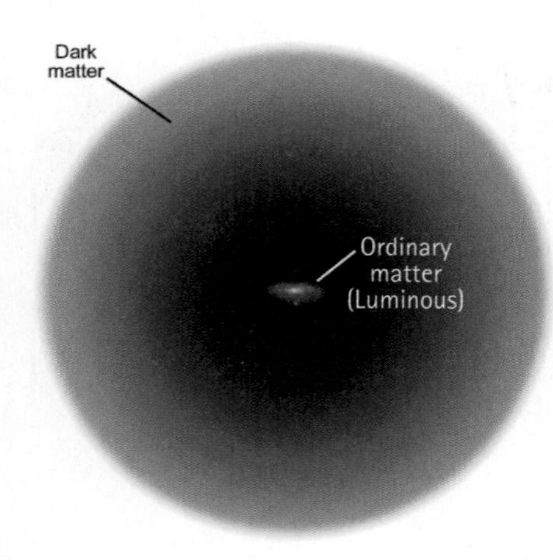

FIGURE 29.43
As large as a galaxy is, its diffuse halo of dark matter is much larger. This halo may measure up to 10 times the diameter of the luminous galaxy and be about six times as massive.

For a solar system, planets orbit as they do because most of the solar system's mass is concentrated within the central sun. For a galaxy such as the Milky Way or Andromeda, it sure looks as though most of the mass is concentrated within the central bulge. The measured orbital speeds of stars, however, tell us that the bulk of the galaxy's mass lies outside the galaxy itself within a diffuse yet massive invisible halo many times the diameter of the visible galaxy, as shown in Figure 29.43. We know it's invisible because all of our telescopes see right through it! But something must be there affecting stellar orbital speeds.

Another bit of evidence for dark matter comes from measuring the speeds of galaxies as they orbit one another within clusters. The measured speeds tell us that the masses of these galaxies are many times greater than the total mass of all their stars.

Last, we know that the path of light is bent by gravity much as it is bent by an optical lens. A cluster of galaxies, therefore, can bend the light from an even farther cluster lying directly behind it—we say the foreground cluster behaves as a *gravitational lens*. Such a gravitational lensing effect was shown in Figure 29.34. The degree to which the light from the distant cluster bends is a function of the mass of the foreground cluster. Once again, the degree of light bending tells us that the mass of the closer cluster far exceeds the mass we would expect based solely on the cluster's luminosity. So, by carefully studying the bending of light from distant galaxies, we can build a map of the dark matter's distribution. Such a study using the Hubble Space Telescope is illustrated in Figure 29.44.

FYI An alternative theory to dark matter is Modified Newtonian Dynamics (MOND), proposed by the physicist Mordehai Milgrom in the early 1980s. According to MOND, Newton's equation $a = F/m$ fails when the force is exceedingly weak, perhaps because of quantum effects. A modified version of this equation was thus created to account for the observed orbital velocities of stars within galaxies. The theory is controversial. To date, most astronomers find the dark matter theory to be more acceptable.

So the evidence for dark matter is strong. Our current problem, however, is trying to figure out exactly what dark matter is made of. It's clear that dark matter is not simply ordinary matter, such as expired stars, that has gotten so cold that it emits no light. Although numerous theories abound about the fundamental nature of dark matter, no dark matter particles have been detected. Until that happens, we are left with yet another fascinating mystery of our universe.

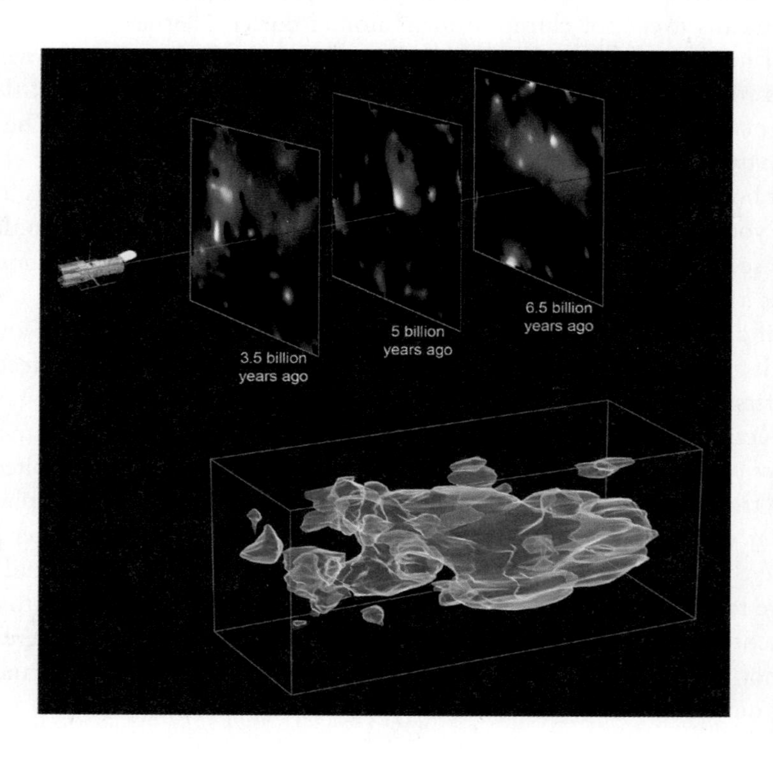

FIGURE 29.44
Images of dark matter, shown in blue, were created through the Hubble Space Telescope's Cosmic Evolution Survey. Of course, the dark matter is not blue. This graphic, however, shows dark matter's distribution over a narrow region of the sky back to about 6.5 billion years ago.

Galaxy Formation

We can speculate that when the universe formed, ordinary matter plus an even greater amount of dark matter was produced. Held together by gravity, the ordinary matter and dark matter would have been strewn outward in a clumpy fashion. Within a clump, ordinary and dark matter may initially have been uniformly mixed together. These two forms of matter differ significantly in that when ordinary matter collides with ordinary matter, energy is released. With this loss of energy, the ordinary matter loses orbital speed and thus falls closer to the center of the clump. Over time, while all dark matter stayed distributed throughout the clump, the ordinary matter became concentrated at the center (Figure 29.45). This concentration of ordinary matter at the center of the clump allowed for the formation of stars. Also, as ordinary matter congregated toward the center, the rate of rotation would increase—angular momentum would have been conserved. If the original clump of ordinary and dark matter was just barely spinning, then the stars forming at the center would take on the form of an elliptical galaxy. If the original clump was spinning a bit faster, then the new stars would be spinning fast enough to flatten the galaxy, much like a rapidly spinning ball of pizza dough. The resulting disk would take on the form of a spiral galaxy. So from a clump of ordinary and dark matter, ordinary matter condensed to form a central galaxy. The dark matter remained diffuse, forming an invisible halo surrounding the newly formed galaxy.

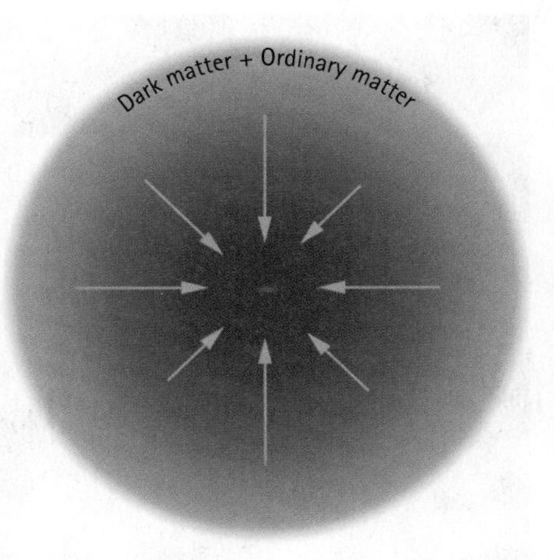

FIGURE 29.45
Ordinary matter condensed out of a mixture of dark and ordinary matter.

Dark Energy

In the years just before Hubble's discovery of the expansion of the universe, Einstein was struggling to understand why gravity wasn't causing the universe to collapse in a Big Crunch. He was thinking of the universe as static, neither collapsing nor expanding. But, in order for the universe to remain static against the inward force of gravity, there would need to be another fundamental outward force counteracting gravity. In other words, if gravity is the "pull inward," there should be a phenomenon that creates a "push outward." To allow for such a balance, he introduced into his equations a factor that he called the *cosmological constant*. He had no proof for the existence of such a phenomenon. Rather, he just postulated it to account for the apparent stability of the universe.

About 10 years later, Hubble announced that the universe was *not* stable but very dynamic and expanding. Einstein later remarked that his own failure to predict a dynamic universe was the "greatest blunder of his life." Subsequent workings of Einstein's equations showed that a static universe would not be stable. Just the slightest push this way or that would cause it to either collapse or expand. Einstein abandoned his notion of a cosmological constant, which for the next 75 years remained a historical curiosity.

Then in the 1990s, some 40 years after Einstein's death, two teams of astronomers made a startling discovery. High-resolution data from very distant galaxies showed that space, beginning about 7.5 billion years ago, started to accelerate in its expansion. Galaxies are not simply coasting away from each other and slowing down. Rather, some unknown form of energy is causing an increase in the rate at which galaxies are receding. Here was a phenomenon that acted as the opposite of gravity, possible evidence of Einstein's once-proposed cosmological constant.

There remains much speculation about how these findings affect the fate of the universe. This form of unknown energy is generally described as

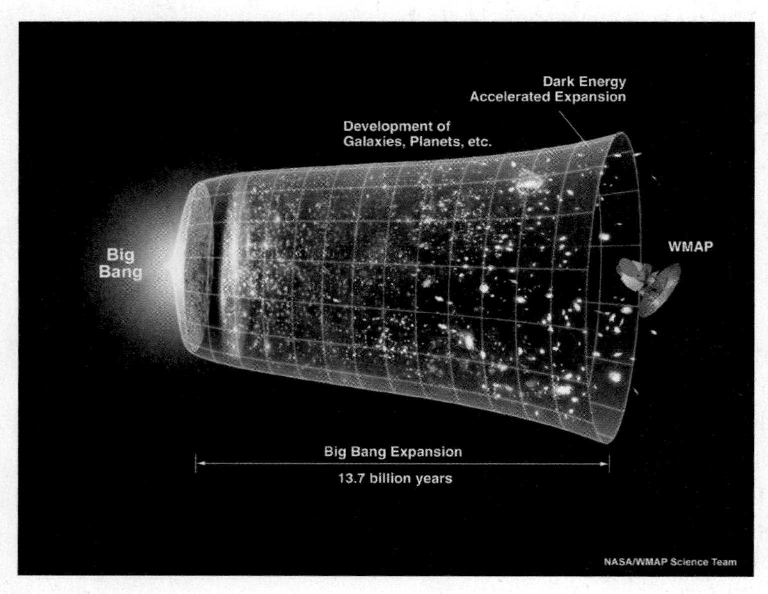

FIGURE 29.46
The expansion of space started to accelerate about 7.5 billion years ago, which is shown on this diagram as a gradual widening just after the development of galaxies. The cause of this accelerated expansion has been given the name *dark energy*. But just because we can name something doesn't mean we understand it. Such is the case with dark energy.

dark energy (Figure 29.46). Theorists have various ideas about the nature of dark energy, including the possibility that the effects of dark energy may be embodied by Einstein's famed cosmological constant.

Some current models suggest that dark energy originates within the emptiness of space. Matter has the effect of pulling space together so that it contracts. The classic example is what happens upon the formation of a black hole—space contracts to a point of zero volume (and infinite density). In the absence of matter—within a perfect vacuum—empty space seethes with an energy that creates an opposite curvature, which allows space to expand. As more empty space is created, the *vacuum energy* becomes more predominant, which accelerates the formation of even more empty space. Distant galaxies are thus seen to be accelerating from each other. It is as though gravity and dark energy are diametrically opposed. When gravity gains full rein, the result is infinite density—a black hole. When dark energy gains full rein, the result may be an infinite vacuum—a region of truly empty space.

Many other intriguing models attempt to explain the nature of dark energy. Which model is best can be determined only after more evidence is collected. Stay tuned for science news reports. In particular, the European Space Agency's *Planck Surveyor* may answer many of our current questions. However, this powerful space telescope and its many successors will, no doubt, also raise more questions than they answer. One thing is for sure: The universe holds no shortage of mysteries.

FYI The space through which our planets orbit contains about 100 hydrogen atoms per liter. The space between the stars of our galaxy contains about 2 hydrogen atoms per liter. If you want really empty space, you'll need to travel to the vast voids that separate the superclusters, as was shown in Figure 29.36. It's only from these regions that dark energy appears to be taking hold. The space within and between our local galaxies is too dense!

The Fate of the Universe

The universe is expanding. Matter, however, has the effect of reversing this expansion. Might enough matter exist within the universe to halt or even reverse this expansion? Before the discovery of dark energy and dark matter, astronomers had calculated that the ordinary matter in the universe was only 4% of the mass needed to halt the expansion. Dark matter was then discovered to be about six times as abundant as ordinary matter, making up about 23% of the mass needed to halt the expansion. If the remaining $100\% - (4\% + 23\%) = 73\%$ of matter could be accounted for, then the mass of the universe would be sufficient to one day halt the expansion.

Then came the discovery of dark energy. Recall that matter and energy are related by Einstein's equation $E = mc^2$. Both dark matter and dark energy, therefore, need to be included in tabulations of the total composition of the universe. The abundance of dark energy then fits the bill for making up the remaining 73% of the composition of the observable universe, as shown in Figure 29.47. There's a twist, however, in that dark energy causes the *expansion* of space, not contraction. The current thinking of most cosmologists, therefore, is that our universe is destined for an eternal expansion. This gives rise to a select number of possible scenarios for the fate of our universe.

MasteringPhysics®
TUTORIAL: Fate of the Universe

In one scenario, called **heat death**, the universe will continue to expand, approaching absolute zero and a state of maximum entropy. After 10^{14} years, all stars will have exhausted all possible fuel. The universe will be fully dark. After 10^{16} years, planets and stars will be flung from their orbits because of random collisions, most then falling into supermassive black holes and the rest forming scattered stellar debris. After 10^{40} years, all protons and neutrons will have decayed, leaving behind gamma radiation and *leptons*, of which the electron is an example. From this time to about 10^{100} years in the future, supermassive black holes will likely be the dominant form of mass in the universe. But by the passing of 10^{150} years, they too will depart as they evaporate into photons and leptons. From then to perhaps 10^{1000} years in the future, the wavelengths of photons as well as all other remaining particles will be stretched to the lowest energy states possible. Entropy will have won supreme victory.

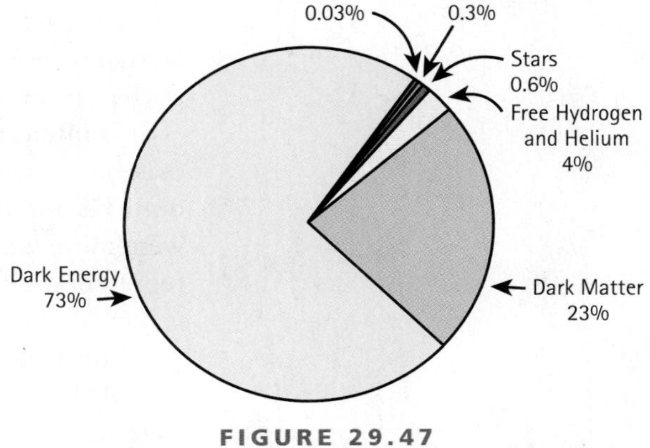

FIGURE 29.47
Ordinary matter, the stuff from which we and the galaxies we live in are made, makes up not more than 4% of the composition of the universe. The remainder is primarily dark matter (23%) and dark energy (73%), both of which we know very little about.

A second scenario, known as the **Big Rip**, recognizes that the influence of dark energy may grow stronger over time (Figure 29.48). As the universe expands exponentially, clusters of galaxies will be pulled farther apart, past the point of being visible to each other. Neighboring galaxies will then be pulled out of each other's sight. About 60 million years before the end, stars within galaxies will fly off in every direction. Some three months before the end, solar systems will disperse. In the last few minutes, stars and planets will become unbound. Then in the last instant, all atoms will be ripped apart, followed by their subatomic particles. The timeframe for the Big Rip is estimated to be about 35 billion years after the Big Bang, which would be about 21 billion years from now. And where will all the matter fly off to? Good question. Perhaps to subsequent Big Bangs?

Although *dark energy* and *dark matter* both begin with the word *dark*, they are uniquely different. One is a form of matter; the other is a form of energy. Both are still mysterious and their existence may yet be disproven by new evidence or alternative explanations. By the next edition of this textbook, the picture may look quite different!

Alan Guth, one of the early developers of modern cosmology, supports the idea that the universe will not end everywhere at once. Some 14 billion years ago, a tiny patch of primordial material inflated to form our own observable universe. But this inflation will keep on going for other patches of this primordial material and will continue to do so eternally. So, while our region of the universe expands

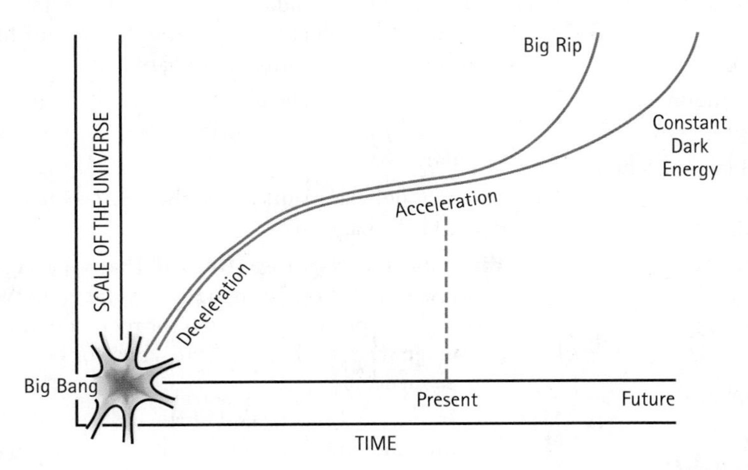

FIGURE 29.48
If the strength of dark energy remains constant, we can expect our universe to suffer heat death. If dark energy gains strength, then the ultimate demise of our universe may be the Big Rip.

to infinity, other regions are just being born. This scenario, in which observable universes are spawned on a perpetual basis, is known as **eternal inflation**. With eternal inflation, the ultimate fate of our own observable universe is not the same as that of the universe as a whole.

It is interesting how the focus of our speculations has been narrowing down over our history. In the beginning, anything seemed possible. But as we opened our eyes and minds to the natural universe, we learned that some speculations were more worthy than others. We once viewed the constellations as heavenly representations of gods. Of course, it was once thought that Earth was the center of the universe. Then we realized that Earth was just a planet orbiting the Sun, which itself was a medium-sized star among many. Our universe was the Milky Way galaxy until Hubble pointed out otherwise. When galaxies were found to be receding, many hypothesized that gravity might be strong enough to pull them back together in a Big Crunch. This possibility has since been discredited. Now we are at the point of wondering what might happen after eternal expansion. Our current speculations are just that—speculations. But they are highly refined speculations based on a great deal of collected evidence. As we continue to look at the natural universe with ever more powerful telescopes and a host of other devices, we can expect that our speculations will become ever more refined. This is the mind-opening art of science, which seeks to learn the nature of the universe for what it is—not for what we might wish it to be. Stay tuned.

SUMMARY OF TERMS (KNOWLEDGE)

Active galactic nucleus The central region of a galaxy in which matter is falling into a supermassive black hole and emitting huge amounts of energy.

Big Bang The primordial creation and expansion of space at the beginning of time.

Big Rip A model for the end of the universe in which dark energy grows stronger over time and causes all matter to rip apart.

Black hole The remains of a supergiant star that has collapsed into itself. It is so dense, and has a gravitational field so intense, that light itself cannot escape from it.

Black-hole singularity The object of zero radius into which the matter of a black hole is compressed.

Celestial sphere An imaginary sphere surrounding Earth to which the stars are attached.

Cosmic background radiation The faint microwave radiation emanating from all directions that is the remnant heat of the Big Bang.

Cosmology The study of the overall structure and evolution of the universe.

Dark energy An unknown form of energy that appears to be causing an acceleration of the expansion of space; thought to be associated with the energy exuded by a perfect vacuum.

Dark matter Invisible matter that has made its presence known so far only through its gravitational effects.

Eternal inflation A model of the universe in which cosmic inflation is not a one-time event but rather progresses to continuously spawn an infinite number of observable universes in its wake.

Event horizon The boundary region of a black hole from which no radiation can escape. Any events within the event horizon are invisible to distant observers.

Galaxy A large assemblage of stars, interstellar gas, and dust, usually catagorized by its shape: elliptical, spiral, or irregular.

Giant stars Cool giant stars above the main-sequence stars on the H–R diagram.

H–R diagram (Hertzsprung–Russell diagram) A plot of luminosity versus surface temperature for stars. When so plotted, stars' positions take the form of a main sequence for average stars, with exotic stars above or below the main sequence.

Heat death A model for the end of the universe in which all matter and energy disperse to the point of maximum entropy.

Hubble's law The farther away a galaxy is from Earth, the more rapidly it is moving away from us: $v = H \times d$.

Light-year The distance light travels in one year.

Local Group Our immediate cluster of galaxies, including the Milky Way, Andromeda, and Triangulum spiral galaxies plus a few dozen smaller elliptical and irregular galaxies.

Local Supercluster The cluster of galactic clusters in which our Local Group resides.

Main sequence The diagonal band of stars on an H–R diagram; such stars generate energy by fusing hydrogen to helium.

Neutron star A small, extremely dense star composed of tightly packed neutrons formed by the welding of protons and electrons.

Nova An event in which a white dwarf suddenly brightens and appears as a "new" star.

Ordinary matter Matter that responds to the strong nuclear, weak nuclear, electromagnetic, and gravitational forces. This is matter made of protons, neutrons, and electrons, which includes the atoms and molecules that make us and our immediate environment.

Planetary nebula An expanding shell of gas ejected from a low-mass star during the latter stages of its evolution.

Pulsar A celestial object (most likely a neutron star) that spins rapidly, sending out short, precisely timed bursts of electromagnetic radiation.

Quasar The core of a distant galaxy early in its life span when its central black hole has not yet swept much matter from its vicinity, leading to a rate of radiation higher than that from entire older galaxies.

Starburst galaxy A galaxy in which stars are forming at an unusually fast rate.

Supernova The explosion of a massive star caused by gravitational collapse with the emission of enormous quantities of matter and radiation.

White dwarf A dying star that has collapsed to the size of Earth and is slowly cooling off; located at the lower left on the H–R diagram.

READING CHECK (COMPREHENSION)

29.1 Observing the Night Sky

1. What are constellations?
2. Why does an observer at a given location see one set of constellations in the winter and a different set of constellations in the summer?
3. Why do the stars appear to turn on an imaginary north–south axis about once every 24 hours?
4. Is the light-year a measurement of time or distance?

29.2 The Brightness and Color of Stars

5. Which is hotter—a red star or a blue star?
6. What is the difference between apparent brightness and luminosity?

29.3 The Hertzsprung–Russell Diagram

7. What is an H–R diagram?
8. Where are the great majority of stars plotted on an H–R diagram?
9. Where does our Sun reside on an H–R diagram?
10. Among stars originating from the main sequence, which are larger—red stars or yellow stars?

29.4 The Life Cycles of Stars

11. What event changes a protostar into a full-fledged star?
12. What are the outward and inward forces that act on a star?
13. Is the lifetime of a high-mass star longer or shorter than that of a low-mass star?
14. What is the relationship between the heavy elements that we find on Earth today and supernovae?
15. What is the relationship between a supernova and a neutron star?

29.5 Black Holes

16. What is the relationship between a supergiant star and a black hole?
17. Why don't we think the Sun will eventually become a black hole?
18. If black holes are invisible, what is the evidence for their existence?
19. Is a black hole's event horizon a physical or mathematical boundary?

29.6 Galaxies

20. What type of galaxy is the Milky Way?
21. What is a starburst galaxy?
22. How many spiral galaxies are in the Local Group?
23. Name three galactic clusters found in our Local Supercluster.

29.7 Looking Back in Time

24. Is the universe in space or is space in the universe?
25. Which depends on distance—a star's brightness or its luminosity?
26. What is the approximate age of the universe?

29.8 Dark Matter and Dark Energy

27. If we can't see dark matter, how do we know it is there?
28. Is dark matter found mostly within or just outside a galaxy?

29. According to recent evidence, how long ago did the expansion of the universe start accelerating?
30. What does WMAP stand for?
31. Which is more abundant—dark matter or ordinary matter?
32. What does the Big Rip scenario assume about dark energy?

THINK INTEGRATED SCIENCE

29A—The Drake Equation

33. What is SETI? Was it a mistake for Congress to cut its funding? Why or why not?
34. The total number of stars in the universe is much greater than the number of grains of sand on all the beaches of Earth. Given this, do you think it's possible that there are other civilizations in the universe?

THINK AND DO (HANDS-ON APPLICATION)

35. To observe the daily diurnal motion of the stars, go star watching tonight. Pick a star or constellation that lines up with a stationary landmark such as a tree or house. Then come back in an hour or so and you will see that the star has moved away from the landmark but remains in place relative to the other stars. In what direction has that star moved? To the east? West? Southwest? Northwest? Where will the star be when the Sun rises? Will it be in the same location in 24 hours?
36. To observe the revolutionary motion of Earth around the Sun, go star watching and make note of the stars directly above you. Sketch their pattern on a piece of paper and write down the date and time at which you make this observation. If these stars are not already a well-known constellation, make up a new constellation and give it a creative name. After a month has passed, look for this same constellation at the same time of night. Why isn't it still directly above you? In what direction has your constellation shifted? Why?

THINK AND COMPARE (ANALYSIS)

37. Rank the appearance of the North Star in order of increasing height from the horizon as seen from (a) Alaska, (b) Florida, (c) Vermont.
38. Rank the objects in order of increasing intrinsic motion as viewed from Earth: (a) the Moon, (b) Venus, (c) the North Star.
39. Rank the following stars in order of increasing radius:

	Star A	Star B	Star C
Surface temperature (K):	6,000	4,000	30,000
Luminosity (solar units):	1	100	0.01

40. Rank these stages of stellar development from earliest to latest: (a) white dwarf, (b) nova, (c) red giant.
41. Rank the nuclear fuels in order of being consumed, from first to last: (a) carbon, (b) helium, (c) hydrogen.
42. Rank in order of increasing size: (a) solar system, (b) Local Group, (c) galaxy.
43. Rank these elements in order of increasing abundance in the universe: (a) helium, (b) hydrogen, (c) carbon.
44. Rank the following in order of increasing abundance in the universe: (a) dark matter, (b) ordinary matter, (c) dark energy.

THINK AND SOLVE (MATHEMATICAL APPLICATION)

45. Suppose Star A is four times as luminous as Star B. If these stars are both 500 light-years away from Earth, how will their apparent brightness compare? How will the apparent brightness of these stars compare if Star A is twice as far away as Star B?
46. If you were to travel straight up from the core of our galaxy and then look back, you would have a grand view of the Milky Way's spiral shape. If the distance from the core to the outer edges was 50,000 light-years, how much surface area are you looking at? Assume the galaxy is a circle whose area can be found by the equation Area $= \pi r^2$.
47. Assume the Milky Way contains 100 billion stars evenly distributed with none concentrated toward the center. What would be the surface area density of stars? Use the equation Surface area density $=$ number of stars/surface area.

48. Use the information in the preceding problem to figure out the average area around a single star in units of AU. (Note: 1 light-year = 63,000 AU.)
49. From your answer to the preceding problem, would it be possible for two galaxies with stars evenly distributed to pass right through each other?

THINK AND EXPLAIN (SYNTHESIS)

50. Is any star bright enough for us to see on a sunny day?
51. On the Moon, stars other than the Sun can be seen during the daytime. Why?
52. We see the constellations as distinct groups of stars. Discuss why they would look entirely different from some other location in the universe, far distant from Earth.
53. Distinguish between the diurnal and intrinsic motions of celestial objects.
54. Which moves faster from horizon to horizon—the Sun or the Moon? Explain.
55. What is the relationship between a planetary nebula and a white dwarf?
56. What do the outward and inward forces acting on a star have to do with its size?
57. What is the relationship between a white dwarf and a nova?
58. What does the color of a star tell you about the star?
59. What is expected to happen to the Sun in its old age?
60. In what sense are we all made of star dust?
61. Would you expect metals to be more abundant in old stars or in new stars? Defend your answer.
62. What is the evidence for believing our Sun is a relatively young star in the universe?
63. Why are massive stars generally shorter lived than low-mass stars?
64. Why are supermassive stars relatively rare?
65. What does the spin rate of a star have to do with whether or not it has a system of planets?
66. With respect to stellar evolution, what is meant by the statement, "The bigger they are, the harder they fall"?
67. Why isn't the Sun able to fuse carbon nuclei in its core?
68. Which has the highest surface temperature—a red star, white star, or blue star?
69. In terms of the life cycle of the Sun, explain why life on Earth cannot last forever.
70. What happens to the radial distance of the event horizon as more and more mass falls into the black hole? Explain.
71. Will the Sun become a supernova? A black hole? Defend your answer.
72. Are there galaxies other than the Milky Way that can be seen with the unaided eye? Discuss.
73. Does the Milky Way galaxy contain an active galactic nucleus?
74. When was most of the helium in the universe created?
75. What does the expansion of space do to light passing through it?
76. What is the relationship between dark energy and Einstein's cosmological constant?
77. The average temperature of the universe right now is about 2.73 K. Will this temperature likely go up or down over the next billion years?
78. If the initial universe remained hotter for a longer period of time, would there likely be more or less helium?
79. Are astronomers able to point their telescopes in the direction of where the Big Bang occurred?
80. True or false: A helium balloon here on Earth pops, releasing direct remnants of the Big Bang. Explain.
81. If we are made of stardust, what are stars made of?
82. Why doesn't dark matter clump together as effectively as ordinary matter?
83. What is one important difference between dark matter and dark energy?
84. Why isn't dark energy called the dark force?
85. If there are so many stars and galaxies, why do we see so much darkness in the clear night sky?
86. If we can't even predict the weather, how can we ever expect to predict the fate of the universe?

THINK AND DISCUSS (EVALUATION)

87. Compare and contrast astronomy and astrology.
88. Project what human civilization would be like if our Sun were hidden in a dusty part of the galaxy such that no stars were ever visible to us at night.
89. It takes an infinite amount of time to watch an object fall through a black hole's event horizon. In what sense, therefore, can it be said that black-hole singularity does not exist?
90. Why is it important to have a science-based understanding of the structure of our universe? Try condensing your answer into a single philosophical sentence.

READINESS ASSURANCE TEST (RAT)

If you have a good handle on this chapter, if you really do, then you should be able to score 7 out of 10 on this RAT. If you score less than 7, you need to study further before moving on.

Choose the BEST answer to each of the following:

1. Summer and winter constellations are different because
   (a) of the spin of Earth about its polar axis.
   (b) the night sky faces in opposite directions in summer and winter.
   (c) of the tilt of Earth's polar axis.
   (d) the universe is symmetrical and harmonious.
2. Polaris is always directly over
   (a) the North Pole.
   (b) any location north of the equator.
   (c) the equator.
   (d) the South Pole.
3. What is the star nearest Earth?
   (a) Proxima Centauri
   (b) Polaris
   (c) Mercury
   (d) the Sun
4. The property of a star that relates to the amount of energy per unit time it is producing is its
   (a) luminosity.
   (b) apparent brightness.
   (c) color.
   (d) volume.
   (e) mass.
5. The longest-lived stars are those of
   (a) low mass.
   (b) high mass.
   (c) intermediate mass.
   (d) infinite mass.
6. After our Sun burns all the hydrogen in its core, it will become a
   (a) white dwarf.
   (b) black dwarf.
   (c) black hole.
   (d) red giant.
   (e) blue giant.
7. The shape of an active starburst galaxy tends to be
   (a) elliptical.
   (b) spiral.
   (c) irregular.
   (d) all of these
8. Scientists estimate the age of our universe to be about
   (a) 5000 years old.
   (b) 1 billion years old.
   (c) 14 billion years old.
   (d) 42 billion years old.
9. Which of the following is not accepted evidence for the Big Bang?
   (a) cosmic background radiation
   (b) homogeneity of the temperature of the universe
   (c) the abundance of helium
   (d) dark energy
10. Dark matter is
    (a) ordinary matter that is no longer emitting light.
    (b) affecting the orbits of stars.
    (c) a dense form of dark energy.
    (d) repelled by ordinary matter.

Answers to RAT

1. b, 2. a, 3. d, 4. a, 5. a, 6. d, 7. c, 8. c, 9. d, 10. b

APPENDIX A

On Measurement and Unit Conversion

Two major systems of measurement prevail in the world today: the United States Customary System (USCS, formerly called the British system of units), used primarily in the United States, and the Système International (SI) (known also as the international system and as the metric system), used in most of the world. Each system has its own standards of length, mass, and time. The units of length, mass, and time are fundamental units because other quantities can be measured in terms of them.

United States Customary System

The USCS is familiar to everyone in the United States. It uses the foot as the unit of length, the pound as the unit of weight or force, and the second as the unit of time. The USCS is presently being replaced by the SI—rapidly in science and technology and some sports (track and swimming) but very slowly in other areas. Americans are wedded to their 100-yard football fields.

Measurements of time are the same in the two systems. In SI the only time unit is the second (s, not sec) with prefixes; but in general, minute, hour, day, year, and so on, with two or more lettered abbreviations (h, not hr), are accepted in the USCS.

Système International

Table A.1 lists SI units and their symbols. SI is based on the metric system, which originated in France after the French Revolution in 1791. The metric system branches into two systems of units: the *meter-kilogram-second* (mks) system, preferred

TABLE A.1 SI UNITS

Quantity	Unit	Symbol
Length	meter	m
Mass	kilogram	kg
Time	second	s
Force	newton	N
Energy	joule	J
Current	ampere	A
Temperature	kelvin	K

TABLE A.2 CONVERSIONS BETWEEN DIFFERENT UNITS OF LENGTH

Unit of Length	Kilometer	Meter	Centimeter	Inch	Foot	Mile
1 kilometer =	1	1000	100,000	39,370	3280.84	0.62140
1 meter =	0.00100	1	100	39.370	3.28084	6.21×10^{24}
1 centimeter =	1.0×10^{25}	0.0100	1	0.39370	0.032808	6.21×10^{26}
1 inch =	2.54×10^{25}	0.02540	2.5400	1	0.08333	1.58×10^{25}
1 foot =	3.05×10^{24}	0.30480	30.480	12	1	1.89×10^{24}
1 mile =	1.60934	1609.34	160,934	63,360	5280	1

in physics, and the *centimeter-gram-second* (cgs) system, favored in chemistry due to its smaller values. The cgs and mks units are related to each other as follows: 100 centimeters equal 1 meter; 1000 grams equal 1 kilogram. Table A.2 shows how several units of length relate to one another.

Note that the metric system uses the decimal system, in which all units are related by dividing or multiplying by 10. The prefixes listed in Table A.3 are commonly used to show the relationships among units.

TABLE A.3 SOME PREFIXES

Prefix	Definition
micro-	One-millionth: a microsecond is one-millionth of a second
milli-	One-thousandth: a milligram is one-thousandth of a gram
centi-	One-hundredth: a centimeter is one-hundredth of a meter
kilo-	One thousand: a kilogram is 1000 grams
mega-	One million: a megahertz is 1 million hertz

Meter

The standard of length of the metric system originally was defined as the distance from the North Pole to the equator. This distance was thought at the time to be close to 10,000 kilometers. One ten-millionth of this, the meter, was carefully determined and marked off by means of scratches on a bar of platinum-iridium alloy, kept at the International Bureau of Weights and Measures in France. The meter is now defined as the length of the path traveled by light in a vacuum during a time interval of 1/299,792,458 of a second.

FIGURE A.1
The standard kilogram.

Kilogram

The standard unit of mass, the kilogram, is a block of platinum, also preserved at the International Bureau of Weights and Measures in France (Figure A.1). The kilogram equals 1000 grams. A gram is the mass of 1 cubic centimeter (cc) of water at a temperature of 4°C. (The standard pound is defined in terms of the standard kilogram: The mass of an object that weighs 1 pound is equal to 0.4536 kilogram.)

Second

The official unit of time in both the USCS and the SI is the second. Until 1956, the second was defined in terms of the mean solar day, which was divided into 24 hours. Each hour was divided into 60 minutes and each minute into 60 seconds. Thus, there were 86,400 seconds per day, and the second was defined as 1/86,400 of the mean solar day. This definition proved unsatisfactory, however, because the rate of Earth's rotation is gradually becoming slower. In 1964, the second was officially defined as the time that a cesium-133 atom takes to make 9,192,631,770 vibrations.

Newton

One newton is the force required to accelerate 1 kilogram at 1 meter per second per second. This unit is named after Sir Isaac Newton.

Joule

One joule is equal to the amount of work done by a force of 1 newton acting over a distance of 1 meter. In 1948, the joule was adopted as the unit of energy by the International Conference on Weights and Measures. Therefore, the specific heat of water at 15°C is now given as 4185.5 joules per kilogram Celsius degree. This figure is always associated with the mechanical equivalent of heat—4.1855 joules per calorie (in this text we round this to 4.19 J/cal).

Ampere

The ampere is defined as the amount of electric charge flowing in each of two long parallel conductors 1 m apart, which results in a force of exactly 2×10^{-7} N/m of length of each conductor. We treat current in this text as being the rate of flow of 1 coulomb of charge per second, where 1 coulomb is the charge of 6.25×10^{18} electrons.

Kelvin

The fundamental unit of temperature is named after the scientist William Thomson, Lord Kelvin. The kelvin is defined to be 1/273.15 the thermodynamic temperature of the triple point of water (the fixed point at which ice, liquid water, and water vapor coexist in equilibrium). The former name *degree Kelvin* (°K) has been changed to *kelvin* (K). The temperatures of melting ice and boiling water at atmospheric pressure are 273.15 K and 373.15 K, respectively.

Area

The unit of area is a square that has a standard unit of length as a side. In the USCS, it is a square with each side 1 foot in length, called 1 square foot and

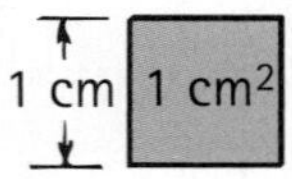

FIGURE A.2
Unit square.

written 1 ft^2. In the SI, it is a square with sides 1 meter in length, making a unit of area of 1 m^2. In the cgs system it is 1 cm^2. Formulas for the surface areas of other shapes can be found on the Web.

Volume

The volume of an object is the space it occupies. The unit of volume is the space taken up by a cube that has a standard unit of length for its edge. In the USCS, one unit of volume is 1 cubic foot, written 1 ft^3. In the metric system, it is 1 m^3 or 1 cm^3 (or cc). The volume of a given space is specified by the number of cubic feet, cubic meters, or cubic centimeters that will fill it.

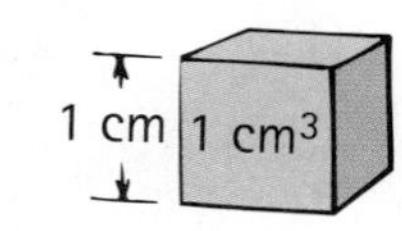

FIGURE A.3
Unit volume.

In the USCS, volumes can also be measured in quarts, gallons, and cubic inches as well as in cubic feet. There are 1728 ($12 \times 12 \times 12$) cubic inches in 1 ft^3. A U.S. gallon is a volume of 231 in^3. Four quarts equal 1 gallon. In the SI, volumes are also measured in liters. A liter is equal to 1000 cm^3.

Unit Conversion

Often in science, and especially in a laboratory setting, it is necessary to convert from one unit to another. To do so, you need to multiply the given quantity by the appropriate *conversion factor*. All conversion factors are ratios in which the numerator and denominator represent the equivalent quantity expressed in different units. Because any quantity divided by itself is equal to 1, all conversion factors are equal to 1. So, multiplying by a conversion factor doesn't affect the quantity itself.

For example, let's find the number of joules in 400 calories: We multiply by a *conversion factor*, in which, as said, the quantity in the numerator is equal to the quantity in the denominator, but expressed in different units. We know there are 4.19 joules in 1 calorie. So we multiply 400 calories by the conversion factor with calories in the denominator (so they'll cancel). Then we find

$$400 \text{ calories} \times \frac{4.19 \text{ joules}}{1 \text{ calories}} = 1676 \text{ joules}$$

As another example, suppose we wish to convert 8 km/s, the orbital speed in near Earth orbit, to miles per hour. By expressing conversion factors with numerators and denominators that result in the cancellation of units, we see whether to multiply or divide quantities:

$$8 \text{ km/s} = \frac{8 \cancel{\text{km}}}{1 \cancel{\text{s}}} \times \frac{1 \text{ mi}}{1.6 \cancel{\text{km}}} \times \frac{60 \cancel{\text{s}}}{1 \cancel{\text{min}}} \times \frac{60 \cancel{\text{min}}}{1 \text{ h}} = 1800 \text{ mi/h}$$

There are many other ways to do the same conversion. For example, we could have used 3600 seconds = 1 hour. Units guide the process.

APPENDIX B

Linear and Rotational Motion

Advanced Concepts of Motion

When we describe the motion of something, we say how it moves relative to something else—to a *reference frame* (see Chapter 2). A reference frame with zero acceleration is called an *inertial frame*. In an inertial frame, forces produce acceleration of objects in accord with Newton's second law. But when the frame of reference itself is accelerating, we observe fictitious forces and motions. Observations from a carousel, for example, are different when it is rotating and when it is at rest. Our description of motion and force depends on our "point of view."

We distinguish between speed and velocity (see Chapter 2). Speed is how fast something moves, or the time rate of change of position (excluding direction); it is a scalar quantity. Velocity includes the direction of motion; it is a vector quantity whose magnitude is speed. Objects moving at constant velocity move the same distance in the same time in the same direction.

But here's the subtlety: There is a distinction between speed and velocity that has to do with the difference between distance and net distance, or *displacement*. Distance is a scalar quantity, while displacement is a vector quantity. Speed is distance per duration, while velocity is displacement per duration. After completing one circuit on a racetrack, for example, a car has zero displacement, since it ends where it started (it's change in position is zero). For motion in a straight line, linear motion, there is no need to distinguish between distance and displacement. They are, for practical purposes, the same. In this book we simply use distance.

As you learned in Chapter 2, acceleration is the rate at which velocity changes. This can be a change in speed only, a change in direction only, or both. Negative acceleration is often called *deceleration*.

Finally, in Newtonian space and time, space has three dimensions—length, width, and height—each with two directions. We can go, stop, and return in any of them. Time has one dimension with two directions—past and future. We cannot stop or return in time, only go. In Einsteinian spacetime, these four dimensions merge (but this *very* advanced topic may await you in a follow-up course).

Computing Velocity and Distance Traveled on an Inclined Plane

A staple of any physics course is the study of motion on an inclined plane. We develop this topic here to help you sharpen your analytical concepts of motion. Recall from Chapter 2 Galileo's experiments with inclined planes. We considered a plane tilted such that the speed of a rolling ball increases at the rate of 2 meters per second each second—an acceleration of 2 m/s^2. So, at the instant the ball starts moving, its velocity is zero, and 1 second later it is rolling at 2 m/s, at the end of the next second 4 m/s, the end of the next second 6 m/s, and so on. The velocity of the ball at any instant is: Velocity = acceleration × time. Or, in shorthand notation, $v = at$. (It is customary to omit the multiplication sign, ×, when expressing relationships in mathematical form. When two symbols are written together, such as *at* in this case, it is understood that they are multiplied.)

How fast the ball rolls is one thing; how far it rolls is another. To understand the relationship between acceleration and distance traveled, we must first investigate the relationship between *instantaneous* velocity (velocity at a particular point in time) and *average* velocity (velocity computed over some extended time interval). If the ball shown in Figure B.1 starts from rest, it will roll a distance of 1 meter in the first second. What will be its average speed? The answer is 1 m/s (it covered 1 meter in the interval of 1 second). But we have seen that the instantaneous velocity at the end of the first second is 2 m/s. Since the acceleration is uniform, the average in any time interval is found the same way we usually find the average of any two numbers: Add them and divide by 2. (Be careful not to do this when acceleration is not uniform!) So, if we add the initial speed (zero in this case) and the final speed of 2 m/s and then divide by 2, we get 1 m/s for the average velocity.

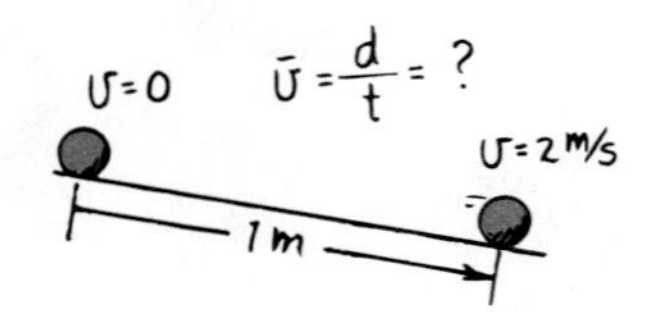

FIGURE B.1

The ball rolls 1 meter down the incline in 1 s and reaches a speed of 2 m/s. Its average speed, however, is 1 m/s. Do you see why?

In each succeeding second we see the ball roll a longer distance down the same slope in Figure B.2. Note that the distance covered in the second time interval is 3 meters. This is because the average speed of the ball in this interval is 3 m/s. In the next 1-second interval the average speed is 5 m/s, so the distance covered is 5 meters. It is interesting to see that successive increments of distance increase as a sequence of odd numbers. Nature clearly follows mathematical rules!

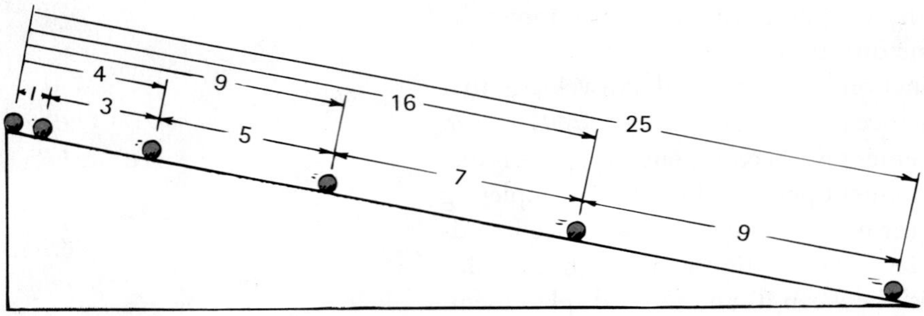

Investigate Figure B.2 carefully and note the total distance covered as the ball accelerates down the plane. The distances go from zero to 1 meter in 1 second, zero to 4 meters in 2 seconds, zero to 9 meters in 3 seconds, zero to 16 meters in 4 seconds, and so on in succeeding seconds. The sequence for total distances covered is the *squares* of the time. Now we'll investigate the relationship between the distance traveled and the square of the time for constant acceleration more closely in the case of free fall.

FIGURE B.2

If the ball covers 1 m during the first second, then it will cover the odd-numbered sequence of 3, 5, 7, 9 m in each successive second. Note that distance increases as the square of time.

Computing Distance When Acceleration Is Constant

How far will an object released from rest fall in a given time? To answer this question, let us consider the case in which it falls freely for 3 seconds, starting at rest. Neglecting air resistance, the object will have a constant acceleration of 10 meters per second each second (more precisely 9.8 m/s^2, but we want to make the numbers easier to follow):

Velocity at the *beginning* = 0 m/s

Velocity at the *end* of 3 seconds = (10 × 3) m/s

$$\begin{aligned}\text{Average velocity} &= \tfrac{1}{2} \text{ the sum of these two velocities}\\ &= \tfrac{1}{2} \times (0 + 10 \times 3) \text{ m/s}\\ &= \tfrac{1}{2} \times 10 \times 3 = 15 \text{ m/s}\\ \text{Distance traveled} &= \text{average velocity} \times \text{time}\\ &= (\tfrac{1}{2} \times 10 \times 3) \times 3\\ &= \tfrac{1}{2} \times 10 \times 3^2 = 45 \text{ m}\end{aligned}$$

We can see from the meanings of these numbers that

$$\text{Distance traveled} = \tfrac{1}{2} \times \text{acceleration} \times \text{square of time}$$

This equation is true for an object falling not only for 3 seconds but for any length of time, as long as it falls from a rest position and the acceleration is constant. If we let d stand for the distance traveled, a for the acceleration, and t for the time, the rule may be written, in shorthand notation, as

$$d = \tfrac{1}{2}at^2$$

This relationship was, in effect, first deduced by Galileo, who reasoned that the total distance fallen should be proportional to the square of the time.

In the case of objects in free fall, it is customary to use the letter g rather than a to represent the acceleration (g because acceleration is due to gravity). Although the value of g varies slightly in different parts of the world, it is nearly equal to 10 m/s^2 (in laboratory use 9.8 m/s^2 for more accurate measurements). If we use g for the acceleration of a freely falling object (with negligible air resistance), the equations for falling objects starting from a rest position become

$$v = gt$$

$$d = \tfrac{1}{2}gt^2$$

When Chelcie Liu releases both balls simultaneously, he asks, "Which will reach the end of the equal-length tracks first?" (*Hint*: On which track is the average speed of the ball greater?)

When the object doesn't start from rest but has an initial speed v_0,

$$v = v_0 + gt$$

$$d = v_0t + \tfrac{1}{2}gt^2$$

Much of the difficulty in learning physics, like learning any discipline, has to do with learning the language—the many terms and definitions. Speed is somewhat different from velocity, and acceleration is vastly different from speed or velocity.

CHECK YOURSELF

1. A car starting from rest has a constant acceleration of 4 m/s^2. How far will it go in 5 s?
2. How far will an object released from rest fall in 1 s? Assume acceleration is $g = 10$ m/s^2.
3. If it takes 4 s for an object to freely fall to the water when released from the Golden Gate Bridge, how high is the bridge?

CHECK YOUR ANSWERS

1. Distance $= \frac{1}{2} \times 4 \times 5^2 = 50$ m
2. Distance $= \frac{1}{2} \times 10 \times 1^2 = 5$ m
3. Distance $= \frac{1}{2} \times 10 \times 4^2 = 80$ m

Circular Motion

Linear speed is what we have been calling simply *speed*—the distance traveled in meters or kilometers per unit of time. A point on the perimeter of a merry-go-round or turntable moves a greater distance in one complete rotation than a point nearer the center. Moving a greater distance in the same time means a greater speed. The speed of something moving along a circular path is called **tangential speed** because the direction of motion is tangent to the circle.

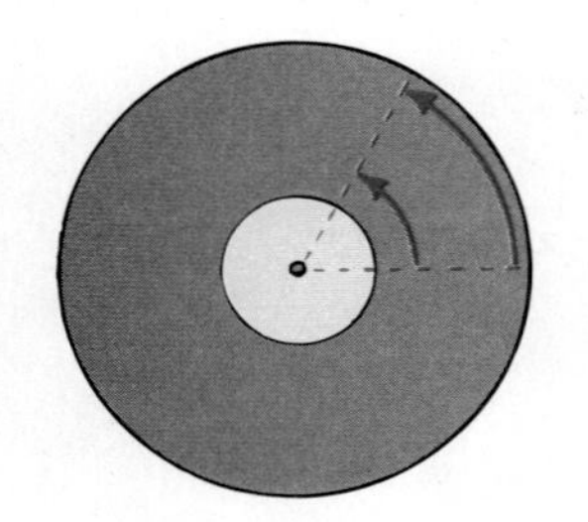

FIGURE B.3

When a phonograph record turns, a ladybug farther from the center travels a longer path in the same time and has a greater tangential speed.

Rotational speed (sometimes called angular speed) is the number of rotations or revolutions per unit of time. All parts of the rigid merry-go-round turn about the axis of rotation *in the same amount of time*. All parts share the same rate of rotation, or *number of rotations or revolutions per unit of time*. It is common to express rotational speeds in revolutions per minute (rpm).* Phonograph records that were common some years ago rotate at 33⅓ rpm. A ladybug sitting anywhere on the surface of the record revolves at 33⅓ rpm (Figures B.3 and B.4).

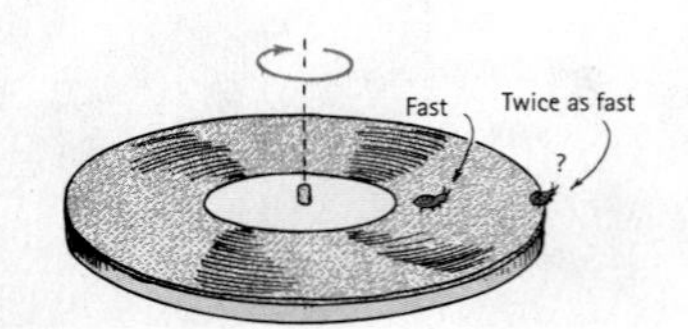

FIGURE B.4

The entire disk rotates at the same rotational speed, but ladybugs at different distances from the center travel at different tangential speeds. A ladybug twice as far from the center moves twice as fast.

Tangential speed is *directly proportional* to rotational speed. Unlike rotational speed, tangential speed depends on the distance from the axis. Something at the center of a rotating platform has no tangential speed at all and merely rotates. But, as it approaches the edge of the platform, its tangential speed increases. Tangential speed is directly proportional to the distance from the axis (for a given rotational speed). Twice as far from the rotational axis, the speed is twice as great. Three times as far from the rotational axis, there is three times as much tangential speed. When a line of people locked arm in arm at the skating rink makes a turn, the motion of "tail-end Charlie" is evidence of this greater speed. So, tangential speed is directly proportional both to rotational speed and to radial distance.†

CHECK YOURSELF

On a rotating platform similar to the disk shown in Figure B.4, if you sit halfway between the rotating axis and the outer edge and have a rotational speed of 20 rpm and a tangential speed of 2 m/s, what will be the rotational and tangential speeds of your friend who sits at the outer edge?

CHECK YOUR ANSWER

All parts of the rigid platform have the same rotational speed, so your friend also rotates at 20 rpm. Tangential speed is a different story; since she is twice as far from the axis of rotation, she moves twice as fast—4 m/s.

* Physics types usually describe rotational speed in terms of the number of "radians" turned in a unit of time, for which they use the symbol ω (the Greek letter omega). There's a little more than 6 radians in a full rotation (2π radians, to be exact).

† When customary units are used for tangential speed v, rotational speed ω, and radial distance r, the direct proportion of v to both r and ω becomes the exact equation $v = r\omega$. So the tangential speed will be directly proportional to r when all parts of a system simultaneously have the same ω, as for a wheel, disk, or rigid wand. (The direct proportionality of v to r is not valid for the planets because planets don't all have the same ω.)

Torque

Whereas force causes changes in velocity, *torque* causes changes in rotation. To understand torque (rhymes with *dork*), hold the end of a meterstick horizontally with your hand, as in Figure B.5. If you dangle a weight from the meterstick near your hand, you can feel the meterstick twist. Now if you slide the weight farther from your hand, the twist you feel is greater, although the weight is the same. The force acting on your hand is the same. What's different is the torque:

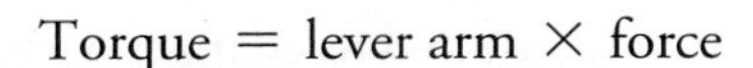

$$\text{Torque} = \text{lever arm} \times \text{force}$$

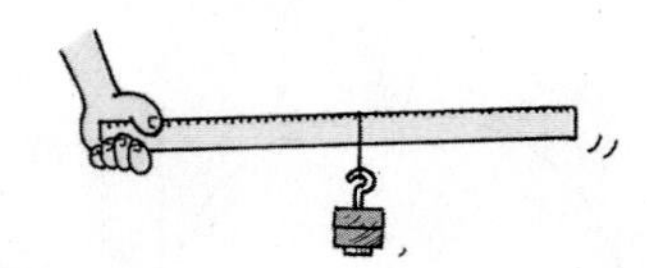

FIGURE B.5

If you move the weight away from your hand, you can feel the difference between force and torque.

Lever arm is the distance between the point of application of the force and the axis of rotation—the axis about which turning occurs. The lever arm is the shortest distance between the applied force and the rotational axis. Torques are intuitively familiar to youngsters playing on a seesaw. Kids can balance a seesaw even when their weights are unequal. Weight alone doesn't produce rotation. Torque does, and children soon learn that the distance they sit from the pivot point is every bit as important as weight (Figure B.6). When the torques are equal, which makes the net torque zero, no rotation is produced.

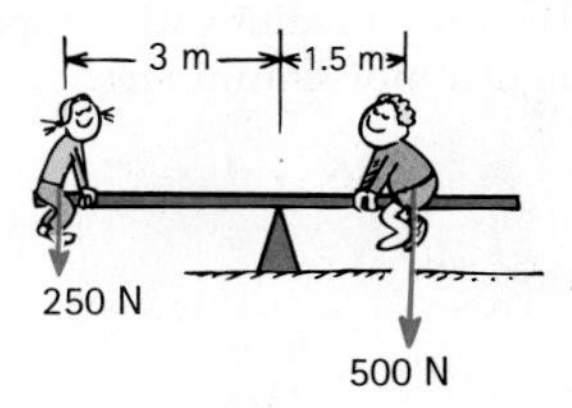

FIGURE B.6

No rotation is produced when the torques balance each other.

Recall the equilibrium rule from Chapter 2: The sum of the forces acting on a body or any system must equal zero for mechanical equilibrium: that is, $\Sigma F = 0$. We now see an additional condition. The net torque on a body or on a system must also be zero for mechanical equilibrium. Things in mechanical equilibrium have no acceleration—neither linearly nor rotationally.

Suppose that the seesaw is arranged so that the half-as-heavy girl is suspended from a 4-meter rope hanging from her end of the seesaw (Figure B.7). She is now 5 meters from the fulcrum, and the seesaw is still balanced. We see that the lever-arm distance is 3 meters, not 5 meters. The lever arm about any axis of rotation is the perpendicular distance from the axis to the line along which the force acts. This will always be the shortest distance between the axis of rotation and the line along which the force acts.

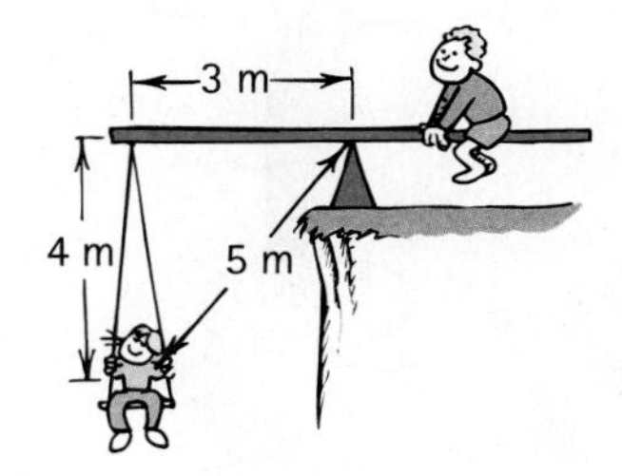

FIGURE B.7

The lever arm is still 3 m.

This is why we can turn the stubborn bolt shown in Figure B.8 more easily if we apply force perpendicular to the handle, rather than at an oblique angle as shown in the first figure. In the first figure, the lever arm is shown by the dashed line, and it is shorter than the length of the wrench handle. In the second figure, the lever arm is equal to the length of the wrench handle. In the third figure, the lever arm is extended with a pipe to provide more leverage and a greater torque.

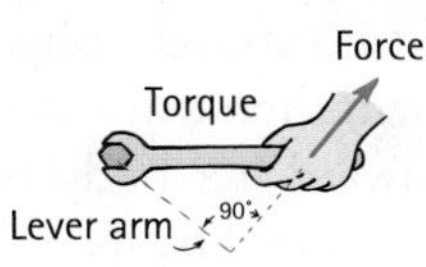

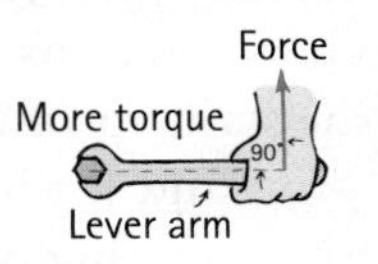

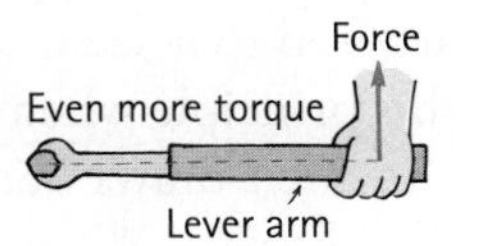

FIGURE B.8

Although the magnitudes of the force in each case are the same, the torques are different.

Angular Momentum

Things that rotate—whether a cylinder rolling down an incline, an acrobat doing a somersault, or the cloud of gas and dust that would become the solar system (see Chapter 28)—keep on rotating until something stops them. A rotating object has "inertia of rotation." Recall from Chapter 4 that all moving objects have "inertia of motion" or momentum—the product of mass and velocity. This kind of momentum is **linear momentum**. Similarly, the "inertia of rotation" of rotating objects is called **angular momentum**.

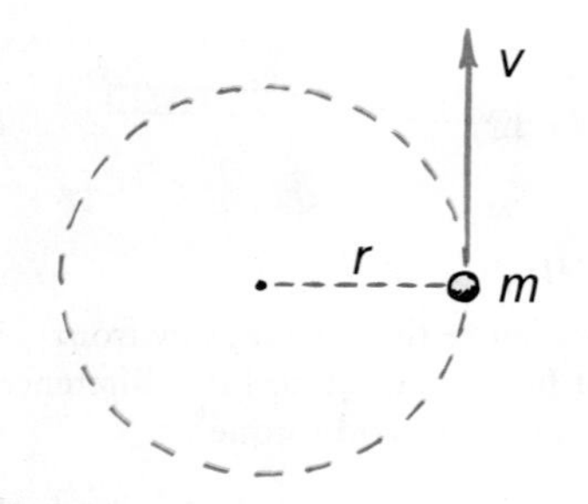

FIGURE B.9
A small object of mass m whirling in a circular path of radius r with a speed v has angular momentum mvr.

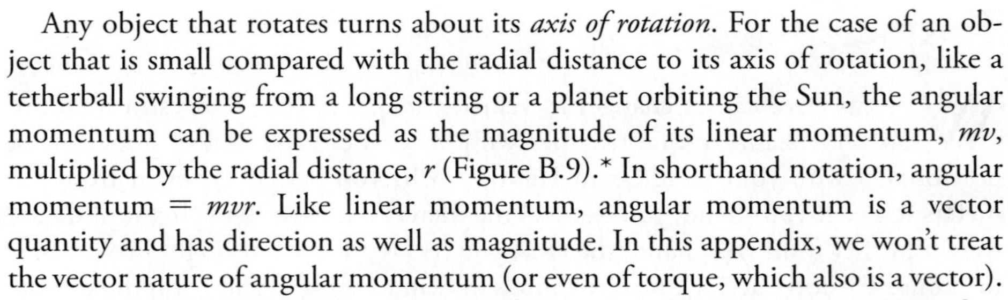

Any object that rotates turns about its *axis of rotation*. For the case of an object that is small compared with the radial distance to its axis of rotation, like a tetherball swinging from a long string or a planet orbiting the Sun, the angular momentum can be expressed as the magnitude of its linear momentum, mv, multiplied by the radial distance, r (Figure B.9).* In shorthand notation, angular momentum = mvr. Like linear momentum, angular momentum is a vector quantity and has direction as well as magnitude. In this appendix, we won't treat the vector nature of angular momentum (or even of torque, which also is a vector).

Just as an external net force is required to change the linear momentum of an object, an external net torque is required to change the angular momentum of an object. We can state a rotational version of Newton's first law (the law of inertia):

An object or system of objects will maintain its angular momentum unless acted upon by an unbalanced external torque.

We can see an application of this rule when we look at a spinning top. If the friction is low so that the torque is also low, the top tends to remain spinning. Earth and the planets spin in torque-free regions, and once they are spinning, they remain so.

Angular momentum is a vector. However, in this book we will not treat the vector nature of momentum.

Conservation of Angular Momentum

Just as the linear momentum of any system is conserved if no net forces act on the system, angular momentum is conserved if no net torque acts on the system. In the absence of an unbalanced external torque, the angular momentum of that system remains constant. This means that its angular momentum at any one time will be the same as at any other time.

Conservation of angular momentum is shown in Figure B.10. The man stands on a low-friction turntable with his arms and weights extended. To simplify, consider only the weights in his hands. When he is slowly turning with his arms extended, much of the angular momentum is due to the distance between the weights and the rotational axis. When he pulls the weights inward, the distance is considerably reduced. What is the result? His rotational speed increases!** This example is best appreciated by the turning person, who feels changes in rotational speed that seem to be mysterious. But it's straight physics! A figure skater uses this principle when she starts to whirl with her arms and perhaps a leg extended and then draws her arms and leg in toward her body to obtain a greater rotational speed. Whenever a rotating body contracts, its rotational speed increases.

The law of conservation of angular momentum is seen in the motions of the planets and the shape of the galaxies. When a slowly rotating ball of gas in space gravitationally contracts, the result is an increase in its rate of rotation. The conservation of angular momentum is far-reaching.

FIGURE B.10
Conservation of angular momentum. When the man pulls his arms and the whirling weights inward, he decreases the radial distance between the weights and the axis of rotation, and the rotational speed increases correspondingly.

* For rotating bodies that are large compared with radial distance—for example, a planet rotating about its own axis—the concept of rotational inertia must be introduced. Then angular momentum is rotational inertia × rotational speed. See any of Paul Hewitt's *Conceptual Physics* textbooks for more information.

** When a direction is assigned to rotational speed, we call it rotational velocity (often called *angular velocity*). By convention, the rotational velocity vector and the angular momentum vector have the same direction and lie along the axis of rotation.

APPENDIX C

More on Vectors

Vectors and their components are briefly treated in Chapter 3 and in the *Conceptual Integrated Science Practice Book*. Here we consider some instructive examples of vector applications.

1. Ernie Brown applies a force that pushes a lawnmower forward and also against the ground. In Figure C.1, we consider only a single force ***F*** applied by Ernie. We can separate this force into two components. The vector ***D*** represents the downward component, and ***S*** is the sideways component, the force that moves the lawnmower forward. If we know the magnitude and direction of the vector ***F***, we can estimate the magnitude of the components from the vector diagram.

FIGURE C.1

2. Is it easier to push or pull a wheelbarrow over a step? Figure C.2 shows a vector diagram for each case. When you push a wheelbarrow, part of the force is directed downward, which makes it harder to get over the step. When you pull, however, part of the pulling force is directed upward, which helps to lift the wheel over the step. Note that the vector diagram suggests that pushing the wheelbarrow may not get it over the step at all. Do you see that the height of the step, the radius of the wheel, and the angle of the applied force determine whether the wheelbarrow can be pushed over the step? We see how vectors help us analyze a situation so that we can see just what the problem is!

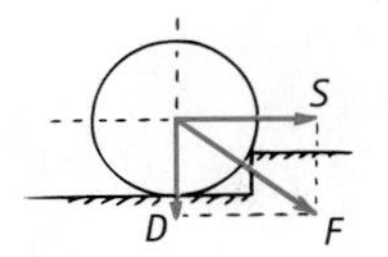

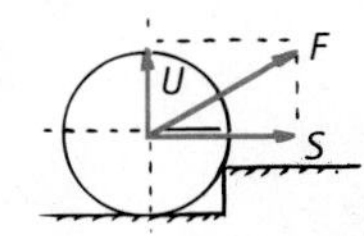

FIGURE C.2

3. If we consider the components of the weight ***W*** of a barrel rolling down an incline, we can see why its pickup in speed depends on the angle (Figure C.3).
   Note that the steeper the incline, the greater the component ***S*** becomes and the faster the barrel picks up speed. When the incline is vertical, ***S*** becomes equal to the barrel's weight and pickup in speed is a maximum—the acceleration of free fall. There are two more forces that are not shown: the normal force ***N***, which is equal and oppositely directed to ***D***, and the friction force ***f***, which is oppositely directed to ***S***.

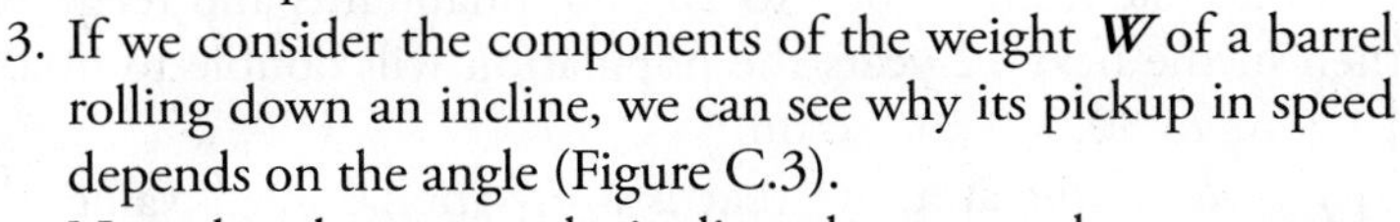

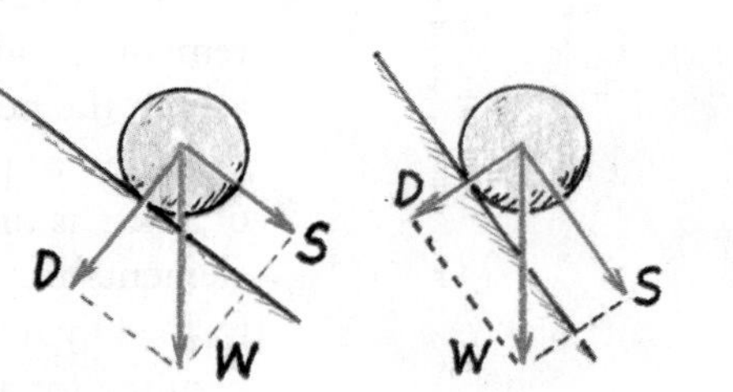

FIGURE C.3

4. Consider Monkey Lil, suspended at rest in a zoo by holding on to a rope with one hand and the side of the cage with the other (Figure C.4). Is the tension in the rope less than, equal to, or greater than Lil's weight? A diagram of the forces acting on Lil is instructive. Vector ***T*** is the rope tension, ***W*** is Lil's weight, and ***S*** is the sideways pull on the cage. Since Lil is in equilibrium, these forces must sum to zero (the equilibrium rule). Applying the parallelogram method, we see that ***T*** and ***S*** describe a parallelogram in which the resultant, the dashed vector, is equal and opposite to ***W***. Vector ***T*** is clearly greater than ***W***, so the tension in the rope exceeds Lil's weight.

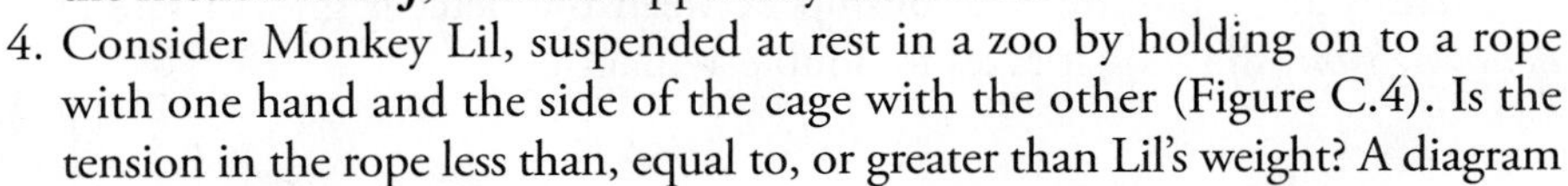

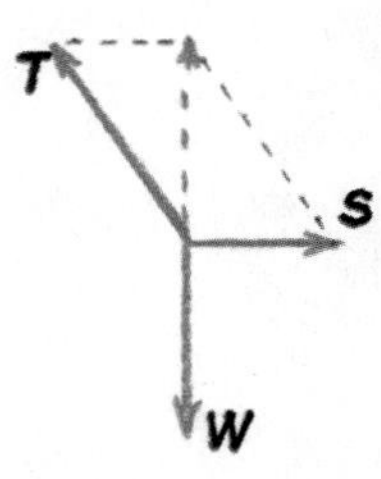

FIGURE C.4

APPENDIX D

Exponential Growth and Doubling Time*

Try to fold a piece of paper in half and then fold it upon itself again and again nine times. You won't be able to do it. It gets too thick to fold. Even if you could fold a fine piece of tissue paper upon itself 50 times, it would be more than 20 kilometers thick! The continual doubling of a quantity builds up exponentially. Double one penny 30 times, so that you begin with one penny, then have two pennies, then four, and so on, and you'll accumulate a total of $10,737,418.23! One of the most important things we seem unable to perceive is the process of exponential growth. If we could, we could "see" why compound interest works the way it does, prices of goods rise the way they do, and populations and pollution proliferate out of control.

When a quantity such as money in the bank, population, or the rate of consumption of a resource steadily grows at a fixed percent per year, we say the growth is *exponential*. Money in the bank may grow at 2% per year; electric power generating capacity in the United States grew at about 7% per year for the first three-quarters of the 20th century. The important thing about exponential growth is that the time required for the growing quantity to double in size (increase by 100%) is also constant. For example, if the population of a growing city takes 12 years to double from 10,000 to 20,000 inhabitants and its growth remains steady, then in the next 12 years the population will double to 40,000 and in the next 10 years to 80,000 and so on.

When a quantity is *decreasing* at a rate that is proportional to its value, that quantity is undergoing *exponential decay*. As discussed in Chapter 10, radioactive elements are subject to radioactive decay, meaning that "parent" isotopes are transformed by nuclear processes into "daughter" isotopes (see Figure 10.17). The time required for an exponentially decaying quantity to be reduced to half of its initial value is called its *half-life*. Just as doubling time is constant for an exponentially growing quantity, half-life is constant for an exponentially decaying quantity.

There is an important relationship between the percent growth rate and its *doubling time*, the time it takes to double a quantity:

$$\text{Doubling time} = \frac{69.3}{\text{percent growth per unit time}} \sim \frac{70}{\%}$$

So, to estimate the doubling time for a steadily growing quantity, we divide the number 70 by the percent growth rate. For example, the 7% growth rate of electric power generating capacity in the United States means that in the past the capacity has doubled every 10 years ($\frac{70\%}{7\%/\text{year}} = 10$ years). A 2% growth rate for

*This appendix is drawn from material by University of Colorado physics professor Albert A. Bartlett, who strongly asserted, "The greatest shortcoming of the human race is our inability to understand the exponential function."

world population means that the population of the world doubles every 35 years ($\frac{70\%}{2\%/\text{year}} = 35$ years). A city planning commission that accepts what seems like a modest 3.5% growth rate may not realize that this means that doubling will occur in 70/3.5 or 20 years; that's double the capacity for such things as water supply, sewage-treatment plants, and other municipal services every 20 years (Figure D.1).

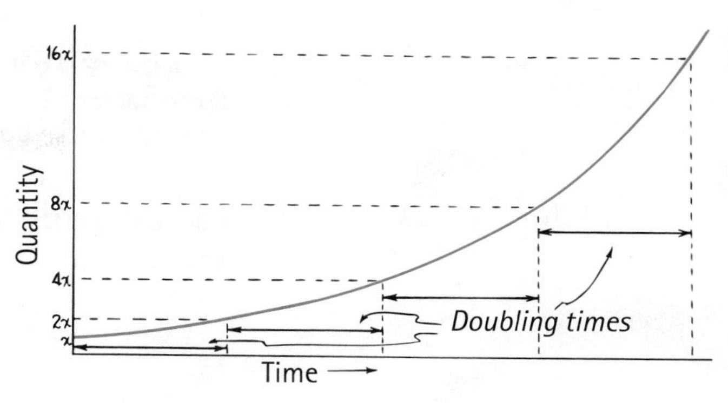

FIGURE D.1

An exponential curve. Notice that each of the successive equal time intervals noted on the horizontal scale corresponds to a doubling of the quantity indicated on the vertical scale. Such an interval is called the *doubling time.*

What happens when you put steady growth in a finite environment? Consider the growth of bacteria that grow by division, so that one bacterium becomes two, the two divide to become four, the four divide to become eight, and so on. Suppose the division time for a hypothetical strain of very tiny bacteria is 1 minute. This is then steady growth—the number of bacteria grows exponentially with a doubling time of 1 minute. Furthermore, suppose that one bacterium is put in a bottle at 11:00 A.M. and that growth continues steadily until the bottle becomes full of bacteria at 12 noon. Consider seriously the following question.

CHECK YOURSELF

When was the bottle half full?

CHECK YOUR ANSWER

11:59 A.M.!

It is startling to note that at 2 minutes before noon the bottle was only $\frac{1}{4}$ full. Table D.1 summarizes the amount of space left in the bottle in the last few minutes before noon. If you were an average bacterium in the bottle, at which time would you first realize that you were running out of space? For example, would you sense there was a serious problem at 11:55 A.M., when the bottle was only 3% filled ($\frac{1}{32}$) and had 97% of the space open (just yearning for development)? The point here is that there isn't much time between the minute when the effects of growth become noticeable and when they become overwhelming.

TABLE D.1 THE LAST MINUTES IN THE BOTTLE

Time	Part Full (%)	Part Empty
11:54 A.M.	1/64 (1.5%)	63/64
11:55 A.M.	1/32 (3%)	31/32
11:56 A.M.	1/16 (6%)	15/16
11:57 A.M.	1/8 (12%)	7/8
11:58 A.M.	1/4 (25%)	3/4
11:59 A.M.	1/2 (50%)	1/2
12:00 noon	Full (100%)	None

Suppose that at 11:58 A.M. some farsighted bacteria see that they are running out of space and launch a full-scale search for new bottles. Luckily, at 11:59 A.M. they discover three new empty bottles, three times as much space as they had ever known. This quadruples the total resource space ever known to the bacteria, for they now have a total of four bottles, whereas before the discovery they had only one. Further suppose that, thanks to their technological proficiency, they are able to migrate to their new habitats without difficulty. Surely, it seems to most of the bacteria that their problem is solved—and just in time.

CHECK YOURSELF

If the bacteria growth continues at the unchanged rate, what time will it be when the three bottles are filled to capacity?

CHECK YOUR ANSWER

12:02 P.M.!

TABLE D.2 EFFECTS OF THE DISCOVERY OF THREE NEW BOTTLES

Time	Effect
11:58 A.M.	Bottle 1 is 1/4 full
11:59 A.M.	Bottle 1 is 1/2 full
12:00 noon	Bottle 1 is full
12:01 P.M.	Bottles 1 and 2 are both full
12:02 P.M.	Bottles 1, 2, 3, and 4 are all full

We can see from Table D.2 that quadrupling the resource extends the life of the resource by only two doubling times. In our example the resource is space—but it could just as well be coal, oil, uranium, or any nonrenewable resource.

Continued growth and continued doubling lead to enormous numbers. In two doubling times, a quantity will double twice ($2^2 = 4$; quadruple) in size; in three doubling times, its size will increase eightfold ($2^3 = 8$); in four doubling times, it will increase sixteenfold ($2^4 = 16$); and so on. This is best illustrated by the story of the court mathematician in India who years ago invented the game of chess for his king. The king was so pleased with the game that he offered to repay the mathematician, whose request seemed modest enough. The mathematician requested a single grain of wheat on the first square of the chessboard, two grains on the second square, four on the third square, and so on, doubling the number of grains on each succeeding square until all the squares had been used (Figure D.2). At this rate there would be 2^{63} grains of wheat on the 64th square. The king soon saw that he could not fill this "modest" request, which amounted to more wheat than had been harvested in the entire history of Earth!

FIGURE D.2

A single grain of wheat placed on the first square of the chessboard is doubled on the second square, this number is doubled on the third, and so on, presumably for all 64 squares. Note that each square contains one more grain than all the preceding squares combined. Does enough wheat exist in the world to fill all 64 squares in this manner?

It is interesting and important to note that the number of grains on any square is 1 grain more than the total of all grains on the preceding squares. This is true anywhere on the board. Note from Table D.3 that when 8 grains are placed on the fourth square, the eight is 1 more than the total of 7 grains that were already on the board. The 32 grains placed on the sixth square are 1 more than the total of 31 grains that were already on the board. We see that in one doubling time we use more than all that had been used in all the preceding growth!

TABLE D.3 FILLING THE SQUARES ON THE CHESSBOARD

Square Number	Grains on Square	Total Grains Thus Far
1	1	1
2	2	3
3	4	7
4	8	15
5	16	31
6	32	63
7	64	127
.	.	.
.	.	.
.	.	.
64	2^{63}	$2^{64} - 1$

So, if we speak of doubling energy consumption in the next however many years, bear in mind that this means that in these years we will consume more energy than has been consumed during the entire preceding period of steady growth.

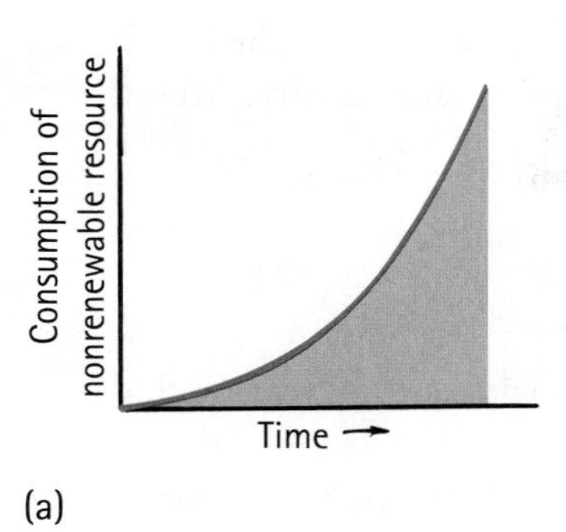

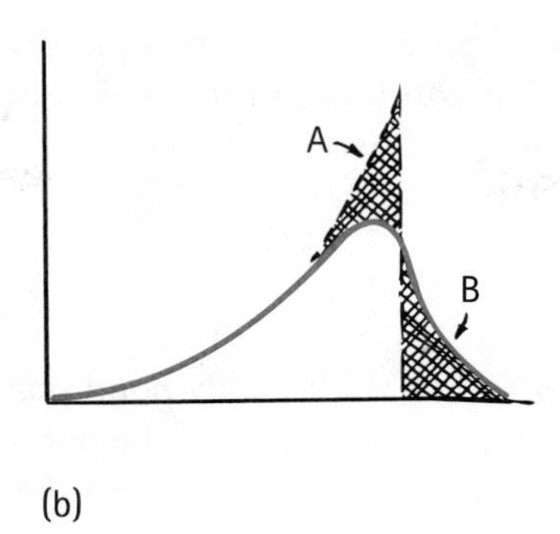

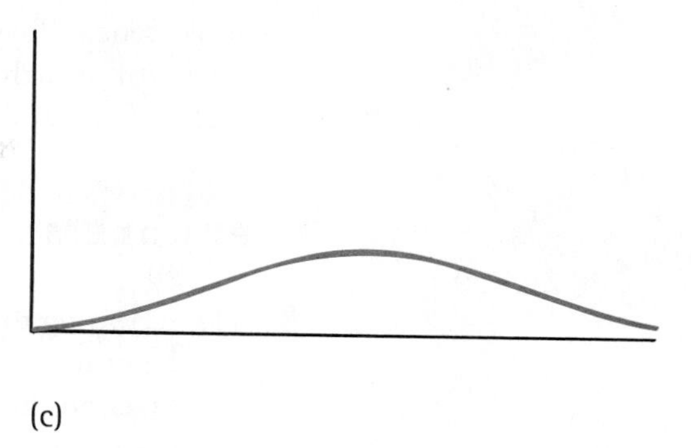

FIGURE D.3

(a) If the exponential rate of consumption for a nonrenewable resource continues until the resource is depleted, consumption falls abruptly to zero. The shaded area under this curve represents the total supply of the resource. (b) In practice, the rate of consumption levels off and then falls less abruptly to zero. Note that the crosshatched area A is equal to the crosshatched area B. Why? (c) At lower consumption rates, the same resource lasts a longer time.

And, if power generation continues to use predominantly fossil fuels, then except for some improvements in efficiency, we will burn up in the next doubling time a greater amount of coal, oil, and natural gas than has already been consumed by previous power generation, and except for improvements in pollution control, we can expect to discharge even more toxic wastes into the environment than the millions upon millions of tons already discharged over all the earlier years of industrial civilization. We would also expect more human-made calories of heat to be absorbed by Earth's ecosystem than have been absorbed in the entire past! At the previous 7% annual growth rate in energy production, all this would occur in one doubling time of a single decade. If over the coming years the annual growth rate remains at half this value, 3.5%, then all this would take place in a doubling time of two decades. Clearly this cannot continue!

The consumption of a nonrenewable resource cannot grow exponentially for an indefinite period because the resource is finite and its supply finally expires. The most drastic way this could happen is shown in Figure D.3a, where the rate of consumption, such as barrels of oil per year, is plotted against time—say, in years. In such a graph the area under the curve represents the supply of the resource. We see that when the supply is exhausted, the consumption ceases altogether. This sudden change is rarely the case because the rate of extracting the supply falls as it becomes more scarce. This is shown in Figure D.3b. Note that the area under the curve is equal to the area under the curve in part (a). Why? Because the total supply is the same in both cases. The principal difference is the time taken to finally extinguish the supply. History shows that the rate of production of a nonrenewable resource rises and falls in a nearly symmetrical manner, as shown in Figure D.3c. The time during which production rates rise is approximately equal to the time during which these rates fall to zero or near zero.

Production rates for all nonrenewable resources decrease sooner or later. Only production rates for renewable resources, such as agricultural or forest products, can be maintained at steady levels for long periods of time (Figure D.4), provided such production does not depend on waning supplies of nonrenewable resources such as petroleum. Much of today's agriculture is so dependent on petroleum that it can be said that modern agriculture is simply the process whereby land is used to convert petroleum into food. The implications of petroleum scarcity go far beyond the rationing of gasoline for cars or fuel oil for home heating.

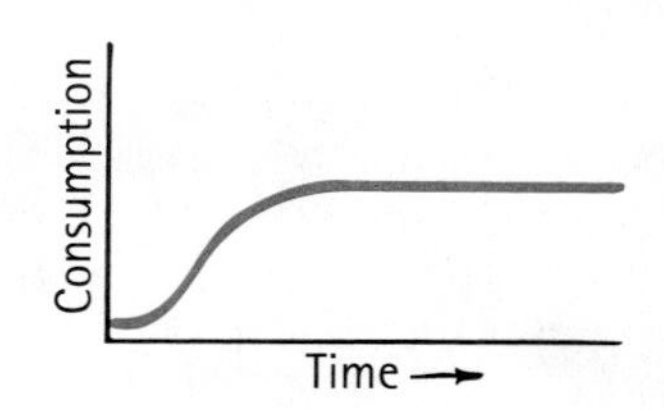

FIGURE D.4

A curve showing the rate of consumption of a renewable resource such as agricultural or forest products, where a steady rate of production and consumption can be maintained for a long period of time, provided this production is not dependent on the use of a nonrenewable resource that is waning in supply.

The consequences of unchecked exponential growth are staggering. It is important to ask: Is growth really good? In answering this question, bear in mind that human growth is an early phase of life that continues normally through adolescence. Physical growth stops when physical maturity is reached. What do we say of growth that continues in the period of physical maturity? We say that such growth is obesity—or worse, cancer.

PROBLEMS

1. According to a French riddle, a lily pond starts with a single leaf. Each day the number of leaves doubles, until the pond is completely covered by leaves on the 30th day. On what day was the pond half-covered? One-quarter covered?
2. In an economy that has a steady inflation rate of 7% per year, in how many years does a dollar lose half its value?
3. At a steady inflation rate of 7%, what will be the price every 10 years for the next 50 years for a theater ticket that now costs $30? For a coat that now costs $300? For a car that now costs $30,000? For a home that now costs $300,000?
4. If the sewage-treatment plant of a city is just adequate for the city's current population, how many sewage-treatment plants will be necessary 42 years later if the city grows steadily at a rate of 5% annually?
5. If world population doubles in 50 years and world food production also doubles in 50 years, how many people will be starving each year then compared to now?
6. Suppose you get a prospective employer to agree to hire your services for wages of a single penny for the first day, 2 pennies for the second day, and double each day thereafter, providing the employer keeps to the agreement for a month. What will be your total wages for the month? How will your wages for only the 30th day compare to your total wages for the previous 29 days?

APPENDIX E

Physics of Fluids

Liquids and gases have the ability to flow; hence they are called *fluids*. In order to discuss the physics of fluids properly, we first need to understand two concepts—*density* and *pressure*.

Density

As introduced in Chapter 2, a basic property of materials—whether in the solid, liquid, or gaseous phase—is the measure of compactness, or **density**:

$$\text{Density} = \frac{\text{mass}}{\text{volume}}$$

A loaf of bread has a certain mass, volume, and density. When the loaf is squeezed, the volume decreases and its density increases—but its mass remains the same. Mass is measured in either grams or kilograms and volume in either cubic centimeters (cm^3) or cubic meters (m^3). Another unit of volume is the liter, which is 1000 cm^3. One gram per cubic centimeter = 1 kg per liter (Figure E.1). The densities of some materials are listed in Table E.1.

FIGURE E.1
A liter of water occupies a volume of 1000 cm^3, has a mass of 1 kg, and weighs 9.8 N. Its density may therefore be expressed as 1 kg/L and its weight density as 9.8 N/L. (Seawater is slightly denser, about 10 N/L.)

A quantity known as *weight density* is commonly used when discussing liquid pressure:

$$\text{Weight density} = \frac{\text{weight}}{\text{volume}}$$

Weight density is common to British units, in which 1 cubic foot of fresh water (almost 7.5 gallons) weighs 62.4 pounds. So, fresh water has a weight density of 62.4 lb/ft^3. Saltwater is a bit denser, 64 lb/ft^3.

Pressure

Pressure is defined as the force exerted over a unit of area, such as a square meter or square foot:

$$\text{Pressure} = \frac{\text{force}}{\text{area}}$$

We see in Figure E.2 that force and pressure are different from each other. In the figure we see pressure due to the weight of a solid. Pressure occurs in fluids as well:

$$\text{Liquid pressure} = \text{weight density} \times \text{depth}$$

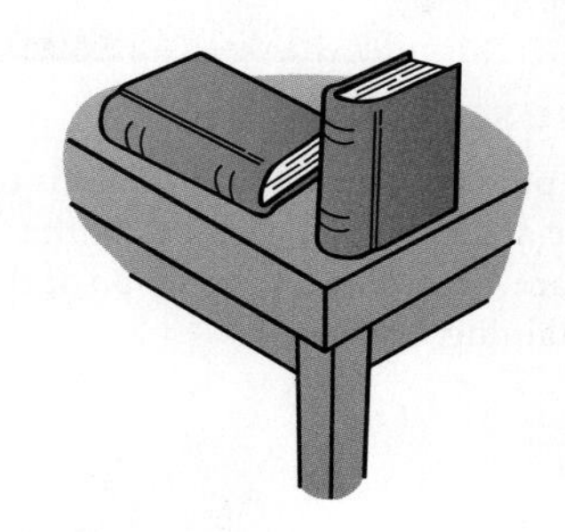

FIGURE E.2
Although the weight of both books is the same, the upright book exerts greater pressure against the table.

It is important to note that pressure depends not on the amount of liquid but on its depth. You feel the same pressure a meter deep in a small pool as you do a meter deep in the middle of the ocean. The pressure is the same at the bottom of

TABLE E.1 DENSITIES OF SOME MATERIALS		
Material	Grams per Cubic Centimeter	Kilograms per Cubic Meter
Liquids		
Mercury	13.60	1,360
Glycerin	1.26	1,260
Seawater	1.03	1,025
Water at 4°C	1.00	1,000
Benzene	0.90	899
Ethyl alcohol	0.81	806
Solids		
Osmium	22.5	22,480
Platinum	21.5	21,450
Gold	19.3	19,320
Uranium	19.0	19,050
Lead	11.3	11,344
Silver	10.5	10,500
Copper	8.9	8,920
Brass	8.6	8,560
Iron	7.8	7,800
Tin	7.3	7,280
Aluminum	2.7	2,702
Ice	0.92	917
Gases (atm pressure, at sea level)		
Dry air		
0°C	0.00129	1.29
10°C	0.00125	1.25
20°C	0.00121	1.21
30°C	0.00116	1.16
Hydrogen at 0°C	0.00090	0.090
Helium at 0°C	0.00178	0.178

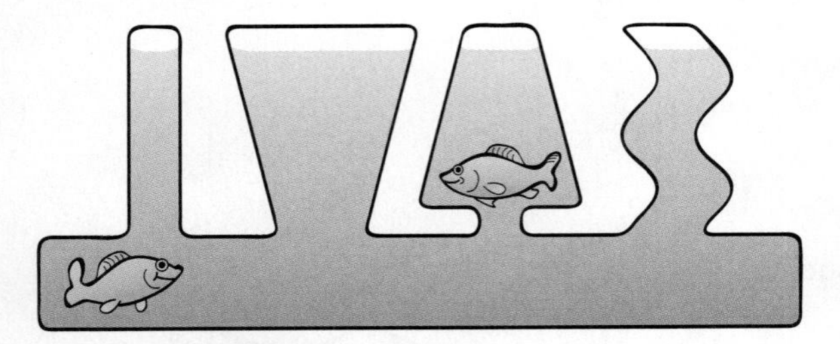

FIGURE E.3

The pressure exerted by a liquid is the same at any given depth below the surface, regardless of the shape of the containing vessel.

each of the connected vases in Figure E.3, for example. Depth, not volume, is the key to liquid pressure.

Pressure in a liquid at any point is exerted in equal amounts in all directions. For example, if you are submerged in water, no matter which way you tilt your head, you feel the same amount of water pressure on your ears. When liquid presses against a surface, there is a net force directed perpendicular to the surface (Figure E.4). If there is a hole in the surface, the liquid spurts out at right angles to the surface before curving downward due to gravity (Figure E.5). At greater depths, the pressure is greater and the speed of the escaping liquid is greater.

Buoyancy in a Liquid

When an object is submerged in water, the greater pressure on the bottom of the object results in an upward force called the **buoyant force**. We see why in Figure E.6. The arrows represent the forces at different places due to water

FIGURE E.4
The forces that produce pressure against a surface add up to a net force that is perpendicular to the surface.

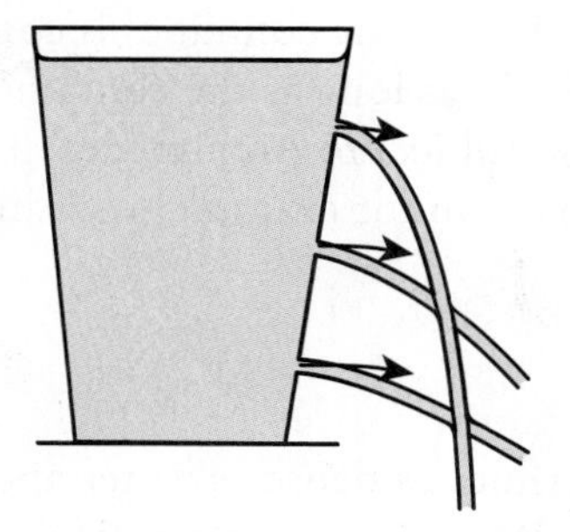

FIGURE E.5
Water pressure pushes perpendicularly against the sides of a container and increases with increasing depth.

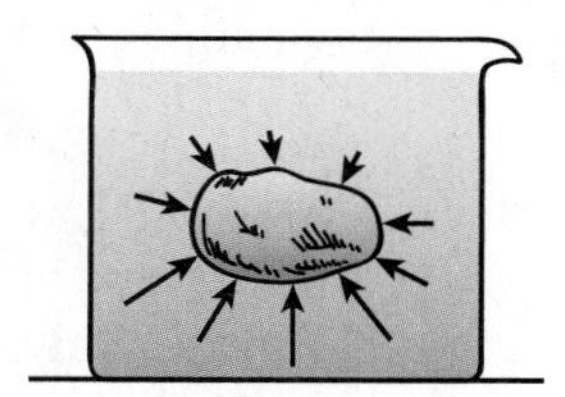

FIGURE E.6
The pressure against the bottom of a submerged object produces an upward buoyant force.

pressure. Forces that produce pressures against opposite sides cancel one another because they are at the same depth. Pressure is greater against the bottom of the object because the bottom is deeper (more pressure). Because the upward forces against the top are less, the forces do not cancel, and there is a net force upward. This net force is the buoyant force.

If the weight of the submerged object is greater than the buoyant force, the object sinks. If the weight is equal to the buoyant force, the object remains at any level, like a fish. If the buoyant force is greater than the weight of the completely submerged object, the object rises to the surface and floats.

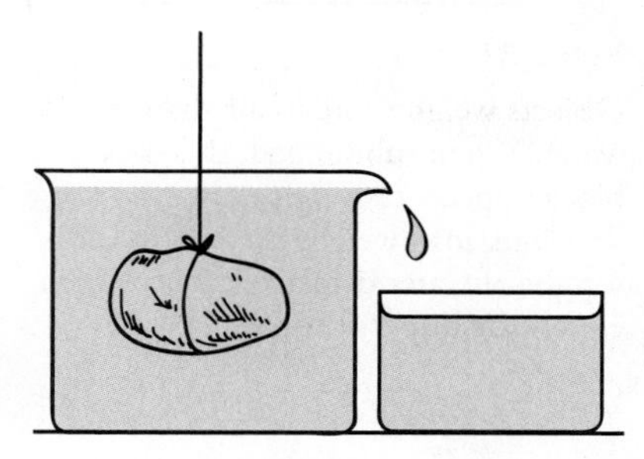

FIGURE E.7
When a stone is submerged, it displaces a volume of water equal to the volume of the stone.

Understanding buoyancy requires understanding the meaning of the expression "volume of water displaced." If a stone is placed in a container that is brimful of water, some water will overflow (Figure E.7). Water is *displaced* by the stone. A little thought tells you that the *volume of the stone*—that is, the amount of space (cubic centimeters) it takes up—is equal to the volume of water displaced. Place any object in a container partially filled with water, and the level of the surface rises (Figure E.8). By how much? By exactly the same amount as if we had added a volume of water equal to the volume of the immersed object. This is a good method for determining the volume of irregularly shaped objects: *A completely submerged object always displaces a volume of liquid equal to its own volume.*

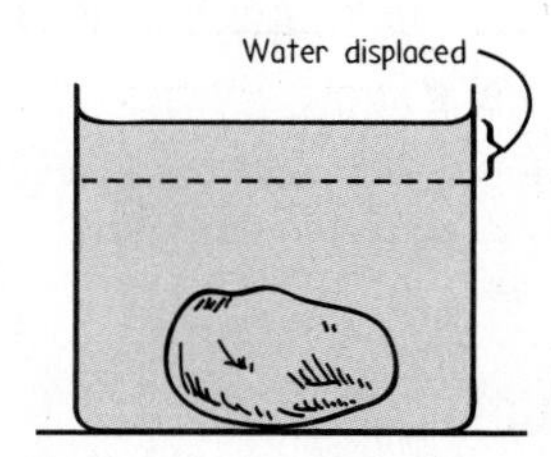

FIGURE E.8
The increase in the water level is the same as that which would occur if, instead of placing the stone in the container, we had poured in a volume of water equal to the stone's volume.

Archimedes' Principle

The relationship between buoyancy and displaced liquid was first discovered in the third century B.C. by the Greek scientist Archimedes. It is stated as follows:

An immersed body is buoyed up by a force equal to the weight of the fluid it displaces.

This relationship is called **Archimedes' principle**; it is true of all fluids, both liquids and gases (Figure E.9). By *immersed*, we mean either *completely* or *partially submerged*. If we immerse a sealed 1-liter container halfway into a tub of water, it displaces a half-liter of water and is buoyed up by a force equal to

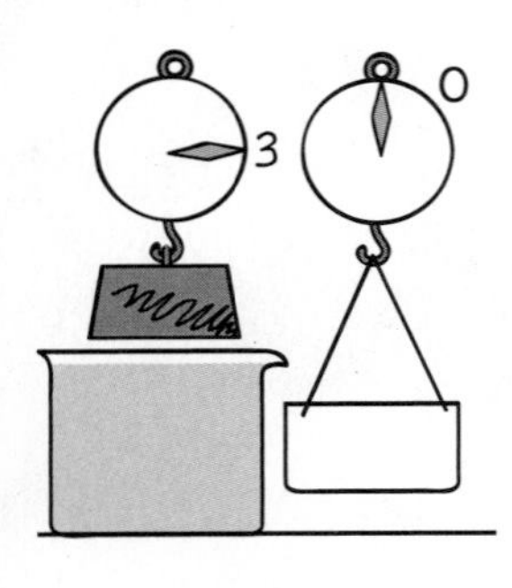

FIGURE E.9

Objects weigh more in air than in water. When submerged, this 3-N block appears to weigh only 1 N. The "missing" weight is equal to the weight of water displaced, 2 N, which equals the buoyant force.

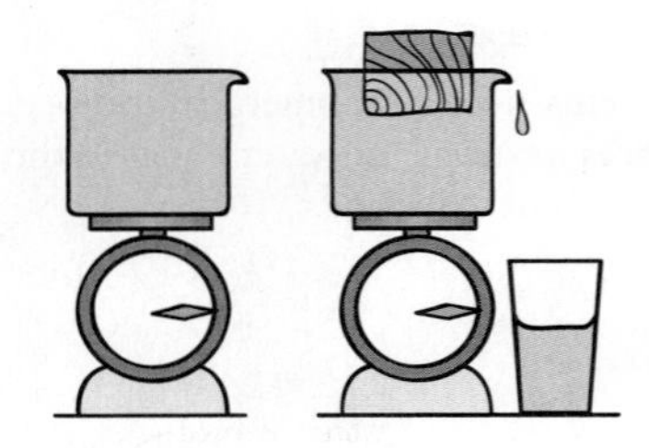

FIGURE E.10

A floating object displaces a weight of fluid equal to its own weight.

the weight of a half-liter of water. If we immerse it completely (submerge it), a force equal to the weight of a full liter or 1 kilogram of water (which is 9.8 newtons) buoys it up. Unless the container is compressed, the buoyant force is 9.8 newtons at *any* depth, as long as the container is completely submerged. A gas is a fluid and is also subject to Archimedes' principle. Any object less dense than air, a gas-filled balloon for example, rises in air.

Flotation

Iron is nearly eight times as dense as water and therefore sinks in water, but an iron ship floats. Why? Consider a solid 1-ton block of iron. When submerged, it doesn't displace 1 ton of water because it's eight times more compact than water. It displaces only $\frac{1}{8}$ ton of water—certainly not enough to make it float. Suppose we reshape the same iron block into a bowl. It still weighs 1 ton but, when placed in water, it displaces a greater volume of water than before. Its greater volume displaces more water. The deeper it is immersed, the more water it displaces and the greater the buoyant force acting on it. When the buoyant force equals 1 ton, it sinks no farther. It floats. When any object displaces a weight of water equal to its own weight, it floats (Figure E.10). This is the **principle of flotation**:

A floating object displaces a weight of fluid equal to its own weight.

A 500-N friend floating in a swimming pool displaces 500 N of water. To accomplish this, your friend must be slightly less dense than water (which may or may not involve the use of a life preserver). Any floating object is less dense than the fluid it floats upon. The density of ice, for example, is 0.9 that of water, so icebergs float in water. Interestingly, mountains are less dense than the semimolten mantle beneath them, so they float. Just as most of an iceberg is below the water surface (90%), most of a mountain (about 85%) extends into the mantle. If you could shave off the top of an iceberg, the iceberg would be lighter and would be buoyed up to nearly its original height before its top was shaved. Similarly, when mountains erode and wear away, they become lighter and are pushed up from below to float to nearly their original heights. So, when a kilometer of mountain erodes away, about 0.85 kilometer of mountain pops up from below. That's why it takes so long for mountains to weather away. Floating objects, whether mountains or ships, must displace a weight of fluid equal to their own weight. Thus, a 10,000-ton ship must be built wide enough to displace 10,000 tons of water before it sinks too deep in the water. The same holds true for vessels in the air. A dirigible or huge balloon that weighs 100 tons displaces at least 100 tons of air. If it displaces more, it rises; if it displaces less, it falls. If it displaces exactly its weight, it hovers at constant altitude.

Pressure in a Gas

There are similarities and there are differences between gases and liquids. The primary difference between a gas and a liquid is the distance between molecules. In a gas, the molecules are far apart and free from the cohesive forces that dominate their motions when in the liquid and solid phases. The motions of gas molecules are less restricted. A gas expands, fills all the space available to it, and exerts a pressure against its container. Only when the quantity of gas is very large, such as the Earth's atmosphere or a star, do gravitational forces limit the size or determine the shape of the mass of gas.

Dalton's Law

The pressure in a gas depends on only the number of gas particles in a given volume and their kinetic energy. The type of atom or molecule is unimportant. Many gases—for example, air—are mixtures. If we know the pressure that each type of gas in a mixture exerts, we can add the individual pressures together to get the total pressure. The pressure exerted by each type of gas is called the *partial pressure* exerted by that gas. Stated mathematically, this is **Dalton's law**:

$$P_{\text{Total}} = P_1 + P_2 + P_3$$

Dalton's law tells us that, at constant volume and temperature, the total pressure exerted by a mixture of gases is equal to the sum of the partial pressures.

Boyle's Law

Pressure and volume in a confined gas are nicely related. "Pressure × volume" for a quantity of gas at any specified time is equal to any "different pressure × different volume" at any other time. In shorthand notation,

$$P_1V_1 = P_2V_2$$

where P_1 and V_1 represent the original pressure and volume, respectively, and P_2 and V_2 the second pressure and volume. This relationship is called **Boyle's law**. Boyle's law applies to ideal gases. An *ideal* gas is one in which the disturbing effects of the forces between molecules and the finite size of the individual molecules can be neglected. Air and other gases under normal pressure approach ideal-gas conditions.

The Ideal Gas Law

The **ideal gas law** relates all four variables that are used to measure changes in a gas—pressure (P), volume (V), temperature (T), and number of moles n. The law is stated as

$$PV = nRT$$

where R is a proportionality constant, whose value depends on the units you are using for pressure and volume.

The Combined Gas Law

If the number of gas particles is a constant—for example, for an enclosed gas—another useful gas law is the **combined gas law**. It is stated as

$$\frac{P_1V_1}{T_1} = \frac{P_2V_2}{T_2}$$

where T_1 and T_2 represent the initial and final *absolute* temperatures, measured in kelvins (see Chapter 6). The combined gas law is often useful—for example, when determining the volume of a gas at standard temperature and pressure (STP).* Thus far we have treated pressure only as it applies to stationary fluids. Motion produces an additional influence.

Bernoulli's Principle

When a fluid flows through a narrow constriction, its speed increases. This is easily demonstrated by the increased speed of water that spurts from a garden hose when you narrow the opening of the nozzle. The fluid must speed up in the constricted region if the flow is to be continuous.

The Swiss scientist Daniel Bernoulli experimented with fluids in the 18th century. He wondered how the fluid gained this extra speed and reasoned that it is acquired at the expense of a lowered internal pressure. His discovery, now called **Bernoulli's principle**, states:

When the speed of a fluid increases, pressure in the fluid decreases.

Bernoulli's principle is a consequence of the conservation of energy. In a steady flow of fluid, there are three kinds of energy: kinetic energy due to motion, gravitational potential energy due to elevation, and work done by pressure forces. In a steady fluid flow where no energy is added or removed, the sum of these forms of energy remains constant. If the elevation of the flowing fluid doesn't change, then an increase in speed means a decrease in pressure, and vice versa. Bernoulli's principle is accurate only for steady flow. If the speed is too great, the flow may become turbulent and follow a changing, curling path known as an *eddy*. This type of flow exerts friction on the fluid and causes some of its energy to be transformed into heat. Then Bernoulli's principle does not hold. Hold a sheet of paper in front of your mouth. When you blow across the top surface, the paper rises. This is because the pressure of the moving air against the top of the paper is lower than the pressure of the air at rest against the lower surface.

If we imagine the rising paper as an airplane wing, we can better understand the lifting force that supports a heavy airliner. A blend of Bernoulli's principle and Newton's laws account for the air flight we see today. Quite awesome!

*To study temperature and pressure changes in a gas, it is useful to specify a set of standard conditions that can be used for comparison. These conditions are known as standard temperature and pressure and they have the values 0°C and 1000 atm.

APPENDIX F

Chemical Equilibrium

The arrow of a chemical equation indicates the direction of the reaction. In the equation for the formation of water, for example, the arrow indicates that hydrogen and oxygen combine to form water:

$$2\,H_2 + O_2 \rightarrow 2\,H_2O$$

The reverse reaction is also possible; that is, oxygen and hydrogen can be formed from water:

$$2\,H_2 + O_2 \leftarrow 2\,H_2O$$

Most chemical reactions are *reversible*, but the extent of the reverse reaction depends very much on conditions. Under usual conditions, for example, water does not readily convert to oxygen and hydrogen. When electricity passes through water, however, the water does decompose to O_2 and H_2.

Nitrogen dioxide, NO_2, a brown gas and one of the toxic components of air pollution, provides a good example of a reaction that is fairly easy to reverse. At room temperature nitrogen dioxide molecules pair off to become colorless dinitrogen tetroxide, N_2O_4:

$$NO_2 + NO_2 \rightarrow N_2O_4$$

Once dinitrogen tetroxide molecules are formed, however, they break apart to re-form nitrogen dioxide:

$$N_2O_4 \rightarrow NO_2 + NO_2$$

Of course, once the nitrogen dioxide molecules are re-formed, they can get together to re-form the dinitrogen tetroxide. The net result is two competing reactions, which we depict using double arrows:

$$NO_2 + NO_2 \rightleftarrows N_2O_4$$

Initially, we may start with pure nitrogen dioxide. In time, however, the percentage of nitrogen dioxide decreases and the percentage of dinitrogen tetroxide increases. This shift in amounts present continues to a point where the two reactions balance each other and the percentages of NO_2 and N_2O_4 settle to constant values (at room temperature, about 31% NO_2 and 69% N_2O_4). At this point, the forward and reverse reactions have achieved what is called **chemical equilibrium**, where the rate at which products are formed is equal to the rate at which they are converted back into reactants (Figure F.1).

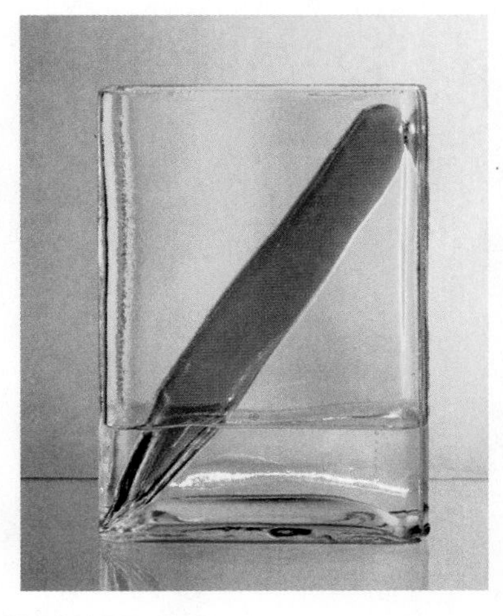

FIGURE F.1
The NO_2 and N_2O_4 in this ampoule are in equilibrium at 16% NO_2 and 84% N_2O_4. Although the nonchanging brown color seems to indicate no activity, individual molecules are continuously converting back and forth with a balance between forward and reverse reactions.

Chemists do not usually express the balance of reactants and products at equilibrium in terms of percentages. Rather, they use equilibrium constants. An equilibrium constant is a number that relates the amounts of reactants to products at equilibrium. For example, consider the reaction in which *a* moles of reactant A and b moles of reactant B react to give c moles of product C and d moles of product D at equilibrium:

$$aA + bB \rightleftharpoons cC + dD$$

Because the reaction is at equilibrium, there is no net change in the amounts of chemical reactants and products—A, B, C, and D—at any time. Therefore, we can use a number—an equilibrium constant, *Keq*—with the following mathematical form to express the relative concentrations of reactants and products at equilibrium:

$$Keq = \frac{[C]^c \times [D]^d}{[A]^a \times [B]^b}$$

Equilibrium constants provide valuable information in a shorthand format. Can you see that, since an equilibrium constant is a ratio of products to reactants, a value of *Keq* greater than 1 means products are favored over reactants? Can you also see that if *Keq* is less than 1, reactants are favored at equilibrium?

Now, to help with your understanding of chemical equilibrium, let's go back and use an analogy to develop this concept. Chemical equilibrium is like a store in which there is a steady flow of customer traffic. Think of people inside the store as product molecules and those outside as reactant molecules. The steady customer flow means that the number of people entering the store in any given time interval equals the number of people leaving in the time interval. As a result, the number of people inside the store never changes. The individual people in the store change as Ms. X and Mr. Y enter and Mr. A and Ms. B leave, but the total number inside remains constant. In the same way, the number of N_2O_4 molecules in the ampoule of Figure F.1 stays constant because the rate at which N_2O_4 molecules convert to NO_2 molecules equals the rate at which N_2O_4 molecules are formed from NO_2 molecules.

Chemical equilibrium does not mean that the amounts of reactants and products are equal. Just as the equilibrium point in our store analogy might be 10 people in the store and 100 on the sidewalk, the NO_2/N_2O_4 reaction reaches an equilibrium point when the NO_2:N_2O_4 ratio is 31:69. The important thing to remember is that chemical equilibrium is a dynamic state. Although the concentrations of products and reactants remain constant, individual molecules are continuously changing back and forth.

If the store holds a big sale, the equilibrium of customers is affected. At the beginning of the sale, the rate at which people enter the store will be greater than the rate at which they leave. Hence the number of people inside the store increases. There is a limit to the number of customers that the store can hold, however. Eventually, a doorman is hired to allow people to enter only at the rate people leave. Thus, a new equilibrium is established. In this case, however, there are now more people in the store than there were before the sale.

Likewise, we find that chemical equilibria are affected by prevailing conditions. The formal statement of this is known as **LeChatelier's principle:**

If a stress is applied to a system in a dynamic equilibrium, the system changes to relieve the stress.

For example, when dinitrogen tetroxide is exposed to energy, it tends to form nitrogen dioxide. Increasing the temperature of a flask of nitrogen dioxide and dinitrogen tetroxide increases the rate of nitrogen dioxide formation until a new equilibrium is reached. At this higher-temperature equilibrium, nitrogen dioxide predominates and the overall color of the reaction mixture is dark brown. This is one of the reasons air pollution becomes particularly noticeable on warm days; the higher the temperature, the greater the concentration of brown nitrogen dioxide (Figure F.2).

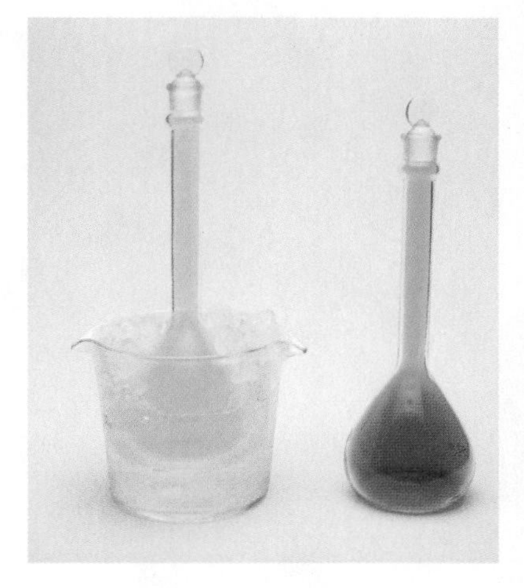

FIGURE F.2

Higher temperatures favor the formation of brown nitrogen dioxide, and the mixture in the flask is dark brown (right). Lower temperatures favor the formation of colorless dinitrogen tetroxide, and the mixture is a lighter brown (left).

CHECK YOURSELF

After a mixture of nitrogen dioxide and dinitrogen tetroxide initially at room temperature is brought to 100°C and kept at that temperature for a few minutes, which is greater—the rate of nitrogen dioxide formation or the rate of dinitrogen tetroxide formation?

CHECK YOUR ANSWER

Neither! After a few minutes a new equilibrium is achieved. It's like the store holding a sale. Initially, equilibrium is lost as more customers pile in. Eventually, however, the store becomes so packed that the rate at which people come in matches the rate at which they go out. Similarly, with the application of heat, the rate of NO_2 formation will initially be greater than the rate of N_2O_4 formation. As the N_2O_4 becomes depleted, however, the rate of NO_2 formation decreases to the point that the forward and reverse reactions become matched as they were before.

Pressure as well as temperature can affect the chemical equilibrium of reacting gases. As we saw in Appendix E, we can increase the pressure of a gas by decreasing the volume of its container. This brings the gaseous molecules closer together, which has the effect of increasing the number of collisions among the molecules (Figure F.3).

Take a container filled with an equilibrium mixture of NO_2 gas and N_2O_4 gas and squeeze it to a smaller volume. The rate at which the molecules collide with one another increases. This is advantageous for the formation of N_2O_4, which requires the coming together (collision) of two NO_2 molecules. Under higher pressures, therefore, the rate of N_2O_4 formation becomes greater than the rate of NO_2 formation and the gaseous mixture becomes lighter in color, as shown in Figure F.2. As more and more N_2O_4 is formed at high pressure, however, the supply of NO_2 dwindles, reducing the number of NO_2 collisions, which decreases the rate of N_2O_4 formation. Eventually a new equilibrium is reached where the rates of NO_2 formation and N_2O_4 formation are once again the same. The relative concentration of N_2O_4 at this new equilibrium, however, remains greater than it is at lower pressures. At twice the pressure, for example, the ratio of NO_2:N_2O_4 changes to 24:76 (Figure F.4).

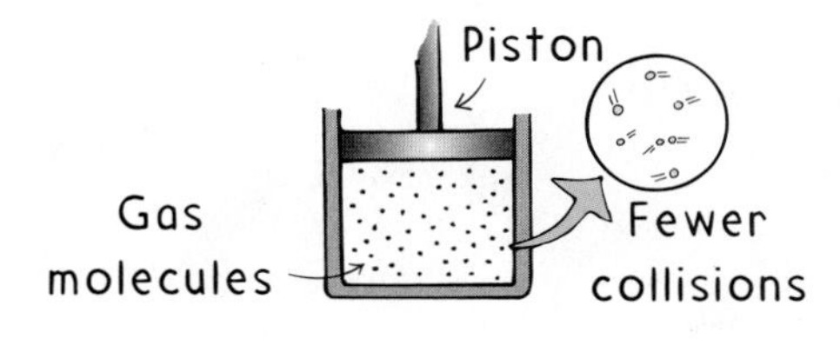

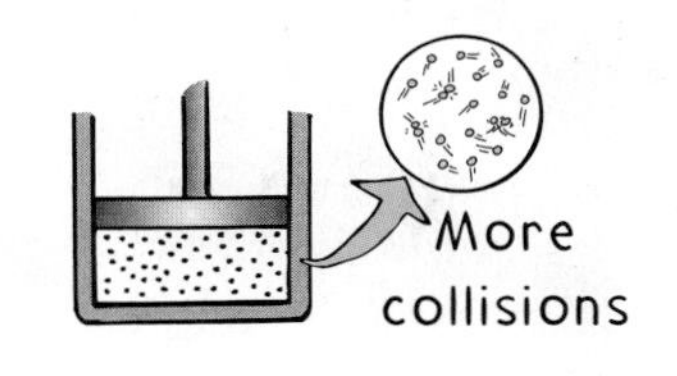

FIGURE F.3

Gas particles at higher pressures are closer together, which increases the rate at which they collide with one another.

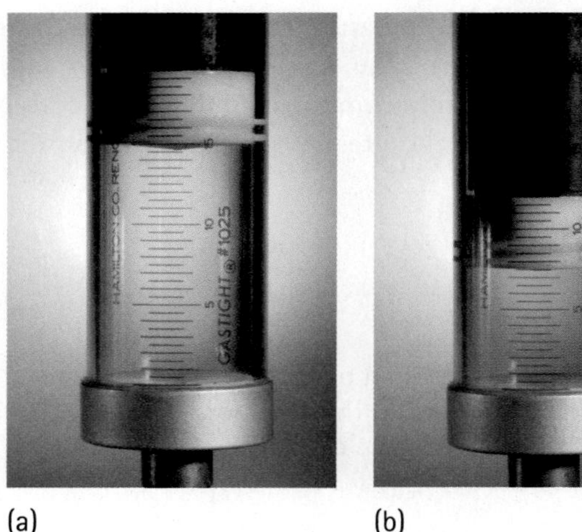

(a) (b) (c)

FIGURE F.4

(a) A syringe containing an equilibrium mixture of nitrogen dioxide and dinitrogen tetroxide. (b) Compressing the gas initially shows a deepening of the brown color as brown NO_2 molecules are bunched together. (c) The higher proportion of NO_2 molecules, however, favors a greater number of collisions among these molecules, which upon colliding convert back to the colorless N_2O_4 molecules. This is evidenced by a rapid lightening of the brown color soon after the NO_2 molecules have been compressed.

CHECK YOURSELF

Would an increase in pressure favor the formation of gaseous sulfur trioxide, SO_3, from gaseous sulfur dioxide, SO_2, and oxygen, O_2?

$$2\ SO_2(g) + O_2(g) \rightleftharpoons 2\ SO_3$$

CHECK YOUR ANSWER

Yes. An increase in pressure favors a greater number of collisions, which has a greater benefit for the formation of SO_3. Incidentally, SO_2 is an industrial pollutant. In the atmosphere it transforms into SO_3, as shown above. The SO_3 then reacts with water to form sulfuric acid, H_2SO_4—a prime component of acid rain.

Odd-Numbered Solutions

Chapter 1

1. The era of modern science was launched when Galileo revived the Copernican view of the heliocentric universe, using experiments to study nature's behavior. 3. We mean that it must be capable of being proved wrong. 5. Galileo showed the falseness of Aristotle's claim with a single experiment—dropping heavy and light objects from the Leaning Tower of Pisa. 7. In everyday speech, a theory is the same as a hypothesis—a statement that hasn't been tested. In science a theory is a synthesis of a large body of verified information. 9. The term *supernatural* literally means "above nature." But science works within nature, not above it. 11. Science is concerned with gathering knowledge and organizing it. Technology enables humans to use that knowledge for practical purposes, and it provides the instruments scientists need to conduct their investigations. 13. Biology is more complex than the physical sciences because it involves matter that is alive and therefore engaged in complex biochemical processes. 15. The disciplines of biology and chemistry were needed to understand the behavior of the Antarctic amphipods. 17. McClintock and Baker hypothesized that amphipods carry sea butterflies to produce a chemical that deters a predator of the amphipod. This is a scientific hypothesis because it would be proven wrong if the chemical secreted by sea butterflies were found not to deter amphipod predators. 19. This is an example of a direct proportion. 21. Answers will vary. Sample answer: The more practice at basketball, the fewer misses in making a basket. $P \sim 1/\text{misses}$. 23. Just as the printing press resulted in more people having more information, so it is with the Internet—only more so! 25. Experimentation can show that a hypothesis is wrong; this is less likely with philosophical discussions. Experiments are tests for truth. 27. Religious beliefs, the core of which is faith, are outside the realm of science and therefore cannot be proven wrong. A person's understanding of a scientific concept, based on experimental evidence, can be proven wrong.

Chapter 2

1. Aristotle classified motion into *natural motion* and *violent motion*. 3. Galileo discredited Aristotle's ideas that heavy objects fall faster than light ones and that a force is necessary to maintain motion. 5. The property is inertia. 7. Your weight would be greater on Earth because of its stronger force of gravity. Mass, however, is the same everywhere. 9. One kg would weigh less on the Moon. 11. The net force on the box is 10 N to the right. 13. The force is tension. 15. $\Sigma F = 0$ means that the vector sum of all the forces that act on an object in equilibrium equals zero. The forces cancel. 17. The forces are the same. You actually read the support force from the scale, which is the same as your weight when the scale is stationary. 19. Since the crate slides in equilibrium (constant velocity), we know that the friction must be equal and opposite to our push. That way the forces cancel and the crate slides without changing velocity. 21. Friction is to the left. 23. Speed is distance per time: $v = d/t$. 25. The speedometer shows the instantaneous speed. 27. Acceleration = change in velocity/unit of time; $a = \Delta v/\Delta t$. 29. The unit of time appears once for the unit of velocity and again for the time during which the velocity changes. 31. Synovial fluid is a lubricant. It protects the bones against the wearing effects of friction. The bones rub against the lubricating synovial fluid instead of against each other. 33. Friction is one of the causes of earthquakes because it prevents rock from moving when pushed. As a result, elastic strain builds up in the rock. More about this in Chapter 22. 35. Length of legs and strength of muscles affect an animal's jumping ability. 37. Answers will vary. The speeds are calculated by measuring the distance and the time to cover that distance. 39. Average speed = $(1.0\text{ m})/(0.5\text{ s}) = 2\text{ m/s}$. 41. Acceleration = $(100\text{ km/h})/(10\text{ s}) = 10\text{ km/h·s}$. 43. C, B, A 45. (a) B, A, C, D (b) B, A, C, D 47. (a) 30 N + 20 N = 50 N. (b) 30 N − 20 N = 10 N, in the direction of the 30-N force. 49. From $\Sigma F = 0$, friction equals weight; $mg = (100\text{ kg})(10\text{ m/s}^2) = 1000\text{ N}$ (or 980 N using $g = 9.8\text{ m/s}^2$).

51. $a = \dfrac{\text{change in velocity}}{\text{time interval}} = \dfrac{-100\text{ km/h}}{10\text{ s}} = -9\text{ km/h·s}$. (The vehicle decelerates at 9 km/h·s.)

53.

Time (*in seconds*)	Velocity (*in meters/second*)	Distance (*in meters*)
6	60	180
7	70	245
8	80	320
9	90	405
10	100	500

55. The Leaning Tower experiment discredited the idea that heavy things fall proportionally faster. The incline plane experiments discredited the idea that a force is needed for motion. 57. A dieter loses mass. To lose weight, the person could go to the top of a mountain where the force of gravity is less. But the amount of matter and therefore the mass would be the same. 59. The density of aluminum is $(5.4\text{ g})/(2\text{ cm}^3) = 2.7\text{ g/cm}^3$. Density is a ratio of weight or mass per volume. This ratio is greater for any amount of lead than for any amount of aluminum, so 5 kg of lead has a greater density than 10 g of aluminum. 61. From $\Sigma F = 0$, the upward forces are 400 N, and the downward forces are 250 N + weight of the staging. So the staging must weigh 150 N. 63. Yes, the forces are equal and opposite and cancel to zero, thus putting the person in equilibrium. But the two forces don't make up an action–reaction pair because they both act on the same object. The reaction to the downward pull of gravity (the world pulling down on the person) is the person pulling up on the world. 65. No, because the force of gravity acts on the object. Its motion is undergoing change, as should be evident a moment later. 67. The impact speed is 2 km/h. 69. More than 2 hours, because you cannot maintain an average speed of 60 mi/h without exceeding the speed limit. You begin at zero and end at zero, so even if there's no slowing down along the way, you'll have to exceed 60 mi/h to average 60 mi/h. So the trip will take you more than 2 hours. 71. The distance increases as the square of the time, so each successive distance covered is greater than the preceding distance covered. 73. Acceleration is 10 m/s^2, constant all the way down. (Velocity, however, is 50 m/s at 5 seconds and 100 m/s at 10 seconds.) 75. No, an object cannot be in mechanical equilibrium if only one force acts on it because there would then be a non-zero net force. It would undergo a change in its motion. 77. In the left figure, Harry is supported by two strands of rope that share his weight (like the little girl in Exercise 76). So each strand supports only 250 N, less than the breaking point. The total force upward supplied by ropes equals the weight acting downward, which gives a net force of zero and no acceleration. In the right figure, Harry is supported by only one strand, which for Harry's well-being requires that the tension be 500 N. Since this is greater than the breaking point of the rope, the rope breaks. The net force on Harry is then only his weight, giving him a downward acceleration of g. The sudden return to zero velocity changes his vacation plans. 79. Ball B will finish first because its average speed along the lower part of the track as well as on the downward and upward slopes is greater than the average speed of ball A.

Chapter 3

1. Every object continues in a state of rest, or in a state of motion in a straight line at constant speed, unless acted upon by a net force. 3. The acceleration produced by a net force on an object is directly proportional to the net force, is in the same direction as the net force, and is inversely proportional to the mass of the object. 5. There is no change in acceleration. 7. When in free fall, the ratio of weight to mass is the same for all objects. 9. Air resistance depends on the speed and frontal area of the falling object. 11. Two forces are required. 13. The boxer can't exert any more force on the tissue paper than the tissue paper can exert on the boxer. The tissue paper has insufficient inertia to absorb a great force. 15. The reaction force is the force of the ball against the bat. 17. Each of the equal forces acts on a different mass. 19. Force, velocity, and acceleration are vector quantities. Time, speed, and volume are scalar quantities. 21. The diagonal represents the resultant vector. 23. Yes, each has less magnitude. 25. The amount of air resistance a falling object

encounters depends on the object's surface area. The more air resistance an animal encounters, the slower and more controllable is its fall. 27. In animal locomotion, an animal typically pushes against some medium (the ground, water, or air) that pushes back on it, providing the force the animal needs to accelerate. 29. The force of friction between your back foot and the floor pushes you forward. 31. The spool will roll in the direction of the pull, to the right. (If the string is pulled vertically, the spool will roll to the left. Other angles will also roll the spool to the left, but when pulled horizontally to the right, the spool rolls to the right!) 33. $a = \Delta v/\Delta t = (6.0 \text{ m/s})/(1.2 \text{ s}) = 5.0 \text{ m/s}^2$. 35. $a = F_{net}/m = (10 \text{ N})/(1 \text{ kg}) = 10 \text{ N/kg} = 10 \text{ m/s}^2$. 37. (a) D, A = B = C. (b) A = C, B = D. 39. (a) A = B = C. (b) C, A, B. 41. C, A = B. 43. The given pair of forces produces a net force of 200 N forward, which accelerates the cart. To make the net force zero, an oppositely directed force of 200 N must be exerted on the cart. 45. Acceleration $a = F_{net}/m = (40 \text{ N} - 12 \text{ N})/4 \text{ kg} = 28 \text{ N}/4 \text{ kg} = 7 \text{ m/s}^2$. 47. $F = ma = (100{,}000 \text{ kg})(2 \text{ m/s}^2) = 200{,}000 \text{ N}$. 49. Acceleration $a = F_{net}/m = (100{,}000 \text{ N} - 90{,}000 \text{ N})/30{,}000 \text{ kg} = 0.3 \text{ N/kg} = 0.3 \text{ m/s}^2$. When the air resistance builds up to 100,000 N, the acceleration will be zero and cruising speed is reached. 51. The wall pushes on you with the same force, 30 N. Acceleration $a = F_{net}/m = 30 \text{ N}/60 \text{ kg} = 0.5 \text{ m/s}^2$. 53. Acceleration $a = F_{net}/m = \sqrt{[(3 \text{ N})^2 + (4 \text{ N})^2]}/5 \text{ kg} = 1 \text{ m/s}^2$. 55. By the Pythagorean theorem, $V = \sqrt{[(120 \text{ m/s})^2 + (90 \text{ m/s})^2]} = 150 \text{ m/s}$. 57. Newton's first law is most applicable because the bag in motion tends to continue in motion, which results in a squashed hand. 59. Let Newton's second law guide the answer: $a = F/m$. As m gets less (much of the mass is the fuel), acceleration a increases for a constant force. 61. Only on the hill in the middle does the acceleration along the path decrease with time because the hill becomes less steep as motion progresses. When the hill levels off, acceleration will be zero. On the first hill, acceleration is constant. On the third hill, acceleration increases as the hill becomes steeper. In all three cases, speed increases. 63. Air resistance on the thrown object decreases the net force on it ($mg - R$), making its acceleration less than that of free fall. 65. (a) Two force pairs act: Earth's pull on the apple (action) and the apple's pull on Earth (reaction). Your hand pushes the apple upward (action), and the apple pushes your hand downward (reaction). (b) If air drag can be neglected, one force pair acts: Earth's pull on the apple and the apple's pull on Earth. If air drag counts, then air pushes upward on the apple (action) and the apple pushes downward on air (reaction). 67. The friction on the crate is 200 N, which cancels your 200-N push on the crate to yield the zero net force that accounts for the constant velocity (zero acceleration). Although the friction force is equal and oppositely directed to the applied force, the two do *not* make an action–reaction pair of forces. That's because both forces act on the same object—the crate. The reaction to your push on the crate is the crate's push back on you. The reaction to the friction force of the floor on the crate is the opposite friction force of the crate on the floor. 69. The forces on each are the same in magnitude, and their masses are the same, so their accelerations are the same. Each person slides an equal distance of 6 m to meet at the midpoint. 71. The winning team pushes harder against the ground. The ground then pushes harder on them, producing a net force in their favor. 73. Before the falling body reaches terminal velocity, the weight is greater than the air resistance. After it reaches terminal velocity, both the weight and the air resistance are the same. Then the net force and the acceleration are both zero. 75. The terminal speed attained by the falling cat is the same whether it falls from 50 stories or 20 stories. Once terminal speed is reached, falling extra distance does not affect the speed. (The low terminal velocities of small creatures enable them to fall without harm from heights that would kill larger creatures.) 77. A hammock stretched tightly has more tension in the supporting ropes than one that sags. The tightly stretched ropes are more likely to break. 79. (a) See the drawing. (b) Yes. (c) The stone is in equilibrium.

81. The force-vector diagram would be the same except that the upward vector would be absent. Only the downward mg vector acts.

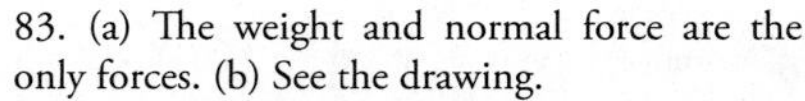

83. (a) The weight and normal force are the only forces. (b) See the drawing.

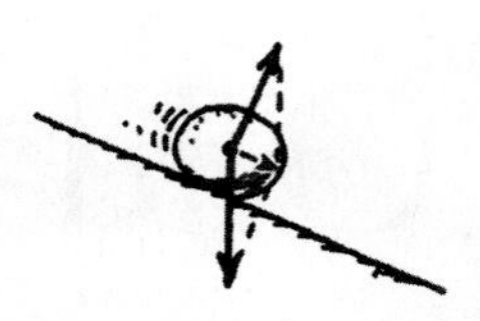

85. Newton's first law applies again: When you jump down on your feet, you compress the disks in your spine. At night, while you are lying prone, relaxation undoes the compression of the disks that has occurred during the day, and you're a little taller in the morning! 87. The sudden stop involves a large acceleration. So, in agreement with $a = F/m$, a large a means a large F. Ouch! 89. No, because in action–reaction fashion, the tail also wags the dog. How much depends on the relative masses of the body of the dog and its tail. 91. Air resistance, however small, builds up with speed, so the air resistance of a ball dropped from the Leaning Tower of Pisa is not really negligible for so high a drop. The result is that the heavier ball does strike the ground first. But, although a twice-as-heavy ball strikes the ground first, it falls only a little faster, not twice as fast, which is what followers of Aristotle believed. Galileo recognized that the small difference is due to air friction and would not be present if there were no air friction. 93. The heavier tennis ball will strike the ground first for the same reason the heavier parachutist in Figure 3.10 strikes the ground first. Note that although the air resistance on the heavier ball is smaller relative to the weight of the ball, it is actually greater than the air resistance that acts on the other ball. Why? Because the heavier ball falls faster, and air resistance is greater for greater speed.

Chapter 4

1. A moving skateboard has greater momentum; anything at rest has *no* momentum. 3. Force is a push or pull, whereas impulse is the product of force and time. 5. A cannonball has more momentum coming from the long cannon because the force acts over a longer time. 7. Case (c) requires the greatest impulse—twice the momentum if the speed thrown is the same as the speed caught. 9. It means that the quantity remains unchanged in a process. 11. B has same initial speed as A. 13. Energy is most evident when it undergoes a change—as when it is being transferred or transformed. 15. Work and energy are measured in joules. 17. Four watts are expended. 19. The heavier car has twice the potential energy. 21. The evidence is whether or not the object experiences a change in energy. 23. With no external work input or output, the energy of a system doesn't change. Energy cannot be created or destroyed. 25. Yes. Yes. No! (There is *no* way to increase energy without doing work.) 27. No way! If a machine has an efficiency greater than 100%, then new physics is afoot! It would violate the conservation of energy. 29. According to Newton's third law, the forces are equal. Only the resilience of the human hand and the training Cassy has undergone to toughen her hand allow her to perform this feat without harm or broken bones. 31. The "burning" that goes on in cells differs from the burning or combustion of a log on a campfire in that the cellular process is much slower and more controlled. 33. You impart energy to the sand when you shake it. Its temperature will increase. 35. Momentum $= mv = (10 \text{ kg})(3 \text{ m/s}) = 30 \text{ kg·m/s}$. 37. Impulse $= Ft = (10 \text{ N})(5 \text{ s}) = 50 \text{ N·s}$. 39. $W = Fd = (2 \text{ N})(3 \text{ m}) = 6 \text{ N·m} = 6 \text{ J}$. 41. $P = W/t = Fd/t = (1 \text{ N})(2 \text{ m})/1 \text{ s} = 2 \text{ N·m/s} = 2 \text{ W}$. 43. $\text{PE} = mgh = (1.5 \text{ kg})(10 \text{ N/kg})(2 \text{ m}) = 30 \text{ J}$. 45. $\text{KE} = \frac{1}{2}mv^2 = \frac{1}{2}(1 \text{ kg})(6 \text{ m/s})^2 = 18 \text{ J}$. 47. $W = \Delta\text{KE} = 6000 \text{ J}$. 49. (a) B, A, C. (b) C, B, A. (c) C, B, A. 51. (a) D, B, C, E, A. (b) D, B, C, E, A. (c) A, E, C, B, D. 53. Let m be the mass of the freight car, $4m$ the mass of the diesel engine, and v the speed after they couple together. Before the collision, the total momentum is due to only the diesel engine, $4m(5 \text{ km/h})$, because the momentum of the freight car is 0. After the collision, the combined mass is $(4m + m)$ and the combined momentum is $(4m + m)v$. By the conservation of momentum equation:

$$\text{Momentum}_{\text{before}} = \text{momentum}_{\text{after}}$$

$$4\text{m}(5 \text{ km/h}) + 0 = (4m + m)v$$

$$v = \frac{(20m\cdot\text{km/h})}{5m} = 4 \text{ km/h}$$

(Note that you don't have to know m to solve the problem.) 55. By momentum conservation, asteroid mass × 800 m/s = Superman's mass × v. Since the asteroid's mass is 1000 times Superman's mass, we have $(1000m)(800 \text{ m/s}) = mv$, so $v = 800{,}000 \text{ m/s}$. This is nearly 2 million miles per hour! 57. At three times the speed, the car has nine (3^2) times the KE and will skid nine times as far—135 m. Since the frictional force is about the same in both cases, the distance has to be nine times as great for nine times as much work done by the pavement on the car. 59. From

$W = Fd$, $d = 25\% W/F = (0.25)(40{,}000{,}000 \text{ J/L})/500 \text{ N} = 20{,}000$ m/L. That's 20 km per liter of fuel (which is about 47 mi/gal). 61. (a) From PE = KE, we have $mgh = \frac{1}{2}mv^2$, so $v^2 = 2gh$ and $v = \sqrt{(2gh)}$. (b) Twice v is twice$\sqrt{(2gh)}$, or four times h. Check: $\sqrt{[2g(4h)]} = 2\sqrt{[2gh]}$. 63. Supertankers are so massive that even at modest speeds their motional inertia, or *momentum*, is enormous. This means enormous impulses are needed to change their motion. How can large impulses be produced with modest forces? By applying modest forces over long periods of time. Hence the force of the water resistance over the time it takes to coast 25 km sufficiently reduces the momentum. 65. When the moving egg makes contact with a sagging sheet, the time it takes to stop is extended. More time means less force, so the egg is less likely to be broken. 67. Oops, you overlooked the conservation of momentum. Your momentum forward equals (approximately) the momentum of the recoiling raft. 69. When a cannon with a long barrel is fired, more work is done as the cannonball is pushed through the longer distance. A greater KE is the result of the greater work, so, of course, the cannonball emerges with a greater velocity. (It might be mentioned that the force acting on the cannonball is not constant but decreases with increasing distance inside the barrel. So we think in terms of average force.) 71. If something has KE, then it must have momentum—because it is moving. But something can have potential energy without being in motion and therefore without having momentum. And every object has "energy of being"—stated in the celebrated equation $E = mc^2$. So, whether something moves or not, it has some form of energy. If it has KE, then with respect to the frame of reference in which its KE is measured, it also has momentum. 73. The pro gives 25 times more energy (speed is squared for kinetic energy). 75. The KE of a pendulum bob is at a maximum where it moves fastest, at the lowest point; the PE is at a maximum at the uppermost points. When the pendulum bob swings by the point that marks half its maximum height, it has half its maximum KE and its PE is halfway between its minimum and maximum values. If we define PE = 0 at the bottom of the swing, then the place where KE is half its maximum value is also the place where the PE is half its maximum value, and KE = PE at this point. (In agreement with energy conservation, total energy = KE + PE.) 77. An engine that is 100% efficient would not be warm to the touch, nor would its exhaust heat the surrounding air, nor would it make any noise, nor would it vibrate. This is because all these transfers of energy cannot happen if all the energy given to the engine is transformed to useful work. No fuel would remain unused. 79. When a boxer hits his opponent, the opponent contributes to the impulse that changes the momentum of the punch. When punches miss, no impulse is supplied by the opponent—all the effort that goes into reducing the momentum of the punches is supplied by the boxer himself. This tires the boxer. 81. The momenta of both bug and bus change by the same amount because both the amount of force and the time, and therefore the amount of impulse, are the same on each. Momentum is conserved. Speed is another story. Because of the huge mass of the bus, its reduction of speed is very tiny—too small for the passengers to notice. 83. Whether or not momentum is conserved depends on the system. If the system in question is you as you fall, then there is an external force acting on you (gravity) and your momentum increases and is therefore not conserved. But, if you enlarge the system to be you and Earth, which pulls on you, then momentum *is* conserved because the force of gravity on you is internal to the system. Your momentum of fall is balanced by the equal but opposite momentum of Earth coming up to meet you! 85. Cars brought to a rapid halt by sticking together experience a change in momentum and a corresponding impulse. A greater change in momentum occurs if the cars rebound upon contact, with correspondingly greater impulse and therefore greater damage. Less damage results if the cars stick upon impact than if they rebound. 87. The game ends after the first toss. Here's why: We assume the equal strengths of the astronauts means that each throws with the same speed. Since the masses are equal, when the first throws the second, both the first and second move away from each other at equal speeds. Assume the thrown astronaut moves to the right with velocity V and the first recoils with velocity $-V$. When the third makes the catch, both she and the second move to the right at velocity $V/2$ (twice the mass moving at half the speed, like the freight cars in Figure 4.11). When the third makes her throw, she recoils at velocity V (the same speed she imparts to the thrown astronaut), which is added to the $V/2$ she acquired in the catch. So her velocity is $V + V/2 = 3V/2$, to the right—too fast to stay in the game. Why? Because the velocity of the second astronaut is $V/2 - V = -V/2$, to the left—too slow to catch up with the first astronaut who is still moving at $-V$. The game is over. Both the first and the third astronauts got to throw the second astronaut only once! 89. The design is impractical. Note that the summit of each hill on the roller coaster is the same height, so the PE of the car at the top of each hill would be the same. If no energy were spent in overcoming friction, the car would get to the second summit with as much energy as it starts with. But in practice there is considerable friction, and the car would not roll to its initial height and have the same energy. So the maximum height of succeeding summits should be lower to compensate for friction. 91. Both balls will have the same speed because both have the same PE at the ends of the track—and therefore the same KE. This is a relatively easy question to answer because it asks for *speed*, whereas the similar question in Chapter 2 asked for which ball got to the end sooner. That question asked for *time*—which meant first establishing which ball had the greater average speed. 93. Momentum is conserved, but kinetic energy is not. If one ball pops out with twice the speed, it has four times as much kinetic energy. So the reason the two-balls/one-ball-at-twice-the-speed scenario never happens is that energy conservation would be violated. Nature doesn't operate that way!

Chapter 5

1. Newton realized that both a falling apple and the Moon are under the influence of Earth's gravity. 3. The Moon falls beneath the straight line it would follow if no forces acted on it. 5. The force is one-fourth as much, in accord with the inverse-square law. 7. Nowhere! The gravitational force approaches zero at great distances but never actually reaches zero. (At Earth's center, however, gravitation *cancels* to zero!) 9. The gravitational force is 9.8 N (or, rounded off, 10 N). 11. Oops, almost a trick question: The scale readings would be unchanged. Readings change only with *changes* in velocity—acceleration. 13. Your weight is greater than mg when you're in an elevator that accelerates upward (or a rotating habitat with excess rotation). 15. The direction of the force is inward, toward the center of the circle. 17. A projectile is any object projected by some means that continues in motion by its own inertia. 19. The same, $\frac{1}{2}gt^2$, which is 5 m in the first second. 21. The other angle of launch is 15° (the complementary angle to 75°). 23. Both height and range are less. 25. The speed of 8 km/s will persist only in the absence of air drag. Hence the motion must be above Earth's atmosphere. 27. The gravitational force varies because the distance between gravitationally attracting masses varies. 29. The sense organs that allow people to sense gravity are called the vestibular organs and are located in the inner ear. 31. The tower's center of gravity lies within its base. 33. The minimum speed is 8 km/s, and the maximum speed is 11.2 km/s. An object projected from Earth at a speed greater than 11.2 km/s will escape Earth's gravitational influence and no longer orbit Earth. 35. For most observers, the two hands look much the same size, with maybe the nearer hand looking slightly bigger. Seen through the eyes of the inverse-square law, the farther hand is twice as tall and twice as wide—that's four times the area! We see through knowledge! 37. With your back to the wall, bending over moves your center of gravity beyond your toes—beyond your support base—and you topple. Doing this without the wall "in the way" is successful because you move your lower body backward so that your center of gravity is above your support base. 39. The water *does* fall into the swinging bucket, just as satellites in Earth orbit fall as they circle Earth. In the case of the water bucket, the bucket falls at least as fast as the water falls. In the case of the satellite, Earth curves away at the same distance the satellite drops beneath each straight-line segment.

41. $F = G\frac{m_1 m_2}{d^2} = 6.67 \times 10^{-11} \text{ N·m}^2/\text{kg}^2 \times \frac{(1 \text{ kg})(6 \times 10^{24} \text{ kg})}{[2(6.4 \times 10^6 \text{ m})]^2} = 2.5 \text{ N}.$

43. $F = G\frac{m_1 m_2}{d^2} = 6.67 \times 10^{-11} \text{ N·m}^2/\text{kg}^2 \times \frac{(6.0 \times 10^{24} \text{ kg})(2.0 \times 10^{30} \text{ kg})}{(1.5 \times 10^{11} \text{ m})^2} = 3.5 \times 10^{22} \text{ N}$

45. $F = G\frac{m_1 m_2}{d^2} = 6.67 \times 10^{-11} \text{ N·m}^2/\text{kg}^2 \times \frac{(3.0 \text{ kg})(100 \text{ kg})}{(0.5 \text{ m})^2} =$ 8.0×10^{-8} N. So the obstetrician exerts twice the force of gravity on the baby! 45. C, B, A 47. (a) A, B, C. (b) C, B, A. 49. From $F = G\frac{m_1 m_2}{d^2}$,

three times d squared is $9d^2$, which means the force is one-ninth of the surface weight. 51. From $F = G\frac{m_1 m_2}{d^2}$, $10d$ squared is $100d^2$, with a force 100 times smaller.

53. From $F = G\frac{m_1 m_2}{d^2}$, five times d squared is $(1/25)d^2$, with a force 25 times greater.

55. From $F = G\frac{2m_1 2m_2}{(2d)^2} = G\frac{4m_1 m_2}{4d^2} = G\frac{m_1 m_2}{d^2}$, with the same force of gravity.

57. $F = G\frac{m_1 m_2}{d^2} = [(6.67 \times 10^{-11}\ \text{N·m}^2/\text{kg}^2) \times$ $(6 \times 10^{24}\ \text{kg})(2 \times 10^{30}\ \text{kg})]/(1.5 \times 10^{11}\ \text{m})^2 = 3.6 \times 10^{22}\ \text{N}$.

59. In accord with the law of inertia, the Moon would move in a straight-line path instead of circling both the Sun and Earth. 61. The force of gravity is the same in both cases because the masses are the same, as Newton's equation for gravitational force verifies. When dropped, the crumpled paper will fall faster only because it will encounter less air drag than the sheet. 63. Gravitational force is indeed acting on a person who falls off a cliff and on a person inside an orbiting space vehicle. Both are falling under the influence of gravity. 65. Yes, the planets are falling around the Sun just as Earth satellites fall around Earth. That's what orbital motion is—falling around a host body. 67. The crate will not hit the car but will crash a distance beyond it that is determined by the height and speed of the plane. 69. There are no forces acting horizontally (neglecting air resistance), so there is no horizontal acceleration; hence, the horizontal component of velocity doesn't change. Gravitation acts vertically, which is why the vertical component of velocity changes. 71. Both balls have the same range (see Figure 5.28). The ball with the initial angle of 30°, however, is in the air for a shorter time and hits the ground first. 73. The initial vertical climb gets the rocket through the denser, retarding part of the atmosphere most quickly. Eventually the rocket must acquire enough tangential speed to remain in orbit without thrust, so it must tilt until finally its path is horizontal. 75. You are momentarily weightless because no support force acts on you. Gravity is acting on you all the time—before, during, and at the end of your jump. How fast you fall also depends on the air drag you encounter. People may be confused because they think motion cancels gravity. Not so! 77. Your friend is confusing the d in the gravity equation. It's the distance between you and the center of Earth, not the distance to the ground floor. The force on you on any floor is, for all practical purposes, very nearly the same. 79. The dart will fall beneath the projected line of the barrel. If the dart leaves at the moment the monkey begins its fall, the monkey will be hit because both objects fall the same vertical distance in the same time. 81. Only when the gravitational force on the cannonball is perpendicular to the ball's velocity will it not increase its speed. In that case there is no component of force in the direction of the moving cannonball. If the cannonball is fired slower, then there is a force and the speed increases. 83. Astronauts in orbit are being pulled by everything in the universe, as they are when at Earth's surface. But these pulls are negligible. Orbit is achieved when the speed is great enough that the vehicle falls around the curved Earth rather than into it.

Chapter 6

1. Thermal energy consists of the translational, vibrational, and rotational motions of particles. 3. The temperatures for freezing water are 0°C and 32°F; the temperatures for boiling water are 100°C and 212°F. 5. Both the thermometer and whatever it measures reach a common temperature (thermal equilibrium). Hence a thermometer measures its own temperature. 7. Zero energy can be taken from the system. 9. (a) Temperature is a measure of the average translational kinetic energy per molecule. (b) Heat is the thermal energy transferred from one thing to another due to a temperature difference. 11. Both calorie and Calorie are units of energy; 4.19 J = 1 cal. 13. When mechanical work is done on a system, its thermal energy increases and its temperature increases. 15. Silver, which has a lower specific heat, warms up faster. 17. The specific heat capacity of water is much higher. 19. Ice has open-structured crystals. So, when water freezes, the crystals occupy more space, and more volume means lower density. 21. Loose electrons conduct energy by collisions throughout a substance. 23. Wood is a poor conductor, even when it is red hot. So very little heat conducts from the coal to your feet. 25. Convection transfers heat by currents of heated fluid. 27. During the day the shore is warmed more than the water, so winds blow from water toward shore. At night the reverse occurs: The shore cools more than the water, and winds blow in the opposite direction. 29. The frequency of radiant energy is directly proportional to the absolute temperature of the source. 31. When energy is transformed, some of it is converted into heat. Heat is the least useful form of energy because it is the least concentrated form. 33. No, the reaction is not thermodynamically favored because the products have less entropy than the reactants. This reaction will not proceed without an energy input. 35. North Atlantic water cools and releases energy into the air, which blows over Europe. 37. In winter months, when the water is warmer than the air, the air is warmed by the water to produce a warmer climate of nearby land. In summer months, when the air is warmer than the water, the air is cooled by the water to produce a cooler climate of nearby land. This is why seacoast communities and especially islands do not experience the high and low temperature extremes that characterize inland locations. 39. Paper's poor conductivity is nicely shown. The high conductivity of the iron soaks up the thermal energy of the flame and prevents the ignition of the paper. 41. $Q = cm\Delta T = (1\ \text{cal/g·°C})(50\ \text{g})(100°\text{C}) = 5000\ \text{cal}$. 43. b, a, c 45. a, c, b 47. The work that the hammer does on the nail is given by Fd, and the temperature change of the nail can be found from $Q = cm\Delta T$. First, we get everything into more convenient units for calculating: 5 g = 0.005 kg and 6 cm = 0.06 m. Then $Fd = (500\ \text{N})(0.06\ \text{m}) = 30\ \text{J}$, and $30\ \text{J} = (0.005\ \text{kg})(450\ \text{J/kg·°C})(\Delta T)$, which we can solve to get $\Delta T = 30/[(0.005)(450)] = 13.3°\text{C}$. (You will notice a similar effect when you remove a nail from a piece of wood. The nail that you pull out is noticeably warm.) 49. Raising the temperature of 10 g of copper by 1°C takes (10)(0.092) = 0.92 calorie, and raising it through 100°C takes 100 times as much, or 92 calories. By formula, $Q = cm\Delta T = (0.092\ \text{cal/g·°C})\ (10\ \text{g})(100°\text{C}) = 92\ \text{cal}$. Heating 10 g of water through the same temperature difference would take 1000 calories, more than ten times more than for the copper—another reminder that water has a large specific heat capacity. 51. We have $0.5mgh = cm\Delta T$, so $\Delta T = 0.5mgh/cm = 0.5gh/c = (0.5)(9.8\ \text{m/s}^2)(100\ \text{m})/(450\ \text{J/kg}) = 1.1°\text{C}$. Note that the mass cancels, so the same temperature would hold for a ball of any mass, assuming half the heat generated goes into warming the ball. The units check because $1\ \text{J/kg} = 1\ \text{m}^2/\text{s}^2$. 53. The hot coffee has a higher temperature, but not a greater amount of thermal energy. Although the iceberg has less thermal energy per mass, its enormously greater mass gives it a greater total energy than that in the small cup of coffee. (For a smaller volume of ice, the fewer more energetic molecules in the hot cup of coffee may constitute a greater total amount of internal energy—but not compared to an iceberg.) 55. The gas pressure increases when the can is heated and decreases when it is cooled. The pressure that a gas exerts depends on the average kinetic energy of its molecules and therefore on its temperature. 57. Different substances have different thermal properties due to differences in the way energy is stored internally in the substances. When the same amount of heat produces different changes in temperatures in two substances of the same mass, we say that they have different specific heat capacities. Each substance has its own characteristic specific heat capacity. Temperature measures the average kinetic energy of random motion but not other kinds of energy. 59. Water is an exception. At temperatures below 4°C, water expands when cooled. 61. Every part of a metal ring expands when it is heated—not only the thickness but also the outer and inner circumferences. Hence the ball that normally passes through the hole when the temperatures are equal will more easily pass through the expanded hole when the ring is heated. (It is interesting that the hole will expand as much as a disk of the same metal undergoing the same increase in temperature. Blacksmiths mounted metal rims on wooden wagon wheels by first heating the rims. When the rims cooled, the contraction resulted in a snug fit.) 63. The gap in the ring will become wider when the ring is heated. Try this: Draw a couple of lines on a ring where you pretend a gap is. When you heat the ring, the lines will be farther apart—the same amount as if a real gap were there. Every part of the ring expands proportionally when heated uniformly—thickness, length, gap, and all. 65. No, the coat is not a source of heat; it merely keeps the thermal energy of the wearer from leaving rapidly. 67. A good reflector is a poor radiator of heat, and a poor reflector is a good radiator of heat. 69. Inanimate things such as tables, chairs, and furniture have the same temperature as the surrounding air (assuming they are in thermal equilibrium with the air—that is, no sudden gush of different-temperature air or something else). People and other mammals, however, generate their own heat and have body temperatures that are normally higher than air temperature. 71. Only a small percentage of the electrical energy that goes into lighting a lamp becomes light. The rest is thermal energy.

But even the light is absorbed by the surroundings and also ends up as thermal energy. So, by the first law, all the electrical energy is ultimately converted into thermal energy. By the second law, organized electrical energy degenerates to the more disorganized form, thermal energy. 73. The brick would cool off too fast and you'd be cold in the middle of the night. A jug of hot water, with its higher specific heat, would keep you warm through the night. 75. Water has the greatest density at 4°C; therefore, either cooling or heating at this temperature will result in an expansion of the water. A small rise in the water level at 4°C would be ambiguous and make a water thermometer impractical in this temperature region. 77. Air is an excellent insulator. Fiberglass is a good insulator principally because of the vast amount of air spaces trapped in it. 79. Air is a poor conductor, whatever the temperature. So, holding your hand in hot air for a short time is not harmful because the air conducts very little heat to your hand. If you touch the hot conducting surface of the oven, however, heat readily conducts to your hand—ouch! 81. When water vapor in the can cools and condenses, vapor pressure within the can is greatly decreased. So condensation inside greatly lowers the vapor pressure inside. The greater pressure of the atmosphere on the outside crushes it.

Chapter 7

1. The proton has a positive charge, and the electron has a negative charge. 3. Both are inverse-square laws; gravity only attracts, whereas electric charges both attract and repel. 5. Gravitational field and electric field (and magnetic field) are common force fields. 7. Electric potential energy is energy, measured in joules; electric potential is energy per charge, measured in volts. 9. A conductor freely conducts, an insulator resists conduction, and semiconductor has properties of both a conductor and an insulator. 11. A temperature difference is necessary for heat energy to flow. An electric potential difference is necessary for electric charge to flow. 13. Electrons are loosely attached to atomic nuclei; protons are locked in the atomic nucleus. 15. A thin wire has more resistance. A long wire has more resistance. 17. The current is doubled when the voltage is doubled. No change in current occurs if both voltage and resistance are doubled. 19. Parallel is better because a short-circuit in one branch doesn't affect the other branches. 21. Electric power = current × voltage. 23. In both interactions, opposites attract and likes repel. 25. A magnetic field is produced by moving electrons. 27. The direction of force on a moving charged particle is at right angles to its velocity and at right angles to the magnetic field. 29. Voltage can be induced by moving a wire loop into a magnetic field, moving the field near the loop of wire, or changing the current in the loop of wire. 31. High voltage across different parts of a body produces shock. 33. When a current passes from one hand to the other, the heart is in the route—bad! A current that flows from the hand to the elbow is not so dangerous. 35. Convection currents of electric charge in Earth's interior likely produce its magnetic field. 37. This activity nicely illustrates a simple magnet.

39. $F = k\dfrac{q_1q_2}{d^2} = (9 \times 10^9\ \text{N·m}^2/\text{C}^2)\dfrac{(1\ \text{C})(1\ \text{C})}{0.1\ \text{m}^2}$
$= 9 \times 10^9$ N.

41. $I = \dfrac{V}{R} = \dfrac{6\ \text{V}}{1000\ \Omega} = 0.006$ A.

43. $P = IV = (0.5\ \text{A})(120\ \text{V}) = 60$ W. 45. A, C, B 47. A = B = C

49. Electric field is force divided by charge: $E = \dfrac{F}{q} = \dfrac{3.2 \times 10^{-4}\ \text{N}}{1.6 \times 10^{-10}\ \text{C}}$
$= 2 \times 10^6$ N/C. (The unit N/C is the same as the unit V/m, so the field can be expressed as 2 million volts per meter.)

51. From $I = V/R$, we find $R = V/I = (120\ \text{V})/(20\ \text{A}) = 6\ \Omega$. 53. First, 100 W = 0.1 kW. Second, there are 168 hours in one week (7 days × 24 hours/day = 168 hours). So 168 hours × 0.1 kilowatt = 16.8 kilowatt-hours, which at $0.20 per kWh comes to $3.36. 55. Gravitational force and electrical force both vary in accord with the inverse-square law. They differ in that gravitational forces between masses are always attractive, whereas electrical forces between charges can both attract and repel. 57. The electrons in a penny are attracted to the protons in the atoms of the penny. 59. At twice the distance, the force is one-quarter as great, in accord with the inverse-square law. 61. This process illustrates the attraction of the paint particles via charge polarization. 63. Auto headlights are wired in parallel, so that when one burns out the other remains functional. 65. The resistors should be connected in series. 67. Large wingspans can span parallel wires on power poles and short them out. Ouch! 69. The clustering of iron atoms into magnetic domains gives rise to magnets. If iron atoms are not aligned in domains, an iron object is not a magnet. 71. Newton's third law applies here: The paper clip and the magnet exert equal forces on each other. 73. The same sign of charge is on both leaves, so they repel each other. 75. Here we must distinguish between the ratio of a quantity and the quantity of energy. Although the temperature (energy per molecule) and voltage (energy per charge) are high, the energies themselves may be quite low. A small amount of energy on a very, very small molecule or charge is not harmful, even though the ratios are high. 77. Only circuit 5 lights the bulb. Note that the other circuits do not connect the bottom of the battery (negative terminal) to one end of the bulb filament and the top of the battery (positive terminal) doesn't connect to the opposite end of the bulb filament. (Look at Figure 7.17 to see where connections should be made.) 79. The physics of both is similar. In one case we have charge inducing charge; in the other we have magnets inducing magnetization in neighboring iron nails. 81. The aluminum ring experiences a force due to magnetic induction because of a sudden magnetic field produced by a hidden coil. 83. CFLs are more efficient because they do not glow via incandescence but by excited gases. Hence, more energy input becomes light rather than heat. 85. Agree! A battery is a voltage source, which gives energy to electrons in the circuit.

Chapter 8

1. Period is the time for a complete cycle to occur; amplitude is the maximum displacement from the midpoint of a wave to the crest (or trough); wavelength is the distance between successive crests, troughs, or identical parts of a wave; and frequency is the number of vibrations per unit time. 3. Energy 5. Wave speed = frequency × wavelength. 7. A compression is a region of compressed air, while a rarefaction is a region of rarefied air. Both travel in the same direction, however. 9. No, sound cannot travel because in a vacuum there is no medium to vibrate. 11. Forced vibrations occur when a surface is forced to vibrate. When the forcing frequency matches the natural frequency of the surface, the result is resonance. 13. The frequency of the wave is the same as the frequency of the vibrations that produce it. 15. Yes, but the law holds only on small segments of a surface. 17. (a) The energy of ultraviolet light is absorbed. (b) The energy of visible light cascades from atom to atom in a transparent medium. 19. Low frequencies appear red, high frequencies appear violet, and intermediate frequencies make up the intermediate colors of the rainbow. 21. The slowing of light causes the bending in refraction. 23. Although a single drop disperses all the colors, a viewer is in a position to view only one color per drop. 25. Diffraction is more pronounced for shorter waves. 27. All kinds of waves exhibit interference. 29. Waves are compressed when coming from an approaching source, which means they have a higher frequency. 31. Photons of violet light are more successful at dislodging electrons because the higher frequency of violet has more energy per photon. 33. The higher the frequency of sound, the higher the pitch. 35. Overlapping of the three colors we see produces the wide spectrum we observe. 37. The Doppler effect is evident when the buzzer changes motion relative to you. When its motion relative to you doesn't change, no change in frequency is heard. 39. The sound heard is truly impressive, like chimes in an auditorium. 41. The pinhole mirror works because light travels in straight lines. This is a must-do activity!

43. Frequency $= \dfrac{1}{\text{period}} = \dfrac{1}{3\ \text{s}} = \dfrac{1}{3}$ Hz.
45. Wave speed $= f\lambda = (1.5/\text{s})(3\ \text{m}) = (1.5\ \text{Hz})(3\ \text{m}) = 4.5$ m/s.
47. $v = f\lambda = (256\ \text{Hz})(1.33\ \text{m}) = (256/\text{s})(1.33\ \text{m}) = 340$ m/s.
49. A, C, B 51. B, C, A 53. (a) $f = 1/T = 1/0.10\ s = 10$ Hz; (b) $f = 1/T = 1/5\ \text{s} = 0.2$ Hz; (c) $f = 1/T = 1/(1/60\ \text{s}) = 60$ Hz.

55. From $c = f\lambda$, $\lambda = \dfrac{c}{f} = \dfrac{3.00 \times 10^8 \text{m/s}}{6 \times 10^{14}\ \text{Hz}} = 5 \times 10^{-7}$ m, or 500 nm.
This is 5000 times larger than the size of an atom, which is about 0.1 nm. (The nanometer is a common unit of length in atomic and optical physics.)

57. (a) Frequency $= \dfrac{\text{speed}}{\text{wavelength}} = \dfrac{3.00 \times 10^8\ \text{m/s}}{0.03\ \text{m}} =$
1.0×10^{10} Hz = 10 GHz. (b) Distance = speed × time, so time =
$\dfrac{\text{distance}}{\text{speed}} = \dfrac{10{,}000\ \text{m}}{3.00 \times 10^8\ \text{m/s}} = 3.3 \times 10^{-5}$ s.

59. For a transverse wave, shake the horizontal Slinky up and down; for a longitudinal wave, shake it back and forth. 61. Light travels faster than sound—hence the time delay. 63. The fundamental source is vibrating electrons. 65. Radio waves are electromagnetic waves and travel at speed *c*. 67. UV light cannot penetrate glass, but it can penetrate clouds. 69. Yellow-green is the color our eyes are most sensitive to, which is why fire engines are often yellow-green in color rather than the traditional red. 71. Cowboy Joe should aim at the assailant. The bullet will ricochet as light does from a shiny surface. 73. The two observers see rainbows slightly displaced from each other, the amount of which depends on how far apart the observers are from each other. 75. The shadow of the airplane will be at the center of the circular rainbow. 77. Longer waves undergo more diffraction than shorter waves. 79. Overlapping of rays doesn't occur for small pinholes, so sharpness is the same at any distance. If the distance is so great that some overlapping does occur, then the pinhole should be smaller. 81. From $v = f\lambda$, since frequency doesn't change, a lower speed means a shorter wavelength. The speed is lower in water than in air, so wavelengths are shorter in water. 83. (a) The frequency increases as the locomotive increases speed. (b) The wavelength of the sound correspondingly decreases. (c) The speed of sound in the air, however, doesn't change. 85. Aim the spear below where the fish appears (to counter refraction). In the case of light, however, aim directly at the fish because the laser light will correspondingly refract the same. 87. The distance to the origin is the difference between the speeds of the waves multiplied by their times. If a second location records corresponding differences, then "triangulation" can occur as the two paths intersect at the origin of the disturbance. 89. Electrons dislodged by ultraviolet light produce a current. Any measure of current activates an electric eye, and the relative amounts of current are useful in light meters and sound tracks.

Chapter 9

1. An element consists of only one type of atom. 3. As shown in the periodic table, the atomic symbol for cobalt is Co. 5. The atomic number is the number of protons in the nucleus. The mass number is the total number of protons and neutrons in the nucleus. 7. Most of the known elements are metallic. 9. Across any period, the properties of the elements gradually change. 11. Atoms give off light as they are subjected to energy. 13. A beam of light consists of zillions of small, discrete packets of energy. Each packet is called a quantum. 15. No, Bohr's model merely illustrated the different energy levels of an electron in an atom. 17. The electron microscope is the practical application. 19. The noble gas shell model explains the organization of the elements in the periodic table. From the shell model, we can understand how the atoms of elements get larger as we move down a group and how they get smaller as we move from left to right across a period. 21. The maximum number of electrons each shell can hold corresponds to the number of elements in the period. 23. No, atoms are smaller than the wavelengths of visible light, so they cannot be seen using any visible light microscope. 25. The atoms in the baseball would be the size of Ping-Pong balls if the baseball were the size of Earth. 27. A physical model attempts to replicate an object on a different scale. A conceptual model describes a system. 29. This is a demonstration of quantization. 31. c, b, a 33. a, b, c 35. a, b, c. Sodium loses one electron, magnesium loses two electrons, and aluminum loses three. The reason for this can be seen with the shell model. Note that sodium has only one electron in its outermost shell. Magnesium has two electrons, and aluminum has three. So why can't sodium lose a second electron? If it did, this second electron would have to come from the second shell. These electrons in the second shell are closer to the nucleus, which means the electrical force that holds them is stronger. This is Coulomb's law, which tells us that the electrical force is stronger over shorter distances. 37. The class average is 80%. One person scoring 100% would slightly raise this average to 81%. Similarly, the atomic mass of an element is the average mass of all the various isotopes of that element. The heavier isotopes have the effect of slightly raising the average. Carbon, for example, has an atomic mass of 12.011, which is slightly higher than 12.000 because of the few but heavier carbon-13 isotopes found in naturally occurring carbon. 39. A 50:50 mix of Br-80 and Br-81 would result in an atomic mass of about 80.5, while a 50:50 mix of Br-79 and Br-80 would result in an atomic mass of about 79.5. Neither of these is as close to the value reported in the periodic table as is a 50:50 mix of Br-79 and Br-81, which would result in an atomic mass of about 80.0. The answer is (c). 41. A body would have no odor if all its molecules remained part of the body. A body has odor only if some of its molecules enter a nose. 43. The carbon atoms that make up coauthor Leslie's hair or anything else in this world originated in the fires of ancient stars. 45. The remaining nucleus is that of carbon-12. 47. The iron atom is electrically neutral when it has 26 electrons to balance its 26 protons. 49. Carbon-13 atoms have more mass than carbon-12 atoms. Therefore, a given mass of carbon-13 contains fewer atoms than the same mass of carbon-12. Look at it this way: Golf balls have more mass than Ping-Pong balls. So, which contains more balls—1 kg of golf balls or 1 kg of Ping-Pong balls? Because Ping-Pong balls are so much lighter, you need many more of them to get 1 kg. 51. Protons, which are nearly 2000 times more massive than electrons, contribute much more to the mass of an atom. Protons, however, are held within the atomic nucleus, which is only a tiny, tiny fraction of the volume of the atom. The size of the atom is determined by the electrons, which sweep through a relatively large volume of space surrounding the nucleus. 53. From the periodic table we see that a carbon atom has a mass of about 12 amu. The carbon dioxide molecule, CO_2, consists of one carbon atom bonded to two oxygen atoms, for a total mass of about 44 amu. 55. The air we exhale is somewhat more massive than the air we inhale because it contains a greater proportion of the heavier carbon dioxide molecules. Interestingly, the added water vapor in our exhaled breath is less massive than the oxygen it replaced, but not massive enough to counteract the much greater mass of the carbon dioxide. Breathing causes you to lose weight. 57. The diagram on the far right, where the nucleus is not visible, is the best representation. 59. Based on its location in the periodic table, gallium, Ga, is more metallic than germanium, Ge. This means that gallium should be a better conductor of electricity. Computer chips manufactured from gallium, therefore, operate faster than chips manufactured from germanium. (Gallium has a low melting point of 30°C, which makes it impractical for use in the manufacture of computer chips. Mixtures of gallium and arsenic, however, have found great use in the manufacture of ultrafast, though relatively expensive, computer chips.) 61. Here is a list of 18 elements. Aluminum (as in aluminum foil); tin (as in tinfoil and tin cans); carbon (as in graphite and diamond); helium (as in a helium balloon); nitrogen (which makes up about 78% of the air we breathe); oxygen (which makes up about 21% of the air we breathe); argon (which makes up about 1% of the air we breathe); silicon (as in integrated circuits for computers and calculators); sulfur (a mineral used for many industrial processes); iron (as in most metal structures); chromium (as in chromium bumpers on cars); zinc (as in the coating of any galvanized nail or the insides of any post-1982 copper penny); copper (as in copper pennies); nickel (as in nickel nickels); silver (as in jewelry and old silver coins); gold (as in jewelry); platinum (as in jewelry); and mercury (as in mercury thermometers). 63. Figure 9.19 shows gallium and arsenic atoms as red and green, respectively. These colors are merely an artificial computer rendering. Atoms do not have color in the same sense that a macroscopic object may have color. Atoms, however, can be a source of light, which is what we observe as we look at an element's atomic spectrum. 65. The scanning probe microscope shows us only the relative sizes and positions of atoms. It does this by detecting the electrical forces between the tip of the microscope needle and the outer electrons of the atom. The atom itself is made of mostly empty space, so the best "image" of the inside of an atom would be a picture of nothing. Therefore, it doesn't make sense to talk about taking an "image" of the inside of an atom. Instead, we develop models that give us a visual handle on how the components of atoms behave. 67. Many objects or systems may be described just as well by a physical model as by a conceptual model. In general, the physical model is used to replicate an object or system of objects on a different scale. The conceptual model, by contrast, is used to represent abstract ideas or to demonstrate the behavior of a system. A gold coin, a car engine, and a virus may be adequately described using a physical model. A dollar bill (which represents wealth but is really only a piece of paper), air pollution, and the spread of a sexually transmitted disease may be described using a conceptual model. 69. The greater the frequency of a photon of light, the more energy is packed into that photon. 71. As the electron vibrates within the atom, it generates electromagnetic waves. The electron vibrates back and forth between different energy levels. 73. The blue frequency is a higher frequency and therefore corresponds to a greater energy transition. 75. The dimensions of the car and the nature of the materials of the car dictate that there will be certain frequencies that reinforce themselves upon vibration. When the vibration of the tires matches the car's "natural frequency," the result is a resonance, which is self-reinforcing waves. The wave nature of the car, however, is simply due to the back-and-forth vibrations of the materials of the car. The wave nature of the electron is entirely different. For the electron moving at very high speeds, some of its mass is converted into energy, which is manifest in its wave nature. Given the dimensions of the atom, certain frequencies of the electron will also be

"natural"—that is, self-reinforcing. The vibrating car, therefore, is analogous to one of the energy levels of the electron, which is the point at which the electron forms a self-reinforcing standing wave.

77.

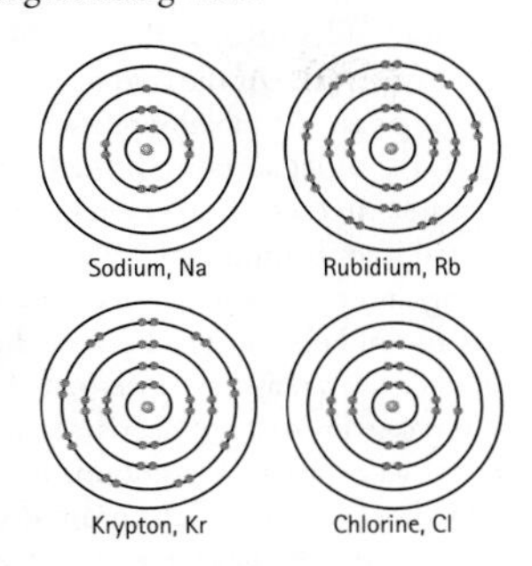

79. Neon's outermost shell is already filled to capacity with electrons. Any additional electrons would have to occupy the next shell, where the electrical attraction from the atomic nucleus is much weaker. 81. You are composed of atoms that at one time were part of not only every person around you but also every person who has ever lived on Earth! 83. The outsides of the atoms of the chair are made of negatively charged electrons, as are the outsides of the atoms that make up your body. Atoms don't pass through one another because of the repulsive forces that occur between electrons. 85. Curious? You need to research this question on your own because we need to draw the line somewhere between what to cover and not to cover in this textbook. Briefly, in the same way that Rutherford was able to deduce the atomic nucleus by bombarding atoms with alpha particles, evidence for the existence of even smaller subatomic particles has been obtained by bombarding the atom with highly energetic radiation. This research over the past century has evolved into what is known as the "standard model of fundamental particles," which places all constituents of matter in one of two categories: quarks and leptons. There are six quarks, and they have the whimsical names *up*, *down*, *charm*, *strange*, *top*, and *bottom*. The proton is made of two up quarks and one down quark, while the neutron is made of two down quarks and one up quark. There are six leptons, and they all appear to be pointlike particles without internal structure. The best-known lepton is the electron.

Chapter 10

1. Alpha radiation decreases the atomic number of the emitting element by 2 and the atomic mass number by 4. Beta radiation increases the atomic number of an element by 1 and does not affect the atomic mass number. Gamma radiation does not affect the atomic number or the atomic mass number. So, alpha radiation results in the greatest change in atomic number, and hence charge, and mass number as well. 3. The attractive strong nuclear force exerted by all nucleons is able to overcome the repulsive electric force of protons. 5. Neutrons act as "nuclear cement" to hold the nucleus together. 7. Emission of an alpha particle decreases the atomic number of the nucleus by 2. Emission of a beta particle increases the atomic number of the nucleus by 1. 9. The half-life of a radioactive sample is the amount of time it takes for half of the original quantity of a radioactive element to decay. 11. An elongated uranium-235 nucleus splits in half because the strong nuclear force between distant nucleons quickly diminishes. 13. A nuclear reactor and a fossil-fuel power plant both boil water to produce steam for turbines. The main difference between them is the amount of fuel involved. Nuclear reactors are much more efficient—1 kg of uranium-235 yields more than 30 freight-car loads of coal. 15. Nucleons have the least mass in iron. 17. Most of the radiation we encounter is natural background radiation that originated in Earth and in space. 19. No, radioactivity has been around longer than the human race and has been part of the environment for as long as the Sun, rain, and soil. 21. Carbon-14 lost to decay is replenished with carbon-14 from the atmosphere. When a living thing dies, replenishment stops and the percentage of carbon-14 decreases at a constant rate. 23. The mass is less after fusion because the mass difference has been converted into energy. 25. Thermonuclear fusion is responsible for sunshine. 27. The number of pennies in the mix should not affect the activity of the dimes. The set of coins with only 2 dimes left has gone through a greater number of tosses. The set with 10 dimes left is the more "radioactive" substance. Remember, according to this simulation, radiation is emitted every time a dime lands heads-up. The set with only 2 dimes is analogous to a sample of once-living ancient material. 29. A nuclear chain reaction spreads in three dimensions within a lump of fissionable material, such as U-235. The first mouse trap triggered by the ball would fly up and trigger other mouse traps, which would fly up and trigger more mouse traps in an exponential manner. The effect is made more dramatic by placing two Ping-Pong balls on each of the mouse traps. 31. (a) N-14, C-14 = C-12 (b) C-14, N-14, C-12 (c) N-14 = C-14, C-12. 33. b, a, c. Splitting uranium-235 into three equal fragments brings the resulting nuclei closest to iron. Splitting uranium-235 into 92 equal fragments creates 92 hydrogen atoms, which requires the input of energy. 35. At the end of 2 weeks, one-fourth is left; at the end of 3 weeks, one-eighth is left; at the end of 4 weeks, one-sixteenth is left. 37. Any isotope of uranium has atomic number 92. U-238 plus one neutron becomes U-239. When it emits a beta particle, a neutron becomes a proton, making the element atomic number 93. So, it is an element with atomic number 93 and mass 239. That's Np-239. 39. The protons repel each other by the electric force and attract each other by the strong nuclear force. The strong force predominates. (If it didn't, there would be no atoms beyond hydrogen!) If the protons are separated to the point where the longer-range electric force overcomes the shorter-range strong force, they fly apart. 41. Because like charges repel and because protons have the same sign of charge (positive) as the target atomic nuclei, the protons must be driven into the target area with enormous energies if they are to bombard the nuclei. Lower-energy protons would be easily repelled by any nuclei they approach. 43. As rocks age, the uranium they contain decays to lead. Therefore, the higher the ratio of lead to uranium, the older the rock. Likewise, a lower ratio of lead to uranium indicates a younger rock. 45. The elements that are lighter than uranium with short half-lives exist as the product of the radioactive decay of uranium. As long as uranium is decaying, their existence is assured. 47. The irradiated food does not become radioactive as a result of being zapped with gamma rays. Gamma rays are simply high-energy light. The high energy breaks down the molecules in the biological tissues, including those of microorganisms, which are thus killed. 49. No, not for materials that are a few years old. Too small a fraction of the carbon-14 has decayed. You couldn't tell the difference between a few years and a few dozen or even a hundred years. To the second question, yes, in a few thousand years a significant fraction of the carbon-14 has decayed. Carbon dating gives the best results for ages not so very different from the half-life of the isotope. To the third question, no, not for materials a few million years old. Essentially all of the carbon-14 will have decayed, and there will be none left to detect. You wouldn't be able to distinguish among a million or 10 million or 100 million years. 51. Stone tablets cannot be dated by carbon dating. Nonliving stone does not ingest carbon and transform that carbon by radioactive decay. Carbon dating is used with organic materials. 53. Strong radioactivity comes from this relatively pure uranium ore, which makes mining this ore particularly hazardous. However, naturally occurring uranium contains more than 99% of the U-238 isotope, which does not readily undergo nuclear fission. 55. Nuclear fission is a poor prospect for powering automobiles primarily because of the massive shielding that would be required to protect the occupants and others from the radioactivity as well as the problem of radioactive waste disposal. However, automobiles can be powered indirectly by the electricity generated by a nuclear fission power plant. 57. The mass per nucleon in uranium is greater than the mass per nucleon of the fission fragments of uranium. 59. Because plutonium generates more triggering neutrons per atom, a smaller mass will produce the same neutron flux as a somewhat larger mass of uranium. So plutonium-239 will have a smaller critical mass than a similar shape of uranium-235. 61. If uranium were split into three parts, the segments would be nuclei with smaller atomic numbers—more toward iron on the graph in Figure 10.31. The resulting mass per nucleon would be less, and more mass would be converted into energy in such a process. 63. No, U-235 (with its shorter half-life) undergoes radioactive decay six times faster than U-238 (half-life 4.5 billion years), so natural uranium in an older Earth would contain a much smaller percentage of U-235, not enough for a critical reaction to occur without enrichment. Conversely, in a younger Earth, natural uranium would contain a higher percentage of U-235 and would more easily sustain a chain reaction. Interestingly, there is strong evidence that 2 billion years ago, when the percentage of U-235 in uranium ore was higher, there was a natural reactor in Gabon, West Africa. 65. Although more energy is released in the fission of a single uranium nucleus than in the fusing of a pair of deuterium nuclei, the much greater number of lighter deuterium atoms in a gram of matter compared to the smaller number of heavier uranium atoms in a gram of matter results in more energy liberated per gram for the fusion of deuterium. 67. The radioactive decay of radioactive elements found under Earth's surface warms Earth from the inside and is responsible for the molten lava that spews from volcanoes.

The nuclear fusion in the Sun is responsible for warming everything on our planet's surface. 69. Energy from the Sun is our chief source of energy, which itself is the energy of fusion. Harnessing that energy on Earth has proven to be a formidable challenge. 71. Two carbon nuclei would fuse to atomic number 12 and the beta emission would increase it by 1, so the element produced would have atomic number 13, which is aluminum. 73. If easily accessible fossil fuels were much more abundant, then perhaps it would take us much longer to respond to the environmental hazards that they pose simply because they are relatively inexpensive. If centuries from now we have sources of clean and abundant energy, our descendants will view the burning of fossil fuels for energy as deadly, thoughtless, and primitive. 75. Agree with Paul, who attributes helium to radioactive decay. It's true; alpha particles emitted by radioactive isotopes in the ground slow down and stop, capture two electrons, and become helium atoms. Our supplies of helium come from underground. Any helium in the atmosphere is soon dissipated into space. Your friend Alison will encounter more radioactivity from the granite outcroppings than she will get from the nearby nuclear power plant. Furthermore, at a high altitude she will be treated to increased cosmic radiation. But the radiation encountered in the vicinity of the plant, on the granite outcropping, or at a high altitude is not appreciably different from the radiation one encounters in the "safest" situation. Advise your friend to enjoy life anyway! Tell Michele that Earth's natural energy heating the water in the hot spring is the energy of radioactive decay. Just as a piece of radioactive material is warmer than its surroundings because of thermal agitation from radioactive decay, Earth's interior is similarly warmed. The intense radioactivity in Earth's interior therefore heats the water but doesn't make the water itself radioactive. The warmth of hot springs is one of the "nicer effects" of radioactive decay.

Chapter 11

1. Chemistry touches all of the sciences. 3. Members of the American Chemistry Council have pledged to manufacture products without causing environmental damage. 5. Molecules are made of atoms. Atoms link together to form larger but still small basic units of matter called molecules. 7. One gram of water vapor, the gaseous phase of water, occupies the greatest volume. 9. Sublimation is the formation of a gas from a solid; evaporation is the formation of a gas from a liquid. 11. 335 joules of heat are needed to melt 1 g of ice. 13. A physical property is a physical attribute of a substance, such as color, hardness, density, texture, and phase. Chemical properties characterize the tendency of a substance to react with or transform into other substances. 15. During a chemical reaction there is a change in the way atoms are bonded together. 17. Both physical and chemical changes involve changes in appearance. 19. If, by returning to the original set of conditions you return to the original physical appearance, then the change was physical. If the substance that has undergone the change has new physical properties, then the change was chemical. 21. An element has only one type of atom. A compound has combinations of different types of atoms. 23. The chemical formula tells us the ratio in which atoms join together to form one unit of a particular substance. 25. The element closer to the left side of the periodic table is given first. A good example is sodium chloride, NaCl. 27. The formula for titanium dioxide is TiO_2. 29. Commercial products developed through nanotechnology, from sunscreens to computer hard drives, are already available. 31. Nature is the ultimate expert at nanotechnology. 33. The dye should become dispersed uniformly within the hot water first. The higher the temperature, the greater the average kinetic energy of the molecules. Because the molecules in the hot water are moving faster, they have a quicker effect on the dye of the Kool-aid crystals. Furthermore, the hot water tends to have more convection currents that help to distribute the dye throughout the water. 35. Water vapor condenses into water droplets on the outside of the pot. This water vapor is a product of the reaction of the natural gas with oxygen. Ice water in the pot would favor the condensation of more liquid water from the water vapor. As the pot gets warmer, the condensed water evaporates. Physical changes include the condensation and evaporation of water. The water vapor from the flame results from chemical changes occurring within the flame. 37. c, b, a 39. c, b, a 41. You simply don't yet understand the language. Likewise, don't assume science is difficult because you are not yet familiar with its language. 43. Physics is the most fundamental science because it lays the foundation for chemistry, which is the study of the physics of the atom. Chemistry, in turn, lays the foundation for biology, the most complex science. 45. Chemistry is the careful study of matter and can take place at a number of different levels, including submicroscopic, microscopic, and macroscopic. 47. The 50 mL plus 50 mL do not add up to 100 mL because, within the mix, many of the smaller water molecules are able to fit inside the pockets of empty space in the 50 mL of larger alcohol molecules. This is an example where molecules help to explain observed phenomena. 49. He could smell the cinnamon molecules because they readily leaked through the microscopic pores of the balloon. 51. At the cold temperatures of your freezer, water molecules in the vapor phase are moving relatively slowly, which makes it easier for them to stick to the inner surfaces of the freezer or to other water molecules. 53. When water freezes, heat is released to the surroundings. 55. Box (b). The trail lines behind each circle in this diagram indicate that the particles are moving faster. Box (a) shows the particles congregated to one side of the box. This is an unlikely scenario because the motions of gaseous particles are randomly oriented. You may have interpreted box (a) to mean that the gas is hotter, and as everyone knows, hot air rises. Fair enough, but notice that the trail lines do not indicate any faster motion. Box (c) shows fatter gas particles. When heat is added, gas particles don't grow any fatter. Instead, they move faster, which is to say that they have a greater average kinetic energy. 57. The alcohol molecules evaporate into the gaseous phase, which is a physical change. 59. The greater warmth of the tabletop provides more energy for the alcohol molecules to go into the gaseous phase. The rate at which the alcohol evaporates would be greater. As a result there would be more alcohol molecules in the air more quickly, which would register in your nose as more noticeable. 61. Boiling down the maple syrup involves the evaporation of water. As the syrup hits the snow, it warms the snow, causing it to melt while the syrup becomes more viscous. These are all physical changes. As the maple syrup is boiled, the sugar in the syrup begins to caramelize, which is a chemical change. 63. The changes that occur as we age involve the chemical reformation of our biomolecules. These are chemical changes. 65. All are physical properties. 67. (a) Chemical, (b) chemical, (c) physical, (d) chemical. 69. All the oxygen in water is bonded to hydrogen atoms to make water molecules. Water is uniquely different from the elements oxygen, O_2, and hydrogen, H_2, from which it is made. The oxygen our bodies are designed to breathe is gaseous molecular oxygen, O_2. We drown when we breathe in water because it contains so little O_2. Although both contain oxygen, gaseous oxygen, O_2, and water, H_2O, are vastly different materials. 71. Water use to be classified as an element but that was before people recognized that the basic building blocks of matter are tiny particles called atoms. Today an element is defined as a material that consists of only one kind of atom. 73. Nothing. The properties of a compound are uniquely different from the properties of the elements used to make that compound. Copper sulfide, CuS, is a black powder. 75. The chemical name is tribarium dinitride or, more simply, barium nitride because no ratio other than three bariums to two nitrogens is possible. 77. The name is strontium phosphate. For a polyatomic ion, the convention is to leave out the *tri-* or *di-* suffix. So, to a chemist, the name "tristrontium diphosphate" would sound a little weird and overdone, but he or she would still know what it means. 79. The scanning probe microscope provides us with images of the nanoscopic world, which is where individual atoms begin to appear. The optical microscope allows us to see at the microscopic level, where individual atoms are too small to be seen. Also, the scanning probe microscope collects images by dragging an ultrasharp needle across the surface. The optical microscope works by collecting and focusing visible light. 81. Chemistry and nanotechnology are similar in that both focus on the world of the atom. They are different in that chemistry focuses on how to combine atoms in bulk to create novel molecules, also in bulk. Nanotechnology works by manipulating individual atoms or small groups of individual atoms. Great things will no doubt be achieved through nanotechnology—but as a complement to the great things also being achieved through novel chemical reactions. 83. You probably should not spend too much time showing politicians your analysis of the data. If your credentials are good and you are respected among your peers, then politicians will likely trust your assessments. Politicians will be much more interested in the impact climate change will have on the general public, especially in terms of the economy. Your time would be well spent focusing on potential solutions to the problem. All the better if your solutions offer new jobs for the politician's constituents. John Ashton also said, "The more effort scientists put into how their message might be heard, how it might be manipulated and made mischief, the better. If we want to successfully respond to climate change we have to form it as part of the solution to the economic crisis." 85. We marvel at a new technology for several years and then the appeal begins to fade. Later, instead of marveling at this technology, people come to take it for granted. This happened when plastics were introduced in the 1940s and 1950s. Plastics were known as a wonder material. Now we do our best to recycle plastics so they don't pollute our roadways and overfill

our garbage dumps. Life is no doubt more comfortable and convenient with our warm GORE-TEX® jackets and cell phones. But, are we necessarily happier? Many think that happiness arises from within the person. To them it is no surprise that comfort and convenience beyond what is necessary for our basic needs do little to improve one's sense of well-being.

Chapter 12

1. Two electrons fit into the first shell. Eight electrons fit into the second shell. 3. The electron-dot structures have the same number of valence electrons. 5. A gain of electrons causes a negative ion. 7. Nonmetallic elements tend to form covalent bonds. 9. A negatively charged oxygen atom forms only one bond and has three lone pairs as a consequence. 11. Fluorine has the greatest electronegativity; francium has the least electronegativity. 13. A nonpolar substance tends to have weak attractions to itself, which causes a low boiling point. 15. Oil and water do not mix because water molecules are more attracted to themselves than to oil molecules. 17. A hydrogen bond is an extremely strong dipole–dipole attraction. 19. The volume of the solution gradually increases as more sugar is dissolved in it. 21. Liters of solution is the typical unit used for the concentration of a solution. 23. The crushed sugar has a larger surface area in contact with the water. 25. Metals lose electrons more readily. 27. Almost all the metals we work with are manufactured. A native metal is one that occurs in nature. Only a few metals, such as gold and silver, are native metals. 29. Atoms and molecules are very small. If one atom or molecule out of a trillion is different, then the sample is no longer pure. 31. The potassium chloride crystals are more rounded at the edges because they are softer. The crystals are softer because the ionic bond between potassium and chlorine ions is weaker than the ionic bond between sodium and chlorine ions. The ionic bond is weaker because the potassium ions are larger than the sodium ions, which means that potassium ions are not able to get as close to the oppositely charged chloride ions. When it comes to electrical attractions, the farther away, the weaker the force. 33. The tiny bubbles come from the air that dissolved in the water. These bubbles contain mostly air, along with some residual water vapor. Gases do not dissolve well in hot liquids. Air that is dissolved in room-temperature water, for example, quickly bubbles out when the water is heated. Thus you can speed up the formation of air bubbles by using warm water. You'll find that boiling *de-aerates* the water—that is, removes the atmospheric gases. Chemists sometimes need to use de-aerated water, which is made by allowing boiled water to cool in a sealed container. Why don't fish live very long in de-aerated water? 35. Interesting crystals can also be made from supersaturated solutions of Epsom salts ($MgSO_4 \cdot 7\ H_2O$) and alum ($KAl(SO_4)_2 \cdot 12\ H_2O$), which is used for pickling and is available in the spice section of some grocery stores. Compare the crystal shapes of Epsom salts, alum, and sugar. 37. a, c, b. Fluorine's boiling point is −188°C, hydrogen chloride's is −85°C, and hydrogen fluoride's is +19°C. Fluorine, F_2, is a nonpolar compound and so should have the lowest boiling point—the attraction between fluorine molecules is so weak. Fluorine has a greater electronegativity than chlorine, which means that the H—F bond is more polar than the H—Cl bond. That HF is the most polar suggests that it has the highest boiling point. 39. c, b, a 41. Hexanol, butanol, ethanol. The carbon–hydrogen structures are nonpolar. The structure for hexanol, therefore, is the most nonpolar of these three molecules and hence it has the lowest solubility in water. Another way to look at this is as follows: The O—H bond is a polar bond, and this is needed for good solubility in water. Only a small percentage of the hexanol molecule is made up of the O—H bond. A greater percentage of the ethanol molecule (about 33%) is made up of the O—H bond. Water, therefore, has an easier time being attracted to an ethanol molecule than to a hexanol molecule. 43. The formula is $MgCl_2$. Two single negatively charged chlorine ions are needed to balance the one double positively charge magnesium ion. A shortcut way of answering these sorts of questions is to take the charge of one ion and make it the subscript of the opposite ion. For example, take the 2+ charge of the magnesium and make it the subscript of the chlorine. Then take the 1− charge on the chlorine and make that the subscript of the magnesium. Because subscripts of 1 are implied when not written, we have $MgCl_2$, not Mg_1Cl_2. 45. Mass = concentration × volume = 3.0 g/L × 15 L = 45 g. 47. There is room for only one additional electron. 49. Two inner shells of electrons shield the valence electron from the nucleus. Consider that the charge of sodium's nucleus is +11. There are ten inner-shell electrons (−10) that shield the outermost-shell electron from this positive charge. This means that the outer-shell electron experiences the nucleus as though it were +1, which is significantly less than the actual nuclear charge of +11. 51. Magnesium has only two electrons in its valence shell, and these two electrons are well shielded from the nuclear charge by inner shells of electrons. 53. Sulfuric acid loses two protons to form the sulfate ion, SO_4^{2-}. A water molecule loses a single proton to form OH^-. 55. The electrical force of attraction weakens with increasing distance. For this reason, the force of attraction between the smaller sodium ion and the chloride ion is stronger than the force of attraction between the larger potassium ion and the chloride ion. This explains, in part, why potassium chloride crystals are weaker (softer) than sodium chloride crystals. 57. The mold should be bigger than 6 inches because the metal will shrink (contract) as it cools. Unfortunately, however, as the molten metal cools and shrinks, it also pulls away from the inside of the mold while it is still soft, which means that it doesn't retain the form of the mold very well. Some exceptions are low-melting-point metals, such as lead. In general, the best practice is to stamp or machine metals to the desired dimensions. 59. The valence electrons of a potassium atom are weakly held by the nucleus. The potassium atom has a hard enough time holding onto its one valence electron, let alone a second one, which is what would happen if the potassium joined in a covalent bond. Potassium atoms are joined together by the metallic bond. 61. When bonded to an atom with low electronegativity, such as any group 1 element, the nonmetal atom pulls the bonding electrons so close to itself as to form an ion. 63. The atoms closer to the lower left corner of the periodic table carry the greater positive charge: (a) hydrogen, (b) bromine, (c) carbon, (d) neither! 65. Sometimes true and sometimes false! Within any one period of the periodic table, it is true that electronegativity increases with increasing nuclear charge. The nuclear charge of a fourth-period potassium atom, however, is much greater than the nuclear charge of a second-period fluorine atom, but its electronegativity is much less. 67. In the top pair, the left molecule with the two chlorines on the same side is more polar and will thus have a higher boiling point. In the bottom pair, the chlorine atoms have a relatively strong electronegativity that pulls electrons away from the carbon. In the left molecule, $COCl_2$, this tug of the chlorines is counteracted by a relatively strong tug of the oxygen, which tends to defeat the polarity of this molecule. The hydrogens in the molecule on the right, $C_2H_2Cl_2$, have an electronegativity that is less than that of carbon, so they actually assist the chlorines in allowing electrons to be yanked toward one side, which means that this molecule is more polar. The right molecule therefore has the higher boiling point. 69. The covalent bonds within a molecule are many times stronger than the attractions between two neighboring molecules. We know this because, while two molecules can move away from each other (as occurs in the liquid or gaseous phase), the atoms within a molecule remain stuck together as a single unit. To pull the atoms apart requires some form of chemical change. 71. The magnitude of the electric charge associated with an ion is much greater. 73. Motor oil is much thicker (viscous) than gasoline. This suggests that molecules of motor oil are more strongly attracted to one another than are molecules of gasoline. A material consisting of molecules of Structure A has greater induced dipole–induced dipole molecular interactions than a material consisting of molecules of Structure B. Motor oil molecules, therefore, are best represented by Structure A, while gasoline molecules are best represented by Structure B. 75. Salt, sodium chloride, is a compound. Stainless steel, a mix of iron and carbon, is a mixture. Tap water, dihydrogen oxide plus impurities, is a mixture. Sugar, chemical name sucrose, is a compound. Vanilla extract, butter, and maple syrup, all natural products, are mixtures. Aluminum, a metal, is an element in pure form. (It is sold commercially as a mixture of mostly aluminum with trace metals, such as magnesium.) Ice, dihydrogen oxide, is a compound in pure form; it is a mixture when made from impure tap water. Milk, a natural product, is a mixture. Cherry-flavored cough drops, a pharmaceutical, is a mixture. 77. Box A shows oval molecules dissolved in smaller circle molecules. 79. The 100 mL of fresh sparkling seltzer water should weigh more than the 100 mL of fresh water because of the carbon dioxide that is still dissolved in the water. The 100 mL of flat seltzer water at 20°C should weigh more because the seltzer water warmed to 80°C has less dissolved carbon dioxide, since the solubility of gases decreases with increasing temperature. 81. All water exposed to the atmosphere contains oxygen, O_2. There's enough oxygen in the water for fish to survive because their gills are especially equipped to absorb what little oxygen is found there. Exposing water to 100% oxygen gas would increase the amount of oxygen in the water because the oxygen would displace the atmospheric nitrogen, N_2, which is also in the water. Still, the relative amount of O_2 in the water is quite small. When we ingest this water, the amount of dissolved O_2 that the gut is able to absorb is orders of magnitude smaller than the amount of O_2 that we routinely absorb through our lungs. Save your money. Buy carbonated water and enjoy the effervescence. If you desire oxygen, just take a deep breath.

You'll find all the oxygen you need—not too much, not too little. And it's free! 83. A high boiling point means that the substance interacts with itself strongly. Put yourself in the position of a water molecule, and you will see that you are attracted to both ends of 1,4-butanediol. In fact 1,4-butanediol is infinitely soluble in water! 85. A saturated solution of $NaNO_3$ is more concentrated. 87. Although oxygen gas, O_2, is not very soluble in water, many other gases are very soluble in water. Hydrogen chloride is a polar molecule and therefore is very soluble in water by virtue of the dipole–dipole attractions between the HCl and H_2O molecules. 89. There are many obstacles to recycling. There's confusion about what materials can and can't be recycled. There's concern about how clean a container must be before it can be recycled. Some people don't know or understand the value of recycling. Others may just be lazy. Campaigns to educate the general public can help overcome obstacles such as these. An easy-to-read pamphlet describing the dos and don'ts about recycling is most effective. These pamphlets are typically sponsored by local governments in cooperation with local recyling companies. State and federal governments can also play a role in educating the general public and also in setting the packaging standards for companies that use recyclable materials to package their goods. Many states offer refunds for empty containers. Some communities even charge fines against individuals who do not sort plastics correctly. Other communities ask their citizens not to sort at all, thinking that the task of sorting inhibits people's tendency to recycle. Ultimately, recycling is a global issue; peoples of all nations should be encouraged not to waste material resources.

Chapter 13

1. Coefficients show the ratios in which reactants combine and products form in a chemical reaction. 3. A chemical equation must be balanced because the law of conservation of mass says that mass can be neither created nor destroyed. There must be the same number of each atom on both sides of the equation. 5. Energy is released by an exothermic reaction. 7. Reactants must collide in a certain orientation with enough energy in order to react. 9. Reactant molecules that move the fastest are the first to pass over the energy barrier. 11. Dissolving an acid in water causes the formation of a hydronium ion. 13. The concentration of hydronium ions in neutral water is 0.0000001 M, so, relatively speaking, there are very few hydronium ions in neutral water. 15. The pH indicates the acidity of the solution as judged by the concentration of hydronium ions. 17. Metallic elements at the left of the periodic table, especially the lower left side, are the best reducing agents because of the ease with which they can lose electrons. 19. Zinc is the coating on galvanized nails. 21. The oxygen atoms become more negatively charged, so they are gaining electrons. 23. A catalyst lowers the activation energy for a reaction. 25. The product of CO_2 and H_2O is carbonic acid, H_2CO_3. 27. Carbon dioxide reacts with water to form carbonic acid, which reacts with carbonate ion to form bicarbonate ions. 29. Fuel cells are similar to batteries. Batteries, however, run down, while fuel cells can keep going indefinitely so long as fuel is supplied. 31. This is an exothermic process, as evidenced by the warmth generated. At the molecular level, alcohol and water molecules join together, forming dipole–dipole attractions. Note that this mixing is a physical process, not a chemical process. In other words, no chemical bonds are formed. The energy that comes out of this process arises from the formation of the dipole–dipole attractions. 33. The bubbling occurs as the result of a reaction between the Alka-Seltzer tablet and the water. The corn syrup–water mixture has a lower proportion of water molecules and therefore a slower rate of reaction. In terms of molecular collisions, with fewer water molecules, there is a lower probability of collisions between the molecules of Alka-Seltzer and water. 35. Notice that the cabbage itself is purple when you buy it from the store. This color indicates that it is only slightly acidic, so the juices in the cabbage are far less acidic than vinegar. The baking soda would react with the vinegar to form gaseous carbon dioxide. The two should neutralize each other so that the color moves toward purple. 37. Copper ions are positively charged. Each copper atom in Cu_2O carries a 1+ charge. The oxygen carries a 2− charge. The copper ion must gain electrons in order to transform into copper metal. The iron atoms release electrons to the copper ions so that they transform back into metallic copper, which happens on the surface from which these electrons came. 39. c, a, b. The endothermic reaction (c) will likely be slower than the exothermic reaction (a) because it requires a decrease in entropy. The fastest of these reactions will be (b), which has no energy of activation. 41. c, b, a. An oxidizing agent causes other materials to lose electrons because of its tendency to gain electrons. Atoms with high electronegativity tend to have a strong attraction for electrons and, therefore, are also strong oxidizing agents. Na has the weakest electronegativity, which means it is the weakest oxidizing agent of these three elements. (It is, however, the strongest reducing agent of these three elements.) S is in between, while Cl, with its high electronegativity, is the strongest oxidizing agent. 43. Ethane, ethanol, acetaldehyde, acetic acid. How is it that acetaldehyde is more oxidized than ethanol when both structures have just one oxygen atom? Think of it from the point of view of the carbon atom bonded to the oxygen. Oxygen is a highly electronegative atom and so it tends to pull electrons toward itself. In ethanol, the carbon has only a single bond (two electrons) in a tug-of-war with oxygen. These are the two electrons that the oxygen will "win" over the carbon. In acetaldehyde, however, the carbon shares four electrons (within the double bond) with the oxygen. This carbon has more electrons to lose, which it does. The carbon of the acetaldehyde, therefore, is more oxidized.

45. (a)

Energy to break bonds	Energy released from bond formation
N—N = 159 kJ	H—H = 436 kJ
N—H = 389 kJ	H—H = 436 kJ
N—H = 389 kJ	N—N = 946 kJ
N—H = 389 kJ	
N—H = 389 kJ	
Total = 1715 kJ absorbed	Total = 1818 kJ released

Net = 1715 kJ absorbed − 1818 released
= −103 kJ (released, exothermic)

(b)

Energy to break bonds	Energy released from bond formation
O—O = 138 kJ	O=O = 498 kJ
H—O = 464 kJ	H—O = 464 kJ
H—O = 464 kJ	H—O = 464 kJ
O—O = 138 kJ	O—H = 464 kJ
H—O = 464 kJ	O—H = 464 kJ
H—O = 464 kJ	
Total = 2132 kJ absorbed	Total = 2354 kJ released

Net = 2132 kJ absorbed − 2354 kJ released
= −222 kJ (released, exothermic)

47. The concentration of hydronium ions in the pH=1 solution is 0.1 M. Doubling the volume of the solution with pure water means that its concentration is cut in half. So, the new concentration of hydronium ions after the addition of 500 mL of water is 0.05 M. To calculate the pH: pH $= -\log [H_3O^+] = -\log (0.05) = -(-1.3) = 1.3$. 49. (a) 4, 3, 2; (b) 3, 1, 2; (c) 1, 2, 1, 2; (d) 1, 2, 1, 2. (Remember that, by convention, 1s are not shown in the balanced equation.) 51. Both equations are balanced. 53. Molecule C has an excess of one diatomic molecule. This extra molecule is not needed for the chemical reaction, which is why it also appears with the products. 55. $CH_4 + H_2O \rightarrow CO + 3\,H_2$ 57. The chemical reactions in a disposable battery are exothermic, as evidenced by the electric energy they release. To recharge a rechargeable battery requires the input of electric energy, so the reactions that proceed during the recharging process are endothermic. 59. When the carbon-based fuel combusts, it gains mass as it combines with the oxygen from the atmosphere to form carbon dioxide, CO_2, which comes out in the exhaust. 61. In pure oxygen there is a greater concentration of one of the reactants (oxygen) for the chemical reaction (combustion). The greater the concentration of reactants, the faster the reaction. 63. Photosynthesis produces oxygen, O_2, which migrates from Earth's surface up into the stratosphere, where it is converted by the energy of ultraviolet light into ozone, O_3. Plants and all other organisms living on Earth's surface benefit from this ozone because of its ability to shade the planet's surface from ultraviolet light. 65. Hydrogen chloride, HCl, does not stay in the atmosphere for long periods of time because it is quite soluble in water, as can be deduced from its polarity. Thus, atmospheric hydrogen chloride mixes with atmospheric moisture and precipitates with the rain. 67. The base accepts the hydrogen ion, H^+, and thus gains a positive charge. The base thus forms the positively charged ion. Conversely, the acid donates a hydrogen ion and thus loses a positive charge. The acid thus forms the negatively charged ion. 69. Squeeze lemon juice onto the fish. The citric acid in lemon juice reacts with the smelly alkaline compounds to form less smelly salts. The smell and taste of the lemon also help to mask any additional undesirable fishy odors. 71. (a) pH is a measure of the hydronium ion concentration. The higher the hydronium ion concentration, the lower the pH. According to the information given in this exercise, as water warms,

the hydronium ion concentration increases, albeit only slightly. Thus, pure water that is hot has a slightly lower pH than pure water that is cold. (b) As water warms up, the hydronium ion concentration increases, but so does the hydroxide ion concentration—and by the same amount. Thus, the pH decreases and yet the solution remains neutral because the hydronium and hydroxide ion concentrations are still equal. At 40°C, for example, the hydronium and hydroxide ion concentrations of pure water are both 1.71×10^{-7} mole per liter. The pH of this solution is the negative log of this number, or 6.77. This is why most pH meters need to be adjusted for the temperature of the solution being measured. Except for this exercise, which probes your powers of analytical thinking, this textbook ignores the slight role that temperature plays in pH. Strictly speaking, the pH of neutral water is 7.0 only at a temperature of 24°C. 73. This solution has a hydronium ion concentration of 10^3 M, or 1000 moles per liter. The solution is impossible to prepare because only so much acid can dissolve in water before the solution is saturated and no more will dissolve. The greatest concentration possible for hydrochloric acid, for example, is 12 M. Beyond this concentration, any additional HCl, which is a gas, added to the water simply bubbles back out into the atmosphere. 75. Consider this analogy: You are at a bottle cap convention where you hope to sell your bottle caps. You soon realize, however, that you're a lousy salesperson. Furthermore, all the other vendors at the convention are excellent salespeople. Not only do they steal away all your potential customers, but they're so good that you end up buying bottle caps from them as well. In the end, you have given away no bottle caps to others. Likewise, carbonic acid is unable to give away any hydrogen ions in a concentrated solution of such a strong acid. If it did, an HCl molecule would come along and quickly give one right back. Adding carbon dioxide gas will not lower the pH even though the carbon dioxide readily transforms into carbonic acid. 77. When fossil fuels burn, they react with the oxygen, O_2, in the air to form gaseous carbon dioxide, CO_2, and water vapor, H_2O. The carbon dioxide reacts with atmospheric moisture to form carbonic acid, which lowers the pH of rainwater. As the acidic rainwater precipitates into the ocean, it neutralizes alkaline ocean salts, such as calcium carbonate. This, in turn, lowers the pH of the ocean. 79. The tin ion, Sn^{2+}, is the oxidizing agent because it causes the silver, Ag, to lose electrons. Meanwhile, the silver atoms, Ag, are the reducing agents because they cause the tin ion to gain electrons. 81. The unsaturated fatty acids are gaining hydrogen atoms, so they are reduced. 83. As the carbon of propane, C_3H_8, forms carbon dioxide, CO_2, it is losing hydrogen and gaining oxygen, which tells us that the carbon is oxidized. 85. The oxygen is chemically bonded to hydrogen atoms to make water, which is a completely different molecule from the oxygen, O_2, required for combustion. Another way to answer this question is to say that the oxygen in water is already "reduced" in the sense that it has gained electrons from the hydrogen atoms to which it is attached. Being already reduced, this oxygen atom no longer has a great attraction for additional electrons. 87. The body must oxidize the elemental iron, Fe, into the iron 2+ ion, Fe^{2+}. 89. Putting more ozone into the atmosphere to replace the ozone that has been destroyed would be like throwing more fish into a pool of sharks to replace the fish that have been eaten. The solution is to remove the CFCs that destroy the ozone. Unfortunately, CFCs degrade slowly and the ones up there now will remain there for many years to come. Our best bet is to stop the current production of CFCs and hope that we haven't already caused too much damage.

Chapter 14

1. Structural isomers have different arrangements of their carbon atoms, but the number of carbon atoms in each is the same. 3. The boiling points of hydrocarbons are used in fractional distillation. 5. A carbon atom in a saturated hydrocarbon is bonded to four atoms. 7. A hydrocarbon must have at least one multiple (double or triple) bond to be classified as unsaturated. 9. A heteroatom is any atom other than carbon or hydrogen in an organic molecule. 11. Small alcohols are soluble in water because they have polar oxygen–hydrogen bonds similar to water. Like dissolves like. 13. An alcohol has a hydroxyl (oxygen bonded to hydrogen) group, but an ether contains an oxygen atom bonded to two carbon atoms. 15. Amines tend to be basic. The lone electron pair in nitrogen atoms makes them basic because this lone pair is able to accept a hydrogen ion. 17. Morphine and caffeine are two examples of alkaloids. 19. Both ketones and aldehydes contain carbonyl groups. Ketones have the carbonyl carbon bonded to two adjacent carbon atoms; aldehydes have the carbonyl carbon bonded to either one hydrogen and one carbon or two hydrogens. 21. Aspirin is made from salicylic acid. 23. In the formation of a condensation polymer, a small molecule (i.e., water or hydrochloric acid) is released from each monomer. 25. A copolymer is a polymer composed of two or more different monomers. 27. Intermolecular attractions, such as hydrogen bonding, hold a drug to its receptor site. 29. Two carbons connected by a double bond are not able to rotate relative to each other. The structure of butane, as seen in Figure 14.7, has many more conformations than that of 2-butene, which cannot rotate around the double bond. Atoms connected by a triple bond are similarly unable to rotate relative to each other. 31. This is a good demonstration of the destructive action of isopropyl alcohol. 33. a = c, b. Cyclobutane and 2-butene both have eight hydrogen atoms, while butane has ten. 35. a, b, c. Recall that the hydroxyl group, $-OH$, is very polar. The more hydroxyl groups a compound has, the greater its polarity, which means the greater its solubility in water. Compound (a) is nonpolar and therefore insoluble in water. Compound (b) has a single hydroxyl group, which makes it somewhat soluble in water. Compound (c) has two hydroxyl groups, so it is very soluble in water. 37. Carbon atoms are unique in their ability to form strong chemical bonds repeatedly with other carbon atoms, which permits the formation of countless possible structures.

39.

Hexane 2-Methylpentane 3-Methylpentane

2, 3-Dimethylbutane 2, 2-Dimethylbutane

41. Both compounds have the same chemical formula. Because they have the same chemical formula but different chemical structures, they are structural isomers. 43. The pressure may be greater at the bottom of the tower because of the higher temperature and the greater number of vaporized molecules. 45. Heavier hydrocarbons have a higher proportion of carbon, which is why, on a gram-to-gram basis, they produce more carbon dioxide. Natural gas containing primarily methane, CH_4, produces the least amount of carbon dioxide. Also, natural gas contains very little of the pollution-causing impurities, such as sulfur compounds, that are in petroleum and coal. This is the reason for the claim that natural gas, CH_4, is a cleaner-burning fuel. 47. These are two different conformations (configurations) of the same structure. To understand this, find the longest chain of carbon atoms in each structure and then note that the side groups coming off of this longest chain are the same.

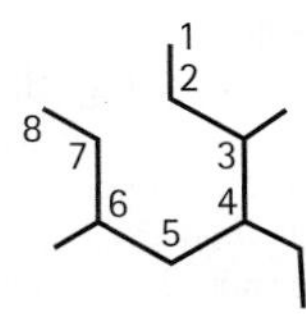

49. There is no hydrogen bonding in an ether. The weaker intermolecular forces in ethers give them lower boiling points than alcohols. 51. This compound is a phenol because the hydroxyl groups it contains are connected to the benzene ring. 53. As shown in Table 14.3, the structure is three ethyl groups surrounding a central nitrogen atom.

55.

Caffeine (free base) $+ H_2O + Na^+H_2PO_4^-$

57. Don't let the name fool you. Look at the structure to deduce its physical or chemical properties. Because of the nitrogen in this alkaloid molecule, it behaves as a base. 59. At an acidic pH, structure (b) is more likely. The amine group behaves as a base, which means that it reacts with the hydronium ions (which are abundant at low pH) to form the positively charged nitrogen ion. 61. Aspirin's chemical name is acetyl salicylic acid. It is the acidic nature of aspirin that gives rise to its sour taste. Recall that the sour taste of vinegar is from the acetic acid it contains. 63. The products are water and the salt sodium benzoate, which is a common food preservative. 65. The aspirin

doesn't "know" to go to your head rather than to your big toe. In fact, once the aspirin enters the bloodstream it is distributed throughout the body, which is why aspirin is good for alleviating headaches as well as toe aches. 67. Polypropylene consists of a polyethylene backbone with methyl groups attached to every other carbon atom. This side group interferes with the close packing that could otherwise occur among the molecules. As a consequence, we find that polypropylene is actually less dense than polyethylene—even low-density polyethylene. 69. Note the similarities between the structures of SBR and polyethylene and polystyrene, all of which possess no heteroatoms. SBR is an addition polymer made from the monomers 1,3-butadiene and styrene combined in a 3:1 ratio. Notably, SBR is the key ingredient that allows the formation of bubbles from bubble gum. 71. Initially, the beta-carotene structure looks complex. Upon careful examination, however, we see that this molecule is many smaller isoprene units joined together. Note that there are 40 carbon atoms in beta-carotene and 5 carbon atoms in isoprene. Take 40 divided by 5 to get that beta-carotene is made from 8 isoprene molecules joined together, with a somewhat different organization of double bonds. 73. The HCl would react with the free base to form the water-soluble, but diethyl ether-insoluble hydrochloric acid salt of caffeine. With no water available to dissolve this material, it would precipitate out of the diethyl ether as a solid that may be collected by filtration. 75. Many medicines are flushed down drains or toilets either directly or after passing through the human body and into the urine. These drugs end up in downstream water supplies. By the time they reach the downstream consumer, their concentration is at a relatively low level. But this level is easily measured. Perhaps of greatest concern is the effect these low doses might have over the long term. Drinking a single glass of drug-tainted water might not have a noticeable effect. But what if this water is consumed continually by an individual over his or her lifetime?

Chapter 15

1. All living organisms use energy. They take energy from the environment and convert it into other forms of energy for their own use. Also, all living organisms develop and grow. They maintain themselves; generate structures, such as stems and leaves or skin and bones; and repair damage done to those structures. Living organisms have the capacity to reproduce. Finally, they are part of populations that evolve. Populations do not remain constant from one generation to the next but change over time. 3. Prokaryotes include two major lineages, the bacteria and the archaea. Eukaryotes include all animals, plants, fungi, and protists. 5. Eukaryotic cells have their DNA in a distinct nucleus, a feature that distinguishes them from prokaryotes. In addition, the DNA of eukaryotic cells is found in linear, rather than circular, chromosomes. 7. Mitochondria are organelles that break down organic molecules to obtain energy in a form that cells can use. Ribosomes are organelles that assemble proteins. Lysosomes are the garbage disposals of a cell. They break down organic materials, such as damaged or worn-out organelles. In plants, organelles called chloroplasts capture energy from sunlight and use it to build organic molecules. 9. Phospholipids have hydrophilic "heads" and hydrophobic "tails." The heads are naturally drawn to the watery environment inside and outside the cell, whereas the tails naturally try to avoid it. The result is that the phospholipids form a double layer, or bilayer, with the hydrophobic tails pointing in and the hydrophilic heads pointing out. You can think of the phospholipids as making a sandwich, with the heads as two slices of bread and the tails as the peanut butter inside. 11. Hydrophobic molecules, such as the gases oxygen and carbon dioxide, can pass directly through the double layer of hydrophobic tails. Some small hydrophilic molecules—including water—can also cross the cell membrane this way. 13. Passive transport requires no energy from the cell. In active transport, a transport protein moves molecules *against* a concentration gradient—from an area of low concentration to an area of high concentration. In this case, the second law of thermodynamics tells us that energy is required. 15. In animals and plants, plasmodesmata allow very local messages to pass directly from one cell to an adjacent cell. Plasmodesmata are slender threads of cytoplasm that link adjacent plant cells. 17. No, receptors are extremely specific about the molecules they bind to. This is because a message molecule and its receptor fit together like a key in a lock—only the right combination will work. 19. During prophase, the normally loosely packed chromosomes condense and the membranes surrounding the nucleus break down. When the chromosomes condense, it becomes clear that each consists of two identical sister chromatids attached at a point called the centromere. The mitotic spindle also forms during prophase. The mitotic spindle, which consists of a series of fibers that attach to the duplicated chromosomes, is responsible for splitting the genetic material between the two daughter cells. During metaphase, the chromosomes line up at the equatorial plane, which passes through the imaginary "equator" of the cell. During anaphase, the two sister chromatids are pulled apart by the shortening of the mitotic spindle fibers and move to opposite poles of the cell. During telophase, new nuclear membranes form around each set of chromosomes, and the chromosomes return to their loosely packed state. 21. Many essential chemical reactions have a very high activation energy. Some important reactions would take 100 years to happen on their own. Because of this, cells rely on catalysts to lower the activation energy of reactions and allow them to happen more quickly. The catalysts in cells are large, complex proteins called enzymes. 23. The antibiotic penicillin kills bacteria by inhibiting an enzyme that bacteria need to build cell walls. 25. During the light-dependent reactions, energy is captured from sunlight. During the light-independent reactions, carbon is fixed—that is, carbon atoms are moved from atmospheric carbon dioxide to the organic molecule glucose. 27. Glycolysis takes place in the cell cytoplasm. During glycolysis, the six-carbon glucose molecule is split into two molecules of pyruvic acid, each of which contains three carbon atoms. Two molecules of ATP are produced in the process. 29. The pyruvic acid from glycolysis is broken down into ethanol and carbon dioxide. 31. The protein keratin provides structure for living organisms in the form of skin, hair, and feathers. Insulin is a protein that acts as a hormone, allowing one type of cell in the body to communicate with other types. Actin and myosin allow muscles to contract. Hemoglobin, a protein found in red blood cells, transports oxygen to body tissues. Antibodies protect the body from disease. And proteins known as digestive enzymes break down food during digestion. 33. Nucleic acids are made up of strands of smaller units called nucleotides. A nucleotide includes a sugar molecule, a phosphate group, and a nitrogenous base. DNA consists of two nucleic acid strands twisted into a spiral, which is why it is sometimes called a double helix. DNA contains four kinds of nucleotides—adenine, cytosine, guanine, and thymine (or A, C, G, and T). 35. With a resolving power of 10^{-6} m, light microscopes enable us to view cells and to make out the larger features inside them, such as the nucleus and mitochondria. However, they do not really allow us to see organelles and other cellular structures in detail. 37. Cells get energy from ATP when one of its phosphate groups is removed, leaving adenosine diphosphate, or ADP. 39. The sodium-potassium pump is a transport protein in the cell membrane. It behaves like a swinging door, shuffling ions into and out of the cell. In its default state, the protein is open to the inside of the cell. There, it binds three sodium ions. An ATP reaction, in which a molecule of ATP transfers a phosphate group to the protein, causes the protein to shift and open to the outside of the cell. (This is the step where ATP is required.) The sodium ions are released, and two potassium ions are bonded, causing the phosphate group to be released from the transport protein. The loss of the phosphate group causes the protein to shift back to its original position and to release the potassium ions inside the cell. 41. This exploration examines the process of diffusion. At first, the water is darker closer to the tea bag because molecules take a longer time to diffuse farther away. If you remove the teabag, the water will eventually become uniformly colored. This is because diffusion continues to move molecules from areas of higher concentration to areas of lower concentration until the concentration is the same throughout. 43. c, b, a 45. b, a, c 47. About 38 molecules of ATP are obtained from the breakdown of one glucose molecule. So, 10,000,000/38 is about 263,158 molecules of glucose. 49. A chain of two amino acids can include 20 amino acids for the first position and 20 amino acids for the second position, so the number of possibilities is $(20)(20) = 400$. A chain of three amino acids can have any of 20 amino acids in the first position, any of 20 amino acids in the second position, and any of 20 amino acids in the third position, so the number of possibilities is $(20)(20)(20) = 8000$. For 10 amino acids, the number of possibilities is $(20)(20)(20)(20)(20)(20)(20)(20)(20)(20) = 20{,}480{,}000{,}000{,}000$, practically countless. 51. This is an example of asexual reproduction because a bacteria that divides in two is reproducing all by itself. 53. The four nucleotides in DNA can be strung together in a huge number of ways to code for many unique genes. It's like making lots of different English sentences using only the 26 letters in the alphabet—you can string the letters together in so many different ways. 55. Chloroplasts are found only in plant cells; they are responsible for photosynthesis. Through photosynthesis, plants are able to make their own organic molecules that they can then break down for energy. 57. Cell walls typically help to protect a cell. Cell membranes separate the inside of a cell from the outside and control the transport of molecules into and out of the cell. Human cells have cell membranes but not cell walls. 59. The specific fit is important so that a transport protein transports only a specific molecule. This allows cells to control which molecules

enter and exit them. 61. There is a higher concentration of glucose molecules outside cells than inside cells because the glucose molecules inside cells are quickly broken down to make ATP (through cellular respiration). 63. Highly branched roots increase the amount of available surface area for absorption of water (and other soil nutrients). 65. Both gap junctions and plasmodesmata allow messages to pass directly from one cell to an adjacent cell. So, they have similar functions. However, they are different structurally. Gap junctions are tiny channels in the cell membrane surrounded by specialized proteins. Plasmodesmata are slender threads of cytoplasm that link adjacent plant cells. 67. The cell will have two nuclei and twice the genetic material of a normal cell. 69. This is competitive inhibition, since it binds at the enzyme's active site, where the substrate normally binds. 71. Plants remove carbon dioxide from the atmosphere during photosynthesis. 73. The bubbles come from carbon dioxide released during alcoholic fermentation. 75. Answers will vary. Dear Grandma: Humans are made up of a huge number of different kinds of cells—skin cells, muscle cells, liver cells, brain cells, and so on. Embryonic stem cells come from human embryos that have yet to differentiate into distinct types of cells. As a result, embryonic stem cells have the capacity to develop into all the different kinds of cells in the body. The hope of stem-cell research is to develop techniques for growing populations of healthy cells from embryonic stem cells and to use these cells to replace defective or diseased cells in the body. Because embryonic stem cells are easily grown in the lab, this source of healthy cells is potentially quite large. Embryonic stem cells provide great promise for treating a variety of conditions, particularly ones where only one or a few cell types are affected, including diabetes, Parkinson's disease, heart disease, Alzheimer's disease, arthritis, strokes, spinal-cord injuries, and burns. 77. The parts of a membrane protein that are near the phospholipid heads are hydrophilic, as are the parts that stick into or out of the cell. The parts of the membrane protein that are inside the cell membrane, adjacent to the phospholipid tails, are hydrophobic. 79. By stirring the air, the fan brings the smell of perfume to you more quickly. However, even without the fan, the perfume would eventually diffuse all around the room. 81. Yeast obtain a lot more ATP from glucose when they are in aerobic conditions and use cellular respiration—about 38 ATP per glucose molecule. When they are kept in anaerobic conditions and obtain ATP through fermentation, each molecule of glucose yields only 2 ATP (during glycolysis). This is why yeast need more sugar under anaerobic conditions—though only about 19 times as much.

Chapter 16

1. A gene is a section of DNA that contains the instructions for building a protein. 3. Each chromosome consists of a long DNA molecule wrapped around small proteins called histones. 5. The two strands of DNA are separated as if the spiral ladder were unzipped down the middle. Each strand then serves as a template for building a new partner. This is possible because of the way the nitrogenous bases pair—A always pairs with T, and G always pairs with C. As free nucleotides pair up with nucleotides on the template strand, they are attached to the new DNA strand by enzymes called DNA polymerases. 7. RNA differs from DNA in three ways: RNA has only one strand instead of two, RNA uses the sugar ribose instead of deoxyribose, and RNA uses the nitrogenous base uracil (U) instead of thymine (T). 9. During RNA processing, the RNA transcript becomes a mature messenger RNA (mRNA) molecule that is ready for translation. First, a "cap" and "tail" are added to the beginning and end of the molecule. The cap and tail allow the cell to recognize the molecule as mRNA. Second, stretches of nucleotides that are not needed to build the protein, called introns, are removed. The stretches of nucleotides that are used to build the protein, called exons, remain. 11. A tRNA molecule consists of a sequence of three nucleotides called an anticodon and carries a single, specific amino acid. During translation, the mRNA molecule binds to a ribosome. The codon being translated is positioned at a specific site. A tRNA molecule with the appropriate anticodon binds to the codon. The binding of anticodon to codon follows the usual rules—A pairs with U, and G pairs with C. The amino acid carried by the tRNA is then added to the growing protein. 13. Nucleotides can be inserted into or removed from DNA. If this happens in a gene, the codons that are "read" during translation can be shifted, producing a frameshift mutation. Frameshift mutations completely change a protein's amino acid sequence and usually result in nonfunctional proteins. 15. In meiosis, one diploid cell, with two of each kind of chromosome, divides into four haploid cells, each with only one of each kind of chromosome. 17. The inheritance of one trait is independent of the inheritance of a second trait. 19. Independent assortment occurs when genes on different chromosomes separate independently during meiosis. The genes for Mendel's pea traits sorted independently because each was found on a different chromosome. If two genes are on the same chromosome, however, they are often inherited together—though not quite all the time because crossing over may shuffle them up. In fact, the closer two genes are to each other on a chromosome, the more likely they are to be inherited together. 21. Humans have about 22,000 genes. 23. No, a mutation in a single gene is not enough to produce cancer—mutations in many important genes are required. 25. A transgenic organism contains a gene from another species. 27. A mammal is cloned through a process called nuclear transplantation, in which the nucleus of a cell from the animal being cloned is placed into an egg cell that has had its own nucleus removed. The resulting embryo is then implanted into a surrogate mother. 29. DNA is made up of two strands twisted into a spiral or helix. 31. The four nitrogenous bases are adenine (A), guanine (G), cytosine (C), and thymine (T). Adenine always pairs with thymine (A–T), and guanine always pairs with cytosine (G–C). 33. Cells that divide frequently have less time to repair DNA damage before that DNA is replicated and mutations are passed on. As a result, frequently dividing cells are especially vulnerable to ionizing radiation. 35. Environmental factors that increase the risk of cancer risk include smoking, poor diet, radiation, ultraviolet light, chemicals, and infection by certain viruses and bacteria. 37. Answers will vary. Remember that the *A* and *B* alleles are both dominant to *O*. The *A* and *B* alleles are also codominant. Some heterozygotes will be able to tell which allele they inherited from which parent, but others will not be able to tell. 39. The first 30 nucleotides are:

Mouse:	agcatcctag	agtccagatt	ccaaactgct
Human:	aggatcccaa	ggcccaactc	cccgaaccac

There are differences at positions #3, 8, 10, 11, 13, 17, 18, 20, 23, 24, 26, 27, 28, 29, and 30. Many of these also contribute to differences in the amino acid sequence of the two species. 41. $A_TA_TB_TB_TC_TC_T$, $A_TA_TB_TB_TC_TC_S$, $A_SA_TB_TB_TC_TC_S$, $A_SA_TB_TB_SC_SC_T$, $A_SA_TB_SB_TC_SC_S$, $A_SA_TB_SB_SC_SC_S$, $A_SA_SB_SB_SC_SC_S$ 43. Its haploid cells will have $64/2 = 32$ chromosomes. 45. For RNA, G, C, A, and T pair with C, G, U, and A, respectively (that is, RNA uses U instead of T). So the opposite strand will be GACUCCAGUCCU. 47. UGA is a stop codon, so a mutation from G to A in the ninth position of the sequence—that is, to AGUCGUUGACAGGAAGUA—will produce a nonsense mutation. (Other answers may be possible.) 49. No, the two possibilities are *RR* and *Rr*, since round seeds are dominant. If you let an *RR* plant self-fertilize, you expect all the offspring to be *RR* and have round seeds. If you let an *Rr* plant self-fertilize, you expect offspring with round seeds as well as offspring with wrinkled seeds in a 3:1 ratio. 51. Your finger is made of diploid cells, like most of your body except for your sex cells. (Although, in women, eggs do not actually complete meiosis and become haploid until they are fertilized!) 53. Every new DNA molecule has one old and one new strand because during DNA replication, the original DNA molecule is unzipped and each strand is used as a template for putting together a new strand. 55. In both processes, DNA is unzipped and one strand is used as a template for building a new nucleic acid strand using base pairing. During transcription, a molecule of RNA is made, whereas during replication, a strand of DNA is made. Thus, differences include the sugar in the new strand (ribose for RNA and deoxyribose for DNA) as well as the nucleotides in the strand (RNA has uracil in place of thymine). 57. No, some codons are stop codons that tell the ribosome there are no more amino acids in the protein being assembled. 59. No. Recall that each pair of homologous chromosomes separates independently during meiosis I. Even without crossing over and recombination, each egg would receive either the chromosome the woman inherited from her mother or the chromosome she inherited from her father. One egg could receive chromosomes 1, 3, 4, 5, 7, 10, 13, and so on from her mother and chromosomes 2, 6, 8, 9, 11, 12, and so on from her father. A second egg the woman produces is almost certain to receive a different set of chromosomes—perhaps chromosomes 2, 3, 5, 6, 7, and so on from her mother and chromosomes 1, 4, 8, 9, and so on from her father. The independent separation of homologous chromosomes alone produces a huge number of possible egg cells. Crossing over and recombination only expand the possibilities. 61. This person has Down syndrome, which is a result of trisomy 21 (having three copies of chromosome 21 rather than two). Down syndrome is characterized by mental retardation and defects of the heart and respiratory system. Chromosomal abnormalities such as trisomies usually result from mistakes that occur during meiosis that result in an egg or sperm having two copies of a chromosome rather than the usual single copy. (The third copy is added at fertilization.) 63. A pea plant with round seeds is either *RR* or *Rr* (*R* stands for round in this case), since having round seeds is dominant. A red snapdragon must be *RR* (*R* stands for red in this case), since flower color in snapdragons shows incomplete dominance. 65. Not necessarily. If you do not have dimples, your genotype is *dd* and you will pass a *d* allele to your child. However, your child may receive a *D* allele from your spouse. Then your child's genotype will be *Dd*, and he or she will have dimples. 67. Anyone can receive type O blood, so this is the type you should be given. The red blood cells in type O blood have neither A nor B molecules that could cause the cells to be rejected and attacked. People with type O blood are called universal donors because they can donate blood to anyone. 69. This could be caused by pleiotropy, in which a single gene causes both blue eyes and deafness. It could also be caused by linked genes, in which the blue-eye gene and the deafness gene are found close together and alleles for blue eyes and deafness are associated with each other more frequently than either is with alleles for non-blue eyes and non-deafness. 71. Because cystic fibrosis is recessive, children are at risk only if both parents carry a recessive allele (that is, are carriers or are ill with the disease). So, the couple's children are not at risk for cystic fibrosis. However, they could be carriers if they inherit the carrier parent's disease allele. 73. The insertion of a single nucleotide is more likely to disrupt protein function because it will throw the codon reading frame off—that is, produce a frameshift mutation that drastically affects the amino acid sequence. The insertion of three nucleotides causes the insertion of an extra amino acid, but the rest of the amino acid sequence will not be changed. 75. No, crossing over may shuffle them. 77. The mutations that cause cancer are, in most cases, not inherited from one's parents. Instead, cancer-causing mutations occur in some cells in the body during the lifetime of the individual. 79. Messenger RNA (mRNA) is produced during transcription. The mRNA molecule carries information from DNA to ribosomes, where the mRNA molecule is translated to build a protein. Ribosomal RNA (rRNA) is one component of ribosomes, the organelles that translate mRNA to build a protein. Transfer RNA (tRNA) carries amino acids to the protein being assembled. Each tRNA molecule has an anticodon that binds to a codon in the mRNA molecule. 81. A mistake during meiosis can result in an egg or sperm that has two copies of chromosome 21. At fertilization, a third copy is added, resulting in the fertilized egg having three copies of chromosome 21—trisomy 21. 83. Genes that are on the same chromosome will be inherited together unless crossing over occurs at a point between their two locations. The closer two alleles are on a chromosome, the less likely it will be for crossing over to occur between them. 85. A pink snapdragon has the genotype *RW*. If you breed two pink snapdragons, you will get $\frac{1}{4}$ *RR*, $\frac{1}{2}$ *RW*, and $\frac{1}{4}$ *WW* offspring. In other words, $\frac{1}{4}$ of the offspring will be red, $\frac{1}{4}$ will be pink, and $\frac{1}{4}$ will be white. 87. Yes, and in fact some scientists are trying to develop a genetically engineered version of artemisinin now. For various reasons, the process is more difficult for artemisinin than for other molecules such as insulin and human growth hormone, but scientists may get there one day!

Chapter 17

1. Pasteur designed a flask that kept out dust and other airborne particles, filled the flask with sterile meat broth, let it sit, and waited for life to emerge. It never did, and Pasteur concluded that life does not arise from nonlife. 3. Liposomes have double membranes similar to cell membranes. Although they are not alive, liposomes sometimes behave like living cells—they grow, shrink, and divide. Liposomes also run chemical reactions inside their membranes and control what molecules enter and exit them, two key features of living cells. 5. The evolution of autotrophs was crucial because, without autotrophs, heterotrophs would have eaten through their food supply and died out. 7. Scientists believe mitochondria and chloroplasts evolved from prokaryotes that started to live inside the earliest eukaryotic cells. 9. Darwin's finches showed remarkable variation in the size and shape of their beaks, with each beak being suited to, and used for, a different diet. Darwin wrote, "Seeing this gradation and diversity of structure in one small, intimately related group of birds, one might really fancy that from an original paucity of birds in this archipelago, one species had been taken and modified for different ends." 11. Lyell, a geologist, argued that Earth's geologic features were created not by major catastrophic events—the favored theory of the time—but by gradual processes that produced their effects over long time periods. For example, the formation of a deep canyon did not require a cataclysmic flood; it could be the result of a river's slow erosion of rock over millennia. Darwin realized that Lyell's ideas could apply to biological organisms as well: The accumulation of gradual changes over long time periods could produce all the diversity of living organisms as well as all of their remarkable features. 13. Many traits are determined at least partly by genes and so are heritable—that is, they are passed from parents to offspring. 15. Survival is usually an important first step in successful reproduction. 17. Parental care evolved because natural selection favored organisms that were able to help their offspring survive and thrive. 19. Genetic drift is the evolution of populations due to chance. Imagine a half-polluted town where light moths and dark moths are equally successful—neither allele is advantageous. Now suppose a sudden storm wiped out part of the town's peppered moth population. It might happen—just by chance—that more dark moths survive. In this case, the frequency of the dark allele would increase. Notice that genetic drift produces evolution (heritable changes in a population over time) but that this evolution is not the result of natural selection—the dark allele was not advantageous. Genetic drift can also occur when, just by chance, more alleles of one type are transmitted to the next generation than alleles of the other type. For example, even if light and dark moths have equal fitness, light moths might just happen to leave more offspring (and therefore more light alleles) one year. 21. Genetic mutations constantly create new variation within populations. For example, when a genetic mutation changes the amino acids in a protein, it may produce a new allele for a given gene. Sexual reproduction also contributes to variation by bringing together alleles for different traits in new combinations. 23. Prezygotic reproductive barriers prevent individuals of different species from mating in the first place or prevent fertilization from occurring if they do mate. There are many examples of prezygotic barriers—organisms may differ in when they breed, in where they breed, or in the details of their courtship rituals. Their sex organs may not fit together properly, preventing successful sperm transfer, or other factors may prevent fertilization even if sperm is transferred. Postzygotic reproductive barriers act after fertilization has taken place. Mating may produce hybrids that either don't survive or are sterile—unable to breed themselves. The mule, the offspring of a horse and a donkey, is sterile and cannot reproduce. 25. Adaptive radiations are spectacular examples of allopatric speciation, where many new species, each adapted to a distinct way of life, evolve from a single ancestor. Adaptive radiations have commonly occurred on island archipelagos, which have abundant opportunities for geographic isolation. Adaptive radiations often occur after a new habitat is colonized. 27. If each of these mammals had originated independently, we would expect their limbs to look completely different. Each limb would have been designed from scratch to best perform its function. But, despite the different functions of human hands, cat legs, whale flippers, and bat wings, all these limbs show the same arrangement of bones. This suggests that the limbs were inherited from a common ancestor and then modified through natural selection for different functions. 29. The bones of Lucy's pelvis make it clear that she walked upright on two legs. In fact, older *Australopithecus* fossils show that an upright posture dates to at least 4 million years ago. 31. The earliest fossils of modern humans, *Homo sapiens sapiens*, were found in Ethiopia and are 195,000 years old. 33. Life may have found its way to Earth in Martian dust that was set adrift in space when a comet collided with Mars. 35. The heat an animal produces depends on its volume. The heat it loses to its environment depends on its surface area because heat is lost through the body's surface. As a result, animals are better able to lose heat if they have a high surface-area-to-volume ratio and better able to retain heat if they are have a low surface-area-to-volume ratio. 37. The tropical species is likely to be smaller, with longer limbs and larger ears. The Arctic goat is likely to be larger, with shorter limbs and smaller ears. 39. *Archaeopteryx*, the famous 150-million-year-old fossil bird, shows intermediate traits in the evolution of birds from their dinosaur ancestors. *Archaeopteryx* has many birdlike features, such as feathers, wings, and a wishbone. However, it also has dinosaur-like features absent in modern birds, including claws on its wings, bones in its tail, and teeth. 41. Answers will vary, but there should be plenty of organisms and adaptations to notice! 43. A, B, C 45. Yes, this is natural selection because color is a variable heritable trait and brown and green individuals have different fitness. What is happening here is that, over time, the population is shifting toward a higher proportion of brown individuals.

Generation	Brown	Green	Proportion Brown
1	2	2	0.50
2	4	2	0.67
3	8	2	0.80
4	16	2	0.89
5	32	2	0.94
6	64	2	0.97
7	128	2	0.98
8	256	2	0.992
9	512	2	0.996
10	1024	2	0.998

47. If there is only a single individual in the population and that individual has genotype *RR*, then the frequency of the *R* allele is 1 and the frequency of the *W* allele is 0. Yes, this is an example of genetic drift. 49. By building a simple model of early Earth, Miller and Urey were able to produce large numbers of complex organic molecules—the building blocks of life. This suggests that early Earth might well have been full of organic molecules that could eventually come together in specific ways to make living cells. 51. Answers will vary. Traits that do not show variation include having a four-chambered heart, five fingers on each hand, two arms, two legs, and one nose. Traits that show variation include height, weight, arm length, foot length, eye color, and hair color. Heritable traits include having a four-chambered heart, five fingers on each hand, two arms, two legs, and one nose, as well as height, weight, arm length, foot length, eye color, and hair color. Nonheritable traits include hairstyle, hair length, clothing, scars, tattoos, pierced ears, and language spoken. 53. In both cases, the administration of a lethal agent (antibiotics or the myxoma virus) led to natural selection for resistant individuals in the population. These resistant individuals survived and reproduced. As a result, over time, populations evolved resistance to the lethal agent. 55. It could be evolution, but the change is more likely due to better nutrition. 57. Bright colors on certain male birds help them attract female mates. However, bright colors also make these males more visible to predators. It is adaptive for females to be less colorful because they don't need the bright colors to attract mates and their duller feathers make it harder for predators to see them. 59. Alternative explanations are genetic drift, migration into or out of the population, and mutation pressure. To determine whether natural selection is responsible, you could compare the fitness (number of offspring left) of red individuals versus yellow individuals. If this turned out to be difficult, you could compare their survival or ability to acquire mates. 61. This is a postzygotic barrier, since it occurs after fertilization. 63. Probably, your observations suggest that there is a prezygotic reproductive barrier. 65. Artificial selection occurs in the breeding of dogs as well as in the breeding of corn from teosinte. Both artificial selection and natural selection are driven by differences in reproductive success (imposed by human beings in the case of artificial selection and by the environment in the case of natural selection). Artificial selection and natural selection are different in that humans can impose whatever conditions they like on artificial selection, and perhaps produce forms that wouldn't do very well in the wild. Humans can also make artificial selection much more stringent than natural selection, insofar as they can choose two specific racehorses and breed them. As a result, evolution by artificial selection can occur surprisingly quickly compared to evolution by natural selection. 67. Humans are primates. We belong to a group of mammals that also includes monkeys and apes. This doesn't mean that humans are descended from any existing species of monkey or ape. It just means that we're more closely related to monkeys and apes than we are to other living things, such as dogs, lizards, or plants. 69. On average, the beaks of the finches will become larger and stronger. This is exactly what happened in certain species of Galápagos finches studied by Peter and Rosemary Grant after a drought. 71. Your friend is probably correct—poisonous animals tend to advertise their toxicity with bright colors so that you notice them before you bite down on them. 73. In the grassy environment, the green individuals may have greater fitness because they are better camouflaged. The population is likely to evolve to have more green individuals. Because of migration, sandy genes will constantly enter the grassy population, and green genes will constantly enter the sandy population. Migration and gene flow make it harder for each population to adapt to its environment. 75. Evolution is falsifiable. In the case of biogeography, it is possible to imagine a situation where the same species is found in many different isolated places on Earth, where islands have more biodiversity than mainlands, where islands are full of terrestrial animals that would have had trouble crossing over from the mainland, where closely related species are scattered over Earth, and other scenarios. There are rare examples; however, those patterns can usually be explained by plate tectonics and the history of the movements of Earth's continents! 77. Yes, the use of narrow-spectrum antibiotics could slow the evolution of antibiotic resistance because narrow-spectrum antibiotics affect fewer species of bacteria, and so they select for resistance in fewer species of bacteria. A broad-spectrum antibiotic selects for resistance in all the (many) species it affects.

Chapter 18

1. At every level in the Linnaean system, species are grouped together based on shared similarities. For example, all species in the class Mammalia (all mammals) have hair and nurse their young with milk. 3. A clade is a group of species that includes an ancestor and all of its descendants. 5. Eukaryotes belong to the domain Eukarya, living things whose cells have a nucleus. 7. Bacteria typically reproduce asexually by dividing. However, most species exchange genetic material at least occasionally—when they take up small pieces of naked DNA from the environment, when bacterial viruses inadvertently transfer DNA between organisms, or when two bacteria join together and one passes DNA to the other. 9. The ribosomes of archaeans are like those of eukaryotes, their genes contain introns like those of eukaryotes, and their DNA is associated with histone proteins, like that of eukaryotes. 11. Protists are eukaryotes that are not plants, animals, or fungi. 13. All the different kinds of seaweeds are multicellular photosynthetic protists. Kelp forms huge oceanic forests that are home to many unique species. Red algae are the source of some of the seaweed we eat, including Japanese nori. Green algae are multicellular protists that likely gave rise to terrestrial plants. 15. Malaria is caused by *Plasmodium* protists, which divide their life cycle between mosquitoes and humans. Humans contract the disease when infected mosquitoes bite them. The protists then move into our red blood cells and reproduce in huge numbers. The synchronized emergence of protists from host red blood cells causes chills, fever, and vomiting. 17. In mosses, the gametophyte is much larger than the sporophyte. When you see a moss in the forest, you are looking at a haploid gametophyte. 19. Fungi are heterotrophs that obtain food from other organisms. Fungi release digestive enzymes over organic matter and then absorb the nutrients. 21. Fungi are essential to most plants because, in most plant species, the roots form close associations with fungi called mycorrhizae. The fungus receives nutrients from the plant while helping the roots absorb water and minerals from the soil. 23. All arthropods have an external skeleton called an exoskeleton that protects and supports the body. The exoskeleton is incapable of growth and must be shed periodically as the animal grows. The bodies of arthropods are divided into different segments, and their legs have bendable joints. Some arthropod legs have been modified during evolution to function as mouthparts, antennae, and reproductive organs. Arthropods have a brain and well-developed sense organs. Many species pass through a distinct larval stage during their growth and development. 25. Amphibians are restricted to moist habitats because their skins are made of living cells that dry out. Amphibian eggs do not have shells and also require moisture. 27. Monotremes, such as the platypus and spiny echidna, are mammals that lay eggs. 29. Prions are proteins that are incorrectly folded. (Remember that proteins are folded strings of amino acids. In prions, the folding has somehow gone wrong.) Prions infect cells and then "reproduce" by converting normal proteins into the incorrectly folded form. 31. Cohesion and adhesion maintain the continuous column of water molecules in the xylem. 33. Coral bleaching is most often triggered by an increase in seawater temperature. 35. The extra hydrogen ions react with carbonate, reducing the amount of carbonate in the water. Many marine animals—including corals, echinoderms, crustaceans, and mollusks—need carbonate to build their calcium carbonate shells. More acidic seawater also increases the rate of dissolution of their shells. 37. In order to generate lift, birds must move forward, which they do by flapping their wings. When a bird pulls its wings downward, it pushes air down and back. Recall Newton's third law: If a bird pushes air backward, the air pushes the bird forward. 39. Answers will vary, depending on the kind of habitat you take your walk in. Enjoy! 41. Answers will vary. Enjoy! 43. domain, kingdom, phylum, class, order, family, genus, species 45. The population doubles every 20 minutes, so:

8:00	1	9:40	32	11:00	512
8:20	2	10:00	64	11:20	1024
8:40	4	10:20	128	11:40	2048
9:00	8	10:40	256	12:00	4096
9:20	16				

47. Under Linnaean classification, different orders are grouped together into a class. This means that two species that belong to the same order must also belong to the same class. However, these two species do not have to belong to the same genus because an order can contain one or more families, and each family can contain one or more genera. So, two species that are in the same order could be in different genera. For example, the order Primates includes many different genera. 49. A chemoautotroph is an autotroph that makes food and organic molecules using chemical energy. Like plants, chemoautotrophs are autotrophs—they convert inorganic molecules into food and organic molecules. However, plants use energy from sunlight to do this, and chemoautotrophs use chemical energy. 51. Decomposition, the breaking down of dead organic matter, frees up resources that would otherwise be trapped in the corpses of dead organisms. Important decomposers include fungi and bacteria among other groups. 53. Mosses are most dependent on living in a moist environment because they have no vascular systems. Instead, every part of a moss plant receives water directly from the environment. In addition, mosses have swimming sperm that require moisture in the environment in order to travel to and fertilize moss eggs. Seed plants are least dependent on living in a moist environment because they use pollen rather than swimming sperm during sexual reproduction, and they have vascular systems. (The third group, ferns, have vascular systems but also use swimming sperm, so they are more moisture-dependent than seed plants.) 55. Seed plants produce pollen. Pollen is carried to the female flower or cone by wind or by animals. Most flowering plants make use of animal pollinators, particularly insects. 57. Wind-pollinated plants are more likely to cause allergies because they make larger quantities of pollen. (Wind blows pollen all over the place, so releasing lots of pollen helps to ensure that some of the pollen will reach its destination.) 59. Fungal spores function in reproduction and also help fungi disperse to new locations. 61. The muscles of roundworms all run longitudinally (from head to tail) down the body. As a result, roundworms move like flailing whips as muscles on alternate sides of the body contract. The muscles of earthworms are arranged in both circular (around the body) and longitudinal (from head to tail) orientations, allowing for great flexibility of motion. Unlike roundworms, for example, earthworms are able to contract one part of the body while keeping the rest of the body still. 63. Birds are endotherms and need a lot of food to maintain their high, stable body temperatures. Snakes are also unusual in that they have adaptations for eating very large prey; this is part of the reason they don't need to eat as frequently. 65. Hollow bones make birds lighter. Since birds have to produce enough lift to counteract the effects of gravity, a lighter body makes it easier for a bird to stay in the air. 67. Answers will vary. Dear Grandpa: Global warming increases seawater temperatures, which has the potential to affect many species. We already know that corals in particular are very sensitive to temperature increases because their dinoflagellates have stringent temperature requirements. If seawater temperatures become too high, the corals kick out their dinoflagellates and turn white—this is called coral bleaching. Without their dinoflagellates, corals starve to death. Because coral reefs are extremely diverse ecosystems, the loss of corals would affect many other species that live in and around the reefs as well. A second way global warming threatens marine life is through ocean acidification. As excess carbon dioxide in the atmosphere is dissolved in ocean water, it reacts with the water to form carbonic acid. Carbonic acid then reacts with carbonate minerals such as calcium carbonate, reducing the amount of calcium carbonate available for shelled animals to build their shells. Seawater that is more acidic also increases the rate of dissolution of the shells. 69. Certain bacteria naturally consume oil; however, bacteria can consume oil only on the edges of oil slicks. By breaking up a large oil slick into tiny droplets, chemical dispersants expose much more surface area to hungry bacteria. The result is that the bacteria can consume the same volume, or amount, of oil much more quickly. Chemical dispersants have serious drawbacks as well: They are highly toxic, and the damage they do to aquatic and terrestrial environments may be far worse than the benefits they offer. 71. Genetic exchange results in genetic diversity among the offspring. Nature doesn't put all its eggs into one (genetic) basket. Since future environmental conditions are impossible to predict, at least some offspring are likely to do well no matter what happens. 73. Carbon dioxide enters plants when it diffuses into leaves through the stomata. Diffusion is a passive process; it does not require energy from the plant. 75. An air bubble in the xylem would break the continuous column of water molecules that extends from the leaves to the roots. The plant would no longer be able to obtain water through the transpiration-cohesion-tension mechanism in that part of the xylem. 77. Because lift comes from air pressure under the wings, the amount of lift a bird generates depends on the surface area of its wings. Lift must counteract gravity, which is proportional to the volume of the bird. Since the surface-area-to-volume ratio is lower in larger animals, larger birds have more difficulty generating the necessary lift. This explains why large birds have relatively larger wings than small birds. It also explains why small birds are better fliers—think of the maneuverability of a hummingbird! Many of the largest birds actually spend most of their time soaring on thermals.

Chapter 19

1. The four types of tissues are epithelial, connective, muscle, and nervous. Epithelial tissue consists of sheets of tightly packed cells that cover the internal and external surfaces of the body—skin is an example. Connective tissue consists of cells scattered within an external matrix—bone, cartilage, and blood are examples. Muscle tissue consists of cells that are able to contract or shorten—skeletal, smooth, and cardiac muscles are examples. Nervous tissue transmits information from one place in the body to the other—examples are found in the brain, spinal cord, and nerves. 3. Multiple organs make up an organ system, which is responsible for a particular bodily function. 5. Oxygen supply and body temperature are two of the many variables the body carefully regulates. The amount of water in the body, the concentration of nutrients and waste products in the blood, the concentrations of important ions inside and outside cells, and blood pH are all carefully regulated as part of maintaining homeostasis. 7. The cerebellum controls balance, posture, coordination, and fine movements. 9. The frontal lobes deal with reasoning, voluntary movements, and speech. The parietal lobes take in sensory information about temperature, touch, taste, and pain. The occipital lobes process what you see—that is, visual information. The temporal lobes deal with sound and help you comprehend language. 11. Dendrites receive information from other neurons or cells, the cell body contains the neuron's nucleus and organelles, and the axon transmits information to other neurons or cells. 13. Motor neurons in the somatic nervous system control voluntary actions, such as running to class. These neurons provide instructions to your voluntary muscles, such as the muscles of your arms, legs, and fingers. Motor neurons in the autonomic nervous system control involuntary actions. These neurons control organs and involuntary muscles (such as heart or stomach muscle). 15. An influx of positively charged ions causes the membrane potential to spike and become positive. This spike is called an action potential. 17. When an action potential begins, sodium ions enter the end of the axon that is closest to the cell body. These ions diffuse into adjacent areas along the axon. Because the sodium ions are positively charged, they cause the local membrane potential to increase. When the local membrane potential reaches threshold, a new action potential, farther along the axon, begins. The process is similar to the way a row of dominoes falls—the first domino knocks down the next one, which knocks down the next one, and so forth. In this way, the action potential travels down the entire axon. 19. In a chemical synapse, a neuron communicates with a target cell using molecules of neurotransmitter. When an action potential arrives at the end of an axon, vesicles containing neurotransmitter fuse to the neuron's cell membrane. Neurotransmitter molecules are released into the narrow space between the axon and the target cell, and they diffuse to the target cell. There, the neurotransmitter molecules bind to receptors on the target cell's membrane. The binding causes ion channels to open, and ions enter the target cell. The membrane potential of the target cell may increase or decrease, depending on the type of ion that enters. An excitatory signal increases the target cell's membrane potential and makes the target cell more likely to fire. An inhibitory signal decreases the target cell's membrane potential and makes it less likely to fire. 21. Sound waves move through the air to the outer ear, or pinna, the cartilaginous flap on the side of your head. The pinna funnels the waves in, and they move toward a thin membrane of skin—the eardrum. Sound waves make the eardrum vibrate, just the way blowing on a piece of paper makes it shake. The eardrum's vibrations move three middle ear bones—the hammer, anvil, and stirrup—in sequence. These bones amplify the vibrations, making them more powerful. The stirrup then transfers the vibrations to the fluid-filled inner ear. In the inner ear, sound vibrations enter the cochlea, a coiled tube containing the organ of Corti. The organ of Corti contains the sensory cells responsible for hearing. Fluid vibrations in the inner ear move the organ of Corti's basilar membrane, causing sensory "hairs" embedded in it to brush against an overlying membrane and bend. This bending causes ion channels to open, initiating action potentials that are transmitted to the brain. 23. Proprioceptors in your muscles, tendons, and joints tell you where different parts of your body are. 25. Many of the hormones of the anterior pituitary regulate the activity of other endocrine organs. For example, anterior pituitary hormones tell the thyroid, sex organs, and adrenal glands to

release their hormones. 27. The adrenal glands, located above the kidneys, make epinephrine (also called adrenaline) and norepinephrine. Epinephrine and norepinephrine are involved in the "fight or flight" response. 29. Enzymes released from the heads of many sperm eat away the zona pellucida. 31. All of the embryo's major organs and body parts develop during the first trimester. 33. A muscle contracts when it receives a signal from a motor neuron. An action potential in the motor neuron arrives at a chemical synapse connecting the neuron to a muscle cell. The action potential triggers the release of the neurotransmitter acetylcholine at the synapse. Acetylcholine then binds to receptors on the muscle cell's membrane, initiating an action potential in the muscle cell. This action potential causes calcium ions to be released from the muscle cell's endoplasmic reticulum. Calcium ions allow a series of "heads" on the myosin fibers to attach to actin. The myosin heads attach and pivot, pulling on the actin filaments. Each pull shortens the length of the sarcomere a tiny bit—about 10 nm—and, consequently, the length of the muscle as a whole. After pulling, the myosin heads release, extend, attach, and pull again. This cycle repeats until the signal to contract ends or until the muscle has fully contracted. 35. Action potentials travel faster if the sodium ions that produce them move more quickly down an axon. How quickly sodium ions move down an axon is determined by Ohm's law: Current = voltage/resistance. The lower the resistance, the higher the current and the faster the action potential. Like any other material, an axon has lower resistance if it is thicker. This is because a thick axon resists current less than a thin axon, the same way a thick pipe resists water flow less than a thin pipe. So, one way to get a fast action potential is to build a thick axon. 37. In a myelinated axon, the action potential is not regenerated at every point along the axon. Instead, the action potential "jumps" from one gap in the myelin sheath to the next, thus saving time. 39. Endorphins bind to opiate receptors on neurons, the same receptors used by drugs such as morphine, codeine, opium, and heroin. 41. The dot disappears when it is in your blind spot. Your brain fills in that space with what is visible around the blind spot—in this case, white paper. 43. cell, tissue, organ, organ system 45. (a) Rods are better able to see in dim light than cones. (b) Cones are better able to make out fine details than rods. (c) Cones are better able to distinguish different colors than rods. 47. We have a total of $125 + 6.5 = 131.5$ million rods and cones, so $131{,}500{,}000/1000 \text{ mm}^2 = 131{,}500$ sensory cells per mm^2. 49. The volume of a human egg $= 4/3\pi r^3 = 4/3\pi(50)^3 = 4/3\pi(125{,}000)$, and the volume of a human sperm $= 4/3\pi r^3 = 4/3\pi(2)^3 = 4/3\pi(8)$. So a human egg is $125{,}000/8 = 15{,}625$ times bigger in volume than a human sperm. 51. The cerebellum is also involved: It controls balance, posture, coordination, and fine movements. It also controls all the automated motions you perform "without thinking." 53. A motor neuron goes to your biceps muscle and tells you to bend your elbow. A sensory neuron transmits information from your feet regarding how they feel. 55. The action potential would travel much more slowly along the axon and not "jump" from one gap in the myelin sheath to the next. 57. A target cell responds to a neurotransmitter if it has receptors for that neurotransmitter. 59. Your cones are responsible for making out fine details, such as the letters in a book. 61. It is a steroid hormone. Steroid hormones have receptors in the cytoplasm or nucleus, whereas protein hormones have receptors on the cell membrane. 63. Meiosis is used to produce eggs and sperm. After fertilization, mitosis allows for growth and development. 65. The fertilized egg also makes the membranes that surround and protect the embryo, including the amnion and the embryonic contribution to the placenta. 67. Yes, in feedback regulation the level of a body variable initiates a process that comes back to affect that variable. When body temperature is low, the body initiates activities such as shivering or putting on more clothes that help to increase body temperature. When body temperature is high, sweating and other activities help to lower it. Both are examples of feedback regulation. 69. You use the temporal lobes of your cerebrum to understand spoken language and the frontal lobes of your cerebrum to talk to your friend—to speak. 71. When you are looking at the night sky, there is very little light, so you use your rods to see. Rods see only in black and white, so you cannot see the colors of the stars. This is why stars look much more colorful in photographs than to the naked eye. 73. Your body's internal clock is regulated by the hormone melatonin. When you fly across many time zones, your internal clock ends up out of sync with day and night at your destination. The result is that you may be wide awake during the night and sleepy during the day. Over time, light cues from your eyes help the pineal gland change the times it releases melatonin. (Melatonin is released during the night.) This allows you to adjust to your new time zone. There is no such thing as "train lag" or "bicycle lag" because trains and bicycles do not go fast enough to take you across many time zones over the course of a short period of time. 75. Only one sperm fertilizes an egg, but many sperm are involved in the process of getting through the zona pellucida that surrounds the egg. Enzymes released from the heads of many sperm help to penetrate the zona pellucida. 77. You decide to wiggle your toe. Wiggling is a voluntary movement, so the instruction originates in the frontal lobe of your cerebrum (the left frontal lobe if you wiggle your right toe, the right frontal lobe if you wiggle your left toe). An action potential in a motor neuron that goes to your toe signals muscles there to contract. The action potential arrives at a chemical synapse connecting the neuron to the muscle cells and causes the neurotransmitter acetylcholine to be released. Acetylcholine binds to receptors on the muscle cell's membrane, initiating an action potential in the muscle cell. This causes calcium ions to be released from the endoplasmic reticulum, and the calcium ions in turn allow myosin heads to attach to actin. When the myosin heads pivot, pulling on the actin filaments, the sarcomeres shorten and the muscle contracts. After pulling, the myosin heads release, extend, attach, and pull again. This cycle repeats, and the muscle becomes shorter and shorter. The shortened muscle pulls on the toe bone it is attached to, and your toe bends one way. A similar process causes opposing muscles to contract, bending your toe the other way. And there you are, wiggling your toe!

Chapter 20

1. The respiratory system brings oxygen into the body, and the circulatory system transports the oxygen to the tissues. 3. Heart muscle does not need a neuron to tell it to contract the way voluntary muscles do. Instead, the heart contracts on its own. Each heartbeat begins on its own in a part of the right atrium called the sinoatrial node, or pacemaker. 5. Capillaries are tiny, thin-walled blood vessels through which molecules are exchanged between blood and body tissues. 7. The three types of blood cells are red blood cells, white blood cells, and platelets. Red blood cells transport oxygen to the body's tissues; they contain numerous molecules of the protein hemoglobin, which binds to oxygen. White blood cells are part of the immune system and help defend your body against disease. Platelets are involved in blood clotting. 9. Both the alveolus and the surrounding capillaries have extremely thin walls, consisting of only a single flattened cell. This allows gas exchange to happen through diffusion. 11. During digestion, food is broken down into small organic molecules that can be absorbed and used by the body. 13. A small flap of cartilage at the back of your tongue—the epiglottis—covers the trachea when you swallow. 15. The small intestine's inner surface is covered with fingerlike projections called villi. The villi are covered with even tinier projections called microvilli. Both the villi and microvilli increase the surface area of the small intestine. 17. From food, you obtain calcium for bones and teeth, phosphorus for ATP, and iron for hemoglobin. 19. Urea is one of the most important wastes found in urine. When amino acids are broken down to make ATP, ammonia, a nitrogen-containing waste, is produced. Because ammonia is highly toxic, it is immediately converted into urea, a less-toxic waste, by the liver. The liver releases urea back into the bloodstream for excretion. 21. The proximal tubule is like a sorting machine. "Good" molecules in the filtrate—such as ions, glucose, vitamins, and amino acids—are transported back into the blood so that the body can keep them. "Bad" waste molecules are transported from the blood into the filtrate. 23. In the presence of antidiuretic hormone, more water is absorbed from the filtrate and retained by the body. As a result, urine becomes more concentrated. 25. The skin is a crucial barrier to pathogens. It is a tough outer layer that is hard to penetrate when intact. In addition, skin cells are shed frequently, making it hard for pathogens to establish a foothold. Hair follicles in skin also secrete special enzymes and acids that kill bacteria and fungi. 27. An antigen is a molecule or part of a molecule that belongs to a pathogen. Most often, antigens are parts of foreign proteins. 29. Memory cells stay in the body for a long time—years or even a lifetime. If the body encounters the same pathogen again, the memory cells initiate a quick, aggressive attack. Pathogens may be eliminated from your body before you even develop symptoms. This is why you catch many diseases only once and then are immune for life. 31. Each hemoglobin molecule can carry up to four molecules of oxygen. 33. If you use up more Calories than you take in, then your body uses stored energy—such as fat—to support its activities. 35. People on low-carb diets actually lose weight for the same reason that people on low-Calorie diets lose weight—they consume fewer total Calories. In addition, low-carb diets cause people to retain less water. This water is used during excretion to flush out the extra proteins consumed. 37. Answers will vary. 39. Answers will vary.

41. (25,000,000,000,000 red blood cells) × (300,000,000 hemoglobin molecules/red blood cell) × (4 molecules of oxygen/hemoglobin molecule) = 30,000,000,000,000,000,000,000 molecules of oxygen, or 3×10^{22}. 43. For heartbeat:
(70 beats per minute) × (60 minutes per hour) = 4200 beats per hour
(4200 beats per hour) × (24 hours per day) = 100,800 beats per day
(100,800 beats per day) × (365 days per year) = 36.8 million beats per year
For breaths:
(12 breaths per minute) × (60 minutes per hour) = 720 breaths per hour
(720 breaths per hour) × (24 hours per day) = 17,280 breaths per day
(17,280 breaths per day) × (365 days per year) = 6.3 million breaths per year
45. Heart muscle has the ability to contract on its own, without nervous system control. This is true not only in developing embryos but throughout our whole lives. 47. The contractions of voluntary muscles help move blood back toward the heart. 49. The body controls the diameters of the arterioles that go to different parts of the body. 51. The trachea is stiffened by rings of cartilage that keep it open. 53. This arrangement enables you to breathe through your mouth. Air can enter your mouth, go into the pharynx, and then go down your trachea. 55. Carbon dioxide and nitrogen-containing wastes (in the form of ammonia, which is then quickly converted into urea) are the waste products produced during cellular respiration. Carbon dioxide is removed from the body by the respiratory system. Nitrogenous wastes are removed by the excretory system. 57. Elimination is associated with the digestive system, excretion with the excretory system. Elimination removes substances in food that cannot be absorbed or used by the body. (Feces are composed of primarily living and dead bacteria and indigestible materials such as plant cellulose.) Excretion helps control the amount of many substances in the body, but it helps us get rid of urea in particular, which is a product of breaking down proteins during cellular respiration. 59. Tears contain enzymes that kill bacteria. This prevents many pathogens from entering the body through the eyes. 61. The acquired immune system is very specific in its response to pathogens and other foreign substances, whereas the innate immune system has defenses that work against a wide variety of potential pathogens. Each cell of the acquired immune system has receptors that respond to a single antigen, whereas innate immune receptors respond to features that characterize multiple pathogens. The acquired immune response is much slower than that of the innate immune system, usually taking three to five days to reach full strength; the innate immune response is immediate. The acquired immune system retains a "memory" of pathogens it has encountered in the past, so that subsequent responses to the same pathogen can be faster and more aggressive. The innate immune system has no such "memory." 63. B cells respond to antigens in bodily fluids. T cells respond to antigens that infected body cells display (like SOS flags) on their surface. 65. Hearing requires the functioning of our skin (which makes up part of the eardrum) and skeleton (the cartilaginous outer ear as well as the middle ear bones). Smelling requires the respiratory system, which brings air in to be "sampled" by our sensory cells for smell. Tasting requires saliva, part of the digestive system, because molecules must be dissolved in liquid in order to be sensed. 67. No. Gas exchange occurs through diffusion, a form of passive transport that does not require energy from your cells. 69. Saliva contains a digestive enzyme that breaks down the starches found in crackers. Meat is largely protein, and the digestive enzymes that break down proteins are not found in your mouth; they are farther down the digestive tract. 71. Breaking down proteins to make ATP results in the production of ammonia. The liver is responsible for converting ammonia into urea, and the kidney is responsible for excreting urea. So both organs are particularly taxed when the diet includes large amounts of protein. 73. The ammonia molecule is released into the bloodstream and moves through the circulatory system until it eventually reaches the liver. The liver converts it into urea and releases it back into the bloodstream. From there, the urea molecule remains in the blood until it eventually reaches the kidneys, where it enters the filtrate and is ultimately excreted in urine. 75. Answers will vary. It is interesting (and perhaps not surprising) that the leading causes of death in the United States are mostly "lifestyle" diseases; they are not caused by a specific pathogen but by a lifetime of accumulated harm to the body. Many of the leading causes of death in low-income countries are related to issues of health care availability associated with poverty.

Chapter 21

1. Ecology is the study of how organisms interact with their environments. 3. A community consists of all the organisms that live in a specific area. An ecosystem consists of all the organisms that live in a specific area *and* all the abiotic features of their environment. 5. Favorable conditions start exponential growth. When resources become scarce, the population crashes. 7. Type I organisms have low death rates early in life, with most individuals surviving until fairly late in life. Type III organisms have high death rates early in life, with few individuals surviving until late in life. Type II organisms fall between Types I and III. Type II organisms experience a steady death rate throughout life—individuals are as likely to die early in life as late in life. 9. The age structure describes the distribution of people's ages within the population. One thing we can learn from the age structure is how a population is growing. A pyramid-shaped age structure, such as Africa's, indicates a rapidly growing population; most of the population is young. The age structure of the United States is much more even; its population is growing slowly. 11. The name is food chain or food web. 13. A decomposer obtains food by eating dead organic matter. Fungi and bacteria are important decomposers in many communities. 15. Parasitism is a relationship that benefits one member of the interaction and harms the other. Fleas, ticks, and tapeworms are parasites that obtain nutrients from their hosts. This relationship benefits the parasite, which gets nutrients and a home, but harms the host. Pathogens such as bacteria or viruses are parasites as well. 17. The type of biome found in a certain habitat depends mainly on climate, including factors such as temperature, rainfall, and the severity of seasonal variation. 19. Fires prevent tree growth. 21. Organisms that live in the intertidal zone, such as barnacles, sea anemones, and starfish, must be able to deal with exposure to air, temperature fluctuations, and waves. Special adaptations allow them to live in this environment. For example, many intertidal species have thick shells or hide in crevices to keep from drying out. In addition, all species can attach firmly to rocks or other surfaces so that they do not get washed up onto the beach. 23. Sunlight energy enters the biotic world when plants and other organisms use it to build organic molecules during photosynthesis. 25. Primary succession occurs when bare land, devoid of soil, is colonized by successive waves of living organisms. Primary succession may begin when new land is formed by volcanic activity or when a glacier's retreat reveals bare rock. Secondary succession occurs when a disturbance destroys the existing life in a habitat but leaves the soil intact. Secondary succession may begin after a fire or when old farmland is abandoned. 27. During the entire process of succession, the total biomass of the ecosystem typically increases, as does the number of species present. 29. Living organisms are made up of many different substances—water, carbon, nitrogen, phosphorus, sodium, calcium, sulfur, chlorine, and others. All these substances move around Earth in a series of biogeochemical cycles, going back and forth between the tissues of living organisms and the abiotic world. The word *biogeochemical* emphasizes that substances cycle between living organisms (*bio*) and Earth (*geo*)—specifically, Earth's atmosphere, crust, and waters. 31. Most of Earth's nitrogen exists as nitrogen gas in the atmosphere. However, nitrogen gas is not a form of nitrogen that most living things can use. Bacteria help convert nitrogen into a usable form. Nitrogen-fixing bacteria in soil convert nitrogen gas into ammonium (NH_4), and then nitrifying bacteria convert ammonium into nitrates (NO_3). Plants absorb nitrogen primarily in the form of nitrates, although they make some use of ammonium as well. Nitrogen then moves from plants up the food chain. 33. The chemical reaction for burning glucose is

$$C_6H_{12}O_6 + 6\,O_2 \rightarrow 6\,CO_2 + 6\,H_2O + 673\text{ kcal/mole}$$

That is, glucose and oxygen react to form carbon dioxide and water, releasing 673 kcal of energy per mole in the process. If this reaction were perfectly efficient in organisms, the entire 673 kcal per mole would be captured as ATP. Is it? We know that about 38 molecules of ADP are converted into ATP as the result of burning a single glucose molecule. ATP then provides 7 kcal per mole when it is broken down into ADP and phosphate during cellular processes. But, 38 × 7 = 266, much less than 673. Clearly, a lot of energy is missing! It is lost to the environment as heat. 35. On day 10, you will have 512 pennies. Yes, your graph should look like exponential growth. 37. Answers will vary, but most humans (except for strict vegetarians) are primary consumers and secondary consumers as well as top predators. 39. desert, coniferous forest, temperate forest, tropical forest

41.

Year	Individuals Alive
2010	100
2011	90
2012	80
2013	70
2014	60
2015	50
2016	40
2017	30
2018	20
2019	10
2020	0

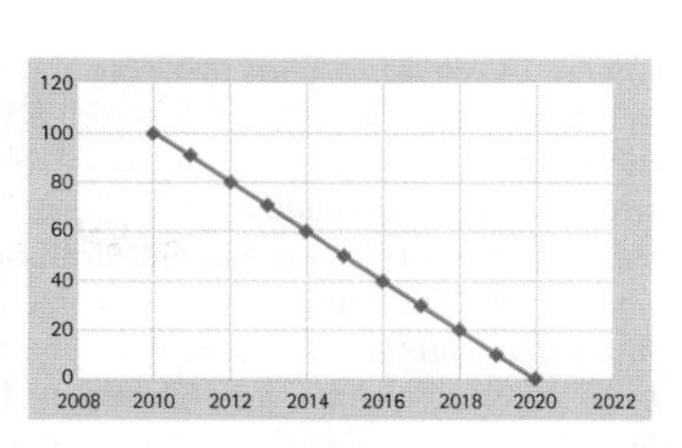

This is Type II survivorship.

43. The carnivores use a total of (5 carnivores) × (2000 kcal/day) = 10,000 kcal/day. If 10% of the energy available at each level of the food chain moves up to the next level and the carnivores are secondary consumers, then 10% × (energy used by primary consumers) = 10,000 kcal/day, so the energy used by primary consumers must be 100,000 kcal/day. Also, 10% × (energy used by producers) = 100,000 kcal/day, so the energy used by producers must be 1,000,000 kcal/day. 45. This is a community-level study because it considers interactions among different species that occupy the same area. 47. For population growth to slow, either the birth rate has to decrease or the death rate has to increase. When a population is large, growth might slow because of more competition for resources such as food and space. This could make it more difficult for organisms to get enough resources to reproduce, causing birth rates to fall. Or, crowded conditions could increase the death rate. 49. In 1999 there was higher infant and early mortality in India than in England. This may have been because poverty affected more people in India and more people were unable to get proper medical care. There is a huge difference between 1999 and 17th-century survivorship curves in England. The 17th-century curves resemble Type III curves, with very high infant and child mortality. 51. All else being equal, a meat-eater will have a larger ecological footprint than a vegetarian. By eating higher on the food chain, a meat-eater ends up using more of Earth's resources. This is because it takes more area to grow the grain to feed animals for consumption (such as cows or chickens) than it takes to supply a person with a vegetarian diet. This goes back to the "10% rule" regarding energy flow through an ecosystem: only about 10% of the energy available at each level of the food chain moves up to the next level. 53. This is mutualism because both species benefit. The insects receive food and the plants are pollinated, a key step in their reproduction. 55. No. The strips of tundra occur in high altitudes, where climatic factors such as extreme cold result in tundra. 57. Humans are heterotrophs, and we get all our carbon from the food we eat. Ultimately, all this carbon came from plants or aquatic producers such as diatoms, seaweeds, and other species. 59. Flying insects and birds had the easiest time getting to the island from the mainland. 61. Yes. According to the intermediate disturbance hypothesis, regular disturbances, if not too extreme, actually contribute to biodiversity because different species make use of different habitats, and periodic disturbances guarantee that there will always be habitat at varying stages of recovery. However, a habitat that received regular, extreme disturbances would probably always be in the early stages of succession and be less diverse. 63. The population that has the higher rate of increase r has a steeper curve and a population that grows more quickly over time. What could account for this difference? Any difference in resource availability could result in one population having a higher rate of increase than the other. For example, more rainfall, richer soils, or more nutritious plants in the community could result in more plentiful food. More plentiful food could in turn allow the rabbits to have larger litters, resulting in faster exponential population growth. 65. Answers will vary. Climate is one obvious variable that can affect carrying capacity. Years with good rainfall may increase the carrying capacity for producer populations as well as for populations of species higher up the food chain. Drought may lead to a reduced carrying capacity for many populations in the habitat. The introduction of new species into a community can also affect the carrying capacity, if this new species competes with or preys on the population of interest. Earth's carrying capacity for humans has increased over time as a result of technological advances such as the origin of agriculture and the green revolution. However, as human activities cause global warming, Earth's carrying capacity for humans may decrease. 67. In areas where both species are found, the salamanders experience interspecific competition for food. Individuals of each species that happen to differ more in size from members of the other species are more successful because they experience less competition for food. That is, the larger *Plethodon hoffmani* are more successful at surviving and reproducing because they compete less with *Plethodon cinereus*. Similarly, the smaller *Plethodon cinereus* are more successful at surviving and reproducing because they compete less with *Plethodon hoffmani* salamanders. Over time, both size and diet evolve in both species. 69. The energy pyramid can't be inverted because of conservation of energy and the second law of thermodynamics. Energy that enters an ecosystem in the form of producer biomass can be converted only into other forms of energy. And, in fact, much of this energy is lost as heat. An upside-down energy pyramid would require that energy be created from nothing. 71. Mammals and birds are endotherms. They use a lot of the energy they obtain from food to maintain their stable, high body temperatures. As a result, they have much less energy available for building biomass.

Chapter 22

1. Answers will vary. Meteorology is the study of weather and climate, paleontology is the study of Earth's early history, hydrology is the study of Earth's waters, and volcanology is the study of volcanoes. 3. Earth, like a hard-boiled egg, has three layers, including a thin and brittle outer layer. 5. Earth's structural layers are the lithosphere, which is cool, rigid rock; the asthenosphere, which is plastic rock; the lower mantle, which is strong, rigid rock; the outer core, which is hot and liquid; and the inner core, which is very hot and solid and under extreme pressure. 7. The large pieces of lithosphere are called tectonic plates. 9. Scientists did not know of any mechanism for the continents to move. 11. Lithosphere is created at spreading centers and destroyed at subduction zones. 13. The rate of subduction is about equal to the rate of seafloor spreading. 15. Tectonic plates move in different directions at different speeds, ranging from 2 cm to about 15 cm per year. 17. The interior of a plate is a more geologically stable region. Much geologic activity occurs at plate boundaries because this is where plates collide, pull apart, and rub beside one another. 19. In continental–continental convergence, plates with continental crust at their leading edges meet and produce crumpled rock. In oceanic–oceanic convergence, plates capped by continental crust collide. The older and denser plate subducts beneath the younger, less dense plate. Partial melting of the subducting plate can produce volcanic island arcs. In oceanic–continental convergence, continental and oceanic plates meet. The dense oceanic plate subducts beneath the more buoyant continental plate. A coastal mountain range may develop over time from partial melting of the subducting plate. 21. Iron and oxygen make up 98% of Earth by weight. 23. Oxygen and silicon constitute about 75% of Earth's solid surface. 25. The P-wave shadow reveals that the mantle–core boundary lies at a depth of 2900 ft. The S-wave shadow reveals that the outer portion of Earth's core is liquid. 27. A magnetic reversal is the switching of the positions of Earth's north and south magnetic poles. Magnetic reversals are recorded in Earth's rock by magnetite crystals that act like compasses. 29. Magnetic stripes show that the youngest rock occurs nearest a divergent boundary while the oldest rock is farthest away. This suggests that, like a conveyor belt, rock is created at a divergent boundary and then moves away from it, thereby expanding the seafloor. 31. The two main factors are the transfer of heat through convection and gravity, which pulls heavy plates downward into subduction zones. 33. Microbes obtain energy through chemosynthesis—the metabolism of energy-rich molecules such as hydrogen sulfide. 35. No, not all the locations in ocean trenches that support life are hydrothermal vents. Extremophiles, whose adaptations are not well understood, can live in the cold, black, nutrient-poor crevasses of ocean trenches. 37. Density = mass/volume = $5.6\ \text{g}/2\ \text{cm}^3 = 2.8\ \text{g/cm}^3$. 39. $d = rt = (3.5\ \text{cm/yr})(20\ \text{yr}) = 70\ \text{cm}$. 41. No, the inner core is solid and the outer core is liquid because the greater weight of the overlying material makes the pressure greater at the core. The pressure packs the atoms of the inner core too tightly to allow them to be in the liquid state. 43. Earth would become larger or more mountainous over time. This is not observed to be true. 45. Oceanic–continental convergent boundaries and oceanic–oceanic

convergent boundaries feature subduction zones. 47. An island arc, or an arc-shaped region of volcanic islands, forms where there is oceanic–oceanic convergence of plates. The subducting plate partially melts as it descends, thereby creating magma. This magma, as well as magma from the overlying plate, rises toward the surface and erupts. The erupted magma cools and accumulates to form a volcanic mountain. This process is repeated in various locations along the subduction zone so that a string of volcanic mountains forms. 49. The speed and direction of seismic waves depend on the density and physical state of the medium. The increases and decreases in the speed of P waves and the inability of S waves to penetrate the outer core led scientists to deduce the presence of Earth's internal layers. 51. The crust "floats" on the mantle because the crust is less dense. 53. The evidence comes largely from seafloor spreading, which is suggested by the ages of rocks, magnetic striping, and the existence of midocean ridges, trenches, and other signs of underwater geologic activity. Correlation of intense geologic activity along plate boundaries also supports the movement of plates. Finally, evidence for continental drift (the migration of ancient ice sheets, and the matching of fossils and rock formations) is evidence that the tectonic plates move. 55. Seafloor spreading is faster in the Pacific. 57. New crust is generated as lava at diverging plate boundaries. The lava cools to form rock. Volcanic rock is formed in this way. Sedimentary rock is made of bits of preexisting rock, which usually is ultimately volcanic rock. 59. Yes. Plate tectonics incorporates information from many scientific disciplines. For example, plate tectonics relies on geologic information such as the types of rock found on Earth as well as on physics concepts such as heat transfer, gravity, and density. Fossils (biology) provide evidence for the theory, and astronomical instruments (radio telescopes) are used to measure the actual motions of the plates. 61. Either the climate changed or the regions themselves migrated southward, away from the poles. As it turns out, the latter explanation is correct. 63. Moving charged particles make up an electric current, and an electric current produces a magnetic field. Iron and nickel in Earth's outer core supply charged particles, and Earth's rotation and convection keep the particles moving, which creates an immense magnetic field. 65. Answers will vary. Sample letter: Dear Grandma, Plate tectonics is an amazing theory that explains why Earth's continents fit together like a jigsaw puzzle! Basically, plate tectonics says that the continents move and change shape over geologic time because they are riding on tectonic plates—big slabs of brittle rock resting on flowing, plastic rock. Magma seeps up from Earth's mantle and erupts. The erupting magma separates the plates at divergent plate boundaries. On the seafloor, the process is called *seafloor spreading.* Because plates get pushed apart at divergent boundaries they are always colliding elsewhere. And that makes living on Earth's surface a bumpy ride sometimes! 67. Scientists cannot travel to Earth's core. The crust is so dense at depth that drilling equipment breaks down before the mantle can be reached. The temperature and pressure are far too high for scientists to journey to the core. And, how could a nuclear explosion set the core spinning? The movie doesn't provide the answers. 69. The theory of plate tectonics provides a mechanism for continental drift (seafloor spreading). 71. Yes and no. The continents are continually being moved and reshaped by tectonic forces, so their locations and sizes are temporary. On the other hand, the continental crust is buoyant, so it isn't subducted and subject to recycling as oceanic crust is. Continental crust is therefore much older—much more permanent—than oceanic crust. Ultimately nothing about Earth is truly permanent, however. One reason is that the Sun will run out of nuclear fuel and die someday just as all stars do. 73. The heat comes from radioactivity, the "Great Bombardment," and the gravitational potential energy of material that sank to Earth's interior through differentiation. 75. One problem is that subduction zones are inherently unstable and unpredictable, so it would be difficult to know how to fortify containers of nuclear waste against all foreseeable geologic events. A second problem is that subduction zones are typically very deep. Practical problems include transporting waste to an appropriate depth, determining where suitable dump sites might be, and drilling into the basalt to secure waste once a suitable site was identified.

Chapter 23

1. They are different minerals because they have different crystal structures. 3. In a mineral, the atoms, ions, and molecules are arranged in regular, repeating patterns. 5. Some of the properties of minerals are crystal form, specific gravity, color, hardness, streak, luster, cleavage, radioactivity, toxicity, magnetism, and fluorescence. 7. There are about 4000 known minerals classified into silicates and nonsilicates. 9. Silicates are more common. 11. Pressure from overlying rock can keep rock that would melt at Earth's surface solid. 13. Much rock is hidden under soil, plants, and pavement. 15. Extrusive igneous rocks solidify above Earth's surface. Intrusive igneous rocks solidify below the surface. 17. Sedimentary rocks form through weathering, erosion, and lithification (compaction and/or cementation). 19. The heat and pressure come from plate tectonics (regional metamorphism) and the movement of magma (contact metamorphism). 21. No, because rocks are transformed from one kind into another, as the rock cycle illustrates. 23. All of the most common minerals in Earth's crust contain the silicate tetrahedron. 25. Coal is not made of rock fragments or dissolved minerals. 27. Coal is made of ancient plant remains. 29. The iridium layer suggests that a large asteroid crashed into Earth at the end of the Cretaceous, the time the dinosaurs virtually disappeared. The asteroid explosion could have triggered wildfires that altered climate. 31. Enjoy your observation of the salt cubes. 33. Answers will vary. For a class of 30 students: (4700 lb/student) $\times$ (30 students/class) = 141,000 lb/class. That amount of concrete weighs as much as about six school busses. 35. The texture is porphyritic. The magma that produced the rock crystallized over a broad range of temperatures; the large crystals cooled slowly, but the small ones cooled quickly. 37. A slowly cooling crystal has plenty of time to grow before it has to compete with its neighbors for space and available atoms. In other words, the rate of nucleation is slow compared to the rate of crystal growth. Two examples are granite and gabbro. 39. Coesite and quartz do not have the same physical properties because properties depend on physical structure as well as chemical composition. 41. Some other natural cycles are the hydrologic cycle, the carbon cycle, the nitrogen cycle, and the biogeochemical cycle. 43. The impression is in sedimentary rock because fossils exist only in sedimentary rock. 45. The plate is subjected to tremendous pressure, which induces metamorphism. 47. Mica is a mineral, limestone is a rock, and plastic is neither rock nor mineral. 49. Intrusive igneous rocks solidify from magma that cools slowly underground. Extrusive igneous rocks cool and solidify quickly at Earth's surface. Slow cooling produces large crystals because each seed crystal can grow by attaching liquid atoms drifting by without competing with so many neighboring seed crystals. 51. No. Fossils exist only in sedimentary rock because they cannot withstand the heat and pressure that are involved in the formation of metamorphic and igneous rock. Marble is a metamorphic rock. 53. The formula of coesite is SiO_2. 55. Sediments, from which sedimentary rocks are made, are continually pulled downhill by gravity. So there is a great deal of sediment under the sea from which sedimentary rock can form. 57. Basaltic magma is less viscous and richer in the dark minerals iron and magnesium but poorer in silica compared to rhyolitic magma. 59. Hardness relates to the strengths of bonds in a crystal. Quartz and diamond have stronger internal bonds than gold has. 61. Answers will vary. Possible arguments for mining regulation could involve controlling toxic waste and consumption of water, fossil fuels, land, and other resources. Possible arguments for deregulation could involve free-market economic philosophy. Many students will favor a middle ground. 63. It is a sedimentary rock because it consists of sediments (sand grains) held together by cement.

Chapter 24

1. Oceans cover 70% of Earth. 3. (a) Compressional stress is the pushing together of masses of rock, tensional stress is the pulling apart of rock, and shear stress is produced by rock blocks sliding past one another. (b) Rock can break, deform elastically, or deform plastically in response to applied stress. 5. In a reverse fault, the hanging wall and footwall are pushed together, which results in the rocks in the hanging wall being pushed above the footwall. In a normal fault, rocks in the hanging wall and footwall are pulled apart. The rocks in the hanging wall drop down relative to the rocks in the footwall. In a strike-slip fault, rock masses have side-to-side motion with respect to one another with no vertical displacement. 7. Folded mountains: the Appalachians, Rockies, and Himalayas. Upwarped mountains: the Black Hills of South Dakota and the Adirondack Mountains. Fault-block mountains: the Tetons in Wyoming and the Sierra Nevada in California. (4) Volcanic mountains: Mauna Loa, Mt. Fuji, and Sunset Crater. Examples may vary. 9. Most volcanoes are in the Ring of Fire in the Pacific Ocean. 11. Plateaus are flat areas elevated more than 600 m above sea level; plains are broad, flat areas that do not rise far about sea level. 13. (a) Most of Earth's water (97.6%) is in the oceans. (b) Most of Earth's fresh water is in ice caps and glaciers. 15. (a) The three parts of a passive continental margin are the continental shelf, an underwater extension of the continent; the continental slope, a steep, narrow boundary between

continental and oceanic crust; and the continental rise, a mound of sediment built up at the base of the continental slope. (b) An active continental margin consists of a continental slope and an accretionary wedge, a mass of continental crust donated to the overlying plate by the descending plate. 17. About 80% of Earth's fresh water is frozen in ice caps and glaciers, about 20% is in groundwater, and less 1% is in streams and lakes. 19. Yes, a watershed is the land that drains into a stream. The water where you live drains somewhere, so everyone lives in a watershed. 21. The water table is not flat but roughly approximates surface contours. 23. A large percentage of precipitation does not become runoff—it infiltrates into the ground and slowly moves until it empties into stream channels, which convey it to the surface. 25. About 90% is in the Antarctic Ice Sheet. 27. The salt removed from ocean water (by precipitation, subduction, biological mechanisms, etc.) is in equilibrium with the salt that is added (by eroded continental rock, dead marine organisms, volcanism, etc.). 29. The salt comes mainly from weathered rock, but other sources, such as dead organisms and volcanism, contribute too. 31. Answers will vary. See Table 24.4. 33. The sample with greatest permeability will allow the water to pass through it most quickly. The sample with greatest porosity may or may not be the sample with greatest permeability. 35. From Figure 24.18, 2.4% of Earth's water is fresh water, but 1.9% is frozen, so $2.4\% - 1.9\% = 0.50\%$. 37. A stress is an applied force. A rock "cracks up" (breaks) or folds when the stress is sufficient. 39. All the world's oceans are interconnected. 41. Answers will vary. Some major landforms are the Mississippi Delta, the Black Hills, the San Andreas Fault, the Great Plains, the Appalachian Mountains, and the Grand Canyon. 43. There are active and passive continental margins. Active margins are sites of plate boundaries, whereas passive margins are located away from plate boundaries. Their structural features vary as well. For example, the continental shelf along a passive margin typically has a gradual slope. An active margin often features an ocean trench with an accretionary wedge. 45. Yes, because groundwater can take thousands or millions of years to recharge. 47. The area changes because sea level rises and falls, thereby exposing more or less of the edge of the continental crust. 49. A well must be drilled down past the level of the water table. 51. If groundwater removal is stopped, land subsidence will stop. The ground will not return to its original level, however. The weight of overlying sediments prevents the clays from expanding. Once the aquifer has been compacted, it cannot expand to its original level. 53. Fresh water provides sustenance for life, including water for drinking, agricultural uses, sanitation, and transportation. 55. Runoff flows toward the Amazon River. 57. Kelp must photosynthesize in order to live. Photosynthesis requires light, but there is no light in the abyssal plains. 59. The infiltration is greater on gentle, sandy slopes because sandy materials have a high porosity and because runoff is greater on steeper slopes. 61. The order from longest to shortest residence time is deep groundwater, Atlantic Ocean, Lake Tahoe, the air over Miami. (b) A water drop evaporates from the Atlantic ocean; from deep groundwater, a water drop flows to the ocean or is incorporated in magma and erupts at Earth's surface as lava; from Lake Tahoe, a water drop evaporates or flows to the ocean; a water drop leaves the air as precipitation.

Chapter 25

1. Surface processes originate at or near Earth's surface, while tectonic processes originate in Earth's interior. 3. The two types of weathering are mechanical and chemical. 5. Erosion involves lifting and transporting sediment. 7. The size of a stream depends on the size of its drainage basin. 9. Dissolved load is sediments carried in solution—for example, the minerals sodium and calcium. Suspended load is sediments carried in suspension that make the stream look cloudy. Bed load is the largest sediments that are dragged and bounce along the streambed. 11. Answers will vary. Stream channels, V-shaped valleys, and the outside of a meander are erosional landforms produced by running water. 13. Large sediments drop out of the water first as a stream slows. 15. Glaciers move as ice crystals slip beside one another and as their weight melts their base, which allows them to slide. 17. Glaciers dump large accumulations of debris. A moraine is produced by glacial deposition. 19. Trees acquire a bent trunk—the "pistol butt" shape—because of creep along slopes. As sloping land erodes, a tree would tilt at an increasingly sharp angle if it did not bend to turn upward, seeking maximum sunlight. 21. A flow is the chaotic movement of unconsolidated material that mixes together as it moves, like a fluid. 23. Mass wasting events are usually caused by the absorption of rainwater. 25. Groundwater erosion produces caves, sink holes, and karst topography. 27. Karst typography is topography where the effects of groundwater erosion can be seen above ground. The Guilin region in China is an example. 29. Both stalagmites and stalactites are formed by the gradual precipitation of calcium calcite and other minerals from groundwater in caves. A stalagmite forms on the floor of a cave as groundwater drips down from above; a stalactite forms on the roof as groundwater drips down. 31. Sea caves, sea arches, sea stacks, and headlands are produced by erosion from ocean waves. 33. Currents are another agent of change. 35. Headlands form from rock formations that originally featured both hard and soft rock, but the softer rock eroded away first. 37. Wind is less powerful because air is so much less dense than water. 39. Sand dunes are a depositional landform produced by wind. 41. Mechanical weathering breaks up rock through physical processes; chemical weathering acts through chemical means. 43. Water seeps into cracks in rock while it is liquid. When the water freezes, its expansion pushes outward on the cracks, which breaks up the rock. 45. Two end products are clay and quartz. 47. Soil is composed in part of weathered rock. The acidification of rainwater accelerates the rate of chemical weathering. 49. Humus provides essential plant nutrients, which are minerals as well as organic compounds. 51. Silty soil and sandy soil differ in particle size and therefore in properties, including the ability to retain water. 53. Answers will vary. Fertile soil is topsoil rich in humus, which is derived from living things. Fungi, bacteria, and other microorganisms also play a role as decomposers of organisms that have died. 55. Sketch the streams you see, and make notes about the stream properties you observe. 57. clay loam, silty loam, sandy loam, sand 59. Discharge (Q) = channel cross-section (A) $\times$ average speed (s) $= 30\ \text{m}^2 \times 5\ \text{m/s} = 150\ \text{m}^3/\text{s}$. 61. Volume of the mountain $= 4\ \text{km} \times 3\ \text{km} \times 4\ \text{km} = (4000\ \text{m})(3000\ \text{m})(4000\ \text{m}) = 4.8 \times 10^{10}\ \text{m}^3$. Assume the rate of erosion is about $146\ \text{m}^3$ per year as stated in the Math Connection. Then the duration of the mountain = (volume of mountain)/(rate of erosion) $= 3.29 \times 10^8$ years or 329 million years 63. Photo (a) shows a V-shaped valley. V-shaped valleys are produced by erosion from running water. Photo (b) shows a U-shaped valley. U-shaped valleys are produced by glaciers. 65. The smallest particles make up the dissolved load and are carried in solution. Medium-sized sediments are suspended in running water and move with it as the suspended load. Large sediments bounce and drag along the stream floor as the bed load. 67. Sand dunes drift by wind erosion. They move in the direction of the wind. 69. Soil can erode slowly—or creep—from under the foundation of a house built on a slope. The steeper the slope, the faster the erosion (all other things being equal). 71. Ice, because it can incorporate and transport huge blocks of rock through the process of plucking. 73. Rainwater is primarily an agent of of chemical weathering. As it flows across the surface and into cracks in rock, it chemically weathers rock by dissolving minerals and by reacting with the rock. Rainwater can erode rock as well as weather it because it can carry away sediments, including dissolved minerals, from the rock. 75. Faster moving streams carry more particles and larger particles. 77. It is probably an old river because meanders develop gradually over time. 79. Silt and soil are composed of medium-sized particles, so the muskrat increases the stream's suspended load. 81. Wind can carry small particles such as sand grains. Sandstorms and migrating sand dunes are possible environmental hazards for people who live in the desert. 83. Answers will vary. Discussions should include the angle of the slope, factors related to weathering, and the amount of vegetation.

Chapter 26

1. Weather is the state of the atmosphere at a particular time and place; climate is the average weather over a period of time. 3. Nitrogen and oxygen make up more than 99% of the atmosphere. 5. The stratosphere has a high temperature because of the ozone it contains; ozone is a good absorber of solar radiation. The mesosphere has a low temperature because it absorbs little solar radiation. 7. Climate zones with lower average temperatures are generally located at lower latitudes. 9. The winter solstice is the shortest day of the year and the first day of winter. The summer solstice is the longest day of the year and the first day of summer. The equinoxes are the days when night and day are of equal length. 11. Wind blows from regions of higher pressure to regions of lower pressure. 13. The trade winds blew European trading ships westward toward the southeastern American colonies and eastward from the northeastern colonies toward Europe. 15. Surface currents affect both water and air temperature. 17. Clouds form by the condensation of water vapor. 19. Two other names are "lows" and "cyclones." They are associated with stormy weather. 21. We don't feel atmospheric pressure because there is no net force on our skin; atmospheric pressure is equal inside and outside

the body. 23. Earth absorbs about two-billionths of the radiation emitted by the Sun. The portion of intercepted solar radiation that's absorbed by the ground is 50%. 25. A greenhouse gas is an atmospheric gas that absorbs infrared radiation, emits the radiation, reabsorbs it, and then re-emits it numerous times, thereby warming the atmosphere. Examples are carbon dioxide, water vapor, and methane. 27. Wind blows air that has been warmed by the body away from the body. 29. The Coriolis effect turns global winds and ocean currents relative to the rotating Earth. 31. This is a good illustration of cloud formation. 33. Raindrops vary in size. 35. Troposphere, stratosphere, mesosphere, thermosphere, exosphere 37. Standard atmospheric pressure $= 14.7\ \text{lb/in}^2$. We can convert this to units of feet: $(14.7\ \text{lb/in}^2)(144\ \text{in}^2/\text{ft}^2) = 2100\ \text{lb/ft}^2$. Pressure ($P$) is defined as force ($F$) per area ($A$), so $F = P \times A$. Then the total force that the air inside the house exerts upward on the ceiling is $P \times A = (2100\ \text{lb/ft}^2)(2000\ \text{ft}^2) = 4.2 \times 10^6$ lb of force. 39. The air inside the house is at standard atmospheric pressure; from Table 25.1 this is $14.7\ \text{lb/in}^2$. The outside air is at 85% of this pressure or $(0.85)(14.7\ \text{lb/in}^2) = 12.5\ \text{lb/in}^2$. Thus, the net pressure pushing the door outward is $(14.7\ \text{lb/in}^2) - (12.5\ \text{lb/in}^2) = 2.2\ \text{lb/in}^2$. The area of the door is 6.5 ft by 3 ft, or $(78\ \text{in})(36\ \text{in}) = 2808\ \text{in}^2$. To find the outward force on the door, we use the equation Force = pressure $\times$ area. So the force $= (2.2\ \text{lb/in}^2)(2808\ \text{in}^2) = 6178$ lb. Rounding to significant figures, we find that the outward force on the door due to the tornado is 6200 lb. It's easy to see that the door will be flung far in the storm. 41. No, the summer solstice is the day that Earth is tilted most directly toward the Sun and therefore receives the most intense solar radiation for the longest period of time. It's not the day that Earth is closest to the Sun. 43. Updrafts carry the water droplets upward against the force of gravity. 45. The cumulus clouds shown in the photo are clouds of vertical development. Like other clouds, they form from water vapor in rising air that expands, cools, and condenses. They are associated with fair weather. 47. The picture is not accurate because it is not complete. The global pattern of circulation is broken up into six convection cells because of the Coriolis effect and other forces. 49. Because air moves from higher pressure to lower pressure, high atmospheric pressure blocks a storm from blowing in. On the other hand, if air pressure is low, it provides a path of least resistance for a storm. 51. Westerly winds move heat inland on the West Coast but offshore on the East Coast. 53. Air pressure is lower at higher altitudes. It takes time for the pressure inside your body to match the atmospheric pressure outside. So the air inside your body pushes outward against the membranes in your ears, producing that popping feeling. 55. Earth's major climate zones are tropical, temperate, and polar. The tropical zones are the latitudes between 0° and 30° north and south, where the temperature varies little and averages around 64°F (18°C). The temperate zones are the latitudes between 30° and 60° north and south, where the temperature varies widely, with average monthly temperatures ranging between 10°C and 18°C (50°F and 64°F). The polar zones are the latitudes beyond 60° north and south, where there is little seasonal variation and monthly temperatures are usually below 10°C (50°F) even in the warmest months. 57. Cold air advancing into a region occupied by warm air is a cold front. 59. There is very little oxygen on Mt. Everest—not because the percentage of oxygen in the air has decreased but because the density of all air is low at that altitude. 61. Relative humidity = (amount of water vapor in air)/(amount of water vapor air can possibly hold at a given temperature). As the temperature decreases, the denominator in this equation decreases, so the relative humidity increases. 63. Without the natural greenhouse effect, Earth's average temperature would be a frigid −18°C, so the natural greenhouse effect is a good thing for life on Earth. The anthropogenic greenhouse effect is a bad thing for life, however, because it causes climate to change quickly and significantly. 65. The radiation emitted by the Earth is terrestrial radiation, which is in the infrared range. Infrared radiation can be felt as heat when absorbed by the body, but it cannot be seen with the naked eye.

Chapter 27

1. The Richter scale measures how much the ground shakes during an earthquake. 3. Earthquakes mostly occur along faults and along plate boundaries. 5. Both a tsunami and a ripple are waves that carry energy away from a disturbance. A tsunami is caused by a large disturbance, whereas ripples are triggered by tiny ones. 7. On December 6, 2004, a tsunami was triggered by an earthquake off the coast of Sumatra. It was costly because the wave struck many settlements throughout the Pacific and moved far inland. On March 11, 2011, a tsunami in Honshu, Japan, swept over Fukushima nuclear power plant's seawall, producing a radiation leak that displaced tens of thousands of people. 9. Shield volcanoes erupt quietly. Composite cones erupt most explosively. 11. A hurricane's energy is liberated by the condensation of water vapor over warm ocean water. (This energy of a state change is often called "latent heat.") 13. The costliest U.S. natural disaster was Hurricane Katrina. 15. Yes; an example is the breakup of Pangaea, which opened up seaways between landmasses and allowed cold-water currents to cool the land. 17. "Global warming" is the name given to the observed recent spike and projected future increase in average global temperature that is caused by an enhanced greenhouse effect. 19. Deforestation also causes carbon dioxide buildup. 21. Greenhouse gas emissions from burning fossil fuels have caused the increased temperature. 23. Permian (due to extensive volcanic eruptions); Cretaceous (due to asteroid impact triggering wildfires that filled the atmosphere with smoke and ash). 25. Without the natural greenhouse effect, Earth's average temperature would be a frigid −18°C, so the natural greenhouse effect is a good thing for life on Earth. The anthropogenic greenhouse effect is a bad thing for life, however, because it causes climate to change quickly and significantly. 27. Melting glaciers and sea ice release liquid water into the sea and so the volume of seawater increases due to thermal expansion. 29. Burning the plant removes the plant from the environment so that it can no longer absorb atmospheric carbon dioxide. Also, combustion reactions produce carbon dioxide as a product. 31. This is a good illustration of motion along a transform fault. 33. Answers will vary. The average footprint for U.S. citizens is about 20 tons. 35. Hurricanes die because they have no energy source over land. They run on the condensation of the large amounts of water vapor that reside over warm oceans. 37. Rock stores elastic potential energy before an earthquake. 39. Cities tend to be hotter than other environments because pavement absorbs solar radiation so well. 41. Burning fossil fuels causes air pollution. Also, consumption of fossil fuels should be reduced because the supply is finite and dwindling quickly. 43. See Table 27.5. 45. Plants absorb carbon dioxide, which reduces the greenhouse effect. 47. A volcano is under pressure and explodes when the top is removed. 49. The Ring of Fire encircles much of the Pacific Ocean basin. It is named for the frequent volcanic eruptions that occur there. 51. Down-going crust is consumed at a subduction zone and driven into the mantle. So, waste buried in a subduction zone could perhaps be conveyed into the mantle, far from the surface air, soil, water, organisms, and other resources that people depend on. However, subduction zones are inherently unstable and unpredictable. It would be hard to know what sorts of geologic activity the buried waste would be subject to over the thousands of years that it would remain radioactive. 53. Moving because of possible stress is a personal decision that should be based on the understanding that earthquake prediction is inexact. 55. Graph (b) represents the response. 57. Paving reduces albedo. 59. Answers will vary. A landslide is a preventable natural disaster. An earthquake is a nonpreventable natural disaster. 61. Edith's checkerspot butterfly traveled north seeking a colder climate; Arctic species have nowhere to go to find a colder climate. 63. Hurricanes require warm air and warm water. 65. The type of magma affects how explosive the eruption is, the nature of the ejected material, the viscosity of the lava, and the hazards posed by the eruption. 67. Responses should state that Earth's climate can change naturally (due to many causes—for example, Milankovitch cycles) and because of human activities (principally due to an enhanced greenhouse effect). The spike in average global temperature over the past 100 years coincides with increased atmospheric carbon dioxide concentrations over the same period and is considered by a consensus of climate scientists to be anthropogenic in origin. 69. Discussion of this question may be lively! Encourage students to support their opinions with evidence, to cite scientific studies from credible sources, and to discuss the issue in depth.

Chapter 28

1. There are eight known planets in our solar system. 3. The outer planets are gaseous and much larger than the inner planets, which are solid. 5. Accretion of the planets began before the Sun ignited. 7. Sunspots are relatively cool regions on the solar surface that move with the Sun's rotation and are created by strong magnetic fields. 9. Because the Sun is a fluid, equatorial regions spin faster than regions at higher latitudes. 11. Mercury is close to the Sun and has very long days and nights due to its slow rotation. 13. The blue oceans dominate Earth's surface. 15. Dry ocean beds and channels cut from water are evidence of Mars's wetter past. 17. Saturn's inner rings move faster (like the greater speed of inner planets or any close-orbiting satellites).

19. Neptune is bluer because it contains greater amounts of methane and ammonia. 21. The Sun is directly in back of you. 23. Eclipses occur only when the plane of the Moon's orbit intersects the plane of Earth's orbit about the Sun, which occurs only rarely. 25. The asteroid belt is between the orbits of Mars and Jupiter. 27. The Kuiper belt is a disk-shaped region beyond Neptune that is populated by many icy bodies; it is considered the source of short-period comets. 29. A falling star is a meteor visible in the sky as it burns in the atmosphere. 31. Each carbon atom is able to form bonds to four different atoms. This allows for an unlimited number of different molecules that can be built based on carbon. Carbon is thus the most versatile atom of the periodic table. 33. This is a good illustration of moonlight. 35. Consider how Earth and the Moon move. 37. Toward the equator, the plane of our solar system appears perpendicular to the horizon. When viewed from the North Pole, it appears more parallel to the horizon. 39. c, b, a. The closer a planet is to the Sun, the faster it orbits and less far it has to travel in order to orbit once. 41. b, a, c (as obtained from Table 26.1) 43. b, a, c. A lunar eclipse is visible to everyone on the night side of planet Earth (assuming clear skies). A solar eclipse traces a narrow path across the surface of Earth. To see a solar eclipse—partial or total—most people need to travel to a distant and often remote location. The new Moon is visible only during a *total* solar eclipse, which is when the new Moon is not obscured by the Sun's glare. The new Moon is just barely visible as it is illuminated by the light of a "full Earth." 45. Use the formula Speed = distance/time, and solve for time. Then convert from units of seconds to minutes:

$$(300{,}000 \text{ km/s}) = \frac{(150{,}000{,}000 \text{ km})}{\text{time}}$$

$$\text{Time} = \frac{(150{,}000{,}000 \text{ km})}{(300{,}000 \text{ km/s})} = 500 \text{ s}$$

$$(500 \text{ s})(1 \text{ min}/60 \text{ s}) = 8.3 \text{ min}$$

47. In AU, the diameter of our solar system including the Oort cloud would be twice 50,000 AU, or about 100,000 AU. We can use the given conversion factor to convert to units of light-years: (100,000 AU)(1 light-year/63,000 AU) = 1.6 light-years. 49. From Exercise 47 we are given that 1 light-year equals 63,000 AU. We can use this relationship to calculate the distance to Alpha Centauri in units of AU: (4.4 light-years)(63,000 AU/1.0 light-year) = 277,000 AU. If, according to the analogy, 1 AU corresponds to 110 m, then 277,000 AU corresponds to (277,000 AU)(110 m/1 AU) = 30.5 million meters. We then divide by 1000 to get about 30,000 km. For perspective, this is about a tenth of the way to the Moon, which is about 300,000 km away. 51. As a nebula contracts and spins faster, it flattens out into a disc. 53. The ball has a greater surface area in the disk shape, which allows it to radiate more energy. 55. If the Earth didn't spin on its axis and only revolved, the Sun would appear to move eastward across the sky. Earth's rotation is also eastward (counterclockwise when viewed from the North Pole). This eastward rotation causes the Sun to appear to move toward the west. If the Earth rotated once for every one revolution (year), then the two motions would balance each other out and the Sun would appear to hover at a single point in the sky—just as Earth appears to remain in a fixed position when viewed from the lunar surface. The Earth, however, rotates about 365 times in a single revolution and this causes the Sun to appear to move toward the west. The eastward moving effect of revolution, however, is enough to push the Sun eastward by 3 minutes and 57 seconds every day. One solar day (24 hours), therefore, is actually longer than one full rotation (23 hours, 56 minutes, 3 seconds). In other words, it takes Earth less than a day to spin around once! 57. Unequal heating of the surface and therefore the atmosphere causes winds. 59. No, because the water from Venus has already been lost. 61. The jovian planets are large, gaseous, low-density worlds that have rings. The terrestrial planets are rocky and have no rings. 63. Adding more mass results in stronger gravitational forces that compress the volume. A more massive object, therefore, can have a smaller volume. An interesting fact is that black holes, which are very, very massive, are quite small. 65. The impacting object must have hit the young Earth not dead-on but askew, which gave Earth a fast rotation. 67. The gravitation at the Moon's surface is too small. The escape velocity at the Moon's surface is less than the speeds that molecules of gas would have at regular Moon temperatures, so any gases on the Moon escape. 69. In both photos the Sun is off to the right, at about 2:30 o'clock, slightly in back. If the two photos were taken on the same day, then the Moon would necessarily appear behind the Ping-Pong ball. So, unless the Ping-Pong ball is exactly covering the Moon, then these photos were likely taken on different days. 71. An observer on the back side of the Moon would never see Earth. To see Earth, the observer would need to travel to the front side of the Moon. As she did so, she would finally reach a point where she would see Earth coming up from the horizon. She would then be midway between the back and front sides. If she stopped at this point for a picnic, would Earth continue to rise? No, it wouldn't. In fact, on the Moon, as long as you don't move about, Earth remains in the same place in the sky for the same reason we always see only one face of the Moon. When the astronauts walked on the Moon, they could count on Earth being in the same place in the sky during their entire stay. What does change, however, is Earth's phase, which takes a month to cycle from full Earth to new Earth and back to full Earth. 73. A lunar eclipse occurs when all three bodies are aligned, with Earth between the Sun and Moon. 75. It would be nighttime on the Moon because you would be located directly between Earth and the Sun. Earth would be so bright, however, that you could clearly see your shadow. From Earth, people would look up and see a dark new Moon. But it would be daytime on Earth, and the sky would be so bright and the new Moon so dim that people wouldn't see a thing—unless, of course, it was during a spectacular solar eclipse. 77. The Sun would always appear in the same location in the sky. One side of Earth would have constant light, while the other side would have constant dark. 79. A lunar eclipse is in view of the whole hemisphere of Earth facing the Moon. But a solar eclipse is in view of only a small part of the hemisphere that faces the Sun at that time, so few see it. 81. When close to the Sun, the Moon is necessarily in a crescent phase. Frame C shows the proper orientation of this crescent Moon. 83. The sailboat is sailing closer to the equator, as evidenced by the perpendicular orientation of the crescent Moon. 85. Comets are icy bodies that orbit the Sun. When they get close to the Sun, they lose volatile compounds, such as water, which escape and appear as a tail. Pluto is an ice body. If it were somehow knocked off its orbit (unlikely because it's so massive), then it too would show a tail as it came closer to the Sun. Because of its great mass, it would be a most spectacular and potentially dangerous tail for those viewing from Earth. 87. Meteorites are more easily found in Antarctica because so many are imbedded in ice. On regular ground, they are not so obvious. 89. There would be almost no consequence except for spectacular meteor showers high in the atmosphere. 91. If Earth had no Moon, then the rate of Earth's rotation would be much slower and the planet might not be amenable to life. Assuming the same habitability, however, we can evaluate how human civilization might be different by considering how the Moon has influenced our history and worldview. For starters, the year would likely not be divided into 12 months. Consider also the role our Moon has played in the timing of the seasons, navigation, nighttime activities, religions, and our perspective of our place in the universe.

Chapter 29

1. Constellations are visible groups of stars in the nighttime sky. 3. This motion of the stars is due to the rotation of Earth. 5. A blue star is hotter than a red star. 7. An H–R diagram is a plot of intrinsic brightness versus surface temperature for stars. When plotted, the stars' positions take the form of a main sequence for average stars, with exotic stars above or below the main sequence. 9. Our Sun is along the middle of the main sequence on an H–R diagram. 11. The ignition of nuclear fuel and the subsequent thermonuclear fusion change a protostar into a star. 13. High-mass stars have shorter lifetimes than low-mass stars. 15. A neutron star, enormously dense, is what remains after a supernova. 17. The Sun has too little mass for the collapse necessary to become a black hole. 19. Anything can pass into the black hole through its event horizon. Once it passes by this mathematical boundary, however, there is no possible return. 21. A starburst galaxy is a galaxy in which stars are forming at an unusually high rate. 23. Three galactic clusters in our Local Supercluster are the Local Group and the Virgo and Eridanus Clusters. 25. The luminosity of a star is the amount of energy it puts out per second. The brightness of the star, however, diminishes with distance. 27. We can detect dark matter by its gravitational effects. 29. The expansion of the universe started to accelerate about 7.5 billion years ago. 31. Dark matter is about six times more abundant than ordinary matter. 33. SETI stands for search for extraterrestrial intelligence, and it is now a privately funded organization. Whether it was a good move or a mistake for Congress to cut its government funding depends on your political point of view. Perhaps the money is better spent on more realistic goals. Perhaps it is worth our tax dollars to support such an exciting endeavor. 35. The star has moved to the west. When the Sun rises, the star will have already set and is obscured by the ground. After 24 hours, it will be slightly more to the west because of the revolution of Earth around

the Sun. 37. b, c, a. The higher the latitude, the higher the North Star appears in the sky. 39. Star C, Star A, Star B. Star A is similar in size to our own Sun. Star B is slightly cooler, but its energy output is 100 times greater. The only way for this to occur is if the star is also very large. The temperature and small luminosity of Star C are consistent with a white dwarf. 41. c, b, a. Hydrogen is the initial fuel that ignites when a star is formed. Once the hydrogen is consumed, helium atoms are fused, followed by carbon. 43. c, a, b 45. If these stars are the same distance from Earth, their apparent brightness depends on only luminosity, and Star A will be four times as bright as Star B. However, if Star A is twice as far away as Star B, the stars will have equal apparent brightness because apparent brightness is related to distance through the inverse-square law. 47. Surface area density $= (1.00 \times 10^{11}$ stars$)/7.85 \times 10^{9}$ ly$^2 = 12.7$ stars/ly^2. 49. The stars within a galaxy are spaced so far apart that the collision of two galaxies is analogous to the collision of two swarms of mosquitoes—it is possible for the swarms to pass right through each other with no mosquito–mosquito collisions. The stars within a galaxy, however, are more concentrated toward the center, so if the galactic collision were dead-center, then a greater number of stellar collisions would occur. The combination of the cores of the two galaxies would most likely result in an active galactic nucleus. Mostly what happens when two galaxies collide is that they sideswipe each other. They pass through each other but then gravity tugs them back into a single irregular galaxy, which eventually transforms into an elliptical galaxy. As the two galaxies merge, there is a mixing of the interstellar gases and dust. This allows for the creation of many new stars, which is why most starburst galaxies are galaxies in the process of colliding. 51. On Earth, the Sun lightens up our atmosphere to the point that we are not able to see the relatively dim light of stars. On the Moon, however, there is no atmosphere. The sky therefore appears black and stars are visible. 53. Diurnal motion is the westward motion of celestial objects due to the daily counterclockwise (eastward) rotation of Earth on its axis. Diurnal motion, for example, gives us sunsets and moon rises. Intrinsic motion is the motion of celestial objects relative to the stars. From our point of view on Earth, intrinsic motion is much less apparent than diurnal motion. The intrinsic motion of stars, for example, is difficult to detect because these stars are so far away. Hundreds or thousands of years will pass before the intrinsic motion of stars will cause a noticeable shift in the shape of a constellation. The planets are much closer, which means that their intrinsic motion can be noted over the course of weeks or months. The intrinsic motion of the Moon can be noted from one night to the next. 55. A planetary nebula is a shell of interstellar material put forth by a medium-mass star in the process of transforming into a white dwarf. 57. A nova is the thermonuclear fusion of material collected by a white dwarf from a neighboring star. 59. The Sun will expand into a red giant before ultimately turning into a white dwarf. 61. Since all the heavy elements are manufactured in supernovae, the newer the star, the greater the percentage of heavy elements available for its construction. Very old stars were made when heavy elements were less abundant. 63. The more massive a star, the greater the gravitational forces. This translates into a hotter core temperature, which helps the fusion reactions to burn faster. So, although a more massive star begins with more fuel, it burns that fuel at a faster rate; hence, its lifetime tends to be shorter. 65. Just as a spinning skater slows down when her or his arms are extended, a spinning star in formation similarly slows down when the material that forms planets is extended. Thus, a slow-spinning star is more likely to have a system of planets. 67. There is insufficient gravitational pressure within the Sun to initiate carbon fusion, which requires greater squashing than hydrogen to fuse. 69. The Sun will eventually run out of hydrogen nuclei for the fusion synthesis of helium. At this point, gravity will cause the Sun to collapse. This, in turn, will provide energy to allow the fusion of helium nuclei into carbon. As this happens, the Sun will grow in size to become a red giant, which will bake the surface of Earth and cause it to lose its atmosphere and hydrosphere. No air or water means no life. 71. The Sun does not have sufficient mass to become either a supernova or a black hole. 73. There is likely a supermassive black hole at the core of most, if not all, spiral galaxies, including the Milky Way. When relatively large amounts of matter are falling into that black hole, large amounts of energy are produced and we classify the core as an active galactic nucleus. The recent discovery of Fermi bubbles radiating from the core of the Milky Way suggests an active galactic nucleus. The degree of activity, however, is likely small compared to the activity in quasars. 75. The light waves are stretched, which means the frequency of the light is lowered. 77. As the universe continues to expand, the average temperature of the universe should continue to decrease. 79. No matter where astronomers point their telescopes, they are looking in the direction of where the Big Bang occurred, which is all around us. The whole universe originated within the Big Bang. 81. Stars are dense collections of hydrogen, which originated from the Big Bang. 83. Although *dark energy* and *dark matter* both begin with the word *dark*, they are different. One is a form of matter, and the other is a form of energy. Both are still mysterious. 85. The darkness of the clear night sky was a mystery to astronomers. Many early astronomers believed the universe consisted of a finite number of stars enclosed within some sort of huge black wall. As late as the beginning of the 20th century, many astronomers believed the universe consisted of a finite number of stars bunched together within an infinite space. Today, the Big Bang theory tells us that the universe is changing. There are zillions of galaxies that we don't see because the universe is not old enough for their light to have yet reached us. 87. Astronomy is a science dedicated to the study of celestial objects. The mission of astronomy is to learn about the nature and origins of these objects so that we can better understand the natural universe in which we live. Astrology is an ancient pseudoscience founded in the belief that the positions of celestial objects influence our physical, mental, and social well-being. Astronomy relies greatly on discoveries made using advanced technologies, such as space telescopes. As a field it has matured greatly over the past 100 years, but it is still in a golden era of new and astounding discoveries. Astrology, by contrast, relies on nonconfirmable anecdotal evidence and, though very popular, has not changed significantly during the past 100 years. 89. From our point of view outside of the black hole, the singularity does not exist at present, only in the infinite future and therefore outside the realm of our observable universe.

Glossary

Absolute zero The lowest possible temperature that a substance may have—the temperature at which molecules of the substance have their minimum kinetic energy.

Abyssal plains The flat regions of the ocean basin.

Acceleration The rate at which velocity changes with time; the change in velocity may be in magnitude, or in direction, or in both. It is usually measured in m/s^2.

Acid A substance that donates hydrogen ions.

Acidic Description of a solution in which the hydronium ion concentration is higher than the hydroxide ion concentration.

Acquired immune system A set of highly specific body defenses that respond to specific features of pathogens.

Action potential A signal from a neuron or other cell that occurs when the cell's membrane potential becomes positive.

Activation energy The minimum energy required in order for a chemical reaction to proceed.

Active galactic nucleus The central region of a galaxy in which matter is falling into a supermassive black hole and emitting huge amounts of energy.

Active transport Transport across the cell membrane that requires energy from the cell.

Adaptations Evolved traits that make organisms well suited to living and reproducing in their environments.

Addition polymer A polymer formed by the joining together of monomer units with no atoms being lost as the polymer forms.

Air mass A large pool of air that has similar temperature and moisture characteristics throughout.

Air resistance The force of friction acting on an object due to its motion through air.

Albedo The fraction of solar radiation arriving at a surface and reflected by it.

Alcohol An organic molecule that contains a hydroxyl group bonded to a saturated carbon.

Aldehyde An organic molecule that contains a carbonyl group, the carbon of which is bonded either to one carbon atom and one hydrogen atom or to two hydrogen atoms.

Allele A version of a gene.

Allopatric speciation Speciation that occurs after a geographic barrier divides a group of organisms into two isolated populations.

Alloy A mixture of two or more metallic elements.

Alpha particle A subatomic particle made up of the combination of two protons and two neutrons ejected by a radioactive nucleus. The composition of an alpha particle is the same as that of the nucleus of a helium atom.

Alternating current (ac) An electric current that repeatedly reverses its direction; the electric charges vibrate about relatively fixed positions, 60 Hz in the United States.

Alternation of generations The plant life cycle, which alternates between haploid and diploid stages.

Alveoli The tiny sacs in the lungs where gas exchange occurs.

Amide An organic molecule that contains a carbonyl group, the carbon of which is bonded to a nitrogen atom.

Amine An organic molecule that contains a nitrogen atom bonded to one or more saturated carbon atoms.

Amphoteric Description of a substance that can behave as either an acid or a base.

Amplitude For a wave or a vibration, the maximum displacement on either side of the equilibrium (midpoint) position.

Animals A clade of multicellular, heterotrophic eukaryotes that take food into their bodies for digestion.

Antibody A large, Y-shaped protein that binds to a pathogen's antigen, disabling the pathogen or marking it for destruction.

Anticyclone (*high-pressure system* or *high*) A weather system organized around an area of high pressure.

Antigen A molecule or part of a molecule belonging to a pathogen that is recognized by cells of the acquired immune system.

Applied research Research that focuses on developing applications of knowledge gained through basic research.

Aquifer A zone of water-bearing rock through which groundwater can flow.

Archaea One of the three domains of life, consisting of many different kinds of prokaryotic organisms that are more closely related to eukaryotes than bacteria; some archaea are adapted to extreme environments.

Aromatic compound Any organic molecule that contains a benzene ring.

Arteries Blood vessels that carry blood away from the heart.

Artificial selection The selective breeding of organisms with desirable traits in order to produce offspring with the same traits.

Asteroid belt A region between the orbits of Mars and Jupiter that contains small, rocky, planet-like fragments that orbit the Sun. These fragments are called *asteroids* ("small star" in Latin).

Asthenosphere One of Earth's structural layers—a layer of weak, warm rock that flows slowly over geologic time.

Astronomical unit (AU) The average distance between Earth and the Sun; about 1.5×10^8 km (about 9.3×10^7 mi).

Atmosphere The thin envelope of gases, tiny solid particles, and liquid droplets surrounding the solid planet.

Atmospheric pressure The weight of all the air molecules in the atmosphere pressing down on Earth's surface.

Atomic mass The mass of an element's atoms; listed in the periodic table as an average value based on the relative abundance of the element's isotopes.

Atomic nucleus The dense, positively charged center of every atom.

Atomic number A count of the number of protons in the atomic nucleus.

Atomic spectrum The pattern of frequencies of electromagnetic radiation emitted by the atoms of an element; considered to be an element's fingerprint.

Atomic symbol An abbreviation for an element or atom.

ATP Adenosine triphosphate, the molecule that provides the energy for most cellular processes.

Autotrophs Living organisms that convert inorganic molecules into food and organic molecules.

Bacteria One of the three domains of life, consisting of an extremely diverse array of prokaryotic organisms.

Base A substance that accepts hydrogen ions.

Basic Description of a solution in which the hydroxide ion concentration is higher than the hydronium ion concentration; also sometimes called *alkaline*.

Basic research Research that leads us to a better understanding of how the natural world operates.

Beta particle An electron emitted during the radioactive decay of a radioactive nucleus.

Big Bang The primordial creation and expansion of space at the beginning of time.

Big Rip A model for the end of the universe in which dark energy grows stronger over time and causes all matter to rip apart.

Biogeochemical cycles The movement of substances such as water, carbon, and nitrogen between the tissues of living organisms and the abiotic world.

Biology The study of life and living organisms.

Biomass The amount of organic matter in an ecosystem.

Biomes Earth's major types of terrestrial ecosystems.

Black hole The remains of a supergiant star that has collapsed into itself. It is so dense, and has a gravitational field so intense, that light itself cannot escape from it.

Black-hole singularity The object of zero radius into which the matter of a black hole is compressed.

Boiling Evaporation in which bubbles form beneath the liquid surface.

Bond energy The amount of energy required to pull two bonded atoms apart, which is the same as the amount of energy released when the two atoms are brought together into a bond.

Calorie The amount of heat needed to change the temperature of 1 gram of water by 1 Celsius degree.

Capillaries Tiny, thin-walled blood vessels from which molecules are exchanged between blood and body tissues.

Carbohydrates Organic molecules that consist of a single simple sugar or chains of simple sugar molecules.

Carbon-14 dating The process of estimating the age of once-living material by measuring the amount of radioactive carbon-14 present in the material.

Carbonyl group A carbon atom double-bonded to an oxygen atom; found in ketones, aldehydes, amides, carboxylic acids, and esters.

Carboxylic acid An organic molecule that contains a carbonyl group, the carbon of which is bonded to a hydroxyl group.

Carrying capacity The maximum number of individuals or the maximum population density that a habitat can support.

Catalyst Any substance that increases the rate of a chemical reaction without itself being consumed by the reaction.

Celestial sphere An imaginary sphere surrounding Earth to which the stars are attached.

Cell membrane The structure that separates the inside of the cell from the outside of the cell.

Cells The basic units of life that make up all living organisms.

Cellular respiration The aerobic breakdown of glucose to produce ATP.

Centripetal force Any force that is directed at right angles to the path of a moving object and that causes the object to follow a circular path.

Chain reaction A self-sustaining reaction in which the products of one reaction event initiate further reaction events.

Chemical bond The force of attraction between two atoms that holds them together.

Chemical change The formation of a new substance(s) by the rearrangement of the atoms of the original material(s).

Chemical equation A representation of a chemical reaction in which reactants are written to the left of an arrow that points toward the products on the right.

Chemical formula A notation that indicates the composition of a compound, consisting of the atomic symbols for the elements in the compound and numerical subscripts indicating the ratio in which the atoms combine.

Chemical property A property that characterizes the ability of a substance to react with other substances or to change into a different substance under specific conditions.

Chemical reaction A term synonymous with chemical change.

Chemical rocks Sedimentary rocks formed from minerals precipitated out of solution.

Chemical weathering Change in the chemical structure of the minerals in rock.

Chemistry The study of matter and the transformations it can undergo.

Chemoreception A way of sensing that occurs when chemicals bind to receptors on sensory cells, initiating action potentials.

Chloroplasts The organelles within plant cells in which photosynthesis occurs.

Chromosomes The DNA-containing structures within cells.

Clade A group of species that includes an ancestor and all of its descendants.

Clastic rocks Sedimentary rocks that are composed of pieces of preexisting rocks.

Climate The general pattern of weather that occurs over a period of years.

Climate change (global warming) A long-term rise in average global temperature that results from an increase in the greenhouse effect.

Cloud A place in the atmosphere that contains enough water droplets or ice crystals to be visible.

Codon A sequence of three nucleotides in an mRNA molecule that either stands for an amino acid or ends translation.

Combustion A rapid exothermic oxidation–reduction reaction between a material and molecular oxygen.

Comet A body composed of ice and dust that orbits the Sun, usually in a very eccentric orbit, and that casts a luminous tail produced by solar radiation pressure when close to the Sun.

Community All the organisms that live in a specific area.

Compound A structure in which atoms of different elements are bonded to one another.

Concentration A quantitative measure of the amount of solute in a solution.

Conceptual model A representation of a system that helps us predict how the system behaves.

Condensation The transformation of a gas to a liquid.

Condensation polymer A polymer formed by the joining together of monomer units accompanied by the loss of small molecules, such as water.

Conduction The transfer of thermal energy by molecular and electronic collisions within a substance (especially within a solid).

Conductor Any material having free charged particles that easily flow through it when an electrical force acts on them.

Configuration The way the atoms within a molecule are connected. For example, two structural isomers have the same numbers and the same kinds of atoms, but in different configurations.

Conformation One of a wide range of possible spatial orientations of a particular configuration.

Conservation of energy In the absence of external work input or output, the energy of a system remains unchanged. Energy cannot be created or destroyed.

Conservation of energy and machines The work output of any machine cannot exceed the work input. In an ideal machine, where no energy is transformed into heat,

$$\text{work}_{\text{input}} = \text{work}_{\text{output}}$$

and

$$(Fd)_{\text{input}} = (Fd)_{\text{output}}$$

Conservation of momentum In the absence of an external force, the momentum of a system remains unchanged. Hence, the momentum before an event involving only internal forces is equal to the momentum after the event:

$$mv_{(\text{before})} = mv_{(\text{after})}$$

Consumer An organism that obtains food by eating other organisms.

Continental drift The hypothesis that the world's continents move slowly over Earth's surface.

Continental margin The transition zone between dry land and the deep ocean bottom.

Control A test that excludes the variable being investigated in a scientific experiment.

Convection The transfer of thermal energy in a gas or liquid by means of currents in the heated fluid. The fluid flows, carrying energy with it.

Convergent boundaries Places where tectonic plates come together.

Core Earth's innermost layer, which is mostly iron and includes the inner core and outer core.

Coriolis effect The tendency of moving bodies not attached to Earth (such as air molecules) to move to their right in the Northern Hemisphere and to their left in the Southern Hemisphere.

Corrosion The deterioration of a metal, typically caused by atmospheric oxygen.

Cosmic background radiation The faint microwave radiation emanating from all directions that is the remnant heat of the Big Bang.

Cosmology The study of the overall structure and evolution of the universe.

Coulomb The SI unit of electric charge. One coulomb (symbol C) is equal in magnitude to the total charge of 6.25×1018 electrons.

Coulomb's law The relationship among force, charge, and distance:

$$F = k\frac{q_1 q_2}{d^2}$$

If the charges are alike in sign, the force is repelling; if the charges are unlike, the force is attractive.

Covalent bond A chemical bond in which atoms are held together by their mutual attraction for two or more shared electrons.

Covalent compound A substance, such as an element or a chemical compound, in which atoms are held together by covalent bonds.

Critical mass The minimum mass of fissionable material needed for a sustainable chain reaction.

Crust Earth's surface layer, consisting of oceanic and continental crust.

Cyclone (*low-pressure system* or *low*) A weather system organized around an area of low pressure.

Cytoplasm The part of a cell that is inside the cell membrane but outside the nucleus.

Dark energy An unknown form of energy that appears to be causing an acceleration of the expansion of space; thought to be associated with the energy exuded by a perfect vacuum.

Dark matter Invisible matter that has made its presence known so far only through its gravitational effects.

Decomposer An organism that obtains food by eating dead organic matter.

Delta A mass of sediment that is deposited where a river slows down as it enters a large body of water such as a lake or the ocean.

Density A measure of mass per volume for a substance.

Deoxyribonucleic acid (DNA) The cell's genetic material: a double-stranded molecule, consisting of sugar–phosphate backbones attached by pairs of matched nitrogenous bases, in the form of a double helix.

Deposition In the context of chemistry, the transformation of a gas to a solid. In Earth science, the process of laying down eroded material in new locations.

Dew point The temperature at which the air becomes saturated.

Diaphragm The sheet of muscle that covers the bottom of the thoracic cavity; its contraction and relaxation cause inhalation and exhalation.

Differential weathering Weathering that is varied in rate and degree due to climate, rock geometry, and other factors.

Differentiation The process by which Earth formed layers according to density.

Diffraction Any bending of light by means other than reflection and refraction.

Diffusion The tendency of the particles of a material, such as molecules, to move from an area of high concentration to one of low concentration.

Digestion The breakdown of food into small organic molecules that can be absorbed and used by the body.

Diploid Description of a cell that has two of each kind of chromosome.

Dipole A separation of charge that occurs in a chemical bond because of differences in the electronegativities of the bonded atoms.

Direct current (dc) An electric current that flows in one direction only.

Discharge The volume of water that a stream transports in a given amount of time.

Dispersion The separation of light into colors arranged by frequency.

Dissolving The process of mixing a solute in a solvent to produce a homogeneous mixture.

Divergent boundaries Places where tectonic plates pull apart.

Divide An imaginary line that is the highest ground separating two drainage basins.

DNA replication The process through which a DNA molecule is copied in order for cells to divide and reproduce.

Dominant Description of an allele that is expressed in a heterozygote.

Doppler effect The change in frequency of a wave due to the motion of the source (or due to the motion of the receiver).

Drainage basin The area that supplies a stream with the water that flows through it.

Dwarf planet A relatively large icy body, such as Pluto, that originated within the Kuiper belt.

Earth science The study of the history, structure, and natural processes of planet Earth.

Earthquake The shaking of the ground that results when rock under Earth's surface moves or breaks.

Earthquake The sudden ground motion and energy release that occur when rocks slip along a fault.

Ecliptic The plane of Earth's orbit around the Sun. All major objects of the solar system orbit roughly within this same plane.

Ecological succession Changes over time in the community of species living in an ecosystem.

Ecology The study of how organisms interact with their environments.

Ecosystem All the organisms that live in a specific area and all the abiotic features of their environment.

Ectotherm An organism that uses behavior to regulate its body temperature, which may vary to some degree.

Efficiency The percentage of the work put into a machine that is converted into useful work output. (More generally, efficiency is useful energy output divided by total energy input.)

Elastic collision A collision in which colliding objects rebound without lasting deformation or the generation of heat.

Electric current A flow of charged particles that transports energy from one place to another, measured in amperes, where 1 A is the flow of 6.25×1018 electrons per second, or 1 coulomb per second.

Electric field Defined as force per unit charge, it can be considered an energetic aura surrounding charged objects. About a charged point, the field decreases with distance according to the inverse-square law, like a gravitational field. Between oppositely charged parallel plates, the electric field is uniform.

Electric potential The electric potential energy per amount of charge, measured in volts, and often called *voltage*:

$$\text{Electric potential} = \frac{\text{electric potential energy}}{\text{charge}}$$

Electric potential energy The energy a charge possesses by virtue of its location in an electric field.

Electric power The rate of energy transfer, or rate of doing work; the amount of energy per unit time, which can be measured by the product of current and voltage:

$$\text{Power} = \text{current} \times \text{voltage}$$

Electric power is measured in watts (or in kilowatts).

Electrical resistance The property of a material that resists the flow of electric charge through it; measured in ohms (Ω).

Electrically polarized Description of an atom or molecule in which the charges are aligned so that one side has a slight excess of positive charge and the other side a slight excess of negative charge.

Electromagnet A magnet whose field is produced by an electric current. It is usually in the form of a wire coil with a piece of iron inside the coil.

Electromagnetic induction The induction of voltage when a magnetic field changes with time.

Electromagnetic spectrum The continuous range of electromagnetic waves that extends in frequency from radio waves to gamma rays.

Electromagnetic wave An energy-carrying wave produced by an oscillating electric charge.

Electron An extremely small, negatively charged subatomic particle found in a cloud outside the atomic nucleus.

Electron-dot structure A shorthand notation of the noble gas shell model of the atom, in which valence electrons are shown as dots surrounding an atomic symbol. The electron-dot structure for an atom or ion is sometimes called a *Lewis dot symbol*, and the electron-dot structure for a molecule or polyatomic ion is sometimes called a *Lewis structure*.

Electronegativity The ability of an atom to attract a bonding pair of electrons to itself when bonded to another atom.

Element Any material that is made up of only one type of atom.

Elemental formula A notation that uses the atomic symbol and (sometimes) a numerical subscript to denote how many atoms are bonded in one unit of an element.

Ellipse A closed curve of oval shape for which the sum of the distances from any point on the curve to two internal focal points is a constant.

Endoskeleton An internal skeleton, such as that found in echinoderms and chordates.

Endotherm An organism that maintains a constant, high body temperature by breaking down food.

Endothermic Description of a chemical reaction in which there is a net absorption of energy.

Energy The property of a system that enables it to do work.

Entropy The measure of the energy dispersal of a system. Whenever energy freely transforms from one form into another, the direction of transformation is toward a state of greater disorder and, therefore, toward greater entropy.

Enzyme A complex protein that catalyzes a chemical reaction in a living organism.

Epicenter The point at Earth's surface directly above the focus of an earthquake.

Equilibrium rule The vector sum of forces acting on a nonaccelerating object equals zero: $\Sigma F = 0$.

Erosion The removal and transport of pieces of preexisting rock.

Escape speed The speed that a projectile, space probe, or similar object must reach in order to escape the gravitational influence of Earth or of another celestial body to which it is attracted.

Ester An organic molecule that contains a carbonyl group, the carbon of which is bonded to one carbon atom and one oxygen atom bonded to another carbon atom.

Eternal inflation A model of the universe in which cosmic inflation is not a one-time event but rather progresses to continuously spawn an infinite number of observable universes in its wake.

Ether An organic molecule that contains an oxygen atom bonded to two carbon atoms.

Eukarya One of the three domains of life, consisting of eukaryotic organisms whose cells have a nucleus: animals, fungi, plants, and protists.

Eukaryotes Organisms whose cells have a true nucleus, such as protists, animals, plants, and fungi.

Evaporation The transformation of a liquid to a gas.

Event horizon The boundary region of a black hole from which no radiation can escape. Any events within the event horizon are invisible to distant observers.

Evolution Inherited changes in populations of organisms over time.

Evolutionary tree A diagram that shows the relationships among a set of organisms.

Exoskeleton An external skeleton, such as that found in arthropods.

Exothermic Description of a chemical reaction in which there is a net release of energy.

Exponential growth A model of population growth in which a population grows at a fixed rate per amount of time.

Extrusive igneous rock An igneous rock that forms at Earth's surface.

Fact A phenomenon about which competent observers can agree.

Faraday's law An electric field is induced in any region of space in which a magnetic field is changing with time. The magnitude of the induced electric field is proportional to the rate at which the magnetic field changes. The direction of the induced field is at right angles to the changing magnetic field.

Fault A fracture through a rock along which sections of the rock have moved.

Fault-block mountain A mountain that forms from tension and that has at least one side bounded by a normal fault.

Feedback loop A path between the parts of a system such that one part affects a second part and the second part in turn affects the initial part. Regarding climate change, positive feedback accelerates climate change, while negative feedback decelerates it.

Fermentation The anaerobic breakdown of glucose that results in the production of ethanol and carbon dioxide gas (alcoholic fermentation) or lactic acid (lactic acid fermentation).

Fertilization The joining of an egg and a sperm to form a diploid cell that can develop into a new organism.

First law of thermodynamics A restatement of the law of energy conservation, usually as it applies to systems involving changes in temperature: Whenever heat flows into or out of a system, the gain or loss of thermal energy equals the amount of heat transferred.

Fitness The number of offspring an organism produces in its lifetime compared to other organisms in the population.

Floodplain A flat area along a stream that is covered by water only during a flood.

Flower The reproductive structure of flowering plants.

Fold A layer or layers of bent (plastically deformed) rock.

Folded mountain A mountain that features extensive folding of rock layers.

Foliated rocks Rocks that contain crystals arranged in parallel bands.

Force Simply stated, a push or a pull.

Force vector An arrow drawn to scale so that its length represents the magnitude of a force and its direction represents the direction of the force.

Forced vibration The setting up of vibrations in an object by a vibrating source.

Fossils The remains or traces of ancient life forms.

Free fall Motion under the influence of gravitational pull only.

Freezing The transformation of a liquid to a solid.

Frequency For a vibrating body, the number of vibrations per unit time. For a wave, the number of crests that pass a particular point per unit time.

Friction The resistive force that opposes the motion or attempted motion of an object through a fluid or past another object with which it is in contact.

Front (*weather front*) A boundary along which air masses meet.

Fruit In flowering plants, a structure surrounding the seeds that typically helps to spread the seeds.

Full Moon The phase of the Moon when its sunlit side faces Earth.

Functional group A specific combination of atoms that behaves as a unit in an organic molecule.

Fungi A clade of heterotrophic eukaryotes that obtain food by secreting digestive enzymes over organic matter and then absorbing the nutrients.

Galaxy A large assemblage of stars, interstellar gas, and dust, usually catagorized by its shape: elliptical, spiral, or irregular.

Gamma rays High-frequency electromagnetic radiation emitted by radioactive nuclei.

Gas Matter that has neither a definite volume nor a definite shape, always filling any space available to it.

Gene A section of DNA that contains the instructions for building a protein.

Gene flow The evolution of a population due to the movement of alleles into or out of the population.

Genetic drift The evolution of a population due to chance.

Genetic mutation A change in the nucleotide sequence of an organism's DNA.

Genome The total genetic content of an organism.

Genotype The genetic makeup of an organism.

Geology The Earth science that is concerned with the composition and structure of Earth.

Geothermal gradient The increase of temperature with depth in Earth's interior.

Giant stars Cool giant stars above the main-sequence stars on the H–R diagram.

Glacier A mass of dense ice that forms when snow on land is subjected to pressure from overlying snow, so that it is compacted and recrystallized.

Gradient (of a stream) The slope; equivalently, the change in a stream's elevation over a particular distance.

Gravitational force The attractive force between objects due to mass.

Greenhouse effect The process by which certain gases warm the atmosphere by trapping infrared radiation that is emitted by Earth's surface.

Groundwater The water that resides in a saturation zone.

Group A vertical column in the periodic table; also known as a family of elements.

H–R diagram (Hertzsprung–Russell diagram) A plot of luminosity versus surface temperature for stars. When so plotted, stars' positions take the form of a main sequence for average stars, with exotic stars above or below the main sequence.

Half-life The time required for half the atoms in a sample of a radioactive isotope to decay.

Half reaction One half of an oxidation–reduction reaction, represented by an equation showing electrons as either reactants or products.

Haploid Description of a cell that has one of each kind of chromosome.

Heat The thermal energy that flows from a substance of higher temperature to a substance of lower temperature, commonly measured in calories or joules.

Heat death A model for the end of the universe in which all matter and energy disperse to the point of maximum entropy.

Heat of fusion The amount of energy needed to change any substance from a solid to liquid or liquid to solid phase.

Heat of vaporization The amount of energy needed to change any substance from a liquid to gas or gas to liquid phase.

Hemoglobin A protein found in red blood cells that binds to and transports oxygen.

Heritable Description of traits that are passed from parents to offspring because they are at least partially determined by genes.

Heteroatom Any atom other than carbon or hydrogen in an organic molecule.

Heterogeneous mixture A mixture in which the different components can be seen as individual substances.

Heterotrophs Living organisms that obtain energy and organic molecules from other living organisms or other organic materials.

Heterozygote An organism that has two different alleles for a given gene.

Homeostasis The maintenance of a relatively stable internal environment.

Homogeneous mixture A mixture in which the components are so finely mixed that any one region of the mixture has the same ratio of substances as any other region; the components cannot be seen as identifiable individual substances.

Homozygote An organism that has two identical alleles for a given gene.

Hormone A chemical molecule that gives instructions to the body; it is produced in one place in the body, released into the blood, and received by target cells elsewhere in the body.

Hot spot A place on Earth's surface that sits directly above a rising column of magma arising from an exceptionally hot region usually near the mantle–core boundary.

Hubble's law The farther away a galaxy is from Earth, the more rapidly it is moving away from us: $v = H \times d$.

Humidity The amount of water vapor in the air.

Humus The decayed remains of animal and plant life in soil.

Hurricane A large rotating tropical storm with wind speeds of at least 119 km/h.

Hydrocarbon A chemical compound that contains only carbon and hydrogen atoms.

Hydrogen bond An unusually strong dipole–dipole attraction between molecules that have a hydrogen atom covalently bonded to a highly electronegative atom, usually nitrogen, oxygen, or fluorine.

Hydrologic cycle The cycle of evaporation and precipitation that controls the distribution of Earth's water.

Hydronium ion A polyatomic ion made by adding a proton (hydrogen ion) to a water molecule.

Hydroxide ion A polyatomic ion made by removing a proton (hydrogen ion) from a water molecule.

Hypothesis An educated guess or a reasonable explanation. When the hypothesis can be tested by experiment, it qualifies as a *scientific hypothesis*.

Ice age A time period that is cold enough so that glaciers collect at high latitudes and move toward lower latitudes.

Igneous rock Rock that forms from magma that cools.

Impulse The product of the force acting on an object and the time during which it acts.

Impulse–momentum relationship Impulse is equal to the change in the momentum of the object upon which the impulse acts. In symbolic notation,

$$Ft = \Delta mv$$

Induced dipole A temporarily uneven distribution of electrons in an otherwise nonpolar atom or molecule.

Inelastic collision A collision in which the colliding objects become distorted, generate heat, and possibly stick together.

Inertia The property of things to resist changes in motion.

Infiltration Absorption of water by the ground.

Innate immune system A set of nonspecific body defenses that work against many different pathogens.

Inner core One of Earth's structural layers—a solid sphere of hot metal, mostly iron, at the center of Earth.

Inner planets The four planets orbiting within 2 AU of the Sun: Mercury, Venus, Earth, and Mars—all are rocky and known as the *terrestrial* planets.

Insulator Any material with very little or no free charged particles and through which current does not easily flow.

Interaction Mutual action between objects during which one object exerts an equal and opposite force on the other object.

Interference The combined effect of two or more waves overlapping.

Interglacial period A period within an ice age that is warm enough so that glaciers melt back to polar latitudes.

Intrusive igneous rock Igneous rock that forms from magma that cools underground.

Inverse-square law A law relating the intensity of an effect to the inverse square of the distance from the cause:

$$\text{Intensity} \sim \frac{1}{\text{distance}^2}$$

Ion An atom that has a net electric charge because of either a loss or gain of electrons.

Ionic bond A chemical bond created by an electrical attraction between two oppositely charged ions.

Ionic compound A chemical compound containing ions.

Isotope Any member of a set of atoms of the same element whose nuclei contain the same number of protons but different numbers of neutrons.

Joule The SI unit of energy and work, equivalent to a newton-meter.

Ketone An organic molecule that contains a carbonyl group, the carbon of which is bonded to two carbon atoms.

Kilogram The unit of mass. One kilogram (symbol kg) is the mass of 1 liter (symbol L) of water at 4°C.

Kinetic energy The energy of motion, described by the relationship

$$\text{Kinetic energy} = \tfrac{1}{2}mv^2$$

Kuiper belt The disk-shaped region of the sky beyond Neptune populated by many icy bodies and a source of short-period comets.

Latent heat The energy absorbed or released in order to cause a change of phase.

Lava Magma that erupts at Earth's surface.

Law A general hypothesis or statement about the relationship of natural quantities that has been tested over and over again and has not been contradicted; also known as a *principle*.

Law of mass conservation Matter is neither created nor destroyed during a chemical reaction; atoms are merely rearranged, without any apparent loss or gain of mass, to form new molecules.

Law of reflection The angle of reflection equals the angle of incidence.

Law of universal gravitation Every body in the universe attracts every other body with a mutually attracting force. For two bodies, this force is directly proportional to the product of their masses and inversely proportional to the square of the distance separating them:

$$F = G\frac{m_1 m_2}{d^2}$$

Light-year The distance light travels in one year.

Lipids Hydrophobic organic molecules, many of which include fatty acids as a primary component.

Liquid Matter that has a definite volume but not definite shape, assuming the shape of its container.

Lithosphere Earth's outermost structural layer, consisting of cool, rigid rock.

Load The amount and type of sediment that a stream carries.

Local Group Our immediate cluster of galaxies, including the Milky Way, Andromeda, and Triangulum spiral galaxies plus a few dozen smaller elliptical and irregular galaxies.

Local Supercluster The cluster of galactic clusters in which our Local Group resides.

Lock-and-key model A conceptual model explaining how drugs function by interacting with receptor sites on proteins in the body.

Logistic growth A model of population growth in which growth slows as the population approaches the habitat's carrying capacity.

Longitudinal wave A wave in which the medium vibrates in a direction parallel (longitudinal) to the direction in which the wave travels.

Lower mantle One of Earth's structural layers—the lowest portion of the mantle, a zone of strong, rigid rock.

Lunar eclipse The phenomenon in which the shadow of Earth falls on the Moon, producing the relative darkness of the full Moon.

Lymph A clear fluid inside lymph vessels that contains large numbers of immune cells.

Magma Molten rock in Earth's interior.

Magnetic domains Clustered regions of aligned magnetic atoms. When these regions themselves are aligned with one another, the substance containing them is a magnet.

Magnetic field The region of magnetic influence around a magnetic pole or around a moving charged particle.

Magnetic force (1) Between magnets, it is the attraction of unlike magnetic poles for each other and the repulsion between like magnetic poles. (2) Between a magnetic field and a moving charge, it is a deflecting force due to the motion of the charge; the deflecting force is perpendicular to the velocity of the charge and perpendicular to the magnetic field lines.

Main sequence The diagonal band of stars on an H–R diagram; such stars generate energy by fusing hydrogen to helium.

Mantle The layer of thick, hot rock between Earth's crust and core.

Mass The quantity of matter in an object. More specifically, mass is a measure of the inertia or sluggishness that an object exhibits in response to any effort made to start it, stop it, deflect it, or change its state of motion in any way.

Mass movement The downward movement of rock or soil due to gravity.

Maxwell's counterpart to Faraday's law A magnetic field is induced in any region of space in which an electric field is changing with time.

Mechanical weathering The breakdown of rock into smaller pieces through physical means.

Meiosis A form of cell division in which one diploid cell divides to produce four haploid cells.

Melting The transformation of a solid to a liquid.

Messenger RNA (mRNA) An RNA molecule made during transcription that carries genetic information from DNA to the ribosomes.

Metallic bond A chemical bond in which positively charged metal ions are held together within a "fluid" of loosely held electrons.

Metamorphic rock Rock that forms by the alteration of preexisting rock deep in Earth's interior due to the effects of temperature, pressure, or reactive chemicals.

Meteor The streak of light produced by a meteoroid burning in Earth's atmosphere; a "shooting star."

Meteorite A meteoroid, or a part of a meteoroid, that has survived passage through Earth's atmosphere to reach the ground.

Meteoroid A small rock in interplanetary space, which can include a fragment of an asteroid or comet.

Midocean ridge A global system of underwater mountains created by seafloor spreading.

Milankovitch theory The idea that slow variations in Earth's orbit, the tilt of its axis, and the path of its wobble affect the way sunlight strikes Earth and therefore alter its temperature.

Mineral A naturally occurring, solid, inorganic, crystal material with a specific chemical composition.

Mitochondria Eukaryotic organelles that break down organic molecules to produce ATP.

Mitosis A form of cell division in which one cell divides into two daughter cells, each of which contains the same genetic information as the original cell.

Mixture A combination of two or more substances in which each substance retains its own properties.

Molarity A common unit of concentration equal to the number of moles of a solute per liter of solution.

Mole A large number equal to 6.02×10^{23}; usually used in reference to the number of atoms, ions, or molecules within a macroscopic amount of a material.

Molecule The fundamental unit of a chemical compound, which is a group of atoms held together by covalent bonds.

Momentum The product of the mass of an object and its velocity.

Monomers The small molecular units from which a polymer is formed.

Moon phases The cycles of change of the "face" of the Moon, changing from *new*, to *waxing*, to *full*, to *waning*, and back to *new*.

Nanotechnology Technology on the scale of a nanometer, where materials are engineered by the manipulation of individual atoms or molecules.

Natural frequency A frequency at which an elastic object naturally tends to vibrate.

Natural selection The process in which organisms with heritable, advantageous traits leave more offspring than organisms with other traits, causing these advantageous traits to become more common in a population over time.

Nebular theory The idea that the Sun and planets formed together from a cloud of gas and dust, a *nebula*.

Nephron The functional unit of a kidney; it filters waste molecules from blood to produce urine.

Net force The combination of all forces that act on an object.

Neuron A cell that receives and transmits electrical signals from one part of the body to another.

Neutral Description of a solution in which the hydronium ion concentration is equal to the hydroxide ion concentration.

Neutralization A reaction between an acid and a base.

Neutron An electrically neutral subatomic particle in the atomic nucleus.

Neutron star A small, extremely dense star composed of tightly packed neutrons formed by the welding of protons and electrons.

New Moon The phase of the Moon when darkness covers the side facing Earth.

Newton The scientific unit of force.

Newton's first law of motion Every object continues in a state of rest, or of uniform speed in a straight line, unless acted on by a nonzero force.

Newton's second law of motion The acceleration produced by a net force on an object is directly proportional to the net force, is in the same direction as the net force, and is inversely proportional to the mass of the object.

Newton's third law of motion Whenever one object exerts a force on a second object, the second object exerts an equal and opposite force on the first object.

Niche The total set of biotic and abiotic resources a species uses within a community.

Noble gas shell A region of space around the atomic nucleus within which electrons may reside.

Nonbonding pairs Two paired valence electrons that do not participate in a chemical bond.

Nonpolar Description of a chemical bond or molecule that has no dipole; the electrons are distributed evenly.

Nova An event in which a white dwarf suddenly brightens and appears as a "new" star.

Nuclear fission The splitting of the atomic nucleus into two smaller halves.

Nuclear fusion The combining of nuclei of light atoms to form heavier nuclei.

Nucleic acids Organic molecules that are composed of chains of nucleotides and store genetic information.

Nucleon Any subatomic particle found in the atomic nucleus; another name for either a proton or a neutron.

Nucleus A structure within eukaryotic cells that is surrounded by a double membrane and that contains the cell's DNA.

Ocean basin The deep portion of the sea floor between continental margins and the midocean ridge.

Ocean trenches Long, deep, steep troughs in the seafloor where an oceanic plate sinks beneath an overlying plate.

Ohm's law The amount of current in a circuit varies in direct proportion to the potential difference or voltage and inversely with the resistance:

$$\text{Current} = \frac{\text{voltage}}{\text{resistance}}$$

Oort cloud The region beyond the Kuiper belt populated by trillions of icy bodies and a source of long-period comets.

Opaque Description of materials that absorb light without reemission.

Ordinary matter Matter that responds to the strong nuclear, weak nuclear, electromagnetic, and gravitational forces. This is matter made of protons, neutrons, and electrons, which includes the atoms and molecules that make us and our immediate environment.

Ore Mineral deposits that are large enough and pure enough to be mined for profit.

Organ A structure in the body that has a specific function.

Organ system A set of organs that work together to perform a particular bodily function.

Organelles Structures within the cytoplasm of eukaryotic cells that perform specific functions in the cell.

Organic chemistry The study of carbon-containing compounds.

Outer core One of Earth's structural layers—a shell of hot, liquid metal beneath the mantle and above the inner core.

Outer planets The four planets orbiting beyond 2 AU of the Sun, Jupiter, Saturn, Uranus, and Neptune—all are gaseous and known as the *jovian* planets.

Ovulation The release of a mature egg cell that occurs once during each menstrual cycle.

Oxidation The process by which a reactant loses one or more electrons.

Parabola The curved path followed by a projectile near Earth under the influence of gravity only.

Parallel circuit An electric circuit with two or more devices connected in such a way that the same voltage acts across each one, and any single one completes the circuit independently of all the others.

Passive transport Transport across the cell membrane that does not require energy from the cell.

Pathogens Disease-causing agents such as bacteria, viruses, or other organisms.

Period In the context of physics, the time required for a vibration or a wave to make one complete cycle. In chemistry, a horizontal row in the periodic table.

Periodic table A chart in which all known elements are listed in order of atomic number.

Peristalsis A moving wave of muscular contractions that moves food down the digestive tract.

Permeability The ease with which fluids flow through the pore spaces in sediments and rocks.

pH A measure of the acidity of a solution; equal to the negative logarithm of the hydronium ion concentration.

Phenol An organic molecule in which a hydroxyl group is bonded to a benzene ring.

Phenotype The traits of an organism.

Photosynthesis The process in plants and some other organisms in which light energy from the Sun is converted into chemical energy in organic molecules.

Physical change A change in which a substance's physical properties are altered but with no change in its chemical identity.

Physical model A representation of an object on some convenient scale.

Physical property Any physical attribute of a substance, such as color, density, or hardness.

Placenta The organ that allows nutrients and wastes to be exchanged between a pregnant woman and a developing embryo.

Planetary nebula An expanding shell of gas ejected from a low-mass star during the latter stages of its evolution.

Planets The major bodies orbiting the Sun that are massive enough for their gravity to make them spherical and small enough to avoid having nuclear fusion in their cores. They also have successfully cleared all debris from their orbital paths.

Plants A clade of autotrophic, multicellular, terrestrial eukaryotes that make food through photosynthesis.

Plate tectonics The theory that Earth's lithosphere is divided into large plates that move slowly around the globe.

Platelets Blood cells that function in blood clotting.

Polar Description of a chemical bond or molecule that has a dipole; the electrons are congregated on one side, which makes that side slightly negative while the opposite side (lacking electrons) becomes slightly positive.

Pollen In seed plants, immature male gametophytes (reproductive cells) wrapped in protective coatings.

Polyatomic ion A molecule that carries a net electric charge.

Polymer A long organic molecule made up of many repeating units.

Population A group of individuals of a single species that lives in a specific area.

Porosity The percentage of the volume of a rock or sediment that consists of open space.

Potential difference The difference in electric potential between two points, measured in volts; often called *voltage difference.*

Potential energy The stored energy that a body possesses because of its position.

Power The time rate of work:

$$\text{Power} = \frac{\text{work}}{\text{time}}$$

(More generally, power is the rate at which energy is expended.)

Precipitate A solute that has come out of solution.

Precipitation Water in the liquid or solid state that returns to Earth's surface from the atmosphere.

Principle of falsifiability For a hypothesis to be considered scientific, it must be testable—it must, in principle, be capable of being proven wrong.

Producer An organism that makes organic molecules using inorganic molecules and energy.

Products The new materials formed in a chemical reaction.

Projectile Any object that moves through the air or through space under the influence of gravity.

Prokaryotes Single-celled organisms, such as bacteria and archaea, whose cells lack a nucleus.

Proteins Organic molecules composed of folded chains of amino acids.

Protists A miscellaneous group of eukaryotic organisms that includes all the eukaryotes that are not plants, animals, or fungi.

Proton A positively charged subatomic particle in the atomic nucleus.

Pseudoscience A theory or practice that is considered to be without scientific foundation but purports to use the methods of science.

Pulsar A celestial object (most likely a neutron star) that spins rapidly, sending out short, precisely timed bursts of electromagnetic radiation.

Pure The state of a material that consists of only a single element or compound.

Pyroclastics Lava and rock fragments ejected into the atmosphere during a violent volcanic eruption.

Quantum A small, discrete packet of light energy.

Quantum number *n* An integer that specifies the quantized energy level of an atomic orbital.

Quasar The core of a distant galaxy early in its life span when its central black hole has not yet swept much matter from its vicinity, leading to a rate of radiation higher than that from entire older galaxies.

Rad A quantity of radiant energy equal to 0.01 J absorbed per kilogram of tissue.

Radiation The transfer of energy by means of electromagnetic waves.

Radioactivity The high-energy particles and electromagnetic radiation emitted by a radioactive substance.

Reactants The reacting substances in a chemical reaction.

Reaction rate A measure of how quickly the concentration of products in a chemical reaction increases or the concentration of reactants decreases.

Recessive Description of an allele that is not expressed in a heterozygote.

Recombination The production of new combinations of genes that differ from combinations found in the parental chromosomes resulting from crossing over during meiosis.

Red blood cells Blood cells that transport oxygen to body tissues.

Reduction The process by which a reactant gains one or more electrons.

Reflection The returning of a wave to the medium from which it came when it hits a barrier.

Refraction The bending of waves due to a change in wave speed.

Relative humidity The ratio of the water vapor actually in the air to the maximum water vapor the air can hold at that temperature.

Rem The unit for measuring radiation dosage in humans based on harm to living tissue.

Reservoir Each of the places where water is stored as it moves through the water cycle.

Residence time The length of time water exists in a reservoir.

Resonance A dramatic increase in the amplitude of a wave that results when the frequency of forced vibrations matches an object's natural frequency.

Resultant The net result of a combination of two or more vectors.

Ribonucleic acid (RNA) A single-stranded molecule consisting of a sugar–phosphate backbone attached to a series of nitrogenous bases.

Ring of Fire A horseshoe-shaped zone of frequent earthquakes and volcanic eruptions that encircles the basin of the Pacific Ocean.

Rock A solid, cohesive mixture of one or more kinds of mineral crystals.

Rock cycle A series of processes by which rock is formed, breaks down, and re-forms into different kinds of rock.

Runoff Precipitation not absorbed by the ground that moves over Earth's surface.

Salt An ionic compound commonly formed from the reaction between an acid and a base.

Sarcomeres The contractile units of muscle cells.

Satellite A projectile or small body that orbits a larger body.

Saturated hydrocarbon A hydrocarbon that contains no multiple covalent bonds, with each carbon atom bonded to four other atoms.

Saturated solution A solution containing the maximum amount of solute that will dissolve in its solvent.

Scalar quantity A quantity, such as mass, volume, speed, and time, that can be completely specified by its magnitude.

Scanning probe microscope A tool of nanotechnology that detects and characterizes the surface atoms of materials using an ultrathin probe tip, which is detected by laser light as it is mechanically dragged over the surface.

Science The collective findings of humans about nature, and the process of gathering and organizing knowledge about nature.

Scientific method An orderly method for gaining, organizing, and applying new knowledge.

Seafloor spreading The process by which new lithosphere is created at midocean ridges as older lithosphere moves away.

Seamount An undersea volcanic peak.

Second law of thermodynamics Heat never spontaneously flows from a lower-temperature substance to a higher-temperature substance. Also, all systems tend to become more and more disordered as time goes by.

Sedimentary rock Rock that forms over time as bits and pieces of preexisting rock are consolidated through compaction and/or cementation.

Sediments Fragments of weathered rock or living organisms.

Seed In seed plants, a structure consisting of a plant embryo, a food supply, and a tough outer coating.

Seismic waves Mechanical waves produced by an earthquake that propagate through Earth.

Seismology The study of seismic waves, waves that travel through Earth as a result of an earthquake or other disturbance.

Semiconductor A material that can be made to behave sometimes as an insulator and sometimes as a conductor.

Series circuit An electric circuit with devices connected in such a way that the same electric current exists in all of them.

Silicates Minerals that contain silicon and oxygen and make up more than 75% of Earth's crust.

Soil A mixture of weathered rock plus air, water, and organic matter that supports the growth of plants.

Solar eclipse The phenomenon in which the shadow of the Moon falls on Earth, producing a region of darkness in the daytime.

Solar intensity Solar radiation per area.

Solar radiation Electromagnetic energy given off by the Sun.

Solid Matter that has a definite volume and a definite shape.

Solubility The ability of a solute to dissolve in a given solvent.

Soluble Capable of dissolving to an appreciable extent in a given solvent.

Solute Any component in a solution that is not the solvent.

Solution A homogeneous mixture in which all components are in the same phase.

Solvent The component in a solution that is present in the largest amount.

Speciation The formation of new species.

Species A group of organisms whose members can breed with one another but not with members of other species.

Specific heat capacity The quantity of heat required to raise the temperature of a unit mass of a substance by 1 degree Celsius.

Spectroscope A device that uses a prism or diffraction grating to separate light into its color components.

Speed The distance traveled per unit of time.

Starburst galaxy A galaxy in which stars are forming at an unusually fast rate.

Strata Parallel layers of sedimentary rock.

Stress A force applied to rock.

Strong nuclear force The attractive force between all nucleons, effective at only very short distances.

Structural isomers Molecules that have the same molecular formula but different chemical structures.

Subduction The sinking of oceanic lithosphere into the mantle.

Subduction zone The region where an oceanic plate sinks into the asthenosphere at a convergent plate boundary.

Sublimation The transformation of a solid to a gas.

Submicroscopic The realm of atoms and molecules, where objects are smaller than can be detected by optical microscopes.

Sunspots Temporary, relatively cool and dark regions on the Sun's surface.

Supernova The explosion of a massive star caused by gravitational collapse with the emission of enormous quantities of matter and radiation.

Support force The force that supports an object against gravity; often called the normal force.

Surface current A wind-driven, shallow ocean current.

Surface processes Geologic processes that have causes and effects at or near Earth's surface.

Suspension A homogeneous mixture in which the various components are thoroughly mixed but remain in different phases.

Sympatric speciation Speciation that occurs without geographic isolation.

Synapse A connection between a neuron and a target cell.

Tangential velocity Velocity that is parallel (tangent) to a curved path.

Technology The means of solving practical problems by applying the findings of science.

Tectonic plates Separate pieces of lithosphere that move on top of the asthenosphere.

Temperature A measure of the hotness or coldness of substances, related to the average translational kinetic energy per molecule in a substance; measured in degrees Celsius, in degrees Fahrenheit, or in kelvins.

Terminal speed The speed at which the acceleration of a falling object terminates when air resistance balances weight.

Terrestrial radiation Infrared radiation emitted by Earth's surface.

Texture (of soil) The size of the particles in soil.

Theory A synthesis of a large body of information that encompasses well-tested hypotheses about certain aspects of the natural world.

Thermal energy The total energy (kinetic plus potential) of the submicroscopic particles that make up a substance (often called *internal energy*).

Thermodynamics The study of heat and its transformation into different forms of energy.

Thermonuclear fusion Nuclear fusion brought about by high temperatures.

Third law of thermodynamics No system can reach absolute zero.

Tissue A group of similar cells that performs a certain function.

Topsoil The upper layer of soil, rich in humus, that supports plant life.

Transcription The first step in building a protein, in which a molecule of RNA is assembled from information contained in DNA.

Transfer RNA (tRNA) An RNA molecule that transfers an amino acid to a growing protein during translation.

Transform boundaries Places where tectonic plates slide along beside one another as they move.

Translation The assembly of a protein based on information contained in an mRNA molecule.

Transmutation The changing of an atomic nucleus of one element into an atomic nucleus of another element through a decrease or increase in the number of protons.

Transparent Description of materials that allow light to pass through them in straight lines.

Transverse wave A wave in which the medium vibrates in a direction perpendicular (transverse) to the direction in which the wave travels.

Tributary A smaller stream that flows into a larger one.

Tsunami A huge seismic sea wave.

Universal constant of gravitation, *G* The proportionality constant in Newton's law of gravitation.

Unsaturated hydrocarbon A hydrocarbon that contains at least one multiple covalent bond.

Unsaturated solution A solution that is capable of dissolving additional solute.

Upwarped mountain A dome-shaped mountain produced by a broad arching of Earth's crust.

Valence electron The electrons in the outermost occupied shell of an atom.

Valence shell The outermost occupied shell of an atom.

Variation Differences in a trait from one individual to another.

Vector An arrow whose length represents the magnitude of a quantity and whose direction represents the direction of the quantity.

Vector components Parts into which a vector can be separated and that act in different directions from the vector.

Vector quantity A quantity that specifies direction as well as magnitude.

Veins Blood vessels that carry blood toward the heart.

Velocity A vector quantity that specifies both the speed of an object and its direction of motion.

Virus A small piece of genetic material wrapped in a protein coat that infects and reproduces within a host cell.

Volcano A hill or mountain formed by the extrusion of lava, ash, and rock fragments.

Water table The upper boundary of the saturated zone.

Watershed The area of land that drains into a stream.

Wave A disturbance that travels from one place to another transporting energy, but not necessarily matter, along with it.

Wave-particle duality The exhibition of both a wave nature and a particle nature of light.

Wavelength The distance from the top of one crest to the top of the next crest or, equivalently, the distance between successive identical parts of the wave.

Weather The state of the atmosphere at a particular time and place.

Weathering The disintegration or decomposition of rock at or near Earth's surface.

Weight Simply stated, the force of gravity on an object. More specifically, the gravitational force with which a body presses against a supporting surface.

Weightlessness A condition encountered in free fall in which a support force is lacking.

White blood cells Blood cells that are part of the immune system.

White dwarf A dying star that has collapsed to the size of Earth and is slowly cooling off; located at the lower left on the H–R diagram.

Wind Air flowing horizontally from an area of high pressure to an area of lower pressure.

Work The product of the force and the distance through which the force moves:

$$W = Fd$$

Work–energy theorem The work done on an object equals the change in kinetic energy of the object:

$$\text{Work} = \Delta KE$$

Credits

Text and Art Credits

Page 5: ("Science is a way to teach . . .") Source: James Gleick, *Genius: The Life and Science of Richard Feynman.* 295: ("There has to be . . .") Source: John Ashton. 306: (Fig 12.13) Suchocki, John A., *Conceptual Chemistry 4th Ed.*, © 2011. Reprinted and electronically reproduced by permission of Pearson Education, Inc., Upper Saddle River, New Jersey. 359: (Fig. 13.26) Hewitt, Paul G., Suchocki, John A., Hewitt, Leslie A., *Conceptual Physical Science, 5th Ed.*, © 2012. Reprinted and electronically reproduced by permission of Pearson Education, Inc., Upper Saddle River, New Jersey. 476: ("pervasively contaminated . . .") Source: Margaret Mellon & Jane Rissler, *Gone To Seed: Transgenic Contaminants In The Traditional Seed Supply.* 507: (Fig. 17.36a) UC Museum of Paleontology's Understanding Evolution (evolution .berkeley.edu). Reprinted by permission of the University of California at Berkeley. 564: (top image) Reprinted by permission of the Boston Retinal Implant Project. 602: (map) UNAIDS Report on the Global AIDS Epidemic 2010. 616: (graph) Global Footprint Network. Number of Planets Scenario 2008. www.footprintnetwork.org. 636: ("Some people have . . .") Wengari Maathai www.greenbeltmovement .org. 737: (Fig. 25.13a) Reprinted by permission of The McGraw-Hill Companies. From Carla Montgomery, *Environmental Geology*. 769: (Fig. 26.25) Paul G. Hewitt, Suzanne Lyons, John Suchocki, Jennifer Yeh, *Conceptual Integrated Science: Explorations, NASTA First Edition* © 2010. Reprinted and electronically reproduced by permission of Pearson Education, Inc., Upper Saddle River, New Jersey. 814: (Fig. 27.36) Source: http://alternativestandrews.blogspot.com/p/carbon-footprint.html. 832: (Fig. 28.17) Hewitt, Paul G., Suchocki, John A., Hewitt, Leslie, A., *Conceptual Physical Science 5th Ed.*, © 2012. Reprinted and electronically reproduced by permission of Pearson Education, Inc. Upper Saddle River, NJ. 837: (Fig. 28.22) Hewitt, Paul G., Suchocki, John A., Hewitt, Leslie, A., *Conceptual Physical Science 5th Ed.*, © 2012. Reprinted and electronically reproduced by permission of Pearson Education, Inc. Upper Saddle River, NJ. 846: (Fig. 28.42) Hewitt, Paul G., Suchocki, John A., Hewitt, Leslie, A., *Conceptual Physical Science 5th Ed.*, © 2012. Reprinted and electronically reproduced by permission of Pearson Education, Inc. Upper Saddle River, NJ. 847: (Fig. 28.44) Hewitt, Paul G., Suchocki, John A., Hewitt, Leslie, A., *Conceptual Physical Science 5th Ed.*, © 2012. Reprinted and electronically reproduced by permission of Pearson Education, Inc. Upper Saddle River, NJ. 860: (Fig. 29.10) Hewitt, Paul G., Suchocki, John A., Hewitt, Leslie, A., *Conceptual Physical Science 5th Ed.*, © 2012. Reprinted and electronically reproduced by permission of Pearson Education, Inc. Upper Saddle River, NJ. A-12: ("The greatest shortcoming . . .") Reprinted by permission of Albert A. Bartlett, University of Colorado.

Photo Credits

Part One Page 1: Courtesy of Enrique Baez, 2: Reed Kaestner/Corbis, 13: (top) Jim McClintock, 13: (bottom) Jim Maestro, 17: Paul G. Hewitt, 18: Natali Glado/Shutterstock.com, 20: Panos Karapanagiotis/Shutterstock.com, 21: Sustermans, Justus (1597-1681), 25: Paul G. Hewitt, 31: robcocquyt/Shutterstock.com, 32: (top) Suzanne Lyons, 32: (bottom) Dirk Freder/Vetta Collection/iStockphoto.com, 33: Alan Schein Photography/CORBIS, 34: Paul G. Hewitt, 37: ZUMA Press/Newscom, 44: Larry Sargeant/Comstock Images, 47: World History Archive/Newscom, 51: Fundamental Photographs, NYC, 52: (a) Stephen Dalton/Photo Researchers, Inc., 52: (b) Ocean/Corbis, 58: Paul G. Hewitt, 62: iStockphoto.com/Duncan Walker, 67: Paul G. Hewitt, 68: Paul G. Hewitt, 69: John Wilhelmsson/StockShot/Alamy, 72: Edward Knisman/Photo Researchers, Inc., 73: Paul G. Hewitt, 77: (top) Paul G. Hewitt, 77: (bottom) ZUMA Press/Newscom, 79: Paul G. Hewitt, 81: Paul G. Hewitt, 82: (a–b) Michael Vollmer, 83: Shutterstock, 86: smereka/Shutterstock.com, 88: Paul G. Hewitt, 94: Paul G. Hewitt, 95: NASA, 103: NASA Earth Observing System, 111: Addison Wesley Longman, Inc./San Francisco, 114: NASA Marshall Space Flight Center Collection, 115: NASA/Goddard Space Flight Center, 121: bierchen/Shutterstock.com, 125: Paul G. Hewitt, 126: Paul G. Hewitt, 128: Paul G. Hewitt, 133: (top) AP Photo/Associated Press, 133: (bottom left) LU Engineers, 133: (bottom right) Hu Meidor, 134: Claude Nuridsany & Marie Perennou/Photo Researchers, Inc., 137: (top) Tracy Suchocki, 137: (b) Don Hynek/Paul G. Hewitt, 138: (top) Paul G. Hewitt, 138: (bottom) Paul G. Hewitt, 140: David J. Green/Alamy, 146: Paul G. Hewitt, 147: iStockphoto.com/Josemaria Toscano, 154: John Downer/Nature Picture Library, 155: Pearson Education, 157: Tim Ridley/Dorling Kindersley Media Library, 160: (top) Pearson Education, 160: (bottom) Pearson Education, 161: (top) Pearson Education, 161: (bottom) Pearson Education, 163: (top, bottom) Richard Megna/Fundamental Photographs, NYC, 165: (a–c) Richard Megna/Fundamental Photographs, NYC, 166: John Downer/Nature Picture Library, 168: (left, right) Pearson Education, 176: (left, right) Paul G. Hewitt, 178: Paul G. Hewitt, 180: Dudarev Mikhail/Shutterstock.com, 185: (top) Paul G. Hewitt, 185: (bottom) AP Photo/Associated Press, 188: Paul G. Hewitt, 189: (left) Bill Bachman/Alamy, 189: (right) Institute of Paper Science & Technology, 191: Paul G. Hewitt, 192: Suzanne Lyons, 193: Hu Meidor/Paul G. Hewitt, 195: Paul G. Hewitt, 199: Paul G. Hewitt, 200: (a–c) 1972 George Resch - Fundamental Photographs, 201: Ken Kay/Fundamental Photographs, NYC, 202: Richard Megna/Fundamental Photographs, NYC, 210: Paul G. Hewitt, 211: Paul G. Hewitt.

Part Two Page 213: John Suchocki, 214: IBM Corporate Archives, 215: IBM Corporate Archives, 216: John Suchocki, 217: (left) Rachel Epstein/PhotoEdit Inc., 217: (middle) MarcelClemens/Shutterstock.com, 217: (right) Joseph Sibilsky/Alamy, 222: (Ag) PhotoDisc/Getty Images, 222: (Ti) PhotoDisc/Getty Images, 222: (C) Shutterstock, 222: (He) Photodisc/Getty Images, 222: (Hg) MarcelClemens/Shutterstock.com, 222: (Zn) Eric Schrader/Pearson Science, 222: (Si) Geostock/PhotoDisc/Getty Images, 222: (Br) Richard Megna/Fundamental Photographs, NYC, 222: (bottom) NASA, 225: Pearson Education/Pearson Science, 226: (baseball) CLM/Shutterstock.com, 226: (Earth) dundanim/Shutterstock.com, 227: (a) National Institute of Standards and Technology, 227: (b–c) IBM Corporate Archives, 228: (a) Geoff Brightling/Dorling Kindersley Media Library, 228: (b) NOAA, 229: (a) GIPhotoStock/Photo Researchers, Inc., 229: (b) John Suchocki/Paul G. Hewitt, 230: (top) John Suchocki/Paul G. Hewitt, 230: (strontium) Tom Bochsler/Pearson Education, 230: (potassium, barium, copper) Richard Megna/Fundamental Photographs, NYC, 234: (a) John Suchocki/Paul G. Hewitt, 234: (b) Dartmouth Electron Microscope Facility, 235: (a–c) John Suchocki/

Paul G. Hewitt, 237: Tom Hollyman/Getty Images, 239: (left, right) Pearson Education, 240: (top) Pearson Education/Pearson Science, 240: (bottom) Paul G. Hewitt, 243: iStockphoto.com/RelaxFoto.de, 246: (top) International Atomic Energy Agency, 246: (middle) Steve Cole/Getty Images, Inc.- Photodisc./Royalty Free, 247: iStockphoto.com/Mark Kostich, 248: Jerry Nulk and Sra Joshua Baker, 249: Iris Christiani/Bionerd via Flickr, 252: (a–b) Saint-Gobain Crystals & Detectors, 254: Lawrence Berkeley National Laboratory, 260: iStockphoto.com/Michael Utech, 266: International Thermonuclear Experimental Reactor (ITER), 268: John Suchocki/Paul G. Hewitt, 272: John Suchocki/Paul G. Hewitt, 273: (top) NASA/Goddard Institute for Space Studies, 273: (bottom) Getty Images, Inc.-Photodisc./Royalty Free, 274: Used with the permission of the American Chemistry Council (ACC). Responsible Care (R) and the "Hands" program logo are registered service marks of the ACC., 275: (dune) Galyna Andrushko/Shutterstock.com, 275: (glass) Pearson Education, 276: (a–c) Paul G. Hewitt, 278: yuris/Shutterstock, 279: (top) Epcot Images/Alamy, 279: (bottom) Steve Warmowski/Journal-Courier/The Image Works, 280: (top left) iStockphoto.com/murat boylu, 280: (top middle) Irene Springer/Pearson Education, 280: (top right) Pearson Education, 280: (bottom) Getty Images, Inc.-Photodisc./Royalty Free, 281: (top left) Steve Allen/Brand X Picture/Jupiter Images PictureArts Corporation, 281: (top middle) Eric Schrader/Pearson Science, 281: (top right) blickwinkel/Alamy, 282: Charles D. Winters/Photo Researchers, Inc., 283: (top) Sue Smith/Shutterstock.com, 283: (bottom) Paul G. Hewitt, 284: (top, bottom) John Suchocki/Paul G. Hewitt, 286: (left) Richard Megna/Fundamental Photographs, NYC, 286: (middle and right) Pearson Education, 289: (left) Victor Habbick Visions/Photo Researchers, Inc., 289: (right) Mondolithic Studio, 290: (left) IBM Corporate Archives, 290: (right top) European Communities, 1995-2006, 290: (right bottom) Dr. Georg Fritz/Team Nanotec GmbH, 293: (top, bottom) Paul G. Hewitt, 297: Eric Schrader/Pearson Science, 299: Eric Schrader/Pearson Science, 303: (ruby) iStockphoto/Laurie Knight, 303: (sapphire) marcel/Fotolia, 304: Dee Breger/Photo Researchers, Inc., 305: US Mint, 306: Shutterstock, 307: Theodore Gray, 308: (a) Pearson Education/Pearson Science, 308: (b) Shutterstock, 309: Shutterstock, 313: David Taylor/Photo Researchers, Inc., 315: SPANISH NAVY PRESS/AFPphotos/NewsCom, 318: Pearson Education/Pearson Science, 320: REUTERS/Daniel Wallis, 321: Getty Images Inc./RF, 322: (top) Don Geddis, 322: (a left) Colin Keates/Dorling Kindersley Media Library, 322: (a middle) iStockphoto, 322: (a right) Getty Images, Inc.- Photodisc., 322: (b left, middle) Shutterstock, 322: (b right) John Suchocki/Paul G. Hewitt, 323: Shutterstock, 331: Paul G. Hewitt, 332: Paul G. Hewitt, 334: Paul G. Hewitt, 336: Getty Images - Digital Vision, 339: (left, right) Shutterstock, 342: NASA, 347: Associated Press [Wide World Photos], 349: Peter Arnold Images/Photolibrary/Getty Images, 350: (top) egd/Shutterstock, 350: (a) Corbis/Alamy, 350: (b–d) Pearson Education/Pearson Science, 351: (a, d) Eric Schrader/Pearson Education/Pearson Science, 351: (b–c) Pearson Education/Pearson Science, 353: Pearson Education/Pearson Science, 357: (a) Richard Megna/Fundamental Photographs, NYC, 357: (b) Eric Schrader/Pearson Science, 359: (a left) Library of Congress Prints and Photographs Division Washington, D.C. 20540 USA, 359: (a right) Will McIntyre/Photo Researchers, Inc., 359: (b) Photoshot Holdings Ltd/Alamy, 361: Richard Megna/Fundamental Photographs, NYC, 364: Eric Schrader/Pearson Science, 365: Ballard Power Systems, 366: (top) John Suchocki/Paul G. Hewitt, 366: (bottom) Pearson Education/Pearson Science, 367: Angelika Bentin/Fotolia, 370: (top, bottom) John Suchocki/Paul G. Hewitt, 371: Pearson Education/Pearson Science, 376: John Suchocki/Paul G. Hewitt, 380: Pearson Science/Pearson Education, 382: Casey Kelbaugh/AGEfotostock, 392: (a) Eduardo Jose Bernardino/istockphoto, 392: (b) James Stuart Griffith/Shutterstock, 399: John Suchocki/Paul G. Hewitt, 400: Pearson Education/Pearson Science, 402: Fujitsu Ltd., HO/AP Wide World Photos, 403: Pearson Education/Pearson Science.

Part Three Page 409: Suzanne Lyons, 410: Jeff Rotman/Alamy, 411: (a) Pakhnyushcha/Shutterstock, 411: (b) Noam Wind/Shutterstock, 411: (c) thinkdo/Shutterstock, 412: (a) David Wrobel, 412: (b) Stefan Lundgren/Johner Images/Alamy, 414: Cathy Keifer/Shutterstock, 416: (left) Juan Gaertner/Shutterstock, 416: (right) Mediscan/Medical-on-Line/Alamy, 417: Bettmann/CORBIS, 418: (a) Dimarion/Shutterstock, 418: (b) Eye of Science/Photo Researchers, Inc., 421: Daniel S. Friend, 423: Heather Angel/Natural Visions/Alamy, 425: Don W. Fawcett/Photo Researchers, Inc., 427: (a) Don W. Fawcett/Photo Researchers, Inc., 427: (b) Biophoto Associates/Photo Researchers, Inc., 428: (a) WaterFrame/Alamy, 428: (b) Prof. Jeff Hardin/U of WI-Madison, 428: (c) David McCarthy/Photo Researchers, Inc., 430: Pascal Goetgheluck/Photo Researchers, Inc., 435: Dr. Jeremy Burgess/Photo Researchers, Inc., 443: (left) Michael Abbey/Photo Researchers, Inc., 443: (right) Clément Philippe/Arterra Picture Library/Alamy, 445: J Emilio Flores/CORBIS, 447: (top) Biophoto Associates/Photo Researchers, Inc., 447: (bottom) CNRI/Photo Researchers, Inc., 448: Driscoll, Youngquist & Baldeschwieler, CALTECH/Photo Researchers, Inc., 449: (left) Science Source/Photo Researchers, Inc., 449: (middle) A. Barrington Brown/Photo Researchers, Inc., 449: (right) Science Source/Photo Researchers, Inc., 450: USDA/APHIS Animal And Plant Health Inspection Service, 461: Library of Congress, 468: (a) vadim kozlovsky/shutterstock, 468: (b) Colin Cuthbert/SPL/Photo Researchers, Inc., 469: Tomasz Markowski/Shutterstock, 470: Science Photo Library/Alamy, 472: Jennifer Yeh, 473: ITAR-TASS Photo Agency/Alamy, 474: (top) FLPA/Mark Newman/AGE Fotostock, 474: (middle) AquaBounty/MCT/Newscom, 474: (bottom) Trinity Mirror/Mirrorpix/Alamy, 476: (top) Cobbing/Hoffman Photo Library/David Hoffman Photo Library/Alamy, 476: (bottom) Everett Collection, 482: demarfa/Fotolia, 483: Robert Thom/Pfizer, Inc., 484: (left) F. M. Menger and Kurt Gabrielson, 484: (right) Stanley Awramik/Biological Photo Service, 485: NASA, 486: Francois Gohier/Photo Researchers, Inc., 487: Braun, C./Arco Images GmbH/Alamy, 488: (top) Science Source/Photo Researchers, Inc., 488: (a) Stubblefield Photography/Shutterstock, 488: (b) David Kennedy/Alamy, 488: (c) Tierbild Okapia/Photo Researchers, Inc., 490: (a–b) Australian Department of the Environment and Heritage, 491: Winfried Wisniewski/Corbis Bridge/Alamy, 492: Kim Taylor/Nature Picture Library, 493: (top left) Manuela Herrmann/bilderlounge/Corbis, 493: (top right) David Fleetham/Alamy, 493: (middle left) Michael S. Yamashita/CORBIS, 493: (middle right) Keza/Fotolia.com, 493: (bottom) Amazon-Images/Alamy, 494: Bartomeu Borrell/AGE Fotostock America, Inc., 495: (a) Terry Whittaker/Photo Researchers, Inc., 495: (b) erectus/Fotolia, 496: (a) Neilld/Dreamstime.com, 496: (b) moodboard Premium/Fotolia, 500: (top) plusphoto/iStockphoto, 500: (bottom) Dave Welling/Jaynes Gallery/Danita Delimont, Agent/Alamy, 501: (top left) Bradley Ireland/Danita Delimont/Alamy, 501: (top right) Pete Oxford/Danita Delimont/Alamy, 501: University of Florida, 502: (a) Farmyard Nurseries, 502: (b) Jason Rick and Loren Rieseberg, 503: (top) Chris Collins/Corbis Bridge/Alamy, 503: (bottom) Chris Collins/CORBIS-NY, 504: Dante Fenolio/Photo Researchers, Inc., 505: (left) Goinyk Volodymyr/Fotolia LLC, 505: (right) outdoorsman/Fotolia LLC, 506: Honolulu Zoo, 508: Jason Edwards/National Geographic Image Collection/Alamy, 509: The Cleveland Museum of Natural History, 510: AP Photo/Martin Meissner, 512: (left, right) Michael Willmer Forbes Tweedie/Photo Researchers, Inc., 513: (left) Rick & Nora Bowers/Alamy, 513: (right) Juniors Bildarchiv/Alamy, 514: (left) Marcio Jose Bastos Silva/Shutterstock, 514: (middle) Rich Leighton/Fotolia LLC, 514: (right) AfriPics.com/Alamy, 516: Mark Conlin/Alamy, 521: Penn State University/Photo Researchers, Inc., 522: STEPHEN LEHMANN/AFP/Getty Images/Newscom, 523: Arco

Images GmbH/Alamy, 524: (top) Jim Zuckerman/Corbis Flirt/Alamy, 524: (middle) Robert Pickett/Papilio/Alamy, 524: (bottom) Gerd Guenther/Photo Researchers, Inc., 525: (top) Tony Brain/SPL/Photo Researchers, Inc., 525: (middle) Ron Boardman/Frank Lane Picture Agency/CORBIS-NY, 525: (bottom) Robert Cable/Design Pics Inc./Alamy, 527: (top) HP Canada/Alamy, 527: (bottom) Andrzej Tokarski/Fotolia, 528: BOB GIBBONS/Science Photo Library/Alamy, 529: (a) Valery Seleznev/Fotolia, 529: (b) Mayskyphoto/Shutterstock, 529: (c) Dr. Merlin D. Tuttle/Photo Researchers, Inc., 530: MICHAEL REYNOLDS/EPA/Newscom, 532: Picture Hooked/Malcolm Schuyl/Alamy, 534: (top) Corbis RF, 534: (bottom) eddie kidd/Fotolia, 535: Georgette Douwma/SPL/Photo Researchers, Inc., 536: (top) Steve Gschmeissner/Photo Researchers, Inc., 536: (bottom) Eye of Science/SPL/Photo Researchers, Inc., 537: (a) Heather Angel/Natural Visions/Alamy, 537: (b) Steve Bower/Shutterstock, 537: (c) Gary K Smith/Alamy, 537: (bottom) Martin Strmiska/Alamy, 538: (top) Astrid & Hanns-Frieder Michler/Photo Researchers, Inc., 538: (bottom) WaterFrame/Alamy, 539: (top) Darktgbe A. Murawski/National Geographic Image Collection, 539: (middle) Tom McHugh/Photo Researchers, Inc., 539: (bottom left) Christian Ziegler/Danita Delimont/Alamy, 539: (bottom right) Michael P. Fogden/BRUCE COLEMAN INC./Alamy, 540: (a) David Woods/Corbis Bridge/Alamy, 540: (b) MartinMaritz/Shutterstock, 540: (bottom) Sam Yeh/Getty Images, Inc. - Agence France Presse, 541: Nicole Duplaix/Getty - National Geographic Society, 543: (top) NIBSC/Science Photo Library/Photo Researchers, Inc., 543: (middle) Daniel Lai/Aurora Photos/Alamy, 543: (bottom) AP Photo/Jacquelyn Martin, 546: (top left) Lee Dalton/Alamy, 546: (top right) Hans Dieter Brandl/Frank Lane Picture Agency/CORBIS- NY, 546: (bottom) Scott Camazine/Alamy, 548: Ruby/Alamy, 549: (a) Michael Abbey/Photo Researchers, Inc., 549: (b) Klaus Guldbrandsen/Photo Researchers, Inc., 549: (c) leonello calvetti/Alamy, 552: Paul Hewitt, 553: Daniel D. Langleben, 554: Steve Gschmeissner/Photo Researchers, Inc., 555: AP Photo/Fabian Bimmer, 558: Steve Gschmeissner/Photo Researchers, Inc., 560: Edwin R. Lewis, Y. Y. Zeevi and T. E, Everhart, University of California at Berkeley, 562: Ralph C. Eagle Jr./Photo Researchers, Inc., 563: SPL/Photo Researchers, Inc., 564: (inset sign) Henryk Sadura/Fotolia, 567: SPL/Photo Researchers, Inc., 569: Khéops27/Fotolia, 571: David M. Phillips/Photo Researchers, Inc., 572: (a–c) Scanpix Sweden AB, 575: Don W. Fawcett/Photo Researchers, Inc., 580: PCN Photography/Alamy, 582: Lennart Nilsson/Albert Bonniers Forlag AB. The Body Victorious, Dell Publishing Company/Scanpix, 584: Eye of Science/SPL/Photo Researchers, Inc., 585: Science Source/Photo Researchers, Inc., 586: CNRI/Science Photo Library/Photo Researchers, Inc., 589: (top) WG/Science Photo Library/Photo Researchers, Inc., 589: (a) Eye of Science/SPL/Photo Researchers, Inc., 589: (b) Steve Gschmeissner/Photo Researchers, Inc., 590: ChooseMyPlate.gov, 591: John Suchocki/Paul Hewitt, 596: (top) NIBSC/Science Photo Library/Getty Images USA, Inc., 596: (bottom) Eye of Science/Photo Researchers, Inc., 602: (a–b) Eye of Science/Photo Researchers, Inc., 608: Krys Bailey/Alamy, 609: Darren Irwin/Jennifer Yeh, 610: (a) Anita Patterson Peppers/Shutterstock, 610: (b) Keren Su/China Span/Alamy, 610: (c) Photoshot Holdings Ltd/Alamy, 610: (d) Beverly Joubert/National Geographic Society/Getty Images USA, Inc., 616: (Earth) Sergej Khackimullin/Fotolia, 619: Martin Harvey/Alamy, 620: (a) arrxxx/Fotolia LLC, 620: (b) Undersea Discoveries/Shutterstock, 620: (c) scubaluna/Shutterstock, 621: James F. Lubner, University of Wisconsin Sea Grant Institute., 622: David Kjaer/Nature Picture Library, 623: (top) aodaodaodaod/Shutterstock, 623: (bottom) Dhoxax/Shutterstock, 624: (top) cecoffman/Shutterstock, 624: (middle) John Schwieder/Alamy, 624: (bottom left) Oleg Znamenskiy/Shutterstock, 624: (bottom middle) David R. Frazier Photolibrary, Inc./Alamy, 624: (bottom right) Mariusz Jurgielewicz/Melastmohican/Dreamstime, 625: Jerl71/Dreamstime, 626: (top) Matt Tilghman/Shutterstock, 626: (bottom) Cary Kalscheuer/Shutterstock, 630: Hugh Spencer/Photo Researchers, Inc., 636: SIMON MAINA/AFP/GETTY IMAGES/Newscom.

Part Four Page 643: Ariel Skelley/Blend Images/Alamy, 644: Courtesy of MODIS Science Team, Goddard Space Flight Center, NASA and NOAA, 645: (left) G. Brad Lewis/Photo Resource Hawaii/Alamy, 645: (top right) Robbie Shone/Alamy, 645: (bottom right) USDA/Nature Source/Photo Researchers, Inc., 645: (middle) Mark Burnett/Alamy, 646: (a) Doug Perrine/Peter Arnold/Getty Images USA, Inc., 646: (b) Kenneth Sponsler/Fotolia LLC, 646: (c) Jim Lundgren/Alamy, 651: G. Glatzmaier, LANL/P. Roberts, UCLA/Nature/Photo Researchers, Inc., 654: (left) akg-images/Newscom, 654: (a) Colin Keates/Dorling Kindersley, Courtesy of the Natural History Museum, London, 654: (b) Sinclair Stammers/Photo Researchers, Inc., 656: (top) Mike Agliolo/Photo Researchers, Inc., 656: (left) Simon Fraser/Photo Researchers, Inc., 658: edobric/Shutterstock, 661: Steve Shott/Dorling Kindersley Limited, 664: Leo & Mandy Dickinson/Nature Picture Library, 665: (top) U.S. Geological Survey, Denver, 665: (bottom) Bernhard Edmaier/Photo Researchers, Inc., 667: (top) David Shale/Nature Picture Library, 667: (bottom) David A. Hardy/Photo Researchers, Inc., 672: Jacques Jangoux/Photo Researchers, Inc., 673: (right) The Natural History Museum/Alamy, 673: (bottom left) Prehistoric/The Bridgeman Art Library/Getty Images USA, Inc., 673: (bottom right) Mauro Fermariello/Photo Researchers, Inc., 675: (a) Rémi BORNET/Fotolia, 675: (b) Dorling Kindersley, 675: (c) Lee Bottin/Paul G. Hewitt, 675: (d) Arnold Fisher/Photo Researchers, Inc., 675: (e) Harry Taylor/Dorling Kindersley Limited, 675: (f) National Institute for Occupational Safety & Health, 676: (a) Paul Silverman/Fundamental Photographs, 676: (b) Fundamental Photographs, 676: (bottom) Dave King/Dorling Kindersley, Courtesy of The Science Museum, London, 677: (top) Corbin17/Alamy, 677: (middle) M. Claye/Jacana/Photo Researchers, Inc., 677: (bottom) Mark A. Schneider/Photo Researchers, Inc., 678: (top) style67/Fotolia, 678: (augite, sodium feldspar, calcium feldspar, calcite, pyrite) Stan Celestian, Glendale Community College, 678: (clay minerals) michal812/Fotolia, 678: (gold) Stefan Schorn, 679: (c) Melissa Carroll/iStockphoto, 679: (d) Andre Jenny/Alamy, 679: (a) Mark A. Schneider/Photo Researchers, Inc., 679: (b) Charles D Winters/SPL/Photo Researchers, Inc., 680: (a) Janhofmann/Dreamstime, 680: (b) PjrStudio/Alamy, 680: (c) Fotolia, 682: (inset, top) Charles D. Winters/Photo Researchers, Inc., 682: (inset, left) Martin Land/Photo Researchers, Inc., 682: (inset, bottom) Harry Taylor/Dorling Kindersley Limited, 682: (inset, right) Mark A. Schneider/Photo Researchers, Inc., 683: (a) Jay Brousseau/Getty Images USA, Inc., 683: (b) Don Wilke/iStockphoto, 684: Calvin Larsen/Photo Researchers, Inc., 685: (granite, left) Harry Taylor/Dorling Kindersley Limited, 685: (quartz) Fribus Ekaterina/Shutterstock, 685: (hornblende, feldspar) PjrStudio/Alamy, 685: (a, left) Harry Taylor/Dorling Kindersley, 685: (a, right) michal812/Fotolia, 685: (b, left) Andrew J. Martinez/Photo Researchers, Inc., 685: (b, right) Superstock, 685: (b, left) John Foxx/Getty Images, 685: (c, right) geoz/Alamy, 686: (left) Paul Hewitt, 686: (a) Michael Szoenyi/Photo Researchers, Inc., 686: (b) Georg Lehnerer/Fotolia, 687: EROS Data Center, U.S. Geological Survey, 688: (a) Stan Celestian, Glendale Community College, 688: (b) Mark A. Schneider/Photo Researchers, Inc., 688: (c) Harry Taylor/Dorling Kindersley, 688: (d) Harry Taylor/Dorling Kindersley Limited, 689: (top) Bob Betenia/Alamy, 689: (bottom) Roderick Chen/All Canada Photos/Alamy, 690: (top, a, c) Gary Ombler/Dorling Kindersley Limited, 690: (top, b) Joyce Photographics/Photo Researchers, Inc., 690: (top, d) Harry Taylor/Dorling Kindersley Limited, 690: (bottom, a) Lyroky/Alamy, 690: (bottom, b) Dorling Kindersley Limited, 690: (bottom, c) Pecold/Shutterstock, 691: (a) Leo Lintang/Fotolia, 691: (b) Lee Prince/Shutterstock, 693: ricknoll/Fotolia, 693: (a) Andreas Einsiedel/Dorling Kindersley Media Library, 694: (shale) U.S. Geological

Survey, Denver, 694: (slate) Jeffliu1216/Dreamstime, 694: (schist) Nancy Gill/Shutterstock, 694: (gneiss) Daveallenphoto/Dreamstime, 695: (top) Docsub/Dreamstime, 695: (middle right) Caitlin Mirra/Shutterstock, 695: (bottom right) Carroteater/Dreamstime, 695: (bottom left) Leene/Shutterstock, 695: (middle left) Andreas Einsiedel/Dorling Kindersley Limited, 696: (top) Dr. Samuel A. Bowring/MIT, 696: (bottom left) Lynette Cook/Photo Researchers, Inc., 696: (a) John Reader/Science Photo Library/Photo Researchers, Inc., 696: (b) Christian Jegou Publiphoto Diffusion/SPL/Photo Researchers, Inc., 697: petra b./Fotolia, 698: (top, left) Detlev van Ravenswaay/Photo Researchers, Inc., 700: Colin Keates/Dorling Kindersley Media Library, 702: JJAVA/Fotolia, 703: (top) Suerob/Dreamstime, 703: (bottom) Jsuyk/Shutterstock, 704: (top) Andy/Fotolia, 704: (a–c) Digital Globe, Eurimage/Photo Researchers, 705: (top) Dirk Wiersma/Photo Researchers, 705: (bottom) Bernhard Edmaier/SPL/Photo Researchers, 708: Indian Pictures/Alamy, 709: Charles Bolin/Dreamstime, 711: (a) Paul Dix/PNI/Cascades Volcano Observatory, U.S. Geological Survey, 711: (b) Cascades Volcano Observatory, U.S. Geological Survey, 711: (c) U.S. Geological Survey, 712: (a) Tetra Images/Alamy, 712: (b) Henrik Winther Ander/Fotolia, 716: Michael Schmeling/Alamy, 717: totophotos/Shutterstock, 718: George Steinmetz/Corbis, 719: (a) Mikael Svensson/Glow Images, 719: (b) e51/Fotolia, 719: (c) Crobard/Fotolia, 723: (a) Garyhartz/Dreamstime, 723: (b) Geoeye/SPL/Photo Researchers, 729: Phillip Minnis/Shutterstock.com, 730: (a) Prisma Bildagentur AG/Alamy, 730: (b) Jacques Jangoux/Photo Researchers, Inc., 730: (c) Michael Wald/Alamy, 730: (d) Impala/Fotolia.com, 731: (a–b) Dr. Jeremy Burgess/Science Photo Library/Photo Researchers, Inc., 731: (bottom) Design Pics Inc./Alamy, 732: Niebrugge Images/Alamy, 733: (top) Dr. Rob Stepney/SPL/Photo Researchers, Inc., 733: (middle) Fotofeeling/Westend61 GmbH/Alamy, 733: (bottom) Images of Africa Photobank/Alamy, 734: (top) Michael Szoenyi/Photo Researchers, Inc., 734: (bottom) Eric Nathan/Alamy, 735: Mariusz Blach/Fotolia.com, 737: (b) creativenature.nl/Fotolia.com, 737: (c) fairyseasky/Fotolia.com, 737: (bottom) Gary Hincks/Photo Researchers, Inc., 738: (top) F1online digitale Bildagentur GmbH/Alamy, 738: (bottom) evron.info/Fotolia.com, 740: (left) Exactostock/SuperStock, 740: (right) Jon Arnold Images Ltd/Alamy, 741: (top) Nigel Cattlin/Alamy, 741: (bottom left) B.A.E. Inc./Alamy, 741: (bottom right) imagebroker/Alamy, 742: imagebroker/Alamy, 743: (top) John A. Anderson/Shutterstock.com, 743: (a) Chuck Haney/DanitaDelimont.com "Danita Delimont Photography"/Newscom, 743: (b) Elenathewise/Fotolia.com, 743: (c) Bergfee/Fotolia.com, 744: ReinhardT/Fotolia.com, 745: (a) keepics/Alamy, 745: (b) Suzanne Lyons, 745: (c) AP Photo/Carlos Osorio, 745: (bottom) Suzanne Lyons, 746: Michael Nichols/National Geographic Image Collection, 747: (a) Celeste Scholz/Fotolia.com, 747: (b) Michael J. Thompson/Shutterstock.com, 747: (middle) AP Photo/Atlanta Journal-Constitution, Vino Wong, 747: (bottom) Pat on stock/Fotolia.com, 749: (a) Markus Koller/Fotolia.com, 749: (b) Goodshot /Jupiter Images Royalty Free, 749: (c) Ian Danbury/Fotolia.com, 749: (d) james entwistle/Fotolia.com, 750: (top) Photo 24/Jupiter Images PictureArts Corporation/Brand X Royalty Free, 750: (bottom) Everett Collection Inc/Alamy, 753: (a) Kevin Kucks/Alamy, 753: (b) Emily Riddell/Alamy, 753: (right) DLILLC/Corbis, 755: NASA Archive/Alamy, 757: lena zajchikova/Fotolia, 759: NASA, 761: Roman Krochuk/Fotolia, 766: (a) GlowImages/Alamy, 766: (b) BA LaRue/Alamy, 766: (c) Thomas Barrat/Shutterstock, 776: (top, a) Corbis Images RF, 776: (top, b) CAN BALCIOGLU/Shutterstock, 776: (top, c) vencav/Fotolia, 776: (top, d) martin phelps/Alamy, 776: (bottom, a) Thierry Maffeis/Fotolia, 776: (bottom, b) Hewitt, Paul, 776: (bottom, c) Konstantin Sutyagin/Fotolia, 777: (a) hide-mori/a.collectionRF/amana images inc./Alamy, 777: (b) iStockphoto, 777: (c) ril/Fotolia, 777: (d) Dean Baird, 778: (left) Kichigin/Shutterstock, 778: (a) Warren Faidley/Corbis RF, 778: (b) blickwinkel/Held/Alamy, 781: NASA Headquarters, 782: NASA, 785: Paul G. Hewitt, 787: Barry Lewis/Alamy, 788: AP Photo/Paul Sakuma, 790: North Wind Picture Archives via AP Images, 792: Guy Croft SciTech/Alamy, 793: AP Photo/Gemunu Amarasinghe, 795: (a) Frans Lanting/CORBIS, 795: (b) Corbis Images, 795: (c) INTERFOTO/History/Alamy, 797: (top) Kimberly Johnson/Alamy Creativity/Alamy, 797: (bottom) Walter Bibikow/Jon Arnold Images Ltd/Alamy, 798: (a) National Oceanic and Atmospheric Administration, 798: (b) Reuters America, 799: (top) NASA Earth Observing System, 799: (middle) Courtesy Dept. of Library Services, American Museum of Natural History., 799: (bottom) Lynette Cook/SPL/Photo Researchers, Inc., 803: Detlev van Ravenswaay/Photo Researchers, Inc., 805: (a) Bob Gibbons/Alamy, 805: (b) ZUMA Press/Newscom, 812: (top) Tomas Sereda/Fotolia, 812: (bottom) Ian Wood/Alamy, 813: (a) seaskylab/Shutterstock, 813: (b) alephnull/Fotolia, 815: oliveromg/Shutterstock.

Part Five Page 819: Paul Hewitt, 820: NASA, 822: NASA, ESA, M. Robberto (Space Telescope Science Institute/ESA) and the Hubble Space Telescope Orion Treasury Project Team, 826: (top left) SOHO (ESA & NASA), 826: (top right) Jerry Lodriguss/Photo Researchers, Inc., 826: (bottom) Chris Butler/Photo Researchers, Inc., 827: NASA, 828: NASA, 829: NASA Earth Observing System, 830: (a–b) NASA Earth Observing System, 832: (top) NASA Earth Observing System, 832: (middle) David A. Hardy, Futures: 50 Years In Space/Photo Researchers, Inc., 833: (top) NASA Earth Observing System, 833: (bottom left, middle) NASA, 833: (bottom right) NASA/Jet Propulsion Laboratory, 834: NASA, ESA, L. Sromovsky and P. Fry (University of Wisconsin), H. Hammel (Space Science Institute), and K. Rages (SETI Institute), 835: John Suchocki/Paul G. Hewitt, 838: (top) NASA Earth Observing System, 838: (bottom) Lick Observatory Publications Office, 839: NASA/Goddard Institute for Space Studies, 841: Detlev van Ravenswaay/Photo Researchers, Inc., 842: (a–c) Paul G. Hewitt, 843: Dennis DiCicco/Paul G. Hewitt, 845: (top) NASA/Goddard Institute for Space Studies, 845: (middle) ESO/S. Guisard CC-By-3.0. http://www.eso.org/public/images/potw1033a/, 845: (bottom) David J. Roddy/U.S. Geological Survey, Denver, 846: Dr. R. Albrecht, ESA/ESO Space Telescope European Coordinating Facility; NASA, 847: (top left) The Australian National University, 847: (top right) NASA, 848: (a) NASA/Science Photo Library/Photo Researchers, Inc., 848: (b) NASA Headquarters, 852: (a–b) Paul Hewitt, 854: NASA, 856: Rick Whitacre/Shutterstock, 858: NASA/Jet Propulsion Laboratory, 862: NASA, 865: (top) P. Harrington/Space Telescope Science Institute, 865: (bottom) Mark Garlick/Photo Researchers, Inc., 866: NASA, ESA, J. Hester (Arizona State University), 867: (top) Space Telescope Science Institute, 867: (bottom) Lick Observatory Publications Office, 868: Julian Baum/New Scientist/JPL, 872: GALEX, JPL-Caltech, NASA, 873: (top) Gordon Garradd/Jeffrey Bennett, 873: (bottom) Russell Croman/Photo Researchers, Inc., 874: (top left) NASA, ESA, and the Hubble Heritage Team (STScI/AURA), 874: (top right, bottom left) NASA/Jet Propulsion Laboratory, 874: (bottom right) NASA/WikiSky, 875: (a) National Optical Astronomy Observatories, 875: (b) A. Nota (ESA/STScI) et al., ESA, NASA, 875: (bottom) NASA, 876: NASA, 877: (top) NASA's Goddard Space Flight Center, 877: (bottom) NASA, 878: NASA, 884: Emilio Segre Visual Archive/American Institute of Physics, 885: Emilio Segre Visual Archive/American Institute of Physics, 886: NASA Headquarters, 888: NASA, 890: NASA.

Page A-1: Paul G. Hewitt, A-7: Paul G. Hewitt, A-23: © 1990 Richard Megna, Fundamental Photographs, NYC, A-25: © 1983 Chip Clark, Fundamental Photographs, NYC, A-26: (a–c) © 2006 Richard Megna, Fundamental Photographs, NYC.

Index

Note: Page numbers followed by n indicate footnotes; those preceded by A indicate Appendices.

D

E

F

G

H

M

N

Q

R

S

T

U

V

T

U

V